ROSENBERG'S MOLECULAR AND GENETIC BASIS OF NEUROLOGICAL AND PSYCHIATRIC DISEASE

FIFTH EDITION

ELSEVIER science & technology books

Companion Web Site:

http://store.elsevier.com/product.jsp?&isbn=9780124105294

Rosenberg's Molecular and Genetic Basis of Neurological and Psychiatric Disease, 5e
Roger N. Rosenberg, Juan M. Pascual Editors

Available Resources:

- All figures from the book available in .tif, .pdf, and PowerPoint presentation formats
- Print book Table of Contents
- Abstract for each chapter
- All tables from the volume in .pdf format

ROSENBERG'S MOLECULAR AND GENETIC BASIS OF NEUROLOGICAL AND PSYCHIATRIC DISEASE

FIFTH EDITION

Edited by

ROGER N. ROSENBERG
The Abe (Brunky), Morris and William Zale Distinguished Chair in Neurology
Department of Neurology and Neurotherapeutics
Department of Physiology
Head, Section of Cognitive and Memory Disorders
Director, Alzheimer's Disease Center
The University of Texas Southwestern Medical Center
Dallas, TX
USA

JUAN M. PASCUAL
The Once Upon a Time Foundation Professorship in Pediatric Neurologic Diseases
Director, Rare Brain Disorders Program
Department of Neurology and Neurotherapeutics
Department of Physiology
Department of Pediatrics
Eugene McDermott Center for Human Growth & Development/Center for Human Genetics
Division of Pediatric Neurology
The University of Texas Southwestern Medical Center
Dallas, TX
USA

AMSTERDAM • BOSTON • HEIDELBERG • LONDON
NEW YORK • OXFORD • PARIS • SAN DIEGO
SAN FRANCISCO • SINGAPORE • SYDNEY • TOKYO
Academic Press is an imprint of Elsevier

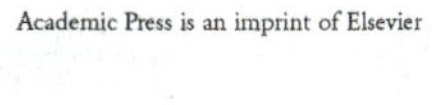

Academic Press is an imprint of Elsevier
32 Jamestown Road, London NW1 7BY, UK
225 Wyman Street, Waltham, MA 02451, USA
525 B Street, Suite 1800, San Diego, CA 92101-4495, USA

Fifth edition

British Library Cataloguing-in-Publication Data
A catalogue record for this book is available from the British Library

Library of Congress Cataloging-in-Publication Data
A catalog record for this book is available from the Library of Congress

ISBN: 978-0-12-410529-4

For information on all Academic Press publications
visit our website at www.store.elsevier.com

Typeset by SPI

Printed and bound in United States of America

15 16 17 18 19 10 9 8 7 6 5 4 3 2

ELSEVIER
Book Aid International
Working together
to grow libraries in
developing countries
www.elsevier.com • www.bookaid.org

Dedications

We dedicate this text to our colleagues, who, by perseverance and dedication, have provided essential new scientific knowledge about the molecular and genetic basis of neurologic and psychiatric disorders, and, in so doing, have conceptualized important insights into disease causation and therapies for the future.

Roger N. Rosenberg and Juan M. Pascual

I wish to dedicate this work to my parents, Cora and Sol Rosenberg, and to my wife, Adrienne. They have been an inspiration to me and have provided me with their care and love to maintain my focus and resilience throughout my life and career, for which I will forever be grateful.

Roger N. Rosenberg

Juan M. Pascual dedicates this work to the memory of his father, Juan Pascual Toledo, magister, who traversed his life and ours loyal, unswerving, and serene, and awaits:

> *"Venisti tandem, tuaque exspectata parenti*
> *vicit iter durum pietas?*
> *datur ora tueri,*
> *nate, tua et notas audire et reddere voces?"*

Contents

4. Pharmacogenomic Approaches to the Treatment of Sporadic Alzheimer Disease using Cholinomimetic Agents

JUDES POIRIER, JUSTIN MIRON, AND CYNTHIA PICARD

5. Application of Mouse Genetics to Human Disease: Generation and Analysis of Mouse Models

TERESA M. GUNN AND BRENDA CANINE

6. DNA Sequencing and Other Methods of Exonic and Genomic Analyses

JUN MITSUI, HIROYUKI ISHIURA, AND SHOJI TSUJI

7. Association, Cause and Causal Association: Means, Methods and Measures

WALTER A. KUKULL

8. Gene Therapy for Neurological Disease

THEODORE FRIEDMANN

9. Direct Induction of Neural Stem Cells from Somatic Cells

WADO AKAMATSU AND HIDEYUKI OKANO

10. Neuroimaging in Dementias

PRASHANTHI VEMURI, MELISSA E. MURRAY, AND CLIFFORD R. JACK, JR.

11. Cognitive Enhancers and Mental Impairment: Emerging Ethical Issues

FABRICE JOTTERAND, JENNIFER L. McCURDY, AND BERNICE ELGER

12. Genetic Counseling

WENDY R. UHLMANN

II NEUROLOGIC DISEASES

13. Cerebral Malformations

WILLIAM D. GRAF AND SHIHUI YU

14. Global Developmental Delay and Intellectual Disability

MYRIAM SROUR AND MICHAEL SHEVELL

15. Down Syndrome

ALLISON CABAN-HOLT, ELIZABETH HEAD, AND FREDERICK SCHMITT

16. An Overview of Rett Syndrome

KRISTEN L. SZABLA AND LISA M. MONTEGGIA

17. Fragile X-Associated Disorders

REYMUNDO LOZANO, EMMA B. HARE, AND RANDI J. HAGERMAN

18. Autism Spectrum Disorders: Clinical Considerations

PATRICIA EVANS, SAILAJA GOLLA, AND MARY ANN MORRIS

19. Metabolic and Genetic Causes of Autism

SAILAJA GOLLA AND PATRICIA EVANS

20. Angelman Syndrome

CHARLES A. WILLIAMS AND JENNIFER M. MUELLER

21. Prion Diseases

JAMES A. MASTRIANNI

III NEUROMETABOLIC DISORDERS

MITOCHONDRIAL DISORDERS

LYSOSOMAL DISORDERS

26. Gaucher Disease: Neuronopathic Forms

RAPHAEL SCHIFFMANN

27. The Niemann–Pick Diseases

EDWARD H. SCHUCHMAN AND ROBERT J. DESNICK

28. G_{M2}-Gangliosidoses

GREGORY M. PASTORES AND GUSTAVO H.B. MAEGAWA

29. Metachromatic Leukodystrophy and Multiple Sulfatase Deficiency

FLORIAN S. EICHLER

30. Krabbe Disease: Globoid Cell Leukodystrophy

DAVID A. WENGER AND PAOLA LUZI

31. The Mucopolysaccharidoses

REUBEN MATALON, KIMBERLEE MICHALS MATALON, AND GEETHA L. RADHAKRISHNAN

32. The Mucolipidoses

REUBEN MATALON AND KIMBERLEE MICHALS MATALON

33. Disorders of Glycoprotein Degradation: Sialidosis, Fucosidosis, α-Mannosidosis, β-Mannosidosis, and Aspartylglycosaminuria

WILLIAM G. JOHNSON

34. β-Galactosidase Deficiency: G_{M1} Gangliosidosis, Morquio B Disease, and Galactosialidosis

WILLIAM G. JOHNSON

35. Acid Ceramidase Deficiency: Farber Lipogranulomatosis and Spinal Muscular Atrophy Associated with Progressive Myoclonic Epilepsy

MICHAEL BECK, HUGO W. MOSER, AND KONRAD SANDHOFF

36. Wolman Disease

ISAAC MARIN-VALENCIA AND JUAN M. PASCUAL

37. Lysosomal Membrane Disorders: LAMP-2 Deficiency

KAZUMA SUGIE AND ICHIZO NISHINO

38. Fabry Disease: α-Galactosidase A Deficiency

ROBERT J. DESNICK

39. Schindler Disease: Deficient-N-Acetylgalactosaminidase Activity

DETLEV SCHINDLER AND ROBERT J. DESNICK

METAL METABOLISM DISORDERS

49. Disorders of Pyridoxine Metabolism

CLARA VAN KARNEBEEK AND SIDNEY M. GOSPE, JR.

LIPID METABOLISM DISORDERS

50. Disorders of Lipid Metabolism

STEFANO DI DONATO AND FRANCO TARONI

51. Lipoprotein Disorders

MARY J. MALLOY AND JOHN P. KANE

52. Cerebrotendinous Xanthomatosis

VLADIMIR M. BERGINER, GERALD SALEN, AND SHAILENDRA B. PATEL

OTHER METABOLIC DISORDERS

53. Organic Acid Disorders

MARGRETTA REED SEASHORE

54. Glycogen Storage Diseases

SALVATORE DIMAURO AND HASAN ORHAN AKMAN

55. Disorders of Galactose Metabolism

GERARD T. BERRY

56. Inborn Errors of Amino Acid Metabolism

WILLIAM L. NYHAN AND RICHARD HAAS

57. Urea Cycle Disorders

NICHOLAS AH MEW, MARIA BELEN PAPPA, AND ANDREA L. GROPMAN

58. Glucose Transporter Type I Deficiency and Other Glucose Flux Disorders

JUAN M. PASCUAL, DONG WANG, AND DARRYL C. DE VIVO

59. Maple Syrup Urine Disease: Clinical and Therapeutic Considerations

DAVID T. CHUANG, R. MAX WYNN, RODY P. COX, AND JACINTA L. CHUANG

60. Congenital Disorders of *N*-Linked Glycosylation

MARC C. PATTERSON

61. Disorders of Glutathione Metabolism

KOJI AOYAMA AND TOSHIO NAKAKI

IV DEGENERATIVE DISORDERS

67. Alzheimer Disease

DENNIS J. SELKOE

68. Genetics of Parkinson Disease and Related Diseases

JILL S. GOLDMAN AND STANLEY FAHN

69. Frontotemporal Dementia

SHUNICHIRO SHINAGAWA AND BRUCE L. MILLER

70. The Neuronal Ceroid-Lipofuscinoses (Batten Disease)

SARA E. MOLE AND MATTI HALTIA

V MOVEMENT DISORDERS

71. The Inherited Ataxias

ROGER N. ROSENBERG AND PRAVIN KHEMANI

72. Friedreich Ataxia

MASSIMO PANDOLFO

73. Ataxia-Telangiectasia

SHUKI MIZUTANI

74. Dystonia

KATJA LOHMANN AND CHRISTINE KLEIN

75. Huntington Disease

ANDREW J. McGARRY, KEVIN BIGLAN, AND FREDERICK MARSHALL

76. Non-Parkinsonian Movement Disorders

STANLEY FAHN AND JILL S. GOLDMAN

77. Hereditary Spastic Paraplegia

JOHN K. FINK

VI NEURO-ONCOLOGY

78. Glioblastoma

ELIZABETH A. MAHER AND ROBERT M. BACHOO

VII NEUROCUTANEOUS DISORDERS

79. Neurofibromatoses

ADAM P. OSTENDORF AND DAVID H. GUTMANN

80. Tuberous Sclerosis Complex

MONICA P. ISLAM AND E. STEVE ROACH

81. Sturge–Weber Syndrome

ANNE M. COMI, DOUGLAS A. MARCHUK, AND JONATHAN PEVSNER

82. Hemangioblastomas of the Central Nervous System

ANA METELO AND OTHON ILIOPOULOS

83. Incontinentia Pigmenti

A. YASMINE KIRKORIAN AND BERNARD COHEN

VIII EPILEPSY

84. The Genetic Epilepsies

ROBERT L. MACDONALD AND MARTIN J. GALLAGHER

IX WHITE MATTER DISEASES

85. Multiple Sclerosis

STEPHEN L. HAUSER, JORGE R. OKSENBERG, AND SERGIO E. BARANZINI

86. Vanishing White Matter Disease

ORNA ELROY-STEIN AND RAPHAEL SCHIFFMANN

X NEUROPATHIES AND NEURONOPATHIES

87. Amyotrophic Lateral Sclerosis

JEMEEN SREEDHARAN AND ROBERT H. BROWN, JR.

88. Peripheral Neuropathies

STEVEN S. SCHERER, KLEOPAS A. KLEOPA, AND MERRILL D. BENSON

89. Spinal Muscular Atrophy

BAKRI H. ELSHEIKH, W. DAVID ARNOLD, AND JOHN T. KISSEL

90. Pain Genetics

WILLIAM RENTHAL

XI MUSCLE AND NEUROMUSCULAR JUNCTION DISORDERS

91. Dystrophinopathies

ERIC P. HOFFMAN

92. Limb-Girdle Muscular Dystrophy

WEN-CHEN LIANG AND ICHIZO NISHINO

93. The Congenital Myopathies

HEINZ JUNGBLUTH, CAROLINE SEWRY, AND FRANCESCO MUNTONI

94. The Distal Myopathies

AMI MANKODI, BJARNE UDD, AND ROBERT C. GRIGGS

95. Hereditary Inclusion-Body Myopathies

ALDOBRANDO BROCCOLINI AND MASSIMILIANO MIRABELLA

96. The Myotonic Dystrophies

RICHARD T. MOXLEY, III, JAMES E. HILBERT, AND GIOVANNI MEOLA

97. Facioscapulohumeral Dystrophy

RABI TAWIL

98. Muscle Channelopathies: Periodic Paralyses and Nondystrophic Myotonias

JEFFREY RALPH AND LOUIS PTÁČEK

99. Congenital Myasthenic Syndromes

ANDREW G. ENGEL

XII STROKE

100. Cerebral Vasculopathies

MICHAEL M. DOWLING

101. Coagulopathies

FENELLA J. KIRKHAM

102. Sickle Cell Disease

FENELLA J. KIRKHAM

XIII PSYCHIATRIC DISEASE

103. Depression

STEVEN T. SZABO AND CHARLES B. NEMEROFF

104. Bipolar Disorder

SCOTT C. FEARS AND VICTOR I. REUS

105. Schizophrenia

DAVID W. VOLK AND DAVID A. LEWIS

106. Obsessive–Compulsive Disorder

MICHAEL H. BLOCH, JESSICA B. LENNINGTON, GABOR SZUHAY, AND PAUL J. LOMBROSO

107. Tourette Syndrome

JESSICA B. LENNINGTON, MICHAEL H. BLOCH, LAWRENCE D. SCAHILL, GABOR SZUHAY, PAUL J. LOMBROSO, AND FLORA M. VACCARINO

108. Addiction

SCOTT D. PHILIBIN AND JOHN C. CRABBE

XIV A NEUROLOGIC GENE MAP

109. A Neurologic Gene Map

SAIMA N. KAYANI, KATHLEEN S. WILSON, AND ROGER N. ROSENBERG

Preface to the Fifth Edition

We are publishing the fifth edition of the *Molecular and Genetic Basis of Neurological and Psychiatric Disease*. The first edition appeared in 1993 followed by editions in 1997, 2003, and 2008. We are most grateful for the foresight, dedication, and authorship of our former editors for the success of the first four editions. They are Stanley B. Prusiner, Salvatore DiMauro, Robert L. Barchi, Louis M. Kunkel, Henry L. Paulson, Louis Ptáček and Eric J. Nestler. The fifth edition is edited by Roger N. Rosenberg and Juan M. Pascual.

There are several major new aspects to the fifth edition: The text now includes well over 100 chapters and 200 contributors. Every chapter has been thoroughly updated either by previous contributors or by new experts in the field, all of which are of international renown. A standard, unified chapter format has been followed as much as possible. Most illustrations are new or have been newly drawn, and color has been used wherever helpful throughout the text. The book is available both in print and in up-to-date electronic format. Additional new chapters in this edition cover the following topics: DNA sequencing and other methods of exonic and genomic analysis; pharmacogenomics; causation and association; stem cells and therapeutic development; neuroimaging; genetic counseling; the ethics of cognitive enhancement and mental impairment; cerebral malformations; global developmental delay and intellectual disability; neurodegeneration with brain iron accumulation; pantothenate kinase deficiency; Wilson disease; Menkes disease and other ATP7A disorders; disorders of manganese transport; aceruloplasminemia; neurotransmitter disorders; frontotemporal dementias; dystonia; glioblastoma; tuberous sclerosis; von Hippel–Lindau disease; Sturge–Weber syndrome; incontinentia pigmenti; channelopathies; vanishing white matter disease; pyruvate metabolism and Krebs cycle disorders; pain; vasculopathies; coagulopathies; sickle cell disease; and autism. Clearly, neurogenics/neurogenomics has advanced rapidly and is now poised to develop in the next decade effective targeted neurotherapeutics.

In the 21 years spanning the five editions of our book, molecular genomic analyses of the human genome have been implemented seeking the genetic basis for natural selection providing biological fitness and also risk of developing disease. Genome-wide association studies (GWAS) seeking gene variations, single nucleotide polymorphisms (SNP), causal of several human diseases have been conducted in recent years including autism, schizophrenia, obesity, diabetes and heart disease.

Several GWAS for risk association with neurological diseases, neuromic studies, have been reported. An increased risk for amyotrophic lateral sclerosis (ALS), Alzheimer disease (AD), restless leg syndrome (RLS), and multiple sclerosis have been associated with polymorphisms in specific genes. These observations have advanced an understanding of the causation of inherited, complex polygenetic, multifactorial neurological diseases. They have been made possible by the publication of the human genome and haplotype studies (HapMap analyses).

The hope with neurome-wide association studies has been that the complete complement of variant genes will be identified causal of the major neurodegenerative diseases. Then, pharmaconeuromic therapy would not be far behind. GWAS has provided new and important data of the major genes responsible for major human traits and common diseases. GWAS has provided insights into gene variations in low penetrant genes causal for polygenetic, multifactorial neurological disease, such as Alzheimer disease. Overall, about 400 genetic variants have been identified that contribute to human traits and diseases including neurological diseases.

Sequencing candidate genes for disease including their surrounding regions in thousands of people will be needed to discover more associations with disease. SNPs are turning out not to be a stringent enough level of analysis seeking genetic risks for disease. The change in mindset is going from seeking analyses of common, low-penetrance variants causal of common diseases to seeking rare low- or moderate-penetrance variants that have been missed by GWAS. It may be necessary to move beyond sequencing candidate genes and surrounding regions for disease association and begin sequencing whole genomes to find the missing heritability. Francis Collins, Director of the National Institutes of Health, has suggested that the 1000 genomes project, designed to sequence the genomes of at least 1000 people from all over the world, would provide a powerful approach to finding the hidden heritability.

The genetic explanations that would be of primary interest to find the missing heritability for genetic neurological disease missed by GWAS include copy-number variation (CNV), epistatic effects, and epigenetics. CNV refers to regions of DNA that are up to hundreds of base pairs long that are deleted or duplicated between individuals.

There are strong CNV associations between schizophrenics compared to normals and they may arise *de novo* in persons without a family history of the mutation. Epistasis, where one or more modifying genes reduce or enhance the effect of another gene, may be an important genetic mechanism at work to explain heritability not found by GWAS. Epigenetics is another vital area to be explored. It refers to changes in gene expression that are inherited but not caused by alteration in the sequence of the gene. We now know that gene expression is altered by methylation or acetylation, and also by inhibition of messenger RNA expression by iRNA or microRNA binding.

The 21,000 protein-coding genes in the human genome make up less than 1.2% of the human genome. Analysis of the remaining 98.8% of the human genome and its role in the causation of human neurological diseases, both inherited and acquired, is a formidable challenge yet unexplored to any degree. RNA transcripts and their effects on regulation and levels of gene expression is one of the next frontiers for neuromics.

Then there is the issue that natural selection only functions before or during the reproductive years and not afterwards, when Alzheimer disease and Parkinson disease occur. Natural selection has as its major biological function to select for fitness allowing for reproduction and maintenance of a lineage or species. Aging and neurodegenerative diseases seem to have escaped the forces of natural selection by occurring after the reproductive years. On the other hand, perhaps evolution has actually selected for aging and neurodegenerative diseases as a means to maintain the limits of a finite lifespan. Clearly, neuromics must address the molecular basis of brain aging and why the aging process provides a permissive environment to allow the opportunistic neuromic program causal of late-onset neurodegenerative diseases to be expressed.

The cause of Alzheimer disease is due both to genetic polymorphisms and environmental stimuli. In this view, environmental stimuli, to be determined, influence the production of an abnormal pattern of gene expression causal of Alzheimer disease. So, we will have to understand the process of natural selection in the context of the selection pressures from the environments that we inhabit. Darwin emphasized adaptation to a changing environment as the principal selective influence for evolution. This principle is valid studying the interaction of environmental stimuli and the genetic factors causal of neurodegenerative diseases.

Deriving induced pluripotential stem cells from late-onset Alzheimer disease patients and differentiating them into neuroblasts would be one way to screen compounds to see if an abnormal pattern of gene expression is produced compared to derived neuroblasts from normal controls. Here would be a method to link environment to the genetic program causal of Alzheimer disease. It would also be a means to screen potential therapeutic agents that correct an abnormal pattern of gene expression seen in AD patients as a prelude to a clinical trial.

The 200 years since Charles Darwin's birth, 150 years since the publication of *On the Origin of Species,* and the 20 years of the publication of the four editions of this book, is a brief time in human experience. The fifth edition builds on the development of neurogenetics during the past 20 years and documents the advances in genome sequencing, CNV, epistasis, epigenetics, RNA regulation of gene expression, and stem cell applications to decipher how mutations in these genetic functions are causal of neurological diseases.

We look forward to future editions of the book and wish to express our gratitude to our many loyal colleagues who have participated in all five editions, and thank our new authors for their contributions to maintain the book's scientific rigor and excellence. Whereas we have made every effort towards comprehensiveness and clarity, many omissions and imprecisions are bound to remain. To that effect, we will welcome comments and suggestions at Rosenberg5ed@gmail.com. We have retained the names of Hugo W. Moser and John H. Menkes through the kindness of their families to honor their memory. The outstanding editorial contributions of Kristi Anderson, project manager, Mica Haley, publisher for neuroscience, and Julia Haynes, book production project manager, are most gratefully acknowledged. We are also thankful to our families, patients, colleagues and trainees both for interactions and for lost time while we were working on the fifth edition. While a textbook on the human experience of neurological or psychiatric patients has not yet been written, we hope that ours will assist in the understanding of one important dimension of their existence.

Roger N. Rosenberg
Juan M. Pascual
Editors

Contributors

Nicholas Ah Mew The Center for Neuroscience and Behavioral Medicine, Department of Genetics and Metabolism, Children's National Medical Center, Washington, DC, USA

Wado Akamatsu Department of Physiology, School of Medicine, Keio University, Tokyo, Japan

Hasan Orhan Akman Department of Neurology, Columbia University Medical Center, New York, NY, USA

Koji Aoyama Department of Pharmacology, Teikyo University School of Medicine, Tokyo, Japan

W. David Arnold Deparment of Neurology, The Ohio State University, Wexner Medical Center, Columbus, OH, USA

Rafael Artuch Hospital Sant Joan de Déu, Barcelona, Spain; CIBERER (Network for Research in Rare Diseases), Instituto de Salud Carlos III, Madrid, Spain

Robert M. Bachoo Departments of Internal Medicine and Neurology & Neurotherapeutics, Annette G. Strauss Center for Neuro-Oncology, Simmons Cancer Center, The University of Texas Southwestern Medical Center, Dallas, TX, USA

Sergio E. Baranzini Department of Neurology, University of California, San Francisco, CA, USA

Michael Beck Children's Hospital, University Medical Center, University of Mainz, Mainz, Germany

Merrill D. Benson Department of Pathology and Laboratory Medicine, Indiana University School of Medicine, Indianapolis, IN, USA

Vladimir M. Berginer Department of Neurology, Soroka Medical Center, Beer Sheva, Israel

Gerard T. Berry Boston Children's Hospital, Division of Genetics and Genomics, Harvard Medical School, Boston, MA, USA

Kevin M. Biglan Department of Neurology, University of Rochester, Rochester, NY, USA

Thomas D. Bird Departments of Neurology and Medicine, University of Washington and VA Medical Center, Seattle, WA, USA

D. Montgomery Bissell National Institutes of Health (NIH)-supported Liver Center, Division of Gastroenterology, UCSF Medical Center, San Francisco, CA, USA

Michael H. Bloch Child Study Center, Yale University School of Medicine, New Haven, CT, USA

Aldobrando Broccolini Institute of Neurology, Department of Geriatrics, Neurosciences and Orthopedics, Catholic University, Rome, Italy

Robert H. Brown, Jr. Department of Neurology, University of Massachusetts Medical School, Worcester, MA, USA

Allison Caban-Holt Department of Behavioral Science, Sanders-Brown Center on Aging, University of Kentucky, Lexington, KY, USA

Brenda Canine McLaughlin Research Institute, Great Falls, MT, USA

C. Thomas Caskey Department of Molecular and Human Genetics, Baylor College of Medicine, Houston, TX, USA

Patrick F. Chinnery Department of Neurology, Institute of Genetic Medicine, Newcastle University, Newcastle NIHR Biomedical Research Centre, Newcastle upon Tyne, UK

David T. Chuang Departments of Biochemistry and Internal Medicine, The University of Texas Southwestern Medical Center, Dallas, TX, USA

Jacinta L. Chuang Department of Biochemistry, The University of Texas Southwestern Medical Center, Dallas, TX, USA

Bernard A. Cohen Dermatology and Pediatrics, Johns Hopkins Children's Center, Baltimore, MD, USA

Anne M. Comi Neurology and Pediatrics, Kennedy Krieger Institute, Johns Hopkins School of Medicine, Hunter Nelson Sturge-Weber Center, Baltimore, MD, USA

Rody P. Cox Department of Internal Medicine, The University of Texas Southwestern Medical Center, Dallas, TX, USA

John C. Crabbe Department of Behavioral Neuroscience, Department of Veterans Affairs Medical Center Director, Portland Alcohol Research Center, Oregon Health & Science University, Portland, OR, USA

Marie Y. Davis Department of Neurology, University of Washington, Seattle, WA, USA

Darryl C. De Vivo SMA Clinical Research Center, Motor Neuron Center, Colleen Giblin Laboratories for Pediatric Neurology, Columbia University Medical Center, New York, NY, USA

Robert J. Desnick Department of Genetics and Genomic Sciences, Icahn School of Medicine at Mount Sinai, New York, NY, USA

Stefano Di Donato Fondazione IRCCS Istituto Neurologico "Carlo Besta," Milan, Italy

Salvatore DiMauro Department of Neurology, Columbia University Medical Center, The Neurological Institute of New York, New York, NY, USA

Michael M. Dowling Departments of Pediatrics, Neurology and Neurotherapeutics, The University of Texas Southwestern Medical Center, Dallas, TX, USA

David A. Dyment Children's Hospital of Eastern Ontario Research Institute, University of Ottawa, Ottawa, ON, Canada

Florian S. Eichler Department of Neurology, Massachusetts General Hospital, Harvard Medical School, Boston, MA, USA

Ramyiadarsini Elangovan Functional Genomics Unit, Department of Physiology, Anatomy and Genetics and Medical Research Council, University of Oxford, Oxford, UK

Bernice Elger Institute of Biomedical Ethics, University of Basel, Basel, Switzerland

Sara Elrefai Department of Medical Genetics, Henry Ford Hospital, Detroit, MI, USA

Orna Elroy-Stein Department of Cell Research and Immunology, Faculty of Life Sciences, Tel Aviv University, Tel Aviv, Israel

Bakri H. Elsheikh Saudi Aramco, Dhahran, Saudi Arabia

Andrew G. Engel Mayo Clinic College of Medicine, Department of Neurology, Mayo Clinic, Rochester, MN, USA

Patricia Evans Department of Neurology and Pediatrics, The University of Texas Southwestern School of Medicine, Dallas, TX, USA

Stanley Fahn Movement Disorder Division, Department of Neurology, Neurological Institute, Columbia University Medical Center, New York, NY, USA

Scott C. Fears Ronald Reagan UCLA Medical Center, Stewart and Lynda Resnick Neuropsychiatric Hospital at UCLA, Los Angeles, CA, USA

John K. Fink Department of Neurology, University of Michigan, Geriatric Research Education and Care Center, Ann Arbor Veterans Affairs Medical Center, Ann Arbor, MI, USA

Theodore Friedmann Department of Pediatrics, UCSD School of Medicine, La Jolla, CA, USA

Martin J. Gallagher Department of Neurology, Vanderbilt University, Nashville, TN, USA

Àngels García-Cazorla Department of Neurology, Hospital Sant Joan de Déu, Barcelona, Spain; CIBERER (Network for Research in Rare Diseases); Instituto de Salud Carlos III, Madrid, Spain

Jill S. Goldman Taub Institute, Columbia University Medical Center, New York, NY, USA

Sailaja Golla Neurodevelopmental Pediatrics, Division Of Pediatric Neurology, The University of Texas Southwestern Medical Center, Children's Medical Center, Dallas TX, USA

Sidney M. Gospe, Jr. Departments of Neurology and Pediatrics, University of Washington, and Seattle Children's Hospital, Seattle, WA, USA

William D. Graf Department of Pediatrics, Department of Neurology, Yale School of Medicine, New Haven, CT, USA

Robert C. Griggs Departments of Neurology, Pathology and Laboratory Medicine and Pediatrics, University of Rochester Medical Center, Rochester, NY, USA

Andrea L. Gropman Division of Neurogenetics and Developmental Pediatrics, Department of Neurology and Pediatrics, Children's National Medical Center and the George Washington University of the Health Sciences, Washington, DC, USA

Yian Gu Taub Institute on Alzheimer's Disease and the Aging Brain, Department of Neurology, Columbia University Medical Center, New York, NY, USA

Teresa M. Gunn McLaughlin Research Institute, Great Falls, MT, USA

David H. Gutmann Department of Neurology, Washington University Neurofibromatosis Center, Washington University School of Medicine, St. Louis, MO, USA

Richard Haas Departments of Neurosciences and Pediatrics, University of California, San Diego, La Jolla, CA, USA

Randi J. Hagerman MIND Institute, UC Davis Health System, Sacramento, CA, USA

Matti J. Haltia Department of Pathology, Children's Hospital, University of Helsinki, Helsinki, Finland

Emma B. Hare MIND Institute, UC Davis Medical Center, Sacramento, CA, USA

Tamar Harel Department of Molecular and Human Genetics, Baylor College of Medicine, Houston, TX, USA

Stephen L. Hauser Department of Neurology, University of California, San Francisco, CA, USA

Elizabeth Head Department of Molecular and Biomedical Pharmacology, Sanders-Brown Center on Aging, University of Kentucky, Lexington, KY, USA

James E. Hilbert Department of Neurology, Neuromuscular Disease Center, University of Rochester School of Medicine and Dentistry, Rochester, NY, USA

Eric P. Hoffman Research Center for Genetic Medicine, Children's Research Institute, Department of Integrative Systems Biology, George Washington University, Washington, DC, USA

Othon Iliopoulos Harvard Medical School, Massachusetts General Hospital Cancer Center, Boston, MA, USA

Hiroyuki Ishiura Department of Neurology, Graduate School of Medicine, The University of Tokyo, Tokyo, Japan

Monica P. Islam Department of Clinical Pediatrics, The Ohio State University College of Medicine, Department of Pediatric Neurology, Nationwide Children's Hospital, Columbus, OH, USA

Clifford R. Jack, Jr. Aging and Dementia Imaging Laboratory, Department of Radiology, Mayo Clinic, Rochester, MN, USA

William G. Johnson Laboratory of Molecular Neurogenetics, Rutgers Robert Wood Johnson Medical School, New Brunswick, NJ, USA

Fabrice Jotterand Department of Health Care Ethics, Regis University, Denver, CO, USA; Institute of Biomedical Ethics, University of Basel, Basel, Switzerland

Heinz Jungbluth Department of Clinical Neuroscience, King's College London, Guy's & St. Thomas' Hospital NHS Foundation Trust, London, UK

John P. Kane Departments of Medicine, Biochemistry and Biophysics, Cardiovascular Research Institute, UCSF Medical Center, San Francisco, CA, USA

Clara van Karnebeek Department of Pediatrics, University of British Columbia, Vancouver, BC, Canada

Saima N. Kayani Department of Neurology and Neurotherapeutics, and Pathology, The University of Texas Southwestern Medical Center, Dallas, TX, USA

Pravin Khemani Department of Neurology and Neurotherapeutics, The University of Texas Southwestern Medical Center, Dallas, TX, USA

Fenella J. Kirkham Department of Paediatric Neurology, Neurosciences Unit, Institute of Child Health, University College London, London, UK

A. Yasmine Kirkorian Division of Pediatric Dermatology, Department of Dermatology, Johns Hopkins University School of Medicine, Baltimore, MD, USA

John T. Kissel Department of Neurology and Pediatrics, Wexner Medical Center, The Ohio State University, Columbus, OH, USA

Christine Klein Institute of Neurogenetics and, Department of Neurology, University of Lübeck, Lübeck, Germany

Kleopas A. Kleopa Department of Clinical Neurosciences, The Cyprus Institute of Neurology and Genetics, Nicosia, Cyprus

Satoshi Kono First Department of Medicine, Hamamatsu University School of Medicine, Handayama, Hamamatsu, Japan

Michael C. Kruer University of South Dakota Sanford School of Medicine, Sanford Children's Specialty Clinic, Sanford Children's Research Center, Sioux Falls, SD, USA

Walter A. Kukull Department of Epidemiology, University of Washington, Seattle, WA, USA

Jessica B. Lennington Child Study Center, Yale University School of Medicine, New Haven, CT, USA

David A. Lewis Department of Psychiatry, University of Pittsburgh, Western Psychiatric Institute and Clinic, Pittsburgh, PA, USA

Wen-Chen Liang Department of Pediatrics, Kaohsiung Medical University Hospital, Department of Pediatrics, School of Medicine, College of Medicine, Kaohsiung Medical University, Kaohsiung, Taiwan

Katja Lohmann Institute of Neurogenetics, University of Lübeck, Lübeck, Germany

Paul J. Lombroso Child Study Center, Yale University School of Medicine, New Haven, CT, USA

Reymundo Lozano MIND Institute, UC Davis Medical Center, Sacramento, CA, USA

James R. Lupski Department of Pediatrics, Department of Molecular and Human Genetics, Baylor College of Medicine, Texas Children's Hospital, Houston, TX, USA

Paola Luzi Department of Neurology, Assistant Director, Lysosomal Diseases Testing Laboratory, Jefferson Medical College, Philadelphia, PA, USA

Qian Ma Rare Brain Disorders Program, Department of Neurology and Neurotherapeutics, The University of Texas Southwestern Medical Center, Dallas, TX, USA

Robert L. Macdonald Department of Neurology, Vanderbilt University, Nashville, TN, USA

Gustavo H.B. Maegawa McKusick-Nathans Institute of Genetic Medicine, Department of Pediatrics, Johns Hopkins University School of Medicine, Baltimore, MD, USA

Elizabeth A. Maher Departments of Internal Medicine and Neurology & Neurotherapeutics, Annette G. Strauss Center for Neuro-Oncology, Simmons Cancer Center, The University of Texas Southwestern Medical Center, Dallas, TX, USA

Mary J. Malloy* Departments of Medicine and Pediatrics, Cardiovascular Research Institute, UCSF Medical Center, San Francisco, CA, USA

Ami K. Mankodi Neurogenetics Branch, National Institute of Neurological Disorders and Stroke, National Institutes of Health, Bethesda, MD, USA

Douglas A. Marchek Department of Molecular Genetics and Microbiology, Duke University School of Medicine, Durham, NC, USA

Isaac Marin-Valencia Rare Brain Disorders Program, Department of Neurology and Neurotherapeutics, Department of Pediatrics, Division of Pediatric Neurology, The University of Texas Southwestern Medical Center, Dallas, TX, USA

Frederick J. Marshall Geriatric Neurology Unit, University of Rochester School of Medicine and Dentistry, Rochester, NY, USA

James A. Mastrianni Center for Comprehensive Care and Research on Memory Disorders, Department of Neurology, The University of Chicago, Chicago, IL, USA

Reuben Matalon Department of Pediatrics, The University of Texas Medical Branch (UTMB), Galveston, TX, USA

Richard Mayeux Gertrude H. Sergievsky Center, Taub Institute on Alzheimer's Disease and the Aging Brain, Columbia University Medical Center, New York, NY, USA

Jennifer L. McCurdy Department of Health Care Ethics, Rueckert-Hartman College of Health Professions, Regis University, Denver, CO, USA

Andrew J. McGarry Department of Neurology, Cooper University Health Care, Cherry Hill, NJ, USA

John H. Menkes* Cedars-Sinai Medical Center, Los Angeles, CA, USA

Giovanni Meola Department of Neurology, University of Milan, IRCCS Policlinico San Donato, Milan, Italy

Ana Metelo Faculty of Science and Technology, Coimbra University, Coimbra; Portugal; Harvard Medical School, Massachusetts General Hospital Cancer Center, Boston, MA, USA

Kimberlee Michals Matalon Health and Human Performance, The University of Houston, Houston, TX, USA

Bruce L. Miller Neurology, Memory and Aging Center, University of California, San Francisco, CA, USA

Massimiliano Mirabella Institute of Neurology, Department of Geriatrics, Neurosciences and Orthopedics, Catholic University, Rome, Italy

Justin Miron Department of Neuroscience, McGill University, Montreal, QC, Canada

Jun Mitsui Department of Neurology, Graduate School of Medicine, The University of Tokyo, Tokyo, Japan

Hiroaki Miyajima First Department of Medicine, Hamamatsu University School of Medicine, Handayama, Hamamatsu, Japan

Shuki Mizutani Department of Pediatrics and Developmental Biology, Tokyo Medical and Dental University, Graduate School of Medicine, Yushima, Tokyo, Japan

Sara E. Mole MRC Laboratory for Molecular Cell Biology, UCL Institute of Child Health, University College London, London, UK

Lisa M. Monteggia Department of Neuroscience, The University of Texas Southwestern Medical Center, Dallas, TX, USA

*Deceased

Hugo W. Moser Department of Neurogenetics, Johns Hopkins University, Kennedy Krieger Institute, Baltimore, MD, USA

Mary Ann Morris Department of Neurology and Neurotherapeutics, The University of Texas Southwestern School of Medicine, Dallas, TX, USA

Richard T. Moxley, III Department of Neurology and Pediatrics, Neuromuscular Disease Center, University of Rochester School of Medicine and Dentistry, Rochester, NY, USA

Jennifer M. Mueller Division of Genetics and Metabolism, Department of Pediatrics, University of Florida, Gainesville, FL, USA

Francesco Muntoni Department of Paediatric Neurology, Dubowitz Neuromuscular Centre, UCL Institute of Child Health, London, UK

Melissa E. Murray Department of Neuroscience, Mayo Clinic, Jacksonville, FL, USA

Toshio Nakaki Department of Pharmacology, Teikyo University School of Medicine, Tokyo, Japan

Charles B. Nemeroff Department of Psychiatry and Behavioral Sciences, Leonard M. Miller School of Medicine, University of Miami, Miami, FL, USA

Ichizo Nishino Department of Neuromuscular Research, National Institute of Neuroscience, National Center of Neurology and Psychiatry (NCNP), Tokyo, Japan

William L. Nyhan Department of Pediatrics, University of California, San Diego, La Jolla, CA, USA

Hideyuki Okano Department of Physiology, School of Medicine, Keio University, Tokyo, Japan

Jorge R. Oksenberg Department of Neurology, University of California, San Francisco, CA, USA

Adam P. Ostendorf Department of Neurology, Washington University School of Medicine, St. Louis, MO, USA

Massimo Pandolfo Université Libre de Bruxelles, ULB Institute of Neurosciences (UNI), Department of Neurology, Hôpital Erasme, Brussels, Belgium

Maria Belen Pappa Division of Neurogenetics and Developmental Pediatrics, Department of Neurology and Pediatrics, Children's National Medical Center and the George Washington University of the Health Sciences, Washington, DC, USA

Carmen Paradas Unidad de Enfermedades Neuromusculares, Servicio de Neurologia, Hospital Universitario Virgen del Rocio/Instituto de Biomedicina de Sevilla, Universidad de Sevilla, Seville, Spain

Juan M. Pascual Rare Brain Disorders Program, Departments of Neurology and Neurotherapeutics, Physiology and Pediatrics and Eugene McDermott Center for Human Growth & Development/ Center for Human Genetics. Division of Pediatric Neurology, The University of Texas Southwestern Medical Center, Dallas, TX, USA

Gregory M. Pastores Adult Metabolic Service, Department of Medicine, National Centre for Inherited Metabolic Disorders Mater Misericordiae University Hospital, Dublin, Ireland

Shailendra B. Patel Division of Endocrinology, Metabolism and Clinical Nutrition, Clement J Zablocki Veterans Medical Center, Medical College of Wisconsin, Milwaukee, WI, USA

Marc C. Patterson Division of Child and Adolescent Neurology, Departments of Neurology, Pediatrics and Medical Genetics, Mayo Clinic Children's Center, Rochester, MN, USA

Davut Pehlivan Department of Molecular and Human Genetics, Baylor College of Medicine, Houston, TX, USA

Scott D. Philibin Department of Psychiatry, The University of Texas Southwestern Medical Center, Dallas, TX, USA

Cynthia Picard Department of Neuroscience, McGill University, Montreal, QC, Canada

Judes Poirier Centre for Studies on the Prevention of Alzheimer's Disease, Douglas Mental Health University Institute, Montreal, QC, Canada

Louis J. Ptáček Department of Neurology, Howard Hughes Medical Institute, University of California, San Francisco, San Francisco, CA, USA

Geetha L. Radhakrishnan Department of Pediatrics, The University of Texas Medical Branch (UTMB), Galveston, TX, USA

Jeffrey W. Ralph Department of Neurology, University of California, San Francisco, CA, USA

Sreeram V. Ramagopalan Functional Genomics Unit, Department of Physiology, Anatomy and Genetics and Medical Research Council, University of Oxford, Oxford, UK

Gerald V. Raymond Department of Neurology, University of Minnesota, Minneapolis, MN, USA

William Renthal Department of Neurology and Neurotherapeutics, The University of Texas Southwestern Medical Center, Dallas, TX, USA

Victor I. Reus Department of Psychiatry, University of California San Francisco School of Medicine, San Francisco, CA, USA

E. Steve Roach Department of Pediatrics and Neurology, The Ohio State University College of Medicine, Nationwide Children's Hospital, Columbus, OH, USA

Roger N. Rosenberg Department of Neurology and Neurotherapeutics, Department of Physiology, Section of Cognitive and Memory Disorders, Alzheimer's Disease Center, The University of Texas Southwestern Medical Center, Dallas, TX, USA

David S. Rosenblatt Departments of Human Genetics, Medicine, Pediatrics and Biology, McGill University, Montreal, QC, Canada

Gerald Salen Division of Gastroenterology, Rutgers, New Jersey Medical School, Newark, NJ, USA

Konrad Sandhoff LIMES, Kekulé-Institut, Bonn University, Bonn, Germany

Lawrence D. Scahill Department of Pediatrics, Marcus Autism Center, Emory University School of Medicine, Atlanta, GA, USA

Steven S. Scherer Department of Neurology, Perelman School of Medicine at the University of Pennsylvania, Philadelphia, PA, USA

Raphael Schiffmann Institute of Metabolic Disease, Baylor Research Institute, Dallas, TX, USA

Detlev Schindler Department of Human Genetics, University of Würzburg, Würzburg, Germany

Frederick Schmitt Department of Neurology, Sanders-Brown Center on Aging, University of Kentucky, Lexington, KY, USA

Susanne A. Schneider Department of Neurology, University of Kiel, Kiel, Germany

Eric A. Schon Department of Neurology, Columbia University Medical Center, The Neurological Institute of New York, New York, NY, USA

Edward H. Schuchman Icahn School of Medicine at Mount Sinai, New York, NY, USA

Nicole Schupf The Gertrude H. Sergievsky Center, Columbia University Medical Center, New York, NY, USA

Margretta Reed Seashore Departments of Genetics, Laboratory Medicine and Pediatrics, Yale University School of Medicine, New Haven, CT, USA

Dennis J. Selkoe Center for Neurologic Diseases, Department of Neurology, Brigham and Women's Hospital/Harvard Medical School, Boston, MA, USA

Caroline Sewry Dubowitz Neuromuscular Centre, UCL Institute of Child Health and Great Ormond Street Hospital for Children, London, UK; RJAH Orthopaedic Hospital, Oswestry, UK

Michael Shevell Departments of Pediatrics and Neurology/Neurosurgery, McGill University, Montreal Children's Hospital/McGill University Health Centre, Montreal, QC, Canada

Shunichiro Shinagawa Department of Psychiatry, The Jikei University School of Medicine, Tokyo, Japan

Jemeen Sreedharan Babraham Institute, Cambridge, UK

Myriam Srour Departments of Pediatrics and Neurology/Neurosurgery, McGill University, Division of Pediatric Neurology, Montreal Children's Hospital/McGill University Health Centre, Montreal, QC, Canada

Kazuma Sugie Department of Neurology, Nara Medical University School of Medicine, Nara, Japan, Department of Neuromuscular Research, National Institute of Neuroscience, National Center of Neurology and Psychiatry (NCNP), Tokyo, Japan

Kristen L. Szabla Department of Neuroscience, The University of Texas Southwestern Medical Center, Dallas, TX, USA

Steven T. Szabo Department of Psychiatry and Behavioral Sciences, Division of Translational Psychiatry, Duke University Medical Center, Durham, NC, USA

Gábor Szuhay Child Study Center, Yale University School of Medicine, New Haven, CT, USA

Franco Taroni Fondazione IRCCS Istituto Neurologico "Carlo Besta," Milan, Italy

Rabi Tawil Department of Neurology, University of Rochester Medical Center, Rochester, NY, USA

Mireia Tondo Rare Brain Disorders Program, Department of Neurology and Neurotherapeutics, The University of Texas Southwestern Medical Center, Dallas, TX, USA

Shoji Tsuji Department of Neurology, Neuroscience Program, Graduate School of Medicine, Medical Genome Center, The University of Tokyo Hospital, The University of Tokyo, Tokyo, Japan

Bjarne Udd Department of Neurology, Neuromuscular Research Center, University and University Hospital of Tampere, Tampere, Finland, Folkhalsan Institute of Genetics, University of Helsinki, Helsinki, Finland; Vasa Central Hospital, Department of Neurology, Vasa, Finland

Wendy R. Uhlmann Departments of Internal Medicine and Human Genetics, University of Michigan Medical School, Ann Arbor, MI, USA

Flora M. Vaccarino Child Study Center, Program in Neurodevelopment and Regeneration, Yale Kavli Institute for Neuroscience, Yale University School of Medicine, New Haven, CT, USA

Prashanthi Vemuri Aging and Dementia Imaging Laboratory, Department of Radiology, Mayo Clinic, Rochester, MN, USA

Charles P. Venditti Organic Acid Research Section, National Human Genome Research Institute, National Institutes of Health, Bethesda, MD, USA

David W. Volk Department of Psychiatry, University of Pittsburgh, Western Psychiatric Institute and Clinic, Pittsburgh, PA, USA

Dong Wang Department of Neurology, Southern Regional Medical Center, Riverdale, GA, USA

David Watkins Department of Human Genetics, McGill University, Montreal, QC, Canada

David A. Wenger Department of Neurology, Director, Lysosomal Diseases Testing Laboratory, Jefferson Medical College, Philadelphia, PA, USA

Charles A. Williams Division of Genetics and Metabolism, Department of Pediatrics, University of Florida, Gainesville, FL, USA

Kathleen S. Wilson Department of Pathology and the McDermott Center for Human Growth and Development, Cytogenomic Microarray Analysis Laboratory, The University of Texas Southwestern Medical Center, Dallas, TX, USA

Golder N. Wilson Department of Pediatrics, Texas Tech University Health Science Centers, Amarillo and Lubbock (Pediatrics), KinderGenome Genetic Practice, Medical City Hospital, Dallas, TX, USA

Barry Wolf Department of Medical Genetics, Henry Ford Hospital, Center for Molecular Medicine and Genetics, Wayne State University School of Medicine, Detroit, MI, USA

R. Max Wynn Departments of Biochemistry and Internal Medicine, The University of Texas Southwestern Medical Center, Dallas, TX, USA

Shihui Yu Department of Pathology and Laboratory Medicine, Seattle Children's Hospital and University of Washington School of Medicine, Seattle, WA, USA

SECTION I

GENERAL CONCEPTS AND TOOLS

SECTION I

GENERAL CONCEPTS AND TOOLS

CHAPTER

1

Mendelian, Non-Mendelian, Multigenic Inheritance, and Epigenetics

Tamar Harel[*], *Davut Pehlivan*[*], *C. Thomas Caskey*[*], *and James R. Lupski*[*,†]

[*]Baylor College of Medicine, Houston, TX, USA
[†]Texas Children's Hospital, Houston, TX, USA

INTRODUCTION

Genetic influence on neurologic disease expression can include the contribution of a highly penetrant Mendelian variant (HPMV) and be the most prominent and perhaps singular factor required to manifest a disease phenotype, or it can be a genetic modifier and one relatively minor component of many different disease-associated factors. Perhaps the best example of the former is monogenic Mendelian disorders with complete penetrance wherein a mutation in a single disease-causing gene usually results in a relatively uniform disease phenotype. The latter pattern can be observed in many common diseases in which genetic factors contribute a portion of the risk and may play a role in either increasing or decreasing disease susceptibility. Between these extremes of genetic pathophysiology, however, there is a continuum of genetic influence on disease pathophysiology.

Mendelian traits represent the most basic and simple pattern of inheritance. Mutations in a gene encoded on an autosome or sex chromosome result in specific inheritance patterns. Non-Mendelian traits reveal some complexity in their mode of inheritance, in which the classic pattern of inheritance may not always apply, and epigenetic factors are often associated with disease mechanisms. Furthermore, in some diseases one gene is not sufficient to cause the clinical phenotype, but when two or more genes are involved, a particular disease becomes apparent. This latter mechanism is usually referred to as multigenic inheritance, and termed oligogenic inheritance when only a small number of genes are involved and digenic inheritance when variation or mutation in two genes is a prerequisite to disease trait manifestation. Finally, complex traits can involve multiple genes as susceptibility or protective factors but also require internal factors, including other health conditions, as well as external factors such as environment, lifestyle, diet, accident, infection, and drug exposure.

Regardless of the mode of inheritance, defining specific genetic factors that are associated with certain diseases and their functional role in phenotypic manifestations is important for patient management and genetic counseling, as well as for understanding disease mechanisms at the molecular level and ultimately developing new therapeutic approaches. Molecular diagnostics contribute to patient management by establishing an accurate diagnosis, by enabling presymptomatic or prenatal diagnosis, by providing prognostic information, and by further refining or subclassifying more general diagnostic labels. It is estimated that approximately one-third of all human genes are expressed in the nervous system; thus, neurogenetic phenotypes are common.[1] This chapter provides an update to the corresponding chapter by Shiga et al.[2] in the previous edition of this book; here we review the modes of inheritance that can be observed in various human neurologic and psychiatric diseases, and how genetics and more recently genomics is increasing our molecular understanding of neurological disease.

Molecular and Genetic Basis of Neurological and Psychiatric Disease
http://dx.doi.org/10.1016/B978-0-12-410529-4.00001-2

MENDELIAN TRAITS

Mendel's Laws

The basic rules of inheritance were delineated from first principles by Gregor Mendel based upon his observation of the segregation of traits in the common garden pea, *Pisum sativum*.[3] Mendel's first law, the principle of independent segregation, referred to the ability of genes, which he called factors, to segregate independently during the formation of gametes or sex cells. Mendel's second law, the principle of independent assortment, was derived from his observations using peas that differed by more than one characteristic or trait. Mendel postulated that only one factor from each pair was independently transmitted to the gamete during sex-cell formation and that any one gamete contains only one type of inherited factor from each pair. There is no tendency for genes arising from one parent to stay together. Of course, we now know that this latter principle is true only for unlinked genes. Genes or loci that are linked, or physically located in close proximity on the same chromosome, do not assort independently. The closer these loci are, the more frequently they will cosegregate. Linkage analysis is a quantitative measurement of this cosegregation (expressed as a LOD score or $\log_{10}$ of the odds ratio for cosegregation vs. independent assortment)[4] and has been a powerful tool in human genetics to map genes for disease traits to particular regions in the human genome.

Chromosomes and Genes

The chromosomal theory of heredity expounded by Walter Sutton emphasized that the diploid chromosome group consists of a morphologically similar set, a homolog pair, for each chromosome and that during meiosis every gamete receives only one chromosome of each homolog pair. This observation was used to explain Mendel's results by assuming that genes, or factors, were part of the chromosome. Genes are arranged in a linear order on the chromosome, each having a specific position or locus. There are two copies for each gene at a given locus, one on each chromosome homolog. These two copies, or alleles, may be identical, or homozygous, at the specific autosomal locus. Alternatively, the two gene copies at a particular locus may be different and represent heterozygous alleles. When only one copy is physically present, either because of deletion of a specific genomic region on the other homolog or because of the special circumstances of the X chromosome in XY males, this condition is referred to as hemizygous. The genes are passed to the next generation through parental gametes, which contain only one of the two alternative gene copies. A particular gamete may contain alleles from different chromosome homologs because of chromosome crossover and recombination of alleles that occur during meiosis.

Mendelian Inheritance

Mendelian inheritance refers to an inheritance pattern that follows the laws of segregation and independent assortment in which a gene inherited from either parent segregates into gametes at an equal frequency. Three major patterns of Mendelian inheritance for disease traits are described: autosomal dominant, autosomal recessive, and X-linked (Figure 1.1). Mendelian inheritance patterns refer to observable traits, not to genes. Some alleles at a specific locus may encode a trait that segregates in a dominant manner, whereas another allele may encode the same or a similar trait, but instead it segregates in a recessive manner.

Autosomal dominant alleles exert their effect despite the presence of a corresponding normal allele on the homologous chromosome. A vertical transmission pattern is observed in the pedigree, with the trait manifested in approximately half of the individuals in each generation (Figure 1.1A). An affected individual will have a 50% chance of transmitting the disease to each independent offspring, which is a reflection of whether a mutant or a normal allele is segregated in the gamete involved in fertilization. Usually, unaffected members of the family do not carry the mutant allele; thus they cannot transmit a disease allele to the next generation. If an affected male transmits the disease to his son, this is considered proof of autosomal dominant inheritance. Male-to-male transmission is inconsistent with X-linked inheritance because a father contributes the Y chromosome but no X chromosome to all his sons.

In autosomal recessive inheritance, both alleles must be abnormal for the disease trait to be expressed. The unaffected parents of an affected child are obligate heterozygote carriers for the recessive mutant allele. Affected children may be homozygous for a specific recessive mutant allele, as is more commonly observed with consanguineous matings, or they may be compound heterozygotes for two different mutations. Couples who are heterozygous carriers of a recessive mutant allele have a 25% risk of having an affected child with each pregnancy. The pattern of transmission observed in the pedigree is horizontal, with multiple members of one generation affected (Figure 1.1B). The unaffected siblings have a 67% (two-thirds) chance of being a carrier for the mutant allele.

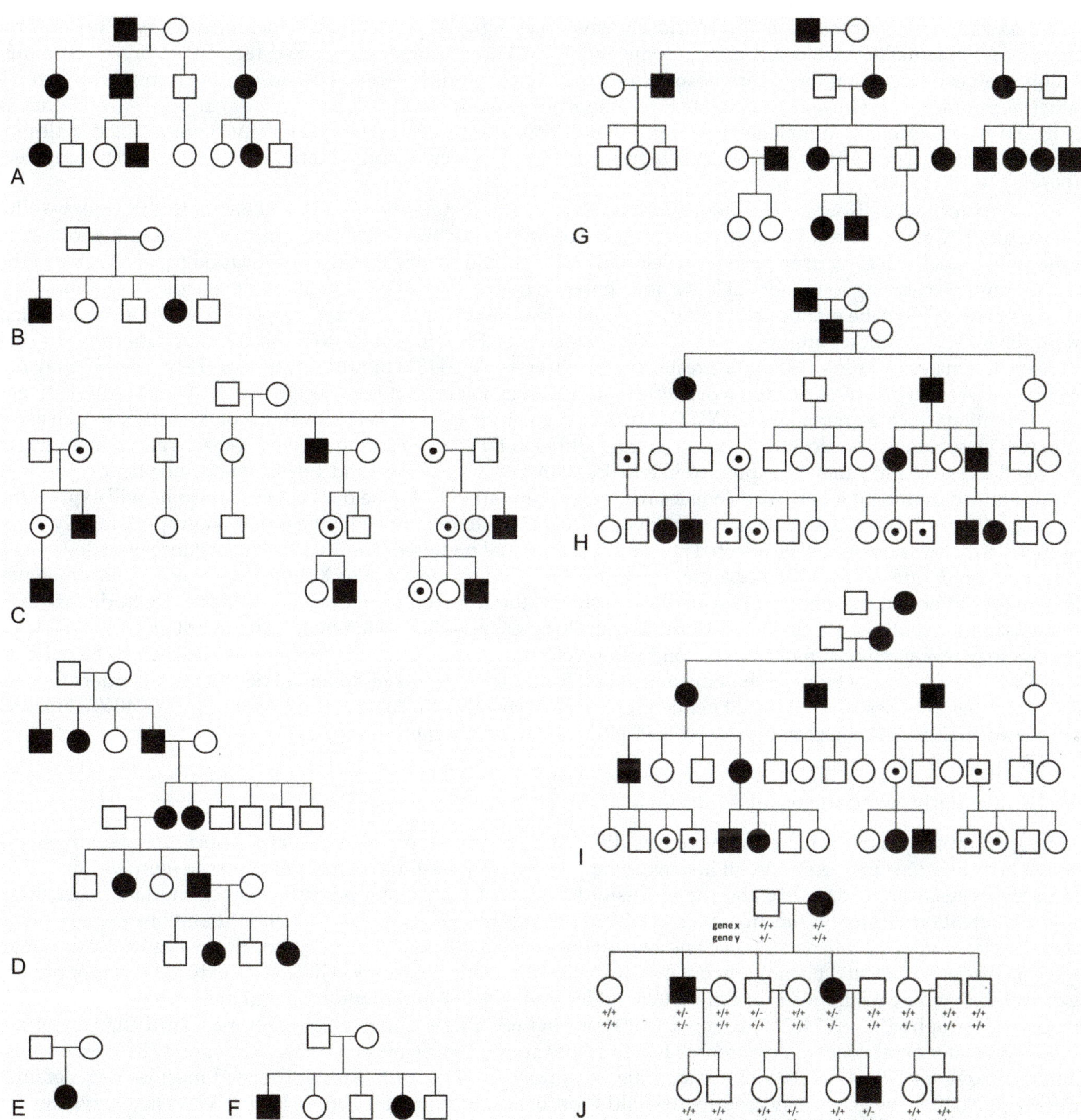

FIGURE 1.1 **Pedigrees representing different patterns of inheritance that can be observed for disease traits in families.** Unaffected males (open squares) and females (open circles) are shown. Filled symbols represent individuals manifesting the trait. Inheritance patterns include **(A)** autosomal dominant; **(B)** autosomal recessive; **(C)** X-linked recessive showing individuals who carry the mutant gene but do not manifest the disease trait (filled-in symbols with large dot in center); **(D)** X-linked dominant; **(E)** sporadic or new mutation or autosomal recessive; **(F)** new mutation with gonadal mosaicism or autosomal recessive; **(G)** maternal or mitochondrial inheritance; **(H)** dominant maternally imprinted showing individuals who carry the mutant imprinted gene on the non-expressing allele and do not manifest the disease trait (filled-in symbols with small dot in center); **(I)** dominant paternally imprinted; and **(J)** digenic inheritance showing wild-type (plus signs) and mutant (minus signs) alleles at disease loci X and Y.

X-linked inheritance patterns reflect special circumstances regarding sex chromosomes. Females have two X chromosomes, while males have one X chromosome and one Y chromosome. In X-linked recessive inheritance, a mutation in a gene located on the X chromosome may not express itself in females because of the normal copy on the other X chromosome. However, all males who inherit the mutant allele will be affected. An important feature of X-linked inheritance is that male-to-male transmission never occurs, but all female offspring of affected males inherit the abnormal gene. Therefore, affected fathers will have genetically normal sons and obligate carrier daughters (Figure 1.1C).

X-linked recessive disorders may sometimes be observed in females because of a skewing in the process of lyonization or X-inactivation.[5] Usually, the expression of one of the two X chromosomes in females is suppressed randomly in each cell early in embryonic development. If nonrandom or skewed X-inactivation occurs, such that the X chromosome carrying a mutant allele for an X-linked recessive trait is predominantly the active X chromosome, then a phenotype will be manifested in a carrier female. In normal females, a "bell-shaped" or gaussian distribution represents X-inactivation patterns with a 50:50 average ratio. Therefore, a skewed inactivation pattern is not rare in normal females. A ratio of 80:20 or greater can be observed in 5–10% of normal females. This skewing may not result in a phenotype unless one has a deleterious mutation in a gene that is subject to X-inactivation. Alternatively, rare conditions such as mutations in *XIST* (a master regulatory gene of X-inactivation) or X-autosome translocation (which separates translocated genes from the regulation by *XIST*) result in skewed X-inactivation. In addition, X-linked recessive traits may be expressed in females with a 45,X karyotype and Turner syndrome phenotype.

In X-linked dominant inheritance, females who carry a mutation in a gene on the X chromosome will express the disease phenotype but usually will have a milder clinical course than males with the mutation. Approximately twice as many females as males will be affected in a multigenerational pedigree. There will be no instance of male-to-male transmission (Figure 1.1D). Whether a trait is considered X-linked recessive or X-linked dominant may sometimes be a matter of how the phenotype is scored. For instance, the X-linked form of Charcot–Marie–Tooth disease may manifest either subtle or no clinical features on neurologic examination of a female patient, but electrophysiologic studies may reveal reduced motor nerve conduction velocities. Some X-linked dominant disorders may be lethal in males and therefore may be observed only in females. Examples of X-linked dominant neurologic disorders that are usually lethal in males include incontinentia pigmenti (MIM# 308300), Aicardi syndrome (MIM# 304050), bilateral periventricular nodular heterotopia (MIM# 300049), and Rett syndrome (MIM# 312750).

Molecular Pathomechanisms of Mutations

Specific mutations in each gene may behave differently in disease pathogenesis and associated phenotypic expression.[6] Mutations may result in an inactive gene product and are thus referred to as *loss-of-function* mutations. Null alleles result from complete absence or loss-of-function of the protein product, whereas mutations resulting in some retention of protein function are considered *hypomorphic alleles*. Loss-of-function mutations explain many recessive traits in which phenotypes become evident only when both alleles are mutated. Heterozygous carriers may have no disease-associated phenotypes but may present with subtle biochemical defects or reduced levels of protein expression that may result in susceptibility to a milder trait. *ABCA4* mutations are identified in recessive Stargardt macular dystrophy (MIM# 248200) patients. Mutations in both alleles can give Stargardt macular dystrophy or the milder and later onset cone–rod dystrophy; if both alleles are null, early onset retinitis pigmentosa (RP) can be seen, and heterozygous *ABCA4* mutations are identified in a fraction of patients with age-related macular degeneration (MIM# 153800). Some genes may require two wild-type alleles for normal function. In such case, heterozygotes for loss-of-function mutations may reveal phenotypes that can transmit as dominant traits because of the insufficient amount of gene products (*haploinsufficiency*).

Two other mutational mechanisms can explain a dominant inheritance pattern. When a translated mutant protein interferes with the function of a normal (wild-type) protein that is produced from the normal allele, a *dominant-negative* effect occurs (*antimorphic mutations*). Dominant-negative alleles occur when the encoded proteins compose a subunit structure (homodimer, heterodimer, or other multimeric complex formation) or when they interact with other proteins (ligand–receptor) or DNA (transcription factors). In contrast, *gain-of-function* is the mechanism in which mutant proteins abnormally enhance the normal function, or acquire novel functions that are toxic to cells without interfering with the wild-type allele function (also referred to as *neomorphic alleles*). Many neurodegenerative disorders with dominant inheritance are likely associated with gain-of-function mutations (e.g., polyglutamine diseases, prion disorders) that may prevent proteins from proper cellular processing, such as folding, transport, or degradation.

The identical or similar trait can often be caused by mutations in different genes at different loci; this is known as *genetic* or *locus heterogeneity*. The molecular basis of genetic heterogeneity is diverse but may be explained by abnormalities in different genes that function in the same biologic process. This may involve genes that encode various enzymes in the same metabolic pathway (e.g., G_{M2} gangliosidoses), abnormalities in genes that code for discrete subunits of a functional protein complex (e.g., leukodystrophy with vanishing white matter, MIM# 603896), or a macromolecular structure such as myelin (Charcot–Marie–Tooth disease type 1). Different types of mutations in a single gene may result in the same clinical disease phenotypes. This phenomenon is referred to as *allelic heterogeneity*. Generally, disorders within a single family represent neither genetic nor allelic heterogeneity. It has been recognized more recently that different mutations in the same gene may give distinct clinical phenotypes—*allelic affinity* (e.g., Duchenne/Becker muscular dystrophies; PCWH [peripheral demyelinating neuropathy, central dysmyelinating leukodystrophy, Waardenburg syndrome and Hirschsprung disease]/Waardenburg–Shah syndrome;[7] and Yunis–Varon syndrome/Charcot–Marie–Tooth disease type 4J).[8]

The molecular basis for at least one example of allelic affinity has been clarified by functional determination of the effects of mutations using exogenous mutant gene transfer and endogenous gene silencing *in vitro* and *in vivo*. Mutations in the coding exons can be missense, nonsense, small insertions/deletions, or nonsynonymous alterations, whereas other mutations can be found in noncoding regions, such as intronic splicing junctions, regulatory elements upstream or downstream of exons. Variations in clinical phenotypes conveyed by these mutants have been considered to be a direct result of the function of mutant proteins. However, the *in vivo* consequences of mutations can be complex, and factors other than protein structure/function effects, such as mRNA instability or posttranslational modifications, may be important to ultimate clinical outcome. For example, mutations causing premature termination codons (PTCs) by nonsense, frameshift, or splice-junction mutations can be processed differently during translation depending on the location of PTCs. The nonsense-mediated decay (NMD) surveillance pathway typically degrades transcripts containing PTCs in 5′ exons to prevent translation of aberrant transcripts, resulting in a loss-of-function allele. Evidence suggests a model for NMD in which PTCs in the last exon or distal to 50–55 bp of the penultimate exon can escape NMD, subsequently being translated into truncated mutant proteins, which can act as a dominant-negative, gain-of-function, or hypomorphic allele.[9] Thus, the distribution of PTC mutations can often be correlated with the mode of inheritance.[10] Furthermore, either triggering or escaping NMD can result in distinct neurological disease phenotypes, as shown in *SOX10* and *MPZ* mutations; PTCs in *SOX10* that trigger NMD cause a milder disease, WS4 (Waardenburg–Shah syndrome type IV), whereas those escaping NMD result in a severe neurocristopathy, referred to as PCWH. Likewise, PTCs in *MPZ* that trigger NMD result in an adult-onset neuropathy (CMT1B), whereas those escaping NMD cause childhood-onset or congenital neuropathies.[7]

Factors That Modify Classic Mendelian Inheritance Patterns

New Mutations, Mosaicism, and Somatic Mutations

In some dominant diseases, new mutations or *de novo* mutations may occur frequently (e.g., tuberous sclerosis [MIM# 191100], neurofibromatosis type 1 [MIM# 162200], Alexander disease [MIM# 203450], Charcot–Marie–Tooth disease type 1A [MIM# 118220]). Such diseases may present sporadically in families (Figure 1.1E) and may not be recognized as involving hereditary factors if the phenotypes resemble nongenetic diseases. New point mutations result from DNA replication or repair errors and frequently occur in germ cells of men at an advanced age (e.g., paternal age effect observed in achondroplasia [MIM# 100800]). The *de novo* mutation (DNM) rate for single nucleotide variations in the paternal germline is about four times greater than that in the maternal germline, and increases linearly by about two DNMs per year in line with spermatogonial stem cell turnover after puberty.[11] New mutations appear to play a prominent role in intellectual disability[12] and other neurodevelopmental and neuropsychiatric traits such as autism spectrum disorders. This is not unexpected because individuals with these disorders are less likely to bear offspring, placing the disease-causing mutations under strong negative selection. Because affected people rarely transmit the mutation to children, the presence of disease reflects the ongoing appearance of new mutations.[13]

Mosaicism refers to the mixture of two or more different cell populations carrying either heterozygous mutant or homozygous normal alleles in the somatic cells generated by *de novo* mutations during postzygotic mitosis, and is perhaps much more common within multicellular organisms than our limited genomic assays have detected thus far.[14] Our ability to detect a pathogenic somatic mutation by using current clinical methods depends on how abundant it is in the leukocytes. However, in some cases of autism, epilepsy, and perhaps other neuropsychiatric conditions, somatic mutations affecting a specific lineage of neurons may be overlooked by conventional genetic testing of leukocyte-derived DNA.[13]

Somatic mutations are clinically significant when they occur early in organogenesis and the organs comprise a reasonable ratio of mutant cells. Some mutations that are not compatible with embryonic development might be found only as somatic mosaic and not as inherited mutations. Examples include somatic activating mutations in the

GNAQ gene, encoding guanine nucleotide-binding protein q polypeptide, that are associated with Sturge–Weber syndrome; and somatic activating mutations in the gene *AKT3*, encoding PKBγ in the mTOR pathway, which lead to hemimegalencephaly. The latter condition is characterized by enlargement and extensive malformation of an entire cerebral hemisphere with functional preservation of the other hemisphere.[13,15]

In other cases, somatic mosaicism may not result in a phenotype, but its presence in germline cells (gonadal mosaicism) may cause recurrence of affected offspring despite the parents having no detectable mutation in DNA isolated from blood (Figure 1.1F). In the case of Duchenne muscular dystrophy (MIM# 310200), somatic mosaicism, in this case germline mosaicism, can account for a recurrence rate up to approximately 15%. A single somatic mutation may result in a tumor-associated phenotype when it arises in the context of an existing cellular recessive mutation on the other allele ("two-hit" model).[16] *NF1* and *RB1* mutations in neurofibromatosis type 1 (MIM# 162200) and retinoblastoma (MIM# 180200), respectively, are examples of cellular recessive mutations in dominantly inherited disorders that require somatic mutation events to manifest the phenotype.

Penetrance and Expressivity

We often observe differences in the severity of clinical manifestation within a pedigree (intrafamilial variability) or between different pedigrees (interfamilial variability). *Penetrance* describes the proportion of individuals with the disease allele that have any manifestations of the disorder. If one observes individuals who have genotypes that usually result in disease but are present with no sign of disorder over a lifetime, the condition is referred to as nonpenetrant. A family with nonpenetrant individuals may represent reduced penetrance.

The use of these terms is sometimes confusing. *Expressivity* illustrates the range of phenotypic expression in individuals who carry an identical mutation. Variation in age of onset may be included in the expressivity, but it may also be considered as age-dependent penetrance.[2] In the strictest sense, penetrance is qualitative (step function), whereas expressivity is quantitative and reflects variability in degree (continuous function), but when referring to a population of patients/subjects the term variable penetrance has been used.

REPEAT EXPANSION DISORDERS

Repeat expansion disorders are those in which unstable expansion of tandem nucleotide repeats, ranging from tri-, tetra-, penta- to dodecanucleotide repeats, result in distinct diseases. Such a mechanism is currently responsible for more than 40 neurological or neuromuscular phenotypes.[17] Pedigrees segregating repeat expansion disorders reveal a Mendelian inheritance pattern; however, unlike static mutations in Mendelian diseases, the repeat mutation process is dynamic. Repeats continue to expand in subsequent generations and even within tissues of the same individual. Table 1.1 shows several examples of neurological disease that are caused by trinucleotide repeat expansions.

First, repeat expansions show different genetic properties depending on the locations of the repeats. Nucleotide repeats can be found in untranslated regions (UTRs) of genes (3′ distal UTR, 5′ proximal UTR, introns, and antisense

TABLE 1.1 Neurological Diseases Associated with Trinucleotide Repeat Expansions

Disease	MIM #	Gene	Trinucleotide repeat expansion	Phenotype
Myotonic dystrophy	160900	*DMPK*	CTG	Myotonia, muscular dystrophy, cataracts, hypogonadism
Huntington disease	143100	*HTT*	CAG	Chorea, dystonia, incoordination, cognitive decline, behavioral difficulties
Spinocerebellar ataxia 1	164400	*ATXN1*	CAG	Cerebellar ataxia, opthalmoplegia, peripheral neuropathy, dementia
Dentatorubral-pallidoluysian atrophy	125370	*ATN1*	CAG	Myoclonic epilepsy, dementia, ataxia, choreoathetosis
Oculopharyngeal muscular dystrophy	602279	*PABPN1*	GCG	Dysphagia, progressive ptosis of eyelids
Friedreich ataxia	229300	*FXN*	GAA	Ataxia, limb muscle weakness, dysarthria
Fragile X syndrome	300624	*FMR1*	CGG	Mental retardation, macro-orchidism, long face, large ears, prominent jaw

sequence) or in coding sequences, such as CAG polyglutamine repeats observed in Huntington disease and in spinocerebellar ataxias[18] and polyalanine repeats in congenital malformation syndromes.[19] Diseases conveying repeats within untranslated regions usually confer greater repeat instability and multisystemic involvement.

Second, affected individuals in successive generations often present with a more severe disease phenotype and an earlier age of onset compared to their affected predecessors (*anticipation*). The molecular basis for anticipation is an increasing number of repeats in subsequent generations, whereby nucleotide repeats tend to increase in number through the transmission from a parent to offspring. Phenotypic manifestations may dramatically change between generations with anticipation, as observed in myotonic dystrophy type 1 (MIM# 160900) and Huntington disease (MIM# 143100).

Third, in some clinically unaffected individuals, the number of repeats appears to exceed beyond the normal range of the general population in a state called *premutation*. Premutations show both somatic and germline instability and often can be found in the phenotypically normal antecedent of patients in a pedigree. Premutations often develop into full mutations that contain longer repeat expansions, resulting in disease phenotypes in progeny. One example is fragile X syndrome (MIM# 309550), in which a CGG repeat in the 5′ untranslated region of the *FMR1* gene ranges in size from approximately 5 to 40 repeat units in the normal population. The premutation range for *FMR1* is between 55 and 200 repeats, whereas full mutations exceed 230 repeats. A subset of individuals with premutations displays a unique neurologic phenotype showing late-onset progressive intention tremor, ataxia, and cognitive decline.[20]

The fourth facet of nucleotide repeat disorders is parent-of-origin effects on anticipation wherein the repeat instability is influenced by the transmitting parent. Overall, paternal intergenerational instability is common, whereas fragile X syndrome, Friedreich ataxia, and myotonic dystrophy type 1 usually reveal greater repeat instability with maternal transmission.

Finally, in some repeat expansion disorders, repeat alleles often associate with a distinct haplotype on the disease chromosome more frequently than expected by chance, indicating linkage disequilibrium (LD). Myotonic dystrophy type 1 shows complete LD both in white and Japanese individuals, suggesting a common Eurasian founder mutation. Huntington disease reveals multiple founder alleles that result in strong LD in different populations. Many sporadic cases of repeat expansion disease are not new mutations; rather, they are likely new full mutations developed from inherited founder chromosome that conveyed premutations. This feature is in sharp contrast to genomic disorders wherein numerous sporadic cases represent *de novo* mutations and no evidence for founder mutations exist.

NON-MENDELIAN INHERITANCE

Non-Mendelian inheritance refers to an inheritance pattern that does not follow the law of segregation in which a gene inherited from either parent segregates into germline cells at an equal probability. Non-Mendelian inheritance includes *mitochondrial inheritance*, wherein maternal transmission of mitochondrial DNA is the rule; *imprinting*, in which only one parental allele is transcribed due to parental-origin-dependent methylation of CpG dinucleotide sites on DNA; *uniparental disomy*, in which an individual receives both copies of a homologous chromosome pair or of a specific chromosomal region from one parent; and *digenic* and *oligogenic* traits.

Mitochondrial Inheritance

In addition to the nuclear genome, mitochondria contain DNA that transmits genetic information to subsequent generations. Because of the cytoplasmic localization and high copy number of mitochondria, mitochondrial DNA (mtDNA) or the mitochondrial genome has a unique inheritance pattern. Mitochondrial DNA is a circular genome of approximately 16.6 kilobase pairs (kb) located within the mitochondrial matrix in the cytoplasm of the cell. It encodes 13 polypeptides of subunits of the mitochondrial respiratory chain and oxidative phosphorylation system, two rRNAs, and 22 tRNAs. Each human cell contains hundreds of mitochondria, and each mitochondrion contains 5–10 mitochondrial genomes.

Traits that result from mtDNA mutations show a specific segregation pattern in a pedigree referred to as *maternal inheritance* (Figure 1.1G). This is due to the fact that the ovum supplies the total complement of mtDNA, while there are effectively no mitochondria that can be transmitted from sperm. Disorders resulting from mtDNA mutations can have tremendous variability in clinical expression because of *heteroplasmy*, wherein different tissues may have a different percentage of a mutant mitochondrial genome. During mitotic growth, the large number of mtDNA can result in asymmetric distribution of mutant versus normal copies (replicative segregation), leading to a change of the heteroplasmy ratio within a cell or organ over time. Moreover, different tissues have varying requirements for the energy generated by oxidative phosphorylation, which contributes further to the clinical heterogeneity of mitochondrial disorders. These variant features present a significant challenge for both clinical diagnosis and genetic counseling in such families. Nevertheless, some mitochondrial diseases can be due to mutations in nuclear encoded genes wherein the proteins are either important to mitochondrial function or to nuclear–mitochondrial communication. mtDNA depletion syndromes (MDS) are a group of autosomal recessive disorders characterized by a severe reduction in mtDNA content leading to impaired energy production in affected tissues and organs. They are due to defects in mtDNA maintenance caused by mutations in nuclear genes that function in either mitochondrial nucleotide synthesis or mtDNA replication. Phenotypic presentation may be myopathic, encephalomyopathic, hepatocerebral, or neurogastrointestinal. An example of the latter is mitochondrial neurogastrointestinal encephalopathy (MNGIE; MIM #603041), which presents with progressive gastrointestinal dysmotility and peripheral neuropathy.[21]

Imprinting

Imprinting is an epigenetic marking placed on certain genes or genomic regions as a result of passage through male or female gametogenesis. Imprinting clusters contain 3–12 genes spread over 20–3700 kb of DNA. Each cluster has a discrete *imprinting control region* (ICR) that exhibits parent-of-origin-specific epigenetic modifications including allele-specific DNA methylation at the cytosine residue in CpG dinucleotides and posttranslational histone modifications. Functionally, DNA methylation inhibits the transcriptional machinery from accessing DNA, leading to decreased transcription of genes with high levels of promoter methylation. Deletion of ICRs results in loss of imprinting of multiple genes within the cluster. Recently, long noncoding RNAs (lncRNAs) within the imprinted cluster have emerged as contributing to transcriptional gene silencing in *cis*, by a variety of proposed mechanisms including transcriptional interference of adjacent imprinted genes by a sense–antisense physical overlap or direct recruitment of repressive chromatin proteins to the imprinted cluster.[22,23]

An inheritance pattern in which expression of a disease depends on the specific parent who transmitted the mutant allele (i.e., imprinting) is referred to as a *parent-of-origin effect*. Pedigrees in Figures 1.1H and 1.1I show families with defects in a maternally or paternally imprinted gene, respectively. In the family H, the gene is only expressed when inherited from the father, but silenced when inherited from the mother. Thus, the disease phenotype is only apparent when the mutant allele is inherited from a father. Family I represents the reciprocal situation, where the gene is expressed only when inherited from the mother. Of note, a woman who inherited a paternally imprinted allele will switch this allele to a maternal imprint when she transmits it to her offspring. Likewise, a man who inherits a maternally imprinted allele will switch to the paternal imprint when he transmits it to his offspring. This reflects the fact that although imprinting is strictly maintained through numerous somatic cell divisions, it is reset during gametogenesis, when the imprint from the previous generation is erased and a new gender-appropriate imprint is established. Of concern, children conceived by assisted reproductive technology (ART) have an increased incidence of rare epigenetic disorders such as Angelman syndrome (MIM# 105830) and Beckwith–Wiedemann syndrome (MIM# 130650), with most of these patients exhibiting loss of DNA methylation at ICRs. This disruption of normal epigenetic programming may be due to embryo culture, embryo transfer, or hormonal treatments and is currently under investigation.[22]

Uniparental Disomy

According to Mendel's first law, only one of two factors in a parent is transmitted to the next generation. The laws of chromosome inheritance also state that only one parental chromosome from the homologous pair is transmitted to the offspring. Uniparental disomy (UPD) is an exception to this segregation rule, and it is defined as inheritance of both copies of homologous chromosomes from one parent.[24,25] Because UPD maintains numerical and structural features of diploid chromosomes, it may not always be associated with a clinical phenotype unless additional genetic components (such as a recessive allele or imprinting) occur on the same chromosome. Thus, UPD may be more frequent than observed because it may not produce a recognizable clinical phenotype. Four distinct mechanisms were originally proposed and each recognized to result in UPD:[25] trisomy rescue, monosomy rescue, gamete complementation, and postfertilization errors.

When UPD results from two copies of a single chromosome homolog that was present in one parent, this condition is referred to as *uniparental isodisomy*. In contrast, *heterodisomy* is the inheritance of the two different chromosomes for one homologous pair derived from one parent. Consequently, in the isodisomic condition, a homozygous allele for a recessive mutation can be transmitted from a heterozygous carrier parent, resulting in a non-Mendelian inheritance pattern for a recessive mutation. At least 40 diseases have been reported in which UPD results in reduction to homozygosity for a recessive mutation.[26] These include UPD 5 and UPD X isodisomy causing spinal muscular atrophy (MIM# 253300) and Duchenne muscular dystrophy, respectively.[27]

Imprinting, UPD, and Genetic Disorders

Prader–Willi syndrome (PWS; MIM# 176270) and Angelman syndrome (AS; MIM# 105830) are phenotypically distinct neurobehavioral syndromes that result from mutations in chromosome 15q11.2-q13, a region in which a number of imprinted genes are localized. PWS is characterized by hypotonia and failure to thrive in infancy, small hands and feet, hypogonadism, mild mental retardation, and obesity. AS is characterized by developmental delay, including absence of speech, severe mental retardation, ataxia, hyperactivity, seizures, aggressive behavior; and excessive, inappropriate laughter. PWS results from loss of the paternal contribution of genes in the 15q11.2-q13 region, whereas AS is caused by failure of a maternal contribution of the same region. Approximately 60–70% of cases with either disease result from a common 4–4.5-Mb chromosomal deletion that comprises a 2-Mb imprinted domain. A deletion of the paternally transmitted chromosomal segment results in PWS, whereas a deletion of the maternally derived segment causes AS. In approximately 10% of patients with AS, the disease results from maternally derived intragenic mutations in the *UBE3A* gene, which encodes an E3 ubiquitin ligase. Recent evidence suggests that paternally derived deficiency of the small nucleolar RNA (snoRNA) *SNORD116* cluster (HBII-85) causes the key characteristics of the PWS phenotype, although other genes in the region may make more subtle phenotypic contributions.[28,29] In addition, in approximately 5% of PWS and AS patients, the disease results from imprinting defects in which uniparental DNA methylation and gene expression occur despite biparental chromosome inheritance; submicroscopic deletions of the *imprinting center* were identified in some of these cases.

The remainder of PWS and AS cases (20–30% of PWS and 5% of AS cases) are caused by UPD of the 15q11.2-q13 region. UPD of maternal chromosome 15 results in a lack of its paternal contribution, causing PWS. Conversely, UPD of paternal chromosome 15 results in a lack of a maternal contribution, causing AS. Rare balanced chromosomal translocations involving paternally derived chromosome 15 also cause PWS. UPD of imprinted loci also results in other genetic disorders, including transient neonatal diabetes (pUPD6), Russell–Silver syndrome (MIM# 180860; mUPD7), and Beckwith–Wiedemann syndrome (MIM# 130650; pUPD11). UPD14 presents with a different phenotype depending on the parental origin: paternal UPD14 yields short stature and precocious puberty, whereas maternal UPD14 results in dwarfism, skeletal dysplasia, and thoracic narrowing.[27]

CHROMOSOMAL AND GENOMIC DISORDERS

In contrast to Mendelian diseases in which alteration or mutation at a single locus, usually within a single gene, is responsible for the disease phenotype, chromosomal and genomic disorders may involve loss or gain of an entire chromosome or a portion of a chromosome that usually contains multiple genes. A human somatic cell comprises 23 pairs of different chromosomes: 22 autosomes numbered 1 to 22 (from longest to shortest), and a divergent pair of sex chromosomes, X and Y. Abnormalities in segregation of each member of the pair in either meiosis or mitosis may result in chromosomal disorders that display distinct modes of inheritance. Constitutional chromosome rearrangements are usually observed in 100% of the cells when they are inherited from parents or occur in gametes. Alternatively, mosaicism of two or more different karyotypes may be observed in an individual when the chromosomes are rearranged in somatic cells during mitosis in early embryogenesis.

Chromosomal and genomic disorders can be classified into four major groups: aneuploidy; translocations; rearrangements including deletion, duplication, and inversion; and isochromosome formation. In the following sections, we first review these categories of chromosomal abnormalities, and then describe molecular mechanisms underlying submicroscopic genomic rearrangements.

Aneuploidy

Chromosome aneuploidy, defined as an abnormal number of chromosomes, is the most common identified chromosome abnormality, occurring in at least 10% of recognized pregnancies. Most aneuploid conceptuses spontaneously abort, which makes this the leading genetic cause of pregnancy loss. Trisomy refers to three copies of a single chromosome or the one extra chromosome. The most common trisomy in live births is for chromosome 21, which is associated with a Down syndrome phenotype (MIM# 190685; karyotype 47,XX+21 in females or 47,XY+21 in males). Trisomy 21 occurs in approximately 1:660 live births, making this a much more frequently observed mutational mechanism than alteration at a single locus.

Monosomy refers to the absence of one chromosome from the pair, commonly observed in the 45,X karyotype associated with Turner syndrome. Since most of 45,X conceptuses are spontaneously aborted, the incidence of this aneuploidy is only 1:5000. Other sex chromosome aneuploidies, such as 47,XXY and 47,XXX, are relatively widespread, occurring at a frequency of approximately 1:1000 live births. Marker chromosomes are small additional chromosomes that are structurally abnormal and are also referred to as supranumerary marker chromosomes or SMC,[30] often derived from chromosomes X, 15, and 22.

Most chromosome aneuploidy results from nondisjunction of allelic chromosomes in maternal meiosis. Meiosis consists of one chromosome replication ($2n$ to $4n$) and two contiguous cell divisions ($4n$ to $2n$, then $2n$ to $1n$). The first division separates homologous chromosomes from each other (meiosis I), followed by the second division that separates sister chromatids (meiosis II), resulting in a haploid set of 23 chromosomes. These unique divisions are preceded by a unique meiotic prophase, during which homologous chromosomes synapse at chiasmata (sites of recombination) and undergo recombination. Most aneuploidy formation occurs in meiosis I with complete nondisjunction of both homologous chromosomes, resulting in a 4:0 segregation.

Maternal age has long been recognized to be a significant risk factor for trisomies. Potential etiologies include: events occurring in the fetal ovary that influence the prophase interactions between homologous chromosomes; that the long prophase arrest in females may contribute because of age-dependent decay of components of the meiotic machinery; and that environmental effects may act at several different stages of oogenesis to influence the likelihood of mistakes.[31]

Isochromosomes

An isochromosome is a mirror-image abnormal chromosome consisting of two copies of either a short arm or a long arm, often observed for X and acrocentric (13, 14, 15, 21, and 22) chromosomes. Isochromosome X is the most common (approximately 1:13,000) and accounts for more than 15% of cases of Turner syndrome. Isochromosome 21 can result in a trisomy 21 and Down syndrome. Both dicentric (two centromeres) and monocentric (one centromere) formations have been found. A U-type strand exchange between sister chromatids at the short arm (dicentric) or misdivision of centromere (monocentric) is likely the mechanism for the formation of isochromosomes.

Translocations

Translocation is an exchange of chromosomal segments or whole arms between two (or more, in rare cases of complex chromosomal rearrangement [CCR]) different chromosomes. The most frequent type is robertsonian translocations. These are whole-arm exchanges between the acrocentric chromosomes, which are found in 1:1000 individuals. In contrast, reciprocal translocations, a segmental exchange between two chromosomes, are mostly unique and can involve any of the chromosomes. Translocation carriers who have the proper amount of genetic information (balanced translocation) have no abnormal clinical phenotype unless the translocation breakpoint disrupts a gene or genes that result in a phenotype. Such patients with a balanced translocation and an associated disease phenotype are extremely rare but potentially enable the identification of disease genes (e.g., identification of the *DMD* gene). Translocation may also cause a single-gene disease through a position effect by disturbing the normal regulation of a gene close to the breakpoint.[26] Typical examples are translocations with breakpoints at a 15 to 250-kb distance from the sonic hedgehog (*SHH*) gene that may suppress *SHH* transcription by a *position effect* and result in holoprosencephaly (MIM# 142945). Others include campomelic dysplasia (MIM# 114290; *SOX9*), cleidocranial dysplasia (MIM# 119600; *CBFA1*), and Saethre–Chotzen syndrome (MIM# 101400; *TWIST*). Most of the genes involved in position effects are developmental regulatory genes that convey complex spatiotemporal expression profiles, requiring multiple *cis*-acting elements.

Balanced translocation carriers are frequent in the population, occurring in about 1:250 individuals. These individuals are at a high risk of transmitting a chromosome with missing or excess genetic information and conceiving a child with an unbalanced karyotype.[32,33] This mechanism should be clinically considered when a couple has had three or more spontaneous early abortions or a history of infertility.

Intrachromosomal Rearrangements

Deletion, duplication, triplication, and inversion are intrachromosomal rearrangements of particular segments within a single chromosome, resulting from strand exchange between either two homologous chromosomes or two sister chromatids. Deletions can be terminal (at the end of the chromosome) or interstitial. Large deletions are usually lethal; only microdeletions are clinically relevant. Well-recognized terminal deletion syndromes include monosomy 1p [del(1)(p36.3)], Wolf–Hirschhorn syndrome [MIM# 194190, del(4)(p16)], Cri-du-chat syndrome [MIM# 123450, del(5)(p15)], Jacobsen syndrome [MIM# 147791, del(11)(q23)], and Miller–Dieker lissencephaly syndrome [MIM# 247200, del(17)(p13.3)]. Moreover, as mentioned earlier, microdeletions of telomeric termini appear to play a role in the etiology of idiopathic mental retardation. Terminal deletions have been identified for each chromosome, and the size of each deletion appears to vary in different individuals. Some terminal deletions occur more frequently than others.

In contrast, interstitial deletions and their reciprocal duplications appear to occur recurrently in specific segments of human chromosomes and have been associated with a number of syndromes presenting behavioral and neurologic phenotypes (Table 1.2). Common disorders associated with interstitial deletion include Williams–Beuren syndrome [WBS; MIM# 194050, del(7)(q11.23q11.23)], PWS [pat del(15)(q11.2q13)], AS [mat del(15)(q11.2q13)], Smith–Magenis syndrome [SMS; MIM# 182290, del(17)(p11.2p11.2)], hereditary neuropathy with liability to pressure palsy [HNPP; MIM# 162500, del(17)(p12p12)], and DiGeorge syndrome/velocardiofacial syndrome [DG/VCFS; MIM# 188400, del(22)(q11.2q11.2)]. Interstitial duplications, some of which have been recognized recently as the reciprocal to the deletion syndromes, have been observed in disorders such as Russell–Silver syndrome [dup(7)(p12p13)], autistic syndrome [mat invdup(15)(q11.2q13)], Potocki–Lupski syndrome [PLS; dup(17)(p11.2p11.2)], Charcot–Marie–Tooth disease type 1A [MIM# 118220, dup(17)(p12p12)], microduplication 22q11.2 [MIM# 608363, dup(22)(q11.2q11.2)], Pelizaeus–Merzbacher disease [MIM# 312080, dup(X)(q22.2q22.2)], and a mental retardation (MR)–seizure disorder associated with the duplication of the *MECP2* gene (point mutations and deletions of *MECP2* cause Rett syndrome).[32] Different sized (0.2–2.6 Mb) microduplications involving the *MECP2* gene encoding X-linked methyl-CpG-binding protein 2 in Xq28 appear to be the most common nonrecurrent pathogenic subtelomeric microduplication. Males with *MECP2* duplications manifest developmental delay, infantile hypotonia, absent speech, and a history of recurrent infections, in addition to less consistent signs such as genital or digital abnormalities and seizures. Most of the microduplications are inherited from the mothers, and female carriers of *MECP2* duplications can manifest susceptibility to neurobehavioral and psychiatric symptoms.[34]

In general, chromosome microdeletion of a genomic segment has more profound phenotypic consequences than duplication of the same segment. One exception to this trend is CMT1A and HNPP, where increased gene dosage of the *PMP22* gene by CMT1A duplication results in a progressive demyelinating neuropathy, whereas haploinsufficiency of *PMP22* by HNPP deletion results in an episodic, asymmetric, and usually milder neuropathy. Similarly, a more severe phenotype is observed in Pelizaeus–Merzbacher disease when patients have *PLP1* duplications rather

TABLE 1.2 Neuropsychiatric Phenotypes in Genomic Disorders Caused by Deletions and their Reciprocal Duplications

Syndrome	Deletion/duplication	Dosage sensitive gene	Neuropsychiatric phenotype
Williams–Beuren	del(7)(q11.23q11.23)	CGS including *ELN*	Gregarious and loquacious behavior, impaired visuospatial recognition
dup(7)q11.23	dup(7)(q11.23q11.23)	CGS including *ELN*	MR, speech delay, growth retardation
Smith–Magenis	del(17)(p11.2p11.2)	CGS including *RAI1*	Self-injurious and aggressive behavior, self-hugging, attention seeking, onychotillomania, polyembolokoilamania
Potocki–Lupski	dup(17)(p11.2p11.2)	*RAI1*	Hypotonia, autistic features, poor balance, ADHD
Velocardiofacial/DiGeorge	del(22)(q11.2q11.2)	CGS including *TBX1*	ADHD, OCD, cognitive deficits
dup(22)q11.2	dup(22)(q11.2q11.2)	CGS including *TBX1*	Motor delay, cognitive deficits, ADHD
Rett	del(X)(q28q28)	*MECP2*	Stereotypic behavior, ataxia, cognitive impairment, autism, spastic paraparesis, epilepsy
Rett-like	dup(X)(q28q28)	*MECP2*	Hypotonia, spasticity, MR, absent language/speech, stereotypic behaviors
Pelizaeus–Merzbacher disease	del(X)(q22.2q22.2)	*PLP1*	Developmental delay, neuropathy, spastic quadriplegia
Pelizaeus–Merzbacher disease	dup(X)(q22.2q22.2)	*PLP1*	Nystagmus, spastic quadriplegia, ataxia, dysarthria, MR, developmental delay

Abbreviations: del, deletion; dup, duplication; ADHD, attention-deficit hyperactivity disorder; CGS, contiguous gene syndrome; MR, mental retardation; OCD, obsessive–compulsive disorder.

than *PLP1* deletions; interestingly, whereas duplication gives a central nervous system (CNS) disorder, deletion results in a milder CNS disorder with a peripheral neuropathy.

The genomic segments that are deleted or duplicated often contain multiple genes; thus, the phenotype observed in chromosomal rearrangements was thought to result from the additive effect of several genes, hence the term contiguous gene syndromes (CGS). However, recent studies have revealed that the disease-specific phenotypes are often associated with a small number of genes, or even a single gene, which are sensitive to dosage alterations (dosage-sensitive genes)[35] and lead to the phenotype via haploinsufficiency or excess dosage in deletions and duplications, respectively. From a genomics perspective, a CGS in which two genes contribute to the syndromic phenotype might be considered a digenic inheritance of two physically linked genes.

Inversion is an intrachromosomal rearrangement generated by chromosomal breaks at two locations and "flipping" of the internal chromosomal segment. Inversions that involve the centromere are called *pericentric inversions*, whereas those occurring within single chromosomal arms are called *paracentric inversions*. Both types of inversions are usually balanced and thus result in no phenotypic manifestation. Homologous pairs of chromosomes with an inversion may have a greater chance of generating a gamete with an unbalanced chromosome by misalignment at the inverted segment in meiosis followed by recombination within the inverted segments.

Mechanisms for Formation of Chromosomal Rearrangements

Molecular mechanisms for interstitial chromosomal rearrangements (deletions, duplications, and inversions) are perhaps the best characterized inherited chromosomal disorders to date, and include the recombination mechanisms of nonalleleic homologous recombination (NAHR) and nonhomologous end joining (NHEJ), and the replication-based mechanism of fork stalling and template switching/microhomology-mediated break-induced replication (FoSTeS/MMBIR).

NONALLELIC HOMOLOGOUS RECOMBINATION (NAHR)

Low-copy repeats (LCRs), constituting up to 5% of the reference haploid human genome, are DNA fragments >1 kb in size that contain highly homologous sequences and are present in two or more copies in the genome. When two copies of paralagous LCRs are located at a distance less than 10 Mb from each other, they can lead to misalignment of chromosomes or chromatids and unequal crossing over, resulting in deletion or duplication of the unique genome segment between the LCRs (Figure 1.2). This is termed nonallelic homologous recombination (NAHR) and results in recurrent rearrangements, i.e., rearrangements that include the same genomic interval occurring independently in unrelated individuals.[34,36]

If paralog LCRs are present in an inverted orientation, NAHR results in an inversion of the unique sequence genomic segment flanked by these LCRs. An inversion does not alter the copy number of the rearranged genomic segment, but it may disrupt a gene that spans the recombination breakpoints, cause a position effect, or result in susceptibility to deletion rearrangement in progeny (e.g., WBS, PWS, and 17q21.31).[37,38] Some LCRs found in the human genome show a complex structure. They contain multiple modules with some units in a direct orientation and other units in an inverted orientation. Depending on which modules are used as substrates for NAHR, deletion/duplication (between tandem modules) or inversion (between inverted modules) may occur (Figure 1.2A).[39]

Studies of recombination breakpoints within a few LCRs revealed clustering of the strand exchange to small regions of perfect sequence identity (>300 bp), referred to as recombination *hotspots*.[40] Specifically, a degenerate 13-bp motif is crucial for recruiting crossovers in 40% of all human hotspots based on studies from historical recombinants.[41] These observations suggest not only that substantial regions of homology are required to facilitate NAHR, but also that a certain length of perfect sequence match and hotspot regions that may reflect genomic instability and susceptibility to double-strand breaks may be associated with efficient NAHR.[35] The entire human genome is punctuated by recombination hotspots of normal allelic homologous recombination (AHR) during meiosis that share many features with and can be coincident with NAHR hotspots. Thus, common architectural features of the human genome predispose to genomic rearrangements.[42,43]

NONHOMOLOGOUS END JOINING (NHEJ)

Nonrecurrent rearrangements, in which junction fragments differ in size between families and breakpoints do not cluster, is thought to result from a *nonhomologous end-joining* (NHEJ) mechanism. In this error-prone mechanism, double-strand breaks are detected and then both broken DNA ends are bridged, modified, and finally ligated. In contrast to NAHR, NHEJ does not require LCRs or minimal efficient processing segments, but may be stimulated by other components of genomic architecture.[34,36]

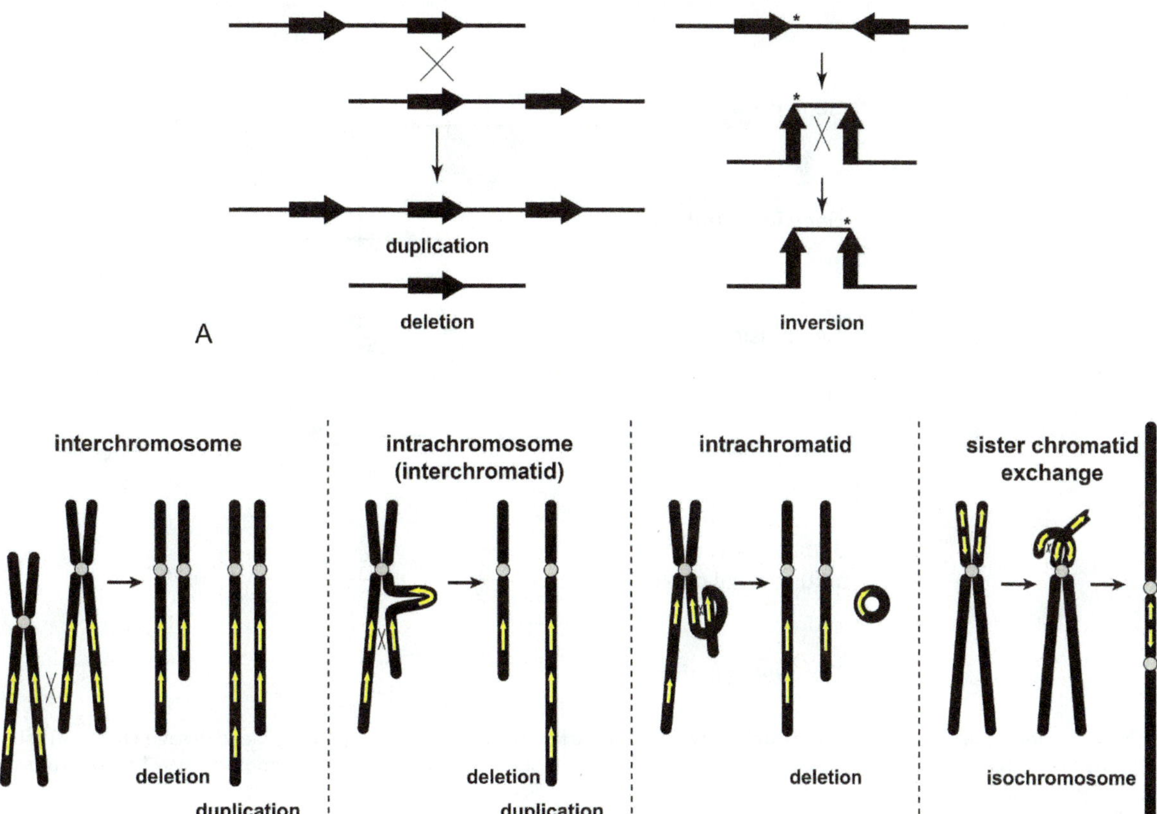

FIGURE 1.2 **Nonalleleic homologous recombination (NAHR) as the mechanism for recurrent genomic rearrangements.** (**A**) Ectopic crossing-over between directly oriented repeats *in trans* can lead to deletion and reciprocal duplication; whereas ectopic crossing-over between inversely oriented repeats *in cis* can result in an inversion. (**B**) NAHR can produce deletion or duplication in three ways: interchromosomal crossover, intrachromosomal (or interchromatidal) crossover, and intrachromatidal crossover. Note that intrachromatidal recombination can only produce deletion, not duplication. NAHR between inverted LCRs on sister chromatids can also result in isochromosome formation. Adapted from[36].

FORK STALLING AND TEMPLATE SWITCHING/MICROHOMOLOGY-MEDIATED BREAK-INDUCED REPLICATION (FoSTeS/MMBIR)

Fork stalling and template switching (FoSTeS) is based on a replication-based mechanism of DNA repair, and plays a role in complex nonrecurrent rearrangements such as deletions and/or duplications interrupted by either normal copy number or triplicated genomic segments. In this model, the DNA replication fork stalls and the lagging strand disengages from the original template and anneals to another replication fork in physical proximity, by virtue of microhomology at the 3′ end, "priming" or reinitiating DNA synthesis. An additional model proposed to be responsible for nonrecurrent and complex genomic rearrangements based on experimental observations in model organisms is termed microhomology-mediated break-induced replication (MMBIR). While FoSTeS is a template-switching model resulting from fork stalling, MMBIR is induced by a collapsed fork (i.e., one-ended, double-strand break in the DNA) and extensive processing of the end resulting in a 3′ overhang that can be used as a primer for DNA replication. Both replicative mechanisms result in a template driven juxtaposition of discrete genomic segments and can lead to complex genomic rearrangements. Depending on the location of the new fork (upstream or downstream), a deletion or duplication, respectively, will occur.[34,44]

How Chromosomal Rearrangements Confer Phenotypes

Deletions, duplications, triplications, and insertions can result in a *copy number variation* (CNV), defined as a segment of DNA ranging from one kilobase to several megabases in size that is present at a variable copy number in comparison with a reference genome. CNVs have been found to represent a major source for human genetic variation and genomic diversity, but while many CNVs might represent benign polymorphisms, others can convey clinical phenotypes. The term *genomic disorders* refers to a group of genetic diseases caused by genomic rearrangements

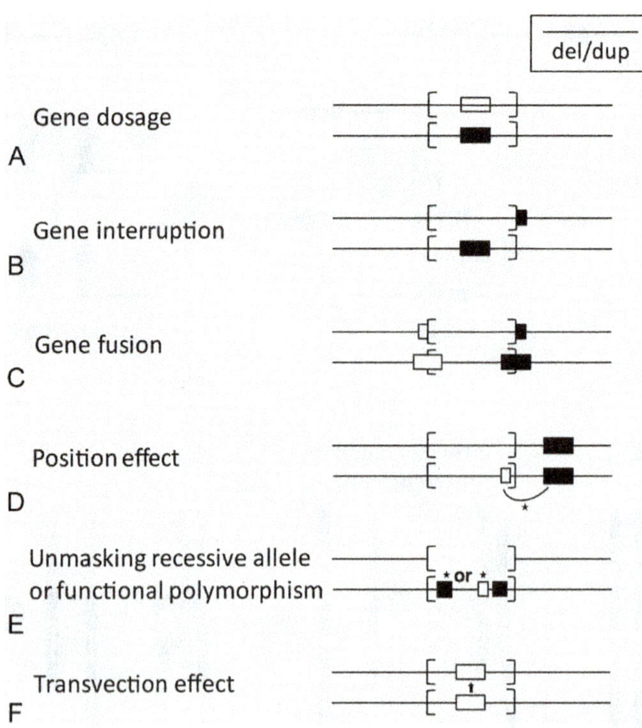

FIGURE 1.3 **Molecular mechanisms by which copy number variation confers phenotypes.** (**A**) Gene dosage effect; (**B**) interruption of a gene; (**C**) gene fusion; (**D**) position effect; (**E**) unmasking of either recessive mutations or functional polymorphism; (**F**) possible transvection effect. Adapted from[37].

that result from structural changes of the genome often resulting from genomic instability due to unique genome architectural features.[45]

Interstitial chromosomal rearrangements can cause a phenotype by several molecular mechanisms (Figure 1.3).[37] The first mechanism is the alteration of copy number (i.e., copy number variation [CNV]) of a gene (or genes) that is sensitive to a dosage effect (Figure 1.3A), as exemplified by *PMP22*, which is associated with CMT1A/HNPP, and *RAI1*, which is responsible for SMS. Second, the breakpoint of the rearrangement may interrupt a gene, thus resulting in a loss of function by inactivating the gene (Figure 1.3B), as exemplified by disruption of *NEMO* that causes incontinentia pigmenti (MIM# 308300). Alternatively, a fusion gene can form at the breakpoint, generating a gain-of-function mutation (Figure 1.3C); this mechanism is prominent among cancers associated with specific chromosomal translocations in somatic tissues. Rearrangements can also manifest through a position effect (Figure 1.3D). Such position effects have been documented for apparently balanced translocations that even exert their influence when the breakpoints map as far as 1 Mb away either upstream or downstream from the culprit gene. Position effects have been observed both with deletion and duplication rearrangements that occur outside the intact gene. One such example is a microduplication located downstream of the *PLP1* gene conferring a phenotype of spastic paraplegia type 2,[46] and duplication upstream of *PMP22* resulting in CMT1A.[47] Other mechanisms for genomic rearrangements may result from the reciprocal relationship between the rearranged segment on one chromosome and the corresponding allele on the other chromosome. These mechanisms include unmasking either recessive mutations or functional polymorphic alleles when a deletion occurs (Figure 1.3E), as well as potential effects on transvection (communication between alleles on homologous chromosomes) wherein regulatory elements required for communication between alleles are disrupted (Figure 1.3F).

Assays for Chromosomal and Genomic Disorders

The development and implementation of genomic microarray technology have ushered in a new era in medical genetics and genomics. Previously, Giemsa or G-banded chromosome analysis had been the standard for over 35 years. Chromosome analysis detects numerical and structural chromosomal abnormalities through counting chromosomes and analyzing chromosomal banding patterns in metaphase (classical karyotype) or prometaphase (high-resolution karyotype) cells. To be detectable by chromosome analysis, the structural changes must be greater than 3–10 Mb in size.[48]

Fluorescence *in situ* hybridization (FISH) is a powerful cytogenetic technique used to investigate specific genomic regions of chromosomes in greater detail by hybridizing a fluorescent-labeled DNA probe of interest. FISH is gaining widespread clinical applicability, although this technology requires prior knowledge and targeting of the specific region that might be abnormal. Telomere FISH can simultaneously examine the presence of subtelomeric deletions using a set of probes specific to the telomere for each chromosome arm. Telomere FISH identifies abnormalities in 2–5% of samples from patients with mental retardation whose previous G-banding karyotypes were normal. Detection of a balanced translocation carrier in such a family enables accurate assessment of recurrence risk and prenatal diagnosis.

High-resolution genome analysis technologies include *comparative genomic hybridization* (CGH), which was developed as a genome-wide screening strategy for detecting DNA copy number imbalances. In CGH, DNA from a test and a reference genome are labeled with distinct fluorescent dyes, followed by competitive hybridization that results in different colors depending on either loss or gain of a certain genomic interval. Microarray-based formats for CGH (array CGH, or chromosomal microarray analysis [CMA]) have been increasingly used in different clinical settings and have tremendous implications for neurological and psychiatric diseases. Notably, screening patients with mental retardation of unknown etiologies using array CGH has uncovered novel microdeletions or duplications that have not been identified previously.[37] Although chromosomal microarray analysis should be the first-tier test for clinical diagnosis of chromosome abnormalities, there is still value for traditional chromosome analysis in the detection of mosaicism and delineation of chromosomal structural rearrangements.[48]

MULTIGENIC INHERITANCE

Mendelian traits segregate with mutations in one or both alleles of a single gene that in itself is sufficient to express the disease phenotype. In contrast, complex traits appear to result from multifactorial interaction of various susceptibility genes and environmental exposures; thus, the role of a single gene may not be fundamental. The concept of multigenic inheritance was solidified when it appeared that some diseases have primarily a genetic etiology, but involve two or more genes in their pathophysiology.

Digenic Inheritance

Digenic inheritance refers to a mode of inheritance in which coexistence of mutations in two independent genes appears to be required to manifest a phenotype (Figure 1.1J). Early evidence for digenic inheritance was found in families with retinitis pigmentosa (RP; MIM# 268000), where a combination of heterozygous mutations in two unlinked genes with photoreceptor-specific expression, *PRPH2* (RDS/peripherin 2) and *ROM1*, results in retinitis pigmentosa, while neither mutation by itself is sufficient to cause symptoms.[49] Subsequent studies have found that interaction of the encoded proteins is critical for stable photoreceptor disc formation.[50] A second example of digenic inheritance concerns autosomal recessive nonsyndromic deafness (DFNB1; MIM# 220290), where a subset of patients have a mutation in the gap-junction protein connexin 26 gene (*GJB2*) in addition to a heterozygous genomic deletion containing the connexin 30 gene (*GJB6*). While mutations in *GJB2* and deletions in *GJB6* can independently be inherited as monogenic autosomal recessive traits, digenic compound heterozygous mutations of *GJB2* and *GJB6* seem to also play a major role in the etiology of recessive nonsyndromic deafness.[51]

A more complex digenic model with triallelic inheritance was identified in studies of Bardet–Biedl syndrome (BBS; MIM# 209900), a genetically heterogeneous multisystemic ciliopathy associated with retinitis pigmentosa, obesity, polydactyly, and mental retardation. In some families, mutations in three alleles at two loci are apparently necessary to manifest a BBS phenotype.[52,53] Subsequent work has provided strong evidence that epigenetic interactions contribute to both penetrance and variable phenotypic expression across the ciliopathy spectrum.[54]

Recently, facioscapulohumeral muscular dystrophy type 2 (FSHD2) was shown to result from digenic inheritance of an allele of the D4Z4 microsatellite array on chromosome 4, which is permissive for the expression of the embedded *DUX4* gene, and single-nucleotide variation at the *SMCHD1* locus. The point mutation in *SMCHD1* (encoding structural maintenance of chromosomes flexible hinge domain containing 1) was shown to act as an epigenetic modifier of the D4Z4 allele.[54,55]

Modifier Genes

After the identification of Mendelian disease genes, enormous mutation analyses have been carried out in search of genotype–phenotype correlations enabling prognostic information and potentially influencing management decisions. Except for a limited number of disease alleles, the vast majority of Mendelian diseases have failed to establish robust and reproducible precise genotype–phenotype correlations. These findings have suggested that, even in a disease that is inherited as a simple Mendelian trait, phenotypic manifestations are likely modified by other factors.

Cystic fibrosis (MIM# 219700) has several known modifier loci and genes, among them low-expressing mannose-binding lectin gene (*MBL*), endothelial receptor type A (*EDNRA*), tumor necrosis factor-α gene (*TNFA*), transforming growth factor B1 gene (*TGFB1*), interferon-related developmental regulator 1 gene (*IFRD1*), and IL-8 for the pulmonary phenotype; nitric oxide synthase 1 (*NOS1*) for microbial infection; SNVs near the transcription factor 7-like 2 (*TCF7L2*) gene for type 2 diabetes; genes of the CEACAM family at the *CFM1* locus on 19q13.1oci for meconium ileus susceptibility; and mucin for gastrointestinal aspects.[56,57]

Familial amyotrophic lateral sclerosis (ALS) accounts for about 10% of all ALS patients, and can be associated with mutations in several genes including superoxide dismutase 1 (*SOD1*) and genes encoding RNA-processing proteins (*TDP43, FUS*), proteostatic proteins (*UBQLN2, OPTN*), and cytoskeleton/cellular transport proteins (*VAPB*), to name a few.[58] Recently, an expanded hexanucleotide repeat (GGGGCC) in the first intron of C9orf72 has been associated with both ALS and frontotemporal dementia,[59] and likely accounts for 40–50% of all familial ALS.[58] Significant variability in the age of onset and disease severity of patients harboring *SOD1* mutations is suggestive of modifying factors. A patient with a particularly severe phenotype of early onset was found to harbor a homozygous mutation in the ciliary neurotrophic factor gene (*CNTF*) in addition to the *SOD1* mutation.[60] *CNTF* was suggested to be a modifier of disease onset in ALS, and transgenic mice with the human *SOD1*-G93A mutation crossed with *Cntf*-deficient mice were shown to have an earlier onset of disease as compared to *Cntf* wild-type mice. Furthermore, human patients with sporadic ALS who are homozygous for *CNTF* mutations show earlier onset than those with wild-type *CNTF*. Each of these observations suggests that *CNTF* is likely a modifier for age of onset of familial ALS and possibly of sporadic ALS.[60]

Charcot–Marie–Tooth (CMT) disease has been attributed to genetic alterations in more than 50 distinct genes or loci, including *PMP22, MPZ, GJB1, EGR2*, and *PRX*. In rare instances, a combination of two mutations in two different CMT genes (a *PMP22* duplication plus a missense mutation of *LITAF*[61] or *GJB1*,[62] and a *EGR2* mutation plus a *GJB1* mutation[63]) have been shown to result in a more severe or earlier-onset neuropathy phenotype, compared to those resulting from mutations of a single gene mutation, wherein one CMT gene can be considered as a modifying factor for the other gene in terms of disease severity.

One last example of modifier genes involves spinal muscular atrophy (SMA), wherein patients exhibit a homozygous functional loss of the survival motor neuron 1 gene (*SMN1*), but also carry one or more *SMN2* copies that modulate the severity of the disease. Additional modifiers include plastin 3 (*PLS3*), which can fully protect against SMA; and exogenous factors such as deficient nutrition and hypoxia, which exacerbate SMA. Studies in animal models suggest that overexpression of *PLS3* can rescue the SMA phenotype. Furthermore, SMA mice exhibit a significant increase in active RhoA (RhoA-GTP), a major upstream regulator of the actin cytoskeleton, and treatment with fasudil, a Rho A inhibitor, significantly improves their motor abilities and survival. This illustrates the importance of modifiers for the understanding of disease pathology and the development of strategies to overcome the main causative gene defect.[64]

COMPLEX TRAITS

In contrast to Mendelian traits, *complex traits* refer to a group of diseases that are influenced by interactions between multiple genetic loci and environmental factors. Complex traits can have significant genetic components. Twin studies and familial aggregation studies have revealed evidence for heritability in many complex traits including autism, schizophrenia, bipolar disease, multiple sclerosis, and Alzheimer disease.[65] For instance, a first-degree relative of a patient with multiple sclerosis (MS) has 3.0–5.0% chance of having the disease (15–25 times higher than the general population), and the monozygotic female twin of a MS patient has a 34% chance of having MS (170 times greater than the general population).[66] However, the susceptibility loci are challenging to map and isolate, and initial positive findings are often not replicated due to relatively modest genetic effects of each contributing locus and diverse genetic backgrounds in different populations. In this section, we first characterize genetic features of complex traits, many of which are common diseases, and then briefly summarize methods to analyze genetic loci and provide examples of susceptibility genes for such diseases.

Genetic Features of Complex Traits

Common diseases that are complex traits share various features. The genetic models of inheritance cannot usually be specified, since parameters such as mode of transmission, penetrance, and allele frequency may vary for any given susceptibility locus and are influenced by environmental factors. Furthermore, genetic heterogeneity, whereby each disease phenotype may result from the contribution of multiple loci in the genome, is common and the contribution of a particular locus to a phenotype may vary between different families and populations. Phenotypic manifestations of common traits are often variable among individuals and may show a continuous spectrum of quantitative values from normality to full-blown disease state (as seen in body mass index, blood pressure, IQ), posing difficulties in defining affected versus unaffected individuals and further complicating the analysis. The genetic loci that are tightly related to these quantitative values and affect natural variation in complex traits are called *quantitative trait loci* (QTL).

Assessing Variation in the Human Genome

Previously, strategies to map susceptibility loci in complex disease traits included linkage analyses and association studies. Linkage analysis elucidates the relationship between a disease phenotype and the location of a specific genetic marker from a marker set encompassing the entire genome. Classical association studies, in contrast, unravel the relationship between a disease phenotype and genetic variations within specific candidate loci predicted by biological functions.

The advent of genome-wide association studies (GWAS), in which hundreds of thousands of single-nucleotide polymorphisms (SNPs) are tested for association with a disease in hundreds or thousands of persons, has revolutionized the search for genetic influences on complex traits. Such studies have identified SNPs implicating hundreds of robustly replicated loci for common traits. However, GWAS has several limitations. The identified variants generally have low associated risks and account for little heritability, and SNP associations identified in one population are frequently not transferable to other populations. Also, few of the SNPs identified have clear functional implications relevant to the mechanisms of disease. It remains to be determined how to use the data obtained in GWAS to screen for and predict disease and to improve the processes of drug selection and dosing.[67] Moreover, since SNPs are defined as variants with a frequency >1% of the population, GWAS do not capture information about rare variants and have limited statistical power to detect small gene–gene and gene–environment interactions.[68] While GWAS were driven by the "common disease/common variant" (CDCV) hypothesis, which presupposes that different combinations of common alleles aggregate in specific individuals to increase disease risk, research efforts have recently shifted to exploration of less frequent variants in common disorders. There are emerging examples wherein specific loci that cause "Mendelian disease" are contributing to the background risk for a parallel common disorder.[69]

In the era of personal genomics and with the decreasing cost of massively parallel sequencing technologies, whole-exome sequencing (which involves targeted sequencing of only the protein-coding DNA sequences of the human genome) is emerging as a popular approach to test for association of rare coding variants with complex phenotypes. In contrast to GWAS that use linkage disequilibrium patterns between common markers, exome-sequencing studies enable the direct identification of causal variants. However, focusing exclusively on the exome sequence is a limitation in complex trait genetics, where noncoding genetic variation is believed to have a larger role than in Mendelian genetics.[70] Whole-genome sequencing will likely produce more complete assessment of genetic variation contributing to personal health.[69]

Genetic Variation and Complex Traits

Genomic variations in the human haploid genome range from *single nucleotide polymorphisms* (SNPs) to large (>5 Mb) microscopically visible chromosome anomalies. In between lie submicroscopic CNVs, variable number of tandem repeats, and microsatellites (di-, hi-, tetra-, etc., nucleotide repeats). *Single nucleotide variation* (SNV) refers to variations in nucleotide sequence where one of the four nucleotides is substituted for another (e.g., T for G), as well as small indels (insertions and deletions). Single nucleotide variants with a minor allele frequency of 1% or greater are termed SNPs and are a major contributor to the variation in the human genome. At least 11 million SNPs occur in the human genome, with an average of 1 every 300 nucleotides among the human genome of 3 billion bases. SNPs may account for a proportion of susceptibility to common diseases, individual responses to drugs, and other normal phenotypic variation.[71,72] As documented for the ε4 allele of *APOE4* and predispositions to Alzheimer disease, one SNP or a combination of SNPs in a coding exon or a promoter region of a gene can confer a susceptibility to a disease or complex traits. This has also been observed for promoter mutations in *APP* and *SNCA* for Alzheimer and Parkinson disease, respectively.[73,74]

Another class of genetic diversity is *copy number variation* (CNV), defined as a copy number change involving a DNA fragment that is 1 kb or larger. Recent analyses revealed 11,700 CNVs overlapping over 1000 genes, thus accounting for about 13% of the human genome. CNVs can be inherited or sporadic; large *de novo* CNVs are considered more likely to be disease causative. However, the phenotypic effects of CNVs are sometimes unclear and depend mainly on whether dosage-sensitive genes or regulatory sequences are affected by the genomic rearrangement. CNVs have been implicated in several common complex traits, including subsets of patients with autism spectrum disorders, schizophrenia, epilepsy, developmental delay/mental retardation, Parkinson disease (PD), and Alzheimer disease. Specifically, the most common cytogenetic abnormality in autism spectrum disorders is maternally derived duplication of 15q11q13, found in 1–3% of cases. Susceptibility to autism has also been associated with a microdeletion and reciprocal microduplication at 16p11.2,[75] and schizophrenia has been associated with 15q13.3 deletions and 1q21.2 deletions and duplications. It is estimated that up to 40% of epilepsies are genetically determined, with various modes of inheritance. Most genes implicated in monogenic epilepsies encode subunits of neuronal voltage- or ligand-gated ion channels or encode proteins related to neuronal maturation and migration during embryonic development. Inactivating point mutations and genomic deletions on 2q24.3 leading to haploinsufficiency of the *SCN1A* gene have been found in patients with generalized epilepsy with febrile seizures plus (GEFS+) and severe myoclonic epilepsy of infancy (SMEI) or Dravet syndrome.[34] Variable-sized deletion CNVs of the *CDKL5* gene were found in three females with early-onset seizures, while point mutations in *CDKL5* have been reported in 10% of females with a Rett-syndrome-like phenotype and early-onset seizures.[76]

APP (encoding amyloid beta A4 precursor protein; MIM# 104760) exemplifies the various mechanisms by which genetic variation confers susceptibility to a complex disease. *APP* copy number and/or expression can be altered by trisomy 21,[77] submicroscopic duplication causing CNV,[78] or SNPs in the gene promoter;[73] all are associated with susceptibility to Alzheimer disease. Another example of possible gene dosage effect in common disease is *SNCA* (α-synclein), a gene responsible for autosomal dominant Parkinson disease. Genomic gains in *SNCA* copy number have recently been shown to be associated with the severity of Parkinson disease phenotype in a dose-dependent manner. The clinical phenotype associated with *SNCA* duplications is consistent with late-onset idiopathic PD,[79,80] whereas *SNCA* triplication results in a more severe form of PD.[81] Consistent with the hypothesis that increased gene dosage is causative for Parkinson disease, susceptibility to Parkinson disease is associated with SNPs in *SNCA* promoter regions.[74]

Examples of Susceptibility Genes for Complex Traits

Rare Mendelian diseases can sometimes display a phenotype similar to a common disease. Thus, genes for such Mendelian traits can be strong candidates for susceptibility genes for the complex trait. One such example is Wolfram syndrome, a rare autosomal recessive disorder characterized by juvenile diabetes mellitus and progressive optic atrophy caused by mutation in the *WFS1* gene. Patients with Wolfram syndrome often present with severe psychiatric manifestations. Heterozygous carriers for *WFS1* mutations have a 26-fold increased risk for hospitalization due to psychiatric illness, especially for depression. A second example is ataxia-telangiectasia, an autosomal recessive disorder characterized by cerebellar ataxia, telangiectases, immune defects, and a predisposition to malignancy of different types caused by mutations in the *ATM* gene. Heterozygous carriers for *ATM* mutations have an estimated relative risk of 5:1 to develop breast cancer. A third example includes age-related macular degeneration (AMD). Heterozygous *ABCA4* mutations are associated with AMD, whereas homozygous mutations of the same gene are identified in patients with recessive Stargardt macular dystrophy.[82] Recently, the association between β-glucocerbridase (*GBA*) mutations and parkinsonism has been defined. Gaucher disease is an autosomal recessive lysosomal storage disorder caused by mutations in *GBA*. Heterozygous *GBA* mutations confer increased susceptibility to parkinsonism, presumably via interactions between β-glucocerbridase and α-synuclein.[83,84] In each of these four examples, heterozygous mutations of rare recessive genes appear to be associated with increased risk of a common complex trait (Table 1.3).

EPIGENETICS

The science of epigenetics describes how gene expression and function are controlled in individual cells and tissues and how gene–gene and gene–environmental interactions are mediated during development and adult life. The major epigenetic mechanisms that have been described include DNA methylation and hydroxymethylation, histone modifications and higher order chromatin remodeling, and noncoding RNA (ncRNA) regulation.

DNA Methylation and Hydroxymethylation

DNA methylation involves the covalent modification of cytosine residues in DNA, which occurs in gene regulatory regions, such as promoter element CpG dinucleotides and also at other genomic sites. DNA methyltransferase

TABLE 1.3 Mendelian Disorders and Heterozygous Predisposition to Multifactorial Disease

Monogenic disease	MIM#	Gene	Multifactorial disease
Familial hypercholesterolemia	143890	*LDLR*	Coronary artery disease
Cystic fibrosis	219700	*CFTR*	Pancreatic insufficiency, chronic rhinosinusitis, idiopathic bronchiectasis
Ataxia-telangiectasia	208900	*ATM*	Breast cancer
α_1-antitrypsin deficiency	107400	*AAT*	Chronic obstructive lung disease
Hyperlipoproteinemia	238600	*LPL*	Ischemic heart disease
Stargardt disease	248200	*ABCR(ABCA4)*	Age-related macular degeneration
Progressive familial intrahepatic cholestasis	171060	*ABCB4*	Intrahepatic cholestasis of pregnancy
Gaucher disease	231000	*SGA*	Parkinson disease

(DNMT) enzymes catalyze *de novo* methylation and maintain methylation "marks." DNA methylation is thought to inhibit the transcriptional machinery from accessing DNA, leading to decreased transcription of genes with high levels of promoter methylation. Proteins that recognize methylated DNA (e.g., the methyl-CpG-binding domain [MBD] family of proteins) can recruit a range of additional epigenetic modulatory factors to these loci. DNA methylation is generally associated with transcriptional repression and long-term gene silencing, and it also plays a role in the establishment and maintenance of higher order epigenetic states (e.g., X chromosome inactivation and genomic imprinting, as previously discussed). Hydroxymethylation counterbalances the effect of methylation by inhibiting the binding of MBD proteins.[23,85]

Mutations in the *DNMT1* gene interferes with proper methylation and causes hereditary forms of neurodegeneration with central and peripheral manifestations including a sensory neuropathy, dementia, and hearing loss syndrome (MIM# 614116) and a cerebellar ataxia, deafness, and narcolepsy syndrome.[86] Mutations in the *SNCA* gene and alterations in the dosage of the wild-type gene are, respectively, associated with familial and sporadic forms of Parkinson disease.[87] In addition, other Parkinson disease risk associated genes also exhibit differential levels of DNA methylation.[88] Furthermore, characteristic DNA methylation profiles at the *frataxin* (*FXN*) gene locus, which harbors an expansion GAA repeat mutation in Friedreich ataxia, have been correlated with mutant FXN expression levels, age of onset of symptoms, and clinical disease severity rating scores.[23,89]

Histone Modifications and Higher Order Chromatin Remodeling

DNA exists as a highly compact structure within the cell nucleus that is referred to as chromatin, comprised of DNA wrapped around an octamer of histone proteins. In loosely packaged chromatin, DNA sequences are relatively accessible to the diverse range of factors present in the nucleus, including the machinery responsible for transcription and DNA replication and repair. The architecture of chromatin is dynamic, evolving with the lifecycle of the cell and in response to environmental cues. Posttranslational modifications (i.e., histone acetylation and methylation) are regulated by histone acetylases/deacetylases and histone methyltransferases/demethylases.

Deficiency of the ataxia-telangectasia mutated (ATM) protein leads to accumulation of the histone deacetylase protein HDAC4 in the neuronal cell nucleus, which promotes neurodegeneration.[90] Altered expression levels of *HDAC11* and *HDAC2* mRNA compared with controls has been demonstrated in postmortem brain and spinal cord specimens of patients with ALS.[91] The pathogenesis wherein mutant ataxin-1 protein causes spinocerebellar atrophy 1 (SCA1) also involves histone acetylation and epigenetic regulatory complexes.[23]

The increased understanding of the role of epigenetics in the pathogenesis of neuropsychiatric disorders extends to therapeutic implications. Small molecule inhibitors of histone deacetylases (HDACs) are being researched and developed as drugs for cancer and neurological disorders, such as Rubinstein–Taybi syndrome, Friedreich ataxia, and fragile X syndrome. Mutations in the *GRN* gene encoding progranulin are a frequent cause of familial frontotemporal dementia, a currently untreatable progressive neurodegenerative disease. A HDAC inhibitor, suberoylanilide hydroxamic acid (SAHA), was shown to increase *GRN* mRNA expression and protein levels in haploinsufficient cells from human subjects, holding promise for potential therapeutic intervention.[92] It has been shown in Alzheimer disease that cognitive capacities in the neurodegenerating brain are constrained by an epigenetic blockade of gene transcription that is potentially reversible. This block is mediated by HDACs, and research is underway for the development of selective HDAC inhibitors to potentially relieve this epigenetic blockade.[93,94]

Noncoding RNA Regulation

The vast majority of human genomic DNA is non-protein-coding, leading to formation of large numbers of noncoding RNAs (ncRNAs). Among the various functional classes of ncRNAs are microRNAs (miRNAs) and long ncRNAs (lncRNAs), which have emerging roles in controlling the expression and function of individual genes and large gene networks through transcriptional, posttranscriptional, and epigenetic mechanisms. miRNAs are 19–22 nucleotide single-stranded RNAs that regulate target messenger RNAs (mRNAs) via sequence-specific interactions. They generally bind to the 3′ UTRs of these mRNAs, inhibiting their translation. miRNA dysregulation has been implicated in multiple sclerosis;[95] CNS injury;[96] and in Parkinson disease, wherein mutant LRRK2-mediated dopaminergic neuronal degeneration is mediated in part by dysregulation of miR-184.[97] lncRNAs have been described above in mechanisms of imprinting, and have a diverse and emerging spectrum of functions in epigenetic regulation. Deletions of a small noncoding differentially methylated region at 16q24.1, including lncRNA genes, have recently been implicated in a lethal lung developmental disorder, alveolar capillary dysplasia with misalignment of pulmonary veins (ACD/MPV). Perturbation of lncRNA-mediated chromatin interactions may, in general, be responsible for position effect phenomena and potentially cause many disorders of human development.[98]

Major international efforts such as the Human Epigenome Project are underway in an attempt to catalog and interpret epigenetic profiles in health and disease. The epigenomic data must ultimately be integrated with genomic and other phenomic (e.g., transcriptomic, proteomic, and metabolomic) profiles in order to build a comprehensive understanding of neurodegenerative diseases and the process of neurodegeneration.[23] One of the most intriguing aspects of disorders that involve monoallelically expressed genes is the prospect for therapy that involves derepressing the silenced allele in situations where the expressed allele of an imprinted gene is deleted or contains a loss-of-function mutation. A recent success was reported for Angelman syndrome, where a screen revealed that small molecule topoisomerase inhibitors reactivated the silenced *UBE3A* gene and repressed the imprinting control region (ICR)-associated antisense RNA.[99]

THE HUMAN GENOME: HIGH-THROUGHPUT TECHNOLOGIES

The 2001 draft sequence of the human genome, by the Human Genome Project (HGP),[100] marked a turning point for human genetics and the starting point for human genomics. Technical development during the decade following the HGP enabled massively parallel next-generation sequencing (NGS), commencing with a series of key examples of individual genomes and continuing with rapid progression into routine research and clinical settings. The 1000 Genomes Project (TGP) aimed to characterize human variation of all types by high-throughput and unbiased sequencing of 1000+ human genomes from diverse populations, with a low-coverage sequencing (4–6× average depth) of the entire genome and subsequent deep resequencing of exons in over 1000 samples in order to capture most of the "normal" coding variation.[101,102] Resequencing of personal human genomes provided further insight into the extent of both simple nucleotide variation (SNV) and structural variation.[103,104] Notably, ~35% of the genes in the human genome are encompassed either totally or partially by a CNV that can alter their expression or even their structure, possibly giving rise to novel fusion transcripts.[105]

The exome comprises the coding sequences of all annotated protein-coding genes (~23,000) and is equivalent to ~1% of the total haploid genomic sequence. Targeted-capture methodologies were developed using arrays and later beads in solution that hybridize to the exonic sequences being captured.[106,107] Several examples of the utility of exome sequencing for gene discovery in Mendelian disorders have been published, including *DHODH* in Miller syndrome[108] (MIM# 263750), *SETBP1* in Schinzel–Giedion syndrome[109] (MIM# 269150), *MLL2* in Kabuki syndrome[110] (MIM# 147920), *STAMBP* in microcephaly-capillary malformation syndrome (MIM# 614261),[111] and *SRCAP* in Floating-Harbor syndrome (MIM# 136140).[112] The NHLBI GO Exome Sequencing Project (ESP) set a goal to discover novel genes and mechanisms contributing to heart, lung, and blood disorders through exome sequencing with a high average coverage (>100× depth). Variants are deposited on the publically available Exome Variant Server (EVS) (http://evs.gs.washington.edu/EVS).

Whole-exome sequencing (WES) has become more widely used for genetic diagnosis and gene discovery because it is less costly than whole-genome sequencing (WGS). However, a limitation is that WES assesses nucleotide variation in only ~1% of the genome, while unappreciated variation in the remaining 99% of the human genome may be

TABLE 1.4 Comparison of Whole-Genome Sequencing (WGS) and Exome Sequencing Approaches For Disease–Gene Identification

	Whole-genome sequencing	Exome sequencing
Cost	Still costly, but decreasing rapidly	Reduced cost
Technical	No capture step, automatable	Capture step, technical bias
Variation	Uncovers all genetic and genomic variation (SNVs and CNVs) Discovery of functional coding and noncoding variation ~3.5 million variants	Focuses on ~1% of genome Limited to coding and splice-site variants in annotated genes ~20,000 variants
Disease	Suitable for Mendelian and complex trait gene identification, as well as sporadic phenoytpes caused by *de novo* SNVs or CNVs	Good for highly penetrant Mendelian disease gene identification

Modified from[106].

particularly important in complex, heterogeneous, or more subtle phenotypes. Whole genome sequencing enables assessment of variation in the entire genome (coding and noncoding). Table 1.4 lists the strengths and weaknesses of WGS compared to exome-sequencing approaches for disease-gene identification. The usefulness of WGS for genetic diagnosis was evaluated by Lupski et al.[113] in a patient with Charcot–Marie–Tooth disease. Compound heterozygous causative alleles in the *SH3TC2* gene were identified, in addition to other clinically relevant variants that provide diagnostic information and may help direct care of patients.[113] Other examples of WGS in medical diagnosis include identification of mutations in the *SPR* gene in a pair of fraternal twins with dopa-responsive dystonia[114] (DRD, MIM# 605407), and mutations in *ABCG5* in an infant with a clinical diagnosis of hypercholesterolemia, redefined as sitosterolemia by the WGS molecular diagnosis.[115] In both cases, medically actionable variants were identified and the applied medical treatment resulted in amelioration of symptoms, marking a true landmark in personal medical genomics.

Probably the most immediate applicability of genomic sequencing in clinical practice, in addition to reaching an accurate genetic diagnosis, is in the field of pharmacogenomics. It is now possible to identify the genome-wide totality of potential clinically relevant pharmacogenomic variants and ascertain if an individual is a fast or slow metabolizer of a certain drug, allowing individualized dosage adjustment to maximize therapeutic effect and minimize side effects.[105,116]

The true challenge for personalized genomics remains to identify the disease-causing or susceptibility-conferring mutations from amongst the 3.0–3.5 million SNVs and ~1000 CNVs on average in any given human diploid genome. Multiple bioinformatic prediction tools are available to aid in discerning deleterious from benign variation, based on parameters such as the conservation of an amino acid at a particular location across species and the biochemical properties of the reference versus variant amino acid.

Exome or genome sequencing is helpful in detection of mutations that might not be anticipated based on clinical phenotype. For many genetic conditions, the phenotypes are either nonspecific (e.g., intellectual disability without congenital anomalies), very obscure (due to the rarity of the condition), or variable from patient to patient, making it difficult to choose a particular gene to test. The clinical utility is the ability to establish a diagnosis, often bringing peace of mind to the family, avoiding further expensive and fruitless testing, providing a basis for genetic counseling, informing surveillance for complications, and even suggesting possible avenues of therapy. However, a sometimes unintended consequence of high throughput sequencing is the identification of secondary or incidental findings and interpretation of those findings.[117,118] Thus, the era of genomics brings about many legal, ethical, and social issues. Given the potential clinical implications of much genomic information, decisions about querying, interpreting, and delivering information arise.[119] Moreover, new risks to privacy arise where the benefits of broad data sharing must be balanced against the imperative to respect and protect individual patients and families.[120]

CONCLUSIONS

Recognizing the mode of inheritance in a family with a genetic disease is not only crucial for accurate diagnosis but also essential for providing appropriate counseling for recurrent risk, prognosis, medical management, and possible therapeutic choices. Human disease traits display many different patterns of inheritance. Because the mode of inheritance is recognized as observed traits in pedigrees, obtaining a thorough family history is important and should be performed in great detail in all patients.[2] At the same time, genetics may play an important role in the etiology of diseases that present as sporadic cases such as *de novo* point mutations or genomic rearrangements. Thus, because of new mutations and variability of expression as well as reduced penetrance, transmission or inheritance of a disease trait is not always readily observed in genetic disorders.

The rapid advances in human and medical genetics in combination with the Human Genome Project have contributed to major progress in the discovery of genes and genetic mechanisms responsible for neurologic and psychiatric diseases. The International HapMap Project has clarified details of genetic variations in four distinct populations, providing crucial resources for genome-wide approaches to complex traits. In addition to SNPs, emerging evidence has revealed that CNV and epigenetic factors are significant components of genetic and epigenetic variations, each of which could potentially play a role in complex traits. Data accumulated from genome-wide studies supports a unified picture wherein previously distinct entities or categories of human diseases, chromosomal syndromes, genomic disorders, Mendelian traits, and common diseases or complex traits, can now be considered part of one continuum, whereby common and rare variants including *de novo* mutations in the context of environmental influences result in perturbation of the biological balance of a restricted set of networks activating final common pathways that ultimately cause disease.[70] We anticipate that further progress into the molecular and genetic bases of neurologic diseases will provide the practitioner with novel tools for diagnosis, management, and treatment of neuropsychiatric disorders.

References

1. Beaudet A, Scriver C, Sly W, Valle D. Genetics, Biochemistry, and Molecular Bases of Variant Human Phenotypes: The Metabolic and Molecular Basis of Inherited Diseases. McGraw-Hill; 2001.
2. Shiga K, Inoue K, Lupski JR. Mendelian, nonmendelian, and multigenic inheritance and complex traits. In: Rosenberg RN, ed. The Molecular and Genetic Basis of Neurological and Psychiatric Disease. 4th ed. Philadelphia: Lippincott Williams and Wilkins; 2008:14–34.
3. Vogel F, Motulsky A, eds. Human Genetics: Problems and Approaches. Berlin: Springer; 1996.
4. Ott J, ed. Analysis of Human Genetic Linkage. Baltimore: Johns Hopkins University Press; 1999.
5. Willard H, ed. The Sex Chromosomes and X Chromosome Inactivation. New York: McGraw-Hill; 2001, The Metabolic and Molecular Basis of Inherited Diseases.
6. Muller H. *Further studies on the nature and causes of gene mutations.* Paper presented at: Proceedings of the 6th International Congress of, Genetics 1932.
7. Inoue K, Khajavi M, Ohyama T, et al. Molecular mechanism for distinct neurological phenotypes conveyed by allelic truncating mutations. *Nat Genet*. 2004;36(4):361–369.
8. Campeau PM, Lenk GM, Lu JT, et al. Yunis-Varon syndrome is caused by mutations in FIG4, encoding a phosphoinositide phosphatase. *Am J Hum Genet*. 2013;92(5):781–791.
9. Khajavi M, Inoue K, Lupski JR. Nonsense-mediated mRNA decay modulates clinical outcome of genetic disease. *Eur J Hum Genet*. 2006;14(10):1074–1081.
10. Ben-Shachar S, Khajavi M, Withers MA, et al. Dominant versus recessive traits conveyed by allelic mutations - to what extent is nonsense-mediated decay involved? *Clin Genet*. 2009;75(4):394–400.
11. Hurles M. Older males beget more mutations. *Nat Genet*. 2012;44(11):1174–1176.
12. Vissers LE, de Ligt J, Gilissen C, et al. A *de novo* paradigm for mental retardation. *Nat Genet*. 2010;42(12):1109–1112.
13. Poduri A, Evrony GD, Cai X, Walsh CA. Somatic mutation, genomic variation, and neurological disease. *Science*. 2013;341(6141):1237758.
14. Lupski JR. Genetics. Genome mosaicism—one human, multiple genomes. *Science*. 2013;341(6144):358–359.
15. Lee JH, Huynh M, Silhavy JL, et al. De novo somatic mutations in components of the PI3K-AKT3-mTOR pathway cause hemimegalencephaly. *Nat Genet*. 2012;44(8):941–945.
16. Knudson AJ. Mutation and cancer: statistical study of retinoblastoma. *Proc Natl Acad Sci*. 1971;68:820–823.
17. Pearson CE, Nichol Edamura K, Cleary JD. Repeat instability: mechanisms of dynamic mutations. *Nat Rev Genet*. 2005;6(10):729–742.
18. Cummings CJ, Zoghbi HY. Trinucleotide repeats: mechanisms and pathophysiology. *Annu Rev Genomics Hum Genet*. 2000;1:281–328.
19. Albrecht A, Mundlos S. The other trinucleotide repeat: polyalanine expansion disorders. *Curr Opin Genet Dev*. 2005;15(3):285–293.
20. Jacquemont S, Hagerman RJ, Leehey M, et al. Fragile X premutation tremor/ataxia syndrome: molecular, clinical, and neuroimaging correlates. *Am J Hum Genet*. 2003;72(4):869–878.
21. El-Hattab AW, Scaglia F. Mitochondrial DNA depletion syndromes: review and updates of genetic basis, manifestations, and therapeutic options. *Neurotherapeutics*. 2013;10(2):186–198.
22. Lee JT, Bartolomei MS. X-inactivation, imprinting, and long noncoding RNAs in health and disease. *Cell*. 2013;152(6):1308–1323.
23. Qureshi IA, Mehler MF. Epigenetic mechanisms governing the process of neurodegeneration. *Mol Aspects Med*. 2013;34(4):875–882.
24. Engel E. A new genetic concept: uniparental disomy and its potential effect, isodisomy. *Am J Med Genet*. 1980;6(2):137–143.
25. Spence JE, Perciaccante RG, Greig GM, et al. Uniparental disomy as a mechanism for human genetic disease. *Am J Hum Genet*. 1988;42(2):217–226.
26. Engel E. A fascination with chromosome rescue in uniparental disomy: mendelian recessive outlaws and imprinting copyrights infringements. *Eur J Hum Genet*. 2006;14(11):1158–1169.
27. Engel E, Antonarakis S, eds. *Genomic Imprinting and Uniparental Disomy in Medicine: Clinical and Molecular Aspects*. New York: Wiley-Liss; 2002.
28. Sahoo T, del Gaudio D, German JR, et al. Prader–Willi phenotype caused by paternal deficiency for the HBII-85 C/D box small nucleolar RNA cluster. *Nat Genet*. 2008;40(6):719–721.
29. Duker AL, Ballif BC, Bawle EV, et al. Paternally inherited microdeletion at 15q11.2 confirms a significant role for the SNORD116 C/D box snoRNA cluster in Prader–Willi syndrome. *Eur J Hum Genet*. 2010;18(11):1196–1201.

30. Liehr T, Claussen U, Starke H. Small supernumerary marker chromosomes (sSMC) in humans. *Cytogenet Genome Res*. 2004;107(1–2):55–67.
31. Nagaoka SI, Hassold TJ, Hunt PA. Human aneuploidy: mechanisms and new insights into an age-old problem. *Nat Rev Genet*. 2012;13(7):493–504.
32. Van Dyke DL, Weiss L, Roberson JR, Babu VR. The frequency and mutation rate of balanced autosomal rearrangements in man estimated from prenatal genetic studies for advanced maternal age. *Am J Hum Genet*. 1983;35(2):301–308.
33. Hook EB, Schreinemachers DM, Willey AM, Cross PK. Inherited structural cytogenetic abnormalities detected incidentally in fetuses diagnosed prenatally: frequency, parental-age associations, sex-ratio trends, and comparisons with rates of mutants. *Am J Hum Genet*. 1984;36(2):422–443.
34. Stankiewicz P, Lupski JR. Structural variation in the human genome and its role in disease. *Annu Rev Med*. 2010;61:437–455.
35. Inoue K, Lupski JR. Molecular mechanisms for genomic disorders. *Annu Rev Genomics Hum Genet*. 2002;3:199–242.
36. Liu P, Carvalho CM, Hastings PJ, Lupski JR. Mechanisms for recurrent and complex human genomic rearrangements. *Curr Opin Genet Dev*. 2012;22(3):211–220.
37. Lupski JR, Stankiewicz P. Genomic disorders: molecular mechanisms for rearrangements and conveyed phenotypes. *PLoS Genet*. 2005;1(6):e49.
38. Lupski JR. Genome structural variation and sporadic disease traits. *Nat Genet*. 2006;38(9):974–976.
39. Shaffer LG, Lupski JR. Molecular mechanisms for constitutional chromosomal rearrangements in humans. *Annu Rev Genet*. 2000;34:297–329.
40. Reiter LT, Hastings PJ, Nelis E, De Jonghe P, Van Broeckhoven C, Lupski JR. Human meiotic recombination products revealed by sequencing a hotspot for homologous strand exchange in multiple HNPP deletion patients. *Am J Hum Genet*. 1998;62(5):1023–1033.
41. Myers S, Freeman C, Auton A, Donnelly P, McVean G. A common sequence motif associated with recombination hot spots and genome instability in humans. *Nat Genet*. 2008;40(9):1124–1129.
42. Lupski JR. Hotspots of homologous recombination in the human genome: not all homologous sequences are equal. *Genome Biol*. 2004;5(10):242.
43. Lindsay SJ, Khajavi M, Lupski JR, Hurles ME. A chromosomal rearrangement hotspot can be identified from population genetic variation and is coincident with a hotspot for allelic recombination. *Am J Hum Genet*. 2006;79(5):890–902.
44. Zhang F, Carvalho CM, Lupski JR. Complex human chromosomal and genomic rearrangements. *Trends Genet*. 2009;25(7):298–307.
45. Lupski JR. Genomic disorders: structural features of the genome can lead to DNA rearrangements and human disease traits. *Trends Genet*. 1998;14(10):417–422.
46. Lee JA, Madrid RE, Sperle K, et al. Spastic paraplegia type 2 associated with axonal neuropathy and apparent PLP1 position effect. *Ann Neurol*. 2006;59(2):398–403.
47. Zhang F, Seeman P, Liu P, et al. Mechanisms for nonrecurrent genomic rearrangements associated with CMT1A or HNPP: rare CNVs as a cause for missing heritability. *Am J Hum Genet*. 2010;86(6):892–903.
48. Bi W, Borgan C, Pursley AN, Hixson P, et al. Comparison of chromosome analysis and chromosomal microarray analysis: what is the value of chromosome analysis in today's genomic array era? *Genet Med*. 2012;15:450–457.
49. Kajiwara K, Berson EL, Dryja TP. Digenic retinitis pigmentosa due to mutations at the unlinked peripherin/RDS and ROM1 loci. *Science*. 1994;264(5165):1604–1608.
50. Goldberg AF, Molday RS. Defective subunit assembly underlies a digenic form of retinitis pigmentosa linked to mutations in peripherin/rds and rom-1. *Proc Natl Acad Sci U S A*. 1996;93(24):13726–13730.
51. del Castillo I, Villamar M, Moreno-Pelayo MA, et al. A deletion involving the connexin 30 gene in nonsyndromic hearing impairment. *N Engl J Med*. 2002;346(4):243–249.
52. Katsanis N, Ansley SJ, Badano JL, et al. Triallelic inheritance in Bardet-Biedl syndrome, a Mendelian recessive disorder. *Science*. 2001;293(5538):2256–2259.
53. Eichers ER, Lewis RA, Katsanis N, Lupski JR. Triallelic inheritance: a bridge between Mendelian and multifactorial traits. *Ann Med*. 2004;36(4):262–272.
54. Lupski JR. Digenic inheritance and Mendelian disease. *Nat Genet*. 2012;44(12):1291–1292.
55. Lemmers RJ, Tawil R, Petek LM, et al. Digenic inheritance of an SMCHD1 mutation and an FSHD-permissive D4Z4 allele causes facioscapulohumeral muscular dystrophy type 2. *Nat Genet*. 2012;44(12):1370–1374.
56. Dipple KM, McCabe ER. Modifier genes convert "simple" Mendelian disorders to complex traits. *Mol Genet Metab*. 2000;71(1–2):43–50.
57. Knowles MR, Drumm M. The influence of genetics on cystic fibrosis phenotypes. *Cold Spring Harb Perspect Med*. 2012;2(12):a009548.
58. Robberecht W, Philips T. The changing scene of amyotrophic lateral sclerosis. *Nat Rev Neurosci*. 2013;14(4):248–264.
59. DeJesus-Hernandez M, Mackenzie IR, Boeve BF, et al. Expanded GGGGCC hexanucleotide repeat in noncoding region of C9ORF72 causes chromosome 9p-linked FTD and ALS. *Neuron*. 2011;72(2):245–256.
60. Giess R, Holtmann B, Braga M, et al. Early onset of severe familial amyotrophic lateral sclerosis with a SOD-1 mutation: potential impact of CNTF as a candidate modifier gene. *Am J Hum Genet*. 2002;70(5):1277–1286.
61. Meggouh F, de Visser M, Arts WF, De Coo RI, van Schaik IN, Baas F. Early onset neuropathy in a compound form of Charcot–Marie–Tooth disease. *Ann Neurol*. 2005;57(4):589–591.
62. Hodapp JA, Carter GT, Lipe HP, Michelson SJ, Kraft GH, Bird TD. Double trouble in hereditary neuropathy: concomitant mutations in the PMP-22 gene and another gene produce novel phenotypes. *Arch Neurol*. 2006;63(1):112–117.
63. Chung KW, Sunwoo IN, Kim SM, et al. Two missense mutations of EGR2 R359W and GJB1 V136A in a Charcot–Marie–Tooth disease family. *Neurogenetics*. 2005;6(3):159–163.
64. Wirth B, Garbes L, Riessland M. How genetic modifiers influence the phenotype of spinal muscular atrophy and suggest future therapeutic approaches. *Curr Opin Genet Dev*. 2013;23(3):330–338.
65. Shih RA, Belmonte PL, Zandi PP. A review of the evidence from family, twin and adoption studies for a genetic contribution to adult psychiatric disorders. *Int Rev Psychiatry*. 2004;16(4):260–283.
66. Dyment DA, Ebers GC, Sadovnick AD. Genetics of multiple sclerosis. *Lancet Neurol*. 2004;3(2):104–110.
67. Manolio TA. Genomewide association studies and assessment of the risk of disease. *N Engl J Med*. 2010;363(2):166–176.

68. Frazer KA, Murray SS, Schork NJ, Topol EJ. Human genetic variation and its contribution to complex traits. *Nat Rev Genet.* 2009;10(4):241–251.
69. Lupski JR, Belmont JW, Boerwinkle E, Gibbs RA. Clan genomics and the complex architecture of human disease. *Cell.* 2011;147(1):32–43.
70. Kiezun A, Garimella K, Do R, et al. Exome sequencing and the genetic basis of complex traits. *Nat Genet.* 2012;44(6):623–630.
71. The international HapMap project. *Nature.* 2003;426(6968):789–796.
72. A haplotype map of the human genome. *Nature.* 2005;437(7063):1299–1320.
73. Theuns J, Brouwers N, Engelborghs S, et al. Promoter mutations that increase amyloid precursor-protein expression are associated with Alzheimer disease. *Am J Hum Genet.* 2006;78(6):936–946.
74. Maraganore DM, de Andrade M, Elbaz A, et al. Collaborative analysis of alpha-synuclein gene promoter variability and Parkinson disease. *JAMA.* 2006;296(6):661–670.
75. Weiss LA, Shen Y, Korn JM, et al. Association between microdeletion and microduplication at 16p11.2 and autism. *N Engl J Med.* 2008;358(7):667–675.
76. Erez A, Patel AJ, Wang X, et al. Alu-specific microhomology-mediated deletions in CDKL5 in females with early-onset seizure disorder. *Neurogenetics.* 2009;10(4):363–369.
77. Rumble B, Retallack R, Hilbich C, et al. Amyloid A4 protein and its precursor in Down's syndrome and Alzheimer's disease. *N Engl J Med.* 1989;320(22):1446–1452.
78. Rovelet-Lecrux A, Hannequin D, Raux G, et al. APP locus duplication causes autosomal dominant early-onset Alzheimer disease with cerebral amyloid angiopathy. *Nat Genet.* 2006;38(1):24–26.
79. Chartier-Harlin MC, Kachergus J, Roumier C, et al. Alpha-synuclein locus duplication as a cause of familial Parkinson's disease. *Lancet.* 2004;364(9440):1167–1169.
80. Ibanez P, Bonnet AM, Debarges B, et al. Causal relation between alpha-synuclein gene duplication and familial Parkinson's disease. *Lancet.* 2004;364(9440):1169–1171.
81. Singleton AB, Farrer M, Johnson J, et al. alpha-Synuclein locus triplication causes Parkinson's disease. *Science.* 2003;302(5646):841.
82. Allikmets R, Shroyer NF, Singh N, et al. Mutation of the Stargardt disease gene (ABCR) in age-related macular degeneration. *Science.* 1997;277(5333):1805–1807.
83. Trinh J, Farrer M. Advances in the genetics of Parkinson disease. *Nat Rev Neurol.* 2013;9:445–454.
84. Swan M, Saunders-Pullman R. The association between ss-glucocerebrosidase mutations and parkinsonism. *Curr Neurol Neurosci Rep.* 2013;13(8):368.
85. Auclair G, Weber M. Mechanisms of DNA methylation and demethylation in mammals. *Biochimie.* 2012;94(11):2202–2211.
86. Winkelmann J, Lin L, Schormair B, et al. Mutations in DNMT1 cause autosomal dominant cerebellar ataxia, deafness and narcolepsy. *Hum Mol Genet.* 2012;21(10):2205–2210.
87. Corti O, Lesage S, Brice A. What genetics tells us about the causes and mechanisms of Parkinson's disease. *Physiol Rev.* 2011;91(4):1161–1218.
88. A two-stage meta-analysis identifies several new loci for Parkinson's disease. *PLoS Genet.* 2011;7(6):e1002142.
89. Evans-Galea MV, Carrodus N, Rowley SM, et al. FXN methylation predicts expression and clinical outcome in Friedreich ataxia. *Ann Neurol.* 2012;71(4):487–497.
90. Li J, Chen J, Ricupero CL, et al. Nuclear accumulation of HDAC4 in ATM deficiency promotes neurodegeneration in ataxia telangiectasia. *Nat Med.* 2012;18(5):783–790.
91. Janssen C, Schmalbach S, Boeselt S, Sarlette A, Dengler R, Petri S. Differential histone deacetylase mRNA expression patterns in amyotrophic lateral sclerosis. *J Neuropathol Exp Neurol.* 2010;69(6):573–581.
92. Cenik B, Sephton CF, Dewey CM, et al. Suberoylanilide hydroxamic acid (vorinostat) up-regulates progranulin transcription: rational therapeutic approach to frontotemporal dementia. *J Biol Chem.* 2011;286(18):16101–16108.
93. Graff J, Rei D, Guan JS, et al. An epigenetic blockade of cognitive functions in the neurodegenerating brain. *Nature.* 2012;483(7388):222–226.
94. Sung YM, Lee T, Yoon H, et al. Mercaptoacetamide-based class II HDAC inhibitor lowers Abeta levels and improves learning and memory in a mouse model of Alzheimer's disease. *Exp Neurol.* 2013;239:192–201.
95. de Faria Jr O, Moore CS, Kennedy TE, Antel JP, Bar-Or A, Dhaunchak AS. MicroRNA dysregulation in multiple sclerosis. *Front Genet.* 2012;3:311.
96. Bhalala OG, Srikanth M, Kessler JA. The emerging roles of microRNAs in CNS injuries. *Nat Rev Neurol.* 2013;9(6):328–339.
97. Gehrke S, Imai Y, Sokol N, Lu B. Pathogenic LRRK2 negatively regulates microRNA-mediated translational repression. *Nature.* 2010;466(7306):637–641.
98. Szafranski P, Dharmadhikari AV, Brosens E, et al. Small noncoding differentially methylated copy-number variants, including lncRNA genes, cause a lethal lung developmental disorder. *Genome Res.* 2013;23(1):23–33.
99. Huang HS, Allen JA, Mabb AM, et al. Topoisomerase inhibitors unsilence the dormant allele of Ube3a in neurons. *Nature.* 2012;481(7380):185–189.
100. Lander ES, Linton LM, Birren B, et al. Initial sequencing and analysis of the human genome. *Nature.* 2001;409(6822):860–921.
101. Abecasis GR, Altshuler D, Auton A, et al. A map of human genome variation from population-scale sequencing. *Nature.* 2010;467(7319):1061–1073.
102. Abecasis GR, Auton A, Brooks LD, et al. An integrated map of genetic variation from 1,092 human genomes. *Nature.* 2012;491(7422):56–65.
103. Sebat J, Lakshmi B, Troge J, et al. Large-scale copy number polymorphism in the human genome. *Science.* 2004;305(5683):525–528.
104. Iafrate AJ, Feuk L, Rivera MN, et al. Detection of large-scale variation in the human genome. *Nat Genet.* 2004;36(9):949–951.
105. Gonzaga-Jauregui C, Lupski JR, Gibbs RA. Human genome sequencing in health and disease. *Annu Rev Med.* 2012;63:35–61.
106. Albert TJ, Molla MN, Muzny DM, et al. Direct selection of human genomic loci by microarray hybridization. *Nat Methods.* 2007;4(11):903–905.
107. Bainbridge MN, Wang M, Burgess DL, et al. Whole exome capture in solution with 3 Gbp of data. *Genome Biol.* 2010;11(6):R62.
108. Ng SB, Buckingham KJ, Lee C, et al. Exome sequencing identifies the cause of a mendelian disorder. *Nat Genet.* 2010;42(1):30–35.
109. Hoischen A, van Bon BW, Gilissen C, et al. *De novo* mutations of SETBP1 cause Schinzel-Giedion syndrome. *Nat Genet.* 2010;42(6):483–485.

110. Ng SB, Bigham AW, Buckingham KJ, et al. Exome sequencing identifies MLL2 mutations as a cause of Kabuki syndrome. *Nat Genet.* 2010;42(9):790–793.
111. McDonell LM, Mirzaa GM, Alcantara D, et al. Mutations in STAMBP, encoding a deubiquitinating enzyme, cause microcephaly-capillary malformation syndrome. *Nat Genet*. 2013;45(5):556–562.
112. Hood RL, Lines MA, Nikkel SM, et al. Mutations in SRCAP, encoding SNF2-related CREBBP activator protein, cause Floating-Harbor syndrome. *Am J Hum Genet*. 2012;90(2):308–313.
113. Lupski JR, Reid JG, Gonzaga-Jauregui C, et al. Whole-genome sequencing in a patient with Charcot-Marie-Tooth neuropathy. *N Engl J Med*. 2010;362(13):1181–1191.
114. Bainbridge MN, Wiszniewski W, Murdock DR, et al. Whole-genome sequencing for optimized patient management. *Sci Transl Med*. 2011;3(87):87re83.
115. Rios J, Stein E, Shendure J, Hobbs HH, Cohen JC. Identification by whole-genome resequencing of gene defect responsible for severe hypercholesterolemia. *Hum Mol Genet*. 2010;19(22):4313–4318.
116. Wang L, McLeod HL, Weinshilboum RM. Genomics and drug response. *N Engl J Med*. 2011;364(12):1144–1153.
117. Korf BR, Rehm HL. New approaches to molecular diagnosis. *JAMA*. 2013;309(14):1511–1521.
118. Green RC, Lupski JR, Biesecker LG. Reporting genomic sequencing results to ordering clinicians: incidental, but not exceptional. *JAMA*. 2013;310(4):365–366.
119. McGuire AL, McCullough LB, Evans JP. The indispensable role of professional judgment in genomic medicine. *JAMA*. 2013;309(14):1465–1466.
120. Rodriguez LL, Brooks LD, Greenberg JH, Green ED. Research ethics. The complexities of genomic identifiability. *Science*. 2013;339(6117):275–276.

CHAPTER

2

Genotype–Phenotype Correlations

Thomas D. Bird and Marie Y. Davis

University of Washington, Seattle, WA, USA

INTRODUCTION

Neurogenetics is experiencing a fascinating interplay between the fields of molecular biology and clinical neurology. The clinical description of diseases leads to the discovery of underlying genes, which, in turn, leads to a more accurate classification of the diseases, which then allows the discovery of additional genes. This circle of understanding has really just begun and will continue for many years. This also has led to a tremendous effort to tease out the mechanisms at the biochemical and physiologic levels connecting gene mutations to disease manifestation. This field of genotype–phenotype correlations, although only in its infancy, is already an area of widespread investigation that cannot possibly be fully documented here. Instead, our aim is to highlight a few models that are likely to generate new knowledge in the near future and will serve as common ground for communication and collaboration between bench and clinical scientists.

SINGLE PHENOTYPE: MULTIPLE GENES

Many relatively common neurological phenotypes have proved to have multiple genetic etiologies. This is the phenomenon of nonallelic genetic heterogeneity. Figure 2.1 indicates seven neurogenetic phenotypes and the diverse array of genes thus far discovered to underlie them.

The most impressive examples of this phenomenon are the autosomal dominant spinocerebellar ataxias (SCAs).[1] The common characteristics of this syndrome are progressive cerebellar atrophy associated with unsteady gait and poor coordination. Mutations in no less than 36 different genes/loci have been associated with this general phenotype. The genes discovered to date probably represent about 75% of the affected families.[2] Thus, there are additional genes that remain to be discovered. This situation reflects the complexity of cerebellar circuits involving several different types of neurons and their connections. The number of proteins that must be involved in these neuronal circuits suggests that a large number of genes could potentially produce this phenotype. It is of interest that many of the known mutations producing cerebellar ataxia are trinucleotide repeat expansions. The association of this particular type of mutation with diseases of the nervous system represents a fruitful field for further investigation. In general, the ataxia phenotype does not predict the genotype. A careful examination of the ataxia phenotype, however, leads to some consistent clinical differences among these diseases, such as the association of retinopathy with SCA 7, peripheral neuropathy with SCA 4, and dementia or mental retardation with several of the others.[3]

A similar story occurs with the various genetic forms of epilepsy.[4] The common phenotype is recurrent seizures associated with abnormalities in the electroencephalogram (EEG). The phenotypes sometimes can be subdivided on the bases of age of onset (infantile vs. childhood), location of seizure focus (frontal, temporal, or occipital lobes) or other features (nocturnal or febrile). Several of the identified genes encode ion channel proteins. Many genes remain to be discovered and are indicated only by regional chromosomal location.

Alzheimer disease (AD) is of special interest because there is both a pathologic and clinical phenotype. The clinical disease is progressive dementia and the neuropathologic characteristics are diffuse cerebral cortical atrophy with neuronal loss, neuritic amyloid plaques, amyloid angiopathy, and neuronal neurofibrillary tangles. Mutations in

Rosenberg's Molecular and Genetic Basis of Neurological and Psychiatric Disease
http://dx.doi.org/10.1016/B978-0-12-410529-4.00002-4

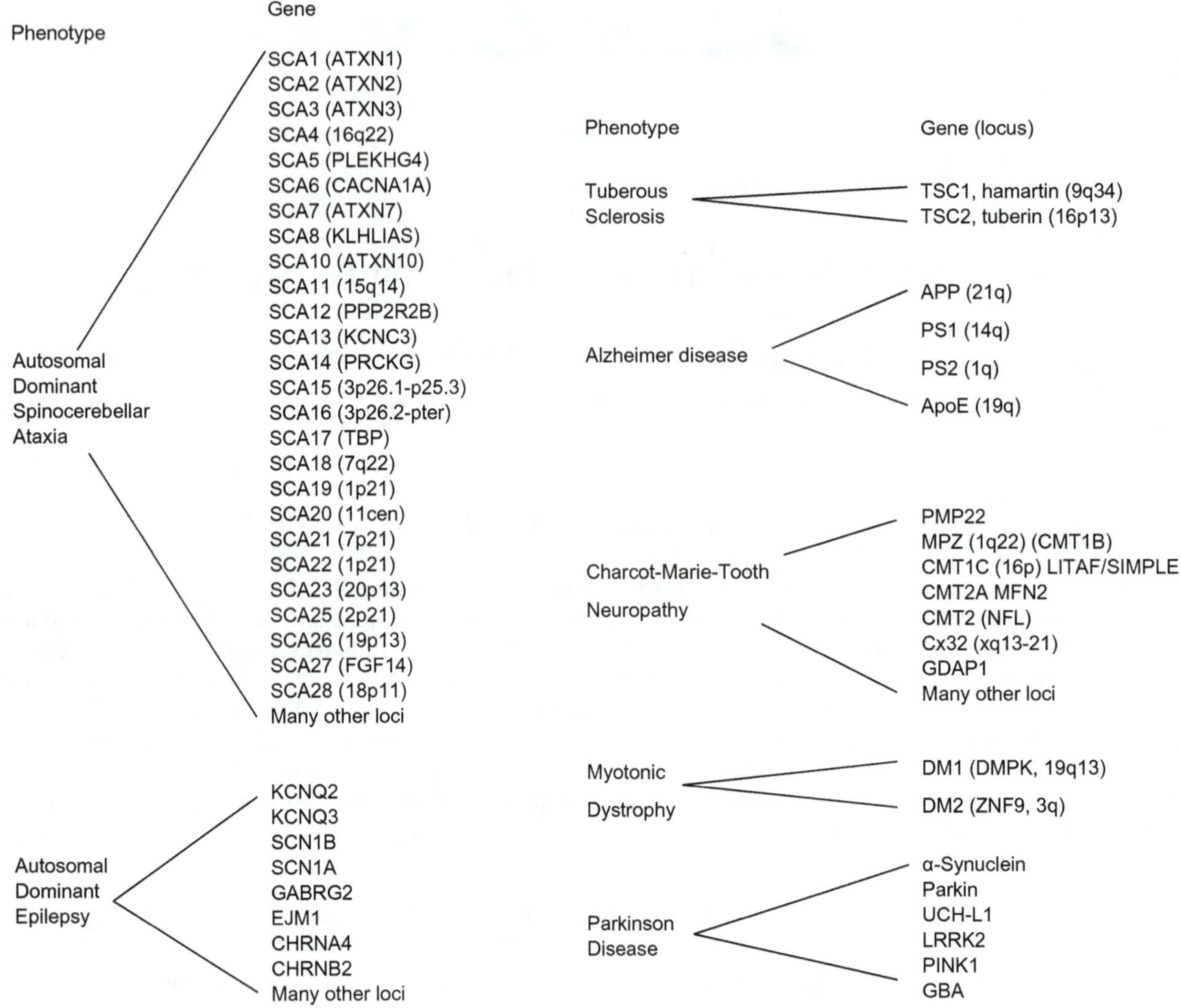

FIGURE 2.1 **Multiple genetic etiologies of common phenotypes.**

three distinct genes (amyloid precursor protein [*APP*], presenilin 1 [*PS1*], presenilin 2 [*PS2*]) all produce essentially an identical clinical and pathologic phenotype: early onset familial AD.[5] The phenotypes are so similar that neither an astute clinician nor an experienced neuropathologist can accurately predict the underlying genotype in any single case. The three genes all influence the same biochemical pathway: the production of Aβ amyloid. Apolipoprotein E also seems to influence this amyloid system. This knowledge has led to an intense investigation of how this accumulation results in the disease phenotype and how the presenilin proteins may be involved in amyloid metabolism.

The phenotype of Parkinson disease is caused by mutations in more than six identified genes (*α-synuclein*, *parkin*, *LRRK2*, *PINK1*, *GBA*, *DJ-1* and *UCH-L1*) and unknown genes at several other loci.[6] Several of the known genes are all involved with the ubiquitin–proteosome complex. Persons with *parkin* mutations are characterized by earlier age of onset and lack of Lewy body pathology whereas those with mutations in *LRRK2* have later onset and Lewy body pathology. Animal models have demonstrated that *PINK1* and *parkin* are important for mitochondrial quality control, adding to growing evidence that Parkinson disease may be due to mitochondrial dysfunction.[7]

The multigenerational Charcot–Marie–Tooth (CMT) hereditary neuropathy syndrome can be caused by mutations in more than 10 different genes in addition to many autosomal recessive genes.[8,9] Myotonic muscular dystrophy (DM) and tuberous sclerosis (TS) have each been associated with mutations in two separate genes.[10,11] The phenotypes within each of these three categories are so similar and overlapping that the genotype cannot be predicted consistently from the individual clinical presentation. Just as in the cerebellar ataxias and the epilepsies, this indicates multiple mechanisms affecting a final common pathway, such as peripheral nerve function in CMT and loss of tumor suppression in TS.

SINGLE GENE: MULTIPLE PHENOTYPES

Conversely, there are many examples of mutations in a single gene producing different and sometimes quite distinct phenotypes (Figure 2.2). A voltage-dependent calcium channel gene (*CACNA1A*) on chromosome 19p13 is associated with three different phenotypes.[12,13] A small CAG trinucleotide repeat expansion in the coding region of this gene produces an autosomal dominant, slowly progressive, late-onset cerebellar ataxia (SCA 6). Point mutations resulting in protein truncation produce an episodic ataxia (EA 2) in which individuals are clinically normal between episodes. Missense mutations are associated with familial hemiplegic migraine (FHM). Some persons with FHM eventually develop cerebellar atrophy and ataxia. Although different, each of these three phenotypes involves a cerebellar theme. This genotype–phenotype association has stimulated investigations into the involvement of ion channels in more common forms of migraine and cerebellar physiology.

Mutations in the peripheral myelin protein 22 gene (*PMP22*) can produce three different forms of familial neuropathy.[8,9,14] Duplication of this gene causes autosomal dominant CMT with slow nerve conductions (CMT1A), whereas deletion of *PMP22* results in hereditary neuropathy with liability to pressure palsy (HNPP). Thus, there is a dosage effect in which too much or too little of the protein produces peripheral nerve dysfunction. Also, missense point mutations in this gene have been associated with severe, early onset neuropathy of infancy or childhood called Dejerine–Sottas syndrome (DSS).

Mutations in the microtubule-associated protein tau can produce a variety of phenotypes, including frontotemporal dementia (FTD), atypical psychosis, and progressive aphasia.[15] At the neuropathologic level, mutations in exon 10 of tau are associated with an accumulation of tau proteins in neurons and glial cells, whereas mutations in exons 12 and 13 produce tau accumulation only in neurons. Some mutations in this gene cause tau to abnormally interact with microtubules, whereas others cause a change in the proportions of two tau isoforms (three-repeat and four-repeat tau). A specific haplotype of tau is also a risk factor for progressive supranuclear palsy (PSP), another degenerative neurological phenotype with abnormal neuronal tau inclusions. In addition, a hexanucleotide repeat expansion on chromosome 9 (*C9ORF72*) can be associated with FTD or amyotrophic lateral sclerosis (ALS) or a combination of FTD with motor neuron disease.[16]

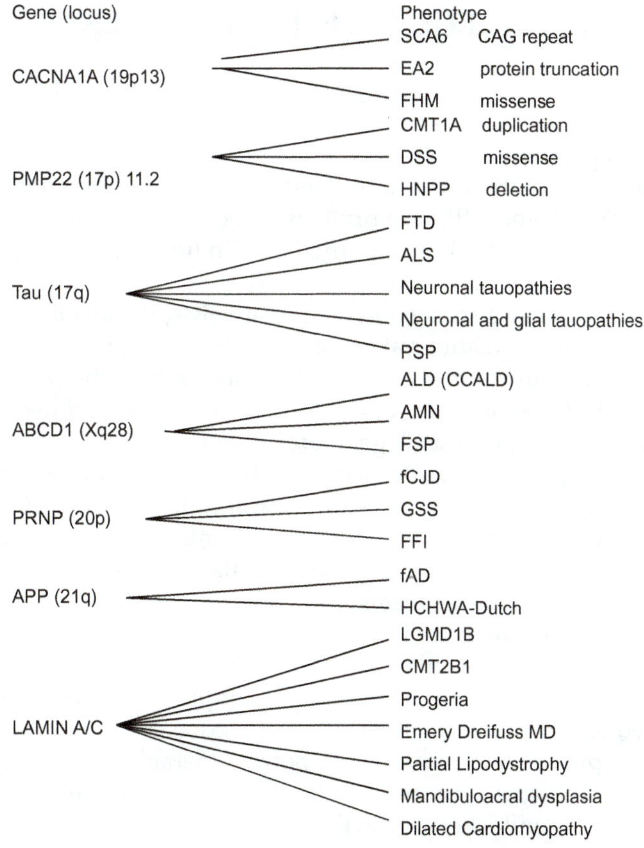

FIGURE 2.2 Multiple phenotypes in single genes.

Mutations in the *ABCD1* gene on the X chromosome affect a peroxisomal ABC half transporter involved in the import of very long-chain fatty acids.[17] For reasons that remain obscure, different mutations in this gene can result in severe, early onset adrenoleukodystrophy or milder, later onset adrenomyeloneuropathy, or a relatively uncomplicated spastic paraplegia.

Lesch–Nyhan disease provides an excellent model of phenotypic variability associated with different mutations in a single gene (*HPRT1*) often related to differential effects on the resultant enzyme activity.[18] The phenotype ranges from asymptomatic overexcretion of uric acid in the urine to a severe neurobehavioral syndrome that may include self-mutilation and HPRT1 enzyme activity that may be only mildly reduced to completely absent.

Mutations in the prion protein gene (*PRNP*) on chromosome 20 produce the severe, rapidly progressive, dementing syndrome of familial Creutzfeldt–Jakob disease (fCJD).[19] Some mutations in the same gene produce a somewhat less rapidly progressive disorder often associated with ataxia and cerebral cortical prion-containing plaques known as Gerstmann–Sträussler–Scheinker (GSS) disease. A missense mutation at codon 178 may produce the unusual phenotype of familial fatal insomnia (FFI). Whether the mutation at codon 178 results in CJD or FFI depends on the presence or absence of methionine at codon 129. Thus the interaction between a mutation and a normal polymorphism in the same gene can alter the resulting phenotype through mechanisms that are not understood.

Missense mutations in the *APP* gene usually result in familial, early onset Alzheimer disease. However, a specific mutation at codon 693 occurring in a large Dutch family produces primarily an amyloid angiopathy associated with severe hemorrhagic stroke (HCHWA-D).[20] Both phenotypes have the common phenomenon of amyloid accumulation, but it is unclear why the Dutch mutation primarily directs the accumulation to cerebral blood vessels.

Mutations in lamin A/C produce some of the most striking varieties of clinical phenotypes: at least eight different clinical presentations affecting nerve, muscle, fat or growth. The critical role of lamin A/C in the nuclear envelope and the envelope's central role in cellular function must be related to this variety of manifestations.[21]

Mutations in the glucocerebrosidase (*GBA*) gene produce an interesting phenomenon in which the initial phenotype is Gaucher disease (hepatosplenomegaly and bone involvement), but this may be followed much later in adult life with a parkinsonian disorder (in both homo- and heterozygotes).[22]

NEURONAL/CELLULAR SELECTIVE VULNERABILITY

An important issue in moving from genotype to phenotype is why only certain cell types become dysfunctional, degenerate, and die, while thousands of others remain healthy. The vulnerable cell type will obviously direct the clinical phenotype and suggests potential targets for clinical treatment. Factors determining cellular vulnerability must vary from disease to disease and are not well understood for many disorders. A few are obvious and fairly straightforward. For example, PMP22 and MPZ are proteins associated with myelin in the peripheral nervous system. Mutations disrupting these proteins result in abnormal myelin function, peripheral nerve disease, and the syndrome of hereditary neuropathy (CMT).

Examples of several other neurogenetic diseases and their relatively specific cell type targets are listed in Figure 2.3. For each disease, there are ongoing investigations attempting to determine precisely how each mutation is directed to disrupt function in only selected neurons. For example, normal ataxin 7, the protein relevant to SCA 7, has been found to be widely expressed in brain, retina, and other peripheral tissues, and even in the central nervous system (CNS) it is not limited to areas in which neurons degenerate.[23,24] In general, differential cellular expression of genes rarely seems to explain the selective phenotypic effects on specific cell populations. Ataxin 7 has been shown to have a functional nuclear localization signal at codons 378 to 393 and localizes preferentially to the nuclear compartment of cells in the cerebellum, pons, inferior olive, and retinal photoreceptor cells.[25] LaSpada and coworkers have further demonstrated in transgenic mice that polyglutamine-expanded ataxin 7 can suppress the transactivation of CRX, a cone-rod homeobox protein.[26] This interference seems to account for the retinal degeneration observed in the phenotype in SCA 7 and provides an explanation for how cell-type specificity is achieved in this polyglutamine-repeat disease. There is also non-cell autonomous degeneration in Purkinje cells related to surrounding Bergmann glia.[27] Similar approaches are being taken for the other diseases. For example, it needs to be explained why the mutation in chorea acanthocytosis is limited to certain neurons as well as red blood cell membranes.[28,29] These two very different cell types must have a common protein that makes them both vulnerable to mutations in this gene. Likewise, the special vulnerability of dorsal root ganglia sensory neurons and cardiac muscle in Friedreich ataxia must be related to the involvement of iron transport in mitochondria in this disease.[30] It is clear that mutations in glial fibrillary acidic protein (GFAP) cause dysfunction in astrocytes and a resultant leukodystrophy (Alexander disease).[31] Dipeptide

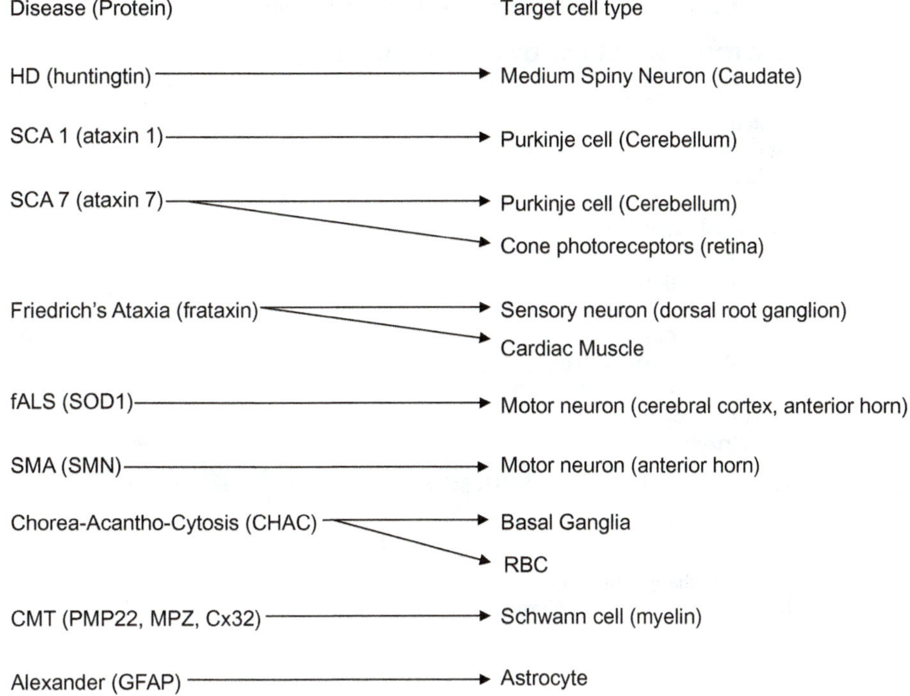

FIGURE 2.3 **Neurogenetic diseases and their specific cell type targets.**

repeat proteins may play an important role in the neurodegeneration (ALS/FTD) associated with the C9ORF72 hexanucleotide repeat mutation.[32]

Facioscapulohumeral muscular dystrophy (FSHD) has been found to be caused by released expression of the *DUX4* gene on chromosome 4q.[33] It primarily affects muscles (although sometimes retina), and first preferentially affects face, shoulder girdle and distal leg muscles.[34] The reason for this muscle selectivity is unknown, as is the explanation for why the muscle weakness is often asymmetrical.

HIGHLY VARIABLE SYSTEMIC PHENOTYPES

Although mutations in some genes result in abnormal function in only single or a limited number of cell types, mutations in other genes produce a remarkable variety of abnormalities in many different organs and cell types. Three examples are shown in Table 2.1, namely myotonic muscular dystrophy (DM1), tuberous sclerosis (TS), and neurofibromatosis type 1 (NF1). DM1 is associated with a CTG repeat expansion in the 3′ untranslated region (UTR) of the *DMPK* gene on chromosome 19q. The disease affects skeletal, smooth and cardiac muscle, skin, ocular lens, endocrine glands, skull, and brain. The molecular mechanism appears to include expanded CUG repeats in RNA, having a toxic gain-of-function effect on splicing of other genes controlling cell metabolism in a variety of organs.[10,35–37]

The two types of TS and NF1 have many phenotypic characteristics in common. Both are neurocutaneous syndromes affecting the nervous system and skin. They are associated with a wide variety of tumors and cancers.[38,39] They also have genetic molecular similarities in that they are caused by mutations in various documented or presumed tumor suppressor genes.[40] The highly variable expression fits the Knudson two-hit hypothesis.[41] Affected persons are heterozygotes who have inherited a single tumor suppressor gene mutation from one parent. Over a lifetime, random environmentally acquired mutations in the homologous gene on the other chromosome (the second hit) produce a loss of heterozygosity, loss of tumor suppression, and unregulated cell growth. Furthermore, early somatic mutations in these genes may produce individuals who are mosaics and have only partial or anatomically restricted phenotypes.[42,43] The two different TS genes (hamartin and tuberin) affect the same molecular pathway through direct interaction with each other.[44]

Because mitochondria play a critical role in cellular energy metabolism, it is not surprising that mutations in mitochondrial DNA result in widespread systematic symptoms. MERRF (myoclonic epilepsy with ragged red fibers)

TABLE 2.1 Single Genes with Wide-Ranging Systemic Phenotypes

MYOTONIC MUSCULAR DYSTROPHY (DM1)
Muscle weakness
Myotonia
Cataract
Cardiac arrhythmia
Diabetes
Intestinal dysmotility
Baldness
Gall bladder disease
Thick skull/large sinuses
Cognitive deficits
Cerebral white matter changes
Sleep apnea
Testicular atrophy
Pilomatrixomas
TUBEROUS SCLEROSIS (TSC1/TSC2)
Mental retardation
Epilepsy
Cutaneous changes (angiofibroma, hypomelanotic macules, shagreen patches, ungual fibroma)
Brain tumors/nodules
Kidney tumors
Cardiac rhabdomyomas
Pulmonary lymphangiomyomatosis
Retinal nodules
NEUROFIBROMATOSIS (NF1)
Neurofibromas
Café au lait spots
Lisch nodules (iris)
Skeletal anomalies
Brain gliomas
Mental retardation/learning disabilities
Seizures
Vascular dysplasia
Malignant tumors
Axillary freckling
Optic glioma
Hypertension
Leukemia
MELAS (MITOCHONDRIAL DNA MUTATIONS)
Seizures
Headache
Stroke
Vomiting
Hearing loss
Short stature
Impaired cognition
Diabetes
Muscle weakness
Cardiac conduction defect
Visual impairment

and MELAS (mitochondrial encephalomyopathy, lactic acidosis, and stroke-like episodes) are two examples of this systemic phenotype.[45] Also there is increasing evidence that multiple neurodegenerative disease, such as Parkinson, Huntington, Alzheimer, ALS and several ataxias have defects in mitochondrial function.[46] Further research is necessary to understand whether mitochondrial dysfunction is a common endpoint in a neurodegenerative pathway or a primary tissue-specific dysfunction in mitochondria in each of these disease processes.

PENETRANCE AND AGE OF ONSET

Penetrance refers to the proportion of gene mutation carriers who express any aspect of the phenotype. Penetrance is age- and test-dependent. That is, at a young age only a small proportion of mutation carriers may manifest the phenotype, whereas at a later age a much larger proportion will do so. An example of test dependence would be a person carrying an epilepsy gene mutation that may never have a clinical seizure but may have an abnormal EEG with electrical seizure discharges. The penetrance is only revealed by the more sensitive test. Likewise, some persons with the DM1 mutation may have only subclinical cataract on slit lamp examination or persons with a TS mutation may have no symptoms or signs other than intracranial calcification on imaging studies. The genetic factors influencing age of onset and incomplete penetrance are poorly understood. Several possible mechanisms are presented in Tables 2.2 and 2.3.

The rate at which a cell or organ accumulates a toxic substance certainly correlates with the age at onset of symptoms. The best examples are the genetic lipid storage diseases. Mutations result in defective enzymatic metabolism of a substance that is toxic to the cell, such as the toxic accumulation of sulfatide in metachromatic leukodystrophy, glucocerebroside in Krabbe leukodystrophy and nonesterified cholesterol in Niemann–Pick type C disease. Some mutations result in a complete absence of enzyme activity, relatively rapid toxin accumulation, and early onset. Other mutations allow partial, very low enzyme activity, slower toxin accumulation, and later onset. Some storage disorders such as Niemann–Pick type C have a wide range in age of onset without a clear-cut genotype–phenotype correlation.[47] Slow accumulation of a toxic protein may play a role in many other neurodegenerative diseases such as AD (Aβ amyloid), FTD (tau), Huntington disease (HD), and other CAG repeat (polyglutamine) expansion disorders.

Duchenne and Becker muscular dystrophies are overlapping clinical syndromes caused by a variety of mutations in the same dystrophin gene.[48] The earlier onset Duchenne muscular dystrophy (DMD) phenotype is associated with mutations that result in the absence of the dystrophin protein. In Becker syndrome, the less severe phenotype, other mutations allow a small amount of protein to be produced, enable partial functional integrity of dystrophin, and result in a later onset.

Alzheimer disease has a very wide range in age of onset. Many of the early onset familial types are caused by single gene mutation in *APP, PS1,* and *PS2*. Nevertheless, the sporadic, late onset variety of AD has an at least 40-year range of onset (60 to 100 years). Apolipoprotein E (ApoE) influences this age of onset variability.[49] Through unclear mechanisms, the ApoE 4 allele is a risk factor for AD that lowers the average age of onset in a dose-dependent manner for persons carrying one or two copies of the E4 allele. The E2 allele may have a protective influence and delay the age of onset. Other unidentified genetic factors must also play a role in AD. For example, in the Volga German familial Alzheimer disease (FAD) kindreds carrying the same *PS2* mutation, there is a 35-year difference in the age of onset (40 to 75 years). Therefore, the ApoE genotype is responsible for only a fraction of the variation in age of onset of nonmonogenic forms of AD.

TABLE 2.2 Examples of Factors Affecting Penetrance and Age of Onset

ACCUMULATION OF A TOXIN OR PROTEIN
Lipid storage diseases
Polyglutamine disorders (?)
Alzheimer disease
Parkinson disease
COMPLETE VS. PARTIAL ABSENCE OF A PROTEIN/ENZYME
Duchenne muscular dystrophy vs. Becker muscular dystrophy
Infantile vs. juvenile metachromatic leukodystrophy
Infantile vs. adult Tay–Sachs disease
GENE/ENVIRONMENT INTERACTIONS
Acute intermittent porphyria
Knudson two-hit phenomenon/tumor suppression
UNKNOWN MECHANISMS
Alzheimer disease/apolipoprotein E
Huntington disease
Torsion dystonia (DYT1)
Adrenoleukodystrophy (ALD)
Frontotemporal dementia (progranulin)

TABLE 2.3 Correlation of Trinucleotide Repeat Expansion Size with Disease Penetrance and Severity in Huntington Disease and Myotonic Dystrophy

	HD (CAG)	DM1 (CTG)
Normal	2–26	5–37
Premutation	27–35	38–49
Decreased penetrance	36–39	50–100
Full penetrance	40+	100+
Severe early onset	70+	1000+

The trinucleotide-repeat expansion disorders have provided additional insights into the phenomena of age of onset and penetrance.[50] Both onset age and severity of disease correlate with the size of the repeat expansion. This is especially well illustrated with the CAG repeat in HD and the CTG repeat in DM1 (Table 2.3). The CAG expansion in HD has a normal size seen in the general population that is not associated with disease, a premutation range that is also never associated with clinical disease but may expand to a larger size in the next generation, a decreased penetrance range in which persons may or may not show clinical signs during a normal life span, and a full penetrance range in which persons always develop clinical signs during a normal life span.[51,52] The expansion size is also correlated with age of onset and accounts for up to 70% of variance in age onset.[53] HD individuals with 70 or greater CAG expansions often have both early onset and severe, more rapidly progressive disease.

A similar phenomenon occurs in DM1. Small CTG expansions may fall into a clearly abnormal range but be associated with minimal clinical expression (only cataracts), whereas expansions of 1000 or greater may be associated with severe, infantile onset (congenital myotonic dystrophy).[10,54]

Type 1 torsion dystonia (DYT1) is associated with a GAG deletion in codon 302 of the torsin A gene.[55] For unknown reasons, only 30–40% of mutation carriers ever manifest symptoms (30–40% penetrance). Other genetic or environmental factors, or both, must play a role.

The authors have seen an unpublished family with a mutation in the *TSC1* (hamartin) gene. Two persons in this family have had severe classic signs of tuberous sclerosis. However, another person in the family carrying the identical mutation displays no symptoms, signs, or manifestations of the phenotype on extensive imaging studies. Perhaps this family represents the wide range in random occurrence of second-mutation hits in the relevant tumor suppressor gene.

There has been an interesting report of monozygotic twins carrying the same mutation for X-linked adrenoleukodystrophy (ALD),[56] with a surprising and dramatic difference in disease severity between the two twins.

Certainly some of the differences in penetrance, age of onset, and expression of neurogenetic disease must be related to an extensive variety of environmental influences. It is easy to speculate on the nature of these environmental factors (smoking, alcohol, bacteria, viruses, chemicals, and so on), but it is much more difficult to document and establish their impact. Clearly, various drugs and medications may unmask the phenotype in a previously unrecognized gene mutation carrier. A well-known example is acute intermittent porphyria (AIP) in which the medical use of barbiturates or sulfa drugs may result in serious neurological symptoms in otherwise asymptomatic carriers of mutations in the gene encoding hydroxymethylbilane (HMB) synthase.[57] These drugs inhibit activity in the porphyrin–heme pathway that is already compromised by reduced activity of the HMB synthase. Genetic and environmental interactions such as these represent yet another area for extensive investigation into the mechanisms of genotype–phenotype correlations.

Finally, there is no doubt that epigenetic factors such as methylation and histone modification of chromatin play important roles in phenotypic variability.[58–60]

CONCLUSION AND FUTURE DIRECTIONS

The relationships between mutations in genes (genotype) and the associated clinical characteristics (phenotype) represent a fascinating, complicated, and developing area of investigation. In this chapter, we have given a brief overview of several aspects of this area, including: 1) single phenotypes caused by multiple genes; 2) single genes associated with multiple phenotypes; 3) neuronal and cellular selective vulnerability; 4) genes related to highly variable systemic phenotypes; and 5) issues concerning penetrance and age of onset. The next decade will experience an increased understanding of these phenomena and how our genetic constitution initiates and modifies both common and rare diseases of the nervous system.

References

1. Bird T, Jayadev S. Hereditary ataxia. Overview. *Genet Med.* 2013;15(9):673–683. doi:10.1038/gim.2013.28.
2. Durr A. Autosomal dominant cerebellar ataxias: polyglutamine expansions and beyond. *Lancet Neurol.* 2010;9:885–894.
3. Maschke M, Oehlert G, Xie TD, et al. Clinical feature profile of spinocerebellar ataxia type 1-8 predicts genetically defined subtypes. *Mov Disord.* 2005;20:1405–1412.
4. Helbig I, Scheffer IE, Mulley JC, Berkovic SF. Navigating the channels and beyond: unraveling the genetics of the epilepsies. *Lancet Neurol.* 2008;7:231–245.
5. Querfurth HW, LaFerla FM. Alzheimer's disease. *N Engl J Med.* 2010;362:329–344.
6. Alcalay RN, Caccappolo E, Mejia-Santana H, et al. Frequency of known mutations in early-onset Parkinson disease. *Arch Neurol.* 2010;67:1116–1122.
7. de Vries RL, Przedborski S. Mitophagy and Parkinson's disease: be eaten to stay healthy. *Mol Cell Neurosci.* 2013;55:37–43.
8. Scherer S. Finding the causes of inherited neuropathies. *Arch Neurol.* 2006;63:812–816.
9. Murphy SM, Laura M, Fawcett K, et al. Charcot–Marie–Tooth disease: frequency of genetic subtypes and guideline for genetic testing. *J Neurol Neurosurg Psychiatry.* 2012;83:706–710.
10. Udd B, Krahe R. The myotonic dystrophies: molecular, clinical, and therapeutic challenges. *Lancet Neurol.* 2012;11:891–905.
11. Jones AC, Shyamsundar MM, Thomas MW, et al. Comprehensive mutation analysis of TSC1 and TSC2 and phenotypic correlations in 150 families with tuberous sclerosis. *Am J Hum Genet.* 1999;64:1305–1315.
12. Romaniello R, Zucca C, Tonelli A, et al. A wide spectrum of clinical, neurophysiological and neuroradiological abnormalities in a family with a novel *CACNA1A* mutation. *J Neurol Neurosurg Psychiatry.* 2010;81:840–843.
13. Rajakulendran S, Schorge S, Kullmann DM, Hanna MG. Dysfunction of the $Ca_v2.1$ calcium channel in cerebellar ataxias. *F1000 Biol Rep.* 2010 Jan 18;2.
14. Li J. Inherited neuropathies. *Semin Neurol.* 2012;32:204–214.
15. Seelar H, Kamphorest W, Rosso SM, et al. Distinct genetic forms of frontotemporal dementia. *Neurology.* 2008;71:1220–1226.
16. DeJesus-Hernandez M, Mackenzie IR, Boeve BF, et al. Expanded GGGGCC hexanucleotide repeat in noncoding region of C9ORF72 causes chromosome 9p-linked FTD and ALS. *Neuron.* 2011;72:245–256.
17. Wang Y, Busin R, Reeves C, et al. X-linked adrenoleukodystrophy: ABCD1 *de novo* mutations and mosaicism. *Mol Genet Metab.* 2011;104:160–166.
18. Fu R, Ceballos-Picot I, Torres RJ, et al. Genotype–phenotype correlation in neurogenetics: Lesch–Nyhan disease as a model disorder. *Brain.* 2013 Aug 22. doi:10.1093/brain/awt202.
19. Kovacs GG, Trabattoni G, Hainfellner JA, et al. Mutations of the prion protein genes: phenotypic spectrum. *J Neurol.* 2002;249:1567–1582.
20. Natte R, Maat-Schieman MLC, Haan J, et al. Dementia in hereditary cerebral hemorrhage with amyloidosis-Dutch type is associated with cerebral amyloid angiopathy but is independent of plaques and neurofibrillary tangles. *Ann Neurol.* 2001;50:765–771.
21. Smith ED, Kudlow BA, Frock RL, et al. A-type nuclear lamins, progerias and other degenerative disorders. *Mech Ageing Dev.* 2005;126:447–460.
22. Sidransky E, Lopez G. The link between the GBA gene and parkinsonism. *Lancet Neurol.* 2012;11:9896–9898.
23. Cancel G, Duyckaerts C, Holmberg M, et al. Distribution of ataxin-7 in normal human brain and retina. *Brain.* 2000;123:2519–2530.
24. Lindenberg KS, Yvert G, Muller K, Landwehrmeyer GB. Expression analysis of ataxin-7 mRNA and protein in human brain: evidence for a widespread distribution and focal protein accumulation. *Brain Pathol.* 2000;10:385–394.
25. Kaytor MD, Duvick LA, Skinner PJ, et al. Nuclear localization of the spinocerebellar ataxia type 7 protein, ataxina-7. *Hum Mol Genet.* 1999;8:1657–1664.
26. LaSpada AR, Fu Y-H, Sopher BL, et al. Polyglutamine-expanded ataxin-7 antagonizes CRX function and induces cone–rod dystrophy in a mouse model of SCAT. *Neuron.* 2001;31:1–15.
27. Garden GA, LaSpada AR. Molecular pathogenesis and cellular pathology of spinocerebellar ataxia type 7 neurodegeneration. *Cerebellum.* 2008;7:138–149.
28. Danek A, Jung HH, Melone MA, et al. Neuroacanthocytosis: new developments in a neglected group of dementing disorders. *J Neurol Sci.* 2005;229–230:171–186.
29. Rampoldi L, Dobson-Stone C, Rubio JP, et al. A conserved sorting-associated protein is mutant in chorea-acanthocytosis. *Nat Genet.* 2001;28:119–120.
30. Pandolfo M, Pastore A. The pathogenesis of Friedreich ataxia and the structure and function of frataxin. *J Neurol.* 2009;256:9–17.
31. Li R, Johnson AB, Salomons G, et al. Glial fibrillary acidic protein mutations in infantile, juvenile, and adult forms of Alexander disease. *Ann Neurol.* 2005;57:310–326.
32. Mori K, Weng SM, Arzberger T, et al. The C9orf72 GGGGCC repeat is translated into aggregating dipeptide-repeat proteins in FTLD/ALS. *Science.* 2013;339:1335–1338.
33. van der Maarel S, Tawail R, Tapscott SJ. Facioscapulohumeral muscular dystrophy and DUX4: breaking the silence. *Trends Mol Med.* 2011;17:252–258.
34. Tawil R, van der Maarel SM. Facioscapulohumeral muscular dystrophy. *Muscle Nerve.* 2006;34:1–15.
35. Savkur RS, Philips AV, Cooper TA. Aberrant regulation of insulin receptor alternative splicing is associated with insulin resistance in myotonic dystrophy. *Nat Genet.* 2001;29:40–47.
36. Mankodi A, Logigian E, Callahan L, et al. Myotonic dystrophy in transgenic mice expressing an expanded CUG repeat. *Science.* 2000;289:1769–1772.
37. Todd PK, Paulson HL. RNA-mediated neurodegeneration in repeat expansion disorders. *Ann Neurol.* 2010;67:291–300.
38. Dabora SL, Jozwiak S, Franz DN, et al. Mutational analysis in a cohort of 224 tuberous sclerosis patients indicates increased severity of TSC2, compared with TSC1, disease in multiple organs. *Am J Hum Genet.* 2001;68:64–80.
39. Creange A, Zellar J, Rostaing-Rigattieri S, et al. Neurological complications of neurofibromatosis type 1 in adulthood. *Brain.* 1999;122:473–481.

40. Miloloza A, Rosner M, Nellist M, et al. The TSC1 gene product, hamartin, negatively regulates cell proliferation. *Hum Mol Genet.* 2000;9:1721–1727.
41. Gottfried ON, Viskochil DH, Couldwell WT. Neurofibromatosis Type 1 and tumorigenesis: molecular mechanisms and therapeutic implications. *Neurosurg Focus.* 2010;28:E8.
42. Au KS, Williams AT, Gambello MJ, Northrup H. Molecular genetic basis of tuberous sclerosis complex: from bench to bedside. *J Child Neurol.* 2004;19:699–709.
43. Ruggieri M, Huson SM. The clinical and diagnostic implications of mosaicism in the neurofibromatoses. *Neurology* 2001;56:1433–1443.
44. Au KS, Williams AT, Roach ES, et al. Genotype/phenotype correlation in 325 individuals referred for a diagnosis of tuberous sclerosis complex in the United States. *Genet Med.* 2007;9:88–100.
45. Kaufmann P, Engelstad K, Wei Y, et al. Protean phenotypic features of the A3243G mitochondrial DNA mutation. *Arch Neurol.* 2009;66:85–91.
46. Johri A, Beal MF. Mitochondrial dysfunction in neurodegenerative diseases. *J Pharmacol Exp Ther.* 2012;342(3):619–630.
47. Patterson MC, Mengel E, Wijburg FA, et al. Disease and patient characteristics in NP-C patients: findings from an international disease registry. *Orphanet J Rare Dis.* 2013;8:12.
48. Takeshima Y, Yagi M, Okizuka Y, et al. Mutation spectrum of the dystrophin gene in 442 Duchenne/Becker muscular dystrophy cases from one Japanese referral center. *J Hum Genet.* 2010;55:379–388.
49. Breitner JCS, Wyse BW, Anthony JC, et al. APOE-epsilon4 count predicts age when prevalence of AD increases, then declines. *Neurology.* 1999;53:321–331.
50. Everett CM, Wood NW. Trinucleotide repeats and neurodegenerative disease. *Brain.* 2004;127:2387–2405.
51. Brinkman RR, Mezei MM, Theilmann J, et al. The likelihood of being affected with Huntington disease by a particular age, for a specific CAG size. *Am J Hum Genet.* 1997;60:1202–1210.
52. Langbehn DR, Hayden MR, Paulsen JS, et al. CAG-repeat length and the age of onset in Huntington disease (HD): a review and validation study of statistical approaches. *Am J Med Genet.* 2010;153B(2):397–408.
53. Weydt P, Soyal SM, Gellera C, et al. The gene coding for PGC-1alpha modifies age at onset in Huntington's Disease. *Mol Neurodegener.* 2009;4:3.
54. Arsenault ME, Prevost C, Lescault A, et al. Clinical characteristics of myotonic dystrophy type 1 patients with small CTG expansions. *Neurology.* 2006;66:1248–1250.
55. Kabacki K, Hedrich K, Leung JC, et al. Mutations in DYT1: extension of the phenotypic and mutational spectrum. *Neurology.* 2004;62:395–400.
56. Sobue G, Ueno-Natsukari I, Okamoto H, et al. Phenotypic heterogeneity of an adult form of adrenoleukodystrophy in monozygotic twins. *Ann Neurol.* 1994;36:912–915.
57. Anderson KE, Bloomer JR, Bonkovsky HL, et al. Recommendations for the diagnosis and treatment of the acute porphyrias. *Ann Intern Med.* 2005;142:439–450.
58. Chen T, Dent SYR. Chromatin modifiers and remodelers: regulators of cellular differentiation. *Nat Rev Genet.* 2013;doi:10.1038/nrg3607.
59. Kato T, Iwamoto K. Comprehensive DNA methylation and hydroxymethylation analysis in the human brain and its implication in mental disorders. *Neuropharm.* Jan 4, 2014;doi:10.1016/j.neuropharm.2013.12.019 epub.
60. Yu CE, Cudaback E, Foraker J, et al. Epigenetic signature and enhancer activity of the human APOE gene. *Hum Mol Genet.* 2013;22:5036–5047.

CHAPTER

3

Immunogenetics of Neurological Disease

Ramyiadarsini I. Elangovan, Sreeram V. Ramagopalan*, and David A. Dyment†*

*University of Oxford, Oxford, UK

†Children's Hospital of Eastern Ontario Research Institute, Ottawa, ON, Canada

INTRODUCTION

There are few neurological diseases with both a genetic and immunological basis and none have been as extensively studied as multiple sclerosis (MS). As such, MS serves as the prototypic immunogenetic disease of the nervous system (CNS). It is thought to occur as a result of an inappropriate immune response to self-antigens in genetically predisposed individuals.[1,2] Evidence for an inherent susceptibility to MS has come largely from family-based investigations and decades of association studies with genetic polymorphisms; in particular, the major histocompatibility complex (MHC) that operates as a significant contributor to MS risk.[3–5] The immune basis of MS is implicated by several lines of evidence including the observation of lymphocytes and macrophages within CNS lesions, the presence of oligoclonal bands in the cerebrospinal fluid, the autoimmune encephalomyelitis (EAE) animal model of the disease, the effectiveness of current immunotherapies, and the aforementioned genetic associations that implicate genes of the immune system.[6,7] This chapter will focus on the evidence for a significant genetic contribution to MS and will highlight the known immune-related genes and the complex interactions of these genes with the environment.

EPIDEMIOLOGICAL EVIDENCE FOR GENETIC SUSCEPTIBILITY

Multiple sclerosis is a chronic and debilitating condition characterized by episodes of CNS inflammation with a concomitant and progressive axonal degeneration.[8] MS is common and affects 1 in 1000 adults in Europe, Canada and the Northern United States.[9,10] Worldwide there are significant differences in prevalence that likely reflect a region's background genetic make-up in addition to endogenous broad-based environmental factors. For example, the condition is virtually absent in those of non-Caucasian ancestry living in Africa and parts of Asia.[11] And yet, African-Americans have a risk estimated to be half that of Americans of Northern European descent,[12,13] although this appears to be changing rapidly.[14] This may be explained by an environmental factor present, or perhaps the degree of admixture resulting in an increased liability. There are several other ethnic groups, such as the Lapps,[15] Romani,[16] Amerindians,[17,18] and the Hutterites,[19] that appear to have increased protection to MS despite living in relatively high-risk regions.

There is a well-described increase in MS prevalence with distance from the equator.[20] This is seen in North America, and is correlated with both the distance from the equator and the proportion of individuals reporting "Scandinavian" ancestry.[21] In Europe, where there is a similar north-south cline in MS prevalence, there is also a gradation in the frequency of another risk factor, the MHC *HLA-DRB1*15* allele (discussed below). As such, the relative contributions of the environment and genetics can be difficult to delineate. However, a clear example of the broad-based environmental contribution to risk can be best seen in the changes in MS prevalence in Australia. The prevalence of MS in the south (Hobart) is 75.6 per 100,000 compared with a prevalence of 11 per 100,000 in northern Queensland, and the differences occur within the context of a relatively uniform genetic background.[22,23]

Rosenberg's Molecular and Genetic Basis of Neurological and Psychiatric Disease
http://dx.doi.org/10.1016/B978-0-12-410529-4.00003-6

More complex patterns of disease distribution also exist. In Norway, for instance, MS prevalence does not increase with latitude but instead appears to correlate with proximity to coastal fishing areas and fish consumption.[10] Exposure to ultraviolet (UV) light is also shown to correlate with MS prevalence, as observed in France. Vitamin D may be the common element when considering latitude, UV exposure and fish consumption.[24] Epidemiological evidence, therefore, supports the actions of both genes and environmental factors.

GENETICS OF MS: FAMILY-BASED INVESTIGATIONS

Family studies of MS show a striking familial aggregation of the disease. Fifteen to twenty percent of MS patients have a first, second, or third degree relative with MS.[25] The empiric risks in family members of MS patients correlates with degree of relatedness. For example, the risk to a full-sibling of an MS patient is 3–5%, while the risk to a half-sibling of an MS patient is 1–2%[26,27] (Table 3.1). Not only do these risks provide useful information for counseling purposes, but the changes in risk can be used to estimate the number of genes contributing to liability. While this number can vary with the models tested, the background prevalence, and assumptions made, this drop in risk would be in keeping with several dozens of susceptibility genes of small-effect size versus monogenic inheritance of an MS gene with a large-effect size.

Classically, the concordance rate in twins can be used to demonstrate a genetic contribution to a trait or disease.[28] If there is a greater concordance in monozygotic twins versus the concordance rate in dizygotic twins, a genetic contribution can be inferred. Moreover, the concordance estimates in twins are often used to estimate heritability. In MS there is a clear difference in monozygotic (MZ) twin concordance (25%) versus dizygotic twin concordance (5%).[29] Heritability estimates for MS depends on the study (and associated biases) but tends to range from 25–75%.[30] It is assumed that the increase in MZ concordance is due to shared genes. However, the increased concordance is likely not so straightforward; twins concordant for MS are almost always female–female pairs.[29] Whilst we might expect an increase in female pairs, given that females are more often affected with MS, the female identical-twin concordance is much greater than expected (11 female pairs to 1 male pair) and cannot be readily explained.

The increase in females affected with MS in twin studies is not operating in isolation. The female-to-male ratio of MS has been increasing over the last century, as demonstrated in large population-based studies in Denmark and Canada.[31,32] In the latter, the sex ratio of MS patients born in the 1930s was less than 2 females to 1 male and has increased to more than 3 females for each male born in recent decades. Such a change is too rapid for a purely genetic explanation and instead suggests that environmental factors are operating in a sex-specific manner. Additional work needs to be conducted to explore this trend though changes in smoking habits, oral contraception, and sun exposure are potential contributors.[31]

In relation to MS, the study of twins also permits a comparison between concordance in dizygotic twins and nontwin siblings. Though not significantly higher than the full-sibling risks, there is a trend for the dizygotic twin

TABLE 3.1 Recurrence Risk to Relatives of MS Patients

Relative	Risk (95% confidence interval)	Kinship	Reference
Identical twin	25.3% (16.7–33.9)	100%	Willer et al.[29]
Female–female MZ twin	34% (22.8–45.2)	100%	Willer et al.[29]
Male–male MZ twin	6.5% (0–15.5)	100%	Willer et al.[29]
Dizygotic twin	5.4% (0–10.8)	50%	Willer et al.[29]
Nontwin sibling	3–5%	50%	Sadovnick et al.[25]
Half-sibling	1.89% (1.36–2.41)	25%	Ebers et al.[26]
Maternal half-sibling	2.35% (1.57–3.13)	25%	Ebers et al.[26]
Paternal half-sibling	1.31% (0.65–1.96)	25%	Ebers et al.[26]
AUNN	0.81% (0.54–1.33)	25%	Robertson et al.[99]
Cousins	0.88% (0.56–1.32)	12.5%	Robertson et al.[99]
Population	0.1–0.2%	–	Beck et al.[100]

Abbreviations: MZ, monozygotic; AUNN, aunt/uncle–niece/nephew.

risk to be greater than the nontwin risk.[29] This suggests the presence of an *in utero* or "maternal" effect. Other investigations corroborate this observation, such as the half-sibling risks, where the maternal half-sibling risk is nearly double (2.4%) that of paternal half-siblings (1.3%) despite both sharing a quarter of their genetic constitution with the MS patient.[26] Evidence from a Dutch MS population also shows an increase in maternal transmissions for those with MS,[33] and there are more individuals connected by a matrilineal inheritance when assessing aunt/uncle and niece/nephew concordant pairs.[34]

A family-based study that convincingly demonstrates the importance of shared genes to explain MS familial aggregation was performed in the 1990s. Adoptive relatives living in the same household as the MS patient were shown to have a similar risk as the background Canadian population (1/1000) and the risk was not comparable to that seen in first-degree relatives.[35] While the adoption study has not been replicated to date, it has been corroborated in other family studies and continues to provide compelling evidence in support of a genetic explanation for familial risk.[36,37]

THE ROLE OF MAJOR HISTOCOMPATIBILITY COMPLEX GENES

Researchers have sought to identify the molecular causes for the increased familial risk of MS for decades and much of this risk appears to be localized to the major histocompatibility complex (MHC) region on chromosome 6p21 (Figure 3.1). This region has proven to have the most consistent and strongest associations with MS, as demonstrated by numerous family-based, candidate case-control, and genome-wide association studies (GWAS) in different populations.[38] The first positive association between MS and the MHC was reported in the early 1970s and was an association to antigens of the MHC Class I.[39,40] This was later refined to a stronger association with antigens within the MHC Class II,[41] and then refined to the extended *DQA1*0102-DQB1*0602-DRB1*1501-DRB5-0101* haplotype.[42] The MHC is large, encompassing over 4 Mb and containing over 240 genes, most of which have an immune-related function. The extensive polymorphism and linkage disequilibrium makes the region ideal for association studies,

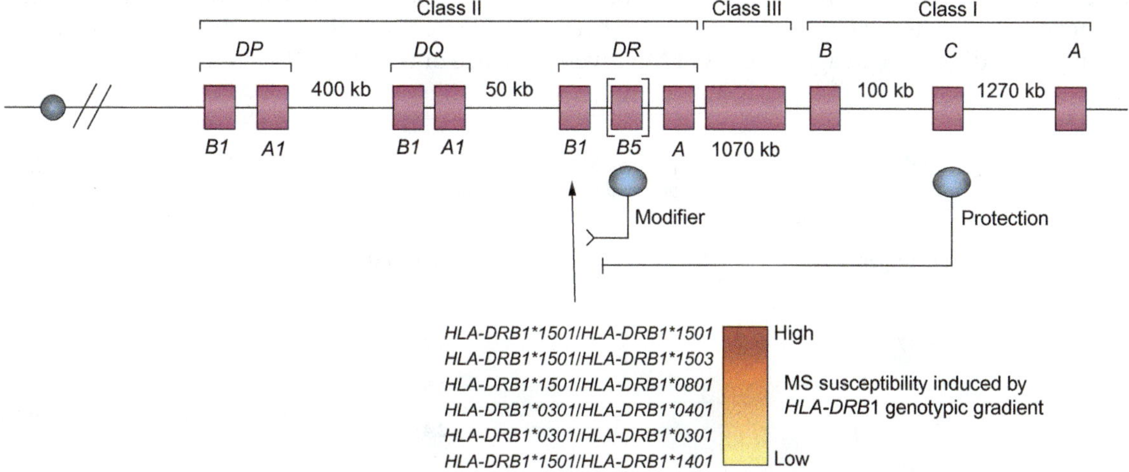

FIGURE 3.1 **The major histocompatibility complex.**[98] The human leukocyte antigen (HLA) gene complex is located on the short arm of chromosome 6 at p21.3, spanning almost 4000 kb of DNA. The full sequence of the region was completed and reported in 1999. From 224 identified loci, 128 are predicted to be expressed and about 40% to have an immune response function. There are two major classes of HLA-encoding genes involved in antigen presentation. The telomeric stretch contains the class I genes, whereas the centromere proximal region encodes HLA-class II genes. The HLA-class II gene HLA-DRB5 (in square brackets) is only present in the DR51 haplotypic group (*HLA-DRB1*15* and *HLA-DRB1*16* alleles). HLA class I- and class II-encoded molecules are cell-surface glycoproteins, the primary role of which, in an immune response, is to display and present short antigenic peptide fragments to specific T-cells, which can then be activated by a second stimulatory signal and initiate an immune response. In addition, HLA molecules are present on stromal cells on the thymus during development, helping to determine specificity of the mature T-cell repertoire. A third group of genes, collectively known as class III, cluster between class I and II regions and include genes coding for complement proteins, 21αhydroxylase, tumor necrosis factor and heat shock proteins. This super-locus contains at least one gene (*HLA-DRB1*, indicated by an arrow) that influences susceptibility to multiple sclerosis. *HLA-DRB1* allelic copy number and *cis* and/or *trans* effects have been detected, suggesting a disease association gradient, ranging from high vulnerability (*HLA-DRB1-1501* homozygotes and *HLA-DRB1*1501/HLA-DRB1*0801* compound heterozygotes) to moderate susceptibility (*HLA-DRB1*03/HLA-DRB1*04* heterozygotes and *HLA-DRB1*03* homozygotes) and resistance (*HLA-DRB1*1501/HLA-DRB1*1401* compound heterozygotes; in this genotypic configuration, the presence of HLA-*DRB1*1401* nullifies the susceptibility effects of *HLA-DRB1*1501*). Emerging data also suggest a modifier role for *HLA-DRB5* and HLA class I alleles. *Permission has been given to reproduce figure from Nature Publishing Group (ref: 3000450365).*

although identifying any additional genes within the MHC can be problematic given this extensive linkage disequilibrium and requires large study samples.

The *HLA-DRB1*1501* haplotype increases risk by 2–3 fold.[43] Moreover, if an individual carries two copies of *HLA-DRB1*15*, the risk of MS is more than 6-fold greater. As such, this genotype acts as the risk factor with the largest impact on MS susceptibility.[44] The mechanism by which *HLA-DRB1*15* increases MS risk is likely due to an inappropriate T-cell recognition of self-antigens presented in the context of *HLA-DRB1*15*.[2] The peripherally activated T-cells cross the blood–brain barrier and evoke an episode of inflammation against self-antigens within the CNS.[1] The human leukocyte antigens (HLA) are therefore central players for proper functioning immune responses.[45] A humanized animal model of MS affirms this mechanism of antigen presentation leading to disease. Researchers generated a mouse model that includes a human *HLA DRB1*15* allele expressed in conjunction with a specific T-cell receptor (TCR).[46] This TCR recognizes epitopes derived from myelin basic protein in the context of *HLA-DRB1*15*. These mice spontaneously develop a severe, progressive experimental autoimmune encephalomyelitis (EAE).[46,47]

The *HLA-DRB1*1501* allele does not appear to act in a simple, additive manner and there are other alleles of the *HLA-DRB1* gene that act to modulate risk, independently, or together with *HLA-DRB1*1501*. *HLA-DRB1*17* is one of the *DRB1* alleles that has also been shown to increase the risk of MS independently, albeit to a lesser extent than *HLA-DRB1*15*; *DRB1*17* has a relative risk of 1.4 and is a common HLA allele in the Northern European population.[48] When the susceptibility alleles are considered together, there is a recognized hierarchy of *DRB1* disease alleles, with some alleles such as *HLA-DRB1*14* playing a role in protecting against MS.[44] To complicate matters, the allele hierarchy is better viewed by genotype rather than by single allele; for example, if *HLA-DRB1*15* and *HLA-DRB1*14* are co-inherited, any increase in risk associated with inheriting the *HLA-DRB1*15* allele is nullified by the presence of the protective *DRB1*14* allele[44] (Table 3.2). Of note, the frequency of the *HLA-DRB1*14* allele is relatively high in Asia and the prevalence of MS in Asia is low. While the *HLA-DRB1*14* allele appears "dominantly" protective, other protective alleles such as *HLA-DRB1*01* appear to reduce risk only when co-inherited with the *HLA-DRB1*15* allele. This interaction between *DRB1*15* and *DRB1*0101* is illustrated in a study of familial transmission that shows DRB1*0101 is transmitted less often than expected by chance when the individual has already inherited *DRB1*15* from the other transmitting parent.[48] An additional observation highlights phenotypic complexity, as *DRB1*0101* also appears to be associated with a more benign clinical outcome compared to non-*DRB1*0101* carriers,[37] although this is not a universal finding. As for *HLA-DRB1*15*, it has been investigated for effects on clinical course and there is a small, but detectable, difference in age of onset of MS given presence of the *DRB1*15* allele.[49]

TABLE 3.2 DRB1 Genotypes and MS Risk
DRB1 genotypes presented in order of increasing risk observed in a Canadian MS sample.[44]

DRB1 genotype	Odds ratio
15/15	5.4
15/17	3.2
15/X	2.9
15/11	2.4
17/17	2.1
X/X	1.0
17/11	1.0
14/15	1.0
11/X	0.7
11/14	0.7
11/14	0.7
14/17	0.60
11/11	0.45
14/X	0.4

*Note: No DRB1*14/14 genotypes were observed in the MS population.*

The effect size of the various *HLA-DRB1* alleles shows substantial heterogeneity between populations with regards to their impact on the development of MS. For example, the *HLA-DRB1*03*, *HLA-DRB1*04*, and *HLA-DRB1*13* alleles increase the risk of MS in the Sardinian population,[50,51] while protection is exerted by *HLA-DRB1*01*, *HLA-DRB1*10*, *HLA-DRB1*11*, and *HLA-DRB1*14* in those of Northern European ancestry[48,52] and *HLA-DRB1*09* in the Japanese.[53] This may speak to the broad-based environmental factors at work and an interaction between these multiple factors (e.g., viruses) and the MHC.

In another humanized mouse model, the *HLA-DRB1*1501* antigen is expressed with the adjacent *HLA DRB5*0101* antigen.[54] These two genetic variants are in tight linkage disequilibrium in humans and there is a clear *epistatic* interaction between these two different genes as observed in this animal model. The mice express both the HLA antigens in addition to a TCR specific for myelin basic protein. In this scenario the mice have a relapsing–remitting disease and a later onset.[54] This is in contrast to those with just DRB1*1501, mentioned above, that present with severe, early, and progressive disease and, as such, the two genes may be responsible for the human relapsing–remitting disease. There are *HLA-DRB1*15*-positive, *DRB5*-null individuals in African-American populations and this group of MS patients have been reported to have a more progressive-type disease which may be due to the absence of *DRB5*0101*, among other factors.[55] However, this does not explain why progressive Northern European MS patients have the same frequency of the *HLA-DRB1*15* haplotype as relapsing–remitting patients of the same ancestry, suggesting that additional influences (likely nongenetic) exist in influencing disease course. Epistatic interactions at HLA Class II have also been shown in human studies. As mentioned, *DRB1*1501* is inherited as a "cassette" with *DQA1*0102* and *DQB1*0602* in Northern Europeans. In MS patients who have inherited this allelic cassette from one parent, the transmission of alleles and haplotypes from the other parent was tested. The transmissions showed that *DQA1*0102* was over-transmitted and suggests an interaction *"in trans"* between the classic *DRB1*15*-bearing haplotype and *DQA1*0102* present on the other co-inherited haplotype.[56]

There have been several association studies to implicate the HLA Class I region of the MHC. Recently a GWAS confirmed that the HLA Class I allele *A*0201* was protective against MS as had been previously identified in the Scandinavian population.[57,58] The HLA Class I presents antigens to CD8 T-cells and, in addition to the predominance of CD8-activated T-cells in MS lesions, suggests that these T-cells are also involved in MS.[2] There are other studies that show *HLA Class I A*03* potentially acts to increase risk.[57,59] Showing these genetic effects in an animal model is a powerful method to validate association studies and this has been performed for these HLA Class I alleles.[60–62] In a humanized mouse model that incorporates the HLA Class I antigen, there is an increased frequency of disease when the mice express PLP epitopes within the context of *HLA-A*0301*.[60] Initially the disease is mild, but becomes severe and dependent on CD4-activated T-cells over time.[2,60] However, this remains a model as the self-antigen presented to activated T-cells has not been conclusively identified, although MOG, MBP and PLP have all been implicated.[2]

The *DRB1*1501* promoter is known to carry a vitamin D-responsive element (VDRE). This sequence is comprised of two 6-base pair motifs that bind the cognate nuclear receptor of vitamin D, the vitamin D receptor (VDR). In an *in vitro* experiment, vitamin D bound to the HLA promoter and increased the expression of *DRB1*15*.[63] This provides a hypothetical mechanism for a gene–environment interaction with the HLA.[64] It is proposed that lower levels of vitamin D may result in less efficient self-antigen presentation during immune development. As a result, T-cells able to recognize self-antigen could potentially escape clonal deletion. However, much work is required to convincingly prove this model.

To summarize, the notion that *HLA-DRB1*15* is the sole determinant of risk is now yielding to a highly complex story that incorporates a hierarchy of independent HLA Class I and Class II susceptibility and resistance genotypes, gene–gene interactions, and gene–environment interactions (such as vitamin D) that contribute to MS risk.

OTHER IMMUNE-RELATED GENES

Beyond the MHC genes, there are dozens of other genetic variants that have recently been shown to increase (or decrease) MS risk, albeit modestly. Historically, these other genetic variants have been exceedingly difficult to identify and candidate gene approaches have failed to replicate initially reported positive findings. To overcome the challenges of the candidate gene approach, genome-wide linkage studies were performed in the 1990s, although they typically did not have the statistical power to detect susceptibility genes of small effect size ($OR<1.5$).[65–67] In 2007, these difficulties were overcome with an MS GWAS.[68] GWAS of adequate sample size provide well-powered investigations into complex traits and typically significance values of 5×10^{-8} are required in an effort to avoid false-positive results due to the number of markers being tested. They are based on linkage disequilibrium versus linkage and are for the most part "hypothesis-free." The first GWAS for MS was in 2007 and involved the genotyping of over 300,000 single nucleotide polymorphisms (SNPs) spanning the genome and close to 1000 MS patients. The results were then followed-up in a tiered approach to validation. The study confirmed the well-known association with *HLA*

*DRB1*1501* and convincingly showed an association with *IL7RA* and *IL2RA*.[68] These variants exert a small effect on risk with odds ratios of 1.3 and 1.2, respectively, and could each explain less than 0.2% of the variance in risk. Several GWAS in MS have since been performed.[58,68–72] Recently, the International Multiple Sclerosis Genetics Consortium (IMSGC) and the Wellcome Trust Case Control Consortium 2 (WTCCC2) have performed a high-powered GWAS including more than 9000MS patients and identified a total of 57 significantly associated variants – in addition to the MHC associations with MS susceptibility[58] (Table 3.3). A gene-ontology analysis of the genes located within the

TABLE 3.3 Susceptibility Genes Identified by Genome-Wide Association Studies

SNP	Gene	Location	Gene function	Effect size
rs4648356	*TNFRSF14*	1p36.32	T-cell activation	1.16
rs11810217	*EVI5*	1p22.1	Cell cycle regulation	1.15
rs11581062	*VCAM1*	1p21.2	Adhesion of monocytes and lymphocytes	1.07
rs1335532	*CD58*	1p13.1	Immune recognition of APC and Th cells	1.18
rs1323292	*RGS1*	1q31	G-protein signaling in B cells	1.12
rs7522462	*KIF21B*	1q31-q32	Intracellular transport along axons and dendritic microtubules	1.11
rs12466022	X	2p21	n/a	1.16
rs7595037	*PLEK*	2p14-p13	Substrate for PKC in platelets	1.15
rs17174870	*MERTK*	2q13	Tyrosine kinase	1.15
rs10201872	*SP140*	2q37	Myeloid cell development	1.15
rs11129295	*EOMES*	3p24.1	CD8+ T-cell differentiation	1.09
rs2028597	*CBLB*	3q13.11	T-cell development	1.13
rs2293370	*TMEM39A*	3q13	Unknown	1.16
rs9282641	*CD86*	3q21	T-cell induction	1.2
rs2243123	*IL12A*	3q25	Subunit of Il12 and Il35 and modulates immune response	1.09
rs669607	*intergenic*	3p24	n/a	1.15
rs228614	*NFKB1*	4q24	Pleiotropic effects including immune and inflammatory response	1.09
rs6897932	*IL7R*	5p13.2	Development and regulation of T-cells	1.11
rs4613763	*PTGER4*	5p13.1	G-protein receptor that activates T-cells	1.21
rs2546890	*ILI2B*	5q33.3	Induction of Th1 development	1.15
rs12212193	*BACH2*	6q15	Transcriptional regulation	1.08
rs802734	*THEMIS*	6q22.33	T-cell development	1.13
rs11154801	*MYB*	6q23.3	Modifies lymphocyte development	1.09
rs170660096	*IL22RA2*	6q23.3	Soluble cytokine receptor	1.14
rs1738074	*TAGAP*	6q25.3	T-cell activation	1.14
rs13192841	*OLIG3*	6q23.3	Neuronal development	1.1
rs354031	*ZNF746*	7q36.1	Dopaminergic neuron degeneration	1.14
rs1520333	*unknown*	8q21	Unknown	1.11
rs4410871	*MYC*	8q24	Cell proliferation and transformation	1.09
rs2019960	*PVT1*	8q24	Activator of MYC	1.16
rs2150702	*MLANA*	9p24	Melanosome development	1.16
rs3118470	*IL2RA*	10p15	Immune homeostasis	1.12

TABLE 3.3 Susceptibility Genes Identified by Genome-Wide Association Studies—cont'd

SNP	Gene	Location	Gene function	Effect size
rs7090512	*IL2RA*	10p15	Immune homeostasis	1.21
rs1250542	*ZMIZI*	10q22	Co-regulator of androgen receptor	1.15
rs7923837	*HHEX*	10q23	Myeloid transcriptional repressor	1.09
rs650258	*CD6*	11q12	T-cell activation	1.12
rs630923	*CXCR5*	11q23	B-cell migration	1.13
rs1800693	*TNFRSF1A*	12p13	Activate NF-κB, mediate apoptosis, and regulator of inflammation	1.12
rs104668829	*CLECL1*	12p13	Modifies cytokine expression	1.12
rs12368653	*CYP27B1*	12q14	Vitamin D metabolism	1.11
rs949143	*MPHOSPH9*	12q24	Cell cycle signaling	1.08
rs9596270	*Intergenic*	13q14	n/a	1.35
rs4902647	*ZFP36L1*	14q24	Erythroid cell differentiation	1.13
rs2300603	*BATF*	14q24	Promotes Th17 differentiation	1.08
rs2119704	*GALC*	14q31	Galactosylceramide catabolism	1.12
rs7200786	*CLEC16A*	16p13	Unknown function, expressed in B cells, APCs and NK cells	1.15
rs7191700	*TNP2*	16p13	Chromatin structure regulation	1.15
rs13333054	*IRF8*	16p24	Bone metabolism	1.12
rs9891119	*STAT3*	17q21	Transcription factor	1.1
rs8070463	*KPNB1/TBKBP1/ TBX21*	17q21	Protein transport	1.15
rs180515	*RPS6KB1*	17q23	Phosphorylation of ribosomal proteins	1.05
rs7238078	*MALT1*	18q21	T-cell activation and proliferation	1.14
rs1077667	*TNFSF14*	19p13	T-cell proliferation	1.14
rs8112449	*TYK2*	19p13	Th1 differentiation and repression of Th2 cells	1.1
rs10411936	*EPS15L1*	19p13	Receptor mediated endocytosis	1.16
rs874628	*MPV17L2*	19p13	Mitochondrial membrane protein	1.07
rs2303759	*DKKL1*	19q13	Extracellular signaling	1.11
rs6074022	*CD40*	20q13	B-cell function	1.15
rs2248359	*CYP24A1*	20q13	Vitamin D metabolism	1.11
rs6062314	*TNFRSF6B*	20q13	T-cell dependent induction of apoptosis	1.14
rs2283792	*MAPK1*	22q11	Intracellular signaling	1.12
rs140522	*SCO2*	22q13	Mitochondrial function	1.12

Abbreviations: APC, antigen-presenting cell; NK cells, natural killer cells; PKC, protein kinase C.
Adapted from[3,58,97].

MS-associated regions showed a substantial over-representation of immune-related processes, further confirming the immunological nature of the disease[58] (Figure 3.2).

As with the case of the MHC, it is important to understand how the variants modify the normal functioning of the immune system. This provides key information with regards to pathogenesis and, ultimately, potential targets for treatment. This can be a challenge, as many genes throughout the genome are not fully characterized. Furthermore, the associated variants from GWAS may be present within an intron or between genes, and a causal explanation may

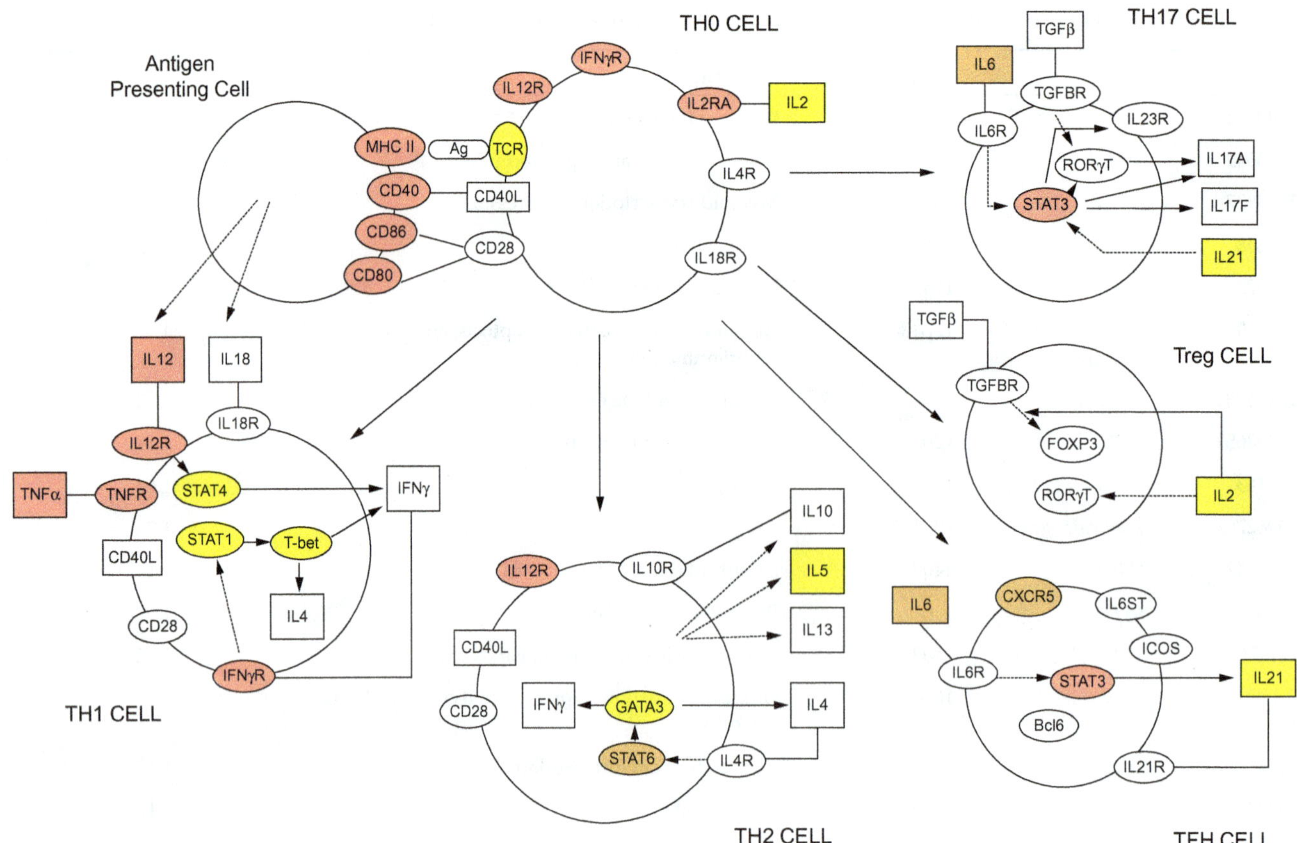

FIGURE 3.2 **Graphical representation of the T-helper cell differentiation pathway.** Figure derived from an image by Ingenuity Pathway Analysis (IPA) software version 8.8 (Ingenuity Systems Inc., Redwood City, CA, USA). Alphanumeric labels indicate the individual genes and gene complexes (nodes) included in the pathway (note some are included more than once). Colored nodes are those containing a gene implicated by proximity to a single nucleotide polymorphism (SNP) showing evidence of association. Red: in bold or gray was highlighted in Sawcer et al.[58] (plus MHC class II region and TNF alpha). Orange: other loci in Sawcer et al.[58] or discover P value $<1\times10^{-4.5}$ and consistent replication data. Yellow: discovery P value $<1\times10^{-3}$. Other molecules (proteins, vitamins, etc.) may also be of relevance in these processes but are not included here as they are not currently listed as being part of this particular pathway in the IPA database. This was originally published in Sawcer et al.[58] *Permission has been given to reproduce figure by Nature Publishing Group (ref: 3000450365).*

be unclear. The mechanism of action of the 57 genes is currently work in progress although insight into several of the associations has been gathered to date.

For example, the interleukin 7 (IL-7) receptor gene (*IL7R*) was implicated in the first GWAS and in previous association studies.[68,73,74] The gene has been known to be involved in T-cell development and homeostasis as somatic mutations are seen in T-cell acute lymphoblastic leukemias.[75] The initial association with MS was with a variant that affects splicing of exon 6 within *IL7R*. When the risk variant is present, there is an increased frequency of exon 6 "skipping," resulting in the increased expression of the soluble form of IL7R versus a membrane-bound form of IL7R.[73,74] This is in keeping with the observation that increased soluble IL7R promotes IL-7 activity and hence T-cell proliferation and autoimmunity as seen in the EAE model and also in MS.[76] Another gene identified in the original GWAS of 2007 was *IL2Ra* (CD25). The IL-2/IL-2RA (CD25) pathway plays an essential role in regulating immune responses. IL-2 is central for both the expansion and apoptosis of T-cells. This is an important reminder that a degree of complexity is likely to be present in other genes as it is in the MHC. The CD25 pathway has been targeted in a treatment for MS, as a monoclonal immunoglobulin G1 (IgG1) antibody (daclizumab) is able to bind the IL2RA and inhibit binding of IL-2 and there have been several trials on its efficacy.[77] A recent clinical trial showed that use of daclizumab results in decreased MS lesion load, decreased number of relapses and a decreased rate of progression of disease.[78]

TYK2 (tyrosine kinase 2) shows evidence of a protective allele. The initial case-control studies and GWAS gave an odds ratio of 0.6–0.8.[71,79–81] The protective allele results in lower kinase activity and T-lymphocytes tend to differentiate into Th2 cells in preference to the autoimmunity-inducing Th1 cells.[80] A different variant in *TYK2* was identified as a susceptibility allele in a single family with several members with MS. In this family, four affected family members were sequenced for over 22,000 genes by next-generation sequencing. No novel MS genes were identified but when the investigators filtered for the genes identified by GWAS, and within regions of haplotype sharing, the susceptibility variant at *TYK2* was observed.[82] Next-generation sequencing is a powerful method to identify rare variants causing disease and can be used in conjunction with GWAS to identify other variants in heterogeneous diseases like MS.

A gene without an overt immunological function that was highlighted in a GWAS is *CYP27B1*.[58] The gene encodes the 1-α-hydroxylase enzyme and mutations are known to cause vitamin D-dependent rickets.[83] The enzyme converts 25-hydroxyvitamin D to the active 1,25-dihydroxyvitamin D (calcitriol). Researchers have also performed next-generation sequencing in 43 families with more than four individuals affected with MS. There were no shared, rare, variants among the 43 representative MS patients, but there were rare variants observed within the GWAS-identified genes and only *CYP27B1* was validated.[84] Follow-up studies by other research groups have not identified rare variants at this gene.[85,86] The involvement of *CYP27B1* (and hence vitamin D) in multiple sclerosis is unclear at present; however, vitamin D does appear to have a role in other non-MHC, GWAS-identified genes. The previously mentioned vitamin-D responsive elements (VDREs) were mapped across the genome by chromatin immunoprecipitation experiments followed by next-generation sequencing. VDR-bound regions across the genome correlated with known MS associations and were found to be present together much more often than expected by chance.[87] As in the case of the VDRE driving *HLA-DRB1*1501*, the presence of the VDRs located near MS-associated genes provides additional mechanisms for gene–environment interactions.

Despite the large number of identified risk alleles in GWAS, taken together, these variants in total account for a small proportion of MS genetic risk, with a great proportion of this risk arising from the MHC variants. The remaining risk has been hypothesized to arise from several different mechanisms including epigenetic and gene–environment interactions and, potentially, may be a result of overestimation of genetic risk. Beyond this, a role for rare and highly penetrant variants has also been proposed to explain the basis of families with several affected members although, to date, the evidence has been lacking.[88]

THE ENVIRONMENT AND IMMUNE-RELATED GENES

As described, there is ample evidence to link vitamin D to MS.[89,90] From an immunogenetic perspective, vitamin D is required for a healthy immune system and the VDR is expressed in B, T, and antigen-presenting cells. Vitamin D has been shown to inhibit B- and T-cell proliferation and it can shift T-cell development to a Th17 response as well as induce T regulatory cells.[91] Given the previously mentioned north–south gradient correlating with UV exposure, the evidence for rare variants in *CYP27B1* and the potential effects of VDR binding throughout the genome, vitamin D appears to play a central role in an individual's genetic risk and to explain the overall epidemiology of MS.

Another environmental risk factor that may interact with immune-related genes is smoking. This behavior was identified as a risk factor in several prospective studies and the increased risk is $\approx$1.5-fold.[92,93] There have been genetic studies that show an interaction with the HLA, and smokers who were *DRB1*1501* positive and *HLA*A2* negative had an odds ratio of 13.5 compared to those nonsmokers who were *DRB1*1501* negative and *HLA*A2* positive.[94] One hypothesis as to how smoking influences MS risk is that smoking can modify proteins and render them autoantigenic to the immune system.[95]

Lastly there is a well-known association with Epstein–Barr virus (EBV) and MS. Virtually all MS patients (>99%) are seropositive for EBV versus approximately 90% of controls.[10] Studies have looked into an interaction with smoking, *HLA-DRB1*15* and EBV seropositivity, and have found evidence for an interaction between EBV and smoking, but *HLA-DRB1*15* appears to operate independently.[96] The mechanism as to how EBV influences MS risk remains elusive, but may involve molecular mimicry between EBV peptides and those in the CNS.

CONCLUSION

MS is an immunogenetic disease with both genes and the environment acting in concert to increase one's risk. There has been a tremendous leap forward with genome-wide association studies identifying non-MHC-related MS genes after decades of searching. However, the impact of each gene identified is modest and mechanisms of disease causation are still unclear and need to be considered in the milieu of other genes, the environment, and the interactions between these two complex systems.

References

1. Nylander A, Hafler DA. Multiple sclerosis. *J Clin Invest*. 2012;122:1180–1188.
2. Goverman J. Autoimmune T cell responses in the central nervous system. *Nat Rev Immunol*. 2009;9:393–407.
3. Fugger L, Friese MA, Bell JI. From genes to function: the next challenge to understanding multiple sclerosis. *Nat Rev Immunol*. 2009;9:408–417.
4. Oksenberg JR, Baranzini SE. Multiple sclerosis genetics—is the glass half full, or half empty? *Nat Rev Neurol*. 2010;6:429–437.
5. Gourraud PA, Harbo HF, Hauser SL, Baranzini SE. The genetics of multiple sclerosis: an up-to-date review. *Immunol Rev*. 2012;248:87–103.
6. Constantinescu CS, Gran B. Multiple sclerosis: autoimmune associations in multiple sclerosis. *Nat Rev Neurol*. 2010;6:591–592.
7. Weissert R. The immune pathogenesis of multiple sclerosis. *J Neuroimmune Pharmacol*. 2013;8(4):857–866.
8. Paty D, Ebers G. *Multiple sclerosis*. Philadelphia: FA Davis; 1997.
9. Ebers GC, Daumer M. Natural history of MS. *Eur J Neurol*. 2008;15:881–882.
10. Ebers GC. Environmental factors and multiple sclerosis. *Lancet Neurol*. 2008;7:268–277.
11. Kurtzke JF. Geography in multiple sclerosis. *J Neurol*. 1977;215:1–26.
12. Rosati G. The prevalence of multiple sclerosis in the world: an update. *Neurol Sci*. 2001;22:117–139.
13. Kurtzke JF, Beebe GW, Norman JE. Epidemiology of multiple sclerosis in U.S. veterans: 1. Race, sex, and geographic distribution. *Neurology*. 1979;29:1228–1235.
14. Wallin MT, Culpepper WJ, Coffman P, et al, Veterans Affairs Multiple Sclerosis Centers of Excellence Epidemiology Group. The Gulf War era multiple sclerosis cohort: age and incidence rates by race, sex and service. *Brain*. 2012;135:1778–1785.
15. Grønning M, Mellgren SI, Schive K. Optic neuritis in the two northernmost counties of Norway. A study of incidence and the prospect of later development of multiple sclerosis. *Arctic Med Res*. 1989;48:117–121.
16. Kálmán B, Takács K, Gyódi E, et al. Sclerosis multiplex in gypsies. *Acta Neurol Scand*. 1991;84:181–185.
17. Flores J, González S, Morales X, Yescas P, Ochoa A, Corona T. Absence of multiple sclerosis and demyelinating diseases among lacandonians, a pure Amerindian ethnic group in Mexico. *Mult Scler Int*. 2012;2012:292–631.
18. Warren S, Svenson LW, Warren KG, Metz LM, Patten SB, Schopflocher DP. Incidence of multiple sclerosis among first nations people in Alberta, Canada. *Neuroepidemiology*. 2007;28:21–27.
19. Hader WJ, Seland TP, Hader MB, Harris CJ, Dietrich DW. The occurrence of multiple sclerosis in the Hutterites of North America. *Can J Neurol Sci*. 1996;23:291–295.
20. Davenport C. Multiple sclerosis from the standpoint of geographic distribution amd race. In: *Association for research in nervous and mental diseases*. New York: Herber; 1921:8–19.
21. Bulman D, Ebers G. The geography of MS reflects genetic susceptibility. *J Trop Geogr Neurol*. 1992;2:66–72.
22. Hammond SR, de Wytt C, Maxwell IC, et al. The epidemiology of multiple sclerosis in Queensland, Australia. *J Neurol Sci*. 1987;80:185–204.
23. Hammond SR, McLeod JG, Millingen KS, et al. The epidemiology of multiple sclerosis in three Australian cities: Perth, Newcastle and Hobart. *Brain*. 1988;111(Pt 1):1–25.
24. Ramagopalan SV, Hanwell HE, Giovannoni G, et al. Vitamin D-dependent rickets, HLA-DRB1, and the risk of multiple sclerosis. *Arch Neurol*. 2010;67:1034–1035.
25. Sadovnick AD, Baird PA, Ward RH. Multiple sclerosis: updated risks for relatives. *Am J Med Genet*. 1988;29:533–541.
26. Ebers GC, Sadovnick AD, Dyment DA, Yee IM, Willer CJ, Risch N. Parent-of-origin effect in multiple sclerosis: observations in half-siblings. *Lancet*. 2004;363:1773–1774.
27. Sadovnick AD, Baird PA. The familial nature of multiple sclerosis: age-corrected empiric recurrence risks for children and siblings of patients. *Neurology*. 1988;38:990–991.
28. Galton F. *Hereditary genius: an inquiry into its laws and consequences*. London: Macmillan; 1869.
29. Willer CJ, Dyment DA, Risch NJ, Sadovnick AD, Ebers GC, CCS Group. Twin concordance and sibling recurrence rates in multiple sclerosis. *Proc Natl Acad Sci U S A*. 2003;100:12877–12882.
30. Watson CT, Disanto G, Breden F, Giovannoni G, Ramagopalan SV. Estimating the proportion of variation in susceptibility to multiple sclerosis captured by common SNPs. *Sci Rep*. 2012;2:770.
31. Willer CJ, Dyment DA, Sadovnick AD, et al. Timing of birth and risk of multiple sclerosis: population based study. *BMJ*. 2005;330:120.
32. Templer DI, Trent NH, Spencer DA, et al. Season of birth in multiple sclerosis. *Acta Neurol Scand*. 1992;85:107–109.
33. Hoppenbrouwers IA, Liu F, Aulchenko YS, et al. Maternal transmission of multiple sclerosis in a dutch population. *Arch Neurol*. 2008;65:345–348.
34. Herrera BM, Ramagopalan SV, Lincoln MR, et al. Parent-of-origin effects in MS: observations from avuncular pairs. *Neurology*. 2008;71:799–803.
35. Ebers GC, Sadovnick AD, Risch NJ. A genetic basis for familial aggregation in multiple sclerosis. Canadian Collaborative Study Group. *Nature*. 1995;377:150–151.
36. Dyment DA, Yee IM, Ebers GC, Sadovnick AD, CCS Group. Multiple sclerosis in stepsiblings: recurrence risk and ascertainment. *J Neurol Neurosurg Psychiatry*. 2006;77:258–259.
37. DeLuca GC, Ramagopalan SV, Herrera BM, et al. An extremes of outcome strategy provides evidence that multiple sclerosis severity is determined by alleles at the HLA-DRB1 locus. *Proc Natl Acad Sci U S A*. 2007;104:20896–20901.
38. Ramagopalan SV, Knight JC, Ebers GC. Multiple sclerosis and the major histocompatibility complex. *Curr Opin Neurol*. 2009;22:219–225.
39. Jersild C, Svejgaard A, Fog T. HL-A antigens and multiple sclerosis. *Lancet*. 1972;1:1240–1241.
40. Naito S, Namerow N, Mickey MR, Terasaki PI. Multiple sclerosis: association with HL-A3. *Tissue Antigens*. 1972;2:1–4.
41. Winchester R, Ebers G, Fu SM, Espinosa L, Zabriskie J, Kunkel HG. B-cell alloantigen Ag 7a in multiple sclerosis. *Lancet*. 1975;2:814.

42. Fogdell A, Hillert J, Sachs C, Olerup O. The multiple sclerosis- and narcolepsy-associated HLA class II haplotype includes the DRB5*0101 allele. *Tissue Antigens*. 1995;46:333–336.
43. Tiwari JL, Terasaki PI. HLA-DR and disease associations. *Prog Clin Biol Res*. 1981;58:151–163.
44. Ramagopalan SV, Morris AP, Dyment DA, et al. The inheritance of resistance alleles in multiple sclerosis. *PLoS Genet*. 2007;3:1607–1613.
45. Dyer P, McGilvray R, Robertson V, Turner D. Status report from 'double agent HLA': health and disease. *Mol Immunol*. 2013;55:2–7.
46. Madsen LS, Andersson EC, Jansson L, et al. A humanized model for multiple sclerosis using HLA-DR2 and a human T-cell receptor. *Nat Genet*. 1999;23:343–347.
47. Goverman J, Woods A, Larson L, Weiner LP, Hood L, Zaller DM. Pillars article: transgenic mice that express a myelin basic protein-specific T cell receptor develop spontaneous autoimmunity. Cell. 1993. 72: 551–560. *J Immunol*. 2013;190:3018–3027.
48. Dyment DA, Herrera BM, Cader MZ, et al. Complex interactions among MHC haplotypes in multiple sclerosis: susceptibility and resistance. *Hum Mol Genet*. 2005;14:2019–2026.
49. Masterman T, Ligers A, Olsson T, Andersson M, Olerup O, Hillert J. HLA-DR15 is associated with lower age at onset in multiple sclerosis. *Ann Neurol*. 2000;48:211–219.
50. Marrosu MG, Sardu C, Cocco E, et al. Bias in parental transmission of the HLA-DR3 allele in Sardinian multiple sclerosis. *Neurology*. 2004;63:1084–1086.
51. Cocco E, Murru R, Costa G, et al. Interaction between HLA-DRB1-DQB1 haplotypes in Sardinian multiple sclerosis population. *PLoS One*. 2013;8:e59790.
52. Ghabanbasani MZ, Gu XX, Spaepen M, et al. Importance of HLA-DRB1 and DQA1 genes and of the amino acid polymorphisms in the functional domain of DR beta 1 chain in multiple sclerosis. *J Neuroimmunol*. 1995;59:77–82.
53. Yoshimura S, Isobe N, Yonekawa T, et al, SJMSG Consortium. Genetic and infectious profiles of Japanese multiple sclerosis patients. *PLoS One*. 2012;7:e48592.
54. Gregersen JW, Kranc KR, Ke X, et al. Functional epistasis on a common MHC haplotype associated with multiple sclerosis. *Nature*. 2006;443:574–577.
55. Weinstock-Guttman B, Ramanathan M, Hashmi K, et al. Increased tissue damage and lesion volumes in African Americans with multiple sclerosis. *Neurology*. 2010;74:538–544.
56. Lincoln MR, Ramagopalan SV, Chao MJ, et al. Epistasis among HLA-DRB1, HLA-DQA1, and HLA-DQB1 loci determines multiple sclerosis susceptibility. *Proc Natl Acad Sci U S A*. 2009;106:7542–7547.
57. Fogdell-Hahn A, Ligers A, Grønning M, Hillert J, Olerup O. Multiple sclerosis: a modifying influence of HLA class I genes in an HLA class II associated autoimmune disease. *Tissue Antigens*. 2000;55:140–148.
58. Sawcer S, Hellenthal G, Pirinen M, et al. Genetic risk and a primary role for cell-mediated immune mechanisms in multiple sclerosis. *Nature*. 2011;476:214–219.
59. Harbo HF, Lie BA, Sawcer S, et al. Genes in the HLA class I region may contribute to the HLA class II-associated genetic susceptibility to multiple sclerosis. *Tissue Antigens*. 2004;63:237–247.
60. Friese MA, Jakobsen KB, Friis L, et al. Opposing effects of HLA class I molecules in tuning autoreactive CD8+ T cells in multiple sclerosis. *Nat Med*. 2008;14:1227–1235.
61. Simmons SB, Pierson ER, Lee SY, Goverman JM. Modeling the heterogeneity of multiple sclerosis in animals. *Trends Immunol*. 2013;34(8):410–422.
62. Friese MA, Montalban X, Willcox N, Bell JI, Martin R, Fugger L. The value of animal models for drug development in multiple sclerosis. *Brain*. 2006;129:1940–1952.
63. Ramagopalan SV, Maugeri NJ, Handunnetthi L, et al. Expression of the multiple sclerosis-associated MHC class II Allele HLA-DRB1*1501 is regulated by vitamin D. *PLoS Genet*. 2009;5:e1000369.
64. Disanto G, Sandve GK, Berlanga-Taylor AJ, et al. Vitamin D receptor binding, chromatin states and association with multiple sclerosis. *Hum Mol Genet*. 2012;21:3575–3586.
65. Sawcer S, Jones HB, Feakes R, et al. A genome screen in multiple sclerosis reveals susceptibility loci on chromosome 6p21 and 17q22. *Nat Genet*. 1996;13:464–468.
66. Ebers GC, Kukay K, Bulman DE, et al. A full genome search in multiple sclerosis. *Nat Genet*. 1996;13:472–476.
67. Haines JL, Ter-Minassian M, Bazyk A, et al. A complete genomic screen for multiple sclerosis underscores a role for the major histocompatability complex. The Multiple Sclerosis Genetics Group. *Nat Genet*. 1996;13:469–471.
68. Hafler DA, Compston A, Sawcer S, et al. Risk alleles for multiple sclerosis identified by a genomewide study. *N Engl J Med*. 2007;357:851–862.
69. Sanna S, Pitzalis M, Zoledziewska M, et al. Variants within the immunoregulatory CBLB gene are associated with multiple sclerosis. *Nat Genet*. 2010;42:495–497.
70. Jakkula E, Leppä V, Sulonen AM, et al. Genome-wide association study in a high-risk isolate for multiple sclerosis reveals associated variants in STAT3 gene. *Am J Hum Genet*. 2010;86:285–291.
71. Australia and New Zealand Multiple Sclerosis Genetics Consortium (ANZgene). Genome-wide association study identifies new multiple sclerosis susceptibility loci on chromosomes 12 and 20. *Nat Genet*. 2009;41:824–828.
72. De Jager PL, Baecher-Allan C, Maier LM, et al. The role of the CD58 locus in multiple sclerosis. *Proc Natl Acad Sci U S A*. 2009;106:5264–5269.
73. Lundmark F, Duvefelt K, Iacobaeus E, et al. Variation in interleukin 7 receptor alpha chain (IL7R) influences risk of multiple sclerosis. *Nat Genet*. 2007;39:1108–1113.
74. Gregory SG, Schmidt S, Seth P, et al. Interleukin 7 receptor alpha chain (IL7R) shows allelic and functional association with multiple sclerosis. *Nat Genet*. 2007;39:1083–1091.
75. Zenatti PP, Ribeiro D, Li W, et al. Oncogenic IL7R gain-of-function mutations in childhood T-cell acute lymphoblastic leukemia. *Nat Genet*. 2011;43:932–939.
76. Lundström W, Highfill S, Walsh ST, et al. Soluble IL7Rα potentiates IL-7 bioactivity and promotes autoimmunity. *Proc Natl Acad Sci U S A*. 2013;110:E1761–E1770.

77. Wiendl H, Gross CC. Modulation of IL-2Rα with daclizumab for treatment of multiple sclerosis. *Nat Rev Neurol*. 2013;9:394–404.
78. Gold R, Giovannoni G, Selmaj K, et al. Daclizumab high-yield process in relapsing-remitting multiple sclerosis (SELECT): a randomised, double-blind, placebo-controlled trial. *Lancet*. 2013;381:2167–2175.
79. Ban M, Goris A, Lorentzen AR, et al. Replication analysis identifies TYK2 as a multiple sclerosis susceptibility factor. *Eur J Hum Genet*. 2009;17:1309–1313.
80. Couturier N, Bucciarelli F, Nurtdinov RN, et al. Tyrosine kinase 2 variant influences T lymphocyte polarization and multiple sclerosis susceptibility. *Brain*. 2011;134:693–703.
81. Mero IL, Lorentzen AR, Ban M, et al. A rare variant of the TYK2 gene is confirmed to be associated with multiple sclerosis. *Eur J Hum Genet*. 2010;18:502–504.
82. Dyment DA, Cader MZ, Chao MJ, et al. Exome sequencing identifies a novel multiple sclerosis susceptibility variant in the TYK2 gene. *Neurology*. 2012;79:406–411.
83. Kitanaka S, Takeyama K, Murayama A, et al. Inactivating mutations in the 25-hydroxyvitamin D3 1alpha-hydroxylase gene in patients with pseudovitamin D-deficiency rickets. *N Engl J Med*. 1998;338:653–661.
84. Ramagopalan SV, Dyment DA, Cader MZ, et al. Rare variants in the CYP27B1 gene are associated with multiple sclerosis. *Ann Neurol*. 2011;70:881–886.
85. Barizzone N, Pauwels I, Luciano B, et al. No evidence for a role of rare CYP27B1 functional variations in multiple sclerosis. *Ann Neurol*. 2013;73:433–437.
86. Ban M, Caillier S, Mero IL, et al. No evidence of association between mutant alleles of the CYP27B1 gene and multiple sclerosis. *Ann Neurol*. 2013;73:430–432.
87. Ramagopalan SV, Heger A, Berlanga AJ, et al. A ChIP-seq defined genome-wide map of vitamin D receptor binding: associations with disease and evolution. *Genome Res*. 2010;20:1352–1360.
88. Hunt KA, Mistry V, Bockett NA, et al. Negligible impact of rare autoimmune-locus coding-region variants on missing heritability. *Nature*. 2013;498:232–235.
89. Ascherio A, Munger KL. Environmental risk factors for multiple sclerosis. Part II: noninfectious factors. *Ann Neurol*. 2007;61:504–513.
90. Ascherio A, Marrie RA. Vitamin D in MS: a vitamin for 4 seasons. *Neurology*. 2012;79:208–210.
91. Tiosano D, Wildbaum G, Gepstein V, et al. The role of vitamin D receptor in innate and adaptive immunity: a study in hereditary vitamin D-resistant rickets patients. *J Clin Endocrinol Metab*. 2013;98:1685–1693.
92. Hernán MA, Olek MJ, Ascherio A. Cigarette smoking and incidence of multiple sclerosis. *Am J Epidemiol*. 2001;154:69–74.
93. Hernán MA, Jick SS, Logroscino G, Olek MJ, Ascherio A, Jick H. Cigarette smoking and the progression of multiple sclerosis. *Brain*. 2005;128:1461–1465.
94. Hedström AK, Sundqvist E, Bäärnhielm M, et al. Smoking and two human leukocyte antigen genes interact to increase the risk for multiple sclerosis. *Brain*. 2011;134:653–664.
95. Doyle HA, Mamula MJ. Posttranslational protein modifications: new flavors in the menu of autoantigens. *Curr Opin Rheumatol*. 2002;14:244–249.
96. Simon KC, van der Mei IA, Munger KL, et al. Combined effects of smoking, anti-EBNA antibodies, and HLA-DRB1*1501 on multiple sclerosis risk. *Neurology*. 2010;74:1365–1371.
97. Patsopoulos NA, Esposito F, Reischl J, et al. Genome-wide meta-analysis identifies novel multiple sclerosis susceptibility loci. *Ann Neurol*. 2011;70:897–912.
98. Oksenberg JR, Baranzini SE, Sawcer S, Hauser SL. The genetics of multiple sclerosis: SNPs to pathways to pathogenesis. *Nat Rev Genet*. 2008;9:516–526.
99. Robertson NP, Fraser M, Deans J, Clayton D, Walker N, Compston DA. Age-adjusted recurrence risks for relatives of patients with multiple sclerosis. *Brain*. 1996;119(Pt 2):449–455.
100. Beck CA1, Metz LM, Svenson LW, Patten SB. Regional variation of multiple sclerosis prevalence in Canada. *Mult Scler*. 2005;11(5):516–519.

CHAPTER

4

Pharmacogenomic Approaches to the Treatment of Sporadic Alzheimer Disease using Cholinomimetic Agents

Judes Poirier, Justin Miron, and Cynthia Picard

Douglas Mental Health University Institute, Montreal, QC, Canada
McGill University, Montreal, QC, Canada

INTRODUCTION

Pharmacogenomics can be summarized by the analysis of the interplay between drug responses versus genetics or side effects versus genetic susceptibility. It is generally not characterized by the nature of the chemical entity itself nor by the biochemistry of the response, but by the fact that a response may lack uniformity and that this lack of uniformity has a genetic origin.[1] Traditionally, pharmacogenetics has focused on the interaction between specific genetic polymorphisms in drug-metabolizing enzymes such as the P450 system, drug-transport enzymes, drug response and, to a certain extent, toxicity. In recent years, pharmacogenomics has refocused on genetic polymorphisms in the drug target-related genes such as cell surface and nuclear receptors, as well as catalytic enzymes. More recently, genes underlying the disease risk, process, onset, progression and pathology have been added to the list, including variants modulating responses to specific therapies.[2–6] This chapter will examine the pharmacogenomic role of specific, yet prevalent, genetic risk factors associated with the common form of Alzheimer disease (AD) and of variants known to be involved in the regulation of specific neurotransmitter pathways.

GENETIC RISK FACTORS AND SPORADIC ALZHEIMER DISEASE

AD is a progressive neurodegenerative disease that has a strong genetic basis.[7] The disease is characterized by the progressive loss of memory, functional decline, behavioural symptoms, and finally, death. It is estimated that by 2030 there will be some 10 million AD patients in North America alone and more than 70 million around the world. The treatment of this disease represents a large financial burden on the healthcare system, which has been estimated to be in the order of 600 billion US$ (World Alzheimer Report 2010). In 1993, apolipoprotein E type 4 (apoE4) was found to act as a major susceptibility gene, being associated with 40–65% of cases of familial and sporadic forms of the disease.[8–10] The initial observations were rapidly confirmed and now more than 30 different genome-wide association studies (GWAS: http://www.alzgene.org/largescale.asp) have replicated the original association in the sporadic form of the disease. AD patients born with at least one E4 allele tend to exhibit: a) an earlier age of onset,[11] b) higher amyloid plaque and tangle densities,[10,12,13] c) lower apoE levels in the blood, brain tissues, and CSF,[14–17] d) greater cerebrovascular amyloid deposition, and more importantly, e) marked reductions in synaptic density in multiple brain areas[18] and in multiple neurotransmitter systems[2,19,20] when compared to non-E4 subjects.

ApoE has been extensively studied in non-nervous tissues as one of several proteins that regulate lipid transport and metabolism. ApoE facilitates cholesterol, phospholipid and vitamin E transport between different cell types and

http://dx.doi.org/10.1016/B978-0-12-410529-4.00004-8

different organs. It binds to large lipid–protein particles of the high-density lipoprotein (HDL) family in the brain. This binding increases the ability of large lipid complexes to transport cholesterol and phospholipids in the blood and the parenchyma of the brain. The mature form of apoE found in human plasma and brain is a single glycosylated 37-kDa polypeptide containing 299 amino acids. It was shown to coordinate the mobilization and redistribution of cholesterol in repair, growth, and maintenance of myelin and neuronal membranes during development or after neuronal cell injury or neurodegenerative conditions.[21] In the brain, apoE coordinates the redistribution of cholesterol and phospholipids during membrane remodeling associated with synaptic plasticity and dendritic remodeling (Figure 4.1). In the absence of apoE in the brain, such as in apoE-knockout mice, cortical synaptic density declines steadily with normal aging, and synaptic remodeling and plasticity become progressively compromised; cognitive deficits emerge in late adulthood and the cholinergic system suffers from severe attrition.[22–26]

In contrast to rodents, human apoE is encoded by a four-exon gene (3.6 kb) on the long arm of chromosome 19 and three major isoforms of apoE (E4, E3 and E2) differing by a single unit of net charge. These isoforms are expressed from multiple alleles at a single apoE genetic locus, giving rise to three homozygous phenotypes (E4/4, E3/3 and E2/2) and three heterozygous phenotypes (E4/3, E4/2 and E3/2).[14]

Although polymorphism associations in more than 700 genes have been studied in sporadic AD (http://www.alzgene.org), only a small number of genetic links have been thoroughly replicated worldwide and a handful of genetic variants were found to affect drug responsiveness or adverse side effect profiles in sporadic AD. In addition

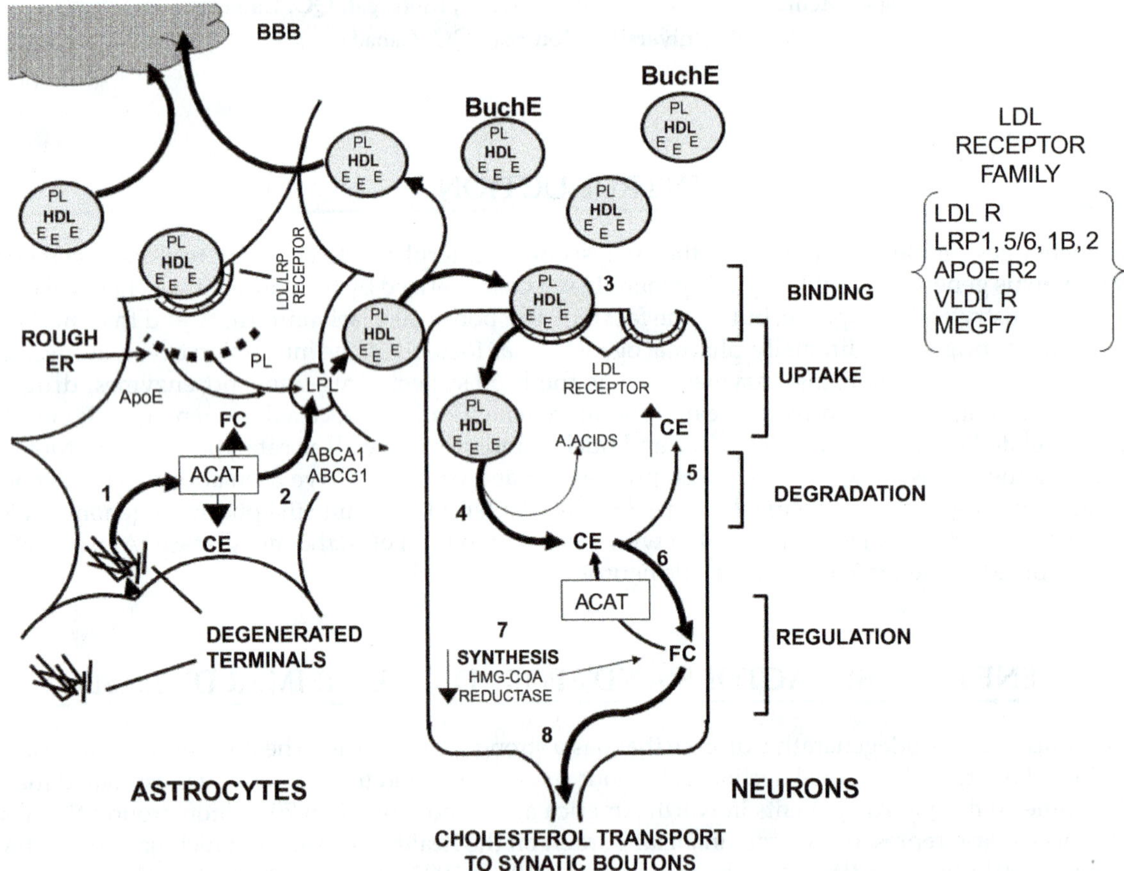

FIGURE 4.1 **Apolipoprotein E and the cholesterol recycling cascade in response to neurodegeneration in the adult CNS.** Degenerating neurons and terminals are initially internalized and degraded by astrocytes and/or microglia. The nonesterified cholesterol released during breakdown of terminals is used as free cholesterol (FC) (1) for the assembly of an extracellular apoE–cholesterol–lipoprotein complex (2) or converted into cholesterol esters (CE) for storage purposes. The newly formed apoE–cholesterol–lipoprotein complexes are then directed toward: a) the blood brain barrier and transferred into the blood stream, presumably through the ependymal cells surrounding the ventricles and/or b) to specific brain cells requiring lipids. ApoE complexes then bind to the different members of the low-density lipoprotein (LDL) receptor family (3), but it is through the LDL receptor pathway that the complex is internalized (4), the cholesterol is then transiently stored as an ester (CE; 5) or converted into free cholesterol (FC; 6), which is then mobilized for dendritic proliferation and/or synaptogenesis (8). As a consequence of the internalization process, cholesterol synthesis in neurons (via the HMG-CoA reductase pathway) becomes progressively repressed (7). BBB, blood–brain barrier; E, ApoE; PL, phospholipids; A. Acids, amino acids; CE, cholesterol ester; FC, free cholesterol; HDL, high-density lipoprotein; ACAT, cholesterol acetyltransferase; ER, endoplasmic reticulum; LPL: Lipoprotein lipase.

to apoE, butyrylcholinesterase (BuChE), paraoxonase (PON), acetylcholinesterase (AchE), interleukin-1 (IL-1), and the insulin degrading enzyme (IDE) genes exhibit common polymorphic variants that have been associated with both risk and drug responsiveness in sporadic AD and related dementias.

GENETIC RISK FACTORS, CHOLINERGIC DYSFUNCTION, AND ALZHEIMER DISEASE

The biological basis for the pharmacogenomic response in sporadic AD stems from the postmortem analyses of autopsy-confirmed control and age-matched AD subjects, in which key cholinergic markers were examined in relation to specific disease-related polymorphisms. Choline acetyltransferase (ChAT) activity, the rate limiting factor in acetylcholine synthesis, in the hippocampus and the temporal cortex of AD cases was reported to be inversely proportional to the apoE4 allele dose,[19,27] i.e., as apoE4 allele copy number increases, ChAT activity decreases in the CNS. Similar gene dose–response associations were reported with other cholinergic pre-synaptic markers, such as acetylcholinesterase (AChE) activity,[2,20,28] nerve growth factor (NGF) receptor,[20] and nicotinic receptor density.[2] Cohen and colleagues reported a marked effect of apoE4 allele on the distribution volume of pre-synaptic M2 receptor sites in the living human brain using positron emission tomography [18 F]FP-TZTP tracing techniques.[29,30] Altogether, these studies indicate that the cholinergic pre-synaptic structures are preferentially damaged in the brain of apoE4 allele carriers (ChAT, nicotinic receptor, NGF receptor, AChE and M2 receptor), whereas post-synaptic markers are clearly spared in the brain of autopsied AD subjects, irrespective of the apoE genotype. These results also indicate that the AD patients who have the most to gain in terms of cholinergic benefit (and cognitive performance) belong to the apoE4 carrier group.

The apoE4 allele–cholinergic dysfunction association was shown to be weakened in older patients, as the E4-linked risk level becomes nonsignificant in subjects aged 80 years and older.[31,32] In contrast, strong pharmacogenomic responses have been reported in younger mild cognitively impaired (MCI) subjects receiving AChE-specific cholinomimetic medications.[33–35]

ApoE4 AND CHOLINOMIMETIC DRUGS IN ALZHEIMER DISEASE

A 1995 retrospective pilot analysis of the clinical data of the so-called 30-week pivotal double-blind, randomized, tacrine clinical trial (a dual acetylcholinesterase/butyrylcholinesterase inhibitor) led to the breakthrough discovery of a potential link between the presence of the apoE4 allele and the clinical outcome in subjects with mild-to-moderate AD treated with cholinomimetics.[2] The analysis revealed an apoE4 genotype mediated effect on the Caregiver-Rated Global Impression of Change (CGIC) scale and the Alzheimer's Disease Assessment Scale-Cognitive subscale (ADAS-Cog). However, both the apoE4 and non-E4 subjects treated at the highest dose of tacrine (160 mg/day) showed significant improvement of symptoms when compared to the placebo group.[3] Closer examination of the tacrine dataset revealed that most of the pharmacogenomic effects stemmed mainly from the placebo group, where the apoE4 allele was shown to markedly modulate the rate of decline, consistent with previous and subsequent publications on the natural progression of AD.[36–38] There are also reports of opposite findings in open-labeled trials with no placebo arms,[39–41] suggesting genotype profile differences when comparing outcome results versus baseline or to a parallel placebo arm. More importantly, it was shown that the action of apoE4 on disease progression varies significantly with the stages of the disease,[33,34,42,43] as well as the actual loss of cholinergic activity from the early to more severe stages of the disease (see Figure 4.2).[44–46]

This initial series of analyses gave a first indication of a possible pharmacogenomic interaction between apoE gene polymorphisms and existing pharmacological treatments for sporadic AD. This observation spawned many follow-up studies that confirmed that an individual's apoE4 allele carrier status may have to be considered by the physician in choosing the most appropriate cholinergic treatment for mild-to-moderate AD cases. But more importantly, the nature of the cholinergic agent (mono vs. dual cholinesterase inhibitors vs. muscarinic agonist) was found to directly impact on the clinical outcome as a function of genetic predisposition involving one or several variants.

Intent-to-treat (ITT) analyses of the influence of apoE genotype on tacrine drug response in mild-to-moderate AD in the whole cohort of subjects replicated the original pilot observation.[47] MacGowan et al. (1998) independently confirmed apoE pharmacogenomic effects on the tacrine response in an independent 12-month study. Using quantitative electroencephalographic responses to tacrine administration, Riekkinen and colleagues reported similar apoE genotype preference in AD subjects treated with the dual BuChE/AChE inhibitor tacrine.[48] Furthermore, a similar

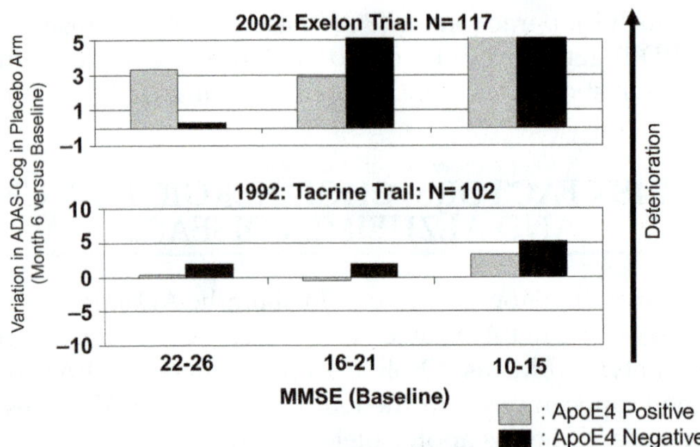

FIGURE 4.2 **Effect of apoE4 on the rate of decline in mild-to-moderate Alzheimer disease (AD) as a function of baseline disease severity.** Analysis of the impact of apoE genotype on the rate of decline using the ADAS-Cog scale in placebo arms of two independent 30-week randomized placebo-controlled clinical trials in mild-to-moderate AD cases. Only observed cases (OC) who completed the study were analyzed. Following diagnosis, rate of decline seems modulated by the presence of the apoE4 allele; the latter displaying the slowest decline. In the moderate stages, the rate of decline of both placebo groups is significantly more pronounced in the apoE4-negative subjects when compared to apoE4 allele carriers. *Adapted from* [80] *and* [81].

TABLE 4.1 Cholinesterase Inhibitor Selectivity in the Mammalian Brain

Compound	IC50 (nM) AChE	IC50 (nM) BuChE	Ratio BuChE/AChE
AChE-specific inhibitors			
Donepezil	5.7	7139	1252
Galantamine	0.35	19	53
BuChE/AchE dual inhibitors			
Rivastigmine	48000	54000	1.1
Tacrine	190	47	0.25

Adapted from [83].

pharmacogenomic relationship was reported with the topical response to tropicamide, a selective cholinergic agonist, in healthy cognitively normal elderly subjects carrying different apoE genotypes. The authors reported that the cholinergic antagonist drug[49] caused a marked dilatation of the pupil area in apoE4 carriers consistent with a neuronal hypersensitivity caused by the preferential damage of the local cholinergic neurons due to normal aging or neurodegenerative diseases.

Tacrine, like the more recent and safer rivastigmine, inhibits both BuChE and AChE activities (Table 4.1). In contrast, galantamine, metrifonate, and donepezil, three mono-specific acetylcholinesterase inhibitors that exhibit little inhibitory effect toward BuChE, were found to display different pharmacogenomic profiles in subjects with AD or amnestic MCI.

Imaging studies have found that ApoE4 may accelerate the progression of the hippocampal atrophy in prodromal and early AD,[50,51] but once an individual is advanced in age or in the progression of the disease, any influence of ApoE4 on the rate of progression is lost.[50] Actually, the rate of cerebral atrophy in AD subjects with a mean age of 70 years may be slower in association with ApoE4 relatively to other genotypes, whereas in older patients with a mean age of 80 years,[50] the progression was not different between genotypes. This is supported by clinical data that demonstrate that, during the prodromal phase of AD, carriers of an ApoE4 allele progress faster than noncarriers.[33,34,42,43]. Figure 4.2 summarizes the situation observed in the placebo arms of two double-blind clinical trials performed at a 10-year interval in mild-to-moderate AD subjects. In mild disease, progression may be comparable or slightly faster, but in more advanced stages, progression was shown to be consistently slower in ApoE4 carriers relative to noncarriers.

However, the story is somewhat different in patients treated with mono-specific AChE inhibitors that exhibit little or no effect on BuChE activity, such as metrifonate, galantamine, and donepezil. Farlow and coworkers (1999) examined the influence of apoE genotype on a short-term metrifonate treatment (6 months) in mild-to-moderate AD patients.[52] In contrast to tacrine and rivastigmine, metrifonate is much more selective toward AChE than to BuChE, although it is still categorized as a weak dual inhibitor. In this particular study, data pooled from four double-blind placebo-controlled clinical trials were analyzed retrospectively for the possible interaction between apoE genotype and cognitive response to metrifonate treatment after 26 weeks for the entire group (n=959); a trend that did not reach significance suggested a possible interaction between apoE genotype and treatment effect.[52] A subsequent metrifonate study that used DNA samples from a 3-year open-labeled clinical trial in mild-to-moderate AD revealed a progressive pharmacogenomic effect of the apoE4 allele emerging after 24 months of treatment.[3] Consistent with the previous meta-analysis report, no significant difference in the Mini Mental State Examination (MMSE) scores was detected after 26 weeks of treatment in E4 versus non-E4 groups. However, the longer follow-up assessments indicated a clear progressive dissociation of the apoE4 and non-E4 groups over time, particularly noticeable after 120 weeks and highly significant after 240 weeks.[3]

As we examine the cholinesterase inhibitors that are orders of magnitude more selective toward AChE, a slightly different picture emerges: a stronger than expected effect of the apoE4 allele on efficacy parameters. McGowan and colleagues first reported the effect of galantamine on cognitive performance in 84 mild-to-moderate AD patients over the course of 6 months of treatment and found that the best responders belonged to the apoE4 homozygotes subgroup.[53] More recently, Wilcock and collaborators and Raskind and collaborators examined galantamine efficacy in two large international multicentre placebo-controlled randomized clinical trials. They reported that while subjects with and without apoE4 allele exhibited a significant improvement on the ADAS-Cog scale, sub-analyses clearly identified the apoE4/4 group as the responder cohort with an average 6.6 points improvement on the ADAS-Cog scale.[54] Similar findings were reported by an independent research team using patients from Eastern Europe.[55]

At the highest dose (32 mg/day) of galantamine, Raskind and collaborators reported a strong apoE4-dependent improvement on the ADAS-Cog scale when compared to the non-apoE4 group – a pharmacogenomic effect that progressively disappears at lower doses in mild-to-moderate AD cases.[56] Consistent with this observation, lead investigators at Johnson and Johnson presented detailed evidence from a 2-year clinical trial with 1600 MCI patients that revealed a strong apoE4-dependent response in galantamine-treated subjects when compared to placebo.[34] Figure 4.3

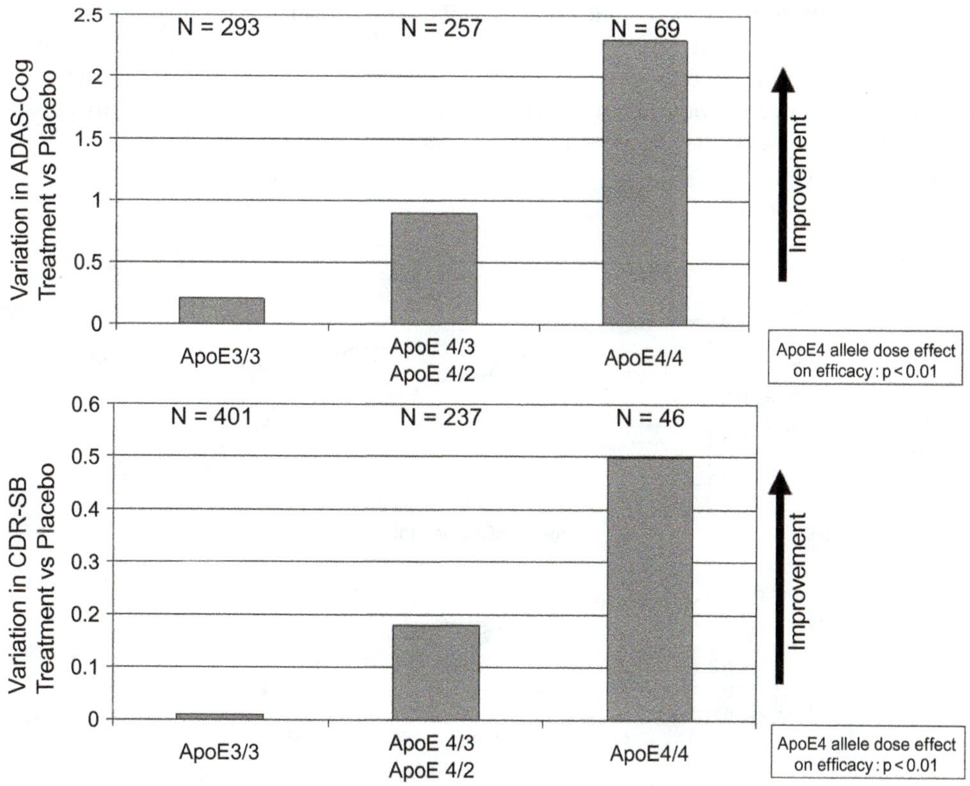

FIGURE 4.3 **Efficacy of a 24-month galantamine treatment in mild cognitively impaired (MCI) subjects as a function of apoE4 allele dose.** Mean changes from placebo in ADAS-Cog/MCI (Study 1) and CDR-SB (Study 2) scores in galantamine-treated subjects as a function of apoE genotype. The allele dose–response analysis reveals a significant ($p<0.01$) association between ADAS-Cog-MCI and CDR-SB score improvement and apoE4 allele copy number. *Adapted from* [34].

summarizes the results of the first of the two parallel studies where ADAS-Cog variations were contrasted with apoE4 allele dose in subjects exposed to either galantamine or placebo for a period of 2 years.[34] The efficacy profile of the drug was found to be tightly associated with apoE4 allele dose, the subjects carrying the apoE4/4 genotype exhibiting outstanding responses when compared to all other genotypes.

The first report suggesting a possible pharmacogenomic effect of the apoE4 allele on donepezil drug response was published by Lucotte's team in France using an open-labeled clinical trial in mild-to-moderate AD subjects.[57] Results clearly showed that the donepezil response was restricted to subjects carrying the apoE4 allele as opposed to the dual inhibitors tacrine and rivastigmine. This pilot observation was subsequently replicated by a small, randomized crossover study using donepezil in mild-to-moderate AD.[58] Although the focus of the study was on global clinical improvement, a clear apoE4-dependent effect was documented on the ADAS-Cog scale. Several investigators followed up with open trial designs with no placebo arm and either succeeded or failed to replicate the original observation with donepezil.[59–61]

Only two very large randomized placebo-controlled studies examined this pharmacogenomic association with the proper statistical power; the first one, performed by Windblad and coworkers in North European countries, failed to detect any apoE4 contribution to donepezil efficacy in mild-to-moderate AD subjects (N=198).[62] In contrast, Lendon and coworkers in England reported a significant contribution of the apoE genotype on donepezil efficacy at the level of cognition and activities of daily living in mild-to-moderate AD cases (N=785).[63]

While the bulk of the findings suggests a better donepezil response in apoE4 allele carriers with sporadic AD, only recently has the observation been extended to MCI populations using double-blind, randomized, placebo-controlled designs.[33] In a 3-year prospective study of the rate of conversion from MCI to sporadic AD, the authors reported that donepezil was not particularly effective at altering the rate of conversion of MCI during the 3-year period, except for the first 12 months of the study. However, when subjects were stratified by their apoE genotype, results clearly showed a significant effect of donepezil on the rate of conversion in apoE4 allele carriers during the entire 3-year period, while E4-negative subjects did not benefit from the 3-year treatment (Figure 4.4).

Figure 4A illustrates the effect of apoE4 allele on the conversion from MCI case to diagnosed AD as opposed to apoE4-negative case (n=424), whereas Figure 4B illustrates the marked and significant impact of donepezil on the conversion rate from MCI to AD in apoE4 allele carrier subjects only.

These results, combined with those of large-scale clinical trials with various AChE inhibitors, strongly support the notion that the cholinergic system, which is particularly affected by the presence of the E4 allele in AD,[2] becomes a target of choice for cognitive-enhancing drugs designed to modulate cholinergic activity in the CNS. The selective damage of the pre-synaptic cholinergic compartment in human apoE4 allele carriers is certainly consistent with a hypersensitivity of the post-synaptic cholinergic sites and the expected enhanced cholinergic drug response observed in apoE4 carriers with MCI and AD.[3]

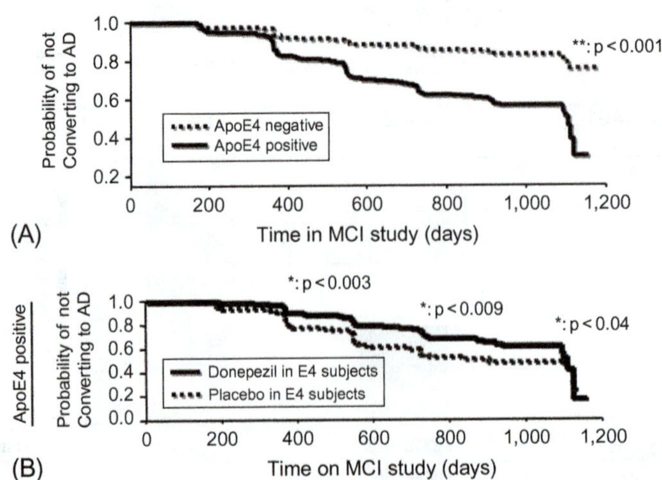

FIGURE 4.4 Kaplan–Meier estimates of the rate of progression from mild cognitive impairment (MCI) to Alzheimer disease (AD). Panel A shows the effect of apoE4 carrier status on the rate of progression to AD, whereas panel B illustrates the long-term beneficial effect of donepezil treatment on the rate of progression in the apoE4 allele-carrier population only. There was no significant difference between the placebo and donepezil groups in apoE4-negative subjects. *Adapted from* [33].

EXPERIMENTAL DRUGS AND THEIR RELATIONSHIP TO THE ApoE4 ALLELE

In recent years, the implementation of phase III clinical trials in the field of dementia has led to careful analyses of the apoE genotype, as well as many other genes believed to affect drug response in dementing illnesses. However, most of the pharmacogenomic analyses performed in clinical trials with MCI, mild-to-moderate, and moderate-to-severe AD were done in a retrospective manner to minimize interference with the commercialization strategy of the products under development. While this approach has led to some controversy as to the full public disclosure and publication of both positive and negative pharmacogenomic results,[64–66] a few pharmaceutical companies have chosen to publish or present some of their clinical trial results at international meetings.

The pharmacogenomic profile of xanomeline (Eli Lilly) was assessed in a phase II drug trial in mild-to-moderate AD. This compound is a M1-specific cholinergic agonist that bypasses the pre-synaptic cholinergic terminals and directly stimulates the post-synaptic receptor sites in the brain. Since the M1 site densities are not affected by apoE4 allele in postmortem brains of AD subjects,[2] it was postulated that xanomeline efficacy would be less affected by apoE4 allele than current cholinesterase inhibitor treatments. Patients exposed to a 75-mg dose of xanomeline were monitored over a period of 6 months using the ADAS-Cog as the primary outcome variable. Figure 4.5 illustrates the observed apoE4 allele-dependent dose response after xanomeline administration with a near-complete absence of response in the apoE4/4AD subject population, suggesting that some pre-synaptic cholinergic components must be present for xanomeline to exert its post-synaptic effect. A second follow-up xanomeline drug trial (phase IIb) was implemented by Eli Lilly shortly after and the data analysis of more than 180 mild-to-moderate AD patients revealed a clear apoE4-dependent pharmacogenomic profile, which replicated the original observation (Figure 4.5A),[82] with apoE4/4 showing no improvement after 6 months of treatment relative to E3 carriers. The development of this agent for the treatment of AD was subsequently abandoned due to side effect profile and the observed pharmacogenomic effects.

Richard and collaborators examined the influence of apoE genotype on the responsiveness of the experimental vasopressinergic/noradrenergic drug S12024 (Servier) in mild-to-moderate AD (Figure 4.5B).[67] While no significant overall benefit of the drug was observed in the AD versus placebo group, stratification by apoE genotype revealed a significant E4-dependent improvement on both ADAS-Cog and Clinician Interview-Based Impression of Change (CIBIC) scales. Citicoline (cytidine 5′-diphosphocholine), an endogenous intermediate in the synthesis of membrane lipids and acetylcholine, has been used in the experimental treatment of neurodegenerative disorders like AD. Alvarez and coworkers (1999) published findings of a double-blind placebo-controlled study in mild-to-moderate AD patients who received citicoline for 12 weeks. The authors reported that citicoline significantly improved cognitive function versus placebo, but only in patients carrying at least one apoE4 allele.

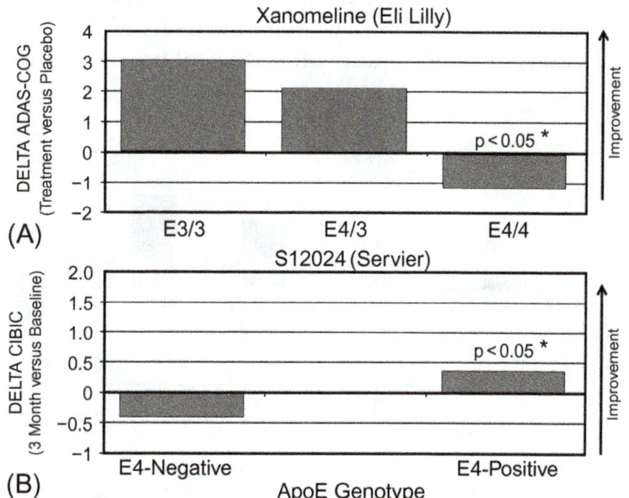

FIGURE 4.5 **Effect of xanomeline (A) and S12024 (B) on disease progression in mild-to-moderate Alzheimer disease (AD) cases stratified by apoE genotype.** Xanomeline, a M1 cholinergic receptor agonist developed by Eli Lilly was administered over 6 months, whereas S12024, a noradrenergic/vasopressinergic agent from Servier, was tested for only 3 months. Results are expressed as mean differences between Clinician Interview-Based Impression of Change (CIBIC) or Alzheimer's Disease Assessment Scale-Cognitive subscale (ADAS-Cog) scores in treated versus placebo subjects. *Adapted from* [67] *and* [82].

ACETYLCHOLINESTERASE AND BUTYRYLCHOLINESTERASE GENETIC VARIANTS IN DEMENTIA

In humans, BuChE (EC3.1.1.8) is present in plasma and in most tissues, including certain regions of the brain.[68] Despite today's extensive knowledge about allelic BuChE variants and their pharmacogenomic impact, its physiological function still remains largely unclear. The 574-amino-acid glycoprotein is coded by a single-copy gene on chromosome 3q26.1–26.2. Several dysfunctional BuChE mutations have been characterized, both at the phenotypic and genetic level.[69] Since 1965, several studies have reported the co-localization of the enzyme with senile plaques and neurofibrillary tangles, the hallmarks of the pathology of AD, and the severe loss of cholinergic neurons in AD brains has been found to be accompanied by higher than normal levels of BuChE.[70,71] The K-variant has been identified as the most frequent functional mutation of BuChE by far.[68] The K-polymorphism is found in various ethnic populations with homozygote frequencies between 1 and 4%. This is consistent with the fact that the BuChE-K genetic variant, which has also been associated with a higher risk of developing AD, is characterized by higher amyloid plaque density relatively to BuChE wild-type cases.[72,73] This genetic association was recently replicated using a GWAS designed to identify polymorphisms that catalyze amyloid deposition *in vivo* using PIB-positron emission tomography scans in the so-called ADNI cohort.[74] The BuChE-K polymorphism has been reported to occur with a higher frequency in late onset AD patients, especially in carriers of the apoE4 allele.[65,75] For all of these reasons, BuChE was deemed to act as a potential pharmacogenomic modulator of cholinomimetic therapies in AD and other related disorders.

Patients suffering from moderate-to-severe AD, Parkinson disease dementia, and dementia with Lewy bodies, harboring the BuChE-K allele, which encodes for lower expression of the enzyme, decline less rapidly than those with wild-type BuChE.[76,77] These preliminary observations prompted the analysis of the BuChE genetic variants in several clinical trials involving the dual AChE/BuChE inhibitor rivastigmine. BuChE wild-type carriers younger than 75 years showed differential efficacy to cholinesterase inhibitor therapies: patients receiving rivastigmine displayed significantly greater treatment responses over 2 years than patients receiving donepezil. In contrast, BuChE-K variant carriers experienced similar long-term treatment effects with both agents, although adverse events were more frequent in rivastigmine-treated patients.[78] Figure 4.6 illustrates the results obtained in a *post-hoc* analysis of the subjects with advanced symptoms (moderate AD) as a function of their combined apoE/BuChE polymorphisms. As shown previously for most forms of dementia (Lewy body, Parkinson disease and moderate-to-severe AD cases), BuChE-K carriers display rate of deterioration markedly slower in more advanced dementia than those expressing the wild-type variant of the BuChE gene,[79] and the clinical response to the dual inhibitor rivastigmine was found to reach significance only in BuChE K-negative (or wild-type) cases. Stratification of subjects with Parkinson dementia as a function of BuChE genotype reveals a similar, but more pronounced, improvement of cognitive performance on the Clinical Dementia Rating scale (CDR-CRT) in subjects treated with rivastigmine who are K-allele negative. Fortunately, this subgroup represents the vast majority of the patients enrolled in the trial.

However, it is in the MCI population that the pharmacogenomic effect of BuChE-K allele was deemed to be the most obvious in a randomized placebo-controlled double-blind clinical trial with rivastigmine which lasted 4 years.

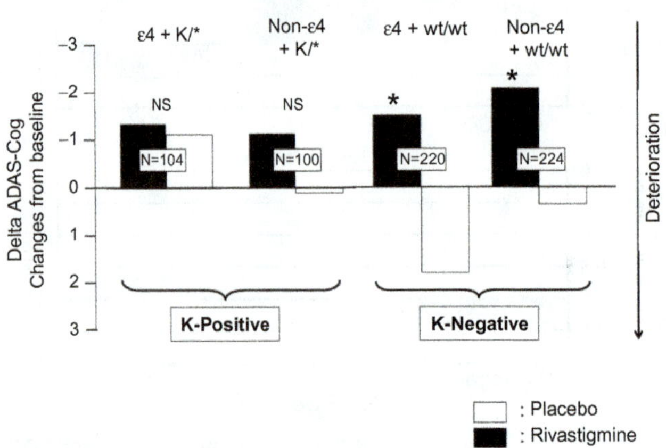

FIGURE 4.6 **ApoE4–butyrylcholinesterase K interaction in a 6-month, randomized, placebo-controlled clinical trial in subjects with moderate Alzheimer disease (AD) receiving rivastigmine.** Cognitive decline (ADAS-Cog) compared to placebo and response to rivastigmine (a dual inhibitor of butyrylcholinesterase (BuChE) and acetylcholinesterase (AChE)) treatment according to genotypes over 6 months in moderate AD patients. *(Adapted from [35]). *, p<0.05, intent-to-treat/last observation carried forward population of subjects.*

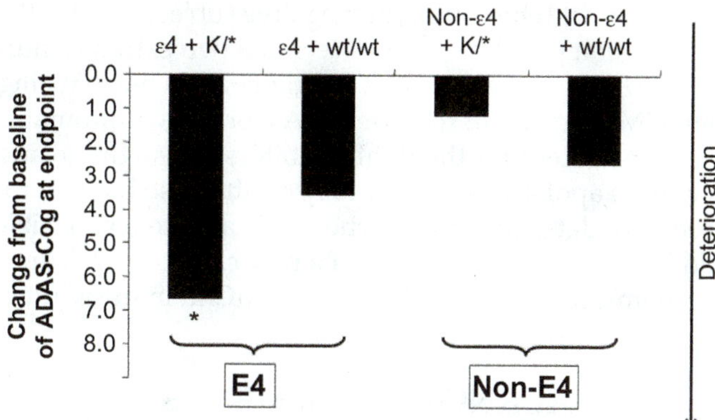

FIGURE 4.7 **ApoE4–butyrylcholinesterase K interaction in rivastigmine-treated mild cognitively impaired (MCI) subjects enrolled in a 4-year, randomized, placebo-controlled clinical trial.** Cognitive decline measured over 3–4 years in subjects with MCI (intent-to-treat) who were genotyped for apoE and BuChE polymorphisms. ADAS-Cog (Alzheimer's Disease Assessment Scale for Cognition) intent-to-treat/last observation carried forward scores. *Adapted* [35] *and* [84].

The combined effect of ApoE and BuChE pharmacogenomic effects on efficacy parameters was quite impressive, particularly in the E4/K carriers, to the point of markedly impacting on the conversion rate from MCI to AD status. Results are summarized in Figure 4.7.

In summary, pharmacogenomic advances in the past 15 years have clearly identified specific genomic markers, such as apoE and BuChE, that significantly impact on both the therapeutic outcome and the molecular biological process that underlies the progression of prodromic and established dementias. A flurry of new pharmacogenomic markers has been published recently in the field of dementia (choline acetyltransferase, acetylcholinesterase, interleukin-1B, paraoxonase-1, insulin degrading enzyme, 5-HT2A, BDNF, CYT2D6), but replication remains an important issue.

Figure 4.8 illustrates the model of disease progression in the pre- and post-diagnostic phases of the disease as a function of apoE and BuChE polymorphisms. In the prodromal (MCI) phase, subjects with the fast rate of cognitive decline and conversion belong to the E4/K allele population. In contrast, E4- and K-negative MCI subjects typically exhibit low rate of cognitive decline and modest annual conversion rates. As a direct consequence, the latter population displays much later ages of onset. As had been well established, the E4 carriers are characterized by early onsets of the disease and the addition of the K allele definitively exacerbates this phenomenon.[8,9,65]

Interestingly, once the diagnosis of dementia is established, the pattern is inverted: the E4-negative/K-negative carriers who exhibit later onsets display a fast rate of decline, which, for certain individuals, is simply catastrophic. On the other hand, E4/K carriers who are characterized by an earlier age of onset display a much slower rate of

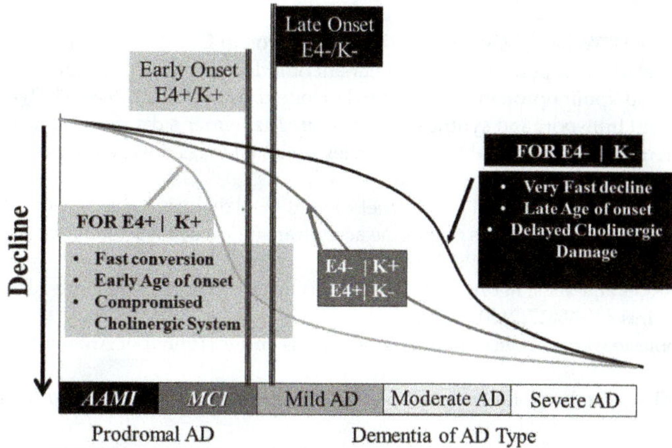

FIGURE 4.8 **Working hypothesis explaining the dual impact of ApoE and BuChE genotypes on the prodromal progression of cognitive deficits, the conversion, the age of onset, and the rate of progression following formal diagnostic of Alzheimer disease (AD).** This model combined the results of multiple double-blind, placebo-controlled randomized clinical trials performed both in mild cognitive impairment (MCI) and in sporadic AD over the past 15 years. AAMI, age-associated memory deficit; ADAS-Cog, Alzheimer's Disease Assessment Scale-Cognitive subscale.

decline that contrasts quite significantly with the genotype-negative subjects. As for those subjects with mixed genotypes, the resulting onset and progression profiles span between the two extremes illustrated in Figure 4.8.

Cholinomimetic drugs, as a group, definitively exhibit strong cognitive-enhancing properties in apoE4 and/or BuChE-K carriers in the prodromal (MCI) phase of the disease. As for the symptomatic phase of the disease, esterase inhibitors work best in apoE4 carriers, except for the dual inhibitors rivastigmine and tacrine, which display solid cognitive-enhancing properties in both apoE4 and non-E4 subjects that also happen to be BuChE-K negative.

Further investigations are needed to determine the biochemical cascade responsible for these pharmacogenomic effects. We believe that this could lead to the development of a new class of AD drugs that would be specifically designed to modify the pharmacogenomic impact of apoE4 and/or BuChE in the adult brain.

ACKNOWLEDGEMENTS

We wish to thank Mrs L. Théroux, D. Dea, V. Leduc for their technical contributions and Drs H. Weibush, K. Scheppert, P. Amouyel and P. Sévigny for their continuous insights onto this exciting emerging field of research. We also wish to acknowledge the support of the Canadian Institute Health Research (J.P.) and the FRSQ (J.P.).

References

1. Kalow W. Life of a pharmacologist or the rich life of a poor metabolizer. *Pharmacol Toxicol*. 1995;76(4):221–227.
2. Poirier J, Delisle MC, Quirion R, et al. Apolipoprotein E4 allele as a predictor of cholinergic deficits and treatment outcome in Alzheimer disease. *Proc Natl Acad Sci U S A*. 1995;92(26):12260–12264.
3. Poirier J. Apolipoprotein E: a pharmacogenetic target for the treatment of Alzheimer's disease. *Mol Diagn*. 1999;4(4):335–341.
4. Farlow MR, Lahiri DK, Poirier J, Davignon J, Schneider L, Hui SL. Treatment outcome of tacrine therapy depends on apolipoprotein genotype and gender of the subjects with Alzheimer's disease. *Neurology*. 1998;50(3):669–677.
5. Nebert DW, Ingelman-Sundberg M, Daly AK. Genetic epidemiology of environmental toxicity and cancer susceptibility: human allelic polymorphisms in drug-metabolizing enzyme genes, their functional importance, and nomenclature issues. *Drug Metab Rev*. 1999;31(2):467–487.
6. McCarthy JJ, Hilfiker R. The use of single-nucleotide polymorphism maps in pharmacogenomics. *Nat Biotechnol*. 2000;18(5):505–508.
7. Lambert JC, Amouyel P. Genetics of Alzheimer's disease: new evidences for an old hypothesis? *Curr Opin Genet Dev*. 2011;21(3):295–301.
8. Poirier J, Davignon J, Bouthillier D, Kogan S, Bertrand P, Gauthier S. Apolipoprotein E polymorphism and Alzheimer's disease. *Lancet*. 1993;342(8873):697–699.
9. Strittmatter WJ, Saunders AM, Schmechel D, et al. Apolipoprotein E: high-avidity binding to beta-amyloid and increased frequency of type 4 allele in late-onset familial Alzheimer disease. *Proc Natl Acad Sci U S A*. 1993;90(5):1977–1981.
10. Rebeck GW, Reiter JS, Strickland DK, Hyman BT. Apolipoprotein E in sporadic Alzheimer's disease: allelic variation and receptor interactions. *Neuron*. 1993;11(4):575–580.
11. Corder EH, Saunders AM, Strittmatter WJ, et al. Gene dose of apolipoprotein E type 4 allele and the risk of Alzheimer's disease in late onset families [see comments]. *Science*. 1993;261(5123):921–923.
12. Schmechel DE, Saunders AM, Strittmatter WJ, et al. Increased amyloid beta-peptide deposition in cerebral cortex as a consequence of apolipoprotein E genotype in late-onset Alzheimer disease. *Proc Natl Acad Sci U S A*. 1993;90(20):9649–9653.
13. Beffert U, Poirier J. Apolipoprotein E, plaques, tangles and cholinergic dysfunction in Alzheimer's disease. *Neurobiology of Alzheimer's Disease*. 1996;777:166–174.
14. Utermann G, Langenbeck U, Beisiegel U, Weber W. Genetics of the apolipoprotein E system in man. *Am J Hum Genet*. 1980;32:339–347.
15. Beffert U, Cohn JS, Petit-Turcotte C, et al. Apolipoprotein E and beta-amyloid levels in the hippocampus and frontal cortex of Alzheimer's disease subjects are disease-related and apolipoprotein E genotype dependent. *Brain Res*. 1999;843(1–2):87–94.
16. Poirier J. Apolipoprotein E, cholesterol transport and synthesis in sporadic Alzheimer's disease. *Neurobiol Aging*. 2005;26(3):355–361.
17. Cruchaga C, Kauwe JS, Nowotny P, et al. Cerebrospinal fluid APOE levels: an endophenotype for genetic studies for Alzheimer's disease. *Hum Mol Genet*. 2012;21(20):4558–4571.
18. Arendt T. Disturbance of neuronal plasticity is a critical pathogenetic event in Alzheimer's disease. *Int J Dev Neurosci*. 2001;19(3):231–245.
19. Soininen H, Kosunen O, Helisalmi S, et al. A severe loss of choline acetyltransferase in the frontal cortex of Alzheimer patients carrying apolipoprotein epsilon 4 allele. *Neurosci Lett*. 1995;187(2):79–82.
20. Arendt T, Schindler C, Bruckner MK, et al. Plastic neuronal remodeling is impaired in patients with Alzheimer's disease carrying apolipoprotein epsilon 4 allele. *J Neurosci*. 1997;17(2):516–529.
21. Poirier J. Apolipoprotein E, and cholesterol metabolism in the pathogenesis and treatment of Alzheimer's disease. *Trends Mol Med*. 2003;9(3):94–101.
22. Masliah E, Mallory M, Ge N, Alford M, Veinbergs I, Roses AD. Neurodegeneration in the central nervous system of apoE-deficient mice. *Exp Neurol*. 1995;136(2):107–122.
23. Veinbergs I, Masliah E. Synaptic alterations in apolipoprotein E knockout mice [comment]. *Neuroscience*. 1999;91(1):401–403.
24. Champagne D, Dupuy JB, Rochford J, Poirier J. Apolipoprotein E knockout mice display procedural deficits in the Morris water maze: analysis of learning strategies in three versions of the task. *Neuroscience*. 2002;114(3):641–654.
25. Krzywkowski P, Ghribi O, Gagne J, et al. Cholinergic systems and long-term potentiation in memory-impaired apolipoprotein E-deficient mice. *Neuroscience*. 1999;92(4):1273–1286.

26. Gordon I, Grauer E, Genis I, Sehayek E, Michaelson DM. Memory deficits and cholinergic impairments in apolipoprotein E- deficient mice. *Neurosci Lett*. 1995;199(1):1–4.
27. Allen SJ, MacGowan SH, Tyler S, et al. Reduced cholinergic function in normal and Alzheimer's disease brain is associated with apolipoprotein E4 genotype. *Neurosci Lett*. 1997;239(1):33–36.
28. Soininen H, Lehtovirta M, Helisalmi S, Linnaranta K, Heinonen O, Riekkinen Sr. P. Increased acetylcholinesterase activity in the CSF of Alzheimer patients carrying apolipoprotein epsilon4 allele. *Neuroreport*. 1995;6(18):2518–2520.
29. Cohen RM, Podruchny TA, Bokde AL, et al. Higher *in vivo* muscarinic-2 receptor distribution volumes in aging subjects with an apolipoprotein E-epsilon4 allele. *Synapse*. 2003;49(3):150–156.
30. Cohen RM, Carson RE, Filbey F, Szczepanik J, Sunderland T. Age and APOE-epsilon4 genotype influence the effect of physostigmine infusion on the *in-vivo* distribution volume of the muscarinic-2-receptor dependent tracer [18 F]FP-TZTP. *Synapse*. 2006;60(1):86–92.
31. Svensson AL, Warpman U, Hellstrom-Lindahl E, Bogdanovic N, Lannfelt L, Nordberg A. Nicotinic receptors, muscarinic receptors and choline acetyltransferase activity in the temporal cortex of Alzheimer patients with differing apolipoprotein E genotypes. *Neurosci Lett*. 1997;232(1):37–40.
32. Reid RT, Sabbagh MN, Thal LJ. Does apolipoprotein E Apo-E. genotype influence nicotinic receptor binding in Alzheimer's disease. *J Neural Transm*. 2001;108(8–9):1043–1050.
33. Petersen RC, Thomas RG, Grundman M, et al. Vitamin E and donepezil for the treatment of mild cognitive impairment. *N Engl J Med*. 2005;352(23):2379–2388.
34. Gold M, Franke S, Nye JS, Goldstein HR, Fijal B, Cohen N. Impact of ApoE genotype on the efficacy of galantamine for the treatment of mild cognitive impairment. *Neurobiol Aging*. 2004;24:521.
35. Lane R, Feldman HH, Meyer J, et al. Synergistic effect of apolipoprotein E epsilon4 and butyrylcholinesterase K-variant on progression from mild cognitive impairment to Alzheimer's disease. *Pharmacogenet Genomics*. 2008;18(4):289–298.
36. Frisoni GB, Govoni S, Geroldi C, et al. Gene dose of the epsilon 4 allele of apolipoprotein E and disease progression in sporadic late-onset Alzheimer's disease. *Ann Neurol*. 1995;37(5):596–604.
37. Dal FG, Rasmusson DX, Brandt J, et al. Apolipoprotein E genotype and rate of decline in probable Alzheimer's disease. *Arch Neurol*. 1996;53(4):345–350.
38. Stern Y, Brandt J, Albert M, et al. The absence of an apolipoprotein epsilon4 allele is associated with a more aggressive form of Alzheimer's disease. *Ann Neurol*. 1997;41(5):615–620.
39. Murphy Jr. GM, Taylor J, Kraemer HC, Yesavage J, Tinklenberg JR. No association between apolipoprotein E epsilon 4 allele and rate of decline in Alzheimer's disease. *Am J Psychiatry*. 1997;154(5):603–608.
40. Craft S, Teri L, Edland SD, et al. Accelerated decline in apolipoprotein E-epsilon4 homozygotes with Alzheimer's disease. *Neurology*. 1998;51(1):149–153.
41. Slooter AJ, Houwing-Duistermaat JJ, van Harskamp F, et al. Apolipoprotein E genotype and progression of Alzheimer's disease: the Rotterdam Study. *J Neurol*. 1999;246(4):304–308.
42. Jonker C, Schmand B, Lindeboom J, Havekes LM, Launer LJ. Association between apolipoprotein E epsilon4 and the rate of cognitive decline in community-dwelling elderly individuals with and without dementia. *Arch Neurol*. 1998;55(8):1065–1069.
43. Farlow MR, He Y, Tekin S, Xu J, Lane R, Charles HC. Impact of APOE in mild cognitive impairment. *Neurology*. 2004;63(10):1898–1901.
44. Davis KL, Mohs RC, Marin D, et al. Cholinergic markers in elderly patients with early signs of Alzheimer disease. *JAMA*. 1999;281(15):1401–1406.
45. DeKosky ST, Scheff SW. Synapse loss in frontal cortex biopsies in Alzheimer's disease: correlation with cognitive severity. *Ann Neurol*. 1990;27(5):457–464.
46. Gilmor ML, Erickson JD, Varoqui H, et al. Preservation of nucleus basalis neurons containing choline acetyltransferase and the vesicular acetylcholine transporter in the elderly with mild cognitive impairment and early Alzheimer's disease. *J Comp Neurol*. 1999;411(4):693–704.
47. Farlow MR, Lahiri DK, Poirier J, Davignon J, Hui S. Apolipoprotein E genotype and gender influence response to tacrine therapy. *Apolipoprotein E Genotyping in Alzheimer's Disease*. 1996;802:101–110.
48. Riekkinen Jr. P, Soininen H, Partanen J, Paakkonen A, Helisalmi S, Riekkinen Sr. P. The ability of THA treatment to increase cortical alpha waves is related to apolipoprotein E genotype of Alzheimer disease patients. *Psychopharmacology (Berl)*. 1997;129(3):285–288.
49. Higuchi S, Matsushita S, Hasegawa Y, Muramatsu T, Arai H, Hayashida M. Apolipoprotein E epsilon 4 allele and pupillary response to tropicamide. *Am J Psychiatry*. 1997;154(5):694–696.
50. Bigler ED, Lowry CM, Anderson CV, Johnson SC, Terry J, Steed M. Dementia, quantitative neuroimaging, and apolipoprotein E genotype. *AJNR Am J Neuroradiol*. 2000;21(10):1857–1868.
51. de Leon MJ, Convit A, Wolf OT, et al. Prediction of cognitive decline in normal elderly subjects with 2-[F-18]fluoro-2-deoxy-D-glucose/positron-emission tomography FDG/PET. *Proc Natl Acad Sci U S A*. 2001;98(19):10966–10971.
52. Farlow MR, Cyrus PA, Nadel A, Lahiri DK, Brashear A, Gulanski B. Metrifonate treatment of AD: influence of APOE genotype. *Neurology*. 1999;53(9):2010–2016.
53. MacGowan SH, Wilcock GK, Scott M. Effect of gender and apolipoprotein E genotype on response to anticholinesterase therapy in Alzheimer's disease. *Int J Geriatr Psychiatry*. 1998;13(9):625–630.
54. Wilcock GK, Lilienfeld S, Gaens E. Efficacy and safety of galantamine in patients with mild to moderate Alzheimer's disease: multicentre randomised controlled trial. Galantamine International-1 Study Group. *BMJ*. 2000;321(7274):1445–1449.
55. Babic T, Mahovic LD, Sertic J, Petrovecki M, Stavljenic-Rukavina A. ApoE genotyping and response to galanthamine in Alzheimer's disease—a real life retrospective study. *Coll Antropol*. 2004;28(1):199–204.
56. Raskind MA, Peskind ER, Wessel T, Yuan W. Galantamine in AD: A 6-month randomized, placebo-controlled trial with a 6-month extension. The Galantamine USA-1 Study Group. *Neurology*. 2000;54(12):2261–2268.
57. Oddoze C, Michel BF, Lucotte G. Apolipoprotein E epsilon 4 allele predicts a better response to donepezil therapy in Alzheimer's disease. *Alzheimers Reports*. 2000;3(4):213–216.
58. Greenberg SM, Tennis MK, Brown LB, et al. Donepezil therapy in clinical practice: a randomized crossover study. *Arch Neurol*. 2000;57(1):94–99.

59. Bizzarro A, Marra C, Acciarri A, et al. Apolipoprotein E epsilon4 allele differentiates the clinical response to donepezil in Alzheimer's disease. *Dement Geriatr Cogn Disord*. 2005;20(4):254–261.
60. Rigaud AS, Traykov L, Latour F, Couderc R, Moulin F, Forette F. Presence or absence of at least one epsilon 4 allele and gender are not predictive for the response to donepezil treatment in Alzheimer's disease. *Pharmacogenetics*. 2002;12(5):415–420.
61. Borroni B, Colciaghi F, Pastorino L, et al. ApoE genotype influences the biological effect of donepezil on APP metabolism in Alzheimer disease: evidence from a peripheral model. *Eur Neuropsychopharmacol*. 2002;12(3):195–200.
62. Winblad B, Engedal K, Soininen H, et al. A 1-year, randomized, placebo-controlled study of donepezil in patients with mild to moderate AD. *Neurology*. 2001;57(3):489–495.
63. Lendon CL, Hills R, Sellwood E, Bentham P, Gray R. Determinant of response to anticholinesterase therapies in the treatment of Alzheimer's disease. In: *Proceedings of the 8th international conference on Alzheimer's disease and related disorders*, Stockholm, Sweden, 20–25 July 2002; 2002.
64. Sinha G. Drug companies accused of stalling tailored therapies. *Nat Med*. 2006;12(9):983.
65. Wiebusch H, Poirier J, Sevigny P, Schappert K. Further evidence for a synergistic association between APOE epsilon 4 and BCHE-K in confirmed Alzheimer's disease. *Hum Genet*. 1999;104(2):158–163.
66. Hedgecoe A. Pharmacogenetics as alien science: Alzheimer's disease, core sets and expectations. *Soc Stud Sci*. 2006;36(5):723–752.
67. Richard F, Helbecque N, Neuman E, Guez D, Levy R, Amouyel P. APOE genotyping and response to drug treatment in Alzheimer's disease [letter]. *Lancet*. 1997;349(9051):539.
68. Bartels CF, Jensen FS, Lockridge O, et al. DNA mutation associated with the human butyrylcholinesterase K-variant and its linkage to the atypical variant mutation and other polymorphic sites. *Am J Hum Genet*. 1992;50(5):1086–1103.
69. Primo-Parmo SL, Bartels CF, Wiersema B, van der Spek AF, Innis JW, La Du BN. Characterization of 12 silent alleles of the human butyrylcholinesterase BCHE. gene. *Am J Hum Genet*. 1996;58(1):52–64.
70. Gomez-Ramos P, Moran MA. Ultrastructural localization of butyrylcholinesterase in senile plaques in the brains of aged and Alzheimer disease patients. *Mol Chem Neuropathol*. 1997;30(3):161–173.
71. Perry EK, Perry RH, Blessed G, Tomlinson BE. Changes in brain cholinesterases in senile dementia of Alzheimer type. *Neuropathol Appl Neurobiol*. 1978;4(4):273–277.
72. Lehmann DJ, Nagy Z, Litchfield S, Borja MC, Smith AD. Association of butyrylcholinesterase K variant with cholinesterase-positive neuritic plaques in the temporal cortex in late-onset Alzheimer's disease. *Hum Genet*. 2000;106(4):447–452.
73. Ghebremedhin E, Thal DR, Schultz C, Braak H. Age-dependent association between butyrylcholinesterase K-variant and Alzheimer disease-related neuropathology in human brains. *Neurosci Lett*. 2002;320(1–2):25–28.
74. Ramanan VK, Risacher SL, Nho K, et al. APOE and BCHE as modulators of cerebral amyloid deposition: a florbetapir PET genome-wide association study. *Mol Psychiatry*. 2013.
75. Lehmann DJ, Johnston C, Smith AD. Synergy between the genes for butyrylcholinesterase K variant and apolipoprotein E4 in late-onset confirmed Alzheimer's disease. *Hum Mol Genet*. 1997;6(11):1933–1936.
76. O'Brien KK, Saxby BK, Ballard CG, et al. Regulation of attention and response to therapy in dementia by butyrylcholinesterase. *Pharmacogenetics*. 2003;13(4):231–239.
77. Holmes C, Ballard C, Lehmann D, et al. Rate of progression of cognitive decline in Alzheimer's disease: effect of butyrylcholinesterase K gene variation. *J Neurol Neurosurg Psychiatry*. 2005;76(5):640–643.
78. Blesa R, Bullock R, He Y, et al. Effect of butyrylcholinesterase genotype on the response to rivastigmine or donepezil in younger patients with Alzheimer's disease. *Pharmacogenet Genomics*. 2006;16(11):771–774.
79. Lane R, Farlow M. Lipid homeostasis and apolipoprotein E in the development and progression of Alzheimer's disease. *J Lipid Res*. 2005;46:949–968.
80. Farlow MR, Lahiri D, Hui S, Davignon J, Poirier J. Apolipoprotein E genotype predicts response to tacrine in Alzheimer's disease. *Neurology*. 1996;46(2):14002.
81. Farlow M, Lane R, Kudaravalli S, He Y. Differential qualitative responses to rivastigmine in APOE epsilon 4 carriers and noncarriers. *Pharmacogenomics J*. 2004;4(5):332–335.
82. Altstiel L, Mohs R, Marin D, Bodick N, Poirier J. ApoE genotype and clinical outcome in Alzheimer's disease. *Neurobiol Aging*. 1998;18:S33–S34.
83. Giacobini E. *Cholinesterases and Cholinesterase Inhibitors. Basic, Preclinical and Clinical Aspects*. London: Martin Dunitz; 2000.
84. Ferris S, Nordberg A, Soininen H, Darreh-Shori T, Lane R. Progression from mild cognitive impairment to Alzheimer's disease: effects of sex, butyrylcholinesterase genotype, and rivastigmine treatment. *Pharmacogenet Genomics*. 2009;19(8):635–646.

CHAPTER

5

Application of Mouse Genetics to Human Disease: Generation and Analysis of Mouse Models

Teresa M. Gunn and Brenda Canine

McLaughlin Research Institute, Great Falls, MT, USA

INTRODUCTION

The laboratory mouse has a well-deserved reputation as the leading mammalian model system for studies into the genetic and molecular basis of neurologic diseases. As mice are mammals, they are physiologically, anatomically and genetically similar to humans. Their relatively short lifespan (2–3 years) makes studies of diseases associated with aging feasible. Large numbers can be maintained at reasonable cost under controlled environmental conditions. Genetically identical mice are available (inbred strains), as well as a plethora of genetic variants that have resulted from natural variation due to spontaneous and induced mutations. The fact that their genome can be manipulated in a variety of ways allows functional annotation of the genome using gene- or phenotype-driven approaches. There are physiological, anatomical and inflammatory response differences between the mouse and human brain, however, and behavioral traits can be difficult to translate between mice and humans.

It has been surprisingly difficult to generate accurate mouse models of age-dependent neurodegenerative disorders. In most cases, mice carrying mutations that cause neurodegeneration in humans do not fully recapitulate the human disease. For example, loss of function mutations in *PARKIN* cause Parkinson disease in humans, but *parkin*-null mutant mice only develop mild mitochondrial dysfunction and subtle defects in dopamine handling, with no other characteristics of Parkinson disease.[1] Familial forms of Parkinson disease can also be caused by mutation of an alanine residue to threonine at position 53 of α-synuclein, but threonine is the naturally occurring residue in mice and other nonprimate animals. Mice carrying mutations in the *amyloid precursor protein* gene (*App*) that is known to cause early onset Alzheimer disease in humans may develop amyloid plaques and learning deficits, but neuronal loss and neurofibrillary tangles associated with the human disease are not observed in these mouse models.[2] These differences could be due to one or more of a number of factors. It is possible, if not likely, that modifier genes influence the penetrance and expressivity of most neurological phenotypes, but inbred mice have no background variation. Another factor is the short lifespan of mice. The relationship between cellular aging and actual time is not clear. For example, are the cumulative effects of oxidative damage or the presence of toxic protein species in mouse neurons over 2–3 years equivalent to those in human neurons over 80 or more years? There are also significant differences in the inflammatory response in the central nervous system of mice and humans,[3] which could play a role. Some studies suggest compensatory changes that protect against the effect of some loss-of-function mutations in mice may not occur in humans. For instance, constitutive ablation of *parkin* does not cause neurodegeneration, but deleting *parkin* from adult mouse neurons does.[4] Similarly, conditional deletion of *glial derived neurotrophic factor* (*Gdnf*) led to a profound degeneration of catecholaminergic neurons that was not observed when the gene was deleted during embryogenesis.[5] Identifying the factors that appear to protect mice against neurodegeneration is necessary to generate better models – ones that will be useful for understanding the mechanistic basis of disease,

Rosenberg's Molecular and Genetic Basis of Neurological and Psychiatric Disease
http://dx.doi.org/10.1016/B978-0-12-410529-4.00005-X

identifying new therapeutic targets, and providing more accurate predictions for clinical trials. The need for mouse models is now perhaps greater than ever as new technologies, coupled with the availability of the complete human genome sequence, have led to a rapid increase in the identification of disease-associated variants. Mice provide an ideal system for confirming that these variants cause disease, investigating causative mechanisms, identifying disrupted pathways, and testing therapeutic strategies. This chapter will describe the different ways in which the mouse genome can be manipulated to generate disease models and the various resources and tools available to study them.

CREATING MOUSE MODELS

The mouse genome can be modified in a multitude of ways. Specific human disease-associated mutations can be recapitulated, and new genes and pathways that contribute to a disease of interest can be identified using random mutagenesis and phenotype-based screens. This section will describe the basic concepts underlying the methods most commonly used to create new mouse models and discuss the advantages and disadvantages associated with each (summarized in Table 5.1).

Transgenesis

Dominant human disorders caused by known genetic variants can most easily be modeled by generating a transgenic mouse. A transgene is, in effect, a minigene consisting of a characterized promoter that drives expression in an appropriate tissue or cell type, an intron (to improve expression), the cDNA for the gene to be expressed, and a polyA tail (Figure 5.1). The cDNA can encode the wild-type protein, if the main goal is simply over-expression, or a mutant protein, to create a model for a dominant disorder. The transgene is linearized and injected into one of the pronuclei of a fertilized mouse egg. Injected embryos are transferred into surrogate mothers. A percentage of pups born will be founders (have integrated the transgene DNA into their genomic DNA) and should express the transgene. Each founder is bred to create independent lines for further analysis. Transgene expression can be influenced by the chromatin conformation of the integration site, however, and in some cases the transgene will be silent or expressed in a mosaic pattern. Most transgenes insert randomly into the genome as a head-to-tail tandem array containing multiple copies of the transgene. This can result in very high expression levels, but since the integration site is different in every founder, expression levels can vary substantially between founder lines. More recently, systems have been developed to direct integration to a specific location, within a ubiquitously expressed gene.[6] These transgenes typically insert as a single copy, resulting in consistent levels and pattern of expression, even between different founder lines.

Transgenes are generally used to create mouse models of dominant diseases that act through a gain-of-function mechanism (for example, amyotrophic lateral sclerosis (ALS),[7] spinocerebellar ataxia type 1 (SCA1),[8] and prion disease[9]) although a few studies have used transgenes expressing dominant negative proteins[10] or shRNAs[11] to disrupt function or expression of the endogenous gene product. Transgenes can also be used to express regulatory enzymes that control the expression of other transgenes. Tet-On and Tet-Off are the most commonly used binary transgenic systems (Figure 5.2). Both regulate expression of the target transgene (expressing the gene of interest) by exposing the animals to varying concentrations of tetracycline or, more commonly, tetracycline derivatives such as doxycycline (Dox). One transgene expresses a recombinant tetracycline-controlled transcription factor (the tetracycline-controlled transactivator protein [*tTA*] for Tet-Off and the reverse tetracycline-controlled transactivator protein [*rtTA*] for Tet-On) under control of a promoter that drives expression in the tissue or cell type of interest. The other transgene expresses the gene of interest under control of a minimal promoter that is regulated by the tetracycline-responsive element (TRE, which consists of multiple copies of the TetO operator). In the Tet-Off system, Dox prevents *tTA* binding to the TRE, leading to expression of the target transgene when Dox is absent and silencing when it is present. In the Tet-On system, the opposite is true: *rtTA* requires Dox to bind the TRE and activate transcription, leading to expression of the target transgene only when the mice are given Dox. Regulatable transgenes have proven particularly useful for investigating whether specific effects of neurodegeneration-related mutant genes are reversible. For example, Tet-Off transgenic mice expressing aggregation-prone human Tau showed synaptic and neuronal loss, impaired memory, and loss of long-term potentiation (LTP).[12] When the Tau transgene was silenced by Dox treatment, starting at ~10 months of age and continuing for ~4 months, memory and LTP recovered, there was a moderate reduction in Tau aggregates and their composition changed from human and mouse Tau to mouse Tau only. Although neuronal loss persisted, synaptic plasticity was partially recovered. These findings indicate that Tau pathology correlates with β-structure rather than the presence of aggregates, and that many of the adverse effects caused by mutant Tau are reversible.

TABLE 5.1 Advantages and Disadvantages of Methods used to Generate Mouse Genetic Models

Type of mutation	Advantages	Disadvantages
Transgene – random	• Relatively quick and easy to generate • Very high expression levels possible • Express normal or mutant allele • Ubiquitous or cell/tissue-specific expression	• Generally only models gain-of-function disorders • Can insert into and disrupt an endogenous gene • Variable expression between founder lines • Each founder line must be characterized
Transgene – targeted	• Relatively quick and easy to generate • Equivalent expression between founder lines • Express normal or mutant allele • Ubiquitous or cell/tissue-specific expression	• Generally only models gain-of-function disorders • Tissue-specific expression requires insertion of floxed transcriptional termination sequence upstream of cDNA and mating to Cre line
Regulatable transgenes (Tet-Off, Tet-On)	• Temporal and tissue specific controls • Reversible • Low background expression • Many activator lines readily available	• Bigenic system (two lines of mice needed) • Dox treatment is costly • Dox treatment does not alter expression quickly • Side-effects of long-term Dox exposure
Knockout (nonconditional)	• Insight into gene function • Models recessive and haploinsufficient disorders • Gene expression pattern assayed using reporter gene	• Time and labor intensive • Cannot model dominant gain-of-function diseases • Disease-causing gene must be known • Phenotype may not fully reflect gene function (functionally redundant)
Conditional knockout	• Assess gene function in specific tissues, even if constitutive knockout is lethal • Spatial and temporal control of gene deletion • Gene expression pattern assayed using reporter gene • Many Cre transgenic lines readily available	• Time and labor intensive • Cannot model dominant gain-of-function diseases • Disease-causing gene must be known • Phenotype may not fully reflect gene function (functionally redundant)
Tamoxifen-inducible knockout (CreERT)	• Temporal control of gene deletion	• Tamoxifen treatment *in utero* causes embryo loss • Expression can be leaky
Knock-in	• Precise recapitulation of human alleles • Models recessive or dominant disorders • Uses endogenous promoter	• Time and labor intensive • Disease-causing mutation must be known
Nuclease-mediated gene targeting (ZFNs, TALENs and CRISPR/Cas9 systems)	• Relatively quick and easy to generate • Models loss- or gain-of-function mutations • Modular assembly kits openly available • CRISPR/Cas9 system can target multiple genes at once (models multigenic disorders) • Can modify genome of species without ESC	• Target sites may not be present where desired • Activity and specificity must be verified • Target gene/mutation must be known
Gene trap	• Can create null or partial loss-of-function (hypomorphic) mutations • Relatively quick and easy to generate • Constitutive/conditional (vector dependent) • Unbiased, phenotype-based annotation of gene function • Gene expression pattern assayed using reporter gene • Vector tag identifies disrupted gene	• Single exon genes cannot be trapped • Genes not expressed in ESC unlikely to be trapped • Insertions near the end of a gene may not disrupt protein expression or function
Transposon-mediated mutagenesis	• Unbiased, phenotype-based annotation of gene function • Transposon tag identifies disrupted gene	• Targets may be limited by sequence preference of transposon • Does not usually model gain-of-function
ENU mutagenesis	• Unbiased method to identify mutations that cause specific phenotypes • Can create null, hypomorphic or dominant-negative alleles • Can model multigenic disorders	• Mutation must be identified by mapping and/or sequencing • Requires large cohort of animals • Time intensive

Abbreviations: CRISPR, clustered regularly interspaced short palindromic repeats; Dox, doxycycline; ENU, N-ethyl-N-nitrosourea; ESC, embryonic stem cells; TALENs, transcriptional activator-like effectors coupled to a nuclease; ZFNs, zinc finger nucleases.

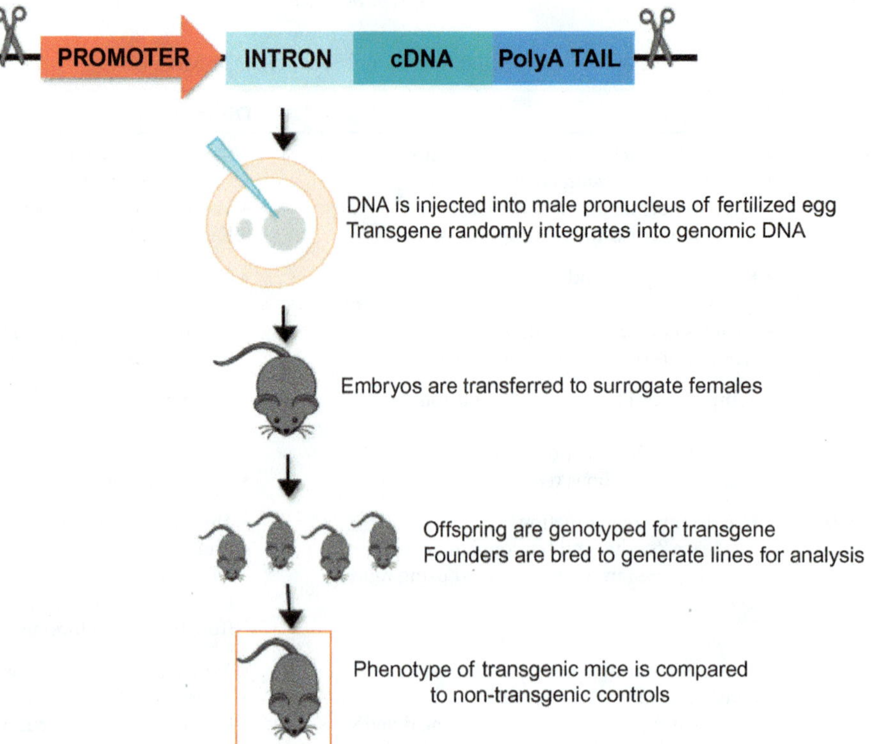

FIGURE 5.1 **Transgenic mice.** Transgene constructs are made using molecular cloning techniques. The main components are a promoter that drives gene expression, an intron to enhance expression, the cDNA for the coding region for the gene of interest, and a polyA tail. The transgene is cut out of the vector in which it is created, purified, and the DNA injected into the male pronucleus of a fertilized egg. Transgene DNA randomly integrates into the embryonic genomic DNA. Embryos are transferred to surrogate mothers and viable offspring genotyped for the transgene to identify founders, which are mated to establish transgenic lines. Each line must be characterized to verify appropriate transgene expression and phenotypic consequences. The site of insertion using this method is unknown but can be determined by sequencing or mapping methods.

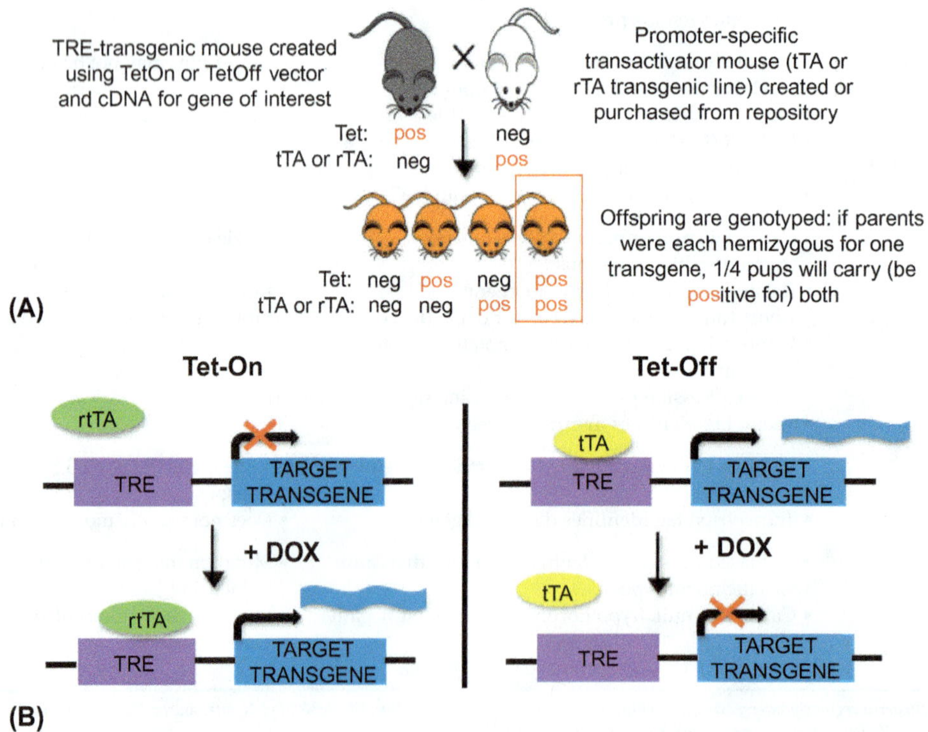

FIGURE 5.2 **Regulatable bigenic transgenic models.** (**A**) In the Tet-On or Tet-Off bigenic transgenic systems, two transgenic mice are created. The first is a TRE-transgenic mouse that expresses the cDNA for the gene of interest under control of the Tet-On or Tet-Off tetracycline-responsive element (TRE). The second mouse carries a transgene that expresses the appropriate tetracycline-controlled transcription factor (transactivator, *tTA*, or reverse transactivator, *rtTA*). Mating these mice together produces pups carrying both transgenes at an expected frequency of 25%. (**B**) In the Tet-On system, rtTA does not bind to the TRE promoter unless Dox is present, resulting in transgene expression only if Dox is administered to the mice. In the Tet-Off system, tTA binds the TRE and the transgene is expressed unless Dox is present, in which case tTA does not bind the TRE and the transgene is not expressed.

Gene Targeting

The most commonly used method for generating mouse models of recessive disorders is gene targeting (Figure 5.3). Briefly, the desired modification is first made in a targeting construct, which also contains long regions ("arms") of homology immediately adjacent to the altered DNA. The targeting construct is introduced into mouse embryonic stem cells (ESC) by electroporation and, in a small proportion of cells, homologous recombination occurs between the arms of homology in the construct and the endogenous locus to replace the original DNA with the modified sequences. Correctly targeted ESC will carry an antibiotic resistance gene to enable selection of correctly targeted ESC clones. Appropriately modified clones are injected into mouse blastocysts. When the ESC incorporate into the developing embryo, a chimeric mouse is produced that contains cells derived from both the donor blastocyst and the ESC. If the ESC contribute to the germline, this mouse will produce progeny heterozygous for the modification, and intercrossing those animals will produce homozygotes. Various types of modifications can be made using this approach: an entire gene can be deleted in a constitutive or conditional manner; one or more exons can be deleted to create a null (knockout) allele by causing a frameshift and premature termination of translation; or the normal sequence can be replaced by a specific mutation (point mutation, deletion, insertion, etc.) to create a knock-in allele. The knock-in approach can also be used to generate "humanized" alleles that provide new insights into disease mechanisms.[13–18] Knockouts can be used to study mutations with recessive (loss-of-function) effects or dominant mutations that act by causing haploinsufficiency. Knock-ins can be used to model dominant or recessive diseases.

Since many gene knockouts are lethal or cause phenotypes outside the tissue of interest, new methodologies were developed to control tissue specificity and timing of gene deletion. These conditional knockouts use the same basic approach described above for traditional knockouts, but the targeting construct contains extra sequences (Figure 5.4). LoxP sites are 34 base pair sequences recognized by the bacterial Cre enzyme, which promotes recombination between loxP sites. If loxP sites are in the same orientation within the genomic DNA, sequences between them will be deleted following Cre-mediated recombination. If they are in the opposite orientation, the sequences between them will be inverted. In a typical conditional knockout targeting construct, one or more exons that encode an essential

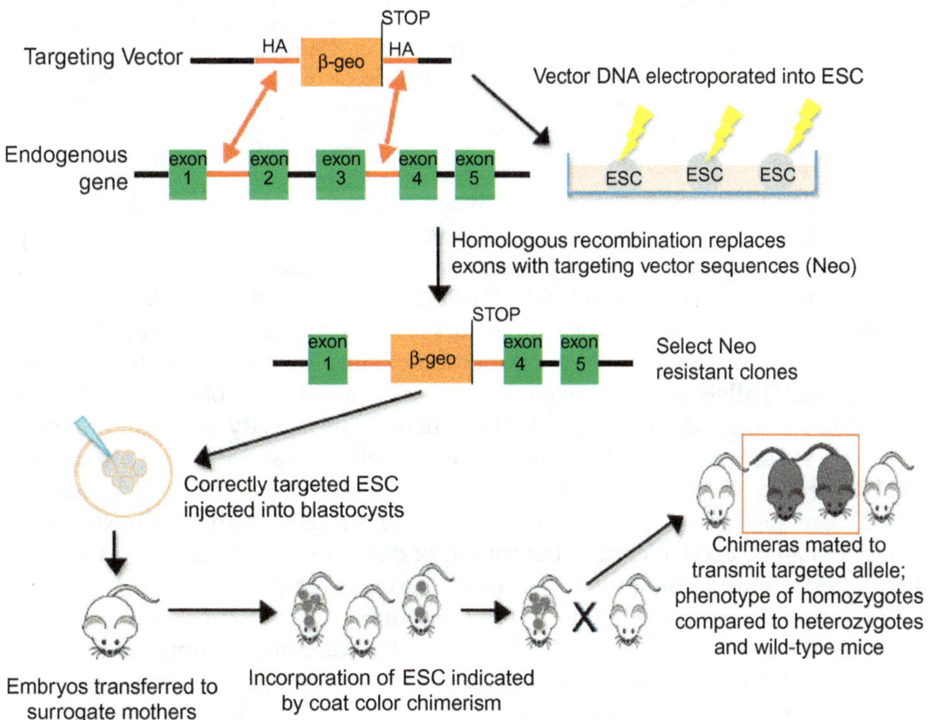

FIGURE 5.3 **Gene targeting.** A targeting construct is generated that contains regions of homology (HA) adjacent to the DNA sequence to be modified. To create a knockout allele, the modifications usually include insertion of a reporter gene fused to a selectable marker (usually β-geo, which is the β-galactosidase [LacZ] gene fused to the Neomycin resistance gene) and creation of a premature stop codon. The targeting construct is electroporated into embryonic stem cells (ESC), where recombination can occur between the regions of homology and the corresponding sequences in the endogenous gene. Homologous recombination replaces the sequences between the regions of homology in the endogenous gene with modified sequences in the construct. Correctly targeted ESC are injected into blastocysts, which are transferred to surrogate mothers. Chimeric offspring are mated to transmit the targeted allele.

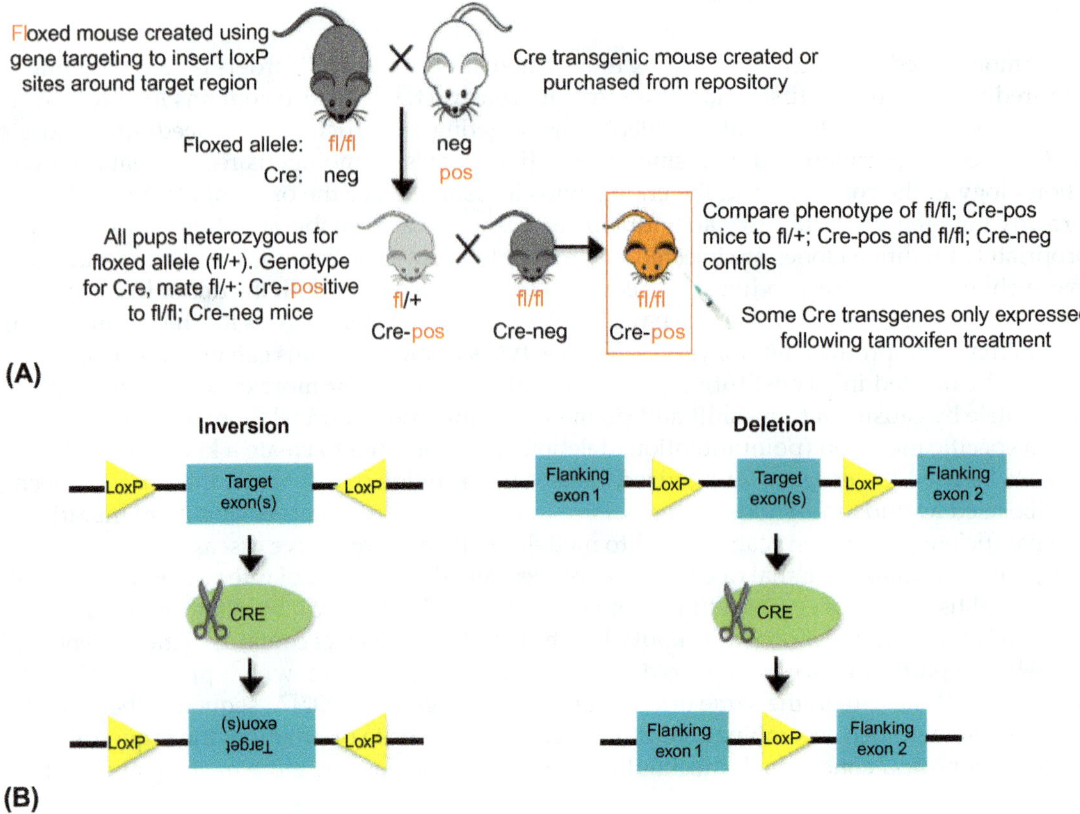

FIGURE 5.4 **Conditional gene targeting.** (A) Conditional knockouts use the gene targeting approach to introduce modifications that include the insertion of loxP sites around essential sequences within the target gene. Targeted mice are crossed to Cre transgenic mice, which express the bacterial Cre enzyme. Cre promotes recombination between loxP sites. (B) LoxP sites in opposite orientation will result in an inversion of the intervening sequences, while loxP sites in an aligned orientation will result in deletion of essential sequences within the target gene. The tissue or cell type in which the target gene sequences are deleted is determined by the promoter of the Cre transgene, which may drive expression in a specific tissue or cell type, and may require activation by administration of tomoxifen.

portion of the gene product and will cause a frameshift when deleted, are floxed (flanked by loxP sites). Following Cre-mediated recombination, the floxed exon(s) will be deleted and the resulting transcript will produce a nonfunctional protein or no protein at all (if the frameshift leads to nonsense-mediated decay). Cre activity is supplied in one of two ways. The first is to directly inject a lentiviral (LV) or adeno associated-virus (AAV) vector that expresses Cre. The viral vector can be injected stereotaxically into a specific brain region, or intra-cerebrally (for mice up to approximately 8 weeks of age). The second and more common method of supplying Cre activity is to mate mice carrying the conditional knockout ("floxed") allele to Cre transgenic mice and intercross to obtain mice homozygous for the targeted allele that carry the Cre transgene. Tissue specificity is dictated by activity of the Cre transgene promoter. Many Cre transgenic lines (expressing Cre using different tissue- or cell-type-specific promoters) have been generated and are readily available (see Cre-X-Mice: A Database of Cre Transgenic Lines [http://nagy.mshri.on.ca/cre_new/]; The Jackson Laboratory Cre Repository [http://cre.jax.org/]; or MouseCre [http://www.ics-mci.fr/mousecre/]). Activatable Cre transgenes can be used to control the timing of deletion. In Cre/ERT and Cre/ERT2 transgenes, Cre is fused to a mutated ligand-binding domain of the human estrogen receptor. The fusion protein only translocates to the nucleus to mediate recombination when bound to tamoxifen. Gene disruption in conditional knockout mice mated to Cre/ERT or Cre/ERT2 mice is thus only induced once the mice are administered tamoxifen (usually by intraperitoneal injection). Some CreERT/2 transgenic lines express a reporter gene from the same promoter, to identify cells in which the floxed allele has been deleted.[19]

One of the biggest drawbacks to gene targeting is that it is time and labor intensive. Additionally, homologous recombination in ESC and obtaining chimeras that transmit the targeted allele to their offspring are low-frequency events. It typically takes at least 6 months to obtain a mouse heterozygous for a targeted allele. This time can be shortened in one of several ways, described below. Targeting constructs and targeted ESC lines for many genes have been created and are available through the International Knockout Mouse Consortium (IKMC). The goal of the IKMC is to mutate all protein-coding genes in the mouse using gene trapping and gene targeting in C57BL/6 mouse ESC.[20]

Their website (http://www.knockoutmouse.org/) can be searched to determine a) whether a gene of interest is being or has been targeted, b) the types of projects underway (nonconditional or conditional-ready targeted or trapped alleles, or deletion of the entire gene), and c) the status of any given project. A gene that is not being targeted can easily be nominated.

Not every mouse gene has been successfully targeted by IKMC, however, and even a conditional knockout will not provide an answer to every experimental question or model human diseases caused by gain-of-function mutations. Other approaches are therefore still needed to create desired mutant alleles. Several groups have developed new approaches that target nucleases to specified DNA sequences to induce double-stranded DNA breaks (Figure 5.5). These breaks can be repaired via one of two mechanisms to create modified alleles. The first, nonhomologous end-joining (NHEJ), is error-prone and typically results in a small deletion at the break site. If the deletion disrupts the reading frame of a transcript, this will usually create a loss-of-function mutation. For the second mechanism, an exogenous sequence containing the desired change is provided as a template for homologous recombination. Large or small deletions or insertions (reporter genes, floxed exon(s), etc.) can be introduced by this method, as well as specific point mutations. Three methods currently exist for targeting a nuclease to the DNA site to be modified. Zinc finger nucleases (ZFNs) and transcriptional activator-like effectors coupled to a nuclease (TALENs) are engineered proteins that couple a customizable DNA binding protein to a nuclease.[21,22] The clustered regularly interspaced short palindromic repeats (CRISPRs) and CRISPR-associated (Cas) system uses a short, complementary single-stranded RNA (CRISPR RNA or crRNA) to guide the Cas9 nuclease to the DNA target. More than one DNA sequence can be targeted (mutated) at once with this system.[23,24] A key advantage to these nuclease-based systems is that mRNAs encoding the ZFNs, TALENs or CRISPR-Cas components can be injected directly into fertilized eggs. The pups born from these injections are often compound heterozygotes for deletions created by NHEJ or homozygous for the desired change(s) created by homologous recombination and can themselves be assessed for phenotypic alterations.

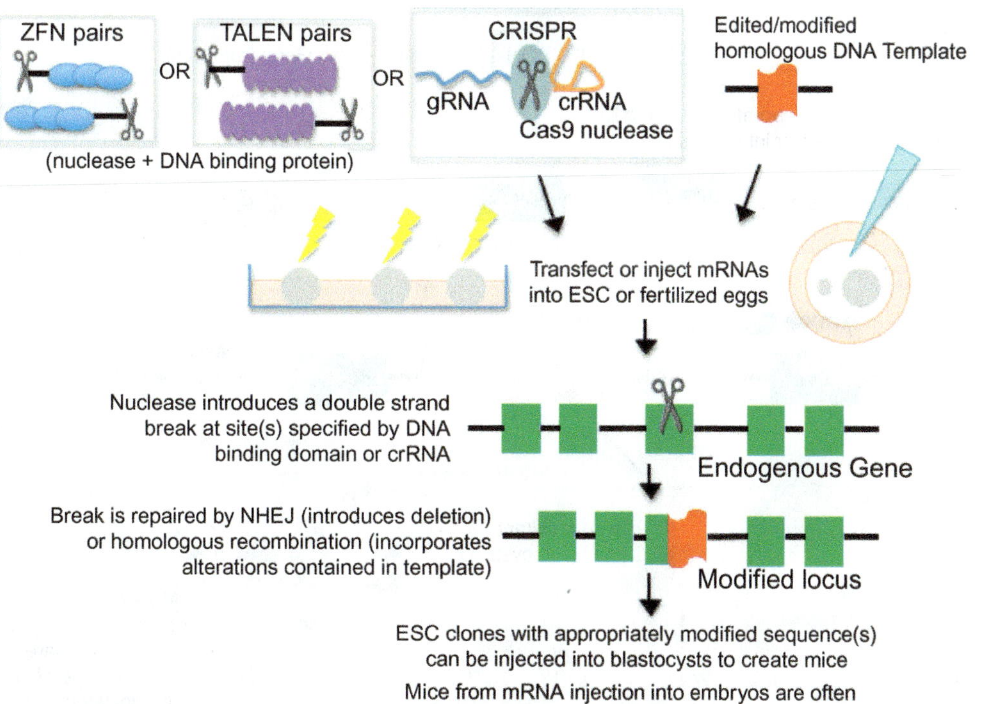

FIGURE 5.5 **Targeted nucleases.** Zinc finger nucleases (ZFNs) and transcriptional activator-like effectors coupled to a nuclease (TALENs) use a DNA-binding protein to target a nuclease to a specific DNA sequence. The CRISPR-Cas system uses a complementary RNA (crRNA). The nuclease creates a double-stranded DNA break that can be repaired by nonhomologous end-joining (NHEJ), which is error-prone and often causes a small deletion at the break site, or homologous recombination if a DNA template is provided. The template can include modified sequences between the regions of homology. ZFNs, TALENs and CRISP-Cas component mRNAs can be transfected into embryonic stem cells (ESC) or directly injected into one-cell embryos to modify their DNA. Modified ESC can be injected into blastocysts to create mice (as in Figure 5.3). Pups born from embryo injections may be analyzed for phenotypic changes, as they are often homozygous for the desired modification.

Random Mutagenesis

The methods described above are typically used to generate mouse models of human disorders when the causative gene and/or mutations have been identified. When the gene (or genes) is unknown, random mutagenesis provides an unbiased, phenotype-driven approach to generating models of disease. Random mutagenesis can also be used to identify genetic modifiers of disease, which may represent novel therapeutic targets.

One strategy for random mutagenesis is gene trapping.[25] This uses an approach similar to gene targeting but uses a gene trap cassette that randomly inserts into genomic DNA in ESC (Figure 5.6) instead of a targeting construct that inserts into a specific location by homologous recombination. The gene trap cassette contains an intron, strong splice acceptor, a reporter gene fused with a selectable (antibiotic resistance) marker, a stop codon and a poly-adenylation signal. When it inserts into an intron, the endogenous transcript will splice into and terminate in the gene trap cassette, most often creating either a hypomorphic or a null allele. The trapped ESC, identified by expression of the reporter gene (usually lacZ, which can be assayed by X-galactosidase staining) and antibiotic resistance (Neo), are injected into blastocysts to produce chimeric mice, which are mated to produce animals carrying the trapped allele. Heterozygous and homozygous mice are subjected to phenotypic screens to identify phenodeviants of interest, and the gene that is disrupted can be identified using gene trap sequences for 5′ rapid amplification of cDNA ends (RACE) and sequencing the products. Alternatively, 5′RACE can be performed on ESC mRNA to identify the gene(s) trapped in each clone and mice produced only from cells in which a specific gene has been trapped. Transposon-based systems such as Sleeping Beauty[26] and PiggyBAC[27] can also be used to randomly insert defined DNA sequences into the genome, followed by phenotypic and molecular assays to identify mice with insertion events that disrupt genes and cause an aberrant phenotype.

Random mutagenesis using chemicals or radiation can also generate interesting animal models. Presently, the most commonly used mutagen is *N*-ethyl-*N*-nitrosourea (ENU),[28] an alkylating agent that typically causes point mutations (Figure 5.7). It is administered by intraperitoneal injection to male mice to target their spermatagonial stem cells. Once the mice have recovered fertility, they are mated and their first, second, or third generation offspring are subjected to phenotypic screens to identify phenodeviants, which are likely to carry one or more mutations. ENU mutations can cause a range of effects, including complete or partial loss of function, novel, or exaggerated function, and dominant

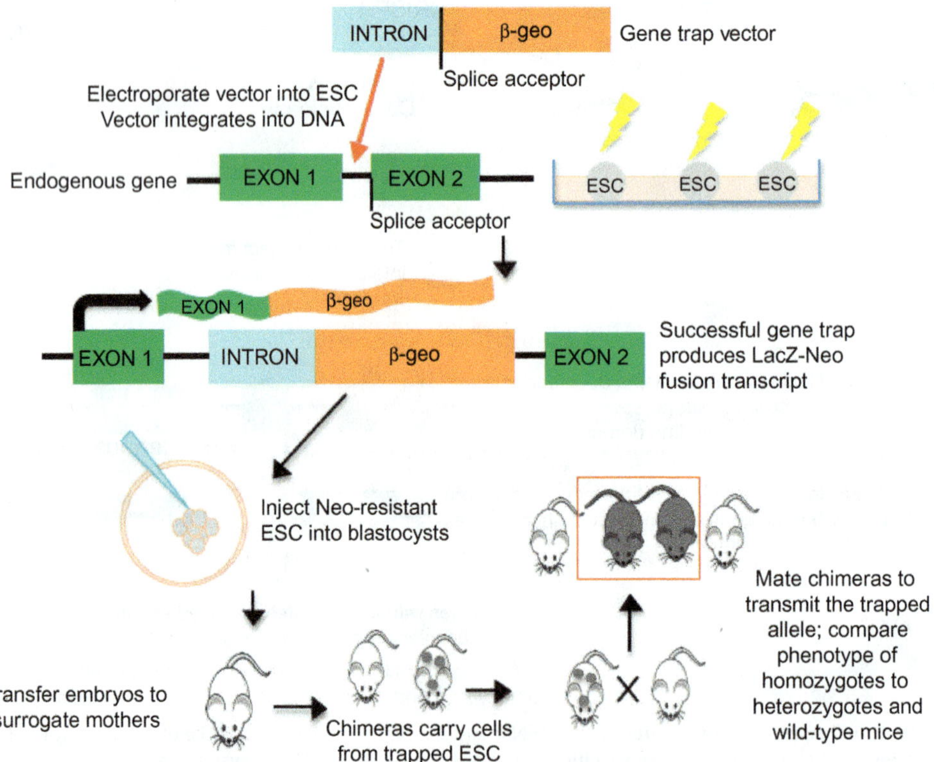

FIGURE 5.6 **Gene trapping.** Gene trapping is a method for random mutagenesis. A gene trap cassette containing an intron, strong splice acceptor, and β-geo (reporter and selectable marker) is transfected into embryonic stem cells (ESC) and randomly inserts into the genomic DNA. If it inserts into an intron, transcripts are expected to splice from the endogenous locus into the gene trap cassette. The resulting protein will contain sequences encoded by exons upstream of the trap site fused to β-geo. ESCs expressing β-geo are injected into blastocysts to generate chimeric mice. The location of the gene trap can be determined by 5′RACE.

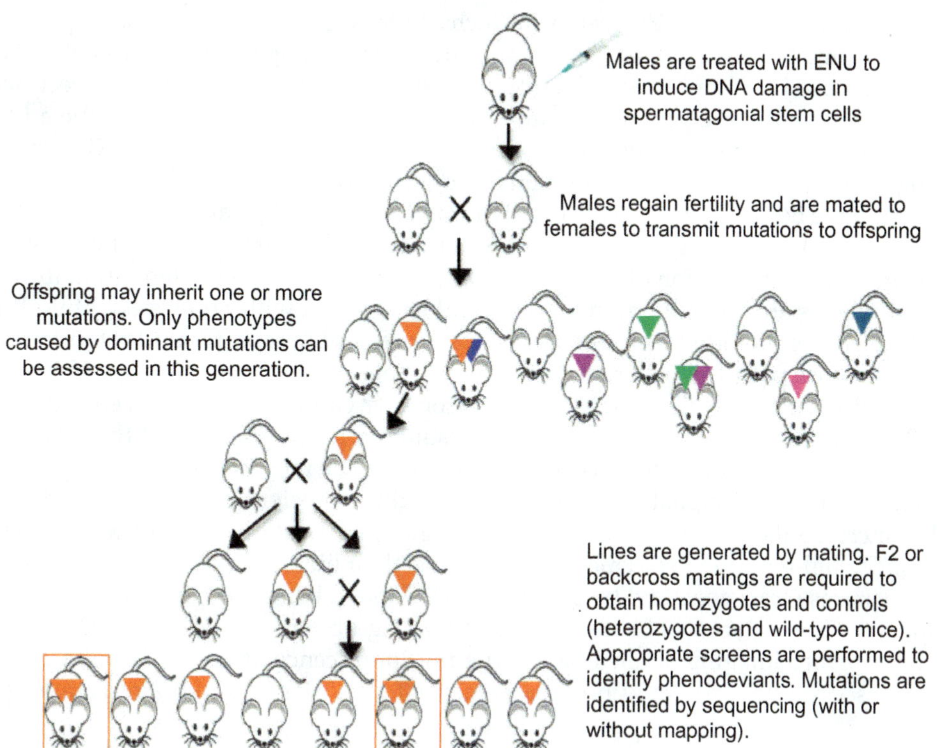

FIGURE 5.7 **Chemical mutagenesis.** An alternative approach to random mutagenesis is to administer a chemical mutagen, typically *N*-ethyl-*N*-nitrosourea (ENU), to male mice. Their first, second, and/or third generation offspring may carry one or more ENU-induced mutations. They are subjected to phenotypic screens to identify mice with mutations (represented by colored triangles) that cause phenotypes of interest. ENU can be used to model dominant, recessive, or multigenic disorders. Mutations must be identified by whole genome or exome sequencing or mapping followed by candidate gene analysis.

negative gain of function. Different mutations within the same gene can have distinct phenotypic consequences and provide unique insights into protein function. For example, mice homozygous for one ENU-induced missense mutation in the *Disrupted in schizophrenia-1* gene (*Disc1*) exhibited schizophrenic-like behavior, while mice homozygous for a different ENU-induced missense mutation showed depressive-like behavior.[29] The main disadvantage is that the mutations must be identified. Previously, this required a positional cloning approach of mapping the mutation and screening candidate genes within the identified candidate interval. "Next-generation" sequencing methodologies provide a new, rapid approach to identifying phenotype-associated ENU mutations, however, and their ever-dropping cost now makes it feasible to select a small number of mice from the same line with the same mutation and to sequence their entire exome or genome to identify shared sequence variants. This approach also makes it possible to use ENU mutagenesis to model multigenic traits. Many diseases are caused by the combined effects of sequence variants in multiple genes. Since each ENU-treated male will carry multiple mutations (approximately one visible dominant mutation per 100 offspring), their first generation progeny may carry mutations in multiple genes that only cause a specific phenotype when they are all inherited together. Those mutations can be identified by whole genome or exome sequencing. In addition, many disease-associated mutations are sensitive to genetic background effects, suggesting they are influenced by modifier loci, but these effects are difficult to identify and interfere in studies to map modifier mutations. Now that genetic background can be held constant and ENU-induced mutations identified by genome sequencing, outcrossing mice to map the mutation is not necessary, making more powerful modifier screens feasible.

PHENOTYPIC ANALYSIS OF MOUSE MODELS

Generating a mouse model is only half the battle: careful phenotype analysis is critical to obtaining valuable information, as is the use of appropriate controls. Neurologic diseases that cause pathology are much easier to analyze than those that cause behavioral alterations. Histology coupled with immunohistochemistry to identify specific cell types or proteins is the most common method used to assess neuropathological changes such as neuronal loss, gliosis, protein aggregates, or other gross changes. Mouse models on inbred strain backgrounds are advantageous for these studies as they are genetically identical: unless there are environmental or stochastic factors that influence disease pathogenesis, the brain of any mouse at any given time-point should have the same pathology as any age-matched animal of the same strain.

In vivo imaging may be used for noninvasive longitudinal studies by using the same subject over the course of disease, allowing one to follow disease progression in individual mice. This approach has the added advantage of using fewer mice to generate more data-points. X-ray, ultrasound, computed tomography (CT), single-photon emission computed tomography (SPECT), positron emission tomography (PET), magnetic resonance imaging (MRI) and optical imaging systems have been developed specifically for live imaging of small animals. Each includes a method to keep animals anesthetized during imaging, and many systems have high throughput options for imaging multiple animals at once. The anatomical resolution of optical imaging is limited by signal attenuation through tissue, but it is highly sensitive and images are obtained more quickly and easily than with other modalities. It requires a light-emitting reporter, which can be provided by a transgene that expresses the reporter under the control of a promoter that is activated by the disease pathway being studied. For example, since gliosis is observed in most neurodegenerative diseases, the glial fibrillary acidic protein (GFAP) promoter can be used to drive expression of a reporter that will be activated when astrocytosis is induced.[30] High-resolution *in vivo* imaging of the mouse brain is possible using fluorescent reporters and two- or three-photon microscopy.[31,32] The most commonly used fluorescent reporters are GFP and TdTomato, with better signal-to-noise ratios at higher wavelengths. New far-red reporters continue to be developed that increase the power of fluorescence imaging, particularly multiplex imaging using different wavelength reporters. The light-emitting enzyme luciferase is used for bioluminescence assays. This system requires that animals carry a luciferase transgene and be given the luciferase substrate luciferin (usually by intraperitoneal injection) prior to imaging. The high cost of luciferin is a disadvantage, but the wavelength of light emitted (>600 nm) is the most sensitive for *in vivo* imaging applications and auto-fluorescence of tissues is not a concern.[33] Figure 5.8 shows two methods of assessing gliosis: immunohistochemistry for GFAP on fixed brain sections (Figure 5.8A) and

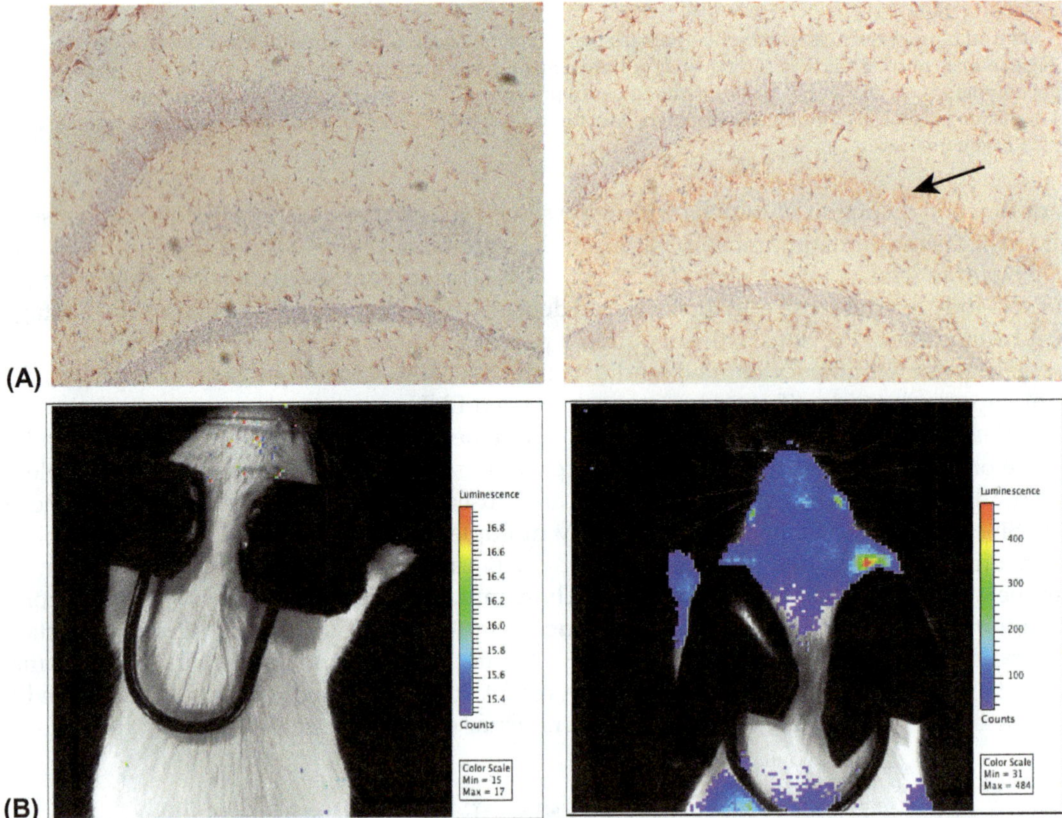

FIGURE 5.8 Imaging GFAP activation. Two methods for assessing gliosis, based on activation of glial fibrillary acidic protein (GFAP) expression. (**A**) Immunohistochemistry using an antibody against GFAP on 10 μm sections of paraffin-embedded brain, showing region around the dentate gyrus. Image on left is from a control mouse; image on right is from a mutant mouse model of neurodegeneration. GFAP signal is reddish brown. Sections were counterstained with hematoxylin (blue). GFAP is expressed in both samples but elevated around the CA3 region of the hippocampus (arrow) in the mutant brain. (**B**) *In vivo* imaging of control (left) and prion-inoculated (right) mice that carry a GFAP promoter-driven luciferase transgene. Luminescence reflects luciferase expression level, which is used as an indicator of prion disease progression. Images were captured using a PerkinElmer (Caliper Life Sciences) IVIS Lumina imaging system.

in vivo imaging of transgenic mice that express luciferase using the GFAP promoter (Figure 5.8B). The former provides more detailed information about the brain regions that are affected, but the latter allows noninvasive tracking of disease progression.

Models of many neurologic and psychiatric disorders are best characterized using cognitive and behavioral assays, although some aspects of psychiatric diseases are almost impossible to model in most animals, for example, depression or suicidal tendencies. Because genetic background, age and sex can have profound effects on the expression of many phenotypic traits, especially behavior, the selection of appropriate control animals is a critical aspect of any behavioral or cognitive study.[34] Furthermore, there can be significant differences in testing protocols and results between laboratories, or even between different research personnel within a laboratory, due to differences in housing conditions or animal handling.[35] It is important to first assess animals for gross phenotypes that might affect the interpretation of behavioral assays. For instance, mice with an inner ear defect will not be able to complete the Morris water maze since they cannot swim. An initial screen such as SHIRPA[36] is often used to assess gross appearance and basic behaviors (arousal, respiration, aggression, general activity level, gait, startle response, reflexes, etc.). Further tests, such as those summarized in Table 5.2, will depend on the outcome of the preliminary screen and the expected phenotype of the mutant mouse, based on the disease being modeled. A mouse model of autism, for example, would be examined for deficits in social interaction, communication and repetitive disorders. Parkinson disease causes

TABLE 5.2 Common Mouse Behavioral Assays

Test	Measurement	Reference
Prepulse inhibition/startle response	Sensory–motor gating, hearing, startle	Valsamis and Schmid, 2011[37]
Latent inhibition test	Latent inhibition	Barad et al., 2004[38]
Rotarod	Motor coordination, motor learning	Deacon, 2013; Moretti et al., 2005[39,40]
Activity chamber, wheel running	Activity, circadian rhythm	Moretti et al., 2005; Bailey and Crawley, 2009[40,41]
Open field test	Locomotor activity, anxiety-like behavior	Moretti et al., 2005; Bailey and Crawley, 2009[40,41]
Light/dark transition	Anxiety-like behavior	Bailey and Crawley, 2009; Takao and Miyakawa, 2006[41,42]
Elevated plus maze	Anxiety	Komada et al., 2008[43]
Crawley's sociability and preference for social novelty test	Social behavior	Kaidanovich-Beilin et al., 2011[44]
24-h home cage monitoring	Social behavior	Moretti et al., 2005[40]
Porsolt forced swim test	Depression	Can et al., 2012[45]
Tail suspension test	Depression-like behavior	Can et al., 2012[46]
Radial arm maze	Working memory, reference memory, perseveration	Wenk, 2004[47]
T- or Y-maze	Working memory, reference memory, perseveration	Shoji et al., 2012[48]
Morris water maze	Reference memory, perseveration, working memory	Wenk, 2004; Nunez, 2008[47,49]
Barnes' circular maze	Reference memory, perseveration, working memory	Larson et al., 2012[50]
Object recognition test	Reference memory	Antunes and Biala, 2012[51]
Cued and contextual fear conditioning test	Context memory	Curzon et al., 2009[52]
Conditioned place preference/aversion test	Reward/addiction	Cunningham et al., 2006[53]
Self-administration test	Reward/addiction	Kmiotek et al., 2012[54]
Olfaction test (buried food)	Olfactory defects	Moretti et al., 2005[40]

motor and nonmotor phenotypes, such as reduced sense of smell, sleep abnormalities, gastrointestinal disturbances, anxiety, depression, and impaired cognition. Activity, gait and coordination can be assessed using a locomotor activity chamber, gait analysis, rotarod and pole test. Olfactory disturbances (hyposmia and anosmia) can be assayed using the buried pellet test, novel scent test, or social olfactory discrimination (block test). Tests for anxiety and depression include the elevated plus maze, open field, light–dark exploration, forced swim test, and tail suspension test. Cognitive deficits are commonly evaluated using the Morris water maze, T- and Y-maze or the radial arm maze. Some of these tests would also be appropriate for models of other neurodegenerative disorders, particularly those that assess cognition. Ideally, characterization of a novel mouse model of a neurologic or psychiatric disorder will combine multiple behavioral tests with imaging, histology, molecular, and/or biochemical assays.

SUMMARY

Mouse models have led – and are certain to continue to lead – to significant breakthroughs in identifying genes, mechanisms, and pathways that underlie human neurologic diseases. Mice are also ideal for testing therapeutic approaches, something we are likely to see more of in the coming years. New methodologies have increased the speed and accuracy with which new mouse models can be generated, and technological advances have led to improved tools to analyze them. Models of multigenic disorders remain scarce. This is primarily because it is difficult to identify the variants that cause these traits, and most mouse models are presently generated using gene targeting, which requires the causative loci be known. Random mutagenesis and thorough phenotypic analysis (including behavioral studies) of existing mutants may reveal subtle and/or unexpected traits, and will complement other, ongoing projects aimed at discovering disease-associated variants in human populations. There is much excitement over the ability to reprogram fibroblasts or other patient-derived cells into induced pluripotent stem cells (iPSC), and the ability to differentiate those iPSC into neuronal stem cells allows for the analysis of those cells in culture. Injecting these cells into the mouse brain will create a new class of mouse models that will provide insight into the *in vivo* behavior of patient-derived cells in the mammalian nervous system. Combining these models with existing genetic models and reporter mice will create a powerful system for analyzing the pathogenesis of neurological disorders.

References

1. Palacino JJ, Sagi D, Goldberg MS, et al. Mitochondrial dysfunction and oxidative damage in parkin-deficient mice. *J Biol Chem*. 2004;279:18614–18622.
2. Braidy N, Munoz P, Palacios AG, et al. Recent rodent models for Alzheimer's disease: clinical implications and basic research. *J Neural Transm*. 2012;119:173–195.
3. Seok J, Warren HS, Cuenca AG, et al. Genomic responses in mouse models poorly mimic human inflammatory diseases. *Proc Natl Acad Sci U S A*. 2013;110:3507–3512.
4. Shin JH, Ko HS, Kang H, et al. PARIS (ZNF746) repression of PGC-1alpha contributes to neurodegeneration in Parkinson's disease. *Cell*. 2011;144:689–702.
5. Pascual A, Hidalgo-Figueroa M, Piruat JI, Pintado CO, Gomez-Diaz R, Lopez-Barneo J. Absolute requirement of GDNF for adult catecholaminergic neuron survival. *Nat Neurosci*. 2008;11:755–761.
6. Tasic B, Hippenmeyer S, Wang C, et al. Site-specific integrase-mediated transgenesis in mice via pronuclear injection. *Proc Natl Acad Sci U S A*. 2011;108:7902–7907.
7. Gurney ME, Pu H, Chiu AY, et al. Motor neuron degeneration in mice that express a human Cu, Zn superoxide dismutase mutation. *Science*. 1994;264:1772–1775.
8. Burright EN, Clark HB, Servadio A, et al. SCA1 transgenic mice: a model for neurodegeneration caused by an expanded CAG trinucleotide repeat. *Cell*. 1995;82:937–948.
9. Prusiner SB, Scott M, Foster D, et al. Transgenetic studies implicate interactions between homologous PrP isoforms in scrapie prion replication. *Cell*. 1990;63:673–686.
10. Saito H, Yamamura K, Suzuki N. Reduced bone morphogenetic protein receptor type 1A signaling in neural-crest-derived cells causes facial dysmorphism. *Dis Model Mech*. 2012;5:948–955.
11. Hitz C, Steuber-Buchberger P, Delic S, Wurst W, Kuhn R. Generation of shRNA transgenic mice. *Methods Mol Biol*. 2009;530:101–129.
12. Sydow A, Van der Jeugd A, Zheng F, et al. Tau-induced defects in synaptic plasticity, learning, and memory are reversible in transgenic mice after switching off the toxic Tau mutant. *J Neurosci*. 2011;31:2511–2525.
13. van den Maagdenberg AM, Pietrobon D, Pizzorusso T, et al. A Cacna1a knockin migraine mouse model with increased susceptibility to cortical spreading depression. *Neuron*. 2004;41:701–710.
14. Lin CH, Tallaksen-Greene S, Chien WM, et al. Neurological abnormalities in a knock-in mouse model of Huntington's disease. *Hum Mol Genet*. 2001;10:137–144.
15. Lorenzetti D, Watase K, Xu B, Matzuk MM, Orr HT, Zoghbi HY. Repeat instability and motor incoordination in mice with a targeted expanded CAG repeat in the Sca1 locus. *Hum Mol Genet*. 2000;9:779–785.

16. Price MG, Yoo JW, Burgess DL, et al. A triplet repeat expansion genetic mouse model of infantile spasms syndrome, Arx(GCG)10+7, with interneuronopathy, spasms in infancy, persistent seizures, and adult cognitive and behavioral impairment. *J Neurosci*. 2009;29:8752–8763.
17. van den Broek WJ, Nelen MR, Wansink DG, et al. Somatic expansion behaviour of the (CTG)n repeat in myotonic dystrophy knock-in mice is differentially affected by Msh3 and Msh6 mismatch-repair proteins. *Hum Mol Genet*. 2002;11:191–198.
18. Bontekoe CJ, Bakker CE, Nieuwenhuizen IM, et al. Instability of a (CGG)98 repeat in the Fmr1 promoter. *Hum Mol Genet*. 2001;10:1693–1699.
19. Guo F, Ma J, McCauley E, Bannerman P, Pleasure D. Early postnatal proteolipid promoter-expressing progenitors produce multilineage cells *in vivo*. *J Neurosci*. 2009;29:7256–7270.
20. Skarnes WC, Rosen B, West AP, et al. A conditional knockout resource for the genome-wide study of mouse gene function. *Nature*. 2011;474:337–342.
21. Carbery ID, Ji D, Harrington A, et al. Targeted genome modification in mice using zinc-finger nucleases. *Genetics*. 2010;186:451–459.
22. Wefers B, Meyer M, Ortiz O, et al. Direct production of mouse disease models by embryo microinjection of TALENs and oligodeoxynucleotides. *Proc Natl Acad Sci U S A*. 2013;110:3782–3787.
23. Cong L, Ran FA, Cox D, et al. Multiplex genome engineering using CRISPR/Cas systems. *Science*. 2013;339:819–823.
24. Wang H, Yang H, Shivalila CS, et al. One-step generation of mice carrying mutations in multiple genes by CRISPR/Cas-mediated genome engineering. *Cell*. 2013;153:910–918.
25. Joyner AL, Auerbach A, Skarnes WC. The gene trap approach in embryonic stem cells: the potential for genetic screens in mice. *Ciba Found Symp*. 1992;165:277–288, discussion 288–297.
26. Horie K, Yusa K, Yae K, et al. Characterization of Sleeping Beauty transposition and its application to genetic screening in mice. *Mol Cell Biol*. 2003;23:9189–9207.
27. Ding S, Wu X, Li G, Han M, Zhuang Y, Xu T. Efficient transposition of the piggyBac (PB) transposon in mammalian cells and mice. *Cell*. 2005;122:473–483.
28. Brown SD. Mouse models of genetic disease: new approaches, new paradigms. *J Inherit Metab Dis*. 1998;21:532–539.
29. Clapcote SJ, Lipina TV, Millar JK, et al. Behavioral phenotypes of Disc1 missense mutations in mice. *Neuron*. 2007;54:387–402.
30. Zhu L, Ramboz S, Hewitt D, Boring L, Grass DS, Purchio AF. Non-invasive imaging of GFAP expression after neuronal damage in mice. *Neurosci Lett*. 2004;367:210–212.
31. Horton NG, Wang K, Kotbat D, et al. In vivo three-photon microscopy of subcortical structures within an intact mouse brain. *Nature Photon*. 2013;7:205–209.
32. Spires-Jones TL, de Calignon A, Meyer-Luehmann M, Bacskai BJ, Hyman BT. Monitoring protein aggregation and toxicity in Alzheimer's disease mouse models using *in vivo* imaging. *Methods*. 2011;53:201–207.
33. Negrin RS, Contag CH. In vivo imaging using bioluminescence: a tool for probing graft-versus-host disease. *Nat Rev Immunol*. 2006;6:484–490.
34. Rogers DC, Jones DN, Nelson PR, et al. Use of SHIRPA and discriminant analysis to characterise marked differences in the behavioural phenotype of six inbred mouse strains. *Behav Brain Res*. 1999;105:207–217.
35. van der Staay FJ, Steckler T. Behavioural phenotyping of mouse mutants. *Behav Brain Res*. 2001;125:3–12.
36. Rogers DC, Fisher EM, Brown SD, Peters J, Hunter AJ, Martin JE. Behavioral and functional analysis of mouse phenotype: SHIRPA, a proposed protocol for comprehensive phenotype assessment. *Mamm Genome*. 1997;8:711–713.
37. Valsamis B, Schmid S. Habituation and prepulse inhibition of acoustic startle in rodents. *J Vis Exp*. 2011;e3446.
38. Barad M, Blouin AM, Cain CK. Like extinction, latent inhibition of conditioned fear in mice is blocked by systemic inhibition of L-type voltage-gated calcium channels. *Learn Mem*. 2004;11:536–539.
39. Deacon RM. Measuring motor coordination in mice. *J Vis Exp*. 2013 May 29;(75):e2609.
40. Moretti P, Bouwknecht JA, Teague R, Paylor R, Zoghbi HY. Abnormalities of social interactions and home-cage behavior in a mouse model of Rett syndrome. *Hum Mol Genet*. 2005;14:205–220.
41. Bailey KR, Crawley JN. Anxiety-related behaviors in mice. In: Buccafusco JJ, ed. *Methods of behavior analysis in neuroscience*. 2nd ed. Boca Raton, FL: CRC Press; 2009, Chapter 5, Frontiers in neuroscience.
42. Takao K, Miyakawa T. Light/dark transition test for mice. *J Vis Exp*. 2006;1:104.
43. Komada M, Takao K, Miyakawa T. Elevated plus maze for mice. *J Vis Exp*. 2008;22:1088.
44. Kaidanovich-Beilin O, Lipina T, Vukobradovic I, Roder J, Woodgett JR. Assessment of social interaction behaviors. *J Vis Exp*. 2011;48:2473.
45. Can A, Dao DT, Arad M, Terrillion CE, Piantadosi SC, Gould TD. The mouse forced swim test. *J Vis Exp*. 2012;(59):e3638.
46. Can A, Dao DT, Terrillion CE, Piantadosi SC, Bhat S, Gould TD. The tail suspension test. *J Vis Exp*. 2012;59:e3769.
47. Wenk GL. Assessment of spatial memory using the radial arm maze and Morris water maze. *Curr Protoc Neurosci*. 2004, Chapter 8:Unit 8 5A.
48. Shoji H, Hagihara H, Takao K, Hattori S, Miyakawa T. T-maze forced alternation and left-right discrimination tasks for assessing working and reference memory in mice. *J Vis Exp*. 2012 Feb 26;(60):pii, 3300.
49. Nunez J. Morris water maze experiment. *J Vis Exp*. 2008 Sep 24;(19):pii, 897.
50. Larson ME, Sherman MA, Greimel S, et al. Soluble alpha-synuclein is a novel modulator of Alzheimer's disease pathophysiology. *J Neurosci*. 2012;32:10253–10266.
51. Antunes M, Biala G. The novel object recognition memory: neurobiology, test procedure, and its modifications. *Cogn Process*. 2012;13:93–110.
52. Curzon P, Rustay NR, Browman KE. Cued and contextual fear conditioning for rodents. In: Buccafusco JJ, ed. *Methods of behavior analysis in neuroscience*. 2nd ed. Boca Raton, FL: CRC Press; 2009.
53. Cunningham CL, Gremel CM, Groblewski PA. Drug-induced conditioned place preference and aversion in mice. *Nat Protoc*. 2006;1:1662–1670.
54. Kmiotek EK, Baimel C, Gill KJ. Methods for intravenous self administration in a mouse model. *J Vis Exp*. 2012;e3739.

CHAPTER

6

DNA Sequencing and Other Methods of Exonic and Genomic Analyses

Jun Mitsui, Hiroyuki Ishiura, and Shoji Tsuji

Graduate School of Medicine, The University of Tokyo, Tokyo, Japan

DNA SEQUENCING TECHNOLOGIES

DNA sequencing is determination of the order of nucleotides (deoxyadenine, deoxyguanine, deoxycytosine, and deoxythymine) in a strand of DNA molecules.[1] Furthermore, DNA sequencing has been used not only for sequencing genomic DNA molecules, but also for sequencing that of complementary DNA (cDNA) prepared from RNA molecules and analyses of epigenetic modifications of genomic DNA molecules.

In the late 1970s, two DNA sequencing techniques were developed independently, namely the Maxam–Gilbert method[2] and the Sanger method.[3] Briefly, in the Maxam–Gilbert method, DNA fragments labeled at the 5′-end with radioisotopes are subjected to random cleavage at specific nucleotide positions using chemical reactions. This generates sets of fragments from two chemical cleavage reactions: one cleaves DNA molecules at deoxyguanine (G) and deoxyadenine (A) and the other cleaves DNA molecules at deoxycytosine (C) and deoxythymine (T). The first reaction can be modified to cleave DNA molecules at G only, and the second can be modified to cleave DNA molecules at C only. The products of these four reactions are then electrophretically separated through a denaturing polyacrylamide gel. The Sanger method, also called the dideoxynucleotide chain terminator method, employs DNA polymerase and dideoxynucleotides (ddNTPs) for the generation of DNA fragments terminated at specific nucleotides by incorporating corresponding dideoxynucleotides. It requires a single-stranded DNA molecule as the template, a DNA primer (a short oligonucleotide with a DNA sequence complementary to the template DNA), a DNA polymerase, deoxynucleotides (dNTPs), and ddNTPs that terminate DNA strand elongation. Briefly, a primer is annealed to a specific region on the template DNA, which provides a starting point for DNA synthesis in the presence of DNA polymerases. The ddNTPs lack the 3′-hydroxyl group of dNTP, which is required for the phosphodiester bond formation between one nucleotide and the following nucleotide during strand elongation. When a ddNTP is incorporated into the elongating strand, it inhibits further strand extension. The reaction is conducted in four tubes, each containing the appropriate amounts of one of the four ddNTPs and the four dNTPs. All the generated fragments have the same 5′-end, whereas the residue at the 3′-end is determined by the specific ddNTP used in the reaction. DNA fragments are detected on the basis of radioactivity by labeling them with a radioactive dNTP (for example, [^{35}S]dCTP) in the reaction mixture. After the four reactions were completed, the mixture of different sized DNA fragments was resolved by electrophoresis through a denaturing polyacrylamide gel, in four parallel lanes. The pattern of bands on the autoradiograms showed the distribution of the specific nucleotides with termination in the synthesized strand of DNA and thus nucleotide sequences could be read out by autoradiography (Figure 6.1).

The original Sanger sequencing method has been subjected to several important improvements and evolved markedly over the couple of decades. First, although cloning of DNA fragments into plasmid vectors was necessary in Sanger sequencing, the application of polymerase chain reaction (PCR)[4] for the amplification of specific DNA fragments *in vitro* has explosively broadened in the field of Sanger sequencing. Second, as an alternative to radioisotope labeling ([^{32}P] when first developed and then replaced with [^{35}S]),[5] the development of the technique of labeling of the chain terminator ddNTPs with four different fluorescent dyes permits sequencing in a single

Rosenberg's Molecular and Genetic Basis of Neurological and Psychiatric Disease
http://dx.doi.org/10.1016/B978-0-12-410529-4.00006-1

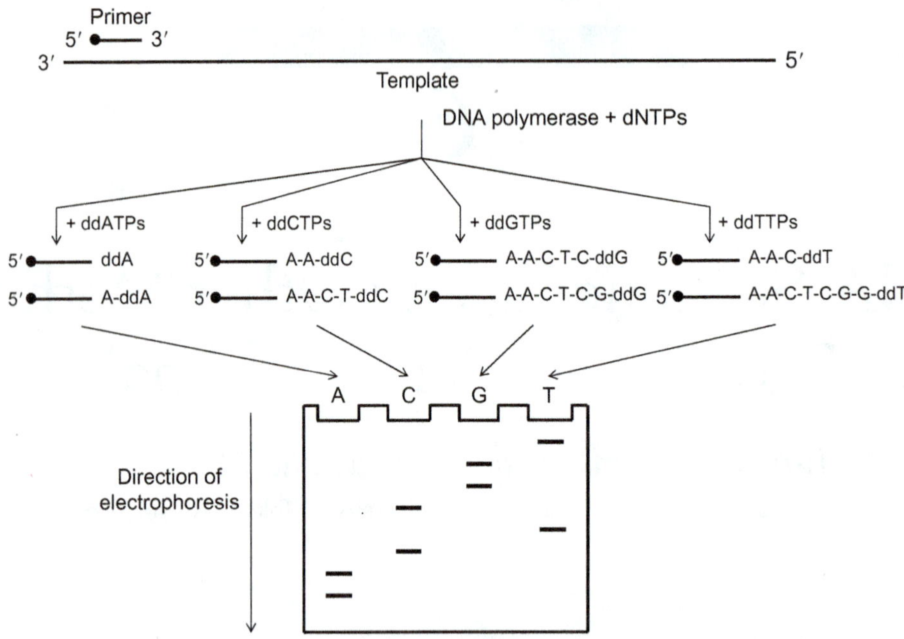

FIGURE 6.1 **Schematic representation of sequencing process in Sanger method.** The Sanger method employs dideoxynucleotides that terminate newly synthesized DNA fragments at specific bases (either A, C, T, or G). The resulting fragments are then resolved by electrophoresis through a denaturing polyacrylamide gel in four parallel lanes, and the DNA sequence can be read.

reaction, and also sets the stage for the use of automated DNA-sequencing instruments.[6] Lastly, technological advances offered by capillary electrophoresis along with highly sensitive detectors and its parallelization have markedly improved the throughput of Sanger sequencing.[7,8] The Human Genome Project was only possible due to the innovative technological advances described above.[9] Automated DNA-sequencing instruments can sequence up to 96 samples in a single batch and automatically carry out capillary electrophoresis for size separation, detection, and recording of dye fluorescence, and data output as fluorescent peak trace chromatograms. Lengths of DNA fragments that Sanger sequencing can determine are approximately 500–900 base pairs (bp) of a DNA fragment.

More recently, next-generation sequencing (NGS) technologies together with improvements in the algorithm of short read alignment and assembly have evolved and rapidly replaced Sanger sequencing, particularly for large-scale sequencing projects, such as whole-exome sequencing (WES; Figure 6.2) or whole-genome sequencing (WGS; Figure 6.3). In combination with hybridization-based target sequence-enrichment technologies, WES is an efficient strategy to selective sequencing of the coding regions of the human genome. Because exonic sequences represent nearly 2% of the whole human genome and the majority of disease-causing mutations involve exonic sequences or splice sites, comprehensive sequencing of exonic sequences and flanking sequences after enrichment is considered to be an efficient method.[10] In principle, the concept underlying NGS technologies is similar to that of Sanger sequencing in that the nucleotide sequences of a DNA fragment are sequentially identified from signals emitted. NGS technologies extend this sequencing process across from millions to billions of reactions. A genomic sequence is first fragmented into a library of small segments, which are subsequently sequenced in a massively parallel manner (Figure 6.4). The newly identified strings of nucleotides are then aligned using a known reference genome as a scaffold (Figure 6.5). For NGS technologies, various platforms have been introduced. Methodologically, they can be classified in terms of the template preparation methods and sequence detection systems used. Methods of template preparation are divided into two groups: preparing clonally amplified templates originating from single DNA molecules (emulsion PCR[11] and solid-phase amplification) and single-molecule sequencing free from amplification of DNA molecules. Sequence detection systems can be divided into two types: detection of fluorescent dyes (e.g., pyrosequencing,[12] reversible dye terminator,[13,14] oligonucleotide probe ligation,[15] and phospholinked fluorescent nucleotides[16]) and fluorescent dye-free methods (e.g., ion semiconductor sequencing[17] and nanopore sequencing[18]). For the currently available second-generation sequencers (sequencing of clonally amplified DNA molecules employing fluorescent nucleotides),[15] millions to billions of amplified DNA molecules can be sequenced, albeit the read lengths are relatively short (100–500bp). Regarding the third-generation sequencers, the PacBio RS system, commercialized in 2010 by Pacific Biosciences, Inc., is the first single molecule real-time DNA sequencer.[16] Although it depends on the size of the library fragments and the time of data

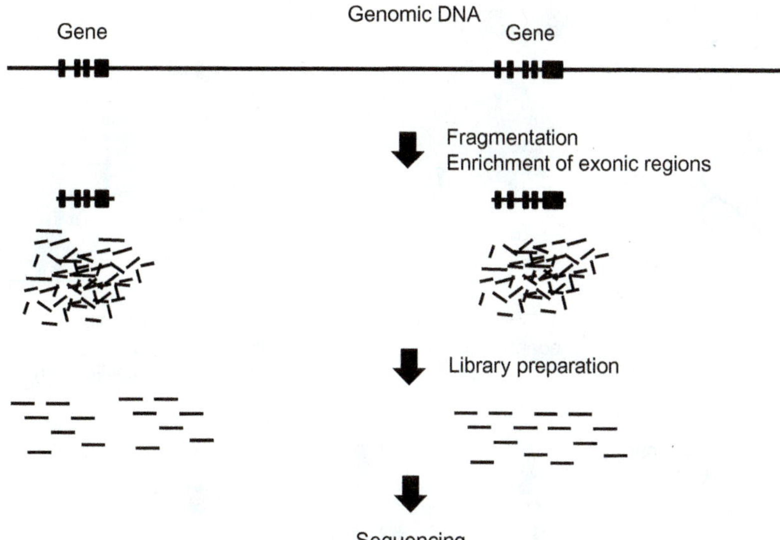

FIGURE 6.2 **Concept of whole-exome sequencing.** Hybridization-based target sequence-enrichment technologies allow one to selectively capture genomic regions of interest from a DNA sample prior to sequencing.

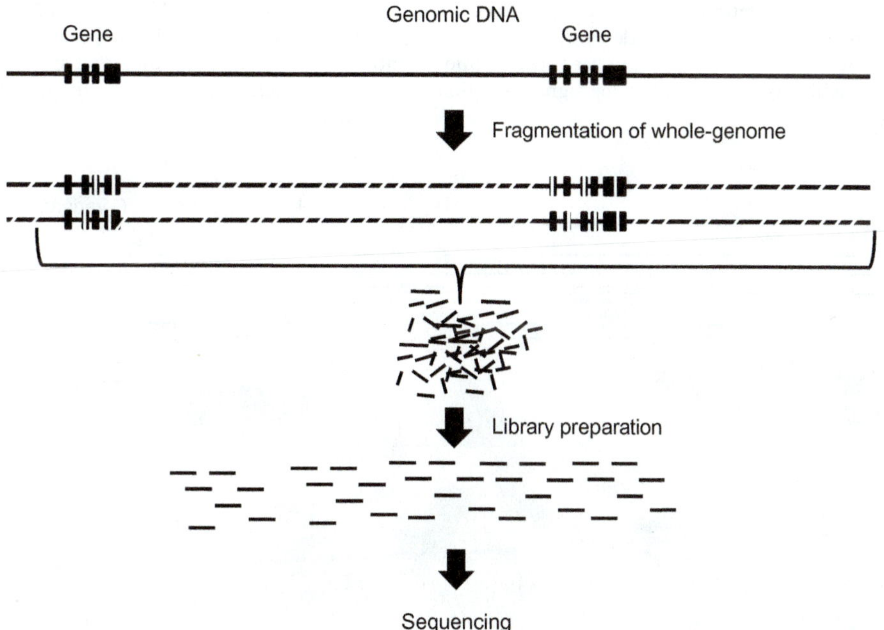

FIGURE 6.3 **Concept of whole-genome sequencing.**

collection, the read lengths can be quite long (up to 20,000 bp), compared with those of the second-generation sequencers. The error rate is quite high (15%) on a per read basis, yet the rate of accuracy obtained from multiple passes on a single molecule can exceed 99%.

NGS FOR ELUCIDATING MENDELIAN-TRAIT DISEASES

NGS platforms are currently revolutionizing the field of genomic research, enabling diverse and flexible sequencing applications of genomes, transcriptomes, and epigenomes for various purposes. In particular, genome-wide screening approaches employing WGS/WES have demonstrated excellent competence to not only identify causative genes for Mendelian-trait diseases,[19] but also identify causative mutations in the context of

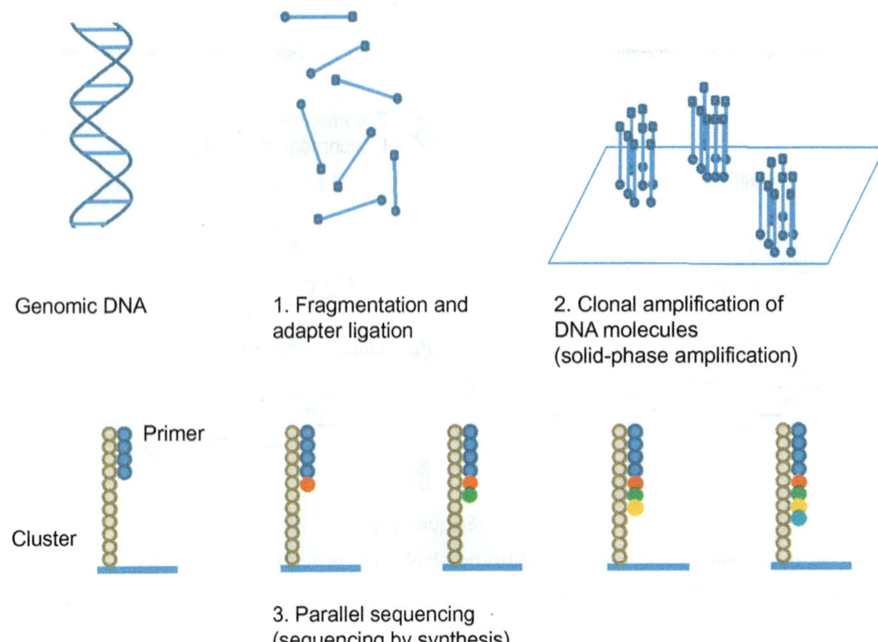

FIGURE 6.4 Schematic representation of sequencing process of second-generation sequencer. 1) Genomic DNA is sheared into small fragments and adaptor sequences are ligated. 2) A single molecule of DNA is attached to a flat surface and is amplified to form a cluster (solid-phase amplification). 3) DNA polymerase and a mix of four base-specific fluorescently labeled reversible chain terminators are added. A primer directs incorporation of one base, which is decoded following signal detection. The process is iterated, enabling addition and detection of one base on each template per cycle.

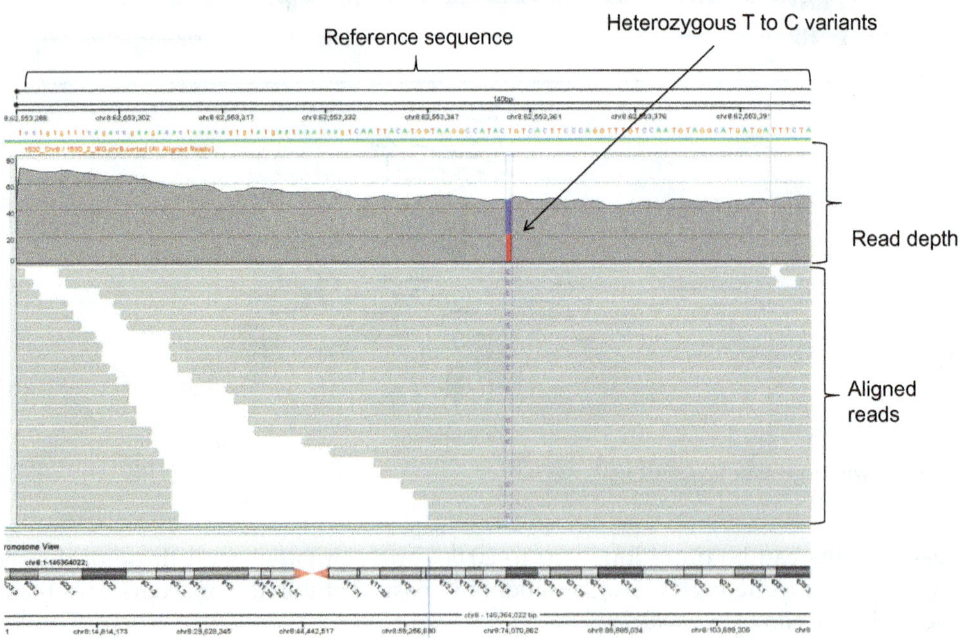

FIGURE 6.5 Example of short-read mapping and variant call.

"clinical sequencing." Compared with WGS, the WES approach has limited power to identify variants in noncoding regions that may be important for transcriptional regulation or splicing and structural variants. Nevertheless, the coding regions of the genome have been considered to contain approximately 85% of the disease-causing mutations,[20] and numerous WES studies have demonstrated the power to identify the causal genetic variants for Mendelian-trait diseases.[21–27]

NGS FOR ELUCIDATING MOLECULAR BASES OF DISEASES WITH MENDELIAN TRAIT

NGS for the Discovery of Causative Genes (Positional Cloning)

Until the era of NGS, the search for causative genes for diseases with Mendelian trait required many laborious steps: a) collection of family members including affected as well as unaffected individuals; b) linkage analyses to narrow down the candidate regions on the basis of genome-wide genotyping of single-nucleotide or microsatellite polymorphisms; and c) Sanger sequencing of the genes located within the candidate regions identified by linkage analyses, and confirmation of the causality of the mutated genes by demonstrating cosegregation of the mutations with the disease and functional evaluation of the mutant gene products. These approaches are called positional cloning.[28] When the sizes of the families are small or the number of families is limited, the candidate regions cannot be sufficiently narrowed down, making it extremely laborious to conduct Sanger sequencing of a tremendous number of genes located in the candidate genes.

Availability of NGS has markedly simplified these laborious steps. Now, WES/WGS allows us to identify eventually all the variants located in the exonic or genomic sequences. Even when the candidate regions are not sufficiently narrowed down, we can identify all the variants in the candidate regions, and we can apply various approaches including cosegregation of the mutations with the disease and functional assays. Thus, many new genes for diseases with Mendelian trait are expected to be discovered by NGS technologies.[21–27]

Search for *De Novo* Mutations Based on WES/WGS Analyses of Trios

Genome-wide identification of *de novo* mutations (new germline mutations arising from the gametes of parents) was not feasible until the arrival of NGS technologies. Family-based WES/WGS has been shown to be highly efficient in identifying *de novo* mutations that are responsible for the diseases with autosomal dominant traits characterized by early onset and/or impaired reproductive fitness.[29] Because the number of *de novo* mutations in exonic sequences is estimated to be one or two per generation at most, *de novo* mutations identified in the proband of a trio are good candidates as the disease-causing mutations.[30–34]

NGS FOR ELUCIDATING MOLECULAR BASES OF COMPLEX-TRAIT DISEASES

The majority of the patients with neurodegenerative diseases such as Alzheimer disease, Parkinson disease, and amyotrophic lateral sclerosis lack obvious family history and are considered to have complex traits where genetic and environmental factors play roles as the molecular basis.[35] To elucidate the genetic factors underlying the complex-trait diseases, genome-wide association studies (GWASs) employing common single-nucleotide polymorphisms (SNPs) have been intensively conducted to identify genomic variations associated with complex-trait diseases.[36] The theoretical framework of GWAS is based on the "common disease–common variants" hypothesis, in which common diseases are attributable, in part, to relatively common allelic variants present in the population. Although GWASs have successfully revealed numerous susceptibility genes, the odds ratios associated with these risk alleles are generally low and account for only a small proportion of estimated heritability. Given these results, it is expected that low-frequency variants, particularly functional variants in coding sequences, with relatively large effect sizes may be involved in the pathogenesis.[37]

Because NGS technologies allow identification of all the variants in the human genome irrespective of allele frequencies, it is anticipated that WGS/WES will be applied to elucidating low-frequency variants with relatively large effect sizes for development of complex-trait diseases. To accomplish this, however, there are growing concerns about the research paradigm employing whole-exome-association studies. Owing to the complexity and magnitude of variants identified by WGS/WES, much larger sample sizes may be required to obtain the statistical significance, particularly for low-frequency variants, compared with conventional GWAS employing common SNPs.[38,39] To date, successes in identification of rare variants with high odds ratios have been due to the use of multiplex families, as best illustrated by the discoveries of multiple rare variants of *GBA* in Parkinson disease,[40,41] those of *COQ2* in multiple system atrophy,[42] and those of *PLD3* in Alzheimer disease.[43] Since genetic variants with large effect sizes are expected to underlie intrafamilial clustering of patients, the search for genetic variants associated with the disease in multiplex families is a highly efficient approach to identifying low-frequency variants with large effect sizes. Furthermore, identification of such variants allows association studies to focus on specific variants, which substantially increase the power of statistical analyses bypassing the correction for multiple comparisons. Detailed evaluation will be needed to estimate the required sample size that will provide sufficient power to detect disease-associated variants taking the allele frequencies and odds ratios associated with the variants into account.

NGS FOR CLINICAL SEQUENCING

As an example of the successful application of WGS for establishing molecular diagnosis,[44] a patient with Charcot–Marie–Tooth neuropathy was subject to WGS analysis. As a result, as many as 3,420,306 single-nucleotide variants (SNVs) were identified. Among these, 2,255,102 of the SNVs were in extragenic regions and 1,165,204 SNVs were within gene regions. Of the intragenic SNVs, 9069 were predicted to result in nonsynonymous codon changes, and 121 of them were nonsense mutations. A search for mutations in 40 genes associated with hereditary neuropathy led to the identification of 54 SNVs, in which two mutations were located at the *SH3TC2* locus.[45] They included one novel missense mutation (Y169H) and one nonsense mutation (R954X), which were previously reported in patients with Charcot–Marie–Tooth type 4C. The study demonstrated the role of WGS in identifying causative mutations with Mendelian traits. Indeed, NGS technologies have intensively been applied to the molecular diagnosis of diseases.

Recently, an example of the clinical use of WES using samples from 250 consecutive patients for the diagnosis of Mendelian-trait diseases has been reported.[46] Among the 250 patients, approximately 80% were children with phenotypes related to neurologic conditions: 60 had neurologic disorders including developmental delay, speech delay, autism spectrum disorder, and intellectual disability; 140 had neurologic disorders and other organ–system disorders; 13 had specific neurological disorders such as ataxia or seizure; and 37 had non-neurological disorders. Variants were compared computationally with the list of reported mutations from the Human Gene Mutation Database (HGMD).[47] Variants in this database with a minor allele frequency of less than 5%, as determined on the basis of either the 1000 Genomes Project[48] or ESP5400 data (http://evs.gs.washington.edu/EVS), were retained. For changes that are not in the HGMD, synonymous variants, intronic variants that were more than 5bp from exon boundaries, and common variants (minor allele frequency, >1%) were discarded. Retained variants were classified as deleterious mutations (potentially pathogenic variants), variants of unknown clinical significance (VUS), or benign variants, in accordance with the interpretation guidelines of the American College of Medical Genetics and Genomics (ACMG).[49,50] Deleterious mutations and VUS that are potentially related to the patient's phenotype were further examined to determine whether they were consistent with ACMG category 1 (previously reported and recognized cause of the disorder) or category 2 (previously unreported and of the type that is expected to cause the disorder).[49] All putative causative alleles were then subjected to extensive literature and database searches, and the results were discussed in roundtable sessions by laboratory directors and physicians with appropriate clinical expertise. Applying the above criteria, the overall rate of a positivity in molecular diagnosis was 25%, including 33 patients with autosomal dominant disease, 16 with autosomal recessive disease, and nine with X-linked disease. These reports have highlighted the diagnostic power of WGS/WES in the context of genetically heterogeneous Mendelian-trait diseases, for which it is difficult to focus on particular genes for the mutational analysis depending solely on phenotypes.

In the context of clinical sequencing, most WES/WGS are conducted typically with a specific clinical purpose. The results are, for example, screened for pathogenic variants in specific candidate genes. However, because WES/WGS generates an unprecedented amount of genetic data, there is a possibility that medically actionable findings, which put individuals at risk for life-threatening but preventable or treatable diseases, are obtained beyond the scope of the purpose. There is ongoing debate on how to deal with incidental findings and we need to pay attention to such findings during the interpretation of the overall results of WES/WGS (http://bioethics.gov/node/3183).[50]

OTHER METHODS OF EXONIC AND GENOMIC ANALYSIS

Structural variants are certain types of genomic alterations, comprising unbalanced forms of variation, including deletions, multiplications (leading to copy-number variation), repeat expansions, and insertions; and balanced forms of variation, such as inversions and translocations.[51] Small genomic alterations including SNVs and indels (addition or deletion of several bases) are detectable by short-read (100–500bp) sequencing methods, because short-read sequences containing such a variant can be easily aligned to a reference genome, sorting out what an alteration is. In contrast, short-read sequences derived from the regions with structural variants are difficult to align to a reference genome, particularly in WES analysis.

In particular, certain genes responsible for neuromuscular diseases are rich in structural variants. For example, deletions or duplications with the sizes of 10 kb to 1 Mb are frequently observed in autosomal recessive juvenile parkinsonism (*PARK2*) or Duchenne muscular dystrophy (*DMD*),[52] where deletions or duplications affect some exons within the gene, but not the entire gene. Gene multiplications (increased copy number of the entire gene) are causative in CMT1A (*PMP22*),[53] CMT1B (*MPZ*),[54] Pelizaeus–Merzbacher disease (*PLP1*),[55] autosomal dominant leukodystrophy (*LMNB1*),[56] autosomal dominant Parkinson disease type 4 (*SNCA*),[57] and autosomal dominant early-onset Alzheimer disease with cerebral amyloid angiopathy (*APP*).[58] Repeat expansions are causative mutations in Huntington disease (*HTT*), spinal-bulbar muscular atrophy (*AR*), certain types of autosomal dominant cerebellar ataxia (SCA1, SCA2, Machado–Joseph disease/SCA3, SCA6, SCA7, SCA8, SCA10, SCA12, SCA17, SCA31, SCA36, and DRPLA), 9p-linked amyotrophic lateral sclerosis (*C9ORF72*), myotonic dystrophy type 1 (*DMPK*), and fragile X syndrome, among others.[59] Because the read length of commonly used DNA sequencing platforms is about 100–500 bp, genomic alterations with size exceeding these limits are generally difficult to analyze currently. There is a possibility that the sequencing technology that offers long read lengths will enable us to comprehensively detect all types of variant including SNVs, indels, structural variants, and repeat expansions.[60]

Other methods of exonic and genomic analysis include fluorescence *in situ* hybridization (FISH) analysis, microarray-based comparative genomic hybridization (array CGH) analysis, PCR-based methods (quantitative PCR analysis, multiplex ligation-dependent probe amplification [MLPA] analysis,[61] PCR fragment analysis, and repeat-primed PCR analysis[62]), and Southern blot hybridization analysis. FISH analysis enables the detection of unbalanced forms of structural variation (deletion and multiplications) with the size of more than a hundred kilobases (e.g., *PMP22* and *PLP1*). Using multi-fluorescent probes, FISH analysis enables the detection of balanced forms of variations, such as inversions and translocations. Using high-resolution microarrays in conjunction with CGH techniques, researchers have developed array CGH analysis with a high resolution of several kilobases, which has become the first-line approach for unbalanced forms of structural variations (Figure 6.6).[63]

The different PCR-based approaches are quantitative PCR and MLPA analyses, which also enable a locus-by-locus measurement of copy number. Compared with array CGH analysis, they are relatively rapid and low cost and can easily be multiplexed. MLPA analysis has been widely used for mutational analysis of several genes, such as *DMD* and *PARK2*.[64,65] One of the limitations of quantitative PCR and MLPA analyses is that we can only obtain information on a restricted number of genomic loci. Note that array CGH, quantitative PCR, and MLPA analyses cannot detect inversions and translocations. PCR fragment analysis is a robust method of demonstrating the presence of the increased size of PCR products containing a repeat-expansion allele. When the repeat-expansion allele is too long and

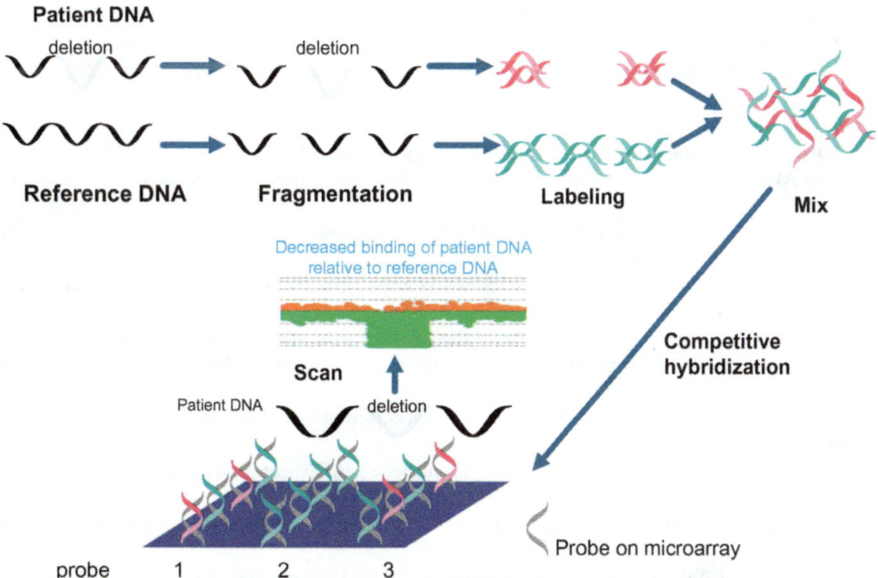

FIGURE 6.6 Schematic representation of array comparative genomic hybridization (CGH) method. Array comparative genomic hybridization consists of a set of DNA fragments from a patient and reference DNA samples labeled with different fluorescent dyes (red and green, respectively). These fragments are mixed and applied to complementary genomic DNA fragments (probes) on a microarray. Copy number loss of a DNA fragment from the patient leads to a decreased binding of the patient DNA (red) relative to the reference DNA (green), resulting in green fluorescence. The copy number gain of a DNA fragment from the patient leads to an increased binding of the patient DNA (red) relative to the reference DNA (green), resulting in red fluorescence.

difficult to be reliably amplified, the alternative approaches are repeat-primed PCR and/or Southern blot hybridization analysis. Repeat-primed PCR analysis is a method using a locus-specific unique primer flanking the repeat together with a paired primer composed of several repeat units that can bind anywhere in the repeat region, thus creating PCR amplicons of varying sizes.[62] A limitation of repeat-primed PCR analysis is that this approach cannot provide a precise estimate of the size of the repeat expansion. Southern blot hybridization analysis is a method of detecting a specific DNA sequence in a genomic DNA sample, which involves the transfer of DNA fragments separated by gel electrophoresis onto a membrane and subsequent fragment detection by hybridization using a specific probe labeled with a fluorescent dye or radioisotopes. This method is used for detecting DNA fragments encompassing the region of repeat expansion, allowing estimation of the repeat number. When there is a variation in the repeat number in a sample, a smear is observed, making it difficult to correctly determine the expanded alleles.[66] Southern blot hybridization analysis is also useful for determining complex rearrangements, such as recombinant alleles involving pseudogenes, novel sequence insertions, or translocations.[67]

References

1. França LT, Carrilho E, Kist TB. A review of DNA sequencing techniques. *Q Rev Biophys*. 2002;35(2):169–200.
2. Maxam AM, Gilbert W. A new method for sequencing DNA. *Proc Natl Acad Sci U S A*. 1977;74(2):560–564.
3. Sanger F, Nicklen S, Coulson AR. DNA sequencing with chain-terminating inhibitors. *Proc Natl Acad Sci U S A*. 1977;74(12):5463–5467.
4. Saiki RK, Gelfand DH, Stoffel S, et al. Primer-directed enzymatic amplification of DNA with a thermostable DNA polymerase. *Science*. 1988;239(4839):487–491.
5. Biggin MD, Gibson TJ, Hong GF. Buffer gradient gels and 35S label as an aid to rapid DNA sequence determination. *Proc Natl Acad Sci U S A*. 1983;80(13):3963–3965.
6. Smith LM, Sanders JZ, Kaiser RJ, et al. Fluorescence detection in automated DNA sequence analysis. *Nature*. 1986;321(6071):674–679.
7. Ruiz-Martinez MC, Salas-Solano O, Carrilho E, Kotler L, Karger BL. A sample purification method for rugged and high-performance DNA sequencing by capillary electrophoresis using replaceable polymer solutions. A development of the cleanup protocol. *Anal Chem*. 1998;70(8):1516–1527.
8. Huang XC, Quesada MA, Mathies RA. DNA sequencing using capillary array electrophoresis. *Anal Chem*. 1992;64(18):2149–2154.
9. Consortium IHGS. Finishing the euchromatic sequence of the human genome. *Nature*. 2004;431(7011):931–945.
10. Hodges E, Xuan Z, Balija V, et al. Genome-wide in situ exon capture for selective resequencing. *Nat Genet*. 2007;39(12):1522–1527.
11. Williams R, Peisajovich SG, Miller OJ, Magdassi S, Tawfik DS, Griffiths AD. Amplification of complex gene libraries by emulsion PCR. *Nat Methods*. 2006;3(7):545–550.
12. Margulies M, Egholm M, Altman WE, et al. Genome sequencing in microfabricated high-density picolitre reactors. *Nature*. 2005;437(7057):376–380.
13. Bentley DR. Whole-genome re-sequencing. *Curr Opin Genet Dev*. 2006;16(6):545–552.
14. Bentley DR, Balasubramanian S, Swerdlow HP, et al. Accurate whole human genome sequencing using reversible terminator chemistry. *Nature*. 2008;456(7218):53–59.
15. Mardis ER. Next-generation DNA, sequencing methods. *Annu Rev Genomics Hum Genet*. 2008;9:387–402.
16. Eid J, Fehr A, Gray J, et al. Real-time DNA sequencing from single polymerase molecules. *Science*. 2009;323(5910):133–138.
17. Rothberg JM, Hinz W, Rearick TM, et al. An integrated semiconductor device enabling non-optical genome sequencing. *Nature*. 2011;475(7356):348–352.
18. Yang Y, Liu R, Xie H, et al. Advances in nanopore sequencing technology. *J Nanosci Nanotechnol*. 2013;13(7):4521–4538.
19. Bamshad MJ, Ng SB, Bigham AW, et al. Exome sequencing as a tool for Mendelian disease gene discovery. *Nat Rev Genet*. 2011;12(11):745–755.
20. Choi M, Scholl UI, Ji W, et al. Genetic diagnosis by whole exome capture and massively parallel DNA sequencing. *Proc Natl Acad Sci U S A*. 2009;106(45):19096–19101.
21. Ng SB, Bigham AW, Buckingham KJ, et al. Exome sequencing identifies MLL2 mutations as a cause of Kabuki syndrome. *Nat Genet*. 2010;42(9):790–793.
22. Ng SB, Buckingham KJ, Lee C, et al. Exome sequencing identifies the cause of a mendelian disorder. *Nat Genet*. 2010;42(1):30–35.
23. Bilgüvar K, Oztürk AK, Louvi A, et al. Whole-exome sequencing identifies recessive WDR62 mutations in severe brain malformations. *Nature*. 2010;467(7312):207–210.
24. Otto EA, Hurd TW, Airik R, et al. Candidate exome capture identifies mutation of SDCCAG8 as the cause of a retinal-renal ciliopathy. *Nat Genet*. 2010;42(10):840–850.
25. Haack TB, Danhauser K, Haberberger B, et al. Exome sequencing identifies ACAD9 mutations as a cause of complex I deficiency. *Nat Genet*. 2010;42(12):1131–1134.
26. Gilissen C, Arts HH, Hoischen A, et al. Exome sequencing identifies WDR35 variants involved in Sensenbrenner syndrome. *Am J Hum Genet*. 2010;87(3):418–423.
27. Ng SB, Turner EH, Robertson PD, et al. Targeted capture and massively parallel sequencing of 12 human exomes. *Nature*. 2009;461(7261):272–276.
28. Collins FS. Positional cloning: let's not call it reverse anymore. *Nat Genet*. 1992;1(1):3–6.
29. Ku CS, Polychronakos C, Tan EK, et al. A new paradigm emerges from the study of de novo mutations in the context of neurodevelopmental disease. *Mol Psychiatry*. 2013;18(2):141–153.
30. Gibson WT, Hood RL, Zhan SH, et al. Mutations in EZH2 cause Weaver syndrome. *Am J Hum Genet*. 2012;90(1):110–118.

31. Harakalova M, van Harssel JJ, Terhal PA, et al. Dominant missense mutations in ABCC9 cause Cantú syndrome. *Nat Genet*. 2012;44(7):793–796.
32. Michaelson JJ, Shi Y, Gujral M, et al. Whole-genome sequencing in autism identifies hot spots for de novo germline mutation. *Cell*. 2012;151(7):1431–1442.
33. Simons C, Wolf NI, McNeil N, et al. A de novo mutation in the β-tubulin gene TUBB4A results in the leukoencephalopathy hypomyelination with atrophy of the basal ganglia and cerebellum. *Am J Hum Genet*. 2013;92(5):767–773.
34. Saitsu H, Nishimura T, Muramatsu K, et al. De novo mutations in the autophagy gene WDR45 cause static encephalopathy of childhood with neurodegeneration in adulthood. *Nat Genet*. 2013;45(4):445–449 9e1.
35. Tsuji S. Genetics of neurodegenerative diseases: insights from high-throughput resequencing. *Hum Mol Genet*. 2010;19(R1):R65–R70.
36. Manolio TA. Genomewide association studies and assessment of the risk of disease. *N Engl J Med*. 2010;363(2):166–176.
37. Manolio TA, Collins FS, Cox NJ, et al. Finding the missing heritability of complex diseases. *Nature*. 2009;461(7265):747–753.
38. Goldstein DB, Allen A, Keebler J, et al. Sequencing studies in human genetics: design and interpretation. *Nat Rev Genet*. 2013;14(7):460–470.
39. Foo JN, Liu JJ, Tan EK. Whole-genome and whole-exome sequencing in neurological diseases. *Nat Rev Neurol*. 2012;8(9):508–517.
40. Sidransky E, Nalls MA, Aasly JO, et al. Multicenter analysis of glucocerebrosidase mutations in Parkinson's disease. *N Engl J Med*. 2009;361(17):1651–1661.
41. Goker-Alpan O, Schiffmann R, LaMarca ME, Nussbaum RL, McInerney-Leo A, Sidransky E. Parkinsonism among Gaucher disease carriers. *J Med Genet*. 2004;41(12):937–940.
42. Mitsui J, Matsukawa T, Ishiura H, et al. Mutations in COQ2 in familial and sporadic multiple-system atrophy. *N Engl J Med*. 2013;369(3):233–244.
43. Cruchaga C, Karch CM, Jin SC, et al. Rare coding variants in the phospholipase D3 gene confer risk for Alzheimer's disease. *Nature*. 2013;505:550–554.
44. Lupski JR, Reid JG, Gonzaga-Jauregui C, et al. Whole-genome sequencing in a patient with Charcot–Marie–Tooth neuropathy. *N Engl J Med*. 2010;362(13):1181–1191.
45. Senderek J, Bergmann C, Stendel C, et al. Mutations in a gene encoding a novel SH3/TPR domain protein cause autosomal recessive Charcot–Marie–Tooth type 4C neuropathy. *Am J Hum Genet*. 2003;73(5):1106–1119.
46. Yang Y, Muzny DM, Reid JG, et al. Clinical whole-exome sequencing for the diagnosis of mendelian disorders. *N Engl J Med*. 2013;369(16):1502–1511.
47. Stenson PD, Mort M, Ball EV, et al. The human gene mutation database: 2008 update. *Genome Med*. 2009;1(1):13.
48. Abecasis GR, Auton A, Brooks LD, et al. An integrated map of genetic variation from 1,092 human genomes. *Nature*. 2012;491(7422):56–65.
49. Richards CS, Bale S, Bellissimo DB, et al. ACMG recommendations for standards for interpretation and reporting of sequence variations: revisions 2007. *Genet Med*. 2008;10(4):294–300.
50. Green RC, Berg JS, Grody WW, et al. ACMG recommendations for reporting of incidental findings in clinical exome and genome sequencing. *Genet Med*. 2013;15(7):565–574.
51. Sharp AJ, Cheng Z, Eichler EE. Structural variation of the human genome. *Annu Rev Genomics Hum Genet*. 2006;7:407–442.
52. Mitsui J, Takahashi Y, Goto J, et al. Mechanisms of genomic instabilities underlying two common fragile-site-associated loci, PARK2 and DMD, in germ cell and cancer cell lines. *Am J Hum Genet*. 2010;87(1):75–89.
53. Lupski JR, de Oca-Luna RM, Slaugenhaupt S, et al. DNA duplication associated with Charcot–Marie–Tooth disease type 1A. *Cell*. 1991;66(2):219–232.
54. Maeda MH, Mitsui J, Soong B-W, et al. Increased gene dosage of myelin protein zero causes Charcot–Marie–Tooth disease. *Ann Neurol*. 2012;71(1):84–92.
55. Inoue K, Osaka H, Sugiyama N, et al. A duplicated PLP gene causing Pelizaeus-Merzbacher disease detected by comparative multiplex PCR. *Am J Hum Genet*. 1996;59(1):32–39.
56. Padiath QS, Saigoh K, Schiffmann R, et al. Lamin B1 duplications cause autosomal dominant leukodystrophy. *Nat Genet*. 2006;38(10):1114–1123.
57. Singleton AB, Farrer M, Johnson J, et al. alpha-Synuclein locus triplication causes Parkinson's disease. *Science*. 2003;302(5646):841.
58. Rovelet-Lecrux A, Hannequin D, Raux G, et al. APP locus duplication causes autosomal dominant early-onset Alzheimer disease with cerebral amyloid angiopathy. *Nat Genet*. 2006;38(1):24–26.
59. Mirkin SM. Expandable DNA, repeats and human disease. *Nature*. 2007;447(7147):932–940.
60. Jiao X, Zheng X, Ma L, et al. A benchmark study on error assessment and quality control of CCS reads derived from the PacBio RS. *J Data Min Genom Proteomics*. 2013;4(3):16008.
61. Stuppia L, Antonucci I, Palka G, Gatta V. Use of the MLPA assay in the molecular diagnosis of gene copy number alterations in human genetic diseases. *Int J Mol Sci*. 2012;13(3):3245–3276.
62. Warner JP, Barron LH, Goudie D, et al. A general method for the detection of large CAG repeat expansions by fluorescent PCR. *J Med Genet*. 1996;33(12):1022–1026.
63. Korf BR, Rehm HL. New approaches to molecular diagnosis. *JAMA*. 2013;309(14):1511–1521.
64. Janssen B, Hartmann C, Scholz V, Jauch A, Zschocke J. MLPA analysis for the detection of deletions, duplications and complex rearrangements in the dystrophin gene: potential and pitfalls. *Neurogenetics*. 2005;6(1):29–35.
65. Scarciolla O, Brancati F, Valente EM, et al. Multiplex ligation-dependent probe amplification assay for simultaneous detection of Parkinson's disease gene rearrangements. *Mov Disord*. 2007;22(15):2274–2278.
66. Buchman VL, Cooper-Knock J, Connor-Robson N, et al. Simultaneous and independent detection of C9ORF72 alleles with low and high number of GGGGCC repeats using an optimised protocol of Southern blot hybridisation. *Mol Neurodegeneration*. 2013;8:12.
67. Wafaei JR, Choy FY. Glucocerebrosidase recombinant allele: molecular evolution of the glucocerebrosidase gene and pseudogene in primates. *Blood Cells Mol Dis*. 2005;35(2):277–285.

CHAPTER

7

Association, Cause and Causal Association: Means, Methods and Measures

Walter A. Kukull

University of Washington, Seattle, WA, USA

The motivation to examine associations between single or multiple factors and the occurrence or progression of health events has often been driven by the desire to better the health of individuals or to affect the public health in general. As such, the enterprise is equally at home in the basic science laboratory as it is in observing population level phenomena and chronicling their occurrences. The approaches to establish "causal criteria" to evaluate such associations have, in large part, been pragmatic rather than philosophical. However, philosophical logic is woven into the pragmatic fabric of the task of characterizing observed associations as potentially "causal" ones.

Russell considered that a definition of causality might be stated as a principle: "Given any event e_1 and a time-interval t, such that whenever e_1 occurs, e_2 follows after an interval t."[1] (p. 4). He argues, however, that the range of events and the variability of time interval length make the statement as a "law of causality," somewhat absurd, or at least impractical. Evidence of this is seen in his often-quoted remark that, "the law of causality, like much of what passes muster among philosophers, is a relic of a bygone age, surviving, like the monarchy, only because it is supposed to do no harm"[1] (p. 1). Usually, however, as medical researchers we are not called to the metaphysics involved in constructing immutable laws that apply to the dynamically elegant universe. Our lot, although perhaps more mundane, is actually no less important. Our charge is to design and conduct scientific studies in a rigorous enough fashion that from the evidence we amass we might infer factors that might help us explain or modify the occurrence and course of disease and health events. We, collectively, tend to navigate that path with the equivalent of a parkinsonian shuffle, unfortunately.

LEARNING FROM INFECTIOUS DISEASE

The efforts to describe causal associations or to ascribe cause to a particular agent or factor have seemed to diverge based on whether studies were focused on putatively infectious agents or examination of "chronic diseases" (perhaps those for which no infectious basis has yet been found). Koch's Postulates, as they are frequently called, are attributed to both Robert Koch and Jakob Henle.[2-4] These postulates were developed primarily to aid the search for and correct recognition of microbial infectious agents, as well as to "convince skeptics that microorganisms could cause disease …".[3] Briefly, there are three postulates: "1) The parasite occurs in every case of the disease in question and under circumstances which can account for pathological changes and clinical course of the disease. 2) It occurs in no other disease as a fortuitous and nonpathogenic parasite. 3) After being fully isolated from the body and repeatedly grown in pure culture, it can induce the disease anew."[2] While the postulates were highly regarded, they suffered some criticism, notably because it did not allow for a carrier state, or did not account for use of an inappropriate animal model in which to reproduce the infection (i.e., the agent would not induce disease in all models).[2] However, over the years, with some occasional clarifications or additions, Koch's Postulates have remained quite central to describing pathogens that cause disease.

Rosenberg's Molecular and Genetic Basis of Neurological and Psychiatric Disease
http://dx.doi.org/10.1016/B978-0-12-410529-4.00007-3

With the evolution of new technology making it possible to characterize not only gross pathogens but also their molecular and genetic constituents, Falkow offered his "Molecular Koch's Postulates" in 1988,[3] to correspond directly to Koch and Henle's original ones

> "1. The phenotype or property under investigation should be associated with pathogenic members of a genus or pathogenic strains of a species;
>
> 2. Specific inactivation of the gene(s) associated with the suspected virulence trait should lead to a measurable loss in the pathogenicity of virulence, or the gene(s) associated with the supposed virulence trait should be isolated by molecular methods. Specific inactivation or deletion of the gene(s) should lead to loss of functions in the clone;
>
> 3. Reversion or allelic replacement of the mutated gene should lead to restoration of the pathogenicity, or the replacement of the modified gene(s) for its allelic counterpart in the strain of origin should lead to loss of function and loss of pathogenicity or virulence. Restoration of the pathogenicity should accompany the reintroduction of the wild-type gene(s)" ***Falkow, 2004[3] (p. 68).***

In revisiting his Molecular Koch's Postulates, Falkow described also how the postulates might apply or be adapted to establishing specific causes through the use of "knockout" mice or to expand them to DNA microarrays and studies or RNA interference (RNAi) (i.e., blocking of gene expression by double-stranded RNA). The point here is not to show all the specific instances where Koch's Postulates or their modifications might be in use today, but rather to counter the notion often harbored by introductory epidemiology textbooks that Koch's Postulates are a relic of the past, important in their time, as was John Snow, but deficient in their ability to describe the occurrence or causes of (chronic) disease. At the basic science/molecular level these postulates are, in fact, active and consistent with modern technology, and useful in helping scientists to describe very specific causal relationships that will project to the large stage of public health and advance our understanding of not only known infectious diseases, but also cancers, and potentially other diseases not currently recognized as due to infectious agents.

CAUSAL "GUIDELINES" AND OBSERVATIONAL VS. EXPERIMENTAL DESIGNS

The applicability of Koch's Postulates and their continuing application in the molecular genetic arena are likely due to experimental design. In the laboratory, experiments may be accomplished in which the investigator has control of a specific experimental maneuver as well as its assignment to his unwitting subjects (who for the most part may be small animals, cells, or viruses). However, the public's trust of scientists is not so great that they would volunteer in great numbers to undergo a toxic exposure, which could result in a fatal disease, at the request of a scientific investigator. It happens, however, that the public routinely exposes itself, often in great numbers, to all manners of potentially toxic agents and behaviors. Epidemiologists have recognized this odd fact for some time and have endeavored to design studies that will allow inferences about "causes" to be drawn from them. The arena in which they must work is much more crowded and chaotic than is the basic science laboratory. Each person may have a large complex of "exposures" (including genetic and environmental, behavioral and occupational, social and demographic), and some of the data points of interest may occur in greater frequency among persons who develop a given disease. Some exposures may play a biological role in the initiation or promotion of a disease process, while others may be more frequent but actually have no role at all in the disease process. Causal factors and noncausal ones may be highly correlated or significantly associated, adding further analytic confusion. Correct identification of valid causal factors could lead to interventions that would modify the occurrence of disease. Some factors (e.g., age and sex) may not be modifiable in a way that would modify disease occurrence, but they still may be important in order to identify high-risk groups for more intense, focused study. Galloping into this arena, on aging draft horses, come the epidemiologists, hoping to bring some order to the chaos, and perhaps also some manner of benefit to the public health.

Observational studies are perhaps no more difficult to do than experimental studies, but they are more difficult to do well, and to do in a way that will allow causal conclusions to be drawn from them. Experiments (e.g., clinical trials, lab studies) have the great benefit of "randomization" in the assignment of the exposure, treatment or maneuver under study, thus tending to create comparison groups that have an equal mix of other personal characteristics and prior exposures, except for the experimental treatment. In an experimental setting, it is more logical and less risky than in an observational design to ascribe differences in outcome to the experimental treatment. In experimental studies, randomization is often viewed as the cure-all for the potential effects of both measured and unmeasured confounders, and in an ideal world it can be. Frequently, lack of compliance with the experimental treatment assigned, as characterized by "always-takers, never-takers and defiers (those who take other than the assigned treatment)"[5] causes the "complier average causal effect" of a treatment to become distorted in comparison with the result of the usual intent-to-treat analysis. Little and Rubin show that an unbiased estimate of the "complier average causal effect" can be obtained by dividing the "intent-to-treat" estimate by the difference in proportions actually adopting the experimental treatment, in treatment and control groups. This technique is often said to be a formulating an "instrumental variable" estimator.[5] Instrumental variable estimators can be created to adjust for confounding in both experimental and observational studies. These instrumental variables can ideally be a powerful tool in sharpening the estimate of causal effect, but they must be applied with a great deal of caution and understanding of mathematical assumptions associated with them.[6]

Basic observational designs include: 1) "case-control" studies, where groups of persons a) with the disease in question, and, b) without the disease, are compared for their history of an exposure of interest, occurring prior to disease onset; and 2) the "Cohort study" design, where groups with and without exposures of interest are determined, among persons *without* the disease, and then exposure groups are followed longitudinally to observe the occurrence of the disease. Naturally there are also much more complicated hybrids from these basic designs incorporating matching and other techniques,[7] but these should serve for the purpose of discussion here. Both attempt to emulate experimental designs in that they strive to equalize extraneous factors between comparison groups, both in the design and in the analysis phases, but obviously cannot randomize subjects to disease or exposure groups. Because the way subjects are ascertained, agree to participate, provide information, and are followed may be associated with the exposures and outcomes of interest, the resulting bias and confounding can obscure a true association or produce a spurious one. Thus, inferring that a greater exposure frequency observed among cases indicates that the exposure is causal becomes quite tenuous.

Around the time that many scientists were attempting to determine whether cigarette smoking might contribute to causing lung cancer, Hill was acutely aware of the practical shortcomings of causal inference ability from observational designs, as well as the lack of direct applicability of Koch's Postulates to the problem.[8] He provided not a philosophical discussion, but some pragmatic guidelines against which the results of observational studies might be measured and then judged with respect to the causal nature of an association observed. Hill's "criteria" are shown in Table 7.1.

Hill adds that "No formal tests of significance can answer those questions. Such tests can, and should, remind us of the effects that the play of chance can create, and they will instruct us in the likely magnitude of those effects. Beyond that they contribute nothing to the 'proof' of our hypothesis."[8] (p. 299). The implication is that significance tests tell us little about the validity or causality of the result in an observational setting. Rather, careful evaluation of how the design, conduct, and analysis of the study were carried out, coupled with an examination on the guidelines posited, should allow us to infer whether an observed association might be causal. Hill's criteria have persisted for almost 50 years now, though they have also been adopted and adapted by many researchers involved in observational studies. When considering diseases with many potential causes, perhaps through different mechanisms or pathways, and causes that may be independent, synergistic, or antagonistic, necessary or sufficient or not, Hill's guidelines still provide some method for our inference of causality.

While some adaptations and interpretations of Hill's criteria have led to more complex and convoluted scenarios, Koepsell and Weiss offer a simplified version giving preference to data from human beings and placing experimental studies ahead of nonrandomized studies, as associations more likely seen to be "causal" in nature.[9] But for "nonrandomized studies in human beings" that show an association between an exposure and disease, they instruct that before inferring that the observed association may be causal, one should evaluate evidence concerning the following: "1) The suspected cause precedes the presence of disease; 2) the association is strong; 3) there is no

TABLE 7.1 Hill's Criteria: Aspects of an Association that Should be Considered Before Interpreting it to be Causal[8]

1. Strength of the association (e.g., magnitude of the relative risk estimate)
2. Consistency (Has a similar association occurred in other independent studies?)
3. Specificity of the association
4. Temporality (Did the putative exposure actually occur before biologic onset of the disease?)
5. Biological gradient (Is there evidence of a dose–response relationship?)
6. Plausibility (From what is known of the biology, is it plausible the exposure may have led to disease – obviously dependent on biological knowledge, hence tenuous)
7. Coherence (i.e., does not conflict with generally known facts of disease biology)
8. Experiment (Is there evidence from an experimental study, e.g., randomized controlled trial?)
9. Analogy (Is there similar evidence from an analogous situation?)

plausible noncausal explanation that would account for the entirety of the association, and a plausible explanation for the association's being a causal one; 4) the magnitude of the association is strongest when it is expected to be so"[9] (p. 182). This reformulation also carries with it some important generalizations about causes. First, it relaxes the priority of direct over indirect causes of disease by saying that an exposure should be considered a cause if except "for the presence of the exposure, some additional cases of the disease would not occur." Thus, the causal web may be large, and if the exposure were accountable for at least some disease, then removal of that "cause" would potentially reduce disease occurrence. This notion is consistent with Joffe's explanation: "A causal relationship is one that has a mechanism which by its operation makes a difference."[10] Koepsell and Weiss continue that such causal exposures inferred to be causal need not be present in every case, nor is it necessary that a single such exposure be capable of producing the disease on its own – i.e., the exposure presence may contribute to disease occurrence that requires other exposures as well does not detract from its appreciation as a cause.[9]

Unfortunately, the simple comparison of an association against any of the causal guidelines or criteria already mentioned will not in itself allow us to always and directly infer causality. As with most academic settings, there are always prerequisites. In the case of observational studies (and experimental studies, as mentioned earlier), great care must be taken in the design, conduct, and analysis. Problems with selection of subjects (including self-selection and loss to follow-up) can generate a spurious association between exposure and disease, is often of substantial magnitude and statistical significance. Thus, critical appraisal of the design and analysis should always be in the forefront of the researcher's mind. Further, how accurately and comparably the information is obtained from all subjects can also create bias. No matter how "exposure" or treatment data are collected, they should be scrutinized for the possibility that more complete or valid data might be gathered from one group than the other. Inaccurate data could be obtained due to subject recall, but could also result from variability of biological measurements due to different assays, diurnal variation, or perhaps how biologic specimens were obtained differentially by group. Such study conduct "mistakes" produce invalid, although often strong, associations. These mistakes are often called bias, and in epidemiology they tend to fall into two categories: 1) selection bias and 2) information bias. Biases such as these are considered to be active only to the extent that they would change the magnitude of the association from its true value. Both selection and information biases can often be quite subtle, but are serious threats to study validity because they can seldom be "adjusted for" in the analysis phase.

More familiar in most research parlance is "confounding," often confused with the biases mentioned above but fundamentally different. Confounding is based on valid data collection – not mistakes as with the biases mentioned above. Confounding factors are legitimate other potential risk factors or causes of the disease or health event being studied; confounding cannot occur due to factors that are not also risk factors for disease.[11] Because these confounders have their own association with disease, if they are also unequally distributed among the comparison groups, their presence can change and obscure the true association between the exposure of interest and disease, and thereby alter our assessment of potential causality. The damage that could be caused by reckless oversight of confounders can be avoided in the design phase of the study. If potential confounders are recognized and captured, they can be statistically adjusted for in the analysis – so long as accurate and valid data on the potential confounders are obtained. Classic methods of restriction and stratification have given way to regression modeling as means of removing or adjusting-for the effects of confounding. However, these techniques should not be applied without the investigator's careful appraisal of the potential mode of association between the supposed confounder and both the exposure and disease. Factors associated with only exposure *or* disease cannot confound the association of interest; adjusting for such factors is inefficient and unnecessary. Adjusting for factors that might be in the causal chain between exposure and disease should not be done, as it will reduce the association of interest toward the null. Purely statistical criteria for entering variables into a model may sometimes result in noncausal factors being substituted for causal ones in the modeling process, because of co-linearity. Thus, there is no effective substitute for careful thought in the examination of potential confounders by the researcher.

Frequently applied methods of adjusting for confounding include the well-known multivariable regression models, such as logistic regression, proportional hazards regression, and Poisson regression. Becoming more common is the use of "propensity scores"[12] (derived from regressing the "exposure" on the other covariates or potential confounders). The derived "score" can then be used in the analysis as a weight or possibly to form strata.[5,7] Linear and nonparametric structural equations in combination with graphical Bayesian networks and counterfactual methods[13,14] have also been used to represent causal effects and to create "instrumental variables" that claim to facilitate adjustment for both measured and unmeasured confounders – both in experimental and observational studies.[5,6,15] As Hernan cautions, however, the claims for such an ideal confounding adjustment through the use of instrumental variable estimation in observational studies may be relatively exaggerated.[6] While all of these methodological tools provide us with additional information on which to base our inference of causality, none apparently does so automatically without our own judgment of the evidence.

An extension of the instrumental variable approaches mentioned above, which is thought to alleviate the confounding ills of observational studies through incorporation of genetic data, has been termed "Mendelian randomization."[16–21] The general idea behind the concept of Mendelian randomization came from Katan who, in about 1986, was struggling with ways to answer the question of whether low serum cholesterol was in fact a risk factor for cancer.[16] The standard approaches to the question seemed to be caught in the loop of confounding and the potential for reverse causation (i.e., low cholesterol levels came about because of cancer). He recognized that the apolipoprotein E gene had three functional allelic forms and that each of those forms produced different levels of serum cholesterol. Assuming then that in a relatively large population APOE gene would assort roughly according to Mendelian laws, and at the same time be unaffected by confounding factors or reverse causation (which confused the study of actual serum cholesterol levels), he opined that using APOE cholesterol phenotype (due to allelic genotype) as an instrumental variable in a case-control study with which to associate cancer would provide an "unconfounded" test of the hypothesis that would "be just as good as measuring it in a large number of newborns and following them to see who developed the disease and who did not"[16] (p.10). He did not complete the study, however, perhaps due to the lack of modern technology as well as the large number of subjects who would need to be processed. Nevertheless, the concept survived, and was distinguished from the usual "candidate gene"–disease association studies in that the functional forms of the genes were seen as a proxy measure for levels of circulating gene products. The inference then was on the projected levels rather than the gene itself. This would imply a potential for intervention on the circulating level if it were associated with disease and an appropriate modifier were available.

Exemplary of recent Mendelian randomization studies is one by Holmes et al.[20] They conducted a study of high-density lipoprotein cholesterol (HDL-C) and triglycerides on coronary heart disease (CHD). Rather than the single-gene system imagined by Katan, Holmes et al. used a meta-analysis of 17 studies that included over 62,000 subjects with 12,099 CHD events, and included single-nucleotide polymorphisms (SNPs) for HDL-C (48 SNPs), triglycerides (67 SNPs), and low-density lipoprotein cholesterol (LDL-C) (42 SNPs). Where the SNPs were associated with more than one lipid, both independent and nonindependent SNP analyses were conducted, and while they were able to report a potential causal association for CHD with LDL-C and triglycerides (as have been shown by numerous other study designs), they were unable to do more than suggest a "possible" effect of (low) HDL-C. It is quite obvious then that the complexity and volume of the required analysis is quite large, although this may now be typical of the Mendelian randomization approach and perhaps also necessary for it to provide any results. There is no doubt that studies requiring the technological advances employed by Holmes et al.[20] would not have been possible relatively few years ago, and will likely be replaced by even greater advances in the near future. Large subject numbers and high levels of technology (genotyping) appear to be a prerequisite for the success of this method. It may well be a useful complement to the classically designed randomized trials and epidemiologic studies with which we are most familiar.

THE FUTURE?

To quote Ebrahim, "The ability of observational datasets to generate spurious associations of nongenetic exposures and outcomes is extremely high, reflecting the correlated nature of many variables and the temptation to publish such findings must rise as the p-values for the associations get smaller"[22] (p. 363). Obviously this would also have implications for our ability to infer causal associations as well. The above view should serve to motivate those who do observational studies to do better, or it may just as well motivate those who fund such studies to throw the baby out with the bathwater. There are two very important articles concerned with strengthening reporting of observational studies in epidemiology (STROBE)[23] and also with strengthening reporting for genetic association studies (STREGA).[24] STROBE shows a checklist of about 22 items that should be addressed in each section of a research manuscript (e.g., introduction, methods, results) to provide the evidence on which a reader might base a judgment concerning the validity as well as the causal nature of an association. STREGA expands the STROBE checklist by including genetics-specific items in each section. Author adherence to these checklists when writing papers should be of great value to readers if the checklists are widely used. However, the checklists really address only the reporting of studies, not the design, conduct, and analysis – except that it may be embarrassing for authors when writing their checklist-based manuscripts to have blanks by many items! Causal inference, regardless of the design and technical details, still requires sound evidence on which to make that judgment.

The legacy of John Snow (the anesthesiologist-turned-epidemiologist who correctly deduced the "cause" of London's cholera epidemics in the mid-1800s) was the subject of a recent conference reported in The Lancet.[25] Scientists from various disciplines considered how Snow's dogged determination to compile rather hum-drum statistics led to a solid causal inference, even before the organism *Cholera vibrio* was characterized. Today we still benefit by "better" data collection. Today also there is still debate over which underlying philosophy best serves the application of "causal inference," without apparent resolution. However, there does appear to be some recognition of the gaps as well as the interactions between Koch's Postulates[2] and Hill's criteria[8] when considering diseases we face today. Failure to effect change in the public health despite solid causal inference concerning the basic mechanisms and transmission of a disease such as cholera is evident in Haiti, as well as in Africa for human immunodeficiency virus (HIV) diseases.[25] Perhaps the studies and models we have been striving to accomplish are less effective at making a difference than we usually hope or even care to believe.

We have discussed the molecular and the pragmatic, the philosophical and the mathematical approaches to causal inference earlier in this chapter. There are at once distinct and overlapping parts to each, with no clear resolution that one method or tool is clearly superior. Against a similar backdrop, Glass et al. discuss causal inference needs and projections in public health, placing in context many of the tools we have mentioned, but at the same time aiming toward a more global approach to interventions that may have the power to effect needed changes for the better in public health.[26] An ongoing challenge for implementing the knowledge we obtain through epidemiologic studies concerns how to intervene when the characteristics we find as causal risk factors either are unable to be manipulated or reflect larger societal ills. If we find that higher cancer rates or obesity rates are causally associated with poverty, what should the intervention be? It would be nice to eradicate poverty, of course. But perhaps we should continue to examine within poverty to see if there are components of it more proximal to disease and perhaps also ones that can be modified by intervention. Certainly this fits with our earlier consideration of component causes that "make a difference"[10] or "but for their presence, some additional cases would not occur."[9] Small changes and larger societal changes are all consistent with the reformulated framework described by Glass et al.[26] Their framework would span geographic and timescales, locating agents, processes, and outcomes within the multidimensional schema – a "complex systems approach." Through this framework and approach, Glass argues that we should re-examine the ability to develop effective interventions for those potentially causal associations we discover, and to make the intervention the focus of our discussion of causal inference. In order to effectively address this broad goal, researchers must still apply the best and most rigorous means of conducting both observational and experimental studies. Further, researchers should beware of narrowing their study data collection excessively, as often the additional detail and richness afforded by casting a broader net may not only result in unexpected new factors but also could serve to capture the previously unmeasured confounders always seen to be "a troll under the bridge" to causal associations.

A broader approach is also valuable for reasons described by Dacks et al. in a recent article.[27] As we discussed earlier in reviewing causal criteria, experimental studies, also known as randomized controlled trials (RCT), are usually taken to provide a higher level of evidence for causal association when compared to observational studies.[8,9] In fact, the usual framework for the design of observational studies is to make them emulate experiments. Unfortunately, not all potential risk factors, especially deleterious ones for humans or the more global processes such as diet or education, can ethically or feasibly be assigned to Subjects in a funded RCTs, so we must rely on other designs for that information. Sometimes too, perhaps in the case of big pharma, because RCTs are quite expensive, the results might need to suggest a next step as a marketable intervention. Suppose that the observational evidence presented by Doll and Hill[28] had been required to be substantiated by an RCT before any intervention on smoking could have taken place. Would we have had the Surgeon General's Report,[29] the changes in cigarette labeling, restrictions on smoking in the workplace, or other smoking interventions? Well, it does sound unlikely, in retrospect. The key was that the observational studies really amassed the evidence with strength of association, temporality, consistency, dose–response, etc., from observational studies that were well designed and analyzed. Ultimately, even if ethical, it would have been essentially unnecessary to conduct a large and lengthy RCT. The causal evidence was just too strong for science or the public to think otherwise (but then, I guess, the same could be said about global warming, and look where we are with that!). Longitudinal studies and data sharing among scientists should also add significantly to the benefits realized from nonrandomized studies. Clearly, many of today's genome-wide association studies and other meta-analyses (e.g., Holmes et al.[20]) could not have been accomplished without collaboration and data sharing. Causal inference is dependent on many things and has many views; truly a distal perspective is necessary to properly appreciate all of its components.

References

1. Russell B. On the notion of cause. *Proc Aristot Soc.* 1912;13:1–26.
2. Evans AS. Causation and disease: the Henle-Koch postulates revisited. *Yale J Biol Med.* 1976;49(2):175.
3. Falkow S. Molecular Koch's postulates applied to bacterial pathogenicity—a personal recollection 15 years later. *Nat Rev Microbiol.* 2004;2(1):67–72.
4. Inglis TJJ. Principia ætiologica: taking causality beyond Koch's postulates. *J Med Microbiol.* 2007;56(11):1419–1422.

5. Little RJ, Rubin DB. Causal effects in clinical and epidemiological studies via potential outcomes: concepts and analytical approaches. *Annu Rev Public Health*. 2000;21(1):121–145.
6. Hernán MA, Robins JM. Instruments for causal inference: an epidemiologist's dream? *Epidemiology*. 2006;17(4):360–372.
7. Stuart EA. Matching methods for causal inference: a review and a look forward. *Stat Sci*. 2010;25(1):1.
8. Hill AB. The environment and disease: association or causation? *Proc R Soc Med*. 1965;58(5):295.
9. Koepsell TD, Weiss NS. *Epidemiologic Methods: Studying the Occurrence of Illness*. Oxford: Oxford University Press; 2003.
10. Joffe M. The gap between evidence discovery and actual causal relationships. *Prev Med*. 2011;53(4):246–249.
11. Greenland S, Morgenstern H. Confounding in health research. *Annu Rev Publ Health*. 2001;22(1):189–212.
12. Rosenbaum PR, Rubin DB. The central role of the propensity score in observational studies for causal effects. *Biometrika*. 1983;70(1):41–55.
13. Balke A, Pearl J. *Probabilistic Evaluation of Counterfactual Queries*. 2011.
14. Dawid AP. Causal inference without counterfactuals. *J Am Stat Assoc*. 2000;95(450):407–424.
15. Kleinberg S, Hripcsak G. A review of causal inference for biomedical informatics. *J Biomed Informat*. 2011;44(6):1102–1112.
16. Katan MB. Commentary: mendelian randomization, 18 years on. *Int J Epidemiol*. 2004;33(1):10–11.
17. Lawlor DA, Harbord RM, Sterne JA, Timpson N, Davey Smith G. Mendelian randomization: using genes as instruments for making causal inferences in epidemiology. *Stat Med*. 2008;27(8):1133–1163.
18. Smith GD, Ebrahim S. 'Mendelian randomization': can genetic epidemiology contribute to understanding environmental determinants of disease? *Int J Epidemiol*. 2003;32(1):1–22.
19. Didelez V, Sheehan N. Mendelian randomization as an instrumental variable approach to causal inference. *Stat Methods Med Res*. 2007;16(4):309–330.
20. Holmes MV, Asselbergs FW, Palmer TM, et al. Mendelian randomization of blood lipids for coronary heart disease. *Eur Heart J*. 2014; eht571.
21. Smith GD, Timpson N, Ebrahim S. Strengthening causal inference in cardiovascular epidemiology through Mendelian randomization. *Ann Med*. 2008;40(7):524–541.
22. Ebrahim S. Improving causal inference. *Int J Epidemiol*. 2013;42(2):363–366.
23. von Elm E, Altman DG, Egger M, Pocock SJ, Gøtzsche PC, Vandenbroucke JP. The Strengthening the Reporting of Observational Studies in Epidemiology (STROBE) statement: guidelines for reporting observational studies. *J Clin Epidemiol*. 2008;61(4):344–349.
24. Little J, Higgins JPT, Ioannidis JPA, et al. Strengthening the reporting of genetic association studies (STREGA)—an extension of the Strengthening the Reporting of Observational Studies in Epidemiology (STROBE) statement. *J Clin Epidemiol*. 2009;62(6):597–608.e594.
25. Fine P, Victora CG, Rothman KJ, et al. John Snow's legacy: epidemiology without borders. *Lancet*. 2013;381(9874):1302–1311.
26. Glass TA, Goodman SN, Hernán MA, Samet JM. Causal inference in public health. *Annu Rev Publ Health*. 2013;34:61–75.
27. Penny A. Evidence needs to be translated, whether or not it is complete. *JAMA*. 2014;71(2):137.
28. Doll R, Hill AB. The mortality of doctors in relation to their smoking habits. *Br Med J*. 1954;1(4877):1451.
29. Brawley OW, Glynn TJ, Khuri FR, Wender RC, Seffrin JR. The first surgeon general's report on smoking and health: the 50th anniversary. *CA Cancer J Clin*. 2014;64(1):5–8.

CHAPTER

Gene Therapy for Neurological Disease

Theodore Friedmann

The University of California, San Diego School of Medicine, La Jolla, CA, USA

INTRODUCTION AND RECENT PROGRESS

It has now been a little more that four decades since the first formal published statement of the need and potential for gene therapy for human disease, and it is clear that the field has made impressive and incontrovertible progress at levels of both basic research and clinical application. In some discussions, that period of time has been described to be very long and the apparent delays in implementing research concepts into clinical reality have often been attributed not only to the difficulty of the task but also to a number of mis-steps and reversals in the field of gene therapy resulting from unrealizable and exaggerated early expectations as well as ethical and logistical lapses. From a different vantage point, however, such a long development time between initial concepts, ensuing advances, setbacks and serious adverse events, and finally convincing clinical successes is not unique to gene therapy and has characterized a number of other areas of biomedicine that have become indispensible tools of modern medicine.

For instance, the earliest studies of antimetabolite-based cancer chemotherapy were carried out by Sidney Farber and his colleagues as early as 1948[1,2] with children suffering from acute lymphoblastic leukemia, but it was not until well into the 1980s that survival and even cure rates with chemotherapy for that disease began to reach the 80–90% levels that are now routinely achieved. Of course, the discovery of cancer chemotherapy came to be applied to many other forms of cancer, with variable degrees of success. In addition, the earliest human applications of bone marrow transplantation occurred in 1957 through the work of Donnal Thomas and his colleagues,[3] and for the following 20–30 years the procedure was considered by many to be destined to fail, a feeling that seemed to be supported by many early failures and clinical adverse events and setbacks. Nevertheless, to this day, bone marrow transplantation remains an indispensible cornerstone tool in the management of hematological malignancies and other blood diseases, despite the fact that the procedure is difficult and harsh and fails in many cases.

The concept of gene therapy was first formally proposed in 1972 and earliest *in vitro* proof-of-principle studies appeared in the 1980s.[4–7] Because of the enormous hope and promise inherent in this field of medicine, human clinical studies quickly began in 1989/1990. Results from initial clinical trials generally showed poor levels of gene transfer and no clinical benefit, despite a high level of expectation and a great deal of exaggerated early claims of therapeutic benefit. However, during the early years of the 21st century, gene therapy studies began to show the first incontrovertibly positive clinical results in gene therapy studies in patients suffering from inherited forms of immunodeficiency. The earliest rigorous demonstrations of efficacy of gene therapy in such disorders came through the work of Alain Fischer and his colleagues in Paris, who demonstrated that the introduction of a normal copy of a gene encoding the common γ-chain of a number of interleukin receptors into bone marrow stem cells of patients the X-linked severe combined immunodeficiency diseases (SCID) followed by transplantation resulted in reconstitution of a largely normalized immune system and resulting correction of the immune deficiency phenotype of life-threatening infections.[8–12] In some patients, the therapeutic effect has now been stably maintained for 10–12 years and most patients have remained clinically well and immunologically intact.

Rosenberg's Molecular and Genetic Basis of Neurological and Psychiatric Disease
http://dx.doi.org/10.1016/B978-0-12-410529-4.00008-5

Unfortunately, the mechanisms of molecular and clinical efficacy were associated with the development of lymphocytic leukemia in some of the patients because of the integration of the gene transfer vectors into or near oncogenes in the patients' genome.[13,14] The system is being studied very actively to learn how to avoid such insertional mutagenesis events. To make up for these problems, a conceptually similar approach has been taken with children suffering from a version of SCID that results from deficiency of the enzyme adenosine deaminase (ADA), and in a large number of such patients, functionally complete correction of the immune deficiency has been accomplished without the appearance of leukemia or any other serious adverse events.

Although gene therapy for SCID is still considered experimental, it is safe to say that most of these SCID patients have been very effectively treated by gene therapy and that this new therapeutic method has begun to rival and, even in the case of ADA-SCID, surpass existing therapy.[15–17] Some would even be forgiven for arguing that in communities where appropriate research institutions are available to carry out such procedures, gene therapy may even be considered to have come very close to becoming the "treatment of choice."

Another clinical setting in which gene therapy has proven itself very effective at the experimental level is in the area of degenerative retinal disease and some forms of congenital blindness. The most impressive therapeutic result of gene therapy in eye disease has come with a form of congenital blindness called disease Leber amaurosis. There are many clinically similar but genetically distinct forms of this disorder,[18] one form of which has provided one of the most impressive examples of truly successful gene therapy. This version of Leber amaurosis is caused by deficiency in retinal pigmented epithelial cells of the gene that encodes RPE65 and whose expression is required to convert all-*trans* retinol to 11-*cis* retinal during phototransduction, which is then used to regenerate visual pigment in photoreceptor cells. If that process is defective, photoreceptor function is impaired, photoreceptor cells degenerate, thereby causing blindness. Gene therapy in this disorder has been carried out, initially by J. Bennett and colleagues at Children's Hospital of Philadelphia, and additionally in several other clinics by using an adenovirus-associated gene transfer vector to introduce a normal copy of the RPE65 gene directly into the RPE layer of the retina.[19–21] The procedure successfully restored RPE65 expression and normal visual pigment turnover in sufficient numbers of photoreceptors to lead to the restoration of useful vision and to protect photoreceptor cells from progressive photoreceptor degeneration. Without doubt, the same approach will be taken for many of the other forms of Leber amaurosis and for many additional retinal degenerative disorders such as retinitis pigmentosa, macular degenerative syndromes, and others. Even more so than in the case of the SCID syndromes described above, successful treatment and vision restoration in these diseases represent the only therapeutic option available to these patients and, therefore, should rapidly become standard of care, provided of course that potential safety concerns are thoroughly examined and resolved.

This review is intended not as an exhaustive summary of all current approaches to gene therapy of central nervous system (CNS) disease, but rather to highlight some recent studies that have been particularly informative about the possibility that gene therapy will fill a niche in the collection of therapies for some developmental, degenerative, and neoplastic diseases of the CNS.

PROGRESS IN GENE THERAPY FOR NEURODEVELOPMENTAL AND NEURODEGENERATIVE DISEASE

It is paradoxical that what seems to have been an enormous growth in our understanding of neurogenetics, neurophysiology, and brain function during the past several decades has not be accompanied by a comparable increase in the efficacy of treatment of most forms of CNS disease. The reasons for such a discrepancy include many factors, including the fact that the brain and other parts of the nervous system are immensely complicated structures, more so than other organs in the body, and while we now have a superficial understanding of that complexity, we do not yet have very many genetic, pharmacological, or neurophysiological tools available to modify and correct most aberrant neural functions to treat neurodevelopmental, neurodegenerative, or neuropsychiatric disorders. Fortunately, these common and devastating neurological diseases exist in rare familial Mendelian forms in which responsible single genes are well recognized. It seems likely that eventually these genetically "simpler" variants of disease offer many opportunities for a better understanding of pathogenesis and potential therapy.

Nevertheless, because of the dire nature of most of these diseases and the lack of effective alternatives, gene therapy clinical trials have been carried out in patients with a variety of neurodevelopmental and neurodegenerative diseases. In some cases, results, even in early phase I/II studies, have been encouraging and give reason to expect that future technical advances will finally make even these very difficult and intractable disorders available for therapy. A number of reviews have been published recently on overall progress toward gene therapy for neurological diseases.[22,23]

Mendelian Neurodevelopmental and Metabolic Diseases

Because of their relative genetic simplicity, single-gene Mendelian disorders often offer the best opportunities to understand genetic mechanisms underlying disease pathogenesis, and recent clinical gene therapy trials with several metabolic neurological disorders have provided convincing evidence of clinical benefit through gene therapy.

Canavan Disease

Canavan disease was one of the first CNS disorders subjected to gene therapy clinical trials and differs from some of the later studies in that it has been approached by direct *in-vivo* delivery of the potentially therapeutic gene transfer vector rather than by an *ex-vivo* approach followed by grafting genetically corrected cells into the circulation or into the CNS.[24–27] This gene therapy model has given evidence that such a direct genetic reconstitution approach can be safe but, despite early claims of clinical benefits in treated patients, the therapeutic consequence of this form of gene therapy in this disease still awaits rigorous confirmation. Canavan disease is a spongiform white matter degenerative leukodystrophy and psychomotor disorder caused by mutations in the aspartoacylase gene (ASPA), defective expression of ASPA enzymatic activity, and resulting accumulation of the enzyme substrate *N*-acetyl-aspartate (NAA). The disease has been an early target for *in-vivo* gene therapy, i.e., direct injection of an adenovirus-associated virus vector expressing the normal ASPA gene (AAV2-ASPA) into the brain. Clinical trials have been interpreted to demonstrate the safety of such a procedure but the quality of clinical improvement has been somewhat more difficult to corroborate. Injection of gene transfer vector into multiple affected sites was reported to lead to decreased levels of the accumulated NAA in the brain and imaging studies have suggested slowed progression of brain atrophy. Furthermore, there have been suggestions of reduced seizure activity and improved overall clinical status. These studies await confirmation.

X-Linked Adrenoleukodystrophy

X-linked adrenoleukodystrophy (X-ALD) is a demyelinating childhood lysosomal storage disorder caused by a deficiency of the ATP-binding cassette transporter ALD protein encoded by the ABCD1 gene. The only current proven effective therapy for this disorder is allogenic bone marrow transplantation, which can, in some cases, stabilize, but not reverse, the clinical disorder by halting progression of the demyelination process. In the past several years, a small clinical study of gene therapy has demonstrated what seems to be a clinically significant stabilization of the clinical status of two patients resulting from an apparent lack of progression of demyelination. In this study, a lentivirus vector was used to introduce a normal copy of the ABCD1 cDNA into autologous peripheral blood CD34+ cells from two ALD patients, and the genetically corrected cells were returned to the circulation of the patients after myeloablation to make room for the incoming cells.[28,29] Both patients demonstrated reduced and possibly arrested demyelination up to 14–16 months following treatment, presumably through the re-population of the CNS with genetically corrected hematopoietic stem cell-derived brain microglia. The degree of arrested demyelination was comparable to that obtained by bone marrow transplantation. While the study is very small and very early in its course, and while long term safety and efficacy remain to be established, this study has represented a very important demonstration of the value of gene therapy for treatment of what previously has been a completely intractable disease for patients not eligible for bone marrow transplantation and its only partial therapeutic effect. Of course, this kind of genetic reconstitution method would ideally be extended to an earlier stage of development of CNS damage and ideally to a method that could reverse and not merely arrest the disease phenotype.

Metachromatic Leukodystrophy

Metachromatic leukodystrophy (MLD) is an autosomal recessive lysosomal storage disease in which mutations in the gene encoding the arylsulfatase A (ARSA) gene lead to accumulation of cerebroside sulfate and other metabolites, with resulting destruction of myelin and severe CNS disease.[30] It can present with severe infantile, juvenile or even adult forms, with varying degrees of neurological impairment ranging from rapid progression of convulsions, blindness, paralysis, and death in the most common infantile form, to slower progression in the juvenile form, and dementia and psychiatric manifestations and death in the adult form. A recent gene therapy clinical trial was carried out using an approach similar to that described for the ALD study.[31] A lentivirus vector expressing the normal ARSA gene was used to genetically correct hematopoietic stem cells from diagnosed but pre-symptomatic patients with infantile MLD. The resulting cells were infused into the circulation of the patients who had undergone partial myeloablation and, as in the case of ALD, marrow-derived glial cells re-populated the CNS glial compartment. In three pre-symptomatic patients, none of the usual and expected signs or symptoms of CNS damage appeared for up to 21 months, far past the time when symptoms would be expected to appear. While this does not demonstrate correction of a disease phenotype, it does argue very strongly for the efficacy of genetic prophylactic intervention in an otherwise inevitably devastating neurodegenerative disease.

Batten Disease

Batten disease is a collection of untreatable CNS disorders called neuronal ceroid lipofuscinoses (NCLs) and is characterized by lysosomal accumulation in the brain of lipoprotein complexes called lipofuscins that cannot be degraded because of deficiency of one of a number of enzymes required for metabolic degradation and removal from lysosomes. The disease presents clinically with a variety of symptoms, including slow learning, seizures, progressive loss of sight, dementia, and death, often by the late teen or early adult years. Batten disease comprises at least eight genetically distinct disorders and presents in infantile, juvenile, or adult forms, depending on precisely which gene among the eight known and additional presumed genes is mutated.

Like other lysosomal diseases, the various forms of Batten disease present attractive targets for gene therapy. Results have been reported of a clinical gene therapy trial of the late infantile CLN2 variant of the disease *CLN 2*, associated with defects in the gene that encodes the acid protease tripeptidyl peptidase 1 (TPP1).[32] In that study, patients received stereotactically guided injections into multiple brain locations of an adeno-associated virus vector CLN2/AAV2 expressing the normal TPP1 cDNA. Despite some serious adverse events, the causes of which were not fully explained, and despite the fact that the study was not blinded or randomized and did not include contemporaneous placebo/sham controls, neurologic rating scale tests showed a modest but significantly reduced rate of disease progression and clinical decline compared with control subjects. Despite these modestly encouraging clinical findings, this set of diseases will probably continue to present seriously difficult challenges to therapy, including gene therapy.

Huntington Disease

Enormous progress has been made in our understanding of the molecular genetic mechanisms underlying this untreatable and devastating neurodegenerative disease since the discovery of the responsible underlying gene and the CAG trinucleotide repeat mechanism that causes neuronal degeneration principally in the basal ganglia.[33,34] Nevertheless, despite such molecular progress in the past three decades, there has been little improvement in therapy, although new therapeutic concepts have emerged that are showing potentially useful therapeutic effects *in vitro* and, in some cases, in animal models of Huntington disease (HD). These developments are occurring in two major areas: 1) the use of neurotrophic or neuroprotective factors to prevent neuronal degeneration,[35] and the use of small, noncoding interfering RNAs (siRNA) to interfere with the transcriptional expression of the huntingtin gene and thereby prevent the production of the responsible aberrant huntingtin protein and its resulting neurotoxicity.[36] While results are encouraging, they have not yet led to human clinical trials.

The potential usefulness of neurotrophic factors is based on their known function of preventing cell death in degenerative processes and their ability to enhance growth and function of striatal neurons that are affected in HD. Neurotrophins such as glial cell line-derived neurotrophic factor family members and ciliary neurotrophic factor have been shown to exert neuroprotective effects on striatal neurons and thereby to help maintain the integrity of the corticostriatal pathway. It seems reasonable to hope that these and possibly other neurotrophic factors may be useful neuroprotective agents in neurodegenerative disorders of the basal ganglia.

Common Neurodegenerative Diseases

As important as gene therapy will be for the Mendelian metabolic disorders associated with CNS dysfunction, the most pressing medical and societal need for effective genetic and other forms of therapy are for the very common neurodegenerative diseases in our aging modern society – Parkinson and Alzheimer diseases.

Parkinson Disease

The basic neurological aberrations underlying most sporadic cases of Parkinson disease (PD) is the progressive loss of the neurotransmitter dopamine due to progressive degeneration and loss of dopamine neurons in the substantia nigra pars compacta of the basal ganglia. The prevalence of the disorder and the lack of long-lasting effective therapy represent a major medical and social problem. Although the traditional replacement therapy with the dopamine precursor levodopa is very useful for many patients, loss of efficacy resulting from progressive cellular loss and the development of adverse neurological side effects of levodopa underscores the need for more effective and more stable and long-lasting forms of therapy.

A number of clinical trials have been undertaken for PD,[37,38] with special attention to three major areas of potential therapy: restoration of dopamine production, neuroprotection of substantia nigra neurons, and enhancement of the inhibitory neurotransmitter γ-aminobutyric acid (GABA). One popular approach has involved the increase of dopamine production by over-expressing amino acid decarboxylase (AADC) by adeno-associated virus-mediated gene transfer.[39,40] Although detailed results have not emerged from these studies, some evidence has been reported for beneficial clinical effect. An alternative approach has involved the use of an AAV vector to over-express the glutamate decaraboxylase (GAD) cDNA, the enzyme responsible for conversion of glutamate to the inhibitory neurotransmitter GABA. A placebo-controlled clinical trial of this method via stereotactic-guided vector introduction has suggested some symptomatic improvement, although not accompanied by reduced dopaminergic neuronal loss.[41] Another approach involves neuroprotection of dopamine neuronal degeneration through the AAV vector-mediated introduction of the neurotrophic factor neurturin into the putamen or the substantia nigra of PD patients.[42,43] Putamen injections resulted in no apparent clinical improvement and further studies are currently underway.

Alzheimer Disease

Alzheimer disease (AD) exemplifies very well the difficulty in moving from extensive molecular and cellular advances to effective clinical application. The past few decades have witnessed extensive characterization of molecular genetic mechanisms suspected of being responsible for the underlying pathology and the possible role of a number of genes in the etiology of AD, including amyloid precursor protein (APP), apolipoprotein E4, presenilin-1 (Psen-1) and presenilin-2 (Psen-2) and in the associated accumulation of fibrillary aggregates of the neurotoxic product amyloid-β. These and other mechanisms have been extensively reviewed.[44]

These findings have pointed to potential pharmacological or genetic targets for therapy but, to date, no effective therapies have emerged. One clinical trial was aimed at examining the possible ability of a nerve growth factor (NGF) transgene to protect cholinergic neurons from degeneration. Loss of these cells is a signature cellular defect in AD and NGF is known to be protective for these cells in animal studies. In a phase I uncontrolled study, autologous fibroblasts genetically modified to express NGF from a Moloney leukemia retrovirus vector were grafted into the forebrain of patients.[45] The investigators reported a significant improvement in positron emission tomography scan cortical fluorodeoxyglucose uptake and a 36–51% reduction in the rate of cognitive decline in treated patients, but these results have not been confirmed by other investigators.

Another approach has been tested in animal models of AD and has utilized lentivirus vectors to deliver a cDNA encoding brain-derived neurotrophic factor (BDNF) to the brain. BDNF gene delivery was reported to reverse synapse loss, improve cell signaling, improve age-related aberrations of gene expression and cell signaling, restore learning and memory, and prevent lesion-induced death of entorhinal cortical neurons.[46] All these effects were independent of any amyloid-related mechanisms. No comparable human clinical trials have yet been reported.

A different approach was reported in 2004 involving the use of an AAV2-based gene transfer vector for the direct *in-vivo* delivery of an NGV-delivered NGF transgene to the affected brain of rat and nonhuman primate models of AD.[47] The investigators concluded that such a direct gene transfer approach with an AAV vector could protect degeneration of basal forebrain cholinergic neurons and might therefore be a useful therapy for AD. Clinical results of such a study have not been reported.

A Comment on Neurodegenerative Diseases

Overall, one might conclude that although current approaches to gene therapy for Parkinson and Alzheimer disease are based on solid molecular and cellular understanding of the underlying pathogenic defects, clinical results have not been startlingly positive. These approaches have taken too little advantage of the clues that the uncommon familial Mendelian variants of these disorders can provide to understanding the etiology and therapy. A deep understanding of the genetic and metabolomic effects of these cleaner genetic aberrations will almost certainly open new conceptual and therapeutic doors to the far more common sporadic version of the disorders. We have learned repeatedly that Mendelian disease can identify unexpected pathogenic mechanisms and potential therapeutic targets and is therefore an entrée to more complex, multigenic, and multifactorial forms of disease. Too often, funding and publication objections are raised to the rare Mendelian variants of common diseases as being "not directly related."

References

1. Farber S, Pinkel D, Sears EM, Toch R. Advances in chemotherapy of cancer in man. *Adv Cancer Res*. 1956;4:1–71.
2. Farber S. Approaches to the chemotherapy of cancer. *Trans Stud Coll Physicians Phila*. 1955;23(2):74–82.
3. Thomas ED. *Nobel Lecture: Bone Marrow Transplantation – Past, Present and Future*. Nobelprize.org. Nobel Media AB 2013 Jul 2013. http://www.nobelprize.org/nobel_prizes/medicine/laureates/1990/thomas-lecture.html.
4. Friedmann T, Roblin R. Gene therapy for human genetic disease? *Science*. 1972;175:949–955.
5. Jolly DJ, Okayama H, Berg P, et al. Isolation and characterization of a full length, expressible cDNA for human hypoxanthine guanine phosphoribosyl transferase. *Proc Natl Acad Sci U S A*. 1983;80:477–481.
6. Willis RC, Jolly DJ, Miller AD, et al. Partial phenotypic correction of human Lesch–Nyhan (HPRT-deficient) lymphoblasts with a transmissible retroviral vector. *J Biol Chem*. 1984;259:7842–7849.

7. Rosenberg MB, Friedmann T, Robertson RC, et al. Grafting of genetically modified cells to the damaged brain: restorative effects of ngf gene expression. *Science*. 1988;242:1575–1578.
8. Cavazzana-Calvo M, André-Schmutz I, Fischer A. Haematopoietic stem cell transplantation for SCID patients: where do we stand? *Br J Haematol*. 2013;160(2):146–152. doi:10.1111/bjh.12119 Epub 2012 Nov 20.
9. Ferrua F, Brigida I, Aiuti A. Update on gene therapy for adenosine deaminase-deficient severe combined immunodeficiency. *Curr Opin Allergy Clin Immunol*. 2010;10:551–556.
10. Fischer A, Hacein-Bey-Abina S, Cavazanna-Calvo M. Gene therapy for primary immunodeficiencies. *Immunol Allergy Clin North Am*. 2010;30:237–248.
11. Gaspar HB, et al. Hematopoietic stem cell gene therapy for adenosine deaminase-deficient severe combined immunodeficiency leads to long-term immunological recovery and metabolic correction. *Sci Transl Med*. 2011;3:97ra80.
12. Hacein-Bey-Abina S, et al. Efficacy of gene therapy for X-linked severe combined immunodeficiency. *N Engl J Med*. 2010;363:355–364.
13. Hacein-Bey-Abina S, et al. Insertional oncogenesis in 4 patients after retrovirus-mediated gene therapy of SCID-X1. *J Clin Invest*. 2008;118:3132–3142.
14. Dave UP, et al. Murine leukemias with retroviral insertions at Lmo2 are predictive of the leukemias induced in SCID-X1 patients following retroviral gene therapy. *PLoS Genet*. 2009;5:e1000491.
15. Aiuti A, et al. Gene therapy for immunodeficiency due to adenosine deaminase deficiency. *N Engl J Med*. 2009;360:447–458.
16. Candotti F, et al. Gene therapy for adenosine deaminase-deficient severe combined immune deficiency: clinical comparison of retroviral vectors and treatment plans. *Blood*. 2012;120:3635–3646.
17. Cavazzana-Calvo M, Fischer A, Hacein-Bey-Abina S, Aiuti A. Gene therapy for primary immunodeficiencies: part 1. *Curr Opin Immunol*. 2012;24:580–584.
18. Online Mendelian Inheritance in Man (OMIM). *Leber Congenital Amaurosis, Type I*; LCA1–204000. http://omim.org/entry/204000.
19. Maguire AM, Simonelli F, Pierce EA, et al. Safety and efficacy of gene transfer for Leber's congenital amaurosis. *N Engl J Med*. 2008;358(21):2240–2248.
20. Bainbridge JW, Smith AJ, et al. Effect of gene therapy on visual function in Leber's congenital amaurosis. *N Engl J Med*. 2008;358(21): 2231–2239. doi:10.1056/NEJMoa0802268.
21. Simonelli F, Maguire AM, Testa F, et al. Gene therapy for Leber's congenital amaurosis is safe and effective through 1.5 years after vector administration. *Mol Ther*. 2010;18(3):643–650.
22. Simonato M, Bennett J, Boulis NM, et al. Progress in gene therapy for neurological disorders. *Nat Rev Neurol*. 2013;9(5):277–291.
23. Nagabhushan Kalburgi S, Khan NN, Gray SJ. Recent gene therapy advancements for neurological diseases. *Discov Med*. 2013;15(81):111–119.
24. Leone P, Shera D, McPhee SW, et al. Long-term follow-up after gene therapy for Canavan disease. *Sci Transl Med*. 2012;4(165):165ra163. doi:10.1126/scitranslmed.3003454.
25. Janson C, Mcphee S, Bilaniuk L, et al. Clinical protocol. Gene therapy of Canavan disease: AAV-2 vector for neurosurgical delivery of aspartoacylase gene (ASPA) to the human brain. *Hum Gene Ther*. 2002;13(11):1391–1412.
26. Kumar S, Mattan NS, De Vellis J. Canavan disease: a white matter disorder. *Ment Retard Dev Disabil Res Rev*. 2006;12(2):157–165.
27. Leone P, Janson CG, Mcphee SJ, During MJ. Global CNS gene transfer for a childhood neurogenetic enzyme deficiency: Canavan disease. *Curr Opin Mol Ther*. 1999;1(4):487–492.
28. Cartier N, Hacein-Bey-Abina S, Bartholomae CC, et al. Lentiviral hematopoietic cell gene therapy for X-linked adrenoleukodystrophy. *Methods Enzymol*. 2012;507:187–198. doi:10.1016/B978-0-12-386509-0.00010-7.
29. Cartier N, et al. Hematopoietic stem cell gene therapy with a lentiviral vector in X-linked adrenoleukodystrophy. *Science*. 2009;326:818–823.
30. Online Mendelian Inheritance in Man (OMIM). http://www.omim.org/entry/250100.
31. Biffi A, Montini E, Lorioli L, et al. Lentiviral hematopoietic stem cell gene therapy benefits metachromatic leukodystrophy. *Science*. 2013;341(6148):1233158.
32. Worgall S, Sondhi D, Hackett NR, et al. Treatment of late infantile neuronal ceroid lipofuscinosis by CNS administration of a serotype 2 adeno-associated virus expressing CLN2 cDNA. *Hum Gene Ther*. 2008;19(5):463–474.
33. Gusella JF, Wexler NS, Conneally PM, et al. A polymorphic DNA marker genetically linked to Huntington's disease. *Nature*. 1983;306(5940):234–238.
34. The Huntington's Disease Collaborative Research Group. A novel gene containing a trinucleotide repeat that is expanded and unstable on Huntington's disease chromosomes. *Cell*. 1993;72(6):971–983.
35. Alberch J, Pérez-Navarro E, Canals JM. Neurotrophic factors in Huntington's disease. *Prog Brain Res*. 2004;146:195–229.
36. McBride JL, Pitzer MR, Boudreau RL, et al. Preclinical safety of RNAi-mediated HTT suppression in the rhesus macaque as a potential therapy for Huntington's disease. *Mol Ther*. 2011;19(12):2152–2162. doi:10.1038/mt.2011.219 Epub 2011 Oct 25.
37. Feng LR, Maguire-Zeiss KA. Gene therapy in Parkinson's disease: rationale and current status. *CNS Drugs*. 2010;24(3):177–192. doi:10.2165/11533740-000000000-00000.
38. Douglas MR. Gene therapy for Parkinson's disease: state-of-the-art treatments for neurodegenerative disease. *Expert Rev Neurother*. 2013;13(6):695–705. doi:10.1586/ern.13.58.
39. Christine CW, et al. Safety and tolerability of putaminal AADC gene therapy for Parkinson disease. *Neurology*. 2009;73:1662–1669.
40. Muramatsu S, et al. A phase I study of aromatic L-amino acid decarboxylase gene therapy for Parkinson's disease. *Mol Ther*. 2010;18:1731–1735.
41. LeWitt PA, et al. AAV2-GAD gene therapy for advanced Parkinson's disease: a double-blinded, sham surgery controlled, randomized trial. *Lancet Neurol*. 2011;10:309–319.
42. Marks Jr. W, et al. Safety and tolerability of intraputaminal delivery of CERE-120 (adeno-associated virus serotype 2-neurturin) to patients with idiopathic Parkinson's disease: N open-label phase I trial. *Lancet Neurol*. 2008;7:400–408.
43. Marks Jr. W, et al. Gene delivery of AAV1-neurturin for Parkinson's diseases: a double blind, randomized, controlled study. *Lancet Neurol*. 2010;9:1164–1172.
44. Tanzi RE. The genetics of Alzheimer disease. *Cold Spring Harb Perspect Med*. 2012;2(10).

45. Tuszynski MH, Thal L, Pay M, et al. A phase 1 clinical trial of nerve growth factor gene therapy for Alzheimer disease. *Nat Med*. 2005;11(5):551–555 Epub 2005 Apr 24.
46. Nagahara AH, et al. Neuroprotective effects of brain-derived neurotrophic factor in rodent and primate models of Alzheimer's disease. *Nat Med*. 2009;15:331–337.
47. Bishop KM, Hofer EK, Mehta A, et al. Therapeutic potential of CERE-110 (AAV2-NGF): targeted, stable, and sustained NGF delivery and trophic activity on rodent basal forebrain cholinergic neurons. *Exp Neurol*. 2008;211(2):574–584. doi:10.1016/j.expneurol.2008.03.004 Epub 2008 Mar 19.

CHAPTER

9

Direct Induction of Neural Stem Cells from Somatic Cells

Wado Akamatsu and Hideyuki Okano

School of Medicine, Keio University, Tokyo, Japan

INTRODUCTION

Neural stem cells (NSCs) are defined as self-renewing stem cells, which give rise to multiple types of neural cells. These NSCs are present in the mammalian central nervous system (CNS) and can be isolated using a floating culture system to form a neurosphere.[1] This system is known as a neurosphere assay, and each single NSC forms a neurosphere under the presence of fibroblast growth factor-2 (FGF-2). These FGF-2-dependent NSCs first appear on embryonic day 8.5 (E8.5) in the neural plate of fetal mice, as shown using neurosphere cultures.[2] We previously reported that these NSCs derived from fetal neuroepithelium were effective for treatment in rat[3] and marmoset[4] spinal cord injury (SCI) models. NSCs can also be derived from pluripotent stem cells, which can differentiate into multiple types of somatic cells.[5] We previously reported that transplantations of neurospheres derived from mouse,[6] marmoset (Okano et al., unpublished results) and human (Okano et al., unpublished results) embryonic stem (ES) cells can improve the recovery outcomes in spinal cord injury models. These NSCs can be expanded in the pluripotent ESC state, and this unique property provides significant promise for cell-based therapies to restore damaged central nervous systems. In 2006, induced pluripotent stem cells (iPSCs) were established from mouse fibroblasts using four reprogramming factors (Oct4, Sox2, Klf4, and cMyc).[7] We previously reported that mouse[8] and human[9] iPS cells give rise to NSCs to form neurospheres, which were transplanted into the spinal cord of host animals 9 days after the SCI. The iPSC-derived cells differentiated into multiple neural lineages, including neurons, astrocytes, and oligodendrocytes, without forming tumors and were involved in remyelination and axonal regrowth. Thus, the therapeutic effects of the iPSC-derived NSCs were very similar to those of embryonic- and ESC-derived NSCs. Given the results of a preclinical study using a marmoset SCI model,[10] a clinical trial is currently being prepared.[11,12] However, several issues remain to be resolved for iPSC-based autologous transplantation. First, the induction of NSCs from human iPS cells is time consuming. In the preclinical studies that we have reported, more than 3–4 months were required to induce NSCs via embryoid body formation from undifferentiated human iPSCs.[9] We have also reported that immature NSCs, which give rise to mainly neurons, did not improve motor function in transplanted animals.[6,8] Such immature NSCs can be developed into mature giliogenic neural stem cells using long-term cultivation in vitro. Second, various factors, including cell origin and methods of reprogramming, are associated with the property of iPSC clones. We previously reported that the frequency of safe iPS clones, which do not form tumors *in vivo*, was approximately 10% (5/55) when they were generated from adult mouse tail-tip fibroblasts.[13] However, we have found that the transplantation of NSCs is effective only during the subacute phase following SCI, but not during the acute or chronic phase.[3] In patients with SCI, the subacute phase is thought to transition to the chronic phase at 4–6 weeks after injury. These findings lead us to conclude that the cells used for clinical trials should be induced and prepared as neural stem cells prior to transplantation. These cells should be derived from a genetically nonidentical donor and should be carefully tested for their safety and differentiation properties. The first clinical trial of allogenic human iPSC therapies for spinal cord injury patients is likely to begin within 5 years.

Rosenberg's Molecular and Genetic Basis of Neurological and Psychiatric Disease
http://dx.doi.org/10.1016/B978-0-12-410529-4.00009-7

DIRECT INDUCTION OF NSCs FROM SOMATIC CELLS

We sought to overcome the current technical limitations in the production of human NSCs by establishing a novel method to induce neural stem cells rapidly. Mouse fibroblasts were transduced with Oct4, Sox2, Klf4 and c-Myc using retroviral vectors for reprogramming, and these cells were cultured for 4 days to advance reprogramming. Without clonal isolation of the iPSCs, partially reprogrammed fibroblasts were directly dissociated in serum-free suspension cultivation under the presence of leukemia inhibitory factor (LIF) and FGF-2 to form neurospheres. Two weeks later, floating neurospheres were observed in the medium (Figure 9.1). Each neurosphere was thought to be generated from a single neural stem cell, suggesting that less than 0.001% of the fibroblasts became NSCs.[14]

Differentiation Properties of Directly Induced NSCs

These neurospheres were differentiated by the removal of growth factors and analyzed using immunostaining for neuronal and glial markers. Interestingly, these directly induced NSCs gave rise to both neurons and glial cells even at primary neurospheres, while primary neurospheres derived from pluripotent stem cells (ESCs)/iPSCs differentiated mainly into neurons. These directly induced neural stem cells (diNSCs) in the secondary and tertiary neurospheres were EGF-dependent mature NSCs, which mainly produce glial cells. Therefore, we were able to remove the residual pluripotent cells and to expand the EGF-dependent mature neural stem cells by removing FGF-2 and LIF from the culture medium. Using this modification of the culture medium, the frequency of Nanog-expressing residual pluripotent cells below was reduced to 0.01%, even in the primary neurospheres directly generated from adult fibroblasts. Recently, another group reported a similar phenomenon in hepatic cell induction.[15] Although hepatocyte-like cells can be induced from human iPSCs/ESCs using directed differentiation protocols, these hepatocyte-like cells expressed cytochrome P450 (CYP450) enzyme incompletely and failed to restore the human serum albumin level when transplanted into mice. Therefore, the expansion and maturation of these iPSC-derived hepatocyte-like cells was thought to be insufficient. In this report, they reprogrammed human fibroblasts with retroviruses expressing OCT4, SOX2, and KLF4 and cultured the cells in a medium optimized for endoderm differentiation. Then, they established multipotent progenitor cells, which do not express pluripotent markers (e.g., Nanog, Oct4) and are self-renewing. These cells were differentiated into hepatocytes using small molecules and were transplanted into an immune-deficient mouse model of human liver failure. These transplanted hepatocytes proliferated extensively and exhibited a hepatocyte function similar to those of adult hepatocytes *in vivo*. These results suggest that the differentiation properties of directly induced cells, which bypassed complete reprogramming into iPSCs, have characteristics distinct from similar cells differentiated from iPSCs/ESCs. Currently, the reason why these cells have distinct differentiation properties is unknown. A detailed comparison between directly differentiated cells and iPSC-derived cells may reveal this mechanism.

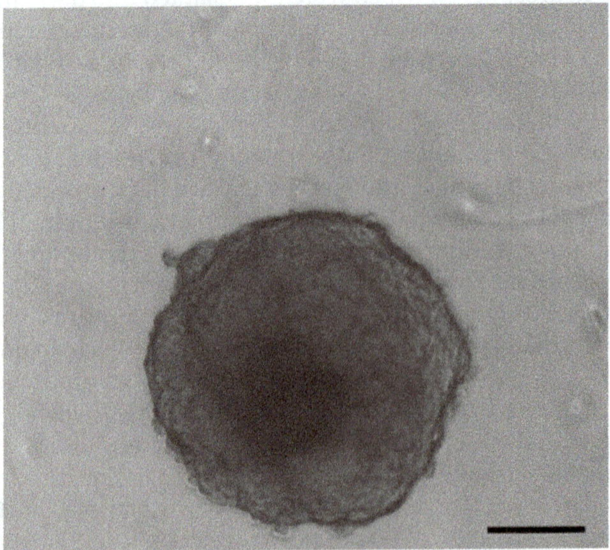

FIGURE 9.1 **A neurosphere derived from mouse fibroblasts infected with Oct4-, Sox2-, Klf4-, and c-Myc-expressing retroviruses, followed by adherent and suspension culture.** Scale bar = 100 μm.

COMPARISON OF DIRECT INDUCTION INTO NSCS

The first report of direct reprogramming in neural cells was made by Wernig and colleagues in 2010. Using three transcription factors (Brn2, Mytl1, and Ascl1) specifically expressed in neurons, they directly induced functional neurons (iN cells) from mouse,[16] marmoset,[17] and human[18] fibroblasts. However, these iN cells were postmitotic and were impossible to expand *in vitro* once they were converted into neurons. In 2011 and 2012, several groups, including the authors', reported the successful induction of multipotent neural cells from somatic cells. The first report of the generation of neural precursor-like cells was achieved by the transient expression of Yamanaka factors using doxycycline-inducible lentiviral vectors and a subsequent neural culture under the presence of FGF2, FGF4, and EGF.[19] Although these cells formed rosette-like structures similar to neural progenitors that had differentiated from pluripotent stem cells, whether these cells had a self-renewing property was not determined. Another group reported that self-renewing neural progenitors were induced from mouse fibroblasts by the transduction of Yamanaka factors. They delivered Oct4 into the cells as a protein preventing the reactivation of the endogenous Oct4 locus.[20] These cells were multipotent and gave rise to neurons, astrocytes, and oligodendrocytes as well as clonally derived neural stem cells *in vivo*. Furthermore, transplantation experiments revealed that these cells survived and differentiated into three lineages in the host animals. The first successful report of the induction of neural progenitors was achieved by the forced expression of nine transcription factors (Ascl1, Ngn2, Hes1, Id1, Pax6, Brn2, Sox2, c-Myc, and Klf4) in Sertoli cells derived from testes.[21] Although the self-renewing ability of the reprogrammed cells was not tested, these cells were able to differentiate into neurons, astrocytes, and oligodendrocytes. Wernig and colleagues generated self-renewing and multipotent neural progenitors by inducing Sox2 and FoxG1 in mouse fibroblasts.[22] Although one of the Yamanaka factors, Sox2, was used for the reprogramming, the fact that only two factors were sufficient to induce clonal and self-renewing neural progenitors with NSC-markers was surprising. These cells were able to differentiate into neurons and astrocytes, but not oligodendrocytes. Interestingly, additional Brn2 overexpression changed the differentiation potential of these cells. The cells reprogrammed by Sox2, Brn2, and FoxG1 were self-renewing, and had the ability to differentiate into all three lineages as well as NSCs *in vivo*. However, these proliferating neural precursor-like cells lost their stem cell-like ability with the removal of the transgenes, while these transgenes were necessary once the cells had differentiated into their progeny. These observations suggested that these neural precursor-like cells were incompletely reprogrammed and that these three factors, Sox2, Brn2, and FoxG1, are not sufficient to induce complete epigenetic reprogramming into NSCs. Schöler and colleagues induced another type of neural precursor-like cells from mouse fibroblasts by transduction with Brn4 and E47/Tcf3 in addition to Sox2, Klf4 and c-Myc.[23] Although three of the Yamanaka factors (Sox2, Klf4 and c-Myc) were used for reprogramming, these three factors were not sufficient to fully reprogram the cells into iPSCs. These neural precursor-like cells were tripotent and could be maintained for a long time. Interestingly, the retroviral transgenes were silenced in these cells, and the expressions of endogenous Sox2 and Brn4 were induced. These results suggested that the endogenous transcriptional network of these cells was similar to that of NSCs *in vivo* through the induction of these five factors. Another group, Huang and colleagues, tried to establish multipotent neural precursor cells from mouse embryonic fibroblasts by the forced expression of NSC-specific factors, including Sox2, Bmi-1, TLX, Hes1 and Oct1.[24] Consequently, Huang and colleagues were able to narrow down the reprogramming factors to Sox2 alone. Furthermore, they succeeded in making these multipotent neural precursor cells from human fetal fibroblasts using Sox2. The expression of retrovirally induced Sox2 was gradually silenced in these cells, suggesting that Sox2 reprogrammed the endogenous transcriptional network of these cells.

DIRECT INDUCTION OF NEURAL STEM CELL IN REGENERATIVE MEDICINE

As described above, various kinds of neural stem-like cells have been established so far. These methods can be classified into two categories based on the set of transgenes: the four Yamanaka factors[14,19,20] and others.[21–24] While the neural stem-like cells induced by the four Yamanaka factors resembled NSCs derived from pluripotent cells or mammalian CNS in terms of their differentiation potential and self-renewing ability, the properties of cells induced by other factors varied. Direct reprogramming by tissue-specific transcriptional factors has been reported in several types of cells including neurons,[16] hepatocytes,[25] and cardiomyocytes.[26] These directly reprogrammed cells are relatively mature and can be reprogrammed within several days. Recently, Abeliovich and colleagues reported that direct reprogramming could be used for disease modeling. They reported that hiN cells derived from familial Alzheimer disease (FAD) patients were able to exhibit a cell type-selective pathology in FAD.[27] Although these properties may facilitate a rapid preparation of desired cells in regenerative medicine, these cells are not self-renewing and impossible to expand *in vitro*. While the risk of teratoma in these cells is thought to be zero without exogenous Yamanaka factors, whether such post-mitotic cells can be used for cell therapy remains unclear. So far, in all the protocols that induce self-renewing neural stem cell-like cells, one of the Yamanaka factors, Sox2, has always been used. Therefore, the safety of these cells must be examined for their use in cell therapy. "Real" induced NSCs, which are reprogrammed by NSC-specific transcriptional factors and are identical to those *in vivo*, should be established for future regenerative medicine.

References

1. Reynolds BA, Weiss S. Generation of neurons and astrocytes from isolated cells of the adult mammalian central nervous system. *Science*. 1992;255:1707–1710.
2. Tropepe V, Sibilia M, Ciruna BG, Rossant J, Wagner EF, van der Kooy D. Distinct neural stem cells proliferate in response to EGF and FGF in the developing mouse telencephalon. *Dev Biol*. 1999;208:166–188.
3. Ogawa Y, Sawamoto K, Miyata T, et al. Transplantation of in vitro-expanded fetal neural progenitor cells results in neurogenesis and functional recovery after spinal cord contusion injury in adult rats. *J Neurosci Res*. 2002;69:925–933.
4. Iwanami A, Kaneko S, Nakamura M, et al. Transplantation of human neural stem cells for spinal cord injury in primates. *J Neurosci Res*. 2005;80:182–190.
5. Tropepe V, Hitoshi S, Sirard C, Mak TW, Rossant J, Van Der Kooy D. Direct neural fate specification from embryonic stem cells: a primitive mammalian neural stem cell stage acquired through a default mechanism. *Neuron*. 2001;30:65–78.
6. Kumagai G, Okada Y, Yamane J, et al. Roles of ES cell-derived gliogenic neural stem/progenitor cells in functional recovery after spinal cord injury. *PLoS One*. 4(11):e7706.
7. Takahashi K, Yamanaka S. Induction of pluripotent stem cells from mouse embryonic and adult fibroblast cultures by defined factors. *Cell*. 2006;126:663–676.
8. Tsuji O, Miura K, Okada Y, et al. Therapeutic potential of appropriately evaluated safe-induced pluripotent stem cells for spinal cord injury. *Proc Natl Acad Sci U S A*. 2010;107:12704–12709.
9. Nori S, Okada Y, Yasuda A, et al. Grafted human-induced pluripotent stem-cell-derived neurospheres promote motor functional recovery after spinal cord injury in mice. *Proc Natl Acad Sci*. 2011;108:16825–16830.
10. Kobayashi Y, Okada Y, Itakura G, et al. Pre-evaluated safe human iPSC-derived neural stem cells promote functional recovery after spinal cord injury in common marmoset without tumorigenicity. *PLoS One*. 2012;7(12):e52787.
11. Okano H, Nakamura M, Yoshida K, et al. Steps toward safe cell therapy using induced pluripotent stem cells. *Circ Res*. 2013;112:523–533.
12. Okano H, Yamanaka S. iPS cell technologies: significance and applications to CNS regeneration and disease. *Mol Brain*. 2014;7(1):22.
13. Miura K, Okada Y, Aoi T, et al. Variation in the safety of induced pluripotent stem cell lines. *Nat Biotechnol*. 2009;27:743–745.
14. Matsui T, Takano M, Yoshida K, et al. Neural stem cells directly differentiated from partially reprogrammed fibroblasts rapidly acquire gliogenic competency. *Stem Cells*. 2012;30(6):1109–1119.
15. Zhu S, Rezvani M, Harbell J, et al. Mouse liver repopulation with hepatocytes generated from human fibroblasts. *Nature*. 2014; 508(7494):93–97.
16. Vierbuchen T, Ostermeier A, Pang ZP, Kokubu Y, Südhof TC, Wernig M. Supplementary data - Direct conversion of fibroblasts to functional neurons by defined factors. *Nature*. 2010;463:1035–1041. doi:10.1038/nature08797.
17. Zhou Z, Ibata K, Kohda K, et al. Reprogramming non-human primate somatic cells into functional neuronal cells by defined factors. *Mol Brain*. 2014;7(1):24.
18. Pang ZP, Yang N, Vierbuchen T, et al. Induction of human neuronal cells by defined transcription factors. *Nature*. 2011;476:220–223.
19. Kim J, Efe JA, Zhu S, et al. Direct reprogramming of mouse fibroblasts to neural progenitors. *Proc Natl Acad Sci U S A*. 2011;108(19):7838–7843.
20. Thier M, Wörsdörfer P, Lakes YB, et al. Direct conversion of fibroblasts into stably expandable neural stem cells. *Cell Stem Cell*. 2012;10:473–479.
21. Sheng C, Zheng Q, Wu J, et al. Direct reprogramming of Sertoli cells into multipotent neural stem cells by defined factors. *Cell Res*. 2012;22:208–218.
22. Lujan E, Chanda S, Ahlenius H, Sudhof TC, Wernig M. Direct conversion of mouse fibroblasts to self-renewing, tripotent neural precursor cells. *Proc Natl Acad Sci*. 2012;109:2527–2532.
23. Han DW, Tapia N, Hermann A, et al. Direct reprogramming of fibroblasts into neural stem cell by defined factors. *Cell Stem Cell*. 2012;10:465–472.
24. Ring KL, Tong LM, Balestra ME, et al. Direct reprogramming of mouse and human fibroblasts into multipotent neural stem cells with a single factor. *Cell Stem Cell*. 2012;11:100–109.
25. Sekiya S, Suzuki A. Direct conversion of mouse fibroblasts to hepatocyte-like cells by defined factors. *Nature*. 2011;475:390–393.
26. Ieda M, Fu JD, Delgado-Olguin P, et al. Direct reprogramming of fibroblasts into functional cardiomyocytes by defined factors. *Cell*. 2010;142:375–386.
27. Qiang L, Fujita R, Yamashita T, et al. Directed conversion of Alzheimer's disease patient skin fibroblasts into functional neurons. *Cell*. 2011;146:359–371.

CHAPTER

10

Neuroimaging in Dementias

Prashanthi Vemuri, Melissa E. Murray*†*, and Clifford R. Jack, Jr.**

*Mayo Clinic, Rochester, MN, USA

†Mayo Clinic, Jacksonville, FL, USA

INTRODUCTION

Late-onset dementia is usually a multifactorial disease wherein cumulative pathological brain insults (of more than one pathology) results in progressive cognitive decline that ultimately leads to impairment in ability to function at work and/or perform usual activities/tasks. Alzheimer disease (AD) is the most common cause of dementia in the elderly; however, AD is the sole cause of dementia in approximately 45% of all dementia cases. There are other cerebrovascular and degenerative conditions that contribute to the burden of dementia on a population basis. These latter include dementia with Lewy bodies (DLB) and frontotemporal dementia (FTD). DLB is considered to be the second most common cause of late-onset dementia (≥65 years of age)[1,2] whereas FTD is considered to be the second most common cause of young onset dementia (<65 years of age).[3,4] At present, there can be considerable uncertainty in the antemortem clinical diagnosis of these pathologic syndromes because of clinical heterogeneity (wherein the dementia underlying pathology does not match the clinical syndrome), subtle symptoms early in the disease process, and frequent occurrence of mixed dementias. Until recently, postmortem examination has been the only way to accurately determine the underlying dementia pathology. However, with the recent emergence of advanced imaging technologies, imaging indicators of disease that closely reflect the underlying pathology have been found to be very useful in aiding the prediction of the underlying dementia pathology. In this chapter we will cover the three common neurodegenerative dementias—AD, DLB and FTD. For each of these dementia subtypes we will discuss the characteristic clinical manifestation of the disease when it exists in a pathologically pure form, the pathologies underlying the disease, and the neuroimaging correlates that can be observed using the currently available imaging methods. Additionally we will also discuss the imaging of vascular disease that is extremely useful in determining if vascular factors contribute to the clinical symptoms of dementia.

NEUROIMAGING TECHNOLOGIES

At the current time there are primarily two technologies that are available for imaging of neurodegenerative dementias—positron emission tomography (PET) and magnetic resonance imaging (MRI).

PET Imaging

In PET a radioactive tracer is injected into a subject intravenously and is chemically incorporated into a biologically active molecule of interest. The PET cameras capture the gamma rays that are produced when the tracer decays and undergoes an annihilation reaction. The common radioactive tracers use ^{11}C, ^{15}O and ^{18}F isotopes.[5–7] The three main PET methods that are used for imaging dementias are: amyloid imaging (^{11}C- or ^{18}F-based), fluorodeoxyglucose (^{18}F-FDG) imaging, and tau (^{18}F-based) imaging.

Amyloid Imaging

Pittsburgh compound B (PiB) accumulation as detected by PiB-PET imaging is the most commonly used *in vivo* imaging measure of amyloid burden, which is a hallmark feature of AD.[8] Excellent concordance exists between

Rosenberg's Molecular and Genetic Basis of Neurological and Psychiatric Disease
http://dx.doi.org/10.1016/B978-0-12-410529-4.00010-3

PiB accumulation in different regions of the brain and fibrillary amyloid-β (Aβ) deposition in subjects who have undergone antemortem PiB-PET imaging and autopsy.[9,10] Other amyloid imaging compounds include florbetapir and flutemetamol (^{18}F based).

FDG-PET

Fluorodeoxyglucose (^{18}F-FDG) PET is used to measure net brain metabolism, which, although it includes many neural and glial functions, largely indicates synaptic activity.[11,12] Brain glucose metabolism measured with FDG-PET is correlated with postmortem measures of the synaptic structural protein synaptophysin.[13] The regional and global loss of brain metabolism is seen in all dementias.

Tau Imaging

Tau is a hallmark feature of AD as well as of a spectrum of sporadic FTD subtypes. Although not discussed in this chapter, the recent advancements in the development of PET ligands for imaging tau (^{18}F-based T807, T808, THK523) have shown considerable promise for *in vivo* visualization of tau in dementias.

MRI Imaging

MRI imaging is based on the principles of nuclear magnetic resonance of the atomic nuclei. Due to the biological abundance of ^{1}H (in water) in the human body, the present MRI scanners use the signal from ^{1}H for imaging. Different MR imaging techniques (structural MRI, diffusion tensor imaging [DTI] and functional MRI [fMRI]) have specialized methods to acquire the signal underlying different tissue properties.

Structural MRI

Structural MRI (sMRI) is the most common clinical imaging technology used to visualize neurodegeneration in dementias. Cerebral atrophy detectable by sMRI is the macroscopic manifestation of loss of neurons, synapses, and dendritic dearborization that occurs on a microscopic level in neurodegenerative diseases. While sMRI is not a direct measure of neuronal pathology, strong correlations have been shown between volume measured on MRI and histology based neuronal numbers.[7]

Diffusion Tensor Imaging (DTI)

DTI measures the diffusion properties of water molecules in the brain and therefore is useful in visualizing the white matter tracts that connect different regions of the brain. There are two primary measures of DTI that are used to measure microstructural changes in dementias due to the degeneration—fractional anisotrophy (FA), which measures degree of directionality of the diffusion process (the FA of highly directional normal white matter tracts is one and FA decreases with white matter degeneration), and mean diffusivity (MD), which measures the magnitude of the diffusivity (there is increased MD with gray and white matter loss due to unrestricted motion of the water molecules).

Task-Free Functional MRI (TF-fMRI)

Task-free or resting state fMRI measures low-frequency fluctuations (0.1–0.01 Hz) in the blood oxygen level-dependent signal during undirected task-free paradigms and has been shown to be specific to gray matter. TF-fMRI can be used to identify the spatial extent and statistical strength of temporally correlated networks of structural and functional connectivity activity within the brain, and thus can measure connectivity breakdown due to dementia. Specific regions of the brain (posterior cingulate, precuneus, medial prefrontal cortex, and the hippocampus) have high intrinsic connectivity when thought is directed inward. This network is suppressed during any task-induced state[14] and is thought to be part of a default mode network (DMN) of the brain. This is the most studied functional connectivity network in dementia.

ALZHEIMER DISEASE

Clinical Manifestation of Alzheimer Disease

Episodic memory loss characterizes the cognitive disorder in AD; however, nonamnestic presentations do occur.[15–17] According to the new diagnostic guidelines set forth by the partnership of the National Institute on Aging and the Alzheimer's Association (NIA–AA), the core clinical criteria incorporate the common amnestic presentation and/or the less common nonamnestic presentations for the diagnosis of probable AD.[18] Using the National Institute of Neurological and Communicative Disorders and Stroke (NINCDS) and the Alzheimer's Disease and Related Disorders Association (ADRDA) criteria,[19] the NIA–AA set forth criteria for all-cause dementia and criteria for dementia caused by AD. According to the recommendations by the NIA-AA guidelines: a patient should meet criteria for "all-cause" dementia, insidious onset of symptoms, progressive worsening of cognitive deficits (not attributable to a major psychiatric disorder or delirium); and these deficits evidenced by history or objective cognitive assessment should present as either an amnestic presentation or nonamnestic presentation.[18] Whether an amnestic or nonamnestic presentation is apparent, an observed deficit in other cognitive domains should be evident. Nonamnestic presentations include language disorders (prominently in word-finding),[17] visuospatial disorder,[15] and executive or behavioral disorder.[20] A diagnosis of probable AD is not recommended if any of the following are evident: significant coexisting cerebrovascular disease; core features of DLB other than dementia; prominent features of behavioral variant frontotemporal dementia (bvFTD); prominent features of nonfluent/agrammatic variant primary progressive aphasia (PPA) or semantic variant primary progressive aphasia; or evidence for another concomitant, active neurological disease, or a non-neurological medical comorbidity or medication use that could have a significant effect on cognition. Importantly, the NIA–AA workgroup incorporates causative AD genetic mutations and biomarker evidence of AD pathology into their recommended guidelines that were previously unknown or underappreciated at the time of the original criteria established in 1984.[19]

Pathologies Underlying Alzheimer Disease

Macroscopic inspection of a brain with a postmortem diagnosis of AD often reveals generalized atrophy predominantly involving medial temporal lobe structures (amygdala, hippocampal formation, and fusiform gyrus) and association cortices. Sequential sections through the supratentorial tissues often reveals marked enlargement of the frontal and temporal horns of the lateral ventricle. The cortical gray mantle is often thinned, but normal in distribution. The hippocampal formation and amygdala typically have marked atrophy.[21] The substantia nigra should have visible pigmentation, but the locus coeruleus will likely appear depigmented depending on the length of disease duration. The classic hallmark neuropathology observed upon microscopic inspection includes extracellular accumulation of Aβ deposits (senile plaques) and intracellular accumulation of hyperphosphorylated tau (neurofibrillary tangles). The stereotypic spread of neurofibrillary tangles (NFTs; "stages") from the transentorhinal to limbic and finally neocortex was first described by Braak and Braak in 1991.[22] This topographic spread of neurofibrillary tangles can be distinguished into six stages and has become a widely accepted method. Evidence supports that there are ~25% of cases that do not conform to this stereotypic progression, with tangle pathology either relatively sparing the hippocampus (i.e., hippocampal-sparing AD) or remaining relatively confined to transentorhinal or limbic regions (i.e., limbic-predominant AD).[16] Senile plaques can be observed in a wide variety of focal deposits, including lake-like plaques and diffuse/primitive plaques, compact plaques, cored plaques, and neuritic plaques. The NIA–AA has also proposed new guidelines for neuropathologic assessment of AD, which incorporates a currently established method of assessing neuritic plaques and proposes the incorporation of a method that evaluates progressive accumulation of any Aβ plaques. The Consortium to Establish a Registry for AD (CERAD) neuropathologic criteria is an established method for assessing the age-related severity of neuritic plaques to evaluate histologic findings associated with AD.[23,24] Thal and colleagues have proposed a topographic classification scheme ("phases") for Aβ plaques.[25] Previously, NIA–Reagan criteria recommended CERAD's age-adjusted four-point scale in combination with Braak neurofibrillary tangle stage to derive a probability statement that dementia was caused by AD pathology.[26] In the most recent NIA–AA criteria, they recommend an "Amyloid, Braak, CERAD (ABC) score," where Thal amyloid phase is now recommended to be incorporated into the AD neuropathologic classification guidelines.

Imaging in Alzheimer Disease

Based on the biomarker literature, the deposition of amyloid pathology in the brain is hypothesized to be the first change in the largely sequential pathologic cascade of AD.[27] Neuronal pathology (tauopathy, neuronal injury, and neurodegeneration) appears to occur relatively later compared to amyloid and neuronal pathology and are key determinants of cognitive impairment in AD.[28,29] The major PET and MRI findings in AD are described here.

Amyloid Imaging

Using cross-sectional and serial amyloid imaging in cognitively normal and clinically impaired subjects, recent studies have found that the deposition of Aβ is a slow and protracted process that occurs over a couple of decades

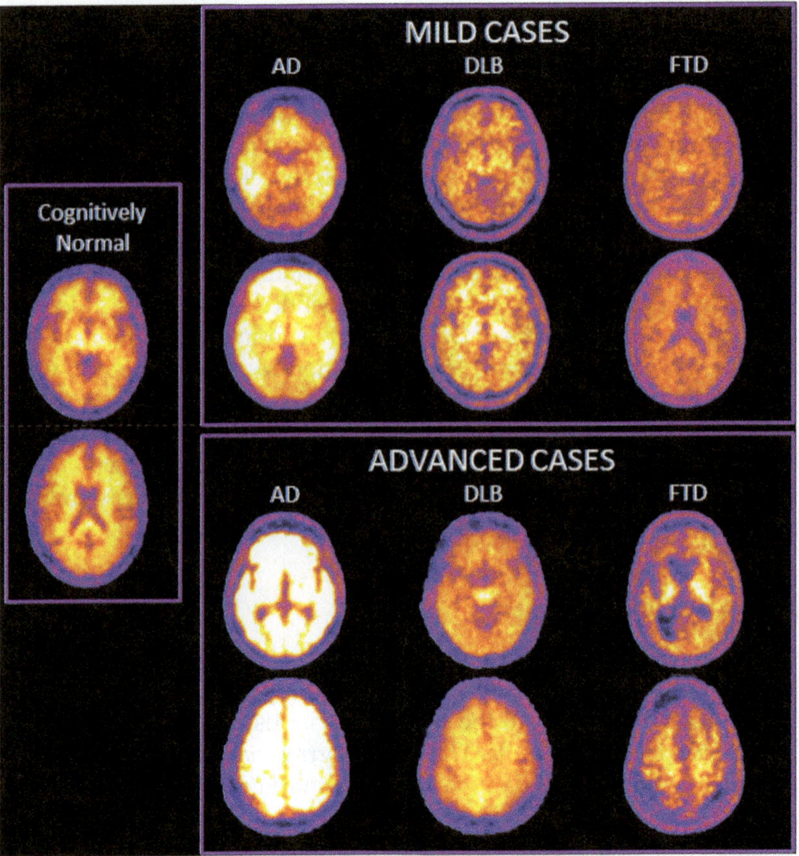

FIGURE 10.1 **Amyloid imaging (PiB-PET) scans in cognitively normal elderly, mild cases of dementia and advanced cases of dementia.** Each image is scaled by the cerebellum accumulation to standardize the scale across images.

typically starting in the 60–70s.[27] AD-related changes in amyloid imaging can be seen very distinctly even before the onset of symptoms. About 30% of cognitively normal elderly, 60% of mild cognitively impaired subjects and about 80–90% of Alzheimer disease patients aged above 70 are amyloid positive (i.e., have significant amyloid deposition as seen on an amyloid scan; this corresponds to a value of 1.5 or more for PiB-PET scans). Figure 10.1 illustrates the PiB scans of a cognitively normal elderly individual as well as a mild and severe case of AD. As can be observed from the figure, even the mild case of AD has a significant amount of amyloid deposition.

FDG-PET Imaging

Normal metabolism shown by FDG-PET accumulation in cognitively normal elderly is shown in Figure 10.2. The hypometabolism seen in AD subjects on FDG-PET scans reflects metabolic deficits due to synaptic dysfunction and (probably) tau-mediated neuronal injury. The hypometabolism patterns are mainly seen in the posterior cingulate and parietal lobes. Unlike amyloid imaging, the hypometabolic changes seen on FDG-PET imaging are subtler in mild AD cases, becoming more dramatic and widespread in severe AD cases because FDG-PET captures changes in neuronal pathology.

MRI Imaging

The pattern of atrophy or neurodegeneration seen using sMRI is similar to the progression of neurofibrillary pathology (Braak stages) described above. It begins typically in the medial temporal lobe and becomes more widespread covering the entire neocortex later in the disease process. Visual and quantitative assessment of the degree of atrophy in the medial temporal lobe, specifically the hippocampus, is used as a metric to measure and track AD progression. Figure 10.3 illustrates the degree of atrophy seen in mild cases of AD where there is subtle atrophy in the medial temporal lobe before severe neocortical atrophy is seen in severe cases of AD.

Diffusion changes seen in DTI are typically seen in the same regions as the sMRI changes and add to the diagnostic information provided by sMRI.[30] Mean diffusivity increases in the medial temporal structures and fractional

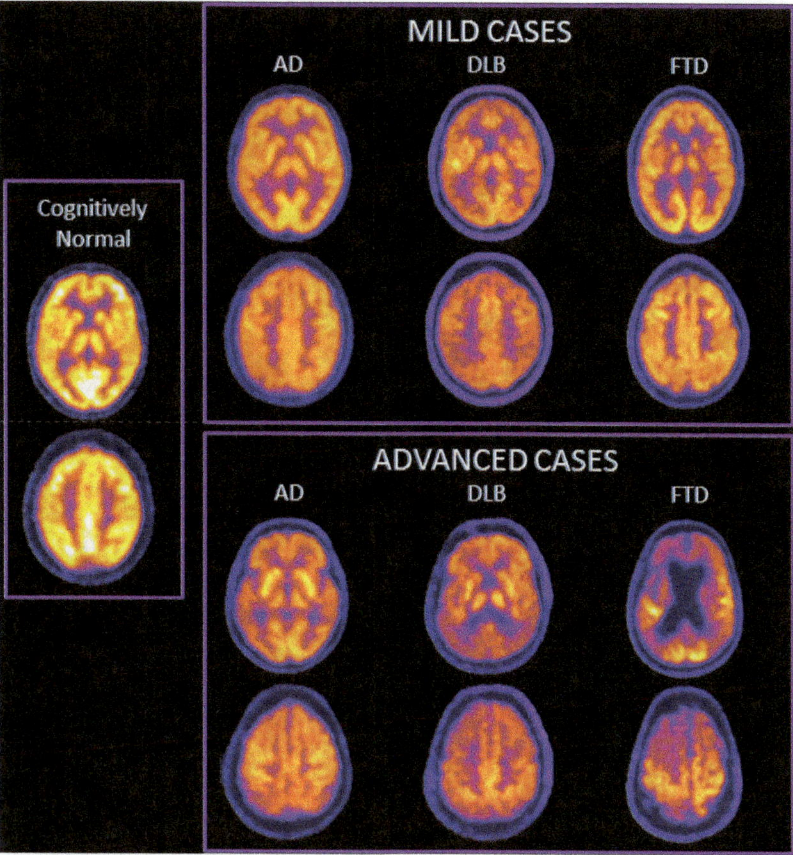

FIGURE 10.2 **Metabolic accumulation as measured by fluorodeoxyglucose (FDG)-PET imaging in cognitively normal elderly, mild cases of dementia, and advanced cases of dementia.** Each image is scaled by the pons accumulation to standardize the scale across images.

anisotrophy decreases in the uncinate fasciculus and fornix are commonly seen in AD. Connectivity changes seen on TF-fMRI in AD appear to be similar to changes of accelerated aging.[31] Since the DMN regions in TF-fMRI are related to episodic memory, the breakdown or decrease in functional connectivity of this network specifically in the posterior parts of the DMN is the most common finding in TF-fMRI AD studies.

DEMENTIA WITH LEWY BODIES

Clinical Manifestation of DLB

Dementia with Lewy bodies (DLB) is a clinical dementia syndrome characterized by cognitive impairment at presentation and three core features: spontaneous parkinsonism, fluctuations, and visual hallucinations.[32] According to the third report from the DLB consortium, two of the core features are necessary for a clinical diagnosis of probable DLB and one is necessary for possible DLB. Unlike AD, DLB patients often do not have episodic memory impairment early in their disease course. Cognitive domains that are typically affected early in DLB include: attention, visual perception, and visuospatial performance. According to research criteria, a "1-year rule" describes the onset of dementia 1 year prior to parkinsonism, versus Parkinson disease dementia whereupon parkinsonism precedes dementia by 1 year.[32] Parkinsonian features include bradykinesia, rigidity, postural instability, and tremor. Resting tremors, however, are less common in DLB patients.[33] Fluctuations can be described as daytime drowsiness, daytime sleeping for >2hours, "staring off," and/or disorganized speech.[34] Visual hallucinations can be an early feature of DLB and often involve seeing nonthreatening animals or people. The temporal onset of visual hallucinations relative to dementia onset has been found to distinguish DLB from AD.[35] Suggestive features were added to the third report to enhance the diagnosis of probable/possible DLB,[32] including: rapid eye movement (REM) sleep behavior disorder (RBD),[36] severe sensitivity to neuroleptics,[37] and low basal ganglia dopamine transporter accumulation as demonstrated by single-photon emission computed tomography (SPECT) or PET imaging.[38]

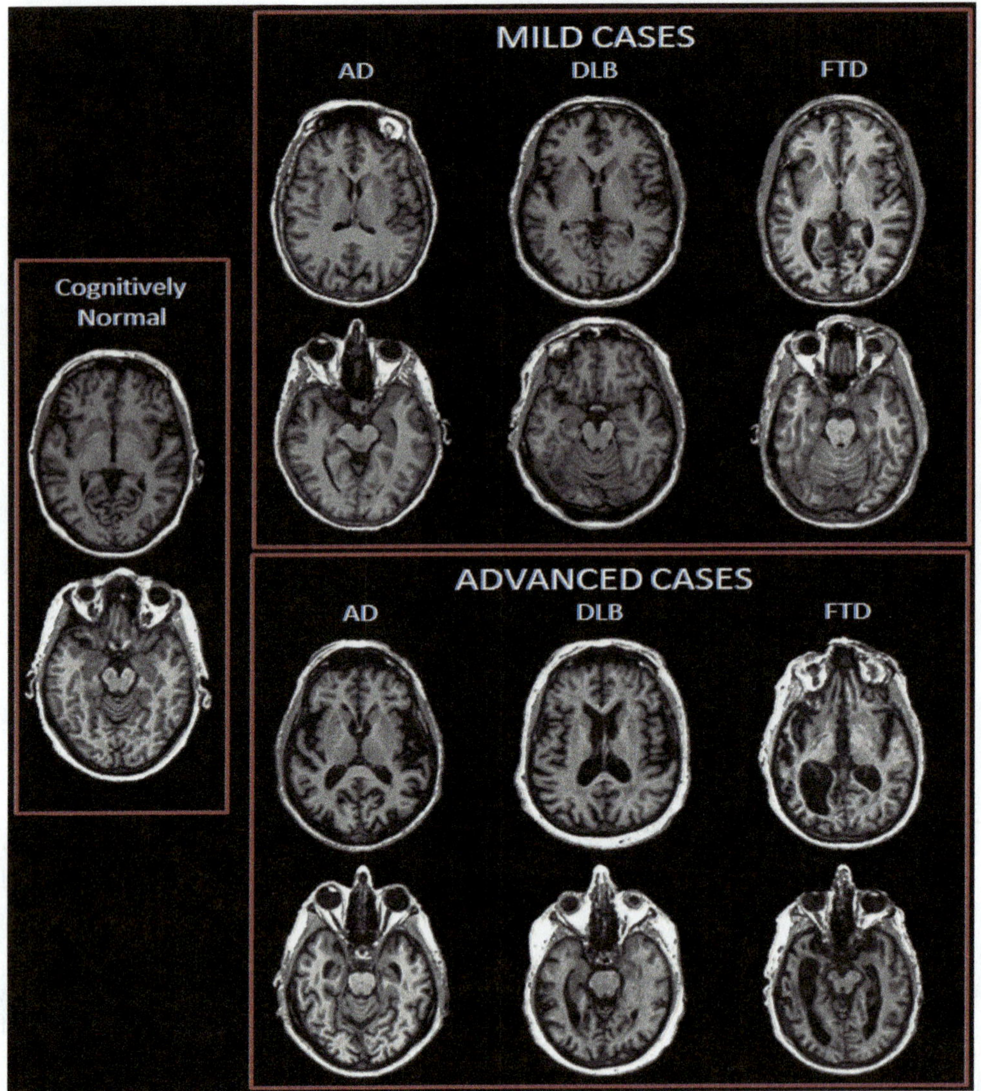

FIGURE 10.3 **Structural MRIs in typical cognitively normal elderly, mild cases of dementia, and advanced cases of dementia.**

Pathologies Underlying DLB

Macroscopic inspection of a brain with a clinical diagnosis of DLB is not typically associated with cerebral atrophy, although diffuse atrophy of frontal, temporal, and parietal lobes can be present. Sequential sections through the supratentorial tissues do not often reveal significant enlargement of the frontal or temporal horn of the lateral ventricle. The cortical gray mantle is usually normal in thickness and distribution. The substantia nigra and locus coeruleus will typically appear depigmented depending on the length of disease duration. The hippocampal formation is typically unremarkable. The range in macroscopic appearance likely reflects the extent of concomitant AD pathology, which has been reported in over half of patients with Lewy body disease.[36,39–41] Given these previous observations, the third report of the DLB consortium proposed a probability statement that incorporates the degree of AD pathology (i.e., Braak neurofibrillary tangle stage[22] or NIA–Reagan criteria[26]) versus distribution of Lewy body disease,[42] to assess the "likelihood" of DLB that is accounted for by neuropathologic findings.[32] Microscopic inspection of the brain should reveal α-synuclein pathology, which can be classified based on the progression of Lewy body disease pathology from the brainstem to limbic regions (hippocampus and amygdala) and to neocortices.[42] There exists a unique Alzheimer variant of Lewy body disease where the α-synuclein pathology is relatively limited to the amygdala, but otherwise consists of plaques and tangles associated with a pathologic diagnosis of AD.[43] A brain with amygdala predominant Lewy body disease is not considered in the probability statement for DLB likelihood.

In limbic and cortical brain regions with significant α-synuclein pathology, microvacuolation of the neuropil is often observed.[44] Microscopic inspection of the midbrain reveals severe neuronal loss in the ventrolateral cell group of the substantia nigra with Lewy bodies in residual neurons, which is considered to underlie a history of parkinsonism.[45] Attributing α-synuclein pathology to specific clinical syndromes in DLB is made difficult by concomitant AD pathology, which may interact synergistically or competitively with α-synuclein.[46]

Imaging in DLB

One of the key problems in the field of dementia diagnosis is the antemortem separation of DLB patients from AD. DLB patients are often misclassified as AD and the sensitivity of DLB diagnosis in different studies varies between 0% and 100%.[47] AD and Lewy body disease pathology both increase in prevalence dramatically with age. The pathological hallmarks of AD (NFTs and neuritic plaques) and Lewy body disease (Lewy bodies and Lewy neurites) often coexist, and there is conflicting evidence of hippocampal involvement in both diseases.[48,49] For this reason, amyloid imaging could play an important role in separating these two dementia subtypes.

Amyloid Imaging

About 60–80% of all DLB cases have medium–high load of amyloid, indicating concomitant AD.[50] Therefore amyloid imaging can be extremely useful in detecting the extent of coexisting AD to benefit the differential diagnosis and consideration of treatment options. In a recent study, clinical DLB subjects who showed reliable clinical improvement to the treatment of acetylcholinesterase inhibitors had significantly lower coexistent AD pathology based on amyloid and sMRI.[51] Figure 10.1 illustrates the PiB scans of a mild and severe case of DLB. As can be observed from the figure, there appears to be a very low extent of AD pathology in these DLB patients suggesting that the underlying Lewy body pathology may be the cause of the dementia symptoms.

FDG-PET

Glucose hypometabolism in the primary visual cortex is one of the key hallmarks of DLB and is associated with the clinical features in DLB. In some cases it is believed to be a prodromal sign of DLB. Sparing of the posterior cingulate relative to the precuneus has been proposed as a specific and sensitive FDG-PET imaging feature of DLB, termed the "cingulate island sign".[52,53] Figure 10.2 shows the hypometabolism patterns in a mild and severe case of DLB and there is considerable posterior as well as global hypometabolism in DLB patients.

MR Imaging

Since amygdala is one of the first cortical regions to be affected in DLB, there is significant sMRI atrophy seen in the amygdala. The extent of atrophy in DLB is much smaller in the medial temporal lobe compared to the atrophy seen in AD.[39,54,55] There are also significant decreases in gray matter density observed in the dorsal pontomesencephalic junction area with LBD.[56] Figure 10.3 illustrates the typical sMRI scan of a mild DLB patient (which is not very different when compared to a cognitively normal individual sMRI scan) and sMRI scan of an advanced DLB patient where there is more widespread atrophy observed along with some ventricular enlargement. DTI changes are also seen in the amygdala even in the absence of tissue loss seen on sMRI indicating possible microvacuolation.[57] While it has been shown that normal DMN connectivity as measured by TF-fMRI is disrupted in DLB patients, there is no clear consensus in the literature on the exact changes that are expected due to Lewy body pathology.

FRONTOTEMPORAL DEMENTIA

Clinical Manifestation of FTD

Broadly, the clinical syndromes of FTD can be divided into two categories: PPA and bvFTD.[58] bvFTD is the preferred clinical diagnostic term that describes the progressive decline in personality and social cognition.[59] Common clinical presentations associated with bvFTD can include apathy, disinhibition, new-onset pathologic compulsive gambling, repetitive or perseverative behaviors, hoarding, proclivity toward sweet food, or changes in emotional expression.[60–62] In an effort to enhance the specificity of clinical diagnosis, an international bvFTD criteria consortium was assembled.[63] It was agreed that probable bvFTD required significant functional deterioration, neuroimaging markers consistent with bvFTD, and three of the six core criteria. Core criteria include: early disinhibition, early apathy, early loss of sympathy, early perseveration, hyperorality/changes in eating habits, and a neuropsychological profile consistent with bvFTD (i.e., executive dysfunction, preservation of episodic memory and visuospatial skills).[63] Suspected bvFTD patients can be excluded if a non-neurodegenerative or medical disorder can account for the pattern of deficits, psychiatric diagnoses can explain behavioral changes, or there is biomarker evidence of AD or other neurodegenerative disease. The insidious deterioration of language skills characterizes PPA, which can be further classified into: nonfluent/agrammatic PPA (PPA-G), semantic dementia or semantic PPA (PPA-S), or logopenic PPA (PPA-L).[60,64,65] PPA-G is characterized by relative preservation of single-word comprehension when a failure to comprehend word order or other aspect of grammar (written or spoken) is impaired. The patient's speech can be effortful and hesitant with an observed impairment in fluency. PPA-S is characterized by relative preservation of grammar and fluency when single-word comprehension and naming is impaired. PPA-L is characterized by phonemic paraphasis, sporadic word-finding hesitations, and impairments in spelling. Naming can be impaired PPA-L, but not to the extent observed in PPA-S.[66]

Pathology Underlying FTD

Frontemporal lobar degeneration (FTLD) encompasses a long list of proteinopathies that can be found in sporadic patients, as well as unknown proteinopathies found in familial forms of FTLD. Macroscopic examination can reveal mild-to-severe cortical atrophy of the frontal, temporal and, to a lesser extent, parietal lobe. Greater than 70% of FTLD variant with TAR DNA-binding protein 43 (FTLD-TDP) brains will have coexisting hippocampal sclerosis, which underlies medial temporal lobe atrophy.[67] Sequential sections through the supratentorial tissues can reveal marked enlargement of the lateral ventricles dependent upon the extent of cortical involvement. The cortical gray mantle may appear thinned, but be normal in distribution. Microscopic examination of affected cortical regions will typically reveal neuronal loss, reactive astrogliosis, activated microglia, and superficial (laminar) spongiosis. A consensus of the FTLD consortium updated the FTLD neuropathologic diagnostic criteria in 2007.[68] FTLD types are described based on the predominance of abnormal accumulation of proteins. Given the breadth of proteins discovered, the readers are directed to the consortium criteria,[68] as well as an in-depth review that provides a logical investigation in order to demystify FTLDs.[69] Briefly, proteins implicated in FTLD include: tau (FTLD-tau [corticobasal degeneration, progressive supranuclear palsy, multiple system tauopathy, Pick disease] and FTLD with mutations in microtubule associated protein tau); TAR DNA-binding protein 43 (FTLD-TDP); and fused in sarcoma (FTLD-FUS [neuronal intermediate filament inclusion disease, basophilic inclusion body disease]). FTLD inclusions labeled by ubiquitin proteasome system (FLD-UPS) that are negative for tau, TDP-43, and FUS have an as yet unidentified protein, but have been found associated with mutations (*CHMP2B* and *C9ORF72*).

Imaging in FTD

Amyloid Imaging

Most of the clinically diagnosed FTD cases do not show any amyloid deposition related to AD. Figure 10.1 illustrates the typical amyloid scans in mild and severe cases of FTD.

FDG-PET and MRI

There is a significant literature that has demonstrated progressive degeneration of the frontal and temporal lobes, with relative sparing of the parietal and occipital lobes, in frontotemporal dementia.[70–73] Figures 10.2 and 10.3 illustrate the extent of hypometabolism and atrophy seen in a typical FTD patient, which becomes more severe in advanced FTD patients. The specific patterns of hypometabolism and atrophy typically vary according to the specific clinical syndrome.[74] Here we discuss the hypometabolism and atrophy patterns in the two common pathologies underlying FTD—FTLD-tau and FTLD-TDP.

FTLD-TAU PATHOLOGIES

With FTLD-tau pathologies, the following are the primary clinicopathologic subtypes and the corresponding atrophy and hypometabolism patterns:

PROGRESSIVE SUPRANUCLEAR PALSY (PSP) PSP is a sporadic neurodegenerative disorder with prominent hyperphosphorylated tau aggregates in the brain accompanied by neuronal loss and gliosis. PSP is characterized by significant atrophy and metabolic changes in the brainstem with additional involvement of cortical regions specifically the medial frontal regions.[75] Atrophy of the midbrain on a midsagittal MRI, described as the "hummingbird sign," is a useful predictor of PSP.[76] Visual assessment or quantification of atrophy in the superior cerebellar peduncle on MRI significantly increases accuracy of the clinical diagnosis.[77]

CORTICOBASAL DEGENERATION (CBD) CBD is a rare neurodegenerative disorder classified as a primary 4R tauopathy due to neuronal and glial aggregates of hyperphosphorylated tau in both gray and white matter of the neocortex, basal ganglia, thalamus, and, to a lesser extent, the brainstem of these patients.[78] CBD is characterized by significant focal atrophy and metabolic changes that are typically asymmetric and are observed in the frontoparietal regions with involvement of subcortical structures.[79,80] Additionally the rates of global atrophy observed in CBD are significantly higher than other neurodegenerative disorders.[81]

PICK DISEASE (PiD) PiD is a rare form of FTLD-tau that is associated with severe circumscribed cortical atrophy of frontal and temporal lobes, described as "knife-edge" atrophy of cortical gyri. PiD is associated with widespread metabolic abnormality and atrophy in the frontal regions and to a lesser extent in the temporal lobe regions.[82,83]

FTLD-TDP PATHOLOGIES

With FTLD-TDP pathologies, there are four distinct subtypes of the disease. Since the recent reclassification of the pathological staging, there are only a few studies that have characterized the imaging changes associated with FTLD-TDP pathologies. Patients with FTLD-TDP type 1 show widespread atrophy in the temporal, frontal and parietal lobes; patients with FTLD-TDP type 2 show anteromedial temporal atrophy; and patients with FTLD-TDP type 3 predominantly have frontal lobe atrophy.[83]

IMAGING VASCULAR DISEASE

Cerebrovascular disease (CVD) is the second most prevalent pathology that is associated with age related cognitive decline.[84,85] The hallmarks of CVD are the presence of microvascular changes (white matter hyperintensities) and macrovascular changes (subcortical and cortical infarcts). It is important to account for the presence and severity of CVD because CVD worsens the effect of pathologies discussed so far on the clinical expression of dementia.[86–89] Fluid-attenuated inversion recovery (FLAIR)-MRI is considered a reasonable proxy for tissue damage due to CVD. White matter hyper intensity (leukoaraiosis) on FLAIR-MRI in elderly persons is suspected to be a direct manifestation of microvascular ischemic injury in the distribution of the penetrating arteriolar vessels. Figure 10.4 illustrates a FLAIR-MRI image in an elderly subject with a large cortical infarct, subcortical infarcts, and white hyper intensities.

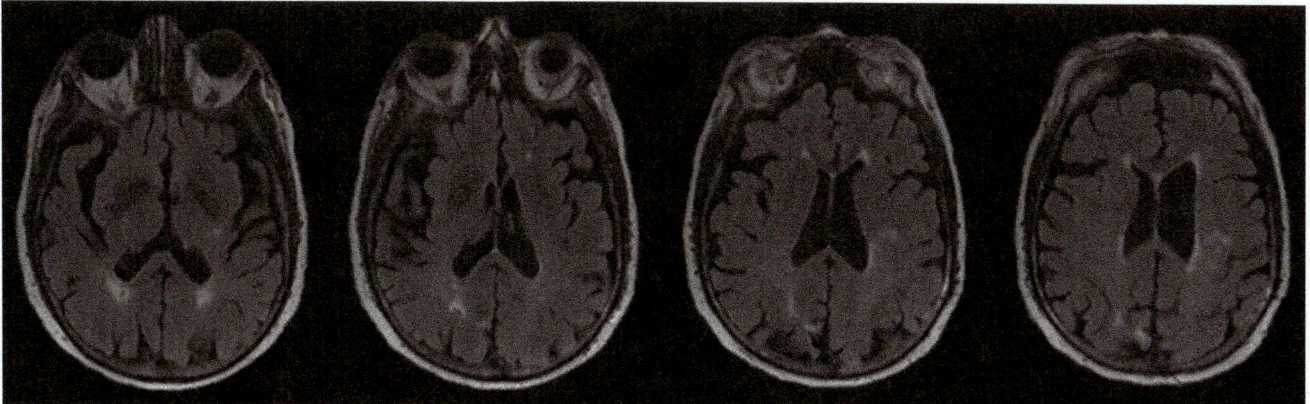

FIGURE 10.4 **Fluid-attenuated inversion recovery (FLAIR)-MRI images depicting the extent of vascular disease in the elderly patient.** Subcortical infarcts are seen on the first two slices and cortical infarct on the last two slices. Also seen are white hyperintensities indicating microvascular changes.

References

1. Aarsland D, Rongve A, Nore SP, et al. Frequency and case identification of dementia with Lewy bodies using the revised consensus criteria. *Dement Geriatr Cogn Disord*. 2008;26(5):445–452.
2. Jellinger KA, Wenning GK, Seppi K. Predictors of survival in dementia with Lewy bodies and Parkinson dementia. *Neurodegener Dis*. 2007;4(6):428–430.
3. Ratnavalli E, Brayne C, Dawson K, Hodges JR. The prevalence of frontotemporal dementia. *Neurology*. 2002;58(11):1615–1621.
4. Rosso SM, Donker Kaat L, Baks T, et al. Frontotemporal dementia in the Netherlands: patient characteristics and prevalence estimates from a population-based study. *Brain*. 2003;126(9):2016–2022.
5. Gómez-Isla T, Hollister R, West H, et al. Neuronal loss correlates with but exceeds neurofibrillary tangles in Alzheimer's disease. *Ann Neurol*. 1997;41(1):17–24.
6. Zarow C, Vinters HV, Ellis WG, et al. Correlates of hippocampal neuron number in Alzheimer's disease and ischemic vascular dementia. *Ann Neurol*. 2005;57(6):896–903.
7. Bobinski M, de Leon MJ, Wegiel J, et al. The histological validation of post mortem magnetic resonance imaging-determined hippocampal volume in Alzheimer's disease. *Neuroscience*. 2000;95(3):721–725.
8. Klunk WE, Engler H, Nordberg A, et al. Imaging brain amyloid in Alzheimer's disease with Pittsburgh Compound-B. *Ann Neurol*. 2004;55(3):306–319.
9. Ikonomovic MD, Klunk WE, Abrahamson EE, et al. Post-mortem correlates of in vivo PiB-PET amyloid imaging in a typical case of Alzheimer's disease. *Brain*. 2008;131(Pt 6):1630–1645.
10. Bacskai BJ, Frosch MP, Freeman SH, et al. Molecular imaging with Pittsburgh Compound B confirmed at autopsy: a case report. *Arch Neurol*. 2007;64(3):431–434.
11. Schwartz WJ, Smith CB, Davidsen L, et al. Metabolic mapping of functional activity in the hypothalamo-neurohypophysial system of the rat. *Science*. 1979;205(4407):723–725.
12. Attwell D, Laughlin SB. An energy budget for signaling in the grey matter of the brain. *J Cereb Blood Flow Metab*. 2001;21(10):1133–1145.
13. Rocher AB, Chapon F, Blaizot X, et al. Resting-state brain glucose utilization as measured by PET is directly related to regional synaptophysin levels: a study in baboons. *Neuroimage*. 2003;20(3):1894–1898.
14. Raichle ME, MacLeod AM, Snyder AZ, et al. A default mode of brain function. *Proc Natl Acad Sci U S A*. 2001;98(2):676–682.
15. Alladi S, Xuereb J, Bak T, et al. Focal cortical presentations of Alzheimer's disease. *Brain*. 2007;130(Pt 10):2636–2645.
16. Murray ME, Graff-Radford NR, Ross OA, et al. Neuropathologically defined subtypes of Alzheimer's disease with distinct clinical characteristics: a retrospective study. *Lancet Neurol*. 2011;10(9):785–796.
17. Rabinovici GD, Jagust WJ, Furst AJ, et al. Abeta amyloid and glucose metabolism in three variants of primary progressive aphasia. *Ann Neurol*. 2008;64(4):388–401.
18. McKhann GM, Knopman DS, Chertkow H, et al. The diagnosis of dementia due to Alzheimer's disease: recommendations from the National Institute on Aging–Alzheimer's Association workgroups on diagnostic guidelines for Alzheimer's disease. *Alzheimers Dement*. 2011;7(3):263–269.
19. McKhann G, Drachman D, Folstein M, et al. Clinical diagnosis of Alzheimer's disease: report of the NINCDS-ADRDA Work Group under the auspices of Department of Health and Human Services Task Force on Alzheimer's Disease. *Neurology*. 1984;34(7):939–944.
20. Swanberg MM, Tractenberg RE, Mohs R, Thal LJ, Cummings JL. Executive dysfunction in Alzheimer disease. *Arch Neurol*. 2004;61(4):556–560.
21. Halliday GM, Double KL, Macdonald V, Kril JJ. Identifying severely atrophic cortical subregions in Alzheimer's disease. *Neurobiol Aging*. 2003;24(6):797–806.
22. Braak H, Braak E. Neuropathological stageing of Alzheimer-related changes. *Acta Neuropathol*. 1991;82(4):239–259.
23. Mirra SS, Heyman A, McKeel D, et al. The Consortium to Establish a Registry for Alzheimer's Disease (CERAD). Part II. Standardization of the neuropathologic assessment of Alzheimer's disease. *Neurology*. 1991;41(4):479–486.
24. Morris JC, Heyman A, Mohs RC, et al. The Consortium to Establish a Registry for Alzheimer's Disease (CERAD). Part I. Clinical and neuropsychological assessment of Alzheimer's disease. *Neurology*. 1989;39(9):1159–1165.
25. Thal DR, Rüb U, Orantes M, Braak H. Phases of A beta-deposition in the human brain and its relevance for the development of AD. *Neurology*. 2002;58(12):1791–1800.
26. Consensus recommendations for the postmortem diagnosis of Alzheimer's disease. The National Institute on Aging, and Reagan Institute Working Group on Diagnostic Criteria for the Neuropathological Assessment of Alzheimer's Disease. *Neurobiol Aging*. 1997;18(4 suppl), S1-2.
27. Jack Jr. CR, Knopman DS, Jagust WJ, et al. Tracking pathophysiological processes in Alzheimer's disease: an updated hypothetical model of dynamic biomarkers. *Lancet Neurol*. 2013;12(2):207–216.
28. DeKosky ST, Scheff SW. Synapse loss in frontal cortex biopsies in Alzheimer's disease: correlation with cognitive severity. *Ann Neurol*. 1990;27(5):457–464.
29. Terry RD, Masliah E, Salmon DP, et al. Physical basis of cognitive alterations in Alzheimer's disease: synapse loss is the major correlate of cognitive impairment. *Ann Neurol*. 1991;30(4):572–580.
30. Kantarci K, Petersen RC, Boeve BF, et al. DWI predicts future progression to Alzheimer disease in amnestic mild cognitive impairment. *Neurology*. 2005;64(5):902–904.
31. Jones DT, Machulda MM, Vemuri P, et al. Age-related changes in the default mode network are more advanced in Alzheimer disease. *Neurology*. 2011;77(16):1524–1531.
32. McKeith IG, Dickson DW, Lowe J, et al. Diagnosis and management of dementia with Lewy bodies: third report of the DLB Consortium. *Neurology*. 2005;65(12):1863–1872.
33. Galasko D, Katzman R, Salmon DP, Hansen L. Clinical and Neuropathological Findings in Lewy Body Dementias. *Brain Cogn*. 1996;31(2):166–175.
34. Ferman TJ, Smith GE, Boeve BF, et al. DLB fluctuations: specific features that reliably differentiate DLB from AD and normal aging. *Neurology*. 2004;62(2):181–187.

35. Ferman TJ, Arvanitakis Z, Fujishiro H, et al. Pathology and temporal onset of visual hallucinations, misperceptions and family misidentification distinguishes dementia with Lewy bodies from Alzheimer's disease. *Parkinsonism Relat Disord*. 2013;19(2):227–231.
36. Ferman TJ, Boeve BF, Smith GE, et al. Inclusion of RBD improves the diagnostic classification of dementia with Lewy bodies. *Neurology*. 2011;77(9):875–882.
37. Aarsland D, Perry R, Larsen JP, et al. Neuroleptic sensitivity in Parkinson's disease and parkinsonian dementias. *J Clin Psychiatry*. 2005;66(5):633–637.
38. McKeith I, O'Brien J, Walker Z, et al. Sensitivity and specificity of dopamine transporter imaging with 123I-FP-CIT SPECT in dementia with Lewy bodies: a phase III, multicentre study. *Lancet Neurol*. 2007;6(4):305–313.
39. Gómez-Isla T, Growdon WB, McNamara M, et al. Clinicopathologic correlates in temporal cortex in dementia with Lewy bodies. *Neurology*. 1999;53(9):2003–2009.
40. Galasko D, Hansen LA, Katzman R, et al. Clinical-neuropathological correlations in Alzheimer's disease and related dementias. *Arch Neurol*. 1994;51(9):888–895.
41. Schneider JA, Arvanitakis Z, Bang W, Bennett DA. Mixed brain pathologies account for most dementia cases in community-dwelling older persons. *Neurology*. 2007;69(24):2197–2204.
42. Kosaka K, Yoshimura M, Ikeda K, Budka H. Diffuse type of Lewy body disease: progressive dementia with abundant cortical Lewy bodies and senile changes of varying degree–a new disease? *Clin Neuropathol*. 1984;3(5):185–192.
43. Uchikado H, Lin WL, DeLucia MW, Dickson DW. Alzheimer disease with amygdala Lewy bodies: a distinct form of alpha-synucleinopathy. *J Neuropathol Exp Neurol*. 2006;65(7):685–697.
44. Hansen LA, Masliah E, Terry RD, Mirra SS. A neuropathological subset of Alzheimer's disease with concomitant Lewy body disease and spongiform change. *Acta Neuropathol*. 1989;78(2):194–201.
45. Dickson DW, Braak H, Duda JE, et al. Neuropathological assessment of Parkinson's disease: refining the diagnostic criteria. *Lancet Neurol*. 2009;8(12):1150–1157.
46. Jellinger KA, Attems J. Prevalence and impact of vascular and Alzheimer pathologies in Lewy body disease. *Acta Neuropathol*. 2008;115(4):427–436.
47. McKeith I, Mintzer J, Aarsland D, et al. Dementia with Lewy bodies. *Lancet Neurol*. 2004;3(1):19–28.
48. Klucken J, McLean PJ, Gomez-Tortosa E, et al. Neuritic alterations and neural system dysfunction in Alzheimer's disease and dementia with Lewy bodies. *Neurochem Res*. 2003;28(11):1683–1691.
49. Kantarci K, Lowe VJ, Boeve BF, et al. Multimodality imaging characteristics of dementia with Lewy bodies. *Neurobiol Aging*. 2012;33(9):2091–2105.
50. Kantarci K. Molecular imaging of Alzheimer disease pathology. *AJNR Am J Neuroradiol*. 2014;35:S31–36S.
51. Graff-Radford J, Boeve BF, Pedraza O, et al. Imaging and acetylcholinesterase inhibitor response in dementia with Lewy bodies. *Brain*. 2012;135(Pt 8):2470–2477.
52. Imamura T, Ishii K, Sasaki M, et al. Regional cerebral glucose metabolism in dementia with Lewy bodies and Alzheimer's disease: a comparative study using positron emission tomography. *Neurosci Lett*. 1997;235(1–2):49–52.
53. Lim SM, Katsifis A, Villemagne VL, et al. The 18F-FDG PET cingulate island sign and comparison to 123I-beta-CIT SPECT for diagnosis of dementia with Lewy bodies. *J Nucl Med*. 2009;50(10):1638–1645.
54. Lippa CF, Johnson R, Smith TW. The medial temporal lobe in dementia with Lewy bodies: a comparative study with Alzheimer's disease. *Ann Neurol*. 1998;43(1):102–106.
55. Hashimoto M, Kitagaki H, Imamura T, et al. Medial temporal and whole-brain atrophy in dementia with Lewy bodies: a volumetric MRI study. *Neurology*. 1998;51(2):357–362.
56. Schmeichel AM, Buchhalter LC, Low PA, et al. Mesopontine cholinergic neuron involvement in Lewy body dementia and multiple system atrophy. *Neurology*. 2008;70(5):368–373.
57. Kantarci K, Avula R, Senjem ML, et al. Dementia with Lewy bodies and Alzheimer disease: neurodegenerative patterns characterized by DTI. *Neurology*. 2010;74(22):1814–1821.
58. Piguet O, Hornberger M, Mioshi E, Hodges JR. Behavioural-variant frontotemporal dementia: diagnosis, clinical staging, and management. *Lancet Neurol*. 2011;10(2):162–172.
59. Kipps CM, Mioshi E, Hodges JR. Emotion, social functioning and activities of daily living in frontotemporal dementia. *Neurocase*. 2009;15(3):182–189.
60. Neary D, Snowden JS, Gustafson L, et al. Frontotemporal lobar degeneration: a consensus on clinical diagnostic criteria. *Neurology*. 1998;51(6):1546–1554.
61. Manes FF, Torralva T, Roca M, et al. Frontotemporal dementia presenting as pathological gambling. *Nat Rev Neurol*. 2010;6(6):347–352.
62. Postiglione A, Milan G, Pappatà S, et al. Fronto-temporal dementia presenting as Geschwind's syndrome. *Neurocase*. 2008;14(3):264–270.
63. Rascovsky K, Hodges JR, Knopman D, et al. Sensitivity of revised diagnostic criteria for the behavioural variant of frontotemporal dementia. *Brain*. 2011;134(9):2456–2477.
64. Hodges JR, Patterson K. Semantic dementia: a unique clinicopathological syndrome. *Lancet Neurol*. 2007;6(11):1004–1014.
65. Grossman M. Primary progressive aphasia: clinicopathological correlations. *Nat Rev Neurol*. 2010;6(2):88–97.
66. Mesulam M, Wieneke C, Rogalski E, et al. Quantitative template for subtyping primary progressive aphasia. *Arch Neurol*. 2009;66(12):1545–1551.
67. Josephs KA, Dickson DW. Hippocampal sclerosis in tau-negative frontotemporal lobar degeneration. *Neurobiol Aging*. 2007;28(11):1718–1722.
68. Cairns NJ, Bigio EH, Mackenzie IR, et al. Neuropathologic diagnostic and nosologic criteria for frontotemporal lobar degeneration: consensus of the Consortium for Frontotemporal Lobar Degeneration. *Acta Neuropathol*. 2007;114(1):5–22.
69. Bigio EH. Making the diagnosis of frontotemporal lobar degeneration. *Arch Pathol Lab Med*. 2013;137(3):314–325.
70. Broe M, Hodges JR, Schofield E, et al. Staging disease severity in pathologically confirmed cases of frontotemporal dementia. *Neurology*. 2003;60(6):1005–1011.
71. Seeley WW, Crawford R, Rascovsky K, et al. Frontal paralimbic network atrophy in very mild behavioral variant frontotemporal dementia. *Arch Neurol*. 2008;65(2):249–255.

72. Kril JJ, Macdonald V, Patel S, Png F, Halliday GM. Distribution of brain atrophy in behavioral variant frontotemporal dementia. *J Neurol Sci.* 2005;232(1–2):83–90.
73. Whitwell JL, Jack Jr. CR, Senjem ML, et al. MRI correlates of protein deposition and disease severity in postmortem frontotemporal lobar degeneration. *Neurodegener Dis.* 2009;6(3):106–117.
74. Rosen HJ, Gorno-Tempini ML, Goldman WP, et al. Patterns of brain atrophy in frontotemporal dementia and semantic dementia. *Neurology.* 2002;58(2):198–208.
75. Eckert T, Tang C, Ma Y, et al. Abnormal metabolic networks in atypical parkinsonism. *Mov Disord.* 2008;23(5):727–733.
76. Kato N, Arai K, Hattori T. Study of the rostral midbrain atrophy in progressive supranuclear palsy. *J Neurol Sci.* 2003;210(1–2):57–60.
77. Paviour DC, Price SL, Stevens JM, Lees AJ, Fox NC. Quantitative MRI measurement of superior cerebellar peduncle in progressive supranuclear palsy. *Neurology.* 2005;64(4):675–679.
78. Dickson DW, Bergeron C, Chin SS, et al. Office of Rare Diseases neuropathologic criteria for corticobasal degeneration. *J Neuropathol Exp Neurol.* 2002;61(11):935–946.
79. Hosakaa K, Ishii K, Sakamotoa S, et al. Voxel-based comparison of regional cerebral glucose metabolism between PSP and corticobasal degeneration. *J Neurol Sci.* 2002;199(1–2):67–71.
80. Josephs KA, Whitwell JL, Boeve BF, et al. Rates of cerebral atrophy in autopsy-confirmed progressive supranuclear palsy. *Ann Neurol.* 2006;59(1):200–203.
81. Whitwell JL, Jack Jr. CR, Parisi JE, et al. Rates of cerebral atrophy differ in different degenerative pathologies. *Brain.* 2007;130(Pt 4):1148–1158.
82. Ishii K, Sakamoto S, Sasaki M, et al. Cerebral glucose metabolism in patients with frontotemporal dementia. *J Nucl Med.* 1998;39(11):1875–1878.
83. Whitwell JL, Josephs KA. Neuroimaging in frontotemporal lobar degeneration–predicting molecular pathology. *Nat Rev Neurol.* 2011;8(3):131–142.
84. Schneider JA, Aggarwal NT, Barnes L, et al. The neuropathology of older persons with and without dementia from community versus clinic cohorts. *J Alzheimers Dis.* 2009;18(3):691–701.
85. Jellinger KA. Understanding the pathology of vascular cognitive impairment. *J Neurol Sci.* 2005;229–230:57–63.
86. Nagy Z, Esiri MM, Jobst KA, et al. The effects of additional pathology on the cognitive deficit in Alzheimer disease. *J Neuropathol Exp Neurol.* 1997;56(2):165–170.
87. Snowdon DA, Greiner LH, Mortimer JA, et al. Brain infarction and the clinical expression of Alzheimer disease. The Nun Study. *JAMA.* 1997;277(10):813–817.
88. Weller RO, Boche D, Nicoll JA. Microvasculature changes and cerebral amyloid angiopathy in Alzheimer's disease and their potential impact on therapy. *Acta Neuropathol.* 2009;118(1):87–102.
89. Zekry D, Duyckaerts C, Moulias R, et al. Degenerative and vascular lesions of the brain have synergistic effects in dementia of the elderly. *Acta Neuropathol.* 2002;103(5):481–487.

CHAPTER

11

Cognitive Enhancers and Mental Impairment: Emerging Ethical Issues

Fabrice Jotterand[*,†], *Jennifer L. McCurdy*[*], *and Bernice Elger*[†]

[*]Regis University, Denver, CO, USA

[†]University of Basel, Basel, Switzerland

INTRODUCTION

In the last 10 years, the use of neurotechnology, including psychopharmacology and neurostimulation, has been increasingly accepted in clinical practice for the treatment of individuals with cognitive disabilities resulting from neurodegenerative disorders such as schizophrenia, Parkinson disease (PD), Alzheimer disease (AD), etc., or traumatic brain injury (TBI).[1] There is evidence that their usage diminishes the effects of neurodegenerative diseases (e.g., cognitive decline) and improves patient quality of life.[2] While their therapeutic benefits provide a strong incentive to an even broader acceptance in the clinical setting, there are ethical challenges arising from the manipulation and modification of brain functions using psychopharmacology and neurostimulation in healthy individuals, which is usually understood in the literature as *enhancement*. This chapter explores a third use of neurotechnology that neither fits squarely into the category of treatment, nor carries some of the moral squeamishness of pure enhancement in healthy persons. This third category is the enhancement of those persons who have baseline cognitive impairments related to a mental disability, but are relatively healthy otherwise. For the remainder of this chapter, these two types of enhancement will be referred to as "enhancement-1" (enhancement of healthy individuals with no cognitive impairments) and "enhancement-2" (enhancement of healthy individuals with cognitive impairments). Specifically, the aim of this chapter is threefold: 1) to delineate between the two types of neuroenhancement (enhancement-1 and enhancement-2); 2) to discuss the issues surrounding personal identity in the use of neurotechnologies; and 3) to outline ethical dimensions of enhancement-2. The question that remains is whether the use of cognitive enhancers to increase functional abilities and improve long-term baseline functioning (in contrast to treating an illness or disease) in the cognitively impaired is morally justified recognizing the potential change of personal identity as a side effect.

In a previous publication, Jotterand and Giordano examine whether changes in personal identity as a side effect of the use of deep brain stimulation (DBS) or transcranial magnetic stimulation (TMS) is ethically acceptable.[3] They conclude that "neurostimulation therapy, as a doctoring act, should be directed, and adherent to goals of restoring and/or preserving, patients' personal identity" if at all possible, and "... any change in personality and/or personal identity that occurs as a consequence of the treatment would need to be regarded as consistent with both [the goal of restoring health] and [the definitions of health and normality] in order to be acceptable."[3] This chapter considers cases where the definitions of health and normality cannot provide the basis for intervention because they do not establish realistic goals (i.e., the restoration of health or attainment of normality). Patients with impairments related to genetics, illness, or injury (i.e., cerebral palsy with accompanying developmental delay) will not, at least in the current state of our knowledge, regain or achieve levels of functioning consistent with the nondisabled. These patients may be limited in "major life activities" such as "caring for oneself, performing manual tasks, seeing, hearing, eating, sleeping, walking, standing, sitting, reaching, lifting, bending, speaking, breathing, learning, reading, concentrating, thinking, communicating, interacting with others, and working."[4] Despite the inability to fully overcome

http://dx.doi.org/10.1016/B978-0-12-410529-4.00011-5

impairments, there is evidence that psychopharmacology and neurostimulation techniques can enhance cognitive functions such as attention, learning, memory, and executive abilities, as well as mood and behavior, and hence improve the quality of life of individuals. The drawback of these interventions is that these patients may incur a change in personal identity as a side effect. But considering that these enhancement procedures could help physically and cognitively impaired patients to reach a greater level of functioning, albeit without achieving full governance of all major life activities, assumptions about the ethical implications of human enhancement need to be challenged.

NEUROETHICS IN CONTEXT

This chapter builds on previous work that has identified many of the ethical implications in the use of cognitive enhancers in persons with mental disabilities. The field of neuroethics has emerged in response to issues raised by rapid advances in neuroscience and the development of novel neurotechnology. Ethical and policy implications have been identified by Arthur Caplan and Martha J. Farah in the fourth edition of *Rosenberg's Molecular and Genetic Basis of Neurological and Psychiatric Disease* with regard to several applications of neurotechnology: diagnostic testing, brain imaging, treatment, and enhancement. Ethical issues they associated with diagnostic testing include: 1) the lack of established standards for user training, patient counseling about the interpretation of the findings, and other clinical applications; 2) concerns about privacy of personal information in relation to third parties; 3) discrimination by health insurance companies, employers, and the undermining of immigration rights; 4) loss of privacy and personal autonomy; and 5) disregard of proper informed consent standards.[5] In addition to ethical issues surrounding diagnostic testing, considerations regarding brain imaging include: 1) the potential stigmatization of individuals with mental illness; 2) hasty interventions based on the probabilistic nature of imaging; and 3) concerns about the autonomy and privacy of the individual.[5]

Caplan and Farah state that "the most intriguing ethical challenge facing those in the clinical neurosciences involves whether and how to set limits on clinical application".[5] They suggest three clinical applications of neurotechnology: treatment of disease, enhancement (which they label cosmetic neurology), and prophylactic use prior to any symptoms. In their analysis of enhancement specifically, they identify several issues which include: 1) a disregard for the value of pain and suffering in attaining success; 2) the raising of standards of "normalcy" and a resultant pressure for using enhancements; 3) the potential presence of an unequal playing field due to an individual's desire to gain an edge for success, 3) an inequality of access to enhancement technology; and 4) the argument that enhancement is simply unnatural. Their discussion of enhancement focuses primarily on the use of neurotechnology in healthy individuals. What follows is the discussion of yet another nuance of the moral implications of the clinical use of neurotechnology not mentioned elsewhere, that of the cognitive enhancement of healthy but cognitively disabled individuals.

THERAPY–ENHANCEMENT: A FALSE DICHOTOMY

Progress in the development of neurotechnology continues to challenge our ability to clearly differentiate between therapy and enhancement. Such distinction is increasingly difficult because in some instances therapeutic interventions require the enhancement of some capacities for individuals to improve their health status (vaccines, prosthetics, etc.). Therapy implies the restoration, partial or complete, of biological functions. Therapeutic interventions may enhance certain traits or abilities beyond the individual's baseline, but the intrinsic goal is to treat a disorder caused by a particular pathology or traumatic event.[6] Enhancement, on the other hand, has usually been referred to as the augmentation of biological capacities beyond what is species-typical. The nature of these interventions focuses on the improvement of the nondisabled and otherwise healthy individual's capacities for nonhealth related purposes.[6] But a third, and until now overlooked, category of the clinical application of neurotechnology resides in the treatment of disabled individuals who have limited functioning but are otherwise stable and healthy.

Health and Disease

Although it is beyond the scope of this chapter to wrestle with the concepts of health, disability, normality, and disease at length, the meaning of health requires clarification in order to frame this challenge to the therapy–enhancement dichotomy. The term is problematic in that it is a slippery concept that is socially constructed. Many have attempted to define health relative to normative standards and according to what is considered species-typical.[7,8] But a definition that proves to be more useful for the disabled population is one that is more contextualized and that defines health relative to the individual's baseline, rather than that of the species. According to Jotterand and Tenzin,

TABLE 11.1 Enhancement-1 vs. Enhancement-2

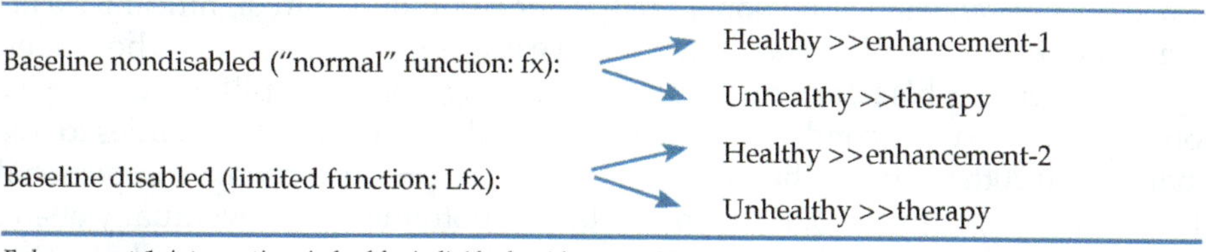

Baseline nondisabled ("normal" function: fx):	Healthy >>enhancement-1
	Unhealthy >>therapy
Baseline disabled (limited function: Lfx):	Healthy >>enhancement-2
	Unhealthy >>therapy

Enhancement-1: interventions in healthy individuals with no cognitive impairments. Enhancement-2: interventions in healthy individuals with cognitive impairments.

"… any conceptualization of health and disease needs a careful explanation of its nature and meaning in order to become relevant to particular contexts and populations…."[9] For example, a child with Down syndrome might have a baseline stable cognitive deficit, but is otherwise free of illness and suffering, and is living a life maximized in every way possible in relation to her baseline.

For this reason, an important distinction between enhancement-1 and enhancement-2 is needed. Enhancement-1 involves the enhancement of healthy individuals with no cognitive impairments whereas enhancement-2 consists of the enhancement of healthy individuals with cognitive impairments (Table 11.1).

Distinguishing between these two types of enhancement avoids conflating two concepts that possess important moral distinctions relevant to the discussion on the use of cognitive enhancers in the cognitively impaired.

The increasing use of enhancers in the clinical context seems to indicate that at least some clinicians disagree that enhancement procedures are intrinsically wrong and therapeutics are necessarily good, as is often portrayed in the therapy–enhancement debate over the last decade. Since the bioconservative camp and the pro-enhancement camp advance their positions without concessions, the debate has become somewhat sterile. Alan Buchanan notes that the low quality of the debate is related to the following five points: 1) the use of rhetoric to justify ill-founded arguments; 2) the disregard of the fundamentals of evolutionary biology and its significance in the enhancement debate; 3) the naïveté in the methodological approach to address complex issues; 4) the inability to communicate clear arguments for and against biomedical enhancement; and 5) the inertia in the current debate which has focused on the same questions in the last two decades.[10]

The failure to make progress in the debate does not mean that cognitive enhancement is morally justifiable due to the lack of convincing arguments against it. Rather the inertia indicates the need to discriminate between instances where enhancement might benefit the human species in general and, in the clinical context, patients in particular. While the idea of enhancing human capacities seems dubious at first, not every enhancement of human capacities is inherently problematic. Specifically, there are procedures that will enhance the aptitudes of individuals with disabilities beyond their *current* baseline biological capacities to potentially achieve higher levels of functioning at the physical, cognitive, behavioral/emotional, and social levels.

COGNITIVE ENHANCERS

In this chapter the term cognitive enhancement is employed not only to signify "the amplification or extension of core capacities of the mind, using augmentation or improvements of our information processing systems," which includes perception (information acquisition), attention, understanding, and memory.[11] Two types of cognitive enhancers are currently under development. Psychopharmaceutical enhancers have the capacities to enhance cognitive functions such as an increased ability to focus and concentrate, to enhance mood, memory, and learning; they include modafinil (Provigil), dextroamphetamine, methylphenidate (Ritalin), fluoxetine (Prozac), donepezil (Aricept), and propranolol (Inderal).[12–16] Their clinical applications encompass a variety of neuropsychiatric disorders such as traumatic brain injury, mood disorders, attention-deficit hyperactivity disorder (ADHD), early stages of schizophrenia, Alzheimer disease, Parkinson disease, and cerebral palsy.[2,17,18] The other type of cognitive enhancers are brain stimulation techniques used to treat or alleviate the symptoms of a variety of neurological and psychiatric disorders. Deep brain stimulation (DBS), transcranial direct current stimulation (tDCS), and transcranial magnetic stimulation (TMS) offer alternative treatment to traditional therapy in psychiatry for conditions such as addiction, severe depression, obsessive–compulsive disorder (OCD), dystonia, Alzheimer disease, and Tourette syndrome.[19–22] In addition to therapeutic applications, there is strong evidence that certain brain stimulation techniques may not only affect personal identity or sense of self of individuals undergoing these procedures,[3,19,23] but also enhance cognition and mood[14] as well as procedural learning tasks, motor learning, attention, visuomotor coordination tasks, and working memory.[11]

The use of cognitive enhancers is not without its shortfalls. There is strong evidence that brain manipulation or the alteration in brain structure can change personal identity or sense of self in significant ways.[3] A few examples are worth mentioning. First, the case of Phineas Gage, a 25-year-old railway worker who underwent a change of personality following the piercing of his left frontal lobe by a tamping iron, exemplifies how the damage to the brain structure can affect one's personality.[25] Second, we should consider the practice of lobotomies to treat psychiatric conditions in the early part of the 20th century. The procedure, meant to produce changes in personality and behavior, consisted of the cutting of neural pathways of the frontal lobes. Lobotomies were eventually abandoned in the early 1950s with the advent of psychopharmacology, particularly with the approval by the Food and Drug Administration (FDA) of Thorazine.[26,27] Lastly, as stated earlier, currently there are various psychopharmacological agents that can affect patients' sense of self and personality[28,29] and at least 13 methods of brain stimulation techniques to treat psychiatric disorders that also produce changes in personal identity.[30–32]

PERSONAL IDENTITY AND MENTAL IMPAIRMENTS

Personal Identity

A comprehensive examination of the concept of personal identity is beyond the scope of this chapter. However, a working definition is necessary in order to situate the nature of the issues arising from the use of cognitive enhancers. Some of the groundwork on the concept of personality in neuroethics has been outlined in Jotterand and Giordano,[3] who adopt a pragmatic interpretation of personal identity that provides enough conceptual depth to frame the subsequent discussion. Personal identity, they assert, "includes the biological and psychosocial aspects of human development and experience. Crucial to this concept is the notion of embodiment. The body of a biological organism constitutes the medium through which the realities of time and place are experienced and interpreted… [t]his means that (under non-pathological conditions) a person experiences, through the medium of the body located in time and space, a variety of events that define his or her sense of self."[3] This definition does not capture the complexities of defining personal identity. That said, it nevertheless offers enough breadth to establish that in the normal development of an individual who reaches adulthood, the following criteria are constitutive of personal identity: 1) consciousness—the ability to recognize oneself located in space and time; 2) embodiment—bodily experience of the outside world and the ability to internalize these experiences; 3) sentience—the capacity to reason, establish beliefs and ideas about the world, and make choices; 4) constructs of self—the capacity to understand oneself as an autonomous self able to relate to others; and 5) agency—the ability to develop character traits and particular aims.[3,33]

Persons with Mental and/or Cognitive Impairment

Mental impairments can exist at different levels and encompass conditions such as dysfunctional personal identity, dysfunctional mood, dysfunctional thought process and perception (or how one would describe psychosis), cognitive dysfunctions as well as mixed variants. Various factors such as traumatic brain injuries, degenerative diseases, neurotechnology devices, and psychopharmacological agents can alter one's personal identity. What follows focuses on individuals who possess a "… mental impairment that substantially limits one or more of the major life activities of such individual" that is related to impaired cognition or dysfunctional personal identity.[4]

Based on the type and seriousness of an individual's cognitive impairment, it becomes more unlikely that these individuals will ever recover full cognitive functioning. However, persons with cognitive, mood, and/or thought process-related mental impairments possess various levels of personal identity, depending on their level of disability, that can influence the person's functioning and quality of life. Identity in the mentally impaired population is a complex issue and can be conceptualized along a continuum. One extreme on the spectrum is the individual with a well-developed self-identity, in contrast to the other extreme, an individual whose identity is fully based on the perception of those with who they are in relationship, and includes their dispositions and behaviors. To clarify, all persons have an identity based on dispositions and behaviors, but not all have a self-identity, by which "a person experiences, through the medium of the body located in time and space, a variety of events that define his or her sense of self …".[3] For instance, those with mild disability will possess a fairly conceptual view of themselves and their location in time and space, thereby forming a clear self-identity. Those disabled persons with moderate levels of disability might not reach full cognitive functioning but nonetheless will retain some basic preferences, interests, relational ability, and a sense of self, albeit less developed. On the opposite end of the disability spectrum, there are persons who are unable to express knowledge of self, but still possess certain dispositions and behaviors such as smiling, irritability, calmness, or lability of mood.

IDENTITY AND ENHANCEMENT-2

Three major aims for cognitive enhancement in the mentally impaired include: 1) to increase quality of life; 2) to increase functioning with a decrease in the need for health care and daily assistance; and 3) to increase the individual's social capacity. The effects of neuroenhancement have the potential to support the disabled person's identity formation in two important ways. For the mildly cognitively disabled individual with self-awareness, it may allow for increased competency, self-sufficiency, and a more active lifestyle. In a study by Anna Kittelsaa, *Self-Presentations and Intellectual Disability,* she concludes that "In their self-presentations, [the cognitively impaired interviewees] were showing that they were active, intentional, and goal-oriented."[34,35] If the use of cognitive enhancers were available to these otherwise healthy persons with varying degrees of cognitive impairment, might they improve their personal identities into more competent and ordinary contributors to society, consistent with Kittelsaa's findings of the same? Mathews, quoting Maura Tumulty, captures well the nature of the issue stating: "'being a self is a job of work'.... Questions about personal identity arise when a person seems to be losing the capacity to engage in that work."[30]

A second use of cognitive enhancers in the cognitively disabled population would be to improve quality of life in the more severely disabled; and when self-identity is not present, at a minimum to maintain the individual's caregiver-perceived identity that allows for maximum expression and happiness regardless of the presence of a high level of self-awareness. The case below of LM, a 24-year-old man with cerebral palsy, illustrates the challenge these individuals encounter in major life activities, which often limits their ability to improve their quality of life.

> LM is a 24-year-old with spastic cerebral palsy and developmental delay secondary to perinatal birth asphyxia. He has severe motor dysfunction with accompanying spasticity and contractures treated with tendon release surgery and ongoing physical and occupational therapy. His primary mode of mobility is a self-operated electric wheelchair, and he requires full assistance with transfers and activities of daily living. LM is unable to swallow and requires a gastric tube for nutrition and hydration. In addition, in the past he experienced frequent hospitalizations for pneumonia and sepsis. LM's cognitive development has remained at the level of a 3-year-old. He can communicate with a limited vocabulary and uses an electronic communication board to relay more complex phrases. He is frequently combative, exhibiting dissatisfaction through crying, temper-tantrums, screaming, and biting. LM's behaviors have discouraged his acceptance into school-based programs as well as sheltered workshops. Apart from weekly speech and physical therapy visits, LM gets little socialization beyond his mother and the occasional caregiver.

As in the case of LM, he has attained a certain level of health in that his physical needs and suffering have been addressed. He has maximized his physical abilities with adaptive devices. His spasticity and contractures have been treated with surgery and ongoing pharmaceutical interventions. He no longer requires frequent hospitalizations for pneumonia. Any physical etiologies for his behavioral traits have been ruled out. His caregiver might report that LM is otherwise healthy, despite his baseline cognitive limitations. Yet, LM possesses certain baseline behavioral traits that, although "normal" for him, preclude him from maximizing his social interactions *in the context of his life*, and *relative to his peers*, not relative to any normativity of the human species. His cognitive limitations, underdeveloped identity, and reactive mood disposition impede his ability to attend school and other social functions. The use of cognitive enhancers could increase his capacity to socialize and be in relationship with others; expand his range of personal preferences, self-expression, and choice; and optimize his engagement in the world.[36] So far, the underlying moral issue has been to determine whether, in cases similar to LM, it would be justifiable to use cognitive enhancers to increase his cognitive functioning toward a "better" quality of life, recognizing 1) no long-term studies have been done on the potential effects of cognitive enhancers, and 2) no moral and legal frameworks have been developed, and broadly accepted, for an evaluation of the alteration of the identity of individuals with cognitive impairment.

The ethical implications of using cognitive enhancers in healthy volunteers in order to increase abilities beyond those of the human species have been debated at length. So far there seems to be no consensus as to whether clinicians should embark on such use of neuroenhancing techniques. The use of these same enhancers on persons with mental impairments raises a new set of questions because, while the intention is to palliate disability, the overarching question is whether this is treatment, enhancement, or both. Clearly, the types of enhancements under consideration have therapeutic implications since they also have the potential to treat any underlying mood and thought process-related mental impairments for the enhancement of the quality of life and social functioning of this patient population. In short, since therapeutic interventions, for the most part, do not raise quandaries and the question of the justified use of enhancers has not been settled, what issues does the use of enhancers raise in healthy persons with mental impairments?

ETHICAL IMPLICATIONS FOR PERSONS WITH MENTAL IMPAIRMENT

The implications of side effects such as personal identity changes for persons with mental impairment share similar ethical considerations as with nonimpaired adults, most importantly, whether changing one's identity for a certain outcome is desirable and/or whether it should be ethically permissible. In the case of therapeutic interventions that affect neurological functions and personal identity, the ultimate determinants that should guide any decision are twofold. First, an assessment should take place concerning whether these changes in personal identity are permanent or long lasting; and if so, what are the risks incurred by the patient? Second, what are the therapeutic benefits and do they outweigh the risks and side effects of the alteration of personal identity?[3]

Cognitive enhancers pose further challenges in their use with the mentally impaired population when compared with their use in nonimpaired adults.

1. The mentally impaired are a vulnerable population requiring further care and protection of their best interests in the testing and application of new treatment/enhancement modalities.
2. An extra layer of protection must be put in place considering the experimental nature of therapeutic enhancers.
3. Mentally impaired persons demonstrate varying degrees of cognitive, behavioral, emotional, and social development and abilities, which require a nuanced assessment of their capacity for informed consent, assent, understanding of their diagnosis, and implications of proposed treatments.[37]
4. Permanently disabled adults will often never realize full autonomy, depending on the level of disability, thereby limiting their right to an open future—one that can only be realized as fully autonomous adults.[38]
5. The mentally impaired person can either be subject to the best interests standard as a means for decision-making if the person never possessed full autonomy (as in the case of perinatal birth asphyxia causing damage early on);[39] or alternatively, if the person had reached adulthood and had previously expressed a set of informed preferences, decisions would be made by his or her proxy using the substituted judgment standard (as in the case of an adult who sustains a head injury). It is important to be sensitive to factors that might interfere with best interest judgments of legal representatives. There are inherent risks for either the impaired person or the legal representatives. An outside-appointed proxy decision-maker lacking a close relationship with a mentally impaired person might not fully appreciate the importance of the effects of neurotechnology on the person's self-identity. When administrating cognitive enhancers, clinicians must always consider whether such procedures truly advance the interests of their patients and do not unduly burden them. In other words, a risks/burden–benefits analysis is crucial before undertaking any brain alterations, which should include consideration of cost and third-party reimbursement.
6. Serious consideration must be given to contemporary approaches to persons with disability that places principles of empowerment, self-determination, choice and self-expression as paramount.[36] This is often embodied in disability advocates' assertion of the social nature of disability. The physical "built environment is implicated in disabled people's marginal status in society and serves to reinforce their incomplete citizenship".[40] In essence, it is imperative that the disabled person is accommodated by a functional environment prior to or concurrent with personal adaptive measures in order to allow for choice and self-determination. Although cognitive enhancers target the individual's body rather than the environment, would cognitive enhancers be morally justifiable if they provided the cognitively impaired person with maximum self-determination, choice, and self-expression, and therefore improved their self-identity? This is an important question.

RECOMMENDATIONS

The intention of the above analysis is not to stigmatize disability and acquiesce to the imperative of "normalcy". Some might argue that the idea of removing or diminishing the effects of mental disabilities using cognitive enhancers might be ethically problematic because it could convey the idea of a new kind of social engineering. Yet, the prospect of being able to enhance the level of functioning of individuals with cognitive impairment, which includes deliberately inducing personality change to positively influence quality of life, should be a strong incentive for the use of cognitive enhancers for mentally impaired individuals in the clinical setting. To be clear, it is not statistical normalcy that is the primary aim, but an improved quality of life as defined by the individual and his or her caregivers within the context of the individual's life. The arguments advanced in this chapter aim to stimulate further critical analysis and develop legal and ethical frameworks grounded on the realities of neuroscience and medical practice. Advances in neuroscience and psychopharmacology could have the potential to provide means to relieve human suffering and enhance the quality of life of individuals with mental impairment. Consequently, the clinical use of enhancement can be ethically permissible, if not recommended, in certain cases if the complexities of each person's preferences and circumstances are accounted for in any decision.

CONCLUSION

The present analysis indicates that the use of cognitive enhancers in persons with mental impairment raises important ethical concerns. Therefore the widespread clinical usage of cognitive enhancers should be restricted until their safety and efficacy is established, and ethical and legal frameworks are developed and adopted to avoid misuse and abuse. However, considering that cognitive-impaired healthy individuals could improve functionality in "major life activities" and enhance their overall quality of life, the use of cognitive enhancers can be morally justified on a case-by-case analysis. A careful examination of the risks involved in the procedure for particular individuals, in consultation with their legal representative of incompetent individuals, warrant a prudent acceptance of enhancement technologies for the mentally impaired.

References

1. Husain M, Mehta MA. Cognitive enhancement by drugs in health and disease. *Trends Cogn Sci*. 2011;15(1):28–36.
2. Sahakian B, Morein-Zamir S. Neuroethical issues in cognitive enhancement. *J Psychopharmacol*. 2011;25:197–204.
3. Jotterand F, Giordano J. Transcranial magnetic stimulation, deep brain stimulation and personal identity: ethical questions, and neuroethical approaches for medical practice. *Int Rev Psychiatry*. 2011;23(5):476–485.
4. Labor, 29 CFR § 1630.2, 2010; available at: www.law.cornell.edu/cfr/text/29/1630.2.
5. Caplan AL, Farah MJ. Emerging ethical issues in neurology, psychiatry, and the neurosciences. In: Rosenberg RN, DiMauro S, Paulson HL, Ptacek L, Nestler EJ, eds. *The Molecular and Genetic Basis of Neurological and Psychiatric Disease*. 4th ed. Philadelphia, PA: Lippincott Williams & Wilkins; 2007.
6. Jotterand F. Beyond therapy and enhancement: the alteration of human nature. *Nanoethics*. 2008;2:15–23.
7. Boorse C. Health as a theoretical concept. *Philos Sci*. 1977;44:542–573.
8. Kass LR. *Toward a More Natural Science: Biology and Human Affairs*. New York: The Free Press; 1985.
9. Jotterand F, Wangmo T. The principle of equivalence reconsidered: assessing the relevance of the principle of equivalence in prison medicine. *Am J Bioeth*. 2014, In press.
10. Buchanan A. *Beyond Humanity*. Oxford: Oxford University Press; 2011.
11. Sanberg A. Cognition enhancement: Upgrading the brain. In: Savulescu J, Meulen TR, Kahane G, eds. *Enhancing human capacities*. Oxford: Blackwell Publishing.
12. Houdsen CR, Morein-Zamir S, Sahakian BJ. Cognitive enhancing drugs: Neuroscience and society. In: Savulescu J, Meulen TR, Kahane G, ed. *Enhancing Human Capacities*. Oxford: Blackwell Publishing.
13. Smith ME, Farah MJ. Are prescription stimulants "Smart Pills"? The epidemiology and cognitive neuroscience of prescription stimulant use by normal health individuals. *Psychol Bull*. 2011;137(5):717–741.
14. Gehring K, Patwardhan SY, Collins R, et al. A randomized trial on the efficacy of methylphenidate and modafinil for improving cognitive functioning and symptoms in patients with a primary brain tumor. *J Neurooncol*. 2012;107(1):165–174.
15. Morein-Zamir S, Robbins TW, Turner D, Sahakian BJ. State-of-science review: SR-E9: pharmacological cognitive enhancement. In: *Mental Capital and Wellbeing: Making the Most of Ourselves in the 21st Century*; 2008:3–16.
16. Morein-Zamir S, Turner DC, Sahakian BJ. A review of the effects of modafinil on cognition in schizophrenia. *Schizophr Bull*. 2007;33:1298–1306.
17. Mohamed AD, Sahakian B. The ethics of elective psychopharmacology. *Int J Neuropsychopharmacol*. 2011;1–13.
18. Meyer DR, Madaan V. Pharmacological cognitive enhancers in children and adolescents. *J Am Acad Child Adolesc Psychiatry*. 2012.
19. Mathews DJH. Deep brain stimulation, personal identity and policy. *Int Rev Psychiatry*. 2011;23(5):486–492.
20. Nitsche MA, Boggio PS, Fregni F, Pascual-Leone A. Treatment of depression with transcranial direct current stimulation (tDCS): a review. *Exp Neurol*. 2009;219(1):14–19.
21. Fregni F, Boggio PS, Nitsche MA, Marcolin MA, Rigonatti SP, Pascual-Leone A. Treatment of major depression with transcranial direct current stimulation. *Bipolar Disord*. 2006;8(2):203–204.
22. Ferrucci R, Priori A. Transcranial cerebellar direct current stimulation (tcDCS): motor control, cognition, learning and emotions. *Neuroimage*. 2014;85(Pt 3):918–923.
23. Glannon W. Stimulating brains, altering minds. *J Med Ethics*. 2009;35:289–292.
24. Gilbert F. Nano-bionic devices for the purpose of cognitive enhancement: toward a preliminary ethical framework. In: Hildt E, Franke AG, eds. *Cognitive Enhancement: An Interdisciplinary Perspective*. Dordrecht: Springer.
25. Rabins PV, Blass DM. Toward a neurobiology of personal identity. In: Mathews DJH, Bok H, Rabins PV, eds. *Personal Identity and Fractured Selves*. Baltimore: Johns Hopkins University Press; 2009:38–49.
26. Shutts D. *Lobotomy: Resort to the Knife*. New York: Van Nostrand Reinhold; 1982.
27. Healy D. *The Creation of Psychopharmacology*. Cambridge, MA: Harvard University Press; 2002.
28. Glannon W. Psychopharmacological enhancement. *Neuroethics*. 2008;1:45–54.
29. Merkel R, Boer G, Fegert J, Galert T, Nuttin B, Rosahl SK. *Intervening in the Brain: Changing Psyche and Society*. Springer: Berlin and Heidelberg; 2007.

30. Mathews DJH, Bok H, Rabins PV. *Personal Identity and Fractured Selves*. Baltimore, MD: Johns Hopkins University Press; 2009.
31. Funkiewiez A, Ardouin C, Caputo E, et al. Long term effects of bilateral subthalamic nucleus stimulation on cognitive function, mood, and behaviour in Parkinson's disease. *J Neurol Neurosurg Psychiatry*. 2004;75:834–839.
32. Gabriëls L, Cosyns P, Nuttin B, Demeulemeester H, Gybels J. Deep brain stimulation for treatment refractory obsessive – compulsive disorder: psychopathological and neuropsychological outcome in three cases. *Acta Psychiatr Scand*. 2003;107:275–282.
33. Perry J. Diminished and fractured selves. In: Mathews DJH, Bok H, Rabins PV, eds. *Personal Identity and Fractured Selves*. Baltimore, MD: Johns Hopkins University Press; 2009:129–162.
34. Kittelsaa AM. Self-presentations and intellectual disability. *Scand J Disabil Res*. 2014;16(1):29–44.
35. Goodly D. Accessing the views of people with learning difficulties. In: Moore M, ed. *Insider Perspectives on Inclusion*. Sheffield: Philip Armstrong; 2000:165–183.
36. Fujiura GT, RRT Expert Panel on Health Measurement. Self-reported health of people with intellectual disability. *Intellect Dev Disabil*. 2012;50(4):352–369.
37. American Academy of Pediatrics Committee on Bioethics. Informed consent, parental permission, and assent in pediatric practice. *Pediatrics*. 1995;95:314–317.
38. Feinberg J. The child's right to an open future. In: Aiken W, LaFollette H, eds. *Whose Child? Children's Rights, Parental Authority, and State Power*. Totowa, NJ: Littlefield, Adams & Co; 1980.
39. Kopelman LM. Using the best-interests standard in treatment decisions for young children. *Pediatric Bioethics*. 2010;22–37.
40. Sherman S, Sherman J. Design professionals and the built environment: encountering boundaries 20 years after the Americans with Disabilities Act. *Disabil Soc*. 2012;27(1):51–64.

CHAPTER

12

Genetic Counseling

Wendy R. Uhlmann

University of Michigan Medical School, Ann Arbor, MI, USA

GENETIC COUNSELING DEFINED AND PROVIDERS

Many neurological and psychiatric conditions either are genetic or have genetic factors that contribute to the condition. As a result, a diagnosis can have implications not only for the affected individual but also for family members whose risk may consequently be increased. Providing genetic counseling for patients and their families can be instrumental for care, identifying at-risk relatives and decisions about testing and management. Genetic counseling is defined as:

> "the process of helping people understand and adapt to the medical, psychological and familial implications of genetic contributions to disease. This process integrates:
> - Interpretation of family and medical histories to assess the chance of disease occurrence or recurrence
> - Education about inheritance, testing, management, prevention, resources and research
> - Counseling to promote informed choices and adaptation to the risk or condition."[1,2]

Traditionally, the genetic counseling approach has been nondirective with presentation of information in an even-handed way and with an emphasis on patient ownership and autonomy in decision-making. The goal of genetic counseling is to provide patients with comprehensive and understandable information so that they can make informed health care and life decisions consistent with their values, beliefs, and needs.[3]

Providers with specific expertise and clinical training in genetics include clinical geneticists, genetic counselors, and clinical nurse specialists in genetics. Clinical geneticists complete a residency in clinical genetics (usually after a residency in pediatrics, obstetrics and gynecology, internal medicine, neurology, or other specialty) and are boarded by the American Board of Medical Genetics (or by boards in other countries). Genetic counselors are master's degree-level trained health care providers who are boarded by the American Board of Genetic Counseling (or by boards in other countries) and are licensed providers in many states. Nurses with special education and training in genetics are bachelor's-level trained genetics clinical nurses (GCN) and master's-level trained advanced practice nurses in genetics (APNG) and are credentialed by the Genetic Nursing Credentialing Commission (GNCC). Genetic counselors and clinical nurse specialists in genetics provide genetic counseling for patients with neurological conditions in general pediatric and adult genetics clinics, and also work in neurology clinics and some in psychiatric clinic settings. Resources for locating genetics clinics and certified genetics providers in the United States are in Table 12.1. For patients residing far away from genetic service providers, there are companies that offer phone genetic counseling that can be ascertained through an Internet search.

WHICH PATIENTS COULD BENEFIT FROM GENETIC COUNSELING?

Patients diagnosed with a hereditary condition will be faced with adjusting to the condition and informing at-risk relatives and can benefit from genetic counseling. They may need to make decisions about genetic testing and other family members may consider testing to determine their status. If a patient is of reproductive age, there may be reproductive

Rosenberg's Molecular and Genetic Basis of Neurological and Psychiatric Disease
http://dx.doi.org/10.1016/B978-0-12-410529-4.00012-7

TABLE 12.1 Resources for Locating Genetics Clinics and Certified Genetics Providers in the United States

Organization	Website
American Board of Genetic Counseling	www.abgc.net
American Board of Medical Genetics	www.abmg.org
American College of Medical Genetics and Genomics	www.acmg.net
GeneTests—Clinic Directory (includes international)	www.genetests.org/clinics/
International Society of Nurses in Genetics	www.isong.org
National Society of Genetic Counselors	www.nsgc.org
Orphanet (international)	www.orpha.net

and prenatal testing decisions to make. Particularly when there is no family history of a genetic condition, the patient has no context for the diagnosis and likely will require more education about a condition, familial implications and support.

The affected patient's relatives, especially siblings and children, can benefit from genetic counseling to learn about their risks and genetic testing options to determine their risk. For asymptomatic individuals at 50% risk for a neurogenetic condition like Huntington disease, there are predictive genetic testing guidelines that specifically require genetic counseling.[4,5] The guidelines describe the testing process, essential information to convey, communication and implications of results, and emphasize that the decision should be voluntary, informed, and made by the patient—not by the physician or anyone else. The guidelines for predictive genetic testing for Huntington disease have been adapted for predictive genetic testing for other hereditary neurological conditions including early-onset Alzheimer disease,[6,7] amyotrophic lateral sclerosis,[8,9] and frontotemporal dementia/autosomal dominant dementias.[7,10]

COMPONENTS OF GENETIC COUNSELING AND CASE PREPARATION

A typical genetic counseling session can last 1 to 1.5 hours. The components of genetic counseling are listed in Table 12.2 and discussed below. There is extensive case preparation that typically needs to be done before a patient is seen,[11] and resources to accomplish this work can be found in Uhlmann and Guttmacher (2008)[12] and Uhlmann (2009).[11] Conducting a literature search is important for ensuring that information is up-to-date given the rapid pace of advances in genetics. In addition to literature searches, key resources for obtaining information about genetic conditions include MedGen (www.ncbi.nlm.nih.gov/medgen), GeneReviews (www.ncbi.nlm.nih.gov/books/NBK1116/) and Online Mendelian Inheritance in Man (OMIM; www.ncbi.nlm.nih.gov/omim). OMIM can also be used to generate differential diagnoses using the search function and providing clinical features. Obtaining a three-generation pedigree from the patient up through grandparents is standard in genetic counseling and is used not only for the basis of risk assessment but also can provide insights about the patient's family dynamics, support and psychosocial issues. Pedigree symbols utilized are summarized in articles by Bennett et al.[13,14]

Like any medical interaction, genetic counseling consists of both obtaining and providing information. What differs is the nature of the interaction, approach and time involved. The National Society of Genetic Counselors has information about genetic counseling for patients and providers and videos of simulated counseling sessions available at

TABLE 12.2 Components of Genetic Counseling

Contracting (ascertaining patient's questions/concerns; providing overview of clinic visit)
Information gathering (family and medical histories)
Physical examination (if indicated)
Risk assessment (determination of likely pattern(s) of inheritance, risks)
Education about genetic condition (clinical features, prognosis, management, inheritance and implications for patient and family members)
Ascertaining and addressing psychosocial issues/concerns
Discussion of genetic tests/other tests/evaluations/referrals to consider
Provision of resources about the genetic condition and contact information for national organizations/support groups
Facilitating decision-making

their website. The American College of Medical Genetics and Genomics, the Genetics in Primary Care Institute (www.geneticsinprimarycare.org) and the National Coalition for Health Professional Education in Genetics/The Jackson Laboratory (www.nchpeg.org) have developed clinical resources and point of care tools for physicians and providers that explain basic genetics concepts and genetic testing.

RISK ASSESSMENT

Establishing the pattern of inheritance of a genetic condition in a family can be critical for making a diagnosis, identifying at-risk relatives and providing an accurate risk assessment. The three-generation pedigree is used to evaluate the family history information to determine likely pattern(s) of inheritance; when indicated and possible, medical records are requested to confirm reported diagnoses. Assessing transmission, generations and genders affected is important for determining inheritance. For X-linked recessive, X-linked dominant and mitochondrial conditions, there can be no male-to-male transmission. Autosomal recessive inheritance is the most likely explanation when individuals are affected in a single generation and/or there is consanguinity or the patient is of an ethnicity with high frequency of carriers of a condition (for example, Tay–Sachs disease in the Ashkenazi Jewish population).

Depending on the pattern of inheritance, gender, and relative(s) affected, recurrence risks can range from general population risks to 100% (Table 12.3). Sometimes a condition can have multiple patterns of inheritance (Table 12.3). Other factors that can impact pedigree interpretation and risk assessment include whether the genetic condition has heterogeneity, reduced penetrance, variable expressivity, anticipation, gonadal or somatic mosaicism. Nonpaternity could impact pedigree interpretation. Bayesian analysis can sometimes be used to modify a patient's risk by factoring

TABLE 12.3 Risks and Examples of Genetic Conditions Associated with Different Patterns of Inheritance

Pattern of inheritance	Risks (for the affected individual)	Examples of conditions
Autosomal dominant	Note: Penetrance impacts risk to be affected. **Children**: 50% risk to be affected **Siblings**: 50% risk to be affected if a parent has gene mutation **Parents**: A parent may have gene mutation or could arise *de novo* in affected individual	Charcot–Marie–Tooth syndrome Hereditary spastic paraplegia Huntington disease Myotonic dystrophy
Autosomal recessive	**Children**: All will be obligate carriers. Risk to be affected generally < 1% (depends on carrier risk in the general population) but if partner is a carrier, then risk is 50% **Siblings**: 25% risk to be affected. If unaffected, ⅔ (67%) risk to be a carrier **Parents**: Both parents would be obligate carriers	Friedreich ataxia Hereditary spastic paraplegia Spinal muscular atrophy Wilson disease
X-linked recessive	**Carrier female**: (Note: some carrier females can be "manifesting carriers" and have mild symptoms due to skewed X-inactivation) Sons: 50% risk to be affected Daughters: 50% risk to be carriers **Affected male**: Sons: All unaffected Daughters: All carriers **Affected female** (can result from affected father and carrier mother): Sons: All affected Daughters: All carriers	Becker/Duchenne muscular dystrophy Charcot–Marie–Tooth syndrome Fragile X syndrome Hereditary spastic paraplegia
X-linked dominant	Few conditions with this pattern of inheritance. Most X-linked dominant conditions are lethal in males. Females are more mildly affected **Female with X-linked dominant condition**: Daughters: 50% risk Sons: 50% risk	Aicardi syndrome Incontinentia pigmenti Rett syndrome
Mitochondrial	Affected status will depend on percentage of mitochondria with mutation that are inherited, 0–100% **Affected female**: Will transmit mitochondria to all of her children. Children may or may not be affected **Affected male**: Will not transmit mitochondria to any of his children	Leber hereditary optic neuropathy (LHON) Mitochondrial encephalopathy, lactic acidosis, strokes (MELAS) Mitochondrial encephalopathy with ragged-red fibers (MERRF)
Complex/multifactorial	Empiric risk studies. Increased risks based on relatedness (1st degree relative > 2nd > 3rd)	Alzheimer disease Bipolar disorder Schizophrenia

in family history, gender, age, clinical status, or test results and is often used to derive the risk modification figures in genetic test reports. For non-Mendelian conditions, empiric risk studies are utilized. One resource that has a summary of empiric risk studies for psychiatric conditions is the National Coalition for Health Professional Education in Genetics/The Jackson Laboratory.

Fundamental to genetic counseling is explaining to a patient how a condition is inherited and the implications for other relatives. Online resources for patient information about patterns of inheritance include: Genetics Home Reference (http://ghr.nlm.nih.gov/), Centre for Genetics Education (www.genetics.edu.au/) and for this information in different languages EuroGentest (www.eurogentest.org/), Genetic Alliance UK (www.geneticalliance.org.uk/) and Centre for Genetics Education.

COUNSELING AND EDUCATION ABOUT THE GENETIC CONDITION

Central to genetic counseling is helping patients understand the genetic condition and educating them about the clinical features, inheritance, and familial implications, prognosis, treatment, management, and testing options. Creation and use of counseling aids (e.g., developing and printing PowerPoint slides, purchase of genetic counseling aids) and use of photos and diagrams from patient brochures and internet searches can be instrumental in clearly presenting complex information and facilitating patient understanding about the genetic condition. Exploring a patient's journey with a diagnosis and/or family history of a genetic condition can help in identifying stressors, challenging issues and difficult family dynamics and assist in normalizing reactions and identifying supportive resources. Educating patients about their genetic condition in understandable terms and providing resources is important both in terms of their adaptation and because frequently they may in turn have to be the ones to communicate this information to other at-risk family members.[15]

GENETIC TESTING

Genetic testing generally is not as simple as just checking off a box on a form. For many conditions, genetic testing is not yet standard of care and there is only a single or few labs that offer testing, particularly if the condition is rare. Uses of genetic testing include diagnostic, predictive, carrier and prenatal applications, which raise different issues and have different implications even though the same analysis from the laboratory standpoint is being performed. Genetics expertise is frequently needed to decide whether genetic testing is indicated, which genetic test to order (e.g., single gene, panel, chromosome microarray analysis, sequencing), select a laboratory, and interpret test results. Depending on the genetic condition, there may be practice guidelines available for genetic testing from the American College of Medical Genetics and Genomics, the National Society of Genetic Counselors, and other professional organizations that can be ascertained by searches of the literature and organization websites and at www.guidelines.gov. For example, the American College of Medical Genetics and Genomics has practice guidelines for Alzheimer disease,[6] autism spectrum disorders,[16] fragile X syndrome,[17] and several other neurological conditions.

Laboratories can differ in the tests offered, methodologies used, cost, and result time, even when testing for the same genetic condition. In the United States, the Genetic Testing Registry (www.ncbi.nlm.nih.gov/gtr/) and GeneTests (www.genetests.org) are key online databases that can be used to determine if genetic testing is available; Orphanet can be used to identify international labs offering testing. Selecting the "right test" and the "right laboratory" are important and, given different testing methodologies and sensitivities, can potentially make a significant difference in making a diagnosis. If genetic testing is costly and/or sensitivity is limited, deferring genetic testing may be preferable. DNA banking is an option to consider if a patient's lifespan is limited and genetic testing is not a reasonable option; generally banking a sample can be done for approximately $100.

Genetic counseling can help patients understand the benefits, risks and limitations of genetic testing and make an informed decision. Pre- and post-test counseling and informed consent are essential components of genetic testing. Topics typically covered during pretest counseling and informed consent are in Table 12.4. Online resources for patients that explain genetic testing and uses include: Genetics Home Reference, National Human Genome Research Institute, Centre for Genetics Education, Genetic Alliance and EuroGentest.

For asymptomatic individuals with a family history of a genetic condition, generally an affected individual needs to be tested first to determine if there is an identifiable mutation(s). In the absence of testing an affected family member, it will not necessarily be known whether an asymptomatic individual who tests negative is in fact a "true negative" for the genetic condition or still at risk (an affected individual may have also tested negative given limitations in testing technology). While genetic testing may not necessarily affect patient care, it can be important for confirming a suspected diagnosis, enable at-risk relatives to have the option of testing and be of use reproductively.

TABLE 12.4 Topics Covered During Pretest Counseling and Informed Consent

Genetic test recommended
Explanation of test and likelihood of finding gene mutation(s)
Benefits, risks and limitations of genetic testing, including psychosocial implications
Significance of a result that is positive, negative or a variant of unknown significance and implications for patient and family members
Laboratory where sample will be sent
Cost and insurance coverage/issues
Timeframe for results and how results will be communicated
Possibility additional testing of patient and/or family members may be indicated

INSURANCE CONSIDERATIONS AND IMPLICATIONS

Insurance coverage of genetic testing and implications results could have for the patient to obtain desired types of insurance are major considerations. Genetic tests can range in cost from a couple of hundred dollars to several thousand dollars. Coverage of genetic testing differs by type of insurance and plan; coverage of genetic testing for patients with Medicaid insurance is particularly difficult to obtain. Insurance generally will not cover the cost of genetic testing if it will not alter a patient's care. Some insurers will provide pre-authorization for genetic testing. There are some laboratories that offer the service of checking whether the genetic test will be covered by the patient's insurance. For many patients, a letter of medical necessity will need to be written after a clinic visit to an insurer that includes clinical indication, test requested, CPT (Current Procedural Terminology) codes, lab and costs in order to determine if genetic testing will be covered.[11]

Patients may be concerned about the potential for stigmatization and discrimination based on their genetic condition or genetic test results. In the United States, the Genetic Information Nondiscrimination Act (GINA), passed in 2008, prohibits discrimination in health insurance and employment based on genetic information. However, this law does not cover life, long-term disability and long-term care insurance. For predictive genetic testing for conditions like Huntington disease, it is particularly important for patients to have desired insurance coverage prior to proceeding with genetic testing. Information about GINA's health insurance and employment protections can be obtained at www.ginahelp.org. The National Human Genome Research Institute maintains a genome statute and legislative database with information about state laws at: www.genome.gov/PolicyEthics/LegDatabase/pubsearch.cfm. HumGen (www.humgen.org) is an international database of laws and policies.

TEST INTERPRETATION

Genetics expertise is often needed in interpreting genetic test results and determining whether additional genetic testing with different methodologies is indicated. Identification of a gene mutation does not necessarily mean that it is causative of the genetic condition and if present in an asymptomatic individual does not necessarily mean he/she will become symptomatic. For example, if there is reduced penetrance, it is possible that an individual with a gene mutation may not become symptomatic. Depending on the gene mutation, it could be an unrelated coincidental finding of no significance or could be disease-causing, a common polymorphism, a rare sequence variant, or a variant of unknown significance (VUS). Use of online databases (e.g., ClinVar, locus specific mutation databases) and conducting a literature review can be important for result interpretation. For some conditions, there are genotype–phenotype correlations. Guidance in determining the significance of a gene mutation and how variants of unknown significance are classified, interpreted and reported can be found in the American College of Medical Genetics and Genomics practice guidelines.[18–20]

Sometimes multiple testing methodologies need to be considered, for example deletion/duplication analysis after negative sequencing results are obtained. Unless the mutation is a known or presumed pathogenic mutation, clarification of test results may require testing additional family members to determine whether the gene mutation(s) are inherited or *de novo* and whether the mutation(s) track with affected family members. The most reassurance that can be given is to an asymptomatic patient who tests negative for a familial gene mutation—a "true negative" result. Understanding these nuances of testing, determining next steps and explaining the complexities of test result interpretation is central to genetic counseling.

ETHICAL ISSUES

Ethical issues can arise in genetic counseling and particularly with genetic testing. In some families, information about genetic conditions is not openly shared or communicated. There can be pressure and coercion by family members to be tested or not tested. While respect for a patient's autonomy is paramount, ethical issues can arise when one relative's right to know conflicts with another relative's right not to know. Given shared genetics, it is possible that an individual's test results could reveal the genetic status of another family member who did not want their status determined—for example, an adult child who has positive predictive genetic test results for Huntington disease and consequently, their at-risk parent, who did not want their status determined, will now be known positive.

Genetics professionals are knowledgeable about ethical issues that can arise and issues to consider. To address and resolve ethical issues, guidelines and position statements from different genetics professional organizations, specific genetic condition guidelines, and hospital ethics committees can be consulted. In general, testing children for adult-onset conditions should be deferred so that the child has an "open future" and can decide for himself/herself whether their status should be determined.[21] Generally, testing a pregnancy should only be considered if it will alter the management of the pregnancy. While patients should be encouraged to share genetic information with other at-risk relatives, which should be documented in clinic visit notes, direct familial disclosure of genetic information by physicians is restricted only to those cases where the potential for harm from nondisclosure is significant and specific criteria have been met.[22] At the present time, genetics professionals do not have a duty to re-contact patients and patients should be encouraged to re-contact a genetics clinic to learn about genetic advances.[23] Additional ethical issues that can arise with genetic testing for neurogenetic conditions are summarized in Uhlmann (2006),[24] Roberts and Uhlmann (2013),[25] and in this book.

IDENTIFYING SUPPORTIVE RESOURCES FOR PATIENTS

Patients leave their doctors' offices and have to face the challenges of living with or being at-risk for a genetic condition. One of the goals of genetic counseling is to help patients adjust to the genetic condition. As a result of having or being at-risk for a genetic condition, a patient can experience sadness, anger, concern, anxiety, guilt, stress, and other emotions. There can be changes in view of self/self-worth, and family dynamics. Linking patients up with supportive resources and helping them connect with other individuals who are facing similar challenges can be of great benefit. Through their experiences of living with a genetic condition, patients may be aware of helpful resources unknown to health care providers that can benefit other affected individuals. Support group/advocacy organizations frequently have resources and online publications about genetic conditions that provide information in understandable terms for patients and may also have publications written for health care professionals. Online resources that can be used to find support groups and resources for patients include: the Genetic Alliance (www.geneticalliance.org), Genetics Home Reference, the International Genetic Alliance (www.intga.org), National Organization for Rare Disorders (www.rarediseases.org) and Orphanet (www.orpha.net). The online directories of genetic testing laboratories discussed above also include links to support groups/resources. Patients can also benefit from referrals to local therapists.

CONCLUSIONS

Genetic advances in identifying gene mutations associated with hereditary neurological and psychiatric conditions will make it possible for more individuals to learn about their risk status. Consulting with health care providers with genetics expertise can be beneficial in establishing/confirming a diagnosis, determining what genetic test to order, selecting a laboratory, and interpreting test results. Genetic counseling can help patients understand and adapt to living with a genetic condition, learn about implications for other relatives, make informed decisions about testing and care, and find out about support groups and resources. Given time constraints of typical clinic visits and potential to feel overwhelmed when confronted with a genetic condition, genetic counseling provides an opportunity for patients to be educated about their condition and discuss the psychosocial implications so that they have the information needed to make informed health care and life decisions.

References

1. National Society of Genetic Counselors. http://nsgc.org/p/cm/ld/fid=175.
2. National Society of Genetic Counselors' Definition Task Force, Resta R, Biesecker BB, Bennett RL, et al. A new definition of genetic counseling: National Society of Genetic Counselors' Task Force Report. *J Genet Couns*. 2006;15:77–83.
3. Walker AP. The practice of genetic counseling. In: Uhlmann WR, Schuette JL, Yashar BM, eds. *A Guide to Genetic Counseling*. 2nd ed. New Jersey: John Wiley & Sons, Inc; 2009:1–35.
4. International Huntington Association (IHA) and the World Federation of Neurology (WFN) Research Group on Huntington's Chorea. Guidelines for the molecular genetics predictive test in Huntington's disease. *Neurology*. 1994;44:1533–1536.
5. MacLeod R, Tibben A, Frontali M, et al. Editorial Committee and Working Group. 'Genetic Testing Counselling' of the European Huntington Disease Network. Recommendations for the predictive genetic test in Huntington's disease. *Clin Genet*. 2013;83:221–231.
6. Goldman JS, Hahn SE, Catania JW, et al. Genetic counseling and testing for Alzheimer disease: joint practice guidelines of the American College of Medical Genetics and the National Society of Genetic Counselors. *Genet Med*. 2011;13:597–605.
7. Goldman JS. New approaches to genetic counseling and testing for Alzheimer's disease and frontotemporal degeneration. *Curr Neurol Neurosci Rep*. 2012;12:502–510.
8. EFNS Task Force on Diagnosis and Management of Amyotrophic Lateral Sclerosis, Andersen PM, Abrahams S, Borasio GD, et al. EFNS guidelines on the clinical management of amyotrophic lateral sclerosis (MALS) – revised report of an EFNS task force. *Eur J Neurol*. 2012;19:360–375.
9. Chiò A, Battistini S, Calvo A, et al. The ITALSGEN Consortium. Genetic counseling in ALS: facts, uncertainties and clinical suggestions. *J Neurol Neurosurg Psychiatry*. 2014;85(5):478–485.
10. Quaid KA. Genetic counseling for frontotemporal dementias. *J Mol Neurosci*. 2011;45:706–709.
11. Uhlmann WR. Thinking it all through: case preparation and management. In: Uhlmann WR, Schuette JL, Yashar BM, eds. *A Guide to Genetic Counseling*. 2nd ed. New Jersey: John Wiley & Sons, Inc; 2009:93–131.
12. Uhlmann WR, Guttmacher AE. Key internet genetics resources for the clinician. *JAMA*. 2008;299:1356–1358.
13. Bennett RL, Steinhaus KA, Uhrich SB, et al. Recommendations for standardized human pedigree nomenclature: Pedigree Standardization Task Force of the National Society of Genetic Counselors. *Am J Hum Genet*. 1995;56:745–752.
14. Bennett RL, French KS, Uhrich SB, Resta RG, Doyle DL. Standardized human pedigree nomenclature: update and assessment of the recommendations of the National Society of Genetic Counselors. *J Genet Counsel*. 2008;17:424–433.
15. Uhlmann WR, Schuette JL, Yashar BM. *A Guide to Genetic Counseling*. 2nd ed. New Jersey: John Wiley & Sons, Inc; 2009.
16. Schaefer GB, Mendelsohn NJ. Professional Practice and Guidelines Committee. Clinical genetics evaluation in identifying the etiology of autism spectrum disorders: 2013 guideline revisions. *Genet Med*. 2013;15:399–407.
17. Sherman S, Pletcher BA, Driscoll DA. Fragile X syndrome: diagnostic and carrier testing. *Genet Med*. 2005;7:584–587.
18. Kearney HM, Thorland EC, Brown KK, Quintero-Rivera F, South ST. Working Group of the American College of Medical Genetics Laboratory Quality Assurance Committee. American College of Medical Genetics standards and guidelines for interpretation and reporting of postnatal constitutional copy number variants. *Genet Med*. 2011;13:680–685.
19. Richards CS, Bale S, Bellissimo DB, et al. Molecular Subcommittee of the ACMG Laboratory Quality Assurance Committee. ACMG recommendations for standards for interpretation and reporting of sequence variations: revisions 2007. *Genet Med*. 2008;10:294–300.
20. Rehm HL, Bale SJ, Bayrak-Toydemir P, et al. Working Group of the American College of Medical Genetics and Genomics Laboratory Quality Assurance Committee. ACMG clinical laboratory standards for next-generation sequencing. *Genet Med*. 2013;15:733–747.
21. Ross LF, Saal HM, David KL, Anderson RR. American Academy of Pediatrics, American College of Medical Genetics and Genomics. Technical report: ethical and policy issues in genetic testing and screening of children. *Genet Med*. 2013;15:234–245.
22. American Society of Human Genetics Social Issues Subcommittee on Familial Disclosure. ASHG statement. Professional disclosure of familial genetic information. *Am J Hum Genet*. 1998;62:474–483.
23. Hirschhorn K, Fleisher LD, Godmilow L, et al. Duty to re-contact. *Genet Med*. 1999;1:171–172.
24. Uhlmann WR. Ethical dilemmas. In: Lynch D, ed. *Neurogenetics: Scientific and Clinical Advances*. New York: Taylor & Francis; 2006:87–106.
25. Roberts JS, Uhlmann WR. Genetic susceptibility testing for neurodegenerative diseases: ethical and practice issues. *Prog Neurobiol*. 2013;110:89–101.

SECTION II

NEUROLOGIC DISEASES

CHAPTER

13

Cerebral Malformations

William D. Graf and Shihui Yu†*

*Yale School of Medicine, New Haven, CT, USA

†Seattle Children's Hospital and University of Washington School of Medicine, Seattle, WA, USA

CLINICAL FEATURES

Historical Overview

Better understanding of cerebral malformations parallels advancements in science and technology. Early neuroscience in the 17th and 18th centuries used simple low-power microscopy for viewing autopsy brain specimens. The 19th century was an era of astute clinical observation combined with improved neuropathological microscopic investigation. By the mid-20th century, intracellular structures were visualized through electron microscopy and a chemical basis of brain morphogenesis was proposed.[1] Over the ensuing decades, numerous textbooks and atlases illustrated the normal structures of the human embryo and fetus, as well as its developmental neuropathology.[2–5] By the mid-1990s, knowledge about cerebral malformations was dominated by neuropathology and the routine clinical application of magnetic resonance imaging (MRI). More recently, knowledge about central nervous system (CNS) malformations has grown exponentially due to the contribution of genetics and genomics. Classification schemes began to organize cerebral–cortical malformations based on the most disruptive step during the prenatal neurodevelopmental process (i.e., cell proliferation, neuronal migration, cortical organization).[6,7] Sarnat and Flores-Sarnat were the first to propose a malformation classification system to correlate molecular genetics to basic embryology and neuropathology.[8] During the last decade, however, the enormous complexity and sheer difficulty of classifying the endless array of genotypes with heterogeneous clinical and neuroimaging phenotypes became widely appreciated as discoveries about the genetic basis of embryology and the many cerebral malformations grew in correlation to more sensitive neuroimaging.[9–11]

Identification of Cerebral Malformations

PRENATAL IDENTIFICATION OF CEREBRAL MALFORMATION

Prenatal imaging using ultrasonography (including high-frequency transvaginal scanning and three-dimensional [3D] sonography) or fetal MRI has allowed the clinical identification of major fetal cerebral malformations (Tables 13.1 and 13.2). Technical issues such as fetal movement, the position of the fetus during the neuroimaging study, and the exact age of the fetus at the time neuroimaging are important in the interpretation of findings. Before counseling parents about a suspected malformation in a fetus, repeat ultrasonography studies with optimal positioning of the fetus are often necessary. Fetal MRI after 20 weeks gestation provides better neuroanatomical resolution compared to fetal ultrasound but milder anatomic variants of uncertain significance may pose ethical questions after the age of fetal viability. Genetic defects causing cerebral malformations can be prenatally identified by established cytogenetic and molecular genetic/genomic techniques both invasively and noninvasively.

POSTNATAL IDENTIFICATION OF CEREBRAL MALFORMATION

The assessment of an individual patient with a cerebral malformation begins with a detailed description of the general clinical circumstances (e.g., prematurity, prenatal ultrasound findings), the physical examination features (e.g., presence of microcephaly or facial dysmorphology) and the specific neuroanatomical anomaly discovered by postnatal MRI. Postnatal malformation investigation is typically initiated because of abnormal neurodevelopment,

Rosenberg's Molecular and Genetic Basis of Neurological and Psychiatric Disease
http://dx.doi.org/10.1016/B978-0-12-410529-4.00013-9

TABLE 13.1 Major Clinically Relevant Malformations Based on Developmental Timeline, Developmental Process, and Molecular Function with Some Gene Examples

Developmental stage (gestational age)	CNS malformation (model disorder)	Exemplary genes (OMIM)	Molecular function
Primary neurulation (20–28 days) Secondary neurulation (27–35 days)	Anencephaly, Myelomeningocele Lipomyelomeningocele	Multifactorial	Neural plate folds into neural tube
Segmentation/formation of prosencephalon, mesencephalon, and rhombencephalon	Neural patterning malformations of forebrain, midbrain, hindbrain, spinal column	*HOX* gene family	Transcription factors expressed in early hindbrain development
Polarity/regionalization (3–5 weeks)	Holoprosencephaly (forebrain cleavage)	*FOXH1* (603621) *GLI2* (165230) *NODAL* (601265) *PTCH1* (601309) *SHH* (600725) *SIX3* (603714) *TGIF1* (602630) *ZIC2* (603073)	Transcription factor for NODAL signaling Transcription factor—hedgehog signaling Release of hedgehog ligands Receptor for hedgehog ligands Ventral central nervous system patterning Forebrain and eye development Transcriptional repressor including retinoids Axis formation and dorsal brain development
Rathke pouch formation (Stage 13/32 days)	Pituitary hypoplasia and septo-optic dysplasia	*HESX1* (601802)	Homeobox protein
Midline formation (3–12 weeks)	Agenesis of the corpus callosum (ACC)	Heterozygous syndromes; >30 loci with gene gain/loss function *SLC12A6* (604878)	Abnormal development of longitudinal callosal fascicles (Probst bundles) Solute carrier family 12 member 6
Axon outgrowth/fascicle formation (mid-trimester)	L1CAM syndrome (X-linked hydrocephalus)	*L1CAM* (308840)	Immunoglobulin superfamily cell adhesion molecules (CAMs) mediate cell-to-cell adhesion
Proliferation (mid-trimester)	Primary microcephaly	*MCPH1* (607117) *ASPM* (605481) *CDK5RAP2* (608201) *CENPJ* (609279)	Microcephalin Abnormal spindle-like microcephaly Cyclin dependant kinase regulatory protein Centromere-associated protein J
Radial migration (mid-trimester)	Heterotopia	*ARFGEF2* (605371) *GPR56* (604110) *FLNA* (300017)	Ventricular epithelium anchoring Pial limiting membrane anchoring Vesicle trafficking and fusion
Neuroblast migration (mid-trimester)	Lissencephaly (classic)	*LIS1* (607432) *TUBA1A* (602529) *MAP3K4* (602425) *ARX* (300382)	Microtubule and nuclear transport/centrosomes Microtubule/postmigrational development Neuroependymal integrity Fetal forebrain interneuron development
Neuroblast migration (mid-trimester)	Lissencephaly (cobblestone)	*ASCL1, CDK5R1, DCX, DRD2, NDN, NEUROG2, NRCAM, NTN1, PAFAH1B1, ROBO1, SLIT2*	Abnormal terminal migration and defects in pial limiting membrane
Neuroblast migration (mid-trimester)	Heterotopia ("double cortex")	*DCX* (300121)	Microtubule stabilization
Neuroblast migration (mid-trimester)	Polymicrogyria Polymicrogyria-like dysplasias	*TUBA8* (605742) *TUBB2B* (612850) *TUBB3* (602661)	Microtubule/postmigrational development

13

TABLE 13.1 Major Clinically Relevant Malformations Based on Developmental Timeline, Developmental Process, and Molecular Function with Some Gene Examples—cont'd

Developmental stage (gestational age)	CNS malformation (model disorder)	Exemplary genes (OMIM)	Molecular function
Neocortical formation	Schizencephaly	*EMX2* (600035)	Empty spiracles/homeobox gene
Axonogenesis (mid–late trimester)		*APBB1, APP, DCX, DRD2, ERBB2, MAP2, NOTCH1, NRCAM, PARD3, POU4F1, S100A6, S100B*	Various mechanisms (e.g., ligand selectivity, calcium binding)
Axon growth and guidance (mid–late-trimester)	Cellular lineage disorders	*NKX2-1* (600635)	Expression of neuropilin-2 guiding interneurons into the developing striatum
Neuronal differentiation (mid–late trimester)		*ASCL1, BDNF, BMP2, BMP4, CDK5R1, CDK5RAP2, HES1, HEYL, MEF2C, NEUROD1, NEUROG1, NEUROG2, NOG, NRCAM, OLIG2, PAFAH1B1, PAX3, POU4F1, RTN4, SOX2*	Various mechanisms
Neuronal cell fate determination (all trimesters)	Congenital hypoventilation Neurite outgrowth Motorneuron–spinal cord Microphthalmia	*ASCL1* (100790) *NTF3* (162660) *OLIG2* (606386) *SOX2* (184429)	Proneural achaete-scute gene/ neurosensory Neurotrophic factor Transcription factor oligodendrocytes Sry-related hmg-box gene 2
Cell differentiation regulators (all trimesters)		*HDAC4, MDK, NRG1, ODZ1, PAX5, PAX6*	
Cerebellum	Cerebellar hypoplasia Dandy–Walker Lissencephaly with cerebellar hypoplasia	*OPHN1* (300127) *ZIC1* (600470)/*ZIC4* (608948) *RELN* (600514)	Oligophrenin-related X-linked ID Zinc finger in cerebellum 1 and 4 Extracellular matrix (ECM) proteins

brain growth (e.g., acquired microcephaly) or seizures. Clinical pattern recognition of the more common malformation syndromes may allow targeted genetic testing and counseling.

Gene Identification

Traditionally, chromosomal aneuploidies of trisomies 13 and 18 and some large structural chromosomal abnormalities leading to cerebral malformations were detected by chromosome karyotyping and fluorescence *in situ* hybridization (FISH). For Mendelian disorders, including some cerebral malformations with well-established genotype–phenotype correlations, conventional molecular diagnostics, such as a combination of polymerase chain reaction (PCR) and Sanger sequencing methods, routinely evaluate selected exons for one or more targeted genes based on the clinical recognition of a specific syndrome.

Advances in genetic technologies have led to higher expectations and standards in clinical practice. Clinical recognition of phenotypes along with a series of "first-line" and "second-line" metabolic and genetic tests provides an approach to diagnose both the type and cause of cerebral malformations. Chromosomal microarray analysis (CMA; including both microarray-based comparative genomic hybridization and single-nucleotide polymorphism microarray) has become the first-line genetic test in the etiological evaluation of CNS malformations identified pre- or postnatally.[12–14] The widespread use and high diagnostic yield of CMA provides the awareness of the unpredictability of human genomic gains and losses as a major cause of cerebral malformations including those disorders initially described as "developmental delay," microcephaly, intellectual disability, and multiple congenital anomalies.[15,16] With wide application of next-generation sequencing (NGS) methods, genes related to cerebral malformations have been identified and more discoveries will be anticipated. In current practice, the combination of "first-line" CMA followed by NGS should provide an etiological diagnosis "yield" of approximately 30–40%. Variability in diagnostic rates depends on the differences in the types of patients included in different studies.[17,18] Thus, CMA followed by NGS testing technology might enable the "genotype first" screening approach ahead of the traditional "phenotype first" patient-examination approach but pretesting patient selection is essential to exclude patients who are not likely to have a genetic condition (see Differential Diagnosis, page 147–148).

TABLE 13.2 Overview of the Proposed Developmental and Genetic Classification of Cerebral Cortical Malformations, after Barkovich et al.

Group	Designation	Basis of classification scheme
I. MALFORMATIONS SECONDARY TO ABNORMAL NEURONAL AND GLIAL PROLIFERATION OR APOPTOSIS		
I.A	Severe congenital microcephaly	Premigrational reduced proliferation or excess apoptosis
I.B	Megalencephalies	Including both congenital and early postnatal
I.C	Cortical dysgenesis without neoplasia	Cortical dysgenesis with abnormal cell proliferation without neoplasia
I.D	Cortical dysplasias with neoplasia	Cortical dysplasias with abnormal cell proliferation and neoplasia
II. MALFORMATIONS DUE TO ABNORMAL NEURONAL MIGRATION		
II.A	Heterotopia	Malformations with neuroependymal abnormalities: periventricular heterotopia
II.B	Lissencephaly	Malformations due to generalized abnormal transmantle migration (radial and nonradial)
II.C	Subcortical heterotopia and sublobar dysplasia	Malformations presumably due to localized abnormal late radial or tangential transmantle migration
II.D	Cobblestone malformations	Malformations due to abnormal terminal migration+defects in pial limiting membrane
III. MALFORMATIONS SECONDARY TO ABNORMAL POSTMIGRATIONAL DEVELOPMENT		
III.A	Polymicrogyria and schizencephaly	Malformations with polymicrogyria or resembling polymicrogyria
III.B	Polymicrogyria (metabolic)	Cortical dysgenesis secondary to inborn errors of metabolism
III.C	Focal cortical dysplasias	Focal cortical dysplasias (without dysmorphic neurons) due to late developmental disturbances
III.D	Severe congenital microcephaly	Postmigrational developmental microcephaly (previously postnatal microcephaly) with birth OFC –3 SD or larger, later OFC below –4 SD and no evidence of brain injury

Abbreviations: OFC, occipitofrontal head circumference; SD, standard deviation.

Early Treatment After Identification of Cerebral Malformation

Advanced prenatal diagnosis may evolve to unrealistic parental expectations or beliefs that a pregnancy should always result in the delivery of a healthy baby. Early identification of a severe cerebral malformation during prenatal life rarely allows the possibility of curative treatment. Prenatal decision-making should be a well-informed process, and all treatment options should be considered respectfully by the physicians and healthcare team members in support of the parents as they attempt to understand the best interests of the child and their family.

Selective Abortion

For a fetus with a severe malformation, selective abortion may be the only clinical option in some circumstances for some parents. Medical ethicists, jurists, philosophers, and laypersons hold widely divergent views about when and whether a pregnancy may be actively terminated. Arguments for and against the selective abortion of a fetus with malformations have been extensively debated, especially over the last 50 years since abortion has become a legal option until the age of viability in many countries. Advanced screening technologies, including fetal MRI and prenatal CMA or NGS, are much more proficient in identifying relatively minor anatomic differences and genetic variations. However, it is difficult to make clinical judgments about subtle fetal anomalies or genetic findings under many circumstances. Ethical decision-making requires careful contemplation of the risks and benefits of any action compared to the alternative decision of no action (i.e., allowing the natural history to evolve).

Prenatal Surgery

Prenatal repair of myelomeningocele before 26 weeks of gestation may result in better neurologic function than postnatal surgical repair; however intrauterine surgery poses higher risks of maternal and fetal morbidity including premature membrane rupture leading to preterm labor and early delivery, placental abruption, and pulmonary edema.[19]

Early-Postnatal Surgery

A few postnatal surgical treatments of malformations are the standard of care. For example, congenital hydrocephalus or large cysts with evidence of mass effects on surrounding brain tissue will be routinely treated by direct ventriculoperitoneal shunting as the best approach to relieve increased intracranial pressure. Endoscopic third ventriculostomy is an alternative surgical procedure that creates an opening in the floor of the third ventricle allowing the cerebrospinal fluid to flow directly to the basal cisterns.

Mode of Inheritance and Prevalence

Primary brain malformations may either be inherited or caused by *de novo* mutations. Of the inherited primary malformations, different types of genetic/genomic alterations from single-nucleotide substitution to whole chromosomes are found with different modes of inheritance, including: 1) single gene (or Mendelian) mutations; 2) numerical or structural chromosomal abnormalities; 3) multigenic or oligogenic effects; 4) mitochondrial gene mutations; and 5) epigenetic aberrations. Single-gene disorders (both autosomal and sex-linked) and chromosomal/subchromosomal abnormalities account for the majority of known genetic/genomic factors causing cerebral malformations.

Natural History

The natural history of cerebral malformations depends on the severity of the individual malformation. Approximately half of all fertilized eggs die spontaneously during the first 2 months of pregnancy, but the exact causes are rarely discovered. Many severe malformations are typically unrecognized until miscarriage (prior to viability) or stillbirth (after viability) occurs. Chromosomal abnormalities account for about half of all miscarriages in the first trimester. Prenatal ultrasound and fetal MRI may reveal major structural malformations including diffuse and focal cortical dysplasias, neuroblast migration abnormalities (e.g., gray matter heterotopias), midbrain–hindbrain abnormalities, and corpus callosum abnormalities (e.g., agenesis), myelination maturation, metabolic disorders, and vascular anomalies. Prenatal counseling is often difficult and must be based on limited knowledge during fetal life. Milder malformations are associated with survival to viability and live birth, and are typically identified postnatally when an infant or child is evaluated because of "delays" in attaining expected neurodevelopmental milestones, abnormalities in growth (e.g., acquired microcephaly or macrocephaly), or neurological symptoms (e.g., seizures). The prognosis and life expectancy for individuals with cerebral malformations is variable, depending mostly on the severity of the malformation and the provision of supportive care.

Age of Onset

The formation of the human cerebral cortex is essentially complete at the end of 40 weeks gestation, and thus virtually all cerebral malformations have a prenatal age of onset. More severe malformations may be clinically evident around the time of birth ("congenital") but milder malformations may be overlooked until a later age when a neuroimaging finding reveals the dysgenesis during the evaluation of a seizure or a significant neurodevelopmental disorder. The postnatal discovery of a cerebral malformation should not prevent the ascertainment of its prenatal onset.

Cerebral Malformation Evolution

Early brain malformation due to genetic defects or environmental factors leads to cell dysfunction and secondary cerebral malformation that can be visualized anatomically. Underlying genetic/environmental mechanisms interfering with neurodevelopment vary, in part, because of the phenomenon of incomplete penetrance and variable expressivity of the genes involved.

Cerebral Malformation Variants and Current Classification Schemes

Cerebral malformation is a group of disorders with great heterogeneity in both causes and clinical findings. The understanding of brain malformation is still insufficient to formulate a single scientifically based classification system. Cerebral malformations are traditionally grouped and organized based on structural changes described through neuroimaging or neuropathology, sometimes with knowledge of concurrent metabolic and immune-mediated disorders. Genetic classification schemes are as imperfect as phenotypic description because of the spectrum of variability.

For example, Barkovich et al. have maintained their basic proposed descriptive schematic organizational categories ("malformations secondary to abnormal neuronal and glial proliferation or apoptosis" and "malformations secondary to abnormal neuronal migration"), but recently changed one category from "malformations secondary to abnormal cortical organization" to "malformations secondary to abnormal post-migrational development," while eliminating the category "malformations of cortical development, not otherwise classified" (Table 13.2).[11]

Mutations of various genes have been associated with microcephaly, lissencephaly, cobblestone cortex, heterotopia, and polymicrogyria have been described (Tables 13.1 and 13.2). The "phenotype–genotype correlation" is the essence of current understanding of ontogenesis and the various malformations of the nervous system. New cerebral malformations will continue to be defined with help of advanced neuroimaging, while the underlying genetic mutations will be revealed by CMA and NGS techniques. Accordingly, cerebral malformation classification schemes must continue to be evidence-based and integrated across scientific disciplines. Terms such as "idiopathic," "cryptogenic," and "multifactorial" will fade from the terminology as better understanding of the genetic etiologies and molecular mechanisms emerge.[20]

End of Life: Mechanisms and Comorbidities

End of life issues in this vast group of individual disorders is highly variable. For very severe malformations such as anencephaly, the natural outcome indicates a median survival time is less than 1 hour, although survival may be up to a few days in a small minority of neonates and up to 10 weeks in one reported case.[21] For relatively minor CNS malformations, a full life expectancy without significant neurodevelopmental disability might be expected.

MOLECULAR GENETICS

Any genes involving any process in neurodevelopment can potentially cause cerebral malformations if these genes are mutated or dysregulated. Genetic/genomic defects involving the induction of the ectoderm or pattern formation of the neural tube in the early stages of embryonic neurodevelopment, or involving regulation of neuronal migration and differentiation in the later stages of fetal neurodevelopment, will generally cause major malformation. Various levels of genetic/genomic alteration from single nucleotide mutation to whole chromosomes are found to cause cerebral malformations with all known genetic inheritances involved. Some examples of genes associated with microcephaly, lissencephaly, cobblestone cortex, heterotopia, and polymicrogyria are listed in Table 13.1.

Genomic Screening by CMA

Genomic copy-number variants (CNV) identified by CMA can be clearly pathogenic or presumably benign, depending on the size, location and genetic content correlating to a certain region of the genome. In addition, CMA may identify some CNVs as "variants of uncertain significance" (VUS) based on insufficient knowledge about the functions and clinical relevance of the gene content. Genome-wide CMA scanning will continue to reveal some novel pathogenic CNVs and some VUS. However, the uncertainty of some VUS will diminish as more is learned about the biological function of human genomic variations and their relevance to clinical phenotypes. By thoroughly scanning the human genome, CMA can frequently identify incidental findings that are discovered unintentionally and are unrelated to the original aim of the test.[22] In general, there are numerous advantages of CMA over conventional chromosome karyotyping such as high sensitivity, high throughput, and high detection rate. CMA has become the first-line genetic test in the etiological evaluation of fetus with ultrasound abnormalities or infants and children with neurodevelopment disorders (NDD) including cerebral malformations.[23,24] CMA has clinical utility for diagnosis and genetic counseling as well as for the potential for patient management. Up to one-third of all pathogenic CNVs may contain genetic information that leads to useful clinical action.[25]

Postnatal CMA

A genomic cause of an NDD will be identified by CMA in 5–20% of affected infants and children depending on the severity of the phenotype and the CMA platforms applied. The detection rate is even higher in children with dysmorphic features and other organ system anomalies. Clinically, the majority of patients with primary microcephaly, especially those with cerebral dysgenesis, multiple organ anomalies, or dysmorphic facial features, have unrecognizable syndromes—even to the most experienced clinicians. Conversely, it is not uncommon that the diagnoses made by CMA frequently involve specific clinical features that may have been present but not apparent or were not yet manifest at the time of testing. These situations lead to the concept of genotype-first over the transitional phenotype-first in clinical practice.

Prenatal CMA

Prenatal diagnosis through CMA using uncultured cells from chorionic-villus sampling or amniocentesis fluid has identified additional, clinically relevant information when compared with the previous standards of chromosome karyotyping.[6] Prenatal CMA is most beneficial after ultrasonographic examination has identified fetal structural anomalies.[26] Expanded prenatal genetic testing involves benefits, limitations, and consequences. Inconclusive prenatal test results (e.g., neuroimaging differences with CNVs of uncertain significance) cause parental anxiety and clinical dilemmas. Uncertainty scenarios underscore the critical need for comprehensive pretest consultation with informed consent, judicious reporting of test results, and access to qualified genetic counselors in the process of enabling parents to make well-informed decisions.

Genomic Screening by NGS Techniques

Although in its adolescent stage, NGS techniques have had a revolutionary impact on clinical genetic/genomic testing, and are gradually making their way into clinical laboratories. NGS will replace conventional candidate gene approaches, in which the recognition of certain features of a known phenotype directs a clinician to a specific gene test in an attempt to confirm a suspected diagnosis. NGS-based genetic/genomic tests include disease-targeted gene panels (NGS-panels), whole-exome sequencing (WES) or whole-genome sequencing (WGS). Both NGS-panels and WES require an enrichment step of the desired genomic regions before NGS can be performed, but this step is not necessary for WGS. WGS techniques uncover virtually all variants within an individual's genome simultaneously, while WES typically evaluates all known genes and NGS-panels target a group of selected genes (from several to hundreds) related to certain diseases or disorders for which both allelic and locus heterogeneity are substantial. Different from conventional "targeted and specific" genetic testing strategies and NGS-panels, which mostly target only known disease genes for diagnosis purposes, WES/WGS is a process of scanning the whole exome/genome for both discovery and diagnostic purposes. In many cases, research discoveries may be directly translated into clinical diagnoses when compelling evidence for establishing causal relationships between novel variants and unique phenotypes exists.

Postnatal NGS Analysis

WES has rapidly become a popular diagnostic test in the characterization of possible genetic causes of nonspecific or unusual disease presentations.[27] Available data indicates that WES identified the underlying genetic defect in 25–30% of patients without a clear clinical diagnosis or for patients who have negative test results for genes known to be associated with suspected disorders.[28] However there are some major limitations of WES clinical utility including: 1) excessive cost; 2) capture methods restricted to "known genes"; 3) the absence of WES capacity to sequence noncoding-regulatory or deep-intronic regions in known genes that may contain the etiologic mutations; 4) the inadequacy of WES to cover 5–10% of coding regions due to the presence of pseudogenes (or repetitive regions including trinucleotide repeats) or GC-rich regions that obscure capture and sequencing procedures; 5) the shortfall of WES technology to detect germline CNV due to biased enrichment of capture procedures (PCR amplicons), uneven genomic distribution of exons, and insufficient software; 6) overwhelming numbers of variants without full understanding of their biological implication to human health or disease; 7) false positive findings due to imperfect data-filtering algorisms; 8) the magnitude of data per sample and the time required for data interpretation prior to clinical use; 9) the lack of tools, standards, regulations and policies to integrate meaningful output for physicians and families; and 10) many unsettled issues relating to ethics, privacy, consent and legal protections.

Instead of targeting all exons of the known genes in WES, NGS-panels target a group of selected genes related to certain diseases or disorders. Many disease-specific diagnostic assays by NGS-panel methods are now commercially available for genetically heterogeneous constitutional disorders. Considering the limitations of WES, NGS-panel strategies are expected to remain the major application of NGS in diagnostic testing for the next few years.[29] NGS-panels can avoid some of the WES limitations by: 1) filling in missing NGS content with supplemental Sanger sequencing and other complementary technologies; 2) supplementing disease-targeted sequencing tests with CNV detection approaches currently missed by WES services; 3) limiting the numbers of VUS and incidental findings that are unrelated to the indication for testing; and 4) providing disease-specific expertise already residing in laboratories that previously carried out disease-targeted testing. Thus, the American College of Medical Genetics and Genomics (ACMG) recommended that WES or WGS should be reserved for those cases in which disease-targeted testing is negative or unlikely to return a positive result in a timely and cost-effective manner.[30]

WGS is currently applied mostly at the research level and few examples have successfully demonstrated utility in patient management for single-gene disorders.[31,32] WGS has some advantages over WES and NGS-panels, such as fewer sample biases during preparation, more comprehensive genome coverage, and easier identification of large deletions/duplications and other genomic abnormalities. When the understanding of noncoding regions improves and sequencing costs decrease, and when software improvement and sequencing technologies are capable of detecting all types of mutations across all genetic loci, it is anticipated that WGS will become the standard clinical method to enable the "genotype-first" screening approach in the future—even in the practice of neonatal or fetal medicine, in an attempt to attain individualized healthcare and predetermine optimal patient management and eventually eliminate the need for CMA as a separate test. Furthermore, with the understanding of the effects of new genetic variabilities, a more comprehensive genetic/genomic testing era will encompass DNA sequencing as well as transcriptomics, proteomics and epigenomics.

Prenatal NGS Analysis

Both invasive and noninvasive methods of prenatal diagnosis by NGS for genetic/genomic abnormalities are available; however, the use of these technologies in the prenatal period raises many ethical and policy questions.[33,34] Similar to prenatal CMA analysis, NGS-based prenatal testing requires chorionic-villus sampling or amniocentesis fluid, from which DNA is extracted. Noninvasive prenatal detection of common fetal aneuploidies with maternal plasma cell-free DNA (cfDNA) using various NGS platforms has been widely applied in a clinical setting.[35] Meanwhile, NGS methods to detect genomic CNVs using cfDNA have been achieved.[36] Recently several groups provided promising solutions for fetal Mendelian diseases noninvasively, especially for couples with a born proband.[37]

Genotype–Phenotype Correlations

Patterns of morphological development are continuously correlated with new data regarding the genetic programming of the CNS. The over-reaching goal would be to create an integrated scheme that would explain brain malformations both in terms of morphogenesis and genetic gradients along the axes of the neural tube and segmentation.

DISEASE MECHANISMS

Both genetic and environmental factors contribute to underlying etiologies of cerebral malformations although the exact pathological mechanisms are usually not known in most individual patients. There is an indirect relationship between genes, genetic transcription, protein translation, and neurodevelopmental outcome. Many gene products affect molecular and cellular functions that are influenced by environmental factors and metabolism.[38] Nonetheless, individual genes and gene products are essential elements interacting to support and guide the complex process of brain development.

Synopsis of Developmental Physiology and Pathophysiology

Neural Induction

Neural induction is a complex process resulting from a combination of graded extracellular signals and transcription factor gradients involving several fields of progenitor cells that lead to later neocortical development. Sonic hedgehog (SHH) signaling from the notochord induces the formation of the floor plate. The embryonic neural plate evolves after neural induction from the ectoderm. This process is regulated by a multifaceted interplay between intrinsic genetic mechanisms and extrinsic information.[20,39]

Neural Tube Formation (Neurulation)

After gastrulation and alignment of neural progenitor cells along the neural plate, the neural tube forms between 20–28 days of gestation. The folding and fusion of the neural tube involves a bidirectional rostral–caudal process. The (rostral) anterior neuropore closes by 25 days and the (caudal) posterior neuropore closes by 28 gestational days. The rostral neural progenitor cells produce the brain and the caudally positioned progenitor cells produce the hindbrain and spinal column. Failure of the embryonic neural tube to close results in developmental defects of the spinal cord itself but also has secondary effects on the surrounding mesodermal and ectodermal structures. The neural ectoderm merges with cells from the endoderm to form the medullary cord. Secondary neurulation involves the development of the caudal cell mass forming the caudal-most portion of the neural tube in spinal segments below L-2. After neural tube closure, disjunction of epithelial ectoderm and neural ectoderm occurs, and mesenchymal cells migrate to form the meninges, neural arches of the vertebrae, and paraspinal muscles. The majority of neurulation malformations

are sporadic fitting a multifactorial polygenic or oligogenic pattern combined with environmental factors, although a strong genetic component is indicated by the high recurrence risk in certain families.[40,41]

Patterning

The neural tube includes four distinct regions during early neurodevelopment. The most anterior portion of the neural tube, the prosencephalon, gives rise to the forebrain. The mesencephalon, just caudal to the prosencephalon, gives rise to the midbrain, while the hindbrain regions evolve from the both the metencephalon and myelencephalon. By the third month of gestation, the spinal cord extends the full length of the embryo. Malformations in the midbrain, hindbrain, and cerebellum are observed in conjunction with cerebral cortical dysgenesis. Specific, genetically defined disorders may lead to failure of differentiation in specific cell types.

Proliferation

Neural cell proliferation originates mainly from the medial and caudal ganglionic eminences and the dorsal subventricular zone of the germinal epithelium before tangential (radial) migration to the developing cerebral cortex. In the subventricular zone, neuroepithelial cells differentiate into radial glial cells by dividing asymmetrically (i.e., daughter cells with different cellular fates) leading to the generation and regulation of intermediate progenitor cells with a balance between cellular self-renewal and progression to the development of the cerebral cortex. Genetic mutations regulating the progenitor cell mitotic cycle are associated with some forms of congenital microcephaly (Table 13.1). The subventricular zone is comprised of an outer subventricular zone and a layer of radially oriented corticocortical, corticothalamic and thalamocortical axons. The outer subventricular zone contains radial glial-like nonepithelial cells and intermediate progenitor cells. The expansive proliferation of progenitor cells in the outer subventricular zone helps to explain the evolutionary expansion of the number of radial glial units, surface area and gyrification in the primate cortex, as these later-born cells are presumed to occupy the outer cortical layers.[42]

Neuroblast Migration

Neuroblasts produced by neural stem cells in the subventricular zone generally migrate to different sites before their neuronal maturation and formation of synaptic connections. The migration of neuroblasts is genetically programmed but the process may be disrupted by environmental factors and injury. Neuroblasts reach the ultimate destinations in cortical cortex along the cellular scaffold of radial glial cells through the intermediate zone out into the developing cortical plate. Migration of neuroblasts into the developing neocortex results in the formation of an ordered six-layered structure with earlier migrating neuroblasts forming the deepest layers of cortex (the preplate) and later migrating neuroblasts forming the outer layers (i.e., inside-out layering). Some neuroblasts travel from the ganglionic eminence through tangential migration along preformed axons rather than radial glial fibers. The mutation of many known genes that facilitate neuroblast migration may cause cortical dysgenesis (Table 13.1). Nonmigrating neurons that move just outside the subventricular region may be described as periventricular nodular heterotopia by postnatal neuroimaging. Neuroblasts that arrest within the white matter are called subcortical laminar heterotopia (or "band heterotopia") whereas overmigrating neuroblasts may form neural cell collections called "glioneuronal heterotopia" within the leptomeninges.

Differentiation

The cerebral cortex may be decreased in size because of deficient cortical layering secondary to errors in proliferation, cell cycle progression, or early neuronal differentiation. Malformations secondary to postmigrational developmental errors (e.g., polymicrogyria) may occur after abnormal expression of late cortical progenitors. Neuronal differentiation is a complex process requiring the regulation of proliferation, cell death (apoptosis), differentiation and maturation. Coding genes and noncoding miRNAs exert major roles in regulating cellular networks but the exact mechanisms remain mostly unknown.

Axon Extension and Synapse Formation

After neuronal differentiation, functional neurons extend axons and dendrites and reach adjacent targets. The genetic regulation of neurite genesis, outgrowth and synapse formation is complex and likewise incompletely understood. Noncoding RNAs may regulate miRNAs activity as a balance of gene expression in the process of normal neuronal development and as an explanation for neurodevelopmental disorders such as autism.[43]

Apoptosis and Neuronal Pruning

The postnatal cerebral cortex decreases in size due to naturally occurring cell death and pruning in a process that involves considerable excess of neuronal production followed by systematic elimination of up to half of all cells.

Most naturally occurring apoptosis occurs prenatally whereas the pruning of neuronal connections generally occurs postnatally. Apoptosis is a genetically regulated process, or intrinsic "suicide" program, that involves the breakdown of nuclear chromatin and cell fragmentation. Although cell death plays an essential role in eliminating cell populations that serve only a transient function in brain development (such as cells of the marginal zone and the subplate), genes involved in these processes are generally not associated with cerebral malformations.

Cerebellum

The protracted development of the cerebellum makes it vulnerable to malformation. Early development begins shortly after neural tube closure and the formation of the primary brain vesicles: prosencephalon (forebrain), mesencephalon (midbrain), and rhombencephalon (hindbrain). The developing cerebellum is derived from the mesencephalon and the metencephalon (later pons and cerebellum) between 4–5 weeks of gestation. After the flexion of the neural tube, the rhombencephalon subdivides into eight rhombomeres and isthmus develops at the junction of the mesencephalon and metencephalon in a genetically regulated patterning process for the midbrain and pontine development.[44] The roof of the fourth ventricle develops into the rudimentary cerebellum in the second month of gestation with the flocculonodular lobe (archicerebellum) and dentate nuclei. The rostral regions give rise to the midline vermis (paleocerebellum) and the caudal regions produce the cerebellar hemispheres (neocerebellum). The vermis is foliated by 4 months and the large cerebellar hemispheres by 5–6 months gestation. Bergmann glial cells that occupy the Purkinje cell layer between Purkinje neurons have long centrifugal processes that also extend to the pial surface of the cerebellum and serve as guide fibers for the external granular cells that need to move into the internal granular layer beneath the row of Purkinje cells. Postnatally, cerebellar growth continues with complete foliation patterns by 7 months of life and complete cellular differentiation of the cerebellar layers, final migration, proliferation, and arborization of cerebellar neurons in humans by about 20 months of life. Disorders with cobblestone lissencephaly and mid–hindbrain abnormalities with cerebellar hypoplasia include autosomal recessive disorders associated with congenital muscular dystrophy and ocular anomalies such as muscle–eye–brain disease, Walker–Warburg syndrome, and Fukuyama congenital muscular dystrophy.

Primary Ciliopathies

Common patterns among many developmental disorders have been recognized including various genetic syndromes with overlapping pathophysiological mechanisms. Primary ciliopathies are genetic disorders resulting in dysfunction of cilia—small nonmotile, hair-like organelles protruding from cell surfaces. Cilia involve sensation of the extracellular environment in the coordination of developmental and homeostatic signaling pathways. Ciliopathies are thought to be the cause of various malformations including distinct cortical and cerebellar dysgenesis syndromes along the Joubert syndrome spectrum of disorders.[45] Other recognized ciliopathies include primary ciliary dyskinesia, Bardet–Biedl syndrome, polycystic kidney and liver disease, nephronophthisis, Alstrom syndrome, Meckel–Gruber syndrome, and some forms of retinal degeneration. Joubert syndrome is characterized by cerebellar vermis hypoplasia with the neuroimaging "molar tooth sign" along with other clinical signs of breathing dysregulation, retinal dystrophy with ocular motor apraxia, hypotonia progressing to ataxia, and diffuse developmental encephalopathy variably associated with renal, liver, and skeletal involvement. Joubert syndrome genes such as *AHI1* are highly expressed in neurons that give rise to the crossing axons of the corticospinal tract and superior cerebellar peduncles.[46] At least 21 genes have been associated with Joubert syndrome. All gene products of the *JBTS* 1–20 genes are related to the function of proteins expressed in the primary cilia or its basal body and centrosome apparatus (Table 13.3).[46]

Current Research

Research on cerebral malformation involves many areas, ranging from basic research (such as identification of genetic causes, environmental factors, or pathological mechanisms) to clinical areas (such as prevention, diagnosis, and treatment). Two active research areas are listed below: one is related to collecting transcriptome data for further understanding of the molecular organization of the developmental brain, and the other takes advantage of NGS techniques for the analysis of maternal plasma cell-free DNA for the diagnosis of genetic/genomic disorders including cerebral malformations.

Anatomical and Functional Maps of the Human Brain

Better understanding of pathological mechanisms of cerebral malformation requires correlation of the regional anatomical and functional changes during development along with the timing of cellular integration and network organization. The availability of new techniques, such as diffusion tensor imaging tractography, allows the study

TABLE 13.3 Genes Involving Primary Ciliopathies Associated with Joubert Syndrome and Other Malformation Syndromes

Joubert loci	Gene (chromosome location)	Protein	Function/mechanism
JBTS1	*INPP5E* (9q34.3)	Inositol polyphosphate-5-phosphatase	Metabolic pathways and phosphatidylinositol signaling
JBTS2	*TMEM216*	Transmembrane protein 216	D-myo-inositol-trisphosphate biosynthesis and inositol phosphate metabolism
JBTS3	*AHI1* (6q23.3)	Jouberin	Primary cilia transition zone/cerebellar–cortical development
JBTS4	*NPHP1* (2q13)	Nephronophthisis 1	Multifunctional/cell division/cell–cell adhesion
JBTS5	*CEP290* (12q21.32)	Centrosomal protein 290 kDa	Centrosome and cilia function in cell division and microtubule assembly retina development
JBTS6	*TMEM67* (8q22.1)	Meckelin	Centriole migration/primary cilium formation
JBTS7	(FTM) *RPGRIP1L* (16q12.2)	RPGRIP1-like	Programmed cell death, craniofacial development, patterning of the limbs, and formation of the left–right axis
JBTS8	*ARL13B*	ADP-ribosylation factor-like 13B	Sonic hedgehog signaling
JBTS9	*CC2D2A* (4p15.32)	Coiled-coil and C2 domain+2A	Coiled-coil/calcium binding domain protein/cilia formation
JBTS10	*CXOrf5* (OFD1)		Associated with oralfaciodigital syndrome 1
JBTS11	*TTC21B*	Tetratricopeptide repeat protein 21B	Retrograde intraflagellar transport in cilia
JBTS12	*KIF7* (15q26.1)	Kinesin family member 7	Regulator of sonic hedgehog (SHH) signaling pathway—associated with Joubert, acrocallosal, and hydrolethalus syndromes
JBTS13	*TCTN1*	Tectonic-1	Tectonic family of secreted and transmembrane proteins
JBTS14	*TMEM237* (2q33.2)	Transmembrane protein 237	Tetraspanin protein involved in WNT signaling
JBTS15	*CEP41*	Centrosomal protein	Ciliogenesis for tubulin glutamylation in cilium
JBTS16	*TMEM237* (2q33)	Transmembrane protein 237	WNT signaling
JBTS17	*TMEM138* (11q12.2)	Transmembrane protein 138	Ciliogenesis
JBTS18	*TCTN3*	Tectonic-3	Hedgehog signal transduction
JBTS19	*ZNF423*	Zinc finger protein 423	Zinc finger DNA-binding transcription factor
	OFD1	Orofaciodigital syndrome 1	Formation of cilia
JBTS20	*TMEM231*	Transmembrane protein 231	Transmembrane component of tectonic-like complex

of functional brain networks alongside structure in an attempt to characterize neuronal connectivity. Longitudinal studies will be needed to understand causality between brain structure, connectivity, plasticity, and intellectual disability in patients with brain malformations.

Improvement of Prenatal Diagnosis

Although fetal MRI and invasive genetic/genomic testing are the current standard for prenatal diagnosis of fetus with cerebral malformation, noninvasive methods using NGS-based fetal DNA analysis from maternal plasma cell-free DNA have become technically feasible and commercially popular for earlier prenatal diagnosis of genetic/genomic disorders. Such testing is expected to lead to earlier prenatal diagnosis.

DIFFERENTIAL DIAGNOSIS

The combination of neuroimaging and genetic testing often fails to provide a precise etiologic diagnosis of malformation even after thorough clinical assessments and genetic counseling. Neurodevelopmental disorders result from a vast range of genetic and environmental causes during the embryological and fetal periods of life.

Traditional *nature-versus-nurture* debates have gradually shifted toward the analysis of *gene–environment* interactions (GxE), which is considered the phenotypic effect of interactions between genes and the environment. Complex adaptive processes and developmental plasticity account for the effects of many epigenetic processes such as maternal nutritional status, certain environmental toxins, maternal illnesses, and genetic disorders, alone or in combination. Cell growth and metabolism alterations during critical periods of prenatal brain development are associated with GxE effects. Further understanding of various determinants of biological-phenotype unpredictability offer promise for better risk-modifying strategies in the prevention of those neurodevelopmental disorders related to GxE interactions.

Neuropathology

Many developmental brain anomalies beneath the resolution of current MRI technology are only identified microscopically in tissue after epilepsy surgery or postmortem at autopsy. Brain tissue is often unavailable for neuropathological diagnostics, especially from patients with milder and clinically stable CNS malformations. Other defects are microscopic or require special immunocytochemical techniques of tissue examination. Thus, some malformation diagnoses are only feasible by postmortem neuropathologic examination, justifying the need to perform an autopsy in a patient who does not survive even though a tentative diagnosis may have been established by neuroimaging and genetic markers. Most brain malformations are true neurodevelopmental *disorders* resulting from changes in genes highly expressed in brain tissues during the prenatal period—rather than neurological *diseases*. Recognizable malformation patterns (e.g., Miller–Dieker syndrome with a *de novo* chromosome 17p13.3 deletion syndrome involving the *PAFAH1B1* gene) might be construed as a "disease" diagnosis because of its progressive postnatal clinical nature.

TESTING

Neuroimaging Examination

MRI is the standard tool for defining structural CNS anomalies and is generally considered the minimum initial investigation for cerebral malformations. More sophisticated neuroimaging methods, such as tractography and diffusion tensor imaging, are valuable to identify subtle structural abnormalities not visible on routine MRI. Techniques such as positron-emission tomography (PET) and single-photon emission computed tomography (SPECT) generate insight into the metabolism and perfusion of regional brain malformations. However, these technologies are not yet routinely employed in clinical settings for evidence-based CNS malformation diagnostics.

Genetic/Genomic Testing

When any cerebral malformations in a fetus are observed prenatally, CMA evaluation using uncultured cells from chorionic-villus sampling or amniocentesis fluid should be recommended to exclude (or identify) any chromosomal numerical/structural abnormalities. Chromosomal karyotype for possible trisomies or locus-specific FISH (e.g., for clinically recognizable Miller–Dieker syndrome) are still useful. DNA sequencing for selected gene(s) relevant to cerebral malformations is also possible if neuroimaging findings point to any specific disorder. Postnatally, it is now routine practice to first evaluate a child with any malformation using CMA. If the CMA study is normal, NGS-panels to evaluate known genes relevant to specific cerebral malformations are recommended. Considering that the majority of patients with primary microcephaly, especially those with cerebral dysgenesis, multiple organ anomalies, or dysmorphic facial features, have unrecognizable syndromes, NGS-panels or WES testing will be used more often for patients with cerebral malformations.

Metabolic Testing

Although abnormal metabolism has been noticed in every disorder of cerebral malformations, metabolic testing using advanced techniques, such as mass spectrometry, are usually not specific for the diagnosis or differential diagnosis of cerebral malformation. Thus, metabolic testing is indicated with cerebral malformations, especially in infants with nonspecific metabolic abnormalities (e.g., hypoglycemia, metabolic acidosis, hyperammonemia) or suspicious physical signs (e.g., lethargy, decreased feeding, vomiting, tachypnea, decreased perfusion, and seizures) or any progressive worsening of tone (hypotonia, hypertonia), posture (fisting, opisthotonos), movements (tongue-thrusting, myoclonus), or consciousness (coma).

MANAGEMENT

Standard of Care

The combination of neuroimaging and genetic screening has become an accepted approach to the prenatal and postnatal evaluation of CNS malformation. Knowledge of the array of genes specific to human brain development is growing but the practical consequences of gene alterations often remain a challenge to clinicians. The concept of diagnostic specificity is important as not all diagnoses have equal certainty or clinical predictability (i.e., estimated prognosis). As innovative technologies become increasingly implemented for the screening of prenatal neurodevelopmental disorders and malformations, many legal and ethical dilemmas will continue to evolve. For many disorders, no definitive guidelines will be available to direct parents about unambiguous reasons for providing or withholding care for a particular infant with signs of severe neurodevelopmental disorder with or without a known cause or mode of effective therapy. However, an accurate diagnosis will remain imperative for family clarification, reassurance, genetic counseling and future prenatal screening.

Postnatal management of infants and children with cerebral malformations is mostly symptomatic and supportive care, including early intervention services (occupational, physical, and speech therapies), rehabilitation services (e.g., wheelchair for mobility, if needed), nutritional supervision (e.g., feeding tubes for individuals with swallowing dysfunction), spasticity management, and epilepsy treatment (e.g., monitoring antiepileptic drugs).

Failed Therapies and Barriers to Treatment Development

Insurmountable barriers to the treatment of many cerebral malformations are underlying injuries to critical neuronal pathways occurring in the early stages of embryogenesis. It is extremely difficult, or even impossible, to treat brain tissue that never formed. Thus, useful treatments for many infants with severe cerebral malformations are very limited and none of current options for treating cerebral malformation should be considered as failed therapies.

Therapies Under Investigation

New treatments for genetically defined malformations such as early gene transfer therapy may be feasible for single-gene disorders (e.g., hematopoietic disorders). Prenatal gene transfer and the prospects of gene therapy are still in experimental stages posing numerous risks for the fetus and the pregnant mother including the potential alteration of normal development, the possibility of inadvertent germline gene transfer, and the danger of oncogenesis.[47] Therapies for malformations arising from multiple genetic contributors with complex inheritance will be much more challenging.

References

1. Turing AM. The chemical basis of morphogenesis. *Phil Trans R Soc Lond B*. 1952;237(641):37–72.
2. Warkany J, Lemire RJ, Cohen Jr. MM. *Mental Retardation and Congenital Malformations of the Central Nervous System*. Chicago: Yearbook; 1981.
3. Friede RL. *Developmental Neuropathology*. 2nd ed. Berlin: Springer; 1989.
4. Sarnat HB. *Cerebral Dysgenesis: Embryology and Clinical Expression*. New York: Oxford University Press; 1992.
5. O'Rahilly R, Muller F. *The Embryonic Human Brain. An Atlas of Developmental Stages*. Hoboken, New Jersey: Wiley; 2006.
6. Wapner RJ, Martin CL, Levy B, et al. Chromosomal microarray versus karyotyping for prenatal diagnosis. *N Engl J Med*. 2012;367:2175–2184.
7. Barkovich AJ, Kuzniecky RI, Jackson GD, Guerrini R, Dobyns WB. Classification system for malformations of cortical development: update 2001. *Neurology*. 2001;57:2168–2178.
8. Sarnat HB, Flores-Sarnat L. Integrative classification of morphology and molecular genetics in central nervous system malformations. *Am J Med Genet*. 2004;126A:386–392.
9. Graf WD, Sarnat HB. Intermediate filament proteinopathies: from cytoskeletons to genes to functional nosology. *Neurology*. 2002;58(10):1451–1453.
10. Jissendi-Tchofo P, Kara S, Barkovich AJ. Midbrain– hindbrain involvement in lissencephalies. *Neurology*. 2009;72:418–426.
11. Barkovich AJ, Guerrini R, Kuzniecky RI, Jackson GD, Dobyns WB. A developmental and genetic classification for malformations of cortical development: update 2012. *Brain*. 2012;135:1348–1369.
12. South ST, Lee C, Lamb AN, Higgins AW, Kearney HM. Working Group for the American College of Medical Genetics and Genomics Laboratory Quality Assurance Committee. ACMG Standards and Guidelines for constitutional cytogenomic microarray analysis, including postnatal and prenatal applications: revision 2013. *Genet Med*. 2013;15(11):901–909.
13. Liao C, Fu F, Li R, et al. Implementation of high-resolution SNP arrays in the investigation of fetuses with ultrasound malformations: five years of clinical experience. *Clin Genet*. 2013 Sep 3; [Epub ahead of print].
14. Southard AE, Edelmann LJ, Gelb BD. Role of copy number variants in structural birth defects. *Pediatrics*. 2012;129(4):755–763.

15. Girirajan S, Rosenfeld JA, Coe BP, et al. Phenotypic heterogeneity of genomic disorders and rare copy-number variants. *N Engl J Med*. 2012;367(14):1321–1331.
16. Mikhail FM, Lose EJ, Robin NH, et al. Clinically relevant single gene or intragenic deletions encompassing critical neurodevelopmental genes in patients with developmental delay, mental retardation, and/or autism spectrum disorders. *Am J Med Genet A*. 2011;155A(10):2386–2396.
17. Gahl WA, Markello TC, Toro C, et al. The National Institutes of Health Undiagnosed Diseases Program: insights into rare diseases. *Genet Med*. 2012;14(1):51–59.
18. Need AC, Shashi V, Hitomi Y, et al. Clinical application of exome sequencing in undiagnosed genetic conditions. *J Med Genet*. 2012;49(6):353–361.
19. Adzick NS, Thom EA, Spong CY, et al. A randomized trial of prenatal versus postnatal repair of myelomeningocele. *N Engl J Med*. 2011;364(11):993–1004.
20. Rakic P, Ayoub AE, Breunig JJ, Dominguez MH. Decision by division: making cortical maps. *Trends Neurosci*. 2009;32:291–301.
21. Poretti A, Meoded A, Ceritoglu E, Boltshauser E, Huisman TA. Postnatal in-vivo MRI findings in anencephaly. *Neuropediatrics*. 2010;41(6):264–266.
22. Green RC, Berg JS, Grody WW, et al. ACMG recommendations for reporting of incidental findings in clinical exome and genome sequencing. *Genet Med*. 2013;15(7):565–574.
23. Miller DT, Adam MP, Aradhya S, et al. Consensus statement: chromosomal microarray is a first-tier clinical diagnostic test for individuals with developmental disabilities or congenital anomalies. *Am J Hum Genet*. 2010;86(5):749–764.
24. Manning M, Hudgins L. Professional Practice and Guidelines Committee. Array-based technology and recommendations for utilization in medical genetics practice for detection of chromosomal abnormalities. *Genet Med*. 2010;12(11):742–745.
25. Ellison JW, Ravnan JB, Rosenfeld JA, et al. Clinical utility of chromosomal microarray analysis. *Pediatrics*. 2012;130(5):e1085–e1095.
26. American College of Obstetricians and Gynecologists Committee on Genetics. Committee Opinion No. 581: the use of chromosomal microarray analysis in prenatal diagnosis. *Obstet Gynecol*. 2013;122(6):1374–1377.
27. Bamshad MJ, Ng SB, Bigham AW, et al. Exome sequencing as a tool for Mendelian disease gene discovery. *Nat Rev Genet*. 2011;12(11):745–755.
28. Yang Y, Muzny DM, Reid JG, et al. Clinical whole-exome sequencing for the diagnosis of Mendelian disorders. *N Engl J Med*. 2013;369(16):1502–1511.
29. Rehm HL. Disease-targeted sequencing: a cornerstone in the clinic. *Nat Rev Genet*. 2013;14(4):295–300.
30. ACMG Board of Directors. Points to consider in the clinical application of genomic sequencing. *Genet Med*. 2012;14(8):759–761.
31. Lupski JR, Reid JG, Gonzaga-Jauregui C, et al. Whole-genome sequencing in a patient with Charcot-Marie-Tooth neuropathy. *N Engl J Med*. 2010;362(13):1181–1191.
32. Bainbridge MN, Wiszniewski W, Murdock DR, et al. Whole-genome sequencing for optimized patient management. *Sci Transl Med*. 2011;3(87):87re3.
33. Yurkiewicz IR, Korf BR, Lehmann LS. Prenatal whole-genome sequencing–is the quest to know a fetus's future ethical? *N Engl J Med*. 2014;370(3):195–197.
34. Donley G, Hull SC, Berkman BE. Prenatal whole genome sequencing: just because we can, should we? *Hastings Cent Rep*. 2012;42(4):28–40.
35. Bianchi DW, Wilkins-Haug L. Integration of noninvasive DNA testing for aneuploidy into prenatal care: what has happened since the rubber met the road? *Clin Chem*. 2014;60(1):78–87.
36. Srinivasan A, Bianchi DW, Huang H, Sehnert AJ, Rava RP. Noninvasive detection of fetal subchromosome abnormalities via deep sequencing of maternal plasma. *Am J Hum Genet*. 2013;92(2):167–176.
37. Kitzman JO, Snyder MW, Ventura M, et al. Noninvasive whole-genome sequencing of a human fetus. *Sci Transl Med*. 2012;4(137):137ra76.
38. Graf WD. Cerebral dysgeneses secondary to metabolic disorders in fetal life. In: Sarnat HB, Curatolo P, eds. *Handbook of Clinical Neurology: Malformations of the Nervous System*. Elsevier B. V; 2007:459–476.
39. Stern CD. Neural induction: old problem, new findings, yet more questions. *Development*. 2005;132(9):2007–2021.
40. Greene NDE, Philip Stanier P, Copp AJ. Genetics of human neural tube defects. *Hum Mol Genet*. 2009;18(2):R113–R129.
41. Carter TC, Pangilinan F, Troendle JF, et al. Evaluation of 64 candidate single nucleotide polymorphisms as risk factors for neural tube defects in a large Irish study population. *Am J Med Genet A*. 2011;155A(1):14–21.
42. Tabata H, Yoshinaga S, Nakajima K. Cytoarchitecture of mouse and human subventricular zone in developing cerebral neocortex. *Exp Brain Res*. 2012;216(2):161–168.
43. Colantuoni C, Lipska BK, Ye T, et al. Temporal dynamics and genetic control of transcription in the human prefrontal cortex. *Nature*. 2011;478(7370):519–523.
44. Garel C, Fallet-Bianco C, Guibaud L. The fetal cerebellum: development and common malformations. *J Child Neurol*. 2011;26(12):1483–1492.
45. Valente EM, Rosti RO, Gibbs E, Gleeson JG. Primary cilia in neurodevelopmental disorders. *Nat Rev Neurol*. 2014;10(1):27–36.
46. Romani M, Micalizzi A, Valente EM. Joubert syndrome: congenital cerebellar ataxia with the molar tooth. *Lancet Neurol*. 2013;12(9):894–905.
47. Coutelle C, Ashcroft R. Risks, benefits and ethical, legal, and societal considerations for translation of prenatal gene therapy to human application. *Methods Mol Biol*. 2012;891:371–387.

CHAPTER

14

Global Developmental Delay and Intellectual Disability

Myriam Srour and Michael Shevell†*

*McGill University, Montreal, QC, Canada

†McGill University Health Centre, Montreal, QC, Canada

CLINICAL FEATURES

Global developmental delay and intellectual disability are complementary chronologically framed entities that encapsulate heterogeneous "symptom complexes" that are frequent reasons for medical evaluation and intervention.[1] Global developmental delay is defined as a significant functional delay in two or more developmental domains (e.g., motor [gross/fine], speech/language [expressive, receptive, mixed], cognition, personal–social, activities of daily living). This delay is evident when compared in a standardized fashion on appropriate norm-referenced evaluations to chronological peers. Typically, the term is used in reference to children less than 5 years of age.[2]

"Intellectual disability" has supplanted the term "mental retardation," which is considered to be pejorative terminology, and thus avoided. It also reflects an alteration in the conceptualization of the construct of disability that increasingly emphasizes contextual factors and adaptive behaviors rather than objective measurement in contrast to a "normative" construct. Originally formulated in strictly psychometric terms as performance greater than 2.5 standard deviations below the mean on intelligence testing, intellectual disability's conceptualization has been extended to include defects in "adaptive behavior as expressed in conceptual, social, practical and adaptive skills". Employing assessments that incorporate sensitivity to cultural and linguistic diversity, limitations for the individual with intellectual disability are apparent within varied environments. These limitations are present from an early age, exist across the lifespan, and require the implementation of systems of support to maximize individual participation in all environments. The use of the term "intellectual disability" is largely restricted to individuals older than 5 years of age.[3]

As "symptom complexes," global developmental delay and intellectual disability are in essence never the same disorder twice. Presentations, underlying etiologies, associated comorbidities, medical challenges, rehabilitation service needs, trajectories, and outcomes vary from case to case.[4] Their relative merits as meaningful constructs rest on a commonality of approach, evaluation, and management principles. Global developmental delay reflects the parental emphasis on the child's attainment of developmental competencies and skills as a prelude to successful integration and autonomy. Intellectual disability reflects our distinctive human capacity to reason, think abstractly, and plan that are preconditions to our ability for learning, solving problems, and truly comprehending our surroundings.[5] These entities are obviously interrelated in that many individuals originally diagnosed with "global developmental delay" will later merit a diagnosis of "intellectual disability," while many individuals with "intellectual disability" were once labeled as having "global developmental delay."

Given the normative population-wide distribution of developmental and intellectual skills, global developmental delay and intellectual disability will, not unexpectedly, affect between 2–3% of the population.[3] Roughly two-thirds of affected individuals will have a mild-to-moderate level of impairment, while one-third will have a severe-to-profound level of impairment. Males are more affected than females and there appears to be an inverse socio-economic status gradient with respect to prevalence, although this is not without controversy. The gender and

Rosenberg's Molecular and Genetic Basis of Neurological and Psychiatric Disease
http://dx.doi.org/10.1016/B978-0-12-410529-4.00014-0

socio-economic bias noted appears to be operative for mild-to-moderate degrees of impairment only. Individuals with global developmental delay or intellectual disability are at an increased risk for a variety of comorbidities, including epilepsy or convulsive disorders, behavioral disturbances, attentional limitations, psychiatric illness, and sensory impairments (i.e., vision, hearing). These comorbidities may be the major burden of care needs challenging the optimization of intrinsic individual potential and adversely impacting on quality of life at both an individual and familial level. The full economic impact of these disorders remains unknown, although a recent study places additional lifetime costs per individual at greater than 1 million US$, with a lifetime additional cost for medical care above 50 billion US$ for the cohort of US children born in the year 2000 with intellectual disability.

Aside from classifying according to the severity of observed impairment, some authors have distinguished between syndromic and nonsyndromic intellectual disability (NSID).[6] Syndromic intellectual disability is said to occur when in addition to intellectual disability a distinct clinical phenotype (e.g., Trisomy 21) may be apparent or comorbidities in addition to intellectual disability are readily evident. The documentation of dysmorphic features or congenital anomalies in non-central nervous system (CNS) organ systems may also suggest a syndromic subclassification. In contra-distinction, NSID is defined by intellectual disability being the sole discernable clinical feature. This classification may sometimes be somewhat arbitrarily assigned, as comorbid features may evolve over time and co-existing clinical features may be subtle and difficult to discern accurately or with certainty.

The etiology of global developmental delay and intellectual disability as befitting their characterization as "symptom complexes" is quite heterogeneous. Causes may be congenital or acquired with prenatal, perinatal, and postnatal pathogenic timing. Over the past decade, various groups have formulated practice parameters and guidelines to assist the clinician in the standardized evidence-based evaluation of these entities.[2,7–9] At present, it appears that between one-quarter to one-half of identified causes are genetic in origin. Genetic etiologies are multiple and pleiotropic, including chromosomal anomaly (e.g., aneuplodies), sub-microscopic deletions/duplications/rearrangements (copy number variant changes) and monogenic disorders.[10] Indeed 450 genes have been implicated in intellectual disability, with 400 attributed to syndromic intellectual disability and 50 to NSID. All manner of Mendelian inheritance (autosomal dominant, autosomal recessive, X-linked) have been documented with the bulk of genes known to result in NSID having an X-chromosome location. Although the number of genes implicated in intellectual disability has increased substantially over the last decade, the majority of suspected genetic causes currently lack a specific molecular diagnosis. At present, only a few specific well-characterized single-gene associations with a highly recognized clinical phenotype (i.e., FMR1-Fragile X, MECP2-Rett syndrome) are tested for at a molecular level during diagnostic evaluation.[9] In addition to the nuclear genome, defects in the mitochondrial genome can give rise to syndromic intellectual disability featuring a maternal pattern of inheritance.

Progress in the identification of genes responsible for global developmental delay and intellectual disability has furthered our understanding of the molecular basis for learning and memory that is fundamental for comprehending cognition and intellect from a neurological perspective.[6,10] Increasing knowledge of molecular pathways will enable the eventual selection of pharmacologic and candidate gene therapeutic approaches. Furthermore, while there may be evidence for a bewildering array of genes involved, there appears to be a merger of action into several discrete networks of functional processes that yield a convergence of phenotypes of intellectual disability. These basic networks include neurogenesis (i.e., neuronal proliferation), neuronal migration, inter-neuronal connectivity (i.e., pre-synaptic vesicle formation, synaptogenesis, synaptic plasticity, dendrite morphogenesis, post-synaptic density), cellular signaling cascades, and regulation of transcription and translation (both genetic and epigenetic in origin).

DIAGNOSIS

Accurate diagnosis of global developmental delay or intellectual disability is an essential pre-condition to initiating a proper evaluation relevant to service referrals, appropriate ongoing management directed at expected comorbidities, and counseling that meets the needs of families.[11,12] This diagnosis is predicated on careful attention to the operational definition of these entities as outlined above. The diagnosis is typically formulated initially on the basis of clinical judgement. The validity of such a diagnosis is related to the degree of direct experience with these individuals by the diagnostician. Validity is increased by direct observation, inputs from reliable third-party informants (e.g., educators), repeated observation over time, and input either concurrently or subsequently from an inter-disciplinary professional team offering complementary skill sets (e.g., physicians, occupational therapists, physiotherapists, speech–language pathologists, psychologists). A contextual sensitivity to social, cultural and linguistic diversity is also especially pertinent. Indeed, such varying contexts may preclude the availability for administration of standardized evaluations that are the hallmark of an objective corroborating diagnosis.

A variety of widely used standardized measures for cognitive function exist.[3] These measures have been extensively used on a variety of populations and demonstrate a normative distribution psychometrically. To qualify for a diagnosis of intellectual disability, performance greater than 2.5 standard deviations below the mean is expected. Those administering a test must be trained and experienced in its application and interpretation requires an awareness of the tests standard error of measure (SEM). Routinely used measures in practice include: the Wechsler Intelligence Scales for Children, 4th edition (WISC-IV), the Wechsler Preschool and Primary Scales of Intelligence, 3rd edition (WPPSI-III), and the Standford-Binet Intelligence Scales, 5th edition (SBS). To merit the diagnosis of intellectual disability, concurrent deficits in adaptive behavior must also be demonstrated. Typically these can be obtained in an indirect way through functional ratings obtained through interviews of a parent or caregiver. An example of one such widely used measure of adaptive behavior is the Vineland Adapted Behavior Scale, 2nd edition (VABS-II).

A diagnosis of global developmental delay is chronologically limited typically to children less than 5 years. Hence, for this diagnosis there must be a reliance on accepted widely used standardized measures of developmental performance and attainment in the young child that are psychometrically robust. Measures generally acknowledged to meet this threshold include the Bayley Scales of Infant Development, 2nd edition (Bayley-II) and the Battelle Developmental Inventory (BDI). An indirect evaluation measure that utilizes third-party reports is the Child Development Inventory (CDI).[3]

An important component of accurate diagnosis for both global developmental delay and intellectual disability include the delineation of any autistic features meriting a possible diagnosis of an autistic spectrum disorder.[13] If autistic features are suspected on the initial clinical assessment, a trio of standardized autism diagnostic tools are readily available in practice. These include: the Autism Diagnosis Observation Scale (ADOS), the Autism Diagnosis Inventory (ADI), and the Childhood Autism Rating Scale (CARS). At present, there is no unifying consensus of opinion to suggest that one evaluation measure is consistently superior.

EVALUATION AND TESTING

The diagnostic evaluation of global developmental delay and intellectual disability begins with a detailed history.[7,9,11,12] Particular attention needs to be directed to potential clues for a genetic or acquired etiology. A genetic etiology may be suspected by family history, prior stillbirths or postnatal deaths of prior offspring. Antenatal history may ascertain adverse toxic or infectious exposures or substantive intra-uterine difficulties. Labor, delivery and neonatal historical details are of essence in identifying a potentially causal perinatal event. Furthermore, developmental progression, possible regression and current developmental and functional status must be determined. Possible co-existing medical conditions must be elicited as these will impact on future management as well as the presence or absence of appropriate rehabilitation service provision.

The requisite physical examination of the affected individual begins with informed indirect observation throughout history taking. Ideally this should be in a setting that enables the formulation of an awareness of the individual's developmental, functional, and cognitive skills. Specific aspects of the physical examination of interest include measurements of height, weight and head circumference, and screening for dysmorphology, hepatosplenomegaly, the cutaneous stigmata of a phakomatosis, spinal dysraphism, as well as the integrity of the hearing and vision apparatus. This needs to be accompanied by a thorough neurological examination that may yield clues to localization and as full as possible developmental, functional, and cognitive assessment as permitted by individual co-operation.

Laboratory testing is directed towards the determination of an underlying etiologic cause for an individual's global developmental delay or intellectual disability, with a particular emphasis on possible treatable causes. Recent consensus papers have assisted the clinician by the formulation of an approach based on a systematic review of the relevant literature. The consensus recommendations incorporate both the use of specific identifying clinical features to suggest disease specific testing based on a heightened pre-test probability and screening investigations with an established greater than 1% yield. Indeed, if subsequent to history and physical examination a specific etiology is suspected, then testing is directed and focused on confirming this diagnostic suspicion.

In the absence of a strong specific etiologic suspicion subsequent to history and physical examination, present consensus opinion suggests the following diagnostic approach:[7,9] Chromosomal microarray (i.e., comparative genomic hybridization [CGH], single nucleotide polymorphism [SNP]) directed at the determination of potentially pathogenic genomic structural variations in DNA copy number is the single test with the highest etiologic yield in this particular clinical setting. "Abnormal" results must be compared to known reference databases as well as the determination of parental copy number variation (CNV) status to establish specific pathogenicity. Further genetic testing that can be conducted on a screening basis include determining a possible FMR-1 triplet expansion that underlies fragile X syndrome in all individuals and MECP 2 analysis for Rett syndrome in moderately to severely impaired females. An X-linked inheritance pattern evident in a particular family will direct testing preferentially towards a group of now identified X chromosome-located genes predominantly involved in synaptic function. Additionally, high-resolution magnetic resonance imaging (MRI) provides the detection of cerebral dysgenesis (some of which have a known genetic relationship) and acquired injuries (cortical and sub-cortical; gray and/or white matter involvement). When paired with proton spectroscopy, there exists a potential for detection of central mitochondrial disorders (i.e., lactate peaks) and disorders of creatine deficiency.

Consideration must also be given to diagnostic testing targeting inborn errors of metabolism, especially those that are amenable to causally beneficial therapy.[14] At present, 81 such disorders have been delineated. A recent review highlights that 65% of such disorders can be identified by first-tier screening tests that are both generally available and inexpensive.[15] Such first-tier testing includes serum ammonia, lactate, copper, ceruloplasmin, homocysteine, plasma amino acids, and urine organic acids, purines, pyrimidines, creatine metabolites, oligosaccharides, and glycosaminoglycan. Such first-tier testing must be considered when no diagnosis is evident following history or physical exam and completion of microarray, imaging, FMR1 and MECP2 testing. The remaining treatable inborn errors of metabolism can only be diagnosed by second-tier testing that are characterized by a highly specific orientation (e.g., cerebrospinal fluid [CSF] neurotransmitter analysis) that feature a "single test for a single disease" yield that are directed primarily by phenotypic recognition. Such recognition is typically dependant on the subspecialty input of a medical or biochemical geneticist. To assist in the diagnosis and management of these treatable inborn errors of metabolism, a recent freely available smart phone app (www.treatable-id.org) has been developed for clinical use.

Rapid advances in genetic diagnostic technology will in the very near future enable the clinical introduction on a widespread basis of next-generation sequencing, whole-genome sequencing and whole-exome sequencing.[9] A particular challenge will be the interpretation of results of particular relevance to a single individual and family. Family-based analysis using trios (affected individual and parents) will likely emerge as the methodology of choice to detect *de novo* point mutations that are linked causally to global developmental delay (GDD) and intellectual disability (ID). The application of such technology and a demonstration of its utility and cost–benefit still requires validation in clinical population samples.

MICRODELETION SYNDROMES

Microdeletion syndromes involve chromosomal deletions that include several genes, but are too small to be detected by karyotype. They are usually *de novo*, and tend to recur in the same regions due to homologous recombination of flanking low-copy repeat gene clusters.[16] These low-copy repeats (duplicons) are prone to deletion, duplication, and inversion. The classical microdeletion syndromes include, amongst others, Angelman syndrome (15q11.2-q13), Prader–Willi syndrome (15q11.2-q13), Williams–Beuren syndrome (7q11.23), Smith–Magenis (17p11.2) and velocardiofacial/DiGeorge syndrome (22q11.2). A list of the common syndromic microdeletion syndromes along with their main clinical characteristics is provided in Table 14.1. Each of these syndromes was first described based on common clinical and phenotypic features, and the underlying associated chromosomal defect was later identified. In the past, the confirmation of a clinically suspected microdeletion was done using fluorescence *in situ* hybridization (FISH), since karyotype did not have the necessary resolution. Chromosomal microarray (CMA) allows the rapid screening of the entire genome for CNVs, thus circumventing the requirement of a hypothesis-driven locus-based approach (such as for FISH studies). CMA has enabled the delineation of new microdeletion syndromes. Although some of these are characterized by recognizable clinical features (e.g., chromosome 17q21[17,18] and 15q24[19–21] microdeletion syndromes), the majority of recurrent pathogenic microdeletions are not, and thus can only be diagnosed by screening. A list of selected microdeletion/duplication syndromes is available in Table 14.2. (For a more complete list, please refer to [22,23].) Furthermore, CNVs can be associated with significant clinical heterogeneity, even within the same family. For example, deletions in 15q13.3 were first described in patients who had ID, epilepsy, and facial dysmorphisms,[24] then subsequently associated with a spectrum of phenotypes including idiopathic generalized epilepsy, attention-deficit and hyperactivity disorder, autism spectrum disorder, speech delay, and bipolar disorder.[25,26] Importantly, CNVs can have incomplete penetrance, as well-described pathogenic microdeletions have been inherited from phenotypically normal parents. The reasons for the variability in phenotype are several-fold. The exact breakpoints of the CNV can vary, along with the genes it encompasses. Additional genomic variants can also have an impact on the expressivity of the phenotype. These may include the "unmasking" of recessive single nucleotide variants or insertion/deletions on the hemizygous allele. For example, mutations in the nondeleted *SNAP29* gene in individuals with 22q11 microdeletions are believed to account for atypical clinical features such as polymicrogyria or skin abnormalities.[27]

TABLE 14.1 Selected Classical Syndromes Associated with GDD and ID

Classical syndrome	Locus	Clinical features	Causal/candidate genes
Angelman syndrome	15q11.2-q13	Severe GDD or ID, gait ataxia, tremulousness of limbs, inappropriate happy demeanor, microcephaly, seizures	*UBE3A*
Prader–Willi syndrome (PW)	15q11.2-q13	Neonatal hypotonia and poor suck, GDD or mild ID, eventual hyperphasia and obesity	Unknown Loss of paternally expressed genes in the PW chromosome region – *SUNRF-SNRPN, MKRN3, MAGEL2, NDN*
Smith–Magenis syndrome	17p11.2	FTT, dysmorphisms, sleep abnormalities, GDD and mild–moderate ID, behavioral abnormalities	*RAI1*
Williams–Beuren syndrome (WBS)	17q11.2	GDD and mild ID, cardiovascular disease (SVAS, PPS), distinctive facies (periorbital fullness, full lips), short stature, endocrine abnormalities (hypercalcemia, hypothyroidism), voice, "cocktail personality"	WBS chromosome region – *LIMK1, GTF2I, STX1A, BAZ1B, CLIP2, GTF2IRD1, NCF1* *ELN* for cardiovascular abnormalities
Fragile X syndrome	Xq27.3	Males: moderate ID, characteristic appearance (large head, long face, prominence chin, protruding ears), post-pubertal macro-orchidism, behavioral abnormalities. Females: mild ID, behavioral abnormalities	*FMR1*
Rett syndrome (classical)	Xq28	Female with normal early psychomotor development followed by regression, microcephaly, loss of purposeful hand movements, ataxia, spasticity	*MECP2*
Rubinstein–Taybi syndrome	16p13.3 (*CREBBP*) and 22q13.2 (*EP300*)	Distinctive facies, broad angulated thumbs and great toes, short stature, moderate to severe ID	*CREBBP* and *EP300*
Velocardiofacial/DiGeorge (DG) syndrome	22q11.2	Congenital heart abnormalities (TOF), palatal abnormalities, characteristic facies, learning difficulties, immune deficiencies, hypocalcemia, schizophrenia	DG Chromosome Region

Abbreviations: FTT, failure to thrive; GDD, global developmental delay; ID, intellectual disability; PPS, peripheral pulmonic stenosis; SVAS, supraventricular aortic stenosis; TOF, tetralogy of Fallot.

This mechanism, however, likely explains only a minority of the variability. Another hypothesis involves the "compound inheritance" model, where a second hit such as a second rare CNV or simple nucleotide variant (SNV) in combination with the microdeletion accounts for the variable clinical presentation. In this model, the microdeletion is considered the primary causal insult, and the "second hit," a modifier of the phenotype. This mechanism can be illustrated by the 16p12 microdeletion syndrome. The 16p12 deletion has been found at a higher frequency in patients with GDD/ID than in control individuals. In 95%, these deletions were inherited from a parent that was either normal or had neuropsychiatric symptoms. Importantly, one-quarter of the probands also possessed a second large CNV, which likely explains the more severe phenotype.[28] In addition, another study revealed that, among 2312 children known to carry a rare CNV associated with intellectual disability and congenital abnormalities, those carrying a second rare CNV were at a 8-fold higher risk of having developmental delay.[29] Finally, epigenetic modifications may contribute to the variability of the phenotype associated with CNVs.

MONOGENETIC CAUSES OF ID

Over the past 15 years, many single-gene causes of syndromic and nonsyndromic ID have been identified, in great part due to the tremendous molecular technological advances. Prior to the advent of massive high-throughput parallel technologies, the identification of ID genes was limited by the great genetic heterogeneity of neurodevelopmental disorders, and the inability to study or lump together affected individuals if they did not share specific recognizable

TABLE 14.2 Selected Microdeletion and Microduplication Syndromes Associated with GDD and ID

Chromosome region	Deletion or duplication	Coordinates in Mb (Hg 18)	Size	Main clinical features	Identified in unaffected parent	Candidate genes	References
1q21.1	Deletion	Chr1:145.0-146.35	1.3	Nonsyndromic; mild ID, microcephaly, mild dysmorphic facial features	Yes	*PRKAB2, FMO5, CHD1L, BCL9, ACP, GJA5, GJA8,* and *GPR89B*	65–67
15q13.3	Deletion	Chr15:28.7-30.2	2.0	ID, epilepsy, schizophrenia, autism, cardiac abnormalities	Yes	*CHRNA7*	9,11,68–70
15q24	Deletion	Chr15:72.2-73.3	1.1	Moderate ID, facial dysmorphisms, FTT, growth retardation, congenital malformations	No	*CYP11A1, SEMA7A, CPLX3, SRTA6, ARID3B, SIN3A, CSK*	4,5,71
16p11.2 (distal)	Deletion	Chr16:29.5-30.1	0.2	ID, ASD, obesity, schizophrenia	Yes	*SH2B1*	72–74
16p11.2 (proximal)	Deletion/ duplication	Chr16:28.7-29.0	0.5–0.6	ID, language delay, ASD, obesity	Yes	*MVP, CDIPT1, SEZ6L2, ASPHD1, KCTD13*	75–77
Xq28	Duplication	ChrX:152.9-123.1	0.2	Males: hypotonia, severe GDD and ID, progressive spasticity, seizures, ASD Females: milder phenotype	No	*MECP2*	78–80

Abbreviations: ASD, autism spectrum disorder; FTT, failure to thrive; GDD, global developmental delay; ID, intellectual disability.

clinical characteristics. The use of classic mapping techniques relied on large families with multiple affected individuals, which was rare in recessive ID, and uncommon in dominant ID.

X-LINKED ID

Given the higher male-to-female ratio in ID (30–50% more common in males),[30–33] the availability of large X-linked pedigrees (since carrier females could reproduce) and the easier mapping of deletion breakpoints in hemizygous chromosomal aberrations, many genetic studies focused on X-linked ID (XLID). There are over 100 genes that are associated with nonsyndromic X-linked ID (NS-XLID).[34] Together with CNVs, these explain the genetic etiology in 60% of males with apparent XLID, and 10–12% of ID in males even in the absence of clear X-linked inheritance.[34,35] In 2009, Tarpey et al. performed a systematic sequencing of 718 genes on the X chromosome in 208 families with XLID and identified nine novel XLID genes.[36] The genes on chromosome X account for part, but not all, of the increased prevalence of ID in males.

The most common etiology of XLID, accounting for 0.5–3% of males with ID, is fragile X syndrome, which is caused by a dynamic CGG triplet-repeat expansion mutation in the *FMR1* gene.[37] Boys with a CGG repeat above 200 generally have moderate intellectual impairment. They have typical facial characteristics such as long face, large ears, prominent forehead, prominent jaw, and develop post-pubertal macro-orchidism. Autistic features are present in 25%. Approximately half of carrier females are phenotypically normal and half have a mild fragile X phenotype. Males with a pre-mutation (55–200 repeats) are at risk of developing fragile X-related tremor and ataxia syndrome, a neurodegenerative disorder that usually presents after age 50 years (75% of pre-mutation carrier males develop the syndrome by age 80). Females with a pre-mutation are at risk of premature ovarian failure (30%). In addition, because the triplet repeat expansion is unstable in females but not males, females with pre-mutations have a high risk of having progeny with a full mutation.[34,38]

Rett syndrome is another important cause of X-linked ID. Mutations in *MECP2*, which encodes methyl-CpG-binding protein 2, result in Rett syndrome in girls, and is usually lethal in boys.[39] Classically, affected girls have a normal psychomotor development during the first 6–18 months, and then develop rapid regression of motor and language skills with microcephaly. They lose purposeful hand movements and develop repetitive, stereotyped hand movements. Other characteristic features include gait ataxia, tremor, autistic features, bruxism, breathing irregularities, inconsolable crying, and peripheral vasomotor disturbances. Atypical variants include a more severe phenotype, where affected girls do not have a period of normal development, or a milder phenotype where girls have milder or no regression, and less severe ID. Boys with severe ID have also been described with a severe neonatal encephalopathic form (for review see [40]).

AUTOSOMAL DOMINANT ID

Again, due to high locus heterogeneity, large reproductive disadvantage of the probands, and the fact that many of the mutations are *de novo*, autosomal dominant causes of ID were previously difficult to explore by classical methods such as linkage mapping. Sequencing of trios that include the proband and both parents has proven a successful strategy for identification of ID genes. The Canadian consortium Synapse to Disease Sanger-sequenced 197 synapse genes in 95 sporadic individuals with NSID and identified 10 *de novo* potentially deleterious mutations.[41] This strategy has also been applied using whole exome and whole-genome sequencing (WES). WES in 10 trios with individuals with sporadic NSID identified nine *de novo* variants in six individuals.[42] The *de novo* variants were in known ID genes in two individuals, and in a plausible candidate gene that needed confirmation in four other individuals. O'Roak et al. used the trio approach in 20 children with autism, and found potentially causative *de novo* events in four patients.[43] Rauch et al. used the WES in trios approach and studied 45 probands with NSID and 14 controls.[44] The *de novo* mutation rate in the control and proband groups was comparable, but the proband group had a much higher incidence of loss-of-function variants, suggesting the important contribution of the *de novo* events to the ID phenotype. Sixteen of the 45 patients carried *de novo* variants in known ID genes, with mutations in multiple patients in *STXBP1*, *SYNGAP1* and *SCN2A*, and loss-of-function mutations in an additional six novel candidate ID genes. It is estimated that mutations in *SYNGAP1*, which encodes a component of the *N*-methyl-D-aspartic acid (NMDA)-receptor complex, cause 3–5% of NSID.[41,45,46] Mutations involving *SHANK2 and SHANK3*, which encode a synaptic scaffolding protein, are another recurrent cause of NSID and Autism.[47–49]

AUTOSOMAL RECESSIVE ID

Autosomal recessive ID is extremely heterogeneous, and common nonsyndromic forms do not exist.

DISEASE MECHANISMS

Common Pathways

As the number of single genes associated with ID is exploding, one way of making sense of the complexity is to consider common pathways in which genes are acting. The structure and function of the synapse is central to ID as it modulates plasticity, cellular signaling and transcriptional regulation.

The Synapse and Ionotropic Glutamate Receptors

The majority of excitatory synapses in the CNS are glutamatergic, and these have long been suspected to have an important role in neurodevelopmental and psychiatric diseases.[50,51] The two main ionotropic glutamatergic channels found in the post-synaptic density are α-amino-3-hydroxy-5-methyl-4-isoxazolepropionic acid (AMPA) and NMDA receptors. Mutations in genes encoding components of glutamateric receptors such as *GRIA3*, *GRIK2*, *GRIN2A*, *GRIN2B*, *NR2A* and *NR2B* have been associated with various degrees of neurodevelopmental impairment.[52–61] SYNGAP1 is a component of the NMDA receptor complex and binds to the NR2B subunit.[46] It is a negative regulator of NMDA-mediated extracellular signal-regulated kinases (ERK) activation and causes inhibition of the Ras/ERK pathway.[62,63]

The membrane-associated guanylate kinase (MAGUK) proteins are multidomain scaffolding proteins involved in the clustering, targeting and anchoring of ionotropic glutamate receptors in the post-synaptic density. Mutations in several MAGUK or MAGUK-associated genes have been involved in ID. Mutations in *IL1RAPL1*, a gene with several known mutations in NSID and autism, result in abnormal localization of DLG4, a MAGUK family protein. IL1RAPL1 also inhibits calcium-dependent exocytosis and neurotransmitter release and thus has an important role in vesicle cycling in the presynaptic membrane. *CASK* encodes a calcium/calmodulin-dependent serine protein kinase and is another MAGUK family protein that is mutated in individuals with ID.[36,64]

Other important ID genes that encode synaptic scaffolding proteins include *SHANK2* and *SHANK3*.

Several ID genes are involved in neurotransmitter release at the pre-synaptic membrane. The SNARE complex has a key role in this process as it mediates membrane fusion of the vesicle with the pre-synaptic membrane.[65] Mutations in *STXBP1*, a subunit of the plasma membrane SNARE complex, result in a severe infantile epileptic encephalopathy but have also been identified in patients with isolated ID.[66,67] *SYN*, another ID gene, encodes synaptophysin, a pre-synaptic transport vesicle membrane protein that interacts with synpatobrevin, a component of the vesicle SNARE complex.[36] The scaffolding proteins also are important for vesicle cycling.

Intracellular Signaling Cascades

Activation of glutamate receptors initiates a number of post-synaptic signaling cascades including the RAS-MAPK signaling pathway. Disruption of this pathway at multiple levels gives rise to a group of conditions termed neuro-cardio-facio-cutaneous conditions (NCFCs). These include Noonan, Noonan-like, Costello, neurofibromatosis, LEOPARD, cardio-facio-cutaneous (CFC) and Legius syndromes. The typical features of the NCFCs are congenital heart defects, short stature and distinctive craniofacial features (for review see [68]). Cognitive impairment is frequent and can be of variable severity. Patients also have an increased risk of developing tumors. NCFCs can result from mutations in the RAS GTPases HRAS and KRAS, the downstream effectors of RAS (RAF1, BRAF, MEK1, MEK2 and RSK2), as well as the regulators of the RAS-MAPK pathway such as SHP2, SOS, NF1, SPRED1 and SHOC. *RAS* and *RAF* have long been known as proto-oncogenes. The individual NCFC syndromes can be each caused by mutations in several genes in the pathway, and inversely, mutations in the same gene can cause several individual NCFCs.

The P13K-mTOR pathway is another major signaling cascade that is disrupted in neurodevelopmental disorders. It has a role in regulating protein translation at the dendrites, participating in dendrite and spine morphogenesis and playing an important role in spine plasticity.[69] Disruption of this pathway has been reported in multiple ID and autism spectrum disorder (ASD) syndromes such as tuberous sclerosis, Cowden syndrome, fragile X syndrome, Rett syndrome, and trisomy 21.

Epigenetic Regulation of Transcription

Epigenetics refer to changes in gene expression that occur by mechanisms that do not permanently alter DNA sequence.[70] The main epigenetic mechanisms are DNA methylation and histone post-translational modification (e.g., acetylation, methylation, phosphorylation) that result in modification of the chromatin structure and modulation of the access of the translational machinery, thus affecting gene expression. Epigenetic regulation of transcription has an important role in synaptic plasticity, memory formation, and consolidation.[71]

Examples of ID disorders that implicate epigenetic regulation of transcription as an underlying mechanism include fragile X, Rett syndrome, Rubinstein–Taybi and Coffin Lowry syndrome.

FMR1, the gene responsible for fragile X syndrome, encodes the fragile X mental retardation protein (FMRP), a member of the nuclear ribonucleoprotein family of RNA-binding proteins that regulates the transport to synapses and represses the translation of a subset of neuronal mRNAs. Approximately one-third of all mRNAs encoding post-synaptic and pre-synaptic proteins are targets of FMRP.[72] Methylation of the CGG expansion results in hypermethylation of a CpG island in a promotor region and a subsequent decrease or silencing of *FMR1* transcription and loss of FMRP. This leads to subsequently dysregulated and exaggerated protein synthesis.[73–75]

Rubinstein–Taybi syndrome is a multiple congenital anomaly syndrome characterized by mental retardation, postnatal growth deficiency, microcephaly, broad thumbs and halluces, and dysmorphic facial features. In 50–70% of cases it is associated with mutations in *CREBBP* (cAMP response element-binding protein), and in 3% of cases with mutations in *EP300*, both transcriptional coactivators and potent histone acetyltransferases.[76,77]

MECP2 encodes methyl CpG-binding protein 2, which is believed to act as a transcriptional modulator by repressing or activating genes through long-range chromatin reorganization by binding methylated CpG DNA.[78]

MANAGEMENT

The comprehensive care of individuals with global developmental delay or intellectual disability is based on an inter-disciplinary patient and family centered approach.[3,80] A heterogeneous mix of challenges and care needs arise that are both individual in their occurrence or severity and fluctuating in their impact and burdens. Non-medical health professionals are an essential component in care provision. These include occupational therapists, physiotherapists, speech language pathologists, psychologists, special educators, nurses, and social workers. Each brings their own particular skill set, professionalism, and complementary orientation that is directed at both minimizing disability and maximizing health and wellbeing. Indeed to minimize impairments and maximize activity and participation, particular attention needs to be paid to the importance of personal and environmental contexts. Such an approach is consistent with the International Classification of Functioning, Disability and Health (ICF) model of health and disease promulgated and endorsed by the World Health Organization.

Requisite physician expertise crosses professional specialty boundaries as well. General pediatricians, developmental pediatricians, pediatric neurologists, geneticists, psychiatrists, family practitioners, internists, physiatrists, neurologists, orthopedic surgeons, and palliative care specialists amongst others each have something tangible and important to contribute. Particular medical issues that frequently arise include epilepsy (i.e., convulsive disorders), movement disorders, behavioral challenges, orthopedic deformities, and sleep disturbances.

Seizures may be a prominent part of the underlying disorder responsible for global developmental delay or intellectual disability or they may incidentally arise as a result of increased risk. Seizure type varies and the seizures themselves may be attributed to medical disturbances or associated cerebral dysgenesis. General principles include the individuation of antiepileptic drug selection according to seizure type, the avoidance where possible of polypharmacy, and the minimal use of sedating or cognitively depressing antiepileptic drugs (i.e., barbiturates, benzodiazepine, topiramate). Valproic acid derivatives are avoided in the context of possible mitochondrial cytopathies, while the use of the ketogenic diet is precluded as a means of effecting seizure control for refractory seizures in disorders of carbohydrate or fatty acid metabolism. The ketogenic diet is, however, an option of first choice for GLUT1 deficiency.

Movement disorders include prominent spasticity. Conservative measures of first resort include physiotherapy and stretching exercises. Spasticity that results in pain or functional limitation will lead to a consideration of intramuscular botox injections (in conjunction with intensive physiotherapy) and oral antispasmodic agents such as baclofen, a GABA-B antagonist. Pronounced burdens of care issue may lead to the administration of baclofen intrathecally through a continuous pump infusion. Disorders affecting deep gray matter structures (i.e., basal ganglia) may lead to prominent extrapyramidal symptoms characterized by dyskinesias such as dystonia and choreoathetosis. For the former, dopamine agonists and anticholinergics are therapeutic options, and for the latter dopamine antagonists should be considered.

Behavioral challenges may be particularly disruptive, with a substantial impact on quality of life. These include disorders of attention and impulsivity (i.e., executive function), agitation, opposition, disinhibition, and aggression towards both others and self. Psychological and psychiatric intervention emphasizing behavioral management approaches is the standard first line. If need be, a pharmacologic approach can then be utilized, employing stimulant (i.e., methylphenidate) and nonstimulant (i.e., amotexine) alternatives for prominent inattention with or without associated hyperactivity. Atypical antipsychotics (e.g., respiradone) for agitation and opposition have been found to be effective in this population. Pharmacologic therapy beyond these first-line agents needs to be individualized and directed at major adverse symptoms and managed by an experienced psychiatrist. Often, sleep disorders are overlooked in this clinical setting, though evidence exists that these disorders are increased in frequency and impact in neurodevelopmental disabilities collectively. Proper sleep hygiene is a first step, followed, if need be, by the judicious use of antihistamines (diphenhdramine), hypnotics (chloral hydrate) and melatonin.

A final aspect of medical management is predicated on the accurate diagnosis of a precise underlying etiology if at all possible. Such a diagnosis may reveal a specific potentially beneficial therapeutic adjunct that is directed at the underlying cause. Treatment paradigms in this instance are individualized depending on the specific cause and include approaches such as substrate restriction, cofactor activation, the promotion of alternative or elimination pathways, deficient product replacement, and protein replacement. The promise of gene therapy "fixes" has not yet been realized due to issues of the development of reliable gene delivery systems that reach the target tissue prior to the time by which the deleterious effect of the underlying gene defect has become irreversible. Advances in this area are both presently being intensively investigated and anxiously awaited.

References

1. Shevell MI. Present conceptualization of early childhood neurodevelopmental disabilities. *J Child Neurol*. 2010;25:120–126.
2. Shevell MI, Ashwal S, Donley D, et al. Practice parameter: evaluation of the child with global developmental delay. *Neurology*. 2003;60:367–379.
3. Sherr EH, Shevell MI. Global developmental delay and mental retardation/intellectual disability. In: Swaiman KF, Ashwal S, Ferriero DM, Schor NF, eds. *In: Pediatric Neurology: Principles and Practice*. 5th ed. Philadelphia: Elsevier Saunders; 2012:554–574.
4. Shevell MI. Global developmental delay and mental retardation/intellectual disability: conceptualization, evaluation and etiology. In: Patel D, ed. *Developmental Disabilities*. 2008:1071–1089. Pediatric Clinics of North America; Vol 55.
5. Shevell MI. Etiology and evaluation of neurodevelopmental disabilities. In: Shevell MI, ed. *Neurodevelopmental Disabilities: Clinical & Scientific Foundations*. London UK: Wiley-Blackwell [MacKeith Press]; 2009:106–118.
6. Kaufman L, Ayub M, Vincent JB. The genetic basis of non-syndromic intellectual disability: a review. *J Neurodev Disord*. 2010;2(4):182–209.

7. Michelson DJ, et al. Evidence report: genetic and metabolic testing on children with global developmental delay: report of the Quality Standards Subcommittee of the American Academy of neurology and the Practice Committee of the Child Neurology Society. *Neurology*. 2011;77:1629–1635.
8. Moeschler JB. Genetic evaluation of intellectual disabilities. *Semin Pediatr Neurol*. 2008;15:2–9.
9. Sherr EH, Michelson DJ, Shevell MI, Moeschler J, Gropman AL, Ashwal S. Neurodevelopmental disorders and genetic testing: current approaches and future advances. *Ann Neurol*. 74(2):164–170.
10. Bokhoven H. Genetic and epigenetic networks in intellectual disabilities. *Annu Rev Genet*. 2011;45:81–104.
11. Shevell MI. Office evaluation of the child with developmental delay. In: Dooley JM, ed. *Common Office Problems in Pediatric Neurology;* 2006:256–261. Seminars in Pediatric Neurology; Vol 13.
12. Shevell MI. Diagnostic approach to developmental delay. In: Maria B, ed. *Current Management in Child Neurology*. 4th ed. Shelton Connecticut: BC Decker Inc; 2009:292–299.
13. Yeargin-Allsopp M, et al. Prevalence of autism in a US metropolitan area. *JAMA*. 2003;289:49–55.
14. Van Karnebeek C, Shevell MI, Zschocke J, Moeschler J, Stockler S. The metabolic evaluation of the child with an intellectual development disorder: diagnostic algorithm for identification of treatable causes and new digital resource. *Mol Genet Med*. 2014;111(4):428–438.
15. van Karnebeek CDM, Stockler S. Treatable inborn errors of metabolism causing intellectual disability: a systematic literature review. *Mol Genet Metab*. 2012;105:368–381.
16. Chen KS, Manian P, Koeuth T, et al. Homologous recombination of a flanking repeat gene cluster is a mechanism for a common contiguous gene deletion syndrome. *Nat Genet*. 1997;17:154–163.
17. Sharp AJ, Hansen S, Selzer RR, et al. Discovery of previously unidentified genomic disorders from the duplication architecture of the human genome. *Nat Genet*. 2006;38:1038–1042.
18. Shaw-Smith C, Pittman AM, Willatt L, et al. Microdeletion encompassing MAPT at chromosome 17q21.3 is associated with developmental delay and learning disability. *Nat Genet*. 2006;38:1032–1037.
19. Andrieux J, Dubourg C, Rio M, et al. Genotype-phenotype correlation in four 15q24 deleted patients identified by array-CGH. *Am J Med Genet A*. 2009;149A:2813–2819.
20. Sharp AJ, Selzer RR, Veltman JA, et al. Characterization of a recurrent 15q24 microdeletion syndrome. *Hum Mol Genet*. 2007;16:567–572.
21. El-Hattab AW, Smolarek TA, Walker ME, et al. Redefined genomic architecture in 15q24 directed by patient deletion/duplication breakpoint mapping. *Hum Genet*. 2009;126:589–602.
22. Carvill GL, Mefford HC. Microdeletion syndromes. *Curr Opin Genet Dev*. 2013;23:232–239.
23. Deak KL, Horn SR, Rehder CW. The evolving picture of microdeletion/microduplication syndromes in the age of microarray analysis: variable expressivity and genomic complexity. *Clin Lab Med*. 2011;31:543–564, viii.
24. Sharp AJ, Mefford HC, Li K, et al. A recurrent 15q13.3 microdeletion syndrome associated with mental retardation and seizures. *Nat Genet*. 2008;40:322–328.
25. Dibbens LM, Mullen S, Helbig I, et al. Familial and sporadic 15q13.3 microdeletions in idiopathic generalized epilepsy: precedent for disorders with complex inheritance. *Hum Mol Genet*. 2009;18:3626–3631.
26. Helbig I, Mefford HC, Sharp AJ, et al. 15q13.3 microdeletions increase risk of idiopathic generalized epilepsy. *Nat Genet*. 2009;41:160–162.
27. McDonald-McGinn DM, Fahiminiya S, Revil T, et al. Hemizygous mutations in SNAP29 unmask autosomal recessive conditions and contribute to atypical findings in patients with 22q11.2DS. *J Med Genet*. 2013;50:80–90.
28. Girirajan S, Rosenfeld JA, Cooper GM, et al. A recurrent 16p12.1 microdeletion supports a two-hit model for severe developmental delay. *Nat Genet*. 2010;42:203–209.
29. Girirajan S, Rosenfeld JA, Coe BP, et al. Phenotypic heterogeneity of genomic disorders and rare copy-number variants. *N Engl J Med*. 2012;367:1321–1331.
30. Leonard H, Wen X. The epidemiology of mental retardation: challenges and opportunities in the new millennium. *Ment Retard Dev Disabil Res Rev*. 2002;8:117–134.
31. David M, Dieterich K, Billette de Villemeur A, et al. Prevalence and characteristics of children with mild intellectual disability in a French county. *J Intellect Disabil Res*. 2013.
32. Westerinen H, Kaski M, Virta LJ, Almqvist F, Iivanainen M. Age-specific prevalence of intellectual disability in Finland at the beginning of new millennium – multiple register method. *J Intellect Disabil Res*. 2013; 58(3):285–295.
33. Gillberg C, Cederlund M, Lamberg K, Zeijlon L. Brief report: "the autism epidemic". The registered prevalence of autism in a Swedish urban area. *J Autism Dev Disord*. 2006;36:429–435.
34. Lubs HA, Stevenson RE, Schwartz CE. Fragile X and X-linked intellectual disability: four decades of discovery. *Am J Hum Genet*. 2012;90:579–590.
35. Ropers HH, Hamel BC. X-linked mental retardation. *Nat Rev Genet*. 2005;6:46–57.
36. Kleefstra T, Hamel BC. X-linked mental retardation: further lumping, splitting and emerging phenotypes. *Clin Genet*. 2005;67:451–467.
37. Tarpey PS, Smith R, Pleasance E, et al. A systematic, large-scale resequencing screen of X-chromosome coding exons in mental retardation. *Nat Genet*. 2009;41:535–543.
38. Crawford DC, Acuna JM, Sherman SL. FMR1 and the fragile X syndrome: human genome epidemiology review. *Genet Med*. 2001;3:359–371.
39. Wijetunge LS, Chattarji S, Wyllie DJ, Kind PC. Fragile X syndrome: from targets to treatments. *Neuropharmacology*. 2013;68:83–96.
40. Amir RE, Van den Veyver IB, Wan M, Tran CQ, Francke U, Zoghbi HY. Rett syndrome is caused by mutations in X-linked MECP2, encoding methyl-CpG-binding protein 2. *Nat Genet*. 1999;23:185–188.
41. Smeets EE, Pelc K, Dan B. Rett syndrome. *Mol Syndromol*. 2012;2:113–127.
42. Hamdan FF, Gauthier J, Araki Y, et al. Excess of *de novo* deleterious mutations in genes associated with glutamatergic systems in nonsyndromic intellectual disability. *Am J Hum Genet*. 2011;88:306–316.
43. Vissers LE, de Ligt J, Gilissen C, et al. A *de novo* paradigm for mental retardation. *Nat Genet*. 2010;42:1109–1112.

44. O'Roak BJ, Deriziotis P, Lee C, et al. Exome sequencing in sporadic autism spectrum disorders identifies severe *de novo* mutations. *Nat Genet*. 2011;43:585–589.
45. Rauch A, Wieczorek D, Graf E, et al. Range of genetic mutations associated with severe non-syndromic sporadic intellectual disability: an exome sequencing study. *Lancet*. 2012;380:1674–1682.
46. Hamdan FF, Gauthier J, Spiegelman D, et al. Mutations in SYNGAP1 in autosomal nonsyndromic mental retardation. *N Engl J Med*. 2009;360:599–605.
47. Berryer MH, Hamdan FF, Klitten LL, et al. Mutations in SYNGAP1 cause intellectual disability, autism, and a specific form of epilepsy by inducing haploinsufficiency. *Hum Mutat*. 2013;34:385–394.
48. Berkel S, Marshall CR, Weiss B, et al. Mutations in the SHANK2 synaptic scaffolding gene in autism spectrum disorder and mental retardation. *Nat Genet*. 2010;42:489–491.
49. Betancur C, Buxbaum JD. SHANK3 haploinsufficiency: a "common" but underdiagnosed highly penetrant monogenic cause of autism spectrum disorders. *Mol Autism*. 2013;4:17.
50. Herbert MR. SHANK3, the synapse, and autism. *N Engl J Med*. 2011;365:173–175.
51. Kaufman L, Ayub M, Vincent JB. The genetic basis of non-syndromic intellectual disability: a review. *J Neurodev Disord*. 2010;2:182–209.
52. Pocklington AJ, Cumiskey M, Armstrong JD, Grant SG. The proteomes of neurotransmitter receptor complexes form modular networks with distributed functionality underlying plasticity and behaviour. *Mol Syst Biol*. 2006;2, 2006.0023.
53. Chiyonobu T, Hayashi S, Kobayashi K, et al. Partial tandem duplication of GRIA3 in a male with mental retardation. *Am J Med Genet A*. 2007;143A:1448–1455.
54. Magri C, Gardella R, Valsecchi P, et al. Study on GRIA2, GRIA3 and GRIA4 genes highlights a positive association between schizophrenia and GRIA3 in female patients. *Am J Med Genet B Neuropsychiatr Genet*. 2008;147B:745–753.
55. Bonnet C, Leheup B, Beri M, Philippe C, Gregoire MJ, Jonveaux P. Aberrant GRIA3 transcripts with multi-exon duplications in a family with X-linked mental retardation. *Am J Med Genet A*. 2009;149A:1280–1289.
56. Casey JP, Magalhaes T, Conroy JM, et al. A novel approach of homozygous haplotype sharing identifies candidate genes in autism spectrum disorder. *Hum Genet*. 2012;131:565–579.
57. Sampaio AS, Fagerness J, Crane J, et al. Association between polymorphisms in GRIK2 gene and obsessive-compulsive disorder: a family-based study. *CNS Neurosci Ther*. 2011;17:141–147.
58. Lesca G, Rudolf G, Bruneau N, et al. GRIN2A mutations in acquired epileptic aphasia and related childhood focal epilepsies and encephalopathies with speech and language dysfunction. *Nat Genet*. 2013; 45(9):1061–1066.
59. Lemke JR, Lal D, Reinthaler EM, et al. Mutations in GRIN2A cause idiopathic focal epilepsy with rolandic spikes. *Nat Genet*. 2013; 45(9):1067–1072.
60. Kim KT, Kim J, Han YJ, Kim JH, Lee JS, Chung JH. Assessment of NMDA receptor genes (GRIN2A, GRIN2B and GRIN2C) as candidate genes in the development of degenerative lumbar scoliosis. *Exp Ther Med*. 2013;5:977–981.
61. Endele S, Rosenberger G, Geider K, et al. Mutations in GRIN2A and GRIN2B encoding regulatory subunits of NMDA receptors cause variable neurodevelopmental phenotypes. *Nat Genet*. 2010;42:1021–1026.
62. Allen AS, Berkovic SF, Cossette P, et al. *De novo* mutations in epileptic encephalopathies. *Nature*. 2013; 501(7466):217–221.
63. Komiyama NH, Watabe AM, Carlisle HJ, et al. SynGAP regulates ERK/MAPK signaling, synaptic plasticity, and learning in the complex with postsynaptic density 95 and NMDA receptor. *J Neurosci*. 2002;22:9721–9732.
64. Tomoda T, Kim JH, Zhan C, Hatten ME. Role of Unc51.1 and its binding partners in CNS axon outgrowth. *Genes Dev*. 2004;18:541–558.
65. Hackett A, Tarpey PS, Licata A, et al. CASK mutations are frequent in males and cause X-linked nystagmus and variable XLMR phenotypes. *Eur J Hum Genet*. 2010;18:544–552.
66. Weber HS, Cyran SE. Transvenous "snare-assisted" coil occlusion of patent ductus arteriosus. *Am J Cardiol*. 1998;82:248–251.
67. Hamdan FF, Piton A, Gauthier J, et al. *De novo* STXBP1 mutations in mental retardation and nonsyndromic epilepsy. *Ann Neurol*. 2009;65:748–753.
68. Hamdan FF, Gauthier J, Dobrzeniecka S, et al. Intellectual disability without epilepsy associated with STXBP1 disruption. *Eur J Hum Genet*. 2011;19:607–609.
69. Zenker M. Clinical manifestations of mutations in RAS and related intracellular signal transduction factors. *Curr Opin Pediatr*. 2011;23:443–451.
70. Troca-Marin JA, Alves-Sampaio A, Montesinos ML. Deregulated mTOR-mediated translation in intellectual disability. *Prog Neurobiol*. 2012;96:268–282.
71. Urdinguio RG, Sanchez-Mut JV, Esteller M. Epigenetic mechanisms in neurological diseases: genes, syndromes, and therapies. *Lancet Neurol*. 2009;8:1056–1072.
72. Jiang Y, Langley B, Lubin FD, et al. Epigenetics in the nervous system. *J Neurosci*. 2008;28:11753–11759.
73. Darnell JC, Van Driesche SJ, Zhang C, et al. FMRP stalls ribosomal translocation on mRNAs linked to synaptic function and autism. *Cell*. 2011;146:247–261.
74. Bear MF, Huber KM, Warren ST. The mGluR theory of fragile X mental retardation. *Trends Neurosci*. 2004;27:370–377.
75. Dolen G, Osterweil E, Rao BS, et al. Correction of fragile X syndrome in mice. *Neuron*. 2007;56:955–962.
76. Osterweil EK, Krueger DD, Reinhold K, Bear MF. Hypersensitivity to mGluR5 and ERK1/2 leads to excessive protein synthesis in the hippocampus of a mouse model of fragile X syndrome. *J Neurosci*. 2010;30:15616–15627.
77. Petrij F, Giles RH, Dauwerse HG, et al. Rubinstein-Taybi syndrome caused by mutations in the transcriptional co-activator CBP. *Nature*. 1995;376:348–351.
78. Roelfsema JH, White SJ, Ariyurek Y, et al. Genetic heterogeneity in Rubinstein–Taybi syndrome: mutations in both the CBP and EP300 genes cause disease. *Am J Hum Genet*. 2005;76:572–580.
79. Della Sala G, Pizzorusso T. Synaptic plasticity and signaling in Rett syndrome. *Dev Neurobiol*. 2014;74:178–196.
80. Shevell MI. Chromosomal disorders, inborn errors of metabolism and pediatric neurodegenerative diseases. In: Noseworthy J, ed. *Neurological Therapeutics: Principles & Practice*. 2nd ed. Abingdon, UK: Informa Health Care; 2006:1856–1866.

CHAPTER

15

Down Syndrome

Allison Caban-Holt, Elizabeth Head, and Frederick Schmitt
University of Kentucky, Lexington, KY, USA

INTRODUCTION

Down syndrome (DS) or trisomy 21 is one of the most common causes of intellectual disability (ID) and recent prevalence estimates suggest that there are 11.8[1] to 14.47[2] per 10,000 live births in the United States with DS. It is estimated that 250,000 people in the United States have DS. In turn, World Health Organization estimates for DS range between 10 to 11 in 10,000 live births worldwide.[3] This chapter summarizes current knowledge on DS risk and prevalence, behavioral, psychiatric, physical and neurological manifestations of the disease, and the effects of aging on cognitive functioning of individuals with DS.

HALLMARKS OF DOWN SYNDROME

There are an array of features that may be present in DS. Ten hallmark symptoms seen in newborns have been identified to assist in postnatal confirmation of DS,[4–6] and include: flat facial profile, slanted palpebral fissures, anomalous ears, hypotonia, poor Moro reflex, dysplasia of midphalanx of the fifth finger, transverse palmar crease, excessive skin at the nape of the neck, hyperflexibility of joints, and dysplasia of pelvis. Other common features that may also present are: small head, short neck, protruding tongue, unusually shaped eyes, Brushfield spots in the eye, relatively short fingers, space between first and second toes (sandal gap), narrow palate, slow growth, shorter stature, late developmental milestones, and ID (typically mild to moderate, with IQ in the 50–70 or 35–50 range, respectively). Symptoms associated with DS may include: heart defect (50% of children), digestive anomalies, hypothyroidism, celiac disease, gastroesophageal reflux, childhood leukemia, and behavioral issues such as attention problems, obsessive/compulsive behavior, stubbornness, tantrums, autism, and dementia (after age 50).

INHERITANCE

Maternal age is a primary risk factor for DS. The chances of a woman having a child with DS increase with age because older eggs have a higher risk of improper chromosome division. By age 35, a woman's risk of conceiving a child with DS is 1 in 400; by age 45 the risk is 1 in 35. However, most children with DS are born to women under the age of 35, as they are responsible for the majority of all births. Having had one child with DS leads to a marginal increase of 1% risk of having another child with DS.[7]

Health of the mother prior to and during pregnancy has also been linked to DS risk. Mothers of DS children tended to have more significant illnesses before conception and more medication ingestion the year before conception.[8] Women who suffered from gestational diabetes were twice as likely to have offspring with chromosomal abnormalities, including DS, than women who did not.[7]

Timing of pregnancy is thought to have some bearing on DS. Short intervals between pregnancies has been linked to increased DS occurrence.[9] Jongbloet proposed that short periods of anovulatory activity followed by conception

Rosenberg's Molecular and Genetic Basis of Neurological and Psychiatric Disease
http://dx.doi.org/10.1016/B978-0-12-410529-4.00015-2

correlate with greater occurrence of DS.[10,11] It is suggested that conceptions during the transition between anovulation and the re-establishment of regular ovulation may be a time vulnerable to maternal meiotic nondisjunction.

Last, some genetic issues may increase risk of DS. One such example is that a parent being a carrier of the genetic translocation for DS increases risk of DS. It has also been shown that families with histories of Alzheimer disease (AD) are more likely to have offspring with DS.[12]

DIAGNOSIS AND TESTING

First-trimester screening involves offering prenatal testing if the maternal age is 35 years or older or if maternal serum markers in combination with ultrasound measures of fetal nuchal translucency are suggestive.[13] Subsequently, the primary method for identifying DS genetically is through a karyotype, which requires 10–14 days for test results as cell cultures are required. Additionally, using FISH (fluorescence *in situ* hybridization using a probe for chromosome 21) or polymerase chain reaction (PCR) with amniotic fluid samples, lymphocytes from blood or buccal mucosa cellular preparations is a more rapid approach and appropriate for establishing mosaicism. Karyotyping and FISH/PCR each have unique advantages and disadvantages, as discussed by Gekas and colleagues,[14] but a consensus is growing that using both FISH and karyotyping provides the highest accuracy.[15]

Gene Identification

DS was initially described by J. Langdon Down in 1866[16] and identified as a chromosome 21 trisomy by Lejeune in 1959.[17] Called nondisjunction trisomy 21, this type of DS is caused by abnormal cell division and results in three copies of chromosome 21. Nondisjunction accounts for 95% of all DS cases. Two additional types of DS have also been identified. Partial trisomy 21 is a rare form of DS (estimates range from 1–5% from a few case reports)[18–20] and involves the triplication of a piece of chromosome 21 when part of chromosome 21 breaks off during cell division and becomes attached (translocated) to another chromosome (usually chromosome 14). The total number of chromosomes in the cells is still 46, however, the additional part of chromosome 21 causes the characteristics of DS. The second additional cause of DS is mosaicism (which affects approximately between 1–4% of individuals), where an extra chromosome 21 is present in some but not all cells in the body and typically the phenotype is milder than full trisomy 21.[21–23] Interestingly, the severity of the DS phenotype is linked to the percentage of mosaicism.[23] Heredity is not a factor in trisomy 21 (nondisjunction) or mosaicism. Yet, in one-third of cases of DS resulting in a translocation, there is a hereditary component.

Disease Identification

Chapman and Hesketh[24] have written a comprehensive review of the behavioral phenotype in DS for children and adolescents that is characterized by ID, specific deficits in expressive language development, impaired speech intelligibility, and impaired verbal short-term memory (STM). Adaptive behavior is consistent with general intelligence. Levels of maladaptive behavior are lower than for comparison groups with ID and do not change significantly with age. In DS adults, the conditions that most affect behavior are depression, hypothyroidism, and dementia.[24]

EARLY INTERVENTION/TREATMENT

At the current time, children with DS are almost always referred for early intervention (EI) programs after birth.[25] The most common EI services for babies with DS are physical therapy (PT) and speech therapy (ST). PT for children with DS focuses on motor development and improving muscle tone. ST is important for children with DS because they frequently have small mouths with enlarged tongues, and, therefore, have trouble speaking clearly. Further, hypotonia also inhibits the movement of the facial muscles required for speech, and hearing problems inhibit the ability to hear and mimic speech. ST focuses on learning to communicate clearly through talking and frequently through sign language.

The effectiveness of EI has been investigated. Connolly and colleagues[26] compared the intellectual and adaptive functioning of children with DS who participated in an EI program to those with DS who did not participate in an EI program. The mean IQ for the EI group was found to be 12 points higher than for the comparison group ($p<.005$, a significant difference). The social quotient (SQ), measured by the Vineland Social Maturity Scale, for the EI group was also found to be significantly higher by 9 points ($p<.05$) than the comparison group. Children in the EI group met developmental milestones earlier than the comparison group, though not as early as children without ID. The authors suggest that EI during early infancy bolsters the development of intellectual and adaptive skills during early childhood. Other research shows that the effects of EI have lasting benefit years after the end of treatment.[27]

EI through medical treatment has likely been a major factor that has helped improve health and quality of life of children with DS. Corrective surgery for heart defects, gastrointestinal irregularities, screening for visual impairment, ear infections, hearing loss, hypothyroidism, and obesity are amongst the early medical treatments individuals with DS can benefit from given the comorbidities associated with their condition.[28]

PREVALENCE

The prevalence of DS births has been increasing. A cross-sectional analysis of live-born infants with DS during 1979–2003 from 10 regions of the United States found the prevalence of DS at birth increased by 31%, from 9.0 to 11.8 per 10,000 live births. DS births over time were found to be increasing in proportion to the increasing rate of mothers aged 35 or older giving birth in the studied regions. These findings are in line with findings from prior research.[1] In the United States, rates of DS vary by race/ethnicity. The prevalence of DS at birth is significantly lower among non-Hispanic blacks and was higher among Hispanics compared to the prevalence rates among non-Hispanic whites. Hispanic infants were found to have higher rates of DS than other infants, even when differences in maternal age were accounted for.[29] Rates for Hispanic, white and African American infants were, 11.8, 9.2 and 7.3, respectively, per 10,000 live births.[29] Differences seen may be due to contrasting use of prenatal screening. Use of prenatal diagnosis and termination of pregnancy significantly reduced the prevalence of DS births among white women, but not among women of other races.[30] However, this finding was not supported in another study.[31]

DISEASE EVOLUTION

Along with longer life expectancy comes a larger population of adults with DS who display premature age-related changes in their health.[32] A key challenge for adults with DS as they age is the increasing risk for developing clinical symptoms of Alzheimer disease (AD). (For a review of risk factors for dementia in DS see [33,34].)

Virtually all adults with DS (full trisomy 21) over the age of 40 years have sufficient amyloid-β plaques and neurofibrillary tangles for a neuropathological diagnosis of AD.[35–37] This is due to the overexpression of the amyloid precursor protein gene on chromosome 21, which is cleaved to produce amyloid-β. The early signs of dementia in people with DS might be a result of dysfunction of the frontal lobe and hippocampus—areas in which amyloid-β first accumulates during the early stages of dementia.[38] However, despite the presence of AD neuropathology, not all adults with DS show clinical signs of dementia. The prevalence of clinical dementia increases with age over 50 years, reaching over 75% in people with DS aged over 65 years.[33,38]

Noetzel[39] suggests that the phases of clinical deterioration in people with DS and dementia can involve memory impairment in the initial stages with reduced verbal output in those with mild–moderate ID. In more severe ID, the first symptoms of dementia may be more behavioral in nature, including apathy, inattention, decreased social interaction, and spatial disorientation. In the middle-severity phase, slowed gait, shuffling, loss of self-help skills, and increased seizure activity can occur.[40] In the final phase of dementia, individuals become nonambulatory and bedridden, with onset of spasticity and pathological release reflexes.[40–44]

Diagnosis of Dementia in DS

Deb and Braganza[45] compared diagnosing dementia in individuals with DS using the International Classification of Diseases, 10th edition (ICD-10) clinician rating, Dementia Scale for Down Syndrome (DSDS), Dementia Questionnaire for Persons with Mental Retardation (DMR), and Mini-Mental State Exam (MMSE). It was found that clinician rating, DSDS and DMR had high levels of concordance in terms of diagnosing dementia. Advanced AD was readily diagnosed in the study. Difficulties with diagnosis arose when individuals with DS were suspected to be in early stages of dementia. In such cases, rating scales and clinician report were discrepant. MMSE did not effectively diagnose dementia in DS in the majority of cases. Strydom et al.[46] found that dementia was common in older individuals with ID, but prevalence differed according to the diagnostic criteria being used. These authors found that the Diagnostic and Statistical Manual of Mental Disorders, 4th edition, (DSM-IV) criteria were more inclusive, while ICD-10 excluded individuals with even moderate dementia.

Current evidence suggests that behavioral and personality changes may be the first and most reliable indicators of cognitive decline in DS. Research by Holland and colleagues[47] demonstrated that when informants first reported symptoms of change in individuals with DS they were predominantly behavioral or psychiatric, consistent with frontal lobe dysfunction. This suggests that functions served by the frontal lobes may be the first compromised with the development of AD-like neuropathology in people with DS.[48] In a 14-year follow-up study of 77 women with DS over the age of 35 years, it was found that amongst the measures used to examine cognitive functioning, the informant-based Dementia Questionnaire for Mentally Retarded Persons (DMR) was found to show decline approximately 5 years prior to diagnosis of dementia.[49] Therefore, honing in on the behavioral observations of caregivers of older individuals with DS to detect change may prove to be important in early identification of cognitive decline. Standardized evaluation of behavioral/psychiatric issues could contribute to the ability to diagnose dementia early in adults with DS. Better understanding of the pattern of progression of dementia in individuals with DS may help to improve diagnostic accuracy and allow for earlier treatment for those affected.[50]

As psychiatric symptoms are prevalent features of dementia in the population with DS and may appear prior to substantial changes in daily functioning, the specific types of psychiatric symptoms common in AD in those with and without DS have been evaluated to examine potential differences in presentation of symptoms in these populations.[50]

The most common types of neuropsychiatric symptoms in AD in the general population are a) delusions, b) hallucinations, c) agitation, d) aggression, e) depression, f) anxiety, and g) apathy. Study of psychiatric and behavioral symptoms in adults with DS has been conducted[51,52,53] and found that generally, adults with DS without dementia had a low prevalence of delusions, hallucinations, behavioral problems, or depression coupled with high levels of apathy. Individuals with DS in the questionable cognitive status group were much more likely to have delusions and exhibit verbal and physical violence than those with no dementia. The authors suggest these aforementioned behaviors may be early indicators of dementia onset.[50] For individuals with DS diagnosed with dementia, delusions, visual hallucinations, misidentifications, and depression were prevalent, persistent and less amenable to correction.

Recently, Lott and colleagues[40] investigated the occurrence of seizures as another early indicator of dementia onset in individuals with DS. Prior research indicates that individuals with DS over the age of 45 years with new-onset seizures are more likely to develop AD and seizures may demarcate the beginning of dementia.[54,55] In a sample of 53 adults with DS and AD, it was found that those with seizures were more likely to become untestable on cognitive measures (used as a surrogate for severe impairment) than those without seizures, showing a more marked cognitive decline. The authors suggest that high brain levels of amyloid-β peptides in individuals with DS and AD are likely interfering with neuronal and synaptic activity, thereby lowering the seizure threshold. Taken together, the presence of seizures in individuals with DS may be an early sign that cognitive decline is impending, and may be useful in diagnosis of dementia in DS.

Nieuwenhuis-Mark[56] reviewed the issues involved with diagnosing AD in DS and provided recommendations which include: annual screening for dementia for all people with DS age 35 and over; modified classification systems to capture early and atypical symptoms of dementia; and developing repeatable standardized assessment batteries to be utilized for diagnosing dementia in DS. Further, assessments of individuals with DS should also include a clinical history and physical examination[57] and be sensitive to psychiatric, behavioral, and mood changes, while taking into account the sensory and physical deficits that are more likely to occur at older ages.[58]

In suggesting a test battery for diagnosing dementia in individuals with ID, Burt and Aylward[59] posit that dementia should be diagnosed only when longitudinal data demonstrate clinically significant declines in functioning. To meet diagnostic criteria for dementia, documented declines on at least one memory test and one other test of cognitive ability are needed. Changes on cognitive tests should also coincide with changes in everyday functioning.

Although a specific benchmark test or test battery that can reliably diagnose dementia in DS across ID levels and situations has yet to be endorsed as the standard, Edgin and colleagues[58] have developed the Arizona Cognitive Test Battery (ACTB) to assess the cognitive phenotype in DS. ACTB utilizes tasks from the Cambridge Neuropsychological Testing Automated Battery (CANTAB)[60] to test general cognitive abilities, prefrontal, hippocampal and cerebellar functions. The developers of the battery have found that the ACTB provided consistent results across contexts, socioeconomic backgrounds and ethnicities, and was positively correlated with informant reports. With further investigation, the ACTB may become a beneficial tool for following changes in individuals with DS by helping to identify the specific brain regions affected as cognition changes over time.

End of Life: Mechanisms and Comorbidities

Common causes of death in the DS population include leukemia (in childhood), respiratory illness, congenital circulatory defects, diseases of the digestive system, dementia and AD.[32] The likelihood of death from a particular comorbidity varies by age. Looking at comorbidity and mortality in DS across the lifespan it has been determined that across all age categories pneumonia and other respiratory infections were the most common cause of death ranging from 23% in adulthood to 40% in senescence.[61] With attention to specific stages:

- Childhood to early adulthood (ages 0–18 years): coronary heart disease (CHD) = 13% of deaths
- Adulthood (ages 19–40 years): CHD = 23% of deaths
- Senescence (ages 40+): coronary artery disease = 10% of deaths; cardiac, renal, and respiratory failure = 9% of deaths

Race has been shown to be a factor in mortality in individuals with DS in the United States.[62,63] Review of death certificates found that the median age at death increased from 25 years in 1983 to 49 years in 1997 for white individuals with DS. However, there were apparent racial disparities in median age at death. For blacks the median age at death was 25 years, and for people of other races only 11 years. This may be due to differences in the manifestations of symptoms in DS in individuals of other races, as it was found that CHD and pulmonary circulation were reported more frequently on the death certificates of people of other races with DS than on those of whites with DS.

With reference to DS adults, age, dementia status, and mobility restrictions were found to be the most important predictors of mortality in a study of 500 people with DS age 45 and older.[64] Studies investigating whether prior levels of functional abilities or declines in functional abilities predict mortality in adults with DS have shown mixed results. Some researchers have found that declines in functional abilities and cognition were related to increased mortality,[64,65] while findings of other studies have not found this relationship.[66–68] Table 15.1 provides a synopsis of the current evidence of the dementia health risks in DS dementia.

TABLE 15.1 Dementia Health Risks and Down Syndrome Dementia

Dementia health risk factor	Risk and putative mechanism	Down syndrome
Hypertension	Stroke; cerebrovascular disease; protein extravasation	Individuals with DS have lower resting heart rates and lower blood pressure than general population
Obesity	High BMI is associated with a 59% increased risk for AD; also a risk for sleep apnea syndrome (see below)	45–79% of males; 56–96% of females are reported to be overweight
Diabetes	May alter Aβ clearance in brain; may promote inflammation	Age of onset ~ 22 years for type 1 diabetes is comparable to general population; preliminary data on type 2 diabetes suggests a lower rate, however
Cardiovascular disease	Promotion of cerebrovascular disease; associated dyslipidemia increases risk of brain plaque pathology	Rate of mitral valve prolapse is high However, lower risk for cardiovascular disease in adults with DS compared to general population; this includes lower rates of hypercholesterolemia and heart disease compared to adults with other intellectual disabilities
Cerebrovascular disease	Direct injury to brain regions involved in cognition; inflammation; hypoperfusion; increased Aβ production	Lower risk for cerebrovascular disease observed in adults with DS compared to general population
Head injury	Aβ and tau pathologies are increased in brain; increased APP production	No available epidemiological reports
Sleep apnea	Lowered oxygen during sleep impacting brain	An estimated 94% of persons with DS, ages 17–56 have obstructive sleep apnea of varying severity
Thyroid dysfunction	May reflect thyroid stimulating hormone on Aβ processing; a cofactor for vascular dementia risk	Seen in 35–40% of adults with incidence increasing with advancing age; Hashimoto thyroiditis may be mistaken for dementia
Seizures	Seen with neurodegeneration, early onset, and ApoEε4 in ~ 2–8% of persons with AD	Possible link to myoclonus epilepsy gene on chromosome 21; rate increases with age in DS from 7–46% (over age 50) up to 84% of persons with DS and dementia

Abbreviations: Aβ, amyloid-β peptide; APP, amyloid precursor protein; BMI, body-mass index.
Reproduced with permission from[68a].

PATHOPHYSIOLOGY

Postmortem observations and volumetric magnetic resonance imaging (MRI) studies show that people with DS have reduced brain volumes and brachycephaly with disproportionately smaller volumes in frontal and temporal areas and the cerebellum compared with healthy individuals. While in contrast, subcortical areas have relatively normal brain volumes. The parahippocampal gyrus appears larger on MRI in people with DS than in healthy individuals, and neuropsychological tests assessing function in parahippocampal and perirhinal regions suggest that these areas are functioning normally.[38] In relation to aging and brain structure, a recent investigation utilized fractional anisotropy (FA) from diffusion tensor imaging (DTI) to detect changes in white matter (WM) integrity between adults with DS without dementia, from adults with DS with dementia and from age-matched non-DS adults.[69] FA in conjunction with scores on cognitive testing (Brief Praxis Test and the Severe Impairment Battery) demonstrate significantly lower WM integrity in DS as compared to non-DS, particularly in the frontal lobes, corpus callosum, and association tracts predominantly within the frontoparietal regions. Comparing DS with and without dementia, DTI shows significant clusters where DS with dementia have lower FA than DS without dementia. The authors suggest that, taken together, developmental reductions in frontal cortex FA in DS that are present prior to dementia leave this area more vulnerable to the effects of brain aging and the progression to dementia.

Morphometric studies of the cortex in people with DS show fewer neurons, decreased neuronal densities, and abnormal neuronal distribution. The synaptic density, synaptic length, and synaptic contact zones are abnormal. These neuroanatomical abnormalities might be associated with the learning and memory deficits.[38]

The cognitive profile in DS shows deficits primarily in morphosyntax, verbal short-term memory (STM), and explicit long-term memory (LTM), while visuospatial STM, associative learning, and implicit LTM are typically preserved.[38]

Other research has shown a relative sparing of basic STM skills in children with DS, a strength that decays when greater demands such as increased memory load or executive requirements are placed onto the children.[70] This investigation also indicated that children with DS were extremely impaired on measures sensitive to hippocampal dysfunction,[71] a structure critically impaired in the DS brain phenotype.[72] Taken together, linking behavioral results to the neurobiological underpinnings of the visual–spatial memory in children with DS, the relative sparing of spatial STM performance could be sustained by the documented preservation of parietal lobe gray matter.[73] The (pre)frontal abnormalities frequently documented in the study of the DS brain phenotype[73] could underpin poor performance in the strategic self-ordered task. Tasks that are sensitive to medial temporal lobe, and in particular to hippocampal functioning (pattern and spatial recognition), show poor performance as a consequence of the disproportionate impairment of this structure in the DS brain.[70,72,73]

In contrast, Pennington and colleagues[72] tested prefrontal and hippocampal functions in a sample of school-aged individuals with DS compared with a sample of typically developing children individually matched on mental age (MA). The group with DS performed worse than MA controls on each hippocampal measure but not on any of the prefrontal measures. These authors suggest the pattern of results provides evidence for dissociation between two neuropsychological functions: hippocampal-mediated LTM and prefrontally mediated working memory. Thus, there is some work that remains to be done with regard to how cognition in DS is measured so that the field can make sense of the discrepant research findings. (For detailed review of neural and cognitive features of DS see [74]).

CONCLUSIONS

DS is one of the most common genetic diseases and its prevalence is increasing with the growing rates of women over the age of 35 years giving birth. DS brings with it a host of implications for the health of affected individuals at all stages of the lifespan. EI at younger ages has shown to benefit DS children by early identification and amelioration of congenital defects, improved sensory capabilities, and development of better communication skills. Improved health in early life results in increased longevity. Thus, there are more DS individuals living into older age, and experiencing age-related problems such as AD. Better understanding of the cognitive, psychiatric, and neuropathological brain changes that occur in older adults with DS will help increase early detection of cognitive decline and ultimately may lead to improved quality of life at the end of the lifespan.

ACKNOWLEDGEMENTS

Funding provided by NIH/DHHS Eunice Kennedy Shriver National Institute of Child Health & Human Development R01 HD064993.

References

1. Shin M, Besser LM, Kicik JE, et al. Congenital Anomaly Multistate Prevalence and Survival (CAMPS) Collaborative. Prevalence of Down syndrome among children and adolescents in 10 regions of the United States. *Pediatrics*. 2009;124:1565–1571.
2. Parker SE, Mai CT, Canfield MA, et al. Updated national birth prevalence estimates for selected birth defects in the United States, 2004–2006. *Birth Defects Res A Clin Mol Teratol*. 2010;88:1008–1016.
3. World Health Organization. www.who.int/genomics/public/geneticdiseases/en/index1.html.
4. Jones KL. Down syndrome. In: *Smith's Recognizable Patterns of Human Malformation*. 6th ed. Philadelphia: Elsevier Saunders; 2006:7–12.
5. Hall B. Mongolism in newborn inants. An examination of the criteria for recognition and some speculations on the pathogenic activity of the chromosomal abnormality. *Clin Pediatr*. 1966;5:4.
6. Ostermaier KK. Clinical features and diagnosis of Down syndrome. In: Rose BD, ed. *UpToDate*. Wellesley, MA: UpToDate; 2010.
7. Moore L, Bradlee M, Singer M, Rotheman K, Milunsky A. Chromosomal anomalies among the offspring of women with gestational diabetes. *Am J Epidemiol*. 2002;155.
8. Murdoch JC, Ogston SA. Characteristics of parents of Down's children and control children with respect to factors present before conception. *J Ment Defic Res*. 1984;28:177–187.
9. Brender J. Down Syndrome Cluster in Pampa, Gray County – 1985. *Internal Report of the Texas Department of Health*; 1986 [unpublished].
10. Jongbloet PH, Mulder A, Hamers AJ. Seasonality of pre-ovulatory nondisjunction and the etiology of Down syndrome. A European collaborative study. *Hum Genet*. 1982;62:134–138.
11. Jongbloet PH, Vrieze OJ. Down syndrome: increased frequency of maternal meiosis I nondisjunction during the transitional stages of the ovulatory seasons. *Hum Genet*. 1985;71:241–248.
12. National Institutes of Health Conference. Alzheimer's disease and Down's syndrome: new insights. *Ann Intern Med*. 1985;103:566–678.
13. Palomaki GE, Lee JE, Canick JA, McDowell GA, Donnenfeld AE. ACMG Laboratory Quality Assurance Committee. Technical standards and guidelines: prenatal screening for Down syndrome that includes first-trimester biochemistry and/or ultrasound measurements. *Genet Med*. 2009;11(9):669–681.
14. Gekas J, van den Berg DG, Durand A, et al. Rapid testing versus karyotyping in Down's syndrome screening: cost-effectiveness and detection of clinically significant chromosome abnormalities. *Eur J Hum Genet*. 2011;19:3–9.
15. Caine A, Maltby AE, Parkin CA, Waters JJ, Crolla JA. UK Association of Clinical Cytogeneticists (ACC). Prenatal detection of Down's syndrome by rapid aneuploidy testing for chromosomes 13, 18, and 21 by FISH or PCR without a full karyotype: a cytogenetic risk assessment. *Lancet*. 2005;366(9480):123–128.
16. Down JL. Observations on an ethnic classification of idiots. *London Hosp Rep*. 1866;3:259–262.
17. Lejeune J, Gautier M, Turpin R. Etude des chromosomes somatiques de neuf enfants mongoliens. *C R Hebd Seances Acad Sci*. 1959;248:1721–1722.
18. Antonarakis SE. 10 years of Genomics, chromosome 21, and Down syndrome. *Genomics*. 1998;51:1–16.
19. Slavotinek AM, Chen XN, Jackson A, et al. Partial tetrasomy 21 in a male infant. *J Med Genet*. 2000;37:E30.
20. Korbel JO, Tirosh-Wagner T, Urban AE, et al. The genetic architecture of Down syndrome phenotypes revealed by high-resolution analysis of human segmental trisomies. *Proc Natl Acad Sci U S A*. 2009;106:12031–12036.
21. Hamerton JL, Giannelli F, Polani PE. Cytogenetics of Down's syndrome (Mongolism). I. Data on a consecutive series of patients referred for genetic counselling and diagnosis. *Cytogenetics*. 1965;4:171–185.
22. Richards BW, Zaremba J, Stewart A. Cytogenetic studies in Down's syndrome with references to familial cases. *Neurol Neurochir Pol*. 1969;3:249–256.
23. Papavassiliou P, York TP, Gursoy N, et al. The phenotype of persons having mosaicism for trisomy 21/Down syndrome reflects the percentage of trisomic cells present in different tissues. *Am J Med Genet A*. 2009;149A:573–583.
24. Chapman RS, Hesketh LJ. Behavioral phenotype of individuals with Down syndrome. *Ment Retard Dev Disabil Res Rev*. 2000;6:84–95.
25. Fergus K. *Treatment of Down Syndrome. A Brief Overview of Medical Care and Therapies Used to Treat Down Syndrome*. About.com; 2009. http://downsyndrome.about.com/od/downsyndrometreatments/a/Treatmentess_ro.htm.
26. Connolly B, Morgan S, Russell FF, Richardson B. Early intervention with Down syndrome children: follow-up report. *Phys Ther*. 1980;60:1405–1408.
27. Myrelid A, Bergman S, Elvik Stromberg M, et al. Late effects of early growth hormone treatment in Down syndrome. *Acta Paediatr*. 2010;99:763–769.
28. Heyn S. *Down syndrome symptoms, causes, diagnosis and treatment*. In: Perlstein D, ed. OnHealth.com: MedicineNet, Inc; 2010.
29. Centers for Disease Control and Prevention. Down syndrome prevalence at birth – United States, 1983–1990. *MMWR Morb Mortal Wkly Rep*. 1994;43:616–622.
30. Krivchenia E, Huether CA, Edmonds LD, May DS, Guckenberger S. Comparative epidemiology of Down syndrome in two United States populations, 1970–1989. *Am J Epidemiol*. 1993;137:815–828.
31. Wilson MG, Chan LS, Herbert WS. Birth prevalence of Down syndrome in a predominantly Latino population: a 15-year study. *Teratology*. 1992;45:285–292.
32. Ebensen AJ. Health conditions associated with aging and end of life adults with Down syndrome. *Int Rev Res Ment Retard*. 2010;39:107–126.
33. Schupf N. Genetic and host factors for dementia in Down's syndrome. *Br J Psychiatry*. 2002;180:405–410.
34. Bush A, Beail N. Risk factors for dementia in people with Down syndrome: issues in assessment and diagnosis. *Am J Ment Retard*. 2004;2:83–97.
35. Mann DMA, Esiri MM. The pattern of acquisition of plaques and tangles in the brains of patients under 50 years of age with Down's syndrome. *J Neurol Sci*. 1989;89:169–179.
36. Wisniewski K, Howe J, Williams G, Wisniewski HM. Precocious aging and dementia in patients with Down's syndrome. *Biol Psychiatry*. 1978;13:619–627.
37. Wisniewski K, Wisniewski H, Wen G. Occurrence of neuropathological changes and dementia of Alzheimer's disease: state of the sciences. *Arch Neurol*. 1985;17:278–282.

38. Lott I, Dierssen M. Cognitive deficits and associated neurological complications in individuals with Down syndrome. *Lancet Neurol.* 2010;9:623–633.
39. Noetzel MJ. Dementia in Down syndrome. In: Morris JC, ed. *Handbook of Dementing Illnesses*. New York: Marcel Dekker; 1994:243–260.
40. Lott I, Doran E, Nguyen VQ, Tournay A, Movsesyan N, Gillen DL. Down syndrome and dementia: seizures and cognitive decline. *J Alzheimers Dis*. 2012;30:1–9.
41. Lai F, Williams RS. A prospective study of Alzhemier's disease in Down syndrome. *Arch Neurol*. 1989;46:849–853.
42. Deb S, Hare M, Prior L. Symptoms of dementia among adults with Down's syndrome: a qualitative study. *J Intellect Disabil Res.* 2007;51:726–739.
43. Cooper SA, Prasher VP. Maladaptive behaviors and symptoms of dementia in adults with Down syndrome compared with adults with intellectual disability of other etiologies. *J Intellect Disabil Res*. 1998;42:293–300.
44. Ball SL, Holland AJ, Hon J, Huppert FA, Treppner P, Watson PC. Personality and behavioral changes mark the early stages of Alzheimer's disease in adults with Down syndrome: findings from a prospective population-based study. *Int J Geriatr Psychiatry*. 2006;21:661–673.
45. Deb S, Braganza J. Comparison of rating scales for the diagnosis of dementia in adults with Down's syndrome. *J Intellect Disabil Res.* 1999;43:400–407.
46. Strydom A, Livingston G, King M, Hassiotis A. Prevalence of dementia in intellectual disability using different diagnostic criteria. *Br J Psychiatry*. 2007;191:150–157.
47. Holland AJ, Hon J, Huppert FA, Stevens F. Incidence and course of dementia in people with Down's syndrome: finding from a population-based study. *J Intellect Disabil Res*. 2000;44:138–146.
48. Nelson LD, Scheibel KE, Ringman JM, Sayer JW. An experimental approach to detecting dementia in Down syndrome: a paradigm for Alzheimer's disease. *Brain Cogn*. 2007;64:92–103.
49. Evenhuis H. The natural history of dementia in Down syndrome. *Arch Neurol*. 1990;47(3):263–267.
50. Urv TK, Zigman WB, Silverman W. Psychiatric symptoms in adults with Down syndrome and Alzheimer's disease. *Am J Intellect Dev Disabil*. 2010;4:265–276.
51. Silverman W, Schupf N, Zigman WB, et al. Dementia in adults with mental retardation: assessment at a single point in time. *Am J Ment Retard*. 2004;109:111–125.
52. Urv TK, Zigman WB, Silverman W. Maladaptive behavior related to dementia status in adults with Down syndrome. *Am J Ment Retard.* 2008;113:73–86.
53. Levy M, Cummings J, Fairbanks L, et al. Apathy is not depression. *J Neuropsychiatry Clin Neurosci*. 1998;10:314–319.
54. Menendez M. Down syndrome, Alzheimer's disease and seizures. *Brain Dev*. 2005;27:246–252.
55. Puri BK, Ho KW, Singh I. Age of seizure onset in adults with Down syndrome. *Int J Clin Pract*. 2001;55:442–444.
56. Nieuwenhuis-Mark RE. Diagnosing Alzheimer's dementia in Down syndrome: problems and possible solutions. *Res Dev Disabil.* 2009;30:827–838.
57. Ball SL, Holland AJ, Treppner P, Watson PC, Huppert FA. Executive dysfunction and its association with personality and behavioral changes in the development of Alzheimer's disease in adults with Down syndrome and mild to moderate learning disabilities. *Br J Clin Psychol*. 2008;47:1–29.
58. Edgin JO, Mason GM, Allman MJ, et al. Development and validation of the Arizona Cognitive Test Battery for Down syndrome. *J Neurodev Disord*. 2010;2:149–164.
59. Burt DB, Aylward EH. Test battery for the diagnosis of dementia in individuals with intellectual disability. *J Intellect Disabil Res.* 2000;44:175–180.
60. Robbins TW, James M, Owen A, Sahakian BJ, McInnes L, Rabbit PM. Cambridge Neuropsychological Test Automated Battery (CANTAB): a factor analytic study of a large sample of normal elderly volunteers. *Dementia*. 1994;5:266–281.
61. Bittles AH, Bower C, Hussain R, Glasson EJ. The four ages of Down syndrome. *Eur J Public Health*. 2006;17:221–225.
62. Frid C, Drott P, Lundell B, Rasmussen F, Annerén G. Mortality in Down's syndrome in relation to congenital malformations. *J Intellect Disabil Res*. 1999;43:234–241.
63. Yang Q, Rasmussen SA, Friedman JM. Mortality associated with Down's syndrome in the USA from 1983 to 1997: a population-based study. *Lancet*. 2002;359:1019–1025.
64. Coppus AMW, Evenhuis HM, Verberne G, et al. Survival in Down syndrome. *J Am Geriatr Soc*. 2008;56:2311–2316.
65. Ebensen AJ, Seltzer MM, Greenberg JS. Factors predicting mortality in midlife adults with and without Down syndrome living with family. *J Intellect Disabil Res*. 2007;52:1039–1050.
66. Glasson EJ, Sullivan SG, Hussain R, Peterson BA, Montgomery PD, Bittles AH. The changing survival profile of people with Down syndrome: implications for genetic counselling. *Clin Genet*. 2002;62:390–393.
67. Strauss D, Eyman RK. Mortality of people with mental retardation in California with and without Down syndrome, 1986–1991. *Am J Ment Retard*. 1996;100:643–653.
68. Powell D, Caban-Holt A, Jicha G, et al. Frontal white matter integrity in adults with Down syndrome with and without dementia. *Neurobiol Aging*. 2014;35(7):1562–1569.
68a. Head E, Powell D, Gold BT, Schmitt FA. Alzheimer's disease in Down syndrome. *Eur J Neurodeg Dis*. 2012;1:353–364.
69. Strauss D, Zigman WB. Behavioral capabilities and mortality risk in adults with and without Down Syndrome. *Am J Ment Retard.* 1996;101:29–281.
70. Visu-Petra L, Benga O, Tincas I, Miclea M. Visual-spatial processing in children and adolescents with Down's syndrome: a computerized assessment of memory skills. *J Intellect Disabil Res*. 2007;51:942–952.
71. Miller LA, Munoz DG, Finmore M. Hippocampal sclerosis and human memory. *Arch Neurol*. 1993;50:391–394.
72. Pennington BF, Moon J, Edgin J, Stedron J, Nadel L. The neuropsychology of Down syndrome: Evidence for hippocampal sclerosis. *Child Dev*. 2003;74:75–93.
73. Pinter JD, Eliez S, Schmitt JE, Capone GT, Reiss AL. Neuroanatomy of Down's syndrome: a high-resolution MRI study. *Am J Psychiatry*. 2001;158:1659–1665.
74. Nadel L. Down's syndrome: a genetic disorder in biobehavioral perspective. *Genes Brain Behav*. 2003;2:156–166.

CHAPTER

16

An Overview of Rett Syndrome

Kristen L. Szabla and Lisa M. Monteggia
The University of Texas Southwestern Medical Center, Dallas, TX, USA

INTRODUCTION

Disease Characteristics

Hallmark Manifestations

Rett syndrome (RTT) is a neurological disorder that primarily affects girls and is one of the leading causes of intellectual disability and autism in women worldwide.[1] Individuals with RTT typically have a normal period of development for the first 6–18 months of age followed by the appearance of a range of symptoms including a slowing of development, loss of purposeful use of the hands, distinctive hand movements, including hand wringing or clapping, slowed brain and head growth, seizures, and often intellectual disability.[2] Autistic features often manifest, including social withdrawal, hypersensitivity to sound, lack of eye-to-eye contact, and indifference to the surrounding environment.[3]

Inheritance

These behavioral traits have been traced to genetic phenotypes that, in a majority of cases, contain mutations in the methyl-CpG binding protein 2 (*MeCP2*) gene.[4] Indeed, at least 95% of individuals with RTT have a mutation in their *MeCP2* gene.[5] Although Rett syndrome is considered a genetic disorder, less than 1% of recorded cases are inherited. In most cases, the mutation in *MeCP2* occurs spontaneously.

Mutations in other genes such as cyclin-dependent kinase like 5 (*CDKL5*) and forkhead box G1 (*FOXG1*) can cause phenotypes overlapping with those seen in RTT; however, several features, such as congenital onset and infantile spasms in *CDKL5*-mutant patients, and congenital onset and hypoplasia of the corpus callosum in *FOXG1*-mutant patients, distinguish these disorders from typical RTT.

Diagnosis Testing

Tests that are commonly used to diagnose RTT include blood tests for genetic screening, urine tests, nerve conduction studies, magnetic resonance imaging (MRI) or computerized tomography (CT) scans, hearing tests, eye and vision exams, and/or electroencephalograms.[6] Some individuals who have lost communication skills are additionally evaluated using the Autism Diagnostic Observation Schedule (ADOS) and the Autism Diagnostic Interview–Revised (ADIR), which are standardized protocols for assessing social and communicative behavior.[6] Physicians use different criteria to diagnose RTT, but all include similar signs and symptoms. One common set of criteria is spelled out in the Diagnostic and Statistical Manual of Mental Disorders, Fourth Edition (DSM-IV), published by the American Psychiatric Association.

The criteria required for a diagnosis of Rett syndrome include:

- Apparently normal development for the first 5 months after birth;
- Normal head circumference at birth, followed by a slowing of the rate of head growth between the ages of 5 months and 4 years;
- Severely reduced language skills;

Rosenberg's Molecular and Genetic Basis of Neurological and Psychiatric Disease
http://dx.doi.org/10.1016/B978-0-12-410529-4.00002-4

- Loss of hand skills and development of repetitive hand movements between the ages of 5 months and 30 months;
- Loss of interaction with others (though this often improves later);
- An unsteady walk or poorly controlled torso movements;
- Severely impaired ability to communicate and move normally.

Current Research

Although MeCP2 is expressed in most tissues, loss of *MeCP2* results primarily in neurological symptoms.[7–9] Several studies have thus propelled the notion that RTT is due exclusively to loss of MeCP2 function in neurons.[8,10,11] While defective neurons clearly underlie the aberrant behaviors, recent work by Mandel and colleagues showed that the loss of *MeCP2* from glia negatively influences neurons in a noncell autonomous fashion,[12] suggesting that *MeCP2* deficiency from glia may also impact neuronal deficits observed in RTT. Several mouse models for RTT have also been developed in recent years in which *MeCP2* has been deleted, both constitutively and in broad forebrain regions during early postnatal development,[8,9,13,14] which recapitulate many of the human symptoms of the disorder. Interestingly, recent studies demonstrate disease reversibility in some RTT mouse models, suggesting that the neurological defects in *MeCP2* disorders are not permanent.[15,16]

The availability of several mouse models of RTT prompted intense searches for MeCP2 target genes.[17,18] The initial attempts to identify such targets were heavily influenced by the model that MeCP2 acted solely as a transcriptional repressor, such that only genes whose expression increased in response to the loss of MeCP2 function were characterized. Indeed, initial studies indicated that MeCP2 functioned as a transcriptional repressor by binding to methylated DNA and recruiting corepressors and chromatin remodeling proteins such as Sin3A and HDACs.[19,20] In fact several genes, such as *Sgk1*, *Fkbp1*, and *Crh*, that have been associated with RTT symptomology were found to be enhanced in mice with loss of function of MeCP2.[7,21,22] However, this initial view of MeCP2 function had to be reconsidered because of the observation that the majority of genes modulated in mice with either MeCP2 loss or gain of function are downregulated in MeCP2 knockout mice and activated in overexpressing mice,[23] suggesting that MeCP2 is also an activator of gene transcription. Regardless of whether MeCP2 is a repressor or an activator, these studies brought forth the view that transcriptional deregulation takes place in a large set of genes when MeCP2 is dysfunctional. However, more recent high-throughput profiling studies using either human postmortem tissue or whole brain tissue from *Mecp2* knockout mice revealed only a few genes with altered transcription, with modest differences to control samples.[24,25] Jordan et al. suggested that significant changes in a region highly relevant to RTT could be diluted in the whole brain samples. Therefore, Jordan and colleagues carried out microarray-based global gene expression studies only in the cerebellum of *Mecp2* knockout mice, a region important for motor coordination and thus relevant to RTT symptoms, and found that several hundred genes were deregulated.[26] Such widespread changes were later confirmed in the hypothalamus, which is a region responsible for autonomic phenotypes present in RTT.[23] Nevertheless, the interpretation of these gene changes may not be straightforward. In previous work, Martinowich et al. found increased synthesis of brain-derived neurotrophic factor (BDNF) in neurons after depolarization, and that increased BDNF transcription involved dissociation of the MeCP2–histone deacetylase–mSin3A repression complex from its promoter, which was then extrapolated as a mechanism by which MeCP2 regulates gene expression.[27] However, a recent study utilizing chromatin immunoprecipitation followed by sequencing (ChIP-Seq) to look for genome-wide DNA–protein associations followed by ChIP-quantitative polymerase chain reaction (ChIP-qPCR) to define specific MeCP2–gene associations[28] found that although MeCP2 bound extensively to genes in cultured cortical neurons, its association with specific target genes was not altered upon neuronal stimulation, suggesting that MeCP2 might not function as a classical transcription factor, but rather as a global regulator of gene transcription. These data suggest a compensatory role for histone H1 in *Mecp2* knockout mice,[29] indicating that the modest changes in gene expression found by Smrt et. al (2007) may be the result of chromatin remodeling due to the loss or gain of a histone-like protein, i.e. MeCP2. However, the apparent discrepancies in how MeCP2 precisely regulates the expression of genes call for future investigations.

CLINICAL FEATURES

Historical Overview

Disease Identification

In 1966, Dr. Andreas Rett published the first description of Rett Syndrome. However, it would not be until 1983 that RTT became recognized in the medical community when Dr. Bengt Hagberg and colleagues reported 35 cases in the English language.[30]

Gene Identification

A turning point for RTT research came in 1999 with the discovery by Dr. Huda Zoghbi and colleagues of mutations in the gene *MeCP2* as the cause of some cases of RTT. In five of 21 sporadic patients, they found three *de novo* missense mutations in the region encoding the highly conserved methyl-binding domain (MBD) as well as a *de novo* frameshift and a *de novo* nonsense mutation, both of which disrupt the transcriptional repression domain (TRD) of MeCP2.[4] Their study was the first report of disease-causing mutations in RTT and pointed to abnormal transcriptional regulation as the mechanism underlying the pathogenesis of RTT.

Mode of Inheritance and Prevalence

Loss of function in *MeCP2*, an X-linked gene, accounts for the majority (95%) of typical RTT cases.[4] In most cases the mutation in *MeCP2* occurs spontaneously in the paternal germline; thus, individuals with RTT are typically females who, owing to X-chromosome inactivation, are somatic mosaics for normal and mutant *MeCP2*. Boys with mutations that cause RTT in females typically die before or soon after birth with a severe encephalopathy.[1] However, in some families of individuals affected by Rett syndrome, there are other female family members who have a mutation of their *MeCP2* gene but do not show clinical symptoms. These females are known as "asymptomatic female carriers."[1] Some evidence suggests that an increased prevalence of skewing of X-chromosome inactivation within affected individuals can help influence the severity of symptoms.

Natural History

A major source of the phenotypic variability in RTT patients is the pattern of X-chromosome inactivation (XCI). In females, only one of the two X chromosomes is active in each cell and the choice of which X chromosome is active is usually random, such that half of the cells have the maternal X chromosome active and the other half have the paternal X chromosome active. A female with a *MeCP2* mutation is typically mosaic,[7] whereby half of her cells express the wild-type *MeCP2* allele and the other half express the mutant *MeCP2* allele. Depending on the extent of skewing of the X-chromosome inactivation, some patients can be mildly affected or are even asymptomatic carriers of *MeCP2* mutations, while others may show the classic symptoms of RTT.

Perhaps one of the more intriguing findings regarding RTT and MeCP2 is the fact that loss of function and gain in *MeCP2* dosage result in clinically similar neurological disorders.[7] Duplications of chromosome Xq28 that span the *MeCP2* locus have been reported in males with progressive neurodevelopmental phenotypes. The patients typically suffer from mental retardation with facial and axial hypotonia, progressive spasticity, seizures, recurrent respiratory infections, and often premature death.[7] In addition, autistic features and RTT phenotypes, including head growth deceleration, motor delay, ataxia, hand stereotypies, teeth grinding, and absence of speech, have been reported in these boys.[31,32] A male patient with a triplication of the locus was also described with a worse early-onset neurological phenotype.[33] Only one duplication case has been identified in females, in a patient with the preserved speech variant of RTT.[34] These observations, together with the finding that *MeCP2* is the only common gene shared among all patients with the duplication syndrome, give credence to the notion that *MeCP2* is the gene within the ~400-kb duplicated region that is responsible for these phenotypes.

End of Life: Mechanisms and Comorbidities

One-quarter of deaths in RTT are sudden and unexpected,[35] and might result from complications of cardiorespiratory dysfunction. However, despite the difficulties with symptoms, many individuals with RTT continue to live well into middle age and beyond. Because the disorder is rare, very little is known about long-term prognosis and life expectancy. While there are women in their 40s and 50s with the disorder, according to the National Institutes of Health, it is currently not possible to make reliable estimates about life expectancy beyond age 40.

MOLECULAR GENETICS

Normal Gene Product

MeCP2 is a member of the methyl-CpG binding protein family[36] and contains three main functional domains: the methyl-binding domain (MBD), the transcriptional repression domain (TRD), and a C-terminal domain, in addition to two nuclear localization signals (NLS). The MBD specifically binds to methylated CpG dinucleotides, with preference for CpG sequences with adjacent A/T-rich motifs.[37] MBD also binds to unmethylated four-way DNA junctions with a similar affinity,[38] implicating a role for the MeCP2 MBD in higher-order chromatin interactions. The more downstream TRD is involved in transcriptional repression through recruitment of corepressors and chromatin remodeling complexes. The *C*-terminus facilitates MeCP2 binding to naked DNA and to the nucleosomal core, and it also contains evolutionarily conserved poly-proline runs that can bind to group II WW domain splicing factors.[39] Although the *C*-terminal region of MeCP2 is not yet well characterized, it is important for protein function as evidenced by the numerous RTT-causing mutations that involve deletion of this domain.

The function of MeCP2 as a transcriptional repressor was first suggested based on *in vitro* experiments in which MeCP2 specifically inhibited transcription from methylated promotors.[40] When MeCP2 binds to methylated CpG dinucleotides of target genes via its MBD, its TRD recruits the corepressor Sin3A and histone deacetylases (HDACs) 1 and 2.[19,20] The transcriptional repressor activity of MeCP2 involves compaction of chromatin by promoting nucleosome clustering, either through recruitment of HDAC and histone deacetylation or through direct interaction between its *C*-terminal domain and chromatin.[41] In addition, the interaction with Sin3A is not stable and appears to be dependent on MeCP2 being DNA-bound.[42] This suggests that Sin3A is not an exclusive partner of MeCP2 and that other factors may interact with MeCP2 to modulate gene expression or other unknown functions. Additional research into alternative functions of MeCP2 suggests that the protein can also influence RNA splicing as well as activate gene transcription.[7,21,43] This range of MeCP2 functions indicates a complex assortment of possible mechanisms leading to neurological dysfunction in RTT.

Although the MeCP2 protein is widely expressed, it is relatively more abundant in the brain, primarily in mature neurons.[44] MeCP2 protein levels are low during embryogenesis and increase progressively during the postnatal period of neuronal maturation.[10,45,46] In the olfactory epithelium, MeCP2 expression coincides with maturation of the olfactory receptor neurons (ORNs) and precedes the onset of synaptogenesis.[46] MeCP2 is nuclear and colocalizes with methylated heterochromatic foci in mouse cells. A recent report suggests that MeCP2 translocates to the nucleus upon neuronal differentiation.[47] Since MeCP2 is expressed in mature neurons and its levels increase during postnatal development, MeCP2 may play a role in modulating the activity or plasticity of mature neurons. Consistent with this, *MeCP2* mutations do not seem to affect the proliferation or differentiation of neuronal precursors.

Abnormal Gene Products

Mutations in *MeCP2* are found in more than 95% of classic RTT cases; most arise *de novo* in the paternal germline and often involve a C-to-T transition at CpG dinucleotides.[5,48] The spectrum of mutation types includes missense, nonsense, and frameshift mutations, with over 300 unique pathogenic nucleotide changes described,[49] as well as deletions encompassing whole exons.[50–52] Eight missense and nonsense mutations account for ~70% of all mutations, while small *C*-terminal deletions account for another ~10%, and complex rearrangements constitute ~6%.

Genotype–Phenotype Correlations

Several genotype–phenotype correlation studies have been reported and while there are no clear correlations, some general conclusions can be made. Because several mutations introduce premature stop codons throughout the gene and are predicted to result in a null allele, RTT is believed to result from loss of MeCP2 function. However, hypomorphic alleles may also occur and result in truncated forms of MeCP2 that retain partial function. Mutations affecting the nuclear localization sequence of MeCP2 or early truncating mutations tend to cause more severe phenotypes than missense mutations, whereas *C*-terminal deletions are associated with milder phenotypes.[53] In addition, the R133C mutation causes an overall milder phenotype,[54,55] while the R270X mutation is associated with increased mortality.[56]

DISEASE MECHANISMS

Pathophysiology

Human Observations

Despite the severity and phenotypic complexity of RTT, the brains of individuals with RTT do not show gross neuropathological changes, nor evidence of neuronal or glial atrophy, degeneration, gliosis, or demyelination, indicating that RTT is not a neurodegenerative disorder.[57,58] Smaller total brain volume and smaller neurons (but with a higher cell density) have been observed in several brain regions, including the cerebral cortex, hypothalamus, and the hippocampus.[59] MRI studies demonstrate that patients with Rett syndrome have global hypoplasia of the brain and progressive cerebellar atrophy increasing with age, which is consistent with increased motor dysfunction with age.[60,61] The size and complexity of dendritic trees are reduced in cortical pyramidal cells,[62,63] and levels of microtubule-associated protein-2 (MAP-2), a protein involved in microtubule stabilization, are lower throughout the neocortex of RTT autopsy material.[64,65] In addition, the density of dendritic spines is lower in pyramidal neurons of the frontal cortex[57,66] and in the CA1 region of the hippocampus.[67]

Animal Models

To uncover the molecular changes that underlie RTT, mouse models with different *MeCP2* mutations have been generated (Table 16.1). The following section provides an overview of some of these mice. *Mecp2* constitutive knockout mice[8,9] develop a stiff, uncoordinated gait, hypoactivity, tremor, hindlimb clasping, and irregular breathing with early postnatal death, typically by 7–10 weeks of age. The brains of these mice are smaller in size and weight than brains of wild-type littermates and, structurally, knockout (KO) mice show reduced neuronal size. Mice as young as 2–4 weeks of age display reduced cortical thickness, increased neuronal density, and immature synapse formation.[68] Female *Mecp2* $^{+/-}$ heterozygous mice have behavioral abnormalities as well, but with a later age of onset.

The severity and specificity of neurological phenotypes promoted the generation of brain-specific *MeCP2* KOs. Deletion of *MeCP2* using the Cre/loxP system of recombination, in which floxed *MeCP2* mice were crossed with a Nestin Cre mouse line, specifically reduced the expression of the gene in neurons and glia as early as embryonic day 11.[8,9] These mice develop normally for the first few weeks and then display similar features as constitutive KO mice, including abnormal gait, hindlimb clasping, and shortened lifespan. When *Mecp2* is deleted in postmitotic neurons using a calcium–calmodulin-dependent protein kinase II (CaMKII) promotor driving Cre-mediated recombination, similar neurological phenotypes are observed with a later age of onset, confirming a critical role for MeCP2 in more mature neurons. These mice display gait ataxia, increased anxiety, deficits in social behavior, and impairments in learning and memory at approximately 16 weeks of age; however, these mice have normal lifespans.[13] These studies suggest that neuronal MeCP2 dysfunction may play a critical role in the pathogenesis of RTT and also highlight the ability to recapitulate aspects of this disorder in mice. There is currently a great deal of work in the field using conditional KO approaches to delete MeCP2 within specific neuronal populations to better assess the role of this factor in the brain as well as to start to delineate the cell populations that may be involved in the pathophysiology of RTT.

Another RTT mouse model, generated by truncating MeCP2 at amino acid 308, results in a hypomorphic allele that retains the MBD, TRD, and NLS.[14] *Mecp2*^{308} male mice appear normal until 6 weeks of age, when they develop progressive neurological phenotypes including motor dysfunction, forepaw stereotypies, hypoactivity, tremor, seizures, social behavior abnormalities, decreased diurnal activity, increased anxiety-related behavior, and learning and memory deficits. Female mice heterozygous for the truncation display milder and more variable features. *In vivo*, the truncated protein maintains normal chromatin localization, but histone H3 is hyperacetylated in the brain, indicating abnormal chromatin architecture.

Mouse models have also been developed to explore the effects of *MeCP2* overexpression and to help further elucidate the role of MeCP2 in the brain. Two-fold overexpression of human *MeCP2* under the control of its endogenous promoter in mice (*MeCP2*^{Tg}) results in the onset of phenotypes around 10 weeks of age. Initially *MeCP2*^{Tg} mice display increased synaptic plasticity, with enhancement in motor and contextual learning abilities.[69] However, at 20 weeks of age, transgenic mice become hypoactive and develop forepaw clasping, aggressiveness, seizures, and motor abnormalities, and die by 1 year of age.[69] In addition, higher levels of MeCP2 expression in other similar transgenic lines are associated with more severe phenotypes. In contrast, overexpression of *MeCP2* selectively in adult neurons under control of the *tau* promoter results in a mouse model with progressive neurological phenotypes that recapitulate key features of *MECP2* duplication syndrome, including motor coordination deficits, heightened anxiety, and impairments in learning and memory that are accompanied by deficits in long-term potentiation and short-term synaptic plasticity.[70] The differences in behavioral and synaptic phenotypes between the two overexpression mouse models may be the result of MeCP2 overexpression throughout the body compared to only in postmitotic neurons. Future work is necessary to explore this and other possibilities. Collectively, data from mouse models have shown that MeCP2 levels must be tightly regulated, even postnatally, and that the slightest perturbation results in deleterious neurological consequences.[69,70]

Current Research

The absence of neuronal degeneration in RTT has begged the question of whether restoring *MeCP2* expression could rescue normal neuronal function and reverse the disease phenotypes. Recent work by Guy et al.[15] and

TABLE 16.1 MeCP2 Mouse Models

	MeCP2$^{-/y}$	*MeCP2* Nestin-Cre	*MeCP2* CaMK-Cre	*MeCP2*^{308y}	*MeCP2*^{Tg}	*MeCP2*^{tau}	*MeCP2*$^{lox\text{-}Stop}$	*MeCP2e2*$^{lox\text{-}Stop\text{-}lox}$
MeCP2 aberration	Constitutive deletion of exon 3 or exons 3 and 4	Conditional deletion of *MeCP2* at embryonic day (E) 11	Conditional deletion of *MeCP2* in postmitotic neurons	Truncation at amino acid 308 resulting in a hypomorphic allele	Overexpression of *MeCP2*	Overexpression of *MeCP2*	Endogenous *MeCP2* silenced by insertion of Lox-Stop cassette and conditionally activated through Cre-mediated deletion of the cassette	*LSL-MeCP2* transgene placed on *MeCP2* null background with the goal of assessing if phenotypic rescue could be obtained upon excision of Stop signal
Neurological phenotype	Severe	Severe	Progressive	Progressive	Progressive	Progressive	Severe*	Severe**
LTP	Reduced	?	?	Reduced	Enhanced	Reduced	?	?
Hypoactivity	✓	✓	✓	✓	✓	✓	✓*	✓
Stereotypies				✓	✓			
Spine curvature				✓	✓			
Spasticity	Hindlimb clasping	Hindlimb clasping	Hindlimb clasping	Hindlimb clasping	Hindlimb clasping		Hindlimb clasping*	Hindlimb clasping
Tremors	✓	✓		✓			✓	✓
Seizures	✓	✓		✓	✓			
Motor dysfunction	✓	✓	✓	✓	✓	✓	✓*	✓**
Breathing abnormalities	✓				?		✓*	✓
Anxiety	Decreased	Decreased	Increased	Increased	Increased	Increased	Decreased	Decreased
Learning and Memory Deficits	✓	✓	✓	✓		✓	✓	✓
Social behavior abnormalities			✓	✓				
Ataxia	✓	✓	✓	✓	✓		✓*	✓**
Age of Death	7–10 wks	7–10 wks	Adulthood	15 months	7–12 months	Adulthood	7–10 wks*	7–10 wks**
References	Guy et al. (2001);[9] Chen et al. (2001)[8]	Guy et al. (2001),[9] Chen et al. (2001)[8]	Gemelli et al. (2006)[13]	Shahbazian et al. (2002)[14]	Collins et al. (2004)[69]	Guy et al. (2001),[9] Na et al. (2012)[70]	Guy et al. (2007)[15]	Giacometti et al. (2007)[16]

Abbreviation: LTP, Long-term potentiation.

Giacometti et al.[16] provide evidence supporting the feasibility of disease reversibility in mouse models of RTT. Guy and colleagues created a mouse in which endogenous *Mecp2* is silenced by insertion of a *Lox-Stop* cassette and can be conditionally activated through Cre-mediated deletion of the cassette. The *Mecp2*$^{lox\text{-}Stop}$ allele behaved as a null mutation, and its activation was controlled by a tamoxifen-inducible (TM-inducible) Cre transgene.[15] Acute TM injections caused sudden activation of *Mecp2* and, somewhat surprisingly, led to complete phenotypic rescue of the null mice.[15] These results indicate that MeCP2-deficient neurons are not permanently damaged, since *Mecp2* activation leads to robust abrogation of advanced neurological defects in both young and adult animals. The authors propose a model in which neuronal MeCP2 target sites are defined by DNA methylation patterns that are preserved in its absence and that guide newly synthesized MeCP2 to its correct chromosomal positions. MeCP2 then resumes its role as interpreter of the DNA methylation signal required for normal neuronal function.[15] Although additional research is needed to further explore the validity of this proposed model and these reversibility results in mice do not provide immediate therapeutic strategies for RTT, they do establish that consequences of MeCP2 loss of function are reversible, and suggest that the neurological defects in RTT are not impervious to therapeutic possibilities.

In an independent study, Giacometti et al.[16] demonstrated partial disease rescue by postnatal reactivation of MeCP2 in mutant animals. The investigators targeted a transgene carrying the mouse *Mecp2-e2* cDNA downstream of a LoxP-Stop-LoxP (LSL) cassette to the *Col1a1* locus, which provides a strong ubiquitous promoter.[16] The LSL-*Mecp2* transgene was then placed on a *Mecp2* null background to determine if it could rescue upon excision of the Stop signal. Four different Cre transgenes, as follows, were tested to activate *Mecp2*: Nestin-Cre in neuronal and glial precursors; *Tau*-Cre in postmitotic neurons during embryogenesis; CaMKII-Cre 93 (C93) in the forebrain, hippocampus, midbrain, and brainstem at postnatal day (P) ~ 0–P15; or CaMKII-Cre 159 (C159) in the forebrain at ~ P15–P30.[16] Although activation of the LSL-*Mecp2* transgene prolonged the lifespan and delayed motor deterioration of *Mecp2*$^{-/Y}$ mice, the extent of rescue directly correlated with the time, level, and site of Cre expression.[16] The most efficient symptomatic rescue was obtained in lines that provided early and wide Cre expression in most neurons (Nestin-Cre and *Tau*-Cre), extending the lifespan of mutant mice to 8 months.[16] Postnatal MeCP2 activation in C93 and C159 lines extended the life span by 4 weeks, with the earlier C93 expression giving a more efficient rescue than the later C159 expression.[16] Inappropriate *Mecp2* expression levels resulting from use of nonendogenous promoters could account for the partial disease rescue obtained in this study, since this was not seen by Guy et al. when *Mecp2* was activated under the control of its endogenous promoter. In fact, MeCP2 levels were much lower than the expected wild-type levels when induced by the C159 line.[16] The findings from the C93 and C159 lines argue that it is critical that MeCP2 levels are restored to the expected wild-type levels and done so in all neurons, or at least in neurons outside of CaMKII domains, to achieve rescue. This conclusion is supported by another study in which induction of MeCP2 expression using a CaMKII promoter failed to rescue the lethality in MeCP2 null mice.[71] Altogether, these data underscore the importance of proper levels of MeCP2 expression throughout the central nervous system (CNS), or at least beyond the CaMKII expression domain, to achieve rescue. What is interesting, however, is that once proper expression of MeCP2 is achieved, symptoms can be reversed even in adult animals, implying that MeCP2-deficient neurons do not experience any permanent functional deficits regardless of age.

Emerging evidence has implicated reduced BDNF levels in MeCP2 based mouse models of RTT. BDNF expression remains unaffected in the early presymptomatic stage of *MeCP2* knockout mice, while it declines with the onset of RTT-like neuropathological and behavioral phenotypes.[72,73] Furthermore, conditional deletion of *BDNF* in postnatal forebrain excitatory neurons results in several phenotypes similar to those of *MeCP2* knockout mice, such as hindlimb clasping, decreased brain weight, and smaller olfactory and hippocampal neurons.[9,72,74] Due to technical limitations to measuring BDNF levels in the human brain, as to whether BDNF expression is impaired in RTT individuals remains unclear. Nevertheless, improving BDNF expression or signaling has received much attention for neurological disorders such as RTT.[6,75] For example, BDNF overexpression in postnatal excitatory forebrain neurons of MeCP2 knockout mice has been shown to significantly extend mouse lifespan, improve locomotor function, increase brain weight, and reverse dampened spontaneous firing of cortical pyramidal neurons.[72] In addition, BDNF overexpression in cultured hippocampal neurons reversed impaired dendritic and axonal complexity caused by either small hairpin RNA (shRNA)-mediated MeCP2 knockdown or expression of RTT-associated MeCP2 mutations.[76]

Pharmacological manipulations that involve the delivery of recombinant mature BDNF expression or its downstream signaling pathways have been attempted in mouse models as they are more amenable alternatives for application in humans than gene therapy. For example, the respiratory dysfunction in *MeCP2* knockout mice that phenocopies the irregular breathing seen in RTT patients is significantly improved by pharmacological manipulations of BDNF signaling.[77] Acute exposure to BDNF reverses neuronal hyperexcitability in the brainstem and the nucleus of solitary tract, which is conveyed to central autonomic pathways and stabilizes cardiorespiratory instability. As an alternative to using recombinant BDNF, the use of AMPAkines have been demonstrated to increase

BDNF expression by preventing the desensitization of AMPA-type glutamate receptors.[78] Indeed, chronic treatment with the AMPAkine CX546 restored normal breathing frequency and minute volume/weight in *MeCP2* KO mice. However, AMPAkine-mediated increases in BDNF expression are likely due to indirect effects as AMPAkines are postulated to change to expression of many molecules, not just BDNF. Another promising compound is the TrkB mimetic LM224A-4, which selectively activates TrkB but no other Trk family members.[79] LM224A-4 administration to *MeCP2* female heterozygous mice increased levels of phosphorylated TrkB and downstream Akt and ERK, as well as restored normal respiration frequency to these animals in comparison to control animals.[80]

In addition to the specific reversal of cardiorespiratory deficits in *MeCP2* knockout mice, other BDNF-related compounds have been examined for the amelioration of RTT-like neurological symptoms. Fringolimod, a compound with sufficient blood–brain barrier permeability, stimulates ERK signaling and leads to enhanced BDNF expression, ultimately resulting in the improvement of RTT-like features in MeCP2 knockout mice.[81] Similarly, oral treatment with cysteamine was found to extend the lifespan and improve locomotor activity in MeCP2 knockout mice,[82] resulting from increased BDNF transport and secretion. The TrkB receptor activator 7,8-dihydroxyflavone (7,8-DHF) has also been suggested to lengthen lifespan and improve locomotor activity, as well as prevent weight loss and breathing pattern irregularities in *MeCP2* knockout mice.[83,84]

Like BDNF, insulin-like growth factor 1 (IGF-1) is widely expressed in the CNS during normal development,[85] strongly promotes neuronal cell survival and synaptic maturation,[85,86] and facilitates the maturation of functional plasticity in the developing cortex. While BDNF stimulates synaptic strengthening via a pathway involving PI3K/pAkt/PSD-95[87] and MAPK signaling,[88] IGF-1 stimulates the same pathways[89] and has been shown to elevate excitatory postsynaptic currents.[90] The biological action of IGF-1 is also regulated by the binding of IGF binding proteins (IGFBP1–6), which may be of significance to RTT and other disorders. IGFBP3, for example, has a binding site for the MeCP2 protein,[91] and MeCP2 null mice and RTT patients express aberrantly high levels of IGFBP3,[92] which can be expected in turn to inhibit IGF-1 signaling. Depressed IGF-1 signaling has indeed been implicated in autism spectrum disorders.[93]

IGF-1 is capable of crossing the blood–brain barrier, particularly in its tripeptide form,[94] where it retains strong neurotrophic efficacy.[89,95] IGF-1 signaling thus offers an attractive target for engaging key molecular pathways to potentially stimulate synaptic maturation and reverse the RTT phenotype, in a format that is amenable to therapeutic administration to RTT patients. Indeed, Tropea and colleagues administered the IGF-1 peptide to *MeCP2* knockout mice and found that the administration of IGF-1 increased lifespan, improved locomotor activity and cardiac and respiratory function in these mice.[96] They also found that IGF-1 partially restored spine density and synaptic amplitude by increasing PSD95 and stabilized cortical plasticity to wild-type levels. Another independent study found that IGF-1 restored spine density 24 hours after administration in *MeCP2* KO mice.[97] Pilot studies have suggested that there are no risks to giving IGF-1 to human patients;[98] clinical trials are currently underway with the goal of assessing the potential beneficial effects of IGF-1 in RTT individuals based on the beneficial effects of IGF1 in *MeCP2* KO mice.

MANAGEMENT

Standard of Care

There are currently no specific treatments that halt or reverse the progression of the disease, and there are no known medical interventions that will change the outcome of patients with RTT. Management is mainly symptomatic and individualized, focusing on optimizing each patient's functional and cognitive abilities.[99] A multidisciplinary approach is usually used, with specialist input from dietitians, physiotherapists, occupational therapists, speech therapists, and music therapists.[99] Regular monitoring for scoliosis and possible heart abnormalities may be recommended. The development of scoliosis (seen in about 87% of patients by age 25 years) and the development of spasticity can have a major impact on mobility and the development of effective communication strategies. Occupational therapy can help children develop skills needed for performing self-directed activities (such as dressing, feeding, and practicing arts and crafts), while physical therapy and hydrotherapy may prolong mobility.[99]

Pharmacological approaches to managing problems associated with RTT include melatonin for sleep disturbances, and several agents for the control of breathing disturbances, seizures, and stereotypic movements.[99] RTT patients have an increased risk of life-threatening arrhythmias associated with a prolonged QT interval, and the avoidance of a number of drugs is recommended, including prokinetic agents, antipsychotics, tricyclic antidepressants, antiarrhythmics, anesthetic agents, and certain antibiotics.

The Challenge for Developing Therapies for RTT

The recent studies demonstrating that neurological deficits resulting from loss of MeCP2 can be reversed upon restoration of gene function are quite exciting. The next phase of research needs to assess how complete the recovery is. Clearly, lethality, level of activity, and hippocampal plasticity are rescued, but are the animals free of any other RTT symptoms such as social behavior deficits, anxiety, and cognitive impairments? Since postnatal rescue results in viability, it will be important to evaluate if these subtler phenotypes of RTT are rescued when protein function is restored postnatally. The genetic rescue data are promising because they show that neurons that have suffered the consequences of loss of MeCP2 function are poised to regain functionality once MeCP2 is restored to the proper expression levels and in the correct spatial distribution. This provides hope for restoring neuronal function in patients with RTT. However, the strategy in humans will require providing the critical factors that function downstream of MeCP2 because of the challenges in delivering the correct MeCP2 dosage only to neurons that lack it, given that the slightest perturbation in MeCP2 level is deleterious. Thus, therapeutic strategies necessitate the identification of X-chromosomal inactivation patterns and the molecular mechanisms underlying individual RTT phenotypes, as well as picking out the candidates that can be therapeutically targeted.

Two viable gene candidates that are currently being investigated for potential therapeutics are BDNF and IGF-1. The recent studies with BDNF and IGF-1 demonstrating that neurological deficits can be reversed by administering either IGF-1- or BDNF-related compounds to MeCP2-deficient animals are exciting in that lifespan, level of activity, and synaptic plasticity are rescued. However, as with the genetic rescue of MeCP2, the next phase of research with these studies is to assess how complete the recovery is and to ascertain whether the animals are free of any other RTT symptoms. The pharmacological rescue data are promising because they show that neurons that have suffered the consequences of loss of MeCP2 function are positioned to regain functionality once a downstream target of MeCP2 is restored to the proper expression levels. This provides hope for restoring neuronal signaling in RTT individuals. While it is conceivable that many other molecules other than BDNF or IGF-1 could potentially serve as therapeutic substrates, it might also prove challenging to restore proper levels of numerous target genes. An alternative approach will be to identify proteins or pathways that suppress MeCP2 dysfunction phenotypes, which might prove easier to target therapeutically. The fact that there are human patients with milder phenotypes in spite of severe mutations[100] argues that some variant in another protein or proteins might subdue the disease. Identifying such modifiers using various mouse models might prove very helpful. The recent data regarding the role of MeCP2 in the CNS and how its dysregulation contributes to RTT has advanced our knowledge on this disorder, with the ultimate hope that rational therapies for the treatment of RTT will be forthcoming.

References

1. Percy AK, Lane JB. Rett syndrome: model of neurodevelopmental disorders. *J Child Neurol*. 2005;20:718–721.
2. Neul JL, Kaufmann WE, Glaze DG, et al. Rett syndrome: revised diagnostic criteria and nomenclature. *Ann Neurol*. 2010;68:944–950. doi:10.1002/ana.22124.
3. Nomura Y, Segawa M. Natural history of Rett syndrome. *J Child Neurol*. 2005;20:764–768.
4. Amir RE, Van den Veyver IB, Wan M, et al. Rett syndrome is caused by mutations in X-linked MECP2, encoding methyl-CpG-binding protein 2. *Nat Genet*. 1999;23:185–188. doi:10.1038/13810.
5. Trappe R, Laccone F, Cobilanschi J, et al. MECP2 mutations in sporadic cases of Rett syndrome are almost exclusively of paternal origin. *Am J Hum Genet*. 2001;68:1093–1101. doi:10.1086/320109.
6. Katz DM, Berger-Sweeney JE, Eubanks JH, et al. Preclinical research in Rett syndrome: setting the foundation for translational success. *Dis Model Mech*. 2012;5:733–745. doi:10.1242/dmm.011007.
7. Chahrour M, Zoghbi HY. The story of Rett syndrome: from clinic to neurobiology. *Neuron*. 2007;56:422–437. doi:10.1016/j.neuron.2007.10.001.
8. Chen RZ, Akbarian S, Tudor M, Jaenisch R. Deficiency of methyl-CpG binding protein-2 in CNS neurons results in a Rett-like phenotype in mice. *Nat Genet*. 2001;27:327–331. doi:10.1038/85906.
9. Guy J, Hendrich B, Holmes M, Martin JE, Bird A. A mouse Mecp2-null mutation causes neurological symptoms that mimic Rett syndrome. *Nat Genet*. 2001;27:322–326. doi:10.1038/85899.
10. Shahbazian MD, Antalffy B, Armstrong DL, Zoghbi HY. Insight into Rett syndrome: MeCP2 levels display tissue- and cell-specific differences and correlate with neuronal maturation. *Hum Mol Genet*. 2002;11:115–124.
11. Luikenhuis S, Giacometti E, Beard CF, Jaenisch R. Expression of MeCP2 in postmitotic neurons rescues Rett syndrome in mice. *Proc Natl Acad Sci U S A*. 2004;101:6033–6038. doi:10.1073/pnas.0401626101.
12. Lioy DT, Garg SK, Monaghan CE, et al. A role for glia in the progression of Rett's syndrome. *Nature*. 2011;475:497–500. doi:10.1038/nature10214.
13. Gemelli T, Berton O, Nelson ED, et al. Postnatal loss of methyl-CpG binding protein 2 in the forebrain is sufficient to mediate behavioral aspects of Rett syndrome in mice. *Biol Psychiatry*. 2006;59:468–476. doi:10.1016/j.biopsych.2005.07.025.

14. Shahbazian M, Young J, Yuva-Paylor L, et al. Mice with truncated MeCP2 recapitulate many Rett syndrome features and display hyperacetylation of histone H3. *Neuron*. 2002;35:243–254.
15. Guy J, Gan J, Selfridge J, Cobb S, Bird A. Reversal of neurological defects in a mouse model of Rett syndrome. *Science*. 2007;315:1143–1147. doi:10.1126/science.1138389.
16. Giacometti E, Luikenhuis S, Beard C, Jaenisch R. Partial rescue of MeCP2 deficiency by postnatal activation of MeCP2. *Proc Natl Acad Sci U S A*. 2007;104:1931–1936. doi:10.1073/pnas.0610593104.
17. Boggio EM, Lonetti G, Pizzorusso T, Giustetto M. Synaptic determinants of Rett syndrome. *Front Synaptic Neurosci*. 2010;2:28. doi:10.3389/fnsyn.2010.00028.
18. Calfa G, Percy AK, Pozzo-Miller L. Experimental models of Rett syndrome based on Mecp2 dysfunction. *Exp Biol Med (Maywood)*. 2011;236:3–19. doi:10.1258/ebm.2010.010261.
19. Jones PL, Veenstra GJ, Wade PA, et al. Methylated DNA and MeCP2 recruit histone deacetylase to repress transcription. *Nat Genet*. 1998;19:187–191. doi:10.1038/561.
20. Nan X, Ng HH, Johnson CA, et al. Transcriptional repression by the methyl-CpG-binding protein MeCP2 involves a histone deacetylase complex. *Nature*. 1998;393:386–389. doi:10.1038/30764.
21. LaSalle JM, Yasui DH. Evolving role of MeCP2 in Rett syndrome and autism. *Epigenomics*. 2009;1:119–130. doi:10.2217/epi.09.13.
22. Nuber UA, Kriaucionis S, Roloff TC, et al. Up-regulation of glucocorticoid-regulated genes in a mouse model of Rett syndrome. *Hum Mol Genet*. 2005;14:2247–2256. doi:10.1093/hmg/ddi229.
23. Chahrour M, Jung SY, Shaw C, et al. MeCP2, a key contributor to neurological disease, activates and represses transcription. *Science*. 2008;320:1224–1229. doi:10.1126/science.1153252.
24. Colantuoni C, Jeon OH, Hyder K, et al. Gene expression profiling in postmortem Rett Syndrome brain: differential gene expression and patient classification. *Neurobiol Dis*. 2001;8:847–865. doi:10.1006/nbdi.2001.0428.
25. Tudor M, Akbarian S, Chen RZ, Jaenisch R. Transcriptional profiling of a mouse model for Rett syndrome reveals subtle transcriptional changes in the brain. *Proc Natl Acad Sci U S A*. 2002;99:15536–15541. doi:10.1073/pnas.242566899.
26. Jordan C, Li HH, Kwan HC, Francke U. Cerebellar gene expression profiles of mouse models for Rett syndrome reveal novel MeCP2 targets. *BMC Med Genet*. 2007;8:36. doi:10.1186/1471-2350-8-36.
27. Martinowich K, Hattori D, Wu H, et al. DNA methylation-related chromatin remodeling in activity-dependent BDNF gene regulation. *Science*. 2003;302:890–893. doi:10.1126/science.1090842.
28. Cohen S, Gabel HW, Hemberg M, et al. Genome-wide activity-dependent MeCP2 phosphorylation regulates nervous system development and function. *Neuron*. 2011;72:72–85. doi:10.1016/j.neuron.2011.08.022.
29. Skene PJ, Illingworth RS, Webb S, et al. Neuronal MeCP2 is expressed at near histone-octamer levels and globally alters the chromatin state. *Mol Cell*. 2010;37:457–468. doi:10.1016/j.molcel.2010.01.030.
30. Hagberg B, Aicardi J, Dias K, Ramos O. A progressive syndrome of autism, dementia, ataxia, and loss of purposeful hand use in girls: Rett's syndrome: report of 35 cases. *Ann Neurol*. 1983;14:471–479. doi:10.1002/ana.410140412.
31. Friez MJ, Jones JR, Clarkson K, et al. Recurrent infections, hypotonia, and mental retardation caused by duplication of MECP2 and adjacent region in Xq28. *Pediatrics*. 2006;118:e1687–e1695. doi:10.1542/peds.2006-0395.
32. Meins M, Lehmann J, Gerresheim F, et al. Submicroscopic duplication in Xq28 causes increased expression of the MECP2 gene in a boy with severe mental retardation and features of Rett syndrome. *J Med Genet*. 2005;42:e12. doi:10.1136/jmg.2004.023804.
33. del Gaudio D, Fang P, Scaglia F, et al. Increased MECP2 gene copy number as the result of genomic duplication in neurodevelopmentally delayed males. *Genet Med*. 2006;8:784–792. doi:10.1097/01.gim.0000250502.28516.3c.
34. Ariani F, Mari F, Pescucci C, et al. Real-time quantitative PCR as a routine method for screening large rearrangements in Rett syndrome: report of one case of MECP2 deletion and one case of MECP2 duplication. *Hum Mutat*. 2004;24:172–177. doi:10.1002/humu.20065.
35. Kerr AM, Armstrong DD, Prescott RJ, Doyle D, Kearney DL. Rett syndrome: analysis of deaths in the British survey. *Eur Child Adolesc Psychiatr*. 1997;6(suppl 1):71–74.
36. Hendrich B, Bird A. Identification and characterization of a family of mammalian methyl-CpG binding proteins. *Mol Cell Biol*. 1998;18: 6538–6547.
37. Klose RJ, Sarraf SA, Schmiedeberg L, et al. DNA binding selectivity of MeCP2 due to a requirement for A/T sequences adjacent to methyl-CpG. *Mol Cell*. 2005;19:667–678. doi:10.1016/j.molcel.2005.07.021.
38. Galvao TC, Thomas JO. Structure-specific binding of MeCP2 to four-way junction DNA through its methyl CpG-binding domain. *Nucleic Acids Res*. 2005;33:6603–6609. doi:10.1093/nar/gki971.
39. Buschdorf JP, Stratling WH. A WW domain-binding region in methyl-CpG-binding protein MeCP2: impact on Rett syndrome. *J Mol Med*. 2004;82:135–143. doi:10.1007/s00109-003-0497-9.
40. Nan X, Campoy FJ, Bird A. MeCP2 is a transcriptional repressor with abundant binding sites in genomic chromatin. *Cell*. 1997;88:471–481.
41. Nikitina T, Ghosh RP, Horowitz-Scherer RA, et al. MeCP2-chromatin interactions include the formation of chromatosome-like structures and are altered in mutations causing Rett syndrome. *J Biol Chem*. 2007;282:28237–28245. doi:10.1074/jbc.M704304200.
42. Klose RJ, Bird AP. MeCP2 behaves as an elongated monomer that does not stably associate with the Sin3a chromatin remodeling complex. *J Biol Chem*. 2004;279:46490–46496. doi:10.1074/jbc.M408284200.
43. Young JI, Hong EP, Castle JC, et al. Regulation of RNA splicing by the methylation-dependent transcriptional repressor methyl-CpG binding protein 2. *Proc Natl Acad Sci U S A*. 2005;102:17551–17558. doi:10.1073/pnas.0507856102.
44. Jung BP, Jugloff DG, Zhang G, et al. The expression of methyl CpG binding factor MeCP2 correlates with cellular differentiation in the developing rat brain and in cultured cells. *J Neurobiol*. 2003;55:86–96. doi:10.1002/neu.10201.
45. Balmer D, Goldstine J, Rao YM, LaSalle JM. Elevated methyl-CpG-binding protein 2 expression is acquired during postnatal human brain development and is correlated with alternative polyadenylation. *J Mol Med*. 2003;81:61–68. doi:10.1007/s00109-002-0396-5.
46. Cohen DR, Matarazzo V, Palmer AM, et al. Expression of MeCP2 in olfactory receptor neurons is developmentally regulated and occurs before synaptogenesis. *Mol Cell Neurosci*. 2003;22:417–429.
47. Miyake K, Nagai K. Phosphorylation of methyl-CpG binding protein 2 (MeCP2) regulates the intracellular localization during neuronal cell differentiation. *Neurochem Int*. 2007;50:264–270. doi:10.1016/j.neuint.2006.08.018.

48. Wan M, Lee SS, Zhang X, et al. Rett syndrome and beyond: recurrent spontaneous and familial MECP2 mutations at CpG hotspots. *Am J Hum Genet*. 1999;65:1520–1529. doi:10.1086/302690.
49. Christodoulou J, Weaving LS. MECP2 and beyond: phenotype-genotype correlations in Rett syndrome. *J Child Neurol*. 2003;18:669–674.
50. Archer HL, Whatley SD, Evans JC, et al. Gross rearrangements of the MECP2 gene are found in both classical and atypical Rett syndrome patients. *J Med Genet*. 2006;43:451–456. doi:10.1136/jmg.2005.033464.
51. Pan H, Li MR, Nelson P, et al. Large deletions of the MECP2 gene in Chinese patients with classical Rett syndrome. *Clin Genet*. 2006;70:418–419. doi:10.1111/j.1399-0004.2006.00694.x.
52. Ravn K, Nielsen JB, Schwartz M. Mutations found within exon 1 of MECP2 in Danish patients with Rett syndrome. *Clin Genet*. 2005;67:532–533. doi:10.1111/j.1399-0004.2005.00444.x.
53. Smeets E, Terhal P, Casaer P, et al. Rett syndrome in females with CTS hot spot deletions: a disorder profile. *Am J Med Genet A*. 2005;132A:117–120. doi:10.1002/ajmg.a.30410.
54. Kerr AM, Archer HL, Evans JC, Prescott RJ, Gibbon F. People with MECP2 mutation-positive Rett disorder who converse. *J Intellect Disabil Res*. 2006;50:386–394. doi:10.1111/j.1365-2788.2005.00786.x.
55. Leonard H, Colvin L, Christodoulou J, et al. Patients with the R133C mutation: is their phenotype different from patients with Rett syndrome with other mutations? *J Med Genet*. 2003;40:e52.
56. Bienvenu T, Chelly J. Molecular genetics of Rett syndrome: when DNA methylation goes unrecognized. *Nat Rev Genet*. 2006;7:415–426. doi:10.1038/nrg1878.
57. Jellinger K, Armstrong D, Zoghbi HY, Percy AK. Neuropathology of Rett syndrome. *Acta Neuropathol*. 1988;76:142–158.
58. Reiss AL, Faruque F, Naidu S, et al. Neuroanatomy of Rett syndrome: a volumetric imaging study. *Ann Neurol*. 1993;34:227–234. doi:10.1002/ana.410340220.
59. Bauman ML, Kemper TL, Arin DM. Microscopic observations of the brain in Rett syndrome. *Neuropediatrics*. 1995;26:105–108. doi:10.1055/s-2007-979737.
60. Murakami JW, Courchesne E, Haas RH, Press GA, Yeung-Courchesne R. Cerebellar and cerebral abnormalities in Rett syndrome: a quantitative MR analysis. *AJR Am J Roentgenol*. 1992;159:177–183. doi:10.2214/ajr.159.1.1609693.
61. Wenk GL. Rett syndrome: neurobiological changes underlying specific symptoms. *Prog Neurobiol*. 1997;51:383–391.
62. Armstrong D, Dunn JK, Antalffy B, Trivedi R. Selective dendritic alterations in the cortex of Rett syndrome. *J Neuropathol Exp Neurol*. 1995;54:195–201.
63. Armstrong DD, Dunn K, Antalffy B. Decreased dendritic branching in frontal, motor and limbic cortex in Rett syndrome compared with trisomy 21. *J Neuropathol Exp Neurol*. 1998;57:1013–1017.
64. Kaufmann WE, MacDonald SM, Altamura CR. Dendritic cytoskeletal protein expression in mental retardation: an immunohistochemical study of the neocortex in Rett syndrome. *Cereb Cortex*. 2000;10:992–1004.
65. Kaufmann WE, Naidu S, Budden S. Abnormal expression of microtubule-associated protein 2 (MAP-2) in neocortex in Rett syndrome. *Neuropediatrics*. 1995;26:109–113. doi:10.1055/s-2007-979738.
66. Belichenko PV, Oldfors A, Hagberg B, Dahlstrom A. Rett syndrome: 3-D confocal microscopy of cortical pyramidal dendrites and afferents. *Neuroreport*. 1994;5:1509–1513.
67. Chapleau CA, Calfa GD, Lane MC, et al. Dendritic spine pathologies in hippocampal pyramidal neurons from Rett syndrome brain and after expression of Rett-associated MECP2 mutations. *Neurobiol Dis*. 2009;35:219–233. doi:10.1016/j.nbd.2009.05.001.
68. Fukuda T, Itoh M, Ichikawa T, Washiyama K, Goto Y. Delayed maturation of neuronal architecture and synaptogenesis in cerebral cortex of Mecp2-deficient mice. *J Neuropathol Exp Neurol*. 2005;64:537–544.
69. Collins AL, Levenson JM, Vilaythong AP, et al. Mild overexpression of MeCP2 causes a progressive neurological disorder in mice. *Hum Mol Genet*. 2004;13:2679–2689. doi:10.1093/hmg/ddh282.
70. Na ES, Nelson ED, Adachi M, et al. A mouse model for MeCP2 duplication syndrome: MeCP2 overexpression impairs learning and memory and synaptic transmission. *J Neurosci*. 2012;32:3109–3117. doi:10.1523/JNEUROSCI.6000-11.2012.
71. Alvarez-Saavedra M, Saez MA, Kang D, Zoghbi HY, Young JI. Cell-specific expression of wild-type MeCP2 in mouse models of Rett syndrome yields insight about pathogenesis. *Hum Mol Genet*. 2007;16:2315–2325. doi:10.1093/hmg/ddm185.
72. Chang Q, Khare G, Dani V, Nelson S, Jaenisch R. The disease progression of Mecp2 mutant mice is affected by the level of BDNF expression. *Neuron*. 2006;49:341–348. doi:10.1016/j.neuron.2005.12.027.
73. Wang H, Chan SA, Ogier M, et al. Dysregulation of brain-derived neurotrophic factor expression and neurosecretory function in Mecp2 null mice. *J Neurosci*. 2006;26:10911–10915. doi:10.1523/JNEUROSCI.1810-06.2006.
74. Chen B, Dowlatshahi D, MacQueen GM, Wang JF, Young LT. Increased hippocampal BDNF immunoreactivity in subjects treated with antidepressant medication. *Biol Psychiatry*. 2001;50:260–265.
75. Gadalla KK, Bailey ME, Cobb SR. MeCP2 and Rett syndrome: reversibility and potential avenues for therapy. *Biochem J*. 2011;439:1–14. doi:10.1042/BJ20110648.
76. Larimore JL, Chapleau CA, Kudo S, et al. Bdnf overexpression in hippocampal neurons prevents dendritic atrophy caused by Rett-associated MECP2 mutations. *Neurobiol Dis*. 2009;34:199–211. doi:10.1016/j.nbd.2008.12.011.
77. Kline DD, Ogier M, Kunze DL, Katz DM. Exogenous brain-derived neurotrophic factor rescues synaptic dysfunction in Mecp2-null mice. *J Neurosci*. 2010;30:5303–5310. doi:10.1523/JNEUROSCI.5503-09.2010.
78. Ogier M, Wang H, Hong E, et al. Brain-derived neurotrophic factor expression and respiratory function improve after ampakine treatment in a mouse model of Rett syndrome. *J Neurosci*. 2007;27:10912–10917. doi:10.1523/JNEUROSCI.1869-07.2007.
79. Massa SM, Yang T, Xie Y, et al. Small molecule BDNF mimetics activate TrkB signaling and prevent neuronal degeneration in rodents. *J Clin Invest*. 2010;120:1774–1785. doi:10.1172/JCI41356.
80. Schmidt HD, Sangrey GR, Darnell SB, et al. Increased brain-derived neurotrophic factor (BDNF) expression in the ventral tegmental area during cocaine abstinence is associated with increased histone acetylation at BDNF exon I-containing promoters. *J Neurochem*. 2012;120:202–209. doi:10.1111/j.1471-4159.2011.07571.x.
81. Deogracias R, Yazdani M, Dekkers MP, et al. Fingolimod, a sphingosine-1 phosphate receptor modulator, increases BDNF levels and improves symptoms of a mouse model of Rett syndrome. *Proc Natl Acad Sci U S A*. 2012;109:14230–14235. doi:10.1073/pnas.1206093109.

82. Roux JC, Zala D, Panayotis N, et al. Unexpected link between Huntington disease and Rett syndrome. *Med Sci (Paris)*. 2012;28:44–46. doi:10.1051/medsci/2012281016.
83. Jang SW, Liu X, Yepes M, et al. A selective TrkB agonist with potent neurotrophic activities by 7,8-dihydroxyflavone. *Proc Natl Acad Sci U S A*. 2010;107:2687–2692. doi:10.1073/pnas.0913572107.
84. Johnson RA, Lam M, Punzo AM, et al. 7,8-dihydroxyflavone exhibits therapeutic efficacy in a mouse model of Rett syndrome. *J Appl Physiol*. 2012;112:704–710. doi:10.1152/japplphysiol.01361.2011.
85. D'Ercole AJ, Ye P, Calikoglu AS, Gutierrez-Ospina G. The role of the insulin-like growth factors in the central nervous system. *Mol Neurobiol*. 1996;13:227–255. doi:10.1007/BF02740625.
86. O'Kusky JR, Ye P, D'Ercole AJ. Insulin-like growth factor-I promotes neurogenesis and synaptogenesis in the hippocampal dentate gyrus during postnatal development. *J Neurosci*. 2000;20:8435–8442.
87. Yoshii A, Constantine-Paton M. BDNF induces transport of PSD-95 to dendrites through PI3K-AKT signaling after NMDA receptor activation. *Nat Neurosci*. 2007;10:702–711. doi:10.1038/nn1903.
88. Carvalho AL, Caldeira MV, Santos SD, Duarte CB. Role of the brain-derived neurotrophic factor at glutamatergic synapses. *Br J Pharmacol*. 2008;153(suppl 1):S310–S324. doi:10.1038/sj.bjp.0707509.
89. Tropea D, Kreiman G, Lyckman A, et al. Gene expression changes and molecular pathways mediating activity-dependent plasticity in visual cortex. *Nat Neurosci*. 2006;9:660–668. doi:10.1038/nn1689.
90. Ramsey MM, Adams MM, Ariwodola OJ, Sonntag WE, Weiner JL. Functional characterization of des-IGF-1 action at excitatory synapses in the CA1 region of rat hippocampus. *J Neurophysiol*. 2005;94:247–254. doi:10.1152/jn.00768.2004.
91. Chang YS, Wang L, Suh YA, et al. Mechanisms underlying lack of insulin-like growth factor-binding protein-3 expression in non-small-cell lung cancer. *Oncogene*. 2004;23:6569–6580. doi:10.1038/sj.onc.1207882.
92. Itoh M, Ide S, Takashima S, et al. Methyl CpG-binding protein 2 (a mutation of which causes Rett syndrome) directly regulates insulin-like growth factor binding protein 3 in mouse and human brains. *J Neuropathol Exp Neurol*. 2007;66:117–123. doi:10.1097/nen.0b013e3180302078.
93. Riikonen R, Makkonen I, Vanhala R, et al. Cerebrospinal fluid insulin-like growth factors IGF-1 and IGF-2 in infantile autism. *Dev Med Child Neurol*. 2006;48:751–755. doi:10.1017/S0012162206001605.
94. Baker AM, Batchelor DC, Thomas GB, et al. Central penetration and stability of N-terminal tripeptide of insulin-like growth factor-I, glycine-proline-glutamate in adult rat. *Neuropeptides*. 2005;39:81–87. doi:10.1016/j.npep.2004.11.001.
95. Guan J, Thomas GB, Lin H, et al. Neuroprotective effects of the N-terminal tripeptide of insulin-like growth factor-1, glycine-proline-glutamate (GPE) following intravenous infusion in hypoxic-ischemic adult rats. *Neuropharmacology*. 2004;47:892–903. doi:10.1016/j.neuropharm.2004.07.002.
96. Tropea D, Giacometti E, Wilson NR, et al. Partial reversal of Rett Syndrome-like symptoms in MeCP2 mutant mice. *Proc Natl Acad Sci U S A*. 2009;106:2029–2034. doi:10.1073/pnas.0812394106.
97. Landi S, Putignano E, Boggio EM, et al. The short-time structural plasticity of dendritic spines is altered in a model of Rett syndrome. *Sci Rep*. 2011;1:45. doi:10.1038/srep00045.
98. Pini G, Scusa MF, Congiu L, et al. IGF1 as a potential treatment for Rett syndrome: safety assessment in six Rett patients. *Autism Res Treat*. 2012;2012:679801. doi:10.1155/2012/679801.
99. Williamson SL, Christodoulou J. Rett syndrome: new clinical and molecular insights. *Eur J Hum Genet*. 2006;14:896–903. doi:10.1038/sj.ejhg.5201580.
100. Dayer AG, Bottani A, Bouchardy I, et al. MECP2 mutant allele in a boy with Rett syndrome and his unaffected heterozygous mother. *Brain Dev*. 2007;29:47–50. doi:10.1016/j.braindev.2006.06.001.

CHAPTER

17

Fragile X-Associated Disorders

Reymundo Lozano, Emma B. Hare, and Randi J. Hagerman
University of California Davis Medical Center, Sacramento, CA, USA

INTRODUCTION

Fragile X-associated disorders include fragile X syndrome (FXS) caused by the full mutation (>200 CGG repeats that are usually methylated) on the front end of the Fragile X Mental Retardation 1 gene (*FMR1*) and premutation disorders (55–200 CGG repeats) including depression, anxiety, fragile X-associated primary ovarian insufficiency (FXPOI; cessation of menses before age 40) and fragile X-associated tremor/ataxia syndrome (FXTAS). Premutation carriers are common in the general population with approximately 1 in 130–250 females and 1 in 250–810 males, whereas those with the full mutation and FXS occur in approximately 1 in 4000 to 1 in 5000[42]. Males with FXS usually have intellectual disability (ID) and behavior problems, including autism spectrum disorder (ASD) in 60%, attention-deficit hyperactivity disorder (ADHD) in 80%, anxiety in 80%, and aggression in 40%. Females are less affected and only 30% have ID because they have a second, normal X chromosome. FXS is the most commonly identified genetic cause of autism or inherited ID. Fragile X DNA testing should be carried out in anyone diagnosed with ID or ASD. The full mutation is always inherited from the mother who may be a premutation carrier or have a full mutation. Males with the premutation will pass on only the premutation to all of their daughters, and their sons will receive the Y chromosome so they will not be affected by the fragile X mutation.

Premutation carriers usually have normal intellectual abilities when the *FMR1* is not methylated and the level of FMRP is usually normal, except in the upper range of the premutation with >120 CGG repeats when FMRP may be mildly deficient. However, the level of *FMR1* messenger RNA (mRNA) is increased in carriers ranging from 2–8 times normal leading to toxic effects in the neurons. Psychiatric problems are common in carriers, including anxiety, which can start in childhood and present as shyness or social anxiety and persist in adulthood, and often worsen with aging because of the onset of neurological problems associated with FXTAS. Depression is also common in adulthood and is often associated with alcohol abuse particularly in males. FXPOI occurs in approximately 20% of carriers and can also be associated with psychiatric problems related to early estrogen deficiency with the early onset of menopause. Female carriers often experience hypertension, hypothyroidism, and/or fibromyalgia in mid-adult life. FXTAS occurs overall in 40% of males and up to 16% of females with an average age of onset at 62 years. FXTAS symptoms include a cerebellar intention tremor, which typically occurs first followed by ataxia, neuropathy pain, autonomic dysfunction, and cognitive decline. Approximately 50% of males with FXTAS develop dementia. Memory problems and executive function deficits are common, typically when individuals present with tremor and/or ataxia. Approximately 30% have parkinsonian symptoms, including resting tremor, so patients are often misdiagnosed with Parkinson disease or benign tremor. Current research is focused on the mechanisms of RNA toxicity that occur in the premutation range and targeted treatments that can reverse the neurobiological abnormalities that occur in FXS and in premutation disorders.

DISEASE CHARACTERISTICS

While males with FXS or fragile X-associated disorders are more affected than females due to the compensatory nature of the second X chromosomes in females, both genders can present with a range of intellectual disability (ID) from learning problems to severe ID. Patients with FXS usually look normal (Figure 17.1), but facial features may

Rosenberg's Molecular and Genetic Basis of Neurological and Psychiatric Disease
http://dx.doi.org/10.1016/B978-0-12-410529-4.00017-6

FIGURE 17.1 **A teenage boy with FXS who has very subtle physical features of FXS without prominent ears, but his behavior is typical for FXS.**

include a long, narrow face, prominent ears with cupping of the ear pinna, and a high arched palate.[1] The muscle tone of the body may be decreased and infants are often diagnosed with hypotonia. Absence of FMRP (the protein produced from the *FMR1* gene) leads to loose connective tissue and manifestations include soft skin, hyperextensible finger joints, double-jointed thumbs, flat feet with pronation, mitral valve prolapse, and recurrent otitis media in early childhood because of collapsible Eustachian tubes.[1] Approximately 60% of boys with FXS have ASD,[2] 80% have ADHD,[3] and almost all have an anxiety disorder.[4] Social anxiety is a hallmark feature of both the pre- and full mutation children so they have difficulty making and maintaining eye contact.[5] Almost all males and approximately 30% of females with FXS have impaired speech,[6] intellectual impairment,[7] and those with a normal IQ usually have executive function deficits.[8] Stereotyped repetitive behaviors such as hand flapping and hand biting are common, along with perseveration and hypersensitivity to sensory stimulation.[1] Seizures occur in 20%,[9] and mood instability with outbursts and aggression[10] occur in approximately 40% and are often the presenting problem to the physician or health care provider.

While there is some overlap with the full mutation, those with the premutation are usually normal but may present with ASD, learning disabilities, mood instability, social anxiety,[11–13] hyperactivity, and obsessive-compulsive behavior.[14] Additional clinical symptoms commonly seen in adults include anxiety, depression,[13] immune mediated disorders such as fibromyalgia and hypothyroidism,[15] restless legs syndrome,[16] hypertension,[17] and migraines.[18] Fragile X-associated disorders that are seen in premutation carriers include fragile X-associated primary ovarian insufficiency (FXPOI) and fragile X-associated tremor/ataxia syndrome (FXTAS). FXPOI is the cessation of menses prior to 40 years of age and affects many premutation carriers.[19] FXTAS is a late-onset, neurodegenerative disorder accompanied by progressive cerebellar ataxia, executive function and memory deficits, autonomic dysfunction, neuropathy and intention tremor.[14] Males are more likely to develop neurological problems including FXTAS and the prevalence increases with age, ranging from approximately 15% in the 50s to 75% in the 80s.[20] Cognitive decline leading to a frontal temporal dementia occurs in approximately 50% of males with FXTAS but it is rare in females.[21] Due to the genetics of the disease, it is important to identify all affected family members when a proband is discovered (Figure 17.2).

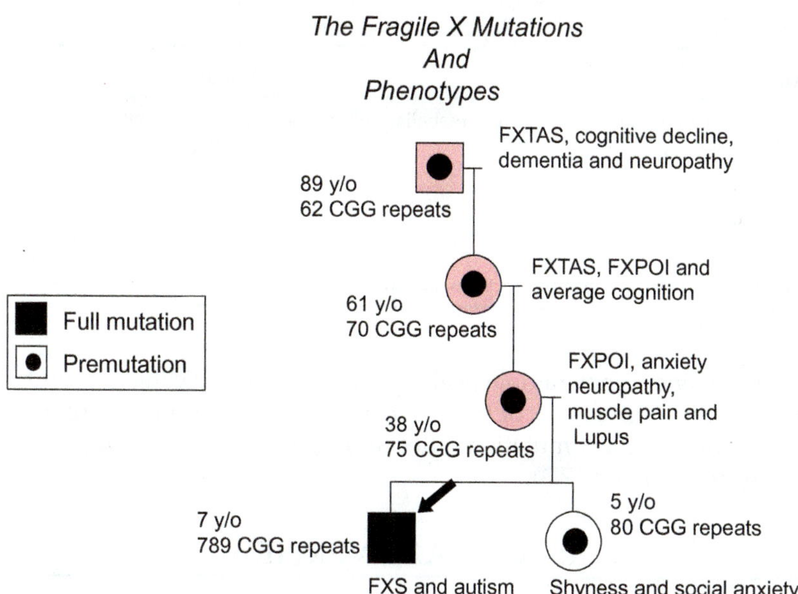

FIGURE 17.2 **A four-generation pedigree with FXS diagnosed initially in the boy with FXS and autism, but each generation demonstrated clinical problems related to either the premutation or the full mutation identified through cascade testing.**

CLINICAL DIAGNOSIS

Early diagnosis is critical to maximize therapeutic interventions. Young children with FXS often have delayed or absent speech, hyperactivity, impulsivity, anxiety, irritability, and repetitive behaviors, which often manifest as hand biting, hand flapping, rocking, and head banging.[1] Many children are diagnosed with ASD before they are identified with a fragile X mutation so all children with ASD should have fragile X DNA testing as a medical workup for their autism. Most children will present to the pediatrician with speech delay at 2 years of age. A lack of eye contact, hyperextensible finger joints, prominent ears, and perseverative behavior should suggest the diagnosis of FXS. Any child or adult with ID also requires a fragile X DNA test. Approximately 2–3% of those with ID and 2–6% of those with ASD will be positive for FXS.[1,22] If there is a family history of ID, particularly in an X-linked pattern, which refers to males with more involvement than females and passing from mother to son, but not inherited from father to son, then the chance of identifying FXS increases to 50%.

The typical time period for diagnosis of young boys is 35–37 months, while girls garner a later diagnosis at around 42 months[23] due to reduced expression in females. Once a diagnosis is confirmed by the *FMR1* DNA test (CGG repeats over 200), the family should meet with a genetic counselor who can review and discuss the possibility of full mutation or premutation involvement in other family members and in future children. The genetic counselor will discuss prenatal diagnosis with the family and can coordinate the testing of other family members—so-called cascade testing. The testing of seemingly normal siblings or other family members at risk for inheriting a mutation will often reveal a mutation that can stimulate a more detailed workup for cognitive or learning problems and subsequent treatment (Figure 17.2). Often the family members who are most resistant to testing will eventually demonstrate the most significant involvement, particularly with psychiatric problems.

FXPOI can be diagnosed in women who have had a cessation of menses prior to age 40 with confirmation of the premutation (55–200 CGG repeats) in the *FMR1* allele. All females with a history of infertility or ovarian dysfunction should be tested for the premutation. The prevalence of the premutation in women with premature ovarian failure is 3–7%, but if there is a family history of early menopause this increases up to 14%.[19]

FXTAS is more difficult to diagnose, but also requires confirmation of the premutation in the *FMR1* allele and can be divided into a possible, probable, and definite FXTAS. Symptoms of FXTAS are divided into major and minor radiological and clinical symptoms (Table 17.1). Major radiological signs include magnetic resonance imaging (MRI) white matter lesions in the middle cerebellar peduncle and/or brain stem. Major clinical signs include intention tremor and gait ataxia. Minor neuroradiologic signs are MRI white matter lesions in the cerebral white matter and moderate-to-severe generalized atrophy. Minor clinical signs include parkinsonism, moderate-to-severe short-term memory deficiency, and executive function deficit. A "definite" diagnosis of FXTAS includes one major clinical symptom and one major radiological criteria or the presence of intranuclear neuronal and astrocytic inclusions.[24] A "probable" diagnosis indicates a major radiological sign and minor clinical sign or two major clinical signs.[25]

TABLE 17.1 Diagnosis of FXTAS: Must Include a CGG Repeat in *FMR1* of 55–200

	Major	Minor
Radiological signs	MRI white matter lesions in middle cerebellar peduncles or brain stem	MRI white matter lesions in cerebral white matter Moderate-to-severe generalized atrophy
Clinical signs	Intention tremor Gait ataxia	Parkinsonism Moderate-to-severe short-term memory deficiency Executive function deficits
Neuropathological signs	Intranuclear neuronal and astrocytic inclusions	

Adapted from[24,25].

A "possible" diagnosis requires one minor radiological sign and one major clinical symptom (Table 17.1).[25] CGG repeat length can also be a significant predictor of the number of intranuclear inclusions and neuropathological involvement in males.[26] In summary, fragile X mutations affect multiple generations and present with multiple clinical presentations in families (Figures 17.2 and 17.3).

HISTORICAL OVERVIEW

In 1943, J. Purdon Martin and Julia Bell first reported X-linked intellectual impairment.[27] In 1969, Herbert Lubs discovered a secondary constriction on the bottom end of the long arm of the X chromosome.[28] However, cytogenetic testing for the fragile site was not initiated until Grant Sutherland discovered that folate-deficient tissue culture media was necessary to identify the fragile site at Xq27.3.[29] Cytogenetic testing was then utilized to diagnose FXS (but could not identify carriers) until the *FMR1* gene was discovered in 1991.[30] Subsequently, a fragile X DNA test has been utilized to confirm the diagnosis of FXS and premutation involvement by determining the exact number of CGG repeats in the 5′ end of the *FMR1* gene. The normal range is 5–44 CGG repeats, the gray zone is 45–54 repeats and this is associated with twice the rate of FXPOI compared to the general population;[31] the premutation range is 55–200 repeats and almost always unmethylated. The full mutation range is >200 repeats and often over 1000 repeats and typically methylated. Both polymerase chain reaction (PCR) and Southern blot testing should be utilized for diagnostic testing, because the former accurately identifies the premutation size and the latter demonstrates full mutation involvement and methylation status. The higher rate of FXPOI in premutation carriers was identified in 1991.[32] The first report of FXTAS occurred in 2001,[33] with diagnostic criteria established in 2003[34] as outlined in Table 17.1. In 2002, the presence of the middle cerebellar peduncle sign of white matter disease was reported[35] and in the same year the discovery of intranuclear eosinophilic inclusions that are tau- and synuclein-negative in neurons and astrocytes were reported, and both findings are considered to be diagnostic of FXTAS and added to the diagnostic criteria (Table 17.1).[24,36] Subsequent reports over the last decade of a variety of medical problems associated with the premutation have significantly expanded the number of conditions that can occur to a greater extent in premutation carriers compared to healthy individuals. These problems relate to the toxicity of the expanded CGG repeat as described below and they include sleep apnea, hypothyroidism, fibromyalgia, migraines, seizures, hypertension, neuropathic pain, ASD, anxiety disorders, depression, and alcohol or other substance abuse.[14] The pain syndromes, such as neuropathic pain and fibromyalgia pain, can lead to excessive use and abuse of narcotics and the psychiatric symptoms, particularly anxiety and depression, can lead to the excessive use and abuse of alcohol.[37] However, in the upper end of the premutation, the level of FMRP can be lowered and this upregulates the metabotropic glutamate receptor 5 pathway (mGluR5), which can also be associated with drug and alcohol abuse.[38,39] The more recent report of FXTAS inclusions throughout the peripheral nervous system in addition to organs such as the heart, islets of Langerhans of the pancreas, adrenals, thyroid, and kidneys facilitates our understanding of additional medical problems such as cardiac arrhythmias, thyroid disease, excessive stress, and cortisol release in premutation carriers;[40] however, further research into understanding the involvement in carriers is needed. For most families identified through a proband, there are multiple individuals throughout the family tree with medical problems associated with the premutation or full mutation (Figures 17.2 and 17.3).

MODE OF INHERITANCE AND PREVALENCE

Due to the X-linked nature of the disease, females have a 50% chance of passing on the premutation or full mutation to any of their children regardless of gender. If the female has the premutation, they will either pass on the premutation or the premutation will expand to the full mutation, particularly if the mother has a premutation with

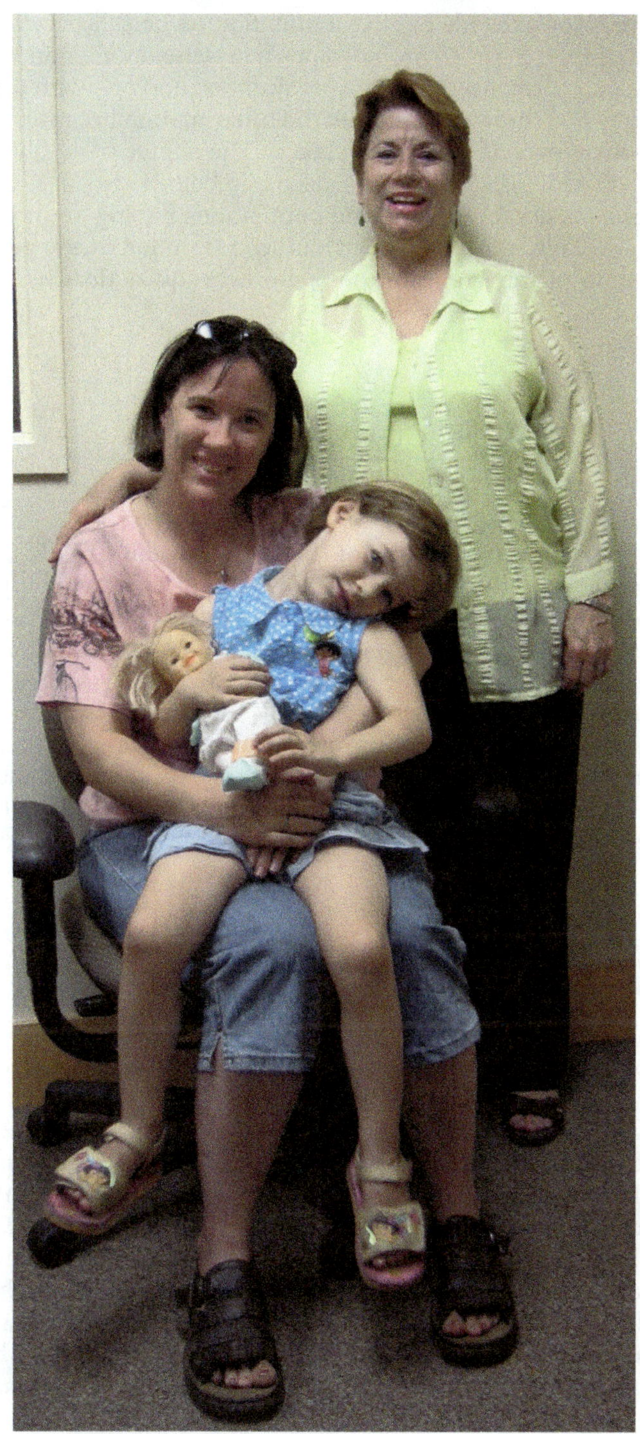

FIGURE 17.3 **A three-generation family with the grandmother standing who has some symptoms of FXTAS (her father died from FXTAS).** The grandmother's daughter has the full mutation and a normal IQ but some learning disabilities in addition to anxiety and depression. The young granddaughter has FXS and mild intellectual impairment, in addition to anxiety and attention deficit hyperactivity disorder (ADHD).

90 or more CGG repeats. Males, conversely, will pass on their X chromosome, which will stay as a premutation to all of their daughters who are obligate carriers. Since carrier males will pass on the Y chromosome to their sons there is no chance they will inherit FXS. The percentage of offspring with FXS of a carrier mother not only depends on the length of her CGG repeat but also on how many AGG anchors she has for every 10 CGG repeats. If the mother has two or three AGG anchors within her CGG repeat sequence she is far less likely to have a child with a full mutation than if she had no anchors within her CGG repeat sequence.[41]

A far less common occurrence is for a female to have a mutation on both X chromosomes. Case studies have reported females with the premutation on both X chromosomes where the mother and father were both carriers, and in some cases females have had an X chromosome with a premutation and one with the full mutation. However, it is impossible for a woman to have two X chromosomes with the full mutation because one of her Xs will always come from her father and this X will never expand to a full mutation. The degree of involvement depends on the level of FMRP and also on the level of elevated *FMR1* mRNA, as discussed below.

The prevalence of FXS varies considerably, with some populations having an incredibly high prevalence of the syndrome such as seen in a small village in Colombia that has been secluded due to geographical and socioeconomic barriers and the early immigration of carriers from the Spanish conquistadors, which is a founder effect to this community.

MOLECULAR GENETICS

Normal Gene Product

The *FMR1* gene is responsible for the production of the *FMR1* protein (FMRP), which is an mRNA-binding and -carrier protein that stabilizes mRNAs but also regulates translation, usually through inhibition of hundreds of other mRNAs important for synapse development and plasticity. It is estimated that FMRP binds and regulates up to 4–8% of total brain mRNAs. Many of the mRNAs produced by genes, that when mutated cause autism (e.g., *Arc*, *NGLN3*, *NGLN4*, *NRXN1*, *SHANK2* and *SHANK3*, *CYFIP*, *PTEN*, *PSD95*) are also regulated by FMRP, which is the basis of the relationship between these two disorders.[43,44] At the synapse, FMRP is stimulated by mGluR5 activation and FMRP inhibits subsequent long-term depression (LTD), which weakens synaptic strength. FMRP is important for the balance of excitatory (glutamate) and inhibitory (γ-aminobutyric acid [GABA]) activity in the central nervous system (CNS). In the absence of FMRP there is upregulation of the mGluR5 system and downregulation of the GABA system, which may occur in many other disorders including autism.[45,46]

FMRP Deficits and Genotype–Phenotype Correlations

Almost all *FMR1* mutations (>99%) leading to FXS are caused by CGG repeat expansions accompanied by abnormal hypermethylation of the gene. Single-base mutations and deletions in the *FMR1* gene account for less than 1% so that sequencing of the *FMR1* gene is rarely needed. Methylation of the CGG expansion is an epigenetic process whereby a methyl group is placed on the backbone of the DNA, which results in decreased or silenced *FMR1* transcription. Therefore, affected males with a deletion of *FMR1* or a full mutation that is fully methylated results in a lack of FMRP and consequently FXS. The degree of involvement from FXS in females results from their activation ratio—the percentage of cells with the normal X active. The level of FMRP correlates directly with IQ.[47] Males who are unmethylated or only partially methylated with a full mutation produce more FMRP than those who are fully methylated. These individuals usually constitute a group of high-functioning males with FXS (IQ higher than 70). Some individuals have some cells with the premutation and some cells with the full mutation, which is called size mosaicism. Those individuals who are mosaic may have a higher IQ than those who are fully methylated if a high percentage of their cells have the premutation. The full mutation cells have a deficit of FMRP and the premutation cells produce an excess of *FMR1* mRNA, leading to RNA toxicity. Individuals with both elevated mRNA and lowered FMRP have a "double hit," meaning pathological involvement from two different mechanisms. We have seen higher rates of psychotic thinking in this population than boys who are fully methylated with FXS.[48]

The association between low FMRP and psychosis is seen in those with schizophrenia without an *FMR1* mutation. Both age of onset and overall IQ correlate with the FMRP level in those with schizophrenia.[49] In addition, the severity of visual-spatial deficits in those with schizophrenia correlates strongly with the level of FMRP.[50] There is wide variability of FMRP levels in the general population.[52] Low FMRP levels in the brain in neuropathology studies have correlated with a variety of neuropsychiatric disorders including bipolar disorder, depression, autism, and schizophrenia in individuals without a fragile X mutation.[53,54]

DISEASE MECHANISMS

Pathophysiology

An imbalance between the excitatory glutamatergic and the inhibitory GABAergic neurotransmission is proposed to cause the cognitive impartments, anxiety, hyperarousal, autism features, and epilepsy in children with FXS.[25–27] Furthermore, dysfunctional amygdala activation is implicated in the social avoidance and anxiety evident in FXS. The decrease of inhibitory input related to the absence of FMRP adds to the phenotype, as does the upregulation of the mGluR pathway and overproduction of a variety of proteins in the hippocampus and elsewhere in the brain.[55] Recent studies also showed that the GABA system is negatively affected at different levels including, synthesis, metabolism and catabolism.[56,57]

The pathophysiology of premutation involvement including FXTAS is very different from FXS and it is based on the excess level of *FMR1* mRNA leading to toxicity in the neuron, astrocyte, and PNS cells. The mRNA has the excess CGG repeats, which form hairpin structures with the CGGs and bind to each other. These hairpin structures are sticky and sequester a variety of proteins that are important to cell survival, including Sam68 that splices mRNAs and Drosha/DGCR8, which are proteins critical for maturing the microRNAs (miRNAs) that regulate translation throughout the body.[58] Neurons with the premutation have upregulation of a variety of proteins, such as heatshock proteins, αB crystallin, and also disruption of lamin A/C that are all signs of toxicity.[59] Neurons with the premutation die earlier in culture compared to normal neurons so their viability is reduced and they appear to be more sensitive to toxins.[60,61] Neurons with the premutation have an enhanced number of spikes and calcium dysregulation[62] that can be predisposed to seizures, which occur in approximately 13% of premutation boys; the presence of seizures in boys with the premutation is also associated with a higher rate of ASD compared to premutation carriers without seizures.[12]

Animal Models

Due to the range, severity, and genetic complexity of FXS, numerous animal models have been developed for the full mutation including the *Drosophila* dfmr1 fly model and the Fmr1 knockout (KO) mouse. Both models exhibit phenotypic features, most of which mimic the human FXS phenotype, and have been reversed with targeted treatments. The *Drosophila* melanogaster model contains a dfmr gene (also called dxfr) that is 35% identical and 60% similar to the human *FMR1* gene and largely expressed in the CNS.[63,64] The phenotype is characterized by altered circadian rhythms, abnormal synaptic morphology, naïve courtship behavior,[65] memory and social paradigm deficits, as well as an overactive grooming behavior.[65,66] The KO mouse also has abnormal synaptic morphology, leading to reduced functional and structural synaptic plasticity and motor skills.[67] The ultrasonic vocalizations are less specific with varied frequency ranges;[68,69] perseveration and hyperactivity are also present,[70] and they have an increased susceptibility to audiogenic seizures,[71] as seen in humans with the full mutation.

For premutation involvement there is an Fmr1 knock-in (KI) mouse carrying a premutation range of CGG repeats or some carrying a set repeat length of 98 CGG repeats[72] as a model for carriers. The KI mice also have abnormal neuronal morphology.[73] While the development and initial follicle pool is normal in the KI mice, increased loss of follicles is observed later.[74] The KI mouse displays elevated mRNA levels, reduced FMRP translation, decreased protein levels and enhanced mGluR LTD, which overlaps with the full mutation model, but unlike the full mutation is dependent on the synthesis of new protein. They also display age-dependent cognitive, neuromotor, and behavioral impairments as are indicative of FXTAS.[75]

Current Research

Current research in animal models of FXS aims to define the pathogenic mechanism of disease and determine effective targeted treatments. Clinical trials in humans are also being conducted to further characterize the phenotype, assess treatment efficacy of currently available medications (both FDA approved and investigational), as well as develop more suitable biomarkers to evaluate molecular improvements in the biochemical and biomolecular profile of affected individuals. Most of the intervention-based studies target those with the full mutation of FXS using pharmaceutical or behavioral interventions, but studies combining the two are needed to evaluate the benefits of multi-faceted treatments, which are already commonly used by clinicians. Considerable research efforts are still needed to better understand the pathogenesis and treatment of the premutation.

DIFFERENTIAL DIAGNOSIS

Developmental Delay/Intellectual Disability/Autism

FXS is the cause of 2–6% of all cases of ID and ASD. Fragile XE syndrome (FRAXE) causes mild ID and nonspecific facial features in males with expanded CCG repeats in the *FMR2* gene at the FRAXE fragile site. Other differential diagnoses include other forms of X-linked ID, when the family inheritance is suggestive, and one can order a panel of tests for these disorders. Microdeletion and microduplication syndromes (Prader–Willi syndrome, Sotos syndrome and others) are responsible for about 15–22% of all cases of ID, therefore a microarray is a first-tier of testing in these individuals. Metabolic syndromes should be considered in the presence of neurological signs and clear signs of regression.

Adult-Onset Neurodegenerative Disorders

These include Parkinson disease (PD), different forms of cerebellar ataxia, Alzheimer disease (AD) and Lewy body dementia. About 1% of individuals with PD will have a premutation and 2–4% of males with ataxia have a premutation.[76,77]

Premature Ovarian Failure

The premutation is one of the most common causes of primary ovarian insufficiency (POI) and therefore genetic testing should be performed in females with compatible history of POI. Other causes include other chromosomal abnormalities (Turner syndrome), exposures (radiation or chemotherapy) and autoimmune disorders.

TESTING

The American College of Medical Genetics recommends ordering fragile X DNA testing for all boys with ASD or ID of unknown etiology and testing for females with the same diagnoses who have a family history of ASD, ID, DD, POI, and/or tremor and ataxia symptoms. Similarly, the American College of Obstetricians and Gynecologists recommend genetic counseling and fragile X DNA testing for women with the same family history characteristics. The fragile X DNA test uses the Southern blot for detection of normal, large premutations and the full mutation. In addition, PCR analysis determines specifically the actual CGG repeat number for normal and premutation alleles. The methylation status can be tested by PCR-based methods or Southern blot analysis. Cascade of testing and genetic counseling is indicated for family members at risk for either a premutation or a full mutation.

MANAGEMENT

Standard of Care

Usually those with FXS have significant anxiety and respond well to a selective serotonin reuptake inhibitor (SSRI), which typically lowers their level of anxiety and sometimes aggression or outburst behavior.[78] Recent studies have shown that sertraline begun at 2 years of age in FXS at a low dose (e.g., 2.5 mg/day) may improve receptive and expressive language,[78,79] but controlled trials are underway to confirm safety and efficacy. Approximately 20% of patients with FXS treated with sertraline or another SSRI may experience hyperarousal, particularly with higher doses by the second or third month of treatment. If this occurs, the dose of sertraline must be lowered or discontinued.[77]

An SSRI can also be very helpful for premutation carriers who have significant depression or anxiety and these agents are also know to stimulate neurogenesis, which can be useful for aging and FXTAS.[14] Once someone is diagnosed with FXTAS or with the premutation, treatment of associated disorders such as hypertension, hypothyroidism, substance abuse, sleep apnea, and migraine headaches is essential if they are present. An exercise regimen can also improve mitochondrial function, stimulate neurogenesis, and improve serotonin levels in the CNS, as a treatment for psychiatric symptoms.[14] The use of antioxidants may also be helpful for both premutation and full mutation disorders, but only animal studies have demonstrated efficacy.[80]

ADHD is a problem for over 80% of boys and approximately 30% of girls with FXS and these problems usually respond well to a stimulant medication by the age of 5 years.[78] If hyperactivity, impulsivity, or aggression is a problem before 5 years of age then use of guanfacine, which is an alpha2A agonist is typically helpful for calming down behavior problems. A related alpha2 agonist, clonidine, is also helpful for nighttime wakefulness, which is very common in young children with FXS. Clonidine can be more sedating than guanfacine, which is why it is helpful at bedtime.[78,81] Melatonin (1–3mg at bedtime) is a sleep hormone that also helps the child or adult initiate sleep and has shown efficacy in children with either FXS or ASD.[82]

Aggression and mood instability is a common problem in boys and occasionally girls with FXS. Although it is present in approximately 30–40% of children it may increase and become harder to control in adolescence, particularly in boys. The use of atypical antipsychotics, such as aripiprazole and risperidone, are the treatment of choice for aggression if an SSRI, stimulant, or guanfacine are not helpful. Sometimes an anticonvulsant is also needed to stabilize mood and valproate is usually the treatment of choice.[78] However, lithium is also helpful because it can directly downregulate the mGluR5 system and it is therefore a targeted treatment for FXS.[83]

Another targeted treatment that can be effective at an early age is minocycline, which is an antibiotic that lowers the level of matrix metalloproteinase 9 (MMP9) that is elevated in FXS.[84,85] High levels of MMP9 interfere with the development of synaptic connections, and a controlled trial of minocycline in children between the ages of 3 and 17 (doses ranging from 25mg per day to 100mg per day) recently demonstrated efficacy on the Clinical Global Impression – Improvement (CGI-I) scale in improving overall behavior and on the Visual Analogue Scale for improving mood and anxiety.[84] Side effects of minocycline include graying of the permanent teeth if given before the age of 8 years and graying or darkening of tissue such as the gums and nailbeds at any age. On rare occasions minocycline can cause a lupus-like syndrome with a rash or swollen joints or pseudotumor cerebri leading to a severe headache, and if these problems occur minocycline should be discontinued immediately. Minocycline is an antibiotic that changes the bacterial flora in the intestines, so the patient should take a daily probiotic or yogurt with active culture when given minocycline. Milk or milk products can interfere with absorption of minocycline if given simultaneously; therefore, waiting 30 minutes to 1 hour before or after the once daily dosing of minocycline is recommended before taking milk or milk products.

Failed Therapies

Additional targeted treatments have been tried in FXS including the $GABA_B$ agonist arbaclofen. In the preliminary study of children and adults with FXS between 6 and 40 years of age it was beneficial in only a subgroup of patients who had ASD or a low sociability score on the Aberrant Behavior Checklist (ABC).[86] However, in a subsequent study of patients with ASD, arbaclofen was not efficacious so the manufacturing company folded.[86] This demonstrates the need to use biological subtyping, particularly in a heterogenous group of patients such as those with ASD. For the subgroup of patients with FXS who did very well with arbaclofen it became a problem to obtain the drug once the company folded and the withdrawal from arbaclofen was very difficult for many families.

A similar fate occurred with fenobam, an mGluR5 antagonist made by Neuropharm. Although an initial open-label safety trial in 12 patients with FXS demonstrated some promising results,[87] the company closed after an unsuccessful study of long-acting fluoxetine in ASD.

Therapies Under Investigation

A number of new, targeted treatments are currently under investigation for FXS. AFQ056, an mGluR5 antagonist made by Novartis, completed a controlled trial in adolescents and adults in Europe demonstrating efficacy in those who had a full mutation that was completely methylated but not in those with incomplete methylation.[88] These results have stimulated multicenter trials in adolescents and adults with FXS utilizing more complete testing of the methylation status, but the results have not yet been published. Roche also has an mGluR5 antagonist, RO4917523, which is currently in multicenter trials for children through adulthood with FXS, and the results are pending.

A number of new treatments are also being studied in clinical trials after promising efficacy studies in the animal models of FXS. These include lovastatin, which lowers extracellular-signal-regulated kinase (ERK) phosphorylation in the mTOR pathway and an insulin-like growth factor-1 (IGF1) analog made by Neuren. Other treatment trials are in the pipeline but have yet to come to clinical trials.

Barriers to Treatment Development

There are a number of barriers to new treatments, including the time and expense of carrying out the toxicity studies in animals and humans and subsequently multicenter human trials to demonstrate efficacy. Although FXS is a single-gene disorder there is significant heterogeneity in clinical involvement and response to treatment. For many trials, approximately 30% respond well but this may not be adequate to demonstrate overall efficacy, which prohibits US Food and Drug Administration (FDA) approval for marketing. The lack of biomarkers that would predict efficacy is greatly needed so that a "likely to respond subgroup" could be identified. For AFQ056 this appears to be those who are fully methylated, but additional biomarkers are needed. In addition, outcome measures that are quantitative and relate to CNS function or molecular changes and do not depend on questionnaires from the family would be useful to decrease the placebo effect. Event-related potential (ERP) paradigms of cognitive processing would be useful and we have seen a positive effect on an oddball paradigm and on a habituation task in children treated with minocycline in a controlled trial.[89]

Only one controlled trial has been carried out in patients with FXTAS. This was a memantine trial that was focused on improving tremor and executive function deficits by using memantine to block glutamate toxicity in patients with FXTAS, and subjects were randomized to 1 year of treatment with memantine or placebo. This was a negative trial with no effect on tremor or severity in the overall group,[90] although a subgroup of patients appeared to do well on the drug.[91] In addition, subsequent analysis of secondary outcome measures utilizing ERP appears to demonstrate a positive effect on language learning and also improvements in a measure of attention in P2 amplitude on the ERP.[92]

The study of new, targeted treatments in FXS and in premutation disorders including FXTAS has a bright future for the many needful patients with these disorders.

ACKNOWLEDGEMENTS

This work was supported by grant R40 MC 22641 from the Maternal and Child Health Research Program, Maternal and Child Health Bureau (Combating Autism Act of 2006, as amended by the Combating Autism Reauthorization Act of 2011), Health Resources and Services Administration, Department of Health and Human Services, National Institute of Health grants HD036071; DOD PR101054, support from the Health and Human Services Administration on Developmental Disabilities grant 90DD05969 and the National Center for Advancing Translational Research UL1 TR000002.

CONFLICTS

Randi Hagerman has received funding from Novartis, Roche, Seaside Therapeutics, and Forest for carrying out treatment studies in fragile X syndrome and autism. She has also served on advisory committees for Roche/ Genentech and Novartis regarding treatment of fragile X syndrome.

References

1. Hagerman RJ, Hagerman PJ. *Fragile X Syndrome: Diagnosis, Treatment, and Research*. Baltimore, MA: JHU Press; 2002.
2. Harris SW, Hessl D, Goodlin-Jones B. Autism profiles of males with fragile X syndrome. *Am J Ment Retard*. 2008;113(6):427–438.
3. Cornish K, Munir F, Wilding J. A neuropsychological and behavioural profile of attention deficits in fragile X syndrome. *Rev Neurol*. 2001;33(suppl 1):S24–S29.
4. Cordeiro L, Ballinger E, Hagerman R, Hessl D. Clinical assessment of DSM-IV anxiety disorders in fragile X syndrome: prevalence and characterization. *J Neurodev Disord*. 2011;3(1):57–67.
5. Schneider A, Hagerman RJ, Hessl D. Fragile X syndrome – from genes to cognition. *Dev Disabil Res Rev*. 2009;15(4):333–342.
6. Finestack LH, Richmond EK, Abbeduto L. Language development in individuals with fragile X syndrome. *Top Lang Disord*. 2009;29(2):133–148.
7. Bennetto L, Pennington B. Fragile X syndrome: diagnosis, treatment, and research. *Neuropsychology*. 2002;3:206–248.
8. Hooper SR, Hatton D, Sideris J, et al. Executive functions in young males with fragile X syndrome in comparison to mental age-matched controls: baseline findings from a longitudinal study. *Neuropsychology*. 2008;22(1):36–47.
9. Berry-Kravis E, Raspa M, Loggin-Hester L, Bishop E, Holiday D, Bailey DB. Seizures in fragile X syndrome: characteristics and comorbid diagnoses. *Am J Intellect Dev Disabil*. 2010;115(6):461–472.
10. Hessl D, Tassone F, Cordeiro L. Brief report: aggression and stereotypic behavior in males with fragile X syndrome – moderating secondary genes in a "single gene" disorder. *J Autism Dev Disord*. 2008;38(1):184–189.
11. Farzin F, Perry H, Hessl D, et al. Autism spectrum disorders and attention-deficit/hyperactivity disorder in boys with the fragile X premutation. *J Dev Behav Pediatr*. 2006;27(2 suppl):S137–S144.
12. Chonchaiya W, Au J, Schneider A, et al. Increased prevalence of seizures in boys who were probands with the FMR1 premutation and co-morbid autism spectrum disorder. *Hum Genet*. 2012;131(4):581–589.
13. Bourgeois JA, Seritan AL, Casillas EM, et al. Lifetime prevalence of mood and anxiety disorders in fragile X premutation carriers. *J Clin Psychiatry*. 2011;72(2):175.
14. Hagerman R, Hagerman P. Advances in clinical and molecular understanding of the FMR1 premutation and fragile X-associated tremor/ ataxia syndrome. *Lancet Neurol*. 2013;12(8):786–798.

15. Winarni TI, Chonchaiya W, Sumekar TA. Immune-mediated disorders among women carriers of fragile X premutation alleles. *Am J Med Genet A*. 2012;158a(10):2473–2481.
16. Summers S, Cogswell J, Goodrich J, et al. Prevalence of restless legs syndrome and sleep quality in carriers of the fragile X premutation. *Clin Genet*. 2013.
17. Hamlin AA, Sukharev D, Campos L. Hypertension in FMR1 premutation males with and without fragile X-associated tremor/ataxia syndrome (FXTAS). *Am J Med Genet A*. 2012;158a(6):1304–1309.
18. Au J, Akins RS, Berkowitz-Sutherland L, et al. Prevalence and risk of migraine headaches in adult fragile X premutation carriers. *Clin Genet*. 2013;84(6):546–551.
19. Sullivan SD, Welt C, Sherman S. FMR1 and the continuum of primary ovarian insufficiency. *Semin Reprod Med*. 2011;29(4):299–307.
20. Jacquemont S, Hagerman RJ, Leehey MA, et al. Penetrance of the fragile X-associated tremor/ataxia syndrome in a premutation carrier population. *JAMA*. 2004;291(4):460–469.
21. Seritan A, Cogswell J, Grigsby J. Cognitive dysfunction in FMR1 premutation carriers. *Curr Psychiatr Rev*. 2013;9(1):78–84.
22. Schaefer GB, Mendelsohn NJ. Genetics evaluation for the etiologic diagnosis of autism spectrum disorders. *Genet Med*. 2008;10(1):4–12.
23. Bailey DB, Raspa M, Bishop E, Holiday D. No change in the age of diagnosis for fragile X syndrome: findings from a national parent survey. *Pediatrics*. 2009;124(2):527–533.
24. Hagerman PJ, Hagerman RJ. The fragile-X premutation: a maturing perspective. *Am J Hum Genet*. 2004;74(5):805–816.
25. Jacquemont S, Hagerman RJ, Leehey M, et al. Fragile X premutation tremor/ataxia syndrome: molecular, clinical, and neuroimaging correlates. *Am J Hum Genet*. 2003;72(4):869–878.
26. Greco CM, Berman RF, Martin RM, et al. Neuropathology of fragile X-associated tremor/ataxia syndrome (FXTAS). *Brain*. 2006;129(Pt 1):243–255.
27. Martin JP, Bell J. A pedigree of mental defect showing sex-linkage. *J Neurol Neurosurg Psychiatry*. 1943;6(3–4):154.
28. Lubs HA. A marker X chromosome. *Am J Hum Genet*. 1969;21(3):231.
29. Sutherland GR. Fragile sites on human chromosomes: demonstration of their dependence on the type of tissue culture medium. *Science*. 1977;197(4300):265–266.
30. Verkerk AJ, Pieretti M, Sutcliffe JS, et al. Identification of a gene (FMR-1) containing a CGG repeat coincident with a breakpoint cluster region exhibiting length variation in fragile X syndrome. *Cell*. 1991;65(5):905–914.
31. Bretherick KL, Fluker MR, Robinson WP. FMR1 repeat sizes in the gray zone and high end of the normal range are associated with premature ovarian failure. *Hum Genet*. 2005;117(4):376–382.
32. Cronister A, Schreiner R, Wittenberger M, Amiri K, Harris K, Hagerman RJ. Heterozygous fragile X female: historical, physical, cognitive, and cytogenetic features. *Am J Med Genet*. 1991;38(2–3):269–274.
33. Hagerman RJ, Leehey M, Heinrichs W, et al. Intention tremor, parkinsonism, and generalized brain atrophy in male carriers of fragile X. *Neurology*. 2001;57(1):127–130.
34. Jacquemont S, Hagerman RJ, Leehey M, et al. Fragile X premutation tremor/ataxia syndrome: molecular, clinical, and neuroimaging correlates. *Am J Hum Genet*. 2003;72(4):869–878.
35. Brunberg JA, Jacquemont S, Hagerman RJ. Fragile X premutation carriers: characteristic MR imaging findings of adult male patients with progressive cerebellar and cognitive dysfunction. *AJNR Am J Neuroradiol*. 2002;23(10):1757–1766.
36. Greco CM, Hagerman RJ, Tassone F, et al. Neuronal intranuclear inclusions in a new cerebellar tremor/ataxia syndrome among fragile X carriers. *Brain*. 2002;125(Pt 8):1760–1771.
37. Kogan CS, Turk J, Hagerman RJ, Cornish KM. Impact of the Fragile X mental retardation 1 (FMR1) gene premutation on neuropsychiatric functioning in adult males without fragile X-associated Tremor/Ataxia syndrome: a controlled study. *Am J Med Genet B Neuropsychiatr Genet*. 2008;147b(6):859–872.
38. Besheer J, Stevenson RA, Hodge CW. mGlu5 receptors are involved in the discriminative stimulus effects of self-administered ethanol in rats. *Eur J Pharmacol*. 2006;551(1–3):71–75.
39. Quintero GC. Role of nucleus accumbens glutamatergic plasticity in drug addiction. *Neuropsychiatr Dis Treat*. 2013;9:1499–1512.
40. Hunsaker MR, Greco CM, Spath MA, et al. Widespread non-central nervous system organ pathology in fragile X premutation carriers with fragile X-associated tremor/ataxia syndrome and CGG knock-in mice. *Acta Neuropathol*. 2011;122(4):467–479.
41. Yrigollen CM, Durbin-Johnson B, Gane L, et al. AGG interruptions within the maternal FMR1 gene reduce the risk of offspring with fragile X syndrome. *Genet Med*. 2012;14(8):729–736.
42. Tassone F. Newborn screening in fragile X syndrome. *JAMA*. 2014; 71(3):355–359.
43. Darnell JC, Klann E. The translation of translational control by FMRP: therapeutic targets for FXS. *Nat Neurosci*. 2013;16(11):1530–1536.
44. Darnell JC, Van Driesche SJ, Zhang C, et al. FMRP stalls ribosomal translocation on mRNAs linked to synaptic function and autism. *Cell*. 2011;146(2):247–261.
45. Bagni C, Oostra BA. Fragile X syndrome: from protein function to therapy. *Am J Med Genet A*. 2013;161a(11):2809–2821.
46. Won H, Mah W, Kim E. Autism spectrum disorder causes, mechanisms, and treatments: focus on neuronal synapses. *Front Mol Neurosci*. 2013;6:19.
47. Loesch DZ, Huggins RM, Hagerman RJ. Phenotypic variation and FMRP levels in fragile X. *Ment Retard Dev Disabil Res Rev*. 2004;10(1):31–41.
48. Schneider A, Seritan A, Tassone F, Rivera SM, Hagerman R, Hessl D. Psychiatric features in high-functioning adult brothers with fragile x spectrum disorders. *Prim Care Companion CNS Disord*. 2013;15(2).
49. Kovacs T, Kelemen O, Keri S. Decreased fragile X mental retardation protein (FMRP) is associated with lower IQ and earlier illness onset in patients with schizophrenia. *Psychiatry Res*. 2013;210(3):690–693.
50. Kelemen O, Kovacs T, Keri S. Contrast, motion, perceptual integration, and neurocognition in schizophrenia: the role of fragile-X related mechanisms. *Prog Neuropsychopharmacol Biol Psychiatry*. 2013;46:92–97.
51. Iwahashi C, Hagerman PJ. Isolation of pathology-associated intranuclear inclusions. *Methods Mol Biol*. 2008;463:181–190.
52. Iwahashi C, Tassone F, Hagerman RJ, et al. A quantitative ELISA assay for the fragile x mental retardation 1 protein. *J Mol Diagn*. 2009;11(4):281–289.

53. Fatemi SH, Kneeland RE, Liesch SB, Folsom TD. Fragile X mental retardation protein levels are decreased in major psychiatric disorders. *Schizophr Res*. 2010;124(1–3):246–247.
54. Fatemi S, Folsom T, Rooney R, Thuras P. mRNA and protein expression for novel GABAA receptors θ and ρ2 are altered in schizophrenia and mood disorders; relevance to FMRP-mGluR5 signaling pathway. *Transl Psychiatry*. 2013;3(6):e271.
55. Qin M, Kang J, Burlin TV, Jiang C, Smith CB. Postadolescent changes in regional cerebral protein synthesis: an *in vivo* study in the FMR1 null mouse. *J Neurosci*. 2005;25(20):5087–5095.
56. D'Hulst C, Heulens I, Brouwer JR, et al. Expression of the GABAergic system in animal models for fragile X syndrome and fragile X associated tremor/ataxia syndrome (FXTAS). *Brain Res*. 2009;1253:176–183.
57. Lozano R, Hare E, Hagerman R. Modulation of the GABAergic pathway for the treatment of fragile X syndrome. *Neuropsychiatr Dis Treat*. under review.
58. Sellier C, Freyermuth F, Tabet R, et al. Sequestration of DROSHA and DGCR8 by expanded CGG RNA repeats alters microRNA processing in fragile X-associated tremor/ataxia syndrome. *Cell Rep*. 2013;3(3):869–880.
59. Hagerman P. Fragile X-associated tremor/ataxia syndrome (FXTAS): pathology and mechanisms. *Acta Neuropathol*. 2013;1–19.
60. Chen Y, Tassone F, Berman RF, et al. Murine hippocampal neurons expressing Fmr1 gene premutations show early developmental deficits and late degeneration. *Hum Mol Genet*. 2010;19(1):196–208.
61. Paul R, Pessah IN, Gane L, et al. Early onset of neurological symptoms in fragile X premutation carriers exposed to neurotoxins. *Neurotoxicology*. 2010;31(4):399–402.
62. Cao Z, Hulsizer S, Cui Y, et al. Enhanced asynchronous Ca(2+) oscillations associated with impaired glutamate transport in cortical astrocytes expressing Fmr1 gene premutation expansion. *J Biol Chem*. 2013;288(19):13831–13841.
63. Wan L, Dockendorff TC, Jongens TA, Dreyfuss G. Characterization of dFMR1, a Drosophila melanogaster homolog of the fragile X mental retardation protein. *Mol Cell Biol*. 2000;20(22):8536–8547.
64. Zhang YQ, Bailey AM, Matthies HJ, et al. Drosophila fragile X-related gene regulates the MAP1B homolog Futsch to control synaptic structure and function. *Cell*. 2001;107(5):591–603.
65. McBride SM, Holloway SL, Jongens TA. Using Drosophila as a tool to identify pharmacological therapies for fragile X syndrome. *Drug Discov Today*. Spring 2013;10(1):e129–e136.
66. Dockendorff TC, Su HS, McBride SMJ. Drosophila lacking dfmr1 activity show defects in circadian output and fail to maintain courtship interest. *Neuron*. 2002;34(6):973–984.
67. Padmashri R, Reiner BC, Suresh A, Spartz E, Dunaevsky A. Altered structural and functional synaptic plasticity with motor skill learning in a mouse model of fragile X syndrome. *J Neurosci*. 2013;33(50):19715–19723.
68. Roy S, Watkins N, Heck D. Comprehensive analysis of ultrasonic vocalizations in a mouse model of fragile X syndrome reveals limited, call type specific deficits. *PLoS One*. 2012;7(9):e44816.
69. Lai JK, Sobala-Drozdowski M, Zhou L, Doering LC, Faure PA, Foster JA. Temporal and spectral differences in the ultrasonic vocalizations of fragile X knock out mice during postnatal development. *Behav Brain Res*. 2014;259:119–130.
70. Kramvis I, Mansvelder HD, Loos M, Meredith R. Hyperactivity, perseveration and increased responding during attentional rule acquisition in the Fragile X mouse model. *Front Behav Neurosci*. 2013;7:172.
71. Musumeci SA, Bosco P, Calabrese G, et al. Audiogenic seizures susceptibility in transgenic mice with fragile X syndrome. *Epilepsia*. 2000;41(1):19–23.
72. Iliff AJ, Renoux AJ, Krans A, Usdin K, Sutton MA, Todd PK. Impaired activity-dependent FMRP translation and enhanced mGluR-dependent LTD in Fragile X premutation mice. *Hum Mol Genet*. 2013;22(6):1180–1192.
73. Berman RF, Murray KD, Arque G, Hunsaker MR, Wenzel HJ. Abnormal dendrite and spine morphology in primary visual cortex in the CGG knock-in mouse model of the fragile X premutation. *Epilepsia*. 2012;53(s1):150–160.
74. Hoffman GE, Le WW, Entezam A, et al. Ovarian abnormalities in a mouse model of fragile X primary ovarian insufficiency. *J Histochem Cytochem*. 2012;60(6):439–456.
75. Van Dam D, Errijgers V, Kooy RF, et al. Cognitive decline, neuromotor and behavioural disturbances in a mouse model for fragile-X-associated tremor/ataxia syndrome (FXTAS). *Behav Brain Res*. 2005;162(2):233–239.
76. Brussino A, Gellera C, Saluto A, et al. FMR1 gene premutation is a frequent genetic cause of late-onset sporadic cerebellar ataxia. *Neurology*. 2005;64(1):145–147.
77. Cellini E, Forleo P, Ginestroni A, et al. Fragile X premutation with atypical symptoms at onset. *Arch Neurol*. 2006;63(8):1135–1138.
78. Hagerman RJ, Berry-Kravis E, Kaufmann WE, et al. Advances in the treatment of fragile X syndrome. *Pediatrics*. 2009;123(1):378–390.
79. Winarni TI, Chonchaiya W, Adams E, et al. Sertraline may improve language developmental trajectory in young children with fragile x syndrome: a retrospective chart review. *Autism Res Treat*. 2012;2012:104317.
80. de Diego-Otero Y, Romero-Zerbo Y, el Bekay R, et al. Alpha-tocopherol protects against oxidative stress in the fragile X knockout mouse: an experimental therapeutic approach for the Fmr1 deficiency. *Neuropsychopharmacology*. 2009;34(4):1011–1026.
81. Hagerman RJ, Rivera SM, Hagerman PJ. The fragile X family of disorders: a model for autism and targeted treatments. *Curr Pediatr Rev*. 2008;4(1):40–52.
82. Wirojanan J, Jacquemont S, Diaz R, et al. The efficacy of melatonin for sleep problems in children with autism, fragile X syndrome, or autism and fragile X syndrome. *J Clin Sleep Med*. 2009;5(2):145.
83. Berry-Kravis E, Sumis A, Hervey C, et al. Open-label treatment trial of lithium to target the underlying defect in fragile X syndrome. *J Dev Behav Pediatr*. 2008;29(4):293–302.
84. Leigh MJ, Nguyen DV, Mu Y, et al. A randomized double-blind, placebo-controlled trial of minocycline in children and adolescents with fragile x syndrome. *J Dev Behav Pediatr*. 2013;34(3):147–155.
85. Dziembowska M, Pretto DI, Janusz A. High MMP-9 activity levels in fragile X syndrome are lowered by minocycline. *Am J Med Genet A*. 2013;161a(8):1897–1903.
86. Berry-Kravis EM, Hessl D, Rathmell B. Effects of STX209 (Arbaclofen) on neurobehavioral function in children and adults with fragile X syndrome: a randomized, controlled, phase 2 trial. *Sci Transl Med*. 2012;4(152):152ra127.

87. Berry-Kravis E, Hessl D, Coffey S, et al. A pilot open label, single dose trial of fenobam in adults with fragile X syndrome. *J Med Genet.* 2009;46(4):266–271.
88. Jacquemont S, Curie A, des Portes V. Epigenetic modification of the FMR1 gene in fragile X syndrome is associated with differential response to the mGluR5 antagonist AFQ056. *Sci Transl Med.* 2011;3(64):64ra61.
89. Schneider A, Leigh MJ, Adams P. Electrocortical changes associated with minocycline treatment in fragile X syndrome. *J Psychopharmacol.* 2013;27(10):956–963.
90. Seritan AL, Nguyen DV, Mu Y. Memantine for fragile X-associated tremor/ataxia syndrome: a randomized, double-blind, placebo-controlled trial. *J Clin Psychiatry.* Dec 10 2013;.
91. Ortigas MC, Bourgeois JA, Schneider A, et al. Improving fragile X-associated tremor/ataxia syndrome symptoms with memantine and venlafaxine. *J Clin Psychopharmacol.* 2010;30(5):642–644.
92. Yang JC, Rodriguez A, Avar M, et al. Memantine effects on attention in FXTAS: a double-blind ERP study [Abstract]. In: *Annual Meeting.* Philadelphia, PA: American Academy of Neurology; 2014.

CHAPTER

18

Autism Spectrum Disorders: Clinical Considerations

Patricia Evans, Sailaja Golla, and Mary Ann Morris
The University of Texas Southwestern Medical Center, Dallas, TX, USA

INTRODUCTION

Autism spectrum disorders (ASD) are a group of neurodevelopmental disabilities that typically first appear in childhood and are characterized by significant difficulties in social, communicative, and behavioral functioning. Although no one single cause has been found, it is suspected that many etiologies have a common final pathway within early neurologic development. Bearing a diagnosis of an ASD typically causes an individual and his/her family difficult financial, emotional, and at times, physical burdens over the individual's lifespan. New burdens are being experienced by state and federal agencies as communities seek ways to medically and financially support the increasing volume of adults with ASD. Consequently, research continues to better understand the complex process of ASD, find disease-specific interventions, and better ways to enable communities to provide resources for families.

OVERVIEW

Identifying individuals with an ASD ideally starts in infancy, and requires two levels of surveillance, each addressing a distinct component of patient management. Routine developmental surveillance and screening specifically for ASD has been recommended to be performed on all children, with subsequent in-depth investigation specific for ASD as needed. Such an approach ensures that all children with risk for any type of atypical development are screened, and within that group, specifically identifying those children with a possible ASD, with appropriate referral. Specific information about the recommended developmental screening and diagnostic tools can be found at the American Academy of Neurology (AAN) website (http://www.aan.com).[1]

The American Academy of Pediatrics as well as the AAN recommend developmental surveillance at all well-child visits from infancy through school age, and at any time afterwards in the context of concerns regarding social acceptance, learning, or behavior.[2–4] A child has failed level one surveillance if he or she fails to meet any of the following specific milestones: no evidence of babbling by 12 months; no gesturing by 12 months; the lack of single words by 16 months; no two-word spontaneous phrases by 24 months; and finally, loss of any previously acquired language or social skills at any age.[3,4] Additionally, siblings of children with an ASD should be carefully monitored for acquisition of social, communication, and play skills, and the occurrence of maladaptive behaviors. Screening should be performed not only for ASD-related symptoms but also for language delays, learning difficulties, social problems, and anxiety or depressive symptoms.

Screening specifically for an ASD should be performed on all children failing routine developmental surveillance procedures using one of the validated instruments. A release for all school records as well as any private, individualized assessment is important to attain pertinent information. A child having failed initial surveillance typically requires assessment by practitioners with particular interest and training in ASD.

Rosenberg's Molecular and Genetic Basis of Neurological and Psychiatric Disease
http://dx.doi.org/10.1016/B978-0-12-410529-4.00018-8

CLINICAL FEATURES AND DIAGNOSTIC EVALUATION

The fifth edition of the Diagnostic and Statistical Manual of Mental Disorders (DSM-5)[2] was released in May 2013. The changes from the previous Diagnostic and Statistical Manual of Mental Disorders, Fourth Edition, Text Revision (DSM-IV-TR)[3] have been as dramatic as they have been controversial: DSM-5 no longer contains the subdiagnoses of Autistic Disorder, Asperger Syndrome, Pervasive Developmental – Not Otherwise Specified, Childhood Disintegrative Disorder, or Rett Disorder. The DSM-IV-TR's autistic symptoms were divided into three areas, specifically, social reciprocity, communicative intent, restricted and repetitive behaviors; while the new DSM-5's diagnostic criteria has been reduced into two areas, specifically, social communication/interaction and restricted/repetitive behaviors. If the diagnostic criteria for ASD are met, then the diagnosis must also specify if it is with or without accompanying intellectual impairment and/or language impairment, as well as noting if the diagnosis is associated with a known medical or genetic condition and/or environmental factor. Additionally, the ASD diagnosis must also be specified by severity levels (Level 1 – Requiring Support, Level 2 – Requiring Substantial Support, and Level 3 – Requiring Very Substantial Support).[2,4] Children who have deficits only in social communication should be considered for the new DSM-5 diagnostic category of Social Communication Disorder (SCD).[4,5]

Clinical Diagnostic Procedures and Methods

Interview Method

The Autism Diagnostic Interview – Revised (ADI-R)[6] is a structured interview conducted with the parent or caretaker of a child and adult with a mental age of at least 2 years old and who has been referred for the evaluation of a possible ASD. The ADI-R is useful for diagnosing ASD, planning treatment, and distinguishing ASD for other developmental disorders.[6–8]

The ADI-R typically takes 1–2 hours and focuses on the child's current behavior or behavior at a certain point in the areas of reciprocal social interaction, communication and language, and patterns of behavior.[7] The interview is divided into five sections: opening questions, communication questions, social development and play questions, repetitive and restricted behavior questions, and questions about general behavior problems. Due to the lengthy interview, the ADI-R is primarily used in clinical or research settings.[8]

Observational Methods

The Autism Diagnostic Observation Schedule, Second Edition (ADOS-2) is a semi-structured set of observations and series of activities involving the referred individual and a trained examiner.[9] This revision improves an instrument already viewed as "the gold standard" for observational assessment of an ASD. With updated protocols, revised algorithms, a new Comparison Score, and a Toddler Module,[10] the ADOS-2 provides a highly accurate picture of current symptoms, unaffected by language. It can be used to evaluate almost anyone suspected of having ASD, as young as a 12 months old with no language to verbally fluent adults.

Like its predecessor, the ADOS,[11] the ADOS-2 is a semi-structured, standardized assessment of communication, social interaction, play, and restricted and repetitive behaviors. It presents various activities that elicit behaviors directly related to a diagnosis of ASD. By observing and coding these behaviors, information can be attained for diagnosis, treatment planning, and educational placement.

In Modules 1 through 4, algorithm scores are compared with cut-off scores to yield one of three classifications: Autism, Autism Spectrum, and Non-Spectrum. The difference between *Autism* and *Autism Spectrum* classifications is one of severity, with the former indicating more pronounced symptoms.[9] In the Toddler Module, algorithms produce "ranges of concern" rather than classification scores.[10]

The Childhood Autism Rating Scale, Second Edition (CARS2) is a widely used autism assessment tool.[12] The revised second edition expands its responsiveness to individuals that are high functioning on the autism spectrum with average and above cognitive abilities, as well as better verbal skills, and more subtle social and behavioral deficits.

The CARS2 includes the Standard Version Rating Booklet, for use with individuals younger than 6 years of age with communication challenges or below-average cognitive abilities, and the High-Functioning Version Rating Booklet, for the assessment of verbally fluent individuals, 6 years of age and older, with cognitive abilities above 80. Rating values for all items are summed to produce a total raw score that is converted to a standard score or percentile rank, as well as guidelines for score interpretation and suggestions for intervention.[12]

Checklist and Rating Scale Methods

The Social Responsiveness Scale, Second Edition (SRS-2) can be completed in approximately 15 minutes and identifies social impairment associated with ASD, as well as quantifying its severity.[13] The SRS-2 has a Preschool Form (for ages 2.5–4.5 years) and the School-Age Form (for ages 4–18 years), which are completed by a parent/guardian or teacher. The Adult (Relative/Other Report) Form is for ages 19 years and up and is completed by a relative or friend and the Adult (Self-Report) Form is completed by the individual and begins at 19 years of age through adulthood.

The SRS-2 generates scores for Social Awareness, Social Cognition, Social Communication, Social Motivation, Restricted Interests, and Repetitive Behavior treatment scales that are useful in determining appropriate treatment and intervention.[13]

The Social Communication Questionnaire (SCQ) is a brief, 35-item, true/false questionnaire, completed by parents or guardians regarding the communication skills and social functioning of children who may be suspected of having ASD.[14] It parallels the ADI-R in content and is used for brief screening to determine the need to conduct a full ADI-R interview and/or the ADOS-2.[7,9] Because the SCQ is brief, quick, and easily completed by a parent or caregiver in less than 10 minutes, it allows clinicians and educators to routinely screen children for ASD, as well as determine the need for comprehensive diagnostic evaluation.

The Gilliam Autism Rating Scale, Third Edition (GARS-3) assists parents/guardians, teachers, and clinicians in identifying ASD in individuals, as well as estimating its severity.[13] The GARS-3 items are based on the DSM-5 diagnostic criteria for ASD.[2] The GARS-3 yields standard scores, percentile ranks, severity levels, and the probability of an ASD. The GARS-3 is normed for ages 3-22 years, and takes approximately 10 minutes to rate. The instrument consists of 56 clearly stated items describing the characteristic behaviors of persons with ASD. The items are grouped into six subscales: Restrictive, Repetitive Behaviors, Social Interaction, Social Communication, Emotional Responses, Cognitive Style, and Maladaptive Speech.[15]

Medical Assessment

A detailed and careful history is imperative, and should include probes into family history with potential risk for consanguinity, postnatal deaths, and perinatal issues.[16] Family prevalence is important: family studies note a 50–100-fold increase in the rate of ASD in first-degree relatives of autistic children.[17] For all referrals of individuals with an ASD, a careful review of psychometric measures is critical, which has been detailed above. Children with ASD have an increased rate of larger head circumferences, often noted over the first few years of life rather than at birth, although only a few will have a true macrocephaly.[1,18,19] There are well-known associations with a range of genetic disorders, including, but not limited to, tuberous sclerosis complex, fragile X syndrome, Angelman Syndrome, and a wide range of DNA genetic variations.[20–25]

The physical exam should include the graphing of height, weight, and head circumference, particularly in young children, as well as simply observing the child's interaction with his family and caregivers during history gathering. In particular, attention should be given to possible dysmorphology, curvature of spine, the shape and size of the head, and the appearance of any birth marks.[1,18,19] The degree to which a child can make and sustain eye contact with family members as well as with the examiners should be specifically described, and adjectives such as fleeting, brief, intermittent, and others, may all be appropriate to the exam. The mental status exam should include a description of the speech patterns, if present; the level of social maturity relative to age; and the capacity to engage with the examiner.[26–28] Attention should be also given to the range and quality of speech; language deficits may range from mutism to verbal fluency, but with other pragmatic and receptive language difficulties.[27,28] The neurologic examination may be quite limited depending upon the anxiety or timidity of the child; however, attempts should be made to assess optic discs, visual fields, strength, tone, at least a general sensory exam, gait, and coordination.[1,19]

Laboratory investigations that are recommended in the context of developmental delay, with or without elements of an ASD, typically include a formal audiology examination as well as lead screening for children who have had an extended oral–motor stage of play, even in the absence of true pica.[4,18] Additional laboratory investigations will be guided by findings on exam; however, genetic testing yields abnormalities in more than 1% of children with an autistic phenotype, including abnormalities in the proximal long arm of chromosome 15q, such as Angelman Syndrome and Prader–Willi syndrome, disorders typically associated with moderate to severe intellectual impairment.[22,23] Less than 5% of all children with an ASD phenotype have inborn errors of metabolism; however, assessing serologies for abnormalities in amino acid, carbohydrate, purine, peptide, and mitochondrial metabolism must be made based on each individual case.[4,18,30] Electrophysiological testing is dictated by the history and physical. The prevalence of epilepsy in autistic children has been estimated at 7%[18] to 14%,[28] whereas the cumulative prevalence by adulthood is estimated at 20–35%.[31,32] Seizure onset peaks in early childhood and again in adolescence. Intellectual disability, with or without motor abnormalities or family history of epilepsy, is a significant risk factor for the development of seizures in individuals with an ASD.[31–33] A higher incidence of epileptiform electroencephalogram (EEG) abnormalities in autistic children whose presentation includes a history of regression has been reported when compared to autistic children without clinical regression.[34,35] Seizures or epileptiform discharges were more prevalent in children with regression who demonstrated cognitive deficits. Regression in cognition and language in adolescence associated with seizure onset has also been observed, but little is known about its cause or prevalence. There may be a causal relationship between a subgroup of children with autistic regression and EEG-defined benign focal epilepsies, although there has not yet been consensus on how to treat.[35] Radiographic images are rarely indicated in children with an autistic phenotype unless there is evidence for focal findings by exam or by EEG. Prevalence of lesions on magnetic resonance imaging (MRI) in children with an ASD is similar to normal control subjects.[1,4,30,36] Computed tomography (CT) and MRI studies of autistic subjects screened to exclude those with disorders other than ASD have confirmed the absence of significant structural brain abnormalities.[36,37] There is no evidence to support a role for functional neuroimaging studies in the clinical diagnosis of an ASD at the present time.[1,4,30]

There is insufficient evidence to support the use of other tests such as hair analysis for trace elements, celiac antibodies, allergy testing for gluten, casein, candida, and other molds, immunologic or neurochemical abnormalities, micronutrients such as vitamin levels, intestinal permeability studies, stool analysis, urinary peptides, mitochondrial disorders, thyroid function tests, or erythrocyte glutathione peroxidase studies.[4]

THERAPEUTIC APPROACHES

Because of the complexity of the ASD condition, reducing the social and emotional discomfort for an individual with an ASD requires a multidisciplinary approach. Such approaches typically include providing supportive environments at home, school, and where applicable, work. Additionally, therapeutics that involve behavior modification and enhancement of effective communication are important. Finally, the use of psychopharmacologic agents can be helpful in reducing troublesome issues of disruptive, aggressive, or anxious behaviors.

Environmental Approaches

After a diagnosis of an ASD has been made, it is optimal to invite all caregivers to participate in a relaxed and supportive conference with the medical team to review specific challenges. Although predictability and consistent routines are often difficult for many 21st century families, such qualities are key to an individual's successful outcome.[38] Evidence continues to reinforce that the earlier and more inclusive the intervention, the more successful individuals are. The individual's environment must include consistency in schedules and physical space as well as supportive relationships between all caregivers.[38–40]

Nonpharmacologic Therapeutic Options

In addition to environmental supports and structure, the addition of formal therapies has been shown to be critically important for many children and teenagers. Although speech and occupational therapies are particularly important for infants and young children, the introduction of appropriate therapies at any point in development, from childhood to the adult years, can yield significant improvement in an individual's independence and capacity for interaction with others. Of these, five are the most commonly used, and include speech and language therapy; occupational therapy; behavior therapy; social skills therapy; and virtual reality therapeutics.

Speech and language therapies include a wide range of techniques and methods, and may include didactic and naturalistic behavioral therapy: verbal behavior, natural language paradigm, pivotal response training, and milieu teaching have been studied.[19,29,38,41] Augmentative and alternative communication modalities, such as American Sign Language, picture and communication boards, and even the child's best gestural commands, can all be effective in building communication skills.[41–44]

Occupational therapy has been the cornerstone of addressing many of the sensory issues found in this population of children. While there are many approaches, the general goals seek to reduce sensory overstimulation and provide improved focus for learning tasks, both cognitively and for visual motor skills.[45–47]

Behavioral and social skills therapies are also important, and can range in goals from better navigating the pragmatics of conversation to simply not being as aggressive when feeling anxious, and include the use of discrete trial teaching (DTT) and applied behavioral analysis (ABA). Certainly the use of practicing scripted dialogue, particularly with improved eye contact, can provide individuals with tools to better understand their social environments.[41,43,44,48]

Finally, individuals with an ASD can be greatly assisted by the emerging field of virtual reality therapeutics. Although often expensive, immersive virtual reality (VR) has the potential to be very helpful in coaching socialization and self-confidence. Some studies have observed an increase in gaze duration, vocalizations, as well as significant reduction in social anxiety. Further research is needed in this very promising field.[49,50]

Pharmacologic Interventions

Patients with ASD typically present with impairments associated with social interaction, communication difficulties, and stereotyped behaviors. These difficulties most often manifest as relative degrees of aggression, self-injurious behaviors, irritability, sleep problems, anxiety, mood disorders, obsessive–compulsive behaviors, inattention, overactivity, and impulsiveness. Such behavioral problems may be responsive to the combination of environmental support, nontherapeutic interventions as reviewed, and adjunctive pharmacological intervention as well. Research into the pharmacotherapy of ASD and related disorders have been ongoing since the middle of the 20th century. But the utility of earlier research is limited by methodological flaws. Behavioral heterogeneity, over interpretation of case studies, and lack of appropriate outcome measures have been some of the major problems.[51] Most current pharmacotherapies target one or more of the several major biochemical pathways, most commonly through the dopaminergic and serotonergic pathways.[52,53] A comprehensive review of psychopharmacology is beyond the scope of this chapter. Readers are referred to the ninth edition of Kaplan and Sadock's *Comprehensive Textbook of Psychiatry* for a detailed discussion of psychopharmacology referenced below.[54]

Neuroleptics (Antipsychotics)

Neuroleptics typically block D2 receptors in the dopaminergic system. In addition to D2 antagonistic effects, neuroleptics also block serotonin receptors, with newer generations of atypical neuroleptics demonstrating greater selectivity of receptors.[54] The use of atypical antipsychotics in children with ASD is based, in part, on their efficacy in the treatment of psychiatric disorders that have behavioral symptoms similar to the core symptoms seen with ASD, as well as their relatively safe adverse effect profile in comparison with older, conventional antipsychotics, and the possible role of dopamine in the etiology of ASD.[55–57]

Typical neuroleptics are less selective than atypical neuroleptics, and because the typical neuroleptics more often block dopamine receptors in the mesocortical pathway, tuberoinfundibular pathway, and the nigrostriatal pathway, typical neuroleptics such as haloperidol provoke frequent side effects, which include sedation, headaches, dizziness, hyperprolactinemia, sexual dysfunction, osteoporosis, weight gain, and orthostasis.[54,58] Neuroleptics also produce anticholinergic effects, including angle-closure glaucoma, constipation, dry mouth, and reduced perspiration. Tardive dyskinesias occur more often for individuals taking high-potency first-generation neuroleptics such as haloperidol, and also appear after chronic, rather than acute, exposure to the drug. Tardive dyskinesias are manifest by repetitive, involuntary, and purposeless movements, most often of the face, lips, legs or torso; they tend to resist treatment and are frequently irreversible.[54] Neuroleptic malignant syndrome is a potentially fatal condition characterized by autonomic instability, which can manifest with tachycardia, nausea, vomiting, diaphoresis, and hyperthermia; mental status alteration; and muscle rigidity, any of which may be associated with laboratory abnormalities, including elevated creatinine kinase, reduced iron plasma levels, and electrolyte derangements.[54] Finally, the use of neuroleptics has also been associated with pancreatitis, pharyngitis, and changes on electrocardiogram, specifically a QT interval prolongation. Seizures have been observed, especially in the context of chlorpromazine and clozapine. Finally, other cardiac complications may include thromboembolism, myocardial infarction, stroke, and torsades de pointes.

Haloperidol, a typical neuroleptic, was one of the first medications studied systematically as a treatment for ASD-related symptoms. The effects of typical neuroleptics, such as haloperidol and chlorpromazine, are primarily mediated through dopamine-receptor (D2) antagonism, and have been shown to be effective in targeting a range of maladaptive behavioral symptoms, withdrawal, and stereotypy, especially in school-age children.[55,56] However, significant adverse effects including dose-related sedative effects and extrapyramidal symptoms such as dystonias and dyskinesias limit the use of haloperidol primarily to treatment-refractory patients.[56] Hence, atypical neuroleptics, with significantly less severe extrapyramidal side effects, have emerged as the first-line drugs in the management of children with ASD.

Atypical neuroleptics, by contrast, are potent serotonin-receptor as well as dopamine-receptor antagonists, and these include aripiprazole, olanzapine, quetiapine, risperidone, and ziprasidone.[54,59] The serotonin receptor antagonism decreases the propensity for tardive dyskinesias and extrapyramidal symptoms. Risperidone is one of the most commonly used medications in this group. In 2003, the US Food and Drug Administration (FDA) approved risperidone for the short-term treatment of the mixed and manic states associated with bipolar disorder. In 2006, the FDA approved risperidone for the treatment of irritability in children and adolescents with autism. The FDA's decision was based, in part, on a study of autistic people with severe and enduring problems of violent meltdowns, aggression, and self-injury. Risperidone is not recommended for autistic people with mild aggression and explosive behavior without an enduring pattern. Weight gain is a major adverse effect of this drug. Hence, it is important to regularly monitor body mass index, lipid panel and blood glucose in patients receiving risperidone.[58–60] Aripiprazole is another commonly used atypical neuroleptic. Weight gain is considered less severe with aripiprazole compared to risperidone, although some of the more recent studies have shown significant weight gain with it.[59,61,62] Ziprasidone is found to be weight neutral, but can potentially cause QT_c prolongation on the electrocardiogram.[63]

Stimulants

Hyperactivity, inattention, and impulsiveness are common traits among individuals with ASD. Stimulants have been shown to decrease these symptoms in a significant proportion of such patients, although these agents can be problematic in this group of patients because of the propensity to exacerbate anxiety, a feature common in ASD.[64,81] Stimulants exert their effects through a number of different pharmacological mechanisms, the most prominent of which include facilitation of norepinephrine, or dopamine, or both; adenosine receptor antagonism; and nicotinic acetylcholine receptor agonism.[65] Commonly used stimulants in the ASD population include amphetamines; methylphenidate; and dopamine and/or norepinephrine reuptake inhibitors.[66]

Amphetamines are a group of phenylethylamine stimulants such as amphetamine, methamphetamine, Dexedrine and levoamphetamine (Adderall). Amphetamines increase the levels of norepinephrine and dopamine in the brain via reuptake inhibition; however, the more important mechanism by which amphetamines cause stimulation is through the direct release of these catecholamines from storage vesicles in cells. Amphetamines are known to cause elevated mood and euphoria as well as rebound depression and anxiety, the latter a particular challenge for many individuals with an ASD.[54,66]

Methylphenidate (MPH; Ritalin, Concerta, Metadate, or Methylin) is a drug approved for treatment of attention-deficit hyperactivity disorder (ADHD), postural orthostatic tachycardia syndrome, and narcolepsy.[65] It belongs to the piperidine class of compounds and increases the levels of dopamine and norepinephrine in the brain through reuptake inhibition of the monoamine transporters. It also increases the release of dopamine and norepinephrine. MPH possesses structural similarities to amphetamine, and, although it is less potent, its pharmacological effects are even more closely related to those of cocaine.[65] A recent meta-analysis reviewed seven randomized-controlled trials looking at the effects of MPH on ADHD symptoms in children with pervasive developmental delay.[64] This review found MPH to be an effective treatment for ADHD symptoms in these children, but noting that weight loss and cardiac toxicity are two of the major side effects associated with its use.

Norepinephrine and/or dopamine reuptake inhibitors (NRIs) also function as stimulants, and inhibit the reuptake of norepinephrine and/or dopamine, resulting in increased extracellular levels and, therefore, enhanced neurotransmission, ultimately producing a stimulant effect. Among individuals with an ASD, the NRI atomoxetine (Strattera) is commonly used in this population of patients. Limited studies have demonstrated its use in ASD as being both effective and safe.[64–66]

Also used is bupropirone (Wellbutrin), which blocks dopamine primarily, and to a much lesser extent, norepinephrine. Both drugs have a considerably lower abuse potential in comparison to other stimulants like the amphetamines.[67,68]

Alpha Adrenergic Agonists

Clonidine and guanfacine are among the most commonly used alpha adrenergic agents in ASD in which inattention and impulsivity prominently feature. The mechanism of action in the treatment of ADHD is to increase noradrenergic tone in the prefrontal cortex (PFC) directly by binding to postsynaptic α_{2A} adrenergic receptors and indirectly by increasing norepinephrine input from the locus coeruleus.[69,70] Alpha 2 agonists have been used as stimulant-treatment extenders in the context of immediate-release methylphenidate or amphetamine, when the goal is to limit both the number of stimulant doses given per day and extend the duration of therapeutic action in combination with a stimulant.[71,72] Also, alpha 2 agonists, with their pronounced sedative side effects, are used to reduce the delay in sleep onset, a problem common in children with ADHD and impulse control problems.[73] In contrast to clonidine, guanfacine is more selective for the alpha 2A receptor,[74] with less alpha 2B activity, making it less sedating than clonidine. The use of these agents as adjunctive sleep aids should be short term at most.[75,76]

Alpha 2 adrenergic agonists are additionally used with a stimulant to enhance the stimulant efficacy. The use of combination of alpha 2 adrenergic agonists with a stimulant is often used in presence of ADHD with comorbid features, specifically oppositional defiant disorder, Giles de Tourette syndrome, and aggressive/impulsive behavior.

The rationale for combination therapy in these cases has been that the primary effects of stimulants and alpha 2 agonists are mediated by different neurotransmitter systems. Functional neuroimaging studies[78] demonstrate that guanfacine produces selective and circumscribed activation of frontal and frontal association areas while at the same time "turning down" or inhibiting striatal activity. Research suggests that stimulants such as methylphenidate improve working memory and regulate attention primarily through alpha 2A receptors to enhance norepinephrine tone in the prefrontal cortex.[79] These effects of methylphenidate on prefrontal cortical function can be prevented by blocking either norepinephrine alpha 2 receptors or dopamine D1 receptors,[79] establishing that both D1 and alpha 2A receptors are essential to the optimum functioning of the prefrontal cortex.

Anxiolytics and Selective Serotonin Reuptake Inhibitors (SSRIs)

Selective serotonin reuptake inhibitors or serotonin-specific reuptake inhibitors (SSRIs) are a class of compounds typically used as antidepressants in the treatment of depression, anxiety disorders, and some personality disorders. SSRIs are believed to increase the extracellular level of the neurotransmitter serotonin by inhibiting its reuptake into the presynaptic cell, increasing the level of serotonin in the synaptic cleft available to bind to the postsynaptic receptor.[80]

Buspirone is a serotonin agonist that is used primarily to treat anxiety disorders but has recently proven to be useful to treat ASD, as these children have alteration in brain serotonin synthesis during critical periods of brain development.[82] Serotonin receptor inhibitors (SRIs) have been beneficial for obsessive–compulsive behaviors and repetitive behaviors with ASD. The most commonly used are fluoxetine, fluoxamine, and clomipramine.[83–92] Further research is needed to better understand mechanisms affected by SSRIs and serotonin agonists during brain development in ASD.

Anticonvulsants

Anticonvulsants such as valproic acid,[93] lamictal,[94,95] carbamazepine and oxcarbamazapine[96] have shown improvement in a range of symptoms in ASD, particularly as mood stabilizers with mixed results. Overall very few studies and limited evidence limits the use of these medications. Some studies demonstrated some improvement in attention span and impulsivity.

Other Medications and Experimental Therapies

AMANTADINE HYDROCHLORIDE

The psychotropic effect of amantadine is related to its antagonism of the *N*-methyl-D-aspartate (NMDA) receptor. By decreasing the toxic effects of the glutamatergic neurotransmitter system, amantadine is thought to ameliorate issues particularly related to ASD. Two randomized controlled trials of amantadine were identified in children and adolescents. One reported beneficial effects in controlling the symptoms of irritability and hyperactivity in ASD and the other described a significant impact in ADHD. Studies in adults, with relevance to children and adolescents, reported effectiveness in resistant depression, obsessive–compulsive disorder and in counteracting side effects of some psychotropic medications. Available data for such use, although promising, require further confirmation.[97]

CHOLINESTERASE INHIBITORS

Cholinesterase inhibitors increase acetylcholine levels in the brain and have been used for a long time in the treatment of Alzheimer disease. Some of the commonly used agents are donepezil,[98] galantamine,[99] and rivastigmine tartrate.[100] Very few studies have been attempted on ASD that have shown some improvement with hyperactivity and irritability.

SECRETIN

Evidence from seven randomized controlled trials supports a lack of effectiveness of secretin for the treatment of ASD symptoms. Secretin has been studied extensively in multiple randomized controlled trials, with clear evidence that it lacks any benefit. Given the high strength of evidence for a lack of effectiveness, secretin as a treatment approach for ASDs warrants no further study.[101] Other therapies like gluten-free diet[102] and hyperbaric oxygen[103] also have not shown any significant beneficial effect for children with ASD.

Treatment and Future Research

The goal of all current therapies in ASD is to relieve the core symptoms, most often anxiety, aggression, impulsivity, and inattention. Many patients with ASD receive some form of pharmacotherapy, much of it off-label, with the primary therapeutic aim being to improve behaviors that interfere with behavioral and educational therapies.[104–106] Because of the difficulties inherent in conducting randomized, placebo-controlled studies in pediatric populations, evidence of the benefits of this strategy is limited, and most such studies have been small and of short duration.[1,30] There remains a great need for research into the long-term benefits and risks of pharmacotherapy in ASD.[1,30]

References

1. Filipek PA, Accardo PJ, Ashwal S, et al. Practice parameter: screening and diagnosis of autism: report of the Quality Standards Subcommittee of the American Academy of Neurology and the Child Neurology Society. *Neurology*. 2000;55(4):468–479. doi:10.1212/wnl.55.4.468.
2. American Psychiatric Association. *Diagnostic and Statistical Manual of Mental Disorders*. 5th ed. Washington, DC: American Psychiatric Association; 2013.
3. American Psychiatric Association. *Diagnostic and Statistical Manual of Mental Disorders - Text Revision*. 4th ed. Washington, DC: American Psychiatric Association; 2000.
4. Filipek PA, Accardo PJ, Ashwal S, et al. Practice parameter: screening and diagnosis of autism: report of the Quality Standards Subcommittee of the American Academy of Neurology and the Child Neurology Society. *Neurology*. 2000;55(4):468–479, PMID:10953176.
5. New DSM-5 includes changes to autism criteria. American Academy of Pediatrics. *AAP News*. Retrieved from www.aapnews.org; June 4, 2013. Accessed 31.10.13.
6. Ozonoff S. Editorial perspective: Autism spectrum disorders in DSM-5 – an historical perspective and the need for change. *J Child Psychol Psychiatry*. 2012;53:1092–1094.
7. Rutter M, LeCouteur A, Lord C. *Autism Diagnostic Interview-Revised (ADI-R)*. Los Angeles, CA: Western Psychological Services; 2003.
8. Kim SH, Lord C. New autism diagnostic interview-revised algorithms for toddlers and young preschoolers from 12 to 47 months of age. *J Autism Dev Disord*. 2012;42(1):82–93. doi:10.1007/s10803-011-1213-1, PMID:21384244.
9. Lord C, Rutter M, DiLavore P, Risi S, Gotham K. *Autism Diagnostic Observation Schedule – Second Edition (ADOS-2): Modules 1 through 4*. Los Angeles, CA: Western Psychological Services; 2012.
10. Lord C, Luyster R, Gotham K, Guthrie W. *Autism Diagnostic Observation Schedule – Second Edition (ADOS-2): Toddler Module*. Los Angeles, CA: Western Psychological Services; 2012.
11. Lord C, Rutter M, DiLavore P, Risi S. *Autism Diagnostic Observation Schedule (ADOS)*. Los Angeles, CA: Western Psychological Services; 1999.
12. Schopler E, Van Bourgondien ME, Wellman GJ, Love SR. *Childhood Autism Rating Scale*. 2nd ed. Los Angeles, CA: Western Psychological Services; 2010.
13. Constantino JM, Gruber CP. *Social Responsiveness Scale* (SRS-2). 2nd ed. Los Angeles, CA: Western Psychological Services; 2012.
14. Rutter M, Bailey A, Lord C. *Social Communication Questionnaire*. Los Angeles, CA: Western Psychological Services; 2003.
15. Gilliam JE. *Gilliam Autism Rating Scale*. 3rd ed. Austin, TX: PRO-ED; 2013.
16. American Academy of Pediatrics. Management of children with autism spectrum disorder. *Pediatrics*. 2007;120(5):1183–1215. Available at http://pediatrics.aappublications.org/content/120/5/1162.full, Accessed 01.03.14.
17. Center for Disease Control. *Autism: Data and Statistics*. Available at http://www.cdc.gov/ncbddd/autism/data.html, Accessed 01.03.14.
18. Dykens EM, Volkmar FR. Medical conditions associated with autism. In: Cohen DJ, Volkmar FR, eds. *Handbook of Autism and Pervasive Developmental Disorders*. 2nd ed. New York, NY: John Wiley & Sons; 1997:388–410.
19. Evans PA, Morris MA. *A Clinical Guide to Autistic Spectrum Disorders*. New York, NY: Lippincott Williams & Wilkins; 2011.
20. Gillberg IC, Gillberg C, Ahlsen G. Autistic behaviour and attention deficits in tuberous sclerosis: a population-based study. *Dev Med Child Neurol*. 1994;36:50–56, PMID:8132114.
21. Piven J, Gayle J, Landa R, Wzorek M, Folstein S. The prevalence of Fragile X in a sample of autistic individuals diagnosed using a standardized interview. *J Am Acad Child Adolesc Psychiatry*. 1991;30:825–830, PMID:1938801.
22. Cook Jr. EH, Lindgren V, Leventhal BL, et al. Autism or atypical autism in maternally but not paternally derived proximal 15q duplication. *Am J Hum Genet*. 1997;60:928–934, PMID:9106540.
23. Schroer RJ, Phelan MC, Michaelis RC, et al. Autism and maternally derived aberrations of chromosome 15q. *Am J Med Genet*. 1998;76: 327–336, PMID:9545097.
24. Weidmer–Mikhail E, Sheldon S, Ghaziuddin M. Chromosomes in autism and related pervasive developmental disorders: a cytogenetic study. *J Intellect Disabil Res*. 1998;42:8–12, PMID:9534109.
25. Steffenburg S, Gillberg CL, Steffenburg U, Kyllerman M. Autism in Angelman syndrome: a population-based study. *Pediatr Neurol*. 1996;14:131–136 PMID:8703225.
26. Stone WL, Ousley OY, Yoder PJ, Hogan KL, Hepburn SL. Nonverbal communication in two-and three-year-old children with autism. *J Autism Dev Disord*. 1997;27:677–696, PMID:9455728.
27. Rapin I. *Preschool Children with Inadequate Communication: Developmental Language Disorder, ASD, Low IQ*. London, UK: MacKeith Press; 1996.
28. Tuchman RF, Rapin I, Shinnar S. Autistic and dysphasic children. II Epilepsy. *Pediatrics*. 1991;88:1219–1225, PMID:1956740.
29. Greenspan SI, Wieger S. *Engaging AutismASD: The Floortime Approach to Helping Children Relate, Communicate, and Think*. Reading, MA: Perseus Book; 2005.
30. Myers S, Johnson CPAmerican Academy of Pediatrics Council on Children with Disabilities. Management of children with autism spectrum disorders. *Pediatrics*. 2007;120(5):1162–1182, PMID:17967921.
31. Rossi PG, Parmeggiani A, Bach V, Santucci M, Visconti P. EEG features and epilepsy in patients with autism. *Brain Dev*. 1995;17:169–174 PMID:7573755.
32. Volkmar FR, Nelson DS. Seizure disorders in autism. *J Am Acad Child Adolesc Psychiatry*. 1990;29:127–129, PMID:2295565.
33. Wong V. Epilepsy in children with autistic spectrum disorder. *J Child Neurol*. 1993;8:316–322, PMID:7693796.
34. Tuchman RF, Rapin I. Regression in pervasive developmental disorders: seizures and epileptiform electroencephalogram correlates. *Pediatrics*. 1997;99:560–566, PMID:9093299.
35. Nass R, Gross A, Devinsky O. Autism and autistic epileptiform regression with occipital spikes. *Dev Med Child Neurol*. 1998;40:453–458, PMID:9698058.
36. Cauda F, Costa T, Palermo S, et al. Concordance of white matter and gray matter abnormalities in autism spectrum disorders: a voxel-based meta-analysis study. *Hum Brain Mapp*. 2013. doi:10.1002/hbm.22313 PMID:23894001.

37. Filipek PA. Neuroimaging in the developmental disorders: the state of the science. *J Child Psychol Psychiatry*. 1999;40:113–128, PMID:10102728.
38. National Research Council, committee on Education interventions for Children with Autism. In: Lord C, McGee JP, eds. *Educating Children with Autism*. Washington, DC: National Academies Press; 2001.
39. Goldstein H. Communication intervention for children with autism: a review of treatment efficacy. *J Autism Dev Disord*. 2002;32:373–396, PMID:12463516.
40. Ventola PE, Oosting D, Anderson LC, Pelphrey KA. Brain mechanisms of plasticity in response to treatments for core deficits in autism. *Prog Brain Res*. 2013;207:255–272. doi:10.1016/B978-0-444-63327-9.00007-2.
41. Paul R, Sutherland D. Enhancing early language in children with autism spectrum disorders. In: Volkmar FR, Paul R, Klin A, Cohen D, eds. *Handbook of Autism and Pervasive Developmental Disorders*; Vol II 3rd ed. Hoboken, NJ: John Wiley & Sons; 2005:946–976.
42. Solomon R, Necheles J, Ferch C, Ruckman D. Pilot study of a parent training program for young children with autism. *Autism*. 2007;11(3):205–224, PMID:17478575.
43. American Speech-Language-Hearing Association, *Ad Hoc* Committee on Autism Spectrum Disorders. *Principles for Speech-Language Pathologists in Diagnosis, Assessment, and Treatment of Autism Spectrum Disorders Across the Life Span*. Available at http://www.asha.org/public/speech/disorders/Autism.htm; 2005. Accessed 01.03.14.
44. DeThorn LS, Johnson CJ, Walder L, Mahurin-Smith J. When "Simon Says" doesn't work: alternatives to imitation for facilitating early speech development. *Am J Speech Lang Pathol*. 2009;18(2):133–145, PMID:18930909.
45. Petrus C, Adamson SR, Block L, Einarson SJ, Sharifenjad M, Harris SR. Effects of exercise interventions on stereotypic behaviors in children with autism spectrum disorder. *Physiother Can*. 2008;60(2):134–145, PMID:20145777.
46. Chen YW, Cordier R, Brown NA. Preliminary study on the reliability and validity of using experience sampling method in children with autism spectrum disorders. *Dev Neurorehabil*. 2013 Dec 4, [Epub ahead of print] PMID:24304202.
47. Wauang Y, Wang C, Huang M, Su CY. The effectiveness of simulated developmental horse-riding. *Adopt Phys Active Q*. 2010;27(2):113–126, PMID:20440.
48. Weiss JA, Viecili MA, Sloman L, Lunsky Y. Direct and indirect psychosocial outcomes for children with autism spectrum disorder and their parents following a parent-involved social skills group intervention. *J Can Acad Child Adolesc Psychiatry*. 2013;22(4):303–309, PMID:24223050.
49. Wang M, Reid D. Using the virtual reality-cognitive rehabilitation approach to improve contextual processing in children with autism. *Scientific World Journal*. 2013;2013(Nov 13):716890. doi:10.1155/2013/716890.
50. Mineo BA, Ziegler W, Gill S, Salkin D. Engagement with electronic screen media among students with autism spectrum disorders. *J Autism Dev Disord*. 2009;39(1):172–187, PMID:18626761.
51. Volkmar FR. Pharmacological interventions in autism: theoretical and practical issues. *J Clin Child Psychol*. 2001;30(1):80–87, PMID:11294081.
52. Aman MG, Farmer CA, Hollway J, Arnold LE. Treatment of inattention, overactivity, and impulsiveness in autism spectrum disorders. *Child Adolesc Psychiatr Clin N Am*. 2008;17(4):713–738, vii. PMID: 18775366.
53. Stigler KA, McDougle CJ. Pharmacotherapy of irritability in pervasive developmental disorders. *Child Adolesc Psychiatr Clin N Am*. 2008;17(4):739–752, vii–viii. PMID:18775367.
54. Sadock BJ, Sadock VA, Ruiz P. *Biological Therapies. Kaplan and Sadock's Comprehensive Textbook of Psychiatry*. 9th ed. New York, NY: Lippincott, Williams & Wilkins; 2009, SBN/ISSN: 9780781768993.
55. Campbell M, Anderson LT, Meier M. A comparison of haloperidol, behavior therapy, and their interaction in autistic children [proceedings]. *Psychopharmacol Bull*. 1979 Apr;15(2):84–86, PMID:373009.
56. Cohen IL, Campbell M, Posner D, Small AM, Triebel D, Anderson LT. Behavioral effects of haloperidol in young autistic children. An objective analysis using a within-subjects reversal design. *J Am Acad Child Psychiatry*. 1980;19(4):665–677, PMID:7204797.
57. Campbell M, Armenteros JL, Malone RP, et al. Neuroleptic-related dyskinesias in autistic children: a prospective, longitudinal study. *J Am Acad Child Adolesc Psychiatry*. 1997;36(6):835–843.
58. Pierre JM. Extrapyramidal symptoms with atypical antipsychotics: incidence, prevention and management. *Drug Saf*. 2005;28(3):191–208, PMID:15733025.
59. Ho J, Panagiotopoulos C, McCrindle B, Grisaru S, Pringsheim T. Management recommendations for metabolic complications associated with second-generation antipsychotic use in children and youth. *Paediatr Child Health*. 2011;16(9):575–580, PMCID: PMC3223901.
60. Scahill L. How do I decide whether or not to use medication for my child with autism? Should I try behavior therapy first? *J Autism Dev Disord*. 2008;38(6):1197–1198. doi:10.1007/s10803-008-0573-7, PMID 18463973.
61. Stigler KA, Posey DJ, McDougle CJ. Aripiprazole for maladaptive behavior in pervasive developmental disorders. *J Child Adolesc Psychopharmacol*. 2004;14(3):455–463, PMID:15650503.
62. Valicenti-McDermott MR, Demb H. Clinical effects and adverse reactions of off-label use of aripiprazole in children and adolescents with developmental disabilities. *J Child Adolesc Psychopharmacol*. 2006;16(5):549–560, PMID: 17069544.
63. Malone RP, Delaney MA, Hyman SB, Cater JR. Ziprasidone in adolescents with autism: an open-label pilot study. *J Child Adolesc Psychopharmacol*. 2007;17(6):779–790, PMID:18315450.
64. Reichow B, Volkmar FR, Bloch MH. Systematic review and meta-analysis of pharmacological treatment of the symptoms of attention-deficit/hyperactivity disorder in children with pervasive developmental disorders. *J Autism Dev Disord*. 2013;43(10):2435–2441, PMID:23468071.
65. Riddle EL, Fleckenstein AE, Hanson GR. Role of monoamine transporters in mediating psychostimulant effects. *AAPS J*. 2005;7(4):E847–E851. doi:10.1208/aapsj070481, PMC 2750953. PMID 16594636.
66. Gibbs TT. Pharmacological treatment of autism. In: Blatt GJ, Gibbs TT, eds. *The Neurochemical Basis of Autism*. 2010:245–267. doi:10.1007/978-1-4419-1272-5_15.
67. Arnold LE, Aman MG, Cook AM, et al. Atomoxetine for hyperactivity in autism spectrum disorders: placebo-controlled crossover pilot trial. *J Am Acad Child Adolesc Psychiatry*. 2006;45(10):1196–1205, PMID:17003665.

68. Lile JA, Nader MA. The abuse liability and therapeutic potential of drugs evaluated for cocaine addiction as predicted by animal models. *Curr Neuropharmacol*. 2003;1:21–46. doi:10.2174/1570159033360566.
69. Jaselskis CA, Cook Jr. EH, Fletcher KE, Leventhal BL. Clonidine treatment of hyperactive and impulsive children with autistic disorder. *J Clin Psychopharmacol*. 1992;12(5):322–327, PMID: 1479049.
70. Posey DJ, Posey DJ, Puntney JI, Sasher TM, Kem DL, McDougle CJ. Guanfacine treatment of hyperactivity and inattention in pervasive developmental disorders: a retrospective analysis of 80 cases. *J Child Adolesc Psychopharmacol*. 2004;14(2):233–241, PMID:15319020.
71. Pliszka SR, Crismon ML, Hughes CW, et al., Texas Consensus Conference Panel on Pharmacotherapy of Childhood Attention Deficit Hyperactivity Disorder. The Texas Children's Medication Algorithm Project: revision of the algorithm for pharmacotherapy of attention-deficit/hyperactivity disorder. *J Am Acad Child Adolesc Psychiatry*. 2006;45:642–657, PMID:16721314.
72. Scahill L, Chappell PB, Kim YS, et al. A placebo-controlled study of guanfacine in the treatment of children with tic disorders and attention deficit hyperactivity disorder. *Am J Psychiatry*. 2001;158:1067–1074, PMID:11431228.
73. Wilens TE, Biederman J, Spencer T. Clonidine for sleep disturbances associated with attention-deficit hyperactivity disorder. *J Am Acad Child Adolesc Psychiatry*. 1994;33:424–426, PMID:8169189.
74. Uhlen S, Wikberg JE. Delineation of rat kidney alpha 2A- and alpha 2B-adrenoceptors with [3H] RX821002 radioligand binding: computer modeling reveals that guanfacine is an alpha 2A-selective compound. *Eur J Pharmacol*. 1991;202:235–243, PMID:1666366.
75. Kugler J, Seus R, Krauskopf R, Brecht HM, Raschig A. Differences in psychic performance with guanfacine and clonidine in normotensive subjects. *Br J Clin Pharmacol*. 1980;10(suppl 1):71S–80S, PMID:6994783.
76. Daviss WB, Patel NC, Robb AS, et al., the CAT STUDY TEAM. Clonidine for attention-deficit/hyperactivity disorder: II. ECG changes and adverse events analysis. *J Am Acad Child Adolesc Psychiatry*. 2008;47:189–198.
77. Biederman J, Melmed RD, Patel A, et al., for the SPD503 Study Group. A randomized, double-blind, placebo-controlled study of guanfacine extended release in children and adolescents with attention-deficit/hyperactivity disorder. *Pediatrics*. 2008;121:e73–e84.
78. Easton N, Shah YB, Marshall FH, Fone KC, Marsden CA. Guanfacine produces differential effects in frontal cortex compared with striatum: assessed by phMRI BOLD contrast. *Psychopharmacology (Berl)*. 2006;189:369–385, PMID:17016709.
79. Arnsten AFT, Dudley AG. Methylphenidate improves prefrontal cortical cognitive function through alpha2 adrenoceptor and dopamine D1 receptor actions: relevance to therapeutic effects in attention deficit hyperactivity disorder. *Behav Brain Funct*. 2005;1:2, PMID: 15916700.
80. Ming X, Gordon E, Kang N, Wagner GC. Use of clonidine in children with autism spectrum disorders. *Brain Dev*. 2008;30:454–460, PMID:18280681.
81. Reichow B, Volkmar FR, Bloch MH. Systematic review and meta-analysis of pharmacological treatment of the symptoms of attention-deficit/hyperactivity disorder in children with pervasive developmental disorders. *J Autism Dev Disord*. 2013;43(10):2435–2441, PMID:23468071.
82. Edwards DJ, Chugani DC, Chugani HT, Chehab J, Malian M, Aranda JV. Pharmacokinetics of buspirone in autistic children. *J Clin Pharmacol*. 2006;46(5):508–514, PMID:16638734.
83. Soorya L, Kiarashi J, Hollander E. Psychopharmacologic interventions for repetitive behaviors in autism spectrum disorders. *Child Adolesc Psychiatr Clin N Am*. 2008;17(4):753–771. doi:10.1016/j.chc.2008.06.003 viii.
84. Oswald DP, Sonenklar NA. Medication use among children with autism spectrum disorders. *J Child Adolesc Psychopharmacol*. Jun 2007;17(3):348–355, PMID:17630868.
85. McPheeters ML, Warren Z, Sathe N, et al. A systematic review of medical treatments for children with autism spectrum disorders. *Pediatrics*. 2011;127(5):e1312–e1321, PMID:21464191.
86. Blier P. Pharmacology of rapid-onset antidepressant treatment strategies. *J Clin Psychiatry*. 2001;62(suppl 15):12–17.
87. Hollander E, Phillips A, Chaplin W, et al. A placebo controlled crossover trial of liquid fluoxetine on repetitive behaviors in childhood and adolescent autism. *Neuropsychopharmacology*. 2005;30(3):582–589, PMID: 15602505.
88. Owley T, Walton L, Salt J, et al. An open-label trial of escitalopram in pervasive developmental disorders. *J Am Acad Child Adolesc Psychiatry*. 2005;44(4):343–348, PMID:15782081.
89. Namerow LB, Thomas P, Bostic JQ, Prince J, Monuteaux MC. Use of citalopram in pervasive developmental disorders. *J Dev Behav Pediatr*. 2003;24(2):104–108, PMID:12692455.
90. Couturier JL, Nicolson R. A retrospective assessment of citalopram in children and adolescents with pervasive developmental disorders. *J Child Adolesc Psychopharmacol*. Fall 2002;12(3):243–248, PMID:12427298.
91. Boyce RD, Handler SM, Karp JF, Hanlon JT. Age-related changes in antidepressant pharmacokinetics and potential drug-drug interactions: a comparison of evidence-based literature and package insert information. *Am J Geriatr Pharmacother*. 2012;10(2):139–150, PMCID: PMC3384538.
92. US Food and Drug Administration. *Celexa (citalopram hydrobromide): Drug Safety Communication – Abnormal Heart Rhythms Associated with High Doses*. Available at http://www.fda.gov/Safety/MedWatch/SafetyInformation/SafetyAlertsforHumanMedicalProducts/ucm269481.htm, Accessed 24.10.12.
93. Hollander E, et al. An open trial of divalproex sodium in autism spectrum disorders. *J Clin Psychiatry*. 2001;62(7):530–534, PMID: 11488363.
94. Rugino TA, Samsock TC. Levetiracetam in autistic children: an open-label study. *J Dev Behav Pediatr*. 2002;23(4):225–230, PMID:12177568.
95. Douglas JF, Sanders KB, Benneyworth MH, et al. Brief report: retrospective case series of oxcarbazepine for irritability/agitation symptoms in autism spectrum disorder. *J Autism Dev Disord*. 2013;43(5):1243–1247. doi:10.1007/s10803-012-1661-2.
96. Uvebrant P, Bauziene R. Intractable epilepsy in children. The efficacy of lamotrigine treatment, including non-seizure-related benefits. *Neuropediatrics*. 1994;25(6):284–289.
97. Hosenbocus S, Chahal R. Amantadine: a review of use in child and adolescent psychiatry. *J Can Acad Child Adolesc Psychiatry*. 2013;22(1):55–60, PMID:23390434.
98. Hardan AY, Handen BL. A retrospective open trial of adjunctive donepezil in children and adolescents with autistic disorder. *J Child Adolesc Psychopharmacol*. 2002;12(3):237–241, PMID:12427297.
99. Nicolson R, Craven-Thuss B, Smith J. A prospective, open-label trial of galantamine in autistic disorder. *J Child Adolesc Psychopharmacol*. 2006;16(5):621–629, PMID:17069550.

100. Chez MG, et al. Treating autistic spectrum disorders in children: utility of the cholinesterase inhibitor rivastigmine tartrate. *J Child Neurol.* 2004;19(3):165–169, PMID: 15119476.
101. Krishnaswami S, McPheeters ML, Veenstra-Vanderweele J. A systematic review of secretin for children with autism spectrum disorders. *Pediatrics.* 2011;127(5):e1322–e1325. doi:10.1542/peds.2011-0428.
102. Dosman C, et al. Complementary, holistic, and integrative medicine: autism spectrum disorder and gluten- and casein-free diet. *Pediatr Rev.* 2013;34(10):e36–e41, PMID: 24085796.
103. Rossignol DA, Rossignol LW, James SJ, Melnyk S, Mumper E. The effects of hyperbaric oxygen therapy on oxidative stress, inflammation, and symptoms in children with autism: an open-label pilot study. *BMC Pediatr.* 2007;7(Nov 16):36, PMID:18005455.
104. Stahl SM. *Stahl's Essential Psychopharmacology: Neuroscientific Basis and Practical Applications.* Cambridge University Press; 2008.
105. Palermo M, Curatolo P. Pharmacologic treatment of autism. *J Child Neurol.* 2004;19:155–164, PMID:15119475.
106. Masi G. Pharmacotherapy of pervasive developmental disorders in children and adolescents. *CNS Drugs.* 2004;18:1031–1052, PMID: 15584771.

CHAPTER

19

Metabolic and Genetic Causes of Autism

Sailaja Golla and Patricia Evans

The University of Texas Southwestern Medical Center, Dallas, TX, USA

INTRODUCTION

Most children with an autism diagnosis rarely have an etiology that can be identified as the cause for their behavior. However, there are well-described syndromes that are clearly associated with the development of the autistic phenotype. For children in whom no clear etiology can be found, the term idiopathic or primary autism is applicable. By contrast, those individuals whose autism is clearly a result of genetic or environmental causes are said to have a secondary autism.[1–3] The rate of associated cognitive deficit in autism has dropped dramatically from approximately 90% before 1990 to less than 50%, due in large part to the improved psychometric measurement of intellectual capacity in individuals with an autistic spectrum disorder (ASD).[4] Subsequently, the comorbid presentation of significantly impaired intellectual capacity in the presence of an autistic spectrum behavior, particularly with associated dysmorphisms on physical examination, is more often consistent with an identifiable disorder.[5–8] Complicating the diagnostic process is the highly variable phenotype seen in ASD. Epidemiology points to multiple genetic lines as causative and potentially causative for this widely varying condition, both for symptoms and severity.[9,10] In this chapter, we provide a table which notes references specific to genomic regions most often associated with the ASD phenotype, followed by a brief discussion of several of the more well known genetic causes for ASD.

The American College of Medical Genetics and Genomics reports that the rate of success for identifying a specific etiologic diagnosis in persons with ASDs remains modest, ranging from 6 to 15%.[9,11,12] However, advances in clinical testing technology have increased the diagnostic yield to as high as 30–40%.[9] Chromosomal abnormalities have consistently been reported in persons with ASDs. However, it is the advent of molecular cytogenetic testing modalities such as chromosomal microarray (CMA) that has dramatically improved the diagnostic power of genetic evaluations. This progress has been so effective that it has largely replaced conventional cytogenetics as a first-tier test.[13,14] Currently two methods of CMA are used in the clinical setting. These two methods, array-comparative genomic hybridization and single-nucleotide polymorphism arrays, use different techniques to scan the genome for copy-number variants (CNVs). With the increased number of CMA studies, new information has emerged regarding the contribution of genome CNVs in ASDs. Estimates of CNV frequencies in unselected populations of individuals with an ASD currently range from 8 to 231%.[15–21] Table 19.1 provides an overview of genetic regions and candidate genes which represent either known or suspected loci associated with ASDs.

FRAGILE X SYNDROME (FRX)

FXS is regarded as the most common cause of genetically inherited ASD, as well as cognitive delay in boys.[91] The mutations in the Fragile X mental retardation 1 gene (FMR1), located at Xp23.7, causes a variety of disorders, depending on the length of the cytosine–guanine–guanine (CGG) repetitive sequence. The normal range has 5–44 CGG repeats with a mean of 30. Carriers who are unaffected intellectually typically will have 55–200 repeats. However, carriers may produce an excessive amount of FMR1 messenger RNA that leads to serious illness in later life.[92] This includes the fragile X-associated tremor ataxia syndrome in approximately 40% of men and 8% of women.[93–96] The full mutation is associated with greater than 200 CGG repeats. Physical features for FXS include: long face; prominent

http://dx.doi.org/10.1016/B978-0-12-410529-4.00019-X

TABLE 19.1 Brief Review of Known and Possible Loci Associated With Autism Spectrum Disorders

Region	Microarray	Candidate genes	Common Associations	References
1p	1p36.13	MECP2	Rett syndrome, metabolic abnormalities, neuroblastoma	(Artuso et al., 2011)[22]
1q	1q21.1	HYDIN	Autism, schizophrenia, tetralogy of Fallot	(Goodbourn et al., 2014; Dolcetti et al., 2013)[23,24]
	1q42.2	DISC1, DISC2, TSNAX MARK 1	Mental retardation, schizophrenia, bipolar disorder, autism and Asperger syndrome	(Crepel et al., 2010; Williams et al., 2009)[25,26]
2p	2p16.3	NRXN1	Autism, schizophrenia	(Walker et al., 2013; Van Winkel et al., 2010) [27,28]
2q	2q34	ERB4 NRG1	Intellectual disability, schizophrenia, epilepsy, developmental delay, urea cycle defects	(Kasnauskiene et al., 2013; Rashidi-Nezhad et al., 2012; Loscalzo et al., 2004)[29–31]
	2q37	HDAC4, PRLH, PER2, TWIST2, CAPN10, KIF1A, FARP2, D2HGDH and PDCD1	Autism, overweight, behavioral problems, brachydactyly, seizures	(Leroy et al., 2013)[32]
3p	3p26.3	CHL1	Nonsyndromic intellectual disability	(Shoukier et al., 2013)[33]
3q	3q24	DLA1, DLAR 1	Autism spectrum, X-linked mental retardation	(Aziz et al., 2011)[34]
	3q29	PAK2, DLG1, DLG9, SLC9A9, FBX045	Intellectual disability (ID), schizophrenia, autism, bipolar disorder, depression, mild facial morphological anomalies/congenital malformations	(Città et al., 2013; Quintero-Rivera et al., 2010)[35,36]
4p	4p16.3	WHSC1, LETM1	Wolf–Hirschhorn syndrome (WHS), facial dysmorphism, mental retardation	(Debost-Legrand et al., 2013; Andersen et al., 2014)[37,38]
	4p12	GABRG1	Autism, dyspraxia, developmental delay, bipolar disorder, ADHD	(Polan et al., 2014; Vincent et al., 2006)[39,40]
4q	4q35.2	TRIML1	Autism, mental retardation	(Wang et al., 2010)[41]
5p	5p14.1	MSNP1	Autism	(Kerin et al., 2012; Ma et al., 2009)[42,43]
5q	5q35.3	DRD1, NSD1	Megalencephaly, perisylvian polymicrogyria, polydactyly and hydrocephalus, Sotos syndrome	(Verkerk et al., 2010)[44]
7q	7q11.23	ELN	Williams–Beuren syndrome	(Cuscó et al., 2008)[45]
	7q31.1	IMMP2L1, WNT2	Tourettes syndrome, retinitis pigmentosa	(Patel et al., 2011; Bowne et al., 2002; Matsuzaki et al., 2012)[46–48]
8p	8p23	CLN8, GATA-4, CGAT, TNKS, MCPH1	Neuronal ceroid lipofuscinosis, cognitive deficit, craniofacial dysmorphism, epilepsy, panic attacks and congenital heart disease, behavioral problems, autism spectrum disorder	(Allen et al., 2012; Aguiar et al., 2013; Ozgen et al., 2009)[49–51]
	8q12.2	CA8, RAB2, RLBP1L1 and CHD7	Duane retraction syndrome, developmental delay	(Amouroux et al., 2012)[52]
9q	9q34.13	COQ4	Coenzyme Q10 deficiency, mental retardation, encephalomyopathy, dysmorphic features	(Salviati et al., 2012)[53]
	9q34.3	EHMT1	Autism spectrum, Kleefstra syndrome, speech delay, neurodevelopmental impairment	(Yatsenko et al., 2012)[54]
10q	10q22.3		Autism, childhood schizophrenia, epilepsy, intellectual disability	(Ahn et al., 2014)[55]
	10q23.31	PTEN	Cowden syndrome, megalencephaly, autism, colorectal cancer	(Eng, 2003)[56]
11p	11p14.1	BDNF, LIN7C	ADHD, autism, neurobehavioral problems, developmental delay, obesity	(Shinawi et al., 2011)[57]

TABLE 19.1 Brief Review of Known and Possible Loci Associated With Autism Spectrum Disorders—cont'd

Region	Microarray	Candidate genes	Common Associations	References
	11p13	BDNF	Lower adaptive behavior, reduced cognitive functioning, WAGR (Wilms tumor, aniridia, genitourinary anomalies, and mental retardation) syndrome	(Han et al., 2013)[58]
12p	12p13.33	ELKS, ERC1	Childhood apraxia of speech	(Thevenon et al., 2013)[59]
13q	13q13.2	Neurobeachin	Idiopathic autism	(Castermans et al., 2003)[60]
15q	15q11 to 15 q 13	UBE3A, GABR B3	Autism, intellectual disability, schizophrenia, epilepsy, speech delay	(Martins-Taylor et al., 2014; Horváth et al., 2013[61,62]
	15q24	*LMANL1*, CHL1, FGFBP3, POUF41	Autism, dysmorphic craniofacial features, intellectual disability, hypotonia, digital and genital abnormalities	(McInnes et al., 2010; Salyakina et al., 2011)[63,64]
16p	16p13.3	TSC2, CREBBP	Autism, tuberous sclerosis, Rubinstein–Taybi syndrome	(Sorte et al., 2013; Mozaffari et al., 2009; Schorry et al., 2008)[65–67]
	16p13.11	NDE1, MYH11, ABCC1, ABCC6	Autism, ADHD, intellectual disability and schizophrenia	(Tropeano et al., 2013)[68]
	16p11.2	PRKCB1	Autism spectrum, scoliosis, vertebral anomalies	(Shishido et al., 2014; Al-Kateb et al., 2014)[69,70]
	16q24.3	C16orf95, ZCCHC14, MAP1LC3B and FBXO31	Autism spectrum disorder, intellectual disability and congenital renal malformation	(Handrigan et al., 2013)[71]
	17p11.2	RAI1	Infantile hypotonia, failure to thrive, cardiovascular malformations, developmental delay, intellectual disability, behavior abnormalities, Smith–Magenis association, Potocki–Lupski syndrome, autism	(Magoulas et al., 2014; Flax et al., 2010)[72,73]
	17q11.2	NF1	Autism, motor speech disorder, neurofibromatosis 1, cognitive delay, attention problems, and inadequate social skills	(Klein-Tasman et al., 2014; Brison et al., 2013)[74,75]
	17q21.31	PARK2	Autistic features, behavioral problems, mild mental retardation, poor social skills, developmental delay, dysmorphic features, hypotonia, seizures	(Grisart et al., 2009; Cooper et al., 2011)[76,77]
	20p12.1	C20orf133	Kabuki syndrome, mental retardation	(Maas et al., 2009)[78]
	Trisomy 21	PSEN1	Down syndrome, autism, cognitive impairment, Alzheimer disease	(Valenti et al., 2014; Anderson et al., 2013)[79,80]
	22q11.21	HIRA, TBX1	Schizophrenia, autism spectrum disorder, intellectual disability, speech delay, developmental delay	(Crespi et al., 2012)[81]
	22q13.1	ADSL	Autism spectrum disorder	(Sorte et al., 2013)[82]
	22q13.33	SHANK3	Phelan–McDermid syndrome, developmental delay, autism	(Betancur et al., 2013)[83]
	Xp22.31	PNPLA4	Ichthyosis, epilepsy, hyperactivity, autism and mental retardation	(Carrascosa-Romero et al., 2012)[84]
	Xp22.13	CDKL5	Rett-like phenotype, mental retardation, epilepsy, Angelman-like phenotype	(Castrén et al., 2011)[85]
	Xp21.3	IL1RAPL1	Intellectual disability, dysmorphic features	(Youngs et al., 2012)[86]
	Xp11.22	IQSEC2, MECP2, FOXG1, CDKL5, MEF2C	Intellectual disability, seizures, plagiocephaly, developmental delay, hypotonia	(Gandomi et al., 2013)[87]
	Xq24	UBE2A, SLC25A43 SLC25A5	Syndromic and nonsyndromic intellectual disability	(Vandewalle et al., 2013)[88]
	Xq27.3	FMR-1	Fragile X syndrome, autism, mental retardation	(Sabaratnam et al., 2003)[89]
	Xq28	MECP2	Intellectual disability, autism, epilepsy, hypotonia	(Scott Schwoerer et al., 2014)[90]

Abbreviation: ADHD, Attention deficit hyperactivity disorder.

and long ear pinnae; high-arched palate; prolapsed mitral valve, dilated aortic arch; flat feet; hyperextensible finger joints in childhood; macroorchidism (testicle volume >30 ml bilaterally in adulthood); soft velvet-like skin.[97,98] The full mutation occurs in 1 in 2500 alleles and the permutation occurs in 1 per 130–250 females and 1 per 250–810 males.[28] Therefore, genetic testing should be carried out in all children and adults with autism, ASD, and intellectual disability of unknown etiology. Once an individual is identified with an FMR1 mutation, a careful family history will often reveal many other family members who are either carriers or significantly affected by FXS. Genetic testing and counseling are therefore recommended for the extended family. Being vigilant when an individual presents with an autistic behavioral phenotype to the possibility of a comorbid diagnosis of fragile X is important, because up to 50% of individuals with genetically confirmed fragile X demonstrate some autistic traits.[99]

NEUROCUTANEOUS SYNDROMES

Two conditions are most commonly associated with the appearance of autistic behavior. The first, tuberous sclerosis (TS)[100–103] is characterized by hypopigmented macules, fibroangiomas, kidney lesions, central nervous system hamartomas, seizures, cognitive deficiency, and autistic behaviors. TS is a dominantly inherited disorder, with two gene locations at 9q and 16p. However, most cases are new mutations and the absence of positive family history should not be a deterrent to examining a child with a Wood's lamp for possible hypopigmented lesions.

Although less commonly associated with ASDs, neurofibromatosis type 1 (NF1) can also present with cognitive impairment as well as autistic features. NF1 is characterized by café-au-lait macules and freckling of the axillary and inguinal regions, neurofibromas, and ocular Lisch nodules, and although autosomal dominant, it is most often associated with new mutations at 17q.[104] NF1 is more commonly associated with learning disabilities rather than autism.

PHENYLKETONURIA

Fortunately phenylketonuria is rare in the United States because of newborn screening; consequently, it is rarely associated with cognitive deficit and autism.[105] However, in children in whom there is reason to suspect missed standard newborn screenings, screening for phenylketonuria should be an important consideration, particularly in the context of cognitive delays, microcephaly, seizures, a musty smell, and autistic features. Early detection permits dietary modification that can markedly improve a child's clinical outcome.

ANGELMAN SYNDROME

Angelman syndrome (AS)[61,62] is a particularly severe disorder in which the individual does not have the maternally expressed gene, called the ubiqitin-protein ligase gene (UBE3A), on 15q through deletion, paternal uniparental disomy, or imprinting errors. Such individuals are typically very delayed, often completely nonverbal, hypotonic initially as children, with subsequent development of ataxia and spasticity. Individuals with AS also typically have seizures that can often be refractory to standard medication. Screening for a deletion of 15q can be accomplished by commercial fluorescence *in situ* hybridization (FISH) testing; however, a negative FISH results in a child who clinical fits an ASD diagnosis should also have a methylation study to assess for uniparental disomy.

RETT SYNDROME

Genetic testing for Rett syndrome (RX),[106–115] a rare disorder, should be considered in all girls who present with a regressive autistic phenotype, and particularly in girls who demonstrate microcephaly, seizures, and hand wringing compulsions. DNA testing at the MECP2 gene is confirmatory in 80% of cases. Although possible to manifest in boys, it is far less common and present with greater variability.

SMITH–LEMLI–OPITZ SYNDROME

Also quite rare, Smith–Lemli–Opitz syndrome (SLOS)[116] is an autosomal recessive disorder caused by errors in cholesterol metabolism, and has a prevalence of not more than 1 in 20,000 children, most often Caucasian or of Scandinavian descent. Children with SLOS have microcephaly, syndactyly of second and third toes, and occasional polydactyly, with malformations often seen in the heart, kidney, gastrointestinal tract, genitalia, with hypotonicity. Because recurrence occurs at 25% within families it is an important consideration in this clinical context.

IN UTERO DRUG EXPOSURE

In utero exposure to drugs has long been associated with neurodevelopmental difficulties; more challenging has been monitoring the effects of drugs as possibly causative to the development of ASD. Fetal exposure to maternal use of recreational drugs, most typically alcohol, cocaine, and cannabis, have been associated with autistic behavior in offspring. Both maternal depression as well as *in utero* exposure to antidepressants also carries a higher risk for ASD in children. Finally, the role of antiepileptics, particularly *in utero* exposure to valproic acid, is associated with significant neurodevelopmental disorders.

Children who are exposed to alcohol during gestation have an increased risk for ASDs as well as other behavioral and cognitive problems.[117] Careful history may yield important information for a possible diagnosis of fetal alcohol syndrome (FAS) and can provide significant help for a child in his or her academic years, because it is associated more often with not just ASDs but mood disorders, learning disabilities, and attention and focusing problems.[118] Physical features may include a smooth philtrum, small palpebral fissures, and thin upper vermillion. Central nervous system complaints include poor memory, cognition, judgment, and executive functions, often with poor self-regulation, hyperactivity, and attention problems. Other organs may be included, such as cardiac, renal, ocular, and dental problems.

In utero exposure to recreational drugs, particularly cocaine, is also associated with autism. In one study of 70 children for whom all mothers used cocaine, significant neurodevelopmental abnormalities were observed, including an 11% rate of autism, and greater than 90% language delays.[119] Children exposed *in utero* to cannabis develop permanent neurobehavioral and cognitive impairments. Psychoactive constituents from cannabis, particularly Δ^9-tetrahydrocannabinol (THC), bind to cannabinoid receptors in the fetal brain and alter neuronal connectivity, subsequently limiting the computational power of neuronal circuitries in affected offspring.[120]

Additionally, it has been suggested that history of maternal, though not paternal, depression is associated with an increased risk of autism spectrum disorders in offspring. Additionally, one study noted an increase of ASDs in women who used antidepressants, regardless of whether selective serotonin reuptake inhibitors (SSRIs) or nonselective monoamine reuptake inhibitors were reported.[121]

Antiepileptic drugs are used both for epilepsy as well as mood disorders. One study followed over 400 children from the newborn age to over 6 years, and found significant risk of neurodevelopmental disorders in children exposed to monotherapy sodium valproate (VPA) as well as those exposed to polytherapy with sodium VPA compared with control children with ASD as the most frequent diagnosis. No significant increase was found among children exposed to carbamazepine (1/50) or lamotrigine (2/30). An accumulation of evidence demonstrates that the risks associated with prenatal sodium VPA exposure include an increased prevalence of neurodevelopmental disorders. Whether such disorders are discrete or represent the severe end of a continuum of altered neurodevelopmental functioning requires further investigation.[122,123]

SECOND-HIT THEORY

Regardless of the mechanism, a review of studies published in the last 50 years revealed convincing evidence that most cases of ASDs result from interacting genetic factors. However, the expression of the autism gene(s) may be influenced by environmental factors. Although still only theoretical, these factors may represent a "second-hit" that primarily occurs during fetal brain development. That is, environmental factors may modulate already existing genetic factors responsible for the manifestation of ASDs in individual children.[124]

SUMMARY

Although known genetic and environmental causes for ASDs are still rare, investigation with clinically driven studies is important. The presence of cognitive impairment as well as dysmorphic features in the context of a child with an ASD should always prompt genetic investigation as well as a thorough history for possible maternal exposures to infection, drugs, or alcohol. When a genetic cause is identified, genetic counseling not only for the parents but for the extended family as well may be valuable.

References

1. Evans PA, Morris MA. *A Clinical Guide to Autistic Spectrum Disorders*. New York, NY: Lippincott Williams & Wilkins; 2011.
2. Fombonne E. Epidemiology of autistic disorder and other pervasive disorders. *J Clin Psychiatry*. 2005;66(suppl 10):3–8.
3. Chakrabarti D, Fombonne E. Pervasive developmental disorders in preschool children: confirmation of high prevalence. *Am J Psychiatry*. 2005;1(62):1133–1141.
4. Myers S, Johnson CP. American Academy of Pediatrics Council on Children with Disabilities. Management of children with autism spectrum disorders. *Pediatrics*. 2007;120(5):1162–1182.
5. Filipek PA, Accardo PJ, Ashwal S. Practice parameter: screening and diagnosis of autism: report of the Quality Standards Subcommittee of the American Academy of Neurology and the Child Neurology society. *Neurology*. 2000;55(4):468–479.
6. http://www.cdc.gov/ncbddd/autism/data.html.
7. http://pediatrics.aappublications.org/content/120/5/1162.full.
8. Dykens EM, Volkmar FR. Medical conditions associated with autism. In: Cohen DJ, Volkmar FR, eds. *Handbook of Autism and Pervasive Developmental Disorders*. 2nd ed. New York, NY: John Wiley & Sons; 1997:388–410.
9. Schaefer GB, Mendelsohn NJ. Clinical genetics evaluation in identifying the etiology of autism spectrum disorders: 2013 guideline revisions. ACMG Practice Guidelines. *Genet Med*. 2013;.
10. http://www.ncbi.nlm.nih.gov/pubmed/20059518.
11. Kosinovsky B, Hermon S, Yoran-Hegesh R, et al. The yield of laboratory investigations in children with infantile autism. *J Neural Trans*. 2005;112:587–596.
12. Battaghlia A, Carey JC. Etiologic yield of autistic spectrum disorders: a prospective study. *Am J Med Genet C Semin Med Genet*. 2006;142C:3–7.
13. Manning M, Hudgins L, Professional Practice and Guidelines Committee. Array-based technology and recommendations for utilization in medical genetics practice for detection of chromosomal abnormalities. *Genet Med*. 2010;12:742–745.
14. Miller DT, Adam MP, Aradhya S. Consensus statement: chromosomal microarray is a first-tier clinical diagnostic test for individuals with developmental disabilities or congenital anomalies. *Am J Hum Genet*. 2010;86:749–764.
15. Couter ME, Miller DT, Harris DJ, et al. Chromosomal microarray testing influences medical management. *Genet Med*. 2011;13:770–776.
16. Sebat J, Lakshmi B, Malhotra D, et al. Strong association of de novo copy number mutations with autism. *Science*. 2007;316:445–449.
17. Rosenfield JA, Ballif BC, Torchia BS, et al. Copy number variations associated with autism spectrum disorders contribute to a spectrum of neurodevelopmental disorders. *Genet Med*. 2010;12:694–702.
18. Schaefer GB, Starr L, Pickering D, Skar G, Dehaai K, Sanger WG. Array evaluation. *J Child Neurol*. 2010;25:1498–1503.
19. Shen Y, Dies KA, Holm IA, et al, Autism consortium clinical Genetics/DNA Diagnostics Collaboration. Clinical genetic testing for patients with autism spectrum disorders. *Pediatrics*. 2010;125:e727–e735.
20. Bremer A, Giacobini M, Eriksson M, et al. Copy number variation characteristics in subpopulations of patients with autism spectrum disorders. *Am J Med Genet B Neuropsychiatr Genet*. 2011;156:115–124.
21. McGrew SG, Peters BR, Crittendon JA, Veenstra-Vanderweele J. Diagnostic yield of chromosomal microarray analysis in an autism primary care practice: which guidelines to implement? *J Autism Dev Disord*. 2012;42:1582–1591.
22. Artuso R, Papa FT, Grillo E, et al. Investigation of modifier genes within copy number variations in Rett syndrome. *J Hum Genet*. 2011;56(7):508–515.
23. Goodbourn PT, Bosten JM, Bargary G, Hogg RE, Lawrance-Owen AJ, Mollon JD. Variants in the 1q21 risk region are associated with a visual endophenotype of autism and schizophrenia. *Genes Brain Behav*. 2014;13(2):144–151.
24. Dolcetti A, Silversides CK, Marshall CR, Lionel AC, Stavropoulos DJ, Scherer SW. Bassett AS.1q21.1 Microduplication expression in adults. *Genet Med*. 2013;15(4):282–289.
25. Crepel A, Breckpot J, Fryns JP, et al. DISC1 duplication in two brothers with autism and mild mental retardation. *Clin Genet*. 2010 Apr;77(4):389–394.
26. Williams JM, Beck TF, Pearson DM, Proud MB, Cheung SW, Scott DA. A 1q42 deletion involving DISC1, DISC2, and TSNAX in an autism spectrum disorder. *Am J Med Genet A*. 2009;149A(8):1758–1762.
27. Walker S, Scherer SW. Identification of candidate intergenic risk loci in autism spectrum disorder. *BMC Genomics*. 2013;14(Jul 24):499.
28. Van Winkel R, Esquivel G, Kenis G, et al. Review: Genome-wide findings in schizophrenia and the role of gene-environment interplay. *CNS Neurosci Ther*. 2010 Oct;16(5):e185–e192.
29. Kasnauskiene J, Ciuladaite Z, Preiksaitiene E, Utkus A, Peciulyte A, Kučinskas V. A new single gene deletion on 2q34: ERBB4 is associated with intellectual disability. *Am J Med Genet A*. 2013;161A(6):1487–1490.
30. Rashidi-Nezhad A, Parvaneh N, Farzanfar F, et al. 2q34-qter duplication and 4q34.2-qter deletion in a patient with developmental delay. *Eur J Med Genet*. 2012;55(3):203–210.
31. Loscalzo ML, Galczynski RL, Hamosh A, Summar M, Chinsky JM, Thomas GH. Interstitial deletion of chromosome 2q32-34 associated with multiple congenital anomalies and a urea cycle defect (CPS I deficiency). *Am J Med Genet A*. 2004;128A(3):311–315.
32. Leroy C, Landais E, Briault S, et al. The 2q37-deletion syndrome: an update of the clinical spectrum including overweight, brachydactyly and behavioural features in 14 new patients. *Eur J Hum Genet*. 2013 Jun;21(6):602–612.
33. Shoukier M, Fuchs S, Schwaibold E, et al. Microduplication of 3p26.3 in nonsyndromic intellectual disability indicates an important role of CHL1 for normal cognitive function. *Neuropediatrics*. 2013;44(5):268–271.
34. Aziz A, Harrop SP, Bishop NE. DIA1R is an X-linked gene related to Deleted In Autism-1. *PLoS One*. 2011;6(1):e14534.
35. Città S, Buono S, Greco D, et al. 3q29 microdeletion syndrome: cognitive and behavioral phenotype in four patients. *Am J Med Genet A*. 2013;161A(12):3018–3022.
36. Quintero-Rivera F, Sharifi-Hannauer P, Martinez-Agosto JA. Autistic and psychiatric findings associated with the 3q29 microdeletion syndrome: case report and review. *Am J Med Genet A*. 2010;152A(10):2459–2467.
37. Debost-Legrand A, Goumy C, Laurichesse-Delmas H, et al. Prenatal ultrasound findings observed in the Wolf-Hirschhorn syndrome: data from the registry of congenital malformations in Auvergne. *Birth Defects Res A Clin Mol Teratol*. 2013 Dec;97(12):806–811.

38. Andersen EF, Carey JC, Earl DL, et al. Deletions involving genes WHSC1 and LETM1 may be necessary, but are not sufficient to cause Wolf-Hirschhorn syndrome. *Eur J Hum Genet*. 2014 Apr;22(4):464–470.
39. Polan MB, Pastore MT, Steingass K, et al. Neurodevelopmental disorders among individuals with duplication of 4p13 to 4p12 containing a GABAA receptor subunit gene cluster. *Eur J Hum Genet*. 2014;22(1):105–109.
40. Vincent JB, Horike SI, Choufani S, et al. An inversion inv(4)(p12–p15.3) in autistic siblings implicates the 4p GABA receptor gene cluster. *J Med Genet*. 2006;43(5):429–434.
41. Wang L-S, D, K, et al. Population-based study of genetic variation in individuals with autism spectrum disorders from Croatia. *BMC Med Genet*. 2010;11:134.
42. Kerin T, Ramanathan A, Rivas K, Grepo N, Coetzee GA, Campbell DB. A noncoding RNA antisense to moesin at 5p14.1 in autism. *Sci Transl Med*. 2012;4(128):128ra40.
43. Ma DQ, Salyakina D, JM I, Whitehead PL, Andersen AL, Hoffman JD. A genome-wide association study of autism reveals a common novel risk locus at 5p14.1. *Ann Hum Genet*. 2009;73(Pt 3):263–273.
44. Verkerk AJMH, Schot R, LV, et al. Unbalanced der(5)t(5;20) translocation associated with megalencephaly, perisylvian polymicrogyria, polydactyly and hydrocephalus. *Am J Med Genet A*. 2010;152A(6):1488–1497.
45. Cuscó I, Corominas R, Bayés M, et al. Copy number variation at the 7q11.23 segmental duplications is a susceptibility factor for the Williams-Beuren syndrome deletion. *Genome Res*. 2008;18(5):683–694.
46. Patel C, Cooper-Charles L, McMullan DJ, Walker JM, Davison V, Morton J. Translocation breakpoint at 7q31 associated with tics: further evidence for *IMMP2L* as a candidate gene for Tourette syndrome. *Eur J Hum Genet*. 2011;19(6):634–639.
47. Bowne SJ, Sullivan LS, Blanton SH, et al. Mutations in the inosine monophosphate dehydrogenase 1 gene (IMPDH1) cause the RP10 form of autosomal dominant retinitis pigmentosa. *Hum Mol Genet*. 2002;11(5):559–568.
48. Matsuzaki H, Iwata K, Manabe T, Mori N. Triggers for autism: genetic and environmental factors. *J Cent Nerv Syst Dis*. 2012;4:27–36.
49. Allen NM, O'hIci B, Anderson G, Nestor T, Lynch SA, King MD. Variant late-infantile neuronal ceroid lipofuscinosis due to a novel heterozygous CLN8 mutation and de novo 8p23.3 deletion. *Clin Genet*. 2012;81(6):602–604.
50. Aguiar P, Cruz D, Ferro Rodrigues R, Araújo F, Ducla Soares JL. Subvalvular aortic stenosis associated with 8p23 deletion. *Rev Port Cardiol*. 2013;32(2):153–157.
51. Ozgen HM, van Daalen E, Bolton PF, et al. Copy number changes of the microcephalin 1 gene (MCPH1) in patients with autism spectrum disorders. *Clin Genet*. 2009;76(4):348–356.
52. Amouroux C, Vincent M, Blanchet P, et al. Duplication 8q12: confirmation of a novel recognizable phenotype with duane retraction syndrome and developmental delay. *Eur J Hum Genet*. 2012 May;20(5):580–583.
53. Salviati L, Trevisson E, Rodriguez Hernandez MA, et al. Haploinsufficiency of COQ4 causes coenzyme Q10 deficiency. *J Med Genet*. 2012;49(3):187–191.
54. Yatsenko SA, Hixson P, Roney EK, et al. Human subtelomeric copy number gains suggest a DNA replication mechanism for formation: beyond breakage-fusion-bridge for telomere stabilization. *Hum Genet*. 2012;131(12):1895–1910.
55. Ahn K, Gotay N, Andersen TM, et al. High rate of disease-related copy number variations in childhood onset schizophrenia. *Mol Psychiatry*. 2014;19(5):568–572.
56. Eng C. PTEN: one gene, many syndromes. *Hum Mutat*. 2003;22(3):183–198.
57. Shinawi M, Sahoo T, Maranda B, et al. 11p14.1 microdeletions associated with ADHD, autism, developmental delay, and obesity. *Am J Med Genet A*. 2011;155A(6):1272–1280.
58. Han JC, Thurm A, Golden Williams C, et al. Association of brain-derived neurotrophic factor (BDNF) haploinsufficiency with lower adaptive behaviour and reduced cognitive functioning in WAGR/11p13 deletion syndrome. *Cortex*. 2013;49(10):2700–2710.
59. Thevenon J, Callier P, Andrieux J, et al. 12p13.33 microdeletion including ELKS/ERC1, a new locus associated with childhood apraxia of speech. *Eur J Hum Genet*. 2013;21(1):82–88.
60. Castermans D, Wilquet V, Parthoens E, et al. The neurobeachin gene is disrupted by a translocation in a patient with idiopathic autism. *J Med Genet*. 2003;40:352–356.
61. Martins-Taylor K, Hsiao JS, Chen PF, et al. Imprinted expression of UBE3A in non-neuronal cells from a Prader–Willi syndrome patient with an atypical deletion. *Hum Mol Genet*. 2014;23(9):2364–2373.
62. Horváth E, Horváth Z, Isaszegi D, et al. Early detection of Angelman syndrome resulting from de novo paternal isodisomic 15q UPD and review of comparable cases. *Mol Cytogenet*. 2013;6(1):35.
63. McInnes LA, Nakamine A, Pilorge M, et al. A large-scale survey of the novel 15q24 microdeletion syndrome in autism spectrum disorders identifies an atypical deletion that narrows the critical region. *Mol Autism*. 2010;1(1):5.
64. Salyakina D, Cukier HN, Lee JM, et al. Copy number variants in extended autism spectrum disorder families reveal candidates potentially involved in autism risk. *PLoS One*. 2011;6(10):e26049.
65. Sorte HS, Gjevik E, Sponheim E, Eiklid KL, Rødningen OK. Copy number variation findings among 50 children and adolescents with autism spectrum disorder. *Psychiatr Genet*. 2013;23(2):61–69.
66. Mozaffari M, Hoogeveen-Westerveld M, Kwiatkowski D, et al. Identification of a region required for TSC1 stability by functional analysis of TSC1 missense mutations found in individuals with tuberous sclerosis complex. *BMC Med Genet*. 2009.
67. Schorry EK, Keddache M, Lanphear N, et al. Genotype-phenotype correlations in Rubinstein-Taybi syndrome. *Am J Med Genet A*. 2008;146A(19):2512–2519.
68. Tropeano M, Ahn JW, Dobson RJ, et al. Male-biased autosomal effect of 16p13.11 copy number variation in neurodevelopmental disorders. *PLoS One*. 2013;8(4):e61365.
69. Shishido E, Aleksic B, Ozaki N. Copy-number variation in the pathogenesis of autism spectrum disorder. *Psychiatry Clin Neurosci*. 2014;68(2):85–95.
70. Al-Kateb H, Khanna G, Filges I, et al. Scoliosis and vertebral anomalies: Additional abnormal phenotypes associated with chromosome 16p11.2 rearrangement. *Am J Med Genet A*. 2014;164(5):1118–1126.
71. Handrigan GR, Chitayat D, Lionel AC, et al. *J Med Genet*. 2013;50(3):163–173.

72. Magoulas PL, Liu P, Gelowani V, et al. Inherited dup(17)(p11.2p11.2): Expanding the phenotype of the Potocki-Lupski syndrome. *Am J Med Genet A*. 2014;164(2):500–504.
73. Flax JF, Hare A, Azaro MA, Vieland VJ, Brzustowicz LM. Combined linkage and linkage disequilibrium analysis of a motor speech phenotype within families ascertained for autism risk loci. *J Neurodev Disord*. 2010;2(4):210–223.
74. Klein-Tasman BP, Janke KM, Luo W, et al. Cognitive and psychosocial phenotype of young children with neurofibromatosis-1. *J Int Neuropsychol Soc*. 2014;20(1):88–98.
75. Brison N, Devriendt K, Peeters H. Genetic counseling for susceptibility loci and neurodevelopmental disorders: the del15q11.2 as an example. *Am J Med Genet A*. 2013;161A(11):2846–2854.
76. Grisart B, Willatt L, Destrée A, et al. 17q21.31 microduplication patients are characterised by behavioural problems and poor social interaction. *J Med Genet*. 2009;46(8):524–530.
77. Cooper GM, Coe BP, Girirajan S, et al. A copy number variation morbidity map of developmental delay. *Nat Genet*. 2011;43(9):838–846.
78. Maas NM, Van de Putte T, Melotte C, et al. The C20 or f133 gene is disrupted in a patient with Kabuki syndrome. *BMJ Case Rep*. 2009;2009: pii: bcr06.2009.1994.
79. Valenti D, de Bari L, De Filippis B, Henrion-Caude A, Vacca RA. Mitochondrial dysfunction as a central actor in intellectual disability-related diseases: an overview of Down syndrome, autism, Fragile X and Rett syndrome. *Neurosci Biobehav Rev*. 2014 Feb 15; pii: S0149-7634(14)00025-6.
80. Anderson JS, Nielsen JA, Ferguson MA, et al. Abnormal brain synchrony in Down Syndrome. *Neuroimage Clin*. 2013;2(May 24):703–715.
81. Crespi BJ, Crofts HJ. Association testing of copy number variants in schizophrenia and autism spectrum disorders. *J Neurodev Disord*. 2012;4(1):15.
82. Sorte HS, Gjevik E, Sponheim E, Eiklid KL, Rødningen OK. Copy number variation findings among 50 children and adolescents with autism spectrum disorder. *Psychiatr Genet*. 2013;23(2):61–69.
83. Betancur C, Buxbaum JD. SHANK3 haploinsufficiency: a "common" but underdiagnosed highly penetrant monogenic cause of autism spectrum disorders. *Mol Autism*. 2013;4(1):17.
84. Carrascosa-Romero MC, Suela J, Alfaro-Ponce B, Cepillo-Boluda AJ. X-chromosome-linked ichthyosis associated to epilepsy, hyperactivity, autism and mental retardation, due to the Xp22.31 microdeletion. *Rev Neurol*. 2012;54(4):241–248.
85. Castrén M, Gaily E, Tengström C, Lähdetie J, Archer H, Ala-Mello S. Epilepsy caused by CDKL5 mutations. *Eur J Paediatr Neurol*. 2011;15(1):65–69.
86. Youngs EL, Henkhaus R, Hellings JA, Butler MG. IL1RAPL1 gene deletion as a cause of X-linked intellectual disability and dysmorphic features. *Eur J Med Genet*. 2012;55(1):32–36.
87. Gandomi SK, Farwell Gonzalez KD, Parra M, et al. Diagnostic exome sequencing identifies two novel IQSEC2 mutations associated with X-linked intellectual disability with seizures: implications for genetic counseling and clinical diagnosis. *J Genet Couns*. 2013 Dec 4.
88. Vandewalle J, Bauters M, Van Esch H, et al. The mitochondrial solute carrier SLC25A5 at Xq24 is a novel candidate gene for non-syndromic intellectual disability. *Hum Genet*. 2013;132(10):1177–1185.
89. Sabaratnam M, Murthy NV, Wijeratne A, Buckingham A, Payne S. Autistic-like behaviour profile and psychiatric morbidity in Fragile X Syndrome: a prospective ten-year follow-up study. *Eur Child Adolesc Psychiatry*. 2003;12(4):172–177.
90. Scott Schwoerer J, Laffin J, Haun J, Raca G, Friez MJ, Giampietro PF. MECP2 duplication: possible cause of severe phenotype in females. *Am J Med Genet A*. 2014 Jan 23.
91. Hagerman PJ, Hagerman RJ. The fragile X premutation: a maturing perspective. *Am J Hum Genet*. 2004;74:805–816.
92. Rogers SJ, Wehner DE, Hagerman R. The behavioral phenotype in fragile X: symptoms of autism in very young children with fragile X syndrome, idiopathic autism, and other developmental disorders. *J Dev Behav Pediatr*. 2001;22:409–417.
93. Hagerman RJ. Fragile X, syndrome and associated disorders in adulthood. *Continuum Lifelong Learning Neurol*. 2009;15(6):32–49.
94. Coffey SM, Cook K, Tartaglia N, et al. Expanded clinical phenotype of women with the FMR1 premutation. *Am J Med Genet*. 2008;52(pt 6):1009–1016.
95. Jacquemaont S, Hagerman RJ, Leehey MA, et al. Penetrance of the fragile X associated tremor /ataxia syndrome in a permutation carrier population. *JAMA*. 2004;291(4):460–469.
96. Hagerman RJ, Hall DA, Coffey S, et al. Treatment of fragile X associated tremor ataxia syndrome (FXTAS) and related neurological problems. *Clin Interv Aging*. 2008;3(2):251–262.
97. Rogers SJ, Wehner DE, Hagerman R. The behavioral phenotype in fragile X: symptoms of autism in very young children with fragile X syndrome, idiopathic autism, and other developmental disorders. *J Dev Behav Pediatr*. 2001;22:409–417.
98. Watson MS, Leckman JF, Annex B, et al. Fragile X in a survey of 75 autistic males. *N Engl J Med*. 1984;310:1462.
99. Hagerman FJ. Physical and behavioral phenotype. In: Hagerman RJ, Hagerman PJ, eds. *Fragile X: Diagnosis, Treatment and Research*. 3rd ed. Baltimore, MD: Johns Hopkins University Press; 2002:3–109.
100. Smalley SL. Autism and tuberous sclerosis. *J Autism Dev Disord*. 1998;28:407–414.
101. Baker P, Piven J, Sato Y. Autism and tuberous sclerosis complex: prevalence and clinical features. *J Autism Dev Disord*. 1998;28:279–285.
102. Curatolo P, Porfirio M, Manzi B, Seri S. Autism in tuberous sclerosis. *Eur J Paediatr Neurol*. 2004;8:327–332.
103. Curatolo P. Tuberous sclerosis: genes, brain, and behavior. *Dev Med Child Neurol*. 2006;48:404.
104. Lauritsen M, Ewald H. The genetics of autism. *Acta Psychiatr Scand*. 2001;103:411–427.
105. Baiele S, Pavone L, Meli C, Fiumara A, Coleman M. Autism and phenylketonuria. *J Autism Dev Disord. Genet*. 2003;40:87–95.
106. Thatcher KN, Peddada S, Yasui DH, LaSalle JM. Homologous pairing of 15q11-13 imprinted domains in brain is developmentally regulated but deficient in Rett and autism samples. *Hum Mol Genet*. 2005;14:785–797.
107. Lopez-Rangel E, Lewis ME. Do other methyl-binding proteins play a role in autism? *Clin Genet*. 2006;69:25.
108. Niemitz EL, Feinberg AP. Epigenetics and assisted reproductive technology: a call for investigation. *Am J Hum Genet*. 2004;74:599–609.
109. Ham AL, Kumar A, Deeter R, Schanen NC. Does genotype predict phenotype in Rett syndrome? *J Child Neurol*. 2005;20:768–778.
110. Kerr AM, Ravine D. Review article: breaking new ground with Rett syndrome. *J Intellect Disabil Res*. 2003;47:580–587.
111. Kerr AM, Prescott RJ. Predictive value of the early clinical signs in Rett disorder. *Brain Dev*. 2005;27(suppl 1):S20–S24.

112. Kerr A. Annotation: Rett syndrome-recent progress and implications for research and clinical practice. *J Child Psychol Psychiatry*. 2002;43:277–287.
113. Einspieler C, Kerr AM, Prechtl HF. Abnormal general movements in girls with Rett disorder: the first four months of life. *Brain Dev*. 2005;27(suppl 1):S8–S13.
114. Ravn K, Nielsen JB, Uldall P, Hansen FJ, Schawartz M. No correlation between phenotype and genotype in boys with a truncating MECP2 mutation. *J Med Genet*. 2003;40:e5.
115. Moog U, Smeets EE, van Roozendall KE, et al. Neurodevelopmental disorders in males related to the gene causing Rett syndrome in females (MECP2). *Eur J Paediatr Neurol*. 2003;7:5–12.
116. Elias E, Giampietro P. *Autism may be caused by Smith-Lemli-Opitz syndrome (SLOS)*. Presented at Annual Clinical Genetics Meeting. Dallas, TX; March 17–20, 2005.
117. Aronson M, Hagberg B, Gillberg C. Attention deficits and autistic spectrum problems in children exposed to alcohol during gestation: a follow-up study. *Dev Med Child Neurol*. 1997;39:583–587.
118. Stevens SA, Nash K, Koren G, Rovet J. Autism characteristics in children with fetal alcohol spectrum disorders. *Child Neuropsychol*. 2013;19(6):579–587.
119. Davis E, Fennoy I, Laraque D, Kanem N, Brown G, Mitchell J. Autism and developmental abnormalities in children with perinatal cocaine exposure. *J Natl Med Assoc*. 1992;84(4):315–319.
120. Tortoriello G, Morris CV, Alpar A, et al. Miswiring the brain: Δ9-tetrahydrocannabinol disrupts cortical development by inducing an SCG10/stathmin-2 degradation pathway. *EMBO J*. 2014;33(7):668–685.
121. Rai D, Lee BK, Dalman C, Golding J, Lewis G, Magnusson C. Parental depression, maternal antidepressant use during pregnancy, and risk of autism spectrum disorders: population based case-control study. *BMJ*. 2013;346(Apr 19):f2059.
122. Bromley RL, Mawer GE, Briggs M, et al. The prevalence of neurodevelopmental disorders in children prenatally exposed to antiepileptic drugs. *J Neurol Neurosurg Psychiatr*. 2013;84(6):637–643.
123. Palac S, Meador KJ. Antiepileptic drugs and neurodevelopment: an update. *Curr Neurol Neurosci Rep*. 2011;11(4):423–427.
124. Lainhart JE, Ozonoff S, Coon H, et al. Autism, regression, and the broader autism phenotype. *Am J Med Genet*. 2002;113(3):231–237.

CHAPTER

20

Angelman Syndrome

Charles A. Williams and Jennifer M. Mueller

University of Florida College of Medicine, Gainesville, FL, USA

INTRODUCTION

Angelman syndrome (AS) is a neurodevelopmental disorder caused by deficiency of the ubiquitin-protein ligase protein, E6-AP, in the brain. AS is seen in 1 in 20,000 individuals in the population and the characteristic features include severe developmental delay and speech impairment, gait ataxia and/or tremulousness of the limbs, and a unique behavior with a happy demeanor that includes smiling, excitability, and frequent laughing. Other common features are microcephaly and seizures. While developmental delays are first noted by around 6 months of age, the unique clinical features usually manifest after the first year of life. The behavioral phenotype is crucial in suspecting the diagnosis, and may not be present during early clinical encounters. Therefore, the timely diagnosis of AS remains challenging, since there may be no obvious dysmorphic features or specific clinical features during infancy that easily direct the clinician to a diagnosis.

UBE3A demonstrates parent-specific differential expression, or imprinting, limited to brain and spinal cord neurons (glial cells are not imprinted). Disruption of the maternal functional allele of *UBE3A* results in AS through several different mechanisms. In contrast, there are no phenotypic effects resulting from the disruption of the functional paternal allele. Analysis of parent-specific DNA methylation imprints in the critical 15q11.2-q13 genomic region identifies 75–80% of all cases, including individuals with cytogenetic deletions, imprinting center (IC) defects, and paternal uniparental disomy (UPD). Analysis of the methylation imprint for AS can be performed by Southern blot, methylation-specific polymerase chain reaction (PCR), or methylation-specific multiplex ligation-dependent probe amplification (MLPA) testing. MLPA, single nucleotide polymorphism (SNP) array, and fluorescence *in situ* hybridization (FISH) can also be used to detect 15q11.2-13 deletions. Detection of UPD is available through SNP array and DNA microsatellite testing. Sequence analysis of *UBE3A* will detect an additional ~10% of cases, but a small remaining percentage of clinically diagnosed cases will have normal genetic studies for AS.

Current research is aimed at identifying relevant *UBE3A* targets and understanding regulatory aspects of *UBE3A* action and gene expression. Although there are several therapeutic approaches for treatment of the syndrome, the strategy that has a possibility for cure is to induce *UBE3A* expression on the paternal allele by modulating the paternally expressed antisense transcript that normally represses or "silences" *UBE3A* transcription.

CLINICAL FEATURES

Historical Overview

In 1965, Dr. Harry Angelman, an English physician, first described three children with characteristics now known as the Angelman syndrome (AS).[1] He noted that all of these patients had a stiff, jerky gait, absent speech, and exhibited excessive laughter and seizures. In 1987, Ellen Magenis identified children with 15q11.2-13 microdeletions who were expected to have the Prader–Willi syndrome (PWS) but had clinical features that were incongruent. It was quickly realized that these children had microdeletions on the maternally derived number 15. In 1991, researchers determined AS can also be caused by two copies of the paternal chromosome 15. This was followed by the discovery

Rosenberg's Molecular and Genetic Basis of Neurological and Psychiatric Disease
http://dx.doi.org/10.1016/B978-0-12-410529-4.00020-6

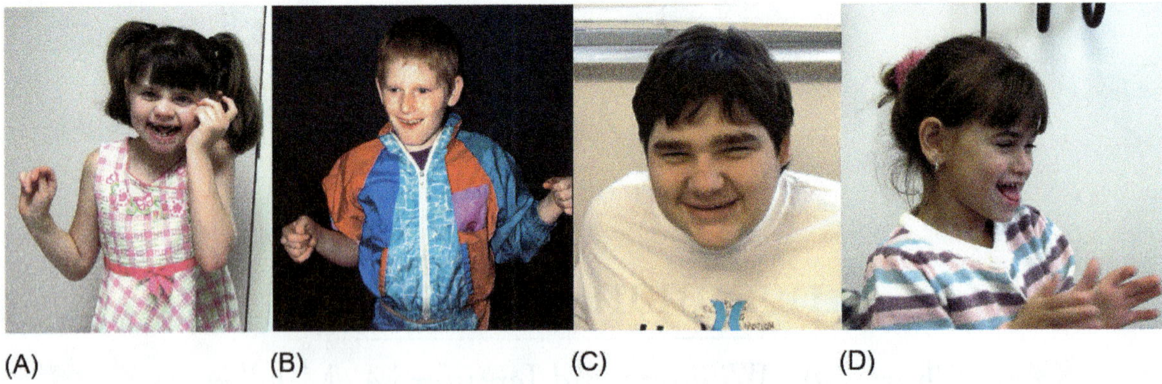

FIGURE 20.1 **Pictured are individuals with genetic test-proven Angelman syndrome.** The *UBE3A* defect identified in each is: 15q11.2-q13 deletion **(A, B)**; *UBE3A* mutation **(C)** and paternal uniparental disomy **(D)**.

of disruptions of the regulatory region (the *imprinting center*) as another causative mechanism in 1993. Finally, in 1997, the AS gene *UBE3A* was isolated. This discovery led to the development of animal models and catalyzed neuroscience research on *UBE3A* and its role in neural development.

Prevalence

Cases of AS are seen worldwide and without apparent racial predilection. While the true incidence remains unknown, it is estimated to be 1 in 12,000 to 1 in 20,000.[2,3]

Clinical Features

Consensus criteria for the clinical diagnosis of AS have been established.[4] Newborns typically have a normal phenotype but developmental delays are first noticeable by 6 months of age. However, the unique clinical features of AS do not become apparent until after 1 year of age. The diagnosis is usually first suspected due to a combination of movement disorder, absent speech, and happy demeanor. Because few individuals exhibit craniofacial dysmorphism, it may take several years before the clinical diagnosis is evident (Figure 20.1).

NATURAL HISTORY

Craniofacial anomalies are present in some individuals with AS, but typically do not represent significant facial dysmorphism. Those with prominent oral-motor behaviors associated with tongue protrusion often have deformational changes, leading to some degree of mandibular prognathism. Individuals with microcephaly may have diminished length of the cranial base, leading to midface retrusion.[5]

Excessive mouthing behaviors are common in infants with AS, with active exploration of objects through manipulation and chewing. Tongue protrusion is seen in 30–50%, associated with drooling. There may be trouble initiating sucking and sustaining breastfeeding. Frequent spitting up may be related to gastroesophageal reflux.

Hyperkinetic movements such as jitteriness or tremulousness of the trunk and limbs may be seen in early infancy.[6] Voluntary movements may be slightly jerky, and uncoordinated coarse movements can prevent walking. Gross motor milestones are delayed and sitting frequently occurs after 12 months of age and walking is often delayed until 3–5 years of age.[7,8] About 10% may fail to achieve walking.[2] The mildly impaired child can have almost normal walking in early childhood, while severely affected children can be extremely shaky and jerky when walking. The legs are kept wide-based, the feet are often flat, and ankles pronated and turned outward. Arms are kept uplifted with flexed elbows and downward-turned hands.

Hypermotoric behaviors in combination with the jerky limb movements and frequent smiling and/or laughter give children with AS a distinctive behavioral phenotype.[9] Some behaviors may suggest an autism spectrum problem but social engagement is typically good. Stereotypical behaviors such as lining up of toys or fascination with spinning objects or flashing lights are rare.[10] Bursts of laughter may occur in up to 70% of older individuals and their apparent happiness and laughter tends to be contextually appropriate.[7,11–13]

Onset of seizures is between 1–3 years of age and includes varied types (generalized tonic-clonic, absence, atonic, complex partial, myoclonic). Infantile spasms are rare. Seizures are associated with specific nonepileptic electroencephalogram (EEG) changes: runs of high-amplitude delta activity with intermittent spike and slow-wave discharges (at times observed as a notched delta pattern); runs of rhythmic theta activity over a wide area; and runs of rhythmic sharp theta activity of 5–6 per second over the posterior third of the head, forming complexes with small spikes.[14] These are usually seen only with eye closure.[15–17] Nonconvulsive status epilepticus can occur.[18] It is felt that seizures are usually well controlled on anticonvulsants, but a questionnaire study by Thibert et al. suggested that the epilepsy is relatively refractory since only 15% of patients respond to the first antiepileptic drug.[19] The brain appears to be structurally normal except for delayed or abnormal myelination and mild atrophy.[20,21] Research by Peters et al. noted abnormalities in diffuser tensor imaging suggestive of dysmyelination.[22]

Children with AS have sleep difficulties including frequently waking up at night.[23,24] Research by Pelc et al. attributes this to abnormal neurodevelopmental functioning of the thalamocortical axis.[25]

Intellectual deficiency is in the severe to profound range of functioning; appropriate use of even one or two words is rare. Receptive language skills are always more advanced than expressive language skills.[26] Effective fluent use of sign language does not occur.[27] Some communication via gestures and communication boards is possible. Accurate developmental testing is challenging due to lack of speech, hyperactivity, and inattentiveness. Overall, psychometric testing estimates that individuals with AS have an upper developmental potential in the 24–30 month range.[26,28,29]

Adult Life

Pubertal onset and development are generally normal and procreation appears possible for both genders. Fertility appears to be normal with reported transmission of an AS deletion to a fetus by the affected mother.[30] Independent living is not possible for adults with AS, but most can live at home or in home-like placements. Lifespan does not appear to be dramatically shortened but may be decreased by 10–15 years. There are reports of AS individuals living beyond 70 years, although at this point there no actual data that estimates lifespan.[31,32]

MOLECULAR GENETICS

UBE3A lies within 15q11.2-q13, and the gene spans approximately 120 kb of genomic DNA and contains 16 exons (Figure 20.2). The coding region is 60 kb and the main three messenger RNA (mRNA) transcripts have 10 exons. They are approximately 5 kb in size and encode three protein isoforms (I, II, and III).[33–36] In addition, there appear to be 8–10 other smaller transcripts of uncertain significance. Isoform I corresponds to the open reading frame for E6-AP, isoform II has an additional 20 amino acids, and isoform III has an additional 23 amino acids at the amino terminus. The functions of these isoforms are unknown but it is predicted they may interact with different substrates in different intracellular regions. The 5′ untranslated region (UTR) has a complex structure with additional exons located upstream of the initiation site. The 3′ UTR extends for about 2.0 kb.[34] *UBE3A* produces an 865 amino acid, E6-associated protein (E6-AP). E6-AP was first recognized as a protein that binds to p53 and mediates its association with human papilloma virus E6 protein. Binding leads to degradation of p53 tumor suppressor via the ubiquitin proteasome pathway, which promotes development of cervical carcinoma.[37,38]

E6-AP facilitates the transfer and covalent linkage of activated ubiquitin (a 76-amino acid protein) to target proteins. E6-AP and other HECT-domain proteins form a thioester bond with ubiquitin before ubiquitin is transferred to substrates, unlike other E3s. E6-AP belongs to the HECT (homologous to E6-AP COOH-terminus) class of E3 enzymes that share a 40 kDa conserved COOH-terminal catalytic domain. The HECT domain (encoded by exons 9

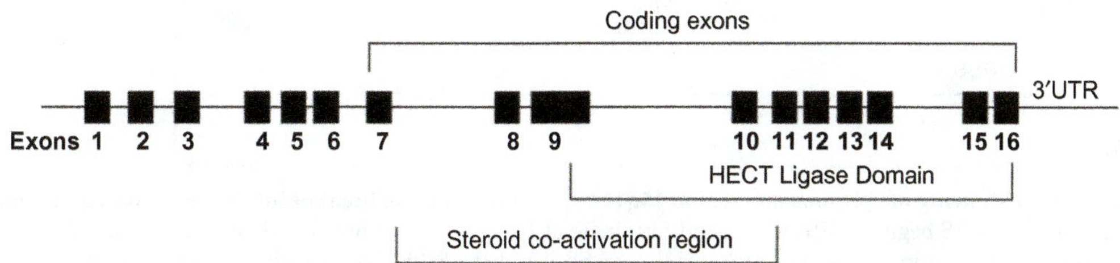

FIGURE 20.2 **Diagram of *UBE3A* illustrating that exons 9–16 constitute the HECT domain.** The steroid coactivation region does not occupy a contiguous genomic region but spans a region that contains several five-amino acid motifs known to be coactivation-interacting motifs.

through 16) of E6-AP is a bilobed structure with a broad catalytic cleft at the junction of the two lobes. The E6-binding site is encoded by exon 9. The active site cysteine residue that accepts ubiquitin from the E2 ubiquitin-conjugating enzyme is encoded within exon 16.[34,36] Mutations within the cleft interfere with ubiquitin-thioester bond formation. Therefore, it makes sense that most AS mutations are due to missense or single amino acid insertion or deletion mutations in the HECT domain map to the catalytic clefts.[37]

A steroid receptor coactivation domain is located upstream of the HECT region, but its role in neuronal development remains uncertain. E6-AP appears to have at least two independent functions since the ligase region and the HECT domain are not required for function of the coactivation domain.[39]

UBE3A displays predominant maternal expression in human fetal brain and adult frontal cortex.[35,40,41] There is widespread and possibly global *UBE3A* allele-specific expression in the mouse and human brain neurons. Primary cell cultures from fetal mouse brains reveal that *UBE3A* imprinting is limited to neurons, whereas glial cells show biallelic expression.[42] *UBE3A* has a large 5′ CpG island but its DNA methylation does not differ between the maternal and paternal alleles (both are unmethylated).[43] Runte et al. have shown that a long *SNURF-SNRPN* sense/*UBE3A* antisense RNA transcript exists in the AS/PWS region, starting from the *SNURF-SNRPN* IC and extending more than 460kb to at least the 5′ end of *UBE3A*.[44] It appears that the paternally active *UBE3A* antisense transcript blocks paternal *UBE3A* transcription.[45]

Genotype/Phenotype Correlations

Those with the large chromosome deletions appear to have more severe symptoms, presumably due to haploinsufficiency of genes adjacent to *UBE3A* such as the downstream *GABA* genes (*GABRB3*, *GABRA5* and *GABRG3*) or those located upstream in the BP1 to BP2 breakpoint region (*NIPA1*, *NIPA2*, *CYFIP1*, and *GCP5*), (Figure 20.3).[43,46–50] Individuals with the large deletions (class I [BP1–BP3] or class II [BP2–BP3]) are more likely to have microcephaly, seizures and severe language impairment as compared to those with *UBE3A* mutations, UPD, or imprinting defects. *OCA2* plays a role in tyrosine metabolism and is important for the development of pigment in the skin, hair, and irides. Most large deletions cause haploinsufficiency of *OCA2* that, in combination with E6-AP haploinsufficency in skin, leads to relatively hypopigmented irides, skin, and hair.[51] Research by Tan et al. presented clinical data from 92 children with a molecular diagnosis of AS established between 5 and 60 months of age.[52] Their study demonstrated that individuals with the larger deletions have diminished weight compared to both the general population and to those with UPD/imprinting defects. This finding may be secondary to diminished muscle mass in individuals with the large deletion.

Individuals with an epigenetic type of imprinting defect associated with some degree of somatic mosaicism have relatively higher verbal speech ability and cognitive functioning, demonstrating up to 50–60 words and use of simple sentences.[47] Language impairment and autism spectrum traits are more frequently seen in individuals with larger class I deletion than those with smaller class II deletions.[53] AS patients with intellectual disabilities or UPD have

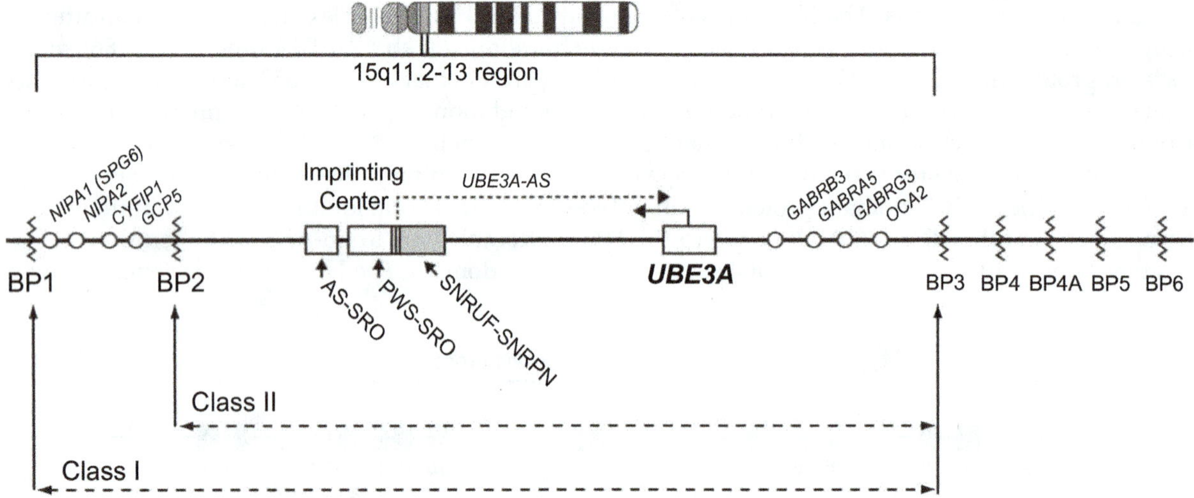

FIGURE 20.3 Schematic drawing of chromosome region 15q11.2-q13 indicating the breakpoint (BP) regions. Approximately 90% of chromosome deletions that result in AS begin at BP1 or BP2 and terminate at BP3. Genes not imprinted and thus biparentally expressed are noted by the open circles. The bipartite imprinting center includes the AS-SRO and the PWS-SRO, which are drawn as open boxes. The shaded box for the *SNRUF-SNRPN* gene is shown with some overlap with the PWS-SRO. The *SNURF-SNRPN* sense/*UBE3A* antisense transcript is labeled *UBE3A*-AS.

relatively higher developmental and language ability, but it is important to recognize there is significant overlap between all of the disease mechanisms.

Individuals with UPD appear to have better physical growth and fewer movement abnormalities, ataxia, and seizures. Further, they are less likely to have microcephaly but the reason for this is unclear.[43,49]

DISEASE MECHANISMS

In normal neurons, *UBE3A* is active only on the maternally derived allele and transcriptionally inactivated (imprinted) on the paternally derived allele of chromosome 15. All other somatic cells have biallelic transcription. AS can occur by four different mechanisms affecting the *maternally* derived chromosome 15: deletion of the gene (e.g., chromosome microdeletion), paternal uniparental disomy (UPD), a defect in the imprinting center that controls *UBE3A* transcription, or intragenic mutation (Figure 20.4).

Chromosome Microdeletions

Almost 70% of cases of AS are due to *de novo* chromosome deletion events. There are three major breakpoints along chromosome 15q11.2-q13, proximal BP1and BP2 and the more distal BP3.[54–56] The BP1–BP3 regions are characterized by low-copy repeats (LCRs), typically in direct orientation, that contain repeats of several pseudogenes and other expressed sequences (Figure 20.3). Class I deletions account for 40% of deletion cases and extend from BP1 to BP3. Class II deletions extend from BP2 to BP3 and account for 50% of cases. Fewer than 10% of individuals with AS may have deletions extending from the BP1/BP2 region to regions more distal at BP4, BP4A, BP5 or BP6.[53]

Paternal Uniparental Disomy 15

Paternal UPD of chromosome 15 causes 3% of AS, which most likely results from a somatic segregation error.[57] Paternal UPD cases of meiotic origin do occur but this mechanism is more commonly seen in maternal UPD cases associated with the PWS. Individuals with UPD should have chromosomal analysis to ensure that they do not have a paternally inherited robertsonian translocation.

Imprinting Defects

Approximately 6% of cases are the direct result of a defect in the resetting/maintaining of imprints during gametogenesis or after fertilization. Genetic (small deletions) and epigenetic (abnormal DNA methylation pattern but no deletion) defects in the AS imprinting center (IC) change the DNA methylation and expression patterns along 15q11.2-q13. Although there is biparental inheritance of chromosome 15, the maternal 15q11.2-q13 region has a paternal epigenotype. The consequence is transcriptional incompetence for the maternal-only expressed *UBE3A*.[58–60]

The IC has a bipartite structure that regulates, in *cis*, the imprint resetting and maintenance within the 15q11.2-q13 imprinted domain.[61–63] About 8–15% of cases with an imprinting defect will have deletions that disrupt the IC and mapping of these deletions, as well as those associated with PWS. There are two smallest regions of deletion overlap

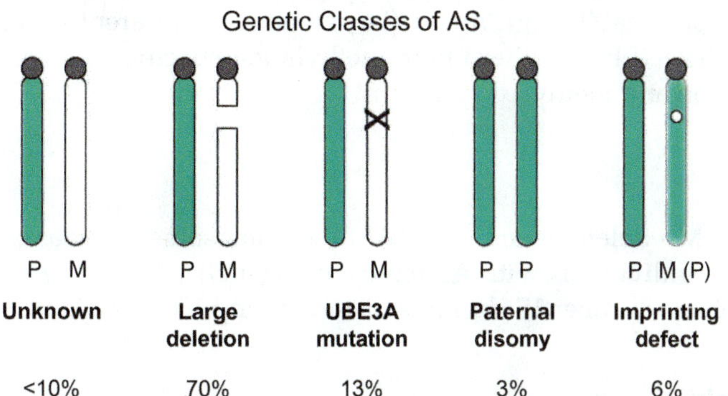

FIGURE 20.4 **A chromosome 15 pair is illustrated for each genetic class of AS.** The P indicates the paternally derived chromosome and the M indicates the maternally derived one. The shaded chromosomes have a paternal pattern of gene functioning while the unshaded chromosomes have a maternal pattern. Percentages indicate the proportion of individuals with the clinical diagnosis of AS who have the indicated mechanism.

(SRO) that define two critical elements in the IC region, the AS-SRO and the PWS-SRO.[63] The PWS-SRO is 4.3 kb in size and overlaps with the *SNURF-SNRPN* exon 1/promoter region.[64] IC deletions in AS patients affect the more centromeric *SNURF-SNRPN* promoter/exon 1 region. The smallest region of overlap in patients with AS and an IC deletion (AS-SRO) is 880 bp in size and maps 35 kb proximal to *SNURF-SNRPN* exon 1.[61,65]

In the vast majority of AS patients (>90%), the imprinting defect represents a primary epimutation without any changes in the DNA sequence.[59,65] IC deletions are only found in a small portion of individuals with imprinting defects.

UBE3A Mutations

The majority of *UBE3A* mutations that cause AS are protein-truncating mutations.[33,43,66,67] To date, more than 60 mutations have been reported and of these, 60–70% involve small deletions and duplications leading to frameshift mutations.[68–70] An estimated 25% involve missense and nonsense mutations. The rest consist of splicing defects, gross deletions and complex rearrangements.[70] All these mutations are predicted to disrupt the HECT ligase domain. Although exons 9 and 16 (which code for part of the HECT domain) account for a high percentage of all mutations, these coding regions are disproportionately large. In actuality, they probably do not represent true hot spots for mutations. Individuals with milder effect mutations, such as certain missense and in-frame deletions/duplications, may show some degree but not all the clinical features associated with AS. Determining pathogenicity of novel missense mutations, especially *de novo* ones, can be a challenge. In these cases, additional testing to establish paternal origin of the mutation through the maternal chromosome will help clarify the pathogenicity of these novel missense mutations.[71] Complete or partial overlapping deletions and intragenic deletions of *UBE3A* are also seen. While these deletions can be missed by sequencing, detection through various methods such as quantitative PCR, real time PCR, and MLPA, etc., is possible.

DIFFERENTIAL DIAGNOSIS

The differential diagnosis for AS is often broad given that infants with AS present with general psychomotor delays and/or seizures. It is not surprising that the differential encompasses entities including cerebral palsy, static encephalopathy, and mitochondrial encephalomyopathy. However, AS should be distinguished from the following chromosome deletion syndromes or single-gene conditions: Mowat–Wilson syndrome (*ZEB2*), Christianson syndrome (*SLC9A6*), Pitt–Hopkins syndrome (*TCF4*), adenylosuccinate lyase deficiency (*ADSL*), Rett syndrome (*MECP2*), *MECP2* duplications, Phelan–McDermid syndrome (22q13.3 deletion), Kleefstra syndrome (9q34.3 deletion, *EHMT1*), *CDKL5* syndrome (*CDKL5*), 2q23.1 deletion (*MBD5*), 17q21.31 deletion (*KANSL1*), *HERC2* deficiency syndrome (*HERC2*), and alpha-thalassemia mental retardation syndrome (*ATRX*).

TESTING

DNA Methylation Analysis

Individuals with AS caused by a 5–6 Mb deletion of 15q11.2-q13, uniparental disomy, or an imprinting defect have only an unmethylated (i.e., "paternal") contribution (i.e., an abnormal parent-specific DNA methylation imprint). Methylation status can be detected by Southern blot, methylation-specific PCR or MLPA. MLPA may be used for concurrent detection of deletion and methylation imprint.

Cytogenetic Analysis

In 70% of individuals, 5–6 Mb deletions are detected by chromosome microarray analysis, MLPA, or less commonly FISH. Less than 1% of individuals with AS have a cytogenetically visible chromosome rearrangement (i.e., translocation or inversion) of one number 15 chromosome involving 15q11.2-q13.

Uniparental Disomy Analysis

UPD can be detected using DNA microsatellite testing, which requires a DNA sample from the proband and both parents or via SNP microarray identification of loss of heterozygosity (LOH).

Imprinting Center Analysis

Approximately 6% of the imprinting defects are caused by microdeletions (6–200 kb) affecting the AS IC, but the great majority of imprinting defects only involve an abnormal DNA methylation imprint without any evidence of a deletion or mutation within the AS IC. Testing for an IC deletion or epigenetic defect is available in only a few clinical laboratories.

UBE3A Sequence Analysis

For those individuals with clinical features of AS and normal DNA methylation analysis, *UBE3A* sequence analysis should be considered.[43,66,72] A few individuals with AS have multiexonic or whole-gene deletions of *UBE3A* not identified by gene sequencing. These deletions may be detected by various methods including some array-comparative genomic hybridization (CGH) platforms and intragenic deletion analysis.[73,74]

Failure to detect AS-causing genetic abnormalities occurs in <10% of cases of clinically diagnosed AS. Possible explanations to consider include incorrect clinical diagnosis, undetected mutations outside of the coding sequence of *UBE3A*, and other unidentified mechanisms or gene(s) involved in disrupting *UBE3A* function.

MANAGEMENT

Newborns with AS can have weak and uncoordinated sucking, which may require special nipples and other feeding strategies. Poor weight gain and emesis are often seen with gastroesophageal reflux, and customary medical treatment is usually effective (i.e., upright positioning, motility drugs). In some cases, surgery may be required. Children with AS often have excessive tongue protrusion causing drooling. Surgical or medication treatments are not generally effective (e.g., surgical reimplantation of the salivary ducts or use of local scopolamine patches).

Many antiepileptic drugs are used to treat seizures in individuals with AS but there is no agreement as to the optimal seizure medication, although valproic acid (Depakote), topiramate (Topamax), lamotrigine (Lamictal), levetiracetam (Keppra), and clonazepam (Klonopin) are more commonly used. Single medication use is preferred but seizure breakthrough is common.

Children with AS usually do not receive drug therapy for their hypermotoric or hyperactive behaviors, although stimulant medications such as methylphenidate can be beneficial in some individuals. Parents and therapists implement behavioral modification for treating undesirable behaviors that are self-injurious or socially disruptive.

Nighttime wakefulness poses a challenge for families and use of diphenylhydramine or clonidine may be helpful. Administration of 0.3 mg melatonin 1 hour before sleep is effective in some children, but should not be given if the child wakes up in the middle of the night.[75] Due to negative side effects, sedating agents are generally not recommended. Many families construct safe but confining bedrooms to help manage nighttime wakefulness.

Orthopedic problems (particularly subluxed or pronated ankles or tight Achilles tendons) may be an issue for some and can be corrected by orthotic bracing or surgery. Individuals with scoliosis may need thoracolumbar jackets or surgical rod stabilization in severe cases. Encouraging ambulation is essential for individuals with ataxia and severe scoliosis to prevent the loss of their ability to walk.

Accessibility to a full range of both educational training and enrichment programs is critical for children with AS. These educational strategies should be both individualized to the child and flexible. Physical therapy from an early age helps both children who are nonambulatory or unstable, while occupational therapy improves fine motor and oral-motor control. Children who are extremely ataxic benefit from special adaptive chairs or positioners. Speech therapy should focus on nonverbal methods of communication. Augmentative communication aids such as picture cards or communication boards are available. Parents and teachers should consider implementing these aids at the earliest appropriate time. Special physical provisions in the classroom and teacher aides or assistants are necessary for effective class integration and safety.

Individuals with AS demonstrate food-related behaviors including eating non-food items, apparent increased appetite, and increased behavioral orientation to food. This may contribute to obesity, although severe (e.g., morbid) obesity is very uncommon.[76] As older adults become less mobile and active, scheduled physical activity is helpful to reduce obesity if it becomes a management concern.

Therapies Under Investigation

Many *UBE3A* target proteins have now been identified in addition to a large repertoire of associated proteins. It remains unclear as to whether there are only a few target proteins accounting for the symptomatology in AS or whether there are many.

It is known that CaMKII activity is diminished in the AS animal model due to inhibitory phosphorylation. Correction of the mouse phenotype has occurred when this inhibitory phosphorylation is prevented.[77] Phosphatases that act on CaMKII are not direct targets for *UBE3A* but it may be that modulating CaMKII's inhibitory phosphatases can provide a therapeutic route.

Individuals with AS have at least one normally functioning paternal allele that has been silenced by the upstream imprinting center (via the antisense transcript). Since it is not possible yet to insert *UBE3A* throughout the brain using viral gene therapy vectors, there has been interest in unsilencing the paternal *UBE3A*. Recently, topoisomerase inhibitors have been shown to do this, perhaps by interfering with the initiation or maintenance of the *UBE3A* antisense RNA.[78] Also, antisense oligonucleotides are being studied that can also interrupt the antisense transcription and thereby allow for *UBE3A* transcription.[79]

Earlier approaches have not been unsuccessful in unsilencing the paternal allele. Several clinical trials used dietary supplements that would increase the methylation pool (e.g., folic acid, B12, betaine and creatine) but those have not shown a proven benefit.[80] Although topoisomerase inhibitors have been shown to unsilence the paternal allele these agents are still under study in animal models and no human clinical trials are yet planned.

References

1. Angelman H. 'Puppet' children. A report of three cases. *Dev Med Child Neurol*. 1965;7:681–688.
2. Clayton-Smith J, Pembrey ME. Angelman syndrome. *J Med Genet*. 1992;29(6):412–415.
3. Steffenburg S, Gillberg CL, Steffenburg U, Kyllerman M. Autism in Angelman syndrome: a population-based study. *Pediatr Neurol*. 1996;14(2):131–136.
4. Williams CA, Beaudet AL, Clayton-Smith J, et al. Angelman syndrome 2005: updated consensus for diagnostic criteria. *Am J Med Genet A*. 2006;140(5):413–418.
5. Frias JL, King GJ, Williams CA. Cephalometric assessment of selected malformation syndromes. *Birth Defects Orig Artic Ser*. 1982;18(1):139–150.
6. Fryburg JS, Breg WR, Lindgren V. Diagnosis of Angelman syndrome in infants. *Am J Med Genet*. 1991;38(1):58–64.
7. Buntinx IM, Hennekam RC, Brouwer OF, et al. Clinical profile of Angelman syndrome at different ages. *Am J Med Genet*. 1995;56(2):176–183.
8. Zori RT, Hendrickson J, Woolven S, Whidden EM, Gray B, Williams CA. Angelman syndrome: clinical profile. *J Child Neurol*. 1992;7(3):270–280.
9. Williams CA. The behavioral phenotype of the Angelman syndrome. *Am J Med Genet C Semin Med Genet*. 2010;154C(4):432–437.
10. Walz NC. Parent report of stereotyped behaviors, social interaction, and developmental disturbances in individuals with Angelman syndrome. *J Autism Dev Disord*. 2007;37(5):940–947.
11. Horsler K, Oliver C. Environmental influences on the behavioral phenotype of Angelman syndrome. *Am J Ment Retard*. 2006;111(5):311–321.
12. Horsler K, Oliver C. The behavioural phenotype of Angelman syndrome. *J Intellect Disabil Res*. 2006;50(Pt 1):33–53.
13. Oliver C, Demetriades L, Hall S. Effects of environmental events on smiling and laughing behavior in Angelman syndrome. *Am J Ment Retard*. 2002;107(3):194–200.
14. Galvan-Manso M, Campistol J, Conill J, Sanmarti FX. Analysis of the characteristics of epilepsy in 37 patients with the molecular diagnosis of Angelman syndrome. *Epileptic Disord*. 2005;7(1):19–25.
15. Boyd SG, Harden A, Patton MA. The EEG in early diagnosis of the Angelman (happy puppet) syndrome. *Eur J Pediatr*. 1988;147(5):508–513.
16. Korff CM, Kelley KR, Nordli Jr. DR. Notched delta, phenotype, and Angelman syndrome. *J Clin Neurophysiol*. 2005;22(4):238–243.
17. Rubin DI, Patterson MC, Westmoreland BF, Klass DW. Angelman's syndrome: clinical and electroencephalographic findings. *Electroencephalogr Clin Neurophysiol*. 1997;102(4):299–302.
18. Pelc K, Boyd SG, Cheron G, Dan B. Epilepsy in Angelman syndrome. *Seizure*. 2008;17(3):211–217.
19. Thibert RL, Conant KD, Braun EK, et al. Epilepsy in Angelman syndrome: a questionnaire-based assessment of the natural history and current treatment options. *Epilepsia*. 2009;50(11):2369–2376.
20. Harting I, Seitz A, Rating D, et al. Abnormal myelination in Angelman syndrome. *Eur J Paediatr Neurol*. 2009;13(3):271–276.
21. Castro-Gago M, Gomez-Lado C, Eiris-Punal J, Rodriguez-Mugico VM. Abnormal myelination in Angelman syndrome. *Eur J Paediatr Neurol*. 2010;14(3):292.
22. Peters SU, Bird LM, Kimonis V, et al. Double-blind therapeutic trial in Angelman syndrome using betaine and folic acid. *Am J Med Genet A*. 2010;152A(8):1994–2001.
23. Bruni O, Ferri R, D'Agostino G, Miano S, Roccella M, Elia M. Sleep disturbances in Angelman syndrome: a questionnaire study. *Brain Dev*. 2004;26(4):233–240.
24. Didden R, Korzilius H, Duker P, Curfs L. Communicative functioning in individuals with Angelman syndrome: a comparative study. *Disabil Rehabil*. 2004;26(21–22):1263–1267.
25. Pelc K, Cheron G, Boyd SG, Dan B. Are there distinctive sleep problems in Angelman syndrome? *Sleep Med*. 2008;9(4):434–441.
26. Trillingsgaard A, Ostergaard JR. Autism in Angelman syndrome: an exploration of comorbidity. *Autism*. 2004;8(2):163–174.
27. Clayton-Smith J. Clinical research on Angelman syndrome in the United Kingdom: observations on 82 affected individuals. *Am J Med Genet*. 1993;46(1):12–15.
28. Didden R, Korzilius H, Kamphuis A, Sturmey P, Lancioni G, Curfs LM. Preferences in individuals with Angelman syndrome assessed by a modified Choice Assessment Scale. *J Intellect Disabil Res*. 2006;50(Pt 1):54–60.
29. Peters SU, Goddard-Finegold J, Beaudet AL, Madduri N, Turcich M, Bacino CA. Cognitive and adaptive behavior profiles of children with Angelman syndrome. *Am J Med Genet A*. 2004;128(2):110–113.
30. Lossie AM, Driscoll DJ. Transmission of Angelman syndrome by an affected mother. *Genet Med*. 1999;1(6):262–266.

31. Bjerre I, Fagher B, Ryding E, Rosen I. The Angelman or "happy puppet" syndrome. Clinical and electroencephalographic features and cerebral blood flow. *Acta Paediatr Scand*. 1984;73(3):398–402.
32. Philippart M, Minassian BA. Angelman syndrome from infancy to old age [abstract]. *Am J Hum Genet*. 2005;79(suppl):605.
33. Kishino T, Lalande M, Wagstaff J. UBE3A/E6-AP mutations cause Angelman syndrome [published erratum appears in *Nat GenetI*. 1997;15(4):411]. *Nat Genet*. 1997;15(1):70–73.
34. Kishino T, Wagstaff J. Genomic organization of the UBE3A/E6-AP gene and related pseudogenes. *Genomics*. 1998;47(1):101–107.
35. Vu TH, Hoffman AR. Imprinting of the Angelman syndrome gene, UBE3A, is restricted to brain [letter]. *Nat Genet*. 1997;17(1):12–13.
36. Yamamoto Y, Huibregtse JM, Howley PM. The human E6-AP gene (UBE3A) encodes three potential protein isoforms generated by differential splicing. *Genomics*. 1997;41(2):263–266.
37. Huang L, Kinnucan E, Wang G, et al. Structure of an E6AP-UbcH7 complex: insights into ubiquitination by the E2–E3 enzyme cascade. *Science*. 1999;286(5443):1321–1326.
38. Huibregtse JM, Scheffner M, Howley PM. A cellular protein mediates association of p53 with the E6 oncoprotein of human papillomavirus types 16 or 18. *EMBO J*. 1991;10(13):4129–4135.
39. Ramamoorthy S, Nawaz Z. E6-associated protein (E6-AP) is a dual function coactivator of steroid hormone receptors. *Nucl Recept Signal*. 2008;6:e006.
40. Herzing LB, Kim SJ, Cook Jr. EH, Ledbetter DH. The human aminophospholipid-transporting ATPase gene ATP10C maps adjacent to UBE3A and exhibits similar imprinted expression. *Am J Hum Genet*. 2001;68(6):1501–1505.
41. Rougeulle C, Glatt H, Lalande M. The Angelman syndrome candidate gene, UBE3A/E6-AP, is imprinted in brain [letter]. *Nat Genet*. 1997;17(1):14–15.
42. Yamasaki K, Joh K, Ohta T, et al. Neurons but not glial cells show reciprocal imprinting of sense and antisense transcripts of Ube3a. *Hum Mol Genet*. 2003;12(8):837–847.
43. Lossie AC, Whitney MM, Amidon D, et al. Distinct phenotypes distinguish the molecular classes of Angelman syndrome. *J Med Genet*. 2001;38(12):834–845.
44. Runte M, Huttenhofer A, Gross S, Kiefmann M, Horsthemke B, Buiting K. The IC-SNURF-SNRPN transcript serves as a host for multiple small nucleolar RNA species and as an antisense RNA for UBE3A. *Hum Mol Genet*. 2001;10(23):2687–2700.
45. Mabb AM, Judson MC, Zylka MJ, Philpot BD. Angelman syndrome: insights into genomic imprinting and neurodevelopmental phenotypes. *Trends Neurosci*. 2011;34(6):293–303.
46. Fridman C, Varela MC, Kok F, Diament A, Koiffmann CP. Paternal UPD15: further genetic and clinical studies in four Angelman syndrome patients. *Am J Med Genet*. 2000;92(5):322–327.
47. Nazlican H, Zeschnigk M, Claussen U, et al. Somatic mosaicism in patients with Angelman syndrome and an imprinting defect. *Hum Mol Genet*. 2004;13(21):2547–2555.
48. Saitoh S, Harada N, Jinno Y, et al. Molecular and clinical study of 61 Angelman syndrome patients. *Am J Med Genet*. 1994;52(2):158–163.
49. Saitoh S, Wada T, Okajima M, Takano K, Sudo A, Niikawa N. Uniparental disomy and imprinting defects in Japanese patients with Angelman syndrome. *Brain Dev*. 2005;27(5):389–391.
50. Smith A, Marks R, Haan E, Dixon J, Trent RJ. Clinical features in four patients with Angelman syndrome resulting from paternal uniparental disomy. *J Med Genet*. 1997;34(5):426–429.
51. Low D, Chen KS. UBE3A regulates MC1R expression: a link to hypopigmentation in Angelman syndrome. *Pigment Cell Melanoma Res*. 2011;24(5):944–952.
52. Tan WH, Bacino CA, Skinner SA, et al. Angelman syndrome: mutations influence features in early childhood. *Am J Med Genet A*. 2011;155A(1):81–90.
53. Sahoo T, Bacino CA, German JR, et al. Identification of novel deletions of 15q11q13 in Angelman syndrome by array-CGH: molecular characterization and genotype–phenotype correlations. *Eur J Hum Genet*. 2007;15(9):943–949.
54. Amos-Landgraf JM, Ji Y, Gottlieb W, et al. Chromosome breakage in the Prader–Willi and Angelman syndromes involves recombination between large, transcribed repeats at proximal and distal breakpoints. *Am J Hum Genet*. 1999;65(2):370–386.
55. Christian SL, Fantes JA, Mewborn SK, Huang B, Ledbetter DH. Large genomic duplicons map to sites of instability in the Prader–Willi/Angelman syndrome chromosome region (15q11-q13). *Hum Mol Genet*. 1999;8(6):1025–1037.
56. Knoll JH, Nicholls RD, Magenis RE, et al. Angelman syndrome: three molecular classes identified with chromosome 15q11q13-specific DNA markers. *Am J Hum Genet*. 1990;47(1):149–155.
57. Robinson WP, Christian SL, Kuchinka BD, et al. Somatic segregation errors predominantly contribute to the gain or loss of a paternal chromosome leading to uniparental disomy for chromosome 15. *Clin Genet*. 2000;57(5):349–358.
58. Buiting K, Barnicoat A, Lich C, Pembrey M, Malcolm S, Horsthemke B. Disruption of the bipartite imprinting center in a family with Angelman syndrome. *Am J Hum Genet*. 2001;68(5):1290–1294.
59. Buiting K, Gross S, Lich C, Gillessen-Kaesbach G, el-Maarri O, Horsthemke B. Epimutations in Prader–Willi and Angelman syndromes: a molecular study of 136 patients with an imprinting defect. *Am J Hum Genet*. 2003;72(3):571–577.
60. Glenn CC, Nicholls RD, Robinson WP, et al. Modification of 15q11-q13 DNA methylation imprints in unique Angelman and Prader–Willi patients. *Hum Mol Genet*. 1993;2(9):1377–1382.
61. Buiting K, Lich C, Cottrell S, Barnicoat A, Horsthemke B. A 5-kb imprinting center deletion in a family with Angelman syndrome reduces the shortest region of deletion overlap to 880bp. *Hum Genet*. 1999;105(6):665–666.
62. Sutcliffe JS, Nakao M, Christian S, et al. Deletions of a differentially methylated CpG island at the SNRPN gene define a putative imprinting control region. *Nat Genet*. 1994;8(1):52–58.
63. Buiting K, Saitoh S, Gross S, et al. Inherited microdeletions in the Angelman and Prader–Willi syndromes define an imprinting centre on human chromosome 15. *Nat Genet*. 1995;9(4):395–400.
64. Ohta T, Buiting K, Kokkonen H, et al. Molecular mechanism of Angelman syndrome in two large families involves an imprinting mutation. *Am J Hum Genet*. 1999;64(2):385–396.
65. Horsthemke B, Buiting K. Genomic imprinting and imprinting defects in humans. *Adv Genet*. 2008;61:225–246.
66. Malzac P, Webber H, Moncla A, et al. Mutation analysis of UBE3A in Angelman syndrome patients. *Am J Hum Genet*. 1998;62(6):1353–1360.

67. Matsuura T, Sutcliffe JS, Fang P, et al. De novo truncating mutations in E6-AP ubiquitin-protein ligase gene (UBE3A) in Angelman syndrome. *Nat Genet*. 1997;15(1):74–77.
68. Abaied L, Trabelsi M, Chaabouni M, et al. A novel UBE3A truncating mutation in large Tunisian Angelman syndrome pedigree. *Am J Med Genet A*. 2009;152A(1):141–146.
69. Camprubi C, Guitart M, Gabau E, et al. Novel UBE3A mutations causing Angelman syndrome: different parental origin for single nucleotide changes and multiple nucleotide deletions or insertions. *Am J Med Genet A*. 2009;149A(3):343–348.
70. Stenson PD, Mort M, Ball EV, Phillips AD, Shaw K, Cooper DN. The human gene mutation database: 2008 update. *Genome Med*. 2009;1(1):13.
71. Horsthemke B, Wawrzik M, Gross S, et al. Parental origin and functional relevance of a *de novo* UBE3A variant. *Eur J Med Genet*. 2010;54(1):19–24.
72. Fang P, Lev-Lehman E, Tsai TF, et al. The spectrum of mutations in UBE3A causing Angelman syndrome. *Hum Mol Genet*. 1999;8(1):129–135.
73. Lawson-Yuen A, Wu BL, Lip V, Sahoo T, Kimonis V. Atypical cases of Angelman syndrome. *Am J Med Genet A*. 2006;140(21):2361–2364.
74. Sato K, Iwakoshi M, Shimokawa O, et al. Angelman syndrome caused by an identical familial 1,487-kb deletion. *Am J Med Genet A*. 2007;143(1):98–101.
75. Zhdanova IV, Wurtman RJ, Wagstaff J. Effects of a low dose of melatonin on sleep in children with Angelman syndrome. *J Pediatr Endocrinol Metab*. 1999;12(1):57–67.
76. Barry RJ, Leitner RP, Clarke AR, Einfeld SL. Behavioral aspects of Angelman syndrome: a case control study. *Am J Med Genet A*. 2005;132(1):8–12.
77. van Woerden GM, Harris KD, Hojjati MR, et al. Rescue of neurological deficits in a mouse model for Angelman syndrome by reduction of alphaCaMKII inhibitory phosphorylation. *Nat Neurosci*. 2007;10(3):280–282.
78. Huang HS, Allen JA, Mabb AM, et al. Topoisomerase inhibitors unsilence the dormant allele of UBE3A in neurons. *Nature*. 2012;481(7380):185–189.
79. Meng L, Person RE, Beaudet AL. UBE3A-ATS is an atypical RNA polymerase II transcript that represses the paternal expression of UBE3A. *Hum Mol Genet*. 2012;21(13):3001–3012.
80. Bird LM, Tan WH, Bacino CA, et al. A therapeutic trial of pro-methylation dietary supplements in Angelman syndrome. *Am J Med Genet A*. 2011;155A(12):2956–2963.

CHAPTER

21

Prion Diseases

James A. Mastrianni

University of Chicago, Chicago, IL, USA

INTRODUCTION

The prion diseases, also known as transmissible spongiform encephalopathies (TSEs), represent a unique family of rare neurodegenerative disorders. They have much in common with other neurodegenerative diseases, but they are distinguished from all others by their highly transmissible nature. In addition, the host range of prion disease (PrD) is unusual in that animals, in addition to humans, can be affected and, as demonstrated by the experience of mad cow disease, prion disease can jump between species. Although the core features of these diseases include progressive dementia, ataxia, and variable neurological and psychiatric symptoms, several clinical and histopathologic phenotypes of disease are recognized. Despite this phenotypic variation, the etiology of all is traced to the misfolded isoform of a single protein known as the prion protein (PrP), a normal constituent of the plasma membrane of cells, especially neurons. There are several fascinating aspects of these diseases, including the origins of its discovery, the nature of its biology, and the future challenges for diagnosis and treatment. From sheep to cannibals to mad cows, the prion diseases have a rich and controversial history. Deciphering the molecular nature of prion generation and propagation will be key to developing a therapeutic approach to these diseases.

ORIGINS OF DISCOVERY

The earliest account of PrD comes from farmers in the 18th century,[1] although it may have origins dating as far back as ancient China.[2] Scrapie is a naturally occurring disease of sheep characterized by irregular behavior, ataxic gait, and a persistent pruritis. Brain pathology reveals diffuse vacuolation of the gray matter, also termed "spongiform degeneration." An infectious basis for scrapie was confirmed when disease was transmitted to a healthy goat by inoculation into its eye of spinal cord homogenate prepared from a scrapie-affected sheep.[3] Scrapie was labeled a transmissible spongiform encephalopathy (TSE) and predicted to have a viral etiology.

In the 1950s, Gajdusek reported kuru, a strange neurodegenerative disease endemic to the Fore tribe in the highlands of New Guinea. Kuru produced progressive ataxia with dementia and rapid progression to death.[4–6] Histologic examination of the brain[7] revealed extensive vacuolation and a small number of amyloid plaques. The clinicopathologic phenotype was noted to be remarkably similar to scrapie, and transmission of kuru to chimpanzees in the 1960s confirmed suspicions that this was also a TSE.[8] Its transmission among humans was determined to be the result of cannibalistic rituals practiced by the Fore. Two years after transmission of kuru, Creutzfeldt–Jakob disease (CJD), an obscure neurological disease with undetermined etiology, but similar neuropathology,[9] was similarly transmitted to chimpanzees. This finding placed new significance to the TSEs. In the late 1990s, the emergence of PrD in cows, in the form of bovine spongiform encephalopathy (BSE), and its transmission to humans as variant CJD (vCJD), catapulted these diseases into the mainstream.

Based on the prolonged incubation times required for disease expression, the etiologic agent was hypothesized to be a "slow virus."[10,11] This presumed virus was indeed unconventional, as it provoked no immune response in the host and was resistant to formalin, ultraviolet (UV), and ionizing radiation treatments that would destroy nucleic acids required

Rosenberg's Molecular and Genetic Basis of Neurological and Psychiatric Disease
http://dx.doi.org/10.1016/B978-0-12-410529-4.00021-8

for viral replication.[9,12–14] Based on this data, it was hypothesized that an infection might induce the replication of a host protein.[15] Stanley Prusiner seized upon this possibility and found that a single protein with relative resistance to proteases was consistently present in the infectious fraction of brain samples.[16] The infectious element was dubbed a prion, to designate its *pro*teinaceous *in*fectious nature, and the protein that composed it was labeled prion protein (PrP). Sequence analysis confirmed that PrP was encoded by a chromosomal gene of the host.[7–19] Based on an immense body of evidence, Prusiner proposed the "protein-only" prion hypothesis, which posits that the etiologic agent of PrD is composed largely, if not entirely, of a misfolded scrapie-associated isoform (PrP^{Sc}) of the normal cellular form (PrP^{C}) of PrP.[17] Although several questions regarding the mechanism of conversion remain unanswered, the prion hypothesis is virtually universally accepted. In fact, a prion-like pathogenic mechanism has been proposed for a variety of other neurodegenerative disease related proteins, such as α-synuclein, tau, and amyloid-β (Aβ), among others (for a review, see [20]).

EPIDEMIOLOGY

PrD in humans is uncommon, with a yearly incidence of 1–1.5 cases per million, worldwide. This incidence reflects three modes of occurrence: sporadic, genetic, and transmissible. There is no obvious gender preference,[21] nor is there a predilection for a particular ethnic group, although certain mutations within the PrP gene (*PRNP*) that cause familial disease may be more restricted to certain cultures, based on the origin of the mutation. Disease progression is rapid in the majority of cases, resulting in an annual incidence rate that is nearly equal to the annual mortality rate. The bulk of cases occur on a sporadic basis, for which no identifiable source of exposure or genetic cause is appreciated. Peak incidence is between 60 and 70 years of age. Up to 15% of cases result from the inheritance of an autosomal dominant mutation in *PRNP*. These patients generally present earlier in life (<55 years) and have a more protracted course of up to several years, with rare case reports of survival beyond 20 years. Environmental exposure and geographical location are not risk factors for sporadic CJD (sCJD) and the disease is not spread by typical modes of transmission.[21] Less than 1% of cases are explained by exposure to exogenously derived prions, such as from contaminated biological therapeutics (growth hormone, gonadotrophic hormone, blood products from vCJD cases), transplantation of infected tissues (dura mater grafts, corneal transplants), the use of improperly sterilized surgical or neurodiagnostic tools and, as evidenced by the emergence of vCJD, the introduction of prions into the food supply.

Bovine spongiform encephalopathy (BSE), a prion disease of cows that is responsible for vCJD in humans, demonstrates the potential for animals to act as a reservoir for prions that can affect humans. Sheep scrapie, chronic wasting disease (CWD) of deer and elk, and some spontaneous forms of BSE are naturally occurring animal-borne diseases, whereas prion diseases detected in several zoo-based animals and domestic cats were likely the result of feed contaminated with BSE.

PATHOLOGIC FEATURES OF PRION DISEASES

Three principal pathologic features associated with PrD include vacuolation (spongiform degeneration), extracellular plaque deposits comprised of PrP, and gliosis. The morphology of plaques and vacuoles, in addition to their relative distribution within the brain, are used to specifically define several PrD subtypes. While these subtypes carry some clinical phenotypic differences among them, it is the histopathologic characteristics that ultimately define them (Figure 21.1). Creutzfeldt–Jakob disease (CJD), the most common PrD subtype, displays widespread spongiform degeneration without PrP-plaque deposits. The vacuoles observed by light microscopy represent focal swellings of axonal and dendritic neuronal processes associated with the loss of synaptic organelles, and accumulation of abnormal membranes by electron microscopy.[22–24] They may be distributed throughout the gray and white matter but are generally most prominent in the gray matter neuropil. They typically range in size from 5 to 25 μm but in advanced cases they may be as large as 100 μm.[25] Vacuolation is typically observed within the cerebral neocortex, subiculum of the hippocampus, caudate, putamen, thalamus, and the molecular layer of the cerebellar cortex.[25] However, subtypes of CJD have been further defined, based on the morphology (large or small vacuoles) and distribution or the spongiform degeneration (see Table 21.2). Reactive gliosis is also consistently present, although inflammatory cells are conspicuously absent. In the fatal insomnia (FI) subtype, gliosis restricted to the thalamus and inferior olives and neuronal cell loss in the absence of vacuolation and amyloid plaques is the characteristic phenotype. However, vacuolation may be observed in some cases, especially those that extend longer than 1 year.

Amyloid plaques are the hallmark of Gerstmann–Sträussler–Scheinker disease (GSS), but they also occur in kuru (no longer observed), variably protease-sensitive prionopathy (VPSPr), and variant CJD (vCJD). In GSS, plaques are

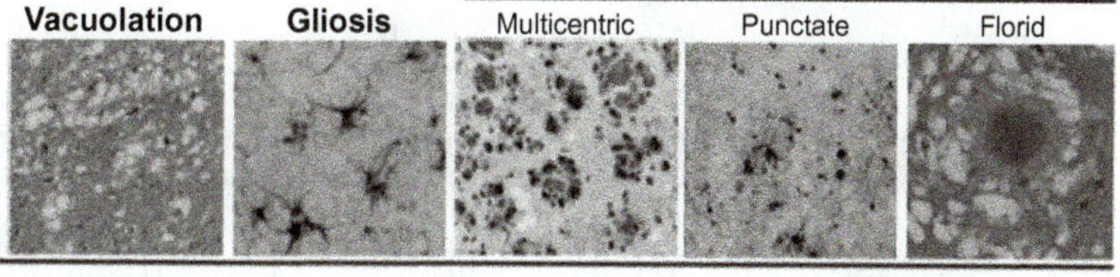

Disease	Vacuolation	Gliosis	PrP-amyloid plaques: Multicentric	Punctate	Florid
CJD	Extensive*	Extensive	Absent	<10% of cases (unicentric)	Absent
vCJD	Extensive	Extensive	Absent	Absent	Prominent*
GSS	Mild	with plaques	Extensive*	Occasional	Absent
FI	Minimal	Thalamus*	Absent	Absent	Absent
VPSPr	Minimal	Minimal	Absent	Prominent*	Absent

*Pathognomonic histopathologic feature of the disease subtype

FIGURE 21.1 **Histopathologic features of prion disease.**

most often concentrated in the cerebellum, whereas in vCJD and VPSPr they are diffusely distributed throughout the brain. Plaque morphology also differs among these diseases. In GSS, they are the "multicentric" type, characterized by a central core of amyloid surrounded by smaller amyloid satellites. The "unicentric" plaque is seen in kuru and in low numbers in 10% of sCJD cases. Florid plaques are pathognomonic of vCJD and are defined by an amyloid core surrounded by a halo of intense spongiform change. In VPSPr, plaques are generally smaller and either globular-like or target-like collections of PrP.[26] PrP plaques are true amyloid detected by Thioflavin S and Congo Red and exhibit green birefringence under polarized light. Analysis of plaque deposits in GSS associated with several *PRNP* mutations (e.g., P102L, A117V, F198S, Q217R) reveal amino- and carboxy-terminally clipped PrP peptides that span residues of 58–150 and 81–150 of the mutant alleles of PrP.[27]

The existence of PrD subtypes suggests differences in the cellular metabolism of PrP and/or the ability of prions to target specific populations of neurons. These different human subtypes form the basis of prion strains, which appear to be defined by the tertiary structure of pathogenic PrP^{Sc} (see page 240).

GENETICS OF PRION DISEASES

Autosomal Dominant Mutations

In humans, PrP is encoded by the prion protein gene (PRNP) on the short arm of chromosome 20 (20pter).[28] The entire coding segment of the gene lies within exon 3 and there is no evidence for differential splicing in humans. More than 30 base pair alterations and insertions of *PRNP* are associated with familial PrD (Figure 21.2). All point mutations result in single amino acid substitutions or, in a handful of cases, amber mutations, resulting in expression of a truncated PrP molecule.[27,29,30] Insertional mutations result from the addition of 1–9 multiples of a stretch of eight amino acids that is normally repeated 5 times, with minor variation (PH(Q)GGG(T/S)WGQ), and situated between residues 51 and 91, known as the octarepeat segment, in addition to a 24-base pair insertion between nucleotides 388 and 389 (residues 129 and 130) that repeats some of the adjacent sequence (LGGLGGYV).[31] Genotype–phenotype correlations have been described, the molecular basis of which is proposed to be the predisposition of different PrP^{Sc} conformers depending on the site and nature of the mutation. Whereas clinical phenotypes may vary considerably among individuals carrying the same *PRNP* mutation, the link between mutation and PrD subtype is more consistent. As such, a group of mutations are associated with either the CJD or GSS subtype, however; only one genotype corresponds to the familial fatal insomnia (FFI) subtype (Figure 21.2).

Risk Associated Polymorphisms

In addition to autosomal dominant mutations, there are several naturally occurring polymorphisms that can modify disease risk or its phenotypic expression. The most significant of these is the polymorphic codon 129.

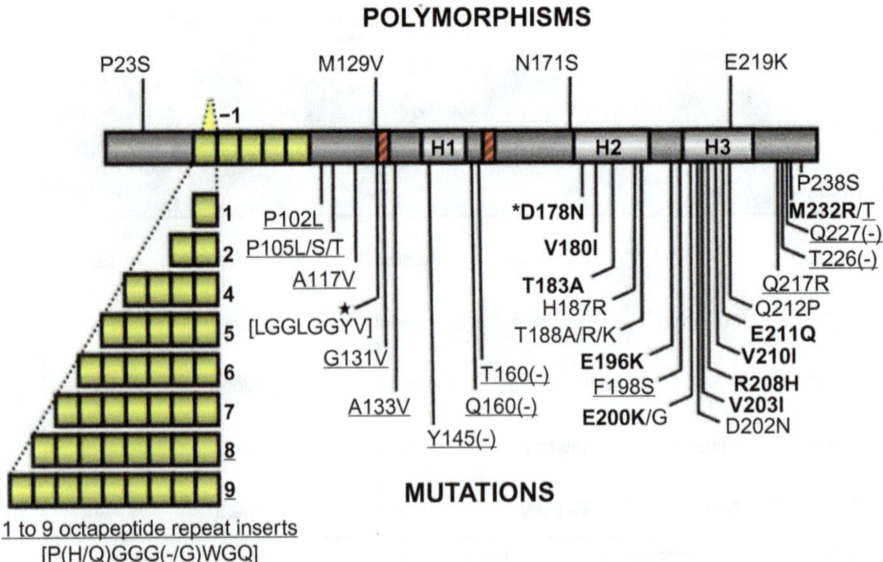

FIGURE 21.2 ***PRNP*** **gene of humans and associated polymorphisms and disease-associated mutations.** The relative positions of polymorphic changes (top) and disease-associated mutations (bottom) are represented along the length of the *PRNP* gene. Reported polymorphisms include single base pair changes, resulting in amino acid substitutions, and deletion of a single octarepeat, whereas disease-associated mutations result from missense mutations, insertions of a variable number of octarepeats or a unique insertion at codon 129 (★), or early stop sequence mutations (–). The letter preceding the residue number indicates the normal sequence and the letter following the residue number is the substitution. The underlined mutations are associated with Gerstmann–Sträussler–Scheinker disease (GSS), those in bold cause Creutzfeldt–Jakob disease (CJD), and the remainder have not been well characterized because of too few examples, or variable presentations. (*) D178N mutation produces CJD when 129 V is allelic (i.e., 178 N/129 V) or FFI when 129 M is allelic (1788 N/129 M). A = alanine, D = aspartate, E = glutamate, F = phenylalanine, G = glycine, H = histidine, I = isoleucine, K = lysine, L = leucine, M = methionine, N = asparagine, P = proline, Q = glutamine, R = arginine, S = serine, T = threonine, V = valine, Y = tyrosine.

Approximately 50% of Caucasians are homozygous for methionine (M) or valine (V) at this position (129MM or 129VV), whereas >85% of patients with CJD are homozygotes, suggesting heterozygosity at this position is protective.[32–36] It has also been shown that homozygotes of either allele have a reduced incubation period for kuru.[37] Interestingly, the distribution of the codon 129 polymorphism differs in Japan, where ~90% of the population is 129MM, thus masking a risk association in that population.[38] The codon 129 polymorphism also appears to influence the phenotype of PrD. The 129MM genotype is typically associated with the onset of dementia and a rapidly progressive disease course, whereas 129MV and 129VV correlate with the onset of ataxia and a slower course.[39] Differences in the patterns of spongiform degeneration and the physicochemical property of PrP^{Sc} appear to be similarly guided by the polymorphic codon 129 (see page 242 and Table 21.2). Homozygosity at codon 129 may also enhance the onset and progression of genetic PrD resulting from a dominant mutation.[40,41]

A striking demonstration of the influence of codon 129 is its role in defining the phenotype of familial PrD caused by the substitution of asparagine (N) for aspartate (D) at codon 178 (D178N) of *PRNP*. This mutation is associated with both familial CJD (fCJD) and fatal familial insomnia (FFI), the clinical and pathologic phenotypes of which are distinctly different (see Human Prion Disease Subtypes section). Goldfarb et al.[42] recognized that patients with FFI carried 129 M on the mutated allele while those with fCJD carried 129 V. This genetic association is even more striking when considered in light of studies that support a difference in conformation of PrP^{Sc} in these two diseases. Structural modeling has since suggested that the amino acid at 129 may interact with the asparagine at 178 to affect the three-dimensional structure of PrP.[43]

In addition to codon 129, ~6% of Japanese encode lysine (K) rather than glutamate (E) at codon 219, and in a small cohort, none of the patients with PrD carried the 219 K codon, suggesting that the lysine substitution is protective.[44] PrP-219 K was subsequently shown to act in a dominant-negative manner to inhibit conversion of PrP^C to PrP^{Sc} in a cell-based prion assay[45] and in Tg mice co-expressing PrP-219 K and PrP^C that were challenged with scrapie.[46] The E219K polymorphism has also been reported to modify the phenotype of the P102L mutation, indicating that in humans, the polymorphism does not prevent conversion of mutated PrP, but may modify it.[47]

A rare serine (S) to asparagine (N) polymorphism and a deletion of a single octarepeat segment, have also been reported, although, with one exception of a family with the D178N mutation and the N171S polymorphism,[48] they have not been shown to influence PrD phenotype or disease risk.[49–51]

CELLULAR PRION PROTEIN BIOLOGY

PrP^C is synthesized within the secretory pathway. It is translated as a 253 amino acid protein that undergoes a variety of posttranslational processing events prior to reaching its final destination at the plasma membrane as a 209 amino acid glycosylated protein. Following translocation into the endoplasmic reticulum (ER), the first 22 amino acids, which function as an ER entry signal, are cleaved and the ~22-amino acid carboxy-terminal glycophosphatidylinositol (GPI) signal peptide is replaced with a GPI anchor.[52,53] A disulfide bond is formed between cysteines at positions 179 and 214 and asparagines at positions 181 and 197 undergo core glycosylation before PrP passes to the Golgi[54] (Figure 21.3), where trimming and modification of the sugars occurs prior to transport to the exofacial membrane of the cell where it resides. The half-life of PrP^C, as assessed in cultured cells, is short (roughly 3–7 hours);[55,56] its metabolic fate largely follows the endocytic pathway, where it undergoes recycling, or it is degraded in the lysosome. It has been proposed that ~10% of PrP undergoes ER-associated degradation by the proteasome. Western blotting of PrP^C reveals three prominent fractions, which correspond to unglycosylated, monoglycosylated, and diglycosylated PrP. In addition, some PrP molecules that retain either the *N*- or *C*-terminal signal sequences, as a result of incomplete or absent ER translocation, may also be detected.[57] This might explain the presence of unglycosylated PrP within the cytosol of cultured cells when artificially expressed.[58] During endocytic recycling, PrP^C may undergo enzymatic cleavage at residue ~110 or 111, generating the N1 fragment PrP23–110/111, and the C1-fragment PrP 111/112-231,[59] which may protect the protein from conversion to PrP^{Sc}. A sheddase, likely ADAM10, may further cleave surface membrane anchored PrP between Gly and Arg, adjacent to the GPI anchor, to release it extracellularly.[60]

PrP^C is expressed in a variety of cell types, although mRNA levels are highest within neurons,[19,61] which provides the basis for PrD being a neurological condition. Initial immunoelectron microscopy studies suggested PrP^C localized primarily within the synapses,[62] but later studies support a more generalized neuronal surface localization that does not predominate at the synapse.[63] PrP^C transcripts are also detected at low levels in glial cells.[64] In addition, PrP^C expression is not exclusive to the central nervous system[65] and more recent quantitative real time reverse transcriptase polymerase chain reaction (RT-PCR) in both sheep and hamsters indicate that mRNA levels in lymph nodes is comparable to brain, whereas levels in spleen, heart, lung, and liver are at moderate levels, and lowest levels are in kidneys.[66] Detection in white blood cells and platelets has also been reported.[67]

PrP^C Function

Initial attempts to define the normal function of PrP^C came from Tg mice constructed with the mouse PrP gene (*Prnp*) disrupted (i.e., *Prnp*$^{0/0}$).[68] Surprisingly, these mice lacked an obvious phenotype, which suggested redundancy

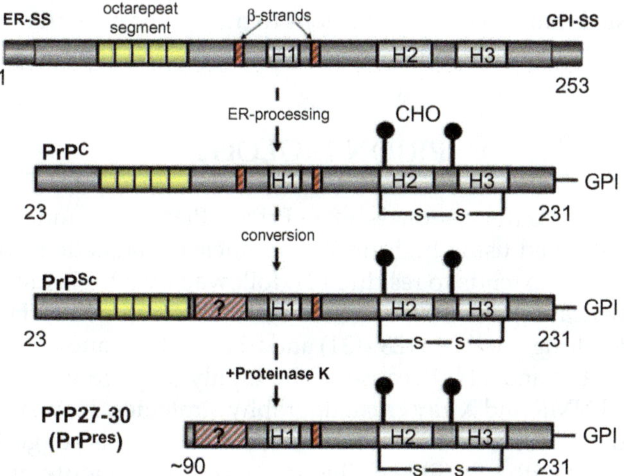

FIGURE 21.3 **Schematic of prion protein (PrP) isoforms.** Nascent PrP contains the ~22 amino acid signal sequences for endoplasmic reticulum (ER) entry (ER-SS) and GPI anchor attachment (GPI-SS). Processing occurs in the ER and Golgi, as described in the text, after which, it is transported to the external surface of the plasma membrane and secured by a GPI anchor, as mature PrP23–231. No sequence difference exists between PrP^C and PrP^{Sc}, as they represent conformational isoforms. The area labeled by '?' signifies the general region predicted to misfold to β-sheet upon conversion to PrP^{Sc}, although the limits of this are not known and appear to extend into H1. Proteinase K (PK) treatment of PrP^{Sc} results in cleavage of the *N*-terminal near residue 90, designated PrP27–30 or PrP^{res}, with associated glycoforms intact. CHO = glycosylations, –S–S– = disulfide bond, H1, 2, and 3 = α-helices. GPI, glycophosphatidyl-inositol anchor.

of function of PrP. Subsequently, it was reported that electrophysiologic studies performed on hippocampal slices from these mice demonstrated impaired GABA receptor-mediated fast inhibition and long term potentiation in CA1 pyramdial neurons,[69] although follow up studies from another group failed to replicate these findings.[70] Two subsequent models of *Prnp*$^{0/0}$ mice were found to exhibit alterations in circadian rhythms and sleep patterns,[71] suggesting a possible role for PrP^C in sleep maintenance. This is intriguing when one considers fatal insomnia, a subtype of PrD that manifests primarily as intractable insomnia. In more recent work, *Prnp*$^{0/0}$ mice were found to display behavioral deficits in hippocampal-dependent spatial learning and a reduction in long-term potentiation in the dentate gyrus, features that were reversed by co-expression of PrP^C.[72] In addition, mice deficient in PrP were shown to develop defects in myelination.[73,74]

PrP^C has been proposed to carry on a variety of functions. A role in signaling was initially proposed when it was demonstrated in a cell culture model of neuronal differentiation that antibody mediated cross-linking of PrP^C caused a caveolin-1-dependent decrease in phosphorylation of the tyrosine kinase Fyn.[75] In addition, *in vivo* administration of antibodies to the hippocampus of normal mice was found to induce dimerization of surface PrP^C that resulted in severe neurotoxicity, supporting the induction of an apoptotic cascade via a signaling function.[76] PrP^C may also possess anti-apoptotic activity; when expressed in human neurons and yeast, it prevents Bax-mediated cell death.[77,78] These and other findings have led to the concept that cell death associated with PrD results, at least in part, from the loss of its protective function.

PrP^C has the potential to bind at least four atoms of copper at histidines within the octarepeat region,[79,80] and it has been noted that the cellular uptake of copper is more efficient in cultured cells expressing PrP^C compared to those lacking it, while the rate of endocytosis of PrP^C from the plasmalemma is enhanced by copper addition to the media.[81,82] In addition, brain homogenates from *Prnp*$^{0/0}$ mice have decreased Cu/Zn superoxide dismutase (SOD) activity,[83] presumably due to the loss of the copper binding potential of PrP^C, since brain homogenates from normal mice immunodepleted of PrP^C also exhibit decreased SOD activity.[84] PrP^C has been proposed to either regulate incorporation of Cu^{2+} into SOD[85] or it possesses its own SOD-like activity,[86] although the latter has been challenged.[87,88]

PrP^C also appears to play a role in neuronal development, differentiation, and neurite outgrowth, a feature that may be related to its interaction with adhesion molecules such as laminin and N-CAM.[89–91] In addition to these molecules, several other binding partners of PrP^C encompassing a range of functions from endocytosis to apoptotic signaling have been detected (see [92,93] for details), although based on the use of *in vitro* systems and the propensity of PrP to aggregate, it is not clear whether many of these binding partners are physiologically relevant.

Of considerable interest in recent years is the possible role of PrP^C in Alzheimer disease (AD). Initial work suggested that PrP^C suppresses beta-secretase (BACE1), a key enzyme leading to the generation of Aβ, suggesting it functions to protect against AD.[94] However, another laboratory reported that PrP^C functions as a specific receptor for soluble Aβ oligomers and mediates toxic effects of Aβ to block long term potentiation and inhibit synaptic plasticity.[95] Further work by the same group reported that PrP^C mediates the synaptic toxicity of Aβ via its interaction with Fyn kinase.[96] Other work has suggested that inoculation of prions to transgenic mice that model AD promotes Aβ plaque deposition,[97] as does the expression and overexpression of PrP^C.[98]

PRION BIOLOGY

The prion hypothesis asserts that a conformational shift of PrP^C to PrP^{Sc} is the underlying basis for induction of PrD. The structure of PrP^C has been resolved using high-resolution nuclear magnetic resonance (NMR) and X-ray.[43,99,100] An unstructured *N*-terminal segment extends to residue 121, followed by a highly structured *C*-terminal region that includes three well-ordered α-helical segments, designated helix 1 (H1; 144–154), H2 (173–194), and H3 (200–228), two antiparallel β-strand regions, designated S1 (128–131) and S2 (161–164), and a loop created by a disulfide bridge formed by cysteines at positions 179 and 214. Because of the highly aggregative nature of PrP^{Sc}, its tertiary structure remains elusive from typical NMR and X-ray crystallography strategies. However, secondary structure analysis from Fourier transform infrared spectroscopy and circular dichroism analysis suggest ~40% of the molecule carries β-pleated sheet structure, compared with <5% in PrP^C (Figure 21.4).[101] This feature underlies the biophysical properties characteristic of PrP^{Sc}, including insolubility in nondenaturing detergents and a relative resistance to proteinase K (PK) digestion. Chromatographic size fractionation of partially disaggregated PrP^{Sc} suggests the most infectious fraction is comprised of oligomers of 14–28 PrP^{Sc} molecules.[102]

The property of PK resistance has been used as a marker for PrP^{Sc} in tissues. If PrP^{Sc} is present, a protease resistant core that is *N*-terminally truncated at about residue 90, designated PrP27–30, based on its rate of migration through an electrophoretic field (also referred to as PrP^{res}), is observed by western blot analysis of a tissue homogenate

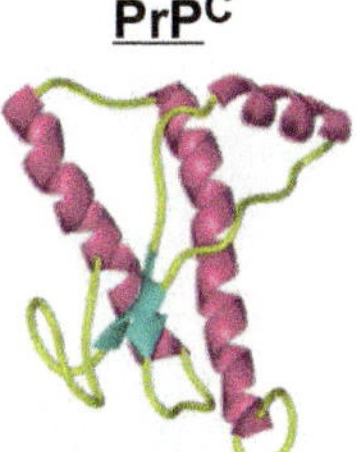
PrPC
• Cellular isoform
• 43% α-helical
• <5% β-sheet
• Cell surface
• Detergent soluble
• PK-sensitive

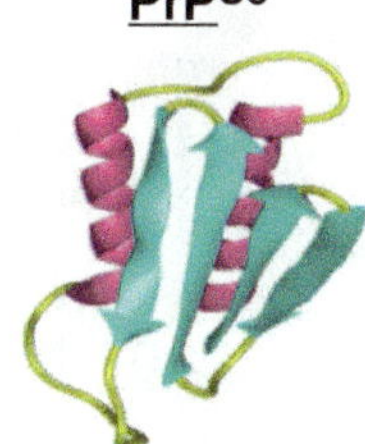
PrPSc
• Scrapie isoform
• 34% α-helical
• 43% β-sheet
• Intra/extra-cellular
• Insoluble
• PK-resistant

FIGURE 21.4 **Comparison of the cellular isoform of prion protein (PrP^C) and scrapie-associated isoform (PrP^{Sc}).** PrP^C and PrP^{Sc} represent structural conformers of each other. The structures presented are based on NMR data for PrP^C and solution secondary structure analysis for PrP^{Sc}. The notable difference between the two structures is the predominance of β-sheet structure of PrP^{Sc} (>40%) compared to the predominantly β-helical PrP^C (<5% β-sheet). This imparts the properties of insolubility, proteinase K (PK) resistance, and presumably, cellular toxicity. Whereas PrP^C is predominantly plasmalemma bound, PrP^{Sc} accumulates intracellularly, but may also deposit in the extracellular space, as with PrP amyloid plaques. It is important to recognize that, although this depicts the major change in conformation of PrP^C to PrP^{Sc}, there are multiple pathogenic conformations of PrP^{Sc}, which forms the basis of prion strains.

(see Figures 21.3 and 21.7). Because the PK-resistant core of PrP^{Sc} includes the two *N*-linked glycosylation sites, all PrP glycoforms display PK resistance, represented as three PrP fractions on western blot that migrate approximately 5–6 kDa faster than its undigested holoprotein, although this may vary depending on prion strain (see page 240) and the presence of specific mutations that may affect its glycosylation.[103] For example, typical appearing PK-resistant PrP is generally absent in humans with GSS and in a Tg mouse model of GSS.[104] Instead of the typical three fractions, a single 7–14-kDa resistant fraction may be present, corresponding to *N*- and *C*-terminal truncated PrP fragments that extend from residue 88 to 153 and which compose the amyloid plaques characteristic of GSS.[105]

The appearance of PrP^{Sc} correlates with progression of PrD in humans, and it appears to predate the development of vacuolation within the gray matter of the brain in experimental scrapie.[106] PrP^{Sc} accumulates primarily within nervous tissues; however, in experimental rodent models of disease, and in vCJD of humans, PrP^{Sc} has also been recovered from lymphoreticular tissues, including tonsils, spleen, and lymph nodes,[107–110] which raised concern early on regarding the safety of blood products, specifically those obtained from individuals who have traveled to Europe during the height of the BSE epidemic, as they may be harboring infectious prions. This concern was subsequently affirmed by reports of secondary transmission of vCJD in four individuals who were exposed to blood products acquired from patients with presymptomatic primary vCJD, who developed disease several months to years later.[111] Prevention of B cell differentiation and activation in mice was found to delay or prevent the onset of PrD following peripheral inoculation, supporting a role of white blood cells in prion neuroinvasion.[112–114]

Other peripheral tissues relevant to the mechanism of transmission of BSE to humans have been shown to harbor prion infectivity. Specifically, quadriceps muscle from scrapie-infected rodents has transmitted disease,[115,116] as did muscle from deer with CWD, using Tg mice expressing a cervid PrP gene sequence as host.[117] Although PK-resistant PrP^{Sc} is documented in nearly all transmissible PrDs, in rare instances, transmissibility has been reported in the absence of western blot detectable PK-resistant PrP.[118,119] In humans, the subtype of prion disease described as variably protease-sensitive prionopathy (VPSPr), is associated with variable degrees, and in some cases, a total lack of, PK-resistance, although whether these diseases are transmissible has not yet been determined.[26,120]

Prion Initiation and Propagation

Three mechanisms lead to the development of PrD: 1) exposure to infectious prions, 2) harboring a mutation of *PRNP*, or 3) an unknown mechanism of spontaneous conversion (Figure 21.5). Prion exposure accounts for less than 1% of all occurrences of PrD, while only ~10–15% are traced to a germline mutation, leaving the vast majority of cases unexplained. Although a somatic mutation of *PRNP* has been suggested to account for the spontaneous development of disease, this is difficult to prove. It is generally accepted that the majority of cases result from the spontaneous conversion of PrP^C to PrP^{Sc}, although the molecular triggers and cellular compartments necessary for this event are unknown.

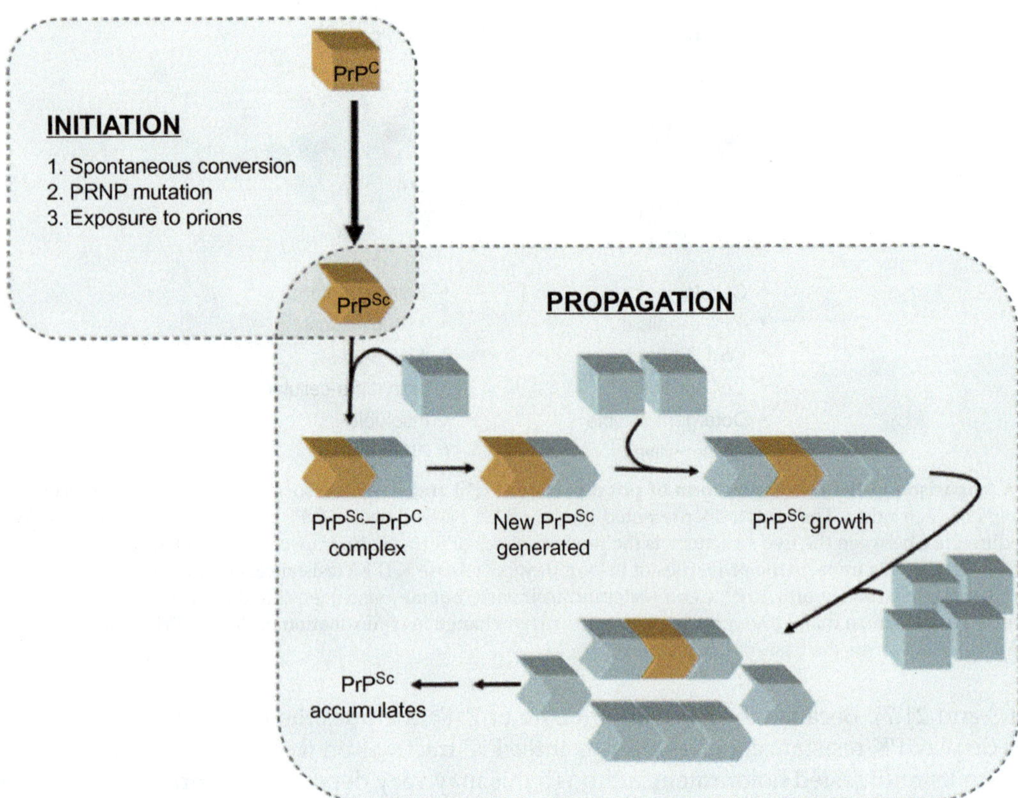

FIGURE 21.5 **Initiation and propagation of scrapie-associated prion protein (PrP^{Sc}).** Initiation of prion disease (PrD) occurs when PrP^{Sc} is generated by one of three mechanisms. Once present at a level that constitutes an "infectious unit," PrP^{Sc} propagation occurs via an autocatalytic process that involves recruiting endogenous cellular form of prion protein (PrP^C) for conversion to PrP^{Sc}. A PrP^{Sc}–PrP^C complex is formed and the misfolded conformation of PrP^{Sc} is templated onto PrP^C. Whether PrP^C binds to the exposed ends of an extending ordered fibril of PrP^{Sc} or some other oligomeric PrP^{Sc} structure, is not clear.

The PrP^{Sc} conformation is thermodynamically more stable than PrP^C, although *in vitro* studies with recombinantly expressed PrP indicate that, at neutral pH, it is folded into the predominantly α-helical PrP^C conformation with rapid kinetics.[121] As such, an activation energy barrier must be overcome for the conversion of PrP^C to PrP^{Sc}.[122] Although a mutation within the coding segment of *PRNP* is thought to destabilize PrP^C sufficiently for it to misfold into an unstable conformation (PrP*), thereby lowering the activation energy barrier, structural prediction models have not fully confirmed this hypothesis.[43]

Many investigators have searched for binding partners of PrP^C, as a way to determine its function and to uncover proteins that may assist in the conversion process. Using yeast two-hybrid, co-immunoprecipitation, and cross-linking strategies, among others, several theoretical partners have been proposed, including GFAP,[123] Bcl-2,[124] caveolin,[125] apoE,[126] several ER chaperones, including calnexin, BiP, and protein disulfide isomerase,[127–129] stress-inducible protein 1,[130] and APLP1,[131] among others. While these proteins may interact with PrP and could feature in the conversion process, direct evidence for their involvement in PrD, is generally lacking. However, the 37-kDa/67-kDa laminin receptor[132] has been reported to mediate PrP^{Sc} propagation in an *in vitro* model of infected neurons,[133–135] and antibody blockade of laminin receptors was shown to prevent uptake of BSE in enterocytes.[136]

One hypothesis proposed for the spontaneous generation of prions is inefficient ER-associated degradation (ERAD).[137,138] In this scenario, PrP misfolds within the ER where resident chaperones fail to refold it to its native state, resulting in retrotranslocation to the cytosol, where it may accumulate under conditions of impaired proteasome function. In support of this hypothesis, PrP-carrying mutations associated with PrD have been reported to undergo prolonged contact with ER chaperones, such as BiP[128] and show delayed exit from the ER.[139,140] In addition, chemical inhibition of the proteasome results in the accumulation of an insoluble fraction of cytosolic PrP in a variety of cultured cell lines expressing either recombinant or endogenous PrP.[137,138,141] However, several lines of evidence suggest cytosolic PrP originates from inefficient translocation of PrP into the ER, rather than active retrotranslocation.[140,142] When PrP expression is directed exclusively to the cytosol, it induces apoptosis in cultured cells, and when expressed in transgenic mice, it produces a phentoype of pruritis, late-onset ataxia, and focal degeneration within the granule cell layer of the cerebellum.[143] Because mice that express only cytosolic PrP are resistant to prion infection, this fraction of PrP may function principally as a toxic, rather than infectious, component, as these mice were found to be resistant to prion inoculation.[144]

Propagation of PrP^{Sc} occurs by a process that may be quite distinct from initiation. In this case, an infectious unit of PrP^{Sc} has already been generated by one of the three mechanisms discussed earlier. To spread and replicate, PrP^{Sc} binds to and transfers its conformation onto resident PrP^{C} (Figure 21.5), thereby generating new PrP^{Sc} molecules with a similar conformation as the parent PrP^{Sc} molecule. This process, labeled as a "template-assisted conversion process," forms the basis for the persistence of prion strain transmission from host to host. Thus, PrP^{Sc} propagation requires the presence of PrP^{C}, a finding initially revealed when mice lacking the PrP gene ($Prnp^{0/0}$) were found to be resistant to prion infection.[68]

Spread of disease within the brain appears to follow neuroanatomical pathways, suggesting axonal transport and release of PrP^{Sc} at the synapse, although release of PrP into the extracellular space that produces PrP amyloid also suggests that diffusible PrP^{Sc} can infect neighboring cells.[145] Evidence suggests that PrP^{Sc} may be packaged into membrane bound exosomes prior to release into the extracellular space,[146] a process which may be common with other neurodegenerative diseases.[147] PrP^{Sc} within the extracellular space presumably interacts directly with PrP^{C} on the plasma membrane of the neuron, and either on the membrane surface or within the early endocytic compartment from lipid-rich rafts or caveolae, PrP^{Sc} converts PrP^{C} to generate new PrP^{Sc}.[148–150] It has been proposed that protein crowding and reduced pH in the endocytic compartments promotes the formation and stabilization of PrP^{Sc}.[151] Because PrP^{C} undergoes constant turnover, there is essentially limitless substrate for conversion, leading to PrP^{Sc} accumulation. Immunoelectron microscopy has suggested lysosomes as a cellular compartment of accumulation.[152] This finding is consistent with recent work that supports a role for autophagy in prion disease. Autophagy is a cell survival mechanism that also functions to isolate cellular protein aggregates within a double membrane vesicle that then fuses with lysosomes, leading to degradation of their cargo. In prion-infected mouse neuroblastoma cells, treatment with autophagy inducers rapamycin or trehalose reduced prion formation,[153] and in Tg mice that model GSS due to the A117V mutation (Tg(PrP-A116V) mice), chronic administration of rapamycin, delayed onset of disease, and prevented PrP plaque pathology.[154]

Synthesizing Prions

It has long been considered that the ultimate proof of the prion hypothesis requires generating PrP^{Sc} *de novo*, either in cells or a test tube, and proving infectivity by bioassay in a receptive animal host. The first effort to reach this goal involved generating *N*-terminally truncated PrP (PrP89–230) via bacterial expression, polymerizing it to amyloid, and inoculating it into Tg mice overexpressing PrP89–230, which led to the development of clinical and pathologic features of PrD, after a significant incubation period.[155] However, it was argued that the Tg(PrP89–230) mice express ~16 times the normal levels of PrP, making them artificially susceptible to developing prion disease, despite the fact that control animals that received nonamyloid PrP did not spontaneously develop the disease after an extended period.

An *in vitro* method of PrP^{Sc} replication, known as protein misfolding cyclic amplification (PMCA), has also contributed to this goal. This process involves adding a small amount of PrP^{Sc} to a sample of normal brain homogenate that provides the PrP^{C} substrate. Using a specified frequency and interval of sonications and incubations, template PrP^{Sc} aggregates/fibrils are disrupted, which increases the number of exposed surfaces to which PrP^{C} can bind and be converted to PrP^{Sc}.[156] Thus, akin to a PCR-type amplification process, PMCA amplifies low levels of PrP^{Sc}. This method has demonstrated potential usefulness as a diagnostic test for tissues with very low levels of PrP^{Sc} in brain and other tissues, including blood.[157] In addition, it has proven useful to identify specific cofactors required in the conversion process to maintain infectivity and strain conformation[158] and it has demonstrated that bacterially expressed PrP^{C} can be spontaneously converted to PrP^{Sc} with proven infectivity to mice, following multiple cycles of PMCA.[159] This finding somewhat dampens the specificity of PMCA as an assay for low level prions in humans or animals, but it provides strong support as proof for the prion hypothesis.

Prion Pathogenesis

PrP^{Sc} accumulates in neurons; however, the actual mechanism of neuronal death in PrD is unclear. Most evidence favors a toxic gain of function; however, the loss of the normal function of PrP^{C} may contribute to cell death. Several arguments can be made in favor of a toxic gain of function. The accumulation of PrP^{Sc} has been linked with apoptotic cell death in animal models[160] and humans.[161] The protease-resistant core of PrP^{Sc} is toxic when applied to cultured cells[162] and a peptide segment of PrP (PrP106–126) is toxic to hippocampal cultures,[163] mixed cerebellar cultures,[164] and primary neuronal cultures. Although the PrP106–126 peptide is not observed in nature, longer peptide segments of PrP extending from residues 85 to 150 have been detected in PrP-amyloid plaques found in the brains of patients with GSS.[105,165] The most compelling data in favor of a toxic function is the finding that *Prnp*$^{0/0}$ mice do not develop spontaneous neurodegeneration.

Two fractions of PrP have been recognized as toxic: 1) an atypical transmembrane topology (^{Ctm}PrP), and 2) a cytosolic fraction. ^{Ctm}PrP was initially detected in Tg mice expressing PrP with mutations within the putative transmembrane domain and in the brain of a patient with GSS caused by the A117V mutation.[166] The fraction of ^{Ctm}PrP was also found to increase with progression of PrD in experimental animal models.[167] PrP carrying mutations that favor ^{Ctm}PrP topology display aberrant trafficking and are retained in the ER and Golgi.[168,169]

Cytosolic PrP was first hypothesized after it was recognized that expression of PrP in the cytosol of yeast produced an insoluble and PK-resistant PrP^{Sc}-like molecule.[170] Expression of PrP23–231, which lacks the signal sequences for ER entry and GPI-anchor attachment, induces apoptotic cell death in mammalian cells, and Tg mice develop a PrD-like phenotype in association with focal cerebellar granule cell layer degeneration.[143] However, neither brain tissue from these mice, nor cells overexpressing cyPrP were transmissible to normal mice, suggesting a toxic but not infectious property.[144] Cytosolic PrP has not been demonstrated in human PrD, but electron microscopy studies have identified low levels of free PrP in the cytosolic compartment.[171]

Several lines of evidence suggest that PrP^{Sc} transmission and toxicity depend on the presence of PrP^{C} on the plasma membrane. This is best modeled by Tg mice that express PrP lacking the GPI anchor [PrP(GPI-)] on a *Prnp*$^{0/0}$ background. When challenged with mouse-adapted scrapie, these mice did not develop clinically apparent disease after several hundred days, despite a dramatic accumulation of extracellular PrP amyloid deposits in their brains, and detection of PK-resistant PrP by western blot.[149] However, when PrP^{C} was co-expressed with PrP(GPI-), the mice showed enhanced susceptibility to prions and displayed severe spongiform degeneration, in addition to amyloid plaque accumulation. This agrees with the demonstration that neurons within the brain of *Prnp*$^{0/0}$ mice are resistant to the toxic effects of PrP^{Sc} produced by grafted normal brain tissue that expresses PrP^{C}.[172]

The conversion of PrP^{C} to PrP^{Sc} during the course of disease, resulting in a relative reduction in total PrP^{C}, might also contribute to the toxicity associated with PrD. PrP^{C} overexpression protects against Bax-induced apoptosis in mammalian cells[77] and when expressed in yeast.[78] In addition, cells derived from *Prnp*$^{0/0}$ mice are hypersensitive to oxidative stress,[83] and Tg mice that express Doppel, a paralog of PrP[173] (discussed later) or *N*-terminally truncated forms of PrP (PrPΔ32–134 and PrPΔ32–121)[174,175] in the absence of PrP^{C} develop neurotoxicity, but not when co-expressed with PrP^{C}. This latter finding has led to the proposal that the *N*-terminal segment of PrP is required to bind to a protective receptor. More recent data suggests that expression of PrP lacking a small, but critical, segment of PrP extending from aa 105–125, known as the toxic peptide segment, results in a clinical and pathological phenotype in mice that is rescued by co-expression of PrP^{C}.[176] When this, or PrP carrying point mutations within the mouse PrP transmembrane domain (aa 111–135) linked to familial prion disease, was expressed in cultured cells, they induced abnormal ionic fluxes consistent with the induction of ion channels.[177] This, and other data, has led to the idea that prions might induce cell death directly via excitotoxicity through the generation of membrane pores or ion channels. Membrane pores have been generated with amyloid-β peptide and in artificial membranes with PrP105–125 peptide.[178] At this time it is difficult to resolve the direct effect of pore formation by PrP105–125 with the ion fluxes observed with PrP lacking the 105–125 segment, since the latter has not been shown to induce amyloid or PK-resistant PrP *in vivo*. In addition, whether mutations outside the transmembrane domain will result in similar changes in channel activity will be an important determination.

Based on these multiple reports and proposed mechanisms, it is evident that a clear theory has not emerged. Could it be that multiple scenarios are possible, each of which depends on the mode of occurrence, albeit sporadic, acquired exogenously, and because of the varied pathologies associated with different mutations, could these depend on the specific position and nature of the mutation and downstream effect on PrP conformation? Until a well-defined mechanism emerges, whether prions induce cell death as a result of a toxic gain of function, a loss of protective function, or a combination of the two, will continue to be debated.

Species Barrier

Humans are not the only hosts to prions. Several animal species develop PrD, including cats, mink, and several zoo animals, the majority occurring through ingestion of prion-contaminated feed. Scrapie is a natural disease of sheep, and this has been used to generate several rodent-passaged isolates that are routinely used in experimental studies. The original cases of BSE are thought to result from the combination of feeding meat and bone meal (MBM) prepared from scrapie-infected sheep, and a reduction in the stringency of the rendering process in the 1970s, resulting in greater survival of prions in MBM.[179] The BSE epidemic highlights the principles of the species barrier to prion transmission. Transmission of prions across species is variable, depending on the species involved, but once

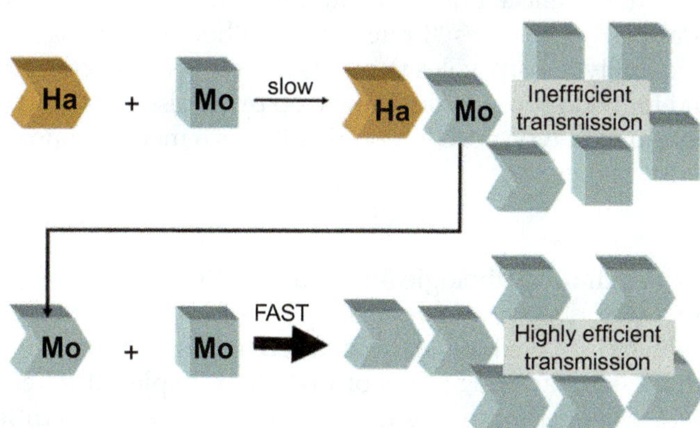

FIGURE 21.6 **The species barrier.** Prions propagate best when they interact with PrP^{C} with a homologous sequence, especially at the PrP^{Sc}–PrP^{C} interface. Initial passage of hamster prions to mice is slow (extended period from inoculation to symptom onset) and inefficient (a small fraction of inoculated animals develop disease) because of differences in amino acid sequences of hamster (Ha) PrP^{Sc} and mouse (Mo) PrP^{C}. Once passed to mice, newly generated PrP^{Sc} now carries mouse sequence, which can be passaged to other mice efficiently, since there is complete sequence homology between PrP^{Sc} and PrP^{C}. Although differences in PrP sequence among species accounts for the majority of the species barrier phenomenon, other factors, such as host proteins involved in prion propagation, and the specific strain passaged, have also been considered to play a role.

passed to the second species, subsequent transmission within that species is highly efficient (Figure 21.6). With BSE, the transmission of sheep prions to cows is inefficient, but when cows with BSE were used in the preparation of MBM that was fed to other cows, disease transmission was highly efficient, resulting in a staggering epidemic that ravaged the cattle population of the UK. A ban placed on MBM feeding, along with improved monitoring, has essentially eliminated this disease. Recently recognized cases of spontaneous BSE that exhibit atypical patterns of PK-resistant PrP, with either high (H-type) or low (L-type) migration rates and/or altered glycosylation, compared with classical BSE (C-type), emphasize the need for continued surveillance, to prevent the possible entry into human food products.[180,181] This is underscored by the demonstration that L-type atypical BSE is more efficiently transmissible than classical BSE to Tg mice expressing either bovine or human PrP,[182,183] and to nonhuman primates, and the pathological features in the latter case comparable to some forms of human sCJD.[184]

Tg animal studies determined that the species barrier is largely defined by the amino acid sequence differences between infecting PrP^{Sc} and host PrP^{C}.[185,186] For example, normal mice are poorly receptive to human and hamster-passaged prions, but Tg mice expressing a chimeric mouse–human or mouse–hamster PrP are highly receptive to each, respectively, as are mice expressing the complete homologous PrP sequence of the prion donor.[187] The receptivity of the host may be quite specific, controlled by several or even a single amino acid. This is best demonstrated by the epidemiology of vCJD and the codon 129 polymorphism. All ~200 documented cases of primary vCJD have the 129MM genotype of *PRNP*. Experimental transmission of vCJD to transgenic mice expressing human (Hu) PrP carrying the 129M haplotype [Tg(HuPrP-129M)] is more efficient and shows a greater fidelity of strain transmission compared with mice expressing HuPrP-129V.[188] A similar enhancement in transmission was observed when sCJD carrying the 129MM or 129VV genotype was transmitted to Tg(HuPrP) mice with the corresponding 129 polymorphism.[189]

Understanding the species barrier is of critical importance, as it relates to future public health concerns, especially with regard to emerging prionoses of animals. Chronic wasting disease (CWD) of deer and elk is a naturally occurring PrD first recognized in Colorado in 1967, but has since displayed an aggressive expansion throughout the US and into Canada. The natural and aggressive horizontal spread of this disease is unique to PrD, although the basis for this facility of transmission is not clear. The host range for CWD currently includes white tailed deer, mule deer, and elk, although moose may be susceptible, while ferrets[190] and squirrel monkeys[191] have proven to be experimentally susceptible to the disease. This has raised concern for the wildlife population and humans that consume venison. Mitigating this concern is a report that Tg mice expressing human PrP were not susceptible to CWD, whereas Tg mice expressing cervid PrP are highly susceptible.[192,193] Interestingly, using PMCA, prions from Tg mice expressing cervid PrP were able to convert human PrP^{C} to PrP^{Sc} after multiple amplification cycles.[194] The sequence(s) within PrP that constitute the species barrier to CWD is not yet known.

In recent years, the bank vole has shown a remarkable susceptibility to prions from several species, including humans.[195–197] The basis for this broad species susceptibility has not been fully elucidated, although specific amino acids within bank vole PrP do seem to alter susceptibility to some, yet not all, species,[198] suggesting a host factor other than the PrP gene in bank voles contributes to this effect. Prior work in transgenic mice supported a hypothetical host factor that participates in the propagation of PrP^{Sc}, labeled as "Protein-X",[199] although this has not been confirmed or disproven. Of note, based on the high susceptibility of bank voles to prions from several species, a transgenic mouse that expresses bank vole PrP was generated and found to spontaneously develop prion disease with spongiform degeneration and PK-resistant PrP^{Sc} that is transmissible to wild-type mice, supporting this as a model of sporadic CJD.[200]

Prion Strains

Prion strains refer to the distinct clinicopathological features of disease following passage of prions to the same host. Strain differences were first described with two isolates of transmissible mink encephalopathy (TME), designated HYPER and DROWSY, which displayed remarkably different behavioral phenotypes, pathologic profiles, and incubation periods, along with PK-resistant fragments of PrP^{Sc} that display different electrophoretic mobilities on western blot.[201,202] The latter feature suggested differential PK cleavage, based on differences in the conformation of PrP^{Sc}, allowing exposure of different sites of PK cleavage. Based on this, and subsequent work, strains are characterized by: 1) the incubation period (i.e., interval between inoculation of the host and disease onset); 2) the distribution and type of associated histopathology; and 3) the PrP^{Sc} type, defined by the migration rate of the PK-resistant PrP^{Sc} fraction and, in some cases, the glycosylation pattern, and/or relative PK resistance.

Several conformations of PrP^{Sc} have been detected in different animal prion strains[203] and this phenomenon is recapitulated in human PrD. Several lines of evidence suggest that prion strains are enciphered in the conformation of PrP^{Sc}, suggesting a tissue or cellular targeting effect, resulting in the observed clinicopathological phenotype. PrP^{Sc} isolated from each of the five major subtypes of human PrD displays a unique electrophoretic mobility pattern (Figure 21.7). In addition, in those that have been transmitted to a susceptible experimental host, the PrP^{Sc}

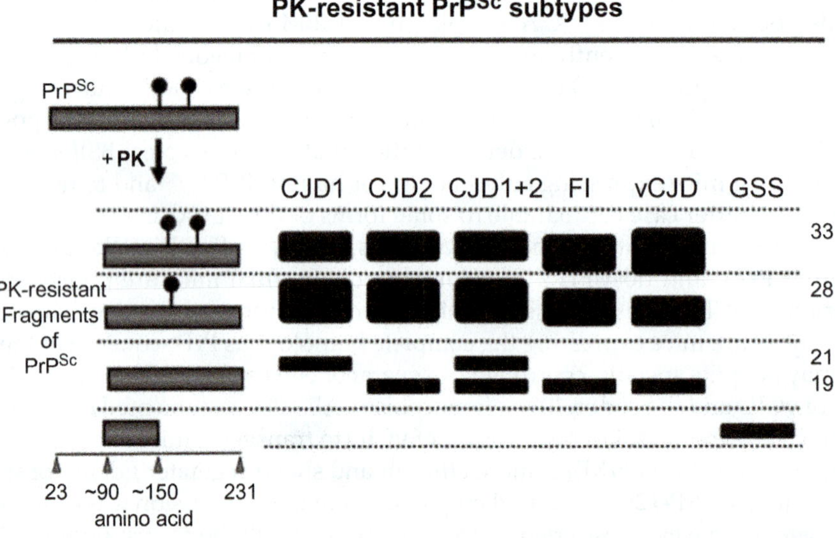

FIGURE 21.7 **Proteinase K (PK)-resistant PrP^{Sc} subtypes.** Diagrammatic representation of the western blot pattern of PK-resistant PrP fragments from each of the major subtypes of prion disease (variably protease-sensitive prionopathy [VPSPr] excluded). The PrP fragments and glycoforms that correspond to the bands on the western blots are represented on the left. The relative position of amino acids as they relate to the cleavage products is displayed below the glycoform fragements. Note there are differences in both migration rate and glycosylation patterns of the subtypes and, in some cases, two PrP^{Sc} subtypes may be present, represented by sCJD1+2. After PK digestion, all remaining fragments are truncated near amino acid 90, but they vary due to the difference in cleavage sites, resulting in different fragment sizes (e.g., type 1 PrP^{Sc} is 21 kDa cleaved primarily at residue ~82, and type 2 is 19 kDa, cleaved at ~97). The uppermost protein band corresponds to diglycosylated PrP, the second band is monoglycosylated, and the lowest band is the unglycosylated fraction. Note that in GSS, the principal, and often only fragment observed, is an unglycosylated fragment that is truncated at both *N*- and *C*-termini (~PrP85/90–150), resulting in the smallest fragments and fastest migration rate. The monoglycosylated fraction is most prominent in typical sporadic Creutzfeldt–Jakob disease (sCJD), as indicated in the diagram, whereas both glycoforms of FI are equally more prominent than the unglycosylated fraction, and in variant Creutzfeldt–Jakob disease (vCJD), the diglycosylated fraction is the most prominent. The migration rates of PK-resistant PrP^{Sc} from fatal insomnia (FI) and vCJD are comparable to type 2 PrP^{Sc} of sCJD whereas that from Gerstmann–Sträussler–Scheinker disease (GSS) runs faster than all because of the additional C-terminal cleavage that occurs. The different migration rates indirectly reflect differences in PrP^{Sc} conformations among the phenotypes. The reason for the predominance of one over another glycoform for each subtype, is less obvious, but likely has to do with PrP conformation, and local cellular environment from which the PrP^{Sc} molecule originated prior to propagating.

conformation propagates and correlates with different histopathologic phenotypes in the experimental host.[187,204] The "protein signature" of PK-resistant PrP (i.e., PrP^{Sc} type) can aid in the classification of PrD, and this feature was instrumental in determining that BSE in cows is the cause of vCJD in humans. BSE shows a characteristic electrophoretic signature that is maintained when experimentally transmitted to several host species, such as hamster, macaque, and mouse,[205–209] and this same protein signature was observed in humans with vCJD. How different PrP^{Sc} conformers can lead to such disparate clinical and pathologic profiles in PrD is not well understood. Full characterization of these different PrP^{Sc} subconformations awaits high-resolution structural information on PrP^{Sc}, and further work is needed to address how these different conformations influence expression of disease.

PRION-RELATED PROTEINS

Doppel

When additional lines of *Prnp*$^{0/0}$ mice were engineered, in contrast to the original line, some developed late onset ataxia with loss of cerebellar Purkinje cells.[210,211] This was subsequently determined to be caused by the deletion of the splice acceptor site of exon 3 of the *Prnp* gene during preparation, which led to the approximation of the *Prnp* promoter with a gene lying 16 kb downstream of *Prnp* in mice. As such, this gene was driven by the PrP promoter. The gene was designated prion-like protein gene (*Prnd*), because it encodes a protein with significant homology of sequence and structure to PrP, appropriately named Doppel (Dpl).[212] The tertiary structure of Dpl in mouse and humans shows significant similarity to PrP^C, although only 25% sequence homology is seen between *C*-terminals of Dpl and PrP^C.[213,214] As in PrP^C, Dpl contains three α helices, the first of which is flanked by two β strands, and two *N*-linked carbohydrate groups. In contrast to PrP, there are two, rather than one, disulfide bonds, no octarepeat segment, and no hydrophobic core. As with PrP^C, however, Dpl follows the secretory pathway and is linked to the plasmalemma by a GPI anchor.[215,216] Dpl expression pattern differs from PrP^C in that its postembryonic expression is generally confined to the testes, although some have reported expression in lymphoid cells.[217,218]

Confirmation that Dpl caused the ataxic phenotype was provided when the original line of *Prnp*$^{0/0}$ mice were constructed to express Dpl, which resulted in ataxia and loss of cerebellar Purkinje cells at 6–12 months of age.[173] No PK-resistant protein or pathology characteristic of PrD was detected in these mice, although Dpl aggregates were observed in cerebellum. When mice co-expressing Dpl and PrP^C were challenged with scrapie, there was no alteration in the properties of scrapie transmission,[173,219] suggesting that Dpl does not play a direct role in PrD. The role of Dpl in PrD is yet unclear, and while evidence suggests it may be directly toxic to neurons,[220] no mutations of *Prnd* have yet been linked to familial PrD,[221] although polymorphisms have been detected and some investigators have argued a risk to sCJD.[221–223] Of note, there is no evidence for elevated expression during the course of PrD in human brain.[224]

Shadoo

Shadoo (Sho), from the Japanese word for "shadow," was identified by scanning public domain protein databases for PrP-like proteins and found to display partial homology with PrP^C to the hydrophobic domain, the octarepeat segment, the *N*-linked glycosylation within the *C*-terminal region, and the GPI anchor.[225] It is expressed by the *Sprn* gene located on chromosome 7 in mice and 10 in humans, and is therefore not linked to the *Prnp*/*Prnd* locus on respective chromosome 2 in mice and 20 in humans. In contrast to Dpl and PrP, Shadoo is expressed only within the brain. Some have considered this to have overlapping function with PrP^C, but Tg mice null for both *Sprn* and *Prnp*[226] did not develop a phenotype or shortened survival compared with wild-type or *Prnp*$^{0/0}$ mice. Interestingly, Sho levels are found to decrease in rodent models of prion infection (see [227,228] for review), speculating this might be an early biomarker of prion disease. While mutations within *SPRN*, the human homolog of *Sprn*, have not been linked to genetic human prion disease, a frameshift mutation was reported in two vCJD cases and a polymorphism in the signal sequence was significantly associated with sCJD.[229]

HUMAN PRION DISEASE SUBTYPES

There are at present five major human subtypes of PrD: CJD, GSS, FI, vCJD, and VPSPr (Table 21.1). CJD and FI occur in sporadic and familial forms, whereas GSS is always found in association with a *PRNP* mutation, and VPSPr is sporadic. Acquired forms of PrD include iatrogenic CJD (iCJD) and vCJD. A brief description of the characteristic features of each phenotype follow.

TABLE 21.1 Summary of Clinical Phenotypes of Current Human Prion Diseases

Phenotype	Onset (years)	Duration (years)	Typical presentation	Pathology	Other clinical data
SPORADIC					
Creutzfeldt–Jakob disease (CJD)	>55	0.5–1	Dementia, ataxia, myoclonus	Diffuse spongiosis	+ periodic EEG, + CSF: 14-3-3 and tau positive in most, + MRI (DWI and FLAIR) in most
Sporadic fatal insomnia (sFI)	40–60	1–1.5	Insomnia, autonomic dysfunction, late dementia, ataxia	Focal thalamic gliosis	Normal or slow wave EEG; MRI not well documented
Variably protease-sensitive prionopathy (VPSPr)	48–81	2–6	Aphasia, ataxia, parkinsonism, and frontal features (impulsivity, apathy, euphoria)	Moderate spongiosis	EEG, MRI, and CSF typically negative
FAMILIAL					
Familial CJD (fCJD)	<55	1–3	As in sCJD	Diffuse spongiosis	EEG and MRI positive in some (*PRNP* mutation dependent)
Gerstmann–Sträussler–Scheinker disease (GSS)	<55	2–7 (up to 20)	Cerebellar dysfunction greater than cognitive in most; some frontal-like presentations	Multicentric and diffuse plaques	Normal or slow wave EEG; MRI generally negative
Fatal familial insomnia	~45	1–1.5	As in sFI	As in sFI	Normal or slow EEG; MRI typically negative
ACQUIRED					
Iatrogenic CJD (iCJD)	<40	<1	Cerebellar signs, minimal dementia	Cerebellar > cortical pathology	Slow EEG; MRI not well documented
Variant CJD (vCJD)	Teens–40s	1–1.5	Personality changes (apathy, depression), painful sensory symptoms, cerebellar signs	Cerebellar and cerebral cortical "florid" plaques	Normal EEG; CSF negative; MRI: "Pulvinar sign" on DWI

Abbreviations: CSF, cerebrospinal fluid; DWI, diffusion-weighted imaging; EEG, electroencephalography; FLAIR, fluid-attenuated inversion recovery; MRI, magnetic resonance imaging.

Sporadic Creutzfeldt–Jakob Disease (sCJD)

This is by far the most common PrD, accounting for roughly 85% of cases. It is generally characterized as a rapidly progressive dementia in association with ataxia and myoclonus. In addition to this typical triad, a host of neurologic signs and symptoms have been reported, including focal weakness, rigidity, bradykinesia, tremor, chorea, alien hand syndrome, and sensory disturbances, among others. Vague complaints of fatigue, headache, sleep disturbance, vertigo, and behavioral changes may precede the development of frank progressive dementia in many cases.[230] Roughly 15% will develop ataxia prior to dementia[231] and about 10% develop cortical blindness as the first symptom, referred to as the Heidenhain variant. Once it begins, the disease is relentless, resulting in a final picture of akinetic mutism, generally within 4 to 6 months of onset. PK-resistant PrP is easily detected in all cases of CJD. Transmission of CJD to non-human primates is relatively efficient (85%)[232] and highly efficient to Tg mice that express human PrP (~100%).[186,199]

Subtypes of sCJD

Several subtypes of sCJD have been designated, based on the molecular typing of PrP^{Sc}, the codon 129 genotype, and the associated histopathology. Two major types of PK-resistant PrP^{Sc} are recognized for sCJD, based on the migration rate of the nonglycosylated fragment on a western blot: type 1 migrates at ~21 kD and type 2 migrates at ~19 kD. Based on these conformational subtypes, and the two possible amino acid substitutions at codon 129, there are six designated sCJD subtypes: 129MM1, 129MM2, 129MV1, 129MV2, 129VV1, and 129VV2. Clinical and pathological features that generally associate with each of the subtypes have been recognized (Table 21.2 and see [233] for review).

Variably Protease-Sensitive Prionopathy (VPSPr)

This is the most recently recognized prion subtype, currently described in only a handful of subjects.[26,120] Onset ranges from 48 to 81 years and duration is 2–6 years. Clinically, these patients are difficult to diagnose, as their

TABLE 21.2 Clinical Phenotypes of sCJD Molecular Subtypes

sCJD molecular subtype	Prominent clinical phenotype	Major histological phenotype	Age at onset	Duration	Positive MRI	Positive CSF	Positive EEG
MM1/MV1 (~40%)	Dementia—rapidly progressive, myoclonus, ataxia	Small (2–10 μm) vacuoles, all layers of cortex	Late 60s (40–90)	~4 months	80%	~90%	60–70%
MM2 (~1%)	Dementia—progressive by 5 months, later other CJD features	Large confluent vacuoles throughout cerebral cortex	Late 60s	~15 months	90%	50–65%	20–50%
VV1 (rare)	Dementia—slowly progressive with behavioral changes, late ataxia and myoclonus	Diffuse spongiform degeneration of cortex and striatum	43 (19–71)	~20 months	100%	~100%	<50%
VV2 (15%)	Ataxic onset, later followed by more typical CJD-like features	Small vacuoles, deep cortex and subcortical nuclei, plaque-like deposits in molecular layer of cerebellum	Late 60s (40–80)	~6 months	60%	>90%	<10%
MV2 (<10%)	Ataxic onset, as in VV2	Like VV2, but with kuru-type plaques	Late 60s	~17 months	~100%	>90%	<10%

Abbreviations: CSF, cerebrospinal fluid; EEG, electroencephalography; MRI, magnetic resonance imaging.

symptoms may be atypical and diagnostic studies such as electroencephalography (EEG), magnetic resonance imaging (MRI), and cerebrospinal fluid (CSF) are most often negative. Common features include aphasia, ataxia, parkinsonian signs, or frontal lobe-like features, such as impulsivity, euphoria, and/or apathy. VPSPr is distinguished from CJD primarily by the presence of PrP^{Sc} that is more sensitive to protease digestion compared to that isolated from the brains of patients affected by CJD. In addition, along with a moderate degree of spongiform degeneration, these patients exhibit punctate and target-like PrP amyloid deposits in their brains, features not typically seen in CJD.

Familial CJD (fCJD)

In general, this is distinguished from sCJD by the earlier onset of disease and a longer duration. Whereas the peak age at onset for sCJD is ~68 years, fCJD usually manifests under 55 years, although this depends somewhat on the mutation involved. Inheritance is autosomal dominant, with nearly 100% penetrance. Those that appear to have reduced penetrance, including especially the V180I and E200K mutations, can occur in later life, up to the ninth decade. Several mutations of *PRNP* are associated with fCJD, which include single base pair alterations throughout the gene, and insertions within the octarepeat segment (Figure 21.2).

Iatrogenic CJD (iCJD)

Approximately 100 individuals have developed CJD after exposure to cadaver-derived human growth hormone (hGH) from at least three separate sources, including the UK, France, and the USA.[234–238] A correlation between duration of therapy and disease risk has been noted.[238] In contrast to sCJD, iCJD typically presents with cerebellar ataxia rather than memory problems.[239] To eliminate this problem, recombinant GH use was initiated in 1985, but because of the prolonged incubation period of up to 30 years, some cases of hGH-associated CJD may still be observed. Iatrogenic CJD has also occurred in over 100 patients exposed to dura mater obtained from a single manufacturer whose preparative procedures were inadequate to decontaminate specimens,[240,241] and as a result of using a pericardium graft to repair a perforated eardrum.[242] Other examples of iCJD include two cases related to corneal transplants,[243] at least five cases in women after receiving human pituitary gonadotropin,[244–246] and two patients exposed to improperly decontaminated brain electrodes.[247] Proof for the latter was provided when the offending electrode produced CJD in a chimpanzee 18 months after brain implantation.[248]

Variant CJD (vCJD)

This disease has been reported in over 175 individuals throughout the UK and Europe since 1995, when it first appeared.[249–251] The clinical distinctions from sCJD include an onset of psychiatric manifestations, especially apathy and depression, occurrence in younger individuals (ages 17–42 years), and a slightly protracted course to beyond a year (Table 21.1). The histopathology of vCJD is distinguished by the presence of florid PrP plaques, which include a central amyloid core surrounded by a halo of spongiform degeneration (Figure 21.1). Spongiform change is typically most prominent in the basal ganglia, and the thalamus shows severe neuronal loss and gliosis, especially in the posterior nuclei.[252] PrP^{Sc} of vCJD has a migration rate comparable to type 2 PrP^{Sc}, but in contrast to a predominance of the monoglycosylated fraction of PrP in sCJD, the diglycosylated fraction is most prominent in vCJD.[206] In contrast to sCJD, vCJD occurs by ingestion of exogenous prions and is absorbed systemically before reaching the brain, supported by detection of PrP^{Sc} in tonsils, spleen, and appendices in vCJD, but not sCJD, although, in one study that used a sensitive method of PrP^{Sc} precipitation with phosphotungstic acid, about one-third of the sCJD cases were found to harbor low levels of PrP^{Sc} in either spleen or muscle.[253] The systemic inoculation of vCJD is further supported by the occurrence of four cases of transfusion related vCJD.[111,254–256]

It is now well established that vCJD is the result of exposure to BSE. Supportive evidence for this includes a temporal correlation with the BSE epidemic that peaked in 1994,[251,257] and the transmission of BSE to macaques and several rodent species that recapitulate both the florid plaque pathology[205] and western blot protein signature of PK-resistant PrP.[109,208,258] In addition, Tg mice that express bovine PrP [Tg(BoPrP)] were susceptible to both vCJD and BSE and they produced the characteristic pathology of vCJD.[259] The number of cases of vCJD has declined significantly since the first cases were identified, with ≤5 cases per year observed in the UK since 2005.

Gerstmann–Sträussler–Scheinker disease (GSS)

In its typical form, as originally described,[260,261] ataxia of gait and dysarthria are the cardinal features of GSS at onset, which is followed by variable degrees of pyramidal and extrapyramidal symptoms and the late development of dementia. Symptom onset typically occurs between the fourth to sixth decade, with progression to death in 2–10 years. Some cases have been described with a 20-year duration. While this is the classic presentation, considerable phenotypic variability, even within families that carry the same mutation, is observed. Some cases have been reported with dementia or spastic paraparesis as primary features, whereas several cases have been reported with personality changes compatible with the clinical features of frontotemporal dementia (FTD). The common thread of all these presentations that defines GSS is the histopathological features, which includes the presence of extracellular PrP plaque deposits, primarily within the cerebellum, but often distributed throughout the CNS. These are usually of the multicentric type, which feature a central amyloid core surrounded by smaller amyloid fragments (Figure 21.1).

The second consistent feature of GSS is the presence of one of several mutations within the *PRNP* gene. These include a number of missense mutations, insertions of 8 or 9 repeats within the octarepeat segment, a unique insertion of LGGLGGYV between residues 129 and 130, and five early stop sequence mutations that result in a *C*-terminally truncated PrP at residues 145, 160, 163, 226, and 227 (Figure 21.2). These latter mutations prevent the attachment of the GPI anchor, which results in secretion of the protein, leading to the extracellular deposition of aggregated PrP in the form of amyloid plaques, as has been modeled in transgenic mice.[149,262]

Transmission of GSS has been demonstrated only in rare cases, in contrast to the highly transmissible nature of CJD,[232] suggesting a fundamental difference in the two diseases. This difference is also reflected in the characteristics of the PK-resistant PrP isolated from the brain of patients with GSS and mice that model this disease, in that it is generally lower in quantity and smaller in size, typically ranging from 7 to 14 kDa, than that from CJD.[105,263] Because *N*- and *C*-terminal clipping of PrP^{Sc} occurs in GSS, this small fragment lacks the two *N*-linked glycosylations.

Fatal Insomnia (FI)

Originally recognized as a genetic prion disease, labeled fatal familial insomnia (FFI), a small number of cases of sporadic FI (sFI) have since been reported.[204,264–269] In contrast to fCJD and GSS, which can each be caused by multiple *PRNP* mutations, the only genotype associated with FFI is a D178N mutation combined with a 129M polymorphism in *cis*.[42] When the D178N mutation is linked to the 129 V polymorphism, fCJD is observed. The disease typically begins with intractable insomnia and dysautonomia (increased blood pressure, sweating, lacrimation, etc.) lasting up to several months before the onset of more typical symptoms of PrD, including ataxia, pyramidal and extrapyramidal features, and myoclonus. Cognitive function may be relatively spared until late in the course and it is typically a "subcortical" presentation, with slowed processing as a primary feature. Age at disease onset ranges from 25 to 61 years (average 48 years), and time to death is generally within 2 years (range 7–33 months).[270,271] In cases where insomnia is not obvious, a sleep study may reveal a shortening of total sleep time.[272,273] Phenotypic heterogeneity has been reported within an Australian family[274] in that some, but not all, members were found to have clinically apparent insomnia early or late in the presentation. Positron emission tomography (PET) with radiolabeled fluorodeoxyglucose (^{18}FDG) shows reduced activity in the thalamus, bilaterally, which corresponds to the prominent neuronal loss and gliosis within the thalamus of patients with FI. The inferior olivary nucleus is similarly involved in most patients, and in those with a prolonged course, the temporal lobes and other cortical regions may show spongiform degeneration and gliosis.

PK resistant PrP^{Sc} is detectable in the brain of affected patients but typically in small amounts and restricted to the thalamus and temporal lobe.[275] Transmission of FFI and sFI to Tg mice susceptible to human prions helped to establish the hypothesis that prion strain is encoded in the conformation of PrP^{Sc}.[187,204] Not only was the pathologic phenotype of FI transmitted to these mice, which displayed the most prominent accumulation of PrP^{Sc} within the thalamus, but the atypical conformation of PrP^{Sc}, as defined by the migration pattern of PK-resistant PrP^{Sc} by western blot, was also efficiently transmitted.

DIAGNOSTIC STUDIES

There is no single diagnostic test for PrD, other than brain biopsy, a procedure that is generally avoided because of the associated risks of human exposure and contamination of surgical equipment and space, necessitating the use of disposable instruments and extensive operating room decontamination. There are established criteria for the diagnosis of sCJD and vCJD, but not for other subtypes (Table 21.1). PrD is a difficult diagnosis to make, especially early in the disease, as the presentation varies and may mimic a variety of neurological conditions. Thus, several etiologies must be considered, including toxic, metabolic, inflammatory, infectious, neoplastic, paraneoplastic, and neurodegenerative. The presence of a mutation of *PRNP* can be used to confirm genetic PrD, once all other possibilities have been ruled out.

Classical methods to diagnose PrD have relied on the history of presentation, neurologic examination, and the EEG. In sCJD, roughly 65% of patients will exhibit generalized periodic sharp wave complexes (PSWCs) with a frequency of 0.5–2 per second[230] (Figure 21.8). However, this feature is less common in fCJD and generally absent in GSS, FI, and vCJD, where the EEG may show only generalized slowing.[257,272] In VPSPr the EEG is most often negative.[26] MRI has proven useful in the diagnostic evaluation, especially of sCJD. Although diffuse atrophy may be noted in some, a more common feature is restricted diffusion within the cortical ribbon and/or the basal ganglia, represented by hyperintense signal on diffusion-weighted imaging (DWI) and fluid-attenuated inversion recovery (FLAIR) sequences[276] (Figure 21.8). This feature is less common in GSS and not well studied in FI. However, in FI, functional imaging (PET or single-photon emission computerized tomography) is most helpful, as it shows a reduction in metabolic activity or blood flow to the thalamus early in the disease.[277] CSF testing is useful for confirming a case of PrD when clinical suspicion is high. Mildly elevated protein, the absence of inflammatory cells, and a significant elevation of the 14-3-3 protein, neuron-specific enolase, or tau protein, are all suggestive of PrD, although tau may have a slight advantage in specificity.[278] CSF testing for these proteins is now commercially available. False positive results and low specificity for PrD caution against using these biomarkers as diagnostic tests in isolation. In general, they appear to be more reliably detected in typical rapidly progressive sCJD, compared with slower progressing sporadic and familial forms of PrD.

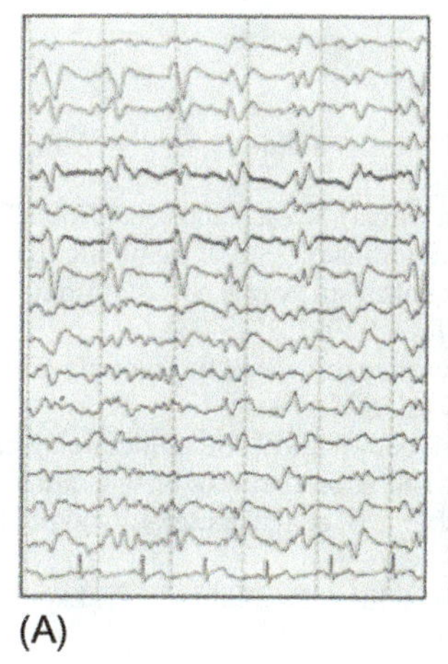

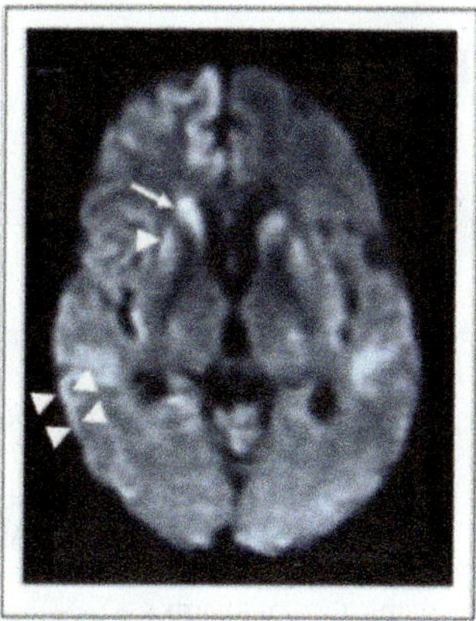

FIGURE 21.8 Clinical tests in prion disease (PrD). (A) Electroencephalogram shows periodic sharp wave complexes (PSWCs) at a frequency of ~1 per second (range, 0.5–2 per second). **(B)** Magnetic resonance imaging shows hyperintensities in basal ganglia (caudate, arrow; putamen, arrowhead) and cortical ribbon (double arrowhead) on diffusion-weighted images. Both features are more characteristic of typical sporadic Creutzfeldt–Jakob disease and less common in other PrD subtypes.

TREATMENT

PrD is invariably fatal. There are currently no proven therapies, although an array of compounds have been tested in several *in vitro* and *in vivo* experimental models of prion disease. Such compounds include sulfated polyanions,[279,280] amphotericin B,[281,282] Congo Red,[283] and tetrapyrroles,[284] among others. Most such compounds display only modest potential to delay the onset of symptoms in mice inoculated with mouse-adapted scrapie, which depended on administering the drug either at the time of inoculation with prions, or prior to it. Pentosan polysulfate, a compound used to treat acquired immunodeficiency syndrome (AIDS)-related Kaposi sarcoma, displayed promise in mouse neuroblastoma cells chronically infected with prions (ScN2a) and in rodents infected with scrapie, when administered intraventricularly,[285] but its poor penetration of the blood–brain barrier limits its potential. A patient with vCJD received intrathecal pentosan and was reported to display some stabilization of the clinical picture, although atrophy of the brain worsened.[286] The acridine quinacrine was the focus of two large clinical trials in the US and UK. These trials were based on its ability to inhibit PrP^{Sc} production in cultured ScN2a cells.[287] Unfortunately, subsequent reports testing the drug in animals as the clinical trial was proceeding did not support its effectiveness *in vivo*,[285,288,289] and the results of the clinical trials came to the same conclusion.[290,291]

Immunotherapy has also been attempted. Anti-PrP antibodies interfere with the association of PrP^{C} and PrP^{Sc} to prevent propagation. Proof of principle was demonstrated by the addition of recombinant Fabs to ScN2a cells, resulting in the clearance of prion infectivity.[292] In addition, active immunization of mice with recombinant mouse PrP delayed the onset of scrapie slightly,[293] whereas passive anti-PrP antibody administration at the time of, or after a significant delay following, peripheral prion inoculation of mice, led to a significant delay in disease onset.[294] Another approach has been to generate transgenic mice that express antiprion antibodies, which were resistant to intraperitoneal PrP^{Sc} inoculation.[295] In addition to using antibodies that target PrP, a beneficial effect has been demonstrated by blocking the laminin receptor, which has been shown to also act as a receptor for PrP.[296] In cell culture, this antibody (pAb W3) prevented propagation of PrP^{Sc},[134] and when the pAb W3 was administered to mice a week prior to prion inoculation, their survival time nearly doubled.[296]

The potential involvement of autophagy in prion disease has led to a handful of studies in cultured cells and *in vivo* models of scrapie infection,[153,297] and in Tg mouse models of genetic prion disease,[154] that support therapies directed at enhancing autophagy to mitigate sporadic disease and delay onset of genetic disease and symptoms.

Another approach that shows great promise is RNA silencing. Silencing of PrP mRNA, leading to reduced levels of PrP^{C}, may provide an option to slow the disease and follow up treatment with a secondary therapy to eliminate accumulated PrP^{Sc}. A proof-of-principle experiment was reported by Mallucci et al.[298] using Tg mice constructed with M*lox*P transgenes that, under Cre recombinase expression, results in the excision of floxed PrP, thereby turning off PrP^{C} expression. This approach was not only effective in halting disease, but evidence for reversal of pathology was presented. This important experiment suggests that some of the pathologic features of disease are, in fact, reversible if PrP^{Sc} production stops. This provides a significant ray of hope for a seemingly inexorable disease.

Finally, a recent report on an orally administered compound, labeled Anle128b, described as an "aggregate modifier," significantly delayed scrapie in rodents, whether administered at the time of inoculation or at the onset of symptoms.[299] Interestingly, this drug appears to work not only for prion disease, but for several neurodegenerative diseases linked to aggregated proteins. At the time writing, it is being tested in transgenic mouse models of familial prion disease.

References

1. Parry HB. Recorded occurrences of scrapie from 1750. In: Oppenheimer DR, ed. *Scrapie Disease in Sheep*. New York: Academic Press; 1983:31–59.
2. Wickner RB. Scrapie in ancient China? *Science*. 2005;309(5736):874.
3. Cuillé J, Chelle PL. Experimental transmission of trembling to the goat. *C R Seances Acad Sci*. 1939;208:1058–1060.
4. Zigas V, Gajdusek DC. Kuru: clinical study of a new syndrome resembling paralysis agitans in natives of the Eastern Highlands of Australian New Guinea. *Med J Aust*. 1957;2:745–754.
5. Gajdusek DC, Zigas V. Degenerative disease of the central nervous system in New Guinea; the endemic occurrence of "kuru" in the native population. *N Engl J Med*. 1957;257:974–978.
6. Gajdusek DC, Zigas V. Clinical, pathological and epidemiological study of an acute progressive degenerative disease of the central nervous system among natives of the eastern highlands of New Guinea. *Am J Med*. 1959;26:442–469.
7. Gibbs Jr. CJ, Gajdusek DC, Asher DM, et al. Creutzfeldt–Jakob disease (spongiform encephalopathy): transmission to the chimpanzee. *Science*. 1968;161:388–389.
8. Beck E, Daniel PM, Alpers M, Gajdusek DC, Gibbs Jr. CJ. Experimental "kuru" in chimpanzees. A pathological report. *Lancet*. 1966;2:1056–1059.

9. Gordon WS. Advances in veterinary research. *Vet Res*. 1946;58:516–520.
10. Sigurdsson B. Rida, a chronic encephalitis of sheep with general remarks on infections which develop slowly and some of their special characteristics. *Br Vet J*. 1954;110:341–354.
11. Chandler RL. Encephalopathy in mice produced by inoculation with scrapie brain material. *Lancet*. 1961;1:1378–1379.
12. Zlotnik I. The pathology of scrapie: a comparative study of lesions in the brain of sheep and goats. *Acta Neuropathol Suppl (Berl)*. 1962;1:61–70.
13. Alper T, Haig DA, Clarke MC. The exceptionally small size of the scrapie agent. *Biochem Biophys Res Commun*. 1966;22:278–284.
14. Alper T, Cramp WA, Haig DA, Clarke MC. Does the agent of scrapie replicate without nucleic acid? *Nature*. 1967;214:764–766.
15. Griffith JS. Self-replication and scrapie. *Nature*. 1967;215:1043–1044.
16. Bolton DC, McKinley MP, Prusiner SB. Identification of a protein that purifies with the scrapie prion. *Science*. 1982;218:1309–1311.
17. Prusiner SB. Novel proteinaceous infectious particles cause scrapie. *Science*. 1982;216:136–144.
18. McKinley MP, Bolton DC, Prusiner SB. A protease-resistant protein is a structural component of the scrapie prion. *Cell*. 1983;35:57–62.
19. Oesch B, Westaway D, Wälchli M, et al. A cellular gene encodes scrapie PrP 27-30 protein. *Cell*. 1985;40:735–746.
20. Polymenidou M, Cleveland DW. Prion-like spread of protein aggregates in neurodegeneration. *J Exp Med*. 2012;209(5):889–893.
21. Brown P, Cathala F, Raubertas RF, Gajdusek DC, Castaigne P. The epidemiology of Creutzfeldt–Jakob disease: conclusion of a 15-year investigation in France and review of the world literature. *Neurology*. 1987;37:895–904.
22. Beck E, Daniel PM, Davey AJ, Gajdusek DC, Gibbs Jr. CJ. The pathogenesis of transmissible spongiform encephalopathy - an ultrastructural study. *Brain*. 1982;105:755–786.
23. Lampert PW, Gajdusek DC, Gibbs Jr CJ. Subacute spongiform virus encephalopathies. Scrapie, kuru and Creutzfeldt–Jakob disease: a review. *Am J Pathol*. 1972;68:626–652.
24. Chou SM, Payne WN, Gibbs Jr CJ, Gajdusek DC. Transmission and scanning electron microscopy of spongiform change in Creutzfeldt–Jakob disease. *Brain*. 1980;103:885–904.
25. DeArmond SJ, Prusiner SB. Prion diseases. In: Lantos P, Graham D, eds. *Greenfield's Neuropathology*. 6th ed. London: Edward Arnold; 1997:235–280.
26. Zou WQ, Puoti G, Xiao X, et al. Variably protease-sensitive prionopathy: a new sporadic disease of the prion protein. *Ann Neurol*. 2010;68(2):162–172.
27. Ghetti B, Piccardo P, Spillantini MG, et al. Vascular variant of prion protein cerebral amyloidosis with t-positive neurofibrillary tangles: the phenotype of the stop codon 145 mutation in *PRNP*. *Proc Natl Acad Sci U S A*. 1996;93:744–748.
28. Liao Y-C, Lebo RV, Clawson GA, Smuckler EA. Human prion protein cDNA: molecular cloning, chromosomal mapping, and biological implication. *Science*. 1986;233:364–367.
29. Finckh U, Muller-Thomsen T, Mann U, et al. High prevalence of pathogenic mutations in patients with early-onset dementia detected by sequence analyses of four different genes. *Am J Hum Genet*. 2000;66(1):110–117.
30. Jansen C, Parchi P, Capellari S, et al. Prion protein amyloidosis with divergent phenotype associated with two novel nonsense mutations in PRNP. *Acta Neuropathol*. 2010;119(2):189–197.
31. Hinnell C, Coulthart MB, Jansen GH, et al. Gerstmann-Straussler-Scheinker disease due to a novel prion protein gene mutation. *Neurology*. 2011;76(5):485–487.
32. Collinge J, Palmer MS, Dryden AJ. Genetic predisposition to iatrogenic Creutzfeldt–Jakob disease. *Lancet*. 1991;337:1441–1442.
33. Owen F, Poulter M, Collinge J, Crow TJ. Codon 129 changes in the prion protein gene in Caucasians. *Am J Hum Genet*. 1990;46:1215–1216.
34. Windl O, Dempster M, Estibeiro JP, et al. Genetic basis of Creutzfeldt–Jakob disease in the United Kingdom: a systematic analysis of predisposing mutations and allelic variation in the *PRNP* gene. *Hum Genet*. 1996;98:259–264.
35. Laplanche J-L, Delasnerie-Lauprêtre N, Brandel JP, et al. Molecular genetics of prion diseases in France. *Neurology*. 1994;44:2347–2351.
36. Salvatore M, Genuardi M, Petraroli R, Masullo C, D'Alessandro M, Pocchiari M. Polymorphisms of the prion protein gene in Italian patients with Creutzfeldt–Jakob disease. *Hum Genet*. 1994;94:375–379.
37. Cervenakova L, Goldfarb LG, Garruto R, Lee HS, Gajdusek DC, Brown P. Phenotype-genotype studies in kuru: implications for new variant Creutzfeldt–Jakob disease. *Proc Natl Acad Sci U S A*. 1998;95(22):13239–13241.
38. Doh-ura K, Kitamoto T, Sakaki Y, Tateishi J. CJD discrepancy. *Nature*. 1991;353:801–802.
39. Parchi P, Castellani R, Capellari S, et al. Molecular basis of phenotypic variability in sporadic Creutzfeldt–Jakob disease. *Ann Neurol*. 1996;39:767–778.
40. Collinge J, Brown J, Hardy J, et al. Inherited prion disease with 144 base pair gene insertion. 2. Clinical and pathological features. *Brain*. 1992;115:687–710.
41. Baker HF, Poulter M, Crow TJ, Frith CD, Lofthouse R, Ridley RM. Amino acid polymorphism in human prion protein and age at death in inherited prion disease. *Lancet*. 1991;337:1286.
42. Goldfarb LG, Petersen RB, Tabaton M, et al. Fatal familial insomnia and familial Creutzfeldt–Jakob disease: disease phenotype determined by a DNA polymorphism. *Science*. 1992;258:806–808.
43. Riek R, Wider G, Billeter M, Hornemann S, Glockshuber R, Wüthrich K. Prion protein NMR structure and familial human spongiform encephalopathies. *Proc Natl Acad Sci U S A*. 1998;95:11667–11672.
44. Shibuya S, Higuchi J, Shin R-W, Tateishi J, Kitamoto T. Codon 219 Lys allele of PRNP is not found in sporadic Creutzfeldt–Jakob disease. *Ann Neurol*. 1998;43:826–828.
45. Zulianello L, Kaneko K, Scott M, et al. Dominant-negative inhibition of prion formation diminished by deletion mutagenesis of the prion protein. *J Virol*. 2000;74(9):4351–4360.
46. Perrier V, Kaneko K, Safar J, et al. Dominant-negative inhibition of prion replication in transgenic mice. *Proc Natl Acad Sci U S A*. 2002;99(20):13079–13084.
47. Tanaka Y, Minematsu K, Moriyasu H, et al. A Japanese family with a variant of Gerstmann-Straussler-Scheinker disease. *J Neurol Neurosurg Psychiatry*. 1997;62(5):454–457.
48. Appleby BS, Appleby KK, Hall RC, Wallin MT. D178N, 129Val and N171S, 129Val genotype in a family with Creutzfeldt–Jakob disease. *Dement Geriatr Cogn Disord*. 2010;30(5):424–431.

49. Fink JK, Peacock ML, Warren JT, Roses AD, Prusiner SB. Detecting prion protein gene mutations by denaturing gradient gel electrophoresis. *Hum Mutat*. 1994;4:42–50.
50. Samaia HB, Mari JJ, Vallada HP, Moura RP, Simpson AJ, Brentani RR. A prion-linked psychiatric disorder [letter]. *Nature*. 1997;390(6657):241.
51. Tsai MT, Su YC, Chen YH, Chen CH. Lack of evidence to support the association of the human prion gene with schizophrenia. *Mol Psychiatry*. 2001;6(1):74–78.
52. Stahl N, Borchelt DR, Hsiao K, Prusiner SB. Scrapie prion protein contains a phosphatidylinositol glycolipid. *Cell*. 1987;51:229–240.
53. Stahl N, Baldwin MA, Beavis R, et al. The search for post-translational modifications of the scrapie prion protein. In: *Prion Diseases in Humans and Animals Conference, London, U.K*; 1991.
54. Harris DA. Trafficking, turnover and membrane topology of PrP. *Br Med Bull*. 2003;66:71–85.
55. Borchelt DR, Scott M, Taraboulos A, Stahl N, Prusiner SB. Scrapie and cellular prion proteins differ in their kinetics of synthesis and topology in cultured cells. *J Cell Biol*. 1990;110:743–752.
56. Caughey B, Neary K, Butler R, et al. Normal and scrapie-associated forms of prion protein differ in their sensitivities to phospholipase and proteases in intact neuroblastoma cells. *J Virol*. 1990;64:1093–1101.
57. Orsi A, Fioriti L, Chiesa R, Sitia R. Conditions of endoplasmic reticulum stress favor the accumulation of cytosolic prion protein. *J Biol Chem*. 2006;281(41):30431–30438.
58. Rane NS, Yonkovich JL, Hegde RS. Protection from cytosolic prion protein toxicity by modulation of protein translocation. *EMBO J*. 2004;23(23):4550–4559.
59. Harris DA, Huber MT, van Dijken P, Shyng S-L, Chait BT, Wang R. Processing of a cellular prion protein: identification of N- and C-terminal cleavage sites. *Biochemistry*. 1993;32:1009–1016.
60. Altmeppen HC, Prox J, Puig B, et al. Lack of a-disintegrin-and-metalloproteinase ADAM10 leads to intracellular accumulation and loss of shedding of the cellular prion protein in vivo. *Mol Neurodegeneration*. 2011;6:36.
61. Kretzschmar HA, Prusiner SB, Stowring LE, DeArmond SJ. Scrapie prion proteins are synthesized in neurons. *Am J Pathol*. 1986;122(1):1–5.
62. Haeberle AM, Ribaut-Barassin C, Bombarde G, et al. Synaptic prion protein immuno-reactivity in the rodent cerebellum. *Microsc Res Tech*. 2000;50(1):66–75.
63. Laine J, Marc ME, Sy MS, Axelrad H. Cellular and subcellular morphological localization of normal prion protein in rodent cerebellum. *Eur J Neurosci*. 2001;14(1):47–56.
64. Ford MJ, Burton LJ, Li H, et al. A marked disparity between the expression of prion protein and its message by neurones of the CNS. *Neuroscience*. 2002;111(3):533–551.
65. Bendheim PE, Brown HR, Rudelli RD, et al. Nearly ubiquitous tissue distribution of the scrapie agent precursor protein. *Neurology*. 1992;42:149–156.
66. Han CX, Liu HX, Zhao DM. The quantification of prion gene expression in sheep using real-time RT-PCR. *Virus Genes*. 2006;33(3):359–364.
67. Perini F, Vidal R, Ghetti B, Tagliavini F, Frangione B, Prelli F. PrP_{27-30} is a normal soluble prion protein fragment released by human platelets. *Biochem Biophys Res Commun*. 1996;223:572–577.
68. Büeler H, Aguzzi A, Sailer A, et al. Mice devoid of PrP are resistant to scrapie. *Cell*. 1993;73:1339–1347.
69. Collinge J, Whittington MA, Sidle KC, et al. Prion protein is necessary for normal synaptic function. *Nature*. 1994;370:295–297.
70. Lledo P-M, Tremblay P, DeArmond SJ, Prusiner SB, Nicoll RA. Mice deficient for prion protein exhibit normal neuronal excitability and synaptic transmission in the hippocampus. *Proc Natl Acad Sci U S A*. 1996;93:2403–2407.
71. Tobler I, Gaus SE, Deboer T, et al. Altered circadian activity rhythms and sleep in mice devoid of prion protein. *Nature*. 1996;380:639–642.
72. Criado JR, Sanchez-Alavez M, Conti B, et al. Mice devoid of prion protein have cognitive deficits that are rescued by reconstitution of PrP in neurons. *Neurobiol Dis*. 2005;19(1–2):255–265.
73. Nishida N, Tremblay P, Sugimoto T, et al. A mouse prion protein transgene rescues mice deficient for the prion protein gene from purkinje cell degeneration and demyelination. *Lab Invest*. 1999;79(6):689–697.
74. Bremer J, Baumann F, Tiberi C, et al. Axonal prion protein is required for peripheral myelin maintenance. *Nat Neurosci*. 2010;13(3):310–318.
75. Mouillet-Richard S, Ermonval M, Chebassier C, et al. Signal transduction through prion protein. *Science*. 2000;289(5486):1925–1928.
76. Solforosi L, Criado JR, McGavern DB, et al. Cross-linking cellular prion protein triggers neuronal apoptosis in vivo. *Science*. 2004;303(5663):1514–1516.
77. Bounhar Y, Zhang Y, Goodyer CG, LeBlanc A. Prion protein protects human neurons against Bax-mediated apoptosis. *J Biol Chem*. 2001;276(42):39145–39149.
78. Li A, Harris DA. Mammalian prion protein suppresses Bax-induced cell death in yeast. *J Biol Chem*. 2005;280(17):17430–17434.
79. Hornshaw MP, McDermott JR, Candy JM. Copper binding to the N-terminal tandem repeat regions of mammalian and avian prion protein. *Biochem Biophys Res Commun*. 1995;207:621–629.
80. Brown DR, Qin K, Herms JW, et al. The cellular prion protein binds copper *in vivo*. *Nature*. 1997;390:684–687.
81. Perera WS, Hooper NM. Ablation of the metal ion-induced endocytosis of the prion protein by disease-associated mutation of the octarepeat region. *Curr Biol*. 2001;11(7):519–523.
82. Pauly PC, Harris DA. Copper stimulates endocytosis of the prion protein. *J Biol Chem*. 1998;273:33107–33110.
83. Brown DR, Schulz-Schaeffer WJ, Schmidt B, Kretzschmar HA. Prion protein-deficient cells show altered response to oxidative stress due to decreased SOD-1 activity. *Exp Neurol*. 1997;146:104–112.
84. Wong BS, Pan T, Liu T, Li R, Gambetti P, Sy MS. Differential contribution of superoxide dismutase activity by prion protein in vivo. *Biochem Biophys Res Commun*. 2000;273(1):136–139.
85. Brown DR, Besinger A. Prion protein expression and superoxide dismutase activity. *Biochem J*. 1998;334:423–429.
86. Brown DR, Wong BS, Hafiz F, Clive C, Haswell SJ, Jones IM. Normal prion protein has an activity like that of superoxide dismutase. *Biochem J*. 1999;344(Pt 1):1–5.
87. Hutter G, Heppner FL, Aguzzi A. No superoxide dismutase activity of cellular prion protein in vivo. *Biol Chem*. 2003;384(9):1279–1285.
88. Jones S, Batchelor M, Bhelt D, Clarke AR, Collinge J, Jackson GS. Recombinant prion protein does not possess SOD-1 activity. *Biochem J*. 2005;392(Pt 2):309–312.
89. Doherty P, Rowett LH, Moore SE, Mann DA, Walsh FS. Neurite outgrowth in response to transfected N-CAM and N-cadherin reveals fundamental differences in neuronal responsiveness to CAMs. *Neuron*. 1991;6:247–258.

90. Graner E, Mercadante AF, Zanata SM, et al. Cellular prion protein binds laminin and mediates neuritogenesis. *Brain Res Mol Brain Res.* 2000;76(1):85–92.
91. Graner E, Mercadante AF, Zanata SM, Martins VR, Jay DG, Brentani RR. Laminin-induced PC-12 cell differentiation is inhibited following laser inactivation of cellular prion protein. *FEBS lett.* 2000;482(3):257–260.
92. Petrakis S, Sklaviadis T. Identification of proteins with high affinity for refolded and native PrPC. *Proteomics.* 2006;6(24):6476–6484.
93. Zafar S, von Ahsen N, Oellerich M, et al. Proteomics approach to identify the interacting partners of cellular prion protein and characterization of Rab7a interaction in neuronal cells. *J Proteome Res.* 2011;10(7):3123–3135.
94. Parkin ET, Watt NT, Hussain I, et al. Cellular prion protein regulates beta-secretase cleavage of the Alzheimer's amyloid precursor protein. *Proc Natl Acad Sci U S A.* 2007;104(26):11062–11067.
95. Lauren J, Gimbel DA, Nygaard HB, Gilbert JW, Strittmatter SM. Cellular prion protein mediates impairment of synaptic plasticity by amyloid-beta oligomers. *Nature.* 2009;457(7233):1128–1132.
96. Um JW, Nygaard HB, Heiss JK, et al. Alzheimer amyloid-beta oligomer bound to postsynaptic prion protein activates Fyn to impair neurons. *Nat Neurosci.* 2012;15(9):1227–1235.
97. Morales R, Estrada LD, Diaz-Espinoza R, et al. Molecular cross talk between misfolded proteins in animal models of Alzheimer's and prion diseases. *J Neurosci.* 2010;30(13):4528–4535.
98. Schwarze-Eicker K, Keyvani K, Gortz N, Westaway D, Sachser N, Paulus W. Prion protein (PrPc) promotes beta-amyloid plaque formation. *Neurobiol Aging.* 2005;26(8):1177–1182.
99. Donne DG, Viles JH, Groth D, et al. Structure of the recombinant full-length hamster prion protein PrP(29-231): the N terminus is highly flexible. *Proc Natl Acad Sci U S A.* 1997;94:13452–13457.
100. Zahn R, Liu A, Luhrs T, et al. NMR solution structure of the human prion protein. *Proc Natl Acad Sci U S A.* 2000;97(1):145–150.
101. Safar J, Roller PP, Gajdusek DC, Gibbs Jr. CJ. Conformational transitions, dissociation, and unfolding of scrapie amyloid (prion) protein. *J Biol Chem.* 1993;268:20276–20284.
102. Silveira JR, Raymond GJ, Hughson AG, et al. The most infectious prion protein particles. *Nature.* 2005;437(7056):257–261.
103. Lehmann S, Harris DA. A mutant prion protein displays aberrant membrane association when expressed in cultured cells. *J Biol Chem.* 1995;270:24589–24597.
104. Hsiao KK, Groth D, Scott M, et al. Neurologic disease of transgenic mice which express GSS mutant prion protein is transmissible to inoculated recipient animals. Paper presented at: Prion Diseases of Humans and Animals Symposium. London; Sept. 2–4 1991 (abstract).
105. Tagliavini F, Lievens PM, Tranchant C, et al. A 7-kDa prion protein (PrP) fragment, an integral component of the PrP region required for infectivity, is the major amyloid protein in Gerstmann-Straussler-Scheinker disease A117V. *J Biol Chem.* 2001;276(8):6009–6015.
106. Jendroska K, Heinzel FP, Torchia M, et al. Proteinase-resistant prion protein accumulation in Syrian hamster brain correlates with regional pathology and scrapie infectivity. *Neurology.* 1991;41:1482–1490.
107. Kuroda Y, Gibbs Jr. CJ, Amyx HL, Gajdusek DC. Creutzfeldt–Jakob disease in mice: persistent viremia and preferential replication of virus in low-density lymphocytes. *Infect Immun.* 1983;41:154–161.
108. Muramoto T, Kitamoto T, Tateishi J, Goto I. Accumulation of abnormal prion protein in mice infected with Creutzfeldt–Jakob disease via intraperitoneal route: a sequential study. *Am J Pathol.* 1993;143:1470–1479.
109. Hill AF, Butterworth RJ, Joiner S, et al. Investigation of variant Creutzfeldt–Jakob disease and other human prion diseases with tonsil biopsy samples. *Lancet.* 1999;353(9148):183–189.
110. Brown P, Rohwer RG, Dunstan BC, MacAuley C, Gajdusek DC, Drohan WN. The distribution of infectivity in blood components and plasma derivatives in experimental models of transmissible spongiform encephalopathy. *Transfusion.* 1998;38:810–816.
111. Wroe SJ, Pal S, Siddique D, et al. Clinical presentation and pre-mortem diagnosis of variant Creutzfeldt–Jakob disease associated with blood transfusion: a case report. *Lancet.* 2006;368(9552):2061–2067.
112. Klein MA, Frigg R, Flechsig E, et al. A crucial role for B cells in neuroinvasive scrapie. *Nature.* 1997;390:687–691.
113. Aguzzi A. Prions and the immune system: a journey through gut, spleen, and nerves. *Adv Immunol.* 2003;81:123–171.
114. Aguzzi A, Heppner FL, Heikenwalder M, et al. Immune system and peripheral nerves in propagation of prions to CNS. *Br Med Bull.* 2003;66:141–159.
115. Thomzig A, Cardone F, Kruger D, Pocchiari M, Brown P, Beekes M. Pathological prion protein in muscles of hamsters and mice infected with rodent-adapted BSE or vCJD. *J Gen Virol.* 2006;87(Pt 1):251–254.
116. Thomzig A, Kratzel C, Lenz G, Kruger D, Beekes M. Widespread PrPSc accumulation in muscles of hamsters orally infected with scrapie. *EMBO Rep.* 2003;4(5):530–533.
117. Angers RC, Browning SR, Seward TS, et al. Prions in skeletal muscles of deer with chronic wasting disease. *Science.* 2006;311(5764):1117.
118. Lasmézas CI, Deslys J-P, Robain O, et al. Transmission of the BSE agent to mice in the absence of detectable abnormal prion protein. *Science.* 1997;275(17):402–405.
119. Shaked GM, Fridlander G, Meiner Z, Taraboulos A, Gabizon R. Protease-resistant and detergent-insoluble prion protein is not necessarily associated with prion infectivity. *J Biol Chem.* 1999;274(25):17981–17986.
120. Gambetti P, Dong Z, Yuan J, et al. A novel human disease with abnormal prion protein sensitive to protease. *Ann Neurol.* 2008;63(6):697–708.
121. Wildegger G, Liemann S, Glockshuber R. Extremely rapid folding of the C-terminal domain of the prion protein without kinetic intermediates. *Nat Struct Biol.* 1999;6(6):550–553.
122. Baskakov IV, Legname G, Prusiner SB, Cohen FE. Folding of prion protein to its native alpha-helical conformation is under kinetic control. *J Biol Chem.* 2001;276(23):19687–19690.
123. Oesch B, Teplow DB, Stahl N, Serban D, Hood LE, Prusiner SB. Identification of cellular proteins binding to the scrapie prion protein. *Biochemistry.* 1990;29:5848–5855.
124. Kurschner C, Morgan JI. The cellular prion protein (PrP) selectively binds to Bcl-2 in the yeast two-hybrid system. *Mol Brain Res.* 1995;30:165–168.
125. Harmey JH, Doyle D, Brown V, Rogers MS. The cellular isoform of the prion protein, PrP^C, is associated with caveolae in mouse neuroblastoma (N_2a) cells. *Biochem Biophys Res Commun.* 1995;210:753–759.

126. Gao C, Lei YJ, Han J, et al. Recombinant neural protein PrP can bind with both recombinant and native apolipoprotein E in vitro. *Acta Biochim Biophys Sin (Shanghai)*. 2006;38(9):593–601.
127. Hammond C, Helenius A. Quality control in the secretory pathway: retention of a misfolded viral membrane glycoprotein involves cycling between the ER, intermediate compartment, and Golgi apparatus. *J Cell Biol*. 1994;126(1):41–52.
128. Jin T, Gu Y, Zanusso G, et al. The chaperone protein BiP binds to a mutant prion protein and mediates its degradation by the proteasome. *J Biol Chem*. 2000;275(49):38699–38704.
129. Capellari S, Zaidi SI, Urig CB, Perry G, Smith MA, Petersen RB. Prion protein glycosylation is sensitive to redox change. *J Biol Chem*. 1999;274(49):34846–34850.
130. Zanata SM, Lopes MH, Mercadante AF, et al. Stress-inducible protein 1 is a cell surface ligand for cellular prion that triggers neuroprotection. *EMBO J*. 2002;21(13):3307–3316.
131. Yehiely F, Bamborough P, Da Costa M, et al. Identification of candidate proteins binding to prion protein. *Neurobiol Dis*. 1997;3:339–355.
132. Rieger R, Edenhofer F, Lasmezas CI, Weiss S. The human 37-kDa laminin receptor precursor interacts with the prion protein in eukaryotic cells. *Nat Med*. 1997;3(12):1383–1388.
133. Gauczynski S, Nikles D, El-Gogo S, et al. The 37-kDa/67-kDa laminin receptor acts as a receptor for infectious prions and is inhibited by polysulfated glycanes. *J Infect Dis*. 2006;194(5):702–709.
134. Gauczynski S, Peyrin JM, Haik S, et al. The 37-kDa/67-kDa laminin receptor acts as the cell-surface receptor for the cellular prion protein. *EMBO J*. 2001;20(21):5863–5875.
135. Leucht C, Simoneau S, Rey C, et al. The 37 kDa/67 kDa laminin receptor is required for PrP(Sc) propagation in scrapie-infected neuronal cells. *EMBO Rep*. 2003;4(3):290–295.
136. Morel E, Andrieu T, Casagrande F, et al. Bovine prion is endocytosed by human enterocytes via the 37 kDa/67 kDa laminin receptor. *Am J Pathol*. 2005;167(4):1033–1042.
137. Yedidia Y, Horonchik L, Tzaban S, Yanai A, Taraboulos A. Proteasomes and ubiquitin are involved in the turnover of the wild-type prion protein. *EMBO J*. 2001;20(19):5383–5391.
138. Ma J, Lindquist S. Wild-type PrP and a mutant associated with prion disease are subject to retrograde transport and proteasome degradation. *Proc Natl Acad Sci U S A*. 2001;98(26):14955–14960.
139. Ivanova L, Barmada S, Kummer T, Harris DA. Mutant prion proteins are partially retained in the endoplasmic reticulum. *J Biol Chem*. 2001;276(45):42409–42421.
140. Drisaldi B, Stewart RS, Adles C, et al. Mutant PrP is delayed in its exit from the endoplasmic reticulum, but neither wild-type nor mutant PrP undergoes retrotranslocation prior to proteasomal degradation. *J Biol Chem*. 2003;278(24):21732–21743.
141. Ma J, Lindquist S. Conversion of PrP to a self-perpetuating PrPSc-like conformation in the cytosol. *Science*. 2002;298(5599):1785–1788.
142. Kang SW, Rane NS, Kim SJ, Garrison JL, Taunton J, Hegde RS. Substrate-specific translocational attenuation during ER stress defines a pre-emptive quality control pathway. *Cell*. 2006;127(5):999–1013.
143. Ma J, Wollmann R, Lindquist S. Neurotoxicity and neurodegeneration when PrP accumulates in the cytosol. *Science*. 2002;298(5599):1781–1785.
144. Norstrom EM, Ciaccio MF, Rassbach B, Wollmann R, Mastrianni JA. Cytosolic prion protein toxicity is independent of cellular prion protein expression and prion propagation. *J Virol*. 2007; 81(6):2831–2837.
145. Borchelt DR, Koliatsis VE, Guarnieri M, Pardo CA, Sisodia SS, Price DL. Rapid anterograde axonal transport of the cellular prion glycoprotein in the peripheral and central nervous systems. *J Biol Chem*. 1994;269:14711–14714.
146. Vella L, Sharples R, Lawson V, Masters C, Cappai R, Hill A. Packaging of prions into exosomes is associated with a novel pathway of PrP processing. *J Pathol*. 2007;211(5):582–590.
147. Schneider A, Simons M. Exosomes: vesicular carriers for intercellular communication in neurodegenerative disorders. *Cell Tissue Res*. 2013;352(1):33–47.
148. Borchelt DR, Taraboulos A, Prusiner SB. Evidence for synthesis of scrapie prion proteins in the endocytic pathway. *J Biol Chem*. 1992;267:16188–16199.
149. Chesebro B, Trifilo M, Race R, et al. Anchorless prion protein results in infectious amyloid disease without clinical scrapie. *Science*. 2005;308(5727):1435–1439.
150. Vey M, Pilkuhn S, Wille H, et al. Subcellular colocalization of the cellular and scrapie prion proteins in caveolae-like membranous domains. *Proc Natl Acad Sci U S A*. 1996;93:14945–14949.
151. Huang L, Jin R, Li J, et al. Macromolecular crowding converts the human recombinant PrPC to the soluble neurotoxic beta-oligomers. *FASEB J*. 2010;24(9):3536–3543.
152. McKinley MP, Taraboulos A, Kenaga L, et al. Ultrastructural localization of scrapie prion proteins in secondary lysosomes of infected cultured cells. *J Cell Biol*. 1990;111:316a.
153. Aguib Y, Heiseke A, Gilch S, et al. Autophagy induction by trehalose counteracts cellular prion infection. *Autophagy*. 2009;5(3):361–369.
154. Cortes CJ, Qin K, Cook J, Solanki A, Mastrianni JA. Rapamycin delays disease onset and prevents PrP plaque deposition in a mouse model of Gerstmann-Straussler-Scheinker Disease. *J Neurosci*. 2012;32(36):12396–12405.
155. Legname G, Baskakov IV, Nguyen HO, et al. Synthetic mammalian prions. *Science*. 2004;305(5684):673–676.
156. Saborio GP, Permanne B, Soto C. Sensitive detection of pathological prion protein by cyclic amplification of protein misfolding. *Nature*. 2001;411(6839):810–813.
157. Saa P, Castilla J, Soto C. Presymptomatic detection of prions in blood. *Science*. 2006;313(5783):92–94.
158. Deleault NR, Walsh DJ, Piro JR, et al. Cofactor molecules maintain infectious conformation and restrict strain properties in purified prions. *Proc Natl Acad Sci U S A*. 2012;109(28):E1938–E1946.
159. Wang F, Wang X, Yuan CG, Ma J. Generating a prion with bacterially expressed recombinant prion protein. *Science*. 2010;327(5969):1132–1135.
160. Chiesa R, Piccardo P, Ghetti B, Harris DA. Neurological illness in transgenic mice expressing a prion protein with an insertional mutation. *Neuron*. 1998;21(6):1339–1351.
161. Gray F, Chretien F, Adle-Biassette H, et al. Neuronal apoptosis in Creutzfeldt–Jakob disease. *J Neuropathol Exp Neurol*. 1999;58(4):321–328.

162. Giese A, Brown DR, Groschup MH, Feldmann C, Haist I, Kretzschmar HA. Role of microglia in neuronal cell death in prion disease. *Brain Pathol*. 1998;8:449–457.
163. Forloni G, Del Bo R, Angeretti N, et al. A neurotoxic prion protein fragment induces rat astroglial proliferation and hypertrophy. *Eur J Neurosci*. 1994;6:1415–1422.
164. Brown DR, Schmidt B, Kretzschmar HA. Role of microglia and host prion protein in neurotoxicity of a prion protein fragment. *Nature*. 1996;380:345–347.
165. Tagliavini F, Prelli F, Verga L, et al. Synthetic peptides homologous to prion protein residues 106–147 form amyloid-like fibrils *in vitro*. *Proc Natl Acad Sci U S A*. 1993;90:9678–9682.
166. Hegde RS, Mastrianni JA, Scott MR, et al. A transmembrane form of the prion protein in neurodegenerative disease. *Science*. 1998;279:827–834.
167. Hegde RS, Tremblay P, Groth D, DeArmond SJ, Prusiner SB, Lingappa VR. Transmissible and genetic prion diseases share a common pathway of neurodegeneration. *Nature*. 1999;402(16):822–826.
168. Stewart RS, Drisaldi B, Harris DA. A transmembrane form of the prion protein contains an uncleaved signal peptide and is retained in the endoplasmic reticulum. *Mol Biol Cell*. 2001;12(4):881–889.
169. Stewart RS, Harris DA. A transmembrane form of the prion protein is localized in the Golgi apparatus of neurons. *J Biol Chem*. 2005;280(16):15855–15864.
170. Ma J, Lindquist S. De novo generation of a PrPSc-like conformation in living cells. *Nat Cell Biol*. 1999;1(6):358–361.
171. Mironov Jr A, Latawiec D, Wille H, et al. Cytosolic prion protein in neurons. *J Neurosci*. 2003;23(18):7183–7193.
172. Brandner S, Isenmann S, Raeber A, et al. Normal host prion protein necessary for scrapie-induced neurotoxicity. *Nature*. 1996;379:339–343.
173. Moore RC, Mastrangelo P, Bouzamondo E, et al. Doppel-induced cerebellar degeneration in transgenic mice. *Proc Natl Acad Sci U S A*. 2001;98(26):15288–15293.
174. Shmerling D, Hegyi I, Fischer M, et al. Expression of amino-terminally truncated PrP in the mouse leading to ataxia and specific cerebellar lesions. *Cell*. 1998;93(2):203–214.
175. Flechsig E, Hegyi I, Leimeroth R, et al. Expression of truncated PrP targeted to Purkinje cells of PrP knockout mice causes Purkinje cell death and ataxia. *EMBO J*. 2003;22(12):3095–3101.
176. Li A, Christensen HM, Stewart LR, Roth KA, Chiesa R, Harris DA. Neonatal lethality in transgenic mice expressing prion protein with a deletion of residues 105–125. *EMBO J*. 2007;26(2):548–558.
177. Solomon IH, Biasini E, Harris DA. Ion channels induced by the prion protein: Mediators of neurotoxicity. *Prion*. 2012;6(1).
178. Kourie JI, Shorthouse AA. Properties of cytotoxic peptide-formed ion channels. *Am J Physiol Cell Physiol*. 2000;278(6):C1063–C1087.
179. Wilesmith JW, Ryan JBM, Atkinson MJ. Bovine spongiform encephalopathy - epidemiologic studies on the origin. *Vet Rec*. 1991;128:199–203.
180. Biacabe AG, Laplanche JL, Ryder S, Baron T. Distinct molecular phenotypes in bovine prion diseases. *EMBO Rep*. 2004;5(1):110–115.
181. Casalone C, Zanusso G, Acutis P, et al. Identification of a second bovine amyloidotic spongiform encephalopathy: molecular similarities with sporadic Creutzfeldt–Jakob disease. *Proc Natl Acad Sci U S A*. 2004;101(9):3065–3070.
182. Buschmann A, Gretzschel A, Biacabe AG, et al. Atypical BSE in Germany–proof of transmissibility and biochemical characterization. *Vet Microbiol*. 2006;117(2–4):103–116.
183. Beringue V, Herzog L, Reine F, et al. Transmission of atypical bovine prions to mice transgenic for human prion protein. *Emerg Infect Dis*. 2008;14(12):1898–1901.
184. Comoy EE, Casalone C, Lescoutra-Etchegaray N, et al. Atypical BSE (BASE) transmitted from asymptomatic aging cattle to a primate. *PLoS One*. 2008;3(8):e3017.
185. Scott M, Foster D, Mirenda C, et al. Transgenic mice expressing hamster prion protein produce species-specific scrapie infectivity and amyloid plaques. *Cell*. 1989;59:847–857.
186. Telling GC, Scott M, Hsiao KK, et al. Transmission of Creutzfeldt–Jakob disease from humans to transgenic mice expressing chimeric human-mouse prion protein. *Proc Natl Acad Sci U S A*. 1994;91:9936–9940.
187. Telling GC, Parchi P, DeArmond SJ, et al. Evidence for the conformation of the pathologic isoform of the prion protein enciphering and propagating prion diversity. *Science*. 1996;274:2079–2082.
188. Wadsworth JD, Asante EA, Desbruslais M, et al. Human prion protein with valine 129 prevents expression of variant CJD phenotype. *Science*. 2004;306(5702):1793–1796.
189. Mallik S, Yang W, Norstrom EM, Mastrianni JA. Live cell fluorescence resonance energy transfer predicts an altered molecular association of heterologous PrPSc with PrPC. *J Biol Chem*. 2010;285(12):8967–8975.
190. Bartz JC, Marsh RF, McKenzie DI, Aiken JM. The host range of chronic wasting disease is altered on passage in ferrets. *Virology*. 1998;251(2):297–301.
191. Marsh RF, Kincaid AE, Bessen RA, Bartz JC. Interspecies transmission of chronic wasting disease prions to squirrel monkeys (Saimiri sciureus). *J Virol*. 2005;79(21):13794–13796.
192. Browning SR, Mason GL, Seward T, et al. Transmission of prions from mule deer and elk with chronic wasting disease to transgenic mice expressing cervid PrP. *J Virol*. 2004;78(23):13345–13350.
193. Kong Q, Huang S, Zou W, et al. Chronic wasting disease of elk: transmissibility to humans examined by transgenic mouse models. *J Neurosci*. 2005;25(35):7944–7949.
194. Barria MA, Telling GC, Gambetti P, Mastrianni JA, Soto C. Generation of a new form of human PrP(Sc) in vitro by interspecies transmission from cervid prions. *J Biol Chem*. 2011;286(9):7490–7495.
195. Agrimi U, Nonno R, Dell'Omo G, et al. Prion protein amino acid determinants of differential susceptibility and molecular feature of prion strains in mice and voles. *PLoS Pathog*. 2008;4(7):e1000113.
196. Heisey DM, Mickelsen NA, Schneider JR, et al. Chronic wasting disease (CWD) susceptibility of several North American rodents that are sympatric with cervid CWD epidemics. *J Virol*. 2010;84(1):210–215.
197. Nonno R, Di Bari MA, Cardone F, et al. Efficient transmission and characterization of Creutzfeldt–Jakob disease strains in bank voles. *PLoS Pathog*. 2006;2(2):e12.

198. Piening N, Nonno R, Di Bari M, et al. Conversion efficiency of bank vole prion protein in vitro is determined by residues 155 and 170, but does not correlate with the high susceptibility of bank voles to sheep scrapie in vivo. *J Biol Chem*. 2006;281(14):9373–9384.
199. Telling GC, Scott M, Mastrianni J, et al. Prion propagation in mice expressing human and chimeric PrP transgenes implicates the interaction of cellular PrP with another protein. *Cell*. 1995;83:79–90.
200. Watts JC, Giles K, Stohr J, et al. Spontaneous generation of rapidly transmissible prions in transgenic mice expressing wild-type bank vole prion protein. *Proc Natl Acad Sci U S A*. 2012;109(9):3498–3503.
201. Bessen RA, Marsh RF. Identification of two biologically distinct strains of transmissible mink encephalopathy in hamsters. *J Gen Virol*. 1992;73:329–334.
202. Bessen RA, Marsh RF. Distinct PrP properties suggest the molecular basis of strain variation in transmissible mink encephalopathy. *J Virol*. 1994;68:7859–7868.
203. Safar J, Wille H, Itri V, et al. Eight prion strains have PrP^{Sc} molecules with different conformations. *Nat Med*. 1998;4(10):1157–1165.
204. Mastrianni JA, Nixon R, Layzer R, et al. Prion protein conformation in a patient with sporadic fatal insomnia. *N Engl J Med*. 1999;340(21):1630–1638.
205. Lasmézas CI, Deslys J-P, Demaimay R, et al. BSE transmission to macaques. *Nature*. 1996;381:743–744.
206. Collinge J, Sidle KCL, Meads J, Ironside J, Hill AF. Molecular analysis of prion strain variation and the aetiology of "new variant" CJD. *Nature*. 1996;383:685–690.
207. Collinge J, Sidle K, Meads J, Ironside J, Hill A. Molecular analysis of prion strain variation and the aetiology of 'new variant' CJD. *Nature*. 1998;383:685–690.
208. Bruce ME, Will RG, Ironside JW, et al. Transmissions to mice indicate that 'new variant' CJD is caused by the BSE agent. *Nature*. 1997;389:498–501.
209. Scott MR, Safar J, Telling G, et al. Identification of a prion protein epitope modulating transmission of bovine spongiform encephalopathy prions to transgenic mice. *Proc Natl Acad Sci U S A*. 1997;94:14279–14284.
210. Sakaguchi S, Katamine S, Nishida N, et al. Loss of cerebellar Purkinje cells in aged mice homozygous for a disrupted PrP gene. *Nature*. 1996;380:528–531.
211. Rossi D, Cozzio A, Flechsig E, et al. Onset of ataxia and Purkinje cell loss in PrP null mice inversely correlated with Dpl level in brain. *EMBO J*. 2001;20(4):694–702.
212. Moore RC, Lee IY, Silverman GL, et al. Ataxia in prion protein (PrP)-deficient mice is associated with upregulation of the novel PrP-like protein doppel. *J Mol Biol*. 1999;292(4):797–817.
213. Mo H, Moore RC, Cohen FE, et al. Two different neurodegenerative diseases caused by proteins with similar structures. *Proc Natl Acad Sci U S A*. 2001;98(5):2352–2357.
214. Luhrs T, Riek R, Guntert P, Wuthrich K. NMR structure of the human doppel protein. *J Mol Biol*. 2003;326(5):1549–1557.
215. Silverman GL, Qin K, Moore RC, et al. Doppel is an N-glycosylated, glycosylphosphatidylinositol-anchored protein. Expression in testis and ectopic production in the brains of Prnp(0/0) mice predisposed to Purkinje cell loss. *J Biol Chem*. 2000;275(35):26834–26841.
216. Peoc'h K, Serres C, Frobert Y, et al. The human "prion-like" protein Doppel is expressed in both Sertoli cells and spermatozoa. *J Biol Chem*. 2002;277(45):43071–43078.
217. Paltrinieri S, Comazzi S, Spagnolo V, Rondena M, Ponti W, Ceciliani F. Bovine Doppel (Dpl) and prion protein (PrP) expression on lymphoid tissue and circulating leukocytes. *J Histochem Cytochem*. 2004;52(12):1639–1645.
218. Paltrinieri S, Spagnolo V, Giordano A, Gelmetti D, Comazzi S. Bovine prion (PrP) and Doppel (Dpl) proteins expression after in vitro leukocyte activation or Dpl/PrP blocking. *J Cell Physiol*. 2006;208(2):446–450.
219. Tuzi NL, Gall E, Melton D, Manson JC. Expression of doppel in the CNS of mice does not modulate transmissible spongiform encephalopathy disease. *J Gen Virol*. 2002;83(Pt 3):705–711.
220. Cui T, Holme A, Sassoon J, Brown DR. Analysis of doppel protein toxicity. *Mol Cell Neurosci*. 2003;23(1):144–155.
221. Mead S, Beck J, Dickinson A, Fisher EMC, Collinge J. Examination of the human prion protein-like gene Doppel for genetic susceptibility to sporadic and variant Creutzfeldt–Jakob disease. *Neurosci Lett*. 2000;290(2):117–120.
222. Jeong BH, Kim NH, Kim JI, Carp RI, Kim YS. Polymorphisms at codons 56 and 174 of the prion-like protein gene (PRND) are not associated with sporadic Creutzfeldt–Jakob disease. *J Hum Genet*. 2005;50(6):311–314.
223. Peoc'h K, Guerin C, Brandel JP, Launay JM, Laplanche JL. First report of polymorphisms in the prion-like protein gene (PRND): implications for human prion diseases. *Neurosci Lett*. 2000;286(2):144–148.
224. Peoc'h K, Volland H, De Gassart A, et al. Prion-like protein Doppel expression is not modified in scrapie-infected cells and in the brains of patients with Creutzfeldt–Jakob disease. *FEBS lett*. 2003;536(1–3):61–65.
225. Premzl M, Sangiorgio L, Strumbo B, Marshall Graves JA, Simonic T, Gready JE. Shadoo, a new protein highly conserved from fish to mammals and with similarity to prion protein. *Gene*. 2003;314:89–102.
226. Daude N, Wohlgemuth S, Brown R, et al. Knockout of the prion protein (PrP)-like Sprn gene does not produce embryonic lethality in combination with PrP(C)-deficiency. *Proc Natl Acad Sci U S A*. 2012;109(23):9035–9040.
227. Watts JC, Drisaldi B, Ng V, et al. The CNS glycoprotein Shadoo has PrP(C)-like protective properties and displays reduced levels in prion infections. *EMBO J*. 2007;26(17):4038–4050.
228. Westaway D, Genovesi S, Daude N, et al. Down-regulation of Shadoo in prion infections traces a pre-clinical event inversely related to PrP(Sc) accumulation. *PLoS Pathog*. 2011;7(11):e1002391.
229. Beck JA, Campbell TA, Adamson G, et al. Association of a null allele of SPRN with variant Creutzfeldt–Jakob disease. *J Med Genet*. 2008;45(12):813–817.
230. Brown P, Cathala F, Castaigne P, Gajdusek DC. Creutzfeldt–Jakob disease: clinical analysis of a consecutive series of 230 neuropathologically verified cases. *Ann Neurol*. 1986;20:597–602.
231. Gomori AJ, Partnow MJ, Horoupian DS, Hirano A. The ataxic form of Creutzfeldt–Jakob disease. *Arch Neurol*. 1973;29:318–323.
232. Brown P, Gibbs Jr. CJ, Rodgers-Johnson P, et al. Human spongiform encephalopathy: the National Institutes of Health series of 300 cases of experimentally transmitted disease. *Ann Neurol*. 1994;35:513–529.
233. Gambetti P, Cali I, Notari S, Kong Q, Zou WQ, Surewicz WK. Molecular biology and pathology of prion strains in sporadic human prion diseases. *Acta Neuropathol*. 2011;121(1):79–90.

234. Anderson JR, Allen CMC, Weller RO. Creutzfeldt–Jakob disease following human pituitary-derived growth hormone administration [Abstr.]. *Neuropathol Appl Neurobiol*. 1990;16:543.
235. Billette de Villemeur T, Beauvais P, Gourmelon M, Richardet JM. Creutzfeldt–Jakob disease in children treated with growth hormone. *Lancet*. 1991;337:864–865.
236. Billette de Villemeur T, Gelot A, Deslys JP, et al. Iatrogenic Creutzfeldt–Jakob disease in three growth hormone recipients: a neuropathological study. *Neuropathol Appl Neurobiol*. 1994;20:111–117.
237. Brown P, Gajdusek DC, Gibbs Jr CJ, Asher DM. Potential epidemic of Creutzfeldt–Jakob disease from human growth hormone therapy. *N Engl J Med*. 1985;313:728–731.
238. Fradkin JE, Schonberger LB, Mills JL, et al. Creutzfeldt–Jakob disease in pituitary growth hormone recipients in the United States. *JAMA*. 1991;265:880–884.
239. Brown P, Preece MA, Will RG. "Friendly fire" in medicine: hormones, homografts, and Creutzfeldt–Jakob disease. *Lancet*. 1992;340:24–27.
240. Otto D. Jacob-Creutzfeldt disease associated with cadaveric dura. *J Neurosurg*. 1987;67:149.
241. Thadani V, Penar PL, Partington J, et al. Creutzfeldt–Jakob disease probably acquired from a cadaveric dura mater graft. Case report. *J Neurosurg*. 1988;69:766–769.
242. Tange RA, Troost D, Limburg M. Progressive fatal dementia (Creutzfeldt–Jakob disease) in a patient who received homograft tissue for tympanic membrane closure. *Eur Arch Otorhinolaryngol*. 1989;247:199–201.
243. Duffy P, Wolf J, Collins G, Devoe A, Streeten B, Cowen D. Possible person to person transmission of Creutzfeldt–Jakob disease. *N Engl J Med*. 1974;290:692–693.
244. Cochius JI, Hyman N, Esiri MM. Creutzfeldt–Jakob disease in a recipient of human pituitary-derived gonadotrophin: a second case. *J Neurol Neurosurg Psychiatry*. 1992;55:1094–1095.
245. Cochius JI, Mack K, Burns RJ, Alderman CP, Blumbergs PC. Creutzfeldt–Jakob disease in a recipient of human pituitary-derived gonadotrophin. *Aust N Z J Med*. 1990;20:592–593.
246. Healy DL, Evans J. Creutzfeldt–Jakob disease after pituitary gonadotrophins. *Br J Med*. 1993;307:517–518.
247. Bernouilli C, Siegfried J, Baumgartner G, et al. Danger of accidental person to person transmission of Creutzfeldt–Jakob disease by surgery. *Lancet*. 1977;1:478–479.
248. Gibbs Jr CJ, Asher DM, Kobrine A, Amyx HL, Sulima MP, Gajdusek DC. Transmission of Creutzfeldt–Jakob disease to a chimpanzee by electrodes contaminated during neurosurgery. *J Neurol Neurosurg Psychiatry*. 1994;57:757–758.
249. Bateman D, Hilton D, Love S, Zeidler M, Beck J, Collinge J. Sporadic Creutzfeldt–Jakob disease in a 18-year-old in the UK (Lett.). *Lancet*. 1995;346:1155–1156.
250. Britton TC, Al-Sarraj S, Shaw C, Campbell T, Collinge J. Sporadic Creutzfeldt–Jakob disease in a 16-year-old in the UK (Lett.). *Lancet*. 1995;346:1155.
251. Will RG, Ironside JW, Zeidler M, et al. A new variant of Creutzfeldt–Jakob disease in the UK. *Lancet*. 1996;347:921–925.
252. Ironside JW, Head MW. Neuropathology and molecular biology of variant Creutzfeldt–Jakob disease. *Curr Top Microbiol Immunol*. 2004;284:133–159.
253. Glatzel M, Abela E, Maissen M, Aguzzi A. Extraneural pathologic prion protein in sporadic Creutzfeldt–Jakob disease. *N Engl J Med*. 2003;349(19):1812–1820.
254. Llewelyn CA, Hewitt PE, Knight RS, et al. Possible transmission of variant Creutzfeldt–Jakob disease by blood transfusion. *Lancet*. 2004;363(9407):417–421.
255. Peden A, McCardle L, Head MW, et al. Variant CJD infection in the spleen of a neurologically asymptomatic UK adult patient with haemophilia. *Haemophilia*. 2010;16(2):296–304.
256. Peden AH, Head MW, Ritchie DL, Bell JE, Ironside JW. Preclinical vCJD after blood transfusion in a PRNP codon 129 heterozygous patient. *Lancet*. 2004;364(9433):527–529.
257. Zeidler M, Stewart GE, Barraclough CR, et al. New variant Creutzfeldt–Jakob disease: neurological features and diagnostic tests. *Lancet*. 1997;350:903–907.
258. Hill AF, Desbruslais M, Joiner S, et al. The same prion strain causes vCJD and BSE. *Nature*. 1997;389:448–450.
259. Scott MR, Will R, Ironside J, et al. Compelling transgenetic evidence for transmission of bovine spongiform encephalopathy prions to humans. *Proc Natl Acad Sci U S A*. 1999;96(26):15137–15142.
260. Gerstmann J. Über ein noch nicht beschriebenes Reflex - phanomen bei einer Erkrankung des zerebellaren Systems. *Wien Med Wochenschr*. 1928;78:906–908.
261. Gerstmann J, Sträussler E, Scheinker I. Über eine eigenartige hereditär-familiäre Erkrankung des Zentralnervensystems zugleich ein Beitrag zur frage des vorzeitigen lokalen Alterns. *Z Neurol*. 1936;154:736–762.
262. Stohr J, Watts JC, Legname G, et al. Spontaneous generation of anchorless prions in transgenic mice. *Proc Natl Acad Sci U S A*. 2011;108(52):21223–21228.
263. Yang W, Cook J, Rassbach B, Lemus A, DeArmond SJ, Mastrianni JA. A new transgenic mouse model of gerstmann-straussler-scheinker syndrome caused by the A117V mutation of PRNP. *J Neurosci*. 2009;29(32):10072–10080.
264. Parchi P, Capellari S, Chin S, et al. A subtype of sporadic prion disease mimicking fatal familial insomnia. *Neurology*. 1999;52(9):1757–1763.
265. Capellari S, Parchi P, Cortelli P, et al. Sporadic fatal insomnia in a fatal familial insomnia pedigree. *Neurology*. 2008;70(11):884–885.
266. Luo JJ, Truant AL, Kong Q, Zou WQ. Sporadic fatal insomnia with clinical, laboratory, and genetic findings. *J Clin Neurosci*. 2012;19(8):1188–1192.
267. Moody KM, Schonberger LB, Maddox RA, Zou WQ, Cracco L, Cali I. Sporadic fatal insomnia in a young woman: a diagnostic challenge: case report. *BMC Neurol*. 2011;11:136.
268. Priano L, Giaccone G, Mangieri M, et al. An atypical case of sporadic fatal insomnia. *J Neurol Neurosurg Psychiatry*. 2009;80(8):924–927.
269. Scaravilli F, Cordery RJ, Kretzschmar H, et al. Sporadic fatal insomnia: a case study. *Ann Neurol*. 2000;48(4):665–668.
270. Medori R, Montagna P, Tritschler HJ, et al. Fatal familial insomnia: a second kindred with mutation of prion protein gene at codon 178. *Neurology*. 1992;42:669–670.
271. Gambetti P. Fatal familial insomnia: a new human prion disease. Paper presented at: Prion Diseases in Humans and Animals Symposium. London; Sept. 2–4 1991.

272. Gambetti P, Parchi P, Petersen RB, Chen SG, Lugaresi E. Fatal familial insomnia and familial Creutzfeldt–Jakob disease: clinical, pathological and molecular features. *Brain Pathol*. 1995;5:43–51.
273. Rancurel G, Garma L, Hauw J-J, et al. Familial thalamic degeneration with fatal insomnia: clinicopathological and polygraphic data on a French member of Lugaresi's Italian family. In: Guilleminault C, Lugaresi E, Montagna P, Gambetti P, eds. *Fatal Familial Insomnia: Inherited Prion Diseases, Sleep, and The Thalamus*. New York: Raven Press; 1994:15–26.
274. McLean CA, Storey E, Gardner RJM, Tannenberg MB, Cervenáková L, Brown P. The D178N (cis-129M) "fatal familial insomnia" mutation associated with diverse clinicopathologic phenotypes in an Australian kindred. *Neurology*. 1997;49:552–558.
275. Parchi P, Castellani R, Cortelli P, et al. Regional distribution of protease-resistant prion protein in fatal familial insomnia. *Ann Neurol*. 1995;38:21–29.
276. Mendez OE, Shang J, Jungreis CA, Kaufer DI. Diffusion-weighted MRI in Creutzfeldt–Jakob disease: a better diagnostic marker than CSF protein 14-3-3? *J Neuroimaging*. 2003;13(2):147–151.
277. Perani D, Cortelli P, Lucignani G, et al. [^{18}F]FDG PET in fatal familial insomnia: the functional effects of thalamic lesions. *Neurology*. 1993;43:2565–2569.
278. Coulthart MB, Jansen GH, Olsen E, et al. Diagnostic accuracy of cerebrospinal fluid protein markers for sporadic Creutzfeldt–Jakob disease in Canada: a 6-year prospective study. *BMC Neurol*. 2011;11:133.
279. Farquhar CF, Dickinson AG. Prolongation of scrapie incubation period by an injection of dextran sulphate 500 within the month before or after infection. *J Gen Virol*. 1986;67(Pt 3):463–473.
280. Farquhar C, Dickinson A, Bruce M. Prophylactic potential of pentosan polysulphate in transmissible spongiform encephalopathies. *Lancet*. 1999;353(9147):117.
281. Pocchiari M, Schmittinger S, Masullo C. Amphotericin B delays the incubation period of scrapie in intracerebrally inoculated hamsters. *J Gen Virol*. 1987;68(Pt 1):219–223.
282. Xi YG, Ingrosso L, Ladogana A, Masullo C, Pocchiari M. Amphotericin B treatment dissociates *in vivo* replication of the scrapie agent from PrP accumulation. *Nature*. 1992;356:598–601.
283. Caughey B, Race RE. Potent inhibition of scrapie-associated PrP accumulation by Congo red. *J Neurochem*. 1992;59:768–771.
284. Priola SA, Raines A, Caughey WS. Porphyrin and phthalocyanine antiscrapie compounds. *Science*. 2000;287(5457):1503–1506.
285. Doh-ura K, Ishikawa K, Murakami-Kubo I, et al. Treatment of transmissible spongiform encephalopathy by intraventricular drug infusion in animal models. *J Virol*. 2004;78(10):4999–5006.
286. Todd NV, Morrow J, Doh-ura K, et al. Cerebroventricular infusion of pentosan polysulphate in human variant Creutzfeldt–Jakob disease. *J Infect*. 2005;50(5):394–396.
287. Korth C, May BC, Cohen FE, Prusiner SB. Acridine and phenothiazine derivatives as pharmacotherapeutics for prion disease. *Proc Natl Acad Sci U S A*. 2001;98(17):9836–9841.
288. Barret A, Tagliavini F, Forloni G, et al. Evaluation of quinacrine treatment for prion diseases. *J Virol*. 2003;77(15):8462–8469.
289. Collins SJ, Lewis V, Brazier M, Hill AF, Fletcher A, Masters CL. Quinacrine does not prolong survival in a murine Creutzfeldt–Jakob disease model. *Ann Neurol*. 2002;52(4):503–506.
290. Collinge J, Gorham M, Hudson F, et al. Safety and efficacy of quinacrine in human prion disease (PRION-1 study): a patient-preference trial. *Lancet Neurol*. 2009;8(4):334–344.
291. Geschwind MD, Kuo AL, Wong KS, et al. Quinacrine treatment trial for sporadic Creutzfeldt–Jakob disease. *Neurology*. 2013;81(23):2015–2023.
292. Peretz D, Williamson RA, Kaneko K, et al. Antibodies inhibit prion propagation and clear cell cultures of prion infectivity. *Nature*. 2001;412(6848):739–743.
293. Sigurdsson EM, Brown DR, Daniels M, et al. Immunization delays the onset of prion disease in mice. *Am J Pathol*. 2002;161(1):13–17.
294. White AR, Enever P, Tayebi M, et al. Monoclonal antibodies inhibit prion replication and delay the development of prion disease. *Nature*. 2003;422(6927):80–83.
295. Heppner FL, Musahl C, Arrighi I, et al. Prevention of scrapie pathogenesis by transgenic expression of anti-prion protein antibodies. *Science*. 2001;294(5540):178–182.
296. Zuber C, Mitteregger G, Pace C, Zerr I, Kretzschmar HA, Weiss S. Anti-LRP/LR antibody W3 hampers peripheral PrPSc propagation in scrapie infected mice. *Prion*. 2007;1(3):207–212.
297. Heiseke A, Aguib Y, Riemer C, Baier M, Schatzl HM. Lithium induces clearance of protease resistant prion protein in prion-infected cells by induction of autophagy. *J Neurochem*. 2009;109(1):25–34.
298. Mallucci G, Dickinson A, Linehan J, Klohn PC, Brandner S, Collinge J. Depleting neuronal PrP in prion infection prevents disease and reverses spongiosis. *Science*. 2003;302(5646):871–874.
299. Wagner J, Ryazanov S, Leonov A, et al. Anle138b: a novel oligomer modulator for disease-modifying therapy of neurodegenerative diseases such as prion and Parkinson's disease. *Acta Neuropathol*. 2013;125(6):795–813.

SECTION III

NEUROMETABOLIC DISORDERS

MITOCHONDRIAL DISORDERS

http://dx.doi.org/10.1016/B978-0-12-410529-4.00022-X

CHAPTER

22

The Mitochondrial Genome

Eric A. Schon

Columbia University, New York, NY, USA

MITOCHONDRIAL ORIGINS

Mitochondria are tiny organelles located in the cytoplasm of almost all mammalian cells. Their small size reflects their evolutionary origins. The now widely accepted endosymbiont hypothesis states that mitochondria were once bacteria that were either "captured" by a proto-eukaryote early in evolution[1] or evolved as the outcome of a symbiotic relationship between two prokaryotes, one of which gave rise to the nucleus while the other gave rise to the mitochondrion.[2] Due to both geochemical processes and the action of primitive photosynthetic organisms, the earth's atmosphere began to shift from a reducing one (rich in ammonia, methane, and hydrogen) to an oxidizing one (rich in nitrogen and oxygen). Bacteria were among the first organisms to evolve systems to deal with a new environmental toxin—oxygen—and convert it into nontoxic or less toxic forms, such as carbon dioxide, nitrate, and water. Over time, the relationship between the two organisms became a symbiotic one, in which one partner (which eventually became the proto-eukaryote) provided food and shelter, while the other (which eventually became the mitochondrion) detoxified the oxygen that threatened the host's survival and, at the same time, provided extra oxidative energy.[2] A bonus that resulted from this evolutionary leap was the increased efficiency of energy utilization afforded by the oxidation of reduced substrates.[2] Two events cemented this symbiotic relationship and eventually converted the bacterium into the mitochondrion we see today: the prokaryote lost 99% of its genes (many of which became incorporated into nuclear DNA [nDNA]), and its translation apparatus evolved so that it could only utilize a modified form of the "universal" genetic code to translate the few remaining messenger RNAs (mRNAs) specified by its DNA (now called mitochondrial DNA [mtDNA]).

There are a number of structural and functional features of the mitochondrion that are still reminiscent of its prokaryotic origin. The most obvious one is morphology: mitochondria are about 1 μm in length, thus approximating the size of most bacteria. Their shapes vary from spherical to rodlike, and can have a network-like appearance. Like many bacteria, they have a two-membrane structure: an *outer membrane* surrounding an *inner membrane,* which is itself invaginated at numerous points to form *cristae.* The *intermembrane space* is located between the outer and inner membranes, while the latter encloses the *matrix* in the organelle's interior. Other similarities to bacteria will become evident as we discuss the unique biochemical and genetic features of mitochondria.

GENOME ORGANIZATION

There is huge diversity in the mitochondrial genomes from different organisms. The malarial parasite *Plasmodium falciparum* contains a tiny linear mtDNA about 6 kilobase pairs (kb) long. The mtDNA in maize is a huge circular genome about 700-kb long, and the mtDNA in some plants can be up to 2500-kb long (about half the size of the circular bacterial genome of *Escherichia coli*). The mtDNA in many fungi (e.g., *Saccharomyces cerevisiae*, or baker's yeast) is about 85-kb long. Among eukaryotes, the mitochondrial genomes of insects and mammals are only 15–20-kb long, which is an example of genetic economy.

The human mitochondrial genome (Figure 22.1) is a double-stranded circle of 16,569 base pairs (bp).[3] It is highly asymmetric in its base composition, with one strand rich in nucleotides G and T (called the heavy strand) and the

Rosenberg's Molecular and Genetic Basis of Neurological and Psychiatric Disease
http://dx.doi.org/10.1016/B978-0-12-410529-4.00022-X

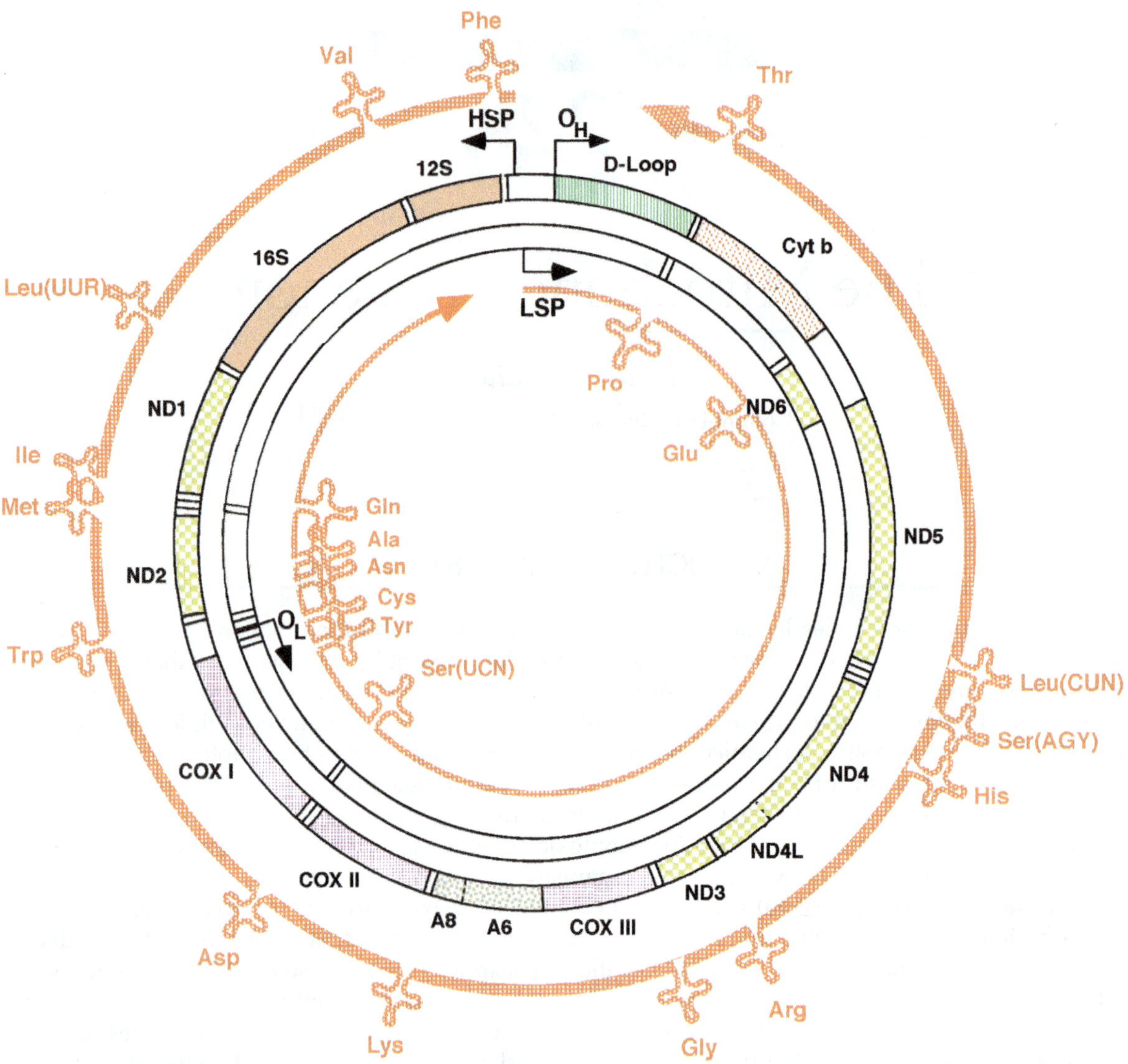

FIGURE 22.1 **Map of the human mitochondrial genome (both the heavy and light strands).** Figure shows the structural genes for the mtDNA-encoded 12S and 16S ribosomal RNAs, the subunits of NADH-coenzyme Q oxidoreductase (ND), cytochrome *c* oxidase (COX), cytochrome *b* (Cyt b), ATP synthetase (A), and 22 tRNAs (3-letter code). The origins of light-strand (O_L) and heavy-strand (O_H) replication, and of the promoters for initiation of transcription from the light-strand (LSP) and heavy-strand (HSP), are shown (arrows). Transcription of the H-strand (outer gray circle) and L-strand (inner gray circle) is shown schematically, with the tRNAs "punctuating" the polycistronic transcripts. See text for further description.

other correspondingly rich in C and A (called the light strand). The "heavy" and "light" nomenclature refers to the differential mobility of the separated strands in alkaline cesium chloride gradients. Of the approximately 4000 genes presumably present in the protomitochondrion, only 37 remain in the modern human mitochondrial genome. About 1700 genes now reside in the nuclear DNA and encode proteins that are synthesized in the cytoplasm and are then imported into mitochondria. Of these, about half are required for "housekeeping" functions necessary for the maintenance of the organelle, such as protein import, nucleic acid metabolism, and protein translation, while the other half are associated with the mitochondrion's specialized functions, such as β-oxidation, the citric acid cycle, the urea cycle, amino acid metabolism, steroid metabolism, and, most importantly, oxidative energy metabolism.

Of the 37 genes on human mtDNA, only 13 specify polypeptides, and like bacterial DNA, none of them contain introns. Remarkably, all 13 encode related functions, as they are all components of the respiratory chain/oxidative phosphorylation system (Figure 22.2). This system is a set of five biochemically related complexes (complexes I–V) located in the mitochondrial inner membrane. Complexes I, III, IV, and V are composed of polypeptides encoded by both nuclear and mitochondrial genes. Nucleus-encoded genes (some of which are tissue-specific, such as the two muscle-specific subunits VIa and VIIa of cytochrome *c* oxidase) are imported into mitochondria, and are

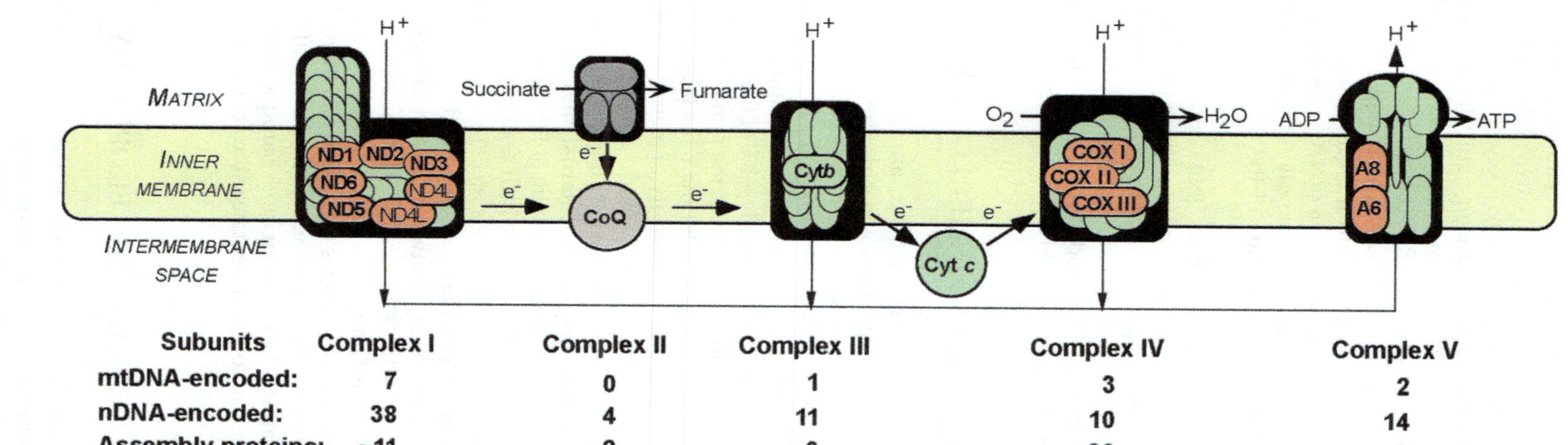

FIGURE 22.2 **The mitochondrial respiratory chain showing nuclear DNA-encoded subunits (green) and mitochondrial DNA-encoded subunits (red; labeled as in Figure 24.1).** Also enumerated are ancillary "assembly factors," all nDNA-encoded, required for the assembly, stability, and integrity of the indicated complexes. Protons (H^+) are pumped from the matrix to the intermembrane space through complexes I, III, and IV. They then flow back to the matrix through complex V, with the concomitant production of ATP. Ubiquinone (Coenzyme Q, or CoQ) and cytochrome *c* (Cyt c) are nucleus-encoded electron (e^-) transfer proteins.

co-assembled with mtDNA-encoded genes into the respective enzyme complexes located in the inner membrane. Complex II, on the other hand, contains no mtDNA-encoded subunits, and is therefore often used as an "internal control" for the presence, amount, and viability of mitochondria in situations where mutations in mtDNA render some or all of the other four respiratory complexes inactive. Notably, two of the four subunits of complex II comprise the activity of succinate dehydrogenase (SDH). Because SDH is also a key enzyme in the citric acid cycle, complex II presumably enables the cell to monitor CoA utilization and reducing equivalents (NADH; $FADH_2$) in the organelle.

The other 24 genes on human mtDNA are required for translation of those 13 polypeptides within the organelle. Two genes specify ribosomal RNAs (called 12S and 16S rRNA); not surprisingly, they are more closely related to bacterial rRNAs than to nucleus-encoded cytoplasmic rRNAs. The remaining 22 genes specify transfer RNAs (tRNAs) that are required for incorporation of amino acids into the growing polypeptide chain as the mtDNA-encoded mRNAs are translated on mitochondrial ribosomes. These tRNA genes are located strategically around the circle, precisely at the borders between the rRNAs and most of the polypeptide-coding genes (see Figure 22.1). This "punctuation" is believed to be crucial to the precise maturation of the tRNAs, rRNA, and mRNAs, as described below in the section on transcription.

Because respiratory chain complexes are derived from both mtDNA- and nDNA-encoded genes, mitochondrial respiratory-chain disorders can be both Mendelian-inherited (e.g., mutations in nDNA-encoded structural subunits and "assembly" proteins) and maternally inherited (e.g., mutations in mtDNA-encoded subunits). Enormous progress has been made in uncovering the molecular bases of maternally inherited disorders, many of which are associated with qualitative and quantitative defects in mtDNA-encoded respiratory chain subunits. In the last few years, significant progress has also been made on the etiology of the Mendelian-inherited gene errors.

MITOCHONDRIAL INHERITANCE

Mammalian mitochondria, and therefore mammalian mtDNAs, are maternally inherited. This means that a mother transmits mitochondria to all of her children, both boys and girls, but only her daughters will transmit their mitochondria to their children. There are at least two competing hypotheses that might explain the limitation of mitochondrial transmission to females. The first hypothesis is based simply on numbers. The typical mammalian sperm contains relatively few mitochondria (albeit large ones, wrapped around the "neck" of the sperm midpiece), whereas the oocyte contains about 100,000 mitochondria. Since only ~1% of the mitochondria present at fertilization will eventually repopulate the embryo (those in the inner cell mass; the remainder become part of the extraembryonic tissues), the chances that a paternal mitochondrion will enter the fetus is extremely low, and even if it entered the fetus, it would be extremely hard to detect amid the vast majority of maternal mitochondria.

The second, and more compelling, hypothesis is based on the finding that during fertilization the sperm enters the ovum in its entirety.[4] The widely held view that only the head of the sperm (which contains the nucleus but no mitochondria) enters the ovum, while the tail (which contains the paternal mitochondria) falls away and never enters the egg, is erroneous. Although paternal mitochondria enter the ovum at fertilization, they are selectively destroyed. Two clearance mechanisms have been proposed: autophagy, in which the entering paternal sperm are engulfed by "autophagolysosomes" (as is the case in the worm *Caenorhabdtis elegans*[5,6]), or proteasomal degradation, in which the sperm mitochondria are "marked" by ubiquitination and are then degraded by the proteasome.[7] Although paternal mitochondria are not inherited in the offspring of matings between mice of the same strains, they can be inherited at high frequency in matings between mice of two different strains,[8] giving rise to a third view, namely, that sperm mitochondria are somehow recognized as foreign "antigens": unknown factors in the ovum recognize and then destroy mitochondria derived from sperm of the same species (the usual situation), but these factors are incapable of recognizing or destroying mitochondria from a different species. In the fruitfly *Drosophila melanogaster*, which has gigantic sperm, paternal mtDNAs are eliminated in an even more baroque fashion: paternal mtDNAs are selectively degraded and are then "swept" into a waste compartment near the sperm tail, so that the sperm are devoid of mtDNA before ever encountering the egg.[9]

SEGREGATION AND HETEROPLASMY

One of the most important concepts in mitochondrial genetics is that it is *population* genetics; this key feature distinguishes it from Mendelian genetics. Depending on the energy requirements of a particular tissue, there may be hundreds or even thousands of mitochondria in a cell, with multiple mtDNAs in each organelle. While the total amount of mtDNA in a cell is tightly regulated,[10] the timing of organellar division and of mtDNA replication are apparently stochastic events unrelated to the cell cycle. Thus, the numbers of mitochondria present in a cell can vary both in space (among cells and tissues) and in time (during development and aging). The numbers can also increase or decrease based on the oxidative energy requirements of the moment (e.g., after prolonged training in athletes or after acclimation at high altitude). This dynamic aspect of mitochondrial transmission from cell to cell is termed *mitotic segregation*.[11]

In Mendelian genetics, a person's nuclear DNA can contain, at most, only two alleles for any particular autosomal gene, one paternally derived (e.g., allele "A") and the other maternally derived (e.g., allele "B"). Thus, an individual may be homozygous (A/A or B/B), heterozygous (A/B), or even hemizygous (A/null or B/null) at a particular gene locus. In mitochondrial genetics, on the other hand, there are theoretically as many alleles for a particular gene in a cell as there are mtDNAs in that cell. To all intents and purposes, however, the mitochondrial genotypes of all the mtDNAs in a normal person are identical. On the other hand, of the 16,569 bp that form the mtDNA genotype, the mtDNA sequence typically differs between two unrelated individuals at approximately 50 positions. These differences are presumably neutral polymorphisms, at least in the normal population. Note that mating does not alter these differences: if a man with mtDNA genotype A marries a woman with mtDNA genotype B, their children will have only genotype B, owing to the effect of exclusive maternal inheritance. Thus, in the normal course of events, individuals are *homoplasmic* (i.e., all their mtDNAs are identical).

The presence of different mitochondrial genotypes begs the question as to why different genotypes exist at all. As with nuclear DNA, mtDNA mutations arise in single molecules spontaneously, probably during DNA replication (discussed below). In mtDNA, these mutations are usually transitions: purines (A or G) replace purines, and pyrimidines (C or T) replace pyrimidines; transversions (purines replacing pyrimidines or *vice versa*) are relatively rare. Once a mutation is fixed, the cell is considered to be *heteroplasmic* (i.e., coexistence of different mtDNA genotypes). If the mutation arises in the female germline, it can be transmitted to the next generation. However, the *bottleneck hypothesis* states that the number of mtDNAs transmitted from mother to child is quite small (estimated at between 5 and 200 segregating units[12]), so even if the germline were heteroplasmic, purification to homoplasmy can take place within a few generations.[13] This would explain why mtDNA genotypes vary among individuals, and yet those individuals are homoplasmic for their particular genotype.

Obviously, a child can be born with heteroplasmic mtDNA if there were heteroplasmy among the few mtDNAs that pass through the mitochondrial bottleneck between mother and child. This is seen most clearly in the maternal inheritance of rare pathogenic mtDNA mutations in which both the clinical phenotype and the associated mtDNA genotype can be tracked in the maternal lineage of a pedigree. In the vast majority of situations, those pathogenic mutations are heteroplasmic (there is a coexistence of normal and mutated mtDNAs). Presumably, homoplasmic pathogenic mtDNA mutations are seen only rarely because they would result either in oocyte failure or in embryonic lethality. The combination of heteroplasmy and mitotic segregation can result in mitochondrial disorders with variable presentation of the pathogenic phenotype among affected members of a pedigree, among different tissues in an affected individual, and during the lifespan of that individual.

MITOCHONDRIAL DNA REPLICATION

The replication of circular human mtDNA is somewhat "prokaryote-like," but has a number of features that make it unusual. In a typical bacterium, such as *E. coli*, the circular genome has a single origin of replication that initiates bidirectional replication from a fixed position on the genome; it has been dubbed "theta" replication because intermediates in such a mode of replication look like the Greek letter theta. Thus, daughter-strand DNA synthesis in *E. coli*, controlled by the combined action of RNA polymerase that synthesizes short RNA primers, and DNA polymerases that use the RNA primers to extend the newly forming daughter DNA using the parental strand as the template, proceeds simultaneously in opposite directions. Notably, because all DNA polymerases synthesize daughter strand DNA only in the 5′ → 3′ direction, one strand (the "leading" strand) is extended in a continuous fashion, whereas the other (the "lagging" strand) is not. In order for the lagging strand to grow in the 3′ → 5′ direction, it is made as a series of fragments ("Okazaki fragments") that are actually synthesized in the 5′ → 3′ direction, albeit discontinuously, with each segment ligated to its adjacent partners. At the end of a round of replication in bacteria, the original parental circle is now a pair of catenated circles, each a double helix consisting of one "old" parental strand and one "new" daughter strand. The circles are then separated into two free-standing decatenated circles by the action of an enzyme called topoisomerase II. (This mode of replication in circular bacterial genomes is quite distinct from that of eukaryotic nuclear DNA, in which thousands of origins of replication, or "eyes," are spaced along the length of the linear chromosome.)

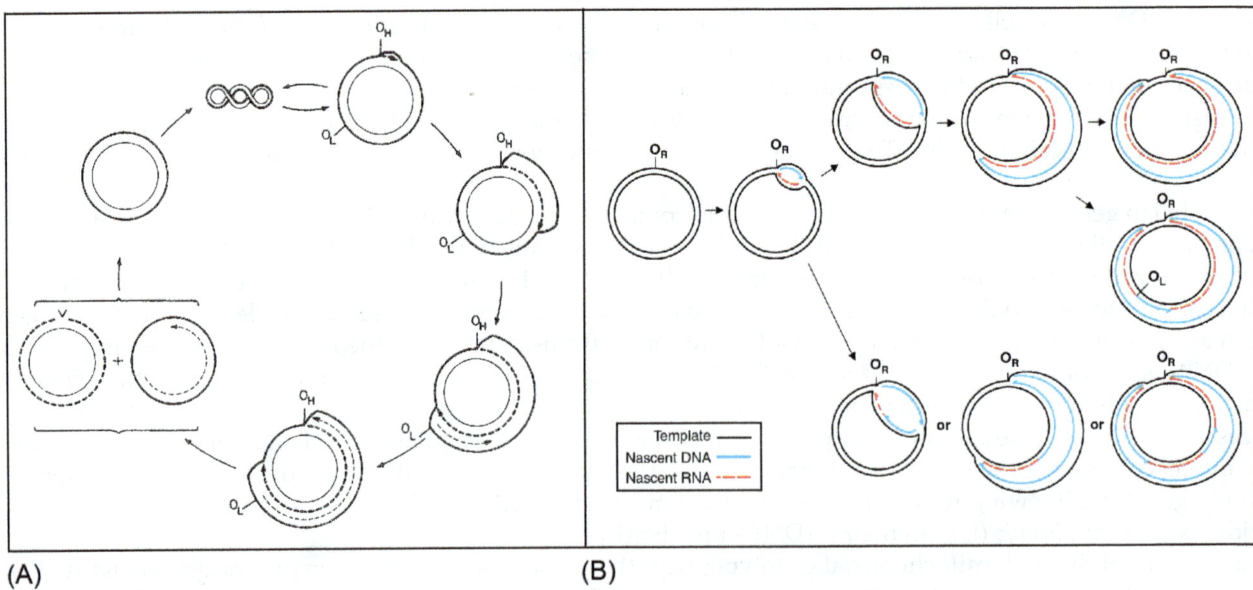

FIGURE 22.3 **Replication of human mtDNA.** **(A)** The "asynchronous, strand displacement" model. Replication of the parental mtDNA (thick solid lines are heavy H strands; thin solid lines are light L strands) proceeds from O_H in the D-loop region (clockwise beginning at "1 o'clock") until daughter strand synthesis (heavy dashed line) reaches O_L, at which point daughter light-strand synthesis (light dashed line) initiates in a counterclockwise direction. *(Adapted from Clayton DA. Replication of animal mitochondrial DNA. Cell 1982; 28: 693. Reproduced with permission from Cell Press, Cambridge, MA).* **(B)** The "synchronous, strand-coupled" model. Replication proceeds bidirectionally on the leading and lagging strands, as in classic bacterial DNA "theta" replication, but with some modifications. Most notably, preformed RNA incorporation throughout the lagging strand (RITOLS) (dashed lines) are "threaded" through the advancing DNA polymerase complex, hybridizing with the displaced H-strand in the 3′–5′ direction, until an upstream RNA processing site (or a transcriptional start site) is encountered. (Adapted from[19] *Reproduced with permission from John Wiley & Sons, Hoboken, NJ*). In both models, following completion of synthesis of the two daughter strands, the two progeny molecules, which are initially catenated circles (not shown), are unlinked; the "relaxed" circles are then supercoiled by topoisomerase. See text for further description.

Human mtDNA also has a single origin of replication, except for one crucial difference: the mtDNA origin has been physically separated into two halves, each controlling synthesis of one of the two daughter DNA strands (Figure 22.3). Synthesis of one strand begins at the *origin of heavy-strand replication* (O_H), which is located at the top of the circle, around map position 200 (nomenclature and numbering of [3]; see Figure 22.1), within a 1123-bp "control region" between the tRNA^{Phe} gene (at map position 577) and the tRNA^{Pro} gene (at map position 16023), and proceeds in a clockwise direction. Synthesis of the other strand begins at the *origin of light-strand replication* (O_L), which is located around "8 o'clock" on the circle (near position 5750), and proceeds in a counterclockwise direction.

This separation of the two origins results in a rather baroque mechanism for overall replication,[14] called the "asynchronous, strand displacement" model (Figure 22.3A).[15,16] As with bacterial DNA, there is a requirement for the synthesis of an RNA primer for initiating DNA synthesis. This RNA primer is synthesized by mitochondrial RNA polymerase as a relatively large precursor molecule starting near one of the three upstream *conserved sequence blocks* (CSB I, II, and III) located between O_H and tRNA^{Phe}. This RNA hybridizes to the L-strand and displaces a portion of the H-strand in the control region, thereby exposing the so-called "D-loop" (D for "displacement"). DNA replication does not commence until the precursor RNA is cleaved at a precise location on the RNA by mitochondrial RNA-processing ribonuclease. RNase MRP is a polypeptide associated with a nucleus-encoded RNA (i.e., it is a ribonucleoprotein) that is imported into mitochondria to form the primer required by the mitochondrial DNA polymerase (called polymerase γ, to distinguish it from polymerases α, β, δ, and ε, which control replication of nuclear DNA).

The nascent daughter H-strand is synthesized from "1 o'clock" toward "8 o'clock," all the while extending the parental D-loop forward as daughter H-strand DNA is laid down on parental L-strand DNA. As the polymerase (and the displaced DNA) passes "8 o'clock," a specific segment of single-stranded parental H-strand is "revealed" at O_L. Owing to the presence of an inverted repeat sequence in this region, the single-stranded DNA around O_L forms a "hairpin" or "stem-loop" structure containing a repeat of Ts in the loop of the hairpin. A second RNA primer, containing a complementary repeat of As, binds to the loop, and becomes a primer for synthesis of daughter L-strand DNA in the opposite (counterclockwise) direction, presumably with the assistance of an O_L-specific "primase." The two opposite-growing strands continue until they both have traversed their respective circles completely. Note that in the strand displacement model, both strands are sythesized continuously in the 5′→3′ direction (i.e., there are no Okazaki fragments), one proceeding clockwise (starting at O_H) and the other counterclockwise (starting at O_L).

In the last few years a second mode of replication, called the "synchronous, strand-coupled" model has been proposed (Figure 22.3B).[17,18] In this model, replication initiates from a single origin and proceeds symmetrically in both a clockwise and counterclockwise direction, similar to the classic "theta replication" mechanism by which circular bacterial genomes divide. However, the strand-coupled model also invokes long tracts of preformed RNA transcripts as replication intermediates on the lagging strand that are "chased" into DNA. Because processed transcripts are successively hybridized to the lagging-strand template while the replication fork advances (dubbed "RNA incorporation throughout the lagging strand," or RITOLS[19]), this mode of replication has also been called the "bootlace" model.[18]

In both models, replication terminates with the creation of a pair of catenated circles (each circle is a double helix containing one "old" parental strand and one "new" daughter strand). Mitochondrial DNA topoisomerase now decatenates the circles, releasing the two daughter molecules. The entire process takes a remarkably long time—approximately 2 hours to replicate the 16,569 bp of mitochondrial DNA (compare this to the 40 minutes required to replicate about 4 million bp of *E. coli* DNA).

There are a number of Mendelian-inherited mitochondrial disorders that are due to errors in the components of the replication machinery or in the regulation of the organellar nucleotide pools. One such disorder is mtDNA depletion, in which infants are born with a severe deficiency in the number of mitochondrial genomes present in specific tissues.[20,21] Another group of disorders are the Mendelian-inherited mtDNA multiple deletion disorders. In these diseases, patients inherit a defective gene that appears to increase the frequency of mtDNA rearrangements, thereby causing pathologic consequences, and are manifested as both autosomal dominant- and autosomal recessive-inherited forms. Mutations causing depletion and multiple deletion syndromes have now been found in genes required for the maintenance and integrity of mtDNA, such as those encoding DNA polymerase γ (both in the catalytic and accessory subunits), mitochondrial DNA/RNA helicase, mitochondrial thymidine kinase, mitochondrial deoxyguanosine kinase, the adenine nucleotide translocator, and cytosolic thymidine phosphorylase.[22] It should be noted that mtDNA rearrangements can also be inherited in a maternal fashion, and can even be sporadic, but the mechanisms responsible for these latter conditions almost certainly do not involve intrinsic defects in the replicative machinery.

TRANSCRIPTION

Like mtDNA replication, transcription of human mtDNA has a prokaryotic "look" about it. In broad view, all 37 genes encoded by human mtDNA are synthesized initially on two huge polycistronic precursor transcripts, one encoded by the L-strand and the other by the H-strand (see Figure 22.1). Of the 37 genes, 28 (two rRNAs, 14 tRNAs, and 12 polypeptide-coding genes) are encoded by the H-strand (note that these RNAs are *encoded* by the H-strand, meaning that they have the *same sequence* as the L-strand). Only eight tRNAs and one mRNA (ND6) are encoded by the L-strand. Since the precursor RNAs are single stranded, the tRNA genes that "punctuate" the circle at strategic positions[23] can now adopt their typical "cloverleaf" conformation (see Figure 22.1). It is thought that these tRNAs become targets for the action of two enzymes, called RNase P (another imported ribonucleoprotein) and RNase Z.[24] These enzymes excise the tRNAs by cleaving them precisely at their 5′ and 3′ edges in the precursor transcript, respectively, thereby releasing not only the tRNAs, but the flanking rRNAs and mRNAs as well. However, there is no tRNA separating ATPase 6 from COX III, and it is not clear how those two messages are processed.

Many details of mitochondrial transcription have been elucidated.[14,25] Human mtDNA contains only two promoters for RNA transcription, both located within the 150-bp region in the D-loop containing the CSBs. One promoter controls transcription of the H-strand (the heavy-strand promoter, or HSP), while the other controls L-strand transcription (the LSP). The LSP is also important because it is required for synthesis of the RNA primer required for replication at O_H. Two *trans*-acting transcription factors, called mtTFA and mtTFB, bind to specific sequences upstream of both the HSP and LSP transcription start sites to promote transcription in the presence of mtRNA polymerase.[26,27]

Because transcription is polycistronic, an interesting arithmetic problem presents itself. Polycistronic transcription means that all of the mature RNAs ultimately generated off the H- strand (or the L-strand) are produced in equimolar amounts. Since each mRNA requires large numbers of rRNAs and tRNAs for its translation, there must be mechanisms whereby the steady-state levels of tRNAs and rRNAs are sufficient to satisfy the translational demands of the mRNAs specifying all 13 polypeptides. One such regulatory mechanism is likely to be differential turnover of classes of RNAs, especially mRNAs as compared to tRNAs.[28]

For rRNAs, however, mitochondria have also evolved at least one elegant solution to the problem. Besides the long 16-kb polycistronic transcript generated off the HSP and encompassing all the H-strand genes, a shorter 3-kb transcript is also synthesized. This transcript, which encompasses only the two rRNA genes and their flanking tRNAs, is synthesized at approximately 25 times the abundance of the long transcript,[29] thereby enabling a sufficient amount of 12S and 16S rRNA to be made for all the ribosomes that the organelle needs for translation. This, of course, raises a new question: how does the transcriptional machinery produce both long and short H-strand transcripts off the same promoter, and in vastly different amounts? It turns out that there is a protein that binds to the $tRNA^{Leu(UUR)}$ gene that is located immediately downstream of the 16S rRNA gene.[30] When this protein, called mitochondrial transcription termination factor (mtTERF), binds to a specific sequence located in the middle of the $tRNA^{Leu(UUR)}$ gene, it blocks the advance of the RNA polymerase, and the short form of the H-strand transcript is released. When mtTERF does not bind (either because it is not present in sufficient amounts or because it is prevented from binding by another factor), the polymerase advances unhindered and the long form of the transcript is made.[31]

Following cleavage of the precursor polycistronic RNAs, other processing events are required for maturation of the RNAs before translation can proceed. The 3′ termini of the mRNAs are polyadenylated (polyadenylation is absent in prokaryotes; thus it is one of the rare instances where a mammalian mitochondrial function is more similar to a process that occurs in eukaryotic, nuclear DNA). The trinucleotide CCA is added to the 3′ termini of the tRNAs (CCA addition is required for aminoacylation of the tRNAs by their cognate aminoacyl synthetases), and specific tRNA bases are chemically modified (e.g., conversion of uridine to pseudouridine).

TRANSLATION

Translation of mitochondrial mRNAs takes place on mitochondrial ribosomes, which consist of mtDNA-encoded 12S and 16S rRNAs together with imported ribosomal proteins.[32] Translation of mRNAs has features unique to mitochondria. The first has to do with the structure of the mRNA. Like bacterial messages, the upstream end of the message (the "5′ end") contains no added "recognition" structure ("cap") for translation initiation. The 5′ end, however, poses a conceptual problem for translation initiation. Ordinarily, the ribosome binds to a specific position in the 5′-untranslated region (the ribosome binding site). The ribosome then travels along the message until it finds the first appropriate initiation codon (usually AUG, specifying methionine). At this point, reading of the codon triplets produces the translated message until the ribosome encounters a "stop" codon (UAA, UAG, or UGA in the universal code) at the other end of the message. In mammalian mitochondria, however, the initiation codon is located at the very beginning of the mature message (there is little or no 5′-untranslated region). In this case, it is unclear how the ribosome recognizes and binds to the message. It is known that the initiation codon is sequestered in a stem structure to which the mitochondrial ribosome binds, but this may not be obligatory. Also, factors similar to those required for translation initiation on prokaryotic and eukaryotic messages are present in mammalian mitochondria.[32]

The downstream end of the message (called the "3′ end") contains its own quirks. Except for COX II, none of the mRNAs has a 3′-untranslated region (a feature present in both bacterial and eukaryotic nuclear mRNAs). The last amino acid-specifying codon is thus located within one or two nucleotides of the end of the message, and the messages often end with a U or a UA. Interestingly, addition of the poly(A) tail to the mRNA converts these "supernumerary" nucleotides to UAA, which is a translational stop codon. Human mitochondria have four stop codons (the standard UAA and UAG, but not UGA [which encodes tryptophan in the mitochondrial code], plus AGA and AGG [which encode arginine in the universal code]). This has raised a conundrum regarding how the mitochondrial translational "release factor" recognizes AGA and AGG, but not UGG. This problem was recently solved after it was noted that terminal AGA and AGG codons are preceded by a U (i.e., UAGA and UAGG), which would create a potential "standard" UAG stop codon, albeit out-of-frame; however, it was then discovered that the mitoribosome stalls at this point and moves backwards by one nucleotide (a "–1 frameshift") prior to recommencing its travel, thereby bringing the UAG stop codons back into frame.[33] In this way, the release factor recognizes both AGA and AGG as UAG, thus terminating translation.[34]

Two of the mRNAs contain overlapping messages, that is, one contiguous piece of mRNA specifies two different polypeptides. In both cases, the pair of overlapping polypeptides belong to the same respiratory complex: one message encodes both ND4L and ND4 of complex I, while the other encodes both ATPase 8 and ATPase 6 of complex V; both overlaps are out-of-frame with each other. It is unclear how both pairs of overlapping messages are translated; perhaps it is by "internal" ribosomal initiation. The discovery of ribosomal frameshifting at termination codons provides a precedent for another mechanism, namely frameshifting to initiate translation internally.

As mentioned earlier, human mitochondria have their own genetic code, which differs from the universal code at five of the 64 triplet positions (the four noted above, plus AUA, which encodes methionine instead of isoleucine). This alteration ensures that only mtDNA-encoded messages can be translated faithfully, even if a "stray" piece of cytoplasmic RNA were to enter the organelle. Conversely, mitochondrial sequences that strayed into the nucleus and integrated into nuclear DNA (rare but documented events; there are ~1000 such mitochondrial "pseudogenes" in the human genome[35]) would not be translated correctly on cytoplasmic ribosomes even if transcribed, and these polypeptides would likely be degraded rapidly.

Errors in mitochondrial translation have been associated with a number of mitochondrial diseases.[11] The most notable diseases are sporadic Kearns–Sayre syndrome and sporadic progressive external ophthalmoplegia (in which all 13 mtDNA-encoded mRNAs are not translated[36]), maternally inherited MELAS (mitochondrial encephalomyopathy, lactic acidosis, and stroke-like episodes; in which the amount of translation products is reduced[37]) and MERRF (myoclonus epilepsy with ragged red fibers; the amount of translation is reduced and a number of aberrant translation products are synthesized[38]).

While many of the details regarding mitochondrial translation have been worked out,[32] much clearly remains to be learned. The recent solution of the structure of the mammalian mitochondrial ribosomal large subunit[39] should go a long way to help addressing these issues.

IMPORTATION

As noted earlier, the vast majority of mitochondrial proteins are nucleus-encoded and imported into mitochondria from the cytosol, and the elucidation of the mechanism of importation of cytosolic proteins into mitochondria has been a major scientific achievement. The process is surprisingly complex, as there are different pathways for the importation and sorting of mitochondrially targeted polypeptides to the various organellar compartments (outer and inner membranes, intermembrane space, and matrix).[40,41] Among the more interesting aspects of this process is the fact that components of the import machinery are members of the family of "heat shock proteins." These "molecular chaperones" are ATP-dependent proteins that unfold and then refold the mitochondrial-targeted polypeptides as they are inserted through the import receptors and are sorted to the appropriate compartments.

A mitochondrial targeting signal (MTS) must be present to address mitochondrial-imported proteins to the correct target (the mitochondria).[42] For most polypeptides destined for the matrix and inner membrane, and for some intermembrane space (IMS) proteins, the MTS is usually located at the *N*-terminus of the polypeptide; once imported, this "leader peptide" is cleaved inside the organelle to release the mature polypeptide. The MTS is usually highly basic, with many arginine and lysine residues and few or no aspartate or glutamate residues. The MTS often contains consensus sequence elements that determine the precise point of cleavage of the presequence inside the organelle (Figure 22.4). Other MTSs are located at the *C*-terminus or are even located internally within the polypeptide; these MTSs are typically not cleaved. Polypeptides destined for the outer membrane, and mitochondrial carriers and transporters that are inserted into the inner membrane, have no obviously recognizable MTSs. For some IMS proteins, the polypeptide is imported in a totally different way: these proteins contain sets of paired cysteine residues (typically CX_3C or CX_9C motifs) that are used, together with specialized chaperones, to import the protein by a "disulfide relay" system in which importation is mediated by oxidization and reduction of the cysteines, which changes the conformation of the polypeptide.[43]

Mutations in leader sequences should result in Mendelian-inherited mitochondrial disorders.[44] A number of such mutations have been identified to date, such as those in the leader peptides of methylmalonyl-CoA mutase (an enzyme of β-oxidation) causing methylmalonic acidemia, and in ornithine transcarbamoylase (an enzyme required in the urea cycle and in arginine biosynthesis). Surprisingly, importation errors can result from mutations in the *mature* portion of the protein, implying that the import machinery recognizes elements downstream of the leader and precursor cleavage site. Mutations in the mature portion of ornithine aminotransferase (OAT) prevent the entry of the OAT precursor into mitochondria, causing gyrate atrophy[45] and deletion of a segment of mature isovaleryl-CoA dehydrogenase (another β-oxidation enzyme) inhibit import, thereby causing isovaleryl acidemia.[46]

Of the nuclear genes encoding mitochondrial-imported polypeptides, a subset of perhaps 50 to 100 genes appears to contain regulatory transcriptional elements that are held in common. This subset includes nuclear genes required for the synthesis of many respiratory chain subunits and some genes required for mtDNA replication. Two good examples of such elements are nuclear respiratory factors 1 and 2 (NRF-1 and NRF-2). These factors have been identified in the promoter region or first intron of a number of such genes.[47] More recently, an entirely new class of regulators of mitochondrial biogenesis has been found, which is associated with the coactivator of the peroxisome proliferator-activated receptor gamma subunit (PPAR-γ).[47] Three such genes have been found, called PGC-1α (PPAR-γ coactivator-1α), PGC-1β (PPAR-γ coactivator-1β), and PRC (PGC1-related coactivator). These regulatory elements are one of the ways in which the nuclear and mitochondrial genomes "communicate" with each other, and are

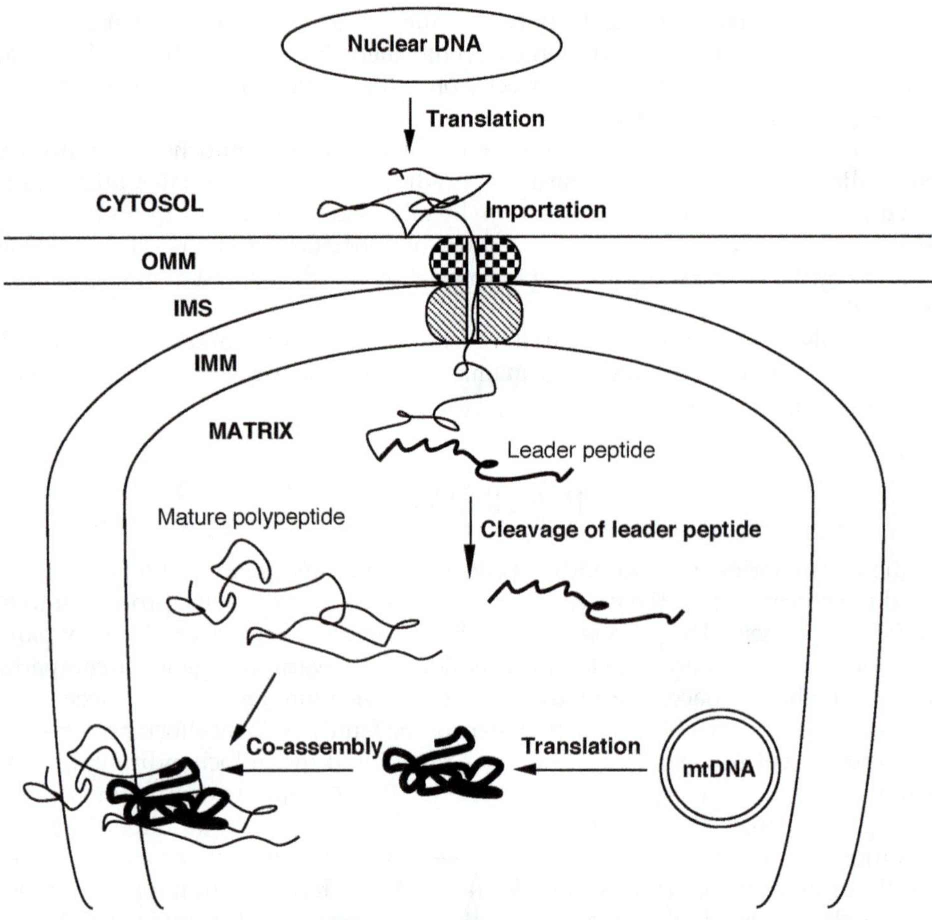

FIGURE 22.4 **Targeting and importation of cytoplasmic polypeptides to mitochondria.** Nucleus-encoded polypeptides are imported into mitochondria through a complex importation machinery (hatched boxes) at import sites where the outer mitochondrial membrane (OMM) and inner mitochondrial membrane (IMM) are apposed. In most cases, a leader peptide at the *N*-terminus is removed by specific proteases after importation into mitochondria. For respiratory-chain complexes I, III, IV, and V, imported nucleus-encoded subunits are coassembled with the appropriate mtDNA-encoded polypeptides and inserted into the IMM, as shown.

fundamental to mechanisms by which the cell can integrate its overall metabolic status and energetic requirements by modulating mitochondrial function and output.[47,48]

ACKNOWLEDGEMENTS

This work was supported by grants from the National Institutes of Health (HD32062), the Department of Defense (W911NF-12-1-0159), the Muscular Dystrophy Association, the Ellison Medical Foundation, and the J. Willard and Alice S. Marriott Foundation.

References

1. Margulis LS. On the origin of mitosing cells. *J Theor Biol*. 1967;14:225–274.
2. Lane N, Martin WF, Raven JA, Allen JF. Energy, genes and evolution: introduction to an evolutionary synthesis. *Philos Trans R Soc Lond B Biol Sci*. 2013;368:20120253.
3. Anderson S, Bankier AT, Barrel BG, et al. Sequence and organization of the human mitochondrial genome. *Nature*. 1981;290:457–465.
4. Shalgi R, Magnus A, Jones R, Phillips DM. Fate of sperm organelles during early embryogenesis in the rat. *Mol Reprod Dev*. 1994;37:264–271.
5. Al Rawi S, Louvet-Vallee S, Djeddi A, et al. Postfertilization autophagy of sperm organelles prevents paternal mitochondrial DNA transmission. *Science*. 2011;334:1144–1147.
6. Sato M, Sato K. Degradation of paternal mitochondria by fertilization-triggered autophagy in *C. elegans* embryos. *Science*. 2011;334:1141–1144.
7. Sutovsky P, Van Leyen K, McCauley T, Day BN, Sutovsky M. Degradation of paternal mitochondria after fertilization: implications for heteroplasmy, assisted reproductive technologies and mtDNA inheritance. *Reprod Biomed Online*. 2004;8:24–33.

8. Kaneda H, Hayashi J-I, Takahama S, Taya C, Fischer Lindahl K, Yonekawa H. Elimination of paternal mitochondrial DNA in intraspecific crosses during early mouse embryogenesis. *Proc Natl Acad Sci U S A*. 1995;92:4542–4546.
9. DeLuca SZ, O'Farrell PH. Barriers to male transmission of mitochondrial DNA in sperm development. *Dev Cell*. 2012;22:660–668.
10. Tang Y, Schon EA, Wilichowski E, Vazquez-Memije ME, Davidson E, King MP. Rearrangements of human mitochondrial DNA (mtDNA): new insights into the regulation of mtDNA copy number and gene expression. *Mol Biol Cell*. 2000;11:1471–1485.
11. DiMauro S, Schon EA, Carelli V, Hirano M. The clinical maze of mitochondrial neurology. *Nat Rev Neurol*. 2013;9:429–444.
12. Shoubridge EA, Wai T. Mitochondrial DNA and the mammalian oocyte. *Curr Top Dev Biol*. 2007;77:87–111.
13. Jenuth JP, Peterson AC, Fu K, Shoubridge EA. Random genetic drift in the female germline explains the rapid segregation of mammalian mitochondrial DNA. *Nat Genet*. 1996;13:146–151.
14. Clayton DA. Transcription and replication of animal mitochondrial DNAs. *Int Rev Cytol*. 1992;141:217–232.
15. Brown TA, Cecconi C, Tkachuk AN, Bustamante C, Clayton DA. Replication of mitochondrial DNA occurs by strand displacement with alternative light-strand origins, not via a strand-coupled mechanism. *Genes Dev*. 2005;19:2466–2476.
16. Brown TA, Tkachuk AN, Clayton DA. Native R-loops persist throughout the mouse mitochondrial DNA genome. *J Biol Chem*. 2008;283:36743–36751.
17. Holt IJ, Lorimer HE, Jacobs HT. Coupled leading- and lagging-strand synthesis of mammalian mitochondrial DNA. *Cell*. 2000;100:515–524.
18. Holt IJ, Reyes A. Human mitochondrial DNA replication. *Cold Spring Harb Perspect Biol*. 2012;4:a012971.
19. Yasukawa T, Reyes A, Cluett TJ, et al. Replication of vertebrate mitochondrial DNA entails transient ribonucleotide incorporation throughout the lagging strand. *EMBO J*. 2006;25:5358–5371.
20. Finsterer J, Ahting U. Mitochondrial depletion syndromes in children and adults. *Can J Neurol Sci*. 2013;40:635–644.
21. Vu TH, Sciacco M, Tanji K, et al. Clinical manifestations of mitochondrial DNA depletion. *Neurology*. 1998;50:1783–1790.
22. Van Goethem G. Autosomal disorders of mitochondrial DNA maintenance. *Acta Neurol Belg*. 2006;106:66–72.
23. Ojala D, Montoya J, Attardi G. tRNA punctuation model of RNA processing in human mitochondria. *Nature*. 1981;290:470–474.
24. Rossmanith W. Of P, and Z: mitochondrial tRNA processing enzymes. *Biochim Biophys Acta*. 1819;2012:1017–1026.
25. Asin-Cayuela J, Gustafsson CM. Mitochondrial transcription and its regulation in mammalian cells. *Trends Biochem Sci*. 2007;32:111–117.
26. McCulloch V, Seidel-Rogol BL, Shadel GS. A human mitochondrial transcription factor is related to RNA adenine methyltransferases and binds S-adenosylmethionine. *Mol Cell Biol*. 2002;22:1116–1125.
27. Parisi MA, Clayton DA. Similarity of human mitochondrial transcription factor 1 to high mobility group proteins. *Science*. 1991;252:965–969.
28. King MP, Attardi G. Post-transcriptional regulation of the steady-state levels of mitochondrial tRNAs in HeLa cells. *J Biol Chem*. 1993;268:10228–10237.
29. Gelfand R, Attardi G. Synthesis and turnover of mitochondrial ribonucleic acid in HeLa cells: the mature ribosomal and messenger ribonucleic acid species are metabolically unstable. *Mol Cell Biol*. 1981;1:497–511.
30. Kruse B, Narasimhan N, Attardi G. Termination of transcription in human mitochondria: identification and purification of a DNA binding protein factor that promotes termination. *Cell*. 1989;58:391–397.
31. Guja KE, Garcia-Diaz M. Hitting the brakes: termination of mitochondrial transcription. *Biochim Biophys Acta*. 1819;2012:939–947.
32. Christian BE, Spremulli LL. Mechanism of protein biosynthesis in mammalian mitochondria. *Biochim Biophys Acta*. 1819;2012:1035–1054.
33. Temperley R, Richter R, Dennerlein S, Lightowlers RN, Chrzanowska-Lightowlers ZM. Hungry codons promote frameshifting in human mitochondrial ribosomes. *Science*. 2010;327:301.
34. Chrzanowska-Lightowlers ZM, Pajak A, Lightowlers RN. Termination of protein synthesis in mammalian mitochondria. *J Biol Chem*. 2011;286:34479–34485.
35. Schon EA, DiMauro S, Hirano M. Human mitochondrial DNA: roles of inherited and somatic mutations. *Nat Rev Genet*. 2012;13:878–890.
36. Nakase H, Moraes CT, Rizzuto R, Lombes A, DiMauro S, Schon EA. Transcription and translation of deleted mitochondrial genomes in Kearns-Sayre syndrome: implications for pathogenesis. *Am J Hum Genet*. 1990;46:418–427.
37. King MP, Koga Y, Davidson M, Schon EA. Defects in mitochondrial protein synthesis and respiratory chain activity segregate with the $tRNA^{(Leu(UUR))}$ mutation associated with mitochondrial myopathy, encephalopathy, lactic acidosis, and strokelike episodes. *Mol Cell Biol*. 1992;12:480–490.
38. Masucci J, Davidson M, Koga Y, DiMauro S, Schon EA, King MP. *In vitro* analysis of mutations causing myoclonus epilepsy with ragged-red fibers in the mitochondrial $tRNA^{Lys}$ gene: two genotypes produce similar phenotypes. *Mol Cell Biol*. 1995;15:2872–2881.
39. Greber BJ, Boehringer D, Leitner A, et al. Architecture of the large subunit of the mammalian mitochondrial ribosome. *Nature*. 2014;505:515–519.
40. Dudek J, Rehling P, van der Laan M. Mitochondrial protein import: common principles and physiological networks. *Biochim Biophys Acta*. 1833;2013:274–285.
41. Paschen SA, Neupert W. Protein import into mitochondria. *IUBMB Life*. 2001;52:101–112.
42. Emanuelsson O, von Heijne G. Prediction of organellar targeting signals. *Biochim Biophys Acta*. 2001;1541:114–119.
43. Herrmann JM, Riemer J. Mitochondrial disulfide relay: redox-regulated protein import into the intermembrane space. *J Biol Chem*. 2012;287:4426–4433.
44. Fenton WA. Mitochondrial protein transport - a system in search of mutations. *Am J Hum Genet*. 1995;57:235–238.
45. Kobayashi T, Ogawa H, Kasahara M, Shiozawa Z, Matsuzawa T. A single amino acid substitution within the mature sequence of ornithine aminotransferase obstructs mitochondrial entry of the precursor. *Am J Hum Genet*. 1995;57:284–291.
46. Vockley J, Nagao M, Parimoo B, Tanaka K. The variant human isovaleryl-CoA dehydrogenase gene responsible for type II isovaleric acidemia determines an RNA splicing error, leading to the deletion of the entire second coding exon and the production of truncated precursor protein that interacts poorly with mitochondrial import receptors. *J Biol Chem*. 1992;267:2494–2501.
47. Scarpulla RC, Vega RB, Kelly DP. Transcriptional integration of mitochondrial biogenesis. *Trends Endocrinol Metab*. 2012;23:459–466.
48. Gupta RK, Rosen ED, Spiegelman BM. Identifying novel transcriptional components controlling energy metabolism. *Cell Metab*. 2011;14:739–745.

CHAPTER

23

Mitochondrial Disorders Due to Mutations in the Mitochondrial Genome

Salvatore DiMauro and Carmen Paradas

Department of Neurology, Columbia University Medical Center, The Neurological Institute of New York, New York, NY, USA

INTRODUCTION

Although mitochondrial DNA (mtDNA) was discovered almost 60 years ago and its biology has been extensively studied (Chapter 22), pathogenic mutations in the mitochondrial genome were not reported until 1988. In that year, Holt et al. described large-scale deletions in patients with mitochondrial myopathies,[1] and Wallace et al. described a point mutation in the gene encoding subunit 4 of complex I (*MTND4*) in a family with Leber hereditary optic neuropathy (LHON).[2] However, these two papers opened up a veritable Pandora's box: the "morbidity map" of mtDNA has gone from the one-point mutation of 1988 to over 200 pathogenic point mutations in 2012 (Figure 23.1).[3]

CLINICAL FEATURES

Mitochondria and mtDNA are ubiquitous, which explains why every tissue in the body can be affected by mtDNA mutations. This is illustrated in Table 23.1, which provides a compilation of all symptoms and signs reported in patients with three different types of mtDNA mutations, including single deletions, point mutations in two distinct transfer RNA (tRNA) genes, and point mutations in a protein-coding gene. As shown by the boxes, certain constellations of symptoms and signs are so characteristic as to make the diagnosis in typical patients relatively easy. On the other hand, because of the peculiar rules of mitochondrial genetics, different tissues harboring the same mtDNA mutation may be affected to different degrees or not at all, which explains the sometimes-puzzling variety of syndromes associated with mtDNA mutations, even within a single family.

Among the maternally inherited encephalomyopathies, four syndromes are more common. The first is MELAS (mitochondrial encephalomyopathy, lactic acidosis, and stroke-like episodes), which usually presents in children or young adults after normal early development and is most commonly associated with the m.3243A>G mutation in the gene (*MTTL*) encoding the $tRNA^{Leu(UUR)}$. Reviews of symptoms in two large series that included both typical patients (with encephalomyopathy, lactic acidosis, and stroke-like episodes) and their matrilinear mutation carriers gave similar results.[4,5] Aside from the acronymic features, the most common symptoms include exercise intolerance, hearing loss, diabetes, migraine, growth failure, ptosis/ophthalmoparesis, gastrointestinal disturbances, and cardiopathy. Magnetic resonance imaging (MRI) of the brain shows "infarcts" that do not correspond to the distribution of major vessels and most commonly affect the occipital and parietal lobes, often causing cortical blindness. The pathogenesis of strokes in MELAS is still uncertain, but there is good pathological evidence that MELAS is ultimately a mitochondrial micro- and macro-angiopathy.[6] Although the most common mtDNA mutation is the m.3243A>G in *MTTL*, about a dozen other mutations have been associated with MELAS, most notably the m.13513G>A in *MTND5*.[7]

The second syndrome is MERRF (myoclonus epilepsy with ragged red fibers). Despite the acronymic priority of myoclonus, in a recent review of 42 Italian patients carrying the m.8344A>G mutation in the gene for $tRNA^{Lys}$ (*MTTK*), the most common symptoms and signs were myopathic weakness and exercise intolerance, generalized

http://dx.doi.org/10.1016/B978-0-12-410529-4.00023-1

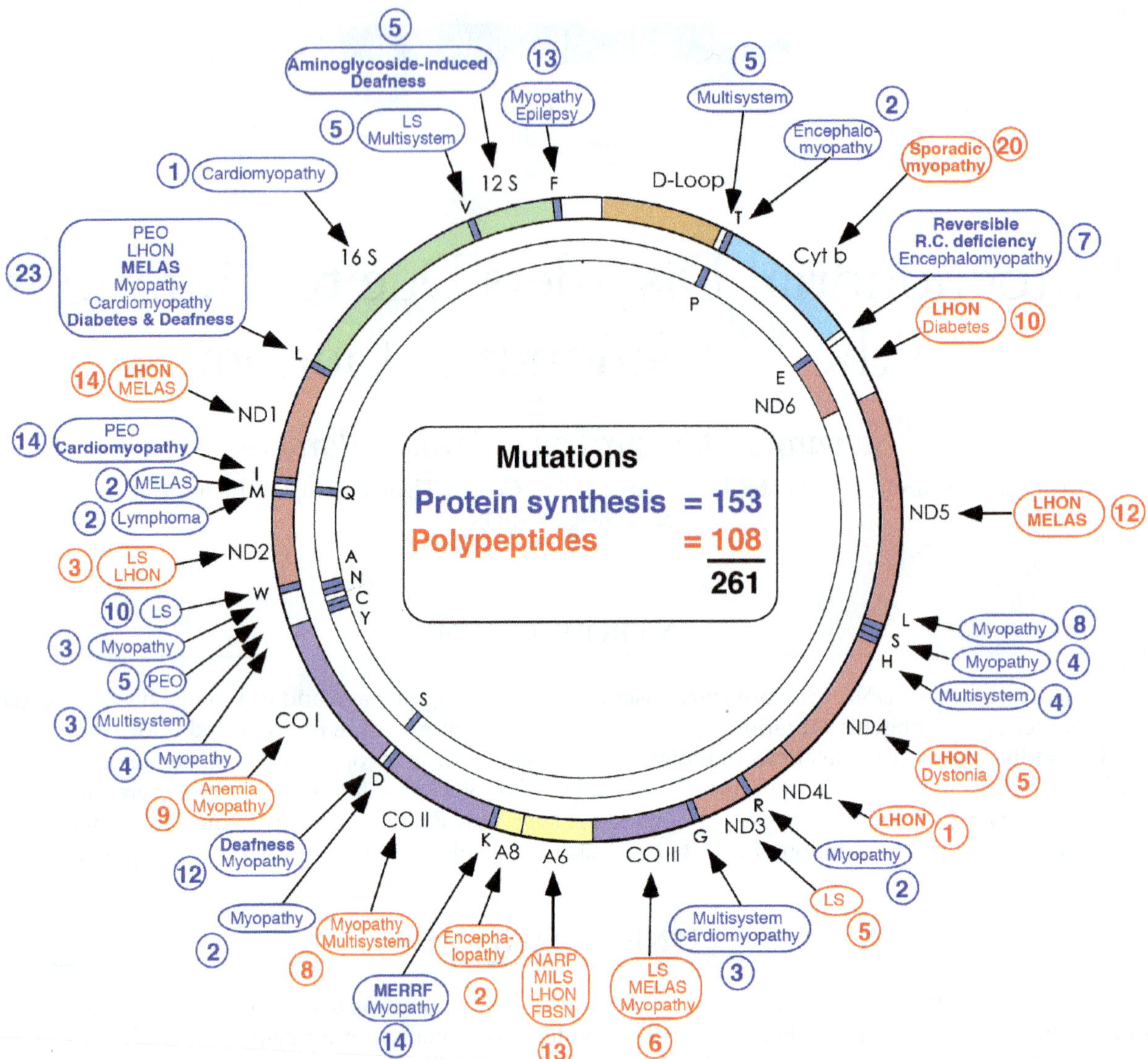

FIGURE 23.1 **Disorders caused by mutations in genes controlling protein synthesis are shown in blue and disorders caused by mutations in protein-coding genes are shown in red.** FBSN, familial bilateral striatal necrosis; LHON, Leber hereditary optic neuropathy; LS, Leigh syndrome; MELAS, mitochondrial encephalomyopathy, lactic acidosis, and stroke-like episodes; MERRF, myoclonus epilepsy and ragged-red fibers; MILS, maternally inherited Leigh symdrome; NARP, neuropathy, ataxia, retinitis pigmentosa; PEO, progressive external ophthalmoplegia. (Modified with permission from DiMauro S, Schon EA. Mitochondrial respiratory-chain diseases. *N Engl J Med*. 2003;348:2656–2668).

seizures, hearing loss, ptosis, and multiple lipomas.[8] Myoclonus was present only in 20% of patients. Less common signs also included dementia, peripheral neuropathy, and cerebellar ataxia. Although the typical mtDNA mutation in MERRF is the m.8344A>G in the *MTTK*, other mutations in the same gene have been reported.[9]

The third syndrome comes in two "flavors." The first is NARP (neuropathy, ataxia, retinitis pigmentosa), which usually affects young adults and causes retinitis pigmentosa, dementia, seizures, ataxia, proximal weakness, and sensory neuropathy. The second is maternally inherited Leigh syndrome (MILS), a more severe infantile encephalopathy with characteristic symmetrical lesions in the basal ganglia and the brainstem. As noted by the late Anita Harding in her original description of this syndrome,[10] the two clinical variants may affect maternal relatives in the same family. The different severity of the two syndromes is explained by the different abundance of the mutation, with mutation loads of about 70% in NARP and around 90% in MILS.[11]

The fourth syndrome, LHON, is characterized by subacute loss of central vision in young adults, more frequently males. The loss of vision is rapid and painless, can start in one or both eyes, and is accompanied by fading of colors (dyschromatopsia). Visual acuity reaches stable residual values at or below 20/200 within months and the visual field shows a large centrocecal absolute scotoma. Fundoscopy typically shows circumpapillary telangiectatic

TABLE 23.1 Clinical Features of Some mtDNA-Related Diseases

		Δ-mtDNA		tRNA		ATPase6	
Tissue	**Symptom/Sign**	***KSS***	**Pearson**	**MERRF**	**MELAS**	**NARP**	**MILS**
CNS	Seizures	–	–	+	+	–	+
	Ataxia	+	–	+	+	+	+/–
	Myoclonus	–	–	+	+/–	–	–
	Psychomor retardation	–	–	–	–	–	+
	Psychomotor regression	+	–	+/–	+	–	–
	Hemiparesis/hemianopia	–	–	–	+	–	–
	Cortical blindness	–	–	–	+	–	–
	Migraine-like headaches	–	–	–	+	–	–
	Dystonia	–	–	–	+	–	+
PNS	Peripheral neuropathy	+/–	–	+/–	+/–	+	–
Muscle	Weakness	+	–	+	+	+	+
	Ophthalmoplegia	+	+/–	–	–	–	–
	Ptosis	+	–	–	–	–	–
Eye	Pigmentory retinopathy	+	–	–	–	+	+/–
	Optic atrophy	–	–	–	–	+/–	+/–
	Cataracts	–	–	–	–	–	–
Blood	Sideroblastic anemia	+/–	+	–	–	–	–
Endocrine	Diabetes mellitus	+/–	–	–	+/–	–	–
	Short stature	+	–	+	+	–	–
	Hypoparathyroidism	+/–	–	–	–	–	–
Heart	Conduction block	+	–	–	+/–	–	–
	Cardiomyopathy	+/–	–	–	+/–	–	+/–
GI	Exocrine pancreas dysfunction	+/–	+	–	–	–	–
	Intestinal pseudo-obstruction	–	–	–	–	–	–

Abbreviations: KSS, Kearns-Sayre syndrome; MELAS, mitochondrial encephalomyopathy, lactic acidosis, and stroke-like episodes; MERRF, myoclonus epilepsy and ragged-red fibers; MILS, maternally inherited Leigh syndrome; NARP, neuropathy, ataxia, retinitis pigmentosa; PNS, peripheral nervous system; tRNA, transfer RNA. Symbols in red highlight the typical symptoms and signs of each syndrome, except MILS, which is defined radiologically or pathologically.
Reproduced from[6], with permission.

microangiopathy and swelling of the nerve fibers around the disc (pseudoedema). Axonal loss in the papillomacular bundle causes temporal pallor of the disc.[12] Mutations in three genes of complex I (ND genes) have been associated with LHON, m.11778G>A in *MTND4*, m.3460G>A in *MTND1*, and m.14484T>C in *MTND6*. Epidemiological studies have shown that LHON is the most common mtDNA-related disorder, with a minimum point prevalence of 3.22 in 100,000 adults (in northeastern England).[13]

Three sporadic conditions are associated with mtDNA single deletions, Kearns–Sayre syndrome (KSS), chronic progressive external ophthalmoplegia (CPEO), and Pearson syndrome (PS). KSS is a multisystem disorder with onset before age 20 and the clinical triad of impaired eye movements (progressive external ophthalmoplegia [PEO]), pigmentary retinopathy, and heart block. Frequent additional signs include ataxia, dementia, and endocrine problems (diabetes mellitus, short stature, hypoparathyroidism). Lactic acidosis and markedly elevated cerebrospinal fluid (CSF) protein (over 100 mg/dL) are typical laboratory abnormalities.[14]

CPEO is a relatively benign disorder characterized by ptosis, PEO, and proximal myopathy that is usually slowly progressive and compatible with a normal lifespan.

Sideroblastic anemia and exocrine pancreatic dysfunction characterize Pearson syndrome, a usually fatal disorder of infancy. Interestingly, those children who do not succumb to the blood dyscrasia often develop KSS later in life.

It is often stated that any patient having multiple organ involvement and evidence of maternal inheritance should be suspected of harboring a pathogenic mtDNA mutation until proven otherwise. While this rule of thumb has some practical value, it is also important to keep in mind that the reverse is not true. That is, patients with involvement of a single tissue and no evidence of maternal inheritance can still have pathogenic mutations of mtDNA. This is especially true for skeletal muscle. We have seen how isolated myopathy with PEO can be due to single large-scale mtDNA deletions. Isolated myopathy without PEO has been associated with point mutations in several tRNA genes.[15]

In addition, we have come to appreciate that exercise intolerance, myalgia, and myoglobinuria can be the sole presentation of respiratory chain defects due to mutations in protein-coding genes (discussed later).

DIAGNOSTIC EVALUATION

The striking clinical heterogeneity of mtDNA-related disorders poses a diagnostic challenge. The following five criteria may be of some help.

Clinical Presentation

As mentioned earlier, the very complexity of the clinical presentation can direct attention to mutations in mtDNA, especially when certain telltale symptoms coexist: short stature, neurosensory hearing loss, PEO, axonal neuropathy, diabetes mellitus, hypertrophic cardiomyopathy, or renal tubular acidosis. On the other hand, it is equally important not to exclude mtDNA mutations in patients with involvement of a single tissue, such as myopathy, cardiomyopathy, or renal disease.

Inheritance

Because most mtDNA-related disorders are maternally inherited but clinical expression is extremely variable, it is crucially important to collect a meticulous family history, with special attention to soft signs in maternal relatives, such as short stature, migraine, deafness, and diabetes. Lack of maternal inheritance, however, does not exclude the diagnosis (such as in cases of KSS and sporadic PEO, discussed previously).

Laboratory

Lactic acidosis is a common finding in defects of the respiratory chain, and the lactate-to-pyruvate ratio is usually elevated (50:1 to 200:1, compared to a normal ratio of 25:1). Information about CSF lactate and pyruvate is important because CSF values can be increased in children with encephalopathy and normal blood lactate. However, lack of lactic acidosis does not exclude the diagnosis. For example, in patients with NARP or MILS, blood lactate and pyruvate can be normal or only mildly elevated. Serum creatine kinase (CK) levels are often normal or modestly increased except, of course, in patients with myopathy during episodes of myoglobinuria. However, in keeping with the frequent finding of myopathic features in MERRF, increased CK was found in 44% of patients.[8] As another exception to this rule, and a useful diagnostic clue, serum CK values can be markedly elevated in children with the myopathic form of mtDNA depletion, such as those with mutations in the gene (*TK2*) encoding thymidine kinase: strictly speaking, however, this is not a primary disorder of mtDNA, as *TK2* is a nuclear gene and TK2 deficiency is an autosomal recessive defect of mtDNA maintenance (see below and Chapter 24). Serum thymidine and deoxyuridine are greatly increased in patients with MNGIE (mitochondrial neurogastrointestinal encephalomyopathy), another autosomal recessive defect of mtDNA maintenance due to mutations in the gene (*TYMP*) encoding thymidine phosphorylase (see Chapter 24). Finally, although fibroblast growth factor 21 (FGF-21) was proposed in 2011 as a valuable new serum or plasma diagnostic biomarker of mitochondrial diseases, especially those affecting muscle,[16] it has not yet been used widely.

Neuroradiology

Bilateral MRI signal hyperintensities in basal ganglia are typical of Leigh syndrome (LS). Stroke-like lesions in the posterior cerebral hemispheres are typically seen in patients with MELAS. Diffuse signal abnormality of the central white matter is characteristic of KSS, whereas basal ganglia calcifications are common in both MELAS and KSS syndrome. Proton magnetic resonance spectroscopy reveals lactate accumulation in the CSF and in specific areas of the brain, where lactate concentration can be compared to the concentration of *N*-acetyl-L-aspartate (NAA), an indicator of cell viability (Figure 23.2).[9]

Exercise Physiology

Impaired oxygen extraction by exercising muscle can be detected by near-infrared spectroscopy, which measures the degree of deoxygenation of hemoglobin. A simpler test is based on the measurement of the partial pressure of

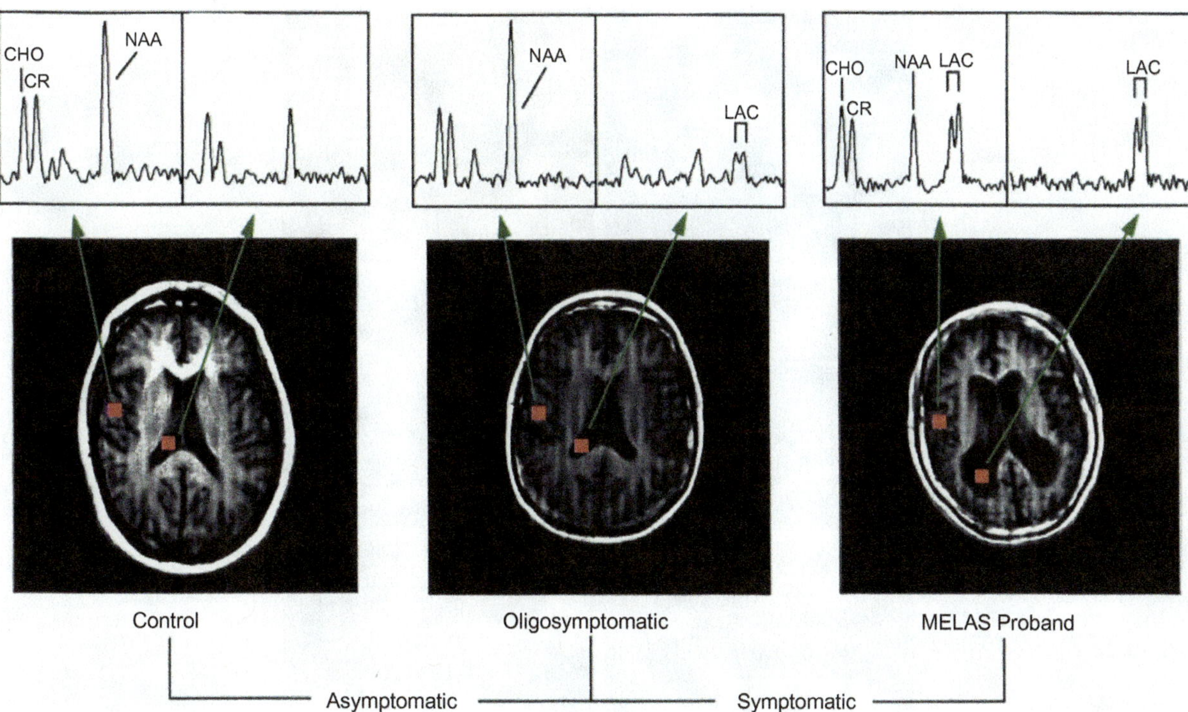

FIGURE 23.2 **T1-weighted magnetic resonance (MR) brain images at the level of the lateral ventricles from three subjects, an asymptomatic control (left panel); an oligosymptomatic mutation carrier (middle panel); and a fully symptomatic MELAS patient (right panel).** Above each image are ^{1}H-MR spectra of a voxel in the left gray matter (left) and from a voxel in the laft lateral ventricle (right). Voxels are represented by red squares. The spectral resonances have been ascribed to: CHO, total choline; CR, total creatine; NAA, *N*-acetylaspartate; and LAC, lactate. All the spectra have been plotted on the same vertical scale. *(Reproduced with permission from Kaufmann P, Shungu DC, Sano MC, et al. Cerebral lactic acidosis correlates with neurological impairment in MELAS. Neurology. 2004;62:1297–1302).*

oxygen (PO_2) in cubital venous blood after forearm aerobic exercise. PO_2 rises paradoxically in patients with mitochondrial myopathy or PEO, and the degree of rise reflects the severity of oxidative impairment. By ^{31}P-magnetic resonance spectroscopy, the ratio of phosphocreatine to inorganic phosphate (PCr to Pi) can be measured in muscle at rest, during exercise, and during recovery. In patients with mitochondrial dysfunction, PCr-to-Pi ratios are lower than normal at rest, decrease excessively during exercise, and return to baseline more slowly than normal.[17]

PATHOLOGY

Skeletal Muscle

Muscle biopsy has a central role in the diagnosis of mtDNA-related disorders. Abnormal mitochondrial proliferation, a hallmark of mitochondrial dysfunction, can be revealed with the modified Gomori trichrome stain ("ragged-red fibers" [RRF])[11] or, better, with the succinate dehydrogenase (SDH) reaction ("ragged blue fibers"). RRF are present in muscle biopsies from patients with most disorders due to mtDNA mutations (Figure 23.3). Rare exceptions to this rule include NARP/MILS and LHON.

The cytochrome *c* oxidase (COX) histochemical reaction typically reveals scattered COX-negative fibers, which usually correspond to—but are not confined to—RRF. RRF are COX-positive in two main conditions: 1) typical MELAS syndrome and 2) mutations in mtDNA genes encoding cytochrome *b* or complex I subunits.[15] To confirm the pathogenicity of an mtDNA mutation, single fibers can be dissected out of thick cross-sections of muscle and used to determine the levels of a given mutation by polymerase chain reaction (PCR). Finding higher levels of the mutation in affected (ragged-red or COX-negative fibers) than in unaffected (non-ragged-red, COX-positive fibers) is strong evidence that the mutation in question is pathogenic.[18]

Electron microscopy has lost much of its historical diagnostic value, but it can reveal focal accumulations of mitochondria or mitochondrial with paracrystalline inclusions in cases in which histochemical results are equivocal.

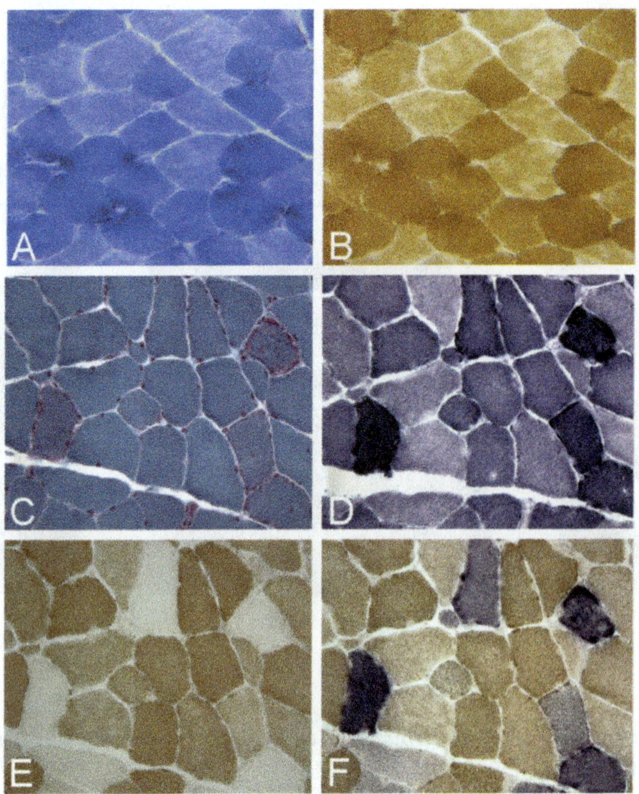

FIGURE 23.3 **Histology of mtDNA-related diseases. Top panels**: in normal human muscle, all fibers stain with the succinate dehydrogenase (SDH) and the cytochrome *c* oxidase (COX) reactions (A and B, respectively). The different intensities correspond to type 1 (darker stains) and type 2 (lighter stains) fibers. **Middle and lower panels**: in a patient with MERRF due to the m.8344A>G mutation, two ragged-red fibers are evident with the modified Gomori trichrome stain (C); the same fibers appear ragged-blue with the SDH stain (D), are COX-negative (E), and appear blue with combined COX/SDH stain (F). *(Reproduced from DiMauro, Tanji, Schon. The many clinical faces of cytochrome c oxidase deficiency. In: Kadenbach ed. Mitochondrial Oxidative Phosphorylation. New York, NY: Springer Science+Business Media; 2012).*

Neuropathology

The basic neuropathology in mitochondrial encephalomyopathies consists of four histologic lesions: spongy degeneration, neuronal loss, gliosis, and demyelination.[19] While these lesions are nonspecific, their distribution patterns vary in patients with different mtDNA-related disorders. Thus, the neuropathology of KSS is characterized by spongiform degeneration, which is usually generalized and affects both gray and white matter. In MELAS, multifocal necrosis predominates, affecting the cerebral cortex or the subcortical white matter, but also the cerebellum, thalamus, and basal ganglia. Calcification of basal ganglia is also common. In MERRF, neuronal degeneration, astrocytosis, and demyelination affect preferentially the cerebellum (especially the dentate nucleus), the brainstem (especially the olivary nucleus), and the spinal cord. In LS, there are symmetrical, bilateral foci of necrosis in the basal ganglia, thalamus, midbrain, and pons. Microscopically, the lesions show vascular proliferation, gliosis, neuronal loss, demyelination, and cystic cavitation.

BIOCHEMICAL FINDINGS

All 13 proteins encoded by mtDNA are components of the mitochondrial respiratory chain. Seven (ND1–ND5, ND4L, ND6) are subunits of complex I, one (cytochrome *b*) is part of complex III, three (COX I, COX II, and COX III) are subunits of complex IV, and two (ATPase 6 and ATPase 8) are subunits of complex V (Figure 23.4). Therefore, disorders owing to mutations of mtDNA are usually associated with biochemical defects of oxidative phosphorylation. Skeletal muscle is the preferred tissue for biochemical analysis, because it is rich in mitochondria, it is invariably affected in multisystem disorders (mitochondrial encephalomyopathies), and it is the most common tissue affected in isolation (mitochondrial myopathies). There is some controversy about whether fresh muscle tissue is needed or frozen tissue suffices. It is the authors' view that—except for the rare cases in which freshly isolated mitochondria

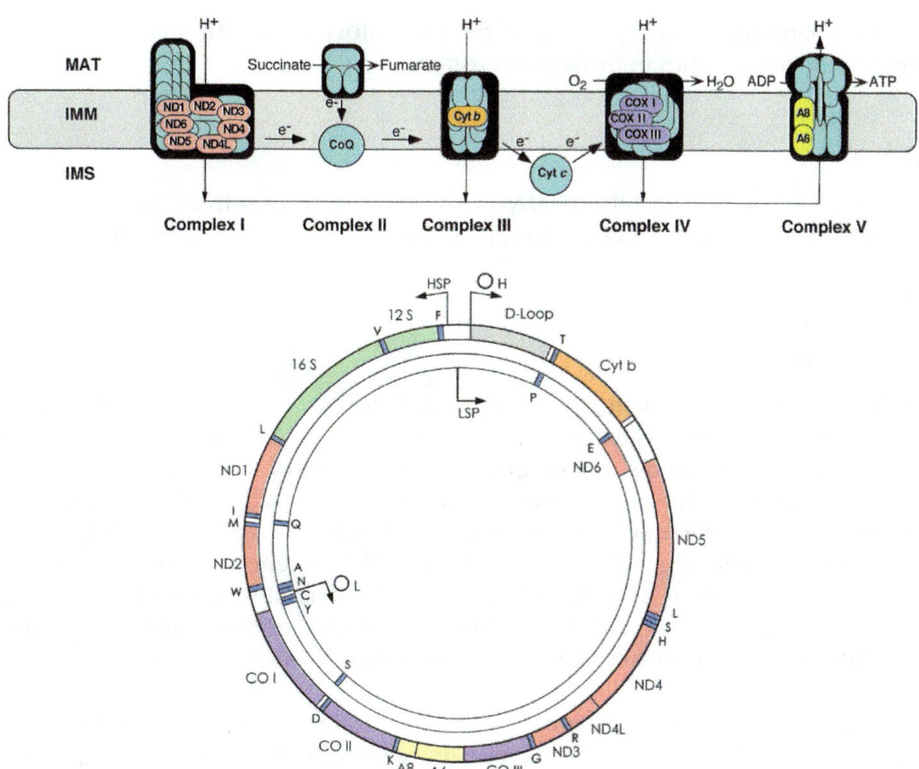

FIGURE 23.4 **Genes and corresponding gene products are similarly color-coded.** ND denote the subunits of NADH-coenzyme Q oxidoreductase (complex I); cyt *b*, cytochrome *b*; subunits of cytochrome *c* oxidase are labeled CO in the mtDNA scheme and COX in the respiratory chain rendition; A6 and A8 indicate subunits 6 and 8 of ATP synthase. The 22 tRNA genes are denoted by one-letter amino acid nomenclature; 12S and 16S denote ribosomal RNAs (rRNAs). O_H and O_L are the origins of heavy- and light-chain replication; HSP and LSP are the promoters of heavy- and light-stranded transcription. ADP, adenosine diphosphate; ATP, adenosine triphosphate; IMM, inner mitochondrial membrane; IMS, intermembrane space; MAT, mitochondrial matrix. *(Reproduced from[6], with permission.)*

for polarographic analyses are required—frozen specimens provide adequate biochemical information. Because mitochondrial proliferation in muscle (RRF) is a common occurrence, activities of respiratory chain complexes should be referred to the activity of citrate synthase, a nuclear encoded enzyme of the mitochondrial matrix that is a good marker of mitochondrial abundance. A vexing problem is the inconsistency of biochemical data among different laboratories. Standardized biochemical procedures have been proposed[20] but not universally adopted.

Biochemical studies in muscle extract from patients with mtDNA-related disorders yield three main types of results. The first pattern consists of partial defects in the activities of multiple complexes containing mtDNA-encoded subunits (I, III, and IV) contrasting with normal or increased activities of SDH (complex II) and citrate synthase. The second pattern is a severe deficiency in the activity of one specific respiratory complex. The third pattern consists of combined defects of complex I, complex II, complex III, and the Krebs cycle enzyme aconitase. As a rule, the first pattern is found in patients with defects of mitochondrial protein synthesis *in toto* (single mtDNA deletions [e.g., KSS], or mutations in tRNA genes [e.g., MELAS, MERRF]). The second pattern is typical of mutations in mtDNA protein-coding genes. For example, mutations in the cytochrome *b* gene (*MTCYB*) cause isolated complex III deficiency.[18] However, mutations in ND genes, which are encountered with increasing frequency in patients with MELAS, LS, LHON, or overlap syndromes, cause inconsistent and often modest impairment of complex I activity.[21] The third pattern suggests a problem with the iron–sulfur (Fe–S) clusters, which are prosthetic groups of respiratory chain complexes I, II, III, and of aconitase. Defects of Fe–S clusters have been associated with various diseases, including isolated myopathy with recurrent myoglobinuria.[22]

MOLECULAR GENETIC FINDINGS

Human mtDNA is a 16,569-bp circle of double-stranded DNA. It is highly compact, and contains only 37 genes (see Figure 23.4): two genes encode ribosomal RNAs (rRNAs), 22 encode transfer RNAs (tRNAs), and 13 encode polypeptides. All 13 polypeptides are components of the respiratory chain (discussed previously). Mitochondrial genetics differs from Mendelian genetics in three major aspects.

Maternal Inheritance

At fertilization, all mitochondria (and all mtDNAs) in the zygote derive from the oocyte. Therefore, a mother carrying an mtDNA mutation is expected to pass it on to all her children, but only her daughters transmit it to their progeny.

Heteroplasmy/Threshold Effect

In contrast to nuclear genes, each consisting of one maternal and one paternal allele, there are hundreds or thousands of mtDNA molecules in each cell. Deleterious mutations of mtDNA usually affect some but not all genomes, such that cells, tissues, or whole individuals, will harbor two populations of mtDNA: normal (wild-type) and mutant, a situation known as heteroplasmy. The situation of normal subjects, in whom all mtDNAs are identical, is called homoplasmy (although supersensitive next-generation sequencing has revealed extremely low-level [0.2–2%] mutations in blood and skeletal muscle of unaffected individuals, a phenomenon dubbed "universal heteroplasmy"[23]).

Not surprisingly, a minimum critical number of mutant mtDNAs must be present before oxidative dysfunction and clinical signs become apparent (threshold effect). Also not surprisingly, the pathogenic threshold will be lower in tissues that are highly dependent on oxidative metabolism than in tissues with higher capacity for anaerobic metabolism.

Pathogenic mutations are usually heteroplasmic, whereas neutral polymorphisms are homoplasmic. In general, this is true, but there are many exceptions and there is an increasing awareness of the possible or documented pathogenicity of homoplasmic mutations. In fact, the first point mutation (m.11778G>A in *MTND4*) associated with a human disease, LHON, was homoplasmic,[2] as are all other mutations causing LHON.[12] Similarly, most nonsyndromic forms of deafness are due to homoplasmic mutations, including m.15555A>G in the 12S rRNA gene, and two mutations in the *MTTS1(UCN)* gene, m.7455A>G, and m.7511 T>C.[21]

In their migration out of Africa, human beings accumulated distinctive variations from the mtDNA of our ancestral "mitochondrial Eve," resulting in several haplotypes characteristic of different ethnic groups.[24] It has been suggested that different mtDNA haplotypes may modulate oxidative phosphorylation, thus influencing the overall physiology of individuals and predisposing them to—or protecting them from—certain diseases.

Mitotic Segregation

At cell division, the proportion of mutant mtDNAs in daughter cells can shift; if and when the pathogenic threshold for a given tissue is surpassed, the phenotype can also change. This explains the time-related variability of clinical features frequently observed in mtDNA-related disorders, including the shift from Pearson syndrome to KSS mentioned above.[25,26]

As already mentioned above, mtDNA mutations fall into two main groups: those affecting mitochondrial protein synthesis and those affecting specific proteins.

Mutations Affecting Mitochondrial Protein Synthesis

Mutations affecting protein synthesis include single deletions (which always encompass one or more tRNA genes), and point mutations in rRNA or in tRNA genes. Clinical experience suggests that these mutations are usually associated with multisystem disorders, lactic acidosis, and massive mitochondrial proliferation in muscle, resulting in the "ragged-red" or "ragged-blue" appearance of fibers in the muscle biopsy. Examples include KSS, MELAS, and MERRF. As far as inheritance is concerned, conditions associated with point mutations, such as MELAS and MERRF, are transmitted maternally, whereas single mtDNA deletions are sporadic events in almost all cases.[27]

While mutations in tRNA genes are usually associated with multisystem disorders, in some cases there is involvement of a single tissue, most commonly skeletal muscle. Family history is negative in some of the patients with pure myopathy, suggesting that the mutations occurred *de novo*. However, in many patients the mutation was also present in blood or cultured skin fibroblasts, implying "skewed heteroplasmy," with preferential accumulation of the mutation in skeletal muscle, where it surpasses the pathogenic threshold.

Mutations in Protein-Coding Genes

The two most common maternally inherited, multisystemic syndromes associated with mutations in mtDNA protein-coding genes are LHON and NARP/MILS. However, we have come to appreciate that sporadic patients with exercise intolerance, myalgia, and myoglobinuria often harbor mutations in protein-coding genes causing defects of complex I, complex III, or complex IV.[18] The sporadic nature and the tissue specificity of these disorders suggests that the underlying mutations are somatic—that is, spontaneous events that occur in muscle and do not affect germline cells.

However, numerous mutations in protein-coding genes have been associated with encephalomyopathies other than LHON and NARP/MILS. In fact, some ND genes, especially *ND5* and *ND3*, are considered hotspots for mutations causing MELAS or LS.[21,28]

Defects of mtDNA Maintenance

These disorders are due to qualitative or quantitative alterations of mtDNA, i.e., multiple mtDNA deletions or mtDNA depletion syndromes.[29] Both integrity and replication of mtDNA are controlled by nuclear DNA and defects of mtDNA maintenance occur when the dialog between the two genomes goes awry. Although these disorders are unequivocally Mendelian, they share many features of mitochondrial genetics, including heteroplasmy and the threshold effect, because the target of the nuclear mutations is the polyploid mtDNA. This is why we mention them here, although they are described in detail in Chapter 24.

Next Generation Genetics

Until recently, mtDNA mutations were identified mainly through the "candidate gene" approach, targeting the most commonly affected gene on the basis of the clinical picture (i.e., the *MTTL* gene in cases of MELAS and the *MTTK* gene in cases of MERRF). Using appropriate primers, all 22 tRNA genes could be sequenced in cases with atypical, usually multisystemic clinical presentations, ragged-blue fibers in the muscle biopsy, and biochemical evidence suggesting a generalized defect of mitochondrial protein synthesis. Nowadays, it is easier to sequence the whole mitochondrial genome through automated microarray-based sequencing and high-throughput sequencing strategies. If a defect of mtDNA maintenance is suspected and obvious candidate nuclear genes have been excluded, either whole-exome sequencing (WES) or mito-exome sequencing, which is targeted to the approximately 1700 nuclear-encoded mitochondrial protein, can be employed. These approaches are both practical (enabling diagnosis of puzzling cases and making prenatal diagnosis possible) and heuristic (leading to the discovery of novel mutant genes and new disease mechanisms).[30]

ANIMAL MODELS

A formidable obstacle to the generation of so-called "mito-mice" was our inability to introduce mutant mtDNA into mitochondria of mammalian cells. To solve this problem, Inoue et al.[31] first prepared synaptosomes from the brain of aging mice, which contained a certain percentage of deleted mtDNAs. They then fused the synaptosomes with mtDNA-less rho^0 cells, thus obtaining hybrid cell lines. One such cell line harboring mtDNA deletions was enucleated, fused with donor embryos, and implanted in pseudopregnant females. Heteroplasmic founder females were bred and mtDNA deletions were transmitted through three generations. Although there are significant differences between these mito-mice and human patients with mtDNA deletions, the fact remains that the animals show mitochondrial dysfunction in various organs.

A slightly more complicated strategy was employed by Sligh et al.[32] to generate mice harboring a point mutation for chloramphenicol resistance (CAP^R). The mutation in homoplasmic or heteroplasmic CAP^R mutants was severe enough to cause death *in utero* or within 11 days of birth, and affected animals showed dilated cardiomyopathy and abnormal mitochondria in both cardiac and skeletal muscle, features often observed in human mtDNA diseases. At the very least, the papers by Inoue et al. and Sligh et al. prove the principle that heteroplasmic mtDNA mutations can be transmitted through germlines and will produce phenotypic abnormalities.

Wallace and coworkers created a "mito-mouse" that recapitulates faithfully human LHON by mutagenesis of LMTK cells, selection of respiratory-defective clones, and identification of a homoplasmic *ND6* mutant line, which was enucleated and fused with female mouse embryonic stem cells.[33] These mito-mice have reduced retinal function and optic atrophy, with partial complex I deficiency and severe oxidative stress.

THERAPY

Therapy of mtDNA-related diseases is still woefully inadequate. Besides palliative pharmacologic and surgical interventions directed at alleviating symptoms, approaches to therapy include[29]: 1) enhancing respiratory chain function through the use of various compounds, most commonly coenzyme Q10 (CoQ10) or its synthetic analogs (idebenone or parabenzoquinone); 2) removing noxious products, such as thymidine and deoxyuridine in patients with MNGIE through allogeneic hematopoietic stem-cell transplantation (AHSCT); 3) shifting heteroplasmy to lower the mutation load to subthreshold levels either by favoring removal of "bad" mitochondria through mitophagy or by introducing "silver bullets" in the form of restriction endonucleases; 4) enhancing mitochondrial fusion and "networking," thus allowing complementation of "bad" and "good" mitochondria and normalization of overall mitochondrial function; and 5) promoting aerobic exercise, which prevents muscle deconditioning and improves functional and biochemical features of muscles harboring pathogenic mtDNA mutations. However, all of these approaches have limitations: some have been successful only in specific disorders (e.g., CoQ10 replacement in primary CoQ10 deficiencies); others are promising but risky (e.g., AHSCT in MNGIE); and others are still at the experimental stage.

An exciting possibility is *preventing* mtDNA-related diseases (especially those due to mutations in tRNA genes, like MELAS and MERRF) through cytoplasmic transfer. In this approach, the nucleus of an *in vitro*-fertilized oocyte from an mtDNA mutation carrier is transferred to an enucleated oocyte from a normal donor: the embryo will have the nuclear DNA of the biological parents but the mtDNA of the normal donor. This technique has proven successful in nonhuman primates[34] and has been effective in fertilized[35] and unfertilized but parthenogenically activated human oocytes.[36] Similar results have been obtained in the UK after pronuclear transfer in abnormally fertilized human oocytes developed to the blastocyst stage.[37] Despite some minor concerns,[38] the stage seems set for approval of this technique for therapeutic application both in the UK and in the USA.

CONCLUSION

Although the mtDNA circle is crowded with pathogenic mutations (see Figure 23.1), there is room for more, and new mutations are still being described, especially in protein-coding genes. In addition, we still do not fully understand the pathophysiology of mtDNA-related diseases, although the advent of "mito-mice" and of knockout mice for nuclear genes will undoubtedly help us in this endeavor. Finally, our therapeutic armamentarium is still totally inadequate, although some interesting novel approaches are being considered.[39] Here, again, the availability of animal models will be of great help.

ACKNOWLEDGEMENTS

Part of the work presented here was supported by NICHD grant PO1HD32062, and by the Marriott Mitochondrial Disorders Clinical Research Fund (MMDCRF).

References

1. Holt IJ, Harding AE, Morgan Hughes JA. Deletions of muscle mitochondrial DNA in patients with mitochondrial myopathies. *Nature*. 1988;331:717–719.
2. Wallace DC, Singh G, Lott MT, et al. Mitochondrial DNA mutation associated with Leber's hereditary optic neuropathy. *Science*. 1988;242:1427–1430.
3. Schon EA, DiMauro S, Hirano K-I. Human mitochondrial DNA: roles of inherited and somatic mutations. *Nat Rev Genet*. 2012;13:878–890.
4. Kaufmann P, Engelstad K, Wei Y-H, et al. Natural history of MELAS associated with mitochondrial DNA m.3243A>G genotype. *Neurology*. 2011;77:1965–1971.
5. Mancuso M, Orsucci D, Angelini C, et al. The m.3243A>G mitochondrial DNA mutation and related phenotypes. A matter of gender? *J Neurol*. 2014;261:504–510.
6. DiMauro S. Mitochondrial encephalomyopathies - Fifty years on. *Neurology*. 2013;281–291.
7. Shanske S, Coku J, Lu J, et al. The G13513A mutation in the ND5 gene of mitochondrial DNA as a common cause of MELAS or Leigh syndrome. *Arch Neurol*. 2008;65:368–372.
8. Mancuso M, Orsucci D, Angelini C, et al. Phenotypic heterogeneity of the 8344A>G mtDNA "MERRF" mutation. *Neurology*. 2013;80:2049–2054.
9. DiMauro S, Hirano M, Kaufmann P, et al. Clinical features and genetics of myoclonic epilepsy with ragged red fibers. In: Fahn S, Frucht SJ, eds. *Myoclonus and Paroxysmal Dyskinesia*. Philadelphia: Lippincott Williams & Wilkins; 2002:217–229.

10. Holt IJ, Harding AE, Petty RK, Morgan Hughes JA. A new mitochondrial disease associated with mitochondrial DNA heteroplasmy. *Am J Hum Genet*. 1990;46:428–433.
11. Tatuch Y, Christodoulou J, Feigenbaum A, et al. Heteroplasmic mtDNA mutation (T>G) at 8993 can cause Leigh disease when the percentage of abnormal mtDNA is high. *Am J Hum Genet*. 1992;50:852–858.
12. Carelli V, Barboni P, Sadun AA. Mitochondrial ophthalmology. In: DiMauro S, Hirano M, Schon EA, eds. *Mitochondrial Medicine*. London: Informa Healthcare; 2006:105–142.
13. Man PY, Griffiths PG, Brown DT, Howell N, Turnbull DM, Chinnery PF. The epidemiology of Leber hereditary optic neuropathy in the Northeast of England. *Am J Hum Genet*. 2003;72:333–339.
14. Hirano M, Kaufmann P, De Vivo DC, Tanji K. Mitochondrial neurology I: encephalopathies. In: DiMauro S, Hirano M, Schon EA, eds. *Mitochondrial Medicine*. London: Informa Healthcare; 2006:27–44.
15. Hays AP, Oskoui M, Tanji K, Kaufmann P, Bonilla E. Mitochondrial neurology II: myopathies and peripheral neuropathies. In: DiMauro S, Hirano M, Schon EA, eds. *Mitochondrial Medicine*. London: Informa Healthcare; 2006:45–74.
16. Suomalainen A, Elo JM, Pietilainen KH, et al. FGF-21 as a biomarker for muscle-manifesting mitochondrial respiratory chain deficiencies: a diagnostic study. *Lancet Neurol*. 2011;10:806–818.
17. Haller RG, Vissing J. Functional evaluation of metabolic myopathies. In: Engel AG, Franzini-Armstrong C, eds. *Myology*. New York: McGraw-Hill; 2004:665–679.
18. Andreu AL, Hanna MG, Reichmann H, et al. Exercise intolerance due to mutations in the cytochrome b gene of mitochondrial DNA. *N Engl J Med*. 1999;341:1037–1044.
19. Tanji K, Kunimatsu T, Vu TH, Bonilla E. Neuropathological features of mitochondrial disorders. *Semin Cell Dev Biol*. 2001;12:429–439.
20. Spinazzi M, Casarin A, Pertegato V, Salviati L, Angelini C. Assessment of mitochondrial respiratory chain enzymatic activities on tissues and cultured cells. *Nat Protocol*. 2012;7:1235–1246.
21. DiMauro S, Davidzon G. Mitochondrial DNA and disease. *Ann Med*. 2005;37:222–232.
22. Mochel F, Knight MA, Tong W-H, et al. Splice mutation in the iron-sulfur cluster scaffold protein ISCU causes myopathy with exercise intolerance. *Am J Hum Genet*. 2008;82:652–660.
23. Payne BAJ, Wilson IJ, Yu-Wai-Man P, et al. Universal heteroplasmy of human mitochondrial DNA. *Hum Mol Genet*. 2013;22:384–390.
24. Wallace DC, Brown MD, Lott MT. Mitochondrial DNA variation in human evolution and disease. *Gene*. 1999;238:211–230.
25. Larsson NG, Holme B, Kristiansson B. Progressive increase of the mutated mitochondrial DNA fraction in Kearns-Sayre syndrome. *Pediatr Res*. 1990;28:131–136.
26. McShane MA, Hammans SR, Sweeney M, et al. Pearson syndrome and mitochondrial encephalopathy in a patient with a deletion of mtDNA. *Am J Hum Genet*. 1991;48:39–42.
27. Chinnery PF, DiMauro S, Shanske S, et al. Risk of developing a mitochondrial DNA deletion disorder. *Lancet*. 2004;364:592–595.
28. Sarzi E, Brown MD, Lebon S, et al. A novel mitochondrial DNA mutation in ND3 gene is associated with isolated complex I deficiency causing Leigh syndrome and dystonia. *Am J Med Genet*. 2007;143A:33–41.
29. DiMauro S, Schon EA, Carelli V, Hirano M. The clinical maze of mitochondrial neurology. *Nat Rev Neurol*. 2013;9:429–444.
30. Coppola G, Geschwind DH. Genomic medicine enters the neurology clinic. *Neurology*. 2012;79:112–114.
31. Inoue K, Nakada K, Ogura A, et al. Generation of mice with mitochondrial dysfunction by introducing mouse mtDNA carrying a deletion into zygotes. *Nat Genet*. 2000;26:176–181.
32. Sligh JE, Levy SE, Waymire KG, et al. Maternal germ-line transmission of mutant mtDNAs from embryonic stem cell-derived chimeric mice. *Proc Natl Acad Sci U S A*. 2000;97:14461–14466.
33. Lin CS, Sharpley MS, Fan W, et al. Mouse mtDNA mutant model of Leber hereditary optic neuropathy. *Proc Natl Acad Sci U S A*. 2012;109:20065–20070.
34. Tachibana M, Sparman M, Sritanaudomchai H, et al. Mitochondrial gene replacement in primate offspring and embryonic stem cells. *Nature*. 2009;461:376–372.
35. Tachibana M, Amato P, Sparman M, et al. Towards germline gene therapy of inherited mitochondrial diseases. *Nature*. 2013;493:632–637.
36. Paull D, Emmanuele V, Weiss KA, et al. Nuclear genome transfer in human oocytes eliminates mitochondrial DNA variants. *Nature*. 2013;493:632–637.
37. Craven L, Tuppen HA, Greggains GD, et al. Pronuclear transfer in human embryos to prevent transmission of mitochondrial DNA disease. *Nature*. 2010;465:82–85.
38. Craven L, Elson JL, Irving L, et al. Mitochondrial DNA disease: new options for prevention. *Hum Mol Genet*. 2011;20:R168–R174.
39. Schon EA, DiMauro S, Hirano M, Gilkerson RW. Therapeutic prospects for mitochondrial disease. *Trends Mol Med*. 2010;16:268–276.

CHAPTER

24

Mitochondrial Disorders Due to Mutations in the Nuclear Genome

Patrick F. Chinnery

Institute of Genetic Medicine, Newcastle University, Newcastle upon Tyne, UK

CLINICAL OVERVIEW AND HISTORY

Ultimately, mitochondrial disorders are due to a defect of the mitochondrial respiratory chain, which is the principal source of intracellular energy in the form of adenosine triphosphate (ATP, Figure 24.1). As a consequence, tissues and organs that are highly dependent on oxygen metabolism are characteristically affected in both mitochondrial DNA (mtDNA)- and nuclear-encoded mitochondrial disorders. Mitochondrial disorders were first recognized in patients with canonical clinical syndromes such as chronic progressive external ophthalmoplegia (CPEO), mitochondrial encephalomyelopathy, lactic acidosis with stroke like episodes (MELAS), and Kearns–Sayre syndrome.[1] However, it has become increasingly evident that most patients do not fit neatly into a particular clinical category or syndrome. Rather, mitochondrial disorders involve an overlapping spectrum of disease, which can either affect a single organ system, or multiple organ systems. Although the field of mitochondrial medicine was initially largely based on the comprehensive investigation of patients with mtDNA-encoded mitochondrial disorders, technological advances led to major insight into nuclear-encoded mitochondrial diseases. It is important to note that there is substantial clinical overlap between the two different genetic etiologies, which is perhaps not surprising given the final common pathway of disease.[2] Having said this, specific phenotypic groups have emerged over the last decade, which point to specific nuclear-genetic etiologies (Table 24.1).

Prevalence

Mitochondrial disorders are amongst the most common inherited neuromuscular diseases. The prevalence of mitochondrial disorders is estimated at 1:5000.[3,4] This is likely to be an underestimate because the subsequent advent of next-generation sequencing has identified several new phenotypic groups presenting in adult life that were not included in the early epidemiological work.

Clinical Presentation

The majority of patients with mitochondrial disease have neurological involvement, although some do have isolated cardiomyopathy, diabetes, or a renal presentation. In some patients, it is possible to identify a defined clinical syndrome. The classical mitochondrial syndromes associated with mtDNA defects are discussed in Chapter 23. The clinically defined syndromes caused by nuclear-mitochondrial genes are shown in Table 24.1. Many patients do not have all of the features listed in Table 24.1, although these may emerge with time.

Clinical Presentation of Defined Syndromes

The clinical classification of mitochondrial disorders is challenging because of the overlapping spectrum of phenotypes. The same genetic defect can cause a range of different clinical syndromes. Conversely, an almost identical clinical syndrome can be caused by a range of different molecular defects of nuclear DNA or mtDNA. The following

Rosenberg's Molecular and Genetic Basis of Neurological and Psychiatric Disease
http://dx.doi.org/10.1016/B978-0-12-410529-4.00024-3

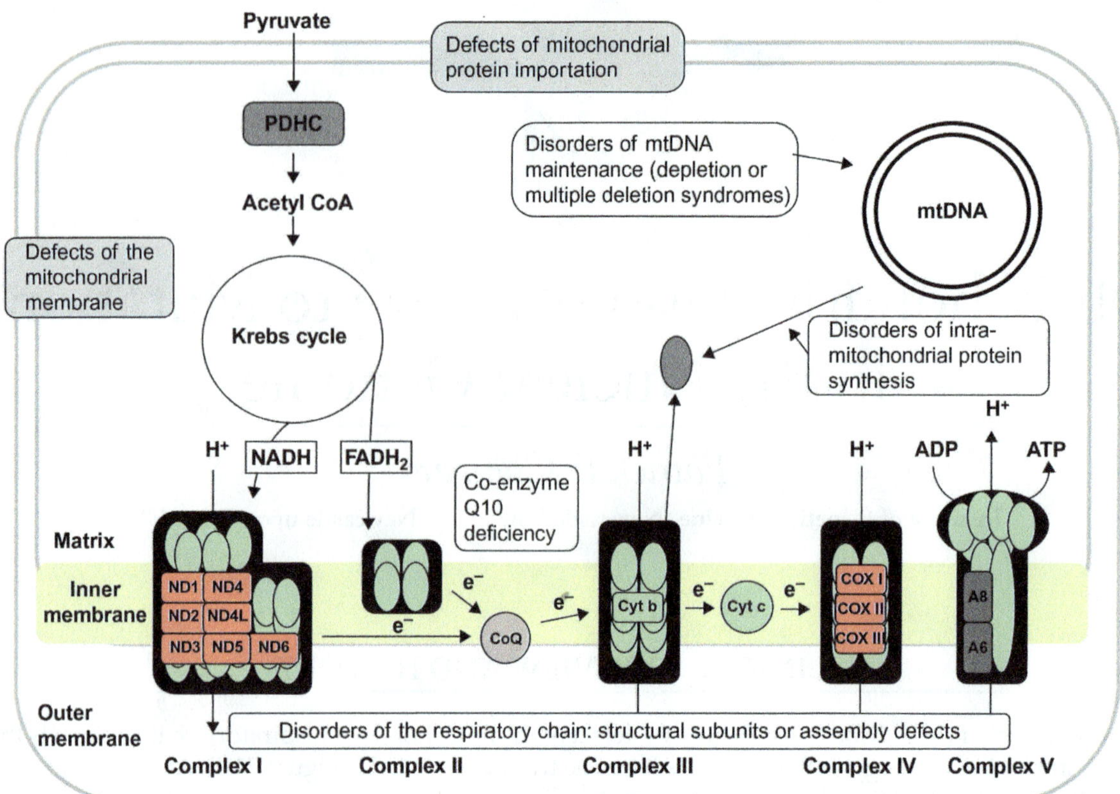

FIGURE 24.1 **Mitochondrial disorders due to mutations in the nuclear genome.** Leigh syndrome can be caused by mutations in mitochondrial DNA (mtDNA), pyruvate dehydrogenase complex (PDHC) deficiency, or mutations in nuclear genes affecting the synthesis, structure or assembly to the mitochondrial respiratory chain. Disorders of mtDNA maintenance are a major cause of disease due to a nuclear gene defect affecting the amount or quality of mtDNA, which causes mtDNA depletion or secondary multiple mtDNA deletions. This group includes mutations in the mtDNA polymerase gene *POLG* and mitochondrial neurogastrointestinal encephalomyopathy (MNGIE). Disorders of intramitochondrial protein synthesis can be due to primary mutations of mtDNA, or a wide range of different nuclear gene defects (see text). Other nuclear-encoded mitochondrial diseases include disorders of mitochondrial protein import (including the Mohr–Tranebjaerg syndrome) and disorders of the lipid membranes (including Barth syndrome), and disorders of coenzyme Q10 biosynthesis. The latter are particularly important to recognize because they are potentially treatable.

classification allows a clinician to identify groups of phenotypically related disorders and groups of genetically related disorders. It should be noted that there is some overlap between these different classifications.

LEIGH SYNDROME

Leigh syndrome was initially described postmortem but now is routinely diagnosed on magnetic resonance imaging (MRI) in an appropriate clinical context. Classical features include a movement disorder, hypotonia, and ataxia related to brainstem and cerebellar involvement. Patients often have a raised lactate in blood or cerebrospinal fluid (CSF), and characteristic brain imaging with high signal in the basal ganglia and stem on MRI (Figure 24.2).[5] Typically the disorder presents with developmental delay, or acutely during a viral infection. This is followed by a relapsing–remitting natural history superimposed on a gradual decline. Patients occasionally have pigmentary retinopathy and can also have liver involvement. Leigh syndrome can be due to a mutation of mtDNA (typically *MTATP6* mutations), X-linked pyruvate dehydrogenase deficiency (Chapter 25), or a mutation in a nuclear-encoded mitochondrial gene. Mutations in *SURF1* are a common cause of Leigh syndrome, causing a biochemical defect of cytochrome *c* oxidase (complex IV), but a large list of nuclear genes have been implicated associated with different biochemical defects (Table 24.1).[6] For this reason patients typically have biochemical analysis performed on skin fibroblasts or a biopsy from an affected tissue to substantiate the mitochondrial etiology and guide subsequent molecular genetic analysis. The different genetic etiologies have different natural histories.[7]

DISORDERS OF MITOCHONDRIAL DNA MAINTENANCE

Disorders of mtDNA maintenance affect either the amount (depletion) or quality (deletions or point mutations) of mitochondrial DNA. The clinical features are primarily due to the mitochondrial DNA abnormality, which is secondary to an underlying nuclear gene defect inherited as either an autosomal dominant or an autosomal recessive

TABLE 24.1 Mitochondrial Clinical Syndromes Due to Mutations in the Nuclear Genome

Syndrome	Clinical features	Inheritance pattern	Gene
Alpers–Huttenlocher syndrome (AHS)	Seizures, developmental delay, hypotonia, hepatic failure	AR	*POLG*
Autosomal dominant optic atrophy (DOA)	Visual failure; 20% have deafness, PEO, ptosis, neuropathy, myopathy	AD	*OPA1*
Chronic progressive external ophthalmoplegia (CPEO)	Ptosis, ophthalmoparesis; proximal myopathy	AD, AR	*POLG, PEO1, ANT1, RRM2B*
Barth syndrome	Cardiomyopathy, myopathy, neutropenia	XLR	*TAZ*
Leigh syndrome (LS)	Diffuse myopathy, or encephalopathy, or hepatocerebral syndrome	AR	*SURF1, SCO1, SCO2, COX10, COX14, TACO1,* respiratory subunit genes
Mitochondrial neurogastrointestinal encephalomyopathy (MNGIE)	PEO, ptosis, gut dysmotility, proximal myopathy, axonal polyneuropathy, leukodystrophy	AR	*TYMP*
Mitochondrial recessive ataxia syndrome (MIRAS)	Encephalopathy precipitated by illness, brainstem and cerebellar dysfunction, neuropathy, cardiomyopathy, epilepsy	AR	*POLG*
Mohr–Tranebjaerg syndrome	Deafness, dystonia	XLR	*TIMM8A*
Sensory ataxic neuropathy with dysphagia and ophthalmoplegia (SANDO)	PEO, dysarthria, sensory neuropathy	AR	*POLG*
Sengers syndrome	Cataract, cardiomyopathy, myopathy, lactic acidosis	AR	*AGK*

Abbreviations: AD, autosomal dominant; AR, autosomal recessive; XLR, X-linked recessive; PEO = progressive external ophthalmoplegia.

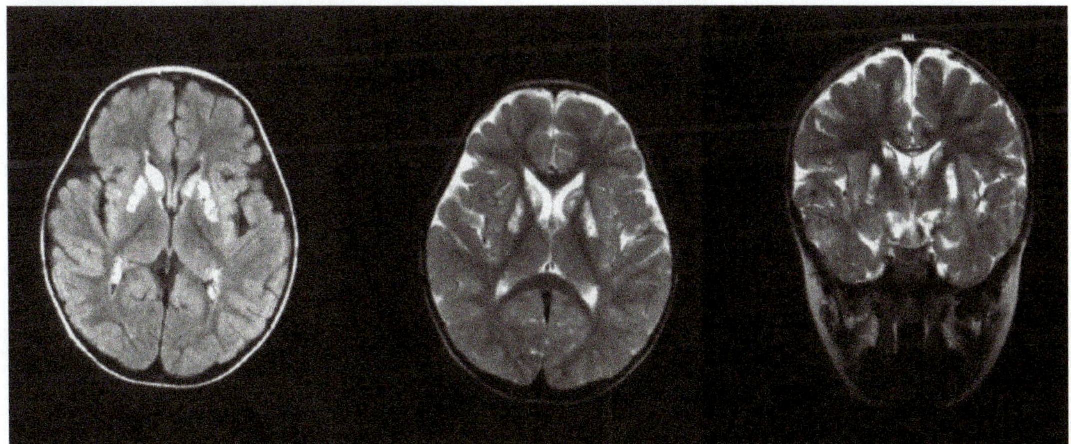

FIGURE 24.2 **MRI findings in Leigh syndrome.** High signal in the caudate and putamen seen on T1- and T2-weighted images in a patient with the homozygous *SCO2* mutation p.E140K. (With thanks to Prof Rita Horvath, Newcastle University).

trait. As a rule of thumb, recessive disorders typically present in childhood, but can also present later in life. Autosomal dominant disorders present in mid-to-late life with a slowly progressive clinical course.

DISORDERS OF MITOCHONDRIAL DNA MAINTENANCE—mtDNA DEPLETION

Mitochondrial DNA depletion disorders fall into two broad categories: a pure myopathic form, and a hepatocerebral syndrome involving a progressive encephalopathy with liver failure. The myopathic cause is formed by mutations in *TK2* and *RRM2B*,[8,9] and hepatocerebral syndromes are caused by mutations in *DGUOK, MPV17, POLG, PEO1, SULCA2* or *SULG1*.[10–13] The Alpers–Huttenlocher syndrome is a well-described mDNA depletion disorder caused by mutations in *POLG*, which typically presents in early childhood with encephalopathy associated with seizures, myoclonus, ataxia, and developmental delay, followed by often irreversible liver failure.[14] Sodium valproate has been shown to cause the acute hepatic crisis in Alpers–Huttenlocher syndrome, and should therefore be avoided in this context.[15]

DISORDERS OF MITOCHONDRIAL DNA MAINTENANCE—MULTIPLE mtDNA DELETION SYNDROMES

A number of mitochondrial DNA deletion syndromes have been described. The first classical presentation was in autosomal dominant and autosomal recessive chronic progressive external ophthalmoplegia.[16] This typically begins in late middle age with ptosis and ophthalmoplegia, followed by mild proximal muscle weakness. Dysphagia can occur later in the disease course, reminiscent of oculopharyngeal muscular dystrophy. Patients typically have multiple cytochrome *c* oxidase (COX)-negative muscle fibers on biopsy and multiple mtDNA deletions in skeletal muscle. Principal genetic causes include mutations in *POLG, POLG2, PEO1, ANT1, RRM2B*.[17–20] Autosomal dominant CPEO associated with optic atrophy is caused by *OPA1* mutations.[21] Classically the visual failure precedes a proximal myopathy, axonal sensory motor neuropathy, and subsequently ophthalmoplegia and ptosis. Mutations in *SPG7* cause autosomal recessive ataxia with spasticity with multiple mtDNA deletions in skeletal muscle, which may explain why some of these patients develop ptosis and PEO later in life.

POLG-RELATED DISORDERS

POLG encodes the catalytic subunit of the mtDNA polymerase. Autosomal recessive homozygous and compound heterozygous mutations have been shown to cause childhood encephalomyopathies sometimes with liver involvement (Alpers–Huttenlocher syndrome, see above)[22] or a late-onset recessive ataxia syndrome (mitochondrial recessive ataxia syndrome; MIRAS),[23] which is highly prevalent in Scandinavia (Figure 24.3). Recessive mutations can also cause sensory ataxic neuropathy with dysphagia and ophthalmoplegia (SANDO).[24,25] Isolated case reports have described phenotypes resembling canonical mitochondrial encephalomyopathy with lactic acidosis and stroke-like episodes (MELAS) and myoclonic epilepsy of the ragged red fibers (MERRF).[25]

DISORDERS OF INTRAMITOCHONDRIAL PROTEIN SYNTHESIS

Disorders of intramitochondrial protein synthesis have emerged as a major cause of nuclear-encoded mitochondrial disease. One-third of patients with mitochondrial disease have a biochemical defect involving multiple respiratory chain enzyme complexes suggesting a defect in intramitochondrial protein synthesis. A minority of these patients have a mutation of mtDNA as the primary cause, but the majority have autosomal recessive disorders.

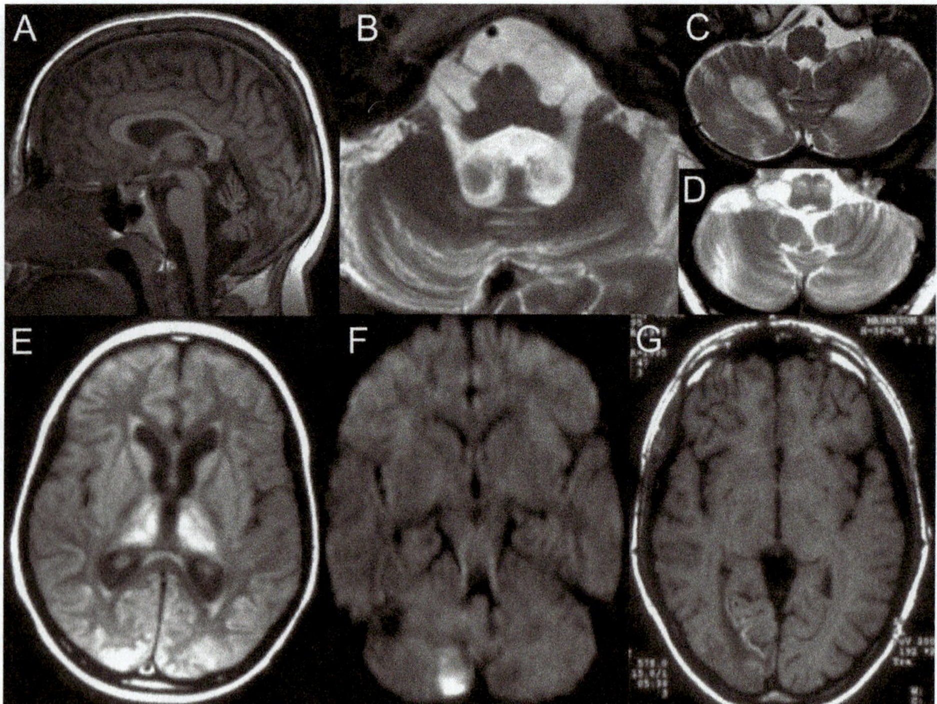

FIGURE 24.3 **MRI findings in *POLG* encephalopathy.** Adapted from [50]. **(A)** Sagittal T1 image showing cerebellar atrophy. **(B)** Axial T2 image showing dentate atrophy (arrow). **(C)** Axial T2 image showing cerebellar white matter hyperintensity. **(D)** Axial T2 image showing bilateral olivary lesions. The olives appear enlarged and hyperintense. **(E)** Axial T2 fluid-attenuated inversion recovery image showing bilateral thalamic and cortical occipital lesions. **(F)** Axial diffusion-weighted image (b = 1000) showing acute stroke-like lesions in the right cerebellar cortex. **(G)** Axial T1 image showing linear, gyriform hyperintensity in the right medial occipital cortex (cortical laminar necrosis). (With thanks to Prof Laurence Bindoff, University of Bergen.)

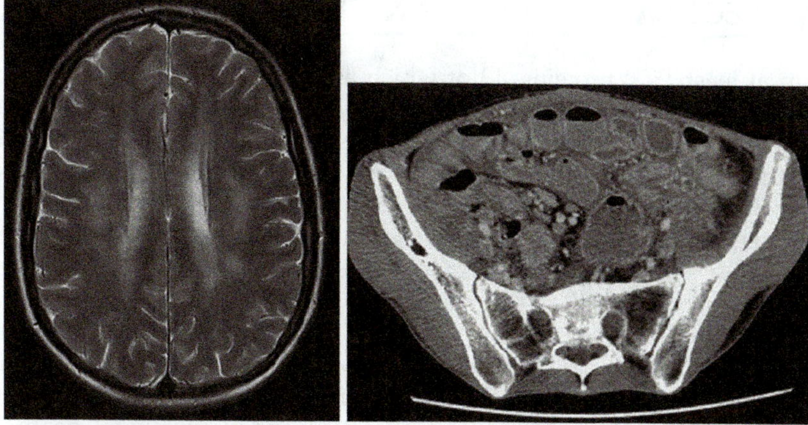

FIGURE 24.4 **Mitochondrial neurogastrointestinal encephalomyopathy (MNGIE).** Adapted from [51]. Left: Axial T2-weighted MRI of the brain showing leukoencephalopathy. Right: intravenous contrast-enhanced computed tomography (CT) of the abdomen and pelvis, revealing multiple loops of small bowel with fluid and gas distension but no transition point, suggestive of ileus/intestinal dysmotility.

The molecular mechanism potentially involves a vast array of different genes affecting mtDNA replication (see above), and the expression of mtDNA, including structural ribosomal proteins, ribosomal assembly proteins, aminoacyl-tRNA synthetases, tRNA-modifying enzymes, tRNA-methylating enzymes and several initiation, elongation, and termination factors of mitochondrial translation. Although there is considerable clinical overlap, specific disorders can have defining characteristics, such as deafness with *MTO1* mutations, or characteristic MR findings with *EARS2* mutations (see below). However, the majority of patients do not have clearly defined features, and present with an overlapping encephalomyelopathy which may involve the heart and liver. In many patients, the disorder is fatal in early life, and the diverse array of genes involved supports the use of early exome sequencing in the diagnosis of this disorder once the biochemical basis has been established.

MITOCHONDRIAL NEUROGASTROINTESTINAL ENCEPHALOMYOPATHY (MNGIE)

MNGIE is an autosomal recessive disorder caused by mutations in *TYMP* leading to thymidine phosphorylase deficiency and an elevation of thymidine detectable in urine and plasma.[26] The disorder typically presents in young adult life with gastrointestinal dysmotility associated with weight loss, a sensory–motor peripheral neuropathy causing profound weakness and ataxia, myopathy and ophthalmoplegia and ptosis.[27] Brain MRI invariably shows leukoencephalopathy (Figure 24.4). Allogenic bone marrow transplantation has been successful in over 30 patients with MNGIE, slowing down progression, and arresting the clinical course in some cases.[28,29] The treatment has been fatal in some cases, largely due to transplantation late in the disease course, or related to severe immune suppression during induction. Alternative approaches include dialysis to reduce circulating thymidine levels, and encapsulated thymidine phosphorylase within autologous erythrocytes. Biosynthetic thymidine phosphorylase is currently being explored as a potential option in parallel to gene therapy approaches.

COENZYME Q10 DEFICIENCY

Coenzyme Q10 deficiency can be primary, or secondary to other inherited neurogenetic disorders.[30] Primary disorders of Q10 biosynthesis fall into four main groups: 1) an encephalopathic form presenting with myeloglobinuria encephalopathy and ragged red fibers on the muscles biopsy; 2) a cerebellar form with prominent cerebellar atrophy on brain MRI; 3) an infantile form with encephalopathy and steroid-unresponsive nephrotic syndrome; and 4) a pure myopathic form with elevated creatine kinase and ragged red fibers. A muscle biopsy may be required to reliably diagnose coenzyme Q10 deficiency. Coenzyme Q10 deficiency is particularly important because of the potential impact of therapies with either coenzyme Q10 or related analogs.[31]

OTHER NUCLEAR-MITOCHONDRIAL DISORDERS

Several additional nuclear-mitochondrial disorders have been described, which are rare. These include X-linked dilated cardiomyopathy and neutropenia (Barth syndrome);[32] X-linked sensorineural hearing loss with dystonia, which can be associated with spasticity, dysphagia, intellectual decline, and cortical blindiness (Mohr–Tranebjaerg syndrome);[33] and X-linked sideroblastic anemia with ataxia. Sengers syndrome presents with congenital cataract, developmental delay, cardiomyopathy, and myopathy.[34]

MOLECULAR GENETICS AND DISEASE MECHANISMS

Mitochondria contain at least 1500 proteins, 13 of which are encoded by small circles of DNA, mitochondrial DNA (mtDNA).[35] The vast majority of mitochondrial proteins are encoded by nuclear DNA. In addition to the majority of respiratory chain subunits, the nuclear-encoded proteins are involved in the maintenance, replication and expression in mitochondrial DNA, the assembly of the respiratory chain, or mitochondrial structural integrity. The respiratory chain itself includes over 100 proteins assembled into five enzyme complexes (I–V). Complexes I and II accept reducing equivalents from intermediary metabolism. Electron flow between the various complexes is linked to the expulsion of protons from the mitochondrial matrix to the intermembrane space. The proton-motive force is harnessed by complex V to synthesize ATP.

Nuclear-encoded mitochondrial disorders can be autosomal dominant, autosomal recessive, or X-linked recessive (Table 24.1). Being exclusively maternally inherited, mtDNA disorders appear as either maternally inherited diseases or sporadic traits (see Chapter 23). The pathophysiology of mitochondrial disorders is ultimately thought to be due to a deficiency of ATP. The bioenergetic failure leads to cellular dysfunction, and ultimately cell death through apoptotic-based mechanisms. Despite mitochondrial disorders being caused by a defect in the same final bioenergetic pathway, they display remarkable clinical heterogeneity. For mtDNA disorders and nuclear disorders of mitochondrial DNA maintenance, differences in the mutation load (heteroplasmy) between cells and tissues provide part of the explanation. However, even with apparently straightforward autosomal recessive respiratory chain enzyme defects, the phenotype can vary from patient to patient. It is currently not clear why one particular respiratory chain complex defect causes one phenotype, and another causes a different clinical presentation. It is particularly intriguing why mutations affecting very similar biosynthetic pathways can cause strikingly distinct phenotypic groups. A good example of this is the predominance of hearing deficit in patients with *MTO1* mutations,[36] and characteristic imaging findings in patients with *EARS2* mutations.[37] Both gene defects cause a defect in intramitochondrial protein synthesis. Understanding tissue specificity is a major focus of current research into the pathophysiology of mitochondrial diseases.

TESTING

Specific phenotypes may point to a precise molecular diagnosis on purely clinical grounds (Table 24.1). Under these circumstances, it may be appropriate to perform specific molecular genetic tests at an early stage. However, phenotypic overlap with other nonmitochondrial disorders and mtDNA disorders necessitates a systematic approach. The first step is to acquire a complete picture of the clinical phenotype. Routine blood tests are rarely helpful, but can reveal a high serum creatine kinase if there is muscle involvement, a high blood or CSF lactate if there is an encephalopathy, and elevated serum FGF21 levels in patients presenting with a severe childhood myopathy.[38] Cardiac investigations are mandatory, with an electrocardiogram and echocardiogram. Brain imaging may reveal basal ganglia lesions, or a white matter high-signal abnormality (Figures 24.2–24.4). Neurophysiological studies may reveal myopathic features or an axonal sensory motor neuropathy. A dorsal root ganglionopathy is also a characteristic of specific nuclear-mitochondrial disorders. Electroencephalography may reveal features of encephalopathy or a seizure predisposition. Some patients with nuclear-mitochondrial disorders develop focal epilepsies, which are intractable to conventional therapy.

Having built up an accurate clinical phenotype, the next step is to establish whether there is an underlying mitochondrial dysfunction. This usually involves a biopsy of an affected tissue (or skin fibroblasts in children), leading to histochemical analysis, or respiratory chain complex biochemistry. The pattern of the biochemistry can point to a specific genetic abnormality, leading to targeted molecular analysis. Failing this, a number of common candidate genes may be screened. Increasingly patients are undergoing early exome and whole-genome sequencing to identify the genetic basis for their mitochondrial disease.

It is worth noting that the mitochondrial dysfunction can occur in several inherited and noninherited disorders as a secondary feature. This presents an additional diagnostic challenge.

MANAGEMENT

A recent Cochrane System Review has shown that there are no treatments of proven benefit for mitochondrial disorders.[39] Clinical management is largely supportive. There are a few exceptions to this, discussed in detail above. These include coenzyme Q10 replacement in disorders of Q10 biosynthesis, and allogenic stem cell transplantation in MNGIE.

Supportive Management

Supportive management involves disease surveillance to enable the early detection of complications. The detection and management of cardiomyopathy is particularly important, as is the early diagnosis of hyperglycemia. Dysphagia is increasingly recognized as an important complication of mitochondrial disease, requiring dietary modification or a gastrostomy in the late stages. Surgical intervention to manage cataract or ptosis can be of enormous benefit to patients. Cochlear implantation has been shown to be highly effective in treating sensorineural deafness in patients with mitochondrial disease.

Specific Therapies

Specific therapies that have been tried include a variety of nutritional supplements, such as creatine, cysteine, dicloracetate, dimethylglycine, lipoic acid, and carnitine. A variety of vitamins and cofactors have also been tried, including vitamin C and K, nicotinamide and dietary manipulations, such as a ketogenic diet (reviewed in [40]). Although isolated case reports describe an improvement, there is no objective evidence that these therapies work based on randomized controlled data.[40] There is emerging evidence that exercise therapy (both endurance and resistance training) may be of benefit to patients with mitochondrial DNA-mediated mitochondrial disorders).[41,42] It remains to be seen whether the same is the case of patients with nuclear-mitochondrial disorders.

Genetic Counseling and Prevention

Establishing a genetic diagnosis is critically important. This will enable accurate genetic counseling, the best guide to prognosis, and potential prenatal or preimplantation genetic diagnosis. It cannot be stressed too strongly that similar clinical phenotypes seen in patients with mitochondrial disorders can have distinctly different genetic implications for the family, and so a biochemical diagnosis is insufficient to provide clear guidance for recurrence risks.

New Approaches to Treatment

There are several new treatment approaches currently being investigated for mitochondrial disorders.[43,44] At the experimental level, a number of gene therapeutic approaches have been studied, with notable success in replacing enzyme defects in mouse models (for example ethylmalonic aciduria, *ETHE1*). Exercise therapy has been evaluated in small open-label clinical studies and shows potential in mtDNA disorders, and may be of benefit in nuclear-mitochondrial disorders. Growing interest from the pharmaceutical industry has led to the first randomized controlled trials for mitochondrial disease.[45] New approaches include targeted antioxidant therapy,[46] agents causing mitochondrial biogenesis,[47] and modified coenzymes and vitamins.[48,49]

References

1. Schon EA, DiMauro S, Hirano M. Human mitochondrial DNA: roles of inherited and somatic mutations. *Nat Rev Genet*. 2012;13:878–890.
2. DiMauro S, Schon EA, Carelli V, Hirano M. The clinical maze of mitochondrial neurology. *Nat Rev Neurol*. 2013;9:429–444.
3. Darin N, Oldfors A, Moslemi AR, Holme E, Tulinius M. The incidence of mitochondrial encephalomyopathies in childhood: clinical features and morphological, biochemical, and DNA anbormalities. *Ann Neurol*. 2001;49:377–383.
4. Skladal D, Bernier FP, Halliday JL, Thorburn DR. Birth prevalence of mitochondrial respiratory chain defects in children. *J Inherit Metab Dis*. 2000;23:138.
5. Rahman S, Blok RB, Dahl HH, et al. Leigh syndrome: clinical features and biochemical and DNA abnormalities. *Ann Neurol*. 1996;39:343–351.
6. Baertling F, Rodenburg RJ, Schaper J, et al. A guide to diagnosis and treatment of Leigh syndrome. *J Neurol Neurosurg Psychiatry*. 2014;85:257–265.
7. Wedatilake Y, Brown R, McFarland R, et al. SURF1 deficiency: a multi-centre natural history study. *Orphane J Rare Dis*. 2013;8:96.
8. Saada A, Shaag A, Mandel H, Nevo Y, Eriksson S, Elpeleg O. Mutant mitochondrial thymidine kinase in mitochondrial DNA depletion myopathy. *Nat Genet*. 2001;29:342–344.
9. Bourdon A, Minai L, Serre V, et al. Mutation of RRM2B, encoding p53-controlled ribonucleotide reductase (p53R2), causes severe mitochondrial DNA depletion. *Nat Genet*. 2007;39:776–780.
10. Sarzi E, Bourdon A, Chretien D, et al. Mitochondrial DNA depletion is a prevalent cause of multiple respiratory chain deficiency in childhood. *J Pediatr*. 2007;150:531–534 4 e1–4 e6.
11. Spinazzola A, Viscomi C, Fernandez-Vizarra E, et al. MPV17 encodes an inner mitochondrial membrane protein and is mutated in infantile hepatic mitochondrial DNA depletion. *Nat Genet*. 2006;38:570–575.
12. Ostergaard E, Christensen E, Kristensen E, et al. Deficiency of the alpha subunit of succinate-coenzyme A ligase causes fatal infantile lactic acidosis with mitochondrial DNA depletion. *Am J Hum Genet*. 2007;81:383–387.
13. Mandel H, Szargel R, Labay V, et al. The deoxyguanosine kinase gene is mutated in individuals with depleted hepatocerebral mitochondrial DNA. *Nat Genet*. 2001;29:337–341.

14. Naviaux RK, Nguyen KV. POLG mutations associated with Alpers' syndrome and mitochondrial DNA depletion. *Ann Neurol.* 2004;55:706–712.
15. Horvath R, Hudson G, Ferrari G, et al. Phenotypic spectrum associated with mutations of the mitochondrial polymerase gamma gene. *Brain.* 2006;129(Pt 7):1674–1684.
16. Zeviani M, Bresolin N, Gellera C, et al. Nucleus-driven multiple large-scale deletions of the human mitochondrial genome: a new autosomal dominant disease. *Am J Hum Genet.* 1990;47:904–914.
17. Van Goethem G, Dermaut B, Lofgren A, Martin J-J, Van Broeckhoven C. Mutation of *POLG* is associated with progressive external ophthalmoplegia characterized by mtDNA deletions. *Nat Genet.* 2001;28:211–212.
18. Longley MJ, Clark S, Yu Wai Man C, et al. Mutant POLG2 disrupts DNA polymerase gamma subunits and causes progressive external ophthalmoplegia. *Am J Hum Genet.* 2006;78:1026–1034.
19. Spelbrink JN, Li FY, Tiranti V, et al. Human mitochondrial DNA deletions associated with mutations in the gene encoding Twinkle, a phage T7 gene 4-like protein localized in mitochondria. *Nat Genet.* 2001;28:223–231.
20. Kaukonen J, Juselius JK, Tiranti V, et al. Role of adenine nucleotide translocator 1 in mtDNA maintenance. *Science.* 2000;289:782–785.
21. Hudson G, Amati-Bonneau P, Blakely EL, et al. Mutation of OPA1 causes dominant optic atrophy with external ophthalmoplegia, ataxia, deafness and multiple mitochondrial DNA deletions: a novel disorder of mtDNA maintenance. *Brain.* 2008;131:329–337.
22. Nguyen KV, Ostergaard E, Ravn SH, et al. POLG mutations in Alpers syndrome. *Neurology.* 2005;65:1493–1495.
23. Hakonen AH, Heiskanen S, Juvonen V, et al. Mitochondrial DNA polymerase W748S mutation: a common cause of autosomal recessive ataxia with ancient European origin. *Am J Hum Genet.* 2005;77:430–441.
24. Van Goethem G, Martin JJ, Dermaut B, et al. Recessive POLG mutations presenting with sensory and ataxic neuropathy in compound heterozygote patients with progressive external ophthalmoplegia. *Neuromuscul Disord.* 2003;13:133–142.
25. Deschauer M, Tennant S, Rokicka A, et al. MELAS associated with mutations in the POLG1 gene. *Neurology.* 2007;68:1741–1742.
26. Nishino I, Spinazzola A, Hirano M. Thymidine phosphorylase gene mutations in MNGIE, a human mitochondrial disorder. *Science.* 1999;283:689–692.
27. Nishino I, Spinazzola A, Hirano M. MNGIE: from nuclear DNA to mitochondrial DNA. *Neuromuscul Disord.* 2001;11:7–10.
28. Hirano M, Marti R, Casali C, et al. Allogenic stem cell transplantation corrects biochemical derangements in MNGIE. *Neurology.* 2006;67(8):1458–1460.
29. Lara MC, Waiss B, Illa I, et al. Infusion of platelets transiently reduces nucleoside overload in MNGIE. *Neurology.* 2006;67(8):1461–1463.
30. Emmanuele V, Lopez LC, Berardo A, et al. Heterogeneity of coenzyme Q10 deficiency: patient study and literature review. *Arch Neurol.* 2012;69:978–983.
31. Hirano M, Garone C, Quinzii CM. CoQ(10) deficiencies and MNGIE: two treatable mitochondrial disorders. *Biochim Biophys Acta.* 1820;2012:625–631.
32. Kirwin SM, Manolakos A, Barnett SS, Gonzalez IL. Tafazzin splice variants and mutations in Barth syndrome. *Mol Genet Metab.* 2014;111:26–32.
33. Tranebjaerg L, Hamel BC, Gabreels FJ, Renier WO, Van Ghelue M. A de novo missense mutation in a critical domain of the X-linked DDP gene causes the typical deafness-dystonia-optic atrophy syndrome. *Eur J Hum Genet.* 2000;8:464–467.
34. Mayr JA, Haack TB, Graf E, et al. Lack of the mitochondrial protein acylglycerol kinase causes Sengers syndrome. *Am J Hum Genet.* 2012;90:314–320.
35. Vafai SB, Mootha VK. Mitochondrial disorders as windows into an ancient organelle. *Nature.* 2012;491:374–383.
36. Ghezzi D, Baruffini E, Haack TB, et al. Mutations of the mitochondrial-tRNA modifier MTO1 cause hypertrophic cardiomyopathy and lactic acidosis. *Am J Hum Genet.* 2012;90:1079–1087.
37. Steenweg ME, Ghezzi D, Haack T, et al. Leukoencephalopathy with thalamus and brainstem involvement and high lactate 'LTBL' caused by EARS2 mutations. *Brain.* 2012;135:1387–1394.
38. Suomalainen A, Elo JM, Pietilainen KH, et al. FGF-21 as a biomarker for muscle-manifesting mitochondrial respiratory chain deficiencies: a diagnostic study. *Lancet Neurol.* 2011;10:806–818.
39. Pfeffer G, Majamaa K, Turnbull DM, Thorburn D, Chinnery PF. Treatment for mitochondrial disorders. *Cochrane Database Syst Rev.* 2012;4:CD004426.
40. Pfeffer G, Horvath R, Klopstock T, et al. New treatments for mitochondrial disease-no time to drop our standards. *Nat Rev Neurol.* 2013;9:474–481.
41. Taivassalo T, Gardner JL, Taylor RW, et al. Endurance training and detraining in mitochondrial myopathies due to single large-scale mtDNA deletions. *Brain.* 2006;129:3391–3401.
42. Murphy JL, Blakely EL, Schaefer AM, et al. Resistance training in patients with single, large-scale deletions of mitochondrial DNA. *Brain.* 2008;131:2832–2840.
43. Hassani A, Horvath R, Chinnery PF. Mitochondrial myopathies: developments in treatment. *Curr Opin Neurol.* 2010;23:459–465.
44. Koopman WJ, Willems PH, Smeitink JA. Monogenic mitochondrial disorders. *N Engl J Med.* 2012;366:1132–1141.
45. Klopstock T, Yu-Wai-Man P, Dimitriadis K, et al. A randomized placebo-controlled trial of idebenone in Leber's hereditary optic neuropathy. *Brain.* 2011;134:2677–2686.
46. Chouchani ET, Methner C, Nadtochiy SM, et al. Cardioprotection by S-nitrosation of a cysteine switch on mitochondrial complex I. *Nat Med.* 2013;19:753–759.
47. Wenz T, Wang X, Marini M, Moraes CT. A metabolic shift induced by a PPAR panagonist markedly reduces the effects of pathogenic mitochondrial tRNA mutations. *J Cell Mol Med.* 2011;15:2317–2325.
48. Enns GM, Kinsman SL, Perlman SL, et al. Initial experience in the treatment of inherited mitochondrial disease with EPI-743. *Mol Genet Metab.* 2012;105:91–102.
49. Hargreaves IP. Coenzyme Q, as a therapy for mitochondrial disease. *Int J Biochem Cell Biol.* 2014;49C:105–111.
50. Tzoulis C, Neckelmann G, Mork SJ, et al. Localized cerebral energy failure in DNA polymerase gamma-associated encephalopathy syndromes. *Brain.* 2010;133:1428–1437.
51. Parry-Jones A, Paine P, Ramdass R, et al. Unexplained gastrointestinal dysmotility: the clue may lie in the brain. *Gut.* 2011;60(758):805.

CHAPTER

25

Pyruvate Dehydrogenase, Pyruvate Carboxylase, Krebs Cycle and Mitochondrial Transport Disorders

Mireia Tondo, Isaac Marin-Valencia, Qian Ma, and Juan M. Pascual

The University of Texas Southwestern Medical Center, Dallas, TX, USA

INTRODUCTION

Considered collectively, energy metabolism disorders spare no organ or tissue and therefore can mimic diseases routinely encountered by primary and specialty care clinicians. In addition to the well-established role of many energy metabolism enzymes and transporters in the formation and dissolution of high-energy molecular bonds and in the supply of fuels to cells, some also serve dual or, possibly, multiple or multifaceted roles. For example, some pyruvate metabolism enzyme mutations impair axonal migration or alter craniofacial configuration; many are linked to neuronal necrosis and apoptosis and to edema (spongiosis) of the cerebral white matter and, paradoxically, most cause enhanced excitation and epilepsy (rather than neuronal hypoexcitability), resulting in further increases in neural energetic demands. These *a priori* unexpected neurological manifestations probably stem from the fact that flux through energy metabolism reactions sustains the synthesis and recycling of neurotransmitters and other signaling molecules by neural cells and circuits. Consequently, the brain often bears the full burden of these diseases, in conjunction with cardiac and skeletal muscle, liver and kidney.

PYRUVATE DEHYDROGENASE DEFICIENCY

The complete aerobic oxidation of glucose and other carbohydrates by the brain requires the sequential action of glycolysis and the citric acid cycle (CAC). The mitochondrial pyruvate dehydrogenase (PDH) complex plays a key role in connecting both sets of reactions by catalyzing the oxidative decarboxylation of pyruvate to form acetyl coenzyme A (acetyl CoA), nicotinamide adenine dinucleotide (NADH), and CO_2[1,2] (Figures 25.1 and 25.2). The PDH complex is a nuclear-encoded multienzymatic assembly present in the mitochondria of most living organisms and is especially prominent in all energy-demanding tissues. Symptoms vary considerably in patients with PDH complex deficiency and, unexpectedly, almost equal numbers of affected males and females have been identified, despite the location of the PDH E1 α subunit gene in the X chromosome, a paradox due to selective female X-inactivation.[3] Thus, the phenotype of PDH deficiency is dictated by mutation severity (especially in males) and by the pattern of X-inactivation in females.[4] Defects in the PDH complex are a frequent cause of lactic acidosis. Neurodevelopmental abnormalities, microcephaly, epilepsy and agenesis of the corpus callosum are characteristic features[5] (Figure 25.3). Infants may exhibit facial features reminiscent of fetal alcoholic syndrome and older children can present with areflexia and intermittent weakness, alternating hemiplegia or Leigh syndrome.[6] Approximately 80% of all cases involve defects in the α subunit of the enzyme (*PDHA1* gene, located in Xp22.12).[7–9] Mutations inherited in autosomal recessive fashion causing defects in other PDH subunits have also been described: E1β (*PDHB* gene, located in

Rosenberg's Molecular and Genetic Basis of Neurological and Psychiatric Disease
http://dx.doi.org/10.1016/B978-0-12-410529-4.00025-5

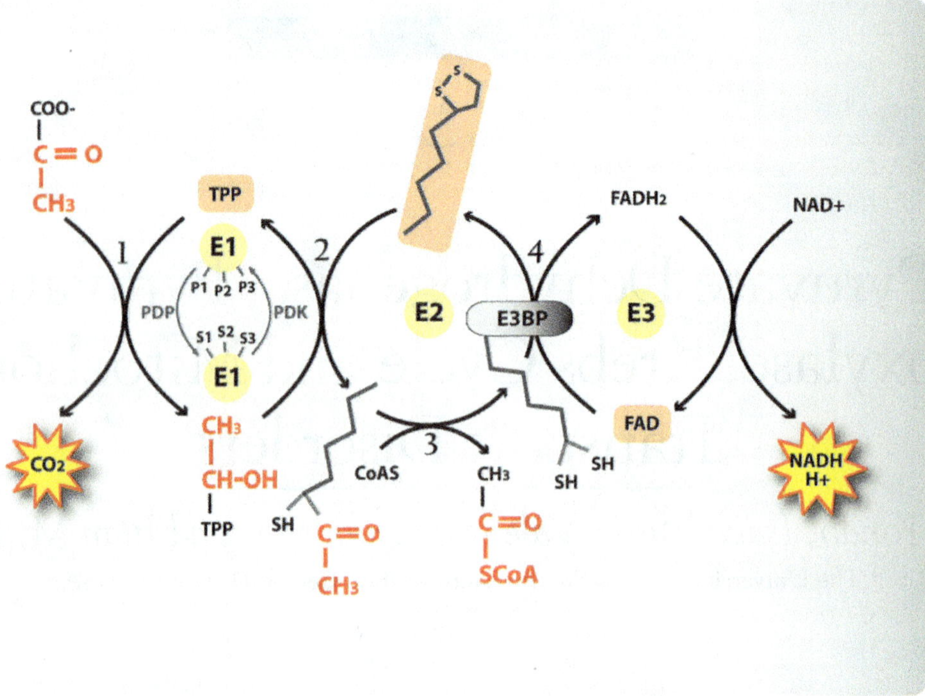

FIGURE 25.1 **Schematic flux diagram illustrating metabolic abnormalities resulting from pyruvate dehydrogenase (PDH) deficiency.** Abbreviations: CAC, citric acid cycle; OAA, oxaloacetate; α-KG, α-ketoglutarate.

3p14.3), E2 (*DLAT* gene, located in 11q23.1), E3 (*DLD* gene, located in chromosome 7q31.1), PDP (*PDP1* gene, located in 8q22.1) and E3BP or X protein (*PDHX* gene, located in 11p13)[1,10] (Table 25.1). Overall, neither gender nor biochemical or clinical features differentiates the various enzymatic or genetic etiologies of the disease.[11] The biochemical diagnosis of these disorders requires measurement of lactate and pyruvate concentrations in plasma and cerebrospinal fluid, analysis of amino acids in plasma and organic acids in urine, as well as neuroradiologic investigations, including magnetic resonance spectroscopy to detect lactate presence or abundance. Enzymatic analysis of fibroblast PDH activity is often performed and molecular diagnosis via DNA sequencing is available for many PDH-related genes. Different therapeutic approaches have been used in PDC deficiency with varying degrees of success[1] owing to relatively modest influence over the disease course.[12] A ketogenic diet is recommended[13] following widely varying standards,[14] as is thiamine supplementation, which can afford substantial benefit in the case mutations that disrupt cofactor affinity, resulting in responsiveness.[15,16] Dichloroacetate is sometimes used to enhance lactate clearance,[17] although it has been associated with neurotoxicity (specifically, neuropathy) in a mitochondrial DNA disorder.[18]

PYRUVATE CARBOXYLASE DEFICIENCY

Pyruvate carboxylase (PC) is a mitochondrial enzyme bound to biotin that catalyzes the conversion of pyruvate to oxaloacetate when abundant acetyl CoA is available, replenishing Krebs cycle intermediates in the mitochondrial matrix. PC is involved in gluconeogenesis, lipogenesis and neurotransmitter synthesis.[15] PC deficiency is a rare cause of lactic acidemia and encephalopathy during the neonatal and infantile periods. Neurological manifestations are often prominent in PC deficiency.[19] PC deficiency manifests three broad degrees of phenotypic severity (Table 25.1): 1) An infantile form (A), characterized by moderate lactic acidosis, mental and motor deficits or delays, hypotonia, pyramidal tract dysfunction, ataxia and seizures and, often, death in infancy. Episodes of vomiting, acidosis and tachypnea can be triggered by metabolic imbalance or infection. 2) A severe neonatal form (B), exhibiting severe lactic acidosis, hypoglycemia, hepatomegaly, depressed consciousness, and severely abnormal development. Abnormal limb and ocular movements are common findings. Brain magnetic resonance imaging (MRI) reveals cystic periventricular leukomalacia. Hyperammonemia and depletion of intracellular aspartate and oxaloacetate can be profound and early death is common. 3) A rare benign form (C), associated with episodic acidosis and moderate mental impairment compatible with survival and near normal neurological performance. PC deficiency is an autosomal recessive disease due to mutation of the *PC* gene, located in chromosome 11 (11q13.4-q13.5), the only gene known to

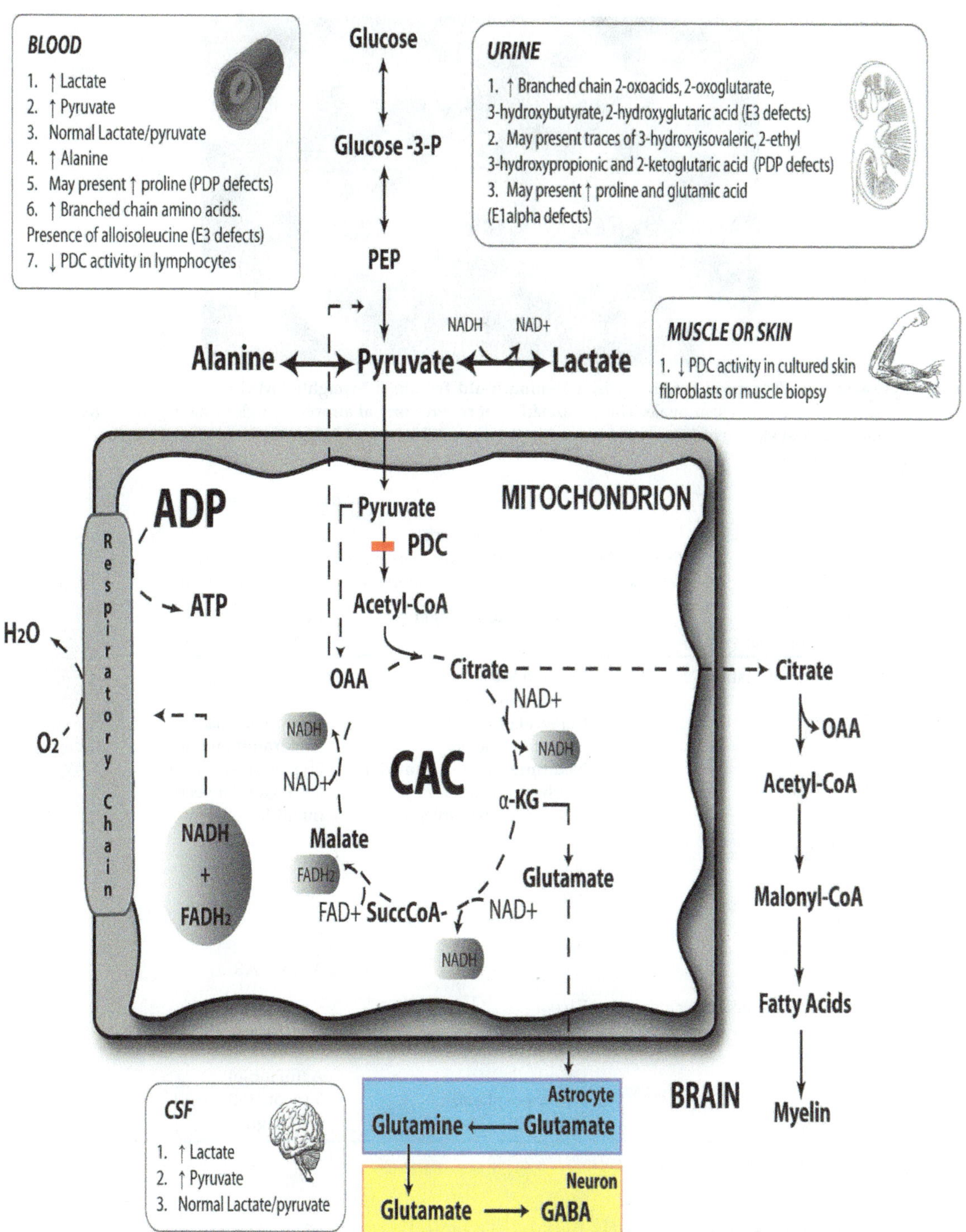

FIGURE 25.2 **Biochemical abnormalities in the principal tissues affected by pyruvate dehydrogenase (PDH) deficiency.**

be associated with PC deficiency.[20] A variety of mutations have been identified, some of which depress PC activity significantly.[21] PC deficiency patients are typically homozygotes, manifesting the disorder with complete penetrance. PC deficiency must be suspected in neonates, infants, or young children presenting with failure to thrive, developmental delay, epilepsy, and chronic or intermittent metabolic acidosis with elevated blood lactate. From an analytical perspective, in addition to lactic acidemia, several other abnormalities such as metabolic acidosis, abnormal plasma and urine amino acids concentration and urine organic acids profile, hypoglycemia, ketonemia and high levels of pyruvate are important clues in the diagnosis of PC deficiency.[19] Structural abnormalities of the brain are frequently present in type A and B patients. The neuroradiological findings reported in these patients include ischemic-like

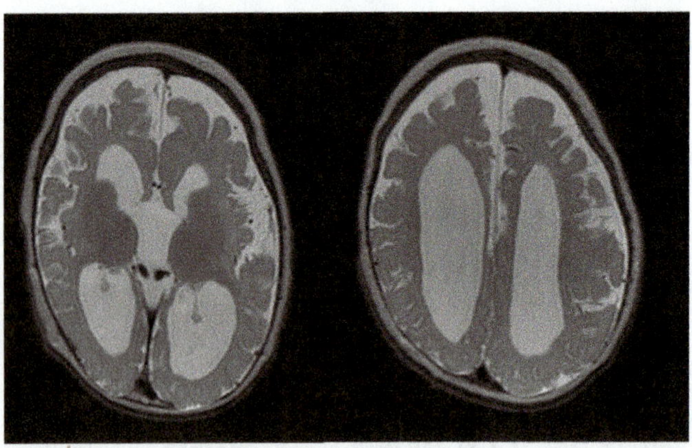

FIGURE 25.3 **Pyruvate dehydrogenase deficiency in a 10-month-old female.** T2-weighted MRI images representing axial sections at the level of the thalamus and of the centrum semiovale. The combination of severe cortical atrophy (predominantly in the frontal lobes and sylvian operculae), white matter underdevelopment (more pronounced in the occipital lobes) and increased water contents, and agenesis of the corpus callosum are characteristic features.

TABLE 25.1 Pyruvate Metabolism Disorders
Pyruvate dehydrogenase (PDH) deficiency and pyruvate carboxylase (PC) deficiency with their clinical variants, inheritance and main manifestations

Disease	Pyruvate dehydrogenase deficiency	Pyruvate carboxylase deficiency
Manifestations	Lactic acidosis	
	Episodic ataxia	
	Cerebral dysgenesis	Lactic acidosis
	Infantile epilepsy	Hyperammonemia
	Neuromuscular dysfunction	Hypercitrullinemia
	Leigh syndrome	Basal ganglia necrosis
	Alternating hemiplegia	Infantile epilepsy
Inheritance	X linked	
	AR	
	AR	
	AR	
	AR	
	AR	Always AR
Biochemical variants	E1α	
	E1β	
	E2	
	E3	A (infantile)
	PDH phosphatase	B (neonatal)
	X protein	C (benign)

Abbreviations: AR, autosomal recessive inheritance.

lesions,[22] ventricular dilatation, periventricular leukomalacia, delayed myelination,[23,24] and subcortical leukodystrophy.[25–27] These findings are usually detected during the neonatal or infantile stages. Enzymatic analysis of fibroblast PC activity can be performed, followed by genotyping. Dietary modification with triheptanoin (a triglyceride) supplementation has been attempted as a means to increase acetyl CoA and anaplerotic propionyl coenzyme A.[28] Liver transplantation has also been performed.[29]

DISORDERS OF THE KREBS CYCLE

Several tricarboxilic acid enzymes are susceptible to mutations that cause severe mitochondrial diseases heritable in autosomal recessive fashion. Of these, aconitase, the E3 component of the α-ketoglutarate dehydrogenase complex (a peptide that is also shared by the PDH complex), succinate dehydrogenase (SDH; which is also part of the mitochondrial respiratory chain and known as complex II) and fumarase (fumarate hydratase; FH) are collectively associated with profound encephalopathy and manifestations that are, again, pleomorphic.

E3 Deficiency

E3 (dihydrolipoamide dehydrogenase) deficiency results in multiple 2-ketoacid dehydrogenase deficiency and can be best understood as a combined syndrome comprising both pyruvate dehydrogenase complex deficiency and a tricarboxylic acid cycle defect. Clinically, E3 deficiency presents with severe progressive hypotonia and failure to thrive, with onset in the first few months of life.[30] Psychomotor retardation, microcephaly and spasticity occur. Some patients develop the typical clinical and radiological manifestations of Leigh syndrome.[31] Biochemically, E3 deficiency causes accumulation of pyruvate, lactate and branched-chain amino acids in plasma, and of branched-chain α-ketoacids in urine.[32] Genetically, the affected gene is *DLD* located on chromosome 7q31-q32 and it presents with a recessive inheritance.

Succinate Dehydrogenase Deficiency

Succinate dehydrogenase (SDH) is part of both the citric acid cycle and respiratory electron transfer chain and it consists of four subunits (named A to D) encoded by the nuclear genome. Clinically, mutations of SDH subunit A cause Leigh syndrome or optic atrophy in the elderly due to progressively necrotic lesions. Mutations in subunits B, C, and D are associated with paraganglioma, and subunits B and D with pheochromocytoma with severely reduced tumor SDH activity (B and D are also associated to papillary and medullary thyroid cancer). Additional mutations in two SDH assembly factors have been associated to infantile leukoencephalopathy (SDHAF1 assembly factor) and paraganglioma (SDH5 assembly factor).[33]

Fumarase Deficiency

Fumaric aciduria due to fumarase deficiency is a rare metabolic disease that may cause structural brain malformations, developmental delay, dysmorphic facial features, neonatal polycythemia, and great accumulation of fumaric acid in urine.[34] Fumarate hydratase mutations are independently associated with uterine and cutaneous leiomyomas and with papillary renal cell cancer. Biochemically, besides a massive excretion of fumaric acid in the urine, additional Krebs cycle intermediates, such as suberic and adipic acids, and succinylpurine derivatives can be found in body fluids.[35] The mode of inheritance of fumarase deficiency is autosomal recessive and the affected gene *FH* is located on chromosome 1q.42.1.

MITOCHONDRIAL TRANSPORTER DISORDERS

Among the first and best-understood functions of mitochondria is the generation of energy in the form of adenosine triphosphate (ATP), which enables cells to perform most energy-consuming processes. Almost one-quarter of patients with documented reduction of substrate oxidation in muscle mitochondria do not exhibit a defect in a mitochondrial enzyme. Hence, renewed emphasis is being placed on the proteins responsible for the transport of substances that transit the inner and outer mitochondrial membranes. Disorders associated with defects in these transporters exemplify an expanding class of human syndromes with a predilection for the central nervous system, skeletal muscle and heart muscle.

Adenine Nucleotide Translocator Deficiency

The electron gradient generated through respiration drives the production of ATP, which is transported to the cytosol by the adenine nucleotide translocator (ANT). This carrier is also responsible for the import of adenosine diphosphate (ADP) into mitochondria.[36] The transporter is encoded by the *ANT1* gene, which is critically important for human muscle function and integrity. Autosomal dominant *ANT1* mutations destabilize mitochondrial DNA maintenance, causing multiple mitochondrial DNA deletions that manifest as progressive external ophthalmoplegia and facioscapulohumeral muscular dystrophy, whereas recessive mutations cause congenital heart defects, cataracts, and lactic acidosis (Senger syndrome).[37]

Malate–Aspartate Shuttle Deficiency

The malate–aspartate shuttle translocates electrons produced during glycolysis across the inner mitochondrial membrane. Shuttle defects can thus impact oxidative phosphorylation. Symptoms include exercise-induced myalgia, pigmenturia, and elevated serum creatine kinase.[38] No neurological impairment has been noted. Nevertheless, one of the components of the malate–aspartate shuttle, the mitochondrial aspartate–glutamate carrier isoform 1 (AGC1), has been associated with global cerebral hypomyelination.[39] This component is specific to neurons and muscle and its role is to supply aspartate to the cytosol thereby contributing to the supply of energy to neurons. Patients manifest arrested psychomotor development, hypotonia and seizures.

VDAC Deficiency

The voltage-dependent anion channels (VDAC) constitute large-diameter pores in the outer mitochondrial membrane that can transport anions, adenine nucleotides, cations, and various uncharged molecules. VDAC defects are associated with an abnormal ion composition of the mitochondrial matrix thereby potentially deteriorating oxidative phosphorylation and other reactions. The principal channel is encoded by the gene *HVDAC1*, located on chromosome Xq13-q21. A case of VDAC deficiency presented with psychomotor retardation, hypotonia, diminished brain myelination, and dysmorphic features. Routine laboratory assays in blood, cerebrospinal fluid and urine were normal despite almost complete VDAC deficiency in skeletal muscle.[40]

Carnitine Transporter Deficiency

The most common carnitine uptake defect is inherited in an autosomal recessive manner as a result of mutation of the *OCTN2* gene, located in 5q31. Over two dozen patients have been reported. The human syndrome is a form of primary carnitine deficiency due to a defect in the high-affinity carnitine transporter OCTN2, a multifunctional polyspecific organic cation channel.[41] The clinical disorder is characterized by carnitine-responsive cardiomyopathy with or without weakness, recurrent hypoglycemic hypoketotic encephalopathy and failure to thrive. Signs of progressive cardiomyopathy and congestive heart failure normally occur in late infancy or early childhood, usually between 1 and 7 years of age. Alternatively, patients may exhibit intermittent hypoglycemic, hypoketotic encephalopathy at a younger age (between 1 month and 5 years).[41] Plasma free and total carnitine levels are extremely low (<5% of normal age-appropriate levels),[42] in association with decreased renal reabsorption of carnitine. Plasma acylcarnitines are normal. In general, the response to oral carnitine supplementation is satisfactory.

ACKNOWLEDGEMENTS

J.M.P. is supported by NIH grants NS077015, NS067015, NS078059, MH084021, RR002584, RR024982 and by the Office of Rare Diseases Research Glucose transporter type 1 deficiency syndrome (G1D) collaboration, education, and test translation (CETT) program for rare genetic diseases.

References

1. Pithukpakorn M. Disorders of pyruvate metabolism and the tricarboxylic acid cycle. *Mol Genet Metab*. 2005;85(4):243–246.
2. Patel MS, Korotchkina LG. Regulation of the pyruvate dehydrogenase complex. *Biochem Soc Trans*. 2006;34(Pt 2):217–222.
3. Lissens W, De Meirleir L, Seneca S, et al. Mutations in the X-linked pyruvate dehydrogenase (E1) alpha subunit gene (PDHA1) in patients with a pyruvate dehydrogenase complex deficiency. *Hum Mutat*. 2000;15(3):209–219.
4. Nissenkorn A, Michelson M, Ben-Zeev B, Lerman-Sagie T. Inborn errors of metabolism: a cause of abnormal brain development. *Neurology*. 2001;56(10):1265–1272.
5. De Vivo DC. Complexities of the pyruvate dehydrogenase complex. *Neurology*. 1998;51(5):1247–1249.
6. De Meirleir L. Disorders of pyruvate metabolism. *Handb Clin Neurol*. 2013;113:1667–1673.
7. Robinson BH. Lacticacidemia. Biochemical, clinical, and genetic considerations. *Adv Hum Genet*. 1989;18:151–179.
8. Patel MS, Naik S, Wexler ID, Kerr DS. Gene regulation and genetic defects in the pyruvate dehydrogenase complex. *J Nutr*. 1995;125(6 suppl):1753S–1757S.
9. De Meirleir L, Lissens W, Benelli C, et al. Pyruvate dehydrogenase complex deficiency and absence of subunit X. *J Inherit Metab Dis*. 1998;21(1):9–16.
10. Brown G. Pyruvate dehydrogenase deficiency and the brain. *Dev Med Child Neurol*. 2012;54(5):395–396.
11. Patel KP, O'Brien TW, Subramony SH, Shuster J, Stacpoole PW. The spectrum of pyruvate dehydrogenase complex deficiency: clinical, biochemical and genetic features in 371 patients. *Mol Genet Metab*. 2012;105(1):34–43.

12. Brown GK, Otero LJ, LeGris M, Brown RM. Pyruvate dehydrogenase deficiency. *J Med Genet*. 1994;31(11):875–879.
13. Wexler ID, Hemalatha SG, McConnell J, et al. Outcome of pyruvate dehydrogenase deficiency treated with ketogenic diets. Studies in patients with identical mutations. *Neurology*. 1997;49(6):1655–1661.
14. Weber TA, Antognetti MR, Stacpoole PW. Caveats when considering ketogenic diets for the treatment of pyruvate dehydrogenase complex deficiency. *J Pediatr*. 2001;138(3):390–395.
15. Robinson BH, MacKay N, Chun K, Ling M. Disorders of pyruvate carboxylase and the pyruvate dehydrogenase complex. *J Inherit Metab Dis*. 1996;19(4):452–462.
16. Duran M, Wadman SK. Thiamine-responsive inborn errors of metabolism. *J Inherit Metab Dis*. 1985;8(suppl 1):70–75.
17. Stacpoole PW, Kurtz TL, Han Z, Langaee T. Role of dichloroacetate in the treatment of genetic mitochondrial diseases. *Adv Drug Deliv Rev*. 2008;60(13–14):1478–1487.
18. Kaufmann P, Engelstad K, Wei Y, et al. Dichloroacetate causes toxic neuropathy in MELAS: a randomized, controlled clinical trial. *Neurology*. 2006;66(3):324–330.
19. Marin-Valencia I, Roe CR, Pascual JM. Pyruvate carboxylase deficiency: mechanisms, mimics and anaplerosis. *Mol Genet Metab*. 2010;101(1):9–17.
20. Walker ME, Baker E, Wallace JC, Sutherland GR. Assignment of the human pyruvate carboxylase gene (PC) to 11q13.4 by fluorescence in situ hybridisation. *Cytogenet Cell Genet*. 1995;69(3–4):187–189.
21. Garcia-Cazorla A, Rabier D, Touati G, et al. Pyruvate carboxylase deficiency: metabolic characteristics and new neurological aspects. *Ann Neurol*. 2006;59(1):121–127.
22. Brun N, Robitaille Y, Grignon A, Robinson BH, Mitchell GA, Lambert M. Pyruvate carboxylase deficiency: prenatal onset of ischemia-like brain lesions in two sibs with the acute neonatal form. *Am J Med Genet*. 1999;84(2):94–101.
23. Pineda M, Campistol J, Vilaseca MA, et al. An atypical French form of pyruvate carboxylase deficiency. *Brain Dev*. 1995;17(4):276–279.
24. van der Knaap MS, Jakobs C, Valk J. Magnetic resonance imaging in lactic acidosis. *J Inherit Metab Dis*. 1996;19(4):535–547.
25. Wexler ID, Kerr DS, Du Y, et al. Molecular characterization of pyruvate carboxylase deficiency in two consanguineous families. *Pediatr Res*. 1998;43(5):579–584.
26. Higgins JJ, Glasgow AM, Lusk M, Kerr DS. MRI, clinical, and biochemical features of partial pyruvate carboxylase deficiency. *J Child Neurol*. 1994;9(4):436–439.
27. Schiff M, Levrat V, Acquaviva C, Vianey-Saban C, Rolland MO, Guffon N. A case of pyruvate carboxylase deficiency with atypical clinical and neuroradiological presentation. *Mol Genet Metab*. 2006;87(2):175–177.
28. Roe CR, Mochel F. Anaplerotic diet therapy in inherited metabolic disease: therapeutic potential. *J Inherit Metab Dis*. 2006;29(2–3):332–340.
29. Nyhan WL, Khanna A, Barshop BA, et al. Pyruvate carboxylase deficiency–insights from liver transplantation. *Mol Genet Metab*. 2002;77(1–2):143–149.
30. Grafakou O, Oexle K, van den Heuvel L, et al. Leigh syndrome due to compound heterozygosity of dihydrolipoamide dehydrogenase gene mutations. Description of the first E3 splice site mutation. *Eur J Pediatr*. 2003;162(10):714–718.
31. Sakaguchi Y, Yoshino M, Aramaki S, et al. Dihydrolipoyl dehydrogenase deficiency: a therapeutic trial with branched-chain amino acid restriction. *Eur J Pediatr*. 1986;145(4):271–274.
32. Patel MS, Harris RA. Mammalian alpha-keto acid dehydrogenase complexes: gene regulation and genetic defects. *FASEB J*. 1995;9(12):1164–1172.
33. Rutter J, Winge DR, Schiffman JD. Succinate dehydrogenase - Assembly, regulation and role in human disease. *Mitochondrion*. 2010;10(4):393–401.
34. Kerrigan JF, Aleck KA, Tarby TJ, Bird CR, Heidenreich RA. Fumaric aciduria: clinical and imaging features. *Ann Neurol*. 2000;47(5):583–588.
35. Allegri G, Fernandes MJ, Scalco FB, et al. Fumaric aciduria: an overview and the first Brazilian case report. *J Inherit Metab Dis*. 2010;33(4):411–419.
36. Bakker HD, Scholte HR, Van den Bogert C, et al. Deficiency of the adenine nucleotide translocator in muscle of a patient with myopathy and lactic acidosis: a new mitochondrial defect. *Pediatr Res*. 1993;33(4 Pt 1):412–417.
37. Sharer JD. The adenine nucleotide translocase type 1 (ANT1): a new factor in mitochondrial disease. *IUBMB Life*. 2005;57(9):607–614.
38. Hayes DJ, Taylor DJ, Bore PJ, et al. An unusual metabolic myopathy: a malate-aspartate shuttle defect. *J Neurol Sci*. 1987;82(1–3):27–39.
39. Wibom R, Lasorsa FM, Tohonen V, et al. AGC1 deficiency associated with global cerebral hypomyelination. *N Engl J Med*. 2009;361(5):489–495.
40. Huizing M, Ruitenbeek W, Thinnes FP, et al. Deficiency of the voltage-dependent anion channel: a novel cause of mitochondriopathy. *Pediatr Res*. 1996;39(5):760–765.
41. Tein I. Carnitine transport: pathophysiology and metabolism of known molecular defects. *J Inherit Metab Dis*. 2003;26(2–3):147–169.
42. Chace DH, Pons R, Chiriboga CA, et al. Neonatal blood carnitine concentrations: normative data by electrospray tandem mass spectometry. *Pediatr Res*. 2003;53(5):823–829.

LYSOSOMAL DISORDERS

CHAPTER

26

Gaucher Disease: Neuronopathic Forms

Raphael Schiffmann

Baylor Research Institute, Dallas, TX, USA

INTRODUCTION

Gaucher disease is one of the most prevalent hereditary lipid storage disorders in humans. Patients are classified into three phenotypes depending on whether the central nervous system (CNS) is involved and on the age of onset of clinical manifestations. This autosomal recessive disorder is caused by mutations in the *GBA1* gene leading to insufficient activity of the hydrolase glucocerebrosidase. This results in accumulation of glucocerebroside in macrophages, and in some cases, in cells of the CNS. The systemic manifestations of the disease are hepatosplenomegaly, anemia, thrombocytopenia and destructive skeletal disease. The neurological manifestations consist of supranuclear gaze palsy, variable cognitive dysfunction, movement disorders and sometimes a progressive myoclonic encephalopathy. Glucocerebrosidase activity in peripheral white blood cells and identification of *GBA1* mutations confirm the diagnosis. Current research is focused on correcting the primary metabolic defect in the CNS and understanding the pathogenesis of the disease. The latter will likely lead to therapeutic approaches designed to counteract the downstream effects of glucocerebrosidase deficiency.

CLINICAL FEATURES

Historical Overview

Disease Identification

Neuronopathic Gaucher disease (NGD) is defined as the presence of neurological involvement in a patient with biochemically proven Gaucher disease, for which there is no explanation other than Gaucher disease.[1] Historically, Gaucher disease has been divided into non-neuronopathic (type 1), acute (type 2), and subacute or chronic (type 3) forms. However, it is increasingly recognized that these divisions are somewhat artificial and do not take into account the full disease spectrum. In the European consensus statement of 2001, the European Working Group Gaucher Disease (EWGGD) Task Force recommended dividing type 2 patients into two subgroups, 2A (milder) and 2B (more severe), based on the severity of symptoms.[2] Traditionally, type 3 patients have been subdivided into 3A, 3B, 3C, and the Norrbottnian variant. However, recently identified intermediate phenotypes suggest that this classification may not comprise all patients and, in fact, type 3C may be the only distinct group within this type.[3,4] The EWGGD Task Force recommended that all patients who do not have type 2 (acute) NGD should be classified as having type 3 (chronic) NGD.[2] More recent publications clearly indicate that the neuronopathic form should be thought of as representing a spectrum of clinical phenotypes, rather than a set of artificial subdivisions.[5]

Gene Identification

The gene for Gaucher disease (MIM 231000, 230900, 231000) was localized to 1q21 in 1983[6] and cloned and sequenced in the following 6 years.

Rosenberg's Molecular and Genetic Basis of Neurological and Psychiatric Disease
http://dx.doi.org/10.1016/B978-0-12-410529-4.00026-7

Early Treatments

Prior to the advent of enzyme replacement therapy (ERT) the main therapeutic approach was to remove the spleen (splenectomy). This procedure does normalize the hemoglobin level and the platelet counts in the short term but often leads to greater skeletal complications.[7] Other supportive therapies such as blood transfusions were used as well. Bone marrow transplantation was found to cure the systemic manifestations of the disease but my clinical experience has shown that it does not ameliorate or prevent neurological complications.[8]

Mode of Inheritance and Prevalence

Gaucher disease is an autosomal recessive disorder. Patients with Gaucher disease type 3 (also called chronic neuronopathic Gaucher disease) constitute about 5% of the population of Gaucher patients in Western countries. Estimated incidence is about 1:100,000. Unlike Gaucher disease type 1, which has a particularly high prevalence among Ashkenazi Jews, the neurological forms are panethnic. However, founder effects have been described in several parts of the world. The most striking example was seen in northern Sweden.[9] The contemporary Swedish index families are found in two geographically distinct clusters with the highest worldwide prevalence of Gaucher type 3 disease. Molecular studies show that the two clusters are compatible with a single founder, who arrived in northern Sweden in or before the 16th century.[9] Another founder effect has been reported in the Mappila Muslims of Kerala state, in Southern India.[10]

However, recent reports suggest that neuronopathic Gaucher disease predominates in countries like Japan, China, Korea and Egypt.[11–16] In the year 2000, less than 1% of the 1698 patients reported to the Gaucher Registry (the largest database of Gaucher disease patients worldwide) had Gaucher type 2 and less than 5% had Gaucher disease type 3.[17] This number likely reflects an ascertainment bias and geographical limitations. The difference in the phenotypic spectrum is explained in part by the absence of alleles that are associated with Gaucher disease type 1 (mostly N370S).

Natural History and Clinical Syndromes

Gaucher Disease Type 2 (Acute Neuronopathic)

Patients with Gaucher disease type 2 commonly present in early infancy with evidence of brainstem dysfunction (supranuclear gaze palsy, strabismus, and dysphagia), visceromegaly, cortical thumbs, dystonic retroflexion of the neck, failure to thrive, and cachexia.[18] The clinical course is a rapid and relentless deterioration, and the child eventually succumbs either to stridor leading to laryngeal obstruction and apnea, or to dyscoordinated swallowing that provokes aspiration. In the original description of 67 patients by Frederickson and Sloan, the average age at death was 8 months. Survival of a typical Gaucher type 2 beyond 2 years without intensive supportive therapy is rare.[19]

However, presentation is varied and some patients may appear alert and attentive, with relatively little spasticity, but with strabismus or paucity of facial movements. The earlier classification into two groups should be dropped.[1] As stated above, milder forms of long-lived Gaucher type 2 patients are in a continuum with severe Gaucher type 3 patients.[5]

LETHAL NEONATAL VARIANT

Congenital Gaucher disease leading to a form of "collodion baby" is associated with the virtual absence of residual GBA activity, an association that was first reported in 1988.[20] These newborns are dead at parturition or die within the first few days of life, at least partly due to the excessive water evaporation caused by the abnormal lipid composition of the epidermis.[21] There have been several reports of this variant in the literature, many associated with hydrops fetalis.[18] The phenotype appears to be analogous to that of the null-allele mouse, which was created by the targeted disruption of the *GBA1* gene.[22]

Gaucher Type 3 (Subacute, Chronic Neuronopathic Gaucher Disease)

Gaucher type 3 is even more heterogeneous than Gaucher type 2. The earlier subclassification into types A, B, C, and Norrbottnian is increasingly considered artificial. The most common clinical form consists of severe systemic involvement and supranuclear gaze palsy, with or without developmental delay, hearing impairment, and other brainstem deficits.[23] Before the availability of ERT, visceral disease accounted for most of the deaths in this group, primarily through lung disease, portal hypertension, and infections. In the vast majority of these patients, the neurological deficit remains stable in patients whose systemic disease is controlled through ERT.[24,25]

A much less common phenotype of Gaucher type 3 exhibits a relatively mild systemic burden of disease and progressive myoclonic encephalopathy, with seizures, dementia, and death.[26] However, there are patients with both severe systemic disease and supranuclear gaze palsy who ultimately develop progressive myoclonic encephalopathy.[25] One of the most consistent clinical presentations is heart valve and aortic calcification, supranuclear gaze palsy, mild hepatosplenomegaly, and bone disease.[27] This variant, sometimes referred to as the type 3C variant, was first described in 1995 by two different groups in Israel and Spain.[27,28] It is almost always associated with homozygosity for the D409H (1342G>C) mutation.[4]

Typical Specific Features

Saccadic Paresis and Other Ocular Features

The most consistent clinical feature is an abnormality of horizontal gaze, first described in the late 1970s.[29,30] This has been mistakenly referred to as oculomotor apraxia, but should more accurately be called supranuclear saccadic gaze palsy. The abnormality consists mainly of saccade (quick eye movement) slowing but also increased saccadic latency with occasional catch-up saccades. It is now widely accepted that this abnormality must be present to make a clinical diagnosis of neuronopathic Gaucher disease. However, the definition of neuronopathic Gaucher as described above does not make saccadic eye movement abnormality a requirement.[1] Supranuclear saccadic gaze palsy can be difficult to detect clinically, especially in early life, but is readily revealed as missed quick-phases during induced optokinetic and vestibular nystagmus.[31] The horizontal component of the saccade is usually involved first, sometimes asymmetrically, and more severely, while vertical movements may be affected later and are almost always less affected. Downward saccades are more affected than upward ones.[32] Smooth pursuit eye movements are typically normal until the end stages of the disease. In severely affected children, vertical movements may be involved from early infancy. These abnormalities can be an early diagnostic sign and are sometimes evident in patients as young as 3–4 months of age, and occasionally even in neonates. It may be the only neurological sign at presentation and, indeed, for many years to come or, conversely, it may develop later in life and even beyond the third decade of life. Older children learn to compensate for their poor saccades by a combination of synkinetic blinking, looping and/or head thrusting. Even so, supranuclear gaze palsy often results in significant functional disability. Children cannot look from side to side quickly, and this has significant implications for safety. Learning is also affected, as the children experience difficulties in reading books and, if the vertical component of the saccade is involved, in looking up to and back down from the blackboard. Adults may have difficulties using a computer or driving.

In patients with Gaucher type 2 and in other progressive or severe forms of Gaucher type 3, saccadic paresis progresses to nuclear paresis and, finally, to complete ocular paralysis. Patients with neuronopathic Gaucher disease type 2 and 3 may develop characteristic bilateral abducens (sixth nerve) palsy, presenting as a convergent squint.

Auditory Dysfunction

The other consistent feature in neuronopathic Gaucher disease is an abnormality of brainstem auditory-evoked potentials. Waveforms are poor and show progressive deterioration over time.[33] Apart from reflecting brainstem dysfunction, the significance of this finding is unclear, as most patients have normal peripheral hearing.

Myoclonus

As mentioned above, some Gaucher type 2 and type 3 patients develop a progressive encephalopathy, characterized by recurrent and sometimes continuous stimulus-sensitive myoclonus.[34] These myoclonic movements are multifocal, usually of neocortical origin, and highly resistant to medical therapy. Some patients have a myoclonic dystonia. Somatosensory-evoked potentials (SEP) in such patients show a greatly increased amplitude of the potential, a so-called "giant potential," which indicates a decreased cortical inhibitory input.[35,36] SEP amplitude is also elevated in Gaucher type 3 patients who do not have myoclonus.[37] In this group, there is an inverse correlation between the SEP amplitude and the patient's intelligence quotient. Therefore, it is likely that the development of overt myoclonus in Gaucher disease is the extreme manifestation of a general cortical pathogenic process that may reflect an intracortical inhibitory deficit.[38]

Seizures

Besides progressive myoclonic epilepsy, Gaucher type 3 patients on long-term ERT may develop partial complex seizures usually of temporal lobe origin that are relatively easy to control. The hippocampal pathological abnormalities found in these patients may explain this phenomenon.[39]

End of Life: Mechanisms and Comorbidities

There is no doubt that overall, patients with neuronopathic Gaucher disease have a shortened lifespan. Most patients with Gaucher type 2 die by age 2 years of progressive brainstem dysfunction, although placement of a tracheostomy can prolong survival. Patients with progressive myoclonic epilepsy die of their neurological disease from epilepsy or respiratory dysfunction. Gaucher disease type 3 patients may survive well into adulthood. However, recently, two patients in the US and one in France died suddenly in their sleep of unknown cause (unpublished data). Gaucher disease type 3 patients who receive ERT at adequate doses do not succumb to the hematologic and visceral complications of their disease.

MOLECULAR GENETICS

Normal Gene Product

The gene coding acid glucocerebrosidase is located on chromosome 1. A pseudogene that has maintained a high degree of homology is approximately 16 kb downstream from the active gene. A number of other active genes have been identified in the flanking regions. These include thrombospondin and metaxin. Liver/red cell pyruvate kinase is located only 71 kb downstream from the glucocerebrosidase gene.[40]

Abnormal Gene Product(s)

There are currently 371 mutations described in the GBA gene (http://www.hgmd.cf.ac.uk/ac/all.php) and of those 286 are missense or nonsense mutations. Others are frameshift mutations, a splicing mutation, deletions, gene fusions with the pseudogene, examples of gene conversions, and total deletions. The naming of the mutations according to the new standard mutation nomenclature that is based on the translation initiation site (the A of the ATG codon) is described in a recent online review of Gaucher disease.[18]

Genotype–Phenotype Correlations

The most common mutation is N370S amino acid substitution. It is empirically regarded as associated exclusively with non-neuronopathic disease, and the clinical manifestations can be relatively mild, particularly in homozygotes.[41] Patients who are either homozygous or compound-heterozygous for certain mutations, such as L444P and D409H, are at high risk of developing CNS involvement. Homozygosity for L444P is the most common neuronopathic Gaucher disease genotype.[42] In Japan, L444P accounts for 41% of all disease alleles.[16] Sixteen consecutive patients with progressive myoclonic encephalopathy had a total of 14 different genotypes.[26] However, mutations V394L, N188S, and G377S are often associated with progressive myoclonic encephalopathy, although each has been seen with other forms of GD.[26] On the other hand, mutation R463C and homozygosity for L444P rarely are associated with myoclonic seizures. The reason for this phenotypical selectivity is unknown.

PATHOPHYSIOLOGY

In Gaucher patients with sufficiently reduced glucocerebrosidase activity,[43] glucosylceramide (Figure 26.1) levels increase above a certain threshold in neurons and possibly also in astrocytes. Once this threshold is crossed, dysfunction and damage develop in susceptible neurons (Figures 26.2 and 26.3).[44] Primary accumulation of glucosylceramide in macrophages (Gaucher cells) does not play a significant role in the neuronopathic process.[39,45,46]

Although the precise mechanism of CNS damage has not yet been elucidated, defective calcium homeostasis may be a mechanism responsible for cerebral pathophysiology in the most severe form of NGD.[45] Cultured neurons were more sensitive to glutamate-induced neuronal toxicity and to toxicity induced via various other cytotoxic agents. Therefore, a major role of glucocerebrosidase in the brain appears to be to keep intracellular glycosphingolipid concentrations below toxic levels.

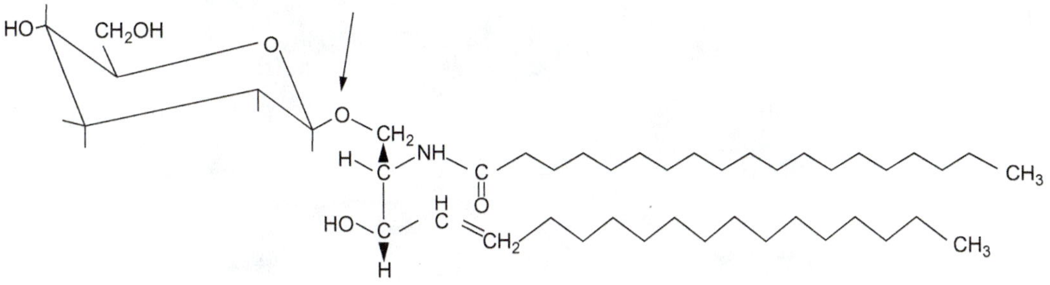

FIGURE 26.1 **The structure of glucosylceramide.** The arrow indicates the point of action of glucocerebrosidase.

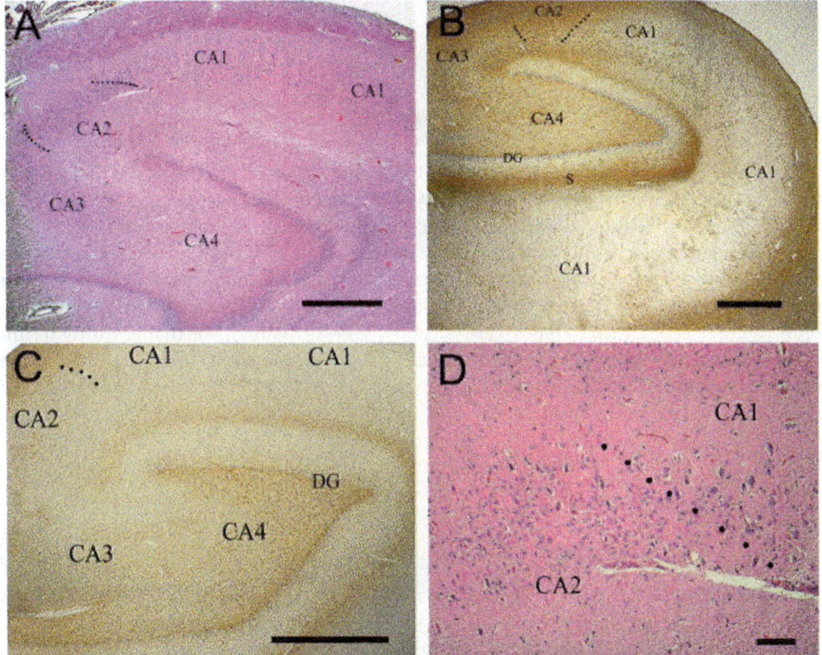

FIGURE 26.2 **Hippocampal CA2–CA4 pyramidal cell neuronal loss and astrogliosis in Gaucher disease. CA1 region is spared.** (**A**) Type 2 Gaucher disease: Hippocampal pyramidal neuron cell loss and astrogliosis are most severe in CA2 (region bracketed between rows of dots), moderate to severe in CA3 and moderate in CA4. The CA1 region of the hippocampus has minimal background gliosis and is largely spared from significant pathology (hematoxylin and eosin [H&E], 40× original magnification). (**B**) Type 1 Gaucher disease: Hippocampal pyramidal neuron cell loss is undetectable and astrogliosis may not be readily detectable, except with glial fibrillary acidic protein staining (GFAP). Hippocampal CA2 region is strongly and densely GFAP immunoreactive. CA3 and CA4 also have moderate GFAP immunoreactivity, whereas the CA1 region only has background scattered perivascular immunoreactivity. "DG" denotes the dentate gyrus. "S" denotes the stratum lacunosum, a region that almost always stains highly GFAP immunoreactive, whether in normal or disease states. (GFAP immunoperoxidase, 40× original magnification). (**C**) Type 3 Gaucher disease: This patient's symptoms included developmental delay, supranuclear gaze palsy and mild dysphagia, but no myoclonus or seizures. The patient died shortly after bone marrow transplantation. CA4 region has mild gliosis and GFAP immunoreactivity that extends into CA3 and proximal CA2. CA1 does not have an appreciable increase in GFAP immunoreactivity. No neuronal loss was visible (GFAP immunoperoxidase, 40× original magnification). (**D**) Type 2 Gaucher disease: Close-up of hippocampal CA2–CA1 interface. The few remaining hippocampal pyramidal cell CA2 neurons are basophilic and shrunken. Astrogliosis with eosinophilic astrocytes and glial processes is prominent. The CA1 region at the interface has mildly affected pyramidal cell neurons (slightly basophilic) and mild astrogliosis. At 150 μm from the CA2–CA1 interface, the CA1 region has no significant changes above background (not shown). (H&E, 100× original magnification). Scale bars: A, B, and C, 1 mm; D, 100 μm. Reproduced from[39].

Observations in Human Subjects

The study of the pathogenesis of neuronopathic Gaucher disease is mostly based on cellular and animal models. The absence of beneficial effect of hematopoietic stem cell transplantation on the course of the neurological manifestations of Gaucher disease suggests that primary neuronal abnormality underlies the disease in the brain. Historically, the most commonly described neuropathological abnormalities have been perivascular accumulation of lipid-laden macrophages (Gaucher cells), neuronal loss and astrogliosis.[47–49] More detailed recent autopsy studies

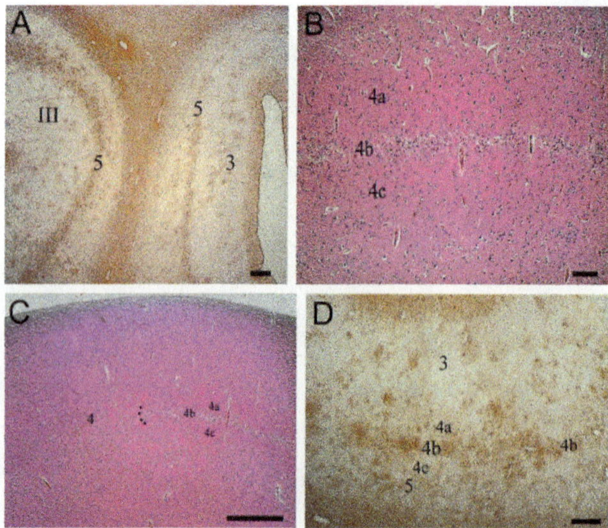

FIGURE 26.3 **Cortical layer-specific neuropathology.** (**A**) Type 1 Gaucher disease, cerebral cortex in parietal lobe: laminar astrogliosis involves cortical layer 5 most consistently. Laminar astrogliosis accentuates the background perivascular astrogliosis along cortical layer 3, but may be diffuse and less organized in other areas (III) (glial fibrillary acidic protein staining [GFAP], 40× original magnification). (**B**) Type 2 Gaucher disease, calcarine cortex, stria of Gennari: a precise, demarcated line of neuronal loss and astrogliosis involves layer 4b, but spares layer 4a and 4c. Pathology involving all of layer 4 in the calcarine cortex was not observed (hematoxylin and eosin [H&E], 40× original magnification). (**C**) Type 2 Gaucher disease, calcarine cortex, termination of V1, stria of Gennari, interface where 4a, 4b, and 4c merge (curved row of dots) into a single layer 4: layer 4b (which corresponds to the stria of Gennari) has severe neuronal loss and astrogliosis that abruptly terminates at the point where layer 4b ends (H&E, 20× original magnification). (**D**) Type 1 Gaucher disease (Patient 4), calcarine cortex, V1, stria of Gennari: in type 1 GD, neuronal cell loss is not identified. A laminar region of astrogliosis along layer 4b, and to a lesser extent, cortical layers 3 and 5 is present (GFAP, 40× original magnification). Scale bars: A, 250 μm; B, 100 μm; C, 1 mm; D, 150 μm. Reproduced from[39].

showed that a characteristic pattern, involving cerebral cortical layers 3 and 5, hippocampal CA2–4, and layer 4b, was involved in all Gaucher disease patients (Figures 26.2 and 26.3).[39] Neuronal loss predominated in both Gaucher type 2 and type 3 patients with progressive myoclonic encephalopathy, whereas patients classified as Gaucher type 1 only had astrogliosis. Adjacent brain regions, including hippocampal CA1 and calcarine cortex lamina 4a and 4c were spared of pathology, highlighting the specificity of the vulnerability of selected neurons (Figures 26.2 and 26.3).[39] Unlike neuronal abnormalities reflecting glucosylceramide storage found in the mouse,[50] typical lysosomal storage is seen very rarely in patients' brains.[39]

Animal Models

Generation of the first genetic mouse model of Gaucher disease was based on production of a null glucocerebrosidase allele (the $gba^{-/-}$ mouse). However, this mouse died soon after birth due to a skin permeability disorder before significant CNS damage occurred and therefore was not useful for studying the long-term effects of glucosylceramide accumulation.[22] Another Gaucher disease mouse model, the L444P mouse, which carries a mutation most commonly leading to neuronopathic Gaucher in humans, did not accumulate significant glucosylceramide levels and did not display CNS pathology.[51,52] Mouse models of neuronopathic Gaucher disease have been particularly useful for elucidating the cascade of events between the primary defect and cellular death. The most useful model has been the $Gba^{flox/flox}$; nestin-Cre mouse.[46] The bone marrow derived cells of this mouse model have normal glucocerebrosidase activity, thus producing a mouse in which glucocerebrosidase deficiency is limited to neurons and astrocytes only, while macrophages and microglia have wild-type enzyme activity. Interestingly, this mouse model survives 3 weeks while a similar model in which the macrophages are enzyme deficient as well survives only two weeks.[46] This confirms the clinical impression that normal glucocerebrosidase activity in macrophages does not appreciably modify the course of neurological Gaucher disease. A nongenetic biochemical manner of obtaining a Gaucher model has been the use of conduritol-B-epoxide (CBE), an irreversible inhibitor of glucocerebrosidase, in cultured cells and in wild-type mice. Various concentrations of CBE lead to correspondingly different residual enzyme activity levels.[53,54]

Current Research

Inflammation and Cellular Death

Cathepsin D elevation in microglia using the nestin-$^{flox/flox}$ neuronopathic Gaucher mouse were found in areas where neuronal loss, astrogliosis, and microgliosis were observed, such as in layer 5 of the cerebral cortex, the lateral globus pallidus, various thalamic nuclei, and other brain regions known to be affected in neuronopathic Gaucher disease.[55] In the same animal model, levels of mRNA expression of interleukin-1 beta (IL-1β), tumor necrosis factor alpha (TNF-α), TNF-α receptor, macrophage colony-stimulating factor, and transforming growth factor beta were elevated by up to -30-fold.[56] The chemokines CCL2, CCL3, and CCL5 were very much elevated and all in a time course that paralleled the disease severity.[56] Blood–brain barrier and excess oxidative stress were also found. Based on these results, the authors suggested that once a critical threshold of glucosylceramide storage is reached in neurons, a signaling cascade is triggered that activates microglia, which in turn release inflammatory cytokines that amplify the inflammatory response, contributing to neuronal death.[51] Neuroinflammation and neuronal loss in susceptible brain areas were detected prior to the noticeable signs linking these to early pathogenic stages.[56]

Using CBE-injected wild-type mice, Vitner et al. showed that neurons and microphage/microglia in a model of severe neuronopathic Gaucher disease die of necroptosis by direct involvement of the Rip kinases pathway.[53] This group also found that the same pathway is involved in the acute neuropathological changes in Krabbe disease and confirmed the findings in the brain autopsy of a patient with Gaucher type 3. Ripk3 elevation in microglia before neuronal loss, together with the improvement in behavioral signs and the attenuation of the pathological injury in peripheral organs in Ripk3$^{-/-}$ mice with Gaucher disease, support the notion that Ripk3 is not only a key activator of necrotic cell death, but also orchestrates inflammation independent of necrosis.[53] Therefore, inhibition of the RIP3 pathway is a potential novel therapeutic approach.

The deacylated form of glucosylceramide, glucosylsphingosine, is often invoked as an offending metabolite in neuronopathic Gaucher disease.[57] However, more recent findings indicate that its levels in brain microsomes appear far too low for it to play a major role in the Ca^{2+} release in Gaucher disease,[58] although it does induce Ca^{2+} release, but via a different mechanism than glucocerebroside.[58] Interestingly, glucosylsphingosine was not elevated in the brain of an adult patient with Gaucher type 3 disease, parkinsonism and dementia.[59] Moreover, unlike for glucosylceramide, there was no correlation between the level of glucosylsphingosine and the pathological changes in susceptible brain regions in the authentic Gaucher mouse model.[50]

DIFFERENTIAL DIAGNOSIS

The two main clinical aspects of neuronopathic Gaucher disease should guide the differential diagnosis. Other disorders that present with liver or spleen involvement are one important group. Besides acquired malignancy or infections such as Epstein–Barr virus-associated infectious mononucleosis, other lysosomal storage disorders such as Niemann–Pick A, B, or C often present with organomegaly, although the liver often is relatively more enlarged than the spleen. The neurological differential diagnosis should include disorders associated with supranuclear gaze palsy such as Niemann–Pick in all its neurologic variants (especially C, in which the vertical movements are predominantly affected), ataxia-telangiectasia, cerebellar hypoplasia such as Joubert syndrome, Huntington disease, as well as acquired diseases.[60] Diseases that are associated with progressive myoclonic epilepsy should be considered in the differential diagnosis of the myoclonic variant of Gaucher disease.[61]

DIAGNOSTIC TESTING

The diagnosis of neuronopathic Gaucher disease is purely clinical and depends on finding neurological abnormalities that cannot be explained by another etiology.[1] The diagnosis of Gaucher disease depends on finding a low glucocerebrosidase activity in peripheral blood cells or other patient cells or tissues. Complete sequencing of the *GBA1* gene is highly recommended in all patients with Gaucher disease.[62] However, neuronopathic Gaucher disease cannot be diagnosed based on a particular level of glucocerebrosidase activity or a particular genotype. A diagnosis of Gaucher type 3 is often overlooked in early stages of the disease or in mild forms. Patients should be observed for the presence of head jerks while reading or for an "eye lag" when turning around when walking. Clinical features, such as early onset of disease, aggressive systemic disease, falling or low IQ scores in a patient with a "high-risk" genotype can also suggest neuronopathic Gaucher disease.[1]

A particular diagnostic challenge is the differentiation between type 2 and type 3 NGD.[5] The striking difference in prognosis between the two neuronopathic forms of Gaucher disease, and the resultant implications for treatment, mean that the clinical distinction at presentation is critical. Severe forms of Gaucher type 2 are relatively easy to identify, but milder cases can be diagnosed with certainty only retrospectively when death occurs before age 4 years. The skin of type 2 patients has an increased ratio of epidermal glucosylceramide to ceramide, as well as extensive ultrastructural abnormalities. It is suggested that these changes might form a reliable basis for discrimination between Gaucher type 2 and type 3.[63]

Clinical and Ancillary Testing: Imaging Techniques

Results of brain imaging tests, such as computed tomography (CT) and magnetic resonance imaging (MRI) are normal except in patients with progressive myoclonic encephalopathy, where diffuse cerebral atrophy is seen in relatively advanced stages. Cerebellar calcification and involvement of the basal ganglia on CT and other abnormalities are rarely found. One small diffusion tensor imaging study in four neuronopathic Gaucher patients and three type 1 patients suggested microstructural abnormalities in both types.[64] Unfortunately, larger prospective longitudinal imaging studies involving cerebral diffusion tensor imaging or volumetric analysis have not been performed to date in neuronopathic Gaucher disease.

EEG

Electroencephalography (EEG) may be normal or show background rhythm slowing or isolated focal epileptogenic activity. In patients with progressive myoclonic encephalopathy, the EEG is more disorganized and associated with multifocal spike and wave activity.[34,65]

Evoked Potentials

Brainstem auditory-evoked potentials are frequently abnormal and often get more so over time even in patients with normal hearing. Somatosensory-evoked potential in patients with progressive myoclonic encephalopathy and other forms of neuronopathic Gaucher disease shows increased cortical amplitude, the so-called "giant potential," which indicates a decreased cortical inhibitory input.[37]

Neuropsychometric Assessment

Cognitive deficits, characterized by visual–spatial dysfunction, are common in Gaucher type 3 patients.[66] Interestingly, patients with normal IQ performed poorly on the Purdue Pegboard test, suggesting selective deficit in eye–hand coordination in Gaucher disease type 3.[32,67]

MANAGEMENT

Standard of Care

The standard of care for neuronopathic Gaucher disease has been summarized in a consensus statement a few years ago and is still valid.[1] Enzyme replacement therapy (ERT) in patients with Gaucher disease type 2 does not result in significant clinical benefit and therefore families should be reassured that every potential therapy has been tried. For patients with Gaucher type 3 disease ERT is the main therapy. However, it has been shown to have no significant effect on the neurological phenotype.[24,25,66] ERT at a dose no lower than 60 IU per kg of body weight is indicated in most cases and especially with those known to have an underlying severe systemic disease, e.g., patients who are homozygous for the L444P mutation.[1] Enzyme doses should however be adjusted according to the clinical response of the individual patient. In addition, standard monitoring is useful, especially when it includes general and neurological examination with particular emphasis on eye movements, hearing test, neurophysiological tests such as EEG and brainstem auditory-evoked potential, and neuropsychological testing.[1]

As mentioned above, bone marrow transplantation (also referred to as hematopoietic stem cell transplantation) with complete engraftment would cure the non-neurological (hematological) manifestations of the disease. However, clinical experience with a number of patients with type 3 Gaucher disease has shown that such an intervention would not appreciably modify the course of the neurological disease and would not prevent neurological deterioration.

Failed Therapies and Barriers to Treatment Development

The biggest obstacle to the delivery of effective therapy for neuronopathic Gaucher disease is the blood–brain barrier. Miglustat, an inhibitor of ceramide glucosyltransferase, a central step in the synthetic process of glycosphingolipids that is able to cross the blood–brain barrier, was tried in a randomized controlled trial but was not shown to be effective.[67] More specific glucosyltransferase inhibitors that cross the blood–brain barrier are being developed and will be tested in the coming years. The development and testing of novel therapies should take into account the heterogeneous natural progression of the disease and the degree to which the neurological abnormalities are reversible or only preventable.

Therapies Under Investigation

In addition to substrate synthesis inhibition,[68] pharmacological chaperones are being developed. These are small molecules that are active site competitive inhibitors that cross the blood–brain barrier, promote normal folding and trafficking to the lysosome of the mutated glucocerebrosidase; they constitute an attractive option. Experiments in the mouse model harboring the most common neuronopathic Gaucher disease mutation (L444P) are promising, but clinical trials have not been initiated thus far.[69] Ambroxol, a drug used to treat airway mucus hypersecretion and hyaline membrane disease in newborns, was identified as a potential pharmacological chaperone.[70,71] This medication has a wide distribution that includes the brain and therefore may be used in future neuronopathic Gaucher disease treatment trials. Modification of glucocerebrosidase that will allow it to cross the intact blood–brain barrier may be an optimal solution. Finally, treatment approaches based on the understanding of neuronal mechanisms by which glucosylceramide causes neuronal dysfunction and death are being developed based on recent findings described above.[53]

References

1. Vellodi A, Tylki-Szymanska A, Davies EH, et al. Management of neuronopathic Gaucher disease: revised recommendations. *J Inherit Metab Dis*. 2009;32(5):660–664.
2. Vellodi A, Bembi B, De Villemeur TB, et al. Management of neuronopathic Gaucher disease: a European consensus. *J Inherit Metab Dis*. 2001;24(3):319–327.
3. Michelakakis H, Skardoutsou A, Mathioudakis J, et al. Early-onset severe neurological involvement and D409H homozygosity in Gaucher disease: outcome of enzyme replacement therapy. *Blood Cells Mol Dis*. 2002;28(1):1–4.
4. Bohlega S, Kambouris M, Shahid M, Al Homsi M, Al Sous W. Gaucher disease with oculomotor apraxia and cardiovascular calcification (Gaucher type IIIC). *Neurology*. 2000;54(1):261–263.
5. Goker-Alpan O, Schiffmann R, Park JK, Stubblefield BK, Tayebi N, Sidransky E. Phenotypic continuum in neuronopathic Gaucher disease: an intermediate phenotype between type 2 and type 3. *J Pediatr*. 2003;143(2):273–276.
6. Barneveld RA, Keijzer W, Tegelaers FP, et al. Assignment of the gene coding for human beta-glucocerebrosidase to the region q21-q31 of chromosome 1 using monoclonal antibodies. *Hum Genet*. 1983;64(3):227–231.
7. Zimran A, Elstein D, Schiffmann R, et al. Outcome of partial splenectomy for type I Gaucher disease. *J Pediatr*. 1995;126(4):596–597.
8. Rappeport JM, Ginns EI. Bone-marrow transplantation in severe Gaucher's disease. *N Engl J Med*. 1984;311(2):84–88.
9. Dreborg S, Erikson A, Hagberg B. Gaucher disease–Norrbottnian type. I. General clinical description. *Eur J Pediatr*. 1980;133(2):107–118.
10. Feroze M, Arvindan KP, Jose L. Gaucher's disease among Mappila Muslims of Malabar. *Indian J Pathol Microbiol*. 1994;37(3):307–311.
11. Jeong SY, Park SJ, Kim HJ. Clinical and genetic characteristics of Korean patients with Gaucher disease. *Blood Cells Mol Dis*. 2011;46(1):11–14.
12. El-Morsy Z, Khashaba MT, Soliman Oel S, Yahia S, El-Hady DA. Glucosidase acid beta gene mutations in Egyptian children with Gaucher disease and relation to disease phenotypes. *World J Pediatr*. 2011;7(4):326–330.
13. Tajima A, Yokoi T, Ariga M, et al. Clinical and genetic study of Japanese patients with type 3 Gaucher disease. *Mol Genet Metab*. 2009;97(4):272–277.
14. Choy FY, Zhang W, Shi HP, et al. Gaucher disease among Chinese patients: review on genotype/phenotype correlation from 29 patients and identification of novel and rare alleles. *Blood Cells Mol Dis*. 2007;38(3):287–293.
15. Wan L, Hsu CM, Tsai CH, Lee CC, Hwu WL, Tsai FJ. Mutation analysis of Gaucher disease patients in Taiwan: high prevalence of the RecNciI and L444P mutations. *Blood Cells Mol Dis*. 2006;36(3):422–425.
16. Ida H, Rennert OM, Iwasawa K, Kobayashi M, Eto Y. Clinical and genetic studies of Japanese homozygotes for the Gaucher disease L444P mutation. *Hum Genet*. 1999;105(1–2):120–126.
17. Charrow J, Andersson HC, Kaplan P, et al. The Gaucher registry: demographics and disease characteristics of 1698 patients with Gaucher disease. *Arch Intern Med*. 2000;160(18):2835–2843.
18. Pastores GM, Hughes DA. Gaucher disease. In: Pagon RA, Adam MP, Bird TD, Dolan CR, Fong CT, Stephens K, eds. *GeneReviews*. 1993 Seattle (WA).
19. Frederickson DS, Sloan HR. Glucosylceramide lipidoses—Gaucher's disease. In: *The Metabolic Basis of Inherited Disease*. 1972:730–759.
20. Lui K, Commens C, Choong R, Jaworski R. Collodion babies with Gaucher's disease. *Arch Dis Child*. 1988;63(7):854–856.

21. Stone DL, Carey WF, Christodoulou J, et al. Type 2 Gaucher disease: the collodion baby phenotype revisited. *Arch Dis Child Fetal Neonatal Ed.* 2000;82(2):F163–F166.
22. Tybulewicz VL, Tremblay ML, LaMarca ME, et al. Animal model of Gaucher's disease from targeted disruption of the mouse glucocerebrosidase gene. *Nature.* 1992;357(6377):407–410.
23. Patterson MC, Horowitz M, Abel RB, et al. Isolated horizontal supranuclear gaze palsy as a marker of severe systemic involvement in Gaucher's disease. *Neurology.* 1993;43(10):1993–1997.
24. Schiffmann R, Heyes MP, Aerts JM, et al. Prospective study of neurological responses to treatment with macrophage-targeted glucocerebrosidase in patients with type 3 Gaucher's disease. *Ann Neurol.* 1997;42(4):613–621.
25. Altarescu G, Hill S, Wiggs E, et al. The efficacy of enzyme replacement therapy in patients with chronic neuronopathic Gaucher's disease. *J Pediatr.* 2001;138(4):539–547.
26. Park JK, Orvisky E, Tayebi N, et al. Myoclonic epilepsy in Gaucher disease: genotype-phenotype insights from a rare patient subgroup. *Pediatr Res.* 2003;53(3):387–395.
27. Abrahamov A, Elstein D, Gross-Tsur V, et al. Gaucher's disease variant characterised by progressive calcification of heart valves and unique genotype. *Lancet.* 1995;346(8981):1000–1003.
28. Chabas A, Cormand B, Balcells S, et al. Neuronopathic and non-neuronopathic presentation of Gaucher disease in patients with the third most common mutation (D409H) in Spain. *J Inherit Metab Dis.* 1996;19(6):798–800.
29. Tripp JH, Lake BD, Young E, Ngu J, Brett EM. Juvenile Gaucher's disease with horizontal gaze palsy in three siblings. *J Neurol Neurosurg Psychiatry.* 1977;40(5):470–478.
30. Sanders MD, Lake BD. Ocular movements in lipid storage disease. Reports of juvenile Gaucher disease and the ophthalmoplegic lipidosis. *Birth Defects Orig Artic Ser.* 1976;12(3):535–542.
31. Harris CM, Taylor DS, Vellodi A. Ocular motor abnormalities in Gaucher disease. *Neuropediatrics.* 1999;30(6):289–293.
32. Benko W, Ries M, Wiggs EA, Brady RO, Schiffmann R, Fitzgibbon EJ. The saccadic and neurological deficits in type 3 Gaucher disease. *PLoS One.* 2011;6(7):e22410.
33. Campbell PE, Harris CM, Vellodi A. Deterioration of the auditory brainstem response in children with type 3 Gaucher disease. *Neurology.* 2004;63(2):385–387.
34. Frei KP, Schiffmann R. Myoclonus in Gaucher disease. *Adv Neurol.* 2002;89:41–48.
35. Rothwell JC, Obeso JA, Marsden CD. On the significance of giant somatosensory evoked potentials in cortical myoclonus. *J Neurol Neurosurg Psychiatry.* 1984;47(1):33–42.
36. Manganotti P, Tamburin S, Zanette G, Fiaschi A. Hyperexcitable cortical responses in progressive myoclonic epilepsy: a TMS study. *Neurology.* 2001;57(10):1793–1799.
37. Garvey MA, Toro C, Goldstein S, et al. Somatosensory evoked potentials as a marker of disease burden in type 3 Gaucher disease. *Neurology.* 2001;56(3):391–394.
38. Hanajima R, Okabe S, Terao Y, et al. Difference in intracortical inhibition of the motor cortex between cortical myoclonus and focal hand dystonia. *Clin Neurophysiol.* 2008;119(6):1400–1407.
39. Wong K, Sidransky E, Verma A, et al. Neuropathology provides clues to the pathophysiology of Gaucher disease. *Mol Genet Metab.* 2004;82(3):192–207.
40. Grabowski GA, Petsko GA, Kolodny EH. Gaucher disease. In: Scriver AL, Beaudet W, Sly S, Valle D, eds. *The Metabolic and Molecular Bases of Inherited Disease.* 9th ed. New York: McGraw-Hill; 2006.
41. Beutler E, Gelbart T. Hematologically important mutations: Gaucher disease. *Blood Cells Mol Dis.* 1997;23(1):2–7.
42. Tylki-Szymanska A, Vellodi A, El-Beshlawy A, Cole JA, Kolodny E. Neuronopathic Gaucher disease: demographic and clinical features of 131 patients enrolled in the International Collaborative Gaucher Group Neurological Outcomes Subregistry. *J Inherit Metab Dis.* 2010;33(4):339–346.
43. Brady RO, Kanfer JN, Bradley RM, Shapiro D. Demonstration of a deficiency of glucocerebroside-cleaving enzyme in Gaucher's disease. *J Clin Invest.* 1966;45(7):1112–1115.
44. Farfel-Becker T, Vitner EB, Pressey SN, Eilam R, Cooper JD, Futerman AH. Spatial and temporal correlation between neuron loss and neuroinflammation in a mouse model of neuronopathic Gaucher disease. *Hum Mol Genet.* 2011;20(7):1375–1386.
45. Pelled D, Trajkovic-Bodennec S, Lloyd-Evans E, Sidransky E, Schiffmann R, Futerman AH. Enhanced calcium release in the acute neuronopathic form of Gaucher disease. *Neurobiol Dis.* 2005;18(1):83–88.
46. Enquist IB, Lo Bianco C, Ooka A, et al. Murine models of acute neuronopathic Gaucher disease. *Proc Natl Acad Sci U S A.* 2007;104(44):17483–17488.
47. Conradi NG, Kalimo H, Sourander P. Reactions of vessel walls and brain parenchyma to the accumulation of Gaucher cells in the Norrbottnian type (type III) of Gaucher disease. *Acta Neuropathol (Berl).* 1988;75(4):385–390.
48. Conradi NG, Sourander P, Nilsson O, Svennerholm L, Erikson A. Neuropathology of the Norrbottnian type of Gaucher disease. Morphological and biochemical studies. *Acta Neuropathol (Berl).* 1984;65(2):99–109.
49. Kaye EM, Ullman MD, Wilson ER, Barranger JA. Type 2 and type 3 Gaucher disease: a morphological and biochemical study. *Ann Neurol.* 1986;20(2):223–230.
50. Farfel-Becker T, Vitner EB, Kelly SL, et al. Neuronal accumulation of glucosylceramide in a mouse model of neuronopathic Gaucher disease leads to neurodegeneration. *Hum Mol Genet.* 2014;23(4):843–854.
51. Vitner EB, Futerman AH. Neuronal forms of Gaucher disease. *Handb Exp Pharmacol.* 2013;216:405–419.
52. Mizukami H, Mi Y, Wada R, et al. Systemic inflammation in glucocerebrosidase-deficient mice with minimal glucosylceramide storage. *J Clin Invest.* 2002;109(9):1215–1221.
53. Vitner EB, Salomon R, Farfel-Becker T, et al. RIPK3 as a potential therapeutic target for Gaucher's disease. *Nat Med.* 2014;20(2):204–208.
54. Pelled D, Shogomori H, Futerman AH. The increased sensitivity of neurons with elevated glucocerebroside to neurotoxic agents can be reversed by imiglucerase. *J Inherit Metab Dis.* 2000;23(2):175–184.
55. Vitner EB, Dekel H, Zigdon H, et al. Altered expression and distribution of cathepsins in neuronopathic forms of Gaucher disease and in other sphingolipidoses. *Hum Mol Genet.* 2010;19(18):3583–3590.

56. Vitner EB, Farfel-Becker T, Eilam R, Biton I, Futerman AH. Contribution of brain inflammation to neuronal cell death in neuronopathic forms of Gaucher's disease. *Brain*. 2012;135(Pt 6):1724–1735.
57. Schueler UH, Kolter T, Kaneski CR, et al. Toxicity of glucosylsphingosine (glucopsychosine) to cultured neuronal cells: a model system for assessing neuronal damage in Gaucher disease type 2 and 3. *Neurobiol Dis*. 2003;14(3):595–601.
58. Lloyd-Evans E, Pelled D, Riebeling C, et al. Glucosylceramide and glucosylsphingosine modulate calcium mobilization from brain microsomes via different mechanisms. *J Biol Chem*. 2003;278(26):23594–23599.
59. Tayebi N, Walker J, Stubblefield B, et al. Gaucher disease with parkinsonian manifestations: does glucocerebrosidase deficiency contribute to a vulnerability to parkinsonism? *Mol Genet Metab*. 2003;79(2):104–109.
60. Leigh RJ, Zee DS. *The Neurology of Eye Movements*. 2nd ed. Philadelphia: F. A. Davis Company; 1991.
61. Franceschetti S, Michelucci R, Canafoglia L, et al. Progressive myoclonic epilepsies: definitive and still undetermined causes. *Neurology*. 2014;82(5):405–411.
62. Hruska KS, LaMarca ME, Scott CR, Sidransky E. Gaucher disease: mutation and polymorphism spectrum in the glucocerebrosidase gene (GBA). *Hum Mutat*. 2008;29(5):567–583.
63. Chan A, Holleran WM, Ferguson T, et al. Skin ultrastructural findings in type 2 Gaucher disease: diagnostic implications. *Mol Genet Metab*. 2011;104(4):631–636.
64. Davies EH, Seunarine KK, Banks T, Clark CA, Vellodi A. Brain white matter abnormalities in paediatric Gaucher Type I and Type III using diffusion tensor imaging. *J Inherit Metab Dis*. 2011;34(2):549–553.
65. Zupanc ML, Legros B. Progressive myoclonic epilepsy. *Cerebellum*. 2004;3(3):156–171.
66. Goker-Alpan O, Wiggs EA, Eblan MJ, et al. Cognitive outcome in treated patients with chronic neuronopathic Gaucher disease. *J Pediatr*. 2008;153(1):89–94.
67. Schiffmann R, Fitzgibbon EJ, Harris C, et al. Randomized, controlled trial of miglustat in Gaucher's disease type 3. *Ann Neurol*. 2008;64(5):514–522.
68. Cabrera-Salazar MA, Deriso M, Bercury SD, et al. Systemic delivery of a glucosylceramide synthase inhibitor reduces CNS substrates and increases lifespan in a mouse model of type 2 Gaucher disease. *PLoS One*. 2012;7(8):e43310.
69. Khanna R, Benjamin ER, Pellegrino L, et al. The pharmacological chaperone isofagomine increases the activity of the Gaucher disease L444P mutant form of beta-glucosidase. *FEBS J*. 2010;277(7):1618–1638.
70. Luan Z, Li L, Higaki K, Nanba E, Suzuki Y, Ohno K. The chaperone activity and toxicity of ambroxol on Gaucher cells and normal mice. *Brain Dev*. 2013;35(4):317–322.
71. Maegawa GH, Tropak MB, Buttner JD, et al. Identification and characterization of ambroxol as an enzyme enhancement agent for Gaucher disease. *J Biol Chem*. 2009;284(35):23502–23516.

CHAPTER

27

The Niemann–Pick Diseases

Edward H. Schuchman and Robert J. Desnick

Icahn School of Medicine at Mount Sinai, New York, NY, USA

INTRODUCTION

Two distinct metabolic derangements are encompassed under the eponym Niemann–Pick disease (NPD). The first is due to the deficient activity of the lysosomal enzyme acid sphingomyelinase (ASM).[1–3] Patients with this enzyme deficiency are classified as having types A and B NPD. Type A NPD patients exhibit hepatosplenomegaly in infancy and profound central nervous system involvement. They rarely survive beyond 2 years of age. Type B patients also have hepatosplenomegaly and pathologic alterations of their lungs, but there are usually no central nervous system signs. The age of onset and rate of disease progression varies greatly among type B patients, and they frequently live into their fourth decade. Patients with phenotypes intermediate between types A and B NPD also have been identified.[4,5] These individuals represent the expected continuum caused by inheriting different mutations in the ASM gene (*SMPD1*).

Patients in the second NPD category are designated as having types C and D NPD. These individuals may have mild hepatosplenomegaly, but the central nervous system is profoundly affected. Impaired intracellular trafficking of cholesterol causes types C and D NPD, and two distinct gene defects have been found. All type D patients originate from a common Nova Scotian ancestry.[6] In this chapter types C and D NPD are referred to as "type C" NPD for simplicity.

CLINICAL FEATURES AND DIAGNOSTIC EVALUATION

Type A NPD patients exhibit profound organomegaly within the first 3 months of life. A cherry-red spot is present in the macula in approximately 50% of these infants. Developmental milestones are rarely attained, and psychomotor deterioration progresses rapidly. Patients become hypotonic and flaccid and rarely survive beyond 2 years of age. In contrast, type B patients have no overt signs of central nervous system involvement, but hepatosplenomegaly may be profound and accompanied by signs of liver failure. Serum triglycerides and low-density lipoprotein (LDL)-cholesterol are often elevated, while high-density lipoprotein (HDL)-cholesterol is low. The lungs are frequently involved in type B NPD, and pulmonary function is often compromised. There is also often a reddish-brown halo surrounding the macula in the eyes of these patients, and in some cases a distinct cherry red spot can be identified. Because insufficient ASM activity is the hallmark of types A and B NPD, measurement of the activity of this enzyme in convenient cells such as circulating leukocytes or cultured skin fibroblasts is the standard confirmatory diagnostic procedure.[7]

Type C NPD is usually suspected in patients with vertical gaze impairment, dysarthria, dementia, ataxia, dystonia, and mild hepatosplenomegaly. Some of these patients also have cataplexy and seizures. The diagnosis is usually confirmed by demonstrating an elevation of free cholesterol in cultured skin fibroblasts by staining with filipin, and/or by determining the rate of cellular cholesterol esterification in these cells.[8]

Rosenberg's Molecular and Genetic Basis of Neurological and Psychiatric Disease
http://dx.doi.org/10.1016/B978-0-12-410529-4.00027-9

RADIOLOGIC AND NEUROPHYSIOLOGIC STUDIES

Magnetic resonance imaging of the brain in patients with types A and C NPD may be normal or reveal cerebral or cerebellar atrophy, sometimes with white matter hyperintensity on T2-weighted imaging. In patients with type C NPD, magnetic resonance spectroscopic imaging reveals a reduction in the *N*-acetyl aspartate: creatine ratio, indicating diffuse brain involvement consistent with the pathologic features of the disease. Brainstem auditory-evoked potentials are often delayed, with absence of acoustic reflexes.

PATHOLOGY

Types A and B Niemann–Pick Disease

Large, lipid-laden cells are present in the liver, spleen lymph nodes, adrenal cortex, lung airways, and bone marrow in types A and B NPD. The cells have a mulberry appearance because of an accumulation of lipid droplets. They are autofluorescent and stain for phospholipids. Some cells are pigmented because of the presence of ceroid. The cells contain concentrically lamellated myelin-like figures. The brain of type A NPD patients is usually atrophic. Ganglion cells are often swollen, and the cytoplasm is pale and vacuolated. Within these cells are membrane-bound inclusions. There is a loss of cells in the cerebral and cerebellar cortices, along with gliosis in both gray and white matter. Some areas of the white matter show demyelination. Foam cells are present in the leptomeninges, tela choroidea, endothelium, and perivascular spaces of cerebral blood vessels. The basal ganglia, brainstem, spinal cord, and autonomic ganglia may also show morphologic alterations. Little is known about structural changes in the brains of patients with type B NPD.

Type C Niemann–Pick Disease

In type C NPD, the spleen is infiltrated with foam cells that stain for cholesterol, phospholipids, and glycolipids. Kupffer cells and, to a lesser extent, hepatocytes in the liver also are involved. In the infantile presentation, extensive hepatic vacuolization is seen along with cholestasis and giant-cell formation. Foam cells are almost always present in the bone marrow. These cells have been called sea-blue histiocytes. The cerebral cortex is atrophic, and ballooned neurons exhibit inclusions. The basal ganglia, thalamus, and brainstem are particularly involved. Cellular atrophy and lipid storage occurs in Purkinje cells and cells in the dentate nucleus of the cerebellum.

BIOCHEMICAL FINDINGS

The principal accumulating lipid in patients with types A and B NPD is sphingomyelin (ceramide-phosphocholine). Sphingomyelin is a major component of the plasma membrane of all cells, and is a principal phospholipid of the myelin sheath. In addition to sphingomyelin, elevated levels of bis(monoacylglycero)phosphate are prominent. Cholesterol, glucocerebroside, lactosylceramide, and gangliosides, particularly ganglioside GM_3, are elevated as well, but not so much as in type C NPD.

The enzyme defective in types A and B NPD is called acid sphingomyelinase because of its highest activity at reduced pH and the fact that it catalyzes the hydrolytic cleavage of sphingomyelin, producing phosphocholine and ceramide (*N*-fatty acylsphingosine) within lysosomes. However, when cells are subjected to stress (e.g., irradiation, heat shock), ASM rapidly translocates from lysosomes to the outer leaflet of the plasma membrane, where it degrades sphingomyelin into ceramide.[9,10] This causes reorganization of raft structures in the plasma membrane and stimulates downstream signaling events. Thus, in addition to lipid storage, the clinical findings in types A and B NPD patients may be due to lipid abnormalities in the plasma membrane. This has been particularly well documented in the brains of ASM knockout mice (see below), where lipid storage in the neuronal membranes leads to abnormal synapse formation and function.[11] It is unclear whether the activity of ASM at the plasma membrane occurs in acidified "micro"compartments, or proceeds at a nonacidic pH. For example, despite the acidic pH optimum of the enzyme *in vitro*, Tabas and coworkers have shown the hydrolysis of sphingomyelin by ASM at physiological pH.[12]

There are two protein abnormalities in type C NPD. The most common (95%) is an alteration of a protein called NPC1, a glycoprotein with 13 transmembrane domains.[13] NPC1 has a high degree of homology with other proteins involved in intracellular cholesterol homeostasis. The second protein that is mutated in type C NPD (NPC2) is a lysosomal protein called HE1, whose precise function has not been identified.[14] Mutations of HE1 are frequently associated with severe clinical manifestations, and NPC2 patients often die of respiratory failure due to extensive pulmonary involvement.

BRAIN IMMUNOCHEMICAL FINDINGS

In type C NPD, neurofibrillary tangles appear in neurons in the orbital gyrus, cingulate gyrus, entorhinal region, basal ganglia, thalamus, hypothalamus, inferior olivary nucleus, and spinal cord. The tangles stain with Alz 50, an antibody that reacts with phosphorylated tau proteins. Amyloid deposition has not been reported. Neurofibrillary tangles have not been reported in the brains of type A NPD patients.

MECHANISM OF DISEASE

NPC1 and HE1 appear to be required for the vesicular shuttling of both lipids and fluid-phase constituents from multivesicular late endosomes to destinations such as the trans-Golgi network. However, the exact function and mode of action of these proteins are unknown. The cholesterol that accumulates in type C NPD arises primarily from extracellular cholesterol associated with low-density lipoprotein. Abnormal cholesterol trafficking leads to numerous secondary metabolic defects in type C NPD patients, including abnormal production of neurosteroids, accumulation of other lipids (e.g., sphingomyelin, gangliosides), abnormal function of Rab proteins, neuroinflammation, and so on. However, how and if these secondary changes contribute to the type C NPD clinical phenotype remains unclear. It is notable that despite the fact that both NPC1 and HE1 are thought to be involved in cholesterol transport, the major accumulating lipids in the brains of type C NPD patients are gangliosides, not cholesterol.

As discussed above, ASM is required for the lysosomal degradation of sphingomyelin, as well as to initiate signal transduction at the plasma membrane. Thus, the clinical findings in types A and B NPD are presumably due to a combination of lipid accumulation within lysosomes/late endosomes and the failure to undergo proper cell signaling at the plasma membrane. The accumulation of sphingomyelin within lysosomes also results in the secondary accumulation of other metabolites, principally cholesterol, which contribute to the disease phenotype. Ganglioside storage also is prominent in the brains of type A NPD patients, as is neuroinflammation.

Of interest, a recent report recently showed that the cholesterol abnormalities in cells from Type C NPD patients can be partially corrected by ASM expression,[15] further revealing the metabolic relatedness of sphingomyelin and cholesterol, and the overlap in biochemical/pathological findings in types A, B and C NPD. These findings also suggest the possibility of common therapeutic solutions.

MOLECULAR GENETICS

ASM is produced from a single gene (*SMPD1*) located within the chromosomal region 11p15.4.[16,17] This region of chromosome 11 is a hotspot for imprinting within the human genome, and studies have shown that the *SMPD1* gene is preferentially expressed from the maternal chromosome (i.e., paternally imprinted).[18] Types A and B NPD are inherited as recessive traits, and thus patients must inherit two mutant alleles. The degree of clinical involvement largely depends on the type of *SMPD1* mutations inherited and the effect of these mutations on the residual ASM polypeptide. In addition, because the *SMPD1* gene is imprinted, the phenotypic variability may also be due, at least in part, to inheritance of specific mutations on the maternal versus paternal alleles. It is of interest that abnormal clinical and laboratory findings have also been reported in heterozygous individuals carrying only one *SMPD1* mutation.[19] This could similarly be due to inheritance of a single, "severe" *SMPD1* mutation on the preferentially expressed maternal chromosome.

To date, more than 120 mutations have been found within the *SMPD1* gene causing types A and B NPD.[20]

These include point mutations (missense and nonsense), small deletions, and splicing abnormalities. No hotspots exist for these mutations. Several polymorphisms also have been found within the *SMPD1* gene of normal individuals, including a varying number of repeated nucleotides within the region encoding the ASM signal peptide. The effects of these polymorphisms on ASM trafficking and function has not been systematically studied. In addition, the existence of a length polymorphism within the signal peptide region of the *SMPD1* gene has led to two different numbering systems for types A and B NPD mutations. For example, the common type B NPD mutation, deltaR608 (see below), is often referred to as deltaR610.

Identification of mutations in type A and B NPD patients has permitted the first genotype–phenotype correlations for this disorder and the first genetic screening efforts. For example, three mutations account for ~90% of Ashkenazi Jewish type A NPD infants. This observation has led to carrier screening of Ashkenazi adults throughout the world and revealed that the carrier frequency for these three mutations in the Ashkenazi population is between 1:80 and 1:100. Another mutation, deltaR608, only occurs in type B NPD patients and is found in 15–20% of NPD type B individuals in Western Europe and North America. Another mutation, Q292K, is associated with an intermediate neurological phenotype. These and other findings have assisted physicians, genetic counselors, and families in predicting the phenotypic outcome in individual NPD patients and in the future may lead to large-scale screening for this disorder in specific populations.

The *NPC1* gene is located on human chromosome 18q11. A variety of alterations have been described in type C NPD patients, including deletions, insertions, and missense mutations.[8] In patients of Western European descent, a relatively common mutation in exon 21 (I1061T) has been described that results in the juvenile and adolescent form of type C NPD. Various reports suggest that this mutation accounts for approximately 15% of the mutant alleles in this population. Several mutations in the *HE1* (NPC2) gene, located on human chromosome 14q24.3, also have been identified.[21]

ANIMAL MODELS

Types A and B Niemann–Pick Disease

Two ASM knockout (ASMKO) mouse models have been produced for types A and B NPD.[22,23] Homozygous animals exhibit progressive lipid storage in reticulodendothelial (RES) organs, as well as in the brain. The principal accumulating lipid is sphingomyelin, but cholesterol and ganglioside storage also has been observed. Progressive loss of Purkinje cells in the cerebellum leads to gait abnormalities, and affected animals die within 6–8 months. The precise cause of death is unknown but is likely secondary to the neurological phenotype.

Profound inflammatory disease has also been documented in the ASMKO mice, and this is a major feature of the pulmonary and neurological phenotypes.[24] Pulmonary lavage has revealed numerous inflammatory cells in the air spaces of ASMKO animals and the release of many inflammatory cytokines into the lavage fluid. These observations and microarray analysis of lung and brain tissue have identified several inflammatory molecules that may be used as biomarkers to monitor progression of the disease and to (potentially) measure the response to treatment (see below).[25]

In addition to the above-noted phenotypes, the ASMKO mice also exhibit a defect in ceramide-mediated signal transduction. This has been studied mostly in the context of stress-induced apoptosis, where the mice are resistant to cell death induced by several treatments (e.g., irradiation, ischemia).[26,27] Although many cell types exhibit this phenotype, it is most evident in endothelial cells. Presumably, the lack of functional ASM inhibits the production of ceramide and reorganization of membrane rafts into platforms, protecting against cell death. ASMKO cells also are resistant to infection by various pathogens, likely also due to defects in membrane reorganization and internalization of the pathogens via receptors.[28] It is also of interest that oocytes from ASMKO mice do not undergo normal apoptosis as they age.[29] This is similarly due to the lack of ASM and failure to produce ceramide in response to normal developmental signals.

In addition to the complete knockout ASMKO models, a transgenic mouse model of type B NPD has been produced.[30] This animal was engineered by introducing a partially functional mouse *spmd1* gene onto the complete knockout (ASMKO) background. The end result was a mouse that produced from 8–15% residual ASM activity in most organs. These mice never developed a neurological phenotype and had normal longevity. However, by about 8–10 months of age, they began to exhibit lipid storage in RES organs. These observations provide direct *in vivo* evidence that low levels of ASM activity in the brain are likely to prevent neurological disease in ASM patients, and they have important implications for the treatment of neurological ASM-deficient NPD (see below).

Lastly, knock-in models of types A and B NPD have recently been constructed.[31] In these models the common type A NPD mutation—R496L, and the common type B mutation—deltaR608 were introduced. The phenotypes of these mice were as expected. R496L mice developed a severe "type A" phenotype indistinguishable from the knockout mice. DeltaR608 developed a later onset, non-neurological "type B" phenotype.

Canine and feline models of type A NPD have been described, but no colonies are currently available.

Type C Niemann–Pick Disease

The naturally occurring mouse model for type C NPD has a well-defined mutation in the murine gene (*npc1*) equivalent to human *NPC1*.[32] Affected mice exhibit a clinical phenotype characterized by progressive loss of Purkinje cell neurons and premature death. Starting at postnatal day 9, mild abnormalities also can be found in the corpus callosum, cerebellar white matter, and nerve fibers. By postnatal day 10, hypomyelination and axonal spheroids are evident in the corpus callosum and subcortical white matter, and by postnatal day 22 activated astrocytes become abundant. Also, neuronal cholesterol accumulation occurs in various regions of the brain. A second mouse model was also identified in Japan with a virtually identical phenotype to that above and, although first described as having "sphingomyelinosis," is now know to have a mutation in the *npc1* gene.[33]

NPC2 mouse models with much lower than normal expression of the NPC2 (HE1) protein (0–4% of normal) also have been produced.[34,35] In terms of disease onset, progression, pathology, and neuronal storage, the phenotypes of the NPC1, NPC2, and NPC1:NPC2 double mutant mice are similar or identical.

Finally, a feline model of type C NPD has been characterized and is phenotypically, morphologically, and biochemically similar to human NPC1.[36] Mutation analysis identified a single base substitution in the feline homolog of the *NPC1* gene.[37] These models have been used to evaluate disease pathogenesis and various approaches to therapy, as described below.

THERAPY

Types A and B Niemann–Pick Disease

Bone marrow transplantation (BMT) has been undertaken in several ASM-deficient NPD patients. Reduction in liver and spleen size has been noted, although complications secondary to the transplant procedure are generally severe.[38] The effect of bone marrow transplantation on the neurological phenotype of type A NPD patients has not been adequately shown, although this procedure has been extensively studied in the ASMKO mouse model.[39] Overall, when transplantation was undertaken in these mice during the newborn period and engraftment levels were high (90%), the positive effects on the RES organs were profound. However, even under such "optimal" transplant conditions, the effects on the central nervous system were intermediate, and animals died secondary to a neurological phenotype. This could be improved by direct intracranial injection of bone marrow-derived cells to provide a local source of ASM in the central nervous system, but even under these conditions the progressive neurological disease was not prevented.[40]

These and other findings suggest that bone marrow transplantation and cord blood transplantation might be considered for non-neurological (type B) ASM-deficient NPD patients. However, BMT is unlikely to have a major impact on the neurological progression in type A NPD. Also, the complications arising from these procedures are likely to be severe and need to be considered carefully. In addition to BMT, liver and amniotic cell transplantation also have been undertaken in ASM-deficient NPD patients, although the outcomes of these procedures have not been fully described.[41–44]

Ultimately, to adequately treat ASM-deficient NPD, ASM activity must be provided to clinically affected organs. The gold standard of achieving such widespread enzyme delivery is enzyme replacement therapy (ERT). To evaluate ERT for ASM-deficient NPD, recombinant human ASM has been produced in Chinese hamster ovary cells and extensively characterized.[45] The recombinant enzyme (rhASM) has been used to treat ASMKO mice, and the effects are well documented.[46] Overall, when rhASM was administered intravenously into young ASMKO mice, lipid storage could be effectively prevented in the RES organs. In addition, the progressive inflammatory disease was prevented in these organs. The effects of the enzyme were dose dependent and were most profound in the liver and spleen, followed by the lung. However, despite the dramatic improvements in the RES organs after ERT, there was no effect on the progression of neurological disease, and the treated ASMKO animals died at the same age as untreated mice.

rhASM also has been used to treat ASMKO mice with established disease.[46] In these studies, which were conducted over a few short weeks (due to the shortened life span of the affected animals), significant reduction in lipid storage and histological improvement of RES organs was noted. Again, this was most profound in the liver and spleen. The lung also responded well, but at a slower rate. This was attributed to the fact that although delivery of rhASM to the lung was rapid after ERT and led to substantial reductions in inflammatory chemokines, the inflammatory cells already present within the airways of these mice were long-lived, and clinical effects could therefore not be seen until these established cells died. Also, the biodistribution of rhASM after intravenous administration indicates that only about 2% of injected dose reaches the lung.

These and other findings regarding ERT in the ASMKO mice have led to the initiation of clinical trials in adult type B NPD patients (by Genzyme/Sanofi). A phase I safety study was first undertaken. Eleven patients were treated with single administrations of rhASM of increasing doses once every 2 weeks. No serious adverse events were observed, and safety findings included only transient elevation of serum cytokines and billirubin. Based on these results the maximum starting dose of rhASM in adult type NPD patients was determined to be 0.6 mg per kg. Next, a phase Ib repeat dosing study was undertaken in five adult type B NPD patients.[47] A dose escalation scheme was used where patients first received several low doses to "debulk" the sphingomyelin load in tissues, followed by escalation to a maximum dose of 3 mg per kg. All patients were successfully dose escalated without serious adverse events. This trial was recently completed and the data is being analyzed. Based on these findings, additional repeat dosing studies will be undertaken in type B NPD adults and children.

Other preclinical therapy studies conducted in the ASMKO mouse model include autologous hematopoietic stem cell gene therapy using retroviral vectors,[48] liver directed gene therapy using AAV vectors,[49] and direct injection of gene therapy vectors and/or rhASM into the brain.[50]

Type C Niemann–Pick Disease

Therapeutic approaches for type C NPD have focused on the downstream metabolic consequences of NPC1 or NPC2/HE1 mutations. Substrate reduction therapy (SRT) is a strategy in which small-molecule inhibitors of glycolipid biosynthesis (miglustat; Zavesca) are used to slow the rate of glycolipid accumulation in patients with glycolipid storage diseases.[51] In type C NPD and other neurodegenerative lysosomal disorders, gangliosides accumulate in the brain, and thus it was suggested that SRT might be used to slow the progression of this disorder as well. SRT was undertaken in type C NPD mice and cats, and modest central nervous system effects were found.[52] Based on these findings and the lack of other effective therapies, a clinical trial was conducted in type C NPD patients,[53] leading to approval of miglustat therapy for type C NPD in Europe. Miglustat has not been approved for NPC in the US by the Food and Drug Administration (FDA).

Another interesting therapeutic approach for type C NPD is the use of cyclodextrans. Cyclodextran therapy came from work evaluating the effects of the allopregnanolone on neurosteroids in the brain of NPC1 mice.[54] NPC1 mice treated with allopregnanolone early in postnatal development exhibited delayed onset of neurological symptoms, increased Purkinje and granule cell survival, reduced gangliosides, and doubling of the lifespan.[55] Subsequently it was found that these positive effects were not due to allopregnanolone, but rather to cyclodextran that was used as an excipient in the drug preparation. Among other properties cyclodextran may be extracting cholesterol from the membranes of neuronal cells. This observation led to subsequent therapeutic studies in NPC1 mice and cats.[56,57] Currently, a phase I study of cyclodextran therapy is being undertaken in type C NPD patients. Due to the high doses required to reach the brain after systemic delivery in NPC1 animals, as well as toxicities observed in NPC cats,[57] in this trial cyclodextran will be delivered directly into the central nervous system, once every 2 weeks or monthly.

Other therapeutic strategies that have been investigated in type C NPD animal models and/or patients include cholesterol-lowering drugs, bone marrow transplantation, and dietary restriction of cholesterol.[58–61] However, none of these approaches substantially improves the neurodegenerative disease course in these individuals.

CONCLUSIONS

The heterogeneity of clinical phenotypes in the NPD group of disorders requires diagnostic sophistication. NPD should be suspected in infants or children who present with hepatosplenomegaly and lipid-storing cells in the bone marrow. The central nervous system may (types A and C) or may not (type B) be involved. The diagnosis of these phenotypes can be established by measuring ASM activity in peripheral blood leukocytes or in cultured skin fibroblasts. Recently, dried blood spot assays have been developed for types A and B NPD based on ASM activity or the elevation of specific lipids (e.g., lyso-sphingomyelin).[62,63] However, while these measurements can reliably rule out type C NPD, they do not adequately predict the occurrence of a neurological or non-neurological phenotype. Further analysis of mutations in the *SMPD1* gene can assist with this analysis in some patients.

Type C NPD is suspected if a patient exhibits mild hepatosplenomegaly, sea-blue histiocytes in the bone marrow, and vertical gaze palsy with or without other signs of central nervous system damage. The diagnosis of this condition can be confirmed by examining intracellular cholesterol esterification. Recent studies also have identified oxysterols as potential biomarkers for NPC, opening a path for population screening for this disorder.[64] Future research directions involve elucidation of specific functions of the products of the *NPC1* and the *NPC2/HE1* genes that are mutated in patients with type C NPD, and the development of specific therapies for patients with all forms of NPD.

References

1. Brady RO, Kanfer JN, Mock MB, et al. The metabolism of sphingomyelin. II. Evidence of an enzymatic deficiency in Niemann–Pick disease. *Proc Natl Acad Sci U S A*. 1966;55:366–369.
2. Schneider PB, Kennedy EP. Sphingomyelinase in normal human spleens and in spleens from subjects with Niemann– Pick disease. *J Lipid Res*. 1967;8:202–209.
3. Schuchman EH, Desnick RJ. Niemann–Pick disease types A and B: acid sphingomyelinase deficiencies. In: Scriver CR, Beaudet AL, Sly WS, et al., eds. *The Metabolic and Molecular Bases of Inherited Diseases*. New York: McGraw-Hill; 2001:3589–3610.
4. Pavlu-Pereira H, Asfaw B, Poupctova H, et al. Acid sphingomyelinase deficiency. Phenotype variability with prevalence of intermediate phenotype in a series of twenty-five Czech and Slovak patients. A multi-approach study. *J Inherit Metab Dis*. 2005;28:203–227.
5. Wasserstein MP, Aron A, Brodie SE, et al. Acid sphingomyelinase deficiency: prevalence and characterization of an intermediate phenotype of Niemann–Pick disease. *J Pediatr*. 2006;149:554–559.
6. Pentchev PG, Comly ME, Kruth HS, et al. A defect in cholesterol esterification in Niemann–Pick disease (type C) patients. *Proc Natl Acad Sci U S A*. 1985;82:8247–8251.
7. Gal AE, Brady RO, Hibbert SR, et al. A practical chromogenic procedure for the detection of homozygotes and heterozygous carriers of Niemann–Pick disease. *N Engl J Med*. 1975;293:632–636.
8. Patterson MC, Vanier MT, Suzuki K, et al. Niemann–Pick disease Type C: a lipid trafficking disorder. In: Scriver CR, Beaudet AL, Sly WS, et al., eds. *The Metabolic and Molecular Bases of Inherited Disease*. New York: McGraw-Hill; 2001:3611–3633.
9. Charruyer A, Grazide S, Bezombes C, et al. UV-C light induces raft-associated acid sphingomyelinase and JNK activation and translocation independently on a nuclear signal. *J Biol Chem*. 2005;280:19196–19204.
10. Falcone S, Perrotta C, De Palma C, et al. Activation of acid sphingomyelinase and its inhibition by the nitric oxide/cyclic guanosine 3,5-monophosphate pathway: key events in Escherichia coli-elicited apoptosis of dendritic cells. *J Immunol*. 2004;173:4452–4463.
11. Ledesma MD, Prinetti A, Sonnino S, Schuchman EH. Brain pathology in Niemann Pick disease type A: insights from the acid sphingomyelinase knockout mice. *J Neurochem*. 2011;116:779–788.
12. Schissel SL, Jiang X, Tweedi-Hardman J, et al. Secretory sphingomyelinase, a product of the acid sphingomyelinase gene, can hydrolyze atherogenic lipoproteins at neural pH. Implications for atherosclerotic lesion development. *J Biol Chem*. 1998;273:2738–2746.
13. Carstea ED, Morris JA, Coleman KG, et al. Niemann–Pick C1 disease gene: homology to mediators of cholesterol homeostasis. *Science*. 1997;277:228–232.
14. Naureckiene S, Sleat DE, Lackland H, et al. Identification of HE1 as the second gene of Niemann–Pick C disease. *Science*. 2000;290:2298–2301.
15. Devlin C, Pipalia NH, Liao X, et al. Improvement in lipid and protein trafficking. *Niemann–Pick C1 cells by correction of a secondary enzyme defect*. 2010;11:601–615.
16. Schuchman EH, Levran O, Pereira LV, et al. Structural organization and complete nucleotide sequence of the gene encoding human acid sphingomyelinase (SMPD1). *Genomics*. 1992;12:197–205.
17. da Veiga Pereira L, Desnick RJ, Adler DA, et al. Regional assignment of the human acid sphingomyelinase gene (SMPD1) by PCR analysis of somatic cell hybrids and in situ hybridization to 11p15.1–p15.4. *Genomics*. 1992;9:229–234.
18. Simonaro CM, Park JH, Eliyahu E, et al. Imprinting at the SMPD-1 gene: Implications for acid sphingomyelinase- deficient Niemann–Pick disease. *Am J Hum Genet*. 2006;78:79–84.
19. Lee CY, Krimbou L, Vincent J, et al. Compound heterozygosity at the sphingomyelin phosphodiesterase-1 (SMPD1) gene is associated with low HDL cholesterol. *Hum Genet*. 2003;112:552–562.
20. Simonaro CM, Desnick RJ, McGovern MM, et al. The demographics and distribution of type B Niemann–Pick disease: novel mutations lead to new genotype/phenotype correlations. *Am J Hum Genet*. 2002;71:1413–1419.
21. Millat G, Chikh K, Naureckiene S, et al. Niemann–Pick disease type C: spectrum of HE1 mutations and genotype/phenotype correlations in the NPC2 group. *Am J Hum Genet*. 2001;69:1013–1021.
22. Otterbach B, Stoffel W. Acid sphingomyelinase-deficient mice mimic the neurovisceral form of human lysosomal storage disease (Niemann–Pick disease). *Cell*. 1995;81:1053–1061.
23. Horinouchi K, Erlich S, Perl DP, et al. Acid sphingomyelinase deficient mice: a model of types A and B Niemann–Pick disease. *Nat Genet*. 2005;10:288–293.
24. Dhami R, He X, Gordon RE, et al. Analysis of the lung pathology and alveolar macrophage function in the acid sphingomyelinase-deficient mouse model of Niemann–Pick disease. *Lab Invest*. 2001;81:987–999.
25. Dhami R, He X, Schuchman EH. Gene expression analysis in acid sphingomyelinase deficient mice. Novel insights into disease pathogenesis and identification of potential bio-markers to monitor Niemann–Pick disease treatment. *Mol Ther*. 2005;13:556–563.
26. Yu ZF, Nikolova-Karakashian M, Zhou D, et al. Pivotal role for acidic sphingomyelinase in cerebral ischemia-induced ceramide and cytokine production, and neuronal apoptosis. *J Mol Neurosci*. 2000;15:85–97.
27. Charruyer A, Grazide S, Bezombes C, et al. UV-C light induces raft-associated acid sphingomyelinase and JNK activation and translocation independently on a nuclear signal. *J Biol Chem*. 2005;280:19196–19204.
28. Utermohlen O, Karow U, Lohler J, et al. Severe impairment in early host defense against Listeria monocytogenes in mice deficient in acid sphingomyelinase. *J Immunol*. 2003;170:2621–2628.
29. Morita Y, Perez GI, Paris F, et al. Oocyte apoptosis is suppressed by disruption of the acid sphingomyelinase gene or by sphingosine-1-phosphate therapy. *Nat Med*. 2000;6:1109–1114.
30. Marathe S, Miranda SR, Devlin C, et al. Creation of a mouse model for non-neurological (type B) Niemann–Pick disease by stable, low level expression of lysosomal sphingomyelinase in the absence of secretory sphingomyelinase: relationship between brain intra-lysosomal enzyme activity and central nervous system function. *Hum Mol Genet*. 2004;9:1967–1976.
31. Jones I, He X, Katouzian F, et al. Characterization of common SMPD1 mutations causing types A and B Niemann–Pick disease and generation of mutation-specific mouse models. *Mol Genet Metab*. 2008;95:152–162.
32. Pentchev PG, Boothe A, Kruth H, et al. A genetic storage disorder in BALB/C mice with a metabolic block in the esterification of exogenous cholesterol. *J Biol Chem*. 1984;259:5784–5791.

33. Sakiyama T, Tsuda M, Kitagawa T, et al. A lysosomal storage disorder in mice: a model of Niemann–Pick disease. *J Inherit Metab Dis.* 1982;5:239–240.
34. Sleat DE, Wiseman JA, El-Banna M, et al. Genetic evidence for nonredundant functional cooperativity between NPC1 and NPC2 in lipid transport. *Proc Natl Acad Sci U S A.* 2004;5886–5891.
35. Nielsen GK, Dagnaes-Hansen F, Holm IE, et al. Protein replacement therapy partially corrects the cholesterol-storage phenotype in a mouse model of Niemann–Pick type C2 disease. *PLoS One.* 2011;6:e27287.
36. Brown DE, Thrall MA, Walkley SU, et al. Feline Niemann–Pick disease type C. *Am J Pathol.* 1994;144:1412–1415.
37. Somers KL, Royals MA, Carstea ED, et al. Mutation analysis of feline Niemann–Pick C1 disease. *Mol Genet Metab.* 2003;79:99–103.
38. Victor S, Coutler JBS, Besley GTN, et al. Niemann–Pick disease type B: 16-year follow-up after allogenic bone marrow transplantation. *J Inherit Metab Dis.* 2003;26:775–785.
39. Miranda SR, Erlich S, Friedrich Jr. VL, et al. Biochemical, pathological, and clinical response to transplantation of normal bone marrow cells into acid sphingomyelinase deficient mice. *Transplantation.* 1998;65:884–892.
40. Jin HK, Schuchman EH. Combined bone marrow and intracerebral mesenchymal stem cell transplantation leads to synergistic visceral and neurological improvements in Niemann–Pick disease mice. *Mol Ther.* 2003;26:775–785.
41. Cerneca F, Andolina M, Simeone R, et al. Treatment of patients with Niemann–Pick disease using repeated amniotic epithelial cells implantation: correction of aggregation and coagulation abnormalities. *Clin Pediatr.* 1997;36:141–146.
42. Daloze P, Delvin EE, Glorieux FH, et al. Replacement therapy for inherited enzyme deficiency: liver orthotopic transplantation in Niemann–Pick disease type A. *Am J Med Genet.* 1977;1:229–239.
43. Scaggiante B, Pineschi A, Sustersich M, et al. Successful therapy of Niemann–Pick disease by implantation of human amniotic membrane. *Transplantation.* 1987;44:59–61.
44. Smanik EJ, Tavill AS, Jacobs GH, et al. Orthotopic liver transplantation in two adults with Niemann–Pick's and Gaucher's disease—implications for the treatment of inherited metabolic disease. *Hepatology.* 1993;17:42–49.
45. He X, Miranda SR, Xiong X, et al. Characterization of human acid sphingomyelinase purified from the media of overexpressing Chinese hamster ovary cells. *Biochim Biophys Acta.* 1999;1432:251–264.
46. Miranda SR, He X, Simonaro CM, et al. Infusion of recombinant human acid sphingomyelinase into Niemann–Pick disease mice leads to visceral, but not neurological correction of the pathophysiology. *FASEB J.* 2000;14:1988–1995.
47. Jones S, Wasserstein M, Soran H, et al. An open-label, multicenter, ascending dose study of the tolerability and safety of recombinant human acid sphingomyelinase (rhASM) in patients with ASM deficiency (ASMD). *Mol Gene Metabol.* 2014;111:557.
48. Miranda SR, Erlich S, Friedrich VL, et al. Hematopoietic stem cell gene therapy leads to marked visceral organ improvements and a delayed onset of neurological abnormalities in the acid sphingomyelinase deficient mouse model of Niemann–Pick disease. *Gene Ther.* 2000;7:1768–1776.
49. Barbon CM, Ziegler RJ, Li C, et al. AAV8-mediated hepatic expression of acid sphingomyelinase corrects the metabolic defect in the visceral organs of a mouse model of Niemann–Pick disease. *Mol Ther.* 2005;12:431–440.
50. Dodge JC, Clarke J, Song A, et al. Gene transfer of human acid sphingomyelinase corrects neuropathology and motor deficits in a mouse model of Niemann–Pick type A disease. *Proc Natl Acad Sci U S A.* 2005;102:17822–17827.
51. Cox TM, Aerts JM, Andria G, et al. The role of iminosugar *N*-butyldeoxynojirimycin (miglustat) in the management of type I (non-neuronopathic) Gaucher disease: a position statement. *J Inherit Metab Dis.* 2003;26:513–526.
52. Zervas M, Somers KL, Thrall MA, et al. Critical role for glycosphingolipids in Niemann–Pick disease type C. *Curr Biol.* 2001;11:1283–1287.
53. Patterson MC, Vecchio D, Prady H, et al. Miglustat for treatment of Niemann–Pick C disease: a randomized controlled study. *Lancet Neurol.* 2007;6:765–772.
54. Compagnone NA, Mellon SH. Neurosteroids: biosynthesis and function of these novel neuromodulators. *Front Neuroendocrinol.* 2000;21:1–56.
55. Griffin LD, Gong W, Verot L, et al. Niemann–Pick type C disease as a model for defects in neurosteroidogenesis. *Nat Med.* 2004;10:704–711.
56. Davidson CD, Ali NF, Micsenyi MC, et al. Chronic cyclodextrin treatment of murine Niemann–Pick C disease ameliorates neuronal cholesterol and glycosphingolipid storage and disease progression. *PLoS One.* 2009;4:e6951.
57. Ward S, O'Donnell P, Fernandez S, et al. 2-hydroxypropyl-beta-cyclodextrin raises hearing threshold in normal cats and cats with Niemann–Pick type C disease. *Pediatr Res.* 2010;68:52–56.
58. Erickson RP, Garver WS, Camargo F, et al. Pharmacological and genetic modifications of somatic cholesterol do not substantially alter the course of CNS disease in Niemann–Pick C mice. *J Inherit Metab Dis.* 2000;23:54–62.
59. Hsu YS, Hwu WL, Huang SF, et al. Niemann–Pick disease type C (a cellular cholesterol lipidosis) treated by bone marrow transplantation. *Bone Marrow Transplant.* 1999;24:103–107.
60. Somers KL, Brown DE, Fulton R, et al. Effects of dietary cholesterol restriction in a feline model of Niemann–Pick type C disease. *J Inherit Metab Dis.* 2001;24:427–436.
61. Boothe AD, Weintroub H, Pentchev PG, et al. A lysosomal storage disorder in the BALB/c mouse: bone marrow transplantation. *Vet Pathol.* 1984;21:432–441.
62. Legnini E, Orsini JJ, Muhl A, et al. Analysis of acid sphingomyelinase activity in dried blot spots using tandem mass spectrometry. *Ann Lab Med.* 2012;32:319–323.
63. Chuang WL, Pacheco J, Cooper S, et al. Lyso-sphingomyelin is elevated in dried blood spots of Niemann–Pick B patients. *Mol Genet Metab.* 2014;111:209–211.
64. Jiang X, Sidhu R, Porter FD, et al. A sensitive and specific LC-MS/MS method for rapid diagnosis of Niemann–Pick C1 disease from human plasma. *J Lipid Res.* 2011;52:1435–1445.

CHAPTER

28

G_{M2}-Gangliosidoses

Gregory M. Pastores[*] *and Gustavo H.B. Maegawa*[†]

[*]Mater Misericordiae University Hospital, Dublin, Ireland
[†]Johns Hopkins University School of Medicine, Baltimore, MD, USA

INTRODUCTION

G_{M2}-gangliosidoses are neurodegenerative lysosomal storage disorders, designated as such because of intraneuronal accumulation of an incompletely metabolized mono-sialoganglioside G_{M2}. Historically, these conditions were referred to as *amaurotic idiocy*, characterized by mental retardation and blindness and the presence of membranous cytoplasmic bodies in distended neurons. Subsequent biochemical analysis of postmortem brains from affected patients led to the recognition of several lipid storage defects, characterized by inherited blocks in the sequential degradation of gangliosides. Gangliosides are complex glycolipids containing ceramide linked to a variable number of monosaccharide and sialic acid residues.

G_{M2}-gangliosides (compound A) and its sialic acid derivative GA2 (compound B) are normally catabolized by the concerted action of the lysosomal enzyme β-hexosamindase A (HexA) and its cofactor G_{M2}-activator protein. The designation G_{M2} indicates a ganglioside (G) with one sialic acid (M) and three monosaccharide residues based on Svennerholm's nomenclature.[1]

Brief Historical Note

In 1881, while working at the Hospital for Children on Hackney Road, Warren Tay, an English physician, examined an infant with progressive neurological dysfunction and "symmetrical changes in the yellow spot in each eye," now known as the cherry-red spot.[2] Independently, in 1886 Bernard Sachs, a neurologist in New York City, had recorded findings in a child with ocular abnormalities and "arrested cerebral development."[3] This child had a similarly affected sibling, which pointed to an inherited basis. Subsequently, it was recognized that Tay and Sachs were describing the same disorder, and the eponymous term *Tay–Sachs disease* (TSD) came into use as an alternative designation for the condition then known as familial amaurotic idiocy. In the 1930s, TSD was identified as a lipid storage disorder by Ernst Klenk,[4] specifically as the accumulation of acid glycosphingolipids, which are normally abundantly found in ganglion cells, hence the term gangliosides. The main storage material in TSD brains, i.e., G_{M2}-ganglioside, was identified by Svennerholm in 1962[5] and its structure elucidated by Makita and Yamakawa in 1969.[6]

In 1968, Robinson and Stirling, separated the total β-*N*-acetyl-hexosaminidase (Hex) isolated from human spleens into two distinct p*I* forms, an acidic and basic β-*N*-acetyl-hexosaminidase A, which were referred to as HexA and HexB, respectively.[7] The biochemical defect resulting in lysosomal G_{M2}-ganglioside was demonstrated by Okada and O'Brien in 1969 to be caused by the HexA form missing in liver, spleen, and brain tissues.[8] Interestingly, in the TSD tissues examined a substantial increase in HexB form was noted. Therefore, it was concluded that HexA and HexB were isoenzymes that derived from two gene products.[9]

In the 1968, Konrad Sandhoff had been isolating the storage material from brains and visceral organs of affected patients from both Jewish and non-Jewish backgrounds. In one of these TSD cases, there was an additional accumulation of globoside in visceral tissues, attributed to complete deficiencies of both isoforms HexA and B, and referred to as the *O* variant.[10] Some of the *O* variant forms were then called Sandhoff disease (SD). Later, the development

http://dx.doi.org/10.1016/10.1016/B978-0-12-410529-4.00028-0

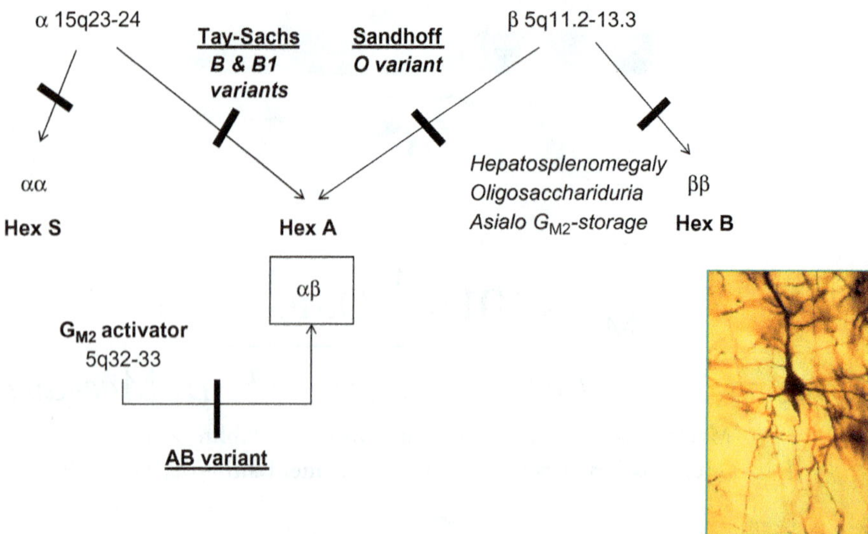

FIGURE 28.1 **G_{M2}-Gangliosidoses variants.** G_{M2}-ganglioside accumulation.

of specific synthetic substrates for hexosaminidase demonstrated that HexB is heat-stable, whereas HexA is heat-labile. These characteristics of the Hex isoenzymes allowed the enzymatic assay to be used for clinical diagnosis of both TSD and SD forms.[11] Subsequently, HexA was shown to be a heterodimer (αβ) and HexB (ββ) a homodimer (Figure 28.1). Additionally, some *O* variant cases studied by Sandhoff had normal levels of HexA and HexB isoenzymes, but lacked total Hex enzymatic activity against the natural substrate; these were then called the AB variant. Ernst Conzelmann demonstrated that the AB variant was caused by deficiency of a protein factor, G_{M2}-activator protein, which normally processes the natural substrate G_{M2}-ganglioside but not the synthetic fluorogenic substrate.[12]

CLINICAL FEATURES

Mutations of the *HEXA* gene encoding the α subunit of HexA are the genetic basis of Tay–Sachs disease, which in its classic form usually presents with onset of symptoms in infancy. Clinical manifestations, starting at 3 months of age onwards, include exaggerated startle (acousticomotor) response, generalized hypotonia and listlessness with failure to follow objects visually. Fundoscopic examination reveals the presence of a cherry-red macula, resulting from lipid deposition in the bipolar ganglion cells of the retina. Affected children subsequently display loss of gained milestones, seizures and myoclonic jerks, and spasticity.[13] There is progression to a vegetative state and cortical blindness, with death usually occurring by age 4 years. This variant was prevalent in the Ashkenazi Jewish community. Carrier screening programs, which in the Orthodox community enabled "suitable" marriage introductions (*shidoth*), combined with prenatal/pre-implantation genetic diagnosis among couples at risk, has led to a reduction in the incidence of TSD in this population.

A retrospective study (N=237), based on patient surveys and review of the literature, revealed that more than half of the infants with G_{M2}-gangliosidoses attained initial motor developmental milestones, and did so within the standard range of normal development.[14] Fifty-five patients who had learned to sit without support lost that ability within 1 year. Those who did not attain early milestones tended not to gain that ability later in life. These findings indicate developmental delay is less a feature of infantile G_{M2}-gangliosidosis than frank regression. In any case, there was no significant difference in lifespan between children with an onset of symptoms before and after 6 months. Early seizures were deemed a marker of disease severity, or at least worse motor developmental outcomes.[14]

Later-onset forms of G_{M2}-gangliosidoses, caused by HexA deficiency, have been identified.[15,16] In contrast to the infantile-onset form, patients with juvenile or adult-onset disease follow a more protracted course and show no ethnic predilection. Onset may be in childhood or later. In the juvenile variant, affected individuals manifest with onset of gait and/or speech problems between ages 2 and 7 years. There is subsequent mental deterioration, with loss of speech and walking ability, and death usually occurring between 10 and 15 years of age. A relatively attenuated form has been described, with insidious onset of progressive proximal muscle weakness during adolescence. Alternatively, patients may develop acute psychosis prior to evident gait problems. Psychiatric symptoms are common and affect as many as 40% of patients with late-onset forms of G_{M2}-gangliosidoses.[16,17] Regardless of presenting complaint, affected individuals develop ataxia with high steppage, and progressive walking difficulty eventually leading to wheelchair use. Dysarthria and communication problems are also common. In these cases, the optic fundi are normal and vision is not impaired, but patients display coarse saccadic pursuit. Brain imaging reveals cerebellar atrophy. Lifespan may extend into the fifth and sixth decade.

Phenotypic characterization was undertaken with 21 adult patients with late-onset G_{M2}-gangliosidosis.[16] The patients (mean age: 27.0 years; range: 14–47) were predominantly male (15/21 individuals) of Ashkenazi Jewish ancestry (15/18 families). Mean age at onset was 18.1 years; balance problems and difficulty climbing stairs were the most frequent presenting complaints. In several cases, the diagnosis was delayed and mean age at diagnosis was 27.0 years. Brain imaging studies revealed marked cerebellar atrophy in all patients (N = 18) tested, regardless of disease stage. A separate report with patients from the same cohort described comprehensive neuropsychological findings in 17 patients (13 men, 4 women), ranging in age from 18 to 56 years and who were at various stages of disease progression. Group mean performance was within the denoted normal range on all measures, except on a task assessing visual sequencing and set shifting. Approximately half of the subjects scored in the impaired range on measures of processing speed, visual sequencing, and set shifting. One-third also scored in the impaired range on measures of delayed verbal recall. Impairment tended to be restricted to a subset of the sample, as 5 of the 14 subjects able to undergo formal testing accounted for 70% of the total number of impaired scores. Overall, 47% of the patients exhibited significant cognitive impairment in at least one cognitive domain.

High-risk populations, due to founder effects for *HEXA* mutations, include Ashkenazi Jews (carrier frequency ~ 1/30), French Canadians, Cajuns, and Pennsylvania Dutch.[18] Carrier frequency in other populations is 1/300. As noted, carrier detection in high-risk populations and related programs have lowered the risk of having affected children.

Sandhoff disease is panethnic; however, there is less published experience on affected individuals. Patients with Sandhoff disease are clinically indistinguishable from patients with the TSD variant. A potential distinguishing feature is the presence of hepatosplenomegaly in the infantile form of Sandhoff disease, although this may be mild and potentially missed on exam. In addition, *N*-acetylglucosamine-containing oligosaccharides may be detected in urine (as only HexB is responsible for cleaving the *N*-acetylglucosamine moiety of the globoside). Cardiomyopathy has also been noted occasionally. In infantile cases, death usually occurs between 2 and 4 years of age. As in TSD, patients presenting with the juvenile or late-onset forms have been reported.[19,20] As with TSD and its variants, ultimate neurologic prognosis is poor.

G_{M2}-activator protein deficiency (AB variant) is a rare condition. Descriptions of a few affected patients report a neurodegenerative course similar to that seen with other clinical subtypes of G_{M2}-gangliosidosis. Cases described to date are patients presenting with an infantile clinical form and have an exaggerated startle response, delayed motor milestones, and increasing weakness, with onset at age 6–9 months.[21,22] Ophthalmologic evaluation showed bilateral macular cherry-red spots. In one case, cerebrospinal fluid (CSF) ganglioside profile revealed a highly elevated total ganglioside content.

DIAGNOSTIC CONFIRMATION

The diagnosis of G_{M2}-gangliosidosis can be established based on demonstration of hexosaminidase deficiency in serum and leukocytes, employing an assay that uses the fluorogenic substrate 4-methlyumbelliferyl-β-*N*-acetyl glucosaminide (4-MU substrate). Total hexosaminidase activity (HexA and B) is obtained, and HexB is measured after heat inactivation of HexA in a second serum aliquot. HexA activity is then calculated as the difference between the total hexosaminidase and HexB activities, with results expressed as the percentage of HexA activity relative to the total hexosaminidase activity.

In TSD and other related variants, total hexosaminidase activity is low, whereas HexB activity is normal or elevated. In Sandhoff disease, HexA and B activities are both low and, in addition, *N*-acetylglucosamine-containing oligosaccharides are detected in urine. Population screening has disclosed the presence of pseudodeficiency (PD) states, when assays use the artificial but not the natural substrate. As a consequence, there can be false-positive results (2% in Ashkenazi Jews, 35% in other populations). Two PD alleles have been identified, R247W and R249W, which can be verified by mutation analysis. False-positive biochemical/enzyme results may also occur when the *serum* of pregnant women and those taking oral contraceptives is used in the assay. This has been attributed to the presence of another form of hexosaminidase (HexP), which is not heat-labile; this leads to an overall reduction in the calculated percentage of HexA activity in these cases. Repeating the assay using white blood cells, which lack HexP, helps clarify the patient's true status.

A form of G_{M2}-gangliosidosis, called the B1 variant, which is clinically indistinguishable from TSD and Sandhoff disease, is associated with normal HexA activity in assays using the nonsulfated fluorogenic substrate. However, HexA activity is decreased in assays that employ the natural or 4-MU sulfated substrate. The B1 variant has been attributed to a specific mutation, p.R178H, in the α subunit of the *HEXA* gene. This mutation has been found among patients in Portugal and the general Mediterranean region.[23–25]

The diagnosis of G_{M2}-activator deficiency, which is also clinically indistinguishable from TSD and Sandhoff disease, may be suspected, based on demonstration of membranous cytoplasmic bodies in axons of the dermal nerves.[26–28] *In vitro*, HexA and B activities are within normal range. Previously, the diagnosis of G_{M2}-activator deficiency was established through G_{M2}-ganglioside loading tests, but this is a labor-intensive assay which has been superceded by mutation analysis and demonstration of a G_{M2}-activator gene defect.

Diagnosis of all G_{M2}-gangliosidosis variants can also be readily confirmed, by mutation analysis.

MOLECULAR GENETICS

The intralysosomal hydrolysis of G_{M2}-ganglioside requires three polypeptides, each encoded by a different gene (Figure 28.1). The α and β subunits come together to form the heteromeric isozyme HexA, responsible for G_{M2}-ganglioside hydrolysis/catabolism. The G_{M2}-activator protein is a small glycolipid binding protein (193 AA) which can form water-soluble complexes with G_{M2}-ganglioside and presents the lipid substrate to the water soluble HexA.[28]

In vitro mutagenesis and protein expression studies have identified certain mutations in the *HEXA* and *HEXB* gene to be severely deleterious, known as null alleles, resulting in no HexA enzymatic activity. Other mutations in these genes can result in a residual HexA enzymatic activity and lead to an attenuated ("mild") phenotype. In general, large deletions and full splice mutations are deleterious alleles associated with early, infantile onset and a rapid neurodegenerative course. Mutations in *HEXA* or *HEXB* associated with a residual HexA enzymatic activity, even in the presence of a deleterious mutation, correlate with later-onset disease and a protracted course, i.e., onset in childhood, with attenuated progression and longer survival, when compared with infantile-onset cases.

To a certain extent, the nature and position of the sequence alteration influences the biochemical activity of the mutant HexA enzyme. Mutations that lead to impairment of the binding of the substrate G_{M2} to the catalytic active site are usually deleterious, as they inactivate or abrogate enzyme function.[29] Missense mutations associated with conformational changes may be "mild" or severe, based on the extent to which they distort the structure of the enzyme, and its ability to interact with the substrate and/or the activator.[29,30]

Over 130 mutations have been described in the three genes encoding components of the HexA system, *HEXA*, *HEXB* and *GM2A*. Mutations of the *HEXA* gene encoding the α-chain are found in patients with TSD. Among patients of Ashkenazi Jewish descent, two common *HEXA* mutations are identified: the 1278insTACT in exon 11 and a G>A transversion at the 5′ splice site of intron 12, accounting for over 95% of disease alleles.[29] In the French Canadian population, two distinct founder mutations in the *HEXA* gene have been identified in patients presenting with the infantile form of TSD: a 7.6-kb deletion in the 5′ end (80% of alleles) and a G>A transition at the +1 position of intron 7.[31,32]

A recent study involving 34 Spanish patients with TSD revealed a high frequency of the c.459+5G>A (IVS4+5G>A) mutation in the *HEXA* gene, accounting for 32.4% of disease alleles.[33] All patients homozygous for this *HEXA* mutation presented with the infantile form of the disease. In this cohort, patients carrying the p.R178H mutation in at least one of the alleles presented with the juvenile form of TSD. Two other mutations, G269S and W494C, were identified in other patients, associated with residual HexA enzymatic activity and later onset and slower but progressive disease course.

Several mutations in the *HEXB* gene encoding the β-chain subunit have been identified in patients with Sandhoff disease.[29] In a study of 14 Spanish patients with Sandhoff disease, a novel deletion, c.171delG, was found to account for 21.4% of disease alleles.[34] All patients with this deletion showed the infantile form of the disease. Compound heterozygosity for the *HEXB* mutations, H235Y and P417L, has been described in an adult form of Sandhoff disease presenting with the motor neuron disease phenotype.

G_{M2}-activator (G_{M2}A) deficiency is rare. A limited number of mutations in the *GM2A* gene have been described.[29] Four disease-associated mutations, ΔAAG264 (ΔLys^{88}), ΔA410 (33 new amino acids, with a loss of 24), and T412→C (Cys^{138}Arg) in exon 3; and G506→C (Arg^{169}Pro) in exon 4, were presumed present in homozygosity in the four respective patients described.[29] None of these mutations resulted in detectable G_{M2}A fragment (cross-reacting material; CRM) in patient cells, although all are associated with normal steady-state levels of G_{M2}A mRNA.[21] In a Laotian patient, sequence analysis revealed a normal-sized G_{M2}A cDNA that contained a single nonsense mutation in exon 2.

DISEASE MECHANISMS

Reduced hexosaminidase enzymatic activity and neuronal accumulation of G_{M2}-ganglioside represent the primary insult to cells. Pathological brain studies reveal widespread neuronal G_{M2}-ganglioside storage throughout the cortex and central nuclear structures, the spinal cord, and autonomic ganglia. Evidence of neuronal storage has been established in fetuses as early as 12–22 weeks of gestation. Intraneuronal G_{M2}-ganglioside storage and reactive gliosis are reflected in the megalencephaly that develops in affected infants, associated with a 20–50% increase in brain weight, compared with age-matched controls.[14] Throughout the neuroaxis, ballooned neurons are encountered with fine granular or vesicular cytoplasm that stains positively with periodic acid–Schiff (PAS) and Luxol fast blue. Older patients show more neuronal loss with marked reduction in cerebellar Purkinje and granule cells. There is loss of white matter accompanied by gliosis with proliferation of protoplasmic astrocytes and microglia. A child with TSD who survives beyond 3–4 years shows cortical atrophy.[14,29]

Gangliosides are implicated in synaptogenesis, and an altered pattern of cellular gangliosides induces ectopic dendritogenesis and potential interference with synaptic transmission.[35,36] Storage in the proximal segment of axons may induce the formation of meganeurites, with additional neurites and aberrant synaptic contacts.

Interestingly, β-hexosaminidase activity has been found to be decreased in Alzheimer disease (AD), pointing to a potential convergent mechanism of disease with other more common disorders seen in the general population.[37–39] Intraneuronal ganglioside-bound amyloid-β peptide (GAβ) immunoreactivity, a proposed prefibrillar aggregate found in AD, has been found to accumulate throughout the frontal cortices of postmortem human TSD and Sandhoff disease brains. Moreover, α-synuclein and amyloid precursor protein carboxy terminal fragments (APP-CTFs) and/or amyloid-β (Aβ) were found to accumulate in different regions of the substantia nigra. Localization of the accumulated intraneuronal Aβ-like immunoreactivity (iAβ-LIR) to endosomes, lysosomes, and autophagosomes, indicates an association with the lysosomal-autophagic turnover of Aβ and fragments of amyloid precursor protein (APP)-containing Aβ epitopes. Altogether, these findings establish an association between the accumulation of gangliosides, autophagic vacuoles, and the intraneuronal accumulation of proteins associated with AD.[40]

It is likely that various cellular events play a contributory role in pathogenesis, although mechanistic links between ganglioside storage and clinical manifestations of G_{M2}-gangliosidoses have not been fully elucidated.[41] Aberrant inflammatory reactions may be relevant. However, no alterations in the differentiation of immunogenic dendritic cells (iDCs) from $CD34^+$ hematopoietic stem cells (HSC), generated by the RNA interference approach, have been observed.[42,43] On the other hand, the knock-down of *HEXA* or *HEXB* genes resulted in a loss of function of iDCs, indicative of an alteration in cell-mediated immune response. The silencing of the *HEXA* gene had a stronger immune inhibitory effect.[43]

In the Sandhoff knockout mouse model, a latent phase during the first few months of life has been noted, followed by a phase of rapid decline that is lethal within 6 weeks of onset. In transgenic mice wherein *HEXB* expression is silenced at 5 weeks of life (as a model for "delayed onset"), stereotypic signs of Sandhoff disease, including tremor, bradykinesia, and hind limb paralysis become evident. As in germline *HEXB* null mice, these neurodegenerative manifestations advanced rapidly. These observations indicate the pathogenesis and progression of Sandhoff disease is not influenced by developmental events in the maturing nervous system.[44]

Splenic B cells isolated from the mouse model for Sandhoff disease and other sphingolipid storage disorders such as NPC1, G_{M1}-gangliosidosis, and Fabry disease showed large (20- to 30-fold) increases in disease-specific glycosphingolipids and up to a 4-fold increase in cholesterol.[41] The glycosphingolipid storage led to an increase in the lysosomal compartment and altered glycosphingolipid trafficking. Specifically, the B cell receptor, CD21 and CD19 had decreased cell surface expression. In contrast, CD40 and MHC II, surface receptors that do not associate with lipid rafts, were unchanged. The clinical significance of these findings is uncertain, although it implicates changes that may be relevant to the subclass of glycosphingolipidosis to which Sandhoff disease belongs.[45]

Studies have been undertaken in two inducible mouse models in which reversible transgenic expression of β-hexosaminidase was directed by two promoters, mouse *HEXB* and human synapsin 1 promoters.[46] This approach enabled disease progression to be modified at predefined ages. Ultimately, late-onset brainstem and ventral spinal cord pathology was observed, associated with increased tone in the limbs of affected mice.

Interestingly, studies of brains from a mouse model of Sandhoff disease showed intraneuronal accumulation of Aβ-like, α-synuclein-like, and phospho-tau-like immunoreactivity.[40] Biochemical and immunohistochemical analyses confirmed that at least some of the iAβ-LIR represents APP-CTFs and/or Aβ.

Lyso-G_{M2}-ganglioside (lyso-G_{M2}), the deacylated form, is increased in the brain and plasma of Sandhoff mice.[40] Elevated levels decreased with intracerebroventricular administration of the modified HexB enzyme. High-performance liquid chromatography in patients with TSD and Sandhoff disease has also revealed elevated lyso-G_{M2}

levels in plasma. Thus, lyso-G_{M2} may be a potential biomarker that can be employed in screening tests and clinical trials as a pharmacodynamic endpoint.[47]

When initially described, the lysolipids detected in other glycosphingolipidoses were believed to exert their toxicity via inhibition of protein kinase C.[48] More recently, studies in animal models have shown that progressive central nervous system (CNS) inflammation (as evident from the expression pattern of several inflammatory markers and cytokines) and alteration in the integrity of the blood–brain barrier occurred coincidentally with the onset of clinical disease signs.[49]

IMAGING

Magnetic resonance imaging (MRI) studies of patients with the various G_{M2}-gangliosidosis variants are indistinguishable, and brain imaging does not enable specific etiologic diagnosis. Findings reflect the pathological features of neuronal cytoplasmic distention, demyelination, and gliosis involving the white matter.[25,50,51]

In TSD (variants B and B1) and Sandhoff disease (*O* variant), the thalami demonstrate symmetrical hyperdensity on computed tomography (CT), and hyper/hyposignal intensity on T1/T2-weighted MR images, respectively.[25] These density/intensity changes, also noted in the mesial temporal lobe gyri, likely reflect the accumulation of calcium and G_{M2}-ganglioside. Mild brain and corpus callosum atrophy in these patients has invariably been attributed to disturbed myelination. The MR spectroscopy findings are nonspecific and indicate progressive neuronal loss and gliosis in the cerebral tissue.[52]

A report on 10 later-onset patients, ranging in age at onset from 3–12 years, revealed diffuse cerebellar atrophy in all patients, and magnetic resonance spectroscopy was characterized by low *N*-acetylaspartate levels in the cerebellum. All patients had abnormal focal supratentorial white matter signal of the corona radiata, while the internal capsule and optic radiation in the peritrigonal region were relatively spared. Fluid-attenuated inversion recovery (FLAIR) imaging signal of the cerebellar cortex was mildly increased in patients with advanced disease and severe cerebellar atrophy.[52]

A detailed review of 300 patients with confirmed cerebellar atrophy on MRI over a 10-year period revealed late-onset G_{M2}-gangliosidosis in 10 patients (4 males, 6 females). Average age at clinical presentation was 9.5 years (range 3–12 years). All presented with behavioral changes or psychotic symptoms and then developed ataxia, seizures, spasticity, and cognitive regression by 17 years. The MRI at initial assessment revealed diffuse cerebellar atrophy in all patients, and magnetic resonance spectroscopy was characterized by low *N*-acetylaspartate levels in the cerebellum. All patients had abnormal focal supratentorial white matter signal of the corona radiata, while the internal capsule and optic radiation in the peritrigonal region were relatively spared. FLAIR imaging signal of the cerebellar cortex was mildly increased in patients with advanced disease and severe cerebellar atrophy.

DIFFERENTIAL DIAGNOSIS

A retrospective study of patients with infantile G_{M2}-gangliosidosis revealed that despite the early and progressive manifestations (average age at first symptom 5.0±3.3 months) the diagnosis was not established until 13.3 months.[13] Misdiagnoses listed in this survey study included cerebral palsy and mitochondrial disorders. It was pointed out that the combination of exaggerated startle response and low tone is relatively uncommon and should prompt the clinician to consider the diagnosis of G_{M2}-gangliosidosis.[13]

As a point of departure from other lysosomal storage disorders associated with neurodegenerative features, patients with G_{M2}-gangliosidoses do not have any of the following signs: dysmorphic features, visceromegaly, and skeletal deformities (e.g., *dysostosis multiplex* as seen in the mucopolysaccharidosis).

Among affected teens and adults with late-onset G_{M2}-gangliosidoses, initial diagnoses given include spinal muscular atrophy (Kugelberg–Welander disease), amyotrophic lateral sclerosis, atypical motor neuron disease, or spinocerebellar degeneration. Clinical signs, disease course, and family history should help in distinguishing one entity from another in this listing.[13,20]

With regard to brain imaging findings in G_{M2}-gangliosidosis patients, differentials include the lesions with CT hyperdense/T2 hypointense thalami such as status marmoratus, neuronal ceroid lipofuscinosis, and Krabbe disease. Again, clinical history and additional signs enable appropriate distinction.[14]

MANAGEMENT

Specific and disease-modifying therapy for G_{M2}-gangliosidosis is not currently available. However, supportive treatment should be undertaken to maintain patients' wellbeing and quality of life.[53] Medications including haloperidol, risperidone, and chlorpromazine are lysosomotropic and impair lysosomal function in cultured cells.[54] In some patients, worsening of neurological symptoms and neuroleptic malignant syndrome have been reported.[55] Acute psychosis may be managed with a combination of lithium salts and lorazepam.[56–58] Over the last few decades, improved symptomatic management, such as use of antibiotics and gastric tube (GT) placement, has had a positive influence on survival.

Hematopoietic stem cell transplantation (HSCT) has been shown to be beneficial for some lysosomal storage diseases (LSDs), such as Hurler syndrome.[59] However, a limited number of patients with infantile G_{M2}-gangliosidosis subjected to HSCT did experience prolonged survival or retention of milestones.[60–62] Presymptomatic HSCT in patients with the juvenile form of G_{M2}-gangliosidosis has not been shown to alter disease course or ultimate prognosis.[63] The presence of inflammatory changes as a putative mechanism of disease and the resultant salutary changes noted in animal models subjected to bone marrow transplantation have generated renewed interest in cellular approaches to treatment.

Enzyme replacement therapy (ERT) is now available as a therapeutic option for some lysosomal storage disorders (e.g., Gaucher and Fabry disease).[64] These treatments involve large molecular-weight molecules (i.e., the recombinant formulation of the cognate enzyme) that are unable to cross the blood–brain barrier (BBB) and reach affected cells within the central nervous system. Thus, intravenously administered ERT is not currently considered as a viable option, although recombinant enzyme given intrathecally is being explored in severe Hunter syndrome (mucopolysaccharidosis type II), an LSD caused by deficient α-L-iduronidase activity.

Small molecule based-therapies are more likely to cross the BBB and treat the neurological symptoms of G_{M2}-gangliosidosis and other LSDs. Two such approaches are under investigation: substrate reduction therapy (SRT) and the use of pharmacologic chaperones. SRT is based on the inhibition of biosynthesis of the substrate of the defective or absent enzyme. In the case of HexA deficiency, the aim is to inhibit glucosylceramide synthase, which catalyzes the first step in the biosynthetic glycosphingolipid pathway.[65] As a consequence, the amount of substrate made will be reduced to a level matched by the residual capacity of the mutant enzyme, thereby preventing or reducing G_{M2}-ganglioside storage.[65,66] One such agent, *N*-butyldeoxynojirimycin (*N*B-DNJ), also known as miglustat (Zavesca, Actelion Pharmaceuticals), was tested in patients with the juvenile form of G_{M2}-gangliosidosis.[67,68] The pharmacokinetics, safety, and tolerability of miglustat in pediatric patients tested were shown to be favorable.[68] However, miglustat failed to ameliorate progressive neurological deterioration. On the other hand, there was no apparent worsening in some areas of cognitive function and brain MRI lesions appeared stable over 24 months of treatment. These findings must be interpreted with care owing to the small sample size and the lack of a control-arm.[67]

Another small-molecule approach is the use of pharmacologic chaperones that can rescue misfolded mutant protein from premature degradation via the endoplasmic reticulum (ER)-associated degradation/ubiquitin pathway.[30,69] By serving as a template for the enzyme, chaperones enable the ultimate delivery of the nascent protein to the lysosome, thereby restoring residual enzyme activity. Pyrimethamine, an antiparasitic drug, has been identified to act as a potential pharmacological chaperone for mutant HexA. Pyrimethamine has been shown to bind to the active site of HexB, which leads to stabilization of the misfolded enzyme. In the lysosome's acidified milieu, the drug disengages and allows the enzyme to process the stored substrate.[70] *In vitro* studies have shown that many mutant α or β subunits of HexA can be partially rescued, following the growth of patients' cells in the presence of the drug.[71] An open-label phase I/II clinical trial of pyrimethamine has been conducted, which enrolled 11 patients (age range 23–50 years) with chronic G_{M2}-gangliosidosis (Tay–Sachs or Sandhoff variants).[72] Pyrimethamine was given, in escalating doses to a maximum of 100 mg, once daily over a 16-week period. In eight evaluable subjects, up to a 4-fold enhancement of HexA activity at doses of 50 mg per day or less was observed.[72] There was marked individual variation in the pharmacokinetics of the drug. Significant adverse effects (AE), such as increased ataxia or incoordination, were experienced by most patients at or above 75 mg of pyrimethamine per day; these AEs remitted with treatment interruption.

Gene transfer when undertaken before neurological signs were manifest, effectively rescued the acute neurodegenerative illness in Sandhoff mice. Intracranial co-injection of recombinant adeno-associated viral vectors (rAAV), serotype 2/1, expressing HexA and B subunits into 1-month-old Sandhoff mice resulted in extended survival (to 2 years) and prevented disease throughout the brain and spinal cord. Classical manifestations of disease, including spasticity—as opposed to tremor/ataxia—were resolved by localized gene transfer to the striatum or cerebellum, respectively.[73] Intracranial transplantation of neural stem cells (NSCs) was examined as an alternative approach, alone or in combination with miglustat, in "juvenile" Sandhoff mice.[74] Compared to untreated and sham-treated controls, NSC transplantation, performed at postnatal day 0, provided a slight increase in Hex activity and significantly decreased GA2 content, but not G_{M2}, at the time of analysis (postnatal day 15). On the other hand, miglustat had no effect on Hex activity but significantly reduced G_{M2} and GA2 content. Combination therapy reduced G_{M2} and GA2 content to a level similar to that of the miglustat-alone treated mice. Thus, no additive or synergistic effect between NSC and SRT was found in these "juvenile" Sandhoff mice.[74]

Using the same AAV gene therapy strategy via bilateral injection of the thalamus, a feline model of Sandhoff disease was shown to increase lifespan and lead to improvement of neurological phenotype.[75] A clinical trial using the AAV gene therapy in patients with G_{M2}-gangliosidosis has been designed and is set to start.

There are ongoing investigations of various therapeutic options for G_{M2}-gangliosidosis; it is hoped this will eventually lead to the introduction of therapy for this recalcitrant lethal condition.

References

1. Svennerholm L. Isolation of gangliosides. *Acta Chem Scand*. 1963;17:239–250.
2. Tay W. Symmetrical changes in the region of yellow spot in each eye of an infant. *Trans Opthal Soc U K*. 1881;1:55–57.
3. Sachs B. A family form of idiocy, generally fatal, associated with early blindness. *J Nerv Ment Dis*. 1896;14:475–479.
4. Klenk E. Uber die Ganglioside des Gehirns bei der infantilen amaurotischen Idiotie vom Typus Tay–Sachs. *Ber Dtsch Chem Ges*. 1942;75:1632–1636.
5. Svennerholm L. The chemical structure of normal brain and Tay–Sachs gangliosides. *Biochem Biophys Res Commun*. 1962;436–441.
6. Makita A, Yamakawa T. The glycolipids of the brain of Tay–Sachs' disease–the chemical structures of a globoside and main ganglioside. *Jpn J Exp Med*. 1963;33:361–368.
7. Robinson D, Stirling JL. *N*-Acetyl-beta-glucosaminidases in human spleen. *Biochem J*. 1968;107(3):321–327.
8. Okada S, O'Brien JS. Generalized gangliosidosis: beta-galactosidase deficiency. *Science*. 1968;160(831):1002–1004.
9. Okada S, O'Brien JS. Tay–Sachs disease: generalized absence of a beta-D-*N*-acetylhexosaminidase component. *Science*. 1969;165(894):698–700.
10. Sandhoff K, Andreae U, Jatzkewitz H. Deficient hexozaminidase activity in an exceptional case of Tay–Sachs disease with additional storage of kidney globoside in visceral organs. *Life Sci*. 1968;7(6):283–288.
11. Kaback MM. Thermal fractionation of serum hexosaminidases: Heterozygote detection and diagnosis of Tay–Sachs disease. *Methods Enzymol*. 1972;28:862.
12. Conzelmann E, Sandhoff K. AB variant of infantile G_{M2} gangliosidosis: deficiency of a factor necessary for stimulation of hexosaminidase A-catalyzed degradation of ganglioside G_{M2} and glycolipid GA2. *Proc Natl Acad Sci U S A*. 1978;75(8):3979–3983.
13. Smith NJW, Winstone AM, Stellitano L, Cox TM. G_{M2} gangliosidosis in a United Kingdom study of children with progressive neurodegeneration: 73 cases reviewed. *Dev Med Child Neurol*. 2012;54(2):176–182.
14. Bley AE, Giannikopoulos OA, Hayden D, Kubilus K, Tifft CJ, Eichler FS. Natural history of infantile $G_{(M2)}$ gangliosidosis. *Pediatrics*. 2011;128(5):e1233–e1241.
15. Neudorfer O, Kolodny EH. Late-onset Tay–Sachs disease. *Isr Med Assoc J*. 2004;6(2):107–111.
16. Neudorfer O, Pastores GM, Zeng BJ, Gianutsos J, Zaroff CM, Kolodny EH. Late-onset Tay–Sachs disease: phenotypic characterization and genotypic correlations in 21 affected patients. *Genet Med*. 2005;7(2):119–123.
17. Navon R. Late-onset G_{M2} gangliosidosis and other hexosaminidase mutations among Jews. *Adv Genet*. 2001;44:185–197.
18. Kaback MM. Screening and prevention in Tay–Sachs disease: origins, update, and impact. *Adv Genet*. 2001;44:253–265.
19. Hendriksz CJ, Corry PC, Wraith JE, Besley GT, Cooper A, Ferrie CD. Juvenile Sandhoff disease—nine new cases and a review of the literature. *J Inherit Metab Dis*. 2004;27(2):241–249.
20. Maegawa GH, Stockley T, Tropak M, et al. The natural history of juvenile or subacute G_{M2} gangliosidosis: 21 new cases and literature review of 134 previously reported. *Pediatrics*. 2006;118(5):e1550–e1562.
21. Schepers U, Glombitza G, Lemm T, et al. Molecular analysis of a G_{M2}-activator deficiency in two patients with G_{M2}-gangliosidosis AB variant. *Am J Hum Genet*. 1996;59(5):1048–1056.
22. Xie B, Kennedy JL, McInnes B, Auger D, Mahuran D. Identification of a processed pseudogene related to the functional gene encoding the G_{M2} activator protein: localization of the pseudogene to human chromosome 3 and the functional gene to human chromosome 5. *Genomics*. 1992;14(3):796–798.
23. Brown CA, Neote K, Leung A, Gravel RA, Mahuran DJ. Introduction of the alpha subunit mutation associated with the B1 variant of Tay–Sachs disease into the beta subunit produces a beta-hexosaminidase B without catalytic activity. *J Biol Chem*. 1989;264(36):21705–21710.
24. dos Santos MR, Tanaka A, sa Miranda MC, Ribeiro MG, Maia M, Suzuki K. G_{M2}-gangliosidosis B1 variant: analysis of beta-hexosaminidase alpha gene mutations in 11 patients from a defined region in Portugal. *Am J Hum Genet*. 1991;49(4):886–890.
25. Grosso S, Farnetani MA, Berardi R, et al. G_{M2} gangliosidosis variant B1 neuroradiological findings. *J Neurol*. 2003;250(1):17–21.
26. Goldman JE, Yamanaka T, Rapin I, Adachi M, Suzuki K. The AB-variant of G_{M2}-gangliosidosis. Clinical, biochemical, and pathological studies of two patients. *Acta Neuropathol (Berl)*. 1980;52(3):189–202.
27. Heng HH, Xie B, Shi XM, Tsui LC, Mahuran DJ. Refined mapping of the G_{M2} activator protein (GM2A) locus to 5q31.3-q33.1, distal to the spinal muscular atrophy locus. *Genomics*. 1993;18(2):429–431.
28. Mahuran DJ. The G_{M2} activator protein, its roles as a co-factor in G_{M2} hydrolysis and as a general glycolipid transport protein. *Biochim Biophys Acta*. 1998;1393(1):1–18.
29. Mahuran DJ. Biochemical consequences of mutations causing the G_{M2} gangliosidoses. *Biochim Biophys Acta*. 1999;1455(2–3):105–138.
30. Tropak MB, Mahuran D. Lending a helping hand, screening chemical libraries for compounds that enhance beta-hexosaminidase A activity in G_{M2} gangliosidosis cells. *FEBS J*. 2007;274(19):4951–4961.

31. De Braekeleer M, Hechtman P, Andermann E, Kaplan F. The French Canadian Tay–Sachs disease deletion mutation: identification of probable founders. *Hum Genet*. 1992;89(1):83–87.
32. Myerowitz R, Hogikyan ND. A deletion involving Alu sequences in the beta-hexosaminidase alpha-chain gene of French Canadians with Tay–Sachs disease. *J Biol Chem*. 1987;262(32):15396–15399.
33. Gort L, de Olano N, Macias-Vidal J, Coll MA. G_{M2} gangliosidoses in Spain: analysis of the HEXA and HEXB genes in 34 Tay–Sachs and 14 Sandhoff patients. *Gene*. 2012;506(1):25–30.
34. Zampieri S, Cattarossi S, Oller Ramirez AM, et al. Sequence and copy number analyses of HEXB gene in patients affected by Sandhoff disease: functional characterization of 9 novel sequence variants. *PLoS One*. 2012;7(7):e41516.
35. Walkley SU, Siegel DA, Dobrenis K. G_{M2} ganglioside and pyramidal neuron dendritogenesis. *Neurochem Res*. 1995;20(11):1287–1299.
36. Walkley SU, Siegel DA, Wurzelmann S. Ectopic dendritogenesis and associated synapse formation in swainsonine-induced neuronal storage disease. *J Neurosci*. 1988;8(2):445–457.
37. Tamboli IY, Tien NT, Walter J. Sphingolipid storage impairs autophagic clearance of Alzheimer-associated proteins. *Autophagy*. 2011;7(6):645–646.
38. Mielke MM, Lyketsos CG. Alterations of the sphingolipid pathway in Alzheimer's disease: new biomarkers and treatment targets? *Neuromolecular Med*. 2010;12(4):331–340.
39. Siegel SJ, Bieschke J, Powers ET, Kelly JW. The oxidative stress metabolite 4-hydroxynonenal promotes Alzheimer protofibril formation. *Biochemistry*. 2007;46(6):1503–1510.
40. Keilani S, Lun Y, Stevens AC, et al. Lysosomal dysfunction in a mouse model of Sandhoff disease leads to accumulation of ganglioside-bound amyloid-beta peptide. *J Neurosci*. 2012;32(15):5223–5236.
41. Sandhoff K, Harzer K. Gangliosides and gangliosidoses: principles of molecular and metabolic pathogenesis. *J Neurosci*. 2013;33(25):10195–10208.
42. Jeyakumar M, Thomas R, Elliot-Smith E, et al. Central nervous system inflammation is a hallmark of pathogenesis in mouse models of G_{M1} and G_{M2} gangliosidosis. *Brain*. 2003;126(Pt 4):974–987.
43. Tiribuzi R, D'Angelo F, Berardi AC, Martino S, Orlacchio A. Knock-down of HEXA and HEXB genes correlate with the absence of the immunostimulatory function of HSC-derived dendritic cells. *Cell Biochem Funct*. 2012;30(1):61–68.
44. Phaneuf D, Wakamatsu N, Huang JQ, et al. Dramatically different phenotypes in mouse models of human Tay–Sachs and Sandhoff diseases. *Hum Mol Genet*. 1996;5(1):1–14.
45. te Vruchte D, Jeans A, Platt FM, Sillence DJ. Glycosphingolipid storage leads to the enhanced degradation of the B cell receptor in Sandhoff disease mice. *J Inherit Metab Dis*. 2010;33(3):261–270.
46. Sargeant TJ, Drage DJ, Wang S, Apostolakis AA, Cox TM, Cachon-Gonzalez MB. Characterization of inducible models of Tay–Sachs and related disease. *PLoS Genet*. 2012;8(9):e1002943.
47. Kodama T, Togawa T, Tsukimura T, et al. Lyso-G_{M2} ganglioside: a possible biomarker of Tay–Sachs disease and Sandhoff disease. *PLoS One*. 2011;6(12):e29074.
48. Huang F, Subbaiah PV, Holian O, et al. Lysophosphatidylcholine increases endothelial permeability: role of PKCalpha and RhoA cross talk. *Am J Physiol Lung Cell Mol Physiol*. 2005;289(2):L176–L185.
49. Hayase T, Shimizu J, Goto T, et al. Unilaterally and rapidly progressing white matter lesion and elevated cytokines in a patient with Tay–Sachs disease. *Brain Dev*. 2010;32(3):244–247.
50. Bano S, Prasad A, Yadav SN, Chaudhary V, Garga UC. Neuroradiological findings in G_{M2} gangliosidosis variant B1. *J Pediatr Neurosci*. 2011;6(2):110–113.
51. Imamura A, Miyajima H, Ito R, Orii KO. Serial MR imaging and 1H-MR spectroscopy in monozygotic twins with Tay–Sachs disease. *Neuropediatrics*. 2008;39(5):259–263.
52. Assadi M, Baseman S, Janson C, Wang DJ, Bilaniuk L, Leone P. Serial 1H-MRS in G_{M2} gangliosidoses. *Eur J Pediatr*. 2008;167(3):347–352.
53. Maegawa G. G_{M2} gangliosidosis: the prototype of lysosomal storage disorders. *Dev Med Child Neurol*. 2012;54(2):104–105.
54. Lullmann H, Lullmann-Rauch R, Wassermann O. Lipidosis induced by amphiphilic cationic drugs. *Biochem Pharmacol*. 1978;27(8):1103–1108.
55. Shapiro BE, Hatters-Friedman S, Fernandes-Filho JA, Anthony K, Natowicz MR. Late-onset Tay–Sachs disease: adverse effects of medications and implications for treatment. *Neurology*. 2006;67(5):875–877.
56. MacQueen GM, Rosebush PI, Mazurek MF. Neuropsychiatric aspects of the adult variant of Tay–Sachs disease. *J Neuropsychiatry Clin Neurosci*. 1998 Winter;10(1):10–19.
57. Rosebush PI, Mazurek MF. Complicating factors in the analysis of acute drug-induced akathisia. *Arch Gen Psychiatry*. 1995;52(10):878–880.
58. Hurowitz GI, Silver JM, Brin MF, Williams DT, Johnson WG. Neuropsychiatric aspects of adult-onset Tay–Sachs disease: two case reports with several new findings. *J Neuropsychiatry Clin Neurosci*. 1993 Winter;5(1):30–36.
59. Prasad VK, Kurtzberg J. Cord blood and bone marrow transplantation in inherited metabolic diseases: scientific basis, current status and future directions. *Br J Haematol*. 2010;148(3):356–372.
60. Bembi B, Marchetti F, Guerci VI, et al. Substrate reduction therapy in the infantile form of Tay–Sachs disease. *Neurology*. 2006;66(2):278–280.
61. Hoogerbrugge PM, Brouwer OF, Bordigoni P, et al. Allogeneic bone marrow transplantation for lysosomal storage diseases. The European Group for Bone Marrow Transplantation. *Lancet*. 1995;345(8962):1398–1402.
62. Martin PL, Carter SL, Kernan NA, et al. Results of the cord blood transplantation study (COBLT): outcomes of unrelated donor umbilical cord blood transplantation in pediatric patients with lysosomal and peroxisomal storage diseases. *Biol Blood Marrow Transplant*. 2006;12(2):184–194.
63. Jacobs JF, Willemsen MA, Groot-Loonen JJ, Wevers RA, Hoogerbrugge PM. Allogeneic BMT followed by substrate reduction therapy in a child with subacute Tay–Sachs disease. *Bone Marrow Transplant*. 2005;36(10):925–926.
64. Grabowski GA, Hopkin RJ. Enzyme therapy for lysosomal storage disease: principles, practice, and prospects. *Annu Rev Genomics Hum Genet*. 2003;4:403–436.
65. Platt FM, Jeyakumar M. Substrate reduction therapy. *Acta Paediatr Suppl*. 2008;97(457):88–93.
66. Platt FM, Butters TD. Substrate deprivation: a new therapeutic approach for the glycosphingolipid lysosomal storage diseases. *Expert Rev Mol Med*. 2000;2(1):1–17.

67. Maegawa GH, Banwell BL, Blaser S, et al. Substrate reduction therapy in juvenile G_{M2} gangliosidosis. *Mol Genet Metab.* 2009;98(1–2):215–224.
68. Maegawa GH, van Giersbergen PL, Yang S, et al. Pharmacokinetics, safety and tolerability of miglustat in the treatment of pediatric patients with G_{M2} gangliosidosis. *Mol Genet Metab.* 2009;97(4):284–291.
69. Fan JQ. A counterintuitive approach to treat enzyme deficiencies: use of enzyme inhibitors for restoring mutant enzyme activity. *Biol Chem.* 2008;389(1):1–11.
70. Bateman KS, Cherney MM, Mahuran DJ, Tropak M, James MN. Crystal structure of beta-hexosaminidase B in complex with pyrimethamine, a potential pharmacological chaperone. *J Med Chem.* 2011;54(5):1421–1429.
71. Maegawa GH, Tropak M, Buttner J, et al. Pyrimethamine as a potential pharmacological chaperone for late-onset forms of G_{M2} gangliosidosis. *J Biol Chem.* 2007;282(12):9150–9161.
72. Clarke JT, Mahuran DJ, Sathe S, et al. An open-label Phase I/II clinical trial of pyrimethamine for the treatment of patients affected with chronic G_{M2} gangliosidosis (Tay–Sachs or Sandhoff variants). *Mol Genet Metab.* 2011;102(1):6–12.
73. Sargeant TJ, Wang S, Bradley J, et al. Adeno-associated virus-mediated expression of beta-hexosaminidase prevents neuronal loss in the Sandhoff mouse brain. *Hum Mol Genet.* 2011;20(22):4371–4380.
74. Arthur JR, Lee JP, Snyder EY, Seyfried TN. Therapeutic effects of stem cells and substrate reduction in juvenile Sandhoff mice. *Neurochem Res.* 2012;37(6):1335–1343.
75. Bradbury AM, Cochran JN, McCurdy VJ, et al. Therapeutic response in feline Sandhoff disease despite immunity to intracranial gene therapy. *Mol Ther.* 2013;21(7):1306–1315.

CHAPTER

29

Metachromatic Leukodystrophy and Multiple Sulfatase Deficiency

Florian S. Eichler

Massachusetts General Hospital, Harvard Medical School, Boston, MA, USA

INTRODUCTION

Disease Characteristics

Sulfatides are a major component of the myelin sheath in the central and peripheral nervous system. Their accumulation leads to clinical symptoms in metachromatic leukodystrophy and multiple sulfatase deficiency.

Hallmark Manifestations

Metachromatic leukodystrophy (MLD) is a lysosomal storage disorder due to arylsulfatase A deficiency.[1] It encompasses three clinical subtypes: a late-infantile, a juvenile and an adult form of MLD. Age of onset within a family is usually similar. The disease is characterized by progressive neurological dysfunction and loss of previously attained milestones. MLD is usually life-limiting, with a disease course that ranges from 3 to 10 or more years in the late-infantile form and up to 20 years or more in the juvenile and adult forms.

Multiple sulfatase deficiency (MSD) is due to faulty processing of an active site cysteine to formylglycine (alanine-semialdehyde), a proenzyme activation step common to most sulfatases.[2] Clinical variability of multiple sulfatase deficiency is great, and features of both MLD and a mucopolysaccharidosis (MPS) may be present.[3] More severe forms of MSD resemble late-infantile MLD. In milder cases, MPS-like features such as coarse facial features and skeletal abnormalities may be evident in infancy and early childhood, with MLD-like symptoms appearing in later childhood.

Inheritance

Both MLD and MSD are inherited in an autosomal recessive manner.[2] The presence of the disease-causing mutations should be determined in both parents so that screening of at-risk relatives can occur. Carrier testing of at-risk family members and prenatal diagnosis for pregnancies at increased risk are possible if both disease-causing mutations have been identified in an affected family member.

Diagnosis/Testing

Characteristic magnetic resonance imaging (MRI) findings in the setting of progressive neurologic symptoms should prompt testing of arylsulfatase A. MLD is suggested by arylsulfatase A (ARSA) enzyme activity in leukocytes that is less than 10% of normal controls. Approximately 0.5–2.0% of the Caucasian population shows a substantial arylsulfatase A deficiency, with a residual enzyme activity of about 10% of normal, without any clinical symptoms related to the deficiency. This phenomenon has been termed arylsulfatase A pseudodeficiency.

Since the assay of ARSA enzymatic activity cannot distinguish between MLD and ARSA pseudodeficiency, molecular genetic testing of ARSA, or urinary excretion of sulfatides must confirm the diagnosis of MLD. At times, confirmation can also arise from metachromatic lipid deposits in nervous system tissue on biopsy samples.

Rosenberg's Molecular and Genetic Basis of Neurological and Psychiatric Disease
http://dx.doi.org/10.1016/10.1016/B978-0-12-410529-4.00029-2

Features characteristic for MSD, such as facial dysmorphia, skeletal deformities, and ichthyosis, may be overlooked. Therefore, once deficiency in arylsulfatase A has been found, total arylsulfatase or another sulfatase should be assayed to prove the specificity of ARSA deficiency.

CLINICAL FEATURES

Historical Overview

Disease and Gene Identification

The first description of a patient with MLD was published in 1910 by Alois Alzheimer.[4] Alzheimer described the metachromatic staining of the nervous system and the clinical symptoms of a patient who today would be classified as having adult onset MLD. In 1921, Witte described a patient not only with metachromatic staining in the brain but also in liver, kidney and testis.[5] In 1963, Austin et al. described the deficiency in arylsulfatase A (ARSA) in MLD;[6] 2 years later, Mehl and Jatzkewitz demonstrated a block in metabolism of sulfatides.[7] The *ARSA* gene was mapped to the long arm of chromosome 22 band q13.[8] Austin first described MSD in 1965.[9]

Mode of Inheritance and Prevalence

The reported prevalence of MLD ranges from 1:40,000 to 1:160,000 in different populations.[1] The disorder seems to occur throughout the world but is more prevalent in particular consanguineous populations:[2] 1:75 in Habbanite Jews in Israel, 1:8000 in Israeli Arabs, 1:10,000 in Christian Israeli Arabs, 1:2500 for the western portion of the Navajo Nation in the US.

Since the first description of MSD in 1965, less than 100 patients have been described. Hence the true incidence and prevalence is not known.

Natural History

Age of Onset, Disease Evolution and Disease Variants

The three clinical subtypes of MLD differ by age of onset: late-infantile MLD, comprising 50–60% of cases; juvenile MLD, approximately 20–30%; and adult MLD, approximately 15–20%. The age of onset within a family is usually similar, but exceptions occur.[10]

LATE-INFANTILE MLD

The majority of patients present with motor and gait abnormalities between ages 1 and 2 years.[11] One-third of the patients may present with seizures. Later signs include inability to stand, difficulty with speech, deterioration of mental function, increased muscle tone, pain in the arms and legs, generalized or partial seizures, compromised vision and hearing, and peripheral neuropathy. In the final stages children have tonic spasms, decerebrate posturing, and general unawareness of their surroundings.

JUVENILE MLD

Most patients present with inattention and difficulties at school between age 4 years and sexual maturity (age 12–14 years). Gait problems, slurred speech, incontinence, and bizarre behaviors also occur. Seizures are less frequent than in the late-infantile form of the disease. Progression is similar to but slower than the late-infantile form.

ADULT MLD

Symptom onset occurs in the fourth or fifth decade. Dementia and behavioral difficulties are the first symptoms.[11] Patients show personality changes, alcohol or drug abuse, poor money management, and emotional lability. In others, weakness, loss of coordination, or seizures occur. Peripheral neuropathy is common. Progression is variable and may be protracted over decades. The final stage is similar to that for the earlier-onset forms.

INFANTILE MSD

The classic form of MSD combines clinical features of late infantile MLD with mild features of MPS. Most children acquire the ability to stand and say a few words. During the second year, children lose acquired abilities and display staring spells, spasticity, blindness, hearing loss, swallowing difficulties, seizures and dementia. Ichthyosis develops around 2–3 years of age. Symptoms of MPS occur at variable stages of the disease and include coarse facial features, stiff joints, growth retardation, skeletal abnormalities, and hepatosplenomegaly. Age at death is usually around 10–18 years of age.

MOLECULAR GENETICS

Normal Gene Product and Abnormal Gene Product(s)

ARSA Enzyme Activity

In MLD, the age of onset is not related to the amount of apparent enzyme activity as usually measured. Age of onset does, however, correlate reasonably well with the ability of cultured fibroblasts to degrade sulfatide added to the culture medium. In early-onset (late-infantile) MLD, affected individuals are usually homozygous or compound heterozygous for I-type ARSA-MLD alleles and make no detectable functional arylsulfatase A enzyme. Later-onset individuals with MLD have one or two A-type ARSA-MLD alleles that encode for an arylsulfatase A enzyme with some functional activity (≤1% when assayed with physiologic substrates).

Genotype–Phenotype Correlations

A genotype–phenotype correlation for MLD has been demonstrated in several independent studies.[2] It can be explained by the varying amount of residual enzyme activity associated with the genotype of the patient.

Patients homozygous for alleles that do not allow for the expression of any enzyme activity (i.e., null alleles) always suffer from the most severe late-infantile form of the disease. The most frequent allele is the IVS459+1A>G splice donor site mutation of exon 2. Most mutations in late-infantile patients, however, are missense mutations. This causes misfolding and trapping of the enzyme in the endoplasmic reticulum, so that no functional enzyme reaches the lysosome.

Many cases of juvenile-onset MLD are associated with heterozygosity for a null allele and a non-null allele.[2] Occasionally this is also found in adult-onset patients. More commonly, though, adult-onset disease shows expression of low amounts of enzyme activity, secondary to alleles being missense mutations. The residual enzyme in these cases can degrade small amounts of sulfatide.

DISEASE MECHANISMS

Pathophysiology

MLD and MSD are disorders of impaired breakdown of sulfatides (cerebroside sulfate or 3-*O*-sulfo-galactosylceramide), sulfate-containing lipids that occur throughout the body and are found in greatest abundance in nervous tissue, kidneys, and testes. Sulfatides are critical constituents of the nervous system, where they comprise approximately 5% of myelin lipids. Nervous system sulfatide accumulation is not restricted to glial cells, however, but also occurs in neurons. Sulfatide accumulation in the nervous system eventually leads to myelin breakdown (leukodystrophy) and a progressive neurologic disorder.[1]

DIFFERENTIAL DIAGNOSIS

Other Leukodystrophies and Lysosomal Storage Diseases

Clinically, MLD is often difficult to differentiate from other progressive degenerative disorders that manifest after initial normal development (Table 29.1). Delayed development in late infancy most often leads to MRI evaluation that can reveal a characteristic pattern of injury. If symmetric confluent white matter changes are present on MRI,

TABLE 29.1 Differential Diagnosis of MLD and MSD

Disorder	Clinical manifestations	Urinary excretion	Enzyme activity
MLD	See above	Elevated sulfatide	Low ARSA enzyme activity
MSD	MLD-like clinical picture, with elevated CSF protein and slowed nerve conduction velocity; MPS-like features, and ichthyosis	Elevated sulfatide and mucopolysaccharides	Very low ARSA enzyme activity; deficiency of most sulfatases in leukocytes or cultured cells
Saposin B deficiency	Similar to MLD	Elevated sulfatide and other glycolipids	ARSA enzyme activity within normal range
Other leukodystrophies	Progressive motor and cognitive decline with variable age of onset	Normal sulfatide and other glycolipids	ARSA enzyme activity within normal range

Adapted from Fluharty AL. Arylsulfatase A Deficiency, Gene Reviews; 2014. Available at: http://www.ncbi.nlm.nih.gov/books/NBK1130/, Accessed May 2014.

other conditions to consider are Krabbe disease and X-linked adrenoleukodystrophy. Hypomyelinating disorders such as Pelizaeus–Merzbacher disease lack the sparing of the subcortical white matter usually seen in metachromatic leukodystrophy.

Although some mucopolysaccharidoses can have a similar presentation to MLD, the most characteristic physical features seen in most mucopolysaccharidoses are not found in individuals with MLD but can be seen in some patients with MSD. The evaluation of appropriate lysosomal enzymes can distinguish these disorders.

It should be kept in mind that low ARSA enzyme activity caused by arylsulfatase pseudodeficiency can be found in association with many disorders. As mentioned above, further genetic and/or urinary sulfatide testing is needed to confirm whether the biochemical changes are disease-causing or not.

TESTING

Neuroimaging

MLD has a characteristic pattern that is quite similar among the late-infantile, juvenile- and adult-onset form of the disease[12] (Figure 29.1). A diffuse sheet-like area of increased T2 signal hyperintensity first involves the frontal and parietal periventricular and central white matter regions. In early disease this can be quite faint and mistakenly thought to be terminal zones of myelination. As severe disease develops the sheet of white matter signal intensity involves the inner half of the subcortical white matter. A tigroid pattern emerges.

MANAGEMENT

Standard of Care

The burden of disease upon both patient and caregiver should not be underestimated. Close collaboration with nursing care helps address the many changing needs of the patient. Supportive therapies to reduce pain and spasticity and optimize hygiene help avoid many end-stage care problems. The 5-year survival has improved since 1970 due to much improved supportive care.

Provision of an enriched environment and an intense but appropriately cautious physical therapy program provides an optimized quality of life at all stages of the disease. The timely use of suction equipment, swallowing aids, feeding tubes, and other supportive measures can improve quality life.

Seizures and contractures should be treated with antiepileptic drugs and muscle relaxants, respectively. Gastroesophageal reflux, constipation, and drooling are common problems that may be helped by specific medications. Many practitioners are not aware of the increased risk of gall bladder stones in MLD. This can be a significant source of pain in a nonverbal child, and detecting and treating this may help relieve pain and discomfort.

Prevention of Primary Manifestations

Hematopoietic stem cell transplantation (HSCT) or bone marrow transplantation (BMT) has long been employed to treat the primary central nervous system manifestations of MLD.[13] Transplantations for leukodystrophies occur worldwide.[14] However, not all individuals with MLD are suitable candidates for these procedures. Overall, children

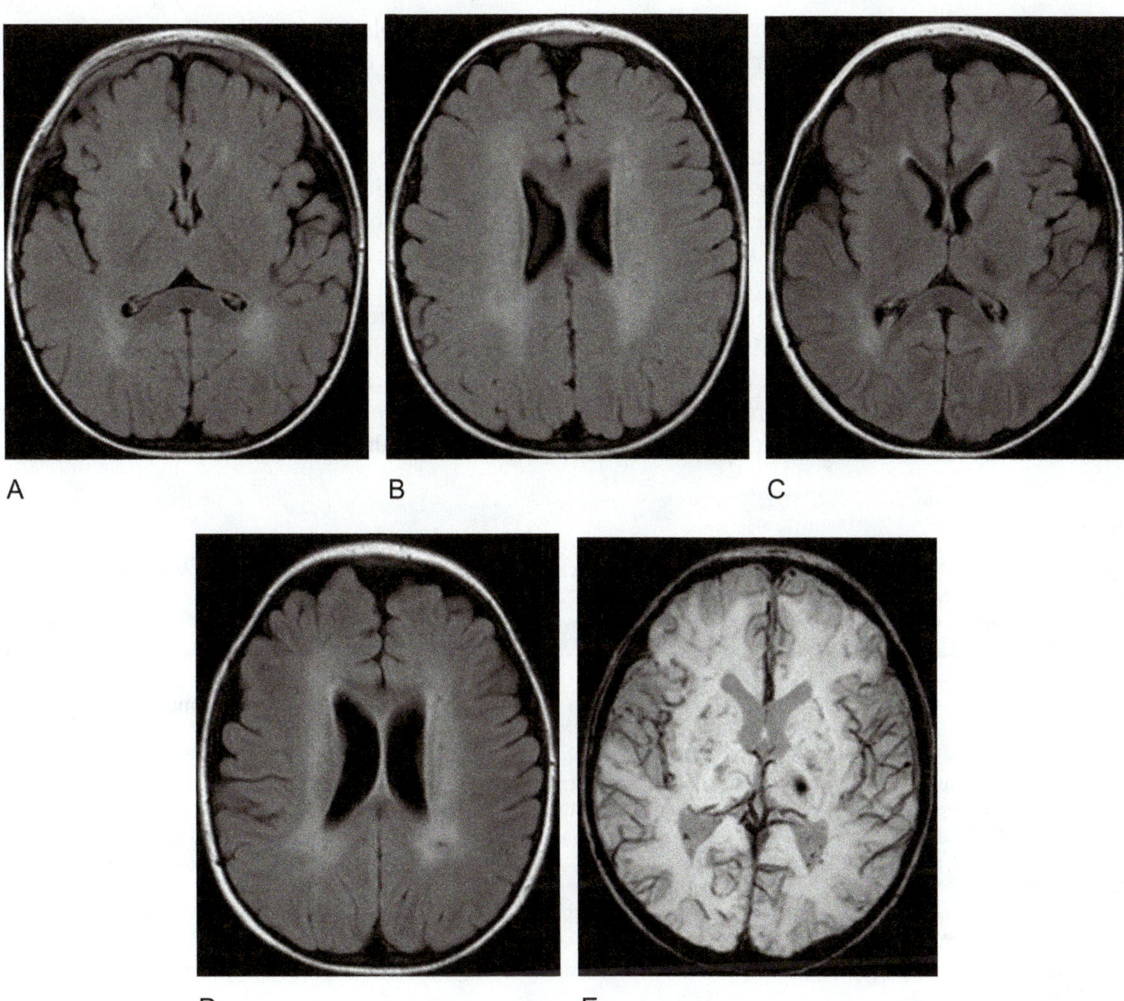

FIGURE 29.1 **MRI of late-infantile metachromatic leukodystrophy in a male child.** The child was born prematurely at 34 weeks gestation due to preeclampsia. At 2 weeks of life, ultrasound (not shown) had illustrated hyperechoic foci in the bilateral posterior parietal periventricular white matter. (**A**, **B**) Age 21 months: Bilateral increased T2-signal abnormality involving the centrum semiovale, corona radiata, periventricular and deep white matter of cerebral hemispheres with relative sparing of the subcortical U fibers. There are multiple foci of signal dropout within the left centrum semiovale and bilateral periventricular white matter, representing foci of remote hemorrhage or calcification. (**C**, **D**) Age 30 months: Dilation of the lateral ventricles and prominence of the sulci have become prominent. As before, there is increased T2 signal in the centrum semiovale and corona radiata white matter. (**E**) Susceptibility-weighted imaging at 30 months of age illustrates scattered hemorrhagic foci with possible mineralization in the centrum semiovale, corona radiata, thalami and globus pallidus. Figure courtesy of Dr. Juan M. Pascual, UT Southwestern Medical Center, with permission.

with juvenile-onset MLD have better outcomes than those with late-infantile MLD.[15] Substantial risk is involved and long-term effects are unclear.

Patients with motor function symptoms at the time of transplant do not improve after transplantation.[16] Brainstem auditory-evoked responses, visual-evoked potentials, electroencephalogram, and/or peripheral nerve conduction velocities stabilize or improve in juvenile patients but continue to worsen in most patients with the late infantile presentation. Pretransplant modified Loes scores are highly correlated with developmental outcomes and predictive of cognitive and motor function. Children who are asymptomatic at the time of transplantation benefit most from the procedure. Children with juvenile onset and minimal symptoms show stabilization or deterioration of motor skills but maintained cognitive skills.

Therapies Under Investigation

Enzyme Replacement Therapy (ERT)

Clinical testing of intravenous recombinant human enzyme was halted after a phase I/II study failed to show substantial improvement. As the major obstacle to ERT has been bypassing the blood–brain barrier, other routes of delivery, such as intrathecal administration, are being pursued. In addition, different forms of human ARSA enzyme are now available, and animal studies suggest that it may be a useful supplement to other therapies.[16] Questions have been raised that the enzymes used in various clinical trials may have had different uptake properties.[17]

Gene Therapy

In the past 10 years much progress has been made in the field of gene therapies for arylsulfatase A. First results of a human gene replacement trial have recently been reported. In three presymptomatic patients, treatment with autologous CD34 cells that had undergone lentiviral gene correction appeared to be safe and effective over a 2-year follow-up period. Some concerns have been raised about the long-term safety of this approach, but the results are encouraging.[18]

Trials are also underway for AAVrh.10 gene delivery in MLD. Different from the lentiviral gene correction mentioned above this is undertaken by direct intracerebral injection.[19] While this necessitates surgery, it may allow for a more rapid delivery in this progressive condition. Colle et al.[20] injected an adeno-associated virus vector containing human ARSA into the brains of nonhuman primates and found that the enzyme was expressed without adverse effects, supporting that a similar approach may be feasible and safe in humans.

References

1. von-Figura KGV, Jaeken J. Metachromatic leukodystrophy. In: Scriver C, Beaudet A, Sly W, Valle D, eds. *The Metabolic & Molecular Bases of Inherited Disease*. 8th ed. New York, NY: McGraw-Hill; 2001:3695–3724.
2. Fluharty AL. *Arylsulfatase A Deficiency, Gene Reviews*; 2014. http://www.ncbi.nlm.nih.gov/books/NBK1130/, Accessed May 2014.
3. Macaulay RJ, Lowry NJ, Casey RE. Pathologic findings of multiple sulfatase deficiency reflect the pattern of enzyme deficiencies. *Pediatr Neurol*. 1998;19:372–376.
4. Alzheimer A: Beitraege zur Kenntnis der pathologischen Neuroglia und ihrer Beziehungen zu den Abbauvorgaengen im Nerven gewebe. Nissl-Alzheimer's Histol Histopathol Arb 3: 1910;401.
5. Witte F. Ueber pathologische Abbauvorgange im Zentralnervensystem. *Munch Med Wochenschr*. 1921;68:69.
6. Austin J. Recent studies in the metachromatic and globoid body forms of diffuse sclerosis. In: Folch-Pi JAB H, ed. *Brain Lipids and Lipoproteins and Leucodystrophies*. 1st ed. New York: Elsevier Publishing Company; 1963.
7. Mehl E, Jatzkewitz H. Evidence for the genetic block in metachromatic leucodystrophy (Ml). *Biochem Biophys Res Commun*. 1965;19:407–411.
8. Phelan MC1, Thomas GR, Saul RA, et al. Cytogenetic, biochemical, and molecular analyses of a 22q13 deletion. *Am J Med Genet*. 1992;43(5):872–876.
9. Austin JH, Balasubramanian AS, Pattabiraman TN, et al. A Controlled Study of Enzymic Activities in three Human Disorders of Glycolipid Metabolism. *J Neurochem*. 1963;10:805–816.
10. Arbour LT, Silver K, Hechtman P, et al. Variable onset of metachromatic leukodystrophy in a Vietnamese family. *Pediatr Neurol*. 2000;23(2):173–176.
11. Mahmood A, Berry J, Wenger D, et al. Metachromatic leukodystrophy: a case of triplets with the late infantile variant and a systematic review of the literature. *J Child Neurol*. 2010;25(5):572–580.
12. Eichler F, Grodd W, Grant E, et al. Metachromatic leukodystrophy: a scoring system for brain MR observations. *AJNR Am J Neuroradiol*. 2009;30(10):1893–1897.
13. Krivit W, Peters C, Shapiro EG. Bone marrow transplantation as effective treatment of central nervous system disease in globoid cell leukodystrophy, metachromatic leukodystrophy, adrenoleukodystrophy, mannosidosis, fucosidosis, aspartylglucosaminuria, Hurler, Maroteaux-Lamy, and Sly syndromes, and Gaucher disease type III. *Curr Opin Neurol*. 1999;12:167–176.
14. Musolino PL, Lund TC, Pan J, et al. Hematopoietic stem cell transplantation in the leukodystrophies: a systematic review of the literature. *Neuropediatrics*. 2014;45(3):169–174.
15. Martin HR, Poe MD, Provenzale JM, et al. Neurodevelopmental outcomes of umbilical cord blood transplantation in metachromatic leukodystrophy. *Biol Blood Marrow Transplant*. 2013;19(4):616–624.
16. Matzner U, Herbst E, Hedayati KK, et al. Enzyme replacement improves nervous system pathology and function in a mouse model for metachromatic leukodystrophy. *Hum Mol Genet*. 2005;14:1139–1152.
17. Schröder S, Matthes F, Hyden P, et al. Site-specific analysis of N-linked oligosaccharides of recombinant lysosomal arylsulfatase A produced in different cell lines. *Glycobiology*. 2010 Feb;20(2):248–259.
18. Biffi A, Montini E, Lorioli L, et al. Lentiviral hematopoietic stem cell gene therapy benefits metachromatic leukodystrophy. *Science*. 2013;341(6148):1233158.
19. Piguet F, Sondhi D, Piraud M, et al. Correction of brain oligodendrocytes by AAVrh.10 intracerebral gene therapy in metachromatic leukodystrophy mice. *Hum Gene Ther*. 2012;23(8):903–914.
20. Colle MA, Piguet F, Bertrand L, et al. Efficient intracerebral delivery of AAV5 vector encoding human ARSA in non-human primate. *Hum Mol Genet*. 2010;19(1):147–158.

CHAPTER

30

Krabbe Disease: Globoid Cell Leukodystrophy

David A. Wenger and Paola Luzi

Jefferson Medical College, Philadelphia, PA, USA

INTRODUCTION

Krabbe disease, or globoid cell leukodystrophy (GLD), is an autosomal recessive disorder affecting white matter in the central and peripheral nervous systems (CNS and PNS). The initial report of infants with "diffuse brain-sclerosis or diffuse gliosis" clearly describes patients we now recognize as having Krabbe disease. In 1916, Krabbe[1] described five patients who had onset of symptoms at about 5 months of age. These included rigidity of musculature, violent tonic spasms, nystagmus, periodic elevations in temperature, progressive paresis, and early death. He noted the depletion of white matter in the cerebellum and degeneration of spinal nerve tracts with replacement by dense fibrillar glia. In 1924, Collier and Greenfield[2] used the term *globoid* to describe these abnormal scavenger cells, which are characteristic for this disorder. Excellent clinical and genetic studies on 32 Swedish patients by Hagberg et al.[3] provide the basis for the clinical delineation of the infantile form of this disorder. Individuals with later-onset forms, onset after infancy to adulthood, are also recognized.

Early research suggested that globoid cells might contain a glycosphingolipid similar to that stored in Gaucher disease. Chemical analysis and production of globoid cells by intracerebral injection of galactosylceramide into brains of experimental animals confirmed this.[4] In 1970 Malone[5] and Suzuki and Suzuki[6] reported that tissue samples from patients with Krabbe disease could not degrade galactosylceramide due to a deficiency of galactocerebrosidase (GALC) activity. Galactosylceramide is produced during the lysosomal degradation of sulfatide. These two glycosphingolipids are important for healthy, stable myelin. The measurement of GALC activity in leukocytes, fibroblasts, and fetal-derived cells can identify patients with Krabbe disease both postnatally and prenatally. In an attempt to identify affected individuals before symptoms occur, newborn screening (NBS) for Krabbe disease has been initiated in several US states.[7] However, the benefits of this program continue to be debated. At this time the treatment of human patients is limited to hematopoietic stem cell transplantation (HSCT) in presymptomatic infants and mildly affected late-onset patients.[8,9] This treatment slows the progression of the disease, but significant deficits remain.

In 1993 GALC was first purified from human urine and brain.[10] This provided amino acid sequence information that was used to clone the human complementary DNA (cDNA)[11] and subsequently the gene.[12] Disease-causing mutations have been identified in human patients and the available animal models.[13,14] In addition, knowing the disease-causing mutations in a patient provides another method for prenatal diagnosis and preimplantation genetic diagnosis for at-risk couples.

Several naturally occurring animal models of globoid cell leukodystrophy are available.[14,15] They are useful to investigate the chemical pathogenesis of GALC deficiency and to explore methods for effective therapy. In addition to HSCT, the mouse and dog models have been used for gene therapy trials, neural stem cell transplantation, chaperone therapy, substrate reduction therapy, and cytokine injections, alone or in various combinations. While no treatments have resulted in a "cure," some treatments have lead to a significant extension of the lives of the treated animals. Although it has been over 40 years since the enzymatic defect in Krabbe disease was described, much more needs to be done before effective therapy for a majority of the patients becomes a reality.

Rosenberg's Molecular and Genetic Basis of Neurological and Psychiatric Disease
http://dx.doi.org/10.1016/B978-0-12-410529-4.00030-9

CLINICAL FEATURES

Hagberg et al.[3] described three stages in the progression of the infantile form of the disease. Stage I is characterized by general irritability, stiffness of limbs with clenched fists, feeding difficulties, arrest of mental and motor development, and episodes of temperature elevation without infection. During the stage II phase, patients have back arching, myoclonic-like jerks of the arms and legs, marked hypertonicity, continued bouts of fever, and regression of any achieved abilities. By stage III, patients are severely decerebrate with no voluntary movements. They are hypotonic and cachectic, and they die of respiratory infections or cerebral hyperpyrexia. In Hagberg's study, the average age of onset was 4 months, and the average age at death was 13 months. In this laboratory we have diagnosed more than 600 patients with Krabbe disease, and about 85% of them present before 6 months of age. With only supportive care, most die before 24 months of age. More recently, Escolar and colleagues described a staging system based on symptoms of disease progression.[16] Stage 1 was characterized by the presence of at least two of the following signs: weak feeding, hypotonia of the shoulder girdle, intermittent thumb clasp, or gastroesophageal reflux. In stage 2, patients developed prolonged periods of irritability, fixed thumb clasp, spasticity of extremities with predominant extensor tone (arching back), severe feeding difficulties with uncoordinated suck and swallowing, and trunk weakness. In stage 3 clinical seizures become evident, deep tendon reflexes are absent, infants show exaggerated startle and other abnormal primitive reflexes, visual deficits including jerky eye movements, poor tracking, and sluggish pupillary responses. In stage 4, infants become unresponsive, severely flaccid or hypotonic, develop blindness and hearing deficit, and have signs of autonomic instability.

While it is difficult to separate patients with later onset into distinct age categories, this group generally includes individuals presenting after 1 year of age. In different publications, patients called "late-onset" will include late infantile, juvenile, adolescent as well as adult individuals. Presenting symptoms in late-onset patients can include vision problems, intention tremor, cerebellar ataxia, spastic hemiparesis, burning paresthesia, peripheral neuropathy, and dementia. Patients diagnosed before 2 years of age are clinically more similar to infantile patients. While some individuals remain stable for long periods of time, others show a steady decline in intellect and develop seizures, severe dysarthria, and blindness, leading to a vegetative state and death. Also, there is considerable variability in age of onset and clinical course between late-onset patients, even those with the same GALC genotype. The reader is referred to a number of papers that describe the clinical and molecular findings in late-onset patients.[17–22]

Recently an infant resembling Krabbe disease with abnormal myelination was found to be homozygous for a mutation in the saposin A region of the prosaposin (PSAP) gene.[23] Saposin A is a heat-stable protein required together with GALC for the enzymatic hydrolysis of galactosylceramide (Figure 30.1).

The incidence of Krabbe disease has been estimated to be 1–100,000 births in the United States population with Northern European ancestry. However, a recent study using death rates estimates the birth prevalence at about 1–250,000.[24] The birth rate for Krabbe disease is higher among certain ethnic groups. As the birth demographics change in the United States population, the incidence of Krabbe disease may change. Also, while about 80–85% of the patients currently diagnosed with this disease have the infantile form, there may be an increase in later-onset and atypical patients.

MOLECULAR GENETICS

The gene for human GALC has been mapped to human chromosome 14 by linkage analysis and to the region 14q31 by *in situ* hybridization using a probe from the cDNA sequence.[25] The cDNA is about 3.8-kb long, including, 2007 bp of open reading frame (coding for 669 amino acids), and 1741 bp of 3′ untranslated sequence.[11] The first 78 nucleotides of the open reading frame code for the 26 amino acid leader peptide. Another potential initiation codon, located 48 nucleotides upstream from the one initially reported, has been identified. This would result in a leader sequence of 42 amino acids. Expression studies utilizing constructs with the two different leader sequences showed no difference in GALC expression (unpublished data). Note that some publications report the location of mutations numbering from the more upstream ATG as the start of the coding region. This would add 48 nucleotides and 16 amino acids to the original numbering system. The remaining 643 amino acids, after cleavage of the leader peptide, are glycosylated at five or six of the potential sites to produce the 80-kDa precursor species. The 80-kDa precursor is processed into the 50–53- and 30-kDa mature forms by the action of a protease probably located in lysosomes. The human GALC gene is nearly 58-kb long and contains 17 exons and 16 introns.[12]

With the available sequence information from the GALC cDNA and gene, mutation analysis on patients with all types of GLD is possible. Over 140 disease-causing mutations have been identified, with a few occurring with an increased frequency (Table 30.1).[13] In addition to disease-causing mutations, several polymorphisms are commonly

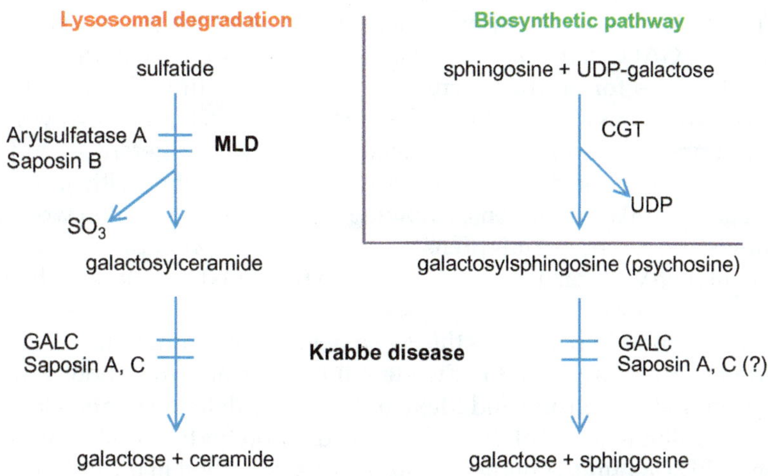

FIGURE 30.1 **Pathway for the lysosomal degradation of sulfatide and galactosylceramide showing the required enzymes and sphingolipid activator proteins (saposins).** In addition, the pathway for the biosynthesis of psychosine and its lysosomal degradation is also shown. Disorders resulting from genetic defects in these pathways (shown by double horizontal lines) are noted. Abbreviations: CGT, ceramide-galactosyltransferase; GALC, galactocerebrosidase; MLD, metachromatic leukodystrophy; UDP, uridine diphosphate.

TABLE 30.1 Mutations Frequently Found in the GALC Gene

Location	Nucleotide change in cDNA	Base change	Effect	Comments[b]
DISEASE-CAUSING MUTATIONS				
Ex 1	c.121G>A (c.169G>A)[a]	GGC>AGC	p.41G>S (p.57G>S)	Mild, southern Italy
Ex 4	c.284G>A (c.332G>A)	GGC>GAC	p.95G>D (p.111G>D)	Severe
Ex 4	c.286A>G (c.334A>G)	ACT>GCT	p.96T>A (p.112T>A)	Mild
Ex 7	c.635 del+ins (c.683 del+ins)	del 12, ins3	del 5 aa+ins 2 aa	Severe, Japanese, Korean
Ex 8	c.809G>A (c.857G>A)	GGC>GAC	p.270G>D (p.286G>D)	Mild
Ex 8	c.860C>T (c.908C>T)	TCC>TTC	p.287S>F (p.303S>F)	Severe
Ex 9	c.908A>G (c.956A>G)	TAT>TGT	p.303Y>C (p.319Y>C)	Mild
In10-end	30 kb deletion	30 kb del	Short mRNA	Severe
Ex 11	c.1138C>T (c.1186C>T)	CGG>TGG	p.380R>W (p.396R>W)	Severe
Ex 13	c.1424 delA (c.1472 delA)	TAAGG>TAGG	FS, PS	Severe
Ex 14	c.1538C>T (c.1586C>T)	ACG>ATG	p.513T>M (p.529T>M)	Severe
Ex 15	c.1652A>C (c.1700A>C)	TAC>TCC	p.551Y>S (p.567Y>S)	Severe
Ex 16	c.1853T>C (c.1901T>C)	TTA>TCA	p.618L>S (p.634L>S)	Mild
COMMON POLYMORPHISMS				
Ex 4	c.502C>T (c.550C>T)	CGT>TGT	p.168R>C (p.184R>C)	The 30-kb deletion always has this polymorphism
Ex 6	c.694G>A (c.742G>A)	GAT>AAT	p.232D>N (p.248D>N)	
Ex 8	c.865A>G (c.913A>G)	ATC>GTC	p.289I>V (p.305I>V)	Japanese
Ex 14	c.1637T>C (c.1685T>C)	ATA>ACA	p.546I>T (p.562I>T)	Very common

[a] *Location of mutations and protein changes in the original numbering system are shown first; those in parentheses are based on a start codon 48 nucleotides longer.*
[b] *The comments reflect the best information available from published and unpublished data.*
Note: Almost all of the disease-causing mutations found in the GALC gene also contain known polymorphisms in the same copy of the gene.
Abbreviations: aa, amino acids; del, deletion; ins, insertion; c., cDNA; p., protein; FS, frameshift; PS, premature stop.

found in the GALC gene (Table 30.1). These missense mutations do affect the measured GALC activity, making carrier identification by measuring GALC activity nearly impossible. The presence of these polymorphisms that lower GALC activity also has implications for newborn screening, which measures GALC activity in dried blood spots. While genotype–phenotype correlations are possible with some mutations, it is not always possible to assign severity to a mutation when it was not previously reported or when it is found with different second alleles. It is also possible that some "mild" mutations would not be disease-causing if inherited together with another "mild" mutation. This may be true for the p.T96A and p.Y303C mutations. Knowing a patient's mutation(s) would improve carrier testing in immediate family members and would permit prenatal diagnosis and preimplantation diagnosis for future pregnancies. However this will not help with carrier assessment of an unrelated individual who is the partner of a known carrier.

The most common mutation found in patients is the 30-kb deletion accounting for 40–50% of the disease-causing alleles. While this mutation probably originated in Sweden, it has spread throughout Europe and is also found in individuals with ancestry from India, Pakistan and Mexico. This large deletion starts within intron 10 and proceeds past the 3′ end of the gene. This eliminates all of the coding information for the 30-kDa subunit plus about 15% of the coding information for the 50-kDa subunit. A number of mutations identified in the authors' laboratory and reported by others occur in only one family. Some mutations only occur within small, defined populations such as in the Druze community of Northern Israel and in two villages of Moslem Arabs located near Jerusalem. Some mutations have only been reported in the Japanese, Korean or Chinese populations. Many patients with a late-onset form of Krabbe disease have at least one copy of the c.G809A (p.G270D) mutation. This missense mutation must result in the production of a small amount of active enzyme that delays the onset of the disease. However, this has never been proven because it is difficult to accurately measure very low enzymatic activity in any tissue sample. This laboratory has identified six families who have the c.G809A mutation on one allele and the 30-kb deletion on the other. Although they all have a late-onset presentation, the age of onset and clinical course are significantly different between and within families. Other genetic and/or environmental factors must be involved in the onset and progression of the disease.

The PSAP (prosaposin) gene has been mapped to 10q21-22 in humans. The expressed prosaposin protein is proteolytically cut into four active saposins, A–D, that have specific abilities to stimulate the lysosomal hydrolysis of certain sphingolipids.

DISEASE MECHANISMS

Biochemical Findings

All patients with Krabbe disease have a deficiency of GALC activity. While almost all patients have mutations in the GALC gene, a very small percentage of patients have mutations in the PSAP gene. GALC has β-galactosidase activity with specificity toward galactosylceramide, psychosine (galactosylsphingosine), galactosylacylalkylglycerol, monogalactosyldiglyceride, and lactosylceramide under specific assay conditions. The metabolism of the first two substrates is important to the pathogenesis of this disorder. It has been shown that globoid cells are composed of high concentrations of galactosylceramide, which comes from the degeneration of myelin. However, the pathogenic lipid may be psychosine, which has been demonstrated to inhibit key enzymes such as protein kinase C and cytochrome *c* oxidase, and induce apoptotic death in oligodendrocytes and Schwann cells. Figure 30.2 shows the probable mechanisms for demyelination and possible avenues for correction. While it is a minor compound in a normal brain, psychosine is stored in relatively large amounts in white matter of humans and animal models with GALC deficiency.[26] GALC is the only enzyme capable of hydrolyzing psychosine. It is synthesized during active myelination by the enzymatic transfer of galactose to sphingosine catalyzed by ceramide-galactosyltransferase (Figure 30.1). It is degraded in people with sufficient GALC activity; however, in tissues of patients with Krabbe disease it accumulates to levels 10–15 times normal.

The role of heat-stable sphingolipid activator proteins (saposins) in stimulating GALC activity has also been investigated. In 1982, we showed that SAP-2 or saposin C could stimulate GALC activity in the presence of acidic lipids, including sulfatide and phosphatidylserine.[27] A mouse model with a point mutation in the saposin A region was generated and these mice have pathologic features resembling a milder form of GLD.[28] A patient with features resembling infantile Krabbe disease was found to be homozygous for a mutation in the saposin A region of the PSAP gene.[23]

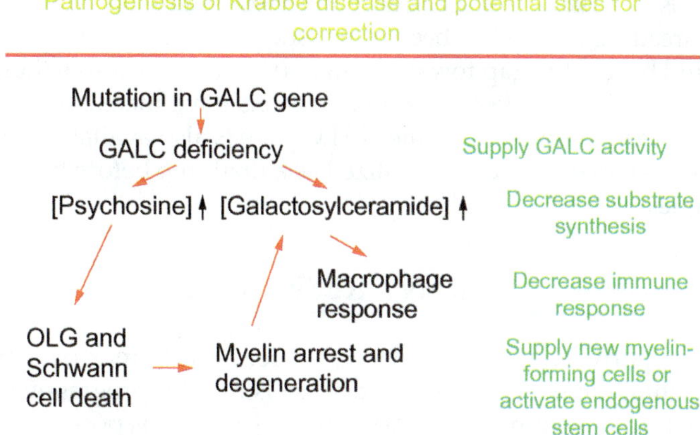

FIGURE 30.2 **Pathogenesis of Krabbe disease and potential sites for correction.**

Pathophysiology

Human

CENTRAL NERVOUS SYSTEM

In patients with the infantile form the white matter is firm, reduced in volume, and whitish-gray in appearance with dilated ventricles. The gray matter appears relatively normal except for a moderate reduction of the cortical thickness. The major histopathologic changes are extensive demyelination, gliosis, and presence of unique macrophages (globoid cells) in the white matter. The fornix, hippocampus, mammillothalamic tract, and nerve fiber bundles in the basal centrum semiovale and the cerebellar white matter are significantly involved. In the spinal cord, the pyramidal tracts are more severely affected than the dorsal columns. In the areas of demyelination, the oligodendroglial cell population is severely diminished and globoid cells are often clustered around blood vessels. Both mononuclear and multinucleated globoid cells are similar ultrastructurally, except for the number of nuclei. They have prominent pseudopods, moderately electron-dense granular cytoplasm containing prominent rough endoplasmic reticulum, many free ribosomes, abundant fine filaments of approximately 9–10nm, and scattered or clustered abnormal cytoplasmic inclusions. The inclusions have moderately electron-dense straight or curved hollow tubular profiles in longitudinal sections and appear irregularly crystalloid in cross sections. Scattered globoid cells with typical inclusions were also found in the spinal cord in affected fetuses of 20–23weeks gestation.

Despite increasing clinical reports describing late-onset cases in recent years, only limited information on their neuropathology has been reported. In late infantile and juvenile cases, neuropathologic changes are similar to those in typical infantile patients. Two identical twin females developed symptoms at about age 18 years. Both received allogeneic bone marrow transplants and both died of severe graft-versus-host disease within 2 months. Neuropathologic changes included degeneration of the frontoparietal white matter and corticospinal tract. The frontoparietal lesion consisted of multiple necrotic foci with calcified deposits and active degeneration of the surrounding white matter with globoid cell infiltration. Globoid cell infiltration was noted in the optic radiations.[17]

PERIPHERAL NERVOUS SYSTEM

In young patients the peripheral nerves tend to be firm and abnormally thick and white on gross inspection. Major pathologic features are marked endoneurial fibrosis, proliferation of fibroblasts, demyelination, and infiltration or perivascular aggregation of histiocytes/macrophages containing periodic acid–Schiff (PAS)-positive materials. Thinly myelinated fibers suggestive of remyelination may be present. Axonal degeneration of varying degrees has been reported. Ultrastructurally, inclusions similar to those in globoid cells in the brain are found in the cytoplasm of histiocytes/macrophages, as well as in the Schwann cells. In a 73-year-old woman sural nerve biopsy revealed a mild loss of myelinated fibers with disproportionately thin myelin sheaths, and Schwann cells contained needle-like inclusions.[17]

Animal Models

There are well-characterized animal models of GLD available for study. Several published reviews describe their clinical and pathologic features, molecular defects, and potential use as models to attempt therapeutic trials.[14,15]

In addition to the mouse models of Krabbe disease, Cairn and West Highland white terriers with GLD have been known for many years, and a breeding colony has been established. This larger animal model has some advantages over the mouse model that could bridge the gap toward human trials once effective therapies are developed. As has been demonstrated in humans with Krabbe disease, animal models have increased psychosine in nervous tissues. A colony of rhesus monkeys that carry a mutation in the GALC gene is also available. Although only a limited number of affected monkeys will be available, they can be utilized as a final step before human trials to investigate safety and effectiveness of new therapies.

CURRENT RESEARCH

While treatment, other than supportive care, is limited to HSCT for some human patients, many therapeutic options are being tried in the available animal models. Since this disease is rapidly progressive in both the infantile human patients and the twitcher mouse model, treatment must be instituted as soon as possible to prevent significant damage to the myelinating cells. Initial trials of bone marrow transplantation (BMT) in the twitcher mouse model were started in 1984.[29] This treatment increased their lifespan, decreased the amount of psychosine, and produced some evidence for remyelination, probably caused by some donor macrophages reaching the brain and donating a small amount of GALC activity to the myelinating cells of the CNS. Many affected mice have been treated by bone marrow transplantation alone and in combination with other therapies, such as substrate reduction, small molecule chemical chaperones and cytokines such as insulin-like growth factor 1 (IGF-1) with clear extension of the lives of the treated mice, but they still die too young with evidence of neurologic disease.[30] As seen in the mice, bone marrow transplantation in the dog model also extends the lives of treated dogs and lowers the psychosine concentration in brain.

More recently, different viral vectors containing the GALC gene have been injected into the brains and blood vessels of affected mice with variable success. While these treatments have resulted in higher GALC activity in the brain than BMT, the extension of life is marginally better. This was due either to low expression of the transgene in critical areas of the CNS and PNS, failure to localize GALC in the lysosomes, possible overexpression of GALC protein that overwhelms the processing system, minimal migration of the vector from the site of injection, and immune reaction to the vector. At this time there is a push to evaluate different adeno-associated viral (AAV) vectors containing the GALC cDNA for their ability to cross the so-called blood–brain barrier, infect cells that can produce GALC activity over a wide area and deliver GALC activity to the cells that need it. In recent studies, affected twitcher mice that received intracerebroventricular, intracerebellar, and intravenous injections of AAVrh10 vector had significant prolongation of their lives, robust GALC expression, improved myelination, retention of normal movement and, in some cases, restored ability of male and female mice to mate and rear their pups.[31] While the untreated twitcher mice live only about 40 days, many of the treated mice live for over 200 days. However, for some unknown reason after a relatively long period of apparent normalcy, the treated mice develop neurologic signs and die within a few weeks. At the time of death they are still expressing high levels of GALC activity and show evidence of myelin repair. The cause for this is being investigated. Meanwhile, studies are underway to inject affected dogs in a similar manner to see if substantial improvement can be seen in a larger animal model.

In addition to HSCT and gene therapy, enzyme replacement therapy (ERT) and stem cell therapy for the treatment of affected mice is being pursued. ERT has the problem of inadequate delivery to the CNS and PNS, although direct delivery to the brains of affected mice has been tried with limited success. As infantile human patients present within the first few months of life and show rapid decline in abilities, there is concern that the myelinating cells are damaged early in the disease. If there is a significant loss of oligodendrocyte progenitor cells, then supplying GALC activity by any means may not be able to restore normal myelination if damage has already occurred. This damage to oligodendrocytes and Schwann cells could be due to psychosine accumulation or release of other harmful chemokines or cytokines. If the neural stem cells are destroyed, it may be necessary to replace these cells by injecting some type of stem or progenitor cells into the brain. Studies injecting virally transduced oligodendrocyte progenitor cells into the brains of twitcher mice appear to show donor cells expressing high GALC activity and transfer of GALC activity to oligodendrocytes that are myelinating axons.[32]

DIFFERENTIAL DIAGNOSIS

A history of normal development for the first few months after birth followed by psychomotor deterioration differentiates Krabbe disease from nonprogressive CNS disorders of congenital or perinatal origin. Differentiation of Krabbe disease from other degenerative diseases purely by clinical features is often difficult. Individuals with later-onset Krabbe disease have a variable age of onset and can have nonspecific presenting symptoms. Individuals of any age with progressive deterioration of the central and/or peripheral nervous systems should be tested for Krabbe disease. The finding of very low GALC activity in the presence of other normal enzymes will readily confirm the diagnosis.

The clinical features of infantile GLD were described earlier in this chapter. Clinical differentiation from other neurodegenerative diseases of early childhood can be challenging. Spongy degeneration of white matter begins in early infancy and is characterized by an enlarged head, initial hypotonia, and a normal cerebrospinal fluid (CSF) protein concentration. Alexander disease is a disorder of cortical white matter affecting mostly infants although neonatal and older-onset forms are also recognized. The infantile form presents in the first 2 years of life typically with megalencephaly, seizures, progressive psychomotor retardation with loss of developmental milestones, and quadriparesis. Affected individuals survive a few weeks to several years. Mutations in the glial fibrillary acidic protein gene are the major cause of Alexander disease. Pelizaeus–Merzbacher disease (PMD) can also occur in the first year of life. The disease can have a wide range of phenotypes with a slowly progressive course, characterized by abnormal involuntary eye movements, which are prominent, hypotonia, and cognitive impairment. The CSF protein concentration is normal and inheritance is X-linked recessive. The finding of mutations in the proteolipid protein 1 gene is diagnostic. The GM2-gangliosidoses, including Tay–Sachs disease and Sandhoff disease, usually manifest themselves in the first year of life. Developmental delay, startle to sharp sounds, seizures, and macular cherry-red spots are commonly present. The initial clinical finding is sluggishness or apathy rather than hyperirritability. Measurement of β-hexosaminidase A and B activity should confirm or rule out this diagnosis. Patients with GM1-gangliosidosis of early onset often present with coarse facial features, hepatosplenomegaly, developmental delay, and radiological findings that may resemble a mucopolysaccharidosis. These individuals are deficient in acid β-galactosidase activity. Canavan disease is characterized by evidence of developmental delay by 5 months of age, with severe hypotonia and failure to achieve independent sitting, ambulation, or speech. Hypotonia evolves into spasticity, and assistance with feeding becomes necessary. However, survival is longer than in patients diagnosed with infantile Krabbe disease. Most individuals with Canavan disease have macrocephaly, a variable finding in individuals with Krabbe disease. Magnetic resonance imaging (MRI) shows prominent involvement of subcortical white matter. The finding of elevated *N*-acetylaspartic acid concentration in urine confirms the diagnosis of Canavan disease. Finally, patients with Gaucher disease and Niemann–Pick disease can present in the first year of life with developmental delay and regression but almost always have hepatosplenomegaly.

It is almost impossible to make a clinical diagnosis of GLD in patients with atypical symptoms and with later onset. Patients with later-onset forms of Krabbe disease may initially be considered to have metachromatic leukodystrophy (MLD), which can begin any time after 1 year of age, with progressive motor weakness, ataxia, and sometimes psychiatric issues. Low arylsulfatase A activity and urinary excretion of sulfatide make the diagnosis. Presenting signs may also be similar to Charcot–Marie–Tooth disease. Patients with the juvenile form of Alexander disease can present with megalencephaly, bulbar/pseudobulbar signs including speech abnormalities, swallowing difficulties, frequent vomiting, lower-limb spasticity, poor coordination (ataxia), gradual loss of intellectual function, and seizures. X-linked adrenoleukodystrophy (X-ALD), caused by mutations in the ABCD1 gene, affects the nervous system white matter and the adrenal cortex. The childhood cerebral form manifests most commonly between 4 and 8 years of age. It can initially resemble attention-deficit disorder with progressive impairment of cognition, behavior, vision, hearing, and motor function, often leading to total disability within 2 years. Adrenomyeloneuropathy (AMN) manifests most commonly in the late 20s as progressive paraparesis, sphincter disturbances, and varying degrees of distal sensory loss. Approximately 50% of carrier females develop neurologic manifestations that resemble adrenomyeloneuropathy, but have later onset (35 years or later) and milder disease than affected males. The plasma concentration of very long-chain fatty acids is elevated in most males with X-ALD of all ages and approximately 85% of female carriers.

TESTING

Biochemical Diagnosis

Initial studies in patients with symptoms suggesting a leukodystrophy usually involve a battery of enzymatic measurements in leukocytes or cultured skin fibroblasts. Other findings are helpful in pointing to the diagnosis of a leukodystrophy. These include elevated protein in CSF (75 to 500 mg/dl), decreased nerve conduction velocities, MRI showing decreased myelination in the brainstem and cerebellum, and computed tomography showing lucencies in the white matter followed by diffuse cerebral atrophy in gray and white matter later in the disease. While these studies can indicate a diagnosis of a leukodystrophy, further testing is required to make a definitive diagnosis. The definitive diagnosis of Krabbe disease rests on the finding of very low GALC activity (less than 5% of our normal mean).

While all individuals of any age with Krabbe disease have very low GALC activity, there is a wide range of GALC activities in noncarrier and carrier individuals. The wide range of values is due to the presence of polymorphisms in the GALC gene. This makes carrier testing by measuring GALC activity in individuals other than immediate family members nearly impossible. Of course, carrier testing in family members is relatively simple when the disease-causing mutation(s) is known. Complete mutation analysis is available in several laboratories.

Prenatal diagnosis of Krabbe disease was first performed in 1971. Initial testing was done using cultured amniotic fluid cells. Since that time, hundreds of pregnancies from at-risk couples have been monitored using cultured amniotic fluid cells and chorionic villus samples. Measurement of GALC activity alone is reliable, but if the mutations in the proband are known, the pregnancy can also be monitored by DNA analysis. It is essential that the fetal sample to be tested is free of maternal contamination. In addition, preimplantation genetic diagnosis is possible if the disease-causing mutations in the family are known.

Newborn Screening

In an attempt to identify patients before the first symptoms occur, New York State has added Krabbe disease to its list of diseases tested for in newborn screening (NBS).[7] The automated screening is done using dried blood spots and a substrate whose hydrolysis can be measured by tandem mass spectrometry.[33] Since its inception in August 2006, over 1.8 million newborns have been screened, and five newborns with very low GALC activity have been predicted to have infantile Krabbe disease based on mutation analysis and neurodiagnostic studies. Four families elected to have HSCT, and two children died of complications from the transplantations. The two other treated individuals are doing better than untreated infantile patients, however, with significant health issues. One family elected not to treat, and that child died. However, in the "high risk" category (based on the measurement in this laboratory of GALC activity below an assigned cutoff) there are eight other individuals who have either mutations not previously reported, only one disease-causing mutation plus polymorphisms in the "normal" allele, mutations only found in patients with late onset or a combination of mutations not previously found in patients confirmed to have Krabbe disease. These individuals are being followed with neurodiagnostic studies and careful observation. However this has created anxiety for the parents who do not know when or if these children will develop Krabbe disease. With the great variability in the age of onset of symptoms and clinical course in later-onset patients, it is difficult to predict when to start therapy. Among the newborns with very low GALC activity less than the expected 80-85% have the infantile form of Krabbe disease. This reflects the high percentage of births in parents with non-Northern European heritage.

MANAGEMENT

Standard of Care

At this time, the only treatment for symptomatic infantile patients is supportive care. The extreme irritability observed in these patients may be controlled by nitrazepan or low-dose morphine. Eleven asymptomatic newborns (ages 12–44 days) with a confirmed diagnosis of Krabbe disease (tested *in utero* or at birth because of a previously affected sibling) received HSCT using umbilical cord blood.[9] Ten of these individuals are living and have out-lived their untreated affected siblings. While cognitive function is preserved, significant deficits in gross motor functioning and expressive language are apparent. It is not known what the long-term outcomes will be in these individuals transplanted before symptoms were predicted to occur. However, the use of umbilical cord blood for transplantation has greatly improved the availability of suitable donors for patients with all lysosomal disorders. The identification of newborns with the potential to develop Krabbe disease by newborn screening (see above) will facilitate the initiation of treatment before neurological damage has occurred. Questions regarding the ideal age to start treatment, prediction of clinical course without treatment, and long-term benefits of treatment remain a concern. Umbilical cord transplantation of symptomatic patients with infantile onset of Krabbe disease did not result in substantive neurological improvement.[9]

HSCT in patients with late-onset Krabbe disease resulted in a slowing of the neurological degeneration, improvement in MRI findings, decrease in spinal fluid protein and improved quality of life.[8] Although it is too early to reach a conclusion regarding the role of HSCT in Krabbe disease, it should be seriously considered in presymptomatic infantile cases and in slowly progressing late-onset cases if a suitable donor is available. *In utero* bone marrow transplantation has been tried without success.

Therapies Under Investigation

It is hoped studies underway in the animal models of Krabbe disease will lead to new approaches to treat this disease. Based on the finding of significant life extension using viral vectors in the animal models, plans to perform the preclinical studies required by the US Food and Drug Administration are in development. However, the rapid progression of the disease and the need for global delivery of GALC activity present significant challenges for the treatment of this disease.

ACKNOWLEDGEMENTS

The authors thank the many coworkers who have helped with the research on this disorder and the families with Krabbe disease who provided samples and support. The research on Krabbe disease was supported in part by grants from the National Institutes of Health and The Legacy of Angels Foundation.

References

1. Krabbe K. A new familial, infantile form of diffuse brain sclerosis. *Brain*. 1916;39:74–114.
2. Collier J, Greenfield J. The encephalitis periaxialis of Schilder: a clinical and pathological study, with an account of two cases, one of which was diagnosed during life. *Brain*. 1924;47:489–519.
3. Hagberg B, Kollberg H, Sourander P, Akesson HO. Infantile globoid cell leukodystrophy (Krabbe's disease): a clinical and genetic study of 32 Swedish cases 1953–1967. *Neuropaediatrie*. 1970;1:74–88.
4. Austin J, Lehfeldt D, Maxwell W. Experimental "globoid bodies" in white matter and chemical analysis in Krabbe's disease. *J Neuropathol Exp Neurol*. 1961;20:284–285.
5. Malone M. Deficiency in degradative enzyme system in globoid leukodystrophy. *Trans Am Soc Neurochem*. 1970;1:56.
6. Suzuki K, Suzuki Y. Globoid cell leukodystrophy (Krabbe's disease): deficiency of galactocerebroside β-galactosidase. *Proc Natl Acad Sci U S A*. 1970;66:302–309.
7. Duffner PK, Caggana M, Orsini JJ, et al. Newborn screening for Krabbe disease: the New York State model. *Pediatr Neurol*. 2009;40:245–252.
8. Krivit W, Shapiro EG, Peters C, et al. Hematopoietic stem-cell transplantation in globoid cell leukodystrophy. *N Engl J Med*. 1998;338:1119–1126.
9. Escolar ML, Poe MD, Provenzale JM, et al. Transplantation of umbilical-cord blood in babies with infantile Krabbe's disease. *N Engl J Med*. 2005;352:2069–2081.
10. Chen YQ, Wenger DA. Galactocerebrosidase from human urine: purification and partial characterization. *Biochim Biophys Acta*. 1993;1170:53–61.
11. Chen YQ, Rafi MA, de Gala G, Wenger DA. Cloning and expression of cDNA encoding human galactocerebrosidase, the enzyme deficient in globoid cell leukodystrophy. *Hum Mol Genet*. 1993;2:1841–1845.
12. Luzi P, Rafi MA, Wenger DA. Structure and organization of the human galactocerebrosidase (GALC) gene. *Genomics*. 1995;26:407–409.
13. Wenger DA, Escolar ML, Luzi L, Rafi MA. Krabbe disease (globoid cell leukodystrophy). In: Valle D, Beaudet AL, Vogelstein B, Kinzler KW, Antonarakis SE, Ballabio A, eds. *The Online Metabolism & Molecular Bases of Inherited Disease*. McGraw Hill; 2013, in press.
14. Wenger DA. Murine, canine and non-human primate models of Krabbe disease. *Mol Med Today*. 2000;6:449–451.
15. Suzuki K, Suzuki K. Genetic galactosylceramidase deficiency (globoid cell leukodystrophy, Krabbe disease) in different mammalian species. *Neurochem Pathol*. 1985;3:53–68.
16. Escolar ML, Poe MD, Martin HR, Kurtzberg J. A staging system for infantile Krabbe disease to predict outcome after unrelated umbilical cord blood transplantation. *Pediatrics*. 2006;118:e879–e889.
17. Kolodny EH, Raghavan S, Krivit W. Late-onset Krabbe disease (globoid cell leukodystrophy): clinical and biochemical features in 15 patients. *Dev Neurosci*. 1991;13:232–239.
18. Luzi P, Rafi MA, Wenger DA. Multiple mutations in the GALC gene in a patient with adult-onset Krabbe disease. *Ann Neurol*. 1996;40:116–119.
19. De Gasperi R, Gama Sosa MA, Sartorato EL, et al. Molecular heterogeneity of late-onset forms of globoid-cell leukodystrophy. *Am J Hum Genet*. 1996;59:1233–1242.
20. Furuya H, Kukita Y, Nagano S, et al. Adult onset globoid cell leukodystrophy (Krabbe disease): analysis of galactosylceramidase cDNA from four Japanese patients. *Hum Genet*. 1997;100:450–456.
21. De Stefano N, Dotti MT, Mortilla M, et al. Evidence of diffuse brain pathology and unspecific genetic characterization in a patient with a typical form of adult-onset Krabbe disease. *J Neurol*. 2000;247:226–228.
22. Duffer PK, Barczykowski A, Kay DM, et al. Later onset phenotypes of Krabbe disease: results of the worldwide registry. *Pediatr Neurol*. 2012;46:298–306.
23. Spiegel R, Bach G, Sury V, et al. A mutation in the saposin A coding region of the prosaposin gene in an infant presenting as Krabbe disease: first report of saposin A deficiency in humans. *Mol Genet Metab*. 2005;84:160–166.
24. Barczykowski AL, Foss AH, Duffner PK, Yan L, Carter RL. Death rates in the U.S. due to Krabbe disease and related leukodystrophy and lysosomal storage diseases. *Am J Med Genet*. 2012;158A:2835–2842.
25. Cannizzaro LA, Chen YQ, Rafi MA, Wenger DA. Regional mapping of human galactocerebrosidase (GALC) to 14q31 by in situ hybridization. *Cytogenet Cell Genet*. 1994;66:244–245.
26. Svennerholm L, Vanier M-T, Månsson JE. Krabbe disease: a galactosylsphingosine (psychosine) lipidosis. *J Lipid Res*. 1980;21:53–64.

27. Wenger PA, Sattler M, Roth S. A protein activator of galactosylceramide β-galactosidase. *Biochim Biophys Acta*. 1982;712:639–649.
28. Matsuda J, Vanier MT, Saito Y, Tohyama J, Suzuki K, Suzuki K. A mutation in the saposin A domain of the sphingolipid activator protein (prosaposin) gene results in a late-onset, chronic form of globoid cell leukodystrophy in the mouse. *Hum Mol Genet*. 2001;10:1191–1199.
29. Yeager AM, Brennan S, Tiffany C, Moser HW, Santos GW. Prolonged survival and remyelination after hematopoietic cell transplantation in the twitcher mouse. *Science*. 1984;225:1052–1054.
30. Luzi P, Rafi MA, Zaka M, et al. Biochemical and pathological evaluation of long-lived mice with globoid cell leukodystrophy after bone marrow transplantation. *Mol Genet Metab*. 2005;86:150–159.
31. Rafi MA, Rao HZ, Luzi P, Curtis MT, Wenger DA. Extended normal life after AAVrh10-mediated gene therapy in the mouse model of Krabbe disease. *Mol Ther*. 2012;20:2031–2042.
32. Luddi A, Volterrani M, Strazza M, et al. Retrovirus-mediated gene transfer and galactocerebrosidase uptake into twitcher glial cells results in appropriate localization and phenotype correction. *Neurobiol Dis*. 2001;8:600–610.
33. Li Y, Scott CR, Chamoles NA, et al. Direct multiplex assay of lysosomal enzymes in dried blood spots for newborn screening. *Clin Chem*. 2004;50:1785–1796.

CHAPTER

31

The Mucopolysaccharidoses

Reuben Matalon[*], *Kimberlee Michals Matalon*[†], *and Geetha L. Radhakrishnan*[*]

[*]The University of Texas Medical Branch, Galveston, TX, USA
[†]The University of Houston, Houston, TX, USA

INTRODUCTION

The mucopolysaccharidoses are a group of inherited disorders caused by specific enzyme deficiencies in the degradation of the glycosaminoglycans (mucopolysaccharides). Enzyme deficiencies result in the accumulation of glycosaminoglycans in lysosomes of various tissues and in the excessive excretion of partially degraded glycosaminoglycans in urine. Clinical manifestations of the mucopolysaccharidoses depend on the specific enzyme deficiency, the end organ affected, and the accumulation of glycosaminoglycans in the affected organs. In diseases in which the brain is not involved, there is no mental retardation. On the other hand, if the brain is affected and other somatic manifestations are minimal, the coarse features that are characteristic of the mucopolysaccharidoses are not as prominent. Specific degradative lysosomal enzyme deficiencies have been identified for all the mucopolysaccharidoses. The glycosaminoglycans that are stored and excreted in the urine of the various mucopolysaccharidoses are dermatan sulfate, heparan sulfate, keratan sulfate, and chondroitin 4/6 sulfates.[1–4]

HISTORY

Hunter in 1917[5] described two brothers that fit the clinical features of the X-linked recessive form of the mucopolysaccharidoses. Hurler in 1919[6] described two unrelated boys with clinical findings that conformed to the syndrome now associated with her name. Initially, all the mucopolysaccharidoses were referred to as Hurler syndrome. In 1936, Ellis et al.[7] suggested the term *gargoylism* because of the coarse facial features of children with these diseases. In 1952, Brante identified storage of mucopolysaccharide in liver of patients with Hurler syndrome.[8] He then coined the term *mucopolysaccharidosis*. Further characterization of the mucopolysaccharides (glycosaminoglycans) in tissues and urine was instrumental for the delineation of the mucopolysaccharidoses.[1–4,9]

The idea that the mucopolysaccharidoses are caused by defects of lysosomal hydrolases was proposed in 1964.[10] The liver from a patient with Hurler syndrome was studied, and distended lysosomes were found, suggesting storage within these organelles. Subsequent studies using cultured skin fibroblasts confirmed that fibroblasts from patients with mucopolysaccharidoses contained large amounts of glycosaminoglycans.[11,12] Studies showed that the accumulation of glycosaminoglycans in fibroblasts was indicative of faulty degradation and that mixing cells from various mucopolysaccharidoses led to cross-correction of the accumulation of glycosaminoglycans, which suggests that each mucopolysaccharide disorder is distinct from the other mucopolysaccharide disorders.[13,14] Matalon et al.[15] discovered the enzyme α-L-iduronidase in human and other mammalian tissues and that deficiency of this enzyme resulted in Hurler syndrome.[16] The discovery of this lysosomal defect led to the subsequent elucidation of the enzyme deficiencies in other mucopolysaccharidoses. Reviews dealing with the biochemistry of the glycosaminoglycans and mucopolysaccharidoses are available.[1–4]

Rosenberg's Molecular and Genetic Basis of Neurological and Psychiatric Disease
http://dx.doi.org/10.1016/B978-0-12-410529-4.00031-0

MANIFESTATIONS OF THE MUCOPOLYSACCHARIDOSES

Hurler syndrome has been the prototype of the entire mucopolysaccharidoses. With the discovery of mucopolysacchariduria, however, delineation of the various mucopolysaccharidoses according to the urinary glycosaminoglycans emerged.[9] Further classification occurred following the enzyme deficiencies assigned for the various mucopolysaccharidoses (Table 31.1). For each enzyme deficiency, there is a spectrum of clinical manifestations. In α-L-iduronidase deficiency, the most severe form is Hurler syndrome (Figure 31.1). The deficiency of α-L-iduronidase leads to the accumulation of dermatan sulfate and heparan sulfate and to the urinary excretion of these glycosaminoglycans.[17–19] In patients with Hurler syndrome, there is a wide occurrence of vacuolated cells containing engorged lysosomes with glycosaminoglycans.

Roentgenographic Findings

The mucopolysaccharidoses have specific skeletal changes referred to as *dysostosis multiplex*. These changes may vary in severity depending on the individual mucopolysaccharide disorder. Sanfilippo syndrome has the least severe radiologic changes. Dysostosis multiplex can also be seen in diseases other than the mucopolysaccharidoses, especially glycoprotein storage diseases such as mannosidosis, aspartylglucosaminuria, and G_{M1} gangliosidosis. The skull is usually dolichocephalic, and the calvarium is thickened with hyperostosis of the cranium. The sella turcica is large and/or boot-shaped. The clavicles are thickened, especially in the middle third. The vertebral bodies are ovoid with beak-like projections on their anterior lower margins. Thoracic vertebra (T-12) tends to be hypoplastic, which may result in a gibbus formation (Figure 31.2). The ribs tend to widen distally, which gives them the shape of a spatula. The iliac bones are flared with shallow acetabula. The hips are deformed, and the head of the femur has changes that resemble aseptic necrosis. The hands show tapering of the terminal phalanges and tapering of the proximal ends of the metacarpals. The fifth metacarpal is usually the first to be affected. The long bones show areas of cortical thinning and irregular widening associated with expansion of the medullary cavity. The radius curves toward the ulna at the distal end, forming a V-shaped deformity (Figure 31.3). The humerus is angulated, and the glenoid fossa is shallow.

MPS I: α-L-Iduronidase Deficiency; Hurler, Hurler–Scheie, and Scheie Syndrome

Before age 1 year, it is difficult to differentiate among the α-L-iduronidase deficiency syndromes clinically or enzymatically. This is important because of experimental therapy and because the milder forms of α-L-iduronidase defects may not require such therapeutic modalities (see section Therapy, page 359). There have been several approaches to resolve this problem, including using enzyme kinetics and monoclonal antibodies for quantitation of α-L-iduronidase.[20] We have used two separate methods to differentiate these variants. The first approach is the analysis of urine for glycosaminoglycans. In Hurler syndrome, there is increased excretion of dermatan and heparan

TABLE 31.1 Enzyme Deficiencies of the Mucopolysaccharidoses

Disease	Enzyme deficiency	Inheritance
Hurler	α-L-iduronidase	Autosomal recessive
Hurler–Scheie	α-L-iduronidase	Autosomal recessive
Scheie	α-L-iduronidase	Autosomal recessive
Hunter (A and B)	Iduronosulfate sulfatase	X-linked recessive
Sanfilippo A	Sulfamidase	Autosomal recessive
Sanfilippo B	α-*N*-acetylglucosaminidase	Autosomal recessive
Sanfilippo C	AcetylCoA:α-glucosaminide *N*-acetyltransferase	Autosomal recessive
Sanfilippo D	*N*-acetylglucosamine-6-sulfate sulfatase	Autosomal recessive
Morquio A	*N*-acetylgalactosamine-6-sulfate sulfatase	Autosomal recessive
Morquio B	β-galactosidase	Autosomal recessive
Maroteaux–Lamy	*N*-acetylgalactosamine-4-sulfate sulfatase	Autosomal recessive
β-Sly	β-glucuronidase	Autosomal recessive

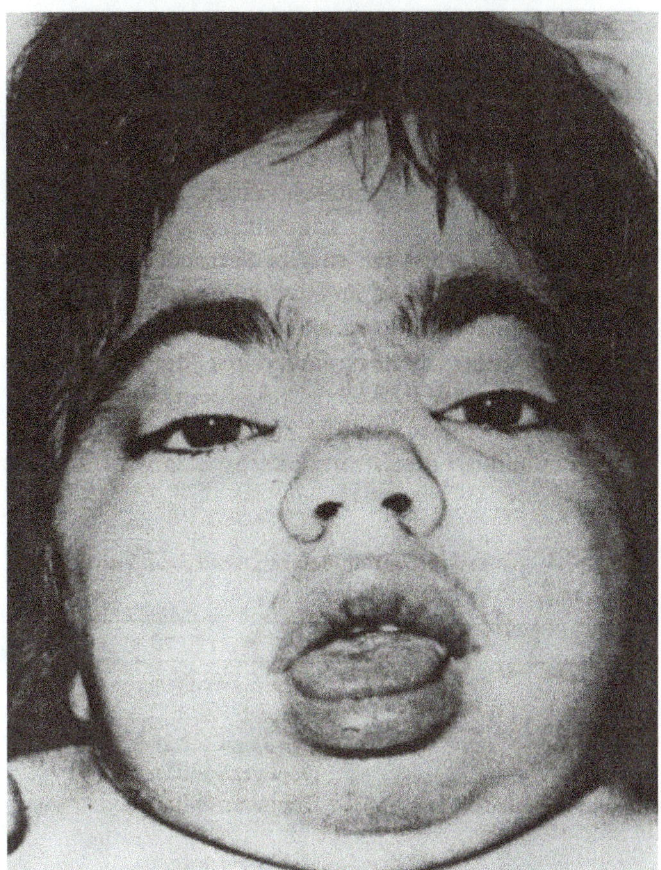

FIGURE 31.1 **Typical facial appearance of a child with Hurler syndrome showing coarse facial features, broad nasal bridge, frontal bossing, clouding of corneas and large tongue.**

sulfate. In the mild forms (Hurler–Scheie and Scheie syndromes), there is excretion of only dermatan sulfate.[18] The second approach is the use of desulfated heparin as a natural substrate for α-L-iduronidase.[19] In Hurler syndrome, iduronic acid is not hydrolyzed from this substrate, whereas in Hurler–Scheie and Scheie syndromes, iduronic acid is hydrolyzed from desulfated heparin. Mutation analysis of the α-L-iduronidase gene may also be informative in some patients.

Hurler Syndrome

The infant with Hurler syndrome appears normal at birth, but after age 6 months, coarse facial appearance can be noted (Figure 31.4). Hepatosplenomegaly can be detected, and umbilical and inguinal hernias may appear. Chronic rhinorrhea may suggest frequent colds. Recurrent upper-airway infection, otitis media, and hypertrophy of tonsils and adenoids may persist beyond early childhood. Hearing impairment may be secondary to such events. When these children attempt to sit, very mild kyphosis can be noticed. As the child grows older, the kyphosis progresses to the gibbus that is typical of Hurler syndrome. Clouding of the corneas leads to impaired vision and can be detected in the first year of life. Children with Hurler syndrome may sit, walk, and develop early language skills, but soon these skills are lost. Severe mental retardation becomes apparent, and these children become bedridden (Figure 31.5).

BIOCHEMICAL AND MOLECULAR STUDIES

Mucopolysaccharidoses are usually diagnosed by the presence of glycosaminoglycans in the urine, and followed by assay for the enzyme diagnosis in cultured fibroblasts, leukocytes, or plasma. The gene for α-L-iduronidase has been localized to the short arm of chromosome 4 (4p16.3) in close proximity to the Huntington gene. DNA based testing starts by examining the parents and the affected child for mutations. If both mutant alleles are identified, then DNA testing can be used for prenatal or carrier testing for family members. Nonsense mutations, including the two most common mutations, W402X and Q70X, in homozygous or compound heterozygous leads to a severe disease with central nervous system involvement. The two most common mutations in milder patients without central

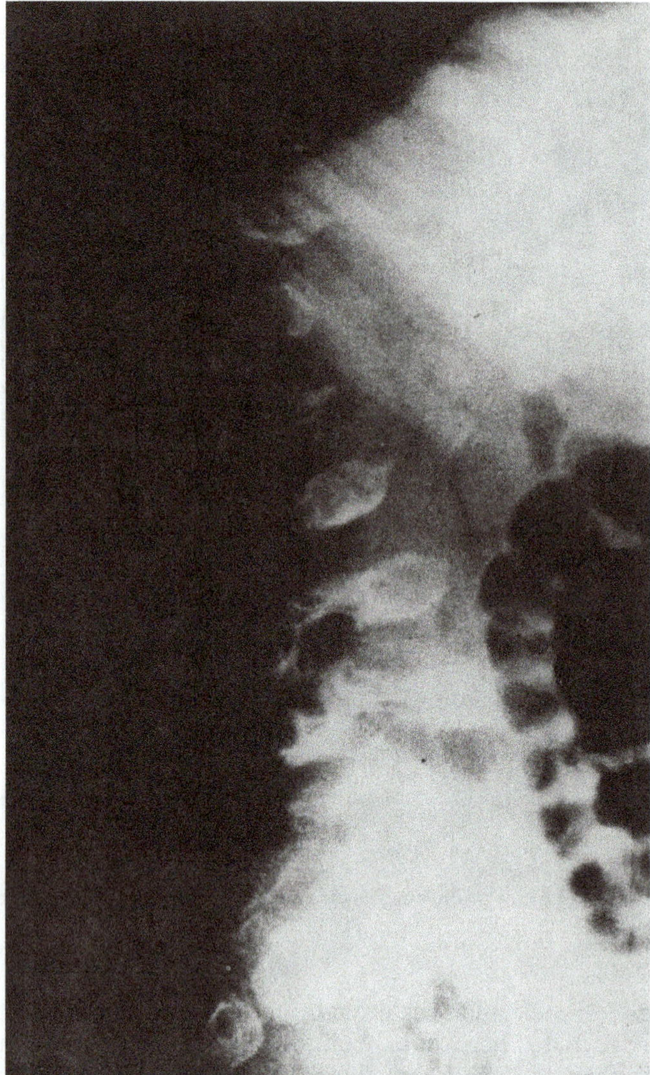

FIGURE 31.2 **X-ray film of lateral spine showing ovoid and beaked vertebrae in a child with mucopolysaccharidosis.**

nervous system involvement are a missense mutation, R89Q and a splice-site mutation, 678-7 g>a. Missense, deletion, insertion or splice site mutations can be found in severe and milder forms of α-L-iduronidase deficiency.

Hurler–Scheie Syndrome

Patients with Hurler–Scheie syndrome have severe joint involvement, short stature, small thorax, hepatosplenomegaly, coarse facial features, and corneal clouding but normal or near-normal mentality (Figure 31.6). The small thorax and the cardiac involvement in this disease, usually mitral valve insufficiency, may be important to recognize because they are the major complicating events that lead to increased mortality.

Scheie Syndrome

Patients with Scheie syndrome usually attain normal height, and their hepatosplenomegaly may be very mild. Stiffness of joints may not be recognized early in life because these children are not retarded and they do not have coarse features. The dysostosis multiplex is very mild. Carpal tunnel syndrome or other joint stiffness may bring these patients to the physician. More commonly, ocular involvement such as corneal clouding and retinal degeneration may lead to the suspicion of mucopolysaccharide disorder. Usually the diagnosis is not made before age 10 years and frequently beyond age 20 years.

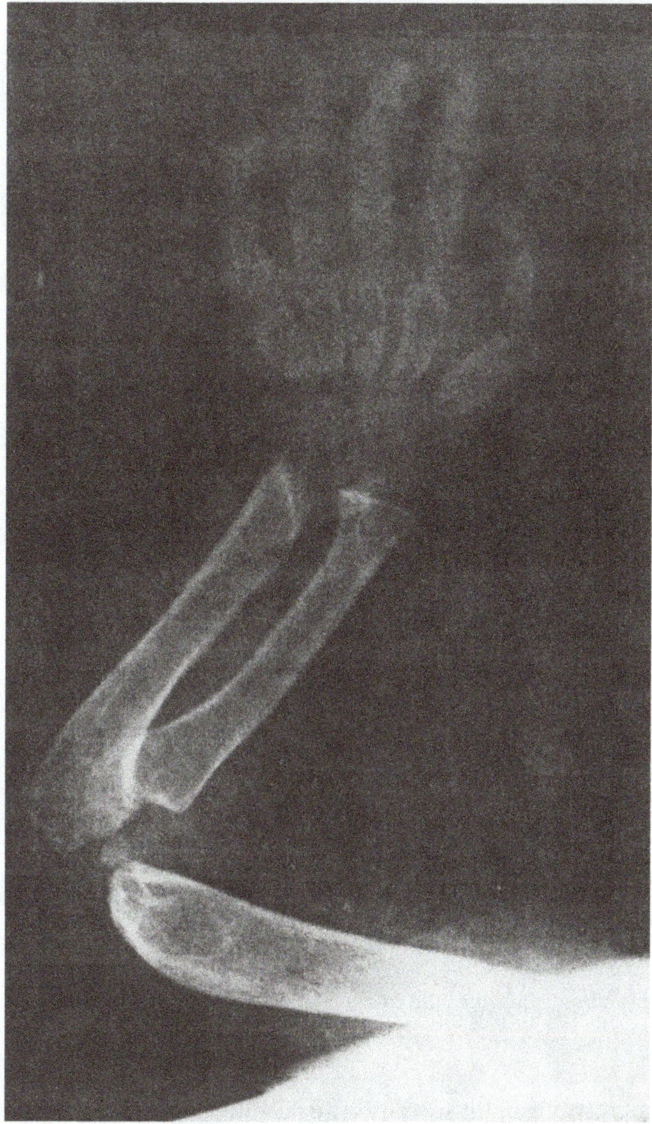

FIGURE 31.3 **X-ray film of hand and forearm showing tapering of the proximal ends of the metacarpals, cortical thinning, and curving of the radius and ulna.**

Biochemical and Molecular Studies

α-L-Iduronidase is required for the hydrolysis of α-L-iduronic acid from the terminal ends of dermatan and heparan sulfates. The gene for α-L-iduronidase has been localized to the short arm of chromosome 4 (4p16.3) in close proximity to the Huntington gene.[21–24]

Mutation analysis indicates that a common allele leading to Hurler syndrome is substitution of the amino acid tryptophan to a stop codon ($Trp_{402} \rightarrow Ter$).[25] Another stop codon involving substitution of glutamine ($Gln_{70} \rightarrow Ter$) and substitution of proline to arginine ($Pro_{533} \rightarrow Arg$) results in a severe phenotype.[26] The $Trp_{402} \rightarrow Ter$ and $Gln_{70} \rightarrow Ter$ mutations account for 72% of alleles of children with Hurler syndrome in Europe.[27] Many other mutations have been described in individual patients and other ethnic groups.[27–29]

The milder forms of iduronidase deficiency are associated with mutations that give some residual activity of iduronidase. The mutation that leads to the substitution of arginine to glutamine ($Arg_{89} \rightarrow Gln$) results in an intermediate phenotype.[30,31] The intronic mutation involving nucleotide in position 678 ($g \rightarrow a$) results in Scheie syndrome.[30]

MPS II: Hunter Syndrome

Hunter syndrome is the only X-linked recessive disease among the mucopolysaccharidoses. Phenotypically there are two forms of Hunter syndrome. One form is with mental retardation (Figure 31.7), and the other is with no retardation.

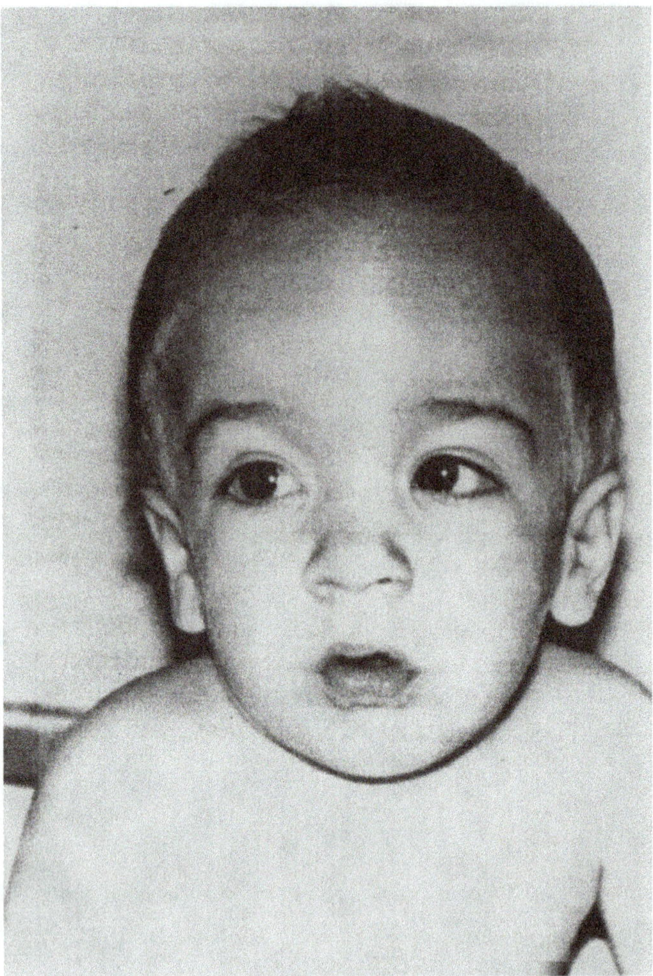

FIGURE 31.4 **Infant with Hurler syndrome showing mild coarse features, presence of metopic sutures, mild corneal clouding, and chronic rhinorrhea.**

This disease may also represent a spectrum of severity. The facial features in both forms are similar, with slightly less coarseness in the mild form. In general, patients with Hunter syndrome look similar to patients with Hurler syndrome with a few exceptions. The corneas are not involved in Hunter syndrome. This can be used as an important distinguishing feature, whereas hearing deteriorates rapidly with Hunter syndrome. In Hunter syndrome, the radiographic finding of dysostosis multiplex is found. The gibbus is not found until the second decade of life, and it is usually very mild. Patients with Hunter syndrome have skin rash over the arms, shoulders, and thighs. The rash is macular and may change in character (Figure 31.7). Patients with Hunter syndrome excrete dermatan and heparan sulfate similar to Hurler syndrome patients. The diagnosis of Hunter syndrome is confirmed by iduronosulfatase deficiency.[31,32]

As they grow older, patients with the mild variant of this disease develop severe hearing problems, carpal tunnel syndrome, and progressive upper-airway obstruction. Death may occur from upper-airway obstruction and heart failure. There are patients with Hunter syndrome who survived beyond the fifth or sixth decade of life.[1] In the α-L-iduronidase deficiency syndromes there is a difference in the urinary glycosaminoglycans between mild and severe forms. In Hunter patients there is no difference.

Biochemical and Molecular Studies

α-L-iduronic acid is sulfated on the C2 position. Iduronosulfatase is required to hydrolyze this ester sulfate as shown in Table 31.1.[31] The gene for iduronosulfatase has been localized to the Xq28 region close to the fragile X site.[33] The gene contains 9 exons and 8 introns.[34] Southern blot analysis of genomic DNA from Hunter patients suggests that a significant number of these patients have gross deletions in the iduronosulfatase gene.[35–38] Patients with complete deletion, partial deletion, or gross rearrangements of the gene all had severe phenotype.[39] Other mutations such as nonsense mutations, small deletions leading to frameshift or premature chain termination, splice mutations,

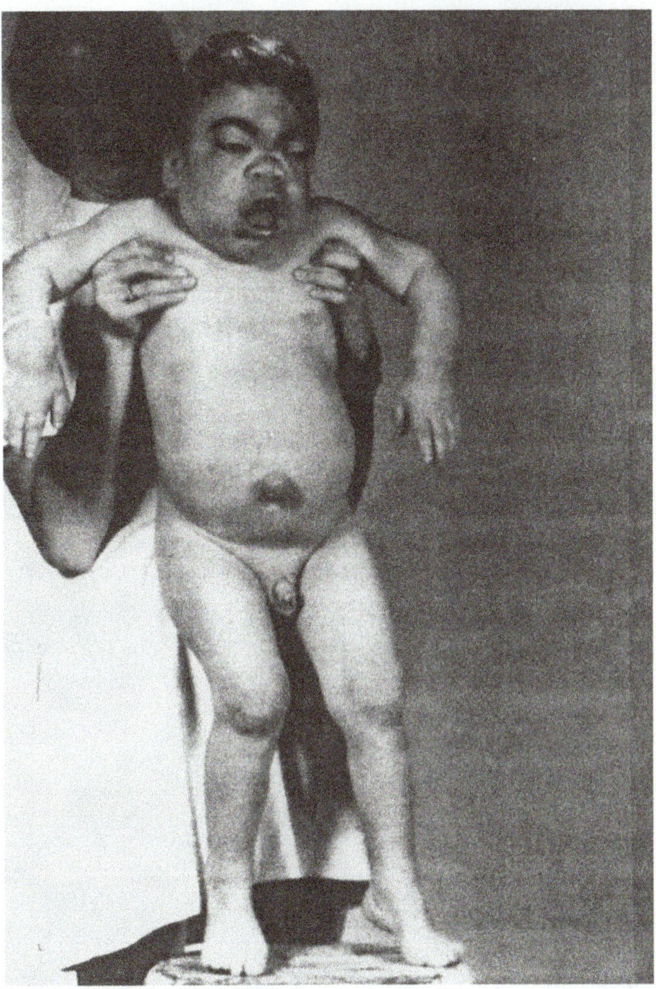

FIGURE 31.5 **Severely retarded child with Hurler syndrome with contracted joints, hepatosplenomegaly, umbilical hernia, and typical facial features.**

and a point mutation eliminating a recognition site have been reported.[39,40] Several missense mutations have been described: $Ala_{68} \rightarrow Glu$, $Ser_{426} \rightarrow Ter$, $Ile_{485} \rightarrow Arg$, $Gln_{293} \rightarrow His$, and $Asp_{478} \rightarrow Gly$.[40] The gross deletion in some patients with Hunter syndrome further suggests that some of the pathology in these severely affected patients may be contributed by the deletion of other neighboring genes.[37] Isolation of the fragile X gene, distal to the iduronosulfatase gene, should make it possible to study whether there is interaction between the two genes.[41]

There have been reports of Hunter syndrome in females. In one such karyotypically normal girl, unbalanced inactivation of the wild-type allele in the maternal chromosome has been suggested.[42] In another case, an X:autosome translocation was reported to span the iduronosulfatase locus.[33] The cDNA can be used to study female Hunter patients and has been used to improve carrier detection for females at risk.[43,44]

MPS III: Sanfilippo Syndromes A–D

There are four types of Sanfilippo syndrome. All have mild dysostosis multiplex and mild coarse facial features (Figure 31.8). Children with Sanfilippo syndrome have hepatosplenomegaly, which is not as pronounced as in Hurler syndrome. As children with this disease grow older, the liver and spleen may resume normal size and hepatosplenomegaly will be missed. Hyperactivity, speech delay, and frank mental retardation are symptoms of this syndrome. By the end of the first decade of life, these children undergo rapid neurologic deterioration and become bedridden and die in their middle teens. Few Sanfilippo patients live beyond the second decade of life.

The mucopolysacchariduria is rather characteristic by increased levels of heparan sulfate only. There is no dermatan sulfate associated with the disease. The specific enzyme assay must be performed to determine which Sanfilippo syndrome the patient has.

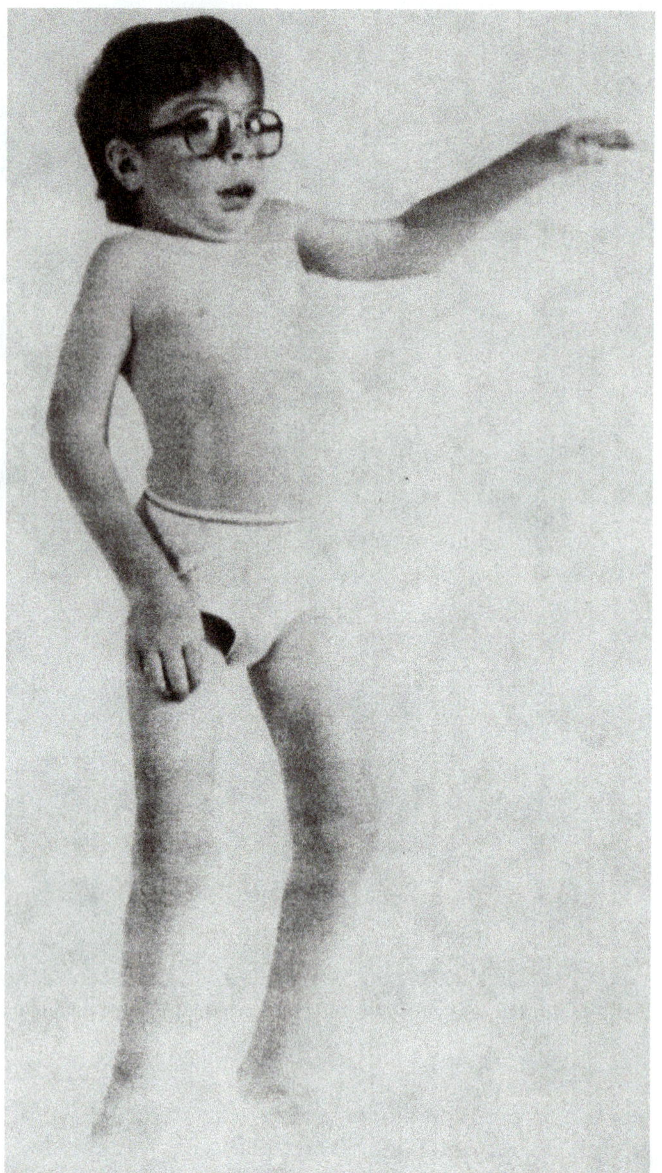

FIGURE 31.6 **Patient with Hurler–Scheie syndrome who is mentally alert with short stature and contracted joints.**

SANFILIPPO A is the most common of the Sanfilippo syndromes. The enzyme deficiency is sulfamidase, which can be assayed on white blood cells or cultured skin fibroblasts (Table 31.1).[45,46]

SANFILIPPO B syndrome is the second in frequency after Sanfilippo A. Clinically the presentation is similar. The enzyme deficiency in this syndrome is α-*N*-acetylhexosaminidase.[46,47]

SANFILIPPO C is probably the mildest of the Sanfilippo syndromes in terms of mental retardation, dysostosis multiplex, and coarse features. Nevertheless, children show symptoms of hyperactivity and mental retardation as they grow older. Joint stiffness is not a prominent feature of Sanfilippo C. Heparan sulfaturia is a feature of this disease, and the enzyme deficiency is acetylCoA:α-glucosamide *N*-acetyltransferase.[48] This enzyme can be assayed on peripheral white blood cells and cultured skin fibroblasts.

SANFILIPPO D is the rarest of the Sanfilippo syndromes. Phenotypically it is usually mild, and the physical findings are typical of the Sanfilippo syndromes. A case of Sanfilippo D was reported in which the patient did not have features of MPS III and presented only with speech delay.[49] The enzyme deficiency of *N*-acetylglucosamine-6-sulfate sulfatase can be assayed on white blood cells and cultured skin fibroblasts.[50] Sanfilippo D patients excrete not only

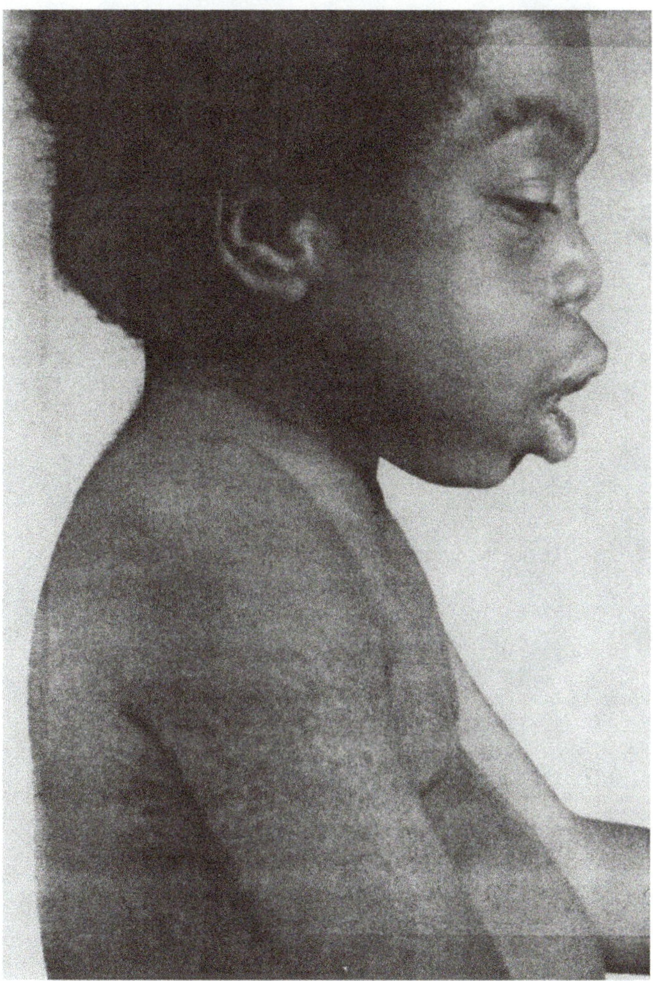

FIGURE 31.7 **Child with Hunter syndrome showing coarse features, depressed nasal bridge, and characteristic rash on the arms and shoulders, frequently seen in Hunter syndrome.**

heparan sulfate but also *N*-acetylglucosamine-6-sulfate. This is caused by the action of hexosaminidase A, which also cleaves sulfated glucosamine.

Biochemical and Molecular Studies

SULFAMIDASE is the enzyme that is specific for sulfate linked to the amino groups of glucosamine. The gene for sulfamidase has been cloned and localized to the long arm of chromosome 17 (17q25.3).[51] The most common mutation among those of European ancestry is $ARG_{245} \rightarrow HIS$.[52] Other mutations have been detected among other ethnic groups.[53–55] The common mutation $Arg_{245} \rightarrow His$ has been described in the Cayman Islands, indicating a founding father effect in a genetic isolate.[56]

α-N-ACETYLGLUCOSAMINIDASE is required for the hydrolysis of *N*-acetylglucosamine residues from heparan sulfate. Deficiency of this enzyme leads to Sanfilippo B syndrome. The gene has been isolated and localized to chromosome 17q21.[57,58] Some mutations have been identified with no predominant mutation.[59–61] Heparan sulfate, which accumulates in the four types of Sanfilippo and in Hunter and Hurler syndromes, is unique among the glycosaminoglycans in two respects: *N*-acetylglucosamine is α-linked in heparan sulfate, and instead of *N*-acetylglucosamine there are *N*-sulfated groups on the glucosamine. Following the action of sulfamidase, the glucosamine moiety has to be cleaved. The free glucosamine to be hydrolyzed has to be acetylated by the specific enzyme acetylCoA:α-glucosaminide *N*-acetyltransferase. This is a lysosomal enzyme that catalyzes the acetylation of the free glucosamine on the polysaccharide terminus. The gene for this enzyme has not been cloned.

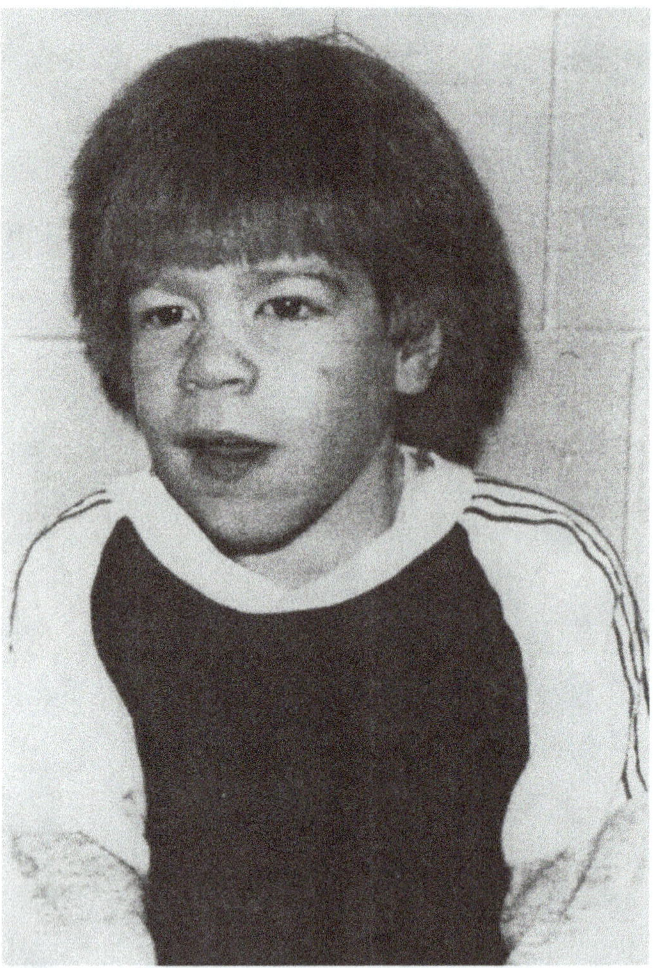

FIGURE 31.8 **Patient with Sanfilippo A syndrome showing mild coarseness of facial features.**

N-ACETYLGLUCOSAMINE-6-SULFATASE hydrolyzes the sulfate on position 6 of the glucosamine of heparan sulfate. This enzyme is deficient in Sanfilippo D. The gene for *N*-acetylglucosamine-6-sulfatase has been cloned and localized to the long arm of chromosome 12 (12q14).[62]

MPS IV: Morquio Syndromes A and B

Morquio[63] and Brailsford[64] described this syndrome with severe skeletal dysplasia and keratan sulfaturia. The skeletal dysplasia can be severe, but also mild forms of Morquio syndrome have been observed. Morquio syndrome usually lacks mental involvement, and is associated with joint laxity, shortness of stature, and pectus carinatum. Skeletal abnormalities include shortened vertebrae (platyspondyly universalis), genu valgum, pes planus, and large joints (Figure 31.9). The neck is short, and the odontoid process of the cervical spine is underdeveloped. This may lead to atlantoaxial subluxation. Corneal clouding is present in 50% of cases. There is usually midfacial hypoplasia and protrusion of the mandible, which makes these children seem like they have a permanent grin. Hepatomegaly may be present, but it is not as prominent as in the other mucopolysaccharidoses. Cardiac involvement includes aortic regurgitation. Tooth enamel is severely affected and is usually very thin. Patients with Morquio syndrome usually survive until middle age. Because of the pectus carinatum and kyphoscoliosis, however, cor pulmonale may develop.

There is a wide range of variability in the phenotypic expression of this disease, and it is often difficult to distinguish Morquio syndrome from other skeletal dysplasias. Keratan sulfaturia, however, is specific for Morquio syndrome.

There are two types of Morquio syndrome: type A caused by deficiency of *N*-acetylgalactosamine-6-sulfate sulfatase[65] and type B caused by deficiency of β-galactosidase.[66] Type B usually lacks enamel hypoplasia. The enzyme deficiency in both conditions can be assayed in white blood cells and cultured skin fibroblasts.

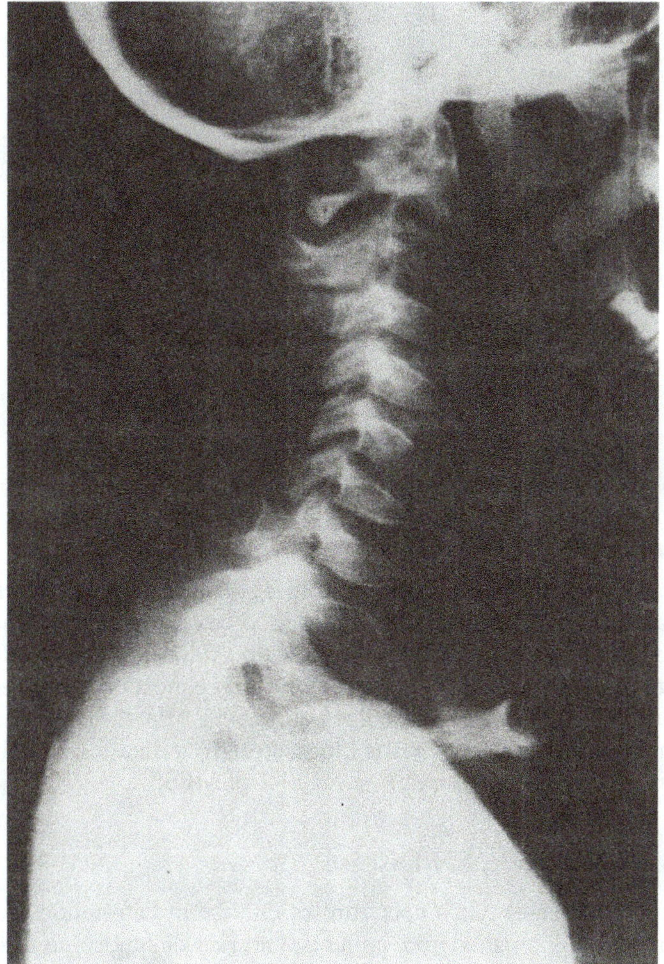

FIGURE 31.9 X-ray film of the cervical spine showing flattened ovoid shaped vertebrae typical of Morquio syndrome.

Biochemical and Molecular Studies

Galactosamine-6-sulfate sulfatase, the enzyme that hydrolyzes sulfate from galactose-6-sulfate and *N*-acetylgalactosamine-6-sulfate, is the enzyme deficient in Morquio type A. The gene for *N*-acetylgalactosamine-6-sulfatase has been cloned and has been localized to the long arm of chromosome 16 (16 g24.3).[67–71] There have been more than 100 mutations reported in Morquio type A patients.[72–83]

β-galactosidase is also required for the sequential degradation of keratan sulfate. Deficiency of this enzyme leads to Morquio type B. The gene has been cloned and assigned to the short arm of chromosome 3 (3p21.33).[84,85] Several point mutations have been reported in the β-galactosidase gene: $Try_{273} \rightarrow$ Leu, $Arg_{201} \rightarrow$ Cys, $Ile_{51} \rightarrow$ Thr, and $Arg_{482} \rightarrow$ His.[86–88] β-galactosidase deficiency can also be caused by a protective protein. Such a disease causes deficiency of sialidase and β-galactosidase (mucolipidosis I).[67]

MPS VI: Maroteaux–Lamy Syndrome

Maroteaux–Lamy syndrome resembles Hurler syndrome but lacks the mental retardation (Figure 31.10). Patients have coarse facial features, corneal clouding, hepatosplenomegaly, and dysostosis multiplex.[89] Urinary mucopolysaccharides show increased excretion of dermatan sulfate, which is caused by the deficiency of *N*-acetylgalactosamine-4-sulfate sulfatase (arylsulfatase B) (Table 31.1).[90,91] Maroteaux–Lamy syndrome also has a spectrum of severity, and in the severe form, children die from constricted chest and upper-airway obstruction. Long-term studies are scarce, however, Brands et al. conducted a prospective open-label follow-up study in 11 Dutch MPS type VI patients (age range 2–18 years old).[92] They showed that enzyme-replacement therapy (ERT) had significant positive effects on cardiac-wall diameters, right and left shoulder flexion, liver/spleen size, urinary glycosaminoglycan excretion and general quality of life.

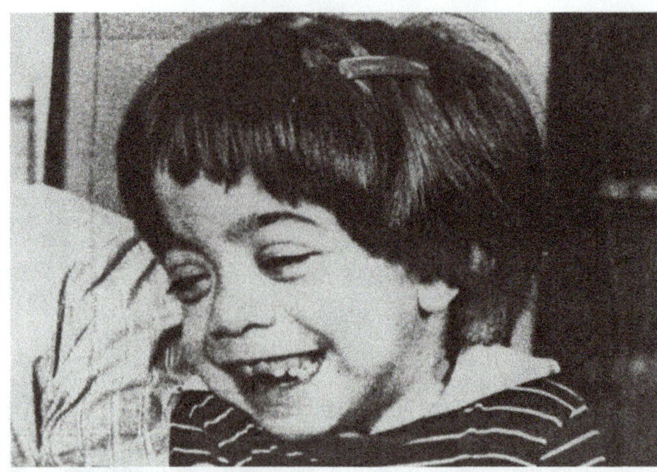

FIGURE 31.10 **Patient with Maroteaux–Lamy syndrome with frontal bossing, large head, and mild coarse features.**

Biochemical and Molecular Studies

N-acetylgalactosamine-4-sulfatase (arylsulfatase B) is a 38–41-kDa monomer enzyme isolated from liver.[93] The gene encoding for arylsulfatase B has been isolated and localized to chromosome 5 (5q13-5q14).[94–96] A feline model of MPS VI has been described and characterized biochemically and pathologically.[97] Mutations have been identified that cause severe, moderate, and mild forms of Maroteaux–Lamy, so genotype and phenotype can be correlated.[98–102] *In vitro* experiments have been performed for nanoparticle bound arylsulfatase B (ASB) for ERT. Adsorption has been demonstrated to be stable for at least 60 minutes in the blood stream, indicating ASB-loaded poly(butyl cyanoacrylate) (PBCA) nanoparticles represent a promising option for ERT of MPS VI.[103]

MPS VII: β-Glucuronidase Deficiency, Sly Disease

β-Glucuronidase deficiency (Sly disease) has a spectrum of clinical manifestations with mild and severe forms.[104] Sometimes these patients cannot be distinguished from patients with Hurler syndrome. The coarse facial features are usually milder, and the gibbus is not pronounced. The mucopolysacchariduria consisted mostly of chondroitin-4/6-sulfate.[105] The enzyme deficiency β-glucuronidase can be documented using white blood cells and cultured skin fibroblasts.

Biochemical and Molecular Studies

Glucuronic acid, found in heparan sulfate, dermatan sulfate, and chondroitin 4/6 sulfate, is hydrolyzed by β-glucuronidase. The gene for β-glucuronidase has been cloned both from human and mouse.[106–112] The exon–intron boundaries between the mouse and human genes are identical. The gene for human β-glucuronidase has been localized to the long arm of chromosome 7 (7q21.11).[113,114]

Canine and mouse models with β-glucuronidase deficiency have been reported.[115–117] A feline colony with the disease has been reported.[118]

A variety of mutations have been identified to cause β-glucuronidase deficiency.[119–122] Two mutations ($Ala_{354} \rightarrow Val$ and $Arg_{611} \rightarrow Trp$) have been identified to result in hydrops fetalis.[123]

Multiple Sulfatase

Multiple sulfatase deficiency has been described where the protective moiety for the hydrolytic enzyme is required and may be deficient in patients with multiple sulfatase. Such patients will store and excrete in their urine various glycosaminoglycans.[124] We have described combined heparan and keratan sulfaturia with glucosamine-6-sulfatase deficiency, which is different than the enzyme deficiency in Sanfilippo D syndrome.[125,126] This unusual mucopolysacchariduria may be a result of a multiple sulfatase defect.

The basic defect for multiple sulfatase is not known. It has been postulated that either a cotranslational or post-translational process results in deficient activity of various sulfatases.[127]

Although multiple sulfatase is not a mucopolysaccharide disease, such patients do accumulate and excrete glycosaminoglycans in addition to other sulfatides. These patients have coarse facial appearance and dysostosis multiplex similar to the mucopolysaccharidoses.

Biochemical and Molecular Studies

Multiple sulfatase deficiency is caused by either homozygous or compound heterozygous mutation in the sulfatase-modifying factor-1 gene (SUMF1) localized to chromosome 3p26.1. Genotype/phenotype show that the most severely affected patient was compound heterozygous for a splice site and a missense mutation. Late infantile patients were homozygous for missense mutations and the highest residual enzyme activity. Patients with intermediate severe infantile form had mutations that compromised stability and caused lower residual enzyme activity.

MPS IX: Hyaluronidase Deficiency

MPS IX is caused by deficiency of hyaluronidase and accumulation of hyaluronan. The single case reported, presented with normal intelligence, short stature and periarticular soft-tissue masses that were associated with painful swelling. The episodes started at the end of the first decade or life and each episode lasted about three days. This condition is inherited in an autosomal recessive manner. The gene has been cloned and localized to chromosome 3p21.3. The patient had two mutations: a missense mutation (c.1412G>A) and a complex rearrangement (c.1361-del137ins14). Imundo et al. discussed complete deficiency of HYAL1 in a consanguineous family that initially presented as familial juvenile idiopathic arthritis.[128] Gene mapping and sequencing identified homozygous deletion of HYAL1, c.104delT, resulting in a premature termination codon in the three children described. Enzymatic analysis confirmed total HYAL1 deficiency and confirmed the diagnosis of MPS IX. In contrast to the previously diagnosed MPS IX patient, these three patients displayed a phenotype limited to the joints. Recently, Natowicz et al. discussed a case report regarding a 14-year-old girl with a normal early past medical history, who then presented with short stature and multiple periarticular soft-tissue masses.[129] She proved to have a storage disease of hyaluronan due to genetic deficiency of hyaluronidase.

Hyaluronidase 2 is a membrane-anchored protein that has been proposed to hydrolyze hyaluronan into smaller fragments that are further internalized for breakdown. Chowdhury et al. have described this using mouse model histological studies of the heart revealing heart valves that were expanded and contained disorganized extracellular matrix.[130] Using electron microscopy, extracellular material was detected throughout the expanded heart valves, which was presumed to be hyaluronan. All HYAL2 knockout mice exhibited increased serum HA as compared to the control mice, supporting a role for HYAL2 in hyaluronan breakdown.

THERAPY FOR THE MUCOPOLYSACCHARIDOSES

Symptomatic treatment is needed to relieve hydrocephalus, cardiac disease, corneal clouding, and upper-airway obstruction. Medication for hyperactivity can be helpful.

Experimental treatment using bone marrow transplants started in 1981 on patients with MPS I. The results included improvement in the coarse facial appearance, corneal clouding, hepatosplenomegaly, and mucopolysacchariduria. Mental function also seemed to improve. Bone marrow treatment was recommended early in life to prevent deterioration in brain function. Bone marrow transplantations were not successful in Hunter, Sanfilippo, or Morquio diseases.

Enzyme-replacement therapy is currently being studied for several mucopolysaccharide disorders. Hurler, Hurler–Scheie and Scheie (MPS I) are using Aldurazyme, produced by Genzyme, to replace the deficiency of α-iduronidase. Results include improvement in forced vital capacity and distance walked. The liver size and glycosaminoglycan level in tissues decreased and urinary excretion of glycosaminoglycans increased. The effect on mental function has not been reported. More than 90% of patients taking Aldurazyme were positive for antibodies against Aldurazyme and adverse reactions including infusion-related reactions and immunogenicity have been reported. A reduced infusion rate and antipyretics and antihistamines can be added to ameliorate symptoms. Enzyme-replacement therapy is being studied on Hunter disease (MPS II) using Elaprase (idursulase), produced by Shire, to replace iduronsulfatase. Clinical trials have documented improvement in distance walked and forced vital capacity. The urinary glycosaminoglycan levels were monitored and normalized at the upper limit of normal and liver and spleen volumes improved. Half of the patients had left ventricular hypertrophy and this normalized in 40% of the patients at the end of the 53-week study. No data is available on the effect of Elaprase on neurological or skeletal manifestations. Clinical studies are underway for Maroteaux–Lamy (MPS VI) disease using Naglazyme (Galsulfase), produced by Biomarin, to replace *N*-acetylhexosamine-4-sulfatase. There were three clinical trials on patients receiving Naglazyme. All trials documented improved functional outcomes; increasing distance walked and improved ability to climb stairs. Some studies reported improved range of motion and lung function and one study reported less pain and stiffness in some patients. Height, weight, cardiac function and bone density did not change during the trials. Adverse reaction to the infusion was the most frequent side effect. A recent study by Horovitz et al. discussed enzyme replacement therapy for children specifically younger than 5 years of age with MPS VI.[131] The prescribed dose of 1 mg per kg intravenously, weekly with galsulfase ERT was shown to be safe and effective in slowing and/or improving certain aspects of the disease process. However, close monitoring of complications is required, especially cardiac valve involvement and spinal cord compression. In February 2014, VIMIZIM (elosulfase alfa) was approved by the US Food and Drug Administration and has been introduced as the only enzyme-replacement therapy for Morquio A (MPS IV).

Experiments with gene therapy have not yet reached human clinical trials.[132] Animal studies have been used for gene therapy and for early stem cell treatment. The cDNA for the canine α-L-iduronidase has been cloned and the canine model can be used to explore gene therapy.[133,134]

References

1. Neufeld EF, Muenzer J. The mucopolysaccharidoses. In: Scriver CR, Beaudet AL, Sly WS, Valle D, eds. *The Metabolic and Molecular Bases of Inherited Disease*. 8th ed. New York: McGraw-Hill; 2001:3421–3452.
2. Matalon R, Kaul R, Michals K. The mucopolysaccharidoses and the mucolipidosis. In: Duckett S, ed. *Pediatric Neuropathology*. Williams & Wilkins: Baltimore; 1995:525–544.
3. Matalon R. Mucopolysaccharidoses. In: Gershwin ME, Robbins DL, eds. *Musculoskeletal Diseases of Children*. New York: Grune & Stratton; 1983:381–445.
4. Kjellen L, Lindahl V. The proteoglycans structures and functions. In: Richardson CC, Abelson JN, Meister A, Walsh CT, eds. *Annual Reviews of Biochemistry*. Annual Reviews: Palo Alto; 1991:443–475.
5. Hunter C. A rare disease in two brothers. *Proc R Soc Med*. 1917;10:104–116.
6. Hurler G. Uber Einen Type multiplier Abortungen, Vorwiegent am Skelet System. *Z Kinderheilkd*. 1919;24:220–234.
7. Ellis RWB, Sheldon W, Capon NB. Gargoylism (chondro-osteo-dystrophy, corneal opacities, hepatosplenomegaly, and mental deficiency). *Q J Med*. 1936;5:119–139.
8. Brante G. Gargoylism: a mucopolysaccharidosis. *Scand J Clin Lab Invest*. 1952;4:43–46.
9. Dorfman A, Lorincz AE. Occurrence of acid mucopolysaccharides in the Hurler syndrome. *Proc Natl Acad Sci U S A*. 1957;43:443–446.
10. Van Hoof F, Hers HG. The ultrastructure of hepatic cells in Hurler's disease (gargoylism). *CR Hebd Seances Acad Sci (D)*. 1964;259:1281–1283.
11. Matalon R, Dorfman A. Biosynthesis of acid mucopolysaccharides in tissue culture. *Proc Natl Acad Sci U S A*. 1966;56:1310–1316.
12. Danes BS, Bearn AG. Hurler's syndrome: demonstration of an inherited disorder of connective tissue in cell culture. *Science*. 1965;149:987–989.
13. Fratantoni JC, Hall CW, Neufeld EF. The defect in Hurler and Hunter's syndromes II: deficiency of specific factors involved in mucopolysaccharide degradation. *Proc Natl Acad Sci U S A*. 1969;64:360–366.
14. Fratantoni JC, Hall CW, Neufeld EF. Hurler and Hunter syndromes: Mutual correction of the defect in cultured fibroblasts. *Science*. 1968;162:570–572.
15. Matalon R, Cifonelli JC, Dorfman A. L-iduronidase in cultured human fibroblasts and liver. *Biochem Biophys Res Commun*. 1971;42:340–345.
16. Matalon R, Dorfman A. Hurler's syndrome: alpha-L-iduronidase deficiency. *Biochem Biophys Res Commun*. 1972;47:959–964.
17. Bach GX, Friedman AX, Weissmann B, Neufeld EF. The defect in the Hurler and Scheie syndromes: deficiency of alpha-L-iduronidase. *Proc Natl Acad Sci U S A*. 1972;69:2048–2051.
18. Matalon R, Deanching M. The enzymic basis for the phenotypic variation of Hurler and Scheie syndromes. *Pediatr Res*. 1977;11:513.
19. Matalon R, Deanching M, Omura K. Hurler, Scheie and Hurler-Scheie "compound" residual activity of alpha-L-iduronidase toward natural substrates suggesting allelic mutations. *J Inherit Metab Dis*. 1983;6:133–134.
20. Clements PR, Brooks DA, McCourt AG, Hopwood JJ. Immunopurification and characterization of human α-L-iduronidase with the use of monoclonal antibodies. *Biochem J*. 1989;259:199–208.
21. Scott HS, Anson DS, Osborn AM, et al. Human α-L-iduronidase: cDNA isolation and expression. *Proc Natl Acad Sci U S A*. 1991;88:9695–9699.
22. Scott HS, Guo XH, Hopwood JJ, Morris CP. Structure and sequence of human α-L-iduronidase gene. *Genomics*. 1992;13:1311–1313.
23. Scott HS, Ashton IJ, Eyre HJ, et al. Chromosomal localization of the human α-L-iduronidase gene (IDUA) to 4P16.3. *Am J Hum Genet*. 1990;47:802–807.
24. MacDonald ME, Scott HS, Whaley WL, et al. Huntington-disease-linked locus D4Siii exposed as the α-L-iduronidase gene. *Somat Cell Mol Genet*. 1991;17:421–425.
25. Scott HS, Litjens T, Hopwood JJ, Morris CP. A common mutation for mucopolysaccharidosis type I associated with a severe Hurler syndrome phenotype. *Hum Mutat*. 1992;1:103–108.
26. Scott HS, Litjens T, Nelson PV, et al. Alpha-L-iduronidase mutations (Q70X and P533R) associate with a severe Hurler phenotype. *Hum Mutat*. 1992;1:333–339.
27. Bunge S, Kleijer WJ, Steglich C, et al. Mucopolysaccharidosis type I: identification of 8 novel mutations and determination of the frequency of the two common α-L-iduronidase mutations (W402X and Q70X) among European patients. *Hum Mol Genet*. 1994;3:861–866.
28. Clarke LA, Nelson PV, Warrington CL, et al. Mutation analysis of 19 North American mucopolysaccharidosis type I patients: identification of two additional frequent mutations. *Hum Mutat*. 1994;3:275–282.
29. Tieu PT, Menon K, Neufeld EF. A mutant stop codon (TAG) in the IDUA gene is used as an acceptor splice site in a patient with Hurler syndrome (MPS IH). *Hum Mutat*. 1994;3:333–336.
30. Scott HS, Litjens T, Nelson PV, et al. Identification of mutations in the alpha-L-iduronidase gene (IDUA) that cause Hurler and Scheie syndromes. *Am J Hum Genet*. 1993;53:973–986.

31. Sjoberg I, Fransson LA, Matalon R, Dorfman A. Hunter's syndrome: a deficiency of L-iduronosulfate sulfatase. *Biochem Biophys Res Commun*. 1973;54:1125–1132.
32. Bach G, Eisenberg Jr F, Cantz M, Neufeld EF. The defect in the Hunter syndrome: deficiency of sulfoiduronate sulfatase. *Proc Natl Acad Sci U S A*. 1973;70:2134–2138.
33. Wilson PJ, Suthers GK, Callen DF, et al. Frequent deletions at xq28 indicate genetic heterogeneity in Hunter syndrome. *Hum Genet*. 1991;86:505–508.
34. Flomen RH, Green EP, Green PM, et al. Determination of the organization of coding sequences within the iduronate sulphate sulphatase (IDS) gene. *Hum Mol Genet*. 1993;2:5–10.
35. Wilson PJ, Morris CP, Anson DS, et al. Hunter syndrome: isolation of an iduronate-2-sulfatase cDNA clone and analysis of patient DNA. *Proc Natl Acad Sci U S A*. 1990;87:8531–8535.
36. Palmieri G, Capra V, Romano G, et al. The iduronate sulfatase gene: isolation of a 1.2-Mb YAC contig spanning the entire gene and identification of heterogeneous deletions in patients with Hunter syndrome. *Genomics*. 1992;12:52–57.
37. Wraith JE, Cooper A, Thornley M, et al. The clinical phenotype of two patients with a complete deletion of the iduronate-L-sulfatase gene (mucopolysaccharidosis II–Hunter syndrome). *Hum Genet*. 1991;87:205–206.
38. Wehnert M, Hopwood JJ, Schroder W, Herrmann FH. Structural gene aberrations in mucopolysaccharidosis II (Hunter). *Hum Genet*. 1992;89:430–432.
39. Hopwood JJ, Bunge S, Morris CP, et al. Molecular basis of mucopolysaccharidosis type II: mutations in the iduronate-2-sulphatase gene. *Hum Mutat*. 1993;2:435–442.
40. Schroder W, Wulff K, Wehnert M, et al. Mutations of the iduronate-2-sulfatase (IDS) gene in patients with Hunter syndrome (mucopolysaccharidosis II). *Hum Mutat*. 1994;4:128–131.
41. Giampietro PF, Haas BR, Matalon R, et al. Fragile X syndrome in two siblings with major congenital malformations. *Am J Hum Genet*. 1994;55:449.
42. Clarke JT, Greer WL, Strasberg PM, et al. Hunter disease (Mucopolysaccharidosis type II) associated with unbalanced inactivation of the X-chromosomes in a karyotypically normal girl. *Am J Hum Genet*. 1991;49:289–297.
43. Schroder W, Petruschka L, Wehnert M, et al. Carrier detection of Hunter syndrome (MPSII) by biochemical and DNA techniques in families at risk. *J Med Genet*. 1993;30:210–213.
44. Gal A, Beck M, Sewell AC, et al. Gene diagnosis and carrier detection in Hunter syndrome by iduronate-2-sulphatase cDNA probe. *J Inherit Metab Dis*. 1992;15:342–346.
45. Matalon R, Dorfman A. The Sanfilippo A syndrome: a sulfamidase deficiency. *Pediatr Res*. 1973;7:156.
46. Matalon R, Dorfman A. Sanfilippo A syndrome: sulfamidase deficiency in cultured skin fibroblasts and liver. *J Clin Invest*. 1974;54:907–912.
47. O'Brien JS. Sanfilippo syndrome: profound deficiency of alpha-acetylglycosaminidase activity in organs and skin fibroblasts from type B patients. *Proc Natl Acad Sci U S A*. 1972;69:1720–1722.
48. Kresse H, von Figura K, Klein U. A new biochemical subtype of the Sanfilippo syndrome: characterization of the storage material in cultured fibroblasts of Sanfilippo C patients. *Eur J Biochem*. 1978;92:333–339.
49. Ozand PT, Thompson JN, Gascon GG, et al. Sanfilippo Type D with acquired language disorder but without features of mucopolysaccharidosis. *J Child Neurol*. 1994;9:408–411.
50. Kresse H, Paschke E, von Figura K, et al. Sanfilippo disease type D: deficiency of *N*-acetylglucosamine-6-sulfate sulfatase required for heparan sulfate degradation. *Proc Natl Acad Sci U S A*. 1980;77:6822–6826.
51. Scott HS, Blanch L, Guo XH, et al. Cloning of the sulphamidase gene and identification of mutations in Sanfilippo A syndrome. *Nat Genet*. 1995;11:465–467.
52. Weber B, Guo XH, Wraith JE, et al. Novel mutations in Sanfilippo A syndrome: implications for enzyme function. *Hum Mol Genet*. 1997;6:1573–1579.
53. Bunge S, Ince H, Steglich C, et al. Identification of 16 sulfamidase gene mutations including the common R74C in patients with mucopolysaccharidosis type IIIa (Sanfilippo A). *Hum Mutat*. 1997;10:479–485.
54. Di Natale P, Balzano N, Esposito S, et al. Identification of molecular defects in Italian Sanfilippo A patients including 13 novel mutations. *Hum Mutat*. 1998;11:313–320.
55. Montfort M, Vilageliu L, Garcia-Giralt N, et al. Mutation 109delC is highly prevalent in Spanish Sanfilippo syndrome type A patients. *Hum Mutat*. 1998;12:274–279.
56. Rady P, Vu AT, Surendran S, et al. Founder mutation, R245H, of Sanfilippo syndrome A in the Cayman Islands. *Genet Test*. 2002;6.
57. Weber B, Blanch L, Clements PR, et al. Cloning and expression of the gene in Sanfilippo B syndrome (mucopolysaccharidosis III B). *Hum Mol Genet*. 1996;5:771–777.
58. Zhao HG, Li HH, Bach G, et al. The molecular basis of Sanfilippo syndrome type B. *Proc Natl Acad Sci U S A*. 1996;93:6101–6105.
59. Whitley CB. The mucopolysaccharidoses. In: Beighton P, ed. *McKusick's Heritable Disorders of Connective Tissue*. 5th ed. St Louis: CV Mosby; 1993:367–499.
60. Schmidtchen A, Greenberg D, Zhao HG, et al. NAGLU mutations underlying Sanfilippo syndrome type B. *Am J Hum Genet*. 1998;62:64–69.
61. Zhao HG, Aronovich EL, Whitley CB. Genotype–phenotype correspondence in Sanfilippo syndrome type B. *Am J Hum Genet*. 1998;62:53–63.
62. Robertson DA, Freeman C, Nelson PV, et al. Human glucosamine-6-sulfatase cDNA reveals homology with steroid sulfatase. *Biochem Biophys Res Commun*. 1988;157:218–224.
63. Morquio L. Surune forme de dystrophie osseuse familiale. *Bull Soc Pediatr (Paris)*. 1929;27:145–152.
64. Brailsford JF. Chondro-osteodystrophy: roentgenographic and clinical features of child with dislocation of vertebrae. *Am J Surg*. 1929;7:404–410.
65. Matalon R, Arbogast B, Justice P, et al. Morquio's syndrome: deficiency of a chondroitin sulfate *N*-acetylhexosamine sulfate sulfatase. *Biochem Biophys Res Commun*. 1974;61:759–765.
66. Arbisser AI, Donnell HA, Scott Jr CI. Morquio-like syndrome with beta galactosidase deficiency and normal hexosamine sulfatase activity: mucopolysaccharidosis IV B. *Am J Med Genet*. 1977;1:195–205.
67. Masuno M, Tomatsu S, Nakashima Y, Hori T. Assignment of the human *N*-acetylgalactosamine-6-sulfate sulfatase (GALNS) gene to chromosome 16q24. *Genomics*. 1993;16:777–778.

68. Baker E, Guo XH, Orsborn AM, et al. The Morquio A syndrome (Mucopolysaccharidosis IVA) gene maps to 16q24.3. *Am J Hum Genet.* 1993;52:96–98.
69. Tomatsu SS, Fukuda M, Masue K, et al. Morquio disease: isolation, characterization and expression of full-length cDNA for human *N*-acetylgalactosamine-6-sulfate sulfatase. *Biochem Biophys Res Commun.* 1991;181:677–681.
70. Morris CP, Guo XH, Apostolou S, et al. Morquio A syndrome: cloning, sequence, and structure of the human N-acetylgalactosamine-6-sulfatase (GALNS) gene. *Genomics.* 1994;22:652–654.
71. Nakashima Y, Tomatsu S, Hori T, et al. Mucopolysaccharidosis IV A: molecular cloning of the human *N*-acetylgalactosamine-6-sulfate gene (GALNS) and analysis of the 5′-flanking region. *Genomics.* 1994;20:99–104.
72. Tomatsu S, Fukuda S, Uchiyama A, et al. Molecular analysis by Southern blot for the *N*-acetylgalactosamine-6-sulphate sulphatase gene causing mucopolysaccharidosis IVA in the Japanese population. *J Inherit Metab Dis.* 1994;17:601–605.
73. Ogawa T, Tomatsu S, Fukuda S, et al. Mucopolysaccharidosis IVA: screening and identification of mutations of the *N*-acetylgalactosamine-6-sulfate sulfatase gene. *Hum Mol Genet.* 1995;4:341–349.
74. Tomatsu S, Fukuda S, Cooper A, et al. Two new mutations, Q473X and N487S, in a Caucasian patient with mucopolysaccharidosis IVA (Morquio disease). *Hum Mutat.* 1995;6:195–196.
75. Tomatsu S, Fukuda S, Cooper A, et al. Mucopolysaccharidosis type IVA: identification of six novel mutations among non-Japanese patients. *Hum Mol Genet.* 1995;4:741–743.
76. Fukuda S, Tomatsu S, Cooper A, et al. Mucopolysaccharidosis IVA (Morquio A): three novel small deletions in the *N*-acetylgalactosamine-6-sulfate sulfatase gene. *Hum Mutat.* 1996;8:187–190.
77. Tomatsu S, Fukuda S, Cooper A, et al. Mucopolysaccharidosis IVA: identification of a common missense mutation I113F in the *N*-acetylgalactosamine-6-sulfate sulfatase gene. *Am J Hum Genet.* 1995;57:556–563.
78. Bunge S, Kleijer WJ, Tylki-Szymanska A, et al. Identification of 31 novel mutations in the *N*-acetylgalactosamine-6-sulfatase gene reveals excessive allelic heterogeneity among patients with Morquio A syndrome. *Hum Mutat.* 1997;10:223–232.
79. Fukuda S, Tomatsu S, Masuno M, et al. Mucopolysaccharidosis IVA: submicroscopic deletion of 16q24.3 and a novel R386C mutation of *N*-acetylgalactosamine-6-sulfate sulfatase gene in a classical Morquio disease. *Hum Mutat.* 1996;7:123–134.
80. Yamada N, Fukuda S, Tomatsu S, et al. Molecular heterogeneity in mucopolysaccharidosis IVA in Australia and Northern Ireland: nine novel mutations including T312S, a common allele that confers a mild phenotype. *Hum Mutat.* 1998;11:202–208.
81. Fukuda S, Tomatsu S, Masue M, et al. Mucopolysaccharidosis type IV A: *N*-acetylgalactosamine-6-sulfate sulfatase exonic point mutations in classical Morquio and mild cases. *J Clin Invest.* 1992;90:1049–1053.
82. Tomatsu S, Fukuda S, Ogawa T, et al. A novel splice site mutation in intron 1 of the GALNS gene in a Japanese patient with mucopolysaccharidosis I VA. *Hum Mol Genet.* 1994;3:1427–1428.
83. Tomatsu S, Fukuda S, Cooper A, et al. Mucopolysaccharidosis IVA: structural gene alterations identified by Southern blot analysis and identification of racial differences. *Hum Genet.* 1995;95:376–381.
84. Oshiwa A, Tsuji A, Nagao Y, et al. Cloning, sequencing and expression of cDNA for human beta-galactosidase. *Biochem Biophys Res Commun.* 1988;157:238–244.
85. Morreau H, Galjart NJ, Gillemans N, et al. Alternative splicing of beta-galactosidase mRNA generates the classic lysosomal enzyme and a beta-galactosidase-related protein. *J Biol Chem.* 1989;264:20655–20663.
86. Oshima A, Yoshida K, Shimmoto M, et al. Human β-galactosidase gene mutations in Morquio B disease. *Am J Hum Genet.* 1991;49:1091–1093.
87. Oshima A, Yoshida K, Itoh K, et al. Intracellular processing and maturation of mutant gene products in hereditary beta-galactosidase deficiency (betagalactosidosis). *Hum Genet.* 1994;93:109–114.
88. Yoshida K, Oshima A, Shimmoto M, et al. β-Galactosidase gene mutations in GM_1-gangliosidase: a common mutation among Japanese adult/chronic cases. *Am J Hum Genet.* 1991;49:435–442.
89. Maroteaux P, Lamy M. La pseudo-polydystrophy de Hurler. *Presse Med.* 1966;74:2889–2892.
90. Matalon R, Arbogast B, Dorfman A. Deficiency of chondroitin sulfate *N*-acetylgalactosamine-4-sulfate sulfatase in Maroteaux–Lamy syndrome. *Biochem Biophys Res Commun.* 1974;61:1450–1457.
91. Stumpf DA, Austin JH, Crocker AC, LaFrance M. Mucopolysaccharidosis type VI (Maroteaux–Lamy syndrome): arylsulfatase B deficiency in tissue. *Am J Dis Child.* 1973;126:747–756.
92. Brands MM, Oussoren E, Ruijter GJ, et al. Up to five years experience with 11 mucopolysaccharidosis type VI patients. *Mol Genet Metab.* 2013;109(1):70–76.
93. Shapira E, Nadler HL. Purification and some properties of soluble human liver arylsulfatase. *Arch Biochem Biophys.* 1975;170:179–187.
94. Schuchman EH, Jackson CE, Desnick RJ. Human arylsulfatase B: MOPAC cloning, nucleotide sequence of a full length cDNA, and regions of amino acid identity with arylsulfatases A and C. *Genomics.* 1990;6:149–158.
95. Peters C, Schmidt B, Rommerskirch W, et al. Phylogenetic conservation of arylsulfatases: cDNA cloning and expression of human arylsulfatase B. *J Biol Chem.* 1990;265:3374–3381.
96. Litgens T, Baker EG, Beckmann KR, et al. Chromosomal localization of ARSB, the gene for human *N*-acetylgalactosamine-4-sulfatase. *Hum Genet.* 1989;82:67–68.
97. Jackson CE, Yuhki N, Desnick RJ, et al. Feline arylsulfatase B (ARSB): isolation and expression of the cDNA, comparison with human ARSB, and gene localization to feline chromosome A1. *Genomics.* 1992;14:403–411.
98. Jin WD, Jackson CE, Desnick RJ, Schuchman EH. Mucopolysaccharidosis type VI: identification of three mutations in the arylsulfatase B gene of patients with the severe and mild phenotypes provides molecular evidence for genetic heterogeneity. *Am J Hum Genet.* 1992;50:795–800.
99. Litgens T, Morris CP, Robertson EF, et al. An *N*-acetylgalactosamine-4-sulfatase mutation (ΔG_{238}) results in a severe Maroteaux–Lamy phenotype. *Hum Mutat.* 1992;1:397–402.
100. Isbrandt D, Arlt G, Brooks DA, et al. Mucopolysaccharidosis VI (Maroteaux–Lamy syndrome): six unique arylsulfatase B gene alleles causing variable disease phenotypes. *Am J Hum Genet.* 1994;54:454–463.

101. Wicker G, Prill V, Brooks D, et al. Mucopolysaccharidosis VI (Maroteaux–Lamy syndrome): an intermediate clinical phenotype caused by substitution of valine for glycine at position 137 of arylsulfatase B. *J Biol Chem*. 1991;266:27386–27391.
102. Voskoboeva E, Isbrandt D, von Figura K, et al. Four novel mutant alleles of the arylsulfatase B gene in two patients with intermediate form of mucopolysaccharidosis VI (Maroteaux-Lamy syndrome). *Hum Genet*. 1994;93:259–264.
103. Muhlstein A, Gelperina S, Kreuter J. Development of nanoparticle-bound arylsulfatase B for enzyme replacement therapy of mucopolysaccharidoses VI. *Pharmazie*. 2013 Jul;68(7):549–554.
104. Sly WS, Quinton BA, McAllister WH, Rimoin DL. Beta-glucuronidase deficiency: report of clinical radiologic and biochemical features of a new mucopolysaccharidosis. *J Pediatr*. 1973;82:249–257.
105. Matalon R, Macias RV, Diekamp U, et al. Beta-glucuronidase deficiency: a mucopolysaccharidosis with chondroitin sulfaturia. *Pediatr Res*. 1985;19(4):316.
106. Nishimura Y, Rosenfeld MG, Kriebich G, et al. Nucleotide sequence of rat preputial gland beta-glucuronidase cDNA and *in vitro* insertion of its encoded polypeptide into microsomal membrane. *Proc Natl Acad Sci U S A*. 1986;83:7292–7296.
107. Oshima A, Kyle JW, Miller RD, et al. Cloning, sequencing and expression of cDNA for human beta-glucuronidase. *Proc Natl Acad Sci U S A*. 1986;84:685–689.
108. Gallaghar PM, D'Amore MA, Lund SD, Ganshow RE. The complete nucleotide sequence of murine beta-glucuronidase mRNA and its deduced polypeptide. *Genomics*. 1988;2:215–219.
109. Powell PP, Kyle JW, Miller RD, et al. Rat liver beta-glucuronidase: cDNA cloning, sequence comparisons and expression of a chimeric protein in COS cells. *Biochem J*. 1988;250:547–555.
110. Miller RD, Hoffmann JW, Powell PP, et al. Cloning and characterization of beta-glucuronidase gene. *Genomics*. 1990;7:280–283.
111. Funkenstein B, Leary SL, Stein JC, Catterall JF. Genomic organization and sequence of the GUS-S alpha allele of the murine beta-glucuronidase gene. *Mol Cell Biol*. 1988;8:1160–1168.
112. Bevilacqua A, Erickson RP. Use of antisense RNA to help identify a genomic clone for 5′ region of mouse beta-glucuronidase. *Biochem Biophys Res Commun*. 1989;160:937–941.
113. Chitayat D, McGillivary BC, Wood S, et al. Interstitial 7q deletion (46, xx, del(7) (Pter-q21.1::q22→qter)) and the location of genes for beta-glucuronidase and cystic fibrosis. *Am J Med Genet*. 1988;31:655–661.
114. Allanson JE, Gemmill RN, Hecht BK, et al. Deletion mapping of beta-glucuronidase gene. *Am J Med Genet*. 1988;29:517–522.
115. Haskins ME, Aguirre GD, Jezyk PE, et al. Mucopolysaccharidosis type VII (Sly syndrome): beta-glucuronidase deficient mucopolysaccharidosis in the dog. *Am J Pathol*. 1991;138:1553–1555.
116. Birkenmeier EH, Davisson MT, Neamer WG, et al. Murine mucopolysaccharidosis type VII: characterization of a mouse with beta-glucuronidase deficiency. *J Clin Invest*. 1989;83:1258–1266.
117. Schuchman EH, Toroyan TK, Haskins ME, Desnick RJ. Characterization of the defective β-glucuronidase activity in canine mucopolysaccharidosis type VII. *Enzyme*. 1989;42:174–180.
118. Gitzelmann R, Bosshard NU, Superti-Furga A, et al. Feline mucopolysaccharidosis VII due to β-glucuronidase deficiency. *Vet Pathol*. 1994;31:435–443.
119. Tomatsu S, Sukegawa K, Ikedo Y, et al. Molecular basis of mucopolysaccharidosis type VII: replacement of ala 619 in beta-glucuronidase with val. *Gene*. 1990;89:283–287.
120. Tomatsu S, Fukuda S, Sukegawa K, et al. Mucopolysaccharides type VII: characterization of mutations and molecular heterogeneity. *Am J Hum Genet*. 1991;48:89–96.
121. Shipley JM, Klinkenberg M, Wu BM, et al. Mutational analysis of a patient with mucopolysaccharidosis type VII, and identification of pseudogenes. *Am J Hum Genet*. 1993;52:517–526.
122. Yamada S, Tomatsu S, Sly WS, et al. Four novel mutations in mucopolysaccharidosis type VII including a unique base substitution in exon 10 of β-glucuronidase gene that creates a novel 5′-splice site. *Hum Mol Genet*. 1995;4:651–655.
123. Wu BM, Sly WS. Mutational studies in a patient with the hydrops fetalis form of mucopolysaccharidosis type VII. *Hum Mutat*. 1993;2:446–457.
124. Kolodny EH, Fluharty AL. Metachromatic leukodystrophy and multiple sulfatase deficiency: sulfatide liposis. In: Scriver CR, Beaudet AL, Sly WS, Valle D, eds. *The Metabolic and Molecular Bases of Inherited Disease*. 7th ed. New York: McGraw-Hill; 1995:2693–2739.
125. Matalon R, Horwitz A, Wappner R, et al. Keratan and heparan sulfaturia: a new mucopolysaccharidosis with *N*-acetylglucosamine-6-sulfatase deficiency. *Pediatr Res*. 1978;12:453.
126. Matalon R, Wappner R, Deanching M, et al. Keratan and heparan sulfaturia: mucopolysaccharidosis with an enzyme defect not previously identified. *J Inherit Metab Dis*. 1982;5:57–58.
127. Rommerskirch W, von Figura K. Multiple sulfatase deficiency: catalytically inactive sulfatases are expressed from retrovirally introduced sulfatase cDNAs. *Proc Natl Acad Sci U S A*. 1992;89:2561–2565.
128. Imundo L, Leduc CA, Guha S, et al. A complete deficiency of hyaluronoglucosaminidase 1 (HYAL1) presenting as familial juvenile idiopathic arthiritis. *J Inherit Metab Dis*. 2011 Oct;34(5):1013–1022.
129. Natowicz MR, et al. Case report: clinical and biochemical manifestations of hyaluronidase deficiency. *N Engl J Med*. 2013;335(14).
130. Chowdhury B, Hemming R, et al. Murine hyaluronidase 2 deficiency results in extracellular hylaronan accumation and severe cardiopulmonary dysfunction. *J Biol Chem*. 2013;288(1):520–528.
131. Horovitz DD, et al. Enzyme replacement therapy with galsulfase in 34 children younger than five years of age with MPS VI. *Mol Genet Metab*. 2013;109(1):62–69.
132. Desnick RJ, Schuchman EH. Human gene therapy: strategies and prospects for inborn errors of metabolism. In: Desnick RJ, ed. *Treatment of Genetic Diseases*. New York: Churchill-Livingstone; 1991:239–259.
133. Stoltzfus LJ, Uhrhammer N, Sosa-Pineda B, et al. Mucopolysaccharidosis I: cloning and characterization of cDNA encoding canine α-L-iduronidase. *Am J Hum Genet*. 1990;47:A167.
134. Menon KP, Tieu PT, Neufeld EF. Architecture of the canine IDUA gene and mutation underlying canine mucopolysaccharidosis I. *Genomics*. 1992;14:763–768.

CHAPTER

32

The Mucolipidoses

Reuben Matalon and Kimberlee Michals Matalon†*

*The University of Texas Medical Branch, Galveston, TX, USA

†The University of Houston, Houston, TX, USA

INTRODUCTION

The mucolipidoses emerged as a separate group of inherited disorders that originally were considered to be part of the mucopolysaccharidoses. The storage material within tissues or cultured skin fibroblasts includes lipids and glycosaminoglycans.[1,2] These disorders are not characterized by mucopolysacchariduria.

MANIFESTATIONS OF THE MUCOLIPIDOSES

The mucolipidoses are a group of diseases that were once phenotypically confused with the mucopolysaccharidoses. The term mucolipidoses was coined because of combined storage of lipids and mucopolysaccharides in tissues and cells of these patients. There are four mucolipidoses that are considered here.

Sialidosis (Mucolipidosis I)

Sialidosis type I, which is due to neuraminidase deficiency as a specific entity that fit mucolipidosis I, was described by Kelly and Graetz in 1977.[3] These patients have myoclonic seizures and cherry-red spot, noted on fundus exam.[4,5] The age of onset is variable, but visual difficulties, myoclonic seizures, and gait disturbances usually appear in the second decade of life. Eye involvement is usually progressive. Neuraminidase deficiency can be detected in cultured skin fibroblasts. The enzyme is unstable after freeze thawing and caution should be exercised in assaying this enzyme. There are severe forms of neuraminidase deficiency that have been referred to as pseudo-Hurler syndrome. There is sometimes a distinction of type I and type II, with type I being the more severe form.

Biochemical and Molecular Studies

Neuraminidase and β-galactosidase exist in a protein complex in association with a protective protein.[6] Deficiency of the protective protein/cathepsin A (PPCA) leads to the secondary deficiency of β-galactosidase and neuraminidase. This is referred to as galactosialidosis or severe form of mucolipidosis I.[7] The protective protein has the enzymatic activity of cathepsin A.[8] The gene for neuraminidase has been cloned, and mutations have been identified.[9,10]

Galactosialidoses

Another form of neuraminidase deficiency, sialidosis type II, is associated with β-galactosidase deficiency in addition to sialidase deficiency.[7,11] Three forms of galactosialidosis have been described. Infantile galactosialidosis is the severe form and late-infantile galactosialidosis is the mild form. Of all these types, the mildest form is the adult type. The onset of symptoms in most patients is early in life, and babies may be born with hydrops fetalis. These children are coarse and have dysostosis multiplex. Periosteal cloaking and hepatosplenomegaly may be present at birth. Foam cells and vacuolated lymphocytes are found in aspirated bone marrow. The storage material is oligosaccharide

Rosenberg's Molecular and Genetic Basis of Neurological and Psychiatric Disease
http://dx.doi.org/10.1016/B978-0-12-410529-4.00032-2

with *N*-acetylneuraminic acid at the nonreducing end. These compounds are excreted in excessive amounts in urine and can be detected using thin-layer chromatography or high-performance liquid chromatography.

Biochemical and Molecular Studies

Galactosidase and neuraminidase make a complex with a protective protein cathepsin A for activity. In sialidosis type II, the deficiency is in the protective protein. The gene for the protective protein was cloned by Galjart et al.[8] Specific mutations have been identified.[12]

I-Cell Disease (Mucolipidosis II)

In 1967, when cell culture for the study of the mucopolysaccharidoses became available, cultured fibroblasts from patients with I-cell disease were thought to exhibit inclusions typical for Hurler fibroblasts.[12–14] Later it was realized that those fibroblasts represented a different disease. Therefore, the letter *I* was used to indicate *inclusions*, and *I-cell* has become an accepted name for this disease.[15] Studies on the storage material in I-cell disease, thought to be a variant of Hurler syndrome, showed the storage of lipids and mucopolysaccharides in these cells.[2] Subsequent studies showed deficiency of multiple lysosomal enzymes in cultured fibroblasts, whereas the level of these enzymes was high in the culture medium.[16] Serum and urine of patients with I-cell disease also show high levels of lysosomal enzymes. These findings led to an important observation on the recognition site and targeting of lysosomal enzymes. Hickman and Neufeld[17] suggested that in I-cell disease the lysosomal enzymes are synthesized normally but lack a recognition marker that targets these enzymes to lysosomes; therefore they are excreted. Subsequent studies showed that mannose-6-phosphate was the recognition marker that targets lysosomal enzymes to the lysosome.[18] The mannose-6-phosphate is added to lysosomal enzymes by UDP-*N*-acetylglucosamine:*N*-acetylglucosaminyl-1-phosphotransferase.[19,20] There are two mucolipidoses associated with this enzyme deficiency: I-cell disease mucolipidosis II, which is the most severe, and mucolipidosis III, which is milder.[21,22]

Coarse features and severe mental retardation characterize I-cell disease (Figure 32.1). Radiologic findings are those of dysostosis multiplex, very similar to those found in Hurler syndrome. Usually coarse facial appearance, umbilical

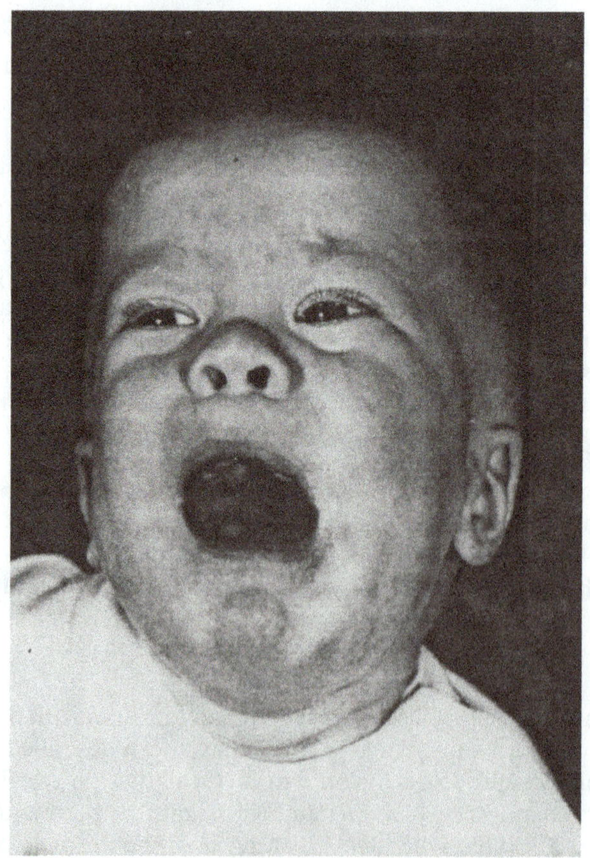

FIGURE 32.1 **Patient with I-cell disease who has mild coarse features and gingival hyperplasia.**

hernias, large liver and spleen, kyphoscoliosis, and lumbar gibbus are typical of I-cell disease. These clinical features are seen earlier than in Hurler syndrome. The thorax is usually small, joint movement is restricted, the tongue is large, and gingival hyperplasia is present. Corneal haziness is also common. Cardiac involvement includes cardiomegaly and aortic insufficiency. These children usually die of cardiopulmonary insufficiency in the first decade of life.

A diagnosis is suggested by the lack of mucopolysacchariduria. Assaying lysosomal enzymes in plasma shows increased activity of lysosomal enzymes. Determining the *N*-acetylglucosaminyl-1-phosphotransferase should confirm the diagnosis.

Mucolipidosis III

Mucolipidosis III (pseudo-Hurler polydystrophy) is a milder form of mucolipidosis II with a late clinical onset, between 2 and 4 years.[21] The enzyme deficiency in mucolipidosis III is a phosphotransferase similar to the one in I-cell disease. Patients with this disease may live to adulthood, and some may not be retarded. Linear growth is severely affected, and they may experience joint stiffness, claw hand deformities, and aortic regurgitation. Coarseness of the face is minimal. Corneal clouding may be associated in some of the patients. Radiologic examination of the skeletal system shows moderate dysostosis multiplex. The biosynthetic pathway of mannose-6-phosphate, which is essential for lysosomal enzymes, showed that the defect in I-cell disease and mucolipidosis III is identical with the exception that in mucolipidosis III there is more residual activity of this enzyme. In general, there is a spectrum of deficiency level of this enzyme activity. In addition to this primary defect, fibroblasts from patients with mucolipidosis III similar to fibroblasts from I-cell disease show deficiency of a large number of lysosomal enzymes, whereas the culture media contain higher activity of these enzymes. Diagnosis can be made by increased activity of lysosomal enzymes in plasma or urine as well as by the direct assay of the *N*-acetylphosphotransferase.[21]

Mucolipidosis IV

Mucolipidosis IV is a disease in which the storage material in the fibroblasts is different to what is observed in I-cell disease.[23] The basic defect has been identified as mucolipin 1 (MCOLN1).[24] It is localized on the short arm of chromosome 19 (19p13.2-13.3).[25] Mucolipin is a membrane protein expressed in all tissues. Mutation analysis identified two founder mutations among Ashkenazi Jews.[26] Random blood samples indicated that 1:100 were carriers for this mutation.[27]

Plasma gastrin concentration is elevated in virtually all individuals with mucolipidoses IV (mean 1507 pg per mL; range 400–4100 pg per mL)(normal 0–200 pg per mL).[28]

Mucolipidosis IV should be considered in any individuals with the following: early-onset developmental delay, whether static (cerebral palsy) or progressive, and dystrophic retinopathy with or without corneal clouding.[28] These patients were also described with psychomotor retardation, lack of facial dysmorphism and no hepatosplenomegaly. Children with mucolipidosis IV share some phenotypic resemblance to other patients with storage diseases. Conjunctival biopsy specimens showed storage material within lysosomes, with lamellar configuration similar to those seen in Tay–Sachs disease.[29,30] The majority of these patients were of Ashkenazi Jewish extraction. All patients had striking findings in the conjunctival biopsy, but no dysostosis multiplex or mucopolysacchariduria was observed. Suggestions of specific neuraminidase deficiency leading to this disease have not been adequately confirmed, and diagnosis relies primarily on the use of conjunctival biopsy.[31–33]

References

1. Hof L, Matalon R, Dorfman A. Gangliosides in human skin fibroblasts and their enrichment in the "Hurler variant" and Krabbe's disease. *Hoppe-Seylers Z Physiol Chem*. 1971;352:1329–1337.
2. Matalon R, Cifonelli HA, Zellweger H, Dorfman A. Lipid abnormalities in a variant of the Hurler syndrome. *Proc Natl Acad Sci U S A*. 1968;59:1097–1102.
3. Kelly TE, Graetz G. Isolated acid neuraminidase deficiency: a distinct lysosomal storage disease. *Am J Med Genet*. 1977;1:31–46.
4. Spranger JW, Gehler J, Cantz M. Mucolipidosis I: a sialidosis. *Am J Med Genet*. 1977;1:21–29.
5. Durand P, Gatti R, Cavalieri S, et al. Sialidosis (mucolipidosis I). *Helv Paediatr Acta*. 1977;32:391–400.
6. Hoogeveen AT, Verheijen FW, Galjaard H. The relation between human lysosomal beta-galactosidase and its protective protein. *J Biol Chem*. 1983;258:12143–12146.
7. Andria G, Del Giudice E, Reuser AJJ. Atypical expression of beta-galactosidase deficiency in a child with Hurler-like features but without neurological abnormalities. *Clin Genet*. 1978;14:16–23.
8. Galjart NJ, Gillemans N, Harris A, et al. Expression of cDNA encoding the human "protective protein" associated with lysosomal beta-galactosidase and neuraminidase: homology to yeast protease. *Cell*. 1988;54:755–764.

9. Bonten E, van der Spoel AV, Fornerod M, et al. Characterization of human lysosomal neuraminidase defines the molecular basis of the metabolic storage disorder sialidosis. *Genes Dev*. 1996;10:3156–3159.
10. Pshezhetsky AV, Richard C, Michaud L, et al. Cloning, expression and chromosomal mapping of human lysosomal sialidase and characterization of mutations in sialidosis. *Nat Genet*. 1997;15:316–320.
11. Wenger DA, Tarby TJ, Wharton C. Macular cherry-red spots and myoclonus with dementia: coexistent neuraminidase and beta-galactosidase deficiencies. *Biochem Biophys Res Commun*. 1978;82:589–595.
12. Zhou XY, van der Spoel A, Rottier R, et al. Molecular and biochemical analysis of protective protein/cathepsin A mutations: correlation with clinical severity in galactosialidosis. *Hum Mol Genet*. 1996;5:1977–1987.
13. DeMars R, Leroy JG. The remarkable cells cultured from a human with Hurler's syndrome: an approach to visual selection for in vitro genetic studies. *In Vitro*. 1967;2:107.
14. Leroy JG, DeMars RI. Mutant enzymatic and cytological phenotypes in cultured human fibroblasts. *Science*. 1967;157:804–806.
15. Leroy JG, Spranger JW, Feingold M, et al. I-cell disease: a clinical picture. *J Pediatr*. 1971;79:360–365.
16. Wiesmann UN, Lightbody J, Vassella F, Herschkowitz NN. Multiple lysosomal enzyme deficiency due to leakage? *N Engl J Med*. 1971;284:109–110.
17. Hickman S, Neufeld EF. Hypothesis for I-cell disease: defective hydrolases that do not enter lysosomes. *Biochem Biophys Res Commun*. 1972;49:992–999.
18. Sly WS, Fischer HD. The phosphomannosyl recognition systems for intracellular and intercellular transport of lysosomal enzymes. *J Cell Biochem*. 1982;18:67–85.
19. Hasilik A, Waheed A, von Figura K. Enzymatic phosphorylation of lysosomal enzymes in the presence of UDP-*N*-acetylglucosamine. Absence of the activity in I-cell fibroblasts. *Biochem Biophys Res Commun*. 1981;98:761–767.
20. Reitman ML, Varki A, Kornfeld S. Fibroblasts from patients with I-cell disease and pseudo-Hurler polydystrophy are deficient in uridine-5-diphosphate-*N*-acetylglucosamine-glycoprotein *N*-acetylglucosaminylphosphotransferase activity. *J Clin Invest*. 1981;67:1574–1579.
21. Varki AP, Reitman ML, Kornfeld S. Identification of a variant of mucolipidosis III (pseudo-Hurler polydystrophy): a catalytically active *N*-acetylglucosaminylphosphotransferase that fails to phosphorylate lysosomal enzymes. *Proc Natl Acad Sci U S A*. 1981;78:7773–7777.
22. Kornfeld S, Sly WS. I-Cell disease and pseudo-Hurler polydystrophy: disorders of lysosomal enzyme phosphorylation and localization. In: Scriver CR, Beaudet AL, Sly WS, Valle D, eds. *The Metabolic and Molecular Bases of Inherited Disease*. 7th ed. New York: McGraw-Hill; 1995:2495–2508.
23. Amir N, Zlotogora J, Bach G. Mucolipidosis type IV: clinical spectrum and natural history. *Pediatrics*. 1987;79:953–959.
24. Bargal R, Avidan N, Ben Asher E, et al. Identification of the gene causing mucolipidosis type IV. *Nat Genet*. 2000;26:118–123.
25. Sun M, Goldin E, Stahl S, et al. Mucolipidosis type IV is caused by mutations in a gene encoding a novel transient receptor potential channel. *Hum Mol Genet*. 2000;9:2471–2478.
26. Bach G. Mucolipidosis type IV. *Mol Genet Metab*. 2001;73:197–203.
27. Edelmann L, Dong J, Desnick RJ, et al. Mucolipidosis type IV in the American Ashkenazi Jewish population. *Am J Hum Genet*. 2002;70:4.
28. Schiffman R, Slaugenhaupt S, et al. Mucolipidoses IV. NCBI bookshelf. *Initial Posting*. Jan 28, 2005, Last updated Jul 20, 2010.
29. Newell FW, Matalon R, Mayer S. A new mucolipidosis with psychomotor retardation, corneal clouding, and retinal degeneration. *Am J Ophthalmol*. 1975;80:440–449.
30. Merin S, Livni N, Berman ER, Yatziv S. Mucolipidosis IV: ocular, systemic, and ultrastructural findings. *Investig Ophthalmol*. 1975;14:437–448.
31. Bach G, Ziegler M, Schaap T, Kohn G. Mucolipidosis type IV: ganglioside sialidase deficiency. *Biochem Biophys Res Commun*. 1979;90:1341–1347.
32. Ben-Yoseph Y, Momoi T, Hahn LC, Nadler HL. Catalytically defective ganglioside neuraminidase in mucolipidosis IV. *Clin Genet*. 1982;21:374–381.
33. Kiedel KG, Zwaan J, Kenyon KR, et al. Ocular abnormalities in mucolipidosis IV. *Am J Ophthalmol*. 1985;99:125–136.

CHAPTER

33

Disorders of Glycoprotein Degradation: Sialidosis, Fucosidosis, α-Mannosidosis, β-Mannosidosis, and Aspartylglycosaminuria

William G. Johnson

Rutgers Robert Wood Johnson Medical School, New Brunswick, NJ, USA

INTRODUCTION

Disorders of glycoprotein degradation were originally recognized as having some clinical resemblance to Hurler syndrome but lacking mucopolysacchariduria. These patients were initially included in the category of mucolipidosis, a somewhat heterogeneous category of disorders with storage of acid mucopolysaccharides, sphingolipids, or glycolipids in visceral and mesenchymal cells and with storage of abnormal amounts of sphingolipids or glycolipids in neural tissue. The mucolipidoses originally included G_{M1} gangliosidosis, fucosidosis, α-mannosidosis, juvenile sulfatidosis–Austin type (now known as mucosulfatidosis), mucolipidosis I (lipomucopolysaccharidosis, now known as sialidosis type II), mucolipidosis II (I-cell disease), and mucolipidosis III (pseudopolydystrophy).

Disorders of glycoprotein degradation have also been included in the category of oligosaccharidoses because they include excessive urinary excretion of oligosaccharides or glycopeptides. This is the basis of useful diagnostic screening tests. Urinary oligosaccharides can be measured by high-performance liquid chromatography, by the simpler and less expensive thin-layer chromatography, or by other methods. Using these methods, abnormal urinary oligosaccharides are found in all disorders discussed in this chapter. In addition, the patterns of oligosaccharides differ from disorder to disorder. Electrospray ionization tandem mass spectrometry has been used[1,2] to determinedisease-specific oligosaccharides in disorders that include sialidosis type II, fucosidosis, α-mannosidosis, aspartylglucosaminuria, with identification of some disorders including α-mannosidosis but having lower sensitivities for sialidosis.[3,4]

Two groups reported the advantages of using a MALDI-TOF/TOF (matrix-assisted laser desorption ionization time-of-flight) mass spectrometric method for screening oligosaccharidoses. Characteristic molecular profiles obtained in patient urine allowed identification of fucosidosis, aspartylglucosaminuria, G_{M1} gangliosidosis, Sandhoff disease, α-mannosidosis, sialidosis and mucolipidoses type II and III in one study[5] and diagnostic urinary patterns for α-mannosidosis, galactosialidosis, mucolipidosis type II/III, sialidosis, α-fucosidosis, aspartylglucosaminuria, Pompe disease, Gaucher disease, and G_{M1} and G_{M2} gangliosidosis in the other study.[6] The oligosaccharidoses are very rare disorders. Birth prevalences in the Netherlands of oligosaccharidoses and mucolipidoses combined were between 0.04 and 0.20 per 100,000 live births for individual diseases, and their combined birth prevalence was 1.0 per 100,000 live births.[7] The diagnosis of oligosaccharidosis is confirmed by assaying the appropriate lysosomal hydrolase enzyme in leukocytes, cultured skin fibroblasts, and, in some cases, G_{M1} gangliosidosis and galactosialidosis. Fifteen protein or metabolite markers evaluated for use in the identification of lysosomal storage diseases in newborn screening showed 100% sensitivity and specificity for the serum. Because oligosacchariduria may occasionally be absent, however, enzymatic testing should be carried out when these disorders are suspected. All of these disorders are inherited in the autosomal recessive pattern. There is no specific therapy, but bone marrow transplant

http://dx.doi.org/10.1016/10.1016/B978-0-12-410529-4.00033-4

has been carried out for a number of lysosomal disorders, including fucosidosis, mannosidosis, and aspartylglucosaminuria and is reported to prolong survival for some patients. Therapy is discussed further in the sections on fucosidosis and aspartylglucosaminuria later in this chapter.

BIOSYNTHESIS AND BIODEGRADATION OF GLYCOPROTEINS

Glycoproteins consist of oligosaccharide chains attached to a protein core. The proteins are synthesized on polysomes in rough endoplasmic reticulum. Oligosaccharide chains are added as the polypeptides pass through smooth endoplasmic reticulum to the Golgi apparatus. Two types of oligosaccharide chains are made there: the *N*-glycosidic type (Figures 33.1 and 33.2), which consists of substrates for the enzymes deficient in the diseases discussed in this chapter, and the *O*-glycosidic type, consisting mostly of those not involved in these diseases. The *O*-glycosidic linkage connects an *N*-acetylgalactosaminide moiety at the reducing end of the oligosaccharide chain through an *O*-glycosidic linkage to a serine or threonine residue of the polypeptide chain. These oligosaccharides are synthesized by sequential transfer of single sugar moieties from sugar–nucleotide intermediates to the end of a growing oligosaccharide chain. The *N*-glycosidic linkage connects an *N*-acetylgalactosaminide moiety at the reducing end of the oligosaccharide chain through an *N*-glycosidic linkage to an asparagine residue of the polypeptide chain. These oligosaccharides are synthesized through the dolichol pathway.[8] A complex oligosaccharide intermediate attached to a lipid moiety, dolichol phosphate, is transferred to an *N*-glycosidic linkage with an asparagine residue of the polypeptide chain. Complex trimming and elongation to give the final *N*-glycosidic oligosaccharide then remodel this oligosaccharide chain. Two basic types of *N*-glycosidic oligosaccharide are important as substrates

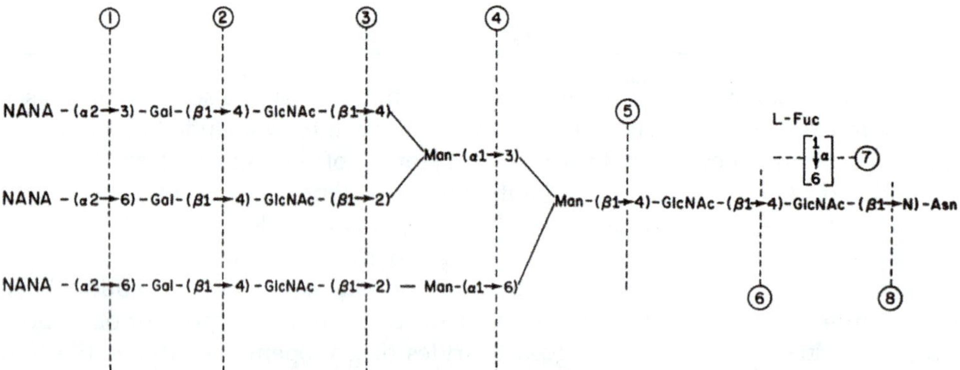

FIGURE 33.1 Structure of a complex-type triantennary oligosaccharide chain of an asparagine-linked glycoprotein consisting of asparagine, neutral sugars, hexosamine, and sialic acid. The mannose-6-phosphate recognition marker is formed by transfer of GlcNAc-1-P to the six hydroxyl groups of the α-linked mannose residues and the subsequent removal of the phosphate-linked GlcNAc residues. The numbers in circles indicate points of cleavage of lysosomal acid hydrolase enzymes that catabolize the oligosaccharide chain. Asn, asparagine; Gal, galactose; GlcNAc, *N*-acetylglucosamine; L-Fuc, L-fucose; Man, mannose; NANA, *N*-acetylneuraminic acid. 1, glycoprotein sialidase; 2, β-galactosidase; 3, hexosaminidases A and B; 4, α-mannosidase; 5, β-mannosidase; 6, endo-β-*N*-acetylglucosaminidase; 7, α-L-fucosidase; 8, aspartylglucosaminidase.

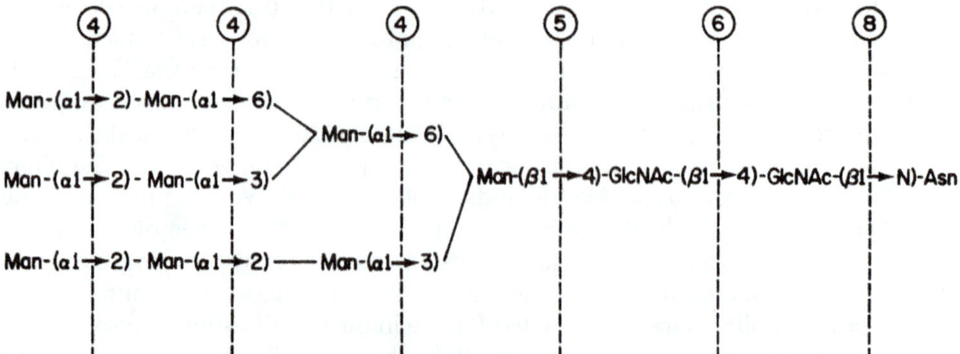

FIGURE 33.2 Structure of a high-mannose-type triantennary oligosaccharide chain of an asparagine-linked glycoprotein. This structure is an intermediate in the synthesis of the complex-type oligosaccharide chain (Figure 33.1). Man, mannose; Asn, asparagine; GlcNAc, *N*-acetylglucosamine. 4, α-mannosidase; 5, β-mannosidase; 6, endo-β-*N*-acetylglucosaminidase; 8, aspartylglucosaminidase.

TABLE 33.1 Genetic Features of Glycoprotein Degradation Disorders

Disease	Deficient enzyme	Gene symbol	Gene locus (human chromosome and band number)
Sialidosis	α-L-*N*-acetylneuraminidase	*NEU, NEU1*	Chromosome 6 (6p21.3)
Fucosidosis	α-L-Fucosidase	*FUCA1*	Chromosome 1 (1p34)
α-Mannosidosis	α-Mannosidase	*MANB, MAN2B1*	Chromosome 19 (19cen–q12)
β-Mannosidosis	β-Mannosidase	*MANB1*	Chromosome 4 (4q22–q25)
Aspartylglucosaminuria	Aspartylglucosaminidase	*AGA*	Chromosome 4 (4q32–q33)

for the diseases in this chapter: the complex-type oligosaccharides (Figure 33.1) and the high-mannose-type oligosaccharides (Figure 33.2). The high-mannose-type oligosaccharide chain is an intermediate in the synthesis of the complex-type oligosaccharide chain.[8]

Various lysosomal enzymes degrade *N*-glycosidic asparagine-linked glycoproteins sequentially (Figures 33.1 and 33.2 and Table 33.1). The polypeptide is catabolized by a variety of lysosomal peptidases. Sialic acid residues and α-L-fucoside residues occur only as terminal moieties and are removed by α-L-*N*-acetylneuraminidase, often called sialidase (Figure 33.1, reaction 1), and β-L-fucosidase (Figure 33.1, reaction 7). Acting next in sequence are β-galactosidase (Figure 33.1, reaction 2), hexosaminidases A and B (Figure 33.1, reaction 3), α-mannosidase (Figures 33.1 and 33.2, reaction 4), and β-mannosidase (Figures 33.1 and 33.2, reaction 5). All of these enzymes are exoglycosidases; that is, they remove a single sugar only when it is the terminal residue at the nonreducing end of the oligosaccharide chain. Endo-β-*N*-acetylglucosaminidase (Figures 33.1 and 33.2, reaction 6) cleaves the chitobiose linkage between two internal β-*N*-acetylglucosaminide residues. This enzyme, an endoglycosidase, can cleave in the middle of the oligosaccharide chain, thus removing the bulk of the oligosaccharide from the protein core. This can be an early step in glycoprotein catabolism. Finally, lysosomal aspartylglucosaminidase (Figures 33.1 and 33.2, reaction 8) cleaves the *N*-glycosidic bond linking β-*N*-acetylglucosamine to the asparagine residue of the peptide.

Besides peptide-linked oligosaccharides, the oligosaccharide portions of some glycolipids are substrates for α-L-fucosidase. Consequently, lipids along with oligosaccharides and glycopeptides are stored in fucosidosis.

SIALIDOSIS

History

This discussion focuses on the true sialidoses. These are genetic disorders in which damage to the gene for glycoprotein sialidase, an α-L-*N*-acetylneuraminidase, results in decreased activity of the enzyme. Patients with this enzyme deficiency were first described under the diagnosis of lipomucopolysaccharidosis and later placed in the category mucolipidosis I before the enzyme deficiency was found.

The true sialidoses, or isolated sialidoses, in which glycoprotein sialidase alone is deficient, have been divided into two groups:[9,10] the nondysmorphic group (sialidosis type I, also known as cherry-red spot–myoclonus syndrome) and the dysmorphic group (sialidosis type II, which includes mucolipidosis I).

Clinical Picture

Sialidosis Type I

Onset of this striking disorder characteristically is between ages 8 and 15 years with myoclonus or decreasing visual acuity. There appears to be a predilection for Italian patients,[9] although patients of various ethnicity, including Japanese, German, Czech, and Saudi Arabian have been described.

Action myoclonus develops, beginning in the limbs, and becomes increasingly debilitating. Patients may become unable to walk or stand and become bedridden. Retinal cherry-red spots are found on funduscopy. Corneas are clear, but punctate lens opacities or lamellar cataracts may be seen. Grand mal seizures may occur. Hyperactive tendon reflexes may be found. Burning pain in the limbs, which is worse in hot weather, is reminiscent of that occurring in Fabry disease. Intelligence is usually normal but is occasionally impaired. Cerebellar ataxia may be present. Renal impairment is not a feature of this disorder, but severe renal involvement has occurred in two siblings with otherwise typical cherry-red spot–myoclonus syndrome. Coarse facial features, organ enlargement, and skeletal dysplasia are not features of this disorder. The cherry-red spot may disappear in the course of the disease; consequently, absence of cherry-red spot is not grounds for excluding this diagnosis. Although there is no treatment for the underlying disorder, 5-hydroxytryptophan has been reported to cause dramatic or mild improvement of the action myoclonus.

Lai et al. reported a longitudinal study of the cherry-red spot–myoclonus syndrome. They studied 17 Taiwanese patients; all carried a 554A->G *NEU1* mutation for which 15 were homozygotes; two had a second novel mutation. Most developed full-blown manifestations within 5 years. All had somatosensory evoked potentials with giant cortical waves. Prolonged P100 peak latency of the visual evoked potentials (VEPs) were found in nearly all patients in the early stage even without visual symptoms. They suggested screening for *NEU1* mutations in patients presenting action myoclonus with abnormal VEPs, even in the absence of macular cherry-red spots.[11]

Sialidosis Type II

Patients with deficiency of glycoprotein sialidase, without deficiency of β-galactosidase, and with clinical features of somatic dysmorphism (facial dysmorphism, skeletal dysmorphism, and organ enlargement) fall into the category of type II sialidosis.[9] There are various clinical phenotypes with onset at birth, in childhood, or later.

Congenital Sialidosis

This is the most severe form of sialidosis known and, in fact, the most severe phenotype of lysosomal disease. Affected neonates are born prematurely and have the clinical appearance of hydrops fetalis. They may be stillborn or may survive a few months. At birth, they are plethoric, hypotonic, and depressed, with generalized skin edema, ascites, hepatosplenomegaly, puffy face and eyelids, and a prominent telangiectatic skin rash. Seizures may occur, and neurologic development is arrested. Skeletal and facial dysmorphism, inguinal hernias, severely dilated coronary arteries, and excessive retinal vascular tortuosity may occur.[12]

Because Rh incompatibility is largely being prevented, it is no longer the most common cause of hydrops fetalis. Although there are many causes of the hydrops fetalis phenotype, metabolic disorders should be considered when such patients are seen. Gaucher disease type II, galactosialidosis, G_{M1} gangliosidosis, mucopolysaccharidosis types IVA and VII, Niemann–Pick type C, Farber disease, mucolipidosis II (I-cell disease), and infantile free sialic storage disease have all produced this phenotype. There is some resemblance to patients reported as having congenital lipidosis or congenital lipidosis of Norman and Wood in whom the biochemical defect was never defined.

Severe Infantile Sialidosis

Patients described as having severe infantile sialidosis show many similarities to patients with congenital sialidosis, but they survive into the second year. These patients have congenital ascites, progressive organomegaly, slowed neurologic development, and progressive renal disease in the second year. This phenotype lacks features present in congenital sialidosis including premature birth, anasarca, skin telangiectasis, vascular abnormalities, optic atrophy, and seizures. Sialidase levels are markedly decreased, and sialic acid content increases in fibroblast sonicates. It is not clear whether severe infantile sialidosis and congenital sialidosis are different manifestations of the same phenotype or different phenotypes. The same problem of distinction is seen if severe infantile sialidosis is compared with nephrosialidosis.

Nephrosialidosis

Nephrosialidosis was defined in studies that focused on the renal component of this disorder. Affected patients appear to have a milder phenotype than those with congenital sialidosis or severe infantile sialidosis. Infants come to medical attention at ages 4 to 6 months because of hernias, hepatosplenomegaly, facial dysmorphism, and psychomotor retardation. Late in the course, the patient develops severe progressive renal disease with proteinuria and cherry-red spot and fine corneal opacities. On autopsy at age 4.5 years, one patient had storage in renal glomeruli and sympathetic ganglia. This was thought to distinguish nephrosialidosis from other sialidoses. The same findings, however, are seen in patients described as having severe infantile sialidosis or congenital sialidosis.

Mucolipidosis I (Lipomucopolysaccharidosis)

The term *mucolipidosis I* is used for the forms of sialidosis that may be noted in infancy but are more slowly progressive and are usually diagnosed in the juvenile period. This disorder was originally defined clinically, and some suspected cases turned out to be α-mannosidosis. The disorder was also termed GAL+disease in the earlier literature because the lysosomal enzyme β-galactosidase was elevated. The term mucolipidosis I, however, now appropriately applies only to patients with the characteristic phenotype and sialidase deficiency without β-galactosidase deficiency.

These patients develop normally for the first 6 months, but motor development then slows, and mental delay is seen. Somatic abnormalities appear in the second or third year but are milder than those in Hurler syndrome. These include short trunk, sometimes with spinal deformity; relatively long limbs; mild coarsening of facial features; mild restriction of joint mobility; and inconsistent enlargement of liver and spleen. Impaired hearing, corneal opacity, and macular cherry-red spots are usually seen, and growth disturbance is common. Neurologic findings include slowly progressive gait ataxia, tremor, cerebellar signs, myoclonic jerks, hypotonia, muscle wasting, peripheral neuropathy, and seizures.

Laboratory Findings

Diagnosis is by the presence of abnormal sialo-oligosaccharides in the urine and demonstration of deficiency in glycoprotein sialidase activity in cultured skin fibroblasts, tissue, leukocytes, amniotic fluid cells, or chorionic villus samples. Patients do not have mucopolysacchariduria, and the amount of urinary uronic acid excretion is comparable to that of controls. Rectal biopsy is a useful confirmatory test because the stored material is visible on ultrastructure of rectal ganglion cells.

Roentgenography

Roentgenographic changes in mucolipidosis I are characteristically mild to moderate dysostosis multiplex. Kyphoscoliosis and hypoplastic odontoid have been described.

Pathology

On pathologic examination of sialidosis type I, the brainstem and spinal cord contain swollen, periodic acid–Schiff (PAS)-positive neurons. PAS-positive material was also seen in the nervous system in swollen dendrites of Purkinje cells, in Bergmann glia, in capillary endothelium, and in macrophages. PAS-positive material was also seen outside the nervous system. Electron microscopy showed atypical membranous cytoplasmic bodies.

Autopsy of a case of congenital sialidosis showed zebra bodies in the neurons of the spinal cord only. Neurons in the cerebral cortex and cerebellar cortex and autonomic ganglion cells showed only membrane-bound vacuoles but no membranous cytoplasmic bodies. Vacuolated cells were also found in the periphery: hepatocytes, endothelial cells, and Kupffer cells in the liver and glomerular and tubular epithelial cells in the kidney.

Vacuolated lymphocytes were plentiful on the peripheral smear, and vacuolated histiocytes were seen in the bone marrow. Foam cells were seen in large numbers in multiple tissues and in the placenta. Fibroblasts contained numerous vacuoles and whorled membranous structures on electron microscopy, although they appeared normal by high-magnification microscopy.

Ultrastructural study of cultured skin fibroblasts in sialidosis showed vacuolar inclusions of low electron density similar to those in α-mannosidosis and the mucopolysaccharidoses. A few myelin-like figures were seen. An unusual type of large, round, membrane-bound inclusion with many concentrically arranged tubules in a fine reticulogranular matrix was also seen. Similar findings in nephrosialidosis by light and electron microscopy have been reported.

Stored Substances in the Sialidoses

Sialic acid content is increased in tissue. In one study of autopsy tissue, water-soluble bound sialic acid was increased between 10-fold and 17-fold in visceral organs but only about two-fold in brain. Lipid-bound sialic acid was increased up to eight-fold in visceral organs but was not elevated in brain. The increase in viscera resulted from elevated amounts of ganglioside G_{D4} and probably G_{M4} and LM1. This pattern of accumulation supported the idea that ganglioside accumulation in the sialidoses was a secondary phenomenon, perhaps induced by inhibition of a ganglioside sialidase by substances stored because of the primary sialidase deficiency. Gangliosides such as G_{M3} and G_{D1a}, which contain only $\alpha(2\rightarrow3)$-linked sialic acid moieties, are apparently not substrates for glycoprotein sialidase but are cleaved by a different enzyme.

Substrates for glycoprotein sialidase include sialyloligosaccharides containing terminal $\alpha(2\rightarrow6)$-linked or $\alpha(2\rightarrow3)$-linked sialic acid residues. Sialyloligosaccharides containing these linkages have been found in urine and cultured fibroblasts. As mentioned earlier, urine oligosaccharides show a characteristic pattern in sialidosis when analyzed by MALDI-TOF mass spectrometry.[5,6]

Urinary glycopeptides have also been found.

Enzyme Deficiency in the Sialidoses

Lysosomal sialidase activity (Figure 33.1, reaction 1), or α-L-*N*-acetylneuraminidase activity, is markedly deficient in both type I and type II sialidoses. Because sialic acid is *N*-acetylneuraminic acid, sialidase is properly equated with α-*N*-acetylneuraminidase, although the terms *sialidase* and *neuraminidase* are often used interchangeably. The sialidase deficient in the sialidoses appears to cleave both α(2→6)-linked and α(2→3)-linked sialyloligosaccharides. Because 80% of the accumulating sialyloligosaccharides have the α(2→6) linkage, it has been suggested that the mutant enzyme retains greater residual activity for α(2→3) linkages. Carrier detection and prenatal diagnosis are possible in this disorder.

Three lysosomal enzymes, sialidase, β-galactosidase, and *N*-acetylaminogalacto-6-sulfate sulfatase, are associated with protective protein/cathepsin A (PPCA; known also as PPGB, protective protein for β-galactosidase, or as cathepsin A, or lysosomal carboxypeptidase). This complex formation affects biogenesis, intracellular sorting, and proteolytic processing and greatly extends the half-life of these enzymes. Association with PPCA is necessary for the expression of sialidase activity.[13]

The α-L-N-acetylneuraminidase Gene

There are three distinct genetic classes of disorders in which glycoprotein sialidase activity is deficient. In the first group, discussed in this chapter, glycoprotein sialidase alone is deficient. These disorders, the sialidoses proper, or isolated sialidase deficiencies, result from mutations of *NEU1*, the structural gene for glycoprotein sialidase that has been localized to chromosome 6 (6p21.3). In the second group, both glycoprotein sialidase activity and β-galactosidase (G_{M1} ganglioside β-galactosidase) activity are secondarily deficient due to primary deficiency of PPCA, which is required for activity of both enzymes and which is coded by a gene on chromosome 20q13.1 (Figure 33.1, reaction 2). PPCA has cathepsin A-like activity, which is distinct from its protective function and is present in the lysosome bound to β-galactosidase and sialidase in a multienzymic complex.[14] This group of disorders is referred to as galactosialidosis. In the third type, sialidase and multiple lysosomal enzymes are decreased because of deficient posttranslational modification. This is the case in mucolipidosis types II and III.

Missense mutations, frameshift mutations, and a splice donor site mutation of the glycoprotein sialidase gene, *NEU1*, have been reported. In 2001, Lukong et al.[15] reported mutations that interfered with binding sialidase to its multienzyme complex. In 2000, Bonten et al.[16] observed that clinical heterogeneity in sialidosis mutant proteins was based on a residual enzyme that was: i) catalytically inactive and not lysosomal (severe infantile type II sialidosis); ii) catalytically inactive, but localized in lysosome (juvenile type II sialidosis); and iii) catalytically active and lysosomal (at least one allele in type I sialidosis). Seyrantepe et al.[17] predicted the impact on the tertiary structure and biochemical properties of the sialidase enzyme for the 34 unique mutations in *NEU1* known in 2003. Mutations found in sialidosis I have been reported.[18]

FUCOSIDOSIS

History

The first case of fucosidosis was described in 1966 and the presence of fucose-rich storage material was later documented in multiple organs. In 1968, the cause of the disorder was found to be absence of α-L-fucosidase activity.

Clinical Picture

Patients with fucosidosis differ greatly in the kind of clinical involvement and the severity of their illness. Some patients (type I) present with severe, progressive neurologic disorder. Others with milder disease (type II) resemble the Hurler phenotype. Some patients have survived into their second, third, or fourth decade. Usually, however, they become symptomatic in childhood. Some authors refer to these adult cases as type II, whereas others recognize a type III. Types I and II may appear in the same family.

In severely affected infants (type I), symptoms develop in the first year of life with psychomotor retardation and hypotonia. Patients then develop hypertonia and, later in the course, spasticity, tremor, and mental deterioration, leading to loss of contact with their environment. Dystonia has been reported. Patients have thick skin and excessive sweating and may lose gallbladder function. They have only mild facial coarsening, radiographic bone changes, and organ enlargement. Cardiac enlargement may occur. Some patients die at age 4 or 5 years. Others (with type II) have onset of symptoms in the second year of life with more markedly coarsened facial features and greater resemblance to Hurler syndrome. These patients also deteriorate mentally and die around age 5 years. Other patients with infantile or juvenile onset have a milder course, with coarsened facial features, skeletal changes, and dwarfing; skin changes resembling the angiokeratoma of Fabry disease; and survival into the third decade. Angiokeratoma corporis diffusum (AKCD) is characteristic of patients with longer survival. Besides Fabry disease and fucosidosis, AKCD has also been reported in sialidosis, α-mannosidosis, β-mannosidosis, aspartylglucosaminuria, galactosialidosis, and other disorders. Tsukadaira et al., presented a striking color photograph of florid angiokeratoma corporis diffusum and recorded its occurrence also in β-galactosidosis, fucosidosis, β-mannosidosis and Kanzaki disease.[19] Recurrent respiratory infections may occur.

In a review of 77 patients, Willems et al.[20] did not find evidence for a rapidly progressive type I and slowly progressive type II fucosidosis but found a wide continuous clinical spectrum. They tabulated the percentage of cases with various clinical signs and symptoms: progressive mental deterioration (95%), motor deterioration (87%), coarse facies (79%), growth retardation (78%), recurrent infections (78%), dysostosis multiplex (58%), angiokeratoma corporis diffusum (52%), visceromegaly (44%), and seizures (38%). Inui et al. reported a case with an interesting late-onset chronic phenotype.[21] Oner et al. reported the occurrence in fucosidosis of diffuse white matter hyperintensity and a pallidal curvilinear streak hyperintensity.[22]

Autti et al. reviewed 117 patients with various lysosomal disease including fucosidosis, mannosidosis, and aspartylglucosaminuria and concluded that decreased T2 signal in the thalami may be a sign of lysosomal storage disease.[23]

Laboratory Findings

All of the phenotypes have marked deficiency of the lysosomal enzyme α-L-fucosidase. Diagnosis requires a high index of suspicion because of the diverse clinical pictures. Abnormal urinary oligosaccharides are present in characteristic pattern.

Diagnosis is based on the presence of abnormal oligosaccharides in urine and severely decreased α-L-fucosidase in serum, leukocytes, and cultured skin fibroblasts. Residual fucosidase in occasional patients may be 60% of control values but with heat-labile enzyme. Therefore, it is important to analyze urine for oligosaccharides because otherwise these patients may be missed. In contrast, false-positive diagnosis of fucosidosis was made in a patient with pseudo-hypoparathyroidism because serum and fibroblast α-fucosidase activity was about 10% of control values.

It is important to realize that 6–11% of the normal population have a polymorphism in which serum α-L-fucosidase but not leukocyte α-L-fucosidase is markedly decreased to 10% of the mean value in controls. Therefore, serum and plasma by themselves are not suitable for enzymatic diagnosis of fucosidosis.

As mentioned earlier, urine oligosaccharides show a characteristic pattern in α-fucosidosis when analyzed by tandem mass spectrometry[4] as well as by MALDI-TOF mass spectrometry.[5,6]

Roentgenography

Roentgenographic changes are those of a mild dysostosis multiplex but are most marked in the pelvis, hips, and spine. These are not in themselves diagnostic but should point to more specific studies. Imaging abnormalities have been reported in thalamus, globus pallidus, internal capsule, and supratentorial white matter.[24]

Pathology

Rectal biopsy specimens have been found to contain three kinds of inclusions in vascular endothelial cells: clear vacuoles, dark vacuoles, and mixed vacuoles. Clear vacuoles and dark vacuoles appeared to contain fucosyl oligosaccharides, and the simultaneous presence of these two types has been considered pathognomonic of fucosidosis. Lectin histochemistry on paraffin-embedded tissue sections can distinguish between α-mannosidosis, β-mannosidosis, fucosidosis, and sialidosis. Fucosidosis tissue stained with Ulex europaeus agglutinin I (UEA-I), whereas the others did not.

Stored Substances in α-L-Fucosidosis

Fucose residues form part of the structure of oligosaccharides, glycoproteins, glycolipids, and keratan sulfate. α-L-Fucoside residues occur in both *N*-glycosidic and *O*-glycosidic types of glycoproteins. Like α-L-*N*-acetyl neuraminic acid residues, α-L-fucoside residues are usually located at the terminal nonreducing position in an oligosaccharide chain.

In fucosidosis, oligosaccharides, glycopeptides, glycolipids, and keratan sulfate may be stored. In the brain, the main stored material is oligosaccharide, chiefly the decasaccharide Fuc-α(1→2)-Gal-β(1→4)-GlcNAc-β(1→2)-Man-[Fuc-β(1→2)-Gal-β(1→4)-GlcNAc-β(1→2)-Man-] α(1→3/6)-Man-β(1→4)-GlcNAc, and the disaccharide Fuc-α(1→6)-GlcNAc. Oligosaccharides are stored in other viscera also. In liver, the glycolipid Fuc-α(1→2)-Gal-β(1→3)-GlcNAc-β(1→4)-Gal-β(1→4)-Glc-ceramide is stored and has H-antigen specificity. Because α-L-fucoside residues are important components of blood-group antigens, the red cells and body fluids of patients with fucosidosis contain increased amounts of blood group-specific compounds as well as abnormal blood-group phenotypes. In the urine, a large number of fucose-containing oligosaccharides and glycopeptides of the *N*-glycosidic type have been recovered.

Enzyme Deficiency in Fucosidosis

In patients with fucosidosis, the lysosomal enzyme α-L-fucosidase is deficient in serum, leukocytes, and cultured skin fibroblasts. The normal enzyme is a tetramer consisting of 50-kDa subunits, which are processed from a 53-kDa precursor. The enzyme is subject to posttranslational processing, and abnormalities of processing have been found in patients with fucosidosis. Eleven patients with severe fucosidase deficiency usually synthesized no precursor, but two synthesized a precursor that was not processed, and one synthesized small amounts of cross-reacting material.

Approximately 6–7% of healthy unaffected individuals have markedly low α-L-fucosidase in serum, but leukocyte and tissue values are within normal limits and they do not have fucosidosis. This trait for low serum fucosidase levels is inherited in Mendelian fashion and is governed by a gene locus on chromosome 6 (6q25-qter).

The α-L-Fucosidase Gene

The α-L-fucosidase gene (structural gene *FUCA1*) has been mapped to human chromosome 1 (1p34). The gene has been cloned and sequenced.[25] Two restriction fragment length polymorphisms have allowed demonstration of genetic linkage between fucosidosis and its structural gene. Willems et al.[26] reviewed 22 mutations detected in fucosidosis. These mutations included missense mutations, large and small deletions, an insertion, a deletion, a splice site mutation, and stop codon mutations. It was not clear how these mutations resulted in the different fucosidosis phenotypes because all appeared to result in nearly total enzyme activity loss. The gene coding for the quantitative polymorphism of serum α-L-fucosidase occurring in normal individuals has been mapped to chromosome 6 (6q25-qter). An α-L-fucosidase pseudogene, *FUCA1P*, has been mapped to chromosome 2 (2q24-q32). Lin et al. reviewed 26 mutations in fucosidosis and reported two more.[27]

Fucosidosis in Other Species

Fucosidosis has been reported in dogs. Bone marrow transplant in canine fucosidosis was not reported to have any beneficial clinical effect but was later reported to show improvement in neurologic signs and tissue storage. Cloned human cDNA encoding for human α-L-fucosidase has been successfully inserted into canine hematopoietic cells using retroviral vectors, and this model may become useful for assessing the effectiveness of gene therapy. α-L-fucosidase has been studied in a number of mammalian species.

Therapy

Bone marrow transplantation in fucosidosis has reportedly increased enzyme levels and improved psychomotor development, evoked potentials, and magnetic resonance imaging (MRI) appearance.[28]

α-MANNOSIDOSIS

History

α-Mannosidosis was first described in 1967. The index patient was tall for his age with large hands and feet, slight hepatosplenomegaly, and mild psychomotor retardation. The patient developed slowed growth, lumbar kyphosis, somewhat coarse features, enlarged tongue, and small, cloudy lenticular opacities (although corneas were clear). Subsequently, the patient developed mental and motor deterioration, hypotonia and hyperreflexia, and Babinski signs. He died at age 5 during an attack of suspected increased intracranial pressure. This patient and others had deficiency of α-mannosidase, and the disorder was called mannosidosis. Subsequently, patients have been found with deficiencies of β-mannosidase, a disorder referred to as β-mannosidosis. Although the term *mannosidosis* still refers to patients with α-mannosidase deficiency, it is clearer to call this disorder α-mannosidosis.

Clinical Picture

The clinical picture of α-mannosidosis is quite varied. More than 70 cases have been reported. These patients may have severe disease, as with the index case just described, or a milder disorder. Patients with a severe form have hepatomegaly, splenomegaly, severe infections, and they die early in life. Severely affected patients have been confused with mucolipidosis I patients.

Other patients have had slower progression of the disorder with greater dysmorphism, corneal opacities, and longer survival. Others have presented primarily with marked mental defects, striking gingival hyperplasia, and survival into the third decade or longer. Facial dysmorphism, skeletal involvement, and organ enlargement have been slight in these patients. This more common mild form shows no clinical abnormalities or nonspecific abnormalities in the first year of life (psychomotor delay, speech delay, or frequent infections), and only later do sensorineural hearing loss, gait and limb ataxia, mild then moderate or severe mental defect, coarse facial features, and dysostosis multiplex develop. Patients develop short stature and may survive into adulthood, at least until age 41, without hepatomegaly. The cognitive and language impairments do not correlate well with the amount of residual α-mannosidase activity measured with an artificial substrate. Pancytopenia with antiplatelet and antineutrophil antibodies may occur. Destructive synovitis may occur with damage at the ankle, knee, or spine. Multiple suture synostosis, macrocephaly, and papilledema with increased intracranial pressure have been reported. Immunodeficiency affecting both cellular and humoral immunity often occurs and may account for the frequent infections.[29]

Malm et al. reviewed the natural history and complications of α-mannosidosis[30] and reported a variety of manifestations including immune deficiency, skeletal abnormalities, hearing impairment, gradual impairment of mental functions and speech, psychosis, muscle weakness, joint abnormalities, and ataxia. Large head, prominent forehead, rounded eyebrows, flattened nasal bridge, macroglossia, widely spaced teeth, prognathism, slight strabismus were apparent with clinical variability, which were slowly progressive over decades. Many patients were over 50 years of age.[31]

Laboratory Findings

Diagnosis requires a high index of suspicion, findings of abnormal urinary oligosaccharides, and demonstration of decreased α-mannosidase in leukocytes and cultured skin fibroblasts. Diagnosis has also been made using transformed lymphoblasts.

Roentgenography

The most notable roentgenographic finding is dense thickening of the calvaria; dysostosis multiplex tends to be mild and in some cases decreases in prominence with the age of the patient. The roentgenographic elements of dysostosis multiplex in α-mannosidosis include flattened vertebral bodies, oar-shaped ribs, hypoplastic ilia, and abnormalities of the small bones of the hands. Poor pneumatization of the sphenoid may be seen. Tight foramen magnum and cervical syrinx have been reported.[32]

Results of computed tomography scans may be normal or may show an increase of subarachnoid spaces and hypodensity of the white matter.

Pathology

The pathology of α-mannosidosis is characterized by the appearance in multiple cell types of enlarged membrane-bound vacuoles that may contain storage material of reticulofibrillar appearance or appear almost empty. These vacuoles appear to be enlarged lysosomes and are not specific for α-mannosidosis. Ultrastructural study of cultured skin fibroblasts in α-mannosidosis showed vacuolar inclusions containing fine reticulogranular material with low electron density similar to those in sialidosis and the mucopolysaccharidoses.

Skeletal muscle is not involved symptomatically in α-mannosidosis; however, muscle biopsy may show ultrastructural abnormalities. A 32-year-old patient had numerous membrane-bound inclusions in biopsied skeletal muscle as well as interstitial fibroblasts and some endothelial cells. The inclusions were small (0.5–5.0 μm) and membrane bound, and they contained dark granules as well as electron-lucent spaces. They were not believed to cause the patient's mild muscle weakness. In another patient, with generalized muscle weakness and spastic paraplegia, muscle pathology may have contributed to the weakness.

Lectin histochemistry on paraffin-embedded tissue sections can distinguish between α-mannosidosis, β-mannosidosis, fucosidosis, and sialidosis. α-Mannosidosis tissue shows staining with concanavalin A (Con A), wheat germ agglutinin (WGA), and succinyl-wheat germ agglutinin (S-WGA). Fucosidosis tissue stains with UEA-I. Sialidosis tissue stains with WGA but not S-WGA. β-Mannosidosis did not stain with Con A, WGA, S-WGA, UEA-I, or peanut agglutinin.

Stored Substances in α-Mannosidosis

In the initial report, mannose-rich storage material was reported in tissues; however, most subsequent studies have examined compounds in the urine. The compounds found have been oligosaccharides, both linear and branching, containing one or more terminal α-mannoside residues at the nonreducing end of the molecule. The most abundant and characteristic oligosaccharide found has been the trisaccharide Man-α(1→3)-Man-β (1→4)-GlcNAc. This trisaccharide may result from partial catabolism of a complex-type oligosaccharide chain (Figure 33.1). After release of the oligosaccharide from the protein core by endo-β-*N*-acetylglucosaminidase (Figure 33.1, reaction 6), stepwise cleavage by exoglycosidases would result in the trisaccharide. This would require removal of an α(1→6)-mannoside; presumably patients with α-mannosidosis have an enzyme that can catabolize this linkage. More than a dozen other oligosaccharides have been isolated from patients' urine, up to decasaccharides in size. Most of these may result from partial catabolism of a complex (Figure 33.1) or high-mannose-type (Figure 33.2) oligosaccharide chain. Some may result from partial catabolism of the dolichol-linked oligosaccharide precursor.[8]

As mentioned earlier, urine oligosaccharides show a characteristic pattern in α-mannosidosis when analyzed by MALDI-TOF mass spectrometry.[5]

Enzyme Deficiency in α-Mannosidosis

Patients with α-mannosidosis have deficiency of lysosomal acid α-mannosidase (Figures 33.1 and 33.2, reaction 4). A variety of α-mannosidases are known, and their genetic origins, relationships, and natural substrate specificities are not well understood. α-Mannosidase activity with acid pH-optimum is found in lysosomes. α-Mannosidase activity with neutral or intermediate pH-optimum is found in cytoplasm and the Golgi apparatus. Tissue α-mannosidase activity has been separated into three forms: A, B, and C. α-Mannosidases A and B are lysosomal forms with a pH-optimum of 4.4; α-mannosidase C is a cytoplasmic form with a pH-optimum of 6.0. An intermediate form of α-mannosidase with pH-optimum 5.5 has also been reported. The neutral cytoplasmic form, α-mannosidase C, is coded for by a gene on chromosome 15, localized to 15q11-q13, and designated *MANA*. The acidic, lysosomal forms, α-mannosidases A and B (which are deficient in α-mannosidosis), are coded for by a gene on chromosome 19, localized to 19 (19cen–q12), and designated *MANB*. The relationship between α-mannosidases A and B is not clear. Both, however, have been reported deficient in α-mannosidosis, and both have been converted to a common, apparently identical, form.

When assayed with artificial substrate 4-methyl umbelliferyl-α-D-mannopyranoside, leukocytes and fibroblasts show marked deficiency of acid α-mannosidase because lysosomal acid α-mannosidases are the largest fraction of the α-mannosidases in these cells. Serum, in contrast, contains a small fraction of acid (lysosomal α-mannosidases) and is not useful for diagnosis of α-mannosidosis.

The α-Mannosidase Gene

The α-mannosidosis gene, *MANB*, has been cloned.[33] Berg et al.[34] reviewed the spectrum of mutations in α-mannosidosis patients from 39 families. One missense mutation, the R750W substitution, accounted for 21% of the disease alleles. Other disease alleles included splicing mutants, nonsense mutations, deletions, insertions, and other missense mutations. There seemed to be no correlation between mutation type and phenotype. Gotoda et al.[35] reviewed the spectrum of mutations in Japanese patients with α-mannosidosis. Kuokkanen characterized 35 *MAN2B1* variants including 29 novel misssense mutations.[36]

Riise Stensland et al. reported that 96 sequence variants were identified in 130 unrelated α-mannosidosis patients from 30 countries. Eighty-three of these were novel, extending the total from 42 to 125. Overall, 256 of the 260 mutant alleles (98.5%) were identified, most unique to a family. One variant, c.2248C>T (p.Arg750Trp) was found in 50 patients, accounting for 27.3% of disease alleles identified.[37]

α-Mannosidosis in Other Species

α-Mannosidosis also occurs in cats. Ingestion of swainsonine, an inhibitor of α-mannosidase, causes an α-mannosidosis-like disorder in cattle, sheep, and horses. Bone marrow transplant in cats with α-mannosidosis has been reported to arrest progression of neurologic signs and prevent neuronal storage; functional acid α-mannosidase activity was demonstrated within neurons and other cells.

Therapy

Yesilipek et al. reported that bone marrow transplantation in two siblings with α-mannosidosis was followed by increased enzyme levels and clinical improvement.[38]

β-MANNOSIDOSIS

History

β-Mannosidosis, deficiency of β-mannosidase, is unusual because the disease was described first in goats.[39] Five years later, the disease was recognized in humans.[40,41]

Clinical Picture

In humans, the disorder presents with mental retardation, angiokeratoma, and tortuousness of conjunctival vessels.[40] Just over a dozen cases have been reported. Hearing loss has been reported in some patients,[41] but not others.[40] Peripheral neuropathy with thenar and hypothenar atrophy has been reported.[42] One patient from consanguineous parents had hypotonia and feeding difficulties in the first few months of life; it is not clear whether the abnormalities of swallowing and esophageal motility and achalasia at age 2 years, which required surgery, were part of her β-mannosidosis. Intrafamilial heterogeneity has been noted. As in Fabry disease, the angiokeratomas are distributed on the buttocks and lower limbs and in males may be seen on the shaft of the penis and scrotum. In 1996, a case was reported of a 22-year-old woman who had had angiokeratomas since age 12 years as her only manifestation. She had β-mannosidase deficiency in serum, leukocytes, and cultured skin fibroblasts, and her parents had intermediate levels consistent with autosomal recessive inheritance. Tsukadaira et al.[19] reported that angiokeratoma corporis diffusum (AKCD) is a feature of a number of lysosomal disorders including Anderson–Fabry disease, β-galactosidosis, fucosidosis, aspartylglucosaminuria, galactosialidosis, β-mannosidosis, and Kanzaki disease, and included an excellent color photo of AKCD. One patient[41] had dysmorphic features, dysostosis multiplex, and sulfamidase deficiency (the enzyme deficient in Sanfilippo syndrome type A), in addition to β-mannosidase deficiency, and may have had two diseases. The other patients did not show dysmorphic features. Seizures or peripheral neuropathy may be prominent. Broomfield et al. reported a severe phenotype with neonatal onset of seizures and communicating hydrocephalus at age 2 years.[43]

Laboratory Findings

Diagnosis of β-mannosidosis is by demonstration of the characteristic oligosacchariduria and by demonstration of the enzyme deficiency in plasma, leukocytes, and fibroblasts. Heterozygotes can be detected, and the disease could presumably be detected during prenatal diagnosis in humans as has been done in goats. Ethanolaminuria has been noted in association with β-mannosidosis.

Lectin histochemistry on paraffin-embedded tissue sections can distinguish between α-mannosidosis, β-mannosidosis, fucosidosis, and sialidosis. α-Mannosidosis tissue showed staining with Con A, WGA, and S-WGA. β-Mannosidosis did not stain with Con A, WGA, S-WGA, UEA-I, or peanut agglutinin.

Pathology

Stored Substances in β-Mannosidosis

Oligosaccharides accumulate in cells and in urine in β-mannosidosis. The major oligosaccharide in the human disease is mannosyl-β(1→4)-*N*-acetylglucosamine, which has been found in leukocytes. Sialyl-α(2→6)-mannosyl-β(1→4)-*N*-acetylglucosamine has also been found. In goats, a trisaccharide accumulates. Oligosacchariduria is present[40,41] and is easily detected by thin-layer chromatography. The disaccharide mannosyl-β(1→4)-*N*-acetylglucosamine, however, requires a special solvent system to distinguish it from lactose.

Enzyme Deficiency in β-Mannosidosis

In β-mannosidosis, the deficient enzyme is lysosomal acid β-mannosidase (Figures 33.1 and 33.2, reaction 5). β-Mannosidase in humans is measurable in plasma, where it varies with age (but not with sex, as in goats), and in leukocytes, fibroblasts, and urine. β-Mannosidase is deficient in all these biologic fractions in β-mannosidosis. β-mannosidase activity in human fibroblasts is 10 times as high as in goat fibroblasts.

The β-Mannosidase Gene

Human β-mannosidase has been mapped to chromosome 4q22–q25 and cloned. Homozygosity for a splice acceptor site mutation was found in one family of Czech Gypsy origin.[44]

Huynh et al. studied β-mannosidase mutations in human, cow, goat and mouse and found a correlation between the genotype and phenotype severity of phenotype.[45]

β-Mannosidosis in Other Species

The disease in goats differs from that in humans.[39] In goats, there is ataxia, pendular nystagmus, inability to stand, dysmorphism, and marked intention tremor. The disorder is present at birth. A major element of the neuropathology is a myelin defect, perhaps caused by an oligodendrocyte defect. In addition, there are axonal lesions with axonal spheroids. In the peripheral nervous system, there are vacuolated Schwann cells and axonal spheroids. Characteristic lysosomal storage vacuoles were seen in all tissues in most cell types. The caprine gene has been cloned and a single base deletion identified that causes a frameshift with premature termination of the protein.

Subsequently, the disorder was diagnosed in Salers cattle. Neuropathology in Salers calves showed hydrocephalus and myelin deficiency in the cerebral hemispheres, cerebellum, and brainstem. On microscopic examination, cytoplasmic vacuolation, myelin deficiency, and axonal spheroids were found. In general, the neuropathologic features in the calves resembled those in goats. In goats, the extent of demyelination did not correlate with the concentration of stored oligosaccharides or the degree of deficiency. The degree of oligosaccharide storage, however, did correlate with the regional β-mannosidase activity levels in control goats. The thyroid gland seems to be the site of severe involvement in goats with β-mannosidosis. Affected neonatal animals have decreased thyroid hormone levels. It has been hypothesized that this hypothyroidism plays a role in the pathogenesis of hypomyelination in affected animals.[45]

ASPARTYLGLYCOSAMINURIA

History

Aspartylglycosaminuria (AGU) was first reported in 1968 in England. Although cases from other countries have been reported, most come from Finland, especially eastern Finland. In eastern Finland, aspartylglycosaminuria is the third most common genetic cause of mental retardation after trisomy 21 and fragile X syndrome.[46,47]

Carrier screening and prenatal diagnosis appear to be effective and are being evaluated in a university-based clinic in eastern Finland.[48]

Clinical Picture

Other than rapid infantile growth, often the first clinical sign of aspartylglycosaminuria, patients present as clinically normal until age 1–5 years, when progressive somatic and mental changes develop. Progressive coarsening of facial features is noted, with depressed nasal bridge, anteverted nostrils, broad nose, broad face, and rosy cheeks. Skeletal changes include thickening of the skull, cranial asymmetry, short neck, joint hypermobility, thinned cortex of the long bones, short stature, and thoracic or lumbar scoliosis. Development is slowed. Diarrhea, frequent respiratory infections, and cutaneous manifestations may be seen. Intellectual deterioration occurs, leading to severe mental defect in the adult. Episodic hyperactivity, psychotic behavior, speech defects, and seizures may occur. Some patients have subcapsular, crystal-like lens opacities. Macroglossia, cardiac murmur, or hernia may occur. Angiokeratoma corporis diffusum (AKCD) and pubertal macro-orchidism have been observed. Tsukadaira et al. note that AKCD occurs in several inherited lysosomal storage disorders, including Anderson–Fabry disease, β-galactosidosis, fucosidosis, aspartylglucosaminuria, galactosialidosis, β-mannosidosis, and Kanzaki disease. They present an excellent color photo of AKCD.[19]

Inheritance is autosomal recessive. Both homozygotes and heterozygotes for AGU disease may develop chronic arthritis.[49]

Laboratory Findings

Diagnosis is by the characteristic clinical and genetic features, findings of aspartylglycosamine in urine and demonstration of deficiency of aspartylglucosaminidase (*N*-aspartyl-β-glucosaminidase). Aspartylglycosamine is easily detected with thin-layer chromatography of urine for oligosaccharides, with liquid chromatography, with the amino acid analyzer, or from urine specimens recovered from absorbent filter paper. Aspartylglycosaminuria has been demonstrated in amniotic fluid and fetal urine.

Aspartylglucosaminidase deficiency can be demonstrated most satisfactorily in cultured skin fibroblasts, although the deficiency is also found in plasma, seminal fluid, and tissue. Elevated serum dolichol levels have been found in aspartylglycosaminuria, as in other neurologic disorders.

Roentgenography

Chief roentgenographic findings include thickening of the calvaria, osteochondrosis of the vertebrae leading to wedge-shaped vertebral bodies with narrowing of the intervertebral spaces, and generalized osteoporosis. Cranial asymmetry may be noted. The cortex of the long bones is thin. Kyphosis or scoliosis may be noted.

Pathology

A wide variety of tissues and cell types show lysosomal storage. The chief ultrastructural finding is enlarged lysosomal vacuoles, which contain fibrillar or granular material. Vacuolated lymphocytes are also seen.

Stored Substances in Aspartylglycosaminuria

Patients excrete in their urine large amounts of aspartyl glycosamine [(2-acetamido-1-β-L-aspartamido)-1,2-dideoxyglucose] and more complex compounds containing this moiety. Aspartylglycosamine is also the chief storage material in tissues. Other glycoasparagines stored in aspartylglycosaminuria include a tetrasaccharide moiety α-D-Man-(1→6)-β-D-Man-(1→4)-β-D-GlcNAc-(1→4)-β-D-GlcNAc-(1→N)-Asn (Man2GlcNAc2-Asn). As mentioned earlier, urine oligosaccharides show a characteristic pattern in aspartylglycosaminuria when analyzed by MALDI-TOF mass spectrometry.[5,6]

Enzyme Deficiency in Aspartylglycosaminuria

The disease results from deficiency of the lysosomal enzyme aspartylglucosaminidase (*N*-aspartyl-β-glucosaminidase), an amidase that cleaves aspartylglycosamine. This bond constitutes the linkage region between the protein portion and saccharide portion of one class of glycoproteins and of keratan sulfate. The enzyme aspartylglucosaminidase has been purified from urine, where it was found to consist of two heavy chains and two light chains. Purified from leukocytes, the enzyme was composed of two nonidentical polypeptides, which gave rise to the tetrameric enzyme protein. Both polypeptides appear to be products of the same structural gene. A 34.6-kDa propeptide is cleaved into a 19.5-kDa α subunit and a 15-kDa β subunit. The enzyme subunits are glycosylated to form the final tetrameric protein.

The Aspartylglucosaminidase Gene

The aspartylglucosaminidase gene has been localized to the long arm of human chromosome 4 (4q32–q33). The gene has been cloned and sequenced. Mutations responsible for aspartylglycosaminuria have been identified. Nearly all Finnish patients have the C163S mutation, but other mutations have been found in patients born outside Finland.[50]

In 2001, Saarela et al.[51] reviewed the consequences of known mutations of the aspartylglucosaminidase gene. They found that some mutations affected the active site, destroying enzyme activity and affecting precursor maturation, but others affected the dimer interface of the protein, thus preventing dimerization.

Two mouse knockout models for aspartylglycosaminuria have been engineered that develop neurologic degeneration, ataxia, progressive motor disease, impaired bladder function, and early death.

Therapy

Bone marrow transplantation for AGU has been beneficial in infancy,[52] but after infancy, the results of bone marrow transplantation for AGU have not been promising.[53] Enzyme replacement therapy in a mouse model greatly reduced concentrations of Man2GlcNAc2-Asn in some tissues.[54] In adenovirus-mediated gene therapy trials using 2 two tissue-specific promoters and the endogenous promoter in the AGU knockout mouse, AGA expression was increased and storage markedly decreased in some cases.[55]

References

1. Ramsay SL, Maire I, Bindloss C, et al. Determination of oligosaccharides and glycolipids in amniotic fluid by electrospray ionisation tandem mass spectrometry: in utero indicators of lysosomal storage diseases. *Mol Genet Metab*. 2004;83:231–235.
2. Ramsay SL, Meikle PJ, Hopwood JJ, et al. Profiling oligosaccharidurias by electrospray tandem mass spectrometry: quantifying reducing oligosaccharides. *Anal Biochem*. 2005;345:30–46.
3. Meikle PJ, Ranieri E, Simonsen H, et al. Newborn screening for lysosomal storage disorders: clinical evaluation of a two-tier strategy. *Pediatrics*. 2004;114:909–916.
4. Sowell J, Wood T. Towards a selected reaction monitoring mass spectrometry fingerprint approach for the screening of oligosaccharidoses. *Anal Chim Acta*. 2011;686(1–2):102–106.
5. Bonesso L, Piraud M, Caruba C, Van Obberghen E, Mengual R, Hinault C. Fast urinary screening of oligosaccharidoses by MALDI-TOF/TOF mass spectrometry. *Orphanet J Rare Dis*. 2014;9(1):19.
6. Xia B, Asif G, Arthur L, et al. Oligosaccharide analysis in urine by MALDI-TOF mass spectrometry for the diagnosis of lysosomal storage diseases. *Clin Chem*. 2013;59(9):1357–1368.
7. Poorthuis BJ, Wevers RA, Kleijer WJ, et al. The frequency of lysosomal storage diseases in the Netherlands. *Hum Genet*. 1999;105:151–156.
8. Kornfeld S, Li E, Tabas I. Synthesis of complex-type oligosaccharides II: characterization of the processing intermediates in the synthesis of the complex oligosaccharide units of the vesicular stomatitis virus G1 protein. *J Biol Chem*. 1978;253:7771–7778.
9. Lowden JA, O'Brien JS. Sialidosis: a review of human neuraminidase deficiency. *Am J Hum Genet*. 1979;31:1–18.
10. O'Brien JS, Warner TG. Sialidosis: delineation of subtypes by neuraminidase assay. *Clin Genet*. 1980;17:35–38.
11. Lai SC, Chen RS, Wu Chou YH, et al. A longitudinal study of Taiwanese sialidosis type 1: an insight into the concept of cherry-red spot myoclonus syndrome. *Eur J Neurol*. 2009;16(8):912–919.
12. Buchholz T, Molitor G, Lukong KE, et al. Clinical presentation of congenital sialidosis in a patient with a neuraminidase gene frameshift mutation. *Eur J Pediatr*. 2001;160:26–30.
13. Pshezhetsky AV, Ashmarina M. Lysosomal multienzyme complex: biochemistry, genetics, and molecular pathophysiology. *Prog Nucleic Acid Res Mol Biol*. 2001;69:81–117.
14. Galjart NJ, Morreau H, Willemsen R, et al. Human lysosomal protective protein has cathepsin A-like activity distinct from its protective function. *J Biol Chem*. 1991;266:14754–14764.
15. Lukong KE, Landry K, Elsliger MA, et al. Mutations in sialidosis impair sialidase binding to the lysosomal multienzyme complex. *J Biol Chem*. 2001;276:17286–17290.
16. Bonten EJ, Arts WF, Beck M, et al. Novel mutations in lysosomal neuraminidase identify functional domains and determine clinical severity in sialidosis. *Hum Mol Genet*. 2000;9:2715–2725.
17. Seyrantepe V, Poupetova H, Froissart R, et al. Molecular pathology of *NEU1* gene in sialidosis. *Hum Mutat*. 2003;22:343–352.
18. Naganawa Y, Itoh K, Shimmoto M, et al. Molecular and structural studies of Japanese patients with sialidosis type 1. *J Hum Genet*. 2000;45:241–249.
19. Tsukadaira A, Hirose Y, Aoki H, Ichikawa K, Sakurai A, Fukuzawa M. Diagnosis of fucosidosis through a skin rash. *Intern Med*. 2005;44(8):907–908.
20. Willems PJ, Gatti R, Darby JK, et al. Fucosidosis revisited: a review of 77 patients. *Am J Med Genet*. 1991;38:111–131.
21. Inui K1, Akagi M, Nishigaki T, Muramatsu T, Tsukamoto H, Okada S. A case of chronic infantile type of fucosidosis: clinical and magnetic resonance image findings. *Brain Dev*. 2000;22(1):47–49.
22. Oner AY1, Cansu A, Akpek S, Serdaroglu A. Fucosidosis: MRI and MRS findings. *Pediatr Radiol*. 2007;37(10):1050–1052.
23. Autti T1, Joensuu R, Aberg L. Decreased T2 signal in the thalami may be a sign of lysosomal storage disease. *Neuroradiology*. 2007;49(7):571–578.
24. Terespolsky D, Clarke JT, Blaser SI, et al. Evolution of the neuroimaging changes in fucosidosis type II. *J Inherit Metab Dis*. 1996;19:775–781.
25. Fukushima H, de Wet JR, O'Brien JS. Molecular cloning of a cDNA for human alpha-L-fucosidase. *Proc Natl Acad Sci U S A*. 1985;82:1262–1265.
26. Willems PJ, Seo HC, Coucke P, et al. Spectrum of mutations in fucosidosis. *Eur J Hum Genet*. 1999;7:60–67.
27. Lin SP, Chang JH, de la Cadena MP, Chang TF, Lee-Chen GJ. Mutation identification and characterization of a Taiwanese patient with fucosidosis. *J Hum Genet*. 2007;52(6):553–556.
28. Miano M, Lanino E, Gatti R, et al. Four year follow-up of a case of fucosidosis treated with unrelated donor bone marrow transplantation. *Bone Marrow Transplant*. 2001;27:747–751.
29. Malm D, Halvorsen DS, Tranebjaerg L, et al. Immuno-deficiency in alpha-mannosidosis: a matched case-control study on immunoglobulins, complement factors, receptor density, phagocytosis and intracellular killing in leucocytes. *Eur J Pediatr*. 2000;159:699–703.
30. Malm D, Riise Stensland HM, Edvardsen Ø, Nilssen Ø. The natural course and complications of alpha-mannosidosis—a retrospective and descriptive study. *J Inherit Metab Dis*. 2014;37(1):79–82.
31. Malm D, Nilssen Ø. Alpha-mannosidosis. *Orphanet J Rare Dis*. 2008;3:21.
32. Patlas M, Shapira MY, Nagler A, et al. MRI of mannosidosis. *Neuroradiology*. 2001;43:941–943.
33. Nebes V, Schmidt MC. Human lysosomal alpha-mannosidase: isolation and nucleotide sequence of the full-length cDNA. *Biochem Biophys Res Commun*. 1994;200:239–245.
34. Berg T, Riise HM, Hansen GM, et al. Spectrum of mutations in alpha-mannosidosis. *Am J Hum Genet*. 1999;64:77–88.
35. Gotoda Y, Wakamatsu N, Kawai H, et al. Missense and non- sense mutations in the lysosomal alpha-mannosidase gene (*MANB*) in severe and mild forms of alpha-mannosidosis. *Am J Hum Genet*. 1998;63:1015–1024.
36. Kuokkanen E, Riise Stensland HM, Smith W, et al. Molecular and cellular characterization of novel α-mannosidosis mutations. *Hum Mol Genet*. 2011;20(13):2651–2661.
37. Riise Stensland HM, Klenow HB, Van Nguyen L, Hansen GM, Malm D, Nilssen Ø. Identification of 83 novel alpha-mannosidosis-associated sequence variants: functional analysis of *MAN2B1* missense mutations. *Hum Mutat*. 2012;33(3):511–520.

38. Yesilipek AM, Akcan M, Karasu G, Uygun V, Kupesiz A, Hazar V. Successful unrelated bone marrow transplantation in two siblings with alpha-mannosidosis. *Pediatr Transplant*. 2012;16(7):779–782.
39. Jones MZ, Dawson G. Caprine beta-mannosidosis. *J Biol Chem*. 1981;256:5185–5188.
40. Cooper A, Sardharwalla IB, Roberts MM. Human beta-mannosidase deficiency. *N Engl J Med*. 1986;315:1231.
41. Wenger DA, Sujansky E, Fennessey PV, et al. Human beta-mannosidase deficiency. *N Engl J Med*. 1986;315:1201–1205.
42. Levade T, Graber D, Flurin V, et al. Human beta-mannosidase deficiency associated with peripheral neuropathy. *Ann Neurol*. 1994;35:116–119.
43. Broomfield A, Gunny R, Ali I, Vellodi A, Prabhakar P. A clinically severe variant of β-Mannosidosis, presenting with neonatal onset epilepsy with subsequent evolution of hydrocephalus. *JIMD Rep*. 2013;11:93–97.
44. Alkhayat AH, Kraemer SA, Leipprandt JR, et al. Human beta-mannosidase cDNA characterization and first identification of a mutation associated with human beta-mannosidosis. *Hum Mol Genet*. 1998;7:75–83.
45. Huynh T, Khan JM, Ranganathan S. A comparative structural bioinformatics analysis of inherited mutations in β-D-Mannosidase across multiple species reveals a genotype–phenotype correlation. *BMC Genomics*. 2011;12(suppl 3):S22.
46. Matilainen R, Airaksinen E, Mononen T, et al. A population-based study on the causes of mild and severe mental retardation. *Acta Paediatr*. 1995;84:261–266.
47. Mononen T, Mononen I, Matilainen R, et al. High prevalence of aspartylglycosaminuria among school-age children in eastern Finland. *Hum Genet*. 1991;87:266–268.
48. Kallinen J, Marin K, Heinonen S, et al. Wide scope prenatal diagnosis at Kuopio University Hospital. 1997–1998: integration of gene tests and fetal karyotyping. *Br J Obstet Gynecol*. 2001;108:505–509.
49. Arvio M, Laiho K, Kauppi M, et al. Carriers of the aspartylglucosaminuria genetic mutation and chronic arthritis. *Ann Rheum Dis*. 2002;61:180–181.
50. Fisher KJ, Aronson Jr NN. Characterization of the mutation responsible for aspartylglucosaminuria in three Finnish patients: amino acid substitution Cys163Ser abolishes the activity of lysosomal glycosylasparaginase and its conversion into subunits. *J Biol Chem*. 1991;266:12105–12113.
51. Saarela J, Laine M, Oinonen C, et al. Molecular pathogenesis of a disease: structural consequences of aspartylglucosaminuria mutations. *Hum Mol Genet*. 2001;10:983–995.
52. Malm G, Mansson JE, Winiarski J, et al. Five-year follow-up of two siblings with aspartylglucosaminuria undergoing allogeneic stem-cell transplantation from unrelated donors. *Transplantation*. 2004;78:415–419.
53. Arvio M, Sauna-Aho O, Peippo M. Bone marrow transplantation for aspartylglucosaminuria: follow-up study of transplanted and non-transplanted patients. *J Pediatr*. 2001;138:288–290.
54. Virta S, Rapola J, Jalanko A, et al. Use of nonviral promoters in adenovirus-mediated gene therapy: reduction of lysosomal storage in the aspartylglucosaminuria mouse. *J Gene Med*. 2006;8:699–706.
55. Kelo E, Dunder U, Mononen I. Massive accumulation of Man2GlcNAc2-Asn in nonneuronal tissues of glycosylasparaginase-deficient mice and its removal by enzyme replacement therapy. *Glycobiology*. 2005;15:79–85.

CHAPTER

34

β-Galactosidase Deficiency: G_{M1} Gangliosidosis, Morquio B Disease, and Galactosialidosis

William G. Johnson

Rutgers Robert Wood Johnson Medical School, New Brunswick, NJ, USA

INTRODUCTION

Genetic deficiency of lysosomal acid β-galactosidase (EC 3.2.1.23) in humans causes G_{M1} gangliosidosis, a neurologic disorder, and Morquio B disease, which is primarily a skeletal–connective tissue disorder.[1] Clinical manifestations of G_{M1} gangliosidosis are highly variable. The classic infantile form is a severe, rapidly progressive neurologic and systemic disorder, and patients rarely survive for more than a few years. Patients with clinically, pathologically, and biochemically less severe juvenile and adult forms, however, have later onset, slower progression, and in many adult patients, little mental involvement. Although it has been clear for many years that the underlying cause of these various clinical phenotypes is genetic deficiency of acid β-galactosidase activity, numerous attempts at their enzymologic delineation have failed.

Availability of cDNA clones coding for the human acid β-galactosidase made it feasible to examine the underlying abnormalities on the gene level in different clinical phenotypes of G_{M1} gangliosidosis. The line between G_{M1} gangliosidosis and Morquio B disease is increasingly blurred because certain mutations have been found to cause intermediate phenotypes with both neurodegenerative and skeletal manifestations. Both G_{M1} gangliosidosis and Morquio B disease can be diagnosed definitively by assay of acid β-galactosidase activity with artificial chromogenic or fluorogenic substrates, provided appropriate precautions are taken to exclude galactosialidosis. Diagnosis based on specific mutations is possible. A mutation-based diagnosis is highly specific and unambiguous, particularly in heterozygote diagnosis. It is, however, often too specific and can never entirely replace enzymatic diagnosis, which tests for the functional abnormality of the enzyme whatever the underlying mutation might be.

Galactosialidosis was initially thought to be a variant of genetic β-galactosidase deficiency. It was later recognized that these patients are deficient in activities of both β-galactosidase and α-neuraminidase.[2] It has since been established that galactosialidosis is genetically distinct from β-galactosidase deficiencies and that the underlying cause is genetic defects of another, unrelated gene, *PPCA*. Three lysosomal enzymes, sialidase, β-galactosidase, and *N*-acetylaminogalacto-6-sulfate sulfatase, are associated with protective protein/cathepsin A (PPCA, known also as PPGB [protective protein for β-galactosidase], or as cathepsin A or lysosomal carboxypeptidase). This complex formation affects the biogenesis, intracellular sorting, and proteolytic processing of the enzymes, and greatly extends their half-life.[3]

G_{M1} GANGLIOSIDOSIS AND MORQUIO B DISEASE

G_{M1} gangliosidosis and Morquio B disease are traditionally classified as entirely different diseases. Their basic relationship is not obvious because their typical clinical phenotypes are so different. Conceptually, however, these two

Rosenberg's Molecular and Genetic Basis of Neurological and Psychiatric Disease
http://dx.doi.org/10.1016/B978-0-12-410529-4.00034-6

diseases merely represent the extremes of phenotypic variations of the same disorder caused by genetic deficiency of acid β-galactosidase activity. The term β-galactosidosis has been suggested as an inclusive nomenclature of both G_{M1} gangliosidosis and Morquio B disease.[1]

G_{M1} GANGLIOSIDOSIS

Clinical Picture and Diagnostic Evaluation

Until the mid-1960s, the only known ganglioside storage disease was the classic Tay–Sachs disease (a G_{M2} gangliosidosis). Existence of another entirely different category of gangliosidoses, G_{M1} gangliosidosis, was firmly established by 1965.[4–6] Advances in the technology of ganglioside analysis were the key for this development because, in retrospect, the disease had been described earlier without definitive chemical evidence. G_{M1} gangliosidosis is transmitted as a Mendelian autosomal recessive disorder. Although the most important clinical manifestations are for the most part neurologic, the disease involves systemic organs, such as the liver and spleen, and the skeletal structure. It is convenient to divide the phenotype into infantile, juvenile, and adult forms according to the time of onset and the rate of progression.

Patients with the infantile form of G_{M1} gangliosidosis generally experience clinical onset by 6 months, although facial and bony abnormalities are often recognized at birth. G_{M1} gangliosidosis is among 10 different lysosomal storage disorders that have been diagnosed in infants with the phenotype of nonimmune hydrops fetalis.[7,8] In the early stages, patients with G_{M1} gangliosidosis appear dull and hypotonic with retarded psychomotor development. Regression soon becomes evident, leading to later manifestations, such as spasticity and seizures. Eventually patients become deaf and blind, totally unresponsive to external stimuli. Macular cherry-red spots are common but not consistent. The entire clinical course rarely exceeds 2 to 3 years, and inevitable death occurs most commonly as a result of concurrent infections. Neurologic manifestations are primarily those of gray matter. White matter and peripheral nerve involvement is relatively minor. Infantile patients show, in addition to the previously mentioned neurologic manifestations, clinical features similar to those in mucopolysaccharidoses, including facial deformity, macroglossia, radiologic bone abnormalities, and visceromegaly.

The late infantile or juvenile form of the disease manifests usually after age 1 year, with milder neurologic signs and slower progression. Systemic involvement, prominent in infantile patients, is less and may be absent. Patients with this phenotype often survive beyond 10 years. A typical case of this form was described early by Derry et al.[9]

The adult or chronic form is very variable clinically. The clinical onset can be any time between a few years to 30 years or older. Perhaps the first clearly delineated patients were described in 1977 by Suzuki et al.[10] This phenotype appears to be relatively common in Japan, but cases are known in other ethnic groups. Slowly progressive dysarthria, gait difficulties, dystonic movements, and other extrapyramidal signs are most prominent. Patients are often slightly or moderately impaired intellectually. Signs prominent in the earlier-onset forms, such as macular cherry-red spots, facial abnormalities, and visceromegaly, are not associated with the adult form. Bony abnormalities are recorded, however, sometimes causing symptoms due to spinal compression. Clinical diagnosis of adult G_{M1} gangliosidosis is difficult, and patients with the above-mentioned clinical picture should be screened using appropriate enzymatic assays. The finding of globus pallidus T2 hypointensity by magnetic resonance imaging (MRI) evaluation suggests the possibility that the patient may have G_{M1} gangliosidosis.[11]

Pathology

Pathology of the infantile and juvenile forms is qualitatively similar, with the infantile form being generally more severe. The gray matter is affected, whereas the white matter and the peripheral nerves are relatively spared. Essentially, all neurons throughout the body are enlarged and contain faintly granular material, which consists of discrete abnormal lamellar bodies (membranous cytoplasmic bodies) that can be detected by electron microscopy. The pathology of cortical neurons is for all practical purposes indistinguishable from that of classic Tay–Sachs disease (G_{M2} gangliosidosis). Many reactive astrocytes also contain abnormal cytoplasmic bodies. White matter often shows mild myelin loss with sudanophilia, presumably secondary to neuronal loss. Many reticuloendothelial organs, such as lung, liver, spleen, lymph nodes, and bone marrow, contain swollen foamy histiocytes with intracellular vacuoles containing strongly periodic acid–Schiff (PAS)-positive materials. This systemic pathology is present even in juvenile patients, who show relatively few clinical signs of systemic organ involvement. Ultrastructure of the systemic organs more closely resembles that of mucopolysaccharidoses and is different from neuropathology. Clear vacuoles in these cells, however, contain interwoven bundles of tubular structures.

The characteristic finding in the neuropathology of adult G_{M1} gangliosidosis is its selective distribution of lesions within the central nervous system. Unlike in the younger phenotypes, cortical neurons tend to be relatively spared, and those in the basal ganglia are primarily affected. Even within the same region, severely swollen neurons can be found next to others that appear relatively normal or pyknotic. The ultrastructural features of affected neurons, however, are qualitatively similar to those in the earlier-onset forms. This selective involvement of the basal ganglia can explain, at least on the phenomenologic level, the predominant extrapyramidal manifestations in the adult disease. Again, systemic organs can show pathology similar to the younger forms even when there is no clinical evidence of their involvement.

Biochemistry and Enzymology

G_{M1} gangliosidosis was originally recognized as a distinct disease when a massive accumulation of G_{M1} ganglioside in the brain was observed. The total concentration of ganglioside in the brain at the terminal stage in the infantile disease can be 3–5 times normal. G_{M1} ganglioside, normally present at 22–25 molar percent of total ganglioside, is present at 90–95 molar percent in patients' gray matter.[12] Thus the increase in G_{M1} ganglioside in the gray matter is actually up to 20 times normal. There is also an abnormal increase in G_{M1} ganglioside in systemic organs, although the absolute amounts remain small. G_{M1} gangliosidosis was named early as generalized gangliosidosis because of the abnormal accumulation of G_{M1} ganglioside in systemic organs. This terminology, however, is a misnomer. Similar degrees of abnormal accumulation of the affected ganglioside, G_{M2}, also occur in systemic organs of patients with Tay–Sachs disease without clinical signs of visceromegaly. There is very limited information available for analytical biochemistry of the adult form of G_{M1} gangliosidosis because of the long survival of patients. In one well-characterized patient, abnormal accumulation of G_{M1} ganglioside paralleled the distribution of pathologic lesions in that the basal ganglia clearly contained a much larger quantity of G_{M1} ganglioside than did the cerebral cortex.[12,13]

The main storage materials in the systemic organs are heterogeneous galactose-rich fragments of varying molecular weights derived from glycoproteins, keratan sulfate, and other carbohydrate-containing materials. The most insoluble of these was first described as keratin sulfate-like based on its behavior in preparative procedures, its electrophoretic properties, and its sugar composition.[14] When all of these materials are combined, the amount in patients' tissues can be more than 20 times normal. Many compounds of similar molecular weight also appear in excess in patients' urine. Several extensive characterizations have been done on these materials, which appear to be for the most part derived from glycoproteins. These materials should, *a priori*, have a terminal β-galactose moiety at the nonreducing end because their accumulation is a consequence of the genetic defect in acid β-galactosidase. Electrospray ionization tandem mass spectrometry has recently been used to determine disease-specific oligosaccharides in disorders that include G_{M1} gangliosidosis, galactosialidosis, sialidosis type II, fucosidosis, α-mannosidosis, and aspartylglucosaminuria.[15,16]

More recently, Bonesso et al. used a matrix-assisted laser desorption ionization time-of-flight (MALDI-TOF/TOF) mass spectroscopic approach to screen oligosaccharidoses in urine. This approach allows identification of G_{M1} gangliosidosis as well as Sandhoff disease, α-mannosidosis, sialidosis fucosidosis, aspartylglucosaminuria, and mucolipidoses type II and III.[17]

The underlying cause of G_{M1} gangliosidosis is a genetic deficiency of the lysosomal acid β-galactosidase.[6,18,19] The defect can be readily demonstrated either with conventional chromogenic or fluorogenic artificial substrate or with the natural substrate, G_{M1} ganglioside. Among several glycolipids that have the terminal β-galactosidase residue, G_{M1} ganglioside, asialo-G_{M1} ganglioside, and lactosylceramide are natural substrates for the enzyme. There is no prominent abnormal accumulation of lactosylceramide in G_{M1} gangliosidosis, however, because the other lysosomal β-galactosidase, galactosylceramidase, also hydrolyzes lactosylceramide. Genetic deficiency of galactosylceramidase is the cause of globoid cell leukodystrophy (Krabbe disease), and this enzyme is normal in G_{M1} gangliosidosis.

MORQUIO B DISEASE

Clinical Picture and Diagnostic Evaluation

Morquio B disease was originally classified as a form of mucopolysaccharidosis because of its clinical and pathologic similarity to other forms of Morquio disease. Unlike G_{M1} gangliosidosis, Morquio B disease is primarily a skeletal disease without neurologic manifestations except those caused by bone deformity. Progressive skeletal dysplasia starts during the first several years of life and is often most prominent in vertebral and pelvic bones. Many patients are of short stature. Severe spinal deformities are common. Odontoid hypoplasia is always present and is the cause of potentially serious complications due to cervical cord compression. Mild corneal opacity, mild organomegaly, and cardiac lesions are frequent manifestations besides bony abnormalities. As in all forms of G_{M1} gangliosidosis, patients with Morquio B disease also excrete excess amounts of galactose-containing materials into urine.

Pathology

Because Morquio B disease was recognized relatively recently as a genetic disorder distinct from other forms of mucopolysaccharidosis IV, information on pathology pertaining specifically to this disease is limited.

Biochemistry and Enzymology

As in other clinical forms of Morquio disease, patients with Morquio B disease excrete keratan sulfate into urine, in addition to galactose-rich fragments of glycoproteins. The first two steps of keratan sulfate degradation are desulfation by *N*-acetylgalactosamine 6-sulfatase (galactose-6-sulfatase) and then cleavage of the β-galactosidic bond by β-galactosidase. A genetic block at the first step causes the classic type A disease, whereas block of the second step causes type B disease. Deficient activities of β-galactosidase can be demonstrated using conventional chromogenic or fluorogenic substrates. The fundamental enzymatic defect underlying Morquio B disease involves the same enzyme that is defective in G_{M1} gangliosidosis. Therefore, it is likely that the mutations underlying the respective disorders differentially affect the specificity of the enzyme toward various natural substrates, resulting in dramatically different clinical pictures. Paschke and Kresse[20] have presented findings consistent with this prediction. They found that the capacity of the β-galactosidase from Morquio B patients to hydrolyze G_{M1} ganglioside could be activated by the natural activator protein, saposin B, up to 20% of the normal activity, whereas it could not be activated to hydrolyze the β-galactose residue from keratan sulfate.

Molecular Genetics of β-Galactosidase Deficiency

Knowledge of the molecular biology of G_{M1} gangliosidosis and Morquio B disease began when cDNA coding for the normal human acid β-galactosidase (GLB1) was cloned and characterized.[21] The genomic organization of the human GLB1 gene has also been characterized.[22] The human *GLB1* gene is located on chromosome 3 at 3p21.33.

The human *GLB1* gene codes by alternate splicing not only for the lysosomal β-galactosidase enzyme but also for elastin binding protein (EBP), which is involved in elastic fiber deposition. While β-galactosidase forms an intralysosomal complex with PPCA, sialidase, and galactosamine 6-sulfate sulfatase, EBP binds to PPCA and sialidase on the cell surface. Impaired elastogenesis in G_{M1} gangliosidosis patients suggests a possible alteration in EBP function.[23] GLB1 mutations may affect both proteins or GLB1 only. Abnormal EBP contributes to specific clinical features of G_{M1} gangliosidosis such as cardiomyopathy and connective-tissue abnormalities.[24] EBP deficiency impairs elastogenesis in G_{M1} gangliosidosis fibroblasts by a primary mechanism in infantile and a secondary mechanism in juvenile patients.[25] Mutations identified in patients with cardiomyopathy were localized in the GLB1 cDNA region common to lysosomal β-galactosidase and EBP. Consequently, both molecules are affected by the mutations, and they may contribute differently to the occurrence of specific clinical manifestations.[26]

Reports on disease-causing mutations have been rapidly increasing in the past several years (Table 34.1). Initially, nine mutations responsible for G_{M1} gangliosidosis were found, by coincidence, among Japanese patients. Within this population and within the relatively limited number of patients examined, one mutation each could be characterized as the cause of the juvenile and adult forms of the disease. It was noteworthy that Ile51→Thr was found in 27 of 32 mutant alleles among 16 Japanese adult patients. Some residual activity of the juvenile mutant enzyme and even higher residual activity in the adult mutant enzyme in an ASVGM1–4 cell expression system are consistent with the hypothesis that each mutation causes less severe, more slowly progressive clinical phenotypes. In contrast to the later-onset forms, mutations responsible for the infantile form of G_{M1} gangliosidosis are highly heterogeneous. The initial series of mutation analyses among Japanese patients was followed by studies with white patients. As expected, most disease-causing mutations found among the white population were different from those among Japanese patients. The mutation Thr82→Met was found among three adult patients from two unrelated families of Scandinavian extraction.

Oshima et al.[21] found a single mutation in three patients with Morquio B disease, although two were siblings. The mutation is an unusual double substitution of two adjacent nucleotides (TGG to CTG). The enzyme protein generated from this mutant gene gave 8% of the normal activity in the ASVGM1–4 cells. This finding is consistent with the

TABLE 34.1 Certain Identified Mutations in Acid β-Galactosidase that Cause G_{M1} Gangliosidosis or Morquio B Disease

Mutation	Phenotype
G_{M1} GANGLIOSIDOSIS	
20-bp Insertion from intron 2, caused by a base insertion in the 5′ donor site	Adult
23-bp Duplication in exon 3 duplication of exons 11 and 12 in mRNA	Infantile
9-bp In-frame insertion in exon 6 (Glu-Asn-Phe)	Adult
9-bp Insertion due to a base change at the acceptor site of intron 6	Juvenile
1-bp Insertion at nucleotide 896	Infantile
1-bp Insertion at nucleotide 1627	Infantile
$Arg^{49} \rightarrow Cys$	Infantile
$Arg^{59} \rightarrow His$	Infantile
$Ile^{51} \rightarrow Thr$	Adult
$Thr^{82} \rightarrow Met$	Very mild adult?
$Arg^{121} \rightarrow Ser$	Infantile
$Gly^{123} \rightarrow Arg$	Infantile
$Arg^{148} \rightarrow Ser$	Infantile
$Arg^{201} \rightarrow Cys$	Juvenile
$Arg^{201} \rightarrow His$	Adult
$Arg^{208} \rightarrow Cys$	Infantile
$Val^{216} \rightarrow Ala$	Infantile
$Val^{240} \rightarrow Met$	Infantile
$Asn^{266} \rightarrow Ser$	Adult
$Tyr^{316} \rightarrow Cys$	Infantile
$Asp^{332} \rightarrow Asn$	Infantile
$Arg^{351} \rightarrow Ter$	Infantile
G_{M1} GANGLIOSIDOSIS—CONT'D	
$Ser^{434} \rightarrow Leu$	Infantile
$Arg^{457} \rightarrow Ter$	Infantile
$Arg^{457} \rightarrow Gln$	Adult/chronic
$Asp^{491} \rightarrow Asn$	Infantile
$Gly^{494} \rightarrow Cys$	Infantile
$Trp^{509} \rightarrow Cys$	Infantile? Found in a heteroallelic patient with Morquio B disease
$Lys^{578} \rightarrow Arg$	Infantile
$Arg^{590} \rightarrow His$	Juvenile or adult
$Thr^{610} \rightarrow Ala$	Juvenile
$Glu^{632} \rightarrow Gly$	Juvenile or adult
$Lys^{655} \rightarrow Arg$	Juvenile
MORQUIO B DISEASE	
32-bp Insertion from intron 2, caused by a mutation at the 3′ acceptor site	Juvenile Morquio B
$Tyr^{83} \rightarrow His$	Juvenile Morquio B
$Arg^{208} \rightarrow His$	GM1/Morquio B intermediate
$Pro^{263} \rightarrow Ser$	GM1/Morquio B intermediate
$Trp^{273} \rightarrow Leu$	Adult Morquio B (one gene dose sufficient for the Morquio B phenotype)
$Asn^{318} \rightarrow His$	GM1/Morquio B intermediate
$Arg^{482} \rightarrow His$	Infantile/Morquio intermediate
$Arg^{482} \rightarrow Cys$	Morquio

The boundary between the phenotypes of G_{M1} gangliosidosis and Morquio B disease is often ill defined, particularly in older patients. Original sources that recorded these mutations are available in Suzuki et al.[1] and Callahan.[52]

relatively mild clinical features compared with severe G_{M1} gangliosidosis. However, it would be of great interest to test the catalytic activity of the enzyme toward different natural substrates to determine whether the enzyme retains its activity toward G_{M1} ganglioside but is inactive toward desulfated keratan sulfate or equivalent oligosaccharide substrates. Such a finding would not only support the hypothesis regarding the enzymology of β-galactosidase mutations, but it also would provide, taken together with the precise knowledge of the primary structure of the mutant enzyme, essential information concerning the functional domains of β-galactosidase. Additional mutations underlying the typical Morquio B phenotype and also a clinical phenotype of combined neurologic manifestations and Morquio-like skeletal abnormalities have been described (Table 34.1). Among 35 patients with G_{M1} gangliosidosis and Morquio disease type B, six patients of Gypsy origin shared not only the same mutation but also a common haplotype.[27]

Exome sequencing is being used as a diagnostic tool for β-galactosidase deficiency diseases, even in cases in which the diagnosis is unknown or even unsuspected.[28] Caciotti et al. have presented a new molecular analysis of 21 unrelated G_{M1} gangliosidosis patients and four Morquio B patients, two of whom were brothers, that showed 27 mutations, nine of which were new: five missense, three microdeletions and a nonsense mutation as well as four new genetics variants with a predicted polymorphic nature.[29] They have also presented a review and analysis of CTSA gene mutations.[30]

As in other genetic disorders, the traditional classification of genetic β-galactosidase deficiency based on the clinical phenotype now faces fundamental as well as pragmatic complications. The fundamental question is how to define a disease or its variant. There are already so many mutations known to cause inactivation of catalytic activity of the gene product. Except for consanguineous cases, compound heterozygosity would be the norm rather than the exception. The number of possible combinations would be enormous. Thus it becomes impossible to define a disease on the basis of the genotype for any given patient. Nevertheless, for pragmatic purposes, one hopes that the traditional clinical classification can be useful. Prediction, however, is not very reassuring. In compound heterozygous individuals, two abnormal alleles contribute to the phenotypic expression. If one makes a reasonable assumption that clinical phenotypes depend on the residual activity of the mutant enzyme protein *in vivo*, a mutant enzyme with higher residual activity tends to override that with less residual activity. Thus it is likely that patients who have at least one allele with the adult mutation are all of the adult phenotype, regardless of the nature of the other allele. In contrast, one juvenile mutant allele will define the juvenile phenotype only if the other allele carries either a juvenile or an infantile mutation. If the other allele carries an adult mutation, the patient will have the adult phenotype. Then patients with the infantile phenotype must carry two infantile mutations. This analysis holds only if there are adult, juvenile, and infantile mutations with discontinuous ranges of residual activities. Probability dictates that it is far more likely that the degree of residual activities of all individual mutations will be continuous between the normal activity to zero activity. Any degree of discontinuity among individual mutations will further be smoothed out by the additive nature of the contributions by the two mutant alleles in autosomal recessive disorders, such as G_{M1} gangliosidosis.

In the case of β-galactosidase, the diverse natural substrates and possible differential effect of any mutation on different substrates further complicate the eventual phenomena. The range of the final outcome of the residual activity contributed by the two mutant genes in any given patient defies any hope for discrete clinical phenotypes and consequently any orderly clinical classification. The traditional line between the primarily neurologic G_{M1} gangliosidosis and the exclusively skeletal Morquio B disease is increasingly blurred. The term β-galactosidosis encompasses all clinical phenotypes, from purely neurologic to purely skeletal. In fact, the argument advanced here finds parallels in all genetic disorders, particularly in those of the autosomal recessive inheritance.

GALACTOSIALIDOSIS

Although included here, galactosialidosis is a genetic disorder distinct from G_{M1} gangliosidosis/Morquio B disease.[2] The acid β-galactosidase gene is not affected in this galactosialidosis. Because patients show deficiencies of both β-galactosidase and sialidase (α-neuraminidase), they were often diagnosed erroneously as having either a variant of G_{M1} gangliosidosis or sialidosis. For example, an early somatic cell hybridization study showed that fibroblasts from a group of patients then classified as a G_{M1} gangliosidosis variant gave genetic complementation when fused with cells from patients with other phenotypes of G_{M1} gangliosidosis, indicating that a different gene is involved in these patients. Wenger et al.[31] was first to point out a combined deficiency of β-galactosidase and sialidase in many of these variant patients. Controversy followed as to which of the two enzyme deficiencies was the primary cause of this disease. Then it was shown that β-galactosidase in fibroblasts from galactosialidosis patients had an abnormally short half-life because of rapid proteolytic degradation *in vivo*. This led to the definitive delineation of this disorder as a genetic PPCA deficiency,[32] as noted earlier. Endothelin has been implicated as an endogenous natural substrate.[33]

Clinical Picture and Diagnostic Evaluation

The mode of inheritance of galactosialidosis is autosomal recessive. The clinical phenotype is heterogeneous, ranging from an early-onset, severe, and rapidly progressive infantile form to a late-onset, slowly progressive adult form. Patients with the infantile form may closely resemble those with infantile G_{M1} gangliosidosis, with severe central nervous system involvement, macular cherry-red spots, visceromegaly, renal insufficiency, coarse facies, and skeletal abnormalities.

Galactosialidosis is among 10 different lysosomal storage disorders that have been diagnosed in infants with the phenotype of nonimmune hydrops fetalis.[7,8,34]

Besides galactosialidosis and lysosomal storage disorders, nonimmune hydrops fetalis has now been associated with more than 75 inborn errors of metabolism, chromosomal aberrations and genetic syndromes.[35]

The late infantile form is essentially a later-onset, milder phenotype of the infantile disease. The juvenile/adult form of the disease appears to have a much higher incidence in Japan. The main clinical manifestations are slowly progressive central nervous system symptoms, including motor disturbance and mental retardation, skeletal abnormalities, dysmorphism, macular cherry-red spots, and angiokeratoma. Patients survive well into adulthood. Galactosialidosis patients of all clinical types excrete excess amounts of a complex mixture of glycopeptide fragments that are rich in sialic acid. Electrospray ionization tandem mass spectrometry has recently been used to determine disease-specific oligosaccharides in disorders including galactosialidosis, G_{M1} gangliosidosis, sialidosis type II, fucosidosis, α-mannosidosis, and aspartylglucosaminuria.[15,16]

Pathology

Relatively little is known about the pathology of patients with verified galactosialidosis. The neuropathology of a 13-year-old Japanese patient was recently described in some detail. Severe neuronal loss was seen in the basal ganglia, lateral geniculate body, and nucleus gracilis, as well as loss of Purkinje cells and retinal ganglion cells. Remaining neurons in the cerebral cortex, motor neurons, spinal cord, and trigeminal and spinal ganglia were distended, containing complex membranous cytoplasmic bodies, lipofuscin-like materials, and cytoplasmic vacuoles.

Biochemistry and Enzymology

Little is known about the analytic biochemistry of tissue constituents of patients with galactosialidosis. Storage materials in cultured fibroblasts and fetal placenta and those excreted into urine, however, have been characterized in detail. They are predominantly sialylated glycopeptides, similar to those found in sialidosis patients.

The enzymatic characteristic of galactosialidosis is the simultaneous deficiency of both β-galactosidase and α-neuraminidase.[27] It was noted earlier that when β-galactosidase was purified, its molecular weight appeared to be approximately 54 kDa, but that another protein component of approximately 32 kDa always copurified with the 54-kDa component. d'Azzo et al.[27] found that, in galactosialidosis, the 32-kDa component was lacking. On the basis of this finding and the earlier finding that the half-life of β-galactosidase in galactosialidosis patients was abnormally short but that it could be prolonged by addition of protease inhibitors, it was proposed that the function of the 32-kDa protein is to protect β-galactosidase and neuraminidase from proteolytic digestion within the lysosome by forming a complex with them and that genetic abnormality in this protective protein, now known as PPCA, is the underlying cause of galactosialidosis.

PPCA associates with three lysosomal enzymes: β-galactosidase, sialidase, and *N*-acetylaminogalacto-6-sulfate sulfatase. As already mentioned, PPCA is also known as protective protein for β-galactosidase, or as cathepsin A or as lysosomal carboxypeptidase. This complex formation affects the biogenesis, intracellular sorting and proteolytic processing of the complexed enzymes and greatly extends their half-life.[3] The hypothesis that PPCA is present both in the lysosome, complexed with β-galactosidase and sialidase, and also at the cell surface, complexed with EBP, and that PPCA is essential for EBP integrity is supported by the observation in a case of galactosialidosis with the fetal hydrops phenotype that the quantity of both GLB1 and EBP proteins were reduced.[36] Through its action as a recyclable chaperone, EBP facilitates tropoelastin secretion. Because EBP is part of a cell surface-targeted molecular complex with PPCA and sialidase, this sialidase may be required for tropoelastin release, facilitating its assembly into elastic fibers.[37] Abnormality of EBP's function as a chaperone for tropoelastin may contribute to impaired elastogenesis in galactosialidosis, G_{M1} gangliosidosis, and sialidosis.[38]

Chaperones are substrate analog competitive inhibitors that can bind to the mutant misfolded enzyme protein and can restore the catalytic activity of a mutant enzyme. These are being studied as therapeutic agents for such disorders as G_{M1} gangliosidosis.[39]

Molecular Genetics of Protective Protein/Cathepsin A Deficiency

The human *PPCA* gene is present on chromosome 20 at 20q13.1. Galjart et al.[40] cloned human and murine cDNA coding for *PPCA* of the respective species. Unexpectedly, the primary structure was homologous with carboxypeptidase.

Consistent with this observation, simultaneous deficiency of carboxypeptidase has been described in galactosialidosis. The exact mechanism by which PPCA stabilizes β-galactosidase and neuraminidase is still not fully understood, in part because of an apparent contradiction: PPCA, the protective protein, which is supposed to protect the two enzymes from proteolytic degradation, is itself a peptidase.

Several specific mutations responsible for human galactosialidosis are known. The common mutation in the late-onset form of the disease prevalent in Japan results in missplicing of the transcript. A substitution of a G at the third base from the 5′ donor site of intron 7 to A appears to cause skipping of exon 7 during processing of the transcript. All nine Japanese patients with the late-onset form had this mutation. A generally clear genotype–phenotype correlation was observed among 19 Japanese patients. No patients with infantile, severe form of the disease had this mutation; all late-onset, mild adult patients were homoallelic; and most patients with an intermediate phenotype were compound heterozygotes with one allele carrying this mutation. More recently, five additional disease-causing mutations have been recorded. Four of them, Gln49→Arg, Trp65→Arg, Ser90→Lys (a double base substitution), and Tyr395→Cys were found among Japanese patients, while Tyr249→Asn was found in a French-German patient. The Tyr395→Cys substitution appeared especially common among infantile/childhood Japanese patients.

ANIMAL MODELS

Naturally occurring animal models of genetic β-galactosidase and PPCA deficiencies are known among domestic and farm animals, including dogs, cats, cattle, and sheep. Sheep appear to have G_{M1} gangliosidosis with low α-neuraminidase activity rather than galactosialidosis because the cells from sick sheep genetically complemented the human galactosialidosis cells. A disease described in Schipperke dog may well be a model for human galactosialidosis. Cloning and characterization of full-length cDNA and the gene coding for mouse β-galactosidase and PPCA opened the way to generate authentic genetic murine models by homologous recombination and transgenic technologies. Two lines of G_{M1} gangliosidosis mouse models and a galactosialidosis mouse line have been successfully produced.

Because the human diseases are rare and because ethical considerations limit experiments on human patients, these animal models provide unique opportunities to study biology and possible treatment of these devastating genetic diseases. The feline and canine models have been used for such experimental purposes. No naturally occurring model, however, is known among small laboratory animals, which would give further advantage for certain types of studies. β-galactosidase knockout mouse lines have been generated in two laboratories independently.[41,42] Similarly, a mouse galactosialidosis model has also been generated.[43] These mouse models are being used actively in studies of the pathogenetic mechanism and of therapeutic trials of these disorders.

STATUS AND FUTURE POSSIBILITY OF THERAPY

Pragmatic, effective therapy is not yet available for human patients. Many laboratories are actively investigating the gene therapy approach, but it remains experimental.[44,45]

Current approaches to treatment for lysosomal diseases including G_{M1} gangliosidosis have been reviewed. Bone marrow transplantation has been disappointing in juvenile G_{M1} gangliosidosis.[46] A new approach called chemical chaperone therapy has been reviewed for β-galactosidase deficiency disorders (G_{M1} gangliosidosis and Morquio B disease) with use of *N*-octyl-4-epi-β-valienamine (NOEV), a potent inhibitor of β-galactosidase.[47] This approach has also been studied in cultured cells and in a mouse model of juvenile G_{M1} gangliosidosis.[48] In a G_{M1} gangliosidosis mouse model, an injected vector encoding β-galactosidase appeared to normalize brain neurochemistry.[49] Potential therapy by neonatal intravenous gene transfer is being studied in a mouse model of G_{M1} gangliosidosis.[50]

In the mouse model of galactosialidosis, lesions in systemic organs could be corrected almost completely using erythroid cells that were made to overexpress the protective protein by *ex vivo* transfection of normal gene.[43]

However, despite the almost complete "cure" of the systemic pathology, brain lesions were not corrected. A conceptually novel approach is activation of mutant enzyme with the use of galactonojirimycin derivatives, which inhibit glycolipid synthesis.[51,52] Substrate reduction therapy using *N*-butyldeoxygalactonojirimycin, an inhibitor of ganglioside synthesis, is being tested in β-galactosidase knockout mice.[53]

Bone marrow transplantation in a mouse model of galactosialidosis ($PPCA^{-/-}$ mice) using transgenic bone marrow overexpressing the corrective enzyme led to complete correction of the disease phenotype with apparently sufficient PPCA expression to delay the onset of Purkinje cell degeneration and to correct the ataxia.[54] A newly discovered lysosomal sialidase, Neu4, ubiquitously expressed and with broad substrate specificity, may be useful for developing new therapies for galactosialidosis and sialidosis.[55]

Future prospects for effective therapy must include the following consideration: If the effectiveness of a treatment can only be demonstrated by well-controlled experiments and perhaps with sophisticated statistical analysis, it is biologically significant but probably not sufficiently effective for clinical therapy. Clinically meaningful therapy requires a different order of magnitude in its effectiveness. From this perspective, pragmatic, effective treatment still seems many years off.

ACKNOWLEDGEMENT

Dr. Kunihiko Suzuki authored this chapter through the first three editions of this book. It has been updated by Dr. William Johnson. The editors are grateful to Dr. Suzuki for authoring this chapter and allowing its inclusion here.

References

1. Suzuki Y, Oshima A, Nanba E. β-Galactosidase deficiency (β-galactosidosis): G_{M1} gangliosidosis and Morquio B disease. In: Scriver CR, Beaudet AL, Sly WS, et al., eds. *The Metabolic and Molecular Bases of Inherited Disease*. 8th ed. New York: McGraw-Hill; 2001:3775–3809.
2. d'Azzo A, Andria G, Strisciuglio P, et al. Galactosialidosis. In: Scriver CR, Beaudet AL, Sly WS, et al., eds. *The Metabolic and Molecular Basis of Inherited Disease*. 8th ed. New York: McGraw-Hill; 2001:3811–3826.
3. Pshezhetsky AV, Ashmarina M. Lysosomal multienzyme complex: biochemistry, genetics, and molecular pathophysiology. *Prog Nucleic Acid Res Mol Biol*. 2001;69:81–114.
4. Jatzkewitz H, Sandhoff K. On a biochemically special form of infantile amaurotic idiocy. *Biochim Biophys Acta*. 1963;70:354–356.
5. Gonatas NK, Gonatas J. Ultrastructural and biochemical observations on a case of systemic late infantile lipidosis and its relationship to Tay-Sachs disease and gargoylism. *J Neuropathol Exp Neurol*. 1965;24:318–340.
6. Okada S, O'Brien JS. Generalized gangliosidosis: beta galactosidase deficiency. *Science*. 1968;160:1002–1004.
7. Stone DL, Sidransky E. Hydrops fetalis: lysosomal storage disorders in extremis. *Adv Pediatr*. 1999;46:409–440.
8. Burin MG, Scholz AP, Gus R, et al. Investigation of lysosomal storage diseases in nonimmune hydrops fetalis. *Prenat Diagn*. 2004;24:653–657.
9. Derry DM, Fawcett JS, Andermann F, et al. Late infantile systemic lipidosis. Major monosialogangliosidosis: delineation of two types. *Neurology*. 1968;18:340–348.
10. Suzuki Y, Nakamura N, Fukuoka K, et al. β-Galactosidase deficiency in juvenile and adult patients: report of six Japanese cases and review of literature. *Hum Genet*. 1977;36:219–229.
11. Vieira JP, Conceição C, Scortenschi E. G_{M1} gangliosidosis, late infantile onset dystonia, and T2 hypointensity in the globus pallidus and substantia nigra. *Pediatr Neurol*. 2013;49(3):195–197.
12. Suzuki K, Suzuki K, Kamoshita S. Chemical pathology of GM1-gangliosidosis (generalized gangliosidosis). *J Neuropathol Exp Neurol*. 1969;28:25–73.
13. Kobayashi T, Suzuki K. Chronic GM1-gangliosidosis presenting as dystonia. II Biochemistry. *Ann Neurol*. 1981;9:476–483.
14. Suzuki K. Cerebral GM1-gangliosidosis: chemical pathology of visceral organs. *Science*. 1968;159:1471–1472.
15. Ramsay SL, Maire I, Bindloss C, et al. Determination of oligosaccharides and glycolipids in amniotic fluid by electrospray ionisation tandem mass spectrometry: in utero indicators of lysosomal storage diseases. *Mol Genet Metab*. 2004;83:231–238.
16. Ramsay SL, Meikle PJ, Hopwood JJ, et al. Profiling oligosaccharidurias by electrospray tandem mass spectrometry: quantifying reducing oligosaccharides. *Anal Biochem*. 2005;345(1):30–46.
17. Bonesso L, Piraud M, Caruba C, Van Obberghen E, Mengual R, Hinault C. Fast urinary screening of oligosaccharososes by MALDI-TOF/TOF mass spectroscopy. *Orphanet J Rare Dis*. 2014;9:19.
18. Sacrez R, Juif JG, Gigonnet JM, et al. La maladie de Landing, ou idiotie amaurotique infantile précose avec gangliosidose généralisée de type G_{M1}. *Pediatrie*. 1967;22:143–162.
19. Seringe P, Plainfosse B, Lautmann F, et al. Gangliosidose généralisée, du type Norman Landing, à GM1. Etude à pro- pos d'un diagnostiqué du vivant du malade. *Ann Pediatr*. 1968;44:165–184.
20. Paschke E, Kresse E. Morquio disease, type B: activation of GM1-beta-galactosidase by GM1-activator protein. *Biochem Biophys Res Commun*. 1982;109:568–575.
21. Oshima A, Tsuji A, Nagano Y, et al. Cloning, sequencing and expression of cDNA for human beta-galactosidase. *Biochem Biophys Res Commun*. 1988;157:238–244.
22. Morreau H, Bonten E, Zhou X-Y, et al. Organization of the gene encoding human lysosomal β-galactosidase. *DNA Cell Biol*. 1991;10:495–504.
23. Caciotti A, Donati MA, Boneh A, et al. Role of beta-galactosidase and elastin binding protein in lysosomal and nonlysosomal complexes of patients with GM1-gangliosidosis. *Hum Mutat*. 2005;25:285–292.
24. Caciotti A, Donati MA, Procopio E, et al. G_{M1} gangliosidosis: molecular analysis of nine patients and development of an RT-PCR assay for GLB1 gene expression profiling. *Hum Mutat*. 2007;28:204.
25. Caciotti A, Donati MA, Bardelli T, et al. Primary and secondary elastin-binding protein defect leads to impaired elastogenesis in fibroblasts from GM1-gangliosidosis patients. *Am J Pathol*. 2005;167:1689–1698.
26. Morrone A, Bardelli T, Donati MA, et al. Beta-galactosidase gene mutations affecting the lysosomal enzyme and the elastin-binding protein in GM1-gangliosidosis patients with cardiac involvement. *Hum Mutat*. 2000;15:354–366.
27. Santamaria R, Chabas A, Coll MJ, et al. Twenty-one novel mutations in the GLB1 gene identified in a large group of GM1-gangliosidosis and Morquio B patients: possible common origin for the prevalent p.R59H mutation among gypsies. *Hum Mutat*. 2006;27:1060.
28. Pierson TM, Adams DA, Markello T, et al. Exome sequencing as a diagnostic tool in a case of undiagnosed juvenile-onset GM1-gangliosidosis. *Neurology*. 2012;79(2):123–126.
29. Caciotti A1, Garman SC, Rivera-Colón Y, et al. G_{M1} gangliosidosis and Morquio B disease: an update on genetic alterations and clinical findings. *Biochim Biophys Acta*. 2011;1812(7):782–790.
30. Caciotti A, Catarzi S, Tonin R, et al. Galactosialidosis: review and analysis of CTSA gene mutations. *Orphanet J Rare Dis*. 2013;8:114.
31. Wenger DA, Tarby TJ, Wharton C. Macular cherry-red spots and myoclonus with dementia: coexistent neuraminidase and beta-galactosidase deficiencies. *Biochem Biophys Res Commun*. 1978;82:589–595.
32. d'Azzo A, Hoogeveen A, Reuser AJJ, et al. Molecular defect in combined beta-galactosidase and neuraminidase deficiency in man. *Proc Natl Acad Sci U S A*. 1982;79:4535–4539.

33. Itoh K, Kase R, Shimmoto M, et al. Protective protein as an endogenous endothelin degradative enzyme in human tissues. *J Biol Chem*. 1995;270:515–518.
34. Kooper AJ, Janssens PM, de Groot AN, et al. Lysosomal storage diseases in non-immune hydrops fetalis pregnancies. *Clin Chim Acta*. 2006;371:176–182.
35. Carvalho S, Martins M, Fortuna A, Ramos U, Ramos C, Rodrigues MC. Galactosialidosis presenting as nonimmune fetal hydrops: a case report. *Prenat Diagn*. 2009;29(9):895–896.
36. Malvagia S, Morrone A, Caciotti A, et al. New mutations in the PPBG gene lead to loss of PPCA protein which affects the level of the beta-galactosidase/neuraminidase complex and the EBP-receptor. *Mol Genet Metab*. 2004;82:48–55.
37. Hinek A, Pshezhetsky AV, von Itzstein M, et al. Lysosomal sialidase (neuraminidase-1) is targeted to the cell surface in a multiprotein complex that facilitates elastic fiber assembly. *J Biol Chem*. 2006;281:3698–3710.
38. Tatano Y, Takeuchi N, Kuwahara J, et al. Elastogenesis in cultured dermal fibroblasts from patients with lysosomal beta-galactosidase, protective protein/cathepsin A and neuraminidase-1 deficiencies. *J Med Invest*. 2006;53:103–112.
39. Suzuki Y. Chaperone therapy update: Fabry disease, GM1-gangliosidosis and Gaucher disease. *Brain Dev*. 2013;35(6):515–523.
40. Galjart NJ, Gillemans N, Meijer D, et al. Mouse "protective protein": cDNA cloning, sequence comparison and expression. *J Biol Chem*. 1990;265:4678–4684.
41. Matsuda J, Suzuki O, Oshima A, et al. β-galactosidase-deficient mouse as an animal model for G_{M1}-gangliosidosis. *Glycoconj J*. 1997;14:729–736.
42. Hahn CN, Martin MD, Schroder M, et al. Generalized CNS disease and massive G_{M1}-ganglioside accumulation in mice defective in lysosomal acid β-galactosidase. *Hum Mol Genet*. 1997;6:205–211.
43. Zhou XY, Morreau H, Rottier R, et al. Mouse model for the lysosomal disorder galactosialidosis and correction of the phenotype with overexpressing erythroid precursor cells. *Genes Dev*. 1995;9:2623–2634.
44. Sena-Esteves M, Camp SM, Alroy J, et al. Correction of acid β-galactosidase deficiency in G_{M1} gangliosidosis human fibroblasts by retrovirus vector-mediated gene transfer: higher efficiency of release and cross-correction by the murine enzyme. *Hum Gene Ther*. 2000;11:715–727.
45. Brady RO. Emerging strategies for the treatment of hereditary metabolic storage disorders. *Rejuvenation Res*. 2006;9:237–244.
46. Shield JP, Stone J, Steward CG. Bone marrow transplantation correcting beta-galactosidase activity does not influence neurological outcome in juvenile G_{M1}-gangliosidosis. *J Inherit Metab Dis*. 2005;28(5):797–798.
47. Suzuki Y. Beta-galactosidase deficiency: an approach to chaperone therapy. *J Inherit Metab Dis*. 2006;29(2–3):471–476.
48. Matsuda J, Suzuki O, Oshima A, et al. Chemical chaperone therapy for brain pathology in $G(_{M1})$-gangliosidosis. *Proc Natl Acad Sci U S A*. 2003;100:15912–15917.
49. Broekman ML, Baek RC, Comer LA, et al. Complete correction of enzymatic deficiency and neurochemistry in the GM1-gangliosidosis mouse brain by neonatal adeno-associated virus-mediated gene delivery. *Mol Ther*. 2007;15:30–37.
50. Takaura N, Yagi T, Maeda M, et al. Attenuation of ganglioside G_{M1} accumulation in the brain of G_{M1} gangliosidosis mice by neonatal intravenous gene transfer. *Gene Ther*. 2003;10(17):1487–1493.
51. Tominaga L, Ogawa Y, Taniguchi M, et al. Galactonojirimycin derivatives restore mutant human β-galactosidase activities expressed in fibroblasts from enzyme-deficient knockout mouse. *Brain Dev*. 2001;23:284–287.
52. Callahan JW. Molecular basis of G_{M1} gangliosidosis and Morquio disease, type B: structure-function studies of lysosomal β-galactosidase and the non-lysosomal β-galactosidase-like protein. *Biochim Biophys Acta Mol Basis Dis*. 1999;1455:85–103.
53. Kasperzyk JL, d'Azzo A, Platt FM, et al. Substrate reduction reduces gangliosides in postnatal cerebrum–brainstem and cerebellum in G_{M1} gangliosidosis mice. *J Lipid Res*. 2005;46:744–751.
54. Leimig T, Mann L, Martin MP, et al. Functional amelioration of murine galactosialidosis by genetically modified bone marrow hematopoietic progenitor cells. *Blood*. 2002;99(9):3169–3178.
55. Seyrantepe V, Landry K, Trudel S, et al. Neu4, a novel human lysosomal lumen sialidase, confers normal phenotype to sialidosis and galactosialidosis cells. *J Biol Chem*. 2004;279:37021–37029.

CHAPTER

35

Acid Ceramidase Deficiency: Farber Lipogranulomatosis and Spinal Muscular Atrophy Associated with Progressive Myoclonic Epilepsy

Michael Beck[*], *Hugo W. Moser*[†], *and Konrad Sandhoff*[‡]

[*]University of Mainz, Mainz, Germany

[†]Deceased, formerly of Johns Hopkins University, Baltimore, MD, USA

[‡]Life and Medical Sciences Institute, University of Bonn, Bonn, Germany

INTRODUCTION

In 1957, Farber et al.[1] reported three patients, two of whom were siblings, who had granulomatous inflammatory lesions of the skin, joints, and larynx associated with neuronal storage of a complex glycosphingolipid. They found this disorder of special interest because it appeared to be a bridge between true metabolic disorders such as Tay–Sachs or Niemann–Pick disease and the inflammatory histiocytoses. Farber et al. named this disorder lipogranulomatosis. In 1967 Prensky et al.[2] demonstrated that this disorder was associated with the abnormal accumulation of ceramide, and in 1972 Sugita et al.[3] demonstrated the deficiency of lysosomal acid ceramidase. In 1996, Koch et al.[4] purified and cloned acid ceramidase. In 1999, Li et al.[5] characterized the gene that codes for acid ceramidase and demonstrated pathogenic mutations in patients with Farber lipogranulomatosis.

CLINICAL PICTURE

Characteristic signs of Farber disease (FD) are painful subcutaneous skin nodules, typically near the joints, and a progressive hoarseness, caused by laryngeal involvement. Often also the central nervous system is affected. Farber disease had been formerly subdivided into seven subtypes that differ in severity and additional organ involvement, like the lungs, nervous system, heart, and lymph nodes.[6] However, one has to take into account that in the last years several patients with a deficient activity of ceramidase have been observed who showed clinical manifestations that do not match the historical classification. As many other lysosomal storage disorders, Farber disease displays rather a continuous phenotypic spectrum than distinct clinical forms.

The occurrence of hydrops fetalis, as described by Kattner at al. and van Lijnschoten et al. represents the most severe form of Farber disease.[7,8] Patients with the neonatal phenotype (also named type 4) are extremely ill during the neonatal period with severe hepatosplenomegaly; they die before the age of 1 year.[9]

The so-called classic form, designated as type 1, seems to be the most common phenotype of Farber disease.[6] The clinical presentation is so characteristic that diagnosis can almost be made at a glance.[1,10] Characteristic features are painful swelling of joints (particularly the interphalangeal, metacarpal, ankle, wrist, knee, and elbow); palpable subcutaneous nodules in relation to affected joints and over pressure points (Figure 35.1) and also in the conjunctiva,

http://dx.doi.org/10.1016/B978-0-12-410529-4.00035-8

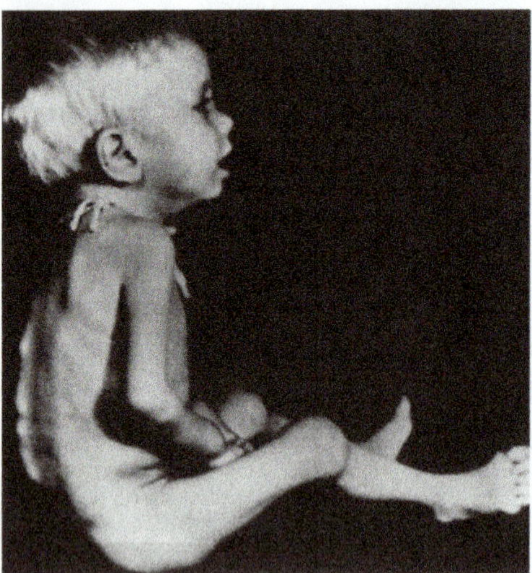

FIGURE 35.1 Farber disease patient at 23 months. Note joint swelling and contractures and subcutaneous nodules over spinous processes. Tracheostomy was performed as a life-saving procedure at age 15 months. (From Zetterstrom R. Disseminated lipogranulomatosis: Farber disease. *Acta Paediatr.* 1958;47:501–510).

external ear, and nostrils; hoarse cry that may progress to aphonia; feeding and respiratory difficulties; poor weight gain; and intermittent fever. Symptoms usually appear between the ages of 2 weeks and 4 months. Disturbances in swallowing, vomiting, and repeated episodes of pulmonary consolidations associated with fever occur frequently, and pulmonary disease is the usual cause of death. Other organs are also involved, and patients may present with hepatosplenomegaly, generalized lymphadenopathy, and cardiac murmurs secondary to valvular involvement. Diminished deep tendon reflexes, hypotonia, and muscular atrophy are due to involvement of anterior horn cells and peripheral neuropathy. Initial psychomotor development may be normal, but seizures and progressive cognitive decline may occur later. Diffuse, grayish opacification of the retina about the foveola, with a cherry-red center, is present in approximately one-third of the patients (although there is no disturbance of visual function, and it is subtler than the cherry-red spot in Tay–Sachs disease).[11] Two siblings, described by Al Jasmi et al. showed clinical features resembling those of type 1 patients, but did not have subcutaneous nodules.[12] This observation clearly demonstrates that in Farber disease a clear distinction between separate phenotypes cannot be made.

Some patients, formerly termed as type 2 and 3, show no or only slight involvement of the central nervous system, but they still suffer from severe signs and symptoms such as joint pain, contractures, hoarseness, failure to thrive, and respiratory involvement.[6] In some patients, assigned to type 5, progressive neurologic deterioration is the most prominent manifestation.[13] Neurologic manifestations in those individuals include loss of speech, ataxia and progressive paraparesis, and macular cherry-red spots.

Zhou et al. have observed patients from three families who were affected by spinal muscular atrophy associated with progressive myoclonic epilepsy (SMA-PME).[14] In the affected children of two families a homozygous missense mutation (c.125C>T [p.Thr42Met]) in exon 2 of ASAH1 has been identified, and the same mutation associated with a deletion of the whole gene in the third family. Expression of the mutant cDNA in Farber fibroblasts generated only 32% of ceramidase activity of that obtained with normal cDNA.[14] In addition, the authors could demonstrate that Morpholino knockdown of the ASAH1 ortholog in zebrafish resulted in a marked loss of motor neuron axonal branching. These observations again reveal a wide clinical spectrum associated with ASAH1 mutations; and it can be speculated that a very low ceramidase activity is responsible for classical Farber disease, and that a higher residual activity may result in spinal muscular atrophy associated with progressive myoclonic epilepsy.

In 1989, Fusch et al. described a patient who in addition to Farber disease also was affected by Sandhoff disease.[15] Subsequent genetic studies indicated that this represented a coincidental combination of two distinct disease entities. Years later it had been detected that in the same family a sib suffered from Sandhoff alone; in a fetus a ceramidase deficiency was found.[16] These cases should not be called type 6 Farber disease anymore. The same applies to a single patient who was classified as type 7, in whom a deficiency of the sphingolipid-activator protein precursor, prosaposin, had been detected.[17] Activities of the lysosomal enzymes acid ceramidase, glucoceramidase, and galactoceramidase were diminished. Clinical manifestations resembled those of Gaucher disease type II, and there was no arthropathy, and subcutaneous nodules were not present.[18] Additional patients with prosaposin deficiency have been described recently.[19–21]

DIAGNOSIS

Clinical diagnosis of classic type 1 FD can be easily made because the triad of subcutaneous nodules, arthropathy, and laryngeal involvement appears to be unique for FD. Diagnostic challenges arise when one or more of these features are missing or mild; in the case described by Al Jasmi et al., for example, subcutaneous nodules were not observed.[12] Joint and/or skin manifestations that resemble FD may occur in juvenile rheumatoid arthritis, multicentric histiocytosis, and fibromatosis hyalin multiplex juvenilis. Acid ceramidase activity is normal in these disorders. Patients with the neonatal forms of FD do not show the classic clinical triad, and diagnosis depends on demonstration of reduced acid ceramidase activity. Other patients in whom the clinical manifestations were dominated by hepatosplenomegaly or by psychomotor deterioration, and in whom joint and skin abnormalities were mild, have been misdiagnosed as having histiocytosis or Tay–Sachs disease.

Deficiency of Acid Ceramidase

In vitro acid ceramidase activity in FD patients is usually $<6\%$ of control values. The test can be applied to white blood cells,[22] cultured skin fibroblasts,[23] amniocytes,[24] and also to plasma[22] and postmortem tissues.[3] *N*-lauroylsphingosine is the preferred substrate.[22] Bedia et al. have synthesized the new fluorogenic substrate Rbm14-12 that allows for measuring ceramidase activity in fibroblasts and lymphoid cell lines using 96-well plates.[25]

Accumulation of Ceramide in Plasma and Urine

Increased levels of ceramide are demonstrable in subcutaneous nodules. Ceramide levels were also increased in the urine of one patient,[26] but this increase was not demonstrable in the urine of four other patients.[6]

Studies of Ceramide Turnover

Impaired degradation of ceramides in FD can also be demonstrated by loading studies with labeled precursors, such as ^{14}C-labeled cerebroside sulfate. Abnormal retention of the ceramide fraction formed has been demonstrated in FD cultured skin fibroblasts,[27,28] but results in FD lymphocytes did not differ from normal.[28] Use of low-density lipoprotein (LDL)-associated ^{3}H sphingomyelin demonstrates impaired degradation of lysosomal ceramides in both FD cultured skin fibroblasts and transformed lymphocytes, and the degree of impairment correlates to some extent with the clinical disease severity.[29] Long-term studies with ^{14}C serine, a substrate for a committed step in the *de novo* synthesis of ceramide and complex sphingolipids,[30] also demonstrated a delayed catabolism of ceramide for FD fibroblasts.

Morphologic Studies

Biopsies of subcutaneous nodules demonstrate granulomas and the presence of lipid cytoplasmic inclusions that are periodic acid–Schiff (PAS)-positive and are extracted by lipid solvents. Under the electron microscope, they have a characteristic curvilinear tubular structure and are referred to as Farber bodies or banana bodies (Figure 35.2).[31]

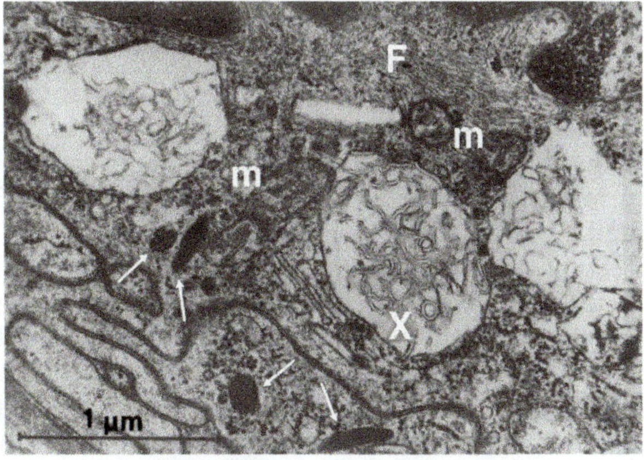

FIGURE 35.2 **Farber disease.** Thin section of an endothelial cell with filaments (F), Wiebel-Palade bodies (arrows), mitochondria (m), and three vacuoles, one of which (X) contains Farber bodies. Reduced from 45,000×. (From[31]).

Identification of Heterozygotes

All of the obligate heterozygotes tested so far had reduced acid ceramidase activity in white blood cells or cultured skin fibroblasts.[6,23] Mutation analysis is used in families in which the molecular defect has been defined.

Prenatal Diagnosis

Prenatal diagnosis has been achieved by measurement of acid ceramidase activity in cultured amniocytes[24] or loading studies in either cultured amniocytes or chorionic villus samples.[16]

PATHOLOGY

Histopathology

Light microscopy shows granulomatous infiltrations in the subcutaneous tissues and joints. The earliest lesions appear to be the accumulation of macrophages or histiocytes. Foam cells are prominent. These lesions have also been found in the larynx, lungs, heart valves, lymph nodes, intestine, liver, spleen, gall bladder, tongue, and thymus.[6] The nervous system is also involved in most cases. The main abnormality is the accumulation of storage material in neuronal cytoplasm. The accumulation is most prominent in the anterior horn cells in the spinal cord, but large cells (nerve cells of the brainstem nuclei, basal ganglia, and cerebellum; retinal ganglion cells; autonomic ganglia; Schwann cells; and, to a lesser extent, cortical neurons) are also involved. The storage material is PAS-positive and is extracted with lipid solvents.[10] Ultrastructural studies show characteristic curvilinear inclusions (Figure 35.2).[31]

Chemical Pathology

Abnormally high ceramide levels, as high as 60-fold over normal, have been found in all FD patients in whom the levels of this lipid have been analyzed. In the subcutaneous nodule, they may make up 20% of total lipids. They are also increased in the kidney. For other tissues, the extent of ceramide excess appears to vary with the severity of the disease. Severely affected patients showed high ceramide levels in the liver, lungs, and brain,[10] whereas mildly affected patients had normal ceramide levels in these tissues.[32] Unlike those of normal subjects, the ceramides of patients with FD may contain significant proportions of 2-hydroxy fatty acids.[10] Levels of gangliosides may also be increased.[10]

Enzyme Deficiency

Acid ceramidase (EC 3.5.1.23), the enzyme that is deficient in FD[3], catalyzes the hydrolysis of ceramide:

$$\text{Ceramide} + H_2O \rightarrow \text{Sphingosine} + \text{Fatty acid}$$

at approximately pH 4.5 and the reverse reaction in a detergent-free assay at approximately pH 5.5.[33]

Acid ceramidase has been purified.[34] It is a heterodimeric glycoprotein with a molecular weight of approximately 50 kDa. It is composed of two subunits. The α subunit has an approximate molecular weight of 13 kDa. The β unit (molecular weight approximately 40 kDa) contains five or six *N*-linked oligosaccharide units, whereas the α unit is not glycosylated. Both subunits arise from a single chain precursor of approximately 55 kDa. Proteolytic processing of the precursor takes place in either the late endosomal and/or lysosomal compartments, and preliminary studies indicate that targeting of the enzyme depends on the mannose-6-phosphate pathway.[35] A complete deficiency of acid ceramidase in mice is embryonic lethal prior to E8.5. Acid ceramidase is essential for the early embryonic survival[36] and improves the outcome of *in vitro* fertilization.[37] Degradation of membrane-bound ceramide by acid ceramidase requires the presence of sphingolipid-activator proteins (SAPs), with Sap-D and Sap-C being most active.[38] As already noted, the formerly called type 7 of Farber disease is due to a genetic defect in the synthesis of the prosaposin precursor. Negatively charged lipids, particularly bis(monoacylglycero)phosphate (BMP), are also required in the ceramide-carrying membranes.[38] It is proposed that the degradation of ceramide by acid ceramidase takes place on luminal lysosomal vesicles that contain the substrate, acid ceramidase, Sap-C, Sap-D, and BMP. Several neutral and alkaline ceramidases have been identified. The activities of these enzymes are not deficient in FD.[6]

Pathogenesis

The ceramide that accumulates in FD is located in the lysosome. Turnover studies show that this ceramide results from the impaired capacity to hydrolyze the ceramide generated during the degradation of complex sphingolipids. The rate of ceramide synthesis is the same as in normal cells, and newly synthesized ceramides in FD fibroblasts are directed to the synthesis of complex sphingolipids as they are in normal cells.[30] The distribution of lesions in FD and the variability of the expression of the disease can only be partially explained. Neuronal storage is not unexpected because ceramide metabolism in brain is known to be active. The striking involvement of subcutaneous tissues may be accounted for by the fact that rather hydrophobic ceramides with fatty acyl chains up to 36 carbon atoms have an important role in normal skin. Ceramides form an essential part of the skin barrier that preserves water impermeability in normal skin.[39–42] Ceramide accumulation appears to be the cause of the granuloma formation. Granulomatous lesions that resemble those in FD were produced by the subcutaneous injection of ceramide in rats.[6] It is of interest that bone marrow transplantation leads to regression of the subcutaneous nodules, possibly by removing this stimulus of granuloma formation.[43]

Ceramide appears to play a critical, but not fully understood, role in many aspects of cell biology,[44,45] such as apoptosis, response to stress, and expression of cytokines. This led to inquiry about whether the ceramide that accumulates in FD could affect these critical processes, but at this time, these roles of ceramide are uncertain. Apoptosis was not demonstrable in FD fibroblasts in spite of the increased ceramide level.[46] This appears to be a consequence of compartmentalization of the ceramide. Biomodulatory actions appear to be exerted by those ceramides located in the inner leaflets of the plasma membrane and at the cell surface in caveolae.[47] The ceramide that accumulates in FD appears to be confined to the lysosome[48] and does not exert these biomodulatory effects. However, investigations in a mouse model disrupting the Asah1 gene led to embryonic death. It was hypothesized that the total loss of acid ceramidase activity resulted in increased pools of ceramide and enhanced propensity to cell death.[49]

CLINICAL GENETICS

The mode of inheritance of FD is autosomal recessive. The prevalence of FD is unknown. The fact that until now only about 80 patients with Farber disease are referred in the literature suggests that it is a very rare condition.[50]

MOLECULAR GENETICS

The human acid ceramidase gene spans about 30 kb and contains a total of 14 exons. It has been mapped to chromosomal region 8p21.3/22.[5] The full-length cDNA contains an 1185-bp open reading frame that encodes 395 amino acids, an 1110-bp 3′ untranslated sequence, and an 18-bp poly (A) tail. To date, 23 different pathogenic mutations, mostly missense mutations, have been described.[50] In addition, a large deletion in the ASAH1 gene has been detected in a severe neonatal form of Farber disease.[50]

ANIMAL MODELS

Using homologous recombination-mediated insertional mutagenesis of the acid ceramidase gene (Asah1), Li et al. developed a mouse model.[49] Embryos homozygous for this deletion died prior to E8.5. A reduction of acid ceramidase activity in heterozygous mice led to elevated ceramide levels and evidence of lipid storage disease. Correlation of these findings with the human disease awaits further studies. A recent murine model of Farber disease with a homozygous Asah1 (P361/P361R) mutation showed acid ceramidase deficiency, accumulated ceramide and a reduced lifespan of 7–13 weeks.[51]

THERAPY

Bone marrow transplant has led to regression of subcutaneous nodules and arthropathy but did not alter neurologic deterioration.[43] Successful hematopoietic stem cell transplantation has been performed in four children who did not have neurologic involvement.[52] The granulomas regressed and mobility improved. Gene therapy may become possible in the future. Introduction of cDNA of human acid ceramidase into FD disease fibroblasts with a retroviral vector restored enzyme activity completely and normalized ceramide levels and lysosomal turnover.[53] Introduction of a single-nucleotide mutation of FD patients into the murine Asah1 gene generated a model of ceramidase deficiency, accumulation of ceramide and reduced lifespan.[51] Treatment of neonates with human ceramidase encoding lentivector decreased ceramide levels and diminished severity of the disease.

CONCLUSION AND FUTURE DIRECTIONS

Farber disease (MIM 22800) is a genetically determined disorder of lipid metabolism associated with the deficiency of lysosomal acid ceramidase and accumulation of ceramide in the lysosome. The disorder presents most commonly during the first few months of life with a unique triad of symptoms: painful and progressively deformed joints, subcutaneous nodules, and progressive laryngeal involvement leading to hoarseness and respiratory impairment. The involved tissues show granulomas and lipid-laden macrophages with characteristic inclusions. Progressive involvement of lungs, heart valves, liver, spleen, peripheral nerves, and brain follow and often lead to death during the first few years of life. Other phenotypes occur, including neonatal, adolescent, and adult forms, in which the nervous system may be spared, and a form in which progressive neurologic deterioration is the main clinical feature. Acid ceramidase (EC 3.5.1.23) has been purified and cloned. The full-length DNA contains an 1185-bp open reading frame. So far, more than 20 different mutations have been identified in FD patients. The ceramide that accumulates is confined to the lysosome and does not appear to contribute to the multiple biomodulatory roles attributed to ceramides in other compartments.

Laboratory diagnosis is achieved by demonstrating reduced acid ceramidase activity in white blood cells, cultured skin fibroblasts, and amniocytes. Prenatal diagnosis is possible. The disease is rare. Data on more than 80 patients in a variety of ethnic groups have been assembled. The mode of inheritance is autosomal recessive. Rarely, ceramide accumulation may also be caused by a deficiency of a sphingolipid-activator protein (prosaposin). There is no effective therapy. Bone marrow transplant can lead to regression of joint manifestations and subcutaneous nodules and relieves the hoarseness, but it does not alter progressive neurologic deterioration. It should be considered for patients without neurologic involvement.

Future directions include the development of additional authentic animal models of acid ceramidase deficiency, purification of the enzyme in sufficient quantity to test the possibility of enzyme replacement therapy, and other therapeutic approaches such as stem cell and gene therapy.

References

1. Farber S, Cohen J, Uzman LL. Lipogranulomatosis; a new lipo-glycoprotein storage disease. *J Mt Sinai Hosp N Y*. 1957;24(6):816–837.
2. Prensky AL, Ferreira G, Carr S, Moser HW. Ceramide and ganglioside accumulation in Farber's lipogranulamatosis. *Proc Soc Exp Biol Med*. 1967;126(3):725–728.
3. Sugita M, Dulaney JT, Moser HW. Ceramidase deficiency in Farber's disease (lipogranulomatosis). *Science*. 1972;178(4065):1100–1102.
4. Koch J, Gärtner S, Li CM, et al. Molecular cloning and characterization of a full-length complementary DNA encoding human acid ceramidase. Identification of the first molecular lesion causing Farber disease. *J Biol Chem*. 1996;271(51):33110–33115.
5. Li CM, Park JH, He X, et al. The human acid ceramidase gene (ASAH): structure, chromosomal location, mutation analysis, and expression. *Genomics*. 1999;62(2):223–231.
6. Moser H, Linke T, Femson A, Levade T, Sandhoff K. Acid ceramidase deficiency: Farber lipogranulomatosis. In: *The Metabolic and Molecular Bases of Inherited Disease*. 8th ed; 2001:3573–3589, chapter 143.
7. Kattner E, Schafer A, Harzer K. Hydrops fetalis: manifestation in lysosomal storage diseases including Farber disease. *Eur J Pediatr*. 1997;156(4):292–295.
8. van Lijnschoten G, Groener JE, Maas SM, Ben-Yoseph Y, Dingemans KP, Offerhaus GJ. Intrauterine fetal death due to Farber disease: case report. *Pediatr Dev Pathol*. 2000;3(6):597–602.
9. Antonarakis SE, Valle D, Moser HW, Moser A, Qualman SJ, Zinkham WH. Phenotypic variability in siblings with Farber disease. *J Pediatr*. 1984;104(3):406–409.
10. Moser HW, Prensky AL, Wolfe HJ, Rosman NP. Farber's lipogranulomatosis. Report of a case and demonstration of an excess of free ceramide and ganglioside. *Am J Med*. 1969;47(6):869–890.
11. Cogan DG, Kuwabara T, Moser H, Hazard GW. Retinopathy in a case of Farber's lipogranulomatosis. *Arch Ophthalmol*. 1966;75(6):752–757.
12. Al Jasmi F. A novel mutation in an atypical presentation of the rare infantile Farber disease. *Brain Dev*. 2012;34(6):533–535.
13. Eviatar L, Sklower SL, Wisniewski K, Feldman RS, Gochoco A. Farber lipogranulomatosis: an unusual presentation in a black child. *Pediatr Neurol*. 1986;2(6):371–374.
14. Zhou J, Tawk M, Tiziano FD, et al. Spinal muscular atrophy associated with progressive myoclonic epilepsy is caused by mutations in ASAH1. *Am J Hum Genet*. 2012;91(1):5–14.
15. Fusch C, Huenges R, Moser HW, et al. A case of combined Farber and Sandhoff disease. *Eur J Pediatr*. 1989;148(6):558–562.

16. Levade T, Enders H, Schliephacke M, Harzer K. A family with combined Farber and Sandhoff, isolated Sandhoff and isolated fetal Farber disease: postnatal exclusion and prenatal diagnosis of Farber disease using lipid loading tests on intact cultured cells. *Eur J Pediatr*. 1995;154(8):643–648.
17. Schnabel D, Schröder M, Fürst W, et al. Simultaneous deficiency of sphingolipid activator proteins 1 and 2 is caused by a mutation in the initiation codon of their common gene. *J Biol Chem*. 1992;267(5):3312–3315.
18. Bradova V, Smid F, Ulrich-Bott B, Roggendorf W, Paton BC, Harzer K. Prosaposin deficiency: further characterization of the sphingolipid activator protein-deficient sibs. Multiple glycolipid elevations (including lactosylceramidosis), partial enzyme deficiencies and ultrastructure of the skin in this generalized sphingolipid storage disease. *Hum Genet*. 1993;92(2):143–152.
19. Elleder M, Jerabkova M, Befekadu A, et al. Prosaposin deficiency—a rarely diagnosed, rapidly progressing, neonatal neurovisceral lipid storage disease. Report of a further patient. *Neuropediatrics*. 2005;36(3):171–180.
20. Kuchar L, Ledvinová J, Hrebícek M, et al. Prosaposin deficiency and saposin B deficiency (activator-deficient metachromatic leukodystrophy): Report on two patients detected by analysis of urinary sphingolipids and carrying novel PSAP gene mutations. *Am J Med Genet A*. 2009;149A(4):613–621.
21. Sikora J, Harzer K, Elleder M. Neurolysosomal pathology in human prosaposin deficiency suggests essential neurotrophic function of prosaposin. *Acta Neuropathol*. 2007;113(2):163–175.
22. Ben-Yoseph Y, Gagne R, Parvathy MR, Mitchell DA, Momoi T. Leukocyte and plasma N-laurylsphingosine deacylase (ceramidase) in Farber disease. *Clin Genet*. 1989;36(1):38–42.
23. Dulaney JT, Milunsky A, Sidbury JB, Hobolth N, Moser HW. Diagnosis of lipogranulomatosis (Farber disease) by use of cultured fibroblasts. *J Pediatr*. 1976;89(1):59–61.
24. Fensom AH, Benson PF, Neville BR, Moser HW, Moser AE, Dulaney JT. Prenatal diagnosis of Farber's disease. *Lancet*. 1979;2(8150):990–992.
25. Bedia C, Camacho L, Abad JL, Fabrias G, Levade T. A simple fluorogenic method for determination of acid ceramidase activity and diagnosis of Farber disease. *J Lipid Res*. 2010;51(12):3542–3547.
26. Sugita M, Iwamori M, Evans J, McCluer RH, Dulaney JT, Moser HW. High performance liquid chromatography of ceramides: application to analysis in human tissues and demonstration of ceramide excess in Farber's disease. *J Lipid Res*. 1974;15(3):223–226.
27. Kudoh T, Wenger DA. Diagnosis of metachromatic leukodystrophy, Krabbe disease, and Farber disease after uptake of fatty acid-labeled cerebroside sulfate into cultured skin fibroblasts. *J Clin Invest*. 1982;70(1):89–97.
28. Levade T, Tempesta MC, Moser HW, et al. Sulfatide and sphingomyelin loading of living cells as tools for the study of ceramide turnover by lysosomal ceramidase–implications for the diagnosis of Farber disease. *Biochem Mol Med*. 1995;54(2):117–125.
29. Levade T, Moser HW, Fensom AH, Harzer K, Moser AB, Salvayre R. Neurodegenerative course in ceramidase deficiency (Farber disease) correlates with the residual lysosomal ceramide turnover in cultured living patient cells. *J Neurol Sci*. 1995;134(1–2):108–114.
30. van Echten-Deckert G, Klein A, Linke T, Heinemann T, Weisgerber J, Sandhoff K. Turnover of endogenous ceramide in cultured normal and Farber fibroblasts. *J Lipid Res*. 1997;38(12):2569–2579.
31. Schmoeckel C, Hohlfed M. A specific ultrastructural marker for disseminated lipogranulomatosis (Farber). *Arch Dermatol Res*. 1979;266(2):187–196.
32. Samuelsson K, Zetterstrom R. Ceramides in a patient with lipogranulomatosis (Farber's disease) with chronic course. *Scand J Clin Lab Investig*. 1971;27(4):393–405.
33. Okino N, He X, Gatt S, Sandhoff K, Ito M, Schuchman EH. The reverse activity of human acid ceramidase. *J Biol Chem*. 2003;278(32):29948–29953.
34. Bernardo K, Hurwitz R, Zenk T, et al. Purification, characterization, and biosynthesis of human acid ceramidase. *J Biol Chem*. 1995;270(19):11098–11102.
35. Ferlinz K, Kopal G, Bernardo K, et al. Human acid ceramidase: processing, glycosylation, and lysosomal targeting. *J Biol Chem*. 2001;276(38):35352–35360.
36. Eliyahu E, Park JH, Shtraizent N, He X, Schuchman EH. Acid ceramidase is a novel factor required for early embryo survival. *FASEB J*. 2007;21(7):1403–1409.
37. Eliyahu E, Shtraizent N, Martinuzzi K, et al. Acid ceramidase improves the quality of oocytes and embryos and the outcome of *in vitro* fertilization. *FASEB J*. 2010;24(4):1229–1238.
38. Linke T, Wilkening G, Sadeghlar F, et al. Interfacial regulation of acid ceramidase activity. Stimulation of ceramide degradation by lysosomal lipids and sphingolipid activator proteins. *J Biol Chem*. 2001;276(8):5760–5768.
39. Doering T, Holleran WM, Potratz A, et al. Sphingolipid activator proteins are required for epidermal permeability barrier formation. *J Biol Chem*. 1999;274(16):11038–11045.
40. Jennemann R, Rabionet M, Gorgas K, et al. Loss of ceramide synthase 3 causes lethal skin barrier disruption. *Hum Mol Genet*. 2012;21(3):586–608.
41. Sandhoff R. Very long chain sphingolipids: tissue expression, function and synthesis. *FEBS Lett*. 2010;584(9):1907–1913.
42. Breiden B, Sandhoff K. The important role of lipids in the epidermis and their role in the formation and maintenance of the cutaneous barrier. *Biochim Biophys Acta*. 2014;1841(3):441–452.
43. Yeager AM, Uhas KA, Coles CD, Davis PC, Krause WL, Moser HW. Bone marrow transplantation for infantile ceramidase deficiency (Farber disease). *Bone Marrow Transplant*. 2000;26(3):357–363.
44. Perry DK, Hannun YA. The role of ceramide in cell signaling. *Biochim Biophys Acta*. 1998;1436(1–2):233–243.
45. Hannun YA, Obeid LM. Many ceramides. *J Biol Chem*. 2011;286(32):27855–27862.
46. Tohyama J, Oya Y, Ezoe T, et al. Ceramide accumulation is associated with increased apoptotic cell death in cultured fibroblasts of sphingolipid activator protein-deficient mouse but not in fibroblasts of patients with Farber disease. *J Inherit Metab Dis*. 1999;22(5):649–662.
47. Liu P, Anderson RG. Compartmentalized production of ceramide at the cell surface. *J Biol Chem*. 1995;270(45):27179–27185.
48. Chatelut M, Leruth M, Harzer K, et al. Natural ceramide is unable to escape the lysosome, in contrast to a fluorescent analogue. *FEBS Lett*. 1998;426(1):102–106.
49. Li CM, Park JH, Simonaro CM, et al. Insertional mutagenesis of the mouse acid ceramidase gene leads to early embryonic lethality in homozygotes and progressive lipid storage disease in heterozygotes. *Genomics*. 2002;79(2):218–224.

50. Alves MQ, Le Trionnaire E, Ribeiro I, et al. Molecular basis of acid ceramidase deficiency in a neonatal form of Farber disease: identification of the first large deletion in ASAH1 gene. *Mol Genet Metab*. 2013;109(3):276–281.
51. Alayoubi AM, Wang JC, Au BC, et al. Systemic ceramide accumulation leads to severe and varied pathological consequences. *EMBO Mol Med*. 2013;5(6):827–842.
52. Ehlert K, Frosch M, Fehse N, Zander A, Roth J, Vormoor J. Farber disease: clinical presentation, pathogenesis and a new approach to treatment. *Pediatr Rheumatol Online J*. 2007;5:15.
53. Medin JA, Takenaka T, Carpentier S, et al. Retrovirus-mediated correction of the metabolic defect in cultured Farber disease cells. *Hum Gene Ther*. 1999;10(8):1321–1329.

CHAPTER

36

Wolman Disease

Isaac Marin-Valencia and Juan M. Pascual

The University of Texas Southwestern Medical Center, Dallas, TX, USA

CLINICAL FEATURES

Wolman disease (WD) and cholesteryl ester storage disease (CESD) embody two distinct clinical phenotypes resulting from either complete or partial deficiency of lysosomal acid lipase (LAL, EC 3.1.1.13), respectively.[1] WD represents the severe form of the spectrum and is characterized by massive lysosomal accumulation of cholesteryl esters and triglycerides, particularly in the liver. This condition was first described in 1956 in a 2-month-old girl presenting with vomiting, distended abdomen, hepatosplenomegaly, and yellowish complexion.[2] Abdominal X-rays showed bilateral calcifications of triangular shape in both adrenal glands, which is a hallmark of the disease. The patient died 3 days after presentation. Since then, approximately 100 cases of WD have been reported. The estimated incidence of WD worldwide is 1/350,000 newborns, although it occurs at higher frequency in Iranian Jews residing in the Los Angeles area,[3] in which the incidence rises to as many as 1/4200 newborns. Overall, children with WD appear normal at birth, but after several weeks or months of life develop severe emesis, steatorrhea, abdominal distension, jaundice, and failure to thrive. Clinical signs include hepatosplenomegaly, anemia, and adrenal gland enlargement with calcification. Although neurological abnormalities are not an invariant component of the disease, lipid inclusions have been identified in the central nervous system (neurons, glia, capillary endothelium, choroid plexus, retinal ganglion cells)[4–7] and in the peripheral nervous system (perineurium, endoneurium, Schwann cells, and sympathetic chain neurons).[7] These children may manifest impaired brain myelination and associated developmental delay.[8] Without treatment, patients rapidly deteriorate and succumb during the first year of life.[9,10] In contrast, CESD patients may present at any age from childhood to early adulthood with hepatic fibrosis, cirrhosis, hyperlipidemia, and accelerated atherosclerosis. The estimated incidence is 2.5/100,000 newborns.[11] Liver disease and cardiovascular failure are the most common causes of death. Because CESD is rare, it may be overlooked or misdiagnosed with nonalcoholic fatty liver disease, as both entities share similar clinical and radiological features.[1]

MOLECULAR GENETICS

Wolman disease and CESD are inherited in autosomal recessive fashion resulting from mutations in the lipase A, lysosomal acid, cholesterol esterase (*LIPA*) gene. The *LIPA* gene maps to chromosome 10q23.2-q23.3, comprises 10 exons, and is approximately 36.5 kb in size.[1] The encoded protein, LAL, varies in size due to alternative cotranslational cleavage sites. For example, human fibroblasts express two different size LAL forms (41 and 49 kDa),[12] which differ from those present in the human liver (41 and 56 kDa).[13]

An extensive variety of mutations in the *LIPA* gene has been associated with LAL deficiency. Some mutations are more commonly associated with WD than with CESD, and *vice versa*, and they generally correlate with residual LAL activity. For instance, deletions, insertions, and nonsense mutations are associated with very low or absent LAL activity, which is typically encountered in patients with WD.[1] On the other hand, missense mutations usually correlate with higher residual enzyme activity (5–10%), typical of CESD patients. The most common mutation carried

Rosenberg's Molecular and Genetic Basis of Neurological and Psychiatric Disease
http://dx.doi.org/10.1016/B978-0-12-410529-4.00036-X

by CESD patients is c.894G>A in exon 8, constituting approximately 54% of the mutant alleles in these individuals.[14] Another reported genetic difference between both phenotypes resides on mutations affecting differentially exon splice variants.[15] For instance, two children with WD carried a homozygous G->A mutation at position +1 of exon 8, preventing the correct splicing of mRNA and therefore skipping exon 8. In contrast, in a patient with CESD, a G->A mutation at position −1 of the exon 8 splice donor site allowed some residual correct splicing (3%), yielding approximately 3% of normal residual enzyme activity.

The genotype–phenotype correlation in WD and CESD is not absolute. For example, patients with nonsense mutations have been reported to manifest either phenotype.[16] This is the case with mutations T22X and D124X linked to WD,[17,18] and R44X associated with CESD,[16] in which complete LAL deficiency is found.[19] A mouse model of LAL deficiency supports this discrepancy:[20] Homozygous null (knockout) mice did not express LAL mRNA, LAL protein, or any residual enzyme activity; yet, they manifested normal development to adulthood, associated with significant accumulation of triglycerides and cholesteryl esters in several organs, such as liver and spleen. Heterozygous mice exhibiting approximately a 50% reduction of enzyme activity did not show lipid accumulation. These findings suggest that additional mechanisms besides LAL deficiency may play a role in the severity of the clinical phenotype.

DISEASE MECHANISMS

Lysosomal acid lipase is an enzyme involved in the intracellular hydrolysis of triglycerides and cholesteryl esters derived from plasma lipoprotein particles (Figure 36.1). Defects of LAL result in failure of hydrolysis and the subsequent accumulation of triglycerides and cholesteryl esters. The residual LAL activity appears to directly depend upon protein conformational changes that result from the underlying mutation(s), such that the more severe structural alterations cause the more severe phenotypes.[21] When residual LAL activity diminishes below 5% of normal, the rate of cholesterol release from the lysosome declines below a critical level, impairing the formation of high-density lipoprotein (HDL) mediated by ATP-binding cassette transporter A1 (ABCA1), leading to hypoalphalipoproteinemia.[1,22] Moreover, it has been postulated that a limited oxidation of low-density lipoproteins (LDL) may be a disease mechanism in WD.[23] In this regard, cultured adrenal cells from WD patients exhibited a higher uptake of mildly oxidized LDL through the LDL-receptor pathway, which is defectively downregulated as a result of limited lysosomal degradation of LDL-cholesteryl esters. The presence of oxidized LDL in adrenal cells leads to a sustained rise in intracellular Ca^{+2}, causing cell damage and death.[23,24] The deposition of calcium in damaged and dead cells in cultured adrenal cells may explain the typical calcifications of the adrenal cortex in WD. Some of these findings have been corroborated in a knockout mouse model of LAL deficiency.[25] Homozygous animals appeared normal at birth, but manifested a significant accumulation of triglyceride and cholesteryl esters in the liver, adrenal glands and small bowel, succumbing approximately at 7–9 months of age.

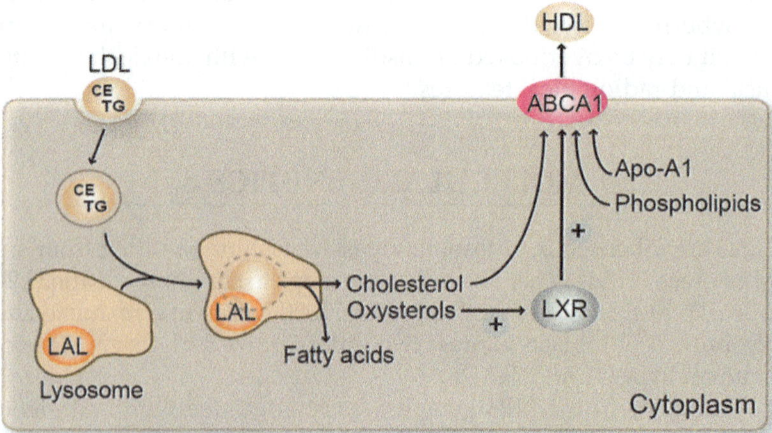

FIGURE 36.1 **Scheme of low-density lipoprotein (LDL) and high-density lipoprotein (HDL) metabolism.** LDL particles are endocytosed by the cell via LDL receptor and metabolized in the lysosome. Lysosomal acid lipase (LAL) hydrolyzes cholesteryl esters (CE) and triglycerides (TG) with the subsequent release of cholesterol, oxysterols and fatty acids. Oxysterols interact with liver X-receptors (LXR), which regulates the activity of ATP-binding cassette transporter 1 (ABCA1) and therefore the formation of HDL particles. (Modified from Figure 1 of Reynolds, 2013.)[1]

DIFFERENTIAL DIAGNOSIS

The differential diagnosis of WD can be extensive on the basis of nonspecific clinical symptoms at presentation. When a child presents with vomiting, diarrhea and hepatosplenomegaly, common conditions such as sepsis, viral infections, intestinal or biliary tract obstructions, and trauma should be ruled out first.[26] The presence of adrenal calcifications in the acute setting without a history of remote adrenal hemorrhage should raise suspicion of WD. Other inborn errors of metabolism (IEM) should be considered if there are additional signs (abnormal skin/hair pigmentation, rash, coarse facial features), family history of metabolic disorders, consanguinity, or if laboratory results are consistent with a metabolic disease. Within this category, lysosomal storage disorders comprise an important component in the differential diagnosis of WD. Niemann–Pick disease encompasses a group of inherited disorders of sphingomyelin metabolism characterized by its accumulation in the liver, spleen and brain. Niemann–Pick type C disease (NPC) resembles WD in that it presents with hepatosplenomegaly, jaundice, and ascites during the first few weeks/months of life or with congenital hydrops fetalis.[10,26] In contrast to WD, severe vomiting and diarrhea are not common in NPC. NPC, in contrast to WD, causes progressive neurological deterioration responsible for the disability and early death of all patients. Other lysosomal storage disorders that present with hepatosplenomegaly and liver dysfunction include mucopolysaccharidosis, sialidosis and galactosialidosis, and Gaucher disease. In these diseases, neurological symptoms are often part of the phenotype, vomiting and diarrhea are usually absent, and most of them (except Gaucher disease) manifest coarse facial features. Specific biochemical testing such as enzyme assay, urine analysis (mucopolysaccharides), and genetic testing are required to confirm the diagnosis. Galactosemia may also resemble WD.[27] In this condition, patients present with vomiting, diarrhea, jaundice, liver dysfunction and poor weight gain during the first few weeks of life and, eventually, develop cataracts. Liver and spleen enlargement can also be observed. Adrenal calcifications, however, are not common. When suspected, urine analysis for reducing substances should be pursued.

The differential diagnosis of CESD includes all conditions associated with nonalcoholic steatohepatitis (obesity, diabetes, hyperlipidemia, drug toxicity), nonalcoholic fatty liver disease (drug action, insulin resistance, iron accumulation, obesity, metabolic disorders, antioxidant deficiencies), hypertriglyceridemia (obesity, diabetes, hypothyroidism, nephrotic syndrome, drug effects, genetic causes), elevated LDL cholesterol (familial hypercholesterolemia, nephrotic syndrome, chronic kidney disease, hypothyroidism, obesity, drug action), and decreased HDL cholesterol (diabetes, chronic kidney disease, cigarette smoking, obesity, drugs).[1,28] Useful diagnostic clues in CESD are the detection of hepatomegaly or elevated transaminases disproportionate to the patient's body mass index (BMI), an enlarged spleen inconsistent with the degree of liver dysfunction, and a history of liver disease in childhood.

TESTING

The diagnosis of WD and CESD relies on clinical suspicion, enzyme assay and genetic testing. Routine blood work-up usually reveals elevated LDL cholesterol, triglycerides, and transaminases (particularly alanine transaminase [ALT]), and reduced HDL cholesterol.[29,30] Enzymatic assay of LAL in lymphocytes or fibroblasts establishes the diagnosis.[1,31–33] Prenatal diagnosis is also available by measuring LAL activity in chorionic villus cells.[34] In general, WD patients manifest a very low or absent LAL activity, in contrast to individuals with CESD, who usually demonstrate higher residual function. Yet, residual activity does not always correlate with phenotypic severity. For example, some patients with CESD manifest very low or absent LAL activity, similar to that encountered in subjects with WD.[14] This could be explained by potential technical limitations, from sample collection and handling to the assay itself. Alternatively, it is possible that LAL activity in peripheral cells (lymphoblasts, fibroblasts) may not correspond to the activity exhibited by other cells involved in cholesterol metabolism, such as hepatocytes.[14] The diagnosis is confirmed by mutation analysis of the *LIPA* gene, which is also used for carrier testing, genetic counseling and prenatal diagnosis. Sequencing and deletion/duplication analysis of the *LIPA* gene are commercially available.

Other diagnostic testing, such as imaging and histology, may provide additional diagnostic information. As a result of the accumulation of lipids in the liver and adrenal glands, the liver appears enlarged and hypodense on computed tomography (CT) and adrenals typically show calcifications (Figure 36.2). Ultrasound may reveal ascites, enlarged liver with normal echogenicity, thickening of bowel loops and adrenal calcifications.[9] Nonenhanced T1-weighted magnetic resonance imaging (MRI) demonstrates increased signal of the adrenal glands with subtle hypointense areas indicative of calcified patches. The spleen may appear hyperintense due to fat deposition.[35] Retroperitoneal lymphadenopathy has also been identified. In contrast to WD, these features are less commonly observed in patients with CESD.[1] Histopathology, on the other hand, provides valuable information to support the diagnosis in some

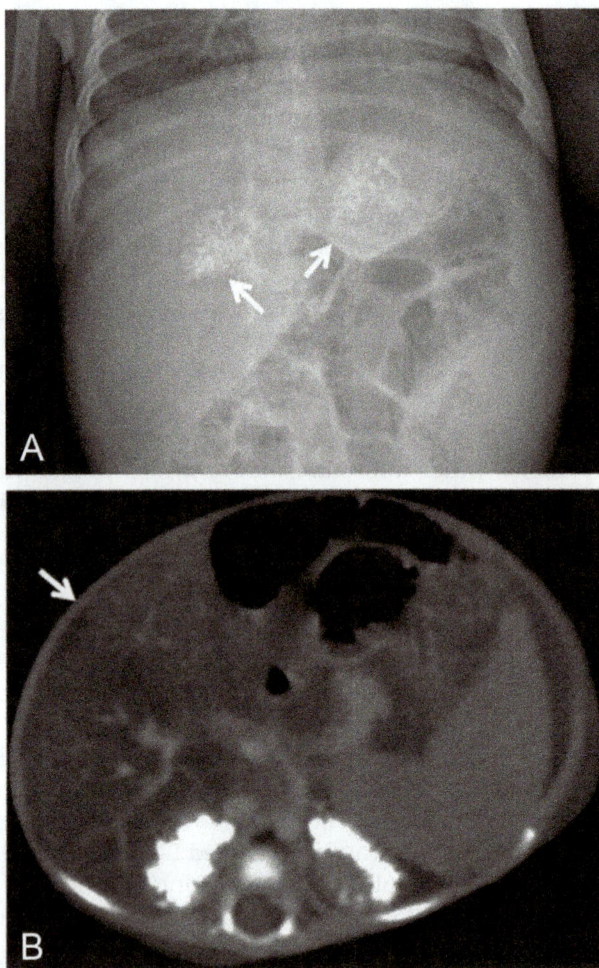

FIGURE 36.2 **Imaging findings in Wolman disease.** (**A**) Plain radiograph of the abdomen from a 5-month-old boy demonstrating bilateral calcification of adrenal glands (arrows). (**B**) Axial CT scan of the abdomen showing bilateral adrenal gland calcification and decreased density of the liver (arrow) consistent with fatty infiltration. (Reproduced with permission from[53].)

cases. Skin biopsies from WD patients demonstrate cytoplasmic accumulation of lipids, which are also present in liver, spleen, gut and adrenal specimens[36] (Figure 36.3). In some WD patients with severe gastrointestinal symptoms, intestinal biopsy has been essential to guide the diagnosis.[37,38] Intestinal villi appear significantly distorted due to infiltration of foam cells into the lamina propria. Patients with CESD manifest moderate fat accumulation in hepatocytes, adrenal cortical cells, bile duct epithelial cells, and Leydig cells.[39,40] The cholesterol content in the liver may exceed 200 times that of normal tissue.[41] Signs of hepatic fibrosis of the periportal region may also be present, which may eventually evolve to cirrhosis.[39]

THERAPEUTIC INTERVENTIONS

Therapeutic strategies in WD aim to enhance LAL activity in order to avert the abnormal accumulation of intracellular lipids. Hematopoietic cell transplantation (HCT), from unrelated umbilical cord blood[8,42] or bone marrow transplantation (BMT),[43] constitutes the only treatment that can prevent hepatic failure and death in WD patients. The rationale is based on the capability of donor peripheral leukocytes to secrete LAL, which can be endocytosed by host cells and used to hydrolyze triglycerides and cholesteryl esters.[42] Graft failure, previous liver injury and post-transplant liver veno-occlusive disease (sinusoidal obstruction syndrome) represent major limitations of this treatment.[44] The first successful long-term remission of a WD patient subjected to BMT was reported by Krivit et al.[43] This resulted in recovery of LAL activity in peripheral leukocytes, resolution of diarrhea, recovery of liver function, and normalization of cholesterol and triglyceride levels in blood. A mild developmental delay, however, remained after

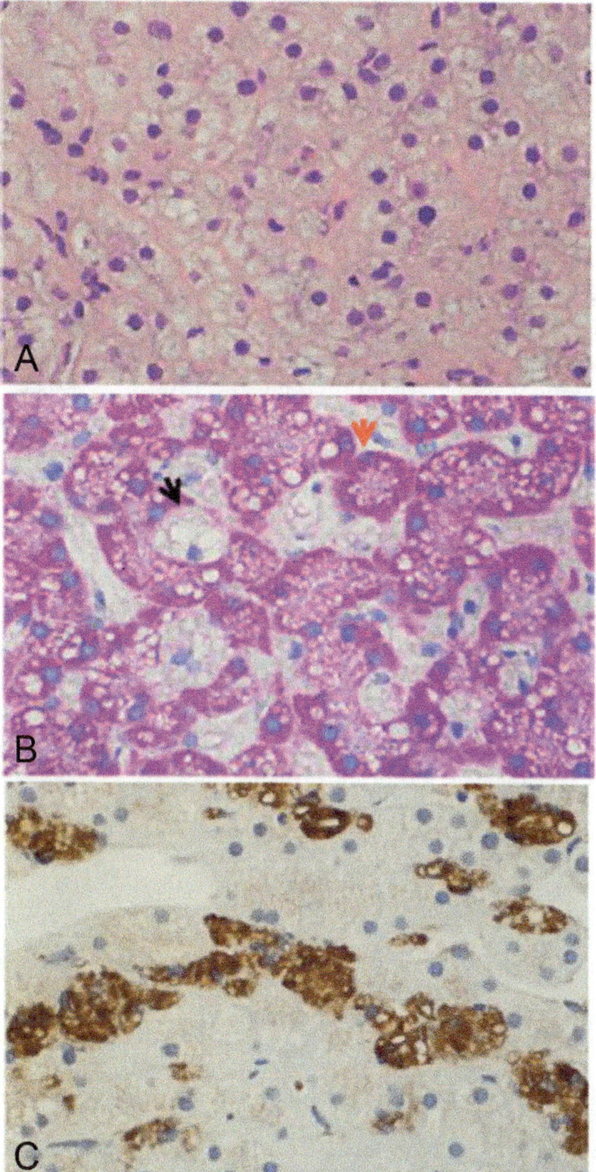

FIGURE 36.3 **Histopathology of Wolman disease.** The panels illustrate the typical liver pathology in an affected patient: microvesicular steatosis, hepatocytes with vacuolated cytoplasm and cholesterol deposits, and foamy histiocytes. (**A**) Hematoxylin and eosin staining illustrating a mixed pattern of hepatocytes and Kupffer cells (histiocytes). (**B**) Periodic acid–Schiff staining showing pale Kupffer cells with vacuolated cytoplasm (black arrow) and hepatocytes containing lipid vacuoles (red arrow). (**C**) CD68 immunohistochemical staining demonstrating Kupffer cells with lipid deposits and needle-shaped crystals of cholesterol. (Reproduced with permission from[42].)

the transplant. On the other hand, Tolar et al. reported the long-term outcomes of four WD infants after HCT. Two of them were long-term survivors, 4 and 11 years at the time of publication.[42] These patients manifested resolution of diarrhea, normalization of liver function, decreased hepatosplenomegaly and, in one patient, adrenal gland function was also restored. The two other children died from complications related to the transplant.

The first successful unrelated cord blood transplant was reported by Stein et al.[8] In this case, umbilical cord blood was chosen due to its immediate availability and reduced propensity for graft-versus-host disease. After the transplant, LAL activity recovered and remained normal for 4 years of follow-up. Liver and spleen normalized in size; the adrenal glands remained calcified, but did not show signs of insufficiency; and the patient continued to thrive in the low end of normal. Her motor and intellectual skills were appropriate for age.

Despite these successful cases of HCT, WD still remains fatal in most patients. Due to its rarity, available data regarding long-term response to transplant are very limited. Only a few reports have published the causes of failed

HCT,[42,44] which limits the design of better therapeutic interventions. Alternative therapeutic options are being explored along with the development of safer conditioning regimens that reduce hepatotoxicity and improve engraftment.[44] For example, enzyme replacement has been used to treat mice lacking LAL.[45,46] Injections of the enzyme decreased liver weight by 36% compared to phosphate-buffered saline solution (PBS)-treated lal(−/−) mice, normalized hepatic color, and reduced the liver accumulation of cholesterol and triglycerides.[45] Currently, several clinical trials with a direct enzyme replacement drug (Sebelipase alfa [SBC-102]) are underway.[47] Moreover, gene transplant targeting the restoration of LAL activity via adenoviral vectors have been tested in lal(−/−) mice and in human fibroblast cells.[48,49] Treated mice showed increased LAL activity in the liver, decreased hepatomegaly, and normalization of histolopathological changes.[48] After infection with adenovirus encoding LAL, fibroblasts from a WD patient manifested a dose-dependent increase in LAL expression and activity, along with a dose-dependent improvement of lipid storage.[49]

The therapeutic battery for CESD differs from that of WD. Treatment strategies are directed to ameliorate hyperlipidemia, hepatic fibrosis and accelerated atherosclerosis. In this regard, lovastatin, a competitive inhibitor of 3-hydroxy-3-methylglutaryl coenzyme A reductase involved in cholesterol synthesis, has been used to reduce serum cholesterol levels in CESD individuals. In two affected sisters, the administration of atorvastatin decreased significantly total serum cholesterol, LDL cholesterol and triglycerides, and increased HDL cholesterol.[50] CT scan showed a significant reduction in hepatic fat content. In one sibling, liver biopsy obtained 6 months after initiation of lovastatin showed 13% less esterified cholesterol compared to prior to treatment. A favorable response to atorvastatin, however, is not always the norm.[39] Treatment with other agents (or combinations of agents) such as simvastatin,[51] lovastatin and ezetimibe,[30] and simvastatin and cholestyramine,[52] have been shown to decrease total serum cholesterol and increase HDL cholesterol. These medications, however, are not always able to normalize hepatic lipid content, and thus the risk for liver failure and cirrhosis persists. Under these circumstances, liver transplantation has been carried out successfully.[52] Ongoing research is focused on devising more effective lipid-lowering drugs and developing novel ways to enhance LAL activity.

References

1. Reynolds T. Cholesteryl ester storage disease: a rare and possibly treatable cause of premature vascular disease and cirrhosis. *J Clin Pathol.* 2013;66:918–923.
2. Abramov A, Schorr S, Wolman M. Generalized xanthomatosis with calcified adrenals. *AMA J Dis Child.* 1956;91:282–286.
3. Valles-Ayoub Y, Esfandiarifard S, No D, et al. Wolman disease (LIPA p.G87V) genotype frequency in people of Iranian-Jewish ancestry. *Genet Test Mol Biomarkers.* 2011;15:395–398.
4. Guazzi GC, Martin JJ, Philippart M, et al. Wolman's disease. Distribution and significance of the central nervous system lesions. *Pathol Eur.* 1968;3:266–277.
5. Wolman M. Involvement of nervous tissue in primary familial xanthomatosis with adrenal calcification. *Pathol Eur.* 1968;3:259–265.
6. Kahana D, Berant M, Wolman M. Primary familial xanthomatosis with adrenal involvement (Wolman's disease). Report of a further case with nervous system involvement and pathogenetic considerations. *Pediatrics.* 1968;42:70–76.
7. Byrd 3rd. JC, Powers JM. Wolman's disease: ultrastructural evidence of lipid accumulation in central and peripheral nervous systems. *Acta Neuropathol (Berl).* 1979;45:37–42.
8. Stein J, Garty BZ, Dror Y, Fenig E, Zeigler M, Yaniv I. Successful treatment of Wolman disease by unrelated umbilical cord blood transplantation. *Eur J Pediatr.* 2007;166:663–666.
9. Ozmen MN, Aygun N, Kilic I, Kuran L, Yalcin B, Besim A. Wolman's disease: ultrasonographic and computed tomographic findings. *Pediatr Radiol.* 1992;22:541–542.
10. Ben-Haroush A, Yogev Y, Levit O, Hod M, Kaplan B. Isolated fetal ascites caused by Wolman disease. *Ultrasound Obstet Gynecol.* 2003;21:297–298.
11. Muntoni S, Wiebusch H, Jansen-Rust M, et al. Prevalence of cholesteryl ester storage disease. *Arterioscler Thromb Vasc Biol.* 2007;27:1866–1868.
12. Sando GN, Rosenbaum LM. Human lysosomal acid lipase/cholesteryl ester hydrolase. Purification and properties of the form secreted by fibroblasts in microcarrier culture. *J Biol Chem.* 1985;260:15186–15193.
13. Anderson RA, Sando GN. Cloning and expression of cDNA encoding human lysosomal acid lipase/cholesteryl ester hydrolase. Similarities to gastric and lingual lipases. *J Biol Chem.* 1991;266:22479–22484.
14. Fasano T, Pisciotta L, Bocchi L, et al. Lysosomal lipase deficiency: molecular characterization of eleven patients with Wolman or cholesteryl ester storage disease. *Mol Genet Metab.* 2012;105:450–456.
15. Aslanidis C, Ries S, Fehringer P, Buchler C, Klima H, Schmitz G. Genetic and biochemical evidence that CESD and Wolman disease are distinguished by residual lysosomal acid lipase activity. *Genomics.* 1996;33:85–93.
16. Redonnet-Vernhet I, Chatelut M, Salvayre R, Levade T. A novel lysosomal acid lipase gene mutation in a patient with cholesteryl ester storage disease. *Hum Mutat.* 1998;11:335–336.
17. Fujiyama J, Sakuraba H, Kuriyama M, et al. A new mutation (LIPA Tyr22X) of lysosomal acid lipase gene in a Japanese patient with Wolman disease. *Hum Mutat.* 1996;8:377–380.
18. Mayatepek E, Seedorf U, Wiebusch H, Lenhartz H, Assmann G. Fatal genetic defect causing Wolman disease. *J Inherit Metab Dis.* 1999;22:93–94.
19. Zschenker O, Jung N, Rethmeier J, et al. Characterization of lysosomal acid lipase mutations in the signal peptide and mature polypeptide region causing Wolman disease. *J Lipid Res.* 2001;42:1033–1040.

20. Du H, Duanmu M, Witte D, Grabowski GA. Targeted disruption of the mouse lysosomal acid lipase gene: long-term survival with massive cholesteryl ester and triglyceride storage. *Hum Mol Genet*. 1998;7:1347–1354.
21. Saito S, Ohno K, Suzuki T, Sakuraba H. Structural bases of Wolman disease and cholesteryl ester storage disease. *Mol Genet Metab*. 2012;105:244–248.
22. Bowden KL, Bilbey NJ, Bilawchuk LM, et al. Lysosomal acid lipase deficiency impairs regulation of ABCA1 gene and formation of high density lipoproteins in cholesteryl ester storage disease. *J Biol Chem*. 2011;286:30624–30635.
23. Fitoussi G, Negre-Salvayre A, Pieraggi MT, Salvayre R. New pathogenetic hypothesis for Wolman disease: possible role of oxidized low-density lipoproteins in adrenal necrosis and calcification. *Biochem J*. 1994;301(Pt 1):267–273.
24. Alomar Y, Negre-Salvayre A, Levade T, Valdiguie P, Salvayre R. Oxidized HDL are much less cytotoxic to lymphoblastoid cells than oxidized LDL. *Biochim Biophys Acta*. 1992;1128:163–166.
25. Du H, Heur M, Duanmu M, et al. Lysosomal acid lipase-deficient mice: depletion of white and brown fat, severe hepatosplenomegaly, and shortened life span. *J Lipid Res*. 2001;42:489–500.
26. Staretz-Chacham O, Lang TC, LaMarca ME, Krasnewich D, Sidransky E. Lysosomal storage disorders in the newborn. *Pediatrics*. 2009;123:1191–1207.
27. Burton BK. Inborn errors of metabolism in infancy: a guide to diagnosis. *Pediatrics*. 1998;102:E69.
28. Bernstein DL, Hulkova H, Bialer MG, Desnick RJ. Cholesteryl ester storage disease: review of the findings in 135 reported patients with an underdiagnosed disease. *J Hepatol*. 2013;58:1230–1243.
29. Lohse P, Maas S, Elleder M, Kirk JM, Besley GT, Seidel D. Compound heterozygosity for a Wolman mutation is frequent among patients with cholesteryl ester storage disease. *J Lipid Res*. 2000;41:23–31.
30. Tadiboyina VT, Liu DM, Miskie BA, Wang J, Hegele RA. Treatment of dyslipidemia with lovastatin and ezetimibe in an adolescent with cholesterol ester storage disease. *Lipids Health Dis*. 2005;4:26.
31. Kuriyama M, Yoshida H, Suzuki M, Fujiyama J, Igata A. Lysosomal acid lipase deficiency in rats: lipid analyses and lipase activities in liver and spleen. *J Lipid Res*. 1990;31:1605–1612.
32. Negre A, Dagan A, Gatt S. Pyrene-methyl lauryl ester, a new fluorescent substrate for lipases: use for diagnosis of acid lipase deficiency in Wolman's and cholesteryl ester storage diseases. *Enzyme*. 1989;42:110–117.
33. Hamilton J, Jones I, Srivastava R, Galloway P. A new method for the measurement of lysosomal acid lipase in dried blood spots using the inhibitor Lalistat 2. *Clin Chim Acta*. 2012;413:1207–1210.
34. van Diggelen OP, von Koskull H, Ammala P, Vredeveldt GT, Janse HC, Kleijer WJ. First trimester diagnosis of Wolman's disease. *Prenat Diagn*. 1988;8:661–663.
35. Fulcher AS, Das Narla L, Hingsbergen EA. Pediatric case of the day. Wolman disease (primary familial xanthomatosis with involvement and calcification of the adrenal glands). *Radiographics*. 1998;18:533–535.
36. Roytta M, Fagerlund AS, Toikkanen S, et al. Wolman disease: morphological, clinical and genetic studies on the first Scandinavian cases. *Clin Genet*. 1992;42:1–7.
37. Castro M, Rosati P, Boldrini R, Lucidi V, Gambarara M, Bosman C. Wolman's disease diagnosed by intestinal biopsy. *Ital J Gastroenterol Hepatol*. 1999;31:610–612.
38. Boldrini R, Devito R, Biselli R, Filocamo M, Bosman C. Wolman disease and cholesteryl ester storage disease diagnosed by histological and ultrastructural examination of intestinal and liver biopsy. *Pathol Res Pract*. 2004;200:231–240.
39. Di Bisceglie AM, Ishak KG, Rabin L, Hoeg JM. Cholesteryl ester storage disease: hepatopathology and effects of therapy with lovastatin. *Hepatology*. 1990;11:764–772.
40. Elleder M, Chlumska A, Ledvinova J, Poupetova H. Testis - a novel storage site in human cholesteryl ester storage disease. Autopsy report of an adult case with a long-standing subclinical course complicated by accelerated atherosclerosis and liver carcinoma. *Virchows Arch*. 2000;436:82–87.
41. Todoroki T, Matsumoto K, Watanabe K, et al. Accumulated lipids, aberrant fatty acid composition and defective cholesterol ester hydrolase activity in cholesterol ester storage disease. *Ann Clin Biochem*. 2000;37(Pt 2):187–193.
42. Tolar J, Petryk A, Khan K, et al. Long-term metabolic, endocrine, and neuropsychological outcome of hematopoietic cell transplantation for Wolman disease. *Bone Marrow Transplant*. 2009;43:21–27.
43. Krivit W, Peters C, Dusenbery K, et al. Wolman disease successfully treated by bone marrow transplantation. *Bone Marrow Transplant*. 2000;26:567–570.
44. Yanir A, Allatif MA, Weintraub M, Stepensky P. Unfavorable outcome of hematopoietic stem cell transplantation in two siblings with Wolman disease due to graft failure and hepatic complications. *Mol Genet Metab*. 2013;109:224–226.
45. Du H, Schiavi S, Levine M, Mishra J, Heur M, Grabowski GA. Enzyme therapy for lysosomal acid lipase deficiency in the mouse. *Hum Mol Genet*. 2001;10:1639–1648.
46. Du H, Cameron TL, Garger SJ, et al. Wolman disease/cholesteryl ester storage disease: efficacy of plant-produced human lysosomal acid lipase in mice. *J Lipid Res*. 2008;49:1646–1657.
47. Clinical Trials. Gov. Field studies of LAL replacement. http://www.clinicaltrials.gov/ct2/results?term=lal+replacement&Search=Search. Accessed 16.03.14.
48. Du H, Heur M, Witte DP, Ameis D, Grabowski GA. Lysosomal acid lipase deficiency: correction of lipid storage by adenovirus-mediated gene transfer in mice. *Hum Gene Ther*. 2002;13:1361–1372.
49. Tietge UJ, Sun G, Czarnecki S, et al. Phenotypic correction of lipid storage and growth arrest in wolman disease fibroblasts by gene transfer of lysosomal acid lipase. *Hum Gene Ther*. 2001;12:279–289.
50. Tarantino MD, McNamara DJ, Granstrom P, Ellefson RD, Unger EC, Udall Jr. JN. Lovastatin therapy for cholesterol ester storage disease in two sisters. *J Pediatr*. 1991;118:131–135.
51. Dalgic B, Sari S, Gunduz M, et al. Cholesteryl ester storage disease in a young child presenting as isolated hepatomegaly treated with simvastatin. *Turk J Pediatr*. 2006;48:148–151.
52. Ferry GD, Whisennand HH, Finegold MJ, Alpert E, Glombicki A. Liver transplantation for cholesteryl ester storage disease. *J Pediatr Gastroenterol Nutr*. 1991;12:376–378.
53. Shenoy P, Karegowda L, Sripathi S, Mohammed N. Wolman disease in an infant. *BMJ Case Rep*. 2014; pii: bcr2014203656.

CHAPTER

37

Lysosomal Membrane Disorders: LAMP-2 Deficiency

Kazuma Sugie,† and Ichizo Nishino†*

*Nara Medical University School of Medicine, Nara, Japan

†National Center of Neurology and Psychiatry, Tokyo, Japan

INTRODUCTION

In normal skeletal muscle, lysosomes have morphologically unremarkable structures. By electron microscopy they are very hard to find and appear to be totally devoid of internal structures compared to other organelles such as mitochondria. Nevertheless, lysosomes have important physiological roles in skeletal muscle, since a number of muscle diseases are accompanied by structurally abnormal lysosomes. Among these disorders, two have been genetically defined, acid maltase deficiency and Danon disease. This chapter will focus on the latter disease, which is due to a primary deficiency of lysosome-associated membrane protein-2 (LAMP-2). To date, this is the only known human muscle disorder caused by a lysosomal membrane protein defect.

DANON DISEASE

In 1981, Danon and colleagues reported two unrelated 16-year-old boys with similar phenotypes[1] characterized by the clinical triad of hypertrophic cardiomyopathy, myopathy, and mental retardation. In addition to these clinical manifestations, the disorder was characterized by a vacuolar myopathy with increased muscle glycogen, resembling glycogen-storage disease type II; however, acid maltase activity was normal. Accordingly, the disease was called "lysosomal glycogen-storage disease with normal acid maltase." However, the disease is not a glycogen-storage disease because glycogen is not always increased and because the defective molecule, LAMP-2, is a structural protein rather than a glycogenolytic enzyme.[2] Therefore, "Danon disease" has been redefined as X-linked vacuolar cardiomyopathy and myopathy due to LAMP-2 deficiency.[2]

CLINICAL FEATURES

Clinically, the disease is characterized by hypertrophic cardiomyopathy, myopathy, and mental retardation.[2,3] All probands were male, but females had milder and later-onset cardiomyopathy; therefore, the disease was thought to be transmitted as an X-linked dominant mode trait, even before the underlying defect was identified. In fact, the causative gene for Danon disease, *LAMP-2*, is on chromosome Xq24.[4]

Pregnancies and deliveries are usually normal. Ages at onset in a study of 20 male and 18 female patients varied from 10 months to 19 years in males and from 12 to 53 years in females.[3] The actual onset can be earlier, but symptoms may remain undetected because of the subacute, insidious nature and slow progression of the disease. For example, two male patients were identified only when isolated increases in serum creatine kinase (CK) were noted, and two

Rosenberg's Molecular and Genetic Basis of Neurological and Psychiatric Disease
http://dx.doi.org/10.1016/10.1016/B978-0-12-410529-4.00037-1

other patients were considered to have Danon disease because they had abnormal electrocardiograms (ECGs) before they showed any cardiac symptoms. Most commonly, onset is in childhood, but one male patient developed dyspnea in the infantile period. Delayed milestones with mild mental retardation have been observed in 15% of male patients.

Sooner or later, all patients develop cardiomyopathy, which is the most severe and life-threatening manifestation. In male patients, cardiac symptoms, such as exertional dyspnea, begin during the teenage years. Hypertrophic cardiomyopathy predominates in males, while dilated cardiomyopathy is more common in females. However, male patients can also show dilated cardiomyopathy, especially later in their lives. Cardiac arrhythmia, especially Wolff–Parkinson–White (WPW) syndrome, is often seen in both males and females. In a study of 38 patients with genetically confirmed Danon disease, ages at death were 19±6 (mean±standard deviation [SD]) years for males and 40±7 years for females, obviously reflecting the milder phenotype in female patients.[3] Among cardiologists, Danon disease is now being increasingly recognized as a cause of hypertrophic cardiomyopathy especially in children.[5–7]

Skeletal myopathy, usually mild, is present in most male patients (90%), but only in a minority of female patients (33%).[3] Patients do not lose the ability to walk despite the myopathy. Weakness and atrophy predominantly affect shoulder-girdle and neck muscles, but distal muscles can also be involved. Interestingly, all male patients show serum CK levels 5- to 10-fold above the upper limits of normal, even in the preclinical state. Therefore, Danon disease should be considered in the differential diagnosis of boys with cardiomyopathy and high serum CK. In contrast, serum CK is elevated in only 63% of female patients. In women, therefore, normal CK does not rule out the diagnosis of Danon disease. Other muscle enzymes including serum aspartate aminotransferase (AST), alanine aminotransferase (ALT), and lactate dehydrogenase (LDH) are also increased. Electrophysiologically, myopathic units have been seen in all male patients studied. In addition, myotonic discharges were recorded in three of 10 male patients.[3]

Although both original patients reported by Danon and colleagues had mental retardation, this manifestation is mild and is absent in 30% of male patients. In the authors' study, there has been only one female patient with mental retardation (1/18, 6%).[3] Brain magnetic resonance imaging (MRI) is usually normal. In two autopsy cases, the authors found vacuolar changes in the cytoplasm of the red nucleus.[8] More recently, the authors' group reported lysosomal storage and advanced senescence in the brain of a Danon disease patient.[9] However, it is still unknown whether or not these abnormalities directly account for the mental retardation.

So far, variable clinical complications have been known in Danon disease. Hepatomegaly was noted in 36% of male patients and splenomegaly was seen in 5% in the authors' study.[3] Ophthalmic abnormalities including retinopathy and cone–rod retinal dystrophy were also reported.[10–12] Retinal impairments have potentially progressive decrease of visual acuity. Foot deformity was sometimes noticed. Nerve conduction studies were usually normal, but sensory and motor neuropathy with pes cavus was rarely reported.[13] Unusually, autism was presented in male children with Danon disease.[14] Cerebrovascular complications have previously been reported.[15] Embolic complications may be relatively common in patients with Danon disease, and early recognition of cardiac arrhythmia and anticoagulant therapy might prevent death or disability because of cerebrovascular events. Restrictive lung problems, possibly related to respiratory muscle weakness, were reported in a few male patients.[16] Clinical features of Danon disease may be broader than previously thought.

In LAMP-2 knockout mice, a wider variety of organs is affected, including liver, kidney, pancreas, small intestine, thymus, and spleen, in addition to heart and skeletal muscle.[17] Similarly, autopsy study showed vacuolar changes in a wide variety of tissues, including cardiac, skeletal, and smooth muscles, and liver.[8] Therefore, Danon disease in mice is essentially a systemic disorder.[18]

Muscle Pathology

Muscle pathology provides an important clue for the diagnosis. As discussed below, the genetic test is now available and can be performed instead of a muscle biopsy; however, considering the rarity of the disease and the absence of a mutation hotspot, it is not cost-effective to screen patients for a mutation in *LAMP-2* without a prior biopsy demonstrating vacuolar myopathy.

In muscle biopsies of Danon disease patients, variation of fiber size is mild to moderate and necrotic fibers are usually not seen. Muscle samples show many scattered intracytoplasmic vacuoles, which, on hematoxylin and eosin staining, often look like solid basophilic granules. These vacuoles are so tiny that they can easily be overlooked (Figure 37.1). Interestingly, the vacuolar membranes have acetylcholinesterase activity and are highlighted by histochemical stains for acetylcholinesterase (Figure 37.1) and nonspecific esterase.[3,19–21] Acetylcholinesterase is present at the neuromuscular junction in specialized sarcolemmal areas called junctional folds. Therefore, the presence of acetylcholinesterase activity indicates that the vacuolar membrane has features of sarcolemma. This characteristic has been confirmed by the immunohistochemical demonstration of other sarcolemma-specific proteins.[21] In fact, virtually all sarcolemmal proteins are

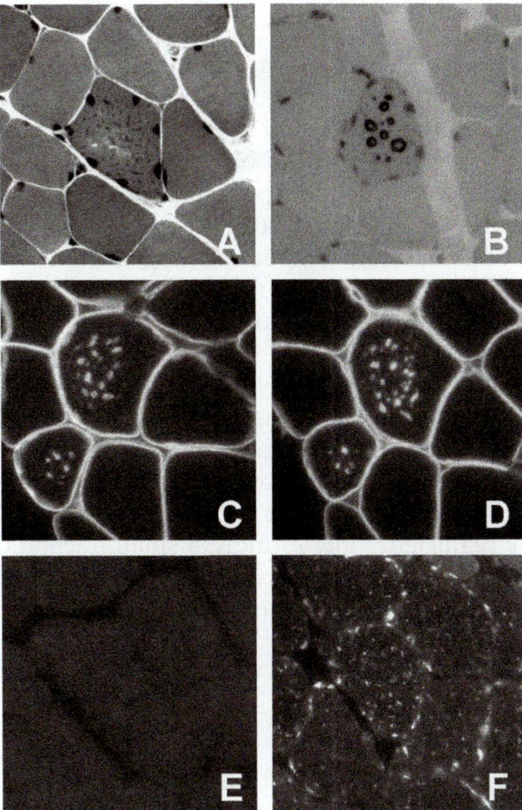

FIGURE 37.1 **Muscle pathology of Danon disease.** With hematoxylin and eosin staining, tiny autophagic vacuoles often look like solid basophilic granules rather than vacuoles, and can be overlooked easily (**A**). Interestingly, the vacuolar membrane has acetylcholinesterase activity (**B**). Immunohistochemical analyses for dystrophin (**C**) and merosin (**D**) show that the vacuolar membrane has features of sarcolemma. Immunostaining for LAMP-2 clearly demonstrates the complete absence of the LAMP-2 protein (**E**) in contrast to control (**F**).

expressed in the vacuolar membrane, including dystrophin, α-, β-, γ-, and δ-sarcoglycans, α- and β-dystroglycans, dystrobrevin, utrophin, dysferlin, perlecan, caveolin-3, collagen IV, and fibronectin (Figure 37.1); hence, these vacuoles are called autophagic vacuoles with sarcolemmal features (AVSF).[21]

By electron microscopy, the intracytoplasmic vacuoles typically contain myelin figures, electron-dense bodies, and various cytoplasmic debris, and, therefore, are considered to be autophagic vacuoles.[3,21] Interestingly, even basal lamina is sometimes seen along the inner surface of autophagic vacuoles, further confirming their AVSF nature.

Occasionally, sarcolemma and vacuolar membranes appear to be connected, suggesting that the unusual vacuolar membrane may arise from indentations of the sarcolemma.[20] However, most vacuoles are not connected to the sarcolemma and they instead form isolated closed spaces, raising the possibility that their peculiar limiting membrane is formed inside the muscle fiber.[21]

Interestingly, the number of AVSF increases with age: whereas only a few AVSF can be observed in patients younger than 2 years, numerous AVSF are seen in older patients.[21,22] However, when autophagic vacuoles in muscle fibers are counted regardless of sarcolemmal features, their total number does not change or even slightly decreases, indicating that most autophagic vacuoles do not have sarcolemmal features early on and that the sarcolemmal structures most likely form later and surround autophagic vacuoles.[21]

Although AVSF is a pathological hallmark of Danon disease, it can be seen in other autophagic vacuolar myopathies, including X-linked myopathy with excessive autophagy (XMEA),[23] infantile autophagic vacuolar myopathy,[24] X-linked congenital autophagic vacuolar myopathy,[25] and adult-onset autophagic vacuolar myopathy with multiorgan involvement.[26] Although these disorders are genetically distinct from Danon disease, these pathological similarities suggest a common pathomechanism. Therefore, autophagic vacuolar myopathies should be categorized as a distinct group of disorders.

By immunohistochemical and western blot analyses, LAMP-2 protein is absent in skeletal muscle regardless of the specific *LAMP-2* gene mutation[2,3,21] (Figure 37.1). Western blot analysis of the cardiac muscle in one patient also showed a complete absence of LAMP-2 protein.[2] In contrast, other lysosomal membrane proteins, such as lysosomal integral membrane protein-I (LIMP-1), are associated with the autophagic vacuoles in Danon disease.[2,3,21]

LAMP-2

LAMP-2 is a type 1 membrane protein with a large luminal domain connected to a transmembrane region and a short cytoplasmic tail. The luminal domain can be divided into two internally homologous domains separated by a hinge region rich in proline, serine, or threonine. Each of the two homologous regions contains four cysteines that are linked in pairs by disulfide bonding between neighboring residues, thus creating two loops in each domain. The luminal domain is heavily glycosylated; most of the potential *N*-linked glycosylation sites are utilized, yielding a molecular mass of 90–120 kDa for the approximately 40 kDa core protein. LAMP-2 is abundantly expressed and is thought to coat the inner surface of the lysosomal membrane together with its autosomal paralog, LAMP-1. The topographical distribution of LAMPs, together with the fact that LAMP-2 is one of the most heavily glycosylated proteins, indicate that LAMPs probably protect lysosomal membrane, and thus also the cytoplasm, from the action of proteolytic enzymes within the lysosomes.[27]

The cytoplasmic tail of LAMP-2 is short, consisting of only 11 amino acids, but has a well-conserved tyrosine residue, which may provide a crucial signal for the transport of LAMP-2 molecules to lysosomes. Moreover, this cytoplasmic tail is thought to function as a receptor for the uptake of certain proteins destined to be degraded into lysosomes (chaperone-mediated autophagy), in association with the 73-kDa heat shock cognate protein.[28]

Whereas LAMP-1 seems constitutively expressed, the expression of LAMP-2 is increased in a variety of situations and, is likely to be specifically regulated.[29] Interestingly, a small fraction (2–3%) of LAMP-2 is present in the plasma membrane,[27] where its expression increases in certain situations, including malignancy[30] and scleroderma.[31] Although the functional significance of LAMP-2 expression at the cell surface is not completely understood, it may be related to the development of the AVSF.

LAMP-2 Gene Mutations

The *LAMP-2* gene is located on Xq24, while the gene for LAMP-1 is on 13q34.[4] The *LAMP-2* open reading frame consists of 1233 nucleotides and encodes 410 amino acids. Exons 1 through 8 and part of exon 9 encode the luminal domain, while the remainder of exon 9 encodes both a transmembrane domain and a cytoplasmic domain. Human exon 9 exists in two forms, 9A and 9B, which are alternatively spliced and produce two isoforms, LAMP-2A and LAMP-2B. LAMP-2A is expressed rather ubiquitously whereas LAMP-2B is expressed specifically in heart and skeletal muscle.[32]

To date, *LAMP-2* gene mutations have been identified in at least 50 ethnically diverse pedigrees, suggesting that this disorder can affect any ethnic group.[2,3,16] Most reported mutations are stop-codon or out-of-frame, and are predicted to truncate the protein and to result in loss of the transmembrane and cytoplasmic domains. Therefore, the mutated products cannot function as lysosomal membrane proteins. The total absence of the LAMP-2 protein in Danon disease muscles suggests that the abnormal proteins are unstable and are rapidly degraded.

An exon-skipping mutation is predicted to cause an in-frame deletion of one of the four loop structures and of several potential glycosylation sites in the luminal domain, resulting in severe structural changes.[2] The patient with this particular mutation also had complete absence of the LAMP-2 protein in skeletal muscle, suggesting that this mutation is as harmful as null mutations.

In one patient harboring a mutation in exon 9B, western blot analysis revealed a trace amount of the LAMP-2 protein.[2] This signal most likely represents LAMP-2A, because the mutation in exon 9B should affect only the LAMP-2B isoform. This particular patient is alive at age 34, suggesting that this mutation causes an exceptionally mild phenotype.[3] Usually, Danon disease is clinically uniform in male patients, without apparent genotype–phenotype variants.[3] The mutation in exon 9B not only supports the idea that LAMP-2B is the major isoform in cardiac and skeletal muscles, but also suggests that a deficiency of LAMP-2B by itself is sufficient to cause the disease, albeit with a milder phenotype.

Although rare, one missense mutation has also been reported.[33] This patient apparently had a milder phenotype with high CK level, exercise intolerance, and hypertrophic cardiomyopathy, but without muscle weakness or mental impairment. However, the LAMP-2 protein was virtually absent in the skeletal muscle.[33]

LAMP-2 Knockout Mouse

LAMP-2 deficient mice produced by a German group provide confirmatory evidence that LAMP-2 deficiency causes Danon disease.[17] About 50% of LAMP-2 deficient mice die between postnatal day 20 and 40, irrespective of sex and genetic background. Surviving mice are smaller and have cardiac hypertrophy, but their lifespan is normal. LAMP-2 deficient mice have autophagic vacuoles in various tissues, including heart and skeletal muscle (analogous to human patients), liver, pancreas, spleen, and kidney. Mice that die early often have stenoses or segmental hemorrhagic infarcts of the small intestine and pancreatic lesions. In addition, apoptotic cell loss is pathologically increased in thymus and the demarcation of white and red pulp is absent. Together with the fact that the *LAMP-2* gene mutations segregate with Danon disease, the findings in LAMP-2 knockout mice clearly demonstrate that Danon disease is primarily a LAMP-2 deficiency. Furthermore, the abnormalities in a wider variety of organs in LAMP-2 deficient mice suggest that more organs could potentially be involved in humans with Danon disease and that patients might develop other symptoms, in addition to the "classical" triad.

Curiously, in contrast to LAMP-2 knockout mice, LAMP-1 deficient mice show normal lysosomal morphology and function and do not develop any symptoms.[34] This is probably due to the compensatory upregulation of LAMP-2 in LAMP-1 knockout mice, contrasting with the fact that LAMP-1 is not upregulated in LAMP-2 knockout mice.[17] This result indicates that the patterns of expression of these highly homologous proteins are regulated differently and bolsters the concept that they may have different functional roles.

Other Autophagic Vacuolar Myopathies

In 1988, Kalimo et al. reported a new type of autophagic vacuolar myopathy, X-linked myopathy with excessive autophagy (XMEA), in a Finnish family.[23] The disease is transmitted in an X-linked recessive manner and is now known to be caused by *VMA21* mutations.[35] VMA21 is an essential assembly chaperone of the V-ATPase, the principal mammalian proton pump complex. Decreased VMA21 raises lysosomal pH, which reduces lysosomal degradative ability and blocks autophagy. XMEA is characterized clinically by slowly progressive muscle weakness and atrophy sparing cardiac and respiratory muscles. Muscle biopsy is characterized by AVSF as in Danon disease. The presence of sarcolemmal proteins, such as dystrophin, in the membrane of autophagic vacuoles in both diseases suggests common or similar molecular pathomechanisms. The distinguishing pathological findings in XMEA, which are not seen in Danon disease, are depositions of complement C5b-9 over the surface of muscle fibers and multilayered basal lamina along the sarcolemma.[3,21,36] Furthermore, the presence of LAMP-2 in XMEA muscle clearly demonstrates that XMEA is distinct from Danon disease.[2]

The list of myopathies characterized by AVSF, aside from Danon disease, is rapidly expanding and includes: 1) infantile autophagic vacuolar myopathy,[24] 2) congenital form of X-linked autophagic vacuolar myopathy,[25] and 3) late-onset autophagic vacuolar myopathy with multiorgan involvement.[26] Interestingly, all these diseases show deposition of complement C5b-9 over the surface of muscle fibers and multiplication of basal lamina, making these myopathies more similar to XMEA than to Danon disease.[24–26]

Despite different clinical and pathologic features, Danon disease, XMEA, and other autophagic vacuolar myopathies can probably be categorized together into a distinct group, because they all show AVSF.[21] Actually, Danon disease and XMEA, both genetically diagnosable autophagic vacuolar myopathies, are primarily due to lysosomal dysfunctions. In contrast, other myopathies characterized by the presence of rimmed vacuoles, such as distal myopathy with rimmed vacuoles, and inclusion body myopathy or hereditary inclusion body myopathy, are secondarily caused by extralysosomal defects. Most likely, there will still be other diseases with AVSF in this group of autophagic vacuolar myopathy, and we expect that the list will continue to expand.[37]

MANAGEMENT

Myopathy is usually mild and can be clinically silent. Symptomatic patients typically had proximal limb weakness, which was very slowly progressive or stable. Myopathic symptoms were noted in only a few female patients and were even milder.[3]

Cardiac symptoms are the dominant clinical features and the most important prognostic factors, because all patients died of cardiac failure. Most male patients developed hypertrophic cardiomyopathy, whereas most female patients showed hypertrophic or dilated cardiomyopathy.[3,16,38] WPW syndrome is more common in male than female patients. The cardiac manifestations occurred approximately 15 years later in female than male patients. Sudden cardiac death is common in patients with Danon disease, especially in female patients.[39] Recently, it was reported that cardiac MRI may be of clinical value for the diagnostic work-up of Danon disease.[40]

Heart transplantation may be the most effective and the only reliable treatment, although implantable cardioverter defibrillators represent one preventive treatment.[3,5–7,38,41] Actually, heart transplantation significantly enhances the survival of patients, although only 17.6% patients undergo heart transplantation.[38] Therefore, we should consider early intervention with heart transplantation once heart failure has been diagnosed and it should be performed as early as possible due to its rapid progression.[38] As in males, cardiomyopathy can be fatal in female patients. This suggests that not only male patients but also female patients with Danon disease should be considered for heart transplantation. In addition, the authors suggest that asymptomatic female relatives of male patients should be investigated for cardiomyopathy and followed closely to detect early signs of a potentially life-threatening condition.[3]

References

1. Danon MJ, Oh SJ, DiMauro S, et al. Lysosomal glycogen storage disease with normal acid maltase. *Neurology*. 1981;31:51–57.
2. Nishino I, Fu J, Tanji K, et al. Primary LAMP-2 deficiency causes X-linked vacuolar cardiomyopathy and myopathy (Danon disease). *Nature*. 2000;406:906–910.
3. Sugie K, Yamamoto A, Murayama K, et al. Clinicopathological features of genetically-confirmed Danon disease. *Neurology*. 2002;58:1773–1778.
4. Mattei M-G, Maaterson J, Chen JW, Williams MA, Fukuda M. Two human lysosomal membrane glycoproteins, h-lamp-1 and h-lamp-2, are encoded by genes localized to chromosome 13q34 and chromosome Xq24-25, respectively. *J Biol Chem*. 1990;265:7458–7551.
5. Charron P, Villard E, Sebillon P, et al. Danon's disease as a cause of hypertrophic cardiomyopathy: a systematic survey. *Heart*. 2004;90:842–846.
6. Arad M, Maron BJ, Gorham JM, et al. Glycogen storage diseases presenting as hypertrophic cardiomyopathy. *N Engl J Med*. 2005;352:362–372.
7. Yang Z, McMahon CJ, Smith LR, et al. Danon disease as an underrecognized cause of hypertrophic cardiomyopathy in children. *Circulation*. 2005;112:1612–1617.
8. Nishino I, Yamamoto A, Tokonami F, Takahashi M, Chino F, Nonaka. Two autopsy cases of Danon disease. *Neuromuscul Disord*. 2001;11:668–669 (Abstr).
9. Furuta A, Wakabayashi K, Haratake J, et al. Lysosomal storage and advanced senescence in the brain of LAMP-2-deficient Danon disease. *Acta Neuropathol*. 2013;125:459–461.
10. Prall FR, Drack A, Taylor M, et al. Ophthalmic manifestations of Danon disease. *Ophthalmology*. 2006;113:1010–1013.
11. Schorderet DF, Cottet S, Lobrinus JA, Borruat FX, Balmer A, Munier FL. Retinopathy in Danon disease. *Arch Ophthalmol*. 2007;125:231–236.
12. Thiadens AA, Slingerland NW, Florijn RJ, Visser GH, Riemslag FC, Klaver CC. Cone-rod dystrophy can be a manifestation of Danon disease. *Graefes Arch Clin Exp Ophthalmol*. 2012;250:769–774.
13. Laforêt P, Charron P, Maisonobe T, et al. Charcot–Marie–Tooth features and maculopathy in a patient with Danon disease. *Neurology*. 2004;63:1535.
14. Burusnukul P, de Los Reyes EC, Yinger J, Boué DR. Danon disease: an unusual presentation of autism. *Pediatr Neurol*. 2008;39:52–54.
15. Spinazzi M, Fanin M, Melacini P, Nascimbeni AC, Angelini C. Cardioembolic stroke in Danon disease. *Clin Genet*. 2008;73:388–390.
16. Boucek D, Jirikowic J, Taylor M. Natural history of Danon disease. *Genet Med*. 2011;13:563–568.
17. Tanaka Y, Guhde G, Suter A, et al. Accumulation of autophagic vacuoles and cardiomyopathy in LAMP-2-deficient mice. *Nature*. 2000;406:902–906.
18. Fanin M, Nascimbeni AC, Fulizio L, Spinazzi M, Melacini P, Angelini C. Generalized lysosome-associated membrane protein-2 defect explains multisystem clinical involvement and allows leukocyte diagnostic screening in Danon disease. *Am J Pathol*. 2006;168:1309–1320.
19. Muntoni F, Catani G, Mateddu A, et al. Familial cardiomyopathy, mental retardation and myopathy associated with desmin-type intermediate filaments. *Neuromuscul Disord*. 1994;4:233–241.
20. Murakami N, Goto Y-I, Itoh M, et al. Sarcolemmal indentation in cardiomyopathy with mental retardation and vacuolar myopathy. *Neuromuscul Disord*. 1995;5:149–155.
21. Sugie K, Noguchi S, Kozuka Y, et al. Autophagic vacuoles with sarcolemmal delineate Danon disease and related myopathies. *J Neuropathol Exp Neurol*. 2005;64:513–522.
22. Sugie K, Koori T, Yamamoto A, et al. Characterization of Danon disease in a man and his affected mother. *Neuromuscul Disord*. 2003;13:708–711.
23. Kalimo H, Savontaus M-L, Lang H, et al. X-linked myopathy with excessive autophagy: a new hereditary muscle disease. *Ann Neurol*. 1988;23:258–265.
24. Yamamoto A, Morisawa Y, Verloes A, et al. Infantile autophagic vacuolar myopathy is distinct from Danon disease. *Neurology*. 2001;57:903–905.
25. Yan C, Tanaka M, Sugie K, et al. A new congenital form of X-linked autophagic vacuolar myopathy. *Neurology*. 2005;65:1132–1134.
26. Kaneda D, Sugie K, Yamamoto A, et al. A novel form of autophagic vacuolar myopathy with late-onset and multiorgan involvement. *Neurology*. 2003;61:128–131.
27. Fukuda M. Biogenesis of the lysosomal membrane. *Subcell Biochem*. 1994;22:199–230.
28. Cuervo AM, Dice JF. A receptor for the selective uptake and degradation of proteins by lysosomes. *Science*. 1996;273:501–503.
29. Sawada R, Jardine KA, Fukuda M. The genes of major lysosomal membrane glycoproteins, lamp-1 and lamp-2. The 5′-flanking sequence of lamp-2 gene and comparison of exon organization in two genes. *J Biol Chem*. 1993;268:9014–9022.
30. Kannan K, Divers SG, Lurie AA, Chervenak R, Fukuda M, Holcombe RF. Cell surface expression of lysosome-associated membrane protein-2 (lamp2) and CD63 as markers of *in vivo* platelet activation in maligancy. *Eur J Heamatol*. 1995;55:145–151.
31. Holcombe RF, Baethge BA, Stewart RM, et al. Cell surface expression of lysosome-associated membrane proteins (LAMPs) in Scleroderma: relationship of lamp2 to disease duration, anti-Scl70 antibodies, serum interleukin-8, and soluble interleukin-2 receptor levels. *Clin Immunol Immunopathol*. 1993;67:31–39.
32. Konecki DS, Foetsch K, Zimmer K-P, Schlotter M, Lichter-Konecki U. An alternatively spliced forma of the human lysosome-associated membrane protein-2 gene is expressed in a tissue-specific matter. *Biochem Biophys Res Commun*. 1995;215:757–767.

33. Musumeci O, Rodolico C, Nishino I, et al. Asymptomatic hyperCKemia in a case of Danon disease due to a missense mutation in Lamp-2 gene. *Neuromuscul Disord*. 2005;15:409–411.
34. Andrejewski N, Punnonen E-L, Guhde G, et al. Normal lysosomal morphology and function in LAMP-1-deficient mice. *J Biol Chem*. 1999;274:12692–12701.
35. Ramachandran N, Munteanu I, Wang P, et al. VMA21 deficiency causes an autophagic myopathy by compromising V-ATPase activity and lysosomal acidification. *Cell*. 2009;137:235–246, Retraction in: *Cell*. 2010;142:984.
36. Villanova M, Louboutin JP, Chateau D, et al. X-linked vacuolated myopathy: complement membrane attack complex on surface membrane of injured muscle fibers. *Ann Neurol*. 1995;37:637–645.
37. Nishino I. Autophagic vacuolar myopathy. *Semin Pediatr Neurol*. 2006;13:90–95.
38. Cheng Z, Fang Q. Danon disease: focusing on heart. *J Hum Genet*. 2012;57:407–410.
39. Miani D, Taylor M, Mestroni L, et al. Sudden death associated with Danon disease in women. *Am J Cardiol*. 2012;109:406–411.
40. Nucifora G, Miani D, Piccoli G, Proclemer A. Cardiac magnetic resonance imaging in Danon disease. *Cardiology*. 2012;121:27–30.
41. Echaniz-Laguna A, Mohr M, Epailly E, et al. Novel Lamp-2 gene mutation and successful treatment with heart transplantation in a large family with Danon disease. *Muscle Nerve*. 2006;33:393–397.

CHAPTER

38

Fabry Disease: α-Galactosidase A Deficiency

Robert J. Desnick

Icahn School of Medicine at Mount Sinai, New York, NY, USA

INTRODUCTION

Fabry disease, an X-linked lysosomal storage disease, results from deficient α-galactosidase A (α-Gal A) activity and the progressive accumulation of globotriaosylceramide (GL-3 or Gb-3) and related glycosphingolipids in the plasma and tissue lysosomes throughout the body (Figure 38.1).[1] There are two major phenotypes, classic and later-onset.[2] Classically affected males have no detectable α-Gal A activity and prominent microvascular GL-3 accumulation. Onset in childhood/adolescence is characterized by severe acroparesthesias, pain crises, gastrointestinal manifestations, angiokeratoma, hypohidrosis, and corneal/lenticular opacities. With advancing age, the progressive GL-3 accumulation in the microvasculature, renal podocytes, and cardiomyocytes leads to premature demise from renal, cardiac, and/or cerebrovascular disease. Later-onset males have residual α-Gal A activity and no vascular endothelial involvement; they develop cardiac and renal disease in maturity. Heterozygous females may be asymptomatic or as severely affected as males. Diagnosis of affected males by α-Gal A enzyme assay is reliable, whereas identification of heterozygous females requires α-Gal A gene mutation analyses. Treatment by enzyme replacement with recombinant human α-Gal A has been shown to be safe and effective in clinical trials,[2–8] and early treatment is required for optimal outcome.

CLINICAL FEATURES AND DIAGNOSTIC EVALUATION

Fabry disease is inherited as an X-linked trait.[1] The α-Gal A gene is fully penetrant; however, different mutations result in variable clinical expressivity in affected males. The disease is panethnic. Based on newborn screening studies,[9–15] the estimated incidence of the classic phenotype is about 1 in 25,000 to 40,000 males,[1] whereas the later-onset phenotype is about 10-fold more frequent.

The Classical Phenotype

Clinical onset in classically affected males, who have absent or severely deficient α-Gal A activity, usually occurs during childhood or adolescence (Table 38.1). Early manifestations include chronic severe pain in the extremities (acroparesthesias), periodic excruciating pain crises, hypohidrosis, angiokeratoma, postprandial cramping and diarrhea, and characteristic corneal and lenticular opacities. With advancing age, progressive vascular glycosphingolipid deposition causes ischemia and infarction leading to cardiac, cerebral, and renal vascular disease, and progressive accumulation in cardiomyocytes and podocytes leads to hypertrophic cardiomyopathy, and renal failure, respectively. Death typically occurs in the fourth or fifth decade of life, most often from renal failure, cardiac failure, or stroke. The mean age of death for 94 affected males prior to hemodialysis, renal transplantation, or enzyme replacement therapy was 41 years.[16] The clinical and pathologic features of the disease have been comprehensively reviewed,[1,2] but are briefly described below.

http://dx.doi.org/10.1016/10.1016/B978-0-12-410529-4.00038-3

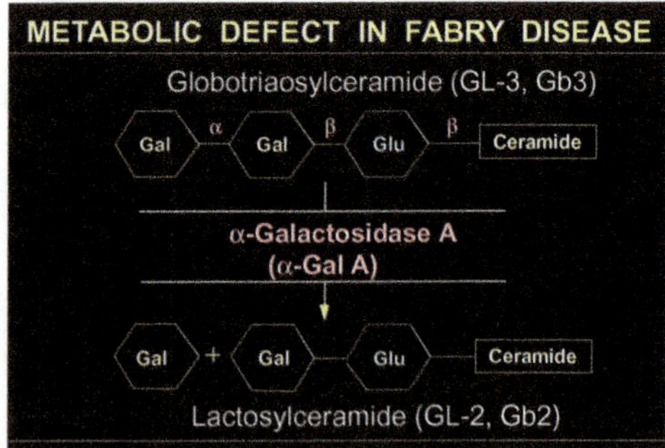

FIGURE 38.1 **The metabolic defect in Fabry disease.** The deficient activity of α-galactosidase A (α-Gal A) results in the accumulation of globotriaosylceramide (GL-3) and other glycoconjugates with terminal α-galactosyl moieties.

TABLE 38.1 Fabry Disease: Major Manifestations in Classical and Later-Onset Phenotypes

		Later-onset phenotypes	
Manifestation	**Classical phenotype**	**Renal***[25]	**Cardiac***[22–24]
Age at onset	4–8 years	>25 years	>40 years
Average age of death	41 years	?	>60 years
Angiokeratoma	+	–	–
Acroparesthesias	+	–	–
Hypohidrosis/anhidrosis	+	–	–
Corneal/lenticular opacity	+	–	–
Heart	Ischemia/LVH/myopathy	LVH	LVH/myopathy
Brain	TIA/strokes	?	–
Kidney	Renal failure	Renal failure	Mild proteinuria
Residual α-Gal A activity	<1%	>1%	>1%

Abbreviations: LVH, left ventricular hypertrophy; TIA, transient ischemic attack; +, present; –, absent.
**The later-onset phenotype presents primarily with renal or heart involvement.*[31–35,38,39]

Acroparesthesias

The single most debilitating early symptom of this disease is pain. Typically, affected males experience episodic crises of excruciating burning pain in the fingers and toes in childhood, which may become more frequent and severe in adolescence. These painful acroparesthesias may last several days to weeks and are associated with low-grade fevers and often elevations of the erythrocyte sedimentation rate. During the second and third decades of life, these recurrent painful episodes may occur less frequently and are usually associated with a fever. In a few patients, however, they may become progressively more frequent and severe, radiate to proximal extremities, and occasionally persist for 1–2 weeks. Affected individuals may be incapacitated for prolonged periods of time with pain that is so severe that suicide has been attempted. Sural nerve biopsies show lipid accumulation in small epineural and endoneural vessels, in perineural cells, in myelinated and unmyelinated axons, and in Schwann cells. The endothelial glycosphingolipid accumulation narrows the vascular lumen; vessel spasms or frank infarction may cause the excruciating pain. Small fiber dysfunction is evident, with Aδ fibers being more severely effected than C fibers, whereas large fibers are less affected.[17] It has been suggested that the etiology of the acroparesthesias is poor nerve perfusion, leading to axonal damage.[18,19]

Angiokeratoma

The cutaneous vascular lesions (angiokeratoma) in most males with the classic phenotype are telangiectases, which usually appear as clusters of individual punctate, dark-red to blue angiectases in the superficial layers of the skin (Figure 38.2). The lesions may be flat or slightly raised and do not blanch with pressure. There may be a slight hyperkeratosis over these lesions. They usually appear during childhood and progressively increase in size and number with age. Characteristically these lesions are most dense between the umbilicus and the knees and have a tendency toward bilateral symmetry. These lesions frequently are found on the oral mucosa and conjunctiva as well as on other mucosal areas. Hypohidrosis is common, and atrophic or sparse sweat and sebaceous glands have been

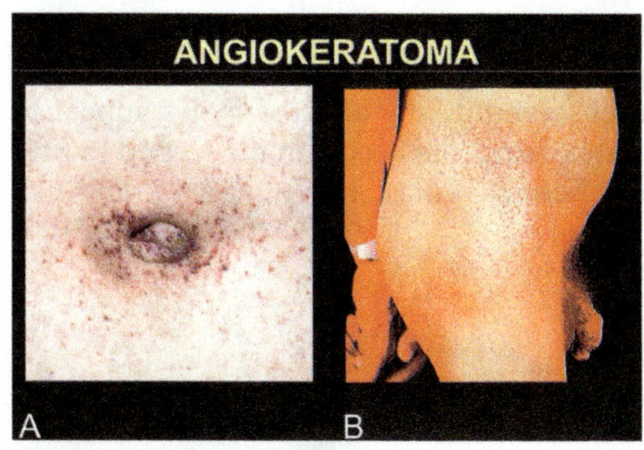

FIGURE 38.2 Clusters of angiokeratomas around the umbilicus (A) and the typical distribution on the torso (B).

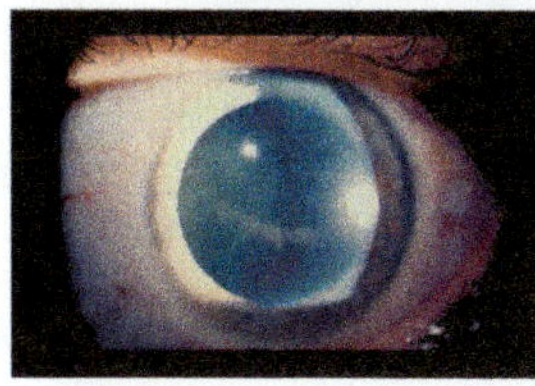

FIGURE 38.3 Corneal opacity in a heterozygous female observed by slit-lamp microscopy.

reported. Heterozygotes from families with mutations causing the classic phenotype, may have angiokeratomas typically on the breasts, flanks, and/or genital areas. Patients with the later-onset phenotype usually do not have angiokeratomas.

Ophthalmologic Features

Ocular manifestations include aneurysmal dilation and tortuosity of conjunctival and retinal vessels, as well as characteristic corneal and lenticular changes (Figure 38.3).[20] The conjunctival and retinal vascular lesions are common and part of the diffuse systemic vascular involvement.[21] The keratopathy is characterized by diffuse haziness or whorled streaks extending from a central vortex in the corneal epithelium. The corneal lesions resemble the changes seen in patients taking chloroquine or amiodarone and must be observed by slit-lamp microscopy; they occur in almost all classically affected males and in most (90%) heterozygous females from classically affected families but are usually absent in affected males and heterozygotes with the later-onset phenotypes (see below). The lenticular changes include a granular anterior capsular or subcapsular deposit in about 30% of classically affected males and a unique linear opacity (termed the Fabry cataract) in classically affected males and some heterozygous females. The cataracts are best observed by retroillumination and appear as whitish, spot-like deposits of fine granular material near the posterior capsule. These lesions do not impair vision.

Renal, Cardiac, and Cerebral Vascular Involvement

With increasing age, the major morbid symptoms result from the progressive involvement of the microvascular system, as well as progressive GL-3 accumulation in renal podocytes and in cardiomyocytes. Early in the course of the disease, casts, red cells, and lipid inclusions with characteristic birefringent Maltese crosses appear in the urinary sediment. Renal microvascular and particularly podocyte glycolipid deposition occurs in classically affected males and heterozygotes in the first decade of life.[22,23] Proteinuria, isosthenuria, gradual deterioration of renal function, and development of azotemia occur in the second to fourth decades of life.[24] Death most often results from uremia. With renal replacement therapy, the mean age of death was about 50 years.[25] Cardiovascular findings may include hypertension, left ventricular hypertrophy (LVH), hypertrophic cardiomyopathy (HCM), anginal chest pain, myocardial ischemia or infarction, and congestive heart failure.[1,2,26] Mitral insufficiency is the most common valvular lesion. Abnormal electrocardiographic and echocardiographic findings are common, and the cardiomyocyte accumulation leads to LVH and then to HCM. Cerebrovascular manifestations result primarily from multifocal small vessel involvement,[27] and early-onset cryptogenic strokes occur in previously unrecognized Fabry disease.

Other Features

Gastrointestinal symptoms, which typically begin in adolescence and may worsen with age, are reported by 70–90% of classically affected males in various studies.[1,2,25,28] The most common symptoms are episodes of postprandial abdominal pain and bloating, followed by multiple bowel movements and diarrhea. Other gastrointestinal symptoms include nausea, vomiting, and early satiety. Other less frequent features include massive lymphedema of the legs and dyspnea. Musculoskeletal system findings have included a permanent deformity of the distal interphalangeal joints of the fingers and avascular necrosis of the head of the femur and talus. Mild normochromic, normocytic anemia, presumably as a result of decreased red cell survival, has been observed. Many affected males appear to have growth retardation or delayed puberty. High-frequency sensorineural hearing loss occurs in the second to fifth decades of life in most affected males, whereas hearing loss occurs in some heterozygotes later in life.[29,30] Tinnitus is also a frequent symptom in affected males (27–40%).[25,29,30]

The Later-Onset Phenotype

Affected males with the later-onset phenotype have residual α-Gal A activity due to missense and some splicing mutations.[1,2] Since later-onset patients do not have microvascular GL-3 accumulation, they lack the early manifestations of classically affected patients, including the angiokeratoma, acroparesthesias, hypohidrosis, gastrointestinal complications, and corneal/lenticular opacities (Table 38.1).[1,2,31,32] The later-onset patients typically present in the fourth to eighth decades of life with cardiac, renal, or cerebrovascular manifestations.[1,2] Many of these patients were identified by screening patients in hemodialysis, transplant, cardiac and stroke clinics.[33–37] Later-onset patients with primary cardiac involvement present with LVH, arrhythmias, mitral insufficiency and/or LVH or HCM and may have mild-to-moderate proteinuria, usually with normal renal function for age. Their renal pathology is limited to glycosphingolipid deposition in podocytes, which is presumably responsible for the proteinuria.[38] Later-onset patients with primary renal involvement also lack the early manifestations of the classic phenotype but develop proteinuria and late-onset end-stage renal disease, often after 50 years of age.[1,2,34,36–39] Most recently, patients presenting with cryptogenic strokes, with or without other manifestations, also have been found to have previously unrecognized Fabry disease.[40,41] Overall, the frequency of previously unrecognized patients with Fabry disease identified by screening hemodialysis, transplantation, cardiac and stroke clinics was $<1\%$ of those screened.

Heterozygous Females

The clinical manifestations in DNA-diagnosed heterozygous females from classically affected families range from asymptomatic throughout a normal lifespan to as severe as affected males, presumably due to skewed X-chromosomal inactivation.[1,42] For example, there are obligate heterozygotes (i.e., daughters of affected males) without any clinical manifestations and with normal levels of leukocyte α-Gal A activity and urinary sediment glycosphingolipids. Also, asymptomatic and symptomatic monozygotic female twins have been described. Although some heterozygotes remain asymptomatic throughout a normal lifespan, many manifest disease symptoms, which are usually less severe than those in affected males.

Most (90%) heterozygotes for the classic phenotype have the whorl-like corneal dystrophy, which does not impair vision, about 10–50% have a few isolated skin lesions, and 50–90% report acroparesthesias, mainly tingling. In addition, heterozygous females may have chronic abdominal pain and diarrhea.[1,2,42,43] With advancing age, some heterozygous females may develop arrhythmias and/or mild to moderate LVH, which may progress to significant cardiomegaly. The occurrence of cerebrovascular disease, including transient ischemic attacks and cerebrovascular accidents, is consistent with the microvascular pathology of the disease. Renal findings in heterozygotes include isothenuria, the presence of erythrocytes, leukocytes, and granular and hyaline casts in the urinary sediment, and especially proteinuria. About 15% of heterozygous females develop renal failure requiring dialysis or transplantation, according to the US and European dialysis and transplant registries. In a retrospective study, the median cumulative survival of heterozygotes was 70 years, 15 years shorter than that of the general female population.[43] To date, limited information is available on the clinical manifestations of heterozygous females in families with the later-onset phenotype.

DIAGNOSTIC EVALUATION

The clinical diagnosis of affected males can be confirmed by the demonstration of deficient α-Gal A activity in plasma, isolated leukocytes, or cultured fibroblasts or lymphoblasts.[1] Classically affected males typically have no detectable or very low α-Gal A activity when the assay is performed using synthetic substrates for α-Gal A, with the addition of α-*N*-acetylgalactosamine in the reaction mixture to inhibit α-galactosidase B activity. The later-onset disease can be detected by the presence of residual α-Gal A activity, which can be up to 10% of mean normal activities. Because the gene encoding α-Gal A undergoes random X inactivation, the expressed level of enzymatic activity in heterozygous females may vary significantly, making accurate carrier detection by measurement of α-Gal A activity unreliable. For example, obligate carriers may have normal plasma and/or leukocyte α-Gal A activities. Note that all daughters of affected males are obligate carriers. Therefore, accurate diagnosis of heterozygous females requires the detection of the family-specific α-Gal A gene mutation.[1]

Prenatal testing can be performed in direct and cultured chorionic villi in the first trimester or in cultured amniocytes in the second trimester using enzymatic and/or molecular techniques.[44]

PATHOLOGY

GL-3 and α-galactosyl-terminal glycosphingolipid substrates progressively accumulate in most cells of classically affected males. However, the major manifestations result from the lysosomal deposition of GL-3 in vascular endothelial cells throughout the body, renal podocytes and tubular cells, and cardiomyocytes. Presumably, the small vessel involvement leads to the renal, cardiac, and cerebrovascular complications in the classically affected patients.

In later-onset males, who do not have the vascular endothelial glycolipid accumulation, the major cardiovascular and renal manifestations result from progressive GL-3 deposition in the cardiomyocytes and renal podocytes, respectively. In addition, vascular ischemia and glycolipid deposition in the perineurium may cause the slowed conduction velocities and distal latency of peripheral nerves. In classically affected males and heterozygous females, glycosphingolipid deposition in nervous tissue appears to be limited to perineural sheath cells of peripheral nerves, neurons of the peripheral and central autonomic nervous system, and certain primary neurons of somatic afferent pathways. Glycosphingolipid deposition has been observed in Schwann cells by some, but not by other investigators. Qualitative and quantitative studies of peripheral sensory neurons in sural nerves and spinal ganglia have shown preferential loss of small myelinated and unmyelinated fibers as well as small cell bodies of spinal ganglia.

Brainstem centers in which lipid deposition has been observed in classically affected males include the nuclei gracilis and cuneatus, dorsal autonomic vagal nuclei, salivary nuclei, nucleus ambiguus, thalamus, reticular substance, mesencephalic nucleus of the fifth nerve, and substantia nigra. Hemisphere involvement has been noted in the amygdaloid, hypothalamic, and hippocampal nuclei. Studies have revealed abnormal lipid deposits in the fifth and sixth cortical layers of the inferior temporal gyrus, the Edinger–Westphal nucleus, the parasympathetic cell column, and the midline nucleus. Lipid storage in neuronal cells of the anterior and posterior lobes of the pituitary also has been described. Immunocytochemical studies using a sensitive antiglobotriaosylceramide monoclonal antibody revealed a highly selective pattern of neuronal involvement. Deposition was observed in selected neurons in the spinal cord and ganglia, brainstem, amygdala, hypothalamus, and entorhinal cortex; however, adjacent areas were spared, including the nucleus basalis, striatum, globus pallidus, and thalamus.

The prominent and progressive vascular endothelial pathology in classically affected patients leads to poor perfusion, dysfunction, and death of peripheral and central autonomic nerve cells, which, together with their endogenous glycolipid accumulation, presumably are responsible for the paresthesias, pain, hypohidrosis, gastrointestinal symptoms such as nausea and diarrhea, and a variety of vague neurologic signs and symptoms.[1] Marked degeneration of the secretory cells and myoepithelial cells of sweat glands has been observed by electron microscopy, suggesting that the hypohidrosis was due to local lipid deposition rather than autonomic nervous system involvement. The observation of a selective decrease in the number of unmyelinated and small myelinated fibers in peripheral nerves has led to the suggestion that selective damage to these fibers may account for the pain production and hypohidrosis in this disorder. Studies of autonomic function revealed sympathetic and parasympathetic dysfunction, particularly in distal cutaneous responses. Histochemical evidence of glycosphingolipid accumulation in neurons and nerve fibers of intestinal nerve plexuses and smooth muscle may account for the uncoordinated intestinal smooth muscle activity, which may lead to complaints of chronic diarrhea or constipation. Glycosphingolipid deposition in the myenteric and submucosal plexuses and a marked decrease in argyrophilic neurons were observed in an involved segment of bowel from an affected male who had jejunal diverticulosis and perforation. Detailed reviews of the neurologic findings are available elsewhere.[1]

BIOCHEMISTRY

The deficient activity of the lysosomal enzyme α-Gal A results in the accumulation of glycosphingolipids and glycoconjugates with terminal α-galactosyl moieties in most visceral tissues and body fluids.[1] The predominant glycosphingolipid accumulated is GL-3 (also called Gb-3), galactosyl-(α1 → 4)galactosyl-(β1 → 4)-glycosyl-(β1 → 1)-ceramide [Gal(α1 → 4) Gal(β1 → 4)Glc(β1 → 1)Cer]. A second neutral glycosphingolipid, galabiosylceramide [Gal(α1 → 4)Gal(β1 → 1) Cer], also accumulates in abnormally high concentrations. Galabiosylceramide deposition appears to be tissue-specific, as this substrate has been detected only in the pancreas, right heart, lung, kidney, and urinary sediment. Recently, the lyso-derivative of GL-3 has been shown to accumulate in classic males, less in classic heterozygotes, and to a much lesser extent in males and heterozygotes with the later-onset phenotype.[45,46] In addition, the blood group B-specific substances, which have terminal α-galactosyl moieties, [Gal(α1 → 3) Gal(2 1αFuc) (β1 → 3)GlcNac(β1 → 3) Gal(β1 → 4) Glc(β1 → 1)Cer] and the blood group B1 glycosphingolipid [Gal(α1 → 3)Gal(2 1αFuc)(β1 → 4)GlcNAc(β1 → 3) Gal(β1 → 4)Glc(β1 → 1)Cer] accumulate in patients who have the blood group B antigen. Thus, affected males and heterozygous females who have blood group B or AB accumulate four major glycosphingolipid substrates, suggesting that they may have a more rapid and severe disease course.

Human α-Gal A is a homodimeric glycoprotein with native and subunit molecular masses of about 101 and 46 kDa. The enzyme has a pH optimum of 4.6 and hydrolyzes the α-linked galactosyl moieties from glycolipids and glycopeptides. Isoelectric focusing studies of the enzyme from various sources reveal multiple forms with pI values ranging from 4.2 to 5.1, due to glycoforms with varying sialylation and phosphorylation.[47] Biosynthetic studies indicated that the α-Gal A subunit in cultured fibroblasts is normally synthesized as a precursor glycopeptide of about 50 kDa. After cleavage of the signal peptide and carbohydrate modifications in the Golgi apparatus and lysosomes, the mature enzyme subunits of 46 kDa form the active, homodimeric enzyme, which contains complex, hybrid, and high-mannose type oligosaccharide moieties.[1,48] A comprehensive review of glycosphingolipid metabolism, the biochemistry of α-Gal A, and the metabolic defect in Fabry disease is available elsewhere.[1]

MOLECULAR GENETICS

α-Gal A Gene and Transcript

Human α-Gal A is encoded by a single housekeeping gene, localized to Xq22.1.[1] The 12-kb α-Gal A gene has been completely sequenced and characterized.[49] The gene contains 7 exons ranging from 92 to 291 bp and 6 introns ranging from 0.2 to 3.8 kb. The full-length 1437-bp α-Gal A cDNA encodes a precursor peptide of 429 amino acids, including a 31-residue signal peptide. The mature 398 amino acid subunit contains four *N*-glycosylation consensus sequences.[49,50]

Characterization of the Molecular Lesions Causing Fabry Disease

More than 750 α-Gal A mutations have been reported (Human Gene Mutation Database; http://www.hgmd.org), including missense and nonsense mutations, large and small gene rearrangements, and splicing defects. All the patients with the later-onset phenotype had missense mutations or cryptic splicing lesions that expressed residual α-Gal A activity. Most of the reported mutations have been private (i.e., confined to a single or a few families). However, several mutations have been found in unrelated families of different ethnic or demographic backgrounds. These include the most common classic mutations, (R227X, A143P, and R342Q) and the most common later-onset mutations (N215S, R112C, IVS4+919G>A). Current efforts to establish genotype–phenotype correlations are underway. Clinical characterization of patients as having classic or later-onset phenotypes by physicians with expertise in Fabry disease are required. In addition, laboratory studies that determine the *in vitro* expression of missense mutations can identify those with essentially no activity (i.e., consistent with classic phenotypes) and those with residual activity (i.e., consistent with the later-onset phenotype). Several missense mutations that occur in certain populations at frequencies of 1 in 1000 or more, including D313Y in Caucasians,[51,52] and E66Q in Japanese and Koreans,[53,54] have been shown to be polymorphic with 50–70% *in vitro* expressed activity and most likely nonpathogenic. These variants have decreased activity at neutral pH, resulting in low plasma α-Gal A activity. These variants have been found in screening patients in hemodialysis, cardiac, and stroke clinics in Europe and the United States.

TREATMENT

Medical Management

Before the availability of enzyme replacement therapy (see below), care of patients with regard to the acroparesthesias and gastrointestinal pain, cardiac, pulmonary, and cerebrovascular manifestations remained nonspecific and symptomatic. In males and heterozygotes with the classical phenotype, the single most debilitating and morbid aspect of the disease is the excruciating pain. Prophylactic administration of low-maintenance dosages of diphenylhydantoin, carbamazepine, or gabapentin may decrease the frequency and severity of the periodic crises of excruciating pain and constant discomfort in affected males and heterozygotes.[55] Narcotic analgesics and nonsteroidal anti-inflammatory drugs should be avoided.

Control of hypertension is essential to minimize renal, cardiovascular, and cerebrovascular disease. Successful heart transplantation in patients with end-stage cardiomyopathy has proved life-saving. Pancrelipase or metoclopramide can help with gastrointestinal symptoms. Patients with microalbuminuria, albuminuria, or proteinuria should be aggressively treated with angiotensin-converting enzyme inhibitors and/or angiotensinogen receptor blockers. Obstructive lung disease has been documented in older affected males and heterozygotes, with more severe impairment in smokers; therefore, patients should be discouraged from smoking. Patients with reversible obstructive airway disease may benefit from bronchodilation therapy. Prophylactic oral anticoagulants are recommended, as these patients are stroke-prone.

Dialysis and Renal Transplantation

Because renal insufficiency is the most frequent complication in patients with this disease, chronic hemodialysis and renal transplantation have become life-saving procedures in patients who develop end-stage renal disease. Patients enter dialysis at an average age of 40 years and have a 3-year survival of about 60%.[56] Successful transplantation will correct renal function, and the normal α-Gal A activity in the allograft will catabolize the turnover of endogenous renal glycosphingolipid substrates. Transplanted patients with Fabry disease do at least as well as transplanted patients with other causes of renal failure.[57,58] Transplantation of kidneys from Fabry heterozygotes should be avoided. Therefore, all potential related donors must be carefully evaluated so that affected males and heterozygous females are excluded.

Enzyme Replacement Therapy

Based on extensive preclinical evaluations in α-Gal A-deficient mice,[59] enzyme replacement therapy with recombinant human α-Gal A has been evaluated in clinical trials.[3–9] Although the two enzymes used in these trials, agalsidase alfa and agalsidase beta, were produced by different techniques, studies comparing the same amount of enzyme (i.e., 1.0 mg of each) revealed similar specific activities, kinetic properties, and biochemical compositions. However, agalsidase beta had had more mannose 6-phosphate residues (i.e., greater lysosomal uptake) and higher sialyation (less hepatic uptake) than agalsidase alfa; these attributes were physiologically and functionally important in studies in cell culture and in Fabry mice.[60–62]

Phase I and I/II Clinical Trials

The phase I/II open-label, dose-escalation trial involving 15 classically affected males evaluated the safety and effectiveness of five doses of agalsidase beta (Fabrazyme, Genzyme Corporation).[4] The enzyme was well tolerated, and rapid and marked reductions in plasma and tissue GL-3 were observed biochemically, histologically, and ultrastructurally. GL-3 deposits were reduced in the vascular endothelium of the kidney, heart, skin and liver by light and electron microscopic evaluation. In addition, patients reported decreased pain and gastrointestinal symptoms and increased ability to perspire.

Phase II and III Clinical Trials and Phase III Extension Study

A single center, double-blind, placebo-controlled phase II trial of agalsidase alfa (Replagal, Shire plc) involved 26 male patients with neuropathic pain who received 0.2 mg per kg every 2 weeks for 22 weeks (12 doses).[6] The primary efficacy end point was pain at its worst, and pain medication was withdrawn before evaluations. The mean (SE) neuropathic pain severity score of the Brief Pain Inventory declined in patients treated with agalsidase alfa (n 14) versus no significant change in the placebo group (n 12; $p=0.02$). Plasma GL-3 levels decreased approximately 50% in patients treated with α-Gal A.

A phase III multinational, multicenter, randomized, placebo-controlled, double-blind trial and 5-year extension study evaluated the safety and effectiveness of enzyme replacement in 58 classically affected patients (56 males, 2 heterozygotes) who received 1.0 mg per kg of agalsidase beta or placebo every 2 weeks for 20 weeks (11 doses).[5] The primary efficacy end point was the percentage of patients whose renal capillary endothelial GL-3 deposits cleared to normal or near normal histologically. Also evaluated were the histological clearance of microvascular endothelial GL-3 deposits in the heart and skin, changes in pain, and in quality of life (Short Form-36 Health Status Survey). In this study, 20 of 29 agalsidase beta-treated patients (69%) cleared the accumulated GL-3 from the renal capillary endothelium versus 0 of 29 placebo-treated patients ($p<0.001$). Compared to the placebo group, agalsidase beta-treated patients also had markedly decreased microvascular endothelial GL-3 in the skin ($p < 0.001$) and heart ($p < 0.001$). Patients receiving agalsidase beta cleared the accumulated GL-3 in plasma to nondetectable levels.[5]

All 58 patients who completed the phase III trial received agalsidase beta in a 5-year extension study.[7,9] After 6 months of the open-label therapy, all 22 former placebo and 20 of 21 α-Gal A-treated patients (42 of 43; 98%) who had a third (optional) biopsy achieved or maintained normal or near-normal renal capillary endothelial histology. Similar results were observed for skin capillary endothelium; 96% of both former placebo patients (22 of 23) and of agalsidase beta patients (23 of 24) who had biopsies achieved normal or near-normal histology. In the capillary endothelium of the heart, histology scores improved with duration of treatment. Among patients who had optional heart biopsies, 10 of the 15 (67%) former placebo patients (6 months of treatment) and 14 of 17 (82%) of the agalsidase beta patients (12 months of treatment) attained normal or near-normal histology. In addition, histologic examination of other renal, cardiac and skin cell types revealed complete or partial clearance of accumulated GL-3. For example, in the kidney, GL-3 clearance to normal or near-normal levels was observed in renal, glomerular, and nonglomerular capillary endothelial cells, mesangial cells, and interstitial cells after 6 and 12 months of treatment. Enzyme replacement was well tolerated; the adverse event incidence and profiles were similar for both treatment groups in the phase III trial, except for mild to moderate infusion reactions to agalsidase beta, which were managed conservatively. Although IgG seroconversion occurred in 51 of 58 agalsidase beta-treated patients, GL-3 clearance was not impaired, and titers decreased with continued treatment.

Based on the phase II and III studies, enzyme replacement therapy for Fabry disease was approved by the European Agency for Evaluation of Medical Products in 2001 (Fabrazyme and Replagal), and by the US Food and Drug Administration in 2003 (Fabrazyme only).

The patients enrolled in the phase III study have been followed in an extension study with reports at 3 and 5 years of enzyme replacement therapy.[7,9] These studies indicated that enzyme replacement therapy continued to be safe and effective after 3 and 5 years of treatment. In particular, patients with normal or near-normal serum creatinine levels remained normal, whereas a few older patients (40 years or older) with significant proteinuria (2.0 g per dL) and more than 50% glomerulosclerosis had increased serum creatinine values. These findings emphasized the importance of initiating enzyme replacement therapy early, optimally in childhood or adolescence, before irreversible renal damage. In addition, aggressive reduction of proteinuria should be undertaken with angiotensin-converting enzyme inhibitors and/or angiotensin receptor blockers.

Since the approval in Europe (2001) and the United States (2003), there have been numerous reports of the clinical benefits of enzyme replacement therapy. These studies have demonstrated the effectiveness of agalsidase beta at 1 mg per kg in stabilizing renal function and improving cardiac function, gastrointestinal manifestations, Fabry neuropathy, sweating, nerve fiber function and quality of life. Also, the enzyme can be administered to patients on hemodialysis and to those who have received transplants.[63,64] A recent study of the effect of dose and cumulative dose in 12 classically affected patients revealed that after 5 years of treatment, patients who had received 1.0 mg per kg every 2 weeks had cleared the accumulated GL-3 in renal podocytes, while treatment with agalsidase alfa at 0.2 mg per kg over 5 years did not.[65]

Phase IV Clinical Trial in Patients with Advanced Disease

The phase IV clinical trial of agalsidase beta at 1.0 mg per kg every 2 weeks was a multinational, multicenter, double-blind, placebo-controlled trial involving 82 patients with classic Fabry disease who had mild-to-moderate renal disease.[8] The patients were randomized 2:1 (agalsidase beta:placebo) at each study site, and the median study period was 18.5 months (35 months total). The primary end point compared the time to the first clinical event (renal, cardiac, cerebrovascular or death) between the two treatment groups. After a clinical event, patients were switched to open-label enzyme. This trial showed that compared to placebo, agalsidase beta at 1 mg per kg every 2 weeks slowed the rate of progression of Fabry disease and substantially reduced the risk of renal, cardiac, and cerebrovascular events together and individually. After the proscribed adjustment for the baseline imbalance in proteinuria between the two treatment groups, patients randomized to agalsidase beta were 53% less likely than the placebo-treated patients to experience a clinically significant renal, cardiac, or cerebrovascular event. Similar to other renal diseases, baseline proteinuria was the most important determinant of outcome. Among the 74 patients who were compliant with the study protocol, after the prespecified adjustment for the proteinuria baseline imbalance between the two treatment groups, the patients who received agalsidase beta were 61% less likely to experience a clinically significant event ($p = 0.034$). The most pronounced benefits of agalsidase beta were seen when therapy was started earlier in the course of the disease (i.e., with less renal dysfunction). These findings emphasize the importance of early treatment with 1 mg per kg of recombinant enzyme.[7]

Pharmacologic Chaperone Therapy

In addition to enzyme replacement therapy, current efforts are focused on the development of other strategies to treat Fabry disease. An attractive approach for lysosomal and other genetic diseases resulting from mutations that cause protein misfolding and/or trafficking is pharmacologic chaperone therapy: the use of small molecules that bind to and stabilize misfolding or unstable proteins, thereby increasing protein function. Using the α-Gal A substrate analog deoxygalactonojirimycin (DGJ), the residual activities of several α-Gal A missense mutations (e.g., Q279E and R301Q) increased in cultured cells as well as in transgenic mice carrying the rescuable R301Q mutation.[66–69] These findings suggested that the active-site inhibitor stabilized the mutant α-Gal A glycoprotein such that more of the enzyme was transported to the lysosome. In addition, clinical proof of concept has been demonstrated in a patient with later-onset Fabry disease.[70] Clinical trials are now underway for pharmacologic chaperone therapy for Fabry disease with the α-Gal A pharmacologic chaperone, 1-deoxygalactonojirimycin (AT1001, Amicus Therapeutics, Inc.). Efforts are also underway to evaluate the safety and effectiveness of coadministration of the chaperone before or with the recombinant enzyme. Studies in the Fabry mouse model demonstrate that cotherapy results in increased enzyme stability and greater tissue uptake.[71]

Substrate Reduction and Gene Therapy

Efforts also are underway to develop substrate reduction therapy for Fabry disease. This strategy uses an inhibitor of glycosphingolipid biosynthesis to reduce the amount of GL-3 accumulation. This approach is in Phase III clinical trials for Gaucher disease, where it has demonstrated its safety and effectiveness.[72]

α-Gal A gene replacement has been undertaken in α-Gal A knockout mice using various vectors to introduce the gene, including retroviral,[73] adeno-associated,[74] and lentiviral vectors.[75] In many of these studies, marked reductions of plasma and tissue GL-3 were demonstrated.[74,75] However, no clinical trials have been reported to date.

SUMMARY

Since the first descriptions by Fabry and Anderson in 1898,[76,77] the characterization of Fabry disease has evolved from a dermatologic curiosity to a treatable inborn error of metabolism with classic and later-onset phenotypes. Diagnosis of this X-linked disease is readily confirmed by demonstration of deficient α-Gal A enzymatic activity in affected males and by α-Gal A mutation detection in heterozygous females. Neurologic manifestations include early onset of chronic pain (acroparesthesias and pain crises) and cerebrovascular complications of small vessel ischemia and infarction. Plasma or leukocyte α-Gal A activity determinations in patients with cryptogenic strokes may reveal previously unrecognized patients with Fabry disease. Enzyme replacement therapy has been shown in randomized, double-blind, placebo-controlled trials to be safe and effective at 1 mg per kg every 2 weeks. Enzyme replacement cleared the vascular endothelial as well as renal podocyte glycolipid deposits and reversed, stabilized, or markedly improved the disease symptoms. Early intervention of classically affected males in childhood is essential to avoid the irreversible renal, cardiac, and cerebrovascular manifestations of the disease.

ACKNOWLEDGEMENTS

The author thanks Kenneth H. Astrin for his assistance with this chapter. This work has been supported in part by grants from the National Institutes of Health, including a research grant (R37 DK34045 Merit Award), a grant UL 1TR 000067 for the Mount Sinai Clinical and Translational Science Award (CTSA) program supported by the National Institutes of Health, and a research grant from the Genzyme Corporation.

References

1. Desnick RJ, Ioannou YA, Eng CM. α-Galactosidase A deficiency: Fabry disease. In: Scriver CR, Beaudet AL, Sly WS, et al., eds. The Metabolic and Molecular Bases of Inherited Disease. 8th ed. New York: McGraw-Hill; 2001:3733–3774.
2. Desnick RJ. Fabry disease. In: Murray MF, Babyatsky MW, Giovanni MA, Alkuraya FS, Stewart DR, eds. Clinical Genomics: Practical Applications in Adult Patient Care. New York: McGraw-Hill; 2014:439–444.
3. Schiffmann R, Murray GJ, Treco D, et al. Infusion of α-galactosidase A reduces tissue globotriaosylceramide storage in patients with Fabry disease. *Proc Natl Acad Sci U S A*. 2000;97:365–370.

4. Eng CM, Banikazemi M, Gordon R, et al. A phase 1/2 clinical trial of enzyme replacement in Fabry disease: pharmacokinetic, substrate clearance, and safety studies. *Am J Hum Genet*. 2001;68:711–722.
5. Eng CM, Guffon N, Wilcox WR, et al. Safety and efficacy of recombinant human α-galactosidase A replacement therapy in Fabry's disease. *N Engl J Med*. 2001;345:9–16.
6. Schiffmann R, Kopp JB, Austin III HA, et al. Enzyme replacement therapy in Fabry disease: a randomized controlled trial. *JAMA*. 2001;285:2743–2749.
7. Wilcox WR, Banikazemi M, Guffon N, et al. Long-term safety and efficacy of enzyme replacement therapy for Fabry disease. *Am J Hum Genet*. 2004;75:65–74.
8. Banikazemi M, Bultas J, Waldek S, et al. Agalsidase beta therapy for advanced Fabry disease: a randomized trial. *Ann Intern Med*. 2007;146:77–86.
9. Germain DP, Waldek S, Banikazemi M, et al. Substained, long-term renal stabilization after 54 months of agalsidase beta therapy in patients with Fabry disease. *J Am Soc Nephrol*. 2007;18:1547–1557.
10. Spada M, Pagliardini S, Yasuda M, et al. High incidence of later-onset Fabry disease revealed by newborn screening. *Am J Hum Genet*. 2006;79:31–40.
11. Lin HY, Chong KW, Hsu JH, et al. High incidence of the cardiac variant of Fabry disease revealed by newborn screening in the Taiwan Chinese population. *Circ Cardiovasc Genet*. 2009;2:450–456.
12. Hwu WL, Chien YH, Lee NC, et al. Newborn screening for Fabry disease in Taiwan reveals a high incidence of the later-onset GLA mutation c.936+919G>A (IVS4+919G>A). *Hum Mutat*. 2009;30:1397–1405.
13. Chien YH, Lee NC, Chiang SC, Desnick RJ, Hwu WL. Fabry disease: incidence of the common later-onset α-galactosidase A IVS4+919G→A mutation in Taiwanese newborns—superiority of DNA-based to enzyme-based newborn screening for common mutations. *Mol Med*. 2012;18:780–784.
14. Wittmann J, Karg E, Turi S, et al. Newborn screening for lysosomal storage disorders in Hungary. *JIMD Rep*. 2012;6:117–125.
15. Liao HC, Huang YH, Chen YJ, et al. Plasma globotriaosylsphingosine (lysoGb3) could be a biomarker for Fabry disease with a Chinese hotspot late-onset mutation (IVS4+919G>A). *Clin Chim Acta*. 2013;426:114–120.
16. Calzavara-Pinton PG, Colombi M, Carlino A, et al. Angiokeratoma corporis diffusum and arteriovenous fistulas with dominant transmission in the absence of metabolic disorders. *Arch Dermatol*. 1995;131:57–62.
17. Dutsch M, Marthol H, Stemper B, et al. Small fiber dysfunction predominates in Fabry neuropathy. *J Clin Neurophysiol*. 2002;19:575–586.
18. Tan SV, Lee PJ, Walters RJ, et al. Evidence for motor axon depolarization in Fabry disease. *Muscle Nerve*. 2005;32:548–551.
19. Üçeyler N, He L, Schönfeld D, et al. Small fibers in Fabry disease: baseline and follow-up data under enzyme replacement therapy. *J Peripher Nerv Syst*. 2011;16:304–314.
20. Sher NA, Letson RD, Desnick RJ. The ocular manifestations in Fabry's disease. *Arch Ophthalmol*. 1979;97:671–676.
21. Nguyen TT, Gin T, Nicholls K, et al. Ophthalmological manifestations of Fabry disease: a survey of patients at the Royal Melbourne Fabry disease treatment centre. *Clin Exp Ophthalmol*. 2005;33:164–168.
22. Tøndel C, Bostad L, Laegreid LM, Houge G, Svarstad E. Prominence of glomerular and vascular changes in renal biopsies in children and adolescents with Fabry disease and microalbuminuria. *Clin Ther*. 2008;30:S42.
23. Najafian B, Svarstad E, Bostad L, et al. Progressive podocyte injury and globotriaosylceramide (GL-3) accumulation in young patients with Fabry disease. *Kidney Int*. 2011;79:663–670.
24. Branton MH, Schiffmann R, Sabnis SG, et al. Natural history of Fabry renal disease: influence of α-galactosidase A activity and genetic mutations on clinical course. *Medicine (Baltimore)*. 2002;81:122–138.
25. MacDermot KD, Holmes A, Miners AH. Anderson-Fabry disease: clinical manifestations and impact of disease in a cohort of 98 hemizygous males. *J Med Genet*. 2001;38:750–760.
26. Linhart A, Lubanda JC, Palecek T, et al. Cardiac manifestations in Fabry disease. *J Inherit Metab Dis*. 2001;24(suppl 2):75–83.
27. Fellgiebel A, Muller MJ, Ginsberg L. CNS manifestations of Fabry's disease. *Lancet Neurol*. 2006;5:791–795.
28. Banikazemi M, Ullman T, Desnick RJ. Gastrointestinal manifestations of Fabry disease: clinical response to enzyme replacement therapy. *Mol Genet Metab*. 2005;85:255–259.
29. Germain DP, Avan P, Chassaing A, et al. Patients affected with Fabry disease have an increased incidence of progressive hearing loss and sudden deafness: an investigation of twenty-two consecutive hemizygous male patients. *BMC Med Genet*. 2002;3:10–20.
30. Ries M, Kim HJ, Zalewski CK, et al. Neuropathic and cerebrovascular correlates of hearing loss in Fabry disease. *Brain*. 2007;130:143–150.
31. Elleder M, Bradova V, Smid F, et al. Cardiocyte storage and hypertrophy as a sole manifestation of Fabry's disease. Report on a case simulating hypertrophic non-obstructive cardiomyopathy. *Virchows Arch A Pathol Anat Histol*. 1990;417:449–455.
32. von Scheidt W, Eng CM, Fitzmaurice TF, et al. An atypical variant of Fabry's disease with manifestations confined to the myocardium. *N Engl J Med*. 1991;324:395–399.
33. Nakao S, Takenaka T, Maeda M, et al. An atypical variant of Fabry's disease in men with left ventricular hypertrophy. *N Engl J Med*. 1995;333:288–293.
34. Nakao S, Kodama C, Takenaka T, et al. Fabry disease: detection of undiagnosed hemodialysis patients and identification of a "renal variant" phenotype. *Kidney Int*. 2003;64:801–807.
35. Sachdev B, Takenaka T, Teraguchi H, et al. Prevalence of Anderson–Fabry disease in male patients with late onset hypertrophic cardiomyopathy. *Circulation*. 2002;105:1407–1411.
36. Spada M, Pagliardini S. Screening for Fabry disease in end-stage nephropathies. *J Inherit Metab Dis*. 2002;25(suppl 1):113.
37. Kotanko P, Kramar R, Devrnja D, et al. Results of a nationwide screening for Anderson-Fabry disease among dialysis patients. *J Am Soc Nephrol*. 2004;15:1323–1329.
38. Meehan SM, Junsanto T, Rydel JJ, et al. Fabry disease: renal involvement limited to podocyte pathology and proteinuria in a septuagenarian cardiac variant. Pathologic and therapeutic implications. *Am J Kidney Dis*. 2004;43:164–171.
39. Rosenthal D, Lien YH, Lager D, et al. A novel α-galactosidase A mutant (M42L) identified in a renal variant of Fabry disease. *Am J Kidney Dis*. 2004;44:85–89.

40. Baptista MV, Ferreira S, Pinho-E-Melo T, et al, PORTuguese Young STROKE Investigators. Mutations of the GLA gene in young patients with stroke: the PORTYSTROKE study—screening genetic conditions in Portuguese young stroke patients. *Stroke*. 2010;41:431–436.
41. Wozniak MA, Kittner SJ, Tuhrim S, et al. Frequency of unrecognized Fabry disease among young European-American and African-American men with first ischemic stroke. *Stroke*. 2010;41:78–81.
42. Opitz JM, Stiles FC, Wise D, et al. The genetics of angiokeratoma corporis diffusum (Fabry's disease) and its linkage relations with the Xg locus. *Am J Hum Genet*. 1965;17:325–342.
43. MacDermot KD, Holmes A, Miners AH. Anderson-Fabry disease: clinical manifestations and impact of disease in a cohort of 60 obligate carrier females. *J Med Genet*. 2001;38:769–775.
44. Desnick RJ. Prenatal diagnosis of Fabry disease. *Prenatal Diagn*. 2007;27:693–694.
45. Aerts JM, Groener JE, Kuiper S, et al. Elevated globotriaosylsphingosine is a hallmark of Fabry disease. *Proc Natl Acad Sci U S A*. 2008;105:2812–2817.
46. van Breemen MJ, Rombach SM, Dekker N, et al. Reduction of elevated plasma globotriaosylsphingosine in patients with classic Fabry disease following enzyme replacement therapy. *Biochim Biophys Acta*. 1812;2011:70–76.
47. Matsuura F, Ohta M, Ioannou YA, et al. Human α-galactosidase A: characterization of the N-linked oligosaccharides on the intracellular and secreted glycoforms overexpressed by Chinese hamster ovary cells. *Glycobiology*. 1998;8:329–339.
48. Lemansky P, Bishop DF, Desnick RJ, Hasilik A, von Figura K. Synthesis and processing of alpha-galactosidase A in human fibroblasts: evidence for different mutations in Fabry disease. *J Biol Chem*. 1987;262:2062–2065.
49. Kornreich R, Desnick RJ, Bishop DF. Nucleotide sequence of the human α-galactosidase A gene. *Nucl Acids Res*. 1989;17:3301–3302.
50. Ioannou YA, Zeidner KM, Grace ME, et al. Human α-galactosidase A: glycosylation site 3 is essential for enzyme solubility. *Biochem J*. 1998;332:789–797.
51. Froissart R, Guffon N, Vanier MT, et al. Fabry disease: D313Y is an α-galactosidase A sequence variant that causes pseudodeficient activity in plasma. *Mol Genet Metab*. 2003;80:307–314.
52. Yasuda M, Shabbeer J, Benson SD, et al. Fabry disease: characterization of alpha-galactosidase A double mutations and the D313Y plasma enzyme pseudodeficiency allele. *Hum Mutat*. 2003;22:486–492.
53. Togawa T, Tsukimura T, Kodama T, et al. Fabry disease: biochemical, pathological and structural studies of the α-galactosidase A with E66Q amino acid substitution. *Mol Genet Metab*. 2012;105:615–620.
54. Lee BH, Heo SH, Kim GH, et al. Mutations of the GLA gene in Korean patients with Fabry disease and frequency of the E66Q allele as a functional variant in Korean newborns. *J Hum Genet*. 2010;55:512–517.
55. Ries M, Mengel E, Kutschke G, et al. Use of gabapentin to reduce chronic neuropathic pain in Fabry disease. *J Inherit Metab Dis*. 2003;26:413–414.
56. Thadhani R, Wolf M, West ML, et al. Patients with Fabry disease on dialysis in the United States. *Kidney Int*. 2002;61:249–255.
57. Mignani R, Gerra D, Maldini L, et al. Long-term survival of patients with renal transplantation in Fabry's disease. *Contrib Nephrol*. 2001;229–233.
58. Tsakiris D, Simpson HK, Jones EH, et al. Report on management of renal failure in Europe, XXVI, 1995. Rare diseases in renal replacement therapy in the ERA-EDTA Registry. *Nephrol Dial Transplant*. 1996;11:4–20.
59. Ioannou YA, Zeidner KM, Gordon RE, et al. Fabry disease: preclinical studies demonstrate the effectiveness of α-galactosidase A replacement in enzyme-deficient mice. *Am J Hum Genet*. 2001;68:14–25.
60. Lee K, Jin X, Zhang K, et al. A biochemical and pharmacological comparison of enzyme replacement therapies for the glycolipid storage disorder Fabry disease. *Glycobiology*. 2003;13:305–313.
61. Sakuraba H, Murata-Ohsawa M, Kawashima I, et al. Comparison of the effects of agalsidase alfa and agalsidase beta on cultured human Fabry fibroblasts and Fabry mice. *J Hum Genet*. 2006;51:180–188.
62. Togawa T, Takada M, Aizawa Y, Tsukimura T, Chiba Y, Sakuraba H. Comparative study on mannose 6-phosphate residue contents of recombinant lysosomal enzymes. *Mol Genet Metab*. 2014;111:369–373.
63. Kosch M, Koch HG, Oliveira JP, et al. Enzyme replacement therapy administered during hemodialysis in patients with Fabry disease. *Kidney Int*. 2004;66:1279–1282.
64. Pisani A, Spinelli L, Sabbatini M, et al. Enzyme replacement therapy in Fabry disease patients undergoing dialysis: effects on quality of life and organ involvement. *Am J Kidney Dis*. 2005;46:120–127.
65. Tøndel C, Bostad L, Larsen KK, et al. Agalsidase benefits renal histology in young patients with Fabry disease. *J Am Soc Nephrol*. 2013;24:137–148.
66. Fan JQ, Ishii S, Asano N, et al. Accelerated transport and maturation of lysosomal α-galactosidase A in Fabry lymphoblasts by an enzyme inhibitor. *Nat Med*. 1999;5:112–115.
67. Fan JQ. A contradictory treatment for lysosomal storage disorders: inhibitors enhance mutant enzyme activity. *Trends Pharmacol Sci*. 2003;24:355–360.
68. Ishii S, Yoshioka H, Mannen K, et al. Transgenic mouse expressing human mutant alpha-galactosidase A in an endogenous enzyme deficient background: a biochemical animal model for studying active-site specific chaperone therapy for Fabry disease. *Biochim Biophys Acta*. 2004;1690:250–257.
69. Khanna R, Soska R, Lun Y, et al. The pharmacological chaperone 1-deoxygalactonojirimycin reduces tissue globotriaosylceramide levels in a mouse model of Fabry disease. *Mol Ther*. 2010;18:23–33.
70. Frustaci A, Chimenti C, Ricci R, et al. Improvement in cardiac function in the cardiac variant of Fabry's disease with galactose-infusion therapy. *N Engl J Med*. 2001;345:25–32.
71. Benjamin ER, Khanna R, Schilling A, et al. Co-administration with the pharmacological chaperone AT1001 increases recombinant human α-galactosidase A tissue uptake and improves substrate reduction in Fabry mice. *Mol Ther*. 2012;20:717–726.
72. Lukina E, Watman N, Arreguin EA, et al. Improvement in hematological, visceral, and skeletal manifestations of Gaucher disease type 1 with oral eliglustat tartrate (Genz-112638) treatment: 2-year results of a phase 2 study. *Blood*. 2010;116:4095–4098.
73. Qin G, Takenaka T, Telsch K, et al. Preselective gene therapy for Fabry disease. *Proc Natl Acad Sci U S A*. 2001;98:3428–3433.

74. Ziegler RJ, Cherry M, Barbon CM, et al. Correction of the biochemical and functional deficits in Fabry mice following AAV8-mediated hepatic expression of α-galactosidase A. *Mol Ther*. 2007;15:492–500.
75. Yoshimitsu M, Higuchi K, Ramsubir S, et al. Efficient correction of Fabry mice and patient cells mediated by lentiviral transduction of hematopoietic stem/progenitor cells. *Gene Ther*. 2007;14:256–265.
76. Anderson W. A case of angiokeratoma. *Br J Dermatol*. 1898;10:113–117.
77. Fabry J. Ein beitrag zur kenntnis der Purpura haemorrhagica nodularis (Purpura papulosa hemorrhagica Hebrae). *Arch Dermatol Syph*. 1898;43:187–200.

CHAPTER

39

Schindler Disease: Deficient-N-Acetylgalactosaminidase Activity

Detlev Schindler and Robert J. Desnick†*

*University of Würzburg, Würzburg, Germany

†Icahn School of Medicine at Mount Sinai, New York, NY, USA

INTRODUCTION

Schindler disease is an autosomal disorder resulting from the deficient activity of α-*N*-acetylgalactosaminidase (α-galactosidase B).[1] The enzymatic defect leads to the cellular accumulation and increased urinary excretion of glycopeptides and oligosaccharides containing α-*N*-acetylgalactosaminyl residues. The disease is clinically heterogeneous, with three major subtypes. Patients with type I disease have early-onset neuroaxonal dystrophy with the characteristic neuropathology,[2,3] whereas type II patients have angiokeratoma corporis diffusum, mild intellectual impairment, and sensory nerve involvement with neuroaxonal degeneration of peripheral nerves.[4–6] Patients with type III disease have an intermediate and variable phenotype with manifestations ranging from seizures and moderate psychomotor retardation in infancy to a milder autistic presentation.[7–9] All three subtypes have very low levels of α-*N*-acetylgalactosaminidase activity, have essentially the same patterns of glycoconjugate accumulation in the urine, and are inherited as autosomal recessive traits. Here, the clinical, pathologic, biochemical, and molecular findings in the severe infantile and milder later-onset forms of this metabolic disease are described.

CLINICAL FEATURES AND DIAGNOSTIC RESULTS

Type I Disease

α-*N*-acetylgalactosaminidase deficiency was first recognized in two German brothers, the offspring of fourth cousins of German descent.[1] The clinical course of these patients was characterized by three stages: i) apparently normal development in the first 9–12 months of life; ii) a period of developmental delay followed by rapid regression starting in the second year of life; and iii) increasing neurologic impairment, resulting by age 3–4 years in profound psychomotor retardation, spasticity, seizures, blindness, and decorticate posturing (Figure 39.1).

Clinical onset of the disease in the older brother was signaled by poor gait coordination, clumsiness, episodes of falling, and startle reactions at 12 months. In the younger brother, grand mal seizures began at 8 months and occurred five times during the next 6 months. Peak development was achieved at about 15 months in both sibs. Thereafter, each experienced a retrogressive course, with the final loss of all mental and motor skills acquired previously. Along with the progressive psychomotor deterioration, both siblings developed strabismus, nystagmus, visual impairment, spasticity, and frequent myoclonic movements. By age 3–4 years, both brothers had profound psychomotor retardation, were immobile and incontinent, and had little or no expressive contact with the environment but had retained some perceptive capability. Since then, these brothers have survived in a vegetative state, dependent on liquid nutrition, intermittent tube feeding, and regular nasopharyngeal suctioning. They remained alive in their teens

Rosenberg's Molecular and Genetic Basis of Neurological and Psychiatric Disease
http://dx.doi.org/10.1016/10.1016/B978-0-12-410529-4.00039-5

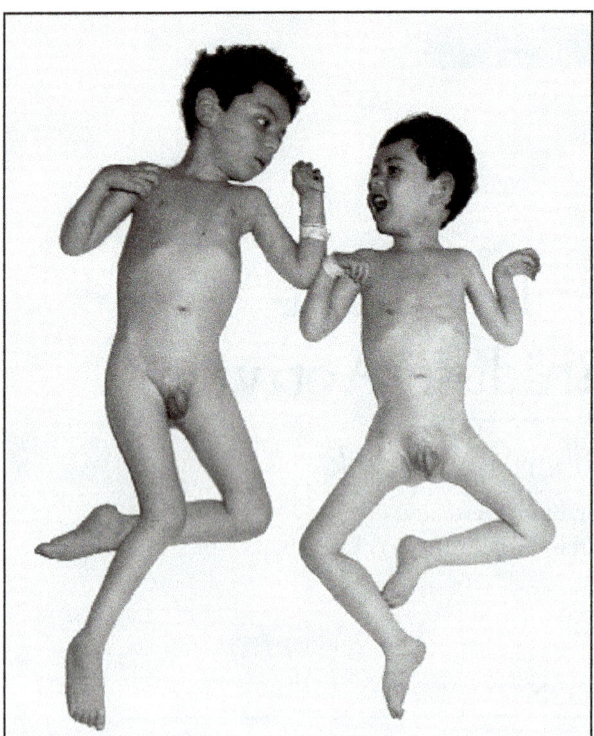

FIGURE 39.1 **The affected German siblings with type I disease at 5 years 3 months (left) and at 4 years (right) of age.** Reproduced with permission from[1] © McGraw-Hill Education LLC.

because of the diligent nursing efforts of their parents. At ages 16 and 17 years, neither sib moved (except for frequent myoclonic jerks and occasional seizures) or responded to optical or acoustic stimuli. They were relatively hypotonic distally and hypertonic proximally. They had bilateral pyramidal tract involvement with symmetric hyperreflexia and clonus. Bilateral optic atrophy was evident. Their developmental skills were at the newborn level. Of note, they did not have ocular, cutaneous, skeletal, or other visceral signs of a storage disease.

A third patient with type I disease, a maternal third and paternal fourth infant cousin of the original sibs, was diagnosed posthumously.[3] This patient experienced several severe grand mal seizures from age 7 months, most occurring with febrile episodes during respiratory tract infections. Developmental progress also was halted from 15 months of age. He died unexpectedly at 18 months from apnea during prolonged convulsions. An autopsy was not performed. However, fibroblasts were preserved, and the diagnosis of α-*N*-acetylgalactosaminidase deficiency was made.

Pertinent Laboratory, Imaging, and Functional Findings

Routine laboratory studies including complete blood counts and blood and cerebrospinal fluid chemistries were within normal limits.[10] Skeletal radiography showed systemic, diffuse, and severe osteopenia and bilateral subluxation of the hips. The electroencephalography (EEG) indicated diffuse brain dysfunction and disclosed irritative features such as multifocal isolated spikes and spike-wave complexes. Quantitative EEG and brain mapping showed marked slowing with substantially increased β and δ activity, in particular over the central and parieto-occipital regions. Brainstem auditory, somatosensory, and visual evoked potentials had low amplitude, delayed responses, or both. These findings were consistent with cortical blindness and deafness, although some residual informational processing was retained. Electroretinography showed increased voltages under all conditions, in a manner consistent with the exclusion of the peripheral visual pathway from the disease process. Nerve conduction velocities were low normal. Computed tomography (CT) and magnetic resonance imaging (MRI) studies showed marked atrophy of the cervical spinal cord, the brainstem, and the cerebellum. The optic tracts and cranial nerves were hypotrophic. Brain atrophy was generalized and also extended to supratentorial regions including the cerebral white matter and cortical structures. These findings were consistent with an overall reduced cerebral glucose metabolism compared to normal values for age, seen on positron emission tomography studies using 2[^{18}F]fluoro-2-deoxy-D-glucose.[11] In addition, the extent of regional glucose hypometabolism was directly correlated with the degree of brain atrophy seen on MRI.

Type II Disease

Type II α-*N*-acetylgalactosaminidase deficiency is an adult-onset disorder characterized by angiokeratoma corporis diffusum, mild intellectual impairment, sensory nerve involvement, and lymphedema. To date, four affected adults in three unrelated consanguineous families have been identified.[4–6] Angiokeratoma corporis diffusum was first noted at about 30 years of age in the unrelated Japanese patients.[4,5] The eruption spread slowly over their bodies, essentially in an appearance and distribution similar to that of affected males with Fabry disease (see Chapter 38). There was no family history of individuals with similar dermatologic findings. Both patients were products of first-cousin marriages.

Dermatologically, these patients had densely peppered, tiny, deep-red to purple maculopapules ranging in diameter from less than 1 mm to 3 mm. The lesions were present on the face and fingers, but they were denser on the axillae, breasts, lower abdomen, groin, buttocks, and upper thighs. Telangiectasias were present on the lips and on the oropharyngeal mucosa. Dilated blood vessels also were present on the ocular conjunctiva and were observed with a corkscrew-like tortuousness in the fundi. There were no retinal hemorrhages, macular changes, or corneal opacities. Cardiac examination and electrocardiographic findings were normal at rest and with exercise. There was no organomegaly, lymphadenopathy, or skeletal abnormality. On neurologic examination, the first Japanese patient had a dull affect and was mentally slow.[4] The second Japanese patient experienced tinnitus and hearing loss that started before age 20 years and recurrent attacks of vertigo from age 25 years.[5] She was diagnosed as having Ménière syndrome and underwent surgery for vestibular dysfunction twice, without success. She appeared to be of normal intelligence.

In 1994, two consanguineous adult sibs of Spanish descent with type II disease were reported.[6] At diagnosis, the brother was 42 years old, and his sister was 38 years old. The affected male had a slightly coarse facies with thick lips and enlarged nasal tip. Both sibs developed angiokeratoma corporis diffusum in adulthood, with the lesions in the male distributed between umbilicus and the upper thighs and those in the female isolated on the abdomen, breasts, and gingival mucosa. Both sibs had tortuous conjunctival vessels, and the male also had tortuous retinal vessels. In the male, massive lymphedema of the lower extremities developed at age 10–14 years, whereas the female had a milder involvement. The male had diffuse haziness in the corneal epithelial layer, which differed from the typical whorl-like opacities seen in Fabry disease, and he had moderate cardiomegaly. His neurologic status was unremarkable.

Pertinent Laboratory, Imaging, and Functional Findings

Routine hematologies and serum and urinary chemistries were normal. Various endocrine studies, chest radiographs, and a skeletal survey also were normal. Psychometric evaluations showed an intelligence quotient (IQ) of 70 in the first Japanese type II patient using the Wechsler adult intelligence scale.[4] Both verbal IQ and performance IQ were below normal. The electroencephalogram was unremarkable. MRI of the brain showed a few small lacunar infarctions; no gross atrophy of the parenchyma was observed. Nerve conduction studies of the right peroneal nerve showed normal conduction amplitude and velocity in the motor fibers. There was, however, a marked decrease in amplitude (0.5 μV; normal range, 1.2–9 μV) with normal velocity in the sensory fibers. Similar results were obtained from the right median nerve. Results of electromyograms of the muscles innervated by these nerves were normal. The second Japanese patient also had markedly decreased conduction amplitude in sensory nerves with normal velocity, whereas the amplitude and velocity in motor nerves were normal.[5] These results were indicative of peripheral neuroaxonal degeneration.

In the Spanish sibs, motor and sensory conduction velocities of the median nerve and motor conduction of the peroneal nerve were within normal ranges. Psychomotor examination of the male showed an IQ of 84 with normal verbal and low performance IQ values.[6]

Type III Disease

Five children with type III α-*N*-acetylgalactosaminidase deficiency from three unrelated families have been described with an intermediate phenotype.[7–9] In a Dutch family, twins were affected.[7] The female was essentially well until she experienced severe asymmetric convulsions at age 11 months. She had a high fever that precipitated uncontrollable grand mal seizures and attacks of apnea and bradycardia that required admission to an intensive care unit. Although the fever and convulsions responded to medication, a few days after detubation she developed pneumonia, a high fever, sepsis, and multiple organ failure from which she slowly recovered. Developmentally, she walked by herself and spoke only single words at age 21 months. On examination at age 2 years, mild psychomotor retardation was evident and strabismus was observed. After age 2.4 years, she had four convulsions, mostly associated with fever, despite treatment with antiepileptic drugs. She did not have dysmorphic features, organomegaly, cutaneous lesions, or vacuolated lymphocytes. A computed tomography scan of the brain appeared normal at age 2 years. The diagnosis was made by the characteristic urinary oligosaccharide pattern and by the demonstration of α-*N*-acetylgalactosaminidase deficiency in fibroblasts and leukocytes.

Her twin brother also was mildly affected. At age 9 months, his developmental performance was reported to be within normal limits, but was probably less advanced than that of his twin sister. At age 4 years, he had some developmental delay, particularly in language skills, but neurologic tests had not been performed. Of interest, the mother was an effectively treated epileptic, and epilepsy also occurred in the father's family.

A third patient with type III disease was recently recognized in France.[9] The proband was the only child of unrelated parents of French, Italian, and Albanian descent. Pregnancy, labor, and delivery were uneventful. The boy's early development was reportedly normal; however, he did not sit alone until age 9 months and did not walk independently until age 18 months. His parents sought medical advice at age 2 years because he did not speak or interact with others. Autism was suspected at 32 months. On physical examination, at age 6 years, he had no dysmorphic features, organomegaly, skeletal, or cutaneous abnormalities or irregular neurologic findings. Weight, height, and head circumference were 0.5 standard deviations below age mean. He was characterized as having a restless attitude, diminished attention span, emotional instability and irritability, and fits of anger with stereotyped and ritualistic behavior. His intellectual development appeared to be impaired, and he only had rudimentary syntax. The behavioral findings led to the diagnosis of autism. At age 8 years, the patient had better social referencing with few compulsions, while he was highly anxious. He did not exhibit seizures or neurologic regression.

Brain MRI and spinal X-ray studies were unremarkable. Psychometric evaluation was difficult because of his behavior; however, his global IQ was markedly decreased (48), the verbal IQ was 58, and the performance IQ was 46. Routine laboratory studies, including complete blood cell counts, blood glucose, serum electrolytes and protein, and liver enzymes were all within normal limits. Vacuolated lymphocytes were not found in blood smears. Serum and urinary amino acid profiles were normal, as were the urinary organic acids.

Two sibs of Moroccan descent with type III disease also have been reported.[8] The index patient was the 3-year-old son of consanguineous parents. He was evaluated for abnormal eye movements at 4 weeks. Urinary oligosaccharide screening showed an abnormal pattern suggestive of α-*N*-acetylgalactosaminidase deficiency, which was subsequently confirmed in leukocytes and skin fibroblasts. He had neuromotor developmental delay at 12 months, which was more prominent at the age of 2 years. At 3 years, he was unable to walk without support but had normal hand coordination and speech development. A brain MRI showed diffuse white matter abnormalities and secondary symmetrical demyelinization, while an ophthalmologic examination showed bilateral cataracts. His 7-year-old brother, identified by urinary oligosaccharide screening and deficient α-*N*-acetylgalactosaminidase activity in leukocytes and fibroblasts similar to the younger sib, had no clinical or neurologic symptoms at that time.[8]

Unclassified Cases

In 2006, Chabás et al. reported an 8-month-old girl referred to an intensive care unit because of high fever, vomiting, and food refusal.[12] Her presenting symptoms were pale skin, tachypnea, hepatomegaly, and cardiomegaly. She died only 12 hours after admission due to dilated cardiomyopathy and cardiac insufficiency. The diagnosis of α-*N*-acetylgalactosaminidase deficiency was made postmortem from cultured fibroblasts and liver tissue. The case was originally interpreted as a patient with infantile-onset (type I) disease because of her age, presenting with visceromegaly and cardiomyopathy. However, this patient's symptoms were consistent with sepsis and heart failure. Her cardiomyopathy and hepatomegaly were not shown to be due to storage histologically. Hence, her α-*N*-acetylgalactosaminidase deficiency should be considered unclassified. The phenotypic spectrum of type I, II, and III Schindler disease is summarized in Table 39.1.

Unpublished cases known to the authors include an adult patient with typical features of type II disease, another adult with ataxia as the guiding symptom and a patient with juvenile disease onset including seizures, consistent with type III α-*N*-acetylgalactosaminidase deficiency.

DIAGNOSTIC EVALUATION

Children with developmental delay and retrogression in the first or second year of life and clinical manifestations compatible with an infantile neuroaxonal dystrophy should be examined for type I disease. Patients with manifestations ranging from seizures and moderate psychomotor retardation in infancy to a milder autistic presentation with speech and language delay and marked behavioral difficulties in early childhood may be investigated for type III

TABLE 39.1 Phenotypic Spectrum of α-*N*-Acetylgalactosaminidase Deficiency

Feature	Type I	Type III	Type II	Unclassified*
Age of clinical onset	Infancy	Childhood	Adulthood	Infancy?
No. of patients/ no. of families	3/1	5/3	4/3	1/1
Major presentation	Global neurodegeneration	Developmental/neurologic/ psychiatric	Dermatological—angiokeratoma	Cardiomyopathy, visceromegaly?
Neuropathology	INAD	Demyelination	?	?
Neurologic severity	+++	++	+	?
Lysosomal storage	+	+	+++	?
Enzyme deficiency	+++	++	+++	+++
Protein defect	+++	++	+++	?
Glycopeptiduria	+++	++	+++	?

**See text for further explanation.*
Abbreviation: INAD, Infantile neuroaxonal dystrophy.

disease. In adolescents and adults with mild intellectual impairment and angiokeratoma, type II disease should be considered. Initial screening for all subtypes can be performed by analyzing urinary oligosaccharide and glycopeptide profiles.[13] Definitive diagnosis of all forms is made by demonstration of deficient α-*N*-acetylgalactosaminidase activity in plasma, isolated leukocytes, or cultured lymphoblasts or fibroblasts using the chromogenic substrate p-nitrophenyl-α-*N*-acetylgalactosaminide or, better, the fluorogenic substrate 4-methylumbelliferyl-α-*N*-acetylgalactosaminide.[10] Prenatal diagnosis can be accomplished by determining the enzymatic activity in chorionic villi obtained at 10 menstrual weeks or in cultured amniocytes obtained by amniocentesis at 15–16 menstrual weeks.

PATHOLOGY

Light microscopic examination of blood, bone marrow, liver, skeletal muscle, skin, peripheral nerve, conjunctiva, and jejunal and rectal wall from the type I patients appeared normal. Electron microscopy of blood leukocytes, secretory cells of eccrine sweat glands, myelinated axons of cutaneous nerves, and cultured fibroblasts from both type I and II patients showed the presence of inclusions with lamellar, fibrillar, vesicular, and granular material in single membrane-bound organelles.[1]

Ultrastructural studies of rectal mucosa from the type I patients showed abnormal tubulovesicular material free in the cytoplasm in only a few preterminal and terminal (intraganglionic) axons in the myenteric plexus.[1,14] Examination of a frontal lobe biopsy revealed the characteristic neuropathology of infantile neuroaxonal dystrophy.[14] There were numerous large, dense axonal swellings, or spheroids, throughout the neocortex, with no apparent laminar distribution; few such formations were observed in axons in the white matter. On light microscopy, the deposits were sharply demarcated, rounded, or polygonal structures that contained prominent angular or curving clefts with a darker, amorphous background. Ultrastructurally, these abnormal formations appeared exclusively within the preterminal and terminal axons and were not observed in the neuronal perikarya, dendrites, axons in white matter, small axon terminals in the cortical neuropil, astrocytes, oligodendrocytes, microglia, endothelial cells, or arachnoid cells. The accumulations were morphologically heterogeneous, comprising dense, labyrinthine membranous tubulovesicular formations, lamelliform membranous arrays, and prominent acicular electron-lucent clefts, all admixed with a few mitochondria, lysosomes, and occasional microtubules. These aggregates were free in the electron-dense axoplasmic matrix and limited by a plasmalemma facing an epithelial interspace. Based on the neocortical findings, the rectal biopsies from the sibs were re-examined, and spheroids were unequivocally identified in the myenteric plexus and ganglionic neuropil.

In Giemsa-stained peripheral blood smears from the probands with type II disease, small cytoplasmic vacuoles were observed in granulocytes, monocytes, and lymphocytes. These vacuoles did not stain positive with periodic acid–Schiff, Alcian blue, or toluidine blue. Histopathologic examination of the skin lesions showed localized hyperkeratosis and dilated, thin-walled blood vessels.[15,16] Dilated lymphatic vessels were observed in the mid-dermis as well as the upper dermis. Ultrastructural examination of both involved and uninvolved skin of type II patients revealed numerous cytoplasmic vacuoles in several cell types, including endothelial cells of blood and lymphatic vessels, pericytes, fibrocytes, fat cells, Schwann cells, axons, arrector smooth-muscle cells, and eccrine sweat gland cells. They were most prominent in vascular endothelial cells and the secretory portion of sweat gland cells. These membrane-lined vacuoles were electron-lucent or contained filamentous material. Electron-dense multilayered structures were observed only in vacuoles of the sweat gland cells. Vascular endothelial cells of the kidney had similar lysosomal vacuolization as those in skin, but the epithelial cells appeared normal. Similar, but smaller, vacuoles also were observed in peripheral leukocytes.

To date, the pathologic studies of type III disease have been limited to peripheral blood cells. In the Dutch sibs, histologic and electron microscopic examination of lymphocytes, granulocytes, and monocytes were normal, and no vacuolization was observed.[3] Similar findings were noted in the French type III patient.[9]

BIOCHEMISTRY

The deficient activity of lysosomal α-*N*-acetylgalactosaminidase is the specific enzymatic defect in types I, II, and III disease.[1] Heterozygotes for both subtypes have approximately half of the normal levels of activity. The enzymatic defect results in the accumulation of glycoconjugates with terminal or internal α-*N*-acetylgalactosaminyl moieties.[1] Structural analysis of the accumulated urinary compounds shows the presence of the glycopeptide, oligosaccharide, and glycosphingolipid structures (Table 39.2). Correlations between residual α-*N*-acetylgalactosaminidase activity and clinical disease severity have been attempted using a physiological substrate.[17]

MOLECULAR GENETICS

Types I, II, and III diseases are inherited as autosomal recessive traits and are very rare. The human α-*N*-acetylgalactosaminidase gene has been mapped to the chromosomal region 22q13.2. The full-length cDNA and complete genomic sequence encoding α-*N*-acetylgalactosaminidase have been characterized.[18,19] The 13,709-bp genomic sequence has 9 exons ranging from 95 to 2028 bp and intronic sequences of 304–2684 bp. Analysis of 1.4 kb of 5′ flanking sequence showed three Sp1 and two CAAT-like promoter elements. The α-*N*-acetylgalactosaminidase gene is transcribed into two transcripts of 2.2 and 3.6 kb, the latter resulting from a second downstream polyadenylation signal in the 3′-untranslated region. Both transcripts have a 1236-bp open reading frame, which encodes 411 amino acids, including a 17-residue signal peptide. The mature 394 amino acid lysosomal polypeptide contains six putative *N*-glycosylation sites.

TABLE 39.2 Accumulated Urinary Compounds in α-*N*-Acetylgalactosaminidase Deficiency

GLYCOPEPTIDES
1 GalNAcα1 → O-Ser/Thr
2 NeuNAα2 → 3Galβ1 → 3GalNAcα1 → O-Ser/Thr
3 Galβ1 → 3(NeuNAcα2 → 6)GalNAcα1 → O-Ser/Thr
4 NeuNAcα2 → 3Galβ1 → 3(NeuNAcα2 → 6)GalNAcα1 → O-Ser/Thr*
5 NeuNAcα2 → 3Galβ1 → 3(Galβ1 → 4G1cNAcβ1 → 6)GalNAcα1 → O-Ser/Thr
6 Galβ1 → 3(NeuNAcα2 → 3Galβ1 → 4GlcNAcβ1 → 6)GalNAcα1 → O-Ser/Thr
7 NeuNAcα2 → 3Galβ1 → 3(NeuNAcα2 → 3Galβ1 → 4G1cNAcβ1 → 6)GalNAcα1 → O-Ser/Thr
8 $(NeuNAc)_2(Gal\beta1 \rightarrow 4GlcNAc)_2Gal\beta1 \rightarrow 3GalNAc\alpha1 \rightarrow$ O-Ser/Thr
OLIGOSACCHARIDE
Blood group A trisaccharide† GalNAcα1 → 3(Fucα1 → 2)Gal
GLYCOSPHINGOLIPID
Blood group A glycolipid A-6-2†
GalNAcα1 → 3(Fucα1 → 2)Galβ1 → 4GlcNAcβ1 → 3Galβ1 → 4Glcβ1 → 1′Cer

**Major accumulated component.*
†Only in patients with blood group A.

TABLE 39.3 Mutations in the α-*N*-Acetylgalactosaminidase Gene Causing Schindler Disease

Disease type	Ancestry	cDNA change*	Genotype	Reference
I	German	973G>A	E325K/E325K	3,19
II	Japanese A	985C>T	R329W/R329W	20
	Japanese B	986G>A	R329Q/R329Q	5
	Spanish	577G>T	E193X/E193X	3
III	Dutch	479C>G/973G>A	S160C/E325K	3
	Moroccan	973G>A	E325K/E325K	8
	French/Albanian/ Italian	973G>A/1099G>A	E325K/E367K	1,9
Unclassified	Spanish	649G>A/973G>A	D217N/E325K	12

**Numbering according to NCBI accession nos. M62783, M29276.1 and M59199.*

Mutations causing the three subtypes of α-*N*-acetylgalactosaminidase deficiency have been determined for each of the published patients (Table 39.3).[3,8,20,21] Causative mutations include homozygosity for E325K (type I disease); homozygosity for R329W, R329Q, or E193X (type II disease); and compound heterozygosity for S160C and E367K or homozygosity for E325K (type III disease). The crystal structure of human α-*N*-acetylgalactosaminidase has been resolved at 1.9 Å and the effect of disease-causing mutations on the three-dimensional structure was analyzed.[22] Apart from D217N of the unclassified case, none of the mutations were localized to the enzyme's active site.[23] Thus, most of the disease-causing mutations presumably impaired the stability of the enzyme, consistent with the biosynthetic studies of the mutant proteins.[1]

RELATION TO OTHER GENE LOCI

A recent gene mapping study identified PLA2G6 mutations in neurodegenerative disorders with high levels of iron in the brain, including a subset of individuals with infantile neuroaxonal dystrophy (INAD; Seitelberger disease).[24] PLA2G6 encodes iPLA2-VI, a calcium-independent phospholipase. A2 phospholipases catalyze the hydrolysis of phospholipids at the sn-2 position. iPLA2 enzymes are critical in the maintenance of membrane phosphatidylcholine levels and cell membrane homeostasis, dysregulated in infantile neuroaxonal dystrophy (INAD). In the study by Morgan et al.,[24] there was no absolute correlation of INAD with PLA2G6 mutations, and the linkage data supported the existence of at least one additional INAD locus. Nonetheless, the fact that PLA2G6 mapped close to the α-*N*-acetylgalactosaminidase locus led to the hypothesis that patients with Schindler disease type I could be homozygous by descent for mutations in two genes, α-*N*-acetylgalactosaminidase and PLA2G6,[25] extending the suggestion by Bakker et al. that type I disease might result from the occurrence of two independent monogenic diseases.[8] To date, there is no published evidence to support this hypothesis.

ANIMAL MODEL

There is no identified naturally occurring animal model for α-*N*-acetylgalactosaminidase deficiency. However, knockout mice with α-*N*-acetylgalactosaminidase deficiency have been generated by targeted gene disruption.[26] Homozygous mice with no detectable α-*N*-acetylgalactosaminidase activity appeared normal, were fertile and bred, lived a normal lifespan, and appeared to have no clinically evident neurologic or other disease. In contrast, there was remarkable cellular pathomorphology. Two main lesions were identified: widespread lysosomal storage of abnormal material in the nervous system and in other organs and focal axonal swellings or spheroids in the central nervous system.[1]

Lysosomal storage was histologically and ultrastructurally observed in numerous cell types throughout the central and peripheral nervous system and in many other organs. Neuronal storage varied from minimal to profound. Perivascular macrophages, either within the neuroparenchyma or in the subarachnoid space, consistently contained the most storage material. Ultrastructurally, the enlarged lysosomes contained flocculent, particulate material, concentric lamellar figures, and multivesicular and dense formations.

The second pathologic finding was neuroaxonal dystrophy. Axonal spheroids were found more frequently in the spinal cord, where they were observed throughout its entire length in a distribution limited to the dorsal gray and white matter, than in the brain. The spheroid content varied from homogeneous to complex. The murine knockout model appears to mimic the characteristic pathologies of the human disease subtypes. However, in absence of autopsied human material, comparisons remain preliminary. Notably, the spheroids in the spinal cord of the knockout mice had remarkable ultrastructural similarity to the cortical spheroids in type I human disease and to those observed in the central nervous system of INAD. Although the definitive distribution of spheroids in type I α-*N*-acetylgalactosaminidase deficiency has not been determined, in INAD spheroids are found in large numbers in the spinal cord and medulla and less so in the cerebrum.

THERAPY

There is no current treatment for this disease. Supportive care should be implemented to optimize patient comfort. Recent studies identify Schindler disease as a typical protein folding disorder.[27] New avenues for therapy of α-*N*-acetylgalactosaminidase-deficiency are open by the findings that the iminosugar 2-acetamido-1,2-dideoxy-D-galactonojirimycin (DGJNAc) can inhibit, stabilize, and chaperone human α-*N*-acetylgalactosaminidase both *in vitro* and *in vivo* and that a related iminosugar, 1-deoxygalactonojirimycin (DGJ), currently in phase III clinical trials for another metabolic disorder, Fabry disease, can also chaperone human α-*N*-acetylgalactosaminidase in Schindler disease.

FUTURE RESEARCH DIRECTIONS

The neuropathology in the murine knockout model, which is similar to the neuroaxonal dystrophy in type I α-*N*-acetylgalactosaminidase-deficient patients, should permit investigation of the role of α-*N*-acetylgalactosaminidase in neuroaxonal transport. Moreover, studies in the knockout mouse should facilitate future evaluation of various neuron-targeted therapeutic strategies, including gene therapy. Screens for PLA2G6 mutations will be subject to further characterization in Schindler disease type I.

ACKNOWLEDGEMENTS

This work was supported in part by a research grant from the National Institutes of Health (Merit Award 5 R37 DK34045).

References

1. Desnick RJ, Schindler D. α-*N*-acetylgalactosaminidase deficiency: Schindler disease. In: Scriver CR, Beaudet AL, Sly WS, et al., eds. *The Metabolic and Molecular Bases of Inherited Disease*. 8th ed. New York: McGraw-Hill; 2001:3483–3505.
2. van Diggelen O, Schindler D, Kleijer W, et al. Lysosomal α-*N*-acetylgalactosaminidase deficiency: a new inherited metabolic disease. *Lancet*. 1987;2:804.
3. Keulemans JLM, Reuser AJJ, Froos MA, et al. Human α-*N*-acetylgalactosaminidase (α-NAGA) deficiency: new mutations and the paradox between genotype and phenotype. *J Med Genet*. 1996;33:458–465.
4. Kanzaki T, Yokota M, Mizuno N, et al. Novel lysosomal glycoaminoacid storage disease with angiokeratoma corporis diffusum. *Lancet*. 1989;1:875–877.
5. Kodama K, Kobayashi H, Abe R, et al. A new case of α-*N*-acetylgalactosaminidase deficiency with angiokeratoma corporis diffusum, with Meniere's syndrome and without mental retardation. *Br J Dermatol*. 2001;144:363–368.
6. Chabas A, Coll MJ, Aparicio M, et al. Mild phenotypic expression of α-*N*-acetylgalactosaminidase deficiency in two adult siblings. *J Inherit Metab Dis*. 1994;17:724–731.
7. de Jong J, van den Berg C, Wijburg H, et al. α-*N*-Acetylgalactosaminidase deficiency with mild clinical manifestations and difficult biochemical diagnosis. *J Pediatr*. 1994;125:385–391.
8. Bakker HD, de Sonnaville ML, Vreken P, et al. Human α-*N*-acetylgalactosaminidase (α-NAGA) deficiency: no association with neuroaxonal dystrophy? *Eur J Hum Genet*. 2001;9:91–96.
9. Blanchon YC, Gay C, Gilbert G, et al. A case of *N*-acetyl-galactosaminidase deficiency (Schindler disease) associated with autism. *J Autism Dev Disord*. 2002;32:145–146.
10. Schindler D, Bishop DF, Wolfe DE, et al. Neuroaxonal dystrophy due to lysosomal α-*N*-acetylgalactosaminidase deficiency. *N Engl J Med*. 1989;320:1735–1740.

11. Rudolf J, Grond M, Schindler D, et al. Cerebral glucose metabolism in type I α-*N*-acetylgalactosaminidase deficiency: an infantile neuroaxonal dystrophy. *J Child Neurol*. 1999;14:543–547.
12. Chabás A, Duque J, Gort L. A new infantile case of α-*N*-acetylgalactosaminidase deficiency. Cardiomyopathy as a presenting symptom. *J Inherit Metab Dis*. 2007;30:108.
13. Schindler D, Kanzaki T, Desnick RJ. A method for therapid detection of urinary glycopeptides in α-*N*-acetylgalactosaminidase deficiency and other lysosomal storage diseases. *Clin Chim Acta*. 1990;190:81–91.
14. Wolfe D, Schindler D, Desnick R. Neuroaxonal dystrophy in infantile α-*N*-acetylgalactosaminidase deficiency. *J Neurol Sci*. 1995;132:44–56.
15. Kanzaki T, Wang AM, Desnick RJ. Lysosomal α-*N*-acetylgalactosaminidase deficiency, the enzymatic defect in angiokeratoma corporis diffusum with glycopeptiduria. *J Clin Invest*. 1991;88:707–711.
16. Kanzaki T, Yokota M, Irie F, et al. Angiokeratoma corporis diffusum with glycopeptiduria due to deficient lysosomal α-*N*-acetylgalactosaminidase activity. *Arch Dermatol*. 1993;129:460–465.
17. Asfaw B, Ledvinová J, Dobrovolńy R, et al. Defects in degradation of blood group A and B glycosphingolipids in Schindler and Fabry diseases. *J Lipid Res*. 2002;43:1096–1104.
18. Wang AM, Bishop DF, Desnick RJ. Human α-*N*-acetylgalactosaminidase: molecular cloning, nucleotide sequence, and expression of a full-length cDNA. *J Biol Chem*. 1990;265:21859–21866.
19. Wang AM, Desnick RJ. Structural organization and complete sequence of the human α-*N*-acetylgalactosaminidase gene: homology with the α-galactosidase A gene proves evidence for evolution from a common ancestral gene. *Genomics*. 1991;10:133–142.
20. Wang AM, Schindler D, Desnick R. Schindler disease: the molecular lesion in the α-*N*-acetylgalactosaminidase gene that causes an infantile neuroaxonal dystrophy. *J Clin Invest*. 1990;86:1752–1756.
21. Wang AM, Kanzaki T, Desnick RJ. The molecular lesion in the α-*N*-acetylgalactosaminidase gene that causes angiokeratoma corporis diffusum with glycopeptiduria. *J Clin Invest*. 1994;94:839–845.
22. Clark NE, Garman SC. The 1.9 Å structure of human α-*N*-acetylgalactosaminidase: The molecular basis of Schindler and Kanzaki diseases. *J Mol Biol*. 2009;393:435–447.
23. Sakuraba H, Matsuzawa F, Aikawa S, et al. Structural and immunocytochemical studies on α-*N*-acetylgalactosaminidase deficiency (Schindler/Kanzaki disease). *J Hum Genet*. 2004;49:1–8.
24. Morgan NV, Westaway SK, Morton JEV, et al. PLA2G6, encoding a phospholipase A2, is mutated in neurodegenerative disorders with high brain iron. *Nat Genet*. 2006;38:752–754.
25. Westaway SK, Gregory A, Hayflick SJ. Mutations in PLA2G6 and the riddle of Schindler disease. *J Med Genet*. 2007;44:64.
26. Wang AM, Stewart CL, Desnick RJ. Schindler disease: generation of a murine model by targeted disruption of the α-*N*-acetylgalactosaminidase gene. *Pediatr Res*. 1994;35:155A.
27. Clark NE, Metcalf MC, Best D, Fleet GW, Garman SC. Pharmacological chaperones for human α-*N*-acetylgalactosaminidase. *Proc Natl Acad Sci U S A*. 2012;109:17400–17405.

METAL METABOLISM DISORDERS

CHAPTER

40

Wilson Disease

Golder N. Wilson

Texas Tech University Health Science Centers, Amarillo and Lubbock TX, USA
Medical City Hospital, Dallas TX, USA

SUMMARY

Disease Characteristics

The hallmark manifestations of Wilson disease are neurodegeneration and cirrhosis, each extremely variable in age, severity, and manner of presentation, as befits a disorder of copper accumulation.[1–8] First characterized by Samuel Kinnear Wilson in 1912, the disorder is also called hepatolenticular degeneration and is typified by the presence of the nonpathognomonic Kayser–Fleischer ring in the ocular limbus (Figure 40.1). Hyperintense signals on head magnetic resonance imaging (MRI) result in characteristic giant panda (midbrain) and cub (pons) shapes (Figure 40.2). Hepatic disease in the second decade is common in children, ranging from asymptomatic elevation of serum transminases to acute hepatic failure.[5–9] Neuropsychiatric abnormalities appear in the third to fourth decade with two major themes:[10–14] 1) neurologic symptoms, including diminished facial expression and movement, tremors, dystonia, and choreoathetosis; and 2) psychiatric symptoms, including personality disorder, depression, cognitive decline, or even psychosis. Symptoms mimicking acute and chronic active hepatitis,[15] parkinsonism,[14] multiple sclerosis, or (rarely) schizophrenia[14] can disguise Wilson disease and delay curative chelation therapy. Pattern recognition is of great help as copper excess can affect other tissues besides its favored liver and basal ganglia, leading to the Kayser–Fleischer ring,[5–9] azure lunules,[16] Fanconi syndrome of the kidney,[17,18] cardiac arrhythmias,[19] arthritis,[20] hemolytic anemia, or dysmenorrhea.[5–9] Suspicion is particularly important in the patient with active cirrhosis or psychiatric symptoms whose inevitable progression of nerve and liver damage can be arrested by therapy.[21–32]

Inheritance

Wilson disease exhibits autosomal recessive inheritance, with a prevalence of 1 in 33,000 that predicts a carrier frequency of about 1 in 90 (2pq in the Hardy–Weinberg equilibrium, where q is approximately the square root of 1/33,000 or 1 in 181).[8,33] The Wilson *WND* or *ATP7B* gene includes 21 exons spanning 60,000 bp (60 Kb) within chromosome band 13q14.3, base pairs 52,506,804 to 52,585,629.[34–40] It encodes a pump- or P-type ATPase belonging to a large family of membrane-bound cation transporters.[3,8,36,40]

Diagnosis/Testing

Clinical findings of hepatorenal and/or neuropsychiatric disease combined with usual metabolic panels may define elevated serum transaminases, hypercalcemia, or hemolytic anemia that justifies specific diagnostic testing.[5–9] Neurology consultation is recommended for those with suspect neuropsychiatric symptoms and head MRI may show hyperintensity of the basal ganglia on T2 imaging.[10–14] Specific testing can begin with serum ceruloplasmin, and acute phase reactant subject to elevation by inflammation is suggestive when concentrations below 20 mg per dL are found. Urinary copper concentrations above 40 μg per ml per 24 hours and the presence of a

Rosenberg's Molecular and Genetic Basis of Neurological and Psychiatric Disease
http://dx.doi.org/10.1016/B978-0-12-410529-4.00040-1

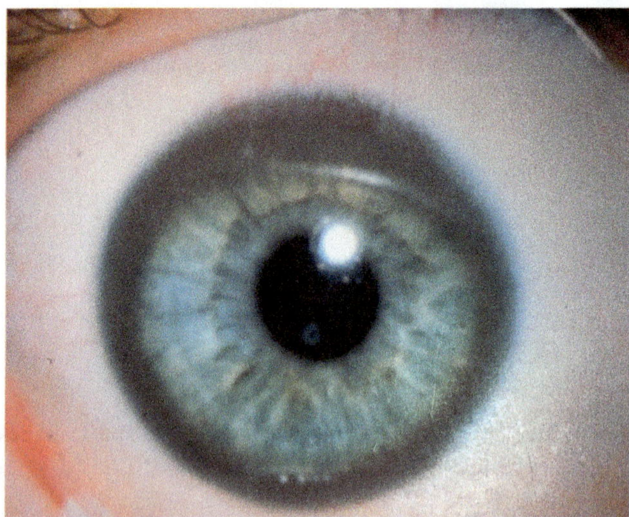

FIGURE 40.1 **Kayser–Fleischer ring.** *Courtesy of Dr. Harold Falls, pioneering ophthalmologist and geneticist at the University of Michigan. From Consultant, July 2007, p. 769, with permission of Consultant/HMP Communications, LLC.*

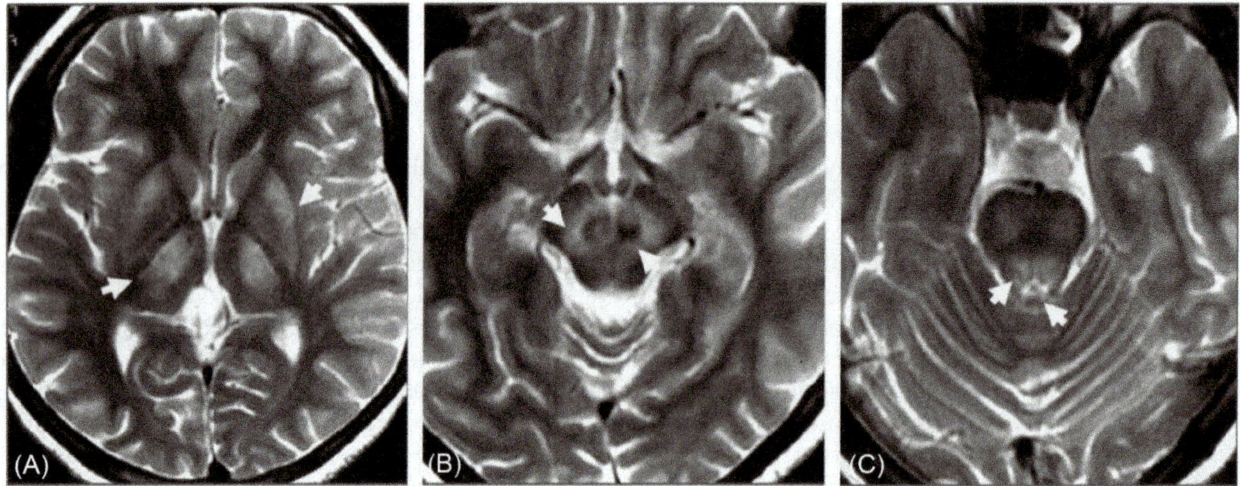

FIGURE 40.2 **T2-weighted axial MRI.** Demonstrates (**A**) symmetric hyperintense signals in the putamen, posterior internal capsule, and thalami (arrows), (**B**) "face of the giant panda" in midbrain with high signal in tegmentum and normal red nuclei (arrows), and (**C**) "face of the panda cub" in pons with hypointensity of central tegmental tracts with hyperintensity of aqueductal opening to fourth ventricle (arrows). *Legend and figure from Shivakumar R and SV Thomas,[11] in Neurology © published by Lippincott Williams & Wilkins, with permission. The primary reference may be viewed at www.neurology.org/content/72/11/e50.long (accessed June, 2013).*

Kayser–Fleischer ring complete the diagnostic triad.[7] Liver biopsy showing typical micronodular cirrhosis and hepatocyte copper concentrations above 250 μg per g of dry weight are useful when one of the prior indicators is borderline, but any cirrhotic process interfering with biliary excretion will elevate hepatic copper.[6,8] DNA analysis of the *WND/ATP7B* gene is available from many academic and commercial laboratories, but the large number of mutations usually requires full gene sequencing costing $1160–2300 as listed on the Next GxDx website (the very useful GeneTests website is no longer linked to genetic disorders found by searching Online Mendelian Inheritance in Man, but is being restructured).[34–37] The high mutation frequency means that most individuals will be compound heterozygotes having two different mutant alleles. Characterization of the mutations in one patient allows targeted mutation analysis for siblings (~$250–500) or consideration of preimplantation genetic/prenatal diagnosis for future pregnancies. DNA sequence analysis of prospective parents for carrier status without affected relatives is too expensive for population screening except in limited regions (e.g., Sardinia) where presumed founder effect has caused many affected individuals to have the same homozygous mutation (an unusual 15-bp deletion in the promoter region).[38]

Current Research

Improved treatment and monitoring strategies are major goals of current research, drawing upon considerable experience with and proven efficacy of chelating agents in Wilson disease.[21–31] Added to penicillamine, zinc, and triene is ammonium tetrathiomolybdate,[31] a new agent that could be useful in patients with neuropsychiatric presentation because it has not caused the neurologic deterioration occasioned by penicillamine.[7,21] Side effects including hepatitis and bone marrow failure complicate the use of this medication, and its powerful diminution of intestinal copper absorption and serum copper concentration limits therapy to a few months before severe copper depletion occurs.[41] Liver transplantation is an effective therapy for individuals with progressive cirrhosis who do not respond to chelation therapy, so targeted hepatic gene therapy is theoretically possible.[6–8] Animal models including the Long–Evans Cinnamon (LEC) rat[42] and the "toxic milk mouse"[43] have respective deletion or mutation of the homologous *ATP7B* gene and are available for gene transfer and expression studies. While Wilson disease ranks high in the short list of therapeutic triumphs over genetic-metabolic disorders, noncompliance, severe disease after therapy resumption, and suicide attempts remain as challenges.[44]

CLINICAL FEATURES

Historical Overview

Samuel A.K. Wilson is no relation to the author and his statement that "the condition was often familial but not congenital or hereditary" in his pioneering paper, is no model for a geneticist.[1,2] He was a New Jersey expatriate who qualified in medicine in Edinburgh at age 24, received a neurologic appointment in London, and achieved fame with his 1912 paper that highlighted the basal ganglia with coinage of the term "extrapyramidal."[1,2] A dramatic lecturer and clinical scholar with little interest in laboratory work, he founded the *Journal of Neurology and Psychopathology* and authored the two-volume *Neurology*. Some saw him as arrogant and insensitive, a view supported by his order to a patient, "See to it that I get your brain when you die!"[2] He would not have liked McKusick's dictum of inappropriate possessives, referring to his condition as "Kinnier Wilson's disease."[2]

Disease Identification

The azure color noted in the limbus by Kayser and Fleischer in 1912 (Figure 40.1),[5–8] later noted in nail lunulae,[16] focused attention on tissue copper accumulation. Measures of decreased copper-binding ceruloplasmin in blood and of increased elemental copper in urine or liver provided objective identification of the disease.[3–9]

Gene Identification

Characterization of mutations in the *ATP7B* copper transporter gene as the cause of Wilson disease linked it to Menkes disease in 1993.[3,4,34–37] Now evidence-based management guidelines[7] and improved genotype–phenotype correlation[33–36] using rapid next generation/parallel DNA sequencing are linking *WND/ATP7B* cation-transporter function with a systems biology for mammalian copper metabolism.[3–5]

Early Treatments

The obvious strategy of removing copper was first tried with British anti-lewisite (dimercatopropanol), then perfected using the less toxic penicillamine by Walshe in 1956.[21,22] Dr. George Brewer at the University of Michigan has contributed greatly to numerous trials that range from zinc exchange[23] to modern combination therapy[7,21–31] and benefit both children[24] and pregnant women.[32]

Mode of Inheritance and Prevalence

Wilson disease exhibits autosomal recessive inheritance with carrier parents having one normal and one abnormal allele, the normal allele allowing sufficient copper transport to prevent deleterious tissue accumulation.[3,8,33] About 1 in 90 individuals will carry an abnormal allele for Wilson disease, corresponding to a prevalence of 1 in 33,000 for affected individuals who have two abnormal and no normal alleles. The later onset of Wilson disease means that many carrier parents will not be aware of their 25% risk until they have completed their family, so diagnosis in one person requires screening of siblings so that preventive therapy can begin. Siblings testing negative have a 2/3 chance of carrying an abnormal Wilson allele, and their risk for an affected child after conception with an unrelated spouse can be dismissed or converted to 1 in 4 by DNA testing.[3,8,34–36] If the family mutation is known, then the sibling with 2/3 risk can be tested relatively inexpensively by polymerase chain reaction (PCR) amplification and sequencing of the mutated Wilson gene region. The unrelated spouse will require full gene sequencing to examine the large Wilson mutation spectrum, and the couple can choose preimplantation genetic or prenatal diagnosis options if both are carriers.

NATURAL HISTORY

Age of Onset

As befits the long trail from copper transport gene mutation to multisystem accumulation and disease, age of onset and severity of Wilson disease is extremely variable.[5–9] Children usually present with liver dysfunction at age 10–13 years, one or two decades before the average affected adult presents with neuropsychiatric symptoms.[6–8] The age and mode of presentation differs even between family members with the same molecular defect, illustrating the role of background genes and environment. Importance of the latter factors and the nonspecific detriment of hepatic copper accumulation are indicated by Indian childhood cirrhosis, where consanguinity and use of copper-containing cooking implements combine to cause accumulation with liver failure.[8]

Disease Evolution and Disease Variants

Wilson disease can present from ages 3–70 with about 40% of patients having hepatic, 40% neurologic, and 20% psychiatric disease.[3] The Kayser–Fleischer ring will be present in less than half of children with their typical hepatic presentation and in over 90% of adults with their typical neuropsychiatric presentation.[3,8] The earlier-onset liver disease usually involves mild elevations of serum transaminases with chronic active hepatitis noted on liver biopsy.[15] On occasion, the presentation mimics autoimmune hepatitis with fatigue, malaise, arthropathy, and rashes.[1,20] Some patients, especially women, can present with acute hepatitis and liver failure that mimics viral disease with the addition of hemolytic anemia caused by outpouring of liver copper.[3,8] Whether cryptic in neuropsychiatric patients, insidious or acute in those with liver disease, the chronic hepatitis from copper accumulation progresses to nodular cirrhosis, fatty liver, and liver failure with release of copper into the bloodstream.[3,6–8]

Later neurologic presentation[10–14] often occurs in the third or fourth decade and has two faces: that of a movement disorder or of dystonia.[3] Movement changes occur earlier, with tremors, incoordination, fine motor changes, such as micrographia with small and cramped handwriting, and chorea with or without athetosis; these changes can mimic multiple sclerosis. Dystonia can include mask facies, rigidity, and gait disturbance that are reminiscent of Parkinson disease.[6–8,14] As Wilson recognized,[1] the basal ganglia are preferentially involved with relative sparing of the sensorimotor cortex, despite copper accumulation throughout the nervous system. Pseudobulbar changes are more common in older persons and can include dysarthria, drooling, and difficulty swallowing.[3,7,8] Psychiatric symptoms can precede or accompany the neurologic or hepatic disease and range from personality change and depression to psychosis suggestive of schizophrenia.[6–8] Suspicion can draw attention to multisystem disease extending beyond the brain, liver, and eye: Fanconi syndrome (obeying the usual correlation of hepatocyte and proximal tubule), renal stones, cardiomyopathy, pancreatitis, anemia, arthritis, rhabdomyolysis, hypoparathyroidism, or menstrual changes emphasize that copper accumulation occurs in all tissues.[15–20]

End of Life: Mechanisms and Comorbidities

Progression of liver disease from chronic active hepatitis to liver failure and neurodegeneration was relentless before chelation and/or zinc therapies were developed.[5–9] Long-term follow-up studies are now available, including a series of 20 patients over 14 years with penicillamine therapy in 90% and zinc in 10%.[26] Median age at diagnosis was 22 years with 50% having hepatic, 25% having neurologic, and 15% having mixed presentation. All had good response, but those on zinc or who switched to zinc because of penicillamine side effects had continued mild elevation of serum transaminases.[26] Another study examined results of penicillamine therapy for 15–27 years in 24 patients of whom 17 (71%) had cirrhosis (11 of these with cirrhosis complications) and 13 (54%) had neuropsychiatric symptoms.[28] Six of the 11 with cirrhotic complications improved or stabilized on penicillamine but three needed liver transplant with one dying. Neuropsychiatric symptoms improved in eight of the 13, stabilized in four, and worsened in one with poor compliance.[28] Therapy with zinc alone can also be effective[23,24,29] as shown by median follow up of

17 symptomatic patients (age range 13–26 years, with half adolescents) over 14 years (range 2–30 years) with hepatic (seven patients), neurologic (five), or mixed (five) presentation.[29] Of the 12 with liver disease, two with decompensated cirrhosis stabilized and two with less severe liver disease worsened; all with milder liver changes stabilized or improved and two with hepatic symptoms developed neurologic disease and *vice versa*. The long-term response to zinc was excellent for those with neurologic disease but "less satisfactory" in hepatic disease because of lesser increase in copper excretion.[29]

Diagnosis and treatment of Wilson disease can provide a normal lifespan with the exception of patients with severe liver disease who do not tolerate transplant. The fulminant liver failure seen in such patients is a function of gene–environment interaction (epigenetic), since particular Wilson *ATP7B* genotypes are not predictive of disease severity or longevity.

MOLECULAR GENETICS

Normal Gene Product

The ATPases causing Wilson (*ATP7B*) and Menkes (*ATP7A*) disease are 55% identical in amino acid sequence, have the same functional motifs, and belong to a family of membrane-bound copper-transporting proteins that occur in all biota. Their amino acid motifs include methionine-X-cysteine-XX-cysteine (MXCXXC) and cysteine-proline-cysteine (CPC) copper-binding/transport sequences at the *N*-terminus plus eight transmembrane domains. The Wilson ATPase gene spans 60 kb at chromosome band 13q41.3 and splices 21 exons into a 7.5-Kb transcript and a 165-kDa protein that are highly expressed in liver and other tissues. The Menkes ATPase gene spans 140 kb at chromosome band Xq13.3 and splices 23 exons into a 178-kDa protein that is expressed in most tissues *except* for liver. Menkes disease is much rarer, with prevalence of 1 in 250,000, and is a tragic opposite of Wilson disease in that its copper deficiency has no treatment.[8]

The copper transporting ATPases mediate transfer of copper between chaperones like *HAH1* to the Golgi secretory pathway, essential for function of many copper proteins.[4,8] Ceruloplasmin is a ferroxidase, also important for iron metabolism in that its rare deficiency causes hemosiderosis with diabetes, dementia, chorea, and ataxia. The superoxide dismutases SOD1 and SOD3 are important for antioxidase defense and dopamine β-hydroxylase converts dopamine to norepinephrine. Important for neurotransmitter metabolism are the peptidylglycine α-amidating monoxygenase and tyrosinase that converts tyrosine (precursor to catecholamines and serotonin) to melanin. Very important in Menkes disease is the copper-dependent cytochrome *c* oxidase that can be mutated to cause mitochondrial disease. On the side of copper cytoplasmic excess from altered transport into the Golgi are metallothioneins encoded by a cluster of 16 genes that seem to have little impact on mouse development but may sequester excess liver copper in Wilson disease.[8]

Abnormal Gene Product(s)

The common histidine-to-glutamine mutation at position 1069 of the Wilson ATPase protein (H1069Q) has been paralleled by site-directed mutagenesis to produce a homologous H1086Q mutation in the Menkes protein, disrupting an evolutionarily conserved residue that is vital for transport function. Such mutations confirm the primary role of disrupted copper transport in the pathogenesis of both diseases and reinforce Garrod's example of lessons from human disease: the Wilson ATPase was the first example of copper transporters that exist in all phyla.[8] Although abnormal transport with copper accumulation/deficiency and its effects on key copper proteins offers an easily appreciated disease mechanism, the systems biology of altered copper metabolism is much more complex with effects on liver transcription, nuclear receptor signaling, stress-response proteins, and RNA processing that perturb lipid metabolism and cell-cycle regulation. These cascading alterations, when added to mutational diversity, explain the complexity of genotype–phenotype correlation for Wilson disease.

Genotype-Phenotype Correlations

Global analysis has defined more than 100 mutations in Wilson disease patients, with the mentioned H1069Q mutation accounting for 40% of alleles in northern Europeans (Germans, Lithuanians, Czechs, Hungarians) and the arginine778leucine (R778L) within a transmembrane domain accounting for 30% of Asian alleles, including Koreans, Chinese, and Japanese—the Asian carrier prevalence is slightly higher at 1 in 51.[3–8,36–38] An expected correlation of increased clinical severity with protein-truncating (nonsense) mutations has been observed, with some 10–25% of mutations altering splice sites or protein reading frames and causing early-onset disease.[3,8] The R778L substitution was thought to have earlier onset with hepatic symptoms, as compared to later onset and neurologic symptoms for the H1069Q mutation, but other studies have refuted these correlations.[3,4,36–38] Corroborating the variable individual expression expected from diverging metabolic effects of altered copper transport and a considerable epigenetic/environmental effect on Wilson disease are the identical twins with homozygous H1069Q mutation, one having asymptomatic liver disease and the other fulminant hepatic failure.[4]

DISEASE MECHANISMS

Pathophysiology

The presence of asymptomatic hepatitis in patients brought to attention by affected family members suggests that toxic effects from copper accumulation initiate pathogenesis. In rodent models, dilation of hepatic cells and nuclei are seen as copper accumulates, progressing to inflammation, necrosis, and animal death from liver failure.[4,42,43] Reflecting the variability in human disease expression, identical affected animal strains survive differently in different laboratories and countries.[4] This progression from elevated liver copper concentration to hepatic inflammation and necrosis is also seen in Indian childhood cirrhosis, but elevation of ceruloplasmin in that disease plus its greater dependence on dietary copper indicates a block distal to entry of copper into the secretory pathway. The fact that nongenetically disposed individuals can consume excess copper without hepatic toxicity indicates the crucial function of *ATP7B* for intracellular copper transport and derived copper-dependent enzyme catalysis..

Consequent effects include downregulation of many nuclear receptors including the multifunctional SREBP-1 and the LXR/RXR receptors involved in oxysterol, retinoic acid, and lipid metabolism plus NF-κB-dependent activation of inflammatory mediators.[4] LEC rats show increased protein degradation as hepatic inflammation progresses, paralleled by increased metallothionein protein synthesis that is seen in Wilson disease patients and all animal models.[4,8] Altered RNA processing also occurs, including decreased production of low-density lipoprotein (LDL) receptor mRNA, but this should not affect Wilson ATPase production itself, since only one isoform of the protein has been observed in rat pineal gland.[4,8] There are general effects on lipid metabolism in LEC rats, with decreased serum cholesterol and increased triglycerides, free cholesterol, and cholesteryl ester in liver. Differences in the rat model include common development of hepatocellular carcinoma that is rare in Wilson disease in humans.[8]

Less is known about neuropsychiatric progression of Wilson disease, but its progression over time is also consistent with copper accumulation. Copper does accumulate in the brain of rodent models, and animal as well as human studies have shown alterations in mitochondrial structure and function. Decreased cytochrome *c* oxidase activity is seen in the copper deficiency of Menkes disease, and corresponding effects on mitochondrial, lipid, and inflammatory pathways documented in liver would certainly be expected to compromise neurons. Copper accumulation in the globus pallidus, putamen, thalamus, and mesencephalon correlates with white matter changes seen on head MRI (Figure 40.2), and suggests that neuropsychiatric symptoms are primary effects rather than secondary to liver disease. However, overflow from this important copper-mediating organ likely drives accumulation in brain and other organs.[4] The correlation of disease with tissue copper levels and its response to chelation therapy argues for a primary role of copper accumulation in the pathophysiology of Wilson disease. The many secondary effects on copper proteins, nuclear receptors, and lipid metabolism then depend on the regulatory constraints of each tissue, producing patterns of disease that vary considerably with individual genetic background and epigenetic circumstance, even in monozygotic twins with the same Wilson mutation.[3,7]

Current Research

Although considerable experience with chelation therapy in Wilson disease has led to standard practice guidelines,[6,7] there is ongoing clinical research that seeks to optimize therapy protocols and evaluate new agents like ammonium tetrathiomolybdate.[31] Important for this endeavor are clinical reports of long-term therapy outcomes in various patient populations[23–31] and the appreciation that chelation therapy can cause side effects from copper depletion.[41] Characterization of the Wilson ATPase gene prompted recognition of copper-transporting proteins in most organisms, and emphasis now is on developing a systems biology approach to copper metabolism[4] that will be aided by finding copper chaperones, transporters, and a COX17 protein that delivers copper to cytochrome *c* oxidase in yeast.[8]

Differential Diagnosis

Copper is elevated in cholestatic liver disease but not in normal individuals exposed to excess copper. When copper accumulation is distal to Golgi transport catalyzed by the Wilson ATPase, ceruloplasmin is always elevated.

Inevitable necrosis caused by excess copper is shown by Indian childhood cirrhosis, where copper accumulates to high levels and progressive liver failure ensues. Here again, there are striking elevations of ceruloplasm that suggest a defect in hepatobiliary excretion that contrasts with the *ATP7B* Golgi transport defect in Wilson disease. The disorder occurs only in India and seems to require genetic predisposition—consanguinity in many affected individuals suggests autosomal recessive influence—plus environmental factors due to traditional use of copper vessels for cooking. Indians have a different Wilson disease mutation, with cysteine at position 271 replaced by a stop codon (C271X), but carrier status does not seem increased in families with Indian childhood cirrhosis.[6] Other rare disorders involving hepatic copper accumulation have been described (idiopathic copper toxicosis, Tyrolean cirrhosis), all having increased ceruloplasmin and normal *ATP7B* mutation analysis that distinguish them from Wilson disease.[8]

TESTING

Wilson disease should be considered in any patient with hepatic dysfunction and unusual symptoms. An algorithm for evaluation of unexplained liver disease begins with serum ceruloplasmin (less than 20mg per dL), 24-hour urinary copper collection (less than 40μg or 0.6μmol per day), and slit lamp examination for the presence of a Kayser–Fleischer ring.[7] This triad of findings establishes the diagnosis of Wilson disease, but liver biopsy for histology and copper quantification should be added if any of the three are equivocal. A liver copper concentration below 50μg per g of dry weight would prompt consideration of other diagnoses, while one above 250 would support the diagnosis of Wilson disease.

Molecular Diagnosis

Because cirrhosis itself can lead to decreased cholestatic excretion with accumulation of copper, and because such accumulation can lead to the presence of a Kayser–Fleischer ring, DNA diagnosis should be performed to confirm a diagnosis of Wilson disease in all situations where the $1000–2000 cost of DNA sequencing is feasible. This will allow much less expensive testing of at-risk family members by targeted PCR amplification and DNA sequencing (~$250–500). Targeted DNA sequencing for alleles frequent in particular ethnic groups (e.g., the H1069Q mutation that comprises ~30% of Wilson disease alleles in Northern European populations) might be considered for uninsured individuals with the clear understanding that a normal result by no means excludes the diagnosis. Finding of one H1069Q mutant allele would confirm carrier status and presumptive disease even though many would have a different companion mutation (compound heterozygotes), a precedent followed in DNA analysis for disorders like spinal muscular atrophy. Analysis for gene duplication or deletions that are common in certain genetic diseases (e.g., Duchenne muscular dystrophy) have less than a 1% yield in Wilson disease. Even complete DNA sequencing is only 98% sensitive because of noncoding base pair changes (in introns or flanking sequences) that cannot be distinguished from normal DNA variation (e.g., single nucleotide polymorphisms [SNPs] that occur every 400–500 base pairs in the average individual).

Analytical Testing

Measures of copper content of tissue and urine are very reliable but require recognition that any form of hepatic cirrhosis can interfere with biliary copper excretion and cause accumulation in liver and other tissues. The selective block of copper entry into the Golgi secretory apparatus due to dysfunction of the *ATP7B* copper transporter in Wilson disease should be distinguished from later interference with biliary excretion in cirrhosis and other genetic or genetically predisposed disorders like Tyrolean or Indian childhood cirrhosis. These diseases have distal blocks in liver copper transport/excretion that lead to increased ceruloplasmin concentrations rather than the low ceruloplasmin that is characteristic of Wilson and Menkes diseases.

Clinical and Ancillary Testing: Imaging, Physiological, Electrodiagnostic

Most patients with Wilson disease will have inflammatory and necrotic changes on liver biopsy that shows micronodular cirrhosis with irregular copper distribution in early stages and chronic hepatitis with nodular regeneration at later stages.[8] Practice guidelines suggest liver biopsy if the triad of low ceruloplasmin, high urine copper excretion, and Kayser–Fleischer ring is not characteristic, with a higher liver copper content differentiating Wilson from other diseases. Head MRI can show white matter loss with the remarkable giant panda and cub shapes in certain planes (Figure 40.2), and the proximal renal tubules may be affected to produce aminoaciduria, glycosuria, and acidosis. Cardiac arrhythmia may be defined by electrocardiogram, hemolytic anemia found on complete blood count (CBC), with occasional findings like rhabdomyolysis complicating the diagnosis.

MANAGEMENT

The primary goal of treatment is to normalize tissue copper levels, particularly in liver with its overflow effects of increasing copper in other sensitive tissues like brain.[25] Chelation therapy is indicated for asymptomatic or mildly symptomatic patients, and treatment is lifelong, including during pregnancy.[3,7–9] If one therapy approach is discontinued, then another must be substituted because cessation leads to refractory hepatic and neurologic disease, even after resumption of therapy.[3,7] For patients with significant liver disease or those who do not respond well to chelation therapy, orthotopic liver transplantation is indicated and has an 80% survival rate in most studies.[7,8]

Standard of Care

Current recommendations emphasize use of chelating agents for patients with asymptomatic or mild disease with hepatic transplantation used for those with advanced inflammatory disease of liver.[7] The largest experience is with penicillamine, but recent studies show better tolerance of trientine, another general chelator that increases urinary copper excretion. Zinc blocks intestinal absorption and induces metallothioneins that enhance cytoplasmic capture of nontransported copper.[23] Tetrathiomolybdate is a more effective chelator that also blocks copper absorption, and there is evidence for greater efficacy in resolving neuropsychiatric symptoms.[31] Studies are available documenting efficacy of each of these agents alone and an increasing number are examining combination therapies.[7,26] Therapy must be individualized because of disease variability and diverse factors influencing therapy, including autoimmune reactions to penicillamine, depression leading to noncompliance, and side effects from unintended copper deficiency.[3,7,8,44]

Traditional therapy begins with a test dose of penicillamine, identified first as a breakdown product of penicillin but similar to the amino acid cysteine with its sulfydryl group functioning as chelator.[7,8] It was originally synthesized as a racemic mixture with one enantiomer interfering with pyridoxine synthesis, but current administration of purer preparations often retains oral dosing with 25–50 mg of vitamin B6. Toleration of the test dose is followed by four divided doses daily, taken 1 hour before or 2 hours after eating to maximize absorption, starting at 250–500 mg per day and increasing incrementally to 1–1.5 g per day.[7,8] The incremental dosing provides an opportunity to identify and minimize side effects and gradual improvement of symptoms should occur over 2–6 months with continued gains up to 1 year.

Some patients will show worsening of neuropsychiatric symptoms, and as many as 30% require cessation of penicillamine therapy due to side effects ranging from sensitivity reactions in the first 1–3 weeks (fever, rash, lymphadenopathy, marrow suppression, proteinuria) to later nephrotoxicity, bone marrow aplasia, or dermatologic changes (progeroid skin changes, pemphigous, aphthous stomatitis).[7] Side effects from restarting the drug in noncompliant patients include myasthenia gravis, polymyositis, serous retinitis, and irreversible neurohepatic degeneration. Therapy is monitored by measures of urinary copper excretion which can reach 1 mg per day and remain at 200–500 μg per day during maintenance therapy (often continued at one-half the dose shown to give a therapeutic response). Low values of urinary copper indicate noncompliance or overtreatment.[3,7,8,27,28]

Trientine (triethylenetetramine dihydrochloride) has similar structure to the common EDTA laboratory agent, chelation reflecting the Latin derivative by its claw-like planar nitrogens.[7,8] Trientine has fewer side effects than penicillamine and a lesser tendency for initial worsening of neurologic symptoms. Trientine is particularly indicated in those with prior renal disease or thrombocytopenia and hypersensitivity is very unusual. Initial dosing at 750–1500 mg per day and administration around meals is similar to penicillamine, with similar monitoring of urinary copper, dose reduction to 750–1000 mg for maintenance, and use of low urinary copper as a sign of noncompliance or overtreatment. Trientine tablets offer the advantage of greater stability in warm climates.[7]

Zinc induces enterocyte and hepatic metallothioneins, which preferentially bind copper over zinc and facilitate fecal excretion. The resulting negative copper balance removes previously stored copper. Administered in three divided doses of 150 mg per day of elemental zinc, the main side effect of gastric irritation may be less with the acetate or gluconate salt than with zinc sulfate. Urinary copper is again monitored to show efficacy and overtreatment is less likely. Because of its low risk for side effects, zinc is often used for treatment of mild or asymptomatic patients.[29,30] It is also preferred for treatment of mildly affected children under the age of 3 years.[24] Other protocols use initial treatment with a chelating agent, switching to zinc after adequate therapy for 1–1.5 years.

The practice guideline of Roberts and Schilsky[7] recommends treatment of presymptomatic patients with chelating agents or zinc but initial treatment of symptomatic patients with chelating agents; trientine or zinc may be used with for the ~30% developing penicillamine side effects.[7,27–30] Treatment should always be complemented by dietary restriction of copper-rich foods that include nuts, liver, chocolate, or shellfish, and avoidance of copper cookware or water with high copper content. Vitamin E as an antioxidant may also be added. For the duration of therapy, monitoring of serum and urinary copper, ceruloplasmin, and liver biochemistries is important plus CBC and urinalysis for those on chelating agents. These measures and a complete physical examination should be performed biannually, with more frequent urinary copper determinations when agent dosage is adjusted or when noncompliance is suspected. Continuation of treatment is extremely important for all individuals, since cessation of therapy can cause irreversible decline, including acute liver failure of pregnancy. Both chelating agents and zinc have been successful for maternofetal therapy, but the former should be reduced in dosage in the third trimester to promote wound-healing.[8,32]

Combined use of chelating agents and zinc can be tried for patients with decompensated cirrhosis that includes hypoalbuminemia, coagulopathy, and/or ascites, but not encephalopathy.[7,28–30] Those with cirrhosis that fail intensive chelation–zinc therapy and those with acute liver failure require transplantation: a scoring system assessing bilirubin and aminotransferase levels plus clotting functions is available to determine which patients will require transplantation to survive.[7] Intensive care strategies such as plasmapheresis, hemofiltration, or dialysis may be used to preserve acutely ill patients for transplant, and even with cadaveric transplants there is a 79–87% survival rate after 1 year.[7,8] Live-donor transplants have been successful using siblings who are normal or carriers for Wilson disease.

Although not mentioned as a standard of care, the author's opinion after review is that psychiatric and medical genetic evaluations should be added to those of gastrointestinal and neurologic experts during the initial management of Wilson disease. The severe consequences of noncompliance and high rates of suicide attempts warrant psychiatric evaluation and counseling analogous to that recommended for patients undergoing presymptomatic testing for Huntington disease. Assessment of resilience and self-image with reinforcement of these coping strategies is critical for patients faced with life-long disease.

Although genetic counseling for an autosomal recessive disease is relatively straightforward, physicians comfortable conveying the 1-in-4 risk for parents to have another affected child or the same risk for untested siblings to be affected may view complex counseling with trepidation. The 2/3 carrier risk for a sibling with negative clinical testing and the ~1-in-90 risk (population prevalence) for an unrelated spouse to be a carrier leads to recurrence risks for affected person/normal spouse marriage ($1 \times 1/90 \times ¼ = 1/360$) for each future child or test-negative sib/normal spouse marriage a lower ($2/3 \times 1/90 \times ¼ = 1/540$) risk. In addition, a seasoned clinical geneticist can weigh the need for expensive DNA testing better than the ward-naïve, masters-level genetic counselors working for and inevitably biased toward genetic testing companies. Effective therapy must be a part of counsel for options such as preimplantation genetic or prenatal diagnosis, a discussion benefited by physician perspective.

Failed Therapies and Barriers to Treatment Development

Failure of therapy includes the approximately 30% of patients who cannot tolerate penicillamine due to early hypersensitivity or later renal, bone marrow, or autoimmune disorders. Problems with resumption of penicillamine therapy in noncompliant patients include irreversible degeneration of hepatic and/or neuropsychiatric status, and these failures occur with other chelating agents and zinc as well. The depression arising from disease dysfunction and need for life-long treatment has led to suicide attempts in as many as 15% of Wilson disease patients, posing considerable challenges for initial and chronic therapy.[3,7,44]

Therapies under Investigation

Ammonium tetrathiomolybdate acts to interfere with intestinal uptake of copper when administered with meals and binds plasma copper when taken between meals.[7,31] It also removes copper from metallothioneins and can form insoluble copper complexes that are deposited in liver. Some studies suggest that this medication is more effective for patients with neuropsychiatric symptoms, and it seems less likely to cause transient increase in these symptoms during therapy initiation. However, its bone marrow toxicity, its powerful copper binding leading to deficiency/neurologic symptoms, and its status as an experimental drug have limited use.[3,7]

The importance of the liver as a primary vehicle for copper accumulation and excretion and the primary role of liver copper accumulation in Wilson disease suggest that targeted gene therapy to the liver would be vehicle for treatment of other symptoms such as neuropsychiatric disease. Transgenic therapy circumvents the barriers to central nervous system therapies that hamper such approaches in other neuropsychiatric diseases,[45] and successful techniques for gene delivery (ultrasound, microvesicle, adenoviral vectors) to the liver have been devised.[46] A clinical trial for introducing bone marrow mesenchymal stem cells into the liver of Wilson disease patients through hepatic vessels is proceeding in Turkey.[47]

References

1. Wilson SAK. Progressive lenticular degeneration: a familial nervous disease associated with cirrhosis of the liver. *Brain*. 1912;34:295–507.
2. Beighton P, Beighton G, Wilson, Samuel AK. (1878–1937). In: *The Man Behind the Syndrome*. Berlin: Springer-Verlag; 1986:200–201.
3. Weiss KH. Wilson disease. *GeneReviews*, updated 5-16-2013; www.ncbi.nlm.nih.gov/books/NBK1512/. Accessed June, 2013.
4. Burkhead JL, Gray LW, Lutsenko S. Systems biology approach to Wilson's disease. *Biometals*. 2011;24:455–466.
5. Rosencrantz R, Schilsky M. Wilson disease: pathogenesis and clinical considerations in diagnosis and treatment. *Semin Liver Dis*. 2011;31:245–259.
6. Ala A, Walker AP, Ashkan K, Dooley JS, Schilsky ML. Wilson's disease. *Lancet*. 2007;369:397–408.
7. Roberts EA, Schilsky ML. A practice guideline on Wilson disease. *Hepatology*. 2003;37:1475–1492.
8. Culotta VC, Gitlin JD. Disorders of copper transport. In: Scriver CR, Beaudet AL, Sly WS, Valle D, eds. *The Metabolic & Molecular Basis of Inherited Disease*. 8th ed. New York: McGraw-Hill; 2001:3105–3126.
9. Gow PJ, Smallwood RA, Angus PW, Smith AL, Wall AJ, Sewell RB. Diagnosis of Wilson's disease: an experience over three decades. *Gut*. 2000;46:415–419.
10. Hedera P, Brewer GJ, Fink JK. White matter changes in Wilson disease. (Letter). *Arch Neurol*. 2002;59:866–867.
11. Shivakumar R, Thomas SV. Teaching Neuro*Images*: face of the giant panda and her cub: MRI correlates of Wilson disease. *Neurology*. 2009;72:e50.
12. van Wassenaer-van Hall HN, van den Heuvel AG, Jansen GH, Hoogenraad TU, Mali WPTM. Cranial MR in Wilson disease: abnormal white matter in extrapyramidal and pyramidal tracts. *AJNR Am J Neuroradiol*. 1995;16:2021–2027.
13. Jung K-H, Ahn T-B, Jeon BS. Wilson disease with an initial manifestation of polyneuropathy. *Arch Neurol*. 2005;62:1628–1631.
14. Burke JF, Dayalu P, Nan B, Askari F, Brewer GJ, Lorincz MT. Prognostic significance of neurologic examination findings in Wilson disease. *Parkinsonism Relat Disord*. 2011;17:551–556.
15. Sternlieb I, Scheinberg IH. Chronic hepatitis as a first manifestation of Wilson's disease. *Ann Intern Med*. 1972;76:59–64.
16. Bearn AG, McKusick VA. Azure lunulae: an unusual change in the fingernails in two patients with hepatolenticular degeneration (Wilson's disease). *JAMA*. 1958;166:904–906.
17. Wiebers DO, Wilson DM, McLeod RA, Goldstein NP. Renal stones in Wilson's disease. *Am J Med*. 1979;67:249–254.
18. Azizi E, Eshel G, Aladjem M. Hypercalciuria and nephrolithiasis as a presenting sign in Wilson disease. *Eur J Pediat*. 1989;148:548–549.
19. Hlubocka Z, Marecek Z, Linhart A, et al. Cardiac involvement in Wilson disease. *J Inherit Metab Dis*. 2002;25:269–277.
20. Menerey KA, Eider W, Brewer GJ, Braunstein EM, Schumacher HR, Fox IH. The arthropathy of Wilson's disease: clinical and pathologic features. *J Rheumatol*. 1988;15:331–337.
21. Walshe JM. Penicillamine, a new oral therapy for Wilson's disease. *Am J Med*. 1956;21:487–495.
22. Walshe JM. Treatment of Wilson's disease: the historical background. *Q J Med*. 1996;89:553–555.
23. Brewer GJ, Yuzbasiyan-Gurkan V, Lee DY, Appelman H. Treatment of Wilson's disease with zinc. VI. Initial treatment studies. *J Lab Clin Med*. 1989;114:633–638.
24. Brewer GJ, Dick RD, Johnson VD, Fink JK, Kluin KJ, Daniels S. Treatment of Wilson's disease with zinc XVI: treatment during the pediatric years. *J Lab Clin Med*. 2001;137:191–198.
25. Weiss KH, Stremmel W. Evolving perspectives in Wilson disease: diagnosis, treatment and monitoring. *Curr Gastroenterol Rep*. 2012;14:1–7.
26. Rodríguez B, Burguera J, Berenguer M. Response to different therapeutic approaches in Wilson disease. A long-term follow up study. *Ann Hepatol*. 2012;11:907–914.
27. Weiss KH, Gotthardt DN, Klemm D, et al. Zinc monotherapy is not as effective as chelating agents in treatment of Wilson disease. *Gastroenterology*. 2011;140:1189–1198.
28. Lowette KF, Desmet K, Witters P, et al. Wilson's disease: long-term follow-up of a cohort of 24 patients treated with D-penicillamine. *Eur J Gastroenterol Hepatol*. 2010;22:564–571.
29. Linn FH, Houwen RH, van Hattum J, van der Kleij S, van Erpecum KJ. Long-term exclusive zinc monotherapy in symptomatic Wilson disease: experience in 17 patients. *Hepatology*. 2009;50:1442–1452.
30. Medici V, Trevisan CP, D'Incà R, et al. Diagnosis and management of Wilson's disease: results of a single center experience. *Clin Gastroenterol*. 2006;40:936–941.
31. Brewer GJ, Askari F, Lorincz MT, et al. Treatment of Wilson disease with ammonium tetrathiomolybdate: IV. Comparison of tetrathiomolybdate and trientine in a double-blind study of treatment of the neurologic presentation of Wilson disease. *Arch Neurol*. 2006;63:521–527.
32. Devesa R, Alvarez A, de las Heras G, de Miguel JR. Wilson's disease treated with trientine during pregnancy. *J Pediatr Gastroenterol Nutr*. 1995;20:102–103.
33. Wilson disease, #277900 in Online Mendelian Inheritance in Man, www.omim.org/entry/277900. Accessed June, 2013.
34. Bennett J, Hahn SH. Clinical molecular diagnosis of Wilson disease. *Semin Liver Dis*. 2011;31:233–238.
35. Schilsky ML, Ala A. Genetic testing for Wilson disease: availability and utility. *Curr Gastroenterol Rep*. 2010;12:57–61.
36. De Bie P, Muller P, Wijmenga C, Klomp LWJ. Molecular pathogenesis of Wilson and Menkes disease: correlation of mutations with molecular defects and disease phenotypes. *J Med Genet*. 2007;44:673–688.
37. Ferenci P. Regional distribution of mutations in the ATP7B gene in patients with Wilson disease: impact on genetic testing. *Hum Genet*. 2006;120:151–159.
38. Loudianos G, Dessi V, Lovicu M, et al. Molecular characterization of Wilson disease in the Sardinian population – evidence of a founder effect. *Hum Mutat*. 1999;14:294–303.
39. Gromadzka G, Schmidt HH-J, Genschel J, et al. Frameshift and nonsense mutations in the gene for ATPase7B are associated with severe impairment of copper metabolism and with an early clinical manifestation of Wilson's disease. *Clin Genet*. 2005;68:524–532.
40. Forbes JR, Cox DW. Copper-dependent trafficking of Wilson disease mutant ATP7B proteins. *Hum Mol Genet*. 2000;9:1927–1935.
41. Harada M, Miyagawa K, Honma Y, et al. Excess copper chelating therapy for Wilson disease induces anemia and liver dysfunction. *Intern Med*. 2011;50:1461–1464.

42. Wu J, Forbes JR, Chen HS, Cox DW. The LEC rat has a deletion in the copper transporting ATPase gene homologous to the Wilson disease gene. *Nat Genet*. 1994;7:541–545.
43. Huang L, Gitschier J. A novel gene involved in zinc transport is deficient in the lethal milk mouse. *Nat Genet*. 1997;17:292–297.
44. Harada M. Wilson disease and its current problems. *Inter Med*. 2010;49:807–808.
45. Simonato M, Bennett J, Boulis NM, et al. Progress in gene therapy for neurological disorders. *Nat Rev Neurol*. 2013;9:277–291.
46. Narmada BC, Kang Y, Venkatraman L, et al. Hepatic stellate cell-targeted delivery of hepatocyte growth factor transgene via bile duct infusion enhances its expression at fibrotic foci to regress dimethylnitrosamine-induced liver fibrosis. *Hum Gene Ther*. 2013;24:508–519.
47. Efficacy of *in vitro* expanded bone marrow derived allogeneic mesenchymal stem cell transplantation via portal vein or hepatic artery or peripheral vein in patients with Wilson Cirrhosis. www.clinicaltrials.gov/ct2/show/NCT01378182?cond=WILSON+DISEASE&rank=6. Accessed June, 2013.

CHAPTER

41

Menkes Disease and Other *ATP7A* Disorders

Juan M. Pascual[*] *and John H. Menkes*[†]

[*]The University of Texas Southwestern Medical Center, Dallas, TX, USA

[†]Deceased

INTRODUCTION

Menkes disease (MD), also known as kinky hair disease, is a multifocal, degenerative disease of gray matter first described in 1962 by Menkes et al.[1] Some 10 years later, Danks et al.[2,3] found that serum copper and ceruloplasmin levels were reduced and suggested that the primary defect in MD involved copper metabolism. In 1993, three groups of workers isolated the gene (*ATP7A*) whose defect is responsible for the disease and found that it encoded a transmembrane copper-transporting P-type ATPase (MNK, or *ATP7A*) and that the disease results from a widespread defect in intracellular copper transport and consequent copper maldistribution. Today, three distinct phenotypes resulting from *ATP7A* defects are recognized: MD, occipital horn syndrome (OHS), and *ATP7A*-related distal motor neuropathy.[4,5] *ATP7A* is located in the trans-Golgi network and encoded by the X chromosome, causing the progressive copper deficiency disorders—Menkes disease and occipital horn syndrome—in addition to a late-adolescence distal motor neuropathy that resembles Charcot–Marie–Tooth disease type 2 and which is not associated with copper deficiency.

MENKES DISEASE

Also known as kinky hair disease, Menkes disease is associated with copper deficiency, in contrast with Wilson disease, which is characterized by a state of copper excess. *ATP7A* allows cellular copper to cross intracellular membranes and to be also translocated from the trans-Golgi network to the plasma membrane in the presence of extracellular copper (Figure 41.1). The fundamental abnormality in this disease is thus the maldistribution of copper, which is unavailable as a cofactor of several enzymes including mitochondrial cytochrome *c* oxidase, lysyl oxidase, superoxide dismutase, dopamine β-hydroxylase and tyrosinase. Thus, the main features of the disease include mitochondrial respiratory chain dysfunction (complex IV deficiency), deficiency of collagen cross-links resulting in hair (pili torti and trichorrhexis nodosa) and vascular abnormalities (elongated cerebral vessels and subdural effusions), neuronal degeneration (markedly affecting Purkinje cells), and deficient melanin production dominate the disease manifestations. Affected neonates present with hypothermia, feeding difficulties and seizures. The infants are pale and exhibit kinky (or, later, no) hair. Serum copper concentration is low and the ratio of urinary homovanillic acid/vanillylmandelic acid ratio is elevated. A variety of minimally symptomatic phenotypes, including ataxia or mental retardation, have been recognized. Intramuscular or subcutaneous administration of copper–histidine affords protection against intellectual deterioration but is less effective in preventing other somatic complications.

Author note (J.M.P.): In all previous editions, this chapter was written by John Hans Menkes, who died on November 22, 2008. I have not been able—nor found incentive—to improve his direct, concise and informative style. Thus, I mostly confine myself to update the scientific aspects of the text, include new disease variants and provide the illustrations. The result, however, can betray two hands at work. And yet, any imprecisions remain my own.

Rosenberg's Molecular and Genetic Basis of Neurological and Psychiatric Disease
http://dx.doi.org/10.1016/B978-0-12-410529-4.00041-3

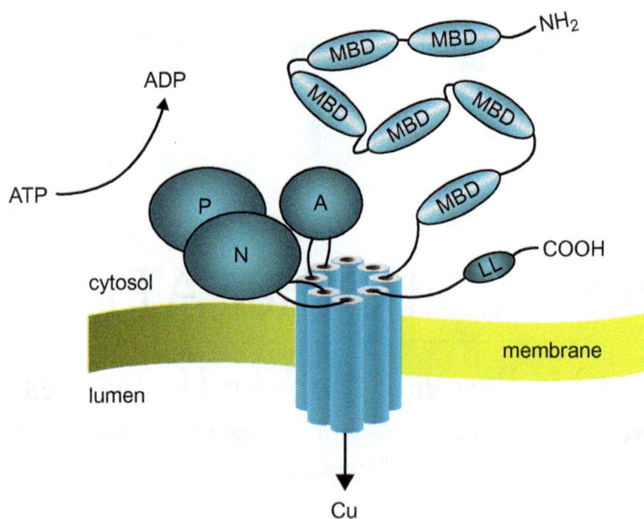

FIGURE 41.1 **Schematic representation of *ATP7A* in the cell membrane.** Figure illustrates amino- (NH_2) and carboxy- (COOH) termini, six metal-binding domains (MBD) for copper (Cu) and a 7-transmembrane domain core highlighting the CPC motif in membrane-spanning domain 6, thought to mediate copper translocation. Other relevant domains are extracellular: activation (A), nucleotide-binding (N), and phosphorylation (P). Reproduced with permission from[5a].

Molecular and Biochemical Pathology

ATP7A, the gene for MD, is located at Xq13. It encodes for MNK, an energy-dependent, copper-transporting P-type membrane ATPase. The structural homology between *ATP7A* and ATP7B is extremely high in the 3′ two-thirds of the genes, but there is considerable divergence between them in the 5′ one-third.[6] *ATP7A* is expressed in all tissue with the exception of liver. It consists of six metal-binding domains, eight transmembrane domains, and domains for phosphatase, phosphorylation, and ATP binding (Figure 41.1). At basal copper levels, the protein is located in the trans-Golgi network, the sorting station for proteins exiting from the Golgi apparatus, where it is involved in copper uptake into its lumen. At increased intra- and extracellular copper concentrations, the MNK protein shifts toward the plasma membrane, presumably to enhance removal of excess copper from the cell.[7]

The characteristic abnormality in copper metabolism as expressed in the human infant is a maldistribution of body copper, a result of defective copper transport across the placenta, the gastrointestinal tract, and the blood–brain barrier.[8] The consequence is a failure of copper incorporation into a variety of essential enzymes.[6,9] Patients absorb little or no orally administered copper; when the metal is given intravenously, they experience a prompt rise in serum copper and ceruloplasmin. As a result of impaired copper efflux, the copper content of cultured fibroblasts, myotubes, and lymphocytes derived from patients with MD is several times greater than that of control cells.[10]

Pathology

Because of defective activity of the metalloenzymes, a variety of pathologic changes are set into motion. In skin, the elastin fibers are reduced in number and consist of thin strands of amorphous elastin associated with numerous microfibrils.[11] Cerebral and systemic arteries are tortuous, with irregular lumens and frayed and split intimal linings. These abnormalities reflect the failure in elastin and collagen cross-linking caused by a decrease in functional activity of copper-dependent lysyl oxidase. Anatomical changes within the brain are believed to result from reduced activity of the various copper-containing enzymes—in particular, from mitochondrial dysfunction, from vascular lesions, or from a combination of the two factors. In addition, copper is specifically protective toward *N*-methyl-D-aspartate (NMDA)-mediated excitotoxic cell death, and *ATP7A* is required to maintain a pool of copper in neurons.[12] Gross examination of the brain discloses diffuse atrophy; unilateral or bilateral subdural hematomas are commonly present. There is extensive focal degeneration of gray matter with neuronal loss and gliosis, and an associated axonal degeneration in white matter.[1] Cellular loss is prominent in the cerebellum. Here Purkinje cells are most affected; many are lost, and others show abnormal dendritic arborization (resembling a "weeping willow") and perisomatic processes. Focal axonal swellings (typically described as "torpedoes") are also observed.[1] On electron microscopy, there are a variety of mitochondrial abnormalities.[13]

Clinical Features

MD is an X-linked disorder; its frequency has been estimated at 0.8–2 per 100,000 live male births. Most of the clinical manifestations can be explained by the low activities of the various copper-containing enzymes. These include cytochrome *c* oxidase, lysyl oxidase, superoxide dismutase, peptidylglycine α-amidating monooxygenase, tyrosinase, dopamine-β-hydroxylase, and ascorbic acid oxidase.

The clinical manifestations of MD are quite variable. In part, the variability can be explained by a large number of mutations. These include mutations that lead to splicing abnormalities, small duplications, nonsense mutations, and missense mutations[14–16] and to aberrant protein fragments that may retain functional activity.[17] To date, all mutations detected have been unique for each given family, and almost all have been associated with a decreased level of the mRNA for the copper-transporting ATPase. As yet there is no good genotype–phenotype correlation and even intrafamilial phenotypic variability can be pronounced.[18] In the allelic variant of MD, occipital horn syndrome, the full-length MNK protein is absent, but the truncated MND protein is expressed and is localized to the endoplasmic reticulum.[19] Baerlocher and Nadal,[20] and Kaler[21] have provided comprehensive reviews of the clinical features. In the classic form, symptoms appear during the neonatal period. Most commonly, there is hypothermia, hypoglycemia, poor feeding, and impaired weight gain. Less often there is a fatal hemorrhagic diathesis and multiple congenital fractures. Cephalhematomas can be prominent in infants born vaginally. The appearance of hair is often unremarkable at birth; newborns can have little or no hair, or normally pigmented hair.[21] Other patients present with seizures, delayed development, or failure to thrive. The most striking finding is the abnormal hair. It is colorless and friable. On examination under the microscope, a variety of abnormalities are evident, most commonly pili torti (a hair shaft which is flattened and twisted 180° on its axis; Figure 41.2). Monilethrix (an elliptical swelling of the hair shafts with intervening tapered constrictions) and trichorrhexis nodosa (small beaded swelling of the hair shaft with fractures at regular intervals) are also seen (Figure 41.3).[1] Other organs can also be involved. Optic disks are pale, and there are microcysts of the pigment epithelium and iris. Hydronephrosis, hydroureter, and diverticula or rupture of the bladder has been reported. Radiographs of long bones show a variety of abnormalities, including osteoporosis, metaphyseal spurring, a diaphyseal periosteal reaction, and scalloping of the posterior aspects of the vertebral bodies. Skull radiographs may reveal the presence of wormian bones in the lambdoidal and posterior sagittal sutures. Neuroimaging studies frequently disclose cerebral atrophy, areas of low density within the cortex, impaired myelination, and tortuous and enlarged intracranial vessels (Figure 41.4). The frequent presence of subdural hematomas has in many instances raised suspicion of intentional trauma.[22]

The course of MD is inexorably downhill, but the rate of neurologic deterioration varies from one patient to the next. There are recurrent infections of the respiratory and urinary tracts. Sepsis and meningitis (Figure 41.5) are fairly common, and there is evidence for dysfunction of the cellular immune responses.

FIGURE 41.2 **Pili torti in the infant from Figures 41.4 and 41.5 shown next to a normal hair.** *Picture courtesy of Dr. Lu Le, Department of Dermatology, UT Southwestern Medical Center.*

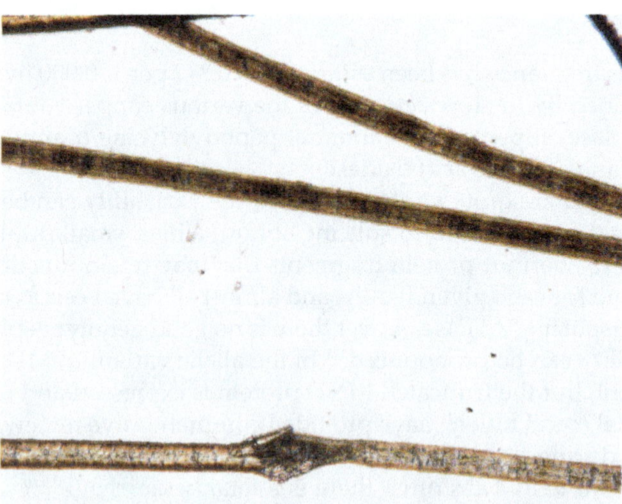

FIGURE 41.3 **Trichorrhexis nodosa in the infant from Figures 41.4 and 41.5, shown next to normal hairs.** *Picture courtesy of Dr. Lu Le, Department of Dermatology, UT Southwestern Medical Center.*

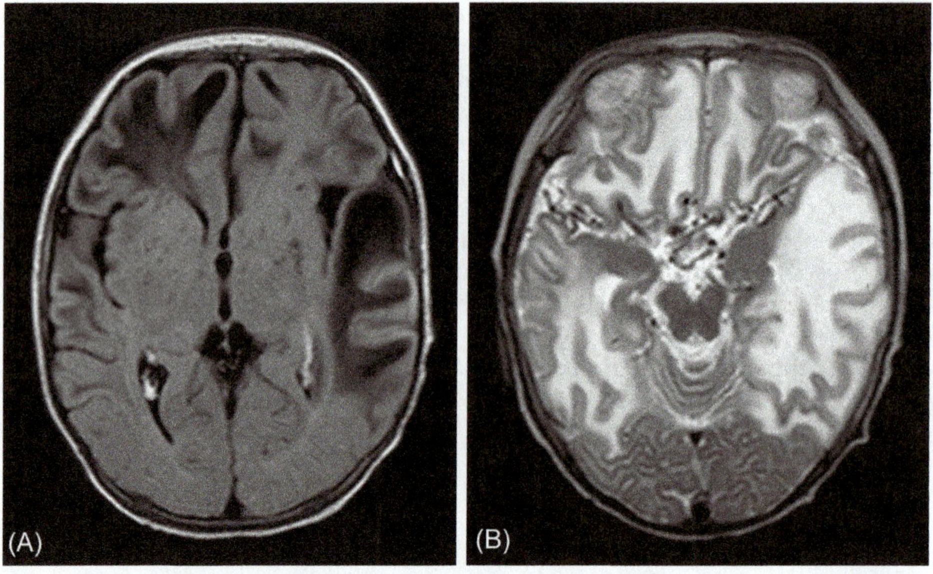

FIGURE 41.4 **MRIs of a male infant with Menkes disease (also shown in Figure 41.5) at age 6 months.** (**A**) Unenhanced, T2-FLAIR axial MRI of the brain: Following resolution of meningitis, he developed multifocal epilepsy, lethargy, and loss of postural control and reflexes. Several infarcts are apparent, as are flow voids in the lenticulostriate vessel territory, which were separately determined to represent tortuous arteries. (**B**) T2-weighted image contemporary with (A) illustrating vascular tortuosity in the circle of Willis and sylvian fissure territory.

Diagnosis

The history of developmental arrest or regression, hypotonia, and seizures, and the appearance of the infant, in particular the unusual hair, should suggest the diagnosis. Reduced levels of serum ceruloplasmin and copper levels are diagnostic of MD. Urinary copper is variable: normal or reduced, and even elevated levels have been reported. This assay is, therefore, of no value for the diagnosis of MD. Heterozygotes are mostly asymptomatic and do not show any biochemical abnormalities. However, areas of pili torti constitute 30–50% of the hair. Prenatal diagnosis has been based on the increased copper content of cultured amniocytes and chorionic villus samples. The determination of abnormal catecholamine metabolites in plasma (for example, dopamine to norepinephrine ratio)[23] and in cerebrospinal fluid resulting from deficiency of dopamine-β-hydroxylase is sensitive and specific, and may allow for earlier patient detection.[24]

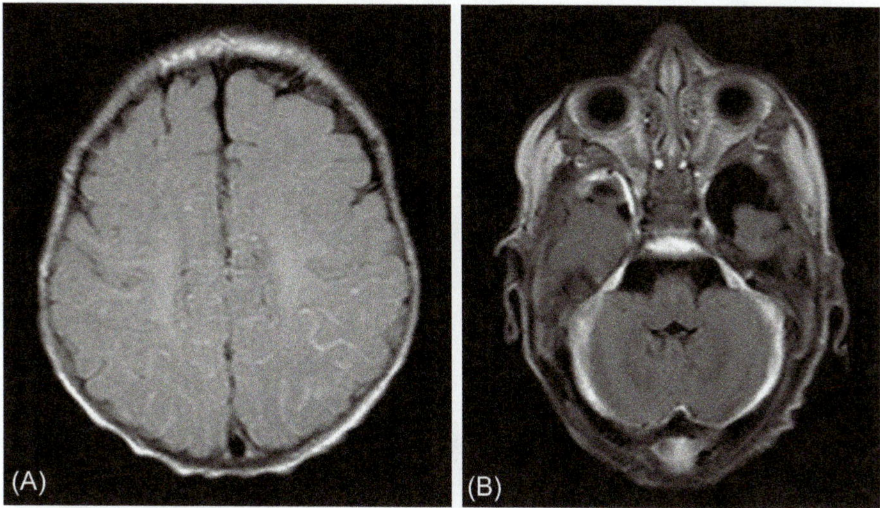

FIGURE 41.5 **Contrast-enhanced, T2-FLAIR axial MRI of the brain of a 4-month male with Menkes disease who presented with meningitis (prostration, acidosis, elevated white cell count, protein and lactate in the cerebrospinal fluid).** (**A**) Leptomeningeal contrast material accumulation is prominent in the parietal and occipital lobes. Note the unremarkable configuration of the rest of the brain. (**B**) Lower axial section illustrating severe meningeal irritation in the cranial base and posterior fossa. The infant harbored a frameshift mutation in *ATP7A*.

Treatment

Because the symptoms of MD are the consequence of impaired activity of the various copper-containing enzymes, copper supplementation would seem to be a rational means of treating this condition. However, neither oral nor parenteral administration of copper has been effective, even though the latter induces a rapid rise of both ceruloplasmin and copper. Copper replacement therapy employing subcutaneous injections of copper histidinate has been suggested, but due to the considerable clinical heterogeneity of patients with MD, the effectiveness of this therapy has been difficult to evaluate. Sheela et al.[25] have reviewed ethical aspects of this treatment. Studies of patients who share the same copper-responsive mutation treated at different ages suggest that early copper injection treatment may be associated with satisfactory outcomes,[26] as also observed in a broader patient group treated neonatally.[27] However, fetal copper injections from the 31st week of uterine life can prove ineffective.[28]

Mouse Model of Menkes Disease

Mottled brindled mouse spontaneous mutants were discovered by 1953,[29] and have served as a testbed for MD after they were confirmed to harbor mutations in *ATP7A* in 1994.[30,31] These mice live less than 14 days and exhibit severe intestinal copper malabsorption associated with tremors and spasms[32] that can be prevented by a subcutaneous injection of copper.[33] The activities of ceruloplasmin oxidase and lysyl oxidase and the pigmentation of skin and fur also normalized.[33] Mottled/brindled mutant (Mo(br/y)) mice exhibit primarily cerebral biochemical abnormalities, including decreased copper content and activity of cytochrome *c* oxidase. Apoptosis is prominent in neocortex and hippocampus and is associated with decreased anti-apoptotic protein Bcl-2 and increased release of cytochrome *c* from mitochondria into the cytosol.[34] Stimulation of noradrenergic function via intraperitoneal injection of L-threo-dihydroxyphenylserine, which is converted into norepinephrine by aromatic-L-amino acid decarboxylase, leads to increases in brain norepinephrine and related metabolites, but does not prevent neocortical or hipoccampal neuronal degeneration.[35] However, neonatal Mo(br/y) mice receiving complementary therapies via lateral ventricle injections of adenovirus harboring a human *ATP7A* DNA construct and copper chloride exhibit prolonged survival and preservation of neural structure.[36]

OCCIPITAL HORN SYNDROME

Occipital horn syndrome is characterized by the presence of protuberant dysostotic lesions located in the cranial base of males, particularly at the insertion of the trapezius and sternocleidomastoid muscles in the occipital bone (Figure 41.6).[37] Occipital horns can be palpated or documented by cranial imaging (Figure 41.5). Patients with OHS exhibit dysautonomia, lax skin and joints, bladder diverticula, inguinal hernias, and vascular tortuosity. Their

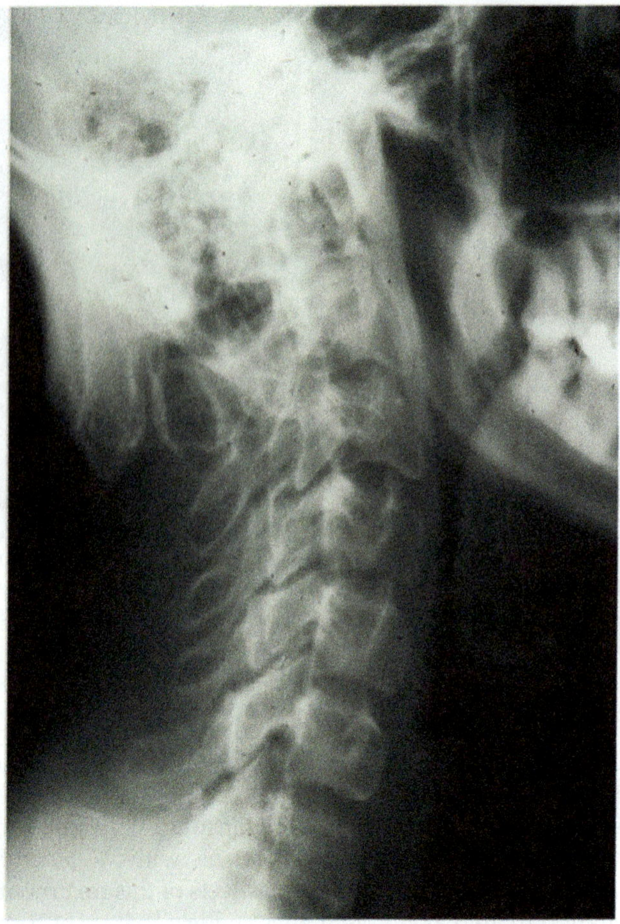

FIGURE 41.6 **Lateral craniospinal radiograph in occipital horn syndrome.** Note the prominent, exostotic areas of insertion of the trapezius muscles well below the level of the first cervical vertebral body. Reproduced with permission from[5a].

intellect is usually normal or mildly impaired. Splice-site and missense mutations endowed with relatively higher residual copper transporter activity are common.[38–40]

ATP7A-RELATED DISTAL MOTOR NEUROPATHY

ATP7A-related distal motor neuropathy is a late-adolescence or young adult-onset motor neuropathy that resembles Charcot–Marie–Tooth disease type 2 in males. Patients can manifest distal muscular atrophy and pes cavus. However, this disorder is biochemically distinct from MD or occipital horn syndrome (Table 41.1). The electromyographic correlate of the disease includes decreased compound motor action potentials, positive waves and fibrillations, normal motor nerve conduction velocities, and minimal or absent sensory nerve involvement. *ATP7A*-related distal motor neuropathy is associated with missense mutation of conserved amino acids in the carboxyl half of *ATP7A* that alter its intracellular translocation without repercussions on protein abundance or copper transport areas of the molecule.[5]

TABLE 41.1 Serum Copper and Ceruloplasmin in Menkes Disease, Occipital Horn Syndrome and *ATP7A*-Related Distal Motor Neuropathy[41]

Serum concentration	Normal	Menkes disease	Occipital horn syndrome	*ATP7A*-related distal motor neuropathy
Copper (μg/dL)	70–150; (0–6 mos: 20–70)	0–55	40–80	Normal (80–100)
Ceruloplasmin (mg/L)	200–450; (0–6 mos: 50–220)	10–160	110–240	Normal (240–310)

MODE OF INHERITANCE OF *ATP7A*-RELATED DISORDERS

The *ATP7A*-related copper transport disorders are inherited in an X-linked recessive manner. Approximately one-third of affected males have no family history of *ATP7A*-related disorders. The risk of transmission of a carrier mother is 50% for each pregnancy. Males who inherit the mutation will be affected and females who inherit the mutation will be carriers. Patients with MD do not reproduce males with occipital horn syndrome or *ATP7A*-related distal motor neuropathy but pass the disease-causing mutation to their daughters, not their sons.

ACKNOWLEDGEMENTS

I would like to thank Mrs. Myrna Menkes for her support of this contribution, which is dedicated to the memory of Dr. John H. Menkes. I am grateful to NIH for grants NS077015, NS067015, NS078059, MH084021, RR002584 and RR024982.

References

1. Menkes JH, Alter M, Steigleder GK, Weakley DR, Sung JH. A sex-linked recessive disorder with retardation of growth, peculiar hair, and focal cerebral and cerebellar degeneration. *Pediatrics*. 1962;29(May):764–779.
2. Danks DM, Campbell PE, Stevens BJ, Mayne V, Cartwright E. Menkes's kinky hair syndrome. An inherited defect in copper absorption with widespread effects. *Pediatrics*. 1972;50(2):188–201.
3. Danks DM, Stevens BJ, Campkell PE, et al. Menkes kinky-hair syndrome. An inherited defect in the intestinal absorption of copper with widespread effects. *Birth Defects Orig Artic Ser*. 1974;10(10):132–137.
4. Kaler SG. ATP7A-related copper transport diseases-emerging concepts and future trends. *Nat Rev Neurol*. 2011;7(1):15–29.
5. Kennerson ML, Nicholson GA, Kaler SG, et al. Missense mutations in the copper transporter gene ATP7A cause X-linked distal hereditary motor neuropathy. *Am J Hum Genet*. 2010;86(3):343–352.

5a. Tümer Z, Møller LB. Menkes disease. *European Journal of Human Genetics*. 2010;18:511–518.

6. Harrison MD, Dameron CT. Molecular mechanisms of copper metabolism and the role of the Menkes disease protein. *J Biochem Mol Toxicol*. 1999;13(2):93–106.
7. Lane C, Petris MJ, Benmerah A, Greenough M, Camakaris J. Studies on endocytic mechanisms of the Menkes copper-translocating P-type ATPase (ATP7A; MNK). Endocytosis of the Menkes protein. *Biometals*. 2004;17(1):87–98.
8. Loudianos G, Gitlin JD. Wilson's disease. *Semin Liver Dis*. 2000;20(3):353–364.
9. Vulpe CD, Packman S. Cellular copper transport. *Annu Rev Nutr*. 1995;15:293–322.
10. Tumer Z, Horn N. Menkes disease: recent advances and new aspects. *J Med Genet*. 1997;34(4):265–274.
11. Pasquali-Ronchetti I, Baccarani-Contri M, Young RD, Vogel A, Steinmann B, Royce PM. Ultrastructural analysis of skin and aorta from a patient with Menkes disease. *Exp Mol Pathol*. 1994;61(1):36–57.
12. Schlief ML, West T, Craig AM, Holtzman DM, Gitlin JD. Role of the Menkes copper-transporting ATPase in NMDA receptor-mediated neuronal toxicity. *Proc Natl Acad Sci U S A*. 2006;103(40):14919–14924.
13. Yoshimura N, Kudo H. Mitochondrial abnormalities in Menkes' kinky hair disease (MKHD). Electron-microscopic study of the brain from an autopsy case. *Acta Neuropathol*. 1983;59(4):295–303.
14. Hsi G, Cox DW. A comparison of the mutation spectra of Menkes disease and Wilson disease. *Hum Genet*. 2004;114(2):165–172.
15. Tumer Z, Birk Moller L, Horn N. Screening of 383 unrelated patients affected with Menkes disease and finding of 57 gross deletions in ATP7A. *Hum Mutat*. 2003;22(6):457–464.
16. Tumer Z. An overview and update of ATP7A mutations leading to Menkes disease and occipital horn syndrome. *Hum Mutat*. 2013;34(3):417–429.
17. Paulsen M, Lund C, Akram Z, Winther JR, Horn N, Moller LB. Evidence that translation reinitiation leads to a partially functional Menkes protein containing two copper-binding sites. *Am J Hum Genet*. 2006;79(2):214–229.
18. Donsante A, Tang J, Godwin SC, et al. Differences in ATP7A gene expression underlie intrafamilial variability in Menkes disease/occipital horn syndrome. *J Med Genet*. 2007;44(8):492–497.
19. Francis MJ, Jones EE, Levy ER, Ponnambalam S, Chelly J, Monaco AP. A Golgi localization signal identified in the Menkes recombinant protein. *Hum Mol Genet*. 1998;7(8):1245–1252.
20. Baerlocher K, Nadal D. Menkes syndrome. *Ergeb Inn Med Kinderheilkd*. 1988;57:77–144.
21. Kaler SG. Menkes disease. *Adv Pediatr*. 1994;41:263–304.
22. Menkes JH. Subdural haematoma, non-accidental head injury or …? *Eur J Paediatr Neurol*. 2001;5(4):175–176.
23. Goldstein DS, Holmes CS, Kaler SG. Relative efficiencies of plasma catechol levels and ratios for neonatal diagnosis of Menkes disease. *Neurochem Res*. 2009;34(8):1464–1468.
24. Kaler SG, Holmes CS. Catecholamine metabolites affected by the copper-dependent enzyme dopamine-beta-hydroxylase provide sensitive biomarkers for early diagnosis of Menkes disease and viral-mediated ATP7A gene therapy. *Adv Pharmacol*. 2013;68:223–233.
25. Sheela SR, Latha M, Liu P, Lem K, Kaler SG. Copper-replacement treatment for symptomatic Menkes disease: ethical considerations. *Clin Genet*. 2005;68(3):278–283.
26. Tang J, Donsante A, Desai V, Patronas N, Kaler SG. Clinical outcomes in Menkes disease patients with a copper-responsive ATP7A mutation, G727R. *Mol Genet Metab*. 2008;95(3):174–181.

27. Kaler SG, Holmes CS, Goldstein DS, et al. Neonatal diagnosis and treatment of Menkes disease. *N Engl J Med*. 2008;358(6):605–614.
28. Haddad MR, Macri CJ, Holmes CS, et al. In utero copper treatment for Menkes disease associated with a severe ATP7A mutation. *Mol Genet Metab*. 2012;107(1–2):222–228.
29. Fraser AS, Sobey S, Spicer CC. Mottled, a sex-modified lethal in the house mouse. *J Genet*. 1953;51:217–221.
30. Levinson B, Vulpe C, Elder B, et al. The mottled gene is the mouse homologue of the Menkes disease gene. *Nat Genet*. 1994;6(4):369–373.
31. Cecchi C, Biasotto M, Tosi M, Avner P. The mottled mouse as a model for human Menkes disease: identification of mutations in the Atp7a gene. *Hum Mol Genet*. 1997;6(3):425–433.
32. Mann JR, Camakaris J, Danks DM. Copper metabolism in mottled mouse mutants: distribution of 64Cu in brindled (Mobr) mice. *Biochem J*. 1979;180(3):613–619.
33. Mann JR, Camakaris J, Danks DM, Walliczek EG. Copper metabolism in mottled mouse mutants: copper therapy of brindled (Mobr) mice. *Biochem J*. 1979;180(3):605–612.
34. Rossi L, De Martino A, Marchese E, Piccirilli S, Rotilio G, Ciriolo MR. Neurodegeneration in the animal model of Menkes' disease involves Bcl-2-linked apoptosis. *Neuroscience*. 2001;103(1):181–188.
35. Donsante A, Sullivan P, Goldstein DS, Brinster LR, Kaler SG. L-threo-dihydroxyphenylserine corrects neurochemical abnormalities in a Menkes disease mouse model. *Ann Neurol*. 2013;73(2):259–265.
36. Donsante A, Yi L, Zerfas PM, et al. ATP7A gene addition to the choroid plexus results in long-term rescue of the lethal copper transport defect in a Menkes disease mouse model. *Mol Ther*. 2011;19(12):2114–2123.
37. Wakai S, Ishikawa Y, Nagaoka M, Minami R, Hayakawa T. Occipital horn syndrome (Ehlers-Danlos syndrome type IX) with severe psychomotor retardation and muscle atrophy–a first Japanese case. *Rinsho Shinkeigaku*. 1991;31(5):534–538.
38. Kaler SG, Gallo LK, Proud VK, et al. Occipital horn syndrome and a mild Menkes phenotype associated with splice site mutations at the MNK locus. *Nat Genet*. 1994;8(2):195–202.
39. Qi M, Byers PH. Constitutive skipping of alternatively spliced exon 10 in the ATP7A gene abolishes Golgi localization of the Menkes protein and produces the occipital horn syndrome. *Hum Mol Genet*. 1998;7(3):465–469.
40. Tang J, Robertson S, Lem KE, Godwin SC, Kaler SG. Functional copper transport explains neurologic sparing in occipital horn syndrome. *Genet Med*. 2006;8(11):711–718.
41. Kaler SG. ATP7A-related copper transport disorders. In: Pagon RA, Adam MP, Bird TD, Dolan CR, Fong CT, Smith RJH, Stephens K, eds. *SourceGeneReviews® [Internet]*. Seattle (WA): University of Washington, Seattle; 1993, 1993–2014. 2003 May 09 [updated 2010 Oct 14].

CHAPTER

42

Neurodegeneration with Brain Iron Accumulation

Susanne A. Schneider

University of Kiel, Kiel, Germany

INTRODUCTION AND CLINICAL FEATURES

Precise regulation of iron metabolism is essential and both iron deficiency and iron overload are associated with disease. The group of syndromes involving neurodegeneration with brain iron accumulation (NBIA) is due to abnormalities in brain iron metabolism with excess iron accumulation in the globus pallidus (GP) and to a lesser degree in the substantia nigra and, sometimes, adjacent areas. NBIA clinically present with a progressive hypo- and/or hyperkinetic movement disorder and excessive iron deposition in the brain detectable by magnetic resonance (MR) neuroimaging, and there is phenotypic overlap with other complicated pallidopyramidal syndromes without iron accumulation. Pathologically NBIA syndromes are characterized by degeneration of both neurons and astrocytes. In addition, spheroid bodies are common to many forms of NBIA and there is overlap with the classification of neuroaxonal dystrophies. Several causative genes underlying NBIA implicated in mitochondrial function, lipid metabolism, and autophagy have already been delineated and rapid developments have prompted the review of this interesting field.

Historical Overview: Disease Identification and Gene Identification

Historical milestones in the delineation of the NBIA disorders include the first explicit clinicopathological description in the 1920s, when high levels of brain iron and spheroids were detected on brain histology in a family affected by a progressive extrapyramidal disorder (onset around 7–9 years, death in the early 20s).[1] For many decades the term Hallervorden–Spatz disease was used but has now largely been replaced by disease names referring to the molecular specifics (see below). The terminology tries to follow a pattern in that the first word(s) (or letters in the abbreviated name) refer to the molecular underpinnings and the last two words stand for "associated neurodegeneration," e.g., pantothenate kinase-associated neurodegeneration (PKAN); PLA2G6-associated neurodegeneration (PLAN).

Another important milestone in the evaluation of these diseases was the development of high-field magnetic resonance imaging (MRI) in the 1980s, which allowed for noninvasive neuroimaging *in vivo* (depicting hypointensity on T2-weighted images or more recently susceptibility-weighted images [SWI]).[2,3] The first gene underlying NBIA disorders was eventually identified by Hayflick and colleagues in 2003, who described mutations in the *PANK2* gene in 49 patients (100%) with a typical presentation and 17 patients (35%) with an atypical presentation. Since then, several other genes have been identified (including *PLA2G6*, *FA2H*, *C19orf12*, *ATP13A2*, *CP* and *FTL*).

Mode of Inheritance and Prevalence

The mode of inheritance of NBIA disorders is autosomal recessive in most subtypes, in particular those with early onset (including PKAN, PLAN, fatty acid hydroxylase-associated neurodegeneration (FAHN) and Kufor-Rakeb disease). In contrast, inheritance is autosomal dominant in neuroferritinopathy and where disease onset usually

Rosenberg's Molecular and Genetic Basis of Neurological and Psychiatric Disease
http://dx.doi.org/10.1016/10.1016/B978-0-12-410529-4.00042-5

occurs in adulthood. Finally, *de novo* mutations occur in X-linked dominant NBIA (beta-propeller protein-associated neurodegeneration [BPAN]).

NATURAL HISTORY

Age of Onset

Age of onset is usually in early childhood in most forms of NBIA. In PLAN, first symptoms occur in infancy (infantile neuroaxonal dystrophy; INAD), although the phenotype of *PLA2G6* mutations is broad and onset may occur in the late teens or early adulthood (atypical PLAN). In PKAN onset is typically around age 3–6 years, but again symptoms may begin later (atypical PKAN). Two forms of NBIA, neuroferritinopathy and aceruloplasminemia, have onset in adulthood (around age 40 and 50), however, this may also vary.

Disease Evolution

Disease progression of NBIA disorders is typically slowly progressive. Loss of ambulation is seen in the majority of PKAN patients with classical disease within 10–15 years of diagnosis. Death is often related to cardiorespiratory complications (aspiration).

Of note, in BPAN disease evolution follows a stepwise progression, with global developmental delay in early childhood, followed by a stable phase. In early adulthood there is a relatively sudden onset and fast deterioration of dystonia-parkinsonism, dementia and spasticity.

Disease Variants

Different genetically defined subforms have been recognized, which will be reviewed in the following. Overall there is great similarity between the clinical presentations of neuroaxonal dystrophies.

PKAN As mentioned above, onset of the classic variant occurs before the age of 6 years in almost 90%,[4,5] often with gait difficulty as the presenting symptom.[5] The clinical picture comprises pyramidal and extrapyramidal features with prominent dystonia, often with predominant oro-lingual-mandibular involvement.[6] Neuropsychiatric[7,8] and visual disturbances due to pigmentary retinopathy are frequent.[9] In late-onset (atypical) PKAN, motor involvement tends to be less severe,[10] while cognitive decline and psychiatric features may be the leading symptoms.[5,11–13]

PLAN In PLAN the early-onset form presents as infantile neuroaxonal dystrophy (INAD) characterized by progressive motor and mental retardation, marked truncal hypotonia, cerebellar ataxia, pyramidal signs, and early visual disturbances due to optic atrophy. Late-onset PLAN has been reported characterized by dystonia-parkinsonism combined with pyramidal signs, eye movement abnormalities, cognitive decline, and psychiatric features.[14] Levodopa-sensitive parkinsonism (the condition was subsequently assigned the PARK14 locus) was characterized by the presence of tremor including a pill-rolling rest component, rigidity, and severe bradykinesia in line with the finding of Lewy body pathology (see below). Cerebellar signs and sensory abnormalities, which are often prominent in the early-childhood variant, were absent.

MPAN The clinical picture of mitochondrial-membrane protein-associated neurodegeneration (MPAN) includes childhood-onset dysarthria and gait difficulty, followed by the development of spastic paraparesis, extrapyramidal features (dystonia and parkinsonism), motor axonal neuropathy, optic atrophy, and psychiatric symptoms. However, a milder late-onset form resembling idiopathic Parkinson disease may occur.[15]

FAHN The phenotype of FAHN is characterized by childhood-onset gait impairment, spastic quadriparesis, severe ataxia, and dystonia. Seizures and divergent strabismus may also be present.

BPAN In BPAN,[16,17] as mentioned above, the phenotype of global developmental delay in early childhood may be unspecific in early stages. In adulthood, there is a relatively sudden onset and fast deterioration of dopa-responsive parkinsonism, dystonia, dementia and spasticity. The phenotype may resemble atypical Rett syndrome or atypical Angelman syndrome.

KUFOR-RAKEB DISEASE The clinical phenotype of Kufor-Rakeb disease[18–20] comprises adolescence-onset parkinsonism, with pyramidal tract signs in some. Eye movement abnormalities with incomplete supranuclear upgaze palsy can be a clue. Oculogyric dystonic spasms, facial-faucial-finger mini-myoclonus and autonomic dysfunction may be present. Psychiatric features include visual hallucinations and dementia.

NEUROFERRITINOPATHY The phenotype of neuroferritinopathy may resemble Huntington disease with prominent extrapyramidal features including chorea, stereotypies, and dystonia[21] with prominent oro-lingual-mandibular dyskinesia and blepharospasm. About 10% present with parkinsonism. Pyramidal involvement and ataxia are usually absent. Cognitive dysfunction, depression, and psychosis may be present.

ACERULOPLASMINEMIA The most common presenting features of aceruloplasminemia are cognitive impairment, retinal degeneration, cerebellar ataxia, and craniofacial dyskinesia. The average age at diagnosis is around 50 years, and ranges from 16–71 years.[22] Diabetes mellitus and microcytic anemia may be a clue and frequently predate neurologic symptoms.

MOLECULAR GENETICS AND GENOTYPE–PHENOTYPE CORRELATIONS

The genes associated with the various NBIA subtypes are listed in Table 42.1. Gene function remains ill understood for most of the forms and in general no obvious genotype–phenotype correlations have emerged yet.

PKAN In PKAN, mutations, mostly missense, have been detected in all seven exons of the *PANK2* gene located at chromosome 20p. Deletions, duplication and splice-site mutations as well as exon deletions have also been reported.[23] Two common mutations account for about one-third of all PKAN cases, that is 1231G>A and 1253C>T. The majority of the remainder cases carry "private mutations,"[10] like the founder mutation 680A>G, found in the Dominican Republic.[10] Some mutations may be associated with milder phenotypes than others.[5,12]

PLAN Numerous different mutations in the *PLA2G6* gene (associated with PLAN) located at chromosome 22q have been reported. No obvious genotype–phenotype correlation has emerged. However, recent functional analysis[24] has suggested that, compared to the wild type, mutant proteins associated with INAD exhibit less than 20% of the specific activity in both lysophospholipase and phospholipase assays, which predicted accumulation of PLA2G6 phospholipid substrates. In contrast, mutations associated with dystonia-parkinsonism did not impair catalytic activity, which may explain the relatively milder phenotype and absence of iron accumulation in at least some cases.

MPAN This is due to mutations in *C19orf12* at chromosome 19q.[15] The gene has three exons and encodes two isoforms with two alternative first exons. All together 28 different mutations (in 55 families) have been described in

TABLE 42.1 Overview of NBIA Conditions and Genes (if known)

Condition (acronym)	Gene	Chromosomal position
PKAN	*PANK2*	20p13
PLAN	*PLA2G6*	22q12
MPAN	*C19orf12*	19q12
FAHN	*FA2H*	16q23
BPAN (SENDA)	*WDR45*	Xp11.23
Kufor-Rakeb disease	*ATP13A2*	1p36
Aceruloplasminemia	*CP*	3q23
Neuroferritinopathy	*FTL*	19q13
Idiopathic late-onset cases	Probably heterogeneous	Probably heterogeneous

Abbreviations: CP, ceruloplasmin; BPAN, beta-propeller associated neurodegeneration; FA2H, fatty acid 2-hydroxylase; FTL, ferritin light chain; MPAN, mitochondrial membrane-associated neurodegeneration; NBIA, neurodegeneration with brain iron accumulation; PANK2; pantothenate kinase 2; PKAN, pantothenate kinase-associated neurodegeneration; PLA2G6, phospholipase A2; PLAN, PLA2G6-associated neurodegeneration; SENDA, static encephalopathy of childhood with neurodegeneration in adulthood.

the literature so far, including frameshift mutations, missense mutations, nonsense mutations and splice-site mutations.[25] A deletion of 11 base pairs leading to a premature stop codon and predicted to cause early truncation of the protein was identified in the majority of patients due to a founder effect in a Polish cohort. A second frequent mutation is the p.Thr11Met mutation, which may be associated with later onset.[25]

FAHN This is due to mutations in *FA2H* located on chromosome 16q. Only a handful of cases, mostly from Asia and southern Europe, have been reported with the phenotype including the allelic disorders leukodystrophy and hereditary spastic paraplegia, type SPG35. No genotype–phenotype correlations have emerged.

BPAN This is due to mutations in *WDR45* located on the X-chromosome. Reported cases carried heterozygous *de novo* mutations, suggesting a dominant mode of inheritance. So far no common mutation has emerged. In the initial report male patients had identical clinical and radiologic features as females, suggesting somatic mosaicism as the most likely mechanism. Little is known about gene function.

KUFOR-RAKEB DISEASE This is due to mutations in the *ATP13A2*[26] gene located on chromosome 1p. The 26 kb-spanning gene contains 29 exons and encodes a lysosomal 5 P-type ATPase. Only a handful of cases have been reported and there is some overlap with neuronal ceroid lipofuscinosis (NCL). Most patients carried homozygous mutations but compound heterozygous cases have also been identified.

NEUROFERRITINOPATHY This is caused by mutations in the *FTL* gene on chromosome 19q. The gene contains four exons, and encodes a single 175 amino-acid protein, the ferritin light chain. To date at least seven pathogenic mutations have been reported including six frameshift mutations and one missense mutation. Six of the seven mutations are located in exon 4. Insertion at position 460 accounts for most cases due to a founder effect (clustering in North England) but private mutations have been observed in patients from France, Canada and Japan.[21,27–30] Most mutations alter the structure of the protein leading to an extended peptide at the site of the pore in the ferritin molecule.[31] Notably, mutations in the iron responsive element (IRE) of the *FTL* gene are associated with hyperferritinemia-cataract syndrome.

ACERULOPLASMINEMIA

This is due to ceruloplasmin gene mutations located at *3q*. The gene contains 20 exons.[32] More than 40 distinct, mostly truncating mutations leading to a premature stop codon have been reported. Most mutations are "private" mutations.[33] No genotype–phenotype association has emerged.[33,34]

DISEASE MECHANISMS

Pathophysiology

Exact mechanisms remain ill understood but abnormal mitochondrial function (PKAN), lipid metabolism (PLAN, MPAN, FAHN), and autophagy (BPAN) have been implicated. For an overview of histological changes in NBIA disorders see Table 42.2.

By catalyzing the phosphorylation of pantothenate (vitamin B5) the associated *PANK2*-encoded protein governs the first regulatory step of coenzyme A synthesis.[35] Coenzyme A is essential for fatty acid synthesis and dysfunction of PANK2 thus likely causes derangement in lipid metabolism. PANK2 is mainly targeted to mitochondria and disease pathophysiology may relate to dysfunction of cellular energy metabolism.[36] Indeed, a recent study examining blood metabolic profiles in PKAN documented elevated levels of lactate suggesting mitochondrial dysfunction and reduced levels of triglycerides, cholesterol metabolites and sphingomyelins confirming the role of PANK2 in lipid metabolism.[37]

The encoded protein associated with PLAN, iPLA2 beta, is a group VIA calcium-independent phospholipase A2 that hydrolyzes the sn-2 acyl chain of phospholipids, thereby generating free fatty acids and lysophospholipids. iPLA2 beta is thought to play a role in remodeling of membrane phospholipids, signal transduction, cell proliferation, and apoptosis. It has been suggested that in the case of loss of iPLA2 function, lipid composition of the plasma membrane, vesicles, or endosomes may be altered.

The underlying pathophysiology of MPAN is ill understood as the function of the orphan protein is still unclear. It is localized predominantly in mitochondria and it is coregulated with genes involved in fatty acid metabolism.

TABLE 42.2 Summary of Pathology Findings in NBIA*

Condition	Gross pathology	Histology	Comment
PKAN	Discoloration of GP No significant atrophy	Neuronal degeneration with loss of GP myelin and axons with perivascular iron deposition and eosinophilic spheroids in GP (and to a lesser degree in SN) Some microglial infiltration, reactive fibrillary astrogliosis Tau-positive neurofibrillary tangles. Large and small spheroid bodies staining positive for amyloid precursor protein	Cerebellum and brainstem relatively spared No significant iron deposition in oligodendrocytes No Lewy bodies
PLAN	Discoloration of GP Variable pallor of SN Diffuse cortical and cerebellar atrophy	Perivascular and intracellular iron accumulation in GP (and variably SN) Widespread spheroids in brain and peripheral nerves Widespread tau and synuclein pathology Mild-to-severe Lewy body pathology. Purkinje cell degeneration	Spheroids also present in skin and selected other peripheral structures
MPAN		Iron deposition, widespread neuronal loss and astrogliosis in GP SN may be affected Widespread spheroids Widespread Lewy bodies (5 times more than in PLAN) and axonal spheroid bodies. Hyperphosphorylated hippocampus tau inclusions. Extracellular deposits of tau and synuclein Only few tau-positive bundles	Cerebellum is largely spared Axonal spheroids on peripheral nerves
Kufor-Rakeb syndrome		No brain pathology available of NBIA patient† Peripheral nerve: reduced myelin. Degenerating axons with axonal loss and perineurial and endoneurial edema‡ Cytoplasmic inclusions were seen in vascular smooth muscle cells	
BPAN§		GP and SN iron accumulation with numerous eosinophilic axonal spheroids Prominent tau and synuclein pathology. Widespread neuronal loss affecting GP, SN and cortical areas Myelin loss and fibrillary astrocytosis in corticospinal and spinocerebellar tracts	Relative sparing of the cerebellum. Rare hippocampal Hirano bodies. Spheroids also present in peripheral nerves.
NFT	Diffuse cerebellar atrophy Cavitary putaminal lesions Discoloration of putamen, GP and SN	Neuronal loss in putamen and GP with intracellular and extracellular GP iron deposition. Prominent loss of oligodendrocytes in paradentate white matter. Extracellular hyaline deposits. Spheroid bodies. Cystic cavitation	
ACP		GP and putaminal neuronal loss Iron deposition (starting from caudate, putamen, thalamus and dentate nucleus, subsequently spreading to the cortex) Widespread astrocytosis Grumose or foamy spheroid bodies (derived from astrocytes)	No significant tau or synuclein pathology

* *Histological data are not available for all NBIA subtypes.*

† *For details on NCL-patient with ATP13A2 mutation, see Bras J, Verloes A, Schneider SA, Mole SE, Guerreiro RJ. Mutation of the parkinsonism gene ATP13A2 causes neuronal ceroid-lipofuscinosis.* Hum Mol Genet. *2012;15;21:2646–2650.*

‡ *Paisán-Ruiz C, Guevara R, Federoff M, et al. Early-onset L-dopa-responsive parkinsonism with pyramidal signs due to ATP13A2, PLA2G6, FBXO7 and spatacsin mutations.* Mov Disord. *2010;25:1791–1800.*

§ *No gene-proven case: data based on clinical case report (see Kruer MC. The neuropathology of neurodegeneration with brain iron accumulation.* Int Rev Neurobiol. *2013;110:165–194.)*

With regards to FAHN, the encoded protein catalyzes hydroxylation at position 2 of the *N*-acyl chain of the ceramide moiety. Glycosphingolipids, which contain a high proportion of 2-hydroxy fatty acid, are important constituents of myelin sheaths.[38]

The encoded protein associated with BPAN belongs to the family of WD40 proteins, which promote protein–protein interactions and play a role in autophagy, cell cycle control, signal and transduction. WDR45 has a beta-propeller tertiary structure and a site allowing interaction with phospholipids. Patient-derived lymphoblast cell lines have a decrease in autophagic flux compared to controls and accumulate early autophagic structures.[17,39]

In ATP13A2-associated neurodegeneration lysosomal dysfunction may be a key mechanism. Histology revealed membrane bound electron-dense material of lamellar appearance with inclusions in Schwann, perineural, and smooth muscle cells of the sural nerve, and skin cells.

In late-onset NBIA, there is altered transport of iron (neuroferritinopathy) and copper (aceruloplasminemia) and storage of iron (neuroferritinopathy) and its mobilization from tissues. Low (neuroferritinopathy) or markedly elevated (aceruloplasminemia) ferritin levels are diagnostic clues. As the name suggests, ceruloplasmin is not detectable in the latter.

Animal Models

The generation of murine and other animal models of NBIA provides clues to the underlying pathophysiology using histology and/or functional assays.

Knockout mouse models and *Drosophila* models of PKAN demonstrate localization of the protein to mitochondria, similar to the human ortholog; and Pank2-defective neurons show an abnormal mitochondrial membrane potential. However, while the mice exhibit retinal degeneration and absence of sperm (azoospermia), the models do not reliably recapitulate the neurological phenotype.[40] Notably, feeding *Drosophila* flies pantethine led to restoration of CoA levels, improvement of mitochondrial function, and enhanced locomotor abilities.[41]

Different mouse models of PLAN, both with *PLA2G6* null mutations and bearing point mutations, have been developed and confirmed presence of Lewy body pathology, in addition to widespread formation of spheroids.[42,43] Functional studies, amongst others, revealed severe disturbance in Ca^{2+} responses to ATP in astrocytes.[44]

Mouse models of FA2H-associated neurodegeneration have also recently been developed.[45,46] In these, marked demyelination and profound axonal loss in the central nervous system could be demonstrated after a period of normal myelin development.[45,46] Axons were abnormally enlarged and there was abnormal cerebellar histology. In contrast, structure and function of peripheral nerves were largely unaffected.

Mouse models of Kufor-Rakeb disease have recently been reported.[47] Behaviorally, Atp13a2(−/−) mice displayed late-onset sensorimotor deficits.[47] While the number of dopaminergic neurons in the substantia nigra and striatal dopamine levels were normal, immunoblot analysis revealed increased insoluble α-synuclein in the hippocampus, but not in the cortex or cerebellum. Accelerated deposition of autofluorescent storage material (lipofuscin) characteristic of neuronal ceroid lipofuscinosis (NCL) was observed in cerebellum, hippocampus and cortex of Atp13a2(−/−) mice in line with the clinical overlap with NCL (see above). In line with this, *ATP13A2* mutations have also been identified in Tibetan terriers clinically affected by NCL.[48,49]

With regards to neuroferritinopathy most animal work was performed in transgenic models harbouring the 498*Ins*TC mutation (found in Cumbrian families). Pathological work-up revealed presence of inclusions in various tissues, including skeletal and cardiac muscle, liver, kidney, the gastrointestinal tract, adrenal glands, and skin. Compared to humans, more cortical involvement was present. Functional studies suggested that over- or underexpression of *FTL* caused relatively little effects in transgenic mice.[50,51]

In aceruloplasminemia, some but not all reported models have increased brain iron and neurological symptoms may inconsistently be present.[52,53] Thus, knockout mice displayed hepatic and reticuloendothelial iron overload but had no neurological phenotype, whereas double knockout mice lacking both ceruloplasmin and hephaestin (a ceruloplasmin homolog) exhibited a neurodegenerative phenotype and retinal degeneration consistent with the human disease.[33,53–55]

So far there are no reports of animal models of MPAN or BPAN.

INVESTIGATION

Molecular Diagnosis

Molecular testing on a diagnostic ground is available for most NBIAs. Knowledge of the pattern of inheritance (i.e., recessive vs. dominant) is helpful to select the most appropriate test. Most recently, panel testing has become available in some countries making it possible to test for all NBIA variants at once rather than in a step-by-step approach. Genetic counseling should be offered before testing and when the result is disclosed to the patient.

Ancillary Testing: The Value of Neuroimaging

Neuroimaging using MRI is an important research and diagnostic tool for NBIA disorders and will often be the clue towards the diagnosis. Although the list of disorders with more or less iron accumulation widens with improved imaging techniques, nevertheless presence of profound iron opens the drawer to this group of conditions to be considered as differential diagnosis. T2*-weighted and SWI images are particularly sensitive to detect iron. Iron deposits may be restricted to the globus pallidus or extend to adjacent areas and this may be somewhat helpful during work-up (Figure 42.1). Importantly, the development of the MRI alterations appears to be a dynamic process

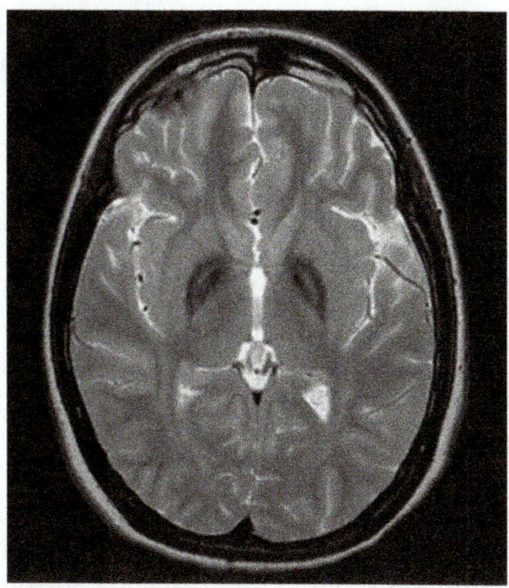

FIGURE 42.1 **Brain MRI demonstrating the "eye of the tiger" sign as typical for pantothenate kinase-associated neurodegeneration (PKAN): There are hypointensities of both globus pallidi representing iron with central hyperintensities.**

(as shown in PKAN) and there is debate regarding the correlation between presence of iron deposition and clinical findings,[56,57] as imaging findings may precede the development of clinical signs[58] (i.e., in asymptomatic carriers of homozygous mutations) but, on the other hand, may also be absent in early disease stages[12,57,59,60]or may alter over time.[57] Thus, repeated MRI may be helpful if unremarkable. Nevertheless, for some NBIA subforms, distinct imaging patterns have been recognized.

Thus, in PKAN the "eye of the tiger" sign (referring to iron accumulation in the anterior-medial part of the globus pallidus (GP) with a central hyperintensity within a surrounding area of hypointensity) is thought to be almost pathognomonic. In some patients hypointensities may extend into the substantia nigra (SN) or internal capsule.[57,61] However, in a recent study[57] only a proportion of patients had the typical eye of the tiger sign and four out of 20 patients did not have any T2-hyperintensity or T1-hypointensity.

In contrast to PKAN, iron accumulation is not a universal feature of PLAN. In PLAN, neuroimaging shows cerebellar atrophy occurring in early stages of INAD, but is absent in late-onset disease. Although half of INAD patients may lack signs of iron accumulation early in the disease course,[62] they usually develop GP hypointensities reflecting iron as the disease progresses.[63] Notably, in contrast to the pattern seen in PKAN, there is no central hyperintensity. Additional iron deposits in the SN may be present.[64,65]

In MPAN iron deposition extends the pallidal region and involves the substantia nigra. Hyperintense streaking of the medial medullary lamina between the globus pallidus interna and externa (which may resemble the eye of the tiger sign) may be present in 20% of patients.[66] Generalized brain atrophy and/or cerebellar atrophy may be present.[25]

In FAHN MRI findings include progressive leukoencephalopathy with cortical, cerebellar, and brainstem atrophy in addition to pallidal iron deposition. Corpus callosum thinning may be a clue.

Neuroimaging in BPAN has demonstrated nigral signal hyperintensity, with a central band of hypointensity and T2-weighted signal hypointensity in the GP and SN, suggesting iron deposition.

Brain imaging in patients with Kufor-Rakeb disease may show diffuse moderate generalized atrophy with iron deposition within the basal ganglia affecting the putamen and caudate in some.

Hypointensities are more widespread in neuroferritinopathy. Thus, iron deposition extends from the globus pallidus to the nigra, cortex, and other nuclei of the basal ganglia. In addition, cavitations depicted as low-T1 and high-T2 signals, sometimes surrounded by a rim of T2 hypointensity are found. Scans may also show generalized cortical and in some cases cerebellar atrophy.[67] Changes were also present in asymptomatic gene mutation carriers.

Iron is also widespread in aceruloplasminemia affecting the caudate nucleus, putamen and pallidum, and the thalamus, as well as red nucleus, and dentate in addition to cerebellar atrophy.[68]

DIFFERENTIAL DIAGNOSIS

The clinical picture of NBIA syndromes is usually complex. Accordingly, the differential diagnosis may be broad, including various inborn errors of metabolism, hereditary spastic paraplegias and other forms of dystonia-parkinsonism.[69] Radiologically, iron deposition may also be found in other non-NBIA disorders,[70–73] including Parkinson disease[74,75] (mainly affecting the substantia nigra) and atypical parkinsonian disorders, Friedreich ataxia, but also multiple sclerosis (for extensive review see [76]), and the list grows with increased use of high-resolution MR imaging (i.e., 7 Tesla MRI).

MANAGEMENT

Standard of Care and Failed Therapies

Treatment for NBIA disorders remains symptomatic. Movement disorders may be alleviated using dopaminergic drugs, anticholinergics, tetrabenazine, baclofen, and other drugs, but have no impact on the long-term prognosis and do not halt progression. Surgical procedures, including stereotactic lesioning such as thalamotomy[77] and pallidotomy[78–80] or deep brain stimulation,[81–84] may also produce benefit in the range of about 30% (for review of reported cases see[85]).

The use of iron chelating agents such as deferiprone has been reported in individual cases with mixed results[86,87] (for review of reported cases see[85]). Recent phase II pilot open trials in PKAN found reduction in GP iron content, ranging from 15–61% but no clinical benefit.[88] A randomized placebo-controlled trial is under way which will hopefully clarify the level of efficacy of chelating agents for the treatment of NBIA disorders.

Therapies under Investigation

A placebo-controlled trial funded by the European Commission Seventh Framework Programme (TIRCON, Treat Iron-Related Childhood-Onset Neurodegeneration) is under way to assess the effect of iron chelating therapy. There are no controlled data regarding pantothenic acid (vitamin B5) for PKAN patients, but individual trials are disappointing (personal communication).

References

1. Hallervorden J, Spatz H. Eigenartige Erkrankung im extrapyramidalen System mit besonderer Beteiligung des Globus pallidus und der Substantia nigra: Ein Beitrag zu den Beziehungen zwischen diesen beiden Zentren. *Z Ges Neurol Psychiat*. 1922;254–302.
2. Schenck JF, Zimmerman EA. High-field magnetic resonance imaging of brain iron: birth of a biomarker? *NMR Biomed*. 2004;17:433–445.
3. Stankiewicz J, Panter SS, Neema M, et al. Iron in chronic brain disorders: imaging and neurotherapeutic implications. *Neurotherapeutics*. 2007;4:371–386.
4. Hayflick SJ. Neurodegeneration with brain iron accumulation: from genes to pathogenesis. *Semin Pediatr Neurol*. 2006;13:182–185.
5. Hayflick SJ, Westaway SK, Levinson B, et al. Genetic, clinical, and radiographic delineation of Hallervorden–Spatz syndrome. *N Engl J Med*. 2003;348:33–40.
6. Schneider SA, Aggarwal A, Bhatt M, et al. Severe tongue protrusion dystonia: clinical syndromes and possible treatment. *Neurology*. 2006;67:940–943.
7. Marelli C, Piacentini S, Garavaglia B, et al. Clinical and neuropsychological correlates in two brothers with pantothenate kinase-associated neurodegeneration. *Mov Disord*. 2005;20:208–212.
8. Thomas M, Hayflick SJ, Jankovic J. Clinical heterogeneity of neurodegeneration with brain iron accumulation (Hallervorden–Spatz syndrome) and pantothenate kinase-associated neurodegeneration. *Mov Disord*. 2004;19:36–42.
9. Egan RA, Weleber RG, Hogarth P, et al. Neuro-ophthalmologic and electroretinographic findings in pantothenate kinase-associated neurodegeneration (formerly Hallervorden–Spatz syndrome). *Am J Ophthalmol*. 2005;140:267–274.
10. Gregory A, Hayflick S. Clinical and genetic delineation of neurodegeneration with brain iron accumulation. *J Med Genet*. 2009;46(2):73–80.
11. Yoon WT, Lee WY, Shin HY, et al. Novel PANK2 gene mutations in Korean patient with pantothenate kinase-associated neurodegeneration presenting unilateral dystonic tremor. *Mov Disord*. 2010;25:245–247.
12. Aggarwal A, Schneider SA, Houlden H, et al. Indian-Subcontinent NBIA: unusual phenotypes, novel PANK2 mutations, and undetermined genetic forms. *Mov Disord*. 2010;25:1424–1431.
13. Chung SJ, Lee JH, Lee MC, et al. Focal hand dystonia in a patient with PANK2 mutation. *Mov Disord*. 2008;23:466–468.
14. Paisan-Ruiz C, Bhatia KP, Li A, et al. Characterization of PLA2G6 as a locus for dystonia-parkinsonism. *Ann Neurol*. 2009;65:19–23.
15. Hartig MB, Iuso A, Haack T, et al. Absence of an orphan mitochondrial protein, c19orf12, causes a distinct clinical subtype of neurodegeneration with brain iron accumulation. *Am J Hum Genet*. 2011;89:543–550.
16. Hayflick SJ, Kruer MC, Gregory A, et al. Beta-propeller protein-associated neurodegeneration: a new X-linked dominant disorder with brain iron accumulation. *Brain*. 2013;136:1708–1717.
17. Saitsu H, Nishimura T, Muramatsu K, et al. *De novo* mutations in the autophagy gene WDR45 cause static encephalopathy of childhood with neurodegeneration in adulthood. *Nat Genet*. 2013;45:445–449.
18. Bruggemann N, Hagenah J, Reetz K, et al. Recessively inherited parkinsonism: effect of ATP13A2 mutations on the clinical and neuroimaging phenotype. *Arch Neurol*. 2010;67:1357–1363.
19. Schneider SA, Paisan-Ruiz C, Quinn NP, et al. ATP13A2 mutations (PARK9) cause neurodegeneration with brain iron accumulation. *Mov Disord*. 2010;25:979–984.

20. Williams DR, Hadeed A, al Din AS, et al. Kufor Rakeb disease: autosomal recessive, levodopa-responsive parkinsonism with pyramidal degeneration, supranuclear gaze palsy, and dementia. *Mov Disord*. 2005;20:1264–1271.
21. Chinnery PF, Crompton DE, Birchall D, et al. Clinical features and natural history of neuroferritinopathy caused by the FTL1 460InsA mutation. *Brain*. 2007;130:110–119.
22. McNeill A, Pandolfo M, Kuhn J, et al. The neurological presentation of ceruloplasmin gene mutations. *Eur Neurol*. 2008;60:200–205.
23. Hartig MB, Hortnagel K, Garavaglia B, et al. Genotypic and phenotypic spectrum of PANK2 mutations in patients with neurodegeneration with brain iron accumulation. *Ann Neurol*. 2006;59:248–256.
24. Engel LA, Jing Z, O'Brien DE, et al. Catalytic function of PLA2G6 is impaired by mutations associated with infantile neuroaxonal dystrophy but not dystonia-parkinsonism. *PLoS One*. 2011;5:e12897.
25. Hartig M, Prokisch H, Meitinger T, et al. Mitochondrial membrane protein-associated neurodegeneration (MPAN). *Inr Rev Neurobiol*. 2013;110:73–84.
26. Ramirez A, Heimbach A, Grundemann J, et al. Hereditary parkinsonism with dementia is caused by mutations in ATP13A2, encoding a lysosomal type 5 P-type ATPase. *Nat Genet*. 2006;38:1184–1191.
27. Chinnery PF, Curtis AR, Fey C, et al. Neuroferritinopathy in a French family with late onset dominant dystonia. *J Med Genet*. 2003;40:e69.
28. Kubota A, Hida A, Ichikawa Y, et al. A novel ferritin light chain gene mutation in a Japanese family with neuroferritinopathy: description of clinical features and implications for genotype-phenotype correlations. *Mov Disord*. 2009;24(3):441–5.
29. Devos D, Tchofo PJ, Vuillaume I, et al. Clinical features and natural history of neuroferritinopathy caused by the 458dupA FTL mutation. *Brain*. 2009;132:e109.
30. Ondo WG, Adam OR, Jankovic J, et al. Dramatic response of facial stereotype/tic to tetrabenazine in the first reported cases of neuroferritinopathy in the United States. *Mov Disord*. 2010;25:2470–2472.
31. Keogh MJ, Jonas P, Coulthard A, et al. Neuroferritinopathy: a new inborn error of iron metabolism. *Neurogenetics*. 2012;13:93–96.
32. Daimon M, Yamatani K, Igarashi M, et al. Fine structure of the human ceruloplasmin gene. *Biochem Biophys Res Commun*. 1995;208:1028–1035.
33. Kono S. Aceruloplasminemia: an update. *Int Rev Neurobiol*. 2013;110:125–151.
34. McNeill A, Pandolfo M, Kuhn J, et al. The neurological presentation of ceruloplasmin gene mutations. *Eur Neurol*. 2008;60:200–205.
35. Kotzbauer PT, Truax AC, Trojanowski JQ, et al. Altered neuronal mitochondrial coenzyme A synthesis in neurodegeneration with brain iron accumulation caused by abnormal processing, stability, and catalytic activity of mutant pantothenate kinase 2. *J Neurosci*. 2005;25:689–698.
36. Lin MT, Beal MF. Mitochondrial dysfunction and oxidative stress in neurodegenerative diseases. *Nature*. 2006;443:787–795.
37. Leoni V, Strittmatter L, Zorzi G, et al. Metabolic consequences of mitochondrial coenzyme A deficiency in patients with PANK2 mutations. *Mol Genet Metab*. 2012;105:463–471.
38. Edvardson S, Hama H, Shaag A, et al. Mutations in the fatty acid 2-hydroxylase gene are associated with leukodystrophy with spastic paraparesis and dystonia. *Am J Hum Genet*. 2008;83:643–648.
39. Haack TB, Hogarth G, Gregory A, et al. BPAN: the only X-linked dominant NBIA disorder. *Int Rev Neurobiol*. 2013;110:85–90.
40. Brunetti D, Dusi S, Morbin M, et al. Pantothenate kinase-associated neurodegeneration: altered mitochondria membrane potential and defective respiration in Pank2 knock-out mouse model. *Hum Mol Genet*. 2012;21:5294–5305.
41. Rana A, Seinen E, Siudeja K, et al. Pantethine rescues a Drosophila model for pantothenate kinase-associated neurodegeneration. *Proc Natl Acad Sci U S A*. 2010;107:6988–6993.
42. Malik I, Turk J, Mancuso DJ, et al. Disrupted membrane homeostasis and accumulation of ubiquitinated proteins in a mouse model of infantile neuroaxonal dystrophy caused by PLA2G6 mutations. *Am J Pathol*. 2008;172:406–416. doi: 10.1002/14651858.CD005045.pub2.
43. Wada H, Kojo S, Seino KI. Mouse models of human INAD by Pla2g6 deficiency. *Histol Histopathol*. 2013;28(8):965–969.
44. Strokin M, Seburn KL, Cox GA, et al. Severe disturbance in the Ca^{2+} signaling in astrocytes from mouse models of human infantile neuroaxonal dystrophy with mutated Pla2g6. *Hum Mol Genet*. 2012;21:2807–2814.
45. Potter KA, Kern MJ, Fullbright G, et al. Central nervous system dysfunction in a mouse model of Fa2h deficiency. *Glia*. 2011;59:1009–1021.
46. Zoller I, Meixner M, Hartmann D, et al. Absence of 2-hydroxylated sphingolipids is compatible with normal neural development but causes late-onset axon and myelin sheath degeneration. *J Neurosci*. 2008;28:9741–9754.
47. Schultheis PJ, Fleming SM, Clippinger AK, et al. Atp13a2-deficient mice exhibit neuronal ceroid lipofuscinosis, limited alpha-synuclein accumulation and age-dependent sensorimotor deficits. *Hum Mol Genet*. 2013;22:2067–2082.
48. Farias FH, Zeng R, Johnson GS, et al. A truncating mutation in ATP13A2 is responsible for adult-onset neuronal ceroid lipofuscinosis in Tibetan terriers. *Neurobiol Dis*. 2011;42:468–474.
49. Wohlke A, Philipp U, Bock P, et al. A one base pair deletion in the canine ATP13A2 gene causes exon skipping and late-onset neuronal ceroid lipofuscinosis in the Tibetan terrier. *PLoS Genet*. 2011;7:e1002304.
50. Cozzi A, Corsi B, Levi S, et al. Analysis of the biologic functions of H- and L-ferritins in HeLa cells by transfection with siRNAs and cDNAs: evidence for a proliferative role of L-ferritin. *Blood*. 2004;103:2377–2383.
51. Cremonesi L, Cozzi A, Girelli D, et al. Case report: a subject with a mutation in the ATG start codon of L-ferritin has no haematological or neurological symptoms. *J Med Genet*. 2004;41:e81.
52. Hineno A, Kaneko K, Yoshida K, et al. Ceruloplasmin protects against rotenone-induced oxidative stress and neurotoxicity. *Neurochem Res*. 2011;36:2127–2135.
53. Texel SJ, Xu X, Harris ZL. Ceruloplasmin in neurodegenerative diseases. *Biochem Soc Trans*. 2008;36:1277–1281.
54. Schulz K, Vulpe CD, Harris LZ, et al. Iron efflux from oligodendrocytes is differentially regulated in gray and white matter. *J Neurosci*. 2011;31:13301–13311.
55. Hahn P, Qian Y, Dentchev T, et al. Disruption of ceruloplasmin and hephaestin in mice causes retinal iron overload and retinal degeneration with features of age-related macular degeneration. *Proc Natl Acad Sci U S A*. 2004;101:13850–13855.
56. Hayflick SJ, Hartman M, Coryell J, et al. Brain MRI in neurodegeneration with brain iron accumulation with and without PANK2 mutations. *AJNR Am J Neuroradiol*. 2006;27:1230–1233.

57. Delgado RF, Sanchez PR, Speckter H, et al. Missense PANK2 mutation without "Eye of the tiger" sign: MR findings in a large group of patients with pantothenate kinase-associated neurodegeneration (PKAN). *J Magn Reson Imaging*. 2012;35(4):788–94 .
58. Hayflick SJ, Penzien JM, Michl W, et al. Cranial MRI changes may precede symptoms in Hallervorden–Spatz syndrome. *Pediatr Neurol*. 2001;25:166–169.
59. Chiapparini L, Savoiardo M, D'Arrigo S, et al. The "eye-of-the-tiger" sign may be absent in the early stages of classic pantothenate kinase associated neurodegeneration. *Neuropediatrics*. 2011;42:159–162.
60. Grandas F, Fernandez-Carballal C, Guzman-de-Villoria J, et al. Treatment of a dystonic storm with pallidal stimulation in a patient with PANK2 mutation. *Mov Disord*. 2011;26:921–922.
61. Fermin-Delgado R, Roa-Sanchez P, Speckter H, et al. Involvement of globus pallidus and midbrain nuclei in pantothenate kinase-associated neurodegeneration: measurement of T2 and T2* time. *Clin Neuroradiol*. 2013;23(1):11–15.
62. Morgan NV, Westaway SK, Morton JE, et al. PLA2G6, encoding a phospholipase A2, is mutated in neurodegenerative disorders with high brain iron. *Nat Genet*. 2006;38:752–754.
63. Kurian MA, Morgan NV, MacPherson L, et al. Phenotypic spectrum of neurodegeneration associated with mutations in the PLA2G6 gene (PLAN). *Neurology*. 2008;70:1623–1629.
64. Gregory A, Polster BJ, Hayflick SJ. Clinical and genetic delineation of neurodegeneration with brain iron accumulation. *J Med Genet*. 2009;46:73–80.
65. Paisan-Ruiz C, Li A, Schneider SA, et al. Widespread Lewy body and tau accumulation in childhood and adult onset dystonia-parkinsonism cases with PLA2G6 mutations. *Neurobiol Aging*. 2012;33:814–823.
66. Hogarth P, Gregory A, Kruer MC, et al. New NBIA subtype: genetic, clinical, pathologic, and radiographic features of MPAN. *Neurology*. 2013;80:268–275.
67. Ohta E, Takiyama Y. MRI findings in neuroferritinopathy. *Neurol Res Int*. 2012;2012:197438.
68. McNeill A, Birchall D, Hayflick SJ, et al. T2* and FSE MRI distinguishes four subtypes of neurodegeneration with brain iron accumulation. *Neurology*. 2008;70:1614–1619.
69. Schneider SA, Bhatia KP, Hardy J. Complicated recessive dystonia parkinsonism syndromes. *Mov Disord*. 2009;24:490–499.
70. Chang MH, Hung WL, Liao YC, et al. Eye of the tiger-like MRI in parkinsonian variant of multiple system atrophy. *J Neural Transm*. 2009;116:861–866.
71. Davie CA, Barker GJ, Machado C, et al. Proton magnetic resonance spectroscopy in Steele-Richardson-Olszewski syndrome. *Mov Disord*. 1997;12:767–771.
72. Molinuevo JL, Munoz E, Valldeoriola F, et al. The eye of the tiger sign in cortical-basal ganglionic degeneration. *Mov Disord*. 1999;14:169–171.
73. Santillo AF, Skoglund L, Lindau M, et al. Frontotemporal dementia-amyotrophic lateral sclerosis complex is simulated by neurodegeneration with brain iron accumulation. *Alzheimer Dis Assoc Disord*. 2009;23:298–300.
74. Zhang J, Zhang Y, Wang J, et al. Characterizing iron deposition in Parkinson's disease using susceptibility-weighted imaging: an *in vivo* MR study. *Brain Res*. 2010;1330:124–130.
75. Rossi M, Ruottinen H, Elovaara I, et al. Brain iron deposition and sequence characteristics in Parkinsonism: comparison of SWI, T(2)* maps, T(2)-weighted-, and FLAIR-SPACE. *Invest Radiol*. 2010;45:795–802.
76. Kell DB. Towards a unifying, systems biology understanding of large-scale cellular death and destruction caused by poorly liganded iron: Parkinson's, Huntington's, Alzheimer's, prions, bactericides, chemical toxicology and others as examples. *Arch Toxicol*. 2010;84:825–889.
77. Tsukamoto H, Inui K, Taniike M, et al. A case of Hallervorden–Spatz disease: progressive and intractable dystonia controlled by bilateral thalamotomy. *Brain Dev*. 1992;14:269–272.
78. Balas I, Kovacs N, Hollody K. Staged bilateral stereotactic pallidothalamotomy for life-threatening dystonia in a child with Hallervorden–Spatz disease. *Mov Disord*. 2006;21:82–85.
79. Kyriagis M, Grattan-Smith P, Scheinberg A, et al. Status dystonicus and Hallervorden–Spatz disease: treatment with intrathecal baclofen and pallidotomy. *J Paediatr Child Health*. 2004;40:322–325.
80. Justesen CR, Penn RD, Kroin JS, et al. Stereotactic pallidotomy in a child with Hallervorden–Spatz disease. Case report. *J Neurosurg*. 1999;90:551–554.
81. Castelnau P, Cif L, Valente EM, et al. Pallidal stimulation improves pantothenate kinase-associated neurodegeneration. *Ann Neurol*. 2005;57:738–741.
82. Mikati MA, Yehya A, Darwish H, et al. Deep brain stimulation as a mode of treatment of early onset pantothenate kinase-associated neurodegeneration. *Eur J Paediatr Neurol*. 2009;13:61–64.
83. Krause M, Fogel W, Tronnier V, et al. Long-term benefit to pallidal deep brain stimulation in a case of dystonia secondary to pantothenate kinase-associated neurodegeneration. *Mov Disord*. 2006;21:2255–2257.
84. Szumowski J, Bas E, Gaarder K, et al. Measurement of brain iron distribution in Hallevorden–Spatz syndrome. *J Magn Reson Imaging*. 2010;31:482–489.
85. Schneider SA, Dusek P, Hardy J, et al. Genetics and pathophysiology of neurodegeneration with brain iron accumulation (NBIA). *Curr Neuropharmacol*. 2013;11:59–79.
86. Li X, Jankovic J. Iron chelation and neuroprotection in neurodegenerative diseases. *J Neural Transm*. 2011;118:473–477.
87. Zhu W, Li X, Luo F, et al. Genetic iron chelation protects against proteasome inhibition-induced dopamine neuron degeneration. *Neurobiol Dis*. 2010;37:307–313.
88. Zorzi G, Zibordi F, Chiapparini L, et al. Iron-related MRI images in patients with pantothenate kinase-associated neurodegeneration (PKAN) treated with deferiprone: Results of a phase II pilot trial. *Mov Disord*. 2011;26:1756–1759.

CHAPTER

43

Pantothenate Kinase-Associated Neurodegeneration

Michael C. Kruer

Sanford Children's Health Research Center, Sioux Falls, SD, USA

INTRODUCTION

Pantothenate kinase-associated neurodegeneration (PKAN) is the cardinal form of neurodegeneration with brain iron accumulation (NBIA) a group of neurodegenerative diseases characterized by iron deposition in the basal ganglia. PKAN (also known as NBIA1) is estimated to account for ~40% of NBIA cases,[1] although there may be some ascertainment bias given the highly distinctive neuroimaging features of PKAN (see below). PKAN was formerly known as Hallervorden–Spatz syndrome. The eponym was bestowed based upon the recognition and characterization of the disease by these two German neuropathologists, and still persists in some literature. However, links to Nazi war crimes[2] led to the redesignation of the disorder as "PKAN" once the causative gene was identified.[3] Subsequently, other forms of NBIA have followed this naming convention based upon the causative gene.

Forms of NBIA share a neurodegenerative course, prominent extrapyramidal symptoms, and the accumulation of iron in the basal ganglia. Ocular pathology and intellectual decline are also frequent features of NBIA. Although NBIA is defined, in part, by the recognition of iron deposits in the brain, it is unclear at the present time whether the progressive iron build-up that is seen substantially contributes to disease pathophysiology or merely represents an epiphenomenon.

Linkage mapping[4] followed by positional cloning of candidate genes in the Old Order Amish led to the identification of mutations in *PANK2* in patients with PKAN.[5] Similar phenotypes and radiographic appearances can result from mutations in *PLA2G6* and *c19orf12*, although important differences exist that can help guide diagnosis. Clinically, PKAN can mimic forms of generalized dystonia and juvenile-onset parkinsonism. Treatments for PKAN are largely symptomatic at this point in time, although recent insights into disease pathogenesis raise the possibility that more targeted therapies may be on the horizon.

CLINICAL FEATURES

Symptoms of PKAN typically begin in childhood, although patients can present with exrapyramidal symptoms and characteristic neuroimaging findings in their 60s (unpublished observation).

Early-Onset PKAN

This form of the disease often manifests in the preschool years. Affected children may exhibit early motor or global developmental delays, followed by gait impairment and progressively worsening falls related to manifesting dystonia, rigidity and/or spasticity. Dystonia, combined with other pyramidal and extrapyramidal features, usually leaves most patients dependent on a wheelchair for locomotion within 10–15 years of onset, sometimes much sooner.

http://dx.doi.org/10.1016/10.1016/B978-0-12-410529-4.00043-7

Patients with PKAN often experience a stepwise decline in function that unfolds over the course of several weeks, followed by prolonged periods of relative stability. Lost function is usually not regained, and clear precipitants have not been able to be identified thus far.

Dystonia

Progressive dystonia is one of the most disabling features of PKAN. Although the dystonia is generalized, many patients are particularly affected by orobuccolingual dystonia and dystonia of the limbs. In PKAN, a task-specific eating dystonia is particularly characteristic. Extremity dystonia is often functionally disabling, but can be severe enough to cause intractable pain and atraumatic fractures (related to chronic stress on long bones and osteopenia). PKAN patients are at risk for episodes of status dystonicus, sometimes after an antecedent infection or surgical procedure. These spells can lead to severe pain, diaphoresis, hyperpyrexia, autonomic hyperactivation, and rhabdomyolysis leading to renal failure.

Parkinsonism

Although dystonia is a probably the most prominent symptom affecting PKAN patients, PKAN can present with juvenile-onset parkinsonism[6] and is often considered in the differential diagnosis for patients presenting with parkinsonian features under 21 years of age. Affected patients often exhibit marked rigidity. However, PKAN patients can also demonstrate freezing of gait (particularly when crossing thresholds), bradykinesia or akinesia, and postural instability contributing to falls. A rest tremor is not typical of PKAN but affected patients can exhibit an action tremor.[7]

Neuro-Ophthalmologic Findings

Abnormal saccades and supranuclear gaze palsy can be observed in PKAN patients. Many patients exhibit signs of Adie's pupil,[8] characterized by sectoral paresis of the iris and irregularity of the pupillary ruff. Funduscopic examination discloses pigmentary retinopathy in most patients with early-onset PKAN, characterized by a "flecked" appearance of the retina at early stages, progressing to a "bull's eye" maculopathy (Figure 43.1). Associated clinical features include night-blindness and peripheral visual field constriction. In some cases, the degenerative retinopathy can progress to blindness, although affected patients who do not have retinopathy at the time of diagnosis do not typically go on to develop this complication. Optic atrophy is not typical of PKAN, although it is common in other forms of NBIA.

Other Clinical Features

Additional features of PKAN include spasticity with associated pyramidal tract findings and dysarthria. A progressive dementia is seen in PKAN,[9] but tends to be milder and more variable than in many other neurodegenerative disorders. PKAN patients can sometimes exhibit chorea or ballism.[10] Seizures are not typical in PKAN.

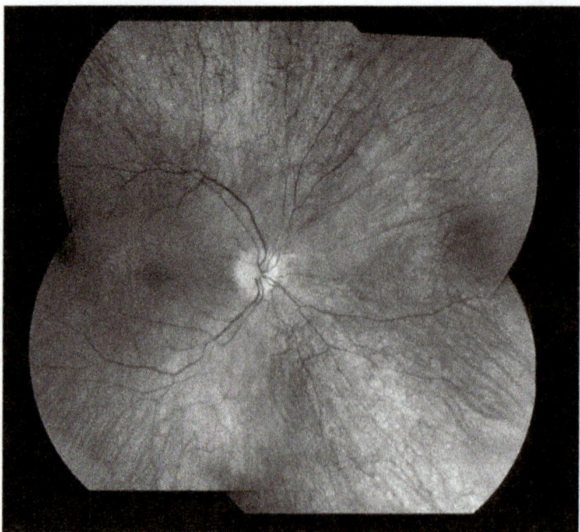

FIGURE 43.1 **Pigmentary retinopathy associated with PKAN.** Montage of right fundus of a patient with PKAN at age 15 years. The optic disk is waxy with mild pallor, the retinal vessels are mildly attenuated with perivenous pigmentary cuffing, and throughout all quadrants there are scattered bone-spicule formations. The reflexes off the inner surface of the retina in the macula suggest retinal swelling or edema. (Reprinted with permission from[8].)

Late-Onset PKAN

Presenting at school age, in adolescence, or early adulthood, patients with late-onset PKAN are more prone to disorders of language production and psychiatric symptoms. Many such patients will present with pallilalia, tachylalia, or tachylogia and dysarthria. Tics are seen not uncommonly in this group, and neuropsychiatric features may be prominent, including depression, anxiety, personality change, apathy, impulsivity, obsessive–compulsive symptoms, and, rarely, psychosis. Those patients with late onset typically have a slower progression of the disease, although they still exhibit dystonia and/or parkinsonism, these symptoms may be more mild. Relative intellectual sparing can be seen in those with late-onset disease.

LABORATORY FINDINGS

The most common laboratory abnormality seen in patients with PKAN is acanthocytosis. This acanthocytosis is most readily identified in fresh blood samples that are diluted 1:1 with heparinized normal saline for 1 hour at room temperature, followed by examination under phase-contrast.[11] In some cases, scanning electron microscopy can be helpful. In addition, in some patients with PKAN, low levels or absence of the pre-beta lipoprotein fraction can be seen on lipoprotein electrophoresis. The HARP syndrome (OMIM #607236), allelic to PKAN, consists of hypobetalipoproteinemia, acanthocytosis, retinopathy, and pallidal degeneration.[12]

NEUROIMAGING FEATURES

The most distinctive neuroimaging feature of PKAN is the "eye of the tiger" (EoT), a radiographic feature owing its characteristic appearance to the marked T2 hyperintensity seen in the central portion of the globus pallidus, surrounded by diffuse hypointensity (indicative of iron deposition) of the remainder of this nucleus (Figure 43.2). The pathophysiology of the EoT has remained cryptic. On T1-weighted images, the globus pallidus appears isointense to mildly hyperintense. The substantia nigra may also demonstrate mild T2 hypointensity in PKAN.

The EoT is remarkably sensitive and specific for PKAN. Although EoT-like presentations have been identified in a variety of other contexts,[13] the EoT in other contexts appears atypical. Interestingly, the EoT can precede symptom onset at times (such as in asymptomatic but mutation-positive siblings of known PKAN patients) or not be seen at early stages of the disease.[7] Furthermore, at later disease stages the EoT can "burn out," with the central hyperintensity overshadowed by accumulating T2 hypointensity related to the paramagnetic effect of iron.[14]

The remainder of the brain looks remarkably normal in PKAN, without evidence of cerebral or cerebellar atrophy, white matter disease, or T2 hyperintensity outside the globus pallidus.

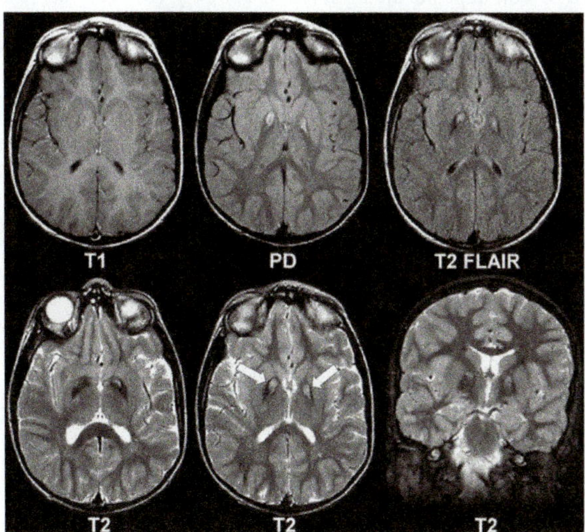

FIGURE 43.2 **Neuroradiologic appearance of PKAN.** T2 hypointensity of the globus pallidus punctuated by a hyperintense center leads to the "eye of the tiger" while the remainder of the brain parenchyma is unremarkable. (Reprinted with permission from[14a].)

DEFINITIVE DIAGNOSIS

DNA sequence analysis of the *PANK2* gene is clinically available and can provide molecular confirmation of clinically suspected cases. Intragenic duplications or deletion that may escape detection by array-based comparative genomic hybridization analyses can be seen in a small number of cases.[15] The causative genes for several other forms of NBIA can also be tested for, and clinical and radiographic features can be used to facilitate diagnosis.[13] Despite molecular investigation, a substantial proportion of NBIA cases remain idiopathic.

NEUROPATHOLOGIC FINDINGS

Neuropathologic findings in PKAN show good agreement with magnetic resonance imaging (MRI) features.[16] Iron deposition is grossly appreciable as a rusty discoloration of the globus pallidus. Histologically, iron is typically detected by staining for Fe^{3+} using Perls' reagent. Iron accumulates intracellularly in both neurons and astrocytes, and a fine "dusting" of the neuropil occurs, along with prominent perivascular staining (Figure 43.3). Histologically, the center of the EoT is a region of profound rarefaction and cell loss (Figure 43.4). There is limited microglial activation with no evidence for exogenous immune cell infiltration.

Eosinophilic neuroaxonal spheroids, 20–70 µm in diamater, are a pathologic hallmark of the NBIA syndromes, and are thought to represent degenerating neurons and/or dystrophic axons.[15] In PKAN, spheroids are found prominently in the globus pallidus and, to a lesser extent, in the substantia nigra pars reticulata (Figure 43.5). Electron microscopic analysis of probable PKAN cases has shown that ultrastructurally, spheroids appear to comprise dense osmiophilic bodies associated with amorphous material and vesicles with an appearance distinct from those associated with infantile neuroaxonal dystrophy.[17] Spheroids are inconsistently positive for tau and amyloid precursor protein, and some spheroids show iron accumulation.

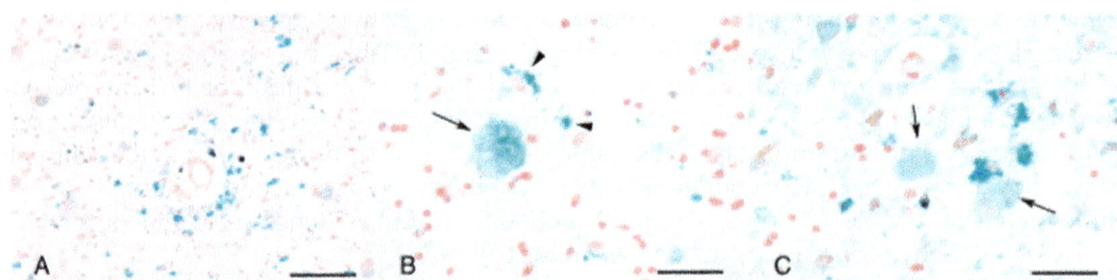

FIGURE 43.3 **Iron deposits in the globus pallidus of a patient with PKAN.** (**A**) Low-magnification view of the globus pallidus in PKAN, stained with Perls' stain for iron. Perivascular iron deposits are seen (scale: 200 µm). (**B**) Perls' stain demonstrating cellular localization of iron in the globus pallidus in PKAN. Degenerating neurons with relative preservation of cytoplasm demonstrate increased cytoplasmic iron staining (arrow). Astrocytes with more intense, dense granular accumulation of cytoplasmic iron (arrowheads) are also noted (scale: 50 µm). (**C**) Perls' stain of the globus pallidus, showing progressive decrease in iron content as neurons degenerate. Iron-positive astrocytes are more conspicuous and greatly outnumber those present in normal globus pallidus. (Reprinted with permission from[15].)

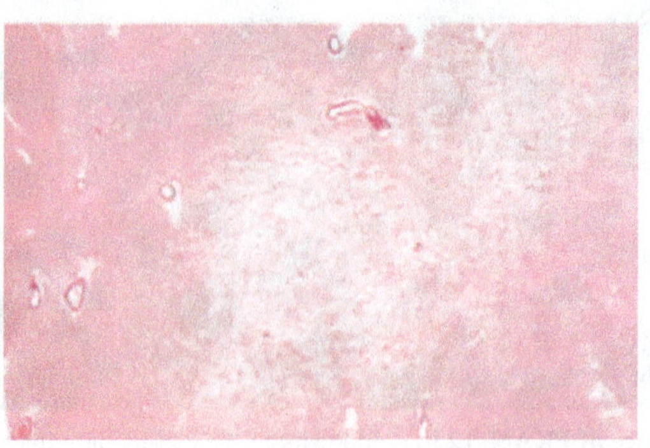

FIGURE 43.4 **Histologic appearance of the eye of the tiger.** Profound rarefaction is seen microscopically. (Reprinted with permission from[15].)

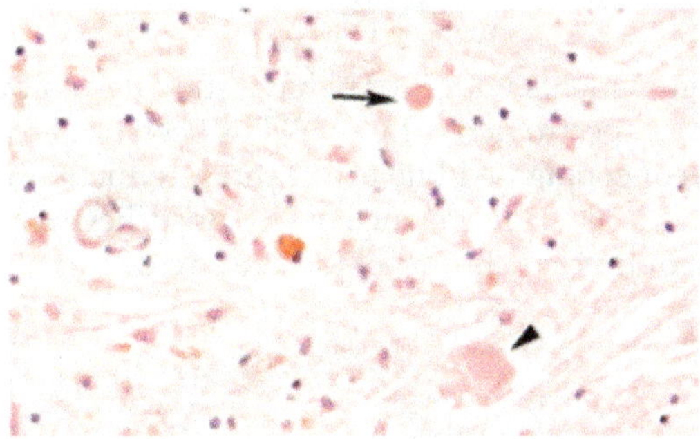

FIGURE 43.5 **Microscopic appearance of neuroaxonal spheroids.** Both small compact spherical structures (arrows) and larger more granular appearing degenerating neurons are observed. (Reprinted with permission from[15].)

PKAN brains show variable tau accumulation, with silver stain-positive neurofibrillary tangles. α-Synuclein accumulation and Lewy body deposition is not seen in PKAN, in distinction to some other forms of NBIA where it is quite prominent.

CURRENT TREATMENT STRATEGIES

Many of the typical oral medications used to treat dystonia and parkinsonism, including baclofen, trihexyphenidyl, benzodiazepines, and dopamine receptor agonists and antagonists can provide some symptomatic relief for patients with PKAN. Levodopa is not typically effective in PKAN, and its use is limited by the development of dyskinesias. However, for many patients with PKAN, their dystonia is poorly controlled by oral medications, and additional measures are necessary.

Injectables and Surgical Interventions

Focal dystonia can respond to botulinum toxin injection, although dystonia is typically widespread in PKAN, limiting the effectiveness of this therapy. Some patients will respond to the intrathecal baclofen pump, albeit often at high settings. At least one patient has been successfully treated with intraventricular baclofen.[18]

Pallidotomy and thalamotomy have been attempted for PKAN, with some improvement seen in affected patients after lesioning. However, lesioning is an irreversible procedure that can be associated with severe side effects, and in some cases, the natural history after this procedure seems to be a gradual return of the underlying dystonia.[19] Deep brain stimulation of the globus pallidus pars interna (Gpi) seems to provide more lasting benefit to many patients with NBIA, including PKAN.[20] However, experience with DBS in PKAN thus far has been limited to a handful of patients from any particular center. In some cases, when patients with PKAN develop status dystonicus, surgical interventions have been emergently applied, sometimes leading to relief.[21]

BIOLOGICAL BASIS OF DISEASE

Molecular Genetics

PANK2 is a pantothenate kinase essential for converting dietary pantothenate into 4′ phosphopantethenic acid, a crucial step in the synthesis of CoA. CoA is an essential cofactor required for ~100 enzymatic reactions including those involved in the citric acid cycle, sterol and bile acid synthesis, heme synthesis, amino acid synthesis, and the β-oxidation of fatty acids. Four PANK isoforms are recognized in mammals (PANK1α, PANK1β, PANK2, and PANK3). PANK4 is not thought to be catalytically active.[22] Although PANKs catalyze the same step in CoA biosynthesis, they are thought to be differentially compartmentalized, and loss of a given PANK is not able to be complemented by the other isoforms.[23] A role for PANKs in the regulation of CoA levels has been proposed, as actual CoA synthesis occurs in the cytoplasm.

Human Cell Culture Models

Studies of fibroblasts cultured from patients with PKAN have shown mild but consistent elevations of lactate. In addition, liquid chromatography–mass spectrometry (LC–MS) has been used for metabolic profiling.[24] These studies have shown reduced tauro- and glycol-conjugated bile acids and a reduction in cholesterol and fatty acid species, putatively related to decreased synthesis as a consequence of CoA insufficiency. In addition, PKAN fibroblasts show an increase in protein carbonylation, indicating chronic oxidative stress.[25]

The Role of Iron in Disease

Despite the close association of iron deposition with PKAN, the role of iron has remained surprisingly enigmatic. Recent studies with *PANK2* fibroblasts have shown that with exposure to exogenous iron, ferritins are inappropriately upregulated leading to increased intracellular iron sequestration while the transferrin receptor 1 (TfR1) is abnormally downregulated.[25] In addition, there was a reduction of membrane-associated mRNA-bound iron regulatory protein 1, with impaired iron responsivity, putatively leading to the observed abnormalities of ferritin and TfR1 expression. Such changes in these crucial iron regulatory proteins were associated with an increase in the bioactive labile iron pool and increased reactive oxygen species levels.

ANIMAL MODELS

Fumble Fly

Mutations in the *Drosophila* PANK homolog (dPANK) lead to infertility and impaired neuromotor function.[26] This mutant, fumble (*dPANK/fbl*), has thus been used to gain insight into fundamental mechanisms of the disease, although there may be differences in subcellular localization between human and *Drosophila* (see PANK2 Knockout Mouse, below). *dPANK/fbl* flies have reduced CoA levels and exhibit a neurodegenerative phenotype.[27] In the *dPANK/fbl* mutant, acetylation of histones and tubulin is impaired, leading to downstream epigenetic effects related to histone hypo-acetylation as well as decreased stability of tubulin-based dynamics.[28] Associated phenotypes include an impaired DNA damage response after irradiation, impaired locomotor function, and diminished survival. Histone acetylation-mediated defects can be partially rescued via treatment with histone deacetylase inhibitors and pantethine (see Burgeoning Therapies and Rationale, page 479).[28] Decreased dPANK and CoA also leads to abnormal elevation of the phosphorylated (active) form of the actin-severing protein cofilin, which impairs normal filamentous actin chain growth and affects neurite extension and morphology in culture.[29]

PANK2 Knockout Mouse

Development of a mouse model for PKAN has been challenging, as germline knockout was initially found to lead to azoospermia and retinal degeneration[30] but no neurological phenotype. Deprivation of dietary pantothenate led to the appearance of a movement disorder in both wild-type and knockout mice, suggesting that this intervention phenocopies the human disease and highlights the role of diet and other subtle factors in the manifestation of disease phenotypes in mice.[31]

Although studies to date have agreed upon a mitochondrial localization for hPANK2, the localization of murine Pank2 remains controversial, with some investigators speculating this may underlie the failure of the mouse model to recapitulate neurological features of the human disease. A cytosolic localization of mPANK2, largely based upon microscopic evidence has been reported,[23] while other investigators have reported a mitochondrial localization of both human and mouse PANK2, largely based on biochemical fractionation.[32] In addition, recent findings have indicated that hPANK2 may be found in the nucleus as well as the mitochondria.[23] Although the significance of this finding is unknown, it is intriguing given the known role of PANK in histone acetylation and DNA repair.

Studies of PANK2 knockout mouse-derived neurons have further characterized mitochondrial dysfunction as a contributor to pathogenesis.[32] Neurons exhibit diminished mitochondrial membrane potential, diminished ATP levels, swollen mitochondrial morphology by electron microscopy, and slightly impaired respiration as measured by oxygen consumption rate.

BURGEONING THERAPIES AND RATIONALE

Although PANK2 catalyzes the crucial initial step in the canonical CoA biosynthesis pathway, recent studies have suggested a role for an ancillary pathway utilizing pantethine as an alternative substrate in CoA synthesis (Figure 43.6). Supplementation of *dPANK/fbl* mutants with pantethine led to improvements in mobility and lifespan in affected flies, rescuing the neurodegenerative phenotype presumably by bypassing the metabolic blockade.[27] Thus far, studies of pantethine as a potential therapeutic have been limited to *Drosophila*, although the compound has been utilized as a lipid-lowering agent in prior human studies.

The efficacy of iron-chelating agents in removing iron deposits from the brain in disorders such as superficial siderosis[33] and Friedreich ataxia[34] has been shown, although linking radiographic changes with clinical improvement has been more challenging. Although the role of iron in the pathogenesis of PKAN remains uncertain, the efficacy of chelators in removing iron from the brain has prompted trials of deferiprone in NBIA (NCT01741532). Limited experience with these agents to date indicates that treatment improves radiographic appearance,[35] but conclusive amelioration of symptoms has yet to be shown.

CONCLUSIONS AND FUTURE DIRECTIONS

The distinct clinical and in particular, neuroimaging features of PKAN lend themselves to the diagnosis of affected patients. Dystonia can be a particularly disabling symptom, and although recent developments have provided patients with new treatment options, therapies that more specifically target the neurodegenerative process are clearly needed. Recent findings from both *in vitro* studies and animal models have begun to shed new light on the mechanisms that lead to iron deposition as well as cellular dysfunction in PKAN, and have suggested new therapeutic avenues for this devastating disease.

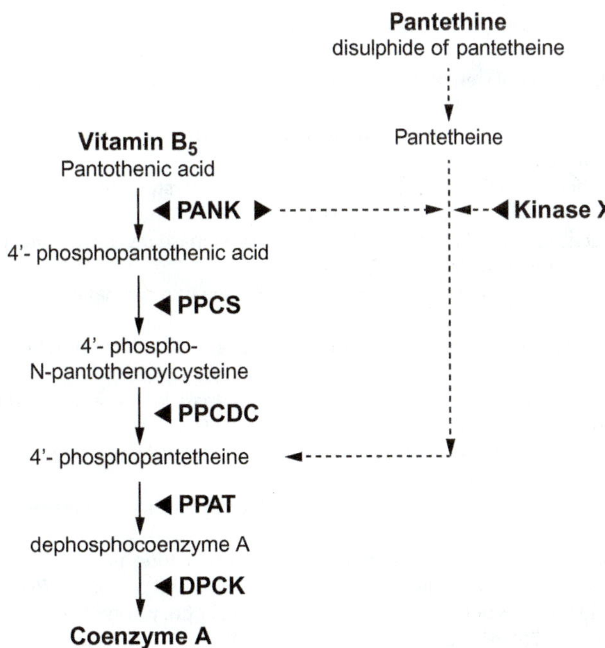

FIGURE 43.6 **Complementary pathways to CoA synthesis.** Pantethine can potentially be used to bypass the metabolic blockade that results from loss of pantothenate kinase activity. (Reprinted with permission from[27].)

References

1. Panteghini C, Zorzi G, Venco P, et al. C19orf12 and FA2H mutations are rare in Italian patients with neurodegeneration with brain iron accumulation. *Semin Pediatr Neurol*. 2012;19(2):75–81.
2. Shevell M. Racial hygiene, active euthanasia, and Julius Hallervorden. *Neurology*. 1992;42(11):2214–2219.
3. Hayflick SJ. Unraveling the Hallervorden–Spatz syndrome: pantothenate kinase-associated neurodegeneration is the name. *Curr Opin Pediatr*. 2003;15(6):572–577.
4. Taylor TD, Litt M, Kramer P, et al. Homozygosity mapping of Hallervorden-Spatz syndrome to chromosome 20p12.3-p13. *Nat Genet*. 1996;14(4):479–481.
5. Zhou B, Westaway SK, Levinson B, Johnson MA, Gitschier J, Hayflick SJ. A novel pantothenate kinase gene (PANK2) is defective in Hallervorden-Spatz syndrome. *Nat Genet*. 2001;28(4):345–349.
6. Healy DG, Abou-Sleiman PM, Wood NW. PINK, PANK, or PARK? A clinicians' guide to familial parkinsonism. *Lancet Neurol*. 2004;3(11):652–662.
7. Aggarwal A, Schneider SA, Houlden H, et al. Indian-subcontinent NBIA: unusual phenotypes, novel PANK2 mutations, and undetermined genetic forms. *Mov Disord*. 2010;25(10):1424–1431.
8. Egan RA, Weleber RG, Hogarth P, et al. Neuro-ophthalmologic and electroretinographic findings in pantothenate kinase-associated neurodegeneration (formerly Hallervorden-Spatz syndrome). *Am J Ophthalmol*. 2005;140(2):267–274.
9. Freeman K, Gregory A, Turner A, Blasco P, Hogarth P, Hayflick S. Intellectual and adaptive behaviour functioning in pantothenate kinase-associated neurodegeneration. *J Intellect Disabil Res*. 2007;51(Pt. 6):417–426.
10. Carod-Artal FJ, Vargas AP, Marinho PB, Fernandes-Silva TV, Portugal D. Tourettism, hemiballism and juvenile Parkinsonism: expanding the clinical spectrum of the neurodegeneration associated to pantothenate kinase deficiency (Hallervorden Spatz syndrome). *Rev Neurol*. 2004;38(4):327–331.
11. Storch A, Schwarz J. Diagnostic test for neuroacanthocytosis: quantitative measurement of red blood cell morphology. In: Danek A, ed. *Neuroacanthocytosis Syndromes* Netherlands: Springer; 2004:71–77 [chapter 9].
12. Ching KH, Westaway SK, Gitschier J, Higgins JJ, Hayflick SJ. HARP syndrome is allelic with pantothenate kinase-associated neurodegeneration. *Neurology*. 2002;58(11):1673–1674.
13. Kruer MC, Boddaert N. Neurodegeneration with brain iron accumulation: a diagnostic algorithm. *Semin Pediatr Neurol*. 2012;19(2):67–74.
14. Chiapparini L, Savoiardo M, D'Arrigo S, et al. The "eye-of-the-tiger" sign may be absent in the early stages of classic pantothenate kinase associated neurodegeneration. *Neuropediatrics*. 2011;42(4):159–162.

14a. Hayflick SJ, Hartman M, Coryell J, Gitschier J, Rowley H. Brain MRI in neurodegeneration with brain iron accumulation with and without PANK2 mutations. *AJNR Am J Neuroradiol*. 2006;27(6):1230–1233.

15. Kruer MC, Hiken M, Gregory A, et al. Novel histopathologic findings in molecularly confirmed pantothenate kinase-associated neurodegeneration. *Brain*. 2011;134(Pt 4):947–958.
16. Li A, Paudel R, Johnson R, et al. Pantothenate kinase-associated neurodegeneration is not a synucleinopathy. *Neuropathol Appl Neurobiol*. 2012 Mar 15;.
17. Malandrini A, Cavallaro T, Fabrizi GM, et al. Ultrastructure and immunoreactivity of dystrophic axons indicate a different pathogenesis of Hallervorden-Spatz disease and infantile neuroaxonal dystrophy. *Virchows Arch*. 1995;427(4):415–421.
18. Albright AL, Ferson SS. Intraventricular baclofen for dystonia: techniques and outcomes. Clinical article. *J Neurosurg Pediatr*. 2009;3(1):11–14.
19. Balas I, Kovacs N, Hollody K. Staged bilateral stereotactic pallidothalamotomy for life-threatening dystonia in a child with Hallervorden–Spatz disease. *Mov Disord*. 2006;21(1):82–85.
20. Timmermann L, Pauls KA, Wieland K, et al. Dystonia in neurodegeneration with brain iron accumulation: outcome of bilateral pallidal stimulation. *Brain*. 2010;133(Pt 3):701–712.
21. Kyriagis M, Grattan-Smith P, Scheinberg A, Teo C, Nakaji N, Waugh M. Status dystonicus and Hallervorden–Spatz disease: treatment with intrathecal baclofen and pallidotomy. *J Paediatr Child Health*. 2004;40(5-6):322–325.
22. Zhang YM, Chohnan S, Virga KG, et al. Chemical knockout of pantothenate kinase reveals the metabolic and genetic program responsible for hepatic coenzyme A homeostasis. *Chem Biol*. 2007;14(3):291–302.
23. Alfonso-Pecchio A, Garcia M, Leonardi R, Jackowski S. Compartmentalization of mammalian pantothenate kinases. *PLoS One*. 2012;7(11):e49509.
24. Leoni V, Strittmatter L, Zorzi G, et al. Metabolic consequences of mitochondrial coenzyme A deficiency in patients with PANK2 mutations. *Mol Genet Metab*. 2012;105(3):463–471.
25. Campanella A, Privitera D, Guaraldo M, et al. Skin fibroblasts from pantothenate kinase-associated neurodegeneration patients show altered cellular oxidative status and have defective iron-handling properties. *Hum Mol Genet*. 2012;21(18):4049–4059.
26. Bosveld F, Rana A, van der Wouden PE, et al. *De novo* CoA biosynthesis is required to maintain DNA integrity during development of the Drosophila nervous system. *Hum Mol Genet*. 2008;17(13):2058–2069.
27. Rana A, Seinen E, Siudeja K, et al. Pantethine rescues a *Drosophila* model for pantothenate kinase-associated neurodegeneration. *Proc Natl Acad Sci U S A*. 2010;107(15):6988–6993.
28. Siudeja K, Srinivasan B, Xu L, et al. Impaired Coenzyme A metabolism affects histone and tubulin acetylation in *Drosophila* and human cell models of pantothenate kinase-associated neurodegeneration. *EMBO Mol Med*. 2011;3(12):755–766.
29. Siudeja K, Grzeschik NA, Rana A, de Jong J, Sibon OC. Cofilin/Twinstar phosphorylation levels increase in response to impaired coenzyme A metabolism. *PLoS One*. 2012;7(8):e43145.
30. Kuo YM, Duncan JL, Westaway SK, et al. Deficiency of pantothenate kinase 2 (Pank2) in mice leads to retinal degeneration and azoospermia. *Hum Mol Genet*. 2005;14(1):49–57.
31. Kovács AD, Pearce DA. Location- and sex-specific differences in weight and motor coordination in two commonly used mouse strains. *Sci Rep*. 2013;3:2116.

32. Brunetti D, Dusi S, Morbin M, et al. Pantothenate kinase-associated neurodegeneration: altered mitochondria membrane potential and defective respiration in Pank2 knock-out mouse model. *Hum Mol Genet*. 2012;21(24):5294–5305.
33. Levy M, Llinas R. Pilot safety trial of deferiprone in 10 subjects with superficial siderosis. *Stroke*. 2012;43(1):120–124.
34. Boddaert N, Le Quan Sang KH, Rötig A, et al. Selective iron chelation in Friedreich ataxia: biologic and clinical implications. *Blood*. 2007;110(1):401–408.
35. Pratini NR, Sweeters N, Vichinsky E, Neufeld JA. Treatment of classic pantothenate kinase-associated neurodegeneration with deferiprone and intrathecal baclofen. *Am J Phys Med Rehabil*. 2013;92(8):728–733.

CHAPTER

44

Disorders of Manganese Transport

Isaac Marin-Valencia

The University of Texas Southwestern Medical Center, Dallas, TX, USA

INTRODUCTION

Manganese is an essential trace metal involved in fundamental biological processes, from photosynthesis to intermediary metabolism. In man, manganese serves as a cofactor for enzymes involved in redox processes, decarboxylation reactions, lipid synthesis, and glucose homeostasis, among other functions. Its levels in the body are tightly regulated, since excessive amounts are toxic to the liver, muscle, and brain. In recent years, an autosomal recessive form of manganese transport disorder has been reported. Patients with this condition manifest hypermanganesemia, dystonia, parkinsonism, liver dysfunction, and polycythemia.

CLINICAL FEATURES

Most of what is known regarding manganese toxicity derives from experience with environmental exposure, which leads to a clinical condition termed manganism. The first cases of manganism were reported in 1837 in five miners exposed to manganese, who developed parkinsonism and aberrant behavior.[1] In more recent times, manganese toxicity has been described under other circumstances; for example in patients on prolonged total parenteral nutrition,[2–4] drug addicts exposed to manganese-contaminated street drugs,[5] and in patients with impaired manganese excretion due to chronic liver disease.[2,6] Regardless of the cause, excess manganese accumulates mainly in the liver, muscle, and brain. Psychological symptoms manifest early in the course of the disease, with psychosis, hallucinations, and behavioral disturbances.[7] Extrapyramidal symptoms soon follow; although in contrast to Parkinson disease, resting tremor is uncommon in manganism. In these patients, blood levels of manganese range between 1000 and 2000 nmol per L (normal < 320 nmol per L).

A genetic form of manganese transport disorder was recently linked to manganese toxicity. This condition is inherited in an autosomal recessive fashion, resulting from mutations of *SCL30A10*. Approximately 20 cases have been reported thus far (Table 44.1). It typically manifests with extrapyramidal symptoms, pica, hyperpigmentation, and, less often, with upper motor neuron signs.[8–10] The reported ages of onset range between 2 and 60 years, with most presenting in childhood and adolescence.[8,10,11] Patients with the childhood-onset form typically exhibit dystonia involving the limbs and tongue, resulting in a "cock-walk" gait, impaired manual dexterity, and dysarthria. Upper motor neuron signs include spastic paraparesis, hyperreflexia, and extensor plantar responses, which has been attributed to the chronic hepatic disease.[8,12] This phenotype is also known as hepatic myelopathy. In general, patients with the childhood-onset form become wheelchair-bound by adolescence. In contrast, individuals with the adult-onset form present with a Parkinson disease-like syndrome, with bradykinesia, hypomimia, hypophonia, and rigidity manifesting after the fourth decade of life. As in manganism, resting tremor is usually absent. Sensorimotor peripheral neuropathy has been also reported in adult-onset patients.[10] In contrast to manganism, cognitive and psychiatric symptoms are uncommon in individuals with defects of manganese transport.

Rosenberg's Molecular and Genetic Basis of Neurological and Psychiatric Disease
http://dx.doi.org/10.1016/10.1016/B978-0-12-410529-4.00044-9

TABLE 44.1 Clinical, Epidemiological and Genetic Characteristics of Patients with Manganese Transport Disorders[8–10,12,100,101]

Presentation	Gender	Age at presentation (years)	Ethnicity	Consanguinity	Clinical manifestations	Polycythemia	Liver damage	Mutation	Exon	Amino acid change
Childhood	F (1), M (2)	2, 10 (F), 14	Dutch	+	Dystonia (both M), hepatomegaly (F)	+	+ (cirrhosis, F)	c.507delG; c.500T>C (both M, F unknown)	1	p.Pro170Leufs*22; p.Phe167Ser
	F (2), M (2)	3 (both F), 5 (both M)	Punjabi (Pakistan)	+	Dystonia (all), hepatomegaly (1)	+ (all)	+ (3)	g.chr11:218.057.426_218.158.564del101139	1–2	
	F (3)	2 (3)	Arabic (Yemen)	+	Dystonia (all)	+ (all)	+ (2)	c.266T>C	1	p.Leu89Pro
	F (1)	2	Acadian (Canada)	–	Dystonia	+	+ (mild elevated ALT)	c.292_402del	1	p.Val98_Phe134del
	F (1), M (1)	2 (M), 11 (F)	Arabic (United Arab Emirates)	+	Dystonia (all), hepatomegaly (all)	+ (F)	+ (cirrhosis)	c.314_322del	1	p.Ala105_Pro107del
	M	14	Caucasian (US)	–	Spastic paraparesis, hepatomegaly	+	+ (cirrhosis)	c.585del	1	p.Thr196Profs*17
	F	11	Indo-Malay (Philippines)	–	Dystonia	+	+	c.765_767del	3	p.Val256del
	F (1), M (1)	2 (M), 3 (F)	Portuguese/ Amerindian (Brazil)	+	Dystonia	+ (F)	+ (cirrhosis)	c.922C>T	3	p.Gln308*
	F	5	Punjabi (India)	–	Dystonia, hepatomegaly	+	+	c.1046T>C	4	p.Leu349Pro
Adult	M (2)	47, 57	Italian	+	Parkinsonism (all), Hepatomegaly (all)	+	+	c.1235delA	4	p.Gln412Argfs*26

Abbreviations: F, female, M, male; number of patients reported in brackets.

MOLECULAR GENETICS

Defects of manganese transport have been associated with mutations of *SCL30A10*.[8,10] Located on chromosome 1q41, *SLC30A10* is comprised of four exons spanning 14,388 nucleotides. It encodes for SLC30A10, a 52-kD protein of 485 amino acids.[10] SLC30A10 is a metal transporter that belongs to the cation diffusion facilitator (CDF) superfamily involved in the efflux of metals (iron, manganese, copper, and zinc) from the cytosol to the extracellular space or to subcellular organelles.[13,14] SLC30A10 is composed of six transmembrane domains, with both *N* and *C* terminals on the cytoplasmic side, and a conserved *C*-terminal domain, named the cation efflux domain.[10] This protein was initially considered a zinc transporter.[15] However, yeast studies revealed that SLC30A10 protects cells from manganese accumulation, indicating a role in transporting this metal as well.[8] In man, SCL30A10 is located in the Golgi apparatus, endosome, and plasma membrane of liver, bile duct, and basal ganglia cells.[10]

As with other CDF transporters, expression of *SLC30A10* is regulated by the concentration of manganese. The mechanisms involved in the expression of *SLC30A10* have been studied in HepG2 hepatocellular carcinoma cells exposed to manganese or zinc.[10] Exposure to manganese led to a significant increase of *SLC30A10* mRNA expression and of SLC30A10 content on immunoblotting analysis. Conversely, zinc exposure led to a small decrease in *SLC30A10* expression. This supports the notion that *SLC30A10* is under tight control by extracellular levels of manganese.[10] To date, ten homozygous mutations have been reported in patients with this disorder, including missense and nonsense mutations, single-base deletions and large deletions (Table 44.1).[8,10] However, due to the small number of patients described thus far, no conclusive genotype–phenotype correlations have been established. It has been postulated that mutations leading to childhood-onset disease have more detrimental effects on SLC30A10 functioning than mutations found in patients with the adult-onset form.[10]

PHYSIOLOGY AND DISEASE MECHANISMS

Manganese homeostasis is tightly regulated, given the potential deleterious effects of its accumulation. The primary source of manganese is the diet. The US Environmental Protection Agency (EPA) recommends a daily intake of 2–5 mg per day in adults and adolescents. Only 1–5% of the ingested manganese is absorbed in the intestinal epithelium.[16] In blood, manganese is predominately in the 2+ oxidation state (>99%). Approximately 80% is bound to β-globulin and albumin, and a small portion to transferrin.[17,18] Manganese (2+) crosses the blood–brain barrier (BBB) by active and facilitated transport systems mediated by, among others, the divalent metal ion transporter 1 (DMT1), *N*-methyl-D-aspartate (NMDA) receptor channel, Zip8, and Zip14.[19–21] On the other hand, transferrin receptors (TfR) are the most efficient transporters for manganese (3+).[22] Manganese is distributed throughout the brain, particularly in the globus pallidus, striatum, and substantia nigra.[23,24] The influx of manganese from the extracellular space to the cytosol is tightly balanced with the efflux mechanisms to prevent manganese accumulation inside the cell. SLC30A10 plays an important role in manganese efflux, and its dysfunction leads to intracellular accumulation of manganese.[8,10]

There is substantial evidence that iron homeostasis is linked to that of manganese via the DMT1, TfR and ferroportin (FPN) transport systems.[25–28] The absorption and excretion of manganese seems to be inversely linked to stored iron, such that depletion of iron stores favors manganese absorption and accumulation through overexpression of *DMT1*, *TfR* and *FPN*,[29–33] whereas iron excess limits manganese absorption via decreased expression of *DMT1* and *FPN*.[34,35]

Within the cell, manganese is stored primarily in the mitochondria, where it serves as a cofactor for numerous enzymes.[36] High levels of manganese inhibit calcium efflux from the mitochondria,[37] leading to the formation of reactive oxygen species (ROS) through the respiratory chain, subsequently limiting oxidative metabolism and ATP synthesis.[38,39] In addition, a high level of manganese directly inhibits mitochondrial complexes I, II and III.[40,41] In turn, oxidative stress favors the opening of the mitochondrial transition pore (MTP), with subsequent shift of protons and molecules across the mitochondrial membrane, loss of inner mitochondrial membrane potential and, ultimately, swelling.[42–44] In addition, manganese-induced mitochondrial toxicity leads to apoptosis by the release of cytochrome *c* and activation of the caspase pathway.[45]

Brain Dysfunction

It has been postulated that the initial detrimental effects of manganese accumulation in the brain occur in astrocytes. This notion is based on the high affinity of the astrocytic uptake system.[46,47] Under physiological conditions, astrocytes accumulate a higher concentration of manganese than neurons (up to 50 times more).[47] The resulting production of ROS leads to activation of microglial cells, which subsequently produce inflammatory cytokines (interleukin 6 [IL-6] and tumor necrosis factor-alpha [TNF-α]). These inflammatory agents induce, in turn, reactive astrocytosis, leading to neuronal damage through the synthesis of nitric oxide and prostaglandins.[48–50] Simultaneously, inflammatory agents downregulate the glutamate transporters GLAST and GLT-1, limiting glutamate uptake by astrocytes.[51–53] It has been shown that nuclear factor kappa-light-chain-enhancer of activated B cells (NF-κB) and mitogen-activated protein kinases (MAPK) signaling pathways mediate TNF-α-induced reduction in GLT-1 expression.[54] Depletion of glutamate transporters favors the accumulation of glutamate in the extracellular space, resulting in excitatory neurotoxicity.

Loss of dopaminergic neurons and reactive gliosis in the globus pallidus and substantia nigra are the most common neuropathological findings in human manganese intoxication.[55,56] The accumulation of manganese leads to misfolding and aggregation of α-synuclein in neurons and glial cells.[57] It has been shown that human α-synuclein-overexpressing cells exposed to manganese manifest activation of NF-κB, p38 MAPK, and apoptotic cascades, leading to cell death, particularly of dopaminergic neurons.[58] On the other hand, manganese can also trigger apoptosis in dopaminergic neurons through the caspase pathway by activation of protein kinase C.[59]

Postsynaptic injury in the dopaminergic pathway is frequently observed in manganese toxicity, with marked reduction of postsynaptic D2 receptor density in the striatum as seen on ^{18}F-methylspiperone positron emission tomography (PET) in a patient with chronic manganism.[60] It is unclear whether dopamine transporters at the presynaptic neuron are also involved in manganese-induced toxicity. On the one hand, PET studies have shown intact dopamine transporter function in presynaptic neurons in the striatum,[61,62] but on the other hand, single photon emission computed tomography (SPECT) studies in patients with chronic manganism revealed decreased binding capacity of ^{123}I-1r-2B-carboxy-methoxy-3B-(4-iodophenyl)-tropane to dopamine transporters in presynaptic cells.[63] Studies of pre- and postsynaptic dopaminergic receptor density and binding capacity have not yet been carried out in patients with manganese transport disorders.

Manganese intoxication can also impair the function of other neurotransmitters. For example, rats exposed to high levels of manganese experienced a two-fold reduction in the content of both protein and mRNA of the α_2-adrenergic receptor in the locus ceruleus and substantia nigra.[64] The resulting impaired noradrenergic transmission is postulated to contribute to the aberrant behavior associated with manganese toxicity. γ-Aminobutyric acid (GABA)ergic transmission might also be affected. Excessive manganese exposure led to increased extracellular concentrations of GABA in the striatum of rats, via altered expression of transport (GAT-1) and receptor proteins ($GABA_A$ and $GABA_B$).[65,66] Dysfunctional GABAergic neurotransmission in the basal ganglia may be associated with the hyperkinesia, ataxia, and dystonia manifested by some of these patients.

Liver Dysfunction

Patients affected with hypermanganesemia of any etiology may manifest hepatic dysfunction to variable degrees. However, little is known about the mechanism of manganese-induced liver toxicity. The liver of rats exposed to manganese exhibited a significant inhibition of superoxide dismutase, glutathione peroxidase, and Na^+/K^+-ATPase activity; decreased levels of glutathione; and increased levels of malondialdehyde, the latter constituting a product from ROS-inducing degradation of polyunsaturated lipids.[67] These changes favor oxidative stress and ion and water shifts across the plasma membrane, ultimately resulting in cell damage. It has been posited that manganese-induced cholestasis occurs within the hepatocyte, not within the canaliculi, since bilirubin conjugation is not required to cause cholestasis in this condition.[68] Physiological dysfunction correlates with histological changes in the livers of these rats, showing mononuclear cell infiltration, enlarged veins and sinusoids, canalicular tight junction alterations, necrotic changes, mitochondrial hyperplasia, swelling, and vacuolization.[67,68] The majority of patients with manganese transport disorders manifest hepatomegaly, elevated transaminases (alanine transaminase [ALT], aspartate transaminase [AST]) and unconjugated hyperbilirubinemia (Table 44.1). Features of liver cirrhosis can be observed by ultrasound. Pathologic features include fibrosis, steatosis, micronodular cirrhosis, and high levels of manganese on rhodamine staining.[8–10]

Polycythemia

Patients with manganese transport disorders and manganism usually have polycythemia.[8,10] Manganese (2+) itself enhances the production of red blood cells by promoting the expression of the erythropoietin gene (*EPO*).[69–71] When cells are exposed to low oxygen levels, or to manganese (2+) or other trace metals such as cobalt (2+) or nickel (2+), hypoxia-inducible factor 1-α (HIF-1α) forms a heterodimer with HIF-1β, resulting in the activation of HIF-1. Subsequently, HIF-1 translocates to the nucleus and promotes the expression of hypoxia-inducible genes, including *EPO*.[69] Erythropoietin is an essential hormone for red blood cell production in the bone marrow, and high levels are associated with polycythemia. In affected individuals, hemoglobin concentrations range between 16 and 23 g per dL (Table 44.1).[8,10]

DIFFERENTIAL DIAGNOSIS

The differential diagnosis of defects of manganese transport includes several extrapyramidal disorders, such as parkinsonian syndromes, inherited and neurodegenerative diseases associated with dystonia, and drug-induced movement disorders. Occasionally, clinical overlap makes distinction among these entities difficult, so that testing becomes crucial to establish a definitive diagnosis. In general, high blood levels of manganese, liver dysfunction, or polycythemia are not observed in other similar conditions. The classic T1 hyperintensities seen in the basal ganglia of individuals with manganese disorders is also helpful in establishing the diagnosis (Figure 44.1). Some of the most common disorders included in the differential diagnosis are listed in Table 44.2.

Parkinson Disease and Associated Syndromes

Parkinson disease (PD) is a common neurodegenerative disorder, characterized by bradykinesia, rigidity, resting tremor, and postural instability. It is a result of dopamine depletion in the substantia nigra (pars compacta) and in the nigrostriatal pathway, leading to increased inhibition of the thalamus and, in turn, reduced excitatory input to the motor cortex. Several clinical features differentiate PD from defects of manganese transport. Patients with manganese disorders do not usually manifest the typical rest tremor of PD, but other cardinal features of parkinsonism (bradykinesia, hypomimia, hypophonia, and rigidity) are usually present in the adult-onset form of manganese transport defects. Dystonia of the limbs and tongue are more common in the childhood-onset form. Defects of manganese transport commonly present in childhood and adolescence, in contrast to PD, which usually manifests after 60 years of age.[72] Also, patients with manganese transport disorders do not respond to dopaminergic medication, in contrast to patients with PD.

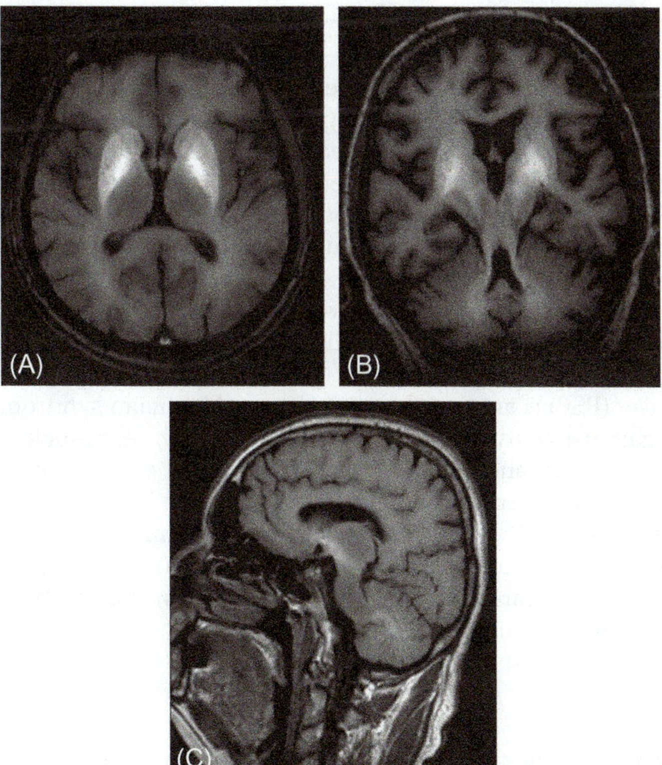

FIGURE 44.1 **Brain MRI from a patient with *SLC30A10* mutations.** Axial (**A**), coronal (**B**) and sagital (**C**) T1-weighted images illustrate the typical bilateral hyperintense signal in globus pallidus, putamen, caudate nucleus, midbrain and cerebellum. Reproduced from [10] with permission of the *American Journal of Human Genetics*.

TABLE 44.2 Conditions Included in the Differential Diagnosis of Manganese Transport Defects

Manganism (acquired hypermanganesemia)
Parkinson disease
Parkinsonian syndromes: Multiple system atrophy Progressive supranuclear palsy Juvenile parkinsonism Vascular parkinsonism Drug-induced parkinsonism
Neurodegenerative conditions associated with dystonia or other extrapyramidal symptoms: Huntington disease Wilson disease Neurodegeneration with brain iron accumulation (pantothenate kinase-associated neurodegeneration (PKAN)), infantile neuroaxonal dystrophy, fatty acid hydroxylase-associated neurodegeneration (aceruloplasminemia, neuroferritinopathy) Organic acidemias Niemann–Pick disease type C Mitochondrial disorders (i.e. pyruvate dehydrogenase complex deficiency)
Inherited forms of dystonia: Early-onset dystonia (DYT1) Dopa-responsive dystonia (Segawa disease) Myoclonus-dystonia Rapid onset dystonia-parkinsonism Hemidystonia
Acquired dystonia: Drug-induced dystonia (antipsychotics or antiemetics) Stroke Demyelinating conditions Infections, abscesses Tumors

Other degenerative parkinsonian syndromes are included in the differential diagnosis. One example is multiple system atrophy (MSA), which results from progressive degeneration in the substantia nigra, striatum, cerebellum, and autonomic nervous system. It presents in middle-aged individuals with bradykinesia and rigidity, but often with minimal or no rest tremor.[73,74] In contrast to manganese disorders, autonomic dysfunction is universally present. At least half of affected individuals have orthostatic hypotension. Other autonomic features include constipation, impotence, dry mouth, loss of sweating, and urinary urgency or retention. Cerebellar ataxia is present in one-third of patients.[74] In contrast to manganese disorders, limb dystonia occurs only occasionally. Brain imaging often reveals putaminal hypointensity on T2-weighted magnetic resonance imaging (MRI), due to iron deposition, as well as pontine and cerebellar atrophy. Similar to manganese disorders, patients with MSA do not typically respond to dopaminergic agents.

Progressive supranuclear palsy (PSP) is another degenerative parkinsonian syndrome, with onset after the sixth decade of life, presenting with gait instability and frequent falls, spasticity, supranuclear visual disturbances, dystonia, slurred speech, dysphagia, dementia, and personality changes.[75] As the disease progresses, affected individuals manifest limitations of extraocular movements, initially vertical and then horizontal saccades, eventually resulting in complete ophthalmoplegia. Dystonia of the face, neck, and limbs is prominent and fairly symmetric. As in manganese disorders, resting tremor is uncommon in PSP. Brain MRI usually shows atrophy of the dorsal mesencephalon, giving the characteristic "Mickey Mouse ears" appearance on axial views, and the "hummingbird sign" on sagittal projections. Dopaminergic medications are usually ineffective.

Huntington Disease

Huntington disease (HD) is an autosomal dominant trinucleotide repeat disorder characterized by choreoathetosis, psychiatric disorders, and dementia. In contrast to disorders of manganese transport, patients with HD initially present with behavioral changes, such as irritability, impulsivity, untidiness, aggressiveness, antisocial behavior, drug abuse, or promiscuity. Depression is common early in the course of the disease, higher suicide risk is frequently associated. Patients invariably develop cognitive impairment. Movement abnormalities are subtle at first, consisting of fidgeting motions in the fingers and hands. This gradually progresses to generalized hyperkinetic movements (chorea or choreoathetosis), in contrast to the hypokinetic movements of manganese disorders. At advanced stages, patients may develop dystonia, bradykinesia, tremor, and rigidity. Juvenile HD (Westphal variant) typically presents in childhood or adolescence with bradykinesia, rigidity, and dystonia. The juvenile form differs from manganese transport defects in that patients manifest behavioral problems, cognitive impairment, cerebellar ataxia, and seizures, in addition to the movement abnormalities.[76] In HD, brain MRI reveals volume loss in the basal ganglia, particularly of the caudate nucleus.[77] Antidopaminergic agents, such as neuroleptics and tetrabenazine, help control chorea.[78–80] In contrast to most parkinsonian disorders, dopaminergic drugs worsen chorea in these patients and may evoke dyskinesias in individuals with the rigid form of the disease.

Wilson Disease

Also known as hepatolenticular degeneration, Wilson disease (WD) is an autosomal recessive disorder of copper transport, resulting in its accumulation in liver and brain tissue. The age of presentation ranges from 3 to over 60 years of age, but like manganese transport defects, most patients present during adolescence and early adulthood.[81] In contrast to manganese disorders, movement abnormalities in WD predominantly consist of tremors, chorea, or choreoathetosis, although dystonia can also be present.[82] Cognitive and psychiatric disturbances are also common in WD, including depression, neurosis, personality disorders, and intellectual impairment. The liver dysfunction, on the other hand, is similar to that of manganese disorders and may present as recurrent jaundice, acute self-limited hepatitis, fulminant liver failure, or chronic liver disease.[83] An additional clinical finding is Kayser–Fleischer rings, copper deposition in Descemet's membrane of the cornea. The diagnosis of WD is suggested by low serum concentration of copper and ceruloplasmin, increased urinary copper excretion, corneal Kayser–Fleischer rings, and/or increased copper concentration on liver biopsy. The diagnosis is confirmed by mutations analysis of *ATP7B*.

Dystonia and Dystonia-Associated Disorders

Dystonia is defined as a sustained contraction of antagonist muscles resulting in repetitive movements or abnormal postures.[84] Inherited forms dystonia may clinically resemble the childhood-onset form of manganese transport defects, although without liver or blood involvement. For example, early-onset primary dystonia (DYT1) presents during childhood and adolescence with involuntary, sustained muscle contractions causing abnormal posturing of upper extremities, lower extremities, or both. A jerky dystonic "tremor" may also be present. The diagnosis is made by genetic testing for *TOR1A* mutations.

"Dystonia-plus" syndromes, including dopa-responsive dystonia, myoclonus-dystonia and rapid-onset dystonia-parkinsonism are also part of the differential diagnosis. Dopa-responsive dystonia is an autosomal dominant disorder resulting from guanosine triphosphate (GTP) cyclohydrolase 1 deficiency. It typically presents in childhood (average 6 years of age) with dystonia of the lower extremities that worsens throughout the day. The condition can progress to generalized dystonia and also to parkinsonism. There is no cognitive impairment.[85,86] Total biopterin and neopterin levels in cerebrospinal fluid (CSF) are characteristically low. In contrast to defects of manganese transport, these patients respond dramatically to the administration of very low doses of levodopa. Myoclonus-dystonia is an autosomal dominant condition that presents during childhood and adolescence with myoclonic jerks of the neck, upper extremities and trunk. Approximately 50% of patients also manifest cervical dystonia and hand dystonia (writer's cramp). Contrary to manganese transport disorders, psychiatric conditions are common in myoclonus-dystonia, including depression, anxiety, addictive behaviors, and personality disorders. Molecular genetic testing of *SGCE* confirms the diagnosis.[87] Rapid-onset dystonia-parkinsonism is a rare autosomal dominant condition characterized by the rapid manifestation (within hours to weeks) of dystonic spasms, bradykinesia, dysarthria and postural instability.[88] It presents during childhood, adolescence, or adulthood. The diagnosis is made based on the clinical presentation of acute-onset parkinsonism, low CSF concentration of homovanillic acid, and genetic analysis of *ATP1A3*. Levodopa does not alleviate extrapyramidal symptoms.

Acquired dystonias result from insults to the basal ganglia due to stroke, demyelinating lesions, infection, tumors, and drugs. The most common acquired dystonia is drug-induced dystonia, which typically manifests as an acute, usually transient, dystonia, although chorea and other extrapyramidal movement disorders may also occur. Antidopaminergic agents such as antipsychotics (e.g., haloperidol) and antiemetics (e.g., phenothiazines and metoclopramide) are the drugs most commonly responsible for acute dystonic reactions. The diagnosis is suggested by acute (except in the case of tardive syndromes) dystonia, temporally associated with use of an antidopaminergic drug. These patients usually respond to anticholinergic medications (e.g., diphenhydramine or trihexyphenidyl).

TESTING

The diagnosis of defects of manganese transport relies primarily on the characteristic neurological presentation, liver dysfunction, and polycythemia, within the context of hypermanganesemia. Manganese levels are high in all biological fluids. In whole blood, manganese concentration correlates with the current exposure and body burden. Whole blood levels are more reliable than urine levels, since manganese is mainly excreted by the biliary system.[89,90] In transport disorders, the levels of manganese in blood are typically higher than in manganism due to environmental exposure, usually >2000 nmol per L.[8] Other distinctive laboratory features include polycythemia, low levels of ferritin, and increased total iron-binding capacity (TIBC); the last two are indicative of depleted iron stores.[8] Liver dysfunction is indicated by elevated transaminases (ALT, AST) and unconjugated hyperbilirubinemia.

Since manganese is paramagnetic, its accumulation in the basal ganglia can be detected on brain MRI (Figure 44.1). Manganese shortens T1 relaxation time, resulting in hyperintensity of the involved regions on T1-weighted imaging.[89] These imaging findings are seen in patients with hypermanganesemia regardless of the underlying cause, and so are present in manganese transport disorders, hypermanganesemia due to environmental exposure, hepatic cirrhosis, prolonged total parenteral nutrition, hemodialysis, and chronic iron deficiency.[91–94] T2-weighted images are generally unremarkable. The hyperintense signal usually resolves about 6–12 months after cessation of the exposure,[95,96] despite the persistence (or progression) of symptoms in untreated patients.[95] Once the condition is suspected, it can be confirmed by molecular genetic testing of *SLC30A10*.

MANAGEMENT

A global assessment of the extent of the disease is required to guide therapy. This includes a general exam, with special emphasis on the neurological evaluation, along with liver and blood assessment. Treatment is aimed at reducing blood manganese levels and consists primarily of chelation and iron supplementation. Chelation therapy with intravenous disodium calcium edetate has led to marked improvement of extrapyramidal symptoms in patients with manganese transport deficiency.[8,10] Potential side effects include hypocalcemia, renal toxicity, trace metal deficiencies, vitamin deficiencies, leukopenia, or thrombocytopenia.[9] In the acute setting, short-term infusions of disodium calcium edetate are administered at 20 mg per kg per dose, twice daily, for 5 days. During the infusion period, it is recommended to check daily plasma and 24-hour urine manganese, along with electrolytes, calcium, phosphate, magnesium, renal and liver function, complete blood count (CBC), and serum levels of trace metals (zinc, copper and selenium). If short-term infusions are clinically effective, then long-term infusions (5 consecutive days per month) are recommended. It has been shown that long-term infusions lower the concentration of manganese and normalize hemoglobin and iron levels.[8,10] As with short-term infusions, it is necessary to monitor the above laboratory studies every 2 months.[97] Other chelating agents have been used where disodium calcium edetate is not available, primarily in developing countries. Two affected siblings diagnosed with manganese transport defect improved clinically on oral dimercaptosuccinic acid (DMSA) and oral iron supplementation.[8] However, two adult patients with environmental manganese intoxication did not improve significantly with DMSA treatment.[98] A paucity of experience with DMSA in patients with hypermanganesemia limits its systematic use.

Iron supplementation is another arm of treatment. Iron competitively inhibits manganese absorption in the intestine, reducing manganese blood levels and improving polycythemia.[8,9] It is recommended to monitor serum iron and total iron-binding capacity every 3–4 months while on oral supplements.[99] If serum iron exceeds 80% of the total iron-binding capacity, it is recommended to discontinue iron supplements. The combination of both chelation therapy and iron supplementation in an affected 12-year-old girl significantly improved motor symptoms, resolved polycythemia, and normalized liver manganese levels.[9] Physical, occupational and speech therapy should be offered as needed, particularly in patients with prominent extrapyramidal symptoms. Patients with manganese transport deficiency do not usually respond to dopaminergic agents or antispasmodics, although one adolescent patient benefited from these agents.[9] Foods with high concentrations of manganese (e.g., wheat, nuts, pumpkin, sesame, dark chocolate, and sunflower seeds) should be avoided.[99] In patients with end-stage liver disease, liver transplant is a consideration, although it has not yet been attempted in patients with this condition.

References

1. Couper J. On the effects of black oxide of manganese when inhaled into the lungs. *Br Ann Med Pharm*. 1837;1:41–42.
2. Chalela JA, Bonillha L, Neyens R, Hays A. Manganese encephalopathy: an under-recognized condition in the intensive care unit. *Neurocritical Care*. 2011;14:456–458.
3. Fell JM, Reynolds AP, Meadows N, et al. Manganese toxicity in children receiving long-term parenteral nutrition. *Lancet*. 1996;347:1218–1221.
4. Hsieh CT, Liang JS, Peng SS, Lee WT. Seizure associated with total parenteral nutrition-related hypermanganesemia. *Pediatr Neurol*. 2007;36:181–183.
5. Sanotsky Y, Lesyk R, Fedoryshyn L, Komnatska I, Matviyenko Y, Fahn S. Manganic encephalopathy due to "ephedrone" abuse. *Mov Disord*. 2007;22:1337–1343.
6. Butterworth RF. Metal toxicity, liver disease and neurodegeneration. *Neurotox Res*. 2010;18:100–105.
7. Olanow CW. Manganese-induced parkinsonism and Parkinson's disease. *Ann N Y Acad Sci*. 2004;1012:209–223.
8. Tuschl K, Clayton PT, Gospe Jr. SM, et al. Syndrome of hepatic cirrhosis, dystonia, polycythemia, and hypermanganesemia caused by mutations in SLC30A10, a manganese transporter in man. *Am J Hum Genet*. 2012;90:457–466.
9. Tuschl K, Mills PB, Parsons H, et al. Hepatic cirrhosis, dystonia, polycythaemia and hypermanganesaemia – a new metabolic disorder. *J Inherit Metab Dis*. 2008;31:151–163.
10. Quadri M, Federico A, Zhao T, et al. Mutations in SLC30A10 cause parkinsonism and dystonia with hypermanganesemia, polycythemia, and chronic liver disease. *Am J Hum Genet*. 2012;90:467–477.
11. Stamelou M, Tuschl K, Chong WK, et al. Dystonia with brain manganese accumulation resulting from SLC30A10 mutations: a new treatable disorder. *Mov Disord*. 2012;27:1317–1322.
12. Gospe Jr. SM, Caruso RD, Clegg MS, et al. Paraparesis, hypermanganesaemia, and polycythaemia: a novel presentation of cirrhosis. *Arch Dis Child*. 2000;83:439–442.
13. Montanini B, Blaudez D, Jeandroz S, Sanders D, Chalot M. Phylogenetic and functional analysis of the Cation Diffusion Facilitator (CDF) family: improved signature and prediction of substrate specificity. *BMC Genomics*. 2007;8:107.
14. Ohana E, Hoch E, Keasar C, et al. Identification of the Zn^{2+} binding site and mode of operation of a mammalian Zn^{2+} transporter. *J Biol Chem*. 2009;284:17677–17686.
15. Sreedharan S, Stephansson O, Schioth HB, Fredriksson R. Long evolutionary conservation and considerable tissue specificity of several atypical solute carrier transporters. *Gene*. 2011;478:11–18.
16. Davidsson L, Almgren A, Hurrell RF. Sodium iron EDTA [NaFe(III)EDTA] as a food fortificant does not influence absorption and urinary excretion of manganese in healthy adults. *J Nutr*. 1998;128:1139–1143.
17. Bowman AB, Kwakye GF, Herrero Hernandez E, Aschner M. Role of manganese in neurodegenerative diseases. *J Trace Elem Med Biol*. 2011;25:191–203.
18. Foradori AC, Bertinchamps A, Gulibon JM, Cotzias GC. The discrimination between magnesium and manganese by serum proteins. *J Gen Physiol*. 1967;50:2255–2266.
19. Au C, Benedetto A, Aschner M. Manganese transport in eukaryotes: the role of DMT1. *Neurotoxicology*. 2008;29:569–576.
20. Itoh K, Sakata M, Watanabe M, Aikawa Y, Fujii H. The entry of manganese ions into the brain is accelerated by the activation of *N*-methyl-D-aspartate receptors. *Neuroscience*. 2008;154:732–740.
21. Aschner M, Gannon M. Manganese (Mn) transport across the rat blood-brain barrier: saturable and transferrin-dependent transport mechanisms. *Brain Res Bull*. 1994;33:345–349.
22. Aschner M, Aschner JL. Manganese transport across the blood–brain barrier: relationship to iron homeostasis. *Brain Res Bull*. 1990;24:857–860.
23. Dorman DC, Struve MF, Marshall MW, Parkinson CU, James RA, Wong BA. Tissue manganese concentrations in young male rhesus monkeys following subchronic manganese sulfate inhalation. *Toxicol Sci*. 2006;92:201–210.
24. Guilarte TR, McGlothan JL, Degaonkar M, et al. Evidence for cortical dysfunction and widespread manganese accumulation in the nonhuman primate brain following chronic manganese exposure: a 1H-MRS and MRI study. *Toxicol Sci*. 2006;94:351–358.
25. Fitsanakis VA, Zhang N, Anderson JG, et al. Measuring brain manganese and iron accumulation in rats following 14 weeks of low-dose manganese treatment using atomic absorption spectroscopy and magnetic resonance imaging. *Toxicol Sci*. 2008;103:116–124.
26. Gunter TE, Gerstner B, Gunter KK, et al. Manganese transport via the transferrin mechanism. *Neurotoxicology*. 2013;34:118–127.
27. Yin Z, Jiang H, Lee ES, et al. Ferroportin is a manganese-responsive protein that decreases manganese cytotoxicity and accumulation. *J Neurochem*. 2010;112:1190–1198.
28. Gunshin H, Mackenzie B, Berger UV, et al. Cloning and characterization of a mammalian proton-coupled metal-ion transporter. *Nature*. 1997;388:482–488.
29. Erikson KM, Shihabi ZK, Aschner JL, Aschner M. Manganese accumulates in iron-deficient rat brain regions in a heterogeneous fashion and is associated with neurochemical alterations. *Biol Trace Elem Res*. 2002;87:143–156.
30. Fitsanakis VA, Zhang N, Garcia S, Aschner M. Manganese (Mn) and iron (Fe): interdependency of transport and regulation. *Neurotox Res*. 2010;18:124–131.
31. DeWitt MR, Chen P, Aschner M. Manganese efflux in Parkinsonism: insights from newly characterized SLC30A10 mutations. *Biochem Biophys Res Commun*. 2013;432:1–4.
32. Erikson KM, Syversen T, Aschner JL, Aschner M. Interactions between excessive manganese exposures and dietary iron-deficiency in neurodegeneration. *Environ Toxicol Pharmacol*. 2005;19:415–421.
33. Aschner M, Shanker G, Erikson K, Yang J, Mutkus LA. The uptake of manganese in brain endothelial cultures. *Neurotoxicology*. 2002;23:165–168.
34. Erikson KM, Pinero DJ, Connor JR, Beard JL. Regional brain iron, ferritin and transferrin concentrations during iron deficiency and iron repletion in developing rats. *J Nutr*. 1997;127:2030–2038.
35. Zoller H, Koch RO, Theurl I, et al. Expression of the duodenal iron transporters divalent-metal transporter 1 and ferroportin 1 in iron deficiency and iron overload. *Gastroenterology*. 2001;120:1412–1419.

36. Gavin CE, Gunter KK, Gunter TE. Manganese and calcium transport in mitochondria: implications for manganese toxicity. *Neurotoxicology*. 1999;20:445–453.
37. Gavin CE, Gunter KK, Gunter TE. Manganese and calcium efflux kinetics in brain mitochondria. Relevance to manganese toxicity. *Biochem J*. 1990;266:329–334.
38. Kruman II, Mattson MP. Pivotal role of mitochondrial calcium uptake in neural cell apoptosis and necrosis. *J Neurochem*. 1999;72:529–540.
39. Kowaltowski AJ, Castilho RF, Vercesi AE. Ca(2+)-induced mitochondrial membrane permeabilization: role of coenzyme Q redox state. *Am J Physiol*. 1995;269:C141–C147.
40. Singh J, Husain R, Tandon SK, Seth PK, Chandra SV. Biochemical and histopathological alterations in early manganese toxicity in rats. *Environ Physiol Biochem*. 1974;4:16–23.
41. Zhang J, Fitsanakis VA, Gu G, et al. Manganese ethylene-bis-dithiocarbamate and selective dopaminergic neurodegeneration in rat: a link through mitochondrial dysfunction. *J Neurochem*. 2003;84:336–346.
42. Gavin CE, Gunter KK, Gunter TE. Mn2+ sequestration by mitochondria and inhibition of oxidative phosphorylation. *Toxicol Appl Pharmacol*. 1992;115:1–5.
43. Yin Z, Aschner JL, dos Santos AP, Aschner M. Mitochondrial-dependent manganese neurotoxicity in rat primary astrocyte cultures. *Brain Res*. 2008;1203:1–11.
44. Zoratti M, Szabo I. The mitochondrial permeability transition. *Biochim Biophys Acta*. 1995;1241:139–176.
45. Gonzalez LE, Juknat AA, Venosa AJ, Verrengia N, Kotler ML. Manganese activates the mitochondrial apoptotic pathway in rat astrocytes by modulating the expression of proteins of the Bcl-2 family. *Neurochem Int*. 2008;53:408–415.
46. Aschner M, Gannon M, Kimelberg HK. Manganese uptake and efflux in cultured rat astrocytes. *J Neurochem*. 1992;58:730–735.
47. Wedler FC, Ley BW, Grippo AA. Manganese(II) dynamics and distribution in glial cells cultured from chick cerebral cortex. *Neurochem Res*. 1989;14:1129–1135.
48. Filipov NM, Seegal RF, Lawrence DA. Manganese potentiates in vitro production of proinflammatory cytokines and nitric oxide by microglia through a nuclear factor kappa B-dependent mechanism. *Toxicol Sci*. 2005;84:139–148.
49. Chen CJ, Ou YC, Lin SY, Liao SL, Chen SY, Chen JH. Manganese modulates pro-inflammatory gene expression in activated glia. *Neurochem Int*. 2006;49:62–71.
50. Hirsch EC, Hunot S, Damier P, Faucheux B. Glial cells and inflammation in Parkinson's disease: a role in neurodegeneration? *Ann Neurol*. 1998;44:S115–S120.
51. Desole MS, Sciola L, Delogu MR, Sircana S, Migheli R, Miele E. Role of oxidative stress in the manganese and 1-methyl-4-(2′-ethylphenyl)-1,2,3,6-tetrahydropyridine-induced apoptosis in PC12 cells. *Neurochem Int*. 1997;31:169–176.
52. Erikson K, Aschner M. Manganese causes differential regulation of glutamate transporter (GLAST) taurine transporter and metallothionein in cultured rat astrocytes. *Neurotoxicology*. 2002;23:595–602.
53. Sitcheran R, Gupta P, Fisher PB, Baldwin AS. Positive and negative regulation of EAAT2 by NF-kappaB: a role for N-myc in TNFalpha-controlled repression. *EMBO J*. 2005;24:510–520.
54. Su ZZ, Leszczyniecka M, Kang DC, et al. Insights into glutamate transport regulation in human astrocytes: cloning of the promoter for excitatory amino acid transporter 2 (EAAT2). *Proc Natl Acad Sci U S A*. 2003;100:1955–1960.
55. Gupta SK, Murthy RC, Chandra SV. Neuromelanin in manganese-exposed primates. *Toxicol Lett*. 1980;6:17–20.
56. Yamada M, Ohno S, Okayasu I, et al. Chronic manganese poisoning: a neuropathological study with determination of manganese distribution in the brain. *Acta Neuropathol*. 1986;70:273–278.
57. Guilarte TR. APLP1, Alzheimer's-like pathology and neurodegeneration in the frontal cortex of manganese-exposed non-human primates. *Neurotoxicology*. 2010;31:572–574.
58. Prabhakaran K, Chapman GD, Gunasekar PG. alpha-Synuclein overexpression enhances manganese-induced neurotoxicity through the NF-kappaB-mediated pathway. *Toxicol Mech Meth*. 2011;21:435–443.
59. Latchoumycandane C, Anantharam V, Kitazawa M, Yang Y, Kanthasamy A, Kanthasamy AG. Protein kinase Cdelta is a key downstream mediator of manganese-induced apoptosis in dopaminergic neuronal cells. *J Pharmacol Exp Ther*. 2005;313:46–55.
60. Kessler KR, Wunderlich G, Hefter H, Seitz RJ. Secondary progressive chronic manganism associated with markedly decreased striatal D2 receptor density. *Mov Disord*. 2003;18:217–218.
61. Shinotoh H, Snow BJ, Hewitt KA, et al. MRI and PET studies of manganese-intoxicated monkeys. *Neurology*. 1995;45:1199–1204.
62. Shinotoh H, Snow BJ, Chu NS, et al. Presynaptic and postsynaptic striatal dopaminergic function in patients with manganese intoxication: a positron emission tomography study. *Neurology*. 1997;48:1053–1056.
63. Kim Y, Kim JM, Kim JW, et al. Dopamine transporter density is decreased in parkinsonian patients with a history of manganese exposure: what does it mean? *Mov Disord*. 2002;17:568–575.
64. Anderson JG, Fordahl SC, Cooney PT, Weaver TL, Colyer CL, Erikson KM. Extracellular norepinephrine, norepinephrine receptor and transporter protein and mRNA levels are differentially altered in the developing rat brain due to dietary iron deficiency and manganese exposure. *Brain Res*. 2009;1281:1–14.
65. Anderson JG, Fordahl SC, Cooney PT, Weaver TL, Colyer CL, Erikson KM. Manganese exposure alters extracellular GABA, GABA receptor and transporter protein and mRNA levels in the developing rat brain. *Neurotoxicology*. 2008;29:1044–1053.
66. Anderson JG, Cooney PT, Erikson KM. Brain manganese accumulation is inversely related to gamma-amino butyric acid uptake in male and female rats. *Toxicol Sci*. 2007;95:188–195.
67. Huang P, Li G, Chen C, et al. Differential toxicity of Mn^{2+} and Mn^{3+} to rat liver tissues: oxidative damage, membrane fluidity and histopathological changes. *Exp Toxicol Pathol*. 2012;64:197–203.
68. Witzleben CL, Boyer JL, Ng OC. Manganese-bilirubin cholestasis. Further studies in pathogenesis. *Lab Invest*. 1987;56:151–154.
69. Ebert BL, Bunn HF. Regulation of the erythropoietin gene. *Blood*. 1999;94:1864–1877.
70. Goldberg MA, Dunning SP, Bunn HF. Regulation of the erythropoietin gene: evidence that the oxygen sensor is a heme protein. *Science*. 1988;242:1412–1415.
71. Goldwasser E, Jacobson LO, Fried W, Plzak LF. Studies on erythropoiesis. V. The effect of cobalt on the production of erythropoietin. *Blood*. 1958;13:55–60.

72. Van Den Eeden SK, Tanner CM, Bernstein AL, et al. Incidence of Parkinson's disease: variation by age, gender, and race/ethnicity. *Am J Epidemiol*. 2003;157:1015–1022.
73. Stefanova N, Bucke P, Duerr S, Wenning GK. Multiple system atrophy: an update. *Lancet Neurol*. 2009;8:1172–1178.
74. Wenning GK, Ben Shlomo Y, Magalhaes M, Daniel SE, Quinn NP. Clinical features and natural history of multiple system atrophy. An analysis of 100 cases. *Brain*. 1994;117(Pt 4):835–845.
75. Litvan I. Update on progressive supranuclear palsy. *Curr Neurol Neurosci Rep*. 2004;4:296–302.
76. Seneca S, Fagnart D, Keymolen K, et al. Early onset Huntington disease: a neuronal degeneration syndrome. *Eur J Pediatr*. 2004;163:717–721.
77. Bamford KA, Caine ED, Kido DK, Cox C, Shoulson I. A prospective evaluation of cognitive decline in early Huntington's disease: functional and radiographic correlates. *Neurology*. 1995;45:1867–1873.
78. Armstrong MJ, Miyasaki JM. Evidence-based guideline: pharmacologic treatment of chorea in Huntington disease: report of the guideline development subcommittee of the American Academy of Neurology. *Neurology*. 2012;79:597–603.
79. Dallocchio C, Buffa C, Tinelli C, Mazzarello P. Effectiveness of risperidone in Huntington chorea patients. *J Clin Psychopharmacol*. 1999;19:101–103.
80. Bonelli RM, Wenning GK. Pharmacological management of Huntington's disease: an evidence-based review. *Curr Pharm Des*. 2006;12:2701–2720.
81. Stremmel W, Meyerrose KW, Niederau C, Hefter H, Kreuzpaintner G, Strohmeyer G. Wilson disease: clinical presentation, treatment, and survival. *Ann Intern Med*. 1991;115:720–726.
82. Lorincz MT. Neurologic Wilson's disease. *Ann N Y Acad Sci*. 2010;1184:173–187.
83. Steindl P, Ferenci P, Dienes HP, et al. Wilson's disease in patients presenting with liver disease: a diagnostic challenge. *Gastroenterology*. 1997;113:212–218.
84. Phukan J, Albanese A, Gasser T, Warner T. Primary dystonia and dystonia-plus syndromes: clinical characteristics, diagnosis, and pathogenesis. *Lancet Neurol*. 2011;10:1074–1085.
85. Trender-Gerhard I, Sweeney MG, Schwingenschuh P, et al. Autosomal-dominant GTPCH1-deficient DRD: clinical characteristics and long-term outcome of 34 patients. *J Neurol Neurosurg Psychiatry*. 2009;80:839–845.
86. Segawa M, Nomura Y, Nishiyama N. Autosomal dominant guanosine triphosphate cyclohydrolase I deficiency (Segawa disease). *Ann Neurol*. 2003;54(suppl 6):S32–S45.
87. Zimprich A, Grabowski M, Asmus F, et al. Mutations in the gene encoding epsilon-sarcoglycan cause myoclonus-dystonia syndrome. *Nat Genet*. 2001;29:66–69.
88. Dobyns WB, Ozelius LJ, Kramer PL, et al. Rapid-onset dystonia-parkinsonism. *Neurology*. 1993;43:2596–2602.
89. Aschner M, Erikson KM, Herrero Hernandez E, Tjalkens R. Manganese and its role in Parkinson's disease: from transport to neuropathology. *Neuromolecular Med*. 2009;11:252–266.
90. Apostoli P, Lucchini R, Alessio L. Are current biomarkers suitable for the assessment of manganese exposure in individual workers? *Am J Ind Med*. 2000;37:283–290.
91. Ohtake T, Negishi K, Okamoto K, et al. Manganese-induced Parkinsonism in a patient undergoing maintenance hemodialysis. *Am J Kidney Dis*. 2005;46:749–753.
92. Herrero Hernandez E, Valentini MC, Discalzi G. T1-weighted hyperintensity in basal ganglia at brain magnetic resonance imaging: are different pathologies sharing a common mechanism? *Neurotoxicology*. 2002;23:669–674.
93. Fitsanakis VA, Zhang N, Avison MJ, Gore JC, Aschner JL, Aschner M. The use of magnetic resonance imaging (MRI) in the study of manganese neurotoxicity. *Neurotoxicology*. 2006;27:798–806.
94. Kim EA, Cheong HK, Choi DS, et al. Effect of occupational manganese exposure on the central nervous system of welders: 1H magnetic resonance spectroscopy and MRI findings. *Neurotoxicology*. 2007;28:276–283.
95. Nelson K, Golnick J, Korn T, Angle C. Manganese encephalopathy: utility of early magnetic resonance imaging. *Br J Ind Med*. 1993;50:510–513.
96. Kim SH, Chang KH, Chi JG, et al. Sequential change of MR signal intensity of the brain after manganese administration in rabbits. Correlation with manganese concentration and histopathologic findings. *Invest Radiol*. 1999;34:383–393.
97. Lamas GA, Goertz C, Boineau R, et al. Design of the trial to assess chelation therapy (TACT). *Am Heart J*. 2012;163:7–12.
98. Angle CR. Dimercaptosuccinic acid (DMSA): negligible effect on manganese in urine and blood. *Occup Environ Med*. 1995;52:846.
99. Tuschl K, Clayton PT, Gospe SM, Mills PB. Dystonia/Parkinsonism, hypermanganesemia, polycythemia, and chronic liver disease. In: Pagon RA, Adam MP, Bird TD, Dolan CR, Fong CT, Stephens K, eds. *GeneReviews*. Seattle (WA): University of Washington; 1993. Available at: http://www.ncbi.nlm.nih.gov/books/NBK100241/.
100. Brna P, Gordon K, Dooley JM, Price V. Manganese toxicity in a child with iron deficiency and polycythemia. *J Child Neurol*. 2011;26:891–894.
101. Sahni V, Leger Y, Panaro L, et al. Case report: a metabolic disorder presenting as pediatric manganism. *Environ Health Perspect*. 2007;115:1776–1779.

CHAPTER

45

Aceruloplasminemia

Satoshi Kono and Hiroaki Miyajima

Hamamatsu University School of Medicine, Hamamatsu, Japan

CLINICAL FEATURES

In 1987, we described the first case of aceruloplasminemia as a form of familial apoceruloplasmin deficiency. The patient was a 52-year-old Japanese female suffering from blepharospasm, retinal degeneration, and diabetes mellitus.[1] Subsequent evaluations revealed the complete absence of serum ceruloplasmin, the presence of mild anemia, a low plasma iron concentration, an elevated plasma ferritin level, and significant iron accumulation in the basal ganglia and liver on T2-weighted magnetic resonance imaging (MRI). Ferrokinetics and a histochemical study showed accumulation of iron, but not copper, in the liver and brain, although ceruloplasmin is a multicopper oxidase harboring six copper ions. Careful family studies in this original case showed that the lack of serum ceruloplasmin was inherited although in an autosomal recessive fashion. A genetic analysis of the ceruloplasmin gene revealed that this patient was homozygous, with a five-base insertion in exon 7, resulting in a frameshift mutation and a truncated open reading frame.[2] The clinical findings and identification of a mutation in the ceruloplasmin gene confirmed that the disorder was a novel disorder of iron metabolism resulting from a lack of ceruloplasmin in the serum. The disorder was termed aceruloplasminemia (MIM 604290). Treatment with the iron chelator desferrioxamine decreased the brain iron stores, prevented the progression of neurological symptoms and reduced the level of plasma lipid peroxidation.[3] The identification of aceruloplasminemia implies that there is a direct connection between iron accumulation in the brain and liver and that ceruloplasmin plays an essential role in the transport and/or metabolism of iron, but not copper, despite its need for copper to complete its functions. An epidemiological study in Japan demonstrated that the prevalence of aceruloplasminemia was estimated to be approximately 1 per 2,000,000 in nonconsanguineous marriages, and subsequent studies have now identified more than 35 affected families from around the world.[4]

The clinical diagnosis should be made based on the complete absence of serum ceruloplasmin and the identification of a mutation in the ceruloplasmin gene. There are two variants associated with mutations in the ceruloplasmin gene. The first variant involves a low amount of ceruloplasmin detected in the serum, as reported in one case.[5] In that case, the patient presented with the typical symptoms of aceruloplasminemia, including asymptomatic hepatic iron overload, retinal degeneration, and diabetes mellitus. MRI of the liver and basal ganglia showed T2-hypointensity signals associated with parenchymal iron accumulation. A gene analysis disclosed a G969S homozygous mutation in the ceruloplasmin gene. An immunoblot analysis of serum ceruloplasmin revealed only apo-form of ceruloplasmin without ferroxidase activity. Another patient with compound heterozygous mutations of R882X and H978Q also presented with typical symptoms, including hepatic and brain iron accumulation.[6] The serum ceruloplasmin level was half the normal value, and apo- and holo-ceruloplasmin were detected in an immunoblot analysis. However, no serum ferroxidase activity was detected in the patient. The patient's mother was heterozygous for the H978Q mutation, and her serum ceruloplasmin level was normal with a level of ferroxidase activity that was half the normal value. Therefore, the serum ceruloplasmin present in H978Q mutation carriers is speculated to be devoid of ferroxidase activity, resulting in the development of aceruloplasminemia in spite of the detection of ceruloplasmin in the serum.

Rosenberg's Molecular and Genetic Basis of Neurological and Psychiatric Disease
http://dx.doi.org/10.1016/10.1016/B978-0-12-410529-4.00045-0

The second variant is symptomatic heterozygous disease. Aceruloplasminemia is an autosomal recessive inherited disease, and heterozygous individuals with a partial ceruloplasmin deficiency may have normal iron metabolism and no clinical symptoms. The first report of symptomatic heterozygous disease showed three Japanese patients from two families with half the normal ceruloplasmin levels in the serum, and who developed cerebellar ataxia from the fourth decade of life.[7] They were all heterozygous for a W858X mutation, and their serum iron concentrations and transferrin saturation levels were normal. At autopsy, pathological and biochemical examinations showed marked loss of Purkinje cells, a large amount of iron deposition in the cerebellum, and small deposits in the basal ganglia, thalamus, and liver. The W858X mutation is frequently detected in Japanese patients; however, most of the carriers heterozygous for the W858X mutation are asymptomatic.[8,9] The second report of a symptomatic heterozygous patient was a young patient who presented with subacute progressive extrapyramidal movement disorders.[10] Although her brain MRI showed no iron accumulation, the iron content in a liver biopsy specimen exceeded the normal expected range. A genetic analysis of her ceruloplasmin gene revealed that she was heterozygous for a R701W mutation. However, her father, who was also heterozygous for the R701W mutation, was asymptomatic and the pathological effects of the mutation on the neurological symptoms are unclear.

These findings have important diagnostic implications, indicating that the presence of ceruloplasmin in the serum of patients with the typical clinical features of aceruloplasminemia requires ceruloplasmin gene analysis before the diagnosis of aceruloplasminemia can be conclusively ruled out.

LABORATORY TESTING

Aceruloplasminemia patients present in the fourth or fifth decade of life with neurological symptoms.[11,12] These neurological features are usually progressive at the time of diagnosis, and are associated with the iron accumulation in the basal ganglia and cerebellum as detected on T2-weighted MRI (Figure 45.1). Ophthalmological examinations usually reveal evidence of peripheral retinal degeneration secondary to iron accumulation and photoreceptor cell loss (Figure 45.2A). Although the neurological features dominate the clinical features in most patients, all individuals have evidence of systemic iron accumulation at the time of diagnosis. The laboratory findings demonstrated microcytic anemia, decreased serum iron content and an increased serum ferritin concentration, usually greater than 1000 ng per ml. T2-weighted MRI of the liver shows low intensity signals associated with the iron accumulation (Figure 45.2B). Liver biopsy samples reveal normal hepatic architecture and histology without cirrhosis or fibrosis; however, they do demonstrate excess iron accumulation (>1200 μg per g dry weight) within hepatocytes and reticuloendothelial cells (Figure 45.2C).

Aceruloplasminemia patients also present with diabetes or evidence of abnormal glucose tolerance. Autopsy studies have revealed significant iron accumulation within the endocrine portion of the pancreas, with marked diminution in the β cell population within the islets of Langerhans.[13–15] Thus, the diagnosis of aceruloplasminemia in a symptomatic individual relies upon the demonstration of the complete absence of serum ceruloplasmin and abnormal laboratory findings, as well as MRI findings suggesting iron overload in both the liver and brain. The neuroimaging studies in aceruloplasminemia patients are strongly supported by the characteristic MRI findings of abnormal low intensities reflecting iron accumulation in the liver and brain, including the basal ganglia, thalamus and dentate nucleus on both T1- and T2-weighted images. Functional neuroimaging studies using fluorodeoxyglucose (FDG)-positron emission tomography (PET) demonstrated hypometabolism in the basal ganglia.[16–18] The clinical characteristics of the patients are summarized in Table 45.1. In general, aceruloplasminemia patients present neurological symptoms, including extrapyramidal signs, in the fourth or fifth decade of life. Although the neurological findings dominate the clinical features in most patients, some patients have been recognized prior to the onset of neurological symptoms due to biochemical abnormalities indicating changes in iron metabolism, the presence of diabetes or evidence of abnormal glucose tolerance and abnormal MRI findings of the liver and the brain.[19–22]

The molecular diagnosis of aceruloplasminemia is usually made based on a sequence analysis of the ceruloplasmin gene using genomic DNA derived from leukocytes. The presence of a processed pseudogene on chromosome 8 encoding the carboxyl-terminal 563 amino acids of this protein must be taken into account when designing polymerase chain reaction (PCR) primers for molecular diagnostic testing.[23] When an intronic mutation that can affect the acceptor or donor splice site is found, a sequence analysis of mRNA using reverse transcriptase PCR should be performed to confirm the production of an aberrant transcript. The use of reverse transcriptase PCR is preferable to conducting assessments of liver samples, the predominant source of the secreted form of ceruloplasmin, because few ceruloplasmin mRNAs are expressed in leukocytes.

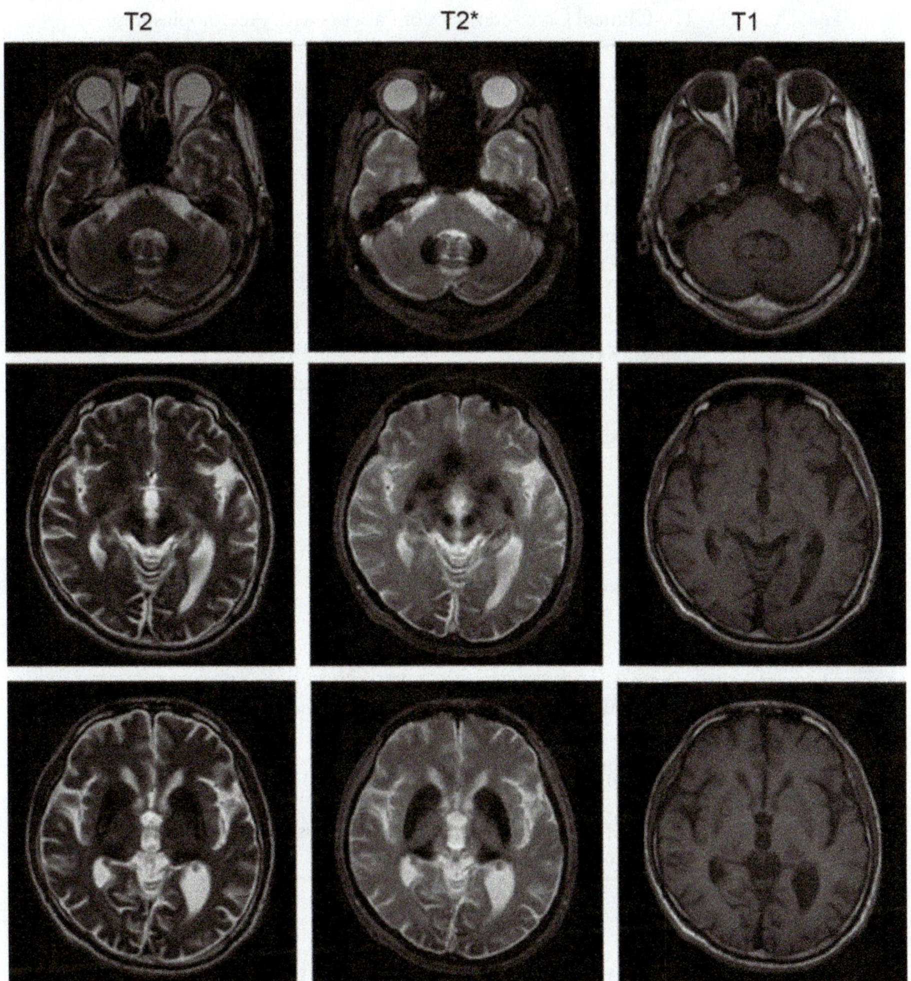

FIGURE 45.1 **Magnetic resonance images (MRI) of an aceruloplasminemia patient.** T1, T2 and T2*-weighted axial images of the brain showed signal attenuation of the dentate nucleus of the cerebellum, globus pallidum, putamen, caudate nucleus, and thalamus.

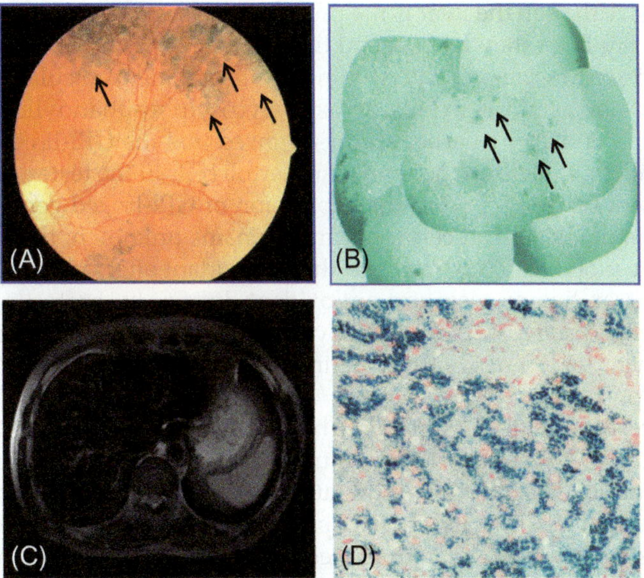

FIGURE 45.2 Ophthalmoscopic findings show several small, yellowish areas of opacity (arrows) scattered over grayish areas of atrophy in the retinal pigment epithelium (**A**). Fluorescein angiography demonstrates window defects (arrows) corresponding to yellowish areas of opacity (**B**). T2-weighted axial images of the liver also showed signal attenuation (**C**). A liver biopsy specimen stained with Perl stain showed iron in the hepatocytes (original magnification ×200) (**D**).

TABLE 45.1 The Clinical Characteristics of Patients with Aceruloplasminemia

Clinical manifestations in 71 patients with aceruloplasminemia:
Anemia (80%)
Retinal degeneration (76%)
Diabetes mellitus (70%)
Neurological symptoms (68%)
1. Ataxia (71%): dysarthria > gait ataxia > limb ataxia
2. Involuntary movement (64%): dystonia (blepharospasm, grimacing, neck dystonia) > chorea > tremors
3. Parkinsonism (20%): rigidity > akinesia
4. Cognitive dysfunction (60%): apathy > forgetfulness

Onset of clinical manifestations:
Diabetes mellitus: under 30 years old, 18%; 30–39 years old, 35%; 40–49 years old, 31%; over 50 years old, 16%
Neurological symptoms: under 40 years old, 7%; 40–49 years old, 38%; 50–59 years old, 42%; over 60 years old, 13%

Laboratory findings:
Undetectable serum ceruloplasmin
Elevated serum ferritin
Decreased serum iron, iron-refractory microcytic anemia
Low serum copper and normal urinary copper levels

MRI (magnetic resonance imaging) findings:
Low intensity on both T1- and T2-weighted MRI in the liver and the basal ganglia, including the caudate nucleus, putamen and pallidum, and the thalamus

Liver biopsy results:
Excess iron accumulation (>1000 μg/g dry weight) within hepatocytes and reticuloendothelial cells
Normal hepatic architecture and histology without cirrhosis or fibrosis
Normal copper accumulation

MOLECULAR GENETICS

Ceruloplasmin is a single-copy gene on chromosome 3 in the human genome. The human ceruloplasmin gene contains 20 exons with total length of about 65 kb.[24] The ceruloplasmin secreted into the plasma is considered to be involved in iron homeostasis. Although the liver is the predominant source of serum ceruloplasmin, the extrahepatic expression of ceruloplasmin has been shown in several tissues, including the central nervous system (CNS).[25] In the brain, ceruloplasmin is expressed in the astrocytes lining the brain microvasculature located in the basal ganglia, where a distinct form of ceruloplasmin is expressed as a glycosylphosphatidylinositol (GPI)-linked form by the alternative splicing of exons 19 and 20.[24,26] Earlier studies showed that the GPI-linked ceruloplasmin was located in leptomeningeal cells, the Müller glial cells in the retina, the Sertoli cells in the testes, and the Schwann cells in peripheral nerves;[27,28] however, recent studies reported the expression of GPI-linked ceruloplasmin in various tissues.[29,30] Although the precise function of GPI-linked ceruloplasmin remains unknown, the GPI-linked ceruloplasmin likely plays an important role in the mobilization of iron and the antioxidant effects in the CNS.[31,32] GPI-linked ceruloplasmin may be associated with iron homeostasis and antioxidant defense by protecting the CNS from iron-mediated free radical injury. The ferroxidase activity of GPI-linked ceruloplasmin is also essential for the stability of cell surface ferroportin.[29,33] The requirement for a ferroxidase to maintain iron transport activity represents a novel mechanism of regulating cellular iron export.

The genetic analyses of aceruloplasminemia patients have identified more than 40 distinct mutations in the ceruloplasmin gene (Figure 45.3).[19,5,6,8–10,20,21,34–51] Most of the mutations detected are unique to specific families, where there is often a history of consanguinity. The majority of mutations are truncated mutations leading to the formation of a premature stop codon. The ferroxidase activity of ceruloplasmin is dependent upon the trinuclear copper cluster, the ligands for which are encoded by exon 18.[52] The truncated mutations identified are predicted to result in the formation of a protein lacking the copper cluster sites presumed to be critical for enzymatic function. The symptoms, onset, and prognosis in single cases have demonstrated that there is no genotype–phenotype association.[12] This finding suggests that unknown genetic or environmental factors may regulate iron accumulation in the brain.

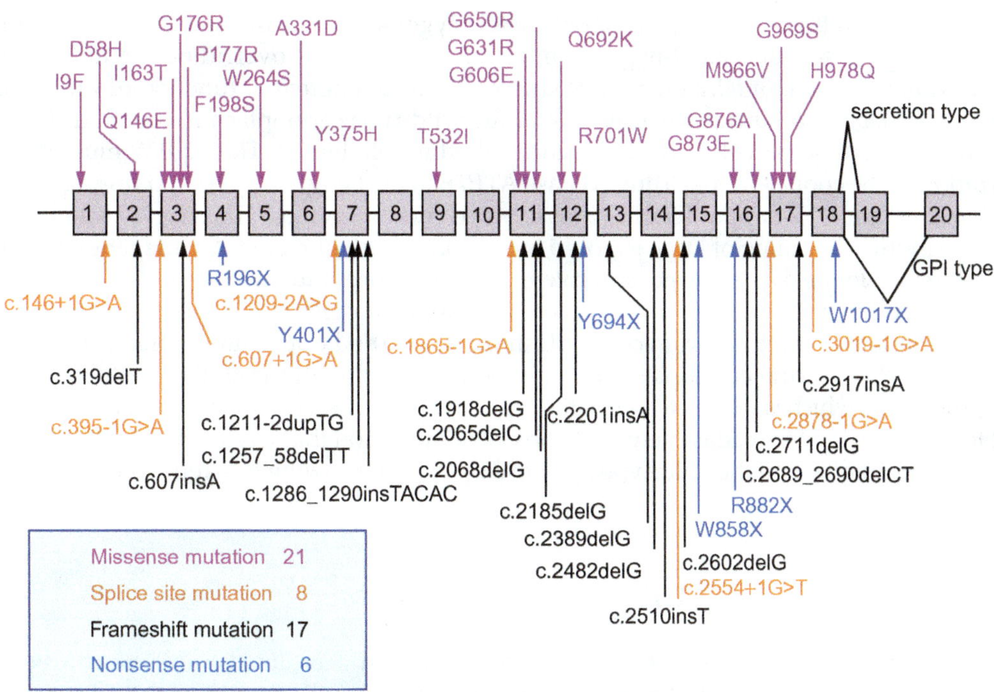

FIGURE 45.3 **Genetic mutations characterized in patients with aceruloplasminemia and their family members.** The indicated mutations are referenced in the text and some mutations are included in our unpublished data.

Biosynthesis studies of mutant ceruloplasmin in mammalian cell culture systems without endogenous ceruloplasmin were performed to investigate the molecular pathogenesis of aceruloplasminemia.[29,37,40,42,49,53] The biosynthesis studies of missense mutants revealed three distinct pathological mechanisms (Figure 45.4). A first group comprising the I9F, G176R, P177R, D58H, F198S, W264S, A331D, G606E and G873E mutants were retained in the endoplasmic reticulum (ER).[29,40,49] The mutants presumably result from the misfolding of ceruloplasmin in the ER. The amino acid sequence of a G(FLI) (LI)GP repeat motif is believed to affect folding during the early secretory pathway.[49] The G873E, G176R, and P177R mutants affect the conserved repeated G(FLI) (LI)GP motif, which is consistent with this hypothesis, whereas the G876A mutant was not retained in the ER.[29] The mutants located beside this motif are speculated to have other molecular mechanisms involved in the cellular trafficking of ceruloplasmin. A second group of mutants, including the G631R, Q692K, M966V and G969S mutants, was synthesized and all were found to be secreted with normal kinetics, but failed to incorporate copper during the late secretory pathway, resulting in apo-ceruloplasmin.[5,29,37] The G631R and G969S mutations are located in the nearby type I copper-binding His637 and His975 sites, respectively. The Q692K and M966V mutations are also located near the type I copper-binding sites of M690 and His975, respectively. A site-directed mutagenesis analysis of the type I copper-binding site indicated that these mutants failed to incorporate copper into the apo-ceruloplasmin.[37] These biochemical studies demonstrated that the type I copper-binding site did not affect either the protein folding for intracellular trafficking from the ER to the Golgi body or the subsequent protein secretion from the cell. However, the copper-binding site may play an essential role in the protein structure for copper incorporation into the apo-ceruloplasmin. The third group containing the Y356H, R701W, and G876A mutants reconstituted both the apo- and holo-proteins, and were secreted extracellularly.[29] However, the mutants had impaired ferroxidase activity, which is required for ferroportin stability. These mutants may have altered iron-binding sites or changes in the trinuclear copper cluster, which are essential for the oxidase activity of the protein. It will be necessary to analyze the crystal structure of the mutant ceruloplasmin proteins in order to obtain insight into the mechanism of ferroxidase activity.

A biogenesis study of nonsense mutations, including Y694X, W858X and R882X, demonstrated that the Y694X and W858X mutants were retained in the ER, while the R882X mutant was secreted.[42] Subsequent site-directed mutagenesis analyses revealed that the truncated mutant containing the cysteine residue at amino acid 881(Cys-881) was able to pass through the ER and was secreted, while the truncated mutant protein without Cys-881 appeared to accumulate in the ER, leading to ER stress, and eventually resulting in cell death. Thus, Cys-881 is necessary for the secretion of almost all of the truncated ceruloplasmin, although a recent biogenesis study of the W1017X mutant with Cys-881 showed that the mutant was exclusively retained in the ER.[54]

The observations made in the cases of symptomatic heterozygous patients indicate that the specific mutations of W858X and R701W may cause a dominant-negative effect of the mutations overriding the influence of ceruloplasmin function in iron metabolism. *In vitro* biogenesis studies showed a potential function of the dominant-negative effect of mutant ceruloplasmin occurring via silencing of the wild-type ceruloplasmin function. The W858X mutant accumulated in the ER, leading to the ER stress, which resulted in cell death.[40] The R701W mutant induced the subcellular relocalization of the copper-transporting ATPase ATP7B in the Golgi complex and fragmentation of the Golgi complex, resulting in a failure of copper loading in wild-type ceruloplasmin.[53] The Arg701 site is located in one of the repeat CX(R/K) motifs consisting of large exposed loops connecting domains. A mutagenesis study of the motifs revealed that the external loops play an important role in copper incorporation.

The clinical phenotype in most patients shows little variation, regardless of the specific mutation.[12] While almost all patients have a complete absence of serum ceruloplasmin, H978Q and G969S mutations were reported in patients who had detectable levels of serum ceruloplasmin, but who presented with the clinical features of aceruloplasminemia. The H978Q mutation, which is located at one of the type I copper-binding sites and constitutes holo-ceruloplasmin, was speculated to be devoid of ferroxidase activity.[6] The G969S mutation that led to apo-ceruloplasmin in the serum was suggested to be less fragile than the wild-type apo-ceruloplasmin, allowing it to remain in circulation for a longer period of time.[5]

DISEASE MECHANISMS

The abnormalities of iron homeostasis observed in patients with aceruloplasminemia can be understood by considering the cellular physiology of systemic iron metabolism. A small amount of total iron is delivered from absorption in enterocytes while a large amount of the iron arising from recycling of the heme iron from aging red blood cells is turned over within the reticuloendothelial system. The recycled iron is released from endothelial cells in the liver and the spleen, and binds to the transferrin in the plasma, resulting in its return to the bone marrow for erythropoiesis. Ceruloplasmin functions as an important factor in the iron cycle by performing iron oxidation, which is required to sustain the iron release and uptake by transferrin (Figure 45.4). The lack of plasma ferroxidase activity in ceruloplasmin results in increased extracellular ferrous iron, which is rapidly taken up into cells. Pathological studies in patients with aceruloplasminemia showed that iron accumulates within hepatocytes, pancreatic endocrine cells, and astrocytes.[13,14,55] Although the mechanism underlying the neurodegeneration in aceruloplasminemia has not been clarified, its pathogenesis is presumably secondary to consistent accumulation of iron within neurons and astrocytes. The electronic properties of iron enable the metal to take part in chemical reactions because the Fenton catalysis of iron plays an important role in cellular redox chemistry by reducing H_2O_2 to the highly cytotoxic hydroxyl (OH•) radical, which may be injurious to neural and other cellular substrates. The antioxidant activity of ceruloplasmin can be mainly ascribed to its ferroxidase activity, which effectively inhibits ferrous ion-stimulated lipid peroxidation and ferrous ion-dependent formation of hydroxyl radicals in the Fenton reaction. A direct role for iron in oxidant-mediated neuronal injury is supported by findings of increased lipid peroxidation and subsequent mitochondrial dysfunction in the brain tissues, cerebral spinal fluid, and erythrocytes of aceruloplasminemia patients.[56–61]

The pathological findings in the brain showed severe iron deposition in both the astrocytes and neurons, and neuronal loss in the same regions associated with the highest iron accumulation and necrosis.[13,14] The neurodegenerative changes were observed in the cerebral cortex, as well as in the basal ganglia, dentate nuclei, and cerebellar cortices. The distribution in order of the iron level is the globus pallidus > putamen > cerebellar cortex and cerebral cortex.[11] The characteristic histopathological findings of the patients were deformed astrocytes and globular structures, which were observed more frequently in the striatum than in the cerebral cortex, which occurred in parallel with significant iron deposition and neuronal loss (Figure 45.5).[55,62] The globular structures were immunologically reactive for a glial marker protein, suggesting that they were ballooned foot processes of astrocytes. The deformed astrocytes were more frequently observed in the basal ganglia in which marked iron deposition was observed. Glial fibrillary acidic protein (GFAP) is one of the proteins most severely modified by oxidative stress in the brains of aceruloplasminemia patients.[63] Intense ferrous iron deposition was demonstrated in the terminal astrocytic processes and the globular structures.[64] The morphological changes of astrocytes may be related to iron-induced tissue damage.

The iron accumulation was observed in neurons as well as astrocytes. This finding indicates that the neurons take up significant amounts of iron due to alternative sources of non-transferrin-bound iron complexed to molecules, such as citrate and ascorbate, because astrocytes without any expression of ceruloplasmin are not able to transport iron to transferrin that binds to transferrin receptor 1 on neurons. A recent pathological study of a murine model of aceruloplasminemia showed that the neuronal cell loss may result from iron deficiency in regions where the iron in

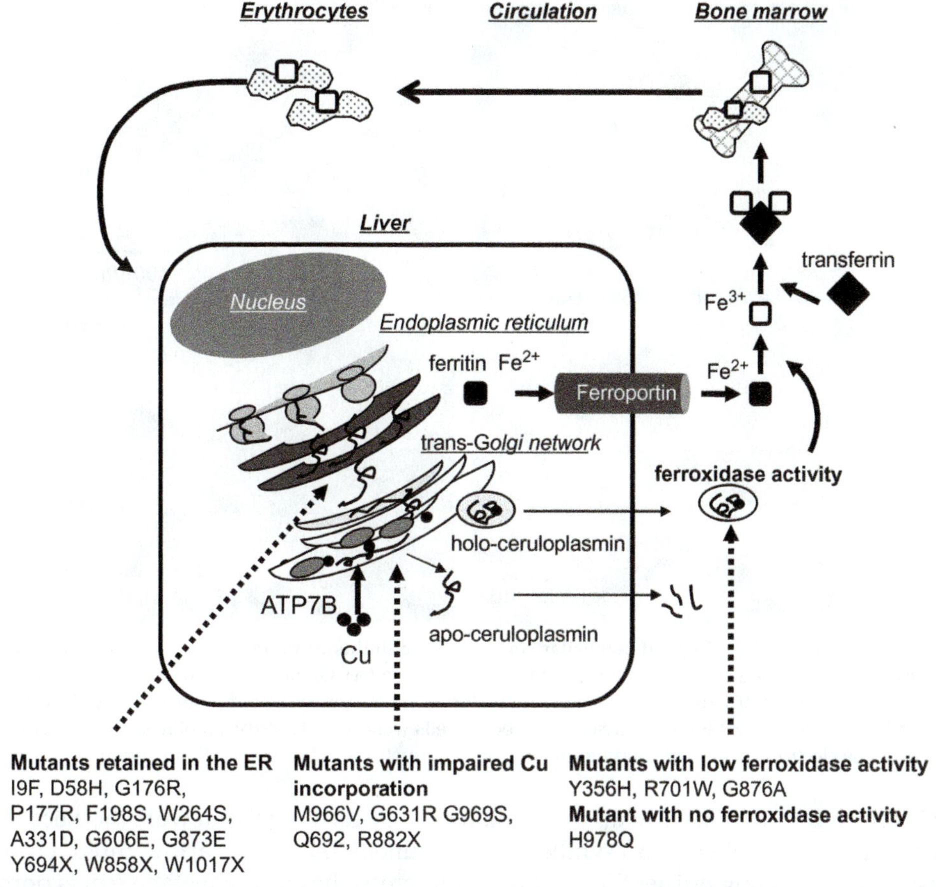

FIGURE 45.4 **A model for the interaction between iron and copper homeostasis in normal subjects and in the aceruloplasminemia patients.** In normal subjects, iron is continuously recycled between the bone marrow and hepatocytes, with serum transferrin acting as a shuttle to deliver iron from hepatocytes to the bone marrow. The role of ceruloplasmin, formed as holo-ceruloplasmin, is as a ferroxidase mediating ferrous iron oxidation and subsequent transfer to transferrin. In normal subjects, copper enters the cell and binds the copper chaperones, which deliver the copper to ATP-7B. The ATP-7B pumps the copper into the trans-Golgi network. Ceruloplasmin is initially synthesized as apo-ceruloplasmin and incorporates the copper into the apo-protein in the Golgi body, resulting in the formation of holo-ceruloplasmin prior to extracellular secretion. Mutant ceruloplasmin biosyntheses were investigated using a cell culture system without endogenous ceruloplasmin. The mutant proteins revealed three distinct pathological mechanisms, including mutants associated with mistrafficking, resulting in the retention of the protein in the ER, mutants altering the intrinsic protein structure, resulting in abrogation of copper incorporation into apo-ceruloplasmin, and mutants with impaired ferroxidase activity.

astrocytes is not able to be mobilized for uptake into neurons, and the excess iron accumulation in astrocytes could also result in oxidative damage to these cells, with subsequent loss of the glial-derived growth factors critical for neurons.[65]

The generation of murine models of aceruloplasminemia provided a critical clue to study the role of ceruloplasmin in iron homeostasis. Three distinct research groups generated ceruloplasmin knockout mice that develop hepatic and reticuloendothelial iron overload. The first knockout mice generated were reported by Harris et al.[66] The mice have an increased iron content, with lipid peroxidation in the brain.[67] However, there is no evidence of neurological symptoms in these mice. Double knockout mice lacking both ceruloplasmin and hephaestin were generated by crossing the ceruloplasmin knockout mice with *sla* mice, which are hephaestin knockout, sex-linked anemic mice. Hephaestin is a ceruloplasmin homolog, and is also a multicopper oxidase with ferroxidase activity, which is abundantly expressed in the neurons in the murine brain, as well as in the enterocytes in the duodenum. The knockout mice lacking both ceruloplasmin and hephaestin exhibited a neurodegenerative phenotype and retinal degeneration consistent with the aceruloplasminemia patients.[67–69] In the mice, hephaestin expression may play a more important role to maintain the cellular redox environment than it does in the CNS in the humans. The iron homeostasis associated with ceruloplasmin was investigated using the ceruloplasmin knockout mice. When the mice were injected with damaged red blood cells in order to induce an increased reticuloendothelial iron overload, or when the mice received a phlebotomy to accelerate reticuloendothelial iron transport to the bone marrow, the mice failed to show

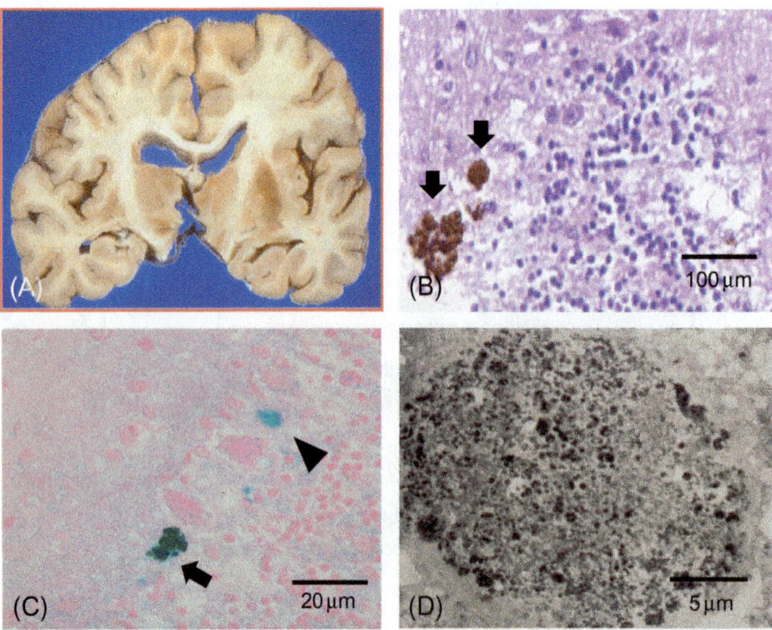

FIGURE 45.5 **The histopathological findings of cerebellar cortices in the brains of patients with aceruloplasminemia.** A coronal section of the brain of the affected patient shows brown pigmentation of the basal ganglia (**A**). Globular structures, indicated by arrows, were seen in the Purkinje cell layer. Many of the globular structures contained brown materials (**B**; hematoxylin and eosin stain). At the cellular layer, iron deposits in Purkinje cells (indicated by an arrowhead) with a decreased number of cells were seen (**C**; Prussian blue stain). The globular structures (arrow) exhibited siderous features. The electron microscopic findings of the globular structures indicated that they contain many electron-dense bodies (**D**).

an increase in their serum iron levels. Only upon injection of holo-ceruloplasmin were the mice able to release iron from a storage compartment for delivery to a synthetic compartment, the bone marrow, thus confirming the essential role for ceruloplasmin in regulating efficient iron efflux. Of interest, the copper metabolism is normal in ceruloplasmin knockout mice, and none of the clinical pathology associated with aceruloplasminemia is secondary to copper deficiency or toxicity.

The second groups of knockout mice were reported by Patel et al.[31] The mice developed a neurodegenerative phenotype and showed increased iron deposition in several regions of the CNS, such as the cerebellum and brainstem. Increased lipid peroxidation due to iron-mediated cellular radical injury was also seen in some regions. Cerebellar neuronal cells from neonatal mice were also more susceptible to oxidative stress *in vitro*. These mice showed deficits in motor coordination that were associated with a loss of brainstem dopaminergic neurons. These results indicate that ceruloplasmin plays an important role in maintaining iron homeostasis in the brain, and in protecting the brain from iron-mediated free radical injury.

The third group of knockout mice were reported by Yamamoto et al.[70] Although the mice showed hepatic iron overload, there was no evidence of iron accumulation in the brain, even after treatment with rotenone, a mitochondrial complex 1 inhibitor that enhances oxidative stress.[71]

Ceruloplasmin plays an essential role in cellular iron efflux by oxidizing the ferrous iron exported from ferroportin. Ferroportin is posttranslationally regulated through internalization triggered by hepcidin binding.[72] Previous studies showed that the ferroxidase activity of GPI-linked ceruloplasmin was essential for the stability of cell surface ferroportin in rat glioma cells lines.[33] The *in vitro* biological analyses suggested that the ceruloplasmin mutants had impaired ferroportin stability on the cell surface, resulting in exacerbated iron accumulation.[29,53] The hepatic expression of ferroportin proteins and the mRNA levels were analyzed to evaluate the involvement of ferroportin in the pathogenesis of aceruloplasminemia.[29] The hepatic ferroportin protein levels were decreased despite the presence of high ferroportin mRNA levels in two aceruloplasminemia patients. Decreased ferroportin protein levels in the liver may be due to degradation due to the absence of ceruloplasmin rather than due to a decreased synthesis of ferroportin at the transcriptional level. Clinical analyses of the hepcidin level in patients with aceruloplasminemia revealed that the serum hepcidin levels and hepatic hepcidin mRNA levels are lower than in control subjects.[29,73] An analysis of ceruloplasmin knockout mice also showed the hepatic hepcidin mRNA levels to decrease in comparison to wild-type and heterozygous mice.[74] The low serum hepcidin levels may induce increased iron absorption in the intestine, where the ceruloplasmin homolog hephaestin retains ferroxidase activity that is involved in basolateral intestinal iron transport. Therefore, the low hepcidin level in the serum and the loss of cell surface ferroportin due to mutant ceruloplasmin may enhance the cellular iron accumulation, contributing to the pathology of aceruloplasminemia.

DIFFERENTIAL DIAGNOSIS

Inherited neurodegenerative disorders termed "neurodegeneration with brain iron accumulation" (NBIA) should be considered in the differential diagnosis of aceruloplasminemia, as the characteristic syndrome of NBIA presents with progressive extrapyramidal symptoms and excessive iron deposition in the brain, particularly affecting the basal ganglia. The main causes of the syndromes are mutations in neuroaxonal dystrophies: panthothenate kinase-associated neurodegeneration (PKAN, formerly known as Hallervorden–Spatz disease, NBIA1) and PLA2G6-associated neurodegeneration (PLAN, NBIA2). Intensive genetic approaches have identified additional genes that cause other NBIA syndromes, including Kufor–Rakeb disease (NBIA3, PARK9), fatty acid hydroxylase-associated neurodegeneration (FAHM), neuroferritinopathy, mitochondrial membrane protein-associated neurodegeneration (MPAN), Woodhouse–Sakati syndrome and beta-propeller protein-associated neurodegeneration.[75,76] Neuroimaging studies using T2* MRI have shown that a wide range of hypointensity regions, including the basal ganglia, caudate nucleus, putamen, pallidum, and thalamus, are typically observed in aceruloplasminemia patients and can be used to distinguish these patients from those with other NBIA syndromes.[77] Among patients with NBIA syndrome, only those with aceruloplasminemia have abnormal serum ceruloplasmin levels. The level of serum ceruloplasmin is usually found to be decreased in Wilson disease patients with progressive extrapyramidal symptoms. In patients with Wilson disease, the inability to transfer copper into the ceruloplasmin precursor protein apo-ceruloplasmin and a decrease in biliary copper excretion results in serum ceruloplasmin deficiency and excess copper accumulation.[78] The presence of Kayser–Fleischer rings on slit-lamp examinations in conjunction with massive accumulation of hepatic copper on liver biopsies is helpful for diagnosing Wilson disease. Increases in the brain level of copper are detectable on MRI relatively early in the course of the disease. T2-weighted images show increased signal intensity with a central core of decreased intensity in the basal ganglia. Neuroimaging may also prove useful for confirming the diagnosis.

MANAGEMENT

Aceruloplasminemia is a fatal disease, and its early diagnosis and early treatment of patients are issues of paramount importance. Iron-mediated lipid peroxidation and oxidative stress are considered to be the main cause of the neuronal degeneration in aceruloplasminemia patients. To reduce the iron accumulation, systemic iron chelation therapy has been introduced in some patients. Desferrioxamine (deferoxamine) is a high-affinity iron chelator that combines with ferric iron. It has been shown to cross the blood–brain barrier and to promote the excretion of excess iron in patients with inherited and acquired forms of iron overload.[79] The administration of desferrioxamine was effective for reducing the hepatic iron overload and leading to a partial improvement of the neurological symptoms and brain iron accumulation, as reported in a single case report.[3] However, subsequent studies showed little effect of desferrioxamine on the central nervous symptoms, despite normalization of the serum ferritin and hepatic iron concentrations and improvement in the insulin requirement and the regional brain iron levels in T2*-weighted MRI.[21,38,80] Desferrioxamine therapy was often discontinued because of a concomitant decrease in hemoglobin and the serum iron level was observed after several months of therapy, suggesting that desferrioxamine sequestered the iron available for erythropoiesis. Combination therapy with fresh frozen plasma for 6 weeks to replenish the blood ceruloplasmin levels and, thereafter, administration of deferoxamine for an additional 6 weeks to deplete ferric iron stores showed unprecedented improvement in neurological symptoms.[81] Deferiprone, which has a lower molecular weight and more lipophilic properties, had no beneficial effects in a patient.[21] Deferasirox, an oral iron-chelating agent, did not lead to any improvement in the neurological symptoms or brain iron accumulation quantified by MRI,[82,83] while deferasirox therapy has been reported to lead to mild improvement in neurological symptoms, including cognitive performance, gait, and balance in an aceruloplasminemia patient who had no response to both desferrioxamine and fresh–frozen plasma therapy.[84] Short-term iron chelation therapy is therefore effective for reducing the hepatic iron overload and improving the diabetic mellitus, but is ineffective for the treatment of neurological symptoms due to brain iron accumulation. In many reports of single cases, the side effects of the iron chelation therapy prohibited the long-term treatment that may be required to mobilize iron from the brain. However, it seems rational to suggest that the therapy should be initiated early in the course of aceruloplasminemia in order to remove the iron before it induces neurodegeneration.

In comparison with iron chelation therapy, oral zinc sulfate therapy (administered for 1.5 years) led to dramatic neurological improvement in a patient with extrapyramidal and cerebellar-mediated movement disorder caused by a heterozygous mutation in the ceruloplasmin gene.[85] Although the patient was bedridden before the zinc treatment, she was able to stand for a short time and walk a few steps after undergoing this treatment. The antioxidant properties of zinc, as well as its effects on iron absorption, are well established.[86,87] While the mechanisms of antioxidation are not fully understood, the induction of metallothionein synthesis is considered to be one relevant aspect. The zinc therapy could be used as an alternative treatment when iron-chelation therapy is discontinued due to side effects or progression of the symptoms, because the zinc therapy shows no side effects and may ameliorate the neurological symptoms in aceruloplasminemia patients.

References

1. Miyajima H, Nishimura Y, Mizoguchi K, Sakamoto M, Shimizu T, Honda N. Familial apoceruloplasmin deficiency associated with blepharospasm and retinal degeneration. *Neurology*. 1987;37(5):761–767.
2. Harris ZL, Takahashi Y, Miyajima H, Serizawa M, MacGillivray RT, Gitlin JD. Aceruloplasminemia: molecular characterization of this disorder of iron metabolism. *Proc Natl Acad Sci U S A*. 1995;92(7):2539–2543.
3. Miyajima H, Takahashi Y, Kamata T, Shimizu H, Sakai N, Gitlin JD. Use of desferrioxamine in the treatment of aceruloplasminemia. *Ann Neurol*. 1997;41(3):404–407.
4. Miyajima H, Kohno S, Takahashi Y, Yonekawa O, Kanno T. Estimation of the gene frequency of aceruloplasminemia in Japan. *Neurology*. 1999;53(3):617–619.
5. Kono S, Suzuki H, Takahashi K, et al. Hepatic iron overload associated with a decreased serum ceruloplasmin level in a novel clinical type of aceruloplasminemia. *Gastroenterology*. 2006;131(1):240–245.
6. Takeuchi Y, Yoshikawa M, Tsujino T, et al. A case of aceruloplasminaemia: abnormal serum ceruloplasmin protein without ferroxidase activity. *J Neurol Neurosurg Psychiatry*. 2002;72(4):543–545.
7. Miyajima H, Kono S, Takahashi Y, Sugimoto M, Sakamoto M, Sakai N. Cerebellar ataxia associated with heteroallelic ceruloplasmin gene mutation. *Neurology*. 2001;57(12):2205–2210.
8. Takahashi Y, Miyajima H, Shirabe S, Nagataki S, Suenaga A, Gitlin JD. Characterization of a nonsense mutation in the ceruloplasmin gene resulting in diabetes and neurodegenerative disease. *Hum Mol Genet*. 1996;5(1):81–84.
9. Daimon M, Kato T, Kawanami T, et al. A nonsense mutation of the ceruloplasmin gene in hereditary ceruloplasmin deficiency with diabetes mellitus. *Biochem Biophys Res Commun*. 1995;217(1):89–95.
10. Kuhn J, Miyajima H, Takahashi Y, et al. Extrapyramidal and cerebellar movement disorder in association with heterozygous ceruloplasmin gene mutation. *J Neurol*. 2005;252(1):111–113.
11. Miyajima H. Aceruloplasminemia, an iron metabolic disorder. *Neuropathology*. 2003;23(4):345–350.
12. McNeill A, Pandolfo M, Kuhn J, Shang H, Miyajima H. The neurological presentation of ceruloplasmin gene mutations. *Eur Neurol*. 2008;60(4):200–205.
13. Morita H, Ikeda S, Yamamoto K, et al. Hereditary ceruloplasmin deficiency with hemosiderosis: a clinicopathological study of a Japanese family. *Ann Neurol*. 1995;37(5):646–656.
14. Kawanami T, Kato T, Daimon M, et al. Hereditary caeruloplasmin deficiency: clinicopathological study of a patient. *J Neurol Neurosurg Psychiatry*. 1996;61(5):506–509.
15. Kato T, Daimon M, Kawanami T, Ikezawa Y, Sasaki H, Maeda K. Islet changes in hereditary ceruloplasmin deficiency. *Hum Pathol*. 1997;28(4):499–502.
16. Miyajima H, Takahashi Y, Kono S, et al. Glucose and oxygen hypometabolism in aceruloplasminemia brains. *Intern Med*. 2002;41(3):186–190.
17. Haemers I, Kono S, Goldman S, Gitlin JD, Pandolfo M. Clinical, molecular, and PET study of a case of aceruloplasminaemia presenting with focal cranial dyskinesia. *J Neurol Neurosurg Psychiatry*. 2004;75(2):334–337.
18. Miyajima H, Takahashi Y, Kono S, Hishida A, Ishikawa K, Sakamoto M. Frontal lobe dysfunction associated with glucose hypometabolism in aceruloplasminemia. *J Neurol*. 2005;252(8):996–997.
19. Hellman NE, Schaefer M, Gehrke S, et al. Hepatic iron overload in aceruloplasminaemia. *Gut*. 2000;47(6):858–860.
20. Hatanaka Y, Okano T, Oda K, Yamamoto K, Yoshida K. Aceruloplasminemia with juvenile-onset diabetes mellitus caused by exon skipping in the ceruloplasmin gene. *Intern Med*. 2003;42(7):599–604.
21. Mariani R, Arosio C, Pelucchi S, et al. Iron chelation therapy in aceruloplasminaemia: study of a patient with a novel missense mutation. *Gut*. 2004;53(5):756–758.
22. Ogimoto M, Anzai K, Takenoshita H, et al. Criteria for early identification of aceruloplasminemia. *Intern Med*. 2011;50(13):1415–1418.
23. Yang F, Naylor SL, Lum JB, et al. Characterization, mapping, and expression of the human ceruloplasmin gene. *Proc Natl Acad Sci U S A*. 1986;83(10):3257–3261.
24. Patel BN, Dunn RJ, David S. Alternative RNA splicing generates a glycosylphosphatidylinositol-anchored form of ceruloplasmin in mammalian brain. *J Biol Chem*. 2000;275(6):4305–4310.
25. Klomp LW, Farhangrazi ZS, Dugan LL, Gitlin JD. Ceruloplasmin gene expression in the murine central nervous system. *J Clin Invest*. 1996;98(1):207–215.
26. Patel BN, David S. A novel glycosylphosphatidylinositol-anchored form of ceruloplasmin is expressed by mammalian astrocytes. *J Biol Chem*. 1997;272(32):20185–20190.
27. Salzer JL, Lovejoy L, Linder MC, Rosen C. Ran-2, a glial lineage marker, is a GPI-anchored form of ceruloplasmin. *J Neurosci Res*. 1998;54(2):147–157.

28. Mittal B, Doroudchi MM, Jeong SY, Patel BN, David S. Expression of a membrane-bound form of the ferroxidase ceruloplasmin by leptomeningeal cells. *Glia*. 2003;41(4):337–346.
29. Kono S, Yoshida K, Tomosugi N, et al. Biological effects of mutant ceruloplasmin on hepcidin-mediated internalization of ferroportin. *Biochim Biophys Acta*. 2010;1802(11):968–975.
30. Mostad EJ, Prohaska JR. Glycosylphosphatidylinositol-linked ceruloplasmin is expressed in multiple rodent organs and is lower following dietary copper deficiency. *Exp Biol Med (Maywood)*. 2011;236(3):298–308.
31. Patel BN, Dunn RJ, Jeong SY, Zhu Q, Julien JP, David S. Ceruloplasmin regulates iron levels in the CNS and prevents free radical injury. *J Neurosci*. 2002;22(15):6578–6586.
32. Jeong SY, David S. Glycosylphosphatidylinositol-anchored ceruloplasmin is required for iron efflux from cells in the central nervous system. *J Biol Chem*. 2003;278(29):27144–27148.
33. De Domenico I, Ward DM, di Patti MC, et al. Ferroxidase activity is required for the stability of cell surface ferroportin in cells expressing GPI-ceruloplasmin. *EMBO J*. 2007;26(12):2823–2831.
34. Yoshida K, Furihata K, Takeda S, et al. A mutation in the ceruloplasmin gene is associated with systemic hemosiderosis in humans. *Nat Genet*. 1995;9(3):267–272.
35. Yazaki M, Yoshida K, Nakamura A, et al. A novel splicing mutation in the ceruloplasmin gene responsible for hereditary ceruloplasmin deficiency with hemosiderosis. *J Neurol Sci*. 1998;156(1):30–34.
36. Bosio S, De Gobbi M, Roetto A, et al. Anemia and iron overload due to compound heterozygosity for novel ceruloplasmin mutations. *Blood*. 2002;100(6):2246–2248.
37. Hellman NE, Kono S, Mancini GM, Hoogeboom AJ, De Jong GJ, Gitlin JD. Mechanisms of copper incorporation into human ceruloplasmin. *J Biol Chem*. 2002;277(48):46632–46638.
38. Loreal O, Turlin B, Pigeon C, et al. Aceruloplasminemia: new clinical, pathophysiological and therapeutic insights. *J Hepatol*. 2002;36(6):851–856.
39. Perez-Aguilar F, Burguera JA, Benlloch S, Berenguer M, Rayon JM. Aceruloplasminemia in an asymptomatic patient with a new mutation. Diagnosis and family genetic analysis. *J Hepatol*. 2005;42(6):947–949.
40. Kono S, Suzuki H, Oda T, et al. Biochemical features of ceruloplasmin gene mutations linked to aceruloplasminemia. *Neuromolecular Med*. 2006;8(3):361–374.
41. Muroi R, Yagyu H, Kobayashi H, et al. Early onset insulin-dependent diabetes mellitus as an initial manifestation of aceruloplasminaemia. *Diabet Med*. 2006;23(10):1136–1139.
42. Kono S, Suzuki H, Oda T, et al. Cys-881 is essential for the trafficking and secretion of truncated mutant ceruloplasmin in aceruloplasminemia. *J Hepatol*. 2007;47(6):844–850.
43. Fasano A, Bentivoglio AR, Colosimo C. Movement disorder due to aceruloplasminemia and incorrect diagnosis of hereditary hemochromatosis. *J Neurol*. 2007;254(1):113–114.
44. Fasano A, Colosimo C, Miyajima H, Tonali PA, Re TJ, Bentivoglio AR. Aceruloplasminemia: a novel mutation in a family with marked phenotypic variability. *Mov Disord*. 2008;23(5):751–755.
45. Kohno S, Miyajima H, Takahashi Y, Inoue Y. Aceruloplasminemia with a novel mutation associated with parkinsonism. *Neurogenetics*. 2000;2(4):237–238.
46. Daimon M, Susa S, Ohizumi T, et al. A novel mutation of the ceruloplasmin gene in a patient with heteroallelic ceruloplasmin gene mutation (HypoCPGM). *Tohoku J Exp Med*. 2000;191(3):119–125.
47. Okamoto N, Wada S, Oga T, et al. Hereditary ceruloplasmin deficiency with hemosiderosis. *Hum Genet*. 1996;97(6):755–758.
48. Shang HF, Jiang XF, Burgunder JM, Chen Q, Zhou D. Novel mutation in the ceruloplasmin gene causing a cognitive and movement disorder with diabetes mellitus. *Mov Disord*. 2006;21(12):2217–2220.
49. Hellman NE, Kono S, Miyajima H, Gitlin JD. Biochemical analysis of a missense mutation in aceruloplasminemia. *J Biol Chem*. 2002;277(2):1375–1380.
50. Bethlehem C, van Harten B, Hoogendoorn M. Central nervous system involvement in a rare genetic iron overload disorder. *Neth J Med*. 2010;68(10):316–318.
51. Harris ZL, Migas MC, Hughes AE, Logan JI, Gitlin JD. Familial dementia due to a frameshift mutation in the caeruloplasmin gene. *QJM*. 1996;89:355–359.
52. Vachette P, Dainese E, Vasyliev VB, et al. A key structural role for active site type 3 copper ions in human ceruloplasmin. *J Biol Chem*. 2002;277(43):40823–40831.
53. di Patti MC, Maio N, Rizzo G, et al. Dominant mutants of ceruloplasmin impair the copper loading machinery in aceruloplasminemia. *J Biol Chem*. 2009;284(7):4545–4554.
54. Hida A, Kowa H, Iwata A, Tanaka M, Kwak S, Tsuji S. Aceruloplasminemia in a Japanese woman with a novel mutation of CP gene: clinical presentations and analysis of genetic and molecular pathogenesis. *J Neurol Sci*. 2010;298(1–2):136–139.
55. Kaneko K, Yoshida K, Arima K, et al. Astrocytic deformity and globular structures are characteristic of the brains of patients with aceruloplasminemia. *J Neuropathol Exp Neurol*. 2002;61(12):1069–1077.
56. Miyajima H, Takahashi Y, Serizawa M, Kaneko E, Gitlin JD. Increased plasma lipid peroxidation in patients with aceruloplasminemia. *Free Radic Biol Med*. 1996;20(5):757–760.
57. Kohno S, Miyajima H, Takahashi Y, Suzuki H, Hishida A. Defective electron transfer in complexes I and IV in patients with aceruloplasminemia. *J Neurol Sci*. 2000;182(1):57–60.
58. Yoshida K, Kaneko K, Miyajima H, et al. Increased lipid peroxidation in the brains of aceruloplasminemia patients. *J Neurol Sci*. 2000;175(2):91–95.
59. Miyajima H, Adachi J, Kohno S, Takahashi Y, Ueno Y, Naito T. Increased oxysterols associated with iron accumulation in the brains and visceral organs of acaeruloplasminaemia patients. *QJM*. 2001;94(8):417–422.
60. Miyajima H, Kono S, Takahashi Y, Sugimoto M. Increased lipid peroxidation and mitochondrial dysfunction in aceruloplasminemia brains. *Blood Cells Mol Dis*. 2002;29(3):433–438.

61. Miyajima H, Fujimoto M, Kohno S, Kaneko E, Gitlin JD. CSF abnormalities in patients with aceruloplasminemia. *Neurology*. 1998;51(4):1188–1190.
62. Kaneko K, Hineno A, Yoshida K, Ohara S, Morita H, Ikeda S. Extensive brain pathology in a patient with aceruloplasminemia with a prolonged duration of illness. *Hum Pathol*. 2012;43(3):451–456.
63. Kaneko K, Nakamura A, Yoshida K, Kametani F, Higuchi K, Ikeda S. Glial fibrillary acidic protein is greatly modified by oxidative stress in aceruloplasminemia brain. *Free Radic Res*. 2002;36(3):303–306.
64. Oide T, Yoshida K, Kaneko K, Ohta M, Arima K. Iron overload and antioxidative role of perivascular astrocytes in aceruloplasminemia. *Neuropathol Appl Neurobiol*. 2006;32(2):170–176.
65. Jeong SY, David S. Age-related changes in iron homeostasis and cell death in the cerebellum of ceruloplasmin-deficient mice. *J Neurosci*. 2006;26(38):9810–9819.
66. Harris ZL, Durley AP, Man TK, Gitlin JD. Targeted gene disruption reveals an essential role for ceruloplasmin in cellular iron efflux. *Proc Natl Acad Sci U S A*. 1999;96(19):10812–10817.
67. Texel SJ, Xu X, Harris ZL. Ceruloplasmin in neurodegenerative diseases. *Biochem Soc Trans*. 2008;36(Pt 6):1277–1281.
68. Hahn P, Qian Y, Dentchev T, et al. Disruption of ceruloplasmin and hephaestin in mice causes retinal iron overload and retinal degeneration with features of age-related macular degeneration. *Proc Natl Acad Sci U S A*. 2004;101(38):13850–13855.
69. Schulz K, Vulpe CD, Harris LZ, David S. Iron efflux from oligodendrocytes is differentially regulated in gray and white matter. *J Neurosci*. 2011;31(37):13301–13311.
70. Yamamoto K, Yoshida K, Miyagoe Y, et al. Quantitative evaluation of expression of iron-metabolism genes in ceruloplasmin-deficient mice. *Biochim Biophys Acta*. 2002;1588(3):195–202.
71. Hineno A, Kaneko K, Yoshida K, Ikeda S. Ceruloplasmin protects against rotenone-induced oxidative stress and neurotoxicity. *Neurochem Res*. 2011;36(11):2127–2135.
72. Nemeth E, Tuttle MS, Powelson J, et al. Hepcidin regulates cellular iron efflux by binding to ferroportin and inducing its internalization. *Science*. 2004;306(5704):2090–2093.
73. Kaneko Y, Miyajima H, Piperno A, et al. Measurement of serum hepcidin-25 levels as a potential test for diagnosing hemochromatosis and related disorders. *J Gastroenterol*. 2010;45(11):1163–1171.
74. Guo P, Cui R, Chang YZ, et al. Hepcidin, an antimicrobial peptide is downregulated in ceruloplasmin-deficient mice. *Peptides*. 2009;30(2):262–266.
75. Kalman B, Lautenschlaeger R, Kohlmayer F, et al. An international registry for neurodegeneration with brain iron accumulation. *Orphanet J Rare Dis*. 2012;7:66.
76. Haack TB, Hogarth P, Kruer MC, et al. Exome sequencing reveals de novo WDR45 mutations causing a phenotypically distinct, X-linked dominant form of NBIA. *Am J Hum Genet*. 2012;91(6):1144–1149.
77. McNeill A, Birchall D, Hayflick SJ, et al. T2* and FSE MRI distinguishes four subtypes of neurodegeneration with brain iron accumulation. *Neurology*. 2008;70(18):1614–1619.
78. Gitlin JD. Wilson disease. *Gastroenterology*. 2003;125(6):1868–1877.
79. Summers MR, Jacobs A, Tudway D, Perera P, Ricketts C. Studies in desferrioxamine and ferrioxamine metabolism in normal and iron-loaded subjects. *Br J Haematol*. 1979;42(4):547–555.
80. Pan PL, Tang HH, Chen Q, Song W, Shang HF. Desferrioxamine treatment of aceruloplasminemia: Long-term follow-up. *Mov Disord*. 2011;26(11):2142–2144.
81. Yonekawa M, Okabe T, Asamoto Y, Ohta M. A case of hereditary ceruloplasmin deficiency with iron deposition in the brain associated with chorea, dementia, diabetes mellitus and retinal pigmentation: administration of fresh-frozen human plasma. *Eur Neurol*. 1999;42(3):157–162.
82. Finkenstedt A, Wolf E, Hofner E, et al. Hepatic but not brain iron is rapidly chelated by deferasirox in aceruloplasminemia due to a novel gene mutation. *J Hepatol*. 2010;53(6):1101–1107.
83. Roberti Mdo R, Borges Filho HM, Goncalves CH, Lima FL. Aceruloplasminemia: a rare disease – diagnosis and treatment of two cases. *Rev Bras Hematol Hemoter*. 2011;33(5):389–392.
84. Skidmore FM, Drago V, Foster P, Schmalfuss IM, Heilman KM, Streiff RR. Aceruloplasminaemia with progressive atrophy without brain iron overload: treatment with oral chelation. *J Neurol Neurosurg Psychiatry*. 2008;79(4):467–470.
85. Kuhn J, Bewermeyer H, Miyajima H, Takahashi Y, Kuhn KF, Hoogenraad TU. Treatment of symptomatic heterozygous aceruloplasminemia with oral zinc sulphate. *Brain Dev*. 2007;29(7):450–453.
86. Powell SR. The antioxidant properties of zinc. *J Nutr*. 2000;130(5S suppl):1447S–1454S.
87. Donangelo CM, Woodhouse LR, King SM, Viteri FE, King JC. Supplemental zinc lowers measures of iron status in young women with low iron reserves. *J Nutr*. 2002;132(7):1860–1864.

VITAMIN DISORDERS

CHAPTER

46

Genetic and Dietary Influences on Lifespan

Yian Gu, Nicole Schupf, and Richard Mayeux

Gertrude H. Sergievsky Center, Taub Institute on Alzheimer's Disease and the Aging Brain, Columbia University Medical Center, New York, NY, USA

INTRODUCTION

Longevity is neither inherent nor inevitable. Longevity, defined as a long duration of life, is the result of a decrease in the cumulative mortality in any population across all ages. The trend toward increasing survival into old age for those living in developed countries has come about as a result of decreasing mortality rates among the older segment of the population. From 1950 to 1995, the mortality rates for women over the age of 80 years declined by 50%.[1] Improvements in environmental hygiene, social welfare, health care systems, and advances in medicine worldwide have no doubt contributed to the current trends in lifespan.

Total lifespan is the result of a complex interaction between genes and the environment. In fact, biological aging results from influences at the genetic and epigenetic level. The finding that normal human and animal cells in culture undergo a finite number of population doublings offered the first clues that aging begins at the cellular level.[2]

Specific pathways involving metabolism have been implicated in longevity,[3] and a large number of studies show that genes may control cell destiny by regulating replication or senescence.[4] For example, variations in genes involved in metabolic control mechanisms in yeast and insulin-signaling pathways in nematodes (*Caenorhabditis elegans*) and in fruit flies (*Drosophila*) significantly extend lifespan. Genes that directly or indirectly modify daf-16, a transcription factor, or daf-2, an insulin and insulin-like growth factor receptor homolog, in *C. elegans* induce diapause, extending the lifespan by preventing maturation from the juvenile state. Mutations in genes resulting in dwarfism can extend lifespan in rodents by affecting growth hormone, which, in turn, affects insulin signaling and metabolism. As might be expected, aging in mammals is more complex than among lower organisms because mutations in genes that modulate apoptosis have also been found to affect lifespan. Some of these same pathways have been investigated in relation to human aging, but the complexity of human aging and longevity show only minimal similarities.

HYPOTHESIS OF LONGEVITY AND SENESCENCE

Aging is physiologically complex. Theories of the biological basis of aging have been developed covering a wide range of molecular systems. The mitochondrial theory of aging implicates oxidative stress within the cell leading to the accumulation of mitochondrial DNA mutations and oxygen-free radicals. The rate of oxidative damage within the mitochondria may represent the biological clock of aging. Alternatively, both germline and mitochondrial mutations may shift the balance between mitochondrial metabolism and ATP-producing pathways that regulate the oxidative burden and determine lifespan. This is done through a complex cell-signaling pathway that regulates glycolytic and mitochondrial ATP generation.[5]

The telomere–telomerase hypothesis of aging proposes that telomere loss results in replicative senescence within the cell[6–10] and that telomerase activation results in immortalization.[11] Telomeres are repetitive DNA consisting of hundreds of concatenated TTAGGG hexanucleotide sequences, located at the end of each human chromosome. Telomeres allow for preservation of the genome during replication and division. In most cells

Rosenberg's Molecular and Genetic Basis of Neurological and Psychiatric Disease
http://dx.doi.org/10.1016/10.1016/B978-0-12-410529-4.00046-2

telomere sequences shorten with each cell replication unless repaired by telomerase,[12] and telomeres may serve as a marker of biological aging.[13–17] Telomere length has been associated with greater longevity in some,[13–16] not all, populations.[16,18,19] Individuals with short telomere length are at an increased risk of age-related diseases (e.g., cardiovascular diseases, diabetes, dementia, cancer) and earlier death compared with similarly aged individuals with longer telomeres.[15,20–25] However, the relationship between telomere length and the risk of these diseases varies across studies,[26] and some investigators have proposed that telomere length may relate more to healthy aging than to survival.[17,19,27]

Telomerase is a reverse transcriptase complex that provides an essential maintenance function for chromosomal ends, or telomeres, during replication. The telomerase complex consists of the enzyme and an RNA template, which is critical for genomic integrity, chromatin assembly, and DNA repair. The telomere is a complex structure composed of both DNA and other proteins and may serve as another type of molecular clock that tallies the number of cell divisions and limits further divisions at a predetermined point. Alterations in telomerase expression have been studied in conditions leading to accelerated aging such as dyskeratosis congenital ataxia telangiectasia, xeroderma pigmentosum, and Werner syndrome.[28] The precise role of telomeres in predicting and limiting cellular lifespan nonetheless remains unclear.[29]

Longevity may also depend on optimal functioning of the immune system. Certain changes in the major histocompatibility complex (MHC), known to control a variety of immune functions, can be associated with the altered lifespan in strains of mice. Human studies have shown contradictory results regarding the association between longevity and variation in the human leukocyte antigen (HLA) system. These molecules are highly polymorphic and, when stimulated, antigenic peptide fragments are taken to the T-lymphocyte receptor. These molecules regulate T-cell responses against specific antigens and are critical for antigen-specific control of the immune response. Therefore, it is not surprising that survival and longevity might be associated with allelic variation that confers immunological resistance or susceptibility to infections and other diseases.[30]

The accumulation of somatic mutations during life has also been proposed as a cause of aging. There is an invariant relationship between lifespan and the number of random somatic mutations. Dietary restriction, which prolongs lifespan in nearly all species, results in slowed accumulation of some somatic mutations in mice. Conversely, senescence-accelerated mice, which have been bred to have a shortened lifespan, show accelerated accumulation of somatic mutations.[31] The accumulation of somatic mutations in the mitochondria results in respiratory chain dysfunction, damage from the creation of the reactive oxygen species, and shortened lifespan.[32]

Lifespan is invariably tied to brain function during senescence.[33] The brains of elderly individuals without dementia show an overall reduction in the brain volume and weight and enlargement of the brain ventricles. These changes are, in part, due to nerve cell loss, but accurate estimates of neuronal loss are difficult to make. There are losses of synapses and dendritic pruning in the aged brain, but these occur in selected areas, rather than globally. Cognitive impairment is a robust predictor of mortality and has been shown to be associated with mortality in both nondemented and demented elderly. In population-based cohorts of the elderly, those with mild as well as severe cognitive impairment have been found to have an increased risk of death.[34–45] Adjustment for a variety of health conditions, lifestyle factors, and sociodemographic characteristics did not decrease the mortality risk associated with poor cognitive function.[34–36,38–40,44–46] These findings suggest that decline in cognitive function with age is a predictor, not simply a surrogate, of rate of aging and mortality risk.[43] Compared with siblings of patients with dementia or those with simply declining cognitive function, siblings of those without dementia or changes in cognitive function are significantly less likely to die early in life, and this effect is more pronounced when the proband is age 75 years or older.[47] Depending on the strain, studies in aged mice also show age-related changes in the sensorimotor performance, spontaneous behavior, and in learning and memory tasks. Complex learning, such as in the Morris water maze task of spatial learning, can be dramatically affected, whereas simple discrimination learning is only impaired in the oldest animals. These changes are not related to sensorimotor or locomotor activity but rather to a loss of synapses in key areas of the hippocampal formation and a decrease in the *N*-methyl-D-aspartate (NMDA)-receptor-mediated response and an alteration of Ca2+ regulation.[48]

CALORIC INTAKE, α-TOCOPHEROL, AND OTHER DIETARY FACTORS

Caloric intake decreases about 20–30% among the elderly because of the decline in physical activity and resting metabolic rate for most people.[49] Other factors such as depression and alterations in taste and odor perception also contribute. In addition to genetic manipulation, dietary modification such as caloric restriction has been found to affect the pathogenesis of many age-associated chronic diseases, slowing aging, and prolonging longevity in laboratory animals. A systematic review[50] confirmed that caloric restriction in adults causes beneficial health effects in a number of parameters, but the authors expressed caution because the precise decrease in calorie intake or body fat mass associated with optimal health and maximum longevity has not been established. Moreover, for certain individuals such a drastic reduction or change in diet could be harmful. Clarification of the underlying biological mechanisms is being intensively investigated. The beneficial effect of caloric restriction on longevity has been investigated in a controlled study of nonhuman primates.[51] In humans, caloric reduction lowers body weight, which in turn lowers blood pressure, and may reduce blood levels of lipoproteins, glucose, and insulin. Combined with exercise, reduced calories can also improve coronary artery disease and extend lifespan. There are other effects such as lower body temperatures and a slower age-related decline in circulating levels of androgenic hormones.

A significant benefit of caloric restriction may be a decreased risk of age-related disorders such as type 2 diabetes, cardiovascular disease, cancer, and dementia, the leading causes for mortality in the US.[52] Caloric restriction increases resistance of neurons to dysfunction and degeneration and improves behavioral outcome in animal models of Alzheimer disease.[53] In contrast, there is also consensus that excessive food and protein intake in early adulthood, and a positive energy balance during adulthood, may contribute to an increased risk of developing several cancer types.[54] A recent study demonstrated that, during 18 years of follow-up, participants aged 50–65 reporting high protein intake had a 75% increase in overall mortality and a four-fold increase in cancer mortality, while high intake of protein reduced cancer and overall mortality among those aged 65 and above, suggesting that low protein intake during middle age (50–65 years) followed by moderate-to-high protein consumption in old adults (over 65 years) may optimize lifespan and longevity.[55]

The mechanisms proposed to explain the effects of caloric restriction center around changes in metabolism. For example, caloric restriction in mice may retard aging by inducing a metabolic shift in protein turnover, which, in turn, decreases macromolecular damage. Lee et al.[56] observed an increase in transcriptional activity for genes related to the upregulation of gluconeogenesis, fatty acid synthesis, and increased synthesis and turnover of proteins. In yeast, *SIR2* (silent information regulator 2a gene), which is associated with lifespan, is affected by calorie restriction. Its ortholog in mammals, *SIRT1*, is activated by caloric restriction, which inhibits the action of peroxisome proliferator-activator receptor-γ, the nuclear receptor that promotes adipogenesis, loss of body fat, and extension of lifespan.[57]

Caloric restriction both enhances and maintains protein turnover and renewal, which, in turn, contributes to the extension of life in laboratory animals. However, each species may have specific effects related to caloric restriction.[58] The Wisconsin National Primate Research Center (WNPRC)[59] and the National Institute on Aging (NIA)[60] studies of caloric restriction in monkeys showed that a 30% caloric restriction without malnutrition prevented cardiovascular disease, cancer, and diabetes incidence,[59,60] as well as protected against age-associated sarcopenia[61] and gray matter volume shrinkage of several key subcortical regions.[59] WNPRC calorically restricted monkeys also had significantly increased survival when considering only age-associated deaths compared with control monkeys, although data to examine caloric restriction on average and maximum longevity will still take a few years to be completed. In a comparison of young and old rhesus monkeys, gene expression analysis revealed selective upregulation of genes involved in inflammation and oxidative stress, and a downregulation of genes involved in mitochondrial electron transport and oxidative phosphorylation.[62]

Drugs that mimic the beneficial effect of caloric restriction, such as 2-deoxyglucose, have also been shown to lower plasma insulin levels and body temperature in rats and nonhuman primates.[63] Caloric restriction or the use of medications that have the same effect in humans will need careful investigation before implementation.[50]

Although there may be advantages to caloric restriction, there may also be disadvantages. For example, with the decrement in caloric intake, there is also a decrease in micronutrient consumption, most notably α-tocopherol (α-TOH), which declines by approximately 30%.[64] The generation of reactive oxygen species in mitochondria, also an important mechanism for aging, and the subsequent effect of oxidative damage to cellular macromolecules is well established. It has been hypothesized that antioxidants from diet may reduce mortality and promote healthy aging. α-TOH is one of the most biologically active antioxidants *in vivo*.[65] α-TOH is a fat-soluble vitamin that, in some forms, is abundant in foods. As a nutrient, α-TOH is the most effective lipid-soluble antioxidant in the biological membrane preventing free radical damage, and it is important for both normal brain development and maintenance during late life. In model systems such as cell culture, α-TOH acts as an oxygen free-radical scavenger and limits lipid peroxidation.[66] Some studies have found telomeric length and the retention of telomerase activity, both related to cellular longevity, were increased in cell culture that includes α-tocopherol.[67] However, studies of supplementing α-TOH in aged rodents showed either no effect or limited effects on lifespan and cognitive performance.[68,69] The National Institute on Aging Interventions Testing Program (ITP) tested agents, including resveratrol, green tea extract, curcumin, oxaloacetic acid, and medium-chain triglyceride oil, on the lifespan of genetically heterogeneous mice. Except for green tea extract, which might diminish the risk of midlife deaths in females, none of the compounds had a statistically significant effect on lifespan of male or female mice.[70]

A large body of observational studies continue to suggest that higher dietary intake of antioxidants, including foods high in α-TOH, is associated with reduced risk of developing some age-related chronic diseases such as cancer, cardiovascular or neurodegenerative diseases, and prolonged longevity.[71] Some studies have also shown promising results of α-TOH supplements. For example, Sano et al.[72] found that α-TOH at doses of 2000 IU per day delayed progression to the more advanced stages of Alzheimer disease over a 10-month period. Recently, use of vitamin E, combined with use of vitamin C supplements, was found to be associated with reduced incidence of Alzheimer's disease in the elderly population.[73] Some studies[74–77] suggested that supplementation of α-TOH, along with other antioxidants, reduced certain cancer mortality or total mortality.

However, the results from clinical trials with α-TOH have been largely negative, and do not support beneficial effects of α-TOH on mortality or disease progression.[78–80] A meta-analysis showed that use of α-TOH did not significantly decrease risk of cardiovascular death or cerebrovascular accident, and did not provide protection from mortality either.[81] The Women's Health Study found α-TOH slightly decreased cardiovascular mortality, but could not lower total mortality.[82] A more recent meta-analysis concludes that α-TOH supplementation including high doses does not affect overall mortality.[80] Furthermore, a large meta-analysis[83] found that α-TOH supplementation at high doses may even increase the all-cause mortality. This finding was also supported by a more recent meta-analysis,[84] although the mechanisms for any adverse effect of a high dose of α-TOH are unknown. In summary, current evidence does not support supplementation of α-TOH or any other antioxidant such as vitamin C, beta-carotene, selenium for primary or secondary prevention of mortality.[78,85,86] There is a need for further work to examine the role of α-TOH in longevity.

Single nutrients with a measurable amount of dosage might be preferable for clinical trials, and the biological roles of the nutrients easy to clarify. Nevertheless, in reality, humans eat meals with complex combinations of nutrients or food items that are likely to be synergistic (or antagonistic) so that the action of the food matrix is different from the individual nutrients or food items.[87] Examining the overall pattern of dietary consumption (dietary pattern analysis) has been suggested.[88,89] The Mediterranean diet, a diet high in plant foods (such as fruits, nuts, legumes, and cereals) and fish, with olive oil as the primary source of monounsaturated fat and low-to-moderate intake of wine, as well as low intake of red meat and poultry, is known to be one of the healthiest dietary patterns in the world due to its protective effects on some chronic diseases.[90] The Mediterranean diet has been associated with a number of healthy outcomes, including reduced risk of cardiovascular disease, cancer, Alzheimer disease, and mortality.[91–95] The PRIDIMED (Prevención con Dieta Mediterránea) randomized clinical trial on the cardiovascular effects of a Mediterranean diet (supplemented with extra-virgin olive oil or nuts) was stopped after a median follow-up of only 4.8 years because this diet significantly reduced the incidence of the combined cardiovascular end points and stroke (but not for myocardial infarction alone).[96]

In summary, caloric restriction remains the major dietary modification that could potentially prolong lifespan. Certain dietary patterns such as the Mediterranean diet, representing a combined higher consumption of healthy foods while avoiding detrimental ones, might also be promising. In contrast, support for antioxidant use is mainly limited to the extension of average lifespan in laboratory animals. Clearly, there is a need for further work to examine preventive interventions.

GENETICS OF AGING AND LIFESPAN

Variations in several groups of genes in specific pathways regulate lifespan in animals. *Age-1* in *C. elegans* encodes phosphatidylinositol-3-kinase in a component of the insulin growth factor pathway. This pathway also includes *DAF16* (FOXO) involved in the regulation of genes related to stress resistance, innate immunity, and metabolic and toxin degradation. Mutations affecting mitochondrial function in *C. elegans*, called *Mit* mutants, result in 20–40% extension of lifespan and also indirectly interact with the insulin and insulin growth factor pathway. Genes such as *SIR2* and *TOR* (target of rapamycin), *MTH* (Methuselah), *INDY* (I'm not dead yet) and *Kl* (KLOTHO) have been studied in rodents and other model organisms but less frequently in humans. Christensen et al.[4] recommend caution in interpretation of these genetic effects because what can extend life in *C. elegans* can cause lethal disruptions in humans.

Studies of lifespan in humans focuses on a number of phenotypes related to survival. Human monozygotic twins are more likely to be concordant for lifespan than are dizygotic twins, particularly among the oldest twins.[97] However, only one-third or less of the variance in longevity in humans is predicted to be attributable to genetic factors. This may be due to the fact that investigating the correlation of age at death between parents and their offspring has proven difficult because secular trends in environmental conditions impede meaningful comparisons. Though a large number of studies imply that genetic variation influences longevity in humans, the extent to which this occurs directly or indirectly is unclear.[4]

Heritability of longevity, defined as the degree to which total years of life or age at death is shared among family members, has been estimated from investigations of human twins, isolated and founder populations, and long-lived families. Lifespan data on all relatives of a cohort of individuals in Utah born between 1870 and 1907 who lived to be at least 65 years of age was used to estimate the influence of family history on the relative risk of longevity. Siblings of probands who reached the 97th percentile of excess longevity (age 95 for men and age 97 for women) were 2.3 times as likely to reach the 97th percentile of longevity as siblings of probands who died at younger ages.[98] Comparison of the excess longevity in near and distant relatives showed that the pattern of familial aggregation of excess longevity was consistent with a relatively simple model of inheritance involving the additive effects of one or a few genes. Families with exceptional longevity also have lower rates of age-related disease. Findings from the Long Life Family Study, in which families with two or more long-lived individuals were recruited, showed that both probands and their offspring had fewer cardiovascular risk factors, better physical function, and less cognitive impairment than spouse controls or members of community-based cohorts not selected for exceptional longevity.[99–101] In the New England Centenarian Study, offspring of centenarians and nonagenarians had an approximately two-fold reduced prevalence, compared with age-matched controls, of cardiovascular disease and cardiovascular risk factors, including myocardial infarction, hypertension, diabetes and stroke[102] and lower all-cause, cancer-specific and coronary heart disease-specific mortality.[103]

Mitchell et al.[104] provided estimates of the heritability of human longevity from several published studies of twins and geographically isolated populations. Estimates of the heritability of lifespan among twins are highly variable, ranging from 10% among twins reared apart, to as high as 50% among same-sex twin pairs. Among 2872 Danish twins, variance was attributed to genetic and environmental factors. The authors modeled both genetic and environmental factors, not shared and shared within the family, and they estimated heritability of longevity to be 25%.[97] A small sex difference was caused by a greater impact of unshared environmental factors among women. Heritability was constant over the three 10-year birth cohorts included. Among Swedish twins, intrapair correlation for lifespan was similar to that for Danish twins, but lower for men than for women. The age at death did not influence the outcome. The authors also compared twins reared apart to twins reared together and concluded that most of the variance in longevity was explained by environmental factors. Over the total age range, 30% of the variance in longevity was attributable to genetic factors, and almost all of the remaining variance was due to unshared individual specific environmental factors.[105] This is also true in other populations.[106] Population-based twin studies in Denmark and Sweden of late-life physical and cognitive functioning demonstrate that genetic factors may become increasingly important at the oldest ages.[107–109] Key aging phenotypes such as physical abilities, grip strength, and cognitive abilities measured as a composite of cognitive tests have heritabilities around 50% among elderly twins.[109,110]

Mitchell et al.[104] also reviewed three large cohort studies involving large, multigenerational families. The highest percentages were found among genealogies of six large New England families. Heritability of lifespan in the older-order Amish indicates that 25% of the variation may be due to genetic factors. Parent and offspring ages at death were highly correlated, as were ages of death among siblings.[104] The strongest effects were for parents and siblings surviving past the age of 75 years. The authors concluded that genetic influences on lifespan might vary at different ages. As noted by these authors, premature death in a parent is a strong risk factor for premature death among offspring. Because susceptibility to disease also increases with age, mortality determinants among older individuals may be very different from those among younger persons.

Studies of centenarians imply much stronger genetic effects. Compared to siblings of individuals who did not survive past age 73 years, siblings of centenarians were four times more likely to live to age 85 years or older,[111] and first-degree relatives of individuals who lived beyond 95 years were twice as likely to survive to the same age as were relatives of controls.[98,112,113] Similarly, mortality rates were 30% lower for first-degree relatives of exceptionally long-lived siblings in the Netherlands.[114]

It may be argued that absence of disease-causing genetic variants throughout life must certainly promote longevity. Alternatively, extreme longevity could be mediated by variations in genes that provide protection against diseases during middle and late life. Evert and colleagues examined morbidity profiles of centenarians and identified three distinct patterns, characterized as "survivors," "delayers," and "escapers."[115] Survivors were centenarians who had a diagnosis of an age-associated illness before age 80 (24% of males and 43% of females) but did not die of these illnesses. Delayers were individuals who showed compressed morbidity and developed age-related disease after age 80 (44% of males and 42% of females), while escapers were individuals who reached the age of 100 without common age-related disease (32% of males and 15% of females).[115] The greater frequency of male compared with female escapers suggests that men may require greater genetic loading of longevity variants to achieve old age. Several recent studies have found that the frequency of disease causing alleles is similar in centenarians and controls from the general population,[116,117] suggesting that centenarians may carry more longevity-enhancing variants, which promote slower aging and delay of age-related disease.[117]

Known Genetic Variants and Longevity

Genes related to cardiovascular disease, growth, and metabolism have been regularly studied for their potential association with longevity on the assumption that certain alleles offer protection and, therefore, prolong lifespan (Table 46.1). Both candidate gene studies and genome-wide association studies (GWAS) have compared genotypes in centenarians and long-lived individuals with those in control groups.

The gene for apolipoprotein E (*APOE*) has been extensively investigated because of its role in lipid metabolism, ischemic cardiovascular disease,[118,119] and Alzheimer's disease,[120,121] and has consistently emerged as a determinant of longevity.[122–125] There are three common alleles of *APOE*: ε2, ε3, and ε4. Although the ε3 allele is the most common allele and is generally present in two copies in about 60–80% of humans, ε4 is considered to be the ancestral allele. The frequency of *APOE* ε4 varies worldwide, from 40.7% among Pygmies to 2–3% among some Asian populations. Variation at the *APOE* locus has also been related to longevity. The ε4 allele has been associated with early mortality, although inconsistently, and the association of the ε4 allele with mortality risk varies by population.[126] Ewbank[127–129] reported that the effect of the ε4 allele on mortality risk diminishes with increasing age, but a recent study in the Danish 1905 cohort showed an increased effect of carrying the ε4 alleles with increasing age.[130] The Finnish Centenarian Study found a decreased association between the ε4 allele and Alzheimer's disease at extreme old ages.[131] The observation of a decreased frequency of the *APOE* ε4 allele amongst Caucasian centenarians is likely due to the phenomenon of demographic selection in which carriers of the allele die of ε4-associated disease, leaving behind a cohort of select survivors without the allele.[132,133] Among offspring of long-lived families, the likelihood of carrying an *APOE* ε4 allele was significantly lower (odds ratio = 0.75) and the likelihood of carrying an *APOE* ε2 allele higher (odds ratio = 1.5) among family members in the offspring generation than among their spouse controls.[134] The finding of a lower *APOE* ε4 allele frequency in a relatively young cohort of offspring of long-lived individuals, before substantial mortality has occurred, suggests that the reduced frequency of the ε4 allele in the oldest old might not be due to early mortality but to a heritable lower frequency of the risk allele, reducing risk for *APOE*-related disease.[134] Recent GWAS in German, Dutch, Danish, and US Caucasian cohorts of long-lived individuals have also identified rs2075650 in *TOMM40* as associated with longevity,[122,124] but close to and in linkage disequilibrium with the rs429358, the *APOE* ε4 allele, and the investigators suggested that rs2075650 might not have an independent effect on longevity. The *APOE* ε2 allele is associated with longevity, although also inconsistently.[133,135,136] Although the mechanism by which the ε2 allele extends life is unknown, Reich et al.[137] have proposed an interaction between ApoE and oxidative stress.

The insulin/insulin-like growth factor 1 (IGF1) signaling pathway influences metabolism and lifespan in *C. elegans* and other model organisms. In *C. elegans*, mutations that decrease insulin/IGF-1 signaling increase lifespan

TABLE 46.1 Partial List of Genetic Variants that have been Associated with Longevity or Lifespan in Humans

Gene	Function	Disease association
APOE (apolipoprotein E)[165,166]	Lipoprotein metabolism	Alzheimer disease, cardiovascular disease
FOX03A (Forkhead Box 03)[138–143]	Insulin–IGF1 signaling	Cardiovascular disease, cancer
APOC3 (apolipoprotein C)[147]	Lipoprotein metabolism	None
ACE (angiotensin 1-converting enzyme)[150]	Renin–angiotensin system	Hypertension, cardiovascular disease
CETP (cholesteryl ester transfer protein)[146]	Transfer of cholesteryl esters	Hypertension, cardiovascular disease, metabolic syndrome, and cognition
HFE (hereditary hemochromatosis)[167]	Regulation of iron homeostasis	Hemochromatosis
HLA class I (histocompatibility antigen)	Immune response	Immune disorders
IGF1R (insulin-like growth factor-1 receptor)[153,168]	Growth factors, metabolism, signal transduction pathway	None
MTHFR (methylene-tetrahydrofolate reductase)[151]	Homocysteine methylation	Cardiovascular disease, cancer susceptibility
MINPP1 (multiple inositol polyphosphate phosphatase 1)[154]	Cellular proliferation	None
IL6 (interleukin-6)[169]	Immunoregulatory cytokine	Arthritis, osteoporosis, Alzheimer disease, type II diabetes
KLOTHO[164]	Microvascular activity	None

by activating the daf-16/FOXO protein and inducing diapause.[138] The Forkhead Box 03 (*FOX03A*), the human homolog of daf-16, has also been consistently associated with human longevity in diverse populations, including Japanese, German, southern Italian, Danish, Ashkenazi Jews and Han Chinese.[138–143] Among German centenarians/nonagenarians, the association of *FOX03A* was stronger in centenarians than in nonagenarians,[140] while in the Danish 1905 longitudinal cohort study, *FOX03a* was associated with survival from younger ages to old age, but was not associated with survival among the oldest old.[143] Among Han Chinese, variants in *FOX03A* were associated with longevity in both men and women, while variants in *FOX01A* were associated with longevity in women only.[141] Willcox and colleagues found that long-lived Japanese men had a reduced prevalence of cancer, cardiovascular disease, high physical and cognitive function, and greater insulin sensitivity compared with controls and several of these phenotypes were associated with the *FOX03A* genotype.[138]

Lipoprotein particle sizes are increased in centenarians and their offspring and related to increased high-density lipoprotein levels. This phenotype is related to a lower prevalence of hypertension, cardiovascular disease, metabolic syndrome, and better cognition.[144–146] Barzilai et al.[146] related this phenotype to an increased frequency of homozygosity for the 1405 V variant in the cholesteryl ester transfer protein (*CETP*) gene, which is involved in regulation of lipoprotein and its particle sizes. A study of centenarians and their offspring have implicated another gene involved in lipid metabolism and lipoproteins.[147] In a follow-up study,[147] it was found that common variants in apolipoprotein C (*APOC3*), located on chromosome 11q23, were not only more frequent among centenarians and their offspring, but also associated with less hypertension and insulin resistance compared with controls.

Thus, cardiovascular and cognitive phenotypes are those most associated with exceptional longevity, and genes that are associated with exceptional longevity affect both phenotypes. While these observations need confirmation, the consistent observation that genes involved in the lipoprotein family, which are related to cardiovascular health, cannot be ignored. Moreover, the biological basis for this relation with longevity in humans is similar to that in laboratory animals and model organisms.

Other regions and genes have been proposed, based on findings from linkage and candidate gene studies and GWAS of long-lived individuals. For example, a genetic linkage study among families of centenarians has identified a locus on chromosome 4. Using a series of 137 families characterized by extreme longevity, defined as survival past the age of 98 years for at least one member of the family, Puca et al.[148] found statistically significant evidence favoring linkage to a region on chromosome 4. Male siblings were 91 years or older, and female siblings were 95 years of age or older. Fine mapping of this locus narrowed the search to microsomal triglyceride transfer protein, MTTP, a protein involved in lipoprotein metabolism,[149] but this finding has not been successfully replicated. Angiotensin I-converting enzyme,[150] and methylenetetrahydrofolate reductase[151] have been shown to promote lifespan, though inconsistently.[152] Among Ashkenazi Jewish centenarians, their offspring and offspring-matched controls, a higher frequency of variants in the IGF1 receptor (IGF1R) gene was found among female centenarians that was associated with high serum IGF1 levels and reduced activity of the IGF1R as measured in transformed lymphocytes.[153] In GWAS using European and American cohorts of the elderly, the Cohorts for Heart and Aging Research (CHARGE) study found a variant near multiple inositol polyphosphate phosphatase 1 (*MINPP1*), a gene that codes inositol phosphate phosphatases and is involved in cellular proliferation, that was associated with living to age 90 or older.[154] A meta-analysis of GWAS involving nine studies from the CHARGE cohorts found variants that predicted death or event-free survival in or near genes that are highly expressed in the brain or genes involved in neural development and function, although no single polymorphism achieved genome-wide significance.[155]

Telomere length is also heritable and associated with longevity[156] and GWAS and meta-analysis across multiple populations have identified several candidate genes for telomere length: *TERC*, telomerase RNA component,[157] located on 3q26[158] and *TERT*, telomerase reverse transcriptase (5p15.33).[158,159] A recent genome-wide family-based association and linkage analysis has identified genetic variants on 17q23.2 and 10q11.21 that contribute to variation in telomere length among families with exceptional longevity.[156] A meta-analysis[160] identified *CTC1* (conserved telomere maintenance component 1, 17p13.1) and *ZNF676* (zinc finger protein 676, 19p12) as candidate genes for telomere homeostasis in humans, and confirmed that minor variants in *OBFC1* on 10q24.33 were associated with shorter telomere length. Although their function is not certain, these genes appear to be involved in maintenance of chromosome structures.

Mutations in a gene encoding a helicase and exonuclease have been identified as the etiology of Werner syndrome, a rare autosomal disorder causing progeria.[161] Severe atherosclerosis and a higher frequency of cancer, as well as rapid aging during young adulthood, are the main features of this disorder.[162,163] However, variant alleles or polymorphisms at the Werner locus have not been found to affect aging in humans. Mice without *Klotho*, a gene that encodes a type I membrane protein and shares homology with glycosides, age prematurely. *Klotho*-deficient mice develop normally during the first month of life, but then rapidly develop growth retardation, become inactive, and die by the second month.[164] Investigators have identified a polymorphism in the human gene on chromosome

13q12 associated with longevity, defined as survival to age 75 years and older. Although they are intriguing, investigations of progeria-like syndromes are limited because they do not always characterize the extent and diverse nature of aging.

CONCLUSION

Longevity is influenced by a complex interaction among genetic, social, and environmental factors. Among the environmental factors, dietary factors may be of greatest interest because of the potential ability to intervene if a rational means becomes available. The ability of dietary restriction to limit the progression of oxidative stress in aged laboratory animals and to slow the aging process needs further investigation but offers promise. Although the genes that influence lifespan are only now being investigated, once they are discovered, new pathways will be identified that may also prove to be important targets for intervention. Current studies point to the role of lipoprotein metabolism, growth factors and metabolism, signal transduction pathways, and immune function. Environmental and behavioral factors, such as diet, may play a role throughout life in the maintenance of good health and longevity.

ACKNOWLEDGEMENTS

This work is from The Taub Institute on Alzheimer's Disease and the Aging Brain, The Gertrude H. Sergievsky Center and The Departments of Neurology and Psychiatry in the College of Physicians and Surgeons. Support was provided by Federal Grants AG/ES18732, U01 AG023749, AG07232, AG08702, K99AG042483, and the Marilyn and Henry Taub Foundation.

References

1. Vaupel JW, Carey JR, Christensen K, et al. Biodemographic trajectories of longevity. *Science*. 1998;280(5365):855–860.
2. Hayflick L. Current theories of biological aging. *Fed Proc*. 1975;34(1):9–13.
3. Finch CE, Ruvkun G. The genetics of aging. *Annu Rev Genomics Hum Genet*. 2001;2:435–462.
4. Christensen K, Johnson TE, Vaupel JW. The quest for genetic determinants of human longevity: challenges and insights. *Nat Rev Genet*. 2006;7(6):436–448.
5. Schieke SM, Finkel T. Mitochondrial signaling, TOR, and lifespan. *Biol Chem*. 2006;387(10–11):1357–1361.
6. Benetos A, Okuda K, Lajemi M, et al. Telomere length as an indicator of biological aging: the gender effect and relation with pulse pressure and pulse wave velocity. *Hypertension*. 2001;37(2 Part 2):381–385.
7. Blackburn EH. Telomere states and cell fates. *Nature*. 2000;408(6808):53–56.
8. Linskens MH, Harley CB, West MD, et al. Replicative senescence and cell death. *Science*. 1995;267(5194):17.
9. von Zglinicki T. Telomeres: influencing the rate of aging. *Ann N Y Acad Sci*. 1998;854:318–327.
10. Zou Y, Sfeir A, Gryaznov SM, et al. Does a sentinel or a subset of short telomeres determine replicative senescence? *Mol Biol Cell*. 2004;15(8):3709–3718.
11. Harley CB. Telomerase is not an oncogene. *Oncogene*. 2002;21(4):494–502.
12. Hodes RJ, Hathcock KS, Weng NP. Telomeres in T and B cells. *Nat Rev Immunol*. 2002;2(9):699–706.
13. Cawthon RM, Smith KR, O'Brien E, et al. Association between telomere length in blood and mortality in people aged 60 years or older. *Lancet*. 2003;361(9355):393–395.
14. Fitzpatrick AL, Kronmal RA, Kimura M, et al. Leukocyte telomere length and mortality in the Cardiovascular Health Study. *J Gerontol A Biol Sci Med Sci*. 2011;66(4):421–429.
15. Honig LS, Kang MS, Schupf N, et al. Association of shorter leukocyte telomere repeat length with dementia and mortality. *Arch Neurol*. 2012;69(10):1332–1339.
16. Martin-Ruiz CM, Gussekloo J, van Heemst D, et al. Telomere length in white blood cells is not associated with morbidity or mortality in the oldest old: a population-based study. *Aging Cell*. 2005;4(6):287–290.
17. Sanders JL, Fitzpatrick AL, Boudreau RM, et al. Leukocyte telomere length is associated with noninvasively measured age-related disease: the Cardiovascular Health Study. *J Gerontol A Biol Sci Med Sci*. 2012;67(4):409–416.
18. Bischoff C, Petersen HC, Graakjaer J, et al. No association between telomere length and survival among the elderly and oldest old. *Epidemiology*. 2006;17(2):190–194.
19. Njajou OT, Hsueh WC, Blackburn EH, et al. Association between telomere length, specific causes of death, and years of healthy life in health, aging, and body composition, a population-based cohort study. *J Gerontol A Biol Sci Med Sci*. 2009;64(8):860–864.
20. Aviv A. Leukocyte telomere length, hypertension, and atherosclerosis: are there potential mechanistic explanations? *Hypertension*. 2009;53(4):590–591.
21. Aviv A. Genetics of leukocyte telomere length and its role in atherosclerosis. *Mutat Res*. 2012;730(1–2):68–74.
22. Epel ES, Blackburn EH, Lin J, et al. Accelerated telomere shortening in response to life stress. *Proc Natl Acad Sci U S A*. 2004;101(49):17312–17315.

23. Jeanclos E, Krolewski A, Skurnick J, et al. Shortened telomere length in white blood cells of patients with IDDM. *Diabetes*. 1998;47(3):482–486.
24. Kaplan RC, Fitzpatrick AL, Pollak MN, et al. Insulin-like growth factors and leukocyte telomere length: the cardiovascular health study. *J Gerontol A Biol Sci Med Sci*. 2009;64(11):1103–1106.
25. Ye S, Shaffer JA, Kang MS, et al. Relation between leukocyte telomere length and incident coronary heart disease events (from the 1995 Canadian Nova Scotia Health Survey). *Am J Cardiol*. 2013;111(7):962–967.
26. Shaffer JA, Epel E, Kang MS, et al. Depressive symptoms are not associated with leukocyte telomere length: findings from the Nova Scotia Health Survey (NSHS95), a population-based study. *PLoS One*. 2012;7(10):e48318.
27. Terry DF, Nolan VG, Andersen SL, et al. Association of longer telomeres with better health in centenarians. *J Gerontol A Biol Sci Med Sci*. 2008;63(8):809–812.
28. Blasco MA. The epigenetic regulation of mammalian telomeres. *Nat Rev Genet*. 2007;8(4):299–309.
29. Stewart SA, Weinberg RA. Senescence: does it all happen at the ends? *Oncogene*. 2002;21(4):627–630.
30. Candore G, Balistreri CR, Listì F, et al. Immunogenetics, gender, and longevity. *Ann N Y Acad Sci*. 2006;1089:516–537.
31. Morley A. Somatic mutation and aging. *Ann N Y Acad Sci*. 1998;854:20–22.
32. Dufour E, Larsson NG. Understanding aging: revealing order out of chaos. *Biochim Biophys Acta*. 2004;1658(1–2):122–132.
33. Drachman DA. Aging of the brain, entropy, and Alzheimer disease. *Neurology*. 2006;67(8):1340–1352.
34. Fried LP, Kronmal RA, Newman AB, et al. Risk factors for 5-year mortality in older adults: the Cardiovascular Health Study. *JAMA*. 1998;279(8):585–592.
35. Bassuk SS, Wypij D, Berkman LF. Cognitive impairment and mortality in the community-dwelling elderly. *Am J Epidemiol*. 2000;151(7):676–688.
36. Bruce ML, Hoff RA, Jacobs SC, et al. The effects of cognitive impairment on 9-year mortality in a community sample. *J Gerontol B Psychol Sci Soc Sci*. 1995;50(6):289–296.
37. Dewey ME, Saz P. Dementia, cognitive impairment and mortality in persons aged 65 and over living in the community: a systematic review of the literature. *Int J Geriatr Psychiatry*. 2001;16(8):751–761.
38. Gussekloo J, Westendorp RG, Remarque EJ, et al. Impact of mild cognitive impairment on survival in very elderly people: cohort study. *BMJ*. 1997;315(7115):1053–1054.
39. Kelman HR, Thomas C, Kennedy GJ, et al. Cognitive impairment and mortality in older community residents. *Am J Public Health*. 1994;84(8):1255–1260.
40. Liu IY, LaCroix AZ, White LR, et al. Cognitive impairment and mortality: a study of possible confounders. *Am J Epidemiol*. 1990;132(1):136–143.
41. Neale R, Brayne C, Johnson AL. Cognition and survival: an exploration in a large multicentre study of the population aged 65 years and over. *Int J Epidemiol*. 2001;30(6):1383–1388.
42. Nguyen HT, Black SA, Ray LA, et al. Cognitive impairment and mortality in older Mexican Americans. *J Am Geriatr Soc*. 2003;51(2):178–183.
43. Schupf N, Tang MX, Albert SM, et al. Decline in cognitive and functional skills increases mortality risk in nondemented elderly. *Neurology*. 2005;65(8):1218–1226.
44. Smits CH, Deeg DJ, Kriegsman DM, et al. Cognitive functioning and health as determinants of mortality in an older population. *Am J Epidemiol*. 1999;150:978–986.
45. Swan GE, Carmelli D, LaRue A. Performance on the digit symbol substitution test and 5-year mortality in the Western Collaborative Group Study. *Am J Epidemiol*. 1995;141(1):32–40.
46. Rozzini R, Franzoni S, Frisoni G, et al. Cognitive impairment and survival in very elderly people. Decreased survival with cognitive impairment seems not to be related to comorbidity. *BMJ*. 1998;316(7145):1674.
47. Schupf N, Costa R, Tang MX, et al. Preservation of cognitive and functional ability as markers of longevity. *Neurobiol Aging*. 2004;25(9):1231–1240.
48. Rosenzweig ES, Barnes CA. Impact of aging on hippocampal function: plasticity, network dynamics, and cognition. *Prog Neurobiol*. 2003;69(3):143–179.
49. Morley JE. Anorexia, sarcopenia, and aging. *Nutrition*. 2001;17(7–8):660–663.
50. Fontana L, Klein S. Aging, adiposity, and calorie restriction. *JAMA*. 2007;297(9):986–994.
51. Mattison JA, Roth GS, Lane MA, Ingram DK. Dietary restriction in aging nonhuman primates. *Interdiscip Top Gerontol*. 2007;35:137–158.
52. Fontana L, Meyer TE, Klein S, et al. Long-term calorie restriction is highly effective in reducing the risk for atherosclerosis in humans. *Proc Natl Acad Sci U S A*. 2004;101(17):6659–6663.
53. Mattson MP. Emerging neuroprotective strategies for Alzheimer's disease: dietary restriction, telomerase activation, and stem cell therapy. *Exp Gerontol*. 2000;35(4):489–502.
54. Longo VD, Fontana L. Calorie restriction and cancer prevention: metabolic and molecular mechanisms. *Trends Pharmacol Sci*. 2010;31(2):89–98.
55. Levine ME, Suarez JA, Brandhorst S, et al. Low protein intake is associated with a major reduction in IGF-1, cancer, and overall mortality in the 65 and younger but not older population. *Cell Metab*. 2014;19(3):407–417.
56. Lee CK, Klopp RG, Weindruch R, et al. Gene expression profile of aging and its retardation by caloric restriction. *Science*. 1999;285(5432):1390–1393.
57. Wolf G. Calorie restriction increases lifespan: a molecular mechanism. *Nutr Rev*. 2006;64(2 Pt 1):89–92.
58. Masoro EJ. Overview of caloric restriction and ageing. *Mech Ageing Dev*. 2005;126(9):913–922.
59. Colman RJ, Anderson RM, Johnson SC, et al. Caloric restriction delays disease onset and mortality in rhesus monkeys. *Science*. 2009;325(5937):201–204.
60. Mattison JA, Roth GS, Beasley TM, et al. Impact of caloric restriction on health and survival in rhesus monkeys from the NIA study. *Nature*. 2012;489(7415):318–321.
61. Colman RJ, Beasley TM, Allison DB, et al. Attenuation of sarcopenia by dietary restriction in rhesus monkeys. *J Gerontol A Biol Sci Med Sci*. 2008;63(6):556–559.

62. Kayo T, Allison DB, Weindruch R, et al. Influences of aging and caloric restriction on the transcriptional profile of skeletal muscle from rhesus monkeys. *Proc Natl Acad Sci U S A*. 2001;98(9):5093–5098.
63. Roth GS, Ingram DK, Lane MA. Caloric restriction in primates and relevance to humans. *Ann N Y Acad Sci*. 2001;928:305–315.
64. Wakimoto P, Block G. Dietary intake, dietary patterns, and changes with age: an epidemiological perspective. *J Gerontol A Biol Sci Med Sci*. 2001;(56 Spec No 2):65–80.
65. Cordero Z, Drogan D, Weikert C, et al. Vitamin E and risk of cardiovascular diseases: a review of epidemiologic and clinical trial studies. *Crit Rev Food Sci Nutr*. 2010;50(5):420–440.
66. Blokhina O, Virolainen E, Fagerstedt KV. Antioxidants, oxidative damage and oxygen deprivation stress: a review. *Ann Bot*. 2003;(91 Spec No):179–194.
67. Tanaka Y, Moritoh Y, Miwa N. Age-dependent telomere-shortening is repressed by phosphorylated alpha-tocopherol together with cellular longevity and intracellular oxidative-stress reduction in human brain microvascular endotheliocytes. *J Cell Biochem*. 2007;102(3):689–703.
68. Navarro A, Gómez C, Sánchez-Pino MJ, et al. Vitamin E at high doses improves survival, neurological performance, and brain mitochondrial function in aging male mice. *Am J Physiol Regul Integr Comp Physiol*. 2005;289(5):R1392–R1399.
69. Sumien N, Forster MJ, Sohal RS. Supplementation with vitamin E fails to attenuate oxidative damage in aged mice. *Exp Gerontol*. 2003;38(6):699–704.
70. Strong R, Miller RA, Astle CM, et al. Evaluation of resveratrol, green tea extract, curcumin, oxaloacetic acid, and medium-chain triglyceride oil on lifespan of genetically heterogeneous mice. *J Gerontol A Biol Sci Med Sci*. 2013;68(1):6–16.
71. Clarke MW, Burnett JR, Croft KD. Vitamin E in human health and disease. *Crit Rev Clin Lab Sci*. 2008;45(5):417–450.
72. Sano M, Ernesto C, Thomas RG, et al. A controlled trial of selegiline, alpha-tocopherol, or both as treatment for Alzheimer's disease. The Alzheimer's Disease Cooperative Study. *N Engl J Med*. 1997;336(17):1216–1222.
73. Zandi PP, Anthony JC, Khachaturian AS, et al. Reduced risk of Alzheimer disease in users of antioxidant vitamin supplements: the Cache County Study. *Arch Neurol*. 2004;61(1):82–88.
74. Blot WJ, Li JY, Taylor PR, et al. Nutrition intervention trials in Linxian, China: supplementation with specific vitamin/mineral combinations, cancer incidence, and disease-specific mortality in the general population. *J Natl Cancer Inst*. 1993;85(18):1483–1492.
75. The Alpha-Tocopherol, Beta Carotene Cancer Prevention Study Group. The effect of vitamin E and beta carotene on the incidence of lung cancer and other cancers in male smokers. *N Engl J Med*. 1994;330(15):1029–1035.
76. Greenwald P, Anderson D, Nelson SA, et al. Clinical trials of vitamin and mineral supplements for cancer prevention. *Am J Clin Nutr*. 2007;85(1):314S–317S.
77. Brinkman MT, Karagas MR, Zens MS, et al. Minerals and vitamins and the risk of bladder cancer: results from the New Hampshire Study. *Cancer Causes Control*. 2010;21(4):609–619.
78. Bjelakovic G, Nikolova D, Gluud C. Antioxidant supplements and mortality. *Curr Opin Clin Nutr Metab Care*. 2014;17(1):40–44.
79. Ernst IM, Pallauf K, Bendall JK, et al. Vitamin E supplementation and lifespan in model organisms. *Ageing Res Rev*. 2013;12(1):365–375.
80. Abner EL, Schmitt FA, Mendiondo MS, et al. Vitamin E and all-cause mortality: a meta-analysis. *Curr Aging Sci*. 2011;4(2):158–170.
81. Vivekananthan DP, Penn MS, Sapp SK, et al. Use of antioxidant vitamins for the prevention of cardiovascular disease: meta-analysis of randomised trials. *Lancet*. 2003;361(9374):2017–2023.
82. Lee IM, Cook NR, Gaziano JM, et al. Vitamin E in the primary prevention of cardiovascular disease and cancer: the Women's Health Study: a randomized controlled trial. *JAMA*. 2005;294(1):56–65.
83. Miller 3rd. ER, Pastor-Barriuso R, Dalal D, et al. Meta-analysis: high-dosage vitamin E supplementation may increase all-cause mortality. *Ann Intern Med*. 2005;142(1):37–46.
84. Bjelakovic G, Nikolova D, Gluud LL, et al. Mortality in randomized trials of antioxidant supplements for primary and secondary prevention: systematic review and meta-analysis. *JAMA*. 2007;297(8):842–857.
85. Bjelakovic G, Nikolova D, Gluud LL, et al. Antioxidant supplements for prevention of mortality in healthy participants and patients with various diseases. *Cochrane Database Syst Rev*. 2012;3: CD007176.
86. Dolara P, Bigagli E, Collins A. Antioxidant vitamins and mineral supplementation, lifespan expansion and cancer incidence: a critical commentary. *Eur J Nutr*. 2012;51(7):769–781.
87. Jacobs Jr. DR, Gross MD, Tapsell LC. Food synergy: an operational concept for understanding nutrition. *Am J Clin Nutr*. 2009;89(5):1543S–1548S.
88. Hu FB. Dietary pattern analysis: a new direction in nutritional epidemiology. *Curr Opin Lipidol*. 2002;13(1):3–9.
89. Gu Y, Scarmeas N. Dietary patterns in Alzheimer's disease and cognitive aging. *Curr Alzheimer Res*. 2011;8(5):510–519.
90. Roman B, Carta L, Martínez-González MA, et al. Effectiveness of the Mediterranean diet in the elderly. *Clin Interv Aging*. 2008;3(1):97–109.
91. Babio N, Bullo M, Salas-Salvado J. Mediterranean diet and metabolic syndrome: the evidence. *Public Health Nutr*. 2009;12(9A):1607–1617.
92. Scarmeas N, Stern Y, Tang MX, et al. Mediterranean diet and risk for Alzheimer's disease. *Ann Neurol*. 2006;59(6):912–921.
93. Sofi F, Cesari F, Abbate R, et al. Adherence to Mediterranean diet and health status: meta-analysis. *BMJ*. 2008;337:a1344.
94. Trichopoulou A, Costacou T, Bamia C, et al. Adherence to a Mediterranean diet and survival in a Greek population. *N Engl J Med*. 2003;348(26):2599–2608.
95. Zazpe I, Sánchez-Tainta A, Toledo E, et al. Dietary patterns and total mortality in a Mediterranean cohort: the SUN project. *J Acad Nutr Diet*. 2014;114(1):37–47.
96. Estruch R, Ros E, Salas-Salvadó J, et al. Primary prevention of cardiovascular disease with a Mediterranean diet. *N Engl J Med*. 2013;368(14):1279–1290.
97. Herskind AM, McGue M, Holm NV, et al. The heritability of human longevity: a population-based study of 2872 Danish twin pairs born 1870–1900. *Hum Genet*. 1996;97(3):319–323.
98. Kerber RA, O'Brien E, Smith KR, et al. Familial excess longevity in Utah genealogies. *J Gerontol A Biol Sci Med Sci*. 2001;56(3):B130–B139.
99. Barral S, Cosentino S, Costa R, et al. Exceptional memory performance in the Long Life Family Study. *Neurobiol Aging*. 34(11):2445–2448.
100. Cosentino S, Schupf N, Christensen K, et al. Reduced prevalence of cognitive impairment in families with exceptional longevity. *JAMA Neurol*. 70(7):867–874.
101. Newman AB, Glynn NW, Taylor CA, et al. Health and function of participants in the Long Life Family Study: a comparison with other cohorts. *Aging (Albany NY)*. 3(1):63–76.

102. Terry DF, Wilcox MA, McCormick MA, et al. Cardiovascular advantages among the offspring of centenarians. *J Gerontol A Biol Sci Med Sci*. 2003;58(5):M425–M431.
103. Terry DF, Wilcox MA, McCormick MA, et al. Lower all-cause, cardiovascular, and cancer mortality in centenarians' offspring. *J Am Geriatr Soc*. 2004;52(12):2074–2076.
104. Mitchell BD, Hsueh WC, King TM, et al. Heritability of lifespan in the Old Order Amish. *Am J Med Genet*. 2001;102(4):346–352.
105. Ljungquist B, Berg S, Lanke J, et al. The effect of genetic factors for longevity: a comparison of identical and fraternal twins in the Swedish Twin Registry. *J Gerontol A Biol Sci Med Sci*. 1998;53(6):M441–M446.
106. Lee JH, Flaquer A, Costa R, et al. Genetic influences on lifespan and survival among elderly African-Americans, Caribbean Hispanics, and Caucasians. *Am J Med Genet A*. 2004;128A(2):159–164.
107. Christensen K, Gaist D, Vaupel JW, McGue M. Genetic contribution to rate of change in functional abilities among Danish twins aged 75 years or more. *Am J Epidemiol*. 2002;155(2):132–139.
108. McClearn GEM, Johansson B, Berg S, et al. Substantial genetic influence on cognitive abilities in twins 80 or more years old. *Science*. 1997;276(5318):1560–1563.
109. McGue M, Christensen K. The heritability of level and rate-of-change in cognitive functioning in Danish twins aged 70 years and older. *Exp Aging Res*. 2002;28(4):435–451.
110. Frederiksen H, Gaist D, Petersen HC, et al. Hand grip strength: a phenotype suitable for identifying genetic variants affecting mid- and late-life physical functioning. *Genet Epidemiol*. 2002;23(2):110–122.
111. Perls TT, Bubrick E, Wager CG, et al. Siblings of centenarians live longer. *Lancet*. 1998;351(9115):1560.
112. Gudmundsson H, Gudbjartsson DF, Frigge M, et al. Inheritance of human longevity in Iceland. *Eur J Hum Genet*. 2000;8(10):743–749.
113. Willcox BJ, Willcox DC, He Q, et al. Siblings of Okinawan centenarians share lifelong mortality advantages. *J Gerontol A Biol Sci Med Sci*. 2006;61(4):345–354.
114. Schoenmaker M, de Craen AJ, de Meijer PH, et al. Evidence of genetic enrichment for exceptional survival using a family approach: the Leiden Longevity Study. *Eur J Hum Genet*. 2006;14(1):79–84.
115. Evert J, Lawler E, Bogan H, Perls T. Morbidity profiles of centenarians: survivors, delayers, and escapers. *J Gerontol A Biol Sci Med Sci*. 2003;58(3):232–237.
116. Beekman M, Nederstigt C, Suchiman HE, et al. Genome-wide association study (GWAS)-identified disease risk alleles do not compromise human longevity. *Proc Natl Acad Sci U S A*. 107(42):18046–18049.
117 Sebastiani P, Perls TT. The genetics of extreme longevity: lessons from the new England centenarian study. *Front Genet*. 3:277.
118. Davignon J, Gregg RE, Sing CF. Apolipoprotein E polymorphism and atherosclerosis. *Arteriosclerosis*. 1988;8(1):1–21.
119. Pablos-Méndez A, Mayeux R, Ngai C, et al. Association of apo E polymorphism with plasma lipid levels in a multiethnic elderly population. *Arterioscler Thromb Vasc Biol*. 1997;17(12):3534–3541.
120. Corder EH, Saunders AM, Strittmatter WJ, et al. Gene dose of apolipoprotein E type 4 allele and the risk of Alzheimer's disease in late onset families. *Science*. 1993;261(5123):921–923.
121. Mayeux R, Stern Y, Ottman R, et al. The apolipoprotein epsilon 4 allele in patients with Alzheimer's disease. *Ann Neurol*. 1993;34(5):752–754.
122. Deelen J, Beekman M, Uh HW, et al. Genome-wide association study identifies a single major locus contributing to survival into old age; the APOE locus revisited. *Aging Cell*. 10(4):686–698.
123. Nebel A, Kleindorp R, Caliebe A, et al. A genome-wide association study confirms APOE as the major gene influencing survival in long-lived individuals. *Mech Ageing Dev*. 132(6–7):324–330.
124. Sebastiani P, Solovieff N, Dewan AT, et al. Genetic signatures of exceptional longevity in humans. *PLoS One*. 7(1):e29848.
125. Beekman M, Blanché H, Perola M, et al. Genome-wide linkage analysis for human longevity: Genetics of Healthy Aging Study. *Aging Cell*. 12(2):184–193.
126. Lee JH, Tang MX, Schupf N, et al. Mortality and apolipoprotein E in Hispanic, African-American, and Caucasian elders. *Am J Med Genet*. 2001;103(2):121–127.
127. Ewbank DC. Mortality differences by APOE genotype estimated from demographic synthesis. *Genet Epidemiol*. 2002;22(2):146–155.
128. Ewbank DC. The APOE gene and differences in life expectancy in Europe. *J Gerontol A Biol Sci Med Sci*. 2004 Jan; 59(1):16–20.
129. Ewbank DC. Differences in the association between apolipoprotein E genotype and mortality across populations. *J Gerontol A Biol Sci Med Sci*. 2007 Aug;62(8)899–907.
130. Jacobsen R, Martinussen T, Christiansen L, et al. Increased effect of the ApoE gene on survival at advanced age in healthy and long-lived Danes: two nationwide cohort studies. *Aging Cell*. 2010;9(6):1004–1009.
131. Sobel E, Louhija J, Sulkava R, et al. Lack of association of apolipoprotein E allele epsilon 4 with late-onset Alzheimer's disease among Finnish centenarians. *Neurology*. 1995;45(5):903–907.
132. Rebeck GW, Perls TT, West HL, et al. Reduced apolipoprotein epsilon 4 allele frequency in the oldest old Alzheimer's patients and cognitively normal individuals. *Neurology*. 1994;44(8):1513–1516.
133. Schächter F, Faure-Delanef L, Guénot F, et al. Genetic associations with human longevity at the APOE and ACE loci. *Nat Genet*. 1994;6(1):29–32.
134. Schupf N, Barral S, Perls T, et al. Apolipoprotein E and familial longevity. *Neurobiol Aging*. 2013;34(4):1287–1291.
135. Hirose N, Homma S, Arai Y, et al. Tokyo Centenarian Study. 4. Apolipoprotein E phenotype in Japanese centenarians living in the Tokyo Metropolitan area. *Nippon Ronen Igakkai Zasshi*. 1997;34(4):267–272.
136. Louhija J, Miettinen HE, Kontula K, et al. Aging and genetic variation of plasma apolipoproteins. Relative loss of the apolipoprotein E4 phenotype in centenarians. *Arterioscler Thromb*. 1994;14(7):1084–1089.
137. Reich EE, Montine KS, Gross MD, et al. Interactions between apolipoprotein E gene and dietary alpha-tocopherol influence cerebral oxidative damage in aged mice. *J Neurosci*. 2001;21(16):5993–5999.
138. Willcox BJ, Donlon TA, He Q, et al. FOXO3A genotype is strongly associated with human longevity. *Proc Natl Acad Sci U S A*. 2008;105(37):13987–13992.
139. Anselmi CV, Malovini A, Roncarati R, et al. Association of the FOXO3A locus with extreme longevity in a southern Italian centenarian study. *Rejuvenation Res*. 2009;12(2):95–104.

140. Flachsbart F, Caliebe A, Kleindorp R, et al. Association of FOXO3A variation with human longevity confirmed in German centenarians. *Proc Natl Acad Sci U S A*. 2009;106(8):2700–2705.
141. Li Y, Wang WJ, Cao H, et al. Genetic association of FOXO1A and FOXO3A with longevity trait in Han Chinese populations. *Hum Mol Genet*. 2009;18(24):4897–4904.
142. Pawlikowska L, Hu D, Huntsman S, et al. Association of common genetic variation in the insulin/IGF1 signaling pathway with human longevity. *Aging Cell*. 2009;8(4):460–472.
143. Soerensen M, Dato S, Christensen K, et al. Replication of an association of variation in the FOXO3A gene with human longevity using both case-control and longitudinal data. *Aging Cell*. 9(6):1010–1017.
144. Arai Y, Hirose N, Yamamura K, et al. Deficiency of choresteryl ester transfer protein and gene polymorphisms of lipoprotein lipase and hepatic lipase are not associated with longevity. *J Mol Med*. 2003;81(2):102–109.
145. Barzilai N, Atzmon G, Derby CA, et al. A genotype of exceptional longevity is associated with preservation of cognitive function. *Neurology*. 2006;67(12):2170–2175.
146. Barzilai N, Atzmon G, Schechter C, et al. Unique lipoprotein phenotype and genotype associated with exceptional longevity. *JAMA*. 2003;290(15):2030–2040.
147. Atzmon G, Rincon M, Schechter CB, et al. Lipoprotein genotype and conserved pathway for exceptional longevity in humans. *PLoS Biol*. 2006;4(4):e113.
148. Puca AA, Daly MJ, Brewster SJ, et al. A genome-wide scan for linkage to human exceptional longevity identifies a locus on chromosome 4. *Proc Natl Acad Sci U S A*. 2001;98(18):10505–10508.
149. Geesaman BJ, Benson E, Brewster SJ, et al. Haplotype-based identification of a microsomal transfer protein marker associated with the human lifespan. *Proc Natl Acad Sci U S A*. 2003;100(24):14115–14120.
150. Frederiksen H, Gaist D, Bathum L, et al. Angiotensin I-converting enzyme (ACE) gene polymorphism in relation to physical performance, cognition and survival – a follow-up study of elderly Danish twins. *Ann Epidemiol*. 2003;13(1):57–65.
151. Todesco L, Angst C, Litynski P, et al. Methylenetetrahydrofolate reductase polymorphism, plasma homocysteine and age. *Eur J Clin Invest*. 1999;29(12):1003–1009.
152. Barzilai N, Shuldiner AR. Searching for human longevity genes: the future history of gerontology in the post-genomic era. *J Gerontol A Biol Sci Med Sci*. 2001;56(2):M83–M87.
153. Suh Y, Atzmon G, Cho MO, et al. Functionally significant insulin-like growth factor I receptor mutations in centenarians. *Proc Natl Acad Sci U S A*. 2008;105(9):3438–3442.
154. Newman AB, Walter S, Lunetta KL, et al. A meta-analysis of four genome-wide association studies of survival to age 90 years or older: the Cohorts for Heart and Aging Research in Genomic Epidemiology Consortium. *J Gerontol A Biol Sci Med Sci*. 65(5):478–487.
155. Walter S, Atzmon G, Demerath EW, et al. A genome-wide association study of aging. *Neurobiol Aging*. 32(11):2109 e15–2109 e28.
156. Lee JH, Cheng R, Honig LS, et al. Genome wide association and linkage analyses identified three loci-4q25, 17q23.2, and 10q11.21-associated with variation in leukocyte telomere length: the Long Life Family Study. *Front Genet*. 2014;4:310.
157. Codd V, Mangino M, van der Harst P, et al. Common variants near TERC are associated with mean telomere length. *Nat Genet*. 2010;42(3):197–199.
158. Soerensen M, Thinggaard M, Nygaard M, et al. Genetic variation in TERT and TERC and human leukocyte telomere length and longevity: a cross-sectional and longitudinal analysis. *Aging Cell*. 2012;11(2):223–227.
159. Hartmann N, Reichwald K, Lechel A, et al. Telomeres shorten while Tert expression increases during ageing of the short-lived fish Nothobranchius furzeri. *Mech Ageing Dev*. 2009;130(5):290–296.
160. Mangino M, Hwang SJ, Spector TD, et al. Genome-wide meta-analysis points to CTC1 and ZNF676 as genes regulating telomere homeostasis in humans. *Hum Mol Genet*. 2012;21(24):5385–5394.
161. Martin GM, Oshima J. Lessons from human progeroid syndromes. *Nature*. 2000;408(6809):263–266.
162. Bohr VA, Brosh Jr. RM, von Kobbe C, et al. Pathways defective in the human premature aging disease Werner syndrome. *Biogerontology*. 2002;3(1–2):89–94.
163. Opresko PL, Otterlei M, Graakjaer J, et al. The Werner syndrome helicase and exonuclease cooperate to resolve telomeric D loops in a manner regulated by TRF1 and TRF2. *Mol Cell*. 2004;14(6):763–774.
164. Arking DE, Krebsova A, Macek Sr. M, et al. Association of human aging with a functional variant of klotho. *Proc Natl Acad Sci U S A*. 2002;99(2):856–861.
165. Bathum L, Christiansen L, Jeune B, et al. Apolipoprotein E genotypes: relationship to cognitive functioning, cognitive decline, and survival in nonagenarians. *J Am Geriatr Soc*. 2006;54(4):654–658.
166. Hayden KM, Zandi PP, Lyketsos CG, et al. Apolipoprotein E genotype and mortality: findings from the Cache County Study. *J Am Geriatr Soc*. 2005;53(6):935–942.
167. Coppin H, Bensaid M, Fruchon S, et al. Longevity and carrying the C282Y mutation for haemochromatosis on the HFE gene: case control study of 492 French centenarians. *BMJ*. 2003;327(7407):132–133.
168. Bonafè M, Barbieri M, Marchegiani F, et al. Polymorphic variants of insulin-like growth factor I (IGF-I) receptor and phosphoinositide 3-kinase genes affect IGF-I plasma levels and human longevity: cues for an evolutionarily conserved mechanism of lifespan control. *J Clin Endocrinol Metab*. 2003;88(7):3299–3304.
169. Ershler WB, Keller ET. Age-associated increased interleukin-6 gene expression, late-life diseases, and frailty. *Annu Rev Med*. 2000;51:245–270.

Further Reading

Zhang J, Asin-Cayuela J, Fish J, et al. Strikingly higher frequency in centenarians and twins of mtDNA mutation causing remodeling of replication origin in leukocytes. *Proc Natl Acad Sci U S A*. 2003;100(3):1116–1121.

Bonafè M, Olivieri F, Mari D, et al. P53 codon 72 polymorphism and longevity: additional data on centenarians from continental Italy and Sardinia. *Am J Hum Genet*. 1999;65(6):1782–1785.

CHAPTER

47

Vitamins: Cobalamin and Folate

David Watkins[*], *Charles P. Venditti*[†], *and David S. Rosenblatt*[*,‡]
[*]McGill University, Montreal, QC, Canada
[†]National Human Genome Research Institute, National Institutes of Health, Bethesda, MD, USA
[‡]McGill University Health Centre, Montreal, QC, Canada

COBALAMIN

Absorption, Transport, and Metabolism

Derivatives of cobalamin (Cbl, vitamin B_{12}) are essential cofactors for two reactions in mammalian cells: 1) mitochondrial methylmalonyl-CoA mutase, which converts methylmalonyl-CoA to succinyl-CoA and requires 5′-deoxyadenosylcobalamin (AdoCbl) for activity; and 2) cytosolic methionine synthase, which catalyzes methylation of homocysteine to form methionine, generating methylcobalamin (MeCbl) during its catalytic cycle. Inherited disorders of transport or metabolism of cobalamin result in deficient activity in either one or both of these enzymes.[1]

Dietary cobalamin is obtained almost exclusively from animal sources. Free dietary cobalamin is initially bound to salivary haptocorrin (HC; R binder, transcobalamin I). Proteolytic hydrolysis of HC occurs in the acidic milieu of the stomach. Subsequently, cobalamin binds in the proximal ileum to intrinsic factor (IF) secreted by the parietal cells of the stomach. The resulting Cbl–IF complex then binds in the distal ileum to "cubam," a specific enterocyte brush border receptor formed by the products of the *CUBN* (cubilin) and *AMN* (amnionless) genes. The cubam–Cbl–IF complex is internalized by endocytosis and undergoes lysosomal degradation, releasing cobalamin. This intracellular cobalamin is released into the portal circulation bound to transcobalamin (TC).[2]

Cobalamin circulates bound to both HC and TC, but only TC-bound cobalamin is taken up by cells outside the liver. The TC–Cbl complex is internalized by carrier-mediated endocytosis mediated by the transcobalamin receptor (TCblR)[3] and is initially processed in the lysosome, where TC undergoes proteolytic degradation. Egress of cobalamin depends on the lysosomal membrane proteins LMBD1 and ABCD4. Upon release from the lysosome, cobalamin associates with the chaperone protein MMACHC, and the upper axial ligand is removed concomitant with the partial reduction of the central cobalt atom. Subsequently, cobalamin may either become associated with methionine synthase in the cytoplasm, with conversion to MeCbl, or be transported into the mitochondria and metabolized into AdoCbl, which becomes associated with methylmalonyl-CoA mutase.[4] The MMADHC protein plays a role in directing cobalamin to either the cytoplasm or mitochondria, but the mechanism is unknown.[5]

Disorders of Absorption, Transport, and Cellular Uptake

These disorders are characterized clinically by the core hematologic finding of megaloblastic anemia.[1] Failure to thrive, developmental delay, or myelopathy may also be seen. These disorders may result from: 1) absent or deficient intrinsic factor; 2) deficient enterocyte uptake and transport of Cbl–IF (Imerslund–Gräsbeck syndrome); or 3) deficiency of TC. An extended duration of the initial illness, often the product of delayed or inadequate treatment, is associated with the prominence of neurologic symptoms. This is particularly true if the anemia is treated with folate alone in the absence of concurrent cobalamin supplementation.

Inheritance for all these disorders is autosomal recessive. Mutations in the *GIF* gene on chromosome 11q13 have been identified in patients with hereditary IF deficiency.[6] Mutations in the *CUBN* gene on chromosome 10p12.1[7] and

Rosenberg's Molecular and Genetic Basis of Neurological and Psychiatric Disease
http://dx.doi.org/10.1016/10.1016/B978-0-12-410529-4.00047-4

the *AMN* gene on chromosome 14q32[8] have been identified in patients with IGS. Mutations in the *TCN2* gene on chromosome 22q12-q13 have been identified in patients with TC deficiency.[9,10] In IF deficiency and IGS, there is low serum cobalamin, normal gastric function and morphology, and absence of autoantibodies to stomach parietal cells and IF. In TC deficiency, serum cobalamin may be normal because the majority of serum cobalamin is bound to HC. Methylmalonic acidemia and hyperhomocysteinemia may be present, but at levels below those encountered in the disorders of intracellular cobalamin metabolism. Proteinuria is often present in patients with IGS.[11] There is phenotypic overlap between IGS and hereditary IF deficiency, and it is not always possible to distinguish between the two on the basis of clinical findings. Traditionally, they could be differentiated using the Schilling test of cobalamin absorption, which was abnormal in both but corrected by the addition of exogenous human IF in IF deficiency, but not in IGS. Since the Schilling test has become increasingly less available, it has been suggested that identification of mutations in the *CUBN*, *AMN* or *GIF* genes represents the best means of determining the correct diagnosis in these patients.[6] For most patients with TC deficiency, serum TC is not detectable by electrophoresis; however, some patients have immunoreactive TC that has abnormal cobalamin-binding properties.[12,13] Molecular genetic studies should also be pursued to search for mutations in the *TCN2* gene.

Treatment of these disorders requires initial therapy with hydroxocobalamin at a daily dose of 1 mg given by the intramuscular route. The dose can be lowered to twice weekly or less with careful monitoring of hematologic and biochemical parameters, particularly the serum methylmalonic acid concentration. In TC deficiency, oral therapy or the use of cyanocobalamin has resulted in a poorer outcomes.[14] A spectrum of outcomes is possible, and reversal of neurologic symptoms has been reported; however, this largely depends on the duration of symptoms before the initiation of treatment. As with patients who suffer nutritional vitamin B_{12} deficiency, peripheral neuropathy can be present and may not resolve despite hematological and biochemical improvement. Folate supplementation will assist in the reversal of hematologic abnormalities but will have no sustained effect without concurrent adequate cobalamin treatment.

Patients with decreased endocytosis of TC-bound cobalamin caused by mutations in the transcobalamin receptor, encoded by the *CD320* gene, have been identified by expanded newborn screening for methylmalonic aciduria.[3,15] These patients had mild elevations in propionylcarnitine, methylmalonic acid and homocysteine, but were without symptoms of metabolic illness. Homozygous mutations in this gene have also been documented in healthy college students,[16] making it uncertain that mutations in *CD320* always cause illness.

Disorders of Intracellular Metabolism

Nine inherited defects in intracellular cobalamin metabolism have been identified by somatic cell complementation analysis (*cblA* to *cblG*, *cblJ*, *cblX*; Figure 47.1).[1,17,18] Affected patients display more severe metabolic disturbances than those encountered in disorders of cobalamin absorption, transport, and uptake. The study of these disorders has provided an improved understanding of many aspects of normal cellular cobalamin metabolism.

Cells from patients with *cblC*, *cblD*, *cblF*, *cblJ*, and *cblX* disorders display decreased synthesis of both AdoCbl and MeCbl, resulting in the elevation of both homocysteine and methylmalonic acid in the body fluids. The *cblC* disorder is the most common of the inborn errors of cobalamin metabolism, with over 550 patients now identified.[19–21] Patients typically present in the first year of life with megaloblastic anemia, feeding difficulties, hypotonia, and lethargy, progressing in the absence of treatment to significant neurologic impairment in which seizures and retinopathy, beginning as a macular "bulls-eye" and progressing into diffuse salt-and-pepper pigmentary changes, are frequent.[22,23] Late-onset *cblC* occurs less frequently and is characterized by mental status changes (delirium, psychosis, executive dysfunction) and long-tract (pyramidal) findings. Fibroblasts from *cblC* patients typically have low intracellular cobalamin content, with decreased synthesis of both AdoCbl and MeCbl. A similar biochemical profile was apparent in the first identified cases of the *cblD* disorder.[24] However, additional *cblD* patients have presented with isolated homocystinuria (*cblD* variant 1) or isolated methylmalonic aciduria (*cblD* variant 2).[5] Clinical and biochemical manifestations in *cblD* variant 1 are virtually identical to those evident in the *cblE* or *cblG* disorders, while those of patients with *cblD* variant 2 greatly resemble those of *cblA* patients, particularly with respect to the *in vivo* responsiveness to hydroxocobalamin.

The *cblF* disorder has been identified in 15 patients to date and the *cblJ* disorder in three. Clinical presentation has been quite variable, with hematological abnormalities, feeding difficulties, skin changes, encephalopathy, hypotonia, and developmental delay.[25,26] A number of patients have had congenital heart defects. Fibroblasts are characterized by increased intracellular cobalamin content, the majority of which is unmetabolized cobalamin trapped in lysosomes. Some have had low levels of serum cobalamin, which has been attributed to impaired intracellular trafficking of cobalamin in the enterocyte caused by the underlying metabolic condition, not dietary insufficiency.

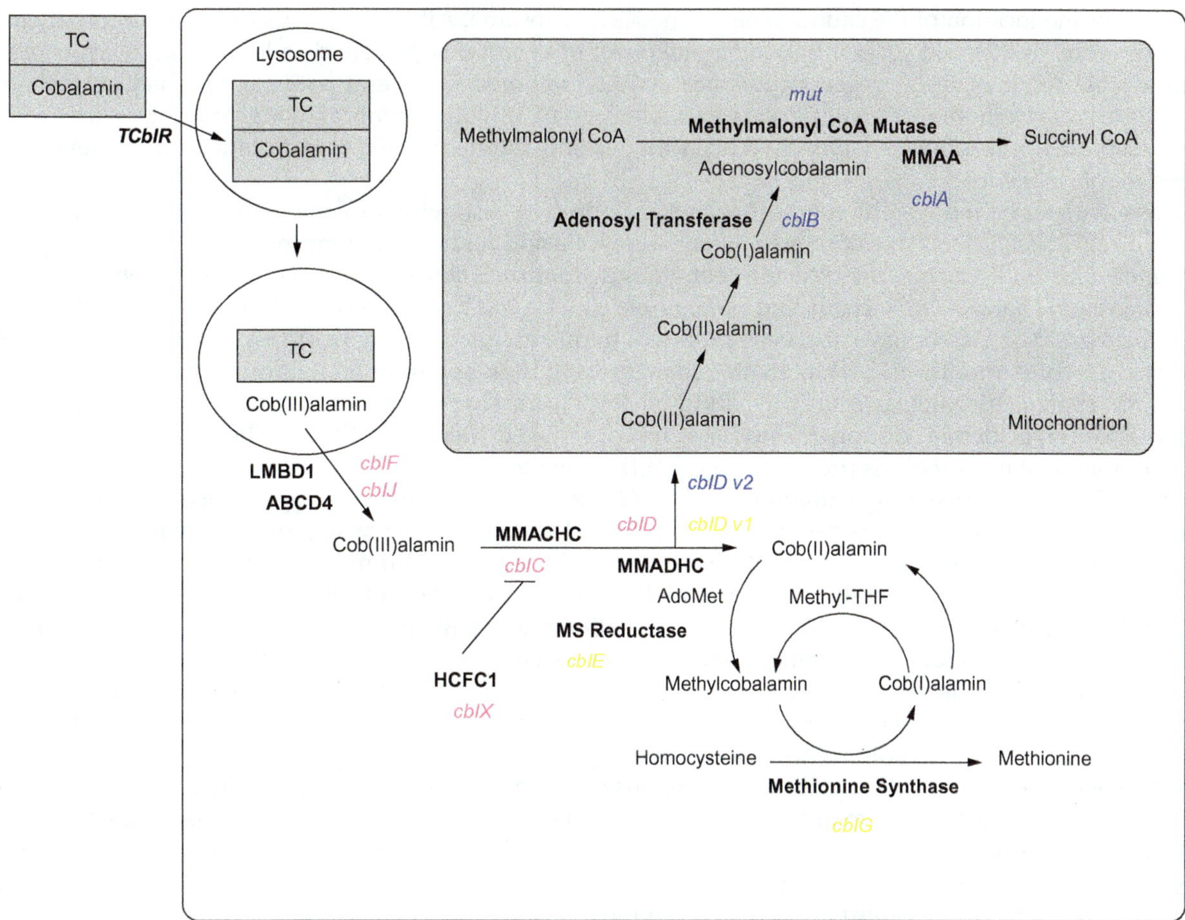

FIGURE 47.1 **Overview of cobalamin metabolism showing the locations of the defects of cellular metabolism.** Circulating transcobalamin–cobalamin complex is internalized via the transcobalamin receptor (TCblR) and egresses from the lysosome in an LMBD1- and ABCD4-dependent fashion to begin cytosolic and mitochondrial metabolism. *cblA-cblX*, location of complementation groups; *cblD v1*, cblD variant 1 (homocystinuria only); *cblD v2*, cblD variant 2 (methylmalonic aciduria only); *mut*, location of the mut complementation class; Cob(III)alamin, cobalamin with cobalt in the Co^{+++} state; Cob(II)alamin, cobalamin with cobalt in the Co^{++} state; Cob(I)alamin, cobalamin with cobalt in the Co^{+} state; MTRR, methionine synthase reductase; MTR, methionine synthase; AdoCbl, 5′ deoxyadenosylcobalamin; MeCbl, methylcobalamin. Disorders in: pink = combined methymalonic acidemia–hyperhomocysteinemia; yellow = isolated hyperhomocysteinemia; blue = isolated methymalonic acidemia. Mutations in HCFC1 (*cblX* disorder) cause decreased expression of MMACHC.

Inheritance of all these disorders is autosomal recessive. More than 75 mutations in the *MMACHC* gene on chromosome 1p34.1 have been identified in over 550 *cblC* patients.[19–21,27] A c.271dupA mutation is particularly common in patients of European descent, and is associated with a severe phenotype. A different mutation, c.609G>A (p.W203X), accounts for over 50% of mutations in the Chinese population, where the c.271dupA mutation is rare.[21] Other correlations between specific *MMACHC* mutations and severity of clinical presentation have been observed.[28] Mutations in the *MMADHC* gene on chromosome 2q23.2 cause *cblD*.[29] MMADHC interacts with the MMACHC protein and appears to be involved in determining whether cobalamin becomes associated with methionine synthase in the cytoplasm or with methylmalonyl-CoA mutase in the mitochondria.[30,31] Mutations affecting its *C*-terminal region are associated with isolated homocystinuria; mutations affecting the *N*-terminal region are associated with isolated methylmalonic aciduria; while truncating mutations are associated with combined homocystinuria and methylmalonic aciduria.[29]

The *cblF* disorder is caused by mutations in the *LMBRD1* gene on chromosome 6q13, while the *cblJ* disorder is caused by mutations of the *ABCD4* gene on chromosome 14q24.3.[17,25] The products of both of these genes are believed to play a role in the egress of cobalamin from lysosomes.

Recently mutations in the *HCFC1* gene on chromosome Xq28 have been identified in 14 male patients that had received a diagnosis of *cblC* on the basis of complementation analysis. In these patients, no *MMACHC* mutations could be identified.[18] The product of the *HCFC1* gene is a transcriptional coregulator.[32] This novel entity was named *cblX* because of the location of the causal gene on the X chromosome. All *cblX* patients came to clinical attention in the first 5 months of life, and several had antenatal onset of symptoms. Severe developmental delay, choreoathetosis, intractable seizures, and congenital microcephaly were noted in several patients; one had polymicrogyria. Homocystinuria/hyperhomocysteinemia were variably present, but all patients exhibited methylmalonic aciduria/acidemia. Typically, metabolic changes have been milder and the neurologic manifestations more severe in *cblX* compared to *cblC* patients.[18]

Disorders with impaired MeCbl synthesis alone include two distinct complementation classes, *cblE* and *cblG*. Patients with these disorders have both megaloblastic anemia and neurologic symptoms.[33] Most patients present at an early age with significant feeding difficulties and failure to thrive. Developmental delay, hypotonia, and seizures often occur. Visual changes are variable and optic nerve disease has been described. Late-onset cases from both of these complementation classes have also been reported, frequently manifesting neuropsychiatric symptoms and/or myelopathy. Cerebral atrophy and white matter changes have been apparent on neuroimaging studies of *cblE* and *cblG* patients. There is megaloblastic anemia, and homocystinuria/hyperhomocysteinemia in the absence of methylmalonic aciduria/acidemia. Cultured fibroblasts reveal reduced MeCbl synthesis, whereas AdoCbl synthesis is normal. Complementation analysis provides the definitive diagnosis.

The *cblG* disorder is caused by mutations in the *MTR* gene on chromosome 1q43, which encodes methionine synthase.[34,35] The most common, a recurrent c.3518C>T (p.P1173L) mutation, was present only in the heterozygous state.[36] The *cblE* disorder is caused by mutations in the *MTRR* gene on chromosome 5p15, which encodes methionine synthase reductase, a protein that maintains methionine synthase-bound cobalamin in its active, fully reduced state.[37] The most common mutations are an intronic mutation that results in inclusion of a 140-bp intronic sequence in MTRR mRNA, and a c.1361C>T mutation that may be associated with mild phenotype when homozygous.[38,39]

Two complementation classes (*cblA* and *cblB*) have been identified in isolated AdoCbl deficiency. Early-onset and late-onset presentations have been described in both. The early-onset form features acutely ill infants with vomiting, lethargy, hyperammonemia, hypotonia and encephalopathy. The later-onset form presents either indolently with developmental delay or acutely with metabolic decompensation that can mimic a toxic ingestion.[40–42] Patients with the *cblA* or *cblB* disorder have methylmalonic aciduria/acidemia without homocystinuria/hyperhomocysteinemia or megaloblastic anemia. Frequent long-term consequences of these disorders include metabolic stroke, predominantly affecting the globus pallidus, and end stage renal disease. All *cblA* and most *cblB* patients are cobalamin-responsive. AdoCbl synthesis is reduced in cultured fibroblasts in the presence of normal MeCbl synthesis. The *cblB* disorder is caused by mutations in the *MMAB* gene on chromosome 12q24.1, which encodes ATP:cob(I)alamin adenosyltransferase;[43,44] this enzyme catalyzes synthesis of AdoCbl from fully reduced cobalamin and ATP, and the gated transfer of AdoCbl to methylmalonyl-CoA mutase.[45] *MMAB* mutations identified in *cblB* patients include a common c.556C>T (p.R186W) mutation found primarily in patients of European background, which is associated with early-onset disease.[46] The *cblA* disorder is caused by mutations at the *MMAA* gene on chromosome 4q31.32.[47] The function of the *MMAA* gene product is unknown. Studies of its bacterial ortholog suggest that it may play a role in maintaining mutase-bound AdoCbl in its active form.[48] The most common *MMAA* mutation identified in *cblA* patients is a stop mutation (c.433C>T, p.R145X).[49]

For all disorders of intracellular cobalamin utilization, treatment consists of systemic pharmacologic cobalamin supplementation. Hydroxocobalamin (OHCbl) is the preferred form of cobalamin and large doses may be needed for optimal biochemical control. For the *cblC* disorder, cyanocobalamin (CNCbl) has been shown to be ineffective. Betaine supplementation has been demonstrated to be helpful in disorders characterized by homocystinuria. Betaine supports conversion of homocysteine to methionine by liver cobalamin-independent homocysteine methyltransferase. Protein restriction is useful in patients with methylmalonic aciduria and carnitine supplementation should be considered. Outcome is variable, depending largely on age of onset, time to diagnosis, and the delay in initiating treatment. Routine metabolic monitoring should include measurements of serum and plasma metabolites, including quantitation of serum methylmalonic acid, plasma total homocysteine, plasma amino acids, plasma carnitine levels and esters, and urinary organic acids. Optimally these patients should be referred to a center with expertise in the management of inherited metabolic disease.

While methylmalonyl-CoA mutase deficiency (the *mut* complementation class) is not strictly a defect in intracellular cobalamin utilization, it will be considered here because of the clinical and biochemical similarity to the *cblA* and *cblB* disorders. A nuclear encoded, intramitochondrial homodimer, methylmalonyl-CoA mutase catalyzes the isomerization of methylmalonyl-CoA to succinyl-CoA, a key step in propionate catabolism. The enzyme has an absolute requirement for AdoCbl as an essential cofactor. Clinical variability in presentation ranges from an acute neonatal crisis, to a subacute indolent course characterized by developmental delay and recurrent ketoacidosis, to asymptomatic cases detected on routine newborn screening.[50] Two subgroups of *mut* cellular phenotypes exist, mut^0 and mut^-.

There is residual mutase activity in cells from *mut*⁻ patients, apparent when cells are grown with high concentrations of hydroxocobalamin in the culture medium. No activity is detectable in *mut⁰* cells under any conditions.

The *mut⁰* patients have an early and often fatal neonatal presentation, whereas the *mut*⁻ patients typically present in later infancy or childhood. Survival is usual for the *mut*⁻ patients; however, there may be significant eventual morbidity, which need not be preceded by episodic acidosis and clinically apparent metabolic imbalance.[50] Inheritance is autosomal recessive. Over 240 mutations in the *MUT* gene, localized to chromosome 6p12, have been identified.[51] Several have been shown to be common within specific populations, including p.G717V in patients of African origin,[52] p.E117X in Japanese patients,[53] and p.R108C in American Hispanic patients.[51] Treatment of mutase deficiency consists of dietary protein restriction, carnitine supplementation and intermittent administration of antimicrobial agents to reduce enteric anaerobic production of propionate. Cobalamin supplementation is not effective in patients with complete absence of mutase activity. Whether *mut*⁻ patients can show an *in vivo* response to cobalamin therapy is uncertain, and individual responses should be assayed.

FOLATE

Absorption, Transport, and Metabolism

Folates are essential cofactors for a variety of one-carbon transfer reactions involved in cellular intermediary metabolism.[1] Folate metabolism is complex (Figure 47.2). One-carbon groups incorporated into reduced tetrahydrofolate (THF) are derived from the metabolism of serine and glycine, and in some tissues, histidine. Derivatives of

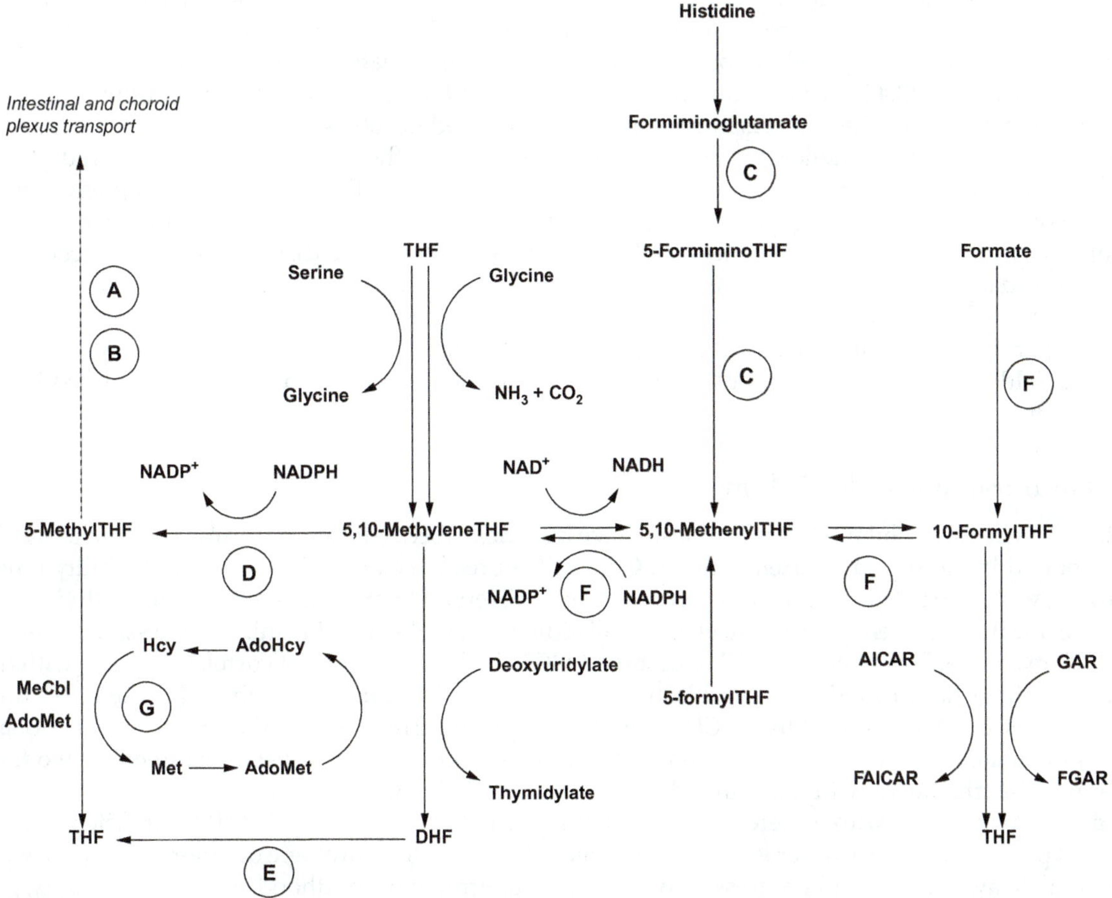

FIGURE 47.2 **Folate metabolism showing sites (A, B, C, D, E, F and G) of the confirmed enzymatic blocks involved in inborn errors of folate metabolism.** A = herrteditary folate malabsorption; B = cerebral folate deficiency; C = Glutamate formiminotransferase deficiency; D = methylenetetrahydrofolate reductase (MTHFR) deficiency; E = dihydrofolate reductase deficiency; F = MTHFD1 deficiency; G = methionine synthase deficiency (cblG) and methionine synthase reductase deficiency (cblE); AdoHcy = S-adenosylhomocysteine; AdoMet = S-adenosylmethionine; AICAR = 5-phosphoribosyl-5-aminoimidazole-4-carboxamide; DHF = dihydrofolate; GAR = 5-phosphoribosylglycinamide; Hcy = homocysteine; Met = methionine; THF = tetrahydrofolate.

THF containing one-carbon groups in different reduction states are required for two steps in the *de novo* synthesis of purines (10-formylTHF), for synthesis of thymidylate (5,10-methyleneTHF), and for synthesis of methionine from homocysteine (5-methylTHF).

Dietary folates are predominantly polyglutamate forms that undergo intestinal hydrolysis to monoglutamates prior to absorption. Intestinal absorption is by a specific transport system, which also appears to mediate folate transport across the choroid plexus.[54] Cellular uptake is also mediated by specific transport systems.

Disorders of Absorption and Transport

Hereditary folate malabsorption (congenital malabsorption of folate) involves a defect in the common transport system for folate at the intestine and choroid plexus levels.[1] Clinical symptoms typically begin in the first few months of life and include megaloblastic anemia, failure to thrive, diarrhea, seizures (frequently intractable), and progressive neurologic deterioration. Infections with unusual organisms associated with hypoimmunoglobulinemia can mimic a severe combined immune deficiency. Intracranial calcifications have been observed on imaging in some patients. Serum, red blood cell, and cerebrospinal fluid folate levels are low. A severe specific abnormality in the intestinal absorption of orally administered folate is apparent. Inheritance is autosomal recessive. Mutations in the gene encoding a high affinity proton-coupled folate transporter (PCFT; the *SLC46A1* gene on chromosome 17q11.2) have been identified in several patients with hereditary folate malabsorption.[54,55] Treatment involves the administration of pharmacologic amounts of reduced folates, usually by parenteral means.

Cerebral folate deficiency in the presence of normal serum folate levels has been identified in a number of patients and has been associated with a specific deficiency of uptake of folate across the blood–brain barrier. Intestinal uptake is unimpaired in these patients. The defect is related to dysfunction of folate receptor α, which is required in addition to PCFT for transport of folate across the blood–brain barrier at the choroid plexus. In many patients, decreased uptake may be the result of acquired antibodies directed against the folate receptor,[56] but mutations in the *FOLR1* gene on chromosome 11q13.4 have been identified in a number of families in which the condition is inherited as an autosomal recessive trait.[57,58] The defect is not manifested in cultured fibroblasts.

Children with cerebral folate deficiency initially appear normal, but there is a deceleration of head growth, irritability, and sleep disturbances starting at approximately 4 months of age.[59,60] Patients develop progressive neurodevelopmental delay, hypotonia, and progressive ataxia. Spasticity, dyskinesia, visual disturbances that can lead to visual loss and optic atrophy, sensorineural deafness, epilepsy, and autism spectrum disorders may also occur.

Oral folinic acid (0.5–1 mg per kg body weight), a stable reduced form of folate that readily crosses the blood–brain barrier, is the most effective treatment for this disorder.[59] Folic acid is less effective and should be avoided. Treatment can result in marked clinical improvement if started at an early age. Neurological recovery may be incomplete in older patients (older than 6 years). Response to therapy needs to be monitored and dosage modified if response is poor.

Disorders of Intracellular Metabolism

Four disorders in intracellular folate metabolism have been described. Glutamate formiminotransferase deficiency has been reported in more than a dozen patients. Clinical features have included cognitive disability ranging from mild to severe, with associated seizures and cortical atrophy observed in more severely affected individuals.[1] Speech delay may be a feature. The enzymatic defect affects histidine metabolism and results in excessive formiminoglutamate (FIGLU) excretion. However, the level of observed FIGLU excretion does not correlate directly with disease severity. Other findings have included megaloblastic anemia, elevated histidine, and normal folate levels. Inheritance is autosomal recessive. Mutations of the *FTCD* gene on chromosome 21q22.3, which encodes a bifunctional enzyme containing glutamate formiminotransferase and cyclodeaminase activities, have been identified in two families.[61] It is not clear whether treatment with folate in order to normalize FIGLU levels is necessary or effective.

More than 100 patients with severe methylenetetrahydrofolate reductase (MTHFR) deficiency are known. Phenotypic expression in this disorder is variable and includes: 1) acute neurologic deterioration in early infancy; 2) developmental delay, seizures, and progressive neurologic impairment in childhood; and 3) myelopathy, psychiatric disease, and recurrent cerebrovascular events in adolescence and adulthood.[62] The enzymatic defect results in deficient formation of methylTHF, causing reduced methylation of homocysteine to yield methionine. Homocystinuria/hyperhomocysteinemia is apparent, although less than that observed in cystathionine β-synthase deficiency. Serum homocysteine is elevated, and serum methionine is low. There is no megaloblastic anemia. It is unclear if the neurologic features of MTHFR deficiency are the result of elevated homocysteine, low methionine, or interference in critical S-adenosylmethionine-dependent methylation reactions. Inheritance is autosomal recessive. Over 55 different mutations at the *MTHFR* gene on chromosome 1p36 have been identified in MTHFR-deficient patients.[63] Nearly all of these appear to be private mutations, restricted to one or two families; an exception is a c.1117C>T mutation that is common among Old Order Amish.[64] Early treatment with betaine (which functions as a substrate for an alternative means of methyl transfer between homocysteine and methionine) is recommended. Folic acid, pyridoxine, methionine, and cobalamin supplementation have also been used but with less success.

A common polymorphism in *MTHFR* (c.677C>T, p.A222V) results in a thermolabile enzyme with about 50% normal enzyme activity.[65] There is evidence that homozygosity for the T allele of this polymorphism, in the presence of inadequate dietary folate intake, results in increased plasma homocysteine levels.[66] While extremely elevated total homocysteine (tHcy) levels, such as those seen in patients with cystathionine β-synthase deficiency or inborn errors affecting homocysteine remethylation, have been shown to increase risk of cardiovascular disorders including stroke, large epidemiologic studies have not shown any effect of lowering tHcy in the general population.[67,68] It is known that periconceptional folate supplementation reduces the risk of offspring with neural tube defects, and that homozygosity for the T allele of the 677C>T polymorphism is a risk factor for neural tube defects.[69] Currently, there is much interest in the effects of additional *MTHFR* polymorphisms, as well as polymorphisms in additional genes implicated in folate and homocysteine metabolism, on susceptibility to a variety of other conditions, including cleft lip and palate, Down syndrome and autism.

Dihydrofolate reductase (DHFR) deficiency, caused by mutations in the *DHFR* gene on chromosome 5q14.1 and inherited as an autosomal recessive trait, has been identified in three families. DHFR is required for reduction of dihydrofolate, which is generated during activity of thymidylate synthase, to tetrahydrofolate. In the absence of DHFR activity, folate is trapped as dihydrofolate and unavailable for cellular metabolism. (Figure 47.2). Severely affected individuals came to medical attention in the first 4 months of life with megaloblastic anemia and pancytopenia, cerebral folate deficiency, and seizures with cerebellar and cerebral atrophy.[70,71] Less severely affected patients presented later, between 2 and 5 years of age, with megaloblastic anemia and seizures; one apparently asymptomatic sibling had macrocytosis and an abnormal electroencephalogram (EEG). Therapy with folinic acid resulted in resolution of biochemical and hematologic symptoms and some improvement of neurological problems, although seizures have continued in some cases, and the more severely affected patients display significant developmental delays. Folic acid therapy resulted in clinical improvement in one patient but another developed serious neurological problems on this regimen.[70,71]

A single patient with MTHFD1 deficiency has been described.[72,73] The *MTHFD1* gene on chromosome 14q22.3 encodes a cytoplasmic trifunctional enzyme that catalyzes synthesis of 10-formyl-THF from formate and THF, and its conversion to 5,10-methenyl-THF and 5,10-methylene-THF. The patient had megaloblastic anemia, severe combined immune deficiency, atypical hemolytic uremic syndrome and seizures, hyperhomocysteinemia, methylmalonic acidemia, and cerebral folate deficiency. Treatment with folinic acid resulted in clinical improvement, although seizures were refractory to treatment and the patient remained developmentally delayed.

References

1. Watkins D, Rosenblatt DS. Inherited disorders of folate and cobalamin transport and metabolism. In: Valle D, Beaudet AL, Vogelstein B, et al., eds. *The Online Metabolic and Molecular Bases of Inherited Disease*; 2011. http://dx.doi.org/10.1036/ommbid.187, Accessed 01.03.14.
2. Kozyraki R, Cases O. Vitamin B12 absorption: mammalian physiology and acquired and inherited disorders. *Biochimie*. 2013;95:1002–1007.
3. Quadros EV, Nakayama Y, Sequeira JM. The protein and gene encoding the receptor for cellular uptake of transcobalamin bound cobalamin. *Blood*. 2009;113:186–192.
4. Gherasim C, Lofgren M, Banerjee R. Navigating the B_{12} road: assimilation, delivery and disorders of cobalamin. *J Biol Chem*. 2013;288:13186–13193.
5. Suormala T, Baumgartner MR, Coelho D, et al. The cblD defect causes either isolated or combined deficiency of methylcobalamin and adenosylcobalamin synthesis. *J Biol Chem*. 2004;279:42742–42749.
6. Tanner SM, Li Z, Perko JD, et al. Hereditary juvenile cobalamin deficiency caused by mutations in the intrinsic factor gene. *Proc Natl Acad Sci U S A*. 2005;102:4130–4133.
7. Tanner SM, Aminoff M, Wright FA, et al. Amnionless, essential for mouse gastrulation, is mutated in recessive hereditary megaloblastic anemia. *Nat Genet*. 2003;33(33):426–429.
8. Aminoff M, Carter JE, Chadwick RB, et al. Mutations in CUBN, encoding the intrinsic factor-vitamin B_{12} receptor, cubilin, cause hereditary megaloblastic anaemia 1. *Nat Genet*. 1999;21:309–313.
9. Li N, Rosenblatt DS, Seetharam B. Nonsense mutations in human transcobalamin II deficiency. *Biochem Biophys Res Commun*. 1994;204:1111–1118.
10. Namour F, Helfer AC, Quadros EV, et al. Transcobalamin deficiency due to activation of an intra exonic cryptic splice site. *Br J Haematol*. 2003;123:915–920.
11. Gräsbeck R. Imerslund-Gräsbeck syndrome (selective vitamin B12 malabsorption with proteinuria). *Orphanet J Rare Dis*. 2006;1:17.

12. Haurani FI, Hall CA, Rubin R. Megaloblastic anemia as a result of an abnormal transcobalamin II (Cardeza). *J Clin Invest*. 1979;64:1253–1259.
13. Seligman PA, Steiner LL, Allen RH. Studies of a patient with megaloblastic anemia and an abnormal transcobalamin II. *N Engl J Med*. 1980;303:1209–1212.
14. Trakadis YJ, Alfares A, Bodamer OA, et al. Update on transcobalamin deficiency: clinical presentation, treatment and outcome. *J Inherit Metab Dis*. 2014;37:1128–1129.
15. Quadros EV, Lai SC, Nakayama Y, et al. Positive newborn screen for methylmalonic aciduria identifies the first mutation in TCblR/*CD320*, the gene for cellular uptake of transcobalamin-bound vitamin B_{12}. *Hum Mutat*. 2010;31:924–929.
16. Pangilinan F, Mitchell A, VanderMeer J, et al. Transcobalamin II receptor polymorphisms are associated with increased risk for neural tube defects. *J Med Genet*. 2010;47:677–685.
17. Coelho D, Kim JC, Miousse IR, et al. Mutations in *ABCD4* cause a new inborn error of vitamin B_{12} metabolism. *Nat Genet*. 2012;44:1152–1155.
18. Yu HC, Sloan JL, Scharer G, et al. An X-linked cobalamin disorder caused by mutations in transcriptional coregulator *HCFC1*. *Am J Hum Genet*. 2013;93:506–514.
19. Lerner-Ellis JP, Anastasio N, Liu J, et al. Spectrum of mutations in *MMACHC*, allelic expression, and evidence for genotype-phenotype correlations. *Hum Mutat*. 2009;30:1072–1081.
20. Nogueira C, Aiello C, Cerone R, et al. Spectrum of *MMACHC* mutations in Italian and Portuguese patients with combined methylmalonic aciduria and homocystinuria, cblC type. *Mol Genet Metab*. 2008;93:475–480.
21. Wang F, Han L, Yang Y, et al. Clinical, biochemical, and molecular analysis of combined methylmalonic acidemia and hyperhomocysteinemia (cblC type) in China. *J Inherit Metab Dis*. 2010;33(suppl 3):S435–S442.
22. Carrillo-Carrasco N, Chandler RJ, Venditti CP. Combined methylmalonic acidemia and homocystinuria, cblC type. I. Clinical presentation, diagnosis and management. *J Inher Metab Dis*. 2012;35:91–102.
23. Carrillo-Carrasco N, Venditti CP. Combined methylmalonic acidemia and homocystinuria, cblC type. II. Complications, pathophysiology, and outcomes. *J Inher Metab Dis*. 2012;35:103–114.
24. Goodman SI, Moe PG, Hammond KB, Mudd SH, Uhlendorf BW. Homocystinuria with methylmalonic aciduria: two cases in a sibship. *Biochem Med*. 1970;4:500–515.
25. Rutsch F, Gailus S, Miousse IR, et al. Identification of a putative lysosomal cobalamin exporter mutated in the cblF inborn error of vitamin B_{12} metabolism. *Nat Genet*. 2009;41:234–239.
26. Rutsch F, Gailus S, Suormala T, Fowler B. *LMBRD1*: the gene for the cblF defect of vitamin B_{12} metabolism. *J Inher Metab Dis*. 2010;33:17–24.
27. Lerner-Ellis JP, Tirone JC, Pawelek PD, et al. Identification of the gene responsible for methylmalonic aciduria and homocystinuria, cblC type. *Nat Genet*. 2006;38:93–100.
28. Morel CF, Lerner-Ellis JP, Rosenblatt DS. Combined methylmalonic aciduria and homocystinuria (*cblC*): phenotype-genotype correlations and ethnic-specific observations. *Mol Genet Metab*. 2006;88:315–321.
29. Coelho D, Suormala T, Stucki M, et al. Gene identification for the cblD defect of vitamin B_{12} metabolism. *N Engl J Med*. 2008;358:1454–1464.
30. Deme JC, Miousse IR, Plesa M, et al. Structural features of recombinant MMADHC isoforms and their interactions with MMACHC, proteins of mammalian vitamin B_{12} metabolism. *Mol Genet Metab*. 2012;107:352–362.
31. Gherasim C, Hannibal L, Rajagopalan D, Jacobsen DW, Banerjee R. The C-terminal domain of CblD interacts with CblC and influences intracellular cobalamin partitioning. *Biochimie*. 2013;95:1023–1032.
32. Michaud J, Praz V, Faresse NJ, et al. HCFC1 is a common component of active human CpG-island promoters and coincides with ZNF143, THAP11, YY1 and GABP transcription factor occupancy. *Genome Res*. 2013;23:907–916.
33. Watkins D, Rosenblatt DS. Functional methionine synthase deficiency (cblE and cblG): clinical and biochemical heterogeneity. *Am J Med Genet*. 1989;34:427–434.
34. Leclerc D, Campeau E, Goyette P, et al. Human methionine synthase: cDNA cloning and identification of mutations in patients of the *cblG* complementation group of folate/cobalamin disorders. *Hum Mol Genet*. 1996;5:1867–1874.
35. Gulati S, Baker P, Li YN, et al. Defects in human methionine synthase in cblG patients. *Hum Mol Genet*. 1996;5:1859–1865.
36. Watkins D, Ru M, Hwang HY, et al. Hyperhomocysteinemia due to methionine synthase deficiency, cblG: structure of the MTR gene, genotype diversity, and recognition of a common mutation, P1173L. *Am J Hum Genet*. 2002;71:143–153.
37. Leclerc D, Wilson A, Dumas R, et al. Cloning and mapping of a cDNA for methionine synthase reductase, a flavoprotein defective in patients with homocystinuria. *Proc Natl Acad Sci U S A*. 1998;95:3059–3064.
38. Homolova K, Zavadakova P, Doktor TK, et al. The deep intronic c.903+469T>C mutation in the *MTRR* gene creates an SF2/ASF binding exonic splicing enhancer, which leads to pseudoexon activation and causes the cblE type of homocystinuria. *Hum Mutat*. 2010;31:437–444.
39. Vilaseca MA, Vilarinho L, Zavadakova P, et al. CblE type of homocystinuria: mild clinical phenotype in two patients homozygous for a novel mutation in the MTRR gene. *J Inher Metab Dis*. 2003;26:361–369.
40. Matsui SM, Mahoney MJ, Rosenberg LE. The natural history of the inherited methylmalonic acidemias. *N Engl J Med*. 1983;308:857–861.
41. Hörster F, Baumgartner MR, Viardot C, et al. Long-term outcome in methylmalonic acidurias is influenced by the underlying defect (*mut⁰*, *mut⁻*, *cblA*, *cblB*). *Pediatr Res*. 2007;62:225–230.
42. Cosson MA, Benoist JF, Touati G, et al. Long-term outcome in methylmalonic aciduria: a series of 30 French patients. *Mol Genet Metab*. 2009;97:172–178.
43. Dobson CM, Wai T, Leclerc D, et al. Identification of the gene responsible for the *cblB* complementation group of vitamin B_{12}-dependent methylmalonic aciduria. *Hum Mol Genet*. 2002;11:3361–3369.
44. Leal NA, Park SD, Kima PE, Bobik TA. Identification of the human and bovine ATP:Cob(I)alamin adenosyltransferase cDNAs based on complementation of a bacterial mutant. *J Biol Chem*. 2003;278:9227–9234.
45. Yamanishi M, Vlasie M, Banerjee R. Adenosyltransferase: an enzyme and an escort for coenzyme B_{12}? *Trends Biochem Sci*. 2005;30:304–308.
46. Lerner-Ellis JP, Gradinger AB, Watkins D, et al. Mutation and biochemical analysis of patients belonging to the *cblB* complementation class of vitamin B_{12}-dependent methylmalonic aciduria. *Mol Genet Metab*. 2006;87:219–225.
47. Dobson CM, Wai T, Leclerc D, et al. Identification of the gene responsible for the *cblA* complementation group of vitamin B_{12}-responsive methylmalonic acidemia based on analysis of prokaryotic gene arrangements. *Proc Natl Acad Sci U S A*. 2002;99:15554–15559.

48. Padovani D, Banerjee R. A G-protein editor gates coenzyme B_{12} loading and is corrupted in methylmalonic aciduria. *Proc Natl Acad Sci U S A*. 2009;106:21567–21572.
49. Lerner-Ellis JP, Dobson CM, Wai T, et al. Mutations in the *MMAA* gene in patients with the *cblA* disorder of vitamin B_{12} metabolism. *Hum Mutat*. 2004;24:509–516.
50. Shevell MI, Matiaszuk N, Ledley FD, Rosenblatt DS. Varying neurological phenotypes among *mut*^{0} and *mut*$^-$ patients with methylmalonylCoA mutase deficiency. *Am J Med Genet*. 1993;45:619–624.
51. Worgan LC, Niles K, Tirone JC, et al. Spectrum of mutations in *mut* methylmalonic acidemia and identification of a common Hispanic mutation and haplotype. *Hum Mutat*. 2006;27:31–43.
52. Adjalla CE, Hosack A, Matiaszuk N, Rosenblatt DS. A common mutation among blacks with *mut*$^-$ methylmalonic aciduria. *Hum Mutat*. 1998;Suppl 1:S248–S250.
53. Ogasawara M, Matsubara Y, Mikami H, Narisawa K. Identification of two novel mutations in the methylmalonyl-CoA mutase gene with decreased levels of mutant mRNA in methylmalonic acidemia. *Hum Mol Genet*. 1994;3:867–872.
54. Qiu A, Jansen M, Sakaris A, et al. Identification of an intestinal folate transporter and the molecular basis for hereditary folate malabsorption. *Cell*. 2006;127:917–928.
55. Shin DS, Mahadeo K, Min SH, et al. Identification of novel mutations in the proton-coupled folate transporter (PCFT-SLC46A1) associated with hereditary folate malabsorption. *Mol Genet Metab*. 2011;103:33–37.
56. Ramaekers VT, Rothenberg SP, Seqeira JM, et al. Autoantibodies to folate receptors in the cerebral folate deficiency syndrome. *N Engl J Med*. 2005;352:1985–1991.
57. Steinfeld R, Grapp M, Kraetzner R, et al. Folate receptor alpha defect causes cerebral folate transport deficiency: a treatable neurodegenerative disorder associated with disturbed myelin metabolism. *Am J Hum Genet*. 2009;85:354–363.
58. Grapp M, Just IA, Linnankivi T, et al. Molecular characterization of folate receptor 1 mutations delineates cerebral folate transport deficiency. *Brain*. 2012;135:2022–2031.
59. Ramaekers VT, Blau N. Cerebral folate deficiency. *Dev Med Child Neurol*. 2004;46:843–851.
60. Serrano M, Pérez-Duenas B, Montoya J, Oramazabal A, Artuch R. Genetic causes of cerebral folate deficiency: clinical, biochemical and therapeutic aspects. *Drug Discov Today*. 2012;17:9–1306.
61. Hilton JF, Christensen KE, Watkins, et al. The molecular basis of glutamate formiminotransferase deficiency. *Hum Mutat*. 2003;22:67–73.
62. Thomas MA, Rosenblatt DS. Severe methylenetetrahydrofolate reductase deficiency. In: Ueland PM, Rozen R, eds. *MTHFR Polymorphisms and Disease*. Georgetown, Texas: Landes Bioscience; 2005:41–53.
63. Leclerc D, Sibani S, Rozen R. Molecular biology of methylenetetrahydrofolate reductase (MTHFR) and overview of mutations/ polymorphisms. In: Ueland PM, Rozen R, eds. *MTHFR Polymorphisms and Disease*. Georgetown, TX: Landes Bioscience; 2005:1–20.
64. Strauss KA, Morton DH, Puffenberger EG, et al. Prevention of brain disease from severe methylenetetrahydrofolate reductase deficiency. *Mol Genet Metab*. 2007;91:165–175.
65. Frosst P, Blom HJ, Milos R, et al. A candidate genetic risk factor for vascular disease: a common mutation in methylenetetrahydrofolate reductase. *Nat Genet*. 1995;10:111–113.
66. Jacques PF, Bostom AG, Williams RR, et al. Relation between folate status, a common mutation in methylenetetrahydrofolate reductase, and plasma homocysteine concentrations. *Circulation*. 1996;93:7–9.
67. Clarke R, Halsey J, Lewington S, et al. Effects of lowering homocysteine levels with B vitamins on cardiovascular disease, cancer, and cause-specific mortality. Meta-analysis of 8 randomized trials involving 37 485 individuals. *Arch Int Med*. 2010;170:1622–1631.
68. Clarke R, Bennett DA, Parish S, et al. Homocysteine and coronary heart disease: meta-analysis of *MTHFR* case-control studies, avoiding publication bias. *PLoS Med*. 2012;9:e1001177.
69. Blom HJ, Shaw GM, van den Heijer M, Finnell RH. Neural tube defects and folate: case far from closed. *Nat Rev Neurosci*. 2006;7:724–731.
70. Banka S, Blom HJ, Walter J, et al. Identification and characterization of an inborn error of metabolism caused by dihydrofolate reductase deficiency. *Am J Hum Genet*. 2011;88:216–225.
71. Cario H, Smith DEC, Blom H, et al. Dihydrofolate reductase deficiency due to a homozygous *DHFR* mutation causes megaloblastic anemia and cerebral folate deficiency leading to severe neurologic disease. *Am J Hum Genet*. 2011;88:226–231.
72. Watkins D, Schwartzentruber JA, Ganesh J, et al. Novel inborn error of folate metabolism: identification by exome capture and sequencing of mutations in the *MTHFD1* gene in a single proband. *J Med Genet*. 2011;48:590–592.
73. Keller MD, Ganesh J, Heltzer M, et al. Severe combined immunodeficiency resulting from mutations in *MTHFD1*. *Pediatrics*. 2013;131:e629–e634.

CHAPTER

48

Disorders of Biotin Metabolism

Sara Elrefai and Barry Wolf*,†*

*Henry Ford Hospital, Detroit, MI, USA

†Wayne State University School of Medicine, Detroit, MI, USA

BIOTIN

Biotin is a water-soluble, B-complex vitamin that consists of a heterocyclic ring structure with an aliphatic carbon side chain (Figure 48.1).[1] Biotin is the coenzyme for four carboxylases in humans: pyruvate carboxylase, which converts pyruvate to oxaloacetate, the initial step in gluconeogenesis; propionyl-CoA carboxylase, which catabolizes several branch-chain amino acids and odd-chain fatty acids; β-methylcrotonyl-CoA carboxylase, which is involved in the catabolism of leucine; and acetyl-CoA carboxylase, which converts malonyl-CoA to acetyl CoA, the first step in the biosynthesis of fatty acids. All four enzymes are found in the mitochondria, although acetyl-CoA carboxylase is mainly found in the cytosol.

The carboxyl group of biotin is covalently attached through an amide bond to the epsilon-amino group of the various apocarboxylases that use it as a coenzyme by HCS (EC 6.3.4.10). This reaction occurs through two partial reactions. The first is the ATP-requiring phosphorylation of biotin resulting in biotinyl 5′-AMP. In the second partial reaction, this AMP intermediate reacts with apocarboxylases, resulting in biotinylated holoenzymes. One of the nitrogens of the ureido portion of the heterocyclic ring (N-1) of biotin is involved with transferring carbon dioxide from various substrates to their respective carboxylated products. After the holocarboxylases are degraded proteolytically, biocytin (biotinyl-ε-lysine) or biotinyl-peptides are cleaved at the amide bond by the enzyme, biotinidase (EC 3.5.1.12), releasing lysine or lysyl-peptides and free biotin; the biotin can then be recycled.

HOLOCARBOXYLASE SYNTHETASE DEFICIENCY

Holocarboxylase synthetase (HCS) deficiency is an autosomal recessive disorder in which the body is unable to covalently bind biotin to the apocarboxylases to form active holocarboxylases.[1] This results in the accumulation of abnormal metabolites that are responsible for biochemical decompensation. Symptomatic individuals with HCS deficiency usually exhibit severe metabolic acidosis, feeding and breathing difficulties, hypotonia, and lethargy during the neonatal period. Diagnosis can be made by testing HCS activity and/or by mutation analysis of the *HCS* gene.

Clinical and Neurological Features

Children with HCS deficiency usually exhibit breathing abnormalities, such as tachypnea, hyperventilation, or apnea; feeding difficulties; hypotonia or hypertonia; seizures; lethargy; and irritability. Many affected children develop skin rash, and several have had alopecia. All affected children have had metabolic ketoacidosis and organic aciduria, and most have varying degrees of hyperammonemia. Abnormal urine odor has been noted in several of these children. These symptoms can result in developmental delay and coma.[1,2] Symptoms usually develop during

Rosenberg's Molecular and Genetic Basis of Neurological and Psychiatric Disease
http://dx.doi.org/10.1016/10.1016/B978-0-12-410529-4.00048-6

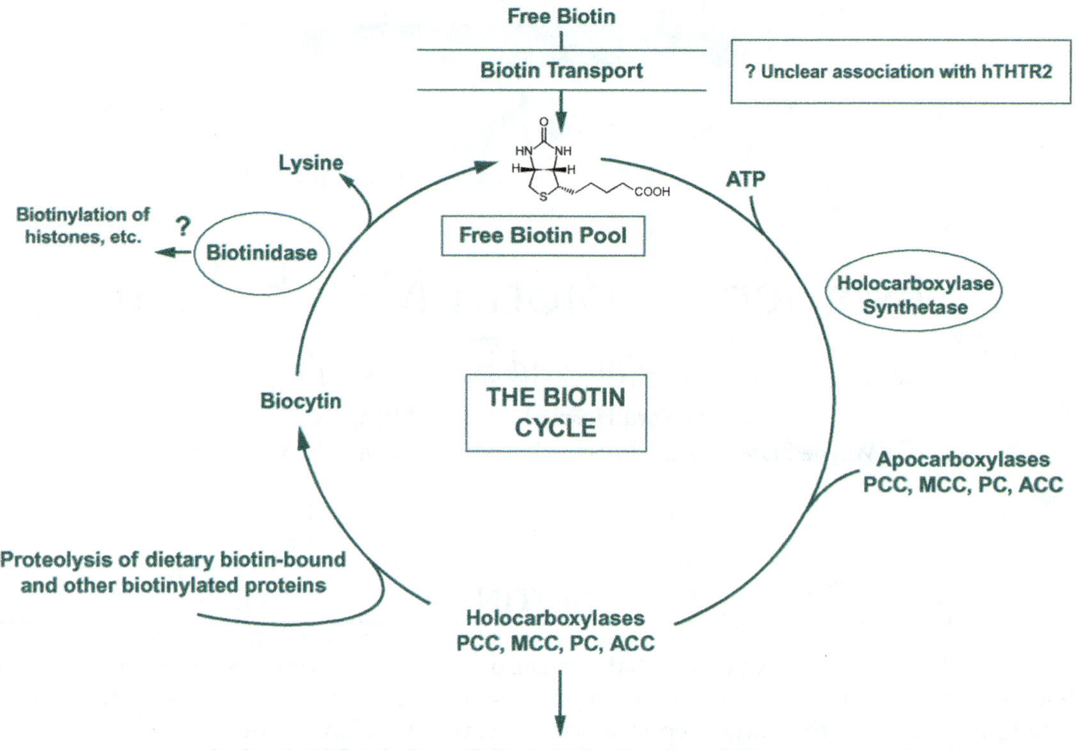

FIGURE 48.1 **The biotin cycle.** Biotin is actively transported across the intestinal and blood–brain barrier to enter the free biotin pool. Free biotin is then covalently attached to the various apocarboxylases that are important for gluconeogenesis, fatty acid synthesis and the catabolism of several branch-chain amino acids. Biotin holocarboxylase synthetase covalently attaches biotin to the various apocarboxylases producing holocarboxylases. The holocarboxylases are degraded proteolytically to biocytin or biotinyl-peptides by biotinidase, thereby recycling biotin that can enter the free biotin pool. In addition, biotinidase has biotinyl-transferase activity that may transfer biotin from biocytin to nucleophilic acceptors, such as histones or other proteins. Biotinidase may also function as a carrier of biotin in plasma. The precise mechanism by which the hTHTR2 gene interacts with biotin metabolism remains to be determined; however, it may be involved with the transport of biotin into various tissues or cells. Abbreviations: PCC: propionyl-CoA carboxylase; MCC: methylcrotonyl-CoA carboxylase; PC: pyruvate carboxylase; ACC: acetyl-CoA carboxylase; hTHTR2: human thiamine transporter 2.

the newborn period, but some have become symptomatic at several months or even years of age.[3] The findings on imaging of the brain are variable, ranging from normal to low-density changes scattered throughout the white matter of both cerebral hemispheres, with loss of demarcation between gray and white matter.[1,4] The electroencephalogram (EEG) abnormalities are also variable, ranging from normal to diffusely abnormal with burst-suppression patterns.[2,4] These aberrant patterns may normalize or resolve with biotin treatment. Siblings of several of the children who had similar symptoms and likely had the same disorder, have become comatose and died before a diagnosis was made or appropriate treatment was instituted. HCS deficiency is inherited as an autosomal recessive trait. Because a limited number of affected individuals have been reported and the disorder is not always ascertained by newborn screening, the gene frequency of the disorder cannot be estimated accurately.

The prognosis for HCS deficiency is good in those who are treated early and do not have severely altered neurological dysfunction or abnormalities. However, if untreated or unrecognized for long periods of time, the outcome can be irreversible or fatal. In acute metabolic decompensation, the affected individual may develop metabolic compromise that can lead to dehydration and seizures progressing to coma and death.

Biochemical Characterization and Molecular Genetics

HCS has been studied in both bacterial and animal systems and is highly specific for biotin.[5–7] The enzyme requires ATP and magnesium for activity. In rats, 70% of the holocarboxylase synthetase activity is located in the cytosol and the remaining 30% is in the mitochondria. The cDNA for the enzyme has been cloned and sequenced.[8,9] The gene for the HCS has been localized to chromosome 21q22.1.[10]

The primary enzyme deficiency was demonstrated by several laboratories at about the same time.[11,12] Studies of the enzymes of multiple individuals with HCS deficiency have revealed elevated K_m values of biotin, ranging from 3–70 times that of normal, and the age of onset of symptoms in these children correlated negatively with the K_m values of biotin.[13] A child with the highest K_m of biotin had a very mild clinical course. Differences were noted in the stability of the mutant enzymes, suggesting biochemical heterogeneity for the disorder.

Multiple mutations have been identified in the HCS gene.[8,14–17] Mutations within the biotin-binding region of the enzyme result in a protein with increased K_m values of biotin, whereas those outside of this region result in enzymes with normal K_m values, but reduced V_{max} values. Children with altered K_m values readily respond to biotin therapy, whereas those with normal K_m values respond to biotin, but usually not as well. One child who did not develop symptoms until 8 years of age has been reported with an abnormal gene-splicing mutation that results in decreased amounts of normal HCS mRNA.[18] Although mutation analyses of children with this disorder may be useful in predicting phenotype, currently there are no clear genotype–phenotype correlations. However, the p.L216R mutant allele, when present in a homozygous state, has been associated with a biotin-unresponsive, severe clinical phenotype.[19]

Pathophysiology

In HCS deficiency, failure to biotinylate the various carboxylases forming holocarboxylases results in the accumulation of organic acid substrates that subsequently cause acidosis and interfere with the urea cycle causing hyperammonemia. This metabolic decompensation causes the neurological dysfunction. Without biotin therapy, the affected individual may become comatose and die. With biotin therapy, the carboxylases can become partially or completely biotinylated and be functional holoenzymes. This reduces the accumulations of abnormal organic acids, resulting in normalization of acid–base status and normal urea cycle function.

Testing

HCS deficiency is usually suspected when high concentrations of the metabolites, β-hydroxyisovalerate, β-methylcrotonylglycine, β-hydroxypropionate, methylcitrate, and/or lactate, indicative of multiple carboxylase deficiency, are found in the urine of a children exhibiting some or all of the above clinical findings.[1] Children with the disorder can be ascertained by elevated organic aciduria using gas–liquid chromatography or by tandem mass spectroscopy, such as used in most newborn screening programs. Prior to biotin treatment, the activities of the various mitochondrial carboxylases are deficient in extracts of the peripheral blood leukocytes. Within hours to days after starting biotin therapy, the carboxylase activities increase to normal or near normal in these cells. Skin fibroblasts from these patients are deficient in the three mitochondrial carboxylase activities when cells are cultured in a medium containing low concentrations of biotin. Upon incubation of these cells in high concentrations of biotin, the carboxylase activities increase to normal or near normal levels. Definitive diagnosis requires the demonstration of deficient activity of holocarboxylase synthetase in peripheral blood leukocytes or cultured skin fibroblasts. A simple assay of the first partial reaction in the HCS reaction has been developed. This activity was deficient in all children previously shown to have HCS deficiency. This assay allows a more rapid confirmation of the diagnosis of this disorder. Mutation analysis of the HCS gene can also be used to make or confirm the diagnosis.

Prenatal Diagnosis

Prenatal diagnosis of HCS deficiency has been performed by demonstrating deficient activities of the mitochondrial carboxylases in cultured amniocytes. When these cells were cultured in a medium supplemented with biotin, the carboxylase activities increased toward normal. In addition, affected fetuses have had elevations of β-hydroxyisovalerate and/or methylcitrate in their amniotic fluid, using stable isotope dilution techniques. Although the latter method is a simpler and more rapid method for prenatal diagnosis, both diagnostic procedures should be performed to confirm the diagnosis, using the same sample of amniotic fluid to avoid the problem of maternal contamination of the amniotic cell sample. Mutation analysis of the HCS gene is the preferred, most accurate method for prenatal diagnosis.

Management

Most individuals with HCS deficiency have improved markedly following oral administration of 10 mg or more of biotin per day. Biochemical abnormalities correct quickly with improvement of many, if not all, of the clinical symptoms. However, a child with one of the highest K_m values of biotin continued to excrete abnormal urinary organic acids even when treated with 60–80mg of biotin per day. Based on the differences in the K_m values of biotin for the synthetase in these patients, the necessity of administering higher concentrations of biotin must be determined on an individual basis. If treatment is delayed, many of the neurologic abnormalities will not be reversible with biotin therapy.

Prenatal treatment of HCS deficiency has been performed in at least two separate at-risk pregnancies. The mothers were treated with 10mg of biotin orally, one from the 34th week of gestation and the other from the 23rd week of gestation. In both cases, the children were asymptomatic at birth, although in the second pregnancy the child became symptomatic after biotin was withheld for a short time. Once biotin treatment was reinstituted, the child's condition returned to normal. The diagnosis was confirmed in both children. Both have remained asymptomatic on biotin therapy. Because biotin treatment initiated at birth apparently prevents development of symptoms in affected newborns, it is unclear if prenatal treatment of affected children is necessary.

BIOTINIDASE DEFICIENCY

Biotinidase deficiency is an autosomal recessively inherited disorder that affects the endogenous recycling of biotin, resulting in multiple carboxylase deficiency. If untreated, individuals develop a variation of symptoms in childhood, including ataxia, seizures, hearing loss, optic atrophy, and developmental delay. Non-neurological manifestations include alopecia, skin rash, and fungal infections. Individuals with profound biotinidase deficiency have less than 10% of mean normal serum biotinidase enzyme activity. Individuals with partial biotinidase deficiency have 10–30% of mean normal serum biotinidase enzyme activity. Both profound and partial biotinidase deficiency are routinely identified by newborn screening in states where such screening is offered. Biotinidase deficiency is caused by mutations in the *BTD* gene and the disorder can be confirmed by mutation analysis.

Clinical and Neurological Features

Symptomatic individuals with biotinidase deficiency often exhibit seizures, skin rash, and/or hypotonia at several months of age.[20,21] Urinary organic acid accumulation is the consequence of multiple carboxylase deficiency due to failure to recycle biotin. It was initially proposed that these children have a defect in the intestinal or renal transport of biotin; however, in 1982 it was shown that these and other children with the late-onset form of multiple carboxylase deficiency actually have a deficiency of serum biotinidase activity.[22]

The most common neurologic features of this disorder are seizures and hypotonia. Seizures occur in about 70% of symptomatic children with profound biotinidase deficiency and are often the initial symptom of the disorder. The most common types of seizures are generalized and tonic/clonic, but some children have exhibited partial, infantile spasms, or myoclonic seizures.[23] Several individuals, especially older children, have developed peripheral neuropathies.[2] Several children have exhibited breathing abnormalities, such as hyperventilation, stridor, and apnea. Frequently they have cutaneous symptoms, including skin rash and varying degrees of alopecia. Sensorineural hearing loss[24] and eye problems, such as optic atrophy,[25] have been described in a large number of untreated children. Several children have had cellular immunologic abnormalities, manifested by fungal infection. These immunologic aberrations are varied and may represent the effects of abnormal organic acid metabolites or biotin deficiency on normal function. Biotin therapy rapidly corrects the immunologic dysfunction. Some with biotinidase deficiency have manifested only one or two of these features, whereas others have exhibited a full spectrum of neurologic and cutaneous findings. The age of onset of symptoms varies from 1 week to adolescents, with a mean age from 3–6 months. As untreated children get older, they may exhibit ataxia and developmental delay. Several children have not developed symptoms until late childhood or adolescence.[19] They exhibited symptoms that were different from those of the younger affected children: spastic paraparesis often with rapid loss of vision with progressive optic neuropathy.[26] The scotomata resolved within weeks to months, whereas the paraparesis markedly improved over months to years.

Adults with partial biotinidase deficiency can develop symptoms when stressed, but the majority will remain asymptomatic. A case report describes an individual with partial biotinidase deficiency who had a 14-month history of persistent vaginal candidiasis despite appropriate medical treatment. Her symptoms resolved after treatment with 20mg of oral biotin.[27]

Biotinidase deficiency is inherited as an autosomal recessive trait with parents exhibiting serum enzyme activities intermediate between that of deficient and normal individuals. The incidence of biotinidase deficiency is about 1:137,000 for profound biotinidase deficiency, 1:110,000 for partial biotinidase deficiency, and 1:61,000 for the combined incidence of profound and partial biotinidase deficiency.

Biochemical Characterization and Molecular Genetics

Biotinidase is found in many prokaryotes and eukaryotes, and in most mammalian tissues, particularly serum, liver, and kidney. Human biotinidase has been purified to homogeneity from serum and plasma by several groups.[28,29] The enzyme is a monomeric glycoprotein with a molecular weight of between 67,000 and 74,000 daltons. Using the protein sequence, the cDNA encoding the biotinidase gene was cloned and isolated[30] and the organization of the entire gene elucidated.[31]

Serum biotinidase is produced mainly in the liver. Biotinidase is localized intracellularly to the rough endoplasmic reticulum. Although the major function of biotinidase is to recycle biotin by cleaving biocytin and biotinyl-peptides, the enzyme has also been shown to biotinylate histones following the cleavage of biocytin. Biotinidase may also function to biotinylate other proteins or to be a biotin-binding protein in serum. The exact site of biocytin cleavage or the metabolism of the enzyme in serum and tissues remains to be determined.

Biotinidase activity is determined by measuring the release of biotin from biocytin or a biocytin analog, such as *N*-biotinyl p-aminobenzoate[28] or biotinyl-aminoquinoline.[32] Other endpoint and kinetic assays for biotinidase activity in serum and tissues are available.

Individuals with profound biotinidase activity have less than 10% mean normal activity. The parents of these children usually have serum enzyme activities intermediate between those of the patients and normal individuals. Deficient biotinidase activity has also been demonstrated in extracts of leukocytes and fibroblasts from some of these patients. At least one patient has also been shown to have deficient biotinidase activity in his liver extract.

In addition to children with profound biotinidase deficiency, a group of patients with partial deficiency, 10–30% of mean normal activity, have been identified by newborn screening. Clinical consequences of partial deficiency were not known until one child, who was identified by newborn screening but was not treated with biotin, developed hypotonia, skin rash, and hair loss at 6 months of age during a bout of gastroenteritis. The child's symptoms rapidly resolved with biotin treatment. It is possible that partial biotinidase deficiency is a problem only when affected individuals are exposed to certain stresses, such as starvation or infection. The full clinical spectrum of partial deficiency remains to be elucidated.

The *BTD* gene is located on chromosome 3p25.1.[33] More than 150 different mutations of the biotinidase gene (*BTD*) have been shown to cause biotinidase and several types of mutations have been found to cause biotinidase deficiency; missense, nonsense, single- and multiple-nucleotide deletions, single- and multiple-nucleotide insertions, deletion/insertions, cryptic splice-site mutations, and compound allelic mutations.[34]

Multiple mutations have been described in symptomatic children with profound biotinidase deficiency.[34] Five mutations are common in children with profound biotinidase deficiency who were identified by newborn screening (p.Cys33Phefs*36, p.Gln456His, p.Arg538Cys, p.Asp444His, and p.[Ala171Thr;Asp444His]).[20,35] These five mutations comprised about 60% of the abnormal alleles. Two mutations, p.Cys33Phefs*36 and p.Arg538Cys, occurred in both symptomatic individuals and children identified by newborn screening, but occurred in the symptomatic population at a significantly greater frequency. The other two common mutations, p.Gln456His and p.[Ala171Thr;Asp444His], occurred mainly in the newborn screening group. There is no clear genotype–phenotype correlation. In one study, children with symptoms of profound biotinidase deficiency with null mutations were more likely to develop hearing loss than those with missense mutations, even if not treated for a period of time.[36] In addition, essentially all children with partial biotinidase deficiency have a specific missense mutation, p.Asp444His, on one allele in combination with a mutation for profound biotinidase deficiency on the other.[37]

Pathophysiology

Biotinidase plays a role in the processing of protein-bound biotin, thereby making the vitamin available to the free biotin pool. Biotinidase also transfers biotin from the cleavage of biocytin to histones.[3]

Serum biotinidase activity is not altered by biotin deficiency. This was demonstrated by the fact that several patients who became biotin deficient while being treated with parenteral hyperalimentation that lacked biotin had normal serum biotinidase activity. Biotinidase appears to play an important role in the processing of protein-bound biotin. This may occur either by secretion into the intestinal tract, where it can release biotin that can subsequently be absorbed, or by cleaving biocytin or biotinyl-peptides in the intestinal mucosa or in the blood.

Biotinidase activity in cerebrospinal fluid and in the brain is very low. This suggests that the brain may not be able to recycle biotin and, therefore, must depend on biotin that is transported across the blood–brain barrier. Several symptomatic children who failed to exhibit peripheral lactic acidosis or organic aciduria have had elevations of lactate and/or organic acids in their cerebrospinal fluid. This compartmentalization of the biochemical abnormalities may explain why neurologic symptoms usually appear before the other symptoms.

The hearing loss in individuals with biotinidase deficiency is usually irreversible, although several young affected children have shown some improvement with therapy. The mechanism causing the hearing loss remains to be determined.

Neuropathology

The brains of two children with biotinidase deficiency have been studied. They have revealed a variety of abnormalities. One child was found to have cerebral degeneration and atrophy with an absence of the Purkinje cell layer with rarification of the granular layer and proliferation of the Bergmann layer. There was gliosis of the white matter and dentate nucleus, but the brainstem and cerebral peduncles were normal. There was focal necrosis with vascular proliferation and infiltration by macrophages, suggesting a subacute necrotizing myelopathy. There was acute meningoencephalitis of the entire central nervous system. In the second child, who died at 3 months of age, the brain revealed defective myelination, focal areas of vacuolization, and gliosis in the white matter of the cerebrum and cerebellum. There was also a mild gliosis in the pyramidal cellular layer of the hippocampus, and there were characteristic changes of viral encephalopathy in the putamen and caudate nucleus.

Computerized axial tomography of the head has been performed in several affected children at different ages, but has failed to reveal consistent findings.[2] The findings are usually normal, but may reveal diffuse cerebral atrophy with or without low attenuation of the white matter, basal ganglia calcifications, ventriculomegaly, subdural effusions, and even cysts. EEG findings range from normal to diffuse slow-wave activities that usually resolve with biotin treatment.

Audiological examinations for sensorineural hearing loss should be performed on an annual basis for individuals with profound biotinidase deficiency and biannually on those with partial biotinidase deficiency. Ophthalmological examinations for optic atrophy and other ocular abnormalities should also be performed on an annual basis for individuals with profound biotinidase deficiency and biannually on those with partial biotinidase deficiency.[38]

Testing

The majority of patients exhibit metabolic ketolactic acidosis and organic aciduria similar to that seen in HCS deficiency. The most commonly elevated urinary organic acid is β-hydroxyisovalerate. These patients may have mild hyperammonemia. There was considerable variability in expression of this disorder even within a family. The absence of organic aciduria or metabolic ketoacidosis in a symptomatic child does not exclude this disorder.

If the degree of enzyme deficiency is equivocal, mutation analysis of the *BTD* gene can be performed to confirm the diagnosis. A repository of known mutations of the *BTD* gene causing biotinidase is available online (http://www.arup.utah.edu/database/BTD/BTD_welcome.php).

Neonatal Screening

Because biotinidase deficiency met many of the criteria for considering inclusion into a neonatal screening program, simple analytical tests for biotinidase activity have been developed to determine biotinidase activity in the same blood spots currently used for other newborn screening tests.[39,40] Although tandem mass spectroscopy can identify the metabolites that are usually elevated in symptomatic children with the disorder, many, if not most, children with the deficiency will have normal organic acids immediately after birth and will not be identified on newborn screening. Therefore, only the enzymatic assay is essential for ascertainment of affected infants. Essentially all the states in the United States and in about 30 countries screen their newborns for biotinidase deficiency.

Prenatal Diagnosis

Biotinidase activity is measurable in cultured amniotic fluid cells and in amniotic fluid. In addition, mutation analysis of DNA from amniocytes is also possible. Therefore, the potential for prenatal diagnosis of biotinidase deficiency exists. Prenatal diagnosis has been performed in several at-risk pregnancies in which amniocentesis was performed because of advanced maternal age. The fetuses were unaffected, and this was confirmed after birth.

Management

All symptomatic children with biotinidase deficiency have improved after treatment with 5–10 mg of biotin per day. Biotin must be administered in the free form, as opposed to the bound form. The biochemical abnormalities and seizures rapidly resolve after biotin treatment, followed by improvement of the cutaneous manifestations. Hair growth returns over a period of weeks to months in the children with alopecia. Optic atrophy and hearing loss seem to be the most resistant to therapy, especially if a long period has elapsed between the time symptoms appeared and the time of diagnosis and initiation of treatment. Some treated children have rapidly achieved developmental milestones, whereas others have continued to show deficits.

Most individuals with biotinidase deficiency excrete large quantities of biocytin in their urine, but there is no evidence of accumulation of this metabolite in the tissues. It remains to be determined whether biotin therapy increases the concentration of biocytin in these children, and results in any adverse effects.

BIOTIN-RESPONSIVE BASAL GANGLIA DISEASE

Biotin-responsive basal ganglia disease, also known as thiamine metabolism dysfunction syndrome-2,[41] was first described in 1998.[42] The two dozen individuals described with the disorder ranged in age from 1 to 23 years and were of different ethnicities, including those of Middle Eastern, Portuguese, Indian and Japanese ancestry.[43]

Individuals with the disorder typically exhibit recurrent subacute episodes of encephalopathy, seizures, ataxia, dystonia, dysarthria, external ophthalmoplegia, and dysphagia during childhood.[44,45] These symptoms can progress to severe quadriparesis and even coma. Episodes can be triggered by febrile illness or trauma. The seizures are usually simple, partial, or generalized. Magnetic resonance imaging of the brains of affected individuals during acute episodes are characterized by bilateral lesions of the caudate nuclei and putamen nuclei. Subsequent imaging studies of these individuals reveals atrophy of the basal ganglia and necrosis.

If untreated, individuals with the disorder can progress to coma and death. However, with high-dose biotin (5–10 mg per kg per day) and thiamine treatment, partial to complete improvement can occur.

Biotin-responsive basal ganglia disease is autosomal recessive and associated with mutations in the *SLC19A3* gene located on chromosome 2q36.3, which encodes human thiamine transporter 2 (hTHTR2).[46] Biotin is not a substrate for hTHTR2 and the mechanism by which biotin improves symptoms of affected patients is still not fully understood.

Several mutations causing the disorder have been identified; however, c.1264 A>G (p.T422A) in exon 5 of the *SLC19A3* gene is the most common and is most frequent in Saudi Arabian individuals.[41] This disorder is likely underdiagnosed and should be considered in individuals with unexplained encephalopathy who show neuroimaging with bilateral edema in the putamen and caudate nuclei, brainstem, and infra- and supratentorial cortex. A trial of biotin and thiamine may be therapeutic and lifesaving.

Differential Diagnosis

Clinical features, such as vomiting, hypotonia, and seizures accompanied by metabolic ketolactic acidosis or mild hyperammonemia, are often observed in inherited metabolic diseases. Neonatal sepsis must be excluded.

Individuals with biotinidase deficiency may exhibit clinical features that are misdiagnosed as other disorders, such as isolated carboxylase deficiency, before they are correctly identified. Other symptoms that are more characteristic of biotinidase deficiency (e.g., skin rash, alopecia) can also occur in children with nutritional biotin deficiency, HCS deficiency, zinc deficiency, or essential fatty acid deficiency.

CONCLUSION

In addition to the biotin-responsive disorders discussed here, there may be other disorders in biotin metabolism that need to be elucidated. Identification of the disease processes will certainly afford a better understanding of the normal intermediary metabolism of this vitamin and perhaps provide clues about the etiologies of other vitamin-responsive syndromes.

References

1. Wolf B. Disorders of biotin metabolism. In: Scriver CR, Beaudet AL, Sly WS, Valle D, eds. *The Metabolic and Molecular Bases of Inherited Disease*. 8th ed. New York: McGraw-Hill; 2001:3935–3962.
2. Wolf B. The neurology of biotinidase deficiency. *Mol Genet Metab*. 2011;104:27–34.
3. Gibson KM, Bennett MJ, Nyhan WL, Mize CE. Late-onset holocarboxylase synthetase deficiency. *J Inherit Metab Dis*. 1996;19:739–742.
4. Wolf B. Disorders of biotin metabolism: treatable neurological syndromes. In: Rosenberg R, Prusiner SB, Di Mauro S, Barchi RL, Kunkel LM, eds. *The Molecular and Genetic Basis of Neurological Disease*. Stoneham, Mass: Butterworth Publishers; 1992:569–581.
5. Chiba Y, Suzuki Y, Narisawa K. Purification and characterization of holocarboxylase synthetase. In: *33rd Meeting of Japanese Society of Inherited Metabolic Disease*. 1991:183.
6. Chiba Y, Suzuki Y, Aoki Y, Ishida Y, Nariawa K. Purification and properties of bovine liver holocarboxylase synthetase. *Arch Biochem Biophys*. 1995;313:8–14.
7. Xia WL, Zhang J, Ahmed F. Biotin holocarboxylase synthetase: purification from rat liver cytosol and some properties. *Biochem Mol Biol Int*. 1994;34:225–232.
8. Suzuki Y, Aoki Y, Ishida Y, et al. Isolation and characterization of mutations in the human holocarboxylase synthetase cDNA. *Nat Genet*. 1994;8:122–128.
9. Leon-Del Rio A, Leclerc D, Akerman B, Wakamatsu N, Gravel R. Isolation of a cDNA encoding human holocarboxylase synthetase by functional complementation of a biotin auxotroph of Escherichia coli. *Proc Natl Acad Sci U S A*. 1995;92:4626–4630.
10. Zhang XX, Leon-Del-Rio A, Gravel RA, Eydoux P. Assignment of holocarboxylase synthetase gene HLCS. to human chromosome band 21q22.1 and to mouse chromosome band 16C4 by *in situ* hybridization. *Cytogenet Cell Genet*. 1997;76:179.
11. Burri BJ, Sweetman L, Nyhan WL. Mutant holocarboxylase synthetase: evidence for the enzyme defect in early infantile biotin-responsive multiple carboxylase deficiency. *J Clin Invest*. 1981;68:1491–1495.
12. Saunders ME, Sherwood WG, Dutchie M, Surh L, Gravel RA. Evidence for a defect of holocarboxylase synthetase activity in cultured lymphoblasts from a patient with biotin-responsive multiple carboxylase deficiency. *Am J Hum Genet*. 1982;34:590–601.
13. Burri BJ, Sweetman L, Nyhan WL. Heterogeneity of holocarboxylase synthetase in patients with biotin-responsive multiple carboxylase deficiency. *Am J Hum Genet*. 1985;37:326–337.
14. Aoki Y, Suzuki Y, Li X, et al. Characterization of mutant holocarboxylase synthetase HCS: a Km for biotin was not elevated in a patient with HCS deficiency. *Pediatr Res*. 1997;42:849–854.
15. Pomponio RJ. *Molecular Identification and Characterization of Mutations that Cause Profound Biotinidase Deficiency in Symptomatic Children: Molecular, Biochemical, and Clinical Correlations in the U.S. and Worldwide Patient Populations*. Doctoral Dissertation 1997.
16. Baumgartner R, Suormala T, Wick H, Geisert J, Lehnert W. Infantile multiple carboxylase deficiency: evidence for normal intestinal absorption but renal loss of biotin. *Helv Paediatr Acta*. 1982;37:499–502.
17. Yang XB, Aoki YB, Li XB, et al. Structure of human holocarboxylase synthetase gene and mutation spectrum of holocarboxylase synthetase deficiency. *Hum Genet*. 2001;109:526–534.
18. Holme E, Jacobson CE, Kristianson B. Biotin responsive carboxylase deficiency in an 8 year old boy with normal serum biotinidase and fibroblast holocarboxylase synthetase activity. *J Inherit Metab Dis*. 1988;11:270–276.
19. Aoki Y, Sakomato O, Hiratsuka M. Identification and characterization of mutations in patients with holocarboxylase synthetase deficiency. *Hum Genet*. 1999;104:143–148.
20. Wolf B. Biotinidase deficiency: if you have to have an inherited metabolic disease, this is the one to have. *Genet Med*. 2012;14:565–575.
21. Sakamoto O, Suzuki Y, Li X, et al. Relationship between kinetic properties of mutant enzyme and biochemical and clinical responsiveness to biotin in holocarboxylase synthetase deficiency. *Pediatr Res*. 1999;46:671–676.
22. Wolf B, Grier RE, Allen RJ, Goodman SI, Kien CL. Biotinidase deficiency: the enzymatic defect in late-onset multiple carboxylase deficiency. *Clin Chim Acta*. 1983;131:273–281.
23. Salbert BA, Pellock JM, Wolf B. Characterization of seizures associated with biotinidase deficiency. *Neurology*. 1993;43:1351–1354.
24. Wolf B, Spencer R, Gleason T. Hearing loss is a common feature of symptomatic children with profound biotinidase deficiency. *J Pediatr*. 2002;140:242–246.
25. Salbert BA, Astruc J, Wolf B. Ophthalmological findings in biotinidase deficiency. *Ophthalmologica*. 1993;206:177–181.
26. Wolf B, Pomponio RJ, Norrgard KJ, et al. Delayed-onset profound biotinidase deficiency. *J Pediatr*. 1998;132:362–365.
27. Strom CM, Levine EM. Chronic vaginal candidiasis responsive to biotin therapy in a carrier of biotinidase deficiency. *Obstet Gynecol*. 1998;92:644–646.
28. Hymes J, Wolf B. Biotinidase and its role in biotin metabolism. *Clin Chim Acta*. 1996;255:1–11.
29. Chauhan J, Dakshinamurti J. Purification and characterization of human serum biotinidase. *J Biol Chem*. 1986;261:4268–4274.
30. Cole H. *Cloning and Characterization of the Human Biotinidase Gene*. Richmond, VA: Doctoral Thesis, Medical College of Virginia/Virginia Commonwealth University; 1994.
31. Knight HC, Reynolds TR, Meyers GA, Pomponio RJ, Buck GA, Wolf B. Structure of the human biotinidase gene. *Mammal Genome*. 1998;9:327–330.
32. Wastell H, Dale G, Bartlett K. A sensitive fluorimetric rate assay for biotinidase using a new derivative of biotin, biotinyl-6-aminoquinoline. *Anal Biochem*. 1984;140:69–73.
33. Cole H, Weremowicz H, Morton CC, Wolf B. Localization of serum biotinidase BTD: to human chromosome 3 in band p25. *Genomics*. 1994;22:662–663.
34. Hymes J, Stanley CM, Wolf B. Mutations in BTD causing biotinidase deficiency. *Hum Mutat*. 2001;200:375–381.
35. Dobrowolski SF, Angeletti J, Banas RA, Naylor EW. Real time PCR assays to detect common mutations in the biotinidase gene and application of mutational analysis to newborn screening for biotinidase deficiency. *Mol Genet Metab*. 2003;78:100–107.
36. Sivri HS, Genc GA, Tokatli A, et al. Hearing loss in biotinidase deficiency: genotype–phenotype correlation. *J Pediatr*. 2007;150:439–442.
37. Swango KL, Demirkol M, Huner G, et al. Partial biotinidase deficiency is usually due to the D444H mutation in the biotinidase gene. *Hum Genet*. 1998;102:571–575.

38. Wolf B. Clinical issues and frequent questions about biotinidase deficiency. *Mol Genet Metab*. 2013;100:6–13.
39. Heard GS, Wolf B, Jefferson LG, et al. Neonatal screening for biotinidase deficiency: results of a 1-year pilot study. *J Pediatr*. 1986;108:40–46.
40. Wolf B, Heard GS. Screening for biotinidase deficiency in newborns: worldwide experience. *Pediatrics*. 1990;85:512–517.
41. Alfadhel M, Alumtashhri, Judah R, et al. Biotin-responsive basal ganglia disease should be renamed biotin-thiamine-responsive basal ganglia disease: a retrospective review of the clinical, radiological and molecular findings of 18 new cases. *Orphanet J Rare Dis*. 2013;8:83.
42. Ozand P, Gascon C, AlEssa M, et al. Biotin-responsive basal ganglia disease: a novel entity. *Brain*. 1998;121:1267–1279.
43. Debs R, Depienne C, Rastetter A, et al. Biotin-responsive basal ganglia disease in ethnic Europeans with novel SLC19A3 mutations. *Arch Neurol*. 2010;67:126–130.
44. Tabarki B, Al-Shafi S, Al-Shahwan S, et al. Biotin-responsive basal ganglia disease revisited: clinical, radiologic, and genetic findings. *Neurology*. 2013;80:261–267.
45. El-Hajj T, Karem P, Mikati M. Biotin-responsive basal ganglia disease: case report and review of the literature. *Neuropediatrics*. 2009;39:268–271.
46. Zeng WQ, Al-Yamani E, Acierno JS, et al. Biotin-responsive basal ganglia disease maps to 2q36.3 and is due to mutations in SLC19A3. *Am J Hum Genet*. 2005;77:16–26.

CHAPTER

49

Disorders of Pyridoxine Metabolism

Clara van Karnebeek[*] *and Sidney M. Gospe, Jr.*[†]

[*]British Columbia Children's Hospital, University of British Columbia, Vancouver, BC, Canada

[†]University of Washington and Seattle Children's Hospital, Seattle, WA, USA

INTRODUCTION

Disorders of pyridoxine metabolism affecting the function of the central nervous system include pyridoxine-dependent epilepsy (PDE), pyridoxal-5′-phosphate (PLP)-responsive epileptic encephalopathy, and tissue nonspecific isoenzyme of alkaline phosphatase (TNSALP) deficiency (Figure 49.1). This chapter will primarily focus on PDE, which is the best-characterized disorder of pyridoxine metabolism, while additional information regarding the other two disorders will be included in the various subsections.

CLINICAL FEATURES

Historical Overview

PDE was first described in 1954 and is considered as the prototypical form of a metabolic epilepsy which is unresponsive to antiepileptic drugs but which is specifically treated with a vitamin or cofactor. Our knowledge of this familial epileptic encephalopathy has increased dramatically over the subsequent six decades, with more than 200 cases having been reported.[1–5] Importantly, many of the clinical reports that have appeared over the past 30 years have discussed atypical phenotypes,[6–8] neurodevelopmental features,[9–13] characteristic electroencephalographic (EEG),[14–19] and imaging findings.[10,19,20]

Patients described in the initial reports of PDE presented with what is commonly thought of as the classic or neonatal form of the disorder. These newborns present with encephalopathy and seizures that are intractable to treatment with antiepileptic drugs (AEDs) and that only come under control after pharmacologic doses of pyridoxine are administered followed by maintenance therapy. In their 1954 report, Hunt et al. first described a newborn with pharmacoresistant seizures that only became controlled after the institution of scheduled treatment with a parenteral multivitamin preparation.[21] With the subsequent elimination of various components of this product, it was determined that pyridoxine (vitamin B_6) was the therapeutic substance that controlled the baby's seizures. As one of the mother's previous pregnancies resulted in a newborn that died from intractable seizures, the familial nature of this disorder was suggested. In 2006, Mills et al. discovered that mutations in the *ALDH7A1* gene, which encodes the protein antiquitin (ATQ), results in the biochemical error that underlies PDE, and this paved the way for laboratory (rather than clinical) confirmation of the disorder.[22–31]

Mode of Inheritance and Prevalence

PDE displays an autosomal recessive mode of inheritance. The *ALDH7A1* gene is located on chromosome 15q and over 80 mutations have been described. While patients with a variety of genotypes consisting of homozygous and compound heterozygous mutations have been reported, only some preliminary genotype–phenotype correlations have been suggested. PDE is a rare condition, hence, few epidemiologic studies have been conducted.[9,10,22,32,33]

Rosenberg's Molecular and Genetic Basis of Neurological and Psychiatric Disease
http://dx.doi.org/10.1016/10.1016/B978-0-12-410529-4.00049-8

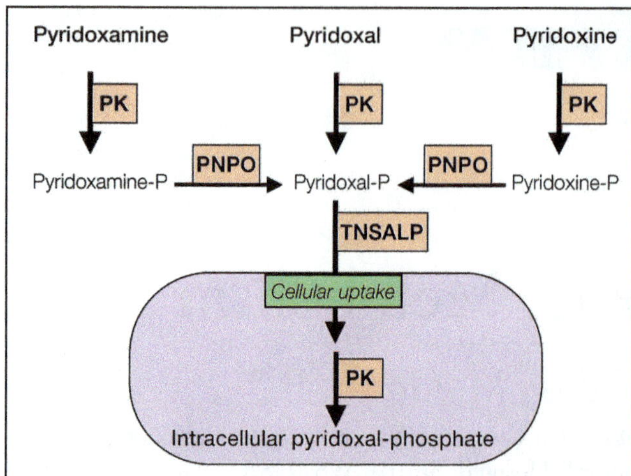

FIGURE 49.1 **Pyridoxal phosphate (PLP) is synthesized from dietary pyridoxal, pyridoxamine, or pyridoxine by pyridoxal kinase and pyridox(am)ine 5′-phosphate oxidase (PNPO); cellular uptake involves membrane-bound tissue nonspecific alkaline phosphatase (TNSALP).**

Research from the United Kingdom and the Republic of Ireland calculated a point prevalence of 1:687,000 for definite and probable cases of PDE,[32] while a somewhat greater birth incidence of 1:396,000 was estimated by a survey conducted in the Netherlands.[22] While these two studies suggest that PDE is uncommon, the disease is probably underdiagnosed and a higher birth incidence is likely. Clinical research from a German center where pyridoxine administration is part of a standard treatment protocol for neonatal seizures reported a birth incidence of probable cases of 1:20,000, which supports this notion,[33] as does a hospital-based study from India where six of 81 children with intractable seizures responded to the vitamin.[34]

NATURAL HISTORY

Patients with PDE typically present within hours or days of birth with encephalopathy and clinical seizures that are resistant to AEDs. In some instances, mothers retrospectively report rhythmic fetal movements suggestive of prenatal intrauterine seizures. Patients with PDE must receive regular pharmacologic doses of pyridoxine, and if not treated in this fashion, death from status epilepticus may follow. This clinical trajectory has been reported retrospectively in babies who expired prior to the birth and subsequent diagnosis of PDE in a younger sibling.[1,5] PDE may present within hours of birth as an epileptic encephalopathy that may mimic hypoxic-ischemic encephalopathy,[1,5,10,32] and neonatal lactic acidosis, hypoglycemia, electrolyte disturbances, hypothyroidism, and diabetes insipidus have been reported.[26,35]

A variety of clinical presentations of PDE have been described. Universally, all patients with PDE eventually develop clinical seizures that, despite treatment with large doses of one or more AEDs, either recur serially or evolve into status epilepticus. In most instances, the institution of either parenteral or oral pyridoxine rapidly results in seizure control, with improvement in the encephalopathy, and the antiepileptic medications that were instituted previously can then be discontinued. While PDE may appear in this fashion at various times during the first 2 months of life, the presentation of PDE after this age is considered to be late-onset and therefore atypical.[5,8,36] Together with the late-onset cases, other atypical forms have been reported, including infants whose seizures first respond to AEDs but then relapse weeks to months later with intractable seizures, and patients whose seizures are not controlled by initial large doses of pyridoxine but which then do respond at a later time to a second trial.[1,6–8,36] Pyridoxine supplementation must continue, or clinical seizures will again develop, typically within days. Critical to the understanding of this disorder is that patients with PDE are not pyridoxine deficient; rather they are metabolically dependent upon high-dose supplementation of the vitamin. Consequently, it is vital to educate parents, primary care providers, therapists, teachers, and others extending services to these patients about this important clinical point.

The constellation of clinical features and seizure semiology in patients with PDE is quite varied. The neonatal epileptic encephalopathy presentation may include gastrointestinal symptoms such as emesis and abdominal distention, sleeplessness, hyperalertness, irritability, paroxysmal facial grimacing, and abnormal eye movements. These striking findings are associated with recurrent partial motor seizures, generalized tonic seizures, or myoclonus. Depending upon the phenotype, response (if any) to AEDs, and timing of the institution of pyridoxine supplementation, complex partial seizures, infantile spasms, and other myoclonic seizures, as well as a mixed seizure pattern and recurrent status epilepticus may develop subsequently.[5,9,10,12,13,18,32] In all instances, after the discontinuation of pyridoxine supplementation there is a high risk of status epilepticus. With adherence to a life-long regimen of daily supplements of pharmacologic doses of pyridoxine, the prognosis for seizure control is generally excellent. Occasional breakthrough seizures may occur, such as during an acute illness when the patient is febrile or has gastroenteritis leading to a temporary reduction in pyridoxine bioavailability. Importantly, some patients taking appropriate weight- and age-based doses of pyridoxine with excellent compliance may still experience recurrent seizures and will therefore require treatment with one or more AEDs. In these instances, a secondary cause of epilepsy, such as mesial temporal sclerosis, hydrocephalus, or other brain dysgenesis such as cortical dysplasia, may be present.[37,38]

Patients with PDE characteristically have associated neurodevelopmental disabilities, including deficiencies in expressive language, as well as nonverbal cognitive deficits, hypotonia and other motor disabilities.[1,5,9] Only a few comprehensive neuropsychologic evaluations of PDE patients have been reported in the literature. A characteristic profile has been suggested by these assessments and includes a reduction in the cognitive/verbal IQ, particularly in measures of expressive language, along with a low normal motor/performance IQ.[1,9–11] In some adolescent and adult PDE patients, behavioral features characteristic of obsessive–compulsive disorder and autistic spectrum disorder have been reported. Cerebral palsy and a significant intellectual disability have been described in some severely affected patients.[9] It is tempting to hypothesize that the early diagnosis and effective treatment of PDE should result in a more favorable neurodevelopmental outcome, and this is frequently the case in kindred where a second affected child is diagnosed and effectively treated at an earlier age than the older affected sibling. Also, individuals with late-onset PDE more frequently have a better prognosis (including a few cases with reportedly normal development). The neurodevelopmental outcome of PDE is probably multifactorial, and is likely affected by: the time of clinical seizure onset; the lag time to diagnosis and effective treatment; compliance with pyridoxine therapy; underlying brain dysgenesis; and the currently unknown correlation between *ALDH7A1* genotype and neurodevelopmental phenotype.[1,5,9,12,13,26,31,39–41]

Other Disorders of Pyridoxine Metabolism

In addition to PDE, two additional disorders of pyridoxine metabolism affecting nervous system function deserve mention (Figure 49.1). PLP-responsive epileptic encephalopathy presents in a similar fashion to PDE with pharmacoresistant neonatal seizures and encephalopathy. Most infants with this disorder present with seizures that do not respond to pyridoxine but do come under control with continuous supplementation of pharmacologic doses of PLP.[42–45] This disorder is due to mutations in the *PNPO* gene, which encodes the enzyme pyridox(am)ine 5′-phosphate oxidase. Similarly, intractable neonatal seizures may occur in patients with the bone mineralization defect hypophosphatasia due to TNSALP deficiency. This deficiency of alkaline phosphatase results in abnormal vitamin B_6 absorption and metabolism, and associated seizures may be treated with pyridoxine supplementation.[46]

MOLECULAR GENETICS: ATQ

Normal Gene Product

ALDH7A1 encodes the enzyme α-aminoadipic semialdehyde dehydrogenase (ALDH7A1; E.C.1.2.1.31) commonly known as ATQ.[27,47] Human ATQ is both mitochondrial and cytosolic, a conclusion based on the presence of a potential mitochondrial targeting sequence and cleavage site, as well as localization studies.[48,49] Located at chromosome 5q31, the human ATQ cDNA (NCBI #NM_001182.3) has an open reading frame of 1620 bp divided among 18 exons.[26] Based on the NCBI predicted sequence (NP_001173.2), the human cDNA codes for a protein of 539 amino acids. Most published mutations in *ALDH7A1* have been named relative to the downstream initiation site (GenBank NM_001201377.1). Recently the Human Gene Mutation Database (HGMD) has revised this practice to name the mutations relative to the upstream site (NM_001182.4), a difference of 84 nucleotides or 28 amino acids. The crystal structure of the fish ATQ (PDB 2JG7) reveals an eight-chain asymmetric unit cell comprised of two tetramers, each a dimer of dimers.[50] Each monomer (~58 kDa mass) has three domains: NAD^+ binding, catalysis, and oligomerization. The human ATQ structure available in the protein database (PDB 2J6L) is superimposable on that of the fish enzyme.[51]

ATQ takes part in lysine catabolism in the brain and in the liver (Figure 49.2). In humans there are two biochemical pathways for lysine catabolism: the saccharopine pathway, which is dominant in the liver and many other tissues,[52] and the pipecolic acid (PA) pathway, which is dominant in brain.[53] While the saccharopine pathway is mitochondrial, the PA pathway has been located mostly in the peroxisomal compartment.[53] Convergence of the

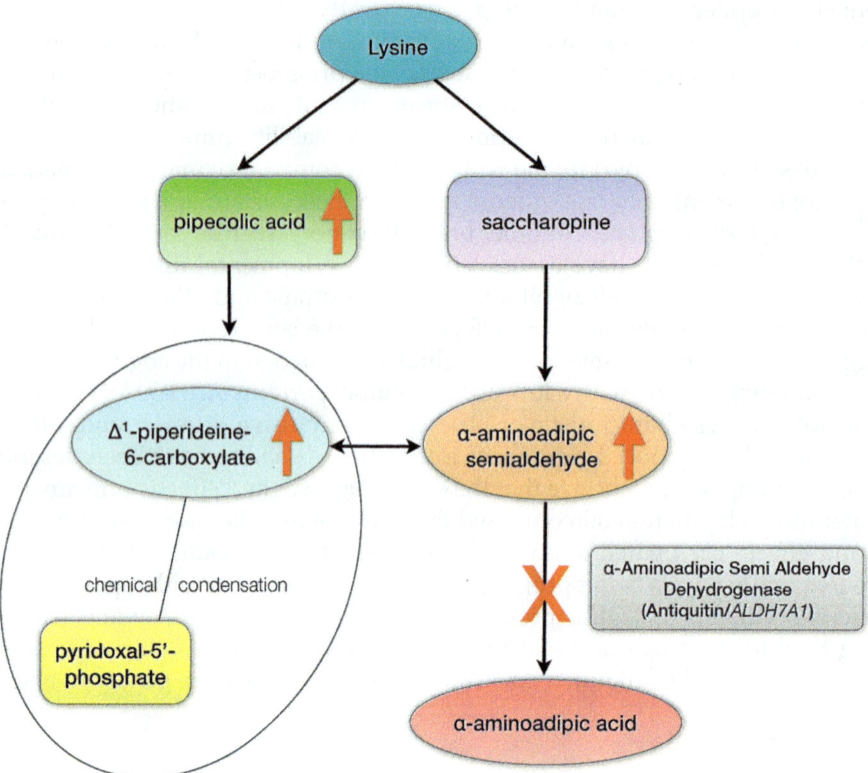

FIGURE 49.2 Schematic overview of the metabolism of L-lysine via the saccharopine and pipecolic acid (PA) pathways and biochemical pathophysiology of antiquitin (ATQ) deficiency. The two pathways converge where L-Δ^1-piperideine 6-carboxylate (P6C), produced via the pipecolic acid pathway, and α-aminoadipic semialdehyde (αAASA), produced via the saccharopine pathway, are in equilibrium. αAASA is then converted to α-aminoadipic acid (αAAA) by ATQ. In ATQ deficiency, P6C and αAASA accumulate due to a block in α-aminoadipic semialdehyde dehydrogenase (antiquitin, *ALDH7A1*). P6C undergoes chemical condensation with pyridoxal phosphate (PLP) resulting in PLP deficiency. PA accumulates due to backpressure from the enzymatic block.

two pathways occurs at the level of α-aminoadipic semialdehyde (αAASA) formation, which occurs in the cytosol. αAASA is in spontaneous equilibrium with Δ^1-piperideine-6-carboxylic acid (P6C) and is oxidized by ATQ (requiring NAD as a cofactor) to α-aminoadipic acid, which is eventually oxidized to produce acetyl CoA.[54] Of note, PA is formed in the PA pathway and modulates the function of γ-aminobutyric acid (GABA), which is a major inhibitory neurotransmitter.[55,56]

Abnormal Gene Product(s)

To date more than 80 mutations have been reported within the 18 exons of *ALDH7A1* on the HGMD website.[12,23,25–27,29–31,40,57–62] Of these, 50–60% are missense mutations, resulting in an altered amino acid sequence. A high number of missense mutations cluster around exons 14, 15, and 16.[31,62] The missense mutation p.Glu399Gln in exon 14 occurs in various populations and accounts for about 30% of published alleles.[26,29] Molecular modeling using human ATQ structure indicates that missense mutations can be divided into three categories: affecting NAD^+ cofactor binding or catalysis, altering the substrate binding pocket, and potentially disrupting dimer or tetramer assembly. The mutations are expected to have different effects on enzyme activity, with the former predicted to have the most significant impact. In at least six patients with the clinical and biochemical phenotypes of ATQ deficiency, only one mutated allele was identified.[23,26,29] Most likely, traditional Sanger sequencing missed a larger deletion or deep intronic mutation; novel technologies such as microarray analysis and whole-genome sequencing might shed light on this in the future.

Genotype–Phenotype Correlations

There are three clinical phenotypes: patients with complete seizure control on pyridoxine, with either normal development (group 1) or developmental delay (group 2), and patients with persistent seizures despite pyridoxine and developmental delay (group 3). Despite extensive studies, correlations between the genotype and these phenotypic categories have not yet been established.[26,31] An example of the different outcomes is shown in the report by Bennett et al. where of three individuals who were homozygous for p.E427Q, two were in the neonatal-onset group and one in the later-onset and clinically milder group.[23]

Because of the urgency of treatment, data on urinary αAASA prior to treatment are limited and so it is difficult to ascertain if there exists any correlation with genotype. However, there appears to be very little correlation between this biomarker in patients on pyridoxine treatment and their genotype.[26] In general, the relation between genotype and biochemical and clinical phenotype is likely to be complex. Aside from the *ALDH7A1* genotype,[31] determinants of brain pyridoxal phosphate level may include dietary lysine intake,[26] anabolic versus catabolic state, pyridoxine intake, maternal prenatal pyridoxine intake,[9,39,47] other environmental factors (e.g., infection), as well as other genetic factors affecting pyridoxine metabolism and lysine catabolism.

MOLECULAR GENETICS: PNPO AND TNSALP

The enzyme encoded by the *PNPO* gene catalyzes the terminal, rate-limiting step in the synthesis of PLP from dietary pyridoxine and pyridoxamine, and *PNPO* gene mutations result in a deficiency of PLP (Figure 49.1). Deficiency of this enzyme has been described in less than 30 patients and typically results in severe neonatal epileptic encephalopathy typically responding only to treatment with PLP; more recently, however, reports have appeared describing PNPO-deficient patients who respond to pyridoxine.[44,45] The full biochemical and clinical spectrum remains to be elucidated and little is known about genotype–phenotype correlations.[43–45,63–66]

ALPL, encoding TNSALP, is the only gene in which mutation is known to cause hypophosphatasia, a mineralization defect. TNSALP is also necessary for proper vitamin B_6 metabolism, and defective TNSALP activity results in hypophosphatasia, and in some cases intractable seizures. Recently, such an infantile case was described by Belachew et al.[46] Most affected individuals have unique mutant alleles, making the prediction of the phenotype difficult. However, there is a good correlation between the severity of the hypophosphatasia phenotype and the residual enzymatic activity produced *in vitro* by the enzyme.[67,68] Specifically for the associated seizures, little is known about genotype–phenotype correlation. According to three-dimensional modeling studies, more than 70% of the mutations affect functional domains of the protein, namely the active site, the calcium-binding site, the crown domain, and the homodimer interface.[68–73]

DISEASE MECHANISMS AND PATHOPHYSIOLOGY

Human Observations and Animal Models

While vitamin B_6 is an essential nutrient, the six vitamers of vitamin B_6 (the alcohol pyridoxine, the aldehyde pyridoxal, the amine pyridoxamine, and their respective 5′-phosphorylated esters) are universally present in both animal- and plant-derived foods, and consequently clinical pyridoxine-deficiency states are rare.[74,75] After the ingestion of pyridoxine vitamers, the phosphorylated pyridoxine esters must first be dephosphorylated by a phosphatase prior to absorption. Subsequently, pyridoxine, pyridoxamine, and pyridoxal may be phosphorylated by a kinase and the six vitamers are systemically distributed (Figure 49.1). PLP is the only biologically active vitamer, and it plays numerous roles as a cofactor in over 140 reactions, which include transamination of amino acids, decarboxylation reactions, modulation of the activity of steroid hormones, and regulation of gene expression.[76,77] Of particular neurologic significance, abnormalities of pyridoxine homeostasis may result in alterations in dopaminergic, serotonergic, glutaminergic and GABAergic neurotransmission.

ATQ dysfunction in PDE patients results in elevations in αAASA in plasma, urine and cerebrospinal fluid (CSF).[14,26,78] As αAASA elevation has also been reported in patients with molybdenum cofactor deficiency and isolated sulfite oxidase deficiency, the presence of this organic acid in body fluids is not specific for PDE. These two disorders can be diagnosed by measuring urinary levels of sulfite, sulfocysteine, and xanthine.[79] Importantly, in the presence of the characteristic clinical scenario, the detection of elevated urinary αAASA would be highly suggestive of a diagnosis of PDE. In patients with PDE, this biochemical finding persists despite years of effective treatment. Elevations of the indirect biomarker PA may also be detected in plasma and CSF, but in some PDE patients it may normalize after effective long-term therapy.[24,27,28,80] αAASA is in equilibrium with P6C, which through a Knoevenagel condensation reaction with PLP results in an inactivation of the cofactor.[27] Pathophysiologically, accumulation of αAASA results in an intracellular reduction in PLP (Figure 49.2). Therefore, while patients with PDE are not systemically pyridoxine deficient, from a cellular metabolic standpoint they are PLP deficient. This PLP deficiency likely affects the function of glutamic acid decarboxylase, the PLP-dependent enzyme that converts the excitatory neurotransmitter glutamic acid into the inhibitory neurotransmitter GABA. An imbalance between these excitatory and inhibitory neurotransmitters may be responsible for the development of encephalopathy and intractable epileptic seizures in these patients. It is likely that dysfunction of other PLP-dependent enzymes also plays a role in the pathophysiology of PDE. Also, elevations of glycine, threonine, and 3-methoxytyrosine in CSF, indicative of dysfunction of glycine-cleavage enzyme threonine hydratase and aromatic L-amino acid decarboxylase, respectively, have been noted in some PDE patients.[26] Lastly, the role of elevated levels of PA in the pathophysiology of PDE needs to be considered. PA can act as a modulator of GABA,[55,56] and build-up of PA in the brain could contribute to an increased risk of seizures and encephalopathy.

ATQ is expressed in radial glial cells in the developing mouse brain and, therefore, some of the findings of brain dysgenesis in patients with PDE may be explained by the absence of this protein in these glial cells that are important in cortical development.[81] To date, an animal model of ATQ deficiency has not been reported.

Current Research

Current research on PDE includes further development of international registries and databases specifically to collect and analyze clinical features, treatment regimens, and natural histories of confirmed PDE patients; study of therapeutic protocols so that the ideal dosing regimen of pyridoxine can be determined, along with the efficacy of lysine restriction to improve the neurodevelopmental outcomes;[47,82,83] the role of ATQ in fetal and early infantile cortical and white matter development;[81,84] development of techniques to screen newborn blood spots for αAASA and P6C;[85] studies of the enzymatic activity of the ATQ protein, in turn leading to an improved understanding of how *ALDH7A1* genotype is related to PDE phenotype;[86] and characterization of pyridoxine vitamer profiles in plasma and CSF.[87,88]

DIFFERENTIAL DIAGNOSIS

The differential diagnosis of neonatal epileptic encephalopathy is quite diverse and includes not only inborn errors of metabolism such as PDE, but also primary disorders of brain dysgenesis, along with acquired disorders such as fetal or neonatal brain injury from maternal or postnatal infection, teratogenesis, fetal hypoxia and/or ischemia, and intracerebral hemorrhage. The clinician caring for an infant with neonatal seizures must consider all of these etiologic categories. As such, early consideration for EEG, brain imaging with magnetic resonance imaging (MRI), evaluation and empiric treatment of presumed infection, and a comprehensive metabolic evaluation including sampling of blood, urine, and CSF are important. Curiously, a few patients with pharmacoresistant neonatal epileptic encephalopathy have been reported to respond to supplementation with folinic acid.[89–92] These patients were subsequently shown to have mutations in *ALDH7A1*, and are, therefore, ATQ deficient.[57] In other words, folinic acid-responsive seizures are allelic to PDE. As patients with PNPO deficiency may present with a similar clinical phenotype,[18] this rare metabolic disorder must also be kept in mind, and for patients with medically intractable neonatal seizures that do not respond to pyridoxine, a trial of PLP should also be instituted.[4,47] Depending upon the region of the world where the patient is located, the ready availability of PLP may be suboptimal. PLP, which can be used to treat both PDE and PNPO deficiency, is the pyridoxine vitamer that is commonly found in Asian formularies, while pyridoxine tends to be stocked in North America and Europe. Similarly, TNSALP deficiency must be considered in young patients with medically refractory neonatal or infantile seizures, particularly when signs of bone demineralization are present.[46]

TESTING

Prior to the discovery in 2006 that PDE is due to mutations in *ALDH7A1*, the diagnosis of this disorder was made on a clinical basis. Neonatologists and neurologists caring for infants with pharmacoresistant seizures would at some point institute a therapeutic trial of pyridoxine. The subsequent clinical diagnosis of PDE was then made by the demonstration of the effectiveness of the vitamin by administering parenteral pyridoxine at a time when the baby was actively experiencing seizures and undergoing continuous EEG monitoring.[1–4] In some patients with PDE, cessation of both clinical and electrographic seizures would be seen, generally within minutes of a single dose of 20–100 mg. However, in some cases, higher doses were needed, and resolution of electrographic seizures did not occur in all patients that were eventually genetically confirmed to have PDE.[93] Therefore, if a patient did not respond to an initial 100-mg dose, up to 500 mg of intravenous pyridoxine would be given in sequential 100-mg doses every 5–10 minutes before concluding that the infant's epilepsy was not immediately responsive to the vitamin. As profound central nervous system depression with associated changes in the EEG have been noted in some PDE patients after the initial treatment with pyridoxine, this trial should take place within an intensive care unit.[1,2,7] An alternate clinical diagnostic approach would be used for patients experiencing frequent, short AED-resistant seizures. Under this clinical scenario, oral pyridoxine (up to 30 mg per kg per day) was administered, and patients with PDE would be expected to have a resolution of clinical seizures within 3–7 days.[1,3,4] Care providers need to be aware that apnea requiring intubation has sometimes been observed in single patients after a first oral administration of pyridoxine.

The definitive clinical confirmation of the diagnosis required the stepwise elimination of the various AEDs that the patient had been receiving, followed by demonstration that the patient had control of seizures on pyridoxine monotherapy, then followed by a withdrawal of the pyridoxine supplementation resulting in the recurrence of seizures, which were again controlled once the vitamin was introduced a second time. Not surprisingly, these confirmatory steps were not acceptable to many clinicians and parents. Consequently, epidemiologic studies and case series conducted prior to 2006 categorized these patients as having possible (rather than definite) PDE.[1,9,32] However, caution was strongly advised in diagnosing all patients treated in this manner with PDE, as certain individuals with seizures that ceased after receiving pyridoxine did not experience a seizure recurrence once the vitamin was discontinued. The term pyridoxine-responsive seizures (PRS) was used to describe this particular condition.[1,32] Using this logic, some patients with possible PDE may actually have PRS.

With the discovery of the biochemical and genetic abnormalities underlying PDE, a fundamental change in our approach to diagnosing this disorder is now possible. Infants with an epileptic encephalopathy resistant to AEDs should be administered pyridoxine as described above (either acutely with one or more intravenous boluses or enterally) and if clinical improvement is noted, biochemical confirmation should then be the next step. Documentation of elevated levels of PA in plasma samples can serve as indirect confirmatory evidence of PDE,[28,80] while the demonstration of elevated levels of αAASA in plasma, CSF or urine represents a more specific biochemical confirmation.[24,26,27,78] The demonstration of either homozygous or compound heterozygous mutations in both *ALDH7A1* alleles will confirm the diagnosis.[23,26,27,29–31,40] Importantly, by taking one or both of these measures, the withdrawal of pyridoxine to clinically prove a PDE diagnosis is no longer necessary. Both biochemical and genetic testing is recommended, as a few individuals with elevated PA and/or αAASA levels have not demonstrated mutations of one or both *ALDH7A1* alleles,[23,26,29] and definite PDE patients treated for an extended period of time may have a normalization of PA levels.[28,29]

In patients with confirmed PDE, a variety of abnormal EEG and neuroimaging findings have been reported, but it is important to emphasize that there are no clear pathognomonic features.[15–19,93,94] In PDE patients, EEGs are commonly performed after several seizures have been observed, as well as after AEDs, typically phenobarbital, have been administered. A PDE-specific pattern is therefore confounded by one or both of these factors. Taking this important point into consideration, abnormal background activity along with a variety of paroxysmal features, including generalized and multifocal epileptiform activity, discontinuous patterns including burst-suppression, bursts of high voltage slow waves, and hypoarrhythmia in patients with infantile spasms have been reported. While electrographic seizures are expected to be recorded during clinical seizures in PDE patients, some behavioral changes considered to be ictal events (such as facial grimacing and abnormal eye movements) may not have an EEG correlate.[18] Importantly, in some untreated patients, as well as in pyridoxine-treated patients, the interictal EEG may be normal or may show only minimal epileptiform activity. In addition, the acute and chronic changes in the EEG in response to pyridoxine treatment are varied. In the classic description of PDE, parenteral pyridoxine led to a rapid resolution of electrographic seizures and normalization of the EEG, correlating with clinical improvement. However, it has recently been shown that this electrographic response to the vitamin may be incomplete or delayed; even a transient deterioration of the EEG may occur.[16,18] Given these inconsistent EEG findings, a definitive diagnosis of PDE should be based on the clinical effectiveness of pyridoxine over hours to days of therapy, together with biochemical and/or genetic confirmation.

Similarly, there are no specific imaging findings in PDE. While normal results are commonly reported when MRI scans are evaluated in a routine qualitative manner, mega cisterna magna, neuronal migration abnormalities, and progressive hydrocephalus requiring shunting have been reported in several patients, and varying degrees of cerebral atrophy have been described in late-diagnosed or inadequately treated patients.[1,10,19,20,26] Thinning of the isthmus

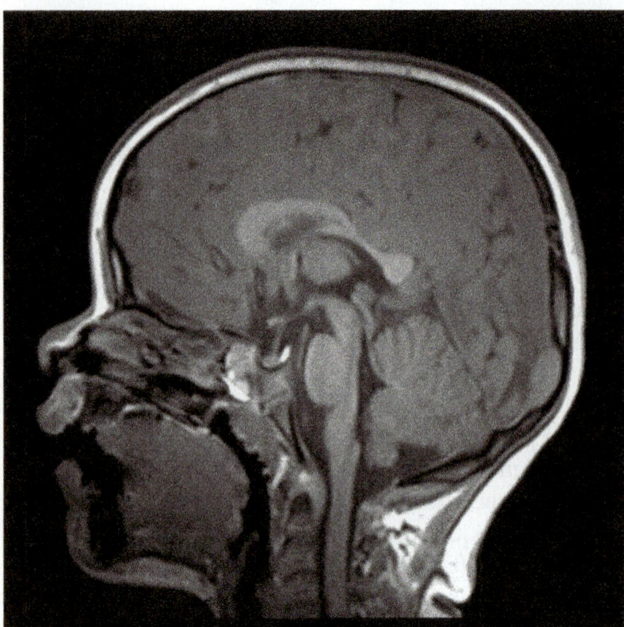

FIGURE 49.3 Midsagittal magnetic resonance imaging scan of a 3-year-old female with PDE. Marked thinning/pinching of the isthmus of the corpus callosum is present.

of the corpus callosum has been suggested to be a common feature in PDE patients (Figure 49.3). Recent studies of callosal cross-sectional area in two cohorts of PDE patients have shown that the entirety of this white matter tract is reduced in size. This change is present in affected infants, children and adults and is not dependent on the timing of the diagnosis and institution of effective therapy.[84]

MANAGEMENT

Management of ATQ Deficiency

An overview of currently practiced treatments, based on supplementation of pharmacological dosages of vitamers, and monitoring is given in Tables 49.1 and 49.2.[47]

Pyridoxine

The standard treatment of ATQ deficiency includes lifelong supplementation of pyridoxine in pharmacologic doses. As described above, in the acutely seizing infant, an initial one or more 100-mg doses of pyridoxine should be given intravenously without delay, or pyridoxine should be given orally/enterally at a dose of up to 30 mg per kg per day. As treatment response may be delayed or masked with concomitant AED use, oral/enteral treatment with pyridoxine should be continued at least for several days, ideally until ATQ deficiency is excluded by negative biochemical or genetic testing.

For long-term treatment, there are no clear-cut dosing recommendations. The recommended daily allowance for pyridoxine, covered by regular diet, is 0.5 mg for infants and 2 mg for adults. In most patients with PDE, therapeutic pyridoxine dosages vary between 15–30 mg per kg per day in infants or up to 200 mg per day in neonates and 500 mg per day in adults. These dosages seem to be safe in long-term treatment. Higher pyridoxine doses may cause sensory and rarely also motor neuropathy, which may be reversible.[38,40] Some patients experience breakthrough seizures during febrile illness. In these cases, higher (e.g., double) dosages may be given during the first 3 days of febrile illnesses. Nerve conduction studies are performed to exclude sensory neuropathy as an adverse effect of pyridoxine treatment. This test is performed upon clinical evidence of neuropathy, or as a regular monitoring (e.g., annually), in particular if high doses (>500 mg per day or >30 mg per kg per day) are used.

Pyridoxal Phosphate

Some patients with medically intractable epilepsy and unresponsiveness to pyridoxine respond to the administration of PLP. In most of them PNPO deficiency has been identified as the underlying genetic condition,[63,95] but idiopathic

TABLE 49.1 Treatment of ATQ Deficiency

Medication	Route	Dosage	Indication	Monitoring	Side effects/ precautions
Pyridoxine	IV	100mg Single dosages	Interruption of initial status epilepticus, or of prolonged breakthrough seizures	EEG is available	May result in respiratory arrest. Administer upon availability of respiratory support
Pyridoxine	Oral/enteral	15–30mg/kg/day divided in up to 3 single dosages Up to 200mg/day in neonates, and 500mg/day in adults	Long-term treatment	Clinical and electrophysiological signs of neuropathy	Continue with dosages above the range only if high dosage has proven essential for effective seizure control
Pyridoxine	Prenatal maternal	100mg/day	Prevention of intrauterine seizures and irreversible brain damage. Start in early pregnancy, continue throughout pregnancy in case of positive prenatal diagnosis or if no prenatal diagnosis has been performed	Monitor of seizures and encephalopathy after delivery in NICU/SCN setting. Consider IV pyridoxine in case of neonatal seizures	Continue oral/ enteral pyridoxine supplementation at 30mg/kg/day immediately after birth and immediately initiate biochemical and molecular genetic investigations to prove or rule out ATQ deficiency
Pyridoxal phosphate (PLP)	Oral/enteral	30mg/kg/day divided in up to 3 single dosages	Interruption of initial status epilepticus: additional to IV pyridoxine in case pyridoxine initially failed to control seizures. Long-term treatment. Alternative to pyridoxine	Same as pyridoxine	Same as pyridoxine
Folinic acid	Oral/enteral	3–5mg/kg/day divided in up to 3 single dosages	Additional therapy if pyridoxine or PLP failed to control seizures	No particular monitoring	None

Abbreviations: IV, intravenous; NICU, neonatal intensive-care unit; SCN, special care nursery.
Adapted from.[47]

TABLE 49.2 Treatment Monitoring: Supplementation of Pyridoxine or PLP

Therapeutic effects	Side effects
Seizures: Clinical and EEG	*Peripheral neuropathy*: NCV
Psychomotor/speech development: <4 years: Bayley Scale >4 years: Kaufman ABC, Wechsler	Upon clinical evidence of neuropathy or as regular monitoring (e.g., annually) in particular if high doses (>500mg/day or>30mg/kg/day) are given

Abbreviations: NCV, nerve conduction velocity; PLP, pyridoxal phosphate.
Adapted from.[47]

PLP response occurs as well.[96] As PLP has the potential to treat both PNPO and ATQ deficiencies, some centers advocate its use (30mg per kg per day divided into three doses) as the first-line form of vitamin B_6, while other centers advocate the consecutive use when pyridoxine, given over 3 consecutive days, has failed to control seizures.[63] In any case, treatment should be continued until results from appropriate metabolic investigations in urine, blood, and CSF are available. Both vitamers (pyridoxine and PLP) are relatively inexpensive. Still, PLP tablets cost 6-10 times more than pyridoxine products, and are less readily available in North America.[74] Therefore, from a practical standpoint, local availability of PLP may determine how this particular vitamer will be used and studied. As shown in animal experiments, high PLP concentrations in the brain can lead to convulsions.[97] Although such effects have not been observed in humans, this should be taken into consideration if there is a deterioration of seizures upon supplementation.

Folinic Acid

Some patients who were later proven to have ATQ deficiency but had an unclear response to pyridoxine, have shown response to folinic acid. Although the mechanism underlying folinic acid responsiveness in ATQ deficiency has not been elucidated, folinic acid (3–5 mg per kg per day) may have potential benefit as an add-on treatment in neonates, especially in the presence of incomplete pyridoxine responsiveness or of breakthrough seizures. In older patients 10–30 mg per day should be tried.[57] It is unknown whether long-term folinic acid is of benefit once the seizures are stabilized. High-dose folinic acid therapy can also exacerbate a seizure disorder, and the clinical benefit has to be closely monitored.

Prophylactic Pre- and Postnatal Treatment

There is a 25% recurrence risk for PDE in subsequent pregnancies. Prenatal treatment of an at-risk fetus with supplemental pyridoxine given to the mother during pregnancy may prevent intrauterine seizures and improve neurodevelopmental outcome.[39] Prenatal treatment with high-dose pyridoxine followed by postnatal treatment was effective to prevent seizures, but did not prevent poor cognitive outcome in two affected offspring in a family with a homozygous stop codon in exon 14 (Y380X).[40] In contrast, a Dutch group reported good developmental outcome after prenatal treatment in three patients homozygous for the missense mutation (E399Q).[14] In these pregnancies, pyridoxine was given to the pregnant women at a dosage of 100 mg per day from early pregnancy. The dosage of 100 mg pyridoxine per day seems to be safe, as it has been used for the treatment of hyperemesis gravidarum without fetal side effects.[32]

Adverse effects, including increased seizure activity have been seen in cases with high pyridoxine or PLP intake.[98,99] A neonate with positive family history was on prophylactic intrauterine treatment with pyridoxine and treatment was continued postnatally until ATQ deficiency was ruled out. On day 15, still on treatment, the patient developed status epilepticus and encephalopathy. His condition only improved after pyridoxine treatment was discontinued after the biochemical testing results were final.[100] Thus, biochemical and genetic testing should be aimed for as soon as possible in order to limit unnecessary high-dose treatment with pyridoxine.

Evidence and Safety of High-Dose Pyridoxine and PLP Treatment

The accumulation of P6C and its condensation with PLP causes chemically induced PLP depletion. Therefore, from a pathophysiological point of view, high pharmacological dosages of pyridoxine are needed to exceed ongoing inactivation of PLP and increase the intracellular proportion of free PLP. This rationale does not explain why a few patients respond to very low dosages of pyridoxine. In the original PDE case,[21] physiologic dosing of pyridoxine was therapeutic if used consistently. Controlled clinical studies to evaluate the optimal dose and shed light into adverse effects of high-dose pyridoxine and PLP treatment have not yet been performed. In addition, there may be substantial interpatient differences in the required effective dose related to the underlying ATQ mutation, other genetic causes, or environmental influences such as the patient's lysine intake.

Novel Treatments: Lysine-Restricted Diet

While treatment with pyridoxine compensates chemical PLP inactivation, the accumulation of substrates from lysine degradation is not sufficiently reduced (Figure 49.2). These potentially neurotoxic compounds could be an explanation of the partial efficacy of pyridoxine, as 75–80% of patients suffer developmental delay or intellectual disability (IQ < 70) despite excellent seizure control in the majority of patients.[41] Standard treatment for inborn errors of metabolism affecting catabolic pathways of essential amino acids consists of substrate reduction to the deficient enzyme activity by dietary modification. Thus, for ATQ deficiency, dietary lysine restriction can reduce the accumulation of lysine-derived substrates and possibly contribute to the improvement of cerebral function (neurodevelopment, cognition, behavior, and seizure control).

Based on this rationale, an open-label observational study was conducted to test the effectiveness and safety of dietary lysine restriction as an adjunct to pyridoxine therapy on chemical biomarkers, seizure control, and developmental outcome in seven children with confirmed ATQ deficiency.[82] The results from the study show that dietary lysine restriction (evidence level IV): 1) is tolerated without adverse events; 2) leads to significant decrease of potentially neurotoxic biomarkers in different body compartments; and 3) has potential benefit for seizure control and neurodevelopmental outcomes. The lysine-restricted diet is thus a promising add-on therapy for ATQ deficiency, with the potential to improve short- and long-term clinical outcomes. However, the evidence is still limited and, as with any diet, it poses a burden on patients and families, often in conflict with social and cultural traditions.[101] Also, such dietary treatment cannot be taken lightly: it requires monitoring by a specialist and metabolic dietician with regular clinical follow-up, dietary protocols, and laboratory testing. Different effects can be expected based on the patient's prenatal history, associated congenital brain abnormalities, genotype and other factors; this requires further study.

Based on the positive outcomes of this study, including the limited evidence level and experience with this burdensome diet, an international PDE Consortium agreed to develop recommendations for implementation of this diet along with monitoring and follow-up.[83] Such standardization will allow a systematic evaluation of the safety and effectiveness of this novel therapy and help generate more solid evidence to substantiate its potential benefits. In summary, all confirmed ATQ-deficient patients are eligible for the lysine-restricted diet as add-on therapy (unless pyridoxine monotherapy has resulted in complete symptom resolution with cessation of seizures, and the establishment of normal behavior and development) regardless of age and gender. Started as early in life as possible, daily lysine intake should be prescribed at an amount that maintains the plasma lysine level within the lower normal age-dependent reference range, preferably in the lower quartile, while allowing for adequate growth and nutrition. To allow for more precise and consistent lysine intake, it is recommended to use a lysine prescription (at least during the initiation of the diet in the first year of life) rather than a simpler protein restriction. The PDE Consortium has adapted the guidelines by Kölker et al. for gluatric aciduria type I,[102] by using both the World Health Organization (WHO) guidelines[103] and those recommended by Yanicelli.[104] In contrast to glutaric aciduria type I, tryptophan restriction is not indicated in the management of PDE. In order to meet the recommended daily protein intake,[103,105] the diet may include commercially available lysine-free amino acid formulas approved for use in conditions affecting lysine metabolism. Lysine-free amino acid formulas are often supplemented with vitamins and minerals in order to provide adequate or significant intakes of these nutrients, but adequate supply of iron, minerals and vitamins must be confirmed by regular laboratory testing. Where lysine-free amino acid formulas are not used, vitamin and mineral supplement preparations must be added. Because limited amounts of high-protein foods from the meat, fish and dairy food groups will be consumed, the use of fats and oils, carbohydrates and special low protein foods are often necessary to provide adequate caloric intake.

Safety monitoring at baseline and thereafter every 1–6 months (frequency depending on age and duration of treatment) includes: lysine and branched-chain amino acids in plasma, albumin, pre-albumin, iron parameters, calcium, phosphate, 25-OH-vitamin D3, zinc, selenium, complete blood count, folic acid, and vitamin B_{12} in serum or plasma. Main outcomes include the biochemical parameters (PA and αAASA levels in body fluids), seizure control, and neurodevelopmental parameters.

Barriers to Development of Novel Treatments

In general, barriers to the implementation and evaluation of treatment for a rare disease such ATQ deficiency include: small patient numbers spread over the world, underdiagnosis due to insufficient awareness, clinical heterogeneity, incomplete insight into the clinical spectrum, and genotype–phenotype correlation; limited financial resources; and diverse specialists treating this metabolic epilepsy. Van Karnebeek et al.[82,83] suggested methods to overcome these roadblocks, including establishment of a task force (PDE Consortium) to develop recommendations for treatment; use of digital tools and databases to unite patients, scientists, clinicians as well as disseminate knowledge; novel trial methodologies with the simultaneous use of different study designs, allowing inclusion of all patients, irrespective of age, severity, symptoms, and interventions. Collaboration and sharing of data is the most important of all.

Therapies under Investigation

In addition to the lysine-restricted diet, two additional therapies are under investigation. Addition of arginine to the lysine-restricted formula, to exploit the transport competition between lysine and arginine over the blood–brain barrier and further reduce cerebral lysine flux is currently being studied (van Hove, Denver, USA; work in progress). A similar treatment strategy has shown promising preliminary results in glutaric aciduria type I.[106] Perez et al. provided a cellular proof-of-concept for antisense therapy application in PDE.[107] This study reported the clinical, biochemical, and genetic analysis of 12 unrelated patients with ATQ deficiency, including seven novel sequence changes. Transcriptional profile analysis showed the silent nucleotide change c.75C>T to be a novel splicing mutation creating a new donor splice site in exon 1. Antisense therapy of the aberrant mRNA splicing in a lymphoblast cell line harboring this mutation was successful, holding the promise that this strategy may be a means of rescuing splice site changes in *ALDH7A1* and preventing neurologic damage.

Management of PNPO Deficiency

AEDs are ineffective in controlling seizures in patients with PNPO deficiency. Instead, individuals with this type of metabolic epileptic encephalopathy are medically treated with oral PLP (30 mg per kg per day in three doses) or in some instances oral pyridoxine (15–30 mg per kg per day).

Management of TNSALP Deficiency

Most recently, enzyme replacement therapy with asfotase alpha has been explored for hypophosphatasia with success in animal models and human clinical trials.[108,109] This bone-targeted, recombinant TNSALP has recently been reported to improve survival, rickets on radiographs, as well as pulmonary function and motor milestones in 11 patients.[109] Belachew et al.[46] reported complete seizure control in their patient (5-months-old when started and 31 months at follow-up) on enzyme replacement therapy, even when antiepileptic medications and pyridoxine were discontinued.

ACKNOWLEDGEMENTS

We gratefully acknowledge the contributions of the following individuals at B.C. Children's Hospital, Vancouver, Canada: Dr. Sylvia Stöckler (Head Division of Biochemical Diseases) for her expert input on this disease, specifically on treatment and monitoring; Mr. Sravan Jaggumantri (PhD student), Mr. Roderick Houben (web designer), and Mrs. Ruth Giesbrecht (administrative assistant).

References

1. Baxter P. Pyridoxine dependent and pyridoxine responsive seizures. In: Baxter P, ed. *Vitamin Responsive Conditions in Paediatric Neurology*. London: MacKeith Press; 2001:109–165.
2. Gospe Jr SM. Current perspectives on pyridoxine-dependent seizures. *J Pediatr*. 1998;132:919–923.
3. Gospe Jr SM. Pyridoxine-dependent seizures: findings from recent studies pose new questions. *Pediatr Neurol*. 2002;26:181–185.
4. Gospe Jr SM. Neonatal vitamin-responsive epileptic encephalopathies. *Chang Gung Med J*. 2010;33(1):1–12.
5. Haenggeli C-A, Girardin E, Paunier L. Pyridoxine-dependent seizures, clinical and therapeutic aspects. *Eur J Pediatr*. 1991;150:452–455.
6. Bankier A, Turner M, Hopkins IJ. Pyridoxine dependent seizures—a wider clinical spectrum. *Arch Dis Child*. 1983;58:415–418.
7. Bass NE, Wyllie E, Cohen B, Joseph SA. Pyridoxine-dependent epilepsy: the need for repeated pyridoxine trials and the risk of severe electrocerebral suppression with intravenous pyridoxine infusion. *J Child Neurol*. 1996;11:422–424.
8. Coker S. Postneonatal vitamin B_6-dependent epilepsy. *Pediatrics*. 1992;90:221–223.
9. Basura GJ, Hagland SP, Wiltse AM, Gospe Jr SM. Clinical features and the management of pyridoxine-dependent and pyridoxine-responsive seizures: review of 63 North American cases submitted to a patient registry. *Eur J Pediatr*. 2009;168:697–704.
10. Baxter P, Griffiths P, Kelly T, Gardner-Medwin D. Pyridoxine-dependent seizures: demographic, clinical, MRI and psychometric features, and effect of dose on intelligence quotient. *Dev Med Child Neurol*. 1996;38:998–1006.
11. Baynes K, Tomaszewski Farias S, Gospe Jr SM. Pyridoxine-dependent seizures and cognition in adulthood. *Dev Med Child Neurol*. 2003;45:782–785.
12. Kluger G, Blank R, Paul K, et al. Pyridoxine-dependent epilepsy: normal outcome in a patient with late diagnosis after prolonged status epilepticus causing cortical blindness. *Neuropediatrics*. 2008;39(5):276–279.
13. Ohtsuka Y, Hattori J, Ishida T, Ogino T, Oka E. Long-term follow-up of an individual with vitamin B_6-dependent seizures. *Dev Med Child Neurol*. 1999;41:203–206.
14. Bok LA, Been JV, Struys EA, Jakobs C, Rijper EA, Willemsen MA. Antenatal treatment in two Dutch families with pyridoxine-dependent seizures. *Eur J Pediatr*. 2010;169(3):297–303.
15. Mikati MA, Trevathan E, Krishnamoorthy KS. Pyridoxine-dependent epilepsy: EEG investigation and long-term follow-up. *Electroencephalogr Clin Neurophysiol*. 1991;78:215–221.
16. Naasan G, Yabroudi M, Rahi A, Mikati MA. Electroencephalographic changes in pyridoxine-dependant epilepsy: new observations. *Epileptic Disord*. 2009;11(4):293–300.
17. Nabbout R, Soufflet C, Plouin P, Dulac O. Pyridoxine dependent epilepsy: a suggestive electroclinical pattern. *Arch Dis Child Fetal Neonatal Ed*. 1999;81:F125–F129.
18. Schmitt B, Baumgartner M, Mills PB, et al. Seizures and paroxysmal events: symptoms pointing to the diagnosis of pyridoxine-dependent epilepsy and pyridoxine phosphate oxidase deficiency. *Dev Med Child Neurol*. 2010;52(7):e133–e142.
19. Shih JJ, Kornblum H, Shewmon DA. Global brain dysfunction in an infant with pyridoxine dependency: evaluation with EEG, evoked potentials, MRI, and PET. *Neurology*. 1996;47:824–826.
20. Gospe Jr SM, Hecht ST. Longitudinal MRI findings in pyridoxine-dependent seizures. *Neurology*. 1998;51:74–78.
21. Hunt AD, Stokes J, McCrory WW, Stroud HH. Pyridoxine dependency: report of a case of intractable convulsions in an infant controlled by pyridoxine. *Pediatrics*. 1954;13:140–145.
22. Been JV, Bok JA, Andriessen P, Renier WO. Epidemiology of pyridoxine-dependent seizures in The Netherlands. *Arch Dis Child*. 2005;90:1293–1296.
23. Bennett CL, Chen Y, Hahn S, Glass IA, Gospe Jr SM. Prevalence of ALDH7A1 mutations in 18 North American pyridoxine-dependent seizure (PDS) patients. *Epilepsia*. 2009;50:1167–1175.
24. Bok LA, Struys E, Willemsen MA, Been JV, Jakobs C. Pyridoxine-dependent seizures in Dutch patients: diagnosis by elevated urinary alpha-aminoadipic semialdehyde levels. *Arch Dis Child*. 2007;92(8):687–689.
25. Kanno J, Kure S, Narisawa A, et al. Allelic and non-allelic heterogeneities in pyridoxine dependent seizures revealed by ALDH7A1 mutational analysis. *Mol Genet Metab*. 2007;91(4):384–389.

26. Mills PB, Footitt EJ, Mills KA, et al. Genotypic and phenotypic spectrum of pyridoxine-dependent epilepsy (ALDH7A1 deficiency). *Brain*. 2010;133(Pt 7):2148–2159.
27. Mills PB, Struys E, Jakobs C, et al. Mutations in antiquitin in individuals with pyridoxine-dependent seizures. *Nat Med*. 2006;12(3):307–309.
28. Plecko B, Hikel C, Korenke G-C, et al. Pipecolic acid as a diagnostic marker of pyridoxine-dependent epilepsy. *Neuropediatrics*. 2005;36:200–205.
29. Plecko B, Paul K, Paschke E, et al. Biochemical and molecular characterization of 18 patients with pyridoxine-dependent epilepsy and mutations of the antiquitin (ALDH7A1) gene. *Hum Mutat*. 2007;28(1):19–26.
30. Salomons GS, Bok LA, Struys EA, et al. An intriguing "silent" mutation and a founder effect in antiquitin (ALDH7A1). *Ann Neurol*. 2007;62(4):414–418.
31. Scharer G, Brocker C, Vasiliou V, et al. The genotypic and phenotypic spectrum of pyridoxine-dependent epilepsy due to mutations in ALDH7A1. *J Inherit Metab Dis*. 2010;33(5):571–581.
32. Baxter P. Epidemiology of pyridoxine dependent and pyridoxine responsive seizures in the UK. *Arch Dis Child*. 1999;81:431–433.
33. Ebinger M, Schutze C, Konig S. Demographics and diagnosis of pyridoxine-dependent seizures. *J Pediatr*. 1999;134:795–796.
34. Ramachandrannair R, Parameswaran M. Prevalence of pyridoxine dependent seizures in south Indian children with early onset intractable epilepsy: a hospital based prospective study. *Eur J Paediatr Neurol*. 2005;9(6):409–413.
35. Mercimek-Mahmutoglu S, Horvath GA, Coulter-Mackie M, et al. Atypical presentation of antiquitin deficiency in a female with neonatal seizures, hypoglycemia, hyperlacticacidemia and intractable myoclonic epilepsy. *J Inherit Metab Dis*. 2010;33(suppl 1):S158.
36. Goutières F, Aicardi J. Atypical presentations of pyridoxine-dependent seizures: a treatable cause of intractable epilepsy in infants. *Ann Neurol*. 1985;17:117–120.
37. McDonald JC, Lavoie J, Cote R, McDonald AD. Chemical exposures at work in early pregnancy and congenital defect: a case referent study. *Br J Ind Med*. 1987;44:527–533.
38. McLachlan RS, Brown WF. Pyridoxine dependent epilepsy with iatrogenic sensory neuronopathy. *Can J Neurol Sci*. 1995;22:50–51.
39. Baxter P, Aicardi J. Neonatal seizures after pyridoxine use [letter]. *Lancet*. 1999;354:2082–2083.
40. Rankin PM, Harrison S, Chong WK, Boyd S, Aylett SE. Pyridoxine-dependent seizures: a family phenotype that leads to severe cognitive deficits, regardless of treatment regime. *Dev Med Child Neurol*. 2007;49(4):300–305.
41. Bok LA, Halbertsma FJ, Houterman S, et al. Long-term outcome in pyridoxine dependent epilepsy. *Dev Med Child Neurol*. 2012;54:849–854.
42. Clayton PT, Surtees RAH, DeVile C, Hyland K, Heales SJR. Neonatal epileptic encephalopathy. *Lancet*. 2003;361:1614.
43. Mills PB, Surtees RAH, Champion MP, et al. Neonatal epileptic encephalopathy caused by mutations in the *PNPO* gene encoding pyridox(am)ine 5′-phosphate oxidase. *Hum Mol Genet*. 2005;14:1077–1086.
44. Pearl PL, Hyland K, Chiles J, McGavin CL, Yu Y, Taylor D. Partial pyridoxine responsiveness in PNPO deficiency. *JIMD Rep*. 2013;9:139–142.
45. Plecko B, Karl P, Mills P, et al. Pyridoxine responsiveness in novel mutations of the PNPO gene. *Neurology*. 2014;82:1425–1433.
46. Belachew D, Kazmerski T, Libman I, et al. Infantile hypophosphatasia secondary to a novel compound heterozygous mutation presenting with pyridoxine-responsive seizures. *JIMD Rep*. 2013;11:17–24.
47. Stöckler S, Plecko B, Gospe Jr SM, et al. Pyridoxine dependent epilepsy and antiquitin deficiency: clinical and molecular characteristics and recommendations for diagnosis, treatment and follow-up. *Mol Genet Metab*. 2011;104(1–2):48–60.
48. Lescuyer P, Strub JM, Luche S, et al. Progress in the definition of a reference human mitochondrial proteome. *Proteomics*. 2003;3(2):157–167.
49. Wong JW, Chan CL, Tang WK, Cheng CH, Fong WP. Is antiquitin a mitochondrial Enzyme? *J Cell Biochem*. 2010;109(1):74–81.
50. Tang WK, Chan CB, Cheng CH, Fong WP. Seabream antiquitin: molecular cloning, tissue distribution, subcellular localization and functional expression. *FEBS Lett*. 2005;579(17):3759–3764.
51. Brocker C, Lassen N, Estey T, et al. Aldehyde dehydrogenase 7A1 (ALDH7A1) is a novel enzyme involved in cellular defense against hyperosmotic stress. *J Biol Chem*. 2010;285(24):18452–18463.
52. Papes F, Kemper EL, Cord-Neto G, Langone F, Arruda P. Lysine degradation through the saccharopine pathway in mammals: involvement of both bifunctional and monofunctional lysine-degrading enzymes in mouse. *Biochem J*. 1999;344 Pt 2:555–563.
53. Cox RP, et al. Errors of lysine metabolism. In: Valle D, Beaudet AL, Vogelstein B, eds. *The Online Metabolic and Molecular Bases of Inherited Disease*. New York: McGraw-Hill; 2006, Updated March 2011.
54. Goodman SI, Freeman FE, et al. Organic acidemias due to defects in lysine oxidation: 2-ketoadipic acidemia and glutaric acidemia. In: Valle D, Beaudet AL, Vogelstein B, eds. *The Online Metabolic and Molecular Bases of Inherited Disease*. New York: McGraw-Hill; 2006, Updated March 2011.
55. Charles AK. Pipecolic acid receptors in rat cerebral cortex. *Neurochem Res*. 1986;11(4):521–525.
56. Gutierrez MC, Delgado-Coello BA. Influence of pipecolic acid on the release and uptake of [3H]-GABA from brain slices of mouse cerebral cortex. *Neurochem Res*. 1989;14:405–408.
57. Gallagher RC, Van Hove JL, Scharer G, et al. Folinic acid-responsive seizures are identical to pyridoxine-dependent epilepsy. *Ann Neurol*. 2009;65:550–556.
58. Kaczorowska M, Kmiec T, Jakobs C, et al. Pyridoxine-dependent seizures caused by alpha amino adipic semialdehyde dehydrogenase deficiency: the first polish case with confirmed biochemical and molecular pathology. *J Child Neurol*. 2008;23(12):1455–1459.
59. Millet A, Salomons GS, Cneude F, et al. Novel mutations in pyridoxine-dependent epilepsy. *Eur J Paediatr Neurol*. 2011;15(1):74–77.
60. Parikh S, Hyland K, Lachhwani DK. Vitamins, not surgery: spinal fluid testing in hemispheric epilepsy. *Pediatr Neurol*. 2009;40(6):477–479.
61. Striano P, Battaglia S, Giordano L, et al. Two novel ALDH7A1 (antiquitin) splicing mutations associated with pyridoxine-dependent seizures. *Epilepsia*. 2009;50(4):933–936.
62. Tlili A, Hamida Hentati N, Chaabane R, Gargouri A, Fakhfakh F. Pyridoxine-dependent epilepsy in Tunisia is caused by a founder missense mutation of the ALDH7A1 gene. *Gene*. 2013;518(2):242–245.
63. Hoffmann GF, Schmitt B, Windfuhr M, et al. Pyridoxal 5′-phosphate may be curative in early-onset epileptic encephalopathy. *J Inherit Metab Dis*. 2007;30(1):96–99.
64. Khayat M, Korman SH, Frankel P, et al. PNPO deficiency: an under diagnosed inborn error of pyridoxine metabolism. *Mol Genet Metab*. 2008;94(4):431–434.

65. Ormazabal A, Oppenheim M, Serrano M, et al. Pyridoxal 5′-phosphate values in cerebrospinal fluid: reference values and diagnosis of PNPO deficiency in paediatric patients. *Mol Genet Metab*. 2008;94(2):173–177.
66. Ruiz A, Garcia-Villoria J, Ormazabal A, et al. A new fatal case of pyridox(am)ine 5′-phosphate oxidase (PNPO) deficiency. *Mol Genet Metab*. 2008;93(2):216–218.
67. Orimo H, Girschick HJ, Goseki-Sone M, Ito M, Oda K, Shimada T. Mutational analysis and functional correlation with phenotype in German patients with childhood-type hypophosphatasia. *J Bone Miner Res*. 2001;16(12):2313–2319.
68. Zurutuza L, Muller F, Gibrat JF, et al. Correlations of genotype and phenotype in hypophosphatasia. *Hum Mol Genet*. 1999;8(6):1039–1046.
69. Brun-Heath I, Lia-Baldini AS, Maillard S, et al. Delayed transport of tissue-nonspecific alkaline phosphatase with missense mutations causing hypophosphatasia. *Eur J Med Genet*. 2007;50(5):367–378.
70. Fukushi M, Amizuka N, Hoshi K, et al. Intracellular retention and degradation of tissue-nonspecific alkaline phosphatase with a Gly317–>Asp substitution associated with lethal hypophosphatasia. *Biochem Biophys Res Commun*. 1998;246(3):613–618.
71. Mornet E, Stura E, Lia-Baldini AS, Stigbrand T, Menez A, Le Du MH. Structural evidence for a functional role of human tissue nonspecific alkaline phosphatase in bone mineralization. *J Biol Chem*. 2001;276(33):31171–31178.
72. Nasu M, Ito M, Ishida Y, et al. Aberrant interchain disulfide bridge of tissue-nonspecific alkaline phosphatase with an Arg433–>Cys substitution associated with severe hypophosphatasia. *FEBS J*. 2006;273(24):5612–5624.
73. Shibata H, Fukushi M, Igarashi A, et al. Defective intracellular transport of tissue-nonspecific alkaline phosphatase with an Ala162–>Thr mutation associated with lethal hypophosphatasia. *J Biochem*. 1998;123(5):968–977.
74. Gospe Jr SM. Pyridoxine-dependent seizures: new genetic and biochemical clues to help with diagnosis and treatment. *Curr Opin Neurol*. 2006;19(2):148–153.
75. Wang HS, Kuo MF. Vitamin B6 related epilepsy during childhood. *Chang Gung Med J*. 2007;30(5):396–401.
76. Bender DA. *Vitamin B6 Nutritional Biochemistry of the Vitamins*. 2nd ed. Cambridge: Cambridge University Press; 2003:232.
77. Percudani R, Peracchi A. A genomic overview of pyridoxal-phosphate-dependent enzymes. *EMBO Rep*. 2003;4(9):850–854.
78. Sadilkova K, Gospe Jr SM, Hahn SH. Simultaneous determination of alpha-aminoadipic semialdehyde, piperideine-6-carboxylate and pipecolic acid by LC-MS/MS for pyridoxine-dependent seizures and folinic acid-responsive seizures. *J Neurosci Methods*. 2009;184(1):136–141.
79. Mills PB, Footitt EJ, Ceyhan S, et al. Urinary AASA excretion is elevated in patients with molybdenum cofactor deficiency and isolated sulphite oxidase deficiency. *J Inherit Metab Dis*. 2012;35(6):1031–1036.
80. Plecko B, Stöckler-Ipsiroglu S, Paschke E, Erwa W, Struys EA, Jakobs C. Pipecolic acid elevation in plasma and cerebrospinal fluid of two patients with pyridoxine-dependent epilepsy. *Ann Neurol*. 2000;48:121–125.
81. Jansen LA, Hevner RF, Roden WH, Hahn SH, Jun S, Gospe Jr SM. Glial localization of antiquitin: implications for pyridoxine-dependent epilepsy. *Ann Neurol*. 2014;75:22–32.
82. van Karnebeek CD, Hartmann H, Jaggumantri S, et al. Lysine restricted diet for pyridoxine-dependent epilepsy: first evidence and future trials. *Mol Genet Metab*. 2012;107(3):335–344.
83. van Karnebeek C, Stockler S, Jaggumantri S, et al. Lysine-restricted diet as adjunct therapy for pyridoxine-dependent epilepsy: the PDE consortium consensus recommendations. *JIMD Rep*. 2014 [Epub April 19].
84. Friedman SD, Ishak GE, Poliachik SL, et al. Callosal alterations in pyridoxine-dependent epilepsy. *Dev Med Child Neurol*. 2014[Epub June 18].
85. Jung S, Tran NT, Gospe Jr SM, Hahn SH. Preliminary investigation of the use of newborn dried blood spots for screening pyridoxine-dependent epilepsy by LC-MS/MS. *Mol Genet Metab*. 2013;110(3):237–240.
86. Coulter-Mackie MB, Li A, Lian Q, Struys E, Stöckler S, Waters PJ. Overexpression of human antiquitin in E. coli: enzymatic characterization of twelve ALDH7A1 missense mutations associated with pyridoxine-dependent epilepsy. *Mol Genet Metab*. 2012;106(4):478–481.
87. Footitt EJ, Clayton PT, Mills K, et al. Measurement of plasma B6 vitamer profiles in children with inborn errors of vitamin B6 metabolism using an LC-MS/MS method. *J Inherit Metab Dis*. 2013;36(1):139–145.
88. Footitt EJ, Heales SJ, Mills PB, Allen GF, Oppenheim M, Clayton PT. Pyridoxal 5′-phosphate in cerebrospinal fluid; factors affecting concentration. *J Inherit Metab Dis*. 2011;34(2):529–538.
89. Frye RE, Donner E, Golja A, Rooney CM. Folinic acid-responsive seizures presenting as breakthrough seizures in a 3-month-old boy. *J Child Neurol*. 2003;18(8):562–569.
90. Hyland K, Buist NR, Powell BR, et al. Folinic acid responsive seizures: a new syndrome? *J Inherit Metab Dis*. 1995;18(2):177–181.
91. Nicolai J, van Kranen-Mastenbroek VH, Wevers RA, Hurkx WA, Vles JS. Folinic acid-responsive seizures initially responsive to pyridoxine. *Pediatr Neurol*. 2006;34(2):164–167.
92. Torres OA, Miller VS, Buist NM, Hyland K. Folinic acid-responsive neonatal seizures. *J Child Neurol*. 1999;14(8):529–532.
93. Bok LA, Maurits NM, Willemsen MA, et al. The EEG response to pyridoxine-IV neither identifies nor excludes pyridoxine-dependent epilepsy. *Epilepsia*. 2010;51(12):2406–2411.
94. Baxter P. Pyridoxine-dependent seizures: a clinical and biochemical conundrum. *Biochim Biophys Acta*. 2003;1647:36–41.
95. Bagci S, Zschocke J, Hoffmann GF, et al. Pyridoxal phosphate-dependent neonatal epileptic encephalopathy. *Arch Dis Child Fetal Neonatal Ed*. 2008;93(2):F151–F152.
96. Wang H-S, Chou M-L, Hung P-C, Lin K-L, Hsieh M-Y, Chang M-Y. Pyridoxal phosphate is better than pyridoxine for controlling idiopathic intractable epilepsy. *Arch Dis Child*. 2005;90:512–515.
97. Kouyoumdjian JC, Ebadi J. Anticonvulsant activity of muscimol and gamma-aminobutyric acid against pyridoxal phosphate-induced seizures. *J Neurochem*. 1981;36:251–257.
98. Clayton PT. B6-responsive disorders: a model of vitamin dependency. *J Inherit Metab Dis*. 2006;29(2–3):317–326.
99. Hammen A, Wagner B, Berkhoff M, Donati F. A paradoxical rise of neonatal seizures after treatment with vitamin B6. *Eur J Paediatr Neurol*. 1998;2:319–322.
100. Hartmann H, Fingerhut M, Jakobs C, Plecko B. Status epilepticus in a newborn treated with pyridoxine due to familial recurence risk for antiqutin deficiency-pyridoxine toxicity? *Dev Med Child Neurol*. 2011;53:1150–1153.
101. Stöckler S, Moeslinger D, Herle M, Wimmer B, Ipsiroglu OS. Cultural aspects in the management of inborn errors of metabolism. *J Inherit Metab Dis*. 2012;35(6):1147–1152.

102. Kölker S, Christensen E, Leonard JV, et al. Diagnosis and management of glutaric aciduria type I–revised recommendations. *J Inherit Metab Dis*. 2011;34(3):677–694.
103. FAO/WHO/UNU. *Energy and Protein Requirements*. Geneva: WHO; 1985 Report No. 724.
104. Yanicelli S. Nutrition management of patients with inherited disorders of organic acid metabolism. In: Acosta PB, ed. *Nutrition Management of Patients with Inherited Metabolic Disorders*. Boston: Jones and Bartlett; 2010:283–341.
105. National Research Council. *Recommended Dietary Allowances*. 10th ed. Washington, D.C.: National Academy Press; 1989.
106. Kölker S, Boy SP, Heringer J, et al. Complementary dietary treatment using lysine-free, arginine-fortified amino acid supplements in glutaric aciduria type I – A decade of experience. *Mol Genet Metab*. 2012;107(1–2):72–80.
107. Perez B, Gutierrez-Solana LG, Verdu A, et al. Clinical, biochemical, and molecular studies in pyridoxine-dependent epilepsy. Antisense therapy as possible new therapeutic option. *Epilepsia*. 2013;54(2):239–248.
108. Millan JL, Narisawa S, Lemire I, et al. Enzyme replacement therapy for murine hypophosphatasia. *J Bone Miner Res*. 2008;23(6):777–787.
109. Whyte MP, Greenberg CR, Salman NJ, et al. Enzyme-replacement therapy in life-threatening hypophosphatasia. *N Engl J Med*. 2012;366(10):904–913.

LIPID METABOLISM DISORDERS

CHAPTER

50

Disorders of Lipid Metabolism

Stefano Di Donato and Franco Taroni

Fondazione IRCCS Istituto Neurologico "Carlo Besta," Milan, Italy

INTRODUCTION

Abnormalities of lipid catabolism as possible causes of human disease were first suggested in the late 1960s by morphological observations of excessive accumulation of lipid droplets within muscle fibers of a young woman who had attacks of muscle weakness lasting from a few weeks to several years.[1] Lipid myopathy associated with muscle pain and cramps and occasional myoglobinuria was later described by Engel (1970) in two twin sisters who, when fed with long-chain fatty acids (LCFAs), showed low ketone production but generated a normal amount of ketones after a medium-chain fatty acid meal.[2] These findings suggested that the patients might suffer from a specific defect in the oxidation of LCFAs. Following these seminal observations, primary lipid myopathy associated with pathogenic carnitine deficiency in muscle was first described by Angelini and Engel (1973),[3] and carnitine palmitoyltransferase (CPT) deficiency was discovered as the first enzyme defect of mitochondrial fatty-acid (FA) oxidation by DiMauro and Melis-DiMauro (1973),[4] in a 29-year-old man who suffered from recurrent episodes of muscle pain and pigmenturia triggered by prolonged exercise. Since this latter description, 18 autosomal recessive defects have been identified, involving almost all enzyme steps in the pathway[5–9] (see Tables 50.1 and 50.2, pages 560 and 562, respectively). With the exception of medium-chain acyl-CoA dehydrogenase (MCAD) deficiency, which has a high frequency (1 in 10,000–30,000 births) among Northern European Caucasians,[10,11] these disorders are uncommon and the prevalence rate is unknown for most of them.

PATHOPHYSIOLOGY

The immediate source of chemical energy for muscle contraction is the hydrolysis of adenosine triphosphate (ATP) to adenosine diphosphate (ADP). ATP can be regenerated from ADP and the high-energy compound phosphocreatine, but during long-term exercise the rephosphorylation of ADP to ATP requires the utilization of other fuels, such as carbohydrate, FA, and ketones. Although anaerobic glycogenolysis in the cytosol can generate ATP up to 100 times faster than aerobic oxidation of glucose, it yields only 2 moles of ATP per mole of glucose as compared to 38 moles of ATP per mole of glucose yielded by mitochondrial oxidative phosphorylation (OXPHOS). Furthermore, it rapidly leads to the accumulation of toxic fatigue-promoting metabolic end products (mainly lactic acid). Therefore, OXPHOS is the primary energy source for the regeneration of ATP during muscle work. Although both carbohydrate and fatty-acid catabolic pathways converge into acetyl-coenzyme A (acetyl-CoA) for final intramitochondrial oxidation through the tricarboxylic acid cycle (TCA) and the respiratory chain (OXPHOS), the pattern of muscle fuel utilization is determined primarily by the intensity and duration of exercise. At rest, most muscle energy is provided by mitochondrial oxidation of LCFA (C_{14}–C_{20}) and the respiratory quotient (respiratory exchange ratio; RER) of resting muscle is close to 0.8, indicating an almost total dependence on the oxidation of FA.[6] During the early phase of exercise (up to ≈45 minutes), energy is derived mainly from catabolism of muscle glycogen stores and blood glucose. After approximately 90 minutes of exercise at an intensity of ≈70% of maximum oxygen uptake (VO_2 max), muscle glycogen stores are depleted and there is a gradual shift from glucose to fatty-acid utilization. After a few hours, about 70% of the skeletal muscle energy requirement is met by the oxidation of fatty acids. Although the mobilization and rate of energy production from fatty acids are slow as compared with those of glycogen, complete oxidation of a fatty-acid molecule is highly exergonic. For example, the oxidation of one molecule of palmitate ($C_{16:0}$) has a net yield of 129 ATPs.[6] Heart is also largely dependent on LCFA oxidation for its functional activity.[6]

http://dx.doi.org/10.1016/10.1016/B978-0-12-410529-4.00050-4

TABLE 50.1 Main Clinical Features of Fatty-Acid β-Oxidation Disorders

	Myopathic symptoms			Hepatic symptoms				
Disorder	Acute[a]	Chronic	Cardiomyopathy	Hypoketotic hypoglycemia	Metabolic encephalopathy[b]	Abnormal organic acids	Other features	MIM No.[c]
LONG-CHAIN FATTY-ACID OXIDATION								
Fatty-acid transport								
CT	–	++	+++	+	+	–	Endocardial fibroelastosis	212140
CPT1	–	–	–	+++	+++	–	Renal tubular acidosis	255120
CACT	?	++	+++[d]	+++	+++	+/–		212138
CPT2, type 1 (muscular)	+++	–	–	–	–	–		255110 600650
CPT2, type 2 (hepatocardio-muscular)	+/–	++	++	++	++	+/–	Recurrent pancreatitis	600649 600650
CPT2, type 3 (lethal neonatal)	–	++	+++	+++	+++	+/–	Brain and kidney dysplasia	600649
β-Oxidation spiral								
VLCAD[e]	+	++	++	++	++	+++		201475
MTP, type 1 (LCHAD)	++	++	++	+++	+++	+++	Retinitis pigmentosa, AFLP, HELLP, lactic acidemia	600890
MTP, type 2 (LCEH/ LCHAD/ LCKT)	++	++	++	+++	+++	+++	Retinitis pigmentosa, peripheral neuropathy, hypoparathyroidism	143450
MEDIUM- AND SHORT-CHAIN FATTY-ACID OXIDATION								
MCAD	+/–	+/–	–	+++	+++	+++		201450
SCAD	–	?[f]	–	+/–[g]	+/–	++	Hypotonia, hypertonia, mental retardation	201470
HAD	–	–	–	++	+/–	+++	Congenital hyperinsulinism	231530
MCKAT	+++	–	–	–	++	+++	Vomiting, hyperammonemia	602199
UNSATURATED FATTY-ACID OXIDATION								
2,4-Dienoyl-CoA reductase	–	++[h]	–	–	–	–[i]	Microcephaly, dysmorphism	222745
ACAD9[j]	+/–	+/–	+	++	++	++[k]	Brain atrophy, cerebellar infarct, chronic thrombocytopenia	611126

TABLE 50.1 Main Clinical Features of Fatty-Acid β-Oxidation Disorders—cont'd

	Myopathic symptoms			Hepatic symptoms				
Disorder	Acute[a]	Chronic	Cardiomyopathy	Hypoketotic hypoglycemia	Metabolic encephalopathy[b]	Abnormal organic acids	Other features	MIM No.[c]
MULTIPLE ACYL-COA DEHYDROGENATION DEFECTS								
ETF or ETF:QO, severe	–	–	–	+++	+++	+++[l]	Congenital anomalies, renal dysplasia, dysmorphism	231680 130410 231675
ETF or ETF:QO, mild	–	+	+/–	+++	+++	+++[l,m]		231680 130410 231675
Riboflavin-responsive MADD[n]	–	+++	–	+++	+	+++	Leukodystrophy, coenzyme Q_{10} deficiency	231680

Abbreviations: CT, carnitine transporter; CPT, carnitine palmitoyltransferase; CACT, carnitine/acylcarnitine translocase; VLCAD, very long-chain acyl-CoA dehydrogenase; MTP, mitochondrial trifunctional protein; LCHAD, long-chain 3-hydroxyacyl-CoA dehydrogenase; LCEH, long-chain 2-enoyl-CoA hydratase; LCKT, long-chain 3-ketoacyl-CoA thiolase; MCAD and SCAD, medium- and short-chain acyl-CoA dehydrogenase, respectively; MCKAT, medium-chain 3-ketoacyl-CoA thiolase; HAD, L-3-hydroxyacyl-CoA dehydrogenase; ACAD9, acyl-CoA dehydrogenase 9; ETF, electron transfer flavoprotein; ETF:QO, ETF:coenzyme Q oxidoreductase; MADD, multiple acyl-CoA dehydrogenation deficiency; AFLP, acute fatty liver of pregnancy; HELLP, hypertension or hemolysis, elevated liver enzymes, and low platelets.

[a]*Myoglobinuria;* [b]*Reye-like episodes;* [c]*Mendelian Inheritance in Man (MIM; McKusick VA. Mendelian Inheritance in Man: A Catalog of Human Genes and Genetic Disorders, 12th ed. Baltimore: Johns Hopkins University Press; 1998); Online MIM database (OMIM™): http://www.ncbi.nlm.nih.gov/omim;* [d]*ventricular arrhythmias in most cases;* [e]*includes cases previously reported as defects of the long-chain acyl-CoA dehydrogenase;* [f]*hypotonia;* [g]*ketotic hypoglycaemia;* [h]*the only patient reported had persistent hypotonia in the neonatal period;* [i]*urinary excretion of the unusual carnitine ester 2-trans,4-cis-decadienoylcarnitine;* [j]*mostly active against unsaturated long-chain acyl-CoA substrates ($C_{16:1}$-, $C_{18:1}$-, $C_{18:2}$-, $C_{22:6}$-CoA);* [k]*abnormal unsaturated long-chain acylcarnitines ($C_{18:1}$ and $C_{18:2}$) in postmortem liver extract;* [l]*glutaric aciduria type II (GAII);* [m]*ethylmalonic-adipic aciduria;* [n]*some patients have mutations in the ETFDH (ETF:QO) gene (see text for details); other patients have been reported to have Coenzyme Q deficiency and mutations in the ETFDH gene (see text for details).*

Mitochondrial oxidation of lipids is a complex process that requires a series of enzymatic reactions[8,10,12,13] (Figure 50.1). Schematically, plasma free FAs delivered into the cytosol are first activated to their corresponding acyl-coenzyme A (CoA) thioesters at the outer mitochondrial membrane by acyl-CoA synthetase(s). Unlike short-chain (C_4–C_6) and medium-chain (C_8–C_{12}) acyl-CoAs, long-chain (C_{14}–C_{20}) acyl-CoAs cannot enter mitochondria directly. The mitochondrial CPT enzyme system, in conjunction with a carnitine/acylcarnitine translocase (CACT), provides the active carnitine-dependent mechanism whereby long-chain acyl-CoAs are transported from the cytosolic compartment into the mitochondrion, where β-oxidation occurs. L-carnitine is supplied for this reaction by a plasma-membrane sodium-dependent carnitine transporter (CT). Once in the mitochondria, FAs are oxidized by repeated cycles of four sequential reactions, acyl-CoA dehydrogenation, 2-enoyl-CoA hydration, L-3-hydroxy-acyl-CoA dehydrogenation, and 3-ketoacyl-CoA thiolysis. The final step of each cycle in the β-oxidation spiral is the release of one molecule of acetyl-CoA and a fatty acyl-CoA, which is two carbon atoms shorter. Each reaction is catalyzed by multiple enzymes that exhibit partially overlapping chain-length specificity.[12]

Complete catabolism of long-chain acyl-CoAs in mitochondria is accomplished by the action of two distinct, albeit coordinated, β-oxidation systems.[14] One is located on the mitochondrial inner membrane and is specifically involved in the oxidation of LCFA. The other system, composed of soluble enzymes located in the mitochondrial matrix, is responsible for the β-oxidation of medium- and short-chain acyl-CoAs (Figure 50.1). Finally, mitochondrial FA β-oxidation is tightly coupled to both the tricarboxylic acid (TCA) and the respiratory chain. Thus, while acetyl-CoA released can enter the TCA cycle, the electrons of the flavin adenine dinucleotide (FAD)-dependent acyl-CoA dehydrogenases and the nicotinamide adenine dinucleotide (NAD^+)-dependent L-3-hydroxy-acyl-CoA dehydrogenases are transferred to the respiratory chain.

Control of FA β-Oxidation and Synthesis

The regulatory control of FA oxidation and synthesis occurs at multiple levels, and with diverse mechanisms in different organs: it is a highly complex interactive system, not yet fully understood, which involves transcriptional and nontranscriptional components.[15] Essential information in the control of lipid metabolism relevant to this chapter is given below.

TABLE 50.2 Molecular Genetics of Fatty-Acid β-Oxidation Disorders

Deficiency	MIM no.[a]	Gene name	Chromosomal localization	Gene structure	cDNA, coding region	Mutations identified	Prevalent mutations
LONG-CHAIN FATTY-ACID OXIDATION							
Fatty-acid transport							
CT	212140	*SLC22A5*	5q33.1	10 exons	1674bp	++	None
CPT1	255120	*CPT1A*	11q13.1-q13.5	20 exons	2322bp	+	None
CACT	212138	*SLC25A2*	3p21.31	9 exons	903bp	+	None
CPT2	255110 600650 600649	*CPT2*	1p32	5 exons	1974bp	++	c.439C>T p.Ser113Leu
β-Oxidation spiral							
VLCAD	201475	*ACADVL*	17p11.2-11.13	20 exons	1968bp	+++	None
MTP, type 1 (LCHAD)	600890	*HADHA*	2p23	20 exons	2289bp	+	c.1528G>C p.Glu474Gln
MTP, type 2 (LCEH/LCHAD/LCKT)	143450	*HADHB*	2p23	16 exons	1422bp	+	None
MEDIUM- AND SHORT-CHAIN FATTY-ACID OXIDATION							
MCAD	201450	*ACADM*	1p31	12 exons	1263bp	+++	c.985A>G p.Lys304Glu
SCAD	201470	*ACADS*	12q22	10 exons	1239bp	++	c.625G>A p.Gly185Ser c.511C>T p.Arg147Trp
HAD	231530	*HADH*	4q22-q26	8 exons	945bp	+	None
MCKAT	602199	–	n.d.	n.d.	n.d.	–	n.d.
UNSATURATED FATTY-ACID OXIDATION							
2,4-Dienoyl-CoA reductase	222745	*DECR1*	8q21.3	10 exons	1008bp	–	n.d.
ACAD9	611126	*ADAD9*	3q26	22 exons	1866bp	+	none
MULTIPLE ACYL-COA DEHYDROGENATION DEFECTS							
ETF α subunit	231680	*ETFA*	15q23-25	12 exons	1002bp	+	c.797C>T p.Thr266Met
ETF β subunit	130410	*ETFB*	19q13.3	6 exons	768bp	+	None
ETF:QO	231675/231680	*ETFDH*	4q32-q35	13 exons	1854bp	+	None
Riboflavin-responsive MADD[b]	231680	*ETFDH*	4q32-q35	13 exons	1854bp	++	None

Abbreviations: CT, carnitine transporter; CPT, carnitine palmitoyltransferase; CACT, carnitine/acylcarnitine translocase; VLCAD, very long-chain acyl-CoA dehydrogenase; MTP, mitochondrial trifunctional protein; LCHAD, long-chain 3-hydroxyacyl-CoA dehydrogenase; LCEH, long-chain 2-enoyl-CoA hydratase; LCKT, long-chain 3-ketoacyl-CoA thiolase; MCAD and SCAD, medium- and short-chain acyl-CoA dehydrogenase, respectively; HAD, L-3-hydroxyacyl-CoA dehydrogenase; MCKAT, medium-chain 3-ketoacyl-CoA thiolase; ETF, electron transfer flavoprotein; ETF:QO, ETF:coenzyme Q oxidoreductase; MADD, multiple acyl-CoA dehydrogenation disorders; n.d., not determined.

[a]*Mendelian Inheritance in Man (MIM; McKusick VA. Mendelian Inheritance in Man: A Catalog of Human Genes and Genetic Disorders, 12th ed. Baltimore: Johns Hopkins University Press; 1998); Online MIM database (OMIM™): http://www.ncbi.nlm.nih.gov/omim;* [b]*some patients have mutations in the* ETFDH *gene (see text for details), other patients have been reported to have coenzyme Q deficiency and mutations in the* ETFDH *gene (see text for details).*

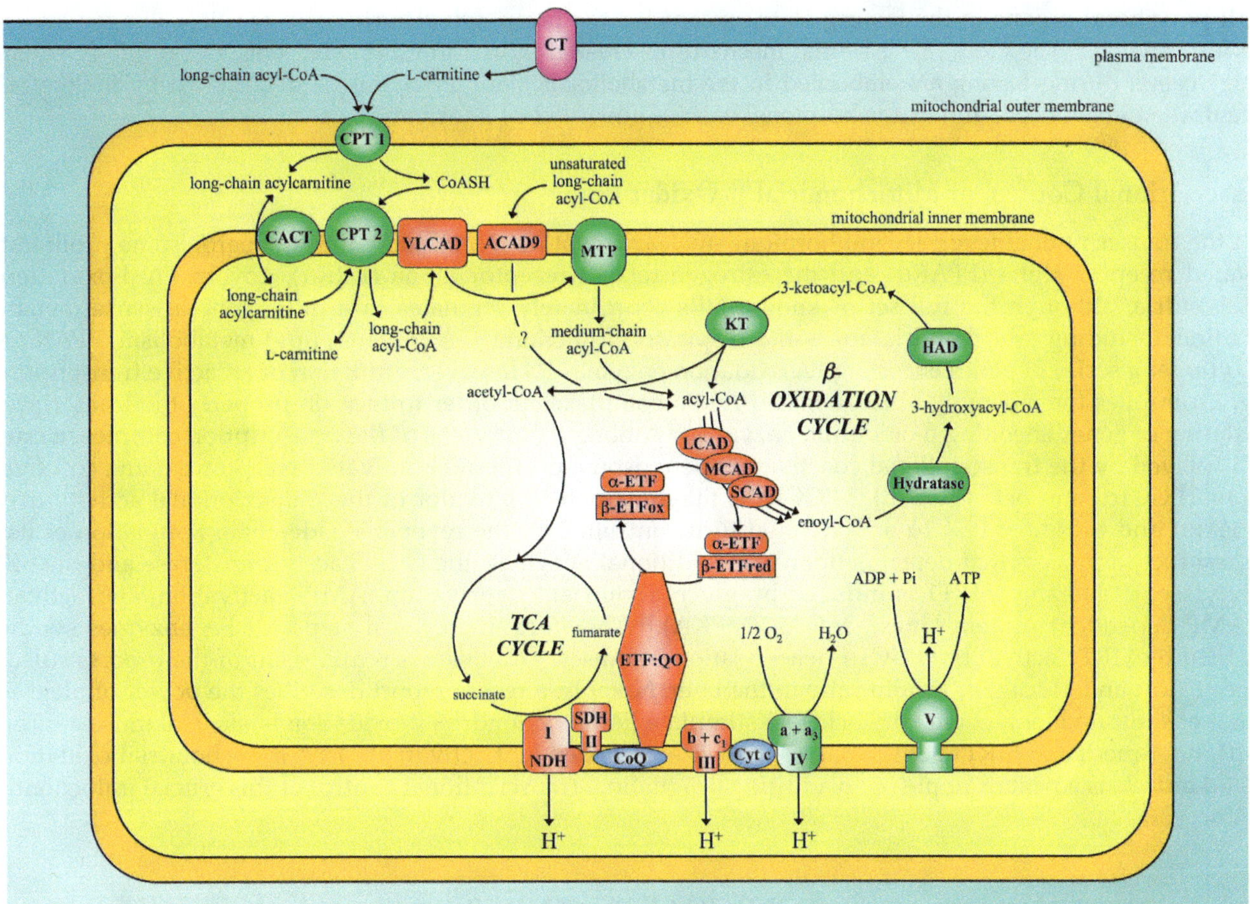

FIGURE 50.1 **Schematic representation of the functional and physical organization of fatty-acid β-oxidation enzymes in mitochondria.** Abbreviations: CT, plasma membrane high-affinity sodium-dependent carnitine transporter (OCTN2); CPT1, carnitine palmitoyltransferase 1; CACT, carnitine/acylcarnitine translocase; CPT2, carnitine palmitoyltransferase 2; VLCAD, LCAD, MCAD, SCAD, very-long-, long-, medium-, and short-chain acyl-CoA dehydrogenase, respectively; ACAD9, acyl-CoA dehydrogenase 9; MTP, mitochondrial trifunctional protein; Hydratase, 2-enoyl-CoA hydratase; HAD, L-3-hydroxyacyl-CoA dehydrogenase; KT, 3-ketoacyl-CoA thiolase; ETF, electron transfer flavoprotein (ox, oxidized; red, reduced); ETF:QO, ETF:coenzyme Q oxidoreductase; I, respiratory chain complex I (NDH, NADH:coenzyme Q reductase); II, respiratory chain complex II (SDH, succinate dehydrogenase); CoQ, coenzyme Q; III, respiratory chain complex III (b, cytochrome *b*; c1, cytochrome *c1*); Cyt c, cytochrome *c*; IV, respiratory chain complex IV (cytochrome *c* oxidase) (a, cytochrome *a*; a3, cytochrome *a3*); V, respiratory chain complex V (ATP synthase). Enzymes which use FAD as a coenzyme are indicated in red.

Hormonal and Allosteric Control of FA Oxidation in Liver

In the fed state, under low glucagon/insulin ratio, the liver avidly takes up glucose from blood. Glucose is partly degraded and oxidized and partly stored as glycogen. FA and triglyceride liver synthesis is high, whereas FA oxidation and ketone body production are shut off because of the high cellular levels of malonyl-CoA, the most potent allosteric suppressor of CPT1 activity. Malonyl-CoA concentration is in turn increased due to the abundance of citric acid and TCA substrates and to the activation of acetyl-CoA carboxylase.[16] In physiological fasting conditions, such as night fasting, or forced nutrient deprivation, glucagon/insulin ratio dramatically increases, and glucagon signaling induces activation of the 5′-AMP-activated protein kinase (AMPK). Activated AMPK phosphorylates acetyl-CoA carboxylase enzymes, promptly turning off their activity. These metabolic events cause a dramatic drop in malonyl-CoA concentration, hence relieving the inhibition on CPT1 and turning on ketone body (acetoacetate and β-hydroxybutyrate) production. Ketone bodies represent the vital energy substrate for all peripheral organs in the fasting state, except for the brain, which primarily requires oxygen and glucose for survival.[16] Therefore, the essential mechanisms created by nature to prevent episodes of life-threatening hypoglycemia during fasting are embodied in the metabolic-allosteric mechanisms determined by malonyl-CoA concentration in liver cells and by the hormonal balance governed by pancreatic β-cells.

Transcriptional Control of Mitochondrial β-Oxidation

At the transcriptional level, several hormone nuclear receptors (NRs), including the peroxisome proliferator-activated receptor alpha (PPARα) and the estrogen-related receptor alpha (ERRα), govern FA β-oxidation in mitochondria, although the full set of known NRs coordinately regulates in a more general sense metabolic activation, including mitochondrial mass and respiratory function.[17,18] As regards lipid metabolism, PPARα and ERRα target a series of genes encoding FA oxidation enzymes.[15] However, induction of an active transcriptional program implies the assembly of large multiprotein complexes in order to turn on properly the transcription–translation of genes encoding β-oxidation enzymes.[17] Among the proteins of the transcription complex, a crucial role is played by the transcriptional coactivators peroxisome proliferator-activated receptor gamma coactivator 1-alpha (PGC1α) and beta (PGC1β).[18] PGC1α is the most potent activator of the transcriptional activity driven by PPARα and ERRα.[17,18] PGC1α activity is in turn controlled by the reversible side-chain acetylation of its lysine residues, a process that depends upon two additional enzymes: the GCN5 acetyltransferase and the SIRT1 deacetylase.[15,17] Notably, SIRT1 is induced by glucose/nutrient deprivation, AMPK activation, and activation of cAMP-dependent protein kinase A (cAMP/PKA).[15] Therefore, the critical steps for FA enzymes synthesis governed by SIRT1 activation/PGC1α deacetylation in muscle and heart is reminiscent of the process ruled by malonyl-CoA and glucagon/insulin ratio in the liver. Notably, a recent report describes the power of oleic acid, but not of saturated long-chain fatty acids, to stimulate FA mitochondrial β-oxidation in skeletal muscle through cAMP/PKA-mediated SIRT1 activation.[15] The oleic acid-dependent activation of FA mitochondrial oxidation in skeletal muscle is a neat example of an additional signaling–transcriptional control of this crucial mitochondrial function.[15]

CLINICAL FEATURES

Defects of FA mitochondrial β-oxidation are autosomal recessive disorders of infancy and childhood, though some patients present later in life. Their classification and main clinical and genetic features are illustrated in Tables 50.1 and 50.2. Overall, the clinical syndromes associated with FA oxidation disorders result from the failure of FA-oxidizing tissues to respond to increased energy demands. Clinical manifestations range from a predominantly myopathic disease, either acute or chronic, to life-threatening systemic metabolic dysfunction (Table 50.3). Symptomatic hypoglycemia, characteristically associated with impaired ketogenesis, is often the earliest clinical manifestation and can be observed in nearly all these disorders.[14] Recurrent episodes of hypoketotic hypoglycemia, with or without concomitant brain involvement, are the most common presentation in the newborn or infant. Nausea and vomiting, hypotonia, drowsiness, and coma are also frequent. Sometimes, attacks are triggered by fasting or minor viral infections. The acute and frequently life-threatening presentation in early infancy requires differential diagnosis from other encephalopathies of infancy because: 1) a few defects of β-oxidation can be effectively cured, such as carnitine deficiency and riboflavin-responsive multiple acyl-CoA dehydrogenase deficiency; and 2) early diagnosis may help to prevent acute metabolic attacks, mental retardation, epilepsy, severe brain damage, and death.[19–21] Some infants survive the acute metabolic attacks but show poor growth, impaired psychomotor development, dystonia, spastic tetraplegia, and intractable seizures. Nervous system involvement is usually secondary to severe acidotic and hypoglycemic attacks, though patients with trifunctional protein deficiency may have retinitis pigmentosa and peripheral neuropathy.[14,22,23] Infants with severe defects of CPT2, ETF, or ETF:QO, however, can present congenital malformations of the brain (microgyria, neuronal heterotopia) and, sometimes, facial dysmorphism reminiscent of Zellweger syndrome, suggesting that LCFA may play a role during human development.[24] In addition to metabolic symptoms, patients often have cardiomyopathy; primary carnitine deficiency, carnitine/acylcarnitine translocase deficiency, CPT2 deficiency, VLCAD deficiency, and MTP deficiency are all associated with various forms of heart disease.[14,21]

Patients with onset in late infancy, childhood, or adulthood tend to have more chronic disorders characterized by progressive myopathy or cardiomyopathy, sometimes associated with mild metabolic symptoms, such as nausea

TABLE 50.3 Clinical Features Associated with Mitochondrial Fatty-Acid β-Oxidation Disorders

HEPATIC SIGNS
Hypoglycemia associated with low ketones (hypoketotic hypoglycemia)
Reye-like syndrome
Steatosis
Acute hepatic failure
Sudden infant death syndrome (SIDS)
MUSCLE SIGNS
Hypotonia
Weakness and wasting
Proximal myopathy with lipid storage
Exercise intolerance and muscle pain with increased levels of creatine kinase
Episodic rhabdomyolysis (with occasional paroxysmal myoglobinuria)
CARDIAC SIGNS
Hypertrophic and dilated cardiomyopathy
Progressive heart failure
Arrhythmias
Cardiac arrest
Sudden infant death syndrome (SIDS)
NERVOUS SYSTEM SIGNS
Permanent brain damage due to hypoglycemia, arrhythmias, or cardiac arrest
Microgyria, cortical atrophy, and neuronal heterotopia
Pigmentary retinopathy
Peripheral sensorimotor neuropathy
MALFORMATIONS
Renal dysplasia and nephromegaly*
Polycystic kidney
Facial dysmorphism
Brain malformations

**Proximal and distal tubulopathy is observed in CPT1 deficiency.*

and drowsiness, or with altered laboratory tests, such as hypoglycemia or poor rise of blood ketone concentrations in provocative tests. Disorders of lipid mitochondrial metabolism may cause two main clinical syndromes in muscle, namely 1) progressive weakness with hypotonia (e.g., carnitine transporter and carnitine/acylcarnitine translocase defects) or 2) acute, recurrent, reversible muscle dysfunction with exercise intolerance and acute muscle breakdown (rhabdomyolysis) with myoglobinuria (e.g., deficiencies of CPT2, very-long-chain acyl-CoA dehydrogenase, or trifunctional protein).[6,7] Approximately 40% of patients affected with all kinds of FA oxidation disorders, except CPT1 and MCAD deficiencies, present with significant muscular involvement.[19] Because of the dependence of heart and skeletal muscle upon LCFA oxidation, cardiomyopathy, typically hypertrophic but sometimes dilated, and skeletal muscle myopathy, either chronic (lipid storage myopathy) or acute (paroxysmal myoglobinuria), are commonly observed in LCFA oxidation defects while they are extremely rare in disorders of medium- and short-chain FA oxidation.[21]

Because clinical presentation is of limited help in differential diagnosis, the only way to reach a definitive diagnosis is to analyze body fluids for accumulating metabolites and to study tissues for specific enzymes of fatty acid metabolism. Because most of the organic acids accumulating in β-oxidation defects are effectively cleared from the blood by the kidneys, gas chromatography–mass spectrometry (GC–MS) analysis of 24-hour urine specimens usually reveals a pattern of metabolites characteristic of a specific disease and is therefore the test of choice.[5] When available, analysis of plasma acylcarnitine profile by tandem (MS/MS) or electrospray mass spectrometry is the most specific and direct approach for the specific diagnosis of most of the FA oxidation disorders.[5] It is very sensitive and can be easily performed on Guthrie cards for newborn screening.[11] Finally, genetic testing is now available for almost all these disorders, and prevalent mutations have been identified in some of them which makes molecular screening feasible and cost-effective.[20,25]

DEFECTS OF MITOCHONDRIAL FATTY-ACID OXIDATION

Carnitine Transporter Deficiency (Primary Carnitine Deficiency)

L-carnitine (β-hydroxy-γ-*N*-trimethylamino-butyrate) is required for the active transport of LCFA into mitochondria. Primary carnitine deficiency (PCD) is characterized by increased urinary carnitine loss and severely decreased carnitine concentration in plasma, heart, and skeletal muscle. The disease is autosomal recessive and has a frequency of 1 : 37,000–1 : 100,000 newborns, as determined by neonatal screening of carnitine levels.[26]

CLINICAL FEATURES

Two major clinical presentations are associated with PCD.[10,26] The most common phenotype is characterized by slowly progressive hypertrophic or dilated cardiomyopathy with lipid storage myopathy (Figure 50.2A and B), occurring between 1 and 7 years of age. A second phenotype, more frequent before 2 years of age, is characterized by acute recurrent episodes of nonketotic hypoglycemic encephalopathy. These two phenotypes are not mutually exclusive, as both metabolic and cardiomuscular presentations have been described in some families.[6]

LABORATORY FINDINGS

PCD has to be distinguished from secondary carnitine deficiency that can be associated with a number of acquired or inherited diseases, including other FA oxidation defects.[6,27] In PCD, carnitine content is very low (<5% of normal) both in tissues (muscle, heart, liver) and in plasma, and analysis of plasma and urine does not show an abnormal

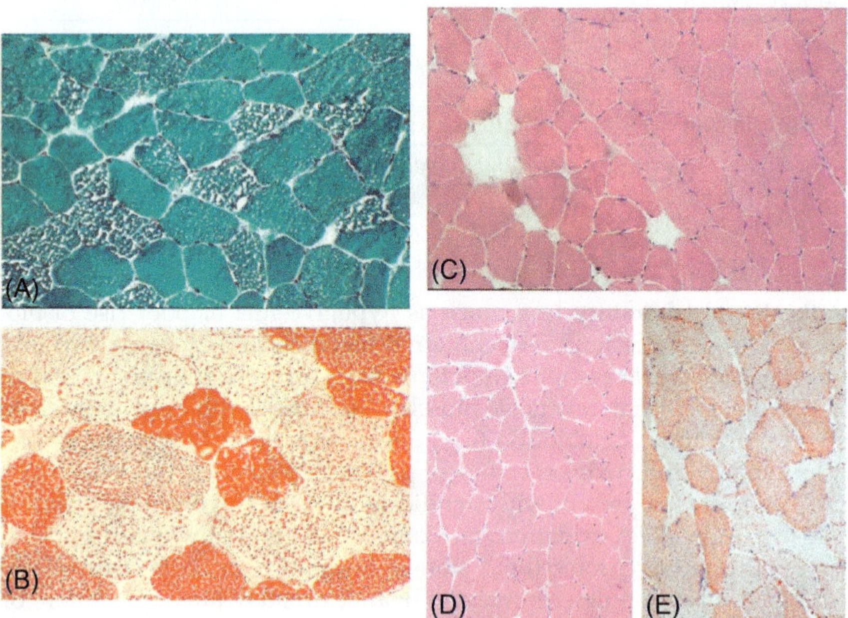

FIGURE 50.2 **Muscle biopsies from patients with fatty-acid oxidation defects.** (**A**) and (**B**) Lipid storage myopathy in a patient with primary carnitine deficiency (PCD) caused by a defect of the high-affinity plasma carnitine transporter (CT); (**A**) modified Gomori trichrome staining showing numerous vacuoles mostly in type 1 fibers. ×160. (**B**) Oil Red O stain showing numerous large lipid droplets within fibers. ×250. (**C**) Recurrent paroxysmal myoglobinuria in a young adult with CPT2 deficiency, harboring the common p.Ser113Leu mutation in the *CPT2* gene. Muscle biopsy performed 10 days after an acute episode shows mild nonspecific morphological alterations. There is evidence of fiber loss and modest variability of fiber diameter. Some fibers show central nuclei. Hematoxylin & eosin, ×160. (**D**) and (**E**) Recurrent paroxysmal myoglobinuria and interictal chronic proximal myopathy in a young woman with VLCAD deficiency. (**D**) Hematoxylin & eosin stain shows mild nonspecific morphological alterations. There is fine vacuolization in some fibers and fiber diameter variability. ×160; (**E**) Oil Red O stain shows signs of mild lipid accumulation with numerous fine droplets within most fibers. Lipid droplets exhibit a subsarcolemmal distribution. ×250.

acylcarnitine profile nor dicarboxylic aciduria which are usually seen in patients with other FA oxidation defects.[6] Once suspected, the transporter defect should be ultimately confirmed by carnitine uptake assay in cultured skin fibroblasts[28] or by molecular analysis.[29]

MOLECULAR GENETICS

PCD is caused by mutations in the *SLC22A5* gene encoding the high-affinity plasma membrane carnitine transporter OCTN2.[30] Most of the mutations are nonsense mutations associated with no residual carnitine transport activity.[26,31] In a few cases, "leaky" missense mutations associated with residual carnitine transport activity have been identified.[29]

THERAPY

If therapy is started before irreversible organ damage occurs, PCD patients respond very well to high-dose oral L-carnitine supplementation (usually 100–600 mg per kg per day),[26,32] which may avoid cardiac transplant. Hypoglycemic episodes also tend to disappear.[26]

Carnitine Palmitoyltransferase Deficiencies

The carnitine palmitoyltransferase (CPT) system is composed of two distinct acyltransferases, CPT1 on the outer mitochondrial membrane, and CPT2 on the inner mitochondrial membrane.[33] CPT1 is expressed in at least three tissue-specific isoforms encoded by distinct genes,[33] whereas CPT2 is present in all tissues in a single form encoded by a gene on chromosome 1.[33,34]

CPT1 Deficiency

This disorder is commonly referred to as the "hepatic" form of CPT deficiency. The disease manifests before the second year of life with encephalopathy, fasting hypoglycemia, hypoketonemia, low plasma insulin concentrations, and *elevated* plasma carnitine levels.[19,33,35] Since the disease is caused by mutations in the *CPT1A* gene on chromosome 11q13.1 encoding the liver isoform of CPT1,[35] there is no cardiomuscular involvement, which makes this defect unique among the disorders of LCFA oxidation[19] (Table 50.1).

CPT2 Deficiency

CLINICAL FEATURES

Three different clinical phenotypes are associated with CPT2 deficiency (Table 50.1): 1) a myopathic form with juvenile–adult onset; 2) an infantile form with hepatic, muscular, and cardiac involvement; and 3) a lethal neonatal form with developmental abnormalities. In all cases, the enzyme defect can be demonstrated in every tissue examined (e.g., skeletal muscle, liver, fibroblasts, platelets, leukocytes).[6,33,36]

The "muscular" form of CPT deficiency is the most common disorder of lipid metabolism in muscle, one of the most common inherited disorders of mitochondrial FA oxidation,[19] and a major cause of hereditary recurrent myoglobinuria in both children and young adults.[9,33,37] The clinical hallmark of the disease is paroxysmal myoglobinuria. Attacks of myoglobinuria are most often precipitated by prolonged exercise (exertional myoglobinuria).[37] Prolonged fasting, infections, usually of viral etiology, and/or fever are the primary precipitating factors in the younger patients.[9,33,37] In approximately 20% of cases, attacks may occur without any apparent cause. True cramps are not a feature, but patients describe instead a feeling of "tightness" and pain in exercising muscles before the appearance of myoglobinuria and weakness. Persistent weakness is very uncommon.[6] The classic "muscular" form of CPT2 deficiency is usually a benign disease with a favorable evolution, provided that acute renal insufficiency, a potential complication of massive myoglobinuria, is adequately managed.[38] There are usually no clinical signs of liver dysfunction. Cardiac involvement is very unusual.[6]

In rare cases, CPT2 deficiency can manifest as a severe life-threatening infantile hepatocardiomuscular form (CPT2 deficiency type 2), characterized by nonketotic hypoglycemia, liver failure, cardiomyopathy, and mild signs of muscle involvement, or a fatal neonatal-onset form (CPT2 deficiency type 3) with acute metabolic decompensation and features of brain and kidney dysgenesis.[2,5,6,13,14,36,39–42]

LABORATORY FINDINGS

Outside episodes of myoglobinuria and at rest, serum creatine kinase (CK) levels are normal. During acute episodes of rhabdomyolysis, there is a massively elevated urinary excretion of myoglobin (≥200 ng per mL) and greatly increased levels of serum CK (20–400-fold) of muscle origin (CK-MM). Prolonged fasting or mild exercise may also provoke an increase in serum CK (2- to 20-fold above normal). Glycemia, ketonemia, ketonuria, urinary organic acid profile and serum and muscle carnitine levels are usually normal. Acute tubular necrosis, a life-threatening condition, may develop in patients excreting more than 1000ng per mL of myoglobin. Following attacks, serum CK levels usually return to normal by 8–10weeks. Between attacks, routine laboratory tests are not contributive to the diagnosis. In most cases, muscle biopsies in interictal periods are normal or may show mild signs of muscle involvement with regenerating fibers (Figure 50.2C). Diagnosis is ultimately made by demonstrating the enzyme defect in muscle or, more conveniently, in peripheral blood leukocytes.[6] However, since only one gene is associated with CPT2 deficiency, molecular genetic testing currently provides the most convenient means for noninvasive, rapid, and specific diagnosis.

MOLECULAR GENETICS

More than 70 mutations in the human *CPT2* gene have been identified.[33,36,42,43] Although most of the mutations are "private," a "common" mutation (p.Ser113Leu) can be identified in approximately 80% of patients with muscular CPT2 deficiency, being present in ≥50% of mutant alleles in patients of different ethnic origins.[6,37,43,44] There is some genotype–phenotype correlation. The muscular form of the disease is always associated with residual CPT2 activity, whereas mutations that abolish enzyme activity are invariably found in patients with the lethal early-onset form.[33,39,44,45]

THERAPY

Effective prevention of attacks may be accomplished by instituting a high-carbohydrate diet with a low amount of long-chain fats and with frequent and regularly scheduled meals, by avoiding the known precipitating factors (fasting, cold, prolonged exercise) and by increasing slow-release carbohydrate intake during intercurrent illness or sustained exercise.[19,33,41,46] More recently, agonists of PPARα such as bezafibrate have been shown to restore CPT2 activity and LCFA oxidation in fibroblasts from patients with the muscular form of CPT2 deficiency and to provide long-term subjective improvement of clinical conditions.[47] These results have not been confirmed by a randomized clinical trial.[48]

Carnitine/Acylcarnitine Translocase Deficiency

Along with infantile and neonatal CPT2 deficiency, CACT deficiency is one of the most severe mitochondrial FA oxidation defects. More than 30 patients have been reported since the first description in 1992.[10,49,50]

CLINICAL FEATURES

Patients exhibit life-threatening episodes in the neonatal period, characterized by neonatal distress with hyperammonemia, variable hypoglycemia, heart beat disorders, and muscle involvement with weakness and high serum CK.[26] The disease is often fatal within the first 2 years of life because of the deleterious combination of energy impairment and the toxic consequences of long-chain acylcarnitine accumulation, which may cause untreatable episodes of arrhythmia.[26,49,50]

LABORATORY FINDINGS

Diagnosis is suspected from the abnormal plasma acylcarnitine profile with low free carnitine and elevated C_{16}–C_{18}.[6,26,51]

MOLECULAR GENETICS

More than 35 mutations in the *SLC25A20* (solute carrier family 25 [carnitine/acylcarnitine translocase], member 20) gene have been reported thus far,[50] most of which are private. Two-thirds of mutations are nonsense, frameshift, or splice-site mutations resulting in premature stop codons (null mutations).[49] Functional analysis of missense mutations has been performed in few cases.[51] Null mutations are associated with rapidly progressive disease whereas hypomorphic mutations cause a milder phenotype with a near normal development with appropriate therapy.[26,51]

THERAPY

Therapy is based on low-LCFA and high-carbohydrate diet in an intensive protocol characterized by frequent or continuous feeding.[6,26] Whether supplementation with carnitine is advisable or potentially hazardous is still to be established, as it could induce an increase in toxic long-chain acylcarnitine production.[50,52]

Very-Long-Chain Acyl-CoA Dehydrogenase Deficiency

CLINICAL FEATURES

Very-long-chain acyl-CoA dehydrogenase (VLCAD) deficiency[53] has been reported in more than 400 cases.[54] It appears to be the most common long-chain fatty-acid oxidation defect, with a disease prevalence of up to 1:30,000.[55] The defect is clinically heterogeneous and can cause three major phenotypes: 1) an acute presentation with exercise-induced rhabdomyolysis and myoglobinuria—myalgia is more severe and episodes more numerous than in CPT2 deficiency;[56] 2) a severe childhood form, with early onset of dilated or hypertrophic cardiomyopathy, recurrent episodes of hypoketotic hypoglycemia, and high mortality rate (50–75%); and 3) a milder childhood form, with later onset of hypoketotic hypoglycemia and dicarboxylic aciduria, low mortality, and rare cardiomyopathy. Overall, acute metabolic decompensation is the most frequent form of presentation in VLCAD-deficient patients and most patients suffer from the severe cardiomyopathic form with early onset and poor outcome.[25]

LABORATORY FINDINGS

In the muscle form, serum CK markedly increases during attacks (20- to>200-fold). However, patients do not exhibit hypoketotic hypoglycemia nor dicarboxylic aciduria and increase of plasma long-chain acylcarnitines is rarely observed.[6] Plasma LCFA profile by GC-MS can be helpful for diagnosis because it may reveal an increase of tetradecenoic ($C_{14:1}$) acid, which persists even after the patient has fully recovered.[57,58] As in CPT2 deficiency, muscle biopsy may not provide any clue to the diagnosis. It may show mild nonspecific morphological alterations with no evidence of lipid accumulation (Figure 50.2D) or may demonstrate a diffuse increase of fat droplets mostly in type 1 fibers[57,58] (Figure 50.2E).

MOLECULAR GENETICS

More than 80 disease-causing mutations have been identified in the *ACADVL* gene, none of which seemed to predominate.[25,53,59,60] There is some genotype–phenotype correlation and mutations that result in some residual enzyme activity are usually found in patients with the milder phenotypes.[25]

THERAPY

VLCAD-deficient patients should be treated with a dietary regimen consisting of avoidance of fasting and a high-carbohydrate, low-LCFA diet. The beneficial effect of medium-chain triglycerides (MCT) is controversial and available evidence indicates that MCT ingestion does not ameliorate exercise performance in VLCAD-deficient myopathic patients.[61] As for CPT2 deficiency, the use of bezafibrate was found to ameliorate the biochemical and cellular phenotype,[62] but a recent randomized clinical trial did not demonstrate its clinical efficacy.[48]

Mitochondrial Trifunctional Protein Deficiency

The mitochondrial trifunctional protein (MTP) is a complex enzyme composed of four α subunits, harboring long-chain 2-enoyl-CoA hydratase (LCEH) and long-chain L-3-hydroxyacyl-CoA dehydrogenase (LCHAD) activities, and four β subunits harboring long-chain 3-ketoacyl-CoA thiolase (LCKT) activity. MTP deficiency is relatively frequent, with more than 80 patients reported thus far.[63,64]

CLINICAL FEATURES

The clinical manifestations of the disease are characteristically associated with urinary excretion of C_6–C_{14} 3-hydroxydicarboxylic acids. Patients can be classified into two groups:[65]

LCHAD DEFICIENCY The vast majority (≥85%) of MTP-deficient patients have an isolated deficiency of LCHAD activity.[65] LCHAD deficiency appears to be a relatively common β-oxidation defect (1 in 50,000 births in Northern Europe).[65] The disease is clinically heterogeneous. In infancy and early childhood, hypoglycemic encephalopathy with or without severe hepatic involvement and cardiomyopathy is the most common presentation. Mortality is high (≈50%). However, cardiomyopathy in patients who survive acute episodes tends to resolve with dietary therapeutic measures.[65,66] Later in childhood, the predominant manifestation is paroxysmal rhabdomyolysis and myoglobinuria. Among the distinctive features of LCHAD deficiency are progressive pigmentary retinopathy[67] and peripheral neuropathy,[68] which are not observed in patients with any other β-oxidation defect. Also characteristic of this disorder is the occurrence of acute fatty liver disease in pregnant women with an affected fetus.[21,65,69]

MTP DEFICIENCY (COMBINED ENZYME DEFICIENCY) In a smaller group of patients, all the three activities harbored by MTP are deficient, albeit to different extents. Clinical manifestations are similar to those observed in patients with isolated LCHAD deficiency, although, in general, the clinical presentation is more severe with a higher mortality rate.[69]

MOLECULAR GENETICS

A prevalent missense mutation (c.1528G>C, p.E510Q) in the LCHAD domain of the α subunit gene (*HADHA*) can be detected in approximately 90% of LCHAD-deficient alleles,[65] thus making molecular screening for the disease quite feasible. No apparent genotype–phenotype correlation has been observed, as patients homozygous for this mutation show widely different phenotypes.[65] Unlike LCHAD deficiency, the molecular basis of MTP deficiency is heterogeneous and different mutations have been identified in both *HADHA* and *HADHB* genes with poor genotype–phenotype correlation.[63,65]

THERAPY

The mainstay of therapy is avoidance of fasting and a high-carbohydrate, low-LCFA diet associated with MCT oil supplementation.[66,70] Deficiency of docosahexaenoic acid (DHA), an essential n-3 polyunsaturated FA necessary for nerve myelination, has been documented in MTP-deficient patients, and encouraging response to cod liver oil extract, high in DHA content, has been observed.[66,71,72]

Medium-Chain Acyl-CoA Dehydrogenase Deficiency

Medium-chain acyl-CoA dehydrogenase (MCAD) deficiency is the most common FA oxidation disorder, with a frequency in the United States of 1:10,000, as determined by newborn screening.[25,73]

CLINICAL FEATURES

Typical symptoms include fasting intolerance, nausea, vomiting, hypoketotic hypoglycemia, lethargy, and coma beginning within the first 2 years of life. Approximately 20% of patients die suddenly at first presentation of the disease because of acute metabolic decompensation in response to either prolonged fasting or intercurrent and common infections.[74] Clinical manifestations, however, are variable, and some patients may be asymptomatic, being recognized through family screening. Skeletal muscle and heart involvement is extremely rare.[6,14]

LABORATORY FINDINGS

Patients have medium-chain dicarboxylic aciduria and secondary carnitine deficiency. The disease is characterized by urinary excretion of C_6–C_{10} dicarboxylic acids (with a characteristic pattern $C_6>C_8>C_{10}$), acylglycine and acylcarnitine conjugates (hexanoylglycine, phenylpropionylglycine, suberylglycine and octanoylcarnitine).[75] C_{12}–C_{14} dicarboxylic acids, the hallmarks of VLCAD deficiency, are absent. Ketones tend to be inappropriately low in plasma.[75] Deficiency of MCAD can be documented in most tissues, including cultured fibroblasts and peripheral blood lymphocytes.[75]

MOLECULAR GENETICS

A prevalent mutation (c.985A>G/p.K329E) in the *ACADM* gene (chromosome 1p31) is found in 90% of patients of Northern European descent.[25] MCAD deficiency is prevalently observed in this Caucasian population, in which the carrier frequency for the common c.985A>G mutation is approximately 1:40.[5] The p.K329E mutation causes impairment of tetramer assembly and instability of the protein.[76] More than 30 mutations account for the remaining alleles[73] and are usually (>90%) present only in compound heterozygous form.[75]

THERAPY

Early diagnosis and treatment of MCAD deficiency can result in good long-term prognosis. Avoidance of fasting and maintenance of adequate caloric intake may prevent life-threatening metabolic attacks.

Multiple Acyl-CoA Dehydrogenase Deficiency (Glutaric Aciduria Type II)

Multiple acyl-CoA dehydrogenation deficiency (MADD) or glutaric aciduria type II (GAII) is an autosomal recessive disorder of FA, amino acid, and choline oxidation, resulting from a generalized defect in intramitochondrial acyl-CoA dehydrogenation due to defective electron transport from the acyl-CoAs to ubiquinone (coenzyme Q; CoQ) in the mitochondrial respiratory chain (Figure 50.1). The function of some 14 FAD-containing dehydrogenases is affected.[21]

CLINICAL FEATURES

Three different phenotypes have been described:[6,77,78] 1) two lethal neonatal forms—with or without multiple congenital anomalies—characterized by hypotonia, hepatomegaly, severe hypoglycemia, and metabolic acidosis; and 2) a milder, late-onset form characterized by potentially life-threatening episodes of metabolic decompensation, ethylmalonic–adipic aciduria, and progressive lipid storage myopathy.

LABORATORY FINDINGS

The disease is characterized by urinary excretion of numerous organic acids (not only glutaric acid, as in glutaric aciduria type I, but also lactic, ethylmalonic, butyric, isobutyric, 2-methyl-butyric, and isovaleric acids) and by multiple elevation of plasma acylcarnitines of different lengths (C_4–C_{16}).[77]

MOLECULAR GENETICS

In most cases, regardless of the phenotype, the disease is due to a defect in the genes encoding the α (*ETFA*) or β subunits (*ETFB*) of electron transfer flavoprotein (ETF) or the electron transfer flavoprotein ubiquinone oxidoreductase (ETF:QO) (*ETFDH*)[6,77–79] (Tables 50.1 and 50.2). Most patients do not respond to riboflavin (vitamin B_2) supplementation.[21]

Riboflavin-Responsive Multiple Acyl-CoA Dehydrogenase Deficiency

Riboflavin is the precursor of the coenzyme flavin adenine dinucleotide (FAD) which is the redox prosthetic group of several flavoproteins including the acyl-CoA dehydrogenases of the β-oxidation system and the electron transfer flavoproteins ETF and ETF:QO[6,80] (Figure 50.1). A subset of MADD patients have been recently characterized who respond to pharmacological doses of riboflavin both clinically and biochemically.

CLINICAL FEATURES

Riboflavin-responsive multiple acyl-CoA dehydrogenase deficiency (RR-MADD) is mostly characterized by impaired oxidation of fatty acids due to multiple deficiencies of short-chain acyl-CoA dehydrogenase (SCAD), MCAD, long-chain acyl-CoA dehydrogenase (LCAD) and VLCAD. There are two major clinical phenotypes: 1) an "infantile form" with nonketotic hypoglycemia, hypotonia, failure to thrive, and acute metabolic episodes reminiscent of Reye syndrome; and 2) a "juvenile form" characterized by progressive proximal lipid storage myopathy.[80]

LABORATORY FINDINGS

There is usually a complex abnormal pattern of urinary excretion of organic acids (glutaric aciduria type II [GAII] or ethylmalonic–adipic aciduria), which indicates a multiple acyl-CoA dehydrogenation defect.[6,21] Activities and protein levels of SCAD, MCAD, and VLCAD are reduced in isolated muscle mitochondria.[80]

MOLECULAR GENETICS

Recessive mutations in the *ETFDH* gene encoding ETF:QO have been identified in some RR-MADD patients presenting with encephalopathy or muscle weakness or a combination of both.[81] Whether *ETFDH* mutations represent a common cause of RR-MADD still remains to be elucidated. Molecular analysis of 23 of our familial patients with RR-MADD has shown a robust prevalence of subjects with a variety of *ETFDH* mutations. Notably, among these patients, one family presented a dominant pattern of transmission (F. Taroni, S. Di Donato, and C. Gellera, unpublished data). *ETFDH* mutatons have also been reported in some patients with Coenzyme Q (CoQ) deficiency presenting with lipid storage myopathy and late-onset GAII.[82,83] In these cases, however, response to therapy was not uniform, with some patients improving following riboflavin or CoQ_{10} (150–500 mg/day) monotherapy, and others requiring the combined therapy. Interestingly, most *ETFDH* mutations in RR-MADD patients are located around the ubiquinone binding pocket.[81,82] Since riboflavin is the precursor of FAD, it has been proposed that riboflavin responsiveness may result from the ability of FAD to act as a chemical chaperone that promotes folding of certain misfolded ETF:QO proteins, thereby ameliorating or normalizing disease symptoms.[84]

THERAPY

The clinical, morphological, and biochemical responses to oral riboflavin supplementation (100–400 mg per day oral riboflavin) are usually dramatic,[6,80] with rapid improvement of muscle weakness and wasting and disappearance of signs of lipid accumulation at muscle biopsy. A prompt response to riboflavin treatment is also observed in encephalopathic patients.[81] Riboflavin supplementation also normalizes the activities of SCAD and MCAD, and restores to normal the amount of protein mass.[6,80] Since riboflavin is the precursor of FAD, and riboflavin responsiveness results from the ability of FAD to act as a chemical chaperone that promotes folding of certain misfolded ETF:QO proteins, the effect is to ameliorate or normalize disease symptoms.[84]

OTHER DISORDERS OF FATTY-ACID β-OXIDATION

Short-Chain Acyl-CoA Dehydrogenase Deficiency

This is a rare disorder which was first reported in 1987.[85] Since then, approximately 25 patients have been reported worldwide, based upon reduced or absent short-chain acyl-coA dehydrogenase (SCAD) activity *in vitro* and the presence of ethylmalonic aciduria (EMA).[86,87] Clinical manifestations range from hypoglycemia and vomiting to hypotonia and seizures accompanied with developmental delay and dysmorphic features.[25,85,87–89] Notably, fasting ketogenesis is not impaired.[89] Laboratory diagnosis is based on ethylmalonic aciduria and elevated plasma C_4-acylcarnitine concentration and is confirmed by enzyme assay in muscle tissue and the detection of disease mutations in the *ACADS* gene.[25,88]

In addition to the small number of patients with SCAD inactivating mutations, a vast number of other patients with predominantly neuromuscular symptoms and EMA were found to carry two common SCAD gene variants (c.511C>T/p.G209S and c.625G>A/p.R171W).[25,89,90] These alleles are not regarded as true disease-causing mutations nor are they polymorphisms, but rather mutations that confer disease susceptibility.[88,90]

Medium-Chain 3-Ketoacyl-CoA Thiolase Deficiency

So far, this defect has been described in only one case, a neonate presenting with vomiting, metabolic acidosis, liver dysfunction, and terminal rhabdomyolysis and myoglobinuria.[91] No information is yet available on the molecular bases of the disorder.

L-3-Hydroxyacyl-CoA Dehydrogenase Deficiency

In earlier reports and literature reviews, the disorder now known as L-3-hydroxyacyl-CoA dehydrogenase (HAD) deficiency was described as short-chain HAD (SCHAD) deficiency.[5] It has now become clear that HAD, rather than SCHAD, provides the majority of 3-hydroxyacyl-CoA dehydrogenase activity for mitochondria.[92] Furthermore, HAD, encoded by the *HADH* gene on chromosome 4q22, has a preference for medium-chain straight 3-hydroxyacyl-CoAs, whereas SCHAD, also known as type 10 17β-hydroxysteroid dehydrogenase, encoded by the *HSD17B10* gene on chromosome Xp11.2, acts on a wide spectrum of substrates, including steroids, cholic acids, and fatty acids, with a preference for short-chain methyl-branched acyl-CoAs.[92]

To date, approximately 10 patients have been reported with missense mutations in the *HADH* gene.[93,94] Presentation of these patients was heterogeneous with either mild late-onset nonketotic hypoglycemia or severe neonatal nonketotic hypoglycemia associated with hyperinsulinism.[93,94] Urine organic acids showed increased dicarboxylic and 3-hydroxydicarboxylic acids with 6–14 carbons. Elevated C_4-hydroxyacylcarnitine was present in plasma.[93,94] Deficient HAD activity was seen in fibroblasts or other tissues.[93,94] HAD deficiency is the only FA oxidation disorder associated with congenital hyperinsulinism. Interestingly, HAD mRNA and activity are particularly high in the pancreas and especially in the islets of Langerhans, which suggests an important role for HAD in insulin secretion, possibly through a novel glucose–fatty acid cycle.[93,94] The hyperinsulinism associated with HAD deficiency is responsive to diazoxide.[93,94]

Defects of Unsaturated-Fatty-Acid Oxidation

2,4-Dienoyl-CoA Reductase Deficiency

2,4-Dienoyl-CoA reductase is an enzyme required in the degradation of unsaturated fatty acids with an even number of double bonds, such as linoleic acid (9-*cis*,12-*cis*-$C_{18:2}$). It converts 2,4-dienoyl-CoA to 3-*trans*-enoyl-CoA. 2,4-Dienoyl-CoA reductase deficiency is a very rare disorder, as it has been described in one female infant only, presenting with hypotonia and fatal respiratory acidosis.[95] She also had microcephaly with a short trunk, arms, and fingers, small feet, and a large face. Organic acid profile was normal but an unusual acylcarnitine species (2-*trans*,4-*cis*-decadienoylcarnitine) was detected in plasma and urine. Enzyme activity was reduced in liver and muscle. No information is available on the molecular basis of the disorder.

Acyl-CoA Dehydrogenase 9 Deficiency

Several acyl-CoA dehydrogenases (ACAD9, ACAD10, and ACAD11) putatively involved in fatty-acid oxidation have been recently described, but their role in the pathophysiology of human disease is not fully elucidated.[21,96] ACAD9, which closely resembles VLCAD, has been shown to have maximum activity with unsaturated long-chain acyl-CoAs.[97] Enzyme defect and mutations in the *ACAD9* gene have been described in three patients presenting with recurrent episodes of acute liver dysfunction and hypoglycemia, cardiomyopathy, and chronic neurologic dysfunction.[97] More recently, however, *ACAD9* mutations have also been found in patients with a defect of complex I of the respiratory chain[98] while evidence indicates a role for ACAD9 in the biogenesis of complex I.[99]

ACKNOWLEDGEMENTS

Part of the original work cited here was made possible by the valuable contribution of Drs. Silvia Baratta, Barbara Castellotti, Patrizia Cavadini, Barbara Garavaglia, Cinzia Gellera, Federica Invernizzi, Eleonora Lamantea, Marco Rimoldi, and Elisabetta Verderio, and the generous support of Telethon-Italia to F.T.

References

1. Bradley WG, Hudgson P, Gardner-Medwin D, Walton JN. Myopathy associated with abnormal lipid metabolism in skeletal muscle. *Lancet*. 1969;i:495–498.
2. Engel WK, Vick NA, Glueck CJ, Levy RI. A skeletal muscle disorder associated with intermittent symptoms and a possible defect of lipid metabolism. *N Engl J Med*. 1970;282:697–704.
3. Engel AG, Angelini C. Carnitine deficiency of human skeletal muscle with associated lipid storage myopathy: a new syndrome. *Science*. 1973;179:899–902.
4. DiMauro S, Melis-DiMauro P. Muscle carnitine palmitoyltransferase deficiency and myoglobinuria. *Science*. 1973;182:929–931.
5. Rinaldo P, Matern D, Bennett MJ. Fatty acid oxidation disorders. *Annu Rev Physiol*. 2002;64:477–502.
6. DiDonato S, Taroni F. Disorders of lipid metabolism. In: Engel AG, Franzini-Armstrong C, eds. *Myology*. New York: McGraw-Hill; 2004:1587–1621.
7. Vissing J, DiDonato S, Taroni F. Metabolic myopathies. Defects of carbohydrate and lipid metabolism. In: Karpati G, Hilton-Jones D, Bushby K, Griggs RC, eds. *Disorders of Voluntary Muscle*. Cambridge, UK: Cambridge University Press; 2009:390–408.
8. Houten SM, Wanders RJ. A general introduction to the biochemistry of mitochondrial fatty acid beta-oxidation. *J Inherit Metab Dis*. 2010;33:469–477.
9. Tein I. Disorders of fatty acid oxidation. In: Aminoff MJ, Boller F, Swaab DF, eds. *Handbook of Clinical Neurology, 3rd Series: Vol. 113: Pediatric Neurology Part III (Dulac O, Lassonde M and Sarnat HN, eds)*. Amsterdam, The Netherlands: Elsevier; 2013:1675–1688.
10. Stanley CA. Carnitine disorders. *Adv Pediatr*. 1995;42:209–242.
11. Lindner M, Hoffmann GF, Matern D. Newborn screening for disorders of fatty-acid oxidation: experience and recommendations from an expert meeting. *J Inherit Metab Dis*. 2010;33:521–526.
12. Kunau WH, Dommes V, Schulz H. β-oxidation of fatty acids in mitochondria, peroxisomes and bacteria: a century of continued progress. *Prog Lipid Res*. 1995;34:267–342.
13. Bartlett K, Eaton S. Mitochondrial beta-oxidation. *Eur J Biochem*. 2004;271:462–469.
14. Taroni F, Uziel G. Fatty-acid mitochondrial β-oxidation and hypoglycemia in children. *Curr Opin Neurol*. 1996;9:477–485.
15. Lim JH, Gerhart-Hines Z, Dominy JE, et al. Oleic acid stimulates complete oxidation of fatty acids through protein kinase A-dependent activation of SIRT1-PGC1alpha complex. *J Biol Chem*. 2013;288:7117–7126.
16. Foster DW. Malonyl-CoA: the regulator of fatty acid synthesis and oxidation. *J Clin Invest*. 2012;122:1958–1959.
17. Rodgers JT, Lerin C, Gerhart-Hines Z, Puigserver P. Metabolic adaptations through the PGC-1 alpha and SIRT1 pathways. *FEBS Lett*. 2008;582:46–53.
18. Scarpulla RC, Vega RB, Kelly DP. Transcriptional integration of mitochondrial biogenesis. *Trends Endocrinol Metab*. 2012;23:459–466.
19. Saudubray JM, Martin D, de Lonlay P, et al. Recognition and management of fatty acid oxidation defects: a series of 107 patients. *J Inherit Metab Dis*. 1999;22:488–502.
20. Wanders RJ, Ruiter JP, IJLst L, Waterham HR, Houten SM. The enzymology of mitochondrial fatty acid beta-oxidation and its application to follow-up analysis of positive neonatal screening results. *J Inherit Metab Dis*. 2010;33:479–494.
21. Olpin SE. Pathophysiology of fatty acid oxidation disorders and resultant phenotypic variability. *J Inherit Metab Dis*. 2013;36:645–658.
22. Tyni T, Paetau A, Strauss AW, Middleton B, Kivela T. Mitochondrial fatty acid beta-oxidation in the human eye and brain: implications for the retinopathy of long-chain 3-hydroxyacyl-CoA dehydrogenase deficiency. *Pediatr Res*. 2004;56:744–750.
23. Spiekerkoetter U, Bennett MJ, Ben-Zeev B, Strauss AW, Tein I. Peripheral neuropathy, episodic myoglobinuria, and respiratory failure in deficiency of the mitochondrial trifunctional protein. *Muscle Nerve*. 2004;29:66–72.
24. Oey NA, den Boer ME, Wijburg FA, et al. Long-chain fatty acid oxidation during early human development. *Pediatr Res*. 2005;57:755–759.
25. Gregersen N, Andresen BS, Corydon MJ, et al. Mutation analysis in mitochondrial fatty acid oxidation defects: exemplified by acyl-CoA dehydrogenase deficiencies, with special focus on genotype-phenotype relationship. *Hum Mutat*. 2001;18:169–189.
26. Longo N, Amat di San Filippo C, Pasquali M. Disorders of carnitine transport and the carnitine cycle. *Am J Med Genet C: Semin Med Genet*. 2006;142:77–85.
27. Stanley CA. Carnitine deficiency disorders in children. *Ann N Y Acad Sci*. 2004;1033:42–51.

28. Garavaglia B, Uziel G, Dworzak F, Carrara F, DiDonato S. Primary carnitine deficiency: heterozygote and intrafamilial phenotypic variation. *Neurology*. 1991;41:1691–1693.
29. Wang Y, Taroni F, Garavaglia B, Longo N. Functional analysis of mutations in the OCTN2 transporter causing primary carnitine deficiency: lack of genotype-phenotype correlation. *Hum Mutat*. 2000;16:401–407.
30. El-Hattab AW. Systemic primary carnitine deficiency. In: Pagon RA, Adam MP, Bird TD, et al., eds. *GeneReviews®* [Internet]. Seattle (WA): University of Washington, Seattle; 1993–2014. Available at: http://www.ncbi.nlm.nih.gov/books/nbk84551/, 2012. Updated 2012 Mar 15.
31. Rose EC, di San Filippo CA, Ndukwe Erlingsson UC, Ardon O, Pasquali M, Longo N. Genotype-phenotype correlation in primary carnitine deficiency. *Hum Mutat*. 2012;33:118–123.
32. Kishimoto S, Suda K, Yoshimoto H, et al. Thirty-year follow-up of carnitine supplementation in two siblings with hypertrophic cardiomyopathy caused by primary systemic carnitine deficiency. *Int J Cardiol*. 2012;159:e14–e15.
33. Bonnefont JP, Djouadi F, Prip-Buus C, Gobin S, Munnich A, Bastin J. Carnitine palmitoyltransferases 1 and 2: biochemical, molecular and medical aspects. *Mol Aspects Med*. 2004;25:495–520.
34. Gellera C, Verderio E, Floridia G, et al. Assignment of the human carnitine palmitoyltransferase II gene (*CPT1*) to chromosome 1p32. *Genomics*. 1994;24:195–197.
35. Bennett MJ, Santani AB. Carnitine palmitoyltransferase 1A deficiency. In: Pagon RA, Adam MP, Bird TD, et al., eds. *GeneReviews®* [Internet]. Seattle (WA): University of Washington, Seattle; 1993–2014. Available at: http://www.ncbi.nlm.nih.gov/books/nbk1527/, 2005. Updated 2013 Mar 07.
36. Wieser T. Carnitine palmitoyltransferase ii deficiency. In: Pagon RA, Adam MP, Bird TD, et al., eds. *GeneReviews®* [Internet]. Seattle (WA): University of Washington, Seattle; 1993–2014. Available at: http://www.ncbi.nlm.nih.gov/books/nbk1253/, 2004. Updated 2011 Oct 06.
37. Deschauer M, Wieser T, Zierz S. Muscle carnitine palmitoyltransferase II deficiency: clinical and molecular genetic features and diagnostic aspects. *Arch Neurol*. 2005;62:37–41.
38. Angelini C, Federico A, Reichmann H, Lombes A, Chinnery P, Turnbull D. Task force guidelines handbook: EFNS guidelines on diagnosis and management of fatty acid mitochondrial disorders. *Eur J Neurol*. 2006;13:923–929.
39. Taroni F, Verderio E, Fiorucci S, et al. Molecular characterization of inherited carnitine palmitoyltransferase II deficiency. *Proc Natl Acad Sci U S A*. 1992;89:8429–8433.
40. Bonnefont JP, Demaugre F, Prip-Buus C, et al. Carnitine palmitoyltransferase deficiencies. *Mol Genet Metab*. 1999;68:424–440.
41. Sigauke E, Rakheja D, Kitson K, Bennett MJ. Carnitine palmitoyltransferase II deficiency: a clinical, biochemical, and molecular review. *Lab Invest*. 2003;83:1543–1554.
42. Isackson PJ, Bennett MJ, Lichter-Konecki U, et al. CPT2 gene mutations resulting in lethal neonatal or severe infantile carnitine palmitoyltransferase II deficiency. *Mol Genet Metab*. 2008;94:422–427.
43. Isackson PJ, Bennett MJ, Vladutiu GD. Identification of 16 new disease-causing mutations in the CPT2 gene resulting in carnitine palmitoyltransferase II deficiency. *Mol Genet Metab*. 2006;89:323–331.
44. Taroni F, Verderio E, Dworzak F, Willems PJ, Cavadini P, DiDonato S. Identification of a common mutation in the carnitine palmitoyltransferase II gene in familial recurrent myoglobinuria patients. *Nat Genet*. 1993;4:314–320.
45. Bonnefont JP, Taroni F, Cavadini P, et al. Molecular analysis of carnitine palmitoyltransferase II deficiency with hepatocardiomuscular expression. *Am J Hum Genet*. 1996;58:971–978.
46. Orngreen MC, Ejstrup R, Vissing J. Effect of diet on exercise tolerance in carnitine palmitoyltransferase II deficiency. *Neurology*. 2003;61:559–561.
47. Bonnefont JP, Bastin J, Laforet P, et al. Long-term follow-up of bezafibrate treatment in patients with the myopathic form of carnitine palmitoyltransferase 2 deficiency. *Clin Pharmacol Ther*. 2010;88:101–108.
48. Orngreen MC, Madsen KL, Preisler N, Andersen G, Vissing J, Laforet P. Bezafibrate in skeletal muscle fatty acid oxidation disorders: a randomized clinical trial. *Neurology*. 2014;82:607–613.
49. Wang GL, Wang J, Douglas G, et al. Expanded molecular features of carnitine acyl-carnitine translocase (CACT) deficiency by comprehensive molecular analysis. *Mol Genet Metab*. 2011;103:349–357.
50. Indiveri C, Iacobazzi V, Tonazzi A, et al. The mitochondrial carnitine/acylcarnitine carrier: function, structure and physiopathology. *Mol Aspects Med*. 2011;32:223–233.
51. Iacobazzi V, Invernizzi F, Baratta S, et al. Molecular and functional analysis of SLC25A20 mutations causing carnitine-acylcarnitine translocase deficiency. *Hum Mutat*. 2004;24:312–320.
52. Liebig M, Gyenes M, Brauers G, et al. Carnitine supplementation induces long-chain acylcarnitine production–studies in the VLCAD-deficient mouse. *J Inherit Metab Dis*. 2006;29:343–344.
53. Leslie ND, Tinkle BT, Strauss AW, Shooner K, Zhang K. Very long-chain acyl-coenzyme A dehydrogenase deficiency. In: Pagon RA, Adam MP, Bird TD, et al., eds. *GeneReviews®* [Internet]. Seattle (WA): University of Washington, Seattle; 1993–2014. Available at: http://www.ncbi.nlm.nih.gov/books/nbk6816/ 2009. Updated 2011 Sep 22.
54. Schiff M, Mohsen AW, Karunanidhi A, McCracken E, Yeasted R, Vockley J. Molecular and cellular pathology of very-long-chain acyl-CoA dehydrogenase deficiency. *Mol Genet Metab*. 2013;109:21–27.
55. Spiekerkoetter U. Mitochondrial fatty acid oxidation disorders: clinical presentation of long-chain fatty acid oxidation defects before and after newborn screening. *J Inherit Metab Dis*. 2010;33:527–532.
56. Laforet P, Acquaviva-Bourdain C, Rigal O, et al. Diagnostic assessment and long-term follow-up of 13 patients with very long-chain acyl-coenzyme A dehydrogenase (VLCAD) deficiency. *Neuromuscul Disord*. 2009;19:324–329.
57. Minetti C, Garavaglia B, Bado M, et al. Very-long-chain acyl-coenzyme A dehydrogenase deficiency in a child with recurrent myoglobinuria. *Neuromuscul Disord*. 1998;8:3–6.
58. Pons R, Cavadini P, Baratta S, et al. Clinical and molecular heterogeneity in very-long-chain acyl-CoA dehydrogenase deficiency. *Pediatr Neurol*. 2000;22:98–105.
59. Gregersen N, Andresen BS, Pedersen CB, Olsen RK, Corydon TJ, Bross P. Mitochondrial fatty acid oxidation defects–remaining challenges. *J Inherit Metab Dis*. 2008;31:643–657.

60. Gobin-Limballe S, McAndrew RP, Djouadi F, Kim JJ, Bastin J. Compared effects of missense mutations in very-long-chain acyl-CoA dehydrogenase deficiency: combined analysis by structural, functional and pharmacological approaches. *Biochim Biophys Acta.* 1802;2010:478–484.
61. Orngreen MC, Norgaard MG, van Engelen BG, Vistisen B, Vissing J. Effects of IV glucose and oral medium-chain triglyceride in patients with VLCAD deficiency. *Neurology.* 2007;69:313–315.
62. Gobin-Limballe S, Djouadi F, Aubey F, et al. Genetic basis for correction of very-long-chain acyl-coenzyme A dehydrogenase deficiency by bezafibrate in patient fibroblasts: toward a genotype-based therapy. *Am J Hum Genet.* 2007;81:1133–1143.
63. Spiekerkoetter U, Sun B, Khuchua Z, Bennett MJ, Strauss AW. Molecular and phenotypic heterogeneity in mitochondrial trifunctional protein deficiency due to beta-subunit mutations. *Hum Mutat.* 2003;21:598–607.
64. Boutron A, Acquaviva C, Vianey-Saban C, et al. Comprehensive cDNA study and quantitative analysis of mutant HADHA and HADHB transcripts in a French cohort of 52 patients with mitochondrial trifunctional protein deficiency. *Mol Genet Metab.* 2011;103:341–348.
65. Olpin SE, Clark S, Andresen BS, et al. Biochemical, clinical and molecular findings in LCHAD and general mitochondrial trifunctional protein deficiency. *J Inherit Metab Dis.* 2005;28:533–544.
66. Spiekerkoetter U, Bastin J, Gillingham M, Morris A, Wijburg F, Wilcken B. Current issues regarding treatment of mitochondrial fatty acid oxidation disorders. *J Inherit Metab Dis.* 2010;33:555–561.
67. Fletcher AL, Pennesi ME, Harding CO, Weleber RG, Gillingham MB. Observations regarding retinopathy in mitochondrial trifunctional protein deficiencies. *Mol Genet Metab.* 2012;106:18–24.
68. Hong YB, Lee JH, Park JM, et al. A compound heterozygous mutation in HADHB gene causes an axonal Charcot-Marie-tooth disease. *BMC Med Genet.* 2013;14:125.
69. Spiekerkoetter U, Khuchua Z, Yue Z, Bennett MJ, Strauss AW. General mitochondrial trifunctional protein (TFP) deficiency as a result of either alpha- or beta-subunit mutations exhibits similar phenotypes because mutations in either subunit alter TFP complex expression and subunit turnover. *Pediatr Res.* 2004;55:190–196.
70. Gillingham MB, Connor WE, Matern D, et al. Optimal dietary therapy of long-chain 3-hydroxyacyl-CoA dehydrogenase deficiency. *Mol Genet Metab.* 2003;79:114–123.
71. Harding CO, Gillingham MB, van Calcar SC, Wolff JA, Verhoeve JN, Mills MD. Docosahexaenoic acid and retinal function in children with long-chain 3-hydroxyacyl-CoA dehydrogenase deficiency. *J Inherit Metab Dis.* 1999;22:276–280.
72. Tein I, Vajsar J, MacMillan L, Sherwood WG. Long-chain L-3-hydroxyacyl-coenzyme A dehydrogenase deficiency neuropathy: response to cod liver oil. *Neurology.* 1999;52:640–643.
73. Grosse SD, Khoury MJ, Greene CL, Crider KS, Pollitt RJ. The epidemiology of medium chain acyl-CoA dehydrogenase deficiency: an update. *Genet Med.* 2006;8:205–212.
74. Andresen BS, Bross P, Udvari S, et al. The molecular basis of medium-chain acyl-CoA dehydrogenase (MCAD) deficiency in compound heterozygous patients: is there correlation between genotype and phenotype? *Hum Mol Genet.* 1997;6:695–707.
75. Matern D, Rinaldo P. Medium-chain acyl-coenzyme A dehydrogenase deficiency. In: Pagon RA, Adam MP, Bird TD, et al., eds. *GeneReviews®* [Internet]. Seattle (WA): University of Washington, Seattle; 1993–2014. Available at: http://www.ncbi.nlm.nih.gov/books/nbk1424/, 2000. Updated 2012 Jan 19.
76. Yokota I, Saijo T, Vockley J, Tanaka K. Impaired tetramer assembly of variant medium-chain acyl-coenzyme A dehydrogenase with a glutamate or aspartate substitution for lysine 304 causing instability of the protein. *J Biol Chem.* 1992;267:26004–26010.
77. Frerman FE, Goodman SI. Defects of electron transfer flavoprotein and electron transfer flavoprotein-ubiquinone oxidoreductase: Glutaric acidemia type II. In: Scriver CR, Beaudet AL, Sly WS, Valle D, Childs B, Kinzler KW, Vogelstein B, eds. *The Metabolic and Molecular Bases of Inherited Disease.* New York: McGraw-Hill; 2001:2357–2365.
78. Olsen RK, Andresen BS, Christensen E, Bross P, Skovby F, Gregersen N. Clear relationship between ETF/ETFDH genotype and phenotype in patients with multiple acyl-CoA dehydrogenation deficiency. *Hum Mutat.* 2003;22:12–23.
79. Schiff M, Froissart R, Olsen RK, Acquaviva C, Vianey-Saban C. Electron transfer flavoprotein deficiency: functional and molecular aspects. *Mol Genet Metab.* 2006;88:153–158.
80. Antozzi C, Garavaglia B, Mora M, et al. Late-onset riboflavin-responsive myopathy with combined multiple acyl coenzyme A dehydrogenase and respiratory chain deficiency. *Neurology.* 1994;44:2153–2158.
81. Olsen RK, Olpin SE, Andresen BS, et al. ETFDH mutations as a major cause of riboflavin-responsive multiple acyl-CoA dehydrogenation deficiency. *Brain.* 2007;130:2045–2054.
82. Gempel K, Topaloglu H, Talim B, et al. The myopathic form of coenzyme Q10 deficiency is caused by mutations in the electron-transferring-flavoprotein dehydrogenase (ETFDH) gene. *Brain.* 2007;130:2037–2044.
83. Liang WC, Ohkuma A, Hayashi YK, et al. ETFDH mutations, CoQ10 levels, and respiratory chain activities in patients with riboflavin-responsive multiple acyl-CoA dehydrogenase deficiency. *Neuromuscul Disord.* 2009;19:212–216.
84. Cornelius N, Frerman FE, Corydon TJ, et al. Molecular mechanisms of riboflavin responsiveness in patients with ETF-QO variations and multiple acyl-CoA dehydrogenation deficiency. *Hum Mol Genet.* 2012;21:3435–3448.
85. Wolfe L, Jethva R, Oglesbee D, Vockley J. Short-chain acyl-CoA dehydrogenase deficiency. In: Pagon RA, Adam MP, Bird TD, et al., eds. *GeneReviews®* [Internet]. Seattle (WA): University of Washington, Seattle; 1993–2014. Available at: http://www.ncbi.nlm.nih.gov/books/nbk63582/, 2011. Updated 2011 Sep 22.
86. Nagan N, Kruckeberg KE, Tauscher AL, Bailey KS, Rinaldo P, Matern D. The frequency of short-chain acyl-CoA dehydrogenase gene variants in the US population and correlation with the C(4)-acylcarnitine concentration in newborn blood spots. *Mol Genet Metab.* 2003;78:239–246.
87. Tein I, Elpeleg O, Ben-Zeev B, et al. Short-chain acyl-CoA dehydrogenase gene mutation (c.319C>T) presents with clinical heterogeneity and is candidate founder mutation in individuals of Ashkenazi Jewish origin. *Mol Genet Metab.* 2008;93:179–189.
88. Pedersen CB, Kolvraa S, Kolvraa A, et al. The ACADS gene variation spectrum in 114 patients with short-chain acyl-CoA dehydrogenase (SCAD) deficiency is dominated by missense variations leading to protein misfolding at the cellular level. *Hum Genet.* 2008;124:43–56.

89. Jethva R, Bennett MJ, Vockley J. Short-chain acyl-coenzyme A dehydrogenase deficiency. *Mol Genet Metab*. 2008;95:195–200.
90. Corydon MJ, Vockley J, Rinaldo P, et al. Role of common gene variations in the molecular pathogenesis of short-chain acyl-CoA dehydrogenase deficiency. *Pediatr Res*. 2001;49:18–23.
91. Kamijo T, Indo Y, Souri M, et al. Medium chain 3-ketoacyl-coenzyme A thiolase deficiency: a new disorder of mitochondrial fatty acid beta-oxidation. *Pediatr Res*. 1997;42:569–576.
92. Yang SY, He XY, Schulz H. 3-Hydroxyacyl-CoA dehydrogenase and short chain 3-hydroxyacyl-CoA dehydrogenase in human health and disease. *FEBS J*. 2005;272:4874–4883.
93. Bennett MJ, Russell LK, Tokunaga C, et al. Reye-like syndrome resulting from novel missense mutations in mitochondrial medium- and short-chain l-3-hydroxy-acyl-CoA dehydrogenase. *Mol Genet Metab*. 2006;89:74–79.
94. Heslegrave AJ, Hussain K. Novel insights into fatty acid oxidation, amino acid metabolism, and insulin secretion from studying patients with loss of function mutations in 3-hydroxyacyl-CoA dehydrogenase. *J Clin Endocrinol Metab*. 2013;98:496–501.
95. Roe CR, Millington DS, Norwood DL, et al. 2,4-Dienoyl-coenzyme A reductase deficiency: a possible new disorder of fatty acid oxidation. *J Clin Invest*. 1990;85:1703–1707.
96. He M, Pei Z, Mohsen AW, et al. Identification and characterization of new long chain acyl-CoA dehydrogenases. *Mol Genet Metab*. 2011;102:418–429.
97. He M, Rutledge SL, Kelly DR, et al. A new genetic disorder in mitochondrial fatty acid beta-oxidation: ACAD9 deficiency. *Am J Hum Genet*. 2007;81:87–103.
98. Haack TB, Danhauser K, Haberberger B, et al. Exome sequencing identifies ACAD9 mutations as a cause of complex I deficiency. *Nat Genet*. 2010;42:1131–1134.
99. Nouws J, Nijtmans L, Houten SM, et al. Acyl-CoA dehydrogenase 9 is required for the biogenesis of oxidative phosphorylation complex I. *Cell Metab*. 2010;12:283–294.

CHAPTER

51

Lipoprotein Disorders

Mary J. Malloy and John P. Kane

University of California, San Francisco School of Medicine and University of California, San Francisco Medical Center, San Francisco, CA, USA

INTRODUCTION

The system of lipoproteins that transports lipids in plasma also interacts with neural tissue. The transport of antioxidant tocopherols is critically dependent on lipoproteins of intestinal and hepatic origin that contain the B apolipoproteins (ApoB-48 and ApoB-100, respectively). Severe impairment of intestinal absorption of tocopherols occurs in disorders such as abetalipoproteinemia (ABL) and in chylomicron retention disease, in which intestinal ApoB-48 is not secreted normally. The second phase of tocopherol transport, from liver to peripheral tissues, is impaired in disorders in which hepatic secretion of ApoB-100 is defective, or with mutations in the α-tocopherol transfer protein. In recessive ABL resulting from mutations in the microsomal triglyceride transfer protein, little or no ApoB-100-containing lipoprotein is secreted, reflecting defects in the assembly and movement of nascent lipoproteins within the hepatocyte. Hypobetalipoproteinemia (HBL) frequently results from mutations that lead to truncation of the ApoB-100 protein. Truncated products, shorter than ApoB-31 in length, cannot be secreted. The neuropathologic features common to these disorders are attributable to the accumulation of oxidized lipids in myelin, leading to peripheral neuropathy and degeneration of posterior columns and spinocerebellar tracts.

Other disorders of lipoproteins that lead to impairment of the centripetal movement of cholesterol and phospholipids with accumulation of abnormal lipoproteins in plasma, including Tangier disease and deficiency of lecithin:cholesterol acyltransferase (LCAT), can cause peripheral neuropathy. In Tangier disease, a syringomyelia-like syndrome has also been observed.

Several apolipoproteins are synthesized in the central nervous system, and others distribute there passively. Some of these play important roles in the sequestration and retrieval of lipids during the regeneration of injured neural tissue. Amyloid-β protein is associated with high-density lipoprotein (HDL) complexes in cerebrospinal fluid and in plasma. The prions of Creutzfeldt–Jakob disease associate with low-density and very low-density lipoproteins.[1]

The relationships so far identified between disorders of lipoprotein metabolism and dysfunction of nerve tissue are chiefly the consequences of alterations in lipid constituents of the neuron and its intimate environment. Notable among these are disorders in which an increase in oxidative modification of lipids takes place due to impaired transport of tocopherol to nerve tissue. Also affecting nerve tissue are disorders of cholesteryl ester metabolism and disorders of the normal centripetal flux of free cholesterol and certain phospholipids. Several apolipoproteins and the lipoprotein-a [Lp(a)] protein are found in central nervous system tissue or in cerebrospinal fluid. Apolipoproteins E, D, J, H, and several members of the ApoL family are expressed in the brain in substantial quantities, and the E, D, and J proteins play important roles in the response of nerve tissue to injury. Apolipoproteins A-I and A-IV, which passively enter the central nervous system, may also contribute to lipid transport and remyelination of nerves. Circulating lipoproteins play a role in the delivery of omega-3 fatty acids to the brain. As the complexities of the lipoprotein system unfold, it is likely that further disorders will be discovered in which alterations of apoprotein or lipoprotein metabolism affect the functions of the central or peripheral nervous systems.

Rosenberg's Molecular and Genetic Basis of Neurological and Psychiatric Disease
http://dx.doi.org/10.1016/10.1016/B978-0-12-410529-4.00051-6

LIPOPROTEIN STRUCTURE AND METABOLISM

The lipid components of most plasma lipoproteins are organized as spherical microemulsions, with varying proportions of the relatively hydrophobic triglycerides and cholesteryl esters in the cores and the amphipathic phospholipids and unesterified cholesterol arrayed in mixed surface monolayers. The protein constituents of lipoproteins, consisting of a number of discrete apoprotein species, are primarily located at the amphipathic monolayer.

The proteins of the lipoprotein system fall into several distinct groups. ApoB-100, a high molecular weight B apolipoprotein, is found in very low-density lipoproteins (VLDL) and in lipoproteins derived from VLDL (intermediate- and low-density lipoproteins; IDL and LDL). It has a molecular weight of 512 kDa. ApoB-48, secreted by the intestine in chylomicrons, is a product of the *APOB* gene but is truncated by RNA editing to yield the *N*-terminal 48% of ApoB sequence. Anomalously truncated *APOB* gene products are likewise designated by their relative lengths using this centile system. Other apolipoproteins are much smaller and transfer among lipoprotein particles.

The B apolipoproteins contain clusters of basic amino acids that bind to heparin, and ApoB-100 includes a ligand domain for the LDL receptor. A number of naturally occurring truncations of apolipoprotein B are recognized that produce abnormal lipoproteins and frequently lead to low levels of VLDL and LDL.[2]

Among the small exchangeable apolipoproteins, ApoA-I is the chief protein of HDL and ApoE has affinity for heparin and for the LDL receptor; others modulate lipoprotein metabolism.

Transport of Dietary Lipid

Ingested triglycerides are hydrolyzed in the intestine by pancreatic lipase in the presence of a protein colipase and bile acids. The monoglycerides and free fatty acids thus formed are absorbed, and triglyceride resynthesis takes place within the enterocyte. Some of the cholesterol absorbed, along with some synthesized *de novo*, is esterified in the enterocyte by an acylCoA acyltransferase (ACAT). ApoB-48 is synthesized within the enterocyte. Association of ApoB-48 with developing microemulsion particles in the endoplasmic reticulum is mediated by a heterodimeric protein that includes the microsomal triglyceride transport protein and a protein disulfide isomerase (MTP-PDI).[3] These particles are secreted via the Golgi apparatus into the lymph spaces. Lipoprotein lipase, situated on capillary endothelium, hydrolyzes up to 80% of the triglyceride content of the particles. As hydrolysis progresses, surface components, including apolipoproteins A-I, A-II, A-IV, and the C apolipoproteins, plus some of the surface lipids, move from the particles to HDL. The ApoB-48-containing particle that results is called a remnant. Chylomicron remnants are removed quantitatively by interaction with LDL receptors and LRP receptors on hepatocytes.

Transport of Endogenous Triglycerides

Triglycerides are formed in the liver from fatty acids derived from lipolysis in adipose tissue and from *de novo* synthesis. One copy of ApoB-100, synthesized in the rough endoplasmic reticulum, joins a growing microemulsion particle as it progresses through the smooth endoplasmic reticulum to the Golgi, catalyzed by microsomal triglyceride transfer protein–protein disulfide isomerase (MTP–PDI). Small amounts of C apolipoproteins, ApoE, and (probably) ApoA-I and A-II are added before the particle is secreted from the Golgi to the blood. Again, lipolysis mediated by lipoprotein lipase produces remnants. In this case, they contain ApoB-100 and ApoE but have lost most of the smaller apolipoproteins and surface lipid to HDL.[2] This transfer appears to require the activity of phospholipid transfer protein. In healthy people, about 50% of VLDL remnants are endocytosed in the liver by a receptor-mediated process involving interaction of ApoE with the LDL receptor. The remaining remnants lose more triglycerides, via the action of hepatic triglyceridase, and also lose ApoE, evolving into LDL. During lipolysis, a conformational change occurs in ApoB-100, in which the ligand domain for the LDL receptor is conformed. LDL catabolism then proceeds through endocytosis via the LDL receptor and endosomal hydrolysis of cholesteryl esters.

Formation of High-Density Lipoproteins and Centripetal Transport

HDL are formed from lipids effluxed from cell membranes and from lipids of the surface monolayers of triglyceride-rich lipoproteins that are shed during lipolysis. Efflux of phospholipid and cholesterol from cell membranes is effected by the ATP-dependent transporter ABCA1, with a small HDL particle (pre-β-1 HDL) as acceptor (Figure 51.1), and by the ABCG1 transporter. In the latter case the principal acceptor HDL species are larger, more mature spherical particles. Phosphatidyl choline and sphingolipids, but not cholesterol, are also effluxed by the ABCA7 transporter to low molecular-weight HDL particles, probably pre-β-1 HDL. This transporter is expressed

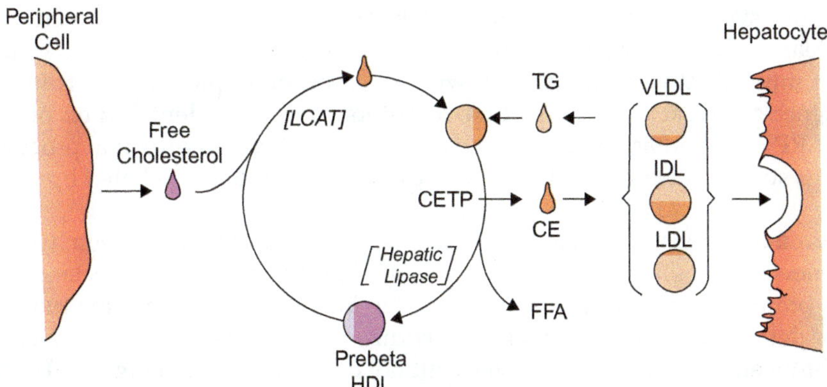

FIGURE 51.1 **Centripetal transport of cholesterol by high-density lipoproteins (HDL).** Free cholesterol (shaded areas) is acquired from cells by pre-β HDL species and is esterified by lecithin:cholesterol acyltransferase. Passage of cholesteryl esters (black areas) to acceptor lipoproteins is mediated by cholesteryl ester transfer protein. The liver takes up a portion of remnant (intermediate-density) lipoproteins and low-density lipoproteins (LDL). Some triglycerides (stippled areas) transfer from very low-density lipoproteins (VLDL) and intermediate-density lipoproteins (IDL) to HDL species, where they are hydrolyzed in part by hepatic lipase. Some HDL particles deliver cholesterol to the liver directly via the SR-BI receptor.

in a number of tissues, including brain.[4] The cholesterol in HDL is esterified by lecithin:cholesterol acyltransferase (which is also found in cerebrospinal fluid). The particles organize into large complexes of 250–450 kDa. Cholesteryl esters are transferred from mature HDL to the liver by two mechanisms: i) they are transferred to ApoB-containing lipoproteins for eventual uptake in the liver, mediated by cholesteryl ester transfer protein, and ii) they are taken up directly from HDL by scavenger receptor type BI (SR-BI) receptors on hepatocytes.

Apolipoproteins in the Nervous System

The realization that apolipoproteins play direct roles in the sequestration and intercellular transfer of lipid in the central nervous system and in peripheral nerves came with the discovery that ApoE is expressed in the brain in amounts higher than would be expected for an ultrafiltered protein. ApoE has been detected immunochemically in association with all forms of astrocytes in a perinuclear distribution associated with the Golgi,[5] and with cell processes that end on the pia or at blood vessels. ApoE is associated with the glia of peripheral nerves, including visceral nerves. It is absent from neurons, oligodendroglia, and other cell types. Astrocytes and microglia secrete ApoE, the most abundant protein in cerebrospinal fluid, which organizes relatively protein-rich lipoprotein particles. These are involved in the efflux of lipid from glia, which appears to require ApoA-I, and in the transport of sulfatides. In addition, they appear to protect against apoptosis of neurons, in a process mediated by the LRP-1 receptor.[6–9] During experimental demyelination, ApoE appears to be secreted by macrophages that enter the tissue. ApoE exists in three common isoforms (E2, E3, and E4) that result from amino acid substitutions. These isoforms influence the distribution of ApoE among lipoproteins, receptor binding, and a number of other biological interactions.

Several classes of lipoprotein complexes exist in cerebrospinal fluid. ApoD, ApoJ, and ApoH (2 glycoprotein I) are synthesized in the brain in addition to ApoE.[10–13] ApoD, a lipocalin, is found in oligodendrocytes, astrocytes, and to a limited extent in some neurons.[14] Its expression increases in inflammatory brain disorders,[15] and it is associated with amyloid plaques.[16] Several members of the newly discovered apolipoprotein L family are also expressed in the brain, including L-I, L-II, and L-III. ApoL-I is expressed in many areas of the brain and in the spinal cord.[17] Apolipoproteins A-I, A-II, and probably A-IV, enter the cerebrospinal fluid by a process of transcytosis that is poorly understood, but ApoA-IV is also transcribed in the hypothalamus in response to neuropeptide Y.[18] ApoA-I appears to play a central role in the efflux of cholesterol from astrocytes in the formation of lipoprotein complexes in the central nervous system.[19] The total lipid content of cerebrospinal fluid is between 1 and 2 mg per dL, consisting primarily of cholesterol and phospholipids. The principal species of cerebrospinal fluid lipoprotein contains apolipoproteins E, A-I, D, H, and J, which possibly function in the redistribution of lipids.[20] A minor particle population contains all of these except ApoA-I. Another population contains only ApoA-I and A-II. Amyloid-β binds to ApoE-containing lipoprotein complexes in cerebrospinal fluid. The E2 and E3 isoforms appear to form these complexes more readily than does E4 (see also Chapter 52).

A number of enzymes and transfer proteins that can modify lipoproteins are found in the brain and may play important roles in lipid transfer there.[21] The phospholipid transfer protein[22] and lipoprotein lipase (LPL)[23,24] are among these. Though fatty acids are not an energy substrate in brain, LPL may be critical to the delivery of omega-3 fatty acids to the central nervous system. The bulk of ApoA-IV is carried on larger particles. The quantity of ApoA-IV in cerebrospinal fluid increases after a fat-rich meal, where it functions as a satiety signal in synergy with the melanocortin system.[25,26] When infused into rat brains, even without associated lipids, it results in decreased food intake. It also appears to influence gastric acid secretion. LDL receptors have been identified on pial cells of the arachnoid and on astrocytes. The LRP receptor has been identified on the cell bodies and proximal processes of cerebral cortical neurons.[27] These findings support the potential importance of apolipoprotein-mediated lipid transport in the central nervous system.

The ApoE in cerebrospinal fluid is in a more highly sialated state than that in plasma, which is chiefly of hepatic origin. In contrast to the small apolipoproteins, virtually no ApoB is normally detectable in cerebrospinal fluid. However, as the blood–brain barrier becomes compromised in infection and with some tumors, LDL and the intact Lp(a) lipoprotein can be detected in cerebrospinal fluid. Oxidized LDL has been demonstrated in the brain following infarction and is apparently endocytosed by astrocytes, but not by neurons or microglia.[28] The mRNA for Apo(a) has been detected in the central nervous system of rhesus monkeys. It is not yet known whether this protein exists in complexes or in a molecular dispersion in the nervous system.

The retina is also an active site for endocytosis and efflux of lipids. The presence of ultraviolet light and a high oxygen tension leads to the formation of hydroperoxidized lipids, which are potentially injurious to the retina.[29] Retinal pigment epithelial (RPE) cells express the ABCA1 and ABCA4 transporters, the SR-BI and -BII scavenger receptors,[30] microsomal triglyceride transfer protein, and possibly apolipoprotein B.[31] RPE cells also actively secrete apolipoprotein E[32] and efflux lipids to HDL analogously to the centripetal transport of lipid from the artery wall.[33]

Studies on regenerating and remyelinating peripheral nerves have revealed a massive increase in local synthesis of ApoE.[5] Synthesis of ApoD also takes place in injured nerves. It appears to originate from astrocytes and oligodendrocytes in the central nervous system and from neurolemma in peripheral nerves.[13,34] Both ApoA-I and A-IV appear in increased amounts in injured nerves. A model is emerging that involves apoproteins in the stabilization and intercellular transfer of lipids derived from degenerating nerve membranes and nerve sheaths. It is likely that ApoE is involved in the storage of lipids, via the LDL receptor, in anticipation of remyelination.

Tocopherols

Tocopherols as Biologic Antioxidants

Tocopherols (vitamin E) have been shown to break free radical chains efficiently because the tocopheroxyl radical is extremely stable. This radical can then annihilate another peroxyl radical. For stereochemical reasons, α-tocopherol cannot neutralize peroxynitrite radicals, whereas γ-tocopherol can. Free-radical chain reactions in biologic systems are promoted by iron and copper, as is the oxidation of LDL. Pathologic states attributed to free radicals have been described in individuals suffering from iron overload. Antioxidant activity is demonstrable in species of HDL that contain transferrin and ceruloplasmin. These complexes inhibit the oxidation of LDL much more effectively than transferrin molecular dispersion. Other natural defenses against free radical chains include catalase, superoxide dismutases, paraoxonase, and glutathione peroxidase, which destroys hydroperoxides of fatty acids. Paraoxonase is carried partially in plasma in several discrete molecular species of HDL containing ApoA-I and ApoJ. It is likely that lipophilic hydroxytyrosyl esters will prove to have significant antioxidant activity in nervous tissue as well.[35]

Tocopherol Metabolism

Natural tocopherols differ with respect to substituents on the chromanol ring. The different arrays are termed α, β, γ, and δ. Of eight possible stereoisomeric arrangements of the isoprenoid side chain, the (2R, 4′R, 8′R) configuration is most potent. The side chain intercalates in lipid monolayers, leading to accumulation in cell membranes and plasma lipoproteins. Tocotrianols are also carried by plasma lipoproteins and may have significant antioxidant activity in the nervous system.[36]

Bile acids and pancreatic enzymes are necessary for tocopherol absorption in the intestine. Ingested tocopherol appears rapidly in chylomicrons. Although some is transferred to tissues during lipolysis, most is endocytosed by liver in chylomicron remnants. Tocopherol then appears in newly secreted VLDL. All isomers of tocopherol appear in chylomicrons, but the liver discriminates via preferential binding of the α species to tocopherol transfer protein when it secretes VLDL. Defects in the transfer protein lead to specific deficiency of tocopherol (familial vitamin E deficiency) that mimic the neurological manifestations of abetalipoproteinemia, including retinitis and spinocerebellar tract disease.[37] During hydrolysis of VLDL triglycerides by lipoprotein lipase, transfer of tocopherol to HDL is mediated by phospholipid transfer protein. LDL formed by lipolysis of VLDL retains the bulk of the tocopherol. Residual tocopherol in LDL then enters cells as LDL and is endocytosed. The decay of tocopherol in LDL corresponds with the

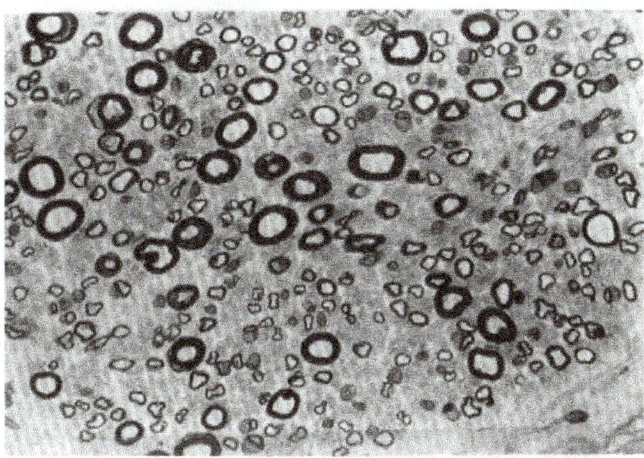

FIGURE 51.2 **Abetalipoproteinemia.** Section of a sural nerve from a 14-year-old patient with homozygous abetalipoproteinemia, showing a marked decrease in the numbers of large-caliber myelinated neurons. (From[37a].)

disappearance of LDL protein. The tocopherol content of peripheral nerves and brain is depleted more slowly than that of other tissues. During depletion, tocopherol levels are partially sustained by efflux from adipocytes via the ABCA1 transporter. Tissue levels of tocopherols are normal in rabbits lacking LDL receptors, indicating that other mechanisms are available. One such mechanism involves the transfer of tocopherol from HDL to endothelial cells via SR-BI receptors associated with caveoli.[38,39] A tocopherol-binding protein widely distributed among tissues may play a critical role in uptake. Levels of tocopherols in plasma correlate with those in cerebrospinal fluid. However, high levels of α-tocopherol suppress those of γ-tocopherol.[40]

Tocopherol Deficiency and Neurologic Disease

Tocopherol deficiency is restricted to conditions in which the transport of the vitamin is interrupted at some point between the intestine and affected tissues. Defects in lipid absorption at the intestinal level due to failure of secretion of ApoB-48 (in ABL), or other defects leading to steatorrhea are predominant causes of deficiency. Tocopherol deficiency in normotriglyceridemic ABL in which chylomicron production is normal, and in the syndrome of isolated tocopherol deficiency,[41] indicates that defects farther along in the transport pathway can also result in deficiency.

The neuropathology of ABL is generally representative of that encountered in tocopherol deficiencies of diverse etiology such as cholestasis and severe fat malabsorption. It includes axonal dystrophy in the posterior columns and in the dorsal and ventral spinocerebellar tracts. Loss of large-caliber myelinated axons is widespread in peripheral sensory nerves (Figure 51.2). In keeping with excess free radical generation, there is an accumulation of lipofuscin in the Schwann cell cytoplasm of peripheral nerves and in dorsal neurons. Positron emission tomography has demonstrated impaired binding of dopa to both the putamen and caudate in ABL, resembling the phenomenon observed in Parkinson disease. Also, studies of mazindol binding reveal a decrease in dopamine terminals in the striatum of tocopherol-deficient rats, further supporting an impact on nigrostriatal function. Electrophysiologic studies demonstrate diminished amplitudes of sensory nerve action potentials with little delayed conduction velocity. Somatosensory-evoked potential studies reveal a central delay in sensory conduction consistent with deficits in function of the posterior columns. A parallel defect is observed in visual evoked potentials. Tocopherol may play a vital role in the embryologic development of the nervous system in addition to maintenance of nerve tissue. This is indicated by the finding that tocopherol, delivered in native lipoproteins, is able to considerably extend the period in which rat brain explants develop in culture.

DISORDERS OF LIPOPROTEINS CONTAINING APOPROTEIN B

Abetalipoproteinemia

Clinical Features and Diagnostic Evaluation

The manifestations of ABL, resulting from the absence of the plasma lipoproteins that contain ApoB, include fat malabsorption, acanthocytosis, retinopathy, and progressive neurologic disease. Malabsorption of fat has been noted as early as the neonatal period with diarrhea, vomiting, and failure to gain weight. Somatic underdevelopment may persist. Malabsorption results from failure of the intestinal cells to secrete chylomicrons. Endoscopy reveals yellowish discoloration of the duodenal mucosa, and biopsy reveals normally formed villi engorged with lipids. Although chylomicrons are not secreted, some essential fatty acids are apparently absorbed and transported to the liver as free fatty acids. In adults with ABL, loss of fatty acids in stools may be as little as 20% of the ingested amount, but this is enough to induce oxalate urolithiasis in some.

Deficiencies of the fat-soluble vitamins A, E, and K result from the malabsorption of lipid. Vitamin D has its own transport protein, and hence absorption is unimpaired by the lack of chylomicron formation. Vitamin A is normally esterified in the enterocyte. Chylomicrons carry vitamin A esters into blood via the intestinal lymphatics and thoracic duct. Levels of vitamin A in the plasma of patients with ABL are low, but normal levels are achieved with supplementation. Low levels of vitamin K in plasma, resulting in prothrombin deficiency, can occur in ABL, and gastrointestinal bleeding has been observed in early childhood. As is the case with vitamin A, modest supplementation of vitamin K maintains normal levels. In contrast, even massive doses of vitamin E do not raise plasma levels to the normal range in some patients.

Acanthocytes account for half or more of circulating erythrocytes in patients with ABL but are not seen in bone marrow. Low sedimentation rates are observed because their structure inhibits rouleaux formation. The red cell envelope has an abnormally high sphingomyelin-to-lecithin ratio, and total phospholipid and cholesterol content is greater than in normal erythrocytes. Maldistribution of lipids between the bilayer leaflets apparently accounts for the acanthocytosis. Erythroid hyperplasia, reticulocytosis, hyperbilirubinemia, and decreased red cell survival have been described. Severe anemia observed in a number of children with ABL probably reflects deficiencies of some nutrients, including iron, folate, and vitamin E, secondary to fat malabsorption.

Pigmentary retinal degeneration, resembling retinitis pigmentosa (the latter is not associated with lipoprotein deficiency), is prominent in ABL. The most severe degeneration occurs in patients who also have severe neurologic symptoms. There is loss of pigment epithelium and photoreceptors, with relative preservation of submacular pigment epithelium. These and other pathologic changes, including abundance of lipofuscin pigment, resemble those of experimental deficiency of tocopherol. The retinopathy of ABL may also be related to deficiency of vitamin A, as partial improvement in electrophysiologic behavior of the retina and in dark adaptation has been noted when some patients receive vitamin A supplements. Vitamin A deficiency, however, is not thought to be a key factor in the retinal degeneration in ABL. Loss of night vision is frequently the earliest symptom. A decrease in visual acuity has occurred in the first decade, but many patients have maintained normal vision until adulthood. Ultimately, blindness can occur. Some patients have nystagmus, probably reflecting loss of visual acuity. Others are unaware of progressing retinal disease because it may involve slowly enlarging annular scotomas with macular sparing. Lenticular opacities, often seen with other forms of retinitis pigmentosa, have not been noted frequently in ABL.

Although some patients with ABL do not develop serious neurologic sequelae until later in life, most experience the onset of symptoms before the end of the second decade. Since the advent of early tocopherol therapy, many patients are remaining free of debilitating symptoms.

The large sensory neurons of the spinal ganglia and their myelinated axons that enter the cord lateral to the posterior funiculus are characteristic sites for the degenerative process. Axonopathy is the pathologic description, and demyelination of the fasciculus cuneatus and fasciculus gracilis has been described.[42,43] Striatonigral involvement has also been recognized.[44]

Deep tendon reflexes may decrease in the first decade, probably reflecting loss of function in posterior columns and spinocerebellar pathways. Ataxic gait and loss of proprioception and vibratory sense are usually progressive. The Romberg sign is often present. Before the advent of vitamin E therapy, patients were frequently unable to stand by the third decade and had severe dysmetric movements and dysarthria. Muscle contractures resulting in equinovarus, pes cavus, and kyphoscoliosis were common. Babinski responses in some patients have been related to pyramidal tract disease, but spastic paralysis has not been described. Head tremor has been reported.

Mental retardation, a feature in some cases, cannot be attributed to the basic metabolic defect of ABL, and evidence of cerebral cortical disease is lacking. Because many cases of this rare recessive disorder involve consanguineous matings, other rare alleles may be responsible for the retardation. Steatorrhea and attendant multiple nutritional deficiencies may underlie the general growth failure and slow neuromuscular development noted in some infants with ABL.

Peripheral neuropathy is an infrequent finding, but hypesthesia in the stocking–glove distribution has been described,[45] with diminished response to local anesthetics. Abnormalities of somatosensory conduction with normal brainstem evoked potentials were seen in 9 of 10 patients studied.[43] Decreased amplitude of sensory potentials with slow conduction velocity has been found in tibial and sural nerves, and evidence of skeletal muscle denervation has been noted on electromyograms.[46] Although denervation of the tongue and oculomotor nerve involvement have been seen, the cranial nerves are usually spared. A loss of large myelinated fibers is found in sural nerves, whereas unmyelinated fibers are relatively unaffected.[46] Paranodal demyelination correlates with age and severity of disease. The lesions bear a striking resemblance to those seen in other malabsorption syndromes that involve vitamin E deficiency.

The deinnervating neuropathy would tend to obscure the myopathy that may be present. The latter was described in a 26-year-old man in whom ceroid pigment was noted in the muscle fibers, resembling that observed in tocopherol-deficient animals. A 10-year-old girl died as a result of cardiomyopathy; the pathologic appearance, again, suggested tocopherol deficiency. The muscle weakness that is a common feature of ABL is probably most often secondary to the deinnervating neuropathy.

Genetic Data

All patients studied have had normal structural genes for ApoB but carry mutations in the microsomal triglyceride transport protein, which plays a critical role in the formation of chylomicrons and VLDL.[3,47]

Hypobetalipoproteinemia

Clinical Features and Diagnostic Evaluation

Patients heterozygous for familial HBL have cholesterol levels ranging from 40 to 180 mg per dL and triglycerides ranging from 15 mg per dL up to the normal range. Lipid composition of the lipoproteins is normal, and patients generally have no symptoms of their disease, with only modest abnormalities in laboratory tests. One patient whose presentation was compatible with heterozygous familial HBL, however, had neurologic findings that resembled olivopontocerebellar atrophy. Homozygous patients resemble those with ABL and generally have no detectable ApoB-containing lipoproteins. Chylomicrons do not appear after consuming fats. The acanthocytosis, gastrointestinal features, retinitis, and neuromuscular manifestations described in ABL are usually present. In one case, a homozygous truncation resulting in an ApoB-50 protein shed further light on the potential effects of truncations. As would be expected, chylomicron secretion appears to be normal. Although a VLDL-like particle is secreted from the liver, LDL are virtually absent from plasma. This patient had ataxia and almost undetectable levels of tocopherols in plasma.[48,49]

Genetic Data

HBL, characterized by very low levels of ApoB and LDL cholesterol in plasma, is caused by mutations in the *APOB* gene that interfere with translation of complete ApoB-100, or *APOB* alleles that produce reduced amounts of normal ApoB-100. A number of different mutations leading to truncations of ApoB proteins have been identified. Truncated B proteins shorter than B-31 apparently cannot be secreted from hepatocytes or enterocytes. Longer truncations are secreted in lipoproteins. Because the ability to incorporate lipid into lipoprotein complexes appears to be a monotonic function of the length of the ApoB chain, many of the secreted lipoproteins have abnormal densities. ApoB chains longer than B-50 are able to organize triglyceride-rich VLDL-like particles, whereas shorter products may appear as hyperdense LDL, and may even be found in the HDL density interval. All the truncations described to date involve the deletion of linear sequence from the carboxyl end of the protein and are attributable to mutations leading to the formation of premature stop codons. Truncated proteins shorter than B-70 would be expected to lack the ligand domain for the LDL receptor. In addition, HBL that cosegregates with the *APOB* locus but that is not associated with truncations has been observed. These may involve regulatory elements for the gene. An additional distinct mechanism for HBL is suggested by a kindred in which the B-48 protein is secreted into plasma but no B-100 is found.[50] Haplotyping studies have established that, unlike most cases of HBL, this disorder is not linked to the *APOB* gene locus. It is possible that this represents a tissue-specific defect in B-100 secretion or perhaps complete editing of B-100 mRNA in liver and intestine.

Chylomicron Retention Disease

Anderson et al.[51] described an infant with fat malabsorption, fat-laden intestinal epithelial cells, and low levels of LDL, HDL, and fat-soluble vitamins. Postprandial chylomicronemia was absent. Subsequently, similar patients who had severe diarrhea and varying degrees of growth retardation have been reported. One patient had mild acanthocytosis, and three developed neurologic symptoms in the second decade that included diminished deep tendon reflexes and vibratory sense. These three patients were also considered to have low or low–normal intelligence. Four of five tested had mildly abnormal retinal function. Results of studies of *in vitro* explants of intestinal biopsy material suggest that ApoB-48 synthesis is normal, but formation and secretion of chylomicrons are impaired.[52] The absence of vertical transmission of this phenotype, its frequency among siblings, and consanguineous relationships in families of affected persons suggest an autosomal-recessive inheritance. Abnormalities of the *APOB* gene locus have been excluded as a cause of this disorder.

Therapy

The intestinal symptoms of ABL and the homozygous form of familial HBL correlate with the amount of dietary fat. Restriction of triglycerides containing long-chain fatty acids therefore is key in management of these patients and those heterozygous for familial HBL who have evidence of malabsorption or oxalate urolithiasis. Medium-chain triglycerides should be used sparingly, if at all, because cirrhosis of the liver has been described with their prolonged use. Supplementation with tocopherol inhibits progression of neurologic sequelae and should be instituted as soon as the diagnosis of ABL or homozygous familial HBL is made. Natural vitamin E is preferred over synthetic preparations or those restricted to α-tocopherol. This is because γ-tocopherol can reduce the peroxynitrite radical and the α form cannot. In patients heterozygous for familial HBL, supplementation is advised because they also can develop neurologic disease. Retinopathy may be prevented, or stabilized once it appears, if vitamin E is given early. Myopathy has also been reversed with tocopherol. Very large doses are well tolerated, and concentrated preparations allow a convenient dosage of 1000–2000 mg per day for infants and up to 10,000–20,000 mg per day for older children or adults. Supplementation with water-soluble preparations of vitamin A are indicated whenever plasma levels are low. Vitamin K should be given if the patient exhibits bleeding or hypoprothrombinemia. Treatment of patients with chylomicron retention disease should include restriction of dietary fat and supplementation with vitamin E and perhaps also vitamin A.

DISORDERS OF HIGH-DENSITY LIPOPROTEINS

Tangier Disease

Clinical Features and Diagnostic Evaluation

Tangier disease is associated with very low levels of HDL, ApoA-I and A-II in plasma, and with lipid-filled histiocytes in many organs. Large, orange-colored tonsils and orange pigment in the rectal mucosa are hallmarks of this disorder. Deposits of cholesteryl esters are also observed in the cornea, thymus, Schwann cells, spinal ganglia, smooth muscle cells of the intestine, ureters, heart valves, bone marrow, and pulmonary arteries. Patients frequently have lymphadenopathy and splenomegaly, with associated thrombocytopenia and increased hemolysis. Cholesterol levels average about 70 mg per dL in homozygotes and about 160 mg per dL in heterozygotes; however, triglycerides tend to be elevated as high as 400 mg per dL. Fasting chylomicronemia is frequent. Levels of ApoA-I and A-II are very low, reflecting increased catabolism.

Neurologic consequences present most commonly as relapsing multiple mononeuropathies, or a syringomyelia-like disorder involving loss of pain and temperature sense.[53] Acute, disabling sensorimotor polyneuropathy has been described.[54] Patients may have loss of corneal sensation and orbicular muscle weakness that can lead to ectropion. Sural nerve biopsy reveals lipid droplets in Schwann cells and interstitial cells, which do not appear to be membrane enclosed. In the syringomyelia-like presentation, lipid inclusions resembling lipofuscin granules have been observed within neurons of the spinal ganglia and cord.[53] In patients who have neuropathies, demyelination and remyelination occur. In the syringomyelia-like syndrome, there is axonal degeneration of both small myelinated and unmyelinated fibers. Spinal cord neurons contain osmiophilic lipid inclusions 1–3 mm in diameter (Figure 51.3). No specific treatment is known for Tangier disease.

Genetic Data

The underlying defect is a marked decrease in the efflux of cholesterol from cells to form HDL, resulting from mutations in the cell-membrane ATP cassette transporter, ABCA1.[55]

Familial Lecithin: Cholesterol Acyltransferase Deficiency

Clinical Features and Diagnostic Evaluation

Familial LCAT deficiency is a rare recessive disorder characterized by extremely low levels of activity of lecithin cholesterol acyltransferase (LCAT).[56] The content of unesterified cholesterol in plasma and all lipoprotein classes is very high, and that of cholesterol esters is low, with a preponderance of palmityl and oleyl esters. Cholesterol-rich lipoproteins with multilamellar and discoidal structures are found in the LDL density interval. Some very large spherical particles appear to be modified chylomicrons. These and a population of large lamellar particles in the LDL density interval, which appear to be modified remnants, diminish when a fat-free diet is consumed. Some LDL particles of

FIGURE 51.3 **Tangier disease.** Electron micrographs of a dorsal root ganglion cell. Top: Low magnification demonstrates the packing of inclusions. Bottom: High magnification demonstrates the presence of an electron-dense granular substance and electron-lucent vacuoles. A distinct membrane surrounds inclusions. Bar = 1 μm. (From[53])

normal diameter but enriched in triglycerides are present. Abnormal HDL particles include at least two discoidal forms containing ApoA-I and ApoE, respectively, and spherical particles of abnormally small diameter (~60 Å).

The structural abnormalities are attributable to two amphipathic lipid species: free cholesterol and lecithin, both substrates for LCAT. Normalization of the morphology of HDL particles in LCAT-deficient plasma occurs with the addition of recombinant enzyme. Erythrocytes in LCAT deficiency contain excess free cholesterol and phosphatidylcholine, and they assume the configuration of target cells. Several functional abnormalities of red cells are present, and they tend to hemolyze, causing anemia.

A variant form of LCAT deficiency known as fish-eye disease has been identified in which the pathophysiologic expression is limited to corneal opacity. Esterification of cholesterol in HDL is blocked, but it does proceed in VLDL and LDL, thus mitigating the accumulation of amphipathic lipids.

Most patients develop nephrosis due to accumulation of lamellar lipid in glomerular capillaries, which often progresses to renal failure. Deposits of C3 complement have been detected in the mesangium. There is a moderately increased incidence of premature atherosclerosis. Though uncommon, peripheral neuropathy occurs in LCAT deficiency, associated with cutaneous xanthomatosis.

Genetic Data

Virtually all cases of familial LCAT deficiency result from mutations in the structural gene for LCAT. Several mutations causing fish-eye disorder have been identified in a specific region of the gene.[56]

Therapy

Treatment for LCAT deficiency involves minimizing the dietary fat intake to reduce the production of lamellar lipoproteins that are derived from chylomicrons.

CONCLUSION

Overt neurologic consequences are recognized in disorders of lipoprotein metabolism that involve deficiency of tocopherol and in those that perturb the lipid composition of nerve tissue. The important roles played by apolipoproteins in the regeneration and remyelination of nerves suggest that abnormalities of these elements may affect the recovery of injured nerve tissue. As knowledge of the function of individual cell populations in the nervous system develops, it is likely that more subtle relationships between lipoproteins and neurologic function will emerge.

ADDENDUM

The molecular activity of apolipoprotein E in the brain is still not completely understood. This is becoming of great importance because of the possibility of employing knockdown of the gene to mitigate Alzheimer disease and other neurodegenerative disorders where the apoE4 variant has a deleterious effect.

In this context a middle-aged patient has been studied at the University of California, San Francisco, who has no apolipoprotein E. A whole-exome analysis has demonstrated homozygosity for a large deletion in the gene, and no apoE protein is detectable in his plasma. This has resulted in severe dyslipidemia, with increased levels of very low-density lipoproteins and their remnant particles in blood, with increased levels of apolipoproteins A-I and A-IV in the lipoproteins that contain apolipoprotein B-100, and decreased levels of apoC-III and C-IV in very low-density lipoproteins, with an increased level of pre-β-1 HDL in plasma. Extensive neurological, cognitive and retinal function studies have revealed no abnormalities. The brain MRI images are normal and the content of amyloid-β and tau proteins in the cerebrospinal fluid are normal. Thus, with respect to function of the central nervous system, the absence of apolipoprotein E appears to have no deleterious effect. This suggests that a knockdown strategy for the gene may offer benefit in disorders associated with the E4 variant.[57]

References

1. Safar JG, Wille H, Geschwind MD, et al. Human prions and plasma lipoproteins. *Proc Natl Acad Sci U S A*. 2006;103:11312–11317.
2. Kane JP, Havel RJ. Disorders of the biogenesis and secretion of lipoproteins containing the B-apolipoproteins. In: Scriver CR, Beaudet AL, Sly WS, et al., eds. *The Metabolic and Molecular Basis of Inherited Disease*. 8th ed. New York: McGraw-Hill; 2001:2705–2716.
3. Wetterau JR, Aggerbeck LP, Bouma LP, et al. Absence of microsomal triglyceride transfer protein in individuals with abetalipoproteinemia. *Science*. 1992;258:999–1001.
4. Wang N, Lan D, Gerbod-Giannone M, et al. ATP-binding cassette transporter A7 (ABCA7) binds apolipoprotein A-I and mediates cellular phospholipid but not cholesterol efflux. *J Biol Chem*. 2003;278:42906–42912.
5. Boyles JK, Pitas RE, Wilson E, et al. Apolipoprotein E associated with astrocytic glia of the central nervous system and with nonmyelinating glia of the peripheral nervous system. *J Clin Invest*. 1985;76:1501–1513.
6. Hayashi H, Campenot RB, Vance DE, et al. Apolipoprotein E-containing lipoproteins protect neurons from apoptosis via a signaling pathway involving low-density lipoprotein receptor-related protein-1. *J Neurosci*. 2007;27:1933–1941.
7. Krul ES, Tang J. Secretion of apolipoprotein E by an astrocytoma cell line. *J Neurosci Res*. 1992;32:227–238.
8. Pitas RE, Boyles JK, Lee SH, et al. Lipoproteins and their receptors in the central nervous system. *J Biol Chem*. 1987;262:14352–14360.
9. Nakai M, Kawamata T, Taniguchi T, et al. Expression of apolipoprotein E mRNA in rat microglia. *Neurosci Lett*. 1996;211:41–44.
10. Koch S, Donarski N, Goetze K, et al. Characterization of four lipoprotein classes in human cerebrospinal fluid. *J Lipid Res*. 2001;42:1143–1151.
11. Caronti B, Calderaro C, Alessandri C, et al. Beta2-glycoprotein I (beta2-GPI) mRNA is expressed by several cell types involved in anti-phospholipid syndrome-related tissue damage. *Clin Exp Immunol*. 1999;115:214–219.
12. Danik M, Chabot JG, Hassan-Gonzalez D, et al. Localization of sulfated glycoprotein-2/clusterin mRNA in the rat brain by in situ hybridization. *J Comp Neurol*. 1993;334:209–227.
13. Patel SC, Asotra K, Patel YC, et al. Astrocytes synthesize and secrete the lipophilic ligand carrier apolipoprotein D. *Neuroreport*. 1995;6:653–657.
14. Navarro A, Del Valle E, Tolivia J. Differential expression of apolipoprotein D in human astroglial and oligodendroglial cells. *J Histochem Cytochem*. 2004;52:1031–1033.
15. del Valle E, Navarro A, Astudillo A, et al. Apolipoprotein D expression in human brain reactive astrocytes. *J Histochem Cytochem*. 2003;51:1285–1290.
16. Desai PP, Ikonomovic MD, Abrahamson EE, et al. Apolipoprotein D is a component of compact but not diffuse amyloid-beta plaques in Alzheimer's disease temporal cortex. *Neurobiol Dis*. 2005;20:574–582.
17. Duchateau PN, Pullinger CR, Cho MH, et al. Apolipoprotein L gene family: tissue-specific expression, splicing, promoter regions; discovery of a new gene. *J Lipid Res*. 2001;42:620–630.
18. Liu M, Shen L, Doi T, et al. Neuropeptide Y and lipid increase apolipoprotein AIV gene expression in rat hypothalamus. *Brain Res*. 2003;971:232–238.

19. Ito J, Li H, Nagayasu Y, Kheirollah A, et al. Apolipoprotein A-I induces translocation of protein kinase C[alpha] to a cytosolic lipid-protein particle in astrocytes. *J Lipid Res*. 2004;45:2269–2276.
20. Messmer-Joudrier S, Sagot Y, Mattenberger L, et al. Injury-induced synthesis and release of apolipoprotein E and clusterin from rat neural cells. *Eur J Neurosci*. 1996;8:2652–2661.
21. Demeester N, Castro G, Desrumaux C, et al. Characterization and functional studies of lipoproteins, lipid transfer proteins, and lecithin:cholesterol acyltransferase in CSF of normal individuals and patients with Alzheimer's disease. *J Lipid Res*. 2000;41:963–974.
22. Albers JJ, Cheung MC. Emerging roles for phospholipid transfer protein in lipid and lipoprotein metabolism. *Curr Opin Lipidol*. 2004;15:255–260.
23. Ben-Zeev O, Doolittle MH, Singh N, et al. Synthesis and regulation of lipoprotein lipase in the hippocampus. *J Lipid Res*. 1990;31:1307–1313.
24. Bessesen DH, Richards CL, Etienne J, et al. Spinal cord of the rat contains more lipoprotein lipase than other brain regions. *J Lipid Res*. 1993;34:229–238.
25. Tso P, Liu M. Apolipoprotein A-IV, food intake, and obesity. *Physiol Behav*. 2004;83:631–643.
26. Gotoh K, Liu M, Benoit SC, et al. Apolipoprotein A-IV interacts synergistically with melanocortins to reduce food intake. *Am J Physiol Regul Integr Comp Physiol*. 2006;290:R202–R207.
27. Wolf BB, Lopes MB, VandenBerg SR, et al. Characterization and immunohistochemical localization of alpha 2-macroglobulin receptor (low-density lipoprotein receptor-related protein) in human brain. *Am J Pathol*. 1992;141:37–42.
28. Shie FS, Neely MD, Maezawa I, et al. Oxidized low-density lipoprotein is present in astrocytes surrounding cerebral infarcts and stimulates astrocyte interleukin-6 secretion. *Am J Pathol*. 2004;164:1173–1181.
29. Hoppe G, O'Neil J, Hoff HF, et al. Accumulation of oxidized lipid-protein complexes alters phagosome maturation in retinal pigment epithelium. *Cell Mol Life Sci*. 2004;61:1664–1674.
30. Duncan KG, Bailey KR, Kane JP, et al. Human retinal pigment epithelial cells express scavenger receptors BI and BII. *Biochem Biophys Res Commun*. 2002;292:1017–1022.
31. Li CM, Presley JB, Zhang X, et al. Retina expresses microsomal triglyceride transfer protein: implications for agerelated maculopathy. *J Lipid Res*. 2005;46:628–640.
32. Ishida BY, Bailey KR, Duncan KG, et al. Regulated expression of apolipoprotein E by human retinal pigment epithelial cells. *J Lipid Res*. 2004;45:263–271.
33. Ishida BY, Duncan KG, Bailey KR, et al. High density lipoprotein mediated lipid efflux from retinal pigment epithelial cells in culture. *Br J Ophthalmol*. 2006;90:616–620.
34. LaDu MJ, Gilligan SM, Lukens JR, et al. Nascent astrocyte particles differ from lipoproteins in CSF. *J Neurochem*. 1998;70:2070–2081.
35. Trujillo M, Mateos R, Collantes de Teran L, et al. Lipophilic hydroxytyrosyl esters. Antioxidant activity in lipid matrices and biological systems. *J Agric Food Chem*. 2006;54:3779–3785.
36. Khanna S, Roy S, Parinandi NL, et al. Characterization of the potent neuroprotective properties of the natural vitamin E alpha-tocotrienol. *J Neurochem*. 2006;98:1474–1486.
37. Mariotti C, Gellera C, Rimoldi M, et al. Ataxia with isolated vitamin E deficiency: neurological phenotype, clinical follow-up and novel mutations in TTPA gene in Italian families. *Neurol Sci*. 2004;25:130–137.
37a. Sokol R. Vitamin E and neurologic function in man. *Free Radic Biol Med*. 1989;6:189.
38. Balazs Z, Panzenboeck U, Hammer A, et al. Uptake and transport of high-density lipoprotein (HDL) and HDL-associated alpha-tocopherol by an in vitro blood–brain barrier model. *J Neurochem*. 2004;89:939–950.
39. Goti D, Hrzenjak A, Levak-Frank S, et al. Scavenger receptor class B, type I is expressed in porcine brain capillary endothelial cells and contributes to selective uptake of HDL-associated vitamin E. *J Neurochem*. 2001;76:498–508.
40. Vatassery GT, Adityanjee, Quach HT, et al. Alpha and gamma tocopherols in cerebrospinal fluid and serum from older, male, human subjects. *J Am Coll Nutr*. 2004;23:233–238.
41. Harding AE, Matthews S, Jones S, et al. Spinocerebellar degeneration associated with a selective defect of vitamin E absorption. *N Engl J Med*. 1985;313:32–35.
42. Sobrevilla LA, Goodman MI, Kane CA. Demyelinating contral nervous system disease, macular atrophy and acanthycytosis (Bassen–Kornzweig syndrome). *Am J Med*. 1964;37:821–828.
43. Brin MF, Nelson MS, Roberts WC, et al. Neurophathology of abetalipoproteinemia: a possible complication of the tocopherol (vitamin E) deficient state. *Neurology*. 1983;33(suppl):142.
44. Dexter DT, Brooks DJ, Harding AE, et al. Nigrostriatal function in vitamin E deficiency: clinical, experimental, and positron emission tomographic studies. *Ann Neurol*. 1994;35:298–303.
45. Schwartz JF, Rowland LP, Eder H, et al. Bassen–Kornzweig syndrome. Neuromuscular disorder resembling Friedreich's ataxia associated with retinitis pigmentosa, acanthocytosis, steatorrhea, and an abnormality of lipid metabolism. *Trans Am Neurol Assoc*. 1961;86:49–53.
46. Wichman A, Buchthal F, Pezeshkpour GH, et al. Peripheral neuropathy in abetalipoproteinemia. *Neurology*. 1985;35:1279–1289.
47. Ricci B, Sharp D, O'Rourke E, et al. A 30-amino acid truncation of the microsomal triglyceride transfer protein large subunit disrupts its interaction with protein disulfide-isomerase and causes abetalipoproteinemia. *J Biol Chem*. 1995;270:14281–14285.
48. Malloy MJ, Hardman DA, Kane JP, et al. Normotriglyceridemic abetalipoproteinemia. absence of the B-100 apolipoprotein. *J Clin Invest*. 1981;67:1441–1450.
49. Hardman DA, Pullinger CR, Hamilton RL, et al. Molecular and metabolic basis for the metabolic disorder normotriglyceridemic abetalipoproteinemia. *J Clin Invest*. 1991;88(5):1722–1729.
50. Naganawa S, Kodama T, Aburatani H, et al. Genetic analysis of a Japanese family with normotriglyceridemic abetalipoproteinemia indicates a lack of linkage to the apolipoprotein B gene. *Biochem Biophys Res Commun*. 1992;182:99–104.
51. Anderson CM, Townley RRW, Freeman JP. Unusual causes of steatorrhea in infancy and childhood. *Med J Aust*. 1961;11:617–622.
52. Levy E, Marcel Y, Deckelbaum RJ, et al. Intestinal apo B synthesis, lipids and lipoproteins in chylomicron retention disease. *J Lipid Res*. 1987;28:1263–1274.
53. Schmalbruch H, Stender S, Boysen G. Abnormalities in spinal neurons and dorsal root ganglion cells in Tangier disease presenting with a syringomyelia-like syndrome. *J Neuropathol Exp Neurol*. 1987;46:533–543.

54. Fazio R, Nemni R, Quattrini A, et al. Acute presentation of Tangier polyneuropathy: a clinical and morphological study. *Acta Neuropathol.* 1993;86:90–94.
55. Oram JF, Lawn RM. ABCA1. The gatekeeper for eliminating excess tissue cholesterol. *J Lipid Res.* 2001;42:1173–1179.
56. Santamarina-Fojo S, Hoeg JM, Assman G, et al. Lecithin:cholesterol acyltransferase deficiency and fish eye disease. In: Scriver CR, Beaudet AL, Sly WS, et al., eds. *The Metabolic and Molecular Basis of Inherited Disease*. 8th ed. New York: McGraw-Hill; 2001:2817–2833.
57. Mak ACY, Pullinger CR, Tang FT, et al. Absence of apolipoprotein E: effects on lipoproteins, neurocognitive and retinal function. *JAMA Neurol.* Published online August 11, 2014, doi: 10.1001/jamaneurol.2014.2011.

CHAPTER

52

Cerebrotendinous Xanthomatosis

Vladimir M. Berginer[*], *Gerald Salen*[†], *and Shailendra B. Patel*[‡]

[*]Soroka Medical Center, Beer Sheva, Israel
[†]Rutgers, New Jersey Medical School, Newark, NJ, USA
[‡]Clement J Zablocki Veterans Medical Center, and Medical College of Wisconsin, Milwaukee, WI, USA

INTRODUCTION

Cerebrotendinous xanthomatosis (CTX; MIM 213700) is a rare, recessively inherited lipid storage disease caused by defective bile acid synthesis. The hallmark manifestations include tendon and tuberous xanthomas, juvenile cataracts, and nervous system dysfunction. Dementia, mental retardation, behavioral and psychiatric problems, pyramidal tract paresis, cerebellar ataxia and peripheral neuropathy are the principal neurologic presentations. Consistent features are epileptic seizures, pathologic electroencephalograms (EEGs) and evoked potentials, and abnormal neuroimaging scans. Chronic diarrhea in children, as well as osteoporosis, bone fractures, and premature atherosclerosis in adults may also be present. The underlying cause of CTX is a block in bile acid synthesis resulting in incomplete oxidation of the cholesterol side chain. This is due to inherited mutations of the mitochondrial enzyme sterol 27-hydroxylase located on chromosome 2. The major biochemical findings include increased concentrations of cholestanol, the 5α-dihydro derivative of cholesterol in every tissue, with greatest enrichment in nervous tissues, cerebrospinal fluid (CSF), xanthomas, and bile, low-to-normal plasma cholesterol with increased tissue concentrations, enhanced expression of low-density lipoprotein (LDL) receptors, abnormal biliary bile acid composition with the absence of chenodeoxycholic acid and the presence of large quantities of C-27 bile alcohol glucuronides in bile, plasma, and urine. Replacement therapy with chenodeoxycholic acid (standard of care) inhibits defective bile acid synthesis to normalize the biochemical abnormalities, in many cases improving neurologic function, and is the mainstay of therapy. Initiation of this therapy early on has been shown to prevent many of the sequelae.

CLINICAL FEATURES

Historical Perspective

The first description of an affected patient with neurologic disease (dementia and ataxia), cataracts, and xanthomas of the tendons and central nervous system with normal serum cholesterol levels was published in 1937 by van Bogaert, Scherer and Epstein.[1] Thirty-one years later, Menkes, Schimschock, and Swanson discovered increased amounts of cholestanol, the 5α-dihydro derivative of cholesterol in the central nervous system (CNS).[2] In 1971, Salen reported that biliary bile acid composition was abnormal with virtually no chenodeoxycholic acid (CDCA; Figure 52.1) detected, but contained large amounts of C-27 bile alcohols.[3] Later studies confirmed markedly reduced primary bile acid synthesis in combination with enhanced cholesterol and cholestanol production.[4] The bile acid synthetic defect was further elucidated by Setoguchi et al., who noted that the excreted bile alcohols contained 27 carbons, indicative of incomplete cleavage of the 8-carbon steroid side-chain, going from cholesterol to cholic acid.[5] As a result, large quantities of cholestanol, cholesterol and bile alcohols are formed and circulate in plasma and are excreted in bile, feces, and urine of CTX subjects.[6–8]

In 1984, Berginer et al. reported that long-term replacement therapy with CDCA, the most deficient biliary bile acid, downregulated abnormal hepatic bile acid synthesis and virtually eliminated C-27 bile alcohols from plasma,

Rosenberg's Molecular and Genetic Basis of Neurological and Psychiatric Disease
http://dx.doi.org/10.1016/10.1016/B978-0-12-410529-4.00052-8

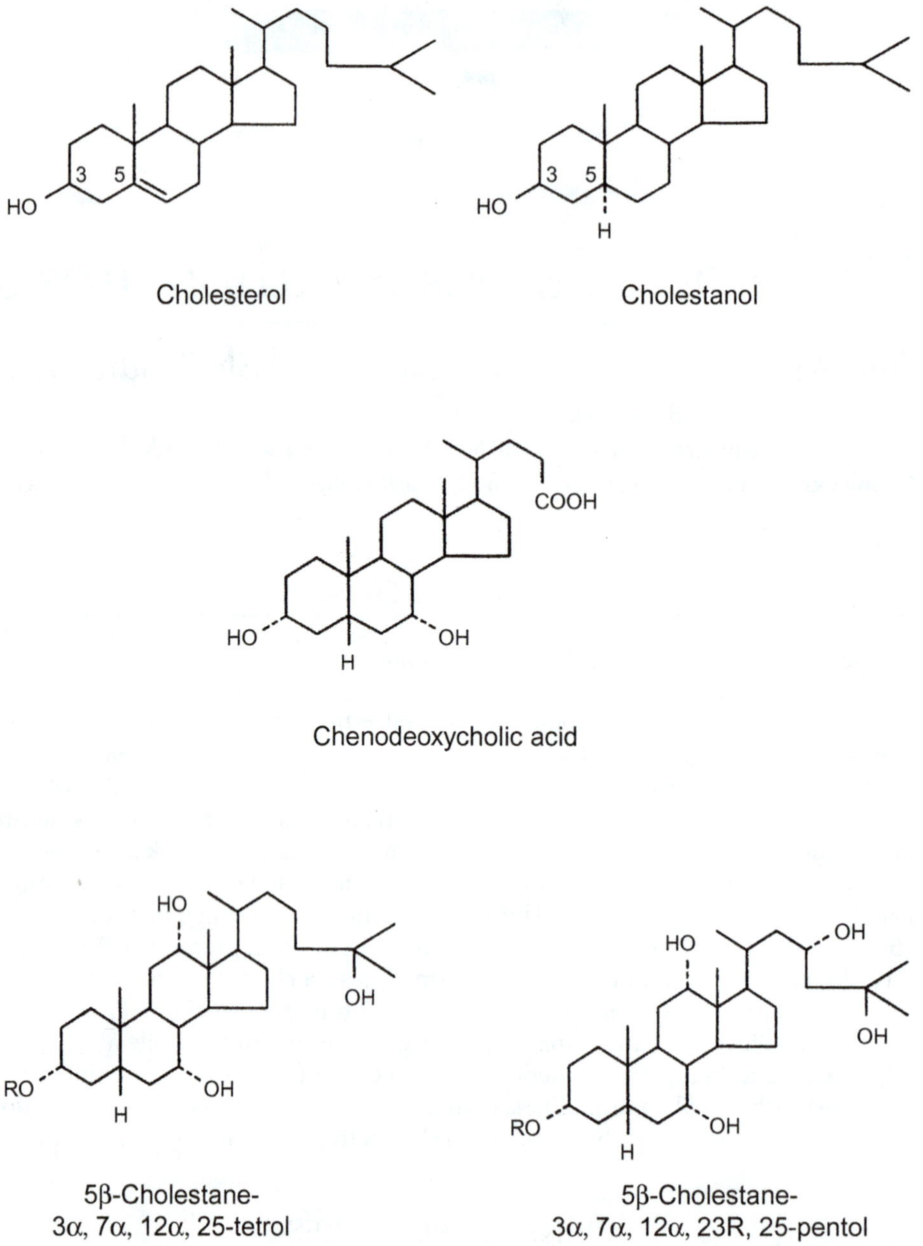

FIGURE 52.1 **Structures of cholesterol, and some of the relevant metabolites identified in CTX.** The structures of cholesterol, cholestanol, chendeoxycholic acid, as well as two bile alcohols (a tetrol and a pentol) are as shown. The 'R' at position 3 (see cholesterol for numbering) is utilized to attach a glucuronic acid moiety to the bile alcohols prior to excretion.

bile and urine.[6,7,9] Elevated cholestanol (Figure 52.1) levels in plasma and cerebrospinal fluid declined and neurologic function improved such that dementia cleared and the progression of other incapacitating neurologic symptoms slowed in most CTX patients.[7]

Natural History

The clinical features develop slowly and may present irregularly in variable combinations in homozygous subjects.[10–13] Table 52.1 lists the clinical findings in 70 patients with CTX, studied by the authors. Tendon xanthomas located in the Achilles tendons were a most consistent feature. It was noteworthy that these xanthomas often enlarged asymmetrically, especially after injury or biopsy (Figure 52.2). The xanthomas most often appeared at around 10 years old and enlarged progressively with age. Tuberous xanthomas located subcutaneously in the elbows, knees and hands also appear and enlarge. Neurologic dysfunction is the most disabling feature followed by juvenile cataracts. The cataracts are detected at an early age and develop through the deposition of cholesterol and cholestanol

TABLE 52.1 Clinical Manifestations in Cerebrotendinous Xanthomatosis

Neurologic manifestations
Dementia
Cerebellar syndrome
Pyramidal: paresis, bulbar palsy
Epileptic seizures
Peripheral neuropathy
Diffuse slow activity on electroencephalogram
Brain and spine atrophy and white matter hypodensity on MRI or CT scans
Pes cavus
Psychiatric disorders
Myopathic-like facies
Juvenile cataracts
Xanthomata
Achilles and other tendons
Brain, spinal cord, lung, skin and sites of trauma
Osteoporosis and bone fractures
Large paranasal sinuses
Chronic diarrhea in children

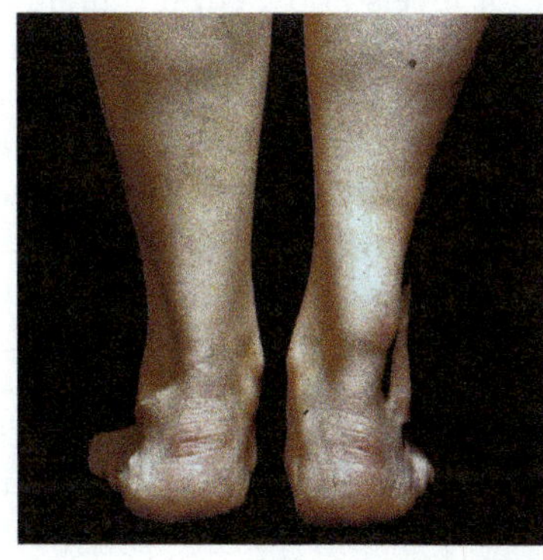
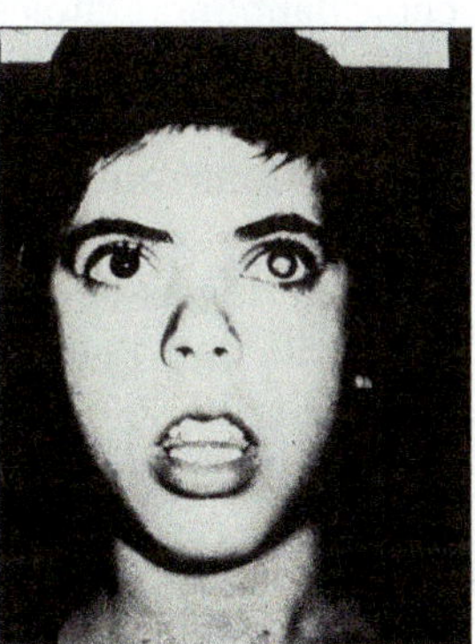

FIGURE 52.2 **Typical clinical features in CTX.** The panel on the left shows asymmetric tendon xanthoma involvement in the Achilles tendon, and is highly suggestive of CTX. The panel on the right shows a typical dense juvenile cataract of the left eye (the right one had been extracted previously), and the open mouth and myopathic facies typically observed in CTX.

in the eye lens (Figure 52.2). As the patients grow older, progressive coronary atherosclerosis, including myocardial infarctions, may appear. Severe osteoporosis that predisposes to bone fractures is present in virtually all subjects. Importantly, unexplained diarrhea in young children is a very early and common childhood symptom and, when present, should direct attention to testing for CTX, especially if juvenile cataracts are also found.[14–16]

Central nervous system abnormalities begin to appear subtly in childhood, associated with the gradual deterioration of intellectual function. Untreated patients usually cannot complete high school. The mental retardation progresses into dementia. Treatment with chenodeoxycholic acid can halt deterioration to preserve existing neurological and intellectual functions. One of our early patients was treated for 40 years with CDCA, from diagnosis in 1972 to her death at age 84; treatment prevented intellectual deterioration. Behavior and psychiatric abnormalities are also common early-in-life symptoms presenting in some patients. Pyramidal tract dysfunction occurs in more than 90% of patients and includes increased deep tendon reflexes, pathologic reflexes, and spastic paraplegia. Cerebellar signs, particularly ataxia, dysarthria and nystagmus are frequently noted. Parkinsonism symptoms (spasticity and resting tremors) were observed in many CTX patients.[17]

Epileptic seizures have occurred in half of symptomatic subjects. In some, the seizure disorder has been the presenting symptom. When juvenile cataracts and/or xanthomas are detected in a young person with seizures, cholestanol and bile alcohol levels should be measured in plasma and urine to establish the diagnosis of CTX.

Peripheral neuropathy due to cholestanol deposition in the nerves with demyelination occurs in more than 50% of CTX subjects when tested. Nerve conduction velocity and visual, auditory and sensory–motor-evolved potentials were abnormally slowed, but improved with CDCA treatment.

EEG abnormalities were found in virtually all CTX patients.[12] Diffuse slowing of background activity presented as poorly organized theta and delta waves and frequent bursts of spikes and high voltage slow delta activity. The abnormal electrical activity improved significantly during treatment with CDCA. However, because abnormal electrical activity can be observed in other lipid storage microencephalopathies, the findings are not specific for CTX.

Neuroimaging scans, including cranial computed tomography (CT) and magnetic resonance imaging (MRI), in patients with CTX demonstrate diffuse brain and spinal atrophy, with brain white matter hypodensity above and especially below the tentorium.[17–20] These findings were attributed, in part, to increased cholestanol and cholesterol infiltration with secondary demyelination. However, they emphasize the possibility that neurologic symptoms, no matter how longstanding, may result from metabolic encephalopathy rather than irreversible destruction of the brain by xanthomas. More than 30 patients with CTX were investigated by CT scan; focal hypodense lesions resembling true xanthomas were detected in only three subjects. MRI scans demonstrated focal brain lesions, located mainly in basal ganglia and mesencephalon areas.

In all CNS examinations, neither surrounding edema nor a significant midline shift was observed, and this was consistent with the absence of symptoms relating to increased intracranial pressure. Treatment with CDCA does not significantly improve neuroimaging atrophy, although in one 13-year old girl, the focal hypodense area (presumed xanthoma) seen by CT scan in the cerebellar hemisphere, disappeared after 1 year of therapy with CDCA. Intracranial calcifications also are infrequently observed in CTX patients.

The CSF from patients is colorless with a normal opening pressure. Cholesterol and cholestanol levels in CSF from untreated patients with CTX were almost 1.5–20 times higher than those in controls.[21] In CSF from untreated patients with CTX, immunoreactive apolipoprotein B fragments were increased 100-fold and albumin about 3.5-fold.[21] These results suggest that the blood–brain barrier permeability is defective and allows the entry of cholestanol and cholesterol attached to LDL from the plasma. During CDCA treatment, the concentration of albumin and apolipoprotein B fragments in the CSF declined and cholestanol levels diminished markedly. It is hypothesized that the high levels of circulating bile alcohols damage the blood–brain barrier to increase permeability with the subsequent rise in brain cholestanol and cholesterol levels and neural damage. Juvenile cataracts are also very common and are found in at least 90% of CTX patients (Figure 52.2). Zonular cortical lens opacities can appear as early as 5–6 years of age or younger, but are almost invariably present by the second decade of life.

Xanthomas are very common and are almost essential features of CTX. They are most often located in the Achilles tendons, but may be present in other muscle extensor tendons and the dermis (tuberous). Interestingly, Achilles tendon xanthomas develop in the proximal portion of the tendon where they attach to the gastrocnemius muscle and usually appear in the second decade of life. Similar xanthomas have been observed in familial hypercholesterolemia when plasma cholesterol concentrations exceed 400 mg per dL, and sitosterolemia, where plasma plant sterol levels are extremely elevated. Although cholestanol and 5α-dihydro plant stanols are increased in this disorder, neither juvenile cataracts nor neurologic symptoms are seen. Cholestanol is increased in sitosterolemia patients' plasma and tissues, but only traces are found in the brain of sitosterolemic patients as evidence of an existing intact blood–brain barrier. Conversely, elevated plasma and tissue phytosterols are not found in CTX plasma or tissues.

Osteoporosis was common in most adult CTX patients and may predispose to and increase the risk of bone fractures. The long bones and vertebra are particularly vulnerable. However, in CTX patients with osteoporosis, serum calcium, inorganic phosphorus, alkaline phosphatase, parathyroid hormone, calcitonin and 1,25-dihydroxy vitamin D3 levels were normal. Serum concentrations of 25-hydroxy vitamin D3 and 24,25-dihydroxy vitamin D3 tend to be low, but not subnormal.[22–25]

Myopathic-like elongated facies, with slightly open mouth (Figure 52.2) and protuberant tongue with prominent paranasal sinuses are commonly observed in CTX patients, who often show serious dental problems.

Intrahepatic pigment and microcrystals are widely distributed in the liver, although their role on pathophysiology has not been established (Figure 52.3).

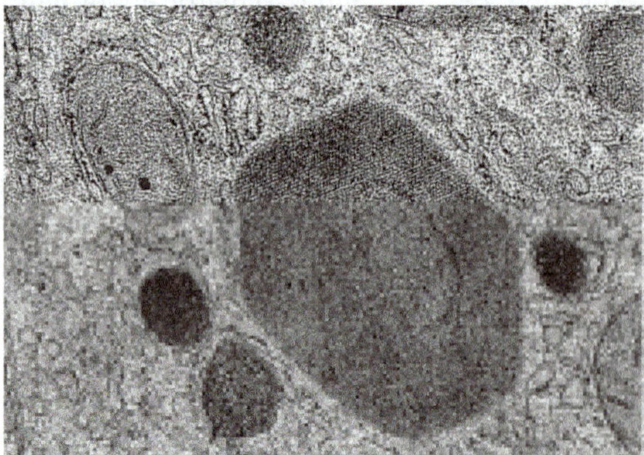
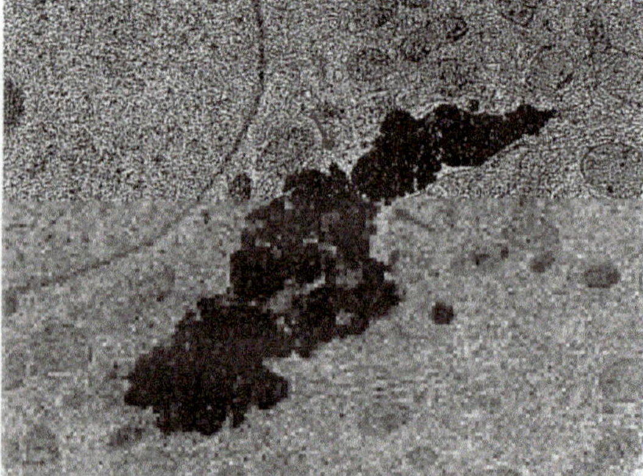

FIGURE 52.3 **Electron-dense deposits and crystal formation in a liver biopsy from a CTX subject.** These microscopic deposits are diffusely located in cytoplasm and may be connected to the overproduction and accumulation of early bile acid precursors that cannot complete the pathway to cholic acid.

MOLECULAR GENETICS

Mode of Inheritance and Prevalence

CTX is an autosomal recessive gene disorder. Most patients have been reported from Japan, the United States, Israel, the Netherlands, Italy, Spain, Pakistan, and China, although CTX cases have been found on all seven continents and have affected all races.[10,26–30] Founder effects in relatively isolated communities underlie the higher prevalence seen in Israel and northern Africa. More than 300 cases worldwide have been reported, but CTX is likely more prevalent, with many more affected subjects undiagnosed or misdiagnosed.

Genetic Defect and Mutational Profiles in CTX

Based upon the biochemical defects identified (see, Historical Perspective, above), sterol 27-hydroxylase (previously known as sterol 26-hydroxlase) was suspected to be defective. Following the purification and sequencing of the rabbit sterol 27-hydroxylase, Cali and Russell isolated the human cDNA.[31] Characterization of the cDNA for sterol 27-hydroxylase showed that it contained a leader signal peptide that allows its targeting and import into the mitochondrion; and comparison of the protein sequence to known cytochrome P450 proteins showed it to have 33% homology. Isolation of a human cDNA clone allowed them to screen two CTX patients and they showed that *CYP27* was mutated in these probands, thus definitively establishing that mutations of *CYP27* caused CTX.[32] The gene, *CYP27*, was localized to chromosome 2q33-ter by somatic cell hybridization[32] and has been subsequently fine-mapped to lie between microsatellite markers D2S1371 and D2S42427. Leitersdorf and colleagues characterized the human *CYP27* gene and showed in a larger cohort of CTX probands that *CYP27* was mutated in all of these cases.[33] Subsequently, various groups have described mutations in probands drawn from almost every part of the world. These are summarized in Figure 52.4. For simplicity, complex mutations, such insertion/deletions, etc., are not depicted. Most mutations result in either introduction of a stop codon, frameshift and premature chain termination or aberrant splicing (mutations indicated below the gene, Figure 52.4). Missense mutations that can potentially be predicted to lead to the synthesis of a full-length polypeptide are depicted above the gene structure

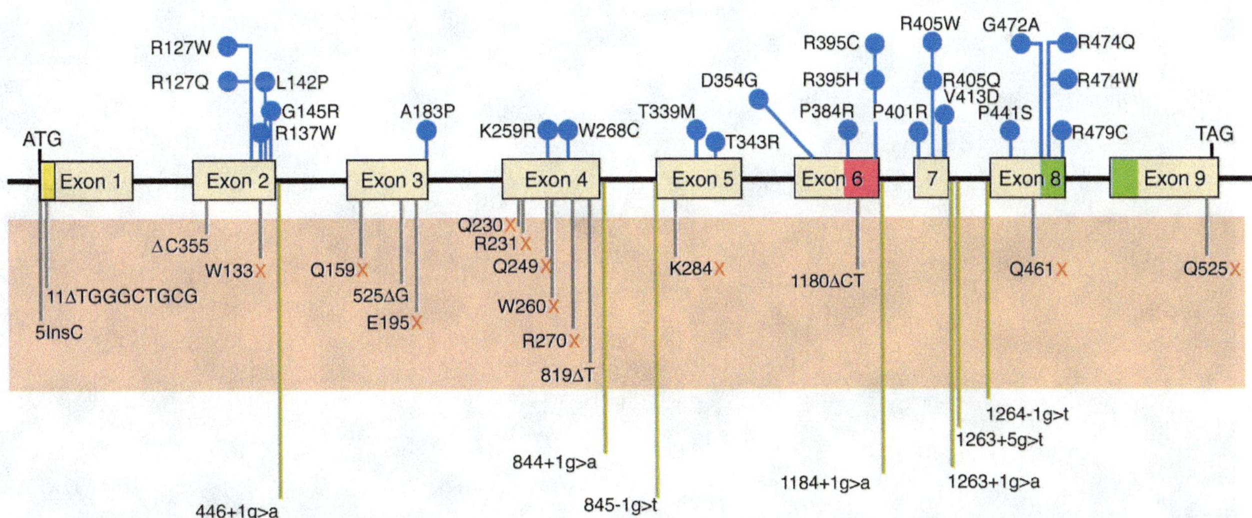

FIGURE 52.4 **Mutations affecting *CYP27* in CTX.** The gene organization, together with the positions of many of the described mutations, is depicted. The gene comprises of 9 exons, and spans >18 Mb of genomic DNA located on human chromosome 2q33. The sequences encoding the mitochondrial signal sequence (exon 1, yellow box), the conserved adrenodoxin (or ferrodoxin) binding domain (exon 6, red box) and the heme binding domain (green boxes, exons 8 and 9) are as shown. The blue circles above the gene structure indicate missense mutations that can be predicted to lead to translation of a full-length protein product. Of these three, R137W, R395C and R395S, may not necessarily lead to protein, as these point mutations are located at the splice boundary regions. In this context, Cali et al. did not report any aberrant splicing for the R395 (1184G>T) mutation. Mutations that can be predicted to lead to truncated proteins are shown below the gene structure. Of these the mutations that lead to premature chain termination or introduction of a stop translation codon are indicated in the orange box. Mutations affecting splicing are shown below this box. These may lead either to exon skipping, frameshift and premature chain termination, or mRNA instability. Insertion/deletion mutations (indicated by "Δ") are not shown as all of these lead to frameshift and premature chain termination, although a few are indicated to show these can be present anywhere in the gene.

(Figure 52.4). Based upon three-dimensional modeling, all but two of the missense mutations are predicted to lead to translation of a peptide map to the critical heme- or adrenodoxin-binding domains, and are thus likely to lead to inactive enzymes (Figure 52.5).[27] However, two of these mutations, affecting peptides Arg127 and Lys259, map to domains outside of these and may indicate other potential functional domains of this enzyme (Figure 52.5).[27] This observation has not been further investigated. The effect some of the missense mutations (affecting R395, R405, R474 and R479) has been shown to lead to an almost complete absence of functional enzyme (either in mutant fibroblasts or by heterologous expression). To date, there does not appear to be a direct relationship between the type of genetic defect and the severity of CTX, suggesting other factors are important in determining the course of this disease (see below). However, efforts are underway to set up a database repository so that such questions may be addressed.[34]

The mRNA for CYP27 is widely expressed with comparable amounts in liver, lung, duodenum, adrenal gland, and lower levels in all tissues including brain; its expression has also been detected in monocytes and in atherosclerotic lesions. Thus, almost all of the cells in the body are capable of converting cholesterol to 27-hydroxycholesterol and cholestanoic acid.[35,36] However, only the liver is capable of transforming these to bile acids; thus, all of these metabolites are secreted into the circulation for clearance by the liver. Björkhem and colleagues have proposed that this pathway may be important in the"reverse" cholesterol transport and thus CYP27 may play a beneficial role in prevention or delay of atherosclerosis, as well as a potential role in maintaining CNS sterol balance.[35,36]

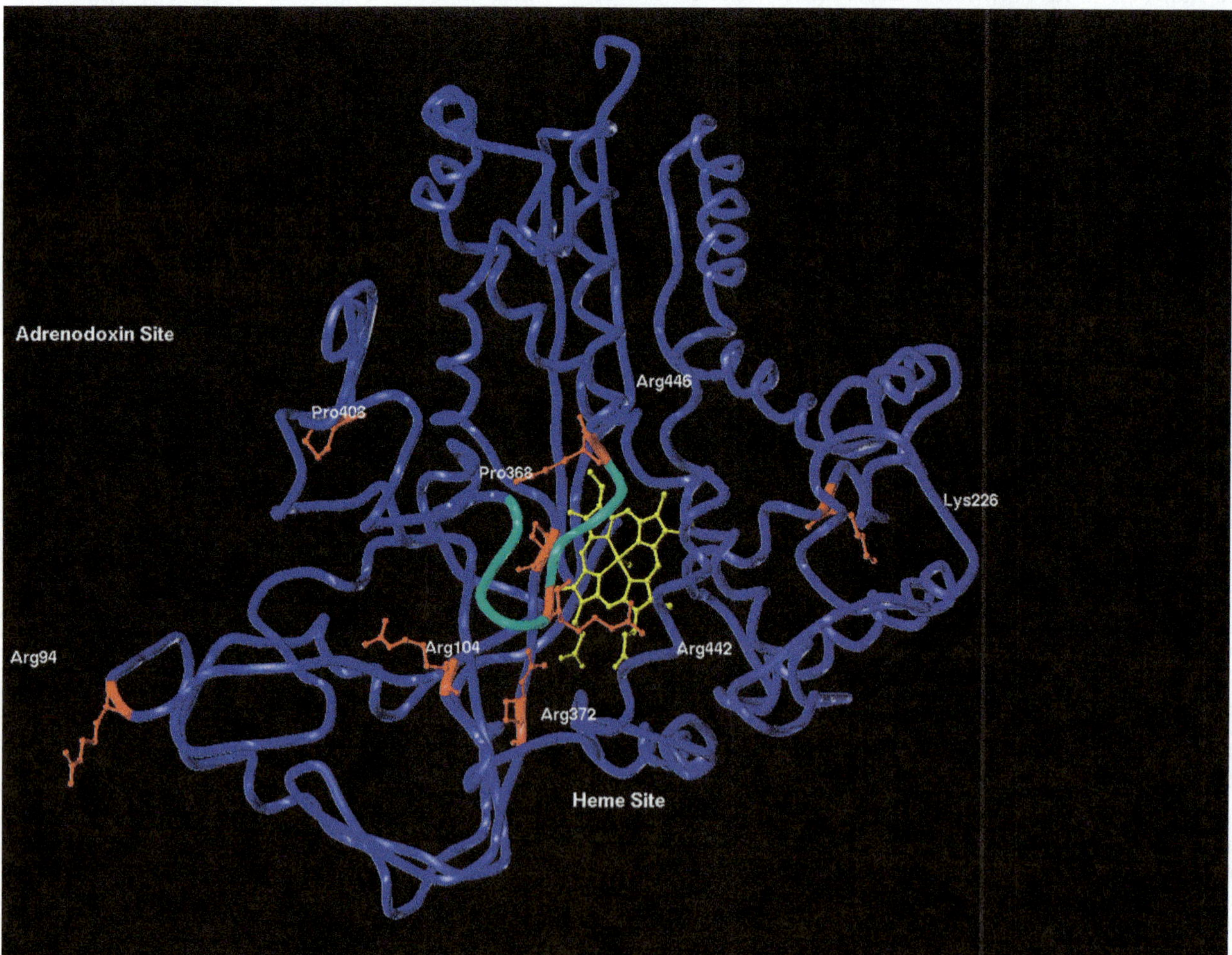

FIGURE 52.5 **Three-dimensional mapping of missense mutations.** Missense mutations shown in Figure 52.3 were mapped to a model of CYP27.[27] This showed that almost all of these, either directly or indirectly, affected the heme and adrenodoxin binding domains. However, two mutations affecting residues R127 and K259 mapped to regions of the protein that do not immediately affect these domains. These sites may lead to an inactive enzyme by means not fully characterized as yet and may reflect important functional domains.

DISEASE MECHANISMS

Biochemical Abnormalities

The major biochemical features that distinguish CTX from other conditions with xanthomas include: 1) increased plasma and tissue cholestanol concentrations with particularly large enrichments in brain, xanthomas, and bile; 2) increased tissue, but low-to-normal plasma cholesterol concentrations associated with enhanced hepatic expression of LDL receptors; 3) defective primary bile acid synthesis with dramatically decreased formation of CDCA with reduced cholic acid in bile, with an abnormally low CDCA-to-cholic acid ratio; 4) increased quantities of C-27 poly-hydroxy bile alcohols with intact 8-carbon side chains conjugated to glucuronic acid at C-3 in bile, feces, urine, and plasma; and 5) disruption of the blood–brain barrier with the accumulation of cholestanol, cholesterol, and apolipoprotein B fragments in CSF. Increased quantities of cholestanol and cholesterol are deposited throughout the brain of CTX patients, but mostly accumulate in white matter. Cholestanol was first discovered in brain tissue by Menkes et al. in 1968.[2] As noted, cholestanol is the 5α-dihydro derivative of cholesterol and normally is present in trace amounts in all tissues and plasma, along with cholesterol. However, in CTX, cholestanol accounts for 2% of plasma and tissue sterols, 10% of biliary and xanthoma sterols, and up to 50% of brain sterols with the remainder being cholesterol.

Cholestanol is transported with cholesterol in the plasma by low-density and high-density lipoproteins. Despite enhanced production of cholesterol and cholestanol in CTX,[4] LDL and HDL sterol levels in plasma are diminished although rates of LDL formation were shown to be increased. LDL catabolism was also enhanced due to the increased hepatic expression of LDL receptors. HDL cholesterol levels tend to be subnormal in most CTX subjects, which may reflect diminished tissue removal and account for the accelerated atherosclerosis and xanthoma formation. Importantly, increased cholestanol and apolipoprotein B fragment concentrations have been found in the cerebrospinal fluid of CTX subjects.[21] It is important to emphasize that apolipoprotein B is neither produced nor catabolized in the central nervous system, so damage to the blood–brain barrier must account for high concentrations in CSF. In addition, the liver, after the diet, is the main source of plasma cholestanol and the abnormal sterol deposits in the brain likely arise from blood to brain transfer when the blood–brain barrier is damaged by the bile alcohols.[37]

Cholestanol formation begins with cholesterol that is overproduced in the liver of CTX patients.[38–40] Two pathways have been proposed in CTX. The classic mechanism begins with the conversion of cholesterol to 4-cholesten-3-one followed by 5α reduction to form cholestanol. Microsomal 3β-hydroxy-steroid dehydrogenase-isomerase that catalyzes the first reaction is 3-fold higher in CTX liver microsomes than controls. However, an alternative mechanism proposed by Skrede et al. showed that 7α-hydroxycholesterol, the initial key intermediate in bile acid synthesis, is overproduced in CTX and can also be partially transformed to cholestanol.[41]

Abnormal Bile Acid Synthesis

An important early observation in CTX subjects was the absence of chenodeoxycholic acid in the bile.[3] There was approximately 14 times more cholic acid relative to chenodeoxycholic acid in bile, which suggests that normal bile acid synthesis was disrupted. This compositional abnormality was confirmed by finding that diminished daily synthesis of cholic acid with virtually no formation of chenodeoxycholic acid combined with the excretion of large quantities of C-27 polyhydroxylated bile alcohol glucuronides (Figure 52.1) in the bile, urine and feces of CTX subjects.

Defective bile acid synthesis became the designated primary inherited biochemical abnormality. Two enzymatic mechanisms seem responsible for the incomplete oxidation of the cholesterol side-chain and subnormal production of bile acids in CTX. Molecular analysis indicated that inherited mutations of sterol 27-hydroxylase, a mitochondrial (P450) enzyme that catalyzes multiple hydroxylation reactions in bile acid synthesis underlie CTX. Deficient mitochondrial CYP27 activity due to an inherited enzyme mutation, in combination with the limited ability to cleave the side-chain via the alternative microsomal side-chain C-25 hydroxylation pathway, results in diminished cholic acid formation and no CDCA production in CTX. As a consequence, increased quantities of C-27 bile alcohols with intact side-chains are produced and excreted as glucuronides (Figure 52.1) mostly in the urine of CTX subjects. It is important to emphasize that CDCA formation does not take place via the alternative microsomal C-25 hydroxylation side-chain pathway. Additionally, while the alternative pathway can lead to cholic acid formation, quantitatively, this seems to be less dominant, since the amounts of cholic acid are also substantially decreased in CTX.

Animal Models

There are currently no adequate animal models that reproduce any of the main features of CTX. A mouse model with a knockout of *Cyp27a1* has been created, but this gene deletion did not lead to accumulation of bile alcohols, nor an elevation in plasma or CNS cholestanol levels.[42] Honda et al. reported that side-chain hydroxylations in bile acid synthesis catalyzed by microsomal CYP3A (part of the alternative 25-hydroxylation pathway that forms only cholic acid) were markedly upregulated in the Cyp27a1 knockout mice, but humans do not seem to manifest this hepatic induction in CTX.[43,44]

Experimental attempts to reproduce the CTX phenotype in mice, rabbits, and rats by feeding cholestanol-enriched diets increased cholestanol levels in liver, heart, and aorta in rabbits and mice (and produced gallstones), but did not increase cholestanol levels in the brain. Damage to the blood–brain barrier, presumably by circulating bile alcohol glucuronides, is therefore necessary to facilitate the uptake of cholestanol and cholesterol in the brain from plasma LDL in subjects with CTX, and may also explain the prevention of neurologic consequences in CTX subjects treated early with CDCA replacement.

DIAGNOSIS

The clinical diagnosis of CTX in symptomatic patients should be entertained when xanthomas (usually located in the Achilles tendons), juvenile cataracts, a history of unexplained childhood diarrhea, and the onset of symptomatic neurologic dysfunction are noted. These abnormalities include behavioral problems, mental retardation and dementia, pyramidal tract and cerebellar dysfunction, epileptic seizures, and peripheral neuropathy. Electrophysiological studies reveal abnormal EEG with diffuse slow wave and occasional bursts of high-voltage activity, slowness of nerve conduction velocity and evoked potentials. Neuroimaging (CT and MRI) scans demonstrate cerebral, cerebellar, and spinal atrophy. X-rays of bones show severe osteoporosis, which accounts for the increased propensity to fracture. Large paranasal sinuses of the skull are noted on X-ray. Poor dental care is also present.

Biochemical diagnosis is established by detection of elevated plasma cholestanol levels (requiring gas or high-performance liquid chromatography) and definitive diagnosis requires the demonstration of bile alcohols in plasma or urine. If the diagnosis remains uncertain (or the biochemistry is equivocal), enhancing endogenous bile acid synthesis by feeding cholestyramine (12 g per day) for 5 days will increase bile acid synthesis and cause plasma cholestanol levels to rise together with increased output of C-27 polyhydroxylated bile alcohols in urine. Interestingly, carriers (one mutated CTX gene) also show a modest increase in urinary bile alcohol excretion but without the rise of plasma cholestanol when treated with cholestyramine. Almost all cases reported to date have also had mutational confirmations performed, presumably because of the rarity of the disease, leading to case reports. Molecular methods are now available to identify mutations in the CYP27A gene to confirm the diagnosis of CTX in homozygotes and to detect heterozygous carriers. Whether this becomes a standard of diagnosis, with the improvements in molecular diagnosis, remains to be seen.

Newer methods that do not involve measuring cholestanol, but detect bile acid precursors also hold promise. DeBarber and colleagues have also suggested that very elevated plasma concentrations of the bile acid precursor 7α-hydroxy-4-cholesten-3-one in a symptomatic subject also is diagnostic of CTX.[45]

MANAGEMENT

Standard of Care

The hypothesis that underlies the treatment of CTX is that increased formation of cholestanol and cholesterol that deposit in the brain, blood vessels, and xanthomas along with the excretion of polyhydroxylated C-27 bile alcohols results from the inherited defect in bile acid synthesis. It further suggests that metabolic encephalopathy due to deposition of cholestanol and cholesterol in nervous tissues is potentially reversible before a critical mass of accumulated sterols and stanols permanently damages nervous tissue.[21] Therefore, it is essential to promptly diagnose CTX and begin replacement therapy with CDCA at a dose of 750 mg per day (calculated to be 33% greater than normal daily bile acid synthesis in a 70 kg individual). This is evidenced by the almost total replacement of CDCA in the enterohepatic circulation, which indicates the suppression of abnormal endogenous bile acid synthesis. Plasma and CSF cholestanol levels decline gradually, associated with small rises in plasma cholesterol concentrations, as cholesterol is no longer needed for endogenous bile acid synthesis.[46] Thus, a normal plasma cholestanol/cholesterol ratio can be reestablished. Selective permeability of the blood–brain barrier will be restored, as noted, by the marked decrease in apolipoprotein B concentrations with the reduction of cholestanol and cholesterol levels in CSF.[21] The rise in plasma cholesterol was attributed to the inhibition of abnormal bile acid synthesis in combination with the reduced expression of LDL receptors. Clinical improvement with CDCA treatment in CTX subjects is impressive. After 1 year, dementia cleared, erratic behavior in many patients improved, and pyramidal and extrapyramidal signs and epileptic seizures disappeared or diminished. EEG showed fewer abnormalities. Nerve conduction velocities and evoked potential improved. The neurologic improvement was better when treatment with CDCA was started at an earlier age and, thus, will prevent the development of permanent neurologic dysfunction in clinically unaffected but biochemically abnormal subjects with CTX.

It is important to emphasize the treatment with CDCA is lifetime replacement therapy. No untoward effects have been encountered; some patients have been treated and remain stable and functional for more than 30 years. Moreover, treatment with CDCA is highly specific, as substituting ursodeoxycholic acid (its 7β-hydroxyepimer) is totally ineffective.[7] It should be noted that the 750 mg per day dose of CDCA acid can compensate for the small conversion of CDCA to ursodeoxycholic acid that occurs in some treated patients.

Failed Therapies and Barriers to Treatment Development

Treatments to lower elevated plasma cholesterol, such as cholestyramine that binds bile acids to promote their elimination by diminishing reabsorption, are clinically ineffective and paradoxically raise plasma cholestanol concentrations. Similarly, diets enriched with unsaturated fats are equally ineffective in halting xanthoma formation or neurological deterioration. Finally, clofibrate, a cholesterol-lowering treatment, when tested in early 1970, did not decrease xanthoma size nor improve biochemical abnormalities.[4]

A brief word of caution about combining CDCA with statin drugs that inhibit HMG-CoA reductase is in order: These statin inhibitors of cholesterol synthesis do not produce additional biochemical benefits in patients with CTX when combined with CDCA,[46] nor have clinical benefits been forthcoming from the combination. Indeed, statin drugs that enhance the expression of LDL receptors to increase the uptake of LDL sterols (cholestanol, cholesterol, and bile alcohols) in tissues actually may be counterproductive. Moreover, LDL apheresis is another suggested treatment to reduce plasma cholestanol levels in CTX. However, this is a temporizing treatment since the moment the apheresis stops, the liver continues to secrete the bile alcohols and cholestanol into the circulation, without benefit of reducing xanthoma size or improving neurologic function.

Surgical removal of xanthomas (except when acute cord compression or pressure relief is needed for emergent therapy) is not recommended. In uncontrolled CTX, xanthomas regrow rapidly and may lead to more clinical problems. However, there are anecdotal reports that surgery may be undertaken without significant regrowth once CTX has been well controlled for a period of time.

References

1. Van Bogaert L, Scherer HJ, Epstein E. *Une forme cerebrale de la cholesterinose generalisse*. Paris, France: Masson et Cie; 1937.
2. Menkes JH, Schimschock JR, Swanson PD. Cerebrotendinous xanthomatosis. The storage of cholestanol within the nervous system. *Arch Neurol*. 1968;19(1):47–53.
3. Salen G. Cholestanol deposition in cerebrotendinous xanthomatosis. A possible mechanism. *Ann Intern Med*. 1971;75(6):843–851.
4. Salen G, Grundy SM. The metabolism of cholestanol, cholesterol, and bile acids in cerebrotendinous xanthomatosis. *J Clin Invest*. 1973;52(11):2822–2835.
5. Setoguchi T, Salen G, Tint GS, et al. A biochemical abnormality in cerebrotendinous xanthomatosis. Impairment of bile acid biosynthesis associated with incomplete degradation of the cholesterol side chain. *J Clin Invest*. 1974;53(5):1393–1401.
6. Batta AK, Shefer S, Batta M, et al. Effect of chenodeoxycholic acid on biliary and urinary bile acids and bile alcohols in cerebrotendinous xanthomatosis; monitoring by high performance liquid chromatography. *J Lipid Res*. 1985;26(6):690–698.
7. Berginer VM, Salen G, Shefer S. Long-term treatment of cerebrotendinous xanthomatosis with chenodeoxycholic acid. *N Engl J Med*. 1984;311(26):1649–1652.
8. Hoshita T, Yasuhara M, Une M, et al. Occurrence of bile alcohol glucuronides in bile of patients with cerebrotendinous xanthomatosis. *J Lipid Res*. 1980;21(8):1015–1021.
9. Batta AK, Salen G, Shefer S, et al. Increased plasma bile alcohol glucuronides in patients with cerebrotendinous xanthomatosis: effect of chenodeoxycholic acid. *J Lipid Res*. 1987;28(8):1006–1012.
10. Verrips A, Hoefsloot LH, Steenbergen GC, et al. Clinical and molecular genetic characteristics of patients with cerebrotendinous xanthomatosis. *Brain*. 2000;123(Pt 5):908–919.
11. Kuriyama M, Fujiyama J, Yoshidome H, et al. Cerebrotendinous xanthomatosis: clinical and biochemical evaluation of eight patients and review of the literature. *J Neurol Sci*. 1991;102(2):225–232.
12. Waterreus RJ, Koopman BJ, Wolthers BG, et al. Cerebrotendinous xanthomatosis (CTX): a clinical survey of the patient population in The Netherlands. *Clin Neurol Neurosurg*. 1987;89(3):169–175.
13. Gallus GN, Dotti MT, Federico A. Clinical and molecular diagnosis of cerebrotendinous xanthomatosis with a review of the mutations in the CYP27A1 gene. *Neurol Sci*. 2006;27(2):143–149.
14. Cruysberg JR. Children with cataract and chronic diarrhoea: cerebrotendinous xanthomatosis. *J Inherit Metab Dis*. 2009;32(2):309.

15. Berginer VM, Gross B, Morad K, et al. Chronic diarrhea and juvenile cataracts: think cerebrotendinous xanthomatosis and treat. *Pediatrics.* 2009;123(1):143–147.
16. van Heijst AF, Wevers RA, Tangerman A, et al. Chronic diarrhoea as a dominating symptom in two children with cerebrotendinous xanthomatosis. *Acta Paediatr.* 1996;85(8):932–936.
17. Su CS, Chang WN, Huang SH, et al. Cerebrotendinous xanthomatosis patients with and without parkinsonism: clinical characteristics and neuroimaging findings. *Mov Disord.* Jan 27 2010;.
18. Colonnese C, Bozzao A, Finocchi V, et al. Role of MR imaging in the diagnosis of cerebrotendinous xanthogranulomatosis, case report and review of literature. *Radiol Med.* 2003;105(1–2):100–103.
19. De Stefano N, Dotti MT, Mortilla M, et al. Magnetic resonance imaging and spectroscopic changes in brains of patients with cerebrotendinous xanthomatosis. *Brain.* 2001;124(Pt 1):121–131.
20. Dotti MT, Federico A, Signorini E, et al. Cerebrotendinous xanthomatosis (van Bogaert-Scherer-Epstein disease): CT and MR findings. *AJNR Am J Neuroradiol.* 1994;15(9):1721–1726.
21. Salen G, Berginer V, Shore V, et al. Increased concentrations of cholestanol and apolipoprotein B in the cerebrospinal fluid of patients with cerebrotendinous xanthomatosis. Effect of chenodeoxycholic acid. *N Engl J Med.* 1987;316(20):1233–1238.
22. Kuriyama M, Fujiyama J, Kubota R, et al. Osteoporosis and increased bone fractures in cerebrotendinous xanthomatosis. *Metabolism.* 1993;42(11):1497–1498.
23. Federico A, Dotti MT, Lore F, et al. Cerebrotendinous xanthomatosis: pathophysiological study on bone metabolism. *J Neurol Sci.* 1993;115(1):67–70.
24. Berginer VM, Shany S, Alkalay D, et al. Osteoporosis and increased bone fractures in cerebrotendinous xanthomatosis. *Metabolism.* 1993;42(1):69–74.
25. Martini G, Mignarri A, Ruvio M, et al. Long-term bone density evaluation in cerebrotendinous xanthomatosis: evidence of improvement after chenodeoxycholic acid treatment. *Calcif Tissue Int.* 2013;92(3):282–286.
26. Pilo-de-la-Fuente B, Jimenez-Escrig A, Lorenzo JR, et al. Cerebrotendinous xanthomatosis in Spain: clinical, prognostic, and genetic survey. *Eur J Neurol.* 2011;18(10):1203–1211.
27. Lee MH, Hazard S, Carpten JD, et al. Fine-mapping, mutation analyses, and structural mapping of cerebrotendinous xanthomatosis in U.S. pedigrees. *J Lipid Res.* 2001;42(2):159–169.
28. Chen W, Kubota S, Kim KS, et al. Novel homozygous and compound heterozygous mutations of sterol 27-hydroxylase gene (CYP27) cause cerebrotendinous xanthomatosis in three Japanese patients from two unrelated families. *J Lipid Res.* 1997;38(5):870–879.
29. Reshef A, Meiner V, Berginer VM, et al. Molecular genetics of cerebrotendinous xanthomatosis in Jews of north African origin. *J Lipid Res.* 1994;35(3):478–483.
30. Leitersdorf E, Safadi R, Meiner V, et al. Cerebrotendinous xanthomatosis in the Israeli Druze: molecular genetics and phenotypic characteristics. *Am J Hum Genet.* 1994;55(5):907–915.
31. Cali JJ, Russell DW. Characterization of human sterol 27-hydroxylase. A mitochondrial cytochrome P-450 that catalyzes multiple oxidation reaction in bile acid biosynthesis. *J Biol Chem.* 1991;266(12):7774–7778.
32. Cali JJ, Hsieh CL, Francke U, et al. Mutations in the bile acid biosynthetic enzyme sterol 27-hydroxylase underlie cerebrotendinous xanthomatosis. *J Biol Chem.* 1991;266(12):7779–7783.
33. Leitersdorf E, Reshef A, Meiner V, et al. Frameshift and splice-junction mutations in the sterol 27-hydroxylase gene cause cerebrotendinous xanthomatosis in Jews or Moroccan origin. *J Clin Invest.* 1993;91(6):2488–2496.
34. Taboada M, Martinez D, Pilo B, et al. Querying phenotype-genotype relationships on patient datasets using semantic web technology: the example of cerebrotendinous xanthomatosis. *BMC Med Inform Decis Mak.* 2012;12:78.
35. Meaney S, Babiker A, Lutjohann D, et al. On the origin of the cholestenoic acids in human circulation. *Steroids.* 2003;68(7–8):595–601.
36. Björkhem I, Diczfalusy U, Lutjohann D. Removal of cholesterol from extrahepatic sources by oxidative mechanisms. *Curr Opin Lipidol.* 1999;10(2):161–165.
37. Bhattacharyya AK, Lin DS, Connor WE. Cholestanol metabolism in patients with cerebrotendinous xanthomatosis: absorption, turnover, and tissue deposition. *J Lipid Res.* 2007;48(1):185–192.
38. Panzenboeck U, Andersson U, Hansson M, et al. On the mechanism of cerebral accumulation of cholestanol in patients with cerebrotendinous xanthomatosis. *J Lipid Res.* 2007;48(5):1167–1174.
39. Salen G, Shefer S, Tint GS. Transformation of 4-cholesten-3-one and 7 alpha-hydroxy-4-cholesten-3-one into cholestanol and bile acids in cerebrotendinous xanthomatosis. *Gastroenterology.* 1984;87(2):276–283.
40. Salen G, Polito A. Biosynthesis of 5α-cholestan-3β-ol in cerebrotendinous xanthomatosis. *J Clin Invest.* 1972;51(1):134–140.
41. Skrede S, Bjorkhem I, Buchmann MS, et al. A novel pathway for biosynthesis of cholestanol with 7 alpha-hydroxylated C27-steroids as intermediates, and its importance for the accumulation of cholestanol in cerebrotendinous xanthomatosis. *J Clin Invest.* 1985;75(2):448–455.
42. Rosen H, Reshef A, Maeda N, et al. Markedly reduced bile acid synthesis but maintained levels of cholesterol and vitamin D metabolites in mice with disrupted sterol 27-hydroxylase gene. *J Biol Chem.* 1998;273(24):14805–14812.
43. Honda A, Salen G, Matsuzaki Y, et al. Side chain hydroxylations in bile acid biosynthesis catalyzed by CYP3A are markedly up-regulated in Cyp27−/− mice but not in cerebrotendinous xanthomatosis. *J Biol Chem.* 2001;276(37):34579–34585.
44. Honda A, Salen G, Matsuzaki Y, et al. Differences in hepatic levels of intermediates in bile acid biosynthesis between Cyp27(−/−) mice and CTX. *J Lipid Res.* 2001;42(2):291–300.
45. DeBarber AE, Connor WE, Pappu AS, et al. ESI-MS/MS quantification of 7 alpha-hydroxy-4-cholesten-3-one facilitates rapid, convenient diagnostic testing for cerebrotendinous xanthomatosis. *Clin Chim Acta.* 2010;411(1–2):43–48.
46. Salen G, Batta AK, Tint GS, et al. Comparative effects of lovastatin and chenodeoxycholic acid on plasma cholestanol levels and abnormal bile acid metabolism in cerebrotendinous xanthomatosis. *Metabolism.* 1994;43(8):1018–1022.

OTHER METABOLIC DISORDERS

CHAPTER

53

Organic Acid Disorders

Margretta Reed Seashore

Yale School of Medicine, New Haven, CT, USA

INTRODUCTION

Most of the disorders termed "organic acid disorders" are disorders of amino acid catabolism whose hallmarks are the abnormal metabolites that accumulate after the amino group is removed from the amino acids derived from protein during catabolism. While most of these compounds are weak acids, they are toxic to cells and may cause damage to organs including liver, brain, kidney, and heart. The phenotype of these disorders may include acidosis, cardiomyopathy, renal damage, central nervous system symptoms and visual symptoms. Diagnosis is based on analysis of blood and urine for the distinctive compounds that characterize each disorder. Age of onset may vary from the first few days of life to early childhood, with occasional onset in adulthood.

For most of these conditions, early symptoms will include acidosis, hyperammonemia, growth failure, and developmental delay. Most of them are inherited in an autosomal recessive manner. While individually quite rare, in the aggregate these conditions have a frequency estimated at between 1:10,000 and 1:30,000 births. With the advent of newborn screening by tandem mass spectrometry, they are being recognized more frequently than these numbers suggest.

Most of these conditions are inherited in an autosomal recessive manner. Affected persons have two abnormal copies of an important gene controlling a specific metabolic pathway. Each parent carries one gene with the mutation and one normal copy of the gene involved. Such carriers do not normally have symptoms. A family history of early infant deaths, serious illness or developmental disability may bring a family to attention.

Standard clinical chemistry testing may show elevated blood ammonia, acidosis reflected by pH and decreased serum bicarbonate, elevated lactate and pyruvate, and elevated plasma ammonia. Diagnosis is usually based initially on an abnormal pattern of metabolites seen in a urine organic acid profile. The urine organic acid profile is normally performed on a spot urine collection. Identification of the abnormal metabolites is based on gas–liquid chromatography and mass spectrum identification of each compound in the urine. Disorders of organic acid metabolism show specific patterns of metabolites depending on the pathway that is disordered. Recognition of these patterns is the critical part of organic acid analysis. Analysis of the glycine derivatives of some metabolites will provide a profile of these acylglycines that are characteristic of some disorders. Follow-up testing will generally include plasma profiles of amino acids and of the carnitine derivatives of the metabolites in an acylcarnitine profile to further elucidate the involved pathway. Molecular testing for specific mutations in the genes controlling the enzymes involved in these pathways is possible for most of these conditions. Such testing enables testing of other family members and may sometimes aid in predicting the severity of the phenotype.

Current research is focused on understanding the variety of clinical presentations, outcome studies directed at developing improved treatment paradigms, and understanding the impact of early diagnosis by newborn screening prior to the onset of symptoms. In addition, since these disorders have been added to the testing done by newborn screening, new previously unrecognized mutations have raised the possibility of milder variants not previously known. Current molecular technology has made possible the identification of the genes involved in controlling the metabolic pathways involved in these disorders. In addition, we need to develop a better understanding of genotype–phenotype correlation and defining the clinical spectrum of the disorders identified by newborn screening. For most of the conditions that are on common newborn screening panels, milder variants are being identified that share biochemical features including plasma acylcarnitine profiles and urine organic acid profiles. The long-term outcome of these patients now being identified is not yet clear.

http://dx.doi.org/10.1016/10.1016/B978-0-12-410529-4.00053-X

CLINICAL FEATURES

Historical Overview

The first organic acid disorder to be completely characterized was isovaleric acidemia (IVA).[1] IVA is a disorder of leucine metabolism characterized clinically by ketoacidosis, sweaty feet odor in body sweat, clothing, and cerumen, pancreatitis and developmental disability. Urine does not have an odor, because the metabolic byproduct in urine, isovaleryl glycine, does not have an odor. Acute neonatal presentation occurs in some infants. Later onset with developmental delay and failure to thrive characterizes other patients. The initial presentation may not predict later clinical course. A broader spectrum of clinical features is reported in patients identified by newborn screening. It was this disorder that drove the development of gas–liquid chromatography with mass spectrometry (GC-MS) as a clinical tool for urine organic acid analysis providing diagnostic specificity. GC-MS is now a critical tool for diagnosis and clinical management of patients with organic acid disorders.

The second organic acid disorder to be well characterized was methylmalonic acidemia. This condition was first reported by Oberholzer and colleagues in 1967.[2] Later Rosenberg and colleagues,[3] using a combination of paper chromatography and thin layer chromatography, identified methylmalonic acid in a urine sample of a young boy admitted in coma after a history of vomiting and acidosis and developmental delay. Based on that diagnosis, treatment with a diet deficient in the amino acid precursors of methylmalonic acid resulted in significant clinical improvement and developmental progress with long-term survival and good clinical outcome. Ultimately the role of vitamin B_{12} as a cofactor for methylmalonyl-CoA mutase was elucidated and methylmalonic acidemia was among the first of the organic acid disorders described with a form amenable to treatment with the vitamin cofactor of the enzyme.

These two disorders established the existence of inherited disorders of organic acid metabolism that resulted in severe metabolic acidosis, developmental disability, seizures and other neurological symptoms (Table 53.1). Further studies established secure diagnostic strategies using GC-MS analysis of urine. Understanding of the disturbance in the underlying metabolic pathway then led the way to the development of successful treatment strategies. Newer molecular genetic methods allowed identification of specific mutations in genes for the proteins involved in these metabolic pathways.

Once the toxic compounds and their precursors were identified, treatment strategies could be devised. Early treatment strategies depended on three principles: definition of the metabolic pathway, identification of the location of the block in the metabolic pathway, and identification of the precursors in the enzyme pathway affected. The important precursors in almost all of these pathways were essential amino acids. Once that knowledge was in hand, dietary strategies could be developed to reduce the intake of those amino acid precursors to just the amount needed for growth and maintenance of skeletal muscle and metabolic well-being. Another effective treatment strategy developed that is based on the observation that some of the enzymes involved had cofactors that could be used to stimulate enzyme activity if residual enzyme activity was present. This approach proved very successful in vitamin B_{12}-responsive methylmalonic acidemia, for example. Removal of the toxic compounds using dialysis was used as well, but cannot be used as a long-term approach.

Even after organic acid disorders were recognized as a category of inborn errors of metabolism, some patients with glutaric aciduria type I (GAI) went unrecognized. GAI was first reported by Goodman in 1975.[4] These patients had macrocephaly and then often had onset of neurologic findings such as seizures and spastic diplegia following an infectious illness. But still this symptom complex was often diagnosed as postencephalitic encephalopathy because a febrile illness seemed to be a prelude to the symptoms. This is one reason why performing a urine organic acid profile in a patient with unexplained neurologic symptoms is important.

TABLE 53.1 Signs and Symptoms of Organic Acid Disorders that Present with Neurologic Findings

Hypotonia or hypertonia
Developmental delay
Failure to thrive
Seizures
Visual disturbances, hallucinations
Loss of developmental milestones
Unusual body odor: musty, "sweaty feet"
Poor feeding, vomiting

Mode of Inheritance and Prevalence

Family history information and the elucidation of the abnormal pathways involved with these conditions led to the conclusion that these are genetic disorders. The identification of the genes that controlled these pathways was a landmark in furthering the understanding of the causes of these disorders. Most of the family histories indicated recessive inheritance: affected siblings with unaffected parents, males and females equally affected and high rates of consanguinity in some affected families. A small number of disorders were identified to be X-linked. An example of an X-linked organic acid disorder is Barth syndrome, 3-methylglutaconic aciduria, type II; (MGCA2). The distinctive organic acid abnormality in that disorder is the presence of 3-methylglutaconic acid in the urine organic acid profile.

The frequency of these disorders varies among different populations. In some areas of the world, there are populations with high rates of consanguinity that result in a significantly higher incidence of organic acid disorders. Examples include: methylmalonic acidemia (MMA), isovaleric acidemia (IVA), and 3-methylcrotonyl-CoA carboxylase deficiency (3-MCC) in Saudi Arabia, and glutaric aciduria type 1 (GA1) in the Old Order Amish in the United States. These disorders occur worldwide, and the clinical presentations vary in severity and age of onset in different populations. Since the advent of newborn screening for these disorders, the numbers that are diagnosed are greater than older incidence data would predict. This observation raises the question of whether there are milder forms of these disorders that might not always be identified by symptoms or indeed might not even cause severe illness.

NATURAL HISTORY

Age of Onset

The most severe forms of the organic acid disorders have a neonatal onset. Symptoms occur within the first few days of life. Early signs and symptoms include irritability, feeding difficulties, and vomiting. As time passes without a diagnosis, the child becomes lethargic and weak and may have seizures. At initial examination the baby may show poor suck, hypotonia, and obtundation. Initial laboratory testing may show acidosis, ketonuria, and hyperammonemia. These signs and symptoms should lead the clinician to consider an inborn error of metabolism. The most effective diagnostic tools to identify an organic acid disorder are the urine organic acid profile and plasma acylcarnitine profile. These will be diagnostic within 24 hours of symptoms.

Later onset is less severe and harder to recognize. Neurologic symptoms may be limited to low muscle tone and slow developmental progress. Poor feeding is often a feature. The child may be less interested in his or her surroundings. Because these symptoms are so nonspecific it is important for the clinician to consider inborn errors of metabolism when poor feeding and slow developmental progress are noted.

PATHOPHYSIOLOGY

The mechanism of brain damage in these disorders is not well understood. Many of them are associated with hyperammonemia, which is well known to cause neuronal damage. Severe acidosis may also be an important pathogenic mechanism. The cause of brain damage in glutaric acidemia type I is poorly understood. Lysine is a precursor of glutaric acid and its transport and metabolism may play a role in the pathophysiology of GA1. Acute striatal necrosis is the cause of the neurologic abnormalities, but the mechanism of this damage is not well understood. The encephalopathy does seem to be age-dependent and does not occur after the age of 5 years or so.

Evidence for the role of genetics in these disorders consists of the observation of multiple cases within sibships and increased frequency of many of these disorders in inbred populations. The Old Order Amish in western Pennsylvania have a high incidence of maple syrup urine disease (MSUD). This population dates back to an early founder population and there is a high degree of consanguinity in the population. Other examples of a high degree of consanguinity correlated with a high incidence of inborn errors of metabolism are found in Saudi Arabia, where a number of organic acid disorders occur with high frequency.[5]

Early in the history of the identification of these disorders, the most severe phenotypes were the ones recognized. Since the introduction of newborn screening for these conditions and the increased diagnostic availability of both biochemical genetic testing and DNA sequencing of genes, milder variants of most of these disorders are being recognized. It is difficult to refrain from treating a child with biochemical and molecular genetic findings consistent with a disorder of organic acid metabolism when the diagnosis is made prior to the onset of symptoms. Once treatment has been instituted, there are few satisfactory criteria for discontinuing treatment. Large studies with clinical correlation between biochemical findings, molecular genetic analysis, and clinical outcome will be needed to make such decisions. However, withdrawing treatment from a patient who is clinically doing well is a very difficult decision to make. Long-term outcome studies will be required to solve these questions.

A few of these pathways involve enzymes for which there is a known cofactor. In those disorders, treatment with that cofactor may be successful. Examples include cobalamin-responsive methylmalonic acidemia, riboflavin glutaryl CoA dehydrogenase deficiency (GA1) and biotin-responsive holocarboxylase synthase deficiency. When unrecognized and thus untreated, the clinical course in these conditions is frequently unrelenting. Acidosis, hyperammonemia, renal failure, and severed developmental disability are common consequences.

DIFFERENTIAL DIAGNOSIS

Other causes of acidosis and hyperammonemia must be excluded when hyperammonemia, acidosis, renal failure, hepatic dysfunction, and developmental delay are observed. Both acquired and other genetic conditions need to be considered. Recognizing that organic acid disorders are an early part of the differential diagnosis is critical, as early treatment makes a big difference in outcome.

MOLECULAR GENETICS

For many of the enzymes involved in inborn errors of metabolism, the normal gene product is known. Current research is focusing on identifying the abnormal gene product(s) and understanding genotype–phenotype correlations.

Gene Identification

Early studies of the identification of the genes involved in these disorders were important in developing treatment strategies for affected individuals and testing strategies for potentially affected family members or members of affected populations. These genes include methylmalonyl mutase, isovaleryl-coenzyme A dehydrogenase, and propionyl-CoA carboxylase.

Protein modeling studies of several enzymes involved in organic acid disorders are being reported.[6] Examples include the molecular structure of isovaleryl dehydrogenase, the enzyme that is deficient in isovaleric acidemia and 3-methylcrotonyl-CoA carboxylase and propionyl-CoA carboxylases.[7] Such modeling is valuable in understanding pathophysiology and developing new treatment strategies such as cofactors that may increase enzyme activity.

TESTING

Once a clinical suspicion is raised of a disorder of organic acid metabolism, a suitable testing strategy needs to be carried out. Clinical laboratory testing will have established the presence of such findings as acidosis, hyperammonemia, and abnormal liver function studies (Table 53.2). Specific biochemical testing will narrow down the diagnosis and provide specific therapeutic guidance. The most useful tests will be the urine organic acid profile and plasma acylcarnitine profile (Table 53.3). A spot urine sample is appropriate for this testing. Fasting is not required for accuracy of the acylcarnitine profile. Carnitine derivatives of the organic acid molecules will be seen in the acylcarnitine profile and may prove diagnostic. These tests should be performed by a laboratory experienced in inherited disorders of metabolism. Some ancillary testing can be considered. Magnetic resonance imaging of the brain may be helpful in defining abnormal anatomy. Some findings are specific to particular disorders, such as lateral hyperdensities of basal ganglia, atrophy of the temporal lobe, or extensive white matter hypodensities in glutaric acidemia

TABLE 53.2 Clinical Laboratory Findings Suggesting an Organic Acid Disorder

Acidosis
Hypoglycemia
Hyperammonemia
Abnormal liver function profile
Elevated plasma lactate and pyruvate

TABLE 53.3 Metabolic Laboratory Studies to Aid in Diagnosis

Urine organic acid profile
Urine acylglycine profile
Plasma acylcarnitine profile
Plasma total and free carnitine
Plasma amino acid profile

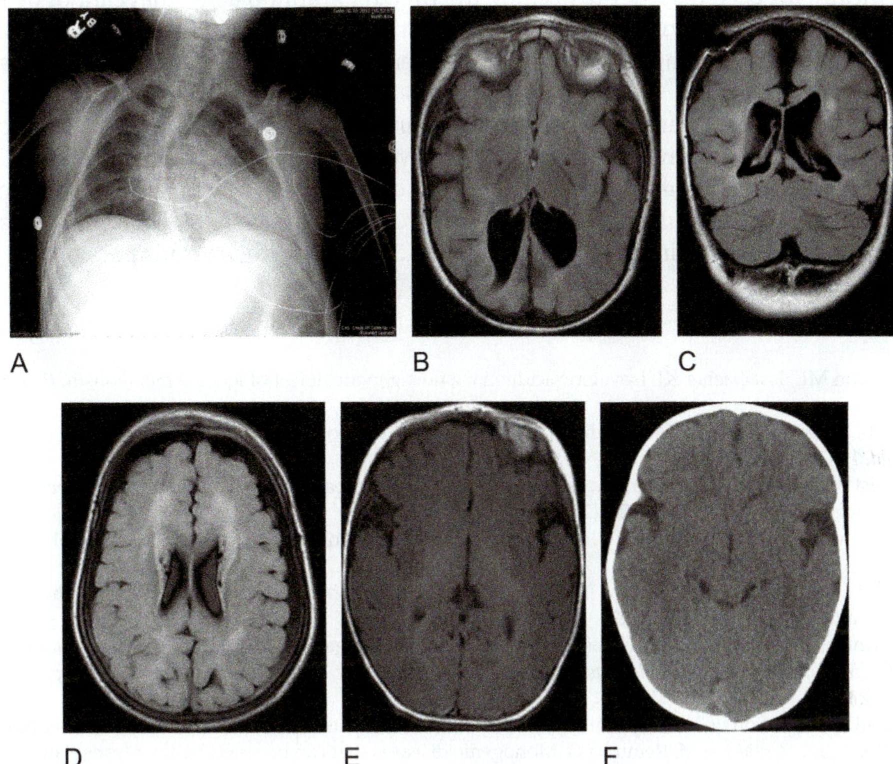

FIGURE 53.1 (**A,B,C**) 2-D-hydroxy glutaric aciduria in a 6-year-old female. (**A**) Chest X-ray illustrating prominent cardiac and mediastinal contours associated with dilated cardiomyopathy. A ventriculoperitoneal shunt catheter and gastrojejunostomy are present, as is patchy perihilar opacification related to chronic lung disease. There is significant scoliosis. (**B,C**) MRI images illustrating periventricular white matter volume loss with extensive gliosis or dysmyelination, occipital lobe encephalomalacia, and cerebral dysgenesis suggested by a simplified gyral pattern and underopercularization. (**D**) 2-D-hydroxy glutaric aciduria in a 7-year-old male. There is microcystic encephalomalacia in the caudate nuclei and in one lentiform nucleus. There is isthmic enlargement of the ventricular atria due to white matter paucity, together with confluent areas of T2 prolongation in the cerebral periventricular and subcortical white matter. Subdural fluid collections probably represent hematohygromas. (**E,F**) Glutaric acidemia type I in a 5-month-old female infant. (**E**) MRI image and (**F**) computerized tomography image. In both, there is slight underopercularization of the insular cortices. Figure courtesy of Dr. Juan M. Pascual, UT Southwestern Medical Center, with permission.

type I. Several representative radiological examples of the broad findings that can be associated with organic acid disorders are illustrated in Figure 53.1. Correlation with the urine organic acid profile is critical, as these findings might be ascribed to a postinfectious encephalopathy. Fasting studies do not have a place in the diagnosis of organic acid disorders.

Molecular diagnosis is available for many of these conditions as the genes are recognized and common mutations are known. DNA testing for a specific known mutation can be performed if there is a family history or if the patient belongs to an ethnic group where specific mutations have been observed. DNA sequencing of the gene is indicated if known mutations are not found, and DNA duplication and deletion studies are also appropriate if a specific mutation is not identified.

In many states in the United States, newborn screening protocols include these conditions. The false negative rate is unknown for these conditions. Newborn screening testing, if performed early before the accumulation of the metabolites occurs, might not identify an organic acid disorder in an affected patient. For this reason, a negative newborn screening test result should never preclude clinical testing for an organic acid disorder in a symptomatic infant.

MANAGEMENT

Standard of care requires significant expertise in these metabolic disorders if treatment is to be effective. Affected infants should be treated in clinics that have physicians and nurses who are experienced in the management of organic acid disorders. Board certification by the American Board of Medical Genetics or comparable certification outside the United States denotes the degree of expertise that is optimal for management of these conditions. The collaboration of a nutritionist with special knowledge of biochemical disorders is crucial. Availability and experience in using the special medical foods and infant formulas that have been developed for the organic acid disorders is critical. These foods are formulated specifically for each disorder to exclude the precursor amino acids that cannot be metabolized and to provide fat and carbohydrate. The use of enzyme cofactors for the pathways involved, such as cobalamin, is essential. Unless identified by newborn screening, symptoms have often been damaging before diagnosis, so diagnosis and treatment as early as possible is critical. Follow-up and monitoring of growth and metabolic parameters is essential.

Liver transplant and kidney transplant have been reported in methylmalonic acidemia and maple syrup urine disease.[8] Liver cell transfer as bridging to liver transplant in urea cycle disorders has been reported in the urea cycle disorders.[9] At present there is no model for this therapy for the organic acid orders. The rarity of these disorders is a barrier to development of treatment models. Populations with known increased frequencies of disorders such as glutaric acidemia type I sometimes are able to participate in clinical research that advances the therapeutic possibilities for these disorders.

References

1. Tanaka K, Budd MA, Efron ML, Isselbacher KJ. Isovaleric acidemia: a new genetic defect of leucine metabolism. *Proc Natl Acad Sci U S A*. 1966;56(1):236–242.
2. Oberholzer VG, Levin B, Burgess EA, Young WF. Methylmalonic aciduria. An inborn error of metabolism leading to chronic metabolic acidosis. *Arch Dis Child*. 1967;42(225):492–504.
3. Rosenberg LE, Lilljeqvist AC, Hsia YE. Methylmalonic aciduria: an inborn error leading to metabolic acidosis, long-chain ketonuria and intermittent hyperglycinemia. *N Engl J Med*. 1968;278(24):1319–1322.
4. Goodman SI, Markey SP, Moe PG, Miles BS, Teng CC. Glutaric aciduria; a "new" disorder of amino acid metabolism. *Biochem Med*. 1975;12:12–21.
5. Moammar H, Cheriyan G, Mathew R, Al-Sannaab N. Incidence and patterns of inborn errors of metabolism in the Eastern Province of Saudi Arabia 1983–2008. *Ann Saudi Med*. 2010;30(4):271–277.
6. Vockley J, Zschocke J, Knerr I, Vockley C, Michael Gibson KK. Branched chain organic acidurias. In: Valle D, Beaudet AL, Vogelstein B, et al., eds. *OMMBID – The Online Metabolic and Molecular Bases of Inherited Diseases*. New York: McGraw-Hill; 2013. http://ommbid.mhmedical.com/content.aspx?bookid=474&Sectionid=45374084. Accessed 23.02.14.
7 Lau EP, Cochran BC, Fall RR. Isolation of 3-methylcrotonyl-coenzyme A carboxylase from bovine kidney. *Arch Biochem Biophys*. 1980;205:352.
8. Fagiuoli S, Daina E, D'Antiga L, Colledan M, Remuzzi G. Monogenic diseases that can be cured by liver transplantation. *J Hepatol*. 2013;59(3):595–612. doi:10.1016/j.jhep.2013.04.004. Epub 2013 Apr.
9. Meyburg J, Hoffmann GF. Liver, liver cell and stem cell transplantation for the treatment of urea cycle defects. *Mol Genet Metab*. 2010;100(suppl 1):S77–S83. doi:10.1016/j.ymgme.2010.01.011. Epub 2010 Jan 29.

Selected Reading

Tanaka K, Matsubara Y, Indo Y, Naito E, Kraus J, Ozasa H. The Acyl-CoA dehydrogenase family: homology and divergence of primary sequence of four acyl-CoA dehydrogenases and consideration of their functional significance. In: Tanaka K, Coates PM, eds. *Fatty Acid Oxidation: Clinical, Biochemical and Molecular Aspects*. New York, NY: Wiley-Liss; 1990.

Matsubara Y, Ito M, Glassberg R, Satyabhama S, Ikeda Y, Tanaka K. Nucleotide sequence of mRNA encoding human isovaleryl-coenzyme A dehydrogenase and its expression in isovaleric acidemia fibroblasts. *J Clin Invest*. 1990;85:1058.

Parimoo B, Tanaka K. Structural organization of the human Isovaleryl-CoA dehydrogenase gene. *Genomics*. 1993;15:582.

Lau EP, Cochran BC, Munson L, Fall RR. Bovine kidney 3-methylcrotonyl-CoA carboxylase and propionyl-CoA carboxylases: each enzyme contains nonidentical subunits. *Proc Natl Acad Sci U S A*. 1979;76:214.

Lau EP, Cochran BC, Fall RR. Isolation of 3-methylcrotonyl-coenzyme A carboxylase from bovine kidney. *Arch Biochem Biophys*. 1980;205:352.

Rao AN, Kavitha J, Koch M, Suresh Kumar V. Inborn errors of metabolism: review and data from a tertiary care center. *Indian J Clin Biochem*. 2009;24:215–222.

Seashore MR. The organic acidemias: an overview. 2001 Jun 27 [Updated 2009 Dec 22], In: Pagon RA, Adam MP, Bird TD, et al., eds. *GeneReviews*™ [Internet]. Seattle (WA): University of Washington, Seattle; 1993–2014. Available from: http://www.ncbi.nlm.nih.gov/books/NBK1134/.

CHAPTER

54

Glycogen Storage Diseases

Salvatore DiMauro and Hasan Orhan Akman

Columbia University Medical Center, New York, NY, USA

INTRODUCTION

The concentration of glycogen in skeletal muscle (about 1 g per 100 g fresh tissue) is second only to that in liver (up to 6 g per 100 g tissue). Liver glycogen serves mainly to keep blood glucose constant, whereas muscle glycogen is exclusively for internal consumption and is one of the major fuels for muscle contraction.[1]

The immediate source of energy for contraction and relaxation derives from the hydrolysis of adenosine triphosphate (ATP), but ATP stores are meager, and ATP resynthesis depends on four metabolic processes: 1) oxidative phosphorylation; 2) anaerobic glycolysis; 3) the creatine kinase (CK) reaction, converting phosphocreatine (PCr) to ATP; and 4) the adenylate kinase reaction, which catalyzes the conversion of two molecules of adenosine diphosphate (ADP) to one molecule of ATP and one of adenosine monophosphate (AMP); this is coupled to the adenylate deaminase reaction converting AMP to inosine monophosphate (IMP).

Quantitatively, oxidative phosphorylation is by far the most important source of energy. Anaerobic glycolysis plays a relatively minor role, essentially limited to conditions of sustained *isometric* contraction, when blood flow and oxygen delivery to exercising muscles are drastically reduced. Conversely, aerobic glycolysis is an important source of energy, especially during the more common, *dynamic* form of exercise, such as walking or running. Accordingly, the pathophysiology of glycogenoses due to defects of glycogenolysis or glycolysis relates more to the impairment of aerobic than anaerobic glycolysis.

There are 15 glycogen storage diseases (GSD) associated with specific defects of glycogen synthesis (glycogen branching enzyme [GSD IV], glycogen synthetase deficiency [GSD 0], and glycogenin deficiency [GSD XV]), glycogen breakdown (GSD I, II, III, V, VI, and VIII), or glycolysis (GSD VII, IX, X, XII, XIII, XIV) (Figure 54.1). Two of these, GSD I and GSD VI, affect mainly the liver, cause neurologic symptoms only indirectly (e.g., hypoglycemic seizures), and will not be considered here. Three new glycogenoses should be added, 5′ AMP-activated protein kinase (AMPK) deficiency, Lafora disease, and RBCK1 deficiency, which do not have numerical definers.[2] Some of these 15 disorders affect only muscle because the genetic defect involves a muscle-specific protein (e.g., myophosphorylase in GSD V) or, more often, the muscle-specific subunit of a multimeric enzyme (e.g., subunit M of phosphofructokinase in GSD VII). The others affect multiple tissues either because the enzyme is a single polypeptide (e.g., acid maltase in GSD II) or because the mutated muscle-specific subunit of a multimeric enzyme is shared with nonmuscle isoforms, causing partial defects in other tissues (e.g., compensated hemolytic anemia in GSD VII).[3]

CLINICAL FEATURES

Glycogen storage diseases cause two main clinical presentations (Figure 54.1), one characterized by exercise intolerance with episodes of cramps and myoglobinuria, the other dominated by fixed weakness.

Rosenberg's Molecular and Genetic Basis of Neurological and Psychiatric Disease
http://dx.doi.org/10.1016/10.1016/B978-0-12-410529-4.00054-1

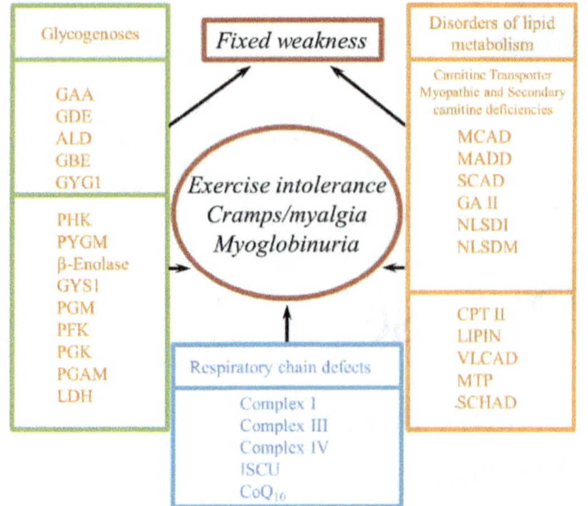

FIGURE 54.1 The two major clinical syndromes associated with metabolic myopathies. Deficient enzymes are denoted by abbreviated gene symbols. GLYCOGENOSES: GAA, acid maltase (acid α-glucosidase, GSD II); GDE, glycogen debranching enzyme (GSD III); ALD, aldolase (GSD IX); GBE, glycogen branching enzyme (GSD IV); GYG1, glycogenin (GSD XV); PHK, phosphorylase *b* kinase (GSD VIII); PYGM, myophosphorylase (GSD V); GYS1, glycogen synthetase (GSD 0); PGM, phosphoglucomutase (GSD XIV); PFK, phosphofructokinase (GSD VII); PGK, phosphoglycerate kinase (GSD IX); PGAM, phosphoglycerate mutase (GSD X); LDH, lactate dehydrogenase (GSD XI). DISORDERS OF LIPID METABOLISM: MCAD, medium-chain acyl-CoA dehydrogenase; MADD, multiple acyl-CoA dehydrogenase; SCAD, short-chain acyl-CoA dehydrogenase; GA II, glutaric aciduria type II; NLSDI, neutral lipid storage disease with ichthyosis; NLSDM, neutral lipid storage disease with myopathy; CPT II, carnitine palmitoyltransferase II; LIPIN, phosphatidate phosphatase; VLCAD, very long-chain acyl-CoA dehydrogenase; MTP, mitochondrial trifunctional protein; SCHAD, short-chain 3-hydroxyacyl-CoA dehydrogenase. RESPIRATORY CHAIN DEFECTS: ISCU, iron sulfur cluster assembly enzyme; CoQ10, coenzyme Q10. (Reproduced, with permission, from[3a]).

Glycogenoses with Exercise Intolerance, Cramps, and Myoglobinuria

The first syndrome is characterized by acute, recurrent, reversible muscle dysfunction, manifesting as exercise intolerance or myalgia, with or without painful cramps (contractures), and often culminating in muscle breakdown (rhabdomyolysis) and myoglobinuria. This is the typical clinical picture of patients with defects in glycogen breakdown (phosphorylase *b*-kinase [PHK] deficiency [GSD VIII] and phosphorylase deficiency [GSD V]) or defects in glycolysis (phosphoglucomutase [PGM] deficiency [GSD XV], phosphofructokinase [PFK] deficiency [GSD VII], phosphoglycerate kinase [PGK] deficiency [GSD IX], phosphoglycerate mutase [PGAM] deficiency [GSD X], β-enolase deficiency [GSD XIII], and lactate dehydrogenase [LDH] deficiency [GSD XI]).

Typically, patients experience myalgia, premature fatigue, and stiffness or weakness of exercising muscles, which is relieved by rest. The type and amount of exercise needed to precipitate symptoms vary from patient to patient, but two types of exertion are more likely to cause problems: brief intense isometric exercise, such as lifting heavy weights, or less intense but sustained dynamic exercise, such as walking uphill. Some patients learn to adjust their activities and lead relatively normal lives. However, when they exceed their limited exercise tolerance, they experience painful cramps and may develop myoglobinuria, with the attendant risk of renal shutdown. Myoglobinuria occurs in about 50% of patients with myophosphorylase deficiency (McArdle disease); it is less common in defects of glycolysis, and least common in PHK deficiency. Uncomplicated episodes of myoglobinuria are followed by complete clinical recovery.

The "second wind" phenomenon is characteristic, in fact pathognomonic, of McArdle disease:[4] when patients slow down or pause briefly at the first appearance of symptoms, they can resume exercising with better endurance. This is related to increased muscle blood flow and increased availability of oxidizable bloodborne fuels (free fatty acids and glucose).

Onset is in childhood, and most patients describe their symptoms as lifelong, but severe cramps and myoglobinuria are rarely reported before adolescence. These disorders appear to be nonprogressive, but fixed weakness tends to set in with increasing age, especially in McArdle disease and in PFK deficiency. This is probably related to chronic muscle damage induced by everyday activities.

Extramuscular symptoms are rare. Jaundice and gouty arthritis may occur in PFK deficiency, reflecting hyperuricemia and hemolytic anemia.[5] Hemolytic anemia, mental retardation, and even Parkinson disease may accompany myopathy in PGK deficiency.[6] Dystocia and dermatological problems were reported in some patients with LDH deficiency.

Rare, atypical phenotypes include fatal infantile myopathies in myophosphorylase or PFK deficiencies.

Exercise intolerance without cramps or myoglobinuria also characterizes the two glycogenoses type 0, so called because muscle shows loss of glycogen rather than glycogen storage. This paradoxical situation makes sense when one considers that glycogenoses type 0 affect the two enzymes that initiate glycogen synthesis, glycogenin and glycogen synthetase.[2] The first mutation in the gene (*GYG1*) encoding the muscle isoform of glycogenin was described in 2010 in a young man who was slower than his peers as a child, suffered from exertional dyspnea, and at age 27 had a life-threatening episode of ventricular fibrillation for which he had to be equipped with a permanent defibrillator.[7] Muscle biopsy showed lack of glycogen both with periodic acid–Schiff (PAS) stain and by electron microscopy whereas heart biopsy showed large accumulation of a poorly structured polysaccharide.

Mutations in the gene (*GYS1*) encoding the muscle/heart isoform of glycogen synthetase were first reported in 2007 in a three Swedish siblings:[8] one boy died of sudden cardiac arrest at age 10, his brother suffered of exercise intolerance and hypertrophic cardiomyopathy, and a younger sister was asymptomatic. Muscle biopsies from the younger siblings showed a conspicuous absence of glycogen and increased numbers of mitochondria. Two more unrelated patients have been reported, an 8-year-old boy who suddenly collapsed and died while running up and down the stairs,[9,10] and a Japanese girl, who, since age 5, had suffered from recurrent and rather stereotypical syncopal episodes after brief exercise. She died of cardiac arrest at age 12. Muscle specimens from both children showed lack of glycogen and excessive numbers of mitochondria. The diagnosis of glycogenosis type 0 is made difficult by the mild muscle symptoms (exercise intolerance), the scarce attention paid by the morphologist to the lack of glycogen, and the misleading clue of the reactive mitochondrial proliferation.

Glycogenoses with Fixed Weakness

The second syndrome is characterized by progressive weakness involving limb and trunk muscles but usually sparing extraocular and facial muscles. This is the typical presentation of branching enzyme deficiency (GSD IV), of the single lysosomal glycogenosis (acid maltase deficiency, GSD II), of AMPK deficiency, debrancher deficiency (GSD III), and aldolase A deficiency (GSD XII). Weakness is also typically associated with two polyglucosan disorders besides GSD IV, Lafora disease and RBCK1 deficiency.

The weakness is severe and generalized in infants with acid maltase deficiency (AMD; α-glucosidase [GAA] deficiency; GSD II; Pompe disease). Some of these infants have isolated myopathy, but most have multisystem involvement, with massive cardiomegaly, less severe hepatomegaly, and, sometimes, macroglossia. Patients with the childhood or adult-onset varieties of AMD have truncal and proximal limb myopathies often simulating muscular dystrophies or polymyositis, and early involvement of respiratory muscles. However, in recent years we have come to realize that late-onset AMD is not confined to muscle, as numerous patients have basilar artery and aortic aneurysms, urinary and fecal incontinence, and dysphagia.[11] Accordingly, a recent postmortem study has shown ultrastructural alterations of the smooth muscle in blood vessels, the gastrointestinal tract, and the bladder.[12]

In debrancher deficiency (GSD III), weakness often becomes manifest in adult life and may be severe and generalized or milder and predominantly distal. Patients with distal weakness are often thought to have Charcot–Marie–Tooth disease or motor neuron disease. However, the apparent paradox that deficiency of an enzyme working hand-in-hand with myophosphorylase should cause weakness rather than a McArdle-like syndrome has been partially clarified by cycle ergometry studies with and without glucose infusion in six patients with only mild or no proximal weakness: their response to exercise and glucose administration was similar to that of patients with McArdle.[13]

Extramuscular involvement is extremely common in this group of glycogenoses and may contribute to the weakness. Cardiomegaly dominates the clinical picture of infantile AMD (Pompe disease) and AMPK deficiency. Cardiomegaly can also be seen in infants and children with brancher deficiency (GSD IV). In debrancher deficiency (GSD III), cardiac involvement is common but cardiomegaly or overt cardiac symptoms are rare.

Hepatosplenomegaly, cirrhosis, and hepatic failure are common causes of death in children with brancher deficiency (GSD IV), but hepatomegaly is also associated with myopathy in severely affected infants. In contrast, hepatomegaly and liver dysfunction often resolve spontaneously in patients with debrancher deficiency, some of whom develop myopathy later in life.

Both central and peripheral nervous systems are involved in infantile AMD (Pompe disease). Peripheral neuropathy is not uncommon in debrancher deficiency. An adult variant of brancher deficiency (adult polyglucosan body disease; APBD) is characterized by progressive upper and lower motor neuron disorder associated with sensory loss, sphincter problems, and, in about half of patients, dementia. Epilepsy, myoclonus, and dementia are the typical clinical triad of Lafora disease.

DIAGNOSTIC EVALUATION

In patients with exercise intolerance, cramps, and myoglobinuria, GSD should be considered in the differential diagnosis, especially when the following clinical features are observed: 1) symptoms occur invariably in association with exercise and the muscles that ache or swell are those that had been exercised; 2) symptoms are triggered by sustained isometric exercise; 3) ischemic exercise causes visible cramps; and 4) resting serum CK is elevated outside

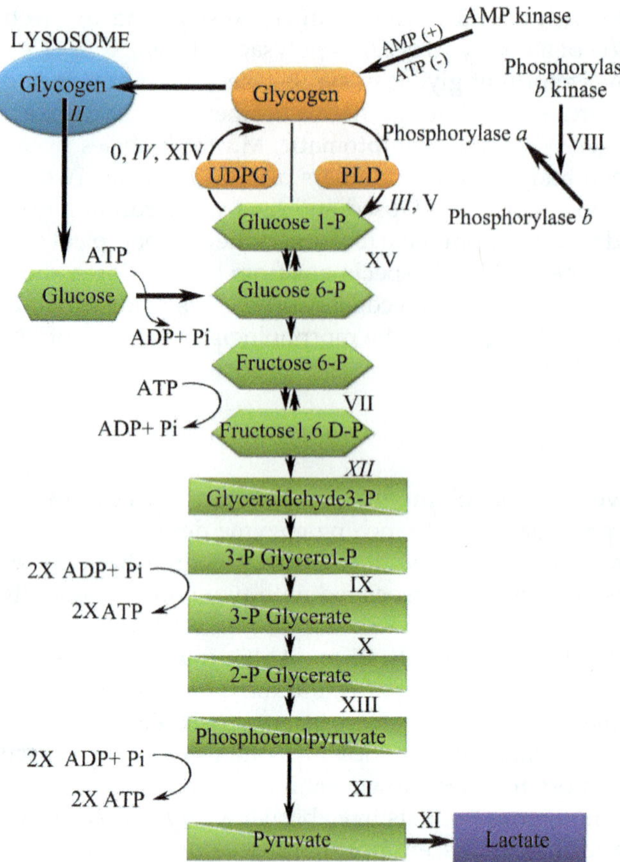

FIGURE 54.2 **Scheme of glycogen metabolism and glycolysis.** Roman numerals indicate enzymes whose deficiencies are associated with glycogen storage diseases (GSD): 0, glycogen synthetase (GYS1); I, glucose-6-phosphatase; II, acid maltase (acid α-glucosidase, GAA); III, debrancher (DBE); IV, branching enzyme (GBE); V, myophosphorylase (PYGM); VII, muscle phosphofructokinase (PFK-M); VIII, phosphorylase *b*-kinase (PHK); IX, phosphoglycerate kinase (PGK); X, phosphoglycerate mutase (PGAM); XI, lactate dehydrogenase (LDH); XII, aldolase-A (ALD-A); XIII, β-enolase (ENO3); XIV, glycogenin (GYG1); XV, phosphoglucomutase (PGM). (Reproduced, with permission, from[3a]).

episodes of myoglobinuria. These features help distinguish GSD from the two other metabolic derangements causing similar symptoms, disorders of lipid metabolism, and defects of the mitochondrial respiratory chain (Figure 54.1). In disorders of lipid metabolism, there are no overt cramps, episodes of myoglobinuria are not triggered exclusively by exercise (prolonged fasting being a major precipitating factor), the offending exercise is prolonged rather than strenuous, and interictal CK is normal. In disorders of the respiratory chain, premature fatigue is the dominating feature, there is myalgia but no cramps, and myoglobinuria is often precipitated by repetitive exercise.[14]

The forearm ischemic exercise—which fails to increase venous lactate when there is a block anywhere in glycogenolysis or glycolysis (Figure 54.2)—is still useful, but is falling out of favor for two main reasons: 1) it depends on the ability and willingness of the patient to exercise vigorously; and 2) it is painful and may provoke local muscle damage with myoglobinuria.

Secure diagnosis of these disorders—of which all but one (PGK deficiency) are due to defects of muscle-specific enzymes—once required muscle biopsy for biochemical (or, less reliably, histochemical) documentation of the enzyme deficiency. Although muscle biopsy is still needed in most cases, it can be bypassed in some patients in whom a probable clinical diagnosis can be confirmed by molecular analysis of genomic DNA isolated from blood cells. For this to happen, however, a few mutations associated with the disease in question must be especially frequent, at least in the ethnic group to which the patient belongs.

In patients with weakness and without involvement of other tissues, the differential diagnosis from muscular dystrophies, neurogenic disorders, and inflammatory myopathies may be very difficult in the absence of a muscle biopsy, in part because serum CK levels can be very high in AMD and debrancher deficiency. The predominant involvement of respiratory muscles in AMD is a useful clue to the diagnosis.

PATHOLOGY

A "mini-atlas" of the muscle pathology in most glycogenoses is provided by a recent review.[2] In GSD with exercise intolerance and myoglobinuria (with the obvious exception of the two glycogenoses type 0), muscle biopsy typically shows subsarcolemmal and less marked intermyofibrillar accumulations of glycogen, best revealed by the PAS reaction in semi-thin plastic-embedded sections. The glycogen storage may be mild enough to escape detection, especially in defects of terminal glycolysis (PGK, PGAM, and LDH deficiencies). Histochemical stains are available for myophosphorylase and PFK. Electron microscopy shows deposits of normal-looking glycogen beta particles in all forms except PFK deficiency.

In PFK deficiency, besides normal glycogen, there is accumulation of an abnormal polysaccharide, which stains intensely with PAS but is not digested by diastase.[15] Ultrastructurally, this polysaccharide is composed of finely granular and filamentous material, similar to the polyglucosan that accumulates in brancher deficiency and in Lafora disease (see later).

Another morphological peculiarity (and a clue to the correct diagnosis) is the presence of tubular aggregates in muscle from more than half of the patients with PGAM deficiency (GSD X): this is not seen in other glycogenoses, but what causes it remains obscure.[2]

In some GSD with progressive weakness, muscle biopsies show massive accumulation of glycogen. This is especially true in both the myopathic and the generalized form of infantile AMD, in which muscle fibers take a "lacework" appearance, and ultrastructure shows both free and intralysosomal glycogen ("glycogenosomes") distorting the contractile system. Severe vacuolar myopathy, with large pools of free glycogen and some reactive glycogenosomes, is also characteristic of debrancher deficiency. In contrast, glycogen storage in muscle was mild in the isolated patient with aldolase A deficiency. In brancher deficiency, the muscle biopsy shows pockets of polyglucosan, whose histochemical and ultrastructural features have been described earlier.

No histochemical reaction is available for acid maltase, but the reaction for acid phosphatase, another lysosomal enzyme, is abnormally intense in patients with AMD.

Neuropathology has shown intralysosomal glycogen accumulation in Pompe disease, especially severe in anterior horn cells of the spinal cord (probably contributing to the flaccid weakness and explaining the tongue fasciculations often observed in these infants). Polyglucosan accumulates to various extents in all tissues of patients with brancher deficiency. In APBD, polyglucosan bodies are present in processes (but not in perikarya) of neurons and astrocytes in both gray and white matter, whereas in Lafora disease polyglucosan bodies are typically seen in neuronal perikarya. In sural nerve biopsies, often used for the diagnosis of APBD, polyglucosan bodies are abundant in the axoplasm of myelinated fibers and less abundant in unmyelinated fibers and Schwann cells. The presence of polyglucosan in muscle biopsies from patients with weakness alone or associated with cardiomyopathy demands sequencing of the *RBCK1* gene, which encodes a ubiquitin ligase[16] (interestingly, malin, one of the two proteins associated with Lafora disease, is also a ubiquitin ligase).

BIOCHEMICAL FINDINGS

Glycogen storage diseases are defined by specific enzyme defects in muscle or other affected tissues. Residual activities are usually below 10% of normal in muscle, and are often explained by the presence of nonmuscle isozymes. For example, in PGAM deficiency, the 5% residual muscle activity is due to the "heart" dimer, HH, and in LDH deficiency the same residual activity is due to the normal "heart" tetramer, H4. Immunochemical studies of muscle from patients with McArdle disease had shown lack of cross-reacting material (that is, enzyme protein) in most cases. The explanation—we now know—is that the vast majority of white patients harbor a nonsense mutation in the first exon of the gene, which impedes protein synthesis.[17]

Why acid maltase deficiency—the defect of a single lysosomal polypeptide—should result in a generalized disorder of infancy or in a mostly myopathic disorder (with onset in infancy, childhood, or adult life) remains unclear, because the enzyme defect is generalized in all forms. A small but crucially important difference in residual activity seems sufficient to protect nonmuscle tissues in myopathic patients.

The differential tissue involvement in some patients with debrancher or brancher deficiency is difficult to explain because both enzymes are single polypeptides encoded by single genes. Accumulating evidence suggests that differential expression of a single gene may give rise to multiple isoforms, only some of which are affected by different mutations.

MOLECULAR GENETIC FINDINGS

The genes encoding all proteins involved in GSD have been cloned, sequenced, and assigned to chromosomal loci. Pathogenic mutations too numerous to be listed here have been identified for each disorder. We will limit this discussion to a few general considerations.

There is striking genetic heterogeneity, often contrasting with relative clinical stereotypy. For example, over 100 mutations in the *PYGM* gene have been identified in patients with typical McArdle disease.[5] Conversely, the few patients with distinct phenotypes, such as fatal infantile myopathy or sudden infant death syndrome, harbored the most common mutation. Thus, genotype–phenotype correlations remain fuzzy, and not just for McArdle disease.

However, some interesting correlations are emerging. In AMD, for which about 200 mutations have been described, there is a reasonably good correspondence between severity of the mutation and severity of the clinical phenotype. Thus, deletions and nonsense mutations are usually associated with infantile AMD, whereas "leaky" mutations are associated with the adult-onset variant of the disease.[18,19] In brancher deficiency, a missense mutation (Y329S) that allows for substantial residual enzyme activity was found in patients with a mild form of hepatopathy, but also in virtually all Ashkenazi Jewish patients with typical APBD, possibly explaining the late onset (although not the brain preference) of this disorder.[20] A good genotype–phenotype correlation was also observed for the neonatal neuromuscular forms of branching enzyme deficiency (GSD IV).[21]

As already mentioned, some mutations predominate in certain ethnic groups, presumably because of founder effects, whereas others, presumably new mutations, are "private."

The ultimate example of genetic heterogeneity is the rare situation in which mutations in the same gene cause two allelic diseases, one of which is a glycogenosis, although the two conditions often coexist in the same patient. This is the case of phosphoglucomutase (PGM) deficiency (GSD XIV), which can cause a pure myopathy with exercise intolerance and myoglobinuria,[22] but more often causes myopathy plus a congenital defect of glycosylation (CDG) with hepatopathy, growth retardation, hypoglycemia, and cardiomyopathy.[23] A second example of allelic disorders is the defect in ubiquitin ligase caused by mutations in *RBCK1*: besides the polyglucosan myopathy/cardiopathy described above, patients may present with failure to thrive, chronic autoinflammation, and recurrent episodes of sepsis.[24] Careful genotype–phenotype correlation may reveal the reason for the different presentations.

ANIMAL MODELS

Spontaneously occurring animal models exist for several GSD. There are two forms of PHK deficiency in the mouse, an X-linked recessive "I" variant (resembling the human myopathic variant), and an X-linked dominant "V" variant.

Myophosphorylase deficiency has been documented in Charolais cattle[25] and in Merino sheep,[26] and molecular defects have been identified in both species. Clinical manifestations include exercise intolerance and myoglobinuria.

English springer spaniel dogs with PFK deficiency suffer from chronic hemolytic anemia and stress-related hemolytic crises, but do not show weakness, premature fatigue, or myoglobinuria. The different clinical phenotype is attributed to the relatively minor role of glycolysis in canine muscle and to the compensatory effect of the PFK-L subunit, which is partially expressed in muscle.[27]

Among the GSD causing weakness, a few animal models have been documented biochemically and genetically for AMD and branching enzyme deficiency: these include Japanese quail, Lapland dog, and cattle for AMD,[18] and the Norwegian forest cat[28] and the American Quarter horse[29] for branching enzyme deficiency. Horses with gain-of-function mutations in glycogen synthetase (*GYS1*) develop polyglucosan storage myopathy[30] by a mechanism similar to that causing polyglucosan storage in GSD VII (see above).

Genetically engineered mouse models have been obtained for AMD,[31] McArdle disease,[32] branching enzyme deficiency,[33] and Lafora disease.[34] Interestingly, in an attempt to boost glycogen accumulation in AMD mice, glycogen synthetase was upregulated, which altered the ratio of synthetase to branching enzyme and resulted in the deposition of polyglucosan[35] (as in human PFK deficiency).

THERAPY

Patients with exercise intolerance and myoglobinuria should be warned about the risks of strenuous exercise, but encouraged to perform some aerobic exercise to avoid deconditioning.[36] A combination of high-protein diet and aerobic exercise has proved useful in patients with McArdle disease[37] and in those with late-onset AMD.[38] Enzyme replacement therapy with recombinant human α-glucosidase (RhGA) has given promising results both in the infantile and in the childhood muscular form of AMD.[39]

Patients with childhood or adult AMD and respiratory insufficiency may need intermittent or permanent mechanically assisted ventilation.

In patients with debranching or branching enzyme deficiency and cirrhosis or liver failure, liver transplantation is an option: of 13 transplanted patients with brancher deficiency, nine had no cardiac or neuromuscular involvement in follow-ups varying from 0.7 to 13.2 years.[40]

CONCLUSION

Although human GSD were among the first inborn errors of metabolism to be described, the field is far from dormant.

Not only are new entities still being discovered, but also previously unsuspected disorders are being unmasked as GSD. For example, the syndrome of hypertrophic cardiomyopathy and pre-excitation due to mutations in *PRKAG2*, the gene for the γ2, regulatory subunit of AMP-activated protein kinase, is, in fact, a cardiac glycogenosis,[41] which, in infants, often underlies a pseudo-PhK deficiency.[42]

More interesting to neurologists is the realization that Lafora disease is a *bona fide* GSD. Clinically, Lafora disease is characterized by seizures, myoclonus, and dementia (and, inconsistently, ataxia, dysarthria, spasticity, and rigidity). Onset is in adolescence and the course is rapidly progressive, with death before age 25 years in most cases. The pathologic hallmarks are the bodies described by Lafora: round, basophilic, strongly PAS-positive intracellular inclusions (3–30 nm in diameter) present only in neuronal perikarya and processes. Ultrastructurally, they are virtually identical to the polyglucosan bodies that accumulate in APBD. However, branching enzyme activity is normal in Lafora disease. Genetic analysis has identified pathogenic mutations in two genes, *EPM2A*, which encodes a dual-specificity phosphatase called laforin, and *EPM2B*, which encodes a putative E3 ubiquitin ligase called malin.[43] Laforin seems to have a protective role in breaking down naturally accumulating polyglucosan, whereas malin appears to modulate the activity of laforin.

Polyglucosan storage diseases are being recognized in increasing numbers. Besides typical GSD IV and its late-onset variant APBD, Lafora disease and, most recently, RBCK1 deficiency have been added to this group of glycogenoses. The creation of mouse models for most GSDs bodes well for the development of therapeutic strategies applicable to humans.

ACKNOWLEDGEMENT

Part of the work described here was supported by the APBD Foundation (APBDF).

References

1. Haller RG, Vissing J. Functional evaluation of metabolic myopathies. In: Engel AG, Franzini-Armstrong C, eds. Myology. New York: McGraw-Hill; 2004:665–679.
2. Oldfors A, DiMauro S. New insights in the field of muscle glycogenoses. *Curr Opin Neurol*. 2013;26:544–553.
3. DiMauro S, Hays AP, Tsujino S. Nonlysosomal glycogenoses. In: Engel AG, Franzini-Armstrong C, eds. Myology. New York: McGraw-Hill; 2004:1535–1558.

3a. DiMauro S, Akman HO, Paradas C. Metabolic myopathies. In: Katjiri B, Kaminski HJ, Ruff RL, eds. *Neuromuscular Disorders in Clinical Practice*. 2nd ed. New York: Springer Science + Business Media; 2014.

4. Vissing J, Haller RG. A diagnostic cycle test for McArdle's disease. *Ann Neurol*. 2003;54:539–542.
5. Mineo I, Kono N, Hara N, et al. Myogenic hyperuricemia. A common pathophysiologic feature of glycogenosis types III, V, and VII. *N Engl J Med*. 1987;317:75–80.
6. Sotiriou E, Greene P, Krishna S, Hirano M, DiMauro S. Myopathy and Parkinsonism in phosphoglycerate kinase deficiency. *Muscle Nerve*. 2010;41:707–710.
7. Moslemi A-R, Lindberg C, Nilsson J, Tajsharghi H, Andersson B, Oldfors A. Glycogenin-1 deficiency and inactivated priming of glycogen synthesis. *N Engl J Med*. 2010;362:1203–1210.
8. Kollberg G, Tulinius M, Gilljam T, et al. Cardiomyopathy and exercise intolerance in muscle glycogen storage disease 0. *N Engl J Med*. 2007;357:1507–1514.
9. Cameron JM, Levandovskiy V, MacKay N, et al. Identification of a novel mutation in GSY1 (muscle-specific glycogen synthase) resulting in sudden cardiac death, that is diagnosable from skin fibroblasts. *Mol Genet Metab*. 2009;98:378–382.
10. Sukigara S, Liang W-C, Komaki H, et al. Muscle glycogen storage disease 0 presenting recurrent syncope with weakness and myalgia. *Neuromuscul Disord*. 2012;22:162–165.
11. Laforet P, Petiot P, Nicolino M, et al. Dilative arteriopathy and basilar artery dolichoectasia complicating late-onset Pompe disease. *Neurology*. 2008;70:2063–2066.
12. Hobson-Webb LD, Proia AD, Thurberg BL, Banugaria S, Prater SN, Kishnani P. Autopsy findings in late-onset Pompe disease: a case report and systematic review of the literature. *Mol Genet Metab*. 2012;106:462–469.

13. Preisler N, Pradel A, Husu E, et al. Exercise intolerance in glycogen storage disease type III: weakness or energy deficiency? *Mol Genet Metab*. 2013;109:14–20.
14. Hays AP, Oskoui M, Tanji K, Kaufmann P, Bonilla E. Mitochondrial neurology II: myopathies and peripheral neuropathies. In: DiMauro S, Hirano M, Schon EA, eds. Mitochondrial medicine. London: Informa Healthcare; 2006:45–74.
15. Agamanolis DP, Askari AD, DiMauro S, et al. Muscle phosphofructokinase deficiency: two cases with unusual polysaccharide accumulation and immunologically active enzyme protein. *Muscle Nerve*. 1980;3:456–467.
16. Nilsson J, Schoser BGH, Laforet P, et al. Polyglucosan body myopathy caused by defective ubiquitin ligase RBCK1. *Ann Neurol*. 2013;74:914–919.
17. Martin MA, Rubio JC, Wevers RA, et al. Molecular analysis of myophosphorylase deficiency in Dutch patients with McArdle's disease. *Ann Hum Genet*. 2003;68:17–22.
18. Hirschhorn R. Glycogen storage disease type II: acid alpha-glucosidase (acid maltase) deficiency. In: Scriver CR, Beaudet AL, Sly WS, Valle D, eds. The Metabolic and Molecular Bases of Inherited Disease. New York: McGraw-Hill; 1995:2443–2464.
19. Kroos M, Hoogeveen-Westerveld M, Van der Ploeg A, Reuser AJJ. The genotype–phenotype correlation in Pompe disease. *Am J Med Genet*. 2012;160C:59–68.
20. Lossos A, Meiner Z, Barash V, et al. Adult polyglucosan body disease in Ashkenazi Jewish patients carrying the Tyr329 Ser mutation in the glycogen-branching enzyme gene. *Ann Neurol*. 1998;44:867–872.
21. Bruno C, van Diggelen OP, Cassandrini D, et al. Clinical and genetic heterogeneity of branching enzyme deficiency (glycogenosis type IV). *Neurology*. 2004;63:1053–1058.
22. Stojkovic T, Vissing J, Petit F, et al. Muscle glycogenosis due to phosphoglucomutase 1 deficiency. *N Engl J Med*. 2009;361:425–427.
23. Tegtmeyer LC, van Scherpenzeel RM, Ng BG, et al. Multiple phenotypes in phosphoglucomutase 1 deficiency. *N Engl J Med*. 2014;370:533–542.
24. Boisson B, Laplantine E, Prando C, et al. Immunodeficiency, autoinflammation and amylopectinosis in humans with inherited HOIL-1 and LUBAC deficiency. *Nat Immunol*. 2012;13:1178–1186.
25. Angelos S, Valberg SJ, Smith BP, et al. Myophosphorylase deficiency associated with rhabdomyolysis and exercise intolerance in six related Charolais cattle. *Muscle Nerve*. 1995;18:736–740.
26. Tan P, Allen JG, Wilton SD, Akkari PA, Huxtable CR, Laing NG. A splice-site mutation causing ovine McArdle's disease. *Neuromuscul Disord*. 1997;7:336–342.
27. Giger U, Argov Z, Schnall M, Chance B. Metabolic myopathy in canine muscle-type phosphofructokinase deficiency. *Muscle Nerve*. 1988;11:1260–1265.
28. Fyfe JC, Giger U, Van Winkle TJ, et al. Glycogen storage disease type IV: inherited deficiency of branching enzyme activity in cats. *Pediatr Res*. 1992;32:719–725.
29. Ward TL, Valberg SJ, Adelson DL, Abbey CA, Binns MM, Mickelson JR. Glycogen branching enzyme (GBE1) mutation causing equine glycogen storage disease IV. *Mamm Genome*. 2004;15:570–577.
30. McCue ME, Valberg SJ, Miller MB, et al. Glycogen synthase (*GYS1*) mutation causes a novel skeletal muscle glycogenosis. *Genomics*. 2008;91:458–466.
31. Raben N, Lu N, Nagaraju K, et al. Conditional tissue-specific expression of the acid alpha-glucosidase (GAA) gene in the GAA knockout mice: implications for therapy. *Hum Mol Genet*. 2001;10:2039–2047.
32. Nogales-Gadea G, Pinos T, Lucia A, et al. Knock-in mice for the R50X mutation in the *PYGM* gene present with McArdle disease. *Brain*. 2012, in press.
33. Akman HO, Sheiko T, Tay SKH, Finegold MJ, DiMauro S, Craigen WJ. Generation of a novel mouse model that recapitulates early and adult onset glycogenosis type IV. *Hum Mol Genet*. 2011;20:4430–4439.
34. Chan EM, Ackerley CA, Lohi H, et al. Laforin preferentially binds the neurotoxic starch-like polyglucosan, which forms in its absence in progressive myoclonus epilepsy. *Hum Mol Genet*. 2004;13:1117–1129.
35. Raben N, Danon MJ, Lu N, et al. Surprises of genetic engineering: a possible model of polyglucosan body disease. *Neurology*. 2001;56:1739–1745.
36. Haller RG. Treatment of McArdle disease. *Arch Neurol*. 2000;57:923–924.
37. Slonim AE, Goans PJ. McArdle's syndrome: improvement with a high-protein diet. *N Engl J Med*. 1985;312:355–359.
38. Slonim AE, Bulone L, Goldberg T, et al. Modification of the natural history of adult-onset acid maltase deficiency by nutrition and exercise therapy. *Muscle Nerve*. 2007;35:70–77.
39. Kishnani P, Beckemeyer AA, Mendelsohn NJ. The new era of Pompe disease: advances in the detection, understanding of the phenotypic spectrum, pathophysiology, and management. *Am J Med Genet*. 2012;160C:1–7.
40. Matern D, Starzl TE, Arnaout W, et al. Liver transplantation for glycogen storage disease types I, III, and IV. *Eur J Pediatr*. 1999;158(suppl 2):S43–S48.
41. Arad M, Benson DW, Perez-Atayde AR, et al. Constitutively active AMP kinase mutations cause glycogen storage disease mimicking hypertrophic cardiomyopathy. *J Clin Invest*. 2002;109:357–362.
42. Burwinkel B, Scott JW, Buhrer C, et al. Fatal congenital heart glycogenosis caused by a recurrent activating R531Q mutation in the gamma2-subunit of AMP-activated protein kinase (*PRKAG2*), not by phosphorylase kinase deficiency. *Am J Hum Genet*. 2005;76:1034–1049.
43. Chan EM, Andrade DM, Franceschetti S, Minassian BA. Progressive myoclonus epilpsies: EPM1, EPM2A, EPM2B. *Adv Neurol*. 2005;95:47–57.

CHAPTER

55

Disorders of Galactose Metabolism

Gerard T. Berry

Harvard Medical School and Boston Children's Hospital, Boston, MA, USA

INTRODUCTION

In man, the Leloir pathway is responsible for galactose metabolism.[1] It consists of three enzymes: 1) galactokinase (GALK); 2) galactose-1-phosphate uridyltransferase (GALT); and 3) UDP-galactose 4′-epimerase (GALE) (Figure 55.1). Each of these cytosolic enzymes is associated with disease that results in hypergalactosemia.[2] Secondary hypergalactosemia may be secondary to liver disease *per se* as the liver is the most important organ for whole body galactose handling. Other secondary causes include a patent ductus venosus and congenital hepatic vascular anomalies (Table 55.1).

The liver is capable of rapid removal of galactose from the circulation and conversion into uridine sugar nucleotides for glycogen synthesis or for the production of complex macromolecules such as the glycan chains of glycoproteins. These enzymes are also present in extrahepatic tissues and probably function primarily in glycoconjugate synthesis and recycling.

CLASSIC GALACTOSEMIA

Classic hereditary galactosemia is an autosomal recessive genetic disorder (OMIM 230400) due to galactose-1-phosphate uridyltransferase (GALT, EC 2.7.7.12) deficiency. In the newborn period, a life-threatening disease with multiorgan involvement emerges, but only in those infants ingesting lactose in breast milk and proprietary formulas containing the ingredients of cow's milk.[2] The first observation of this nutritional toxicity state involving the neonate and breast milk was by von Reuss in 1908.[3] In the subsequent decade, Göppert documented the presence of excess galactose in the urine of a similarly affected infant.[4] The first well-characterized infant with hypergalactosemia and galactosuria who responded to a lactose-restricted diet was by Mason and Turner in 1935.[5] This paper, describing an African American infant with a variant form of galactosemia, served to reveal marked hypergalactosemia as an integral component of GALT deficiency. However, establishment of the abnormal biochemistry awaited the discovery by Schwarz in 1956 that the substrate, galactose-1-phosphate, was elevated in erythrocytes from galactosemic patients exposed to galactose,[6] and the demonstration later that year in the Kalckar laboratory that GALT enzyme activity was absent.[7]

This rare Mendelian disorder with a worldwide frequency of 1/40,000 to 1/60,000 newborn infants entered the modern era of molecular biology when Reichardt and Berg cloned GALT cDNA from a human liver library.[8] Soon thereafter, many causative gene mutations were identified and the structural organization of the GALT gene was delineated.[9–12]

It was clear in the decades following the landmark Mason and Turner paper that restriction of galactose intake in the affected newborn infant would usually permit survival and allow the following early infantile complications to remit, resolve or disappear: poor growth, poor feeding, emesis, jaundice, liver enlargement, cataracts, encephalopathy, including lethargy, irritability and hypotonia, hyperbilirubinemia, liver-associated test abnormalities such as transaminasemia and hypofibrinogenemia, hyperchloremic metabolic acidosis, albuminuria, generalized aminoaciduria, and anemia. Yet, between 1970 and 1990, there was a growing awareness that there are "clouds over

Rosenberg's Molecular and Genetic Basis of Neurological and Psychiatric Disease
http://dx.doi.org/10.1016/10.1016/B978-0-12-410529-4.00055-3

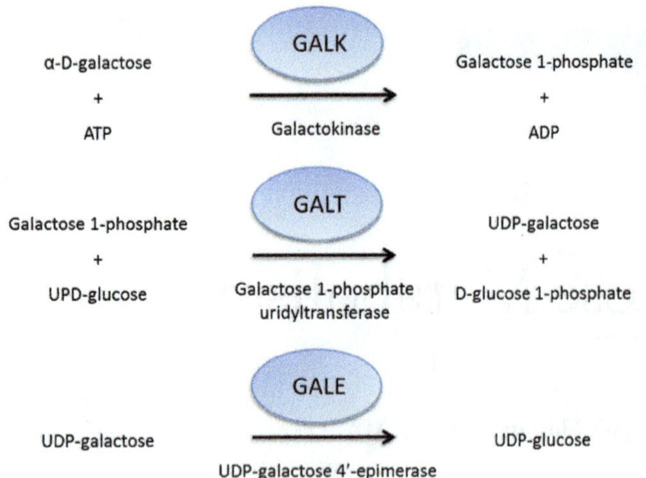

FIGURE 55.1 **The primary enzymes of the galactose pathway: the Leloir pathway enzymes.**

TABLE 55.1 Common GALT Genotypes and Biochemical/Clinical Phenotypes

Classic galactosemia	Clinical variant galactosemia	Biochemical variant galactosemia
Q188R/Q188R	S135L/S135L*	4bp 5′ del+N314D/Q188R†
K285N/K285N		
L195P/L195P		
Δ5.2kb del/Δ5.2kb del		

**The original identification of the S135L mutation was exclusively in African Americans but it is present on occasion in infants without known African American heritage.*

†Known as the very frequent "Duarte-classical galactosemia variant" or more often just the "Duarte variant."

galactosemia,"[13,14] that patients were not faring as well as physicians had expected,[15–21] especially given the fact that lactose restriction largely eliminates death from *Escherichia coli* sepsis.[22] This was all the more poignant as these infants who were treated very early in life, especially after the introduction of newborn screening for galactosemia in the USA in the early sixties, appeared to be indistinguishable from normal, healthy, thriving infants—at least until approximately 18–36 months of age when language acquisition and speech become conspicuous landmarks of nervous system development.

This naïveté came to an end once and for all when Kaufman et al. published their paper in 1981 on the prevalence of primary ovarian insufficiency (POI) in female patients on lactose restricted diets,[23] followed by the 1990 retrospective piece of work of Waggoner et al., summarizing the outcome of 350 patients from around the world.[24] The latter paper is still the most comprehensive of its type. A review of patients' records from many centers indicated that they had suffered developmental delay involving language acquisition and speech defects as well as POI in females even though they had been on a lactose-restricted diet. Furthermore, these late-onset apparent diet-independent complications occurred even if diet therapy began on day 1 of life.

In addition to the above well known diet-independent complications,[25–32] we now know that additional complications include poor growth and/or short stature, reduced bone mineral density, deficits in personality construct, mood disorder, tremor, cerebellar ataxia and extrapyramidal movement disorders.[33–37] In some instances, discrete neurological, imaging and neuropathological findings have been recorded.[38–48] It is still not clear how frequent each of these entities are in patients with classic and variant galactosemia at different stages of maturation including adulthood.

Classic galactosemia is defined as that which occurs in patients with absent or barely detectable GALT enzyme activity in erythrocytes, persistent elevations of galactose-1-phosphate (erythrocyte range of 1–4mg%) and galactitol (urine excretion range of 100–400 μmoles per nmole of creatinine) while on a chronic lactose-restricted diet, and evidence of or the propensity for a neonatal life-threatening multiorgan disorder with the risk of lethal *E. coli* sepsis. Additional defining features include markedly diminished oxidation of [1-^{13}C]galactose to $^{13}CO_2$ in a 2–5 hour breath test,[49–52] elevated serum follicle-stimulating hormone (FSH) and decreased anti-Müllerian hormone (AMH) levels in female infants and children,[53] and brain magnetic resonance imaging (MRI) lesions (white matter hyperintensities, cortical atrophy and cerebellar atrophy). Most of these patients will possess severe GALT gene mutations such as Q188R, K285N, L195P and the Δ5.2-kb deletion either in homozygous or compound heterozygous states. Clinical variant galactosemia, i.e., the patient who may manifest complications such as poor growth, hepatomegaly, jaundice, anemia, developmental delay, and cognitive impairment if exposed to dietary lactose, is best exemplified by the South African S135L/S135L genotype.[54] The S135L allele is present in the majority of African Americans in the USA.[55] These clinically significant variants are in great contradistinction to the largely benign Duarte D_2 compound heterozygous condition that seems to be only a biochemical variant, not a clinical variant. The most common genotype example in the Western world may be the Q188R/4bp deletion-*cis*- N314D.

Genetics

The mode of inheritance is autosomal recessive. The birth incidence of classical galactosemia is 1 in 40,000–60,000. In Ireland it is 1 in 16,476. The GALT gene is on chromosome 9, and over 250 mutations or polymorphisms have been described;[10–12,56] see the following website: http://www.arup.utah.edu/database/galactosemia/GALT_welcome.php.[12] The most prevalent alleles and genotypes are shown in Table 55.1. The Q188R is the most prevalent mutation in European populations, and is particularly common in Ireland and Great Britain where it accounts for over 70% of mutant alleles. The K285N mutation is more common in Eastern Europe. The Ashkenazi Jewish mutation, Δ5.2-kb deletion, is very common in Jerusalem. The S135L is the most frequent mutation in black Africans and African Americans. Some genotype–phenotype matching is available.[27,29,51,57–59] For instance, an unfavorable clinical outcome has been associated with homozygosity for the Q188R mutation,[27,29,59] but outcome is decidedly better in those homozygous for S135L.[58] Many allelic variants associated with a partial enzyme defect have been reported, but the best known is the D_2 Duarte variant due to a N314D2 *GALT* gene polymorphism that exists in *cis* with a small deletion in the 5′ flanking region.[10,11,60]

Variants such as the Q188R/N314D compound heterozygote can be distinguished by *GALT* gene sequencing. The N314D Duarte D_2 variant when combined with the Q188R mutation (D/G defect) is associated with 25% residual GALT activity and is almost always benign. It is relatively common,[10,11,60] so that partial transferase deficiency is more frequent than classical galactosemia.

Diagnostic Tests

Newborn screening (NBS) for galactosemia is undertaken in many countries by measurement of blood total or free galactose, the transferase enzyme, or both using dried blood spots usually collected 24–48 hours after birth. Some programs quantify galactose-1-phosphate in addition to total galactose. Measurement of galactose alone results in a high rate of false positives and in infants with a clinical variant form of galactosemia, may also give false negative results.[61] In infants with classic galactosemia, at the time of discovery by NBS, the first signs may already have appeared, and the infant admitted to a hospital, usually for jaundice.

The diagnosis of galactosemia is confirmed by assaying GALT activity in erythrocytes. An indirect fluorescence blood spot test (the Beutler test) can be used as an initial investigation. An abnormal (positive) result is where there is no fluorescence under ultraviolet (UV) light after 2 hours of incubation of whole blood with UDP-glucose, galactose-1-P, and NADP. False positives occur in infants with glucose-6-phosphate dehydrogenase deficiency. More specific quantitative assays are available where GALT activity is measured in erythrocyte lysates. All assays of red cell GALT activity can give a false negative result if a blood transfusion has been given within the prior 2–3 months. In this situation, an assay of urinary galactitol, red cell galactose-1-phosphate and/or mutation analysis can be helpful. The finding of reduced GALT activity in parental blood may also provide additional information since, in heterozygotes, the enzyme activity in erythrocytes is approximately 50% of normal. Cultured skin fibroblasts can also be used for the enzyme assay. NBS will also detect infants with partial deficiencies, most commonly D/G galactosemia. Although D/G galactosemia appears benign, every newborn with partial GALT deficiency must nevertheless be observed closely in case they have another allelic variant that may be clinically relevant, such as individuals of African descent with a S135L/S135L genotype.[58] Assessment involves quantitation of urine galactitol, of erythrocyte galactose-1-P, GALT enzyme activity, and *GALT* gene sequencing, as well as molecular investigation of the parents. Galactose-tolerance tests are notoriously noxious to the child with classical galactosemia and have no place in evaluating the need for treatment of partial deficiencies. Breath tests with [1-^{13}C]galactose may be helpful in the assessment of patients with very rare genotypes.[51]

Where newborns are not screened for galactosemia or when the results of screening are not yet available, diagnosis rests on clinical awareness; however, some infants are detected from finding a raised blood phenylalanine and tyrosine (as a result of liver disease) assayed as part of the NBS for phenylketonuria.

Prenatal Diagnosis

If the mutation is known within a family, prenatal diagnosis can be performed by analysis of DNA extracted from a chorionic villus biopsy collected at about 11 weeks gestation. Prenatal testing can also be achieved by GALT enzyme activity measured directly in chorionic villus cells, in cultured chorionic villus cells, or in cultured amniotic fluid cells and by measurement of amniotic fluid galactitol.[62] Restricting maternal lactose intake does not interfere with a diagnosis based on galactitol measurements in amniotic fluid.[63]

Treatment and Prognosis

Newborn

Treatment of the newborn with classical galactosemia consists of the exclusion of all lactose from the diet. Breastfeeding or milk-based formulas must be stopped immediately the disorder is suspected clinically or following a positive newborn screening result and before confirmatory diagnostic tests are available. Most newborns are changed to a soy-based formula, although an elemental formula may also be used. Some infants are seriously ill at diagnosis and will require considerable supportive care including the management of coagulopathy and septicemia. When a lactose-free diet is instituted early enough, symptoms disappear promptly, jaundice resolves within days, cataracts may clear, liver and kidney functions return to normal, and liver cirrhosis may be prevented.

Infants and Children

At weaning parents must learn to identify all other sources of lactose and need assistance from the pediatrician and dietitian, who must have recourse to published recommendations. Some Swiss cheeses and certain other hardened cheeses are galactose- and lactose-free, as these sugars are cleared by the fermenting microorganisms.[64,65] Although soy-based infant formulas normally provide sufficient calcium in infancy, at weaning calcium supplement is necessary.

Even with strict adherence to dietary treatment, a completely galactose-free diet is not possible. Galactose is present in a great number of vegetables, fruits and legumes (beans, peas, lentils, etc.),[66] as a component of galactolipids and glycoproteins, in the disaccharide melibiose and in the oligosaccharides raffinose and stachyose.[67,68] The latter two contain galactose in α-galactosidic linkage which is not hydrolyzable by human small intestinal mucosa *in vitro* or *in vivo*.[68] There is no evidence that restriction of other foods, in addition to exclusion of diary products, leads to any improvement in outcome.

Older Children and Adults

Current recommendations are that dietary restriction of galactose needs to be continued for life. Single reports and anecdotal information suggest that children and/or adults may suffer cataracts,[69] liver disease,[70] and organic brain disease with psychiatric manifestations[71] following the ingestion of lactose. Milligram amounts of galactose cause an appreciable rise of galactose-1-phosphate in erythrocytes (e.g., ~500 mg of galactose in a 70 kg adult with Q188R/Q188R genotype will acutely increase galactose-1-phosphate by 30% in 8 hours); it is possible that the same happens in sensitive tissues, such as brain, liver, and kidney. However, it is not possible to define toxic tissue levels of galactose-1-phosphate and, therefore, safe amounts of dietary galactose cannot be determined. Patients with relatively increased alternate metabolic pathway activities should have greater tolerance for galactose. There are reported cases of classic galactosemia, as well as a few anecdotal cases, where stopping dietary restriction after early childhood has resulted in outcomes that are no worse than those of patients who have continued treatment.[72,73] In a few studies, administration of galactose resulted in no adverse effects. Relaxation of the diet in older patients requires further study.

Biochemical Monitoring

Erythrocyte galactose-1-P concentration has been the most common biochemical assay used to monitor treatment. The level is often very high at diagnosis and only falls gradually over weeks and months after initiating treatment. However even with good dietary compliance the concentration remains above normal. Its usefulness is open to question, as it has not been shown to correlate with outcome. Other galactose metabolites including red cell galactitol, urine galactitol, and red cell galactonate are also consistently increased in classic galactosemia and have been suggested as alternative or additional markers.[74] However, there are currently no data available to demonstrate their superiority over galactose-1-P. Analysis of the glycosylation of plasma transferrin from untreated newborn infants indicates both an assembly and processing congenital disorders of glycosylation (CDG)-like pattern.[75,76]

Long-Term Outcome and Complications

Despite the rapid clinical response to lactose exclusion in newborn infants with classic galactosemia long-term complications are common.[19,21,24,26,77] These appear to be largely independent of the severity of any initial illness and the strictness of dietary compliance.[24,48] Mild growth retardation, decreased bone density, delayed speech development, verbal dyspraxia, difficulties in spatial orientation and visual perception, and intellectual deficits have been variably described as complications of treated galactosemia.[19,24,26,32,33,37,78–80] Tremor, ataxia, dystonia, and choreatic movements are also reported. The quality of life in treated patients has been unfavorably compared to phenylketonuria (PKU).[81] Some patients, males in particular, manifest an introverted personality, anxiety and/or depression.[71]

The complete set of sequelae is not necessarily present in every patient, and the degree of handicap appears to vary widely. A minority of patients have developed more severe neurological disease with cerebellar dysfunction. The brain MRI may reveal white matter disease, cerebral and cerebellar atrophy.[42] Fluorodeoxyglucose positron emission tomography (FDG-PET) scanning of the brain has shown altered cerebral glucose metabolism in certain areas although the abnormalities have been highly variable.[82]

Although the majority of these late complications cannot be prevented, early identification of decreased bone mass by dual-energy X-ray absorptiometry (DXA) and advice as to physical activity and ensuring an adequate intake of vitamin D, vitamin K_1, and calcium and estrogen supplementation in girls, may help to reduce the risk of developing osteoporosis.[83,84]

Fertility and Pregnancy

Premature ovarian insufficiency occurs in almost all women with classic galactosemia, and presents clinically with delayed menarche, primary or secondary amenorrhea or oligomenorrhea.[23,30,31] In contrast, male gonadal function has not been shown to be significantly affected.[85] The cause of ovarian damage remains unclear but it is often signaled early in infancy or childhood by hypergonadotropism[23] with a perturbation in granulosa cell function, as evidenced by reduced circulating levels of AMH.[53] Since in female patients the number of expected ovulatory cycles is limited, it may be wise to temporarily suppress cycles by birth-control medication, which is lifted when the young woman wishes to become pregnant. This is not an established form of therapy, in contrast with chronic estrogen and progesterone supplementation. Prescription is hampered by the fact that seemingly all drug tablets contain lactose, providing 100 mg or more per treatment day.[86] However, some female patients have received the birth-control medication containing galactose for many years without any obvious side effects.[71] Women with classic galactosemia, including those with Q188R/Q188R genotype, have experienced one or more successful pregnancies and deliveries; some of them subsequently developed secondary amenorrhea. Galactose metabolite concentrations do not appear to increase significantly during pregnancy, although levels show a transient rise after delivery, peaking within the first week and subsequently falling even in those who chose to breastfeed.[87] Those infants born to mothers with classic galactosemia have been normal.

Dietary Treatment in Pregnant Women at Risk

Based on the presumption that toxic metabolites deriving from galactose ingested by the heterozygous mother accumulate in the galactosemic fetus, mothers may be counseled to refrain from or limit drinking milk for the duration of pregnancy. However, despite dietary restriction by the mother, galactose-1-phosphate and galactitol accumulate in the fetus,[88,89] and in the amniotic fluid,[63] presumably due to endogenous synthesis. Furthermore, the outcome for infants whose mothers restricted milk intake in pregnancy was no better than those who did not.[78]

Management of Partial Transferase Deficiency due to D/G Genotype

Infants with Duarte D_2 galactosemia are frequently detected on NBS. Some centers have adopted a pragmatic approach, prescribing a lactose-free formula to all infants discovered by newborn screening for 1–4 months after birth until erythrocyte galactose-1-phosphate levels normalize and remain so on a regular lactose-containing diet. This transition may be initiated with a lactose challenge. For example, if at the end of a 1-week trial with a daily supplement of formula containing lactose the erythrocyte galactose-1-phosphate level is below 1 mg per dL, the infant will be returned to normal nutrition. Other centers opt for 1 year of treatment and utilize a 1-month challenge with cow's milk. Most clinicians believe that the large majority of evidence indicates that D/G galactosemia is a benign condition: in countries without screening for galactosemia, D/G galactosemia is not reported associated with any clinical symptoms, and a recent study of 30 children between 1 and 6 years of age did not demonstrate any developmental or clinical pathology.[90] However, some controversy still remains; another study has reported an excess of affected children who had specific speech and language problems when compared to the normal population.[91]

Heterozgotes for Classic Galactosemia

Heterozygotes for GALT deficiency have not been shown to be at increased risk from premature menopause, presenile cataracts or other disease manifestations.[92]

Future Treatments

Work is in progress to try and develop a specific and safe inhibitor of the galactokinase (GALK) enzyme.[93] If successful, this may allow classic galactosemia to be transformed into a disease process that would more resemble GALK deficiency. If that were the case, patients may greatly benefit but still be at risk for cataract development. Along with the use of a lactose-restricted diet, patients may benefit from the use of aldose reductase inhibitor (ARI) to block galactose conversion to galactitol.[94]

This is important as the synthesis and accumulation of galactitol within the ocular lens is thought to be the biochemical basis of cataractogenesis.[95]

URIDINE DIPHOSPHATE-GALACTOSE 4′-EPIMERASE DEFICIENCY

Clinical Presentation

GALE deficiency varies from an asymptomatic condition only detected as a result of newborn screening for GALT deficiency, to a severe disorder that presents with a life-threatening illness in the newborn period. The severe form of GALE deficiency is extremely rare, with a total of six patients from two consanguineous Pakistani families and one Indian family known.[96] In those infants exposed to lactose, the initial clinical presentation has been similar to GALT deficiency with jaundice, weight loss, hypotonia and vomiting, generalized aminoaciduria and galactosuria. Learning difficulties, sensorineural hearing loss, growth delay, micrognathia, ligamentous laxity, and contractures of fingers have been present in the majority of patients. One child died from unexplained liver failure at 4 months of age. Unlike GALT deficiency, ovarian function is not affected. A few patients with an intermediate form have also been described with findings ranging from a transient illness with seizures, vomiting, and hypoglycemia in response to lactose ingestion, developmental delay, to juvenile cataracts.[97]

Metabolic Derangement

GALE is essential for the normal biosynthesis of glycoproteins and glycolipids. Complete GALE deficiency is likely to be incompatible with life; even patients with the severe form of the disease have some residual tissue enzyme activity. Asymptomatic individuals have enzyme activity reduced only in red cells (peripheral GALE deficiency) whereas in intermediate and severe forms the reduction is more widespread (generalized GALE deficiency). Red cell GALE activity does not correlate well with that in lymphoblasts and is poor at differentiating between peripheral and generalized forms of the disease. GALE deficiency leads to an accumulation of galactose, galactitol and galactose-1-phosphate but in contrast to GALT deficiency an increase in UDP-galactose and a decrease in UDP-glucose (Figure 55.2). As in GALT deficiency,[75] abnormal glycosylation of proteins, that appears to be dependent, at least in part, on lactose consumption, has been reported in severe GALE deficiency[96] and is thought to be a secondary biochemical complication, not primarily related to the genetic defect.

Genetics

Epimerase deficiency is inherited as an autosomal recessive trait. The true incidence is not known but the peripheral form appears to be approximately 10 times more common in African Americans than other groups. The epimerase gene is on chromosome 1.[98] A number of mutations have been identified[74,97–102] and characterized.

The V94M mutation was present in a homozygous form in all of the patients tested with the severe phenotype[102] whereas other mutations are associated with the intermediate or asymptomatic phenotype.[97,103]

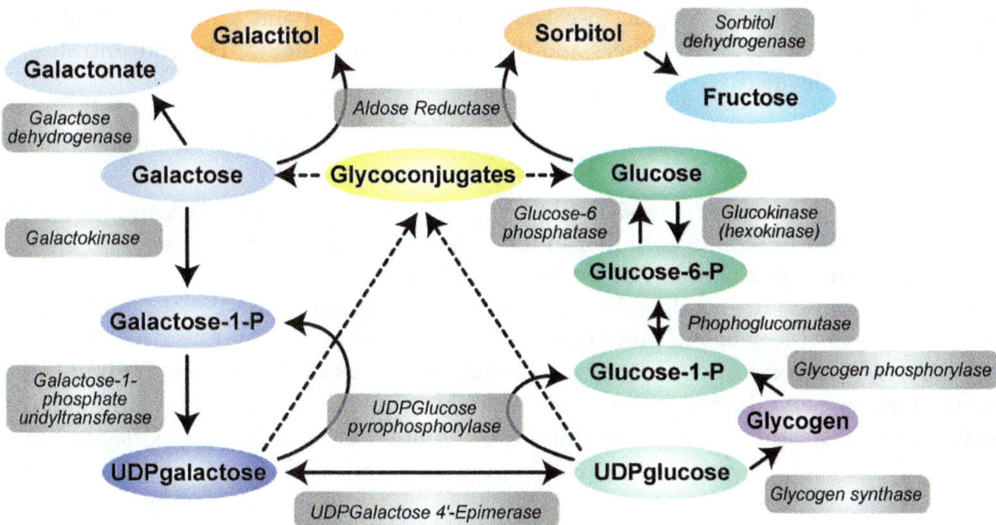

FIGURE 55.2 **An overview of whole-body galactose metabolism.**

Diagnostic Tests

GALE deficiency should be suspected when red cell galactose-1-phosphate is increased while GALT is normal. Newborn screening will give an abnormal result if defined by a raised total blood galactose level with normal GALT activity. Diagnosis is confirmed by the assay of epimerase in erythrocytes. Heterozygous parents have reduced epimerase activity, a finding that can help in the evaluation. Further studies of GALE activity in transformed lymphoblasts and red cell galactose-1-phosphate while on and off dietary galactose may help characterize the disorder further.[97] In those families with the severe form of GALE deficiency, GALE gene sequencing has been the most rapid method of determining whether infants at risk are affected or not.

Treatment and Prognosis

Newborns at risk from the severe form of GALE deficiency must have lactose excluded from their diet until the results of diagnostic tests are available. It is unclear whether infants diagnosed with this form of the disorder require a small amount of exogenous galactose for synthesis of galactoproteins and galactolipids. As with GALT deficiency, red cell galactose-1-phosphate levels have not been particularly helpful for long-term monitoring; concentrations have remained raised on treatment but to a lesser degree than those with GALT deficiency. The oldest patients homozygous for V94M mutation are now in their third decade; they have not shown evidence of progressive disease.

True peripheral GALE deficiency does not require galactose restriction. However since intermediate forms are now recognized, measurement of red cell galactose-1-phosphate and urine reducing substances, while on a normal galactose intake, and monitoring of psychomotor progress, may be advisable.

GALACTOKINASE DEFICIENCY

Clinical Presentation

Cataracts are the only consistent manifestation of untreated galactokinase (GALK) deficiency,[104] though pseudotumor cerebri has been described.[105] Liver, kidney and brain damage are not usually seen.

Metabolic Derangement

Persons with GALK deficiency lack the ability to phosphorylate galactose (Figure 55.2). and consequently accumulate galactose and galactitol, the latter being formed from galactose by aldose reductase. Individuals on a normal galactose intake have high levels of galactose and galactitol in blood and urine. Galactitol also accumulates in the lens,[95] causing osmotic swelling of lens fibers and denaturation of proteins. The concentrations of galactose-1-phosphate in tissues are not elevated.

Genetics

The mode of inheritance is autosomal recessive. In most parts of Europe, in the USA and Japan, birth incidence is in the order of 1 in 150,000 to 1 million. It is higher in the Balkan countries,[106] the former Yugoslavia, Romania, and Bulgaria. In Gypsies, birth incidence was calculated as one in 2,500.

Two genes have been reported to encode galactokinase: GK1 on chromosome 17q24[107] and GK2 on chromosome 15.[108] GK1 mutations cause GALK deficiency and many have now been described.[109–117] The GK1 P28T mutation was identified as the founder mutation responsible for galactokinase deficiency in Gypsies[109,114] and in immigrants from Bosnia in Berlin. The GK2 gene product is primarily a *N*-acetylgalactosamine kinase and although it does have some galactokinase activity this is insufficient to prevent the accumulation of galactose and galactitol in those with GK1 mutations.

Diagnostic Tests

Provided they have been fed breast milk or a lactose-containing formula prior to the test, newborns with the defect are discovered by NBS methods for detecting elevated blood galactose. Every person with nuclear cataracts ought to be examined for GALK deficiency. The final diagnosis is made by assaying GALK activity in red cell lysates.

Treatment and Prognosis

Treatment may be limited to the elimination of milk from the diet. Minor sources of galactose, such as milk products, green vegetables, legumes, drugs in tablet form, etc., can probably be disregarded, since it can be assumed that the small amounts of ingested galactose are either metabolized or excreted before significant amounts of galactitol can be formed. When diagnosis is made rapidly and treatment begun promptly, i.e., during the first 2–3 weeks of life, cataracts can clear. When treatment is late, and cataracts too dense, they will not clear completely (or at all) and must be removed surgically.

As in carriers with GALT deficiency,[118] the speculation that heterozygosity for GALK deficiency predisposes to the formation of presenile cataracts[119] remains unproven.[120] It has been suggested that heterozygotes restrict their milk intake,[119] though scientific proof of the merits of this measure is lacking.

FANCONI–BICKEL SYNDROME

This is a recessively inherited disorder of glucose and galactose transport due to glucose transporter 2 (GLUT2) deficiency and is rare.[121] A few cases have been discovered during newborn screening for galactose in blood. The clinical features of the disorder are those of glycogen storage disease and renal tubular dysfunction. Diagnosis is confirmed by mutation analysis.

PORTOSYSTEMIC VENOUS SHUNTING AND HEPATIC ARTERIOVENOUS MALFORMATIONS

Portosystemic bypass of splanchnic blood via ductus venosus Arantii[122] or intrahepatic shunts[123,124] causes alimentary hypergalactosemia, which is discovered during metabolic newborn screening.

References

1. Fridovich-Keil JL, Walter JH. Galactosemia. In: Valle D, Beaudet AL, Vogelstein B, et al., eds. *OMMBID—The Online Metabolic and Molecular Bases of Inherited Diseases*. New York, NY: McGraw-Hill; 2013.
2. Berry GT, Walter JH. Disorders of galactose metabolism. In: Fernandes J, Saudubray M, Van der Berghe G, Walter JH, eds. *Inborn Metabolic Diseases: Diagnosis and Treatment*. 5th ed. Heidelberg, Germany: Springer-Verlag; 2012.

3. von Reuss A. Zuckerausscheidung im Sauglingsalter. *Wein Med Wochenschr*. 1908;(58):799–801.
4. Göppert F. Galaktosurie Nach Milchzuckergabe Bei Angeborenum, Familiarem Chronicsem Leberleiden. *Klin Wochenschr*. 1917;(54):473–474.
5. Mason HH, Turner ME. Chronic galactosemia. *Am J Dis Child*. 1935;(50):359–363.
6. Schwarz V, Golberg L, Komrower GM, Holzel A. Some disturbances of erythrocyte metabolism in galactosaemia. *Biochem J*. 1956;62:34–40.
7. Isselbacher KJ, Anderson EP, Kurahashi K, Kalckar HM. Congenital galactosemia, a single enzymatic block in galactose metabolism. *Science*. 1956;(3198):635–636.
8. Reichardt JK, Berg P. Cloning and characterization of a cDNA encoding human galactose-1-phosphate uridyl transferase. *Mol Biol Med*. 1988;5(2):107–122.
9. Leslie ND, Immerman EB, Flach JE, Florez M, Fridovich-Keil JL, Elsas LJ. The human galactose-1-phosphate uridyltransferase gene. *Genomics*. 1992;14(2):474–480.
10. Elsas 2nd. LJ, Lai K. The molecular biology of galactosemia. *Genet Med*. 1998;1(1):40–48.
11. Tyfield LA. Galactosemia and allelic variation at the galactose-1-phosphate uridyltransferase gene: a complex relationship between genotype and phenotype. *Eur J Pediatr*. 2000;159(suppl 3):S204–S207.
12. Calderon FR, Phansalkar AR, Crockett DK, Miller M, Mao R. Mutation database for the galactose-1-phosphate uridyltransferase (GALT) gene. *Hum Mutat*. 2007;28(10):939–943.
13. Clouds over galactosaemia. *Lancet*. 1982;2(8312):1379–1380.
14. Holton JB, Leonard JV. Clouds still gathering over galactosaemia. *Lancet*. 1994;344(8932):1242–1243.
15. Komrower GM, Lee DH. Long-term follow-up of galactosaemia. *Arch Dis Child*. 1970;45(241):367–373.
16. Lee DH. Psychological aspects of galactosaemia. *J Intellect Disabil Res*. 1972;16(3):173–191.
17. Fishler K, Donnell GN, Bergren WR, Koch R. Intellectual and personality development in children with galactosemia. *Pediatrics*. 1972;50(3):412–419.
18. Fishler K, Koch R, Donnell GN, Wenz E. Developmental aspects of galactosemia from infancy to childhood. *Clin Pediatr*. 1980;(1):38–44.
19. Komrower GM. Galactosemia: thirty years on. The experiences of a generation. FP Hudson Memorial Lecture. *J Inherit Metab Dis*. 1982;5(suppl 2):96–104.
20. Waisbren SE, Norman TR, Schnell RR, Levy HL. Speech and language deficits in early-treated children with galactosemia. *J Pediatr*. 1983;102(1):75–77.
21. Gitzelmann R, Steinmann B. Galactosemia: how does long-term treatment change the outcome? *Enzyme*. 1984;2(1):37–46.
22. Levy HL, Sepe SJ, Shih VE, Vawter GF, Klein JO. Sepsis due to *Escherichia coli* in neonates with galactosemia. *N Engl J Med*. 1977;297(15):823–825.
23. Kaufman FR, Kogut MD, Donnell GN, Goebelsmann U, March C, Koch R. Hypergonadotropic hypogonadism in female patients with galactosemia. *N Engl J Med*. 1981;304(17):994–998.
24. Waggoner DD, Buist NR, Donnell GN. Long-term prognosis in galactosaemia: results of a survey of 350 cases. *J Inherit Metab Dis*. 1990;13(6):802–818.
25. Nelson CD, Waggoner DD, Donnell GN, Tuerck JM, Buist NR. Verbal dyspraxia in treated galactosemia. *Pediatrics*. 1991;88(2):346–350.
26. Schweitzer S, Shin Y, Jakobs C, Brodehl J. Long-term outcome in 134 patients with galactosaemia. *Eur J Pediatr*. 1993;152(1):36–43.
27. Guerrero NV, Singh RH, Manatunga A, Berry GT, Steiner RD, Elsas 2nd. LJ. Risk factors for premature ovarian failure in females with galactosemia. *J Pediatr*. 2000;137(6):833–841.
28. Robertson A, Singh RH, Guerrero NV, Hundley M, Elsas LJ. Outcomes analysis of verbal dyspraxia in classic galactosemia. *Genet Med*. 2000;2(2):142–148.
29. Webb AL, Singh RH, Kennedy MJ, Elsas LJ. Verbal dyspraxia and galactosemia. *Pediatr Res*. 2003;53(3):396–402.
30. Gubbels CS, Land JA, Rubio-Gozalbo ME. Fertility and impact of pregnancies on the mother and child in classic galactosemia. *Obstet Gynecol Surv*. 2008;63(5):334–343.
31. Berry GT. Galactosemia and amenorrhea in the adolescent. *Ann N Y Acad Sci*. 2008;1135:112–117.
32. Potter NL, Lazarus JA, Johnson JM, Steiner RD, Shriberg LD. Correlates of language impairment in children with galactosaemia. *J Inherit Metab Dis*. 2008;31(4):524–532.
33. Kaufman FR, Loro ML, Azen C, Wenz E, Gilsanz V. Effect of hypogonadism and deficient calcium intake on bone density in patients with galactosemia. *J Pediatr*. 1993;123(3):365–370.
34. Rubio-Gozalbo ME, Hamming S, van Kroonenburgh MJ, Bakker JA, Vermeer C, Forget PP. Bone mineral density in patients with classic galactosaemia. *Arch Dis Child*. 2002;87(1):57–60.
35. Leslie ND. Insights into the pathogenesis of galactosemia. *Annu Rev Nutr*. 2003;23:59–80.
36. Bosch AM. Classical galactosaemia revisited. *J Inherit Metab Dis*. 2006;29(4):516–525.
37. Panis B, Gerver WJ, Rubio-Gozalbo ME. Growth in treated classical galactosemia patients. *Eur J Pediatr*. 2007;166(5):443–446.
38. Crome LA. Case of galactosaemia with the pathological and neuropathological findings. *Arch Dis Child*. 1962;37:415–421.
39. Huttenlocher PR, Hillman RE, Hsia YE. Pseudotumor cerebri in galactosemia. *J Pediatr*. 1970;76(6):902–905.
40. Haberland C, Perou M, Brunngraber EG, Hof H. The neuropathology of galactosemia. A histopathological and biochemical study. *J Neuropathol Exp Neurol*. 1971;30(3):431–447.
41. Jan JE, Wilson RA. Unusual late neurological sequelae in galactosaemia. *Dev Med Child Neurol*. 1973;15(1):72–74.
42. Nelson Jr MD, Wolff JA, Cross CA, Donnell GN, Kaufman FR. Galactosemia: evaluation with MR imaging. *Radiology*. 1992;184(1):255–261.
43. Belman AL, Moshe SL, Zimmerman RD. Computed tomographic demonstration of cerebral edema in a child with galactosemia. *Pediatrics*. 1986;78(4):606–609.
44. Bohles H, Wenzel D, Shin YS. Progressive cerebellar and extrapyramidal motor disturbances in galactosaemic twins. *Eur J Pediatr*. 1986;145(5):413–417.
45. Koch TK, Schmidt KA, Wagstaff JE, Ng WG, Packman S. Neurologic complications in galactosemia. *Pediatr Neurol*. 1992;8(3):217–220.
46. Kaufman FR, McBride-Chang C, Manis FR, Wolff JA, Nelson MD. Cognitive functioning, neurologic status and brain imaging in classical galactosemia. *Eur J Pediatr*. 1995;154(7 suppl 2):S2–S5.
47. Ridel KR, Leslie ND, Gilbert DL. An updated review of the long-term neurological effects of galactosemia. *Pediatr Neurol*. 2005;33(3):153–161.

48. Hughes J, Ryan S, Lambert D, et al. Outcomes of siblings with classical galactosemia. *J Pediatr*. 2009;154(5):721–726.
49. Berry GT, Nissim I, Mazur AT, et al. In vivo oxidation of [^{13}C]galactose in patients with galactose-1-phosphate uridyltransferase deficiency. *Biochem Mol Med*. 1995;56(2):158–165.
50. Berry GT, Nissim I, Gibson JB, et al. Quantitative assessment of whole body galactose metabolism in galactosemic patients. *Eur J Pediatr*. 1997;156(suppl 1):S43–S49.
51. Berry GT, Singh RH, Mazur AT, et al. Galactose breath testing distinguishes variant and severe galactose-1-phosphate uridyltransferase genotypes. *Pediatr Res*. 2000;48(3):323–328.
52. Barbouth DS, Velazquez DL, Konopka S, Wilkinson JJ, Carver VH, Elsas LJ. Screening newborns for galactosemia using total body galactose oxidation to CO_2 in expired air. *Pediatr Res*. 2007;62(6):720–724.
53. Sanders RD, Spencer JB, Epstein MP, et al. Biomarkers of ovarian function in girls and women with classic galactosemia. *Fertil Steril*. 2009;92(1):344–351.
54. Manga N, Jenkins T, Jackson H, Whittaker DA, Lane AB. The molecular basis of transferase galactosaemia in South African negroids. *J Inherit Metab Dis*. 1999;22(1):37–42.
55. Lai K, Langley SD, Singh RH, Dembure PP, Hjelm LN, Elsas 2nd. LJ. A prevalent mutation for galactosemia among black Americans. *J Pediatr*. 1996;128(1):89–95.
56. Tyfield L, Reichardt J, Fridovich-Keil J, et al. Classical galactosemia and mutations at the galactose-1-phosphate uridyl transferase (GALT) gene. *Hum Mutat*. 1999;13(6):417–430.
57. Berry GT, Leslie N, Reynolds R, Yager CT, Segal S. Evidence for alternate galactose oxidation in a patient with deletion of the galactose-1-phosphate uridyltransferase gene. *Mol Genet Metab*. 2001;72(4):316–321.
58. Henderson H, Leisegang F, Brown R, Eley B. The clinical and molecular spectrum of galactosemia in patients from the Cape Town region of South Africa. *BMC Pediatr*. 2002;2:7.
59. Shield JP, Wadsworth EJ, MacDonald A, et al. The relationship of genotype to cognitive outcome in galactosaemia. *Arch Dis Child*. 2000;83(3):248–250.
60. Carney AE, Sanders RD, Garza KR, et al. Origins, distribution and expression of the Duarte-2 (D2) allele of galactose-1-phosphate uridylyltransferase. *Hum Mol Genet*. 2009;18(9):624–1632.
61. Crushell E, Chukwu J, Mayne P, Blatny J, Treacy EP. Negative screening tests in classical galactosaemia caused by S135L homozygosity. *J Inherit Metab Dis*. 2009;32(3):412–415.
62. Jakobs C, Kleijer WJ, Allen J, Holton JB. Prenatal diagnosis of galactosemia. *Eur J Pediatr*. 1995;154(7 suppl 2):S33–S36.
63. Jakobs C, Kleijer WJ, Bakker HD, van Gennip AH, Przyrembel H, Niermeijer MF. Dietary restriction of maternal lactose intake does not prevent accumulation of galactitol in the amniotic fluid of fetuses affected with galactosaemia. *Prenat Diagn*. 1988;8(9):641–645.
64. Steffen C. Enzymatische Bestimmungsmethoden zur Erfassung der Grünungsvorgänge in der milchwirtschaftlichen Technologie. *Lebensm Wiss Technol*. 1975;8:16.
65. Portnoi PA, MacDonald A, Portnoi PA, MacDonald A. Determination of the lactose and galactose content of cheese for use in the galactosaemia diet. *J Hum Nutr Diet*. 2009;22(5):400–408.
66. Acosta PB, Gross KC. Hidden sources of galactose in the environment. *Eur J Pediatr*. 1995;154(7 suppl 2):S87–S92.
67. Wiesmann UN, Rose-Beutler B, Schluchter R. Leguminosae in the diet: the raffinose-stachyose question. *Eur J Pediatr*. 1995;154(7 suppl 2): S93–S96.
68. Gitzelmann R, Auricchio S. The handling of soya α-galactosides by a normal and a galactosemic child. *Pediatrics*. 1965;36:231–235.
69. Beigi B, O'Keefe M, Bowell R, Naughten E, Badawi N, Lanigan B. Ophthalmic findings in classical galactosaemia—prospective study. *Br J Ophthalmol*. 1993;77(3):162–164.
70. Vogt M, Gitzelmann R, Allemann J. Decompensated liver cirrhosis caused by galactosemia in a 52-year-old man. *Schweiz Med Wochenschr*. 1980;110(47):1781–1783.
71. Berry GT. Unpublished observations.
72. Lee PJ, Lilburn M, Wendel U, Schadewaldt P. A woman with untreated galactosaemia. *Lancet*. 2003;362(9382):446.
73. Panis B, Bakker JA, Sels JP, Spaapen LJ, van Loon LJ, Rubio-Gozalbo ME. Untreated classical galactosemia patient with mild phenotype. *Mol Genet Metab*. 2006;89(3):277–279.
74. Yager CT, Chen J, Reynolds R, Segal S. Galactitol and galactonate in red blood cells of galactosemic patients. *Mol Genet Metab*. 2003;80(3):283–289.
75. Pronicka E, Adamowicz M, Kowalik A, et al. Elevated carbohydrate-deficient transferrin (CDT) and its normalization on dietary treatment as a useful biochemical test for hereditary fructose intolerance and galactosemia. *Pediatr Res*. 2007;62(1):101–105.
76. Sturiale L, Barone R, Fiumara A, et al. Hypoglycosylation with increased fucosylation and branching of serum transferrin *N*-glycans in untreated galactosemia. *Glycobiology*. 2005;15(12):1268–1276.
77. Holton JB. Galactosaemia: pathogenesis and treatment. *J Inherit Metab Dis*. 1996;19(1):3–7.
78. Manis FR, Cohn LB, McBride-Chang C, Wolff JA, Kaufman FR. A longitudinal study of cognitive functioning in patients with classical galactosaemia, including a cohort treated with oral uridine. *J Inherit Metab Dis*. 1997;20(4):549–555.
79. Schadewaldt P, Hoffmann B, Hammen HW, Kamp G, Schweitzer-Krantz S, Wendel U. Longitudinal assessment of intellectual achievement in patients with classical galactosemia. *Pediatrics*. 2010;125(2):e374–e381.
80. Bosch AM, Grootenhuis MA, Bakker HD, Heijmans HS, Wijburg FA, Last BF. Living with classical galactosemia: health-related quality of life consequences. *Pediatrics*. 2004;113(5):e423–e428.
81. Bosch AM, Maurice-Stam H, Wijburg FA, Grootenhuis MA. Remarkable differences: the course of life of young adults with galactosaemia and PKU. *J Inherit Metab Dis*. 2009;32(6):706–712.
82. Dubroff JG, Ficicioglu C, Segal S, Wintering NA, Alavi A, Newberg AB. FDG-PET findings in patients with galactosaemia. *J Inherit Metab Dis*. 2008;31(4):533–539.
83. Panis B, van Kroonenburgh MJ, Rubio-Gozalbo ME. Proposal for the prevention of osteoporosis in paediatric patients with classical galactosaemia. *J Inherit Metab Dis*. Nov 2007;30(6):982.

84. Panis B, Vermeer C, van Kroonenburgh MJ, et al. Effect of calcium, vitamins K1 and D3 on bone in galactosemia. *Bone*. 2006;39(5):1123–1129.
85. Rubio-Gozalbo ME, Gubbels CS, Bakker JA, Menheere PP, Wodzig WK, Land JA. Gonadal function in male and female patients with classic galactosemia. *Hum Reprod Update*. 2010;16(2):177–188.
86. Berry GT, Palmieri M, Gross KC, et al. The effect of dietary fruits and vegetables on urinary galactitol excretion in galactose-1-phosphate uridyltransferase deficiency. *J Inherit Metab Dis*. 1993;16(1):91–100.
87. Schadewaldt P, Hammen HW, Kamalanathan L, et al. Biochemical monitoring of pregnancy and breast feeding in five patients with classical galactosaemia—and review of the literature. *Eur J Pediatr*. 2009;168(6):721–729.
88. Irons M, Levy HL, Pueschel S, Castree K. Accumulation of galactose-1-phosphate in the galactosemic fetus despite maternal milk avoidance. *J Pediatr*. 1985;107(2):261–263.
89. Holton JB. Effects of galactosemia *in utero*. *Eur J Pediatr*. 1995;154(7 suppl 2):S77–S81.
90. Ficicioglu C, Thomas N, Yager C, et al. Duarte (DG) galactosemia: a pilot study of biochemical and neurodevelopmental assessment in children detected by newborn screening. *Mol Genet Metab*. 2008;95(4):206–212.
91. Powell KK, Van Naarden Braun K, Singh RH, Shapira SK, Olney RS, Yeargin-Allsopp M. Long-term speech and language developmental issues among children with Duarte galactosemia. *Genet Med*. 2009;11(12):874–879.
92. Knauff EA, Richardus R, Eijkemans MJ, et al. Heterozygosity for the classical galactosemia mutation does not affect ovarian reserve and menopausal age. *Reprod Sci*. 2007;14(8):780–785.
93. Tang M, Wierenga K, Elsas LJ, Lai K. Molecular and biochemical characterization of human galactokinase and its small molecule inhibitors. *Chem Biol Interact*. 2010;188(3):376–385.
94. Berry GT. The role of polyols in the pathophysiology of hypergalactosemia. *Eur J Pediatr*. 1995;154(7 suppl 2):S53–S64.
95. Ai Y, Zheng Z, O'Brien-Jenkins A, et al. A mouse model of galactose-induced cataracts. *Hum Mol Genet*. 2000;9(12):1821–1827.
96. Walter JH, Roberts RE, Besley GT, et al. Generalised uridine diphosphate galactose-4-epimerase deficiency. *Arch Dis Child*. 1999;80(4):374–376.
97. Openo KK, Schulz JM, Vargas CA, et al. Epimerase-deficiency galactosemia is not a binary condition. *Am J Hum Genet*. 2006;78(1):89–102.
98. Daude N, Gallaher TK, Zeschnigk M, et al. Molecular cloning, characterization, and mapping of a full-length cDNA encoding human UDP-galactose 4′-epimerase. *Biochem Mol Med*. 1995;56(1):1–7.
99. Maceratesi P, Daude N, Dallapiccola B, et al. Human UDP-galactose 4′-epimerase (GALE) gene and identification of five missense mutations in patients with epimerase-deficiency galactosemia. *Mol Genet Metab*. 1998;63(1):26–30.
100. Alano A, Almashanu S, Chinsky JM, et al. Molecular characterization of a unique patient with epimerase-deficiency galactosaemia. *J Inherit Metab Dis*. 1998;21(4):341–350.
101. Wohlers TM, Christacos NC, Harreman MT, Fridovich-Keil JL. Identification and characterization of a mutation, in the human UDP-galactose-4-epimerase gene, associated with generalized epimerase-deficiency galactosemia. *Am J Hum Genet*. 1999;64(2):462–470.
102. Wohlers TM, Fridovich-Keil JL. Studies of the V94M-substituted human UDPgalactose-4-epimerase enzyme associated with generalized epimerase-deficiency galactosaemia. *J Inherit Metab Dis*. 2000;23(7):713–729.
103. Wasilenko J, Lucas ME, Thoden JB, Holden HM, Fridovich-Keil JL. Functional characterization of the K257R and G319E-hGALE alleles found in patients with ostensibly peripheral epimerase deficiency galactosemia. *Mol Genet Metab*. 2005;84(1):32–38.
104. Bosch AM, Bakker HD, van Gennip AH, van Kempen JV, Wanders RJ, Wijburg FA. Clinical features of galactokinase deficiency: a review of the literature. *J Inherit Metab Dis*. 2002;25(8):629–634.
105. Litman N, Kanter AI, Finberg L. Galactokinase deficiency presenting as pseudotumor cerebri. *J Pediatr*. 1975;86(3):410–412.
106. Reich S, Hennermann J, Vetter B, et al. An unexpectedly high frequency of hypergalactosemia in an immigrant Bosnian population revealed by newborn screening. *Pediatr Res*. 2002;51(5):598–601.
107. Stambolian D, Ai Y, Sidjanin D, et al. Cloning of the galactokinase cDNA and identification of mutations in two families with cataracts. *Nat Genet*. 1995;10(3):307–312.
108. Lee RT, Peterson CL, Calman AF, Herskowitz I, O'Donnell JJ. Cloning of a human galactokinase gene (GK2) on chromosome 15 by complementation in yeast. *Proc Natl Acad Sci U S A*. 1992;89(22):10887–10891.
109. Kalaydjieva L, Perez-Lezaun A, Angelicheva D, et al. A founder mutation in the *GK1* gene is responsible for galactokinase deficiency in Roma (Gypsies). *Am J Hum Genet*. 1999;65(5):1299–1307.
110. Asada M, Okano Y, Imamura T, Suyama I, Hase Y, Isshiki G. Molecular characterization of galactokinase deficiency in Japanese patients. *J Hum Genet*. 1999;44(6):377–382.
111. Kolosha V, Anoia E, de Cespedes C, et al. Novel mutations in 13 probands with galactokinase deficiency. *Hum Mutat*. 2000;15(5):447–453.
112. Hunter M, Angelicheva D, Levy HL, Pueschel SM, Kalaydjieva L. Novel mutations in the GALK1 gene in patients with galactokinase deficiency. *Hum Mutat*. 2001;17(1):77–78.
113. Okano Y, Asada M, Fujimoto A, et al. A genetic factor for age-related cataract: identification and characterization of a novel galactokinase variant, "Osaka," in Asians. *Am J Hum Genet*. 2001;68(4):1036–1042.
114. Hunter M, Heyer E, Austerlitz F, et al. The P28T mutation in the GALK1 gene accounts for galactokinase deficiency in Roma (Gypsy) patients across Europe. *Pediatr Res*. 2002;51(5):602–606.
115. Timson DJ, Reece RJ. Functional analysis of disease-causing mutations in human galactokinase. *Eur J Biochem*. 2003;270(8):1767–1774.
116. Sangiuolo F, Magnani M, Stambolian D, Novelli G. Biochemical characterization of two GALK1 mutations in patients with galactokinase deficiency. *Hum Mutat*. 2004;23(4):396.
117. Park HD, Bang YL, Park KU, et al. Molecular and biochemical characterization of the GALK1 gene in Korean patients with galactokinase deficiency. *Mol Genet Metab*. 2007;91(3):234–238.
118. Karas N, Gobec L, Pfeifer V, Mlinar B, Battelino T, Lukac-Bajalo J. Mutations in galactose-1-phosphate uridyltransferase gene in patients with idiopathic presenile cataract. *J Inherit Metab Dis*. 2003;26(7):699–704.
119. Stambolian D, Scarpino-Myers V, Eagle Jr RC, Hodes B, Harris H. Cataracts in patients heterozygous for galactokinase deficiency. *Invest Ophthalmol Vis Sci*. 1986;27(3):429–433.
120. Maraini G, Hejtmancik JF, Shiels A, et al. Galactokinase gene mutations and age-related cataract. Lack of association in an Italian population. *Mol Vis*. 2003;9:397–400.

121. Chen Y, Kishnani PS, Koeberl D. Glycogen storage diseases. In: Valle D, Beaudet AL, Vogelstein B, et al., eds. *OMMBID—The Online Metabolic and Molecular Bases of Inherited Diseases*. New York, NY: McGraw-Hill; 2013.
122. Gitzelmann R, Arbenz UV, Willi UV. Hypergalactosaemia and portosystemic encephalopathy due to persistence of ductus venosus Arantii. *Eur J Pediatr*. 1992;151(8):564–568.
123. Matsumoto T, Okano R, Sakura N, et al. Hypergalactosaemia in a patient with portal-hepatic venous and hepatic arterio-venous shunts detected by neonatal screening. *Eur J Pediatr*. 1993;152(12):990–992.
124. Gitzelmann R, Forster I, Willi UV. Hypergalactosaemia in a newborn: self-limiting intrahepatic portosystemic venous shunt. *Eur J Pediatr*. 1997;156(9):719–722.

CHAPTER

56

Inborn Errors of Amino Acid Metabolism

William L. Nyhan and Richard Haas

The University of California, San Diego School of Medicine, La Jolla, CA, USA

PHENYLKETONURIA AND DISORDERS OF BIOPTERIN METABOLISM

Clinical Features

Historical Overview

Phenylketonuria (PKU) is the classic error of amino acid metabolism. It serves as a model for the control of genetic disease because screening programs in every developed country detect infants with elevated blood concentrations of phenylalanine. This permits effective early intervention and therefore prevents mental retardation, the most important consequence of the disease. In untreated PKU, the IQ is usually less than 30. Patients also have fair skin, blond hair, and blue eyes. Itching or an eczematoid rash may occur. Spastic paraplegia and seizures occur in some, and microcephaly in a few.[1] The electroencephalogram (EEG) is abnormal. Neuroimaging studies in patients with untreated PKU have revealed cerebral atrophy or calcification. There is also a family of disorders of biopterin metabolism in which phenylalanine cannot be converted to tyrosine, although the phenylalanine hydroxylase apoenzyme is normal. For these disorders, in which neurologic dysfunction is appreciably worse than in PKU, treatment is more demanding because of abnormalities in biopterin metabolism.

Disease Identification

The diagnosis of PKU is by a positive screening test for phenylalanine on a drop of whole blood collected on filter paper. This must be confirmed by a quantitative test for phenylalanine and tyrosine in the blood. In classic PKU, the phenylalanine is >1200 µmol per L, and the tyrosine is low. The urine gives a positive green-color test for phenylpyruvic acid on the addition of 10% $FeCl_3$. Diagnosis in patients with defective hydroxylase cofactor metabolism may be by the administration of tetrahydrobiopterin (BH_4). Tetrahydrobiopterin reduces phenylalanine levels in patients with biopterin cofactor defects, but it does not change phenylalanine levels in patients with PKU. Another diagnostic test is measurement of pterins in urine and dihydropteridine reductase in blood. Testing for biopterin defects is mandatory in any patient detected by newborn screening.

Disease Pathophysiology and Gene Identification

The gene for human phenylalanine hydroxylase has been cloned and localized to chromosome 12q.[2] The defective enzyme in PKU is phenylalanine hydroxylase, which normally is expressed only in liver. It converts phenylalanine to tyrosine. In its absence, phenylalanine accumulates and is converted to phenylpyruvic acid, phenylacetic acid, and phenylacetylglutamine. Phenylacetic acid accounts for the odor of the patient with untreated PKU.

Tetrahydrobiopterin is an essential cofactor for the phenylalanine hydroxylase enzyme. In the course of the reaction, BH_4 is oxidized to dihydrobiopterin, and it must be reduced with NADH in a reaction catalyzed by dihydropteridine reductase. Patients may have defects in this enzyme or in the synthesis of BH_4 catalyzed by guanosine 5′-triphosphate cyclohydrolase and dihydrobiopterin synthase.[1]

Rosenberg's Molecular and Genetic Basis of Neurological and Psychiatric Disease
http://dx.doi.org/10.1016/10.1016/B978-0-12-410529-4.00056-5

Molecular Genetics, Mode of Inheritance and Prevalence

PKU and the BH_4 variants are autosomal recessive. PKU occurs in about 1 in 10,000 to 1 in 15,000 births. Biopterin abnormalities are found in 1–2% of hyperphenylalaninemic individuals. At least 31 mutations in the human hydroxylase gene have been identified. Two mutations, expressed as zero activity and no cross-reacting material (CRM), accounted for almost half of the northern European patients studied: an arginine to tryptophan change in exon 12 and a splicing mutation in intron 12. Some 9 of 31 abnormal alleles involve cytosinephosphate-guanine dinucleotides (CpG), which are known to be hotspots for mutation.

Management

Clinical manifestations of PKU may be successfully prevented by detection in neonatal screening and restriction of dietary intake of phenylalanine.[3] The authors' clinic aims to maintain blood concentrations at 300 μmol. It is now clear that some patients with PKU respond favorably to BH_4 treatment.[4] These are patients with mutations causing relatively milder phenotypes. Mixtures of large neutral amino acids (PreKUnil, NeoPhe) have been used to compete with phenylalanine for transport into the brain; competition for transport across the intestine may also occur.

Patients with biopterin abnormalities are treated with BH_4.[5] In addition; therapy includes administration of dopamine, 5-hydroxytryptophan, and carbidopa. Carbidopa is necessary to inhibit peripheral decarboxylases that would prevent these compounds from entering the central nervous system.

HEPATORENAL TYROSINEMIA

Clinical Features

Hereditary hepatorenal tyrosinemia is caused by deficient activity of fumarylacetoacetate hydrolase.[6] In its most severe form it presents in early infancy with hepatocellular disease, hepatic failure, and severe coagulopathy. In more indolent forms of the disease renal tubular dysfunction may appear in the first years of life with hypophosphatemic rickets and renal Fanconi syndrome. End results of hepatic disease include advanced cirrhosis and hepatocellular carcinoma. Acute porphyria-like neurologic crises of pain and paresthesia resulting from peripheral neuropathy may lead to respiratory failure and death.[6] Prior to the use of hepatic transplantation there were no long-term survivors.

Gene Identification, Mode of Inheritance and Prevalence

The gene has been cloned[7] and mapped to chromosome 15q23-25. Hepatorenal tyrosinemia is transmitted in an autosomal recessive fashion.[8] Consanguinity has been documented in a number of families. A particularly high frequency of 1.46 per 1000 births has been recorded in a French-Canadian isolate in the Chicoutimi-Lac St. Jean region of northeastern Quebec, where the carrier rate is 1 in 20.[9] An overall incidence of 0.8 per 10,000 births was observed in the French-Canadian population of Quebec.

Testing

The pattern of laboratory findings in this disease is virtually unique. A combination of hypoglycemia, coagulopathy, tyrosinemia, succinylacetone and very high α-fetoprotein is diagnostic. The latter has been employed to monitor for the development of hepatocarcinoma.

Molecular Genetics

Mutations have been identified,[6] including founder mutations in French-Canadian Quebec and in Finland, where the disease is prevalent. The Quebec mutation is a splice mutation IVS12+5 g-a,[10] and that in Finland is W262X.[11]

The clinical course of hepatorenal tyrosinemia has generally followed one of two patterns: an acute or a chronic form. The former has sometimes been referred to as the French-Canadian type and the latter the Scandinavian type, but of course there is considerable overlap. Most patients have had acute presentations. Symptoms develop in early infancy, and they are those of acute hepatic decompensation. Hepatic failure and death occurs usually under 1 year of age. However, some infants with an acute onset of hepatic disease survive to go on to display a chronic disease just like those patients with the chronic form. Until recently, most of these children died at younger than 10 years of age.

The normal gene product is the hepatic fumarylacetoacetic acid hydrolase (fumarylacetoacetase, EC 3.7.1.2). This was originally proposed on the basis of the accumulation of succinylacetone. Deficiency of this enzyme was then documented by assay of activity in liver.

The deficient enzyme is on the catabolic pathway for tyrosine, and this is the cause of the hypertyrosinemia. Fumarylacetoacetone accumulates and is converted to succinylacetoacetate and to succinylacetone. In hepatorenal tyrosinemia, concentrations of tyrosine usually range from 170 to 660 μmol per L (3–12 mg per dL).

Quebec now screens newborns for succinylacetone using a liquid chromatography–tandem mass spectrometry (LC-MS/MS) method. It can also be used for monitoring treatment.[12]

Management

Initial approaches to treatment followed the phenylketonuria model with diets restricted in whole protein and supplements containing an amino acid mixture lacking phenylalanine and tyrosine, such as tyrex. Nevertheless, early death or liver transplantation was the usual result until the development of nitisinone.[13] Treatment with this inhibitor of 4-hydroxyphenylpyruvic acid dioxygenase has abolished the acute complications of the disease. In the extensive Quebec experience, this treatment, along with newborn screening for succinylacetone, has led to an absence of detectable hepatic disease for more than 5 years. Doses approximating 1 mg per kg abolish succinylacetone from the urine and produce levels of the drug approximating 50 μmol per L, and the drug can be given once daily.[14]

NONKETOTIC HYPERGLYCINEMIA

Clinical Features

In nonketotic hyperglycinemia (NKH), the enzyme responsible is usually expressed only in brain and liver, and the diagnosis is made based on accumulation of glycine in the cerebrospinal fluid. Most patients present in the early days of life with life-threatening illness.[15]

Hiccups are common and may occur prenatally. In the first days of life there is rapid progression to deep coma and respiratory arrest. Ventilatory assistance must be provided, or death ensues. A majority of patients die in the neonatal period, presumably many without diagnosis. In those who survive the early crisis, there is little evidence of psychomotor development. Intractable seizures are the rule. Neonates are hypotonic or flaccid. Eventually they become spastic, hyperreflexic, and often opisthotonic. Most are microcephalic.

Disease Variants

NKH is phenotypically heterogeneous. A small number of patients have atypical milder clinical presentations.[16] This disease may produce isolated mild mental retardation. It may also present as a neurodegenerative disease similar to Tay–Sachs or Krabbe disease.

Pathophysiology

Neuropathology reveals spongy changes in the myelin. This is the characteristic finding, particularly in older patients. On electron microscopy, vacuoles are within the myelin sheaths.

Molecular Genetics

The glycine cleavage system is comprised of four proteins, now designated, B, H, T, and L. The P protein contains pyridoxal phosphate and catalyzes the decarboxylation of glycine. The H protein is a lipoic acid-containing protein, which is the amino methyl carrier protein. The T protein is a tetrahydrofolate-requiring flavoprotein that transfers carbon-2 of glycine to tetrahydrofolate. The L protein is a dihydrolipoyl dehydrogenase that catalyzes the oxidation of the lipoic acid of the H protein to its disulfide. The overall reaction yields ammonia, carbon dioxide, and single carbon tetrahydrofolate, which in the presence of another molecule of glycine becomes the hydroxymethyl group of serine. The reaction is localized to the mitochondria.

The gene for the P protein cDNA has been cloned, and a number of mutations has been identified.[16–18] In one patient with a large deletion, no immunoreactive protein could be found. Another classic patient had complete absence of P protein activity, but had immunoreactive P protein as well as mRNA. In this patient, a three-base deletion was found leading to a deletion of the phenylalanine 756. The mutation in the classic disease found commonly in Finland has been identified as a G-to-T point mutation of nucleotide 1691, leading to a serine to leucine change at amino acid 566 of the P protein. In a study of 28 unrelated patients, more than 40 different gene mutations were identified,[18] highlighting the genetic heterogeneity of this disorder.

Testing

Concentrations of glycine in the plasma range from 6 to 12 mg per dL; concentrations in controls are 1 mg per dL. The excretion of glycine in the urine is enormous, often >1 g per day. The ratio of glycine in the cerebrospinal fluid to that of the plasma is used to make the diagnosis. In control individuals, the ratio is <0.02; in NKH, the ratio is >0.10.

A considerable number and variety of mutations have been identified and heterogeneity has been identified.[17,18]

All of the defects in the glycine cleavage reaction are inherited in autosomal recessive fashion. Defects in the P protein have been found in 80% of patients; and in the T protein in 15%; deficiency of the H protein is rare. Phenotypically, the classic neonatal type of NKH is usually associated with virtual absence of the P protein.

Clinical and Ancillary Testing

The classic abnormality of the EEG in NKH is a burst suppression pattern. Hypsarrhythmia has been encountered. Neuroimaging on computed tomography (CT) or magnetic resonance imaging (MRI) shows cerebral atrophy and white matter hypomyelination. The corpus callosum is abnormally thin.

Genotype–Phenotype Correlations

Most patients have had the classic neonatal severe disease.

Management

There is no satisfactory treatment for NKH. Administration of sodium benzoate in doses sufficient to decrease cerebrospinal fluid levels of glycine reduces otherwise intractable seizures.

Early Treatments

Dextromethorphan has been used with less than uniform effect. Benzoate treatment can lower levels of glycine, without clear clinical effect.

MAPLE SYRUP URINE DISEASE

Clinical Features

Natural History

In maple syrup urine disease (MSUD), the classic presentation is early neonatal hypertonicity and opisthotonus,[19] and coma in the first days of life, often before the results of newborn screening become available. Deep tendon reflexes are increased, and there may be clonus. Generalized seizures are common. Lethargy progresses rapidly to coma, apnea, and death, unless intubation and artificial ventilation are initiated promptly. The clinical phenotype is variable. Intermediate, intermittent, and thiamine-responsive variants have been described. All of these variants have more residual enzyme than the classic form. The disease is seldom intermittent; it is virtually always chemically recognizable. Only the acute symptoms are intermittent. Nevertheless, an acute episode in a variant patient may be fatal. It has been reported that milder cases of MSUD can be missed by newborn screening. The 2-ketoacid derivatives of these amino acids may be detected by organic acid analysis of urine.

All forms of branched-chain ketoaciduria result from deficiency of branched-chain α-ketoacid dehydrogenase.

Mode of Inheritance and Prevalence

All forms of the disease are autosomal recessive. The disease is common in Portuguese gypsies and the Mennonites. Incidence was 1 in 290,000 in the New England Newborn Screening Program. It is 1 in 760 in the Pennsylvania Mennonites.

MSUD can be detected in newborn screening by the elevation of leucine in spots of blood.[19] Confirmation of the diagnosis is by quantitative analysis of plasma concentrations of amino acids. Levels of leucine, valine, isoleucine, and alloisoleucine are elevated. A simple screening test on urine is to add dinitrophenylhydrazine, which makes a yellow precipitate in the presence of these ketoacids.

Molecular Genetics

The cDNA of each component has been cloned. The E1 gene is located on chromosome 19q13.1-13.2, and the E1β on 6p21-22. The E2 gene has been localized to chromosome 1. A majority of mutations have been in the E1 (α) and E2 genes.[20]

In the Mennonite population, the common mutation is a T-to-A change that produces a missense Tyr393→Asn change.[21] The same mutation was found in a compound with an eight-base deletion causing a downstream nonsense codon. In an intermediate phenotype patient, there was a compound of a missense and one frameshift in E1. Only one mutation was found in E1β, a deletion in Japanese patients.[22] Eleven mutations have been reported in E2.[23] Two missense mutations have been found in E3.[24] A thiamine-responsive patient had a 17-bp insertion and a Phe215→Cys missense mutation in the E2 mRNA.[25]

Testing

The enzyme complex is composed of three major subunits. Defects have been localized to the E1, E2, or E3 proteins. E3 deficiency also causes lactic acidemia. The activity of the dehydrogenase enzyme is usually measured in lymphocytes, fibroblasts, or lymphoblasts by measuring the conversion of ^{14}C-leucine to $^{14}CO_2$. There is some correlation between clinical severity and amount of residual enzyme activity. Heterozygotes can be detected and prenatal diagnosis accomplished by assay of cultured amniocytes or chorionic villus samples.

Clinical and Ancillary Testing

The EEG in MSUD may show generalized slowing or paroxysmal discharges. Neuroimaging on CT or MRI shows a generalized decrease in attenuation in the white matter, which resolves with treatment and time. There may be cerebral atrophy and, in acute coma, cerebral edema.

Disease Mechanisms

Best correlations with clinical manifestations of the disease are with the concentration of leucine in the plasma.

Management

A few patients have responded to treatment with large doses of thiamine. Therefore, newly diagnosed patients should be tested for thiamine responsiveness. Doses used have ranged between 10 and 300 mg per day.

The treatment of MSUD consists of a diet low in branched-chain amino acids. Preparations of amino acid mixtures free of branched-chain amino acids are helpful. Optimal treatment before damaging ketoacidotic episodes occur can promote normal brain development.

Acute episodes of coma should be treated aggressively. An anabolic approach is to use branched-chain amino acid-free parenteral or enteral nutrition to lay down branched-chain amino acids into protein.[26] Once the acute episode is controlled, protein and branched-chain amino acid intake is limited to provide the minimum quantity of branched-chain amino acids necessary for growth. Patients must be carefully followed by monitoring plasma amino acid concentrations repeatedly.

Liver transplantation has been found to eliminate crises of acute metabolic imbalance in this disease.[27,28] Experience indicates that this not only reduces the risk of death and neurologic disability but also markedly improves quality of life.

References

1. Smith I, Leeming RJ, Cavanagh NP, Hyland K. Neurological aspects of biopterin metabolism. *Arch Dis Child*. 1986;61:130–137.
2. Scriver CR, Waters PJ, Sarkissian C, et al. A locus specific knowledge base. *Hum Mutat*. 2000;15:99–104.
3. National Institute of Child Health Human Development. *Report of the NIH consensus development conference on phenylketonuria (PKU): screening and management*. Bethesda, MD: National Institutes of Health; 2001.

4. Seashore MR. Tetrahydrobiopterin and dietary restriction in mild phenylketonuria. *N Engl J Med*. 2002;347:2094–2095.
5. Smith I, Hyland K, Kendall B. Clinical role of pteridine therapy in tetrahydrobiopterin deficiency. *J Inherit Metab Dis*. 1985;8(suppl 1):39–45.
6. Phaneuf D, Labelle Y, Berube D, et al. Cloning and expression of the cDNA encoding human fumarylacetoacetate hydrolase, the enzyme deficient in hereditary tyrosinemia: assignment of the gene to chromosome 15. *Am J Hum Genet*. 1991;48:525–535.
7. Laberge C. Hereditary tyrosinemia in a French Canadian isolate. *Am J Hum Genet*. 1969;21:36–45.
8. De Braekeleer M, Larocheller J. Genetic epidemiology of hereditary tyrosinemia in Quebec and in Saguenay-Lac-St-Jean. *Am J Hum Genet*. 1990;47:302.
9. Phaneuf D, Lambert M, Laframboise R, Mitchell G, Lettre F, Tanguay RM. Type 1 hereditary tyrosinemia evidence for molecular heterogeneity and identification of a causal mutation in a French Canadian patient. *J Clin Invest*. 1992;90:1185–1192.
10. Grompe M, St-Louis M, Demers S, al-Dhalimy M, Leclerc B, Tanguay RM. A single mutation of the fumarylacetoacetate hydrolase gene in French Canadians with hereditary tyrosinemia type I. *N Engl J Med*. 1994;7:239.
11. Rootwelt H, Hoie K, Berger R, Kvittingen EA. Fumarylacetoacetase mutations in tyrosinemia type I. *Hum Mutat*. 1996;7:239–243.
12. Demers SI, Phaneuf D, Tanguay RM. Hereditary tyrosinemia type I: strong association with haplotype 6 in French Canadians permits simple carrier detection and prenatal diagnosis. *Am J Hum Genet*. 1994;55:327–333.
13. Larochelle J, Alvarez F, Bussieres J-F, et al. Effect of nitisinone (NTBC) treatment on the clinical course of hepatorenal tyrosinemia in Quebec. *Mol Genet Metab*. 2012;107:49–54.
14. Schlune A, Thimm E, Herebian D, Spiekerkoetter U. Single dose NTBC-treatment of hereditary tyrosinemia type I. *J Inherit Metab Dis*. 2012;35:831–836.
15. Tada K, Hayasaka K. Non-ketotic hyperglycinaemia: clinical and biochemical aspects. *Eur J Pediatr*. 1987;146:221–227.
16. Dinopoulos A, Matsubara Y, Kure S. Atypical variants of nonketotic hyperglycinaemia. *Mol Genet Metab*. 2005;86:61–69.
17. Tada K, Kure S. Non-ketotic hyperglycinaemia: molecular lesion, diagnosis and pathophysiology. *J Inherit Metab Dis*. 1993;16:691–703.
18. Conter C, Rolland MO, Cheillan D, Bonnet V, Maire I, Froissart R. Genetic heterogeneity of the GLDC gene in 28 unrelated patients with glycine encephalopathy. *J Inherit Metab Dis*. 2006;29:135–142.
19. Nyhan WL, Barshop BB, Al-Aqeel. *Atlas of Inherited Metabolic Disease*. London: Arnold Hodder; 2012, 164–170.
20. Nellis MM, Kasinski A, Carlson M, et al. Relationship of causative genetic mutations in maple syrup urine disease with their clinical expression. *Mol Genet Metab*. 2003;80:189–195.
21. Matsuda I, Nobukuni Y, Mitsubuchi H, et al. A T-to-A substitution in the E1 alpha subunit gene of the branched-chain alpha-ketoacid dehydrogenase complex in two cell lines derived from Mennonite maple syrup urine disease patients. *Biochem Biophys Res Commun*. 1990;172:646–651.
22. Nobukuni Y, Mitsubuchi H, Akaboshi I, Indo Y, Endo F, Matsuda I. Maple syrup urine disease: clinical and biochemical significance of gene analysis. *J Inherit Metab Dis*. 1991;14:787–792.
23. Fisher CW, Fisher CR, Chuang JL, Lau KS, Chuang DT, Cox RP. Occurrence of a 2-bp (AT) deletion allele and a nonsense (G-to-T) mutant allele at the E2 (DBT) locus of six patients with maple syrup urine disease: multiple-exon skipping as a secondary effect of the mutations. *Am J Hum Genet*. 1993;52:414–424.
24. Liu TC, Kim H, Arizmendi C, Kitano A, Patel MS. Identification of two missense mutations in a dihydrolipoamide dehydrogenase-deficient patient. *Proc Natl Acad Sci U S A*. 1993;90(11):5186–5190.
25. Chuang JL, Wynn RM, Moss CC, et al. Structural and biochemical basis for novel mutations in homozygous Israeli maple syrup urine disease patients: a proposed mechanism for the thiamin-responsive phenotype. *J Biol Chem*. 2004;279(17):17792–17800.
26. Nyhan WL, Rice-Kelts M, Klein J. Treatment of the acute crisis in maple syrup urine disease. *Arch Pediatr Adolesc Med*. 1998;152:593–598.
27. Strauss KA, Mazariegos GV, Sindhi R, et al. Elective liver transplantation for the treatment of classical maple syrup urine disease. *Am J Transplant*. 2006;6:557–564.
28. Khanna A, Hart M, Nyhan WL, Hassanein T, Panyard-Davis J, Barshop BA. Domino liver transplantation in maple syrup urine disease. *Liver Transpl*. 2006;12:876–882.

CHAPTER

57

Urea Cycle Disorders

Nicholas Ah Mew, Maria Belen Pappa, and Andrea L. Gropman

Children's National Medical Center, and the George Washington University of the Health Sciences, Washington, DC, USA

INTRODUCTION

Urea cycle disorders (UCDs) represent a group of rare inherited metabolic disorders resulting from a partial or complete deficiency of one of the urea cycle components, thereby resulting in accumulation of ammonia, as well as other nitrogenous products, including glutamine and alanine. Hyperammonemia results in cerebral edema and neurologic injury. These disorders may present at birth, childhood, or adulthood and may range from a relatively mild encephalopathy to profound developmental disability. Early diagnosis and treatment may prevent some of the adverse outcomes, which are primarily neurological. A schematic of the urea cycle is shown in Figure 57.1.

CLINICAL FEATURES

Historical Overview

The discovery of the urea cycle by Krebs and Heinsleit in 1932[1] was groundbreaking, as it not only described the biochemical pathway by which waste nitrogen is converted into urea, but it was also the first of many biochemical cycles to be identified.

While it was understood well before 1932 that hepatic failure could result in hyperammonemia, it took another 25 years after Krebs' seminal discovery before a congenital deficiency of the urea cycle was observed in humans.[2] Clinical descriptions of the other disorders followed thereafter.[3–9] Nearly another quarter century would pass before the individual genes encoding the enzymes and transporters of the urea cycle were cloned.[10–18] All urea cycle genes have now been identified.

Early Treatments

Historical treatments include exchange transfusion and peritoneal dialysis.[19,20] Exchange transfusion is no longer recommended and peritoneal dialysis is used only when extracorporeal dialysis is not readily available.[21,22]

A seminal event in the treatment of urea cycle disorders was the innovation of employing alternative pathway therapies, such as benzoate or phenylbutyrate, to improve nitrogen disposal.[23] A review of published UCD cases by Shih et al. prior to the advent of such therapies revealed 24 deaths among 35 neonates presenting with hyperammonemia.[24] In contrast, in the initial publication describing the application of alternative pathway therapies in hyperammonemic newborns, only 4 of 26 died of complications related to hyperammonemia.[25]

Incidence, Genetics and Genetic Counseling

The reports of the overall incidence of urea cycle disorders vary between 1/35,000 and 1/46,000.[26–28] The incidence of individual disorders is shown in Table 57.1.

http://dx.doi.org/10.1016/10.1016/B978-0-12-410529-4.00057-7

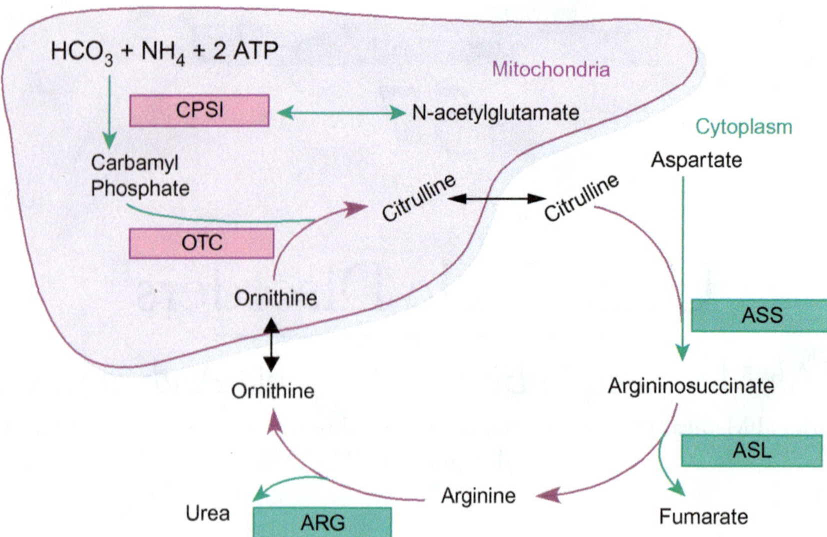

FIGURE 57.1 **The urea cycle.** This figure shows the six enzymes and two transporters involved in the urea cycle.

TABLE 57.1 The Urea Cycle Disorders

Enzyme or transporter deficiency	Incidence[28]	Amino acids (plasma, unless otherwise indicated)	Urine organic acids	IV Sodium phenylacetate/ benzoate loading dose (Over 90–120 minutes)	IV Sodium phenylacetate/ benzoate maintenance dose	Arginine dose
N-acetylglutamate synthetase MIM 237310	<1:2,000,000	↓ citrulline ↓ arginine ↑ glutamine	Unremarkable	250 mg/kg each 5.5 m² if > 20 kg	250 mg/kg each, or 5.5 m² if > 20 kg	200 mg/kg
Carbamyl phosphate synthetase I MIM 237300	1:1,300,000	↓ citrulline ↓ arginine ↑ glutamine	Unremarkable	250 mg/kg each 5.5 m² if > 20 kg	250 mg/kg each, or 5.5 m² if > 20 kg	200 mg/kg
Ornithine transcarbamylase MIM 311250	1:56,500	↓ citrulline ↓ arginine ↑ glutamine	↑↑ orotic acid	250 mg/kg each 5.5 m² if > 20 kg	250 mg/kg each, or 5.5 m² if > 20 kg	200 mg/kg
Argininosuccinate synthetase MIM 215700	1:250,000	↑↑ citrulline ↑ glutamine ↓ arginine	N-to-↑ orotic acid	250 mg/kg each 5.5 m² if > 20 kg	250 mg/kg each, or 5.5 m² if > 20 kg	600 mg/kg
Argininosuccinate lyase MIM 207900	1:218,750	↑ argininosuccinate and anhydrides ↑ glutamine ↑ citrulline ↓ arginine	N-to-↑ orotic acid	250 mg/kg each 5.5 m² if > 20 kg	250 mg/kg each, or 5.5 m² if > 20 kg	600 mg/kg
Arginase MIM 207800	1:950,000	↑ arginine	N-to-↑ orotic acid	250 mg/kg each 5.5 m² if > 20 kg	250 mg/kg each, or 5.5 m² if > 20 kg	Contraindicated
Citrin (SLC25A13) MIM 605814, 603471	<1:2,000,000	↑ citrulline ↑ arginine ↑ methionine ↑ threonine	Unremarkable	250 mg/kg each 5.5 m² if > 20 kg	250 mg/kg each, or 5.5 m² if > 20 kg	Not indicated
Ornithine transporter (SLC25A15) MIM 238970	<1:2,000,000	↑ ornithine ↑ glutamine ↔-to-↓ citrulline ↑ urine homocitrulline	N-to-↑ orotic acid	250 mg/kg each 5.5 m² if > 20 kg	250 mg/kg each, or 5.5 m² if > 20 kg	Not indicated

Abbreviation: IV, Intravenous.

All of the UCDs are inherited as autosomal recessive disorders, with the exception of ornithine transcarbamylase deficiency (OTCD), which is X-linked. A three-generation family history may serve as a valuable diagnostic tool as it may suggest a mode of inheritance. Results of radiologic, biochemical, and genetic testing, as well as dietary preferences should be elicited for any relatives, especially children and young adults for whom neurologic symptoms are reported.

Autosomal recessive conditions usually affect men and women equally and the severity of the symptoms is not sex-related. In these conditions, both parents of an affected individual are assumed to be obligate heterozygotes for a mutation in a UCD-related gene. With each pregnancy, two heterozygotes have a 1-in-4 chance (25%) of having an affected child; however, if only one parent is a carrier, the offspring are at no risk for developing the condition, but have a 50% chance of also being a heterozygote. For this reason, it is not uncommon for the proband to be the only affected individual in the family. In fact, recessive pedigrees are often characterized by multiple affected members of a sibship, rather than in preceding generations. Family-specific characteristics such as small size or lack of information about health status of relatives may hinder this observation. Familial consanguinity should increase diagnostic suspicion of an autosomal recessive disorder.

Given the improved detection rate for some UCDs through extended newborn screening panels, early diagnosis and intervention has helped extend the life expectancy of individuals affected with UCDs into their adulthood and reproductive years. Two affected individuals have a 100% chance of having affected offspring in each pregnancy. This risk is reduced to 50% if one parent is an asymptomatic carrier.

Occasionally, only one parent of an affected homozygote proband is found to be a heterozygote carrier, while no mutation is identified in the other parent. Most frequently this represents laboratory error; however, sometimes this may reflect nonpaternity. Rarely, *de novo* mutations may occur.

Although the previously mentioned factors should be considered in OTCD, its X-linked inheritance raises additional issues. While hemizygote male probands are most likely to have inherited the mutation from a heterozygous mother, *de novo* mutations are common. As a result, for the mother of a hemizygote proband, the *a priori* risk of being a heterozygote carrier is predicted by Bayesian analysis to be 66% when reproductive fitness of the mutation is nil. However, due to the differential rates of spontaneous mutations in sperm versus ova, this risk of being a heterozygote may be as high as 90%.[29]

Although in many X-linked conditions female carriers are asymptomatic, in OTCD female carriers may present with milder symptoms of the condition. This interferes with the typical presentation of an X-linked pedigree in which only female-to-male transmission is noted. Given that female carriers have a 50% chance of having a daughter who is also a carrier (and who may have mild symptoms for the condition) the inheritance pattern may be confused with an autosomal dominant mode of inheritance with decreased penetrance. For this reason, it is important to obtain a detailed description of the extent of the symptomatology in each individual in the family. In contrast to the autosomal recessive forms of UCDs, a more severe symptomatology is expected in affected males than in carrier females.

Overall, although genetic testing may lead to diagnostic confirmation in cases where a UCD is suspected, appropriate counseling should be provided to families at a higher risk for these conditions. A consultation with a genetic counselor may be beneficial to address implications of the predictive testing, including ethical, psychosocial, and emotional issues. Prenatal genetic testing is available clinically; however, families should discuss the ethical implications of this testing with a genetics specialist to ensure proper understanding of the limitations of the testing.

NATURAL HISTORY

Age of Onset

UCDs are "intoxication-type" disorders and, as such, do not typically result in any prenatal phenotype. The fetal–placental circulation clears toxic metabolites such as ammonia, thus preventing any *in utero* sequelae. After delivery, in the absence of placental detoxification, hyperammonemia may result when the defective urea cycle is challenged.

Neonatal Presentation

Children are normal at birth and remain so for at least 24 hours. Symptoms may mimic those of sepsis: poor feeding, vomiting, lethargy, seizures, and coma. Additionally, affected youngsters become tachypneic and develop respiratory alkalosis caused by stimulation of the respiratory center by ammonia. Ammonia levels may rise above 1000 μmol per L. Without treatment, death occurs as a result of irreversible cerebral edema. Severe neonatal hyperammonemia nearly invariably leads to permanent neurologic damage.

A neonatal presentation almost certainly reflects a complete deficiency of a urea cycle enzyme[30,31] resulting in virtually no flux through the urea cycle. Due to their greatly reduced urea cycle flux, children with neonatal-type UCDs typically have the lowest dietary protein tolerance, and are at greatest risk for episodes of hyperammonemia triggered by infections or inadvertent feeding of a high-protein meal.

Late-Onset Presentations

Partial deficiencies of the urea cycle can present at any age. However, this diagnosis is rarely considered in older patients, typically resulting in a delay in diagnosis. The diagnosis of a urea cycle disorder should be considered in patients with chronic encephalopathy, autism, learning disorders, hyperactive and self-injurious behavior, vomiting with changes in level of consciousness, stroke-like episodes and, in teens and adults, psychiatric symptoms including episodic psychosis, bipolar disorder and/or major depression.[32–38]

Because these disorders are rare, most physicians are not familiar with the late onset/adult presentation, which may permit an individual to escape diagnosis often for several decades until an environmental stressor triggers acute hyperammonemia. Some of these stressors include surgery, fractures, severe bleeding, the postpartum period,[39,40] parenteral nutrition with high-nitrogen intake, gastrointestinal bleeding[41–44] and status post gastric bypass surgery.[45] In addition, hyperammonemia may be triggered after a change to a higher protein intake due to a change in dietary habits (e.g., Atkins diet), decreased access to low-protein foods, or a different meal preparation (i.e., summer camp). Medications influencing protein turnover include corticosteroids[46] and chemotherapy. These conditions typically either increase the need for nitrogen clearance or interfere with the enzymes of the urea cycle. Additionally, the underlying urea cycle disorder may be difficult to recognize in the face of another illness. Patient outcome depends on prompt recognition and treatment of hyperammonemia (HA), otherwise the prognosis for these patients is poor.

When approaching an adult with a possible UCD, taking a family history may reveal this, but most families may not be aware of the existence of metabolic abnormalities, and the recessive inheritance pattern of most of these diseases (except OTCD) may make the patient appear to be the only affected individual in the family. A dietary history can be very important in revealing patients who are auto-selective vegetarians (i.e., elective decreased protein intake). In adults, history of behavioral and psychiatric illness that does not fit a clear diagnosis and/or is difficult to treat with usual medications (possibly resulting from chronic low-grade hyperammonemia) or a history of prolonged clinical courses with routine illnesses should also prompt suspicion of a UCD.

Once the diagnosis is made, one should strive to avoid similar triggering events. Therefore, intravenous steroids for asthma or valproic acid[47] for epilepsy or mood stabilization are contraindicated.

Arginase Deficiency

Patients with arginase deficiency do not classically present with hyperammonemic coma. They have milder hyperammonemia, typically in the 200–300 μmol per L range. The most prominent features are progressive spasticity, particularly in the legs, and mild intellectual disability. Some patients have protein intolerance and avoidance, similar to patients with late-onset presentations of the other urea cycle disorders.

Citrullinemia Type II

There are two distinct presentations of citrullinemia type II. Affected adults develop the various neuropsychological symptoms associated with hyperammonemia, as described above.[5] However, neonates present with intrahepatic cholestasis and poor growth.[48] Most patients exhibit spontaneous improvement in the first year of life. A minority of patients continue to have failure to thrive, dyslipidemia, and chronic liver disease.[49]

MOLECULAR GENETICS

Carbamylphosphate Synthetase I (*CPS1*)

Located on the long arm of chromosome 2 (2q35), the *CPS1* gene encodes for the enzyme responsible for catalyzing the entry of ammonia into the urea cycle. Given its role in the urea cycle, CPSI is highly expressed in the liver[50] and the intestine.[51]

Currently, over 200 mutations in the *CPS1* gene have been identified to cause carbamylphosphate synthetase I deficiency, including missense, nonsense, insertion/deletions, and splice-site mutations leading to premature truncation as well as in-frame alterations of amino acid composition.[52] Although large genomic rearrangements and imbalances are more uncommon than single nucleotide variations, they have also been reported in the literature.[53] Interestingly, approximately 90% of the identified mutations have been found to be private in nature, meaning that they are not identified in unrelated families.[53]

Ornithine Transcarbamylase (*OTC*)

The *OTC* gene maps to chromosomal location Xp11.4. Ornithine transcarbamylase plays a role in catalyzing the production of citrulline from ornithine and carbamylphosphate in the liver and small intestine.[54] The 354-amino acid preprotein that results from translation of the human OTC mRNA[15] undergoes posttranslational cleavage after entering the mitochondria, resulting in a mature protein that is composed of 322 amino acids with a predicted molecular weight of 36 kDa.

Mutations of all types (nonsense, missense, and those affecting mRNA splicing) have been associated with OTC deficiency. The pathogenicity of missense mutations, which encompass approximately 68% of mutations in this gene, is directly dependent on the specific amino acid substitution as well as the location of the residue within the folded protein structure.[54–60] Mutations impacting splicing of the mRNA transcript (~14%) can have variable effects on the production of OTC enzyme ranging from decreased production to complete absence.[61,62]

The vast majority of mutations in OTC deficiency are inherited. In affected index male probands, conventional genetics would suggest that two-thirds of mutations are inherited, whereas one-third occur *de novo*. However, due to the differential rates of mutations in sperm versus ova, the proportion of affected hemizygous males with inherited mutations may be as high as 90%.[29]

Argininosuccinate Synthetase (*ASS1*)

Located at chromosomal location 9q34.11, *ASS1* codes for argininosuccinate synthase, a homotetramer of 186 kDa involved in catalyzing the cytosolic condensation of citrulline and aspartate to argininosuccinic acid. *ASS1* comprises 16 exons and its protein product localizes to the cytoplasm after translation. Approximately 100 *ASS1* mutations, both exonic and intronic, have been noted to impact the production of argininosuccinate synthase.[63] More specifically, seven mutations have been associated with a severe phenotype, with p.Arg304Trp, c.421-2A>G, and p.Gly390Arg accounting for most cases of citrullinemia type I.[64]

Argininosuccinate Lyase (*ASL*)

Consisting of 16 exons, the *ASL* gene is located at position 7q11.21, and encodes for the monomer subunit of argininosuccinate lyase. Four identical ASL monomers assemble to form a functional homotetramer of 208 kDa, which is able to breakdown argininosuccinate to yield arginine and fumarate.

Mutations of all types (nonsense, missense mutations, and those affecting mRNA splicing) are distributed throughout the gene with a higher incidence of mutations located on exons 4, 5, and 7.[65,66] In addition, several *ASL* founder mutations have been identified, including the c.1153C>T variant in the Finnish population,[67] and the c.1060C>T and c.346C>T variants in individuals of Arab ancestry.[68]

Arginase (*ARG1*)

Comprising eight exons, the *ARG1* gene localizes at position 6q23.2 and codes for the manganese-dependent enzyme catalyzing the conversion of L-arginine into L-ornithine and urea (arginase 1).

Mutations causing decreased or absent expression of arginase 1 have been identified throughout the coding region of *ARG1*, with a higher incidence of missense mutations in highly conserved regions.[69]

A second isoform of arginase exists—arginase II—and the gene is located on 14q24.1[41]. No human disorder is currently associated with a deficiency of arginase II, and a murine knockout model is completely viable.[70,71]

Ornithine Translocase (*SLC25A15*)

Hyperornithinemia-hyperammonemia-homocitrullinuria (HHH) is caused by mutations in *SLC25A15* (13q14.11), which encodes for the mitochondrial ornithine transporter, a 301-amino acid protein that is inserted in the inner mitochondrial membrane by six-transmembrane domains.[72]

Mutations of all types have been described in the literature, showing a higher frequency of mutations in exons 2–7, corresponding to the gene's coding region.[73,74] The most common mutation (c.562_564delTTC, p.Phe188del), accounting for approximately half of the HHH reported cases results in a truncated unstable protein with 10–15% residual activity.[75,76] This variant is prevalent in individuals of French-Canadian descent. The c.535C>T (p.Arg179*) variant, which is predominantly found in persons of Japanese and Middle Eastern heritage,[77,78] yields a completely nonfunctional protein.[79,80] Other mutations with variable impact on the protein's functionality have been reported with no specific ethnic associations.

Citrin (*SLC25A13*)

Citrin, the carrier of glutamate and aspartate into and out of the mitochondria, respectively, is encoded by the 18 exon gene *SLC25A13*, located at 7q21.3.[81,82]

Although mutations of all types have been reported, two common mutations account for the vast majority (~70%) of citrullinemia type II cases. These mutations, c.1177+1G>A and c.851-854del, have been primarily identified in persons of Japanese descent.[81–86]

N-Acetylglutamate Synthase (NAGS)

N-acetylglutamate synthase conjugates glutamate and acetyl-CoA to generate *N*-acetylglutamate, the obligate activator of the CPS1 enzyme. It is encoded by the 7-exon gene *NAGS* and located at 17q21.31.[10] NAGS deficiency is the rarest of the urea cycle disorders. The majority of the mutations appear to be private.[43]

EXPRESSION OF UREA CYCLE ENZYMES AND NITROGEN METABOLISM

The entire cycle is expressed in hepatocytes. Within the liver, expression is highest in cells near the portal triads. CPS constitutes as much as 30% of mitochondrial protein in the liver. Glutamine synthetase is expressed in hepatocytes surrounding the hepatic veins. Conversely, glutaminase is expressed in cells in the portal triads. Thus, it appears that systemic ammonium is converted to glutamine, a high-affinity, low-capacity process. The glutamine is hydrolyzed and the ammonium converted to urea in the portal cells, a high-capacity, low-affinity process. Conversely, ammonium produced by glutamine degradation or digestive absorption in the intestine reaches the portal triads and is converted into urea. Only a small amount of portal vein-derived ammonia appears to be metabolized to glutamine directly.

Part of the urea cycle is also present in the kidney (the cytosolic enzymes), and in the intestine (the mitochondrial enzymes). Intestine contains CPS and OTC and is capable of producing citrulline from ammonium and ornithine. The gut also contains ornithine aminotransferase (OAT), and is capable of synthesizing ornithine from glutamate. The kidney uses glutamine as the source of ammonium ions as a means of excretion of excess acid into the tubular fluid. Glutamine carbon is converted to glucose. The kidney also takes up citrulline and releases arginine. The source of the citrulline is the intestine. This reaction is a major source of arginine synthesis and explains why arginine is not an essential amino acid in normal humans.

The net result is an active shuttle of glutamine within the liver and between gut, liver, and kidney. The intestine uses glutamine as a fuel, producing ammonium, citrulline, and alanine. The ammonium reaches the liver and is converted either to urea or glutamine. The gut-produced citrulline is not removed by the hepatocytes. Instead, it serves as a source for arginine synthesis in the kidney. Systemically produced ammonia would be preferentially converted to glutamine near the hepatic veins. The glutamine can serve as a source of urinary ammonium (acid excretion) or of urea, depending on whether it was broken down in the kidney or the liver.

In the absence of effective urea cycle activity, only the glutamine-producing reactions would be active in the liver, hence the extreme hyperglutaminemic characterstic of urea cycle disorders. Similarly, a defect in any of the urea cycle enzymes will vitiate the intestinal–renal pathway for arginine synthesis, making arginine an essential amino acid.

DISEASE MECHANISMS/PATHOPHYSIOLOGY

Human Observations: Neuropathology of UCDs

While hyperammonemia can lead to severe consequences in the central nervous system (CNS), the pathophysiology remains unclear. Current theories have focused on: 1) glutamine accumulation, with associated impaired cerebral osmoregulation, and 2) glutamate/*N*-methyl-D-aspartate (NMDA) receptor activation, with resultant excitotoxic injury and energy deficit.[87–90]

Although the plasma ammonia and glutamine levels are typically the parameters by which clinical management decisions are made in the proximal UCDs, the correlation between plasma levels of ammonia and glutamine is quite poor and plasma glutamine levels may not reflect brain levels, which may be higher. Acute hyperammonemia also causes other dynamic changes, particularly increased blood–brain barrier permeability, depletion of intermediates of cell energy metabolism and the disaggregation of microtubules.

Neuropathological findings in patients with urea cycle disorders share similar features with hepatic encephalopathy and hypoxic ischemic encephalopathy, that is, they depend both upon the duration of hyperammonemic coma as well as the interval between coma and death. Evidence from autopsy and imaging studies suggests that OTCD results in white matter injury. Neuroimaging studies performed months later in neonatal coma survivors are consistent with these pathological findings, correlating with hypomyelination of white matter, myelination delay, and cystic changes of the white matter and gliosis of the deep gray matter nuclei.[91–97]

Pathogenic mechanisms may include glutamine-induced cerebral edema, energy failure and neurotransmitter alterations. Ammonia is able to diffuse freely across the blood–brain barrier into the brain in amounts proportional to the arterial blood concentration and blood flow and, as a result, metabolic trapping can occur with the concentration in the brain tending to be higher than in peripheral blood.[88,89] Ammonia in the bloodstream rapidly enters the brain and almost all is converted instantly to glutamine, offering short-term buffering of excess ammonia in patients with hyperammonemia. This is accomplished in the astrocyte via glutamine synthetase (GS). Glutamine is osmotically active and can cause astrocytic swelling, leading to cytotoxic edema.[98–100] Studies in which the GS inhibitor methionine sulfoximine is administered demonstrate reduced ammonia-induced brain edema in both *in vivo* and *in vitro* models.[102] The astrocyte, therefore, is an important intermediate in the interactions of glutamine and ammonia via the glutamate–glutamine cycle (Figure 57.2).

The diagnosis of OTCD occurs after initial metabolic encephalopathy has occurred, unless identification of an affected family member prompts diagnosis in an at-risk relative. The time course of metabolic perturbation in the brain is unknown and not previously studied.

In the acute setting, the early neurological effects of increasing CNS ammonia include anorexia and vomiting as well as changes in mental status. Progressive CNS dysfunction may ensue, reflecting ammonia-induced cell swelling, including lethargy, ataxia, seizures, asterixis, hypothermia, and ultimately coma may follow. Death may occur, or alternatively survival with significant neurological injury ensues. A decreased level of consciousness (lethargy, drowsiness, unresponsiveness, coma, and obtundation), and/or abnormal motor function (slurred speech, tremors, weakness, decreased or increased muscle tone, and ataxia) may be more frequent clinical findings with the first episode rather than with subsequent episodes of acute hyperammonemia.[71]

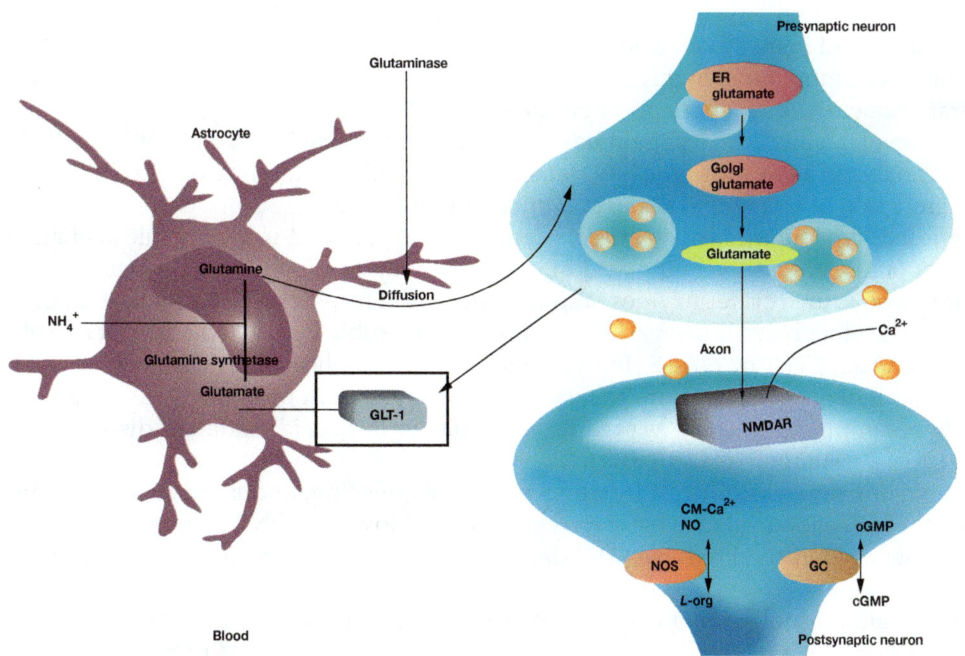

FIGURE 57.2 **The glutamine–glutamate cycle.**

Animal Models

Murine models for all urea cycle defects except HHH syndrome exist.[103–110] The most studied of these include two hypomorphic models of OTC deficiency, sparse fur (*spf*),[103] and *spf-ash*.[104] The *spf*-mouse harbors a mutation, H117N, which affects mRNA processing, resulting in affected males with approximately 10% hepatic OTC activity compared to wild-type littermates.[103] In contrast, the *spf-ash* mutation R129H,[103] which has also been identified in humans,[54] affects the recognition of a splice site. The inefficient splicing results in 5–10% expression of normal OTC.

Complete knockout mouse models of the proximal urea cycle disorders NAGSD, CPSD, and OTCD result in hyperammonemia and neonatal death within 48 hours. Interestingly, the NAGS-deprived mouse can be rescued with *N*-carbamylglutamate and citrulline,[101] and again develops hyperammonemia when this treatment is withdrawn; thus, it may be used as a model of induced hyperammonemia.

DIFFERENTIAL DIAGNOSIS

Conditions that perturb the liver and result in hyperammonemia should be considered in the differential diagnosis of UCDs. Several medications are also known to cause hyperammonemia, including valproic acid[47] and chemotherapy.[111,112] In premature infants, a diagnosis of transient hyperammonemia of the newborn (THAN)[113] should be considered.

Among inherited metabolic disorders, organic acidopathies, in particular propionic and methylmalonic acidemia, should be strongly considered. However, these may initially typically be differentiated from UCDs due to the profound acidosis and ketosis, and later by the diagnostic profile present in urine organic acid analysis. These conditions are hypothesized to interfere with the NAGS[114–118] or CPSI[118] reaction of the urea cycle, thereby resulting in hyperammonemia. This mechanism has also been proposed to explain the uncommon instances of hyperammonemia in other inherited disorders, such as maple syrup urine disease,[119] and fatty acid oxidation disorders.[120,121] Lysinuric protein intolerance[122] may result in a secondary deficiency of the urea cycle,[123] due to urinary excretion of the key intermediates ornithine and arginine.

TESTING

Symptomatic individuals with urea cycle disorders have elevated ammonia levels at the time of diagnosis. Levels in neonates may be well above 1000 μmol per L at presentation; however, adults may be symptomatic with levels above only 100 μmol per L. The presence of tachypnea and respiratory alkalosis in addition to hyperammonemia is virtually pathognomonic for urea cycle disorders. However, this is not a highly sensitive sign, as concurrent infection or sepsis may result instead in metabolic acidosis.

The measurement of ammonia is highly susceptible to artifact, and this should be considered when reviewing causes of hyperammonemia. Improper collection, handling, storage, and analysis may all contribute to a falsely elevated ammonia level. Ideally, blood should be drawn from a free-flowing vessel, into a chilled heparinized tube, placed on ice, and analyzed as soon as possible. Hemolysis or blood left standing may result in in elevated ammonia levels due to deamination of free amino acids, in particular glutamine.

In the undiagnosed hyperammonemic patient, plasma amino acid and urine organic acid analysis should be requested as early as possible. Plasma amino acid quantitation may be diagnostic, in particular when the defect lies within one of the cytosolic urea cycle enzymes or transporters, and results in the accumulation of a diagnostic abnormal amino acid. On the other hand, a decreased citrulline is compatible with a deficiency of one of the mitochondrial urea cycle enzymes, OTC, CPS1 or NAGS. In this instance, the presence of increased orotic acid on urine organic acid analysis would highly suggest OTC deficiency. A urine organic acid profile will discriminate UCDs from other diagnoses with hyperammonemia, such as organic acidemias or fatty acid oxidation disorders. Typical amino acid patterns of each UCD are summarized in Table 57.1.

Diagnostic suspicions can be confirmed by direct genetic sequencing, and is frequently required in order to distinguish between the mitochondrial urea cycle enzyme defects. However, DNA sequencing is not 100% sensitive.[31,69] For instance, in patients confirmed to have OTC deficiency via liver enzyme analysis, a deleterious mutation was found in only 77%.[31]

In OTC deficiency, allopurinol loading may induce orotic aciduria or orotidinuria, but the predictive value is poor, particularly in the absence of a family history. Enzymatic diagnosis of OTC and CPS requires liver tissue. Arginase can be assayed in red blood cells, and argininosuccinate synthetase and argininosuccinate lyase can be assayed in fibroblasts, although these tests are rarely necessary for diagnosis, as the plasma amino acid pattern is typically diagnostic.

MANAGEMENT

There have been two consensus statements reviewing treatment guidelines for the management of urea cycle disorders.[21,22,124]

Acute Hyperammonemia

The primary goal in the treatment of acute hyperammonemia is to reduce the hyperammonemia as rapidly as possible. Extracorporeal detoxification is the most effective[19,21,125] and preferred method of removing ammonia. Because of the greater risk of developing intracranial hypertension and cerebral edema, dialysis should be considered at much lower concentrations of blood ammonia in adults than in children or neonates.[21]

Standard hemodialysis provides the highest clearance rates.[125] However, in infants, there is a high risk of hemodynamic or technical complications.[126] As a result, hemodialysis driven by an extracorporeal membrane oxygenation (ECMO) pump[125] may be preferred, despite the risk of blood vessel damage. Alternatively, continuous renal replacement therapy, although typically achieving slower clearance rates, is usually well tolerated by infants,[21] and is not prone to rebound hyperammonemia, as observed with intermittent forms of dialysis.

Peritoneal dialysis is less effective,[19,22] but may be more accessible to many centers. Exchange transfusion is a historical treatment,[127] and should no longer be used.

Other concurrent measures include halting protein and maximizing caloric intake, typically parenterally via intravenous fluids with 10% dextrose.[21,22] Ammonia scavengers should also be started, even if preparations for dialysis are underway. Initial therapy should include boluses of intravenous sodium benzoate 250 mg per kg and sodium phenylacetate 250 mg per kg over 90 minutes, followed by continuous maintenance intravenous infusions of 250–500 mg per kg day.[21,22] Repeated or high doses have been associated with a greater risk of drug toxicity.[128]

Intravenous arginine may replenish urea cycle intermediates and promote nitrogen excretion in all of the urea cycle disorders, except in arginase deficiency and possibly HHH syndrome. Arginine should be started at a dose of 200 mg per kg per day,[21,124,129] including in patients who do not yet have a definitive diagnosis. In patients with argininosuccinate synthase deficiency (ASD) or argininosuccinate lyase deficiency (ALD), larger doses of up to 600 mg per kg per day may be utilized.[124] NAGS deficiency is the only urea cycle disorder in which normal urea cycle function may essentially be restored with a pharmacological agent. Oral *N*-carbamylglutamate should be started at 100–250 mg per kg per day in patients with a known diagnosis of NAGS deficiency. Additionally, in undiagnosed patients in whom NAGS deficiency remains a diagnostic possibility, a trial of *N*-carbamylglutamate should be attempted.[130]

Long-Term Management

The goals of therapy are to maintain normal growth and development while normalizing ammonia concentrations. This may be achieved by dietary protein restriction and drug therapy.

Diet Management

The diet must be adjusted to provide adequate calories and protein for growth, while minimizing the nitrogen load on the dysfunctional urea cycle. Some patients may require only a low-protein diet, sometimes with a protein-free caloric supplement. References for "safe level of protein intake" have been published by a joint cooperation between the World Health Organization, the Food and Agriculture Organization of the United Nations, and United Nations University. The US Food and Nutrition Board and Institute of Medicine have also published similar references.[131] These recommendations may be used as an initial guide; however, the amount of protein tolerated by each patient must be determined individually and titrated to normal ammonia levels. In general, protein requirements are highest in the neonate (1.5–1.77 g per kg per day) and decrease with age.[131]

If adequate protein cannot be provided from natural sources alone without causing nitrogen accumulation and hyperammonemia, as is often the case with neonatal presentations of UCDs, some of the natural protein can be replaced by products providing "protein" as a mixture of essential amino acids. It is important to avoid deficiencies of essential amino acids or arginine, as an insufficiency of any amino acid will limit growth.

Drug Therapy

In severe urea cycle disorders, optimized dietary therapy is frequently insufficient to prevent hyperammonemia. The development of alternative pathway therapies, drugs that conjugate with nitrogen-containing substances and then are excreted in the urine, has allowed long-term survival.[132]

Oral sodium benzoate, which is not a registered drug, has been used for decades in the urea cycle disorders. It combines with the nonessential amino acid glycine to form hippuric acid. Sodium phenylbutyrate is a registered drug, and has mostly supplanted sodium benzoate as the preferred ammonia scavenger in the United States. Recently, glycerol phenylbutyrate, a triglyceride of sodium phenylbutyrate, was approved in the US. This pre-prodrug is a clear, tasteless liquid as compared to the unpleasant-tasting sodium phenylbutyrate powder, but is equally effective.[133]

Oral arginine and citrulline may help replenish depleted arginine stores, and may improve nitrogen excretion in all urea cycle disorders except arginase deficiency, where this is contraindicated.

N-carbamylglutamate is the treatment of choice in NAGS deficiency. It should be initially started at 100 mg per kg per day, and may be titrated to ammonia levels. Peroral drug dosages for individual urea cycle disorders are shown in Table 57.1.

Transplantation

Orthotopic liver transplantation can correct the hyperammonemia, and eliminates the need for protein restriction or ammonia scavengers in all the urea cycle disorders. However, this procedure is associated with significant morbidity, and necessitates long-term immunosuppression,[134] so the risk–benefit must be carefully evaluated before candidacy is determined. Liver transplantation should be considered in any child affected with hyperammonemia that cannot be controlled medically, particularly in severe CPS1 and OTC deficiency. While liver transplantation will also normalize ammonia in ASD or ALD, the enzymes AS and AL are ubiquitously expressed, and thus it is not yet clear whether there exist other long-term complications of ASS or ASL deficiency in other tissues.[30,135–137] Five-year survival after pediatric hepatic transplantation is nearly 90%.[138] Hepatic transplantation is contraindicated in NAGS deficiency, where complete or near-complete restoration of the urea cycle may be achieved pharmacologically.[139,140]

Outcomes

The two most proximal enzyme defects, namely carbamylphosphate synthetase I (CPSI) deficiency and OTCD, present the highest risk for acute neurological injury. This is especially true for neonatal onset disease in which the outcome, even with early recognition and treatment, has been uniformly poor.[38,141–144]

Late-onset UCDs still carry a potential risk for encephalopathy and neurological damage if not recognized and treated promptly ([38]; Summar et al., personal communication). Prediction of outcome is not straightforward. There is no direct correlation between genotype, age of onset, peak ammonia level, imaging, and/or phenotype.[142–150] Normal intelligence is possible after a hyperammonemic event. Reversible changes in the CNS caused by elevated blood ammonia concentrations may be possible when levels are kept below 200–400 mg per dL; however, accumulating damage results in irreversible impairment.

CURRENT RESEARCH

There are several organizations currently evaluating the natural history of urea cycle disorders. These include the US-based Urea Cycle Disorders Consortium (http://rarediseasesnetwork.epi.usf.edu/ucdc/) and the European Registry and Network for Intoxication-Type Metabolic Disorders (https://www.eimd-registry.org/).

Gene therapy offers the possibility of definitive treatment. The initial gene therapy trial for OTC deficiency was terminated after the death of one of the subjects, apparently due to an adverse reaction to the adenovirus vector.[151] Improved vectors and new techniques to improve expression levels may make gene therapy a viable therapy in the future. Current murine studies employing an adeno-associated vector have been successful.[152] Hepatocyte[153,154] and stem cell[155] transplantation are also promising alternatives.

Current therapies have improved long-term survival and ammonia control, and have, thus, unmasked additional scientific questions. For instance, disruption of the urea cycle may also result in deleterious effects by perturbing nitric oxide production.[135,136] Additionally, future studies may also demonstrate whether urea cycle dysfunction may result in long-term hepatic damage, increasing the risk of liver cancer.[156]

References

1. Krebs HA, Heinsleit. Untersuchungen über die Harnstoffbidung im Tierkörper. *Z Physiol Chem*. 1932;210:33–46.
2. Allan JD, Cusworth DC, Dent CE, Wilson VK. A disease, probably hereditary characterised by severe mental deficiency and a constant gross abnormality of amino acid metabolism. *Lancet*. 1958;1(7013):182–187.
3. Bachmann C, Krähenbühl S, Colombo JP, Schubiger G, Jaggi KH, Tönz O. N-acetylglutamate synthetase deficiency: a disorder of ammonia detoxication. *N Engl J Med*. 1981;304(9):543.
4. Hommes FA, De Groot CJ, Wilmink CW, Jonxis JH. Carbamylphosphate synthetase deficiency in an infant with severe cerebral damage. *Arch Dis Child*. 1969;44(238):688–693.
5. Matsuda I, Anakura M, Arashima S, Saito Y, Oka Y. A variant form of citrullinemia. *J Pediatr*. 1976;88(5):824–826.
6. McMurray WC, Rathbun JC, Mohyuddin F, Koegler SJ. Citrullinuria. *Pediatrics*. 1963;32:347–357.
7. Russell A, Levin B, Oberholzer VG, Sinclair L. Hyperammonaemia. A new instance of an inborn enzymatic defect of the biosynthesis of urea. *Lancet*. 1962;2(7258):699–700.
8. Shih VE, Efron ML, Moser HW. Hyperornithinemia, hyperammonemia, and homocitrullinuria. A new disorder of amino acid metabolism associated with myoclonic seizures and mental retardation. *Am J Dis Child*. 1969;117(1):83–92.
9. Terheggen HG, Schwenk A, Lowenthal A, van Sande M, Colombo JP. Hyperargininemia with arginase deficiency. A new familial metabolic disease. I. Clinical studies. *Z Kinderheilkd*. 1970;107(4):298–312.
10. Caldovic L, Morizono H, Yu X, et al. Cloning and expression of the human *N*-acetylglutamate synthase gene. *Biochem Biophys Res Commun*. 2002;299(4):581–586.
11. Camacho JA, Obie C, Biery B, et al. Hyperornithinaemia-hyperammonaemia-homocitrullinuria syndrome is caused by mutations in a gene encoding a mitochondrial ornithine transporter. *Nat Genet*. 1999;22(2):151–158.
12. Dizikes GJ, Grody WW, Kern RM, Cederbaum SD. Isolation of human liver arginase cDNA and demonstration of nonhomology between the two human arginase genes. *Biochem Biophys Res Commun*. 1986;141(1):53–59.
13. Haraguchi Y, Takiguchi M, Amaya Y, Kawamoto S, Matsuda I, Mori M. Molecular cloning and nucleotide sequence of cDNA for human liver arginase. *Proc Natl Acad Sci U S A*. 1987;84(2):412–415.
14. Haraguchi Y, Uchino T, Takiguchi M, Endo F, Mori M, Matsuda I. Cloning and sequence of a cDNA encoding human carbamyl phosphate synthetase I: molecular analysis of hyperammonemia. *Gene*. 1991;107(2):335–340.
15. Horwich AL, Fenton WA, Williams KR, et al. Structure and expression of a complementary DNA for the nuclear coded precursor of human mitochondrial ornithine transcarbamylase. *Science*. 1984;224(4653):1068–1074.
16. Kobayashi K, Sinasac DS, Iijima M, et al. The gene mutated in adult-onset type II citrullinaemia encodes a putative mitochondrial carrier protein. *Nat Genet*. 1999;22(2):159–163.
17. O'Brien WE, McInnes R, Kalumuck K, Adcock M. Cloning and sequence analysis of cDNA for human argininosuccinate lyase. *Proc Natl Acad Sci U S A*. 1986;83(19):7211–7215.
18. Su TS, Bock HG, O'Brien WE, Beaudet AL. Cloning of cDNA for argininosuccinate synthetase mRNA and study of enzyme overproduction in a human cell line. *J Biol Chem*. 1981;256(22):11826–11831.
19. Donn SM, Swartz RD, Thoene JG. Comparison of exchange transfusion, peritoneal dialysis, and hemodialysis for the treatment of hyperammonemia in an anuric newborn infant. *J Pediatr*. 1979;95(1):67–70.
20. Batshaw ML, Brusilow SW. Treatment of hyperammonemic coma caused by inborn errors of urea synthesis. *J Pediatr*. 1980;97(6):893–900.
21. Häberle J, Boddaert N, Burlina A, et al. Suggested guidelines for the diagnosis and management of urea cycle disorders. *Orphanet J Rare Dis*. 2012;7:32.
22. Summar M. Current strategies for the management of neonatal urea cycle disorders. *J Pediatr*. 2001;138(1 suppl):S30–S39.
23. Brusilow SW, Valle DL, Batshaw M. New pathways of nitrogen excretion in inborn errors of urea synthesis. *Lancet*. 1979;2(8140):452–454.
24. Grisolía S, Báguena R, Mayor F. The urea cycle. New York: John Wiley and Sons; 1976, 579 p.
25. Batshaw ML, Brusilow S, Waber L, et al. Treatment of inborn errors of urea synthesis: activation of alternative pathways of waste nitrogen synthesis and excretion. *N Engl J Med*. 1982;306(23):1387–1392.
26. Keskinen P, Siitonen A, Salo M. Hereditary urea cycle diseases in Finland. *Acta Paediatr*. 2008;97(10):1412–1419.
27. Nagata N, Matsuda I, Oyanagi K. Estimated frequency of urea cycle enzymopathies in Japan. *Am J Med Genet*. 1991;39(2):228–229.
28. Summar ML, Koelker S, Freedenberg D, et al. The incidence of urea cycle disorders. *Mol Genet Metab*. 2013;110(1–2):179–180.
29. Tuchman M, Matsuda I, Munnich A, Malcolm S, Strautnieks S, Briede T. Proportions of spontaneous mutations in males and females with ornithine transcarbamylase deficiency. *Am J Med Genet*. 1995;55(1):67–70.
30. Ah Mew N, Krivitzky L, McCarter R, Batshaw M, Tuchman M, Urea Cycle Disorders Consortium of the Rare Diseases Clinical Research Network. Clinical outcomes of neonatal onset proximal versus distal urea cycle disorders do not differ. *J Pediatr*. 2013;162(2):324–329.e1.
31. McCullough BA, Yudkoff M, Batshaw ML, Wilson JM, Raper SE, Tuchman M. Genotype spectrum of ornithine transcarbamylase deficiency: correlation with the clinical and biochemical phenotype. *Am J Med Genet*. 2000;93(4):313–319.
32. Batshaw ML, Roan Y, Jung AL, Rosenberg LA, Brusilow SW. Cerebral dysfunction in asymptomatic carriers of ornithine transcarbamylase deficiency. *N Engl J Med*. 1980;302(9):482–485.
33. Felig DM, Brusilow SW, Boyer JL. Hyperammonemic coma due to parenteral nutrition in a woman with heterozygous ornithine transcarbamylase deficiency. *Gastroenterology*. 1995;109(1):282–284.
34. Gaspari R, Arcangeli A, Mensi S, et al. Late-onset presentation of ornithine transcarbamylase deficiency in a young woman with hyperammonemic coma. *Ann Emerg Med*. 2003;41(1):104–109.
35. Gilchrist JM, Coleman RA. Ornithine transcarbamylase deficiency: adult onset of severe symptoms. *Ann Intern Med*. 1987;106(4):556–558.
36. Mizoguchi K, Sukehiro K, Ogata M, et al. A case of ornithine transcarbamylase deficiency with acute and late onset simulating Reye's syndrome in an adult male. *Kurume Med J*. 1990;37(2):105–109.
37. Rimbaux S, Hommet C, Perrier D, et al. Adult onset ornithine transcarbamylase deficiency: an unusual cause of semantic disorders. *J Neurol Neurosurg Psychiatry*. 2004;75(7):1073–1075.

38. Smith W, Kishnani PS, Lee B, et al. Urea cycle disorders: clinical presentation outside the newborn period. *Crit Care Clin*. 2005;21 (4 suppl):S9–S17.
39. Arn PH, Hauser ER, Thomas GH, Herman G, Hess D, Brusilow SW. Hyperammonemia in women with a mutation at the ornithine carbamoyltransferase locus. A cause of postpartum coma. *N Engl J Med*. 1990;322(23):1652–1655.
40. Cordero DR, Baker J, Dorinzi D, Toffle R. Ornithine transcarbamylase deficiency in pregnancy. *J Inherit Metab Dis*. 2005;28(2):237–240.
41. Enns GM, O'Brien WE, Kobayashi K, Shinzawa H, Pellegrino JE. Postpartum "psychosis" in mild argininosuccinate synthetase deficiency. *Obstet Gynecol*. 2005;105:1244–1246.
42. Mathias RS, Kostiner D, Packman S. Hyperammonemia in urea cycle disorders: role of the nephrologist. *Am J Kidney Dis*. 2001;37(5):1069–1080.
43. Fenves A, Boland CR, Lepe R, Rivera-Torres P, Spechler SJ. Fatal hyperammonemic encephalopathy after gastric bypass surgery. *Am J Med*. 2008;121(1):e1–e2.
44. Trivedi M, Zafar S, Spalding MJ, Jonnalagadda S. Ornithine transcarbamylase deficiency unmasked because of gastrointestinal bleeding. *J Clin Gastroenterol*. 2001;32(4):340–343.
45. Hu WT, Kantarci OH, Merritt 2nd. JL, et al. Ornithine transcarbamylase deficiency presenting as encephalopathy during adulthood following bariatric surgery. *Arch Neurol*. 2007;64(1):126–128.
46. Lipskind S, Loanzon S, Simi E, Ouyang DW. Hyperammonemic coma in an ornithine transcarbamylase mutation carrier following antepartum corticosteroids. *J Perinatol*. 2011;31(10):682–684.
47. Thakur V, Rupar CA, Ramsay DA, Singh R, Fraser DD. Fatal cerebral edema from late-onset ornithine transcarbamylase deficiency in a juvenile male patient receiving valproic acid. *Pediatr Crit Care Med*. 2006;7(3):273–276.
48. Ohura T, Kobayashi K, Tazawa Y, et al. Neonatal presentation of adult-onset type II citrullinemia. *Hum Genet*. 2001;108(2):87–90.
49. Song YZ, Deng M, Chen FP, et al. Genotypic and phenotypic features of citrin deficiency: five-year experience in a Chinese pediatric center. *Int J Mol Med*. 2011;28(1):33–40.
50. Neill MA, Aschner J, Barr F, Summar ML. Quantitative RT-PCR comparison of the urea and nitric oxide cycle gene transcripts in adult human tissues. *Mol Genet Metab*. 2009;97(2):121–127.
51. Windmueller HG, Spaeth AE. Source and fate of circulating citrulline. *Am J Physiol*. 1981;241(6):E473–E480.
52. Aoshima T, Kajita M, Sekido Y, et al. Novel mutations (H337R and 238-362del) in the CPS1 gene cause carbamoyl phosphate synthetase I deficiency. *Hum Hered*. 2001;52(2):99–101.
53. Loscalzo ML, Galczynski RL, Hamosh A, Summar M, Chinsky JM, Thomas GH. Interstitial deletion of chromosome 2q32-34 associated with multiple congenital anomalies and a urea cycle defect (CPS I deficiency). *Am J Med Genet A*. 2004;128A(3):311–315.
54. Yamaguchi S, et al. Mutations and polymorphisms in the human ornithine transcarbamylase (OTC) gene. *Hum Mutat*. 2006;27(7):626–632.
55. Azevedo L, Soares PA, Quental R, et al. Mutational spectrum and linkage disequilibrium patterns at the ornithine transcarbamylase gene (OTC). *Ann Hum Genet*. 2006;70:797–801.
56. Kim GH, Choi JH, Lee HH, Park S, Kim SS, Yoo HW. Identification of novel mutations in the human ornithine transcarbamylase (OTC) gene of Korean patients with OTC deficiency and transient expression of the mutant proteins in vitro. *Hum Mutat*. 2006;27(11):1159.
57. Ogino W, Takeshima Y, Nishiyama A, et al. Mutation analysis of the ornithine transcarbamylase (OTC) gene in five Japanese OTC deficiency patients revealed two known and three novel mutations including a deep intronic mutation. *Kobe J Med Sci*. 2007;53(5):229–240.
58. Shchelochkov OA, Li FY, Geraghty MT, et al. High-frequency detection of deletions and variable rearrangements at the ornithine transcarbamylase (OTC) locus by oligonucleotide array CGH. *Mol Genet Metab*. 2009;96(3):97–105.
59. Ben-Ari Z, Dalal A, Morry A, et al. Adult-onset ornithine transcarbamylase (OTC) deficiency unmasked by the Atkins' diet. *J Hepatol*. 2010;52(2):292–295.
60. Lin Y, Liu JC, Zhang XJ, et al. Downregulation of the ornithine decarboxylase/polyamine system inhibits angiotensin-induced hypertrophy of cardiomyocytes through the NO/cGMP-dependent protein kinase type-I pathway. *Cell Physiol Biochem*. 2010;25(4–5):443–450.
61. Tuchman M, Jaleel N, Morizono H, Sheehy L, Lynch MG. Mutations and polymorphisms in the human ornithine transcarbamylase gene. *Hum Mutat*. 2002;19(2):93–107.
62. Engel K, Nuoffer JM, Mühlhausen C, et al. Analysis of mRNA transcripts improves the success rate of molecular genetic testing in OTC deficiency. *Mol Genet Metab*. 2008;94(3):292–297.
63. Engel K, Höhne W, Häberle J. Mutations and polymorphisms in the human argininosuccinate synthetase (ASS1) gene. *Hum Mutat*. 2009;30(3):300–307.
64. Gao HZ, Kobayashi K, Tabata A, et al. Identification of 16 novel mutations in the argininosuccinate synthetase gene and genotype–phenotype correlation in 38 classical citrullinemia patients. *Hum Mutat*. 2003;22(1):24–34.
65. Linnebank M, Tschiedel E, Häberle J, et al. Argininosuccinate lyase (ASL) deficiency: mutation analysis in 27 patients and a completed structure of the human ASL gene. *Hum Genet*. 2002;111(4–5):350–359.
66. Trevisson E, Salviati L, Baldoin MC, et al. Argininosuccinate lyase deficiency: mutational spectrum in Italian patients and identification of a novel ASL pseudogene. *Hum Mutat*. 2007;28(7):694–702.
67. Kleijer WJ, Garritsen VH, Linnebank M, et al. Clinical, enzymatic, and molecular genetic characterization of a biochemical variant type of argininosuccinic aciduria: prenatal and postnatal diagnosis in five unrelated families. *J Inherit Metab Dis*. 2002;25(5):399–410.
68. Al-Sayed M, Alahmed S, Alsmadi O, et al. Identification of a common novel mutation in Saudi patients with argininosuccinic aciduria. *J Inherit Metab Dis*. 2005;28(6):877–883.
69. Vockley JG, Goodman BK, Tabor DE, et al. Loss of function mutations in conserved regions of the human arginase I gene. *Biochem Mol Med*. 1996;59(1):44–51.
70. Morris Jr. SM, Bhamidipati D, Kepka-Lenhart D. Human type II arginase: sequence analysis and tissue-specific expression. *Gene*. 1997;193(2):157–161.
71. Shi O, et al. Generation of a mouse model for arginase II deficiency by targeted disruption of the arginase II gene. *Mol Cell Biol*. 2001;21(3):811–813.
72. Camacho JA, Mardach R, Rioseco-Camacho N, et al. Clinical and functional characterization of a human ORNT1 mutation (T32R) in the hyperornithinemia-hyperammonemia-homocitrullinuria (HHH) syndrome. *Pediatr Res*. 2006;60(4):423–429.

73. Debray FG, Lambert M, Lemieux B, et al. Phenotypic variability among patients with hyperornithinaemia-hyperammonaemia-homocitrullinuria syndrome homozygous for the delF188 mutation in SLC25A15. *J Med Genet*. 2008;45(11):759–764.
74. Tessa A, Fiermonte G, Dionisi-Vici C, et al. Identification of novel mutations in the SLC25A15 gene in hyperornithinemia-hyperammonemia-homocitrullinuria (HHH) syndrome: a clinical, molecular, and functional study. *Hum Mutat*. 2009;30(5):741–748.
75. Camacho JA, Obie C, Biery B, et al. Hyperornithinaemia-hyperammonaemia-homocitrullinuria syndrome is caused by mutations in a gene encoding a mitochondrial ornithine transporter. *Nat Genet*. 1999;22(2):151–158.
76. Fiermonte G, Dolce V, David L, et al. The mitochondrial ornithine transporter. Bacterial expression, reconstitution, functional characterization, and tissue distribution of two human isoforms. *J Biol Chem*. 2003;278(35):32778–32783.
77. Miyamoto T, Kanazawa N, Kato S, et al. Diagnosis of Japanese patients with HHH syndrome by molecular genetic analysis: a common mutation, R179X. *J Hum Genet*. 2001;46(5):260–262.
78. Tessa A, Fiermonte G, Dionisi-Vici C, et al. Identification of novel mutations in the SLC25A15 gene in hyperornithinemia-hyperammonemia-homocitrullinuria (HHH) syndrome: a clinical, molecular, and functional study. *Hum Mutat*. 2009;30(5):741–748.
79. Tsujino S, Kanazawa N, Ohashi T, et al. Three novel mutations (G27E, insAAC, R179X) in the ORNT1 gene of Japanese patients with hyperornithinemia, hyperammonemia, and homocitrullinuria syndrome. *Ann Neurol*. 2000;47(5):625–631.
80. Fiermonte G, Dolce V, David L, et al. The mitochondrial ornithine transporter. Bacterial expression, reconstitution, functional characterization, and tissue distribution of two human isoforms. *J Biol Chem*. 2003;278(35):32778–32783.
81. Kobayashi K, Sinasac DS, Iijima M, et al. The gene mutated in adult-onset type II citrullinaemia encodes a putative mitochondrial carrier protein. *Nat Genet*. 1999;22(2):159–163.
82. Sinasac DS, Moriyama M, Jalil MA, et al. Slc25a13-knockout mice harbor metabolic deficits but fail to display hallmarks of adult-onset type II citrullinemia. *Mol Cell Biol*. 2004;24(2):527–536.
83. Yasuda T, Yamaguchi N, Kobayashi K, et al. Identification of two novel mutations in the SLC25A13 gene and detection of seven mutations in 102 patients with adult-onset type II citrullinemia. *Hum Genet*. 2000;107(6):537–545.
84. Ben-Shalom E, Kobayashi K, Shaag A, et al. Infantile citrullinemia caused by citrin deficiency with increased dibasic amino acids. *Mol Genet Metab*. 2002;77(3):202–208.
85. Yamaguchi N, Kobayashi K, Yasuda T, et al. Screening of SLC25A13 mutations in early and late onset patients with citrin deficiency and in the Japanese population: identification of two novel mutations and establishment of multiple DNA diagnosis methods for nine mutations. *Hum Mutat*. 2002;19(2):122–130.
86. Dimmock D, Maranda B, Dionisi-Vici C, et al. Citrin deficiency, a perplexing global disorder. *Mol Genet Metab*. 2009;96(1):44–49.
87. Butterworth RF, Giguère JF, Michaud J, Lavoie J, Layrargues GP. Ammonia: key factor in the pathogenesis of hepatic encephalopathy. *Neurochem Pathol*. 1987;6(1–2):1–12.
88. Ott P, Clemmesen O, Larsen FS. Cerebral metabolic disturbances in the brain during acute liver failure: from hyperammonemia to energy failure and proteolysis. *Neurochem Int*. 2005;47(1–2):13–18.
89. Felipo V, Butterworth R. Mitochondrial dysfunction in acute hyperammonemia. *Neurochem Int*. 2002;40(6):487–491.
90. Butterworth RF. Effects of hyperammonaemia on brain function. *J Inherit Metab Dis*. 1998;21(suppl 1):6–20.
91. Dolman CL, Clasen RA, Dorovini-Zis K. Severe cerebral damage in ornithine transcarbamylase deficiency. *Clin Neuropathol*. 1988;7(1):10–15.
92. Harding BN, Leonard JV, Erdohazi M. Ornithine carbamoyl transferase deficiency: a neuropathological study. *Eur J Pediatr*. 1984;141(4):215–220.
93. Krieger I, Snodgrass PJ, Roskamp J. Atypical clinical course of ornithine transcarbamylase deficiency due to a new mutant (comparison with Reye's disease). *J Clin Endocrinol Metab*. 1979;48(3):388–392.
94. Kornfeld M, Woodfin BM, Papile L, Davis LE, Bernard LR. Neuropathology of ornithine carbamyl transferase deficiency. *Acta Neuropathol*. 1985;65(3–4):261–264.
95. Mattson LR, Lindor NM, Goldman DH, Goodwin JT, Groover RV, Vockley J. Central pontine myelinolysis as a complication of partial ornithine carbamoyl transferase deficiency. *Am J Med Genet*. 1995;60(3):210–213.
96. Yamanouchi H, Yokoo H, Yuhara Y, et al. An autopsy case of ornithine transcarbamylase deficiency. *Brain Dev*. 2002;24(2):91–94.
97. Kreis R, Ross BD, Farrow NA, Ackerman Z. Metabolic disorders of the brain in chronic hepatic encephalopathy detected with H-1 MR spectroscopy. *Radiology*. 1992;182(1):19–27.
98. Batshaw ML, Msall M, Beaudet AL, Trojak J. Risk of serious illness in heterozygotes for ornithine transcarbamylase deficiency. *J Pediatr*. 1986;108(2):236–241.
99. Bender AS, Norenberg MD. Effects of ammonia on L-glutamate uptake in cultured astrocytes. *Neurochem Res*. 1996;21(5):567–573.
100. Norenberg MD. Astrocytic-ammonia interactions in hepatic encephalopathy. *Semin Liver Dis*. 1996;16(3):245–253.
101. Takahashi H, Koehler RC, Brusilow SW, Traystman RJ. Inhibition of brain glutamine accumulation prevents cerebral edema in hyperammonemic rats. *Am J Physiol*. 1991;261(3 Pt 2):H825–H829.
102. Blei AT, Olafsson S, Therrien G, Butterworth RF. Ammonia-induced brain edema and intracranial hypertension in rats after portacaval anastomosis. *Hepatology*. 1994;19(6):1437–1444.
103. DeMars R, LeVan SL, Trend BL, Russell LB. Abnormal ornithine carbamoyltransferase in mice having the sparse-fur mutation. *Proc Natl Acad Sci U S A*. 1976;73(5):1693–1697.
104. Doolittle DP, Hulbert LL, Cordy C. A new allele of the sparse fur gene in the mouse. *J Hered*. 1974;65(3):194–195.
105. Patejunas G, Bradley A, Beaudet AL, O'Brien WE. Generation of a mouse model for citrullinemia by targeted disruption of the argininosuccinate synthetase gene. *Somat Cell Mol Genet*. 1994;20(1):55–60.
106. Reid Sutton V, Pan Y, Davis EC, Craigen WJ. A mouse model of argininosuccinic aciduria: biochemical characterization. *Mol Genet Metab*. 2003;78(1):11–16.
107. Iyer RK, Yoo PK, Kern RM, et al. Mouse model for human arginase deficiency. *Mol Cell Biol*. 2002;22(13):4491–4498.
108. Deignan JL, Livesay JC, Yoo PK, et al. Ornithine deficiency in the arginase double knockout mouse. *Mol Genet Metab*. 2006;89(1–2):87–96.
109. Saheki T, Iijima M, Li MX, et al. Citrin/mitochondrial glycerol-3-phosphate dehydrogenase double knock-out mice recapitulate features of human citrin deficiency. *J Biol Chem*. 2007;282(34):25041–25052.
110. Sinasac DS, Moriyama M, Jalil MA, et al. Slc25a13-knockout mice harbor metabolic deficits but fail to display hallmarks of adult-onset type II citrullinemia. *Mol Cell Biol*. 2004;24(2):527–536.

111. Metzeler KH, Boeck S, Christ B, et al. Idiopathic hyperammonemia (IHA) after dose-dense induction chemotherapy for acute myeloid leukemia: Case report and review of the literature. *Leuk Res*. 2009;33(7):69–72.
112. Mitchell RB, Wagner JE, Karp JE, et al. Syndrome of idiopathic hyperammonemia after high-dose chemotherapy: review of nine cases. *Am J Med*. 1988;85(5):662–667.
113. Hudak ML, Jones Jr. MD, Brusilow SW. Differentiation of transient hyperammonemia of the newborn and urea cycle enzyme defects by clinical presentation. *J Pediatr*. 1985;107(5):712–719.
114. Coude FX, Grimber G, Parvy P, Rabier D. Role of N-acetylglutamate and acetyl-CoA in the inhibition of ureagenesis by isovaleric acid in isolated rat hepatocytes. *Biochim Biophys Acta*. 1983;761(1):13–16.
115. Coude FX, Ogier H, Grimber G, et al. Correlation between blood ammonia concentration and organic acid accumulation in isovaleric and propionic acidemia. *Pediatrics*. 1982;69(1):115–117.
116. Coude FX, Sweetman L, Nyhan WL. Inhibition by propionyl-coenzyme A of N-acetylglutamate synthetase in rat liver mitochondria. A possible explanation for hyperammonemia in propionic and methylmalonic acidemia. *J Clin Invest*. 1979;64(6):1544–1551.
117. Ah Mew N, McCarter R, Daikhin Y, Nissim I, Yudkoff M, Tuchman M. N-carbamylglutamate augments ureagenesis and reduces ammonia and glutamine in propionic acidemia. *Pediatrics*. 2010;126(1):e208–e214.
118. Dercksen M, Ijlst L, Duran M, et al. Inhibition of *N*-acetylglutamate synthase by various monocarboxylic and dicarboxylic short-chain coenzyme A esters and the production of alternative glutamate esters. *Biochim Biophys Acta*. 2013, Epub ahead of print.
119. Kalkan Ucar S, Coker M, Habif S, et al. The first use of *N*-carbamylglutamate in a patient with decompensated maple syrup urine disease. *Metab Brain Dis*. 2009;24(3):409–414.
120. Corvi MM, Soltys CL, Berthiaume LG. Regulation of mitochondrial carbamoyl-phosphate synthetase 1 activity by active site fatty acylation. *J Biol Chem*. 2001;276(49):45704–45712.
121. Costell M, O'Connor JE, Míguez MP, Grisolía S. Effects of L-carnitine on urea synthesis following acute ammonia intoxication in mice. *Biochem Biophys Res Commun*. 1984;120(3):726–733.
122. Lauteala T, Sistonen P, Savontaus ML, et al. Lysinuric protein intolerance (LPI) gene maps to the long arm of chromosome 14. *Am J Hum Genet*. 1997;60(6):1479–1486.
123. Ogier de Baulny H, Schiff M, Dionisi-Vici C. Lysinuric protein intolerance (LPI): a multi organ disease by far more complex than a classic urea cycle disorder. *Mol Genet Metab*. 2012;106(1):12–17.
124. Berry GT, Steiner RD. Long-term management of patients with urea cycle disorders. *J Pediatr*. 2001;138(1 suppl):S56–S60, discussion S60–S61.
125. Summar M, Pietsch J, Deshpande J, Schulman G. Effective hemodialysis and hemofiltration driven by an extracorporeal membrane oxygenation pump in infants with hyperammonemia. *J Pediatr*. 1996;128(3):379–382.
126. Sadowski RH, Harmon WE, Jabs K. Acute hemodialysis of infants weighing less than five kilograms. *Kidney Int*. 1994;45(3):903–906.
127. Kang ES, Snodgrass PJ, Gerald PS. Ornithine transcarbamylase deficiency in the newborn infant. *J Pediatr*. 1973;82(4):64264–64269.
128. Praphanphoj V, Boyadjiev SA, Waber LJ, Brusilow SW, Geraghty MT. Three cases of intravenous sodium benzoate and sodium phenylacetate toxicity occurring in the treatment of acute hyperammonaemia. *J Inherit Metab Dis*. 2000;23(2):129–136.
129. Lanpher BC, Gropman A, Chapman KA, Lichter-Konecki U, Urea Cycle Disorders Consortium, Summar ML. Urea cycle disorders overview. In: Pagon RA, et al., eds. GeneReviews. 1993 Seattle (WA).
130. Guffon N, Schiff M, Cheillan D, Wermuth B, Häberle J, Vianey-Saban C. Neonatal hyperammonemia: the N-carbamoyl-L-glutamic acid test. *J Pediatr*. 2005;147(2):260–262.
131. Trumbo P, Schlicker S, Yates AA, Poos M, Food and Nutrition Board of the Institute of Medicine, The National Academies. Dietary reference intakes for energy, carbohydrate, fiber, fat, fatty acids, cholesterol, protein and amino acids. *J Am Diet Assoc*. 2002;102(11):1621–1630.
132. Batshaw ML, MacArthur RB, Tuchman M. Alternative pathway therapy for urea cycle disorders: twenty years later. *J Pediatr*. 2001;138 (1 suppl):S46–S54 discussion S54–S55..
133. Lichter-Konecki U, Diaz GA, Merritt 2nd. JL, et al. Ammonia control in children with urea cycle disorders (UCDs); phase 2 comparison of sodium phenylbutyrate and glycerol phenylbutyrate. *Mol Genet Metab*. 2011;103(4):32332–32339.
134. Lee B, Goss J. Long-term correction of urea cycle disorders. *J Pediatr*. 2001;138(1 suppl):S62–S71.
135. Erez A, Nagamani SC, Lee B. Argininosuccinate lyase deficiency-argininosuccinic aciduria and beyond. *Am J Med Genet C Semin Med Genet*. 2011;157(1):45–53.
136. Nagamani SC, Campeau PM, Shchelochkov OA, et al. Nitric-oxide supplementation for treatment of long-term complications in argininosuccinic aciduria. *Am J Hum Genet*. 2012;90(5):836–846.
137. Nagamani SC, Erez Alee B. Argininosuccinate lyase deficiency. *Genet Med*. 2012;14(5):501–507.
138. Perito ER, Rhee S, Roberts JP, Rosenthal P. Pediatric liver transplant for urea cycle disorders and organic acidemias: United Network for Organ Sharing (UNOS) data 2002–2012. *Liver Transpl*. 2013; Epub ahead of print.
139. Heibel SK, Ah Mew N, Caldovic L, Daikhin Y, Yudkoff M, Tuchman M. N-carbamylglutamate enhancement of ureagenesis leads to discovery of a novel deleterious mutation in a newly defined enhancer of the NAGS gene and to effective therapy. *Hum Mutat*. 2011;32(10): 1153–1160.
140. Tuchman M, Caldovic L, Daikhin Y, et al. N-carbamylglutamate markedly enhances ureagenesis in *N*-acetylglutamate deficiency and propionic acidemia as measured by isotopic incorporation and blood biomarkers. *Pediatr Res*. 2008;64(2):213–217.
141. Bachmann C, Colombo JP. Increase of tryptophan and 5-hydroxyindole acetic acid in the brain of ornithine carbamoyltransferase deficient sparse-fur mice. *Pediatr Res*. 1984;18(4):372–375.
142. Bachmann C. Long-term outcome of patients with urea cycle disorders and the question of neonatal screening. *Eur J Pediatr*. 2003;162 (suppl 1):S29–S33.
143. Gropman AL, Batshaw ML. Cognitive outcome in urea cycle disorders. *Mol Genet Metab*. 2004;81(suppl 1):S58–S62.
144. Msall M, Batshaw ML, Suss R, Brusilow SW, Mellits ED. Neurologic outcome in children with inborn errors of urea synthesis. Outcome of urea-cycle enzymopathies. *N Engl J Med*. 1984;310(23):1500–1505.
145. Breningstall GN. Neurologic syndromes in hyperammonemic disorders. *Pediatr Neurol*. 1986;2(5):253–262.

146. Nicolaides P, Liebsch D, Dale N, Leonard J, Surtees R. Neurological outcome of patients with ornithine carbamoyltransferase deficiency. *Arch Dis Child*. 2002;86(1):54–56.
147. Bachmann C. Ornithine carbamoyl transferase deficiency: findings, models and problems. *J Inherit Metab Dis*. 1992;15(4):578–591.
148. Enns GM, Berry SA, Berry GT, Rhead WJ, Brusilow SW, Hamosh A. Survival after treatment with phenylacetate and benzoate for urea-cycle disorders. *N Engl J Med*. 2007;356(22):2282–2292.
149. Picca S, Dionisi-Vici C, Abeni D, et al. Extracorporeal dialysis in neonatal hyperammonemia: modalities and prognostic indicators. *Pediatr Nephrol*. 2001;16(11):862–867.
150. Uchino T, Endo F, Matsuda I. Neurodevelopmental outcome of long-term therapy of urea cycle disorders in Japan. *J Inherit Metab Dis*. 1998;21(suppl 1):151–159.
151. Raper SE, Chirmule N, Lee FS, et al. Fatal systemic inflammatory response syndrome in a ornithine transcarbamylase deficient patient following adenoviral gene transfer. *Mol Genet Metab*. 2003;80(1–2):148–158.
152. Wang L, Morizono H, Lin J, et al. Preclinical evaluation of a clinical candidate AAV8 vector for ornithine transcarbamylase (OTC) deficiency reveals functional enzyme from each persisting vector genome. *Mol Genet Metab*. 2012;105(2):203–211.
153. Meyburg J, Hoffmann GF. Liver, liver cell and stem cell transplantation for the treatment of urea cycle defects. *Mol Genet Metab*. 2010;100(suppl 1):S77–S83.
154. Meyburg J, Schmidt J, Hoffmann GF. Liver cell transplantation in children. *Clin Transplant*. 2009;23(suppl 21):75–82.
155. Najimi M, Khuu DN, Lysy PA, et al. Adult-derived human liver mesenchymal-like cells as a potential progenitor reservoir of hepatocytes? *Cell Transplant*. 2007;16(7):717–728.
156. Wilson JM, Shchelochkov OA, Gallagher RC, Batshaw ML. Hepatocellular carcinoma in a research subject with ornithine transcarbamylase deficiency. *Mol Genet Metab*. 2012;105(2):263–265.

CHAPTER

58

Glucose Transporter Type I Deficiency and Other Glucose Flux Disorders

Juan M. Pascual, Dong Wang†, and Darryl C. De Vivo‡*

*The University of Texas Southwestern Medical Center, Dallas, TX, USA

†Southern Regional Medical Center, Riverdale, GA, USA

‡Columbia University Medical Center, New York, NY, USA

OVERVIEW OF GLUCOSE TRANSPORT

Transport of water-soluble molecules across tissue barriers has been a subject of intense interest since 1952.[1] Widdas proposed that the transport of glucose across the erythrocyte membrane required a carrier mechanism to facilitate diffusion driven by concentration. Soon afterwards, Crane postulated a Na^+ glucose cotransport hypothesis to account for the accumulation of glucose in epithelial cells against a concentration gradient.[2] These observations proved correct: Today most cloned genes that encode proteins endowed with significant glucose-transporter capacity follow one of two major protein structural and functional plans: facilitative and concentrative.[3–5] Each of these families includes additional, structurally related molecules of uncertain function, amounting, together with the canonical glucose transporters, to a total of 26 genes. The principal value of these discoveries is that they, in conjunction with *in vitro* structure–function studies, provide important insights into several genetically determined human diseases and help understand pharmacological action.

The first major family of glucose transporters includes the concentrative sodium glucose transporters (SGLT), which are also known as active cotransporters or symporters. These symporters are energized by proton or sodium gradients, which are generated by adenosine triphosphate (ATP)-dependent ion pumps. Twelve types of concentrative SGLT proteins can cotransport glucose with Na^+ ions. The energy required for transport derives from a transmembrane gradient of Na^+ ions that is maintained by a Na^+/K^+-ATPase. SGLT1 is responsible for glucose absorption in the intestinal tract. The sequence of the 664 amino acids of mammalian SGLT1 was inferred in 1987 via expression cloning of its cDNA.[6] The other major family is the glucose transporter (GLUT) gene family. The GLUT family consists of 14 proteins. They differ from one another in their alternative splicing, affinity both for glucose and for a heterogeneous array of other substrates, tissue distribution and type of signals that cause their translocation to the cell membrane. These transporters facilitate passive diffusion of glucose across tissue barriers by energy-independent, stereospecific mechanisms.

CLINICAL FEATURES

SGLT1 Deficiency

Mutations in the *SGLT1* gene cause glucose–galactose malabsorption (OMIM 182380).[7] The glucose–galactose malabsorption syndrome was described in 1962.[8,9] Both reports described a familial entity of early onset characterized by watery, acidic diarrhea and life-threatening dehydration. The symptoms typically begin before the fourth day of life and respond to the elimination of dietary glucose and galactose. This syndrome has no known primary neurologic repercussions.

Rosenberg's Molecular and Genetic Basis of Neurological and Psychiatric Disease
http://dx.doi.org/10.1016/10.1016/B978-0-12-410529-4.00058-9

GLUT2 Deficiency

Two human diseases are associated with mutations of GLUT family proteins. The first is Fanconi–Bickel syndrome (OMIM 227810), and the second is GLUT1 deficiency syndrome (OMIM 606777). The Fanconi–Bickel syndrome was described in 1949[10] and originally named hepatorenal glycogenosis with renal Fanconi syndrome. The first patient presented at age 6 months with failure to thrive, polydipsia, and constipation. Later in childhood, he developed osteopenia, short stature, hepatomegaly, and tubular nephropathy. The nephropathy was associated with glycosuria, phosphaturia, aminoaciduria, and intermittent proteinuria. These urinary metabolites were accompanied by hypophosphatemia and hyperuricemia. The liver was filled with glycogen and fat. Distinctively, ketotic hypoglycemia was evident preprandially, and hyperglycemia was evident postprandially. This distinctive feature is the result of the multifunctional role of GLUT2 in the pancreas and the liver. These patients may manifest intestinal malabsorption and diarrhea, but they do not exhibit any primary neurologic complaints. A mutation in the *GLUT2* gene was identified in the original case by Santer et al.[11] Most patients with Fanconi–Bickel syndrome are homozygous for disease-related mutations, consistent with an autosomal recessive pattern of inheritance.

GLUT1 Deficiency

GLUT1 deficiency syndrome was described in 1991.[12] The phenotypic spectrum of GLUT1 deficiency continues to broaden as more patients are diagnosed. The clinical signature of the most common phenotype is an infantile-onset epileptic encephalopathy associated with delayed neurologic development, deceleration of head growth, acquired microcephaly, incoordination, and spasticity. Diminished brain glucose accumulation can be proven by positron emission tomography of patients using ^{18}F-deoxyglucose as a tracer.[13] Movement disorders (Table 58.1) are almost invariably part of the phenotype, either in isolation or in the context of epilepsy. The two children described by De Vivo et al. in 1991 exhibited these neurologic signs. The presence of hypoglycorrhachia (diminished cerebrospinal glucose concentration) led the investigators to speculate that there was a defect in the transport of glucose across the blood–brain barrier. Seven years later, these speculations were substantiated. The first patient had a large-scale deletion involving one GLUT1 allele causing hemizygosity, and the second patient had a heterozygous nonsense mutation.[14]

As many as 90% of patients exhibit the most common (epileptic) phenotype, which is best characterized a developmental encephalopathy with seizures. In fact, GLUT1 deficiency is a well-recognized cause of absence epilepsy[15–17] and of idiopathic generalized epilepsy.[18] Seizures typically begin within the first 4 months of life. The prenatal and perinatal histories, birth weights, and Apgar scores are normal. The earliest epileptic events include apneic episodes and episodic eye movements simulating opsoclonus. The infantile seizures are clinically fragmented, and the electroencephalographic (EEG) correlate is that of multifocal spike-wave discharges. Seizures become more synchronized with brain maturation and present clinically as generalized events associated with 2.5- to 4-Hz generalized spike and wave discharges electrically. Several seizure types have been described in these patients, including generalized tonic or clonic seizures, myoclonic seizures, atypical absence, atonic, infantile spasms and unclassified seizures.[19,20] The frequency of clinical seizures varies considerably from one patient to another. Some have daily seizures, whereas others have only occasional seizures separated by days, weeks, or months. The average age of onset of epilepsy is 8 months. Seizures are mixed in about 70% of patients: generalized tonic–clonic (53%), absence (49%), complex partial (37%), myoclonic (27%), drop (26%), tonic (12%), simple partial (3%), and spasms (3%). However, about 15% of patients can exhibit a normal EEG. Many of these patients suffer other paroxysmal events, and it remains unclear whether these events are epileptic or nonepileptic. These paroxysmal events include intermittent ataxia, confusion, lethargy or somnolence,

TABLE 58.1 Movement Disorders in GLUT1 Deficiency

Movement disorder	Reference
Spastic ataxia	12
Paroxysmal exercise-induced (kinesigenic) dystonia (DYT18)	22
Paroxysmal choreoathetosis/spasticity (DYT9)	26
Chorea	28
Dystonic tremor	117
Dysarthria	Multiple publications

alternating hemiparesis, abnormalities of movement or posture, total body paralysis, sleep disturbances, and migraines. The frequency of neurologic symptoms can fluctuate and may be influenced by fasting or fatigue.[21] The majority of patients exhibit variable degrees of speech and language impairment. Dysarthria is common. Both receptive and expressive language skills are affected, but expressive language skills are disproportionately affected. Varying degrees of cognitive impairment are also evident, ranging from learning disabilities to severe mental retardation. Social adaptive behavior is a strength in these children: they are remarkably comfortable in the group setting and interact well with their peers and with adults. Acquired microcephaly occurs in 50% of patients, and deceleration of head growth is more common. However, a small number of patients experience normal head growth. Additional neurological signs can relate to the pyramidal, extrapyramidal, and cerebellar systems predominantly and consist of varying degrees of spasticity, dystonia, and ataxia. The limb tone generally is increased, tendon reflexes are brisk, and Babinski signs are present. The gait abnormality is best described as a spastic ataxia. No consistent abnormalities of ocular fundi, cranial nerves, or sensation have been described and the general physical examination is normal.

A second phenotype includes paroxysmal exercise-induced dyskinesia (DYT18) and epilepsy.[22] This syndrome is characterized by choreoathetosis, dystonia, or a combination of both, affecting primarily mainly the lower extremities without intellectual disability. Adolescent or adult-onset cases have been reported.[23] Epilepsy manifests as primary generalized seizures. Some of these patients also exhibit hemolytic anemia with echinocytosis associated with a red cell cation leak that alters intracellular concentrations of sodium, potassium and calcium.[24] The relevance of GLUT1 to hematological function is also highlighted by stomatin-deficient cryohydrocytosis, a distinct, dominantly inherited hemolytic anemia in which the permeability of the erythrocyte membrane to monovalent cations is also pathologically increased.[25] These patients can also harbor GLUT1 mutations in the context of seizures, developmental delay, an episodic movement disorder and cataracts.

A third phenotype is characterized by paroxysmal choreoathetosis/spasticity (DYT9; which includes slowly progressive spastic paraparesis in association with episodic choreoathetosis).[26] Notably, autosomal dominant forms of isolated hereditary spastic paraparesis do not bear association with GLUT1 deficiency.

Additional phenotypes have long been identified and novel variants continue to be periodically reported. One patient has mental retardation and intermittent ataxia without any clinical seizures,[27] one exhibits chorea,[28] and one patient has a movement disorder characterized by choreoathetosis and dystonia.[29] These movement disorders typically respond to a ketogenic diet. These less common phenotypes, thus far observed in about 10% of the GLUT1 deficiency syndrome patient population, indicate that diagnostic studies should be performed on all infants and children with unexplained neurologic symptoms, including epilepsy, mental retardation and movement disorders.

MOLECULAR GENETICS OF GLUT1 DEFICIENCY

The GLUT1 deficiency syndrome generally presents as a new mutation and behaves as a cellular codominant trait with haploinsufficiency. Most patients harbor unique mutations distributed throughout the GLUT1 gene. Several families have been recognized, documenting an autosomal dominant pattern of inheritance over two or more generations. The familial cases all harbor missense mutations that are likely less pathogenic than other loss-of-function mutations.

The first family included five patients in three generations.[30] The affected family members manifested mild to severe seizures, developmental delay, ataxia, hypoglycorrhachia in the three patients undergoing lumbar puncture, and decreased erythrocyte glucose uptake. Other patterns of inheritance also have been identified, including paternal mosaicism, compound heterozygosity and recessivity.[31]

Normal Gene Product

In 1985, Lodish and coworkers first cloned a protein, GLUT1, that facilitated the diffusion of glucose across the hepatic cell membrane.[32] The following year, Birnbaum et al. proceeded to clone and sequence the gene encoding GLUT1 from the rat brain.[33] The gene is localized in human chromosome 1 (1p34.2). GLUT1 and additional GLUT genes (including the proton-driven myoinositol transporter HMIT or GLUT13) constitute a subset within a larger family of transport facilitators designated SLC2A for solute carrier 2A.[34] GLUT1 is highly expressed in erythrocytes and brain. Human GLUT1 has a molecular mass of 54,117 Da, a single *N*-linked oligosaccharide, and 12 transmembrane segments (Figure 58.1). By analogy with XylE, an *Escherichia coli* homolog of GLUT1-4 whose structure has been solved in complex with D-xylose by X-ray crystallography,[35] the GLUT1 transporter is an aqueous sugar translocation pathway through the lipid bilayer via the clustering of several transmembrane helices. The authors and

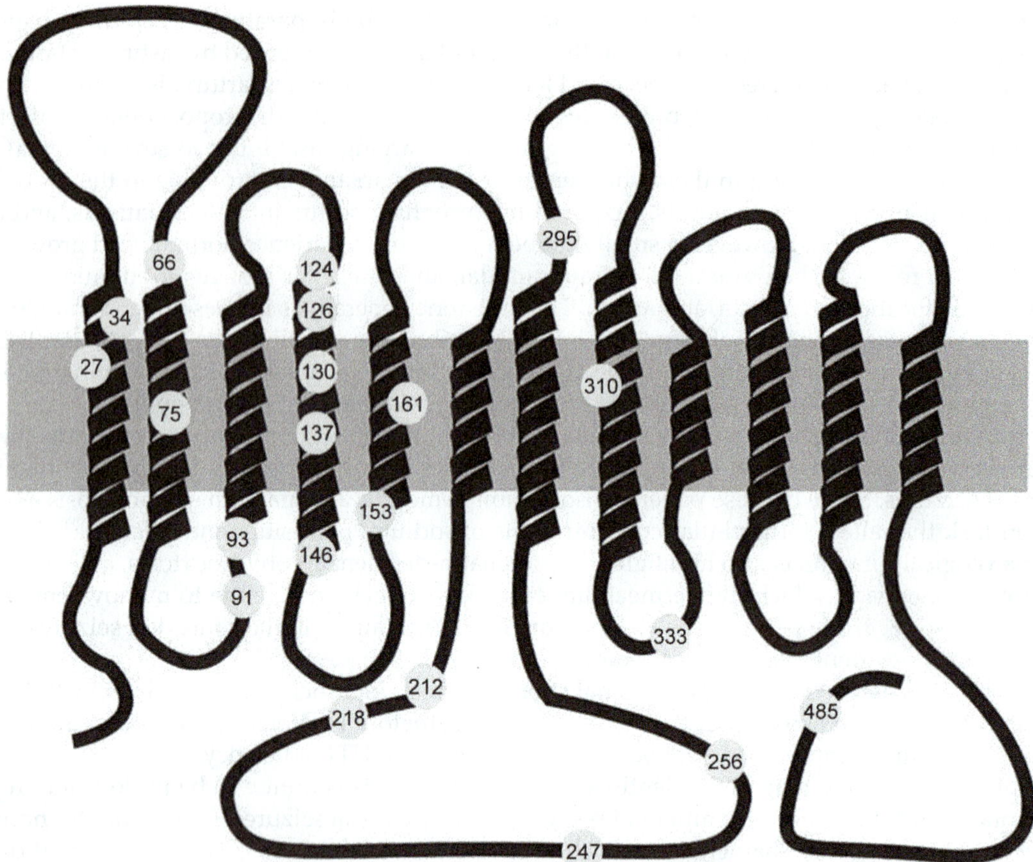

FIGURE 58.1 **GLUT1 topological model.** GLUT1 is represented in the context of the plasma membrane (gray rectangle) and comprising cytoplasmic amino- (left side) and carboxyl-terminal (right side) domains. The extracellular side of the membrane is situated on the top part of the schematic. The relative length of membrane-spanning domains follows the crystal structure of LacY, and the length of the extramembranous amino and carboxyl termini and interhelical loops is also drawn approximately to scale. Numbered circles indicating residue number identify the location of select missense mutations, which are scattered throughout the protein, with a slight predilection for the fourth transmembrane domain. Each mutation may contribute a different molecular pathogenetic mechanism:[36] for example, 218 can be considered a susceptibility allele rather than a deleterious substitution, whereas the disruption caused by the 295 substitution may be temperature dependent.[118]

others have been able to examine the bidirectional flux of glucose, dehydroascorbic acid, and water in the wild-type and mutant GLUT1 proteins using heterologous expression in oocytes and other cells.

Abnormal Gene Product

By 2008, we had identified 100 patients, who harbored an array of mutations in the transporter gene. The structure of the transporter was, at the time, unknown, and at least four conflicting structural models had been put forth by others on the basis of membrane protein homology and/or *in silico* (computational) structural predictions or using site-directed mutagenesis of transporter regions thought to be important for function. In contrast with these approaches, using the location and nature (i.e., the degree of amino acid conservation) of human pathogenic mutations as potential reporters of structurally relevant regions, we constructed a series of site-directed mutants for detailed topological and functional characterization into *Xenopus laevis* oocytes. Then, additional mutagenesis of functionally relevant individual sites identified by this approach was combined with targeted chemical modification (Figure 58.2). The goal of these studies was both topological (i.e., elucidating the location and secondary structure of protein regions) and functional (that is, determining their role in the overall operation of the transporter). This approach yielded several key observations about the structure and function of the transporter and pathogenic mechanisms[36] that were later supported by the solution of the crystal structure of a homologous protein (Figure 58.3).[35] Only a set of conserved amino acids seems essential. These include glycine, capable of inducing helical breaks, and charged residues mainly involved in the stabilization of membrane helices required for conformational changes. These and

FIGURE 58.2 **GLUT1 tridimensional model.** Membrane-spanning (TM) helices are represented in gray except for TM4, indicated by a yellow ribbon, and select helices, labeled and shown in color. The side chains of residues subject to missense mutation are depicted in pink and numbered. Top panel, view normal to the membrane from the extracellular aspect of the transporter. Bottom panel, view parallel to the plane of the membrane. Dotted lines represent the putative boundaries of the plasma membrane and have been arbitrarily placed. IN and OUT denote the extracellular and intracellular spaces, respectively.

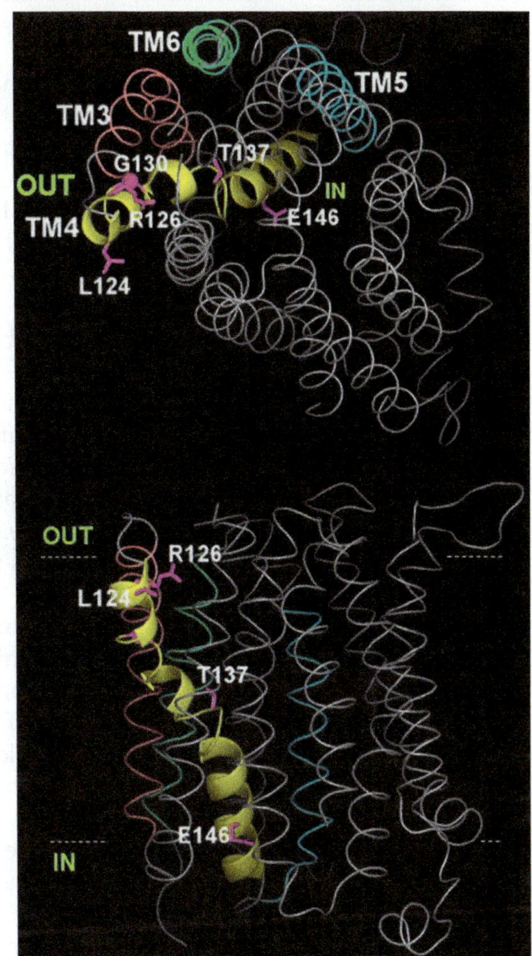

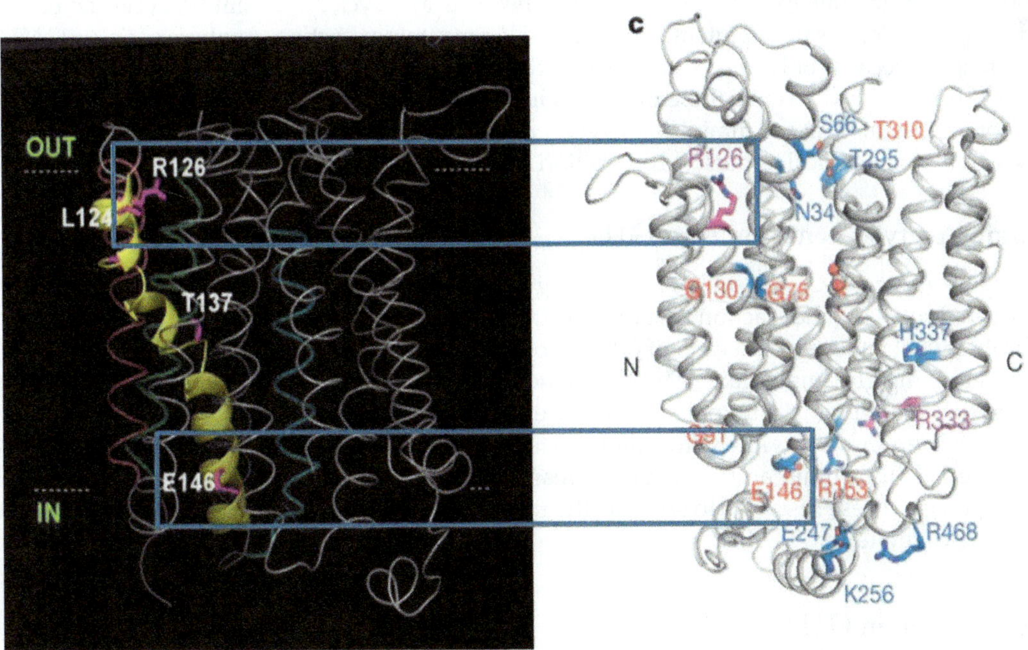

FIGURE 58.3 **Comparison between critical residues in GLUT1 and XylE.** Using the location and nature of human pathogenic mutations as potential reporters of structurally relevant regions, site-directed mutants can be constructed for detailed topological and functional characterization into *X. laevis* oocytes. These studies yield both topological (i.e., elucidating the location and secondary structure of protein regions) and functional (that is, determining their role in the overall operation of the transporter) information.[36] This approach can yield important observations about the structure and function of the transporter and pathogenic mechanisms supported by the solution of the crystal structure of a homologous protein.[35]

similar results, however, do not imply that these or other pathogenic mutations (in the fourth membrane-spanning domain and elsewhere) alter substrate flow through the pore directly. Rather, in light of the available crystallized structures, they identified the fourth membrane-spanning domain (TM4) as a potential transducer of substrate binding. We concluded that TM4 may present a surface of interaction necessary for the intramolecular rearrangement of this and other helices as the substrate is translocated. Further work will be necessary to understand the dynamics of protein conformational changes during the glucose translocation process.

DISEASE MECHANISMS IN GLUT1 DEFICIENCY

Movement disorders and epilepsy are not unexpected manifestations, given the close link between glucose metabolism, the tricarboxylic (TCA) cycle, and the synthesis of glutamate and γ-aminobutyric acid (GABA) and their recycling, which has been observed both *in vitro* and *in vivo* in man and animals.[37] GLUT3 and less abundant homologs allow glucose that has penetrated the brain's extracellular space flow through neuronal and other membranes. Therefore, GLUT1 deficiency may impair *both glial and neuronal metabolism*, including neurotransmitter metabolism as further described below. While GLUT1 haploinsufficiency leads to human and murine epilepsy,[12,38,39] murine GLUT3 haploinsufficiency leads to more modest abnormalities.[40] (GLUT3 deficiency has not been described in man.) Yet, glucose transporter type I deficiency (G1D) remains underexplored, in part because of its apparent rarity, although this is being contested.[18] These disorders also escape conventional neurochemical interpretation. For example, astrocyte–neuron neurotransmitter cycling, postulated to sustain neurotransmission,[41] may not occur in all potentially disease-relevant brain structures.[42–44]

GLUT1 protein quantification alone is insufficient to characterize brain metabolism, as G1D leads to regional overexpression of other transporters[45] (and western blot data illustrating ~0–50% GLUT3 overexpression in mouse cortex, thalamus, striatum or cerebellum, n=8; Pascual, data not shown). After glucose entry, neuronal or glial glycolysis generates most of the brain's pyruvate, which, together with the oxidation of brain fatty acids and of ketones produced in the liver, leads to the synthesis of acetyl-coenzyme A used by the TCA cycle. A fraction of glucose-derived pyruvate is separately used for brain anaplerosis.[46,47] Glia can also generate lactate from glucose, which is transferable to neurons via monocarboxylate transporters for pyruvate generation.[48] Most *de novo* neuronal glutamate synthesis derives from α-ketoglutarate produced in the TCA cycle. Most brain glutamine derives from glutamate via glutamine synthase in glia and can be exchanged during neurotransmission such that neurons release and reuptake glutamate, whereas glia can also take it up and convert it into glutamine for potential transfer to neurons in a cycle.[41,49] A similar cycle can occur in inhibitory synapses,[50] where GABA derives from glutamate[51] via an ATP-consuming reaction. Analysis of tissue GABA metabolism, however, is challenging because only approximately one-fifth of the total glucose consumed in neurotransmitter synthesis is related to GABA synthesis.[52] While these principles invite several testable hypotheses, at the present time, the only established abnormalities in G1D are diminished acetyl-coenzyme A, imbalanced synaptic excitation/inhibition in the cerebral cortex, and thalamocortical hypersynchronization leading to electrical oscillations.[53–55] These are described below.

Deficit of Brain Acetyl-Coenzyme A in G1D

Acetyl-coenzyme A (acetyl-CoA) levels reflect energy production potential in the disease state. If total acetyl-coenzyme A was reduced and the fraction that derives from glucose was also diminished in G1D, then the energetic needs of the brain must be met by oxidation of other substrates capable of yielding acetyl-coenzyme A (such as fatty acids) and of maintaining glutamate flux.[56] This postulate implies that the brain may incur in an anaplerotic (i.e., biosynthetic) deficit that would remain unmatched because of the lack of carbon donors able to sustain citric acid cycle velocity. Mice have been infused with uniformly labeled ^{13}C glucose and total and ^{13}C-labeled acetyl-CoA assayed by mass spectrometry.[53] Total brain acetyl-CoA is decreased in G1D whole brain, as is the fraction of acetyl-CoA that originates from glucose. These results also support the notion that the G1D brain uses alternative substrates to generate acetyl-CoA, glutamate, and GABA.

Synaptic Dysfunction in G1D

The mouse sensory barrel cortex coronal slice is a pertinent preparation because of high-glucose consumption and robust excitatory and inhibitory currents, ideally suited to investigate epileptogenesis by metabolic or pharmacological synaptic manipulation by superfusion.[57–59] At postnatal day 28 (P28), a commonly studied age when electrophysiological slice recordings can be readily performed, GLUT1 expression and cortical glucose metabolism are robust. G1D mice are fully symptomatic (specifically due to ataxia and seizures) at P28, shortly after weaning from maternal (high-fat) milk has been completed[60] and the brain relies primarily on glucose metabolism. Basic neuronal electrical properties are normal in G1D cortex. Generally in epilepsy, preserved neuronal membrane potential leads to consideration of 1) transmitter presynaptic release or 2) postsynaptic action as potential, nonexclusive disease mechanisms assuming that cell structure is preserved (and rapid amelioration by glucose in G1D argues that this is a reasonable assumption). These processes may be disrupted by impaired glucose metabolism. Cortical mouse brain slices maintained in high glucose show a small, significant decrease in mean miniature postsynaptic current (mPSC) amplitude in G1D, indicative of altered spontaneous synaptic transmission onto pyramidal neurons that can be further and rapidly reduced in sustained fashion by decreasing bath glucose closer to physiological concentration.[54] To address mechanisms, excitatory and inhibitory currents can be further separated pharmacologically. Following block of action potentials to facilitate recordings, excitatory currents (mEPSC) are readily measurable. Under these conditions, mEPSC amplitude is reduced in G1D relative to normal. This is exacerbated in low glucose without alteration of mEPSC frequency. On the other hand, inhibitory GABAergic current (mIPSC) amplitude is also but more significantly decreased at high and normal glucose concentrations, with an additional decrease in mIPSC frequency, suggesting *more pronounced inhibitory transmission impairment*. These findings constitute the basis for postulating an excitatory/inhibitory imbalance in G1D cortex.[54,55]

Thalamocortical Synchronization

Spike-wave discharges in G1D recorded by EEG appear generalized without a discernible cortical focus. As part of a return loop, corticothalamic afferents excite the thalamus through the release of glutamate on reticular neurons. The latter are GABAergic inhibitory cells that provide feed forward inhibition onto relay neurons, which are the source of thalamocortical efferents. Most of this inhibition is mediated by ionotropic $GABA_A$ receptors, with a smaller contribution from metabotropic $GABA_B$ receptors.[61] Both thalamic relay and reticular cells generate calcium spikes in response to depolarizarion mediated by T-type channels that are enabled, or primed (i.e., displaced from the inactivated into the closed state), by prior synaptic inhibition. The inactivation state of T channels is responsible for two action potential firing patterns: tonic (T channels are inactivated by steady membrane depolarization), and phasic, which is associated with GABA release capable of activating $GABA_B$ receptors.[62] Phasic enhancement of T-type calcium channels due to reduced GABAergic inhibition leads to thalamocortical synchronization via robust relay cell firing, which translates into intense re-excitation and synchronization of cortical activity and spike-wave epilepsy.[63] However, mechanistic predictions in the thalamus are not as well grounded as in cortex because the effect of perturbations in the thalamocortical network is not easily predictable *a priori*: For reasons that are not fully clear, but probably related to diminished GABA reuptake by the glial transporter GAT-1, the opposite mechanism, enhanced tonic inhibition of relay cells via GABA action on extrasynaptic $GABA_A$ receptors, is a feature of multiple absence spike-wave epilepsies such as GAERS, *stargazer*, *lethargic*, and other pharmacologically-induced models.[64] Therefore, these two distinct mechanisms (decreased GABA action in reticular and relay cells and increased tonic GABA action on relay cells) nucleus constitute the focus of active current research. However, multi-electrode array (MEA) field potential recordings obtained from the somatosensory barrel cerebral cortex[65] of both G1D and normal P28–P35 mice using thalamocortical brain slices unexposed to pharmacological agents illustrate thalamic dysfunction and synchronization. MEA recordings obtained in low glucose show a robust, spontaneous 3-Hz oscillatory activity in the G1D cortex detectable in all the electrodes, indicating hypersynchronization, absent from G1D slices that excluded the thalamus and from normal mouse thalamocortical slices.

ANIMAL MODELS OF GLUT1 DEFICIENCY

Stable G1D mouse lines were first generated by antisense transgene incorporation,[66] followed by a hemizygous line[39] and a similar model.[45] Many informative studies to date have been obtained, except as noted, in a G1D antisense line[67] that expresses about 50% of total brain GLUT1 protein because the hemizygous line manifested a more modest (34%) reduction in brain GLUT1 protein and convulsed only upon fasting.[39] G1D mice display frequent seizures, ataxia, and poor rotarod performance, without incompletely penetrant features. The inheritance pattern is autosomal dominant with homozygous embryonic lethality.

The antisense mouse has allowed addressing a preliminary question central to G1D: Whether brain glucose influx is significantly reduced in the G1D mouse model. This has been confirmed by positron emission tomography (PET;

performed as in [39] with additional normalization to muscle;[68] Pascual, unpublished) and, in greater topographic detail, by emulsion autoradiography via systemic injection of radiolabeled glucose illustrative of deficient cortical and thalamic uptake as described below. The next question addressed was whether: (a) these observations stemmed solely from reduced blood–brain barrier (BBB) glucose penetration; or (b) BBB-independent astrocyte uptake was impaired. The magnitude of astrocyte glucose transport taking place in normal brain tissue is debated, often in relation to uncertainty about the magnitude of neuronal lactate consumption.[69–73] From a therapeutic perspective, disproving (b) would oblige to focus on (a). Consistently with (b), primary G1D astrocyte cultures exhibited diminished glucose uptake (~50% reduction in V_{max}; Pascual, unpublished). Scenario (b) was additionally and more pertinently demonstrated in cortical brain slices (which are devoid of BBB), which exhibited synaptic dysfunction as described above. This dysfunction depended rapidly and reversibly upon bath glucose or acetate concentration. Of note, acetate is an alternative source of acetyl-coenzyme A.

The third question was that of potential TCA cycle precursor depletion impacting the synapse via reduced neurotransmitter and/or energy generation (a direct consequence of TCA cycle intermediate depletion). To render the latter hypothesis testable, the production of acetyl-CoA, glutamate and glutamine that occurred in the cerebral cortex of the mouse was determined. Glutamate and glutamine are uniquely rich in metabolic information related to the TCA cycle[74–77] and can lend themselves to precise multiplet ^{13}C NMR analysis[78,79] and to mass spectrometry, accomplished for the first time by controlled infusion of ^{13}C-labeled substrates in the conscious mouse.[79] Using ^{13}C-glucose, the G1D cortex displayed lower acetyl-CoA abundance and ^{13}C enrichment derived from glucose, which was paralleled by lower ^{13}C-glutamate enrichment. Importantly from a therapeutic point of view, the ^{13}C-glutamate labeling pattern, indicative of TCA cycle integrity, was intact. Therefore, the simplest interpretation is that G1D is associated with both impairment of acetyl-coenzyme A production (due to reduced glucose availability) and reduced glucose-dependent neurotransmitter production, without overall disruption of TCA cycle integrity due to increased consumption of alternative substrates. Consistent with a brain acetyl-coenzyme A deficit in G1D, infused odd-carbon fatty acid (^{13}C-heptanoate) replenish G1D acetyl-CoA and TCA-derived glutamine more effectively than in normal mice.[80]

Complementary investigations in brain slices reveal additional hyperexcitability mechanisms manifested as intrinsic thalamic oscillations. This is the basis for hypothesizing driven (superimposed onto intrinsic cortical) epileptogenesis in G1D. This observation invokes selective neural network vulnerability, an important theme both in epilepsy and metabolic disorders.[81]

The mouse epileptic phenotype is characterized by a spike-wave EEG pattern typical of other models of rodent thalamocortical dysfunction.[82] Several lines of evidence including western blots of GLUT1 and GLUT3 protein, multielectrode array, and electrical field potential recordings from thalamocortical slices, and high-resolution emulsion autoradiography using ^{14}C-2-deoxyglucose (2DG), demonstrate thalamic dysfunction in G1D. 2DG is phosphorylated after uptake and does not significantly progress through glycolysis.[83] Autoradiography was performed in P39 mice by intraperitoneal injection of 2DG in conscious animals: G1D mice exhibit reduced 2DG signal preferentially in the thalamus (53% of normal) and cerebral cortex (full thickness) (61%) relative to other areas such as the cerebellar cortex (70%).[67]

GLUT1 expression (relative to glyceraldehyde-3-phosphate dehydrogenase [GAPDH] as in [36]) in G1D was similarly decreased in thalamus, cortex and whole brain. However, GLUT3/GAPDH expression was significantly increased in whole brain and cortex but not in thalamus, suggesting that neuronal glucose metabolism may be preferentially impaired in the thalamus (Pascual, unpublished).

DIFFERENTIAL DIAGNOSIS OF GLUT1 DEFICIENCY

The differential diagnosis has continued to expand as the GLUT1 deficiency syndrome phenotypic spectrum has broadened. Several patients have been investigated for an occult neuroblastoma because of the opsoclonus-like eye movement abnormalities in early infancy. Other infantile-onset metabolic encephalopathies may be associated with infantile-onset seizures, developmental delay, and deceleration of head growth, including chronic hypoglycemic syndromes, mitochondrial diseases, and disorders of amino acid and organic acid metabolism. Rett syndrome and Angelman syndrome have been considered occasionally, and many patients have been diagnosed initially with cerebral palsy. Occasionally, a misdiagnosis of mitochondrial disease has been rectified by genotyping for GLUT1 deficiency.

TESTING FOR GLUT1 DEFICIENCY

An increasing proportion of patients are diagnosed via single-gene DNA sequencing, Sanger gene sequencing panels, whole-exome DNA sequencing, and comprehensive genomic hybridization (see Chapters 1 and 6). It is expected that the use of relatively unbiased DNA diagnostic methods will result in a better estimation of the prevalence of GLUT1 deficiency and the full phenotypic spectrum.

Examination of the cerebrospinal fluid (CSF) can be diagnostic in the appropriate clinical setting: 99% of patients exhibit hypoglycorrhachia and low lactate concentrations, defined as a CSF glucose level at or below the 10th percentile for age, a CSF to blood glucose ratio at or below the 25th percentile, and a CSF lactate level at or below the 10th percentile. This profile is thus almost specific for GLUT1 deficiency syndrome in the absence of hypoglycemia. CSF glucose concentrations range from 16.2 to 50.5 mg per dL and CSF lactate values range from 5.4 to 13.5 mg per dL.[84]

Brain magnetic resonance imaging (MRI) and computed tomography (CT) images generally are normal. Minor, nonspecific abnormalities have been described in some patients with slight degrees of brain hypotrophy. The PET scan is distinctively abnormal (Figure 58.4). The abnormalities include a global reduction of cerebral glucose accumulation, which is more pronounced in the cerebellum, medial temporal lobes, and thalami. There is a striking, contrasting, increased accumulation of glucose in the basal ganglia.[13,85] The electroencephalographic abnormalities have been described earlier.

The erythrocyte glucose uptake study remains a valuable screening test. The rate of 3-*O*-methyl-D-glucose uptake by freshly isolated, washed erythrocytes *in vitro* is decreased in patients compared to controls. The assay is relatively specific and sensitive for GLUT1 deficiency syndrome, but false negatives have been obtained.[86]

MANAGEMENT OF GLUT1 DEFICIENCY

Standard of Care

The ketogenic diet was introduced in 1991 as a possible treatment for GLUT1 deficiency syndrome before the decrease in glucose transport across tissue barriers was confirmed. A ketogenic diet has led to about two-thirds of patients becoming seizure-free, most of which achieved seizure remission in less than a week. A minority of patients

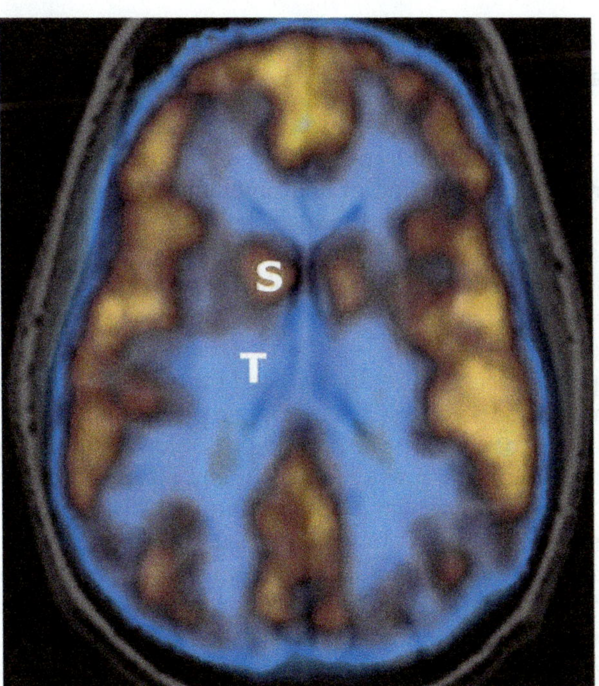

FIGURE 58.4 **^{18}F-deoxyglucose–positron emission tomography (FDG-PET) in human glucose transporter type I deficiency (G1D).** This 17-year old girl harbored a missense mutation. Thalamic (T) hypointensity (blue) contrasts with striatal (S) signal. Cortical FDG signal was also globally depressed. The PET image is superimposed on an equiplanar transaxial T1-weighted MRI image.

achieved epileptic control with anticonvulsants only.[20] As a rule, the clinical seizures respond poorly to antiepileptic drugs and disappear rapidly after beginning a ketogenic diet. Unfortunately, neurological deficits in motor control and cognition tend to persist under the diet, with a significant fraction (about one-third) of patients experiencing recurrent or incompletely treated abnormalities and reduced diet tolerability.[87] Thus, until very recently, most therapeutic efforts for G1D had been limited to early diagnosis and initiation of a ketogenic diet.[20,88] The ketogenic diet, given with the therapeutic intent of stimulating energy metabolism in G1D, may act through several mechanisms,[89,90] including the production of acetyl-coenzyme A that fuels the neural TCA cycle.[91,92] However, ketogenic diets are not universally tolerable[93–97] and do not sufficiently alleviate incapacitating features of G1D, such that relatively few young adults with G1D receiving or who have received a ketogenic diet are employed or subemployed. Older children and adults become increasingly noncompliant, and some choose to resume a regular diet even though their symptoms relapse and seizures recur.

Among all drugs that modulate neuronal excitability, acetazolamide often proves an effective adjuvant in both G1D movement disorders and epilepsy (personal observations and [98]). The modified Atkins diet has also proven effective in G1D movement disorders.[99]

Barbiturates inhibit transport of glucose. Most patients with infantile-onset seizures are treated with phenobarbital, perhaps the most commonly used antiepileptic drug. Parents have reported anecdotally that phenobarbital not only fails to improve seizure control, but also may worsen a child's neurobehavioral state. We have shown that *in vitro* barbiturates aggravate the GLUT1 transport defect in erythrocytes from patients with GLUT1 deficiency syndrome.[100]

We have made similar observations with methylxanthines, which also inhibit transport of glucose by GLUT1.[101]

Barriers to Treatment Development

The mechanistic understanding and treatment of neurometabolic disorders is still in its earliest stages. Consequently, most G1D research has yielded modest benefits. For example, the finding of apoptosis in cultured G1D mouse embryonic cells,[102] led to the search for brain cell death. However, apoptosis is insignificant in the G1D mouse brain[67,103] and human MRI scans are generally unremarkable except for mildly diminished myelin abundance.[85] Analogously, the potential role of specific anticonvulsants in G1D was suggested by its spike-wave EEG pattern[19,20] typical of absence epilepsy,[104] but this approach also proved ineffective.[105] Furthermore, anticonvulsants can worsen epilepsy in G1D and some may inhibit residual glucose transporter function such that, as a result, they are used on a trial and error basis.[20,100,101,106,107]

Therapies Under Investigation

In contrast with these therapies, brain fuels with additional potential to refill TCA cycle intermediates (anaplerosis), such as triheptanoin, may offer superior therapeutic benefit.[46,108] These questions are central for the development of treatments for G1D and for the understanding of neurometabolic disorders. Anaplerotic therapies based on the replacement of part of the daily caloric needs with odd-chain triglycerides or fatty acids supply net carbon to the citric acid cycle and have been biochemically and clinically successful in other energy metabolism defects in which precursor depletion is strongly suspected or documented, such as pyruvate carboxylase deficiency.[109] Dietary triheptanoin (a seven-carbon triglyceride that generates three molecules of the seven-carbon fatty acid heptanoate in the liver[110]) gives rise to plasma heptanoate and to five-carbon ketone bodies.[110] Notably, heptanoate metabolism exerts anticonvulsant effects in epileptic animals and refills depleted brain TCA cycle intermediates.[111] Normally, the principal brain glucose-derived anaplerotic process is the astrocytic pyruvate carboxylase pathway,[112,113] which depends on glycolysis (pyruvate generation) to refill citric acid cycle precursors, including glutamate and GABA.[114] Neurons probably carry out only a modest degree of anaplerosis via the malic enzyme.[115] In contrast, common fats generate even-carbon ketone bodies (β-hydroxybutyrate and acetoacetate),[92] which lead to acetyl-CoA generation and therefore cannot sustain anaplerosis because they are fully converted into water and CO_2 in the citric acid cycle.[46,91] Thus, the ketogenic diet is a limited form of therapy for G1D: even-carbon ketones, generated from common dietary fat or a ketogenic diet, ameliorate seizures, but are not anaplerotic. In fact, the two key metabolic roles of glucose are: (a) energy production by oxidation of acetyl-coenzyme A; and (b) anaplerosis, by providing pyruvate for carboxylation. Ketogenic diets address only (a). In contrast, mouse brain nuclear magnetic resonance (NMR) and mass spectrometry indicate that odd-carbon triheptanoin is anaplerotic, thus fulfilling both roles. G1D patients receiving triheptanoin as a dietary supplement experience increased oxygen cerebral metabolic rate ($CMRO_2$) by MRI, decreased seizures by and improved neuropsychological performance, suggesting that triheptanoin is effective in cerebral hypometabolic states associated with epilepsy.[116]

ACKNOWLEDGEMENTS

JMP and DCD would like to acknowledge the generous support of the Glut1 Deficiency Foundation. JMP is supported by NIH grants NS077015, NS067015, NS078059, MH084021, RR002584, RR024982 and by the Office of Rare Diseases Research Glucose transporter type 1 deficiency syndrome (G1D) collaboration, education, and test translation (CETT) program for rare genetic diseases. DCD is supported by the Colleen Giblin Foundation, the Will Foundation and Milestones for Children.

References

1. Widdas WF. Inability of diffusion to account for placental glucose transfer in the sheep and consideration of the kinetics of a possible carrier transfer. *J Physiol*. 1952;118(1):23–39.
2. Bosackova J, Crane RK. Studies on the mechanism of intestinal absorption of sugars. IX. Intracellular sodium concentrations and active sugar transport by hamster small intestine in vitro. *Biochim Biophys Acta*. 1965;102(2):436–441.
3. Wright EM, Loo DD, Hirayama BA. Biology of human sodium glucose transporters. *Physiol Rev*. 2011;91(2):733–794.
4. Mueckler M, Thorens B. The SLC2 (GLUT) family of membrane transporters. *Mol Aspects Med*. 2013;34(2–3):121–138.
5. Cura AJ, Carruthers A. Role of monosaccharide transport proteins in carbohydrate assimilation, distribution, metabolism, and homeostasis. *Compr Physiol*. 2012;2(2):863–914.
6. Hediger MA, Ikeda T, Coady M, Gundersen CB, Wright EM. Expression of size-selected mRNA encoding the intestinal Na/glucose cotransporter in Xenopus laevis oocytes. *Proc Natl Acad Sci U S A*. 1987;84(9):2634–2637.
7. Turk E, Zabel B, Mundlos S, Dyer J, Wright EM. Glucose/galactose malabsorption caused by a defect in the Na+/glucose cotransporter. *Nature*. 1991;350(6316):354–356.
8. Lindquist B, Meeuwisse GW. Chronic diarrhoea caused by monosaccharide malabsorption. *Acta Paediatr*. 1962;51:674–685.
9. Laplane R, Polonovski C, Etienne M, Debray P, Lods J, Pissarro B. L'intolerance aux sucres a transfert intestinal actif. Ses rapports avec l'intolerance au lactose et le syndrome coeliaque. *Archs fr Pédiat*. 1962;19:895.
10. Fanconi G, Bickel H. Die chronische aminoacidurie (aminosaeurediabetes oder nephrotischglukosurisscher zwergwuchs) ber der glykogenose und cystinkrankheit. *Helv Paediatr Acta*. 1949;4(5):359–396.
11. Santer R, Schneppenheim R, Dombrowski A, Gotze H, Steinmann B, Schaub J. Mutations in GLUT2, the gene for the liver-type glucose transporter, in patients with Fanconi-Bickel syndrome. *Nat Genet*. 1997;17(3):324–326.
12. De Vivo DC, Trifiletti RR, Jacobson RI, Ronen GM, Behmand RA, Harik SI. Defective glucose transport across the blood-brain barrier as a cause of persistent hypoglycorrhachia, seizures, and developmental delay. *N Engl J Med*. 1991;325(10):703–709.
13. Pascual JM, Van Heertum RL, Wang D, Engelstad K, De Vivo DC. Imaging the metabolic footprint of Glut1 deficiency on the brain. *Ann Neurol*. 2002;52(4):458–464.
14. Seidner G, Alvarez MG, Yeh JI, et al. GLUT-1 deficiency syndrome caused by haploinsufficiency of the blood-brain barrier hexose carrier. *Nat Genet*. 1998;18(2):188–191.
15. Arsov T, Mullen SA, Damiano JA, et al. Early onset absence epilepsy: 1 in 10 cases is caused by GLUT1 deficiency. *Epilepsia*. 2012;53(12): e204–e207.
16. Suls A, Mullen SA, Weber YG, et al. Early-onset absence epilepsy caused by mutations in the glucose transporter GLUT1. *Ann Neurol*. 2009;66(3):415–419.
17. Mullen SA, Suls A, De Jonghe P, Berkovic SF, Scheffer IE. Absence epilepsies with widely variable onset are a key feature of familial GLUT1 deficiency. *Neurology*. 2010;75(5):432–440.
18. Arsov T, Mullen SA, Rogers S, et al. Glucose transporter 1 deficiency in the idiopathic generalized epilepsies. *Ann Neurol*. 2012;72(5):807–815.
19. Leary LD, Wang D, Nordli Jr DR, Engelstad K, De Vivo DC. Seizure characterization and electroencephalographic features in Glut-1 deficiency syndrome. *Epilepsia*. 2003;44(5):701–707.
20. Pong AW, Geary BR, Engelstad KM, Natarajan A, Yang H, De Vivo DC. Glucose transporter type I deficiency syndrome: epilepsy phenotypes and outcomes. *Epilepsia*. 2012;53(9):1503–1510.
21. von Moers A, Brockmann K, Wang D, et al. EEG features of glut-1 deficiency syndrome. *Epilepsia*. 2002;43(8):941–945.
22. Suls A, Dedeken P, Goffin K, et al. Paroxysmal exercise-induced dyskinesia and epilepsy is due to mutations in SLC2A1, encoding the glucose transporter GLUT1. *Brain*. 2008;131(Pt 7):1831–1844.
23. Afawi Z, Suls A, Ekstein D, et al. Mild adolescent/adult onset epilepsy and paroxysmal exercise-induced dyskinesia due to GLUT1 deficiency. *Epilepsia*. 2010;51(12):2466–2469.
24. Weber YG, Storch A, Wuttke TV, et al. GLUT1 mutations are a cause of paroxysmal exertion-induced dyskinesias and induce hemolytic anemia by a cation leak. *J Clin Invest*. 2008;118(6):2157–2168.
25. Flatt JF, Guizouarn H, Burton NM, et al. Stomatin-deficient cryohydrocytosis results from mutations in SLC2A1: a novel form of GLUT1 deficiency syndrome. *Blood*. 2011;118(19):5267–5277.
26. Weber YG, Kamm C, Suls A, et al. Paroxysmal choreoathetosis/spasticity (DYT9) is caused by a GLUT1 defect. *Neurology*. 2011;77(10):959–964.
27. Overweg-Plandsoen WC, Groener JE, Wang D, et al. GLUT-1 deficiency without epilepsy–an exceptional case. *J Inherit Metab Dis*. 2003;26(6):559–563.
28. Perez-Duenas B, Prior C, Ma Q, et al. Childhood chorea with cerebral hypotrophy: a treatable GLUT1 energy failure syndrome. *Arch Neurol*. 2009;66(11):1410–1414.
29. Friedman JR, Thiele EA, Wang D, et al. Atypical GLUT1 deficiency with prominent movement disorder responsive to ketogenic diet. *Mov Disord*. 2006;21(2):241–245.

30. Brockmann K, Wang D, Korenke CG, et al. Autosomal dominant glut-1 deficiency syndrome and familial epilepsy. *Ann Neurol.* 2001;50(4):476–485.
31. Klepper J, Scheffer H, Elsaid MF, Kamsteeg EJ, Leferink M, Ben-Omran T. Autosomal recessive inheritance of GLUT1 deficiency syndrome. *Neuropediatrics.* 2009;40(5):207–210.
32. Mueckler M, Caruso C, Baldwin SA, et al. Sequence and structure of a human glucose transporter. *Science.* 1985;229(4717):941–945.
33. Birnbaum MJ, Haspel HC, Rosen OM. Cloning and characterization of a cDNA encoding the rat brain glucose-transporter protein. *Proc Natl Acad Sci U S A.* 1986;83(16):5784–5788.
34. Lam VH, Lee JH, Silverio A, et al. Pathways of transport protein evolution: recent advances. *Biol Chem.* 2011;392(1–2):5–12.
35. Sun L, Zeng X, Yan C, et al. Crystal structure of a bacterial homologue of glucose transporters GLUT1-4. *Nature.* 2012;490(7420):361–366.
36. Pascual JM, Wang D, Yang R, Shi L, Yang H, De Vivo DC. Structural signatures and membrane helix 4 in GLUT1: inferences from human blood-brain glucose transport mutants. *J Biol Chem.* 2008;283(24):16732–16742.
37. Sibson NR, Shen J, Mason GF, Rothman DL, Behar KL, Shulman RG. Functional energy metabolism: in vivo 13C-NMR spectroscopy evidence for coupling of cerebral glucose consumption and glutamatergic neuronal activity. *Dev Neurosci.* 1998;20(4–5):321–330.
38. Pascual JM, Wang D, Lecumberri B, et al. GLUT1 deficiency and other glucose transporter diseases. *Eur J Endocrinol.* 2004;150(5):627–633.
39. Wang D, Pascual JM, Yang H, et al. A mouse model for Glut-1 haploinsufficiency. *Hum Mol Genet.* 2006;15(7):1169–1179.
40. Zhao Y, Fung C, Shin D, et al. Neuronal glucose transporter isoform 3 deficient mice demonstrate features of autism spectrum disorders. *Mol Psychiatry.* 2010;15(3):286–299.
41. Pellerin L, Magistretti PJ. Glutamate uptake into astrocytes stimulates aerobic glycolysis: a mechanism coupling neuronal activity to glucose utilization. *Proc Natl Acad Sci U S A.* 1994;91(22):10625–10629.
42. Kam K, Nicoll R. Excitatory synaptic transmission persists independently of the glutamate-glutamine cycle. *J Neurosci.* 2007;27(34):9192–9200.
43. Bryant AS, Li B, Beenhakker MP, Huguenard JR. Maintenance of thalamic epileptiform activity depends on the astrocytic glutamate-glutamine cycle. *J Neurophysiol.* 2009;102(5):2880–2888.
44. Liang SL, Carlson GC, Coulter DA. Dynamic regulation of synaptic GABA release by the glutamate-glutamine cycle in hippocampal area CA1. *J Neurosci.* 2006;26(33):8537–8548.
45. Ohtsuki S, Kikkawa T, Hori S, Terasaki T. Modulation and compensation of the mRNA expression of energy related transporters in the brain of glucose transporter 1-deficient mice. *Biol Pharm Bull.* 2006;29(8):1587–1591.
46. Brunengraber H, Roe CR. Anaplerotic molecules: current and future. *J Inherit Metab Dis.* 2006;29(2–3):327–331.
47. Mason GF, Petersen KF, de Graaf RA, Shulman GI, Rothman DL. Measurements of the anaplerotic rate in the human cerebral cortex using ^{13}C magnetic resonance spectroscopy and [1-13C] and [2-13C] glucose. *J Neurochem.* 2007;100(1):73–86.
48. Magistretti PJ, Pellerin L. Cellular mechanisms of brain energy metabolism and their relevance to functional brain imaging. *Philos Trans R Soc Lond B Biol Sci.* 1999;354(1387):1155–1163.
49. Mason GF, Gruetter R, Rothman DL, Behar KL, Shulman RG, Novotny EJ. Simultaneous determination of the rates of the TCA cycle, glucose utilization, alpha-ketoglutarate/glutamate exchange, and glutamine synthesis in human brain by NMR. *J Cereb Blood Flow Metab.* 1995;15(1):12–25.
50. Chowdhury GM, Patel AB, Mason GF, Rothman DL, Behar KL. Glutamatergic and GABAergic neurotransmitter cycling and energy metabolism in rat cerebral cortex during postnatal development. *J Cereb Blood Flow Metab.* 2007;27(12):1895–1907.
51. Roberts E, Frankel S. gamma-Aminobutyric acid in brain: its formation from glutamic acid. *J Biol Chem.* 1950;187(1):55–63.
52. Patel AB, de Graaf RA, Mason GF, Rothman DL, Shulman RG, Behar KL. The contribution of GABA to glutamate/glutamine cycling and energy metabolism in the rat cortex in vivo. *Proc Natl Acad Sci U S A.* 2005;102(15):5588–5593.
53. Marin-Valencia I, Good LB, Ma Q, et al. Glycolysis, citric acid cycle flux and neurotransmitter synthesis in mouse brain by ^{13}C NMR spectroscopy. Paper presented at: Society for Neuroscience Annual Meeting 2009; Chicago.
54. Good LB, Espinosa F, Ma Q, Heilig CW, Kavalali ET, Pascual JM. A neuronal excitability defect in a prototypic energy metabolism disorder. In: American Epilepsy Society Annual Meeting. vol. 49. Seattle, WA, USA: Epilepsia; 2008:364 (49), suppl. 7.
55. Good LB, Ma Q, Kavalali ET, Heilig CW, Pascual JM. Diminished synaptic quantal amplitudes in a brain energy metabolic disorder. In: *Society for Neuroscience Annual Meeting.* Chicago: Society for Neuroscience; 2009:330.338/H327.
56. Hertz L, Hertz E. Cataplerotic TCA cycle flux determined as glutamate-sustained oxygen consumption in primary cultures of astrocytes. *Neurochem Int.* 2003;43(4–5):355–361.
57. Cholet N, Pellerin L, Welker E, et al. Local injection of antisense oligonucleotides targeted to the glial glutamate transporter GLAST decreases the metabolic response to somatosensory activation. *J Cereb Blood Flow Metab.* 2001;21(4):404–412.
58. Giaume C, Maravall M, Welker E, Bonvento G. The barrel cortex as a model to study dynamic neuroglial interaction. *Neuroscientist.* 2009;15(4):351–366.
59. Gibson JR, Bartley AF, Hays SA, Huber KM. Imbalance of neocortical excitation and inhibition and altered UP states reflect network hyperexcitability in the mouse model of fragile X syndrome. *J Neurophysiol.* 2008;100(5):2615–2626.
60. Nehlig A. Brain uptake and metabolism of ketone bodies in animal models. *Prostaglandins Leukot Essent Fatty Acids.* 2004;70(3):265–275.
61. Douglas RJ, Koch C, Mahowald M, Martin KA, Suarez HH. Recurrent excitation in neocortical circuits. *Science.* 1995;269(5226):981–985.
62. Kim U, McCormick DA. The functional influence of burst and tonic firing mode on synaptic interactions in the thalamus. *J Neurosci.* 1998;18(22):9500–9516.
63. Ernst WL, Zhang Y, Yoo JW, Ernst SJ, Noebels JL. Genetic enhancement of thalamocortical network activity by elevating alpha 1g-mediated low-voltage-activated calcium current induces pure absence epilepsy. *J Neurosci.* 2009;29(6):1615–1625.
64. Cope DW, Di Giovanni G, Fyson SJ, et al. Enhanced tonic GABAA inhibition in typical absence epilepsy. *Nat Med.* 2009;15(12):1392–1398.
65. Woolsey TA, Van der Loos H. The structural organization of layer IV in the somatosensory region (SI) of mouse cerebral cortex. The description of a cortical field composed of discrete cytoarchitectonic units. *Brain Res.* 1970;17(2):205–242.
66. Heilig CW, Saunders T, Brosius 3rd FC, et al. Glucose transporter-1-deficient mice exhibit impaired development and deformities that are similar to diabetic embryopathy. *Proc Natl Acad Sci U S A.* 2003;100(26):15613–15618.
67. Marin-Valencia I, Good LB, Ma Q, et al. Glut1 deficiency (G1D): epilepsy and metabolic dysfunction in a mouse model of the most common human phenotype. *Neurobiol Dis.* 2012;48(1):92–101.

68. Logan J. Graphical analysis of PET data applied to reversible and irreversible tracers. *Nucl Med Biol*. 2000;27(7):661–670.
69. Barros LF, Bittner CX, Loaiza A, Porras OH. A quantitative overview of glucose dynamics in the gliovascular unit. *Glia*. 2007;55(12):1222–1237.
70. Simpson IA, Carruthers A, Vannucci SJ. Supply and demand in cerebral energy metabolism: the role of nutrient transporters. *J Cereb Blood Flow Metab*. 2007;27(11):1766–1791.
71. Jolivet R, Allaman I, Pellerin L, Magistretti PJ, Weber B. Comment on recent modeling studies of astrocyte-neuron metabolic interactions. *J Cereb Blood Flow Metab*. Dec;30(12):1982–1986.
72. Mangia S, Simpson IA, Vannucci SJ, Carruthers A. The in vivo neuron-to-astrocyte lactate shuttle in human brain: evidence from modeling of measured lactate levels during visual stimulation. *J Neurochem*. 2009;109(suppl 1):55–62.
73. Lund-Andersen H. Transport of glucose from blood to brain. *Physiol Rev*. 1979;59(2):305–352.
74. Hassel B, Sonnewald U, Fonnum F. Glial-neuronal interactions as studied by cerebral metabolism of [2-13C]acetate and [1-13C]glucose: an ex vivo ^{13}C NMR spectroscopic study. *J Neurochem*. 1995;64(6):2773–2782.
75. Brenner E, Sonnewald U, Schweitzer A, Andrieux A, Nehlig A. Hypoglutamatergic activity in the STOP knockout mouse: a potential model for chronic untreated schizophrenia. *J Neurosci Res*. 2007;85(15):3487–3493.
76. Yudkoff M, Daikhin Y, Nissim I, et al. Response of brain amino acid metabolism to ketosis. *Neurochem Int*. 2005;47(1–2):119–128.
77. Bogen IL, Risa O, Haug KH, Sonnewald U, Fonnum F, Walaas SI. Distinct changes in neuronal and astrocytic amino acid neurotransmitter metabolism in mice with reduced numbers of synaptic vesicles. *J Neurochem*. 2008;105(6):2524–2534.
78. Malloy CR, Sherry AD, Jeffrey FM. Carbon flux through citric acid cycle pathways in perfused heart by ^{13}C NMR spectroscopy. *FEBS Lett*. 1987;212(1):58–62.
79. Marin-Valencia I, Good LB, Ma Q, Jeffrey FM, Malloy CR, Pascual JM. High-resolution detection of (13)C multiplets from the conscious mouse brain by ex vivo NMR spectroscopy. *J Neurosci Methods*. 2012;203(1):50–55.
80. Marin-Valencia I, Good LB, Ma Q, Malloy CR, Pascual JM. Heptanoate as a neural fuel: energetic and neurotransmitter precursors in normal and glucose transporter I-deficient (G1D) brain. *J Cereb Blood Flow Metab*. 2013;33(2):175–182.
81. Holopainen IE. Seizures in the developing brain: cellular and molecular mechanisms of neuronal damage, neurogenesis and cellular reorganization. *Neurochem Int*. 2008;52(6):935–947.
82. Huguenard JR, Prince DA. Intrathalamic rhythmicity studied in vitro: nominal T-current modulation causes robust antioscillatory effects. *J Neurosci*. 1994;14(9):5485–5502.
83. Sokoloff L. The deoxyglucose method for the measurement of local glucose utilization and the mapping of local functional activity in the central nervous system. *Int Rev Neurobiol*. 1981;22:287–333.
84. Leen WG, Wevers RA, Kamsteeg EJ, Scheffer H, Verbeek MM, Willemsen MA. Cerebrospinal fluid analysis in the workup of GLUT1 deficiency syndrome: a systematic review. *JAMA Neurol*. 2013;70(11):1440–1444.
85. Pascual JM, Wang D, Hinton V, et al. Brain glucose supply and the syndrome of infantile neuroglycopenia. *Arch Neurol*. 2007;64(4):507–513.
86. Klepper J, Garcia-Alvarez M, O'Driscoll KR, et al. Erythrocyte 3-O-methyl-D-glucose uptake assay for diagnosis of glucose-transporter-protein syndrome. *J Clin Lab Anal*. 1999;13(3):116–121.
87. Klepper J. Glucose transporter deficiency syndrome (GLUT1DS) and the ketogenic diet. *Epilepsia*. 2008;49(suppl 8):46–49.
88. Klepper J, Leiendecker B, Bredahl R, et al. Introduction of a ketogenic diet in young infants. *J Inherit Metab Dis*. 2002;25(6):449–460.
89. Kim do Y, Rho JM. The ketogenic diet and epilepsy. *Curr Opin Clin Nutr Metab Care*. 2008;11(2):113–120.
90. Lutas A, Yellen G. The ketogenic diet: metabolic influences on brain excitability and epilepsy. *Trends Neurosci*. 2013;36(1):32–40.
91. Yudkoff M, Daikhin Y, Melo TM, Nissim I, Sonnewald U, Nissim I. The ketogenic diet and brain metabolism of amino acids: relationship to the anticonvulsant effect. *Annu Rev Nutr*. 2007;27:415–430.
92. Bough K. Energy metabolism as part of the anticonvulsant mechanism of the ketogenic diet. *Epilepsia*. 2008;49(suppl 8):91–93.
93. Kielb S, Koo HP, Bloom DA, Faerber GJ. Nephrolithiasis associated with the ketogenic diet. *J Urol*. 2000;164(2):464–466.
94. Stewart WA, Gordon K, Camfield P. Acute pancreatitis causing death in a child on the ketogenic diet. *J Child Neurol*. 2001;16(9):682.
95. Best TH, Franz DN, Gilbert DL, Nelson DP, Epstein MR. Cardiac complications in pediatric patients on the ketogenic diet. *Neurology*. 2000;54(12):2328–2330.
96. Berry-Kravis E, Booth G, Taylor A, Valentino LA. Bruising and the ketogenic diet: evidence for diet-induced changes in platelet function. *Ann Neurol*. 2001;49(1):98–103.
97. Hoyt CS, Billson FA. Optic neuropathy in ketogenic diet. *Br J Ophthalmol*. 1979;63(3):191–194.
98. Anheim M, Maillart E, Vuillaumier-Barrot S, et al. Excellent response to acetazolamide in a case of paroxysmal dyskinesias due to GLUT1-deficiency. *J Neurol*. 2011;258(2):316–317.
99. Leen WG, Mewasingh L, Verbeek MM, Kamsteeg EJ, van de Warrenburg BP, Willemsen MA. Movement disorders in GLUT1 deficiency syndrome respond to the modified Atkins diet. *Mov Disord*. 2013;28(10):1439–1442.
100. Klepper J, Fischbarg J, Vera JC, Wang D, De Vivo DC. GLUT1-deficiency: barbiturates potentiate haploinsufficiency in vitro. *Pediatr Res*. 1999;46(6):677–683.
101. Ho YY, Yang H, Klepper J, Fischbarg J, Wang D, De Vivo DC. Glucose transporter type 1 deficiency syndrome (Glut1DS): methylxanthines potentiate GLUT1 haploinsufficiency in vitro. *Pediatr Res*. 2001;50(2):254–260.
102. Heilig C, Brosius F, Siu B, et al. Implications of glucose transporter protein type 1 (GLUT1)-haplodeficiency in embryonic stem cells for their survival in response to hypoxic stress. *Am J Pathol*. 2003;163(5):1873–1885.
103. Ullner PM, Di Nardo A, Goldman JE, et al. Murine Glut-1 transporter haploinsufficiency: postnatal deceleration of brain weight and reactive astrocytosis. *Neurobiol Dis*. 2009;36(1):60–69.
104. Posner E. Pharmacological treatment of childhood absence epilepsy. *Expert Rev Neurother*. 2006;6(6):855–862.
105. Pascual JM, Wang D, Vivo DD. Glucose transporter type I deficiency syndrome. *GeneReviews*. 2009; www.genetests.org. Accessed 01.05.09.
106. Klepper J, Florcken A, Fischbarg J, Voit T. Effects of anticonvulsants on GLUT1-mediated glucose transport in GLUT1 deficiency syndrome in vitro. *Eur J Pediatr*. 2003;162(2):84–89.
107. Wang D, Pascual JM, De Vivo D. Glucose transporter type 1 deficiency syndrome. *GeneReviews*. 2002; Resource available on line at, http://www.ncbi.nlm.nih.gov/pubmed/20301603.

108. Marin-Valencia I, Roe CR, Pascual JM. Pyruvate carboxylase deficiency: mechanisms, mimics and anaplerosis. *Mol Genet Metab.* 2010;101(1):9–17.
109. Mochel F, DeLonlay P, Touati G, et al. Pyruvate carboxylase deficiency: clinical and biochemical response to anaplerotic diet therapy. *Mol Genet Metab.* 2005;84(4):305–312.
110. Roe CR, Sweetman L, Roe DS, David F, Brunengraber H. Treatment of cardiomyopathy and rhabdomyolysis in long-chain fat oxidation disorders using an anaplerotic odd-chain triglyceride. *J Clin Invest.* 2002;110(2):259–269.
111. Willis S, Stoll J, Sweetman L, Borges K. Anticonvulsant effects of a triheptanoin diet in two mouse chronic seizure models. *Neurobiol Dis.* 2010;40(3):565–572.
112. Gamberino WC, Berkich DA, Lynch CJ, Xu B, LaNoue KF. Role of pyruvate carboxylase in facilitation of synthesis of glutamate and glutamine in cultured astrocytes. *J Neurochem.* 1997;69(6):2312–2325.
113. Shank RP, Bennett GS, Freytag SO, Campbell GL. Pyruvate carboxylase: an astrocyte-specific enzyme implicated in the replenishment of amino acid neurotransmitter pools. *Brain Res.* 1985;329(1–2):364–367.
114. Kornberg HL. In: Campbell PN, Marshall RD, eds. Essays in Biochemistry. London, UK: Academic Press; 1966:1–31.
115. Hassel B. Carboxylation and anaplerosis in neurons and glia. *Mol Neurobiol.* 2000;22(1–3):21–40.
116. Pascual JM, Good LB, Liu P, et al. Synaptic excitation-inhibition imbalance in glucose transporter I deficiency (G1D) and first treatment of its associated human epilepsy with triheptanoin. Paper presented at: Curing the Epilepsies 2013: Pathways Forward 2013; Bethesda, MD, USA.
117. Roubergue A, Apartis E, Mesnage V, et al. Dystonic tremor caused by mutation of the glucose transporter gene GLUT1. *J Inherit Metab Dis.* 2011;34(2):483–488.
118. Cunningham P, Naftalin RJ. Implications of aberrant temperature-sensitive glucose transport via the glucose transporter deficiency mutant (GLUT1DS) T295M for the alternate-access and fixed-site transport models. *J Membr Biol.* 2013;246(6):495–511.

CHAPTER

59

Maple Syrup Urine Disease: Clinical and Therapeutic Considerations

David T. Chuang, R. Max Wynn, Rody P. Cox, and Jacinta L. Chuang

The University of Texas Southwestern Medical Center, Dallas, TX, USA

INTRODUCTION

The oxidative degradation of branched-chain amino acids (BCAA) leucine, isoleucine and valine begins with reversible transamination in mitochondria by coupling with α-ketoglutarate to give rise to the corresponding branched chain α-ketoacids (BCKA), i.e., α-ketoisocaproate (KIC), α-keto-β-methylvalerate (KMV) and α-ketoisovalerate (KIV). These ketoacids are then irreversibly decarboxylated by a single branched-chain α-ketoacid dehydrogenase complex (BCKDC). The resultant branched-chain acyl-CoAs are further degraded through separate remaining reactions in the BCAA degradative pathways. Acetoacetate and succinyl-CoA from leucine and valine, respectively, serve as fuels via the Krebs cycle, whereas acetyl-CoA produced from leucine or isoleucine is a precursor for fatty acid and cholesterol synthesis. The oxidation of BCKA occurs primarily in the liver, followed by extrahepatic tissues including kidney, muscle, heart, brain and adipose tissues.

Maple syrup urine disease (MSUD) or branched-chain ketoaciduria is an autosomal recessive disorder caused by deficiency in the BCKDC.[1,2] This large mitochondrial enzyme complex contains multiple copies of catalytic and regulatory components,[3] and its activity is regulated through reversible phosphorylation–dephosphorylation[4] (see below). MSUD is a complex disorder since a mutation in any one component of the BCKDC could produce the disease. There are currently five distinct clinical phenotypes: classic, intermediate, intermittent, thiamine-responsive and dihydrolipoamide dehydrogenase (E3)-deficient phenotypes, based on the severity of the symptoms, the age of onset and the protein component affected.[1]

CLINICAL PRESENTATION OF CLASSIC MSUD

Children with classic MSUD appear normal at birth, but within 4–7 days they have the onset of encephalopathy. This presentation is the most common, i.e., in approximately 75% of the patients, and is the most severe form of the disease.[1] The first symptoms are difficulty in feeding with vomiting. The levels of BCAA are markedly increased in blood, cerebrospinal fluid (CSF), and urine. In classic MSUD, the bulk of the BCKA is derived from leucine. The presence of alloisoleucine from L-isoleucine through tautomerization is diagnostic for MSUD. Progressive weight loss and neurologic signs of alternating hypertonic and hypotonic posturing occur. Dystonic extension of the arms resembling decerebrate rigidity is often observed. A maple syrup or burnt sugar odor is present in the diapers.[5] Seizures and coma may ensue, leading to death if not treated. Untreated classic MSUD patients usually die within the first few months of life from metabolic crisis and neurological deterioration, which are often precipitated by infection or other stresses. Surviving patients who initiate dietary therapy late by restricting BCAA levels[6] generally suffer from severe neurological damage including mental retardation, spasticity, or hypotonia, and occasionally cortical blindness. However, there is wide variation in the age of onset and rapidity of neurological damage even among siblings.[7] Physical examination usually shows a bulging frontal, generalized hyperreflexia with spasticity, a Babinsky sign, dystonic posturing, and severe psychomotor retardation.[1,5,7] Cranial nerve palsy with bilateral ptosis, ophthalmoplegia and bilateral facial nerve paralysis has been described in some patients.[8]

http://dx.doi.org/10.1016/10.1016/B978-0-12-410529-4.00059-0

NEUROPATHOLOGY OF MSUD

Edema of the brain was noted in the initial description of the disease, with the brain weighing 650 g rather than the 410 g expected of a normal infant of similar age.[5] The major findings in the brain of MSUD patients occur in the white matter. There was failure of myelination with no signs of demyelination, glial reaction, or neuronal degeneration.[5] Defective myelination also was accompanied by a striking spongy degeneration of the white matter with a decreased number of oligodendroglioma cells. The pyramidal tracts of the spinal cord, the myelin around the dentate nuclei, the corpus callosum, and the cerebral hemispheres are most affected. A moderate but significant degree of astrocytic hypertrophy is present in the white matter of most patients. Impressive alterations are also noted in the cerebellum with necrosis of the granular cell layer, but the molecular and Purkinje cell layers are preserved.[1] In patients treated with restricted diets,[6] the neuropathological findings are similar but to a lesser degree. Chemical analysis of the brain shows reduction in the lipid content. Proteolipids and cerebrosides are particularly affected, indicating reduced amounts of myelin. Figure 59.1 shows a typical cerebral atrophy of an older MSUD patient, who experienced prolonged amino acid imbalances since infancy.[9] The total brain volumes do not differ across the MSUD and the control groups.

Recently, a new neuroradiologic picture resembling Wernicke encephalopathy (WE) was observed during metabolic decompensation in two MSUD patients.[10] Clinical observations and the review of the literature regarding WE and MSUD pathophysiology prompted the authors to hypothesize a pathogenic link between these two disorders. Based on these findings, clinicians and neuroradiologists should be aware of MSUD as a possible predisposing factor of WE in children.

VARIANT TYPES OF MSUD

Milder forms of MSUD have been reported.[1] An intermittent form of the disease presents with episodic ketoacidosis precipitated by infections or excess protein intake. Between these episodes, the subjects are asymptomatic with normal psychomotor development. The BCAA and BCKA levels are also normal between exacerbations but rise dramatically during ketoacidotic episodes. An intermediate form of MSUD was described, in which BCAA and BCKA are moderately elevated but there is no obvious ketoacidosis. Mental retardation and psychomotor delay is present. The diagnosis of MSUD in these patients is usually delayed until the first or second year of life. Institution of dietary therapy often results in some improvement. Fibroblasts from patients with the intermittent and intermediate forms of MSUD have residual decarboxylation activity that is greater than 2% and as high as 25–40% of normal.[1] The

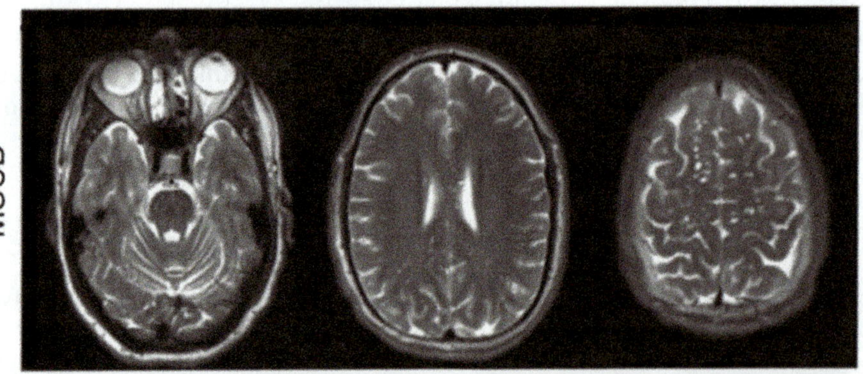

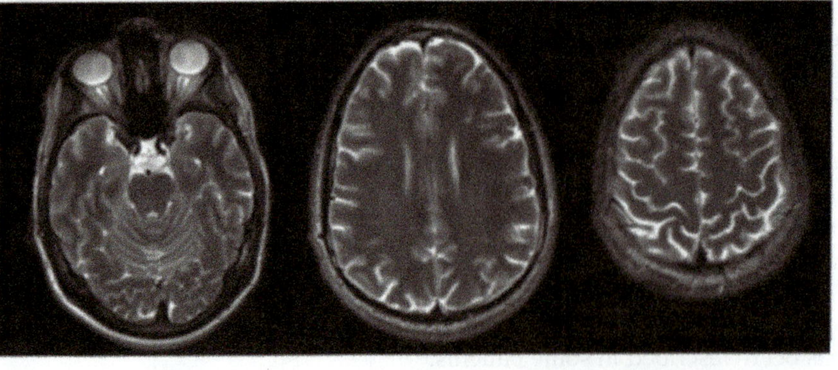

FIGURE 59.1 **Cerebral atrophy in an older MSUD patient.** A series of T2-weighted axial images were taken from an older MSUD patient who experienced poor metabolic control since infancy (**upper panel**) and an age-matched control subject (**lower panel**). There is a general loss of brain tissues at every level of the neuraxis, which is visible as the cortical gyri and cerebellar fossa with prominent sulci. There is an expansion of the perivascular spaces, particularly evident near the cortical surface and temporal lobes. Reproduced with permission from.[9]

residual decarboxylation activity in cells from variant MSUD patients could not be directly related to the phenotype, although it correlated with tolerance for dietary proteins.[11]

A thiamine-responsive type of MSUD was first documented by Scriver and associates.[12] An 11-month female infant had excessive BCAA and BCKA in the urine and exhibited developmental retardation. The patient was placed on a low-protein diet, and 10 mg of thiamine hydrochloride per day was administered. The BCAA in plasma abruptly fell within several days. Withdrawal of thiamine resulted in a prompt rise of plasma BCAA to pre-thiamine treatment levels. Reinstituting thiamine again resulted in a dramatic response. Thiamine responsiveness has been reported in other MSUD subjects, but biochemical improvement required several weeks to months of therapy.[1]

Dihydrolipoamide dehydrogenase (E3) deficiency is a rare disorder.[1] The phenotype is dominated by severe lactic acidosis with relatively modest elevations of BCKA and BCAA.[13] Since E3 is a common component of the pyruvate, α-ketoglutarate, and branched chain α-ketoacid dehydrogenase complexes (see below), its deficiency causes impairment of all three mitochondrial α-ketoacid dehydrogenase complexes. Infants develop persistent lactic acidosis between 8 weeks and 6 months of age. The clinical course is marked by progressive neurological deterioration that includes movement disorders, hypotonia, and seizures. Death usually ensues within the first year of life. The neuropathology shows demyelination and cavitation primarily in the basal ganglia, thalmus and brain stem.

GENETICS AND PREVALENCE

MSUD is clearly a Mendelian recessive disorder as documented by family studies, and more recently by molecular genetic detection of recessive mutations in MSUD families.[1] Prevalence depends upon the population studied. In Mennonites, one in 176 live births is afflicted with classic type MSUD.[14] A large collaborative study of 2.8 million newborns showed an incidence of MSUD at 1:180,000 live births.[15]

COMPONENT ENZYMES AND MACROMOLECULAR ORGANIZATION OF BCKDC

The mammalian BCKDC is a member of the highly conserved mitochondrial α-ketoacid dehydrogenase machines comprising pyruvate dehydrogenase complex (PDC), α-ketoglutarate dehydrogenase complex (α-KGDC) and the BCKDC with similar structure and function.[3] As shown in Table 59.1, the mammalian BCKDC consists of three catalytic components: a heterotetrameric ($\alpha_2\beta_2$) branched-chain α-ketoacid decarboxylase (abbreviated as E1), a homo-24-meric dihydrolipoyl transacylase (E2) and a homodimeric dihydrolipoamide dehydrogenase (E3).

TABLE 59.1 Component Enzymes and Subunit Composition of the Mammalian Branched-Chain α-Ketoacid Dehydrogenase (BCKDC)

Component	Molecular mass (Daltons)	Prosthetic group (P) and cofactor (C)
BCKA decarboxylase (E1)	1.7×10^5 ($\alpha_2\beta_2$)	TPP (C)
α subunit	46,500	Mg^{2+}, K^+ (C)
β subunit	37,200	K^+
Dihydrolipoyl transacylase (E2)	1.1×10^6 (α_{24})	Lipoic acid (P)
Subunit	46,518*	
Dihydrolipoamide dehydrogenase (E3)	1.1×10^5 (α_2)	FAD (C)
Subunit	55,000	
BCKDC kinase	1.8×10^5 (α_4)	Mg^{2+}, K^+ (C)
Subunit	43,000	
BCKDC phosphatase	43,000 (α_1)	Mn^{2+} (C)
Subunit	43,493†	

Abbreviations: BCKA, branched-chain α-ketoacid; FAD, flavin adenine dinucleotide; TPP, thiamine pyrophosphate.

**Calculated from the amino acid composition deduced from a bovine E2 cDNA. The E2 subunit migrates anomalously as a 52-kDa species in sodium dodecyl sulfate polyacrylamide gel electrophoresis (SDS-PAGE).*

†Calculated from the amino acid composition deduced from a human E2 cDNA.

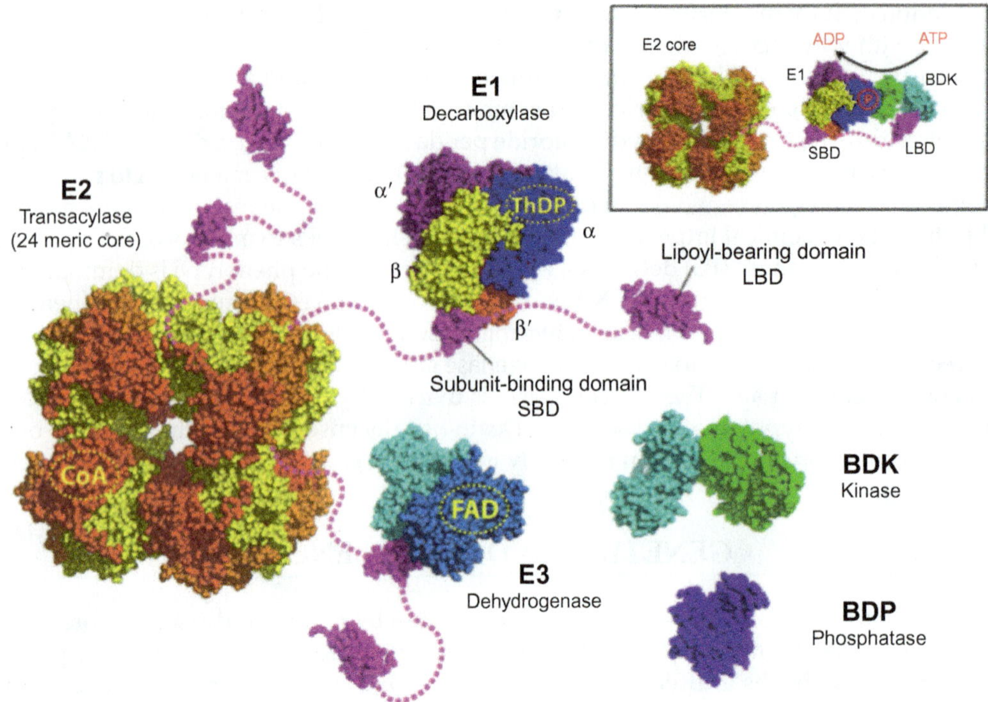

FIGURE 59.2 **Three-dimensional structures and organization of enzyme components in the mammalian BCKDC.** The macromolecular machine (4.5×10^6 Da in size) is organized about a 24-meric cubic core of dihydrolipoyl transacylase (E2), to which multiple copies of branched-chain α-ketoacid decarboxylase (E1) and dihydrolipoamide dehydrogenase (E3), BCKD kinase (BDK) and BCKD phosphatase (BDP) are attached through ionic interactions. Each E2 subunit is made up of three folded domains: lipoyl-binding domain (LBD), subunit-binding domain (SBD), and the E2 core domains that are linked by flexible regions (dashed lines). Only three full-length E2 subunits from the E2 24-mers are shown. E1 $\alpha_2\beta_2$ heterotetramers or E3 homodimers are attached to the SBD. The partial reactions carried out by E1, E2 and E3 are mediated by bound cofactors thiamine diphosphate (ThDP), coenzyme A (CoA) and flavin adenine dinucleotide (FAD), respectively. These partial reactions are coupled through substrate channeling in the BCKDC assembly. **Inset** depicts that the homodimeric BDK binds to the E2 core through the LBD domain of the E2 core and interacts with the E1α subunit to facilitate phosphorylation (the circled letter P), resulting in the inactivation of BCKDC. The space-filling models were based on the coordinates in the following Protein Data Bank accession codes: 1W85 (E1 with SBD), 2II3 (E2 core), 1K8M (LBD), 3RNM (E3 with SBD), 1GKX (BDK) and 4DA1 (BDP).

E1 and E2 components are specific for the BCKDC, whereas the E3 component is common among the three α-ketoacid dehydrogenase complexes.[3] In addition, the mammalian BCKDC contains two regulatory enzymes: the specific kinase and the specific phosphatase that regulate activity of the BCKDC by reversible phosphorylation.[4] The BCKDC is organized around the 24-meric cubic E2 core, to which 12 copies of E1, 6 copies of E3, and unknown numbers of BCKD kinase (BDK) and BCKD phosphatase (BDP) are attached through ionic interactions (Figure 59.2). The molecular mass of the BCKDC multienzyme complex is estimated to be 4.5×10^6 Da. The three catalytic components mediate individual partial reactions that are coupled through substrate channeling in the macromolecular assembly. The overall reaction of the BCKDC is shown below:

$$\text{R-CO-COOH} + \text{CoA-SH} + \text{NAD}^+ \rightarrow \text{R-CO-S-CoA} + \text{CO}_2\uparrow + \text{NADH} + \text{H}^+ \quad (59.1)$$

The mammalian BCKDC is acutely regulated by dynamic phosphorylation (inactivation)/dephosphorylation (activation) cycles.[4] Starvation and diabetes increase BCKDC activity in skeletal muscle by decreasing phosphorylation of the enzyme complex.[16] The state of phosphorylation in BCKDC inversely correlates with the level of BDK in tissues regulated by extracellular stimuli.[17,18] The molecular mechanism for the acute inactivation of BCKDC through phosphorylation by the E2-bound BDK has been elucidated.[19] The presence of a bulky phosphoryl group at Ser292 (phosphorylation site 1) of the E1α subunit renders the phosphorylation loop harboring the phosphoserine residue disordered. The disordered loop conformation interrupts substrate channeling between E1 and E2, leading to the inactivation of BCKDC.[19]

MSUD-causing mutations in catalytic subunits (E1α, E1β, E2 and E3) of the human BCKDC have been documented.[1,20] These mutant alleles are classified as type IA, type IB, type II and type III MSUD, which exhibit the various clinical phenotypes. A novel missense mutation resulting in the inactivation of BDK and activation of BCKDC was recently reported in an autism patient with epilepsy, which is in variance with the traditional MSUD phenotype.[21] Additional missense mutations in BDK resulting in inactivation of the kinase have been reported.[22] As expected,

BCKDC activity is augmented in these patients, which leads to rapid BCAA catabolism and neurobehavioral deficits, which can be reversed by treatment with a protein-rich diet. On the other hand, the *PPM1K* gene that encodes the PP2Cm phosphatase was independently identified by two groups as the BDP phosphatase of BCKDC.[23,24] PP2Cm knockout mice[24] and a homozygous 2-bp deletion[25] resulted in the inactivation of BDP. The absence of a functional BDP in turn causes an inactivation of BCKDC and the intermediate MSUD phenotype.

THE THIAMINE-RESPONSIVE PHENOTYPE IS LINKED TO THE PRESENCE OF MUTANT E2 PROTEINS

The biochemical basis for thiamine-responsive MSUD has been a subject of intense interest and controversies. Thiamine diphosphate (ThDP) is a cofactor of E1, and it was speculated that the mutations in thiamine-responsive patients involved either the α or the β subunit. It was therefore serendipitous when we found that the E2 subunit in the cell extract from Scriver's thiamine-responsive patient WG-34 was much reduced compared to the control samples.[26] The levels of E1α and E1β subunits were normal in WG-34. This was confirmed by a report that showed that the cDNA sequence of the E1α unit of WG-34 were normal.[27] We subsequently identified the two mutant E2 alleles in WG-34, one contained the F215C missense mutation[28] and the other harbored a 3.2-kb deletion in intron 4 of the E2 gene.[29] The 3.2-kb intronic deletion resulted in a null E2 mRNA containing a 17-bp frameshift insertion. The F215C allele produced a functional but unstable E2 protein, which accounts for the relatively high (30–40%) residual BCKDC activity in cultured fibroblasts. Another thiamine-responsive patient was later studied by whole-body [1-^{13}C] leucine oxidation while on a BCAA restricted diet; thiamine supplements at 200 mg per day increased her rate of $^{13}CO_2$ release from undetectable to 14.2% of normal levels.[30] This patient is compound-heterozygous for the K278K and the 15–20-kb deletion alleles.[31] Two additional documented thiamine-responsive patients have also been studied.[30] Both patients were found to carry E2 mutations: one is compound-heterozygous for the P73R and G292R substitutions and the other carries a R223G substitution and the IVSdel[-3.2 kb:-14] deletion. Studies of additional thiamine-responsive patients revealed that they invariably contained at least one missense E2 allele that produces a full-length mutant E2 protein. Most recently, there are two Israeli thiamine-responsive patients that are homozygous for the H391R substitution in the E2 subunit.[32] His-391 is the key catalytic residue serving as a base in the E2 active site; therefore, the H391R mutant E2 protein is devoid of transacylase activity. However, cells from these patients invariably show detectable residual activities for the decarboxylation of BCKA, which are catalyzed by the normal E1 present in these cells. Therefore, there appears to be a tight linkage between the thiamine-responsive phenotype and the presence of at least one allele producing the mutant E2 protein.

The biochemical mechanism for thiamine-responsive MSUD has not been entirely understood. However, *in vitro* studies have shown that the free E1 component, which is not affected in E2-deficient thiamine-responsive MSUD patients, has 5% decarboxylation activity, relative to the overall decarboxylation rate catalyzed by the BCKDC.[33] The binding of the E1 component to the E2 core of the BCKDC further augments E1-catalyzed decarboxylation of BCKA.[32] Based on these results, we suggest that the long-term thiamine supplement increases the mitochondrial ThDP concentration, which prevents the phosphorylation of E1, rendering it fully active *in vivo*.[33] Moreover, the anchoring of normal E1 to the E2 core carrying an MSUD mutation is necessary to stabilize the E1 protein and augment its activity in thiamine-responsive patients. Both mechanisms could account for the favorable clinical outcome in thiamine-responsive MSUD patients receiving dietary thiamine supplements.

ANIMAL MODELS FOR CLASSIC AND INTERMEDIATE MSUD

A severe encephalopathy has been described in association with abnormal BCAA metabolism in Polled Hereford inbred calves in Australia[34,35] and in horned Hereford calves in Canada.[36] This bovine metabolic disorder has many similarities to human MSUD, and it may serve as an animal model. The majority of affected calves are dull at birth; some develop sluggishness, intermittent opisthotonos, and disorganized limb paddling within 2–3 days. Death usually occurs within 5 days of life. The urine has a "bitter sweet odor of burnt sugar" and a positive 2,4-dinitrophenylhydrazine test.

The major neuropathological finding is extensive status spongiosus throughout the white matter.[34,36,37] Electron microscopy shows myelin edema with splitting of the myelin sheath at the intraperiod line.[38] Neurotransmitter studies showed reduced concentrations of the transmitter amino acids glutamate, aspartate, and γ-aminobutyric acid (GABA), as well as a 50% loss in number of postsynaptic GABA receptors as assessed from [^{3}H] diazepam binding.[39]

Markedly increased BCAA were found in serum, cerebrospinal fluid, and brain tissue. A severe deficiency of the BCKDC activity was confirmed in an affected calf by measurement of $^{14}CO_2$ from [1-^{14}C] leucine with intact fibroblasts and from α-[1-^{14}C] KIC using disrupted fibroblasts.[35]

An autosomal recessive inheritance is suggested by breeding experiments. Comparison of this animal model with human MSUD shows that there are some differences. Affected calves were stillborn or born with neurologic symptoms suggesting prenatal metabolic defects. In contrast, patients with MSUD appear normal at birth by both clinical and laboratory examinations and symptoms develop toward the end of the first week of life.

E2 protein knockout mice have been reported, which show the absence of BCKDC activity and a three-fold increase in the circulating concentration of BCAA and BCKA.[40] These phenotypes are remarkably similar to humans with classical MSUD. Reduced neuroactive amino acid levels of alanine, glutamate, and glutamine are likely the cause of reduced neurological function in the brain of E2-deficient mice. They invariably show neonatal lethality, usually resulting in death within 72 hours, likely due to accumulation of BCAA to neurotoxic levels, ketoacidosis, brain edema, dehydration, and malnutrition as observed in MSUD calves[34] and classic MSUD patients.[1] Interestingly, transgenic expression of the "normal" E2-c-*myc* protein in the liver of these E2 knockout mice produced an animal model with intermediate MSUD.[40] BCKDC activity was found to be 5–6% of normal, but was sufficient to allow the animals to survive. This model is interesting in light of the recent X-ray structure of the E2 inner core,[41] whereby deleting the C-terminus exclusively resulted in trimer formation instead of the normal 24-mer assembly. We have since added additional amino acids to the full-length E2 monomers and this extension also results in trimer formation (Wynn et al., unpublished results). Therefore, it is highly likely that the C-terminal addition of the c-*myc* leads to the production of only trimeric E2-c-*myc* proteins. We have found that the trimers only have about 15–20% of the overall activity compared to normal 24-mers, when reconstituted with E1 and E3 (Wynn et al., unpublished results). The above results may explain the intermediate MSUD phenotype presented by the E2-transgenic mice. The availability of this E2-deficient intermediate MSUD model offers an opportunity to investigate its response to thiamine supplements. Positive results, if obtained, may establish the linkage between the presence of mutant E2 proteins and the thiamine-responsive MSUD phenotype.[32,42]

Zebrafish now present a viable option to mice or rat models due to their size and the number of fish that can be produced and subsequently studied. Previous neurological studies with zebrafish have concentrated on central nervous system function, muscle relaxation and other swimming behaviors.[43–46] Recent studies by Friedich et al.[47] revisited the known *que* mutation associated with abnormal postfertilization muscle contraction, and determined that it is a single-point mutation within the E2 gene. The E2 point mutation was shown to be at the exon–intron boundary of exon 6 (AGGT mutated to AG<u>A</u>T), which fails to splice the intron properly for the E2 gene.[47] E2-deficient *que* mutant zebrafish possess abnormally elevated branched-chain amino acid levels, similar to classical MSUD patients, as well as reduced levels of the neurotransmitter glutamate, in both brain and spinal cord. Deficiencies in glutamate levels, most likely lead to severe dystonia in these larval zebrafish and explain their abnormal muscle contractions. The *que* zebrafish model will likely become an important tool to enhance our understanding of the neurological progression for MSUD. In addition, the *que* zebrafish model may present useful therapeutic options for translating pharmacological treatments to MSUD patients.

Another recent MSUD model of interest is the PP2Cm (BDP) knockout mouse. These animals manifested elevated concentrations of BCAA due to the inactivation of BCKDC and produced an intermediate MSUD phenotype.[24] In addition to neuronal tissues, PP2Cm is highly expressed in cardiac muscle, and its expression is diminished in a heart under pathologic stresses. Whereas phenotypic features of heart failure are seen in PP2Cm-deficient zebrafish embryos,[48] cardiac function in PP2Cm-null mice is compromised at a young age and deteriorates faster with mechanical overload. These observations suggest that the catabolism of branched-chain amino acids also has physiologic significance in maintaining normal cardiac function. Defects in PP2Cm (BDP)-mediated catabolism of branched-chain amino acids may be a potentially novel mechanism not only for MSUD, but also for congenital heart diseases and heart failure.

TREATMENTS OF MSUD

The standard dietary therapy, which involves feeding patients a synthetic diet containing reduced BCAA contents, was originally instituted by Snyderman in 1964.[6] When a classic MSUD patient was placed on this dietary regimen for 3 months, the plasma BCAA level was decreased to the normal range, including the disappearance of the alloisoleucine marker for MSUD. Commercial medical diets have since been developed for MSUD patients, based largely on Snyderman's synthetic formula. Single dietary formulations have recently been designed and developed to counteract the metabolic derangements that cause brain disease while also providing a nutritional safety net against essential lipid and micronutrient deficiencies.[49] Orthotopic liver transplantations were originally performed on four MSUD patients for other medical reasons.[1,50] This procedure has proved effective in controlling plasma BCAA concentrations. There was a dramatic drop of plasma leucine concentrations in a classic MSUD patient to the nearly normal level after the liver transplantation while the patient was on a normal diet.[51] This poses an interesting question in terms of inter-organ relationships of BCAA catabolism. The prevailing concept has been that the skeletal muscle is the major site for the decarboxylation of BCAA, due to its large mass and the presence of BCAA aminotransferase activity in the human muscle.[52] The fact that the plasma leucine level returns to normal after the liver transplantation indicates that the transplanted liver alone is capable of degrading more than 90% of BCAA. Other organs, including skeletal muscle, are silent for BCAA decarboxylation, as a result of genetic defects in the BCKDC.

Transplantation of Human Amnion Epithelial Cells

Organ shortage is the primary impetus for clinical cell transplantation. Clinical hepatocyte transplantation has shown promise for a number of inherited metabolic diseases, such as Crigler–Najjar type 1, ornithine transcarbamylase deficiency, citrullinemia, and glycogen storage diseases.[53,54] However, the availability of useful hepatocytes remains a limiting factor of clinical transplantation. Human amnion epithelial cells (hAEC) are easily accessible and free of any ethical and safety concerns common to embryonic stem cells. They can be efficiently cryopreserved and they exhibit many beneficial and immune favored characteristics.[55–58]

Several methods have now been employed to differentiate hAEC into hepatocyte-like cells.[59,60] Undifferentiated hAECs transplanted into livers of mice were found to display hepatic morphology and to express mature liver genes at comparable levels with human adult livers. Transplanted hAECs significantly increased BCKDC enzyme activity in a mouse model of intermediate MSUD resulting in extended animal survival, more normal body weight, and decreased circulating BCAA levels.[61] This study supports the concept that placental-derived stem cells such as hAEC may provide a safe and abundant cell source for the treatment of MSUD, and other liver-based metabolic diseases.

Small Molecule Inhibitors for BDK

Modulation of BDK activity constitutes a major mechanism for BCAA homeostasis *in vivo*,[62] and BDK offers a therapeutic target for ameliorating the accumulation of BCAA and BCKA in disease conditions. BDK is inhibited by KIC from leucine, resulting in the activation of BCKDC in perfused rat hearts.[63] Thus, leucine serves as a "feed-forward" nutritional signal that promotes BCAA disposal through the inhibition of BDK activity. The inhibition of BDK by small molecules such as KIC prompted the development and identification of a series of KIC analogs that function as BDK inhibitors.[63,64] These include KIC analogs α-chloroisocaproate (CIC),[65] phenylpyruvate,[64] clofibric acid,[66] and recently phenylbutyrate (PB).[67] Figure 59.3 shows that plasma BCAA and BCKA concentrations are significantly ($P \leq 0.05$) smaller than reduced after PB treatment on MSUD patients with residual BCKDC activity. The improved MUSD phenotype results from increased mutant BCKDC activity through dephosphorylation. However, these BDK inhibitors are less than robust as BDK inhibitors with reported I_{40} (concentration for 40% inhibition) in the sub-millimolar range (e.g., CIC, phenylpyruvate and clofibric acid). A unique allosteric site has been identified and structure-based design was employed to produce novel BDK inhibitors with improved IC_{50} and binding affinities for BDK.[68] We have shown that one of these newly identified BDK inhibitors robustly augments BCKDC activity and reduces BCAA concentrations in wild-type mice. These novel BDK inhibitors have clinical ramifications for treating hereditary metabolic disorders, such as MSUD, that are associated with the accumulation of BCAA and BCKA.

CONCLUDING REMARKS

MSUD is arguably an underappreciated metabolic disorder because of its rare occurrence in the general population. However, the disease has severe neurological and metabolic consequences caused by the toxicity of the accumulated BCKA. These manifestations often prove fatal or lead to mental retardation in surviving patients. MSUD is, therefore, a dreadful disease to individual patients and their families as well as to certain kindred such as Mennonites, where the incidence of the disease is high as a result of consanguinity.

The six genetic loci associated with MSUD confer the large variations in clinical phenotypes, and complicate the identification and mutational analysis of the affected gene. On the other hand, the mitochondrial BCKDC, which is deficient in MSUD, has been and continues to be a fertile ground for studying how human mutations impede catalysis, assembly and protein-to-protein interactions of the macromolecular multi-enzyme complex. The strong

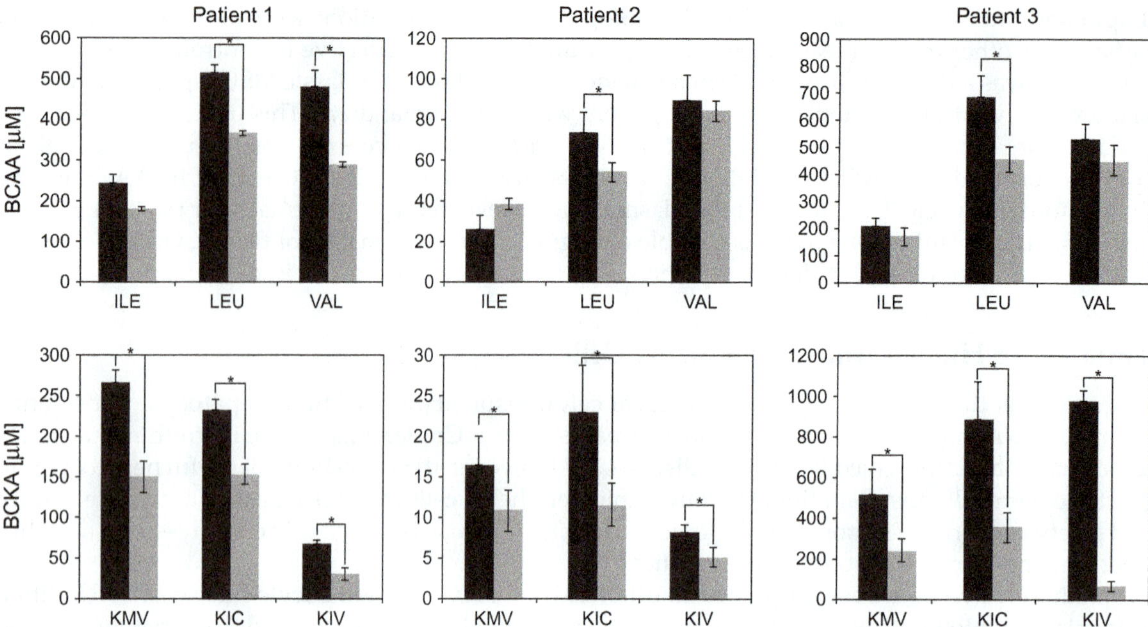

FIGURE 59.3 **Phenylbutyrate treatment reduces plasma BCAA and BCKA levels in MSUD subjects. Upper panel**: shows plasma BCAA concentrations in MSUD patients 1–3 before (black bars) and after (gray bars) the phenylbutyrate treatment. Abbreviations: Ile, isoleucine; Leu, leucine; Val, valine. **P*≤0.05. **Lower panel**: illustrates plasma BCKA concentrations in MSUD patients 1–3 before (black bars) and after (gray bars) the phenylbutyrate treatment. Abbreviations: KMV, α-keto-β-methylvalerate; KIC, α-ketoisocaproate; KIV, α-ketoisovalerate. **P*≤0.05. Patient 1 carried a homozygous V367M missense mutation in the E1α subunit. Patient 2 was a compound heterozygote with a c. 75_76 2-bp deletion (in exon 2) and a R240C missense mutation in the E2 subunit. Patient 3 was a compound heterozygote with a S305P missense mutation and an exon 11 deletion in the E2 subunit. Adapted with permission from.[67]

correlation between E2 mutations and the thiamine-responsive MSUD phenotype suggests a potential role of the vitamin in mitigating mild variants of the disease. Restrictions in BCAA intake is the standard dietary therapy to control BCAA levels in MSUD. Liver transplantation has emerged as an effective treatment to prevent metabolic decompensation.[69] Pharmacological treatments with kinase inhibitors are also under investigation to reduce BCAA concentrations through increased residual BCKDC activity (Figure 59.3). Currently, two murine animal models for MSUD allow for developing new therapies for treatment of MSUD. Human AEC transplantation in the intermediate MSUD mouse model has produced encouraging results.

ACKNOWLEDGEMENTS

This work was supported by grants DK26758, DK62306 and DK92921 from the National Institutes of Health and I-1286 from the Welch Foundation.

References

1. Chuang DT, Shih VE. Maple syrup urine disease (branched-chain ketoaciduria). In: Scriver CR, Beaudet AL, Sly WS, Valle D, Childs B, Kinzler KW, Vogelstein B, eds. *The Metabolic and Molecular Basis of Inherited Disease 1971–2005*. New York, NY: McGraw-Hill; 2001:1971–2005.
2. Dancis J, Hutzler J, Levitz M. Metabolism of the white blood cells in maple-syrup-urine disease. *Biochim Biophys Acta*. 1960;43:342.
3. Chuang DT, Chuang JL, Wynn RM, Song J-L. The branched-chain alpha-ketoacid dehydrogenase complex, human. In: Creighton TE, ed. *Encyclopedia of Molecular Medicine*. vol. 5. New York, NY: John Wiley & Sons; 2001:393–396.
4. Harris RA, Hawes JW, Popov KM, et al. Studies on the regulation of the mitochondrial alpha-ketoacid dehydrogenase complexes and their kinases. *Adv Enzyme Regul*. 1997;37:271–293.
5. Menkes JH, Hurst PL, Craig JM. A new syndrome: progressive familial infantile cerebral dysfunction associated with an unusual urinary substance. *Pediatrics*. 1954;14:462.
6. Snyderman SE, Norton PM, Roitman E, Holt Jr LE. Maple syrup urine disease, with particular reference to dietotherapy. *Pediatrics*. 1964;34:454.
7. Dancis J, Levitz M, Miller S, Westall RG. Maple syrup urine disease. *Br Med J*. 1959;1:91–93.
8. Chhabria S, Tomasi LG, Wong PW. Ophthalmoplegia and bulbar palsy in variant form of maple syrup urine disease. *Ann Neurol*. 1979;6:71–72.

9. Muelly ER, Moore GJ, Bunce SC, et al. Biochemical correlates of neuropsychiatric illness in maple syrup urine disease. *J Clin Invest.* 2013;123:1809–1820.
10. Manara R, Del Rizzo M, Burlina AP, et al. Wernicke-like encephalopathy during classic maple syrup urine disease decompensation. *J Inherit Metab Dis.* 2012;35:413–417.
11. Dancis J, Hutzler J, Snyderman SE, Cox RP. Enzyme activity in classical and variant forms of maple syrup urine disease. *J Paediatr.* 1972;81(2):312–320.
12. Scriver CR, Mackenzie S, Clow CL, Delvin E. Thiamine-responsive maple-syrup-urine disease. *Lancet.* 1971;297:310–312.
13. Munnich A, Saudubray JM, Taylor J, et al. Congenital lactic acidosis, α-ketoglutaric aciduria and variant form of maple syrup urine disease due to a single enzyme defect: dihydrolipoyl dehydrogenase deficiency. *Acta Paediatr Scand.* 1982;71:167–171.
14. Marshall L, DiGeorge A. Maple syrup urine disease in the old order Mennonites. *Am J Hum Genet.* 1981;33:139A.
15. Naylor EW, Guthrie R. Newborn screening for maple syrup urine disease (branched-chain ketoaciduria). *Pediatrics.* 1978;61:262–266.
16. Paul HS, Adibi SA. Role of ATP in the regulation of branched-chain alpha-keto acid dehydrogenase activity in liver and muscle mitochondria of fed, fasted, and diabetic rats. *J Biol Chem.* 1982;257:4875–4881.
17. Zhao Y, Denne SC, Harris RA. Developmental pattern of branched-chain 2-oxo acid dehydrogenase complex in rat liver and heart. *Biochem J.* 1993;290:395–399.
18. Huang Y, Chuang DT. Down-regulation of rat mitochondrial branched-chain 2-oxoacid dehydrogenase kinase gene expression by glucocorticoids. *Biochem J.* 1999;339:503–510.
19. Wynn RM, Kato M, Machius M, et al. Molecular mechanism for regulation of the human mitochondrial branched-chain alpha-ketoacid dehydrogenase complex by phosphorylation. *Structure (Camb).* 2004;12:2185–2196.
20. Chuang DT, Wynn RM, Shih VE. Maple syrup urine disease (branched-chain ketoaciduria) external update online. In: Scriver CR, Beaudet AL, Sly WS, Valle D, Vogelstein B, Childs B, eds. *The Metabolic and Molecular Basis of Inherited Disease.* New York, NY: McGraw-Hill; 2008:1971–2006.
21. Novarino G, El-Fishawy P, Kayserili H, et al. Mutations in BCKD-kinase lead to a potentially treatable form of autism with epilepsy. *Science.* 2012;338:394–397.
22. Garcia-Cazorla A, Oyarzabal A, Fort J, et al. Two novel mutations in the BCKDK gene (branched-chain keto-acid dehydrogenase kinase) are responsible for a neurobehavioral deficit in two pediatric unrelated patients. *Hum Mutat.* 2014;doi:10.1002/humu.22513.
23. Joshi M, Jeoung NH, Popov KM, Harris RA. Identification of a novel PP2C-type mitochondrial phosphatase. *Biochem Biophys Res Commun.* 2007;356:38–44.
24. Lu G, Sun H, She P, et al. Protein phosphatase 2Cm is a critical regulator of branched-chain amino acid catabolism in mice and cultured cells. *J Clin Invest.* 2009;119:1678–1687.
25. Oyarzabal A, Martínez-Pardo M, Merinero B, et al. A novel regulatory defect in the branched-chain α-keto acid dehydrogenase complex due to a mutation in the PPM1K gene causes a mild variant phenotype of maple syrup urine disease. *Hum Mutat.* 2013;34:355–362.
26. Fisher CW, Chuang JL, Griffin TA, Lau KS, Cox RP, Chuang DT. Molecular phenotypes in cultured maple syrup urine disease cells. Complete E1 alpha cDNA sequence and mRNA and subunit contents of the human branched chain alpha-keto acid dehydrogenase complex. *J Biol Chem.* 1989;264:3448–3453.
27. Zhang B, Wapner RS, Brandt IK, Harris RA, Crabb DW. Sequence of the E1α subunit of branched chain α-ketoacid dehydrogenase in two patients with thiamine-responsive maple syrup urine disease. *Am J Hum Genet.* 1990;46:843.
28. Fisher CW, Lau KS, Fisher CR, Wynn RM, Cox RP, Chuang DT. A 17-bp insertion and a Phe215Cys missense mutation in the dihydrolipoyl transacylase (E2) mRNA from a thiamine-responsive maple syrup urine disease patient WG-34. *Biochem Biophys Res Commun.* 1991;174:804–809.
29. Chuang JL, Cox RP, Chuang DT. E2 transacylase-deficient (type II) maple syrup urine disease. Aberrant splicing of E2 mRNA caused by internal intronic deletions and association with thiamine-responsive phenotype. *J Clin Invest.* 1997;100:736–744.
30. Ellerine NP, Herring WJ, Elsas IILJ, McKean MC, Klein PD, Danner DJ. Thiamin-responsive maple syrup urine disease in a patient antigenically missing dihydrolipoamide acyltransferase. *Biochem Med Metab Biol.* 1993;49:363–374.
31. Herring WJ, Litwer S, Weber JL, Danner DJ. Molecular genetic basis of maple syrup urine disease in a family with two defective alleles for branched chain acyltransferase and localization of the gene to human chromosome 1. *Am J Hum Genet.* 1991;48:342–350.
32. Chuang JL, Wynn RM, Moss CC, et al. Structural and biochemical basis for novel mutations in homozygous Israeli maple syrup urine disease patients: a proposed mechanism for the thiamin-responsive phenotype. *J Biol Chem.* 2004;279:17792–17800.
33. Li J, Wynn RM, Machius M, et al. Cross-talk between thiamin diphosphate binding and phosphorylation loop conformation in human branched-chain alpha-keto acid decarboxylase/dehydrogenase. *J Biol Chem.* 2004;279:32968–32978.
34. Harper PA, Healy PJ, Dennis JA. Maple syrup urine disease as a cause of spongiform encephalopathy in calves. *Vet Rec.* 1986;119:62–65.
35. Harper PA, Dennis JA, Healy PJ, Brown GK. Maple syrup urine disease in calves: a clinical, pathological and biochemical study. *Aust Vet J.* 1989;66:46–49.
36. Baird JD, Wojcinski ZW, Wise AP, Godkin MA. Maple syrup urine disease in five Hereford calves in Ontario. *Can Vet J.* 1987;28:505–511.
37. Harper PA, Healy PJ, Dennis JA. Maple syrup urine disease (branched chain ketoaciduria). *Am J Pathol.* 1990;136:1445–1447.
38. Harper PA, Healy PJ, Dennis JA. Ultrastructural findings in maple syrup urine disease in Poll Hereford calves. *Acta Neuropathol (Berl).* 1986;71:316–320.
39. Dodd PR, Williams SH, Gundlach AL, et al. Glutamate and gamma-aminobutyric acid neurotransmitter systems in the acute phase of maple syrup urine disease and citrullinemia encephalopathies in newborn calves. *J Neurochem.* 1992;59:582–590.
40. Homanics GE, Skvorak K, Ferguson C, Watkins S, Paul HS. Production and characterization of murine models of classic and intermediate maple syrup urine disease. *BMC Med Genet.* 2006;7:33.
41. Kato M, Wynn RM, Chuang JL, Brautigam CA, Custorio M, Chuang DT. A synchronized substrate-gating mechanism revealed by cubic-core structure of the bovine branched-chain alpha-ketoacid dehydrogenase complex. *EMBO J.* 2006;25:5983–5994.
42. Chuang DT, Chuang JL, Wynn RM. Lessons from genetic disorders of branched-chain amino acid metabolism. *J Nutr.* 2006;136:243S–249S.
43. Gleason MR, Armisen R, Verdecia MA, Sirotkin H, Brehm P, Mandel G. A mutation in serca underlies motility dysfunction in accordion zebrafish. *Dev Biol.* 2004;276:441–451.

44. Hirata H, Saint-Amant L, Waterbury J, et al. Accordion, a zebrafish behavioral mutant, has a muscle relaxation defect due to a mutation in the ATPase Ca^{2+} pump SERCA1. *Development*. 2004;131:5457–5468.
45. Hirata H, Saint-Amant L, Downes GB, et al. Zebrafish bandoneon mutants display behavioral defects due to a mutation in the glycine receptor beta-subunit. *Proc Natl Acad Sci U S A*. 2005;102:8345–8350.
46. Olson BD, Sgourdou P, Downes GB. Analysis of a zebrafish behavioral mutant reveals a dominant mutation in atp2a1/SERCA1. *Genesis*. 2010;48:354–361.
47. Friedrich T, Lambert AM, Masino MA, Downes GB. Mutation of zebrafish dihydrolipoamide branched-chain transacylase E2 results in motor dysfunction and models maple syrup urine disease. *Dis Model Mech*. 2012;5:248–258.
48. Sun H, Lu G, Ren S, Chen J, Wang Y. Catabolism of branched-chain amino acids in heart failure: insights from genetic models. *Pediatr Cardiol*. 2011;32:305–310.
49. Strauss KA, Wardley B, Robinson D, et al. Classical maple syrup urine disease and brain development: principles of management and formula design. *Mol Genet Metab*. 2010;99:333–345.
50. Wendel U, Saudubray JM, Bodner A, Schadewaldt P. Liver transplantation in maple syrup urine disease. *Eur J Pediatr*. 1999;158(suppl 2):S60–S64.
51. Bodner-Leidecker A, Wendel U, Saudubray JM, Schadewaldt P. Branched-chain L-amino acid metabolism in classical maple syrup urine disease after orthotopic liver transplantation. *J Inherit Metab Dis*. 2000;23:805–818.
52. Suryawan A, Hawes JW, Harris RA, Shimomura Y, Jenkins AE, Hutson SM. A molecular model of human branched-chain amino acid metabolism. *Am J Clin Nutr*. 1998;68:72–81.
53. Strom SC, Bruzzone P, Cai H, et al. Hepatocyte transplantation: clinical experience and potential for future use. *Cell Transplant*. 2006;15 (suppl 1):S105–S110.
54. Fisher RA, Strom SC. Human hepatocyte transplantation: worldwide results. *Transplantation*. 2006;82:441–449.
55. Li H, Niederkorn JY, Neelam S, et al. Immunosuppressive factors secreted by human amniotic epithelial cells. *Invest Ophthalmol Vis Sci*. 2005;46:900–907.
56. Miki T, Strom SC. Amnion-derived pluripotent/multipotent stem cells. *Stem Cell Rev*. 2006;2:133–142.
57. Banas RA, Trumpower C, Bentlejewski C, Marshall V, Sing G, Zeevi A. Immunogenicity and immunomodulatory effects of amnion-derived multipotent progenitor cells. *Hum Immunol*. 2008;69:321–328.
58. Parolini O, Alviano F, Bagnara GP, et al. Concise review: isolation and characterization of cells from human term placenta: outcome of the first international Workshop on Placenta Derived Stem Cells. *Stem Cells*. 2008;26:300–311.
59. Marongiu F, Gramignoli R, Dorko K, et al. Hepatic differentiation of amniotic epithelial cells. *Hepatology*. 2011;53:1719–1729.
60. Miki T, Marongiu F, Ellis EC, et al. Production of hepatocyte-like cells from human amnion. *Methods Mol Biol*. 2009;481:155–168.
61. Skvorak KJ, Dorko K, Marongiu F, et al. Placental stem cell correction of murine intermediate maple syrup urine disease. *Hepatology*. 2013;57:1017–1023.
62. Harris RA, Joshi M, Jeoung NH. Mechanisms responsible for regulation of branched-chain amino acid catabolism. *Biochem Biophys Res Commun*. 2004;313:391–396.
63. Paxton R, Harris RA. Regulation of branched-chain alpha-ketoacid dehydrogenase kinase. *Arch Biochem Biophys*. 1984;231:48–57.
64. Paxton R, Harris RA. Clofibric acid, phenylpyruvate, and dichloroacetate inhibition of branched-chain alpha-ketoacid dehydrogenase kinase in vitro and in perfused rat heart. *Arch Biochem Biophys*. 1984;231:58–66.
65. Harris RA, Paxton R, DePaoli-Roach AA. Inhibition of branched chain alpha-ketoacid dehydrogenase kinase activity by alpha-chloroisocaproate. *J Biol Chem*. 1982;257:13915–13918.
66. Kobayashi R, Murakami T, Obayashi M, et al. Clofibric acid stimulates branched-chain amino acid catabolism by three mechanisms. *Arch Biochem Biophys*. 2002;407:231–240.
67. Brunetti-Pierri N, Lanpher B, Erez A, et al. Phenylbutyrate therapy for maple syrup urine disease. *Hum Mol Genet*. 2011;20:631–640.
68. Tso SC, Qi X, Gui WJ, et al. Structure-based design and mechanisms of allosteric inhibitors for mitochondrial branched-chain alpha-ketoacid dehydrogenase kinase. *Proc Natl Acad Sci U S A*. 2013;110:9728–9733.
69. Strauss KA, Mazariegos GV, Sindhi R, et al. Elective liver transplantation for the treatment of classical maple syrup urine disease. *Am J Transplant*. 2006;6:557–564.

CHAPTER

60

Congenital Disorders of *N*-linked Glycosylation

Marc C. Patterson

Mayo Clinic, Rochester, MN, USA

INTRODUCTION

The congenital disorders of glycosylation (CDG) are a family of anabolic diseases with variable multisystem manifestations. All result from defective activity of enzymes that participate in the modification of proteins and other macromolecules by the addition and processing of oligosaccharide side chains (glycosylation). Disorders of *N*-linked and *O*-linked glycosylation and of glypiation (synthesis of glycosylphosphatidylinositol [GPI] anchors) have been described.[1] Strictly speaking, the galactosemias and mucolipidoses II and III should be included in this group; patients with galactosemia have abnormal *N*-glycan formation,[2] and lysosomal hydrolases are missorted because of impaired addition of mannose-6-phosphate residues to the polypeptide chains in mucolipidoses II and III.[3] These disorders will not be further discussed here because they have traditionally been separately categorized. They are discussed in in Chapters 55 (galactosemias) and 32 (mucolipidoses). Disorders of *O*-linked glycosylation are associated with a variety of phenotypes; the congenital muscular dystrophies (Chapter 91) and limb-girdle dystrophies (Chapter 92) are of most interest to neurologists. The glycoproteinoses (Chapter 33) are the catabolic counterparts of congenital disorders of glycosylation, and result from deficiencies in specific lysosomal hydrolases that catalyze the breakdown of glycoproteins (glycans).

N-glycosylation is localized to the endoplasmic reticulum (ER) and Golgi apparatus in eukaryotes, and is essential to survival.[4] The process occurs in three distinct phases.[5] In the first phase, oligosaccharides are synthesized in a series of reactions catalyzed by enzymes located on the cytoplasmic surface of the ER membrane. Glucose (Glc), fucose (Fuc), and mannose (Man) derived from the diet are activated through linkage to nucleotide bases, and are then sequentially attached to dolichol pyrophosphate (Dol-PP) anchored in the ER membrane to yield a branched-chain structure containing two *N*-acetylglucosamine (GlcNAc) molecules and eight mannose molecules ($GlcNAc_2Man_8$). This complex is then *flipped* (by one or more proteins—including RFT1)[6] from the cytoplasmic to the luminal face of the ER membrane, where additional saccharides are added to form Dol-PP-$GlcNAc_2Man_9Glc_3$ (the lipid-linked oligosaccharide [LLO] precursor). The LLO interacts with the oligosaccharyltransferase complex (OST), and attaches to growing polypeptide chains through amide bonds with asparagine (Asn) residues that are associated with serine (Ser) residues in an Asn-X-Ser motif.[7]

In the second phase, the external glucose residues of the oligosaccharide chains are trimmed by α-glucosidases to yield a $GlcNAc_2Man_9Glc_1$ side chain that enters the calnexin–calreticulin deglycosylation–reglycosylation cycle. Proteins that are appropriately folded are deglycosylated and are able to proceed along the ER–Golgi pathway. Those that are not folded correctly remain in the cycle until they are either appropriately folded or degraded.[8]

In the third phase of *N*-glycosylation, the nascent glycoproteins pass from the ER to the Golgi apparatus, a vesicular transport process requiring the participation of several multiprotein complexes, including the conserved oligomeric Golgi (COG) complex and SNARE (soluble NSF-attachment protein [SNAP] receptor) proteins.[9] The glycoproteins then pass through the Golgi apparatus, where they are exposed to gradually increasing pH and a series of enzymes that first trim the *N*-glycans to a core of mannose residues and then add more saccharides to produce complex, tissue-specific glycoforms.[10] For any given protein, a family of glycoforms is produced, such that the net function of the protein represents the mean activity of the population of glycoforms.

Rosenberg's Molecular and Genetic Basis of Neurological and Psychiatric Disease
http://dx.doi.org/10.1016/10.1016/B978-0-12-410529-4.00060-7

Observations from animal studies suggest that optimal function is achieved when there is a balance between hypoglycosylation and excessive glycosylation.[11] For example, the (hypoglycosylated) *MGATV–/–* mouse is relatively protected from tumor growth and metastasis,[12] but is more susceptible to autoimmune diseases than controls[13] (discussed later).

Two phases of *N*-glycosylation may be conceptualized: a tightly controlled early phase (in the ER) leading to the production of a limited number of glycoproteins of high and consistent quality, followed by the Golgi phase, in which modifiable structural and functional diversity is conferred on glycoproteins through tissue-specific processing of glycans. Impairment of *N*-glycosylation leads to specific cellular dysfunction through impaired activity of the hypoglycosylated glycoconjugates and nonspecific dysfunction by activating the unfolded protein response (UPR).[14]

CLINICAL FEATURES AND DIAGNOSTIC EVALUATION

The *N*-linked (and multiple pathway) CDG syndromes are summarized in Tables 60.1 and 60.2, respectively. Until 2009, the nomenclature of CDG recognized three groups.[15] Group I included those disorders characterized by impaired synthesis of the lipid-linked oligosaccharide and its attachment to the growing polypeptide chain. All members of this group have a type 1 pattern of transferrin glycoforms. Group II included defects in the processing of *N*-glycans. Disorders in groups I and II were given a letter corresponding to their order of description. For example, phosphomannomutase 2 (PMM2) deficiency, the first recognized CDG, was designated CDG 1a, and phosphomannose isomerase (PMI) deficiency, the second type I disorder described was designated CDG 1b, and so on. Group X was established for those patients with evidence of abnormal glycosylation in whom the underlying enzymatic and genetic basis has not yet been determined. This classification superseded previous designations of these disorders as the "diasialotransferrin developmental deficiency" (DDD) syndrome[16] and the "carbohydrate-deficient glycoprotein syndrome"[17] (CDGS), reflecting growing appreciation of the biochemical and clinical spectrum of this family of diseases. By 2009, it was apparent that the rapid growth in the number of CDGs recognized had rendered this classification unworkable, and the current classification was established.[18] This taxonomy delineates four groups: a) defects in protein *N*-glycosylation; b) defects in protein *O*-glycosylation; c) defects in glycosphingolipid and glycosylphosphatidylinositol anchor glycosylation; and d) defects in multiple glycosylation and other pathways. Within these groups, individual disorders are designated by the mutated gene, followed by CDG. Thus, CDG 1a becomes PMM2-CDG, CDG 1a becomes MPI-CDG, and so on.

The most frequently recognized phenotype is PMM2-CDG (PMM2 deficiency); at least 700 affected people are known.[19] The clinical progression of the typical (severe) phenotype has been classified into four phases.[20] The first is the infantile phase, with varying combinations of dysmorphism, abnormal fat distribution (supragluteal and vulval fat pads, focal lipoatrophy),[21] abnormal hair structure,[22] inverted nipples, cryptorchidism, recurrent infections, cardiomyopathy or pericardial effusions, coagulopathies, nephrotic syndrome, hypothyroidism, life-threatening episodes of hepatic failure, and unexplained coma. Infants may exhibit a variety of abnormal eye movements, and esotropia may also become apparent during this phase.[23–25] Mortality may be up to 20% in this phase. In the second phase (the remainder of the first decade), children exhibit evolving facial hypotonia and esotropia, which is often bilateral and fixed, and experience seizures and stroke-like episodes, often precipitated by intercurrent infections. So-called stroke-like episodes may represent ictal or postictal paralysis, focal edema or necrosis or ischemic strokes.[26–30] The third phase (second decade) is marked by slowly progressive cerebellar ataxia and wasting of the legs, accompanied by progressive visual loss secondary to pigmentary retinopathy. In the fourth (adult) phase, the picture is one of moderate intellectual disability with severe ataxia and hypogonadism, with or without skeletal deformities. Long-term follow-up has shown that the phenotype stabilizes after the first decade, with most patients showing stable intellectual disability and motor deficits.[31] As is the case for most inborn errors of metabolism, the availability and application of biochemical tests has led to the identification of milder and partial phenotypes of CDG.[32,33] For example, early reports stressed the ubiquity of cerebellar hypoplasia in CDG-Ia. One girl with CDG-Ia had normal computed tomography (CT) of the head at 9 months, with progressive atrophy on CT at 2 years, and magnetic resonance imaging (MRI) at 9 years. The authors concluded that the (olivoponto)cerebellar hypoplasia reported in infancy in most children with CDG-Ia likely results from atrophy of antenatal onset,[34] rather than hypoplasia. Children with normal development presenting with otherwise nonspecific behavioral disturbances may have PMM2-CDG.[35]

MPI-CDG (MPI deficiency; CDG-Ib) can be effectively treated with oral mannose therapy. At least 20 cases have been recognized.[19] It has no primary neurologic manifestations, the burden of disease falling on the liver, gastrointestinal tract, kidney, and coagulation system. MPI-CDG can present with seizures secondary to severe hypoglycemia,[36] although extraneurologic manifestations usually predominate. The endocrine and gastrointestinal symptoms are usually fully

TABLE 60.1 Classification of CDG: Defects in Protein *N*-Glycosylation

Individual disorder	Gene	Gene product
PMM2-CDG (CDG-Ia)	*PMM2*	Phosphomannomutase 2
MPI-CDG (CDG-Ib)	*MPI*	Mannose-6-phosphate isomerase
ALG6-CDG (CDG-Ic)	*ALG6*	Dolichyl pyrophosphate $Man_9GlcNAc_2$ α-1,3-glucosyltransferase
ALG3-CDG (CDG-Id)	*ALG3*	Dolichyl-P-Man:$Man_5GlcNAc_2$-PP-dolichyl mannosyltransferase
DPM1-CDG (CDG-Ie)	*DPM1*	Dolichol-phosphate mannosyltransferase
MPDU1-CDG (CDG-If)	*MPDU1*	Mannose-P-dolichol utilization defect 1 protein
ALG12-CDG (CDG-Ig)	*ALG12*	Dolichyl-P-Man:$Man_7GlcNAc_2$-PP-dolichyl-α-1,6-mannosyltransferase
ALG8-CDG (CDG-Ih)	*ALG8*	Probable dolichyl pyrophosphate $Glc_1Man_9GlcNAc_2$ α-1,3-glucosyltransferase
ALG2-CDG (CDG-Ii)	*ALG2*	α-1,3-mannosyltransferase ALG2
DPAGT1-CDG (CDG-Ij)	*DPAGT1*	UDP-*N*-acetylglucosamine–dolichyl-phosphate *N*-acetylglucosaminephosphotransferase
ALG1-CDG (CDG-Ik)	*ALG1/HMT-1*	Chitobiosyldiphosphodolichol β-mannosyltransferase
ALG9-CDG (CDG-IL)	*ALG9*	α-1,2-mannosyltransferase ALG9
DOLK-CDG (CDG-Im)	*DOLK (DK1)*	Dolichol kinase
RFT1-CDG (CDG-In)	*RFT1*	Protein RFT1 homolog
DPM3-CDG (CDG-Io)	*DPM3*	Dolichol-phosphate mannosyltransferase subunit 3
ALG11-CDG (CDG-Ip)	*ALG11*	Asparagine-linked glycosylation protein 11 homolog
SRD5A3-CDG (CDG-Iq)	*SRD5A3*	Probable polyprenol reductase
DDOST-CDG (CDG-Ir)	*DDOST*	Dolichyl-diphosphooligosaccharide–protein glycosyltransferase 48-kd subunit
MAGT1-CDG	*MAGT1*	Magnesium transporter protein 1
TUSC3-CDG	*TUSC3*	Tumor suppressor candidate 3
ALG13-CDG	*ALG13*	UDP-*N*-acetylglucosamine transferase subunit ALG13 homolog
PGM1-CDG	*PGM1*	Phosphoglucomutase-1
MGAT2-CDG (CDG-IIa)	*MGAT2*	α-1,6-mannosyl-glycoprotein 2-β-*N*-acetylglucosaminyltransferase
MOGS-CDG (CDG-IIb)	*MOGS (GCS1)*	Mannosyl-oligosaccharide glucosidase
SLC35C1-CDG (CDG-IIc)	*SLC35C1*	GDP-fucose transporter 1
B4GALT1-CDG (CDG-IId)	*B4GALT1*	β-1,4-galactosyltransferase 1
SLC35A2-CDG	*SLC35A2*	UDP-galactose translocator
GMPPA-CDG	*GMPPA*	Mannose-1-phosphate guanyltransferase α
SSR4-CDG	*SSR4*	Translocon-associated protein subunit delta
STT3A-CDG, STT3B-CDG	*STT3A, STT3B*	Dolichyl-diphosphooligosaccharide–protein glycosyltransferase subunit STT3A/STT3B

TABLE 60.2 Classification of CDG: Defects in Multiple Glycosylation and Other Pathways

Individual disorder	Gene	Gene product
COG7-CDG (CDG-IIe)	*COG7*	COG complex subunit 7
SLC35A1-CDG (CDG-IIf)	*SLC35A1*	CMP-sialic acid transporter
COG1-CDG (CDG-IIg)	*COG1*	COG complex subunit 1
COG8-CDG (CDG-IIh)	*COG8*	COG complex subunit 8
COG5-CDG (CDG-IIi)	*COG5*	COG complex subunit 5
COG4-CDG (CDG-IIj)	*COG4*	COG complex subunit 4
TMEM165-CDG (CDG-IIk)	*TMEM165*	Transmembrane protein 165
COG6-CDG (CDG-IIL)	*COG6*	COG complex subunit 6
DPM2-CDG	*DPM2*	Dolichol phosphate-mannose biosynthesis regulatory protein
DHDDS-CDG	*DHDDS*	Dehydrodolichyl diphosphate synthase
MAN1B1-CDG	*MAN1B1*	Endoplasmic reticulum mannosyl-oligosaccharide 1,2-α-mannosidase

responsive to oral mannose, but liver disease may persist.[37] An adult has been described with asymptomatic MPI-CDG, in whom the abnormal transferrin findings were initially attributed to alcohol consumption.[38] Although mannose consumption is usually considered to be harmless, a study in which pregnant hypomorphic Mpi mice were given mannose supplementation showed that embryonic lethality was increased, and that ocular defects were induced in 50% of fetuses.[39] The authors concluded that mannose supplementation should be avoided in pregnant women.

ALG6-CDG (CDG-Ic) results from glucosyltransferase I deficiency. Thirty patients have been recognized.[19] ALG6-CDG generally has a milder phenotype than CDG-Ia, characterized by moderate psychomotor retardation, hypotonia, esotropia, seizures, and ataxia. Severe intellectual disability with seizures and hypotonia has been reported.[40] Abnormalities of glycoproteins are also less consistent and less severe. Protein-losing enteropathy and other hepatic and gastrointestinal manifestations occur in some patients.[41] Skeletal dysplasia[42] and variable hormonal abnormalities, ranging from normal pubertal development to female virilization, have been reported.[43]

A boy with microcephaly, optic atrophy, iris colobomas, epilepsy, spastic quadriparesis, and profound psychomotor delay was initially classified as type IV CDGS.[44] Subsequently, he was found to have deficient mannosyltransferase VI (α-1–3 mannosyltransferase) activity and is now categorized as having ALG3-CDG (CDG-Id).[7,10] A second child with CDG-Id was reported in 2005; the same report also described an affected fetus and noted that several circulating glycoproteins that did not cross the placenta were normally glycosylated, suggesting that unknown maternal mechanisms could partially compensate for the mannosyltransferase deficiency.[45] One child presenting with severe psychomotor delay, primary microcephaly, and optic atrophy showed accumulation of dol-PP-$GlcNAc_2Man_5$ in cultured fibroblasts, suggesting a defect in the asparagine-linked glycosylation 3 homolog (*ALG3*) gene. The patient had segmental maternal isodisomy UPD3 (q21.3-qter) harboring an Arg266 → Cys (796C → T) mutation.[46] Another severely affected infant died at 19 days of age with profound hypotonia, facial dysmorphism, hyperinsulinemic hypoglycemia, islet cell hyperplasia, and a Dandy–Walker malformation. The child was homozygous for a 512G → A mutation in ALG3.[47] Patients with ALG3-CDG have a progressive encephalopathy with seizures and microcephaly. They typically lack the multisystem defects seen in other CDGs, although they may have osteopenia with pathological fractures.[48]

DPM1-CDG (CDG-Ie) was first described as a severe phenotype, with findings including marked psychomotor delay, profound hypotonia, microcephaly, cortical blindness, and intractable seizures associated with elevated creatine kinase (CK), depressed antithrombin III (AT III) and a type 1 pattern of transferrin glycoforms.[49] Some children have dysmorphic features, including down slanting palpebral fissures, flat occiput and nasal bridge, hemangiomas of the occiput and sacrum, a high narrow palate, and mild limb shortening. Dolichol phosphate mannose synthase activity is markedly diminished in all cases.

Two siblings with intronic mutations in the *DPM1* gene have been described.[50] These children had a milder phenotype, comprising ataxia without hepatic disease. One patient has been described with a dystroglycanopathy associated with DPM1 mutations.[51]

Three children have been described with severe psychomotor retardation and variable features, including growth retardation, optic atrophy, ichthyosis, dysmorphism (parietal bossing and thin lips), hypo- or hypertonia, enlarged subarachnoid spaces, thrombocytopenia, transient deficiency of growth hormone and insulin-like growth factor I, and mild elevations of creatine kinase (CK). All have mutations in *MPDU1/Lec35*, leading to deficient function of the Lec35 protein, and are classified as MPDU1-CDG (CDG-If).[52] The Lec35 protein is hypothesized to act as a chaperone for dolichol phosphate in the endoplasmic reticulum (ER) membrane, ensuring appropriate lateral spacing of Dol-P-Man and Dol-P-Glc. In the absence of such spacing, these compounds may form rafts that alter local concentration gradients, impairing synthesis of the LLO and its accessibility to the oligosaccharyltransferase complex.[52] Saudek[53] has suggested that mannose-P-dolichol utilization defect 1 (MPDU1) is a member of a family of transmembrane proteins that function as cargo receptors for vesicular trafficking.

The index patient with ALG12-CDG (CDG-Ig) was born to consanguineous parents, and had congenital hypotonia and facial dysmorphism. Her development was markedly delayed and she had progressive microcephaly, feeding difficulties, and poor growth. She had frequent upper respiratory tract infections that are associated with immunoglobulin A (IgA) deficiency. The activity of dolichol-P-mannose: dolichol mannosyltransferase was severely deficient.[54] Additional patients have been described subsequently, including two in the United States.[55]

Nine patients have been described with ALG8-CDG (CDG-Ih),[56–58] with a range of phenotypes from mild gastrointestinal disease (alive at 3 years) to severe multiorgan failure with neonatal death. Cutaneous manifestations may range from wrinkled skin, cutis laxa, or severe ichthyosis, to abnormal fat distribution.[58] The first reports of ALG8-CDG did not include neurologic defects, but subsequent reports have described children with a variety of neurologic manifestations, including severe developmental delay, ataxia, seizures and visual impairment.[59–61] The deficient enzyme is dolichyl-P-Glc:$Glc_1Man_9GlcNAc_2$-PP-dolichyl α-1,3-glucosyltransferase, associated with mutations in *ALG8*.

ALG2-CDG (CDG-Ii) is the first recognized human molecular defect affecting the transfer of mannosyl residues from GDP-Man to $Man_1GlcNAc_2$-PP-dolichol by the enzyme GDP-Man:$Man_1GlcNAc_2$-PP-dolichol mannosyltransferase (*ALG2*).[62] The affected infant presented with nystagmus and poor vision at 2 months, when bilateral iris colobomas and a cataract were detected. At 4 months, infantile spasms occurred, and subsequent development was severely impaired. Imaging studies showed markedly delayed myelination. Laboratory investigations were otherwise normal but for prolonged activated partial thromboplastin time (APTT) and diminished factor XI. The phenotype of ALG2 has been expanded to include complex neuromuscular disease, with features of a congenital myasthenic syndrome[63] and a limb girdle dystrophy, in which muscle biopsy showed ragged red fibers.[64] Mutations in ALG14 and DPAGT1 have also been associated with congenital myasthenic phenotypes.[63]

DPAGT1-CDG (CDG-Ij) was first described in a girl who presented with infantile spasms at 4 months, 3 days after DPT (diphtheria, pertussis and tetanus) immunization.[65] The child has global developmental delay, severe hypotonia, microcephaly, a high-arched palate, micrognathia, fifth finger clinodactyly, single flexion creases, dimples on the upper thighs and exotropia. Her seizures remained refractory at 6 years. Transferrin isoelectric focusing showed a type I pattern, but other laboratory investigations, including MRI, were normal. Positron emission tomography (PET) showed nonspecific multifocal areas of hypometabolism. This child had deficient activity of UDP-GlcNAc: dolichol phosphate *N*-acetyl-glucosamine-1 phosphate transferase and mutations in *DPAGT1*. Eighteen affected infants from a large consanguineous kindred were homozygous for c.902G>A; all died in infancy with a phenotype characterized by severe hypotonia, global developmental delay, seizures, and microcephaly.[66] MR imaging of the brain showed delayed myelination and muscle biopsy showed fiber-type disproportion. Two more children, homozygous for c.341C>G, also succumbed to severe neurologic disease in infancy.[67] Several subsequent reports have described a limb-girdle congenital myasthenic syndrome associated with DPAGT1 mutations.[68] Mild craniobulbar weakness and tubular aggregates on muscle biopsy may help differentiate DPAGT1-CDG from other congenital muscle diseases.[69]

The first patient with ALG1-CDG (CDG-Ik)[70] was hypotonic and globally delayed from early infancy, and subsequently experienced intractable seizures and was found to be blind. He had episodes of sepsis, bleeding and liver dysfunction in the first year of life. MRI of the head was normal. Mutations were identified in *ALG1*, leading to deficiency of β-1,4 mannosyltransferase. A similarly severely affected infant with a fatal outcome has been reported with compound heterozygosity for c.1145T>C (M382T) and c.1312C>T (R438W).[71] A review of 16 patients expanded the spectrum from infantile onset to mild, adult cases, with neurologic (developmental delay, hypotonia, strabismus, microcephaly, visual loss, and seizures) and systemic features (coagulopathy, abnormal fat distribution) very similar to those of PMM2-CDG.[72] The authors suggested that patients with a PMM2-CDG phenotype with normal PMM2 activity should be screened for ALG1-CDG.

RFT1-CDG[73] (CDG-In) was first described in a child with severe failure to thrive, congenital arthrogryposis, and intractable epilepsy, who died of a pulmonary embolus at 4 years.[74] Sensorineural deafness has been recognized as a salient feature of this disorder.[75] Six children and two young adults have now been reported with RFT1-CDG.[76] As in other CDGs, the phenotype is milder in the adult cases, one of whom has normal hearing, and both of whom have well controlled seizures. Profound intellectual disability appears to be universal.

PGM1-CDG has recently been recognized as a CDG.[77] Nineteen patients from 16 families were found to have deficient activity of phosphoglucomutase 1 activity. This enzymatic defect has previously been associated with glycogenosis type XIV. Patients had liver disease (with elevated transaminases and hypoglycemia), cardiomyopathy (dilated, with or without cardiac arrest) and muscle disease, manifest as weakness, rhabdomyolysis, and decreased exercise intolerance. Most patients had short stature and two experienced malignant hyperthermia with anesthesia.

Twelve patients from nine families with intellectual disability, progressive cerebellar ataxia, and congenital eye malformations associated with visual impairment were found to have mutations in SRD5A3, which encodes steroid 5α-reductase type 3, a key enzyme in the dolichol synthetic pathway.[78] Brothers aged 38 and 40 years have been recognized with a milder phenotype of SRD5A3-CDG.[79]

A number of patients with nonsyndromic intellectual disability have mutations in TUSC3, which encodes a component of the oligosaccharyltransferase complex.[80,81] Both point mutations and deletions have been described, the latter in three families.[82,83] In addition to its role in *N*-linked glycosylation, TUSC3 plays a role in plasma membrane magnesium transport,[84] which, when impaired, may play a role in impaired learning and memory.

Two children have been reported with ALG9-CDG (CDG-Il),[70,85] whose features include severe failure to thrive, developmental delay, hypotonia, progressive microcephaly, and seizures. The heart, liver and kidneys showed functional and structural abnormalities. MR imaging found cerebral and cerebellar atrophy with delayed myelination. Laboratory investigation found low levels of serum cholesterol and proteins, including multiple coagulation factors. Both children had mutations in ALG9, leading to α-1,2-mannosyltransferase deficiency. A balanced chromosomal translocation t(9;11)(p24;q23) was found to cosegregate with bipolar affective disorder in a small family, raising the possibility of a role for ALG9 in this disorder. A subsequent study of several hundred families found no evidence of linkage between ALG9 and bipolar disorder.[86]

Four children have been reported with MGAT2-CDG.[87,88] They were of Belgian, Iranian, and French descent, and all had severe psychomotor delay, acquired microcephaly and growth retardation, and variable combinations of hypotonia, ventricular septal defects, craniofacial dysmorphism (thin lips, hooked nose, large ears, hypertrophied gums, and short neck), stereotypies, and coagulation defects. All showed impaired activity of the Golgi enzyme, *N*-acetylglucosaminyl transferase II (GlcNAc II, GnT II), and an increase in tri- and monosialotransferrin (type II pattern).

The index case of MOGS-CDG (CDG-IIb) was a girl born with marked hypotonia and craniofacial dysmorphism (prominent occiput, scalp alopecia, short palpebral fissures, long eyelashes, broad nose, retrognathia, and a high arched palate).[89] She also had generalized edema, thoracic scoliosis, hypoplastic genitalia, and overlapping fingers. Subsequently she developed progressive hepatomegaly, respiratory failure, and seizures. The child died at 74 days of age despite supportive care. Transferrin glycoforms were normal, but urine contained an abnormal tetrasaccharide whose analysis led to the recognition of glucosidase I deficiency.

SLC35C1-CDG (CDG-IIc) was first described as leukocyte adhesion deficiency, type 2 (LAD II). Affected individuals have persistent marked neutrophilia, recurrent nonpurulent skin infections and periodontitis, short stature, microcephaly and delayed psychomotor development.[90] Fucosylation of glycoproteins is impaired by GDP fucose transporter deficiency.[91] This includes sialyl-Lewis X (sLeX), a key ligand for selectins that act as endothelial adhesion molecules. Transferrin glycoforms are normal in this form of CDG. Therapy with oral fucose is effective in this disorder.[92]

The first case of B4GALT1-CDG (CDG-IId)[93] was a boy with macrocephaly secondary to a Dandy–Walker malformation with hydrocephalus, transient cholestasis, hypotonia, and progressive elevation of CK secondary to myopathy, coagulation abnormalities, and a unique pattern of transferrin glycoforms, with elevation of tri-, di-, mono-, and asialotransferrin. Analysis of the glycosylation pathway led to the finding of β-1,4-galactosyltransferase deficiency.[94] A second patient with mild gastrointestinal disease and normal neurological function has been described.[95]

A number of CDG subtypes have been recognized since 2004 in association with defects in multiple pathways. Most of these are caused by mutations in genes encoding components of the COG complex. Two siblings with a rapidly fatal phenotype and impaired ER–Golgi trafficking were found to have mutations in the COG7 gene.[96] A less severe phenotype was described, whose features included growth retardation, progressive, severe microcephaly, hypotonia, adducted thumbs, feeding problems by gastrointestinal pseudo-obstruction, failure to thrive, cardiac anomalies, wrinkled skin, and episodes of extreme hyperthermia[97] (COG7-CDG; CDG-IIe). COG8 deficiency causes a milder phenotype, albeit one associated with psychomotor retardation, seizures, and wheat and dairy intolerance.[98,99] Cases of COG1-,[100] COG4-,[101] COG6-,[102] COG5-,[103] and TMEM-CDG[104] have been described; the cases are too few to characterize a consistent phenotype for any of these disorders as yet.

An infant who presented with macrothrombocytopenia, neutropenia, spontaneous bleeding into the vitreous humor and skin, and absence of the sialyl-Le(x) antigen was subsequently found to have defective activity of the Golgi cytidine 5′-monophosphate (CMP)–sialic acid transporter.[105] The child's course was marked by recurrent bleeding and infections. Despite aggressive medical management, including bone marrow transplantation, he succumbed at 37 months. Laboratory testing did not show a hypoglycosylation of transferrin or other typical changes of CDG. Investigation confirmed that the disorder results from dysfunction of the CMP–sialic acid transporter encoded by SLC35A1, leading to the designation SLC35A1-CDG.[106]

CDG should be considered in any child (or adult) with otherwise unexplained hypotonia, seizures, and mental retardation or delayed development, as well as the additional features described earlier. Indeed, any child with unexplained multisystem disease and apparently disparate findings could have CDG. The likelihood of CDG is strongest when the neurologic signs are combined with systemic abnormalities, particularly coagulopathies, hepatocellular dysfunction, gastrointestinal disturbances (protein-losing enteropathy and cyclic vomiting), endocrinopathy (hypogonadism, hyperinsulinemic hypoglycemia, and hypothyroidism), and cardiac disease (hypertrophic cardiomyopathy and pericardial effusion).

PATHOLOGY

Biopsy and postmortem studies have been performed in PMM2-CDG (CDG-Ia), and are summarized in a review by Jaeken and coworkers.[107] Findings have included olivopontocerebellar atrophy, with neuronal dropout and gliosis throughout the neuraxis. Dandy–Walker malformations have been reported in PMM2-CDG (CDG-Ia), B4GALT1-CDG (CDG-IId),[93] and ALG3-CDG (CDG-Id).[47] Myelin loss and Schwann cell inclusions have been reported in peripheral nerves. The liver may be enlarged, and show microvesicular steatosis, fibrosis, and inclusions in hepatocytes. Renal cysts, testicular fibrosis, and aging changes in epidermal collagen have also been observed. Dysostosis and gross skeletal deformities (pectus carinatum and kyphoscoliosis) are found in some cases.

Liver biopsy in MPI-CDG (CDG-Ib) has shown fibrosis in some cases and ductal plate hypoplasia in at least one other case.[36] Cryptogenic liver disease has been associated with unclassified CDGs.[108] Postmortem examination of the child with MOGS-CDG (CDG-IIb) found hepatic cholangiofibrosis, macrovesicular steatosis, and enlarged lysosomes. Parenchymal cells contained inclusions with concentric lamellae; similar structures were observed in macrophages. The brain showed increased weight, slightly delayed myelination, and widespread ballooning of neurons that contained empty, membrane-bound vacuoles on electron microscopy.[89]

Biochemical (Enzymatic) Findings

Characterization of transferrin glycoforms is the most widely used screening test for CDG. Group I CDG is associated with a shift in transferrin isoforms from a predominance of tetra- and tri-sialo species to variable increases in a-, mono-, and di-sialo species.[16] The pattern in group II[109] is more variable, with an increase in tri- and monosialotransferrin in MGAT2-CDG (CDG-IIa), a unique pattern in B4GALT1-CDG (CDG-IId), and normal patterns in MOGS-CDG (CDG-IIb) and SLC35C1-CDG (CDG-IIc). False positives occur in genetic variants of transferrin,[110,111] actively imbibing alcoholics,[112] galactosemia, and hereditary fructose intolerance. The hypoglycosylation in galactosemia results from both processing and assembly defects and is corrected by dietary therapy.[113] False negatives have been observed in preterm babies and some patients with PMM2-CDG, whose transferrin isoelectric focusing (IEF) was initially abnormal.[114] Intensification of the classic type 1 pattern has been described in CDG-Ia patients treated with mannose.

Hypoglycosylated transferrin glycoforms can be demonstrated in serum on Guthrie cards cerebrospinal fluid,[115,116] and urine,[117] in addition to blood samples. Isoelectric focusing of apolipoprotein C-III (apoC-III) shows abnormalities in disorders of *O*-linked glycan synthesis,[118,119] but may also be abnormal in some patients with primary disorders of *N*-linked glycosylation. Wopereis et al.[120] were able to classify patients with CDG IIx into six distinct subgroups by combining both transferrin and apoC-III IEF.

Assessment of transferrin glycoforms was originally performed using slab gel immunoelectrophoresis,[115,119] and later by ion-exchange chromatography linked with immunoassay or high-performance liquid chromatography. These methods require substantial time (from hours to days), and the exact nature of the glycoforms measured is uncertain. A single-step technique combining online immunoaffinity-postconcentration-mass spectrometry can determine the relative abundance of transferrin isoforms using less than 5 mL of blood in less than 30 minutes.[121] This technique could be used for large-scale population screening for CDG.

Other approaches designed to allow automation of transferrin screening have been described, including the combination of site-specific glycopeptide analysis with mass spectrometry[122] and protein-chip technology.[123] The sensitivity of transferrin screening for CDG may be increased by studying additional glycoproteins, including thyroid-stimulating hormone,[124] α1-acid glycoprotein, haptoglobin,[125] α1-antitrypsin[110,126] and α1 antichymotrypsin.[126]

Creatine kinase may be elevated, with or without clinical myopathy.[93] Complex abnormalities of coagulation factors have been described in CDG, involving clotting factors, inhibitors,[127] and platelet glycoproteins.[88] Although the plasma profiles are similar, patients with PMM2-CDG have thrombotic tendencies, whereas those with MGAT2-CDG are more likely to bleed excessively, due to differences in platelet glycoproteins.[88]

When a patient has a typical phenotype, direct enzyme measurement or genotyping may be considered. In some patients with PMM2-CDG, the activity of PMM2 has overlapped with that of heterozygotes, mandating genotyping. Common mutations (R141H, F119L) should be screened for first, followed by a search for other mutations if these are not present, or are present on only one allele. Genomic sequencing may be necessary in some cases, where the pathogenic mutations involve splice site donors.

Assays of PMM and PMI activity are readily available in most Western countries. Patients with a type I transferrin glycoform pattern should have PMM and PMI activity assayed first. If these are negative, and genotyping is either not indicated or is negative, direct collaboration with a research glycobiologist is necessary.

MOLECULAR GENETIC DATA

Mutation data for CDGs are increasing rapidly, and may be accessed online (http://www.euroglycanet.org/). As in many other uncommon recessive disorders, patients with CDG are either homozygous for mutations associated with a founder effect (particularly in geographically or genetically restricted populations) or are compound heterozygotes for private mutations. Genomic sequencing has been necessary in some forms of CDG where the mutations occurred in splicing donor sites (PMI and ALG6). Denaturing high-pressure liquid chromatography (DHPLC) was shown to be an efficient and sensitive method for screening in PMM2-CDG, MPI-CDG, ALG6-CDG, ALG3-CDG, DPM1-CDG, and MPDU1-CDG,[128] but rapid progress in the development of gene panels and whole-exome sequencing (WES) is changing the diagnostic approach to uncommon forms of CDG. One study suggested that a gene panel approach was superior to WES, because of inadequate exome coverage by the latter.[129] Some rare forms of CDG have been solved by WES, which would likely have remained unrecognized, or whose solution would have been more arduous using traditional approaches.[64,130]

Most genotyping information is available for PMM2-CDG. PMM2 is the 246-amino-acid product of the *PMM2* gene, localized to chromosome 16p13. Mutations have been described throughout the gene. The most frequent mutation overall is Arg141 → His. In Scandinavian populations, either Arg141 → His or Phe119 → Leu alleles occur in 72% of cases, and Arg141 → His/Phe119 → Leu is the most frequent genotype. Statistical analysis suggests that homozygosity for Arg141 → His should be observed with a frequency of 1 in 20,000, but no such cases have been found. Arg141 → His recombinant protein lacks useful activity, and Jaeken and Matthijs[131] have concluded that the homozygous Arg141 → His is embryonic lethal. The same authors have estimated a frequency of PMM2-CDG as high as 1 in 20,000 based on a frequency of other mutations between 1 in 300 and 1 in 400. The high frequency (approximately 1/72) of the Arg141→ His allele in the Dutch and Danish populations[132] has been attributed to a survival advantage in heterozygotes in whom the mutation may confer resistance to hepatitis B and C viruses.[11] More recent studies suggest that the persistence of this allele, which might normally be expected to vanish over time, results from a transmission ratio distortion. Molecular prenatal diagnosis of 92 pregnancies in 59 PMM2-CDG families found that 34% of fetuses were affected (31/92, p = 0.039), which is higher than the expected 25% predicted by Mendel's second law.

The result is important for genetic counseling in CDG families, implying a recurrence risk close to one third.[133] Genotype–phenotype correlation is not always consistent. For example, Gly691 → Ala has been associated with severe manifestations in several cases, but a child who was a compound heterozygote for Ala647 → Thr/Gly691 → Ala had a relatively benign course.[22] The steps in *N*-glycosylation are tightly interlinked, so that mutations in genes other than that giving rise to the primary syndrome might act as modifiers. Studies in yeast mutants and humans with PMM2-CDG found that the presence of the Thr911 → Cys mutation in one allele of *ALG6* was associated with a more severe phenotype in both cell lines and patients with mutations in both *PMM2* alleles.[134] Most cases of MPI-CDG are associated with missense mutations of the *PMI* gene, involving highly conserved amino acids. Patients homozygous for three of these mutations (Met51 → Thr, Asp131 → Asn, and Arg152 → Gln) have been described. ALG6-CDG was the first of the congenital disorders of glycosylation in which the human biochemistry was deduced from studies in yeast; this has served as a model for many subsequently recognized disorders.

ANIMAL MODELS

Murine models of MGAT 1, 2, 3, and 5 deficiency have been described.[135] The *MGAT1–/–* mouse dies at embryonic day 9.5 with failure of neural tube closure and abnormal vascular development and body symmetry. An *MGAT1–/–* chimera has emphasized the role of *MGAT1* gene products in the normal embryonic development of epithelial tissues. The *MGAT2–/–* mouse is born with multiple anomalies and dies in the neonatal period. Hypoglycosylation reduces the rate of tumor growth by impairing signaling pathways in *MGAT3–/–* and *MGAT5–/–* mice.[12] This protection comes at the expense of an increased risk of autoimmune disease in the *MGAT5–/–* model.[13] These animals show enhanced T-cell receptor (TCR) clustering, perhaps secondary to reduced spatial separation of TCRs by hypoglycosylated galectins. The functional consequence is activation of cell-mediated immunity by what would otherwise be subthreshold stimuli.

A PMI knockout mouse model was embryonic lethal; studies of embryonic fibroblasts found that accumulation of mannose-1-phosphate inhibited glucose metabolism at multiple points and depleted ATP.[136] Targeted disruption of the murine *PMM2* gene produced normal heterozygotes but was embryonic lethal in homozygotes, confirming the essential role of this gene product in early development.[137]

The experience with these models has emphasized the specificity of discrete stages in the pathway of glycosylation commensurate with the variable phenotypes observed in human CDG.

THERAPY

Mannose supplementation partially or completely corrects the hypoglycosylation of multiple conjugates in fibroblast cultures from patients with PMM2-CDG and MPI-CDG, where cellular GDP-mannose levels are low, but not in cells from patients with DPM1-CDG where GDP-mannose levels are normal.[138] Plasma mannose concentrations are lower in PMM2-CDG patients than in controls, and oral mannose supplementation in healthy volunteers and patients boosts plasma concentrations. Despite these encouraging preliminary data, neither oral nor intravenous mannose supplementation produces clinical or biochemical improvement in patients with PMM2-CDG. In some cases, mannose supplementation has been associated with exaggeration of the abnormal pattern of transferrin glycoforms. Studies *in vitro* have shown that hydrophobic, membrane-permeable acylated versions of mannose-1-phosphate (Man-1-P) normalize LLO size and *N*-glycosylation in PMM2-CDG, DPM1-CDG and MPDU1-CDG fibroblasts; no effect was seen in ALG3-CDG and MPDU1-CDG cells.[139] Additional Man-1-P prodrugs for potential treatment of PMM2-CDG have been synthesized.[140]

MPI-CDG (phosphomannose isomerase deficiency) has shown variable response to oral mannose supplementation.[36,141] In some cases, there was virtually complete resolution of clinical manifestations with return of laboratory markers to normal or near normal. Older individuals with irreversible tissue damage are less likely to respond in this fashion.

Oral fucose therapy for CDG-IIc (leukocyte adherence deficiency II) has proved effective in bringing about clinical and laboratory improvement in the single patient treated.[92] The patient's episodes of febrile illness abated, persistent neutrophilia resolved, and psychomotor development improved. In this case, the use of pharmacologic doses of the substrate presumably activated the otherwise minor fucose-salvage pathway, allowing sidestepping of the genetic defect and adequate fucosylation of glycoconjugates.

There are no data on the role of mannose supplementation in the rare phenotypes of CDG, although the experience with CDG-Ia suggests that this is unlikely to be a fruitful approach to therapy.

CONCLUSION

The CDG comprise a diverse and rapidly growing family of diseases whose study promises to yield insights into many fundamental processes that involve the participation of glycoproteins, including fetal development, regulation of immunity, carcinogenesis, and neurodegeneration (through activation of the unfolded protein response [UPR]).[14] Continued progress in the field is predicated on the education and alertness of clinicians, who must consider CDG in patients with a wide range of presentations, and on partnering with glycobiologists to pursue characterization of novel glycosylation defects. EUROGLYCANET,[142] an initiative funded by the European Commission, serves as a model for such collaboration. The network comprises a central database and patient sample repository. Samples are distributed to several expert laboratories, in a process called carousel testing. The network also aims to establish referral laboratories in national centers, with the ultimate goal of providing access to definitive CDG diagnostic testing for all Europeans. Thorough understanding of the consequences of mutations in genes participating in *N*-linked glycosylation requires elucidation of their downstream consequences; a worldwide functional glycomics initiative is underway to accomplish this goal.[143] Progress in DNA sequencing promises to improve the identification of known and novel CDGs, but will still require careful clinical and biochemical evaluation of suspected patients.[1]

Oral mannose and fucose therapy benefit a small number of patients at present. Understanding the discrepancy between the *in vitro* and *in vivo* effects of mannose in PMM2-CDG may lead to better therapy for these patients. It is also likely that knowledge gained from studies of CDG will translate into therapies for other disorders in which glycoproteins play a role, including developmental disorders, cancer, and immune disorders.

References

1. Freeze HH. Understanding human glycosylation disorders: biochemistry leads the charge. *J Biol Chem*. 2013;288(10):6936–6945.
2. Coss KP, Hawkes CP, Adamczyk B, et al. N-glycan abnormalities in children with galactosemia. *J Proteome Res*. 2014;13(2):385–394.
3. Braulke T, Bonifacino JS. Sorting of lysosomal proteins. *Biochim Biophys Acta*. 2009;1793(4):605–614.
4. Aebi M, Hennet T. Congenital disorders of glycosylation: genetic model systems lead the way. *Trends Cell Biol*. 2001;11(3):136–141.
5. Helenius A, Aebi M. Intracellular functions of *N*-linked glycans. *Science*. 2001;291(5512):2364–2369.
6. Haeuptle MA, Hennet T. Congenital disorders of glycosylation: an update on defects affecting the biosynthesis of dolichol-linked oligosaccharides. *Hum Mutat*. 2009;30(12):1628–1641.
7. Kelleher DJ, Gilmore R. An evolving view of the eukaryotic oligosaccharyltransferase. *Glycobiology*. 2006;16(4):47R–62R.
8. Ruddock LW, Molinari M. N-glycan processing in ER quality control. *J Cell Sci*. 2006;119(Pt 21):4373–4380.
9. Laufman O, Hong W, Lev S. The COG complex interacts with multiple Golgi SNAREs and enhances fusogenic assembly of SNARE complexes. *J Cell Sci*. 2013;126(Pt 6):1506–1516.
10. de Graffenried CL, Bertozzi CR. The roles of enzyme localisation and complex formation in glycan assembly within the Golgi apparatus. *Curr Opin Cell Biol*. 2004;16(4):356–363.
11. Freeze HH, Westphal V. Balancing N-linked glycosylation to avoid disease. *Biochimie*. 2001;83(8):791–799.
12. Granovsky M, Fata J, Pawling J, Muller WJ, Khokha R, Dennis JW. Suppression of tumor growth and metastasis in Mgat5-deficient mice. *Nat Med*. 2000;6(3):306–312.
13. Demetriou M, Granovsky M, Quaggin S, Dennis JW. Negative regulation of T-cell activation and autoimmunity by Mgat5 N-glycosylation. *Nature*. 2001;409(6821):733–739.
14. Zhang K, Kaufman RJ. The unfolded protein response: a stress signaling pathway critical for health and disease. *Neurology*. 2006;66(2 suppl 1): S102–S109.
15. Aebi M, Helenius A, Schenk B, et al. Carbohydrate-deficient glycoprotein syndromes become congenital disorders of glycosylation: an updated nomenclature for CDG. First International Workshop on CDGS. *Glycoconj J*. 1999;16(11):669–671.
16. Kristiansson B, Andersson M, Tonnby B, Hagberg B. Disialotransferrin developmental deficiency syndrome. *Arch Dis Child*. 1989;64(1):71–76.
17. Jaeken J, Stibler H, Hagberg B. The carbohydrate-deficient glycoprotein syndrome. A new inherited multisystemic disease with severe nervous system involvement. *Acta Paediatr Scand Suppl*. 1991;375:1–71.
18. Jaeken J, Hennet T, Matthijs G, Freeze HH. CDG nomenclature: time for a change! *Biochim Biophys Acta*. 2009;1792(9):825–826.
19. Sparks SE, Krasnewich DM. Congenital Disorders of *N*-linked Glycosylation Pathway Overview. Seattle: Seattle (WA): University of Washington; 1993–2014. 2014 [updated 2014. Jan 30; cited 2014 Mar 15]; Available from http://www.ncbi.nlm.nih.gov/books/NBK1332/.
20. Hagberg BA, Blennow G, Kristiansson B, Stibler H. Carbohydrate-deficient glycoprotein syndromes: peculiar group of new disorders. *Pediatr Neurol*. 1993;9(4):255–262.
21. Dyer JA, Winters CJ, Chamlin SL. Cutaneous findings in congenital disorders of glycosylation: the hanging fat sign. *Pediatr Dermatol*. 2005;22(5):457–460.
22. Silengo M, Valenzise M, Pagliardini S, Spada M. Hair changes in congenital disorders of glycosylation (CDG type 1). *Eur J Pediatr*. 2003;162(2):114–115.
23. Stark KL, Gibson JB, Hertle RW, Brodsky MC. Ocular motor signs in an infant with carbohydrate-deficient glycoprotein syndrome type Ia. *Am J Ophthalmol*. 2000;130(4):533–535.
24. Jensen H, Kjaergaard S, Klie F, Moller HU. Ophthalmic manifestations of congenital disorder of glycosylation type 1a. *Ophthalmic Genet*. 2003;24(2):81–88.
25. Coorg R, Lotze TE. Child Neurology: a case of PMM2-CDG (CDG 1a) presenting with unusual eye movements. *Neurology*. 2012;79(15):e131–e133.
26. Pearl PL, Krasnewich D. Neurologic course of congenital disorders of glycosylation. *J Child Neurol*. 2001;16(6):409–413.
27. Dinopoulos A, Mohamed I, Jones B, Rao S, Franz D, deGrauw T. Radiologic and neurophysiologic aspects of stroke-like episodes in children with congenital disorder of glycosylation type Ia. *Pediatrics*. 2007;119(3):e768–e772.
28. Arnoux JB, Boddaert N, Valayannopoulos V, et al. Risk assessment of acute vascular events in congenital disorder of glycosylation type Ia. *Mol Genet Metab*. 2008;93(4):444–449.
29. Ishikawa N, Tajima G, Ono H, Kobayashi M. Different neuroradiological findings during two stroke-like episodes in a patient with a congenital disorder of glycosylation type Ia. *Brain Dev*. 2009;31(3):240–243.
30. Freeze HH, Eklund EA, Ng BG, Patterson MC. Neurology of inherited glycosylation disorders. *Lancet Neurol*. 2012;11(5):453–466.
31. Perez-Duenas B, Garcia-Cazorla A, Pineda M, et al. Long-term evolution of eight Spanish patients with CDG type Ia: typical and atypical manifestations. *Eur J Paediatr Neurol*. 2009;13(5):444–451.
32. Casado M, O'Callaghan MM, Montero R, et al. Mild clinical and biochemical phenotype in two patients with PMM2-CDG (congenital disorder of glycosylation Ia). *Cerebellum*. 2012;11(2):557–563.

33. Wolthuis DF, Janssen MC, Cassiman D, Lefeber DJ, Morava-Kozicz E. Defining the phenotype and diagnostic considerations in adults with congenital disorders of *N*-linked glycosylation. *Expert Rev Mol Diagn*. 2014;14(2):217–224.
34. Mader I, Dobler-Neumann M, Kuker W, Stibler H, Krageloh-Mann I. Congenital disorder of glycosylation type Ia: benign clinical course in a new genetic variant. *Childs Nerv Syst*. 2002;18(1–2):77–80.
35. Giurgea I, Michel A, Le Merrer M, Seta N, de Lonlay P. Underdiagnosis of mild congenital disorders of glycosylation type Ia. *Pediatr Neurol*. 2005;32(2):121–123.
36. Babovic-Vuksanovic D, Patterson MC, Schwenk WF, et al. Severe hypoglycemia as a presenting symptom of carbohydrate-deficient glycoprotein syndrome. *J Pediatr*. 1999;135(6):775–781.
37. de Lonlay P, Seta N. The clinical spectrum of phosphomannose isomerase deficiency, with an evaluation of mannose treatment for CDG-Ib. *Biochim Biophys Acta*. 2009;1792(9):841–843.
38. Helander A, Jaeken J, Matthijs G, Eggertsen G. Asymptomatic phosphomannose isomerase deficiency (MPI-CDG) initially mistaken for excessive alcohol consumption. *Clin Chim Acta*. 2014;431C(Feb 6):15–18.
39. Sharma V, Nayak J, Derossi C, et al. Mannose supplements induce embryonic lethality and blindness in phosphomannose isomerase hypomorphic mice. *FASEB J*. 2014;28(4):1854–1869.
40. Ichikawa K, Kadoya M, Wada Y, Okamoto N. Congenital disorder of glycosylation type Ic: report of a Japanese case. *Brain Dev*. 2013;35(6):586–589.
41. Damen G, de Klerk H, Huijmans J, den Hollander J, Sinaasappel M. Gastrointestinal and other clinical manifestations in 17 children with congenital disorders of glycosylation type Ia, Ib, and Ic. *J Pediatr Gastroenterol Nutr*. 2004;38(3):282–287.
42. Drijvers JM, Lefeber DJ, de Munnik SA, et al. Skeletal dysplasia with brachytelephalangy in a patient with a congenital disorder of glycosylation due to ALG6 gene mutations. *Clin Genet*. 2010;77(5):507–509.
43. Miller BS, Freeze HH, Hoffmann GF, Sarafoglou K. Pubertal development in ALG6 deficiency (congenital disorder of glycosylation type Ic). *Mol Genet Metab*. 2011;103(1):101–103.
44. Korner C, Knauer R, Stephani U, Marquardt T, Lehle L, von Figura K. Carbohydrate deficient glycoprotein syndrome type IV: deficiency of dolichyl-P-Man:Man(5)GlcNAc(2)-PP-dolichyl mannosyltransferase. *EMBO J*. 1999;18(23):6816–6822.
45. Denecke J, Kranz C, von Kleist-Retzow J, et al. Congenital disorder of glycosylation type Id: clinical phenotype, molecular analysis, prenatal diagnosis, and glycosylation of fetal proteins. *Pediatr Res*. 2005;58(2):248–253.
46. Schollen E, Grunewald S, Keldermans L, Albrecht B, Korner C, Matthijs G. CDG-Id caused by homozygosity for an ALG3 mutation due to segmental maternal isodisomy UPD3(q21.3-qter). *Eur J Med Genet*. 2005;48(2):153–158.
47. Sun L, Eklund EA, Chung WK, Wang C, Cohen J, Freeze HH. Congenital disorder of glycosylation Id presenting with hyperinsulinemic hypoglycemia and islet cell hyperplasia. *J Clin Endocrinol Metab*. 2005;90(7):4371–4375.
48. Rimella-Le-Huu A, Henry H, Kern I, et al. Congenital disorder of glycosylation type Id (CDG Id): phenotypic, biochemical and molecular characterization of a new patient. *J Inherit Metab Dis*. 2008;31(suppl 2):S381–S386.
49. Kim S, Westphal V, Srikrishna G, et al. Dolichol phosphate mannose synthase (DPM1) mutations define congenital disorder of glycosylation Ie (CDG-Ie). *J Clin Invest*. 2000;105(2):191–198.
50. Dancourt J, Vuillaumier-Barrot S, de Baulny HO, et al. A new intronic mutation in the DPM1 gene is associated with a milder form of CDG Ie in two French siblings. *Pediatr Res*. 2006;59(6):835–839.
51. Yang AC, Ng BG, Moore SA, et al. Congenital disorder of glycosylation due to DPM1 mutations presenting with dystroglycanopathy-type congenital muscular dystrophy. *Mol Genet Metab*. 2013;110(3):345–351.
52. Schenk B, Imbach T, Frank CG, et al. MPDU1 mutations underlie a novel human congenital disorder of glycosylation, designated type If. *J Clin Invest*. 2001;108(11):1687–1695.
53. Saudek V. Cystinosin, MPDU1, SWEETs and KDELR belong to a well-defined protein family with putative function of cargo receptors involved in vesicle trafficking. *PLoS One*. 2012;7(2):e30876.
54. Chantret I, Dupre T, Delenda C, et al. Congenital disorders of glycosylation type Ig is defined by a deficiency in dolichyl-P-mannose: $Man_7GlcNAc_2$-PP-dolichyl mannosyltransferase. *J Biol Chem*. 2002;277(28):25815–25822.
55. Kranz C, Basinger AA, Gucsavas-Calikoglu M, et al. Expanding spectrum of congenital disorder of glycosylation Ig (CDG-Ig): sibs with a unique skeletal dysplasia, hypogammaglobulinemia, cardiomyopathy, genital malformations, and early lethality. *Am J Med Genet A*. 2007;143A(12):1371–1378.
56. Chantret I, Dancourt J, Dupre T, et al. A deficiency in dolichyl-P-glucose:$Glc1Man_9GlcNAc_2$-PP-dolichyl alpha3-glucosyltransferase defines a new subtype of congenital disorders of glycosylation. *J Biol Chem*. 2003;278(11):9962–9971.
57. Grubenmann CE, Frank CG, Hulsmeier AJ, et al. Deficiency of the first mannosylation step in the N-glycosylation pathway causes congenital disorder of glycosylation type Ik. *Hum Mol Genet*. 2004;13(5):535–542.
58. Kouwenberg D, Gardeitchik T, Mohamed M, Lefeber DJ, Morava E. Wrinkled skin and fat pads in patients with ALG8-CDG: revisiting skin manifestations in congenital disorders of glycosylation. *Pediatr Dermatol*. 2014;31(1):e1–e5.
59. Stolting T, Omran H, Erlekotte A, Denecke J, Reunert J, Marquardt T. Novel ALG8 mutations expand the clinical spectrum of congenital disorder of glycosylation type Ih. *Mol Genet Metab*. 2009;98(3):305–309.
60. Vesela K, Honzik T, Hansikova H, et al. A new case of ALG8 deficiency (CDG Ih). *J Inherit Metab Dis*. 2009;32(suppl 1).
61. Sorte H, Morkrid L, Rodningen O, et al. Severe ALG8-CDG (CDG-Ih) associated with homozygosity for two novel missense mutations detected by exome sequencing of candidate genes. *Eur J Med Genet*. 2012;55(3):196–202.
62. Thiel C, Schwarz M, Peng J, et al. A new type of congenital disorders of glycosylation (CDG-Ii) provides new insights into the early steps of dolichol-linked oligosaccharide biosynthesis. *J Biol Chem*. 2003;278(25):22498–22505.
63. Cossins J, Belaya K, Hicks D, et al. Congenital myasthenic syndromes due to mutations in ALG2 and ALG14. *Brain*. 2013;136(Pt 3):944–956.
64. Monies DM, Al-Hindi HN, Al-Muhaizea MA, et al. Clinical and pathological heterogeneity of a congenital disorder of glycosylation manifesting as a myasthenic/myopathic syndrome. *Neuromuscul Disord*. 2014;24(4):353–359.
65. Wu X, Rush JS, Karaoglu D, et al. Deficiency of UDP-GlcNAc:Dolichol Phosphate N-Acetylglucosamine-1 Phosphate Transferase (DPAGT1) causes a novel congenital disorder of glycosylation type Ij. *Hum Mutat*. 2003;22(2):144–150.

66. Imtiaz F, Al-Mostafa A, Al-Hassnan ZN. Further delineation of the phenotype of congenital disorder of glycosylation DPAGT1-CDG (CDG-Ij) identified by homozygosity mapping. *JIMD Rep*. 2012;2:107–111.
67. Wurde AE, Reunert J, Rust S, et al. Congenital disorder of glycosylation type Ij (CDG-Ij, DPAGT1-CDG): extending the clinical and molecular spectrum of a rare disease. *Mol Genet Metab*. 2012;105(4):634–641.
68. Belaya K, Finlayson S, Cossins J, et al. Identification of DPAGT1 as a new gene in which mutations cause a congenital myasthenic syndrome. *Ann N Y Acad Sci*. 2012;1275:29–35.
69. Finlayson S, Palace J, Belaya K, et al. Clinical features of congenital myasthenic syndrome due to mutations in DPAGT1. *J Neurol Neurosurg Psychiatry*. 2013;84(10):1119–1125.
70. Frank CG, Grubenmann CE, Eyaid W, Berger EG, Aebi M, Hennet T. Identification and functional analysis of a defect in the human ALG9 gene: definition of congenital disorder of glycosylation type IL. *Am J Hum Genet*. 2004;75(1):146–150.
71. Rohlfing AK, Rust S, Reunert J, et al. ALG1-CDG: a new case with early fatal outcome. *Gene*. 2014;534(2):345–351.
72. Morava E, Vodopiutz J, Lefeber DJ, et al. Defining the phenotype in congenital disorder of glycosylation due to ALG1 mutations. *Pediatrics*. 2012;130(4):e1034–e1039.
73. Haeuptle MA, Pujol FM, Neupert C, et al. Human RFT1 deficiency leads to a disorder of N-linked glycosylation. *Am J Hum Genet*. 2008;82(3):600–606.
74. Clayton PT, Grunewald S. Comprehensive description of the phenotype of the first case of congenital disorder of glycosylation due to RFT1 deficiency (CDG In). *J Inherit Metab Dis*. 2009;32(suppl 1):S137–S139.
75. Jaeken J, Vleugels W, Regal L, et al. RFT1-CDG: deafness as a novel feature of congenital disorders of glycosylation. *J Inherit Metab Dis*. 2009;32(suppl 1):S335–S338.
76. Ondruskova N, Vesela K, Hansikova H, Magner M, Zeman J, Honzik T. RFT1-CDG in adult siblings with novel mutations. *Mol Genet Metab*. 2012;107(4):760–762.
77. Tegtmeyer LC, Rust S, van Scherpenzeel M, et al. Multiple phenotypes in phosphoglucomutase 1 deficiency. *N Engl J Med*. 2014;370(6):533–542.
78. Morava E, Wevers RA, Cantagrel V, et al. A novel cerebello-ocular syndrome with abnormal glycosylation due to abnormalities in dolichol metabolism. *Brain*. 2010;133(11):3210–3220.
79. Kara B, Ayhan O, Gokcay G, Basbogaoglu N, Tolun A. Adult phenotype and further phenotypic variability in SRD5A3-CDG. *BMC Med Genet*. 2014;15(1):10.
80. Garshasbi M, Hadavi V, Habibi H, et al. A defect in the TUSC3 gene is associated with autosomal recessive mental retardation. *Am J Hum Genet*. 2008;82(5):1158–1164.
81. Molinari F, Foulquier F, Tarpey PS, et al. Oligosaccharyltransferase-subunit mutations in nonsyndromic mental retardation. *Am J Hum Genet*. 2008;82(5):1150–1157.
82. Khan MA, Rafiq MA, Noor A, et al. A novel deletion mutation in the TUSC3 gene in a consanguineous Pakistani family with autosomal recessive nonsyndromic intellectual disability. *BMC Med Genet*. 2011;12:56.
83. Loddo S, Parisi V, Doccini V, et al. Homozygous deletion in TUSC3 causing syndromic intellectual disability: a new patient. *Am J Med Genet A*. 2013;161A(8):2084–2087.
84. Zhou H, Clapham DE. Mammalian MagT1 and TUSC3 are required for cellular magnesium uptake and vertebrate embryonic development. *Proc Natl Acad Sci U S A*. 2009;106(37):15750–15755.
85. Weinstein M, Schollen E, Matthijs G, et al. CDG-IL: an infant with a novel mutation in the ALG9 gene and additional phenotypic features. *Am J Med Genet A*. 2005;136(2):194–197.
86. Baysal BE, Willett-Brozick JE, Bacanu SA, Detera-Wadleigh S, Nimgaonkar VL. Common variations in ALG9 are not associated with bipolar I disorder: a family-based study. *Behav Brain Funct*. 2006;2:25.
87. Cormier-Daire V, Amiel J, Vuillaumier-Barrot S, et al. Congenital disorders of glycosylation IIa cause growth retardation, mental retardation, and facial dysmorphism. *J Med Genet*. 2000;37(11):875–877.
88. Van Geet C, Jaeken J, Freson K, et al. Congenital disorders of glycosylation type Ia and IIa are associated with different primary haemostatic complications. *J Inherit Metab Dis*. 2001;24(4):477–492.
89. De Praeter CM, Gerwig GJ, Bause E, et al. A novel disorder caused by defective biosynthesis of N-linked oligosaccharides due to glucosidase I deficiency. *Am J Hum Genet*. 2000;66(6):1744–1756.
90. Marquardt T, Brune T, Luhn K, et al. Leukocyte adhesion deficiency II syndrome, a generalized defect in fucose metabolism. *J Pediatr*. 1999;134(6):681–688.
91. Lubke T, Marquardt T, von Figura K, Korner C. A new type of carbohydrate-deficient glycoprotein syndrome due to a decreased import of GDP-fucose into the Golgi. *J Biol Chem*. 1999;274(37):25986–25989.
92. Marquardt T, Luhn K, Srikrishna G, Freeze HH, Harms E, Vestweber D. Correction of leukocyte adhesion deficiency type II with oral fucose. *Blood*. 1999;94(12):3976–3985.
93. Peters V, Penzien JM, Reiter G, et al. Congenital disorder of glycosylation IId (CDG-IId) – a new entity: clinical presentation with Dandy–Walker malformation and myopathy. *Neuropediatrics*. 2002;33(1):27–32.
94. Hansske B, Thiel C, Lubke T, et al. Deficiency of UDP-galactose:N-acetylglucosamine beta-1,4-galactosyltransferase I causes the congenital disorder of glycosylation type IId. *J Clin Invest*. 2002;109(6):725–733.
95. Guillard M, Morava E, de Ruijter J, et al. B4GALT1-congenital disorders of glycosylation presents as a non-neurologic glycosylation disorder with hepatointestinal involvement. *J Pediatr*. 2011;159(6):1041–1043 e2.
96. Wu X, Steet RA, Bohorov O, et al. Mutation of the COG complex subunit gene COG7 causes a lethal congenital disorder. *Nat Med*. 2004;10(5):518–523.
97. Morava E, Zeevaert R, Korsch E, et al. A common mutation in the COG7 gene with a consistent phenotype including microcephaly, adducted thumbs, growth retardation, VSD and episodes of hyperthermia. *Eur J Hum Genet*. 2007;15(6):638–645.
98. Kranz C, Ng BG, Sun L, et al. COG8 deficiency causes new congenital disorder of glycosylation type IIh. *Hum Mol Genet*. 2007;16(7):731–741.
99. Vasile E, Oka T, Ericsson M, Nakamura N, Krieger M. IntraGolgi distribution of the Conserved Oligomeric Golgi (COG) complex. *Exp Cell Res*. 2006;312(16):3132–3141.

100. Foulquier F, Vasile E, Schollen E, et al. Conserved oligomeric Golgi complex subunit 1 deficiency reveals a previously uncharacterized congenital disorder of glycosylation type II. *Proc Natl Acad Sci U S A*. 2006;103(10):3764–3769.
101. Reynders E, Foulquier F, Leao Teles E, et al. Golgi function and dysfunction in the first COG4-deficient CDG type II patient. *Hum Mol Genet*. 2009;18(17):3244–3256.
102. Lubbehusen J, Thiel C, Rind N, et al. Fatal outcome due to deficiency of subunit 6 of the conserved oligomeric Golgi complex leading to a new type of congenital disorders of glycosylation. *Hum Mol Genet*. 2010;19(18):3623–3633.
103. Fung CW, Matthijs G, Sturiale L, et al. COG5-CDG with a mild neurohepatic presentation. *JIMD Rep*. 2012;3:67–70.
104. Foulquier F, Amyere M, Jaeken J, et al. TMEM165 deficiency causes a congenital disorder of glycosylation. *Am J Hum Genet*. 2012;91(1):15–26.
105. Martinez-Duncker I, Dupre T, Piller V, et al. Genetic complementation reveals a novel human congenital disorder of glycosylation of type II, due to inactivation of the Golgi CMP-sialic acid transporter. *Blood*. 2005;105(7):2671–2676.
106. Song Z. Roles of the nucleotide sugar transporters (SLC35 family) in health and disease. *Mol Aspects Med*. 2013;34(2–3):590–600.
107. Jaeken J. MGCHVSE. Defects of N-glycan synthesis. In: Valle DBAL, Vogelstein B, Kinzler KW, Antonarakis SE, Ballabio A, Gibson K, Mitchell G, eds. OMMBID – The Online Metabolic and Molecular Bases of Inherited Diseases. New York: McGraw-Hill; 2013.
108. Mandato C, Brive L, Miura Y, et al. Cryptogenic liver disease in four children: a novel congenital disorder of glycosylation. *Pediatr Res*. 2006;59(2):293–298.
109. Jaeken J, Schachter H, Carchon H, De Cock P, Coddeville B, Spik G. Carbohydrate deficient glycoprotein syndrome type II: a deficiency in Golgi localised *N*-acetyl-glucosaminyltransferase II. *Arch Dis Child*. 1994;71(2):123–127.
110. Ziad A, Eliska M, Hubert V, et al. Our experience with diagnostics of congenital disorders of glycosylation. *Acta Medica (Hradec Kralove)*. 2004;47(4):267–272.
111. Albahri Z, Marklova E, Vanicek H, Minxova L, Dedek P, Skalova S. Genetic variants of transferrin in the diagnosis of protein hypoglycosylation. *J Inherit Metab Dis*. 2005;28(6):1184–1188.
112. Golka K, Wiese A. Carbohydrate-deficient transferrin (CDT)—a biomarker for long-term alcohol consumption. *J Toxicol Environ Health B Crit Rev*. 2004;7(4):319–337.
113. Sturiale L, Barone R, Fiumara A, et al. Hypoglycosylation with increased fucosylation and branching of serum transferrin N-glycans in untreated galactosemia. *Glycobiology*. 2005;15(12):1268–1276.
114. Patterson MC. Screening for "prelysosomal disorders": carbohydrate-deficient glycoprotein syndromes. *J Child Neurol*. 1999;14(suppl 1):S16–S22.
115. Stibler H. Direct immunofixation after isoelectric focusing. An improved method for identification of cerebrospinal fluid and serum proteins. *J Neurol Sci*. 1979;42(2):275–281.
116. Marklova E, Albahri Z. Pitfalls and drawbacks in screening of congenital disorders of glycosylation. *Clin Chem Lab Med*. 2004;42(6):583–589.
117. Vakhrushev SY, Mormann M, Peter-Katalinic J. Identification of glycoconjugates in the urine of a patient with congenital disorder of glycosylation by high-resolution mass spectrometry. *Proteomics*. 2006;6(3):983–992.
118. Wopereis S, Grunewald S, Morava E, et al. Apolipoprotein C-III isofocusing in the diagnosis of genetic defects in O-glycan biosynthesis. *Clin Chem*. 2003;49(11):1839–1845.
119. Stibler H, Jaeken J. Carbohydrate deficient serum transferrin in a new systemic hereditary syndrome. *Arch Dis Child*. 1990;65(1):107–111.
120. Wopereis S, Morava E, Grunewald S, et al. Patients with unsolved congenital disorders of glycosylation type II can be subdivided in six distinct biochemical groups. *Glycobiology*. 2005;15(12):1312–1319.
121. Lacey JM, Bergen HR, Magera MJ, Naylor S, O'Brien JF. Rapid determination of transferrin isoforms by immunoaffinity liquid chromatography and electrospray mass spectrometry. *Clin Chem*. 2001;47(3):513–518.
122. Wada Y. Mass spectrometry for congenital disorders of glycosylation, CDG. *J Chromatogr B Analyt Technol Biomed Life Sci*. 2006;838(1):3–8.
123. Mills K, Mills P, Jackson M, et al. Diagnosis of congenital disorders of glycosylation type-I using protein chip technology. *Proteomics*. 2006;6(7):2295–2304.
124. Ferrari MC, Parini R, Di Rocco MD, Radetti G, Beck-Peccoz P, Persani L. Lectin analyses of glycoprotein hormones in patients with congenital disorders of glycosylation. *Eur J Endocrinol*. 2001;144(4):409–416.
125. Ferens-Sieczkowska M, Zwierz K, Midro A, Katnik-Prastowska I. Glycoforms of six serum glycoproteins in a patient with congenital disorder of glycosylation type I. *Arch Immunol Ther Exp (Warsz)*. 2002;50(1):67–73.
126. Fang J, Peters V, Assmann B, Korner C, Hoffmann GF. Improvement of CDG diagnosis by combined examination of several glycoproteins. *J Inherit Metab Dis*. 2004;27(5):581–590.
127. Fiumara A, Barone R, Buttitta P, et al. Haemostatic studies in carbohydrate-deficient glycoprotein syndrome type I. *Thromb Haemost*. 1996;76(4):502–504.
128. Schollen E, Martens K, Geuzens E, Matthijs G. DHPLC analysis as a platform for molecular diagnosis of congenital disorders of glycosylation (CDG). *Eur J Hum Genet*. 2002;10(10):643–648.
129. Jones MA, Rhodenizer D, da Silva C, et al. Molecular diagnostic testing for congenital disorders of glycosylation (CDG): detection rate for single gene testing and next generation sequencing panel testing. *Mol Genet Metab*. 2013;110(1–2):78–85.
130. Hedberg C, Oldfors A, Darin N. B3GALNT2 is a gene associated with congenital muscular dystrophy with brain malformations. *Eur J Hum Genet*. 2014;22(5):707–710.
131. Jaeken J, Matthijs G. Congenital disorders of glycosylation. *Annu Rev Genomics Hum Genet*. 2001;2:129–151.
132. Schollen E, Kjaergaard S, Legius E, Schwartz M, Matthijs G. Lack of Hardy-Weinberg equilibrium for the most prevalent PMM2 mutation in CDG-Ia (congenital disorders of glycosylation type Ia). *Eur J Hum Genet*. 2000;8(5):367–371.
133. Schollen E, Kjaergaard S, Martinsson T, et al. Increased recurrence risk in congenital disorders of glycosylation type Ia (CDG-Ia) due to a transmission ratio distortion. *J Med Genet*. 2004;41(11):877–880.
134. Westphal V, Kjaergaard S, Schollen E, et al. A frequent mild mutation in ALG6 may exacerbate the clinical severity of patients with congenital disorder of glycosylation Ia (CDG-Ia) caused by phosphomannomutase deficiency. *Hum Mol Genet*. 2002;11(5):599–604.
135. Dennis JW, Warren CE, Granovsky M, Demetriou M. Genetic defects in N-glycosylation and cellular diversity in mammals. *Curr Opin Struct Biol*. 2001;11(5):601–607.

136. DeRossi C, Bode L, Eklund EA, et al. Ablation of mouse phosphomannose isomerase (Mpi) causes mannose 6-phosphate accumulation, toxicity, and embryonic lethality. *J Biol Chem*. 2006;281(9):5916–5927.
137. Thiel C, Lubke T, Matthijs G, von Figura K, Korner C. Targeted disruption of the mouse phosphomannomutase 2 gene causes early embryonic lethality. *Mol Cell Biol*. 2006;26(15):5615–5620.
138. Rush JS, Panneerselvam K, Waechter CJ, Freeze HH. Mannose supplementation corrects GDP-mannose deficiency in cultured fibroblasts from some patients with Congenital Disorders of Glycosylation (CDG). *Glycobiology*. 2000;10(8):829–835.
139. Eklund EA, Merbouh N, Ichikawa M, et al. Hydrophobic Man-1-P derivatives correct abnormal glycosylation in type I congenital disorder of glycosylation fibroblasts. *Glycobiology*. 2005;15(11):1084–1093.
140. Hardre R, Khaled A, Willemetz A, et al. Mono, di and tri-mannopyranosyl phosphates as mannose-1-phosphate prodrugs for potential CDG-Ia therapy. *Bioorg Med Chem Lett*. 2007;17(1):152–155.
141. Niehues R, Hasilik M, Alton G, et al. Carbohydrate-deficient glycoprotein syndrome type Ib. Phosphomannose isomerase deficiency and mannose therapy. *J Clin Invest*. 1998;101(7):1414–1420.
142. Matthijs G. Research network: EUROGLYCANET: a European network focused on congenital disorders of glycosylation. *Eur J Hum Genet*. 2005;13(4):395–397.
143. Taniguchi N, Nakamura K, Narimatsu H, von der Lieth CW, Paulson J. Human Disease Glycomics/Proteome Initiative Workshop and the 4th HUPO Annual Congress. *Proteomics*. 2006;6(1):12–13.

CHAPTER

61

Disorders of Glutathione Metabolism

Koji Aoyama and Toshio Nakaki

Teikyo University School of Medicine, Tokyo, Japan

INTRODUCTION

Glutathione (GSH) is the major endogenous low molecular-weight thiol in mammalian cells. GSH plays pivotal roles in antioxidative defense, intracellular redox homeostasis, cysteine carrier/storage, cell signaling, enzyme activity, gene expression, and cell differentiation/proliferation. In particular, GSH is one of the most important antioxidants or antioxidant-related compounds in the brain for neuroprotection. The brain is one of the organs generating large amounts of reactive oxygen species (ROS) because of its high ratio of O_2 consumption (~20% of total body O_2 consumption) to weight of the brain (~2% of body weight) and also generates high levels of lipids with unsaturated fatty acids, leading to oxidative stress. However, the brain contains low antioxidant levels with low antioxidant enzyme activities. ROS can cause lipid peroxidation, DNA damage, mitochondrial dysfunction, and protein oxidation. GSH acts directly as a potent antioxidant by itself or in collaboration with other enzymes to reduce ROS or detoxify xenobiotics.[1] GSH also acts as the major redox buffer to maintain intracellular redox homeostasis. Oxidative damage alters the redox state of the cell, leading to permanent loss of the functions of proteins as enzymes, receptors, and transporters.[2] GSH can preserve protein thiol groups in a reduced state (*S*-glutathionylation) to prevent irreversible protein oxidation.[3] A variety of GSH functions are crucial to protect the cells against the oxidative damage leading to neurodegeneration.

Clinically, inborn errors in the GSH-related enzymes are very rare, while disorders in GSH metabolism are common in neurodegenerative diseases, including Alzheimer disease (AD), Parkinson disease (PD), amyotrophic lateral sclerosis (ALS), and Huntington disease (HD). In this chapter, we focus on disorders of GSH metabolism in neurologic diseases, especially in neurodegenerative diseases related to GSH dysfunction. Additional reviews are presented in the references of this chapter.[4–7]

GSH AND THE γ-GLUTAMYL CYCLE

GSH is a tripeptide constituted of three amino acids: glutamate, cysteine and glycine. GSH synthesis requires two steps involving adenosine triphosphate (ATP)-dependent enzymatic reactions. The first step mediates the reaction between glutamate and cysteine by an enzyme called γ-glutamylcysteine ligase (GCL) or γ-glutamylcysteine synthetase. Glutamate has two carboxyl groups, one of which binds at the γ-position with the amino group of cysteine to form a dipeptide, γ-glutamylcysteine. This step is the rate-limiting reaction for GSH synthesis. GCL is a heterodimer composed of a catalytic (heavy; molecular weight of 73kDa) subunit, GCLC, and a modulatory (light; molecular weight of 28kDa) subunit, GCLM. GCLC, but not GCLM, has all the catalytic function and is subject to feedback inhibition by GSH,[8] while the association of GCLM with GCLC increases the affinity for glutamate.[9] The second step for GSH synthesis is mediated by another enzyme called GSH synthetase (GS), which binds γ-glutamylcysteine and glycine to form GSH, although the precise mechanisms for the regulation of GS activity have not been clarified yet.

GSH reacts nonenzymatically with superoxide, nitric oxide, hydroxyl radical, and peroxynitrite as an antioxidant, while GSH is utilized enzymatically by GSH peroxidase (GPx) and GSH-*S*-transferase (GST) as a reducing agent against oxidative damage. GPx requires GSH as an electron donor to react with H_2O_2 or endogenous hydroperoxides

Rosenberg's Molecular and Genetic Basis of Neurological and Psychiatric Disease
http://dx.doi.org/10.1016/10.1016/B978-0-12-410529-4.00061-9

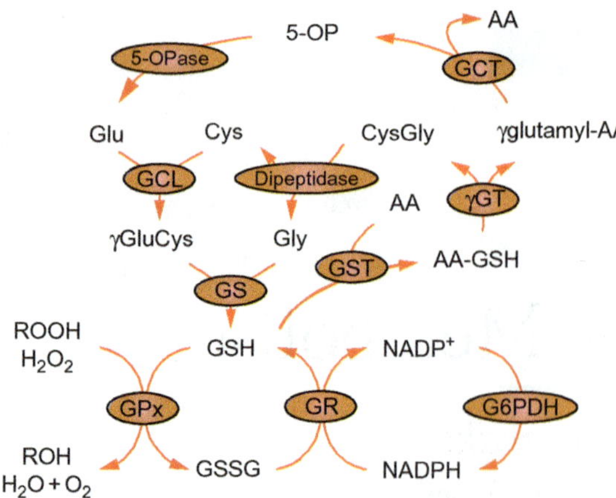

FIGURE 61.1 **The γ-glutamyl cycle.** Abbreviations: AA, amino acids; Cys, cysteine; CysGly, cysteinylglycine; GCL, γ-glutamylcysteine ligase; GCT, γ-glutamyl cyclotransferase; γGT, γ-glutamyl transpeptidase; γGluCys, γ-glutamylcysteine; Glu, glutamate; Gly, glycine; G6PDH, glucose-6-phosphate dehydrogenase; GPx, glutathione peroxidase; GR, glutathione reductase; GS, glutathione synthetase; GSH, glutathione; GSSG, glutathione disulfide; GST, glutathione-*S*-transferase; H_2O_2, hydrogen peroxide, NADPH, nicotinamide adenine dinucleotide phosphate; 5-OP, 5-oxoproline; 5-OPase, 5-oxoprolinase; ROH, alcohol; ROOH, hydroperoxide.

(ROOH), which are formed by the reaction of superoxide with superoxide dismutase (SOD) or lipid peroxidation, respectively. There are four types of selenium-containing GPx. The cytosolic isoform of GPx, named GPx-1, is the most abundant GPx and functions as an important antioxidative enzyme in the brain. In the processes of these GSH-associated antioxidant mechanisms, GSH is oxidized to GSH disulfide (GSSG), which is then reduced back to GSH by the reaction with GSH reductase (GR). This reaction requires nicotinamide adenine dinucleotide phosphate (NADPH), which is produced by glucose-6-phosphate dehydrogenase (G6PDH), as a substrate for supplying electrons to GSSG. Neuronal GR activity is sufficiently active to rapidly regenerate GSH from GSSG.[10] Thus, GSSG is maintained as less than 1% of total GSH under physiological conditions and the GSSG level is doubled (~2%) under oxidatively insulted conditions,[11] although the GSSG increase during oxidative stress is transient because of the reduction by GR. Although catalase can also degrade intracellular H_2O_2 to H_2O and O_2^-, this enzyme can not detoxify other endogenous ROOH and therefore does not work sufficiently for the peroxide detoxification in neurons.[12] GST catalyzes GSH attack on various electrophilic metabolites of xenobiotics to detoxify the compounds and release them from the cell. In mammalian species, seven classes of cytosolic GST have been reported as the isoforms alpha, mu, pi, sigma, theta, omega, and zeta.[13] GSH and GSH-containing molecules, including GSSG and GSH *S*-conjugates, are exported to the extracellular space via the multidrug resistance-associated proteins.[14,15] GSH and its conjugates are then cleaved into the γ-glutamyl moiety and cysteinylglycine by the reaction with γ-glutamyl transpeptidase (γGT), which is a plasma membrane-bound enzyme with its active site on the extracellular side. Subsequently, the γ-glutamyl moiety is cleaved into the corresponding amino acid and 5-oxoproline by the reaction of γ-glutamyl cyclotransferase (GCT), while cysteinylglycine is cleaved into cysteine and glycine in a reaction mediated by membrane-bound dipeptidase. 5-Oxoproline is then converted to glutamate by the reaction of 5-oxoprolinase. These cleaved or converted amino acids are reused for GSH synthesis to form the γ-glutamyl cycle (Figure 61.1).

DISORDERS OF ENZYMES IN THE γ-GLUTAMYL CYCLE

Animal Models

GCLC-deficient mice are embryonic lethal,[16] while GCLM-deficient mice are viable and fertile, although the GSH levels in their organs and plasma are low.[17] Mice deficient in GPx-1, for which the total GPx activity is completely blocked in the brain, show normal growth and fertility and no histological abnormality in the brain or other organs,[18] although they do show increased vulnerability to some neurotoxins.[19] γGT-deficient mice develop glutathionuria, severe growth failure, lethargy, shortened lifespan, hypogonadism, and infertility.[20,21] Mice deficient in some cytosolic GSTs have been developed to show no obvious phenotype with some increased susceptibility to toxic xenobiotics.[13]

Clinical Features and Molecular Genetics

GCL deficiency is a very rare hereditary (autosomal recessive) disease, which has been reported in nine patients in seven families around the world. The patients exhibit hemolytic anemia and, in some cases, neurological symptoms, such as spinocerebellar degeneration, mental retardation, peripheral neuropathy, myopathy, and aminoaciduria. The laboratory data show low GCL activity/levels, and low GSH levels in red blood cells and/or cultured skin fibroblasts. Both human gene loci of GCLC and GCLM are described on chromosome 6p12 and chromosome 1p21, respectively. Mutation analysis of the patients has identified four different mutations in the GCLC subunit in four families. No promising treatment has been established yet. Patients with GCL deficiency should avoid drugs known to cause hemolytic anemia in patients with G6PDH deficiency—e.g., acetyl salicylic acid or sulfonamides.

GS deficiency is also a rare autosomal recessive disease reported in more than 70 patients in more than 50 families. The patients present with hemolytic anemia, metabolic acidosis, 5-oxoprolinuria, progressive neurologic symptoms, such as mental retardation, seizures, spasticity, and ataxia, and recurrent bacterial infections. The human GS gene is localized to chromosome 20q11.2, on which some different mutations or epigenetic modifications have been revealed. For this deficiency as well, the laboratory data indicate low GS activity/levels, and low GSH levels in red blood cells and/or cultured skin fibroblasts. About 25% of the patients die in the neonatal period. Early administration of vitamin C and/or vitamin E improves the long-term clinical outcome. Treatment with bicarbonate, citrate, or trometamol is useful to correct the metabolic acidosis. *N*-acetylcysteine (NAC) administration might be recommended to protect cells from oxidative stress, but accumulated cysteine would be rather neurotoxic in patients with GS deficiency.[22–24] The patients should avoid drugs that could precipitate hemolytic anemia.

There are some other hereditary diseases with enzyme dysfunctions in the γ-glutamyl cycle although all disorders are clinically very rare.[25]

EXCITATORY AMINO ACID TRANSPORTERS (EAATS)

Neurons cannot assimilate extracellular GSH directly into the cell interior, but can directly assimilate cysteine. Most of the neuronal cysteine uptake is mediated by sodium-dependent systems, mainly the excitatory amino acid transporter (EAAT). To date, five EAATs have been reported: glutamate aspartate transporter (GLAST, also termed EAAT1), glutamate transporter-1 (GLT-1, also termed EAAT2), excitatory amino acid carrier 1 (EAAC1, also termed EAAT3), EAAT4, and EAAT5.[26] GLAST and GLT-1 are localized primarily to astrocytes, and EAAC1, EAAT4 and EAAT5 to neurons. EAAT4 and EAAT5 are localized to cerebellar Purkinje cells and the retina, respectively, while EAAC1 is expressed widely throughout the central nervous system (CNS). In the brain, astroglial EAATs, mainly GLT-1, play a central role in removing interstitial glutamate.[27,28] EAATs can transport not only extracellular glutamate but also cysteine into the cells.[29] Notably, EAAC1 preferentially transports extracellular cysteine rather than glutamate into neurons (Figure 61.2). Under normal conditions, EAAC1 is mostly localized in the cytoplasm, with the expression on the plasma membrane accounting for only 20% of the total.[30] Once stimulated by intracellular signaling, such as protein kinase C or phosphatidylinositol 3-kinase activation, EAAC1 translocates to the plasma membrane for glutamate and cysteine uptake, leading to neuronal GSH synthesis.[7] In contrast, the expression of EAAC1 on the plasma membrane is restricted by direct interaction with glutamate transport associated protein 3-18 (GTRAP3-18), which inhibits translocation of

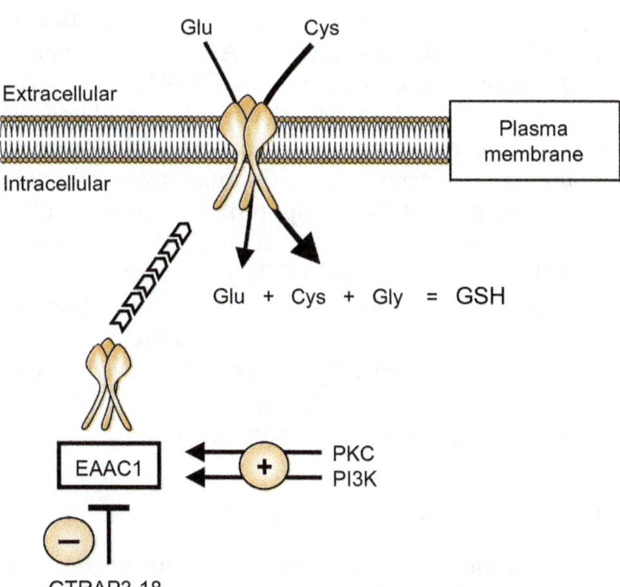

FIGURE 61.2 EAAC1-mediated GSH synthesis in neurons. Abbreviations: Cys, cysteine; EAAC1, excitatory amino acid carrier 1; Glu, glutamate; Gly, glycine; GTRAP3-18, glutamate transport associated protein 3-18; GSH, glutathione; PI3K, phosphatidylinositol 3-kinase; PKC, protein kinase C.

EAAC1 to the plasma membrane.[31] GTRAP3-18 is a member of the prenylated Rab acceptor family, which contains two extensive hydrophobic domains tightly attached to the endoplasmic reticulum (ER) membrane and anchors the target proteins in the ER. GTRAP3-18 negatively regulates the neuronal GSH level by controlling the EAAC1-mediated cysteine uptake.[32–34]

DISORDERS OF EAAC1 LEADING TO GSH DEPLETION

Animal Models

EAAC1-deficient mice develop brain atrophy, spatial learning and memory dysfunction, decreased number of dopaminergic neurons in the substantia nigra (SN), and movement disorder at advanced ages but not at adolescence.[35,36] The total GSH levels in the brain are much lower in EAAC1-deficient mice than in the wild-type mice, while the liver GSH levels are comparable between the groups. The brains of EAAC1-deficient mice show increased oxidant levels and increased vulnerability to oxidative stress, which are alleviated by treatment with NAC to increase GSH synthesis.

Conversely, inhibition of GTRAP3-18 expression increases GSH levels in the brain.[33,34] GTRAP3-18-deficient mice show the increased EAAC1 expression on the plasma membrane, increased cysteine and GSH contents in the brain, resistance to oxidative stress, and facilitated learning and memory functions.

Clinical Features

Several lines of putative evidence have been reported regarding EAAC1 dysfunction leading to GSH depletion in neurodegenerative diseases; however, it is still uncertain whether this finding is the primary cause of neurodegenerative diseases (as discussed below).

NEURODEGENERATIVE DISEASES LEADING TO GSH DEPLETION

Alzheimer Disease

AD is a leading age-related neurodegenerative disease characterized by progressive dementia developed in middle or later life. Pathologically, depositions of amyloid-β plaques and neurofibrillary tangles in the brain are the hallmark of AD. Amyloid-β induces oxidative stress leading to neurofibrillary tangles formation and neuronal death.[37] Changes in GSH metabolism have been reported in AD patients. Polymorphisms in the GPx-1 and GST genes have been identified as possible risk factors for AD.[38,39] GSH levels in erythrocytes are decreased, while plasma oxidation protein products are increased in AD patients.[40] A recent clinical study using magnetic resonance spectroscopy revealed that brain GSH levels are depleted in AD patients as compared to healthy subjects.[41] Postmortem brain tissue samples from AD patients also showed decreased GSH/GSSG ratios with disease progression and decreased GST activities.[42,43] Mild cognitive impairment (MCI), which is considered the earliest stage of AD,[44] has also been associated with decreased GSH levels in erythrocytes, increased oxidation protein products in plasma,[40] and the decreased ratio of GSH/GSSG in the hippocampus.[45] These findings suggest that GSH depletion would precede the onset of the disease.

The ε4 allele of the apolipoprotein E gene (*APOE*) is a major genetic risk factor for late-onset AD. The ApoE4 protein enhances Aβ deposition in the CNS. Brain tissues from AD patients with the *APOE* ε4 allele also show decreased GSH levels with decreased GPx activities as compared to those of age-matched controls or AD patients homozygous for the ε3 allele.[46] α-tocopherol is also a potent antioxidant in the brain. However, the analysis of α-tocopherol levels in the brain revealed no relation with the *APOE* polymorphism. These findings suggest a specific dysregulation of GSH biosynthesis in AD patients. Indeed, in the hippocampus of AD patients, degenerating neurons exhibit aberrant detergent-insoluble EAAC1 accumulation.[47] GSH depletion via EAAC1 dysfunction would lead to neurodegeneration in the hippocampus of patients with AD.

Parkinson Disease

PD is the second most common neurodegenerative disease after AD, and is clinically characterized by resting tremor, rigidity, akinesia, and postural instability. PD is also a progressive, late-onset movement disorder that is affected by dopaminergic neurodegeneration in the SN. Lewy bodies are eosinophilic neuronal inclusions that contain both α-synuclein and ubiquitin as the pathological hallmarks of PD. Oxidative stress leads to α-synuclein aggregation, followed by proteasome dysfunction and neuronal death.[48] Postmortem analysis of normal individuals with incidental Lewy bodies, who would be considered presymptomatic PD subjects, revealed lower GSH levels in the SN than those of age-matched controls without Lewy bodies.[49] As with AD patients, GSH depletion would precede the onset of the disease.[50] Consequently, the brain GSH levels in PD patients have been shown to be depleted in the SN, but not in the other regions, as compared to those of age-matched controls.[51] The activity of the GSH synthetic enzyme GCL was unchanged, while that of the GSH-degrading enzyme γGT was elevated in PD patients.[52] The severity of GSH depletion parallels pathological and/or clinical PD severity.[53] Decreased GSH may be considered not only as an early event in PD progression but even as the primary cause of neurodegeneration.[50,54] Although the primary cause of GSH depletion in PD is still unclear, recent studies suggest that EAAC1 dysfunction is involved in the pathogenesis. Human dopaminergic neurons in the SN express EAAC1.[36,55] Dopaminergic (DA) neurons are more vulnerable to EAAC1 dysfunction than non-DA neurons.[56] Considering these findings, EAAC1 dysfunction would be a possible cause of GSH depletion in PD, although further clinical studies would be required to elucidate the involvement.

Approximately 10% of PD patients present a family history of the Mendelian form of the disease with autosomal dominant or recessive inheritance. To date, there are 18 specific chromosomal regions, designated *PARK1-18*.[57] Mutations in the α-synuclein gene (*SNCA*), formerly termed *PARK1*, were found in autosomal dominant PD. Normal α-synuclein translocates into lysosomes for degradation, while mutated α-synuclein aggregates in neurons to cause neurodegeneration.[58] A mutant A53T model in the α-synuclein gene decreases *de novo* GSH synthesis after treatment with a proteasome inhibitor.[59] Loss-of-function mutations in *parkin* (*PARK2*) are found in autosomal recessive juvenile PD patients. Parkin is an E3 ubiquitin ligase, catalyzing the addition of ubiquitin to specific substrates, including α-synuclein, which targets them for degradation by the ubiquitin–proteasome system. The active sites of parkin are cysteine-rich regions and thereby sensitive to oxidative modification, which alters the protein solubility and its function.[60,61] Mutations in *PINK1* (*PARK6*) cause autosomal recessive early-onset PD. PINK1 is a mitochondrial protein kinase and loss of its function induces mitochondrial dysfunction. In primary fibroblasts from the PD patients, elevated GR and GST activities with increased lipid peroxidation and mitochondrial dysfunction are observed, although the GSH/GSSH ratios are unchanged, compared to controls.[62] *DJ-1* (*PARK7*) is also one of the causative genes for familial PD. Mutations in *DJ-1* cause an autosomal-recessive, early onset familial form of PD. DJ-1 is a redox-dependent molecular chaperone that upregulates GSH synthesis during oxidative stress.[63] Oxidation of a conserved cysteine residue in DJ-1 regulates its chaperone activity against α-synuclein.[64] Conversely, DJ-1 is oxidatively damaged in the brains of idiopathic PD and AD patients.[65]

Amyotrophic Lateral Sclerosis

ALS is a chronic progressive disease characterized by selective degeneration of motor neurons in the spinal cord and motor cortex. Although the precise etiology is still unknown, glutamate neurotoxicity induced by loss of GLT-1 has been reported in the spinal cord and motor cortex of sporadic ALS patients.[66,67] Moreover, the expression of EAAC1 proteins, but not the mRNA, is also slightly downregulated in the spinal cord and motor cortex of sporadic ALS patients.[67,68] GSH depletion has been shown to result in motor neuron death *in vitro* and *in vivo*.[69] GSH metabolism seems to be altered in ALS patients. An early clinical study reported increased protein glutathionylation, which is an important adaptive cellular response to protect crucial protein functions under oxidative stress, in the spinal cords of sporadic ALS patients.[70] A recent study also demonstrated decreased GSH, GR, and G6PDH levels, but increased lipid peroxidation in the erythrocytes of sporadic ALS patients.[71] These changes correlate with the disease progression. The mRNA expression of GST pi is reduced in the spinal cord and motor/sensory cortex in the brain of ALS patients.[72] These changes in GSH metabolism support the clinical evidence implicating oxidative stress in ALS pathogenesis.[73] Approximately 10% of all ALS cases are familial; in turn, ~20% of these familial cases are inherited in an autosomal dominant pattern with mutations in the gene encoding cytosolic Cu/Zn SOD (SOD1). SOD1 mutant mice show decreased GSH levels in the spinal cord and motor neurons.[69]

Huntington Disease

HD is caused by the expansion of *CAG* trinucleotide repeats (in excess of 38 repeats) on chromosome 4 in exon 1 of the gene coding "*huntingtin*" with autosomal dominant inheritance. HD patients show hyperkinetic movement disorders based on basal ganglion dysfunction. The precise mechanisms causing HD are still elusive. However, oxidative stress is considered to be the major cause leading to the neurodegeneration. HD patients have higher plasma lipid peroxidation levels and lower GSH levels than their age and sex-matched controls.[74] In a recent study, depleted GSH levels with elevated ROS levels were found in neurons prepared from a model of HD ($HD^{140Q/140Q}$) in which a human *huntingtin* gene with 140 *CAG* repeats was inserted into the mouse genome.[75] These results were attributable to EAAC1 dysfunction, which impairs cysteine uptake and thereby leads to GSH depletion in neurons. Further clinical studies are needed to clarify the mechanisms of EAAC1 dysfunction in HD patients.

Therapy

Antioxidant therapy may be a potential approach to prevent and treat neurodegenerative diseases. However, oral supplementation with ascorbate and/or α-tocopherol, which are important antioxidants in the brain, have not yet shown definitive benefit in patients with AD and PD. In addition, the levels of ascorbate and α-tocopherol in the CNS do not change in patients with AD and PD compared with controls, while GSH levels in the CNS are consistently reduced in these neurodegenerative diseases, as described above. Epidemiological studies have indicated that incidence of neurodegenerative diseases such as AD and PD is inversely correlated with caffeine consumption or plasma uric acid level.[76,77] These purine derivatives may induce neuronal GSH synthesis by promoting cysteine uptake, leading to neuroprotection.[78] Therefore, the therapeutic strategy of increasing the GSH levels in the brain in neurodegenerative diseases is sound; however, at present there are no therapeutic drugs for increasing brain GSH levels. Cysteine is the rate-limiting substrate for GSH synthesis, but direct administration of cysteine is not recommended due to its neurotoxicity. The plasma half-life of intravenous administered GSH is 2–3 minutes. Orally administered GSH is rapidly degraded in the gut. Moreover, neither cysteine nor GSH in the blood can penetrate the blood–brain barrier (BBB) easily, and these compounds are rapidly oxidized to cystine and GSSG, respectively. NAC is a promising compound as a therapeutic drug for neurodegenerative diseases. NAC acts as both a direct antioxidant and the substrate for GSH synthesis in the brain. NAC can penetrate the BBB and the plasma membrane to supply cysteine into the cells. In the mouse models of neurodegenerative diseases, NAC can increase neuronal GSH levels and reduce neuronal damages induced by oxidative stress.[35,79] Based on the promising results of NAC treatment for AD models *in vitro* and *in vivo*, two clinical trials in the USA (ClinicalTrials.gov identifier: NCT01320527 and NCT01370954) have been initiated and completed to investigate whether a dietary supplement containing NAC maintains or improves cognitive performance in patients with AD or MCI, although these results are not available as of June 2014. For the same reasons as AD, NAC is also a promising agent for the treatment of PD patients. As of June 2014, two clinical studies (ClinicalTrials.gov identifier: NCT01427517 and NCT01470027) are listed in the USA. However, in a randomized, double blind clinical trial with NAC treatments for ALS patients, no significant benefits were found in either survival or disease progression.[80]

An endogenous approach to increase GSH levels in neurons is an alternative strategy against neurodegeneration. Although there is no known drug for clinical use to activate neuronal GSH synthesis in the brain, the compounds facilitating EAAC1 function might be a promising treatment for neurodegenerative diseases in the future.

CONCLUSIONS

Insight into neurological disorders involving GSH metabolism has been obtained from a few rare hereditary diseases with enzymatic dysfunctions of GSH metabolism, as well as from some major neurodegenerative diseases such as AD, PD, ALS, and HD. GSH depletion is an early event in neurodegeneration and is related to disease progression in patients. A strategy to increase neuronal GSH levels would be a promising treatment for patients with these disorders, although it is still under basic and clinical investigations.

References

1. Dringen R. Metabolism and functions of glutathione in brain. *Prog Neurobiol*. 2000;62:649–671.
2. Klatt P, Lamas S. Regulation of protein function by S-glutathiolation in response to oxidative and nitrosative stress. *Eur J Biochem*. 2000;267:4928–4944.
3. Giustarini D, Rossi R, Milzani A, et al. *S*-glutathionylation: from redox regulation of protein functions to human diseases. *J Cell Mol Med*. 2004;8:201–212.
4. Ristoff E, Larsson A. Disorders of glutathione metabolism. In: Rosenberg RN, DiMauro S, Paulson HL, et al., eds. The Molecular and Genetic Basis of Neurologic and Psychiatric Disease. 4th ed. Lippincott Williams & Wilkins/Wolters Kluwer: Boston, MA; 2008:683–688.
5. Aoyama K, Watabe M, Nakaki T. Regulation of neuronal glutathione synthesis. *J Pharmacol Sci*. 2008;108:227–238.

6. Aoyama K, Nakaki T. Inhibition of GTRAP3-18 may increase neuroprotective glutathione (GSH) synthesis. *Int J Mol Sci.* 2012;13:12017–12035.
7. Aoyama K, Nakaki T. Neuroprotective properties of the excitatory amino acid carrier 1 (EAAC1). *Amino Acids.* 2013;45:133–142.
8. Richman PG, Meister A. Regulation of gamma-glutamyl-cysteine synthetase by nonallosteric feedback inhibition by glutathione. *J Biol Chem.* 1975;250:1422–1426.
9. Dickinson DA, Forman HJ. Glutathione in defense and signaling: lessons from a small thiol. *Ann N Y Acad Sci.* 2002;973:488–504.
10. Dringen R, Kussmaul L, Gutterer JM, et al. The glutathione system of peroxide detoxification is less efficient in neurons than in astroglial cells. *J Neurochem.* 1999;72:2523–2530.
11. Maher P. The effects of stress and aging on glutathione metabolism. *Ageing Res Rev.* 2005;4:288–314.
12. Cooper AJ, Kristal BS. Multiple roles of glutathione in the central nervous system. *Biol Chem.* 1997;378:793–802.
13. Hayes JD, Flanagan JU, Jowsey IR. Glutathione transferases. *Annu Rev Pharmacol Toxicol.* 2005;45:51–88.
14. Leier I, Jedlitschky G, Buchholz U, et al. ATP-dependent glutathione disulphide transport mediated by the MRP gene-encoded conjugate export pump. *Biochem J.* 1996;314(Pt 2):433–437.
15. Ballatori N, Krance SM, Marchan R, et al. Plasma membrane glutathione transporters and their roles in cell physiology and pathophysiology. *Mol Aspects Med.* 2009;30:13–28.
16. Dalton TP, Dieter MZ, Yang Y, et al. Knockout of the mouse glutamate cysteine ligase catalytic subunit (Gclc) gene: embryonic lethal when homozygous, and proposed model for moderate glutathione deficiency when heterozygous. *Biochem Biophys Res Commun.* 2000;279:324–329.
17. Yang Y, Dieter MZ, Chen Y, et al. Initial characterization of the glutamate-cysteine ligase modifier subunit Gclm(-/-) knockout mouse. Novel model system for a severely compromised oxidative stress response. *J Biol Chem.* 2002;277:49446–49452.
18. Ho YS, Magnenat JL, Bronson RT, et al. Mice deficient in cellular glutathione peroxidase develop normally and show no increased sensitivity to hyperoxia. *J Biol Chem.* 1997;272:16644–16651.
19. Klivenyi P, Andreassen OA, Ferrante RJ, et al. Mice deficient in cellular glutathione peroxidase show increased vulnerability to malonate, 3-nitropropionic acid, and 1-methyl-4-phenyl-1,2,5,6-tetrahydropyridine. *J Neurosci.* 2000;20:1–7.
20. Harding CO, Williams P, Wagner E, et al. Mice with genetic gamma-glutamyl transpeptidase deficiency exhibit glutathionuria, severe growth failure, reduced life spans, and infertility. *J Biol Chem.* 1997;272:12560–12567.
21. Kumar TR, Wiseman AL, Kala G, et al. Reproductive defects in gamma-glutamyl transpeptidase-deficient mice. *Endocrinology.* 2000;141: 4270–4277.
22. Puka-Sundvall M, Eriksson P, Nilsson M, et al. Neurotoxicity of cysteine: interaction with glutamate. *Brain Res.* 1995;705:65–70.
23. Janaky R, Varga V, Hermann A, et al. Mechanisms of L-cysteine neurotoxicity. *Neurochem Res.* 2000;25:1397–1405.
24. Ristoff E, Hebert C, Njalsson R, et al. Glutathione synthetase deficiency: is gamma-glutamylcysteine accumulation a way to cope with oxidative stress in cells with insufficient levels of glutathione? *J Inherit Metab Dis.* 2002;25:577–584.
25. Ristoff E, Larsson A. Inborn errors in the metabolism of glutathione. *Orphanet J Rare Dis.* 2007;2:16.
26. Danbolt NC. Glutamate uptake. *Prog Neurobiol.* 2001;65:1–105.
27. Tanaka K, Watase K, Manabe T, et al. Epilepsy and exacerbation of brain injury in mice lacking the glutamate transporter GLT-1. *Science.* 1997;276:1699–1702.
28. Holmseth S, Dehnes Y, Huang YH, et al. The density of EAAC1 (EAAT3) glutamate transporters expressed by neurons in the mammalian CNS. *J Neurosci.* 2012;32:6000–6013.
29. Zerangue N, Kavanaugh MP. Interaction of L-cysteine with a human excitatory amino acid transporter. *J Physiol.* 1996;493(Pt 2):419–423.
30. Fournier KM, Gonzalez MI, Robinson MB. Rapid trafficking of the neuronal glutamate transporter, EAAC1: evidence for distinct trafficking pathways differentially regulated by protein kinase C and platelet-derived growth factor. *J Biol Chem.* 2004;279:34505–34513.
31. Ruggiero AM, Liu Y, Vidensky S, et al. The endoplasmic reticulum exit of glutamate transporter is regulated by the inducible mammalian Yip6b/GTRAP3-18 protein. *J Biol Chem.* 2008;283:6175–6183.
32. Watabe M, Aoyama K, Nakaki T. Regulation of glutathione synthesis via interaction between glutamate transport-associated protein 3-18 (GTRAP3-18) and excitatory amino acid carrier-1 (EAAC1) at plasma membrane. *Mol Pharmacol.* 2007;72:1103–1110.
33. Watabe M, Aoyama K, Nakaki T. A dominant role of GTRAP3-18 in neuronal glutathione synthesis. *J Neurosci.* 2008;28:9404–9413.
34. Aoyama K, Wang F, Matsumura N, et al. Increased neuronal glutathione and neuroprotection in GTRAP3-18-deficient mice. *Neurobiol Dis.* 2012;45:973–982.
35. Aoyama K, Suh SW, Hamby AM, et al. Neuronal glutathione deficiency and age-dependent neurodegeneration in the EAAC1 deficient mouse. *Nat Neurosci.* 2006;9:119–126.
36. Berman AE, Chan WY, Brennan AM, et al. *N*-acetylcysteine prevents loss of dopaminergic neurons in the EAAC1-/- mouse. *Ann Neurol.* 2011;69:509–520.
37. Mattson MP. Pathways towards and away from Alzheimer's disease. *Nature.* 2004;430:631–639.
38. Spalletta G, Bernardini S, Bellincampi L, et al. Glutathione *S*-transferase P1 and T1 gene polymorphisms predict longitudinal course and age at onset of Alzheimer disease. *Am J Geriatr Psychiatry.* 2007;15:879–887.
39. Paz-y-Mino C, Carrera C, Lopez-Cortes A, et al. Genetic polymorphisms in apolipoprotein E and glutathione peroxidase 1 genes in the Ecuadorian population affected with Alzheimer's disease. *Am J Med Sci.* 2010;340:373–377.
40. Bermejo P, Martin-Aragon S, Benedi J, et al. Peripheral levels of glutathione and protein oxidation as markers in the development of Alzheimer's disease from Mild Cognitive Impairment. *Free Radic Res.* 2008;42:162–170.
41. Mandal PK, Tripathi M, Sugunan S. Brain oxidative stress: detection and mapping of antioxidant marker 'Glutathione' in different brain regions of healthy male/female, MCI and Alzheimer patients using non-invasive magnetic resonance spectroscopy. *Biochem Biophys Res Commun.* 2012;417:43–48.
42. Lovell MA, Xie C, Markesbery WR. Decreased glutathione transferase activity in brain and ventricular fluid in Alzheimer's disease. *Neurology.* 1998;51:1562–1566.
43. Ansari MA, Scheff SW. Oxidative stress in the progression of Alzheimer disease in the frontal cortex. *J Neuropathol Exp Neurol.* 2010;69:155–167.
44. Petersen RC. Mild cognitive impairment: transition between aging and Alzheimer's disease. *Neurologia.* 2000;15:93–101.

45. Sultana R, Piroddi M, Galli F, et al. Protein levels and activity of some antioxidant enzymes in hippocampus of subjects with amnestic mild cognitive impairment. *Neurochem Res.* 2008;33:2540–2546.
46. Ramassamy C, Averill D, Beffert U, et al. Oxidative insults are associated with apolipoprotein E genotype in Alzheimer's disease brain. *Neurobiol Dis.* 2000;7:23–37.
47. Duerson K, Woltjer RL, Mookherjee P, et al. Detergent-insoluble EAAC1/EAAT3 aberrantly accumulates in hippocampal neurons of Alzheimer's disease patients. *Brain Pathol.* 2009;19:267–278.
48. Dawson TM, Dawson VL. Molecular pathways of neurodegeneration in Parkinson's disease. *Science.* 2003;302:819–822.
49. Dexter DT, Sian J, Rose S, et al. Indices of oxidative stress and mitochondrial function in individuals with incidental Lewy body disease. *Ann Neurol.* 1994;35:38–44.
50. Jenner P. Oxidative damage in neurodegenerative disease. *Lancet.* 1994;344:796–798.
51. Sian J, Dexter DT, Lees AJ, et al. Alterations in glutathione levels in Parkinson's disease and other neurodegenerative disorders affecting basal ganglia. *Ann Neurol.* 1994;36:348–355.
52. Sian J, Dexter DT, Lees AJ, et al. Glutathione-related enzymes in brain in Parkinson's disease. *Ann Neurol.* 1994;36:356–361.
53. Riederer P, Sofic E, Rausch WD, et al. Transition metals, ferritin, glutathione, and ascorbic acid in parkinsonian brains. *J Neurochem.* 1989;52:515–520.
54. Jenner P. Oxidative stress in Parkinson's disease. *Ann Neurol.* 2003;53(suppl 3):S26–S36, discussion S36–S28.
55. Plaitakis A, Shashidharan P. Glutamate transport and metabolism in dopaminergic neurons of substantia nigra: implications for the pathogenesis of Parkinson's disease. *J Neurol.* 2000;247(suppl 2):II25–II35.
56. Nafia I, Re DB, Masmejean F, et al. Preferential vulnerability of mesencephalic dopamine neurons to glutamate transporter dysfunction. *J Neurochem.* 2008;105:484–496.
57. Klein C, Westenberger A. Genetics of Parkinson's disease. *Cold Spring Harb Perspect Med.* 2012;2:a008888.
58. Lee MK, Stirling W, Xu Y, et al. Human alpha-synuclein-harboring familial Parkinson's disease-linked Ala-53→Thr mutation causes neurodegenerative disease with alpha-synuclein aggregation in transgenic mice. *Proc Natl Acad Sci U S A.* 2002;99:8968–8973.
59. Vali S, Chinta SJ, Peng J, et al. Insights into the effects of alpha-synuclein expression and proteasome inhibition on glutathione metabolism through a dynamic in silico model of Parkinson's disease: validation by cell culture data. *Free Radic Biol Med.* 2008;45:1290–1301.
60. Wong ES, Tan JM, Wang C, et al. Relative sensitivity of parkin and other cysteine-containing enzymes to stress-induced solubility alterations. *J Biol Chem.* 2007;282:12310–12318.
61. Meng F, Yao D, Shi Y, et al. Oxidation of the cysteine-rich regions of parkin perturbs its E3 ligase activity and contributes to protein aggregation. *Mol Neurodegeneration.* 2011;6:34.
62. Hoepken HH, Gispert S, Morales B, et al. Mitochondrial dysfunction, peroxidation damage and changes in glutathione metabolism in PARK6. *Neurobiol Dis.* 2007;25:401–411.
63. Zhou W, Freed CR. DJ-1 up-regulates glutathione synthesis during oxidative stress and inhibits A53T alpha-synuclein toxicity. *J Biol Chem.* 2005;280:43150–43158.
64. Zhou W, Zhu M, Wilson MA, et al. The oxidation state of DJ-1 regulates its chaperone activity toward alpha-synuclein. *J Mol Biol.* 2006;356:1036–1048.
65. Choi J, Sullards MC, Olzmann JA, et al. Oxidative damage of DJ-1 is linked to sporadic Parkinson and Alzheimer diseases. *J Biol Chem.* 2006;281:10816–10824.
66. Rothstein JD, Martin LJ, Kuncl RW. Decreased glutamate transport by the brain and spinal cord in amyotrophic lateral sclerosis. *N Engl J Med.* 1992;326:1464–1468.
67. Rothstein JD, Van Kammen M, Levey AI, et al. Selective loss of glial glutamate transporter GLT-1 in amyotrophic lateral sclerosis. *Ann Neurol.* 1995;38:73–84.
68. Bristol LA, Rothstein JD. Glutamate transporter gene expression in amyotrophic lateral sclerosis motor cortex. *Ann Neurol.* 1996;39:676–679.
69. Chi L, Ke Y, Luo C, et al. Depletion of reduced glutathione enhances motor neuron degeneration in vitro and in vivo. *Neuroscience.* 2007;144:991–1003.
70. Lanius RA, Krieger C, Wagey R, et al. Increased [35S]glutathione binding sites in spinal cords from patients with sporadic amyotrophic lateral sclerosis. *Neurosci Lett.* 1993;163:89–92.
71. Babu GN, Kumar A, Chandra R, et al. Oxidant-antioxidant imbalance in the erythrocytes of sporadic amyotrophic lateral sclerosis patients correlates with the progression of disease. *Neurochem Int.* 2008;52:1284–1289.
72. Usarek E, Gajewska B, Kazmierczak B, et al. A study of glutathione *S*-transferase pi expression in central nervous system of subjects with amyotrophic lateral sclerosis using RNA extraction from formalin-fixed, paraffin-embedded material. *Neurochem Res.* 2005;30:1003–1007.
73. Barber SC, Mead RJ, Shaw PJ. Oxidative stress in ALS: a mechanism of neurodegeneration and a therapeutic target. *Biochim Biophys Acta.* 2006;1762:1051–1067.
74. Klepac N, Relja M, Klepac R, et al. Oxidative stress parameters in plasma of Huntington's disease patients, asymptomatic Huntington's disease gene carriers and healthy subjects: a cross-sectional study. *J Neurol.* 2007;254:1676–1683.
75. Li X, Valencia A, Sapp E, et al. Aberrant Rab11-dependent trafficking of the neuronal glutamate transporter EAAC1 causes oxidative stress and cell death in Huntington's disease. *J Neurosci.* 2010;30:4552–4561.
76. Kutzing MK, Firestein BL. Altered uric acid levels and disease states. *J Pharmacol Exp Ther.* 2008;324:1–7.
77. Ribeiro JA, Sebastiao AM. Caffeine and adenosine. *J Alzheimers Dis.* 2010;20(suppl 1):S3–S15.
78. Aoyama K, Matsumura N, Watabe M, et al. Caffeine and uric acid mediate glutathione synthesis for neuroprotection. *Neuroscience.* 2011;181:206–215.
79. Aoyama K, Matsumura N, Watabe M, et al. Oxidative stress on EAAC1 is involved in MPTP-induced glutathione depletion and motor dysfunction. *Eur J Neurosci.* 2008;27:20–30.
80. Louwerse ES, Weverling GJ, Bossuyt PM, et al. Randomized, double-blind, controlled trial of acetylcysteine in amyotrophic lateral sclerosis. *Arch Neurol.* 1995;52:559–564.

CHAPTER

62

Canavan Disease

Reuben Matalon and Kimberlee Michals Matalon†*

*The University of Texas Medical Branch (UTMB), Galveston, TX, USA

†The University of Houston, Houston, TX, USA

INTRODUCTION

Spongy degeneration of the brain, Canavan disease, is an autosomal recessive leukodystrophy prevalent among individuals of Ashkenazi Jewish extraction. Clinically, severe mental retardation, developmental delays, and early death characterize Canavan disease. Deficiency of the enzyme aspartoacylase, causes excessive amounts of *N*-acetylaspartic acid in the urine, brain and body fluids.[1,2] The high levels of *N*-acetylaspartic acid in the urine and body fluids are diagnostic for Canavan disease. Mild or juvenile cases of Canavan disease with mild elevation of *N*-acetylaspartic acid and developmental delay have also been found. The gene for aspartoacylase has been cloned and mutations identified. Ashkenazi Jewish populations have two mutations in 98% of the cases while other ethnic groups have more diverse mutations.

HISTORY

Spongy degeneration of white matter of the brain was described more than 70 years ago.[3–5] Canavan disease had an autosomal recessive mode of inheritance with high prevalence among Ashkenazi Jews.[6–11] Elevated *N*-acetylaspartic acid in the urine and the deficiency of aspartoacylase in cultured skin fibroblasts in Canavan disease were described by Matalon et al. in 1988.[1,2] *N*-acetylaspartic aciduria and the deficiency of aspartoacylase in Canavan disease have since been confirmed by many investigators.[12–17]

BASIC DEFECT

Aspartoacylase hydrolyzes *N*-acetylaspartic acid to acetate and aspartate.[18] Canavan disease is caused by aspartoacylase deficiency with abnormal accumulation of *N*-acetylaspartic acid in the brain and body fluids. Most vertebrates have a high concentration of *N*-acetylaspartic acid in the brain. The concentration in the human brain is 6–7 μmol per gram tissue, with the highest concentration in the cerebral cortex and the lowest in the medulla.[19–21] In spite of its abundance in the brain, the function of *N*-acetylaspartic acid is not clear, and it has been referred to as an inert compound.[22,23] Studies suggest that *N*-acetylaspartic acid is essential for myelin synthesis. The hydrolyzed acetate, in a series of reactions, is used for conversion of lignoceric acid to cerebronic acid, a component of myelin.[24–26] Studies with the knockout mouse for Canavan disease have shown low acetate concentration in the brain.[27] The high concentration of *N*-acetylaspartic acid in the brain may also serve as a molecular water pump. The elevated *N*-acetylaspartic acid in Canavan disease may lead to spongy degeneration of the brain.[28]

Rosenberg's Molecular and Genetic Basis of Neurological and Psychiatric Disease
http://dx.doi.org/10.1016/10.1016/B978-0-12-410529-4.00062-0

CLINICAL FEATURES

Adachi et al.[10] reported three clinical variants of Canavan disease: a congenital form in which the disease is apparent at birth or shortly thereafter; an infantile form, which is most common, in which symptoms manifest after the first 6 months of life; and a juvenile form, in which the disease manifests after the first 5 years of life. Since the discovery of the enzyme deficiency, milder forms of Canavan disease have been documented, indicating that aspartoacylase deficiency leads to a spectrum of clinical and biochemical manifestations.[29–31]

Infantile Form

Infants with Canavan disease do not present the clinical features of the disease at birth. Delayed development, however, may be noted at about 3 months of age. An important characteristic of Canavan disease is persistent hypotonia and head lag. Macrocephaly is another symptom, although head circumference in early infancy may not be remarkably increased and may remain in the upper limits of normal. The triad of hypotonia, head lag, and macrocephaly should suggest Canavan disease in the differential diagnosis. As the infant grows older, hypotonia gives in to spasticity, so some of these children are labeled as having cerebral palsy. Head control remains poor. Seizures, although not a prominent feature, may develop in the second year of life, and optic atrophy will become noticeable. Patients with Canavan disease are irritable, with sleep disturbance and often have fevers of unknown origin. As the disease progresses, problems with gastroesophageal reflux become prominent, leading to feeding difficulties and poor weight gain. Swallowing deteriorates, and many of these children require nasogastric feeding or permanent feeding gastrostomy. Many patients with Canavan disease die in the first decade of life. Improved medical and nursing care have allowed more of these children to survive into the second and third decade of life.[30,31] Unfortunately, these children remain developmentally delayed, unable to walk or talk, in spite of extensive physical and speech therapy. However, they seem to recognize and interact with their parents, smiling and reaching for objects.

Mild/Juvenile Form

Patients with mild/juvenile forms of Canavan disease have been confirmed by biochemical and genetic analysis.[31–35] Some of these patients were found due to mildly elevated urine *N*-acetylaspartic acid during a routine evaluation for developmental delay. These patients do not have sponginess of the brain on magnetic resonance imaging (MRI) but show increased signal intensity in the basal ganglia, bilateral involvement of the globus pallidus or changes in the lentiform and caudate nuclei.[34,35] Most of the children with mild forms of Canavan disease have normal head size, although one case had a large head and retinitis pigmentosa.[32] In spite of developmental delay most of these children attend normal school.

DIAGNOSIS

Excessive levels of *N*-acetylaspartic acid in the urine is the most reliable test for Canavan disease since 1988.[29,30] The levels of *N*-acetylaspartic acid in normal urine, as determined by gas chromatography–mass spectroscopy, are less than 10 μmol per mmol creatinine, whereas in Canavan patients, they are in the range of 3000±1800 μmol per mmol creatinine. High levels of *N*-acetylaspartic acid in plasma, cerebrospinal fluid, and brain tissue can also be detected. Patients with mild Canavan disease have levels of *N*-acetylaspartic acid four or more times the normal.[29] The enzyme, aspartoacylase, is not detected in blood but can be assayed in cultured skin fibroblasts. The enzyme activity in cultured fibroblasts may vary depending on culture conditions, therefore the enzyme activity may not be consistent and the assay is unreliable for the diagnosis. Urine elevation of *N*-acetylaspartic acid remains the diagnostic test for Canavan disease. Molecular testing for mutation analysis or deletion or duplication, can be determined after the diagnosis is confirmed by the elevated urine *N*-acetylaspartic acid.

Computed tomography (CT) or MRI of the brain reveal diffuse white matter degeneration in Canavan disease.[36–38] The involvement is primarily in the cerebral hemispheres, with less involvement in the cerebellum and brainstem. The MRI of a patient with Canavan disease at 9 months of age was thought to be nondiagnostic (Figure 62.1A), whereas at age 2 years, MRI shows the progression of the disease with severe white matter degeneration (Figure 62.1B). Nuclear magnetic resonance (NMR) spectroscopy of the Canavan brain as compared with the normal brain reveals increased signal of *N*-acetylaspartic acid in Canavan disease.[36]

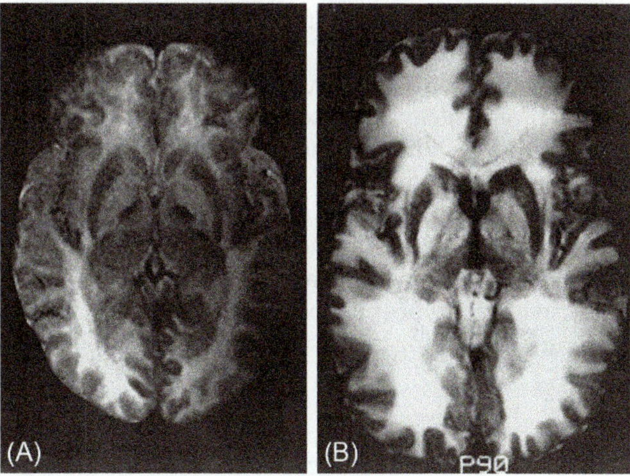

FIGURE 62.1 (A) Axial T2-weighted MRI scan taken of a patient with Canavan disease at 9 months of age was initially interpreted as normal. Closer evaluation, however, reveals moderate expansion of medullary and subcortical white matter. (B) Axial T2-weighted MRI scan taken of a patient with Canavan disease at age 2 years reveals extensive thickening of the white matter radiation.

DIFFERENTIAL DIAGNOSIS

Macrocephaly with white matter disease, characteristic for Canavan disease, can also be found in Alexander disease, benign megalencephaly with leukodystrophy, Krabbe disease, adrenoleukodystrophy, and vacuolating leukodystrophy. Alexander disease is caused by a defect in the synthesis of glial fibrillary acidic protein (GFAP), and this diagnosis can be ruled out by molecular diagnosis on blood lymphocytes. Canavan disease is not a rapidly progressive disease, and often such children are diagnosed with static encephalopathy or cerebral palsy. The child remains relatively stable with severe developmental delays and no obvious deterioration. This leads to a delay in diagnosis in a disease that can be prevented by early diagnosis and counseling.

EPIDEMIOLOGY

Canavan disease is panethnic; however, it is most prevalent among Jews of East European ancestry. Two mutations account for 98% of Canavan disease among Ashkenazi Jews. Screening of healthy Jewish individuals for two mutations revealed a carrier rate of 1:40.[39,40] A carrier rate of 1:58 was found screening the Ashkenazi Jewish population for the three mutations.[41] Other studies documented rates of 1:57 and 1:82 in other Ashkenazi populations.[42,43] Based on these screening results the incidence of Canavan disease is 1:6400 to 1:13,500. Screening for the three most common mutations detects 99% of the carriers among Ashkenazi Jews. Screening is recommended for Ashkenazi Jews of reproductive age because of the high incidence in this population.[44]

Non-Jewish patients have mutations that are more diverse. Canavan disease has been reported among European, Middle-Eastern, Turkish, Gypsy, African American, and Japanese populations. The incidence in the non-Jewish population is about 1:100,000 births.

MOLECULAR BASIS

The gene encoding for aspartoacylase has been isolated, and mutations in patients with Canavan disease have been identified.[45] The human aspartoacylase gene is 20 kb, and the cDNA is comprised of 6 exons. The gene has been mapped to chromosome 17 p-ter. The cDNA for aspartoacylase codes for a 313-amino acid protein with a molecular mass of 36 kDa. The aspartoacylase gene is conserved among species. The coding sequence of the bovine and mouse cDNA shows 92% and 86% identity with the human cDNA, respectively.[46]

Two predominant mutations occur in Jewish patients. The first is a missense mutation on exon 6, p.Glu285Ala, in 86% of the alleles. The other is a nonsense mutation on exon 5, p.Try231X, (termination codon) in 13.6% of the alleles. More than 98% of all Jewish patients with Canavan disease have these two mutations.[47]

In non-Jewish patients the mutations for Canavan disease are more diverse with more than 70 mutations reported.[47–52] The most common mutation in non-Jewish patients is a missense mutation on exon 6, p.Ala305Glu. Sequence analysis is often needed to determine mutations in non-Jewish families. Many case reports document a mutation found in a single family. The diversity of mutations in the non-Jewish patients makes it difficult to offer screening programs.

Patients with mild or juvenile forms of Canavan disease have been compound heterozygotes with mild mutation on one allele and a severe mutation on the other allele. Mild mutations include p.Tyr288Cys, p.Arg71His, p.Pro257Arg, p.Ile143Thr and p.Tyr231Cys.[31]

The phenotype for p.Tyr231X mutation which has no enzyme activity and p.Glu285Ala mutation that has some residual activity is clinically undistinguishable, both mutations lead to a severe phenotype. The common non-Jewish mutation p.Ala305Glu has no residual activity and also leads to a severe phenotype. The mutations associated with the milder phenotypes all have residual enzyme activity that leads to a lower level of *N*-acetylaspartic acid in brain and urine. Genotype–phenotype correlation and aspartoacylase (ASPA) expression indicate that such studies may aid in understanding the disease severity.[53–55]

PREVENTION/PRENATAL DIAGNOSIS

Carrier detection among Jewish individuals can be determined using DNA analysis. The Committee on Genetics of the American College of Obstetrician Gynecologists recommends that Jewish couples be screened for carrier status of Canavan disease.[44] If both parents carry the gene for Canavan, the risk of an affected baby is 25%. Prenatal diagnosis on chorionic villus samples (CVS) or amniocentesis can be offered using DNA analysis. The diagnosis can be readily made in patients of Jewish ancestry because of the two known Jewish mutations. Among non-Jewish individuals, attempts should be made to identify the mutation in the proband so that molecular prenatal diagnosis can be used for future pregnancies.[56]

Other methods of prenatal diagnosis include determination of *N*-acetylaspartic acid in amniotic fluid, which will be increased in an affected pregnancy.[57,58] A specialized center is needed for such determination.

Preimplantation genetics using a single cell to rule out Canavan mutations is new in practice in specialized centers and the Canavan-free fertilized egg can be implanted.[59]

MANAGEMENT

Treatment for the infantile form of Canavan disease is supportive.[29] After the diagnosis, the patients should have developmental and nutritional assessment. Follow-up should be two or three times a year. Canavan children benefit from early intervention and special education programs to maximize abilities. Therapy can enhance communication skills. Physical therapy can minimize contractures, prevent decubiti ulcers, and determine appropriate seating posture to improve nutritional intake and hydration. Seizures and swallowing difficulties may cause aspiration and indicate need for G-tube feedings or a feeding gastrostomy. Seizures should be treated with antiepileptic drugs. Diamox seems to reduce increased intracranial pressure. Botox injections may be used to relieve spasticity. The mild/juvenile form of Canavan disease patients may need special programs for developmental delay, learning difficulties or speech therapy. These children can be monitored once or twice a year.

THERAPY

A knockout mouse with a phenotype similar to that of human Canavan disease, a tremor rat with aspartoacylase deficiency, spongy degeneration of the brain, and increased *N*-acetylaspartic acid, and human Canavan disease patients have been used to investigate pathophysiology, gene therapy, stem cell therapy, enzyme replacement and other modes of treatment.[60–63]

The tremor rat and Canavan disease patients were given lithium citrate to determine if it would reduce osmotic pressure that may be the cause for the spongy degeneration of the brain. The tremor rat showed a mild reduction of *N*-acetylaspartic acid in the brain and the Canavan disease children showed some reduction of *N*-acetylaspartic acid in the basal ganglia and frontal white matter.[64–66] However, there were no clinical improvements and lithium citrate is not recommended for treatment.

Glycerol triacetate supplementation has been used in the knockout mouse, tremor rat, and patients with Canavan disease to determine if the acetate will provide a substrate for myelin synthesis. The knockout mouse had some improved MR imaging with decreased sponginess; however, there was no improvement in the clinical symptoms.

The tremor rat showed an increase in myelin, galactocerebroside, and a modest reduction in brain vacuolation and improved motor performance.[67] Glycerol triacetate given in low and high doses in Canavan disease patients had no clinical improvement.[68,69] Glycerol triacetate is not recommended for treatment of Canavan disease.

Gene therapy trials began in 2000 when two children with Canavan disease had an adeno-associated virus 2 (AAV2) injected into the brain as a vector for aspartoacylase and showed no toxicity.[70,71] Another trial with aspartoacylase AAV2 injected into the brains of 10 patients with Canavan disease showed neutralizing antibodies to AAV2 in 3 out of 10 patients.[72] A 5-year follow-up of gene therapy in Canavan disease showed lack of long-term adverse events with some decrease in the elevation of *N*-acetylaspartic acid in the brain.[73] These studies failed to improve the clinical symptoms.

Enzyme-replacement therapy using native ASPA and pegylated ASPA (i.e., ASPA in which covalent attachment of polyethylene glycol polymer chains masks the enzyme from the host immune system allowing for longer circulation and less renal clearance) were injected into the peritoneum of Canavan disease mice. Results show that the enzyme passed the blood–brain barrier and there was a decrease of *N*-acetylaspartic acid in the brain.[74] In another trial recombinant *ASPA* (rASPA) was injected intraperitoneally with and without hyaluronidase. The brain showed a 25% increase in ASPA activity above baseline with *ASPA* alone, and 50% above baseline with *ASPA* and hyaluronidase.[75]

Stem cell therapy in knockout Canavan disease mice done in collaboration with Genzyme Corporation showed that the stem cells produced some oligodendrocytes but not enough to make myelin.[76] Gene therapy with AAV2 injected intrathecally resulted in localized improvement of sponginess but it did not spread to the entire brain.[77] Novel rAAV serotypes AAV8, 9, and 10 have been found to cross the blood–brain barrier and transduce the central nervous system (CNS). After a single intravenous injection of AAV8 with *ASPA* in knockout Canavan disease mice, the mice improved and lived for 2 years instead of dying at 1 month of age.[78,79]

The supplementation with lithium citrate, glycerol triacetate and the trials with gene therapy, stem cell therapy, and enzyme replacement indicate a continuous effort toward the goal of finding a cure for Canavan disease. The spread of AAV8 and the long-term expression in the Canavan mice hold the promise of successful gene therapy for Canavan disease.

References

1. Matalon R, Michals K, Sebasta D, et al. Aspartoacylase deficiency and *N*-acetylaspartic aciduria in patients with Canavan disease. *Am J Med Genet*. 1988;29:463–471.
2. Matalon R, Kaul RK, Casanova J, et al. Aspartoacylase deficiency: the enzyme defect in Canavan disease. *J Inherit Metab Dis*. 1989;12:329–331.
3. Canavan MM. Schilder's encephalitis periaxialis diffusa. *Arch Neurol Psychiatr*. 1931;25:299–308.
4. Globus JH, Strauss I. Progressive degenerative subcortical encephalopathy: Schilder's disease. *Arch Neurol Psychiatr*. 1928;20:1190–1228.
5. van Bogaert L, Bertrand I. Sur une idiotie familiale avec degerescence sponglieuse de neuraxe (note preliminaire). *Acta Neurol Belg*. 1949;49:572–587.
6. Banker BQ, Robertson JJ, Victor M. Spongy degeneration of the central nervous system in infancy. *Neurology*. 1964;14:981–1001.
7. Buchanan DS, Davis RL. Spongy degeneration of the nervous system: a report of 4 cases with a review of the literature. *Neurology*. 1965;15:207–222.
8. Sacks O, Brown WJ, Aguilar MJ. Spongy degeneration of white matter: Canavan's sclerosis. *Neurology*. 1965;15:165–171.
9. Gamberti P, Mellman WJ, Gonatas NK. Familial spongy degeneration of the central nervous system: van Bogaert–Bertrand disease. *Acta Neuropathol*. 1969;12:103–115.
10. Adachi M, Schneck L, Cazara J, Volk BW. Spongy degeneration of the central nervous system: van Bogaert and Bertrand type; Canavan's disease. *Hum Pathol*. 1973;4:331–346.
11. Banker BQ, Victor H. Spongy degeneration of infancy. In: Goodman RM, Motulsky AG, eds. Genetic Disease among Ashkenazi Jews. New York: Raven Press; 1979:201–217.
12. Kvittingen EA, Guldal G, Borsting S, et al. *N*-acetylaspartic aciduria in a child with a progressive cerebral atrophy. *Clin Chim Acta*. 1986;158:217–227.
13. Hagenfeldt L, Bollgren I, Venizelos N. *N*-acetylaspartic aciduria due to aspartoacylase deficiency: a new etiology of childhood leukodystrophy. *J Inherit Metab Dis*. 1987;10:135–141.
14. Divry P, Viamey-Liaund C, Gay C, et al. *N*-acetylaspartic aciduria: report of three cases in children with a neurological syndrome associating macrocephaly and leukodystrophy. *J Inherit Metab Dis*. 1988;11:307–308.
15. Echeme B, Divry P, Viamey-Liaud C. Spongy degeneration of the neuraxis (Canavan–van Bogaert's disease) and *N*-acetylaspartic aciduria. *Neuropediatrics*. 1989;20:179–181.
16. de Coo IFM, Bakkeren JAJM, Gabreels FJM. Canavan disease: value of *N*-acetylaspartic aciduria? *Neuropediatrics*. 1991;22:3.
17. Michelakakis H, Giouroukos S, Divry P, et al. Canavan disease: findings in four new cases. *J Inherit Metab Dis*. 1991;14:267–268.
18. Birnbaum SM, Levintow L, Kingsley RB, Greenstein JP. Specificity of amino acid acylases. *J Biol Chem*. 1952;194:455–462.
19. Birken DL, Oldendorf WH. *N*-Acetyl-L-aspartic acid: a literature review of a compound prominent in ^{1}H-NMR spectroscopic studies of brain. *Neurosci Biobehav Rev*. 1989;13:23–31.
20. Miyake M, Kakimoto Y, Sorimachi M. A gas chromatographic method for the determination of *N*-acetyl-L-aspartic acid, *N*-acetylalpha-aspartylglutamic acid and beta-citryl-L-glutamic acid and their distributions in the brain and other organs of various species of animals. *J Neurochem*. 1980;36:804–810.

21. Kaul RK, Casanova J, Johnson A, et al. Purification, characterization and localization of aspartoacylase from bovine brain. *J Neurochem.* 1991;56:129–135.
22. Jacobson HB. Studies on the role of *N*-acetylaspartic acid on mammalian brain. *J Gen Physiol.* 1957;43:323–333.
23. McIntosh JM, Cooper JR. Studies on the function of *N*-acetylaspartic acid in the brain. *J Neurochem.* 1965;12:825–835.
24. Shigematsu H, Okamura N, Shimeno H, et al. Purification and characterization of the heat stable factors essential for conversion of lignoceric acid to cerebronic acid and glutamic acid: identification of *N*-acetyl-L-aspartic acid. *J Neurochem.* 1983;40:814–820.
25. Miyake M, Kakimoto Y. Developmental changes of *N*-acetyl-L-aspartic acid, *N*-acetyl-alphaaspartylglutamic acid and beta-citryl-L-glutamic acid in different brain regions and spinal cords of rat and guinea pig. *J Neurochem.* 1981;37:1064–1067.
26. Chakraborty G, Mekala P, Yahya D, et al. Intraneuronal *N*-acetylaspartate supplies acetyl groups for myelin lipid synthesis: evidence for myelin-associated aspartoacylase. *J Neurochem.* 2001;78:736–745.
27. Matalon R, Michals-Matalon K, Surendran S, Tyring SK. Canavan disease: studies on the knockout mouse. *Adv Exp Med Biol.* 2006;576:77–93.
28. Baslow MH. Evidence supporting a role of *N*-acetyl-L-aspartate as a molecular water pump in myelinated neurons in the central nervous system. An analytical review. *Neurochem Int.* 2002;40:295–300.
29. Matalon R, Michals-Matalon K. Canavan disease (August 2011). In: Pagon RA, Bird TD, Colan CR, Stephens K, eds. GeneReviews. Medical Genetics Information Resource. Seattle: University of Washington; 1997–2011. Available at: http://www.genetests.org.
30. Michals K, Matalon R. Canavan disease. In: Raymond GV, Eichler F, Fatemi A, Naidu S, eds. Leukodystrophies. London: Mac Keith Press; 2011:156–169.
31. Matalon K, Matalon R. Canavan disease. *Orphanet J Rare Dis.* December, 2012; http://www.ojrd.com.
32. Surendran S, Bamforth FJ, Chan A, Tyring SK, Goodman SI, Matalon R. Mild elevation of *N*-acetylaspartic acid and macrocephaly: diagnostic problem. *J Child Neurol.* 2003;18:809–812.
33. Janson CG, Kolodny EH, Zeng BJ, et al. Mild-onset presentation of Canavan's disease associated with novel G212A point mutation in aspartoacylase gene. *Ann Neurol.* 2006;59:428–431.
34. Toft PB, Geib-Holtorff R, Rolland MO, et al. Magnetic resonance imaging in juvenile Canavan disease. *Eur J Pediatr.* 1993;152:750–753.
35. Yalcinkaya C, Benbir G, Salomons GS, et al. Atypical MRI findings in Canavan disease: a patient with a mild course. *Neuropediatrics.* 2005;36:336–339.
36. Matalon R, Michals K, Kaul R, Mafee M. Spongy degeneration of the brain: Canavan disease. *Int Pediatrics.* 1990;5:121–124.
37. Brismar J, Brismar G, Gascon G, Ozand P. Canavan disease: CT and MR imaging of the brain. *AJNR Am J Neuroradiol.* 1990;11:805–810.
38. Rushton AR, Shaywitz BA, Dumen CC, et al. Computerized tomography in the diagnosis of Canavan's disease. *Ann Neurol.* 1981;10:57–60.
39. Matalon R, Kaul R, Michals K. Carrier rate of Canavan disease among Ashkenazi Jewish individuals. *Am J Hum Genet.* 1994;55:A908.
40. Kronn D, Oddoux C, Phillips J, Ostrer H. Prevalence of Canavan disease heterozygotes in the New York metropolitan Ashkenazi Jewish population. *Am J Hum Genet.* 1995;5:1250–1252.
41. Sugarman EA, Allitto BA. Carrier testing for seven diseases common in the Ashkenazi Jewish population: implications for counseling and testing. *Obstet Gynecol.* 2001;5:38–39.
42. Fares F, Badarneh K, Abosaleh M, Harari-Shaham A, Diukman R, David M. Carrier frequency of autosomal-recessive disorders in the Ashkenazi Jewish population: should the rationale for mutation choice for screening be reevaluated? *Prenat Diagn.* 2008;28:236–241.
43. Feigenbaum A, Moore R, Clarke J, et al. Canavan disease: carrier-frequency determination in the Ashkenazi Jewish population and development of a novel molecular diagnostic assay. *Am J Med Genet A.* 2004;124A:142–147.
44. ACOG committee opinion. Screening for Canavan disease. Committee on Genetics. American College of Obstetrician Gynecologists, Number 212, November 1998. *Int J Gynaecol Obstet.* 1999;1:91–92.
45. Kaul R, Gao GP, Balamurugan K, Matalon R. Human aspartoacylase cDNA and missense mutation in Canavan disease. *Nat Genet.* 1993;5:118–123.
46. Kaul R, Balamurugan K, Gao GP, Matalon R. Canavan disease: genomic organization and localization of human ASPA to 17p13-ter and conservation of the ASPA gene during evolution. *Genomics.* 1994;21:364–370.
47. Kaul R, Gao GP, Aloya M, et al. Canavan disease: mutations among Jewish and non-Jewish patients. *Am J Hum Genet.* 1994;55:34–41.
48. Elpeleg ON, Shaag A. The spectrum of mutations of the aspartoacylase gene in Canavan disease in non-Jewish patients. *J Inherit Metab Dis.* 1999;4:531–534.
49. Kaul R, Gao GP, Matalon R, et al. Identification and expression of eight novel mutations among non-Jewish patients with Canavan disease. *Am J Hum Genet.* 1996;59:95–102.
50. Shaag A, Anikster Y, Christensen E, et al. The molecular basis of Canavan (aspartoacylase deficiency) disease in European non-Jewish patients. *Am J Hum Genet.* 1995;57:572–580.
51. Sistermans EA, de Coo RF, van Beerendonk HM, et al. Mutation detection in the aspartoacylase gene in 17 patients with Canavan disease: four new mutations in the non-Jewish population. *Eur J Hum Genet.* 2000;7:557–560.
52. Tahmaz FE, Sam S, Hoganson GE, Quan F. A partial deletion of the aspartoacylase gene is the cause of Canavan disease family from Mexico. *J Med Genet.* 2001;9:38.
53. Sommer A, Sass JO. Expression of aspartoacylase (ASPA) and Canavan disease. *Gene.* 2012;505(2):206–210.
54. Zano S, Wijayasinghe YS, Malik R, Smith J, Viola RE. Relationship between enzyme properties and disease progression in Canavan disease. *J Inherit Metab Dis.* 2013;36(1):159–160.
55. Tacke U, Olbrich H, Sass JO, et al. Possible genotype–phenotype correlations in children with mild clinical course of Canavan disease. *Neuropediatrics.* 2005;36:252–255.
56. Matalon R, Michals-Matalon K. Prenatal diagnosis of Canavan disease. *Prenat Diagn.* 1999;7:669–670.
57. Bennett MJ, Gibson KM, Sherwood WG, et al. Reliable prenatal diagnosis of Canavan disease (aspartoacylase deficiency): comparison of enzymatic and metabolite analysis. *J Inherit Metab Dis.* 1993;16:831–836.
58. Kelley RI. Prenatal diagnosis of Canavan disease by measurement of *N*-acetyl-L-aspartate in amniotic fluid. *J Inherit Metab Dis.* 1993;16:918–919.
59. Yaron Y, Schwartz T, Mey-Raz N, Amit A, Lessing JB, Malcov M. Preimplantation genetic diagnosis of Canavan disease. *Fetal Diagn Ther.* 2005;20:465–468.

60. Matalon R, Rady PL, Platt KA, et al. Knock-out mouse for Canavan disease: a model for gene transfer to the central nervous system. *J Gene Med*. 2000;3:165–175.
61. Kumar S, Biancotti JC, Matalon R, de Vellis J. Lack of aspartoacylase activity disrupts survival and differentiation of neural progenitors and oligodendrocytes in a mouse model of Canavan disease. *J Neuroaci Res*. 2009;87:3415–3427.
62. Mattan NS, Ghiani CA, Lloyd M, et al. Aspartoacylase deficiency affects early postnatal development of oligodendrocytes and myelination. *Neurobiol Dis*. 2010;40:432–443.
63. Kitada K, Akimitsu T, Shigematsi Y, et al. Accumulation of *N*-acetyl-L-aspartate in the brain of the tremor rat, a mutant exhibiting absence-like seizure and spongiform degeneration in the central nervous system. *J Neurochem*. 2000;6:2512–2519.
64. Assadi M, Janson C, Wang DJ, et al. Lithium citrate reduces excessive intra-cerebral *N*-acetyl aspartate in Canavan disease. *Eur J Paediatr Neurol*. 2010;14:354–359.
65. Baslow MH, Kitada K, Suckow RF, Hungund BL, Serikawa T. The effects of lithium chloride and other substances on levels of brain *N*-acetyl-L-aspartic acid in Canavan disease-like rats. *Neurochem Res*. 2002;27:403–406.
66. Baslow MH, Guilfoyle DN. Canavan disease, a rare early-onset human spongiform leukodystroph: insights into its genesis and possible clinical interventions. *Biochimie*. 2013;95(4):946–956.
67. Arun P, Madhavarao CN, Moffett JR, et al. Metabolic acetate therapy improves phenotype in the tremor rat model of Canavan disease. *J Inherit Metab Dis*. 2010;33:195–210.
68. Madhavarao CN, Arun P, Anikster Y, et al. Glyceryl triacetate for Canavan disease: a low-dose trial in infants and evaluation of a higher dose for toxicity in the tremor rat model. *J Inherit Metab Dis*. 2009;32:640–650.
69. Segel R, Anikster Y, Zevin S, et al. A safety trial of high dose glyceryl triacetate for Canavan disease. *Mol Genet Metab*. 2011;103(3):203–206.
70. Leone P, Janson CG, Bilaniuk L, et al. Aspartoacylase gene transfer to the mammalian central nervous system with therapeutic implications for Canavan disease. *Ann Neurol*. 2000;1:27–38.
71. Janson C, McPhee S, Bilaniuk L, et al. Clinical protocol. Gene therapy of Canavan disease: AAV-2 vector for neurosurgical delivery of aspartoacylase gene (ASPA) to the human brain. *Hum Gene Ther*. 2002;13:1391–1412.
72. McPhee SW, Janson CG, Li C, et al. Immune responses to AAV in a phase I study for Canavan disease. *J Gene Med*. 2006;8:577–588.
73. Leone P, Shera D, McPhee SW, et al. Long-term follow-up after gene therapy for Canavan disease. *Sci Transl Med*. 2012;4:165ra163.
74. Zano S, Malik R, Szucs S, Matalon R, Viola RE. Modification of aspartoacylase for potential use in enzyme replacement therapy for the treatment of Canavan disease. *Mol Genet Metab*. 2011;102:176–180.
75. Matalon R, Bhatia G, Suzucs S, Michals-Matalon K, Tyring S, Grady J. Aspartoacylase entry to the brain of Canavan mouse with hyaluronidase? *J Inherit Metab Dis*. 2008;31:28.
76. Surendran S, Shihabuddin LS, Clarke J, et al. Mouse neural progenitor cells differentiate into oligodendrocytes in the brain of a knockout mouse model of Canavan disease. *Brain Res Dev Brain Res*. 2004;153:19–27.
77. Matalon R, Surendran S, Rady PL, et al. Adeno-associated virus-mediated aspartoacylase gene transfer to the brain of knockout mouse for Canavan disease. *Mol Ther*. 2003;7:580–587.
78. Ahmed SS, Li H, Cao C, et al. A single intravenous rAVV injection as late as P20 achieves efficacious and sustained CNS gene therapy in Canavan mice. *Mol Ther*. 2013;12:2136–2147.
79. Gao G, Su Q, Michals-Matalon K, Matalon R. Efficacious and safe gene therapy for Canavan disease: a novel approach. *J Inherit Metab Dis*. 2011;34(suppl 3):234.

CHAPTER

63

Neurotransmitter Disorders

Àngels García-Cazorla and Rafael Artuch

Hospital Sant Joan de Déu, Barcelona, Spain; CIBERER (Network for Research in Rare Diseases), Instituto de Salud Carlos III, Madrid, Spain

INTRODUCTION

Disease Characteristics

Neurotransmitter disorders are a group of inherited metabolic diseases that affect the synthesis, catabolism, or transport of the small molecules that neurons use to undergo chemical communication. Chemical transmission is the major means by which nerves communicate with one another in the nervous system. It requires the following steps: 1) synthesis of the neurotransmitter in the presynaptic terminal; 2) storage of the neurotransmitter in secretory vesicles; 3) regulated release of the neurotransmitter on the postsynaptic membrane; and 4) a means for termination of the action of the released neurotransmitter.

Classical neurotransmitter systems involve different biochemical compounds:

- Amino acids: inhibitory (γ-aminobutyric acid [GABA] and glycine) and excitatory (aspartate and glutamate)
- Acetylcholine (cholinergic neurotransmission)
- Biogenic amines: mainly adrenaline, noradrenaline, dopamine, and serotonin
- Purinergic: adenosine and adenosine mono-, di-, and triphosphate
- Other important modulators of the neurotransmission process are neuropeptides, ion channels, and synaptic lipids.

From a clinical point of view, the hallmark manifestations of neurotransmitter disorders can be divided into the following categories:

- Severe early encephalopathies characterized by developmental delay, hypokinetic rigid syndrome, dyskinetic movements, seizures, and dysautonomic signs. These manifestations may be progressive if they are not treated
- Dopaminergic deficiency: parkinsonism (hypokinetic-rigid syndrome), dystonia and other movement disorders, oculogyric crises, and hypersalivation
- Serotonin deficiency: depression, disturbed temperature regulation and intestinal motility, and insomnia
- Hypoglycemia, ptosis, miosis, and low blood pressure are due to noradrenergic deficiency
- Abnormal GABAergic neurotransmission is involved in epilepsy, intellectual disability and developmental delay, episodes of lethargy, abnormal movements, and behavior disturbances
- Pyridoxin and pyridoxal phosphate-dependent seizures.

The inheritance of these disorders is Autosomal recessive for most of them. There are some exceptions: Segawa disease is the dominant form of GTP cyclohydrolase deficiency and monoamine oxidase deficiency-type A is a X-linked disease.

Diagnosis/Testing

The most important diagnostic test is the measurement of neurotransmitter metabolites in the cerebrospinal fluid. As there is a lumbosacral gradient in the concentration of these molecules, measurements should be carried out in a standardized cerebrospinal fluid (CSF) volume fraction in an experienced laboratory. High-pressure liquid chromatography (HPLC) is the most common method used for neurotransmitter detection. Urinary measurements of

http://dx.doi.org/10.1016/10.1016/B978-0-12-410529-4.00063-2

some specific metabolites can be useful only in some particular diseases. Direct enzyme measurement is not always available. Molecular studies confirm the diagnosis.

Current Research

Current research is focused on the establishment of patients' registries within international networks, the development of cellular and animal models, and the study of new treatment options such as genetic therapy.

DISORDERS OF MONOAMINES

These disorders can be divided into defects of synthesis, catabolism, transport, and defects of pterins.

Defects of Synthesis

Tyrosine Hydroxylase Deficiency

CLINICAL FEATURES

Lüdecke et al. identified the disease in 1996 (OMIM *191290). They described a recessively inherited levodopa-responsive parkinsonism in infancy caused by a point mutation (L205P) in the tyrosine hydroxylase (*TH*) gene.[1] TH deficiency is inherited as an autosomal recessive trait. Around 60 cases have been reported worldwide.[1–10] Clinically, it causes a neurological disease with predominant extrapyramidal signs and a variable response to levodopa. Although different phenotypes have been described two main forms can be outlined. Type A: progressive hypokinetic-rigid syndrome (HRS) plus dystonia, with onset in infancy or childhood. Type B: complex encephalopathy with neonatal or early infancy onset (HRS plus developmental delay, a variety of movement disorders and sometimes epilepsy). Nonprogressive mental disability, tremor, chorea, oculogyric crises, ptosis, fluctuation of signs, autonomic dysfunction, and poor response to levodopa, can be present in both groups, but are more likely in type B.[7] Motor and cognitive prognosis is worse in type B. Recently, a family with a new phenotype presenting with levodopa-responsive myoclonus with dystonia, has been described.[8]

MOLECULAR GENETICS

TH converts tyrosine into levodopa, the direct precursor of catecholamine biosynthesis (Figure 63.1), using tetrahydrobiopterin (BH_4) as cofactor. This enzymatic step is rate-limiting in the biosynthesis of the

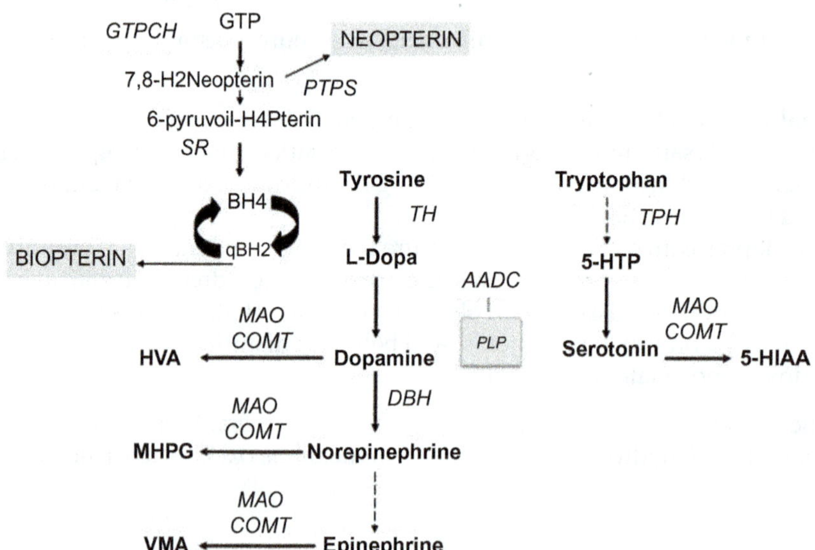

FIGURE 63.1 Metabolic pathways of monoamines. The first step in their formation is catalyzed by amino acid-specific hydroxylases, which require tetrahydrobiopterin (BH_4) as a cofactor. The synthesis of BH_4 comes from guanosine triphosphate (GTP) and it is initiated by the enzyme GTP cyclohydrolase-1 (GTPCH-I), which forms dihydroneopterin triphosphate. Levodopa and 5-hydroxytryptophan (5-HTP) are metabolized by a common B_6-dependent aromatic L-amino acid decarboxylase (AADC) into dopamine and serotonin. Monoamine oxidase A (MAO-A) catabolizes adrenaline and noradrenaline to vanillylmandelic acid (VMA) and 3-methoxy-4-hydroxyphenylethyleneglycol (MHPG). This enzyme is also involved in the catabolism of both dopamine into homovanillic acid (HVA) and serotonin into 5-hydroxyindoleacetic acid (5-HIAA).

catecholamines. The biochemical markers of the disease are low CSF concentrations of homovanillic acid (HVA) and 3-methoxy-4-hydroxyphenylethyleneglycol (MHPG), the catabolites of dopamine (DA) and norepinephrine, respectively, with normal 5-hydroxyindoleacetic acid (5-HIAA) and pterins values. Several mutations including promotor regions and deletions have been described.[5,11] Genotype–phenotype correlations are not well established.

DISEASE MECHANISMS

Chronic low DA brain innervations seem to be the clue to understanding the pathophysiology of this disease. HVA levels tend to be lower in more affected patients.[7] The mouse model showed embryonic lethality with complete DA deficiency. A selective DA deficient mouse was generated by restoring TH function in noradrenergic cells via expression of TH under the noradrenergic-specific dopamine β-hydroxylase promoter. General hypoactivity resulted in early lethality associated with feeding difficulties. Rescue of these animals was achieved with levodopa.[12] Current research is focused on the development of zebrafish and cell models[13,14] to search for new therapeutic approaches in those cases with suboptimal levodopa response.

DIFFERENTIAL DIAGNOSIS

Differential diagnosis should be established with other primary neurotransmitter deficiencies (such as Segawa disease) and levodopa-responsive dystonias, early-onset primary parkinsonism, spastic paraparesis or cerebral palsy, and secondary neurotransmitter deficiencies, especially common in mitochondrial disorders.

TESTING

The most important diagnostic test is the measurement of HVA, MHPG, 5-HIAA, and pterins (neopterin and biopterin) in the CSF.[15] There is no enzyme activity detectable in body fluids, blood cells, and fibroblasts. Molecular analysis of the TH gene is available.

MANAGEMENT

In most cases, TH deficiency can be treated with low-dose levodopa in combination with a levodopa decarboxylase inhibitor. The response is variable, ranging from complete remission (more likely in type A phenotype) to mild improvement. Therapy should be started with low doses. Recommended levodopa initial dose is 1–2 mg per kg per day in four to six divided doses, and only increased over periods of weeks or months, since these patients are especially prone to major side effects even on low doses (mainly dyskinesia).

Aromatic L-Amino Acid Decarboxylase Deficiency

CLINICAL FEATURES

Aromatic L-amino acid decarboxylase (AADC) deficiency (OMIM *107930) was described in 1988.[16] First therapeutic approaches were based on a combination of vitamin B_6, DA agonists, and MAO inhibitors. This autosomal recessive disease has been reported in almost 80 patients worldwide.[17–19] Neonatal symptoms are reported in over half (poor sucking and feeding difficulties, lethargy, increased startle response, hypothermia, ptosis). All patients develop neurological signs within the first 6 months of life. Severe and progressive epileptic encephalopathy may appear.[20,21] Dystonia, ptosis, and autonomic dysfunction (temperature instability with hypothermia, gastrointestinal symptoms, paroxysmal sweating, impaired heart rate, and blood pressure regulation) will also develop. The long-term outcome for the majority of the patients is the development of a severe encephalopathy, cerebral palsy-like associated with movement disorders and dysautonomic signs.

MOLECULAR GENETICS

AADC is implicated in the biosynthesis of catecholamines and of serotonin (Figure 63.1). Its deficiency results in severely reduced concentrations of HVA, 5-HIAA, and MHPG in the CSF, with accumulation of the precursors 3-orthomethyldopa and 5-hydroxytryptophan. The activity of the homodimeric enzyme requires pyridoxal phosphate (PLP) as a cofactor. Diverse mutations in the *AADC* gene have been reported with IVS6+ 4A>T being the most common one.[19] Some mutations such as the pathogenic variants G102S, F309L, S147R and A275T, cause a decreased PLP binding affinity and altered state of the bound coenzyme and of its microenvironment.[22]

DISEASE MECHANISMS

AADC deficiency produces a deficiency of both dopamine and serotonin in the brain, which from early developmental periods can produce devastating consequences in the assembly of neuronal networks. A recent mouse model shows impaired survival. Those who survived grew poorly and exhibited severe dyskinesia.[23] Current research focuses on the effects of some particular mutation variants in AADC function, as well as in the development of gene therapy.[24]

DIFFERENTIAL DIAGNOSIS

Initial suspected diagnoses include other primary neurotransmitter defects, cerebral palsy, epileptic encephalopathy, hyperekplexia, and mitochondrial disorders.

TESTING

Biochemical hallmarks are low concentrations of HVA, 5-HIAA, and MHPG in the CSF and increased concentration of metabolites upstream of the metabolic block: 3-orthomethyldopa, 3-methoxytyrosine, vanillyllactic acid (VLA), and 5-hydroxytryptophan. Often the finding of increased VLA in the urinary organic acid profile is the first important clue towards this diagnosis. In several patients a paradoxical hyperdopaminuria has been noted, probably due to production of dopamine and metabolites in non-neural cells.[25] AADC deficiency can be confirmed by analyzing its enzyme activity in plasma and by sequencing the *AADC* gene.

MANAGEMENT

Treatment in AADC deficiency may be beneficial but the effects are limited and long-term prognosis is poor. The most common therapeutic strategies are cofactor supplementation in the form of vitamin B_6 (PLP), MAO inhibitors (such as tranylcypromine, selegeline or phenelzine), dopamine agonists (pergolide, bromocriptine), high levodopa dose as "substrate therapy," serotoninergic agents (fluoxetine) or combinations of these with anticholinergic drugs (trihexylphenidyl). Transdermal rotigotine has recently been reported to produce some improvements compared with the classical levodopa therapy.[26] Some patients with relatively mild forms improved on a combined therapy with pyridoxine (B_6)/PLP, dopamine agonists, and monoamine oxidase B inhibitors.[19] First assays in gene therapy have shown that this is well tolerated and leads to improved motor function.[24]

Dopamine β-Hydroxylase Deficiency

CLINICAL FEATURES

Dopamine β-hydroxylase (DBH) deficiency (OMIM *609312) is an autosomal recessive disease that was first described in 1987.[27] L-threo-3,4-dihydroxyphenylserine, a synthetic precursor of norepinephrine, has been used as the main therapeutic agent since the first descriptions of the disease. Its clinical hallmark is severe orthostatic hypotension whereas supine blood pressure is normal to low. Most patients complain of exercise intolerance. Symptoms become manifest in early childhood but may worsen in late adolescence. Perinatal hypoglycemia, hypothermia, and hypotension may occur. Smaller brain volumes and temporal attention deficits have recently been described in patients without treatment.[28] Additional symptoms in some patients are ptosis, nasal stuffiness, weak facial musculature, hyperflexible joints, brachydactyly, high palate, sluggish deep tendon reflexes, and mild normocytic anemia.[29] The prognosis on therapy is satisfactory to good.

MOLECULAR GENETICS

DBH converts dopamine into noradrenaline. The patients typically have extremely low plasma noradrenaline and adrenaline values and increased or high–normal concentrations of dopamine. Pathogenic mutations have been found in all known patients with symptomatic DBH deficiency. Interestingly, a noticeable percentage of the population have nearly undetectable DBH activity in plasma with normal concentrations of noradrenaline and adrenaline and without clinical features of DBH deficiency. This is caused by a common allelic variant (–970C-T)[30] at the *DBH* gene.

DISEASE MECHANISMS

DBH is present in sympathetic neurons. A DBH defect should have consequences for (nor)adrenergic neurons as well for the adrenal glands. This defect is characterized by normal parasympathetic and sympathetic cholinergic function but with a lack of sympathetic noradrenergic function. Mice lacking DBH had a very low survival rate due to the absence of noradrenaline *in utero*.[31]

DIFFERENTIAL DIAGNOSIS

Differential diagnosis includes pure autonomic failure/autonomic neuropathy, familial dysautonomia, and Shy–Drager syndrome or central autonomic failure.

TESTING

Tests of autonomic function may provide specific diagnostic information.[29] Orthostatic intraocular pressure and mean arterial blood pressure may be a helpful early screening tool for autonomic dysfunction in children undergoing a ptosis evaluation.[32] The patients have extremely low plasma noradrenaline and adrenaline values and increased or high–normal levels of dopamine. The diagnosis can be confirmed by measuring DBH activity in plasma and by the molecular study.

MANAGEMENT

Therapy with L-dihydroxyphenylserine (L-Dops) is available. Administration of 100–500 mg L-Dops orally twice or three times daily increases blood pressure and restores plasma noradrenaline levels.[29]

Defects Of Catabolism

Monoamine Oxidase-A and -B Deficiency

MAO exists as two isoenzymes (A and B). The genes encoding for both isoenzymes are located on the X chromosome. They metabolize serotonin/5-hydroxytryptamine, catecholamines (dopamine, norepinephrine, epinephrine) and other biogenic amines (Figure 63.1). Patients missing these enzymes have different metabolic and neurologic disturbances. MAOA deficient patients (OMIM *309850) have borderline intellectual deficiency and impaired impulse control. They can manifest aggressive and, sometimes, violent behavior, arson, attempted rape, exhibitionism, and a tendency towards stereotyped hand movements.[33,34] They are deficient in plasma concentrations of catecholamines. MAOB deficient patients (OMIM *309860) exhibit normal clinical characteristics and behavior, as well as normal concentrations of catecholamines. However, patients who are deficient in both MAOA and MAOB due to X-chromosome deletions tend to have the most extreme lower values of these metabolites and may manifest the typical clinical features of Norrie disease[35] (OMIM *300658). These patients manifest congenital blindness, increased excretion of catecholamines, *o*-methylated metabolites, and trace amines, decreased urinary excretion of deaminated metabolites, low levels of MAOB activity in platelets, and increased concentrations of serotonin in platelets. No effective treatment is known at present. Both the borderline mental retardation and the behavioral abnormalities seem to be stable with time.

Defects of Transport

Dopamine Transport Deficiency

Dopamine transport deficiency (DAT1) was described in 2009[36] (OMIM *126455). Homozygous missense *SLC6A3* mutations (p.L368Q and p.P395L), were identified as loss-of-function mutations that severely reduced levels of mature DAT1 (dopamine transporter) while having a differential effect on the apparent binding affinity of dopamine. Eleven patients have been reported so far. Children presented in infancy with either hyperkinesias, parkinsonism, or a mixed hyperkinetic and hypokinetic movement disorder. During childhood, patients developed severe parkinsonism-dystonia associated with eye movement disorders and pyramidal tract signs. All patients had raised HVA/5-HIAA ratio in the CSF. They did not respond to multiple therapeutic agents with no normalization of CSF HVA concentrations.[37]

Vesicular Monoamine Transporter Type 2 Deficiency

Vesicular monoamine transporter type 2 deficiency (OMIM *193001) is a new disorder described in 2013.[38] Eight children of a consanguineous Saudi Arabian family had similar clinical symptoms. They presented an infantile-onset movement disorder (including severe parkinsonism and nonambulation), mood disturbance, autonomic instability, and developmental delay, that was inherited in an autosomal recessive way. Neurotransmitters in the CSF were normal but abnormal in urine (high levels of HVA and 5-HIAA; decreased levels of norepinephrine and dopamine). Genetic studies showed mutation in the *SLC18A2* gene (which encodes vesicular monoamine transporter 2 [VMAT2]). VMAT2 translocates dopamine and serotonin into synaptic vesicles and is essential for motor control, mood, and autonomic function. Treatment with levodopa was associated with worsening, whereas treatment with direct pramipexole (a DA agonist) was followed by near-complete correction of the symptoms.[38]

Defects of Pterins

Guanosine Triphosphate Cyclohydrolase-I Deficiency

Guanosine triphosphate cyclohydrolase-I (GTPCH-I) and sepiapterin reductase (SR) deficiencies are disorders of pterins, in which baseline phenylalanine (Phe) is normal.

CLINICAL FEATURES

Autosomal dominant GTPCH-I deficiency (OMIM *600225) was identified as the cause for dopa-responsive dystonia (DRD) in 1994.[39] Prior to the identification of the gene, this disease was called Segawa syndrome.[40] The clinical spectrum reaches from the classical "DRD" to severe neonatal forms similar to TH deficiency type B. Most patients develop symptoms during the first decade of life. Dystonia in the lower limbs is the most common initial clinical sign. Unless treated with levodopa, the dystonia becomes generalized. Diurnal fluctuation of the symptoms with improvement after sleep is a feature in most patients. Two types of dystonia have been described: the postural type and the action type, with association of vigorous dystonic movements.[41] Paroxysmal exercise-induced dystonia has been described in a family.[42] Adult-onset patients can start with parkinsonism features. Mild cognitive impairment and impulsivity may be associated.[43]

MOLECULAR GENETICS

GTPCH-I deficiency is caused by mutations at the *GCH1* gene, and can be inherited as an autosomal dominant trait with 30% of penetrance. The female:male ratio is approximately 3:1. More than one hundred mutations have been found as a cause for the dominant form of the disease. GTPCH-I is the rate-limiting step in the biosynthesis of BH_4, the essential cofactor of various aromatic amino acid hydroxylases (Figure 63.1) with the highest affinity for TH, but GTPCH deficiency produces defective biosynthesis of both serotonin and catecholamines.

DISEASE MECHANISM

As already mentioned, the enzyme GTPCH-I catalyzes the first and limiting step in the BH_4 biosynthetic pathway, which is now thought to involve up to eight different proteins. GTPCH-I expression is particularly heterogeneous across different populations of human and rodent monoamine-containing neurons.[44] Postural dystonia may be related to low dopamine through the basal ganglia direct pathway and parkinsonism in teens through the indirect pathway. Dysfunction of dopamine in the terminals does not cause degenerative changes.[45]

DIFFERENTIAL DIAGNOSIS

DRD may be caused by TH deficiency, SR deficiency, and the recessive form of GTPCH-I deficiency (which is very rare and also causes hyperphenylalaninemia).

TESTING

Patients with dominant GTPCH-I deficiency have normal Phe levels in body fluids. Some tests may be helpful: measurement of pterines especially in CSF (biopterin and neopterin; both are decreased, from 20–30% of normal levels); measurement of CSF HVA and 5-HIAA (usually, low CSF HVA values in combination with low-normal 5-HIAA concentrations); an oral Phe-loading test (in general, it reveals an increase of phenylalanine/tyrosine ratios during the postloading period); mutation analysis of the *GCH1* gene (presence of deletions is common); measurement of the enzyme activity in fibroblasts. Some patients with the recessive form of the disease may be diagnosed through the hyperphenylalaninemia found on neonatal screening.

MANAGEMENT

The association of levodopa at low doses (4–5mg per kg per day) and a dopa-decarboxylase inhibitor lead to a complete or near-complete response of symptoms. In cases of action dystonia and in late-onset cases, complete normalization is not always reached.[41]

Sepiapterine Reductase Deficiency

Sepiapterine reductase deficiency (SPR; OMIM *182125) is an autosomal recessive disease implicated in the final step of the BH_4 metabolic pathway (Figure 63.1). Around 43 individuals with SRD have been identified.[46] Common, clinical findings are dopa-responsive dystonia, oculogyric crises with diurnal fluctuation and complex motor problems that mimic cerebral palsy. CSF study shows high levels of biopterin and sepiapterin with normal levels of neopterin, and low concentrations of HVA and 5-HIAA (that can be found also in urine). Patients can develop hyperphenylalaninemia under an oral challenge with Phe. Diverse mutations have been described. Treatment with levodopa plus carbidopa and 5-hydroxytryptophan improves dramatically the clinical picture. In these two pterine metabolism defects, BH_4 therapy is not successful, since BH_4 does not easily pass the blood–brain barrier.

DISORDERS OF GABA

Three defects of GABA catabolism have been described: succinic semialdehyde dehydrogenase (SSADH) deficiency (OMIM *610045), GABA transaminase deficiency (OMIM *137150), and homocarnosinosis (OMIM *236130)

(Figure 63.2). We describe here the most common defect: SSADH deficiency. We also consider disorders involving B_6 (pyridoxine and its derivative, pyridoxal phosphate) because they are cofactors required for the synthesis of several neurotransmitters. Defects of GABA receptors or transporters are not included in the chapter.

Succinic Semialdehyde Dehydrogenase Deficiency

CLINICAL FEATURES

Succinic semialdehyde dehydrogenase deficiency (aldhehyde dehydrogenase 5a1, *ALDH5A1* gene) was first reported as γ-hydroxybutyric aciduria (4-hydroxybutyric aciduria) in 1981.[47] It is the most prevalent of the disorders of GABA metabolism. The mode of inheritance is autosomal recessive. Around 450 patients have been reported worldwide. Clinical manifestations include developmental delay and intellectual disability with a marked expressive language disorder. Ataxia, seizures and behavioral disturbances have also been reported.[48] Abnormal brain MRI such as hyperintensity in the globus pallidus, myelination delay, and cerebellar atrophy can be present (Figure 63.3). In general, patients evolve towards a static encephalopathy although behavioral problems or lethargic episodes can appear over time.

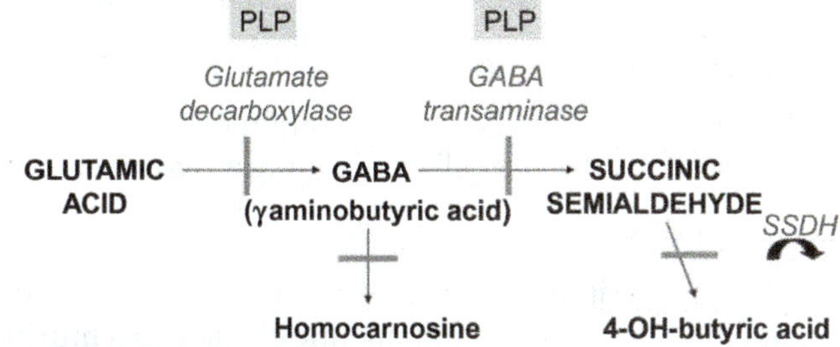

FIGURE 63.2 **Metabolic pathways of γ-aminobutyric acid (GABA).** GABA is formed from glutamic acid by glutamic acid decarboxylase. It is catabolized into succinic acid through the action of two mitochondrial enzymes, GABA transaminase and succinic semialdehyde dehydrogenase (SSDH). Glutamic acid decarboxylase and GABA transaminase have pyridoxal phosphate (PLP) as a coenzyme.

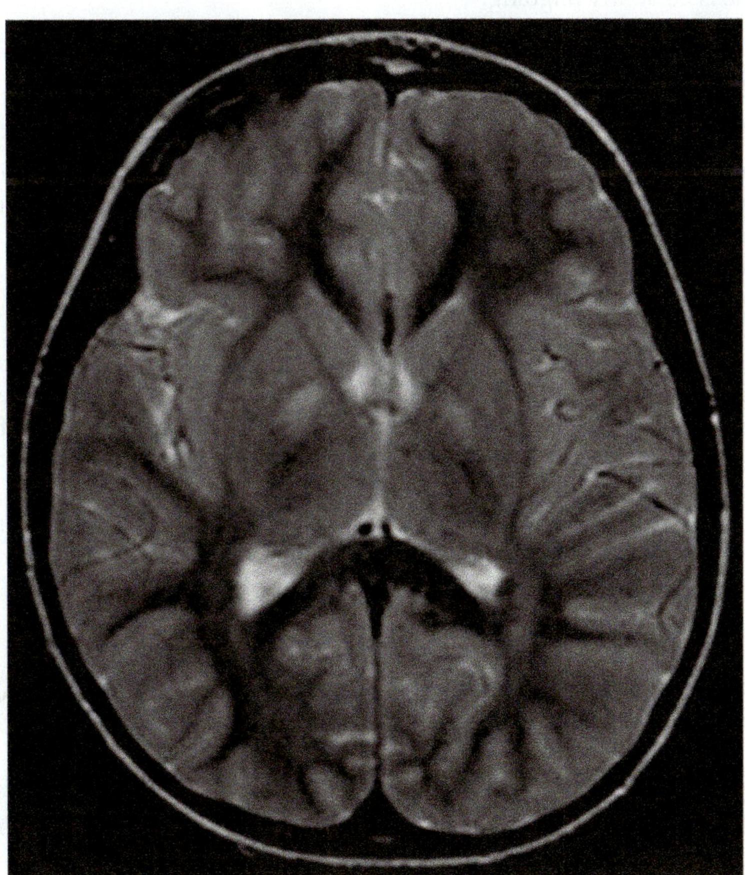

FIGURE 63.3 **Brain MRI of a patient with succinic semialdehyde dehydrogenase (SSDH) deficiency.** High intensity of both nucleus pallidi is observed in a T2 brain MRI image of a patient with SSDH deficiency.

MOLECULAR GENETICS

GABA is formed from glutamic acid by glutamic acid decarboxylase (Figure 63.2). It is catabolized into succinic acid through the sequential action of two mitochondrial enzymes, GABA transaminase and succinic semialdehyde dehydrogenase. SSADH deficiency leads to γ-hydroxybutyrate (GHB) and GABA accumulation. Multiple disease-associated alleles have been identified, but a mutation hotspot has not been detected.

DISEASE MECHANISMS

HyperGABAergic neurotransmission is one of the crucial pathophysiological features reported in both humans and animal models. GHB may exert its action on state and motor control through GABA(B) receptors but also by modulating other neurotransmission pathways. In particular, an inhibition of serotoninergic dorsal raphe neurons and cholinergic laterodorsal tegmentum neurons have been described.[49] GABAergic motor cortex dysfunction in patients[50] points towards reduced long interval intracortical inhibition and reduced cortical silent period. Current research aims to develop new drugs that modulate GABA(B) receptors.

DIFFERENTIAL DIAGNOSIS

Differential diagnosis includes causes of developmental delay or intellectual disability associated with other signs such as lethargic episodes, ataxia, seizures or marked behavioral abnormalities.

TESTING

Diagnosis is made by determination of GHB acid in urine, plasma, and/or CSF. The enzyme deficiency can be demonstrated in white cells employing a fluorimetric assay.[51] Molecular studies confirm the diagnosis.

MANAGEMENT

Therapeutic intervention has traditionally employed vigabatrin with variable results. GABA(B) and GHB receptor antagonists have shown therapeutic efficacy in a murine model,[51] but have yet to be employed clinically in patients. A clinical trial with taurine is currently ongoing.

PYRIDOXINE-RESPONSIVE EPILEPSY

Pyridoxine-responsive epilepsy (PRE) was first reported in 1954.[52] Recently, some quite large series have been reported.[53] Classical presentation refers to onset of refractory convulsions before or within a month of birth with a rapid response to pyridoxine (50–100 mg). Atypical cases may present later (up to the age of 2–3 years) and may require a longer treatment and larger pyridoxine doses before seizures respond.[54] Mutations in the *ALDH7A1* gene (autosomal recessive inheritance) were described as a major cause of PRE in 2006[55] (OMIM *107323). This gene encodes a protein, antiquitin, which is involved in the lysine catabolism pathway in the brain and other organs. This defect leads to a nonenzymatic trapping of PLP by the cyclic form of α-amino adipic semialdehyde, which is causative for the lowered cerebral PLP levels. PLP deficiency leads to decreased GABAergic transmission and seizures. An increased pyridoxine intake is required to compensate for the rate of PLP loss. Biomarkers of the disease are increased levels of plasma and CSF pipecolic acid and α-aminoadipic semialdehyde. Neurotransmitters and GABA values may also be affected as well. In CSF, the presence of an unknown X-peak may be pathognomonic of the disease. The infant should be in an intensive care setting with EEG monitoring before pyridoxine trial since apnea and hypotension may appear. 50–100 mg of intravenous pyridoxine will lead to cessation of seizures within minutes. Permanent control of seizures usually requires a dose of 5–10 mg per kg per day.

PYRIDOXAMINE 5′-PHOSPHATE OXIDASE DEFICIENCY

Pyridoxamine 5′-phosphate oxidase (PNPO) deficiency is an autosomal recessive disease and the first patients were described in 2005[56] (Figure 63.4; OMIM *603287). It causes severe neonatal epileptic encephalopathy. Less than 20 patients have been described so far. This defect leads to an important decrease of pyridoxal phosphate (PLP), the active form of vitamin B_6, and reduced activity of a large number of PLP-dependent enzymes. Main clinical manifestations are fetal seizures, premature birth, seizures resistant to antiepileptic drugs and pyridoxine, hypoglycemia, hyperlactacidemia and a clinical picture suggestive of neonatal hypoxic-encephalopathy. Biomarkers of

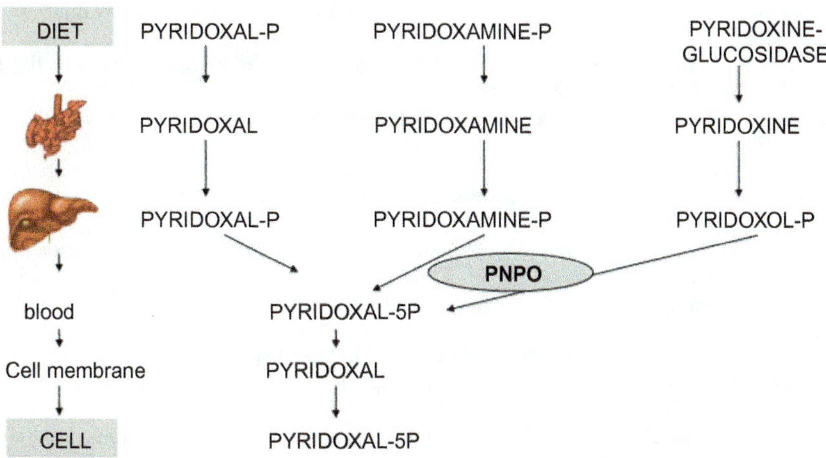

FIGURE 63.4 **Pyridoxal phosphate and pyridoxine.** Conversion of dietary vitamin B_6 (pyridoxine) to intracellular pyridoxal 5′-phosphate cofactor. PNPO: Pyridoxamine 5′-phosphate oxidase.

the disease are low CSF PLP, high 3-methoxy-tyrosine, high plasma glycine and treonine (not constant), HVA and 5-HIAA in the CSF may be low and the urinary excretion of vanillactic acid may be increased. A therapeutic trial of PLP is an important diagnostic tool confirmed by the molecular study of the *PNPO* gene. Treatment should be tried with the infant in an intensive care setting. 50mg of oral PLP has led to cessation of seizures, but was associated with profound hypotonia in some patients. Baseline treatment is usually given at 10mg per kg every 6 hours.[56]

References

1. Lüdecke B, Knappskog PM, Clayton PT, et al. Recessively inherited L-dopa-responsive parkinsonism in infancy caused by a point mutation (L205P) in the tyrosine hydroxylase gene. *Hum Mol Genet*. 1996;5:1023–1028.
2. Hoffmann GF, Assmann B, Bräutigam C, et al. Tyrosine hydroxylase deficiency causes progressive encephalopathy and dopa-nonresponsive dystonia. *Ann Neurol*. 2003;54(suppl 6):S56–S65.
3. Furukawa Y, Kish SJ, Fahn S. Dopa-responsive dystonia due to mild tyrosine hydroxylase deficiency. *Ann Neurol*. 2004;55:147–148.
4. Schiller A, Wevers RA, Steenbergen GC, Blau N, Jung HH. Long-term course of L-dopa-responsive dystonia caused by tyrosine hydroxylase deficiency. *Neurology*. 2004;63:1524–1526.
5. Ribasés M, Serrano M, Fernández-Alvarez E, et al. A homozygous tyrosine hydroxylase gene promoter mutation in a patient with dopa-responsive encephalopathy: clinical, biochemical and genetic analysis. *Mol Genet Metab*. 2007;92:274–277.
6. Pons R, Serrano M, Ormazabal A, et al. Tyrosine hydroxylase deficiency in three Greek patients with a common ancestral mutation. *Mov Disord*. 2010;25:1086–1090.
7. Willemsen MA, Verbeek MM, Kamsteeg EJ, et al. Tyrosine hydroxylase deficiency: a treatable disorder of brain catecholamine biosynthesis. *Brain*. 2010;133:1810–1822.
8. Stamelou M, Mencacci NE, Cordivari C, et al. Myoclonus-dystonia syndrome due to tyrosine hydroxylase deficiency. *Neurology*. 2012;79:435–441.
9. Szentiványi K, Hansíková H, Krijt J, et al. Novel mutations in the tyrosine hydroxylase gene in the first Czech patient with tyrosine hydroxylase deficiency. *Prague Med Rep*. 2012;113:136–146.
10. Chi CS, Lee HF, Tsai CR. Tyrosine hydroxylase deficiency in Taiwanese infants. *Pediatr Neurol*. 2012;46:77–82.
11. Ormazabal A, Serrano M, Garcia-Cazorla A, et al. Deletion in the tyrosine hydroxylase gene in a patient with a mild phenotype. *Mov Disord*. 2011;26:1558–1560.
12. Zhou QV, Palmitter RD. Dopamine-deficient mice are severely hypoactive, adipsic and aphagic. *Cell*. 1995;83:1197–1209.
13. Ren G, Li S, Zhong H, Lin S. Zebra fish tyrosine hydroxylase 2 gene encodes tryptophan hydroxylase. *J Biol Chem*. 2013;288:22451–22459.
14. Díez H, Ortez C, Fernández-Castillo N, et al. Establishing a cell line model to study tyrosine hydroxylase deficiency. Communication presented at the congress "Dopamine 2013", Alghero, Italy, May 24–28, 2013.
15. Marín-Valencia I, Serrano M, Ormazabal A, et al. Biochemical diagnosis of dopaminergic disturbances in paediatric patients: analysis of cerebrospinal fluid homovanillic acid and other biogenic amines. *Clin Biochem*. 2008;41:1306–1315.
16. Hyland K, Surtees RAH, Rodeck C, Clayton PT. Aromatic L-amino acid decarboxylase deficiency: clinical features, diagnosis, and treatment of a new inborn error of neurotransmitter amine synthesis. *Neurology*. 1988;42:1980–1988.
17. Fiumara A, Bräutigam C, Hyland K, et al. Aromatic L-amino acid decarboxylase deficiency with hyperdopaminuria: clinical and laboratory findings in response to different therapies. *Neuropediatrics*. 2002;33:203–208.
18. Swoboda KJ, Saul JP, McKenna CE, Speller NB, Hyland K. Aromatic L-amino acid decarboxylase deficiency. Overview of clinical features and outcomes. *Ann Neurol*. 2003;54 (suppl 6):S49–S55.
19. Brun L, Ngu LH, Keng WT, et al. Clinical and biochemical features of aromatic L-amino acid decarboxylase deficiency. *Neurology*. 2010;75:64–71.
20. Manegold C, Hoffmann GF, Degen I, et al. Aromatic L-amino acid decarboxylase deficiency: clinical features, drug therapy and follow-up. *J Inherit Metab Dis*. 2009;32:371–380.

21. Ito S, Nakayama T, Ide S, et al. Aromatic L-amino acid decarboxylase deficiency associated with epilepsy mimicking non-epileptic involuntary movements. *Dev Med Child Neurol*. 2008;50(11):876–878.
22. Montioli R, Cellini B, Borri Voltattorni C. Molecular insights into the pathogenicity of variants associated with the aromatic amino acid decarboxylase deficiency. *J Inherit Metab Dis*. 2011;34:1213–1224.
23. Lee NC, Shieh YD, Chien YH, et al. Regulation of the dopaminergic system in a murine model of aromatic L-amino acid decarboxylase deficiency. *Neurobiol Dis*. 2013;52:177–190.
24. Hwu WL, Muramatsu S, Tseng SH, et al. Gene therapy for aromatic L-amino acid decarboxylase deficiency. *Sci Transl Med*. 2012;4:134ra61.
25. Wassenberg T, Willemsen MA, Geurtz PB, et al. Urinary dopamine in aromatic L-amino acid decarboxylase deficiency: the unsolved paradox. *Mol Genet Metab*. 2010;101:349–356.
26. Mastrangelo M, Caputi C, Galosi S, Giannini MT, Leuzzi V. Transdermal rotigotine in the treatment of aromatic L-amino acid decarboxylase deficiency. *Mov Disord*. 2013;28:556–557.
27. Man in't Veld AJ, Boomsma F, Moleman P, Schalekamp MA. Congenital dopamine-beta-hydroxylase deficiency. A novel orthostatic syndrome. *Lancet*. 1987;1:183–188.
28. Jepma M, Deinum J, Asplund CL, et al. Neurocognitive function in dopamine-β-hydroxylase deficiency. *Neuropsychopharmacology*. 2011;36:1608–1619.
29. Robertson D, Garland EM. Dopamine beta-hydroxylase deficiency. In: Pagon RA, Bird TC, Dolan CR, Stephens K, eds. *GeneReviews*. Seattle (WA): University of Washington, Seattle; 2005.
30. Deinum J, Steenbergen-Spanjers GC, Jansen M, et al. DBH gene variants that cause low plasma dopamine beta hydroxylase with or without a severe orthostatic syndrome. *J Med Genet*. 2004;41:e38.
31. Thomas SA, Matsumoto AM, Palmiter RD. Noradrenaline is essential for mouse fetal development. *Nature*. 1995;374:643–646.
32. Phillips L, Robertson D, Melson MR, Garland EM, Joos KM. Pediatric ptosis as a sign of treatable autonomic dysfunction. *Am J Ophthalmol*. 2013;156:370–374.
33. Brunner HG, Nelen M, Breakefield XO, Ropers HH, van Oost BA. Abnormal behaviour associated with a point mutation in the structural gene for monoamine oxidase A. *Science*. 1993;262:578–580.
34. Brunner HG, Nelen MR, van Zandvoort P, et al. X-linked borderline mental retardation with prominent behavioural disturbance: phenotype, genetic localisation, and evidence for disturbed monoamine metabolism. *Am J Hum Genet*. 1993;52:1032–1039.
35. O'Leary RE, Shih JC, Hyland K, Kramer N, Asher YJ, Graham Jr. JM. *De novo* microdeletion of Xp11.3 exclusively encompassing the monoamine oxidase A and B genes in a male infant with episodic hypotonia: a genomics approach to personalized medicine. *Eur J Med Genet*. 2012;55:349–353.
36. Kurian MA, Zhen J, Cheng SY, et al. Homozygous loss-of-function mutations in the gene encoding the dopamine transporter are associated with infantile parkinsonism-dystonia. *J Clin Invest*. 2009;119:1595–1603.
37. Kurian MA, Li Y, Zhen J, et al. Clinical and molecular characterisation of hereditary dopamine transporter deficiency syndrome: an observational cohort and experimental study. *Lancet Neurol*. 2011;10:54–62.
38. Rilstone JJ, Alkhater RA, Minassian BA. Brain dopamine-serotonin vesicular transport disease and its treatment. *N Engl J Med*. 2013;368:543–550.
39. Ichinose H, Ohye T, Takahashi E, et al. Hereditary progressive dystonia with marked diurnal fluctuation caused by mutations in the GTP cyclohydrolase I gene. *Nat Genet*. 1994;8:236–242.
40. Segawa M, Hosaka A, Miyagawa F, Nomura Y, Imai H. Hereditary progressive dystonia with marked diurnal fluctuation. *Adv Neurol*. 1976;14:215–233.
41. Segawa M. Autosomal dominant GTP cyclohydrolase I (AD GCH 1) deficiency (Segawa disease, dystonia 5; DYT 5). *Chang Gung Med J*. 2009;32:1–11.
42. Dale RC, Melchers A, Fung VS, Grattan-Smith P, Houlden H, Earl J. Familial paroxysmal exercise-induced dystonia: atypical presentation of autosomal dominant GTP-cyclohydrolase 1 deficiency. *Dev Med Child Neurol*. 2010;52:583–586.
43. López-Laso E, Sánchez-Raya A, Moriana JA, et al. Neuropsychiatric symptoms and intelligence quotient in autosomal dominant Segawa disease. *J Neurol*. 2011;258:2155–2162.
44. Kapatos G. The neurobiology of tetrahydrobiopterin biosynthesis: a model for regulation of GTP cyclohydrolase I gene transcription within nigrostriatal dopamine neurons. *IUBMB Life*. 2013;65:323–333.
45. Segawa M. Dopa-responsive dystonia. *Handb Clin Neurol*. 2011;100:539–557.
46. Friedman J, Roze E, Abdenur JE, et al. Sepiapterin reductase deficiency: a treatable mimic of cerebral palsy. *Ann Neurol*. 2012;71:520–530.
47. Jakobs C, Bojasch M, Monch E, Rating D, Siemes H, Hanefeld F. Urinary excretion of gamma-hydroxybutyric acid in a patient with neurological abnormalities. The probability of a new inborn error of metabolism. *Clin Chim Acta*. 1981;111:169–178.
48. Vogel KR, Pearl PL, Theodore WH, McCarter RC, Jakobs C, Gibson KM. Thirty years beyond discovery–clinical trials in succinic semialdehyde dehydrogenase deficiency, a disorder of GABA metabolism. *J Inherit Metab Dis*. 2013;36:401–410.
49. Kohlmeier KA, Vardar B, Christensen MH. γ-Hydroxybutyric acid induces actions via the GABAB receptor in arousal and motor control-related nuclei: implications for therapeutic actions in behavioral state disorders. *Neuroscience*. 2013;248C:261–277.
50. Reis J, Cohen LG, Pearl PL, et al. GABAB-ergic motor cortex dysfunction in SSADH deficiency. *Neurology*. 2012;79:47–54.
51. Pearl PL, Gibson KM, Cortez MA, et al. Succinic semialdehyde dehydrogenase deficiency: lessons from mice and men. *J Inherit Metab Dis*. 2009;32:343–352.
52. Hunt AD, Stokes J, McCrory WW, Stroud HH. Pyridoxine dependency: report of a case of intractable convulsions in an infant controlled by pyridoxine. *Pediatrics*. 1954;13:140–145.
53. Mills PB, Footitt EJ, Mills KA, et al. Genotypic and phenotypic spectrum of pyridoxine-dependent epilepsy (ALDH7A1 deficiency). *Brain*. 2010;133:2148–2159.
54. Baxter P, ed. *Pyridoxine Dependent and Pyridoxine Responsive Conditions in Paediatric Neurology*. Mac Keith Press for International Child Neurology Association; 2001:109–165.
55. Mills PB, Struys E, Jakobs C, et al. Mutations in antiquitin in individuals with pyridoxine-dependent seizures. *Nat Med*. 2006;12:307–309.
56. Mills PB, Surtees RAH, Champion MP, et al. Neonatal epileptic encephalopathy caused by mutations in the PNPO gene encoding pyridox(am)ine 5′-phosphate oxidase. *Hum Mol Genet*. 2005;14:1077–1086.

CHAPTER

64

Peroxisomal Disorders

Gerald V. Raymond

University of Minnesota, Delaware, MN, USA

INTRODUCTION

The peroxisome is a subcellular organelle whose importance in cellular metabolism was first recognized when the previously identified human disorders Zellweger syndrome, X-linked adrenoleukodystrophy (XALD), and Refsum disease were discovered to be examples of peroxisomal malfunction. Peroxisomes are present in nearly all tissues and most of the peroxisomal disorders have neurologic expression. With the exception of the X-linked disorder adrenoleukodystrophy, all are autosomal recessive conditions. The human peroxisomal disorders are now divided into two major categories: the disorders of peroxisome biogenesis, in which the organelle fails to form normally, and those disorders in which a single peroxisomal enzyme is deficient. This chapter provides a description of the disorders of peroxisome biogenesis, XALD, and Refsum disease, and concludes with a brief description of the remaining single-enzyme disorders.

DISORDERS OF PEROXISOME BIOGENESIS

A key step toward the understanding of the disorders of peroxisome biogenesis (PBD) was the demonstration that patients with Zellweger syndrome (ZS) lacked peroxisomes[1] and had a deficiency of plasmalogens, a class of lipids known to be synthesized in the peroxisome.[2]

Clinical Features

The disorders that are now known to belong to the PBD category were named and described clinically before it was recognized that they are all related to defective formation of the peroxisome. The historical names—Zellweger syndrome, neonatal adrenoleukodystrophy (NALD), infantile Refsum disease (IRD), and rhizomelic chondrodysplasia punctata (RCDP)—are still often used, but studies of the gene defects associated with each have shown that they can be grouped into two categories. The first group is now referred to as the Zellweger spectrum disorders.[3,4] It includes ZS, NALD, and IRD, which differ in respect to clinical severity. RCDP and its variants are the only representatives of the second category.

The Zellweger Spectrum

Table 64.1 lists the main clinical features of patients with ZS, NALD, and IRD; ZS is the most severe, IRD the least severe, and NALD is intermediate. It should be noted, however, that the range of phenotypic expression represents a continuum that is only imperfectly captured by these three designations. Figure 64.1 shows the facial appearance of patients with ZS. Infants with the ZS phenotype rarely live more than a few months due to the severe hypotonia, feeding difficulty, seizures, liver involvement, and apnea. Dysmorphic features are less striking in NALD than in ZS. The clinical course of NALD ranges from that of a severely involved infant who made no psychomotor gains to those who are stable but disabled in their mid-teens. IRD patients have moderate craniofacial features. All patients have had sensorineural hearing loss and pigmentary degeneration of the retina.

Rosenberg's Molecular and Genetic Basis of Neurological and Psychiatric Disease
http://dx.doi.org/10.1016/10.1016/B978-0-12-410529-4.00064-4

TABLE 64.1 Major Clinical Features of Disorders of Peroxisome Biogenesis and Selected Single Enzyme Disorders

Feature	ZS	NALD	IRD	Acyl oxidase deficiency	Bifunctional enzyme deficiency	RCDP	DHAP synthase deficiency	DHAP akyl transferase deficiency
Average age at death or last follow-up (years)	0.76	2.2	6.4	4	0.75	1.0	0.5	?
Facial dysmorphism	2+	+	+	0	73%	2+	2+	2+
Cataract	80%	45%	7%	0	0	72%	+	+
Retinopathy	71%	82%	100%	2+	+	0	0	0
Impaired hearing	100%	100%	93%	2+	?	71%	+	+
Developmental delay	4+	4+	2-4+	2+	4+	4+	4+	?
Hypotonia	99%	82%	52%	+	4+	±	±	?
Neonatal seizures	80%	82%	20%	50%	93%	±	?	?
Neuronal migration defect	67%	20%	±	?	88%	±	?	?
Demyelination	22%	50%	0	60%	75%	0	0	0
Hepatomegaly	100%	79%	83%	0	+	0	?	0
Renal cysts	93%	0	0	0	0	0	0	0
Rhizomelia	3%	0	0	0	0	93%	+	+
Chondrodysplasia punctata	69%	0	0	0	0	100%	+	+
Coronal vertebral cleft	0	0	0	0	0	+	+	+

Abbreviations: DHAP, dihydroxyacetone phosphate; IRD, infantile Refsum disease; NALD, neonatal adrenoleukodystrophy; RCDP, rhizomelic chondrodysplasia punctata; ZS, Zellweger syndrome.

Percentages indicate the known percentage of patients in whom the abnormality is present; 0, abnormality is present; ± up to 4+, degree to which an abnormality is present.

Rhizomelic Chondrodysplasia Punctata

The RCDP phenotype differs significantly from that of ZS.[5] In the most severe cases, children have rhizomelia with striking shortening of their proximal limbs. The facies are abnormal with flattening, frontal bossing, flat nasal bridge, and small nares (Figure 64.2). Bilateral cataracts are present in 75% of patients, and 25% develop ichthyosis after birth. Psychomotor retardation is profound. Endochondrial bone formation is profoundly disturbed. There is stippling of epiphyses, especially at the knees, elbows, hips, and shoulders. Vertebral bodies have coronal clefts that are seen on lateral roentgenograms of the spine. The majority of RCDP patients die in the first year or two of life, but several have survived longer, some into their teens.

Mode of Inheritance and Prevalence

The PBD are inherited in an autosomal recessive manner and in the aggregate occur in approximately 1 in 50,000–100,000 live births.

Molecular Genetics

Peroxisome biogenesis and importation relies on an assembly method that involves steps for the creation and movement of proteins into the nascent peroxisomes. The proteins involved are referred to as peroxins and are encoded by *PEX* genes.[6,7] Peroxisome targeting sequences (PTS) are attached in the endoplasmic reticulum and target proteins for entry into the peroxisomes. PTS1 is employed by most peroxisomal matrix proteins and is an S–K–L sequence at the carboxy terminus of the protein. PEX5 is the receptor for peroxisome targeting signal 1 (PTS1).

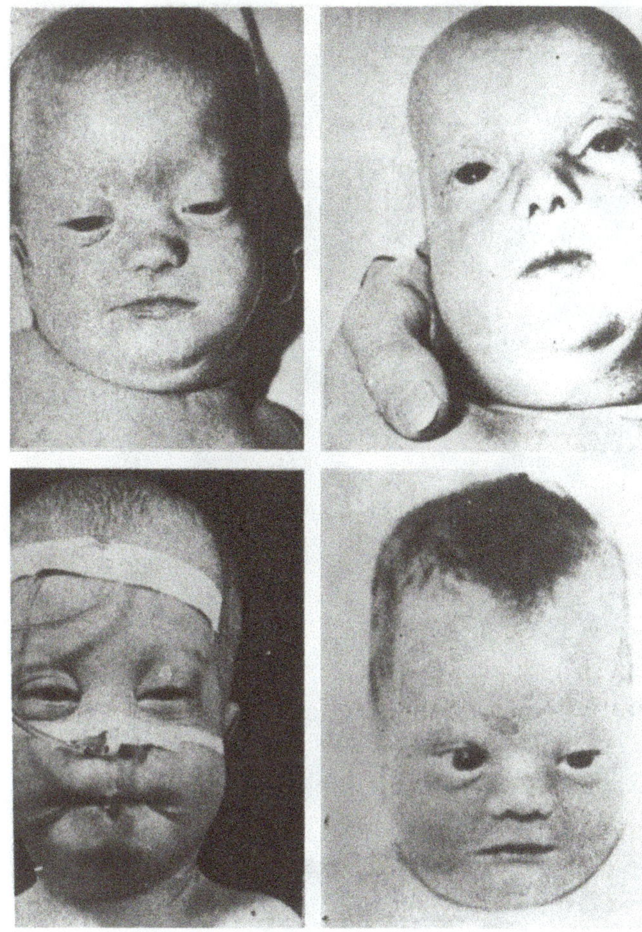

FIGURE 64.1 **Newborn infants with the Zellweger syndrome.** Note prominent forehead, hypertelorism, epicanthal folds, hypoplastic supraorbital ridge, and depressed bridge of nose. (From Lazarow M. Disorders of peroxisome biogenesis. In: Scriver CR, Beaudet AL, Sly WS, et al., eds. *The Metabolic and Molecular Bases of Inherited Disease*. New York, NY: McGraw Hill; 1994:2297.)

A structural model of the PEX5–PTS1 interaction has been proposed and there is recycling of the receptor and other components.

PEX1 deficiency is the most common cause of the PBD.[8] It accounts for about two-thirds of cases. PEX1 interacts closely with PEX6 and both function at late steps in the matrix protein import pathways. The high frequency appears to be caused by presence of two mutated *PEX1* alleles in the general population: one that carries the G843D missense mutation and another that carries the frameshift mutation 2097insT. *In vitro* studies have shown that about 15% of import is maintained in cells with the G843D mutation, whereas, as expected, the frameshift mutation abolishes it completely. These differences are reflected in the phenotype.[8] Patients homozygous for the G843D mutation have the milder phenotypes. Patients homozygous for the 2097insT have the ZS phenotype, and persons who are compounds for the two mutations show phenotypes of intermediate severity. PEX6 accounts for 9% of PBD, while involvement of other PEX genes are rarer. The severity of the phenotype can often be correlated with the degree to which specific mutations impair import *in vitro*.

PEX7 is the receptor for peroxisome targeting sequence-2 (PTS2).[9] The PTS-2 targeting sequence is located near the *N*-terminal and is contained in peroxisomal enzymes that play a role in ether phospholipid (plasmalogen) synthesis and in the α-oxidation of branched-chain fatty acids. PEX7 deficiency has been demonstrated in all patients with the classic RCDP phenotype. The L292X mutation, which completely abolishes import of PTS2-containing proteins, is present in approximately two-thirds of patients with classic RCDP. Other mutations in *PEX7*, which do not abolish import completely, are associated with somewhat milder phenotypes that resemble adult Refsum disease.[10]

Disease Mechanisms

The pathophysiology of the PBD stems from the widespread failure of protein importation and subsequent metabolic failure.[4] With subcellular localization of catalase as the criterion, peroxisomes are reduced or absent in all PBD patients who present as Zellweger spectrum. When peroxisomes are deficient, the catalase is found in the soluble

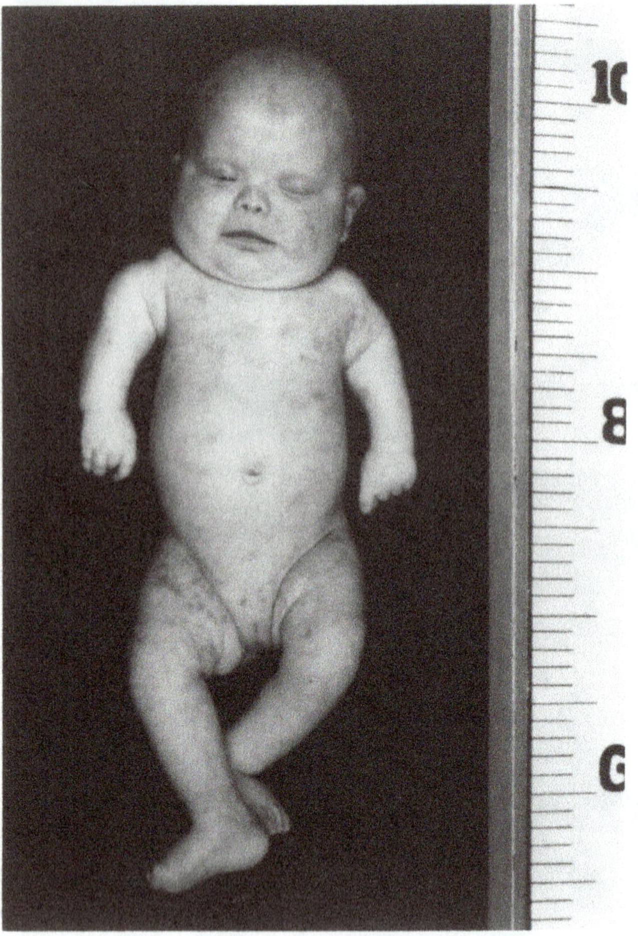

FIGURE 64.2 A newborn female with rhizomelic chondrodysplasia punctata. Note severe shortening of proximal limbs, depressed bridge of nose, hypertelorism, and widespread erythematous and scaling skin lesions. (From Lazarow M. Disorders of peroxisome biogenesis. In: Scriver CR, Beaudet AL, Sly WS, et al., eds. *The Metabolic and Molecular Bases of Inherited Disease*. New York, NY: McGraw Hill; 1994:2309.)

cytosolic fraction. The degree of reduction of peroxisomes correlates roughly with clinical severity.[11] Multiple various biochemical abnormalities result from this failure, including deficits in very long-chain fatty acid oxidation, synthesis of plasmalogens, and phytanic acid oxidation.

Pathological changes are found in many organs in children with ZSD. The liver is enlarged and fibrotic in most patients. Renal cysts are present in nearly all patients with the ZS phenotype, but not in the milder PBD phenotypes. Calcific stippling of the patella and synchondrosis of the acetabulum occurs in 50% of ZS patients. The adrenal glands show striated cells in the reticularis fasciculata inner zone that contain lamellar inclusions. One of the most striking abnormalities is a defect in neuronal migration.[12,13] This leads to characteristic and unique abnormalities that involve the cerebral hemispheres, the cerebellum, and the inferior olivary complex. In the cerebral hemispheres, neurons that are normally destined for outer cortical layers are distributed within the inner cortical layers and in the underlying white matter. This migration failure causes the cerebral convolutions to be abnormally small (microgyria) or thick (pachygyria).

Several murine models with targeted disruption of PEX genes have been described including PEX5, PEX2, PEX11 and PEX13.[14–16] Affected animals showed general intrauterine growth retardation and severe hypotonia and die a few days after birth. These models show a defect in neuronal migration that resembles that seen in ZS patients, and thus they provide an opportunity for determining the mechanism of this syndrome.

Unlike in ZSD, peroxisomes in patients with RCDP do not have demonstrable structural abnormalities, but still lack critical steps in plasmalogen synthesis and phytanic acid oxidation. As a result, plasmalogen levels are low and phytanic acid levels are elevated significantly.[5] The principal pathogenic abnormality appears to be the defect in plasmalogen biosynthesis since single enzymatic abnormalities can reproduce the major features of RCDP.

Study of import-deficient cell lines has provided opportunities for the understanding of the biologic role of the peroxisome and the disease mechanisms of each of the gene defects. Continued studies of animal models will permit us to elucidate the defects in neuronal migration and other manifestations.[16]

TABLE 64.2 Diagnostic Assays for Peroxisomal Disorders

Assay	Zellweger spectrum disorder	RCDP	XALD	Refsum disease	Bifunctional enzyme deficiency	Acyl-CoA deficiency	Racemase deficiency
PLASMA							
VLCFA	Increased	Normal	Increased	Normal	Increased	Increased	Normal
THCA	Increased	Normal	Normal	Normal	Increased	Normal	Increased
Pristanic	Increased	Normal	Normal	Normal	Increased	Normal	Increased
Phytanic	Increased	Increased	Normal	Increased	Increased	Normal	Normal
Pristanic/Phytanic ratio	Normal	Decreased	Normal	Decreased	Increased	Normal	Increased
RED BLOOD CELLS							
Plasmalogen	Decreased	Decreased	Normal	Normal	Normal	Normal	Normal
FIBROBLASTS							
Catalase localization	Cytosol	Peroxisome	Peroxisome	Peroxisome	Peroxisome	Peroxisome	Peroxisome
Plasmalogen biosynthesis	Decreased	Decreased	Normal	Normal	Normal	Normal	Normal
VLCFA	Increased	Normal	Increased	Normal	Increased	Increased	Normal

Differential Diagnosis and Testing

The diagnosis of PBD is suspected on the basis of clinical symptoms. In the newborn, the profound hypotonia may initially raise concerns for a chromosomal abnormality such as Down syndrome. With the more mild ZSD presentations, children may initially be diagnosed with Leber congenital amaurosis or as having Usher syndrome. There are several disorders that may result in bony stippling as seen in RCDP and include the X-linked chondrodysplasia punctata as well as intrauterine warfarin exposure.

The presence or absence of a peroxisomal disorder in affected persons can be determined on the basis of biochemical tests in plasma and red cells (Table 64.2). Definitive designation that the disorder is in the PBD category may require studies of cultured skin fibroblasts because the clinical and biochemical abnormalities in peroxisome single-enzyme defects, particularly D-bifunctional enzyme deficiency, may mimic the Zellweger spectrum PBD disorders. Definition of the gene defect requires mutation analysis.[17] Several reliable methods for prenatal diagnosis are available.

Management

The potential of postnatal treatment is limited by the multiple malformations that originate in fetal life. Based on the hypothesis that some of the biochemical defects secondary to peroxisomal malfunction may lead to additional progressive damage, therapeutic interventions have aimed at correcting one or more of these biochemical abnormalities. These include the oral administration of ether lipids,[18] cholic and deoxycholic acid,[19] and a diet restricted in very long-chain fatty acids and phytanic acid. While it was reported that oral administration of docosahexaenoic acid (DHA) improved clinical status, visual-evoked responses, and brain magnetic resonance imaging (MRI),[20–22] the evaluation of this therapy in a placebo controlled trial demonstrated a lack of efficacy.[23] Symptomatic therapy, often overlooked, has been of benefit to children with ZSD and RCDP.

PEROXISOMAL DISORDERS DUE TO DEFECTS IN SINGLE PEROXISOMAL ENZYMES

Single-enzyme defects have been described for many of the biochemical steps localized to the peroxisome. The two most important disorders are XALD and Refsum disease, which are described in detail.

X-Linked Adrenoleukodystrophy

Although first reported in 1910, the first complete description of what is now referred to as X-linked adrenoleukodystrophy was provided by Siemerling and Creutzfeldt in 1923.[24] They reported a patient with an inflammatory demyelinating disorder similar to the "encephalitis periaxialis diffusa" that had been described by Schilder but in whom the brain pathology was combined with adrenal atrophy. Adrenomyeloneuropathy (AMN), the adult form of the disease, was described in 1977.[25] Lipid storage and the accumulation of saturated very long-chain fatty acids (VLCFA) in the brain and adrenal gland were demonstrated in 1976,[26] and were shown to be the result of the impaired capacity to degrade these substances, a reaction that normally takes place in the peroxisome.[27,28] The defective gene, now referred to as *ABCD1*, was cloned in 1993.[29] It codes for a peroxisomal membrane protein, ALDP, a member of the ATP-binding cassette transporter superfamily of proteins.

Clinical Features

XALD displays a wide range of phenotypic expression (Table 64.3).[30] Three principal phenotypes occur in male patients. The first phenotype comprises the cerebral forms. In the aggregate these forms affect approximately 40% of male XALD patients. The cerebral form is most common in childhood and manifests between the ages of 4 and 10 years. The initial manifestations resemble those of attention deficit disorders or other learning or behavioral problems, with later progression to dementia, impairment of vision, and hearing and motor deficits. The disorder progresses to total disability within 2 to 3 years, and death ensues at varying intervals thereafter. The adolescent and adult cerebral forms are similar in characteristics, presenting with neurobehavioral issues including dementia, may progress more slowly, and also are caused by an inflammatory demyelination. There is a characteristic MRI appearance with bilateral symmetric involvement of the posterior parieto-occipital white matter, with a garland of contrast enhancement seen in approximately 85% of individuals.[31] The remainder is usually frontal in location.

The second phenotype is AMN. Men with AMN present most commonly in their twenties or thirties with a slowly progressive paraparesis with sphincter and sexual disturbances. These myelopathic findings are often misdiagnosed

TABLE 64.3 Presentations in Males with X-linked Adrenoleukodystrophy

Presentation	Description	Estimated frequency
Childhood cerebral	Onset at 4–10 years of age Progressive neurologic deficit Total disability in 2–3 years Inflammatory cerebral demyelination	30–35%
Adolescent	Similar to childhood cerebral Often with seizures and hemiparesis Onset 11–21 years of age Maybe slower in progression	4–7%
Adrenomyeloneuropathy (AMN)	Onset 20–30 years of age Progressive spastic paraparesis over decades Axonopathy with no evidence of inflammation	40–46%
Cerebral AMN	Varying ages in adulthood Development of cerebral inflammatory demyelination 20–40% of men with AMN	
Adult cerebral	Cerebral disease without preceding AMN Rapidly progressive	2–5%
Olivopontocerebellar	Mainly cerebellar and brainstem involvement in late adolescence or adulthood	1–2%
Addison-only	Primary adrenal insufficiency without neurologic involvement	Varies with age; up to 50% in childhood
Asymptomatic	Genetic and biochemical abnormality without demonstrable adrenal or neurologic deficit	Diminishes with age; very common in males <4 years of age

as multiple sclerosis or other forms of hereditary spastic paraparesis. MRI typically shows only spinal cord atrophy without evidence of cerebral involvement. This presentation is consistent with a normal lifespan, but approximately 20% of individuals will develop cerebral disease in adulthood.[32]

The Addison-only phenotype is the third principal phenotype of XALD. Although the majority of neurologically affected individuals will have evidence of primary adrenal insufficiency, the occurrence of the two manifestations is independent.[33] Therefore, 15–20% of patients have primary adrenocortical insufficiency without evidence of neurologic involvement. These patients cannot be distinguished clinically from patients with other forms of adrenal insufficiency. It is estimated that XALD is the cause of adrenal insufficiency in approximately 35% of boys with idiopathic Addison disease who are less than 7.5 years old. Most XALD patients with the Addison-only phenotype later exhibit neurological involvement, so the incidence will vary with age.

Women heterozygous for XALD remain asymptomatic in childhood and early adulthood, but up to 50% develop an AMN-like syndrome in their later adult years.[34] Manifestations include spasticity and bladder involvement. This syndrome is of later onset and milder than in the AMN patients, but occasionally patients require the use of a wheelchair. Neuropathic pain with pronounced dysesthesias is frequently reported in this population. Less than 1% of heterozygous women have cerebral involvement or adrenal insufficiency.

Mode of Inheritance and Prevalence

The mode of inheritance of XALD is X-linked recessive. The frequency of the disorder in the United States is approximately 1 in 21,000 males, and the combined frequency in men and women in the total population is 1 in 16,800.[35] XALD has been reported in all ethnic groups, and there is no evidence that there is significant variation in the frequency among them.

It is important to emphasize that the various phenotypes co-occur in the same family, and phenotypic expression does not correlate with the nature of the molecular defect or with the degree of elevation of VLCFA levels. Therefore, one cannot predict the presentation in other affected family members based on the index case in a family.

Molecular Genetics

The gene located at Xq28 deficient in XALD is now referred to as *ABCD1*. It is composed of 10 exons, and it encodes a protein of 745 amino acids.[36] The amino acid sequence is related to the superfamily of ATP-binding cassette (ABC) transporter proteins. The resultant protein is a peroxisomal membrane protein required for transport of very long-chain fatty acids into the peroxisome. Mutations in the *ABCD1* gene have been identified in nearly all XALD patients and these mutations are updated on a website (www.x-ald.nl).[37] More than 500 different mutations have been identified. Mutations are varied and spread through the gene. The majority of kindreds have private mutations and, again, there is no correlation between the nature of the mutation and the phenotype. While the existence of a modifier has been proposed, extensive studies have not discerned a gene or genes.[38–42]

Disease Mechanisms

The pathology of the cerebral forms of XALD differs in important ways from that of AMN. An inflammatory demyelinating process with cellular infiltration and increased cytokine levels is a characteristic feature of the cerebral forms,[43,44] whereas AMN is a noninflammatory distal axonopathy that clinically involves mainly the dorsal columns and the corticospinal tract.[45] The main biochemical abnormality is the abnormal accumulation of saturated VLCFA, most strikingly in the brain white matter and adrenocortical cells,[26] and to a lesser and varying extent in all other tissues. The VLCFA excess is present in a variety of lipid classes, including cholesterol esters, sphingomyelin, glycerophospatides, and gangliosides. The increase in VLCFA levels appears to be caused by both the impaired capacity to degrade these substances, a reaction that normally takes place in the peroxisome, as well as the continued elongation of very long-chain fatty acids.[46]

The pathogenesis of XALD is not completely understood. VLCFA excess impairs membrane stability, and it has been proposed that this contributes to the axonopathy in AMN. The pathogenesis of the inflammatory reaction may involve an immune-mediated inflammatory reaction to VLCFA excess.[47]

Mouse models of XALD have been produced by targeted inactivation of *ABCD1*.[48,49] VLCFA levels in the brain and adrenal glands are increased as in the human disease, but the animals appear to develop normally. The inflammatory brain disease observed in patients with the cerebral phenotypes has never been observed in the mouse model. The mouse model develops an AMN-like syndrome at the advanced age of 18–24 months.[50] Recent work has demonstrated the role of reactive oxygen species and oxidative stress in the pathogenesis of disease in the animal models. Work continues to attempt to recapitulate the cerebral form of the condition in animal models and to better understand the discrepant presentation that can arise in at-risk males.

Diagnosis

The diagnosis of XALD is suspected on the basis of the clinical symptoms and, in childhood cerebral disease, by the characteristic MRI findings. Confirmation of diagnosis is based on biochemical studies of plasma and, less frequently, on mutation analysis.[51,52] The demonstration of increased levels of saturated VLCFAs in plasma is the most frequently used diagnostic assay. Levels of hexacosanoic acid (C26:0) and the ratios of C26:0 and of tetracosanoic acid (C24:0) to docosahexaenoic acid (C22:0) are increased. This test is reliable for the diagnosis of affected males, irrespective of age. Plasma VLCFA levels are increased to a lesser degree in women who are heterozygous for XALD, but they are normal or borderline in 20% of obligate heterozygotes. Mutation analysis is the most reliable method for heterozygote identification.[53] Prenatal diagnosis is achieved by demonstration of increased VLCFA levels in cultured chorion villus cells or amniocytes and by mutation analysis.[54]

Management

Adrenal steroid replacement therapy is mandatory for all patients with impaired adrenocortical function. It can be life saving and improves general strength and wellbeing, but it does not affect neurologic status.

Hematopoietic stem cell (HSCT) or bone marrow transplantation has provided long-term stabilization and occasionally improvement in boys and adolescents with cerebral involvement that is still mild.[55,56] Key to success is the identification of patients with cerebral involvement that is still in the early stage. This is aided by monitoring asymptomatic patients with MRI. Abnormalities can be detected by neuroimaging before the manifestation of clinical symptoms.[57] Semiannual or annual (depending on age) monitoring can help identify patients who are candidates for this therapy. HSCT carries a significant risk, so it is not recommended for those without cerebral involvement, or for those who already have advanced cerebral involvement. The promising results of HSCT in patients with early cerebral involvement heightens the need to identify asymptomatic or mildly symptomatic patients by increasing awareness of the disease and by screening at-risk family members.[58]

Dietary therapy in which reduced fat intake is combined with the oral administration of a 4:1 mixture of glyceryl trioleate and glyceryl trierucate, also referred to as Lorenzo's oil, leads to rapid normalization of plasma VLCFA levels. Not surprisingly, the effect in patients who are already symptomatic is limited by the speed and severity of active cerebral disease. An open study reported that the administration of Lorenzo's oil and a specially designed low-fat diet to boys who were younger than 6 years old who were neurologically asymptomatic with a normal MRI reduced the incidence of MRI abnormalities in at-risk boys. Limitations of this study included the lack of control group, inability to predict males at risk, and the need for significant reduction of VLCFA for potential benefit.[59,60]

An *ex vivo* gene therapy has demonstrated efficacy in arresting childhood cerebral disease.[61]

Refsum Disease

In 1946, the Norwegian neurologist Sigvald Refsum described a progressive familial disorder that he named heredopathia atactica polyneuritiformis and that has since been referred to as Refsum disease.[62] In 1963, Klenk and Kahike demonstrated that Refsum disease was associated with the accumulation of phytanic acid.[63] In 1967 Steinberg al. demonstrated that the oxidation of phytanic acid involved an unusual initial α-oxidation and that this reaction was deficient in patients with Refsum disease.[64] They also showed that phytanic acid in humans is of dietary origin exclusively and that dietary restriction of phytanic acid leads to clinical improvement.[65] The steps involved in the α-oxidation of phytanic acid are complex, and it was not until 1997 that Jansen et al. demonstrated that phytanoyl-coenzyme A hydroxylase is the enzyme that is deficient in Refsum disease.[66]

Clinical Features

The main clinical features of Refsum disease are retinitis pigmentosa, peripheral neuropathy, neurogenic hearing loss, anosmia, cardiac abnormalities, and ichthyosis.[67] Symptoms begin most commonly in the second decade. Pigmentary degeneration of the retina appears to be present in all patients and is an early manifestation. Night blindness may be present years before other clinical symptoms occur. Over the years, concentric visual field constriction gradually develops until only tubular vision remains. A progressive polyneuropathy is another consistent manifestation but may not be recognized at the start of the illness. It is a mixed motor and sensory neuropathy that affects distal parts of the lower extremities most severely, with weakness, atrophy, and loss of deep tendon reflexes. Vibration and position sense are affected most severely, and peripheral nerves may be palpably enlarged and firm. Both the olfactory and the auditory nerves are affected. Loss of hearing is of the cochlear type and may be almost complete. Cardiomyopathy with cardiomegaly, heart failure, conduction disturbances, and electrocardiographic changes is common. The involvement of the skin is highly variable. It ranges from slightly dry skin to florid ichthyosis, although some patients never develop skin involvement.

Mode of Inheritance and Prevalence

The pattern of inheritance is autosomal recessive and it must be emphasized that Refsum disease is very rare. Most cases seem to have come from Scandinavia, northern France, the British Isles, and Ireland, a distribution that suggested the disease was spread by the Vikings, but the disease has also been observed in other ethnic groups where a connection with the Vikings is unlikely.

Molecular Genetics

Adult Refsum disease results from mutations in either the gene that encodes phytanoyl-CoA hydroxylase (*PHYH*) or *PEX7*, which impacts importation into the peroxisome. The *PHYH* gene spans 21.5 kb on chromosome 10p13 of the human genome and encodes an mRNA of about 1.6 kb. The 1014 nucleotides of the open-reading frame sequence are separated by 8 introns. Nineteen different mutations have been identified in Refsum disease patients.[68] Interesting structure–function relationships have been described by Mukherji and colleagues.[69] The alteration P29S does not abolish PHYH activity *in vitro*, but because it lies within the PTS2 targeting sequence, it may affect transport of the enzyme to the peroxisome. Mukherji et al. have noted that in the United Kingdom only about 45% of adult Refsum disease patients have defects in the function of PHYH. Other cases are associated with defects in PEX7.[70] The A218V *PEX7* mutation is associated with milder phenotypes that resemble adult Refsum disease.

Disease Mechanism

BIOCHEMICAL DEFECT

The metabolism of phytanic acid is complex.[67] An initial beta oxidation is impossible because of the methyl group in the three position. The initial step is the formation of the coenzyme A derivative, followed by the formation of 2-hydroxyphytanoyl-CoA, catalyzed by phytanoyl-CoA hydroxylase. PHYH is deficient in classic Refsum disease, owing to a variety of mutations. It is located in the peroxisome and contains the PTS-2 targeting sequence, therefore requiring the PEX7 receptor for import into the peroxisome. PHYH activity is also deficient in RCDP (PEX7 deficiency) and in ZSD. The deficiency in RCDP results from the fact that PEX7 is the receptor for proteins that contain the PTS-2 so that the enzyme is not targeted to the peroxisome and remains in the cytosolic compartment, where it is unstable.

In humans, phytanic acid is exclusively of dietary origin, a finding that forms the basis for current therapy. The impaired capacity to degrade it leads to striking accumulation of phytanic acid in many tissues. In some tissues, such as the heart, phytanic acid may account for up to 50% of total fatty acids. The fact that lowering phytanic acid levels by dietary restriction leads to clinical improvement suggests that phytanic acid excess contributes to the pathogenesis. Several mechanisms have been proposed. These include the molecular distortion hypothesis, which states that the increased cross-sectional area of methylated fatty acids when compared to that of straight-chain fatty acids could cause membrane instability in myelin. Another hypothesis is that phytanic acid may interfere with the formation of prenylated proteins. Of particular interest are studies that indicate that phytanic acid is a ligand for members of the nuclear receptor hormone receptor superfamily.[71]

Diagnosis

Biochemical and genetic confirmation of the diagnosis is essential. Demonstration of abnormally high concentrations of phytanic acid in plasma is the key diagnostic assay. It is recommended that an assay procedure be used that measures levels of VLCFAs and of pristanic acid concurrently. Plasma concentrations of phytanic acid are always increased except in early infancy. Normal concentration of phytanic acid in a patient who is 1 year or older thus effectively rules out Refsum disease and related disorders, and permits distinction from the many other disorders that can cause retinitis pigmentosa or peripheral neuropathy. The concurrent measurements of VLCFA and pristanic acid permits distinction of Refsum disease from other disorders associated with increased levels of phytanic acid. In ZS, NALD, IRD, and bifunctional enzyme deficiency, VLCFA levels are increased, whereas they are normal in Refsum disease. Pristanic acid levels are increased in racemase deficiency and in bifunctional enzyme deficiency, whereas they are normal or decreased in classic Refsum disease. The phytanic and pristanic acid patterns in RCDP and in PEX7 defects with milder manifestations resemble those in classic Refsum disease. Classic RCDP can be distinguished because its clinical presentation is totally different. Enzymatic assays, mutation analysis, or both are required to distinguish classic Refsum disease from the milder forms of PEX7 deficiency.

Management

The prognosis of untreated patients with Refsum disease is poor. A survey in Norway found that 10 of 11 patients were blind and that half had died before age 30 years. Dietary therapy, based on restriction of phytanic acid intake, has revolutionized the prognosis.[72] This therapy reduces the levels of phytanic acid in plasma from more than 100 mg per dL to 10 mg per dL. The dietary therapy improves the peripheral neuropathy and stabilizes the visual and auditory deficits, cardiomyopathy, and skin manifestations. Because the diet does not appear to reverse deficits that are already present, except for the neuropathy, the therapy should be begun early, hence the need for early diagnosis, including the screening of at-risk relatives. The phytanic acid content is highest in all kinds of dairy products. Certain cautions in respect to dietary management are essential. The adipose tissues may contain an enormous amount of phytanic acid. Release of phytanic acid may occur during periods of reduced food intake, including initiation of dietary therapy, dental or surgical procedures, or stress. It is essential, therefore, to maintain adequate caloric intake. Very high levels of phytanic acid (>100 mg per dL) may precipitate toxic and life-threatening symptoms, such as cardiomyopathy or tetraparesis. Under these circumstances, plasmapheresis has been successful as an emergency measure.

OTHER PEROXISOMAL SINGLE-ENZYME DEFECTS

Overall, this is a rarer group of disorders. They were predominantly defined by their biochemical abnormalities and originally came to attention because of their overlap with the previously described disorders. In this group, bifunctional enzyme deficiency is the most common. At the Kennedy Krieger Institute, 15% of patients with the ZSD phenotype had bifunctional enzyme deficiency. The enzymatic and molecular basis of bifunctional enzyme deficiency has now been determined.[73] It was shown that the first patient reported to have this disorder, who was thought to have a deficiency of L-bifunctional enzyme, actually had a deficiency of the D-bifunctional enzyme and that there is no documented case of human L-bifunctional enzyme deficiency. Furthermore, biochemical re-evaluation of the patient who was originally reported as an example of peroxisomal 3-oxoacyl-CoA thiolase 1 deficiency (pseudo-Zellweger syndrome) demonstrated that this patient also had D-bifunctional enzyme deficiency.[74] The phenotypic expression of D-bifunctional enzyme deficiency ranges from ZS to milder IRD-like phenotypes.

Peroxisomal acyl-CoA oxidase deficiency, also referred to as pseudoneonatal adrenoleukodystrophy, is a rare disorder, and its clinical manifestations are milder than those of bifunctional enzyme deficiency.[75]

The continuous deletion of the X-linked adrenoleukodystrophy gene and DXS1357E (CADDS) syndrome has been described.[76] Patients presented with a combination of cholestasis, hypotonia, and profound psychomotor retardation and died during the first year. Their clinical presentation thus resembled that of a PBD. VLCFA levels were markedly increased and they lacked ALDP, but other peroxisomal functions were normal. Molecular analysis revealed that they had a contiguous deletion of *ABCD1* and of *DXS1357E*. A point of great practical significance is that the mode of inheritance of CADDS is X-linked recessive, whereas that of the PBD is autosomal recessive. Patients with racemase deficiency may present with a sensory motor neuropathy that bears some resemblance to Charcot–Marie–Tooth disease, Refsum disease, or a mild peroxisome biogenesis disorder.[77] As shown in Table 64.2, the laboratory findings are characteristic.

References

1. Goldfischer S, Moore CL, Johnson AB, et al. Peroxisomal and mitochondrial defects in the cerebro-hepato-renal syndrome. *Science.* 1973;182(4107):62–64.
2. Heymans HS, Schutgens RB, Tan R, Van den BH, Borst P. Severe plasmalogen deficiency in tissues of infants without peroxisomes (Zellweger syndrome). *Nature.* 1983;306(5938):69–70.
3. Steinberg SJ, Raymond GV, Braverman NE, Moser AB, Moser HW. Peroxisome Biogenesis Disorders, Zellweger Syndrome Spectrum. Gene Reviews at Gene Tests: Medical Genetics Information Resource [Database online]. Seattle, WA: University of Washington; 2003.
4. Steinberg SJ, Dodt G, Raymond GV, Braverman NE, Moser AB, Moser HW. Peroxisome biogenesis disorders. *Biochim Biophys Acta.* 2006;1763(12):1733–1748.
5. Braverman NE, Moser AB, Steinberg SJ. Rhizomelic chondrodysplasia punctata type 1. 2001 Nov 16 [updated 2012 Sep 13]. In: Pagon RA, Adam MP, Ardinger HH, Bird TD, Dolan CR, Fong CT, Smith RJH, Stephens K, eds. *GeneReviews® [Internet].* Seattle (WA): University of Washington, Seattle; 1993–2014. Available from: http://www.ncbi.nlm.nih.gov/books/NBK1270/.
6. Waterham HR, Ebberink MS. Genetics and molecular basis of human peroxisome biogenesis disorders. *Biochim Biophys Acta.* 2012;1822(9): 1430–1441.
7. Fujiki Y, Yagita Y, Matsuzaki T. Peroxisome biogenesis disorders: molecular basis for impaired peroxisomal membrane assembly: in metabolic functions and biogenesis of peroxisomes in health and disease. *Biochim Biophys Acta.* 2012;1822(9):1337–1342.

8. Reuber BE, GermainLee E, Collins CS, et al. Mutations in PEX1 are the most common cause of peroxisome biogenesis disorders. *Nat Genet.* 1997;17(4):445–448.
9. Braverman N, Steel G, Obie C, et al. Human PEX7 encodes the peroxisomal PTS2 receptor and is responsible for rhizomelic chondrodysplasia punctata. *Nat Genet.* 1997;15(4):369–376.
10. van den Brink DM, Brites P, Haasjes J, et al. Identification of PEX7 as the second gene involved in Refsum disease. *Adv Exp Med Biol.* 2003;544:69–70.
11. Moser AB, Rasmussen M, Naidu S, et al. Phenotype of patients with peroxisomal disorders subdivided into sixteen complementation groups. *J Pediatr.* 1995;127(1):13–22.
12. Gressens P, Baes M, Leroux P, et al. Neuronal migration disorder in Zellweger mice is secondary to glutamate receptor dysfunction. *Ann Neurol.* 2000;48(3):336–343.
13. Volpe JJ, Adams RD. Cerebro-hepato-renal syndrome of Zellweger: an inherited disorder of neuronal migration. *Acta Neuropathol (Berl).* 1972;20(3):175–198.
14. Faust PL, Su HM, Moser A, Moser HW. The peroxisome deficient PEX2 Zellweger mouse: pathologic and biochemical correlates of lipid dysfunction. *J Mol Neurosci.* 2001;16(2–3):289–297.
15. Baes M, Van Veldhoven PP. Generalised and conditional inactivation of Pex genes in mice. *Biochim Biophys Acta.* 2006;1763(12):1785–1793.
16. Baes M. Mouse models for peroxisome biogenesis disorders. *Cell Biochem Biophys.* 2000;32:229–237.
17. Steinberg S, Chen L, Wei L, et al. The PEX Gene Screen: molecular diagnosis of peroxisome biogenesis disorders in the Zellweger syndrome spectrum. *Mol Genet Metab.* 2004;83(3):252–263.
18. Wilson GN, Holmes RG, Custer J, et al. Zellweger syndrome: diagnostic assays, syndrome delineation, and potential therapy. *Am J Med Genet.* 1986;24(1):69–82.
19. Setchell KD, Heubi JE. Defects in bile acid biosynthesis–diagnosis and treatment. *J Pediatr Gastroenterol Nutr.* 2006;43(suppl 1):S17–S22.
20. Noguer MT, Martinez M. Visual follow-up in peroxisomal-disorder patients treated with docosahexaenoic acid ethyl ester. *Invest Ophthalmol Vis Sci.* 2010;51(4):2277–2285.
21. Martinez M, Pineda M, Vidal R, Conill J, Martin B. Docosahexaenoic acid–a new therapeutic approach to peroxisomal-disorder patients: experience with two cases. *Neurology.* 1993;43(7):1389–1397.
22. Martinez M, Vazquez E. MRI evidence that docosahexaenoic acid ethyl ester improves myelination in generalized peroxisomal disorders. *Neurology.* 1998;51(1):26–32.
23. Paker AM, Sunness JS, Brereton NH, et al. Docosahexaenoic acid therapy in peroxisomal diseases: results of a double-blind, randomized trial. *Neurology.* 2010;75(9):826–830.
24. Siemerling E, Creutzfeldt HG. Bronzekrankheit und sklerosierende encephalomyelitis. *Arch Psychiatr Nervenkr.* 1923;68:217–244.
25. Griffin JW, Goren E, Schaumburg H, Engel WK, Loriaux L. Adrenomyeloneuropathy – probable variant of adrenoleukodystrophy. 1. Clinical and endocrinologic aspects. *Neurology.* 1977;27(12):1107–1113.
26. Igarashi M, Schaumburg HH, Powers J, Kishmoto Y, Kolodny E, Suzuki K. Fatty acid abnormality in adrenoleukodystrophy. *J Neurochem.* 1976;26(4):851–860.
27. Singh I, Moser AE, Goldfischer S, Moser HW. Lignoceric acid is oxidized in the peroxisome: implications for the Zellweger cerebro-hepato-renal syndrome and adrenoleukodystrophy. *Proc Natl Acad Sci U S A.* 1984;81(13):4203–4207.
28. Singh I, Moser AE, Moser HW, Kishimoto Y. Adrenoleukodystrophy: impaired oxidation of very long chain fatty acids in white blood cells, cultured skin fibroblasts, and amniocytes. *Pediatr Res.* 1984;18(3):286–290.
29. Mosser J, Lutz Y, Stoeckel ME, et al. The gene responsible for adrenoleukodystrophy encodes a peroxisomal membrane-protein. *Hum Mol Genet.* 1994;3(2):265–271.
30. Moser HW, Mahmood A, Raymond GV. X-linked adrenoleukodystrophy. *Nat Clin Pract Neurol.* 2007;3(3):140–151.
31. Melhem ER, Barker PB, Raymond GV, Moser HW. X-linked adrenoleukodystrophy in children: review of genetic, clinical, and MR imaging characteristics. *Am J Roentgenol.* 1999;173(6):1575–1581.
32. van Geel BM, Bezman L, Loes DJ, Moser HW, Raymond GV. Evolution of phenotypes in adult male patients with X-linked adrenoleukodystrophy. *Ann Neurol.* 2001;49(2):186–194.
33. Dubey P, Raymond GV, Moser AB, Kharkar S, Bezman L, Moser HW. Adrenal insufficiency in asymptomatic adrenoleukodystrophy patients identified by very long-chain fatty acid screening. *J Pediatr.* 2005;146(4):528–532.
34. Jangouk P, Zackowski KM, Naidu S, Raymond GV. Adrenoleukodystrophy in female heterozygotes: underrecognized and undertreated. *Mol Genet Metab.* 2012;105(2):180–185.
35. Bezman L, Moser HW. Invited editorial comment - incidence of X-linked adrenoleukodystrophy and the relative frequency of its phenotypes. *Am J Med Genet.* 1998;76(5):415–419.
36. Sarde CO, Mosser J, Kioschis P, et al. Genomic organization of the adrenoleukodystrophy gene. *Genomics.* 1994;22(1):13–20.
37. Kemp S, Pujol A, Waterham HR, et al. ABCD1 mutations and the X-linked adrenoleukodystrophy mutation database: role in diagnosis and clinical correlations. *Hum Mutat.* 2001;18(6):499–515.
38. Barbier M, Sabbagh A, Kasper E, et al. CD1 gene polymorphisms and phenotypic variability in X-linked adrenoleukodystrophy. *PLoS One.* 2012;7(1):e29872.
39. Brose RD, Avramopoulos D, Smith KD. SOD2 as a potential modifier of X-linked adrenoleukodystrophy clinical phenotypes. *J Neurol.* 2012;259(7):1440–1447.
40. Semmler A, Bao X, Cao G, et al. Genetic variants of methionine metabolism and X-ALD phenotype generation: results of a new study sample. *J Neurol.* 2009;256(8):1277–1280.
41. Maier EM, Mayerhofer PU, Asheuer M, et al. X-linked adrenoleukodystrophy phenotype is independent of ABCD2 genotype. *Biochem Biophys Res Commun.* 2008;377(1):176–180.
42. Holzinger A, Mayerhofer PU, Maier EM, Roscher AA, Berger J. Evidence against the adrenoleukodystrophy-related gene acting as a modifier of X-adrenoleukodystrophy. *Adv Exp Med Biol.* 2003;544:95–96.
43. Powers JM, Liu Y, Moser AB, Moser HW. The inflammatory myelinopathy of adrenoleukodystrophy - cells, effector molecules, and pathogenetic implications. *J Neuropathol Exp Neurol.* 1992;51(6):630–643.

44. Hudspeth MP, Raymond GV. Immunopathogenesis of adrenoleukodystrophy: current understanding. *J Neuroimmunol.* 2007;182(1–2):5–12.
45. Powers JM, DeCiero DP, Ito M, Moser AB, Moser HW. Adrenomyeloneuropathy: a neuropathologic review featuring its noninflammatory myelopathy. *J Neuropathol Exp Neurol.* 2000;59(2):89–102.
46. Kemp S, Theodoulou FL, Wanders RJ. Mammalian peroxisomal ABC transporters: from endogenous substrates to pathology and clinical significance. *Br J Pharmacol.* 2011;164(7):1753–1766.
47. Kemp S, Berger J, Aubourg P. X-linked adrenoleukodystrophy: clinical, metabolic, genetic and pathophysiological aspects. *Biochim Biophys Acta.* 2012;1822(9):1465–1474.
48. Lu JF, Lawler AM, Watkins PA, et al. A mouse model for X-linked adrenoleukodystrophy. *Proc Natl Acad Sci U S A.* 1997;94(17):9366–9371.
49. Forss-Petter S, Werner H, Berger J, et al. Targeted inactivation of the X-linked adrenoleukodystrophy gene in mice. *J Neurosci Res.* 1997;50(5):829–843.
50. Pujol A, Hindelang C, Callizot N, Bartsch U, Schachner M, Mandel JL. Late onset neurological phenotype of the X-ALD gene inactivation in mice: a mouse model for adrenomyeloneuropathy. *Hum Mol Genet.* 2002;11(5):499–505.
51. Moser AB, Kreiter N, Bezman L, et al. Plasma very long chain fatty acids in 3,000 peroxisome disease patients and 29,000 controls. *Ann Neurol.* 1999;45(1):100–110.
52. Moser HW, Moser AB, Frayer KK, et al. Adrenoleukodystrophy - increased plasma content of saturated very long-chain fatty-acids. *Neurology.* 1981;31(10):1241–1249.
53. Boehm CD, Cutting GR, Lachtermacher MB, Moser HW, Chong SS. Accurate DNA-based diagnostic and carrier testing for X-linked adrenoleukodystrophy. *Mol Genet Metab.* 1999;66(2):128–136.
54. Moser AB, Moser HW. The prenatal diagnosis of X-linked adrenoleukodystrophy. *Prenat Diagn.* 1999;19(1):46–48.
55. Peters C, Charnas LR, Tan Y, et al. Cerebral X-linked adrenoleukodystrophy: the international hematopoietic cell transplantation experience from 1982 to 1999. *Blood.* 2004;104(3):881–888.
56. Mahmood A, Raymond GV, Raymond P, Dubey P, Peters C, Moser HW. Survival analysis of haematopoietic cell transplantation for childhood cerebral X-linked adrenoleukodystrophy: a comparison study. *Lancet Neurol.* 2007;6(8):687–692.
57. Loes DJ, Fatemi A, Melhem ER, et al. Disease progression in X-linked adrenoleukodystrophy based on MR imaging: the role of contrast enhancement and location of brain lesion. *Neurology.* 2002;58(7):A38.
58. Bezman L, Moser AB, Raymond GV, et al. Adrenoleukodystrophy: incidence, new mutation rate, and results of extended family screening. *Ann Neurol.* 2001;49(4):512–517.
59. Moser HW, Raymond GV, Lu SE, et al. Follow-up of 89 asymptomatic patients with adrenoleukodystrophy treated with Lorenzo's oil. *Arch Neurol.* 2005;62(7):1073–1080.
60. Moser HW, Moser AB, Hollandsworth K, Brereton NH, Raymond GV. "Lorenzo's oil" therapy for X-linked adrenoleukodystrophy: rationale and current assessment of efficacy. *J Mol Neurosci.* 2007;33(1):105–113.
61. Cartier N, Hacein-Bey-Abina S, Bartholomae CC, et al. Hematopoietic stem cell gene therapy with a lentiviral vector in X-linked adrenoleukodystrophy. *Science.* 2009;326(5954):818–823.
62. Refsum S. Hereditopathia atactica polyneuritiformis. *Acta Psychiat Scan.* 1946;38:1–303.
63. Klenk E, Kahike W. On the presence of 3,7,11,15-tetramethylhexadecanoic acid [phytanic acid] in the cholesterol esters and other lipoid fractions of the organs in a case of a disease of unknown origin [possibly heredopathia atactica polyneuritiformis, Refsum's syndrome]. *Hoppe Seylers Z Physiol Chem.* 1963;333:133–139.
64. Steinberg D, Herndon Jr. JH, Uhlendorf BW, Mize CE, Avigan J, Milne GW. Refsum's disease: nature of the enzyme defect. *Science.* 1967;156(3783):1740–1742.
65. Steinberg D, Mize CE, Herndon Jr. JH, Fales HM, Engel WK, Vroom FQ. Phytanic acid in patients with Refsum's syndrome and response to dietary treatment. *Arch Intern Med.* 1970;125(1):75–87.
66. Jansen GA, Ferdinandusse S, Ijlst L, et al. Refsum disease is caused by mutations in the phytanoyl-CoA hydroxylase gene. *Nat Genet.* 1997;17(2):190–193.
67. Wanders RJA, Waterham HR, Leroy BP, Wanders RJ, Komen JC. Refsum disease peroxisomes, Refsum's disease and the alpha- and omega-oxidation of phytanic acid. *Biochem Soc Trans.* 2007;35(Pt 5):865–869.
68. Jansen GA, Hogenhout EM, Ferdinandusse S, et al. Human phytanoyl-CoA hydroxylase: resolution of the gene structure and the molecular basis of Refsum's disease. *Hum Mol Genet.* 2000;9(8):1195–1200.
69. Mukherji M, Chien W, Kershaw NJ, et al. Structure-function analysis of phytanoyl-CoA 2-hydroxylase mutations causing Refsum's disease. *Hum Mol Genet.* 2001;10(18):1971–1982.
70. Jansen GA, Waterham HR, Wanders RJA. Molecular basis of Refsum disease: sequence variations in phytanoyl-CoA hydroxylase (PHYH) and the PTS2 receptor (PEX7). *Hum Mutat.* 2004;23(3):209–218.
71. Wolfrum C, Ellinghaus P, Fobker M, et al. Phytanic acid is ligand and transcriptional activator of murine liver fatty acid binding protein. *J Lipid Res.* 1999;40(4):708–714.
72. Masters-Thomas A, Bailes J, Billimoria JD, Clemens ME, Gibberd FB, Page NG. Heredopathia atactica polyneuritiformis (Refsum's disease): 1. Clinical features and dietary management. *J Hum Nutr.* 1980;34(4):245–250.
73. van Grunsven EG, van Berkel E, Mooijer PAW, et al. Peroxisomal bifunctional protein deficiency revisited: resolution of its true enzymatic and molecular basis. *Am J Hum Genet.* 1999;64(1):99–107.
74. Ferdinandusse S, van Grunsven EG, Oostheim W, et al. Reinvestigation of peroxisomal 3-ketoacyl-CoA thiolase deficiency: identification of the true defect at the level of D-bifunctional protein. *Am J Hum Genet.* 2002;70(6):1589–1593.
75. Watkins PA, Mcguinness MC, Raymond GV, et al. Distinction between peroxisomal bifunctional enzyme and acyl-CoA oxidase deficiencies. *Ann Neurol.* 1995;38(3):472–477.
76. Corzo D, Gibson W, Johnson K, et al. Contiguous deletion of the X-linked adrenoleukodystrophy gene (ABCD1) and DXS1357E: a novel neonatal phenotype similar to peroxisomal biogenesis disorders. *Am J Hum Genet.* 2002;70(6):1520–1531.
77. Ferdinandusse S, Denis S, Clayton PT, et al. Mutations in the gene encoding peroxisomal alpha-methylacyl-CoA racemase cause adult-onset sensory motor neuropathy. *Nat Genet.* 2000;24(2):188–191.

CHAPTER

65

Disorders of Purine Metabolism

William L. Nyhan

The University of California, San Diego School of Medicine, La Jolla, CA, USA

LESCH–NYHAN DISEASE

Clinical Features

Historical Overview

Lesch–Nyhan disease was first described as a syndrome in 1964 in two brothers aged 4 and 8 years, respectively.[1] The defect is a lack of activity of the enzyme hypoxanthine guanine phosphoribosyltransferase (HPRT) (Figure 65.1).[2] HPRT catalyzes the reaction in which the purines hypoxanthine and guanine are reutilized to form their respective nucleotides, inosinic acid (IMP) and guanylic acid (GMP). The gene for HPRT is located on the X chromosome. Phenotypic heterogeneity in the expression of symptoms is seen and correlates with varying degrees of residual HPRT activity.[3,4] Untreated, the disease usually caused fatal nephropathy early in life. This problem was solved by the development of inhibitors of xanthine oxidase, but treatment of other aspects of the disease remains elusive.

Mode of Inheritance and Prevalence

Best estimates of prevalence are of 1 in 100,000 male births.[4] An X-linked recessive, the disorder occurs almost exclusively in males.

Age of Onset and Disease Evolution

The disease may express as uricosuria in early infancy. In the classic disease, which accounts for most of the patients with HPRT deficiency, affected infants typically develop normally for the first 4–6 months and have no positive findings on neurologic examination. The first sign of the disease is usually a manifestation of crystalluria, often described as "orange sand" in the diapers. Delayed neurologic development is apparent within the first year, and the affected child does not learn to sit alone or loses this ability if sitting has been achieved. The child never learns to stand unassisted or to walk. If securely fastened around the chest, the child is able to sit in a chair. During the first year, extrapyramidal signs develop. In the established disease severe action dystonia is the major motor feature.[5] Involuntary movements begin at 6–12 months and may be relatively minor at rest and most obvious with excitement. Movements may be choreic or athetoid as well as dystonic. Opisthotonic spasms are regularly present. Episodic opisthotonus or truncal arching, reported in 25% of patients,[5] has been uniformly present in the author's experience. This may lead to cervical spinal cord injury. Signs of pyramidal tract involvement develop during the first years of life, and the accompanying spasticity is severe and often leads to dislocation of the hips. Patients have hyperreflexia, and ankle clonus and a positive Babinski response are present in most. Scissoring of the legs is common. Dysarthria (oromandibular and lingual dystonia) and dysphagia are other features of the disease, and the dysarthria can combine with the motor defect and behavior to make proper assessment of the patient's mental capabilities difficult. Most patients eat poorly, and most of them vomit. The vomiting becomes incorporated into their abnormal behavior, so at least some of it appears semivoluntary.

Mental development may be retarded; the IQ is usually in the range of 40–70. However, the behavior and the motor defect make testing difficult. Some patients have had normal cognitive function, and a few have been successful mainstream students. Aggressive, self-injurious behavior is one of the most distinctive features of Lesch–Nyhan

Rosenberg's Molecular and Genetic Basis of Neurological and Psychiatric Disease
http://dx.doi.org/10.1016/10.1016/B978-0-12-410529-4.00065-6

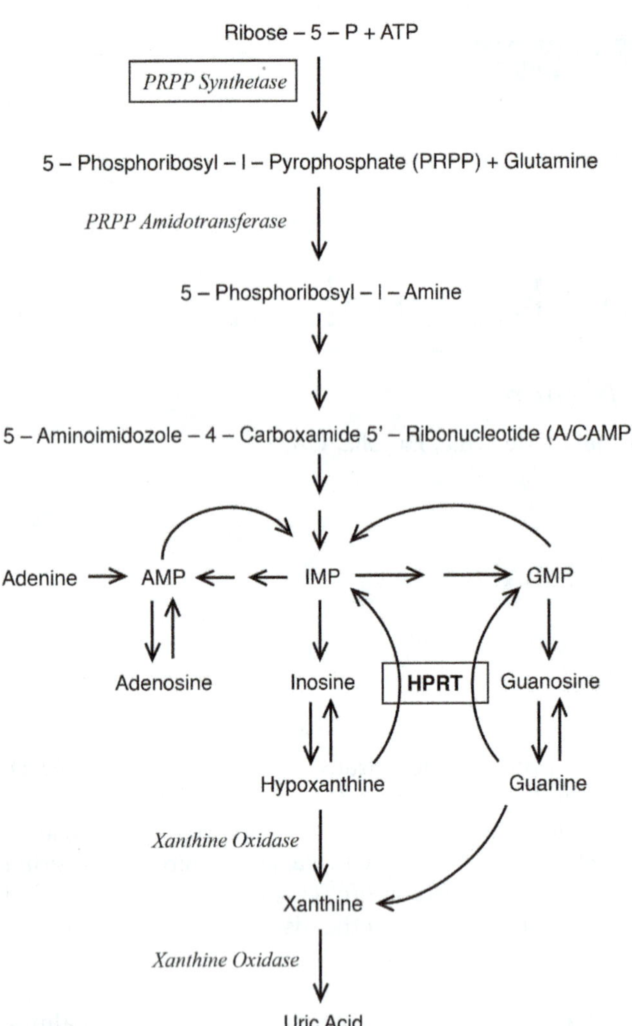

FIGURE 65.1 Metabolic breakdown of purine nucleotides leading to the formation of uric acid. Abbreviations: AMP, adenosine monophosphate; GMP, guanosine monophosphate; HPRT, hypoxanthine guanine phosphoribosyltransferase; IMP, inosine monophosphate.

disease, and in our experience has always been present in patients with the fully developed syndrome. However, there are probably exceptions to every rule. Among 22 patients from Spain,[6] one patient whose relatives displayed the full syndrome, including self-injurious behavior, was observed until age 21 years without displaying this behavior, and another with two typical Lesch–Nyhan-affected uncles had reached 6 years of age without self-injurious behavior.

Patients bite their lips and fingers with resulting loss of tissue, sometimes amputating phalanges of the fingers or severing the tongue. Patients are not insensitive to pain, and they scream in pain when they bite and are usually relieved when restrained to prevent self-injury. Physical restraint and extraction of teeth are the only effective methods of preventing the behavior. Aggressive behavior is also directed to others: they bite, hit, or kick, and verbal aggression becomes common. The abnormal behavior appears to be compulsive and beyond the control of the patient, who is often remorseful about the behavior. Detailed behavioral analysis of 22 patients revealed characteristic behavioral profiles of aggression, distractibility, anxiety, and social problems.[7]

Hyperuricemia is characteristic; serum uric acid levels are usually between 5–10 mg per dL. The excretion of uric acid is 3–4 mg of uric acid per 1 mg creatinine (1.9±0.9 mmol per mol creatinine). Normal children excrete <1 mg uric acid per mg creatinine. A mean of 0.2 mmol per mol creatinine was reported for 20 controls. The consistent finding of an elevated plasma level or urinary uric acid-to-creatinine ratio and their relative ease of measurement usually make these measures useful in initial screening tests for the disease. However, patients may have normal serum levels if they are efficient excreters and particularly if adult male standards are used for normal values; in addition, urinary data for uric acid can be spuriously low as a result of bacterial contamination. This has led us to advise against 24-hour collection of urine. The clinical consequences of hyperuricemia are manifestations shared with gout: arthritis, tophi, painful crystalluria, hematuria, nephrolithiasis, urinary tract infection, and, in untreated patients, renal failure. Patients have also manifested megaloblastic anemia, in some severe enough to require regular blood transfusions.

Macrocytosis or megaloblastic changes in the marrow have also been found in the absence of clinical anemia. A preliminary diagnosis of classic Lesch–Nyhan disease can be made based on the phenotype. A serum concentration in a child of 4–5 mg uric acid per dL and urine uric acid-to-creatinine ratio of 3:4 are supportive. However, a definitive diagnosis requires analysis of HPRT enzyme activity, which is most consistently assayed in erythrocyte lysate. Dried blood spots on filter paper, as in newborn screening programs, provide for a stable source of enzyme for analysis of samples sent from distances.

Disease Variants

Patients have been described in whom HPRT activity is low, who do not have the full clinical phenotype displayed by the classic Lesch–Nyhan patient, and these are usually referred to as variants.[4,8] Patients with HPRT deficiency generally belong to one of three groups: those with all features of Lesch–Nyhan disease, those with the neurologic manifestations and hyperuricemia, and those with hyperuricemia only. In some patients with partial HPRT deficiency, the abnormal enzyme is readily distinguished as different from both control and Lesch–Nyhan HPRT activity. However, no consistent correlation between enzyme activity as measured in the red cell lysate and clinical phenotype has emerged. Nonetheless, a rough inverse correlation between HPRT activity and the severity of the clinical symptoms was found when enzyme activity was measured in intact cultured fibroblasts.[3] Classic Lesch–Nyhan patients exhibited HPRT activities <1.4% of normal. Patients with neurologic symptoms and hyperuricemia had HPRT activities ranging from 1.6–8% of normal. In the hyperuricemic group, the lowest reported activity was 8% of normal, and the highest was around 60%. In partial variants with neurologic disease, hyperuricemia is present together with the typical extrapyramidal and pyramidal symptoms found in the classic Lesch–Nyhan patient; intelligence is normal or near normal, and behavior may appear normal. The findings of neurologic examination of these patients are indistinguishable from that of the patient with classic Lesch–Nyhan disease. A distinct clinical phenotype with partial HPRT deficiency (7.5% of normal enzyme activity) has been reported in a family with four affected members.[9] Each member displayed a phenotype of atypical neurologic disease characterized by spasticity, increased deep tendon reflexes, and a positive Babinski response, but no dystonia or choreoathetosis; mental retardation was mild.

Disease Mechanism: Human Observations

The most significant pathologic feature is the normality of the central nervous system. The accumulation of uric acid in soft tissue tophi yields an amorphous, powdery histologic picture. In joints, an inflammatory response surrounds crystallized needles of sodium urate engulfed by leukocytes. In the renal parenchyma, the inflammatory response results in fibrosis and renal failure. Renal stones and parenchymal urate may be detected by ultrasound or pyelography.

Differential Diagnosis

Overproduction of hyperuricemia is also seen in mutant phosphoribosyl pyrophosphate synthetase (vi). Self-injurious behavior of the degree of severity seen in this disease is unique. None of the other diseases in which self-injurious behavior occurs display this degree of severity.

Animal Models

Animal models of self-injurious behavior resembling that seen in the Lesch–Nyhan patient have been created by the administration of high doses of caffeine or theophylline. The underlying mechanism is not clear in either case, but a role for endogenous adenosine has been proposed in view of the pharmacologic actions of caffeine and theophylline as antagonists at the adenosine receptor. High doses of clonidine have been reported to induce self-injurious behavior in mice, and the mechanism might operate through blocking of adenosine receptors. Amphetamine and amphetamine-related drugs, when given repeatedly or in high doses, also induce self-injurious behavior in mice and rats, and this is probably an effect of increased dopamine release. One of the most well-characterized models for self-injurious behavior was described in rats treated neonatally with 6-hydroxydopamine (6-OHDA). 6-OHDA administered intracisternally in the neonatal brain leads to destruction of dopaminergic nerve fibers. When these animals later are given dopamine or a D1 agonist, self-injurious behavior is induced. This dopamine-triggered behavior is not seen after treatment of adult rats with 6-OHDA. A similar response to dopamine challenge was displayed by monkeys, in which nigrostriatal dopamine neurons were denervated in the neonatal period through surgical lesion of the brainstem. These observations have led to the hypothesis that even late-manifesting self-injurious behavior may be the result of neonatal depletion of dopamine.[10] The first HPRT-deficient animal model engineered was the mouse. Embryonic stem cells were used for the selection of HPRT-deficient cells, which either arose spontaneously or were produced by retroviral insertional mutagenesis. HPRT-negative embryonic stem cells were introduced into mouse blastocysts to produce chimeric mice, which were screened for HPRT deficiency. Heterozygous females were then mated to produce HPRT-deficient males. It has been interesting, but somewhat disconcerting, to learn that in spite of the total absence of HPRT activity in these animals, they seem to develop normally and do not exhibit any of the neurologic manifestations of Lesch–Nyhan disease. Excess uric acid does not accumulate in mice because, unlike humans, mice express the uricase enzyme that metabolizes uric acid to allantoin, which is readily excreted. However, measurements of dopamine levels in the forebrains of these HPRT-negative mice have yielded values of 20–30% depletion compared with normal. A 70–90% depletion of dopamine has been found in the basal ganglia of three Lesch–Nyhan patients studied postmortem. It has been speculated that the mouse might be less dependent than humans on the purine salvage pathway for maintaining adequate intracellular levels of purine nucleotides and that this might be particularly true for cells of the basal ganglia. Nevertheless, the mouse model may be useful in the development of therapy for Lesch–Nyhan disease.

Testing and Molecular Diagnosis

HPRT (EC 2.4.2.8.) catalyzes the reaction of hypoxanthine and guanine with 5-phosphoribosyl-1-pyrophosphate (PRPP) to form their nucleotides (Figure 65.1). In Lesch–Nyhan disease, the activity of HPRT in erythrocytes approximates zero.[2,3] HPRT is ubiquitously expressed in cells of the human body, with the highest levels in the basal ganglia. The subcellular localization is cytoplasmic.

Molecular Genetics: Normal and Abnormal Gene Products and Genotype–Phenotype Correlations

The gene for HPRT is coded on the X chromosome and mapped to the distal part of the long arm at position Xq2.6-Xq2.7. The disease is transmitted in an X-linked recessive fashion. It is almost exclusively a disease of males, but seven affected females have been observed.[11,12] The human gene spans approximately 44 kb of DNA over nine exons, and the sequence of the entire 57-kb gene has been determined. The sequences determined for the cDNA completely matched the 218-amino-acid sequence that had previously been determined for the enzyme protein. The transcriptional promoter region is like that of other housekeeping genes with many copies of the GGGCGG sequence, but lacking CAAT-like and TATA-like sequences characteristic of most mRNA-coding genes. In the majority of patients with HPRT deficiency, there is no gross alteration of the gene.[8,13] Major alteration of the HPRT gene usually results in the classic Lesch–Nyhan phenotype. HPRT Toronto was one of the first HPRT mutants to be examined in this fashion. Amino acid sequence analysis had revealed a glycine-for-arginine substitution. The nucleotide substitution in HPRT Toronto was predicted to result in the loss of a *Taq*I restriction site. This information could then be applied in restriction fragment length polymorphism (RFLP) analysis of DNA from affected and carrier members of this particular family. In most instances of HPRT deficiency, however, RFLP analysis is not informative. The introduction of the polymerase chain reaction (PCR) and the continued refinement of DNA sequence analysis have revolutionized the study of inherited diseases and the characterization of specific mutations.

These approaches have led to the identification of over 400 different HPRT mutations.[8,13] The spectrum is heterogeneous, but most variants represent point mutations. Most families studied have displayed a unique mutation; the same mutation has only rarely been found in unrelated pedigrees. Missense mutations in Lesch–Nyhan patients have generally been nonconservative; for instance, an aspartic acid for a glycine at position 16; a leucine for phenylalanine at position 74; and a tyrosine for an aspartic acid at position 201.[13] A CpG mutational hotspot was identified at arginine 51 of the HPRT protein. This may represent deamination of 5-methylcytosine to thymine. This position has been mutated 11 times among reported patients.[13]

In the neurologic variant group of patients, point mutations have been the rule, and the types of change observed have been relatively conservative. For instance, changing a valine to a glycine would not be expected to make a major difference in protein structure. Others had changes such as an isoleucine to a threonine. In these patients and in the partial or hyperuricemic variants, no deletions, stop codons, or major rearrangements were observed. In HPRT Salamanca,[9] there were two mutations: a T-to-G change at position 128 and a G-to-A change at 130. These changes resulted in the substitution of two adjacent amino acids at positions 43 and 44: methionine to arginine and aspartic acid to asparagine, respectively. These would not appear to be particularly nonconservative, but the phenotype was probably the mildest of the neurologic variants observed.

Prenatal diagnosis can be performed by assay of the enzyme in amniocytes and chorionic villus cells. It is readily done by mutational analysis if the mutation is known.

Management

The excessive uric acid production in Lesch–Nyhan patients is best treated with daily administration of allopurinol. Unaccountably, the marketers of Febuxostat recommend against the use of this xanthine oxidase inhibitor in this disease.

A dose of 20 mg per kg per day promptly leads to normal plasma concentrations of uric acid and prevents the symptoms related to hyperuricemia. The reduction of uric acid is paralleled by increases in urinary levels of hypoxanthine and xanthine. Because these patients cannot reutilize hypoxanthine, the levels of these oxypurines can become quite high. The solubility of hypoxanthine is much higher than uric acid, so this purine will not cause calculi. On the other hand, xanthine is even less soluble than uric acid, so xanthine calculi can be a problem. We measure the concentration of oxypurines to arrive at the optimal dose of allopurinol. The potential complication of xanthine stone formation should not deter treatment with allopurinol. A high fluid intake is recommended. Alkalinization of the urine is not useful. Allopurinol administration does not affect the neurologic or behavioral manifestations of the disease. Effective management of the self-injurious behavior is physical restraint and in some cases extraction of teeth. Numerous trials with different drugs have been tested in the past, but no effective therapy is available for these manifestations.

Allogeneic bone marrow transplant was successfully accomplished in a 22-year-old Lesch–Nyhan patient.[14] No improvement of neurologic or behavioral symptoms was observed, although the HPRT activity in the patient's blood was normal. Hyperuricemia was also not improved. At least five other bone marrow or cord blood stem cell transplants have been carried out in Lesch–Nyhan patients, and all have died of complications of the procedure.

Therapies Under Investigation

Future attempts at therapy will undoubtedly include gene therapy. In several studies, introduction of retroviral based vectors containing HPRT cDNA into HPRT-negative cultured cells has resulted in complete or partial correction of the metabolic defect in these cells. Mouse bone marrow cells expressing human HPRT after retroviral transfer were transplanted into murine hosts, and the recipient animals expressed the human protein in spleen and bone marrow for up to 4 months, but stable long-term expression has yet to be achieved. Deep brain stimulation with electrodes implanted in the brain ganglia in an approach to control dystonia was reported to abolish self-injurious behavior.[15] In that paper, no documentation of deficiency of HPRT was recorded, and other experience has been mixed.

PHOSPHORIBOSYLPYROPHOSPHATE (PRPP) SYNTHETASE ABNORMALITIES

Clinical Features

A familial disorder, in which accelerated activity of PRPP synthetase was associated with overproduction of uric acid and gout, was first reported in 1972 by Sperling and colleagues.[16] A small number of patients has since been identified, establishing the relationship with hyperuricemia.[17,18] In some kindred there has been sensorineural deafness.[18,19]

5-Phosphoribosylpyrophosphate (PRPP) synthetase (EC 2.7.6.2) (Figure 65.1) catalyzes the initial step in the *de novo* synthesis of purines in which ribose-5-P reacts with adenosine triphosphate (ATP) to form PRPP. The formed PRPP provides the substrate for the first rate-limiting step in the ten step reaction. PRPP is coded for by a gene on the X chromosome at Xq.[17–19] At least two mutations have been defined.

Clinical Abnormalities–Modalities

The invariant clinical features of this disease are hyperuricemia and uricosuria. Therefore, a patient is subject to any of the clinical consequences of the accumulation of uric acid in body fluids. Gouty arthritis has been reported with onset as early as 21 years of age.[16] Renal colic has been observed, as well as the passage of calculi.[17] Our patient[18] failed as an infant to cry with tears and was found to have absent lachrymal glands. Other structural anomalies were glandular hypospadias and hypoplastic teeth. The mother also had the abnormal PRPP synthetase and hearing loss.

In addition to hyperuricemia and uricosuria, overproduction of purine *de novo* was documented[18] by measuring the *in vivo* in conversion of ^{14}C-glycine to urinary uric acid; it was 7 times the control level.

Testing and Molecular Diagnosis

PRPP synthetase is coded for by a gene on the long arm of the X chromosome.[20] In some families the abnormality may be fully expressed in the heterozygous female. In others it may be fully recessive.[16] This may reflect different degrees of lyonizaton. On the other hand, it is easier for an overactive enzyme than the more common deficient one encountered in inborn errors to function as an X-linked dominant.

The cDNA encodes transcript of 2.3 kb and a protein of 317 amino acids. In two patients with overactivity of PRPP synthetase, there was an A-to-G transition at nucleotide 341, which led to a change from asparagine to serine at position 113.[20] Another patient had a C-to-G change at nucleotide 547, leading to a change from aspartic acid to histidine at 182.[20]

Management

Allopurinol is the treatment of choice in overproduction of hyperuricemia. Treatment of abnormalities in PRPP synthetase is simpler than that of HPRT deficiency, because in the presence of normal HPRT activity there is extensive reutilization of hypoxanthine leading to a substantial decrease in the overall excretion of oxypurines in the urine.

Hearing should be tested promptly and appropriate intervention provided.

References

1. Lesch M, Nyhan WL. A familial disorder of uric acid metabolism and central nervous system function. *Am J Med*. 1964;36:561–570.
2. Seegmiller JE, Rosenbloom FM, Kelley WN. Enzyme defect associated with a sex-linked human neurological disorder and excessive purine synthesis. *Science*. 1967;155:1682–1684.
3. Page T, Bakay B, Nissinen E, Nyhan WL. Hypoxanthine-guanine phosphoribosyltransferase variants: correlation of clinical phenotype with enzyme activity. *J Inherit Metab Dis*. 1981;4:203–206.
4. Nyhan WL, Barshop BA, Al-Aqeel A. Atlas of Inherited Metabolic Diseases. 3rd ed. London: Hodder/Arnold; 2013, 483–497.
5. Jinnah HA, Visser JE, Harris JC, et al. Delineation of the motor disorder of Lesch–Nyhan disease. *Brain*. 2006;129:1201–1217.
6. Puig JG, Torres RJ, Mateos FA, Ramos T, Buno AS, Arcas J. The spectrum of hypoxanthine-guanine phosphoribosyltransferase (HPRT) deficiency. *Adv Exp Med Biol*. 2000;486:15–21.
7. Schretlen DJ, Ward J, Meyer SM, et al. Behavioral aspects of Lesch–Nyhan disease and its variants. *Dev Med Child Neurol*. 2005;47:673–677.
8. Nyhan WL, O'Neill JP, Jinnah HA, Harris JC. Lesch–Nyhan syndrome. In: *Gene Reviews at Gene Tests: Medical Genetics Information Resource*. 1997–2010. Seattle, Washington. Available at: http://www.genetests.org.
9. Page T, Nyhan WL, Morena de Vega V. Syndrome of mild mental retardation, spastic gait, and skeletal malformations in a family with partial deficiency of hypoxanthine-guanine phosphoribosyltransferase. *Pediatrics*. 1987;79:713–717.
10. Nyhan WL. Dopamine function in Lesch–Nyhan disease. *Environ Health Perspect*. 2000;108:409–411.
11. De Gregorio L, Nyhan WL, Serafin E, Chamoles NA. An unexpected affected female patient in a classical Lesch–Nyhan family. *Mol Genet Metab*. 2000;69:263–268.
12. De Gregorio L, Jinnah HA, Harris JC, Nyhan WL. Lesch–Nyhan disease in a female with a clinically normal monozygotic twin. *Mol Genet Metab*. 2005;85:70–77.
13. Jinnah HA, De Gregorio L, Harris JC, Nyhan WL, O'Neill JP. The spectrum of inherited mutations causing HPRT deficiency: 75 new cases and a review of 196 previously reported cases. *Mutat Res*. 2000;463:309–326.
14. Nyhan WL, Page T, Gruber HE, Parkman R. Bone marrow transplantation in Lesch– Nyhan disease. *Birth Defects Orig Artic Ser*. 1986;22:113–117.
15. Taira T, Kobayashi T, Hori T. Disappearance of self-mutilating behavior in a patient with Lesch–Nyhan syndrome after bilateral chronic stimulation of the globus pallidus internus. Case report. *J Neurosurg*. 2003;98:414–416.
16. Sperling O, Eilam G, Persky-Brosh S. Accelerated erythrocyte 5-phosphoribosyl-1-pyrophosphate synthesis. A familial abnormality associated with excessive uric acid production and gout. *Biochem Med*. 1972;6:310–316.
17. Becker MA, Meyer LJ, Seegmiller JE. Gout with purine overproduction due to increased phosphoribosylpyrophosphate synthetase activity. *Am J Med*. 1973;55:232–242.
18. Nyhan WL, James JA, Teberg AJ, Sweetman L, Nelson LG. A new disorder of purine metabolism with behavioral manifestations. *J Pediatr*. 1969;74:20–27.
19. Simmonds HA, Webster DR, Lingam S, Wilson J. An inborn error of purine metabolism, deafness and neurodevelopmental abnormality. *Neuropediatrics*. 1985;16:106–108.
20. Roessler BJ, Golovoy N, Palella TD, Heidler SK, Becker MA. Identification of distinct PRS1 mutations in two patients with X-linked phosphoribosylpyrophosphate synthetase superactivity. *Adv Exp Med Biol*. 1991;309B:125–128.

CHAPTER

66

The Porphyrias

D. Montgomery Bissell

University of California, San Francisco, CA, USA

INTRODUCTION

Heme is the prosthetic group for essential molecules including hemoglobin and cytochromes. The pathway of heme synthesis starts with formation of δ-aminolevulinic acid (ALA) and proceeds through porphobilinogen (PBG) and a series of porphyrins. At the final step, iron is inserted in protoporphyrin to make heme. Each step is catalyzed by a specific enzyme. Heme synthesis can be compromised by mutations affecting enzymatic activity or by environmental factors such as lead. When this occurs, pathway intermediates preceding the defective step accumulate and appear in blood, urine, or feces in increased amounts.[1] For the hepatic porphyrias, the presentation usually is abdominal pain, but mental changes and seizure also may occur. Although the clinical presentation of porphyria can be highly suggestive of the diagnosis, the symptoms or signs are notoriously nonspecific. Quantitative analysis of porphyrins and porphyrin precursors in urine and feces is required for diagnosis and is definitive. Therapy centers on the administration of heme (as hematin or heme arginate), which compensates for the principal metabolic deficiency of these diseases. Prevention involves screening of family members of an index case and education as to the conditions that increase the risk of an acute attack. Some data suggest that genetic carriers can experience progressive, often subclinical, neurological damage in the absence of acute symptoms. In general, however, when recognized and properly managed, porphyria is not a cause of long-term disability.

Heme and Porphyrin Metabolism

Early clinical observations established the existence of more than one type of porphyria, and their hereditary nature. With elucidation of the pathway of heme synthesis, the biochemical relationship of these diseases to one another was defined. Moreover, the pattern of overproduction associated with an individual type pointed to a block at a specific step in the pathway of heme synthesis, which was confirmed in the 1960s and 1970s by assay of the enzymes of the pathway. By 2000, all of the relevant genes had been cloned. Mutation analysis now provides definitive diagnosis and is essential for screening the relatives of a clinically active case.

Pathway of Heme Synthesis

The building blocks of heme are succinyl CoA and glycine, which combine to form δ-aminolevulinic acid (ALA), the first committed intermediate of the pathway. In subsequent reactions, two molecules of ALA combine to form porphobilinogen (PBG), the pyrrole subunit of the heme ring. Four PBGs are linked, yielding an intermediate linear tetrapyrrole, hydroxymethylbilane, which is cyclized to the initial porphyrin of the pathway, UROgen-I. Because PBG is asymmetric, having one acetyl and one propionyl substituent, four porphyrin isomers are theoretically possible: a regular head-to-tail linkage of the four PBG units (type I), reversal of two adjacent units (type II), reversal of one unit (type III), and reversal of two opposite units (type IV). The only isomers found in nature are types I and III,

http://dx.doi.org/10.1016/10.1016/B978-0-12-410529-4.00066-8

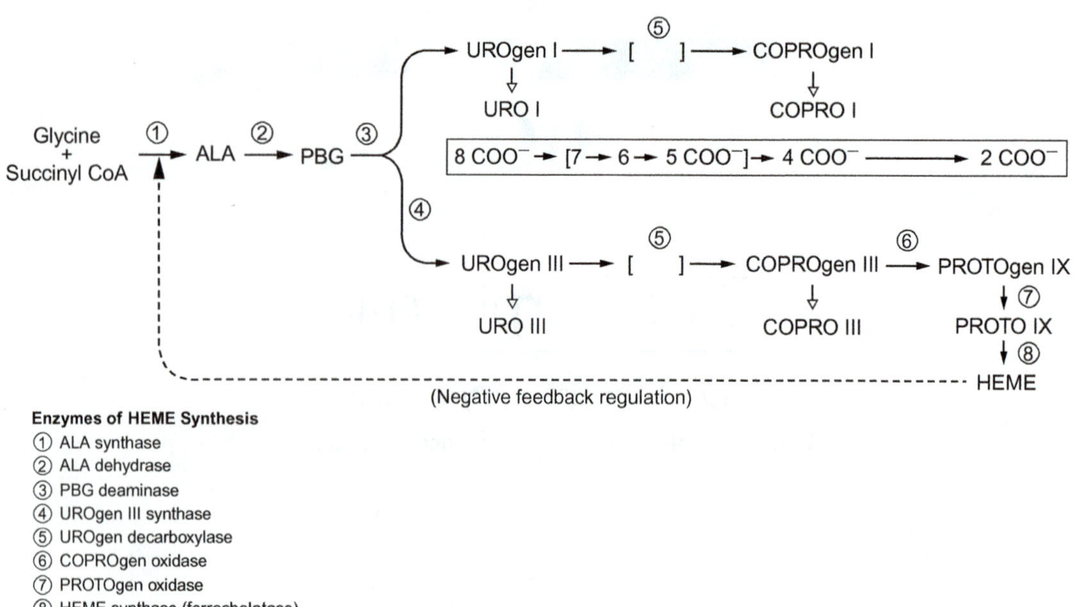

FIGURE 66.1 **The pathway of heme synthesis.** Between the two arms of the pathway, depicting the isomer I and isomer III porphyrin, is a box indicating the number of carboxyl substituents for each porphyrin. ALA, δ-aminolevulinic acid; PBG, porphobilinogen; URO, uroporphyrin; COPRO, coproporphyrin; PROTO, protoporphyrin.

with type III representing the physiologic intermediate in heme synthesis (Figure 66.1). The conversion of UROgen to COPROgen and PROTOgen involves decarboxylation of the side chains, producing intermediates of progressively decreasing water solubility (Figure 66.1). Thus ALA, PBG, and uroporphyrin undergo excretion essentially entirely in urine. Coproporphyrin appears in both urine and feces, and protoporphyrin only in feces.

The true porphyrin intermediates in heme synthesis are uro- and copro*porphyrinogen* together with protoporphyrin. The conversion of UROgen and COPROgen to the corresponding porphyrins represents an irreversible oxidation. Uroporphyrin and coproporphyrin have no biological role and are destined for excretion. Normally, they represent a minute fraction of the total flow of intermediates through the pathway. In porphyria, the levels increase 5- to 200-fold the upper limit of normal.

Regulation of Heme and Porphyrin Synthesis

Each step in heme synthesis is catalyzed by an individual enzyme, all of which have been cloned and characterized. The rate-limiting step for the pathway as a whole is the formation of ALA, which is catalyzed by ALA synthase (ALAS; Figure 66.1). Two forms of ALAS exist and are distinct gene products. The erythroid form (known as ALAS2, human chromosome X) serves hemoglobin synthesis in the bone marrow and is regulated by iron. A sequence in the 5′ untranslated region (UTR) of ALAS2 mRNA recognizes an iron-binding protein. When iron is replete, the iron-loaded binding protein blocks translation of ALAS mRNA and thus limits heme synthesis.[2] The ubiquitous form of ALA (also known as ALAS1, human chromosome 3) is present in all tissues except the erythron.[3] In the liver, it is produced in the cytosol, then translocates into the mitochondrion, where it catalyzes formation of ALA. The end product, heme, regulates the pathway by blocking transfer of ALAS from cytosol to mitochondria (Figure 66.1). This ensures that levels of ALA and subsequent pathway intermediates are appropriate to the need for new heme production. In normal individuals, pathway intermediates are never present in excess.

The pattern of heme precursors in urine or feces for each of the porphyrias is shown in Figure 66.1. Each porphyria can be viewed as resulting from a partial block to the flow of precursors along the pathway, with excess excretion of the intermediates that precede the enzyme deficiency. If the pathway is compromised by a defective enzyme (as in the hereditary porphyrias), a prolonged, exaggerated increase in ALA synthase may occur under inducing circumstances. The result is marked overproduction of intermediates prior to the block, which gives rise to the excretion patterns shown in Figure 66.2. This is the metabolic setting for an acute attack of porphyria.

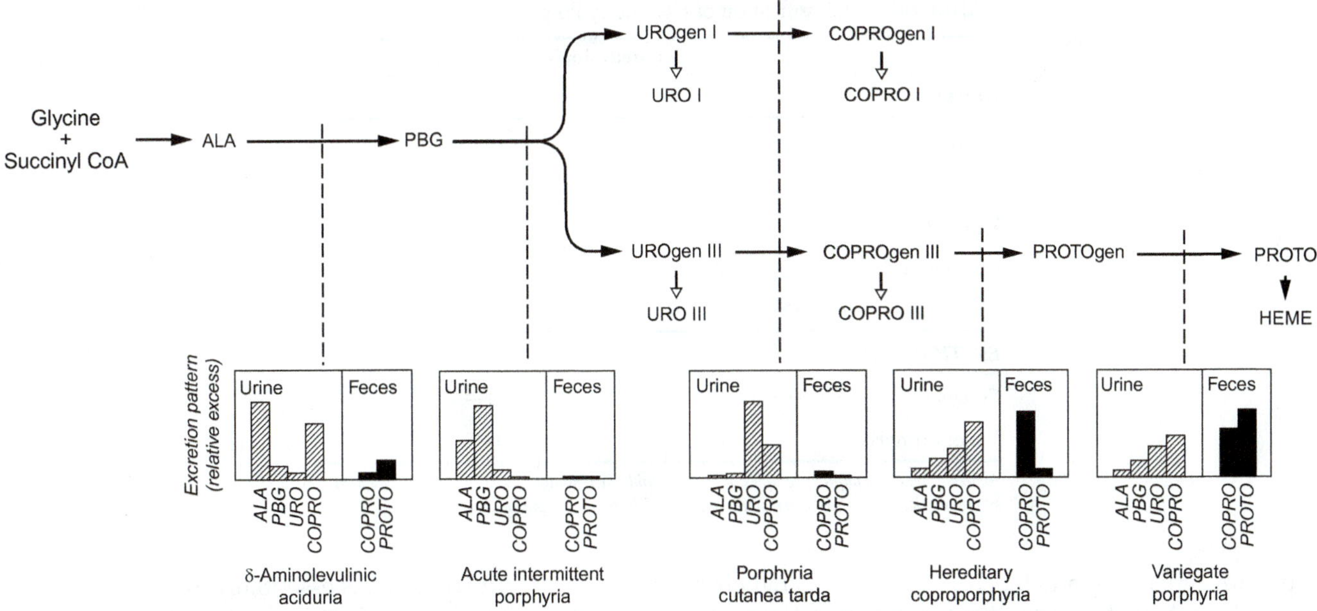

FIGURE 66.2 **The profile of heme precursor excretion associated with the hepatic porphyrias.** The dashed vertical lines indicate the point at which an inherited enzyme deficiency compromises the flow of precursors. For the specific enzymes affected and key to abbreviations, see Figure 66.1.

PORPHYRIA: CLINICAL ASPECTS

Prevalence and Geographic Distribution

Estimates of the incidence of acute attacks of porphyria vary in the range 0.1–1.0 per 100,000 population. It appears, however, that only 10–20% of genetic carriers display symptoms of porphyria. The literature includes only one effort to define the prevalence of mutations for acute intermittent porphyria (AIP), from France. In 3350 healthy blood donors, the prevalence of the genetic carrier state for acute intermittent porphyria was around 1:1650.[4] Although not yet confirmed, the figure suggests that that the number of AIP carriers (who are at risk of symptomatic AIP) is substantially higher than that inferred from the incidence of acute symptoms. Estimates for hereditary coproporphyria and variegate porphyria may be similar, given that these types are more often latent.[5] For reasons of geographic isolation in the past, populations in northern Scandinavia and South Africa have an unusually high prevalence of acute intermittent porphyria and variegate porphyria, respectively. In the case of South Africa, the mutation for variegate porphyria was traced to one of the original settlers who emigrated from the Netherlands to the Cape of South Africa in the late 17th century.[6] Outside of special populations, mutations in porphyria genes appear to be distributed worldwide.[7,8]

Classification of Porphyrias

The porphyrias are grouped according to their predominant manifestation (neurologic or cutaneous) and the tissue in which porphyrin overproduction is most evident (liver or bone marrow; Table 66.1). Liver and bone marrow are sites of relatively high production of heme for synthesis of microsomal cytochromes and hemoglobin, respectively. Thus, some porphyrias are expressed predominantly in liver and others largely or exclusively in bone marrow. In cutaneous porphyria, sun sensitivity correlates with overproduction of porphyrins, which are photosensitizers. In acute porphyria, the relationship between precursor overproduction and symptoms was speculative but is becoming clearer as clinical evidence accumulates implicating ALA in the pathogenesis of acute attacks.

Clinical Manifestations

The porphyrias presenting with acute neurovisceral symptoms comprise acute intermittent porphyria (AIP), hereditary coproporphyria (HCP), variegate porphyria (VP), and δ-aminolevulinic aciduria (ALAD). Although these

TABLE 66.1 Classification of Hereditary Porphyrias

	Acute ("neurologic")	Cutaneous
HEPATIC		
Acute intermittent	+++	0
Coproporphyria	++	+
Variegate	++	++
Cutanea tarda	0	++
ALA uria	+	0
ERYTHROPOIETIC		
Congenital (Gunther's)	0	+++
Protoporphyria	0	++

Abbreviations: 0, symptom complex absent; +, mild when symptomatic; ++, moderate when symptomatic; +++, severe when symptomatic; ALA, δ-aminolevulinic acid.

types are genetically and biochemically distinct, their presentation as acute diseases is essentially identical.[1] Patients typically seek medical attention for abdominal or back pain, which is aching more often than colicky but can suggest inflammation of the gallbladder or bowel. An important differentiating point in porphyric pain is the lack of fever, leukocytosis, or rebound tenderness. Chronic constipation is characteristic and typically worsens with an attack. Abdominal imaging is unremarkable except for signs of ileus.

Mental abnormalities accompany acute attacks in 50–75% of cases and range from depression to delusions; frank psychosis is unusual.[9] King George III of the English house of Stuart (1738–1820) suffered from intermittent psychosis starting around age 50. Although it has been suggested that the cause was acute porphyria,[10] the evidence for this is minimal.[11–13] There is also speculation that the Dutch artist, Vincent van Gogh (1853–1890), who was institutionalized for mental problems before committing suicide at the age of 37, had the same condition.[14–16] If porphyria was present in either person, it was atypical. Most commonly, patients are judged "hysterical" or drug seeking, because their subjective pain is far out of proportion to the physical findings.[17] So striking is this aspect of the acute illness that some patients unfortunately progress to flaccid paralysis or commitment for psychiatric care without a correct diagnosis. The Klüver–Bucy syndrome, consisting of aggressive oral and hypersexual behavior attributed to a temporal lobe lesion, has also been described.[18] Mental changes resolve with treatment of the attack.

Psychosis or seizures can dominate the initial presentation of a porphyric attack. They represent a particular challenge to the neurologist in that many of the widely used long-acting anticonvulsants aggravate the attack and therefore are contraindicated. If present, a history of abdominal pain, nausea or vomiting during the preceding days should suggest porphyria rather than a primary neuropsychiatric disorder. A history of dark urine is helpful but not always present. Hyponatremia may accompany acute attacks and be sufficiently profound to contribute to altered mental status. Indeed, when a young woman presents with seizure and/or abdominal pain and serum sodium <130 meq per ml, acute porphyria should be excluded before any inducing drugs are administered (see Diagnosis, below).

Neuropathy occurs in approximately 40% of acute attacks, generally 1–4 weeks after the onset but in rare cases as late as 11 weeks.[19,20] Weakness is the predominant manifestation and is usually proximal and symmetric. It can involve the upper extremities, the lower extremities, or both (Table 66.2). The rate of neurological progression is highly variable among individual patients but can be rapid, with flaccid tetraplegia and respiratory paralysis appearing within a few days. Cranial nerve involvement, when present, involves nerves X and VII most frequently.[19] Monocular or total blindness can occur and has been attributed to vasospasm; it is usually transient but can be permanent.[21] In patients with altered mental function, electroencephalograms are diffusely abnormal, as in metabolic encephalopathies.[22] Magnetic resonance imaging (MRI) of the brain may reveal posterior lesions. Persistent changes are more frequent in people with a history of acute attack than in genetic carriers who have never had an attack.[23] In some, the presentation resembles posterior leukoencephalopathy syndrome and improves following treatment with hematin.[24] Sensory disturbances occur in as many as 50% of patients in some series and may precede the motor neuropathy.[25] These consist of paresthesias and numbness (usually distal) and are unimpressive in relation to the motor signs.

TABLE 66.2 Symptoms and Signs in Acute Hepatic Porphyria at Presentation

Symptom/sign	Frequency (%)
SYMPTOM	
Abdominal pain	95
Extremity pain	50
Back pain	29
Chest pain	12
Nausea, vomiting	43
Constipation	48
Diarrhea	43
SIGN	
Tachycardia	80
Dark urine	74
Motor deficit	60
Proximal limbs	48
Distal limbs	10
Generalized	42
Altered mentation	40
Hypertension	36
Sensory deficit	26
Seizure	20

Data taken from Ridley[19] and Stein and Tschudy:[108] studies of acute intermittent porphyria.
The findings in hereditary coproporphyria and variegate porphyria are similar.[1,2]

Neuropathology of Acute Porphyria

In early autopsy studies, demyelination was more prominent than axonal degeneration, suggesting a primary disorder of Schwann cells.[26,27] The patchy nature of the process also was emphasized, with damage being present only at intervals along a nerve or affecting some nerves but not others within a bundle, with no relationship to blood vessels or the perineurium.[27] However, subsequent reports, one of which described a sural nerve biopsy during the early phase of an attack, provided evidence for primary axonal degeneration.[28,29] Nerve conduction studies support this, being typically in the normal or low–normal range.[30,31] Also, the time course and pattern of recovery of motor function (proximal to distal) are more consistent with axonal regeneration than with Schwann cell repair.[19]

DIAGNOSIS

In a first presentation of acute porphyria, the principal complaints of abdominal or back pain are not specific, and a delay in diagnosis of weeks to months is not uncommon. Thus, specific laboratory tests are virtually always required for establishing the diagnosis and for characterizing the type of porphyria in an individual case.

A hallmark of the acute porphyrias (other than δ-aminolevulinic aciduria, which manifests as increased ALA rather than PBG) is marked elevation of PBG in urine. This can be established rapidly, if qualitatively, with the two-step Watson–Schwartz test. In the first step, urine is mixed with Ehrlich's aldehyde reagent (*p*-dimethylaminobenzaldehyde in hydrochloric acid), which forms a pink complex with PBG. Although lack of an initial pink reaction product constitutes a negative result, a red complex is not by itself specific for PBG. Urobilinogen, among other substances present in normal urine, reacts with Ehrlich's reagent to give a red/pink color. For a correct interpretation of the test, the pink complex must be tested for solubility in water in a second step, which involves adjusting the solution to pH 4 (with a saturated solution of sodium acetate), then carrying out an extraction with *n*-butanol. The PBG complex remains in the aqueous (lower) layer, whereas color due to urobilinogen and most other chromogens partitions into butanol. When performed in this manner, the Watson–Schwartz test yields few false positives.[32] False negatives can occur if the urine sample is very dilute: the test is insensitive to concentrations of PBG below 10 mg per L. Because of these issues and problems with reproducibility in inexperienced hands, a column test was developed for rapid use (Porphobilinogen Kit, Thermo Scientific TR52001; www.thermoscientific.com). According to two published reports, the performance characteristics of the kit are superior to those of the original Watson–Schwartz test.[33,34] Positive test results with any qualitative method must be confirmed by quantitative column assay,[35] which is widely available; some laboratories are using mass spectrometry.[36] It should be noted that most reference laboratories handle ALA, PBG and fractionated porphyrins as individual tests, which must be ordered separately. A common error is ordering a "porphyrin screen" in a patient with abdominal symptoms suggestive of acute porphyria. This test is for fractionated urine porphyrins, *not* PBG. It is of value mainly in dermatologic diagnosis.

Identifying the type of porphyria requires quantitative assay of urine ALA, PBG, and porphyrins and fecal coproporphyrin and protoporphyrin. As presented in Tables 66.3 and 66.4 and in Figure 66.2, a specific pattern of excretion is associated with each type of acute porphyria. This information can be supplemented by assay of specific enzyme deficiencies. This is particularly important in AIP for the reason that a significant fraction of carriers, varying from 44% to 15% in two European studies, have normal urine analyses when asymptomatic.[37,38] The defective enzyme in AIP, PBG deaminase (PBG-D, also known as hydroxymethylbilane synthase), is present in erythrocytes, making its assay convenient. Because PBG-D does not fluctuate with disease activity, it is a useful adjunct

TABLE 66.3 Normal Values for Porphyrins and Porphyrin Precursors*

	Urine (24 hours)	Feces (dry weight)	Erythrocytes
ALA	<57 pmol (<7.5 mg)	–	–
PBG	<9 pmol (<2 mg)	–	–
URO	12–60 nmol (0.01–0.05 mg)	Trace	–
COPRO	76–382 nmol (0.05–0.25 mg)	<76 nmol/g (<50 pg/g)	<23 nmol/L(<1.5 pg/dl)
PROTO	–	<215 nmol/g (<120 pg/g)	<1.3 pmol/L (<75 pg/dL)

**Column chromatographic and solvent extraction methods are used to quantify ALA and PBG[16] and the porphyrins.[130]*

Abbreviations: ALA, δ-aminolevulinic acid; PBG, porphobilinogen; URO, uroporphyrin; COPRO, coproporphyrin; PROTO, protoporphyrin.

TABLE 66.4 Profile of Heme Precursors in the Porphyrias

		Precursors present in increased amount*			
Type of porphyria	**Enzyme defect**	**Urine**	**Feces**	**Plasma**	**Red blood cells**
Acute intermittent	3[†]	ALA, **PBG**, URO		ALA, PBG	
Hereditary coproporphyria	5	PBG, URO, COPRO	**COPRO**	COPRO	
Variegate porphyria	7	PBG, URO, COPRO	COPRO<**PROTO**	COPRO, PROTO	
δ-Aminolevulinic aciduria	2	**ALA**, COPRO			PROTO
Cutanea tarda	5	**URO**>COPRO	URO	URO	
Congenital erythropoietic	4	URO>>COPRO	URO, COPRO	URO	**URO**
Protoporphyria	8		PROTO	PROTO	**PROTO**
Secondary porphyrinuria	?	(URO), **COPRO**			

**Bold type denotes the precursor that is characteristically increased.*

†For a key to the enzymes of heme synthesis, see Figure 66.1.

Abbreviations: ALA, δ-aminolevulinic acid; PBG, porphobilinogen; URO, uroporphyrin; COPRO, coproporphyrin; PROTO, protoporphyrin.

to urine studies in screening for genetic carriers among the family members of an index case. On average, affected individuals have 50% of normal activity. Problems with this marker include ~20% overlap of the porphyric and normal ranges; in addition, in some families PBG-D activity is reduced in nonerythroid tissues but is normal in erythrocytes. The basis for the latter finding has been determined and is described later in the section on molecular genetics. Patients with subclinical hemolysis also may have false negative tests. PBG-D activity is highest in young erythrocytes and may reach the normal range in a blood sample when red cell turnover is increased. These various circumstances account for a falsely normal PBG-D determination in about 25% of known carriers. Nonetheless, by combining urine studies, PBG-D activity, and pedigree analysis, 80–90% of carriers of AIP can be identified by clinical biochemistry alone.[37,39]

The enzymes that are defective in hereditary coproporphyria and variegate porphyria, respectively, are COPROgen oxidase and PROTOgen oxidase. Although assays for both exist, they require extracts of nucleated cells and are research procedures. Carriers are identified by fecal porphyrin analysis (Figure 66.2), which in the case of variegate porphyria, has a sensitivity of ~75%.[40] A fourth type of acute porphyria is δ-aminolevulinic aciduria, which may be the rarest of the acute porphyrias. It is clinically expressed only in individuals who are homozygous for mutations in PBG synthase.[41,42] Its diagnostic feature is marked elevation of urinary ALA and coproporphyrin, as well as very low activity (<5% of normal) of erythrocyte PBG synthase. The urine pattern resembles that seen in heavy metal intoxication, which should be excluded.

Mutation analysis is the gold standard of diagnosis and is available for all inherited porphyrias. Because of genetic heterogeneity and new mutations, whole-gene sequencing is done in an index case, focused initially on the type of porphyria that seems most likely from clinical and biochemical evaluation. When sequencing is combined with analysis for gene deletions and duplications, the genetic lesion can be identified in 95–99% of cases.[43] Once a family's mutation is defined, screening of relatives is straightforward and can be performed on material from a buccal brush or saliva.

Differential Diagnosis

In patients with progressive weakness due to acute porphyria, the entity most likely to be considered is acute ascending polyneuropathy, the Guillain–Barré syndrome. Several differential points are worth noting. An ascending paralysis (classic in Guillain–Barré, albeit not always present) is rare in acute porphyria.[30] Abdominal pain, constipation, and tachycardia precede the acute neurologic illness in porphyria but either are not present or are coincidental in Guillain–Barré. Also, beyond the first few days of the illness, the protein concentration of the cerebrospinal fluid in Guillain–Barré reaches abnormal levels and mononuclear cells may be present, whereas cerebrospinal fluid usually is normal throughout an attack of porphyria.[44] The most important differential point is the presence in acute porphyria of markedly elevated urine PBG.

Lead intoxication and (in children) hereditary tyrosinemia are considerations in the differential diagnosis, because the symptoms mimic those of acute porphyria. In both, urinary ALA is increased, which is due to inhibition of ALA dehydrase. In lead intoxication, the peripheral neuropathy is predominantly (if not exclusively) motor, as in porphyria, and central nervous system function may be depressed as well. The urine porphyrin pattern (isolated elevation of urine ALA and coproporphyrin) points to the correct diagnosis, which is established with a blood lead determination.

Hereditary tyrosinemia is a disease of tyrosine catabolism involving a mutation in the gene for fumarylacetoacetate hydrolase (FAH).[45] Its inheritance is autosomal recessive. In the liver with defective FAH, fumarylacetoacetate and succinyl acetone accumulate, which have a number of toxic effects including inhibition of ALA dehydrase. Thus, a biochemical hallmark of the disease is high levels of ALA (but not PBG) in blood and urine. Many of the symptoms, including abdominal pain and neurological crises, resemble acute porphyria,[45,46] lending support to the view that ALA is neurotoxic (discussed below). The differentiation of hereditary tyrosinemia from acute porphyria on clinical grounds is straightforward. Tyrosinemia manifests in infancy or early childhood. Acute porphyria, by contrast, is rare before puberty[47] except in individuals who are homozygous for the gene defect (see below). Additionally, hereditary tyrosinemia is associated with vitamin D-resistant rickets and aminoaciduria, neither of which occurs in acute porphyria. Patients with AIP, however—mainly those with a history of acute attacks and/or chronically elevated ALA and PBG—are at risk for slowly progressive renal insufficiency.[48,49]

Secondary Coproporphyrinuria

The evaluation of acute porphyria generally requires quantitative measurement of heme precursors in urine and feces. A frequent finding from screening studies is an isolated increase in urine coproporphyrin. This result is nonspecific, occurring in a wide range of conditions[50,51] and is termed *secondary coproporphyrinuria* (Table 66.4). It should not be construed as evidence of hereditary porphyria.

PATHOGENESIS OF NEUROLOGIC SYMPTOMS

The relationship between altered heme metabolism and neuropsychiatric symptoms, which has challenged investigators for the past 50 years (Table 66.5), is now coming into focus as a result of new studies, both experimental and human, which identify excess circulating ALA as pathogenic. Although PBG is also elevated during acute attacks, its role is in doubt after a 2005 trial in which patients with recurrent attacks of AIP were given recombinant porphobilinogen deaminase (Porphozyme®). The enzyme rapidly reduced the plasma PBG[52] but had no effect on symptoms.

ALA as a Neurotoxin

Two mechanisms of ALA toxicity have been postulated. One is based on the resemblance of ALA to the inhibitory neurotransmitter γ-aminobutyric acid (GABA). GABA is present in the myenteric plexus of the intestine, where it may inhibit peristalsis. Thus ALA, acting as a GABA agonist, could produce the ileus associated with acute porphyria.[53] Several *in vitro* studies exist on the mechanism of GABAergic actions of ALA; in most cases effects were seen only at supraphysiological concentrations of ALA. Although the concentration of ALA at the synapse is unknown, its level in neural tissue appears to be about 10% of that in plasma.[54] In brain, this may be due to both the blood–brain barrier and limited activity of the heme synthetic pathway, which is low at baseline in neural tissue and noninducible by drugs such as barbiturates.[55]

The second hypothesis focuses on ALA as a neurotoxin. In the presence of cuprous or ferrous ion, ALA undergoes oxidation to 4,5-dioxovaleric acid, releasing oxygen radicals,[56,57] which may be directly damaging.[58–62] Experimentally, ALA has been shown to alter mitochondrial permeability and to cause thiol oxidation, actions that are blocked by radical scavengers.[58] However, in two patients with AIP, antioxidant therapy for 8 weeks produced no detectable benefit.[63]

An argument against toxicity of ALA as the primary cause of neural damage in porphyria has been the rather poor correlation between the blood ALA, which can range widely in genetic carriers, and neurological symptoms.[64,65] That said, when an acute attack is treated successfully with heme and ALA drops rapidly to normal or near-normal levels, neurological improvement follows.[66] Also, patients with a history of acute porphyria have more pronounced chronic deficits (motor and sensory) than do asymptomatic carriers,[25] possibly because in many the plasma ALA remains elevated after the symptoms resolve. Administered ALA may cause neurological damage when given as a sensitizer for photodynamic therapy of esophageal adenocarcinoma, as reported in a man age 82 who, after receiving ALA, developed a severe polyneuropathy resembling that of acute porphyria.[67] Infants or children who are homozygous for AIP sustain significant neurological damage, which has been interpreted as evidence for a toxic effect of ALA.[68,69] Finally, patients who undergo liver transplantation for recurrent, severe attacks experience rapid normalization of ALA (and PBG) and durable relief.[70,71]

Heme Deficiency and Altered Hepatocellular Metabolism

Elevation of ALA is but one consequence of an impaired heme synthetic pathway. Cellular heme deficiency may underline some manifestations of an acute attack. Hepatic hemeproteins exhibit varying dependence on readily available heme within the cell. Tryptophan pyrrolase, which catalyzes the conversion of tryptophan to kynurenine,

TABLE 66.5 Postulated Mechanisms for the Neurological Symptoms in Acute Porphyria

Neurotoxicity of ALA
Pro-oxidant
Reduces melatonin production by the pineal gland
ALA as a false neurotransmitter
Structural resemblance to γ-aminobutyric acid (GABA)
Neural heme deficiency
Altered tryptophan metabolism
Increased serotonin in the central nervous system

requires heme as its prosthetic group but binds it with a relatively low affinity. For this reason it is only half-saturated with heme under normal conditions and particularly susceptible to inactivation during heme deficiency. Such inactivation in theory shunts tryptophan from the kynurenine pathway to the tryptophan hydroxylase pathway, which yields 5-OH-tryptamine (serotonin). The predicted result is a serotonergic state. This is consistent with studies in animals and conceivably accounts for some porphyric manifestations, such as a heightened sensitivity to sedatives. Blood tryptophan is increased in some patients with AIP, with serotonin and its urinary metabolite, 5-hydroxyindole acetic acid, both increasing during acute attacks.[72,73] These changes are reversed by administration of heme.[74] It is unclear whether altered tryptophan metabolism contributes to the pain and neurological symptoms in acute porphyria.

Plasma melatonin also decreases in patients with acute porphyria.[75] This is a surprise, given that serotonin, the substrate for melatonin production, is increased. However, unlike the changes in serotonin and other metabolites of tryptophan, the drop in melatonin is unaffected by heme administration[75] and thus does not appear to reflect a state of heme deficiency in the pineal, which is the source of melatonin. The defect in melatonin production occurs also in rats with high circulating ALA.[76] The pineal is outside the blood–brain barrier and thus is likely to be more affected by systemic ALA than are other parts of the brain. Whether the loss of melatonin production represents a direct toxicity of ALA or a GABAergic effect is uncertain. The fact that administered GABA has a similar effect supports the latter possibility.[77] Deficient melatonin (and loss of circadian variation) could account for some of the prodrome of acute porphyria (mood disturbances, insomnia), although this remains to be proven. It is interesting that melatonin is an antioxidant and, in experimental systems, counters the pro-oxidant effects of ALA.[60–62,77] This opens the possibility that a primary loss of melatonin sensitizes the patient to the neurotoxic effects of ALA. Alternatively, loss of melatonin may be secondary to the high levels of ALA associated with acute attacks. Longitudinal studies of patients between and during attacks will be necessary to address this question.

CHEMICAL AND PHYSIOLOGIC INDUCERS OF ACUTE PORPHYRIA

Pharmaceuticals have been linked to many acute attacks of porphyria, and lists of offending drugs have been compiled, which are available on line: American Porphyria Foundation (http://www.porphyriafoundation.com/testing-and-treatment/drug-safety-in-acute-porphyria); and the Norwegian Porphyria Center (NAPOS; http://www.drugs-porphyria.org/index.php). Much of the information is anecdotal, some published, some not. Only the NAPOS site represents an evidence-based approach. The length of some lists creates the mistaken impression that no drug is safe. The most hazardous pharmaceuticals are those that stimulate hepatic heme production, generally by inducing synthesis of apocytochrome P-450, which requires ramping up cellular heme production. The most important class of inducers includes the barbiturates and related compounds such as phenytoin (Dilantin®). Clinical experience confirms the risk of these agents to carriers of acute porphyria (Table 66.6). While this presents a problem for the treatment of seizure disorders, for virtually all other clinical indications, effective but less risky alternatives exist.

Test systems are used to prospectively assess the risk of therapeutic agents. These consist of the rat liver *in vivo*, chick embryo liver *in ovo*, chick embryo hepatocytes in culture, and LMH cells, which are derived from an avian liver carcinoma and in which heme metabolism is remarkably similar to that of hepatocytes *in vivo*.[78] Such systems are helpful for estimating the risk of available drugs for people with acute porphyria, even if the correlation between experimental data and clinical experience is less than perfect. Compounds such as diazepam, which are inducers in chick embryo hepatocytes (probably the most sensitive of the test systems), have been used with impunity in porphyria. Conversely, relatively poor inducers in experimental systems, such as clonazepam, have not been entirely benign in patients.[79] Despite interspecies differences in hepatic metabolism, discordant findings are the exception rather than the rule. A reasonable approach is to categorize all medications that have been linked to at least one well-documented attack of acute porphyria as "hazardous" and to label any chemicals that are active in a test system but not yet implicated as a cause of acute porphyria in humans as "possibly hazardous."[80,81] A third and important category comprises drugs that are not inducers of cytochrome P-450, have never precipitated acute porphyria and are considered safe; it includes some extremely useful agents.

A physiological condition that predisposes humans to acute porphyria is fasting. It is well documented in experimental animals that fasting sensitizes hepatic ALAS to induction by barbiturates and related chemicals and that administered glucose reverses the effect.[82] This is the basis for giving glucose to patients in an acute attack, which tend to occur in the setting of a fast (either presurgical or due to illness). First attacks of acute porphyria have occurred following gastric bypass surgery for morbid obesity.[83] While these observations underline the role of extreme fasting (and fad diets) in the onset of an attack, it does not follow that a diet with excess carbohydrates protects against

TABLE 66.6 Use of Medications in Porphyria

	Unsafe	Believed to be safe
ANTICONVULSANTS		
	Barbiturates Carbamazepine Clonazepam Ethosuximide Felbamate Phenytoin Primidone Tiagabine Valproic acid	Bromides Diazepam Gabapentin Magnesium sulfate
HYPNOTICS/SEDATIVES		
	Barbiturates Chlordiazepoxide Ethchlorvynol Glutethimide Meprobamate Methyprylon	Chloral hydrate Chlorpromazine Diphenhydramine Lithium Lorazepam Meclizine Trifluoperazine
OTHER		
	α-Methyldopa Danazol Ergot preparations Oral contraceptives Griseofulvin Imipramine Pentazocine Pyrazinamide Sulfonamides	Adrenocorticotropic hormone Allopurinol Aminoglycosides Aspirin Codeine Colchicine Furosemide Ibuprofen Insulin Meperidine Morphine Naproxen Warfarin

attacks. Too often, the individual who is "pushing carbs" acquires obesity as an additional health issue. Exercise is another lifestyle choice that can affect caloric balance, but it has never been implicated in an acute attack.

Studies in mice have shed new light on the mechanism of the fasting effect in acute porphyria, which involves the regulation of ALA synthase. The initial observation was that the nuclear protein, peroxisome peroxisome proliferator-activated receptor γ coactivator 1α (PGC-1α), is involved in the metabolic switch that occurs with the change from fed to fasting. Its level is induced with fasting, and it forms a complex with other DNA-binding proteins to stimulate transcription of genes for fatty acid oxidation and hepatic gluconeogenesis. Among the responding genes is ALA synthase. To confirm the functional importance of PGC-1α, mice were generated with deletion of the gene selectively in hepatocytes. In these mice, the effect of fasting on the level of ALA and its induction by porphyrogenic chemicals was significantly blunted.[84]

Spontaneous attacks also occur and can be recurrent. They involve women much more frequently than men, with a peak incidence in the third and fourth decades of life; active porphyria prior to puberty is rare. Attacks in some women are cyclic, occurring just prior to menstruation, which implies changes in ovarian hormone levels. In test systems, progestins may be more porphyrogenic than are estrogens;[85] danazol also has been associated with a worsening clinical picture.[86,87] Acute attacks of hereditary coproporphyria and variegate porphyria occur about equally in males and females, and most are medication related.[5]

MOLECULAR GENETICS

The acute porphyrias originally were viewed as autosomal dominant disorders of variable penetrance. It is recognized now that two forms, δ-aminolevulinic aciduria and harderoporphyria, are recessive in that heterozygotes invariably are asymptomatic. Both are very rare. Fewer than a dozen cases of δ-aminolevulinic aciduria have been reported. Harderoporphyria is a variant of hereditary coproporphyria. The manifestations are markedly elevated urinary porphyrins and photosensitivity; acute neurovisceral attacks do not occur.[88] A number of cases of homozygous, compound heterozygous, or "dual" porphyria have been reported. Homozygous AIP with 2–5% residual PBG deaminase activity is apparent in early childhood, with ataxia, mental retardation and early death.[69,89] In homozygous VP, neurological disease may be present or absent depending on the severity of the mutation. In South Africa, where due to the well-known founder affect (see above) 95% of carriers harbor the same mutation, the homozygous state is manifest as unusually severe disease starting in childhood.[90]

Porphobilinogen Deaminase

PBG-D is the most extensively studied of these enzymes and the one that is mutated in AIP. A single copy located on chromosome 11 has two transcriptional start sites, which encode proteins of 44 and 42 kDa, respectively.[91] The smaller of the two is expressed solely in erythroid cells; the larger is ubiquitous. Each promoter has been examined by DNase-I hypersensitivity mapping and by functional studies of deletion mutants in transgenic mice or erythroleukemia cells. The promoter for the erythroid form is tissue-specific, sharing motifs with other erythroid-specific genes such as globins.[92,93] Erythroid cells also have a mechanism for premature termination of transcription from the upstream promoter, to exclude interference with the tissue-specific promoter.[93] The complete genomic sequence for PBG deaminase has been published and around 300 different mutations documented. Studies of the enzyme from *Escherichia coli* indicate conservation of key functional domains. Several mutations that are clustered in exons 10 and 12 give rise to inactive protein by affecting residues in the active site.[94,95]

Variation in Tissue Expression of Porphobilinogen Deaminase

Before DNA analysis, the first indication of genetic heterogeneity among carriers of AIP was the observation that in rare families, the expected deficiency of PBG-D was present only in nucleated cells, contrary to the usual finding of reduced activity in erythrocytes as well as nucleated cells.[96] Analysis of the transcriptional unit for PBG-D in erythroid and nonerythroid tissues provided an explanation. The mutation in families with AIP but normal erythroid PBG-D lie upstream of the second start site, sparing the erythroid transcript. Most mutations are distal and therefore affect both forms of the enzyme.

A different form of heterogeneity has been inferred by comparison of active PBG-D and total immunoprecipitable enzyme protein in various families with AIP. While in most cases there is a similar decrease in enzyme activity and total enzyme protein, in approximately 15%, the ratio of enzyme activity to total enzyme protein is less than 1, indicating a mutation that results in enzymatically impaired but stable PBG-D.[97]

In most porphyria centers, mutation analysis has not proven to be useful for predicting the likelihood or frequency of acute porphyric attacks. The exception is Sweden, where two HMBS mutations were associated with more active disease.[98] This is consistent with the importance of medications, other environmental factors and modifier genes in the pathogenesis of acute attacks. The latter may include genes that regulate metabolism in the fasting state.[84] Genetic heterogeneity has been documented for hereditary coproporphyria and variegate porphyria as well as AIP.

ANIMAL MODELS

Most models of acute porphyria involve the administration of chemicals that cause rapid degradation of hepatic microsomal heme while blocking a step in the pathway. The result is a disturbance in heme homeostasis that biochemically resembles the human disease. The prototype inducer is allyl isopropylacetamide.[99,100] It undergoes activation on cytochrome P-450 to a radical species, which forms a covalent adduct with one of the pyrrole nitrogens of the enzyme's heme constituent. The modified heme is unstable and undergoes degradation.[101]

While these models have illuminated several important aspects of hepatic heme regulation, none has reproduced the neurological abnormalities of human porphyria. One approach to this question was developing mice with a null mutation in PBG deaminase.[102] As expected, homozygotes for the mutation died *in utero*. Although heterozygotes had 50% of normal hepatic enzyme activity (as in AIP), they exhibited no biochemical or behavioral abnormalities. Even after challenge with phenobarbital, urinary excretion of ALA was normal. Reasoning that the mouse might require a more than a 50% deficiency in PBG deaminase for disease manifestations, the investigators created a second mouse with a partial defect in one allele and crossed it with the mouse line bearing the null allele. PBG deaminase in the compound mutant heterozygote was 30% of normal. Although baseline ALA excretion remained low (less than twofold increase above normal), phenobarbital administration elicited a marked increase in hepatic ALA synthase activity and urinary ALA.[103] Most importantly, over a period of 6–16 months, the animals developed neurological deficits characterized as muscle weakness and loss of motor function; histology showed axonal degeneration. Nerve conduction was essentially unaltered. Because the neurological syndrome proceeds in mice with normal or minimally elevated ALA, the authors concluded that it reflects an underlying neural heme deficiency rather than ALA toxicity. While the model mimics certain aspects of AIP in humans, phenobarbital administration failed to produce acute symptoms such as seizures or rapidly progressive paralysis. Additionally, the neuropathology that developed chronically in the mutant mice appears to be much more severe than that reported for asymptomatic human carriers of AIP.[25,104] Nevertheless, this model lends itself to examining possible mechanisms of neurological damage in porphyria and testing therapeutic interventions.

THERAPY

Acute Attacks

Initial management is directed at pain relief, support for complications such as hyponatremia, and elimination of factors implicated in the attack. Anecdotal reports support the use of propranolol, which counters the tachycardia present in most patients and may also relieve pain.[105] If used, propranolol should be introduced at a low dose, as hypotensive reactions have occurred in some patients.[106] Most cases require opiates (meperidine or morphine) for adequate analgesia, often in large doses. For patients with a clear-cut acute attack, the risk of addiction is minimal. With effective treatment, such as administered heme, pain resolves in less than 5 days,[66] and pain medication can be discontinued at the same time. In known genetic carriers of acute porphyria, opiates are needed only for acute attacks. Their use between attacks should not be necessary and calls the diagnosis into question.

Hyponatremia often accompanies acute attacks of porphyria and may contribute to neurologic abnormalities. Although not always present on admission, hyponatremia nonetheless can emerge rapidly with administration of intravenous fluids (e.g., dextrose in water). Its basis is debated. In some patients, high urine osmolarity suggests the syndrome of inappropriate secretion of antidiuretic hormone (SIADH), whereas in others, hypovolemia may be the underlying problem. Because the presence of SIADH cannot be assumed, water restriction should be avoided in favor of saline administration, with precautions to minimize the risk of brainstem damage.[107]

All nonessential medications, in addition to those implicated as a cause of acute attacks, should be discontinued. For reversing a fasting state (present in most patients because of pain and nausea), carbohydrate is given with the goal of providing approximately 400 g dextrose daily. Although this therapy has not been tested in controlled trials, it is supported by clinical experience and on occasion can result in symptomatic improvement.[108]

Administration of heme, as hematin (hydroxyheme), constitutes definitive therapy, on the premise that the acute porphyrias are diseases of heme deficiency. Initial studies were carried out in gravely ill patients but established that intravenously administered heme was effective in suppressing overproduction of ALA and PBG.[109] Subsequent work from several groups demonstrated the clinical efficacy of this treatment,[66,110,111] leading to the current recommendation that hematin be used early in the course of a first attack, as soon as it can be obtained. In patients without neurologic signs, conservative care involving the measures outlined earlier may be pursued for 24–48 hours, supplemented by hematin after this period if the clinical picture is unchanged or deteriorating. In patients presenting with neurologic signs, hematin is started as early as possible, together with carbohydrates and supportive care.

Heme itself (ferriprotoporphyrin IX) is essentially insoluble in aqueous media but dissolves readily in alkaline solutions (e.g., 0. 1 M sodium carbonate, pH 8.0) as hydroxyheme, or *hematin*. Hematin is unstable in solution[112] and thus has been formulated in dry form for reconstitution immediately prior to infusion (Panhematin, Recordati Rare Diseases, USA). Current information for ordering Panhematin is available on the website of the American Porphyria Foundation (http://www.porphyriafoundation.com/treatment). Panhematin comes with sterile water for preparation. Alternatively it can be dissolved in human albumin solution, which reduces the risk of a chemical phlebitis with a peripheral infusion.[113] Heme may be rendered soluble also by complexing it with the amino acid arginine to form heme arginate, which is stable in a vehicle consisting of 40% 1,2-propanediol and 10% ethanol. Its efficacy is essentially identical to that of hematin.[114] It is marketed in Europe and South Africa as Normosang (Orphan Europe SARL, Paris, France). Orally administered heme is presumably degraded by intestinal heme oxygenase[115] and therefore would be ineffective. The experience with this route of administration is very limited but suggests that oral hematin has little, if any, effect relative to that of intravenous hematin.[110]

Freshly prepared hematin (or heme arginate) is given at a dose of 1.5–2.0 mg per kg body weight by slow intravenous infusion (over 10 minutes). Initially, it was recommended that a dose be given every 12 hours,[116] but a single daily administration is equally effective.[66] Urinary PBG and symptoms are monitored. A rapid drop in PBG (plasma or urine) to less than 20% of the preinfusion level after 48 hours confirms the biochemical efficacy of the infusion. A symptomatic response appears 72–96 hours after the initiation of therapy and typically is dramatic, with marked pain reduction, improved mental status, and discontinuation of opiate use.[66] Objective neurologic signs, however, respond variably depending on the underlying pathology. Motor neuron degeneration is not reversed by hematin, although its progression is slowed or halted.[117]

Treatment failures have been reported.[118,119] Some cases, however, may be attributable to use of decayed hematin. As noted, the drug decays in alkaline aqueous solution, even when refrigerated, and thus must be given directly after its preparation. Another problem is misdiagnosis. In genetic carriers who are acutely ill, it can be difficult to establish with certainty whether symptoms are due to porphyria or to an unrelated problem. If the patient's baseline urinary PBG excretion is known, a substantial increase supports the diagnosis. With a new porphyria attack, the PBG should be more than 25 mg per 24 hours or, for patients with a high baseline PBG, at least twofold increased. In the experience of specialists who are familiar with Panhematin administration and the clinical response, true hematin resistance is rare—limited to patients with frequent attacks and consistently very high excretion of ALA and PBG. Nonresponse in a typical case should raise the question of inactive hematin or a wrong diagnosis.

Adverse effects of hematin are minor and infrequent, provided that the recommended dose is not exceeded. Hematin in large amounts is known to be nephrotoxic; one patient receiving an unusually large dose exhibited renal failure, which was transient.[120] An accidental overdose of heme arginate in another patient (sixfold the recommended dose) resulted in acute liver failure requiring urgent transplant.[121] One case of an anaphylactic reaction to heme arginate has been reported, possibly involving the propanediol component.[122] Chemical phlebitis occurs in approximately 4% of hematin infusions, presumably owing to the alkalinity of the solution. The problem is seen with heme arginate but is much less severe. Hematin phlebitis is minimized by using as large a vein as possible and a slow infusion rate. Preparing the hematin with human albumin solution may help.[113] For patients with small or deep veins, the infusion should be given through a central line. Anticoagulant effects of hematin were noted in early reports and consist of a prolonged prothrombin time and reduced platelets.[123,124] The abnormalities were transient, and clinically significant bleeding was unusual. Anticoagulant effects appear to be attributable to decayed hematin[112,125] and should be minimal with freshly reconstituted material, but caution is advised in a patient with impaired coagulation or a bleeding tendency. Heme arginate has minimal or no effect on coagulation.[126] With chronic administration of either form, iron accumulation is a concern and can be monitored as an increase in serum ferritin. When the ferritin is >1000 ng per ml, the iron burden can be reduced by judicious phlebotomy or administration of a chelator such as deferasirox.

Premenstrual Exacerbations and Pregnancy

Changes in reproductive hormones occurring during menstruation may precipitate attacks of acute intermittent porphyria. The initial approach is standard therapy including hematin, along with elimination of medications that may be hazardous in porphyria (Table 66.6). If these measures are ineffective or if the exacerbations occur monthly, ovulatory suppression with gonadotropin-releasing hormone (GnRH) agonists may be helpful.[127,128] The latter are peptides that have no demonstrable porphyria-inducing activity, in contrast to the synthetic steroids in oral contraceptives, which have been linked to attacks in some patients.[85,129] Administration of hematin, given weekly or timed to the expected onset of symptoms, also has been effective.[130] As noted previously, danazol is hazardous in porphyria.

The effect of pregnancy is unpredictable. In the experience of porphyria specialists, women with porphyria and a history of attacks are more likely to feel better during a pregnancy than worse. Most notice no change. If an attack occurs during gestation, it can be treated with hematin. There are no reports suggesting that the infusion is teratogenic.[85,131]

Treatment Failure and Liver Transplantation

In a few cases, which fortunately are uncommon, attacks become frequent and persist for years. Ovulatory suppression may not be effective, and frequent hematin infusion—even weekly—does not completely control symptoms. Some patients develop neurological signs. In this situation, liver transplant has been performed since

2004.[70] The procedure is followed by rapid normalization of urine chemistry and resolution of acute symptoms. Post-transplant care, however, can be challenging, particularly for patients with severe chronic neurological deficits and/or opiate addiction. Also described has been an unusually high incidence of hepatic artery thrombosis, which occurred in 4 of 10 cases from the UK, detected between 3 days and 9 months after transplant.[71] The sample size is small, and technical factors may have played a role. Nonetheless, the rate of this complication (40% vs. 3–5% for liver transplant overall) is striking and worrisome. The criteria for proceeding to transplant and its timing are open questions.[132] Recurrent attacks may remit spontaneously, even after several years, raising the specter of premature transplant. On the other hand, if the procedure is withheld until neurological deficits are severe, outcomes are likely to be suboptimal.[71] Unfortunately, there is no reliable method for predicting the course of the disease.

Seizures

Seizures, when present, are grand mal and may occur at the onset of an acute attack. They can be suppressed with intravenous diazepam. The use of magnesium sulfate also has been recommended, based on its efficacy in eclamptic seizures,[133] although experience with this agent is limited.[134] In one patient with status epilepticus, magnesium was used in combination with levetiracetam.[135] Gabapentin alone or in combination with propofol also has been effective.[136,137] Propofol has been used for anesthesia of genetic carriers who are not in an acute attack[138] and presumably is safe for treating status epilepticus. Use of sevoflurane also has been reported.[139] Rocuronium[140] and atracurium[141] appear to be safe for neuromuscular blockade. Patients presenting with a seizure should receive hematin or heme arginate, but as already noted, benefit is seen only after a lag period of 3–4 days.

Treatment of chronic seizures in genetic carriers of acute porphyria is singularly difficult, because many of the first-line agents have been implicated in acute attacks.[79,142–144] Barbiturates and related compounds (such as phenytoin) are absolutely contraindicated (Table 66.6). Individual patients may tolerate carbamazepine or valproic acid, although the risk of a porphyric exacerbation is substantial.[79,143,145] Gabapentin has been effective in a few reported cases and without complications.[146] Studies of various antiseizure compounds in cultured liver cells support the use of gabapentin or vigabatrin but suggest that felbamate, lamotrigine, and tiagabine should be avoided.[147] In an *in vivo* model of porphyria (rats with chemical blockade of PROTOgen oxidase), tiagabine and topiramite were porphyrogenic albeit less so than phenobarbital.[148]

Bromides have fallen out of favor as an anticonvulsant but are safe in porphyria and apparently effective.[79,142] Despite their narrow therapeutic index and the necessity for close monitoring, they have been used in the modern era for childhood epilepsy that is resistant to standard therapy, with good effect.[149–151] A conservative approach is to initiate therapy with 10 mg per kg per day, taken in divided doses with food to minimize gastric irritation.[151] A serum level of 80–100 mg per dL is targeted, but this may be adjusted to achieve the desired balance between clinical efficacy and toxicity. Serum bromide is assayed by a colorimetric procedure,[152] which is simple but may not be available in clinical laboratories. Bromide ion interferes with the serum chloride determination by ion-specific electrode, resulting in an apparent hyperchloremia. Side effects occur in virtually all patients who achieve anticonvulsant levels of bromide and consist primarily of sedation, an acneiform skin eruption, and gastrointestinal symptoms. Toxic effects ("bromism") include psychosis and ataxia. Bromide is eliminated through the kidneys. Although its half-life is long (approximately 12 days), excretion can be accelerated by the administration of sodium chloride and diuretics.

Genetic Therapy

The results of liver transplantation suggest that correction of the genetic defect solely in the liver is sufficient for clinical cure, even though all nucleated cells harbor the defective gene. The future for intractable disease, however, is not transplant but gene therapy directed at either knocking down excess ALA production or correcting the mutation itself. The latter may be accomplished through viral vectors that home to hepatocytes or by infusion of autologous hepatocytes in which the porphyria mutation has been corrected. Both strategies, while not ready for clinical use, are under active study. While this area moves forward, it should be kept in mind that the vast majority of genetic carriers have no symptoms provided they avoid circumstances associated with acute attacks. The first line of treatment is prevention, which entails screening of the patient's family for identification of silent genetic carriers and education of all affected members, with the goal of ensuring that they have a long and porphyria-free life.

PROGNOSIS

In a series reported in 1969,[19] 10 of 29 acute porphyric attacks with neuropathy were fatal. The current mortality rate is substantially lower than this, most likely as a result of heightened awareness of the disease and tests for the genetic trait, with screening of possible carriers and early diagnosis.[5] Undoubtedly hematin treatment and improved intensive care also have had a positive impact.

Given that most acute attacks are precipitated by an identifiable factor, repeat episodes are avoidable with appropriate education of the patient and occur in fewer than 20% of those who have had an acute attack. Neurologic deficits, when present, resolve slowly, proceeding from proximal to distal muscles, consistent with axonal regrowth. Although permanent sequelae have been described,[153] complete recovery is the rule over an average time of 10.6 months after the acute episode.[19]

Acute porphyria may manifest as an altered mental state. Given the difficulty of the diagnosis, the question arises as to whether mental institutions have a disproportionate number of people with unrecognized porphyria. Patients hospitalized for mental illness have been surveyed for reduced activity of PBG-D and elevated urinary PBG. The results are mixed. One study found an unexpectedly high prevalence of abnormal tests,[154] although few if any of the patients with positive results had a history suggestive of past episodes of acute porphyria. In another study, the prevalence of positive tests did not differ from that in the control population.[155] For patients with a recognized attack who are given appropriate therapy, the acute mental disturbances resolve along with the pain symptoms. An association with chronic mental illness has not been found.[9,156]

As already mentioned, chronic renal insufficiency is a complication of acute porphyria, which may be due in part to the toxic effect of ALA. The risk appears to be increased in people with a history of acute attacks and in those who are asymptomatic but have substantially elevated circulating ALA.[49] An elevated risk of primary liver cancer (hepatocellular carcinoma, HCC) also has been reported from Sweden.[157] Although it is unclear whether this pertains elsewhere, current recommendations are to perform annual screening for liver cancer, such as imaging (e.g., ultrasound) and serum α-fetoprotein, on all patients over age 50 regardless of a history of attacks. For patients with evidence of significant liver disease, semi-annual screening may be justified.

References

1. Anderson KE, Bloomer JR, Bonkovsky HL, et al. Recommendations for the diagnosis and treatment of the acute porphyrias. *Ann Intern Med.* 2005;142(6):439–450.
2. Dandekar T, Stripecke R, Gray NK, et al. Identification of a novel iron-responsive element in murine and human erythroid delta-aminolevulinic acid synthase mRNA. *EMBO J.* 1991;10(7):1903–1909.
3. Bishop DF. Two different genes encode delta-aminolevulinate synthase in humans: nucleotide sequences of cDNAs for the housekeeping and erythroid genes. *Nucleic Acids Res.* 1990;18(23):7187–7188.
4. Nordmann Y, Puy H, Da Silva V, et al. Acute intermittent porphyria: prevalence of mutations in the porphobilinogen deaminase gene in blood donors in France. *J Intern Med.* 1997;242(3):213–217.
5. Hift RJ, Meissner PN. An analysis of 112 acute porphyric attacks in Cape Town, South Africa: evidence that acute intermittent porphyria and variegate porphyria differ in susceptibility and severity. *Medicine (Baltimore).* 2005;84(1):48–60.
6. Dean G. *The Porphyrias. A Study of Inheritance and Environment.* London: Pitman Medical; 1971.
7. Kondo M, Yano Y, Shirataka M, Urata G, Sassa S. Porphyrias in Japan: compilation of all cases reported through 2002. *Int J Hematol.* 2004;79(5):448–456.
8. Liu YP, Lien WC, Fang CC, et al. ED presentation of acute porphyria. *Am J Emerg Med.* 2005;23(2):164–167.
9. Patience DA, Blackwood DH, McColl KE, Moore MR. Acute intermittent porphyria and mental illness—a family study. *Acta Psychiatr Scand.* 1994;89(4):262–267.
10. Macalpine I, Hunter R, Rimington C. Porphyria in the royal houses of Stuart, Hanover, and Prussia. A follow-up study of George III's illness. *Br Med J.* 1968;1(5583):7–18.
11. Cox TM, Jack N, Lofthouse S, Watling J, Haines J, Warren MJ. King George III and porphyria: an elemental hypothesis and investigation. *Lancet.* 2005;366(9482):332–335.
12. McDonagh AF, Bissell DM. Porphyria and porphyrinology—the past fifteen years. *Semin Liver Dis.* 1998;18:3–15.
13. Peters T. King George III, bipolar disorder, porphyria and lessons for historians. *Clin Med.* 2011;11(3):261–264.
14. Blumer D. The illness of Vincent van Gogh. *Am J Psychiatry.* 2002;159(4):519–526.
15. Bonkovsky HL, Cable EE, Cable JW, et al. Porphyrogenic properties of the terpenes camphor, pinene, and thujone (with a note on historic implications for absinthe and the illness of Vincent van Gogh). *Biochem Pharmacol.* 1992;43(11):2359–2368.
16. Hughes JR. A reappraisal of the possible seizures of Vincent van Gogh. *Epilepsy Behav.* 2005;6(4):504–510.
17. Puy H, Gouya L, Deybach JC. Porphyrias. *Lancet.* 2010;375(9718):924–937.
18. Guidotti TL, Charness ME, Lamon JM. Acute intermittent porphyria and the Kluver–Bucy syndrome. *Johns Hopkins Med J.* 1979;145(6): 233–235.
19. Ridley A. The neuropathy of acute intermittent porphyria. *Q J Med.* 1969;38(151):307–333.

20. Lin CS, Lee MJ, Park SB, Kiernan MC. Purple pigments: the pathophysiology of acute porphyric neuropathy. *Clin Neurophysiol.* 2011;122(12):2336–2344.
21. Lai CW, Hung TP, Lin WS. Blindness of cerebral origin in acute intermittent porphyria. report of a case and postmortem examination. *Arch Neurol.* 1977;34(5):310–312.
22. Pischik E, Kauppinen R. Neurological manifestations of acute intermittent porphyria. *Cell Mol Biol (Noisy-le-grand).* 2009;55(1):72–83.
23. Bylesjo I, Brekke OL, Prytz J, Skjeflo T, Salvesen R. Brain magnetic resonance imaging white-matter lesions and cerebrospinal fluid findings in patients with acute intermittent porphyria. *Eur Neurol.* 2004;51(1):1–5.
24. Utz N, Kinkel B, Hedde JP, Bewermeyer H. MR imaging of acute intermittent porphyria mimicking reversible posterior leukoencephalopathy syndrome. *Neuroradiology.* 2001;43(12):1059–1062.
25. Wikberg A, Andersson C, Lithner F. Signs of neuropathy in the lower legs and feet of patients with acute intermittent porphyria. *J Intern Med.* 2000;248(1):27–32.
26. Gibson JB, Goldberg A. The neuropathology of acute porphyria. *J Pathol Bacteriol.* 1956;71:495–509.
27. Naef RW, Berry RG, Schlesinger MS. Neurologic aspects of porphyria. *Neurology.* 1958;9:313–320.
28. Thorner PS, Bilbao JM, Sima AA, Briggs S. Porphyric neuropathy: an ultrastructural and quantitative case study. *Can J Neurol Sci.* 1981;8(4):281–287.
29. King PH, Petersen NE, Rakhra R, Schreiber WE. Porphyria presenting with bilateral radial motor neuropathy: evidence of a novel gene mutation. *Neurology.* 2002;58(7):1118–1121.
30. Albers JW, Fink JK. Porphyric neuropathy. *Muscle Nerve.* 2004;30(4):410–422.
31. Flugel KA, Druschky KF. Electromyogram and nerve conduction in patients with acute intermittent porphyria. *J Neurol.* 1977;214(4):267–279.
32. Pierach CA, Cardinal R, Bossenmaier I, Watson CJ. Comparison of the Hoesch and the Watson-Schwartz tests for urinary porphobilinogen. *Clin Chem.* 1977;23(9):1666–1668.
33. Deacon AC, Peters TJ. Identification of acute porphyria: evaluation of a commercial screening test for urinary porphobilinogen. *Ann Clin Biochem.* 1998;35(Pt 6):726–732.
34. Vogeser M, Stauch T. Evaluation of a commercially available rapid urinary porphobilinogen test. *Clin Chem Lab Med.* 2011;49(9):1491–1494.
35. Davis JR, Andelman SL. Urinary delta-aminolevulinic acid. *Arch Environ Health.* 1967;15(1):53–59.
36. Zhang J, Yasuda M, Desnick RJ, Balwani M, Bishop D, Yu C. A LC-MS/MS method for the specific, sensitive, and simultaneous quantification of 5-aminolevulinic acid and porphobilinogen. *J Chromatogr B Analyt Technol Biomed Life Sci.* 2011;879(24):2389–2396.
37. Andersson C, Thunell S, Floderus Y, et al. Diagnosis of acute intermittent porphyria in northern Sweden: an evaluation of mutation analysis and biochemical methods. *J Intern Med.* 1995;237(3):301–308.
38. Kauppinen R, von und zu Fraunberg M. Molecular and biochemical studies of acute intermittent porphyria in 196 patients and their families. *Clin Chem.* 2002;48(11):1891–1900.
39. Lamon JM, Frykholm BC, Tschudy DP. Family evaluations in acute intermittent porphyria using red cell uroporphyrinogen I synthetase. *J Med Genet.* 1979;16(2):134–139.
40. von und zu Fraunberg M, Kauppinen R. Diagnosis of variegate porphyria—hard to get? *Scand J Clin Lab Invest.* 2000;60(7):605–610.
41. Bird TD, Hamernyik P, Nutter JY, Labbe RF. Inherited deficiency of delta-aminolevulinic acid dehydratase. *Am J Hum Genet.* 1979;31(6):662–668.
42. de Verneuil H, Doss M, Brusco N, Beaumont C, Nordmann Y. Hereditary hepatic porphyria with delta aminolevulinate dehydrase deficiency: immunologic characterization of the non-catalytic enzyme. *Hum Genet.* 1985;69(2):174–177.
43. Whatley SD, Mason NG, Woolf JR, Newcombe RG, Elder GH, Badminton MN. Diagnostic strategies for autosomal dominant acute porphyrias: retrospective analysis of 467 unrelated patients referred for mutational analysis of the HMBS, CPOX, or PPOX gene. *Clin Chem.* 2009;55(7):1406–1414.
44. Sergay SM. Management of neurologic exacerbations of hepatic porphyria. *Med Clin North Am.* 1979;63(2):453–463.
45. Gentz J, Johansson S, Lindblad B, Lindstedt S, Zetterstrom R. Exertion of delta-aminolevulinic acid in hereditary tyrosinemia. *Clin Chim Acta.* 1969;23(2):257–263.
46. Strife CF, Zuroweste EL, Emmett EA, Finelli VN, Petering HG, Berry HK. Tyrosinemia with acute intermittent porphyria: aminolevulinic acid dehydratase deficiency related to elevated urinary aminolevulinic acid levels. *J Pediatr.* 1977;90(3):400–404.
47. Badcock NR, DA O, Zoanetti GD, Robertson EF, Parker CJ. Childhood porphyrias: implications and treatments. *Clin Chem.* 1993;39(6):1334–1340.
48. Karbownik M, Tan D, Manchester LC, Reiter RJ. Renal toxicity of the carcinogen delta-aminolevulinic acid: antioxidant effects of melatonin. *Cancer Lett.* 2000;161(1):1–7.
49. Marsden JT, Chowdhury P, Wang J, et al. Acute intermittent porphyria and chronic renal failure. *Clin Nephrol.* 2008;69(5):339–346.
50. Doss MO. Porphyrinurias and occupational disease. *Ann N Y Acad Sci.* 1987;514:204–218.
51. Gibson PR, Grant J, Cronin V, Blake D, Ratnaike S. Effect of hepatobiliary disease, chronic hepatitis C and hepatitis B virus infections and interferon-alpha on porphyrin profiles in plasma, urine and faeces. *J Gastroenterol Hepatol.* 2000;15(2):192–201.
52. Sardh E, Rejkjaer L, Andersson DE, Harper P. Safety, pharmacokinetics and pharmocodynamics of recombinant human porphobilinogen deaminase in healthy subjects and asymptomatic carriers of the acute intermittent porphyria gene who have increased porphyrin precursor excretion. *Clin Pharmacokinet.* 2007;46(4):335–349.
53. Brennan MJ, Cantrill RC. Delta-aminolevulinic acid is a potent agonist for GABA autoreceptors. *Nature.* 1979;280(5722):514–515.
54. Paterniti Jr. JR, Simone JJ, Beattie DS. Detection and regulation of delta-aminolevulinic acid synthetase activity in the rat brain. *Arch Biochem Biophys.* 1978;189(1):86–91.
55. Percy VA, Shanley BC. Studies on haem biosynthesis in rat brain. *J Neurochem.* 1979;33(6):1267–1274.
56. Di Mascio P, Teixeira PC, Onuki J, et al. DNA damage by 5-aminolevulinic and 4,5-dioxovaleric acids in the presence of ferritin. *Arch Biochem Biophys.* 2000;373(2):368–374.
57. Douki T, Onuki J, Medeiros MH, Bechara EJ, Cadet J, Di Mascio P. DNA alkylation by 4,5-dioxovaleric acid, the final oxidation product of 5-aminolevulinic acid. *Chem Res Toxicol.* 1998;11(2):150–157.

58. Hermes-Lima M. How do Ca2+ and 5-aminolevulinic acid-derived oxyradicals promote injury to isolated mitochondria? *Free Radic Biol Med*. 1995;19(3):381–390.
59. Costa CA, Trivelato GC, Pinto AM, Bechara EJ. Correlation between plasma 5-aminolevulinic acid concentrations and indicators of oxidative stress in lead-exposed workers. *Clin Chem*. 1997;43(7):1196–1202.
60. Carneiro RC, Reiter RJ. Melatonin protects against lipid peroxidation induced by delta-aminolevulinic acid in rat cerebellum, cortex and hippocampus. *Neuroscience*. 1998;82(1):293–299.
61. Princ FG, Maxit AG, Cardalda C, Batlle A, Juknat AA. *In vivo* protection by melatonin against delta-aminolevulinic acid-induced oxidative damage and its antioxidant effect on the activity of haem enzymes. *J Pineal Res*. 1998;24(1):1–8.
62. Karbownik M, Reiter RJ. Melatonin protects against oxidative stress caused by delta-aminolevulinic acid: implications for cancer reduction. *Cancer Invest*. 2002;20(2):276–286.
63. Thunell S, Andersson D, Harper P, Henrichson A, Floderus Y, Lindh U. Effects of administration of antioxidants in acute intermittent porphyria. *Eur J Clin Chem Clin Biochem*. 1997;35(6):427–433.
64. Ackner B, Cooper JE, Gray CH. Excretion of porphobilinogen and delta-aminolevulinic acid in acute porphyria. *Lancet*. 1961;1:1256–1260.
65. Becker DM, Kramer S. The neurological manifestations of porphyria: a review. *Medicine*. 1977;56(5):411–423.
66. Bissell DM. Treatment of acute hepatic porphyria with hematin. *J Hepatol*. 1988;6(1):1–7.
67. Sylantiev C, Schoenfeld N, Mamet R, Groozman GB, Drory VE. Acute neuropathy mimicking porphyria induced by aminolevulinic acid during photodynamic therapy. *Muscle Nerve*. 2005;31(3):390–393.
68. Roberts AG, Puy H, Dailey TA, et al. Molecular characterization of homozygous variegate porphyria. *Hum Mol Genet*. 1998;7(12):1921–1925.
69. Solis C, Martinez-Bermejo A, Naidich TP, et al. Acute intermittent porphyria: studies of the severe homozygous dominant disease provides insights into the neurologic attacks in acute porphyrias. *Arch Neurol*. 2004;61(11):1764–1770.
70. Soonawalla ZF, Orug T, Badminton MN, et al. Liver transplantation as a cure for acute intermittent porphyria. *Lancet*. 2004;363(9410):705–706.
71. Dowman JK, Gunson BK, Mirza DF, et al. Liver transplantation for acute intermittent porphyria is complicated by a high rate of hepatic artery thrombosis. *Liver Transpl*. 2012;18(2):195–200.
72. Litman DA, Correia MA. Elevated brain tryptophan and enhanced 5-hydroxytryptamine turnover in acute hepatic heme deficiency: clinical implications. *J Pharmacol Exp Ther*. 1985;232(2):337–345.
73. Correia MA, Litman DA, Lunetta JM. Drug-induced modulations of hepatic heme metabolism. Neurological consequences. *Ann N Y Acad Sci*. 1987;514:248–255.
74. Bonkovsky HL, Healey JF, Lourie AN, Gerron GG. Intravenous heme-albumin in acute intermittent porphyria: evidence for repletion of hepatic hemoproteins and regulatory heme pools. *Am J Gastroenterol*. 1991;86:1050–1056.
75. Puy H, Deybach JC, Baudry P, Callebert J, Touitou Y, Nordmann Y. Decreased nocturnal plasma melatonin levels in patients with recurrent acute intermittent porphyria attacks. *Life Sci*. 1993;53(8):621–627.
76. Puy H, Deybach JC, Bogdan A, et al. Increased delta aminolevulinic acid and decreased pineal melatonin production. A common event in acute porphyria studies in the rat. *J Clin Invest*. 1996;97(1):104–110.
77. Qi W, Reiter RJ, Tan DX, Manchester LC, Calvo JR. Melatonin prevents delta-aminolevulinic acid-induced oxidative DNA damage in the presence of Fe2+. *Mol Cell Biochem*. 2001;218(1–2):87–92.
78. Kolluri S, Sadlon TJ, May BK, Bonkovsky HL. Haem repression of the housekeeping 5-aminolaevulinic acid synthase gene in the hepatoma cell line LMH. *Biochem J*. 2005;392(Pt 1):173–180.
79. Bonkowsky HL, Sinclair PR, Emery S, Sinclair JF. Seizure management in acute hepatic porphyria: risks of valproate and clonazepam. *Neurology*. 1980;30(6):588–592.
80. Disler PB, Blekkenhorst GH, Eales L, Moore MR, Straughan J. Guidelines for drug prescription in patients with the acute porphyrias. *S Afr Med J*. 1982;61(18):656–660.
81. Moore MR. International review of drugs in acute porphyria—1980. *Int J Biochem*. 1980;12(5–6):1089–1097.
82. Marver HS, Collins A, Tschudy DP, Rechcigl Jr. M. Delta-aminolevulinic acid synthetase. II induction in rat liver. *J Biol Chem*. 1966;241(19):4323–4329.
83. Bonkovsky HL, Siao P, Roig Z, Hedley-Whyte ET, Flotte TJ. Case records of the Massachusetts general hospital. Case 20-2008. A 57-year-old woman with abdominal pain and weakness after gastric bypass surgery. *N Engl J Med*. 2008;358(26):2813–2825.
84. Handschin C, Lin J, Rhee J, et al. Nutritional regulation of hepatic heme biosynthesis and porphyria through PGC-1alpha. *Cell*. 2005;122(4):505–515.
85. Andersson C, Innala E, Backstrom T. Acute intermittent porphyria in women: clinical expression, use and experience of exogenous sex hormones. A population-based study in northern Sweden. *J Intern Med*. 2003;254(2):176–183.
86. Lamon JM, Frykholm BC, Herrera W, Tschudy DP. Danazol administration to females with menses-associated exacerbations of acute intermittent porphyria. *J Clin Endocrinol Metab*. 1979;48(1):123–126.
87. Hughes MJ, Rifkind AB. Danazol, a new steroidal inducer of delta-aminolevulinic acid synthetase. *J Clin Endocrinol Metabol*. 1981;52(3):549–552.
88. Schmitt C, Gouya L, Malonova E, et al. Mutations in human CPO gene predict clinical expression of either hepatic hereditary coproporphyria or erythropoietic harderoporphyria. *Hum Mol Genet*. 2005;14(20):3089–3098.
89. Hessels J, Voortman G, van der Wagen A, van der Elzen C, Scheffer H, Zuijderhoudt FM. Homozygous acute intermittent porphyria in a 7-year-old boy with massive excretions of porphyrins and porphyrin precursors. *J Inherit Metab Dis*. 2004;27(1):19–27.
90. Corrigall AV, Hift RJ, Davids LM, et al. Homozygous variegate porphyria in South Africa: genotypic analysis in two cases. *Mol Genet Metab*. 2000;69(4):323–330.
91. Chretien S, Dubart A, Beaupain D, et al. Alternative transcription and splicing of the human porphobilinogen deaminase gene result either in tissue-specific or in housekeeping expression. *Proc Natl Acad Sci U S A*. 1988;85(1):6–10.
92. Porcher C, Pitiot G, Plumb M, Lowe S, de Verneuil H, Grandchamp B. Characterization of hypersensitive sites, protein-binding motifs, and regulatory elements in both promoters of the mouse porphobilinogen deaminase gene. *J Biol Chem*. 1991;266(16):10562–10569.

93. Porcher C, Picat C, Daegelen D, Beaumont C, Grandchamp B. Functional analysis of DNase-I hypersensitive sites at the mouse porphobilinogen deaminase gene locus. Different requirements for position-independent expression from its two promoters. *J Biol Chem.* 1995;270(29):17368–17374.
94. Shoolingin-Jordan PM. Porphobilinogen deaminase and uroporphyrinogen III synthase: structure, molecular biology, and mechanism. *J Bioenerg Biomembr.* 1995;27(2):181–195.
95. Brownlie PD, Lambert R, Louie GV, et al. The three-dimensional structures of mutants of porphobilinogen deaminase: toward an understanding of the structural basis of acute intermittent porphyria. *Protein Sci.* 1994;3(10):1644–1650.
96. Grandchamp B, Picat C, Mignotte V, et al. Tissue-specific splicing mutation in acute intermittent porphyria. *Proc Natl Acad Sci U S A.* 1989;86(2):661–664.
97. Desnick RJ, Ostasiewicz LT, Tishler PA, Mustajoki P. Acute intermittent porphyria: characterization of a novel mutation in the structural gene for porphobilinogen deaminase. Demonstration of noncatalytic enzyme intermediates stabilized by bound substrate. *J Clin Invest.* 1985;76(2):865–874.
98. Andersson C, Floderus Y, Wikberg A, Lithner F. The W198X and R173W mutations in the porphobilinogen deaminase gene in acute intermittent porphyria have higher clinical penetrance than R167W. A population-based study. *Scand J Clin Lab Invest.* 2000;60(7):643–648.
99. De Matteis F, Stonard M. Experimental porphyrias as models for human hepatic porphyrias. *Semin Hematol.* 1977;14(2):187–192.
100. Meyer UA, Marver HS. Chemically induced porphyria: increased microsomal heme turnover after treatment with allylisopropylacetamide. *Science.* 1971;171(966):64–66.
101. Correia MA, Farrell GC, Olson S, et al. Cytochrome P-450 heme moiety. The specific target in drug-induced heme alkylation. *J Biol Chem.* 1981;256(11):5466–5470.
102. Lindberg RL, Porcher C, Grandchamp B, et al. Porphobilinogen deaminase deficiency in mice causes a neuropathy resembling that of human hepatic porphyria. *Nat Genet.* 1996;12(2):195–199.
103. Lindberg RL, Martini R, Baumgartner M, et al. Motor neuropathy in porphobilinogen deaminase-deficient mice imitates the peripheral neuropathy of human acute porphyria. *J Clin Invest.* 1999;103(8):1127–1134.
104. Kochar DK, Poonia A, Kumawat BL, Shubhakaran, Gupta BK. Study of motor and sensory nerve conduction velocities, late responses (F-wave and H-reflex) and somatosensory evoked potential in latent phase of intermittent acute porphyria. *Electromyogr Clin Neurophysiol.* 2000;40(2):73–79.
105. Beattie AD, Moore MR, Goldberg A, Ward RL. Acute intermittent porphyria: response of tachycardia and hypertension to propranolol. *Br Med J.* 1973;3(874):257–260.
106. Bonkowsky HL, Tschudy DP. Letter: Hazard of propranolol in treatment of acute prophyria. *Br Med J.* 1974;4(5935):47–48.
107. Laureno R, Karp BI. Myelinolysis after correction of hyponatremia. *Ann Intern Med.* 1997;126(1):57–62.
108. Stein JA, Tschudy DP. Acute intermittent porphyria. A clinical and biochemical study of 46 patients. *Medicine.* 1970;49(1):1–16.
109. Bonkowsky HL, Tschudy DP, Collins A, et al. Repression of the overproduction of porphyrin precursors in acute intermittent porphyria by intravenous infusions of hematin. *Proc Natl Acad Sci U S A.* 1971;68(11):2725–2729.
110. Lamon JM, Frykholm BC, Hess RA, Tschudy DP. Hematin therapy for acute porphyria. *Medicine.* 1979;58(3):252–269.
111. Watson CJ, Pierach CA, Bossenmaier I, Cardinal R. Postulated deficiency of hepatic heme and repair by hematin infusions in the "inducible" hepatic porphyrias. *Proc Natl Acad Sci U S A.* 1977;74(5):2118–2120.
112. Goetsch CA, Bissell DM. Instability of hematin used in the treatment of acute hepatic porphyria. *N Engl J Med.* 1986;315(4):235–238.
113. Anderson KE, Bonkovsky HL, Bloomer JR, Shedlofsky SI. Reconstitution of hematin for intravenous infusion. *Ann Intern Med.* 2006;144(7):537–538.
114. Tenhunen R, Tokola O, Linden IB. Haem arginate: a new stable haem compound. *J Pharm Pharmacol.* 1987;39(10):780–786.
115. Raffin SB, Woo CH, Roost KT, Price DC, Schmid R. Intestinal absorption of hemoglobin iron-heme cleavage by mucosal heme oxygenase. *J Clin Invest.* 1974;54(6):1344–1352.
116. Dhar GJ, Bossenmaier I, Petryka ZJ, Cardinal R, Watson CJ. Effects of hematin in hepatic porphyria. Further studies. *Ann Intern Med.* 1975;83(1):20–30.
117. Bosch EP, Pierach CA, Bossenmaier I, Cardinal R, Thorson M. Effect of hematin in porphyric neuropathy. *Neurology.* 1977;27(11):1053–1056.
118. Herrick AL, McColl KE, Moore MR, Cook A, Goldberg A. Controlled trial of haem arginate in acute hepatic porphyria. *Lancet.* 1989;1(8650):1295–1297.
119. McColl KE, Moore MR, Thompson GG, Goldberg A. Treatment with haematin in acute hepatic porphyria. *Q J Med.* 1981;50(198):161–174.
120. Dhar GJ, Bossenmaier I, Cardinal R, Petryka ZJ, Watson CJ. Transitory renal failure following rapid administration of a relatively large amount of hematin in a patient with acute intermittent porphyria in clinical remission. *Acta Med Scand.* 1978;203(5):437–443.
121. Frei P, Minder EI, Corti N, et al. Liver transplantation because of acute liver failure due to heme arginate overdose in a patient with acute intermittent porphyria. *Case Rep Gastroenterol.* 2012;6(1):190–196.
122. Daimon M, Susa S, Igarashi M, Kato T, Kameda W. Administration of heme arginate, but not hematin, caused anaphylactic shock. *Am J Med.* 2001;110(3):240.
123. Glueck R, Green D, Cohen I, Ts'ao CH. Hematin: unique effects of hemostasis. *Blood.* 1983;61(2):243–249.
124. Morris DL, Dudley MD, Pearson RD. Coagulopathy associated with hematin treatment for acute intermittent porphyria. *Ann Intern Med.* 1981;95(6):700–701.
125. Jones RL. Hematin-derived anticoagulant. Generation in vitro and *in vivo. J Exp Med.* 1986;163(3):724–739.
126. Volin L, Rasi V, Vahtera E, Tenhunen R. Heme arginate: effects on hemostasis. *Blood.* 1988;71(3):625–628.
127. Anderson KE, Spitz IM, Bardin CW, Kappas A. A gonadotropin releasing hormone analogue prevents cyclical attacks of porphyria. *Arch Intern Med.* 1990;150(7):1469–1474.
128. Herrick AL, McColl KE, Wallace AM, Moore MR, Goldberg A. LHRH analogue treatment for the prevention of premenstrual attacks of acute porphyria. *Q J Med.* 1990;75(276):355–363.
129. Perlroth MG, Marver HS, Tschudy DP. Oral contraceptive agents and the management of acute intermittent porphyria. *JAMA.* 1965;94:1037–1042.

130. Lamon JM, Frykholm BC, Bennett M, Tschudy DP. Prevention of acute porphyric attacks by intravenous haematin. *Lancet.* 1978;2(8088):492–494.
131. Wenger S, Meisinger V, Brucke T, Deecke L. Acute porphyric neuropathy during pregnancy–effect of haematin therapy. *Eur Neurol.* 1998;39(3):187–188.
132. Seth AK, Badminton MN, Mirza D, Russell S, Elias E. Liver transplantation for porphyria: who, when, and how? *Liver Transpl.* 2007;13(9):1219–1227.
133. Lucas MJ, Leveno KJ, Cunningham FG. A comparison of magnesium sulfate with phenytoin for the prevention of eclampsia. *N Engl J Med.* 1995;333(4):201–205.
134. Taylor RL. Magnesium sulfate for AIP seizures [letter]. *Neurology.* 1981;31(10):1371–1372.
135. Zaatreh MM. Levetiracetam in porphyric status epilepticus: a case report. *Clin Neuropharmacol.* 2005;28(5):243–244.
136. Krauss GL, Simmons OBE, Campbell M. Successful treatment of seizures and porphyria with gabapentin. *Neurology.* 1995;45(3 Pt 1):594–595.
137. Pandey CK, Singh N, Bose N, Sahay S. Gabapentin and propofol for treatment of status epilepticus in acute intermittent porphyria. *J Postgrad Med.* 2003;49(3):285.
138. Pazvanska EE, Hinkov OD, Stojnovska LV. Uneventful propofol anaesthesia in a patient with acute intermittent porphyria. *Eur J Anaesthesiol.* 1999;16(7):485–492.
139. Evans PR, Graham S, Kumar CM. The use of sevoflurane in acute intermittent porphyria. *Anaesthesia.* 2001;56(4):388–389.
140. Hsieh CH, Hung PC, Chien CT, et al. The use of rocuronium and sevoflurane in acute intermittent porphyria–a case report. *Acta Anaesthesiol Taiwan.* 2006;44(3):169–171.
141. Sneyd JR, Kreimer-Birnbaum M, Lust MR, Heflin J. Use of sufentanil and atracurium anesthesia in a patient with acute porphyria undergoing coronary artery bypass surgery. *J Cardiothorac Vasc Anesth.* 1995;9(1):75–78.
142. Magnussen CR, Doherty JM, Hess RA, Tschudy DP. Grand mal seizures and acute intermittent porphyria. The problem of differential diagnosis and treatment. *Neurology.* 1975;25(12):121–125.
143. Larson AW, Wasserstrom WR, Felsher BF, Chih JC. Posttraumatic epilepsy and acute intermittent porphyria: effects of phenytoin, carbamazepine, and clonazepam. *Neurology.* 1978;28(8):824–828.
144. Schoenfeld N, Greenblat Y, Epstein O, Atsmon A. The influence of carbamazepine on the heme biosynthetic pathway. *Biochem Med.* 1985;34(3):280–286.
145. Morgan CJ, Badawy AA. Effects of acute carbamazepine administration on haem metabolism in rat liver. *Biochem Pharmacol.* 1992;43(7):1473–1477.
146. Zadra M, Grandi R, Erli LC, Mirabile D, Brambilla A. Treatment of seizures in acute intermittent porphyria: safety and efficacy of gabapentin. *Seizure.* 1998;7(5):415–416.
147. Hahn M, Gildemeister OS, Krauss GL, et al. Effects of new anticonvulsant medications on porphyrin synthesis in cultured liver cells: potential implications for patients with acute porphyria. *Neurology.* 1997;49(1):97–106.
148. Krijt J, Krijtova H, Sanitrak J. Effect of tiagabine and topiramate on porphyrin metabolism in an in vivo model of porphyria. *Pharmacol Toxicol.* 2001;89(1):15–22.
149. Livingston S, Pearson PH. Bromides in the treatment of epilepsy in children. *J Dis Child.* 1953;86:717–720.
150. Okuda K, Yasuhara A, Kamei A, Araki A, Kitamura N, Kobayashi Y. Successful control with bromide of two patients with malignant migrating partial seizures in infancy. *Brain Dev.* 2000;22(1):56–59.
151. Woody RC. Bromide therapy for pediatric seizure disorder intractable to other antiepileptic drugs. *J Child Neurol.* 1990;5(1):65–67.
152. Wuth O. Rational bromide treatment. *JAMA.* 1927;88:2013–2017.
153. Sorensen AW, With TK. Persistent pareses after porphyric attacks. *Acta Med Scand.* 1971;190(3):219–222.
154. Tishler PV, Woodward B, O'Connor J, et al. High prevalence of intermittent acute porphyria in a psychiatric patient population. *Am J Psychiatry.* 1985;142(12):1430–1436.
155. Jara-Prado A, Yescas P, Sanchez FJ, Rios C, Garnica R, Alonso E. Prevalence of acute intermittent porphyria in a Mexican psychiatric population. *Arch Med Res.* 2000;31(4):404–408.
156. Millward LM, Kelly P, King A, Peters TJ. Anxiety and depression in the acute porphyrias. *J Inherit Metab Dis.* 2005;28(6):1099–1107.
157. Sardh E, Wahlin S, Bjornstedt M, Harper P, Andersson DE. High risk of primary liver cancer in a cohort of 179 patients with acute hepatic porphyria. *J Inherit Metab Dis.* 2013;36(6):1063–1071.

SECTION IV

DEGENERATIVE DISORDERS

CHAPTER

67

Alzheimer Disease

Dennis J. Selkoe

Harvard Medical School, Brigham and Women's Hospital, Boston, MA, USA

INTRODUCTION

The remarkable rise in life expectancy during the twentieth century has made Alzheimer disease (AD) among the most common disorders of late life. An insidious loss of memory, cognition, reasoning, and behavioral stability leads inexorably to global dementia and premature death of the patient. At autopsy, one finds myriad amyloid plaques and neurofibrillary tangles in the limbic and association cortices and regions that project to them. These lesions have served as the starting point for an intensive biochemical and molecular genetic attack on the disorder. In particular, progress in unraveling the genotype–phenotype relationships of inherited forms of AD has provided substantial support for the hypothesis that cerebral accumulation of the amyloid-β protein (Aβ) is an early, invariant and necessary step in the development of the disease.

In view of concerns about the amyloid—or Aβ—hypothesis of AD, it is useful to review the scientific underpinnings of this concept. At least eight kinds of evidence support a central role for excessive accumulation of Aβ in the initiation and progression of the disease: 1) Immunocytochemical studies on the brains of subjects with Down syndrome, Alzheimer disease, or normal aging all suggest that amorphous, nonfibrillar deposits of Aβ—referred to as diffuse plaques—are the earliest detectable AD-type lesion and precede the development of other neuropathologic features of the disease. 2) Missense mutations immediately surrounding the Aβ region of the amyloid β-protein precursor (*APP*) gene are a specific cause of AD, and each of these mutations has been shown to lead to excess production of amyloidogenic Aβ peptides, particularly $A\beta_{1\text{–}42}$. 3) Altered levels of $A\beta_{1\text{–}42}$ peptide in the cerebrospinal fluid (CSF) can precede the onset of AD symptoms by many years or decades, as observed, for example, in individuals harboring mutations in the presenilin (*PS*) 1 or 2 genes.[1,2] 4) Cortical levels of $A\beta_{1\text{–}42}$ determined by enzyme-linked immunosorbent assay (ELISA) correlate tightly with both the presence and the degree of cognitive impairment.[3] 5) Aggregated Aβ peptides, including soluble oligomers isolated directly from AD cortex, have been shown reproducibly to exert cytotoxicity on neurons, enhance phosphorylated tau protein at AD-specific epitopes,[4] activate astrocytes and microglia *in vitro* and interfere with cognitive function in *in vivo* models.[5,6] 6) Transgenic overexpression of AD-causing mutations in *APP* and/or *PS1* has provided mouse models that, while not perfect, reproduce some of the characteristic neuritic and glial cytopathology of AD and alter the animals' behavior, supporting the concept that Aβ accumulation can initiate neuritic dystrophy, gliosis, synaptic loss and even neurofibrillary tangle formation.[7,8] 7) Treatments that lower cerebral $A\beta_{1\text{–}42}$ levels in mouse models of AD, for example, an Aβ vaccine[9–11] or a modulator of the presenilin/γ-secretase enzyme that generates Aβ,[12] lead to less neuropathology and improved learning behavior in APP transgenic mice. 8) Some evidence is emerging that therapeutics solely directed against Aβ (e.g., the Aβ monoclonal antibody solanezumab) can slow the rate of cognitive decline in patients with mild AD.[13]

This chapter reviews these and other recent molecular advances in the context of the complex clinicopathological phenotype of the Alzheimer syndrome. The influence of the known genetic factors underlying familial forms of AD on the development of Aβ deposition and neuronal dysfunction is discussed. A model of the disease process that links Aβ accumulation to many other features of the AD phenotype, including neurofibrillary tangles, is presented, and emerging points of therapeutic intervention are considered.

Rosenberg's Molecular and Genetic Basis of Neurological and Psychiatric Disease
http://dx.doi.org/10.1016/10.1016/B978-0-12-410529-4.00067-X

CLINICAL FEATURES AND DIAGNOSIS

The Alzheimer Syndrome

Most AD patients present with subtle deficits in episodic declarative memory, particularly retention of the minor details of everyday life—what they did yesterday, whom they telephoned this morning, what was said to them 5 minutes ago, etc. It is usually the family that first becomes concerned about these intermittent lapses in short-term memory, although some patients will notice them and acknowledge that they have memory deficits. Frequently, patients either deny the memory problem or reluctantly admit to minor symptoms but dismiss them as common to older people and not personally worrisome. The patient who comes to the office earnestly worried that he or she is developing specific symptoms of AD may turn out not to have the disorder, although a personal awareness of memory symptoms is becoming recognized as an early indicator of trouble ahead.[14] Most patients show surprising equanimity about their symptoms, making up a variety of excuses for why it should not be shocking that they cannot remember this or that. This apparent lack of interest in one's own mental errors may be explained by the patient's gradual loss of insight and judgment—the inability to sense one's situation in life—that so often accompanies the classical memory deficits of AD.

A salient feature of the clinical presentation is its remarkably insidious onset. Neither patient nor family can pinpoint when it all began. Once they become convinced that there is something wrong, family members search retrospectively for the first clue. Often, past events of major or minor import, perhaps a medical illness, accident or emotional upset, will be viewed by the spouse or children as the starting point. But when one considers the biology of the process, one recognizes that the biochemical and pathological phenotype has been building very slowly over a long time, almost certainly for one to two decades or more. Accordingly, the first symptoms can be so subtle and occur so infrequently that they blend with everyday behavior.

As patients progress over the first few years, they usually develop increasing forgetfulness, disorientation, confusion about mathematical and geographical concepts, decreased abstraction, and word-finding difficulty and other language problems. As these deficits accrue, many patients also develop intermittent behavioral symptoms—agitation, restlessness (e.g., pacing), insomnia, paranoia, angry outbursts, and sometimes frank delusions and hallucinations. But the sensorium remains clear, and alertness is preserved through much of the illness. Motor function usually remains normal until extrapyramidal signs (rigidity, bradykinesia) and slowed gait develop after several or many years of cognitive decline. In later years, global dementia is accompanied by motor impairment, incontinence of bladder and/or bowel, and an increasingly immobile, mute state. Death often comes after one to two decades of progressive symptomatology, with minor respiratory difficulty, aspiration or pneumonia serving as common terminal events.

Much has been written in the popular press about the difficulty that doctors have in diagnosing AD. Because until recently, AD could only be diagnosed definitively at postmortem examination, an assumption has developed that it is challenging to come to an accurate diagnosis during life. This is now no more true for AD than for many commonly diagnosed systemic diseases in which the symptoms, signs, and laboratory tests point clearly to a particular diagnosis but definitive proof requires tissue examination (something that is very rarely done in AD). Experienced clinicians can distinguish AD from most other dementias with a high degree of accuracy, and the advent around 2004 of positron emission tomography (PET) scanning for amyloid deposition[15] has revolutionized the ability to either confirm or deny a suspected clinical diagnosis of AD. Autopsy series from clinics specializing in dementia indicate that the accuracy of a clinical diagnosis of AD can exceed 95%. In the general medical community, this rate is likely to be somewhat lower, perhaps 70–80%.

Until recently, the diagnostic work-up for AD was largely a matter of obtaining negative blood tests and a normal brain imaging study that helped rule out other causes of dementia. Now, positive evidence for the presence of AD has emerged from imaging studies such as fluorodeoxyglucose PET scans and functional magnetic resonance imaging (MRI) scans, and from the measurement of elevated tau and decreased $A\beta_{42}$ protein levels in the CSF.[16] Detailed psychometric tests can solidify a clinical diagnosis and accurately quantify deterioration over time. Psychiatric consultation can help determine whether behavioral symptoms are due to the dementia and suggest how they should be treated pharmacologically. Most excitingly, noninvasive imaging of the cerebral amyloid deposits by PET scanning has emerged as a diagnostic tool with great promise, and a US Food and Drug Administration (FDA)-approved amyloid ligand for PET is being widely used in experimental clinical trials. Careful follow-up evaluations by the clinician can often clarify diagnostic dilemmas and reveal the rather stereotyped clinical progression of AD.

PATHOLOGY

Neuropathology of AD: the Basis for Molecular Advances

In contrast to numerous other hereditary neurologic and psychiatric disorders in which the identification and cloning of causative genes provided the first insights into the biochemical mechanism (e.g., Huntington disease, Parkinson disease), genetic advances in AD derived, in considerable part, from the prior biochemical characterization of the neuropathological phenotype. The morphologic signature provided by the presence of amyloid plaques and neurofibrillary tangles has given rise to countless biochemical, molecular biological and cell biological studies that have cumulatively helped to elucidate the pathogenesis of AD. Based on Alzheimer's original description of the disease more than 100 years ago, as well as formal consensus about its pathodiagnostic criteria, all cases of AD have amyloid-bearing neuritic plaques in densities substantially greater than those found in nondemented subjects of similar age. These plaques are complex, multicellular lesions, the temporal development of which appears to be gradual, as judged by studies of young adults with Down syndrome (see below). The "mature" neuritic plaque generally contains a central extracellular deposit of Aβ that is principally in filamentous form (i.e., skeins of myriad amyloid fibrils of 6–10 nm) and is intimately associated with dilated, ultrastructurally abnormal dendrites and axons (dystrophic neurites). In addition, many neuritic plaques are associated with activated microglia located within the amyloid deposit and reactive fibrillary astrocytes immediately surrounding the lesion.

The other hallmark lesion of AD, the neurofibrillary tangle, usually occurs abundantly in the entorhinal cortex, hippocampus, amygdala, association cortices of the frontal, temporal, and parietal lobes, and certain subcortical nuclei that project to these regions. The tangles are nonmembrane-bound masses of abnormal paired helical filaments (PHF) and straight filaments found in the perinuclear cytoplasm of selected neurons. Smaller bundles of identical PHF are present in many of the dystrophic neurites of the neuritic plaque. In addition to sensitively detecting neurofibrillary tangles and plaque-associated neurites, silver stains also highlight widely scattered dystrophic neurites that contain PHF and are found throughout much of the cortical neuropil in most cases of AD, particularly those with high tangle densities. The occurrence of this diffuse neuritic abnormality in affected brain regions indicates that axonal and dendritic abnormalities are by no means confined to the immediate periplaque region.

The subunit protein of the PHF is the microtubule-associated protein, tau.[17] Biochemical studies reveal that the tau found in the perikaryal tangles and also in many of the dystrophic neurites within and outside of the plaques constitutes hyperphosphorylated, largely insoluble forms of this normally highly soluble cytosolic protein. The insoluble tau aggregates in the tangles are often conjugated with ubiquitin, a feature they share with other intraneuronal proteinaceous inclusions in etiologically diverse disorders such as Parkinson disease and Lewy body dementia. If this ubiquitination represents an attempt to rid the cytoplasm of the tau aggregates by way of proteasomal degradation, it appears to be largely unsuccessful. Tangles also occur in a dozen or more relatively uncommon neurodegenerative diseases in which one finds no Aβ deposits and neuritic plaques. One particularly important example is frontotemporal dementia with parkinsonism linked to chromosome 17 (*FTDP-17*; see Chapters 68 and 69).

In addition to the presence of abundant neuritic plaques, neurofibrillary tangles and dystrophic neurites in virtually all AD cases, shrinkage and loss of many limbic and cortical neurons and, in particular, loss of their synaptic endings are a prominent feature of the pathology. Cortical synaptic loss shows a particularly strong quantitative correlation with the degree of dementia that a patient experienced during life (reviewed in [18]). Robust astrocytosis and microgliosis are observed immediately surrounding and between amyloid-bearing neuritic plaques. Granulovacuolar degeneration in the cytoplasm of selected pyramidal neurons, primarily in the hippocampus, is found in many AD brains.

The original purification and partial sequencing of Aβ by Glenner and Wong[19] and the subsequent development of antibodies to Aβ led to the recognition that diffuse, nonfibrillar Aβ deposits largely lacking neuritic and glial alterations are even more abundant than amyloid-bearing neuritic plaques in AD brain tissue. Moreover, substantial numbers of diffuse plaques can occur in brain areas not obviously implicated in the classical symptomatology of AD, such as the cerebellum and thalamus. The presence of abundant diffuse plaques in these cytopathologically and clinically "unaffected" areas of brain, as well as their very early appearance in the course of the AD pathology of Down syndrome, provides strong neuropathological evidence that Aβ accumulation and deposition can precede local neuritic and glial alteration rather than follow it.

Because some aged, nondemented humans show moderate or even high densities of cortical Aβ deposits at postmortem examination, the relevance of Aβ deposition to the pathogenesis of the clinical disease has been questioned. However, these age-related deposits are often diffuse plaques lacking detectable neuritic and glial pathology. Indeed, if substantial numbers of amyloid-bearing neuritic plaques were present, a pathologic diagnosis of AD would usually be made. Thus, the presence of some, or occasionally many, diffuse Aβ deposits in humans lacking measurable cognitive impairment does not rule against a critical role for Aβ accumulation in the disease but rather suggests that subsequent synaptic, neuritic and glial cytopathology must develop for frank clinical symptoms to appear. The very occurrence of diffuse Aβ deposits in otherwise normal brain tissue of healthy older humans clearly supports the concept that Aβ production and initial deposition do not require pre-existing cellular abnormalities. Biochemical and immunohistochemical quantification of cortical Aβ in normal aged and minimally cognitively impaired individuals has revealed that cerebral Aβ concentrations, particularly of soluble forms of Aβ, are substantially elevated before the first symptoms of dementia appear.[3,20,21]

In addition to many parenchymal Aβ deposits of varying morphologies, almost all AD brains have at least modest and often substantial amounts of fibrillar Aβ deposits in the basement membranes of cerebral capillaries and cerebral and meningeal arterioles and small arteries (i.e., congophilic amyloid angiopathy). Because the amount of vascular amyloid deposition often correlates poorly with the generally high levels of amyloid plaques in AD, the relationship of the processes that lead to these two types of Aβ deposits remains unsettled. The Aβ species that appears to be initially deposited in diffuse plaques and that remains the major constituent of most mature (neuritic) plaques ends at Aβ residue 42, whereas the major Aβ peptide in vascular amyloid ends at residue 40 ($A\beta_{40}$), with some $A\beta_{42}$ species also present in vessels and perhaps depositing initially (discussed later). AD brains often show typical amyloid plaques that are intimately associated with capillaries as well as roughly cylindrical tissue deposits of Aβ immediately surrounding amyloid-bearing cortical vessels ("dyshoric angiopathy"). The mechanistic reasons for this microvascular predilection for Aβ deposits are poorly understood.

BIOCHEMICAL FINDINGS

Biochemistry of Amyloid Deposits Led to Identification and Characterization of APP

Biochemical analyses of isolated amyloid plaques indicate that they contain Aβ peptides ending at Alanine_{42} as their major constituent but show substantial *N*- and *C*-terminal heterogeneity, including species that begin at native or modified residues between Aspartate_{1} and Glutamate_{11} and that end at Valine_{40}, Alanine_{42}, Threonine_{43} or other nearby residues.[22-26] Among the modified *N*-terminal species are peptides that begin at residue 3 with the conversion of glutamate to pyroglutamate or at residue 1 with the conversion of L-aspartate to D-aspartate or to L-isoaspartate, each of which provides a blocked *N*-terminus. The pathogenic significance of this *N*-terminal heterogeneity in AD is unclear, and it is likely that at least some of the modifications occur after years of storage of $A\beta_{1-42}$ and $A\beta_{1-40}$ in plaques. *C*-terminal-specific Aβ enzyme-linked immunosorbent assays (ELISAs) performed on total cortical extracts demonstrate that the major species that accumulates in AD cortex ends at residue 42 but that brains having robust amyloid angiopathy as well as plaques also contain abundant Aβ peptides ending at residue 40.

Immunocytochemistry with antibodies that are specific for particular *N*- or *C*-termini of Aβ confirms the findings obtained by compositional analysis and indicates that Aβ peptides ending at Ala_{42} are the initially deposited species in plaques.[27] Vascular amyloid generally reacts with antibodies specific for Val_{40}, although it is likely that the $A\beta_{42}$ peptide is the initially deposited species in vessels, as it is in brain parenchyma. The heterogeneity found at both ends of Aβ purified from cerebral amyloid deposits correlates in part with the variety of soluble Aβ peptides that are normally secreted by cultured cells (discussed later).

The identification of the protein sequences for vascular and plaque Aβ led to the cloning of their large precursor protein, APP.[25] Aβ was shown to be a proteolytic fragment comprising the last 28 residues of the APP ectodomain just prior to the membrane plus the first 12–15 residues of the transmembrane region (Figure 67.1). APP is a single membrane-spanning, receptor-like glycoprotein that is inserted in part at the plasma membrane but occurs to a great extent in internal vesicular membranes, including the Golgi apparatus and endosomes (reviewed in [28]). As expected from this subcellular distribution, APP has a signal peptide for translocation into the endoplasmic reticulum (ER), after which it is processed through the secretory pathway, undergoing first *N*- and then *O*-linked glycosylation. In addition to its complex posttranslational modification, APP undergoes alternative splicing of several exons in its ectodomain, including a motif homologous to the Kunitz family of serine protease inhibitors that correctly predicts one function of the ectodomain, namely as a secreted protease inhibitor.

Cellular Trafficking of APP: Non-Amyloidogenic and Amyloidogenic Pathways

Intensive studies of the cellular trafficking and proteolytic processing of APP have revealed the alternative metabolic fates that this molecule can undergo.[28] APP is expressed in virtually all neuronal and non-neuronal cells,

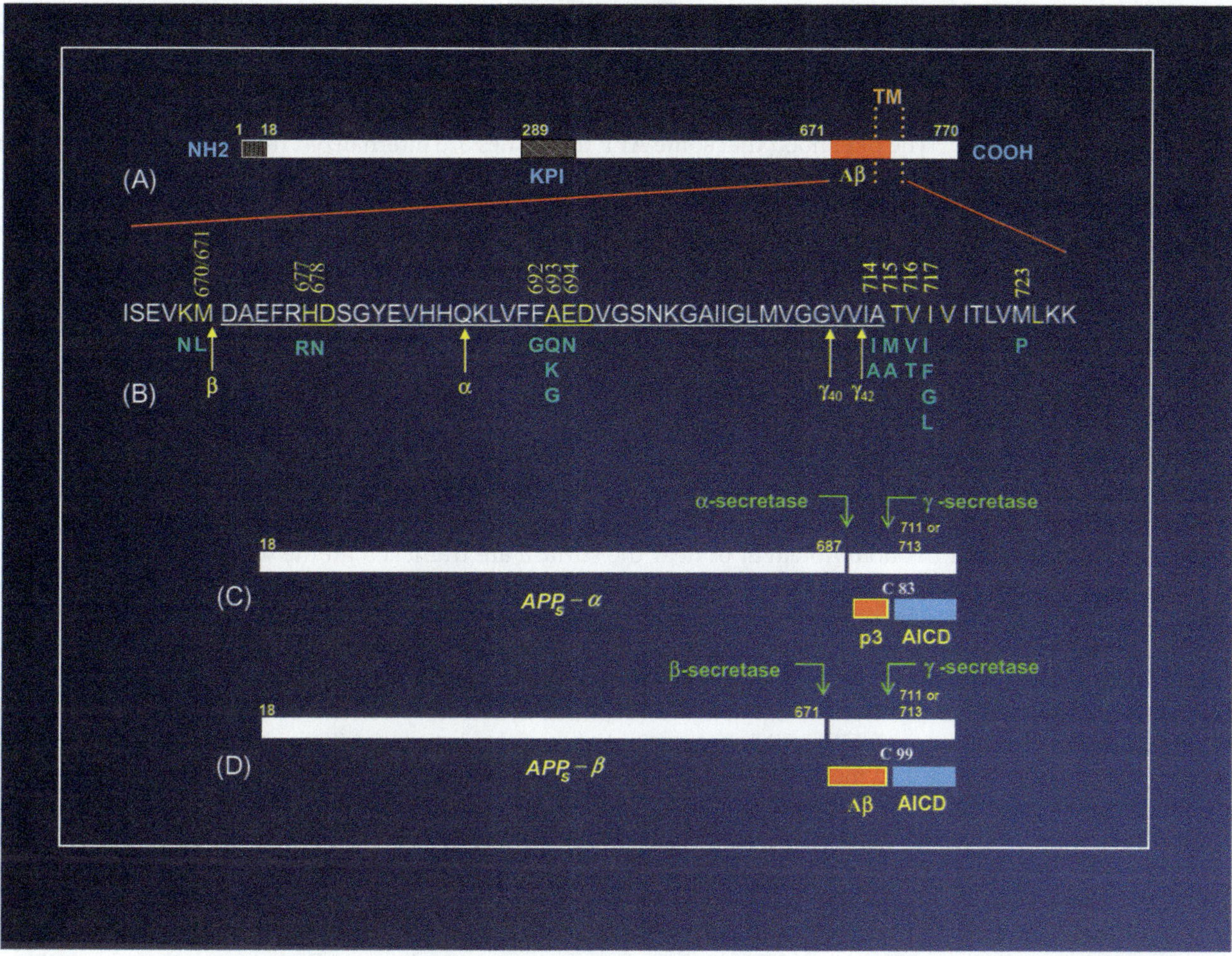

FIGURE 67.1 **Schematic diagrams of the amyloid β-protein precursor (APP) and its principal metabolic derivatives.** **(A)** Diagram showing the largest of the known APP alternate splice forms comprising 770 amino acids. Regions of interest are indicated at their relative positions in the linear sequence. A 17-residue signal peptide occurs at the *N*-terminus (vertical lines). Two alternatively spliced exons of 56 and 19 amino acids are inserted at residue 289; the first contains a serine protease inhibitor domain of the Kunitz type (KPI). A single transmembrane (TM) domain at amino acids 700–723 is shown (vertical dotted lines). The amyloid β-protein (Aβ) fragments include 28 residues just outside the membrane plus the first 12–15 residues of the TM domain. **(B)** Expanded sequence within APP that contains the Aβ and transmembrane regions. The $A\beta_{1-42}$ peptide (underlined residues) is shown. Most of the currently known missense mutations (letters below the wild-type sequence) identified in certain patients with AD and/or hereditary cerebral hemorrhage with amyloidosis are also shown. Three-digit numbers shown just above the sequence represent the codon number (βAPP_{770} isoform). One missense mutation (A673?) has been reported to confer a decrease in β-secretase cleavage of APP at the M671/D672 peptide bond and thus protects against the development of AD. **(C)** Site of the cleavage after residue 687 (arrow) by α-secretase that enables secretion of the large soluble ectodomain of APP (APP_s-α) into the medium, and retention of the 83-residue *C*-terminal fragment (C83) in the membrane. C83 can then undergo cleavage by the γ-secretase complex principally at residue 711 or 713 to release the $p3_{40}$ and $p3_{42}$ peptides. **(D)** Alternative proteolytic cleavage of holoAPP after residue 671 by the β-secretase (BACE-1) that results in the secretion of the slightly truncated APP_s-β molecule and the retention of a 99-residue *C*-terminal fragment (C99). C99 can also undergo cleavage by γ-secretase at either residue 711 or 713 to release the $A\beta_{40}$ and $A\beta_{42}$ peptides. Cleavage of both C83 and C99 by γ-secretase releases the APP intracellular domain (AICD) into the cytoplasm.

although the expression of alternate transcripts shows some cell and tissue differences. Transcripts containing the Kunitz protease inhibitor (KPI) insert in the mid-region of the ectodomain (yielding APP_{751} and APP_{770}) are the most abundantly expressed isoforms in all non-neuronal tissues and in glial, endothelial and smooth muscle cells in the brain. The transcript lacking the KPI insert (APP_{695}) is the major species expressed in neurons, with substantially lower amounts of $APP_{751/770}$ also expressed in these cells. The proteolytic processing of APP varies only modestly as a function of cell type; the findings about to be described are generalized from studies conducted primarily in non-neuronal cells.

Newly synthesized APP molecules are cotranslationally translocated into the ER, with the *N*-terminal ectodomain projecting into the lumen of the vesicle and the *C*-terminus located in the cytoplasm. Following addition of *N*- and *O*-linked sugars in the Golgi, a subset of APP molecules is trafficked via secretory vesicles to the cell surface. The percentage of total cellular APP that exists at the cell surface at any one point is relatively small in most cell types but may be functionally important. Those precursors that arrive at the cell surface can undergo one of two fates: α-secretase cleavage 12 residues *N*-terminal to the beginning of the transmembrane domain to shed the large soluble ectodomain (APP_S) into the extracellular fluid or re-internalization of the holoprotein and trafficking through the endosomal/lysosomal system.[28] The likelihood that some APP molecules are internalized from the cell surface was initially predicted by the observation that APP contains a consensus sequence (Arg-Pro-Xxx-Tyr) for clathrin-mediated endocytosis near the end of its cytoplasmic domain, and this route has been confirmed in several cell types, including primary neurons. Following endocytosis, some APP molecules recycle rapidly to the cell surface. No physiological ectodomain ligand that clearly upregulates APP ectodomain shedding at the surface has yet been confirmed, but several ectodomain ligands (e.g., Reelin, pancortin, Lingo-1) may downregulate this process.[29] Still, it is unclear precisely what signal triggers the internalization, recycling and endoproteolysis of the molecule. The finding that chimeric APP molecules containing a heterologous transactivating domain in the cytoplasmic tail can drive the transcription of reporter genes in the nucleus[30] has suggested that APP functions as a type of signaling receptor, but this may not involve nuclear entry and direct participation in transcriptional regulation,[31] and there has been no consensus on specific downstream genes regulated by the APP intracellular domain (AICD).[32]

Current data suggest that surface-derived APP molecules present in recycling endosomes may serve as the major precursor of Aβ. Some of the endocytosed APP C-terminal fragments (C83 and C99) that arise respectively from α- or β-secretase cleavage of the holoprotein at or near the surface are trafficked all the way to late endosomes and lysosomes, but there is no clear evidence that the C83 and C99 reaching these later compartments actually give rise to p3 and Aβ; rather, they may be directed to lysosomes for complete catabolism. It appears that nonamyloidogenic (α-secretase) and amyloidogenic (β-secretase) processing of APP can occur in part in the secretory pathway, as well. For example, the generation of APP_S-α by α-secretase has been demonstrated intracellularly during the exocytic trafficking of APP, presumably in the trans-Golgi network (TGN) or in secretory vesicles budding from it. Moreover, at least the "Swedish" mutant form of APP, which is a particularly good substrate for β-secretase (see below), appears to undergo some cleavage by this enzyme on the way to the cell surface, whereas the wild-type substrate may do so less.

Although Aβ was first identified as the subunit of amyloid fibrils in AD, it is now known that this fragment is constitutively secreted by APP-expressing cells during normal metabolism throughout life and is present in extracellular fluids such as CSF and plasma (reviewed in [28]). Therefore, the terms *non-amyloidogenic processing* (via α- and γ-secretase) and *amyloidogenic processing* (via β- and γ-secretase) do not imply normal and abnormal processing, respectively. Various influences on the cell, both normal events, such as activation of certain signal transduction cascades, and pathologic events, such as mutations in APP, can shift the relative processing of APP between these two alternate cleavage events. For example, activation of cell-surface receptors for several different first messenger systems, including acetylcholine, bradykinin, vasopressin and other neurotransmitters or modulators that act via the phospholipase C/protein kinase C-dependent pathway, can increase the fraction of APP holoproteins undergoing α-secretory cleavage and thereby presumably reduce Aβ production. Electrical stimulation of brain slices has also been shown to increase APP_S-β release *in vitro*. Such information suggests that certain pharmacologic manipulations may be able to shift APP metabolism away from Aβ production and toward the non-amyloidogenic α-secretase pathway. Of course, these would also increase production of APP_S-α, which appears to have several discrete functions.

The distribution and trafficking of APP shows special features in polarized cells.[28] APP is brought down the axon from the cell body via fast axonal transport. In cultured neurons, the molecule is primarily distributed on the axonal surface in cells that have established mature axonal and somatodendritic compartments. Despite this preferential axonal location, surface APP appears to be sparsely present on axonal growth cones. Surface APP at axonal terminals can be endocytosed and retrogradely transported up the axon. Moreover, some of these molecules are fully transcytosed into the somatodendritic compartment of the neuron. Although the polarized trafficking of APP has been demonstrated in neurons, it has not been possible to specify whether Aβ release occurs principally at axonal, somatic, or dendritic sites. Resolution of this issue would help address whether axonal terminals are the immediate local source of the extracellular Aβ that can alter synaptic function and can lead to neuritic plaques.

MOLECULAR GENETIC ANALYSIS OF ALZHEIMER DISEASE

Genetics

It has long been known that clinically typical AD can cluster in families and sometimes be inherited in an autosomal dominant fashion. Estimates of the fraction of all AD cases determined by genetic factors have varied widely from as low as 5–10% to as high as 50% or more. Some investigators believe that in the fullness of time, a large majority of AD cases will be shown to have underlying genetic determinants, many of which may appear as polymorphic alleles that predispose to the disease but do not invariably cause it. Determining how frequently genetic factors underlie the disease is difficult in a late-onset disorder such as AD, particularly one that was not specifically diagnosed and intensively studied prior to the last three decades. Moreover, the recognition that the ε4 allele of *Apolipoprotein E* predisposes humans to develop typical AD in their 60s and 70s suggests that other polymorphic genes with lesser effect sizes could predispose to the disorder but may be difficult to detect in genetic epidemiologic studies because they do not always produce the disease and thus will not show high penetrance.

Despite the uncertainty about the extent to which AD is accounted for by genetic factors, clinical and neuropathological analyses of familial and nonfamilial ("sporadic") cases make clear that these two forms are highly similar or indistinguishable phenotypically, except for the earlier age of onset of the known autosomal dominant forms. The histologic phenotype of early-onset familial cases is often impossible to distinguish from that of typical late-onset patients. Similarly, the clinical manifestations of dominantly inherited AD are often quite similar or indistinguishable from those of the sporadic cases, although occasional families show distinctive clinical signs (e.g., myoclonus, seizures, early and prominent extrapyramidal signs). This phenotypic similarity strongly suggests that information about the mechanism of dominant forms caused by mutations in the APP and presenilin genes will turn out to be relevant to the pathogenesis of the common, "sporadic" forms.

Missense Mutations in APP

The first specific genetic cause of AD to be identified was the occurrence of missense mutations in APP (reviewed in [33]). Families harboring APP missense mutations that cause AD generally have their onset of symptoms before age 65 years, often in their 50s or earlier. Such mutations have only been observed to date in a very small number of families. Nevertheless, the location of the mutations (Figure 67.1, second line) and the subsequent delineation of their genotype–phenotype relationships have provided critical insights into the pathogenesis of AD. The mutations are strategically located either immediately prior to the β-secretase cleavage site, shortly following the α-secretase site, or just C-terminal to the γ-secretase cleavage site. No AD-causing mutations away from the Aβ region in the large (770-residue) APP protein have been discovered. This finding strongly suggests that the substitutions lead to AD by altering the proteolytic processing at the secretase sites or, in the case of the intra-Aβ mutations, by enhancing the aggregation of the resultant Aβ peptides. This hypothesis has been confirmed by detailed analysis of each mutation, initially in transfected cells and/or primary cells from patients and later in transgenic mouse models.

Alterations in the APP gene can predispose to the development of AD in another way. The overexpression of structurally normal APP owing to elevated gene dosage in trisomy 21 (Down syndrome) leads to the premature occurrence of classical AD neuropathology (neuritic plaques and neurofibrillary tangles) during the middle adult years. A lifelong increase in APP expression owing to duplication of the entire chromosome in trisomy 21 (or in the case of translocation Down syndrome, of a portion of 21q containing the APP gene) results in overproduction of $A\beta_{40}$ and $A\beta_{42}$ peptides beginning prior to birth. This is assumed to be responsible for the highly premature appearance of many $A\beta_{42}$ diffuse plaques, which can occur as early as age 10–12 years. Thus, Down syndrome subjects display diffuse plaques composed principally of $A\beta_{42}$ in their teens and 20s, with accrual of $A\beta_{40}$ peptides onto these plaques and the appearance of associated microgliosis, astrocytosis, and surrounding neuritic dystrophy usually beginning in the late 20s or 30s.[34] This sequence supports the importance of $A\beta_{42}$ accumulation as a seminal event in the development of AD-type pathology. Indeed, $A\beta_{42}$ is the first species of Aβ peptide to be deposited in conventional AD.[35] The appearance of neurofibrillary tangles in Down syndrome patients is also delayed until the late 20s, 30s, or later. The gradual accrual of AD-type neuropathology in these individuals (who are mentally retarded from birth for other reasons) appears to be associated in many cases with progressive loss of cognitive and behavioral function after their late 30s.

Because the entire chromosome 21 is duplicated in the vast majority of Down syndrome subjects, it has been difficult to attribute the AD phenotype that they develop directly to elevated APP gene dosage. However, this issue has been essentially resolved by the evaluation of a patient with translocation Down syndrome in which the obligate Down syndrome region in the distal portion of chromosome 21 was duplicated but the break point was telomeric to the APP gene. The subject bearing this translocation had typical phenotypic features of Down syndrome but did not develop evidence of behavioral deterioration during middle age. At autopsy at age 78, no significant Aβ deposition nor other Alzheimer-type neuropathology was observed.[36] The clinicopathologic findings in this unusual case provide further support for the primacy of Aβ accumulation in the development of AD neuropathology.

The above conclusions have been substantially strengthened by the discovery of rare families harboring microduplications of the region of 21q containing the APP gene.[37] These individuals presented with AD-type dementia and at autopsy had both abundant amyloid plaques and marked amyloid angiopathy in the cortex. The microduplicated regions vary in extent but always contain the APP gene. This striking finding has subsequently been confirmed in other families. Thus, elevated dosage of the wild-type APP gene precipitates clinical AD with amyloid angiopathy in otherwise phenotypically normal individuals, entirely consistent with a causative role for Aβ.

A very important support for the Aβ hypothesis of AD came from the exciting discovery of a missense mutation at the second amino acid of the Aβ region of APP that strongly protects against the development of AD.[38] The authors showed that this mutation can decrease the processing of APP by β-secretase by as much as 40% and thus prevent the build-up of Aβ during aging. Not only AD but also the occurrence of late-life cognitive decline in subjects without a formal diagnosis of AD were lessened by the presence of this mutation.

Missense Mutations in the Presenilins

The realization that autosomal dominant AD is genetically heterogeneous led to intensive searches for loci in the genome besides *APP* that could explain the many families that did not link to chromosome 21. Establishment of a linkage of some families to chromosome 14[39] led to further linkage analyses and positional cloning that ultimately identified a novel gene on chromosome 14q now known as presenilin 1 (*PS1*; reviewed in [40]). Missense mutations in *PS1* are causative of AD in certain families with clinical onset in their 40s and 50s, and sometimes as early as their 20s. A homologous gene (*PS2*) was discovered on chromosome 1, mutations in which explain the early-onset families referred to as the Volga German kindred as well as other rare families.[41] Further genetic surveys have identified over 160 missense mutations in the 467-residue PS1 protein and at least 12 in *PS2* as molecular causes of early-onset AD in hundreds of families worldwide.[40] *PS1* missense mutations cause the earliest and most aggressive form of AD, commonly leading to onset of symptoms before age 45 years and demise of the patient during his or her 60s. These mutations have been instructive for understanding both the role of presenilins (PS) in AD and the normal functions of these fascinating polytopic membrane proteins (see below).

Apolipoprotein ε4: A Major Risk Factor for Late-Onset Alzheimer Disease

Whereas the autosomal dominant mutations in *APP*, *PS1* or *PS2* are very infrequent causes of AD, the discovery that the ε4 allele of apolipoprotein E (ApoE) predisposes to AD provided a major genetic risk factor for the disorder in the typical late-onset period. Studies initiated as a search for proteins in human CSF that bound Aβ peptides led to the identification of ApoE as such an Aβ-binding protein, and its gene was localized to chromosome 19q in a region previously linked to AD in some late-onset families.[42] Further genetic analyses indicated that the ε4 allele of ApoE is overrepresented in subjects with AD compared with the general population, and that inheritance of one or two ε4 alleles heightens the likelihood of developing AD and makes its mean age of onset earlier than in subjects harboring ε2 and/or ε3 alleles.[43] Thus, the presence of the ApoE4 protein helps precipitate the disorder primarily in subjects in their 60s and 70s. Conversely, there is evidence that inheritance of the ε2 allele confers protection against the development of AD. Although inheritance of a single ε4 allele may increase the likelihood of patients developing AD in their 60s and 70s from 2–6-fold, and two ε4 alleles may increase the risk as much as 5–10-fold, it should be emphasized that ApoE4 is a risk factor for, not an invariant cause of, AD. Some humans homozygous for the E4 isoform still show no Alzheimer symptoms in their ninth decade of life and beyond. Conversely, many humans develop AD without harboring ε4 alleles.

Discovery of Tau Mutations Clarifies the Roles of Aβ and Tau in AD

All of the genetic alterations that have been associated with AD to date have been shown to promote the accumulation of Aβ in the brain. The finding of missense and splicing mutations in the tau gene on chromosome 17 in patients with frontotemporal dementia with parkinsonism (*FTDP-17*) helped settle the long-standing debate over whether plaques or tangles have temporal primacy in AD. Mutant tau in *FTDP-17* produces profound neurofibrillary tangle formation and neuronal cell loss, resulting in a severe dementia and the ultimate demise of the host. But no Aβ accumulation is detected in these patients, unless they survive long enough to develop diffuse plaques associated with aging. Therefore, primary Aβ buildup in autosomal dominant AD (e.g., due to APP or presenilin mutations) precedes secondary alterations of wild-type tau, whereas primary defects in tau in *FTDP-17* do not lead to Aβ buildup.

Genotype–Phenotype Conversions in Familial Alzheimer Disease

Experiments to decipher the genotype–phenotype relationships of AD genes have been conducted in cell culture, in transgenic mice and, most importantly, in patients who actually harbor the relevant genetic alterations. For all four AD genes unequivocally confirmed to date (*APP, ApoE, PS1* and *PS2*), inherited alterations in the respective gene products have been credibly linked to increases in the production and/or cerebral deposition of Aβ peptides. Such studies have provided the strongest support for the hypothesis that cerebral accumulation of Aβ is an early, invariant, and necessary event in the genesis of AD.

APP Mutations can Increase the Production of $A\beta_{42}$ Peptides

Most of the missense mutations in APP currently linked to familial AD (Figure 67.1) have been found to increase Aβ production but they do so by somewhat different biochemical mechanisms. A double mutation in the two amino acids immediately preceding the β-secretase cleavage site (often referred to as the "Swedish" APP mutation, based on the ethnic origin of the relevant family) provides a more avid substrate for cleavage by β-secretase, generating more $A\beta_{40}$ and $A\beta_{42}$. The mutations that occur just C-terminal to the γ-secretase cleavage sites enhance the relative production of Aβ species ending at residue 42. Although most of the latter mutations raise the absolute levels of $A\beta_{42}$, a few can apparently act by lowering the amount of cleavage at position 40, raising the ratio of $A\beta_{42}$ to $A\beta_{40}$ peptides and thereby favoring Aβ aggregation. Among the APP mutations located near the middle of the Aβ sequence, it is possible that some may serve to decrease the efficiency of α-secretase and thus allow a relative increase in cleavage by β-secretase. However, there is more evidence that these internal mutations enhance the aggregation propensity of the peptide and thereby lead to cytopathology. Among the APP mutations associated with prominent microvascular deposition, the *E22Q* ("Dutch") Aβ peptide has been shown to aggregate on the surfaces of smooth muscle and endothelial cells and thus confer local toxicity that leads to the hyaline necrosis and vessel rupture seen in the very rare families with hereditary cerebral hemorrhage with amyloidosis of the Dutch type (HCHWA-D).[44] It should be noted that all known patients with APP mutations are heterozygotes. In accord, the resultant microvascular Aβ deposits have been shown to contain mixtures of peptides derived from the mutant and wild-type alleles.

Presenilin Mutations Increase $A\beta_{42}$ Production

Perhaps the most intriguing genotype–phenotype relationships in AD involve the presenilin mutations. When *PS1* and *PS2* were first cloned, the mechanism by which mutations produced the AD phenotype was an open matter and not necessarily expected to involve enhanced Aβ production. However, direct assays of $A\beta_{40}$ and $A\beta_{42}$ in the plasma and cultured skin fibroblast media of humans harboring these mutations soon revealed a selective ~2-fold elevation of $A\beta_{42}$ levels.[1] Extensive modeling of these mutations in cultured cells and transgenic mice has confirmed this finding.[28,45] An important observation has been the finding that crossing mice transgenic for mutant human *APP* with mice expressing a *PS1* missense mutation leads to a substantially accelerated AD-like phenotype in the offspring, with $A\beta_{42}$ plaques (first diffuse and then mature) occurring as early as age 3–4 months.[46] Quantitative immunohistochemistry of brain amyloid deposits in patients with PS mutations has directly demonstrated a 1.5–3-fold increase in the relative abundance of plaques containing $A\beta_{42}$ peptides, as compared to levels observed in sporadic AD.[47] The molecular mechanism by which missense mutations in PS selectively increase the γ-secretase cleavage of C99 (and also C83) to yield relatively more peptides ending at $A\beta_{42}$ will be discussed later, after current knowledge of the biology of presenilin is reviewed.

Inheritance of ApoE4 Alleles Increases Brain Aβ Levels

Even before ApoE4 was recognized as a genetic risk factor for late-onset disease, immunohistochemistry had demonstrated the presence of ApoE protein in a high percentage of Aβ deposits in AD brain tissue. Once the genetic connection between AD and ApoE4 inheritance was made, immunohistochemistry on brains of patients expressing ApoE4 showed that inheritance of this allele was associated with a significantly higher Aβ plaque burden than was observed in patients lacking ApoE4.[48,49] Importantly, studies in nonagenarians who died without showing overt clinical symptoms of AD demonstrated that the ApoE4 genotype was again linked to enhanced amounts of $A\beta_{42}$ plaques in the brain, suggesting that the Aβ-elevating effects associated with ApoE4 inheritance can be observed presymptomatically in hosts who had not necessarily developed clinical AD.

The mechanism by which ApoE4 protein leads to increased Aβ deposition has been difficult to pinpoint. Little evidence has emerged that ApoE4 specifically elevates new Aβ generation. Rather, the ApoE4 protein seems to enhance the steady-state levels of extracellular Aβ peptides, namely by decreasing their clearance from the brain tissue in some way.[50] *In vitro* studies quantifying the degree of Aβ fibrillogenesis using synthetic peptides suggest that the presence of ApoE4 increases the number of fibrils, compared to levels obtained in the presence of ApoE3.[51] Therefore, ApoE4 may serve either as a less effective inhibitor of Aβ fibrillogenesis or as a more potent stimulator than ApoE3 or ApoE2. An alternative mechanism for the AD-promoting effect of ApoE4 inheritance emerges from evidence in transgenic mice expressing either the ε4 or ε3 human protein. Mice expressing ε4 appear to have decreased neuritic outgrowth of cultured neurons and decreased maintenance of established neurites.[52] These studies suggest that the

ApoE4 protein is less supportive of normal neuronal form and function than are the ApoE3 or E2 proteins. However, such a neuronal vulnerability in ApoE4 gene carriers may not be the actual mechanism of the ApoE4 effect on the AD phenotype, given the fact that ApoE4, in a gene dose-dependent fashion, enhances deposition of Aβ into cerebral and meningeal vessels to produce amyloid angiopathy, even in the absence of AD-type neuropathology.[53] The fact that ApoE4 expression enhances Aβ deposition in parenchymal plaque deposits as well as in microvessels outside of the brain parenchyma (including in subjects who do not have AD) potentially separates the effect of ApoE4 in promoting the AD cerebral phenotype (i.e., Aβ-bearing neuritic plaques) from any deleterious effects ApoE4 may have on neuronal/neuritic function in general. Thus, the most parsimonious explanation for ApoE4 effects vis-à-vis AD is that this isoform enhances and decreases the clearance of extracellular Aβ peptides in both the cerebral cortex and its microvasculature.[50]

Such a mechanism is supported by studies in which mice transgenic for mutant human APP are crossed with mice in which the endogenous mouse ApoE gene is deleted. The resultant offspring show substantially decreased Aβ burden compared to that seen in the parental APP line, and the Aβ deposits that develop are overwhelmingly diffuse (nonfibrillar). Moreover, mice lacking endogenous ApoE that transgenically express human APP and either human ApoE3 or E4 initially show even fewer Aβ deposits than mice expressing no ApoE at all. But as the mice develop deposits with age, the presence of the E4 isoform leads to far more fibrillar (neuritic) plaques than does the E3 isoform.[54]

Cell Biology of Presenilin: A Major Molecular Switch in Signaling

How do missense mutations in *PS1* and *PS2* enhance the amyloidogenic processing of APP? The first clue came from deleting the *PS1* gene in mice, which results in markedly altered skeletal and brain development *in utero* and perinatal mortality.[55] Embryonic neurons cultured from *PS1*$^{-/-}$-mice showed a striking 60–70% decrease in total Aβ production, both of $A\beta_{40}$ and $A\beta_{42}$.[56] In addition, a small fraction of APP molecules was found to coimmunoprecipitate with *PS1*, suggesting that the two formed transient complexes.[28] Separate work using peptidomimetic inhibitors of γ-secretase designed to mimic the γ-secretase cleavage site within APP suggested that this unknown enzyme had the properties of an aspartyl protease.[57] Putting these observations together, we hypothesized that presenilin might actually be the γ-secretase.[58] Close inspection of the sequence of presenilin revealed two conserved aspartate residues in adjacent transmembrane regions (TM 6 and 7) of all PS family members (Figure 67.2). Mutation of either aspartate in *PS1* decreased cellular Aβ production by ~60%, similar to the effect of ablating the entire *PS1* gene.[58] Expressing *PS1* and *PS2* in which each had one of its TM aspartates mutated decreased cellular Aβ production to undetectable levels.[59] Moreover, mutation of either TM aspartate in PS1 or PS2 prevented the normal endoproteolysis of these proteins.[58]

These results suggested that the two aspartates constitute the active site of γ-secretase, an unprecedented intramembrane aspartyl protease activated by an autoproteolytic cleavage to create the biologically active heterodimeric form. This interpretation was provocative, given the lack of precedent for a polytopic aspartyl protease with its active site in the membrane. But the findings are analogous to the observation that the metalloprotease (called site 2 protease) responsible for cleaving the sterol regulatory element-binding protein (SREBP) implicated in cholesterol homeostasis had its active site in the membrane. Strong support for the model came from the subsequent finding that γ-secretase inhibitors designed to mimic the transition state of a substrate with an aspartyl protease bound directly to PS heterodimers and no other cellular proteins.[60,61] Moreover, the region immediately around one of the two TM aspartates has homology to part of the active site of a known bacterial aspartyl protease.[62] Further work has confirmed that presenilin does not act alone; three additional membrane proteins (nicastrin, Aph-1 and Pen-2) complex with presenilin to form the active γ-secretase enzyme.[63–65]

The multiple lines of evidence that PS constitutes the active site of γ-secretase are strongly supportive of the "amyloid hypothesis" (or Aβ hypothesis) of AD causation.[28] All of the genetic mutations currently known to cause autosomal dominant forms of AD are found either in the substrate (APP) or the protease (PS) of the reaction that generates Aβ. AD-causing missense mutations in PS may subtly alter the conformation of one or more of the PS transmembrane (TM) domains to alter the interaction of the two TM aspartates with the initial cleavage site in the intramembrane proteolysis of APP, namely the ε-site immediately after either residue 49 or 50 (in the Aβ numbering scheme).[66] The result is an increase in the subsequent cleavage of the APP TM domain at the $A\beta_{42-43}$ peptide bond, thereby releasing relatively more of the highly self-aggregating $A\beta_{42}$ species throughout life.

As the work just reviewed was unfolding, it became apparent that PS had another, biologically more important proteolytic substrate than APP: the Notch family of cell-surface receptors.[67] Shortly after the human presenilins were cloned,[68] the homologous gene in *Caenorhabditis elegans* (*sel-12*) was shown to be a facilitator of signaling by *lin-12*, the worm homolog of the Notch gene.[69] Notch protein is a major regulator of cell fate determination in all metazoans.

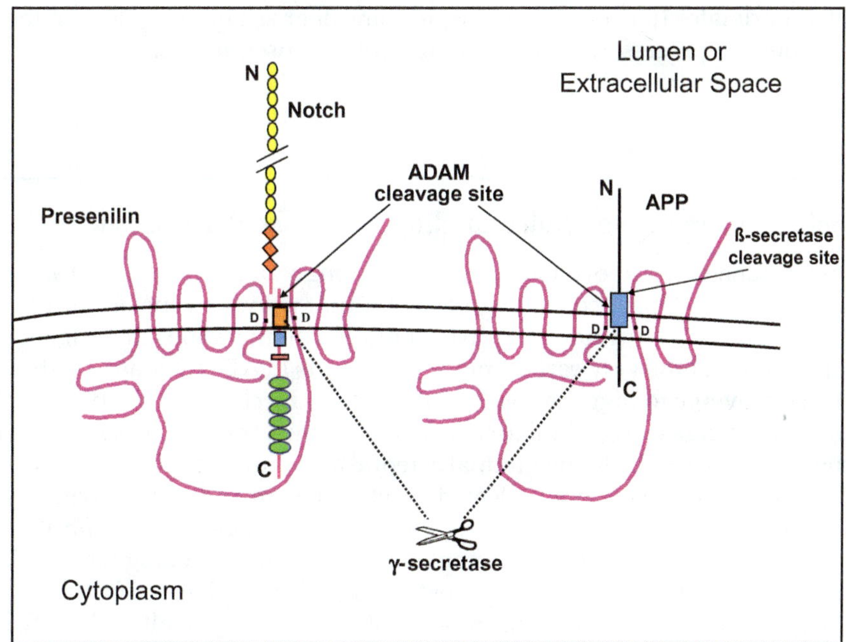

FIGURE 67.2 **Cartoon of the role of presenilin (PS) in Notch and APP processing.** The diagram shows the predicted nine TM domain topology of PS, which occurs principally as a cleaved heterodimer. Notch and APP molecules can form complexes with PS, together with three other protein cofactors (not shown). Two aspartate residues (D) in TM domains 6 and 7 of PS are required for the cleavages of Notch and APP within their TM domains, and these are predicted to align with the respective multiple sites of intramembrane cleavage in the two substrates (and many other type 1 membrane proteins). Ectodomain shedding effected by certain ADAM proteases precedes and is required for the PS-mediated proteolysis of both Notch and APP. In the case of APP (but not Notch), a subset of the holoprotein undergoes alternative ectodomain shedding by β-secretase. Several motifs are depicted in Notch (from top to bottom): EGF-like repeats (vertical ovals), LNG repeats (diamonds), a single TM (rectangle), the RAM23 domain (square), a nuclear localization sequence (narrow rectangle), and six cdc10/ankyrin repeats (horizontal ovals). Following the putative intramembranous cleavage mediated by PS, the Notch intracellular domain is released to the cytoplasm and can enter the nucleus to regulate transcription of target genes. APP contains the Aβ region (rectangle), which is released into the lumen after sequential cleavages of APP by β-secretase and PS/γ-secretase. The APP intracellular domain is released into the cytoplasm as well, and some of it might reach the nucleus in a complex with Fe65 and other proteins.

The binding of protein ligands such as Delta or Jagged to the Notch ectodomain leads to signaling of the Notch cytoplasmic domain by its regulating the transcription of target genes in the nucleus. But how the cytoplasmic domain of Notch reached the nucleus was initially a matter of lively debate. It became clear that it does so as the result of two sequential proteolytic cleavages of full-length Notch: first by ADAM 17 (TNFα-converting enzyme, one of the α-secretases) just 12 amino acids outside the TM domain (at the same relative position as the α-secretase cleavage of APP), and then by a γ-secretase scission within the Notch TM domain that releases the Notch intracellular domain (NICD) (Figure 67.2). Loss-of-function mutations in the *Drosophila melanogaster* presenilin gene produce a phenotype closely resembling that of deleting Notch function, and the generation of the NICD fragment is abrogated.[67] Furthermore, γ-secretase inhibitors lower the proteolytic formation of NICD in a fashion indistinguishable from their effects on Aβ levels (reviewed in [70]). Deletion of both *PS1* and *PS2* in mammalian embryonic stem cells abolishes both NICD and Aβ production. The parallels continue with the discovery that the APP intracellular domain (AICD) can participate in multiprotein complexes that may, in part, enter the nucleus[52] and could help regulate gene transcription, at least in experiments using heterologous reporter systems.[30,31] Thus, one normal function of APP, like Notch, may be as a type of signaling receptor, but its activating ligand(s) and the genes whose transcription it may help regulate remain ill-defined.

Accumulating evidence from numerous laboratories now supports the concept that the presenilins are key molecular switches that affect the proteolytic release of cytoplasmic signaling domains from more than 100 single-TM polypeptides. Thus, work that began with a focus on the pathobiology of AD has uncovered a common mechanism that is required in a large number of different signaling pathways. On this basis, one can postulate that AD arose in the human population because a highly conserved proteolytic machine necessary for developmental decisions in all multicellular organisms has as one of its alternative substrates a protein whose membrane-derived peptide fragment (Aβ) can slowly aggregate and accumulate in the brain during postreproductive life. This age-related accumulation can be detected in most old primates (including humans) to some degree, and mutations in the substrate or the protease can markedly accelerate the process, with devastating clinical consequences.

THERAPY

Therapeutic Approaches Emerge from Understanding the Molecular Cascade of Alzheimer Disease

The unequivocal confirmation of the above four genes that strongly predispose to AD and the accelerating identification of numerous additional AD risk-conferring genes (http://www.alzgene.org) lead to the conclusion that AD is a syndrome that can be initiated by numerous distinct molecular alterations that trigger a common pathologic cascade. Such a model predicts that the very earliest molecular events in AD differ among the distinct genetic forms of the disease but that these converge to trigger a common pathogenic mechanism: a subtle but chronic imbalance between Aβ production and Aβ clearance in cerebral tissue. This imbalance, the degree of which is likely to be regulated by numerous other gene products, results in the gradual cerebral accumulation of $A\beta_{42}$ peptides, a fraction of which undergoes conformational alteration to a more β-pleated sheet-rich form and can aggregate over time into soluble oligomers of a wide range of sizes and, later, insoluble filamentous polymers (amyloid fibrils).

The insidious accrual of Aβ oligomers and their apparent triggering of microgliosis and astrocytosis and the accompanying heightened secretion of inflammatory mediators (e.g., complement factors, cytokines, chemokines, acute phase proteins) may induce subtle—and later more profound—synaptic alterations that produce the symptoms of progressive neuronal dysfunction. Figure 67.3 summarizes certain key pathogenic factors discussed in this chapter. Experimental evidence has emerged that this inflammatory response may not be absolutely required by itself for synaptic dysfunction, i.e., soluble Aβ oligomers may directly interfere with synaptic function via many molecular pathways that are now being worked out (reviewed in [71,72]). The complex pathogenic cascade in neurons prominently includes the dissociation of tau from microtubules, apparently related to its acquisition of excess phosphorylation, and its subsequent aggregation into oligomers and then insoluble PHF within innumerable neurites and neuronal cell bodies. *In vivo* evidence that cerebral Aβ accumulation can drive tau alteration and tangle formation has arisen from the crossing of mutant human tau transgenic mice to mutant human Aβ transgenic mice.[7,8] The ultimate consequences of the neuronal dysfunction induced by the synaptotoxic and inflammatory effects of Aβ oligomers and the accompanying alteration of tau include synaptic and neuronal loss and the multiple neurotransmitter deficiencies of the disease. The precise temporal sequence of the many molecular and cellular alterations that accrue in AD and their relative importance in producing progressive neuronal dysfunction and hence dementia remain vague and may never be completely understood, given the reality that many of the key changes occur virtually simultaneously and have feedforward effects during the decades-long pathogenic cascade.

The Aβ hypothesis of AD has the greatest amount of supporting experimental data among the concepts that have been proposed to explain the etiology of the disorder. The model of Alzheimer disease reviewed here points to new therapeutic strategies aimed at one or more critical steps in its molecular progression (reviewed in [73]): 1) One could partially inhibit one or the other of the proteases (β- and γ-secretases) that liberate Aβ from its precursor. One approach to modulating γ-secretase that has reached human trials is the use of derivatives of certain nonsteroidal anti-inflammatory drugs that lower $A\beta_{42}$ without interfering with the processing of Notch.[12] However, this and other later classes of γ-secretase modulators (e.g.,[74]) have not yet been shown to be sufficiently potent and bioavailable in man to produce clinical efficacy. Certain inhibitors of BACE-1, the β-secretase for APP[75], are in advanced clinical testing at time of writing. 2) One could decrease the production and release of Aβ by a variety of means other than secretase inhibition, for example, by pharmacologically diverting some APP molecules from the amyloidogenic (β-secretase) to the nonamyloidogenic (α-secretase) processing pathway. 3) One could attempt to inhibit the oligomerization of Aβ monomers that precedes fibrillogenesis, but one would presumably need to block the monomer-to-dimer conversion, as blockage solely thereafter could risk increasing the levels of potentially synaptotoxic dimers.[6,76] 4) One could try to clear Aβ peptides (monomers, oligomers and even fibrillar aggregates) from the cortex. This appears to be one of the mechanisms of active Aβ vaccines[9] and passively administered anti-Aβ antibodies,[77] both of which have been successful in AD mouse models but have yet to be proven efficacious in human trials, despite suggestive data in at least one phase III trial.[78] 5) One could interfere with the activities of microglia and astrocytes that contribute to the development of a chronic inflammatory process around Aβ deposits. 6) One could attempt to block the molecules ("Aβ receptors") on the surface of neurons (e.g., certain membrane lipids) or their many intracellular effectors that mediate the neurotoxic effects of Aβ oligomers, although these molecules have not been comprehensively identified as yet, and there are likely to be many of them.

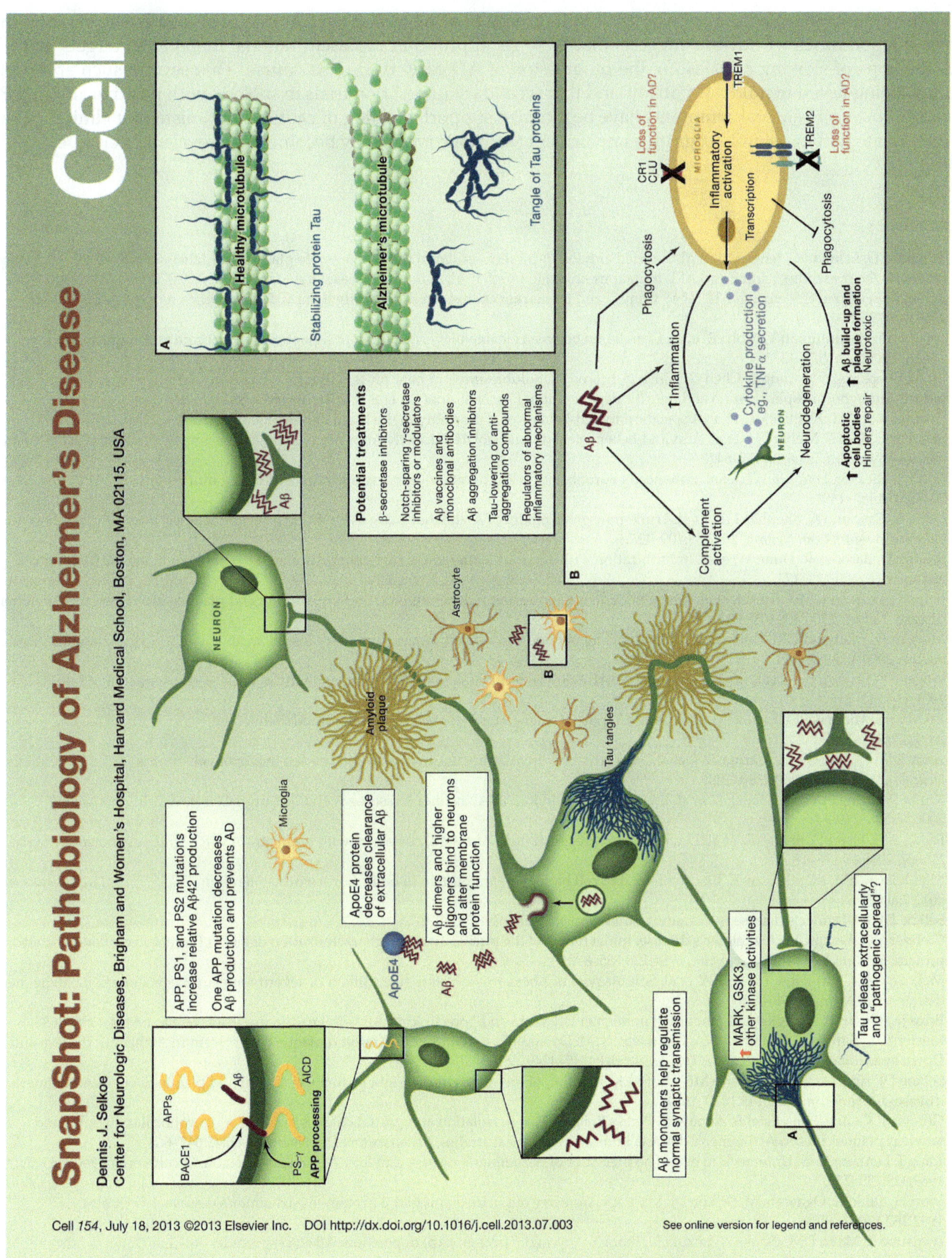

FIGURE 67.3 Summary diagram of many of the principal cellular and molecular factors that contribute to the pathogenesis of Alzheimer disease, based on current information. Reprinted with permission from *Cell*. Drawn by Yvonne Blanco for Cell Press. To view other Snapshots in the collection, please visit snapshots.cell.com.

None of these pharmacological objectives will be easy to achieve. But the confluence of data emerging from hundreds of laboratories worldwide suggests that these six approaches represent rational therapeutic targets that offer the best hope of slowing or arresting the progression of AD early during its course. That several such approaches are now being tested in mild AD patients and that secondary prevention trials in subjects with positive amyloid PET scans but few or no clinical symptoms have begun may support a feeling of cautious optimism that clinical AD and its immediate harbinger, mild cognitive impairment (amnestic type), may become less prevalent over time.

References

1. Scheuner D, Eckman C, Jensen M, et al. Secreted amyloid β-protein similar to that in the senile plaques of Alzheimer's disease is increased *in vivo* by the presenilin 1 and 2 and APP mutations linked to familial Alzheimer's disease. *Nat Med*. 1996;2:864–870.
2. Bateman RJ, Xiong C, Benzinger TL, et al. Clinical and biomarker changes in dominantly inherited Alzheimer's disease. *N Engl J Med*. 2012;367:795–804.
3. Näslund J, Haroutunian V, Mohs R, et al. Correlation between elevated levels of amyloid β-peptides in the brain and cognitive decline. *JAMA*. 2000;283:1571–1577.
4. Jin M, Shepardson N, Yang T, Chen G, Walsh D, Selkoe DJ. Soluble amyloid beta-protein dimers isolated from Alzheimer cortex directly induce Tau hyperphosphorylation and neuritic degeneration. *Proc Natl Acad Sci U S A*. 2011;108:5819–5824.
5. Lesne S, Koh MT, Kotilinek L, et al. A specific amyloid-beta protein assembly in the brain impairs memory. *Nature*. 2006;440:352–357.
6. Shankar GM, Li S, Mehta TH, et al. Amyloid-beta protein dimers isolated directly from Alzheimer's brains impair synaptic plasticity and memory. *Nat Med*. 2008;14:837–842.
7. Lewis J, Dickson DW, Lin WL, et al. Enhanced neurofibrillary degeneration in transgenic mice expressing mutant tau and APP. *Science*. 2001;293:1487–1491.
8. Oddo S, Caccamo A, Shepherd JD, et al. Triple-transgenic model of Alzheimer's disease with plaques and tangles: intracellular Abeta and synaptic dysfunction. *Neuron*. 2003;39:409–421.
9. Schenk D, Barbour R, Dunn W, et al. Immunization with amyloid-β attenuates Alzheimer-disease-like pathology in the PDAPP mouse. *Nature*. 1999;400:173–177.
10. Janus C, Pearson J, McLaurin J, et al. A beta peptide immunization reduces behavioural impairment and plaques in a model of Alzheimer's disease. *Nature*. 2000;408:979–982.
11. Morgan D, Diamond DM, Gottschall PE, et al. A beta peptide vaccination prevents memory loss in an animal model of Alzheimer's disease. *Nature*. 2000;408:982–985.
12. Weggen S, Eriksen JL, Das P, et al. A subset of NSAIDs lower amyloidogenic Aβ42 independently of cyclooxygenase activity. *Nature*. 2001;414:212–216.
13. Doody RS, Thomas RG, Farlow M, et al. Phase 3 trials of solanezumab for mild-to-moderate Alzheimer's disease. *N Engl J Med*. 2014;370:311–321.
14. Amariglio RE, Becker JA, Carmasin J, et al. Subjective cognitive complaints and amyloid burden in cognitively normal older individuals. *Neuropsychologia*. 2012;50:2880–2886.
15. Klunk WE, Engler H, Nordberg A, et al. Imaging brain amyloid in Alzheimer's disease with Pittsburgh Compound-B. *Ann Neurol*. 2004;55:306–319.
16. Fagan AM, Head D, Shah AR, et al. Decreased cerebrospinal fluid Abeta(42) correlates with brain atrophy in cognitively normal elderly. *Ann Neurol*. 2009;65:176–183.
17. Mandelkow EM, Mandelkow E. Biochemistry and cell biology of tau protein in neurofibrillary degeneration. *Cold Spring Harb Perspect Med*. 2012;2:a006247.
18. Selkoe DJ. Alzheimer's disease is a synaptic failure. *Science*. 2002;298:789–791.
19. Glenner GG, Wong CW. Alzheimer's disease: initial report of the purification and characterization of a novel cerebrovascular amyloid protein. *Biochem Biophys Res Commun*. 1984;120:885–890.
20. McLean CA, Cherny RA, Fraser FW, et al. Soluble pool of Abeta amyloid as a determinant of severity of neurodegeneration in Alzheimer's disease. *Ann Neurol*. 1999;46:860–866.
21. Price JL, Morris JC. Tangles and plaques in nondemented aging and "preclinical" Alzheimer's disease. *Ann Neurol*. 1999;45:358–368.
22. Masters CL, Simms G, Weinman NA, Multhaup G, McDonald BL, Beyreuther K. Amyloid plaque core protein in Alzheimer disease and Down syndrome. *Proc Natl Acad Sci U S A*. 1985;82:4245–4249.
23. Selkoe DJ, Abraham CR, Podlisny MB, Duffy LK. Isolation of low-molecular-weight proteins from amyloid plaque fibers in Alzheimer's disease. *J Neurochem*. 1986;146:1820–1834.
24. Gorevic P, Goni F, Pons-Estel B, Alvarez F, Peress R, Frangione B. Isolation and partial characterization of neurofibrillary tangles and amyloid plaque cores in Alzheimer's disease: immunohistological studies. *J Neuropathol Exp Neurol*. 1986;45:647–664.
25. Kang J, Lemaire H-G, Unterbeck A, et al. The precursor of Alzheimer's disease amyloid A4 protein resembles a cell-surface receptor. *Nature*. 1987;325:733–736.
26. Mori H, Takio K, Ogawara M, Selkoe DJ. Mass spectrometry of purified amyloid β protein in Alzheimer's disease. *J Biol Chem*. 1992;267:17082–17086.
27. Iwatsubo T, Mann DM, Odaka A, Suzuki N, Ihara Y. Amyloid β protein (Aβ) deposition: Aβ42(43) precedes Aβ40 in Down syndrome. *Ann Neurol*. 1995;37:294–299.
28. Selkoe DJ. Alzheimer's disease: genes, proteins and therapies. *Physiol Rev*. 2001;81:742–761.
29. Rice HC, Young-Pearse TL, Selkoe DJ. Systematic evaluation of candidate ligands regulating ectodomain shedding of amyloid precursor protein. *Biochemistry*. 2013;52:3264–3277.

30. Cao X, Sudhof TC. A transcriptionally active complex of APP with Fe65 and histone acetyltransferase Tip60. *Science*. 2001;293:115–120.
31. Cao X, Sudhof TC. Dissection of amyloid-beta precursor protein-dependent transcriptional transactivation. *J Biol Chem*. 2004;279:24601–24611.
32. Hebert SS, Serneels L, Tolia A, et al. Regulated intramembrane proteolysis of amyloid precursor protein and regulation of expression of putative target genes. *EMBO Rep*. 2006;7:739–745.
33. Hardy J. The Alzheimer family of diseases: many etiologies, one pathogenesis? *Proc Natl Acad Sci U S A*. 1997;94:2095–2097.
34. Lemere CA, Blustzjan JK, Yamaguchi H, Wisniewski T, Saido TC, Selkoe DJ. Sequence of deposition of heterogeneous amyloid β-peptides and Apo E in Down syndrome: implications for initial events in amyloid plaque formation. *Neurobiol Dis*. 1996;3:16–32.
35. Iwatsubo T, Odaka A, Suzuki N, Mizusawa H, Nukina H, Ihara Y. Visualization of A beta 42(43) and A beta 40 in senile plaques with end-specific A beta monoclonals: evidence that an initially deposited species is A beta 42(43). *Neuron*. 1994;13:45–53.
36. Prasher VP, Farrer MJ, Kessling AM, et al. Molecular mapping of Alzheimer-type dementia in Down's syndrome. *Ann Neurol*. 1998;43:380–383.
37. Rovelet-Lecrux A, Hannequin D, Raux G, et al. APP locus duplication causes autosomal dominant early-onset Alzheimer disease with cerebral amyloid angiopathy. *Nat Genet*. 2006;38:24–26.
38. Jonsson T, Atwal JK, Steinberg S, et al. A mutation in APP protects against Alzheimer's disease and age-related cognitive decline. *Nature*. 2012;488:96–99.
39. Schellenberg GD, Bird TD, Wijsman EM, et al. Genetic linkage evidence for a familial Alzheimer's disease locus on chromosome 14. *Science*. 1992;258:668–671.
40. St George-Hyslop PH. Molecular genetics of Alzheimer's disease. *Biol Psychiatry*. 2000;47:183–199.
41. Sherrington R, Rogaev EI, Liang Y, et al. Cloning of a gene bearing missense mutations in early-onset familial Alzheimer's disease. *Nature*. 1995;375:754–760.
42. Strittmatter WJ, Saunders AM, Schmechel D, et al. Apolipoprotein E: high-avidity binding to β-amyloid and increased frequency of type 4 allele in late-onset familial Alzheimer disease. *Proc Natl Acad Sci U S A*. 1993;90:1977–1981.
43. Corder EH, Saunders AM, Strittmatter WJ, et al. Gene dose of apolipoprotein E type 4 allele and the risk of Alzheimer's disease in late onset families. *Science*. 1993;261:921–923.
44. Van Nostrand WE, Melchor JP. Disruption of pathologic amyloid beta-protein fibril assembly on the surface of cultured human cerebrovascular smooth muscle cells. *Amyloid*. 2001;8(suppl 1):20–27.
45. Price DL, Wong PC, Markowska AL, et al. The value of transgenic models for the study of neurodegenerative diseases. *Ann N Y Acad Sci*. 2000;920:179–191.
46. Holcomb L, Gordon MN, McGowan E, et al. Accelerated Alzheimer-type phenotype in transgenic mice carrying both mutant amyloid precursor protein and presenilin 1 transgenes. *Nat Med*. 1998;4:97–100.
47. Lemere CA, Lopera F, Kosik KS, et al. The E280A presenilin 1 Alzheimer mutation produces increased Aβ42 deposition and severe cerebellar pathology. *Nat Med*. 1996;2:1146–1150.
48. Saunders AM, Strittmatter WJ, Schmechel D, et al. Association of apolipoprotein E allele epsilon 4 with late-onset familial and sporadic Alzheimer's disease. *Neurology*. 1993;43:1467–1472.
49. Rebeck GW, Reiter JS, Strickland DK, Hyman BT. Apolipoprotein E in sporadic Alzheimer's disease: allelic variation and receptor interactions. *Neuron*. 1993;11:575–580.
50. Castellano JM, Kim J, Stewart FR, et al. Human apoE isoforms differentially regulate brain amyloid-beta peptide clearance. *Sci Transl Med*. 2011;3:89ra57.
51. Ma J, Yee A, Brewer Jr. HB, Das S, Potter H. The amyloid-associated proteins α1-antichymotrypsin and apolipoprotein E promote the assembly of the Alzheimer β-protein into filaments. *Nature*. 1994;372:92–94.
52. Nathan BP, Bellosta S, Sanan DA, Weisgraber KH, Mahley RW, Pitas RE. Differential effects of apolipoprotein E3 and E4 on neuronal growth *in vitro*. *Science*. 1994;264:850–852.
53. Greenberg SM, Rebeck GW, Vonsattel JPG, Gomez-Isla T, Hyman BT. Apolipoprotein E ε4 and cerebral hemorrhage associated with amyloid angiopathy. *Ann Neurol*. 1995;38:254–259.
54. Holtzman DM, Bales KR, Tenkova T, et al. Apolipoprotein E isoform-dependent amyloid deposition and neuritic degeneration in a mouse model of Alzheimer's disease. *Proc Natl Acad Sci U S A*. 2000;97:2892–2897.
55. Shen J, Bronson RT, Chen DF, Xia W, Selkoe DJ, Tonegawa S. Skeletal and CNS defects in presnilin-1 deficient mice. *Cell*. 1997;89:629–639.
56. De Strooper B, Saftig P, Craessaerts K, et al. Deficiency of presenilin-1 inhibits the normal cleavage of amyloid precursor protein. *Nature*. 1998;391:387–390.
57. Wolfe MS, Xia W, Moore CL, et al. Peptidomimetic probes and molecular modeling suggest Alzheimer's γ-secretase is an intramembrane-cleaving aspartyl protease. *Biochemistry*. 1999;38:4720–4727.
58. Wolfe MS, Xia W, Ostaszewski BL, Diehl TS, Kimberly WT, Selkoe DJ. Two transmembrane aspartates in presenilin-1 required for presenilin endoproteolysis and γ-secretase activity. *Nature*. 1999;398:513–517.
59. Kimberly WT, Xia W, Rahmati T, Wolfe MS, Selkoe DJ. The transmembrane aspartates in presenilin 1 and 2 are obligatory for gamma-secretase activity and amyloid beta-protein generation. *J Biol Chem*. 2000;275:3173–3178.
60. Esler WP, Kimberly WT, Ostaszewski BL, et al. Transition-state analogue inhibitors of γ-secretase bind directly to presenilin-1. *Nat Cell Biol*. 2000;2:428–434.
61. Li Y-M, Xu M, Lai M-T, et al. Photoactivated γ-secretase inhibitors directed to the active site covalently label presenilin 1. *Nature*. 2000;405:689–694.
62. Steiner H, Kostka M, Romig H, et al. Glycine 384 is required for presenilin-1 function and is conserved in bacterial polytopic aspartyl proteases. *Nat Cell Biol*. 2000;2:848–851.
63. Edbauer D, Winkler E, Regula JT, Pesold B, Steiner H, Haass C. Reconstitution of gamma-secretase activity. *Nat Cell Biol*. 2003;5:486–488.
64. Kimberly WT, LaVoie MJ, Ostaszewski BL, Ye W, Wolfe MS, Selkoe DJ. Gamma-Secretase is a membrane protein complex comprised of presenilin, nicastrin, Aph-1 and Pen-2. *Proc Natl Acad Sci U S A*. 2003;100:6382–6387.
65. Takasugi N, Tomita T, Hayashi I, et al. The role of presenilin cofactors in the γ-secretase complex. *Nature*. 2003;422:438–441.

66. Takami M, Nagashima Y, Sano Y, et al. Gamma-Secretase: successive tripeptide and tetrapeptide release from the transmembrane domain of beta-carboxyl terminal fragment. *J Neurosci*. 2009;29:13042–13052.
67. Struhl G, Greenwald I. Presenilin is required for activity and nuclear access of Notch in Drosophila. *Nature*. 1999;398:522–525.
68. Sherrington R, Rogaev EI, Liang Y, et al. Cloning of a novel gene bearing missense mutations in early onset familial Alzheimer disease. *Nature*. 1995;375:754–760.
69. Levitan D, Greenwald I. Facilitation of *lin-12*-mediated signalling by *sel-12*, a *Caenorhabditis elegans S182* Alzheimer's disease gene. *Nature*. 1995;377:351–354.
70. Haass C, De Strooper B. The presenilins in Alzheimer's disease—proteolysis holds the key. *Science*. 1999;286:916–919.
71. Haass C, Selkoe DJ. Soluble protein oligomers in neurodegeneration: lessons from the Alzheimer's amyloid β-peptide. *Nat Rev Mol Cell Biol*. 2007;8:101–112.
72. Mucke L, Selkoe D. Neurotoxicity of amyloid β-protein: synaptic and network disfunction. In: Selkoe D, Mandelkow E, Holtzman D, eds. *The Biology of Alzheimer Disease*. New York: Cold Spring Harbor Laboratory Press; 2012:317–333.
73. Selkoe DJ, Schenk D. Alzheimer's disease: molecular understanding predicts amyloid-based therapeutics. In: Cho AK, Blaschke TF, Insel PA, Loh HH, eds. *Annual Review of Pharmacology and Toxicology*. Palo Alto, CA: Annual Reviews; 2003:545–584.
74. Kounnas MZ, Danks AM, Cheng S, et al. Modulation of gamma-secretase reduces beta-amyloid deposition in a transgenic mouse model of Alzheimer's disease. *Neuron*. 2010;67:769–780.
75. Vassar R, Bennett BD, Babu-Khan S, et al. Beta-secretase cleavage of Alzheimer's amyloid precursor protein by the transmembrane aspartic protease BACE. *Science*. 1999;286:735–741.
76. Walsh DM, Townsend TM, Podlisny MB, et al. Certain inhibitors of synthetic Aβ fibrillogenesis block oligomerization of natural Aβ and thereby rescue long term potentiation. *J Neurosci*. 2005;25:2455–2462.
77. Bard F, Cannon C, Barbour R, et al. Peripherally administered antibodies against amyloid beta-peptide enter the central nervous system and reduce pathology in a mouse model of Alzheimer disease. *Nat Med*. 2000;6:916–919.
78. Salloway S, Sperling R, Fox NC, et al. Two phase 3 trials of bapineuzumab in mild-to-moderate Alzheimer's disease. *N Engl J Med*. 2014;370:322–333.

CHAPTER

68

Genetics of Parkinson Disease and Related Diseases

Jill S. Goldman and Stanley Fahn†*

*Taub Institute, Columbia University Medical Center, New York, NY, USA

†Department of Neurology, Columbia University Medical Center, New York, NY, USA

INTRODUCTION

Movement disorders can be defined as neurological syndromes in which there is either an excess of movement (hyperkinesias) or a paucity of voluntary and automatic movements, unrelated to weakness or spasticity (hypokinesias). The parkinsonian syndromes are the most common cause of such paucity of movement; other hypokinetic disorders affect only a small group of patients. Genetic research is rapidly revealing the cause of many of the familial parkinsonian diseases and shedding light on mechanisms of disease processes. In this chapter we review the genetic highlights of the parkinsonian syndromes.

PARKINSON DISEASE

Parkinsonism is characterized clinically by any combination of rest tremor, bradykinesia, rigidity, postural instability, freezing phenomena, and flexed posture (Figure 68.1). It is conveniently divided into primary parkinsonism (i.e., Parkinson disease); secondary parkinsonism (insults to the brain); Parkinson-plus syndromes, such as progressive supranuclear palsy (PSP), multiple system atrophy (MSA) and corticobasal degeneration (CBD); and heredodegenerative disorders in which parkinsonism is not the primary clinical feature (e.g., Wilson disease and juvenile Huntington disease). Parkinson disease (PD) is the most common form of parkinsonism and the second most common neurodegenerative disorder. It affects 1–2% of people older than 65 years. The classical neuropathology of PD affects substantia nigra pars compacta with loss of dopaminergic neurons and presence of eosinophilic intracytoplasmic inclusions (Lewy bodies) that contain ubiquitin in the central core and α-synuclein in the halo. The locus ceruleus is similarly affected. PD is differentiated from Parkinson-plus syndromes because the latter have pathology involving additional areas of the basal ganglia and other parts of the brain, producing symptoms beyond those of pure parkinsonism. Additionally the pathology does not manifest Lewy bodies.

The identification of "families with PD" and the discovery that individuals exposed to a potent neurotoxin (1-methyl-4-phenyl-1,2,3,6-tetrahydropyridine, MPTP) developed parkinsonism, suggested both genetic and environmental causes of PD. It is now generally accepted that PD can be caused by genetic and environmental factors in various combinations. Before this was known, PD used to be referred to as idiopathic parkinsonism. However "primary parkinsonism" is the more appropriate term now that numerous genes have been discovered that are definitively associated with monogenic inheritance of PD. A community-based study identified a lifetime risk of PD in a parent or a sibling of an individual with PD as approximately 2%, as opposed to a lifetime risk of PD of 1% in controls, confirming multifactorial inheritance as a mechanism for PD.[1] A twin study found that PD was more likely to be genetic in origin in people with onset of PD under the age of 50 as opposed to over the age of 50.[2] Eight genes have so far been found to be associated with the majority of Mendelian PD, four with autosomal dominant PD, four

Rosenberg's Molecular and Genetic Basis of Neurological and Psychiatric Disease
http://dx.doi.org/10.1016/10.1016/B978-0-12-410529-4.00068-1

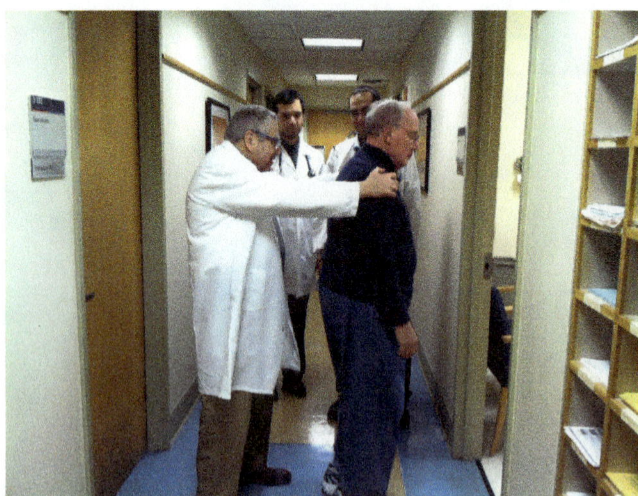

FIGURE 68.1 **The pull test in a patient with Parkinson disease.** The physician is demonstrating the pull test, which evaluates postural instability.

with autosomal recessive PD (discussed below), plus several other recently discovered rare genes, and a number of candidate susceptibility genes. It is possible, though not proven, that several of the Mendelian genes are involved in the development of sporadic PD; regardless, understanding the cause of the familial forms is crucial to uncovering the cause and pathogenesis of sporadic cases.

Autosomal Dominant Parkinson Disease

SNCA (Formerly **PARK1, PARK4)**

The first gene to be associated with PD was the *α-synuclein* (*SNCA*) gene on chromosome 4q22.1 (Table 68.1).[3] Mutations in this gene cause highly penetrant, autosomal dominant Lewy body parkinsonism. Three mutations have been discovered in a small number of families of Italian and Greek (A53T), German (A30P) and Spanish (E46K) ancestry. It is now known that a family with autosomal dominant PD, originally (mis)linked to 4p15 and assigned locus name PARK4, has, in fact, a triplication of a large chromosomal region containing *SNCA* (PARK1).[4] Other triplications and duplications have been reported since then. While copy number variants are very rare, they are more widespread than the point mutations.[5] In fact, *SNCA* duplications may be responsible for 1–2% of autosomal dominant PD.[6] Thus, an excess of normal, not mutated, α-synuclein can also cause PD. The dosage effect of *SNCA* appears to influence both age of onset and phenotype, with triplication carriers presenting earlier than duplication carriers, and often having more severe symptoms. People with missense mutations generally have an earlier age of onset and shorter duration than sporadic PD. Dementia and hallucinations are common. Other features include dysautonomia, speech dysfunction, and behavioral problems. Unlike triplications and missense mutations, *SNCA* duplications have reduced penetrance of 30–50%.[7,8] Lewy body pathology is much more widespread than that found with sporadic PD, and found in both the cerebral cortex and brainstem.

Through numerous genome-wide association studies (GWAS), several polymorphisms in *SNCA* have been associated with increased risk of sporadic PD. The absolute risk may differ depending on the population.

LRRK2 **(PARK8)**

Mutations in *leucine-rich repeat kinase-2 (LRRK2)* are associated with familial (autosomal dominant) and sporadic PD. *LRRK2* encodes the protein dardarin. Mutations in this gene are thought to account for about 5% of all familial autosomal dominant PD, and anywhere from 0.3–3% of sporadic PD.[7] The incidence of *LRRK2* sporadic disease is difficult to define as the gene displays age-dependent penetrance. Mutations appear to be equally common in early- and late-onset PD. Of nearly a dozen different reported mutations, the most common, G2019S, has been reported in various populations, with an increased frequency among Ashkenazi Jews (18.3% of those with PD) and North African Berbers (39% of those with PD—both sporadic and familial).[6,9] Other mutations are much rarer, although the R1441G mutation was found in 8% of Basque familial PD cases.[10]

Estimates of *LRRK2* penetrance varies considerably, depending on the study population and methodology, as well as the particular mutation. Penetrance of the G2019S mutation is age-dependent and studies report a range of 24–74% by age 79.[9,11,12] Not only is penetrance variable, but so is age of onset and phenotype, even among

TABLE 68.1 A List of Monogenic Forms of Parkinson Disease

Disease name/Gene	Inheritance	Location	Protein	Protein function
PARK1/*SNCA*	AD	4q22.1	α-Synuclein	Synaptic protein
PARK2/*PRKN*	AR	6q26	Parkin	Ubiquitin ligase
PARK3/*SPR*	AD	2p13	Sepiapterin reductase	Tetrahydrobiopterin synthesis
PARK4 is the same as PARK1	–	–	–	–
PARK5/*UCHL1*	AD	4p13	Ubiquitin carboxy terminal hydrolase	Ubiquitin hydrolase
PARK6/*PINK1*	AR	1p36.12	PTEN-induced putative kinase 1 (PINK1)	Serine/threonine protein kinase
PARK7/DJ-*1*	AR	1p36.23	DJ-1	Antioxidant scavenger
PARK8/*LRRK2*	AD	12q12	Dardarin and also known as leucine-rich repeat kinase-2	Kinase
PARK9/Kufor-Rakeb	AR	1p36.13	ATP13A2	Lysosomal atpase
PARK10	Susceptibility locus?	1p32	N/I	–
PARK11	Susceptibility locus?	2q37.1	N/I	–
PARK12	Susceptibility locus?	Xq21-q25	N/I	–
PARK13	Susceptibility locus?	2p13.1	HTRA2	Serine protease
PARK14/*PLA2G6*	AR	22q13.1	PLA2G6	Phospholipase A2
PARK15/*FBX07*	AR	22q12.3	F-box only protein 7	Ubiquitin ligase
PARK16	Susceptibility locus?	1q32	N/I	–
PARK17/*VPS35*	AD	16q11.2	Vacuolar protein sorting 35	Endosome-Golgi trafficking
PARK18/*EIF4G1*	AD	3q27.1	Eukarytic translocation initiation factor 4-gamma, 1	Initiation of protein synthesis
PARK19/*DNAJC6*	AR	1p31.3	Auxilin	Clathrin mediated endocytosis
PARK20/*SYNJ1*	AR	21q22.11	Synaptojanin 1	Synaptic vesicle recycling
GBA, Gaucher Disease	Susceptibility locus	1q21	β-Glucocerebrosidase	Lysosomal enzyme
SNCAIP	Susceptibility locus for MSA	5q23.2	Synphilin-1	Binds to α-synuclein, induces the formation of intracellular aggregates
NR4A2 or *NURR1*	Susceptibility locus?	2q24.1	Nuclear receptor subfamily 4, group a, member 2 or nuclear receptor-related 1	
POLG	Susceptibility locus?	15q26.1	Mitochondrial polymerase gamma A subunit	Increases fidelity of mtDNA replication through proofreading

Abbreviations: AD, autosomal dominant; AR, autosomal recessive; MSA, multiple system atrophy; mtDNA, mitochondrial DNA; N/I, not identified.

identical twins. Thus, other genes or epigenetic factors must play a role in *LRRK2* expression. Clinical findings are similar to that of sporadic PD although there are reports that tremor is predominant,[12] where others find more postural instability and gait disorder.[13] Neuropathology is also variable, with Lewy bodies being the most common finding for G2019S. However, tau and ubiquitin have also been found in some cases. Pathology is variable in other mutations.[7]

LRRK2 mutations have been implied in increased risk for nonskin cancers, Crohn disease, and leprosy.[14,15] *LRRK2* mutations appear to be involved with autophagy but the exact mechanism is yet to be determined. Some mutations such as p.R1398H may in fact be protective of PD.[16]

EIF4G1

The *eukaryotic translation initiation factor 4-gamma 1* gene is associated with a rare cause of mild form of autosomal dominant PD with onset between 50 and 80 years of age. Pathology appears to be similar to that of diffuse Lewy body disease. This gene is involved in mRNA translation.[17]

DCTN1

The *dynactin 1* gene (*DCTN1*), first linked to a family with motor neuron disease, has been found to be the causal gene for Perry disease, a rapidly progressive condition with parkinsonism, profound weight loss, and depression. This gene is involved in protein and organelle trafficking. Pathology is a combination of tau and TDP-43 deposits.[18]

VPS35

The *vacuolar protein sorting 35* gene has been reported as another very rare cause of PD (PARK17). Phenotypic findings may include psychiatric symptoms, dementia, or myoclonus, but this may be mutation dependent or an ascertainment bias because of the small number of cases. Penetrance appears to be incomplete.[19]

Autosomal Recessive Parkinson Disease

Parkin (*PRKN*; PARK2)

Mutations in the *parkin* gene on chromosome 6q26 (PARK2) were initially identified in Japanese individuals with juvenile-onset PD. *Parkin* mutations have now been identified pan-ethnically worldwide and were originally reported to be the cause of approximately 50% of familial young-onset PD and up to 10% of sporadic young-onset PD (younger than 45 years).[7] However, a recent study estimates a much lower incidence of 1.4–8.2% of young-onset PD.[20] Over 170 pathologic allelic variants have been described, including exon rearrangements, point mutations, duplications, triplications, insertions, and deletions, many of them recurrent in different populations.[5,6]

Parkin is a very large gene, which contains many polymorphisms. Testing is difficult because of results of unknown significance. The earlier the age of onset of PD, the increased likelihood that *parkin* mutations are identified. However, PARK2 does not appear to be restricted to young-onset PD, and *parkin* mutations have been identified in individuals older than 50 years. Findings suggest that juvenile PD is most often associated with mutations in both *parkin* alleles (homozygous /compound heterozygous), but that later-onset PD, when associated with *parkin*, is associated with a single *parkin* mutation, and thus acts as a susceptibility factor.[6] This association has not been proven.[7]

The frequency and penetrance of *parkin* mutations have yet to be determined. PARK2, especially in juvenile cases, has been characterized by slow progression, sustained response to levodopa, levodopa-induced dyskinesias, dystonia, sleep-benefit, and hyperreflexia, but these findings have not been consistent. The PARK2 phenotype can resemble that of dopa-responsive dystonia (DRD). Some individuals with PARK2, especially with later onset, are clinically indistinguishable from sporadic adult-onset PD. PARK2 is characterized pathologically by loss of neurons in pigmented nuclei, and interestingly, by absence of Lewy bodies, although several case reports identified Lewy bodies in compound heterozygotes.[8] Parkin is an E3 ubiquitin ligase that ubiquinates specific proteins targeted for degradation by the ubiquitin-dependent proteasome system. This information led researchers to suspect that reduced E3 ligase activity may be the mechanism by which *parkin* mutations cause familial PD, possibly by defects in solubility or substrate binding. However, genetic animal models in mice and *Drosophila* (fruit flies) without the *PRKN* gene are associated with mitochondrial impairment, cellular apoptosis, and locomotor dysfunction. Subsequent studies revealed that the parkin protein is selectively recruited to dysfunctional mitochondria. After recruitment, parkin mediates the engulfment of mitochondria by autophagosomes and the selective elimination of impaired mitochondria. Thus, wild-type parkin promotes autophagy of damaged mitochondria. A failure to eliminate dysfunctional mitochondria leads to cell death. *In vitro* studies showed that overproduction of parkin is cytoprotective in certain mitochondrial toxin models and rescues *PINK1* mutant phenotype, indicating parkin works downstream from PINK1 in a common pathway. Other animal studies have demonstrated a neuroprotective effect of parkin against α-synuclein-induced toxicity.[21]

PINK1 (PARK6)

A number of families of European ancestry with autosomal recessive early-onset PD have been linked to the *PTEN-induced putative kinase 1* (*PINK1*) on 1p36.12 (PARK6). *PINK1*-associated PD is characterized by early onset, slow progression, and sustained response to levodopa. Some individuals display psychiatric symptoms.[6] *PINK1* mutations have so far been found to cause 1–9% of early-onset PD, and are thus, the second most common autosomal recessive cause of PD. The mutations are most common in Asians.[20] As in the case with *Parkin*, a heterozygous *PINK1* mutation may be associated with the development of late-onset PD, though some studied have indicated no association.[6] The PINK1 protein (PTEN-induced kinase 1) localizes to mitochondrial membranes and is found throughout the body and throughout the brain. It functions as a mitochondrial protein kinase. Following the depolarization of mitochondria, PINK1 is activated and then phosphorylates parkin on the surface of mitochondria, which results in the ubiquitination of several proteins and activates the removal of defective mitochondria by macroautophagy (mitophagy). In *Drosophila*, *PINK1* knockouts display a phenotype identical to the *parkin* knockout, with mitochondrial impairment and morphological changes that result in cellular apoptosis.

DJ-1 (PARK7)

Mutations in the *DJ-1* gene (PARK7) on chromosome 1p36.23 have been found as a rare cause of autosomal recessive early-onset PD in Europeans, which is characterized by slow progression and a good response to levodopa. Psychiatric symptoms and dystonia have been reported in some families.[7] The DJ-1 protein is ubiquitously expressed and was first described as an oncogene. Mutations in the *DJ-1* gene are associated with early-onset PD in 1–2% of cases.

Wild-type DJ-1 resists oxidative stress. Loss of DJ-1 does not affect mitochondrial respiration or mitochondrial calcium levels but increases reactive oxygen species production, leading to elevated mitochondrial permeability-transition pore opening and reduced mitochondrial transmembrane potential. Wild-type DJ-1 upregulates the vesicular monoamine transporter 3 (VMAT2), protecting cells against dopamine toxicity. Overexpression of DJ-1 increases VMAT2 expression and function, and reduces oxidative stress.[22] Mutations in *DJ-1* increase sensitivity to rotenone, paraquat and hydrogen peroxide.

Mendelian Genes with Parkinsonism and Other Features

ATP13A2: Kufor–Rakeb Syndrome (PARK9)

Kufor–Rakeb syndrome is a rare Parkinson-plus syndrome with autosomal recessive inheritance characterized by juvenile-onset pallidopyramidal degeneration with levodopa-responsive parkinsonism and dystonia, spasticity, supranuclear gaze palsy, and dementia. Mutations in *ATP13A2*, which encode a lysosomal type 5 P-type ATPase *transmembrane active transporter* were first discovered in a consanguineous Jordanian family and a large nonconsanguineous Chilean family. Since that time, about a dozen mutations have been found. Phenotypic variation exists even within families pointing to the presence of modifiers. Additionally, this same gene has been found to be responsible for a type of neurodegeneration with brain iron accumulation (NBIA). Iron accumulation is not a prerequisite for the Kufor–Rakeb presentation. A single copy of this gene has been found in patients with PD and may be a risk factor.[23]

PLA2G6 (PARK14) and *PANK2*

Two NBIA genes, *PLA2G6* and *PANK2*, have been associated with young-onset PD. *PLA2G6* (phospholipase A2, group VI) is involved in fatty acid metabolism and most commonly causes infantile neuroaxonal dystrophy neurodegeneration.[24] The phenotype typically includes dystonia, rigidity, ataxia, spasticity, dysarthria, cognitive impairment, and seizures. This disease can present in infancy or with a juvenile or adult onset. Atypical presentations with later onset and levodopa parkinsonism have been reported in homozygotes and compound homozygotes of *PLA2G6*.[25] Likewise, *PANK2* (pantothenate kinase), which causes infantile- or juvenile-onset pantothenate kinase-associated neurodegeneration (PKAN), can also result in a later-onset, slowly progressive form with parkinsonism. Autopsies of people homozygous for mutations in *PLA2G6* show significant α-synuclein pathology with tau pathology, whereas Lewy bodies are absent from *PANK2* pathology.[7]

FBXO7 (PARK15)

Mutations in the F-box protein 7 (*FBXO7*) result in both spasticity and levodopa-responsive parkinsonism. Inheritance is autosomal recessive.[26] This gene product's function is not well understood, but may be involved with ubiquitin-mediated pathways.[27]

POLG

The DNA *polymerase gamma* gene (*POLG*) is needed for mitochondrial DNA replication and repair. Mutations in *POLG* cause mitochondrial DNA depletion or deletion syndromes with myopathy, progressive external ophthalmoplegia (PEO), neuropathy, ataxia, seizures, and even liver failure and encephalopathy.[28] Inheritance can be autosomal dominant or recessive and can have onset throughout life. *POLG* mutations have also been found in several families with PEO and parkinsonism, as well as parkinsonism without other features. These patients have a loss of dopaminergic cells and some response to anti-Parkinson medications.[29] Within the gene there is a CAG repeat with the common number of repeats being 10 or 11. Reports from Scandinavia have shown an association between non-10/11 repeats and PD.[30]

ATXN2, ATXN3

Expansion of CAG repeats in the *ATXN2* and *ATXN3* genes cause spinocerebellar ataxia types 2 and 3 (SCA 2 and 3), respectively. Several reports describe patients presenting with PD diagnostic criteria. Intermediate number of repeats in *ATXN2* has also been implicated in amyotrophic lateral sclerosis (ALS).[7]

Other Unconfirmed Loci

Numerous loci have been reported to be rare causes of PD in specific populations. Most of these genes have not been confirmed as causal of familial PD. Some of the reported genes include:

- *ubiquitin carboxy terminal hydrolase L1* (*UCH-L1*) (PARK5) on 4p13 in a Northern German family with autosomal dominant PD, and a polymorphism in the gene which may be a risk factor for sporadic PD;[31]
- *sepiapterin reductase* (*SPR*) (PARK3) in a small group of European families with autosomal dominant PD with incomplete penetrance;[32]
- PARK10 on chromosome 1p32 in Icelandic families;[33]
- PARK11 in North American families;[34]
- X-linked PARK12;[35]
- the *Omi/HTRA2* gene, PARK13, in German PD patients;[36]
- mutations in the *NR4A2* gene.[37]

Susceptibility Loci

GBA

Whereas homozygous mutations in the glucocerebrosidase gene (*GBA*) cause the lysosomal storage disease Gaucher disease, heterozygous mutations are associated with increased risk for idiopathic Parkinson disease. This risk is age-related with about a 2–14% risk by age 65 and about an 11–30% by age 85.[38] *GBA* mutations are frequent in the Ashkenazi population with approximately 8.4% carrying a mutation. The common mutation for this population is N370S. About 15% of Ashkenazi PD patients harbor *GBA* mutations and about 3% of non-Ashkenazi patients have a mutation. The overall odds ratio of a PD patient to carry a *GBA* mutation is 5.4.[39]

The clinical presentation of *GBA*-associated PD is similar to that of sporadic PD. However, some reports find earlier onset, and risk of tremor, bradykinesia, psychiatric symptoms, dementia, and a faster progression.[39] Recent studies indicate *GBA* mutations are a significant risk factor for PD dementia and Dementia with Lewy Bodies (DLB) as well as PD.[40–43] PD patients with *GBA* mutations have cortical and brainstem Lewy body pathology.[39]

GWAS

Genome-wide association studies have implicated a number of loci as potential susceptibility genes. Of these, the *SNCA*, *MAPT*, and *LRRK2* loci show the greatest association.[44,45]

Molecular Genetic Testing

Molecular genetic testing is available in the United States and/or Europe for the majority of PD genes. Genetic counseling should be sought to discuss the significance of detecting mutations and implications for family members, especially since mutations in these genes may not act in a purely autosomal dominant or recessive manner, and may in some cases modify risk but not be causal to the development of PD. When testing for *SNCA*, *PARK2*, and *PINK1*, copy number analysis should accompany sequencing. Gene panels are very expensive and should be ordered with caution. Whole-exome or -genome sequencing will reduce the cost of testing but may increase the complexity of interpretation because of the presence of large numbers of variations of unknown significance. Additionally, several direct-to-consumer companies offer testing for *LRRK2*, *parkin*, and *GBA*. Patients should be counseled about the interpretation of these test results as they only reflect a few mutations. Because of the numerous genes that can cause PD, presymptomatic genetic testing should be limited to individuals for whom there is a known family mutation. Additionally, pre- and post-test genetic counseling should be provided.

PARKINSON-PLUS SYNDROMES

Tauopathies

A number of the Parkinson-plus syndromes involve abnormal accumulation of aberrantly phosphorylated microtubule-associated protein tau in the central nervous system. These Parkinson-plus syndromes together with other diseases with characteristic abnormal tau accumulation are collectively described as "tauopathies."

Six major isoforms of tau protein are expressed in the adult brain and are generated by alternate splicing of exons 2, 3, and 10. Alternative splicing of exon 10 leads to a protein containing either three (3R-tau, also called 10–) or four (4R-tau, also called 10+) microtubule binding repeats. Several tauopathies are associated with aberrant splicing of exon 10, which causes an imbalance in the 3R*tau*:4R*tau* ratio. Insoluble tau deposits in different tauopathies have different tau-isoform compositions. For example, in progressive supranuclear palsy (PSP) and corticobasal degeneration (CBD), inclusions contain mostly 4R-tau isoforms, whereas in Pick disease, the classical Pick bodies are made up mostly of 3R-tau isoforms. An extended haplotype that covers the human *tau* gene has been associated with PSP, a Parkinson-plus syndrome that is usually sporadic and is characterized by supranuclear vertical gaze palsy, postural instability, freezing, and falling. Inheriting two copies of the common haplotype (H1) in the *tau* gene is associated with increased risk to develop PSP in Europeans.[46] It is not known at the present time which specific region of the haplotype increases susceptibility to PSP. The same tau haplotype has also been associated with CBD. The finding that the same tau isoform accumulates in PSP and CBD adds weight to the theory that a shared pathway contributes to the development of PSP and CBD.[46] Clinically, PSP and CBD have many overlapping features, although cortical features, such as myoclonus, cortical sensory loss, apraxia, alien limb, unilateral rigidity, and dystonia strongly favor CBD. Specific single-nucleotide polymorphisms (SNPs) within the H1 haplotype have also been associated with risk for PD.[47]

Frontotemporal Degeneration

Frontotemporal degeneration (FTD), formerly called frontotemporal dementia, was first linked to chromosome 17 in 1994 and to the *MAPT* gene in 1998.[48–50] This tau gene acts to stabilize neuronal microtubules. At that time, autosomal dominant FTD was known as frontotemporal dementia with parkinsonism linked to chromosome 17 (FTDP-17). FTD usually presents with behavioral or language changes.

Although up to 50% of FTD cases have a family history, only about 10% of FTD is autosomal dominant. Whereas originally all hereditary cases were attributed to *MAPT*, it is now known that at least seven different genes can lead to autosomal dominant FTD. *MAPT* accounts for approximately 5–20% of familial cases.[51] In 2006 a second gene on chromosome 17 was discovered.[52,53] The progranulin gene *(PGRN, GRN)*, which lies 1.7Mb centromeric to the *tau* gene, encodes a secreted growth factor that is involved in development, wound repair, and inflammation. Unlike cases with *MAPT* mutations, which exhibit tau pathology as do approximately 40% of all FTD cases, *PGRN* mutation carriers exhibit TDP-43 pathology. This is now the most common pathology, and is found in 50% of all FTD cases. *PGRN* accounts for approximately the same amount of familial cases as *MAPT*, but also about 1–5% of sporadic cases.[51] In 2011, the *C9orf72* gene was discovered as the most common cause of both familial FTD and ALS, and accounts for the great majority of families with family history of both diseases; *C9orf72* carriers also exhibit TDP-43 pathology.[54,55] *C9orf72* has an intronic GGGGCC hexanucleotide repeat that expands in mutation carriers. Expanded alleles range from 22 to thousands. Penetrance is not well understood, especially for smaller repeat numbers. Much research is being carried out in studying the function of this gene. Several rare genes have been found to cause FTD. The *valosin containing protein* gene (*VCP*) causes inclusion body myopathy associated with Paget disease of the bone and frontotemporal dementia. This gene is involved in the ubiquitin-dependent protein degradation pathway. The charged multivesicular body protein 2B gene (*CHMP2B*) has ubiquitin-positive, tau- and TDP-43-negative pathology and was described in one large Dutch family. Several cases of FTD were found to harbor two ALS genes, *TARDPB* and *FUS*. However, these seem to be very rare causes of FTD. Strangely, people with *FUS* mutations do not exhibit the third FTD pathology found in 10% of FTD cases.[51]

There is marked inter- and intra-familial variation of disease phenotype, even with the same mutation, suggesting that other factors, genetic or environmental, contribute to the expression of the disease. Parkinsonism is a common finding. Although parkinsonism has been associated with CBD and PSP with *MAPT* mutations, CBD-like symptoms are also present in cases with other mutations.[56,57] Thus the term, corticobasal syndrome (CBS) is now used in reference to these symptoms. Additionally, many cases of CBS will be found to have Alzheimer pathology. New criteria for CBD have been developed to reflect these findings.[58]

Multiple System Atrophy

Multiple system atrophy (MSA) is a sporadic neurodegenerative Parkinson-plus syndrome characterized by a combination of cerebellar symptoms, parkinsonism, autonomic and pyramidal tract dysfunctions. Initially, individual different clinical forms were recognized, including olivopontocerebellar atrophy, Shy–Drager syndrome, and striatonigral degeneration, but all of these are now considered to be components of a single disease entity, MSA, with a common pathology of oligodendroglial inclusions containing α-synuclein, which is characteristic of all clinical variants. There are very few family cases of MSA. MSA, along with PD, is considered a synucleinopathy, although analysis of pathologically confirmed MSA cases failed to find any pathogenic mutations in the *α-synuclein* gene. Oligodendroglial inclusions have been associated with brain specific protein tubulin polymerization promoting protein (TPPP/p25), chaperone protein 14-3-3, heat shock protein 90, and rab3. These are all proteins of interest as the origin of α-synuclein in the oligodendroglial inclusions, and the ways in which they form, remain a mystery. Interestingly, there have been reports of MSA clinical phenotype and pathological features in carriers of *SNCA* duplication and *SCA* expansions. Additionally, GWAS have associated polymorphisms within the *SNCA* as risk factors for MSA.[59] Most recently, mutations in *COQ2* have been associated for increased risk of familial and sporadic MSA.[60]

CONCLUSIONS

Research on genetics of Parkinson disease and related diseases has revealed new information about the etiologies and molecular mechanisms of these diseases, which in turn can provide insights into possible therapies. At the same time, the genetics of these diseases has proved to be much more complex than was originally thought. The diseases display heterogeneity, variable penetrance and expressivity, and genetic and sporadic phenotypes. The interplay between Mendelian genes, susceptibility genes, and epigenetics is yet unknown. Thus, genetic counseling for these conditions can be complicated, but is more necessary than ever for those considering testing.

References

1. Marder K, Tang MX, Mejia H, et al. Risk of Parkinson's disease among first-degree relatives: a community-based study. *Neurology*. 1996;47(1):155–160.
2. Tanner CM, Ottman R, Goldman SM, et al. Parkinson disease in twins: an etiologic study. *JAMA*. 1999;281(4):341–346.
3. Polymeropoulos MH, Higgins JJ, Golbe LI, et al. Mapping of a gene for Parkinson's disease to chromosome 4q21-q23. *Science*. 1996;274(5290):1197–1199.
4. Singleton AB, Farrer M, Johnson J, et al. alpha-Synuclein locus triplication causes Parkinson's disease. *Science*. 2003;302(5646):841.
5. Puschmann A. Monogenic Parkinson's disease and parkinsonism: clinical phenotypes and frequencies of known mutations. *Parkinsonism Relat Disord*. 2013;19(4):407–415.
6. Lesage S, Brice A. Parkinson's disease: from monogenic forms to genetic susceptibility factors. *Hum Mol Genet*. 2009;18(R1):R48–R59.
7. Houlden H, Singleton AB. The genetics and neuropathology of Parkinson's disease. *Acta Neuropathol*. 2012;124(3):325–338.
8. Poulopoulos M, Levy OA, Alcalay RN. The neuropathology of genetic Parkinson's disease. *Mov Disord*. 2012;27(7):831–842.
9. Ozelius LJ, Senthil G, Saunders-Pullman R, et al. LRRK2 G2019S as a cause of Parkinson's disease in Ashkenazi Jews. *N Engl J Med*. 2006;354(4):424–425.
10. Mata IF, Hutter CM, González-Fernández MC, et al. Lrrk2 R1441G-related Parkinson's disease: evidence of a common founding event in the seventh century in Northern Spain. *Neurogenetics*. 2009;10(4):347–353.
11. Clark LN, Wang Y, Karlins E, et al. Frequency of LRRK2 mutations in early- and late-onset Parkinson disease. *Neurology*. 2006;67(10):1786–1791.
12. Healy DG, Falchi M, O'Sullivan SS, et al. Phenotype, genotype, and worldwide genetic penetrance of LRRK2-associated Parkinson's disease: a case-control study. *Lancet Neurol*. 2008;7(7):583–590.
13. Alcalay RN, Mejia-Santana H, Tang MX, et al. Motor phenotype of LRRK2 G2019S carriers in early-onset Parkinson disease. *Arch Neurol*. 2009;66(12):1517–1522.
14. Inzelberg R, Cohen OS, Aharon-Peretz J, et al. The LRRK2 G2019S mutation is associated with Parkinson disease and concomitant non-skin cancers. *Neurology*. 2012;78(11):781–786.
15. Lewis PA, Manzoni C. LRRK2 and human disease: a complicated question or a question of complexes? *Sci Signal*. 2012;5(207):e2.
16. Heckman MG, Elbaz A, Soto-Ortolaza AI, et al. The protective effect of LRRK2 p.R1398H on risk of Parkinson's disease is independent of MAPT and SNCA variants. *Neurobiol Aging*. 2014;35(1):266.e5–266.e14.
17. Chartier-Harlin MC, Dachsel JC, Vilariño-Güell C, et al. Translation initiator EIF4G1 mutations in familial Parkinson disease. *Am J Hum Genet*. 2011;89(3):398–406.
18. Farrer MJ, Hulihan MM, Kachergus JM, et al. DCTN1 mutations in Perry syndrome. *Nat Genet*. 2009;41(2):163–165.
19. Zimprich A, Benet-Pagès A, Struhal W, et al. A mutation in VPS35, encoding a subunit of the retromer complex, causes late-onset Parkinson disease. *Am J Hum Genet*. 2011;89(1):168–175.
20. Kilarski LL, Pearson JP, Newsway V, et al. Systematic review and UK-based study of PARK2 (parkin), PINK1, PARK7 (DJ-1) and LRRK2 in early-onset Parkinson's disease. *Mov Disord*. 2012;27(12):1522–1529.

21. Khandelwal PJ, Dumanis SB, Feng LR, et al. Parkinson-related parkin reduces α-synuclein phosphorylation in a gene transfer model. *Mol Neurodegeneration*. 2010;5:47.
22. Lev N, Barhum Y, Pilosof NS, et al. DJ-1 protects against dopamine toxicity: implications for Parkinson's disease and aging. *J Gerontol A Biol Sci Med Sci*. 2013;68(3):215–225.
23. Di Fonzo A, Chien HF, Socal M, et al. ATP13A2 missense mutations in juvenile parkinsonism and young onset Parkinson disease. *Neurology*. 2007;68(19):1557–1562.
24. Yoshino H, Tomiyama H, Tachibana N, et al. Phenotypic spectrum of patients with PLA2G6 mutation and PARK14-linked parkinsonism. *Neurology*. 2010;75(15):1356–1361.
25. Lu CS, Lai SC, Wu RM, et al. PLA2G6 mutations in PARK14-linked young-onset parkinsonism and sporadic Parkinson's disease. *Am J Med Genet B Neuropsychiatr Genet*. 2012;159B(2):183–191.
26. Shojaee S, Sina F, Banihosseini SS, et al. Genome-wide linkage analysis of a Parkinsonian-pyramidal syndrome pedigree by 500 K SNP arrays. *Am J Hum Genet*. 2008;82(6):1375–1384.
27. Bonifati V. Autosomal recessive parkinsonism. *Parkinsonism Relat Disord*. 2012;18(suppl 1):S4–S6.
28. Horvath R, Hudson G, Ferrari G, et al. Phenotypic spectrum associated with mutations of the mitochondrial polymerase gamma gene. *Brain*. 2006;129(Pt 7):1674–1684.
29. Orsucci D, Caldarazzo Ienco E, Mancuso M, Siciliano G. POLG1-related and other "mitochondrial Parkinsonisms": an overview. *J Mol Neurosci*. 2011;44(1):17–24.
30. Balafkan N, Tzoulis C, Müller B, et al. Number of CAG repeats in POLG1 and its association with Parkinson disease in the Norwegian population. *Mitochondrion*. 2012;12(6):640–643.
31. Maraganore DM, Farrer MJ, Hardy JA, et al. Case-control study of the ubiquitin carboxy-terminal hydrolase L1 gene in Parkinson's disease. *Neurology*. 1999;53(8):1858–1860.
32. Sharma M, Maraganore DM, Ioannidis JP, et al. Role of sepiapterin reductase gene at the PARK3 locus in Parkinson's disease. *Neurobiol Aging*. 2011;32(11):2108.e1–2108.e5.
33. Hicks AA, Petursson H, Jonsson T, et al. A susceptibility gene for late-onset idiopathic Parkinson's disease. *Ann Neurol*. 2002;52(5):549–555.
34. Pankratz N, Nichols WC, Uniacke SK, et al. Genome screen to identify susceptibility genes for Parkinson disease in a sample without *parkin* mutations. *Am J Hum Genet*. 2002;71(1):124–135.
35. Hardy J, Cai H, Cookson MR, et al. Genetics of Parkinson's disease and parkinsonism. *Ann Neurol*. 2006;60(4):389–398.
36. Krüger R, Sharma M, Riess O, et al. A large-scale genetic association study to evaluate the contribution of Omi/HtrA2 (PARK13) to Parkinson's disease. *Neurobiol Aging*. 2011;32(3):548.e9–548.e18.
37. Le WD, Xu P, Jankovic J, et al. Mutations in NR4A2 associated with familial Parkinson disease. *Nat Genet*. 2003;33(1):85–89.
38. Rana HQ, Balwani M, Bier L, Alcalay RN. Age-specific Parkinson disease risk in GBA mutation carriers: information for genetic counseling. *Genet Med*. 2013;15(2):146–149.
39. Sidransky E, Lopez G. The link between the GBA gene and parkinsonism. *Lancet Neurol*. 2012;11(11):986–998.
40. Swan M, Saunders-Pullman R. The association between β-glucocerebrosidase mutations and parkinsonism. *Curr Neurol Neurosci Rep*. 2013;13(8):368.
41. Chahine LM, Qiang J, Ashbridge E, et al. Clinical and biochemical differences in patients having parkinson disease with vs without GBA mutations. *JAMA Neurol*. 2013;70(7):852–858.
42. Alcalay RN, Caccappolo E, Mejia-Santana H, et al. Cognitive performance of GBA mutation carriers with early-onset PD: the CORE-PD study. *Neurology*. 2012;78(18):1434–1440.
43. Nalls MA, Duran R, Lopez G, et al. A multicenter study of glucocerebrosidase mutations in dementia with Lewy bodies. *JAMA Neurol*. 2013;70(6):727–735.
44. Satake W, Nakabayashi Y, Mizuta I, et al. Genome-wide association study identifies common variants at four loci as genetic risk factors for Parkinson's disease. *Nat Genet*. 2009;41(12):1303–1307.
45. Simón-Sánchez J, Schulte C, Bras JM, et al. Genome-wide association study reveals genetic risk underlying Parkinson's disease. *Nat Genet*. 2009;41(12):1308–1312.
46. Houlden H, Baker M, Morris HR, et al. Corticobasal degeneration and progressive supranuclear palsy share a common tau haplotype. *Neurology*. 2001;56(12):1702–1706.
47. Skipper L, Wilkes K, Toft M, et al. Linkage disequilibrium and association of *MAPT* H1 in Parkinson disease. *Am J Hum Genet*. 2004;75(4):669–677.
48. Poorkaj P, Bird TD, Wijsman E, et al. Tau is a candidate gene for chromosome 17 frontotemporal dementia. *Ann Neurol*. 1998;43(6):815–825.
49. Hutton M, Lendon CL, Rizzu P, et al. Association of missense and 5′-splice-site mutations in tau with the inherited dementia FTDP-17. *Nature*. 1998;393(6686):702–705.
50. Spillantini MG, Murrell JR, Goedert M, et al. Mutation in the tau gene in familial multiple system tauopathy with presenile dementia. *Proc Natl Acad Sci U S A*. 1998;95(13):7737–7741.
51. Rademakers R, Neumann M, Mackenzie IR. Advances in understanding the molecular basis of frontotemporal dementia. *Nat Rev Neurol*. 2012;8(8):423–434.
52. Baker M, Mackenzie IR, Pickering-Brown SM, et al. Mutations in progranulin cause tau-negative frontotemporal dementia linked to chromosome 17. *Nature*. 2006;442(7105):916–919.
53. Cruts M, Gijselinck I, van der Zee J, et al. Null mutations in progranulin cause ubiquitin-positive frontotemporal dementia linked to chromosome 17q21. *Nature*. 2006;442(7105):920–924.
54. Mahoney CJ, Beck J, Rohrer JD, et al. Frontotemporal dementia with the *C9ORF72* hexanucleotide repeat expansion: clinical, neuroanatomical and neuropathological features. *Brain*. 2012;135(Pt 3):736–750.
55. Majounie E, Renton AE, Mok K, et al. Frequency of the *C9orf72* hexanucleotide repeat expansion in patients with amyotrophic lateral sclerosis and frontotemporal dementia: a cross-sectional study. *Lancet Neurol*. 2012;11(4):323–330.
56. Rohrer JD, Ridgway GR, Modat M, et al. Distinct profiles of brain atrophy in frontotemporal lobar degeneration caused by progranulin and tau mutations. *Neuroimage*. 2010;53(3):1070–1076.

57. Spina S, Murrell JR, Huey ED, et al. Corticobasal syndrome associated with the A9D Progranulin mutation. *J Neuropathol Exp Neurol.* 2007;66(10):892–900.
58. Armstrong MJ, Litvan I, Lang AE, et al. Criteria for the diagnosis of corticobasal degeneration. *Neurology*. 2013;80(5):496–503.
59. Ahmed Z, Asi YT, Sailer A, et al. The neuropathology, pathophysiology and genetics of multiple system atrophy. *Neuropathol Appl Neurobiol.* 2012;38(1):4–24.
60. Multiple-System Atrophy Research Collaboration. Mutations in COQ2 in familial and sporadic multiple-system atrophy. *N Engl J Med.* 2013;369(3):233–244.

Relevant Websites

http://www.genereviews.org/.
http://www.ncbi.nlm.nih.gov/books/NBK1116/.
http://www.ncbi.nlm.nih.gov/Omim/.
http://www.nsgc.org.

CHAPTER

69

Frontotemporal Dementia

Shunichiro Shinagawa[*,†] *and Bruce L. Miller*[†]

[*]Jikei University School of Medicine, Tokyo, Japan

[†]University of California, San Francisco, CA, USA

HISTORY AND TERMINOLOGY

Frontotemporal dementia (FTD) is a clinically and pathologically heterogeneous syndrome, associated with progressive decline in behavior or language caused by focal degeneration of the frontal and anterior temporal lobes. The FTD clinical syndromes are a common cause of early-onset dementia, with an incidence and prevalence similar to Alzheimer disease (AD),[1] and are likely to be an underestimated cause of dementia both in early- and late-onset patients.

Arnold Pick described the FTD syndrome in a series of clinical and anatomical papers more than 100 years ago. First, he described the presentation of focal language deterioration as a sign of a dementia and helped to introduce the concept of a neurodegenerative disease beginning focally: "Simple progressive brain atrophy can lead to symptoms of local disturbance through local accentuation of the diffuse process".[2] Subsequently, a histopathological substance forming characteristic argyrophilic neuronal inclusions was discovered; these inclusions were called "Pick bodies."[3]

Sadly, these conditions did not attract much research interest until recently. The revival of interest in these syndromes was indicated by a series of seminal papers by Brun and Gustafson, and Mesulam's description of slowly progressive aphasia in 1982:[4] the syndrome with progressive aphasias without dementia in their early stage. Recent progress in neurobiology, histopathology, and molecular genetics have played a key role in helping to understand the complex FTD syndrome.

The selective degeneration of the frontal and temporal regions, which can be related with behavioral and language symptoms, gives rise to the term FTD. First divided into three subtypes by Neary et al. in 1998,[5] FTD is now used to classify three syndromes that can be distinguished based on the early and predominant symptoms: behavioral-variant FTD (bvFTD); semantic-variant primary progressive aphasia (svPPA), and nonfluent-variant primary progressive aphasia (nfvPPA).

The classification of these disorders can be confusing to the person outside of this field, with different terminologies applied in different papers. Additionally, there is a significant clinical, pathological, and genetic overlap between FTD and other neurodegenerative diseases such as amyotrophic lateral sclerosis (ALS), progressive supranuclear palsy (PSP), and corticobasal syndrome (CBS). Furthermore, recent clinicopathological research has revealed that FTD is complex with many potential pathologies including tau, TAR DNA-binding protein 43 (TDP-43), and fused in sarcoma (FUS) pathology.[6] Therefore, in this article we use the term FTD for clinical syndromes including bvFTD, svPPA, and nfvPPA; and frontotemporal lobar degeneration (FTLD) as a pathologically comprehensive term for various pathological conditions.

EPIDEMIOLOGY

FTD typically presents in middle-to-late adulthood; a recent study estimates that 60% of all FTD cases present in patients aged between 45 and 64 years.[7] FTD accounts for roughly 5–10% of all cases of pathologically confirmed

Rosenberg's Molecular and Genetic Basis of Neurological and Psychiatric Disease
http://dx.doi.org/10.1016/10.1016/B978-0-12-410529-4.00069-3

dementia,[8] and is considered to be the third cause of neurodegenerative dementia across all ages, following AD and dementia with Lewy bodies (DLB).

Accurate estimates of the incidence and prevalence of FTD in general population is challenging because of the difficulty with accurate diagnosis in the population-based setting, and the variation in diagnostic criteria. There are several studies regarding the incidence and prevalence of FTD. The incidence of FTD is estimated to be between 2.7 and 4.0 cases per 100,000 person-years, based on epidemiological data from the United States and Europe.[7,9,10] The prevalence of FTD has varied between several research studies. A study in the Netherlands estimated 4 per 100,000;[10] on the other hand, prevalence estimates of 15–22 cases per 100,000 in the United States,[7] and 15 per 100,000 in the United Kingdom[1] have been made. These population studies show nearly equal distribution by gender, which contrasts with many clinical and neuropathology reports that suggest a male predominance for bvFTD.

FTD is frequently familial, with up to 40% of FTD syndrome patients having a suggestive family history, with about 10% of patients showing an autosomal dominant inheritance.[11,12] Nongenetic risk factors have not yet been identified.

Mean survival in FTD has been estimated to range from 6–11 years from symptom onset.[7,13–15] In our center, bvFTD is associated with the shortest survival (median 9 years from onset), svPPA with the longest survival (12 years), and nfvPPA with intermediate survival (9 years)[13] and overall survival is shorter and cognitive and functional decline are more rapid than in AD.

CLINICAL SYNDROMES

In spite of its pathological heterogeneity, FTD can be classified into three clinical syndromes based on the early and predominant symptoms: bvFTD, svPPA and nfvPPA. Each clinical variant is associated with a distinct regional pattern of brain atrophy and, to some degree, a characteristic histopathology. Overlap between the syndromes can occur, particularly later in the course as the disease spreads to involve the frontal and temporal lobes more diffusely.

Behavioral-Variant Frontotemporal Dementia

bvFTD is the most common of clinical syndromes among all FTD subtypes, accounting for approximately 70% of all FTD cases. Patients with bvFTD present with marked changes in personality and behavior. The key features include an insidious onset and gradually progressive course, early behavioral disinhibition, early apathy or emotional blunting, early stereotypic behavior, alterations in eating behavior, loss of sympathy or empathy, and dysexecutive neuropsychological profiles.

Patients with bvFTD show a mixture of disinhibition and apathy.[16] Apathy is characterized by loss of interest in personal affairs and responsibilities, social withdrawal, and loss of awareness of personal hygiene. Disinhibition is manifested by a large amount of socially inappropriate behaviors, such as shoplifting or making hurtful or insensitive remarks to others. Insight is impaired, with either obvious denial of illness or very shallow recognition of their problem. Stereotypic motor behaviors (rubbing, picking, throat clearing, pacing and wandering), idiosyncratic hoarding and collecting, change in eating behavior (overeating and weight gain, loss of table manners) and hyperorality including oral exploration of inedible objects are also common and disease-specific symptoms. Psychotic symptoms such as delusions and hallucinations occur in a small minority of FTD-spectrum disorders.

The most common cognitive symptoms are dysexecutive functions, such as poor judgment, less attention and distractibility, loss of the ability to plan or imagine themselves in the future, and disorganization.[17] Patients show deficits on executive tasks, such as set shifting, mental flexibility, and response inhibition and abstract reasoning on cognitive testing. Visuospatial function is relatively preserved.

Structural and functional neuroimaging studies have highlighted frontal atrophy, hypometabolism, and hypoperfusion in patients with bvFTD[18] (Figure 69.1A). While the dorsolateral prefrontal cortex is often involved, the earliest changes occur in a medial paralimbic network that includes anterior cingulate, orbital frontal and frontoinsular cortices, and atrophy in these regions usually differentiates bvFTD from AD.[18,19] The region of atrophy correlates with the clinical phenotype, with dorsomedial frontal atrophy associated with apathy and aberrant motor behavior, and orbitofrontal atrophy associated with disinhibition.[20]

Although episodic memory and visuospatial function is relatively preserved in contrast to AD patients, several groups reported anterograde amnesia up to 10% of pathology-confirmed bvFTD cases.[21–23] The preponderance of primarily amnestic (versus behavioral) presentations in elderly subjects with bvFTD may be related to hippocampal sclerosis, which was reported in 43% of FTLD cases with late onset cases.[24] It is important to realize the prevalence of

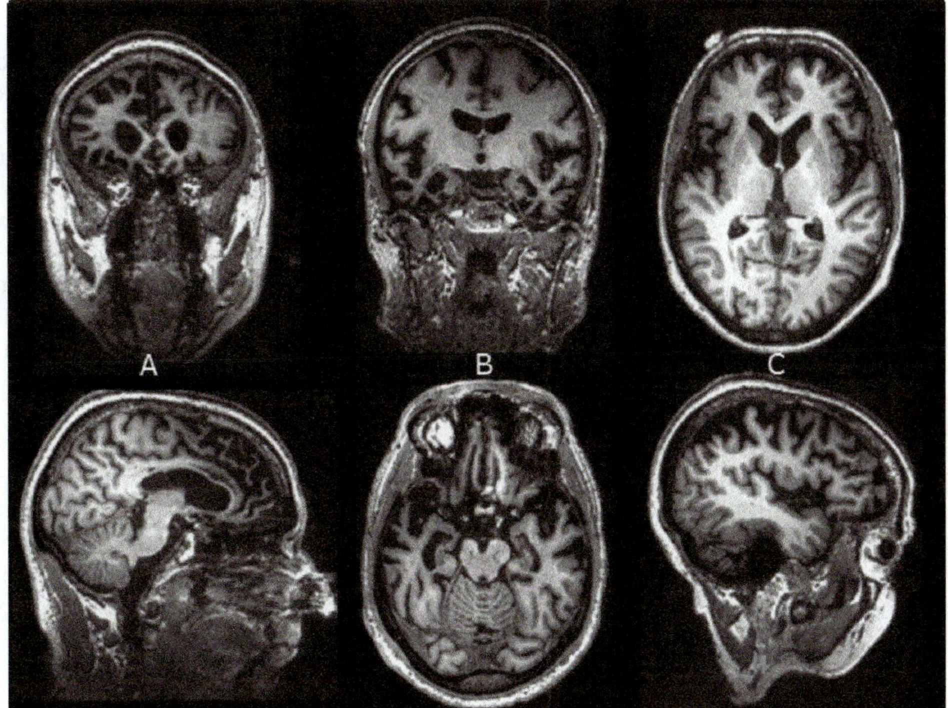

FIGURE 69.1 **MRI findings in frontotemporal dementia (FTD), T1-weighted images.** (**A**) Behavioral-variant frontotemporal dementia (bvFTD); (**B**) semantic-variant primary progressive aphasia (svPPA); **C**: nonfluent-variant primary progressive aphasia (nfvPPA).

memory deficits in many elderly persons, even those without AD, may bias overall clinical impressions by increasing the salience of prominent amnesia while making light of a patient's behavioral symptoms.

Semantic-Variant Primary Progressive Aphasia

svPPA, also called the temporal-variant of FTD or semantic dementia (SD), is characterized by a fluent, anomic aphasia and behavioral changes with remarkable, often asymmetric degeneration of the anterior temporal lobes[25] (Figure 69.1B). Patients with primarily left-sided predominant atrophy present initially with progressive loss of word knowledge and meaning about words, objects and concepts; so called "semantic" knowledge. This is obvious as a fluent aphasia with poor speech content and semantic paraphasic errors, but intact syntax, prosody and motor speech; what they say may not be meaningful. When the disease disproportionately involves the right temporal lobe, deficits in knowledge of facial emotion, diminished recognition of familiar faces, and deficits in empathy for others predominate the clinical syndrome.

Anomia is the most common symptom. The inability to name an object is matched by the patient's inability to give a detailed description of the object. In addition, patients with svPPA have a multimodal agnosia and are unable to recognize word meanings via written, auditory, olfactory and visual modalities.[26] Patients with svPPA present a surface dyslexia while reading, a condition in which the patient has difficulty reading words with irregular pronunciations, for example, yacht is pronounced "ya-ch." These language symptoms are known more generally as "Gogi (word-meaning) aphasia" in Japan. Two systems of letters were used in writing Japanese, Kana (phonogram) and Kanji (ideogram). svPPA patients are not impaired at reading of Kana because of its invariant relationship between orthography and phonology. By contrast, reading of Kanji is impaired in a graded fashion depending on the consistency characteristics of the Kanji target words, with worst performance on words whose component characters take atypical pronunciations, especially if the words are of lower frequency.[27]

Patients with predominant right anterior temporal atrophy present prosopagnosia and behavioral changes similar to bvFTD.[28] Prosopagnosia and associative agnosia are the most emphasized symptoms, but changes in personality and behavior often precede these symptoms by years. There is a clear loss of empathy and interest in other people, and mental rigidity manifested with strict schedules and routines. Left-sided patients are reported more commonly than those with right-sided disease, approximately 3:1 in most centers,[28–30] although this may be referral bias with the right-sided cases never reaching specialized centers.

Nonfluent-Variant Primary Progressive Aphasia

nfvPPA is a progressive disorder of language expression and motor speech.[31] Anatomically, it is associated with atrophy, hypometabolism and hypoperfusion of the left perisylvian area: frontal operculum, premotor and supplementary motor areas, and anterior insula[32] (Figure 69.1C). Patients present with slow, effortful speech, impaired production and comprehension of grammar, and motor speech deficits. Apraxia of speech, defined as difficulty initiating speech, a slow rate of speech or incorrect sequencing or omission of phonemes, is highly characteristic of nfvPPA, while dysarthria is more variably present. Comprehension is spared for single words and for all except the complex syntactic structures. Reading is nonfluent and effortful, while writing is agrammatic and features phonemic paraphasias.

In addition to the aphasia, neuropsychological tests may show mild deficits in executive function, with relatively spared episodic memory and visuospatial function. Behavioral disturbances are less frequent and severe than in bvFTD and svPPA, reflecting less damage in the orbitofrontal areas and the right hemisphere in general.[33]

Other Forms of Primary Progressive Aphasia

There are some patients with progressive language impairment who do not fit into the svPPA or nfvPPA criteria (Table 69.1): a third, more recently defined, subtype of PPA is the logopenic or phonological-variant PPA (lvPPA) or logopenic progressive aphasia (LPA).[31,34] Patients with lvPPA are characterized by a slow rate of speech output, word-finding difficulties, deficits in sentence repetition, and occasional phonemic errors in spontaneous speech and naming, whereas motor speech, expressive grammar, and single-word comprehension are relatively preserved. The underlying pathology of most of LPA cases is AD.

FTD Overlap Syndromes

Amyotrophic lateral sclerosis (ALS) and bvFTD are overlapping syndromes at the clinical, pathological, and genetic level.[35,36] While a proportion of patients presenting with ALS manifest cognitive and behavioral changes that may be severe enough to reach criteria for bvFTD (10–15%), a subgroup of patients with bvFTD develops features of ALS (15%).[37] The molecular pathology and genetics of the bvFTD-ALS spectrum are described later. Because of this overlap between FTD and ALS, all FTD patients should carefully undergo a neuromuscular evaluation with consideration for electromyography and, conversely, all ALS patients should be checked for behavioral changes.

FTD clinical syndromes also show significant overlap with patients in whom the underlying pathology is corticobasal degeneration (CBD) or PSP, even though both of these syndromes were originally classified as movement disorders.[38] Until fairly recently, patients in whom CBD was expected—the clinical syndrome was called CBS—were suspected when there was a syndrome of limb apraxia, axial and limb rigidity, dystonia, postural instability, ocular apraxia, the "alien limb phenomenon" and cortical sensory loss. PSP was described as a syndrome of supranuclear gaze palsies, axial-predominant parkinsonism and profound retropulsion. Pseudobulbar signs such as dysarthria, dysphagia, and pseudobulbar affect were also observed.

More recently, it has been increasingly recognized that many patients with the classical CBS syndrome exhibit AD at neuropathology.[39] Additionally, the prominent behavioral, language and executive deficits associated with both conditions have been increasingly recognized.[40,41] More than one-quarter of all CBD cases show either nfvPPA syndrome or bvFTD, while another third exhibit an executive-motor syndrome with prominent deficits in executive control and parkinsonism. Additionally, PSP often begins as a neuropsychiatric disorder or nfvPPA. Both are associated with prominent tau pathology (see page 779) and these syndromes should be considered within the pathological FTLD spectrum.[42]

TABLE 69.1 Clinical Features, Distribution of Atrophy, and Underlying Pathology of Three Variants of Primary Progressive Aphasia

Variant	Clinical features	Region of cortical atrophy	Common underlying pathology
Semantic-variant PPA	Poor confrontation naming Impaired single-word comprehension	Anterior and ventral temporal lobe	FTLD-TDP
Nonfluent-variant PPA	Effortful speech with speech sound errors Grammatical errors in language production	Left inferior frontal and insula	FTLD-tau
Logopenic-variant PPA	Impaired single-word retrieval Impaired repetition of phrases and sentences	Left posterior temporal and inferior parietal	AD

Abbreviations: AD, Alzheimer disease; FTLD-tau, frontotemporal lobar degeneration with tau-positive pathology; FTLD-TDP, frontotemporal lobar degeneration with TDP43-positive pathology; PPA, primary progressive aphasia.

DIAGNOSTIC CRITERIA

The consensus diagnostic criteria by Neary et al. in 1998 were widely used in research and practice until recently.[5] In 2011, the International Behavioral Variant FTD Criteria Consortium (FTDC) proposed a revision of diagnostic and research criteria for bvFTD, which has higher sensitivity.[43] The details of new consensus criteria for bvFTD are shown in Table 69.2. According to the new criteria, the main feature of bvFTD is progressive deterioration of behavior and/or cognition by observation or history provided by a knowledgeable informant. If this criterion is satisfied, there are three

TABLE 69.2 Diagnostic Criteria for bvFTD by the International Behavioural Variant FTD Criteria Consortium (FTDC)

I. Neurodegenerative disease

The following symptom must be present to meet criteria for bvFTD:
A. Shows progressive deterioration of behavior and/or cognition by observation or history (as provided by a knowledgeable informant)

II. Possible bvFTD

Three of the following behavioral/cognitive symptoms (A–F) must be present to meet criteria. Ascertainment requires that symptoms be persistent or recurrent, rather than single or rare events:
A. Early* behavioral disinhibition (one of the following symptoms [A.1–A.3] must be present):
  A.1. Socially inappropriate behavior
  A.2. Loss of manners or decorum
  A.3. Impulsive, rash or careless actions
B. Early apathy or inertia (one of the following symptoms [B.1–B.2] must be present]:
  B.1. Apathy
  B.2. Inertia
C. Early loss of sympathy or empathy (one of the following symptoms [C.1–C.2] must be present):
  C.1. Diminished response to other people's needs and feelings
  C.2. Diminished social interest, interrelatedness or personal warmth
D. Early perseverative, stereotyped or compulsive/ritualistic behavior (one of the following symptoms [D.1–D.3] must be present):
  D.1. Simple repetitive movements
  D.2. Complex, compulsive or ritualistic behaviors
  D.3. Stereotypy of speech
E. Hyperorality and dietary changes (one of the following symptoms [E.1–E.3] must be present):
  E.1. Altered food preferences
  E.2. Binge eating, increased consumption of alcohol or cigarettes
  E.3. Oral exploration or consumption of inedible objects
F. Neuropsychological profile: executive/generation deficits with relative sparing of memory and visuospatial functions (all of the following symptoms [F.1–F.3] must be present):
  F.1. Deficits in executive tasks
  F.2. Relative sparing of episodic memory
  F.3. Relative sparing of visuospatial skills

III. Probable bvFTD

All of the following symptoms (A–C) must be present to meet criteria:
A. Meets criteria for possible bvFTD
B. Exhibits significant functional decline (by caregiver report or as evidenced by Clinical Dementia Rating Scale or Functional Activities Questionnaire scores)
C. Imaging results consistent with bvFTD (one of the following [C.1–C.2] must be present):
  C.1. Frontal and/or anterior temporal atrophy on MRI or CT
  C.2. Frontal and/or anterior temporal hypoperfusion or hypometabolism on PET or SPECT

IV. Behavioral-variant FTD with definite FTLD pathology

Criterion A and either criterion B or C must be present to meet criteria:
A. Meets criteria for possible or probable bvFTD
B. Histopathological evidence of FTLD on biopsy or at postmortem
C. Presence of a known pathogenic mutation

V. Exclusionary criteria for bvFTD

Criteria A and B must be answered negatively for any bvFTD diagnosis. Criterion C can be positive for possible bvFTD but must be negative for probable bvFTD:
A. Pattern of deficits is better accounted for by other nondegenerative nervous system or medical disorders
B. Behavioral disturbance is better accounted for by a psychiatric diagnosis
C. Biomarkers strongly indicative of Alzheimer disease or other neurodegenerative process

Source: Rascovsky, et al. 2011.[43]

further levels of certainty for bvFTD: possible, probable, or definite. "Possible" bvFTD requires three out of six clinically discriminating features. "Probable" bvFTD meets the criteria of "possible" bvFTD and 1) a significant functional decline by caregiver report or evidenced at neuropsychological testing, and 2) imaging results consistent with bvFTD. "Definite" bvFTD requires the histopathological evidence of FTLD or the presence of a known pathogenic mutation.

In the diagnostic evaluation of a patient with suspected FTD, as with any other neurodegenerative dementia, the clinician should first exclude treatable conditions that can be similar to FTD, such as metabolic disturbances, nutritional deficiencies, central nervous system (CNS) infections, vascular disease, normal or low pressure syndromes, and primary neoplastic conditions. These can be excluded by a careful medical history, laboratory testing, and neuroimaging. Many patients with FTD tend to be misdiagnosed with a psychiatric disorder such as major depression or bipolar affective disorder.[44]

HISTOPATHOLOGY

On autopsy, FTD patients show frontal and anterior temporal lobe atrophy in gross pathological samples. Atrophy can be extreme and circumscribed, with marked sparing of posterior brain regions until the advanced stages of disease.[45] Microscopic examination of the cerebral cortex reveals neuronal loss and microvacuolar degeneration in layers II and III of the frontal and temporal cortex, with a variable degree of cortical gliosis. White matter changes include loss of myelin, and astrocytic gliosis, and neuronal loss in the basal ganglia and substantia nigra may be prominent in some cases[46] (Figure 69.2A).

FTLD is a broader spectrum term for various pathological conditions. Most cases of FTLD can be subdivided into the following three subcategories, which are based on the presence of specific inclusion bodies: (i) FTLD with tau inclusions (FTLD-TAU); (ii) FTLD with TDP-43 inclusions (FTLD-TDP); and (iii) FTLD with FUS inclusions (FTLD-FUS).[6] Additionally a small number of FTLD cases show no pathological aggregates, and it is increasingly realized that other coexisting neurodegenerative diseases, most commonly AD and DLB, are present with microscopy, particularly in older patients.

Recent development of immunohistochemical staining techniques, which use labeled antibodies against the TAU, TDP-43 or FUS protein, make it possible to identify the presence and distribution of pathological protein aggregates and molecular biomarkers of specific neurodegenerative diseases within the FTLD spectrum. Immunohistochemical techniques are more sensitive to underlying molecular pathology; however, immunohistochemical staining does not always correlate with the density and distribution of pathological proteins in FTLD.

Predicting the underlying histopathology of a clinical diagnosis of FTD has been often challenging. On an individual level, it can be difficult to reliably predict the underlying FTLD neuropathology based upon the FTD clinical phenotype, yet there are strong correlations between some clinical syndromes and the neuropathological subtype.[47,48] The underlying histopathology in typical bvFTD is heterogeneous, showing no clear association with one specific

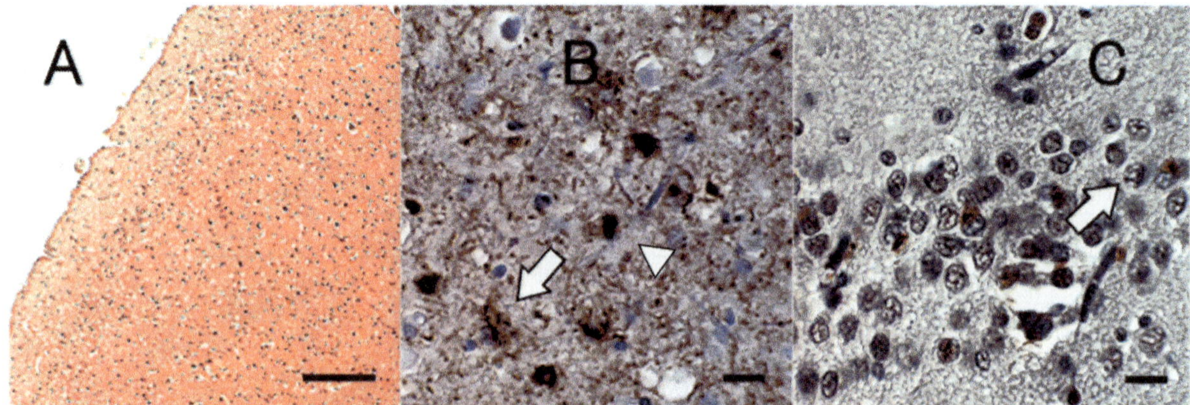

FIGURE 69.2 **Frontotemporal lobar degeneration (FTLD) pathology.** (**A**) Affected regions show neuronal loss, gliosis, and microvacuolation, especially in superficial cortical layers (shown on the left). Hematoxylin and eosin stain. (Scale bar = 200 μm.) (**B**) Pick disease, a tau-positive form of FTLD, shows cytoplasmic hyperphosphorylated tau inclusions within neurons called Pick bodies (arrow head) and in ramified astroglial processes (arrow). Immunohistochemistry for phospho-tau. (Scale bar = 50 μm.) (**C**) In FTLD, due to TAR DNA-binding protein 43 (TDP-43), proteinopathy-affected neurons show cytoplasmic TDP-43 inclusions, dentate gyrus granule cells. Immunohistochemistry for phospho-TDP. (Scale bar = 50 μm.) Images courtesy of L.T. Grinberg, University of California, San Francisco.

pathological subtype. svPPA is associated with predominantly FTLD-TDP type C histopathology, FTD with motor neuron disease (FTD-MND) was associated with predominantly FTLD-TDP type B histopathology. nfvPPA is associated with FTLD-TAU, or TDP type A histopathology.

Tau-Positive FTLD (FTLD-TAU)

Tau is a microtubule-associated protein (MAP). Tauopathies are classified according to the predominant species of tau that accumulates, with tau proteins containing three repeats (3R) or four repeats (4R) of amino acids in the microtubule-binding domain. Tauopathies of the FTLD spectrum include Pick disease, FTD with MAPT mutations CBD, PSP and argyrophilic grain disease.[46]

Pick disease is the prototypical tauopathy of FTLD, and is characterized by the presence of Pick bodies, which are solitary, round, argyrophilic inclusions found in the cytoplasm of neurons located in the limbic system, including the dentate fascia of the hippocampus, entorhinal cortex and amygdala, and in the superficial frontal and temporal neocortex[49] (Figure 69.2B). Pick inclusions are composed primarily of 3R tau. CBD, PSP, and argyrophilic grain disease are tauopathies with predominantly 4R tau inclusions. CBD and PSP are as common as Pick disease in patients presenting with FTD clinical syndrome. The distribution of pathology and pattern of atrophy can distinguish CBD from PSP, and from Pick disease. Patients with FTDP-17 share the common characteristic of filamentous, hyperphosphorylated tau aggregates.[50] Patients with mutations in the MAPT gene may show the pathologic features of 3R, 4R or a combination of 3R and 4R tau.[51]

TDP-43-Positive FTLD (FTLD-TDP)

TDP-43 is a ubiquitously expressed nuclear protein that may function as a transcription repressor, activator of exon skipping or as a transcription regulator. TDP-43 becomes hyperphosphorylated, ubiquinated and cleaved into C-terminal fragments under pathologic conditions.[52] TDP-43 pathology is found in sporadic and familial patients of FTLD involving progranulin (PGRN), C9ORF72 hexanucleotide repeat expansion, and valosin-containing protein (VCP) gene mutations.[53,54]

FTD-TDP is now considered to be the most common neuropathology seen in FTLD. Pathological TDP-43 inclusions are present in the dentate gyrus of the hippocampus, layer II of the frontotemporal cortex and anterior horn of the spinal cord (Figure 69.2C). These inclusions are ubiquitin- and TDP-43-positive, tau-, α-synuclein-, amyloid-β- and neuronal intermediate filament-negative by immunohistochemical method.[46]

Four histological subtypes, A–D, have been recognized according to a revised classification scheme based upon the distribution and morphology of the inclusions.[52] Type A is associated with cortical pathology predominantly in layer II, with progranulin (GRN) mutations, and with some sporadic nfvPPA cases. Type B is found in patients with clinical bvFTD and FTD-MND, and it shows linkage to the C9ORF72 hexanucleotide repeat expansion. Type C is localized primarily to cortical layer II in patients with svPPA and bvFTD. Type D histology is found in all cortical layers and is associated with patients with familial inclusion body myopathy and Paget disease of the bone with frontotemporal dementia and VCP gene mutations. It remains still unclear what differences in underlying pathophysiology determine the distinction between these TDP-43 subtypes.

FUS-Positive FTLD (FTLD-FUS)

The majority of tau-negative/TDP-43-negative, ubiquitin-positive FTLD cases have positive immunohistochemical staining to the FUS protein, thus distinguishing a third category of FTLD neuropathology. The FUS protein contains 526 amino acids. It is as a nuclear protein involved in DNA repair and the regulation of RNA splicing. Mutations in the FUS gene on chromosome 16 emphasize its pathogenetic role in the clinicopathological spectrum of FTD and ALS.[55,56] Clinically, most FTLD-FUS cases are characterized by early-onset FTD (age <50 years), a negative family history, prominent psychiatric features such as delusions and hallucinations, and a caudate atrophy on brain imaging. These TDP-negative FTLD with ubiquitin pathology cases were formerly known as atypical FTLD with ubiquitin-positive inclusions (aFTLD-U), basophilic inclusion body disease (BIBD), and neuronal intermediate filament inclusion disease (NIFID).[57]

GENETICS

Approximately 40% of bvFTD patients report a positive family history of dementia, suggesting a strong genetic component to these disorders, whereas patients with svPPA or nfvPPA have a much lower frequency.[58] An autosomal dominant pattern of inheritance is found in at least 10% of all FTD cases. Genes recognized to play an important role in autosomal dominant FTD include: 1) MAPT, encoding microtubule-associated protein tau, 2) PGRN, encoding the protein progranulin, and 3) C9ORF72, a recently identified hexanucleotide repeat expansion on chromosome 9.[11,12] These gene mutations explain the majority of autosomal dominant FTLD cases with mutations in the VCP, charged multivesicular body protein 2B (CHMP2B), TAR-DNA binding protein (TARDP), and FUS genes are found in less than 5%. There are still families with FTLD in whom no gene has yet to be found.

Microtubule Associated Protein Tau (MAPT)

More than 40 mutations in the MAPT gene have been identified in families with FTD and parkinsonism linked to chromosome 17q (FTDP-17).[59] Most mutations are caused by missense mutations in exons 9–13 affecting the normal function of the tau protein to stabilize microtubules, or in the intronic regions, disproportionately influencing the splicing in of exon 10 at the mRNA level, resulting in a change in ratio of 3R to 4R tau isoforms. MAPT gene mutations account for approximately 20% of all familial FTD cases in our center, although small numbers of cases have been reported at some sites.[60,61]

The most common FTD clinical variant associated with MAPT mutations is bvFTD with prominent behavioral changes including disinhibition and obsessive–compulsive behavior. svPPA and parkinsonian syndromes can also be seen in patients with MAPT mutations. The mean age at onset is 55 years, and the mean duration of illness is approximately 9 years. MAPT mutations are associated with more significant symmetrical anteromedial temporal and orbitofrontal atrophy.[47,62]

PGRN

More than 60 mutations in the PGRN gene on chromosome 17 have been identified. PGRN mutation cases account for approximately 5–10% of all FTD patients, and up to 22% in familial FTD.[12,47] Its frequency is similar to that of MAPT gene mutations as a cause for hereditary forms or FTLD.[63]

The mean age at onset is around 60 years; the mean duration is 8 years.[12] GRN mutations have been most commonly associated with clinical diagnoses of bvFTD and nfvPPA, although AD, Parkinson disease and CBS are also seen. The neuropathology of patients with GRN mutations is characterized by tau-negative, ubiquitin-positive, and TDP-43-positive inclusions. GRN mutations are associated with an asymmetrical frontoparietotemporal pattern of atrophy; the most common behavioral changes are apathy and social withdrawal. Twenty-five percent of patients present with early isolated language dysfunction of an anomic nonfluent type. Hallucinations and delusions are often reported.[64] Extrapyramidal features are frequently seen and can be associated with limb apraxia, and asymmetrical parkinsonism, but FTD-MND is a rare phenotype in patients carrying GRN mutations.[65,66]

C9ORF72 Hexanucleotide Repeat Expansion

In the majority of chromosome 9-associated FTD-MND, it has now been recognized that the cause of disease is an expanded GGGGCC hexanucleotide repeat in a noncoding region of chromosome 9 open reading frame 72 (C9ORF72).[54,67] This mutation is most often associated with bvFTD with or without MND and amyotrophic lateral sclerosis (ALS), while the mutation does not appear to be commonly associated with svPPA or nfvPPA phenotypes. C9ORF72 mutations are reported to pathologically associate with deposition of the FTLD-TDP type B,[68] although type A pathology also occurs at our center. The phenotype associated with C9ORF72 is high prevalence of psychiatric symptoms and prodromal psychiatric features. On structural imaging, in addition to atrophy in dorsolateral frontal, medial orbitofrontal, atrophy in parietal, occipital, cerebellar, and thalamus regions are also observed.[69–71]

Other Hereditary Forms

The genetic heterogeneity of FTD is further emphasized by the rare occurrence of mutations in the VCP, CHMP2B, TARDP, and FUS genes.[59] VCP gene mutations are associated with inclusion body myopathy (90%), Paget disease of the bone (45%), and bvFTD (38%) or sometimes ALS, presenting between the ages of 40 and 60 years. The clinical presentation of CHMP2B gene mutations consists of a frontal lobe syndrome and a more global cognitive impairment, with parkinsonism, dystonia, pyramidal signs, and myoclonus later in the course of the disease. TARDBP gene mutations on chromosome 1 are found in 5% of familial ALS, and occasionally in FTD or FTD-MND cases. Also, FUS

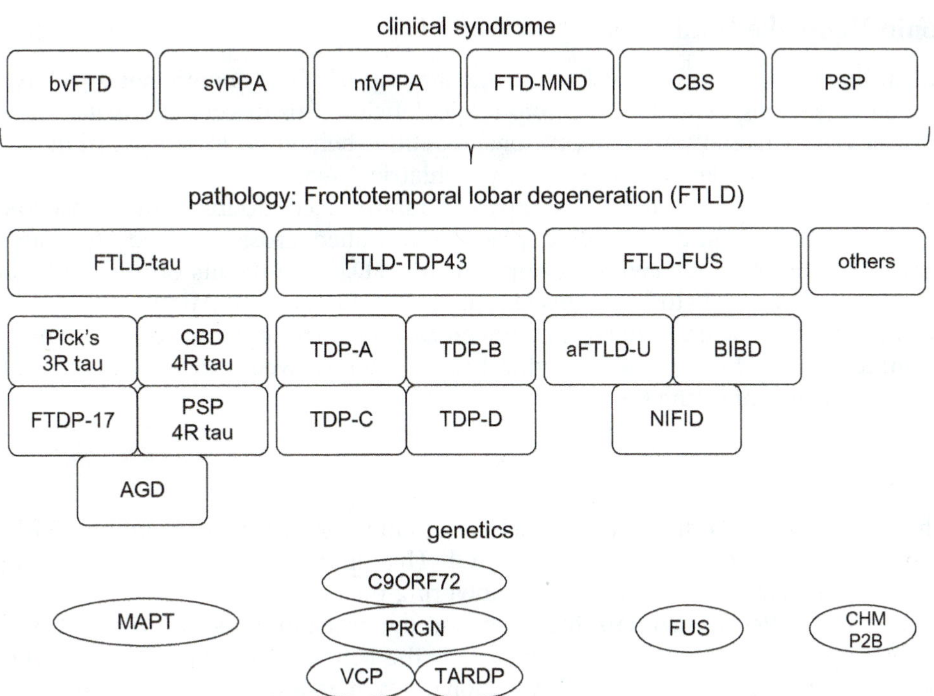

FIGURE 69.3 **Clinical, pathologic, and genetic spectrum of frontotemporal lobar degeneration (FTLD).** Abbreviations: aFTLD-U, atypical frontotemporal lobar degeneration with ubiquitin pathology; BIBD, basophilic inclusion body disease; bvFTD, behavioral-variant frontotemporal dementia; C9ORF72, chromosome 9 open reading frame 72 gene mutation; CBD, corticobasal degeneration; CBS, corticobasal syndrome; CHMP2B, charged multivesicular body protein 2B gene mutation; FTD-MND, frontotemporal dementia with motor neuron disease; FTDP-17, frontotemporal dementia and parkinsonism linked to chromosome 17q; FTLD-FUS, frontotemporal lobar degeneration with FUS pathology; FTLD-tau, frontotemporal lobar degeneration with tau-positive pathology; FTLD-TDP43, frontotemporal lobar degeneration with TDP43-positive pathology; FUS, FUS gene mutation; MAPT, microtubule-associated protein tau gene mutation; nfvPPA, semantic-variant primary progressive aphasia; NIFID, neuronal intermediate filament inclusion disease; PRGN, progranulin gene mutation; PSP, progressive supranuclear palsy; svPPA, semantic-variant primary progressive aphasia; TARDP, TAR-DNA binding protein gene mutation; VCP, valosin-containing protein gene mutation.

gene mutations are found in 5% of the familial ALS cases; in such families bvFTD appears to be rare. An overview of the clinical, pathologic, and genetic spectrum of FTLD is given in Figure 69.3.

TREATMENT

There is no specific disease-modifying treatment for FTD to date. Thus, medications for other types of dementia and neurodegenerative disease are frequently used as off-label symptomatic treatment of FTD. A similar percentage of AD and bvFTD patients have been reported to take AD medications.[72] Current pharmacological strategies for FTD have focused on symptomatic neurotransmitter replacement and modulation for the treatment of behavioral symptoms. These medications include antidepressants, including selective serotonin reuptake inhibitors (SSRIs), atypical antipsychotics, acetylcholinesterase inhibitors (AChEIs) and *N*-methyl-D-aspartate (NMDA) glutamate receptor antagonists. In general, with the exception of SSRIs, medication use is often associated with adverse-effect profiles.

To date, implementation of drug trials has been challenging and most reports of FTD treatments are based on small case series, and few have been large sample, double-blind, randomized control trials. Still, the field is harmed by lack of FTD-specific clinical rating scales. Clinical instruments used to measure efficacy in AD, such as the Mini-Mental States Examination, Alzheimer Disease Assessment Scale—Cognitive, or Clinical Dementia Rating Scale, emphasize memory loss and do not take enough the executive and language deficits seen in FTD patients. There is study-size limitation due to the relatively small number of the patients.[73] With recent advances in the genetics and molecular biology of FTD, pharmacological strategies will be shifting from a symptomatic approach towards treatment of the underlying disease process, such as anti-tau protein compounds.

In addition to drug therapy, there are ranges of nonpharmacological interventions that can ameliorate the impact of this disorder. Most of these are derived from an extensive experience in managing patients with dementia in general; however, some nonpharmacological interventions are specific to FTD. These focus on behavioral management strategies, education, and caregiver support.[74]

Selective Serotonin Reuptake Inhibitors

Of all neurotransmitter-based therapies for FTD, drugs that modify serotonergic neurons have the strongest biological basis, since there is strong evidence for serotonergic deficit in this disorder.[73] Furthermore, many of the behavioral symptoms of FTD, such as apathy, compulsions, repetitive behaviors, stereotypical movements, and eating abnormality respond to SSRIs in patients with primary psychiatric disease.

To date, there are several open-label studies, but only two randomized, double-blind, placebo-controlled studies have been conducted. One is the first double-blind, placebo-controlled, crossover study to evaluate the efficacy of paroxetine (40 mg/day) to target the behavioral symptoms of FTD.[75] The results of this study showed no significant differences between placebo and study groups after 6 weeks of treatment. Another randomized, double-blind, placebo-controlled trial was conducted to determine the efficacy of trazodone (up to 300 mg per day).[76] Trazodone demonstrated a significant reduction in the behavioral score after 12 weeks of treatment, especially irritability, agitation, depression, and aberrant eating behavior.

Atypical Antipsychotics

In patients with severe behavioral disturbances such as agitation or irritability in whom SSRIs are not effective, treatment with atypical antipsychotics may be considered. This application is supported by a single open-label study of olanzapine, which showed similar efficacy in lowering the behavioral score to that reported with SSRIs.[77] However, significant adverse effects often limit their use, and the decision to use antipsychotics in the treatment of FTD should be made with caution. An important concern is the increased susceptibility of FTD patients to extrapyramidal adverse effects of psychotropic drugs.[78] Additionally, a meta-analysis of randomized, placebo-controlled trials determined that treatment of elderly patients with dementia with atypical antipsychotics is associated with a 1.6–1.7-fold increase in mortality secondary to cardiac events or infection, thus prompting the US Food and Drug Administration to place a "black-box warning" on their use.

Acetylcholinesterase Inhibitors and NMDA Glutamate Receptor Antagonists

Acetylcholinesterase inhibitors (AChEIs) have proven to be beneficial in improving the underlying cholinergic system deficits in AD and DLB, they have an important role in treating cognitive and behavioral symptoms of these neurodegenerative disorders. Thus, AChEIs have attracted large interest as a potential treatment in FTD patients. Numbers of studies have investigated the efficacy of AChEIs in the treatment of FTD, however, in contrast to AD and DLB, there is no cholinergic system deficit in FTD patients.[73] In the context of this theory, the efficacy of AChEIs in the treatment of FTD has not been established.

One open-label study found that rivastigmine improved neuropsychiatric symptoms and caregiver burden but did not improve cognitive decline.[79] Another study investigated the efficacy of galantamine in patients with bvFTD and PPA.[80] No significant differences in language or behavioral symptoms were reported between placebo and both treatment groups (combined bvFTD and PPA). However, language functions remained stable in the PPA group compared with the placebo group. It may be because patients in the PPA group include the logopenic-variant patients whose symptoms are associated with underlying Alzheimer's pathology. The third study was an open-label trial with donepezil in bvFTD patients.[81] There was no significant difference in global cognitive function or dementia severity between groups; however, an exacerbation of behavioral symptoms was reported after treatment.

Memantine is a noncompetitive NMDA glutamate antagonist, providing neuronal protection against glutamate-mediated excitotoxicity. Memantine effectively treats agitation in moderate-to-severe AD.[82] There are some reports suggesting that glutamate excitotoxicity may play a role in the pathogenesis of bvFTD; therefore, the therapeutic effects of memantine were thought to be useful in treating the neuropsychiatric features of FTD. A recent randomized, parallel group, double-blind, placebo-controlled trial of memantine in patients with bvFTD could find no benefit.[83] Memantine treatment had no effect on either the neuropsychiatric symptoms or global impression after 26 weeks of treatment. This study confirms the absence of benefit of memantine for treatment of bvFTD.

Future Treatment

Efforts to develop disease-modifying therapies for FTD are focused on eliminating the specific proteins that are implicated in pathogenesis. Patients with CBD, PSP and MAPT mutations would be preferred candidates for a tau-specific drug; upregulation of progranulin is the ideal approach for patients with GRN mutations; and silencing the *C9ORF72* gene is a focus for those individuals who carry this gene. The development of tau- and TDP-43-specific biomarkers will be necessary to further improve prediction of histopathology, particularly in patients with bvFTD, and the advent of tau imaging shows great promise for separating out these subtypes of FTD. The efficacy of candidate drugs should be tested by clinical outcome measures that are sensitive to the changes seen with disease progression in FTD. Collaborative, multicenter trials will be needed in order to recruit sufficient number of subjects to test the promising therapies that are emerging for FTD.

CONCLUSIONS

FTD is recognized as a leading cause of early-onset dementia. Important advances in the research on genetics and molecular mechanisms of FTD over the last two decades have led to increasing clinical recognition of this disease. Detailed clinical information, neuropsychological testing, and neuroimaging results will improve the prediction of the underlying pathology and genetics in clinical FTD syndromes. We need to develop further early and accurate diagnosis with neuropathological prediction in order to develop the therapeutic interventions and to move toward primary prevention for FTD.

References

1. Ratnavalli E, Brayne C, Dawson K, Hodges JR. The prevalence of frontotemporal dementia. *Neurology*. 2002;58(11):1615–1621.
2. Pick A. Über die Beziehungen der senilen Hirnatrophie zur Aphasie. *Prager Med Wochenschr*. 1892 Jul 10;(17):165–167.
3. Alzheimer A. Über eigenartige Krankheitsfälle des späteren Alters. *Zbl Ges Neurol Psych*. 1911;4(Jul 9):356–385.
4. Mesulam M. Slowly progressive aphasia without generalized dementia. *Ann Neurol*. 1982;11(6):592–598.
5. Neary D, Snowden JS, Gustafson L, et al. Frontotemporal lobar degeneration: a consensus on clinical diagnostic criteria. *Neurology*. 1998;1546–1554.
6. Mackenzie IR, Neumann M, Bigio EH, et al. Nomenclature and nosology for neuropathologic subtypes of frontotemporal lobar degeneration: an update. *Acta Neuropathol*. 2010;119(1):1–4.
7. Knopman DS, Roberts ROR. Estimating the number of persons with frontotemporal lobar degeneration in the US population. *J Mol Neurosci*. 2011;45(3):330–335.
8. Snowden JS, Neary D, Mann DMA. Frontotemporal dementia. *Br J Psychiatry*. 2002;180(Feb):140–143.
9. Mercy L, Hodges JR, Dawson K, Barker RA, Brayne C. Incidence of early-onset dementias in Cambridgeshire, United Kingdom. *Neurology*. 2008;71(19):1496–1499.
10. Rosso SM, Kaat LD, Baks T, et al. Frontotemporal dementia in the Netherlands: patient characteristics and prevalence estimates from a population-based study. *Brain*. 2003;126(Pt 9):2016–2022.
11. Rohrer JD, Guerreiro R, Vandrovcova J, et al. The heritability and genetics of frontotemporal lobar degeneration. *Neurology*. 2009;73(18):1451–1456.
12. Seelaar H, Kamphorst W, Rosso SM, et al. Distinct genetic forms of frontotemporal dementia. *Neurology*. 2008;71(16):1220–1226.
13. Roberson ED, Hesse JH, Rose KD, et al. Frontotemporal dementia progresses to death faster than Alzheimer disease. *Neurology*. 2005;65(5):719–725.
14. Rascovsky K, Salmon DP, Lipton AM, et al. Rate of progression differs in frontotemporal dementia and Alzheimer disease. *Neurology*. 2005;65(3):397–403.
15. Kertesz A, Blair MM, McMonagle PP, Munoz DGD. The diagnosis and course of frontotemporal dementia. *Alzheimer Dis Assoc Disord*. 2007;21(2):155–163.
16. Liu W, Miller BL, Kramer JH, et al. Behavioral disorders in the frontal and temporal variants of frontotemporal dementia. *Neurology*. 2004;62(5):742–748.
17. Gregory CA, Serra-Mestres J, Hodges JR. Early diagnosis of the frontal variant of frontotemporal dementia: how sensitive are standard neuroimaging and neuropsychologic tests? *Neuropsychiatry Neuropsychol Behav Neurol*. 1999;12(2):128–135.
18. Rosen HJ, Gorno-Tempini M-L, Goldman WP, et al. Patterns of brain atrophy in frontotemporal dementia and semantic dementia. *Neurology*. 2002;58(2):198–208.
19. Rabinovici GD, Seeley WW, Kim EJ, et al. Distinct MRI atrophy patterns in autopsy-proven Alzheimer's disease and frontotemporal lobar degeneration. *Am J Alzheimers Dis Other Demen*. 2007;22(6):474–488.
20. Rosen HJ, Allison SC, Schauer GF, Gorno-Tempini M-L, Weiner M, Miller BL. Neuroanatomical correlates of behavioural disorders in dementia. *Brain*. 2005;128(Pt 11):2612–2625.
21. Graham A, Davies R, Xuereb J, et al. Pathologically proven frontotemporal dementia presenting with severe amnesia. *Neurology*. 2005; 128(Pt 3):597–605.
22. Boeve BF, Parisi JE, Dickson DW, et al. Antemortem diagnosis of frontotemporal lobar degeneration. *Ann Neurol*. 2005;57(4):480–488.
23. Piguet OO, Hornberger M, Shelley BP, Kipps CM, Hodges JR. Sensitivity of current criteria for the diagnosis of behavioral variant frontotemporal dementia. *Neurology*. 2009;72(8):732–737.
24. Baborie A, Griffiths TD, Jaros E, et al. Pathological correlates of frontotemporal lobar degeneration in the elderly. *Acta Neuropathol*. 2011;121(3):365–371.
25. Hodges JR. Frontotemporal dementia (Pick's disease): clinical features and assessment. *Neurology*. 2001;56(11 suppl 4):S6–S10.
26. Snowden JS, Goulding PJ, Neary D. Semantic dementia: a form of circumscribed cerebral atrophy. *Behav Neurol*. 1989, IOS Press.

27. Fushimi T, Komori K, Ikeda M, Lambon Ralph MA, Patterson K. The association between semantic dementia and surface dyslexia in Japanese. *Neuropsychologia*. 2009;47(4):1061–1068.
28. Seeley WW, Bauer AM, Miller BL, et al. The natural history of temporal variant frontotemporal dementia. *Neurology*. 2005;64(8):1384–1390.
29. Thompson SNA, Patterson K, Hodges JR. Left/right asymmetry of atrophy in semantic dementia: behavioral-cognitive implications. *Neurology*. 2003;61(9):1196–1203.
30. Kashibayashi T, Ikeda M, Komori K, et al. Transition of distinctive symptoms of semantic dementia during longitudinal clinical observation. *Dement Geriatr Cogn Disord* 2010;29(3):224–232.
31. Gorno-Tempini M-L, Dronkers NFN, Rankin KP, et al. Cognition and anatomy in three variants of primary progressive aphasia. *Ann Neurol*. 2004;55(3):335–346.
32. Josephs KA, Duffy JRJ, Strand EAE, et al. Clinicopathological and imaging correlates of progressive aphasia and apraxia of speech. *Brain*. 2006;129(Pt 6):1385–1398.
33. Rosen HJ, Allison SC, Ogar JM, et al. Behavioral features in semantic dementia vs. other forms of progressive aphasias. *Neurology*. 2006;67(10):1752–1756.
34. Gorno-Tempini M-L, Hillis AE, Weintraub S, et al. Classification of primary progressive aphasia and its variants. *Neurology*. 2011;1006–1014.
35. Lomen-Hoerth C. Characterization of amyotrophic lateral sclerosis and frontotemporal dementia. *Dement Geriatr Cogn Disord*. 2004;17(4):337–341.
36. Ringholz GM, Appel SH, Bradshaw M, Cooke NA, Mosnik DM, Schulz PEP. Prevalence and patterns of cognitive impairment in sporadic ALS. *Neurology*. 2005;65(4):586–590.
37. Phukan J, Elamin M, Bede P, et al. The syndrome of cognitive impairment in amyotrophic lateral sclerosis: a population-based study. *J Neurol Neurosurg Psychiatry*. 2012;83(1):102–108.
38. Kertesz A, Martinez-Lage P, Davidson W, Munoz DG. The corticobasal degeneration syndrome overlaps progressive aphasia and frontotemporal dementia. *Neurology*. 2000;55(9):1368–1375.
39. Burrell JR, Hornberger M, Villemagne VL, Rowe CC, Hodges JR. Clinical profile of PiB-positive corticobasal syndrome. *PLoS One*. 2013;8(4):e61025.
40. Murray R, Neumann M, Forman MS, et al. Cognitive and motor assessment in autopsy-proven corticobasal degeneration. *Neurology*. 2007;68(16):1274–1283.
41. Grafman J, Litvan I, Stark M. Neuropsychological features of progressive supranuclear palsy. *Brain Cogn*. 1995;28(3):311–320.
42. Josephs KA, Petersen RC, Knopman DS, et al. Clinicopathologic analysis of frontotemporal and corticobasal degenerations and PSP. *Neurology*. 2006;66(1):41–48.
43. Rascovsky K, Hodges JR, Knopman DS, et al. Sensitivity of revised diagnostic criteria for the behavioural variant of frontotemporal dementia. *Brain*. 2011;134(Pt 9):2456–2477.
44. Woolley JD, Khan BK, Murthy NK, Miller BL, Rankin KP. The diagnostic challenge of psychiatric symptoms in neurodegenerative disease. *J Clin Psychiatry*. 2011;72(02):126–133.
45. Broe MM, Hodges JR, Schofield EE, Shepherd CEC, Kril JJJ, Halliday GM. Staging disease severity in pathologically confirmed cases of frontotemporal dementia. *Neurology*. 2003;60(6):1005–1011.
46. Cairns NJ, Bigio EH, Mackenzie IR, et al. Neuropathologic diagnostic and nosologic criteria for frontotemporal lobar degeneration: consensus of the Consortium for Frontotemporal Lobar Degeneration. *Acta Neuropathol*. 2007;5–22.
47. Rohrer JD, Lashley T, Schott JM, et al. Clinical and neuroanatomical signatures of tissue pathology in frontotemporal lobar degeneration. *Brain*. 2011;134(Pt 9):2565–2581.
48. Josephs KA, Hodges JR, Snowden JS, et al. Neuropathological background of phenotypical variability in frontotemporal dementia. *Acta Neuropathol*. 2011;122(2):137–153.
49. Dickson DW, Kouri N, Murray ME, Josephs KA. Neuropathology of frontotemporal lobar degeneration-Tau (FTLD-Tau). *J Mol Neurosci*. 2011;45(3):384–389.
50. Goedert M, Spillantini MG. Tau mutations in frontotemporal dementia FTDP-17 and their relevance for Alzheimer's disease. *Biochim Biophys Acta*. 2000;1502(1):110–121.
51. Rovelet Lecrux A, Hannequin D, Guillin O, et al. Frontotemporal dementia phenotype associated with MAPT gene duplication. *J Alzheimers Dis*. 2010;21(3):897–902.
52. Mackenzie IR, Neumann M, Baborie A, et al. A harmonized classification system for FTLD-TDP pathology. *Acta Neuropathol*. 2011;122(1):111–113.
53. Neumann M, Mackenzie IR, Cairns NJ, et al. TDP-43 in the ubiquitin pathology of frontotemporal dementia with VCP gene mutations. *J Neuropathol Exp Neurol*. 2007;66(2):152–157.
54. DeJesus-Hernandez M, Mackenzie IR, Boeve BF, et al. Expanded GGGGCC hexanucleotide repeat in noncoding region of C9ORF72 causes chromosome 9p-linked FTD and ALS. *Neuron*. 2011;72(2):245–256.
55. Neumann M, Rademakers R, Roeber S, Baker M, Kretzschmar HA, Mackenzie IR. A new subtype of frontotemporal lobar degeneration with FUS pathology. *Brain*. 2009;132(11):2922–2931.
56. Urwin HH, Josephs KA, Rohrer JDJ, et al. FUS pathology defines the majority of tau- and TDP-43-negative frontotemporal lobar degeneration. *Acta Neuropathol*. 2010;120(1):33–41.
57. Seelaar H, Klijnsma KY, Koning I, et al. Frequency of ubiquitin and FUS-positive, TDP-43-negative frontotemporal lobar degeneration. *J Neurol*. 2010;257(5):747–753.
58. Borroni B, Alberici A, Archetti S, Magnani E, Di Luca M, Padovani A. New insights into biological markers of frontotemporal lobar degeneration spectrum. *Curr Med Chem*. 2010;17(10):1002–1009.
59. Seelaar H, Rohrer JD, Pijnenburg YA, Fox NC, van Swieten JC. Clinical, genetic and pathological heterogeneity of frontotemporal dementia: a review. *J Neurol Neurosurg Psychiatry*. 2011;82(5):476–486.
60. van Swieten JC, Spillantini MG. Hereditary frontotemporal dementia caused by Tau gene mutations. *Brain Pathol*. 2007;17(1):63–73.
61. Boeve BF, Hutton M. Refining frontotemporal dementia with parkinsonism linked to chromosome 17: introducing FTDP-17 (MAPT) and FTDP-17 (PGRN). *Arch Neurol*. 2008;65(4):460–464.

62. Whitwell JL, Weigand SD, Boeve BF, et al. Neuroimaging signatures of frontotemporal dementia genetics: C9ORF72, tau, progranulin and sporadics. *Brain*. 2012;135(3):794–806.
63. Pickering-Brown SM, Rollinson S, Plessis Du D, et al. Frequency and clinical characteristics of progranulin mutation carriers in the Manchester frontotemporal lobar degeneration cohort: comparison with patients with MAPT and no known mutations. *Brain*. 2008;131(3):721–731.
64. Le Ber I, Camuzat A, Hannequin D, et al. Phenotype variability in progranulin mutation carriers: a clinical, neuropsychological, imaging and genetic study. *Brain*. 2008;131(Pt 3):732–746.
65. Yu C-E, Bird TD, Bekris LM, et al. The spectrum of mutations in progranulin A collaborative study screening 545 cases of neurodegeneration. *Arch Neurol*. 2010;67(2):161–170.
66. Chen-Plotkin AS, Martinez-Lage M, Sleiman PMA, et al. Genetic and clinical features of progranulin-associated frontotemporal lobar degeneration. *Arch Neurol*. 2011;68(4):488–497.
67. Renton A, Majounie E, Waite A, et al. A hexanucleotide repeat expansion in C9ORF72 is the cause of chromosome 9p21-linked ALS-FTD. *Neuron*. 2011;72(2):257–268.
68. Dobson-Stone C, Hallupp M, Bartley L, et al. C9ORF72 repeat expansion in clinical and neuropathologic frontotemporal dementia cohorts. *Neurology*. 2012;79(10):995–1001.
69. Mahoney CJ, Beck J, Rohrer JD, et al. Frontotemporal dementia with the C9ORF72 hexanucleotide repeat expansion: clinical, neuroanatomical and neuropathological features. *Brain*. 2012;135(Pt 3):736–750.
70. Sha SJ, Takada LT, Rankin KP, et al. Frontotemporal dementia due to C9ORF72 mutations: clinical and imaging features. *Neurology*. 2012;79(10):1002–1011.
71. Boeve BF, Boylan KB, Graff-Radford NR, et al. Characterization of frontotemporal dementia and/or amyotrophic lateral sclerosis associated with the GGGGCC repeat expansion in C9ORF72. *Brain*. 2012;135(Pt 3):765–783.
72. Hu B, Ross LL, Neuhaus J, et al. Off-label medication use in frontotemporal dementia. *Am J Alzheimers Dis Other Demen*. 2010;25(2):128–133.
73. Huey ED, Putnam KT, Grafman J. A systematic review of neurotransmitter deficits and treatments in frontotemporal dementia. *Neurology*. 2006;66(1):17–22.
74. Mendez MF. Frontotemporal dementia: therapeutic interventions. *Front Neurol Neurosci*. 2009;24:168–178.
75. Deakin JB, Rahman S, Nestor PJ, Hodges JR, Sahakian BJ. Paroxetine does not improve symptoms and impairs cognition in frontotemporal dementia: a double-blind randomized controlled trial. *Psychopharmacology (Berl)*. 2004;172(4):400–408.
76. Lebert F, Stekke W, Hasenbroekx C, Pasquier F. Frontotemporal dementia: a randomised, controlled trial with trazodone. *Dement Geriatr Cogn Disord*. 2004;17(4):355–359.
77. Moretti R, Torre P, Antonello RM, Cazzato G, Griggio S, Bava A. Olanzapine as a treatment of neuropsychiatric disorders of Alzheimer's disease and other dementias: a 24-month follow-up of 68 patients. *Am J Alzheimers Dis Other Demen*. 2003;18(4):205–214.
78. Kerssens CJ, Kerrsens CJ, Pijnenburg YA. Vulnerability to neuroleptic side effects in frontotemporal dementia. *Eur J Neurol*. 2008;15(2): 111–112.
79. Moretti R, Torre P, Antonello RM, Cattaruzza T, Cazzato G, Bava A. Rivastigmine in frontotemporal dementia: an open-label study. *Drugs Aging*. 2004;21(14):931–937.
80. Kertesz A, Morlog D, Light M, et al. Galantamine in frontotemporal dementia and primary progressive aphasia. *Dement Geriatr Cogn Disord*. 2008;25(2):178–185.
81. Mendez MF, Shapira J, McMurtray A, Licht EE. Preliminary findings: behavioral worsening on donepezil in patients with frontotemporal dementia. *Am J Geriatr Psychiatry*. 2007;15(1):84–87.
82. Tariot PN, Farlow MR, Grossberg GT, et al. Memantine treatment in patients with moderate to severe Alzheimer disease already receiving donepezil: a randomized controlled trial. *JAMA*. 2004;291(3):317–324.
83. Boxer AL, Knopman DS, Kaufer DI, et al. Memantine in patients with frontotemporal lobar degeneration: a multicentre, randomised, double-blind, placebo-controlled trial. *Lancet Neurol*. 2013;12(2):149–156

CHAPTER

70

The Neuronal Ceroid-Lipofuscinoses (Batten Disease)

Sara E. Mole[*] *and Matti Haltia*[†]

[*]Medical Research Council, Laboratory for Molecular Cell Biology, University College London, London, UK

[†]Department of Pathology, University of Helsinki, Helsinki, Finland

INTRODUCTION

The neuronal ceroid-lipofuscinoses (NCLs), collectively also called Batten disease, constitute one of the most common groups of inherited neurodegenerative disorders in children, but may also occur in adults.[1] The childhood forms present clinically as progressive mental and motor deterioration and loss of vision, while the rare adult-onset forms are dominated by dementia. To date, 14 genetically distinct forms of NCL have been identified, all characterized by the accumulation of abnormal lipofuscin-like material in the lysosomes of nerve cells, associated with progressive and selective destruction of neurons particularly in the cerebral and cerebellar cortex and, less constantly, in the retina. The distribution of the stored material resembles that of the "wear-and-tear" pigment lipofuscin, which accumulates during normal aging.

HISTORICAL OVERVIEW

Disease Identification

The first clinical description of patients who may have suffered from NCL is usually attributed to Stengel.[2] However, his report (written in Norwegian) remained unnoticed by the contemporary scientific community. Between 1903 and 1913 several authors, including Batten, Vogt, Spielmeyer, Janský, and Bielschowsky described a number of patients with late infantile or juvenile onset of progressive loss of vision and psychomotor retardation in a familial setting (for review see [3]). At neuropathological investigation, all these patients showed intraneuronal accumulation of granular material with lipid-like staining qualities. They thus shared similarities (familial occurrence, progressive loss of vision, psychomotor retardation, and intraneuronal "thesaurismosis" or "storage") with the infantile-onset "amaurotic family idiocy," described by Sachs in 1896, and subsequently known as Tay–Sachs disease. Owing to the superficial similarities, the diseases published by all these authors were gradually grouped together as variants of "amaurotic family idiocy" of either infantile (Tay–Sachs), late infantile (Janský–Bielschowsky), or juvenile (Spielmeyer–Sjögren) onset. In 1925, Kufs published his first paper on adult-onset mental deterioration, associated with similar intraneuronal storage but without evident loss of vision.[3]

Biochemical studies[4,5] gradually led to the identification of the major storage material in Tay–Sachs disease as GM2 ganglioside, and classification of Tay–Sachs disease as a form of gangliosidosis. However, ganglioside accumulation was not found in the juvenile Spielmeyer–Sjögren type of "amaurotic family idiocy." The histochemical and electron microscopic studies of Zeman and his collaborators in the 1960s showed further differences between Tay–Sachs disease and the later-onset forms of "amaurotic family idiocy." The storage cytosomes in the late infantile (Janský–Bielschowsky) and juvenile (Spielmeyer–Sjögren) types were strongly autofluorescent and largely resistant to lipid solvents, in striking contrast to the easily extractable storage granules in the infantile (Tay–Sachs) type. Furthermore, the ultrastructure

Rosenberg's Molecular and Genetic Basis of Neurological and Psychiatric Disease
http://dx.doi.org/10.1016/10.1016/B978-0-12-410529-4.00070-X

of the electron-dense storage material in the Janský–Bielschowsky type (curvilinear pattern) and Spielmeyer–Sjögren type (fingerprint pattern) (Figure 70.1) radically differed from the membranous cytoplasmic bodies, earlier described by Terry and Korey[6] in Tay–Sachs disease. These characteristics and the distribution of the storage granules in the Janský–Bielschowsky and Spielmeyer–Sjögren types were close to those of the "wear-and-tear" pigment lipofuscin, associated with aging, or its pathological counterpart ceroid. Accordingly, in 1969 Zeman and Dyken proposed the new term "neuronal ceroid-lipofuscinosis" (NCL)[7] in order to clearly distinguish the late infantile and juvenile forms of "amaurotic family idiocy" and the histochemically similar adult-onset cases reported by Kufs from Tay–Sachs disease and other gangliosidoses. A few years later, a new infantile type of NCL was described, characterized by electron-dense storage cytosomes with a finely granular ultrastructure (Figure 70.2).[8–10] NCL was first reported in animals, in dogs, in 1970.[11]

During the subsequent decades, the NCL family rapidly expanded. A congenital type, originally reported as "amaurotic family idiocy" in 1941,[12] was now included, in addition to a number of "atypical" or "variant" cases of NCL, many of them with a late infantile or early juvenile onset[3] (Table 70.1). At neuropathological investigation, even an apparently distinct clinical disease "Northern epilepsy" or "progressive epilepsy with mental retardation" emerged as a new form of NCL[3] (Table 70.1). The group of human NCLs now comprises more than 14 distinct entities (see

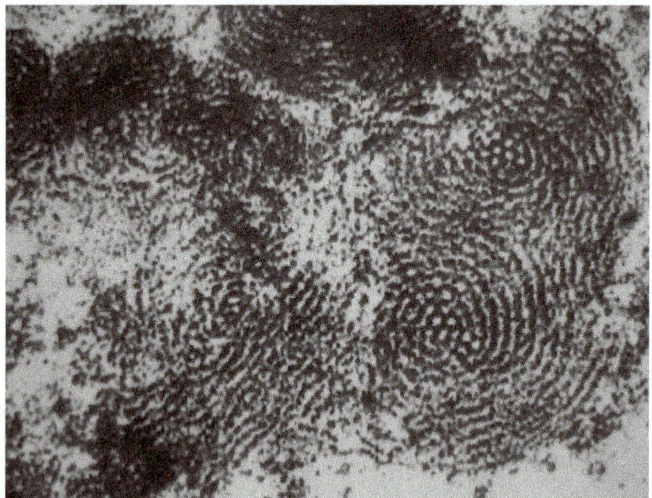

FIGURE 70.1 **Fingerprint patterns are a characteristic ultrastructural feature of the intraneuronal storage cytosomes of patients with juvenile CLN3 disease.** Electron micrograph (original magnification 30 000 x).

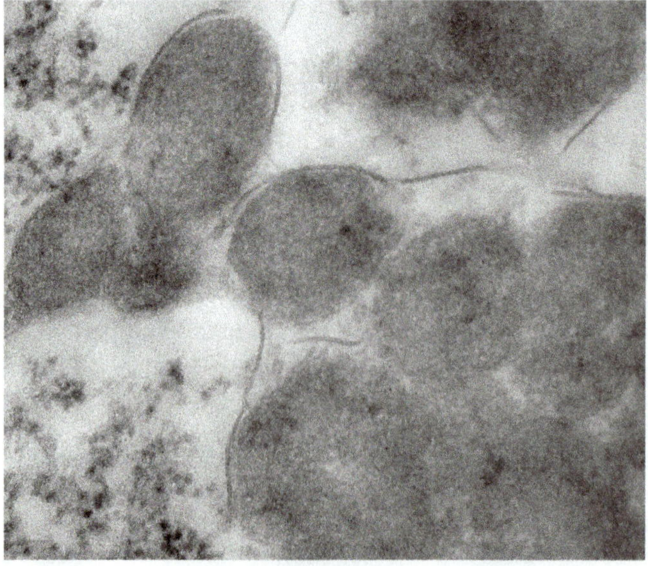

FIGURE 70.2 **The intraneuronal storage material in Children with infantile CLN1 disease is characterized by membrane-bound accumulations of globules with a finely granular internal structure.** Electron micrograph (original magnification 75 000 x).

TABLE 70.1 NCL Diseases Showing the Correlation between Genotype and Phenotype

Gene Protein	No. mutations	Genotype–phenotype correlation*	Widespread common mutations	Country-specific mutations	Gene identification
CLN1/PPT1 Palmitoyl protein thioesterase	64	**Infantile** Late infantile Juvenile Adult	p.Arg122Trp p.Arg151*	p.Thr75Pro and p.Leu10* in Scotland	1
CLN2/TPP1 Tripeptidyl peptidase	105	**Late infantile** Juvenile Protracted	c.509-1G>C p.Arg208*	p.Glu284Val in Canada	2
CLN3 CLN3	57	Juvenile Protracted	1 kb intragenic deletion in Caucasian populations	1 kb deletion in many countries 2.8 kb intragenic deletion in Finland	3
CLN4/DNAJC5 Cysteine string protein α	2	Adult autosomal dominant	p.Leu116del p.Leu115Arg	N/A	4
CLN5 CLN5	36	**Late infantile** Juvenile Protracted Adult	None	p.Tyr392* and p.Trp75* in Finland	5
CLN6 CLN6	68	**Late infantile** Protracted Adult Kufs type A	None	p.Ile154del in Portugal	6
CLN7/MFSD8 Major facilitator superfamily domain 8	31	**Late infantile** Juvenile protracted	None	P.Thr294Lys in Roma Gypsies c.724+2T>A in Eastern Europe	7
CLN8 CLN8	24	**Late infantile** Protracted ***EPMR/Northern epilepsy***	None	p.Arg24Gly in Finland causing EPMR p.Arg204Cys and pTrp263Cys in Turkey	8
CLN9¶	Not known	Juvenile	Not known	Not known	
CLN10/CTSD Cathepsin D	7	**Congenital** Late infantile Juvenile Adult	Not known	Not known	9,10
CLN11/GRN Progranulin	2†	**Adult** *Frontotemporal lobar dementia (when heterozygous)*	N/A	N/A	11
CLN12/ATP13A2 Type 5 P-type ATPase	1‡	**Juvenile** ***Kufor–Rakeb syndrome***	N/A	N/A	12
CLN13/CTSF Cathepsin F	5	**Adult Kufs type B**	N/A	N/A	13,14
CLN14/KCTD7 Potassium channel tetramerization domain-containing protein 7	1§	**Infantile** ***Progressive myoclonic epilepsy-3*** ***Opsoclonus-myoclonus ataxia-like syndrome***	N/A	N/A	13

*__Bold__ = *phenotype caused by complete loss of gene function; Italics = non-NCL disease phenotype that in some cases may be more typically associated with this gene.*

†*Only the mutation that causes NCL when present on both disease alleles is indicated; this mutation, and other mutations in this gene, cause later onset frontotemploral lobar dementia when present in heterozygous form.*

‡*Only the mutation that causes NCL is indicated; this mutation, other mutations cause Kufor–Rakeb syndrome.*

§*Only the mutation that causes NCL is indicated; this mutation, other mutations cause PME-3 or opsoclonus-myoclonus ataxia-like syndrome.*

¶*CLN9 is predicted based on rare clinical observations.*

References cited can be found in the "Literature Cited in Tables" list, pages 806-808.

Table 70.1). In addition to the human forms, numerous new spontaneous forms of NCL were discovered in many breeds of dogs, cats, sheep, goats, cattle, horses, and mice.[3]

Gene Identification

The discovery of subunit c of the mitochondrial ATP synthase[13] and/or saposins A and D[14] as the major protein components of isolated NCL storage granules prompted the search for the primary genomic defects in the corresponding genes, however, with negative results. The first NCL genes were reported in 1995 (Table 70.1). That for the recently described infantile Haltia–Santavuori form of NCL was identified in Leena Peltonen's laboratory by a positional candidate gene approach. This gene, *PPT1*, encodes palmitoyl protein thioesterase (PPT1), a soluble lysosomal enzyme. An international consortium of five research groups succeeded in isolating the gene underlying the juvenile Spielmeyer–Sjögren type of NCL, which had already been linked to the haptoglobin locus on chromosome 16.[15] This novel gene encodes a membrane protein. Since then, 11 additional NCL genes have been discovered (see Table 70.1), with over 400 mutations responsible for further forms of NCL.

Evolution of Classification

The first classifications of the NCLs, now collectively also called Batten disease, were essentially based on the age of clinical onset and the ultrastructure of the storage cytosomes. Zeman's classification of 1976[16] thus included the infantile (Haltia–Santavuori), late infantile (Janský–Bielschowsky), juvenile (Spielmeyer–Sjögren), and adult (Kufs) types, as well as the canine form described by Koppang in English setters. However, the advances in molecular genetics since 1995 made Zeman's original clinicopathological classification problematic, because different mutations in a given gene may give rise to varying clinical phenotypes, including widely different ages of onset. The human NCLs are now classified into 14 genetic forms from CLN1 to CLN14 diseases (Table 70.1). This newest diagnostic classification system[17] derives from the primary genomic defect (i.e., gene name) but takes into account varying phenotype (e.g., late infantile) (Table 70.1).

Early Treatments

Only symptomatic treatments have been available to date, including antiepileptic medication when indicated.[18] In 1975, Zeman and Rider put forward the hypothesis that the basic defect in the NCLs was formation of pathological "lipopigments," possibly due to an increased rate of peroxidation of polyunsaturated fatty acids.[19] However, numerous experimental and clinical studies, including treatment of patients with various antioxidants, did not provide conclusive evidence in favor of their idea.

MODE OF INHERITANCE, INCIDENCE AND PREVALENCE

The NCLs show an autosomal recessive mode of inheritance, with one exception. Penetrance is considered full. The adult-onset CLN4 (Parry) disease follows an autosomal dominant pattern of inheritance. The NCLs have a worldwide distribution, but their reported incidence and prevalence rates differ greatly between countries. The overall incidence rates may vary from 1:67,000 in Italy and Germany to 1:14,000 in Iceland, and the prevalence rates from 1:1,000,000 in some regions to 1:100,000 in the Scandinavian countries.[1] In some populations, a genetic founder effect is well documented,[1] with that for juvenile CLN3 disease being worldwide.[20]

NATURAL HISTORY

Age of Onset

Most forms of NCL have their onset during childhood, either within the infantile, late-infantile or juvenile age range (Table 70.1). Classic CLN10 disease is congenital, whereas CLN4, CLN11, and CLN13 diseases have an adult onset[1] (Table 70.1). However, as indicated before, different mutations in a given gene may give rise to varying clinical phenotypes, including widely different ages of onset. As an example, classic CLN1 disease has an infantile onset but, depending on the mutation, CLN1 disease may not become clinically manifest until later, at the late-infantile, juvenile or even adult age range (Table 70.1). Considerable mutation-dependent variation in the age of onset has also been observed in CLN10 disease, as well as in the usually late infantile-onset CLN5, CLN6, and CLN7 diseases (Table 70.1).

Disease Evolution

Classic congenital CLN10 disease begins *in utero*. The affected fetus may show signs of intrauterine growth retardation and epileptic seizures, which can be confirmed by antenatal real-time ultrasound examination. The newborn infant is microcephalic and presents with intractable seizures, spasticity, and central apnea. The infant usually survives only for a few days. It is likely that affected fetuses may also be aborted or stillborn.[1]

Infants with the classic infantile CLN1 disease seem to develop normally until 6–12 months. Deceleration of head growth and muscular hypotonia usually precede other clinical manifestations. From the age of 12 months the child starts to lose previously acquired skills, including speech, and becomes clumsy. There is gradual loss of vision, hyperexcitability, restlessness, and poor sleep. The child may have characteristic "knitting" hyperkinesias of the hands and upper arms. Typical ophthalmoscopic findings include hypopigmentation of the fundus, optic atrophy, involution of retinal vessels, and brownish discoloration of the macula. After the age of 15–20 months, the deterioration accelerates and leads to microcephaly, truncal ataxia, dystonic features, choreoathetosis, and myoclonic jerks. Although seizures may occasionally be the first sign of the disease, onset of epilepsy is usually by the age of 30 months. Affected children become blind by the age of 2 years and lose all cognitive and active motor skills before the age of 3 years. They can become irritable and distressed, owing to increasing spasticity and pain, and die usually between the ages of 9 and 13 years.[1]

Classic late-infantile CLN2 disease has its clinical onset between 2–4 years of age with clumsiness, ataxia, and deterioration of speech. Rapid decline of motor function leads to complete dependency by the age of 5 years. Epileptic seizures usually start early in the disease, by the age of 3 years. Myoclonus may coexist, and often presents a major problem. Gradual decline in visual ability from around 4 years of age results in blindness by the age of 7 years. Finally, limb spasticity and truncal hypotonia result in total loss of independent motility, and swallowing problems may necessitate nutritional support using a nasogastric or gastrostomy tube. Death usually ensues by the mid-teens.[1]

The variant late infantile CLN5, CLN6, CLN7, and CLN8 diseases typically have their onset between the ages of 2 and 8 years with developmental delay, cognitive and motor decline, and epileptic seizures. Further symptoms include myoclonus, visual failure, speech impairment, spasticity, and swallowing problems. "Northern epilepsy" or progressive epilepsy with mental retardation (EPMR) is a striking mutation-specific phenotype of CLN8 disease, characterized by an early juvenile-onset and very slowly progressive course with epilepsy and cognitive deterioration but no evident loss of vision.[1]

The first clinical manifestation of classic juvenile CLN3 disease is visual failure, noticed by the age of 4–7 years and leading to blindness within 2–4 years. Typical ophthalmological findings include macular degeneration, optic atrophy, thinning of the vessels, and pigment accumulation in the peripheral retina. Learning difficulties become a problem during the early school years, and epileptic seizures often appear by the age of 10. Motor disturbances, including extrapyramidal and less prominent pyramidal involvement, appear around puberty and gradually lead to loss of independent mobility. Motor symptoms are usually associated with speech difficulties. Various psychiatric and sleep problems may occur at any stage of the disease. Those affected usually die during their third decade (mean age at death is 24 years).[1]

An NCL with teenage onset is CLN12 disease, arising from mutation of the same gene that causes Kufor–Rakeb syndrome, with clinical features of Parkinson disease plus spasticity, supranuclear upgaze paresis, and dementia.[21]

The autosomal dominant adult-onset CLN4 disease (Parry disease) usually starts between the ages of 25 and 45 years. The clinical picture is characterized by generalized tonic–clonic seizures, myoclonus, and prominent dementia. Additional features may include speech problems, cerebellar dysfunction and parkinsonism. Adult-onset CLN6 disease (formerly known as Kufs type A) presents with a progressive myoclonus epilepsy syndrome with dementia and ataxia. Pyramidal and extrapyramidal abnormalities appear later. Adult-onset CLN13 disease, which includes some but not all cases previously classified as Kufs type B, shows behavioral changes, dementia, and motor abnormalities, associated with cerebellar and/or extrapyramidal disturbances. In all these adult-onset forms of NCL vision is normal without pigmentary retinal degeneration. In contrast, the young adult-onset CLN11 disease is characterized by rapidly progressive visual loss due to retinal dystrophy, seizures, and cerebellar ataxia. Cognitive decline may also occur. The remaining adult onset NCLs whose genetic cause is known are caused by milder mutations of genes more usually associated with childhood onset.[1]

MOLECULAR GENETICS

Five of the NCL genes encode lysosomal luminal proteins (PPT1, TPP1, CTSD, CTSF, CLN5) and five encode intracellular transmembrane proteins (CLN3, CLN6, MFSD8, CLN8, ATP13A2). Two encode cytoplasmic proteins (DNAJC5, KCTD7) and one is secreted (GRN). Some NCL genes are highly conserved suggesting that they play a fundamental role in eukaryotic cells. There is considerable phenotypic heterogeneity amongst the genes[22,23] (Table 70.1).

CLN1

PPT1 is a 306-amino acid long fatty-acid hydrolase that removes fatty acyl groups from modified cysteines in proteins that are trafficked to the lysosome.[24–26] Mutations that impact the catalytic triad cause infantile disease,[27] with very low levels of enzyme activity remaining. Later onset or a more protracted disease course is usually associated with missense mutations.

CLN2

TPP1 is a 563-amino acid pepstatin-insensitive carboxypeptidase derived from a precursor that cleaves tripeptides from the amino terminus of small polypeptides.[28,29] Most mutations cause classic late infantile CLN2 disease, but some missense changes are associated with more protracted disease or later onset.

CLN3

CLN3 encodes a conserved 438-amino acid polytopic membrane protein product in endosomes/lysosomes and possibly the Golgi compartment and synaptic vesicles. The most common mutation, an intragenic deletion found worldwide, does not completely abolish function.[30] Some missense mutations are associated with more or very protracted disease.[31–33]

CLN4

CLN4 encodes cysteine-string protein alpha (CSPα), a 198-amino acid palmitoylated cytoplasmic protein that attaches to intracellular membranes including synaptic vesicles, and is thought to play a role in synaptic transmission and possibly synaptic preservation.[34,35] There are only two mutations known, both causing dominantly inherited disease with onset in adulthood.

CLN5

CLN5 encodes a soluble lysosomal 407-amino acid glycoprotein that is cleaved at position p.96.[36–38] Most mutations cause variant late infantile CLN5 disease but some cause disease of later onset, even up to adulthood.

CLN6

CLN6 encodes a 311-amino acid transmembrane protein of the endoplasmic reticulum (ER).[39,40] Most mutations cause variant late infantile CLN6 disease, but some cause disease of later onset, notably CLN6 disease, Kufs type A, which is without visual loss.[41–43]

CLN7

CLN7 encodes a 518-amino acid lysosomal protein that belongs to the polytopic major facilitator superfamily,[44] although it is not known what it transports. Most mutations cause disease with late-infantile onset. Rarely there can be later onset and very protracted disease (into the fifth decade).[45]

CLN8

CLN8 encodes a 286-amino acid membrane protein of the ER[46,47] that belongs to the TLC family that have a role in sensing, biosynthesis, and metabolism of lipids, or protection of proteins from proteolysis.[48] Most mutations cause disease with onset in late infancy, occasionally later onset,[49] but there is one notable founder missense mutation originating in Finland, that causes CLN8 disease, EPMR,[46] a protracted disease that is not associated with myoclonus or visual failure.

CLN10

CTSD is an aspartyl protease of the pepsin family, cathepsin D,[50] that cleaves peptide bonds flanked by bulky hydrophobic amino acids, but its activity is not restricted to the lysosome.[51] Mutations that completely abolish enzymatic activity cause congenital disease,[52,53] with milder mutations associated with juvenile or teenage/early-adult onset.[54]

CLN11

GRN is a 593-amino acid active glycoprotein, progranulin, that is proteolytically cleaved once secreted, to produce smaller peptides (granulins), each containing a highly conserved 12-cysteine consensus motif. These have broad activities, being growth factor-like, modulators of immune responses, and neuronal effectors.[55] Compound heterozygous mutations cause adult-onset NCL[56] but heterozygous mutations cause frontotemporal lobar degeneration with TDP-43 inclusions (FTLD-TDP).[57,58]

CLN12

ATP13A2 is a type 5 P-type ATPase polytopic lysosomal membrane protein of 1180 amino acids[59,60] that may regulate intracellular cation homeostasis and neuronal integrity, or protect cells against α-synuclein misfolding and toxicity.[61–64] Mutations in *ATP13A2* generally cause Kufor–Rakeb syndrome (KRS)—neurodegeneration with brain iron accumulation (NBIA).[59,65–67] However, a juvenile-onset NCL presenting with learning difficulties is described in one family.[21]

CLN13

CTSF is a 484-amino acid lysosomal cysteine protease.[68–70] Mutations cause adult-onset NCL (CLN13 disease, adult Kufs type B).[71]

CLN14

KCTD7 is a 289-amino acid cytoplasmic protein containing the potassium channel tetramerization domain-containing protein 7, and part of the BTB/POZ family,[72,73] that may modulate the activity of ion channels. Mutations have been described in three diseases, infantile NCL, infantile progressive myoclonic epilepsy, and opsoclonus-myoclonus ataxia-like syndrome.[74–77]

Other Genetic Types

There remain families that satisfy the clinical and histopathological criteria for a diagnosis of NCL, but which do not have mutations in any of the 13 genes described above, or the genes additionally mutated in some animal models. This group includes a variant juvenile-onset NCL disease subtype termed CLN9, and cases of late-infantile onset and adult onset. The application of exome or whole-genome sequencing technologies in single families is likely to identify further novel genes causing rare types of NCL.

DISEASE MECHANISMS

Human Observations

The molecular genetic heterogeneity of the NCLs is in striking contrast to their remarkably uniform morphological phenotype, the very basis of the NCL concept. All genetically distinct forms share at least the two essential features of NCL: 1) accumulation in the lysosomes of nerve cells of autofluorescent, electron-dense material, containing subunit c of mitochondrial ATP-synthase and/or sphingolipid activator proteins A and D; and 2) progressive and selective loss of neurons, particularly in the cerebral and cerebellar cortex and, less constantly, in the retina.[1,78] The loss of cerebral neurons, their axons and myelin sheaths leads gradually to brain atrophy which in the early-onset

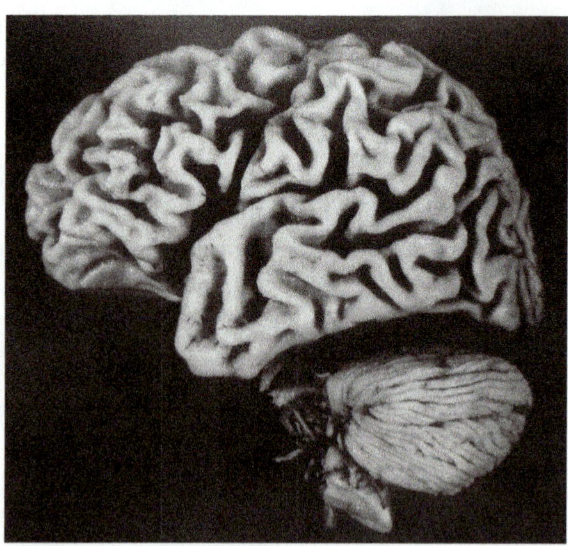

FIGURE 70.3 **All forms of NCL are accompanied by brain atrophy, due to the loss of neurons, their axons and myelin sheaths.** In the early-onset forms the atrophy is generalized and of an extraordinary degree. The brain of this boy with infantile CLN1 disease, who died at the age of 9 years, weighed 420 g. Note the very narrow gyri and the widely gaping sulci.

forms may be striking (Figure 70.3). Most clinical manifestations of the NCLs can be explained by this special pattern of neuronal degeneration and loss. However, it is still not understood how the numerous established genomic defects lead to the characteristic morphological phenotype of the NCLs, nor even whether the accumulation of storage material is detrimental to cells and a major contributing factor to disease. Whether there is interaction between the different NCL proteins or a unifying pathogenetic pathway for this group of diseases remains to be determined.

The mechanism underlying the selective neuronal death is unknown. Study of sheep and mouse models for many of the genetic subtypes has identified where initial cell loss occurs, and these regions can differ between disease models.[79] Certainly, cell loss does not appear to be related to the amount of accumulated storage material. It is preceded by localized glial activation and is accompanied by synaptic pathology. It is not yet determined whether this activation is causative or protective, but neuroinflammation appears to contribute to the pathogenesis, and there may be an immune response contribution. The sensory thalamocortical pathway is particularly affected early in disease progression for most but not all mouse NCL types—for one (CLN5 disease), cell loss begins in the cortex.

The underlying cellular abnormalities are not known, largely because the normal functions of the NCL genes are not fully understood. There may be abnormal lipid metabolism, intracellular trafficking and metabolism, and stress responses.

Animal Models

Animal models of NCL have been and are still essential for delineating the underlying cell biology and disease mechanism, developing and testing out new therapies, including high-throughput screening regimens, as well as new gene identification.[80] There are a range of organisms and mutations available, some occurring naturally and others experimentally derived, to make use of the distinct advantages of different species. Those most used include mouse, dog, sheep, zebrafish, fruit fly and yeast (Table 70.2). Mouse models exist for almost all NCL genes, with other species modeling a subset of genes according to conservation or ease of genetic manipulation (Table 70.2). There exist genetically solved animal models with NCL disease for which no human disease has yet been identified (e.g., *ARSG* in the American Staffordshire dog), as well as many for which the genetic defect is unsolved.

DIFFERENTIAL DIAGNOSIS AND TESTING

Clinical investigations usually begin with neurological and ophthalmological assessment, electrophysiological studies, brain imaging, and analysis of blood and cerebrospinal fluid (CSF). If NCL is suspected, the approach to diagnosis is then guided by the age of onset of symptoms[18] (Table 70.3), although could be supported by the results of tests (e.g., a preschool child with CLN2 disease usually has an occipital spike response to low flash rate photic

TABLE 70.2 Summary of Animal Models for NCL and their Use in Therapeutic Development

Gene	Mouse	Therapy research (in mice)	Reference	Large animal	Therapy research (in large animal)	Reference	Additional models
CLN1	Two	ERT; AAV gene therapy; bone marrow transplantation; stem cells; pharmacological; genetic immunosuppression	15–25	Miniature Dachshund dog	-		Yeasts *Caenorhabditis elegans* *Drosophila* Zebrafish
CLN2	Yes	ERT; AAV gene therapy	26–33	Longhaired Dachshund dog	ERT	34	Zebrafish
CLN3	Four	Pharmacological; immunosuppression	35–38	-	-		Yeasts *Caenorhabditis elegans* *Drosophila* Zebrafish
CLN4	Knockout, rather than dominant mutation	-		-	-		-
CLN5	Yes	-		Borderdale sheep; Devon cattle; Border collie dog	-		-
CLN6	nclf	-		South Hampshire sheep; Merino sheep; Australian Shepherd dog	Hematopoeitic cell transplant	39	Zebrafish
CLN7	Yes	-	52	-	-		Zebrafish
CLN8	mnd	Pharmacological	40–42	English Setter Dog	BMT, Dietary	43,44	-
CLN10	Yes	AAV gene therapy	45	Swedish Landrace sheep; American Bulldog	-		Yeasts *Drosophila*
CLN11	Yes	-		-	-		Zebrafish
CLN12	Yes	-		Tibetan Terrier dog			Zebrafish
CLN13	Yes	-		-	-		-
CLN14	Yes, but model for frontotemporal dementia too	-	46	-	-		-

More detail of animal models in [47], http://www.ucl.ac.uk/ncl/animal.shtml and [48] for therapy development for CLN1 disease.
There are many reports of animals with NCL that have not been genetically defined.
References cited can be found in the "Literature Cited in Tables" list, pp. 806–808.

stimulation). Initial laboratory studies comprise enzymatic tests, light and electron microscopic investigations, and finally DNA analysis. If these are performed in a logical order, they can quickly lead to a molecular genetic diagnosis, which is the ultimate aim.

Thus, for onset in infancy or late infancy, testing for NCL enzyme activities is the first step, followed by electron microscopic examination of pathological material to confirm typical NCL storage if these enzyme activities are normal. For onset in the juvenile age range, particularly if the leading symptom is rapid visual failure, then the first step should be examination of a blood film for vacuolated lymphocytes. For onset in adulthood, enzyme testing followed by examination of tissue specimens for storage material is recommended. In all cases, the final step is DNA analysis to confirm mutation in a candidate gene or sequencing of a specific set of genes to identify which carries a mutation.

TABLE 70.3 Diagnostic Priorities for NCL Diseases

Age of symptom onset	Enzyme and other tests	Additional DNA analysis if typical NCL storage confirmed by EM
Birth	Enzyme: CTSD	*CLN14/KCTD7*
Infant	Enzyme: PPT1, CTSD	–
Late infancy	Enzyme: TPP1, PPT1, CTSD	*CLN5, CLN6, CLN7, CLN8*
Juvenile	Vacuolated lymphocytes and CLN3 DNA	*CLN12/ATP13A2*
Adult	Enzyme: PPT1, TPP1, CTSF, CTSD	*CLN6* if autosomal recessive, *CLN4* if autosomal dominant, *CLN11/GRN*

When NCL is suspected, these are the enzyme tests and subsequent work that should be prioritized according to the age of presenting symptom.

Rarely, NCL may be confirmed pathologically, but no mutation is found in one of the known genes. In this instance, DNA from the family should be passed to a research laboratory to allow further work with the aim of identifying a novel NCL gene. With the advent of massively parallel sequencing technologies, it can be possible to determine the underlying gene in a single family, though this is not yet done routinely.

MANAGEMENT

Standard of Care

There is not yet an established preventative or curative treatment for the NCLs.[18] The focus is, therefore, on palliative care. This is challenging because of the diverse symptoms and because those affected have a limited and decreasing ability to communicate. However, there are clinicians experienced in caring for those with NCL who can be consulted by those with more limited knowledge. In general, epilepsy in NCL is resistant to therapy. It is not realistic to aim for complete absence of seizures or normalization of electroencephalogram (EEG), which should be used mainly for monitoring. Rather, therapy should follow clinical symptoms. Valproate and lamotrigine are the most effective anticonvulsants. Use of carbamazepine, phenytoin, or vigabatrin should be avoided, as these may exacerbate the disease course, particularly myoclonus. Myoclonus may be partially helped by levetiracetam or piracetam, and a reduction in the medication load. Spasticity can be painful and should be treated with physical therapy and medication such as baclofen, tizandine or perhaps tetrahydrocannabinol. Benzodiazepines may also be effective, but can have frequent side effects or a negative effect on the disease course. Episodes of spasticity or pain may be triggered by many factors and must be recognized and carefully addressed. These can result from fractures, renal calculi, or venous thrombosis caused by the lack of mobility. Constipation caused by intestinal hypomotility or malnutrition may lead to abdominal pain. The handling of psychopathological symptoms, such as sleep disturbance, fear, aggressive behavior, depression, and hallucinations can be challenging. The first effort should be managing environmental triggers, rather than medication. For juvenile CLN3 disease, risperidone can be considered for hallucinations or panic attacks and fluvoxamine for anxiety and depression. In all instances, as few drugs as possible should be administered. There are often national family organizations that can offer advice and professional support.

Therapies under Investigation

There are several strategies for the development of therapy for the NCLs, some of which have been applied to the different animal models with promising results. There is awareness that there will be a need for treatment that addresses both the central nervous system (CNS) deficits as well as any consequences outside the CNS, with the eye being targeted separately. Current strategies focus on enzyme replacement therapies for those NCLs caused by deficiencies in lysosomal proteases; gene therapy by stem cell transplantation or by directly targeting cells *in vivo* using a viral vector; pharmacological interventions; and diet modification. Initial focus has been on CLN1[81] and CLN2 diseases, but there has been some work done on other types, primarily using mouse and larger animal models (summarized in Table 70.2). Some approaches have reached clinical trials, and these are summarized in Table 70.4. Other experimental models, from yeast to zebrafish, may provide other avenues for further therapeutic development.

TABLE 70.4 Summary of Human Clinical Trials for Neuronal Ceroid-Lipofuscinoses

NCL disease	Title	Treatment	Start date and status	Sponsor	Country of operation	Clinical Trials Number	Report
CLN1	Cystagon to treat infantile neuronal ceroid-lipofuscinosis	Combination of cysteamine, bitartrate, and N-acetylcysteine	2001–2013 Completed	National Institute of Child Health and Human Development	USA	NCT00028262	[49]
CLN1	Safety and efficacy study of human central nervous system stem cells (HuCNS-SC) in subjects with neuronal ceroid-lipofuscinosis	Stem cells	2010–2011 Withdrawn prior to enrollment	StemCells, Inc	USA	NCT01238315	-
CLN1 and CLN2	Study of human central nervous system stem cells (HuCNS-SC) in patients with infantile or late infantile neuronal ceroid-lipofuscinosis (NCL)	Stem cells	2006–2009 Completed	StemCells, Inc	USA	NCT00337636	[50]
CLN2	Safety study of a gene transfer vector for children with late infantile neuronal ceroid-lipofuscinosis	AAV2CUhCLN2	2005–2019 Active, not recruiting	Weill Cornell Medical College	USA	NCT00151216	[51]
CLN2	Safety study of a gene transfer vector (rh.10) for children with late infantile neuronal ceroid-lipofuscinosis	AAVrh.10CUhCLN2	2010–2016 Recruiting	Weill Cornell Medical College	USA	NCT01161576	-
CLN2	AAVrh.10 administered to children with late infantile neuronal ceroid-lipofuscinosis with uncommon genotypes or moderate/severe impairment	AAVrh.10CUhCLN2	2010–2014 Recruiting	Weill Cornell Medical College	USA	NCT01414985	-
CLN2	Safety and efficacy study of BMN190 for the treatment of CLN2 patients	rhTPP1 BMN190	2013–2016 Recruiting	Biomarin Pharmaceutical	Germany, UK	NCT01907087	-
CLN3	Cellcept for treatment of juvenile neuronal ceroid-lipofuscinosis	Mycophenolate mofetil	2011–2015 Recruiting	University of Rochester	USA	NCT01399047	-

References cited can be found in the "Literature Cited in Tables" list, pp. 806–808.

CONCLUSION

The NCLs remain a heterogenous group of diseases not yet unified by genetic or underlying disease mechanism. Many underlying genes have now been identified, and some disease types are proceeding to therapeutic development and even clinical trials.

ACKNOWLEDGEMENTS

We thank the children and families who keep on contributing to studies, enhancing knowledge of the NCLs, as well as other experts and especially physicians whose experience we have drawn upon. This work was partly supported by funding from the European Union Seventh Framework Programme (FP7/2007-2013) under grant agreement no. 281234.

References

1. Mole SE, Williams RE, Goebel HH, eds. *The Neuronal Ceroid Lipofuscinoses (Batten Disease)*. 2nd ed. Oxford: Oxford University Press; 2011, Contemporary Neurology.
2. Stengel C. Account of a singular illness among four siblings in the vicinity of Røraas. Translated from Eyr (Christiania) 1826; 1:347-352. In: Armstrong D, Koppang N, Rider JA, eds. *Ceroid-lipofuscinosis (Batten´s disease)*. Amsterdam: Elsevier Biomedical Press; 1982:17–19.
3. Haltia M, Goebel HH. The neuronal ceroid-lipofuscinoses: a historical introduction. *Biochim Biophys Acta*. 2013;1832:1795–1800.
4. Klenk E. Beiträge zur Chemie der Lipidosen, Niemann–Pickschen Krankheit und amaurotischen Idiotie. *Hoppe Seylers Z Physiol Chem*. 1939;262:128–143.
5. Svennerholm L. The chemical structure of normal human brain and Tay–Sachs gangliosides. *Biochem Biophys Res Commun*. 1962;9:436–441.
6. Terry RD, Korey SR. Membranous cytoplasmic granules in infantile amaurotic idiocy. *Nature*. 1960;188:1000–1002.
7. Zeman W, Dyken P. Neuronal ceroid-lipofuscinosis (Batten's disease): relationship to amaurotic familial idiocy? *Pediatrics*. 1969;44:570–583.
8. Haltia M, Rapola J, Santavuori P. Infantile type of so-called neuronal ceroid-lipofuscinosis. Histological and electron microscopic studies. *Acta Neuropathol (Berl)*. 1973;26:157–170.
9. Haltia M, Rapola J, Santavuori P, Keränen A. Infantile type of so-called neuronal ceroid lipofuscinosis part 2. Morphological and biochemical studies. *J Neurol Sci*. 1973;18:269–285.
10. Santavuori P, Haltia M, Rapola J, Raitta C. Infantile type of so-called neuronal ceroid lipofuscinosis part 1. A clinical study of 15 patients. *J Neurol Sci*. 1973;18:257–267.
11. Koppang N. Neuronal ceroid-lipofuscinosis in English setters. *J Small Anim Pract*. 1970;10:639–644.
12. Norman RM, Wood N. Congenital form of amaurotic family idiocy. *J Neurol Psych*. 1941;4:175–190.
13. Palmer DN, Martinus RD, Cooper SM, Midwinter GG, Reid JC, Jolly RD. Ovine ceroid lipofuscinosis. The major lipopigment protein and the lipid-binding subunit of mitochondrial ATP synthase have the same NH2-terminal sequence. *J Biol Chem*. 1989;264:5736–5740.
14. Tyynelä J, Palmer DN, Baumann M, Haltia M. Storage of saposins A and D in infantile neuronal ceroid-lipofuscinosis. *FEBS Lett*. 1993;330:8–12.
15. Eiberg H, Gardiner RM, Mohr J. Batten disease (Spielmeyer–Sjögren disease) and haptoglobins (HP): indication of linkage and assignment to chromosome 16. *Clin Genet*. 1989;36:217–218.
16. Zeman W. The neuronal ceroid lipofuscinoses. In: Zimmerman HM, ed. Progress in Neuropathology. vol. III. New York: Grune and Stratton; 1976:207–223.
17. Williams RE, Mole SE. New nomenclature and classification scheme for the neuronal ceroid lipofuscinoses. *Neurology*. 2012;79:183–191.
18. Schulz A, Kohlschutter A, Mink J, Simonati A, Williams R. NCL diseases - clinical perspectives. *Biochim Biophys Acta*. 1832;2013:1801–1806.
19. Zeman W, Rider JA, eds. *The Dissection of a Degenerative Disease*. New York: Elsevier; 1975.
20. Mitchison HM, O'Rawe AM, Taschner PEM, et al. Batten disease (*CLN3*): linkage disequilibrium mapping in the Finnish population and analysis of European haplotypes. *Am J Hum Genet*. 1995;56:654–662.
21. Bras J, Verloes A, Schneider SA, Mole SE, Guerreiro RJ. Mutation of the parkinsonism gene ATP13A2 causes neuronal ceroid-lipofuscinosis. *Hum Mol Genet*. 2012;21:2646–2650.
22. Kousi M, Lehesjoki AE, Mole SE. Update of the mutation spectrum and clinical correlations of over 360 mutations in eight genes that underlie the neuronal ceroid lipofuscinoses. *Hum Mutat*. 2012;33:42–63.
23. Warrier V, Vieira M, Mole SE. Genetic basis and phenotypic correlations of the neuronal ceroid lipofuscinoses. *Biochim Biophys Acta*. 1832;2013:1827–1830.
24. Camp LA, Hofmann SL. Purification and properties of a palmitoyl-protein thioesterase that cleaves palmitate from H-Ras. *J Biol Chem*. 1993;268:22566–22574.
25. Hellsten E, Vesa J, Olkkonen VM, Jalanko A, Peltonen L. Human palmitoyl protein thioesterase: evidence for lysosomal targeting of the enzyme and disturbed cellular routing in infantile neuronal ceroid lipofuscinosis. *EMBO J*. 1996;15:5240–5245.
26. Lyly A, von Schantz C, Salonen T, et al. Glycosylation, transport, and complex formation of palmitoyl protein thioesterase 1 (PPT1)–distinct characteristics in neurons. *BMC Cell Biol*. 2007;8:22.
27. Das AK, Lu JY, Hofmann SL. Biochemical analysis of mutations in palmitoyl-protein thioesterase causing infantile and late-onset forms of neuronal ceroid lipofuscinosis. *Hum Mol Genet*. 2001;10:1431–1439.

28. Vines DJ, Warburton MJ. Classical late infantile neuronal ceroid lipofuscinosis fibroblasts are deficient in lysosomal tripeptidyl peptidase I. *FEBS Lett*. 1999;443:131–135.
29. Warburton MJ, Bernardini F. Tripeptidyl-peptidase I deficiency in classical late-infantile neuronal ceroid lipofuscinosis brain tissue. Evidence for defective peptidase rather than proteinase activity. *J Inherit Metab Dis*. 2000;23:145–154.
30. Kitzmüller C, Haines R, Codlin S, Cutler DF, Mole SE. A function retained by the common mutant CLN3 protein is responsible for the late onset of juvenile neuronal ceroid lipofuscinosis (JNCL). *Hum Mol Genet*. 2008;17:303–312.
31. Munroe PB, Mitchison HM, O'Rawe AM, et al. Spectrum of mutations in the Batten disease gene, *CLN3*. *Am J Hum Genet*. 1997;61:310–316.
32. Lauronen L, Munroe PB, Järvelä I, et al. Delayed classic and protracted phenotypes of compound heterozygous juvenile neuronal ceroid lipofuscinosis. *Neurology*. 1999;52:360–365.
33. Åberg L, Lauronen L, Hämäläinen J, Mole SE, Autti T. A 30-year follow-up of a neuronal ceroid lipofuscinosis patient with mutations in CLN3 and protracted disease course. *Pediatr Neurol*. 2009;40:134–137.
34. Chamberlain LH, Burgoyne RD. Cysteine-string protein: the chaperone at the synapse. *J Neurochem*. 2000;74:1781–1789.
35. Johnson JN, Ahrendt E, Braun JE. CSPalpha: the neuroprotective J protein. *Biochem Cell Biol*. 2010;88:157–165.
36. Schmiedt ML, Bessa C, Heine C, Ribeiro MG, Jalanko A, Kyttala A. The neuronal ceroid lipofuscinosis protein CLN5: new insights into cellular maturation, transport, and consequences of mutations. *Hum Mutat*. 2010;31:356–365.
37. Isosomppi J, Vesa J, Jalanko A, Peltonen L. Lysosomal localization of the neuronal ceroid lipofuscinosis CLN5 protein. *Hum Mol Genet*. 2002;11:885–891.
38. Holmberg V, Jalanko A, Isosomppi J, Fabritius A-L, Peltonen L, Kopra O. The mouse ortholog of the neuronal ceroid lipofuscinosis *CLN5* gene encodes a soluble lysosomal glycoprotein expressed in the developing brain. *Neurobiol Dis*. 2004;16:29–40.
39. Wheeler RB, Sharp JD, Schultz RA, Joslin JM, Williams RE, Mole SE. The gene mutated in variant late infantile neuronal ceroid lipofuscinosis (*CLN6*) and *nclf* mutant mice encodes a novel predicted transmembrane protein. *Am J Hum Genet*. 2002;70:537–542.
40. Mole SE, Michaux G, Codlin S, Wheeler RB, Sharp JD, Cutler DF. CLN6, which is associated with a lysosomal storage disease, is an endoplasmic reticulum protein. *Exp Cell Res*. 2004;298:399–406.
41. Arsov T, Smith KR, Damiano J, et al. Kufs disease, the major adult form of neuronal ceroid lipofuscinosis, caused by mutations in CLN6. *Am J Hum Genet*. 2011;88:566–573.
42. Sharp JD, Wheeler RB, Parker KA, Gardiner RM, Williams RE, Mole SE. Spectrum of *CLN6* mutations in variant late infantile neuronal ceroid lipofuscinosis. *Hum Mutat*. 2003;22:35–42.
43. Cannelli N, Garavaglia B, Simonati A, et al. Variant late infantile ceroid lipofuscinoses associated with novel mutations in CLN6. *Biochem Biophys Res Commun*. 2009;379:892–897.
44. Siintola E, Topcu M, Aula N, et al. The novel neuronal ceroid lipofuscinoses gene MFSD8 encodes a putative lysosomal transporter. *Am J Hum Genet*. 2007;81:136–146.
45. Kousi M, Siintola E, Dvorakova L, et al. Mutations in CLN7/MFSD8 are a common cause of variant late-infantile neuronal ceroid lipofuscinosis. *Brain*. 2009;132:810–819.
46. Ranta S, Zhang Y, Ross B, et al. The neuronal ceroid lipofuscinoses in human EPMR and *mnd* mutant mice are associated with mutations in *CLN8*. *Nat Genet*. 1999;23:233–236.
47. Lonka L, Kyttälä A, Ranta S, Jalanko A, Lehesjoki AE. The neuronal ceroid lipofuscinosis CLN8 membrane protein is a resident of the endoplasmic reticulum. *Hum Mol Genet*. 2000;9:1691–1697.
48. Winter E, Ponting CP. TRAM, LAG1 and CLN8: members of a novel family of lipid-sensing domains? *Trends Biochem Sci*. 2002;27:381–383.
49. Reinhardt K, Grapp M, Schlachter K, Bruck W, Gartner J, Steinfeld R. Novel CLN8 mutations confirm the clinical and ethnic diversity of late infantile neuronal ceroid lipofuscinosis. *Clin Genet*. 2010;77:79–85.
50. Metcalf P, Fusek M. Two crystal structures for cathepsin D: the lysosomal targeting signal and active site. *EMBO J*. 1993;12:1293–1302.
51. Benes P, Vetvicka V, Fusek M. Cathepsin D-many functions of one aspartic protease. *Crit Rev Oncol Hematol*. 2008;68:12–28.
52. Siintola E, Partanen S, Strömme P, et al. Cathepsin D deficiency underlies congenital human neuronal ceroid lipofuscinosis. *Brain*. 2006;129:1438–1445.
53. Tyynelä J, Sohar I, Sleat DE, et al. A mutation in the ovine cathepsin D gene causes a congenital lysosomal storage disease with profound neurodegeneration. *EMBO J*. 2000;19:2786–2792.
54. Steinfeld R, Reinhardt K, Schreiber K, et al. Cathepsin D deficiency is associated with a human neurodegenerative disorder. *Am J Hum Genet*. 2006;78:988–998.
55. Cenik B, Sephton CF, Kutluk Cenik B, Herz J, Yu G. Progranulin: a proteolytically processed protein at the crossroads of inflammation and neurodegeneration. *J Biol Chem*. 2012;287:32298–32306.
56. Smith KR, Damiano J, Franceschetti S, et al. Strikingly different clinicopathological phenotypes determined by progranulin-mutation dosage. *Am J Hum Genet*. 2012;90:1102–1107.
57. Baker M, Mackenzie IR, Pickering-Brown SM, et al. Mutations in progranulin cause tau-negative frontotemporal dementia linked to chromosome 17. *Nature*. 2006;442:916–919.
58. Cruts M, Gijselinck I, van der Zee J, et al. Null mutations in progranulin cause ubiquitin-positive frontotemporal dementia linked to chromosome 17q21. *Nature*. 2006;442:920–924.
59. Ramirez A, Heimbach A, Grundemann J, et al. Hereditary parkinsonism with dementia is caused by mutations in ATP13A2, encoding a lysosomal type 5 P-type ATPase. *Nat Genet*. 2006;38:1184–1191.
60. Schultheis PJ, Hagen TT, O'Toole KK, et al. Characterization of the P5 subfamily of P-type transport ATPases in mice. *Biochem Biophys Res Commun*. 2004;323:731–738.
61. Ramonet D, Podhajska A, Stafa K, et al. PARK9-associated ATP13A2 localizes to intracellular acidic vesicles and regulates cation homeostasis and neuronal integrity. *Hum Mol Genet*. 2012;21:1725–1743.
62. Tan J, Zhang T, Jiang L, et al. Regulation of intracellular manganese homeostasis by Kufor-Rakeb syndrome-associated ATP13A2 protein. *J Biol Chem*. 2011;286:29654–29662.

63. Schmidt K, Wolfe DM, Stiller B, Pearce DA. Cd2+, Mn2+, Ni2+ and Se2+ toxicity to Saccharomyces cerevisiae lacking YPK9p the orthologue of human ATP13A2. *Biochem Biophys Res Commun*. 2009;383:198–202.
64. Gitler AD, Chesi A, Geddie ML, et al. Alpha-synuclein is part of a diverse and highly conserved interaction network that includes PARK9 and manganese toxicity. *Nat Genet*. 2009;41:308–315.
65. Di Fonzo A, Chien HF, Socal M, et al. ATP13A2 missense mutations in juvenile parkinsonism and young onset Parkinson disease. *Neurology*. 2007;68:1557–1562.
66. Eiberg H, Hansen L, Korbo L, et al. Novel mutation in ATP13A2 widens the spectrum of Kufor-Rakeb syndrome (PARK9). *Clin Genet*. 2012;82:256–263.
67. Schneider SA, Paisan-Ruiz C, Quinn NP, et al. ATP13A2 mutations (PARK9) cause neurodegeneration with brain iron accumulation. *Mov Disord*. 2010;25:979–984.
68. Shi GP, Bryant RA, Riese R, et al. Role for cathepsin F in invariant chain processing and major histocompatibility complex class II peptide loading by macrophages. *J Exp Med*. 2000;191:1177–1186.
69. Oorni K, Sneck M, Bromme D, et al. Cysteine protease cathepsin F is expressed in human atherosclerotic lesions, is secreted by cultured macrophages, and modifies low density lipoprotein particles in vitro. *J Biol Chem*. 2004;279:34776–34784.
70. Lindstedt L, Lee M, Oorni K, Bromme D, Kovanen PT. Cathepsins F and S block HDL3-induced cholesterol efflux from macrophage foam cells. *Biochem Biophys Res Commun*. 2003;312:1019–1024.
71. Smith KR, Dahl H-H, Canafoglia L, et al. Cathepsin F mutations cause Type B Kufs disease, an adult-onset neuronal ceroid lipofuscinosis. *Hum Mol Genet*. 2013;22:1417–1423.
72. Stogios PJ, Chen L, Prive GG. Crystal structure of the BTB domain from the LRF/ZBTB7 transcriptional regulator. *Protein Sci*. 2007;16:336–342.
73. Stogios PJ, Downs GS, Jauhal JJ, Nandra SK, Prive GG. Sequence and structural analysis of BTB domain proteins. *Genome Biol*. 2005;6:R82.
74. Blumkin L, Kivity S, Lev D, et al. A compound heterozygous missense mutation and a large deletion in the KCTD7 gene presenting as an opsoclonus-myoclonus ataxia-like syndrome. *J Neurol*. 2012;259:2590–2598.
75. Kousi M, Anttila V, Schulz A, et al. Novel mutations consolidate KCTD7 as a progressive myoclonus epilepsy gene. *J Med Genet*. 2012;49:391–399.
76. Staropoli JF, Karaa A, Lim ET, et al. A homozygous mutation in KCTD7 links neuronal ceroid lipofuscinosis to the ubiquitin-proteasome system. *Am J Hum Genet*. 2012;91:202–208.
77. Van Bogaert P, Azizieh R, Desir J, et al. Mutation of a potassium channel-related gene in progressive myoclonic epilepsy. *Ann Neurol*. 2007;61:579–586.
78. Haltia M. The neuronal ceroid-lipofuscinoses. *J Neuropathol Exp Neurol*. 2003;62:1–13.
79. Palmer DN, Barry LA, Tyynela J, Cooper JD. NCL disease mechanisms. *Biochim Biophys Acta*. 1832;2013:1882–1893.
80. Bond M, Holthaus SM, Tammen I, Tear G, Russell C. Use of model organisms for the study of neuronal ceroid lipofuscinosis. *Biochim Biophys Acta*. 2013;1832(11):1842–1865.
81. Hawkins-Salsbury JA, Cooper JD, Sands MS. Pathogenesis and therapies for infantile neuronal ceroid lipofuscinosis (infantile CLN1 disease). *Biochim Biophys Acta*. 1832;2013:1906–1909.

Literature Cited in Tables

1. Vesa J, Hellsten E, Verkruyse LA, et al. Mutations in the palmitoyl protein thioesterase gene causing infantile neuronal ceroid lipofuscinosis. *Nature*. 1995;376:584–587.
2. Sleat DE, Donnelly RJ, Lackland H, et al. Association of mutations in a lysosomal protein with classical late-infantile neuronal ceroid lipofuscinosis. *Science*. 1997;277:1802–1805.
3. The International Batten Disease Consortium. Isolation of a novel gene underlying Batten disease, *CLN3*. *Cell*. 1995;82:949–957.
4. Noskova L, Stranecky V, Hartmannova H, et al. Mutations in DNAJC5, encoding cysteine-string protein alpha, cause autosomal-dominant adult-onset neuronal ceroid lipofuscinosis. *Am J Hum Genet*. 2011;89:241–252.
5. Savukoski M, Klockars T, Holmberg V, Santavuori P, Lander ES, Peltonen L. *CLN5*, a novel gene encoding a putative transmembrane protein mutated in Finnish variant late infantile neuronal ceroid lipofuscinosis. *Nat Genet*. 1998;19:286–288.
6. Wheeler RB, Sharp JD, Schultz RA, Joslin JM, Williams RE, Mole SE. The gene mutated in variant late infantile neuronal ceroid lipofuscinosis (*CLN6*) and *nclf* mutant mice encodes a novel predicted transmembrane protein. *Am J Hum Genet*. 2002;70:537–542.
7. Siintola E, Topcu M, Aula N, et al. The novel neuronal ceroid lipofusinosis gene MFSD8 encodes a putative lysosomal transporter. *Am J Hum Genet*. 2007;81:136–146.
8. Ranta S, Zhang Y, Ross B, et al. The neuronal ceroid lipofuscinoses in human EPMR and *mnd* mutant mice are associated with mutations in *CLN8*. *Nat Genet*. 1999;23:233–236.
9. Siintola E, Partanen S, Strömme P, et al. Cathepsin D deficiency underlies congenital human neuronal ceroid lipofuscinosis. *Brain*. 2006;129:1438–1445.
10. Steinfeld R, Reinhardt K, Schreiber K, et al. Cathepsin D deficiency is associated with a human neurodegenerative disorder. *Am J Hum Genet*. 2006;78:988–998.
11. Smith KR, Damiano J, Franceschetti S, et al. Strikingly different clinicopathological phenotypes determined by progranulin-mutation dosage. *Am J Hum Genet*. 2012;90:1102–1107.
12. Bras J, Verloes A, Schneider SA, Mole SE, Guerreiro RJ. Mutation of the parkinsonism gene ATP13A2 causes neuronal ceroid-lipofuscinosis. *Hum Mol Genet*. 2012;21:2646–2650.
13. Staropoli JF, Karaa A, Lim ET, et al. A homozygous mutation in KCTD7 links neuronal ceroid lipofuscinosis to the ubiquitin-proteasome system. *Am J Hum Genet*. 2012;91:202–208.
14. Smith KR, Dahl H-H, Canafoglia L, et al. Cathepsin F mutations cause Type B Kufs disease, an adult-onset neuronal ceroid lipofuscinosis. *Hum Mol Genet*. 2013;22:1417–1423.

15. Lu JY, Hu J, Hofmann SL. Human recombinant palmitoyl-protein thioesterase-1 (PPT1) for preclinical evaluation of enzyme replacement therapy for infantile neuronal ceroid lipofuscinosis. *Mol Genet Metab*. 2010;99:374–378.
16. Hu J, Lu JY, Wong AM, et al. Intravenous high-dose enzyme replacement therapy with recombinant palmitoyl-protein thioesterase reduces visceral lysosomal storage and modestly prolongs survival in a preclinical mouse model of infantile neuronal ceroid lipofuscinosis. *Mol Genet Metab*. 2012;107:213–221.
17. Finn R, Kovacs AD, Pearce DA. Treatment of the Ppt1-/- mouse model of infantile neuronal ceroid lipofuscinosis with the *N*-methyl-D-aspartate (NMDA) receptor antagonist memantine. *J Child Neurol*. 2013;28:1159–1168.
18. Macauley SL, Pekny M, Sands MS. The role of attenuated astrocyte activation in infantile neuronal ceroid lipofuscinosis. *J Neurosci*. 2011;31:15575–15585.
19. Macauley SL, Roberts MS, Wong AM, et al. Synergistic effects of central nervous system-directed gene therapy and bone marrow transplantation in the murine model of infantile neuronal ceroid lipofuscinosis. *Ann Neurol*. 2012;71:797–804.
20. Roberts MS, Macauley SL, Wong AM, et al. Combination small molecule PPT1 mimetic and CNS-directed gene therapy as a treatment for infantile neuronal ceroid lipofuscinosis. *J Inherit Metab Dis*. 2012;35:847–857.
21. Wei H, Zhang Z, Saha A, et al. Disruption of adaptive energy metabolism and elevated ribosomal p-S6K1 levels contribute to INCL pathogenesis: partial rescue by resveratrol. *Hum Mol Genet*. 2011;20:1111–1121.
22. Tamaki SJ, Jacobs Y, Dohse M, et al. Neuroprotection of host cells by human central nervous system stem cells in a mouse model of infantile neuronal ceroid lipofuscinosis. *Cell Stem Cell*. 2009;5:310–319.
23. Griffey M, Bible E, Vogler C, et al. Adeno-associated virus 2-mediated gene therapy decreases autofluorescent storage material and increases brain mass in a murine model of infantile neuronal ceroid lipofuscinosis. *Neurobiol Dis*. 2004;16:360–369.
24. Griffey MA, Wozniak D, Wong M, et al. CNS-directed AAV2-mediated gene therapy ameliorates functional deficits in a murine model of infantile neuronal ceroid lipofuscinosis. *Mol Ther*. 2006;13:538–547.
25. Griffey M, Macauley SL, Ogilvie JM, Sands MS. AAV2-mediated ocular gene therapy for infantile neuronal ceroid lipofuscinosis. *Mol Ther*. 2005;12:413–421.
26. Meng Y, Sohar I, Wang L, Sleat DE, Lobel P. Systemic administration of tripeptidyl peptidase I in a mouse model of late infantile neuronal ceroid lipofuscinosis: effect of glycan modification. *PLoS One*. 2012;7:e40509.
27. Xu S, Wang L, El-Banna M, Sohar I, Sleat DE, Lobel P. Large-volume intrathecal enzyme delivery increases survival of a mouse model of late infantile neuronal ceroid lipofuscinosis. *Mol Ther*. 2011;19:1842–1848.
28. Sondhi D, Hackett NR, Peterson DA, et al. Enhanced survival of the LINCL mouse following CLN2 gene transfer using the rh.10 rhesus macaque-derived adeno-associated virus vector. *Mol Ther*. 2007;15:481–491.
29. Sondhi D, Peterson DA, Edelstein AM, del Fierro K, Hackett NR, Crystal RG. Survival advantage of neonatal CNS gene transfer for late infantile neuronal ceroid lipofuscinosis. *Exp Neurol*. 2008;213:18–27.
30. Chang M, Cooper JD, Sleat DE, et al. Intraventricular enzyme replacement improves disease phenotypes in a mouse model of late infantile neuronal ceroid lipofuscinosis. *Mol Ther*. 2008;16:649–656.
31. Cabrera-Salazar MA, Roskelley EM, Bu J, et al. Timing of therapeutic intervention determines functional and survival outcomes in a mouse model of late infantile Batten disease. *Mol Ther*. 2007;15:1782–1788.
32. Passini MA, Dodge JC, Bu J, et al. Intracranial delivery of CLN2 reduces brain pathology in a mouse model of classical late infantile neuronal ceroid lipofuscinosis. *J Neurosci*. 2006;26:1334–1342.
33. Haskell RE, Hughes SM, Chiorini JA, Alisky JM, Davidson BL. Viral-mediated delivery of the late-infantile neuronal ceroid lipofuscinosis gene, TPP-I to the mouse central nervous system. *Gene Ther*. 2003;10:34–42.
34. Vuillemenot BR, Katz ML, Coates JR, et al. Intrathecal tripeptidyl-peptidase 1 reduces lysosomal storage in a canine model of late infantile neuronal ceroid lipofuscinosis. *Mol Genet Metab*. 2011;104:325–337.
35. Finn R, Kovacs AD, Pearce DA. Altered sensitivity of cerebellar granule cells to glutamate receptor overactivation in the Cln3(Deltaex7/8)-knock-in mouse model of juvenile neuronal ceroid lipofuscinosis. *Neurochem Int*. 2011;58:648–655.
36. Kovacs AD, Saje A, Wong A, Ramji S, Cooper JD, Pearce DA. Age-dependent therapeutic effect of memantine in a mouse model of juvenile Batten disease. *Neuropharmacology*. 2012;63:769–775.
37. Kovacs AD, Saje A, Wong A, et al. Temporary inhibition of AMPA receptors induces a prolonged improvement of motor performance in a mouse model of juvenile Batten disease. *Neuropharmacology*. 2011;60:405–409.
38. Seehafer SS, Pearce DA. Spectral properties and mechanisms that underlie autofluorescent accumulations in Batten disease. *Biochem Biophys Res Commun*. 2009.
39. Westlake VJ, Jolly RD, Jones BR, et al. Hematopoietic cell transplantation in fetal lambs with ceroid-lipofuscinosis. *Am J Med Genet*. 1995;57:365–368.
40. Elger B, Schneider M, Winter E, et al. Optimized synthesis of AMPA receptor antagonist ZK 187638 and neurobehavioral activity in a mouse model of neuronal ceroid lipofuscinosis. *ChemMedChem*. 2006;1:1142–1148.
41. Zeman RJ, Peng H, Etlinger JD. Clenbuterol retards loss of motor function in motor neuron degeneration mice. *Exp Neurol*. 2004;187:460–467.
42. Katz ML, Rice LM, Gao C-L. Dietary carnitine supplements slow disease progression in a putative mouse model for hereditary ceroid-lipofuscinosis. *J Neurosci Res*. 1997;50:123–132.
43. Siakotos AN, Hutchins GD, Farlow MR, Katz ML. Assessment of dietary therapies in a canine model of Batten disease. *Eur J Paediatr Neurol*. 2001;5(suppl A):151–156.
44. Deeg HJ, Shulman HM, Albrechtsen D, Graham T, Storb R, Koppang N. Batten's disease: failure of allogeneic bone marrow transplantation to arrest disease progression in a canine model. *Clin Genet*. 1990;37:264–270.
45. Pike LS, Tannous BA, Deliolanis NC, et al. Imaging gene delivery in a mouse model of congenital neuronal ceroid lipofuscinosis. *Gene Ther*. 2011;18:1173–1178.
46. Filiano AJ, Martens LH, Young AH, et al. Dissociation of frontotemporal dementia-related deficits and neuroinflammation in progranulin haploinsufficient mice. *J Neurosci*. 2013;33:5352–5361.

47. Bond M, Holthaus SM, Tammen I, Tear G, Russell C. Use of model organisms for the study of neuronal ceroid lipofuscinosis. *Biochim Biophys Acta*. 2013;1832(11):1842–1865.
48. Hawkins-Salsbury JA, Cooper JD, Sands MS. Pathogenesis and therapies for infantile neuronal ceroid lipofuscinosis (infantile CLN1 disease). *Biochim Biophys Acta*. 1832;2013:1906–1909.
49. Levin SW, Baker EH, Zein WM, et al. Oral cysteamine bitartrate and N-acetylcysteine for patients with infantile neuronal ceroid lipofuscinosis: a pilot study. *Lancet Neurol*. 2014;13:777–787.
50. Selden NR, Al-Uzri A, Huhn SL, et al. Central nervous system stem cell transplantation for children with neuronal ceroid lipofuscinosis. *J Neurosurg Pediatr*. 2013;11:643–652.
51. Worgall S, Sondhi D, Hackett NR, et al. Treatment of late infantile neuronal ceroid lipofuscinosis by CNS administration of a serotype 2 adeno-associated virus expressing CLN2 cDNA. *Hum Gene Ther*. 2008;19:463–474.
52. Damme M, Brandenstein L, Fehr S, et al. Gene disruption of Mfsd8 in mice provides the first animal model for CLN7 disease. *Neurobiol Dis*. 2014;65:12–24.

SECTION V

MOVEMENT DISORDERS

CHAPTER

71

The Inherited Ataxias

Roger N. Rosenberg and Pravin Khemani

The University of Texas Southwestern Medical Center, Dallas, TX, USA

INTRODUCTION

The clinical manifestations and neuropathologic findings of cerebellar disease typically dominate the clinical picture. Depending on the type of ataxia, however, there may also be characteristic changes in the basal ganglia, brainstem, spinal cord, optic nerves, retina, and peripheral nerves. In large families with dominantly inherited disease, clinical features can range from purely cerebellar manifestations to mixed cerebellar and brainstem disorders, cerebellar and basal ganglia syndromes, and spinal cord or peripheral nerve disease. The clinical picture may be consistent within a family with dominantly inherited ataxia, but sometimes many affected family members show one characteristic syndrome, whereas others in the same family manifest a much different phenotype.[1]

To date, many genetically identified dominantly inherited spinocerebellar ataxias (SCAs) and the most common recessively inherited ataxia, Friedreich ataxia (FA), have proved to be caused by dynamic expansions of simple sequence repeats in specific genes.[2] The most common SCAs are caused by expanded CAG repeats that encode a glutamine repeat, or polyglutamine domain, in the disease protein. These polyglutamine ataxias include SCA1,[3,4] SCA2, SCA3 (also known as Machado–Joseph disease),[5,6] SCA6, SCA7, and SCA17. Another genetically identified SCA, SCA8, appears to be caused by a trinucleotide repeat that is transcribed in both directions, yielding an untranslated CTG repeat in one direction and a glutamine-encoding CAG repeat in the other direction. Two other SCAs are caused by dynamic repeat expansions: SCA12 is caused by an untranslated CAG repeat,[7] and SCA10 is caused by an untranslated pentanucleotide repeat.[8]

The clinical phenotypes of the SCAs overlap so one cannot reliably diagnose a particular SCA on clinical grounds alone. A single familial phenotype can be caused by several different genotypes; conversely, a single genotype can manifest as several different familial phenotypes. This phenotypic complexity, observed in so many SCAs, is explained by the dynamic nature of expanded repeats and the tendency for disease features to vary with the size of the expansion.

Although expanded repeats are clearly very important in the dominantly inherited ataxias, the past few years have witnessed the discovery of nonrepeat mutations as the cause of several SCAs and recessive ataxias. For example, SCAs 5,[9] 13,[10] 14, and 27[11] are caused by a variety of mutations in the respective genes. In addition, recent discoveries in the recessively inherited ataxias have widened the range of diseases that the clinician needs to consider as potential diagnoses in sporadic and early-onset cases of progressive ataxias. In particular, mutations that cause two recessive forms of ataxia with ocular apraxia, ataxia with oculomotor apraxia types 1 and 2 (AOA1 and AOA2), have been discovered.[12] These and other recently discovered recessive ataxias highlight an important, if still unexplained role, for DNA repair machinery in normal cerebellar integrity.[13]

Rapidly accumulating knowledge of the genetic basis of many ataxias has settled the formerly vexing issue of classification, and genetic testing has simplified the route to diagnosis in many cases. Here we review the clinical and molecular genetic features of inherited ataxias, discuss current views on mechanisms of pathogenesis with a focus on the expanded repeat SCAs, and describe potential therapeutic approaches to these currently untreatable diseases. We note that not all genetic syndromes manifesting ataxia are included; rather, we focus on classes of hereditary disorders in which ataxia is the primary clinical feature. A classification of the inherited ataxias based on genotype is emphasized in earlier editions of this book.[1]

Rosenberg's Molecular and Genetic Basis of Neurological and Psychiatric Disease
http://dx.doi.org/10.1016/10.1016/B978-0-12-410529-4.00071-1

AUTOSOMAL DOMINANT ATAXIAS

Previously, descriptive neuropathologic terms such as *olivopontocerebellar atrophy* and *cerebellar cortical atrophy* were applied to the class of dominantly inherited ataxias now known as SCAs. The recent explosion of genetic data has revealed that these descriptive syndromes encompass many different genetic disorders with overlapping clinical features. Each of the 36 SCAs described here maps to a different chromosomal region (these loci are indicated in Table 71.1). By the time this book is published, very likely even more will be defined at the mutation level. As presented in Table 71.1, the genetic classification of dominant ataxias currently includes SCA types 1 through 36, as well as dentatorubropallidoluysian atrophy (DRPLA) and several episodic ataxias.[1]

SCA1

SCA1 was the first genetic locus identified for a dominantly inherited ataxia. Although SCA1 may not be as common as several other SCAs, studies of this disease continue to lead the way in our understanding of the entire class of polyglutamine neurodegenerative diseases to which it belongs. Because it is the first and best characterized SCA, we devote more space to it than the other SCAs.

Symptoms and Signs

The clinical syndrome of SCA1, and indeed many other SCAs, is typically characterized by the development in adults of slowly progressive cerebellar ataxia of the trunk and limbs, impairment of equilibrium and gait, slowness of voluntary movements, scanning speech, nystagmus, and tremor of the head and trunk. Dysarthria and dysphagia usually develop in most patients during the course of the illness, and ophthalmoparesis may also occur. Less common extrapyramidal symptoms include rigidity, immobile facies, and parkinsonian tremor. Parkinsonian signs are rather rare in this disease, unlike in SCA3/Machado–Joseph disease. The reflexes are usually normal, but knee and ankle jerks may be lost, and extensor plantar responses may occur. Dementia may occur but is usually mild. Impairment of sphincter function is common late in disease. Cerebellar and brainstem atrophy are evident on magnetic resonance imaging (MRI) in affected individuals (Figure 71.1).

Although the range of neurodegeneration in SCA1 varies, it usually includes prominent cerebellar and brainstem atrophy. Marked shrinkage of the ventral pons, disappearance of the olivary eminence on the ventral surface of the medulla, and atrophy of the cerebellum are evident on postmortem inspection of the brain. Variable loss of Purkinje cells, reduction in the number of cells in the molecular and granular layer, demyelination of the middle cerebellar peduncle and the cerebellar hemispheres, and severe loss of cells in the pontine nuclei and olives are found on histologic examination. Degenerative changes in the striatum, especially the putamen, and loss of the pigmented cells of the substantia nigra may be found in cases with extrapyramidal features. More widespread degeneration in the central nervous system, including involvement of the posterior columns and the spinocerebellar fibers, is often present, but the cerebral hemispheres are typically spared.

Genetics

SCA1 was the first genetically defined SCA. In 1993, Zoghbi et al.[14] determined that SCA1 is due to a CAG repeat expansion in the *SCA1(6p22-p23) gene*. The mutant allele has a CAG repeat of 39 or more repeats in length, whereas alleles from control subjects have 6–44 repeats. Disease-causing expanded repeats are pure, uninterrupted CAG repeats that encode glutamine, whereas normal alleles greater than 20 repeats in length are interrupted by one or more CAT repeat units that encode histidine instead of glutamine. As in many expanded repeat ataxias, there is a strong inverse correlation between repeat size and age of symptom onset: increasingly large repeat sizes lead to a younger age of onset for SCA1, with juvenile-onset cases having the highest repeat numbers. *Anticipation*, the tendency for disease to occur earlier in successive generations, is common in SCA1 families due to further expansion of disease repeats on transmission.

The CAG repeat lies in the protein-coding portion of the *ATXN1* gene, in-frame to encode a pure stretch of glutamine in the disease protein—hence the term *glutamine-repeat* or *polyglutamine* diseases. The SCA1 disease protein, ataxin-1, is a novel protein that regulates transcriptional repression through its association with various nuclear factors.[2] It can shuttle in and out of the cell nucleus but becomes concentrated in the nucleus in diseased neurons. As a protein that can bind RNA, ataxin-1 may also regulate gene expression posttranscriptionally. Ubiquitinated, intranuclear inclusions containing the disease protein have been described in neurons of select brain regions, including the pons and dentate nucleus of the cerebellum.[4]

TABLE 71.1 Genotype Classification of the Spinocerebellar Ataxias

Ataxia	Chromosome location and mutation	Clinical description
SCA1 (autosomal dominant type 1)	6p22-p23 with CAG repeats (exonic); leucine-rich acidic nuclear protein (LANP), region-specific interaction protein Ataxin-1	Ataxia with ophthalmoparesis, pyramidal and extrapyramidal findings; genetic testing is available; 6% of all autosomal dominant (AD) cerebellar ataxia
SCA2 (autosomal dominant type 2)	12q23-q24.1 with CAG repeats (exonic) Ataxin-2	Ataxia with slow saccades and minimal pyramidal and extrapyramidal findings; genetic testing available; 13% of all AD cerebellar ataxia
Machado–Joseph disease/SCA3 (autosomal dominant type 3)	14q24.3-q32 with CAG repeats (exonic); codes for ubiquitin protease (inactive with polyglutamine expansion); altered turnover of cellular proteins due to proteosome dysfunction MJD–ataxin-3	Ataxia with ophthalmoparesis and variable pyramidal, extrapyramidal, and amyotrophic signs; dementia (mild); 23% of all AD cerebellar ataxia; genetic testing available
SCA4 (autosomal dominant type 4)	16q22.1-ter; pleckstrin homology domain-containing protein, family G, member 4; (PLEKHG4; puratrophin-1: Purkinje cell atrophy associated protein-1, including spectrin repeat and the guanine-nucleotide exchange factor, GEF for Rho GTPases)	Ataxia with normal eye movements, sensory axonal neuropathy, and pyramidal signs; genetic testing available
SCA5 (autosomal dominant type 5)	11p12-q12; β-III spectrin mutations; (SPTBN2); stabilizes glutamate transporter EAAT4; descendants of President Abraham Lincoln	Ataxia and dysarthria; genetic testing available
SCA6 (autosomal dominant type 6)	19p13.2 with CAG repeats in –1A-voltage-dependent calcium channel gene (exonic); CACNA1A protein, P/Q type calcium channel subunit	Ataxia and dysarthria, nystagmus, mild proprioceptive sensory loss; genetic testing available
SCA7 (autosomal dominant type 7)	3p14.1-p21.1 with CAG repeats (exonic); ataxin-7; subunit of GCN5, histone acetyltransferase-containing complexes; ataxin-7 binding protein; Cbl-associated protein (CAP; SH3D5)	Ophthalmoparesis, visual loss, ataxia, dysarthria, extensor plantar response, pigmentary retinal degeneration; genetic testing available
SCA8 (autosomal dominant type 8)	13q21 with CTG repeats; noncoding; 3′ untranslated region of transcribed RNA; KLHL1AS	Gait ataxia, dysarthria, nystagmus, leg spasticity, and reduced vibratory sensation; genetic testing available
SCA10 (autosomal dominant type 10)	22q13; pentanucleotide repeat ATTCT repeat; noncoding, intron 9	Gait ataxia, dysarthria, nystagmus; partial complex and generalized motor seizures; polyneuropathy; genetic testing available
SCA11 (autosomal dominant type 11)	15q14-q21.3 by linkage	Slowly progressive gait and extremity ataxia, dysarthria, vertical nystagmus, hyperreflexia
SCA12 (autosomal dominant type 12)	5q31-q33 by linkage; CAG repeat; protein phosphatase 2A, regulatory subunit B, (PPP2R2B); protein PP2A, serine/threonine phosphatase	Tremor, decreased movement, increased reflexes, dystonia, ataxia, dysautonomia, dementia, dysarthria; genetic testing available
SCA13 (autosomal dominant type 13)	19q13.3-q14.4; potassium channel voltage-gated; KCNC3	Ataxia, legs>arms; dysarthria, horizontal nystagmus; delayed motor development; mental developmental delay; tendon reflexes increased; MRI: cerebellar and pontine atrophy; genetic testing available
SCA14 (autosomal dominant type 14)	19q-13.4; protein kinase Cγ (PRKCG), missense mutations including in-frame deletion and a splice site mutation among others; serine/threonine kinase	Gait ataxia; leg>arm ataxia; dysarthria; pure ataxia with later onset; myoclonus; tremor of head and extremities; increased deep tendon reflexes at ankles; occasional dystonia and sensory neuropathy; genetic testing available

Continued

TABLE 71.1 Genotype Classification of the Spinocerebellar Ataxias—cont'd

Ataxia	Chromosome location and mutation	Clinical description
SCA15 (autosomal dominant type 15)	3p24.2-3pter; inositol 1, 4, 5- triphosphate receptor type 1 (ITPRI)	Gait and extremity ataxia, dysarthria; nystagmus; MRI: superior vermis atrophy; sparing of hemispheres and tonsils
SCA16 (autosomal dominant type 16)	8q22.1-24.1	Pure cerebellar ataxia and head tremor, gait ataxia, and dysarthria; horizontal gaze-evoked nystagmus; MRI: cerebellar atrophy; no brainstem changes
SCA17 (autosomal dominant type 17)	6q27; CAG expansion in the TATA-binding protein (*TBP*) gene	Gait ataxia, dementia, parkinsonism, dystonia, chorea, seizures; hyperreflexia; dysarthria and dysphagia; MRI shows cerebral and cerebellar atrophy; genetic testing available
SCA18 (autosomal dominant type 18)	7q22-q32	Ataxia; motor/sensory neuropathy; head tremor; dysarthria; extensor plantar responses in some patients; sensory axonal neuropathy; EMG denervation; MRI: cerebellar atrophy
SCA19 (autosomal dominant type 19)	1p21-q21; KCND3; missense mutations; T352P; M373I; S390N; allelic with SCA22; overlaps with the locus of SCA22	Ataxia, tremor, cognitive impairment, myoclonus; MRI: atrophy of cerebellum
SCA20 (autosomal dominant)	11p13-q11; 260-kb duplication	Dysarthria; gait ataxia; ocular gaze-evoked saccades; palatal tremor; dentate calcification on CT; MRI: cerebral atrophy
SCA21 (autosomal dominant)	7p21.3-p15.1	Ataxia, dysarthria, extrapyramidal features of akinesia, rigidity, tremor, cognitive defect; reduced deep tendon reflexes; MRI: cerebellar atrophy, normal basal ganglia and brainstem
SCA22 (autosomal dominant)	1p21-q23; deletion (in frame); V338E; G345V; T377M; allelic with SCA19; KCND3; Kv4.3 channels	Pure cerebellar ataxia; dysarthria; dysphagia; nystagmus; MRI: cerebellar atrophy
SCA23 (autosomal dominant)	20p13-12.3; prodynorphin (PDYN protein); missense R138S; L211S; R212W; R215C	Gait ataxia; dysarthria; extremity ataxia; ocular nystagmus, dysmetria; leg vibration loss; extensor plantar responses; MRI: cerebellar atrophy
SCA25 (autosomal dominant)	2p15-p21	Ataxia, nystagmus; vibratory loss in the feet; pain loss in some; abdominal pain; nausea and vomiting may be prominent; absent ankle reflexes; sensory nerve action potentials are absent; MRI: cerebellar atrophy, normal brainstem
SCA26 (autosomal dominant)	19p13.3	Gait ataxia; extremity ataxia; dysarthria; nystagmus; MRI: cerebellar atrophy
SCA27 (autosomal dominant)	13q34; fibroblast growth factor 14 protein; mutation F145S; produces reduced protein stability	Tremor extremities and head and orofacial dyskinesia; ataxia of arms > legs, gait ataxia; dysarthria; nystagmus; psychiatric symptoms; cognitive defect; MRI: cerebellar atrophy; genetic testing available
SCA28 (autosomal dominant)	18p11.22-q11.2; ATPase family gene 3- like 2 (AFG3L2 protein) mutations: N432T; S674L; E691K; A694E; R702Q	Extremity and gait ataxia; dysarthria; nystagmus; ophthalmoparesis; leg hyperreflexia and extensor plantar responses; MRI: cerebellar atrophy
SCA30 (autosomal dominant)	4q34.3-q35.1; candidate gene *ODZ3*	Candidate gene *ODZ3*; gait ataxia, dysarthria, saccades; nystagmus, brisk tendon reflexes in legs; MRI: cerebellar atrophy
SCA31 (autosomal dominant)	16q22.1; associated with NEDD4 (BEAN)	Pentanucleotide (TGGAA)n repeat insertions; previously called SCA4; gait ataxia; limb dysmetria; MRI: cerebellar atrophy
SCA32 (autosomal dominant)	7q32-q33	Ataxia, azospermia, mental retardation; absent germ cells on testicular biopsy

TABLE 71.1 Genotype Classification of the Spinocerebellar Ataxias—cont'd

Ataxia	Chromosome location and mutation	Clinical description
SCA35 (autosomal dominant)	20p13;TGM6 protein; transglutaminase 6	Ataxia; ocular dysmetria; upper motor neuron signs; extensor plantars; onset 5th decade
SCA36 (autosomal dominant)	20p13; large intronic expansion of GGCCTG (1500-2500); also phe265leu mutation; RNA gain of function; microRNA; MIR 1292 suppression	Ataxia; onset 5th–6th decades; motor neuron disorder; grouped atrophy (muscle biopsy) fasciculations; increased reflexes; flexor plantars
Prion disease (autosomal dominant)	20p13; Pro102Leu; Ala117Val mutations; proteinase k resistant form PrP27-30 accumulates in brain; eponym: Gerstmann–Straüssler–Scheinker disease	Ataxia; dementia 3rd–7th decades
	Glu200Lys mutation; increased octapeptide repeats; eponym: Creutzfeldt–Jakob disease	Ataxia; dementia; rigidity
Multiple hamartoma syndrome (autosomal dominant)	10q23.31; phosphatase and tensin homolog (PTEN); Cowden; Lhermitte–Duclos syndrome	Skin hamartomas; ataxia; mental retardation; increased intracranial pressure; epilepsy
Cerebellar ataxia, deafness, and narcolepsy (autosomal dominant)	19p13.2; exon 21; missense Ala570Val; Val606Phe mutations	Ataxia; deafness; narcolepsy cataplexy; REM sleep disorder
Cerebellar ataxia (non-progressive) mental retardation (autosomal dominant)	1p36.31-p36.23	Ataxia, mental retardation
Familial dementia with amyloid angiopathy and spastic ataxia (autosomal dominant)	13q14.2; integral membrane protein 2B (ITM2B)	Ataxia; dementia; amyloid angiopathy
Dentatorubropallidoluysian atrophy (autosomal dominant)	12p13.31 with CAG repeats (exonic) Atrophin 1	Ataxia, choreoathetosis, dystonia, seizures, myoclonus, dementia; genetic testing available
Friedreich ataxia (autosomal recessive)	9q13-q21.1 with intronic GAA repeats, in intron at end of exon 1 Frataxin defective; abnormal regulation of mitochondrial iron metabolism; iron accumulates in mitochondria in yeast mutants	Ataxia, areflexia, extensor plantar responses, position sense deficits, cardiomyopathy, diabetes mellitus, scoliosis, foot deformities; optic atrophy; late-onset form, as late as 50 years, with preserved deep tendon reflexes, slower progression, reduced skeletal deformities, associated with an intermediate number of GAA repeats and missense mutations in one allele of frataxin; genetic testing available
Vitamin E deficiency syndrome (autosomal recessive)	8q13.1-q13.3 (α-TTP deficiency)	Same as phenotype that maps to 9q but associated with vitamin E deficiency; genetic testing available
Sensory ataxic neuropathy and ophthalmoparesis (SANDO) with dysarthria (autosomal recessive)	15q25; mutations in DNA polymerase-gamma (*POLG*) gene that leads to mtDNA deletions	Young adult-onset ataxia, sensory neuropathy, ophthalmoparesis, hearing loss, gastric symptoms; a variant of progressive external ophthalmoplegia; MRI: cerebellar and thalamic abnormalities; mildly increased lactate and creatine kinase
von Hippel–Lindau syndrome (autosomal dominant)	3p26-p25	Cerebellar hemangioblastoma; pheochromocytoma
Baltic myoclonus (Unverricht–Lundborg) (recessive)	21q22.3; cystatin B; extra repeats of 12-bp tandem repeats	Myoclonus epilepsy; late-onset ataxia; responds to valproic acid, clonazepam; phenobarbital
Marinesco–Sjögren syndrome (recessive)	5q31; SIL 1 protein, nucleotide exchange factor for the heat-shock protein 70 (HSP70); chaperone HSPA5; homozygous 4-nucleotide duplication in exon 6; also compound heterozygote	Ataxia, dysarthria; nystagmus; retarded motor and mental maturation; rhabdomyolysis after viral illness; weakness; hypotonia; areflexia; cataracts in childhood; short stature; kyphoscoliosis; contractures; hypogonadism
Autosomal recessive spastic ataxia of Charlevoix–Saguenay (ARSACS)	Chromosome 13q12; SACS gene; loss of sacsin peptide activity	Childhood onset of ataxia, spasticity, dysarthria, distal muscle wasting, foot deformity, retinal striations, mitral valve prolapse

Continued

TABLE 71.1 Genotype Classification of the Spinocerebellar Ataxias—cont'd

Ataxia	Chromosome location and mutation	Clinical description
Kearns–Sayre syndrome (sporadic)	mtDNA deletion and duplication mutations	Ptosis, ophthalmoplegia, pigmentary retinal degeneration, cardiomyopathy, diabetes mellitus, deafness, heart block, increased CSF protein, ataxia
Myoclonic epilepsy and ragged red fiber syndrome (MERRF) (maternal inheritance)	Mutation in mtDNA of the $tRNA^{lys}$ at 8344; also mutation at 8356	Myoclonic epilepsy, ragged red fiber myopathy, ataxia
Mitochondrial encephalopathy, lactic acidosis, and stroke syndrome (MELAS) (maternal inheritance)	$tRNA^{leu}$ mutation at 3243; also at 3271 and 3252	Headache, stroke, lactic acidosis, ataxia
Neuropathy; ataxia; retinitis pigmentosa (NARP)	ATPase6 (Complex 5); mtDNA point mutation at 8993	Neuropathy; ataxia; retinitis pigmentosa; dementia; seizures
Episodic ataxia, type 1 (EA-1) (autosomal dominant)	12p13; potassium voltage-gated channel gene, *KCNA1*; Phe249Leu mutation; variable syndrome	Episodic ataxia for minutes; provoked by startle or exercise; with facial and hand myokymia; cerebellar signs are not progressive; choreoathetotic movements; responds to phenytoin; genetic testing available
Episodic ataxia, type 2 (EA-2) (autosomal dominant)	19p-13(*CACNA1A*) (allelic with SCA6) (α_{1A}-voltage-dependent calcium channel subunit); point mutations or small deletions; allelic with SCA6 and familial hemiplegic migraine	Episodic ataxia for days; provoked by stress, fatigue; with down-gaze nystagmus; vertigo; vomiting; headache; cerebellar atrophy results; progressive cerebellar signs; responds to acetazolamide; genetic testing available
Episodic ataxia, type 3 (autosomal dominant)	1q42	Episodic ataxia; 1 min to over 6 h; induced by movement; vertigo and tinnitus; headache; responds to acetazolamide
Episodic ataxia, type 4 (autosomal dominant)	Not mapped	Episodic ataxia; vertigo; diplopia; ocular slow-pursuit defect; no response to acetazolamide
Episodic ataxia, type 5 (autosomal dominant)	2q22-q23; CACNB4β4 protein	Episodic ataxia; hours to weeks; seizures
Episodic ataxia type 6 with seizures, migraine, and alternating hemiplegia (autosomal dominant)	SLC1A3; 5p13; EAAT1 protein; missense mutations; glial glutamate transporter (GLAST); 1047 C to G; proline to arginine	Ataxia, duration 2–4 days; episodic hypotonia; delayed motor milestones; seizures; migraine; alternating hemiplegia; mild truncal ataxia; coma; febrile illness as a trigger; MRI: cerebellar atrophy
Episodic ataxia, type 7 (autosomal dominant)	19q13	Episodic ataxia; vertigo, weakness; less than 24 h
Episodic ataxia with paroxysmal choreoathetosis and spasticity (dystonia-9) (DYT9; CSE) (autosomal dominant)	1p	Ataxia; involuntary movements; dystonia; headache; spastic paraplegia; responds on occasion to acetazolamide
Fragile X tremor/ataxia syndrome (FXTAS) X-linked dominant	Xq27.3; CGG premutation expansion in FMR1 gene; expansions of 55–200 repeats in 5′ UTR of the FMR-1 mRNA; presumed dominant toxic RNA effect	Late-onset ataxia with tremor, cognitive impairment, occasional parkinsonism; males typically affected, although affected females also reported; syndrome is of high concern if affected male has grandson with mental retardation (fragile X syndrome); MRI shows increased T2 signal in middle cerebellar peduncles, cerebellar atrophy, and occasional widespread brain atrophy; genetic testing available
Ataxia telangiectasia (autosomal recessive)	11q22-23; *ATM* gene for regulation of cell cycle; mitogenic signal transduction and meiotic recombination; elevated serum alpha-fetoprotein level; immunoglobulin deficiency	Telangiectasia, ataxia, dysarthria, pulmonary infections, neoplasms of lymphatic system; IgA and IgG deficiencies; diabetes mellitus, breast cancer; genetic testing available; chorea; dystonia
Early-onset cerebellar ataxia with retained deep tendon reflexes (autosomal recessive)	13q11-12	Ataxia; neuropathy; preserved deep tendon reflexes; impaired cognitive and visuospatial functions; MRI: cerebellar atrophy

TABLE 71.1 Genotype Classification of the Spinocerebellar Ataxias—cont'd

Ataxia	Chromosome location and mutation	Clinical description
Ataxia with oculomotor apraxia (AOA1) (autosomal recessive)	9p21; protein is member of histidine triad superfamily, role in DNA repair; elevation of serum LDL cholesterol and low serum albumin level; APTX, aprataxin	Ataxia; dysarthria; limb dysmetria; dystonia; oculomotor apraxia; optic atrophy; motor neuropathy; late sensory loss (vibration); genetic testing available
Ataxia with oculomotor apraxia 2 (AOA2) (autosomal recessive)	9q34; senataxin protein, involved in RNA maturation and termination; helicase superfamily 1; elevated serum alpha-fetoprotein level; SETX, senataxin	Gait ataxia; choreoathetosis; dystonia; oculomotor apraxia; neuropathy, vibration loss, position sense loss, and mild light touch loss; absent leg deep tendon reflexes; extensor plantar response; genetic testing available
Cerebellar ataxia with muscle coenzyme Q10 deficiency (autosomal recessive)	9p13	Ataxia; hypotonia; seizures; mental retardation; increased deep tendon reflexes; extensor plantar responses; coenzyme Q10 levels reduced with about 25% of patients with a block in transfer of electrons to complex 3; may respond to coenzyme 10
Refsum's disease (autosomal recessive)	10pter; elevated serum phytanic acid level; phytanoyl-COH hydroxylase and PEX7	Retinitis pigmentosa; ataxia; sensorineural deafness; demyelinating neuropathy
Cerebrotendinous xanthomatosis (autosomal recessive)	2p33; elevated cholesterol level; CYP27; sterol 27 hydroxylase	Spastic ataxia; mental retardation; dementia; tendon xanthomas; diarrhea; cataracts
Joubert syndrome (autosomal recessive)	9q34.3	Ataxia; ptosis; mental retardation; oculomotor apraxia; nystagmus; retinopathy; rhythmic tongue protrusion; episodic hyperpnea or apnea; dimples at wrists and elbows; telecanthus; micrognathia
Sideroblastic anemia and spinocerebellar ataxia (X-linked recessive)	Xq13; ATP-binding cassette 7 (ABCB7; ABC7) transporter; mitochondrial inner membrane; iron homeostasis; export from matrix to the intermembrane space	Ataxia; elevated free erythrocyte protoporphyrin levels; ring sideroblasts in bone marrow; heterozygous females may have mild anemia but not ataxia
Infantile-onset spinocerebellar ataxia of Nikali et al. (autosomal recessive)	10q23.3-q24.1; twinkle protein (gene); homozygous for Tyr508Cys missense mutations	Infantile ataxia, sensory neuropathy; athetosis, hearing deficit, reduced deep tendon reflexes; ophthalmoplegia, optic atrophy; seizures; primary hypogonadism in females
Hypoceruloplasminemia with ataxia and dysarthria (autosomal recessive)	Ceruloplasmin gene; 3q23-q25 (trp 858 ter)	Gait ataxia and dysarthria; hyperreflexia; cerebellar atrophy by MRI; iron deposition in cerebellum, basal ganglia, thalamus, and liver; onset in the 4th decade
Spinocerebellar ataxia with neuropathy (SCAN1) (autosomal recessive)	Tyrosyl-DNA phosphodiesterase-1 (TDP-1) 14q31-q32	Onset in 2nd decade; gait ataxia, dysarthria, seizures, cerebellar vermis atrophy on MRI, dysmetria
Cerebellar ataxia type 1 (autosomal recessive)	6p25 SCAR8; SYNE1; spectrin repeats-nuclear envelope 1	Pure ataxia
Cerebellar ataxia type 2 (autosomal recessive)	1q42; ADCK3 (CABC1); aarf-domain containing kinase 3; elevation of serum lactate and decreased coenzyme Q10 level	Ataxia; mental retardation; myoclonus; epilepsy; exercise intolerance; stroke or transient ischemic-like episodes
Niemann–Pick type C disease	18q11; NPCI; NPCH1 and 2; skin biopsy (filipin staining)	Ataxia; vertical supra-nuclear ophthalmoplegia splenomegaly; dystonia; impaired cognition

Abbreviations: CSF, cerebrospinal fluid; CT, computed tomography; EMG, electromyogram; MRI, magnetic resonance imaging.

Modified from Rosenberg et al., eds. The Molecular and Genetic Basis of Neurologic and Psychiatric Disease. 4th edition. Philadelphia, PA: Lippincott, Williams & Wilkins, 2007.

Studies have shed light on many molecular features of SCA1 and other polyglutamine diseases. As discussed later, pathogenesis in this and other polyglutamine diseases seems to be mediated by a novel toxic property of the expanded disease protein. In the case of ataxin-1, pathogenesis requires nuclear localization and is modulated by sequence elements in the protein outside of the polyglutamine domain. Although the mechanistic details are still being defined, increasing evidence implicates protein misfolding and transcriptional dysregulation in the pathogenesis of this and other polyglutamine diseases.

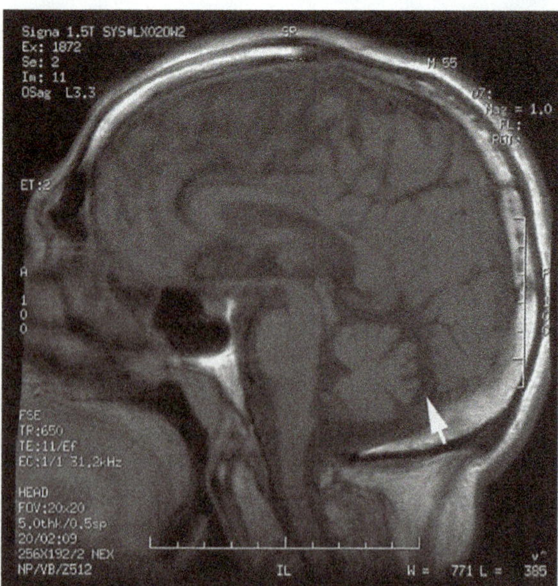

FIGURE 71.1 Sagittal MRI of the brain of a 60-year-old man with SCA1 showing cerebellar and brainstem atrophy.

SCA2

A second ataxic disorder, SCA2, was initially described in a large Cuban population that may represent the largest homogeneous group of ataxic patients described. Since the discovery of its genetic defect, SCA2 has been found to be one of the three most common forms of dominantly inherited ataxia in the United States.

The age of onset ranges from 2 to 65 years, and there is considerable clinical variability within various families (a common refrain in the SCAs). As with all SCAs, the core clinical features reflect cerebellar and brainstem degeneration. These include progressive gait and limb ataxia, dysarthria and, eventually, dysphagia. In SCA2, common additional features are remarkably slow saccades, prominent neuropathy, and occasional parkinsonism. Dementia is also relatively common.[1]

After the SCA2 locus was mapped to chromosome 12, three separate groups identified the disease gene. The mutation proved to be an expanded CAG repeat that encodes polyglutamine in the disease protein, ataxin-2. Normal alleles contain 15–32 repeats, whereas mutant alleles have 35–77 repeats. There is a zone of reduced penetrance (32–34 repeats), within which not all individuals develop signs of disease in a normal lifespan. Occasionally, sporadic cases of SCA2 are identified with no family history. In such cases, an intermediate-sized allele from an asymptomatic parent presumably underwent further expansion when transmitted to the affected individual.[1]

The precise function of ataxin-2 is not yet known. Ataxin-2 was recently shown to assemble with polyribosomes, suggesting that it may help regulate protein translation.

SCA3/Machado–Joseph Disease

SCA3, or Machado–Joseph disease (MJD) was first described among the Portuguese and their descendants in New England and California. Subsequently, SCA3/MJD has been found in families from Portugal, Australia, Brazil, Canada, China, England, France, India, Israel, Italy, Japan, Spain, Taiwan, and the United States.[5,6] In many populations, it is the most common autosomal dominant ataxia.[1]

The phenotypic spectrum in SCA3/MJD is remarkably broad, ranging from early-onset dystonia to late-onset ataxia with neuropathy. This heterogeneity reflects the variable size of the repeat mutation. The broad clinical spectrum has led some to classify SCA3/MJD in three major clinical types. In type 1 SCA3/MJD, neurologic deficits appear in the first two decades and involve weakness and spasticity of extremities, especially the legs, often with dystonia of the face, neck, trunk, and extremities. Hyperreflexia and ankle clonus are common, as are extensor plantar responses. In type 1 SCA3/MJD, the gait is more spastic than ataxic, though ataxia is still part of the clinical spectrum. Pharyngeal weakness and spasticity cause difficulty with speech and swallowing. Ophthalmoparesis, ocular prominence, and facial and lingual fasciculations are common manifestations of disease. In type 2 SCA3/MJD, the classic ataxic type which is the most common form of disease, signs and symptoms begin in the second to fifth decades. True cerebellar deficits predominate, including dysarthria and gait and limb ataxia, along with corticospinal and extrapyramidal signs such as spasticity, rigidity, and dystonia. Ophthalmoparesis, upward vertical gaze deficits, and facial and lingual fasciculations are also common in type 2 SCA3/MJD. Type 3 SCA3/MJD (ataxic-amyotrophic type) occurs in the fifth to the seventh decade with a cerebellar disorder that includes dysarthria and gait and limb ataxia. Distal sensory loss involving pain, touch, vibration, and position senses and distal atrophy are prominent, indicating the presence of peripheral neuropathy. The deep tendon reflexes are depressed to absent, and no corticospinal or extrapyramidal findings occur. Neurologic deficits progress and lead to death from debilitation within 15 years of onset, especially in patients with type 1 and 2 disease. Usually patients retain full intellectual function, but individuals with the earliest onset can suffer more widespread central nervous system dysfunction that reflects involvement of brain regions above the foramen magnum.

The major pathologic findings include variable loss of neurons and glial replacement in the corpus striatum and severe loss in the zona compacta portion of the substantia nigra. A moderate loss of neurons occurs in the dentate nucleus of the cerebellum, the red nucleus, and cranial nerve motor nuclei. Purkinje cell loss and granule cell loss are found in the cerebellar cortex. Relative sparing of the inferior olives distinguishes SCA3/MJD from several other dominantly inherited ataxias.[5,6]

Two groups identified the mutation in families with MJD as a CAG repeat.[15–17] The genetic defect in what had been thought to be a separate dominant ataxia (SCA3), was found to be the same mutation (14q24.3-q32).[16] Hence, SCA3 and MJD are genetically the same disorder, here called SCA3/MJD. The CAG repeat in normal control subjects is between 12 and 40 CAG repeats and is expanded in disease to between 55 and 84 CAG repeats, with most disease-causing repeats being greater than 60. The expanded CAG repeat encodes an expanded polyglutamine tract in the disease protein, ataxin-3, and earlier age of onset is associated with longer repeats. In diseased brain, aggregates of ataxin-3 have been observed in neuronal nuclei in the pons, midbrain, and several other regions undergoing neuropathologic degeneration.[1,18] The disease protein is a deubiquitinating enzyme that likely participates in protein quality-control pathways in the cell, including the ubiquitin–proteasome degradation pathway.[19–21] This biochemical function for ataxin-3 is intriguing in light of the fact that perturbations in protein quality-control may contribute to pathogenesis in the entire class of polyglutamine diseases. The normal function of ataxin-3 may even counter the toxicity of expanded polylgutamine proteins.

SCA4

A familial disorder characterized by progressive ataxia, pyramidal tract deficits, and prominent sensory axonal neuropathy has been mapped to chromosome 16q22.[1] A form of dominantly inherited ataxia also maps to 16q22 in Japanese families,[22] where it has been associated with a mutation in the upstream region of the *PLEKHG4* gene encoding puratrophin.[23] Whether these two represent the same genetic form of ataxia is currently not known.

SCA5

Another ataxic family was reported with two major branches, both descended from the paternal grandparents of Abraham Lincoln. Thus, this relatively pure form of slowly progressive, dominant cerebellar ataxia is sometimes referred to as the Lincoln family ataxia. The defective gene was recently discovered to be the *SPTBN2* gene encoding β-III spectrin (11p12).[24] How the various described nonrepeat mutations in β-III spectrin cause disease is uncertain, but spectrin proteins are implicated in stabilizing and facilitating communication among the extracellular matrix, plasma membrane, and underlying cytoskeleton. β-III spectrin normally stabilizes glutamate transporters at the Purkinje cell surface, suggesting that mutations in this protein could lead to dysregulation of glutamate signaling in the cerebellum.

SCA6

Along with SCA2 and SCA3/MJD, SCA6 represents one of the more common dominantly inherited ataxias. It is a milder, later-onset disease characterized by slowly progressive ataxia, dysarthria, nystagmus, and sensory loss. It is nearly always a pure cerebellar ataxia without prominent spinal cord, brainstem, or extrapyramidal involvement. In some cases, SCA6 occurs sporadically with no family history of similar disease. SCA6 is caused by a relatively small CAG repeat expansion (20–33 triplets in patients vs. 3–18 triplets in normal subjects, with a repeat of 19 being of questionable significance) in a voltage-dependent calcium channel subunit gene (*CACNA1A4*) (19p13.2). Other nonrepeat mutations in the same gene cause two additional neurologic disorders. Episodic ataxia type 2 (EA-2) is usually caused by nonsense mutations leading to truncation of the channel protein. A few missense mutations in *CACNA1A4* have also been reported in EA-2. More commonly, however, missense *CACNA1A4* mutations result in a different disease, familial hemiplegic migraine. Some patients with familial hemiplegic migraine also develop progressive ataxia with cerebellar atrophy.[25]

SCA7

SCA7, an autosomal dominant ataxia, is distinguished from all other SCAs by the presence of retinal pigmentary degeneration. The visual abnormalities first appear as blue–yellow color blindness and proceed to frank visual loss with macular degeneration. In almost all other respects, SCA7 resembles several other SCAs in which ataxia is accompanied by various noncerebellar findings, including ophthalmoparesis, visual loss, and extensor plantar responses.

The genetic defect is an expanded CAG repeat in the *ATXN7* gene (3p14.1-p21.1). Repeats on normal alleles contain 19 or fewer CAG repeats, while disease alleles range from 36 to 400 CAG repeats. The expanded repeat size in SCA7 is highly variable. Consistent with this, the severity of clinical findings varies from essentially asymptomatic to mild late-onset symptoms to severe, aggressive disease in childhood with rapid progression. Marked anticipation has been recorded, more so than in the other SCAs, especially with paternal transmission. The disease protein, ataxin-7, forms aggregates in nuclei of affected neurons, as has been described for the other polyglutamine SCAs (1, 2, 3, 6, and 17).[26–28] Ataxin-7 is a nuclear protein that participates in gene transcription.[29] Indeed, most polyglutamine disease proteins are directly or indirectly implicated in transcriptional regulation.

SCA8

SCA8 is caused by a CTG/CAG repeat expansion in an intriguing gene known as *ATXN8* (13q21). There is marked maternal bias in transmission, perhaps reflecting contractions of the repeat during spermatogenesis. Normal CTG alleles are between 15 and 50 repeats, whereas affected alleles are typically 80 repeats, and usually 100. The mutation is not fully penetrant, as there are some individuals with expansions, including very large expansions, who do not develop disease. Symptoms include dysarthria and gait and truncal ataxia, beginning at a mean age of about 40 years with a range of 20–65 years. Other features include nystagmus, leg spasticity, and reduced vibratory sensation. Patients tend to have slowly progressive disease; severely affected individuals are nonambulatory by their 40s to 60s. MRI shows cerebellar atrophy consistent with the clinical features.[30] The molecular mechanism in SCA8 remains uncertain. It was initially thought to involve a dominant toxic effect occurring at the RNA level via the CTG expansion, as occurs in myotonic dystrophy. Recent findings, however, suggest that the expansion is also transcribed in the opposite direction, encoding a CAG repeat that encodes an essentially pure, expanded polyglutamine domain.

SCA10

SCA10, originally reported in families of Mexican origin, has since been found in Brazilian families, suggesting it may have arisen in the Native American population. In Mexico, it is the second most common SCA after SCA2. SCA10 manifests with gait ataxia, dysarthria, and nystagmus. A distinctive feature of SCA10 is that seizures occur in a substantial segment of the affected patient population.[31] These can be generalized motor or complex partial seizures. The genetic defect in SCA10 is an extremely large expansion of a pentanucleotide repeat (ATTCT) in an intron of the *ATXN10* (22q13) gene. Normal repeats are 10–29 triplets in length, while in disease alleles the repeat is expanded from 800 to 4500 triplets in length. A few cases with repeat lengths of 280–370 suggest there may be intermediate alleles of reduced penetrance. The inverse correlation between repeat length and age of symptom onset is less clear for SCA10 than it is for the polylgutamine SCAs, suggesting that other genetic or environmental factors influence disease severity. The pathogenic mechanism in SCA10 is uncertain, but the CAG expansion, which resides in an intron, and thus does not encode protein, is not thought to act at the protein level.

SCA11

Two British families have been described with a relatively benign, slowly progressive form of gait and limb ataxia that maps to chromosome 15. This SCA represents a relatively pure cerebellar syndrome with mild pyramidal signs. The mean age at onset is roughly 25, with a normal life expectancy.[32]

SCA12

In SCA12, affected persons develop a progressive ataxic syndrome between the ages of 8 and 55 years that is typically accompanied by an action tremor. In addition, hyperreflexia, decreased movement including parkinsonism, ophthalmoparesis, focal dystonia, dysautonomia, and dementia can occur. SCA12 is found in the United States as well as in India. In India, SCA12 is less often associated with tremor and parkinsonism but may include sensory neuropathy. Cerebellar and cerebral atrophy are present on MRI studies. The disease is caused by a CAG repeat expansion in the 5′ untranslated region of the *PPP2R2B* gene, which encodes the protein phosphatase 2A regulatory subunit B (5q31-q33). The repeat, normally 4–32 triplets in length, is expanded to 51 repeats in affected persons. The mechanism of pathogenesis is still uncertain; although it is a CAG repeat disorder, SCA12 does not appear to be a polyglutamine disease.[33]

SCA13

SCA13 has a widely varied age of onset, including childhood, and delayed motor development and mental retardation can be part of the clinical picture. The ataxia is accompanied by dysarthria, nystagmus, and, occasionally, hyperreflexia. MRI usually shows cerebellar and pontine atrophy. Recently it was discovered that SCA13 is caused by various mutations in the *KCNC3* gene, which encodes a voltage-gated potassium channel subunit (19q.13.3-q14.4).[10]

SCA14

SCA14, which is not an expanded repeat ataxia, nevertheless shows phenotypic variability.[34] Most affected individuals develop slowly progressive ataxia with dysarthria in early adulthood. In late-onset cases, SCA14 can manifest as a relatively pure cerebellar ataxia. In earlier onset cases, however, the ataxia can be accompanied by facial myokymia, hyperreflexia, axial myoclonus, dystonia, and vibratory sensory loss. It is usually compatible with a normal lifespan, although affected persons can be wheelchair-bound late in life.

SCA14 is caused by various nonrepeat mutations in the *PRKCG* gene, which encodes a serine-threonine protein kinase (PKC-γ) (19q-13.4). This calcium-activated and phospholipid-dependent kinase is highly expressed in the cerebellum and in Purkinje cells. Most pathogenic mutations are amino acid deletions or nonconservative point mutations, but precisely how they lead to cerebellar dysfunction and atrophy is unknown. Genetic testing is available for SCA14. Unlike in the expanded repeat ataxias, however, SCA14 gene testing requires sequencing of all exons and flanking sequences in the disease gene.

SCA15

SCA15 is an autosomal dominant, slowly progressive, pure cerebellar ataxia described in an Australian family (3p24.2-3pter).[35]

SCA16

This is a slowly progressive dominant cerebellar ataxia described in a four-generation Japanese family between ages 20 and 66 years. One-third of the patients had a head tremor. Dysarthria, horizontal gaze-evoked nystagmus, and impaired smooth movement of the eyes were also present in some patients. Recently, a single mutation in the *CNTN4* gene encoding contactin 4 was reported in SCA16 (9q22.1-24.1).[36,37] Genetic testing is not currently available for this rare SCA.

SCA17

More than any other SCA described to date, SCA17 manifests with widespread cerebral as well as cerebellar dysfunction. Affected persons typically present in early to mid-adulthood with progressive gait and limb ataxia that is usually accompanied by dementia, psychiatric symptoms, and varying extrapyramidal features including parkinsonism, tremor, dystonia, and occasional chorea. In some cases, ataxia is not the predominant feature. Seizures have been reported in some patients. Consistent with the more global neurological phenotype, the MRI findings include diffuse cerebral and cerebellar atrophy.[38]

Nakamura et al.[38] discovered that the genetic defect was an abnormal CAG expansion in the *TBP* gene that encodes TATA-binding protein, a component of the basal transcription machinery that drives expression of most protein-coding genes (6q27). The expansion results in an abnormally long glutamine repeat of 43–66 CAG repeats in *TBP*, whereas the normal repeat number ranges from 25 to 42. Rare repeats of 43–48 may show reduced penetrance. The expanded TBP protein likely alters the normally tightly regulated expression of genes in neurons.[38]

SCA18

Also known as sensorimotor ataxia with neuropathy (SMNA), SCA18 has been reported in a single, five-generation American family of Irish ancestry with variable expressivity and severity. Onset of progressive symptoms has been reported in the second to third decades of life. Gait ataxia, dysmetria, and nystagmus were present in all affected patients; additional symptoms have included sensory loss, pyramidal tract signs, and muscle weakness. Nerve conduction studies were consistent with a sensorimotor axonal neuropathy, and a brain MRI showed mild cerebellar atrophy.[39] The phenotype essentially combines both progressive ataxia (SCA) and hereditary motor/sensory neuropathy (HMSN).

Spinocerebellar Ataxias 19 through 36

SCAs 19 through 36 are uncommon, dominantly inherited ataxias.[40–44] Clinical features for the small numbers of families in which these SCAs have been characterized are described in Table 71.1. More genetically defined dominant ataxias surely will come to light over the next decade.

Dentatorubropallidoluysian Atrophy

A disorder of variable clinical presentation, DRPLA is characterized by progressive ataxia, choreoathetosis, dementia, seizures, myoclonus, and dystonia. DRPLA shares many features of the SCAs and thus is grouped with them here. DRPLA is due to a glutamine-encoding CAG repeat expansion in the *DRPLA* gene encoding the disease protein, atrophin (12p-13.31). As in other CAG repeat diseases, larger expansions cause more severe disease manifesting earlier in life, and anticipation occurs frequently, particularly with paternal transmission. Patients with onset before 20 years of age nearly always have seizures and display more of a progressive myoclonic epilepsy phenotype. In contrast, individuals with older onset typically develop ataxia with choreoathetosis and dementia. The number of repeats in patients with DRPLA is between 48 and 93, whereas repeats in control individuals are between 6 and 39. DRPLA is most prevalent in Japan and quite rare in the United States. One well-characterized African American family in North Carolina has a phenotypic variant known as the Haw River syndrome, in which seizures and cerebral calcifications often accompany the ataxia.[1]

Episodic Ataxias

Two well-established forms of dominantly inherited episodic ataxia, EA-1 and EA-2, are caused by mutations in a voltage-gated potassium channel and calcium channel, respectively. In EA-1, which is due to mutations in the *KCNA1* gene, affected individuals typically have brief episodes of ataxia lasting only minutes, in some cases precipitated by stress, exercise, or sudden change in posture. Myokymia is common in EA-1. In EA-2, which is due to mutations in the *CACNA1A4* gene (and thus is allelic with SCA6 and familial hemiplegic migraine); patients usually have longer episodes of ataxia that can last for hours or days, often precipitated by stress, exercise, or fatigue. Acetazolamide may be therapeutic for either ataxia but more commonly is beneficial for EA-2.[1,25] Indeed, in all cases of episodic ataxia, regardless of genetic cause, a trial of acetazolamide is warranted.

Several other EAs mapping to different chromosomal loci have recently been described, though they may be less common than EA-1 and EA-2. In the small number of patients in whom EA-3 and EA-4 have been described, vestibular symptoms, including vertigo, are prominent during the episodes.[45] The genetic defects underlying EA-3 and EA-4 are still unknown. Three other episodic ataxias, EA-5, EA-6, and EA-7 have been described.

EA-5 is caused by a mutation in a gene encoding a calcium-channel β-subunit,[46] whereas EA-6 is caused by a mutation in a gene encoding a glutamate transporter. The latter patient also had migraine with alternating hemiplegia.[47] This case, together with the overlap between the allelic disorders EA-2 and familial hemiplegic migraine, illustrates the close link between episodic ataxia and migraine. Clearly, ion channel dysfunction contributes mechanistically both to certain forms of headache and to ataxia.[1]

AUTOSOMAL RECESSIVE ATAXIAS

The description of recessive ataxias has lagged behind that of the dominant ataxias. Friedreich ataxia, which was genetically defined in 1996, remains the most commonly diagnosed recessive ataxia in the United States. In the last 5 years, however, many more genetic causes of rare recessive ataxia have been discovered. This advance is leading to a clearer description of the wide range of ataxic phenotypes that can occur in recessive syndromes. In general, recessive ataxias tend to manifest earlier than the SCAs, typically in childhood. Given the recessive inheritance pattern, many cases occur in the absence of a defined family history of progressive ataxia. Thus, the underlying genetic nature of disease may be missed unless the clinician remembers to keep this possibility high on the differential diagnosis. Here we emphasize Friedreich ataxia, because it is relatively common and because new insights into pathogenic mechanisms hold promise for preventive therapies in the near future.

Friedreich Ataxia

FA, a recessively inherited disorder, is the most common form of inherited ataxia. It is typically characterized by ataxia and lower limb areflexia beginning in late childhood. Since the discovery of the underlying genetic defect, it has become clear that FA also manifests as an atypical adult-onset form that differs in many respects from the classic presentation.

Symptoms and Signs

FA classically occurs before 25 years of age with progressive ataxic gait and titubation. The lower extremities are more severely involved. Patients occasionally present with dysarthria. Progressive scoliosis, foot deformity, nystagmus, or cardiomyopathy rarely can be initial signs.[48]

In classic FA, the neurologic examination reveals nystagmus and square wave jerks, slowing of saccadic eye movements, truncal titubation, dysarthria, dysmetria, and gait and limb ataxia. Extensor plantar responses (with normal tone in trunk and extremities), absent deep tendon reflexes in the legs, and weakness (greater distally than proximally) are usually found. Loss of vibratory and proprioceptive sensation occurs. The median age of death is approximately 35 years.

Cardiac involvement occurs in 90% of patients with classic FA. Cardiomegaly, symmetrical hypertrophy, murmurs, and conduction defects are reported. Moderate mental retardation or psychiatric syndromes are present in a small percentage of patients. A high incidence of diabetes (20%) is found and is associated with insulin resistance, pancreatic beta-cell dysfunction, and type 1 diabetes. Musculoskeletal deformities are common, including pes cavus, equinovarus and scoliosis. MRI of the spinal cord shows significant spinal cord atrophy in affected patients (Figure 71.2).

Onset before age 25 years was once a hallmark of FA. Now, however, genetic testing shows that atypical adult-onset disease is surprisingly common, accounting for perhaps as many as one-quarter of patients. Adult-onset disease is less severe, progresses more slowly, and is less frequently accompanied by foot deformities, muscle wasting, areflexia,

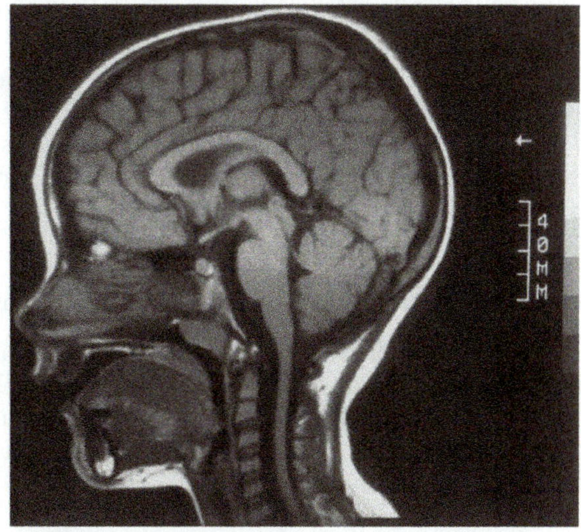

FIGURE 71.2 **Sagittal MRI of the brain and spinal cord showing significant spinal cord atrophy in a 6-year-old girl with Friedreich ataxia.**

and cardiomyopathy. In any adult with unexplained progressive ataxia, FA should be among the first diseases considered in the differential diagnosis, especially if the brain MRI does not show cerebellar atrophy.[48,49]

The primary sites of pathology are the spinal cord, dorsal root ganglion cells, and peripheral nerves. Sclerosis and degeneration are seen predominantly in the spinocerebellar tracts, lateral corticospinal tracts, and posterior columns. Degeneration of the glossopharyngeal, vagus, hypoglossal, and deep cerebellar nuclei has been described. Slight atrophy of the cerebellum and cerebral gyri may occur. The cerebral cortex is histologically normal, except for loss of Betz cells in the precentral gyri. Peripheral nerves are extensively involved, with a loss of large myelinated fibers. The density of small myelinated fibers is normal, but axonal size and myelin thickness are diminished.

Cardiac pathology consists of myocytic hypertrophy and fibrosis, focal vascular fibromuscular dysplasia with subintimal or medial deposition of periodic acid–Schiff-positive material, myocytopathy with unusual pleomorphic nuclei, and focal degeneration of myelinated and unmyelinated nerves and cardiac ganglia.

Genetics

All cases of FA are due to mutations in the *FRDA* gene on chromosome 9q13-q21. In more than 95% of patients, the mutation is an expanded GAA repeat occurring in the first intron of the *FRDA* gene and present on both alleles. In other words, most patients are homozygous for the expansion. A few patients, however, are compound heterozygotes, having an expansion in one allele and a point mutation or deletion in the second allele. Normal alleles have 7–34 GAA repeats, and disease alleles have approximately 120–1700 GAA repeats. Patients with FA have reduced levels of frataxin mRNA and protein; thus the fundamental problem in this disease is insufficient frataxin expression. Frataxin is an essential, iron-binding mitochondrial protein involved in iron homeostasis. Studies of disease tissue have confirmed a deficiency in iron/sulfur cluster-containing mitochondrial enzymes, decreased mitochondrial ATP production, and accumulation of iron in the heart. Thus, FA can be thought of as a nuclear-encoded, mitochondrial disorder in which iron-mediated oxidative stress, or perhaps other iron-dependent sequelae of frataxin deficiency, plays an important role in pathogenesis.[48,50,51]

Given what is now suspected about FA pathogenesis, two potential approaches to therapy in FA are, first, to counter toxic effects of presumptive oxidative stress with free radical scavengers such as idebenone[52] or mitochondrial enhancers such as coenzyme Q10,[53] and, second, to boost expression of frataxin, including use of histone deacetylase inhibitors that reverse gene silencing[54] or small molecules that unblock the sticky DNA triplex that is thought to form at the site of the repeat in the transcribed *FRDA* gene.

Genetic Causes of Vitamin E Deficiency

Classic FA should be compared and contrasted with the similar ataxic presentation associated with vitamin E deficiency. Two forms of recessively inherited ataxias associated with abnormalities in vitamin E (α-tocopherol) have been identified: familial isolated deficiency of vitamin E (FIVE), which is also known as *ataxia with vitamin E deficiency*, and abetalipoproteinemia, which is also known as *Bassen–Kornzweig syndrome*. The clinical features of FIVE are indistinguishable from those of classic FA, but the presence of retinopathy helps distinguish abetalipoproteinemia from FA. FIVE is due to mutations in the gene for α-tocopherol transfer protein (α-TTP) on chromosome 8 (8q13). These patients have an impaired ability to incorporate vitamin E into the very low-density lipoprotein (VLDL) produced and secreted by the liver, resulting in a deficiency of vitamin E in peripheral tissues. Abetalipoproteinemia is caused by mutations in the gene coding for a subunit of the microsomal triglyceride transfer protein (MTP). Defects in MTP result in impaired formation and secretion of VLDL in the liver. This defect results in deficient delivery of vitamin E to tissues, including the nervous system, as VLDL is the transport molecule for vitamin E and other fat-soluble substitutes. Hence, either impaired incorporation of vitamin E into VLDL (FIVE) or the absence of VLDL (abetalipoproteinemia) causes an ataxic syndrome similar to FA.[55,56]

Ataxia Telangiectasia and other DNA Repair Ataxias

Recent genetic discoveries highlight the importance of DNA repair to cerebellar integrity. Precisely why mutations in various DNA repair proteins tend to preferentially affect the nervous system with its postmitotic neurons remains a mystery.[13] Of these disorders, ataxia telangiectasia (AT) is best characterized. Some of these disorders, including AT, are caused by mutations in components of double-strand DNA break repair, whereas others are caused by mutations in components of single-strand DNA break repair.

Ataxia Telangiectasia

Although less common than FA, AT has received considerable attention, largely because of its association with cancer. Patients usually present in the first decade of life with telangiectasias associated with cerebellar ataxia and nystagmus. The neurologic manifestations are similar to those seen in FA, except that extrapyramidal features (including myoclonic jerks, dystonia, and athetosis) are more common and areflexia is less common. There is a high incidence of recurrent pulmonary infections and neoplasms of the lymphatic and reticuloendothelial system in patients with AT. Lymphomas, Hodgkin disease, and acute leukemias are the most common cancers. Chromosomal translocations occur frequently, particularly between chromosomes 7 and 14. Thymic hypoplasia with cellular and humoral immunodeficiencies (particularly IgA and IgG-2), premature aging, and endocrine disorders, such as insulin-dependent diabetes mellitus, can also occur. Elevated levels of α-fetoprotein and carcinoembryonic antigen are useful markers of disease. Fibroblasts can be screened for increased X-ray sensitivity and radioresistant DNA synthesis, two hallmarks of AT. The mean age of death is approximately 20 years, usually from infection or cancer.

The most striking neuropathologic changes include loss of Purkinje, granule, and basket cells in the cerebellar cortex and of neurons in the deep cerebellar nuclei. The inferior olives of the medulla also may have neuronal loss. There is a loss of anterior horn neurons in the spinal cord and of dorsal root ganglion cells associated with posterior column spinal cord demyelination. A poorly developed or absent thymus gland is the most consistent defect of the lymphoid system.

Hundreds of different mutations have been found in the *ATM* gene (ataxia-telangiectasia, mutated) on chromosome 11q22-23, which are broadly distributed throughout the gene. Most mutations are null mutations; rarer mutations that act in a dominant negative manner may be especially prone to predispose heterozygous carriers to cancer. Evidence from cellular studies and knockout mice indicates that the ATM protein, a very large protein kinase, serves as a master regulator of the cell cycle checkpoint in response to double-strand breaks in DNA. By phosphorylating various cell-cycle proteins, DNA repair proteins, and kinases, ATM controls signaling pathways activated during genotoxic stress.[57,58]

Other ataxic syndromes similar to AT result from defects in double-strand break repair. For example, the rare AT-like disorder, so named because of its close resemblance to AT, is caused by mutations in *MRE11*.[59] Mre11 is one of three proteins composing the Mre11/Rad50/Nbs1 (MRN) complex that senses double-strand breaks and recruits and activates ATM at these sites of DNA damage. Defects in a second protein of this complex, Nbs1 or nibrin, cause Nijmegen breakage syndrome, a developmental disorder that shares many features with AT, although the brain impairment occurs earlier and is more widespread.

Ataxia with Ocular Apraxia Type 1

After FA, AOA1 is the most common early-onset recessive ataxia in many populations.[12] In Japan, where FA is virtually absent, AOA1 is the most common recessive ataxia. Affected children develop ataxia between 2 and 14 years of age. The ataxia is usually accompanied by ocular apraxia that progresses to ophthalmoplegia. Axonal neuropathy with areflexia is always seen, and chorea and/or dystonia are observed in approximately half of patients. Cognitive function is impaired to a varying degree, reflecting the various different mutations underlying the disease. AOA1 must be considered as a possible diagnosis in any early-onset ataxia in which FA has been excluded and in which the systemic features of AT are absent (immunodeficiency, telangiectasias, cancer predisposition). Many affected persons will have low serum albumin (a constant feature in Japan, where the disorder has even been called AOA with hypoalbuminemia). Cholesterol may also be elevated. The MRI shows cerebellar atrophy.

AOA1 is caused by loss-of-function mutations in the *APTX* gene (9p21). Various missense (inactivating) and nonsense (truncating) mutations have been reported. The gene product, aprataxin, participates in DNA single-strand break repair. When aprataxin is defective, unrepaired DNA strand breaks may accumulate in the brain due to a failure in religation of strand breaks.

Ataxia with Ocular Apraxia Type 2

AOA2 is a second, early-onset ataxia with similar features to AOA1.[60,61] It tends to manifest at slightly older ages (10–22 years), and only about 50% of affected persons develop ocular apraxia. Severe sensorimotor neuropathy develops in about 90% of patients, but extrapyramidal symptoms are less frequent than in AOA1. Elevated serum α-fetoprotein, seen in most patients, is a clue to diagnosis; although α-fetoprotein is also increased in AT, AOA2 patients do not develop the characteristic, systemic features of AT. As in AOA1, cholesterol may be elevated (9q34).

AOA2 is caused by mutations in the *SETX* gene, which encodes senataxin; patients have two inactive alleles. A single family has been described in which a heterozygous mutation apparently causes an adult-onset form of ataxia with tremor with dominant inheritance;[62] *SETX* mutations are unlikely to be a common cause of adult-onset ataxia, however. Sentaxin contains a DNA/RNA helicase domain and likely functions in RNA processing and transcription-coupled DNA repair. Certain mutations in *SETX* can cause a dominantly inherited, juvenile-onset form of amyotrophic lateral sclerosis (ALS4). In this instance, heterozygous mutation causes early-onset, slowly progressive motor neuron degeneration compatible with a normal lifespan.[63] In contrast, AOA2 is caused by loss of function in both alleles.

Spinocerebellar Ataxia and Axonal Neuropathy

Spinocerebellar ataxia and axonal neuropathy (SCAN1) is a rare, autosomal recessive ataxia characterized by ataxia and severe sensorimotor axonal neuropathy without systemic features.[64] It is caused by an inactivating mutation in the topoisomerase I-dependent DNA damage repair enzyme, tyrosyl phosphodiesterase 1 (Tdp1). This protein was recently found to function in a novel single-strand DNA break–repair pathway. Why its mutation leads to selective damage to neurons is unclear. One possibility is that, in SCAN1, neurons accumulate single-strand DNA breaks that impede neuronal transcription, thereby compromising neuronal integrity.

Xeroderma Pigmentosum

Xeroderma pigmentosum is a rare autosomal recessive neurocutaneous disorder caused by mutations in enzymes necessary for nucleotide excision repair. In addition to skin lesions, patients may show progressive mental deterioration, microcephaly, ataxia, spasticity, choreoathetosis, and hypogonadism. Nerve deafness, peripheral neuropathy (predominantly axonal), electroencephalographic abnormalities, and seizures are reported. Neuronal death is noted in pyramidal cells, cerebellar Purkinje cells, the deep nuclei of the cerebellum, the brainstem, the spinal cord, and peripheral nerves.

Cockayne Syndrome

Cockayne syndrome is a rare autosomal recessive disorder first described by Cockayne in 1936. Like xeroderma pigmentosum, Cockayne syndrome is due to mutations in enzymes of nucleotide excision repair. Clinical features are mental retardation, optic atrophy, dwarfism, neural deafness, hypersensitivity of skin to sunlight, cataracts, and retinal pigmentary degeneration. Cerebellar, pyramidal, and extrapyramidal deficits and peripheral neuropathy may occur, with bird-headed facies and normal-pressure hydrocephalus. Skin fibroblasts show defective DNA repair when exposed to ultraviolet light.[65]

OTHER RECESSIVE ATAXIAS

Autosomal Recessive Spastic Ataxia

Autosomal recessive spastic ataxia was originally reported in the Charlevoix-Saguenay region of Quebec and hence is known as ARSACS. This childhood-onset, progressive spastic ataxia is also characterized by retinal hypermyelination (striations visible on funduscopic examination), dysarthria, distal muscle wasting, foot deformities, absence of somatosensory evoked potentials in the lower limb, and the frequent presence of mitral valve prolapse. ARSACS is caused by a mutation in the *SACS* gene on chromosome 13q12, resulting in loss of expression of the novel disease protein sacsin.[66] Since its discovery in Quebec, ARSACS has been found worldwide.[67]

The recent discovery of additional mutations in a previously unrecognized upstream exon of the gene may lead to a broadening of the phenotype, with retinal hypermyelination not seen in all cases.

Marinesco–Sjögren Syndrome

Marinesco–Sjögren syndrome is a rare, early-onset disorder in which progressive cerebellar deficits occur in association with other features (cataracts, mental retardation, multiple skeletal abnormalities, hypogonadotropic hypogonadism). Mutations in the *SIL1* gene, which encodes a nucleotide exchange factor for the heat-shock protein 70, were recently discovered to be the cause (5q31).[68]

Autosomal Recessive Cerebellar Ataxia Type 1 and 2

Autosomal recessive cerebellar ataxia type 1 (ARCA1) represents the first pure cerebellar ataxia to be described in which the inheritance is autosomal recessive. In the one reported family, onset of ataxia was between 17 and 46 years of age. Gait and limb ataxia is accompanied by dysarthria and jerky ocular pursuits but without signs of extracerebellar involvement. The MRI shows cerebellar atrophy. ARCA1 is caused by mutations in the *SYNE1* gene, which encodes a very large spectrin-like protein, SYNE1 (6p25). The disease protein is highly expressed in the cerebellum. ARCA2 includes ataxia, myoclonus, impaired cognition, epilepsy, exercise intolerance, and ischemic episodes. It maps to 1q42 and is associated with elevated serum lactate and decreased coenzyme Q10 levels.

MITOCHONDRIAL ATAXIAS

As outlined in Table 71.1, several ataxic syndromes are caused by mutations in mitochondrial DNA (mtDNA). The names of these syndromes, all maternally inherited and highly variable in severity and phenotype, provide clues to the major features in each disease: myoclonic epilepsy and ragged red fiber syndrome (MERRF); mitochondrial encephalopathy, lactic acidosis, and stroke syndrome (MELAS); and neuropathy, ataxia, and retinitis pigmentosa (NARP). Of these three, NARP is most commonly associated with ataxia. Point mutations in distinct mitochondrial genes are associated with each disorder. Straightforward genetic tests for these and other mtDNA mutations are readily available. It is important to recognize that other mitochondrial defects can also manifest with ataxia. In individuals with ataxia associated with hearing loss, retinopathy, myopathy, neuropathy, and/or diabetes, especially when there is evidence of possible maternal inheritance, the clinician should seriously consider a primary mitochondrial disorder as the cause. Additional laboratory clues to mitochondrial disorders include elevated serum and cerebrospinal fluid lactate, elevated serum creatine kinase, and evidence of enzymatic and histochemical abnormalities on muscle biopsy.

Classic mitochondrial disorders display maternal inheritance because the genetic defect resides in the mitochondrial organelle. Many mitochondrial proteins, however, are encoded by the nuclear genome. Thus, mutations in nuclear-encoded genes vital for mitochondrial function can cause similar neurodegenerative disease. In these cases, disease may occur sporadically or in an autosomal dominant or recessive manner. For example, in Kearns–Sayre syndrome, a progressive disorder in which mtDNA deletions result in various organ systems being affected, the inheritance can be autosomal dominant. In contrast, a recessively inherited, early-onset ataxia known as infantile-onset spinocerebellar ataxia is caused by mutations in a nuclear gene encoding a mitochondrial DNA helicase, Twinkle.[69] Friedreich ataxia, discussed above, is another nuclear mitochondrial disorder that is recessively inherited.

MOLECULAR GENETICS

DNA Repeat Expansions as a Common Cause of Hereditary Ataxia

Since the identification of the first gene causative of a spinocerebellar ataxia, SCA1, in 1993, several gene loci have been discovered that allow the classification of the SCAs based on their molecular genetics. The single most commonly identifiable cause of the inherited dominant ataxias is DNA repeat expansions either in the coding or noncoding portions of the affected gene. SCAs 1, 2, 3, 6, 7, 17 and DRPLA are caused by CAG or polyglutamine (polyQ) repeat expansions in the exons of the gene, while SCAs 8, 10, 12, 31, and 36 are due to intronic repeat expansions[70] (Figure 71.3). Fragile X-associated tremor/ataxia syndrome (FXTAS) deserves mention as an X-linked ataxia due to a noncoding CGG repeat expansion in the *FMR1* gene resulting in late-onset progressive ataxia and tremor primarily in men; CGG repeats greater than 200 cause the well-known fragile-X mental retardation syndrome.[71] Although DNA expansions are the most common cause of autosomal dominant ataxic disorders as a group, most dominant ataxias are a result of conventional mutations, while gene loci are yet unknown for the remainder of these disorders.

With a few exceptions, namely SCAs 10, 31, 36, genetic ataxias caused by repeat expansions are a result of abnormal trinucleotide repeats, and of these the majority is from pathologically expanded and transcribed CAG repeats. Genotype–phenotype correlations are more predictable for polyQ disorders than for other dominant ataxias due to common mutational mechanisms that dictate clinical course, not unlike other triplet repeat disorders such as Huntington disease, Kennedy disease, and myotonic dystrophy. Intergenerational variability of phenotype due to repeat size expansion, a phenomenon called anticipation, is a hallmark clinical feature of polyQ diseases; longer

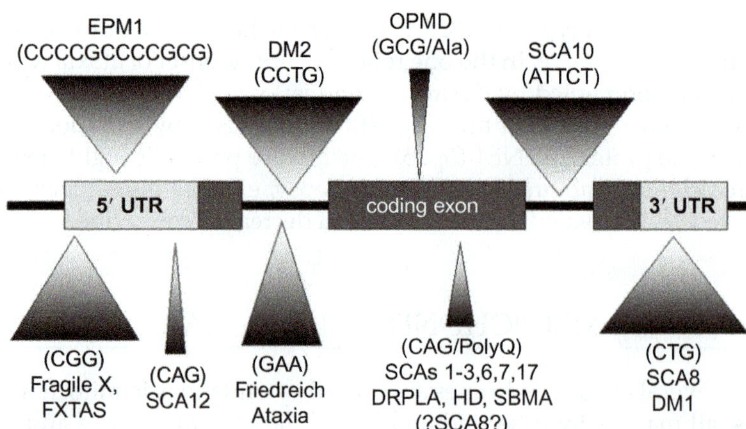

FIGURE 71.3 **Schematic of a gene indicating the diseases and repeat sequence and location within the gene.** All but one CAG repeat (SCA12) are in the protein-coding region, encoding polyglutamine in the disease protein. Other repeats are untranslated and cause disease through several different mechanisms at the transcriptional or posttranscriptional level. The repeat in spinocerebellar ataxia (SCA) 8 may be transcribed in both directions, yielding a CUG in noncoding RNA in one direction and a polyglutamine-coding CAG repeat in the other. Depending on repeat length, the FMR1 repeat can cause fragile-X mental retardation syndrome (full-size expansions) or the fragile-X associated tremor/ataxia syndrome (FXTAS; premutation expansions). Hereditary ataxias are underlined. EPM1, progressive myoclonic epilepsy of Unverricht–Lundborg; DM1, DM2, myotonic dystrophies types 1 and 2; OPMD, oculopharyngeal muscular dystrophy: DRPLA, dentatorubropallidoluysian atrophy; HD, Huntington disease; SBMA, spinobulbar muscular atrophy. Reproduced from chapter 33 of Rosenberg et al., eds. *The Molecular and Genetic Basis of Neurologic and Psychiatric Disease*. 4th edition. Philadelphia, PA: Lippincott, Williams & Wilkins, 2007.

repeats in successive generations cause earlier disease onset and typically a more severe disorder. Additionally, larger repeats are associated with multisystemic neurological manifestations due to more widespread neuronal degeneration than nonrepeat ataxias, which cause primarily cerebellar and brainstem symptoms.[70] Other determinants of phenotype common to polyglutamine disorders include: paternal versus maternal transmission, the former being associated with more prominent anticipation due to triplet expansions during male gametogenesis; interrupted versus uninterrupted CAG repeats; and somatic mosaicism in the central nervous system.[72]

General Molecular Features of Polyglutamine Repeat Ataxias

The disruption of biological processes at the cellular levels in SCAs are complex and not completely understood, but significant progress has been made in illustrating some of these mechanisms and their clinical implications since the identification of CAG repeat-related diseases in the early 1990s. Human and animal cell cultures, transgenic mice, and *Caenorhabditis elegans* and *Drosophila* models of SCA diseases have allowed scientists to identify pathways that may contribute to common biological and clinical features of these disorders.

1. The selective vulnerability of neurons, particularly those of the cerebellum, is intriguing given the genes affected in the SCAs are widely transcribed. One theory is that these genes affect cellular machinery in vulnerable neurons in a similar fashion. Recent studies involving cDNAs of proteins from various ataxia causing genes including those from polyQ SCAs showed that the protein products of mutant genes affected similar cellular processes: transcription regulation, nuclear translocation of these peptides, and RNA modification.[73] The sites of action of these peptides overlapped and included the nucleoplasm, nuclear membrane, and the spliceosome complex, as did their molecular functions, which involved transcription factor activity and kinase inhibition. Other factors that influence regional selectivity and neuronal specificity of polyQ diseases may have to do with the intrinsic electrical property, high metabolic demands, and resultant oxidative stress in the affected cells reminiscent of pathogenesis proposed in Parkinson and Alzheimer disease.
2. It is well recognized that longer CAG repeats cause an earlier onset of disease and a more severe phenotype, with very large repeats having multisystemic neurological symptoms beyond cerebellar and brainstem deficits. Moreover, the typical repeat size at which disease is usually manifested is above a threshold of approximately 35–40 repeats with the exception of SCA6, which is caused by smaller repeat lengths (>19). Repeat size variability among individuals affected with SCAs informs phenotypic variability along with other factors that influence disease expression including size of the unaffected allele, brain region-specific tract length and gene

transcription.[70] For instance, retinal degeneration is characteristic of SCA7, but with smaller CAG repeats, individuals may manifest ophthalmological symptoms only much later in the disease course. It is speculated that progressively larger CAG repeats confer toxicity to regions of the central nervous system that would not be otherwise affected with smaller repeats, the result being diverse symptomatology such as extrapyramidal features and abnormal movements, peripheral neuropathy, cardiomyopathy, and dementia.

3. The destructive effects of polyQ diseases due to a positive correlation between repeat size and disease severity is explained by a toxic gain-of-function mechanism, whereby the abnormal gene products aggregate and form intraneuronal inclusions; whether these inclusions themselves are pathological has been a matter of much discussion and has provided the impetus to look at the various cellular processes affected by the polyglutamine tract and its derivatives. Recent studies have implicated polyQ peptides in cellular functions involving signal transduction and RNA splicing, RNA metabolism and translation, modulation of the ubiquitin–proteosome pathway, and regulation of transcription.[73] In SCA1, mutant protein ataxin-1 (ATXN1), translocates into the nucleus but cannot be transported back into the cytosol. Transgenic mice models have demonstrated disruption of RNA splicing due to an abnormal affinity of mutant ATXN1 for splicing factor RBM17 within the nucleus as the polyQ tract lengthens. The peptide affected in SCA2, ataxin-2 (ATXN2), has been shown by way of its homolog proteins in *C. elegans* and *Drosophila* models to regulate mRNA translation; it is speculated that abnormal polyQ expansions adversely affect translation. SCA3, also known as Machado–Joseph disease, is caused by a polyglutamine repeat expansion in ataxin-3 (ATXN3). Recent studies have confirmed the altered protein's role in regulation of transcription and protein–protein interactions, the primary site of pathogenesis being the nucleus. Specifically, the localization of normal and mutant ATXN3 to the nucleus is increased under heat shock and oxidative stress conditions pointing to its potential role in regulation of genes that modulate the cellular stress response. In SCA3 animal models, ATXN3's inability to function effectively as a deubiquinating enzyme compromises the neuroprotective functions of the ubiquitin–proteasome system, which deals with misfolded proteins and transcription regulation. SCA6 is widely thought to be a channelopathy caused by CAG expansions in strategic portions of the P/Q-type voltage-gated calcium channel *CACNA1A* and more recent studies have demonstrated the nuclear translocation of the polyQ containing transmembrane portion of the protein as a contributor to its pathogenicity. SCA7 and SCA17 are caused by mutations in critical genes that regulate transcription in selective neuronal populations, the former affecting retinal as well as cerebellar cells. ATXN7 in SCA7 modifies the function of a key transcription factor in retinal cells by attaching to it, whereas in SCA17 the expanded polyQ tract occurs in the TATA-binding protein, which alters its ability to transcribe DNA.
4. Anticipation, or earlier and more severe disease in subsequent generations harboring the mutant CAG repeat gene due to repeat size expansion, is a well-established attribute of polyQ diseases. Interactions between polyglutamine tracts, abnormally conformed into noncanonical DNA structures (hairpins, triplexes, quadruplexes, etc.), and the DNA repair machinery contribute to repeat instability and expansion in both germline and somatic tissues.[74] Epigenetic factors including the chromatin context of the triplet repeat also influence stability and, hence, expansion. DNA demethylation, which is a key process during reprogramming in the developing zygote, and which involves DNA repair, has shown to be strongly linked to repeat instability. For instance, SCA1 mice with only one copy of a DNA methylating enzyme passed on the polyQ expansion to their offspring 3–4 times more frequently than mice with both copies of the enzyme. Another mechanism potentially destabilizing CAG repeats is antisense transcription through disease loci generating double-stranded RNA molecules that impact instability through heterochromatin formation.
5. Toxic gain of function leading to protein aggregation may be the leitmotif of the neuropathology of polyglutamine ataxias, but incontrovertible proof of a causal relationship between peptide inclusions and cell death is lacking. Several cellular processes, upstream of inclusion formation, have been implicated in converging towards untimely neuronal demise and they include: direct activation of proapoptotic enzymes; aberrant mitochondrial energetics; formation of toxic polyQ oligomers; impaired ubiquitin–proteasome pathways; and, disruption of cAMP response element-binding protein (CREB)-mediated gene transcription.[75]

Recent experiments have highlighted the role of toxic mRNA gain of function in FXTAS pathology and in the multirepeat ataxias SCA10 and SCA36.[76]

THERAPEUTIC STRATEGIES IN GENETIC ATAXIAS

Although our comprehension of all aspects of SCA pathogenesis remains incomplete, elucidation of the key pathological mechanisms, particularly of the polyglutamine repeat diseases, has been made possible through transgenic knock-in mouse models. Similar animal models have been utilized when researching cell-based therapies for these disorders. These include but are not limited to: decreasing the mutant protein by attenuating transcription using siRNA or short-hairpin RNA; stimulating degradation of the polyQ peptide through rapamycin-related pathways; minimizing aggregation of abnormal proteins by increasing molecular chaperones; modifying disturbed transcription with histone deacetylase inhibitors; alleviating mitochondrial oxidative stress with antioxidants and free-radical scavengers; utilizing antiapoptotic agents such as caspase inhibitors to increase the lifespan of affected cells; modulating ion channel dysfunction (in channelopathies such as SCA6); and, enhancing neurogenesis with nerve growth factor stimulators.[75,77] However, most of these therapies have not stood up to scientific scrutiny as significant breakthroughs in the clinical treatment of ataxias for a variety of reasons, including the observation that a single agent might be inadequate in effectively halting a complex pathological process that may warrant a rational combination of effective drugs. Two drugs that have shown some clinical benefit in recent human trials deserve mention: i) riluzole, a potassium-channel regulator that is thought to reduce neuronal hyperexcitability,[78] and ii) varenicline, a partial agonist of the $\alpha 4\beta 2$ nicotinic agonist, whose mechanism of improving ataxia in SCA3-affected individuals is not fully understood.[79]

Sophisticated molecular therapies will likely enter the mainstream of SCA therapeutics after we have uncovered the exact pathomechanism of SCAs; the rationale will be to precisely and strategically manipulate key steps in the pathological cascade either to arrest the disease or reverse damage and ameliorate symptoms. Initiating therapy early to modify risk of disease in presymptomatic individuals can be facilitated by genetic tests and the continuous discovery of mutations implicated in the genesis of SCAs. The larger dilemma of precisely and safely manipulating the affected genome and its derivatives remains to be conquered.

References

1. Rosenberg RN, Paulson HL. The Inherited Ataxias (Chapter 35). In: Rosenberg RN, Prusiner SB, DiMauro S, et al. eds. *The Molecular and Genetic Basis of Neurologic and Psychiatric Disease.* Philadelphia: Butterworth Heinemann; 2003, 369–382.
2. Gatchel JR, Zoghbi HY. Diseases of unstable repeat expansion: mechanisms and common principles. *Nat Rev Genet*. 2005;6(10):743–755.
3. Orr HT, Chung MY, Banfi S, et al. Expansion of an unstable trinucleotide CAG repeat in spinocerebellar ataxia type 1. *Nat Genet.* 1993;4:221–226.
4. Klement IA, Skinner PJ, Kaytor MD, et al. Ataxin-1 nuclear localization and aggregation: role in polyglutamine-induced disease in SCA1 transgenic mice. *Cell*. 1998;95:41–53.
5. Rosenberg RN, Nyhan WL, Bay C, et al. Autosomal dominant striatonigral degeneration. A clinical, pathologic, and biochemical study of a new genetic disorder. *Neurology*. 1976;26:703–714.
6. Rosenberg RN, Nyhan WL, Coutinho P, et al. *Joseph's Disease: An Autosomal Dominant Neurological Disease in Portuguese of the United States and Azores Islands*. New York: Raven Press; 1978.
7. O'Hearn E, Holmes SE, Calvert PC, et al. SCA-12: tremor with cerebellar and cortical atrophy is associated with a CAG repeat expansion. *Neurology*. 2001;56:299–303.
8. Wakamiya M, Matsuura T, Liu Y, et al. The role of ataxin 10 in the pathogenesis of spinocerebellar ataxia type 10. *Neurology*. 2006;67:607–613.
9. Ranum LP, Schut LJ, Lundgren JK, et al. Spinocerebellar ataxia type 5 in a family descended from the grandparents of President Lincoln maps to chromosome 11. *Nat Genet*. 1994;8:280–284.
10. Waters MF, Minassian NA, Stevanin G, et al. Mutations in voltage-gated potassium channel KCNC3 cause degenerative and developmental central nervous system phenotypes. *Nat Genet*. 2006;38:447–451.
11. van Swieten JC, Brusse E, de Graaf BM, et al. A mutation in the fibroblast growth factor 14 gene is associated with autosomal dominant cerebellar ataxia [corrected]. *Am J Hum Genet*. 2003;72:191–199.
12. Le Ber I, Brice A, Durr A. New autosomal recessive cerebellar ataxias with oculomotor apraxia. *Curr Neurol Neurosci Rep*. 2005;5:411–417.
13. Paulson HL, Miller VM. Breaks in coordination: DNA repair in inherited ataxia. *Neuron*. 2005;46(6):845–848.
14. Zoghbi HY, Orr HT. Glutamine repeats and neurodegeneration. *Annu Rev Neurosci*. 2000;23:217–247.
15. Kawaguchi Y, Okamoto T, Taniwaki M, et al. CAG expansions in a novel gene for Machado-Joseph disease at chromosome 14q32.1. *Nat Genet*. 1994;8:221–228.
16. Twist EC, Casaubon LK, Ruttledge MH, et al. Machado–Joseph disease maps to the same region of chromosome 14 as the spinocerebellar ataxia type 3 locus. *J Med Genet*. 1995;32:25–31.
17. Maciel P, Gaspar C, DeStefano AL, et al. Correlation between CAG repeat length and clinical features in Machado–Joseph disease. *Am J Hum Genet*. 1995;57:54–61.
18. Paulson HL, Perez MK, Trottier Y, et al. Intranuclear inclusions of expanded polyglutamine protein in spinocerebellar ataxia type 3. *Neuron*. 1997;19:333–344.
19. Berke SJ, Chai Y, Marrs GL, et al. Defining the role of ubiquitin-interacting motifs in the polyglutamine disease protein, ataxin-3. *J Biol Chem*. 2005;280:32026–32034.
20. Berke SJ, Paulson HL. Protein aggregation and the ubiquitin proteasome pathway: gaining the UPPer hand on neurodegeneration. *Curr Opin Genet Dev*. 2003;13:253–261.
21. Burnett BG, Pittman RN. The polyglutamine neurodegenerative protein ataxin 3 regulates aggresome formation. *Proc Natl Acad Sci U S A*. 2005;102:4330–4335.
22. Onodera Y, Aoki M, Mizuno H, et al. Clinical features of chromosome 16q22.1 linked autosomal dominant cerebellar ataxia in Japanese. *Neurology*. 2006;67:1300–1302.

23. Ishikawa K, Toru S, Tsunemi T, et al. An autosomal dominant cerebellar ataxia linked to chromosome 16q22.1 is associated with a single-nucleotide substitution in the 5′ untranslated region of the gene encoding a protein with spectrin repeat and Rho guanine-nucleotide exchange-factor domains. *Am J Hum Genet*. 2005;77:280–296.
24. Ikeda Y, Dick KA, Weatherspoon MR, et al. Spectrin mutations cause spinocerebellar ataxia type 5. *Nat Genet*. 2006;38:184–190.
25. Zhuchenko O, Bailey J, Bonnen P, et al. Autosomal dominant cerebellar ataxia (SCA6) associated with small polyglutamine expansions in the alpha 1A-voltage-dependent calcium channel. *Nat Genet*. 1997;15:62–69.
26. Holmberg M, Duyckaerts C, Durr A, et al. Spinocerebellar ataxia type 7 (SCA7): a neurodegenerative disorder with neuronal intranuclear inclusions. *Hum Mol Genet*. 1998;7:913–918.
27. Kaytor MD, Duvick LA, Skinner PJ, et al. Nuclear localization of the spinocerebellar ataxia type 7 protein, ataxin-7. *Hum Mol Genet*. 1999;8:1657–1664.
28. David G, Abbas N, Stevanin G, et al. Cloning of the SCA7 gene reveals a highly unstable CAG repeat expansion. *Nat Genet*. 1997;17:65–70.
29. Palhan VB, Chen S, Peng GH, et al. Polyglutamine-expanded ataxin-7 inhibits STAGA histone acetyltransferase activity to produce retinal degeneration. *Proc Natl Acad Sci U S A*. 2005;102:8472–8477.
30. Koob MD, Moseley ML, Schut LJ, et al. An untranslated CTG expansion causes a novel form of spinocerebellar ataxia (SCA8). *Nat Genet*. 1999;21:379–384.
31. Matsuura T, Yamagata T, Burgess DL, et al. Large expansion of the ATTCT pentanucleotide repeat in spinocerebellar ataxia type 10. *Nat Genet*. 2000;26:191–194.
32. Worth PF, Giunti P, Gardner-Thorpe C, et al. Autosomal dominant cerebellar ataxia type III: linkage in a large British family to a 7.6-cM region on chromosome 15q14–21.3. *Am J Hum Genet*. 1999;65:420–426.
33. Holmes SE, O'Hearn EE, McInnis MG, et al. Expansion of a novel CAG trinucleotide repeat in the 5′ region of PPP2R2B is associated with SCA12. *Nat Genet*. 1999;23:391–392.
34. Chen DH, Cimino PJ, Ranum LP, et al. The clinical and genetic spectrum of spinocerebellar ataxia 14. *Neurology*. 2005;64:1258–1260.
35. Subramony SH, Filla A. Autosomal dominant spinocerebellar ataxias ad infinitum? *Neurology*. 2001;56:287–289.
36. Miyoshi Y, Yamada T, Tanimura M, et al. A novel autosomal dominant spinocerebellar ataxia (SCA16) linked to chromosome 8q22.1–24.1. *Neurology*. 2001;57:96–100.
37. Miura S, Shibata H, Furuya H, et al. The contactin 4 gene locus at 3p26 is a candidate gene of SCA16. *Neurology*. 2006;67:1236–1241.
38. Nakamura K, Jeong SY, Uchihara T, et al. SCA17, a novel autosomal dominant cerebellar ataxia caused by an expanded polyglutamine in TATA-binding protein. *Hum Mol Genet*. 2001;10:1441–1448.
39. Brkanac Z, Fernandez M, Matsushita M, et al. Autosomal dominant sensory/motor neuropathy with Ataxia (SMNA): linkage to chromosome 7 q22–q32. *Am J Med Genet*. 2002;114:450–457.
40. Yu GY, Howell MJ, Roller MJ, et al. Spinocerebellar ataxia type 26 maps to chromosome 19p13.3 adjacent to SCA6. *Ann Neurol*. 2005;57:349–354.
41. Verbeek DS, van de Warrenburg BP, Wesseling P, et al. Mapping of the SCA23 locus involved in autosomal dominant cerebellar ataxia to chromosome region 20p13–12.3. *Brain*. 2004;127(Pt 11):2551–2557.
42. Stevanin G, Broussolle E, Streichenberger N, et al. Spinocerebellar ataxia with sensory neuropathy (SCA25). *Cerebellum*. 2005;4:58–61.
43. Cagnoli C, Mariotti C, Taroni F, et al. SCA28, a novel form of autosomal dominant cerebellar ataxia on chromosome 18p11.22-q11.2. *Brain*. 2006;129(Pt 1):235–242.
44. Schelhaas HJ, van de Warrenburg BP. Clinical, psychological, and genetic characteristics of spinocerebellar ataxia type 19 (SCA19). *Cerebellum*. 2005;4:51–54.
45. Cader MZ, Steckley JL, Dyment DA, et al. A genome-wide screen and linkage mapping for a large pedigree with episodic ataxia. *Neurology*. 2005;65:156–158.
46. Escayg A, De Waard M, Lee DD, et al. Coding and noncoding variation of the human calcium-channel beta4-subunit gene CACNB4 in patients with idiopathic generalized epilepsy and episodic ataxia. *Am J Hum Genet*. 2000;66:1531–1539.
47. Jen JC, Wan J, Palos TP, et al. Mutation in the glutamate transporter EAAT1 causes episodic ataxia, hemiplegia, and seizures. *Neurology*. 2005;65:529–534.
48. Durr A, Cossee M, Agid Y, et al. Clinical and genetic abnormalities in patients with Friedreich's ataxia. *N Engl J Med*. 1996;335:1169–1175.
49. Pandolfo M, Montermini L. Molecular genetics of the hereditary ataxias. *Adv Genet*. 1998;38:31–68.
50. Campuzano V, Montermini L, Molto MD, et al. Friedreich's ataxia: autosomal recessive disease caused by an intronic GAA triplet repeat expansion. *Science*. 1996;271:1423–1427.
51. Seznec H, Simon D, Bouton C, et al. Friedreich ataxia: the oxidative stress paradox. *Hum Mol Genet*. 2005;14(4):463–474.
52. Rustin P, Bonnet D, Rotig A, et al. Idebenone treatment in Friedreich patients: one-year-long randomized placebo-controlled trial. *Neurology*. 2004;62:524–525.
53. Hart PE, Lodi R, Rajagopalan B, et al. Antioxidant treatment of patients with Friedreich ataxia: four-year follow-up. *Arch Neurol*. 2005;62(4):621–626.
54. Herman D, Jenssen K, Burnett R, et al. Histone deacetylase inhibitors reverse gene silencing in Friedreich's ataxia. *Nat Chem Biol*. 2006;2:551–558.
55. Ouahchi K, Arita M, Kayden H, et al. Ataxia with isolated vitamin E deficiency is caused by mutations in the alpha-tocopherol transfer protein. *Nat Genet*. 1995;9:141–145.
56. Sharp D, Blinderman L, Combs KA, et al. Cloning and gene defects in microsomal triglyceride transfer protein associated with abetalipoproteinaemia. *Nature*. 1993;365:65–69.
57. Shiloh Y. ATM and ATR: networking cellular responses to DNA damage. *Curr Opin Genet Dev*. 2001;11:71–77.
58. Savitsky K, Bar-Shira A, Gilad S, et al. A single ataxia telangiectasia gene with a product similar to PI-3 kinase. *Science*. 1995;268:1749–1753.
59. Stewart GS, Maser RS, Stankovic T, et al. The DNA double-strand break repair gene hMRE11 is mutated in individuals with an ataxia-telangiectasia-like disorder. *Cell*. 1999;99:577–587.
60. Moreira MC, Klur S, Watanabe M, et al. Senataxin, the ortholog of a yeast RNA helicase, is mutant in ataxia-ocular apraxia 2. *Nat Genet*. 2004;36:225–227.
61. Asaka T, Yokoji H, Ito J, et al. Autosomal recessive ataxia with peripheral neuropathy and elevated AFP: novel mutations in SETX. *Neurology*. 2006;66:1580–1581.

62. Bassuk AG, Chen YZ, Batish SD, et al. In cis autosomal dominant mutation of Senataxin associated with tremor/ataxia syndrome. *Neurogenetics*. 2007;8:45–49.
63. Chen YZ, Bennett CL, Huynh HM, et al. DNA/RNA helicase gene mutations in a form of juvenile amyotrophic lateral sclerosis (ALS4). *Am J Hum Genet*. 2004;74:1128–1135.
64. Takashima H, Boerkoel CF, John J, et al. Mutation of TDP1, encoding a topoisomerase I-dependent DNA damage repair enzyme, in spinocerebellar ataxia with axonal neuropathy. *Nat Genet*. 2002;32:267–272.
65. Nance MA, Berry SA. Cockayne syndrome: review of 140 cases. *Am J Med Genet*. 1992;42:68–84.
66. Engert JC, Berube P, Mercier J, et al. ARSACS, a spastic ataxia common in northeastern Quebec, is caused by mutations in a new gene encoding an 11.5-kb ORF. *Nat Genet*. 2000;24:120–125.
67. Mrissa N, Belal S, Hamida CB, et al. Linkage to chromosome 13q11–12 of an autosomal recessive cerebellar ataxia in a Tunisian family. *Neurology*. 2000;54:1408–1414.
68. Anttonen AK, Mahjneh I, Hamalainen RH, et al. The gene disrupted in Marinesco–Sjogren syndrome encodes SIL1, an HSPA5 cochaperone. *Nat Genet*. 2005;37:1309–1311.
69. Nikali K, Suomalainen A, Saharinen J, et al. Infantile onset spinocerebellar ataxia is caused by recessive mutations in mitochondrial proteins Twinkle and Twinky. *Hum Mol Genet*. 2005;14:2981–2990.
70. Durr A. Autosomal dominant cerebellar ataxias: polyglutamine expansions and beyond. *Lancet Neurol*. 2010;9:885–894.
71. Leehey MA, Hagerman PJ. Fragile X-associated tremor/ataxia syndrome. *Handb Clin Neurol*. 2012;103:373–386.
72. Verbeek DS, van de Warrenburg BP. Genetics of the dominant ataxias. *Semin Neurol*. 2011;31:461–469.
73. Orr HT. Cell biology of spinocerebellar ataxia. *J Cell Biol*. 2012;197:167–177.
74. Dion V, Wilson JH. Instability and chromatin structure of expanded trinucleotide repeats. *Trends Genet*. 2009;25:288–297.
75. Bauer PO, Nukina N. The pathogenic mechanisms of polyglutamine diseases and current therapeutic strategies. *J Neurochem*. 2009;110:1737–1765.
76. Nelson DL, Orr HT, Warren ST. The unstable repeats—three evolving faces of neurological disease. *Neuron*. 2013;77:825–843.
77. Perlman SL. Treatment and management issues in ataxic diseases. *Handb Clin Neurol*. 2012;103:635–654.
78. Ristori G, Romano S, Visconti A, et al. Riluzole in cerebellar ataxia: a randomized, double-blind, placebo-controlled pilot trial. *Neurology*. 2010;74:839–845.
79. Zesiewicz TA, Greenstein PE, Sullivan KL, et al. A randomized trial of varenicline (Chantix) for the treatment of spinocerebellar ataxia type 3. *Neurology*. 2012;78:545–50.

CHAPTER

72

Friedreich Ataxia

Massimo Pandolfo

Université Libre de Bruxelles, Hôpital Erasme, Brussels, Belgium

CLINICAL FEATURES

Rigorous diagnostic criteria for Friedreich ataxia (FRDA) were established in the late 1970s and early 1980 by the Quebec Collaborative group[1] and by Anita Harding[2] (Table 72.1). Since the gene was discovered in 1996, the phenotype has been extended beyond the limits fixed by these criteria.[3,4]

The typical age of onset of FRDA is around puberty, but it may be earlier or later, even in late adult life. Age of onset may dramatically vary within a sibship, a phenomenon now in part explained by the dynamic nature of the underlying mutation.

Gait instability or generalized clumsiness are the most frequent symptoms at onset. Typically, the patient is a child who recently started to sway when walking and falls easily. The child may be active in sports and have shown no clues of the impending neurological illness, but more commonly the child had been considered clumsy for some time before the overt appearance of symptoms. Scoliosis, often considered to be idiopathic, may precede the onset of ataxia. Rare patients (5%) are diagnosed with idiopathic hypertrophic cardiomyopathy and treated as such for up to 2–3 years before the appearance of neurological symptoms.

Mixed cerebellar and sensory ataxia is the hallmark of the disease. It begins as truncal ataxia causing swaying, imbalance, and falls. At the very beginning of onset, the balance and gait abnormalities are subtle and may be revealed only as a difficulty in tandem gait and in standing on one foot. Subsequently, gait becomes frankly ataxic, with irregular steps, veering, and difficulty in turning. Loss of upright stability becomes evident when standing with feet close together and is worsened by eye closure (positive Romberg sign). With further progression, gait becomes broad based, with frequent loss of balance, requiring at first intermittent support (furniture, walls, an accompanying person's arm), then a cane, and then a walker. Patients become completely unable to stand with feet close together, then need support even with their feet apart. Eventually, on average 10–15 years after onset, patients lose the ability to walk, stand, and sit without support. Evolution is variable, however, with mild cases in which patients are still ambulatory decades after onset.

Limb ataxia appears after truncal ataxia. Fine motor skills become impaired, with increasing difficulty in activities such as writing, dressing, handling utensils. Dysmetria and intention tremor are evident. Ataxia is progressive and unremitting, though periods of relative stability are frequent at the beginning of the illness.

With rare exceptions, dysarthria appears within 5 years from clinical onset. It consists of slow, jerky speech with sudden utterances, and progresses until speech becomes almost unintelligible. Dysphagia, particularly for liquids, appears only with advanced disease. Modified foods, then a nasogastric tube or gastrostomy feedings are eventually required.

Loss of position and vibration sense is invariable, but may not be evident at onset. Perception of light touch, pain, and temperature, initially normal, tends to decrease with advancing disease. Loss of tendon reflexes in the lower limbs was considered essential for the diagnosis. However, a minority of patients with a positive molecular test for FRDA, referred to as "FRDA with retained reflexes" (FARR), have deep tendon reflexes, which may sometimes be brisk and associated with spasticity. Extensor plantar responses are found in most patients. Muscular weakness becomes severe only with advanced disease. Ataxia and not weakness is the primary cause for loss of ambulation: even when patients become wheelchair bound, they still maintain on average 70% of their normal strength in lower limbs.

Rosenberg's Molecular and Genetic Basis of Neurological and Psychiatric Disease
http://dx.doi.org/10.1016/10.1016/B978-0-12-410529-4.00072-3

TABLE 72.1 Diagnostic Criteria for Friedreich Ataxia According to Harding

Autosomal recessive inheritance
Onset before age 25
Within 5 years from onset:
Limb and truncal ataxia showing evolution in a 2-year period
Absent tendon reflexes in the legs
Extensor plantar responses
Motor nerve conduction velocity ~40 m/s in upper limbs with small or absent sensory-evoked action potentials
After 5 years since onset:
As above plus dysarthria
Additional criteria, not essential for diagnosis, present in ~67% of cases:
Scoliosis
Pyramidal weakness of the legs
Absent reflexes in upper limbs
Distal loss of joint position and vibration sense in lower limbs
Abnormal ECG
Other features, present in ~50% of cases:
Nystagmus
Optic atrophy
Deafness
Distal weakness and wasting
Pes cavus
Diabetes

Reproduced with permission from[2].

In general, the combination of sensory neuropathy and of pyramidal tract degeneration results most often in the typical picture of areflexia associated with extensor plantar responses, but sometimes one component prevails. Such partial pictures are usually observed in milder cases. Muscle tone is often normal at onset. However, with advancing pyramidal involvement, and particularly when gait becomes severely impaired, many patients complain of spasms in the lower limbs, mostly nocturnal. Despite limited involvement of lower motor neurons, distal amyotrophy in the lower limbs and in the hands is frequent, even early in the course of the disease. When patients are wheelchair bound, disuse atrophy occurs.

The typical oculomotor abnormality of FRDA is fixation instability with square-wave jerks. Various combinations of cerebellar, vestibular, and brainstem oculomotor signs may be observed, but gaze-evoked nystagmus is uncommon, and ophthalmoparesis does not occur. About 30% of patients develop optic atrophy, with or without visual impairment.[5] Sensorineural hearing loss affects about 20% of the patients, becoming evident with advanced disease. However, auditory pathway dysfunction with a normal tonal audiometry occurs early and is a universal feature of the disease, leading in particular to impaired speech perception.[6] Optic atrophy and sensorineural hearing loss tend to be associated with each other, both frequently occurring in older patients with severe neurological deficits.

Cognitive functions are well preserved, though there are reports of slowed information processing speed not accompanied by major prefrontal cortex and mood disorders.[7,8] Accordingly, on one hand, intellectual and learning disabilities do not seem to be more prevalent in patients with FRDA than in the general population. On the other hand, FRDA, as any disease causing substantial physical disability, may have a substantial impact on academic, professional, and personal development if not properly managed.

Shortness of breath (40%) and palpitations (11%) are the most common symptoms of heart involvement. Electrocardiography (ECG) shows inverted T-waves in essentially all patients, ventricular hypertrophy in most, conduction disturbances in about 10%, occasionally supraventricular ectopic beats, and atrial fibrillation. Repeated ECG recordings are the most sensitive test to detect the FRDA cardiomyopathy. Echocardiography and Doppler-echocardiography demonstrate concentric hypertrophy of the ventricles (62%) or asymmetric septal hypertrophy (29%), with diastolic function abnormalities.

About 10% of FRDA patients develop diabetes mellitus, and 20–40% have carbohydrate intolerance. The mechanisms are complex, involving β-cell dysfunction and loss as well as peripheral insulin resistance.[9] Some cases may be initially controlled by oral hypoglycemia drugs, but insulin is eventually necessary. Diabetes may increase the burden of disease, complicate the neurological picture, and promote potentially fatal complications. Kyphoscoliosis may cause pain and cardiorespiratory problems. Pes cavus and pes equinovarus may further affect ambulation. Autonomic disturbances, most commonly cold and cyanotic legs and feet, become increasingly frequent as the disease advances. Parasympathetic abnormalities such as decreased heart rate variability parameters have been reported. Urgency of micturition does occur with advancing disease.

Variant Phenotypes

A positive molecular test for FRDA is found in up to 10% of patients with recessive or sporadic ataxia who do not fulfill the classical clinical diagnostic criteria for the disease. A reanalysis of cases revealed that no clinical finding or combination of findings characterizes exclusively or is necessarily present in positive cases (Table 72.2). Even neurophysiological evidence of axonal sensory neuropathy is rarely absent in patients with a proven molecular diagnosis. Overall, however, the classical features of FRDA, including cardiomyopathy, are highly predictive of a positive test. Absence of cardiomyopathy and moderate to severe cerebellar cortical atrophy, demonstrated by magnetic resonace imaging (MRI), appear to be the best predictors of a negative test. A molecular test for FRDA is indicated in the initial work-up of all patients with chronic progressive ataxia where neuroimaging and other investigations have excluded structural lesions and show no (or very mild) cerebellar atrophy, regardless of whether or not they fulfill the diagnostic criteria.[10]

Neurophysiological Investigations

Sensory nerve action potentials are severely reduced or, most often, lost. Motor and sensory (when measurable) nerve conduction velocities, in contrast, are within or just below the normal range. These findings clearly distinguish

TABLE 72.2 Frequency of Signs and Symptoms Observed in 59 Consecutive Patients with a Positive Molecular Test for Friedreich Ataxia

Sign/symptom	Frequency
Ataxia	100%
Deep sensory loss in lower limbs	89%
Dysarthria	83%
Cardiomyopathy	79%
Lower limb areflexia	75%
Square-wave jerks	75%
Upper limb areflexia	72%
Scoliosis	68%
Babinski sign	62%
Pes cavus	58%
Weakness of lower limbs	46%
Distal amyotrophy	26%
Decreased muscle tone in lower limbs	22%
Carbohydrate intolerance	18%
Optic atrophy	13%
Cerebellar atrophy at MRI	11%
Increased muscle tone in lower limbs	7%
Hearing loss	6%

an early case of FRDA from a case of hereditary demyelinating sensorimotor neuropathy. Degeneration of both peripheral and central sensory fibers also results in dispersion and delay of somatosensory evoked potentials. Brainstem auditory-evoked potentials, beginning from the most rostral component, wave V, progressively deteriorate in all patients.[11] Visual-evoked potentials are commonly (50–90%) reduced in amplitude with a normal P-100 latency. Central motor conduction velocity, determined by cortical magnetic stimulation, is slower than normal, and, contrary to the sensory involvement, this slowing becomes more severe with increasing disease duration.[12,13]

Biochemical Investigations

Commonly tested biochemical parameters in blood, urine, and cerebrospinal fluid are normal in FRDA. There is an ongoing search for biomarkers of disease status, progression and response to treatments. Measures of iron metabolism and oxidative stress have been investigated first. Blood iron, iron binding capacity and ferritin are normal; an increase in circulating transferrin receptors has been reported but is controversial.[14] Markers of oxidative stress in FRDA include increased urinary 8-hydroxy-2′-deoxyguanosine (8OH2′dG), a marker of oxidative DNA damage,[15] increased plasma malondialdheyde, a marker of lipid peroxidation,[16] and decreased free glutathione. However, measuring oxidative stress markers has proven difficult in clinical practice, and an attempt to use 8OH2′dG as an endpoint in a clinical trial has failed.[17] Proteomic studies of plasma and urine are under way. Gene expression profiling in peripheral blood mononuclear cells (PBMCs) has detected a set of genes whose expression is significantly and consistently different in FRDA patients versus carriers and controls. In *ex vivo* experiments, some of these genes respond to treatments that normalize frataxin levels.[18]

Neuroimaging

The characteristic neuroimaging finding in FRDA is the thinning of the cervical spinal cord. It can be detected on sagittal and axial MRI, along with signal abnormalities in the posterior and lateral columns. Brainstem, cerebellum, and cerebrum are apparently normal, but quantitative studies showed that all these structures are smaller in FRDA patients than in healthy controls. The extent of this diffuse atrophy correlates with clinical severity. Vermian and lobar cerebellar atrophy only occurs in more severe and more advanced cases and is usually mild. In fact, marked cerebellar atrophy early in the course of the disease is a strong predictor of a negative molecular test for FRDA.[10] However, Tc-HMPAO single-positron emission tomography (SPECT) reveals a decrease in cerebellar blood flow, more than expected for the limited degree of atrophy.

More recently, advanced imaging techniques have been used to detect structural changes in FRDA, which have been confirmed to extend beyond the cerebellum and spinal cord. Diffusion-based MRI studies showed significantly lower fractional anisotropy, higher mean diffusivity and increased radial diffusivity in dentatorubral, dentatothalamic and thalamocortical tracts, correlating with clinical severity.[19,20] Altered cerebellocerebral white matter connectivity, as shown by diffusion MRI and tractography,[21] is the likely basis of abnormalities detected by functional imaging studies, in particular decreased brain activation during specific cognitive tasks.[22,23]

Metabolic imaging studies have also revealed widespread abnormalities. Positron emission tomography (PET) scans with fluorodeoxyglucose as tracer reveal an increased glucose uptake in the brain of FRDA patients who are still ambulatory. Glucose consumption decreases as the disease progresses and as patients lose their ability to walk, with glucose consumption eventually becoming subnormal.[24] Although not yet fully explained, this abnormality may relate to mitochondrial dysfunction. Proton magnetic resonance spectroscopy (1H-MRS) shows a decrease in the neuronal marker *N*-acetyl-aspartate (NAA) and in the antioxidant glutathione (GSH) in the cerebellar hemispheres that appear to be proportional to the size of the smaller GAA expanded repeat.[25] Phosphorus MRS analysis of skeletal muscle and heart showed a reduced rate of adenosine triphosphate (ATP) synthesis after exercise, which is inversely correlated to GAA expansion sizes[26] and is improved by coenzyme Q10 treatment.

Prognosis

On average, typical FRDA patients become wheelchair bound 15 years after onset. Patients with milder cases may remain ambulatory much longer, and those with severe cases may lose ambulation in childhood. Early onset and left ventricular hypertrophy predict a faster rate of progression. The burden of neurological impairment, cardiomyopathy, and diabetes, if present, shortens life expectancy. In a retrospective study of FRDA patients to determine cause of death, followed by a case-control analysis comparing characteristics of deceased patients with living, age- and sex-matched FRDA controls, cardiac dysfunction was the most frequent cause of death, most commonly congestive heart failure or arrhythmia.[27-29] In this study, median age of death due to cardiac cause was 17 years, while median age of death due to non-cardiac cause was 29 years.[28]

Survival may be significantly prolonged with appropriate treatment of cardiac symptoms, particularly arrhythmias, and of diabetes and by preventing and managing complications resulting from prolonged neurological disability. Carefully assisted patients may live several more decades.

Diagnosis

Nonhereditary diseases may cause early-onset progressive ataxia with a chronic course. MRI will reveal posterior fossa tumors and malformations, including platybasia and basilar impression, as well as the lesions of multiple sclerosis. Autoimmune disorders may also cause ataxia, although some autoimmune causes such as antiglutamic decarboxylase antibodies are rare in the young age group. Detection of antigliadin and antiendomysial antibodies points to a diagnosis of gluten sensitivity, which may cause a progressive and treatable cerebellar syndrome, even in the absence of gastrointestinal symptoms and signs of malabsorption.

Other recessive ataxias may sometimes resemble typical or atypical FRDA. Recessive ataxias are a large and heterogeneous group of diseases. They include metabolic disorders such as abetalipoproteinemia, Refsum disease, late-onset G_{M2} gangliosidoses, cerebrotendinous xanthomatosis, coenzyme Q10 deficiency, and several leukodystrophies. Other recessive ataxias may be due to primary mitochondrial dysfunction, such as those associated with mutations in DNA polymerase and in the twinkle/twinky mitochondrial helicases. DNA repair abnormalities are found in ataxia-telangiectasia and related disorders, including ataxia-oculomotor apraxia types 1 and 2. Early prominent cerebellar atrophy is found only in these non-FRDA cases, some of which are also characterized by typical clinical, laboratory, and imaging features. An FRDA-like clinical picture occurs in cases of isolated vitamin E deficiency, an autosomal recessive disease due to defective conjugation of this vitamin to lipoproteins. Differentiating features of isolated vitamin E deficiency include a less severe peripheral neuropathy, the common occurrence of a nodding head tremor and of pigmentary retinopathy, and the lack of cardiomyopathy. A similar neurological picture may occur when vitamin E deficiency is secondary to a genetic disorder of lipoproteins (e.g., in abetalipoproteinemia) or to malabsorption.

DNA testing is the gold standard to confirm a clinical diagnosis of FRDA, for carrier detection, and for prenatal diagnosis (see section "Clinical and Molecular Genetics," below).

PATHOLOGY

Central Nervous System

Degeneration of the posterior columns of the spinal cord is the hallmark of FRDA.[30] Atrophy is more severe in the Goll than in the Burdach tract; therefore, the longest fibers originating more caudally are most affected. The spinocerebellar tracts, the dorsal more than the ventral, and their neurons of origin, the Clarke column, also become atrophic. In the brainstem, atrophy involves the gracilis and cuneate nuclei and the medial lemnisci, particularly in their ventral portion deriving from the gracile nuclei, as well as the cranial nerve sensory nuclei and entering roots of the V, IX, and X nerves, the descending trigeminal tracts, the solitary tracts, and the accessory cuneate nuclei, corresponding to the Clarke column in the spinal cord. Overall, FRDA severely affects the sensory systems providing information to the brain and cerebellum about the position and speed of body segments. The motor system is directly affected as well: the long crossed and uncrossed corticospinal motor tracts are atrophic, more so distally, suggesting a dying-back phenomenon. There is a variable loss of pyramidal neurons in the motor cortex, whereas motor neurons in the brainstem and in the ventral horns of the spinal cord are less affected. In the cerebellum, cortical atrophy occurs only late and is mild, whereas atrophy of the dentate nuclei and of superior cerebellar pedunculi is prominent. Quantitative analysis of synaptic terminals indicates a loss of contacts over Purkinje cell bodies and proximal dendrites. The auditory system and the optic nerves and tracts are variably affected. The external pallidus and subthalamic nuclei may show a moderate cell loss. Finally, many patients with FRDA die as a consequence of heart disease, and corresponding widespread hypoxic changes and focal infarcts are often found in the central nervous system.

In the peripheral nervous system, the major abnormalities occur in the dorsal root ganglia, where a loss of large primary sensory neurons is observed, accompanied by proliferation of satellite cells. Loss of large myelinated sensory fibers is prominent in peripheral nerves, but the fine, unmyelinated fibers are well preserved. Interstitial connective tissue is increased. This sensory axonal neuropathy is an early event in the course of FRDA and, according to some authors, is scarcely progressive.

Heart

In the heart, ventricular walls and interventricular septum are thickened. Hypertrophic cardiomyocytes are found early in the disease and are intermingled with normal-appearing ones. Subsequently, atrophic, degenerating, even necrotic fibers progressively appear, there is diffuse and focal inflammatory cell infiltration, and connective tissue increases. In the late stage of the disease, with extensive fibrosis, the cardiomyopathy becomes dilated. A variable number of cardiomyocytes, from 1–10%, show intracellular iron deposits.[31] This is a specific finding in FRDA and a direct consequence of the basic biochemical defect.

Other Organs

More than three-quarters of the patients have kyphoscoliosis, mostly as a double thoracolumbar curve resembling the idiopathic form. Pes cavus, pes equinovarus, and clawing of the toes are found in about half the cases. Patients with diabetes (about 10%) often show a loss of islet cells in the pancreas, which is not accompanied by the autoimmune inflammatory reaction found in type I diabetes.

CLINICAL AND MOLECULAR GENETICS

Gene Structure and Expression

The FRDA gene (*FXN*) is localized in the proximal long arm of chromosome 9 and is composed by seven exons spread over 95 Kb of genomic DNA.[32] The major, and probably only, functionally relevant mRNA has a size of 1.3 kb and corresponds to the first five exons, numbered 1 to 5a. The encoded protein, predicted to contain 210 amino acids, is called frataxin. In adult humans, frataxin mRNA is most abundant in the heart and spinal cord, followed by liver, skeletal muscle, and pancreas. In mouse embryos, expression starts in the neuroepithelium at embryonic day 10.5 (E10.5), then reaches its highest level at E14.5 and into the postnatal period, with the highest levels being found in the spinal cord, particularly at the thoracolumbar level, and in the dorsal root ganglia. The developing brain is also very rich in frataxin mRNA, which is abundant in the proliferating neural cells in the periventricular zone, in the cortical plates, and in the ganglionic eminence.[33]

Intronic GAA Triplet Repeat Expansion

FRDA is the consequence of frataxin deficiency. Most patients are homozygous for the expansion of a GAA triplet repeat sequence (TRS) in the first intron of the gene, which results in inhibited frataxin transcription; 5% are heterozygous for a GAA expansion and a point mutation or a deletion in the frataxin gene. Repeats in normal chromosomes contain up to about 40 triplets, whereas there are 90–1000 triplets in FRDA chromosomes.[34] Expanded alleles show meiotic and mitotic instability.

In vitro and in bacterial plasmids, lengths of the GAA TRS corresponding to pathological FRDA alleles tend to fold back onto themselves and adopt a non-B DNA structure, consisting of an intramolecular triple helix formed by two GAAs and a CTT strand, with an unpaired CTT strand. Two triplex segments may associate to form a novel DNA structure called sticky DNA.[35] Triplexes and sticky DNA strongly inhibit transcription by blocking and possibly sequestering the advancing RNA polymerase complex.[36] According to one hypothesis, the expanded GAA TRS is supposed to oscillate between the triplex and the standard B-DNA structure, which would be responsible for the residual frataxin transcription observed in FRDA patients, with the equilibrium shifting toward the triplex as the length of the repeat increases. However, no direct demonstration of the triplex structure in the cell nucleus has been provided so far.

The current model for transcriptional silencing by the expanded GAA TRS involves the induction of chromatin condensation. This was first based on the observation, made in transgenic mice, that expanded GAA repeats cause the transcriptional silencing of a nearby gene in a way that closely resembles position variegation effect, a phenomenon occurring when a gene is in the vicinity of a heterochromatic region.[37] Notably, the expanded repeating sequence in FRDA (GAA) is similar to a type of heterochromatin-associated satellite DNA in *Drosophila* (GAGAA). Furthermore, studies in cells from FRDA patients and in animal models carrying long GAA repeats have provided evidence in the vicinity of the repeats of decreased acetylation of several lysine residues in core histones and of increased DNA methylation, which are epigenetic marks of transcriptionally silent chromatin.[38] The key modification appears to be the deacetylation of lysine 9 of histone H3 (H3K9) and its trimethylation. Trimethylated H3K9 binds heterochromatin protein 1 (HP1), which is responsible for chromatin condensation and transcriptional silencing. Counteracting H3K9 deacetylation is being intensively explored as a therapeutic approach for FRDA.[39]

The triplex and chromatin hypotheses are not mutually exclusive, as chromating condensation may be triggered by the abnormal structure of this DNA segment.

As expected by the fact that smaller expansions allow a higher residual expression of the frataxin gene, the severity and age of onset of the disease are in part determined by the size of the expanded GAA repeat, in particular of the smaller allele. A number of studies have shown a direct correlation between the size of GAA repeats and earlier age of onset, earlier age when confined in wheelchair, more rapid rate of disease progression, and presence of nonobligatory disease manifestations indicative of more widespread degeneration. However, differences in GAA expansions account for only about 50% of the variability in age of onset. Other factors that influence the phenotype may include somatic mosaicism for expansion sizes,[40] variations in the frataxin gene itself, modifier genes, and environmental factors.

Frataxin Point Mutations

About 2% of FRDA chromosomes carry GAA repeat of normal length, but have missense, nonsense, or splice site mutations ultimately affecting the frataxin coding sequence.[3,41] FRDA patients carrying a frataxin point mutation are in all cases compound heterozygotes for a GAA repeat expansion. Therefore, as GAA expansion homozygotes, they express a small amount of normal frataxin, which is thought to be essential for survival (see section "Animal Models" below). Missense mutations have so far been identified only in the *C*-terminal portion of the protein corresponding to the mature intramitochondrial form of frataxin. Nonsense and most missense mutations result in a typical FRDA phenotype, and a few missense mutations are associated with milder atypical phenotypes with slow progression, suggesting that the mutated proteins preserve some residual function. The p.Gly130Val mutation in particular is associated with early onset but slow progression, no dysarthria, mild limb ataxia, and retained reflexes. For unclear reasons, optic atrophy is more frequent in patients with point mutations of any kind (50%).[41]

DNA Testing

Polymerase chain reaction (PCR) or Southern blot can be used to directly detect the intronic GAA expansion. PCR requires only a small amount of DNA, but high-quality DNA is extremely important. Ambiguous results, particularly in heterozygote detection, can be resolved by subsequent hybridization of PCR products with an oligonucleotide probe containing a GAA or CTT repeat. Southern blot requires more DNA and is less accurate for determining the number of repeats, but it will not miss a heterozygote when heterozygosity detection is crucial.

The question of the lower limit for an expansion to be pathological, so far reported to be 66 triplets, is not so critical in FRDA as in dominant unstable repeat diseases. In any case, the diagnosis is strongly supported when the other chromosome carries a fully expanded GAA repeat or a frataxin point mutation. *In vitro* data indicate that repeats exceeding about 50 triplets start to adopt a triplex structure and may form sticky DNA, suggesting that they may be pathological. Length and sequence of the repeat must both be taken into account in uncertain cases. Interruptions of the GAA triplet repeat sequence prevent the formation of a triplex structure and sticky DNA *in vitro*. *In vivo*, interrupted or variant repeats are occasionally found and may be nonpathological up to the equivalent of 130 triplets.[42]

Testing for point mutation is done in research labs. No mutation in the coding sequence could be identified in up to 40% of compound heterozygous patients in some series. Some of these patients may actually have a different disease and carry a GAA expansion by chance, given its high frequency (1:90) in whites. One such case has been documented.

Individuals with a FRDA-like phenotype but no GAA expansion most likely have mutations at a different locus. Homozygosity for frataxin point mutations has never been reported. However, individuals with progressive degenerative ataxia have been reported with low levels of frataxin in blood cells, but no expanded GAA repeat and no identifiable frataxin point mutations.[43] Further studies will indicate if such individuals carry a different kind of mutation in the *FXN* gene.

Frataxin Function

Frataxin is highly conserved during evolution, with homologs in mammals, invertebrates, yeast, and plants. It is imported into the mitochondrial matrix, where an *N*-terminal peptide of about 40 amino acids is removed generating the mature, functional protein. A homolog of frataxin is also found in Gram-negative bacteria and is called CyaY in *Escherichia coli*. Structural studies have shown that mature frataxin is a compact, globular protein containing an *N*-terminal α-helix, a middle β-sheet region composed of seven β strands, a second α-helix, and a *C*-terminal coil. The α-helices are folded upon the β-sheet, with the *C*-terminal coil filling a groove between the two α-helices. On the outside a ridge of negatively charged residues and a patch of hydrophobic residues are highly conserved.

A number of functions have been attributed to frataxin,[44] with studies suggesting a role in iron storage and detoxification, stimulation of oxidative phosphorylation, activation of antioxidant defenses, regulation of iron metabolism, and protection against pro-apoptotic stimuli. Some of these postulated functions remain controversial; others may represent downstream effects rather than a primary activity of the protein. Genetic, biochemical, and structural data indicate that frataxin is a component of a multiprotein complex that assembles iron–sulfur (Fe–S) clusters in the mitochondrial matrix. Within this complex, frataxin appears to act as an allosteric activator of a key enzyme in Fe–S cluster synthesis, the cysteine desulfurase Nsf1, so that Fe–S cluster synthesis proceeds at a very low rate in the absence of frataxin.[45] Fe–S clusters are cofactors for proteins with a variety of functions and subcellular localizations, including the Krebs cycle enzyme aconitase and several subunits of respiratory complexes I, II and III in the mitochondria, enzymes of amino acid and purine metabolism in the cytosol, and DNA replication and repair factors in the nucleus.[46] In mammalian cells, Fe–S cluster assembly factors are also present in the cytosol, but to be active they need an essential substrate exported from mitochondria and assembled by the mitochondrial Fe–S cluster biogenesis machinery.[46]

ANIMAL AND CELLULAR MODELS

Early embryonic lethality in frataxin knockout mice[47] has complicated the effort to generate a vertebrate animal model of the disease.[48] Viable mouse models were first obtained through a conditional gene-targeting approach by crossing animals with a LoxP-flanked frataxin exon 2 with transgenic animals expressing Cre recombinase under the control of different promoters.[49] Striated muscle-, neuron-, pancreas-, and liver-restricted exon 2 deletions have been obtained this way. Mutants with neuron-specific enolase promoter-directed Cre expression have a low birth weight and develop a progressive neurological phenotype with the average onset of ataxia at 12 days, hunched stance, and loss of proprioception.[49] Mutants with muscle creatine kinase (MCK) promoter-directed Cre expression show cardiac hypertrophy with thickening of the walls of the left ventricle and myocardial degeneration with cytoplasmic vacuolization in the myocytes, evidence of necrosis and postnecrotic fibrosis. Loss of activity of iron-sulfur cluster-containing enzymes is an early finding in these models. The MCK mutants accumulate iron in heart mitochondria at later stages.[49] Using a similar conditional knockout approach but with a tamoxifen-inducible Cre recombinase under the control of a neuron-specific prion protein promoter, two different lines that exhibit a progressive neurological phenotype with slow evolution that recreates the neurological features of the human disease were developed.[50] An autophagic process was detected in the dorsal root ganglia, leading to removal of mitochondrial debris and apparition of lipofuscin deposits.

To have a more faithful model of the human disease, knock-in mice that carry a $(GAA)_{230}$ repeat in the endogenous frataxin gene (*Fxn*) were engineered.[51] This repeat, when homozygous, only leads to a decrease in frataxin expression to 70% of wild-type levels. To further decrease frataxin expression, frataxin knock-in–knockout mice (Fxn/230GAA) were engineered, expressing approximately 25–30% of wild-type frataxin levels, a reduction associated with mild but clinically evident FRDA in humans.[51] These mice have no frank pathology and only minor motor abnormalities, but they show clear changes in gene expression in the central nervous system, heart, and muscle that may represent a basic response to frataxin deficiency not yet blurred by secondary changes due to degeneration and cell loss. Genes identified are involved in nucleic acid and protein metabolism, signal transduction, stress response, and nucleic acid binding, with an over-representation of mitochondria-related transcripts.

Two lines of YAC transgenic mice were obtained carrying the human frataxin gene with an expanded GAA repeat (~200 triplets) and were crossbred with heterozygous frataxin knockout mice.[52] The resultant mice express only low levels of human frataxin mRNA and protein, show decreased aconitase activity and oxidative stress, and develop progressive neurodegenerative and cardiac pathological phenotypes. Coordination deficits are present, as measured by accelerating rotarod analysis, together with a progressive decrease in locomotor activity and increase in weight. Large vacuoles are detected within primary sensory neurons of the dorsal root ganglia, predominantly within the lumbar regions in 6-month-old mice, but spreading to the cervical regions after 1 year of age. Secondary demyelination of large axons is also detected within the lumbar roots of older mice. Lipofuscin deposition is increased in both primary sensory neurons and cardiomyocytes, and iron deposition is detected in cardiomyocytes after 1 year of age.

Cellular models range from yeast cells with deleted or downregulated frataxin homolog, to cells derived from animal models and cells from FRDA patients, in particular fibroblasts and peripheral blood mononuclear cells (PBMCs).

Other models involve transfection with antisense or inhibitory RNA (RNAi)-expressing constructs, or expression of mutated human frataxin on a mouse knockout background. Again, all these models have been useful to investigate some aspects of pathogenesis and to test therapeutics, but none of them consists of specifically vulnerable cells carrying the GAA expansion mutation in human disease.[48] Recently, however, several groups have obtained induced pluripotent stem cells (iPSCs) from an individual with FRDA, which are being used to derive neurons and cardiomyocytes. These cells represent the most faithful *in vitro* model for the disease and are being increasingly used to investigate pathogenesis, specific vulnerability, and the effect of potential therapeutics. Initial studies have demonstrated that iPSCs and iPSC-derived differentiated cells from FRDA patients show GAA repeat instability, epigenetic changes at the *FXN* locus with reduced frataxin expression, and mitochondrial dysfunction.[53]

PATHOGENESIS OF FRIEDREICH ATAXIA

By impairing Fe–S cluster biogenesis, frataxin deficiency results in energy deficit, altered iron metabolism, and oxidative damage. Fe–S cluster proteins in all cellular compartments show reduced activity, but aconitase and respiratory complexes I, II and III in the mitochondria are particularly affected. Impairment of the respiratory chain results in reduced mitochondrial membrane potential, energy deficit, and leaking of electrons that may react with molecular oxygen and form reactive oxygen species (ROS), in particular superoxide (O_2^-).[54] A homeostatic response aiming to restore Fe–S cluster levels, a situation usually caused by iron deficiency, leads to increased cellular iron uptake and import into mitochondria. Iron responsive protein 1 (IRP1) is a cytosolic Fe–S cluster protein functioning as an aconitase when its Fe–S cluster is fully assembled and, when its Fe–S cluster is partially or fully disassembled, as an RNA-binding protein that recognizes a sequence motif on mRNAs for proteins involved in iron metabolism. IRP1 stabilizes mRNAs for iron import proteins, such as the transferrin receptor (TfR), and blocks translation of mRNAs for proteins that need iron, such as aconitase, or that store iron, such as ferritin. Excessive cellular iron import and IRP1 RNA binding activity[55] have indeed been demonstrated in frataxin deficient cells. Other mechanisms must play a role, however, particularly in directing the flow of imported iron into mitochondria. When this response is induced by iron deficiency in a normal cell, increased mitochondrial iron import restores Fe–S cluster levels. In FRDA, because of impaired Fe–S cluster synthesis, iron cannot be efficiently utilized and progressively accumulates in mitochondria.[56] Iron and ROS in mitochondria are the substrates for the Fenton reaction, which generates the toxic hydroxyl radical (OH•) that can damage nucleic acids, proteins and lipids. Excess ROS also lead to the depletion of small antioxidant molecules, in particular glutathione, which is largely consumed by mitochondrial glutathione peroxidase.

Interestingly, several other genes whose mutations lead to defective Fe–S cluster biogenesis have been identified, including *ISCU*, *NFU1*, and others, but the corresponding phenotypes are entirely different from FRDA. The reasons for such heterogeneity are still unknown, as is the basis of specific cell vulnerability in FRDA.

THERAPY

Attempts to treat the disease may involve increasing frataxin production, which in principle could be obtained: i) through gene replacement therapy by introducing a frataxin gene without the GAA expansion into the patient cells; ii) by directly administering frataxin, modified in such a way to be able to reach the nerve cells affected by the disease and the mitochondria inside these cells; or iii) by intervening at the transcriptional level with drugs that decondense chromatin as histone deacetylase inhibitors. Though still in their infancy, these approaches are under study. In experimental models, encouraging results have been obtained with gene replacement therapy, using adeno-associated virus and lentivirus vectors, and with histone deacetylase inhibitors.[57] Therapy may also be directed to the oxidative stress, mitochondrial dysfunction, and iron abnormalities occurring in FRDA. Accordingly, removal of excess mitochondrial iron and/or antioxidant treatment has been attempted. The short chain coenzyme Q10 analog idebenone and the orally administered iron chelator deferiprone have been tested in randomized controlled trials (RCTs), but results have been negative or inconclusive.[58] However, newer, more potent antioxidants and respiratory chain activators are now undergoing RCTs.[59]

Cellular therapies, in particular the use of stem cells, may be explored to promote regeneration and function recovery. The widespread nature of neurodegeneration in FRDA would require diffuse delivery of cells in the central nervous system of the patients—a major difficulty at present, but the object of intense research.

In conclusion, thanks to the remarkable progress in understanding the pathogenesis of FRDA since the responsible gene was discovered in 1996, the perspective of a treatment for this incurable neurodegenerative disease now appears a realistic goal.

References

1. Geoffroy G, Barbeau A, Breton G, et al. Clinical description and roentgenologic evaluation of patients with Friedreich's ataxia. *Can J Neurol Sci*. 1976;3(4):279–286.
2. Harding AE. Friedreich's ataxia: a clinical and genetic study of 90 families with an analysis of early diagnostic criteria and intrafamilial clustering of clinical features. *Brain*. 1981;104(3):589–620.
3. Parkinson MH, Boesch S, Nachbauer W, Mariotti C, Giunti P. Clinical features of Friedreich's ataxia: classical and atypical phenotypes. *J Neurochem*. 2013;126(suppl 1):103–117.
4. Pandolfo M. Friedreich ataxia: the clinical picture. *J Neurol*. 2009;256(suppl 1):3–8.
5. Seyer LA, Galetta K, Wilson J, et al. Analysis of the visual system in Friedreich ataxia. *J Neurol*. 2013;260(9):2362–2369.
6. Rance G, Corben L, Barker E, et al. Auditory perception in individuals with Friedreich's ataxia. *Audiol Neurootol*. 2010;15(4):229–240.
7. Fielding J, Corben L, Cremer P, Millist L, White O, Delatycki M. Disruption to higher order processes in Friedreich ataxia. *Neuropsychologia*. 2010;48(1):235–242.
8. Corben LA, Georgiou-Karistianis N, Fahey MC, et al. Towards an understanding of cognitive function in Friedreich ataxia. *Brain Res Bull*. 2006;70(3):197–202.
9. Cnop M, Igoillo-Esteve M, Rai M, et al. Central role and mechanisms of β-cell dysfunction and death in Friedreich ataxia-associated diabetes. *Ann Neurol*. 2012;72(6):971–982.
10. Pandolfo M, Manto M. Cerebellar and afferent ataxias. *Continuum (Minneap Minn)*. 2013;19(5, Movement Disorders):1312–1343.
11. Vanasse M, Garcia-Larrea L, Neuschwander P, Trouillas P, Mauguière F. Evoked potential studies in Friedreich's ataxia and progressive early onset cerebellar ataxia. *Can J Neurol Sci*. 1988;15(3):292–298.
12. Santoro L, Perretti A, Lanzillo B, et al. Influence of GAA expansion size and disease duration on central nervous system impairment in Friedreich's ataxia: contribution to the understanding of the pathophysiology of the disease. *Clin Neurophysiol*. 2000;111(6):1023–1030.
13. Santoro L, De Michele G, Perretti A, et al. Relation between trinucleotide GAA repeat length and sensory neuropathy in Friedreich's ataxia. *J Neurol Neurosurg Psychiatr*. 1999;66(1):93–96.
14. Wilson RB, Lynch DR, Farmer JM, Brooks DG, Fischbeck KH. Increased serum transferrin receptor concentrations in Friedreich ataxia. *Ann Neurol*. 2000;47(5):659–661.
15. Schulz JB, Dehmer T, Schöls L, et al. Oxidative stress in patients with Friedreich ataxia. *Neurology*. 2000;55(11):1719–1721.
16. Emond M, Lepage G, Vanasse M, Pandolfo M. Increased levels of plasma malondialdehyde in Friedreich ataxia. *Neurology*. 2000;55(11):1752–1753.
17. Di Prospero NA, Baker A, Jeffries N, Fischbeck KH. Neurological effects of high-dose idebenone in patients with Friedreich's ataxia: a randomised, placebo-controlled trial. *Lancet Neurol*. 2007;6(10):878–886.
18. Coppola G, Burnett R, Perlman S, et al. A gene expression phenotype in lymphocytes from Friedreich ataxia patients. *Ann Neurol*. 2011;70(5):790–804.
19. Akhlaghi H, Yu J, Corben L, et al. Cognitive deficits in Friedreich ataxia correlate with micro-structural changes in dentatorubral tract. *Cerebellum*. 2014;13(2):187–198.
20. Lund C, Bremer A, Lossius MI, Selmer KK, Brodtkorb E, Nakken KO. Dravet syndrome as a cause of epilepsy and learning disability. *Tidsskr Nor Laegeforen*. 2012;132(1):44–47.
21. Zalesky A, Akhlaghi H, Corben LA, et al. Cerebello-cerebral connectivity deficits in Friedreich ataxia. *Brain Struct Funct*. 2014;219(3):969–981.
22. Pandolfo M. Drug Insight: antioxidant therapy in inherited ataxias. *Nat Clin Pract Neurol*. 2008;4(2):86–96.
23. Georgiou-Karistianis N, Akhlaghi H, Corben LA, et al. Decreased functional brain activation in Friedreich ataxia using the Simon effect task. *Brain Cogn*. 2012;79(3):200–208.
24. Gilman S, Junck L, Markel DS, Koeppe RA, Kluin KJ. Cerebral glucose hypermetabolism in Friedreich's ataxia detected with positron emission tomography. *Ann Neurol*. 1990;28(6):750–757.
25. Iltis I, Hutter D, Bushara KO, et al. (1)H MR spectroscopy in Friedreich's ataxia and ataxia with oculomotor apraxia type 2. *Brain Res*. 2010;1358:200–210.
26. Lodi R, Cooper JM, Bradley JL, et al. Deficit of in vivo mitochondrial ATP production in patients with Friedreich ataxia. *Proc Natl Acad Sci U S A*. 1999;96(20):11492–11495.
27. Dürr A, Cossée M, Agid Y, et al. Clinical and genetic abnormalities in patients with Friedreich's ataxia. *N Engl J Med*. 1996;335(16):1169–1175.
28. Tsou AY, Paulsen EK, Lagedrost SJ, et al. Mortality in Friedreich ataxia. *J Neurol Sci*. 2011;307(1–2):46–49.
29. Montermini L, Richter A, Morgan K, et al. Phenotypic variability in Friedreich ataxia: role of the associated GAA triplet repeat expansion. *Ann Neurol*. 1997;41(5):675–682.
30. Koeppen AH, Mazurkiewicz JE. Friedreich ataxia: neuropathology revised. *J Neuropathol Exp Neurol*. 2013;72(2):78–90.
31. Lamarche JB, Côté M, Lemieux B. The cardiomyopathy of Friedreich's ataxia morphological observations in 3 cases. *Can J Neurol Sci*. 1980;7(4):389–396.
32. Campuzano V, Montermini L, Moltò MD, et al. Friedreich's ataxia: autosomal recessive disease caused by an intronic GAA triplet repeat expansion. *Science*. 1996;271(5254):1423–1427.
33. Jiralerspong S, Liu Y, Montermini L, Stifani S, Pandolfo M. Frataxin shows developmentally regulated tissue-specific expression in the mouse embryo. *Neurobiol Dis*. 1997;4(2):103–113.
34. Montermini L, Andermann E, Labuda M, et al. The Friedreich ataxia GAA triplet repeat: premutation and normal alleles. *Hum Mol Genet*. 1997;6(8):1261–1266.

35. Sakamoto N, Chastain PD, Parniewski P, et al. Sticky DNA: self-association properties of long GAA.TTC repeats in R.R.Y triplex structures from Friedreich's ataxia. *Mol Cell*. 1999;3(4):465–475.
36. Sakamoto N, Ohshima K, Montermini L, Pandolfo M, Wells RD. Sticky DNA, a self-associated complex formed at long GAA*TTC repeats in intron 1 of the frataxin gene, inhibits transcription. *J Biol Chem*. 2001;276(29):27171–27177.
37. Saveliev A, Everett C, Sharpe T, Webster Z, Festenstein R. DNA triplet repeats mediate heterochromatin-protein-1-sensitive variegated gene silencing. *Nature*. 2003;422(6934):909–913.
38. Herman D, Jenssen K, Burnett R, Soragni E, Perlman SL, Gottesfeld JM. Histone deacetylase inhibitors reverse gene silencing in Friedreich's ataxia. *Nat Chem Biol*. 2006;2(10):551–558.
39. Gottesfeld JM, Rusche JR, Pandolfo M. Increasing frataxin gene expression with histone deacetylase inhibitors as a therapeutic approach for Friedreich's ataxia. *J Neurochem*. 2013;126(suppl 1):147–154.
40. Montermini L, Kish SJ, Jiralerspong S, Lamarche JB, Pandolfo M. Somatic mosaicism for Friedreich's ataxia GAA triplet repeat expansions in the central nervous system. *Neurology*. 1997;49(2):606–610.
41. Cossée M, Dürr A, Schmitt M, et al. Friedreich's ataxia: point mutations and clinical presentation of compound heterozygotes. *Ann Neurol*. 1999;45(2):200–206.
42. Ohshima K, Sakamoto N, Labuda M, et al. A nonpathogenic GAAGGA repeat in the Friedreich gene: implications for pathogenesis. *Neurology*. 1999;53(8):1854–1857.
43. Deutsch EC, Santani AB, Perlman SL, et al. A rapid, noninvasive immunoassay for frataxin: utility in assessment of Friedreich ataxia. *Mol Genet Metab*. 2010;101(2–3):238–245.
44. Pastore A, Puccio H. Frataxin: a protein in search for a function. *J Neurochem*. 2013;126(suppl 1):43–52.
45. Tsai C-L, Barondeau DP. Human frataxin is an allosteric switch that activates the Fe–S cluster biosynthetic complex. *Biochemistry*. 2010;49(43):9132–9139.
46. Rouault TA. Biogenesis of iron–sulfur clusters in mammalian cells: new insights and relevance to human disease. *Dis Model Mech*. 2012;5(2):155–164.
47. Cossée M, Puccio H, Gansmuller A, et al. Inactivation of the Friedreich ataxia mouse gene leads to early embryonic lethality without iron accumulation. *Hum Mol Genet*. 2000;9(8):1219–1226.
48. Perdomini M, Hick A, Puccio H, Pook MA. Animal and cellular models of Friedreich ataxia. *J Neurochem*. 2013;126(suppl 1):65–79.
49. Puccio H, Simon D, Cossée M, et al. Mouse models for Friedreich ataxia exhibit cardiomyopathy, sensory nerve defect and Fe–S enzyme deficiency followed by intramitochondrial iron deposits. *Nat Genet*. 2001;27(2):181–186.
50. Simon D, Seznec H, Gansmuller A, et al. Friedreich ataxia mouse models with progressive cerebellar and sensory ataxia reveal autophagic neurodegeneration in dorsal root ganglia. *J Neurosci*. 2004;24(8):1987–1995.
51. Miranda CJ, Santos MM, Ohshima K, et al. Frataxin knockin mouse. *FEBS Lett*. 2002;512(1–3):291–297.
52. Al-Mahdawi S, Pinto RM, Varshney D, et al. GAA repeat expansion mutation mouse models of Friedreich ataxia exhibit oxidative stress leading to progressive neuronal and cardiac pathology. *Genomics*. 2006;88(5):580–590.
53. Hick A, Wattenhofer-Donzé M, Chintawar S, et al. Induced pluripotent stem cell derived neurons and cardiomyocytes as a model for mitochondrial defects in Friedreich's ataxia. *Dis Model Mech*. 2013;6(3):608–621.
54. González-Cabo P, Palau F. Mitochondrial pathophysiology in Friedreich's ataxia. *J Neurochem*. 2013;126(suppl 1):53–64.
55. Lobmayr L, Brooks DG, Wilson RB. Increased IRP1 activity in Friedreich ataxia. *Gene*. 2005;354(Jul 18):157–161.
56. Pandolfo M. Iron and Friedreich ataxia. *J Neural Transm Suppl*. 2006;70:143–146.
57. Rai M, Soragni E, Jenssen K, et al. HDAC inhibitors correct frataxin deficiency in a Friedreich ataxia mouse model. *PLoS One*. 2008;3(4):e1958.
58. Kearney M, Orrell RW, Fahey M, Pandolfo M. Antioxidants and other pharmacological treatments for Friedreich ataxia. *Cochrane Database Syst Rev*. 2012;4: CD007791.
59. Lynch DR, Willi SM, Wilson RB, et al. A0001 in Friedreich ataxia: biochemical characterization and effects in a clinical trial. *Mov Disord*. 2012;27(8):1026–1033.

CHAPTER

73

Ataxia-Telangiectasia

Shuki Mizutani

Tokyo Medical and Dental University, Tokyo, Japan

CLINICAL FEATURES

Ataxia-telangiectasia (A-T) was first documented in 1926.[1] Individuals of all races and ethnicities are affected, but the incidence may vary and is estimated to be between 1 in 40,000 and 1 in 100,000 people.[2,3]

The severity of the immune deficiency disorder varies from patient to patient. Clinical laboratory characteristics show elevated serum α-fetoprotein levels in all cases, and lymphocytes with chromosomal translocations involving chromosome 7 and 14 are often seen. Malignant tumors develop in 15–30% of patients, with the rate of lymphoid tumors about 100–250 times higher than that of normal individuals.

The susceptibility to infection due to immunodeficiency is a serious problem in A-T. Respiratory infections and infections of the paranasal sinus and middle ear are frequently seen. Bacterial and viral infections are most common, while very few fungus and protozoan infections are reported. A chronic respiratory infection associated with the bronchiectasis is often the cause of death. Dysphagia is a main cause of aspiration-related pneumonia, which is worsened by a concomitant immunodeficiency state.

The most prominent clinical manifestation is characterized by a progressive cerebellar ataxia. Ataxia develops at around 4 years of age and results in a walk that resembles a drunken state that progresses slowly. Dystonia and intention tremor appears later. The voluntary movement is slow, and the deep tendon reflex is diminished, but without sensory disturbance. The cerebellar symptoms are prominent at younger ages, and extrapyramidal symptoms become remarkable at older ages. Other features include a characteristic ocular movement called "oculomotor apraxia" and pathological degeneration and disappearance of Purkinje cells and granule cells.

The ocular telangiectasia begins at one of the bulbar conjunctivas at around 3–5 years of age and progresses symmetrically (Figure 73.1). Telangiectasia develops on the skin such as in areas of a V-neck of a sweater, face, hands and soles of the feet. The mechanism for the telangiectasia is not known.

Additional possible manifestations include atrophy of the testicles in boys, menoxenia and sexual involution in girls, and delayed puberty accompanied by growth failure of the ovaries. A drop in glucose tolerance has been reported in 14% of patients. The abnormal differentiation of the bone and fat cells has also been noted.

The high sensitivity to ionizing radiation may lead to malignant tumor due to deficiencies in DNA-damage response in A-T patients. For this reason, X-rays or computed tomography (CT) inspection should be avoided where possible.

MOLECULAR PATHOLOGY

The gene responsible for A-T was first localized to chromosome 11q22.3-23.1 by Gatti et al.[4] and specifically to the *ATM* (*Ataxia Telangiectasia Mutated*) gene by Savitsky et al. as the causal locus responsible for A-T.[5] ATM protein is a high molecular-weight (350 kDa) protein kinase, a member of the phosphoinositidyl 3-kinase-related protein kinase (PIKK) family including the DNA-dependent protein kinase catalytic subunit (DNA-PKcs), A-T and Rad3-related protein (ATR), mammalian target of rapamycin (mTOR)/FKBP-rapamycin-associated protein (mTOR/FRAP),

http://dx.doi.org/10.1016/10.1016/B978-0-12-410529-4.00073-5

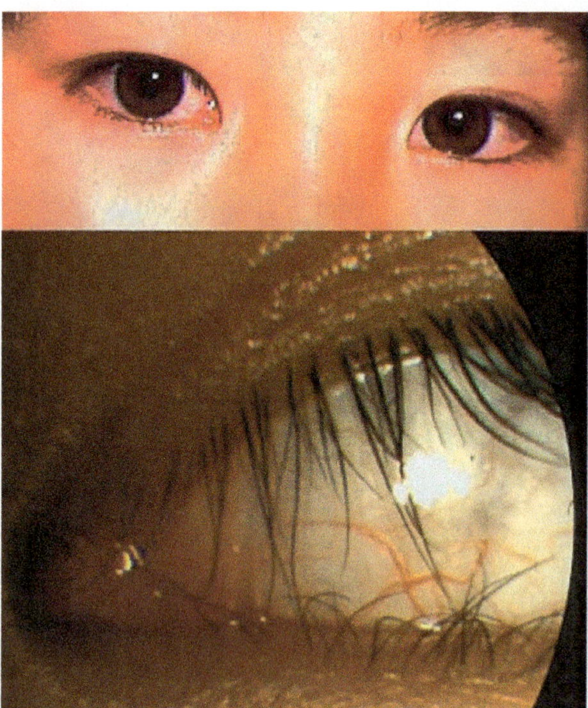

FIGURE 73.1 **Ocular telangiectasia in A-T patient.** The ocular telangiectasia usually begins at one of the bulbar conjunctivas at around 3–5 years of age and progresses symmetrically.

and ATX/SMG. These proteins have similar domains for: 1) kinase activity; 2) FRAP-ATM-TRRAP (FAT); and 3) *C*-terminal FAT (FATC) and binding sites for p53, BLM, and other ATM-binding proteins.

ATM plays a key role in maintaining genomic stability by coordinating the DNA-damage response, including cell-cycle checkpoints and apoptosis (Figure 73.2). In steady-state cells, ATM exists as an inactive dimer/tetramer that can be rapidly activated and recruited to sites of double-strand breaks (DSBs) by the Mre11-Rad50-NBS1 (MRN complex) proteins. Once bound and retained by the MRN complex at a DSB site, ATM undergoes autophosphorylation at a critical residue, Ser1981, leading to the formation of active monomers/dimers. The sustained activation of ATM is modulated by a number of accessory proteins, including MDC1, 53BP1, BRCA1, the protein phosphatases PP2A and WIP1, and TIP60, the histone acetyl-transferase that acetylates ATM on lysine residue (K3016) during its autophosphorylation at Ser1981. Once activated, ATM controls cell cycle-checkpoint signaling, leading to the proper maintenance of independent steps between DNA repair and cell division. These steps should be coordinately regulated and should not occur simultaneously. ATM coordinates: 1) the G1-S cell-cycle checkpoint by phosphorylating p53 at Ser15; 2) the G2/M checkpoint by phosphorylating the protein kinase Chk2 at Thr68 (phosphorylated Chk2 further phosphorylates p53, by which p53 binding with HDM2 is inhibited; HDM2 binding with p53 strengthens the step for p53 degradation); and 3) the intra-S cycle checkpoint by phosphorylating SMC1 at Ser957 and Ser966. In the presence of severe DNA damage, cells are eliminated by ATM/p53 dependent apoptosis. In the absence of ATM, cell-cycle control in the presence of DNA damage is disturbed. One of such historical hallmarks in A-T patient cells has been called "radioresistant DNA synthesis" (RDS). The activity of BRCA1, identified in familial breast cancer, is controlled by ATM in response to radiation exposure. In addition, FANCD2, responsible for Fanconi anemia, is phosphorylated by ATM and is required at the checkpoint for S phase.

The defect in cell cycle regulation and DNA repair leads to defective DNA recombination of antigen receptor genes, which causes primary immunodeficiency and might lead to cell death within the nervous system, particularly Purkinje cells and granule cells in the cerebellum, though the details are not known yet.

Telomeric shortening is also seen in A-T cells. ATM binds to Pin2/TRF1 and regulates telomeric maintenance, and telomeric destabilization in the absence of ATM leads to chromosomal instability and tumor susceptibility.

ATM participates in cell differentiation such as those for osteoblasts and adipocytes, which may partly explain its effect on growth retardation and the glucose tolerance in A-T patients. The RAG-dependent breaks/translocations of the TCRδ chain gene occurs in DN2 to 3 phases of double negative (DN) T cell development in the thymus of ATM deficiency.[6]

Two-thirds of patients have a complete loss of ATM and the remaining one-third show low-level activity. In the latter patients, neurological symptoms are slowly progressive, often manifesting only in adulthood, making early stage

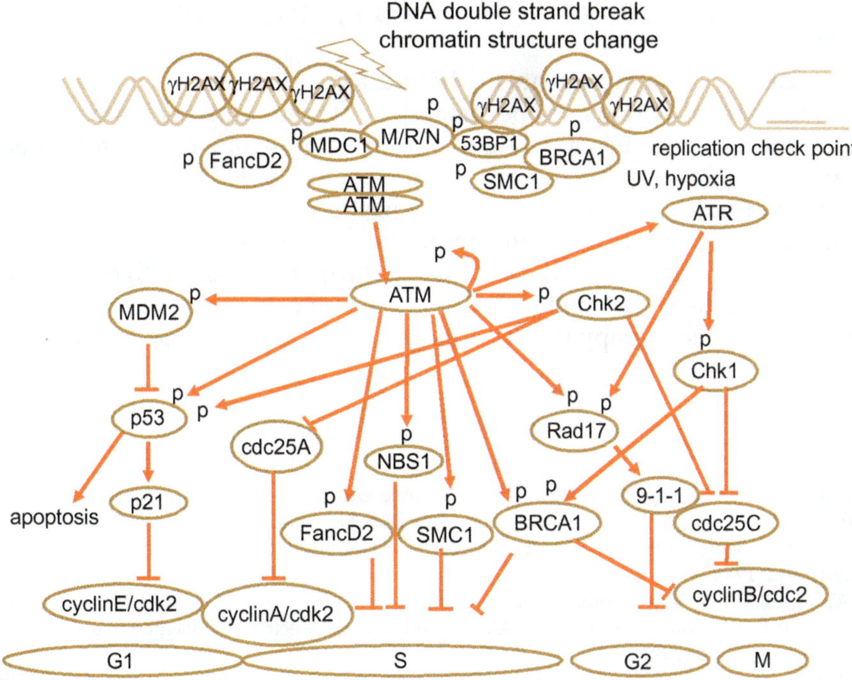

FIGURE 73.2 **ATM-dependent regulation of cell-cycle checkpoints and apoptosis following DNA double-strand breaks.** In steady-state cells, ATM exists as an inactive dimer/tetramer that can be rapidly activated and recruited to sites of double-strand breaks (DSBs) by the Mre11-Rad50-NBS1 (MRN complex) proteins. In response to DNA DSB, ATM undergoes autophosphorylation at residue Ser1981, forming active monomers/dimers. Once activated, ATM controls cell-cycle checkpoint signaling, leading to the proper maintenance of independent steps between DNA repair and cell division. ATM coordinates: 1) the G1-S cell-cycle checkpoint by phosphorylating the tumor suppressor protein p53 at Ser15; 2) the G2/M checkpoint by phosphorylating the protein kinase Chk2 at Thr68; 3) the intra-Scycle checkpoint by phosphorylating SMC1 at Ser957 and Ser966. In the presence of severe DNA damage, cells are eliminated by ATM/p53-dependent apoptosis. In the absence of ATM, cell-cycle control in the presence of DNA damage is disturbed and a condition called radioresistant DNA synthesis (RDS) occurs in A-T patient cells (see text for more details).

diagnosis more difficult. The tumor types may also be dependent on the remaining ATM kinase activity in the patients. These findings suggest that the partial reactivation of the residual kinase activity may help treat A-T patients.

Compound heterozygosity of the ATM gene was discovered in a family of idiopathic torsion dystonia of unidentified type. ATM protein expression is decreased to 20–40%, but the kinase activity is reportedly undetectable in these patients.[7]

Recently, it was indicated that fractions of ATM are also localized in mitochondria. In cells with ATM deficiency, an increase in mitochondrial quantity was shown, and ATM is speculated to play an important role in mitophagy. According to this finding, A-T is disorganized in cerebellar function possibly due to a mitochondrial disorder, which might be difficult to account for by the obstacle of DNA-damage response and cell-cycle checkpoints.[8,9] Thus, it is likely that ATM has separate functions in cytoplasmic-signaling pathways in the presence of oxidative stresses. The disruption of intracellular distribution of ATM might contribute to the clinical pleiotropy of A-T.

DIAGNOSIS AND DIFFERENTIAL DIAGNOSIS

A-T is easy to diagnose if patients show cerebellar ataxia, ocular telangiectasia, and compromised signs within the respiratory system. The last definitive diagnosis is made by genetic analyses. The bedside screen for a conventional diagnosis can be employed, including the detection of ATM protein expression or phosphorylation.[10] Tumor suppressor protein p53 coexists with γ tubulin in a centrosome in normal interphase cells (MCL, mitotic centrosomal localization), but this is reportedly disrupted in ATM-deficient cells.[11]

Friedreich ataxia and oculomotor apraxia 1 (aprataxin deficiency), oculomotor apraxia 2 (senataxin deficiency), and A-T-like disorder (ATLD) needs to be differentially diagnosed. Other diseases characterized by the hypersensitivity to ionizing radiations should be ruled out including the so-called XCIND syndromes such as Nijmegen breakage syndrome, KU70/80 deficiency, DNA-PKcs deficiency, Artemis deficiency, DNA Lig IV deficiency, XRCC4 deficiency, Cernunnos/XLF deficiency, RNF168 deficiency, and RAD50 deficiency.

TREATMENT AND PROGNOSIS

Generally, the prognosis is poor and patients are bound to a wheelchair from the early stage. About half of patients die at around the age of 20 years. Proper treatment of symptoms including the control of infection by appropriate antibiotics and intravenous gamma globulin infusion is one of few recommendations that can be made. X-ray examination and CT scan for diagnosis are not recommended except for unavoidable circumstances. Clinical trials are currently in progress to examine strategies to delay the progression of neurologic symptoms, such as the use of antioxidizer treatment or lead compounds (SMRT: small molecule readthrough compounds) skipping a nonsense mutation.[12,13] Patient-derived induced pluripotent stem (iPS) cells or nerve cell-derived stem cells will be helpful for future discovery of effective treatments. Furthermore, clinical trials evaluating low dose betamethasone or dexamethasone therapy using autologous red blood cells as a carrier (EryDEX) are currently ongoing.[14,15]

References

1. Syllaba L, Henner K. Contribution à l'indépendence de l'athétose double idiopathique et congénitale. Atteinte familiale, syndrome dystrophique, signe du reséau vasculaire conjonctival, intégrité psychique. *Rev Neurol*. 1926;1:541–562.
2. Swift M, Morrell D, Cromartie E, Chamberlin AR, Skolnick MH, Bishop DT. The incidence and gene frequency of ataxia-telangiectasia in the United States. *Am J Hum Genet*. 1986;39:573–583.
3. Shiloh Y, Kastan MB. ATM: genome stability, neuronal development, and cancer cross paths. *Adv Cancer Res*. 2001;83:209–254.
4. Uhrhammer N, Concannon P, Huo Y, Nakamura Y, Gatti RA. A pulsed-field gel electrophoresis map in the ataxia-telangiectasia region of chromosome11q22.3. *Genomics*. 1994;20(2):278–280.
5. Savitsky K, Bar-Shira A, Gilad S, et al. A single ataxia telangiectasia gene with a product similar to PI-3 kinase. *Science*. 1995;268:1749–1753.
6. Isoda T, Takagi M, Piao J, et al. Process for immune defect and chromosomal translocation during early thymocyte development lacking ATM. *Blood*. 2012;120(4):789–799.
7. Meissner WG, Fernet M, Couturier J, et al. Isolated generalized dystonia in biallelic missense mutations of the ATM gene. *Mov Disord*. 2013;28(13):1897–1899.
8. Valentin-Vega YA, Maclean KH, Tait-Mulder J, et al. Mitochondrial dysfunction in ataxia-telangiectasia. *Blood*. 2012;119(6):1490–1500.
9. Valentin-Vega YA, Kastan MB. A new role for ATM: regulating mitochondrial function and mitophagy. *Autophagy*. 2012;8(5):840–841.
10. Honda M, Takagi M, Chessa L, Morio T, Mizuatni S. Rapid diagnosis of ataxia-telangiectasia by flow cytometric monitoring of DNA damage-dependent ATM phosphorylation. *Leukemia*. 2009;23(2):409–414.
11. Prodosmo A, De Amicis A, Nisticò C, et al. p53 centrosomal localization diagnoses ataxia-telangiectasia homozygotes and heterozygotes. *J Clin Invest*. 2013;123(3):1335–1342.
12. Lavin MF, Gueven N, Bottle S, Gatti RA. Current and potential therapeutic strategies for the treatment of ataxia-telangiectasia. *Br Med Bull*. 2007;81–82:129–147.
13. Nakamura K, Du L, Tunuguntla R, et al. Functional characterization and targeted correction of ATM mutations identified in Japanese patients with ataxia-telangiectasia. *Hum Mutat*. 2012;33(1):198–208.
14. Broccoletti T, Del Giudice E, Cirillo E, et al. Efficacy of very-low-dose betamethasone on neurological symptoms in ataxia-telangiectasia. *Eur J Neurol*. 2011;18(4):564–570.
15. Zannolli R, Buoni S, Betti G, et al. A randomized trial of oral betamethasone to reduce ataxia symptoms in ataxia telangiectasia. *Mov Disord*. 2012;27(10):1312–1316.

CHAPTER

74

Dystonia

Katja Lohmann and Christine Klein

Institute of Neurogenetics, University of Lübeck, Lübeck, Germany

DEFINITION

Dystonia is a movement disorder characterized by sustained or intermittent muscle contractions causing abnormal, often repetitive, movements, postures, or both. Examples of dystonic postures are shown in Figure 74.1. Dystonic movements are typically patterned, twisting, and may be tremulous. Dystonia is often initiated or worsened by voluntary action and associated with overflow muscle activation.[1]

CLASSIFICATION

There are two main axes of classification for dystonia including phenotypic and etiologic features. While the phenotypic classification has recently been revised,[1] a task force has been implemented by the Movement Disorder Society to further evaluate a new concept for a novel genetic nomenclature overcoming the shortcomings of the existing system.[2] On clinical grounds, the updated dystonia classification proposes characterization by age of onset (infancy, childhood, adolescence, early and late adulthood), body distribution (focal, segmental, multifocal, and generalized), temporal pattern (static or progressive disease course; persistent, action-specific, diurnal or paroxysmal presentation), and association with additional features (isolated vs. combined with another movement disorder).[1] Formerly, isolated dystonia was referred to as "primary dystonia" and combined dystonia as "dystonia-plus." Syndromes in which dystonia is part of a complex phenotype and not a main and/or consistent finding are termed complex dystonia (formerly "secondary dystonia"). The main clinical classification is depicted in Figure 74.2.

With respect to the etiological axis, we here focus on known underlying genetic causes of dystonia. The current genetic classification is based on the mutated gene (or at least the genetic locus to which a specific form has been linked). It includes information on the mode of inheritance and, if known, the impaired cellular pathway. This list currently comprises 25 "DYTs" (Tables 74.1A and B) but actually represents an assortment of clinically and genetically heterogeneous disorders, with duplicated loci, as well as unconfirmed, and thus uncertain, genetic causes.[2] A new system has recently been proposed which includes only confirmed genes and avoids numbering of "DYTs." Rather, the "DYT" prefix is suggested to be followed by the gene name, for example, "DYT-TOR1A" replaces DYT1.[2] The new genetic classification of the "DYTs" that includes the isolated and combined dystonias is represented in Table 74.1A.

GENETIC CAUSES

Hermann Oppenheim described the first case of genetic (likely DYT-TOR1A) dystonia more than 100 years ago.[3,4] He coined the term "dystonia" and recognized dystonia as an organic disorder. More than seven decades later, the first inherited cause was identified in a combined dystonia (dopa-responsive dystonia) as a mutation in the *GCH1* gene encoding GTP cyclohydrolase I.[5] Within the past two decades, genetic linkage studies in large families and candidate gene approaches have led to the elucidation of many additional dystonia genes (Tables 74.1 and 74.2). Since

Rosenberg's Molecular and Genetic Basis of Neurological and Psychiatric Disease
http://dx.doi.org/10.1016/10.1016/B978-0-12-410529-4.00074-7

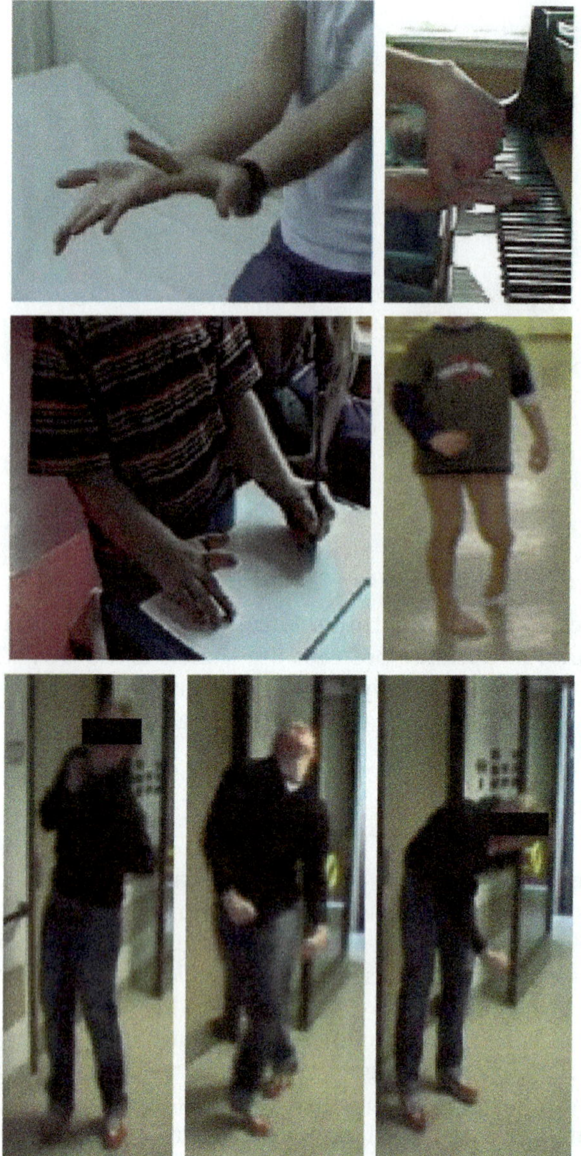

FIGURE 74.1 **Clinical presentation of hereditary dystonias.** Upper left: isolated bibrachial dystonia in a patient with DYT-TOR1A (DYT1); upper right: patient with familial musician's dystonia; middle left: dystonic posturing of the right hand when writing with the left hand in a patient with myoclonus-dystonia (DYT-SGCE; DYT11); middle right: generalized dystonia in a patient with dopa-responsive dystonia (DYT-GCH1; DYT5); lower panel: dystonia combined with ballism, chorea, and athetosis in a patient with paroxysmal kinesigenic dyskinesia (DYT-PRRT2; DYT10).

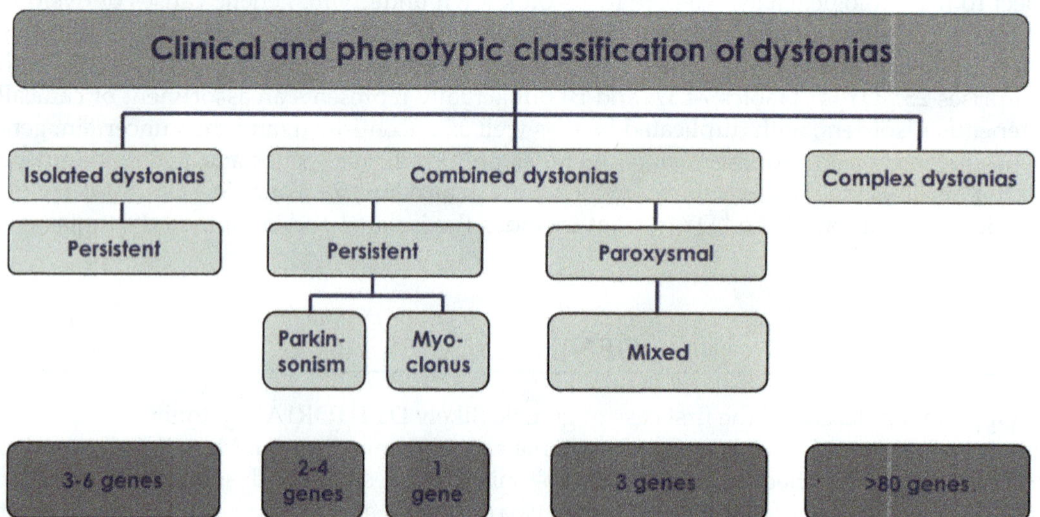

FIGURE 74.2 **Phenotypic classification of the dystonias.** The number of known genetic forms is indicated.

TABLE 74.1A New Classification of Isolated and Combined Forms of Dystonia with Confirmed Genetic Cause

New designation and phenotypic subgroup	Additional phenotypic notes	Inheritance pattern	Locus symbol
ISOLATED DYSTONIAS			
DYT-TOR1A	Early-onset generalized dystonia	AD	*DYT1*
DYT-THAP1	Adolescent-onset dystonia of mixed type	AD	*DYT6*
DYT-TUBB4A	Adult-onset dystonia with prominent spasmodic dysphonia*	AD	*DYT4*
DYT-GNAL	Adult-onset cranial-cervical dystonia	AD	*DYT25*
COMBINED DYSTONIAS			
Dystonia plus parkinsonism			
DYT-GCH1	Dopa-responsive dystonia	AD	*DYT5a*
DYT-ATP1A3	Rapid-onset dystonia-parkinsonism	AD	*DYT12*
DYT-TAF1†	Dystonia-parkinsonism	X-linked	*DYT3*
Dystonia plus myoclonus			
DYT-SGCE	Myoclonus-dystonia	AD	*DYT11*
Paroxysmal dystonia plus other dyskinesia			
DYT-PRRT2	Paroxysmal kinesigenic dyskinesia	AD	*DYT10, DYT19*
DYT-MR-1	Paroxysmal nonkinesigenic dyskinesia	AD	*DYT8*
DYT-SLC2A1	Paroxysmal exertion-induced dyskinesia	AD	*DYT18, DYT9*

**The phenotype of DYT-TUBB4A may be more complex, and may thus justify grouping DYT-TUBB4A under the complex dystonias.*
†Due to a founder effect, genetic testing is possible. The pathogenicity of the TAF1 *gene is not confirmed; however, testing of selected variants in this gene is sufficient for the diagnosis.*
Abbreviations: AD, autosomal dominant; AR, autosomal recessive.

TABLE 74.1B Unconfirmed or Unidentified Genetic Forms of Isolated and Combined Dystonia

Symbol	Gene locus/ Gene	Disorder	Inheritance	Remarks
DYT2	Unknown/ Unknown	Autosomal recessive dystonia	AR	No genetic information, No specified phenotype
DYT5b	11p15.5/ *TH*	Dopa-responsive dystonia with developmental delay	AR	Phenotype is complex, should rather be grouped under complex dystonia
DYT7	18p/ Unknown	Adult-onset focal dystonia	AD	No gene has been identified, Locus unconfirmed since 1996
DYT13	1p36/ Unknown	Multifocal/segmental dystonia	AD	No gene has been identified, Locus unconfirmed since 2001
DYT14	11p15 (false)/ *GCH1*	Dopa-responsive dystonia	AD	Erroneous linkage (*GCH1* mutation)
DYT15	18p11/ Unknown	Myoclonus-dystonia	AD	No gene has been identified, Locus unconfirmed since 2001
DYT16	2q31.2/ *PRKRA*	Young-onset dystonia-(parkinsonism)	AR	No add. homozygous/compound heterozygous mutation since 2008
DYT17	20p-q/ Unknown	Autosomal recessive primary dystonia	AR	No gene has been identified, Locus unconfirmed since 2008
DYT20	2q/ Unknown	Paroxysmal nonkinesigenic dyskinesia (PNKD2)	AD	No gene has been identified, Clinical overlap with/locus close to DYT8
DYT21	2q/ Unknown	Late-onset pure dystonia	AD	No gene has been identified, Locus unconfirmed since 2011
DYT23	9q34/ *CIZ1*	Adult onset cranial-cervical dystonia	AD	No independent confirmation, Variants frequent in controls
DYT24	11p/ *ANO3*	Adult onset cranial-cervical dystonia	AD	No independent confirmation, Variants frequent in controls

Abbreviations: AD, autosomal dominant; AR, autosomal recessive; DYT22 is not reported.

TABLE 74.2 Genetic Forms of Complex Dystonias

Main symptom/ typical presentation	Mode of inheritance	Disease/syndrome	Mutated gene	Dysfunction due to the mutation
Ataxia	AD	SCA1	*ATXN1*	Transcriptional
		SCA2	*ATXN2*	?
		SCA3 (Machado–Joseph disease)	*ATXN3*	Protein degradation/ transcrip.
		SCA6	*CACNA1A1*	Transport
		SCA14	*PRKCG*	?
		SCA17	*TBP*	Transcriptional
	AR	Spastic staxia 5	*AFG3L2*	Mitochondrial
		Ataxia-telangiectasia	*ATM*	DNA repair
		Ataxia-oculomotor apraxia-1	*APTX*	DNA repair
		Ataxia-ocular apraxia-2	*SETX*	Transcrip.+ DNA repair
		DYTCA	*COX20* (1 fam.)	Mitochondrial
		Ataxia with isolated vitamin E deficiency	*TTPA*	?
Parkinsonism	AD	PARK17	*VPS35*	Vesicular transport
	AR	PARK2	*Parkin*	Mitochondrial
		PARK6	*PINK1*	Mitochondrial
		PARK7	*DJ-1*	Mitochondrial?
		PARK9 (Kufor–Rakeb syndrome)	*ATP1A3*	Lysosomal
		PARK15	*FBXO7*	?
		PARK20	*SYNJ1*	Metabolic
		Infantile parkinsonism-dystonia	*SLC6A3* (DAT1)	Metabolic
Spasticity	AR	SPG11	*SPG11*	?
		SPG26	*B4GALNT1*	Metabolic
		SPG35	*FA2H*	Metabolic
	X-linked	Pelizaeus–Merzbacher disease	*PLP1*	Structural
Chorea	AD	Huntington disease	*HTT*	Transcriptional
Dementia	AD	Frontotemporal dementia	*CHMP2B*	?
		Dentatorubral-pallidoluysian atrophy (DRPLA)	*ATN1*	Transcriptional?
Neurodegeneration with brain iron accumulation	AD	NBIA3	*FTL*	Metabolic
	AR	NBIA1	*PANK2*	Metabolic
		NBIA2	*PLA2G6*	Metabolic
		NBIA4	*C19orf12*	?
	X-linked	NBIA5	*WDR45* (*de novo*)	Protein degradation
Idiopathic basal ganglia calcification	AD	IBGC	*SLC20A1*	Transport
		IBGC4	*PDGFRB* (1 case)	Transport?
		IBGC5	*PDGFB*	Transport?
Muscular dystrophy	AR	Limb-girdle muscular dystrophy type 2S	*TRAPPC11*	Vesicle trafficking
		G_{M2} gangliosidosis	*HEXA*	Metabolic

TABLE 74.2 Genetic Forms of Complex Dystonias—cont'd

Main symptom/ typical presentation	Mode of inheritance	Disease/syndrome	Mutated gene	Dysfunction due to the mutation
Deafness	AD	Syndrome with deafness, malformations, and dystonia	*ACTB*	Structural
	AR	Woodhouse–Sakati syndrome	*DCAF17* (C2orf37)	?
		MEGDEL syndrome	*SERAC1*	Mitochondrial
		Mito. DNA depletion syndrome 5	*SUCLA2*	Mitochondrial
	X-linked	Mohr–Tranebjaerg syndrome	*TIMM8A*	Mitochondrial
		DDCH (syndrome with deafness, dystonia, cerebral hypomyelination)	*BCAP31*	Protein degradation
Optic atrophy	AR	Costeff syndrome	*OPA3* (1 family)	Metabolic
Mitochondrial phenotype (without deafness)	AR	Mito. complex II deficiency	*SDHA* or *SDHAF1*	Metabolic
		Mito. complex III deficiency	*TTC19*	Mitochondrial
		Combined oxidative phosphorylation deficiency 10	*MTO1* (1 case)	Mitochondrial
		Combined oxidative phosphorylation deficiency 13	*PNPT1* (1 family)	Transport
	Mitochondrial	LHON (Leber hereditary optic neuropathy) complex I subunits	*MTND1, 3, 4, 6*	Metabolic
		MELAS-like	*MT-TC*	Mitochondrial
Hepatic dysfunction	AR	Wilson disease	*ATP7B*	Transport
		Hypermanganesemia with dystonia, polycythemia, and cirrhosis	*SLC30A10*	Transport
		Mito. DNA depletion syndrome-6	*MPV17*	Transport
Intellectual disability/ developmental delay	AR	Mito. complex III deficiency	*UQCRQ* (1 family)	Metabolic
		Pyruvate dehydrogenase E2 deficiency	*DLAT*	Metabolic
		Pyruvate dehydrogenase E3-binding protein deficiency	*PDHX* (PDX1)	Metabolic
		Combined oxidative phosphorylation deficiency-12	*EARS2*	Metabolic
		Cystic leukoencephalopathy without megalencephaly	*RNASET2*	Transport?
		Homocystinuria	*CBS*	Metabolic
		Dopa-responsive dystonia with developmental delay 1	*GCH1*[*]	Metabolic
		Dopa-responsive dystonia with developmental delay 2	*TH*	Metabolic
		Dopa-responsive dystonia with developmental delay 3	*SPR*	Metabolic
	X-linked	Lesch–Nyhan syndrome	*HPRT1*	Metabolic
		Allan–Herndon–Dudley syndrome	*SLC16A2*	Transport
		X-linked syndromic mental retardation, Christianson type	*SLC9A6* (1 family)	Transport
		Pyruvate dehydrogenase E1-deficiency	*PDHA1*	Metabolic
Epileptic encephalopathy	AD	Rett-like syndrome	*FOXG1*	Transcriptional
		Epileptic encephalopathy, early infantile, 17	*GNAO1* (1 case) (*de novo*)	Signaling
	AR	Cerebroretinal microangiopathy with calcifications and cysts	*CTC1*	?
		Thiamine metabolism dysfunction syndrome 2	*SLC19A3*	Metabolic
	X-linked	Rett syndrome	*MeCP2*	Transcriptional
		Partington syndrome	*ARX*	Transcriptional

Continued

TABLE 74.2 Genetic Forms of Complex Dystonias—cont'd

Main symptom/ typical presentation	Mode of inheritance	Disease/syndrome	Mutated gene	Dysfunction due to the mutation
Other encephalopathy	AR	Glutaricaciduria, type I	*GCDH*	Metabolic
		Aicardi–Goutieres syndrome-1	*TREX1*	?
		Thiamine metabolism dysfunction syndrome 5	*TPK1*	Metabolic
Skeletal abnormalities	AR	G_{M1} gangliosidosis	*GLB1*	Protein degradation/ Signaling
Other (complex and/ or variable)	AD	Porencephaly 1	*COL4A1* (1 family)	?
	AR	Niemann–Pick disease	*NPC1*	Storage (lipids)
		Choreoacanthocytosis	*VPS13A*	?
		Aromatic L-amino acid decarboxylase deficiency	*DDC*	Metabolic
		Combined malonic and methylmalonic aciduria	*ACSF3*	Metabolic
		Methylmalonic aciduria	*MUT*	Metabolic
		Leukoencephalopathy with dystonia, motor neuropathy	*SCP2* (1 case)	Metabolic
Ectopic lenses	AR	Sulfite oxidase deficiency	*SUOX*	Metabolic
Symmetric dyschromatosis	AD	Dyschromatosis symmetrica hereditaria	*ADAR*	RNA stability

[a]The much more common heterozygous GCH1 *mutations are the cause of dopa-responsive dystonia (combined dystonia; Table 74.1A).*
Abbreviations: AD, autosomal dominant; AR, autosomal recessive; MEGDEL, 3-methylglutaconic aciduria with sensorineural deafness, encephalopathy, and Leigh-like syndrome; MELAS, mitochondrial encephalomyopathy, lactic acidosis, and stroke-like episodes.

2011, the field of dystonia genetics has been evolving very rapidly due to the advent of next generation sequencing technology, which has resulted in the description of another 14 genes including five for possibly isolated or combined forms (*PRRT2, CIZ1, ANO3, TUBB4/TUBB4A, GNAL*).

In general, a monogenic cause—i.e., a mutation in a single gene—is mostly found in early-onset dystonia where symptoms affect more than one body part. Notably, these forms represent only a minority of patients with dystonia. In contrast, late-onset isolated dystonia accounts for about 90% of all cases with dystonia, which has a prevalence of about 16 : 100,000.[6]

Isolated Dystonias

Currently, mutations in three genes have undoubtedly been shown to cause isolated dystonia: *TOR1A*,[7] *THAP1*,[8] *GNAL*.[9] These genetic forms are all inherited in an autosomal dominant fashion and are found in about 1–3% of unselected patients.[10–12] Mutations in *TUBB4/TUBB4A* have recently been shown to cause "whispering dysphonia" (DYT4 dystonia) in the original DYT4 family from Australia.[13,14] In addition, mutations in *CIZ1*[15] and *ANO3*[16] have been suggested to cause dystonia but could not yet be confirmed in independent studies and appear to also occur in controls at considerable frequencies.[17,18] Therefore, only the first four forms will be described in more detail.

DYT-TOR1A: Early-Onset Generalized Dystonia; Oppenheim Dystonia (DYT1)

The onset of DYT-TOR1A dystonia is usually in childhood (mean age 13 years) with dystonia in an extremity; the symptoms later progress to other limbs and the torso but usually spare the face and neck.[19,20] The only confirmed mutation is a specific 3-bp deletion in the *TOR1A* gene (c.904_906delGAG; p.302delGlu). Due to a founder effect, this mutation explains up to 80% of patients with early-onset generalized dystonia in the Ashkenazi Jewish population.[21] Interestingly, the penetrance of this mutation is reduced to 30% and a missense variant in the *TOR1A* gene (p.D216H) has been identified as a modifier, explaining the reduced penetrance in a subset of patients.[22] The encoded protein TorsinA is a member of the AAA+superfamily (ATPases associated with a variety of cellular activities) and therefore considered to function as a molecular chaperone. How the mutation leads to dystonia has not yet been fully elucidated and an impact on the nuclear envelope, endoplasmatic reticulum, and cytoskeletal dynamics has been suggested.[23]

DYT-THAP1: Adolescent Onset Dystonia with Mixed Phenotype (DYT6)

DYT-THAP1 dystonia features focal, segmental, and generalized dystonia and in contrast to DYT-TOR1A has a slightly later onset (mean 19 years) and more prominent craniocervical involvement, with dysphonia being a predominant disease manifestation.[24] Mutations in the *THAP1* (*THAP domain containing, apoptosis associated protein 1*) gene were identified to underlie this form of dystonia.[9] To date, more than 80 different mutations have been reported in *THAP1*.[11,25] Notably, the penetrance of *THAP1* mutations is also reduced but may be higher than in DYT-TOR1A.[26] *THAP1* encodes a transcription factor with a DNA-binding THAP domain at the *N*-terminus and a nuclear localization signal (NLS) towards the *C*-terminus. THAP1 represses the expression of *TOR1A in vitro* and several THAP1 mutations disrupt this repression, providing a molecular link between these two forms of dystonia.[27,28] Additional targets of the transcription factor THAP1 remain to be identified, as well as the pathogenic role of mutations outside the DNA binding domain and the NLS.

DYT-GNAL: Adult-Onset Segmental Dystonia (DYT25)

Mutations in the *guanine nucleotide-binding protein (G protein), alpha activating activity polypeptide, olfactory type* gene cause cervical or cranial dystonia with a mean onset in the thirties (range 7–54 years).[9] The number of mutational studies for this very recently identified dystonia gene are limited but 15 different mutations have already been reported and are spread over the entire gene. *GNAL* mutations seem to be highly but not fully penetrant.[29] *GNAL* encodes the stimulatory alpha subunit, $G_{\alpha olf}$, which links G protein-coupled receptors to downstream effector molecules. It is important for dopamine type 1 receptor (D1R) function and odorant signal transduction.[30]

DYT-TUBB4A: Adult-Onset Generalized Dystonia with Whispering Dysphonia (DYT4)

A unique dystonia phenotype (known as DYT4 dystonia or whispering dysphonia) was described in a large Australian family.[31] This family was recently re-evaluated and it became apparent that the phenotype was broader ranging from isolated spasmodic dysphonia to severe generalized dystonia and included extrusional tongue dystonia and a unique "hobby horse gait," as well as mild spasticity. Notably, the symptoms were alcohol-responsive.[32] Genetic linkage analysis combined with exome or genome sequencing resulted in the identification of an Arg2Gly mutation within the autoregulatory domain of tubulin 4 (TUBB4 or TUBB4A) as the disease cause.[13,14] This mutation resulted in lower *TUBB4* expression in a mutation carrier in different tissues.[14] TUBB4 is a highly conserved protein with only little variation in the general population. Among 400 unrelated dystonia patients, one additional mutation carrier was found who had a positive family history and presented with spasmodic dysphonia.[14] Based on the additional clinical features in the original DYT4 family, it might be that DYT-TUBB4A does not belong to the isolated dystonias but rather should be grouped under the complex forms. However, before any conclusion can be reached, more mutation carriers need to be identified to fully evaluate the phenotypic spectrum. Given that TUBB4 is a neuronally expressed tubulin, this implies abnormal microtubule function as another disease mechanism in dystonia.

Combined Dystonias

By definition, in combined dystonias, dystonia is combined with another movement disorder, most commonly parkinsonism or myoclonus. In the paroxysmal dyskinesias, chorea, or ballism are also commonly seen. There is considerable phenotypic variability and the associated movement disorder may in some cases be more prominent than the dystonia or even be the sole disease manifestation. Persistent and paroxysmal forms both belong to the group of combined dystonias. Among the persistent, combined dystonias, there are two confirmed genes (*GCH1*,[5] *ATP1A3*[33]) and two suggested genes (*TAF1*,[34] *PRKRA*[35]) for dystonia-parkinsonsim and one confirmed gene for myoclonus dystonia (*SGCE*[36]). Since mutations in *PRKRA* have not independently been reported in additional families with recessively inherited dystonia, this form has to be considered as unconfirmed and will not be described in further detail (Table 74.1B). Additional genes are known to cause prominent parkinsonism with dystonia and are therefore classified as complex dystonia (Table 74.2). For paroxysmal dyskinesias, mutations in three different genes have been reported as disease cause (see below).

Persistent Dystonia Combined with Parkinsonism

DYT-GCHI: DOPA-RESPONSIVE DYSTONIA; SEGAWA SYNDROME (DYT5A)

Dopa-responsive dystonia (DRD) is characterized by childhood-onset of dystonia, diurnal fluctuation of symptoms, and a dramatic response to levodopa therapy.[37] Parkinsonian features often co-occur. The most frequent form of DRD is dominantly inherited and usually caused by heterozygous mutations in the *GTP cylohydrolase 1* (*GCH1*) gene.[38] To date, more than 100 different mutations, spread across the entire *GCH1* coding region, have been reported and include missense, nonsense, and splice-site mutations, as well as whole-exon or whole-gene deletions. *GCH1* mutation carriers show a high degree of phenotypic variability and reduced penetrance. While penetrance is lower among men than women, the underlying mechanisms affecting penetrance are not yet resolved. Likewise, nonmotor features (sleep disturbances, mood disorders, migraine) are present in a considerable subset of patients and are probably due to involvement of the serotoninergic system.[38] GTP cylohydrolase 1 represents the rate-limiting enzyme in the biosynthesis of dopamine. DRD may also be recessively inherited and is then characterized by a much more severe clinical phenotype than the dominant form, usually including developmental delay and infantile onset.[39] These forms are caused by homozygous or compound-heterozygous mutations in *GCH1*, the *tyrosine hydroxylase* (*TH, DRD5b*) gene, or the *sepiapterin reductase* (*SPR*) gene (Table 74.2).[39–41] Notably, tyrosine hydroxylase as well as sepiapterin reductase are involved in the same pathway as GTP cyclohydrolase 1, namely the biosynthesis of dopamine. This explains the often-dramatic response to levodopa therapy. Recently, it was shown that prenatal treatment with levodopa in a homozygous *GCH1* mutation carrier can prevent development of a severe phenotype.[42]

DYT-ATP1A3: RAPID-ONSET DYSTONIA-PARKINSONISM (DYT12)

DYT-ATP1A3 dystonia has a characteristic sudden onset within hours to weeks, typically in adolescence or young adulthood (but as late as 55 years), in response to physical or mental stress such as fever or injuries. Dystonic symptoms usually involve the bulbar region and are accompanied by symptoms of parkinsonism.[43] It is inherited in an autosomal dominant manner with reduced penetrance. About 10 different mutations were identified in the *ATP1A3* gene, which encodes Na^+/K^+ATPase alpha 3, the catalytic subunit of an ionic pump that uses ATP hydrolysis to exchange Na^+ and K^+ across the cell membrane to maintain ionic gradients.[33]

DYT-TAF1: X-LINKED DYSTONIA-PARKINSONISM (LUBAG, DYT3)

Another well-described, although rare form of dystonia-parkinsonism is X-linked dystonia-parkinsonism.[44] This condition is endemic to the Philippines due to a founder effect and inherited in an X-linked recessive fashion and thus mainly affects males. Disease onset typically occurs in the mid-thirties with complete penetrance. Symptoms start as focal dystonia and progress to generalized dystonia within the first phase of the disease (about 5 years). Later in the disease course, parkinsonism predominates.[44] There is clear evidence for neurodegeneration[45] in this form of combined dystonia and deep brain stimulation has been used as an effective treatment in a few patients.[46] The disease has undoubtedly been linked to chromosome Xq, and several disease specific changes have been reported including five different single base pair changes, a retrotransposon insertion, and a 48-bp deletion.[34,47] All of these changes are located within or near the TATA-binding protein-associated factor 1 (encoded by the *TAF1* gene) and at least some of these variants may affect the expression of a neuron-specific transcript of *TAF1* and other genes.[47,48] Despite these uncertainties of the exact genetic cause, genetic testing for the disease-linked haplotype is possible due to the founder effect.

Persistent Dystonia Combined with Myoclonus

DYT-SGCE: MYOCLONUS-DYSTONIA (DYT11)

Myoclonus-dystonia (M-D) is characterized by a combination of myoclonus and dystonia. Symptom onset is usually in childhood or early adolescence. The disease is inherited as an autosomal dominant trait with reduced penetrance. The action-induced, alcohol-responsive myoclonic jerks typical of M-D are brief, lightning-like movements most often affecting the neck, trunk, and upper limbs. Approximately half of the affected individuals have focal or segmental dystonia that presents as cervical dystonia and/or writer's cramp. Nonmotor features such as psychiatric disease seem to be part of the phenotype of M-D.[49,50] Loss-of-function mutations in the *epsilon-sarcoglycan* gene (*SGCE*) are the cause of M-D in approximately 20–25% of patients.[51,52] More than 70 different mutations have been reported in *SGCE* including large deletions involving adjacent genes resulting in additional, seemingly unrelated clinical features, such as joint problems.[51,53] Reduced penetrance upon maternal transmission of the disease allele is caused by maternal imprinting of the *SGCE* gene leading to expression of only the paternal allele.[54] Due to the imprinting, the M-D phenotype also occurs in patients with maternal uniparental disomy of chromosome 7 who have additional, characteristic, phenotypic features.[55] epsilon-Sarcoglycan has a transmembrane domain and is a member of the sarcoglycan family. Other sarcoglycans form the dystrophin–glycoprotein complex that links the cytoskeleton to the extracellular matrix and mutations in these genes cause limb-girdle muscular dystrophies.[56] However, the role of SGCE in this complex has not yet been elucidated.

Dystonia Combined with Paroxysmal Dyskinesia

Paroxysmal dyskinesia manifest as episodic movement disorders. Based on their respective triggers, three types of paroxysmal dyskinesias are differentiated: paroxysmal kinesigenic dyskinesia (PKD, after sudden movements), paroxysmal nonkinesigenic dyskinesia (PNKD, after intake of a stressor such as caffeine), and paroxysmal exercise-induced dyskinesia (PED, after extended exercise, such as jogging with high energy demand). For each of the three forms, two gene loci have (erroneously) been reported and only one causative gene has yet been identified for each of the forms (Tables 74.1A and B).

DYT-PRRT2: PAROXYSMAL KINESIGENIC DYSKINESIA (DYT10, DYT19)

First attacks of PKD usually occur in childhood or adolescence after sudden movements. Attacks are characterized by dystonic and choreoathetotic movements, usually last several minutes, and may appear up to 100 times per day. Several missense and truncating mutations in the *proline-rich transmembrane protein 2* (*PRRT2*) gene have been identified as the cause of PKD.[57] PRRT2 may function at the synaptosomal membrane[58] but the exact role of the PRRT2 protein needs to be further investigated.

DYT-MR-1: PAROXYSMAL NONKINESIGENIC DYSKINESIA (DYT8)

The first attacks of PNKD usually start in childhood or adolescence and comprise dystonia, chorea, athetosis, or ballism. In addition to caffeine, attacks of PNKD can be precipitated by alcohol, stress, hunger, fatigue, and tobacco. They usually last from minutes to hours, and may occur several times per day or rarely within a year. Three missense mutations in the *myofibrillogenesis regulator 1* (*MR-1*) gene are the cause of the disease and were reported in ethnically diverse families.[59] All mutations affect the mitochondrial targeting sequence and might lead to disturbed mitochondrial import of a specific splice form.[60] The function of MR-1 may be related to glyxalase hydroxyacylglutathione hydrolase which is known to detoxify methylglyoxal, a compound found in coffee and alcohol.[59]

DYT-SLC2A1: PAROXYSMAL EXERTION-INDUCED DYSKINESIA (DYT18, DYT9)

First attacks in PED occur in childhood and may become more prominent during adulthood.[61] Attacks include a combination of chorea, athetosis, and dystonia, last from a few minutes to an hour, and occur after prolonged physical exercise.[62] The legs are most frequently affected. PED is inherited in an autosomal dominant pattern with almost complete penetrance. PED is caused by mutations in the *SLC2A1* gene encoding the glucose transporter GLUT1 that transports glucose into the brain.[63,64] This explains why the attacks occur when energy (glucose) demand is high. A ketogenic diet or variants thereof are effective therapeutic options.[61]

Complex Dystonias

As the name implies, this group of dystonias represent an assortment of different syndromes for which, at least in a subset of patients, dystonia is observed. There is no firm sub-classification of this group, although sorting according to disturbed pathways or prominent clinical features or a mixture thereof has been used.[65] In Table 74.2, more than 80 genes linked to complex dystonias are listed and grouped in accordance with the main symptom of the respective disease. This information is based on descriptions in original papers and summaries thereof in Online Mendelian Inheritance in Man (OMIM; at www.omim.org). Notably, the phenotypic spectrum is often broad and many of the complex dystonias have several prominent features, such as for instance chorea and dementia in Huntington disease. Further, there are two problems related to the description of dystonic features in complex dystonias: First, since these patients are often not followed by movement disorder specialists, the observed abnormal movements may erroneously be described as dystonia but rather represent, for example, spasticity or ataxia. Second, due to the complex nature of the respective syndromes, mild dystonia may be overlooked or simply not reported in some patients, making it difficult to judge the actual prevalence of dystonic features in a given complex dystonia. Similar to the isolated and combined dystonias, many novel genes have recently been identified for complex dystonias in individual families. However, further confirmation of the variants by additional screening efforts is necessary to evaluate whether dystonia is truly part of the phenotype.

PLEIOTROPY

Pleiotropy is a phenomenon increasingly recognized by the usage of hypothesis-free exome sequencing rather than candidate gene approaches to identify novel disease genes. It refers to the effect that mutations in one and the same gene may lead to multiple, seemingly unrelated diseases. Interestingly, even among the 16 genes linked to isolated or combined forms of dystonia, four genes show pleiotropy: *TUBB4A*, *ATP1A3*, *PRRT2*, *SLC2A1*. Specifically, a particular *de novo* mutation (Asp249Asn) in *TUBB4A* has been identified as the cause of hypomyelination with atrophy of the basal ganglia and cerebellum (H-ABC).[66] This is an extremely rare leukoencephalopathy. Notably, subtle changes in brain structure have also been detected in the index patient of the DYT4 family (unpublished data). Further, *de novo* mutations in *ATP1A3* cause alternating hemiplegia of childhood (AHC).[67,68] AHC is a severe neurodevelopmental syndrome characterized by recurrent hemiplegic episodes and distinct neurological manifestations. Very recently, a third phenotype characterized by cerebellar ataxia, areflexia, pes cavus, optic atrophy, and sensorineural hearing loss (CAPOS syndrome) has been shown to be caused by a recurrent mutation (Glu818Lys) in *ATP1A3*.[69] Notably, there seem to be no mutational overlap in *ATP1A3* between rapid-onset dystonia-parkinsonism, AHC, and CAPOS syndrome. Two genes for paroxysmal forms of dystonia, *PRRT2* and *SLC2A1*, have been linked to different forms of epilepsy (which is also an episodic disorder) and other phenotypes. Specifically, the phenotypic spectrum of *PRRT2* includes benign familial infantile seizures (BFIS) and the syndrome of rolandic epilepsy, while *SLC2A1* mutations have been found in cases with GLUT1 deficiency characterized by epilepsy, hemolytic anemia, and/or migraine. The mechanism underlying the phenomenon of pleiotropy and the factors influencing the specific phenotypic outcome have not yet been elucidated. Shedding light on this conundrum may provide further insights into the yet overall poorly understood disease mechanism(s) of dystonia.

SUSCEPTIBILITY GENES

In addition to the monogenic forms of dystonia, in which a mutation in a single gene is causative of the dystonic phenotype, a contribution of genetic risk factors to dystonia has been postulated. Dystonia is a condition with an overall high heritability; however, only a minority of patients can be and will ever be explained by monogenic causes. In the past two decades, a number of common genetic variants (polymorphisms) within genes from the dopamine pathway,[70,71] brain-derived neurotrophic factor (BDNF),[72] and genes involved in monogenic forms of dystonia[73] have been investigated as genetic risk factors in candidate gene association studies. However, the results of these studies were overall inconclusive. Recently, two genome-wide association studies in homogeneous subgroups of dystonia, namely musician's dystonia and cervical dystonia, have suggested the first two genetic risk factors for dystonia in the arylsulfatase G (*ARSG*) gene[74] and in a sodium leak channel (*NALCN*),[75] respectively. Based on the function of the encoded proteins, dysfunction of either protein is a plausible cause of dystonia.[74,75] Replication in independent phenotypically similar and diverse cohorts is needed to confirm these findings.

ACKNOWLEDGEMENT

C.K. is supported by a career development award from the Hermann and Lilly Schilling Foundation and by the German Research Foundation (SFB936). K.L. is supported by the German Research Foundation (LO 1553/3-2; LO 1553/8-1) and the Dystonia Coalition.

References

1. Albanese A, Bhatia K, Bressman SB, et al. Phenomenology and classification of dystonia: a consensus update. *Mov Disord*. 2013;28:863–873.
2. Marras C, Lohmann K, Lang A, Klein C. Fixing the broken system of genetic locus symbols: Parkinson disease and dystonia as examples. *Neurology*. 2012;78:1016–1024.
3. Klein C, Fahn S. Translation of Oppenheim's 1911 paper on dystonia. *Mov Disord*. 2013;28:851–862.
4. Oppenheim H. Ueber eigenenartige Krampfkrankheit des kindlichen und jugendlichen Alters (Dysbasia lordotica progressiva, Dystonia Musculorum Deformans). *Neurol Centrabl*. 1911;30:1090.
5. Ichinose H, Ohye T, Takahashi E, et al. Hereditary progressive dystonia with marked diurnal fluctuation caused by mutations in the GTP cyclohydrolase I gene. *Nat Genet*. 1994;8:236–242.
6. Steeves TD, Day L, Dykeman J, Jette N, Pringsheim T. The prevalence of primary dystonia: a systematic review and meta-analysis. *Mov Disord*. 2012;27:1789–1796.
7. Ozelius LJ, Hewett JW, Page CE, et al. The early-onset torsion dystonia gene (DYT1) encodes an ATP-binding protein. *Nat Genet*. 1997;17:40–48.
8. Fuchs T, Gavarini S, Saunders-Pullman R, et al. Mutations in the THAP1 gene are responsible for DYT6 primary torsion dystonia. *Nat Genet*. 2009;41:286–288.
9. Fuchs T, Saunders-Pullman R, Masuho I, et al. Mutations in GNAL cause primary torsion dystonia. *Nat Genet*. 2012;45:88–92.
10. Grundmann K, Laubis-Herrmann U, Bauer I, et al. Frequency and phenotypic variability of the GAG deletion of the DYT1 gene in an unselected group of patients with dystonia. *Arch Neurol*. 2003;60:1266–1270.

11. Lohmann K, Uflacker N, Erogullari A, et al. Identification and functional analysis of novel THAP1 mutations. *Eur J Hum Genet.* 2012;20:171–175.
12. Dufke C, Sturm M, Schroeder C, et al. Screening of mutations in GNAL in sporadic dystonia patients. *Mov Disord.* 2014;doi:10.1002/mds.25794.
13. Hersheson J, Mencacci NE, Davis M, et al. Mutations in the autoregulatory domain of beta-tubulin 4a cause hereditary dystonia. *Ann Neurol.* 2013;73:546–553.
14. Lohmann K, Wilcox RA, Winkler S, et al. Whispering dysphonia (DYT4 dystonia) is caused by a mutation in the TUBB4 gene. *Ann Neurol.* 2013;73:537–545.
15. Xiao J, Uitti RJ, Zhao Y, et al. Mutations in CIZ1 cause adult onset primary cervical dystonia. *Ann Neurol.* 2012;71:458–469.
16. Charlesworth G, Plagnol V, Holmstrom KM, et al. Mutations in ANO3 cause dominant craniocervical dystonia: ion channel implicated in pathogenesis. *Am J Hum Genet.* 2012;91:1041–1050.
17. Ma L, Chen R, Wang L, Yang Y, Wan X. No mutations in CIZ1 in twelve adult-onset primary cervical dystonia families. *Mov Disord.* 2013;28:1899–1901.
18. Zech M, Gross N, Jochim A, et al. Rare sequence variants in ANO3 and GNAL in a primary torsion dystonia series and controls. *Mov Disord.* 2014;29(1):143–147.
19. Bressman SB, Sabatti C, Raymond D, et al. The DYT1 phenotype and guidelines for diagnostic testing. *Neurology.* 2000;54:1746–1752.
20. Kabakci K, Hedrich K, Leung JC, et al. Mutations in DYT1: extension of the phenotypic and mutational spectrum. *Neurology.* 2004;62:395–400.
21. Ozelius LJ, Bressman SB. Genetic and clinical features of primary torsion dystonia. *Neurobiol Dis.* 2011;42:127–135.
22. Risch NJ, Bressman SB, Senthil G, Ozelius LJ. Intragenic Cis and Trans modification of genetic susceptibility in DYT1 torsion dystonia. *Am J Hum Genet.* 2007;80:1188–1193.
23. Granata A, Warner TT. The role of torsinA in dystonia. *Eur J Neurol.* 2010;17(suppl 1):81–87.
24. Bressman SB, Raymond D, Fuchs T, Heiman GA, Ozelius LJ, Saunders-Pullman R. Mutations in THAP1 (DYT6) in early-onset dystonia: a genetic screening study. *Lancet Neurol.* 2009;8:441–446.
25. Blanchard A, Ea V, Roubertie A, et al. DYT6 dystonia: review of the literature and creation of the UMD Locus-Specific Database (LSDB) for mutations in the THAP1 gene. *Hum Mutat.* 2011;32:1213–1224.
26. Cheng FB, Ozelius LJ, Wan XH, et al. THAP1/DYT6 sequence variants in non-DYT1 early-onset primary dystonia in China and their effects on RNA expression. *J Neurol.* 2012;259:342–347.
27. Gavarini S, Cayrol C, Fuchs T, et al. Direct interaction between causative genes of DYT1 and DYT6 primary dystonia. *Ann Neurol.* 2010;68:549–553.
28. Kaiser FJ, Osmanoric A, Rakovic A, et al. The dystonia gene DYT1 is repressed by the transcription factor THAP1 (DYT6). *Ann Neurol.* 2010;68:554–559.
29. Vemula SR, Puschmann A, Xiao J, et al. Role of Gα(olf) in familial and sporadic adult-onset primary dystonia. *Hum Mol Genet.* 2013;22:2510–2519.
30. Jones DT, Reed RR. Golf: an olfactory neuron specific-G protein involved in odorant signal transduction. *Science.* 1989;244:790–795.
31. Parker N. Hereditary whispering dysphonia. *J Neurol Neurosurg Psychiatry.* 1985;48:218–224.
32. Wilcox RA, Winkler S, Lohmann K, Klein C. Whispering dysphonia in an Australian family (DYT4): a clinical and genetic reappraisal. *Mov Disord.* 2011;26:2404–2408.
33. de Carvalho Aguiar P, Sweadner KJ, Penniston JT, et al. Mutations in the Na^+/K^+-ATPase alpha3 gene ATP1A3 are associated with rapid-onset dystonia parkinsonism. *Neuron.* 2004;43:169–175.
34. Nolte D, Niemann S, Muller U. Specific sequence changes in multiple transcript system DYT3 are associated with X-linked dystonia parkinsonism. *Proc Natl Acad Sci U S A.* 2003;100:10347–10352.
35. Camargos S, Scholz S, Simon-Sanchez J, et al. DYT16, a novel young-onset dystonia-parkinsonism disorder: identification of a segregating mutation in the stress-response protein PRKRA. *Lancet Neurol.* 2008;7:207–215.
36. Zimprich A, Grabowski M, Asmus F, et al. Mutations in the gene encoding epsilon-sarcoglycan cause myoclonus-dystonia syndrome. *Nat Genet.* 2001;29:66–69.
37. Segawa M, Hosaka A, Miyagawa F, Nomura Y, Imai H. Hereditary progressive dystonia with marked diurnal fluctuation. *Adv Neurol.* 1976;14:215–233.
38. Tadic V, Kasten M, Bruggemann N, Stiller S, Hagenah J, Klein C. Dopa-responsive dystonia revisited: diagnostic delay, residual signs, and nonmotor signs. *Arch Neurol.* 2012;1–5.
39. Opladen T, Hoffmann G, Horster F, et al. Clinical and biochemical characterization of patients with early infantile onset of autosomal recessive GTP cyclohydrolase I deficiency without hyperphenylalaninemia. *Mov Disord.* 2011;26:157–161.
40. Willemsen MA, Verbeek MM, Kamsteeg EJ, et al. Tyrosine hydroxylase deficiency: a treatable disorder of brain catecholamine biosynthesis. *Brain.* 2010;133:1810–1822.
41. Friedman J, Roze E, Abdenur JE, et al. Sepiapterin reductase deficiency: a treatable mimic of cerebral palsy. *Ann Neurol.* 2012;71:520–530.
42. Brüggemann N, Spiegler J, Hellenbroich Y, et al. Beneficial prenatal levodopa therapy in autosomal recessive guanosine triphosphate cyclohydrolase 1 deficiency. *Arch Neurol.* 2012;69:1071–1075.
43. Brashear A, Dobyns WB, de Carvalho Aguiar P, et al. The phenotypic spectrum of rapid-onset dystonia-parkinsonism (RDP) and mutations in the ATP1A3 gene. *Brain.* 2007;130:828–835.
44. Lee LV, Rivera C, Teleg RA, et al. The unique phenomenology of sex-linked dystonia parkinsonism (XDP, DYT3, "Lubag"). *Int J Neurosci.* 2011;121(suppl 1):3–11.
45. Waters CH, Faust PL, Powers J, et al. Neuropathology of lubag (x-linked dystonia parkinsonism). *Mov Disord.* 1993;8:387–390.
46. Patel AJ, Sarwar AI, Jankovic J, Viswanathan A. Bilateral pallidal deep brain stimulation for X-linked dystonia parkinsonism. *World Neurosurg.* 2013;82(1):241.e1–241.e4.
47. Makino S, Kaji R, Ando S, et al. Reduced neuron-specific expression of the TAF1 gene is associated with X-linked dystonia-parkinsonism. *Am J Hum Genet.* 2007;80:393–406.

48. Herzfeld T, Nolte D, Grznarova M, Hofmann A, Schultze JL, Muller U. X-linked dystonia parkinsonism syndrome (XDP, lubag): disease-specific sequence change DSC3 in TAF1/DYT3 affects genes in vesicular transport and dopamine metabolism. *Hum Mol Genet.* 2013;22:941–951.
49. Weissbach A, Kasten M, Grunewald A, et al. Prominent psychiatric comorbidity in the dominantly inherited movement disorder myoclonus-dystonia. *Parkinsonism Relat Disord.* 2013;19:422–425.
50. Peall KJ, Waite AJ, Blake DJ, Owen MJ, Morris HR. Psychiatric disorders, myoclonus dystonia, and the epsilon-sarcoglycan gene: a systematic review. *Mov Disord.* 2011;26:1939–1942.
51. Grünewald A, Djarmati A, Lohmann-Hedrich K, et al. Myoclonus-dystonia: significance of large SGCE deletions. *Hum Mutat.* 2008;29:331–332.
52. Carecchio M, Magliozzi M, Copetti M, et al. Defining the epsilon-sarcoglycan (SGCE) gene phenotypic signature in myoclonus-dystonia: a reappraisal of genetic testing criteria. *Mov Disord.* 2013;28:787–794.
53. Asmus F, Salih F, Hjermind LE, et al. Myoclonus-dystonia due to genomic deletions in the epsilon-sarcoglycan gene. *Ann Neurol.* 2005;58:792–797.
54. Müller B, Hedrich K, Kock N, et al. Evidence that paternal expression of the epsilon-sarcoglycan gene accounts for reduced penetrance in myoclonus-dystonia. *Am J Hum Genet.* 2002;71:1303–1311.
55. Guettard E, Portnoi MF, Lohmann-Hedrich K, et al. Myoclonus-dystonia due to maternal uniparental disomy. *Arch Neurol.* 2008;65:1380–1385.
56. Hack AA, Groh ME, McNally EM. Sarcoglycans in muscular dystrophy. *Microsc Res Tech.* 2000;48:167–180.
57. Chen WJ, Lin Y, Xiong ZQ, et al. Exome sequencing identifies truncating mutations in PRRT2 that cause paroxysmal kinesigenic dyskinesia. *Nat Genet.* 2011;43:1252–1255.
58. Lee HY, Huang Y, Bruneau N, et al. Mutations in the novel protein PRRT2 cause paroxysmal kinesigenic dyskinesia with infantile convulsions. *Cell Rep.* 2012;1:2–12.
59. Lee HY, Xu Y, Huang Y, et al. The gene for paroxysmal non-kinesigenic dyskinesia encodes an enzyme in a stress response pathway. *Hum Mol Genet.* 2004;13:3161–3170.
60. Ghezzi D, Viscomi C, Ferlini A, et al. Paroxysmal non-kinesigenic dyskinesia is caused by mutations of the MR-1 mitochondrial targeting sequence. *Hum Mol Genet.* 2009;18:1058–1064.
61. Leen WG, Taher M, Verbeek MM, Kamsteeg EJ, van de Warrenburg BP, Willemsen MA. GLUT1 deficiency syndrome into adulthood: a follow-up study. *J Neurol.* 2014;261:589–599.
62. Bhatia KP, Soland VL, Bhatt MH, Quinn NP, Marsden CD. Paroxysmal exercise-induced dystonia: eight new sporadic cases and a review of the literature. *Mov Disord.* 1997;12:1007–1012.
63. Weber YG, Storch A, Wuttke TV, et al. GLUT1 mutations are a cause of paroxysmal exertion-induced dyskinesias and induce hemolytic anemia by a cation leak. *J Clin Invest.* 2008;118:2157–2168.
64. Weber YG, Kamm C, Suls A, et al. Paroxysmal choreoathetosis/spasticity (DYT9) is caused by a GLUT1 defect. *Neurology.* 2011;77:959–964.
65. LeDoux MS. The genetics of dystonia. *Adv Genet.* 2012;79:35–85.
66. Simons C, Wolf NI, McNeil N, et al. A de novo mutation in the beta-tubulin gene TUBB4A results in the leukoencephalopathy hypomyelination with atrophy of the basal ganglia and cerebellum. *Am J Hum Genet.* 2013;92:767–773.
67. Rosewich H, Thiele H, Ohlenbusch A, et al. Heterozygous de-novo mutations in ATP1A3 in patients with alternating hemiplegia of childhood: a whole-exome sequencing gene-identification study. *Lancet Neurol.* 2012;11:764–773.
68. Heinzen EL, Swoboda KJ, Hitomi Y, et al. De novo mutations in ATP1A3 cause alternating hemiplegia of childhood. *Nat Genet.* 2012;44:1030–1034.
69. Demos MK, van Karnebeek CD, Ross CJ, et al. A novel recurrent mutation in ATP1A3 causes CAPOS syndrome. *Orphanet J Rare Dis.* 2014;9:15.
70. Groen JL, Simon-Sanchez J, Ritz K, et al. Cervical dystonia and genetic common variation in the dopamine pathway. *Parkinsonism Relat Disord.* 2013;19:346–349.
71. Sibbing D, Asmus F, Konig IR, et al. Candidate gene studies in focal dystonia. *Neurology.* 2003;61:1097–1101.
72. Svetel MV, Djuric G, Novakovic I, et al. A common polymorphism in the brain-derived neurotrophic factor gene in patients with adult-onset primary focal and segmental dystonia. *Acta Neurol Belg.* 2013;113:243–245.
73. Newman JR, Sutherland GT, Boyle RS, et al. Common polymorphisms in dystonia-linked genes and susceptibility to the sporadic primary dystonias. *Parkinsonism Relat Disord.* 2012;18:351–357.
74. Lohmann K, Schmidt A, Schillert A, et al. Genome-wide association study in musician's dystonia: A risk variant at the arylsulfatase G locus? *Mov Disord.* 2014;29:921–927.
75. Mok KY, Schneider SA, Trabzuni D, et al. Genome-wide association study in cervical dystonia demonstrates possible association with sodium leak channel. *Mov Disord.* 2013;29:245–251.

CHAPTER

75

Huntington Disease

Andrew J. McGarry*, Kevin Biglan†, and Frederick Marshall†

*Cooper University Hospital, Cherry Hill, NJ, USA

†University of Rochester School of Medicine & Dentistry, Rochester, NY, USA

CLINICAL FEATURES

Historical Overview

While descriptions of hereditary chorea had been made in the 1830s[1] and 1840s,[2] the seminal event in documenting and recognizing the disease that would ultimately bear his name was George Huntington's treatise "On Chorea," presented in 1872.[3] In a relatively brief but highly descriptive document, Huntington imparted in poignant terms:

> "The hereditary chorea, as I shall call it, is confined to certain and fortunately a few families, and has been transmitted to them, an heirloom from generations away back in the dim past. It is spoken of by those in whose veins the seeds of the disease are known to exist, with a kind of horror, and not at all alluded to except through dire necessity, when it is mentioned as 'that disorder.' It is attended generally by all the symptoms of common chorea, only in an aggravated degree, hardly ever manifesting itself until adult or middle life, and then coming on gradually but surely, increasing by degrees, and often occupying years in its development, until the hapless sufferer is but a quivering wreck of his former self."

With growing identification of phenotypic variation, refined classification schemes, and emphasis on the autosomal dominant inheritance pattern, Huntington disease became of worldwide interest through the 20th century. Numerous researchers have studied groups of high prevalence throughout the world, including cohorts in Venezuela, Sweden, and Scotland.[4] The causative gene, "interesting transcript 15" (IT15), was identified by the Huntington's Disease Collaborative Research Group in 1993.[5] Since then, advancements in molecular biology of Huntington disease (HD) pathogenesis, imaging, and the search for biomarkers have continued at a vigorous pace.

Mode of Inheritance and Prevalence

HD is inherited in an autosomal dominant fashion, with offspring carrying a 50% chance of inheriting the abnormal allele, which is completely penetrant. A small percentage of cases are thought to be the result of a spontaneous mutation; however, the presence of an intermediate allele of 30–38 repeats that has potential to expand, particularly when inherited from fathers of advanced age, may account for many of these seemingly *de novo* cases.[6] Prevalence varies substantially, with subjects of European extraction manifesting disease in 30–100 per one million people.[7] African American and Asian populations have substantially lower prevalence, a phenomenon suspected to be secondary to different polymorphisms adjacent to the HD gene and repeat lengths fewer in number compared to normal values in European populations.[8,9] Elevated prevalence in various locales worldwide such as Lake Maracaibo and South Wales[7] is thought secondary to founder effects with subsequent inbreeding in relatively isolated regions.

Natural History

As an insidious, progressive neurodegenerative disorder, phenoconversion—the point at which individuals show clinically evident manifestations of the underlying genetic defect—can be quite subtle and difficult to pinpoint. The reported mean age of onset varies considerably, from early 30s to mid 50s.[7] A juvenile onset variant of HD, accounting

Rosenberg's Molecular and Genetic Basis of Neurological and Psychiatric Disease
http://dx.doi.org/10.1016/10.1016/B978-0-12-410529-4.00075-9

for 10–15% of symptomatic cases, may occur at any time before the age of 21.[10] Initial manifestations in adults are often mild, nonspecific, or otherwise assigned to common elements of human behavior (stress, inattention, depression, anxiety, nonspecific "age related" cognitive change). Similarly, calculating the duration of the disease is equally challenging, thought in general to last 10–15 years.[11] Average age of death for adult subjects is in the early to mid-50s.[12]

Behavioral disturbances and cognitive impairment are often the initial elements of disease presentation, even though chorea is the disease manifestation that typically attracts initial clinical attention. Upwards of 80% of HD subjects manifest irritability, labile mood, anxiety, depression, anhedonia, poor judgment, and impulsivity.[13] Cognitive impairment can be especially subtle but eventually dominates the phenotype. Premanifest gene-positive subjects demonstrate impairment in visuospatial and executive performance.[14] Short- and long-term memory are also impaired,[15] and overall intelligence testing scores are lower in a CAG length-dependent fashion.[16] Abstraction, calculation, and synthetic frontal functions gradually decline further, with global dementia characterized by reduction in verbal fluency more so than agnosia or apraxia.[17]

Neurologically, chorea is usually the earliest conspicuous movement abnormality in adults, although juvenile cases often present with an akinetic-rigid state.[11] Subtle chorea—often brief, isolated, jerking movements in incipient form—is often mistaken for akathisia or psychomotor agitation in the setting of anxiety. Motor impersistence with posture, such as nonsustained hand closure ("milkmaid's grip") can be observed, as well as choreic intrusions adding inefficiency to otherwise coordinated voluntary movements. With progression chorea becomes more diffuse (including the face, trunk, and all limbs), of higher amplitude, takes on a more flowing quality, and is often continuous. Chorea of the diaphragm and thoracic muscles can significantly alter cadence and continuity of speech, and chorea of facial and pharyngeal muscles can resemble tics. Gait and balance is ultimately influenced by numerous choreic intrusions of the limbs, trunk, head, and impairment of normal stride and arm swing.

While chorea is often the abnormal movement that initiates clinical consideration, numerous other movement disorders have been described in HD. It has been long recognized that HD is both a hyperkinetic and hypokinetic disorder. As HD progresses, hypokinetic elements including dystonia and parkinsonism may become increasingly prominent.[11] Dystonia is reported in upwards of 95% in some series, with common postures including internal rotation of the shoulder, hand closure, flexion at the knee, and inversion of the foot.[18] Bradykinesia becomes more prominent as HD progresses, variably associated with prominent rigidity; gait can become more parkinsonian over time, and in some instances an akinetic-mute state characterizes the terminal phase of the disorder. Cerebellar signs are infrequently found in HD but may include intention tremor, appendicular ataxia, and dysdiadochokinesia. Chorea can confound accurate ascertainment of true cerebellar signs. Eye movement abnormalities are prominent in HD. In premanifest subjects, occulomotor abnormalities may be the first movement manifestation of the disorder.[19] Slowed, reduced-amplitude saccades with initiation by eye blink or head thrust are seen, as well as saccadic intrusions in smooth pursuit movements, fixation difficulty, and poor convergence.[20]

An important variation on the classic phenotype of Huntington disease is juvenile-onset HD, referred to as Westphal or rigid HD.[21] Juvenile HD is thought to account for 5–10% of HD cases, with a prevalence of 0.5–1 per 100,000.[22] The phenotype can be quite different to adult-onset HD, with rigidity, bradykinesia, and spasticity dominating the clinical picture rather than chorea. Hand tremor with posture or action, as well as magnetic gait reminiscent of parkinsonism, can be seen. Cerebellar features and epilepsy are more common in this cohort,[23] with dementia and psychiatric disturbances also thought to be more prominent.[24] There appears to be a significant relationship between age of onset and severity of striatal degeneration,[25] although there is conflicting data regarding the effect of age of onset on clinical progression.[26,27] Ravina and colleagues observed that longer CAG repeat length, when adjusted for age, was associated with greater clinical progression, particularly with respect to executive function, working memory, and dystonia.[28] However, substantial variability seen in this study was not associated with CAG length, suggesting currently unknown genetic and environmental factors also influence clinical progression.

Other clinical features of HD include dysarthria, dysphagia, autonomic dysfunction (cardiovascular, sudomotor, urological), frontal release signs, impaired olfaction, dream enactment behavior, and weight loss.[29]

MOLECULAR GENETICS

Huntington disease is a hereditary trinucleotide repeat expansion disease, resulting from an aberrant CAG-encoded polyglutamine stretch. The site of this expansion is located on chromosome 4 at 4p16.3, a locus termed "interesting trasnscript 15," or IT15, which codes for a 348-kDa protein called *huntingtin*,[11] bearing little homology to other known proteins. Structurally, IT15 consists of 210 kb and 67 exons; the HD expansion occurs in the first exon, beginning at the 18th position. Unaffected individuals have between 10 and 29 repeats. Repeat lengths of 36–39 appear to involve reduced penetrance, though repeats greater than 40 invariably lead to phenoconversion. Abnormal HD genes are inherited in autosomal dominant fashion, with offspring of affected individuals carrying a 50% risk of inheriting the pathological allele. CAG repeats may contract or expand. In general expansions tend to increase in size over subsequent generations, thus demonstrating anticipation, particularly if the gene is inherited paternally.[6,30] Repeat length influences age of onset, with longer repeats (50 or more) associated with juvenile onset, although repeat length only accounts for half to two-thirds of variability in age at phenoconversion across the HD population.[30] Modifying genes are hypothesized to contribute to variability, though this is not well understood.[31] Polymorphisms in PGC-1a, a protein known to broadly influence bioenergetics, have been shown to modify age of onset.[32] CAG repeat length in adult-onset HD appears to have no bearing on clinical features or neuropathological findings.[33,34]

DISEASE MECHANISMS

Neuropathology

The pathological hallmark of HD is selective, severe degeneration of the striatum. The caudate is affected earliest, followed by the putamen, particularly medium-sized γ-aminobutyric acid (GABA)ergic spiny neurons that constitute the efferent projections from the striatum.[11] Degeneration tends to progress laterally and ventrally, with the most severe changes in posteromedial and posterodorsal regions.[35] Neurons involved in the indirect pathway of the basal ganglia are generally affected before neurons of the direct pathway. The external segment of the globus pallidus is also heavily affected, with changes also seen in the ventrolateral and centromedian nuclei of the thalamus.[36,37] Cortical atrophy is prominent, particularly frontally in layers 5 and 6 that project to the striatum.[38] White matter also demonstrates axonal damage and demyelination, with the recent suggestion that this may be among the earliest changes in HD pathology.[39]

Molecular Biology

Normal nonexpanded *huntingtin* is ubiquitously expressed and critically important to normal cellular homeostasis. In the brain it is involved in cellular trafficking (endocytosis) and regulation of brain-derived neurotrophic factor (BDNF), a regulator of striatal survival, as well as directly downregulating apoptotic mechanisms.[40,41] Mutant *huntingtin* (mHtt) activates caspase-3, a pro-apoptotic regulator, which fragments *huntingtin*; the unfragmented mutant protein and cleaved elements translocate to the nucleus, where they aggregate and are suspected to interfere with gene transcription.[42] Notably, mHtt binds cAMP-response element binding protein (CREB), specificity protein 1 (Sp1), and p53, interfering broadly with mechanisms promoting cell survival and bioenergetics.[43,44] mHtt or its *N*-terminal fragment containing the polyglutamine expansion accumulates in neurons, microglia, and astrocytes to promote cytotoxicity. In the brain, mHtt inhibits proteosome function and normal transcription of a number of genes, including many with essential roles in energetics such as complex II, III, and IV. Mitochondrial dysfunction subsequent to abnormal gene regulation is thought to promote free-radical formation and an ATP-deficient environment, which in turn destabilizes the resting potential of neurons. This change promotes glutamate activation of *N*-methyl-D-aspartate (NMDA) receptors—which are especially prevalent in medium spiny striatal neurons—resulting in abnormal calcium influx and subsequent promotion of apoptosis. Mitochondrial fragmentation and breakdown of homeostatic trafficking pathways ensue, which compound and accelerate dysfunction in the compromised organelle.

A number of specific molecules have generated interest for their role in HD dysfunction. Peroxisome proliferator-activated receptor gamma coactivator 1-alpha (PGC-1a) is a master respiratory regulator via its role as a transcriptional coactivator.[45,46] It upregulates uncoupling protein 2 and nuclear respiratory factors, which in turn upregulate mitochondrial transcription factor A towards replication and transcription of mitochondrial DNA. It is highly regulated via acetylation, phosphorylation, methylation, and sumoylation, and serves a multiplicity of important roles for estrogen, thyroid, glucocorticoid, mineralcorticoid, glucose, and lipid metabolism. In HD mouse models, PGC-1a is downregulated because of mHtt interference with CREB and TAF4 on the PGC-1a promoter, with subsequent reduction in expression of PGC-1a-facilitated genes in the caudate.[47] Inhibition of PGC-1a also interferes with myelination in oligodendrocytes, consistent with an emerging recognition of white matter abnormalities in HD.[48] PGC-1a knockout mice demonstrate a phenotype reminiscent of HD with accompanying striatal degeneration. Of note, expression of PGC-1a ameliorates neurodegeneration in R6/2 HD mice.[47] As such, PGC-1a may play an integral role in HD neurodegeneration. Creatine kinase (CK) has also garnered attention in HD, given its high sensitivity to oxidative stress and role in maintenance of cellular ATP stores.[44] CK activity has been reported as reduced in HD subjects as well as murine models, with reduction of CK transcripts in the R6/2 murine brain implicating mHtt as a transcriptional interference. It is suspected that reactive oxidation species resulting from mitochondrial dysfunction alter creatine/phosphocreatine balance, with subsequent aberrancy in cellular energy reserves. In R6/2 mice, creatine was associated with improved survival, reduced atrophy, and reduction in striatal huntingtin aggregations.[49]

DIFFERENTIAL DIAGNOSIS

The differential diagnosis of HD centers on other conditions associated with adult-onset chorea. Considerations include neuroacanthocytosis, dentatorubral-pallidoluysian atrophy (DRPLA), spinocerebellar ataxia 17 (SCA-17), autoimmune causes of chorea (systemic lupus erythematosus [SLE], antiphospholipid antibody disease), neuroferritinopathy, pantothenate kinase-associated neurodegeneration (PKAN), and the Huntington disease-like (HDL1-3) disorders.

Neuroacanthocytosis an autosomal recessive disorder characterized by oral dystonia and lip/mouth biting (particularly tongue protrusion), chorea, neuropathy, muscle wasting, cognitive change, seizures, and ophthalmoparesis.[50] The primary laboratory abnormality is greater than 3% acanthocytes on peripheral blood smear (although this can vary), and elevated serum CK. A variety of possible mutations, including nonsense, missense, frameshift, and deletion have been described in the *VPS13A* gene, which codes for the chief malfunctioning protein, chorein.[51]

Like HD, DRPLA is an autosomal dominant, progressive trinucleotide repeat disease, with abnormal expansions in the *ATN1* (atrophin-1) gene numbering from the upper 40s to the 90s.[52] Age of onset varies widely, with phenotypic features that vary accordingly: 20 years and younger typically demonstrate ataxia, seizures (including progressive myoclonic epilepsy), myoclonus, and cognitive disability; while older subjects manifest with ataxia, chorea, and dementia. With an average onset of about 30 years of age, DRPLA can appear similar to HD in a similarly aged cohort, though ataxia is typically more prominent than chorea.

SCA-17 is also an autosomal dominant, progressive disorder caused by a CAG/CAA repeat expansion of more than 42 repeats in the *TBP* gene, which codes for TATA-box binding protein.[53] Ataxia, dementia, and parkinsonism are main manifestations, though dystonia and chorea have been described.

Neuroferritinopathy results from a mutation in the ferritin light chain gene (*FTL*), resulting in abnormal accumulation of ferritin and iron in the brain.[54] These abnormalities, as well as cystic degeneration, are visible on imaging.[55] This adult-onset autosomal dominant disease includes limb dystonia and chorea that becomes progressive; a typical feature is orofacial dystonia provoked by action, as well as oro-buccal-lingual dyskinesias. Cognitive deficits are typically not as prominent as HD, but become more problematic over time.[56]

PKAN (also known as neurodegeneration with brain iron accumulation, or NBIA-1, and formerly known as Hallervorden–Spatz disease) typically onsets before 10 years of age, but approximately 25% of individuals have a later onset with slower progression of symptoms including dystonia, dysarthria, dysphagia, psychiatric manifestations and dementia, as well as chorea and spasticity.[57] The disease is autosomal recessive, due to a mutation in the *PANK2* gene, coding for pantothenate kinase 2, which leads to neuronal toxicity due to the accumulation of *N*-pantothenoyl-cysteine and pantetheine.[58] Magnetic resonance imaging (MRI) shows iron deposits in the basal ganglia known as the "eye-of-the-tiger" sign.

HDL-1 is an autosomal dominant disorder involving eight octapeptide repeats in the *PRNP* gene, coding for prion protein.[59] This mutation was found in a family demonstrating signs similar to HD, which is notably different than phenotypes associated with prion protein mutations (relatively rapid onset dementia, ataxia, and myoclonus). HDL-2 involves expansion in the *JPH3* gene of 41 repeats or greater, with a variable phenotype that should be suspected in patients of African ancestry.[60] JPH-3 codes for junctophilin-3, a protein thought to interface the cytoplasmic membrane with endoplasmic reticulum. Earlier onset, in the 20s to 40s, tends to involve weight loss, rigidity, bradykinesia, tremor, dysarthria, dystonia, psychiatric abnormalities, and relatively mild chorea. Later onset (typically in the 40s or older) shows saccadic abnormalities and more chorea, reminiscent of HD. MRI tends to be similar to HD, with caudate and cortical atrophy. HDL-3 is an autosomal recessive disorder of early childhood onset characterized by a severe phenotype of spasticity, intellectual decline, seizures, dystonia, chorea, and ataxia.[61]

Benign hereditary chorea is caused by an autosomal dominant mutation in the thyroid transcription factor (*TTF*) gene that may manifest from infancy to childhood, with a predilection for movements of the hands, arms, cervical muscles, and face, as well as potential abnormalities in pulmonary and thyroid function.[62] In a younger person, benign hereditary chorea can be differentiated from HD by the absence of features typically found in juvenile HD (parkinsonism, dystonia, relative paucity of chorea), as well as the general absence of significant disability over time.

Autoimmune causes can elicit chorea reminiscent of HD, including SLE and antiphospholipid antibody syndrome.[11] Chorea can be the initial manifestation of SLE in as much as 25% of affected individuals in some series.[63]

TESTING

Testing for subjects whose phenotype is suspected to be HD is performed using a commercially available assay, either polymerase chain reaction (PCR)-based or Southern blot, identifying CAG expansions in the *IT15* gene. *IT15* is the only gene known to be associated with Huntington disease, with other genes implicated in HD phenocopies (see Differential Diagnosis). With respect to test interpretation, a normal allele consists of 26 or fewer repeats. Individuals with 27–35 repeats, or "intermediate" alleles, are notable for their tendency towards instability of the expansion during meiosis and anticipation phenomena, particularly if the allele in question is inherited paternally. HD-causing alleles begin at 36 repeats, although repeats of 36–39 demonstrate incomplete penetrance and may not manifest symptoms during a normal lifespan despite enhanced risk.[27] Repeat lengths of 40 or greater are associated with full penetrance of the phenotype. In gene-positive subjects with two gene-negative parents, spontaneous mutation, non-paternity, or previously unknown adoption may explain such findings.

In families in whom a HD gene-positive person is known, at-risk testing may be undertaken for adults without clinical symptoms, i.e., presymptomatic testing, for risk stratification. Prenatal testing is also possible in fetuses of at-risk individuals through amniocentesis or chorionic villus sampling, as is preimplantation genetic diagnosis, a process that involves *in vitro* fertilization and implantation of embryos without the CAG expansion.[64] The presence of HD-positive embryos identified in this process may or may not be disclosed to the parents.

Because of the mode of transmission and devastating phenotype, genetic counseling is an exceptionally important element of HD patient care. Implications for family members are significant with respect to their own family planning, career trajectory, and lifestyle expectations. These are sensitive issues, all the more so given prevalence of depression, impulsivity, and suicidal ideation in HD. Difficult emotions, including guilt and anger, are understandably possible when at-risk family members decide to undergo testing. Clinical care requires a thorough and detailed conversation regarding possible outcomes of testing and implications for these scenarios. In persons younger than 18 years, testing is not recommended given social and ethical ramifications for an individual who is typically not fully independent. Visitation with a genetic counselor should precede any at-risk presymptomatic testing.

Imaging is not specifically diagnostic for HD, though characteristic caudate and cortical atrophy may be seen on computed tomography and MRI.[65]

MANAGEMENT

Management of HD focuses on symptomatic treatment of movement disorders, psychiatric disturbances, and cognitive deficits. There are currently no disease-modifying agents employed in the management of HD.

Movement Disorders

Chorea is thought to arise from a relative imbalance of striatal degeneration, involving indirect pathway neurons malfunctioning before direct pathway counterparts.[11] This results in a relative reduction in inhibitory tone and subsequent increased probability of involuntary movement. As such, strategies center on reduction of dopaminergic tone in the relatively disinhibited, movement-facilitating direct pathway. Chorea is often present relatively early and may not require pharmacological intervention, though it is often responsive to treatment when treatment is decided upon. Notably, while chorea can be mitigated with dopamine blockade or depletion, motor function progressively worsens by way of bradykinesia and incoordination. Mainstays of reducing chorea include typical and atypical antipsychotics (including haloperidol, chlorpromazine, risperidone, quetiapine, ziprasidone, clozapine, olanzapine). A number of potential side effects including parkinsonism, glucose intolerance, arrhythmias, hypotension, somnolence, and seizures must be considered depending on the extent of the HD phenotype in each patient. Tetrabenazine, an agent that depletes presynaptic monoamines, demonstrated significant reduction in chorea in a 2006 Huntington Study Group trial.[66] It is US Food and Drug Administration (FDA)-approved for the treatment of disabling chorea in HD. Tetrabenazine also depletes serotonin, necessitating adequate management of depression in advance of its use. Currently there are no approved treatments for bradykinesia, although some juvenile HD cases dominated by parkinsonism have improved with levodopa.[67] Dystonia is best treated with botulinum toxin and judicious use of dopamine blocking agents. Anticholinergic agents, while theoretically useful for dystonia, are discouraged given potential for adverse cognitive effects, sedation, and possible worsening of overall movement. Physical therapy is essential for accommodating movement changes, minimizing contractures, and maintaining balance and gait as much as possible. Speech therapy also plays an important role in maximizing communication and safety of swallowing.

Cognitive Issues

No specific therapies exist for the amelioration of cognitive dysfunction in HD. Agents from Alzheimer disease treatment have demonstrated no benefit on dementia. A study of rivastigmine conducted over 24 months suggested a trend toward improvement in cognitive function but no significant overall benefit.[68] Donepezil likewise demonstrated no benefit over 12 weeks in a double-blind, placebo-controlled trial.[69] Similar to approaches in other conditions with dementia, a supportive context that maximizes safety and dignity is essential. Occupational therapy services to investigate potential pitfalls in the home environment can be useful. Provisions to minimize caregiver burnout can be a sensitive issue but are integral to long-term care strategies.

Psychiatric Issues

Behavioral disturbances are very common, including depression, anxiety, irritability, paranoia, apathy, obsessive–compulsive behaviors, aggression, delusions, libido abnormalities, and suicidality. Standard approaches involving selective serotonin reuptake inhibitors (SSRIs), serotonin–norepinephrine reuptake inhibitors (SNRIs), benzodiazepines, and sedatives are generally thought to be useful, although robust clinical trial data is lacking. Venlafaxine has proven effective for major depression, albeit with side effects, in a small trial of 26 HD patients.[70] Counseling can be effective if the cognitive burden of disease is not prohibitive, and is highly recommended at the time of diagnosis while adjusting to its implications. Many subjects maintain ongoing counseling for support and monitoring in the setting of their disease. Impulsivity and suicidal ideation deserves particular mention in this context, as suicide risk is 7–12-fold higher than the general population,[71,72] and can occur with little or no warning, even with routine provider contact. Paulsen et al. found suicidality in HD patients was highest immediately surrounding formal diagnosis and as independence diminishes.[73] Maximizing safety includes attention to screening for intent and planning, access to firearms, and monitoring of life stressors that are bound to accumulate with progression of disease. Social services to minimize impact of changes in lifestyle, such as driving and employment, are important in easing alterations in independence.

References

1. Stevens DL. The history of Huntington's chorea. *J R Coll Physicians Lond.* 1972;6:271–282.
2. Waters CO. Huntington's Chorea 1872-1972. *Adv Neurol.* (1):29–30.
3. Huntington G. On Chorea. *Med Surg Rep Wkly J.* 1872;26(15):317–321.
4. Biglan KM, Shoulson I. Huntington's Disease. Therapeutics of Parkinson's Disease and other Movement Disorders. West Sussex: Wiley-Blackwell; 2008.
5. The Huntington's Disease Collaborative Research Group. A novel gene containing a trinucleotide repeat that is expanded and unstable on Huntington's disease chromosomes. *Cell.* 1993;72(6):971–983.
6. Goldberg YP, Kremer B, Andrew SE, et al. Molecular analysis of new mutations for Huntington's Disease: intermediate alleles and sex of origin effects. *Nat Genet.* 1993;5(2):174–179.
7. Hayden MR. Huntington's Chorea. Berlin: Springer; 1981.
8. Rubensztein DC, Amos W, Leggo J, et al. Mutational bias provides a model for the evolution of Huntington's disease and predicts a general increase in disease prevalence. *Nat Genet.* 1994;4:525–530.
9. Squitieri F, Andrew SE, Goldberg YP, et al. DNA haplotype analysis of Huntington disease reveals clues to the origins and mechanisms of CAG expansion and reasons for geographic variations of prevalence. *Hum Mol Genet.* 1994;3:2103–2114.
10. Rasmussen A, Macias R, Yescas P, Ochoa A, Davila G, Alonso E. Huntington disease in children: genotype-phenotype correlations. *Neuropediatrics.* 2000;31(4):190–194.
11. Fahn S, Jankovic J. Principles and Practice of Movement Disorders. Philadelphia: Elselvier; 2007.
12. Brothers RCD. Huntington's chorea in Victoria and Tasmania. *J Neurol Sci.* 1964;1:405–420.
13. Paulsen JS, Ready RE, Hamilton JM, et al. Neuropsychiatric aspects of Huntington's disease. *J Neurol Neurosurg Psychiatry.* 2001;71:310–314.
14. Jason GW, Pajurkova EM, Suchowersky O, et al. Presymptomatic neuropsychological impairment in Huntington's disease. *Arch Neurol.* 1988;45:769–773.
15. Butters N, Sax D, Montgomery K, et al. Comparison of the neuropsychological deficits associated with early and advanced Huntington's disease. *Arch Neurol.* 1978;35:585–589.
16. Fouroud T, Siemers E, Kleindorfer D. Cognitive scores of Huntington's disease gene carriers compared with non-carriers. *Ann Neurol.* 1995;37:657–664.

17. Shelton PA, Knopman DS. Ideomotor apraxia in Huntington's disease. *Arch Neurol*. 1991;48:35–41.
18. Louis ED, Lee P, Quinn L, et al. Dystonia in Huntington's disease: prevalence and clinical characteristics. *Mov Disord*. 1999;14:95–101.
19. Kirkwood SC, Siemers E, Stout JC, et al. Longitudinal cognitive and motor changes among presymptomatic Huntington disease gene carriers. *Arch Neurol*. 1999;56:563–568.
20. Leigh RJ, Zee DS. The Neurology of Eye Movements. New York: Oxford; 2006.
21. Quarrell O, O'Donovan KL, Bandmann O, Strong M. The prevalence of juvenile Huntington's disease: a review of the literature and meta-analysis. *PLOS Curr*. 2012 Jul 20;doi:10.1371/4f8606b742ef3, Edition 1.
22. van Dijk G, Van der Velde EA, Roos RA, Bruyn G. Juvenile Huntington's disease. *Hum Genet*. 1986;73:235–239.
23. Cloud LJ, Rosenblatt A, Margolis RL, et al. Seizures in juvenile Huntington's disease: frequency and characterization in a multicenter cohort. *Mov Disord*. 2012;27(14):1797–1800.
24. Brackenridge CJ. Factors influencing the dementia and epilepsy in in Huntington's disease of early onset. *Acta Neurol Scand*. 1980;62:305–311.
25. Meyers RH, Vonsattel JP, Stevens TJ, et al. Clinical and neuropathological assessment of severity in Huntington's disease. *Neurology*. 1988;38:341–347.
26. Mahant N, McCusker EA, Byth K, Graham S. Huntington's disease: clinical correlates of disability and progression. *Neurology*. 2003;61:1085–1092.
27. Feigin A, Kieburtz K, Bordwell K, et al. Functional decline in Huntington's disease. *Mov Disord*. 1995;10:211–214.
28. Ravina B, Romer M, Constantinescu R, et al. The relationship between CAG repeat length and clinical progression in Huntington's Disease. *Mov Disord*. 2008;23(9):1223–1227.
29. Donaldson IM, Marsden CD, Schneider SA, Bhatia KP. Marsden's Book of Movement Disorders. New York: Oxford; 2012.
30. Gusella JF, MacDonald ME, Ambrose CM, et al. Molecular genetics of Huntington's disease. *Arch Neurol*. 1993;50:1157–1163.
31. Myers RH. Huntington's disease genetics. *NeuroRx*. 2004;1:255–262.
32. Weydt P, Soyal SM, Gellera C, et al. The gene coding for PGC-1a modifies age of onset in Huntington's Disease. *Mol Degener*. 2009;4:3. doi:10.1186/1750-1326-4-3.
33. MacMillan JC, Snell RG, Tyler A, et al. Molecular analysis and clinical correlations of the Huntington's disease gene mutation. *Lancet*. 1993;432:954–958.
34. Neal JW, Fenton I, MacMillan JC, et al. A study comparing mutation triplet repeat size and phenotypes in patients with Huntington disease. *Neurodegeneration*. 1994;3:73–77.
35. Sieradzan K, Mann DM, Dodge, et al. The selective vulnerability of nerve cells in Huntington's disease. *Neuropathol Appl Neurobiol*. 2001;27:1–21.
36. Dom R, Malfroid M, Baro F. Neuropathology of Huntington's chorea: cytometric studies of the ventrobasal complex of the thalamus. *Neurology*. 1976;26:64–68.
37. Simma K. Die subcorticalen veranderungen bei pickscher krankheit im vergleich zur chorea Huntington. *Mschr Psychiat Neurol*. 1952;123:205–238.
38. Sotrel A, Paskevich PA, Kiely DK, et al. Morphometric analysis of the prefrontal cortex in Huntington's disease. *Neurology*. 1991;41:1117–1123.
39. Ciarmiello A, Cannella M, Lastoria S, et al. Brain white matter volume loss and glucose hypometabolism precede the clinical symptoms of Huntington's disease. *J Nucl Med*. 2006;47(2):215–222.
40. Zuccato C, Ciammola A, Rigamonti D, et al. Loss of huntingtin-mediated BDNF gene transcription in Huntington's disease. *Science*. 2001;293:493–498.
41. Sawa A, Tomoda T, Bae BI. Mechanisms of neuronal cell death in Huntington's disease. *Cytogenet Genome Res*. 2003;100:287–295.
42. Schilling G, Becher MW, Sharp AH, et al. Intranuclear inclusions and neuritic aggregates in transgenic mice expressing a mutant N-terminal fragment of huntingtin. *Hum Mol Genet*. 1999;8:397–407.
43. Steffan JS, Kazantsev A, Spasic-Boskovic O, et al. The Huntington's disease protein interacts with p53 and CREB-binding protein and represses transcription. *Proc Natl Acad Sci U S A*. 2000;97:6763–6768.
44. Dunah AW, Jeong H, Griffin A, et al. Sp1 and TAFII130 transcriptional activity disrupted in early Huntington's disease. *Science*. 2002;296:2238–2243.
45. Tz-Chuen J, Yow-Sien L, Chern Y. Energy dysfunction in Huntington's disease: insights from PGC-1a, AMPK, and CKB. *Cell Mol Life Sci*. 2012;69:4107–4120.
46. Ehrlich M. Huntington's Disease and the striatal medium spiny neuron: cell-autonomous and non-cell-autonomous mechanisms of disease. *Neurotherapeutics*. 2012;9:270–284.
47. Cui L, Jeong HK, Borovecki F, Parkhurst CN, Tanese N, Krainc D. Transcriptional repression of PGC-1a by mutant huntingtin leads to mitochondrial dysfunction and neurodegeneration. *Cell*. 2006;127:59–69.
48. Xiang Z, Valenza M, Cui L, et al. Peroxisome-proliferator-activated receptor gamma coactivator 1a contributes to dysmyelination in experimental models of Huntington's disease. *Neurobiol Dis*. 2011;31(26):9544–9553.
49. Ferrante RJ, Andreassen OA, Jenkins BG, et al. Neuroprotective effects of creatine in a transgenic mouse model of Huntington's disease. *J Neurosci*. 2000;20(12):4389–4397.
50. Danek A, Walker RH. Neuroacanthocytosis. *Curr Opin Neurol*. 2005;18(4):386–392.
51. Dobson-Stone C, Rampoldi L, Bader B, et al. Chorea-acanthocytosis. 2002 Jun 14, Updated 2011 Aug 18, In: Pagon RA, Adam MP, Bird TD, et al., eds. GeneReviews™ Internet. Seattle, WA: University of Washington; 1993–2013. Available from: http://www.ncbi.nlm.nih.gov/books/NBK1387/.
52. Tsuji S. DRPLA. 1999 Aug 6, Updated 2010 Jun 1, In: Pagon RA, Adam MP, Bird TD, et al., eds. GeneReviews™ Internet. Seattle, WA: University of Washington; 1993–2013. Available from: http://www.ncbi.nlm.nih.gov/books/NBK1491/.
53. Toyoshima Y, Onodera O, Yamada M, et al. Spinocerebellar ataxia type 17. 2005 Mar 29, Updated 2012 May 17, In: Pagon RA, Adam MP, Bird TD, et al., eds. GeneReviews™ Internet. Seattle, WA: University of Washington; 1993–2013. Available from: http://www.ncbi.nlm.nih.gov/books/NBK1438/.

54. Curtis AR, Fey C, Morris CM, et al. Mutation in the gene coding ferritin light polypeptide causes dominant adult-onset basal ganglia disease. *Nat Genet*. 2001;28:350–354.
55. Chinnery PF, Curtis AR, Fey C, et al. Neuroferritinopathy in a French family with late-onset dominant dystonia. *J Med Genet*. 2003;40:e69.
56. Chinnery PF. Neuroferritinopathy. 2005 Apr 25, Updated 2010 Dec 23, In: Pagon RA, Adam MP, Bird TD, et al., eds. GeneReviews™ Internet. Seattle,WA: University of Washington; 1993–2013. Available from: http://www.ncbi.nlm.nih.gov/books/NBK1141/.
57. Zhou B, Westaway SK, Levinson B, et al. A novel pantothenate kinase gene (PANK2) is defective in Hallervorden-Spatz syndrome. *Nat Genet*. 2001;28:345–349.
58. Pellecchia MT, Valente EM, Cif L, et al. The diverse phenotype and genotype of pantothenate kinase-associated neurodegeneration. *Neurology*. 2005;64:1810–1812.
59. Moore RC, Xiang F, Monaghan J, et al. Huntington disease phenocopy is a familial prion disease. *Am J Hum Genet*. 2001;69:1385–1388.
60. Margolis RL. Huntington disease-like 2. 2004 Jan 30, Updated 2012 Apr 26, In: Pagon RA, Adam MP, Bird TD, et al., eds. GeneReviews™ Internet. Seattle,WA: University of Washington; 1993–2013. Available from: http://www.ncbi.nlm.nih.gov/books/NBK1529/.
61. Al-Tahan AY, Divakaran MP, Kambouris M, et al. A novel autosomal recessive 'Huntington's disease like' neurodegenerative disorder in a Saudi family. *Saudi Med J*. 1999;20:85–89.
62. Kleiner-Fisman G. Benign hereditary chorea. *Handb Clin Neurol*. 2011;100:199–212.
63. Donaldson I, Espiner EA. Disseminated lupus erythematosis presenting as chorea gravidarum. *Arch Neurol*. 1971;25:240–244.
64. Warby SC, Graham RK, Hayden MR. Huntington disease. 1998 Oct 23, Updated 2010 Apr 22, In: Pagon RA, Adam MP, Bird TD, et al., eds. GeneReviews™ Internet. Seattle,WA: University of Washington, Seattle; 1993–2013. Available from: http://www.ncbi.nlm.nih.gov/books/NBK1305/.
65. McGarry A, Biglan KM. Imaging in Huntington's disease and other choreas. Neuroimaging of Movement Disorders. New York: Springer; 2013.
66. The Huntington Study Group. Tetrabenazine as antichorea therapy in Huntington disease: a randomized controlled trial. *Neurology*. 2006;66(3):366–372.
67. Racette BA, Pearlmutter JS. Levodopa responsive parkinsonism in an adult with Huntington's disease. *J Neurol Neurosurg Psychiatry*. 1998;65:577–579.
68. de Tommaso M, Difruscolo O, Sciruicchio V, et al. Two years' follow up of rivastigmine treatment in Huntington's disease. *Clin Neuropharmacol*. 2007;30:43–46.
69. Cubo E, Shannon KM, Tracy D, et al. Effect of donepezil on motor and cognitive dysfunction in Hunrtington's disease. *Neurology*. 2006;67:1268–1271.
70. Holl AK, Wilkinson L, Painold A, et al. Combating depression in Huntington's disease: effective antidepressive treatment with venlafaxine XR. *Int Clin Psychopharmacol*. 2010;25:46–50.
71. National Institute of Mental Health. In Harm's Way: Suicide in America. Available at: http://www.amhc.org/9-suicide/article/8892-in-harms-way-suicide-in-america.
72. Farrer LA. Suicide and attempted suicide in Huntington disease: implications for preclinical testing of persons at risk. *Am J Med Genet*. 1986;24:305–311.
73. Paulsen JS, Hoth KF, Nehl C, et al. Critical periods of suicide risk in Huntington's disease. *Am J Psychiatry*. 2005;162:725–731.

CHAPTER

76

Non-Parkinsonian Movement Disorders

Stanley Fahn* and Jill S. Goldman†

*Department of Neurology, Columbia University Medical Center, New York, NY, USA
†Taub Institute, Columbia University Medical Center, New York, NY, USA

INTRODUCTION

Movement disorders with excess of movement are commonly referred to as *hyperkinesias* (excessive movements), *dyskinesias* (unnatural movements), and *abnormal involuntary movements*. We commonly use the term dyskinesias, but all are interchangeable. The five major categories of dyskinesias, in alphabetical order, are: chorea, dystonia, myoclonus, tics, and tremor. The number of genes associated with these disorders has multiplied over the last few years.

ESSENTIAL TREMOR

Essential tremor (ET) is a progressive neurological condition characterized by a 4–12 Hz kinetic tremor of the hands, head, or vocal cords that can severely affect basic daily activities (Figure 76.1). ET is the most common form of tremor, and in some age groups has been estimated to be up to 20 times more prevalent than Parkinson disease (PD). In addition to action tremor, patients with ET often have postural tremor, and when very severe, rest tremor can be present. Late in the disease some individuals develop tandem gait abnormalities. Tremor commonly lessens remarkably following alcohol consumption. There are familial and nonfamilial forms of ET, and a number of autosomal dominant families have been described. Sporadic and familial forms of ET are clinically indistinguishable.

ET is clinically and genetically heterogeneous. Segregation analysis of familial ET is usually consistent with an autosomal dominant inheritance. Three genetic loci have been identified on chromosomes 3q13.3 (*ETM1*), 2p25-p22 (*ETM2*), and 6p23 (*ETM3*). No discreet genes have been identified for any of these loci. A polymorphism G312A in the *DRD3* gene has been suggested as a cause of ET in some ETM1 families. A polymorphism in the *HS1-BP3* gene has been found in two families linked to the ETM2 locus, but not all analyzed families had identifiable sequence variants in these genes.[1]

A genome-wide association study (GWAS) identified two polymorphisms within the leucine-rich repeat and immunoglobulin (Ig) domain containing one gene (*LINGO1*) on chromosome 15q that were associated with increased risk for ET and replicated in different populations.[2] However, because no association was found in other studies, the significance of *LINGO* is still controversial. More recently, several autosomal dominant families were found to have mutations in the *FUS* gene, one of the genes causing familial amyotrophic lateral sclerosis (ALS).[3] Some studies have failed to replicate this finding.[4] The identification of genes for ET is complicated by phenocopies and misdiagnoses, and may require the use of endophenotypes.

DYSTONIA

Dystonia refers to twisting movements that tend to be sustained at the peak of the movement, are frequently repetitive, and often progress to prolonged abnormal postures. In contrast to chorea, dystonic movements repeatedly involve the same group of muscles (i.e., they are patterned). Agonist and antagonist muscles contract simultaneously

http://dx.doi.org/10.1016/10.1016/B978-0-12-410529-4.00076-0

FIGURE 76.1 Drawing of spiral by a patient with essential tremor.

(cocontraction) to produce the sustained quality of dystonic movements. The speed of the movement varies widely from slow (athetotic dystonia) to shock-like (myoclonic dystonia). When the contractions are very brief (e.g., less than a second), they are referred to as dystonic spasms. When they are sustained for several seconds, they are called dystonic movements. And when they last minutes to hours, they are known as dystonic postures. When present for weeks or longer, the postures can lead to permanent fixed contractures.

There are numerous types of dystonia that can present in different regions of the body. Dystonia can involve a single body part (focal dystonia), can be segmental or generalized, and can present early or late in life. Many focal dystonias have specific names based on the body part involved, such as blepharospasm, oromandibular dystonia, spasmodic dysphonia, spasmodic torticollis, and occupational cramps or writer's cramp. In the clinical classification, one distinguishes isolated dystonia (formerly called primary dystonia), in which there is no degenerative pathology detected and no other neurological features except dystonia and sometimes kinetic tremor and combined dystonia, in which other neurological features are present, such as parkinsonism or myoclonus.[5] In the etiologic classification, one divides them into inherited or acquired and whether the pathology reveals neurodegeneration, static lesions, or no structural lesion.[5] Genetics has allowed classification of the different types of dystonia into over 20 subtypes, designated DYT1 to DYT25 including both persistent dystonias and paroxysmal dyskinesias, which often include dystonic and choreic movements. It should be emphasized, though, that the vast majority of individuals with dystonia are sporadic, nonfamilial, and with adult onset.

DYT1: Early-Onset Primary Torsion Dystonia

Oppenheim dystonia, named after Hermann Oppenheim, who described dystonia in 1911, is an autosomal dominant condition with onset almost always before age 26 years and usually starting in a limb. It is the most common form of early-onset primary dystonia. Oppenheim dystonia is caused by a 3-bp GAG deletion in the DYT1 gene, *TOR1A* (Table 76.1). DYT1 encodes an ATP-binding protein called TorsinA. The GAG deletion removes a glutamic acid residue in a conserved region of the ATP-binding domain.[6] DYT1 is highly expressed in the dopamine neurons of the substantia nigra pars compacta, but many other regions of the brain also express this protein, including the cholinergic intrastriatal neurons. Intellect is normal. It is rare for DYT1 carriers to develop symptoms after their mid-20s, and diagnostic guidelines for genetic testing reflect this finding, recommending DYT1 testing for individuals with onset of limb dystonia under age 26 years.[7] DYT1 has been found in all ethnic groups, worldwide, with the same GAG deletion. One out of 2000 Ashkenazi Jews are gene carriers; the increased carrier frequency is due to a founder effect. In individuals with a DYT1 phenotype, 10% of Ashkenazi Jewish individuals and 50% with non-Jewish ancestry have no identifiable mutation, suggesting genetic heterogeneity. Only 30–40% of individuals who carry the DYT1 GAG deletion manifest dystonia, and symptoms can vary both between and within families and can manifest as focal limb dystonia, hand cramps (Figure 76.2), or other manifestations.[8] The variability of symptoms and incomplete penetrance of DYT1 suggests a role for modifier genes and possibly

TABLE 76.1 A List of Hereditary Forms of Dystonia

Name (Gene)	Clinical features	Inheritance pattern	Location	Protein	Mutations identified
DYT1, (*TOR1A*), early onset-torsion dystonia	Limb onset, may generalize	AD, incomplete penetrance, increased incidence in Ashkenazi Jews	9q34.11	TorsinA	3-bp deletion (GAG)
DYT2	Torsion dystonia	AR, possible increased incidence in Jews	N/I	N/I	–
DYT3/Lubag (*TAF1*)	Usually generalized, with parkinsonism	XLR, increased incidence in the Philippines, mosaic gliosis in striatum	Xq13.1	Transcription factor IID	–
DYT4 (*TUBB4A*)	Torsion dystonia	AD	19p13.3	Beta-tubulin	–
DYT5A/GTPCH1-deficient dopa-responsive dystonia (*GCH1*)	Childhood-onset dystonia, responsive to levodopa	AD, incomplete penetrance; F>M	14q22.2	GTP cyclohydrolase I	>100
DYT5B (*TH*) childhood onset dopa-responsive dystonia	Onset in infancy results in severe disability	AR	11p15.5	Tyrosine hydroxylase	–
DYT6, variable-onset, mixed type (*THAP1*)	Cranial–cervical dystonia, can generalize	AD, incomplete penetrance, in Amish–Mennonites of German ancestry	8p11.21	THAP1	–
DYT7, adult-onset focal dystonia	Focal dystonia-torticollis, adult onset	AD, incomplete penetrance, in Germans	18p	N/I	–
DYT8, paroxysmal nonkinesigenic dyskinesia (*PNKD*)	Prolonged attacks of dystonia, chorea and athetosis that do not follow sudden voluntary movements	AD	2q35	Myofibrillo-genesis regulator-1	2
DYT9, choreoathetosis/spasticity with episodic ataxia (CSE)	Episodic involuntary movements, dystonic posturing, imbalance, dysarthria, paresthesia	AD	now recognized to be same as DYT18	now recognized to be same as DYT18	–
DYT10, paroxysmal kinesigenic dyskinesia (PKD) (*PRRT2*)	Brief attacks of dystonia, chorea and athetosis, that are induced by sudden voluntary movements	AD	16p11.2	Proline-rich transmembrane protein 2	–
DYT11 (SGCE) myoclonus-dystonia	Myoclonus or dystonia or both	AD, imprinting effect	7q21.3	ε-sarcoglycan	>15
DYT12, rapid-onset dystonia-parkinsonism (RDP) (*ATP1A3*)	Rapid-onset dystonia-parkinsonism; below the age of 10 presents as alternating hemiplegia	AD, incomplete penetrance	19q13.2	alpha3 subunit of Na,K ATPase	>6
DYT13	Cranial/cervical dystonia	AD, incomplete penetrance	1p36.32-p36.13	N/I	–
DYT14	Same as DYT5A				
DYT15	Myoclonus dystonia seen in one large Canadian kindred	AD, incomplete penetrance	18p11	N/I	–
DYT16 (*PRKRA*)	Young-onset dystonia-parkinsonism	AR	2q31.2	Protein kinase	–
DYT17	Young-onset craniocervical dystonia that can generalize	AR	20p11.2-q13.12	N/I	–

Continued

TABLE 76.1 A List of Hereditary Forms of Dystonia—cont'd

Name (Gene)	Clinical features	Inheritance pattern	Location	Protein	Mutations identified
DYT18 (*SLC2A1*), GLUT1 deficiency, paroxysmal exercise-induced dystonia	Episodic involuntary movements, dystonic posturing, imbalance, dysarthria, paresthesia	AD	1p34.2	Glucose transporter 1	–
DYT19 (*PKD2*)	Paroxyxmal kinesigenic dyskinesia	AD	16q13-q22.1	N/I	–
DYT20 (*PNKD2*)	Paroxysmal nonkinesigenic dyskinesia	AD	2q31	N/I	–
DYT21	Generalized torsion dystonia; Swedish family	AD	2q14.3-q21.3	N/I	–
DYT23 (*CIZ1*)	Cervical dystonia	AD	9q34.11	Cip1- interacting zinc finger protein	–
DYT24 (*ANO3*)	Jerky torticollis	AD	11p14.2	Anoctamin 3	–
DYT25 (*GNAL*)	Craniocervical dystonia	AD	18p11.21	G-alpha-olf	–

Abbreviations: AD, autosomal dominant; AR, autosomal recessive; XLR, X-linked recessive; N/I, not identified.

environmental interactions. In fact, a polymorphism at codon 216 of DYT1 is reported to modify the phenotype.[9] No consistent pathology has been found, but fluorodeoxyglucose (FDG) positron emission tomography (PET) scans reveal high lenticular and reduced thalamic regional metabolism.

DYT6

DYT6 presents both as young- and adult-onset isolated, nondegenerative dystonia, usually starting in the cranial region and spreading to become generalized. It is transmitted in an autosomal dominant pattern with incomplete penetrance. It is characterized by focal or generalizable dystonia, with cranial, cervical, or limb involvement. Mutations in the thanatos-associated domain-containing apoptosis-associated protein gene, *THAP1*, are responsible for this condition.[10] Although this gene was first discovered in a few Amish–Mennonite families of German ancestry, mutations have been found across many ethnicities. Penetrance appears to be about 60%.[8] *THAP1* may be involved in regulating transcription of *TOR1A*.[11]

Other Torsion Dystonias

Autosomal dominant adult-onset focal dystonia (cervical and laryngeal dystonia with or without postural tremor) with reduced penetrance has been linked to chromosome 18p in individuals of German ancestry (DYT7); however,

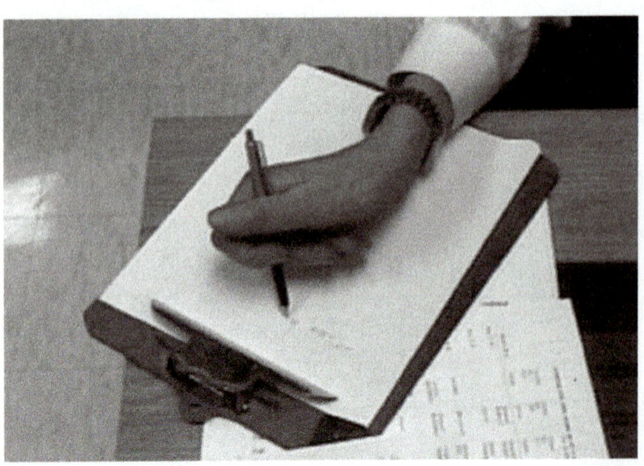

FIGURE 76.2 Oppenheim dystonia manifested as difficulty in writing (writer's cramp).

this association is in question.[12] DYT13 has been linked to 1p36.32-p36.13 in an Italian family with cranial–cervical or upper limb onset dystonia and incomplete penetrance.[13] The disease is characterized by juvenile or early-adult onset, a mild course, and occasional generalization. Next generation sequencing has resulted in an explosion of new dystonia genes. Mutations in the *CIZ1* gene, DYT23,[14] and the *ANO3* gene, DYT24,[15] have been found to cause an autosomal dominant craniocervical dystonia. Most recently, *GNAL* on chromosome 18p has been found to cause another cervical and laryngeal dystonia, DYT25, with variable ages of onset.[16] All of these genes have incomplete penetrance. Mutations in *TUBB4A* have been found to cause DYT4, which results in a generalized dystonia with laryngeal dysphonia, originally described as whispering dystonia, but now recognized as spasmodic dysphonia. This gene appears to be one of the only fully penetrant dystonia genes.[17,18]

DYT3: Lubag Disease

Lubag disease is a rare neurodegenerative, variable-onset, highly penetrant X-linked recessive form of dystonia-parkinsonism characterized by generalized dystonia and levodopa-unresponsive parkinsonism. Mean age of onset is near 40. When onset is during adulthood, focal cranial dystonia (jaw/tongue) is common. DYT3 is located on Xq13.1 and is restricted almost exclusively to males whose ancestors originated from the island of Panay in the Philippines, presumably due to a founder effect.[19] Female carriers are usually asymptomatic, but may occasionally manifest mild chorea or other abnormal movements. Lubag has been linked to a retrotransposon insertion in the *TAF1 (TATA box-binding protein-associated factor 1)* gene.[20]

DYT5: Dopa-Responsive Dystonia

Dopa-responsive dystonia (DRD) typically manifests as a childhood-onset lower-limb dystonia/gait disorder, often with marked diurnal fluctuation. It may be accompanied by parkinsonian features, such as hypomimia, bradykinesia, and loss of postural reflexes, and it demonstrates a dramatic and sustained resolution of symptoms with low doses of levodopa. The phenotype, however, is highly variable, with very severe infantile onset to a nonprogressive adult form. The dystonia usually slowly generalizes.

Three types of DRD are: autosomal dominant GTP cyclohydrolase 1 (GTPCH1)-deficient DRD, also known as Segawa syndrome or DYT5A and caused by mutations in *GCH1*; an autosomal recessive tyrosine hydoxylase (TH)-deficient DRD (often referred to as DYT5B); and, more rarely, autosomal recessive mutations in the sepiapterin reductase gene *(SPR)*. Both sepiapterin reductase and GTPCH1 are enzymes in the biosynthesis of tetrahydrobiopterin, the essential cofactor for TH. TH is the rate-limiting enzyme in the synthesis of dopamine.

Clinical testing for the *GTP cyclohydrolase 1* (*GCH1*) gene detects mutations in approximately 60% of families with DRD. This may be due to genetic heterogeneity or the presence of duplications or deletions. Duplication/deletion analysis adds 5–10% to the detection rate.[21] More than 100 mutations in *GCH1* have been identified. Interestingly, GTPCH1-deficient DRD is more penetrant in females (87%) than in males (38%).

The *TH* gene is located on chromosome 11p15.5. Parents of an affected child are obligate heterozygotes and carry one mutant allele. Usually asymptomatic, carriers in rare instances can have a subtle phenotype, such as exercise-induced stiffness, which is responsive to levodopa. The mutation detection rate in clinical testing for TH-deficient DRD is unknown. Mutations in *TH* also cause a severe progressive infantile encephalopathy. The severity of *TH* disorders is dependent on the amount of enzymatic activity.

A leading differential diagnosis for DRD is PARK2, which can present with parkinsonism or dystonia. The differential diagnosis is resolved by ^{18}F-fluorodopa (FDOPA) PET scans, which show no deficit in DRD, but a loss of FDOPA uptake in PARK2. Additionally mutations in the sepiapterin reductase gene (*SPR*) have been considered to be the cause of PARK3. What was originally thought to be a separate gene causing DRD and named DYT14 was shown to be a deletion within the *GCH1* gene.[22]

DYT11, DYT15: Myoclonus-Dystonia

Myoclonus-dystonia is an autosomal dominant movement disorder characterized by sudden jerking movements and dystonia of variable expression and penetrance. The disease is almost twice as penetrant in women than men (80%:40%).[23] Typical age at onset is in the first decade. It is caused by mutations in either the *epsilon sarcoglycan* gene (*SGCE*) on chromosome 7q21.3 (DYT11), which encodes one member of the dystrophin glycoprotein complex, or on a gene on a locus mapped to chromosome 18p11 (DYT15). For DYT11, maternal imprinting leads to penetrance mainly when inherited through the paternal line. Only about 5% of patients with symptoms have inherited the gene from their mother. Psychiatric abnormalities such as obsessive–compulsive disorder are associated with the disease in some individuals. Alcohol alleviates symptoms of myoclonus, leading some individuals to alcohol addiction. A mutation in the *D2 dopamine receptor* gene has also been associated with myoclonus-dystonia in a large family but this result has not been replicated. Mutations in DYT11 are detected in approximately 30–40% of symptomatic individuals. Heterogeneity and copy number variants may be responsible for the low detection rate. Thus, copy number variations (CNVs) should be analyzed if sequencing is negative.[23]

DYT12, DYT16: Rapid-Onset Dystonia-Parkinsonism

Rapid-onset dystonia-parkinsonism (RDP) is a rare autosomal dominant disease with incomplete penetrance. Onset is sudden and symptoms progress within weeks to a plateau state; it usually occurs in adolescence, but onset has been reported from childhood to adulthood. Penetrance by age 40 is about 85%.[19] Some cases of RDP (DYT12) are caused by mutations in the *Na^+/K^+-ATPase alpha3* subunit gene *(ATP1A3)* implicating a malfunction of the Na^+/K^+ pump. DYT12 is an autosomal dominant disease, but *de novo* mutations seem to occur frequently.[24] Neither levodopa or deep brain stimulation (DBS) relieve symptoms of RDP.[24]

DYT16 is an autosomal recessive, young-onset dystonia-parkinsonism due to mutations in the *protein kinase interferon-inducible double-stranded RNA-dependent activator (PRKRA)* gene. Patients usually present with limb dystonia, which then generalizes. Dysphagia and delays of motor and speech development have been reported in some cases.[19,24]

Neurodegeneration with Brain Iron Accumulation Disorders (NBIAs)

The NBIAs are a group of hereditary degenerative disorders in which iron accumulates in the basal ganglia and commonly in the cerebellum and cerebrum. Features of these diseases include dystonia, parkinsonism, spasticity, neuropsychiatric impairment, retinal degeneration, and optic atrophy. Both the phenotype and age at onset within and among diseases is highly variable. Although the majority of these diseases are autosomal recessive, both X-linked and autosomal dominant diseases also occur.

Pantothenate Kinase-Associated Neurodegeneration (PKAN)

PKAN, formerly called Hallervorden–Spatz disease, is an autosomal recessive disease with typical childhood onset caused by mutations in the *pantothenate kinase gene 2* (*PANK2*) on chromosome 20p13. Missense mutations, CNVs, and large deletions have been found causing a wide variation in phenotype.[25] Clinical features include dystonia, progressive extrapyramidal dysfunction, chorea, dysarthria, parkinsonism, spasticity, dementia, oculomotor abnormalities, and retinal degeneration. Some individuals present with milder, mainly parkinsonian, symptoms at a later age of onset. Histological findings reveal iron deposits in the basal ganglia visualized as brown discolorations. Increased iron in the globus pallidus is usually seen with magnetic resonance imaging (MRI), often with a central atrophic region called the "eye-of-the tiger" sign. *PANK2* is involved in coenzyme A production, which has a crucial role in fatty acid metabolism. Mutations in *PANK2* also cause hypoprebetalipoproteinemia, acanthocytosis, retinitis pigmentosa, and pallidal degeneration (HARP), a syndrome now considered part of the PKAN disease spectrum.

PLA2G6-Associated Neurodegeneration (PLAN)

Neuronal brain iron accumulation genes (NBIA) *PLA2G6* and *PANK2* have been associated with young-onset PD. *PLA2G6*, phospholipase A2, Group VI, is involved in fatty acid metabolism and most commonly causes infantile neuroaxonal dystrophy neurodegeneration.[26] The phenotype typically includes dystonia, rigidity, ataxia, spasticity, dysarthria, cognitive impairment, and seizures. This disease can present in infancy or with a juvenile or adult onset. Atypical presentations with later onset and levodopa-responsive parkinsonism have been reported in homozygotes and compound homozygotes of *PLA2G6*.[27] Autopsies of people homozygous for mutations in *PLA2G6* show significant α-synuclein pathology with tau pathology.

Mitochondrial Membrane-Associated Neurodegeneration (MPAN)

MPAN is an autosomal recessive disorder with spasticity as its primary symptom and dystonia, motor neuron disease, optic atrophy, and cognitive dysfunction. Onset is from childhood to adulthood with a variable rate of progression. A single mutation in the *C190rf12* gene appears to be causal.[26,28]

Beta-Propeller Protein-Associated Neurodegeneration (BPAN)

BPAN is an X-linked dominant disorder caused by *de novo* mutations in the WD repeat domain phosphoinositide-interacting protein 4 gene (*WDR45*). This gene, probably an embryonic lethal in males, produces a syndrome similar to Rett syndrome. Childhood developmental delay is followed by a sudden onset of dystonia, parkinsonism, and dementia in adolescence or adulthood.[29]

Fatty Acid Hydroxylase-Associated Neurodegeneration (FAHN)

FAHN is a rare autosomal recessive disease caused by mutations in the fatty acid 2-hydroxylase gene (*FA2H*). The syndrome consists of dystonia, ataxia, dysarthria, optic atrophy, cognitive decline, and leukodystrophy. It has previously been reported as a spastic paraparesis, SPG35.[30]

Kufor–Rakeb Disease

Kufor-Rakeb disease can result in a rare autosomal recessive NBIA or a juvenile-onset Parkinson-plus syndrome with pallidopyramidal degeneration with levodopa-responsive parkinsonism and dystonia, spasticity, supranuclear gaze palsy, and dementia. Mutations in *ATP13A2,* encoding a transmembrane lysosomal P5-type ATPase, are causal for Kufor–Rakeb. Phenotype is highly variable both in a family and between families.[30]

Neuroferritinopathy

Neuroferritinopathy is an autosomal dominant disease of the basal ganglia caused by mutations in the gene encoding ferritin light polypeptide (*FTL1*) on chromosome 19q13.33. Neuroferritinopathy is characterized by extrapyramidal features with symptoms of both Huntington disease and parkinsonism, and typically presents with progressive adult-onset chorea or dystonia and subtle cognitive deficits. Histological findings reveal abnormal deposition of ferritin and iron in the basal ganglia, detected by MRI. Iron accumulation has been associated with a number of neurodegenerative diseases and occurs naturally with age. Molecular genetic testing is available.[30]

Aceruloplasminemia

Aceruloplasminemia is an autosomal recessive disease due to mutations in the ceruloplasmin gene (*CP*). The disease has an adult onset and results in cognitive impairment, retinal degeneration, diabetes, chorea, ataxia, dystonia, dysarthria, and tremor. Iron accumulates in the liver as well as the brain.[30]

CHOREA

Huntington Disease

Huntington disease (HD) is a neurodegenerative autosomal dominant disease caused by CAG triplet repeat expansions in exon 1 of the *HTT* (*IT15*) gene on chromosome 4p16.3 coding for the huntingtin protein.[31] Abnormal involuntary movements, cognitive decline, and psychiatric disturbance characterize HD. Mean age of onset is 35–44 years. The CAG triplet repeat expansion results in a gain-of-function mutation causing substantial loss of spiny neurons in the striatum. The elucidation of the cause of HD in 1993 was one of the first genetic discoveries in movement disorders. HD research has since led the way as a model of the polyglutamine diseases. In addition, genetic testing protocols first developed for HD are now templates for other adult-onset diseases.[32,33] We have, therefore, placed special emphasis on genetic testing, clinical features, and pathogenesis of HD.

Genetics and Genetic Testing

The HD triplet repeat expansion translates into an expanded polyglutamine tract close to the *N*-terminus of the huntingtin protein. In normal individuals, the number of repeats in the *HTT* gene is 26 or less. Triplet repeat lengths of 27–35 repeats, "the intermediate range," do not cause HD, but may expand into the "HD" range during gametogenesis, placing children of carriers in this range at risk to develop HD. Thirty-six to 39 CAG repeats (reduced penetrance alleles) cause HD with variable penetrance—an individual with a repeat length in this range may or may not develop symptoms of HD, and may be at risk to have offspring with an expansion in the "HD" range. Although the likelihood of developing symptoms and the age of onset tend to correspond to the number of repeats, there have been case reports of symptomatic HD in the intermediate range so these categories cannot be considered firm.[34] Finally, individuals with 40 or more repeats develop fully penetrant HD.

There is a loosely inverse relationship between repeat length and age of onset. In fact, only 50–70% of variation in age of onset can be attributed to repeat length.[35] However, it is clear that factors other than repeat length also determine age of onset. No association between repeat length and clinical features or rate of progression has been consistently reported. Twenty-five percent of individuals with HD develop symptoms after age 50 years. Seven percent of individuals with HD have the juvenile form with onset before age 20 years and with large expansions (over 55, up to 121 repeats). Individuals with later onset HD may have a more slowly progressing disease when compared to those with juvenile-onset HD. Eighty percent of individuals with juvenile HD inherit an expanded paternal allele. This phenomenon is called "anticipation" and produces ever-larger HD expansions from generation to generation, often resulting in earlier ages of onset in successive generations. An unknown mechanism specific to spermatogenesis causes CAG repeat instability and expansion in HD. The children of females with HD have smaller shifts in triplet repeat lengths, and these may take the form of contractions or expansions. Approximately 10% of patients with symptoms of HD do not have a family history. After discounting nonpaternity, early death of family members, and other causes of incomplete family history, it has been suggested that the remaining individuals may represent new expansions, possibly from a parent with a repeat length in the intermediate range.[35]

Genetic testing for individuals with HD and other neurodegenerative genetic diseases is a lengthy and complex process that involves pre- and post-counseling sessions with a genetic counselor or geneticist, who is trained to address not only the complex details of inheritance, but also the psychosocial issues that frequently occur both with the affected individual and at-risk family members. The pre-test protocol also involves evaluation by a psychiatrist and neurologist to look for early symptoms. Presymptomatic testing of minors is strongly discouraged due to issues of informed consent. In addition, there is currently no clinical benefit to knowing presymptomatic status. When ordering genetic tests for HD, it should be noted that very large CAG repeat lengths (more common in juvenile HD) may not be detected by conventional diagnostic polymerase chain reaction (PCR) techniques and may only be detected by Southern blot analysis. Southern blot should be performed when homozygous repeat numbers are found to rule out the possibility of a large expansion.

The prevalence of HD is as high as 5–10/100,000 in some European populations, and as low as 0.1/100,000 in some African and Asian populations. This finding may be related to the observation that alleles with high–normal numbers of repeats are higher in Europeans than in Africans or Asians. In certain relatively isolated regions of the world, the prevalence is as high as 52 (Lake Maracaibo, Venezuela) and 560 (Moray Firth, Scotland) per 100,000.[36]

Clinical Symptoms of HD

The symptoms of HD can vary greatly, and although chorea, and psychiatric and cognitive changes are the hallmarks of the disease, some patients do not develop dementia or chorea. The majority of patients present with movement disorders such as chorea, clumsiness, dropping of objects due to motor impersistence (negative chorea), or impaired coordination of voluntary movements, but often report prior depression or difficulty planning or concentrating. The Prospective Huntington At Risk Observational Study (PHAROS) by the Huntington Study Group followed at-risk individuals to determine signs of HD before onset of abnormal movements.[37] It found that 92.3% of at-risk individuals were judged to have no or nonspecific motor abnormalities, 6.7% were found to have possible or probable motor signs, and 1% had unequivocal HD.

Gait disturbance (prancing, stuttering, unsteady gait) (Figure 76.3), dysarthria, dysphagia, bradykinesia, rigidity, or dystonia often appear further on in the course of the disease. Psychiatric and cognitive changes such as psychosis, agitation, apathy, impulsiveness, depression, mania, paranoia, delusions, sexual disorders, hallucinations, memory loss, and poor judgment may occur. HD is often fatal between 15 and 20 years after onset of symptoms.[38]

Juvenile HD (age 20 or younger) is characterized by progressive parkinsonism (bradykinesia and rigidity), dementia, ataxia, and seizures, and can present as declining cognitive function, behavioral disturbance (such as violence or suicidal ideation), rigidity of limb or trunk, and seizures. Chorea can be absent, and an akinetic-rigid syndrome is usually the presentation. Juvenile HD is often associated with a family history of HD (usually paternal). The constellation of features that characterize juvenile HD differ from adult-onset HD and can lead to misdiagnosis. Juvenile HD is not only characterized by loss of neurons in the striatum but may also show cerebellar atrophy.

Symptoms of HD may overlap with those of other genetic diseases such as dentatorubropallidoluysian atrophy (DRPLA), Wilson disease, neuroacanthocytosis, benign hereditary chorea, familial ataxia, HD-like 2 disease, familial prion disease, and nongenetic causes of chorea-like illnesses (such as tardive dyskinesia, thyrotoxicosis, cerebral lupus and polycythemia).

Therapy for HD includes antidepressants, antipsychotic and anti-anxiety medication, neuroleptics (but these may induce tardive dyskinesia), and tetrabenazine.[39] A recent randomized trial suggests the use of pridopidine.[40] Supportive therapy, including occupational and physical therapy, has proven very effective in managing symptoms.

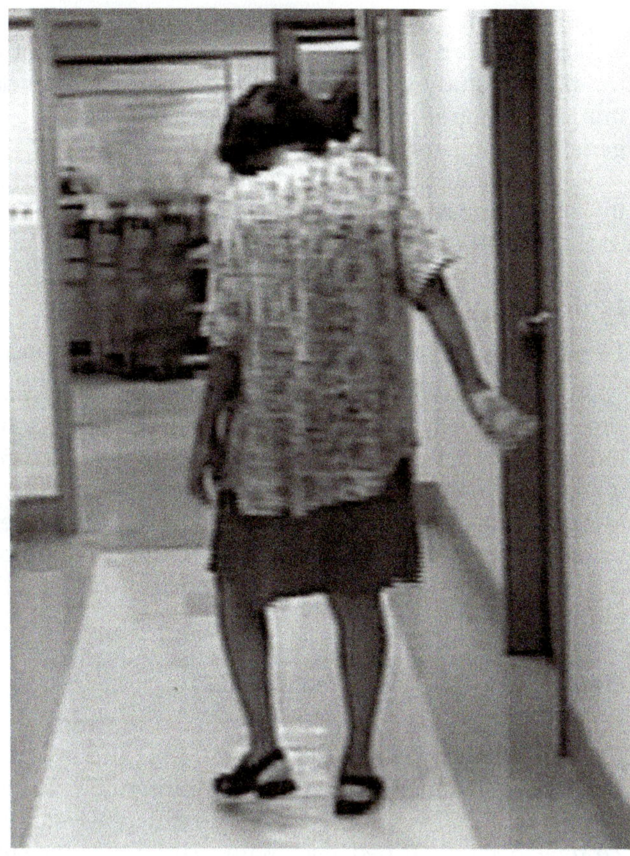

FIGURE 76.3 Stuttering, uncoordinated gait due to chorea, and right arm posturing owing to chorea/dystonia in a patient with Huntington disease.

Pathogenesis of HD

It is still not clear how the polyglutamine expansion causes neurodegeneration. Mice transgenic for human exon 1 containing the expansion, develop symptoms of HD, supporting a gain-of-function role. Observations on the normal role of huntingtin and its mutated version also give clues as to the pathogenesis of the disease. Normal and expanded huntingtin are both expressed in the brain and body. However, the brains of individuals with HD contain large intracellular aggregates or inclusions containing huntingtin and also ubiquitin. The length of huntingtin and its polyglutamine tract is directly proportional to the frequency of intracellular aggregates.[41] The aggregates contain only the part of the huntingtin protein that includes the expanded polyglutamine tract, cleaved from the rest of the protein by caspases. The presence of the abnormal tract may induce apoptosis and it is hypothesized that normal huntingtin may have an anti-apoptotic role. Alternatively, formation of the intracellular aggregates may be a compensatory mechanism, sequestering the expanded huntingtin. This is supported by the observation that aggregates are often found in surviving neurons. In order to try to tease apart the causal mechanism, minocycline, a caspase inhibitor, was administered to mouse models of HD, and was found to decrease HD progression. A pilot study involving patients taking minocycline over a 6-month period did not, however, result in a significant decrease in HD progression. This study was conducted on a small population to determine safety of minocycline use, so further studies are needed to determine the efficacy of caspase inhibitors.

Another theory is that the expanded polyglutamine tract may interact abnormally with a number of transcription factors containing short polyglutamine tracts, thereby altering gene expression. One such protein is the transcriptional coactivator cAMP-responsive element-binding protein (CREB)-binding protein (CBP), which is thought to bind the expanded polyglutamine tract, thus reducing the normal acetyl transferase activity of CBP, decreasing acetylation of histones, and leading to reduced gene transcription, possibly affecting cell survival. Since histone deacetylase (HDAC) inhibitors have been successful in improving symptoms in transgenic mice, trials have now started in HD patients. Another trial is studying the safety and tolerability of phenylbutyrate.[42]

Mutant huntingtin binds to neuronal synaptic vesicles. This adversely affects glutamate levels in the striatum resulting in mitochondrial calcium abnormalities and high levels of free radicals. It was postulated that antioxidant therapy to reduce levels of free radicals might benefit HD patients. A clinical trial demonstrated evidence that coenzyme Q10, an antioxidant, might slow the progression of HD. A trial has recently been completed to determine the highest and safest dosage of coenzyme Q10 so that further studies will have the power to detect whether coenzyme Q10 can be used as a therapy.[43] (For the most current information on clinical trials for HD patients can visit www.huntington-study-group.org.) More recently, studies have implicated altered DNA methylation, RNA splicing, immune cell migration, transcription and trafficking of brain-derived neurotrophic factor (BDNF) as possible pathogenic mechanisms.[44–47]

Polyglutamine tract expansions are now known to cause a number of neurodegenerative diseases. It is hoped that the expanding research in the field of HD may help shed light on how triplet repeat expansions cause disease.

Huntington Disease-Like Phenotypes

Familial Prion Disease (formerly Huntington Disease-Like 1 [HDL1])

Autosomal dominant familial prion disease, characterized by cognitive difficulties, myoclonus and ataxia, can present atypically as an early-onset, slowly progressing disease with features overlapping those of HD. Affected individuals present with adult-onset psychiatric or behavioral disturbance and develop cognitive decline, personality change, motor disturbance with chorea, dysarthria, and ataxia, together with atrophy of the basal ganglia. Some affected family members also develop epileptic seizures, which is atypical for adult-onset HD. This atypical form of prion disease (initially described as HDL1) is caused by eight extra octapeptide repeats in the coding region of the prion protein (*PRNP*) gene on chromosome 20pter-p12.[48] It is sometimes referred to as "early-onset prion disease with prominent psychiatric features."

Huntington Disease-Like 2 (HDL2)

HDL2 is caused by a CTG expansion in exon 2A of the *junctophilin-3* (*JPH3*) gene on chromosome 16q24.2.[49] *JPH3* is primarily expressed in the brain, and appears to be involved in the formation of junctional membrane structures, specifically in modulating internal calcium flux. Exon 2A is variably spliced, and does not appear in the full-length *JPH3* transcript. Three different versions of a truncated transcript containing only exon 1 and exon 2A have been identified, each with a different splice junction that alters the reading frame, so that the CTG repeat may encode polyalanine or polyleucine, or fall in the 3′ untranslated region. The normal repeat is 7–28 triplets in length, with expansions ranging in length from 40–57 triplets; repeats in the 40–45 range may cause incomplete penetrance. Shorter disease-causing expansions may be associated with more prominent chorea and a slower rate of progression, while individuals with longer repeat lengths may initially present with rigidity and dystonia, resembling juvenile HD. It is unknown precisely how the expansion causes disease but it is thought to involve toxic RNA species, and perhaps a loss of normal JPH3 expression. Clinically and pathologically, HDL2 is nearly indistinguishable from HD, and, like HD, is characterized by selective striatal degeneration and neuronal intranuclear inclusions.

HDL2 is almost exclusively limited to people of African ethnicity, or populations with an admixture of African ancestry. Overall, about 1% of individuals with HD-like symptoms but without the HD expansion may have HDL2 if the phenotype is broadly defined. The prevalence is likely to be much higher among individuals with a more classic HD phenotype, a family history, and African ancestry. Gene testing for the expansion is clinically available.

Huntington Disease-Like 3 (HDL3)

HDL3 is an autosomal recessive neurodegenerative disorder segregating in a consanguineous family from Saudi Arabia. It has been mapped to 4p15.3, in close proximity to the HD locus at 4p16.3.[50] The disease manifests at approximately 3–4 years of age and is characterized by both pyramidal and extrapyramidal abnormalities, including chorea, dystonia, ataxia, gait instability, spasticity, seizures, autism, and intellectual impairment. Brain MRI findings include progressive frontal cortical atrophy and bilateral caudate atrophy.

Huntington Disease-Like 4 (HDL4)

HDL4 is otherwise known as SCA17 and is due to a CAG/CAA expansion in *TBP*, encoding the TATA-box-binding protein. This syndrome has a wide age of onset and usually begins with psychiatric symptoms and ataxia. Chorea, dystonia, parkinsonism, and dementia can arise as the disease progresses.[51] The normal range of repeats is 21–40, reduced penetrance alleles are 41–48, and fully penetrant alleles are 49 or more. Anticipation is found far less often than with HD and may require the loss of the CAA interruption. Pathology shows atrophy of the striatum, cerebellum, and sometimes thalamus and cerebrum.

Dentatorubral-Pallidoluysian Atrophy

Dentatorubral-pallidoluysian atrophy (DRPLA) is a progressive neurodegenerative autosomal dominant disease, caused by an expansion of a CAG triplet repeat in exon 5 of *ATN1* (*DRPLA*) which codes for Atrophin-1 on chromosome 12p13.31.[52,53] It is characterized by ataxia, choreoathetosis, progressive dementia, and cognitive decline. The triplet-repeat expansion causes combined systemic degeneration of the dentatofugal and pallidofugal pathways visualized histologically by ubiquitinated neuronal intranuclear inclusions. The relative quantities of DRPLA mRNA and protein in individuals with DRPLA and without are the same, suggesting that the *DRPLA* expansion acts as a gain-of-function mutation possibly by interfering with CREB-dependent transcriptional activation resulting in neuronal toxicity.

Unaffected individuals have 6–35 CAG repeats in the *DRPLA* gene. Individuals with 36–47 repeats do not have symptoms of DRPLA, but alleles in this range are unstable and can expand on transmission, placing offspring at risk to have an expanded allele that is disease-causing. Repeats of 48 and over cause DRPLA. DRPLA demonstrates anticipation, especially when inherited through the paternal line. Gene testing is widely available and involves measuring the number of CAG repeats.

Age of onset varies from 1 to 62 years, with an average of 30 years. CAG triple repeat length has been shown to be inversely proportional to age, with earlier age of onset usually accompanying a greater expansion. Two main phenotypes have been described: The first is the progressive myoclonus epilepsy phenotype (62–79 repeats), which has an age of onset of 21 years and younger, and is usually caused by a paternal expansion. In addition to classical symptoms of DRPLA, individuals with young-onset DRPLA may exhibit myoclonus, epilepsy, and mental retardation. The second phenotype is characterized by nonprogressive myoclonus epilepsy (54–67 repeats) and a later age at onset (older than 21 years); it is clinically similar to HD and spinocerebellar ataxias, but the chorea component of the DRPLA phenotype can mask the presence of the ataxia. Individuals with very small pathogenic expansions (49–55) can exhibit ataxia only, also complicating the diagnostic process.

DRPLA is rare. The prevalence of both DRPLA and HD in the Japanese population is estimated to be 0.2–0.7/100,000. An expansion in *DRPLA* is believed to be responsible for approximately 0.25% of European families demonstrating autosomal dominant forms of cerebellar ataxia.[54] A neurodegenerative disease in an African American family described as "Haw River Syndrome" was later found to be caused by a *DRPLA* expansion.[55] Interestingly, the increased prevalence of DRPLA in the Japanese population corresponds to the increased frequency of triplet repeats in the high–normal range of the *DRPLA* gene in Japanese when compared to Europeans.

Neuroacanthocytosis

Neuroacanthocytosis (chorea acanthocytosis) is a rare neurodegenerative autosomal recessive disorder characterized by chorea, buccal dyskinesias, tics, cognitive decline, seizures, dementia, parkinsonism, absent tendon reflexes, and occasional myopathy.[56] Epilepsy is the initial manifestation in half of individuals with this disease. The mean age of onset is 35 years, but both young- and late-onset cases have been described. Computerized tomography (CT) and MRI reveal atrophy of the caudate nuclei with dilatation of the anterior horns of the lateral ventricles. Acanthocytes are present in up to 50% of red blood cells of affected individuals. Infrequently they are absent altogether, and in some patients they are present only in the latter stages of the disease. Myopathy is usually accompanied by increased serum concentration of muscle creatine phosphokinase. Neuroacanthocytosis is caused by mutations in the vacuolar protein sorting-associated protein 13A gene (*VPS13A*) on chromosome 9q21, which codes for the protein chorein. So far, 17 mutations have been described, dispersed throughout the gene, including missense, deletion, and frameshift mutations. Between 200 and 250 people with neuroacanthocytosis have been identified worldwide, 20% of which are Japanese—suggesting a founder effect in this population.

Although neuroacanthocytosis is a rare disease, it is noticeable for its clinical overlap with McLeod syndrome, abetalipoproteinemia, hypobetalipoproteinemia, and HD. In addition, it can mimic Tourette syndrome with vocal tics, obsessive–compulsive behavior, and impaired impulse control.

Genetic testing is not regularly used to clinically diagnose neuroacanthocytosis, and diagnosis rests on assays for acanthocytes in blood smears and western blot analysis to determine whether the chorein protein is expressed.

Benign Hereditary Chorea

Benign hereditary chorea is an autosomal dominant condition characterized by nonprogressive, benign, early-onset chorea with an absence of mental deterioration, dystonia, myoclonic jerks, or dysarthria.[57] It is associated with mutations in the gene encoding thyroid transcription factor 1 (*TITF1*), whose gene is called *NKX2-1*. A greater penetrance in males than females has been reported, and there is both intra- and interfamilial variability. Onset of chorea usually begins by 5 years of age, and stabilizes between 10 and 20 years of age. For some, symptoms abate during adulthood. MRI has not detected major cerebral changes. Choreoathetosis, congenital hypothyroidism, and respiratory distress are more severe manifestations of mutations in the *NKX2-1* gene.

Haplotype analysis has found no suggestion of a common founder for families with mutations in the *NKX2-1* gene, and most families have a "private" mutation. A number of families with benign hereditary chorea do not have mutations in *NKX2-1*, suggesting that mutations in other genes cause benign hereditary chorea. These families, in addition to symptoms of chorea, tend to have more heterogeneous symptoms than do *NKX2-1* families (such as ataxia, dysarthria, or mental disturbance) and a wider range of age of onset.

Wilson Disease

Wilson disease (WD) is an autosomal recessive neurodegenerative disease with hepatic, renal, neurological and psychiatric manifestations. It is caused by mutations in the *ATP7B* gene on chromosome 13q14.3 leading to defective copper transport and accumulation of copper.[58] *ATP7B* is homologous to the cation-transporting P-type adenosine-triphosphatase (ATPase) gene family, including the gene responsible for Menkes syndrome.

Symptoms of WD are caused by defective biliary copper excretion from hepatocytes. In addition, copper fails to be incorporated into ceruloplasmin. These defects result in copper deposition at toxic levels in the brain, liver, and kidney. Diagnosis of WD is aided by the following findings: low serum ceruloplasmin concentration and increased urinary and hepatic copper concentration. However, a normal serum ceruloplasmin concentration is found in at least 5% of patients with WD with neurologic symptoms and in up to 40% of patients with hepatic symptoms. Some heterozygotes may also have some of these laboratory findings, but are asymptomatic. In addition, Kayser–Fleischer (K–F) rings in the outer circumference of the cornea are caused by copper deposition in Descemet's membrane of the corneal endothelium and can be detected by slit-lamp examination. K–F rings are observed in approximately 50% of individuals with WD presenting with hepatic disease and in approximately 90% of those presenting with neurological or psychiatric manifestations. However, K–F rings are not unique to WD.

Age of onset of WD ranges from 6 to 50 years, and symptoms are variable. Liver disease is the first manifestation in 40% of individuals with WD, and is usually the first manifestation in early-onset WD. Initial presentation of neurological or psychiatric symptoms often occurs at a later age of onset. Interestingly, mutations that eliminate gene function (null mutations) appear to correlate with young-onset WD, presenting with liver disease. Neurological findings can include incoordination, tremor of any type including wing-beating postural tremor, dystonia (Figure 76.4), rigidity, chorea, and a risus sardonicus dystonic facial grin with open mouth. Presentation can also be unilateral. Psychiatric symptoms can include depression, cognitive declines and compulsive behavior. The range of symptoms and variability of initial symptoms can lead to differential diagnoses as wide ranging as PD, juvenile PD, HD, DRPLA, ataxia, or Niemann–Pick type C disease. Early diagnosis of symptomatic and presymptomatic individuals is crucial, since copper chelation therapy has proven very effective, with reversible symptoms in the early stages of the disease.

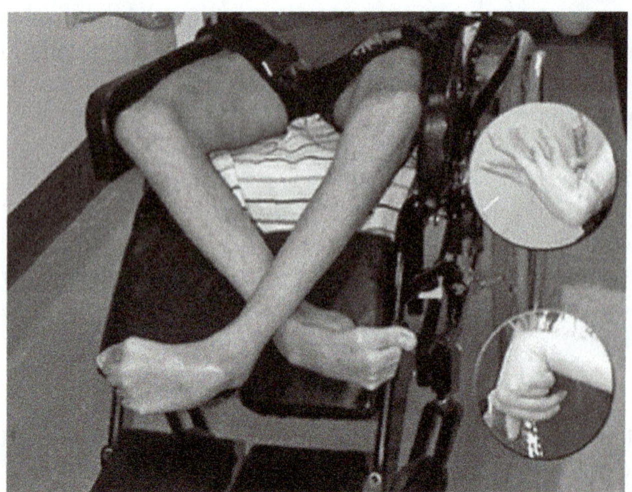

FIGURE 76.4 Dystonic postures in Wilson disease.

The prevalence of WD is approximately 1/30,000 (with a 1/90 carrier frequency), but is higher in Asia and Europe. A single mutation (His 1069 Glu) is thought to account for more than 30% of European cases of WD. More than 200 other mutations have been described across the world, most of which are rare. As a result, most patients are compound heterozygotes (they have a different mutation on each allele). Molecular genetic testing for mutations in the *ATP7B* gene is available; however, the variety of mutations in this gene makes genetic testing a difficult and expensive option. Atypical inheritance and incomplete penetrance may also play a role in some cases.[59] In fact, recently cases with uniparental disomy were described.[59] Both European and Asian populations appear to have, to some degree, individual founding mutations, thus aiding gene analysis. If a specific mutation cannot be found, it is still possible for an individual with WD to undergo linkage studies so that the disease status of other family members can be determined. This can be useful for prenatal diagnosis to identify asymptomatic family members so that prophylactic copper chelation therapy can be administered to offset symptoms. A number of genotype–phenotype correlations have been found. For example, the most severe mutation (that completely prevents functioning of the gene) results in symptoms before the age of 12 years, frequently with liver manifestations. Whole-exome sequencing may be another option for these families.

A naturally occurring rat model of WD, called the Long–Evans Cinnamon (LEC) rat, has played a large role in the elucidation of the underlying defect of WD. The rat homolog of the human *ATP7B* gene is abnormal in the LEC rat. The LEC rat has been used successfully to investigate gene therapy of WD using adenoviral gene transfer.

MYOCLONUS EPILEPSY

The myoclonus epilepsies are a group of progressive neurodegenerative diseases that display the combination of myoclonus and epilepsy in addition to ataxia, dysarthria, psychiatric symptoms, and dementia. The two most common types are discussed.

Myoclonus Epilepsy of Unverricht–Lundborg Type

Of the several genetic causes of progressive myoclonus epilepsy, the most common is myoclonus epilepsy of Unverricht–Lundborg disease (EPM1). This autosomal recessive disease is caused by a dodecamer (12-mer) minisatellite repeat expansion (CCC CGC CCC GCG) located about 70 nucleotides upstream of the transcription start site nearest to the 5′ end of the *Cystatin B* gene *(CSTB)* on chromosome 21q22.3.[60] *Cystatin B* codes for a cysteine protease inhibitor, which has a crucial role in the programming of cell apoptosis and is widely expressed in neurons but not glial cells. The expansion results in loss of expression of *CSTB* leading to uncontrolled cell apoptosis and neuronal cell death (without inclusion bodies). Onset is usually between the ages of 6 and 15 years and is characterized by stimulus-sensitive myoclonus, tonic-clonic seizures and typical electroencephalogram (EEG) findings. Neurodegenerative changes such as ataxia and dementia are common, and cognitive function typically declines 10–20 years after onset of symptoms.[61]

Expansion of the dodecamer repeat in *CSTB* was also found to cause Baltic Mediterranean progressive myoclonus epilepsy, proving that these two diseases are in fact a single disease entity. Founder effects are thought to be responsible for the increased incidence of EPM1 that has been reported in both Finland (where the incidence is 1/20,000) and North Africa. No other genes have been associated with EPM1.

Normal alleles of the *CSTB* gene contain two or three copies of this dodecamer repeat whereas mutant alleles contain more than 30 repeats. Premutation repeats with 12–17 repeats show marked instability when transmitted to offspring. Evidence of the instability of the repeat is also provided by the observation that affected siblings can have widely differing repeat lengths. While the majority of EPM1 mutated alleles contain expansions of the dodecamer repeat (85–90%, 99% in Finland), the remainder are caused by smaller mutations, such as point mutations in the *CSTB* gene. *CSTB* genetic testing is available and analyzes the dodecamer repeat expansion and three of the most common disease-causing mutations.[62] The exclusion of other rarer mutations from the testing procedures results in lower detection rates. No correlation has been found between the size of the repeat and age of onset or severity.

Progressive Myoclonus Epilepsy, Lafora Type (Lafora Body Disease)

Lafora body disease is a rapidly progressive neurodegenerative disorder with adolescent onset. Myoclonus and/or seizures develop; in addition, ataxia, spasticity, dysarthria, visual hallucinations and dementia are common symptoms.[63] Two genes are associated with this disease: *EPM2A* or *NHLRC1* (EPM2B). Phenotype is variable.[64] Both genes are autosomal recessive.

Neuronal Ceroid-Lipofuscinoses

The neuronal ceroid-lipofuscinoses (NCLs) are a group of neurodegenerative genetic lysosomal-storage diseases characterized by seizures, progressive deterioration of cognition (dementia and speech abnormalities) and motor function (involuntary movement, ataxia and spasticity), blindness, and the accumulation of autofluorescent lipopigment material in neurons and other cell types, probably caused by an underlying defect in protein degradation. Characteristic inclusion bodies representing lysosomal storage material such as saposins A and D, and subunit c of mitochondrial ATP are detectable on electron microscopy for some forms of the disease. The lipopigment pattern seen most often in CLN1 is referred to as granular osmiophilic deposits (GROD).[65]

Collectively, this group of diseases account for the most common cause of childhood onset neurodegeneration, with an incidence of approximately 1/14,000 to 1/100,000, but with wide geographical variation, and they are most common in Northern Europe.[66] Most of the NCLs are inherited in an autosomal recessive manner. The NCLs were initially named according to age of onset and symptom presentation, but now are usually classified by their genetic cause. Moreover, the same gene mutation can result in different ages of onset and phenotypes. The discovery of the causative genes has helped to delineate this group of disorders. There is significant overlap in presentation between the NCLs. For example, individuals with mutations in the *CLN1* gene present mainly as infantile NCL, but can also present as late-infantile, juvenile and adult NCL (Table 76.2). This is probably due to genotype–phenotype

TABLE 76.2 Classification of the Neuronal Ceroid-Lipofuscinoses

Name(s)	Inheritance pattern	Gene/location	Protein	Mutations identified (n)
Infantile NCL/Santavuori–Haltia *CLN1* mutations also cause classic late-infantile, juvenile, and adult NCL	AR	*CLN1*/1p34.2	Palmitoyl-protein thioesterase	>38; 3 common
Classic late infantile/Jansky–Bielschowsky. *CLN2* mutations also cause Juvenile NCL	AR	*CLN2*/11p15.4	Tripeptidyl-peptidase 1	>40; 2 common
Juvenile/Batten Disease/Spielmeyer-Vogt-Sjogren. *CLN3* mutations also cause adult NCL	AR	*CLN3*/16p11.2	Battenin (membrane protein)	>31; 1 common
Adult NCL/Kufs disease	AR	*CLN4A*/15q23	Mutations in CLN6 gene	–
Adult NCL/Kufs disease	AD	*CLN4B*/*DNAJC5* 20q13.33	Cysteine string protein	–
Late-infantile Finnish variant	AR	*CLN5*/13q22.3	Membrane protein	>5, Northern Europe only
Late infantile/early juvenile variant/Roma gypsy/Indian	AR	*CLN6*/15q23	Putative membrane protein	>6, increased in Southern Mediterranean
Turkish variant (may be allelic to *CLN8*)	AR	*CLN7*/4q28.2	Major facilitator superfamily domain-containing protein-8	–
Northern epilepsy (NE)/progressive	AR	*CLN8*/8p23.3	Putative membrane protein	1, in Finland
Late infantile	AR	*CLN9*/?	N/I	–
Congenital	AR	*CLN10*/*CTSD* 11p15.5	Cathepsin D	–
Adult onset	AR	*CLN11*/*GRN* 17q21.31	Progranulin	–
Juvenile	AR	*CLN12*/*ATP13A2* 1p36.13	Same protein as in Kufor–Rakeb (see the section on NBIAs above and also Chapter 68)	–
Adult onset	AR	*CLN13*/*CTSF* 11q13.2	Cathepsin F	–
Infantile	AR	*CLN14*/*KCTD7* 7q11.21	Potassium channel	–

Abbreviations: AD, autosomal dominant; AR, autosomal recessive; N/I, not identified.

interactions. Several new genes have been discovered in the last few years leading to 14 different NCL-causal genes. All but some forms of NCL are autosomal recessive, with autosomal dominant inheritance in *CLN4*.[67] Some of these genes are in doubt, so that to date, there are nine accepted types of NCL.[68]

Infantile NCL (INCL) is characterized by normal early development, onset of retinal blindness and seizures by age 2 years, myoclonic jerks, progressive mental deterioration, and death between the ages 8 and 11 years. INCL is caused by mutations in *palmitoyl-protein thioesterase 1 (PPT1)*. Myoclonic epilepsy manifesting between 2 and 4 years of age is usually the first sign of late-infantile NCL (LINCL). Loss of motor and mental milestones and vision, dementia, ataxia, and extrapyramidal and pyramidal signs, with death between age 6 and 40 years are other characteristics of this disease. Finnish, Gypsy/Indian and Turkish variants of LINCL are caused by mutations in *CLN5, CLN6* and *CLN7* respectively.[65]

Juvenile NCL (JNCL), or Batten disease, usually begins with retinal visual failure between age 4 and 10 years. Other symptoms include epilepsy with generalized clonic-tonic seizures, complex-partial or myoclonic seizures, behavioral problems, extrapyramidal signs, and death in the late teens. Adult NCL (ANCL), or Kufs disease, occurs around age 30 years and is characterized by ataxia, dementia, athetosis, dyskinesias, pyramidal and extrapyramidal signs, and progressive myoclonic epilepsy, with death approximately 10 years from onset. ANCL is not characterized by vision loss. Northern epilepsy (NE) manifests with onset of epilepsy between age 5 and 10 years and is characterized by progressive epilepsy with mental retardation, no vision loss, and a protracted course.

Differential diagnoses include other progressive neurodegenerative diseases such as Tay–Sachs, Krabbe disease, Canavan disease, Rett syndrome, metachromatic leukodystrophy, Niemann–Pick A and B, and other lysosomal storage disorders. Treatment usually involves management of symptoms and control of seizures. The diagnosis of NCL is usually based on assay of enzyme activity and/or genetic testing coupled with clinical findings and skin biopsy. Enzyme assay for palmitoyl-protein thioesterase 1, tripeptidyl-peptidase 1 (TPP-1), and cathepsin D (CTSD) and gene testing for *PPT1* (encoding palmitoyl-protein thioesterase 1), *TPP1* (encoding tripeptidyl-peptidase 1), *CLN3, CLN5, CLN6, MFSD8, CLN8,* and *CTSD* (encoding cathepsin D) are available on a clinical basis. A gene for CLN4 (*DNAJC5*) is responsible for some adult types of NCL.[69] Genetic testing either involves targeted mutations analysis for the most common mutations or sequence analysis. Some genotype/phenotype correlation exists which may help the testing strategy. Animal models, both naturally occurring (with mutations in *CLN6* and *CLN8* homologs) and genetically manipulated (*CLN2* and *CLN3*) are helping researchers to understand how mutations in NCL genes cause disease. Enzyme replacement therapy has not been effective in treating the NCLs.[65]

Rett Syndrome

Rett syndrome is an X-linked dominant neurodegenerative disease characterized by normal presentation at birth with loss of acquired skills beginning between 6 and 18 months of age, gradual deceleration of head growth and the development of all or some of the following features: ataxia, seizures, stereotyped hand movements, gait apraxia, tremors and dystonia. Rett syndrome is usually embryonic lethal in males. It is caused by mutations in the methyl-CpG binding protein 2 (*MECP2*) gene on Xq28 and acts as an architectural chromatin-binding protein. *MECP2* is the only gene known to be associated with Rett syndrome.[70]

Molecular genetic testing for Rett syndrome leads to identification of a mutation in *MECP2* in approximately 80% of cases.[71] Nonsense, missense and frameshift mutations have all been described. Due to the large number of mutations, bidirectional sequencing of the entire *MECP2* coding region is usually undertaken alongside quantitative PCR to detect deletions. Mutations in *MECP2* also cause atypical Rett syndrome (either a more severe phenotype with early onset and infantile spasms, or a more benign phenotype), as well as autism, behavioral disorders, and mental retardation with tremor/spasticity. Rarely, afflicted males have severe neonatal hypotonia or encephalopathy or clinically suspected but molecularly unconfirmed Angelman syndrome. Most cases are due to mosaicism, reducing the severity of the phenotype. A small number of males with classical Rett syndrome have been reported who also have Klinefelter syndrome (XXY).

The prevalence of classic Rett syndrome is estimated to be 1/10,000 to 1/15,000, and nearly all cases are sporadic. However, multiple siblings with Rett syndrome have been described. This can be due to gonadal (germline) mosaicism. Alternatively, the mother may be a mutation carrier, but due to favorable (random) X inactivation, have mild or no symptoms. It is therefore advisable to offer the mother of a child with a known *MECP2* mutation genetic testing to determine carrier status. If negative, the family should appreciate that there is still a possible recurrence risk owing to gonadal mosaicism.

The function of *MECP2* is to downregulate transcription of other genes by binding to methylated CpG dinucleotides and recruiting histone deacetylases. Mutations in *MECP2* cause disease due to loss of normal gene function and hence overexpression of those genes. There is considerable phenotypic variation in Rett syndrome.[72] Missense mutations resulting in only partial loss of function have been reported to cause a less severe phenotype than nonsense/frameshift mutations that obliterate gene function. In addition, a number of males with mental retardation, cognitive delay, and movement disorders such as tremor and bradykinesia have *MECP2* mutations that have not been described in females with classic Rett syndrome.

The clinical and developmental features of Rett syndrome suggest that dysfunction of subcortical regulator systems, brainstem, basal forebrain nuclei, and basal ganglia cause abnormal development of the cortex in late infancy. Immaturity of the respiratory regulator is thought to be a crucial step in the development of the disease. A number of diseases share features with Rett syndrome and must be considered in the differential diagnosis, such as infantile neuronal ceroid-lipofuscinosis and Angelman syndrome. A typical EEG pattern is associated with Rett syndrome, but is not unique to the disease. Current management of patients with Rett syndrome focuses on supportive and symptomatic therapy.

MECP2 mouse knockouts have been generated by gene targeting.[72] The mice are born normal and develop symptoms of tremor, behavioral change, and deceleration of head growth from about the age of 6 weeks, and typically die between 8 and 12 weeks of age. Research has shown that mutations in *MECP2*, a widely expressed gene, appear to selectively affect neurons and together with other genes—including the brain-derived neurotrophic factor pathway and glucocorticoid response genes—may affect formation and maturation of synapses or synaptic function in post-mitotic neurons.

PAROXYSMAL DYSKINESIAS

The majority of hyperkinetic movement disorders have symptoms that are continuously present. There is a rare group of hyperkinetic movement disorders that are characterized by transient symptoms of varying frequency, severity, duration, and aggravating factors. Symptoms that manifest as dyskinesias of the choreoathetotic and dystonic type are called "paroxysmal" and those of the ataxic type are described as "episodic."

Episodic ataxia-1 (EA1) is a rare autosomal dominant condition characterized by episodes of ataxia and dysarthria, without vertigo, lasting seconds to minutes with symptoms of myokymia (rippling of muscles, diagnosable by electromyography) during and between episodes, with near normal neurological function in between attacks. It is triggered by exercise, startle, or stress. It is caused by mutations in a voltage-gated (delayed rectifier) potassium channel gene *KCNA1* on chromosome 12p13.32. Individuals with EA1 may also have epilepsy. Recently there has been a report of hearing loss associated with EA1.[73] Genetic testing involves sequencing of the entire coding region.[73]

Episodic ataxia-2 (EA2) is also a rare autosomal dominant condition. EA2 often manifests with vertigo and is associated with episodic ataxia lasting hours with nystagmus between episodes. It is caused by mutations in the calcium channel, voltage-dependent, P/Q type, alpha 1A subunit gene (*CACNA1A*), which is highly expressed in the cerebellum and is located on chromosome 19p13.2.[74] More than 50% of patients have vertigo, nausea, and vomiting during attacks and about half have migraine headaches. Like EA1, EA2 is triggered by exercise, startle, or stress. EA2 is responsive to acetazolamide treatment. Familial hemiplegic migraine is also caused by mutations in *CACNA1A*, and *SCA6*, with a triplet repeat in the 3′ end of *CACNA1A*. Spinocerebellar ataxia can be distinguished from EA2 by progressive rather than episodic ataxia and a later age of onset, although there is an overlap of symptoms. Genetic testing also involves sequencing of the entire coding region, as well as deletion analysis. Phenotypic variations may be due to both mutational and epigenetic differences.

Other rare episodic ataxias have been identified, including the following. EA3 has been mapped to chromosome 1q42 and appears to be unique to a Mennonite family; EA4 is the locus name reserved for a unique periodic ataxia identified in two families; EA5 is caused by a mutation in the *calcium-channel beta4-subunit* gene on chromosome 2q22.3; and EA6 is caused by a mutation in the *SLC1A3* gene on chromosome 5p13.2.[75]

Paroxysmal Kinesigenic Dyskinesia (PKD)

PKD is a rare disease characterized by recurrent, brief attacks of dystonia, choreic, ballistic, or athetoid involuntary movements induced by sudden voluntary movements. Approximately 27–72% of cases are familial, showing autosomal dominant inheritance.[76] In recent years, mutations in the proline-rich transmembrane protein gene (*PRRT2*) on chromosome 16p11.2 have been identified as a cause of PKD (DYT10). The protein interacts with SNAP25 at the synaptic terminal. Mutations may impair the ability of the protein to fuse synaptic vesicles to the plasma membrane. This gene accounts for the majority of autosomal dominant PKD cases. Mutations have also been found in sporadic cases. Duplications, deletions, and point mutations have been identified. However, some PKD families do not show linkage to this region, suggesting further genetic heterogeneity.

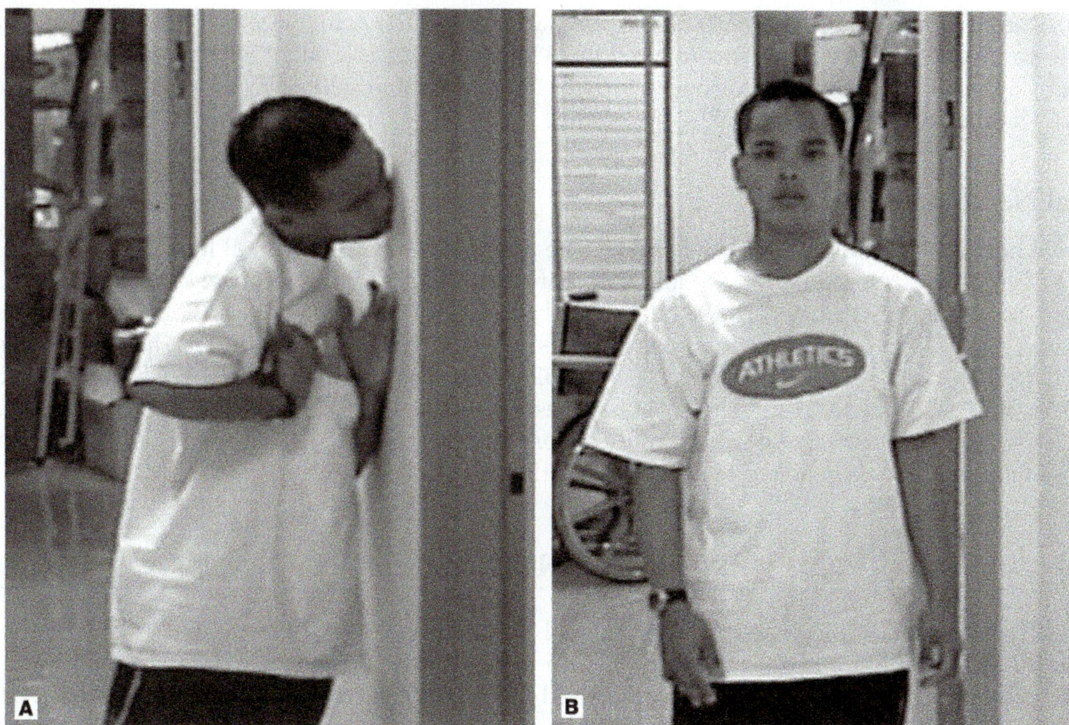

FIGURE 76.5 **During (A) and immediately after (B) an attack in paroxysmal kinesigenic dyskinesia.**

Penetrance is estimated to be between 80–90%. Episodes often begin in late childhood and can last seconds to minutes and occur up to 100 times a day (Figure 76.5). Some patients with PKD have a history of infantile afebrile convulsions with a favorable outcome. EEG, neuroimaging and pathology are normal in individuals with PKD. Individuals with PKD are very responsive to antiepileptics, such as carbamazepine. Additionally mutations in *PRRT2* have been reported in patients with hemiplegic migraine and episodic ataxia.[76]

What was once called benign familial infantile convulsions and paroxysmal choreoathetosis (ICCA) is now known to be due to the same gene that causes PKD, *PRRT2*. Three other genes have been shown to cause benign familial epilepsies, the potassium channel genes, *KCNQ2* and *KCNQ3*, and the sodium channel gene, *SCN2A*.[77]

Paroxysmal Nonkinesigenic Dyskinesias (PNKD)

Autosomal dominant PNKD (dystonic choreoathetosis, familial paroxysmal dyskinesia, or Mount–Reback syndrome) is caused by mutations in the *myofibrillogenesis regulator-1* gene (*MR1*, but also called the *PNKD* gene) on 2q35[78] in the midst of a cluster of ion channel genes. Penetrance is estimated to be greater than 98%.[79] Other loci have been associated with this condition, though genes have not been identified. PNKD differs from PKD in that movements are characterized by attacks of involuntary movements, such as dystonia, chorea and athetosis, that last up to several hours. These begin in childhood or adolescence and are triggered by caffeine or alcohol consumption, stress, or hormonal changes, but do not follow sudden voluntary movements. Treatment with benzodiazepines can sometimes be effective.

Paroxysmal Exercise-Induced Dyskinesias (PED)

PED is a condition in which prolonged exercise induces choreoathetoid and dystonic movements that last 15–30 minutes. The autosomal dominant PED accompanied by epilepsy was found to be due to glucose transporter 1 (GLUT1) deficiency with mutations on the *SLC2A1* gene on chromosome 1p34.2. The syndrome also includes mild developmental delay, reduced cerebrospinal fluid glucose levels, hemolytic anemia with echinocytosis, and altered erythrocyte iron concentrations. The clinical spectrum of GLUT1 deficiency was evaluated by videotape in 57 patients, and a wide range of movement disorders are seen.[80] These include abnormal gait (ataxic-spastic and ataxic), action limb dystonia (86%), mild chorea (75%), cerebellar tremor (70%), myoclonus (16%), and dyspraxia (21%). Nonepileptic paroxysmal events were encountered in 28% of patients, and included episodes of ataxia, weakness, parkinsonism, and PED. An autosomal recessive form of PED associated with rolandic epilepsy and writer's cramp (RE-PED-WC) has been linked to 16p12-p11.2.

Paroxysmal Nocturnal Dyskinesia

Paroxysmal nocturnal dyskinesia is now known to be a form of nocturnal frontal lobe epilepsy with autosomal dominant (ADNFLE) and sporadic presentation. It is characterized by childhood-onset partial epilepsy causing frequent, violent, brief seizures at night. The seizures mainly consist of motor elements, which can be dystonic, tonic, or hyperkinetic. ADNFLE is caused by mutations in genes encoding the *cholinergic receptor, nicotinic, alpha polypeptide subunit 4 (CHRNA4)* or *beta polypeptide subunit 2 (CHRNB2)* of the *neuronal nicotinic acetylcholine receptor* on chromosome 20q13.33 and 1q21.3, respectively. Clinical genetic testing reveals mutations in *CHRNA4* or *CHRNB2* in approximately 20–30% of individuals with a positive family history and fewer than 5% of individuals with a negative family history. Other families have been linked to chromosome 15q24.

HEREDITARY HYPEREKPLEXIA

Hereditary hyperekplexia, otherwise known as familial startle disease (STHE) or stiff baby syndrome, is caused by mutations in the alpha1 subunit of the inhibitory glycine receptor gene (*GLRA1*) on chromosome 5q33.1. GLRA1 is a ligand-gated chloride channel composed of three ligand-binding alpha1 subunits and two structural beta subunits that are clustered on the postsynaptic membrane of inhibitory glycinergic neurons. Dominant and rare recessive and X-linked forms of the disease have been described. STHE can begin in the neonatal period. The exaggerated startle reflex (such as when performing the Moro reflex or tapping the infant on the nose, or in response to unexpected acoustic or tactile stimuli) can result in sustained myoclonus that produces continuous increased muscle tone. Occasionally, the startle reflex results in fatal apneic attacks. Muscle tone usually returns to normal during the first years of life, but the startle reflex continues to be manifested. Additional symptoms may include excessive hypnic jerks (including periodic limb movements in sleep) and exaggerated head retraction reflex. Onset can be later in childhood, with the excessive startle reflex causing the patient to become stiff and unable to maintain a standing position, leading to falls. Sudden infant death has occurred in conjunction with this disease. Although intelligence is usually normal, learning disabilities have occasionally been reported. Some individuals with STHE benefit from clonazepam.

Five different genes have been found to cause this syndrome. The most common causal gene is *GLRA1*, which encodes the alpha1 subunit of the glycine receptor. A small number of individuals with "sporadic" STHE were found to be homozygotes (due to consanguinity). A compound heterozygote has also been reported. Mutations appear to be clustered in exons 7 and 8. Dominant transmission has also been documented. The mouse mutant "spasmodic," which exhibits a phenotype similar to STHE, is caused by mutations in the mouse homolog of *GLRA1*.

The normal function of GLRA1 involves mediating neuronal inhibition in the brainstem and spinal cord resulting in the control of muscle tone. GLRA1 belongs to an ion channel family that also includes γ-aminobutyric acid (GABA), serotonin, and nicotinic acetylcholine receptors. Like other members of the family, GLRA1 has both *N*- and *C*-terminal extracellular components on either side of four transmembrane regions.

Mutations in the *beta subunit of the glycine receptor* gene and *SLC6A5*, the gene encoding the presynaptic glycine transporter 2 (GLYT2) also cause autosomal recessive hyperekplexia.[81] Individuals with mutations in *SLC6A5* present with hypertonia, an exaggerated startle response to tactile or acoustic stimuli, and life-threatening neonatal apnea episodes. Hyperekplexia is also caused by mutations affecting other postsynaptic glycinergic proteins including the beta subunit of the glycine receptor. However, many cases of sporadic hyperekplexia do not have mutations in the genes discovered so far suggesting further genetic heterogeneity. Sequencing and deletion/duplication analysis for *GLRA1*, *SLC6A5*, and the 3 rare genes, *GLRB*, *GPHN*, and *ARHGEF9* are available clinically.

TOURETTE SYNDROME

Tourette syndrome (TS) is a neurologic condition characterized by motor and phonic tics and behavioral symptoms (such as loss of impulse control), which usually manifest in childhood. TS is usually familial and is often accompanied by attention-deficit hyperactivity disorder (ADHD), obsessive–compulsive disorder (OCD), and other behavioral problems. Current theories suggest that TS is a gender-influenced autosomal dominant disease with nearly complete penetrance in males and lower penetrance in females (56%). Twin studies showing high concordance for TS in monozygotic twins provide evidence for a genetic involvement in TS. It has also been proposed that TS is a semi-dominant, semi-recessive disease supported by evidence that parents with partial symptoms of TS are at increased risk to have children manifesting full-blown TS. Interestingly, it appears that individuals with TS undergo assortative mating, which may contribute to this phenomenon (affected individuals nonrandomly choosing likewise affected partners). Recurrence risk for first-degree relatives is 10–15%. Empiric risk to children of an affected person is about 20% for TS and 50% or more for a tic disorder.[82]

Inter- and intrafamilial variability of TS symptoms has been widely reported. OCD and ADHD are often seen in family members of individuals with TS, and can appear alongside symptoms of TS. The spectrum of conditions, including other behavioral disorders that can occur under the "umbrella" of TS, can make diagnosis a confusing task. The lack of clear definition as to what constitutes TS may also contribute to the difficulties of locating susceptibility genes and may adversely affect linkage analyses. Numerous susceptibility loci have so far been tested for linkage to TS (including dopamine D1, D2, D3, D4, D5 receptors, dopamine beta hydroxylase, tyrosinase, and tyrosine hydroxylase), but none have consistently been associated with the disease and more than 95% of the genome has so far been excluded. However, it is still believed that there is one major locus. Interestingly, there is evidence for TS susceptibility loci on 2p11, 8q22, and 11q23 in French-Canadian and Afrikaner family studies. A candidate region on chromosome 7q31 is also being investigated due to the observation of an individual with TS who has a chromosomal rearrangement in this region. It is hoped that examining the breakpoint may help locate genes causal to TS. A study investigating a three-site haplotype spanning three genes on chromosome 17q25 in a number of TS families resulted in highly significant association results. More studies are needed to confirm whether 17q25 is a susceptibility region for TS. Finally, a gene called *Slit and Trk-like 1 (SLITRK1)* has been studied as a candidate gene on chromosome 13q31.1. A frameshift mutation was suggested to be associated with TS. However it remains to be determined if this is a sequence variant in linkage disequilibrium with the causal gene or whether it is causal to the development of TS. It is likely that there are numerous causative genes. Rare copy number variants are a possible cause of at least some of TS.[83–85]

FRAGILE X-ASSOCIATED TREMOR/ATAXIA SYNDROME

Fragile X syndrome is X-linked and is caused by an expansion of over 200 CGG repeats in the *FMR1* gene on chromosome Xq27.3 and results in aberrant methylation of the *fragile X mental retardation 1* gene (*FMR1*). It is characterized by mental retardation with or without behavioral abnormalities, connective tissue findings, large testes, and characteristic facies in affected boys, and mild mental retardation in affected females. The normal range of CGG repeats is 5–40 repeats. It has been discovered that males with a premutation in the *FMR1* gene (from 55 to 200 repeats) are at increased risk to develop fragile X-associated tremor/ataxia syndrome (FXTAS), which presents with late-onset symptoms of progressive cerebellar ataxia and/or intention tremor. Other neurologic findings can include executive cognitive dysfunction and memory deficits, parkinsonism, peripheral neuropathy, lower-limb proximal muscle weakness, and autonomic dysfunction. Some of the symptoms may respond to medications used for other movement disorders. Frontosubcortical dementia is the main neuropsychiatric manifestation of FXTAS. Cognitive and functional impairment increases with increasing CGG repeats in the premutation range, although number of repeats does not appear to be associated with age at onset of either tremor or ataxia. MRI findings include decreased cerebellar volume, increased ventricular volume, and increased white matter hyperdensities. Males with a premutation have about a 40–45% risk of developing symptoms during their lifetime.[86] The daughters of men with FXTAS are obligate carriers of the premutation and have a 50% chance of passing on the premutation, and are at risk for having children with fragile X syndrome, as the premutation may expand into the full mutation range when inherited maternally (a process called anticipation). FMR1-related premature ovarian failure (POF) is associated with 20% of women with an *FMR1* premutation and 8–16% of FXTAS may present in females with expansions in the premutation range.[86] Genetic testing to measure the number of CGG repeats in the *FMR1* gene is clinically available, as is methylation analysis to detect aberrant methylation in the *FMR1* gene caused by CGG expansions.

MRI findings of males with FXTAS include an enhanced T2 signal in the middle cerebellar peduncles and/or brainstem. The pathologic hallmark of FXTAS is the ubiquitin-positive intranuclear inclusion found in neurons and astrocytes in broad distribution throughout the brain. However the protein component of the inclusions makes it unlikely that FXTAS involves breakdown in proteasomal degradation of nuclear proteins.

Ongoing research has revealed that the phenotype of premutation carriers is highly variable. Cognitive symptoms including ADHD, autism spectrum disorder, and increased risk of seizures have been reported. Likewise, adults can present with parkinsonism rather than classic FXTAS, especially if they carry a low premutation repeat.[87–90] In fact the premutation range is in question; symptoms of the premutation syndrome have been seen in repeat sizes below 55.[91]

MOVEMENT DISORDERS, GENETICS, MULTIDISCIPLINARY CARE, AND THE FUTURE

The practice of neurology has entered a new era. Confirmatory genetic testing has changed the way we diagnose neurogenetic conditions. But this gain in diagnostic power means that neurologists will need a multidisciplinary team, including genetic counselors and social workers, who can assist patients with the added burden of genetic disease. Genetic counseling for neurogenetic conditions includes not only discussions of mechanisms of inheritance, risk to relatives, and psychosocial counseling, but also encompasses issues such as confidentiality of genetic information and the unknown risk of discrimination by life and health insurance companies. In addition, asymptomatic individuals at 50% risk to inherit a neurogenetic disease can now choose to learn of their future disease status through genetic testing. The enormously difficult predicament that these individuals experience requires lengthy and complex pre- and post-test genetic counseling sessions to enable a fully informed decision. (For more details, see http://www.nsgc.org.)

Genetic discovery has greatly expanded our knowledge of movement disorders and will continue to change the way we approach the understanding of disease, and hopefully in the future, treatment. The ethical, legal, and social implications of genetics, along with its diagnostic power, will change our medical practice in a way that is unparalleled in the history of medicine.

References

1. Jasinska-Myga B, Wider C. Genetics of essential tremor. *Parkinsonism Relat Disord*. 2012;18(suppl 1):S138–S139.
2. Stefansson H, Steinberg S, Petursson H, et al. Variant in the sequence of the LINGO1 gene confers risk of essential tremor. *Nat Genet*. 2009;41(3):277–279.
3. Merner ND, Girard SL, Catoire H, et al. Exome sequencing identifies FUS mutations as a cause of essential tremor. *Am J Hum Genet*. 2012;91(2):313–319.
4. Parmalee N, Mirzozoda K, Kisselev S, et al. Genetic analysis of the FUS/TLS gene in essential tremor. *Eur J Neurol*. 2013;20(3):534–539.
5. Albanese A, Bhatia K, Bressman SB, et al. Phenomenology and classification of dystonia: a consensus update. *Mov Disord*. 2013;28(7):863–873.
6. Ozelius LJ, Hewett JW, Page CE, et al. The early-onset torsion dystonia gene (DYT1) encodes an ATP-binding protein. *Nat Genet*. 1997;17:40–48.
7. Bressman SB, Sabatti C, Raymond D, et al. The DYT1 phenotype and guidelines for diagnostic testing. *Neurology*. 2000;54:1746–1752.
8. Bressman SB. Genetics of dystonia: an overview. *Parkinsonism Relat Disord*. 2007;13(suppl 3):S347–S355.
9. Risch NJ, Bressman SB, Senthil G, Ozelius LJ. Intragenic Cis and Trans modification of genetic susceptibility in DYT1 torsion dystonia. *Am J Hum Genet*. 2007;80:1188–1193.
10. Fuchs T, Gavarini S, Saunders-Pullman R, et al. Mutations in the THAP1 gene are responsible for DYT6 primary torsion dystonia. *Nat Genet*. 2009;41:286–288.
11. Kaiser FJ, Osmanoric A, Rakovic A, et al. The dystonia gene DYT1 is repressed by the transcription factor THAP1 (DYT6). *Ann Neurol*. 2010;68(4):554–559.
12. Winter P, Kamm C, Biskup S, et al. DYT7 gene locus for cervical dystonia on chromosome 18p is questionable. *Mov Disord*. 2012;27(14):1819–1821.
13. Bentivoglio AR, Ialongo T, Contarino MF, et al. Characterization of DYT13 primary torsion dystonia. *Mov Disord*. 2004;19(2):200–206.
14. Xiao J, Uitti RJ, Zhao Y, et al. Mutations in CIZ1 cause adult onset primary cervical dystonia. *Ann Neurol*. 2012;71(4):458–469.
15. Charlesworth G, Plagnol V, Holmström KM, et al. Mutations in ANO3 cause dominant craniocervical dystonia: ion channel implicated in pathogenesis. *Am J Hum Genet*. 2012;91(6):1041–1050.
16. Fuchs T, Saunders-Pullman R, Masuho I, et al. Mutations in GNAL cuase primary torsion dystonia. *Nat Genet*. 2013;45(1):88–92.
17. Lohmann K, Wilcox RA, Winkler S, et al. Whispering dysphonia (DYT4 dystonia) is caused by a mutation in the TUBB4 gene. *Ann Neurol*. 2013;73(4):537–545.
18. Wilcox RA, Winkler S, Lohmann K, Klein C. Whispering dysphonia in an Australian family (DYT4): a clinical and genetic reappraisal. *Mov Disord*. 2011;26(13):2404–2408.
19. Asmus F, Gasser T. Dystonia-plus syndromes. *Eur J Neurol*. 2010;17(suppl 1):37–45.
20. Makino S, Kaji R, Ando S, et al. Reduced neuron-specific expression of the TAF1 gene is associated with X-linked dystonia-parkinsonism. *Am J Hum Genet*. 2007;80(3):393–406.
21. Clot F, Grabli D, Cazeneuve C, et al. Exhaustive analysis of BH4 and dopamine biosynthesis genes in patients with Dopa-responsive dystonia. *Brain*. 2009;132(Pt 7):1753–1763.
22. Wider C, Melquist S, Hauf M, et al. Study of a Swiss dopa-responsive dystonia family with a deletion in GCH1: redefining DYT14 as DYT5. *Neurology*. 2008;70(Pt 2):1377–1383.
23. Carecchio M, Magliozzi M, Copetti M, et al. Defining the epsilon-sarcoglycan (SGCE) gene phenotypic signature in myoclonus-dystonia: a reappraisal of genetic testing criteria. *Mov Disord*. 2013;28(6):787–794.

24. Charlesworth G, Bhatia KP, Wood NW. The genetics of dystonia: new twists in an old tale. *Brain*. 2013;136(Pt 7):2017–2037.
25. Schneider SA, Bhatia KP. Syndromes of neurodegeneration with brain iron accumulation. *Semin Pediatr Neurol*. 2012;19(2):57–66.
26. Yoshino H, Tomiyama H, Tachibana N, et al. Phenotypic spectrum of patients with PLA2G6 mutation and PARK14-linked parkinsonism. *Neurology*. 2010;75(15):1356–1361.
27. Lu CS, Lai SC, Wu RM, et al. PLA2G6 mutations in PARK14-linked young-onset parkinsonism and sporadic Parkinson's disease. *Am J Med Genet B Neuropsychiatr Genet*. 2012;159B(2):183–191.
28. Hartig MB, Iuso A, Haack T, et al. Absence of an orphan mitochondrial protein, c19orf12, causes a distinct clinical subtype of neurodegeneration with brain iron accumulation. *Am J Hum Genet*. 2011;89(4):543–550.
29. Hayflick SJ, Kruer MC, Gregory A, et al. β-Propeller protein-associated neurodegeneration: a new X-linked dominant disorder with brain iron accumulation. *Brain*. 2013;136(Pt 6):1708–1717.
30. Schneider SA, Dusek P, Hardy J, Westenberger A, Jankovic J, Bhatia KP. Genetics and Pathophysiology of Neurodegeneration with Brain Iron Accumulation (NBIA). *Curr Neuropharmacol*. 2013;11(1):59–79.
31. The Huntington's Disease Collaborative Research Group. A novel gene containing a trinucleotide repeat that is expanded and unstable on Huntington's disease chromosomes. *Cell*. 1993;72(6):971–983.
32. International Huntington Association (IHA) and the World Federation of Neurology (WFN) Research Group on Huntington's Chorea. Guidelines for the molecular genetics predictive test in Huntington's disease. *Neurology*. 1994;44(8):1533–1536.
33. MacLeod R, Tibben A, Frontali M, et al., Huntington Disease Network. Recommendations for the predictive genetic test in Huntington's disease. *Clin Genet*. 2013;83(3):221–231.
34. Squitieri F, Jankovic J. Huntington's disease: how intermediate are intermediate repeat lengths? *Mov Disord*. 2012;27(14):1714–1717.
35. Ross CA, Tabrizi SJ. Huntington's disease: from molecular pathogenesis to clinical treatment. *Lancet Neurol*. 2011;10(1):83–98.
36. Pringsheim T, Wiltshire K, Day L, et al. The incidence and prevalence of Huntington's disease: a systematic review and meta-analysis. *Mov Disord*. 2012;27(9):1083–1091.
37. Huntington Study Group PHAROS Investigators. At risk for Huntington disease: the PHAROS (Prospective Huntington At Risk Observational Study) cohort enrolled. *Arch Neurol*. 2006;63(7):991–996.
38. Sturrock A, Leavitt BR. The clinical and genetic features of Huntington disease. *J Geriatr Psychiatry Neurol*. 2010;23(4):243–259.
39. Huntington Study Group. Tetrabenazine as antichorea therapy in Huntington disease: a randomized controlled trial. *Neurology*. 2006;66:366–372.
40. The Huntington Study Group HART Investigators. A randomized, double-blind, placebo-controlled trial of pridopidine in Huntington's disease. *Mov Disord*. 2013 Feb 28.
41. Finkbeiner S. Huntington's disease. *Cold Spring Harb Perspect Biol*. 2011;3(6).
42. Hogarth P, Lovrecic L, Krainc D. Sodium phenylbutyrate in Huntington's disease: a dose-finding study. *Mov Disord*. 2007;22(13):1962–1964.
43. Huntington Study Group Pre2CARE Investigators, et al. Safety and tolerability of high-dosage coenzyme Q10 in Huntington's disease and healthy subjects. *Mov Disord*. 2010;25(12):1924–1928.
44. Sathasivam K, Neueder A, Gipson TA, et al. Aberrant splicing of HTT generates the pathogenic exon 1 protein in Huntington disease. *Proc Natl Acad Sci U S A*. 2013;110(6):2366–2370.
45. Ng CW, Yildirim F, Yap YS, et al. Extensive changes in DNA methylation are associated with expression of mutant huntingtin. *Proc Natl Acad Sci U S A*. 2013;110(6):2354–2359.
46. Kwan W, Träger U, Davalos D, et al. Mutant huntingtin impairs immune cell migration in Huntington disease. *J Clin Invest*. 2012;122(12):4737–4747.
47. Xie Y, Hayden MR, Xu B. BDNF overexpression in the forebrain rescues Huntington's disease phenotypes in YAC128 mice. *J Neurosci*. 2010;30(44):14708–14718.
48. Moore RC, Xiang F, Monaghan J, et al. Huntington disease phenocopy is a familial prion disease. *Am J Hum Genet*. 2001;69(6):1385–1388.
49. Holmes SE, O'Hearn E, Rosenblatt A. A repeat expansion in the gene encoding junctophilin-3 is associated with Huntington disease-like 2. *Nat Genet*. 2001;29(4):377–378.
50. Kambouris M, Bohega S, Al-Tahan A, Meyer BF. Localization of the gene for a novel autosomal recessive neurodegenerative Huntington-like disorder to 4p15.3. *Am J Hum Genet*. 2000;66(2):445–452.
51. Rolfs A, Koeppen A, Bauer I, et al. Clinical features and neuropathology of autosomal dominant spinocerebellar ataxia (SCA17). *Ann Neurol*. 2003;54(3):367–375.
52. Koide R, Onodera O, Ikeuchi T, et al. Atrophy of the cerebellum and brainstem in dentatorubral pallidoluysian atrophy. Influence of CAG repeat size on MRI findings. *Neurology*. 1997;49(6):1605–1612.
53. Nagafuchi S, Yanagisawa H, Sato K, et al. Dentatorubral and pallidoluysian atrophy expansion of an unstable CAG trinucleotide on chromosome 12p. *Nat Genet*. 1994;6(1):4–18.
54. Le Ber I, Camuzat A, Castelnovo G, et al. Prevalence of dentatorubral-pallidoluysian atrophy in a large series of white patients with cerebellar ataxia. *Arch Neurol*. 2003;60(8):1097–1099.
55. Burke JR, Wingfield MS, Lewis KE, et al. The Haw River syndrome: dentatorubropallidoluysian atrophy (DRPLA) in an African-American family. *Nat Genet*. 1994;7(4):521–524.
56. Walker RH, Jung HH, Dobson-Stone C, et al. Neurologic phenotypes associated with acanthocytosis. *Neurology*. 2007;68(2):92–98.
57. Gras D, Jonard L, Roze E, et al. Benign hereditary chorea: phenotype, prognosis, therapeutic outcome and long term follow-up in a large series with new mutations in the TITF1/NKX2-1 gene. *J Neurol Neurosurg Psychiatry*. 2012;83(10):956–962.
58. Lorincz MT. Recognition and treatment of neurologic Wilson's disease. *Semin Neurol*. 2012;32(5):538–543.
59. Coffey AJ, Durkie M, Hague S. A genetic study of Wilson's disease in the United Kingdom. *Brain*. 2013;136(Pt 5):1476–1487.
60. Joensuu T, Kuronen M, Alakurtti K. Cystatin B: mutation detection, alternative splicing and expression in progressive myclonus epilepsy of Unverricht-Lundborg type (EPM1) patients. *Eur J Hum Genet*. 2007;2:185–193.
61. Genton P. Unverricht-Lundborg disease (EPM1). *Epilepsia*. 2010;51(suppl 1):37–39.
62. Kälviäinen R, Khyuppenen J, Koskenkorva P, et al. Clinical picture of EPM1-Unverricht-Lundborg disease. *Epilepsia*. 2008;49(4):549–556.

63. Turnbull J, Girard JM, Lohi H, et al. Early-onset Lafora body disease. *Brain*. 2012;135(Pt 9):2684–2698.
64. Singh S, Ganesh S. Phenotype variations in Lafora progressive myoclonus epilepsy: possible involvement of genetic modifiers? *J Hum Genet*. 2012;57(5):283–285.
65. Bennett MJ, Rakheja D. The neuronal ceroid-lipofuscinoses. *Dev Disabil Res Rev*. 2013;17(3):254–259.
66. Haltia M, Goebel HH. The neuronal ceroid-lipofuscinoses: a historical introduction. *Biochim Biophys Acta*. 2013;1832(11):1795–1800.
67. Warrier V, Vieira M, Mole SE. Genetic basis and phenotypic correlations of the neuronal ceroid lipofusinoses. *Biochim Biophys Acta*. 2013;1832(11):1827–1830.
68. Mink JW, Augustine EF, Adams HR, Marshall FJ, Kwon JM. Classification and natural history of the neuronal ceroid lipofuscinoses. *J Child Neurol*. 2013;28(9):1101–1105.
69. Anderson GW, Goebel HH, Simonati A. Human pathology in NCL. *Biochim Biophys Acta*. 2013;1832(11):1807–1826.
70. Amir RE, Zoghbi HY. Rett syndrome: methyl-CpG-binding protein 2 mutations and phenotype-genotype correlations. *Am J Med Genet*. 2000;97(2):147–152.
71. Van den Veyver IB, Zoghbi HY. Mutations in the gene encoding methyl-CpG-binding protein 2 cause Rett syndrome. *Brain Dev*. 2001;23(suppl 1):S147–S151.
72. Ricceri L, De Filippis B, Laviola G. Rett syndrome treatment in mouse models: searching for effective targets and strategies. *Neuropharmacology*. 2013;68:106–115.
73. Tomlinson SE, Rajakulendran S, Tan SV, et al. Clinical, genetic, neurophysiological and functional study of new mutations in episodic ataxia type 1. *J Neurol Neurosurg Psychiatry*. 2013;84(10):1107–1112.
74. Jung J, Testard H, Tournier-Lasserve E, et al. Phenotypic variability of episodic ataxia type 2 mutations: a family study. *Eur Neurol*. 2010;64(2):114–116.
75. Conroy J, McGettigan P, Murphy R, et al. A novel locus for episodic ataxia:UBR4 the likely candidate. *Eur J Hum Genet*. 2013 Aug 28;doi:10.1038/ejhg.2013.173.
76. Becker F, Schubert J, Striano P, et al. PRRT2-related disorders: further PKD and ICCA cases and review of the literature. *J Neurol*. 2013;260(5):1234–1244.
77. Zara F, Specchio N, Striano P, et al. Genetic testing in benign familial epilepsies of the first year of life: clinical and diagnostic significance. *Epilepsia*. 2013;54(3):425–436.
78. Lee HY, Xu Y, Huang Y, et al. The gene for paroxysmal non-kinesigenic dyskinesia encodes an enzyme in a stress response pathway. *Hum Mol Genet*. 2004;13(24):3161–3170.
79. Brockmann K. Episodic movement disorders: from phenotype to genotype and back. *Curr Neurol Neurosci Rep*. 2013;13(10):379.
80. Pons R, Collins A, Rotstein M, Engelstad K, De Vivo DC. The spectrum of movement disorders in Glut-1 deficiency. *Mov Disord*. 2010;25(3):275–281.
81. Carta E, Chung SK, James VM, et al. Mutations in the GlyT2 gene (SLC6A5) are a second major cause of startle disease. *J Biol Chem*. 2012;287(34):28975–28985.
82. Keen-Kim D, Freimer NB. Genetics and epidemiology of Tourette syndrome. *J Child Neurol*. 2006;8:665–671.
83. Hooper SD, Johansson AC, Tellgren-Roth C, et al. Genome-wide sequencing for the identification of rearrangements associated with Tourette syndrome and obsessive-compulsive disorder. *BMC Med Genet*. 2012;13:123.
84. Bertelsen B, Debes NM, Hjermind LE, et al. Chromosomal rearrangements in Tourette syndrome: implications for identification of candidate susceptibility genes and review of the literature. *Neurogenetics*. 2013 Aug 29.
85. Yasmeen S, Melchior L, Bertelsen B, et al. Sequence analysis of SLITRK1 for var321 in Danish patients with Tourette syndrome and review of the literature. *Psychiatr Genet*. 2013;23(3):130–133.
86. Rodriguez-Revenga L, Madrigal I, Pagonabarraga J, et al. Penetrance of FMR1 premutation associated pathologies in fragile X syndrome families. *Eur J Hum Genet*. 2009;17(10):1359–1362.
87. Hagerman R, Hagerman P. Advances in clinical and molecular understanding of the FMR1 premutation and fragile X-associated tremor/ataxia syndrome. *Lancet Neurol*. 2013;12(8):786–798.
88. Hagerman P. Fragile X-associated tremor/ataxia syndrome (FXTAS): pathology and mechanisms. *Acta Neuropathol*. 2013;126(1):1–19.
89. Hall DA, O'keefe JA. Fragile x-associated tremor ataxia syndrome: the expanding clinical picture, pathophysiology, epidemiology, and update on treatment. *Tremor Other Hyperkinet Mov (N Y)*. 2012;2.
90. Hall DA, Howard K, Hagerman R, Leehey MA. Parkinsonism in FMR1 premutation carriers may be indistinguishable from Parkinson disease. *Parkinsonism Relat Disord*. 2009;15(2):156–159.
91. Hall D, Tassone F, Klepitskaya O, Leehey M. Fragile X-associated tremor ataxia syndrome in FMR1 gray zone allele carriers. *Mov Disord*. 2012;27(2):296–300.

Selected Reading

Fahn S, Jankovic J, Hallett M. *Principles and Practice of Movement Disorders*. 2nd ed. Edinburgh: Saunders Elsevier; 2011.

Relevant Websites

http://www.genereviews.org/.
http://www.ncbi.nlm.nih.gov/Omim/.
http://www.nsgc.org.

CHAPTER

77

Hereditary Spastic Paraplegia

John K. Fink

University of Michigan, Ann Arbor Veterans Affairs Medical Center, Ann Arbor, MI, USA

INTRODUCTION

Hereditary spastic paraplegia (HSP) is a syndromic designation for inherited neurologic disorders in which the predominant symptom is lower extremity spastic weakness.[1–5] HSP syndromes are clinically diverse and genetically heterogeneous with more than 70 genetic types described.[6,7] The prevalence of HSP (estimated to range from 1.27 to 9.6 per 100,000[8–10]) is similar to that of amyotrophic lateral sclerosis (ALS; approximately 1.6 per 100,000).[11] This summary is an update from the previous edition.[12] See [6] for additional references.

GENETIC AND SYNDROMIC CLASSIFICATIONS

HSP is classified 1) by mode of inheritance (autosomal dominant, recessive, and X-linked); 2) by the genetic locus and causative gene mutation when known (Table 77.1); and 3) by the presence or absence of additional, syndrome-specific features. "Uncomplicated" HSP refers to inherited syndromes of lower extremity spasticity and weakness, typically associated with subtle decrease in vibratory sensation in the toes and urinary urgency. In addition to these signs, "complicated" HSP syndromes exhibit other neurologic or systemic abnormalities such as mental retardation, ataxia, peripheral neuropathy, deafness, cataracts, or muscle atrophy (reviewed in [1]).

SYMPTOMS, SIGNS, AND COURSE OF UNCOMPLICATED HSP

The predominant symptom of uncomplicated HSP is difficulty walking due to lower extremity spasticity and weakness. Urinary urgency is common and occasionally may be an early symptom. Symptom onset may begin at any age. HSP symptoms that begin in infancy may manifest as "toe walking" and be very similar to spastic diplegic cerebral palsy (CP). Like spastic diplegic CP, when HSP symptoms begin in infancy there is often little worsening even over two decades.[13]

In contrast, when HSP symptoms begin after infancy, there is insidiously progressive worsening over many years (decades). Nonetheless, symptoms may not worsen inexorably even in subjects in whom the disorder is initially progressive. After a number of years (e.g., 3–7) of insidious worsening, many subjects appear to reach a functional plateau after which the rate of functional decline in the ability to ambulate slows considerably and may be similar to the rate of functional gait decline attributed to age and similar degrees of exercise. Annual neurologic examinations of such subjects show relatively little change.

Neurologic examination of uncomplicated HSP subjects demonstrates increased tone in the lower extremities (particularly, quadriceps, hamstrings, adductors, and ankles); weakness (particularly in hamstrings, tibialis anterior, and iliopsoas muscles); hyperreflexia; extensor plantar responses (almost always); and mild diminution of vibratory sensation in the toes. The relative proportions of spasticity and weakness are variable. For example, some individuals have significant spasticity but have no or only minimal weakness. Muscle relaxing medications (such as Lioresal) are most beneficial for individuals who have prominent spasticity and are fully strong or have only slight weakness.

Rosenberg's Molecular and Genetic Basis of Neurological and Psychiatric Disease
http://dx.doi.org/10.1016/10.1016/B978-0-12-410529-4.00077-2

TABLE 77.1 Genetic Types of HSP, their Encoded Proteins, and Clinical Syndromes (Updated from [6])

Spastic gait (SPG) locus and OMIM no.	Protein	Clinical syndrome	References
AUTOSOMAL DOMINANT HSP			
SPG3A 182600	Atlastin	Uncomplicated HSP: symptoms usually begin in childhood (and may be nonprogressive); symptoms may also begin in adolescence or adulthood and worsen insidiously. Genetic nonpenetrance reported. *De novo* mutation reported presenting as spastic diplegic cerebral palsy	41–43
SPG4 182601	Spastin	Uncomplicated HSP, symptom onset in infancy through senescence, single most common cause of autosomal dominant HSP (~40%); some subjects have late-onset cognitive impairment	44–48
SPG6 600363	Not imprinted in Prader–Willi/Angelman 1 (NIPA1)	Uncomplicated HSP: prototypical late-adolescent, early-adult onset, slowly progressive uncomplicated HSP. Rarely complicated by epilepsy or variable peripheral neuropathy. One subject with uncomplicated HSP later died from amyotrophic lateral sclerosis	49–55
SPG8 603563	KIAA0196 (Strumpellin)	Uncomplicated HSP	56–60
SPG9 601162	Unknown	Complicated: spastic paraplegia associated with cataracts, gastroesophageal reflux, and motor neuronopathy	61,62
SPG10 604187	Kinesin heavy chain (KIF5A)	Uncomplicated HSP or complicated by distal muscle atrophy	63,64
SPG12 604805	Reticulon 2 (RTN2)	Uncomplicated HSP	57,65
SPG13 605280	Chaperonin 60 (heat shock protein 60, HSP60)	Uncomplicated HSP: adolescent and adult onset	66–68
SPG17 270685	BSCL2/seipin	Complicated: spastic paraplegia associated with amyotrophy of hand muscles (Silver syndrome)	15,69,70
SPG19 607152	Unknown	Uncomplicated HSP	71
SPG29 609727	Unknown	Complicated: spastic paraplegia associated with hearing impairment; persistent vomiting due to hiatal hernia inherited	72
SPG31 610250	Receptor expression enhancing protein 1 (REEP1)	Uncomplicated HSP or occasionally associated with peripheral neuropathy	73–75
SPG33 610244	ZFYVE27/protrudin	Uncomplicated HSP	76
SPG36 613096	Unknown	Onset age 14–28 years, associated with motor sensory neuropathy	77
SPG37 611945	Unknown	Uncomplicated HSP	78
SPG38 612335	Unknown	One family, five affected subjects, onset age 16–21 years. Subjects had atrophy of intrinsic hand muscles (severe in one subject at age 58)	79
SPG40	Unknown	Uncomplicated spastic paraplegia, onset after age 35, known autosomal dominant HSP loci excluded	80
SPG41 613364	Unknown	Single Chinese family with adolescent onset, spastic paraplegia associated with mild weakness of intrinsic hand muscles	81
SPG42 612539	Acetyl CoA transporter (SLC33A1)	Uncomplicated spastic paraplegia reported in single kindred, onset age 4–40 years, possibly one instance of incomplete penetrance	82–84
SPG72 615625	REEP2	Uncomplicated spastic paraplegia manifesting as childhood onset (infancy to age 8 years) of toe walking or progressive spastic gait. Two of 10 subjects had upper extremity spasticity. Cognition, speech, and vision were normal. REEP2 mutations may cause both autosomal dominant HSP (due to heterozygous mutation); or autosomal recessive HSP due to compound heterozygote mutations	20

TABLE 77.1 Genetic Types of HSP, their Encoded Proteins, and Clinical Syndromes (Updated from [6])—cont'd

Spastic gait (SPG) locus and OMIM no.	Protein	Clinical syndrome	References
615290	BICD2	Two mutation-associated phenotypes have been reported: autosomal dominant, congenital spinal muscular atrophy; and autosomal dominant adult onset, progressive spastic paraplegia	85
AUTOSOMAL RECESSIVE HSP			
SPG5 270800	CYP7B1	Uncomplicated or complicated by axonal neuropathy, distal or generalized muscle atrophy, and white matter abnormalities on MRI	86–92
SPG7 602783	Paraplegin	Uncomplicated or complicated: variably associated with mitochondrial abnormalities on skeletal muscle biopsy and dysarthria, dysphagia, optic disk pallor, axonal neuropathy, and evidence of "vascular lesions," cerebellar atrophy, or cerebral atrophy on cranial MRI	93,94
SPG11 604360	Spatacsin (KIAA1840)	Uncomplicated or complicated: spastic paraplegia variably associated with thin corpus callosum, mental retardation, upper extremity weakness, dysarthria, and nystagmus; may have Kjellin syndrome: childhood onset, progressive spastic paraplegia accompanied by pigmentary retinopathy, mental retardation, dysarthria, dementia, and distal muscle atrophy; juvenile, slowly progressive amyotrophic lateral sclerosis reported in subjects with SPG11 HSP; 50% of autosomal recessive HSP is considered to be SPG11	95,96
SPG14 60 5229	Unknown	Single consanguineous Italian family, 3 affected subjects, onset age ~30 years; complicated spastic paraplegia with mental retardation and distal motor neuropathy (sural nerve biopsy was normal)	97
SPG15 270700	Spastizin/ZFYVE26	Complicated: spastic paraplegia variably associated with pigmented maculopathy, distal amyotrophy, dysarthria, mental retardation, and further intellectual deterioration (Kjellin syndrome)	98,99
SPG18 611225	Endoplasmic reticulum, lipid raft associated protein 2 (ERLIN2)	Two families described with spastic paraplegia complicated by mental retardation and thin corpus callosum. ERLIN2 mutations also identified in subjects with juvenile primary lateral sclerosis	100–102
SPG20 607211	Spartin	Complicated: spastic paraplegia associated with distal muscle wasting (Troyer syndrome)	14,103–106
SPG21 248900	Maspardin	Complicated: spastic paraplegia associated with dementia, cerebellar and extrapyramidal signs, thin corpus callosum, and white matter abnormalities (Mast syndrome)	107
SPG23 270750	Unknown	Complicated: childhood onset HSP associated with skin pigment abnormality (vitiligo), premature graying, characteristic facies; Lison syndrome	108
SPG24 607584	Unknown	Complicated: childhood onset HSP variably complicated by spastic dysarthria and pseudobulbar signs	109
SPG25 608220	Unknown	Consanguineous Italian family, four subjects with adult (30–46 years) onset back and neck pain related to disk herniation and spastic paraplegia; surgical correction of disk herniation ameliorated pain and reduced spastic paraplegia. Peripheral neuropathy also present	110
SPG26 609195	Unknown	Single consanguineous Bedouin family with five affected subjects. Complicated: childhood onset (7–8 years), progressive spastic paraparesis with dysarthria and distal amyotrophy in both upper and lower limbs, nerve conduction studies were normal; mild intellectual impairment, normal brain MRI	111

Continued

TABLE 77.1 Genetic Types of HSP, their Encoded Proteins, and Clinical Syndromes (Updated from [6])—cont'd

Spastic gait (SPG) locus and OMIM no.	Protein	Clinical syndrome	References
SPG27 609041	Unknown	Complicated or uncomplicated HSP. Two families described. In one family (seven affected subjects) uncomplicated spastic paraplegia began between ages 25 and 45 years. In the second family (three subjects described) the disorder began in childhood and included spastic paraplegia, ataxia, dysarthria; mental retardation, sensorimotor polyneuropathy, facial dysmorphism and short stature	62,112
SPG28 609340	DDHD1	Uncomplicated: pure spastic paraplegia, onset in infancy, childhood, or adolescence, either as an uncomplicated spastic paraplegia syndrome; or variable associated with axonal neuropathy, distal sensory loss, and cerebellar eye movement disturbance	113,114
SPG29 609727	Unknown	Uncomplicated HSP, childhood onset	
SPG30 610357	KIF1A	Complicated: spastic paraplegia, distal wasting, saccadic ocular pursuit, peripheral neuropathy, mild cerebellar signs	115
SPG32 611252	Unknown	Mild mental retardation, brainstem dysraphia, clinically asymptomatic cerebellar atrophy	
SPG35 612319	Fatty acid 2-hydroxylase (FA2H)	Childhood onset (6–11 years), spastic paraplegia with extrapyramidal features, progressive dysarthria, dementia, seizures. Brain white matter abnormalities and brain iron accumulation; an Omani and a Pakistani kindred reported	116–118
SPG39 612020	Neuropathy target esterase (NTE)	Complicated: spastic paraplegia associated with wasting of distal upper and lower extremity muscles	119
SPG43 615043	C19orf12	Two sisters from Mali, symptom onset 7 and 12 years, progressive spastic paraplegia with atrophy of intrinsic hand muscles and dysarthria (one sister)	120
SPG44 613206	Gap junction protein GJA12/GJC2, also known as connexin47 (Cx47)	Allelic with Pelizaeus–Merzbacher-like disease (PMLD; early-onset dysmyelinating disorder with nystagmus, psychomotor delay, progressive spasticity, ataxia). GJA/GJC2 mutation I33M causes a milder phenotype: late-onset (first and second decades), cognitive impairment, slowly progressive, spastic paraplegia, dysarthria, and upper extremity involvement. MRI and MR spectroscopy imaging consistent with a hypomyelinating leukoencephalopathy	121
SPG45 613162	Unknown	Single consanguineous kindred from Turkey, five subjects described: affected subjects had mental retardation, infantile onset lower extremity spasticity and contractures, one subject with optic atrophy, two subjects with pendular nystagmus; MRI in one subject was normal	122
SPG46 614409	Unknown	Dementia, congenital cataract, ataxia, thin corpus callosum	123
SPG47 614066	AB4B1	Two affected siblings from consanguineous Arabic family with early-childhood onset slowly progressive spastic paraparesis, mental retardation, and seizures; one subject had ventriculomegaly; the other subject had thin corpus callosum and periventricular white matter abnormalities	124
SPG48 613647	KIAA0415	Analysis of *KIAA0415* gene in 166 unrelated spastic paraplegia subjects (38 recessive, 64 dominant, 64 "apparently sporadic") and control subjects revealed homozygous mutation in two siblings with late onset (6th decade) uncomplicated spastic paraplegia; and heterozygous mutation in one subject with apparently sporadic spastic paraplegia	125

TABLE 77.1 Genetic Types of HSP, their Encoded Proteins, and Clinical Syndromes (Updated from [6])—cont'd

Spastic gait (SPG) locus and OMIM no.	Protein	Clinical syndrome	References
SPG49 615031	TECPR2	Five subjects from three apparently unrelated families (Jewish Bukharian ancestry) had infantile onset hypotonia, developmental delay with severe cognitive impairment, dysmorphic features (short stature, bradymicrocephaly, oral, facial, dental, nuchal abnormalities). Spastic, ataxic, rigid gait developed in childhood; additional features included gastroesophageal reflux, recurrent apneic episodes, mild dysmorphic features. Two subjects had epilepsy and MRI of two subjects showed thin corpus callosum and cerebellar atrophy	126
SPG50 612936	AP4M1	Five subjects from one consanguineous Moroccan family exhibited infantile onset, nonprogressive spastic quadriplegia with severe cognitive impairment; variably associated with adducted thumbs. Ventriculomegaly, white matter abnormalities and variable cerebellar atrophy noted on neuroimaging. Neuroaxonal abnormalities, gliosis, and reduced myelin noted on postmortem examination	127,128
SPG51 613744	AP4E1	Two siblings from a consanguineous Palestinian Jordanian family and two siblings from a consanguineous Syrian family exhibited microcephaly, hypotonia, psychomotor delay, spastic tetraplegia, marked cognitive impairment with severe language impairment, facial dysmorphic features; abnormal brain MRI showed (including atrophy and diffuse white matter loss). Seizures were variably present.	128–130
SPG52 614067	AP4S1	Five subjects from a consanguineous Syrian kindred exhibited delayed motor development, and severe cognitive impairment. Neonatal hypotonia was followed by progressive spastic gait with contractures. Dysmorphic features included short stature, microcephaly, and facial abnormalities	130–133
SPG53 614898	VSP37A	Nine subjects from two Arab Moslem families exhibited developmental delay, progressive lower extremity spasticity, and subsequently progressive upper extremity involvement; associated with skeletal dysmorphism (kyphosis and pectus carinatum); mild to moderate cognitive impairment; and variable hypertrichosis and impaired vibration sensation	134
SPG54 615033	DDHD2	Affected subjects reported from four unrelated families exhibited psychomotor delay, cognitive impairment, progressive spasticity (onset before age 2 years), thin corpus callosum, periventricular white matter abnormalities. Additional clinical features include foot contractures, dysarthria, dysphagia, strabismus, optic hypoplasia	135,136
SPG55 615035	C12ORF65	Two Japanese brothers from consanguineous parents exhibited early onset spastic paraplegia variably associated with reduced visual acuity (with central scotoma and optic atrophy), reduced upper extremity strength and dexterity, lower extremity muscle atrophy, and motor sensory neuropathy	137,138
SPG56 615030	CYP2U1	Five unrelated families were reported with early-childhood onset spastic paraplegia, variable upper extremity involvement, upper extremity dystonia, cognitive impairment, thin corpus callosum, brain white matter disturbance, axonal neuropathy, basal ganglia calcifications	114
SPG57 602490	Trk-fused gene (*TFG*)	Two siblings with early-childhood onset spastic paraplegia, optic atrophy (at age 2.5 years), and wasting of hand and leg muscles due to axonal-demyelinating sensorimotor neuropathy; normal intelligence	139

Continued

TABLE 77.1 Genetic Types of HSP, their Encoded Proteins, and Clinical Syndromes (Updated from [6])—cont'd

Spastic gait (SPG) locus and OMIM no.	Protein	Clinical syndrome	References
SPG58 611302	KIF1C	Spastic-ataxia. Subjects are described from four, unrelated consanguineous kindreds. Following normal development, progressive ataxia began between ages 1 and 17 years and was associated with lower extremity spasticity that was nonprogressive. MRI in one subject showed some evidence of demyelination of posterior limb of internal capsule and occipital cortex; and cerebellar and cortical atrophy in another, unrelated subject. Subjects in one family had short stature and microcephaly. Subjects in another family had chorea, developmental delay, and cognitive impairment	7,140
SPG59 603158	USP8	Uncomplicated spastic gait reported in two subjects from one consanguineous family. Toe walking noted at age 20 months. When examined in late childhood, subjects were ambulatory with spastic gait	7
SPG60 612167	WDR48	Early-childhood onset gait impairment associated with increased lower extremity muscle tone and reflexes, nystagmus, peripheral neuropathy, and mild learning disability	7
SPG61 616685	ARL6IP1	Two affected subjects are reported from one consanguineous family, with onset at age 14 months. Subjects were nonambulatory at ages 11 and 12 years. Examination demonstrated increased patellar and absent ankle deep tendon reflexes; diffuse motor and sensory neuropathy with loss of digits and acromutilation; normal intelligence	7
SPG62	ERLIN1	Two affected subjects from one consanguineous family are reported. Symptom onset (toe walking) at ages 2 and 3 years; spastic gait described at ages 16 and 20 years. Subjects were ambulatory despite having flexion contractures at the knees. Cognition was normal cognition. One subject had absent lower extremity reflexes	7
SPG63 615686	AMPD2	Two affected subjects from one consanguineous family are reported: delayed walking milestone, ambulatory with scissoring gait, and normal cognition. One subject had periventricular white matter and corpus callosum abnormalities	7
SPG64 615683	ENTPD1	Two affected subjects from one consanguineous family are reported with childhood onset complicated spastic paraplegia. Symptom onset between ages 3 and 4 years; examination showed lower extremity spasticity, hyperreflexia, amyotrophy (one subject), microcephaly, aggressiveness, and delayed puberty	7
SPG65 (same disorder as SPG45)	NT5C2	Consanguineous kindred from Turkey with five subjects described: affected subjects had mental retardation, infantile onset lower extremity spasticity and contractures, one subject with optic atrophy, two subjects with pendular nystagmus; MRI in one subject was normal. Four additional families reported with infantile onset spasticity, thin or dysplastic corpus callosum, learning difficulty	7,122
SPG66 610009	ARSI	One affected subject from a consanguineous family is described with early-childhood onset of abnormal gait. At 3.5 years the subject was not ambulatory, had lower extremity spasticity but was areflexic; and had severe sensorimotor polyneuropathy, corpus callosum and cerebellar hypoplasia, colpocephaly, and borderline intelligence	7
SPG67 611655	PGAP1	Two affected subjects from one consanguineous family are described. Onset in infancy with global delay, abnormal hand movements, spasticity, borderline intelligence (one subject). MRI scan was abnormal, showing prominent cortical sulci in one subject and corpus callosum agenesis, vermis hypoplasia, and defective myelination in one subject	7

TABLE 77.1 Genetic Types of HSP, their Encoded Proteins, and Clinical Syndromes (Updated from [6])—cont'd

Spastic gait (SPG) locus and OMIM no.	Protein	Clinical syndrome	References
SPG68 604806	FLRT1	Two affected subjects from one consanguineous family are described: early-childhood onset gait impairment with hyperreflexia (patellar clonus) but no spasticity; nystagmus, optic atrophy, foot drop, peripheral neuropathy, normal brain MRI and normal intelligence	7
SPG69	RAB3GAP2	One subject from a consanguineous family is described: infantile onset, global developmental delay, lower extremity spasticity, dysarthria, deafness, cataract, and intellectual impairment	7
SPG70 156560	MARS	Four subjects from one consanguineous family are described: infantile onset, delayed motor milestones, lower extremity spasticity, amyotrophy, borderline intelligence	7
SPG71 615635	ZFR	One affected subject from a consanguineous family is described with symptom onset at age 1 year. At age 3 years had lower extremity spasticity and flexion contractures at the knees. Brain MRI showed thin corpus callosum. Normal intelligence	7
SPG72 615625	REEP2	Uncomplicated spastic paraplegia manifesting as childhood onset (infancy to age 8 years) toe walking or progressive spastic gait. Two of 10 subjects had upper extremity spasticity. Cognition, speech, and vision were normal. REEP2 mutations may cause both autosomal dominant HSP (due to heterozygous mutation) or autosomal recessive HSP due to compound heterozygote mutations	20
603513	GAD1	Four siblings in consanguineous Pakistani family with spastic cerebral palsy, and moderate-to-severe mental retardation	141–143
"SPOAN" syndrome 609541	Unknown	Complicated spastic paraplegia associated with optic atrophy, neuropathy (SPOAN)	144
256840	Epsilon subunit of the cytosolic chaperonin-containing t-complex peptide-1 (Cct5)	Complicated spastic paraplegia associated with mutilating sensory neuropathy	145,146
X-LINKED HSP			
SPG1 303350	L1 cell adhesion molecule (L1CAM)	Complicated: associated with mental retardation, and variably, hydrocephalus, aphasia, and adducted thumbs	147
SPG2 300401	Proteolipid protein	Complicated: variably associated with MRI evidence of CNS white matter abnormality; may have peripheral neuropathy	148–151
SPG16 300266	Unknown	Uncomplicated; or complicated: associated with motor aphasia, reduced vision, nystagmus, mild mental retardation, and dysfunction of the bowel and bladder	152,153
SPG22 300523	Monocarboxylate transport 8 (MCT8)	Complicated (Allan–Herndon–Dudley syndrome): congenital onset, neck muscle hypotonia in infancy, mental retardation, dysarthria, ataxia, spastic paraplegia, abnormal facies	154–156
SPG34 300750		Uncomplicated, onset 12–25 years	157
MATERNAL (MITOCHONDRIAL) INHERITANCE HSP			
No SPG designation	Mitochondrial *ATP6* gene	Adult onset, progressive spastic paraplegia, mild-to-severe symptoms, variably associated with axonal neuropathy, late-onset dementia, and cardiomyopathy	158

Neurologic findings are relatively symmetrical in uncomplicated HSP (although it is not unusual for subjects to report that one leg is stronger or more flexible than the other, particularly early in the course). Pes cavus (high-arched feet) and hammer toe deformation are common in HSP, although may be absent in definitely affected subjects. Although it is common for upper extremity deep tendon reflexes to be brisk (3+) and for Hoffman and Tromner signs to be present, upper extremity muscle tone, strength, and dexterity remain normal in uncomplicated HSP.

HSP DIAGNOSIS

For most individuals, HSP is a clinical diagnosis. Laboratory confirmation is increasingly available but limited primarily by the cost of testing. Diagnostic criteria for HSP are: 1) symptomatic gait disturbance due to spastic weakness affecting both legs approximately symmetrically that may be either essentially nonprogressive (when symptom-onset occurs in infancy) or insidiously progressive (with later symptom onset); 2) neurologic signs of bilateral, typically symmetric, lower extremity spasticity, hyperreflexia, variable degrees of weakness (from full strength to marked weakness) in an upper motor neuron pattern, extensor plantar responses (occasionally flexor), often accompanied by mildly decreased vibration sensation in the toes; and 3) careful exclusion of alternate disorders (Table 77.2).

Family history, though useful when present, is not required for diagnosing HSP. Family history may be absent in autosomal recessive or X-linked forms of HSP; those with incomplete genetic penetrance or wide variability of age-of-symptom onset (e.g., a child may be affected before a parent becomes symptomatic); and in individuals with *de novo* mutation or nonpaternity.

The diagnosis of uncomplicated HSP should be questioned when the course is atypical (abrupt symptom onset, step-wise course, or rapid worsening over several months); when neurologic involvement is unilateral or markedly asymmetric; when spasticity and weakness involve upper extremities (beyond asymptomatic upper extremity hyperreflexia) or bulbar muscles; in the presence of spinal sensory level; or when there is significant muscle atrophy and fasciculations. Though significant muscle atrophy is not a typical feature of uncomplicated HSP, it is characteristic of many types of complicated HSP including Troyer syndrome (SPG20) and Silver syndrome (SPG17),[14,15] and has been reported in SPG3A, SPG4, SPG7, SPG10, SPG14, SPG15, and SPG26 HSP (Table 77.3 and [16]). Symmetrical, distal, and slowly progressive muscle weakness that occurs in some complicated forms of HSP is quite different than that which occurs in amyotrophic lateral sclerosis (generally asymmetric, relatively rapidly progressive, often involving proximal and bulbar muscles).

HSP Gene Analysis

HSP gene analysis, both through direct testing of panels of HSP genes and by whole-exome sequencing, can confirm the diagnosis and identify the genetic cause of HSP for the majority of individuals. Unfortunately, the costs of this testing often limits its clinical utility. HSP gene analysis can be used for genetic counseling and prenatal testing (including preimplantation genetic diagnosis).

TABLE 77.2 Differential Diagnosis of Hereditary Spastic Paraplegia (Reviewed in [3])

Category	Example
Structural abnormality of the brain and spinal cord	Tethered cord syndrome, spinal cord compression from degenerative spondylosis or neoplasm
Leukodystrophy	B_{12} deficiency, multiple sclerosis, adrenomyeloneuropathy, Krabbe disease, metachromatic leukodystrophy, mitochondrial disorder
Infectious diseases	Tropical spastic paraplegia due to HTLV1 infection (which may be familial) and pachymeningitis from tertiary syphilis
Other motor neuron disorders	Amyotrophic lateral sclerosis, primary lateral sclerosis, distal hereditary motor neuropathy
Other degenerative neurologic disorders	Friedreich ataxia (which may have spasticity rather than arreflexia)[159] spinocerebellar ataxia type III (Machado–Joseph disease), spinal cord arteriovenous malformation
Environmental toxins	Lathyrism, Konzo, and Cycad poisoning, and organophosphate-induced delayed neuropathy
Other	Spastic diplegic cerebral palsy, dopa-responsive dystonia

TABLE 77.3 Examples of HSP Proteins Participating in Shared Cellular–Molecular Functions

Cellular–molecular process	Representative types of HSP and mutant genes
Axon transport	SPG30/KIF1A, SPG4/spastin, SPG58/KIF1C
Disturbance in endoplasmic reticulum (ER) morphology	SPG12/reticulon 2, SPG3A/atlastin, SPG4/spastin, SPG31/REEP1
Mitochondrial abnormality	SPG13/chaperonin 60/heat shock protein 60, SPG7/paraplegin, mitochondrial ATP6
Primary myelin abnormality	SPG2/proteolipid protein, SPG42/connexin 47, SGP17/BSCL2 (seipin), CcT5
Abnormal protein conformation leading to ER-stress response, ER protein degradation disturbance	SPG8/strumpellin, SPG62/ERLIN1, SPG18/ERLIN2
Disturbance in vesicle formation and membrane trafficking including selective uptake of proteins into vesicles	SPG47/AP4B1, SPG48/KIAA0415, SPG50/AP4M1, SPG51/AP4E1, SPG52/AP4S1, SPG53/VPS37A
Disturbance of lipid metabolism	SPG28/DDHD1, SPG54/DDHD2, SPG39/NTE, SPG56/CYP2U1, SPG35/FA2H, Sjogren–Larsson/FALDH
Purine nucleotide metabolism	AMPD2, ENTPD1, and NT5C2 are involved in purine nucleotide metabolism

It is important to consider a few caveats regarding HSP gene analysis. First, although testing of HSP genes by gene-panel analysis and by whole-exome sequencing can identify mutations in the majority of affected subjects, it is still incomplete. Recently, whole-exome sequencing of 55 consanguineous kindreds with HSP identified pathogenic mutations that segregated with the disorder in 40 kindreds. In 15 families, whole-exome sequencing did not identify a single genetic cause.

Second, results of HSP gene analysis must be interpreted in the clinical context. In subjects whose clinical syndrome conforms to HSP, the occurrence of a gene variation predicted to result in significant alteration in a known HSP protein can be generally interpreted as the likely cause of disease. Much less certainty can be placed on the significance of either HSP gene variations that do not predict significant alteration in the encoded protein's sequence or secondary structure, or which are identified in subjects whose clinical syndromes do not conform to HSP. Development of functional evaluation of the impact of HSP gene variations *in vitro* or *in vivo* is underway in research laboratories. This will help understand the relevance of HSP gene variations currently "of uncertain significance" to the subject's clinical syndrome.

Third, mutations for dominantly inherited HSP (e.g., SPG4/spastin) are identified in approximately 10% of subjects with no previous family history of HSP. This has important implications for genetic counseling (for example, risk of disease occurrence in progeny could be as high as 50%).

And finally, mutations in the *SPG7*/paraplegin gene[17–19] and SPG72/REEP2[20] can manifest as either autosomal recessive HSP or autosomal dominant HSP. The extent to which dominant-versus-recessive inheritance is related to specific alleles (i.e., particular genotype–phenotype correlation) and the occurrence of both dominant and recessive transmission in other genetic types of HSP is not known.

COMPLICATED HSP

Only limited generalizations may be drawn about the features of the numerous types of "complicated" HSPs since each represents a unique clinicopathologic entity (reviewed in [21]). From a clinical perspective, it is useful to recognize four clinical paradigms of HSP: 1) uncomplicated HSP; 2) spastic paraplegia associated with peripheral motor neuropathy and/or distal wasting (e.g., Troyer and Silver syndromes [Table 77.1]); 3) spastic paraplegia associated with cognitive impairment (e.g., SPG11 which is often associated with thin corpus callosum and considered to represent 50% of recessively inherited HSP); and 4) spastic paraplegia associated with additional neurologic and systemic abnormalities.

Some genetic types of HSP manifest as either "complicated" or "uncomplicated" HSP syndromes. For example, although most individuals with SPG3A and SPG4 HSP exhibit "uncomplicated" spastic paraplegia, some have distal muscle atrophy (SPG3A and SPG4) and cognitive impairment (SPG4).[22–29] Conversely, families with complicated HSP (e.g., those with SPG7 (paraplegin gene mutation), SPG11 (spatacsin gene mutation) or SPG10 (*KIF5A* gene mutation) may have members affected with either complicated or uncomplicated HSP.

CLINICAL VARIABILITY AND GENOTYPE–PHENOTYPE CORRELATION

The age at which symptoms begin, the severity, and degree of disability may vary significantly between affected individuals in a family who share precisely the same HSP gene mutation; between families with different mutations in the same HSP gene; and between different genetic types of HSP. For most types of HSP, there is little correlation between the specific HSP gene mutation and clinical severity or age of symptom onset. Although some types of HSP typically begin in childhood (e.g., SPG3A) and others typically begin in adulthood (e.g., SPG6), it is usually not possible to distinguish one type of uncomplicated HSP from another by clinical parameters alone.

Phenotype variation includes not only variation in severity, rate of progression, and age-of-symptom onset, but, as noted above, may extend to the occurrence of additional neurologic disturbance (i.e., in some genetic types of HSP, both complicated and uncomplicated HSP syndromes occur in family members who inherit precisely the same HSP gene mutation). This phenotype variation between subjects with the same pathogenic HSP gene mutation indicates the importance of gene modification and potentially environmental factors in determining the clinical syndrome.

TREATMENT

Treatment for HSP is presently limited to reducing muscle spasticity through daily, muscle stretching routines and through medications (e.g., oral or intrathecal Lioresal); reducing urinary urgency through medication (e.g., oxybutyinin); reducing toe-dragging through ankle–foot orthotic devices; and improving cardiovascular fitness, endurance, and lower extremity strength through daily physical exercise.

PROGNOSIS

Diagnosing uncomplicated HSP carries with it the following prognoses: 1) lower extremity spasticity and weakness and gait disturbance may worsen insidiously over many years; 2) upper extremity strength and dexterity, and muscles of speech, swallowing, and respiration will not be involved; and 3) longevity will not be affected. As noted above, anecdotal clinical experience shows that after a number of years (e.g., 3–7), many individuals with progressive worsening seem to reach a functional plateau beyond which the rate of worsening is similar to that attributed to age and similar degrees of physical exercise. In view of the clinical variability between and within different genetic types of HSP, it is not possible to accurately predict the extent of an individual's disability. A cautious "wait and see" approach is advised.

NEUROPATHOLOGY[30–37]

Postmortem studies of genetically defined types of HSP are quite limited. In general, such studies of uncomplicated HSP show axon degeneration limited to the central nervous system affecting primarily the distal ends of the longest descending motor fibers (corticospinal tracts, with maximal involvement in the thoracic spinal cord); and the distal ends of the longest ascending fibers (fasciculus gracilis fibers, with maximal involvement in the cervicomedullary region).[16,31,32,34–37] Demyelination of fibers undergoing degeneration in uncomplicated HSP is considered to be the consequence axonal degeneration in uncomplicated HSP and not the primary process. Decreased numbers of cortical motor neurons and anterior horn cells have been observed.[34,35]

Distal degeneration of long central nervous system motor and sensory axons, demonstrated in some types of uncomplicated HSP, is parallel to Charcot–Marie–Tooth type 2 in which distal motor and sensory axon degeneration is limited to the peripheral nervous system. Indeed there is evidence that at least some forms of HSP and CMT-II share similar pathophysiologic mechanisms. SPG10 HSP is due to mutation in kinesin heavy chain (KIF5A),[38] a molecular motor involved in axonal transport. Mutations in another kinesin (KIF1B) cause Charcot–Marie–Tooth type 2A1.[39]

MOLECULAR BASIS OF HSP

58 HSP genes have been discovered. Analysis of the functions of these HSP proteins indicates first that a large diversity of proteins is implicated. Most of the HSP proteins are not members of an extended gene family. A few,

however share some sequence similarities: spastin and paraplegin share an AAA domain; spartin's amino-terminal region is similar to that of spastin's. Some HSP genes encode members of a functional protein complex (e.g., *AP4B1* [SPG47], *KIAA0415* [SPG48], *AP4M1* [SPG50], *AP4E1* [SPG51], *AP4S1* [SPG52], and *VPS37A* [SPG53]).

Though unique, most HSP proteins can be grouped into one or more shared, functional pathways including axon transport, control of endoplasmic reticulum morphology, mitochondrial function, myelination, protein folding and degradation, vesicle transport and membrane trafficking, lipid metabolisms, and purine nucleotide metabolism. (See Table 77.3 for representative examples and [6,7,40] for additional examples and references). The implication of the large diversity of HSP proteins and the number of implicated cellular and molecular processes of HSP proteins is that the hallmark sign of corticospinal tract impairment (ranging from developmental disturbance to progressive degeneration), in different genetic types of HSP, often occurring in the absence of other nervous system or systemic disturbance, can be caused by different biochemical abnormalities.[6,7]

CONCLUSIONS

Table 77.4 provides a summary of HSPs. The HSPs are an extremely large group of disorders that share the primary clinical feature of lower extremity spastic weakness. Although motor symptoms predominate, distal axon degeneration in many types of HSP involves both motor (corticospinal) and sensory (dorsal column) fibers. In this regard, HSP can be considered a "central nervous system homolog" of Charcot–Marie–Tooth disease type 2

TABLE 77.4 HSP Summary

Genetics	More than 70 different genetic types: dominant, recessive, and X-linked forms. SPG4 HSP (due to spastin gene mutation) is the most common type of autosomal dominant HSP. SPG3A HSP (due to atlastin gene mutation) is the most common type of childhood onset, autosomal dominant HSP. Genetic penetrance for autosomal dominant HSP is age-dependent and high (~85% for SPG4) but often incomplete
Clinical variability	Variability between and within different genetic types. Individuals with severe and mild forms may coexist in the same family
"Uncomplicated" and "complicated HSP"	"Uncomplicated HSP": gait disturbance due to lower extremity spastic weakness; subtle vibration impairment in the toes; urinary urgency. "Complicated HSP": symptoms and signs of uncomplicated HSP plus additional systemic or neurologic involvement (such as muscle wasting, peripheral neuropathy, cataracts, ataxia, mental retardation)
Neuropathology	Axon degeneration maximally affecting the distal ends of the longest central nervous system motor (corticospinal tracts) and sensory (fasciculus gracilis) fibers. Complicated HSP types have additional, syndrome-specific neuropathology
Molecular basis	Many different molecular processes underlie various forms of HSP. These include disturbance in microtubule dynamics, axonal transport, mitochondria, corticospinal tract development, and myelination, endoplasmic reticulum morphology, protein folding and degradation, vesicle and membrane trafficking, purine nucleotide synthesis, lipid metabolism
Treatment	Exercise, gait and balance training, spasticity reducing medications (e.g., oral or intrathecal Lioresal), ankle–foot orthotic devices, medication to reduce urinary urgency (e.g., oxybutyin)
Prognosis	Infantile onset HSP often shows little worsening for the first two decades. Childhood- and later-onset HSP typically worsen insidiously over years and decades. After a number of years of insidious worsening, many subjects appear to reach a functional plateau. Assistive devices (including wheelchairs) may be needed. Uncomplicated HSP does not involve upper extremity, speech, bulbar, or respiratory muscles or shorten life expectancy
Diagnosis and genetic testing	Diagnostic criteria: signs and symptoms of HSP; exclusion of other disorders; family history is important but may be absent (e.g., recessive HSP, new mutation, nonpaternity). Genetic testing of HSP gene panels or whole-exome sequencing can confirm the diagnosis for the majority of subjects with HSP syndromes
Differential diagnosis	Structural disorders affecting brain or spinal cord; spinal cord arteriovenous malformation; leukodystrophies (including B_{12}, adrenomyeloneuropathy, Krabbe, metachromatic leukodystrophy, multiple sclerosis), dopa-responsive dystonia, amyotrophic lateral sclerosis, primary lateral sclerosis, HTLV1

(in which distal, motor–sensory axon degeneration is limited to the peripheral nervous system). More than 70 different genetic types of HSP have been identified. "Uncomplicated" HSP syndromes may not be distinguishable by clinical parameters alone. Syndrome-specific neurologic or systemic findings help diagnose various "complicated" forms of HSP. Genetic testing can diagnose the vast majority of subjects with HSP. Nonetheless, because genetic testing is often cost-prohibitive, HSP is a clinical diagnosis for most subjects. The differential diagnosis includes treatable disorders (e.g., B_{12} deficiency, cervical spondylosis, dopa-responsive dystonia, multiple sclerosis), as well as disorders whose prognosis is quite different than HSP (e.g., ALS and primary lateral sclerosis). At present, there is no treatment to reverse or retard progressive disability in HSP. Treatment for HSP is symptomatic and includes physical therapy and the use of medications to reduce spasticity and urinary urgency.

ACKNOWLEDGEMENTS

This research is supported by grants from the Veterans Affairs Merit Review, the National Institutes of Health (NINDS R01 NS069700), the Spastic Paraplegia Foundation, the Paul and Lois Katzman Family, and the Susan Parkinson Foundation. We are grateful for the participation of HSP patients and family members, without whom our investigations would not be possible.

References

1. Fink JK, et al. Hereditary spastic paraplegia. In: Rimoin D, Connor JM, Pyeritz RE, eds. *Emery and Rimoin's Principles and Practice of Medical Genetics*. 5th ed. Philadelphia: Churchill Livingstone Elsevier; 2007:2771–2801.
2. Fink JK. Hereditary spastic paraplegias. In: Schapira AHV, ed. *Neurology and Clinical Neurosciences*. Philadelphia: Mosby Elsevier; 2007:899–910.
3. Fink JK, Heiman-Patterson T, Bird T, et al. Hereditary spastic paraplegia: advances in genetic research. *Neurology*. 1996;46:1507–1514.
4. Fink JK, Hedera P. Hereditary spastic paraplegia: genetic heterogeneity and genotype-phenotype correlation. *Semin Neurol*. 1999;19:301–310.
5. Fink JK. Hereditary spastic paraplegia. *Curr Neurol Neurosci Rep*. 2006;6:65–76.
6. Fink JK. Hereditary spastic paraplegia: clinico-pathologic features and emerging molecular mechanisms. *Acta Neuropathol*. 2013;126:307–328.
7. Novarino G, Fenstermaker AG, Zaki MS, et al. Exome sequencing links corticospinal motor neuron disease to common neurodegenerative disorders. *Science*. 2014;343:506–511.
8. Polo AE, Calleja J, Combarros O, et al. Hereditary ataxias and paraplegias in Cantabria, Spain: an epidemiological and clinical study. *Brain*. 1991;114:855–856.
9. Filla A, DeMichele G, Marconi R, et al. Prevalence of hereditary ataxias and spastic paraplegias in Molise, a region of Italy. *J Neurol*. 1992;239:351–353.
10. Ruano L, Melo C, Silva MC, et al. The global epidemiology of hereditary ataxia and spastic paraplegia: a systematic review of prevalence studies. *Neuroepidemiology*. 2014;42:174–183.
11. Hirtz D, Thurman DJ, Gwinn-Hardy K, et al. How common are the "common" neurologic disorders? *Neurology*. 2007;68:322–323.
12. Fink JK. The hereditary spastic paraplegias. In: Rosenberg R, ed. *Molecular and Genetic Basis of Neurologic and Psychiatric Disease*. 4th ed. Lippincott Williams & Wilkins; 2007.
13. Rainier S, Sher C, Reish O, et al. *De novo* occurrence of novel *SPG3A*/atlastin mutation presenting as cerebral palsy. *Arch Neurol*. 2006;63:445–447.
14. Cross HE, McKusick VA. The Troyer syndrome. A recessive form of spastic paraplegia with distal muscle wasting. *Arch Neurol*. 1967;16:473–485.
15. Patel H, Hart PE, Warner TT, et al. The silver syndrome variant of hereditary spastic paraplegia maps to chromosome 11q12-q14, with evidence for genetic heterogeneity within this subtype. *Am J Hum Genet*. 2001;69:209–215.
16. Fink JK. Hereditary spastic paraplegia: nine genes and counting. *Arch Neurol*. 2003;60:1045–1049.
17. Brugman F, Scheffer H, Wokke JHJ, et al. Paraplegin mutations in apparently sporadic adult-onset upper motor neuron syndromes. *Neurology*. 2008;71:1500–1505.
18. Arnoldi A, Tonelli A, Crippa F, et al. A clinical, genetic, and biochemical characterization of SPG7 mutations in a large cohort of patients with hereditary spastic paraplegia. *Hum Mutat*. 2008;29:522–531.
19. McDermott CJ, Dayaratne RK, Tomkins J, et al. Paraplegin gene analysis in hereditary spastic paraparesis (HSP) pedigrees in northeast England. *Neurology*. 2001;56:467–471.
20. Esteves T, Durr A, Mundwiller E, et al. Loss of association of REEP2 with membranes leads to hereditary spastic paraplegia. *Am J Hum Genet*. 2014;94:268–277.
21. Fink JK. Hereditary spastic paraplegia. In: Rimoin D, Connor JM, Pyeritz RE, et al., eds. *Emery and Rimoin's Principles and Practice of Medical Genetics*. 6th ed. Philadelphia: Churchill Livingstone Elsevier; 2011.
22. Heinzlef O, Paternotte C, Mahieux F, et al. Mapping of a complicated familial spastic paraplegia to locus SPG4 on chromosome 2p. *J Med Genet*. 1998;35:89–93.
23. Lizcano-Gil LA, Garcia-Cruz D, Bernal-Beltran MDP, et al. Association of late onset spastic paraparesis and dementia: probably an autosomal dominant form of complicated paraplegia. *Am J Med Genet*. 1997;68:1–6.

24. Webb S, Coleman D, Byrne P, et al. Autosomal dominant hereditary spastic paraparesis with cognitive loss linked to chromosome 2p. *Brain*. 1998;121:601–609.
25. Byrne PC, Webb S, McSweeney F, et al. Linkage of AD HSP and cognitive impairment to chromosome 2p: haplotype and phenotype analysis indicates variable expression and low or delayed penetrance. *Eur J Hum Genet*. 1998;6:275–282.
26. Reid E, Grayson C, Rubinsztein DC, et al. Subclinical cognitive impairment in autosomal dominant 'pure' hereditary spastic paraplegia. *J Med Genet*. 1999;36:797–798.
27. Tedeschi G, Allocca S, DiCostanzo A, et al. Multisystem involvement of the central nervous system in Strumpell's disease. A neurophysiological and neuropsychological study. *J Neurol Sci*. 1991;103:55–60.
28. McMonagle P, Byrne P, Hutchinson M. Further evidence of dementia in SPG4-linked autosomal dominant hereditary spastic paraplegia. *Neurology*. 2004;62:407–410.
29. Tallaksen CME, Gomez EG, Verpillat P, et al. Subtle cognitive impairment but no dementia in patients with spastin mutations. *Arch Neurol*. 2003;60:1113–1118.
30. Schwarz GA. Hereditary (familial) spastic paraplegia. *AMA Arch Neurol Psychiatry*. 1952;68:655–682.
31. Schwarz GA, Liu C-N. Hereditary (familial) spastic paraplegia. Further clinical and pathologic observations. *AMA Arch Neurol Psychiatry*. 1956;75:144–162.
32. Deluca GC, Ebers GC, Esiri MM. The extent of axonal loss in the long tracts in hereditary spastic paraplegia. *Neuropathol Appl Neurobiol*. 2004;30:576–584.
33. White KD, Ince PG, Lusher M, et al. Clinical and pathologic findings in hereditary spastic paraparesis with spastin mutation. *Neurology*. 2000;55:89–94.
34. Behan W, Maia M. Strumpell's familial spastic paraplegia: genetics and neuropathology. *J Neurol Neurosurg Psychiatry*. 1974;37:8–20.
35. Harding AE. Hereditary spastic paraplegias. *Semin Neurol*. 1993;13:333–336.
36. Sack GH, Huether CA, Garg N. Familial spastic paraplegia: clinical and pathologic studies in a large kindred. *Johns Hopkins Med J*. 1978;143:117–121.
37. Buge A, Escourolle R, Rancurel G, et al. Strumpell-Lorrains familial spasmodic paraplegia - anatomical and clinical review and report on a new case. *Rev Neurol (Paris)*. 1979;135:329–337.
38. Reid E, Kloos M, Ashley-Koch A, et al. A kinesin heavy chain (KIF5A) mutation in hereditary spastic paraplegia (SPG10). *Am J Hum Genet*. 2002;71:1189–1194.
39. Zhao C, Takita J, Tanaka Y, et al. Charcot–Marie–Tooth disease type2A caused by mutation in a microtubule motor KIF1B-beta. *Cell*. 2001;105:587–597.
40. Blackstone C. Cellular pathways of hereditary spastic paraplegia. *Annu Rev Neurosci*. 2012;35:25–47.
41. Zhao X, Alvarado D, Rainier S, et al. Mutations in a novel GTPase cause autosomal dominant hereditary spastic paraplegia. *Nat Genet*. 2001;29:326–331.
42. Hazan J, Lamy C, Melki J, et al. Autosomal dominant familial spastic paraplegia is genetically heterogeneous and one locus maps to chromosome 14q. *Nat Genet*. 1993;5:163–167.
43. Paternotte C, Rudnicki D, Fizames C, et al. Quality assessment of whole genome mapping data in the refined familial spastic paraplegia interval on chromosome 14q. *Genome Res*. 1998;8:1216–1227.
44. Hazan J, Fontaine B, Bruyn RPM, et al. Linkage of a new locus for autosomal dominant familial spastic paraplegia to chromosome 2p. *Hum Mol Genet*. 1994;3:1569–1573.
45. Roll-Mecak A, Vale RD. The drosophila homologue of the hereditary spastic paraplegia protein, spastin, severs and disassembles microtubules. *Curr Biol*. 2005;15:650–655.
46. Charvin D, Fonknechten N, Cifuentes-Diaz C, et al. Mutations in SPG4 are responsible for a loss of function of spastin, an abundant neuronal protein localized to the nucleus. *Am J Hum Genet*. 2003;12:71–78.
47. Evans KJ, Gomes ER, Reisenweber SM, et al. Linking axonal degeneration to microtubule remodeling by Spastin-mediated microtubule severing. *J Cell Biol*. 2005;168:599–606.
48. Hentati A, Pericak-Vance MA, Lennon F, et al. Linkage of the late onset autosomal dominant familial spastic paraplegia to chromosome 2p markers. *Hum Mol Genet*. 1994;3:1867–1871.
49. Rainier S, Chai J-H, Tokarz D, et al. NIPA1 gene mutations cause autosomal dominant hereditary spastic paraplegia (SPG6). *Am J Hum Genet*. 2003;73:967–971.
50. Fink JK, Wu C-TB, Jones SM, et al. Autosomal dominant familial spastic paraplegia: tight linkage to chromosome 15q. *Am J Hum Genet*. 1995;56:188–192.
51. Fink JK, Sharp G, Lange B, et al. Autosomal dominant hereditary spastic paraparesis, type I: clinical and genetic analysis of a large North American family. *Neurology*. 1995;45:325–331.
52. Chen S, Song C, Guo H, et al. Distinct novel mutations affecting the same base in the NIPA1 gene cause autosomal dominant hereditary spastic paraplegia in two Chinese families. *Hum Mutat*. 2005;25:135–141.
53. Martinez-Lage M, Molina-Porcel L, Falcone D, et al. TDP-43 pathology in a case of hereditary spastic paraplegia with a NIPA1/SPG6 mutation. *Acta Neuropathol*. 2012;124:285–291.
54. Svenstrup K, Moller RS, Christensen J, et al. NIPA1 mutation in complex hereditary spastic paraplegia with epilepsy. *Eur J Neurol*. 2013;18:1197–1199.
55. Du J, Hu YC, Tang BS, et al. Expansion of the phenotypic spectrum of SPG6 caused by mutation in NIPA1. *Clin Neurol Neurosurg*. 2011;113:480–482.
56. Hedera P, Rainier S, Alvarado D, et al. Novel locus for autosomal dominant hereditary spastic paraplegia on chromosome 8q. *Am J Hum Genet*. 1999;64:563–569.
57. Reid E, Dearlove AM, Osborn M, et al. A locus for autosomal dominant "Pure" hereditary spastic paraplegia maps to chromosome 19q13. *Am J Hum Genet*. 2000;66:728–732.
58. Valdmanis PN, Meijer IA, Reynolds A, et al. Mutations in the KIAA0196 gene at the SPG8 locus cause hereditary spastic paraplegia. *Am J Hum Genet*. 2007;80:152–161.

59. Bian X, Klemm RW, Liu TY, et al. Structures of the atlastin GTPase provide insight into homotypic fusion of endoplasmic reticulum membranes. *Proc Natl Acad Sci U S A*. 2011.
60. Hedera P, DiMauro S, Bonilla E, et al. Phenotypic analysis of autosomal dominant hereditary spastic paraplegia linked to chromosome 8q. *Neurology*. 1999;53:44–50.
61. Seri M, Cusano R, Forabosco P, et al. Genetic mapping to 10q23.3-q24.2, in a large Italian pedigree, of a new syndrome showing bilateral cataracts, gastroesophageal reflux, and spastic paraparesis with amyotrophy. *Am J Hum Genet*. 1999;64:586–593.
62. Meijer IA, Cossette P, Roussel J, et al. A novel locus for pure recessive hereditary spastic paraplegia maps to 10q22.1-10q24.1. *Ann Neurol*. 2004;56:579–582.
63. Reid E, Dearlove AM, Rhodes M, et al. A new locus for autosomal dominant 'pure' hereditary spastic paraplegia mapping to chromosome 12q13 and evidence for further genetic heterogeneity. *Am J Hum Genet*. 1999;65:757–763.
64. Fichera M, Lo Giudice M, Falco M, et al. Evidence of kinesin heavy chain (KIF5A) involvement in pure hereditary spastic paraplegia. *Neurology*. 2004;63:1108–1110.
65. Montenegro G, Rebelo AP, Connell J, et al. Mutations in the ER-shaping protein reticulon 2 cause the axon-degenerative disorder hereditary spastic paraplegia type 12. *J Clin Invest*. 2012;122:538–544.
66. Fontaine B, Davoine C-S, Durr A, et al. A new locus for autosomal dominant pure spastic paraplegia, on chromosome 2q24-q34. *Am J Hum Genet*. 2000;66:702–707.
67. Hansen JJ, Durr A, Cournu-Rebeix I, et al. Hereditary spastic paraplegia SPG13 is associated with a muatation in the gene encoding the mitochondrial chaperonin Hsp60. *Am J Hum Genet*. 2002;70:1328–1332.
68. Bross P, Naundrup S, Hansen J, et al. The Hsp60-(p.V98I) mutation associated with hereditary spastic paraplegia SPG13 compromises chaperonin function both *in vitro* and in vivo. *J Biol Chem*. 2008;283:15694–15700.
69. Auer-Grumbach M, Schlotter-Weigel B, Lochmuller H, et al. Phenotypes of the N88S Berardinelli-Seip congenital lipodystrophy 2 mutation. *Ann Neurol*. 2005;57:415–424.
70. Windpassinger C, Auer-Grumbach M, Irobi J, et al. Heterozygous missense mutations in BSCL2 are associated with distal hereditary motor neuropathy and Silver syndrome. *Nat Genet*. 2004;36:271–276.
71. Valente EM, Brancati F, Caputo V, et al. Novel locus for autosomal dominant pure hereditary spastic paraplegia (SPG19) maps to chromosome 9q22-q34. *Ann Neurol*. 2002;51:681–685.
72. Ashley-Koch A, Kail ME, Nance M, et al. A new locus for autosomal dominant hereditary spastic paraplegia (SPG29) maps to chromosome 2p12. *Am J Hum Genet*. 2005.
73. Zuchner S, Wang G, Tran-Viet KN, et al. Mutations in the novel mitochondrial protein REEP1 cause hereditary spastic paraplegia type 31. *Am J Hum Genet*. 2006;79:365–369.
74. Zuchner S, Kail ME, Nance M, et al. A new locus for dominant hereditary spastic paraplegia maps to chromosome 2p12. *Neurogenetics*. 2006;7:127–129.
75. Beetz C, Schule R, Deconinck T, et al. REEP1 mutation spectrum and genotype/phenotype correlation in hereditary spastic paraplegia type 31. *Brain*. 2008;131:1078–1086.
76. Mannan AU, Krawen P, Sauter SM, et al. ZFYVE27 (SPG33), a novel spastin-binding protein, is mutated in hereditary spastic paraplegia. *Am J Hum Genet*. 2006;79:351–357.
77. Schule R, Bonin M, Durr A, et al. Autosomal dominant spastic paraplegia with peripheral neuropathy maps to chr12q23-24. *Neurology*. 2009;72:1893–1898.
78. Hanein S, Durr A, Ribai P, et al. A novel locus for autosomal dominant 'uncomplicated' hereditary spastic paraplegia maps to chromosome 8p21.1-q13.3. *Hum Genet*. 2007;122:261–273.
79. Orlacchio A, Patrono C, Gaudiello F, et al. Silver syndrome variant of hereditary spastic paraplegia: a locus to 4p and allelism with SPG4. *Neurology*. 2008;70:1959–1966.
80. Subramony SH, Nguyen TV, Langford L, et al. Identification of a new form of autosomal dominant spastic paraplegia. *Clin Genet*. 2009;76:113–116.
81. Zhao GH, Hu ZM, Shen L, et al. A novel candidate locus on chromosome 11p14.1-p11.2 for autosomal dominant hereditary spastic paraplegia. *Chin Med J (Engl)*. 2008;121:430–434.
82. Lin P, Li J, Liu Q, et al. A missense mutation in SLC33A1, which encodes the acetyl-CoA transporter, causes autosomal-dominant spastic paraplegia (SPG42). *Am J Hum Genet*. 2008;83:752–759.
83. Lin P, Mao F, Liu Q, et al. Prenatal diagnosis of autosomal dominant hereditary spastic paraplegia (SPG42) caused by SLC33A1 mutation in a Chinese kindred. *Prenat Diagn*. 2010;30:485–486.
84. Schlipf NA, Beetz C, Schule R, et al. A total of 220 patients with autosomal dominant spastic paraplegia do not display mutations in the SLC33A1 gene (SPG42). *Eur J Hum Genet*. 2010;18:1065–1067.
85. Oates EC, Rossor AM, Hafezparast M, et al. Mutations in BICD2 cause dominant congenital spinal muscular atrophy and hereditary spastic paraplegia. *Am J Hum Genet*. 2013;92:965–973.
86. Hentati A, Pericack-Vance MA, Hung W-Y, et al. Linkage of the "pure" recessive familial spastic paraplegia to chromosome 8 markers and evidence of genetic locus heterogeneity. *Hum Genet*. 1994;3:1263–1267.
87. Muglia M, Criscuolo C, Magariello A, et al. Narrowing of the critical region in autosomal recessive spastic paraplegia linked to the SPG5 locus. *Neurogenetics*. 2004;5:49–54.
88. Tang BS, Chen X, Zhao GH, et al. Clinical features of hereditary spastic paraplegia with thin corpus callosum: report of 5 Chinese cases. *Chin Med J (Engl)*. 2004;117:1002–1005.
89. Wilkinson PA, Crosby AH, Turner C, et al. A clinical and genetic study of SPG5A linked autosomal recessive hereditary spastic paraplegia. *Neurology*. 2003;61:235–238.
90. Tsaousidou MK, Ouahchi K, Warner TT, et al. Sequence alterations within CYP7B1 implicate defective cholesterol homeostasis in motor-neuron degeneration. *Am J Hum Genet*. 2008;82:510–515.
91. Criscuolo C, Filla A, Coppola G, et al. Two novel CYP7B1 mutations in Italian families with SPG5: a clinical and genetic study. *J Neurol*. 2009;256:1252–1257.

92. Biancheri R, Ciccolella M, Rossi A, et al. White matter lesions in spastic paraplegia with mutations in SPG5/CYP7B1. *Neuromuscul Disord*. 2009;19:62–65.
93. DeMichele G, DeFusco M, Cavalcanti F, et al. A new locus for autosomal recessive hereditary spastic paraplegia maps to chromosome 16q24.3. *Am J Hum Genet*. 1998;63:135–139.
94. Garner CC, Garner A, Huber G, et al. Molecular cloning of microtubule-associated protein 1 (MAP1A) and microtubule-associated protein 5 (MAP1B): identification of distinct genes and their differential expression in developing brain. *J Neurochem*. 1990;55:146–154.
95. Martinez-Murillo F, Kobayashi H, Pegoraro E, et al. Genetic localization of a new locus for recessive spastic paraplegia to 15q13-15. *Neurology*. 1999;53:50–56.
96. Winner B, Uyanik G, Gross C, et al. Clinical progression and genetic analysis in hereditary spastic paraplegia with thin corpus callosum in spastic gait gene 11 (SPG11). *Arch Neurol*. 2004;61:117–121.
97. Vazza GZM, Boaretto F, Micaglio GF, et al. A new locus for autosomal recessive spastic paraplegia associated with mental retardation and distal motor neuropathy SPG14, maps to chromosome 3q27-q28. *Am J Hum Genet*. 2000;67:504–509.
98. Hughes CA, Byrne PC, Webb S, et al. SPG15, a new locus for autosomal recessive complicated HSP on chromosome 14q. *Neurology*. 2001;56:1230–1233.
99. Hanein S, Martin E, Boukhris A, et al. Identification of the SPG15 gene, encoding spastizin, as a frequent cause of complicated autosomal-recessive spastic paraplegia, including Kjellin syndrome. *Am J Hum Genet*. 2008;82:992–1002.
100. Al-Yahyaee S, Al-Gazali LI, De JP, et al. A novel locus for hereditary spastic paraplegia with thin corpus callosum and epilepsy. *Neurology*. 2006;66:1230–1234.
101. Alazami AM, Adly N, Al Dhalaan H, et al. A nullimorphic ERLIN2 mutation defines a complicated hereditary spastic paraplegia locus (SPG18). *Neurogenetics*. 2011;12:333–336.
102. Al-Saif A, Bohlega S, Al-Mohanna F. Loss of ERLIN2 function leads to juvenile primary lateral sclerosis. *Ann Neurol*. 2012;72:510–516.
103. Crosby AH, Patel H, Patton MA, et al. Spartin, the Troyer syndrome gene, suggests defective endosomal trafficking underlies some forms of hereditary spastic paraplegia. *Am J Hum Genet*. 2002;71:516.
104. Patel H, Cross H, Proukakis C, et al. SPG20 is mutated in Troyer syndrome, an hereditary spastic paraplegia. *Nat Genet*. 2002;31:347–348.
105. Proukakis C, Cross H, Patel H, et al. Troyer syndrome revisited. A clinical and radiological study of a complicated hereditary spastic paraplegia. *J Neurol*. 2004;251:1105–1110.
106. Lu J, Rashid F, Byrne PC. The hereditary spastic paraplegia protein spartin localises to mitochondria. *J Neurochem*. 2006;98:1908–1919.
107. Simpson MA, Cross H, Proukakis C, et al. Maspardin is mutated in Mast Syndrome, a complicated form of hereditary spastic paraplegia associated with dementia. *Am J Hum Genet*. 2003;73:1147–1156.
108. Blumen SC, Bevan S, Abu-Mouch S, et al. A locus for complicated hereditary spastic paraplegia maps to chromosome 1q24-q32. *Ann Neurol*. 2003;54:796–803.
109. Hodgkinson CA, Bohlega S, Abu-Amero SN, et al. A novel form of autosomal recessive pure hereditary spastic paraplegia maps to chromosome 13q14. *Neurology*. 2002;59:1905–1909.
110. Zortea M, Vettori A, Trevisan CP, et al. Genetic mapping of a susceptibility locus for disc herniation and spastic paraplegia on 6q23.3-q24.1. *J Med Genet*. 2002;39:387–390.
111. Wilkinson PA, Simpson MA, Bastaki L, et al. A new locus for autosomal recessive complicated hereditary spastic paraplegia (SPG26) maps to chromosome 12p11.1-12q14. *J Med Genet*. 2005;42:80–82.
112. Ribai P, Stevanin G, Bouslam N, et al. A new phenotype linked to SPG27 and refinement of the critical region on chromosome. *J Neurol*. 2006;253:714–719.
113. Bouslam N, Benomar A, Azzedine H, et al. Mapping of a new form of pure autosomal recessive spastic paraplegia (SPG28). *Ann Neurol*. 2005;57:567–571.
114. Tesson C, Nawara M, Salih M, et al. Alteration of fatty-acid-metabolizing enzymes affects mitochondrial form and function in hereditary spastic paraplegia. *Am J Hum Genet*. 2012;91:1051–1064.
115. Klebe S, Azzedine H, Durr A, et al. Autosomal recessive spastic paraplegia (SPG30) with mild ataxia and sensory neuropathy maps to chromosome 2q37.3. *Brain*. 2006;129:1456–1462.
116. Dick KJ, Al-Mjeni R, Baskir W, et al. A novel locus for an autosomal recessive hereditary spastic paraplegia (SPG35) maps to 16q21-q23. *Neurology*. 2008;71:248–252.
117. Dick KJ, Eckhardt M, Paisan-Ruiz C, et al. Mutation of FA2H underlies a complicated form of hereditary spastic paraplegia (SPG35). *Hum Mutat*. 2010;31:E1251–E1260.
118. Kruer MC, Paisan-Ruiz C, Boddaert N, et al. Defective FA2H leads to a novel form of neurodegeneration with brain iron accumulation (NBIA). *Ann Neurol*. 2010;68:611–618.
119. Rainier S, Bui M, Mark E, et al. Neuropathy target esterase gene mutations cause motor neuron disease. *Am J Hum Genet*. 2008;82: 780–785.
120. Meilleur KG, Traore M, Sangare M, et al. Hereditary spastic paraplegia and amyotrophy associated with a novel locus on chromosome 19. *Neurogenetics*. 2010;11:313–318.
121. Orthmann-Murphy JL, Salsano E, Abrams CK, et al. Hereditary spastic paraplegia is a novel phenotype for GJA12/GJC2 mutations. *Brain*. 2009;132:426–438.
122. Dursun U, Koroglu C, Kocasoy OE, et al. Autosomal recessive spastic paraplegia (SPG45) with mental retardation maps to 10q24.3-q25.1. *Neurogenetics*. 2009;10:325–331.
123. Boukhris A, Feki I, Elleuch N, et al. A new locus (SPG46) maps to 9p21.2-q21.12 in a Tunisian family with a complicated autosomal recessive hereditary spastic paraplegia with mental impairment and thin corpus callosum. *Neurogenetics*. 2010;11:441–448.
124. Blumkin L, Lerman-Sagie T, Lev D, et al. A new locus (SPG47) maps to 1p13.2-1p12 in an Arabic family with complicated autosomal recessive hereditary spastic paraplegia and thin corpus callosum. *J Neurol Sci*. 2011;305:67–70.
125. Slabicki M, Theis M, Krastev DB, et al. A genome-scale DNA repair RNAi screen identifies SPG48 as a novel gene associated with hereditary spastic paraplegia. *PLoS Biol*. 2010;8:e1000408.

126. Oz-Levi D, Ben-Zeev B, Ruzzo E, et al. Mutation in TECPR2 reveals a role for autophagy in hereditary spastic paraparesis. *Am J Hum Genet*. 2012;91:1065–1072.
127. Verkerk AJ, Schot R, Dumee B, et al. Mutation in the AP4M1 gene provides a model for neuroaxonal injury in cerebral palsy. *Am J Hum Genet*. 2009;85:40–52.
128. Najmabadi H, Hu H, Garshasbi M, et al. Deep sequencing reveals 50 novel genes for recessive cognitive disorders. *Nature*. 2011;478:57–63.
129. Moreno-De-Luca A, Helmers SL, Mao H, et al. Adaptor protein complex-4 (AP-4) deficiency causes a novel autosomal recessive cerebral palsy syndrome with microcephaly and intellectual disability. *J Med Genet*. 2011;48:141–144.
130. Abou-Jamra R, Philippe O, Raas-Rothschild A, et al. Adaptor protein complex 4 deficiency causes severe autosomal-recessive intellectual disability, progressive spastic paraplegia, shy character, and short stature. *Am J Hum Genet*. 2011;88:788–795.
131. Abou Jamra R, Philippe O, Raas-Rothschild A, et al. Adaptor protein complex 4 deficiency causes severe autosomal-recessive intellectual disability, progressive spastic paraplegia, shy character, and short stature. *Am J Hum Genet*. 2011;88:788–795.
132. Dell'Angelica EC, Mullins C, Bonifacino JS. AP-4, a novel protein complex related to clathrin adaptors. *J Biol Chem*. 1999;274:7278–7285.
133. Hirst J, Bright NA, Rous B, et al. Characterization of a fourth adaptor-related protein complex. *Mol Biol Cell*. 1999;10:2787–2802.
134. Zivony-Elboum Y, Westbroek W, Kfir N, et al. A founder mutation in Vps37A causes autosomal recessive complex hereditary spastic paraparesis. *J Med Genet*. 2012;49:462–472.
135. Al-Yahyaee S, Al-Gazali LI, De Jonghe P, et al. A novel locus for hereditary spastic paraplegia with thin corpus callosum and epilepsy. *Neurology*. 2006;66:1230–1234.
136. Schuurs-Hoeijmakers J, Geraghty M, Kamsteeg EJ, et al. Mutations in DDHD2, encoding an intracellular phospholipase A1, cause a recessive form of complex hereditary spastic paraplegia. *Am J Hum Genet*. 2012;91:1073–1081.
137. Shimazaki H, Takiyama Y, Ishiura H, et al. A homozygous mutation of C12orf65 causes spastic paraplegia with optic atrophy and neuropathy (SPG55). *J Med Genet*. 2012;49:777–784.
138. Antonicka H, Oÿstergaard E, Sasarman F, et al. Mutations in C12orf65 in patients with encephalomyopathy and a mitochondrial translation defect. *Am J Hum Genet*. 2010;87:115–122.
139. Beetz C, Johnson A, Schuh AL, et al. Inhibition of TFG function causes hereditary axon degeneration by impairing endoplasmic reticulum structure. *Proc Natl Acad Sci*. 2013;110:5091–5096.
140. Dor T, Cinnamon Y, Raymond L, et al. KIF1C mutations in two families with hereditary spastic paraparesis and cerebellar dysfunction. *J Med Genet*. 2014;51:137–142.
141. Mitchell S, Bundey S. Symmetry of neurological signs in Pakistani patients with probable inherited spastic cerebral palsy. *Clin Genet*. 1997;51(1):7–14.
142. McHale DP, Mitchell S, Bundey S, et al. A gene for autosomal recessive symmetrical spastic cerebral palsy maps to chromosome 2q24-25. *Am J Hum Genet*. 1999;64:526–532.
143. Lynex C, Carr I, Leek J, et al. Homozygosity for a missense mutation in the 67 kDa isoform of glutamate decarboxylase in a family with autosomal recessive spastic cerebral palsy: parallels with Stiff-Person Syndrome and other movement disorders. *BMC Neurol*. 2004;4:20.
144. Macedo-Souza LI, Kok F, Santos S, et al. Spastic paraplegia, optic atrophy, and neuropathy is linked to chromosome 11q13. *Ann Neurol*. 2005;57:730–737.
145. Bouhouche A, Benomar A, Bouslam N, et al. Autosomal recessive mutilating sensory neuropathy with spastic paraplegia maps to chromosome 5p15.31-14.1. *Eur J Hum Genet*. 2006;14:249–252.
146. Bouhouche A, Benomar A, Bouslam N, et al. Mutation in the epsilon subunit of the cytosolic chaperonin-containing t-complex peptide-1 (Cct5) gene causes autosomal recessive mutilating sensory neuropathy with spastic paraplegia. *J Med Genet*. 2006;43:441–443.
147. Jouet M, Rosenthal A, Armstrong G, et al. X-linked spastic paraplegia (SPG1), MASA syndrome and X-linked hydrocephalus result from mutations in the L1 gene. *Nat Genet*. 1994;7:402–407.
148. Kobayashi H, Hoffman EP, Marks HG. The rumpshaker mutation in spastic paraplegia. *Nat Genet*. 1994;7:351–352.
149. Hudson LD. Pelizaeus-Merzbacher disease and spastic paraplegia type 2: two faces of myelin loss from mutations in the same gene. *J Child Neurol*. 2003;18:616–624.
150. Saugier-Veber P, Munnich A, Bonneau D, et al. X-linked spastic paraplegia and Pelizaeus-Merzbacher disease are allelic disorders at the proteolipid protein locus. *Nat Genet*. 1994;6:257–262.
151. Cambi F, Tang XM, Cordray P, et al. Refined genetic mapping and proteolipid protein mutation analysis in X-linked pure hereditary spastic paraplegia. *Neurology*. 1996;46:1112–1117.
152. Steinmuller R, Lantingua-Cruz A, Carcia-Garcia R, et al. Evidence of a third locus in X-linked recessive spastic paraplegia [letter]. *Hum Genet*. 1997;100:287–289.
153. Tamagaki A, Shima M, Tomita R, et al. Segregation of a pure form of spastic paraplegia and NOR insertion into Xq11.2. *Am J Med Genet*. 2000;94:5–8.
154. Marx J. Alzheimer's research moves to mice. *Science*. 1991;253:266–267.
155. Allan W, Herndon CN, Dudley FC. Some examples of the inheritance of mental deficiency: apparently sex-linked idiocy and microcephaly. *Am J Ment Defic*. 1944;48:325–334.
156. Bialer MG, Lawrence L, Stevenson RE, et al. Allan-Herndon-Dudley syndrome: clinical and linkage studies on a second family. *Am J Med Genet*. 1992;43:491–497.
157. Macedo-Souza LI, Kok F, Santos S, et al. Reevaluation of a large family defines a new locus for X-linked recessive pure spastic paraplegia (SPG34) on chromosome Xq25. *Neurogenetics*. 2008;9:225–226.
158. Verny C, Guegen N, Desquiret V, et al. Hereditary spastic paraplegia-like disorder due to a mitochondrial ATP6 gene point mutation. *Mitochondrion*. 2011;11:70–75.
159. Matsumura R, Takayanagi T, Fujimoto Y, et al. The relationship between trinucleotide repeat length and phenotypic variation in Machado-Joseph disease. *J Neurol Sci*. 1996;139:52–57.

SECTION VI

NEURO-ONCOLOGY

CHAPTER

78

Glioblastoma

Elizabeth A. Maher and Robert M. Bachoo

The University of Texas Southwestern Medical Center, Dallas, TX, USA

INTRODUCTION

Disease Characteristics

Glioblastoma (GBM) is a disease with a median survival of 15 months that strikes without warning, often manifested by a single seizure, new onset of a focal neurological deficit, or a few days of progressive severe headache. Initial evaluation with magnetic resonance imaging (MRI) usually reveals a heterogeneously gadolinium-enhancing mass on T1-weighted images with surrounding T2-weighted fluid-attenuated inversion recovery (FLAIR) signal.[1] It is glial in origin and classified by the World Heath Organization (WHO) as grade IV/IV glioma. Diagnosis is made surgically, with an open craniotomy and tumor sampling, or a stereotactic biopsy if surgical resection is deemed unsafe because of location of the tumor or the patient's clinical status. Since extent of resection may independently impact overall survival and recurrence rates,[2,3] an attempt is usually made at the time of surgery to remove as much tumor as can be taken without causing a neurological deficit. The potential for a gross total resection depends on the location of the tumor and often a subtotal resection is achieved because of the safety concerns.

Diagnosis/Testing

Histological analysis of the tumor reveals hypercellularity, paucity of neurons and glial cells, microvascular proliferation, and usually the hallmark characteristic of GBM, pseudopalisading necrosis in which necrotic foci are typically surrounded by "pseudopalisading" cells.[1] Testing for methylation status of the DNA repair enzyme, O^6-methylguanine methyltransferase (MGMT) is done routinely as a prognostic marker in academic centers[4] because methylation has been shown to be associated with better overall survival.[5] In addition, immunohistochemical analysis for mutated isocitrate dehydrogenase 1 (IDH1) is performed and, if negative, sequencing of commonly mutated regions of IDH1 and 2 are performed in some centers, since the presence of the mutation identifies a good prognostic subgroup of patients.[6]

Management

The current standard therapy for GBM, regardless of clinical or molecular subgroup, is external beam radiation combined with daily oral temozolomide for 6 weeks, followed by 12 months of temozolomide on a 5 out of 28 day schedule.[7] Time to tumor recurrence is highly variable and patients are generally followed with serial MRI scans every 2–3 months until recurrence. Treatment for recurrent GBM generally includes use of the vascular endothelial growth factor (VEGF) inhibitor, bevacizumab (Genentech), alone or in combination with a cytotoxic agent or molecular targeted therapy.[8] Duration of response is generally short and thus treatment on a clinical trial of a new agent or combination is encouraged.

Rosenberg's Molecular and Genetic Basis of Neurological and Psychiatric Disease
http://dx.doi.org/10.1016/10.1016/B978-0-12-410529-4.00078-4

Current Research

Extensive genomic, transcriptional, and epigenomic data has identified the landscape of molecular characteristics of GBM; yet, to date, treatments based on targeting specific genes and pathways have not lead to improvements in overall survival. Research is focused on expanding the range of therapeutic targets, including vaccine therapy[9] and metabolic reprogramming.[10]

CLINICAL FEATURES

Historical Overview

The etiology of GBM remains elusive. Risk related to chemical exposure, in particular sarin nerve gas in Gulf War veterans has been documented,[11] although in the vast majority of cases no etiologic factors can be determined. Signs and symptoms associated with brain tumors occur as a result of either increased intracranial pressure and/or focal cerebral dysfunction. The early signs of increased pressure include nausea, vomiting, headache, and change in mental status, and are sometimes accompanied by clinical findings of papilledema and loss of retinal venous pulsations. The late symptoms of uncal or cerebellar-foramen magnum herniation, associated with clinical findings of hemiplegia, hemianopia, pupillary dilatation and posturing constitute a surgical emergency for tumor debulking and relief of increased pressure. For cerebral hemispheric tumors, the symptoms reflect the neurological functions of the area affected by tumor. For example, temporal lobe tumors in the dominant hemisphere are heralded by progressive development of contralateral hemiparesis and expressive aphasia, while tumors in the parietal lobe frequently manifest contralateral neglect, sensory inattention, dysesthesia, dyslexia, and dysgraphia. Frequently, temporal lobe tumors that have no overt focal neurological symptoms produce clinically subtle brief focal or partial seizures that may include olfactory or gustatory hallucinations, transient sensations of déjà vu or fear that prior to diagnosis may be misdiagnosed as panic attacks.

A major impact of the development and widespread use of MR imaging has been the earlier detection of GBM, such that late presentation with uncal herniation or complete hemiplegia from progressive symptoms has become much less common. Thus, earlier detection preserves neurological function and quality of life, making aggressive treatment more feasible although it does not alter survival, since the major barrier to effective therapy is diffuse infiltration of highly resistant tumor cells. Historically, the treatment for GBM was debulking surgery followed by a 6-week course of external beam radiation. In 2005, the addition of the alkylating agent, temozolomide, to radiation therapy increased median survival from 9 months to 14.6 months.[12] With the results of two large phase III studies published in 2014,[13,14] no further increase in overall survival has been achieved with any single agent or combination therapy.

Mode of Inheritance and Prevalence

Neurofibromatosis and Li–Fraumeni syndromes predispose to gliomas,[15] although most of the tumors are WHO grade II, not GBMs. More than 98% of all GBMs are sporadic. Rare families with multiple GBMs have been described, although with no clearly identified common genetic mutations.[16]

The most comprehensive reporting of the prevalence of GBM is from the Central Brain Tumor Registry in the United States (CBTRUS), which combines data from the Center for Disease Control's (CDC) National Program of Cancer Registry (NPCR) and the National Cancer Institute's (NCI) Surveillance Epidemiology and End Results (SEER) Program. Most recently reported are data from 2006–2010, during which 50,872 GBMs were identified, accounting for 15.6% of all brain tumors and 45.2% of all malignant brain tumors.[17] The age-adjusted incidence of GBM during that time period was 3.19 cases per 100,000 people, with a 1.57 : 1 increased incidence in males versus females and 2.07 : 1 white : black.

Natural History

Age of Onset

The median age at diagnosis for GBM is 64 years with the age-adjusted incidence increasing from 0.14 per 100,000 in pediatric patients (under 20 years), 0.4 in 20–44-year-olds, to 1.2 in 45–54-year-old patients.[17] This is followed by a large increase to 8.08 per 100,000 in 55–64-year-old patients, then 13.09 in 65–74-year-olds, and then remains relatively stable in the 75–84-year age group (14.93 per 100,000).[17] GBMs are far less frequent in the pediatric population. Among pediatric patients, gliomas represent 53% of all brain tumors in patients under 15 and 36% of brain tumors in 15–19-year-olds. Of these, GBMs occur far less commonly than in adults, accounting for only 2.6% of all gliomas in patients under age 15 and 2.9% in 15–19-year-olds.[17] A recent study comparing the molecular features of pediatric versus adult GBMs supports the notion that these tumors are derived along distinct genetic/epigenetic pathways leading to a common histopathologic endpoint.[18]

Disease Evolution and End-of-Life Mechanisms

GBM recurs after treatment with a variable time to progression ranging from a few months to a few years in a small number of patients (median ~6 months). Approximately 80% recur within 2 cm of the tumor resection margin and 20% recur at a site remote from the initial tumor location.[19] Except in very rare cases, GBMs do not metastasize outside the central nervous system (CNS). Aggressive treatment[20] of recurrent disease can result in stable disease for a period of time but inevitably the tumor grows leading to progressive neurological impairment with symptoms related to the tumor location, as well as progressive cognitive impairment, leading to coma and death. The comorbidities seen with progressive disease include a high incidence of deep vein thrombosis and pulmonary emboli, steroid-induced myopathies and diabetes, and seizures.

Disease Variants

GBMs develop along two distinct clinical and pathological pathways. Ninety percent are primary or *de novo* GBMs, which develop with a few month prodrome and no evidence of an antecedent precursor tumor. Ten percent of GBMs develop from a WHO grade II (low grade) astrocytoma or oligoastrocytoma that is generally very slow growing for several years before transforming into a "secondary" GBM (Figure 78.1). Low-grade gliomas occur most commonly in young patients, with a median age of 30, and have a greater than 80% risk of transformation to GBM.[21] Secondary GBMs are histologically and clinically indistinguishable from primary GBM although have distinct genetic profiles.[6,22] Patients with low-grade gliomas that do not progress to GBM die of widely infiltrating tumor.

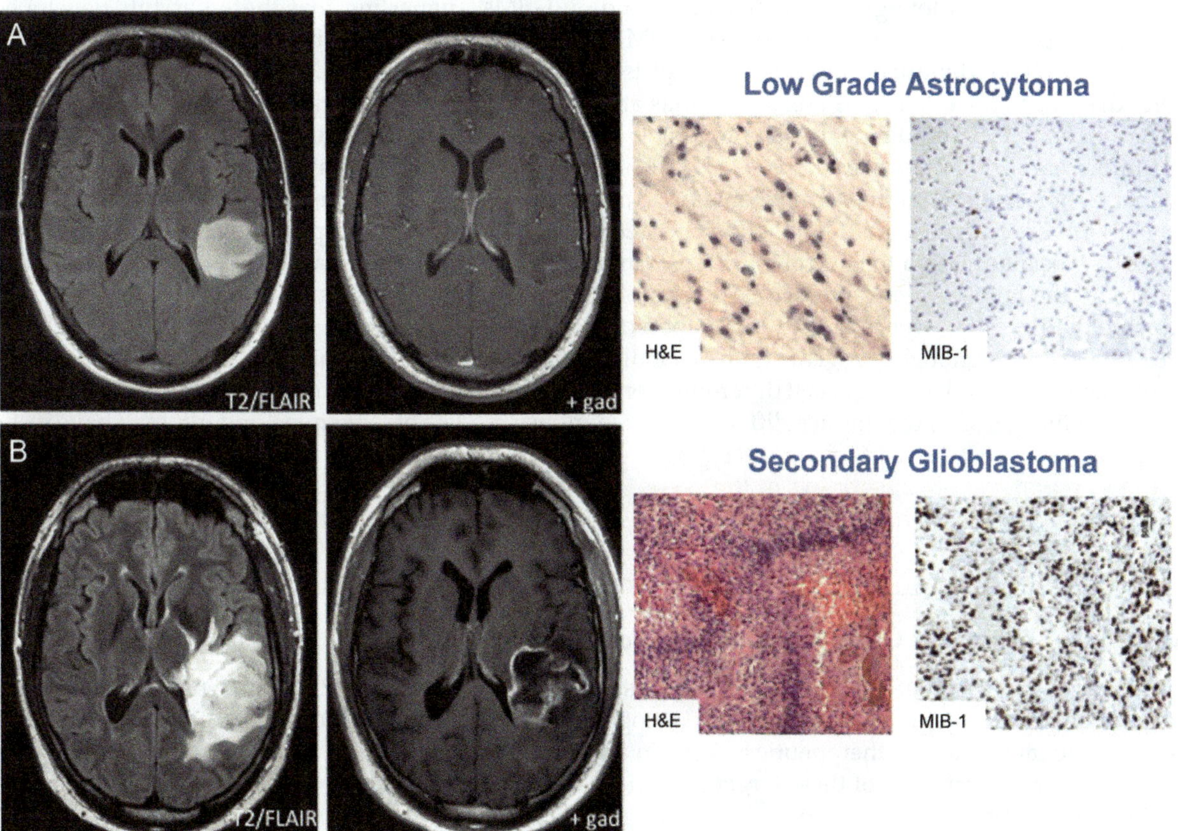

FIGURE 78.1 **Progression from low-grade astrocytoma (A) to glioblastoma (B) in the same patient over a 3-year period.** Images left to right for upper and lower panel series: MRI (T2/FLAIR left panel and with gadolinium, right panel), low power, hematoxylin and eosin, MIB-1. The MIB-1 for low-grade astrocytoma was 2% and increased to 64% in glioblastoma.

MOLECULAR GENETICS

Insights from Comprehensive Molecular Analysis of GBM

Recent large-scale genomic and proteomic studies have provided new insights into the molecular complexity of GBM and identified recurrent gain-of-function mutations, which may serve as driver mutations.[23,24] The Cancer Genome Atlas project performed molecular analysis of 206 GBMs and identified common mutations in the major cancer-associated genetic pathways that impinge on critical activities of cell growth and survival. P53 signaling was altered in 87% of tumors (due to amplification of MDM2 or MDM4, homozygous deletion of CDKN2A, or mutation/deletion in TP53), leading to cellular proliferation, survival and translation. RB signaling was altered in 78% (due to deletions in CDKN2A, 2B, and 2C, amplification of CDK23, CCND2, and CDK6, or deletion of RB1) leading to G1/S progression. The receptor tyrosine kinase (RTK), RAS, PI3Kinase pathway was altered in 88% of tumors (due most commonly to amplification of EGFR and/or loss of PTEN but other members of the pathway were also altered including amplification of PDGFRα, MET, AKT, mutation in ERBB2, or PI3K.[23] Analysis of an additional 250 GBMs has further delineated these pathways and identified novel mutated genes in these and other signaling pathways.[24] The complexity in GBM is further amplified by intratumoral heterogeneity by which different cells harbor different combinations of mutations. For example, it has been shown that individual GBMs can harbor two more activating mutations or amplification of RTKs in the same cell or even in different populations of cells within the same tumor.[25] This has been raised as a possible explanation for a number of ineffective clinical trials that target a single specific RTK, suggesting that either the inhibitor can kill only the cells harboring the one specific RTK or that these are not the oncogenic drivers in the tumors.

Identification of IDH Mutations in the Secondary GBM Pathway

For many years, low-grade gliomas have been characterized by loss of p53 in approximately 70% of cases, with amplification/overexpression of PDGFRα in fewer than 20% of cases.[26] No other RTKs have been reported to be amplified or mutated in low-grade gliomas or secondary GBMs, suggesting that these variants developed along distinct genetic pathways.[22,27] Transformation to GBM is commonly associated with loss of PTEN. Recently, however, mutations have been identified in two isoforms of isocitrate dehydrogenase (IDH1 in the cytosol and IDH2 in the mitochondria) in >70% of grade II and III gliomas and secondary GBMs.[6,28,29] However, fewer than 5% of primary GBMs harbor an IDH1/2 mutation. In gliomas, approximately 90% of the IDH mutations are found in IDH1 (R132H). Other tumors harboring these mutations in IDH1/2 include acute myelogenous leukemia,[30] chondrosarcoma,[31] and intrahepatic cholangiocarcinoma.[32] For gliomas, the presence of an IDH mutation has been shown to be associated with a better overall survival than IDH wild-type (IDH-WT) tumors when compared grade for grade (Figure 78.2), and also is associated with a longer time to transformation in low-grade gliomas when compared with tumors expressing only the wild-type enzymes.[6,29,33] Acquired somatically, IDH1 and 2 mutations are activating mutations in the enzyme binding site, which leads to production of high levels of 2 hydroxyglutarate (2HG) from α-ketoglutaric acid (αKG).[34] The functional oncogenic and metabolic consequences of 2HG production include alteration of the epigenetic landscape and a block in cell differentiation.[35] The field has progressed rapidly. Drugs targeting IDH1 and 2 (Agios, Cambridge, MA) that induce differentiation in mouse models of leukemia[36] and glioma[37] have just entered clinical trial.

DISEASE MECHANISMS

GBM Cell-of-Origin and Genetically Engineered Mouse Models

Identification of the cell-of-origin of GBM remains an unanswered question but is essential for the ultimate elucidation of the pathway(s) that are affected by oncogenic mutations early in tumor development. Such mutations may also be the most relevant therapeutic targets. In addition, defining the biochemical and biological changes that accompany the transformation of these target cells may provide clues to improvements in the treatment of GBM, to eradicate or halt tumor progression.[38]

GBM may arise from malignant transformation of terminally differentiated astrocytes, glial restricted progenitors, or multipotent neural stem cells. A number of reports have supported the role of a restricted population of neural stem cells found within the subventricular region as the cell-of-origin for gliomas.[39–44] However, the studies on which

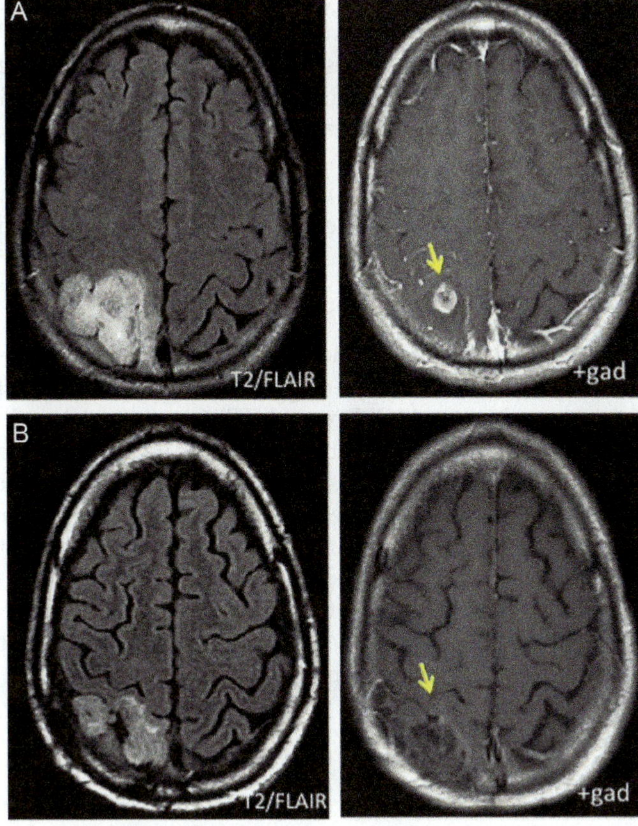

FIGURE 78.2 **IDH1 mutated, methylated MGMT glioblastoma.** (**A**) Preoperative magnetic resonance scans show location of tumor in the region of T2/FLAIR (left) and corresponding region after gadolinium administration. Yellow arrow denotes enhancing tumor. (**B**) Images from the corresponding regions 7 years after initial resection and treatment with concurrent radiation and temozolomide followed by 12 cycles of adjuvant temozolomide. Note posttreatment FLAIR signal and minimal enhancement at the resection margin, which has not changed in the 7 years of follow-up.

these reports are based are limited by reliance on a constitutively active promoter (hGFAP, 2.2 kb) that drives Cre expression during brain development resulting in activation of mutations in multiple lineages.[45,46] Although this strategy is efficient in generating high-grade glioma models, it is a poor representation of human adult GBMs, which have a median age of onset in the sixth decade. Thus, while these studies have identified progenitor cells as potential sources for glioma formation, they neither address nor preclude alternative mechanisms that could lead to glioma formation from a non-neurogenic niche precursor or more differentiated cell types. To circumvent the limitations of constitutive Cre-driver mice, next-generation inducible forms of Cre-Lox-P system have been developed. These rely on Cre recombinase fused to a mutated estrogen receptor, which is entrapped in the cytoplasm by binding to HSP90. Treating the mouse with tamoxifen at low doses allows the Cre-ERT^2 fusion to transiently enter the nucleus and excise lox-P-flanked DNA regions. Such genetic models will be essential for addressing the GBM cell-of-origin question. Using neural stem cell (nestin) and mature astrocyte-specific (PLA2G7) Cre-ERT^2 driver mice crossed to conditional oncogene and tumor suppressor mice has shown that both cell types are equally permissive for generating high-grade gliomas (Bachoo, unpublished observations). This latest generation of genetically engineered mouse models, together with use of Cre-inducible lentiviral vectors, have shown that both stem/progenitor cells as well as terminally differentiated cells in the adult brain are susceptible to transformation and can give rise to gliomas.[47,48]

Cancer Stem Cells

There is growing evidence that therapeutic failure in cancer is due to the relative refractoriness of rare populations of cells to all modes of treatment. Functionally identified as cancer stem cells or tumor-initiating cells, the identification of this cell type has become a major focus in cancer, and specifically for GBM.[49] The implication is that to achieve durable therapeutic responses, successful treatment regimens must target and successfully eliminate the cancer stem cell population within a growing tumor. To date, a major obstacle to eliminating GBM stem cells is the lack of methods for detecting and quantifying them for further studies.

Stem cells are uncommon cells defined by two exceptional features: the capacity for multilineage differentiation and self-renewal. These properties allow stem cells to generate an unlimited supply of progeny and are best exemplified by hematopoietic stem cells (HSCs) and embryonic stem (ES) cells. Observations from embryonic development and cancer biology have given rise to two popular concepts that link these fields: 1) tumors arise from stem cells; and 2) a small population within a tumor—so-called "cancer stem cells"—constitutes an inexhaustible source for tumor cells in the mass lesion. The identification of the cell type(s) from which GBM originates and through which its growth is sustained will be critical for a comprehensive understanding of tumor biology. Specifically, tumors arise and thrive within particular *permissive* cellular contexts that enable them to develop. Characterizing the key elements that define these contexts—namely cellular differentiation states, active signaling pathways, and/or developmental stages—will facilitate the understanding of the mechanism by which transforming mutations cause GBM and allow us to conceive of novel therapies and early detection strategies.

Patient-Derived Tumor Models

Although the latest generation of genetically engineered mouse models are very useful for testing specific hypotheses related to the cell-of-origin (astrocyte vs. neural stem cell) and assessing the relative importance of particular driver oncogenes (EGFR vs. c-MET, for example) and tumor suppressor genes (PTEN vs. p53, for example), these models lack the complexity and the natural history of clonal selection associated with clinical tumors. Furthermore, they are unable to address the existence of a cancer stem cell pool. The recent development of GBM patient-derived xenograft models in mice provide a better approximation of human tumors.[50,51] Despite the drawback due to reliance on an immune-compromised host environment, these models have been shown to preserve the genomic, transcriptomic, and proteomic profiles of the patient's original tumor.[51] In addition, the xenografts preserve critical physiological pathways of the human tumor, such as the tumor's metabolic phenotype.[50] By maintaining GBM cells in a murine host microenvironment, without ever adapting the cells to culture conditions, these models represent the wide variety of genetic, epigenetic, and transcriptional heterogeneity of the disease and, as such, can be very useful in basic investigation and preclinical evaluation of new therapeutics.

TESTING

Molecular Diagnosis

Molecular diagnosis currently focuses on the key genetic changes that have prognostic importance. IDH1/2 sequencing and assessment of methylation status of MGMT are becoming more routine in clinical practice.[4,52] Assessment of EGFR amplification and the presence of the EGFR VIII mutant are done when the specific EGFR-targeted therapies are being considered. This is not standard in clinical practice because the anti-EGFR therapies have not proven to be effective.

Noninvasive Imaging of 2-Hydroxyglutarate as a Clinical Biomarker

The demonstration of high levels of 2HG in glioma cells with mutations in IDH1 or 2 has provided an unexpected and highly valuable biomarker for noninvasive brain tumor imaging. Methods using MR spectroscopy (MRS) at 3Tesla (T) have been developed to detect and quantitate 2HG.[53,54] Since the presence of an IDH mutation is associated with a better prognosis in GBM, imaging with 2HG can be diagnostic and prognostic, as well as potentially important as a dynamic biomarker in the ongoing follow-up of patients with IDH-mutated gliomas. We have demonstrated that the concentration of 2HG is reproducible between MR scans in the same patient, is stable when the tumor is stable, increases when the tumor is growing, and decreases when the tumor is responding to treatment (Maher, unpublished observations). In addition to its clinical value, 2HG represents the first direct link between a genetic mutation and noninvasive imaging biomarker, which is a significant breakthrough in cancer diagnostics and imaging.

MANAGEMENT

Approach to management of GBMs relies first on maximal safe surgical resection, although due to the infiltrative nature of gliomas, none are curable by surgery alone. Postoperative radiation therapy has been shown to significantly improve survival for high-grade astrocytomas.[55,56] It has been established that external beam radiation to the tumor and adjacent brain is as effective as whole brain radiation and is less morbid.[57] Based on these data, the standard of care for GBM until 2005 (see below) was postoperative radiation (54–60 Gy over 6 weeks) to the tumor and a 2-cm margin.

Addition of chemotherapy, either in combination with or following radiation, did not alter overall survival until the development of temozolomide (TMZ, Merck), a second-generation alkylating agent with excellent bioavailability and blood–brain penetration. TMZ was approved for the treatment of recurrent high-grade astrocytomas in 1999 and further development of this drug included investigation in GBM. A landmark phase III study conducted by the European Organization for Research and Treatment of Cancer (EORTC) was published in 2005 that showed that addition of TMZ to radiation in newly diagnosed GBM patients improved median overall survival by 2.5 months.[7] In addition, the 1-year survival increased from 9% to 27%, and 2-year survival from 2% to 11% compared with radiotherapy alone.[7] Based on this study, the US Food and Drug Administration (FDA) approved TMZ for the treatment of newly diagnosed GBM in 2005 and the new standard of care regimen was established as postoperative treatment with concurrent radiation plus daily TMZ (75 mg per m^2 per day) for 6 weeks followed by adjuvant TMZ (150–200 mg per m^2 per day $\times$ 5 out 28 days) for 6–12 months. In a subset of patients from this study tumor was available for correlative studies. Methylation of the DNA repair gene O^6-methylguanine-DNA methyltransferase (MGMT) was found to be correlated with better overall survival[5] (Figure 78.2), thought to be a result of the tumor cell's inability to repair DNA after treatment with the alkylating agent, TMZ, leading to apoptosis. Based on these data, it was hypothesized that depleting MGMT in tumors that were not methylated, and therefore had active MGMT, would achieve the same goal of "inactivating" this mechanism. Several regimens of TMZ treatment were investigated and the regimen of 21 days treatment followed by 7 days off was shown to effectively deplete intracellular levels. A phase III randomized study for newly diagnosed GBM was conducted by the Radiation Therapy Oncology Group (RTOG) in which the "dose intensive regimen" (21 days on/7 days off) was compared with standard dose (5 days on/23 days off) in the adjuvant phase after concurrent radiation and TMZ was completed. It also prospectively evaluated the status of MGMT in predicting response and/or survival benefit. A total of 833 were randomized between the two arms. No statistically significant difference was observed for median overall survival (16.6 vs. 14.9 months, respectively) or progression free survival (5.5 vs. 6.7 months).[58] MGMT methylation was associated with a marked increase in overall survival (21.2 vs. 14 months) as well as response to treatment[58] and therefore has been identified as an important prognostic factor in GBM.

In 2009, the FDA approved bevacizumab (Genentech), a vascular endothelial growth factor (VEGF) inhibitor for the treatment of recurrent GBM, based on two phase II clinical trials that showed responses in ~20% of patients, with effects lasting, on average, approximately 4 months.[59] There has been considerable controversy as to whether bevacizumab is effective in combination with other agents due to variability in results of small phase II clinical trials. A major interest was in determining whether the addition of bevacizumab to radiation/TMZ and adjuvant TMZ would improve overall survival. Two large phase III studies have been completed and were published in 2014.[13,14] Each study randomized ~1000 patients to either the standard regimen with placebo or the standard regimen with bevacizumab given intravenously every 2 weeks throughout the full course of radiation/TMZ and adjuvant TMZ. Neither study was able to show an improvement in median survival with the addition of bevacizumab.[13,14] Moreover, there is some indication that the treatment was associated with increased neurocognitive toxicity.[13] While bevacizumab is an active drug in GBM but not effective in the upfront treatment setting, more work needs to be done to determine how best to use it.[60]

The Future of GBM Therapeutics

The ability to do large, multicenter, international studies in GBM with correlative tissue collection and biomarker analysis has been established. This is undoubtedly a major advantage for the development of future therapeutics. However, the lack of efficacy of a wide range of treatments including the molecularly targeted therapies, alone or in combination with bevacizumab and TMZ, has raised the question as to where the next improvement in median survival will come from. There is a large effort directed at developing vaccines and harnessing the immune system for therapeutic benefit in GBM.[61] To date, vaccines have been shown to be safe with no demonstrable autoimmune activity in the brain. However, efficacy is currently lacking and future studies will be directed at optimizing antigen choices and minimizing immunosuppression due to the tumor itself and treatment with dexamethasone.

It is becoming clear that improved treatments are highly dependent on developing better mouse models of GBM for preclinical testing, as discussed above. Currently, almost no preclinical testing is done and drugs are taken straight from development to clinical trials. This is a very expensive, time consuming, and overwhelmingly disappointing and inefficient process. The patient-derived xenograft models that recapitulate the key features of the human disease are critical for future drug development. Using the models to assess treatment response, develop biomarkers, and understand resistance will mark a new era in GBM therapeutics and are the best chance for finally improving overall survival and heading for cure in this devastating disease.

References

1. Maher EA, MCKee AC. Neoplasms of the central nervous system. In: Skarin AT, ed. 3rd ed. *Diagnostic Atlas of Oncology*. vol. 1. Edinburgh: Elsevier Science Limited; 2002:395–434.
2. Chaichana KL, Cabrera-Aldana EE, Jusue-Torres I, et al. When gross total resection of a glioblastoma is possible, how much resection should be achieved? *World Neurosurg*. Feb 6 2014; pii:S1878-8750(14)00083-7.
3. Chaichana KL, Jusue-Torres I, Navarro-Ramirez R, et al. Establishing percent resection and residual volume thresholds affecting survival and recurrence for patients with newly diagnosed intracranial glioblastoma. *Neuro Oncol*. 2014;16(1):113–122.
4. Cankovic M, Nikiforova MN, Snuderl M, et al. The role of MGMT testing in clinical practice: a report of the association for molecular pathology. *J Mol Diagn*. 2013;15(5):539–555.
5. Hegi ME, Diserens AC, Gorlia T, et al. MGMT gene silencing and benefit from temozolomide in glioblastoma. *N Engl J Med*. 2005;352(10):997–1003.
6. Parsons DW, Jones S, Zhang X, et al. An integrated genomic analysis of human glioblastoma multiforme. *Science*. 2008;321(5897):1807–1812.
7. Stupp R, Mason WP, van den Bent MJ, et al. Radiotherapy plus concomitant and adjuvant temozolomide for glioblastoma. *N Engl J Med*. 2005;352(10):987–996.
8. Norden AD, Young GS, Setayesh K, et al. Bevacizumab for recurrent malignant gliomas: efficacy, toxicity, and patterns of recurrence. *Neurology*. 2008;70(10):779–787.
9. Virasch N, Kruse CA. Strategies using the immune system for therapy of brain tumors. *Hematol Oncol Clin North Am*. 2001;15(6):1053–1071.
10. DeBerardinis RJ, Lum JJ, Hatzivassiliou G, Thompson CB. The biology of cancer: metabolic reprogramming fuels cell growth and proliferation. *Cell Metab*. 2008;7(1):11–20.
11. Barth SK, Kang HK, Bullman TA, Wallin MT. Neurological mortality among U.S. veterans of the Persian Gulf War: 13-year follow-up. *Am J Ind Med*. 2009;52(9):663–670.
12. Stupp R, Dietrich PY, Ostermann Kraljevic S, et al. Promising survival for patients with newly diagnosed glioblastoma multiforme treated with concomitant radiation plus temozolomide followed by adjuvant temozolomide. *J Clin Oncol*. 2002;20(5):1375–1382.
13. Gilbert MR, Dignam JJ, Armstrong TS, et al. A randomized trial of bevacizumab for newly diagnosed glioblastoma. *N Engl J Med*. 2014;370(8):699–708.
14. Chinot OL, Wick W, Mason W, et al. Bevacizumab plus radiotherapy-temozolomide for newly diagnosed glioblastoma. *N Engl J Med*. 2014;370(8):709–722.
15. Piscatelli N, Batchelor T. *Epidemiology of Brain Tumors*. Wellesley: UpToDate; 2000.
16. Sadetzki S, Bruchim R, Oberman B, et al. Description of selected characteristics of familial glioma patients - results from the Gliogene Consortium. *Eur J Cancer*. 2013;49(6):1335–1345.
17. Dolecek TA, Propp JM, Stroup NE, Kruchko C. CBTRUS statistical report: primary brain and central nervous system tumors diagnosed in the United States in 2005–2009. *Neuro Oncol*. 2012;14(suppl 5):v1–v49.
18. Sturm D, Bender S, Jones DT, et al. Paediatric and adult glioblastoma: multiform (epi)genomic culprits emerge. *Nat Rev Cancer*. 2014;14(2):92–107.
19. Sherriff J, Tamangani J, Senthil L, et al. Patterns of relapse in glioblastoma multiforme following concomitant chemoradiotherapy with temozolomide. *Br J Radiol*. 2013;86(1022):20120414.
20. Szerlip NJ, Pedraza A, Chakravarty D, et al. Intratumoral heterogeneity of receptor tyrosine kinases EGFR and PDGFRA amplification in glioblastoma defines subpopulations with distinct growth factor response. *Proc Natl Acad Sci U S A*. 2012;109(8):3041–3046.
21. Louis DN, Ohgaki H, Wiestler OD, et al. The 2007 WHO classification of tumours of the central nervous system. *Acta Neuropathol*. 2007;114(2):97–109.
22. Maher EA, Brennan C, Wen PY, et al. Marked genomic differences characterize primary and secondary glioblastoma subtypes and identify two distinct molecular and clinical secondary glioblastoma entities. *Cancer Res*. 2006;66(23):11502–11513.
23. Cancer Genome Atlas Research Network. Comprehensive genomic characterization defines human glioblastoma genes and core pathways. *Nature*. 2008;455(7216):1061–1068.
24. Brennan CW, Verhaak RG, McKenna A, et al. The somatic genomic landscape of glioblastoma. *Cell*. 2013;155(2):462–477.
25. Stommel JM, Kimmelman AC, Ying H, et al. Coactivation of receptor tyrosine kinases affects the response of tumor cells to targeted therapies. *Science*. 2007;318(5848):287–290.
26. Kleihues P, Ohgaki H. Primary and secondary glioblastomas: from concept to clinical diagnosis. *Neuro Oncol*. 1999;1(1):44–51.
27. Tso CL, Freije WA, Day A, et al. Distinct transcription profiles of primary and secondary glioblastoma subgroups. *Cancer Res*. 2006;66(1):159–167.
28. Hartmann C, Meyer J, Balss J, et al. Type and frequency of IDH1 and IDH2 mutations are related to astrocytic and oligodendroglial differentiation and age: a study of 1,010 diffuse gliomas. *Acta Neuropathol*. 2009;118(4):469–474.
29. Yan H, Parsons DW, Jin G, et al. IDH1 and IDH2 mutations in gliomas. *N Engl J Med*. 2009;360(8):765–773.
30. McKenney AS, Levine RL. Isocitrate dehydrogenase mutations in leukemia. *J Clin Invest*. 2013;123(9):3672–3677.
31. Amary MF, Bacsi K, Maggiani F, et al. IDH1 and IDH2 mutations are frequent events in central chondrosarcoma and central and periosteal chondromas but not in other mesenchymal tumours. *J Pathol*. 2011;224(3):334–343.
32. Grassian AR, Pagliarini R, Chiang DY. Mutations of isocitrate dehydrogenase 1 and 2 in intrahepatic cholangiocarcinoma. *Curr Opin Gastroenterol*. 2014;30(3):295–302.
33. Von Deimling A, Korshunov A, Hartmann C. The next generation of glioma biomarkers: MGMT methylation, BRAF fusions and IDH1 mutations. *Brain Pathol*. 2011;21:74–87.
34. Dang L, White DW, Gross S, et al. Cancer-associated IDH1 mutations produce 2-hydroxyglutarate. *Nature*. 2009;462(7274):739–744.

35. Losman JA, Kaelin Jr. WG. What a difference a hydroxyl makes: mutant IDH, (R)-2-hydroxyglutarate, and cancer. *Genes Dev.* 2013;27(8):836–852.
36. Wang F, Travins J, DeLaBarre B, et al. Targeted inhibition of mutant IDH2 in leukemia cells induces cellular differentiation. *Science.* 2013;340(6132):622–626.
37. Rohle D, Popovici-Muller J, Palaskas N, et al. An inhibitor of mutant IDH1 delays growth and promotes differentiation of glioma cells. *Science.* 2013;340(6132):626–630.
38. Furnari FB, Fenton T, Bachoo RM, et al. Malignant astrocytic glioma: genetics, biology, and paths to treatment. *Genes Dev.* 2007;21(21):2683–2710.
39. Singh SK, Hawkins C, Clarke ID, et al. Identification of human brain tumour initiating cells. *Nature.* 2004;432(7015):396–401.
40. Zhu Y, Guignard F, Zhao D, et al. Early inactivation of p53 tumor suppressor gene cooperating with NF1 loss induces malignant astrocytoma. *Cancer Cell.* 2005;8(2):119–130.
41. Jackson EL, Garcia-Verdugo JM, Gil-Perotin S, et al. PDGFR alpha-positive B cells are neural stem cells in the adult SVZ that form glioma-like growths in response to increased PDGF signaling. *Neuron.* 2006;51(2):187–199.
42. Lim DA, Cha S, Mayo MC, et al. Relationship of glioblastoma multiforme to neural stem cell regions predicts invasive and multifocal tumor phenotype. *Neuro Oncol.* 2007;9(4):424–429.
43. Quinones-Hinojosa A, Chaichana K. The human subventricular zone: a source of new cells and a potential source of brain tumors. *Exp Neurol.* 2007;205(2):313–324.
44. Kwon CH, Zhao D, Chen J, et al. Pten haploinsufficiency accelerates formation of high-grade astrocytomas. *Cancer Res.* 2008;68(9):3286–3294.
45. Groszer M, Erickson R, Scripture-Adams DD, et al. Negative regulation of neural stem/progenitor cell proliferation by the Pten tumor suppressor gene in vivo. *Science.* 2001;294(5549):2186–2189.
46. Zhuo L, Theis M, Alvarez-Maya I, Brenner M, Willecke K, Messing A. hGFAP-cre transgenic mice for manipulation of glial and neuronal function in vivo. *Genesis.* 2001;31(2):85–94.
47. Friedmann-Morvinski D, Bushong EA, Ke E, et al. Dedifferentiation of neurons and astrocytes by oncogenes can induce gliomas in mice. *Science.* 2012;338(6110):1080–1084.
48. Marumoto T, Tashiro A, Friedmann-Morvinski D, et al. Development of a novel mouse glioma model using lentiviral vectors. *Nat Med.* 2009;15(1):110–116.
49. Yan K, Yang K, Rich JN. The evolving landscape of glioblastoma stem cells. *Curr Opin Neurol.* 2013;26(6):701–707.
50. Marin-Valencia I, Yang C, Mashimo T, et al. Analysis of tumor metabolism reveals mitochondrial glucose oxidation in genetically diverse human glioblastomas in the mouse brain in vivo. *Cell Metab.* 2012;15(6):827–837.
51. Joo KM, Kim J, Jin J, et al. Patient-specific orthotopic glioblastoma xenograft models recapitulate the histopathology and biology of human glioblastomas in situ. *Cell Rep.* 2013;3(1):260–273.
52. Cankovic M, Mikkelsen T, Rosenblum ML, Zarbo RJ. A simplified laboratory validated assay for MGMT promoter hypermethylation analysis of glioma specimens from formalin-fixed paraffin-embedded tissue. *Lab Invest.* 2007;87(4):392–397.
53. Choi C, Ganji SK, DeBerardinis RJ, et al. 2-hydroxyglutarate detection by magnetic resonance spectroscopy in IDH-mutated patients with gliomas. *Nat Med.* 2012;18(4):624–629.
54. Andronesi OC, Rapalino O, Gerstner E, et al. Detection of oncogenic IDH1 mutations using magnetic resonance spectroscopy of 2-hydroxyglutarate. *J Clin Invest.* 2013;123(9):3659–3663.
55. Walker MD, Alexander Jr. E, Hunt WE, et al. Evaluation of BCNU and/or radiotherapy in the treatment of anaplastic gliomas. A cooperative clinical trial. *J Neurosurg.* 1978;49(3):333–343.
56. Walker MD, Green SB, Byar DP, et al. Randomized comparisons of radiotherapy and nitrosoureas for the treatment of malignant glioma after surgery. *N Engl J Med.* 1980;303(23):1323–1329.
57. Shapiro WR, Green SB, Burger PC, et al. Randomized trial of three chemotherapy regimens and two radiotherapy regimens in postoperative treatment of malignant glioma. Brain Tumor Cooperative Group Trial 8001. *J Neurosurg.* 1989;71(1):1–9.
58. Gilbert MR, Wang M, Aldape KD, et al. Dose-dense temozolomide for newly diagnosed glioblastoma: a randomized phase III clinical trial. *J Clin Oncol.* 2013;31(32):4085–4091.
59. Friedman HS, Prados MD, Wen PY, et al. Bevacizumab alone and in combination with irinotecan in recurrent glioblastoma. *J Clin Oncol.* 2009;27(28):4733–4740.
60. Fine HA. Bevacizumab in glioblastoma–still much to learn. *N Engl J Med.* 2014;370(8):764–765.
61. Reardon DA, Wucherpfennig KW, Freeman G, et al. An update on vaccine therapy and other immunotherapeutic approaches for glioblastoma. *Expet Rev Vaccine.* 2013;12(6):597–615.

SECTION VII

NEUROCUTANEOUS DISORDERS

CHAPTER

79

Neurofibromatoses

Adam P. Ostendorf and David H. Gutmann

Washington University School of Medicine,
Washington University Neurofibromatosis Center, St. Louis, MO, USA

INTRODUCTION

Tumors of the nervous system typically occur in a sporadic fashion without a pre-existing familial predisposition. However, the common tumors that arise in the context of a tumor predisposition syndrome can provide valuable insights into the critical genetic events important for tumorigenesis in the nervous system. Two such syndromes, neurofibromatosis type 1 (NF1) and neurofibromatosis type 2 (NF2) are autosomal dominant disorders characterized by an increased susceptibility to both benign and malignant tumors. The genes responsible for NF1 (neurofibromin) and NF2 (merlin/schwannomin) were identified by positional cloning and encode proteins that function as negative growth regulators. This chapter will focus on the nervous system tumors in NF1 and NF2.

CLINICAL FEATURES

NF1 was previously known as "peripheral neurofibromatosis" or von Recklinghausen disease in honor of Frederick von Recklinghausen who first described the condition in 1882. It is a relatively common autosomal dominant disorder with complete penetrance and an incidence of 1 in 2500 individuals worldwide.[1] Sporadic cases comprise about half of all cases, with affected males exhibiting approximately half the fitness as affected females.[2,3]

Most individuals with NF1 present in infancy or early childhood with hyperpigmented lesions, including café-au-lait macules, skinfold freckling and Lisch nodules (iris hamartomas). Moreover, NF1 predisposes affected people to the development of brain tumors (optic pathway gliomas in children, malignant gliomas in adults), distinctive bony defects (tibial pseudarthrosis, sphenoid wing dysplasia), and Schwann cell tumors (neurofibromas and plexiform neurofibromas). Based on the increased incidence of these findings in the context of NF1, the diagnostic criteria for NF1 were originally formulated by the National Institutes of Health Consensus Panel in 1988 (Table 79.1). In order for an individual to be given the diagnosis of NF1, two of these seven diagnostic criteria must be present. While the features that comprise the diagnostic criteria represent the most common abnormalities encountered in children and adults with NF1, other overrepresented clinical features include short stature, macrocephaly, cognitive deficits, aqueductal stenosis, seizures, renal and cerebral artery stenosis, and other bony abnormalities (decreased bone mineral density, dystrophic scoliosis).

Café-au-lait macules occur in 99% of individuals with NF1, and are frequently the earliest manifestation of the disorder (Figure 79.1, panel A). Café-au-lait macules are most often noted at birth, and tend to increase in number and size during the first 2 years of life, only to fade later in adulthood. Typically, these macules have smooth regular borders and demonstrate homogenous pigmentation. Histopathologically, café-au-lait macules contain increased numbers of large melanosomes (macromelanosomes). While one or two café-au-lait macules can be seen in unaffected individuals in the general population, the likelihood of an eventual diagnosis of NF1 increases with a greater number (greater than five) and typical morphologic appearance, especially in young children.[4]

Like café-au-lait macules, freckling is also an early diagnostic sign, found in ~50% of children with NF1 by 10 years of age.[3,5] Freckling typically occurs in regions where skinfolds are in frequent opposition, such as the axilla, inguinal region, and beneath the chin. In women, freckling may be seen under the breasts. Lisch nodules are benign pigmented nodules of

Rosenberg's Molecular and Genetic Basis of Neurological and Psychiatric Disease
http://dx.doi.org/10.1016/10.1016/B978-0-12-410529-4.00079-6

TABLE 79.1 Diagnostic Criteria for Neurofibromatosis 1 and Neurofibromatosis 2

NEUROFIBROMATOSIS TYPE 1
Two or more of the following features are required:
1. Six or more café-au-lait spots with diameters greater than 0.5 mm before puberty or 1.5 cm after puberty
2. Two or more neurofibromas or a single plexiform neurofibroma
3. Freckling in the axillary or inguinal regions
4. Optic pathway tumor
5. Lisch nodules (hamartomas of the iris)
6. A distinctive bony lesion: dysplasia or the sphenoid bone or dysplasia/thinning of long bone cortex
7. A first-degree relative diagnosed with NF1
NEUROFIBROMATOSIS TYPE 2
Definite NF2:
1. Bilateral vestibular schwannomas (VS) or
2. Family history of NF2 plus either
A) Unilateral VS at <30 years old, or
B) Any two of the following: meningioma, glioma, schwannoma, juvenile posterior subcapsular lenticular opacities/juvenile cortical cataract
Probable NF2:
1. Unilateral VS at <30 years old, plus at least one of any of the following: meningioma, glioma, schwannoma, juvenile posterior subcapsular lenticular opacities/juvenile cortical cataract, or
2. Multiple meningiomas (two or more) plus unilateral VS at <30 years old, or
3. Any two of the following: glioma, schwannoma, juvenile posterior subcapsular lenticular opacities/juvenile cortical cataract

the iris that are not clinically significant, other than for diagnostic purposes. While only about 30–50% of people with NF1 have Lisch nodules by age 6, nearly all do so by adulthood. Lisch nodules likely develop in response to light exposure, as they are most frequently observed in the lower half of the iris.[6] Slit-lamp examinations are required to accurately identify these lesions.

The most common tumor seen in individuals with NF1 is the neurofibroma, a benign tumor composed of Schwann cells, fibroblasts, perineural cells, collagen, macrophages, and mast cells (Figure 79.1, panel B). These tumors may appear as defined cutaneous or subcutaneous lesions or as deep tumors associated with nerves anywhere within the body. The number of dermal neurofibromas typically increases as a function of age, with tumors first appearing around the time of puberty. Neurofibromas may be a source of skin irritation when they occur in particular locations, and may become a major cosmetic problem when abundant.

Plexiform neurofibromas are more extensive neurofibromas, arising from larger nerve trunks and involving more than a single nerve. These tumors likely represent congenital lesions, found in approximately 30–50% of individuals with NF1 (Figure 79.1, panel C).[7] While often considered congenital lesions that present in infancy, they may arise from deeper structures and remain clinically silent until later in life. Plexiform neurofibromas can grow anywhere along the length of major peripheral nerves, and may result in bony abnormalities, airway compromise, functional impairment, and increased mortality when symptomatic.[8] In this regard, their presence may stimulate underlying bone growth to result in leg length discrepancy or scoliosis, or erode associated bone to create sphenoid wing dysplasia, often resulting in intracranial extension and exophthalmos.

Unlike the discrete neurofibroma, plexiform neurofibromas may transform into malignant peripheral nerve sheath tumors (MPNSTs) in about 5–10% of individuals.[9] Unfortunately, reliable signs of malignant transformation are lacking; however, the two most predictive clinical features suggestive of malignant transformation are pain and neurological deficit.[10] Individuals with clinical signs suspicious for malignancy require prompt and aggressive management, as these cancers are highly aggressive and are associated with dismal 5-year survival rates. A sensitive and specific tool for the evaluation of a suspected MPNST is 18-fluorodeoxyglucose positron emission tomography (FDG-PET) (Figure 79.1, panel D).[11,12] Small biopsies are frequently unhelpful due to sampling error, and open biopsy or wide surgical resection is recommended.[13] Following diagnosis, prompt referral to a medical oncologist is necessary to plan and initiate radiation treatment and chemotherapy. While local radiotherapy after wide surgical excision of an MPNST may delay the time to local recurrence, it does not improve the overall long-term survival.[13] Similarly, MPNSTs are not particularly sensitive to conventional chemotherapy, prompting investigational chemotherapy protocols currently in evaluation. MPNSTs also have a propensity for distant metastasis to bone and lung. New onset of bone pain or fracture in a person with NF1 and a MPNST warrants a search for metastases.

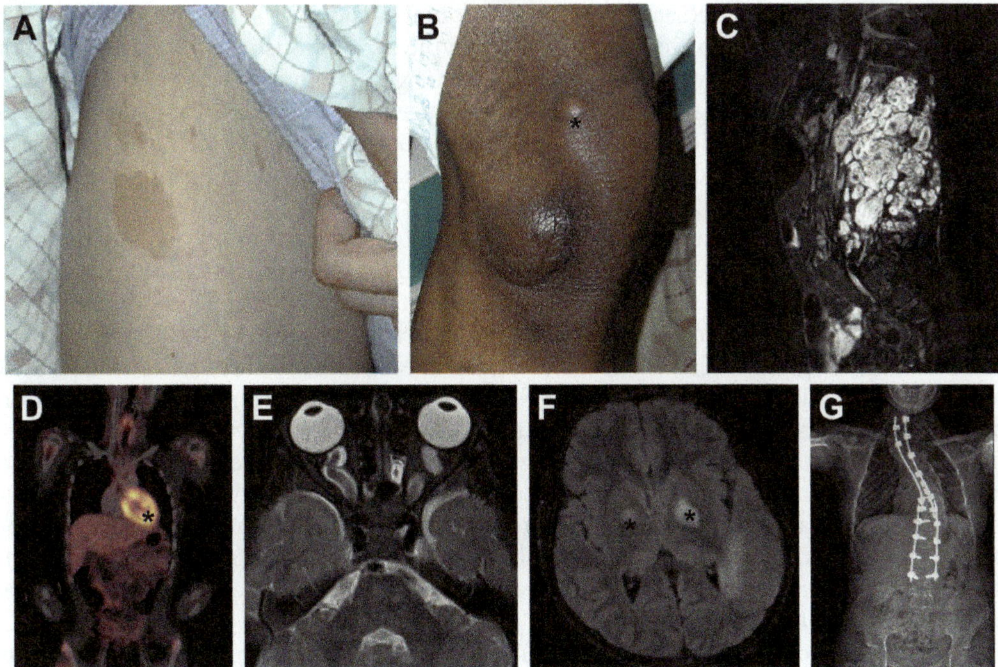

FIGURE 79.1 **Clinical Features of NF1.** (**A**) A representative café-au-lait macule on an individual with NF1. (**B**) Dermal neurofibroma and (**C**) a plexiform neurofibroma in an adult with NF1. (**D**) FDG-PET demonstrating increased uptake in a malignant peripheral nerve sheath tumor. (**E**) Optic pathway glioma predominantly involving the right optic nerve of a young child with NF1 is evident on cranial MRI. (**F**) Brain MRI revealing bilateral T2 hyperintensities in the basal ganglia in a child with NF1. (**G**) Radiographs reveal severe scoliosis following spinal fusion surgery.

The second most frequent tumor in NF1 is the optic pathway glioma (OPG), occurring in 15% of individuals with NF1 (Figure 79.1, panel E). These tumors are almost entirely restricted to childhood, with a mean age at presentation of 4.5 years.[14] Histologically, NF1-associated OPGs are almost always low-grade (World Health Organization [WHO] grade I) pilocytic astrocytomas with little propensity for malignant progression. These tumors can arise anywhere along the optic pathway, from the retrobulbar optic nerve to the postchiasmatic optic radiations.[15] They may also extend into the hypothalamus to cause neuroendocrine dysfunction, such as precocious puberty. Many of these tumors are detected incidentally on brain magnetic resonance imaging (MRI), and only about one-third of these are clinically symptomatic. The decision to treat is largely based on visual decline, and all children with NF1 should be evaluated annually by an ophthalmologist expert in NF1.[16] While no absolute risk factors for optic glioma progression have been identified to date, young children (<2 years of age) and those individuals with postchiasmatic tract involvement are more likely to require treatment.[17]

NF1-associated gliomas may also involve the brainstem in children, accounting for 10–15% all of brain tumors in this population. These tumors tend to arise in slightly older children, are usually low-grade gliomas, and warrant close follow-up with serial MR imaging. Individuals with NF1 are also at a higher risk for other malignancies, including breast cancer, gastrointestinal stromal tumors, pheochromocytoma, and juvenile chronic myeloid leukemia.

In addition to brain tumors, more than 50% of all children with NF1 harbor high signal intensity lesions on T2-weighted images (T2-HSI lesions), previously referred to as unidentified bright objects (UBOs; Figure 79.1, panel F). These nonenhancing well-circumscribed lesions are most often found in the brainstem, thalamus, cerebellum, and basal ganglia. Although there are few pathological studies, these lesions may represent regions of vacuolar or spongiotic change.[18] T2-HSI lesions have no known potential for neoplastic degeneration, but may enlarge and be misdiagnosed as gliomas. The presence of mass effect, diffuse infiltration of surrounding parenchyma, and contrast enhancement should raise suspicion for a glioma. Multiple studies have investigated a potential correlation between the number of these lesions and cognitive deficits, yet no consensus exists as to their clinical significance.[19,20] Notably, most UBOs disappear by late adolescence and early adulthood.[21]

Individuals with NF1 are also at higher risk for specific bony abnormalities. Scoliosis occurs in 10–20% of individuals with NF1 (Figure 79.1, panel G). Scoliosis in children with NF1 may resemble the idiopathic form encountered in the general population, or may present earlier in life when related to vertebral dysplasia or secondary to paravertebral neurofibromas. Scoliosis resulting from a paravertebral neurofibroma may manifest with a dramatic abrupt angle curvature, unusual for non-NF1-associated cases. In addition, tibial pseudarthrosis and sphenoid wing dysplasia are both osseous lesions relatively specific to children with NF1, typically obvious in early infancy. Bowing of the long bones may be the result of cortical thinning or dysplastic bony changes, and may lead to pathological fracture in NF1. Repeated fracture and incomplete healing produces the appearance of a "false joint" or pseudarthrosis. Early recognition is essential to avoid limb amputation. Sphenoid wing dysplasia will usually manifest as an orbital deformity. Skull bone radiographs should readily identify this bony defect. Finally, adults with NF1 have a higher risk for osteopenia and lower serum concentrations of 25-hydroxy-vitamin D.[22,23]

Vascular abnormalities, including renal artery stenosis, which may cause hypertension, and moya-moya, a cerebral vessel abnormality that can result in stroke, are unusual complications of NF1, found in less than 1% of affected individuals.

Learning deficits and developmental delays are reported in >60% of individuals with NF1, and are the most frequently reported complication.[24] The most common affected domains include executive function and attention, visual–spatial and visual–perceptual skills.[25,26] Fine motor, gross motor, and math/premath delays are the most common developmental abnormalities reported.[27] In population studies, there is a left shift in mean IQ scores from 100 in the general population to 85 in children with NF1, although 6–7% of children have IQ scores less than 70.[28]

Individuals with NF1 are also more likely to have seizures, with an incidence of 4–9%.[3,29–33] Focal seizures are the most common type, and the refractory nature of the seizures has been debated (A.P. Ostendorf, D.H. Gutmann and J.L. Weisenberg, unpublished data).[31–34] Seizures are more likely to occur in individuals with an intracranial tumor, which should prompt clinicians to obtain brain neuroimaging (A.P. Ostendorf, D.H. Gutmann and J.L. Weisenberg, unpublished data).[32] In refractory cases, individuals may respond favorably to lesionectomy (A.P. Ostendorf, D.H. Gutmann and J.L. Weisenberg, unpublished data).[35]

One of the hallmark features of NF1 is phenotypic variability. There is typically as much variation in the spectrum of clinical features between individuals within a single family who all harbor an identical germline *NF1* mutation as there is between individuals from different families. Monozygotic twin studies have shown concordance for numbers of café-au-lait macules, cutaneous neurofibromas, and cognitive function, yet were discordant for plexiform neurofibromas and MPNSTs.[36] Parents with few clinical features of NF1 (e.g., café-au-lait macules and skinfold freckles) can have children with more clinically serious manifestations (e.g., MPNSTs and OPGs). Evidence has been conflicting regarding parent of origin effects on phenotypic variability in individuals with NF1,[37] and currently no consensus exists. Phenotypic variability in family members with the same mutation suggests that factors, other than the germline *NF1* mutation, determine the clinical expression of the disorder.

An increasing number of individuals have been identified with segmental NF1, affecting only a single region of the body.[38] Some of these individuals are genetically mosaic for mutations in the *NF1* gene, and therefore are known to harbor the mutation in only a subset of their cells. Mosaicism, resulting from a somatic mutation occurring at a later stage in fetal development, is more likely to affect fewer cells and yield a more restricted somatic distribution of clinical abnormalities. Given that mosaicism can potentially affect gonadal tissues, the probability of transmission of NF1 may be as high as 50%, as seen in typical cases of NF1, but is generally lower. Although the parent may have a mosaic or segmental pattern of NF1, affected offspring will have the generalized NF1 phenotype, as they would have inherited a germline *NF1* mutation in all cells.

Life expectancy is decreased by 8–15 years in individuals with NF1, typically related to the development of a MPNST or glioma.[39,40] Children with NF1 report significant impact on quality of life related to orthopedic manifestations, learning disabilities, and the presence of at least two plexiform neurofibromas.[41,42]

NF2 is much less common than NF1, with an incidence of 1 in 35,000.[1] It is characterized by the formation of multiple schwannomas affecting cranial (Figure 79.2, panel A), spinal, and peripheral nerves as well as meningiomas (Figure 79.2, panel B), ependymomas, and juvenile cataracts (Table 79.1). The hallmark of the disorder is the development of bilateral vestibular schwannomas, which most individuals develop by their third decade. In addition to nervous system tumors, children and young adults may develop posterior subcapsular lenticular opacities (cataracts).

The majority of adult individuals with NF2 present with unilateral hearing loss, often associated with imbalance and tinnitus. Children, who have more relentless disease course, due to an early onset and more rapidly progressive clinical complications, are more likely to present with tumors involving other cranial and peripheral nerves. In some families with NF2, the onset and severity are similar in all affected individuals, suggesting a stronger relationship between the phenotype and the underlying genetic mutation than observed in NF1.[43]

The most common tumor found in NF2 is the schwannoma, composed of neoplastic Schwann cells. Clinically, schwannomas from individuals with NF2 are not different from their sporadic counterparts, except that they are often multiple and typically arise at an earlier age. Despite the misnomer "acoustic neuroma," schwannomas most frequently affect the vestibular branch of the eighth cranial nerve, although other cranial nerves may be affected. Since these tumors may be small, fine continuous MR imaging through the internal auditory canal is required.

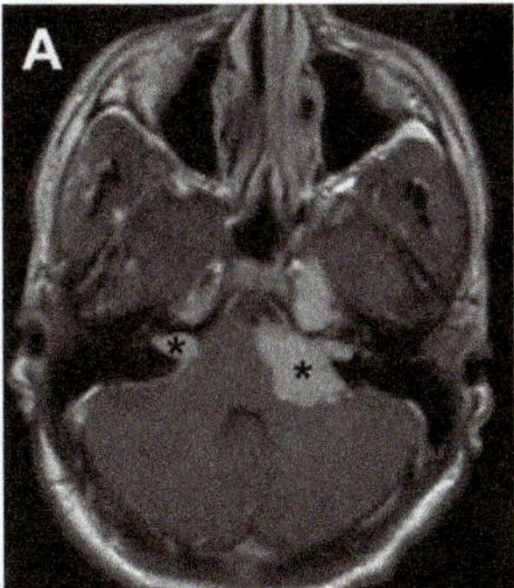

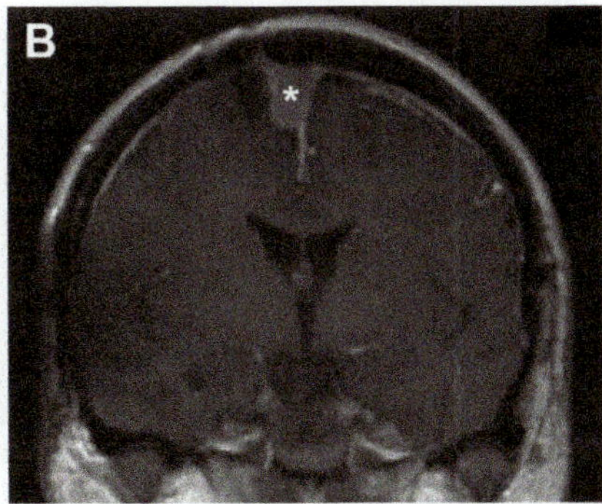

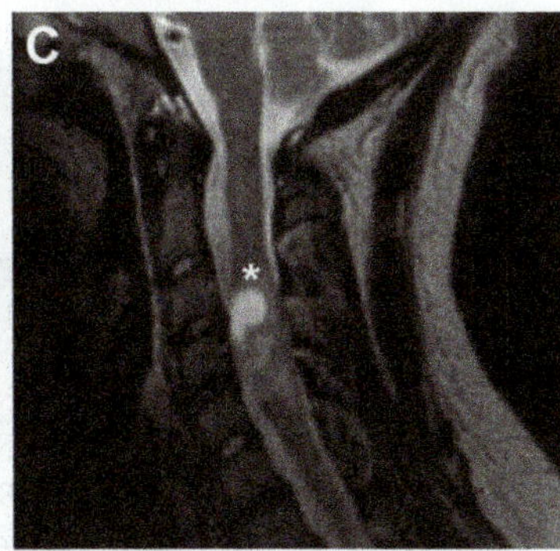

FIGURE 79.2 **Clinical Features of NF2.** (**A**) Bilateral vestibular schwannomas in an individual with NF2, with mass effect on the left. (**B**) MRI revealing a meningioma near the falx of an individual with NF2. (**C**) T2 hyperintensity on spinal MRI demonstrates a spinal schwannoma in an individual with NF2.

Fusiform subcutaneous schwannomas may be palpated in individuals with NF2, but are never present in the hundreds, as may occur with cutaneous neurofibromas in people with NF1. Extramedullary spinal schwannomas (Figure 79.2, panel C) may arise at any vertebral level, and are identified in nearly 80–90% of individuals with NF2 using gadolinium-enhanced MRI. Neuropathies have also been recognized as a feature of NF2, including a mononeuropathy more common in childhood and a progressive polyneuropathy in adulthood.[44–46]

Meningioma is the second most common tumor in NF2, and occurs both intracranially and within the spinal cord. Most are supratentorial, occurring in the parasagittal region, cerebral convexities, or sphenoid ridge. Like schwannomas, meningiomas are also common sporadic tumors; however, only 1% of individuals with a single meningioma meet criteria for NF2. The recognition of multiple meningiomas in a child or young adult should prompt a search for other signs of NF2, as meningiomas are typically found in older adults in the general population.

The initial diagnostic workup for individuals at risk for NF2 includes full cranial and spinal MRI to evaluate for schwannomas, meningiomas, and intramedullary tumors. Brain MRI, neurologic examinations, ophthalmologic and audiologic assessments are recommended, and should be dictated by clinical findings.[47,48] Brainstem auditory evoked responses may also be helpful in assessing eighth cranial nerve function. Brain MRI also serves as an effective screening exam for asymptomatic individuals, as a normal MRI scan by age 30 renders a diagnosis of NF2 extremely unlikely. Once the diagnosis is established, brain MRI should be obtained every 2 years for those under 20 years of age and every 3–5 years thereafter. If tumors are present, annual brain MRI should be obtained. In those individuals with spinal tumors, spinal MRI should be obtained every 3–10 years even if the neurologic exam is normal.[48] Importantly, the frequency of neuroimaging studies should be dictated by the specific clinical manifestations present.

There are also rare individuals with schwannomatosis, a disorder defined by multiple schwannomas in the absence of bilateral vestibular schwannomas, meningiomas, or ependymomas.[49] These individuals tend to develop schwannomas later in life, and therefore have a longer life expectancy. In contrast to people affected with NF2, individuals with schwannomatosis do not harbor germline mutations in the *NF2* gene, but rather harbor mutations in the *SMARCB1* gene.[50–54]

Another subset of individuals with NF2 has disease-related tumors restricted to one region of the nervous system, such as localized meningiomas and unilateral vestibular schwannomas.[38] These individuals have segmental or mosaic NF2, characterized by somatic mutations in the *NF2* gene occurring later during fetal development, much as described above for segmental NF1. Mosaicism is much more common in NF2 than in NF1, accounting for 20–30% of individuals with no previous family history.[55,56]

MOLECULAR GENETICS

Both NF1 and NF2 are considered tumor suppressor disorders. The mechanism by which autosomal dominant tumor suppressor genes result in the formation of tumors is explained by the Knudson "two hit" hypothesis, whereby affected individuals start life with a germline mutation in the *NF1* (or *NF2*) gene in all cells of their body.[57] Random somatic mutation of the one remaining functional *NF1* or *NF2* gene ("second hit") is required for loss of tumor suppressor expression and tumor formation. These second mutations are stochastic and may be affected by other genetic or epigenetic factors, resulting in differential expression of the disorder, even in individuals within the same family. In this respect, while the disorder is inherited as an autosomal dominant trait, tumor formation is autosomal recessive at the cellular level, because both copies of the gene need to be inactivated to result in dysregulated cell growth.

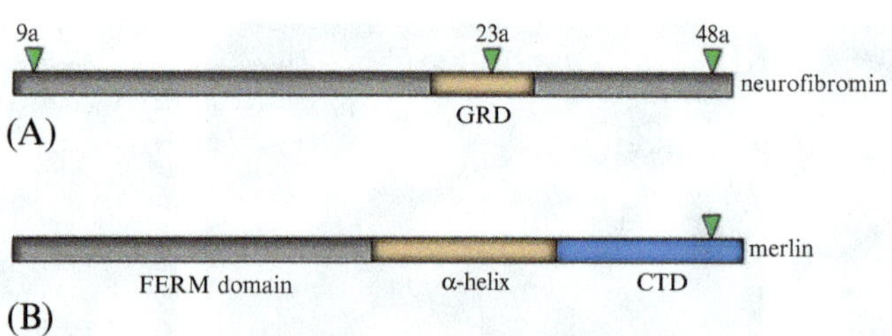

FIGURE 79.3 ***NF1* and *NF2* gene products.** (**A**) The *NF1* gene product, neurofibromin, contains three alternatively spliced exons (9a, 23a, and 48a). The central region of the protein contains a GAP-related domain (GRD) with structural and functional similarity to GTPase-activating proteins (GAPs) that accelerate the conversion of RAS from its active GTP-bound form to an inactive GDP-bound form. Active RAS provides a mitogenic signal for many cell types. (**B**) The *NF2* gene product, merlin, contains three regions of structural significance, including an amino terminal FERM domain (residues 1–302), a central alpha helical region (residues 303–478), and a unique carboxyl terminal domain (CTD; residues 479–595) lacking the conventional actin-binding sequences in other ERM proteins.

The *NF1* gene on the long arm of chromosome 17 was identified by positional cloning, and encodes the tumor suppressor protein neurofibromin.[58–60] The *NF1* gene codes for an mRNA of 11–13 kb with at least 60 exons and a predicted cytoplasmic protein of 2818 amino acids (Figure 79.3A). Neurofibromin, the 220–250 kDa *NF1* gene product, is most abundantly expressed in neurons, oligodendrocytes, and Schwann cells, but is also detected in astrocytes, microglia, macrophages, leukocytes and the adrenal medulla.

Sequence analysis revealed homology between a small portion of neurofibromin and a family of molecules, termed GTPase activating proteins (GAPs), involved in negative regulation (inhibition) of RAS activity (Figure 79.4A).[61] RAS is a low molecular-weight GTPase protein involved in stimulating cell growth and promoting tumor formation. RAS exists in an active form when bound to guanosine triphosphate (GTP) and an inactive form when guanosine diphosphate (GDP)-bound. GAP proteins, like neurofibromin, inhibit RAS by stimulating its intrinsic GTPase activity, and accelerate the conversion of active GTP-bound RAS to inactive GDP-bound RAS. Failure to inactivate RAS, as a result of loss of neurofibromin function due to *NF1* gene inactivation, leads to increased cell proliferation and tumor formation.[62–64] Neurofibromin also increases adenylyl cyclase activity to increase intracellular cyclic adenosine monophosphate (cAMP) levels.[65–68] Finally, neurofibromin has also been shown to regulate the mammalian target of rapamycin (mTOR) pathway in a RAS-dependent manner.[69–72]

Multiple alternatively spliced *NF1* transcripts have been identified as a result of the alternative use of three exons (9a, 23a, and 48a). These three alternatively spliced exons are differentially expressed, and may contribute to potentially unique functions of neurofibromin in specific tissues during development. Exon 48a is expressed only in muscle tissues, whereas exon 9a expression is restricted to forebrain neurons in a developmentally regulated manner. Exon 23a encodes an additional 21 amino acids within the GAP-related domain, such that neurofibromin containing exon 23a is less effective at RAS regulation than neurofibromin lacking this alternatively spliced exon.

The *NF2* gene was identified by positional cloning in 1993, and is located on the long arm of chromosome 22.[73] The *NF2* gene encodes a 2.2-kb messenger RNA and a predicted 595 amino acid protein named merlin (or schwannomin). Merlin is expressed in neurons, lens fibers, blood vessels, leptomeningeal cells, astrocytes, gonadal tissue and Schwann cells. Merlin is a member of the protein 4.1 family whose members link the actin cytoskeleton to cell surface glycoproteins.[74] Merlin consists of an amino-terminal FERM domain followed by a predicted alpha helical region, both of which are conserved in the ERM (ezrin, radixin, and moesin) subclass of protein 4.1 molecules (Figure 79.3B).

Over the two decades since the identification of the *NF2* gene, numerous studies have implicated merlin in a large number of key growth regulatory processes (Figure 79.4B). In this regard, merlin has been shown to regulate Rac GTPase activity,[75–77] receptor tyrosine kinase signaling,[78–80] Yes-associated protein (YAP) and Hippo pathway signaling,[81] and the mammalian target of rapamycin.[82] Further research will be required to determine whether these diverse mechanisms of action reflect tissue-specific functions for merlin.

DISEASE MECHANISM

In an effort to learn more about the molecular pathogenesis of NF-associated clinical features and to develop preclinical models for NF1 and NF2, many research groups have generated mice with targeted disruptions of the *Nf1* and *Nf2* genes.

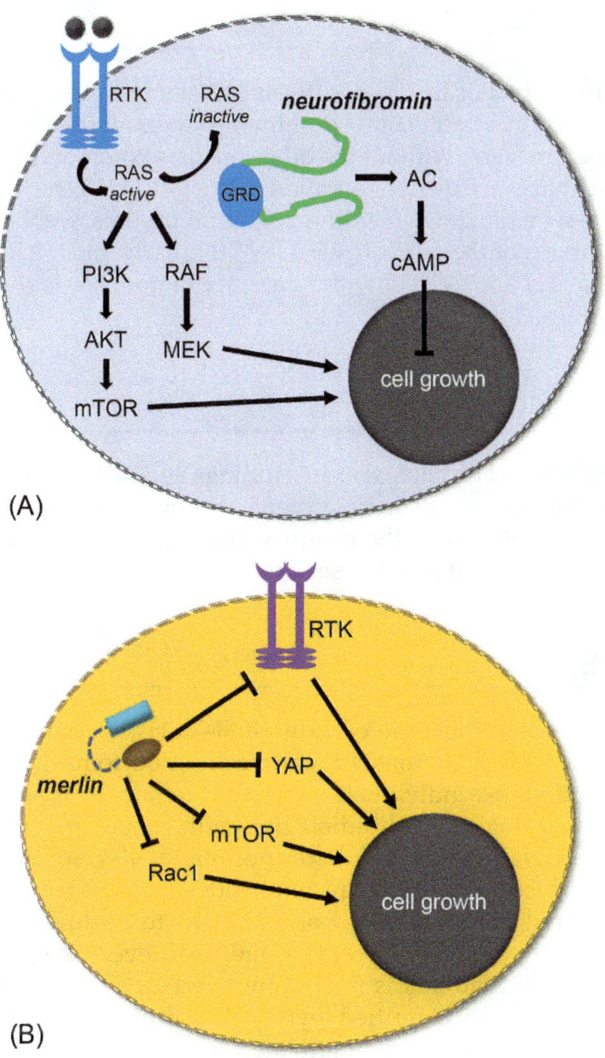

FIGURE 79.4 Neurofibromin and merlin signaling pathways. (**A**) Neurofibromin functions to regulate cell growth and development through both RAS and cAMP pathway regulation. In cells with reduced or absent neurofibromin expression, RAS function is increased, leading to increased PI3K, AKT/mTOR, RAF/MEK signaling and cell growth. Adenylate cyclase (AC) is decreased in cells with reduced or absent neurofibromin expression, resulting in attenuated cAMP levels. (**B**) Absent merlin expression leads to activation of several different receptor tyrosine kinases (RTKs), such as EGFR and ErbB2. In addition, mTOR, Rac1 and YAP activity can be elevated in *NF2*-deficient cells, leading to increased cell growth.

While homozygous deletion of the *Nf1* gene results in embryonic lethality by gestational day 14 due to a cardiac developmental abnormality, mice heterozygous for a mutation in the *Nf1* gene (*Nf1*+/− mice) manifest cognitive deficits in executive, attention and spatial domains similar to those found in individuals with NF1.[83–86] In addition, by 18–24 months of age, *Nf1*+/− mice developed myeloid leukemia and pheochromocytoma, but do not develop neurofibromas or astrocytomas.

To develop more refined models of NF1-associated tumors, Parada and colleagues bypassed the embryonic lethality of *Nf1*-null mice by selectively ablating the gene in specific cell types using the Cre/LoxP system.[87] Astroglial progenitor *Nf1* deletion or activation of *KRas* in *Nf1*+/− mice results in optic gliomas.[88,89] Similarly, *Nf1*+/− mice with Schwann cell precursor *Nf1* inactivation develop plexiform neurofibromas.[90,91] Multiple models of high-grade glioma have also been developed utilizing conditional *Nf1* knockout mice in combination with *Pten* and *p53* gene inactivation in neuroglial progenitor cells.[92–94] These preclinical models of NF1 provide valuable platforms for therapeutic testing.

Mice in which both copies of the *Nf2* gene are inactivated die by embryonic day 7, due to failure to complete gastrulation as a result of an abnormal extraembryologic ectoderm. *Nf2*+/− mice develop highly malignant and metastatic cancers during the second year of life, including malignant sarcomas, hepatocellular carcinomas, and fibrosarcomas.[95] As employed to model NF1-associated tumors, Giovannini and colleagues generated mice by tissue-specific disruption of the *Nf2* gene only in Schwann cells (schwannomas) or meningeal cells (meningiomas).[96–98]

DIFFERENTIAL DIAGNOSIS

Individuals with NF1 may share features found in other disorders. For instance, café-au-lait macules are found in several other less common genetic disorders, or may be inherited in an autosomal dominant fashion. Individuals with multiple CAL macules, macrocephaly and axillary/inguinal freckling without the other features commonly associated with NF1, may have Legius syndrome with an associated loss of function mutation in the *SPRED1* gene.[99,100] Noonan syndrome, characterized by ocular hypertelorism, low-set ears, down-slanting palpebral fissures, webbed neck and pulmonic stenosis, is sometimes associated with mutations of the *PTPN11* gene.[101,102] Similarly, LEOPARD syndrome is associated with mutations of the *PTPN11* gene, and is characterized by the presence of multiple lentigines (freckles), sensorineural deafness, and cardiac abnormalities.

GENETIC TESTING

The diagnosis of NF1 can usually be made accurately and early based on clinical exam findings and history alone (Table 79.1).[47] In rare cases where diagnostic criteria are not met or couples are seeking prenatal counseling to make reproductive decisions, *NF1* mutation analysis is 95% sensitive.[103,104] Similarly, the majority of cases of NF2 can be established on clinical grounds. NF2 mutation analysis is also available, but is not as sensitive.

MANAGEMENT

The mainstay of treatment is anticipatory management.[47] Careful examination of individuals at risk for specific age-dependent features, and the prompt recognition of specific signs and symptoms suggestive of astrocytomas, neurofibromas, meningiomas, and schwannomas is required to identify those individuals who might require treatment. Multidisciplinary teams including surgeons, ophthalmologists, oncologists, and radiologists with experience with these disorders are mandatory to avoid unnecessary treatment. Given the natural history of many NF-associated tumors, a conservative approach with close clinical and radiographic surveillance is recommended.

Routine management of individuals with NF1 involves annual ophthalmologic assessments to evaluate for visual abnormalities suggestive of a symptomatic OPG in children up to 12 years of age, then every 2 years until 18 years of age.[16] Routine brain imaging in asymptomatic individuals is not recommended, as the decision to treat is based on clinical symptoms.[105,106] Once an OPG has been identified, eye examinations should be performed every 3 months for the first year after diagnosis, with subsequently lengthened intervals. Treatment currently entails traditional chemotherapy, typically with carboplatin and vincristine, and radiation therapy is not recommended due to the significant risk of secondary malignancy.[107] The treatment of plexiform neurofibromas presents significant challenges. Surgical removal of plexiform neurofibromas is often difficult due to the significant vascularity of the tumors and the multiple tissue planes that they cross. While radiation therapy has a limited role, due to the risk of malignant transformation, there are currently numerous clinical trials using biologically based treatments. Plexiform neurofibromas should be closely monitored for the development of sudden pain, new neurological deficit, or rapid growth, which may indicate malignant transformation.[10] Growth in children should be closely followed, and signs of precocious puberty should be promptly evaluated. Developmental progress and school performance should be monitored for signs of learning disabilities and attention deficits. Scoliosis should be screened for clinically and tibial bowing should be referred to an orthopedic surgeon. Routine determinations of blood pressure are recommended to recognize potential complications including renal artery stenosis and pheochromocytoma.

The management of individuals with NF2 requires close surveillance for cranial and spinal tumors.[47,108] Complete cranial and spinal magnetic resonance imaging is indicated at diagnosis. Annual physical examinations and repeat cranial and spinal imaging studies are recommended to identify and follow tumors as clinically indicated. Audiometry and brainstem auditory evoked responses aid in the detection of clinically significant vestibular schwannomas. Affected individuals are best followed in centers that have a multidisciplinary team familiar with NF2. Some individuals may benefit from hearing aids, cochlear implants, or brainstem implants, depending on the particular situation. Symptomatic vestibular schwannomas are often treated with complete surgical resection or debulking, depending on the size and local tissue destruction. Radiosurgery has been utilized as an alternative, although this may increase the risk for secondary malignancy. Surgery remains the mainstay of symptomatic meningioma treatment.

The ability to design more rational therapies for NF-associated tumors is heavily dependent on an improved basic understanding of the mechanisms by which merlin and neurofibromin regulate cell growth.[109] For example, multiple studies are presently underway to evaluate the mTOR inhibitors everolimus and sirolimus for the treatment of individuals with low-grade gliomas or plexiform neurofibromas.[110–114] Inhibition of MEK has demonstrated promise in the treatment of NF1-associated peripheral nerve tumors and myeloproliferative disease in mice,[115,116] which prompted a newly launched clinical trial for NF1-associated plexiform neurofibromas.[117]

Nontumor phenotypes of NF1 are also the target of current clinical trials research. Methylphenidate is currently being tested in individuals with NF1 and ADD/ADHD[118] after promising results in a mouse model of NF1-associated attention deficit.[67] Learning deficits in children with NF1 have similarly been linked to inhibition of long-term potentiation due to elevated RAS activity, which in mice can be reversed by treatment with lovastatin, a 3-hydroxy-3-methylglutaryl coenzyme A (HMG-CoA) reductase inhibitor.[86] Although initial human trials with simvastatin did not reveal a significant benefit,[119] further evaluation of lovastatin is underway.[120–122]

In addition, the discovery of multiple pathways through which merlin mediates cell growth, including ErbB and VEGF, has spurred ongoing studies using the inhibitors lapatinib[123,124] and bevacizumab.[125,126] Other potential therapies include PI3K/Akt inhibition[127] and heat shock protein 90 inhibition.[128]

FUTURE DIRECTIONS

The ability to design rational therapies for NF-associated tumors is heavily dependent on an improved basic understanding of the mechanisms by which merlin and neurofibromin regulate cell growth. While preliminary insights into these mechanisms have been achieved over the past decade, future targeted therapies will likely derive from a more detailed appreciation of the molecular and cellular biology of the tissues involved in NF-associated tumorigenesis and the consequence of *NF* gene loss on these cell types. The combined study of human tumor tissue, preclinical animal models, and tumor cell lines will facilitate these fundamental advances.

References

1. Friedman J, Gutmann D, MacCollin M, Riccardi V, eds. *Neurofibromatosis: Phenotype, Natural History, and Pathogenesis*. 3rd ed. Baltimore: Johns Hopkins University Press; 1999.
2. Crowe FW, Schull W, Neel J. *A Clinical, Pathological, and Genetic Study of Multiple Neurofibromatosis*. Springfield, Ill: Thomas; 1956.
3. Huson SM, Harper PS, Compston DA. Von Recklinghausen neurofibromatosis. A clinical and population study in south-east Wales. *Brain*. 1988;111(Pt 6):1355–1381.
4. Nunley KS, Gao F, Albers AC, Bayliss SJ, Gutmann DH. Predictive value of café au lait macules at initial consultation in the diagnosis of neurofibromatosis type 1. *Arch Dermatol*. 2009;145(8):883–887.
5. DeBella K, Szudek J, Friedman JM. Use of the national institutes of health criteria for diagnosis of neurofibromatosis 1 in children. *Pediatrics*. 2000;105(3 Pt 1):608–614.
6. Boley S, Sloan JL, Pemov A, Stewart DR. A quantitative assessment of the burden and distribution of Lisch nodules in adults with neurofibromatosis type 1. *Invest Ophthalmol Vis Sci*. 2009;50(11):5035–5043.
7. Mautner V-F, Asuagbor FA, Dombi E, et al. Assessment of benign tumor burden by whole-body MRI in patients with neurofibromatosis 1. *Neuro Oncol*. 2008;10(4):593–598.
8. Prada CE, Rangwala FA, Martin LJ, et al. Pediatric plexiform neurofibromas: impact on morbidity and mortality in neurofibromatosis type 1. *J Pediatr*. 2012;160(3):461–467.
9. Evans DGR, Baser ME, McGaughran J, Sharif S, Howard E, Moran A. Malignant peripheral nerve sheath tumours in neurofibromatosis 1. *J Med Genet*. 2002;39(5):311–314.
10. King AA, Debaun MR, Riccardi VM, Gutmann DH. Malignant peripheral nerve sheath tumors in neurofibromatosis 1. *Am J Med Genet*. 2000;93(5):388–392.
11. Ferner RE, Golding JF, Smith M, et al. [18 F]2-fluoro-2-deoxy-D-glucose positron emission tomography (FDG PET) as a diagnostic tool for neurofibromatosis 1 (NF1) associated malignant peripheral nerve sheath tumours (MPNSTs): a long-term clinical study. *Ann Oncol*. 2008;19(2):390–394.
12. Derlin T, Tornquist K, Münster S, et al. Comparative effectiveness of 18 F-FDG PET/CT versus whole-body MRI for detection of malignant peripheral nerve sheath tumors in neurofibromatosis type 1. *Clin Nucl Med*. 2013;38(1):e19–e25.
13. Ferner RE, Gutmann DH. International consensus statement on malignant peripheral nerve sheath tumors in neurofibromatosis. *Cancer Res*. 2002;62(5):1573–1577.
14. Listernick R, Charrow J, Greenwald M, Mets M. Natural history of optic pathway tumors in children with neurofibromatosis type 1: a longitudinal study. *J Pediatr*. 1994;125(1):63–66.
15. Kleihues P, Cavenee W. *Tumors of the Nervous System*. Lyon: IARC Press; 2000.
16. Listernick R, Ferner RE, Liu GT, Gutmann DH. Optic pathway gliomas in neurofibromatosis-1: controversies and recommendations. *Ann Neurol*. 2007;61(3):189–198.

17. Fisher MJ, Loguidice M, Gutmann DH, et al. Visual outcomes in children with neurofibromatosis type 1-associated optic pathway glioma following chemotherapy: a multicenter retrospective analysis. *Neuro Oncol*. 2012;14(6):790–797.
18. DiPaolo DP, Zimmerman RA, Rorke LB, Zackai EH, Bilaniuk LT, Yachnis AT. Neurofibromatosis type 1: pathologic substrate of high-signal-intensity foci in the brain. *Radiology*. 1995;195(3):721–724.
19. Chabernaud C, Sirinelli D, Barbier C, et al. Thalamo-striatal T2-weighted hyperintensities (unidentified bright objects) correlate with cognitive impairments in neurofibromatosis type 1 during childhood. *Dev Neuropsychol*. 2009;34(6):736–748.
20. Goh WHS, Khong P-L, Leung CSY, Wong VCN. T2-weighted hyperintensities (unidentified bright objects) in children with neurofibromatosis 1: their impact on cognitive function. *J Child Neurol*. 2004;19(11):853–858.
21. Gill DS, Hyman SL, Steinberg A, North KN. Age-related findings on MRI in neurofibromatosis type 1. *Pediatr Radiol*. 2006;36(10):1048–1056.
22. Dulai S, Briody J, Schindeler A, North KN, Cowell CT, Little DG. Decreased bone mineral density in neurofibromatosis type 1: results from a pediatric cohort. *J Pediatr Orthop*. 2007;27(4):472–475.
23. Petramala L, Giustini S, Zinnamosca L, et al. Bone mineral metabolism in patients with neurofibromatosis type 1 (von Recklinghausen disease). *Arch Dermatol Res*. 2012;304(4):325–331.
24. North KN, Riccardi V, Samango-Sprouse C, et al. Cognitive function and academic performance in neurofibromatosis. 1: consensus statement from the NF1 Cognitive Disorders Task Force. *Neurology*. 1997;48(4):1121–1127.
25. Mautner V-F, Kluwe L, Thakker SD, Leark RA. Treatment of ADHD in neurofibromatosis type 1. *Dev Med Child Neurol*. 2002;44(3):164–170.
26. Dilts CV, Carey JC, Kircher JC, et al. Children and adolescents with neurofibromatosis 1: a behavioral phenotype. *J Dev Behav Pediatr*. 1996;17(4):229–239.
27. Soucy EA, Gao F, Gutmann DH, Dunn CM. Developmental delays in children with neurofibromatosis type 1. *J Child Neurol*. 2012;27(5):641–644.
28. Hyman SL, Shores A, North KN. The nature and frequency of cognitive deficits in children with neurofibromatosis type 1. *Neurology*. 2005;65(7):1037–1044.
29. Friedman JM, Birch PH. Type 1 neurofibromatosis: a descriptive analysis of the disorder in 1,728 patients. *Am J Med Genet*. 1997;70(2):138–143.
30. Riccardi VM. Von Recklinghausen neurofibromatosis. *N Engl J Med*. 1981;305(27):1617–1627.
31. Kulkantrakorn K, Geller TJ. Seizures in neurofibromatosis 1. *Pediatr Neurol*. 1998;19(5):347–350.
32. Hsieh H-Y, Fung H-C, Wang C-J, Chin S-C, Wu T. Epileptic seizures in neurofibromatosis type 1 are related to intracranial tumors but not to neurofibromatosis bright objects. *Seizure*. 2011;20(8):606–611.
33. Korf BR, Carrazana E, Holmes GL. Patterns of seizures observed in association with neurofibromatosis 1. *Epilepsia*. 1993;34(4):616–620.
34. Vivarelli R, Grosso S, Calabrese F, et al. Epilepsy in neurofibromatosis 1. *J Child Neurol*. 2003;18(5):338–342.
35. Barba C, Jacques T, Kahane P, et al. Epilepsy surgery in neurofibromatosis type 1. *Epilepsy Res*. 2013;105(3):384–395.
36. Rieley MB, Stevenson DA, Viskochil DH, Tinkle BT, Martin LJ, Schorry EK. Variable expression of neurofibromatosis 1 in monozygotic twins. *Am J Med Genet A*. 2011;155A(3):478–485.
37. Johnson KJ, Fisher MJ, Listernick RL, et al. Parent-of-origin in individuals with familial neurofibromatosis type 1 and optic pathway gliomas. *Fam Cancer*. 2012;11(4):653–656.
38. Ruggieri M, Huson SM. The clinical and diagnostic implications of mosaicism in the neurofibromatoses. *Neurology*. 2001;56(11):1433–1443.
39. Evans DGR, O'Hara C, Wilding A, et al. Mortality in neurofibromatosis 1: in North West England: an assessment of actuarial survival in a region of the UK since 1989. *Eur J Hum Genet*. 2011;19(11):1187–1191.
40. Rasmussen SA, Yang Q, Friedman JM. Mortality in neurofibromatosis 1: an analysis using U.S. death certificates. *Am J Hum Genet*. 2001;68(5):1110–1118.
41. Kodra Y, Giustini S, Divona L, et al. Health-related quality of life in patients with neurofibromatosis type 1. A survey of 129 Italian patients. *Dermatology*. 2009;218(3):215–220.
42. Wolkenstein P, Rodriguez D, Ferkal S, et al. Impact of neurofibromatosis 1 upon quality of life in childhood: a cross-sectional study of 79 cases. *Br J Dermatol*. 2009;160(4):844–848.
43. Parry DM, MacCollin MM, Kaiser-Kupfer MI, et al. Germ-line mutations in the neurofibromatosis 2 gene: correlations with disease severity and retinal abnormalities. *Am J Hum Genet*. 1996;59(3):529–539.
44. Evans DG, Birch JM, Ramsden RT. Paediatric presentation of type 2 neurofibromatosis. *Arch Dis Child*. 1999;81(6):496–499.
45. Sperfeld AD, Hein C, Schröder JM, Ludolph AC, Hanemann CO. Occurrence and characterization of peripheral nerve involvement in neurofibromatosis type 2. *Brain*. 2002;125(Pt 5):996–1004.
46. Schulz A, Baader SL, Niwa-Kawakita M, et al. Merlin isoform 2 in neurofibromatosis type 2-associated polyneuropathy. *Nat Neurosci*. 2013;16(4):426–433.
47. Gutmann DH, Aylsworth A, Carey JC, et al. The diagnostic evaluation and multidisciplinary management of neurofibromatosis 1 and neurofibromatosis 2. *JAMA*. 1997;278(1):51–57.
48. Evans DGR. Neurofibromatosis 2 [Bilateral acoustic neurofibromatosis, central neurofibromatosis, NF2, neurofibromatosis type II]. *Genet Med*. 2009;11(9):599–610.
49. MacCollin M, Chiocca EA, Evans DG, et al. Diagnostic criteria for schwannomatosis. *Neurology*. 2005;64(11):1838–1845.
50. Hulsebos TJM, Plomp AS, Wolterman RA, Robanus-Maandag EC, Baas F, Wesseling P. Germline mutation of INI1/SMARCB1 in familial schwannomatosis. *Am J Hum Genet*. 2007;80(4):805–810.
51. Hadfield KD, Newman WG, Bowers NL, et al. Molecular characterisation of SMARCB1 and NF2 in familial and sporadic schwannomatosis. *J Med Genet*. 2008;45(6):332–339.
52. Boyd C, Smith MJ, Kluwe L, Balogh A, Maccollin M, Plotkin SR. Alterations in the SMARCB1 (INI1) tumor suppressor gene in familial schwannomatosis. *Clin Genet*. 2008;74(4):358–366.
53. Sestini R, Bacci C, Provenzano A, Genuardi M, Papi L. Evidence of a four-hit mechanism involving SMARCB1 and NF2 in schwannomatosis-associated schwannomas. *Hum Mutat*. 2008;29(2):227–231.

54. Rousseau G, Noguchi T, Bourdon V, Sobol H, Olschwang S. SMARCB1/INI1 germline mutations contribute to 10% of sporadic schwannomatosis. *BMC Neurol*. 2011;11:9.
55. Evans DG, Wallace AJ, Wu CL, Trueman L, Ramsden RT, Strachan T. Somatic mosaicism: a common cause of classic disease in tumor-prone syndromes? Lessons from type 2 neurofibromatosis. *Am J Hum Genet*. 1998;63(3):727–736.
56. Kluwe L, Mautner V, Heinrich B, et al. Molecular study of frequency of mosaicism in neurofibromatosis 2 patients with bilateral vestibular schwannomas. *J Med Genet*. 2003;40(2):109–114.
57. Knudson AG. Hereditary cancer: two hits revisited. *J Cancer Res Clin Oncol*. 1996;122(3):135–140.
58. Cawthon RM, Weiss R, Xu GF, et al. A major segment of the neurofibromatosis type 1 gene: cDNA sequence, genomic structure, and point mutations. *Cell*. 1990;62(1):193–201.
59. Wallace MR, Marchuk DA, Andersen LB, et al. Type 1 neurofibromatosis gene: identification of a large transcript disrupted in three NF1 patients. *Science*. 1990;249(4965):181–186.
60. Viskochil D, White R, Cawthon R. The neurofibromatosis type 1 gene. *Annu Rev Neurosci*. 1993;16:183–205.
61. Cichowski K, Jacks T. NF1 tumor suppressor gene function: narrowing the GAP. *Cell*. 2001;104(4):593–604.
62. Basu TN, Gutmann DH, Fletcher JA, Glover TW, Collins FS, Downward J. Aberrant regulation of ras proteins in malignant tumour cells from type 1 neurofibromatosis patients. *Nature*. 1992;356(6371):713–715.
63. DeClue JE, Papageorge AG, Fletcher JA, et al. Abnormal regulation of mammalian p21ras contributes to malignant tumor growth in von Recklinghausen (type 1) neurofibromatosis. *Cell*. 1992;69(2):265–273.
64. Bollag G, Clapp DW, Shih S, et al. Loss of NF1 results in activation of the Ras signaling pathway and leads to aberrant growth in haematopoietic cells. *Nat Genet*. 1996;12(2):144–148.
65. Tong J, Hannan F, Zhu Y, Bernards A, Zhong Y. Neurofibromin regulates G protein-stimulated adenylyl cyclase activity. *Nat Neurosci*. 2002;5(2):95–96.
66. Dasgupta B, Dugan LL, Gutmann DH. The neurofibromatosis 1 gene product neurofibromin regulates pituitary adenylate cyclase-activating polypeptide-mediated signaling in astrocytes. *J Neurosci*. 2003;23(26):8949–8954.
67. Brown JA, Gianino SM, Gutmann DH. Defective cAMP generation underlies the sensitivity of CNS neurons to neurofibromatosis-1 heterozygosity. *J Neurosci*. 2010;30(16):5579–5589.
68. Hegedus B, Dasgupta B, Shin JE, et al. Neurofibromatosis-1 regulates neuronal and glial cell differentiation from neuroglial progenitors in vivo by both cAMP- and Ras-dependent mechanisms. *Cell Stem Cell*. 2007;1(4):443–457.
69. Dasgupta B, Yi Y, Chen DY, Weber JD, Gutmann DH. Proteomic analysis reveals hyperactivation of the mammalian target of rapamycin pathway in neurofibromatosis 1-associated human and mouse brain tumors. *Cancer Res*. 2005;65(7):2755–2760.
70. Johannessen CM, Reczek EE, James MF, Brems H, Legius E, Cichowski K. The NF1 tumor suppressor critically regulates TSC2 and mTOR. *Proc Natl Acad Sci U S A*. 2005;102(24):8573–8578.
71. Banerjee S, Crouse NR, Emnett RJ, Gianino SM, Gutmann DH. Neurofibromatosis-1 regulates mTOR-mediated astrocyte growth and glioma formation in a TSC/Rheb-independent manner. *Proc Natl Acad Sci U S A*. 2011;108(38):15996–16001.
72. Lee DY, Yeh T-H, Emnett RJ, White CR, Gutmann DH. Neurofibromatosis-1 regulates neuroglial progenitor proliferation and glial differentiation in a brain region-specific manner. *Genes Dev*. 2010;24(20):2317–2329.
73. Gusella JF, Ramesh V, MacCollin M, Jacoby LB. Merlin: the neurofibromatosis 2 tumor suppressor. *Biochim Biophys Acta*. 1999;1423(2):M29–M36.
74. Sun C-X, Robb VA, Gutmann DH. Protein 4.1 tumor suppressors: getting a FERM grip on growth regulation. *J Cell Sci*. 2002;115(Pt 21):3991–4000.
75. Mack NA, Whalley HJ, Castillo-Lluva S, Malliri A. The diverse roles of Rac signaling in tumorigenesis. *Cell Cycle*. 2011;10(10):1571–1581.
76. Xiao G-H, Beeser A, Chernoff J, Testa JR. p21-activated kinase links Rac/Cdc42 signaling to merlin. *J Biol Chem*. 2002;277(2):883–886.
77. Kissil JL, Johnson KC, Eckman MS, Jacks T. Merlin phosphorylation by p21-activated kinase 2 and effects of phosphorylation on merlin localization. *J Biol Chem*. 2002;277(12):10394–10399.
78. Curto M, Cole BK, Lallemand D, Liu C-H, McClatchey AI. Contact-dependent inhibition of EGFR signaling by Nf2/Merlin. *J Cell Biol*. 2007;177(5):893–903.
79. Hennigan RF, Moon CA, Parysek LM, et al. The NF2 tumor suppressor regulates microtubule-based vesicle trafficking via a novel Rac, MLK and p38(SAPK) pathway. *Oncogene*. 2013;32(9):1135–1143.
80. Houshmandi SS, Emnett RJ, Giovannini M, Gutmann DH. The neurofibromatosis 2 protein, merlin, regulates glial cell growth in an ErbB2- and Src-dependent manner. *Mol Cell Biol*. 2009;29(6):1472–1486.
81. Striedinger K, VandenBerg SR, Baia GS, McDermott MW, Gutmann DH, Lal A. The neurofibromatosis 2 tumor suppressor gene product, merlin, regulates human meningioma cell growth by signaling through YAP. *Neoplasia*. 2008;10(11):1204–1212.
82. James MF, Han S, Polizzano C, et al. NF2/merlin is a novel negative regulator of mTOR complex 1, and activation of mTORC1 is associated with meningioma and schwannoma growth. *Mol Cell Biol*. 2009;29(15):4250–4261.
83. Costa RM, Federov NB, Kogan JH, et al. Mechanism for the learning deficits in a mouse model of neurofibromatosis type 1. *Nature*. 2002;415(6871):526–530.
84. Cui Y, Costa RM, Murphy GG, et al. Neurofibromin regulation of ERK signaling modulates GABA release and learning. *Cell*. 2008;135(3):549–560.
85. Shilyansky C, Karlsgodt KH, Cummings DM, et al. Neurofibromin regulates corticostriatal inhibitory networks during working memory performance. *Proc Natl Acad Sci U S A*. 2010;107(29):13141–13146.
86. Li W, Cui Y, Kushner SA, et al. The HMG-CoA reductase inhibitor lovastatin reverses the learning and attention deficits in a mouse model of neurofibromatosis type 1. *Curr Biol*. 2005;15(21):1961–1967.
87. Zhu Y, Romero MI, Ghosh P, et al. Ablation of NF1 function in neurons induces abnormal development of cerebral cortex and reactive gliosis in the brain. *Genes Dev*. 2001;15(7):859–876.
88. Bajenaru ML, Hernandez MR, Perry A, et al. Optic nerve glioma in mice requires astrocyte Nf1 gene inactivation and Nf1 brain heterozygosity. *Cancer Res*. 2003;63(24):8573–8577.

89. Dasgupta B, Li W, Perry A, Gutmann DH. Glioma formation in neurofibromatosis 1 reflects preferential activation of K-RAS in astrocytes. *Cancer Res*. 2005;65(1):236–245.
90. Zhu Y, Ghosh P, Charnay P, Burns DK, Parada LF. Neurofibromas in NF1: Schwann cell origin and role of tumor environment. *Science*. 2002;296(5569):920–922.
91. Yang F-C, Ingram DA, Chen S, et al. Nf1-dependent tumors require a microenvironment containing Nf1+/– and c-kit-dependent bone marrow. *Cell*. 2008;135(3):437–448.
92. Zhu Y, Guignard F, Zhao D, et al. Early inactivation of p53 tumor suppressor gene cooperating with NF1 loss induces malignant astrocytoma. *Cancer Cell*. 2005;8(2):119–130.
93. Kwon C-H, Zhao D, Chen J, et al. Pten haploinsufficiency accelerates formation of high-grade astrocytomas. *Cancer Res*. 2008;68(9):3286–3294.
94. Chen J, Li Y, Yu T-S, et al. A restricted cell population propagates glioblastoma growth after chemotherapy. *Nature*. 2012;488(7412):522–526.
95. McClatchey AI, Saotome I, Mercer K, et al. Mice heterozygous for a mutation at the Nf2 tumor suppressor locus develop a range of highly metastatic tumors. *Genes Dev*. 1998;12(8):1121–1133.
96. Giovannini M, Robanus-Maandag E, van der Valk M, et al. Conditional biallelic Nf2 mutation in the mouse promotes manifestations of human neurofibromatosis type 2. *Genes Dev*. 2000;14(13):1617–1630.
97. Kalamarides M, Niwa-Kawakita M, Leblois H, et al. Nf2 gene inactivation in arachnoidal cells is rate-limiting for meningioma development in the mouse. *Genes Dev*. 2002;16(9):1060–1065.
98. Kalamarides M, Stemmer-Rachamimov AO, Niwa-Kawakita M, et al. Identification of a progenitor cell of origin capable of generating diverse meningioma histological subtypes. *Oncogene*. 2011;30(20):2333–2344.
99. Brems H, Chmara M, Sahbatou M, et al. Germline loss-of-function mutations in SPRED1 cause a neurofibromatosis 1-like phenotype. *Nat Genet*. 2007;39(9):1120–1126.
100. Pasmant E, Sabbagh A, Hanna N, et al. SPRED1 germline mutations caused a neurofibromatosis type 1 overlapping phenotype. *J Med Genet*. 2009;46(7):425–430.
101. Colley A, Donnai D, Evans DG. Neurofibromatosis/Noonan phenotype: a variable feature of type 1 neurofibromatosis. *Clin Genet*. 1996;49(2):59–64.
102. Carcavilla A, Pinto I, Muñoz-Pacheco R, Barrio R, Martin-Frías M, Ezquieta B. LEOPARD syndrome (PTPN11, T468M) in three boys fulfilling neurofibromatosis type 1 clinical criteria. *Eur J Pediatr*. 2011;170(8):1069–1074.
103. Messiaen LM, Callens T, Mortier G, et al. Exhaustive mutation analysis of the NF1 gene allows identification of 95% of mutations and reveals a high frequency of unusual splicing defects. *Hum Mutat*. 2000;15(6):541–555.
104. Valero MC, Martín Y, Hernández-Imaz E, et al. A highly sensitive genetic protocol to detect NF1 mutations. *J Mol Diagn*. 2011;13(2):113–122.
105. Segal L, Darvish-Zargar M, Dilenge M-E, Ortenberg J, Polomeno RC. Optic pathway gliomas in patients with neurofibromatosis type 1: follow-up of 44 patients. *J AAPOS*. 2010;14(2):155–158.
106. King A, Listernick R, Charrow J, Piersall L, Gutmann DH. Optic pathway gliomas in neurofibromatosis type 1: the effect of presenting symptoms on outcome. *Am J Med Genet A*. 2003;122A(2):95–99.
107. Sharif S, Ferner R, Birch JM, et al. Second primary tumors in neurofibromatosis 1 patients treated for optic glioma: substantial risks after radiotherapy. *J Clin Oncol*. 2006;24(16):2570–2575.
108. Evans DGR, Baser ME, O'Reilly B, et al. Management of the patient and family with neurofibromatosis 2: a consensus conference statement. *Br J Neurosurg*. 2005;19(1):5–12.
109. MacCollin M, Gutmann DH, Korf B, Finkelstein R. Establishing priorities in neurofibromatosis research: a workshop summary. *Genet Med*. 2001;3(3):212–217.
110. University of Alabama at Birmingham. Study of RAD001 (Everolimus) for children with NF1 and chemotherapy-refractory radiographic progressive low grade gliomas. In: *ClinicalTrials.gov [Internet]*. Bethesda (MD): National Library of Medicine (US); 2000- [cited 2013 Jun 13]. Available from: http://clinicaltrials.gov/show/NCT01158651.
111. Novartis Pharmaceuticals. Efficacy and safety of RAD001 in treating plexiform neurofibromas (PN) associated with neurofibromatosis (NF1). In: *ClinicalTrials.gov [Internet]*. Bethesda (MD): National Library of Medicine (US); 2000- [cited 2013 Jun 13]. Available from: http://clinicaltrials.gov/show/NCT01365468.
112. Assistance Publique - Hôpitaux de Paris. Use of RAD001 as monotherapy in the treatment of neurofibromatosis 1 related internal plexiform neurofibromas. In: *ClinicalTrials.gov [Internet]*. Bethesda (MD): National Library of Medicine (US); 2000- [cited 2013 Jun 13]. Available from: http://www.clinicaltrials.gov/show/NCT01412892.
113. National Cancer Institute. Sirolimus to treat plexiform neurofibromas in patients with neurofibromatosis type I. In: *ClinicalTrials.gov [Internet]*. Bethesda (MD): National Library of Medicine (US); 2000- [cited 2013 Jun 13]. Available from: http://www.clinicaltrials.gov/show/NCT00652990.
114. University of Alabama at Birmingham. A phase II study of the mTOR inhibitor sirolimus in neurofibromatosis type 1 related plexiform neurofibromas. In: *ClinicalTrials.gov [Internet]*. Bethesda (MD): National Library of Medicine (US); 2000- [cited 2013 Jun 13]. Available from: http://www.clinicaltrials.gov/show/NCT00634270.
115. Jessen WJ, Miller SJ, Jousma E, et al. MEK inhibition exhibits efficacy in human and mouse neurofibromatosis tumors. *J Clin Invest*. 2013;123(1):340–347.
116. Chang T, Krisman K, Theobald EH, et al. Sustained MEK inhibition abrogates myeloproliferative disease in Nf1 mutant mice. *J Clin Invest*. 2013;123(1):335–339.
117. National Cancer Institute. AZD6244 hydrogen sulfate for children with nervous system tumors. In: *ClinicalTrials.gov [Internet]*. Bethesda (MD): National Library of Medicine (US); 2000- [cited 2013 Jun 13]. Available from: http://www.clinicaltrials.gov/show/NCT01362803.
118. Hospices Civils de Lyon. NF1-attention: study of children with neurofibromatosis type 1 treated by Methylphenidate. In: *ClinicalTrials.gov [Internet]*. Bethesda (MD): National Library of Medicine (US); 2000- [cited 2013 Jun 13]. Available from: http://www.clinicaltrials.gov/show/NCT00169611.

119. Krab LC, de Goede-Bolder A, Aarsen FK, et al. Effect of simvastatin on cognitive functioning in children with neurofibromatosis type 1: a randomized controlled trial. *JAMA*. 2008;300(3):287–294.
120. Acosta MT, Kardel PG, Walsh KS, Rosenbaum KN, Gioia GA, Packer RJ. Lovastatin as treatment for neurocognitive deficits in neurofibromatosis type 1: phase I study. *Pediatr Neurol*. 2011;45(4):241–245.
121. University of Alabama at Birmingham. A randomized placebo-controlled study of lovastatin in children with neurofibromatosis type 1. In: *ClinicalTrials.gov [Internet]*. Bethesda (MD): National Library of Medicine (US); 2000- [cited 2013 Jun 13]. Available from: http://www.clinicaltrials.gov/show/NCT00853580.
122. University of California, Los Angeles. Trial to evaluate the safety of lovastatin in individuals with neurofibromatosis type I. In: *ClinicalTrials.gov [Internet]*. Bethesda (MD): National Library of Medicine (US); 2000- [cited 2013 Jun 13]. Available from: http://www.clinicaltrials.gov/show/NCT00352599.
123. Karajannis MA, Legault G, Hagiwara M, et al. Phase II trial of lapatinib in adult and pediatric patients with neurofibromatosis type 2 and progressive vestibular schwannomas. *Neuro Oncol*. 2012;14(9):1163–1170.
124. Sidney Kimmel Comprehensive Cancer Center. Concentration and activity of lapatinib in vestibular schwannomas. In: *ClinicalTrials.gov [Internet]*. Bethesda (MD): National Library of Medicine (US); 2000- [cited 2013 Jun 13]. Available from: http://www.clinicaltrials.gov/show/NCT00863122.
125. Plotkin SR, Merker VL, Halpin C, et al. Bevacizumab for progressive vestibular schwannoma in neurofibromatosis type 2: a retrospective review of 31 patients. *Otol Neurotol*. 2012;33(6):1046–1052.
126. University of Alabama at Birmingham. Phase 2 study of bevacizumab in children and young adults with neurofibromatosis 2 and progressive vestibular schwannomas. In: *ClinicalTrials.gov [Internet]*. Bethesda (MD): National Library of Medicine (US); 2000- [cited 2013 Jun 13]. Available from: http://www.clinicaltrials.gov/show/NCT01767792.
127. Lee TX, Packer MD, Huang J, et al. Growth inhibitory and anti-tumour activities of OSU-03012, a novel PDK-1 inhibitor, on vestibular schwannoma and malignant schwannoma cells. *Eur J Cancer*. 2009;45(9):1709–1720.
128. Tanaka K, Eskin A, Chareyre F, et al. Therapeutic potential of HSP90 inhibition for neurofibromatosis type 2. *Clin Cancer Res*. 2013;19(14):3856–3870.

CHAPTER

80

Tuberous Sclerosis Complex

Monica P. Islam and E. Steve Roach

Nationwide Children's Hospital, The Ohio State University College of Medicine, Columbus, OH, USA

INTRODUCTION

Tuberous sclerosis complex (TSC) is a neurocutaneous disorder of cellular differentiation and proliferation that affects the brain, skin, kidneys, heart, and other organs. Many of the clinical manifestations of TSC result from hamartomatous malformations in the affected organs, and abnormal neuronal migration plays a major additional role in neurological dysfunction.[1,2]

Von Recklinghausen's description in 1862 of a child with cardiac and brain lesions generally is cited as the first description of TSC. Once considered extremely rare, estimates of the frequency of TSC have risen dramatically in recent years with the recognition of milder phenotypes. Difficulty identifying individuals with mild manifestations of TSC makes an exact incidence hard to pinpoint, but population-based studies suggest a prevalence of 1 in 6000 to 15,000 individuals.[3–5]

As with many autosomal dominant disorders, the rate of spontaneous mutation of TSC is high, and phenotypic heterogeneity is common.[6] Variability of clinical expression occurs even among affected members of the same family,[7] and discordant homozygous twins have been described.[8] Given this highly variable clinical picture, diagnosis can be difficult in patients and family members with subtle findings, and gene testing becomes particularly valuable for these individuals.

CLINICAL MANIFESTATIONS

Cutaneous Findings

Similar to other neurocutaneous disorders, the cutaneous manifestations of TSC play an important role in establishing the diagnosis. Skin lesions include hypomelanotic macules, facial angiofibromas, the shagreen patch, and ungual fibromas. A slightly raised, plaque-like lesion on the forehead is a less common but early cutaneous lesion of TSC.

Hypomelanotic macules (ash leaf spots) are found in at least 90% of patients (Figure 80.1A). The lesions may not be present at birth[9] and often are difficult to see in the newborn even with an ultraviolet light. A small patch of white scalp hair is sometimes noted. Individuals without TSC can have one or two hypomelanotic macules; thus, they are not unique to TSC.[10]

Facial angiofibromas (adenoma sebaceum) are hamartomatous nodules containing vascular and connective tissue elements.[11] These lesions develop in about three-quarters of TSC patients. While facial angiofibromas correlate strongly with the presence of TSC, rare examples of families with these lesions without other clinical or molecular evidence of TSC have been documented. They may not become apparent for several years after birth, often appearing after the diagnosis has been established by other means. Typically, the lesions begin during the preschool years as a few small red papules on the malar region. The lesions gradually become larger and more numerous, sometimes extending down the nasolabial folds or onto the chin (Figure 80.1B).

Rosenberg's Molecular and Genetic Basis of Neurological and Psychiatric Disease
http://dx.doi.org/10.1016/10.1016/B978-0-12-410529-4.00080-2

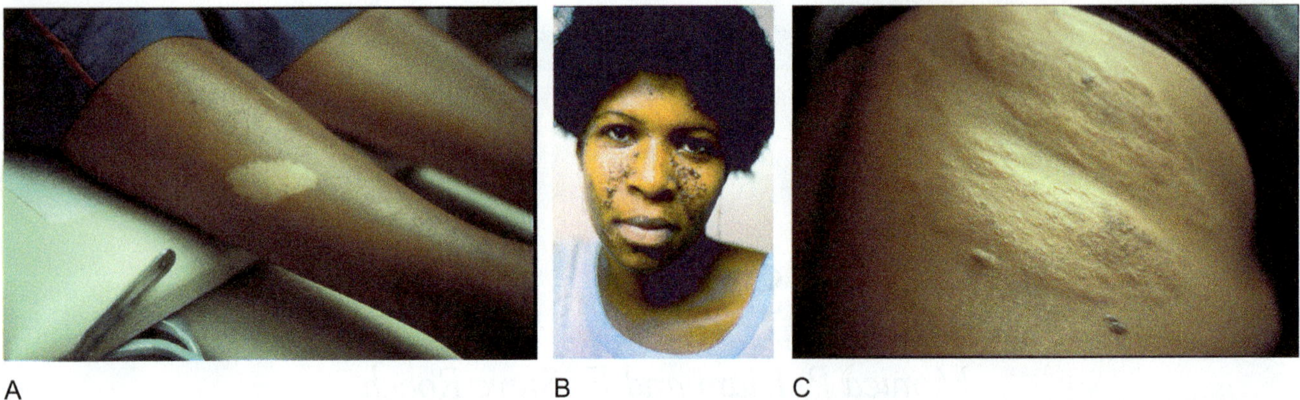

FIGURE 80.1 **Classic cutaneous manifestations of tuberous sclerosis.** (**A**) A hypomelanotic macule, or ash leaf spot. (**B**) Facial angiofibromas. (**C**) Shagreen patch. Reprinted with permission from Roach ES. Diagnosis and management of neurocutaneous syndromes. *Semin Neurol*. 1988;8: 83–96, © 2014 Thieme Publishing Group.

Most commonly found on the back or flank area, the shagreen patch is an irregularly shaped, slightly raised or textured skin lesion measuring several centimeters in diameter (Figure 80.1C). The lesion is found in nearly half of TSC patients but may not develop until after puberty. This finding is not pathognomonic.[12]

Ungual fibromas are nodular or fleshy lesions that arise adjacent to or from underneath the nails. These occasionally develop as singular lesions after trauma in non-TSC individuals, but otherwise they appear specific to TSC. Ungual fibromas are seen in about 15–20% of unselected TSC patients and are more likely to develop in adolescence or later.

Ophthalmologic Findings

Ophthalmologic lesions occur in the form of retinal astrocytic hamartomas and may be difficult to identify without pupillary dilatation and fundoscopy, particularly in uncooperative children. Thus, the frequency of retinal hamartomas in TSC patients has varied from almost one-third to 87% of patients.[13] The appearance of the retinal lesions varies from classic mulberry lesions adjacent to the optic disc (Figure 80.2) to more common plaque-like hamartomas. Retinal hamartomas occur in individuals with other disorders, so they are suggestive of TSC but not pathognomonic. Pigmented iris defects are not as common as retinal lesions.

Most retinal hamartomas remain asymptomatic. Occasional patients have visual impairment caused by a large hamartoma in the macular region, and there are rare instances of visual loss following retinal detachment, vitreous hemorrhage, or hamartoma enlargement.[14] For the most part, treatment of the retinal lesions and repeated ophthalmologic examinations are unnecessary. Like the cutaneous manifestations, the significance of the iris lesion lies mainly in its implication for establishing a diagnosis of TSC.

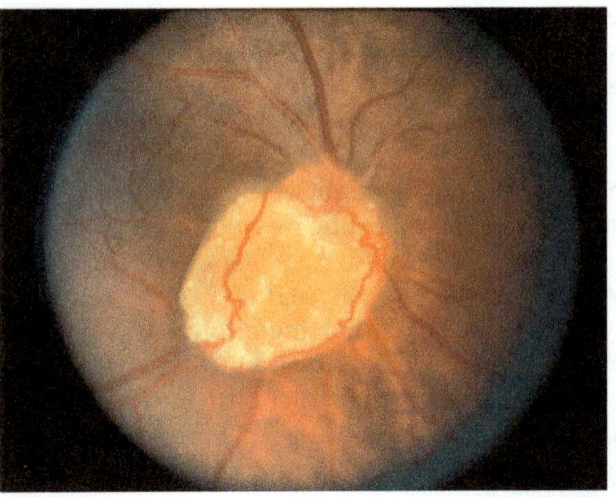

FIGURE 80.2 **This mulberry astrocytoma adjacent to the optic nerve is typical of those found with tuberous sclerosis.** Reprinted with permission from Roach ES. Neurocutaneous syndromes. *Pediatr Clin North Am*. 1992;39: 591–620, © 1992 Elsevier.

Neurological Manifestations

The extent of neurological involvement appears related to the burden of identifiable cortical tubers, cortical dysplasias and subependymal nodules, which can be identified on neuroimaging. Seizures and intellectual disability are the most important neurological manifestations of TSC, although patients with little or no neurological impairment are common. Focal neurological deficits such as hemiparesis sometimes are present.

Seizures occur in about 90% of patients. Infantile spasms are particularly common and may be the first sign that a child has TSC. An infant without known TSC presenting with infantile spasms may be diagnosed by a careful skin examination or by neuroimaging that reveals the pathognomonic lesions. Generalized tonic–clonic seizures, focal seizures, tonic, atonic, myoclonic, and atypical absence seizures also occur. The choice of antiepileptic drugs should be tailored to the patient's age and seizure type; in the case of infantile spasms, the medication of choice has been vigabatrin.[15] Even in the presence of multiple cortical tubers, the dysplasia associated with or surrounding a single cortical tuber may be responsible for all or most of the seizures at a given time—and therefore amenable to surgical resection for treatment.[16]

Intellectual disability has a high association with uncontrolled epilepsy, although many individuals with epilepsy are cognitively normal. Clearly some of the impairment in these children stems from poorly controlled seizures. There is also some evidence that individuals with severe cerebral disruption from TSC tend to have a greater likelihood of both uncontrolled seizures and intellectual disability, whereas patients with only a few small cerebral lesions typically have better intellectual function and control.[17,18] In addition, patients with infantile spasms have more cerebral lesions than children with other seizure types.[19] Of course, a more disabling mutation could lead to more severe intellectual disability and a greater brain lesion burden via different processes. The age at which seizures begin could be merely an epiphenomenon of the severity of the brain lesions, because more cerebral lesions occur in children who develop seizures before age 1 year. The eventual level of the neurological function is probably determined by a combination of the severity of the seizures and the number, size, and location of the cerebral lesions.[18]

Approximately 60% of TSC patients exhibit significant intellectual disability,[20] although the degree of cognitive impairment ranges from mild to profound. Children whose seizures begin early in life are more likely to have intellectual impairment,[21] and children whose seizures continue unchecked despite treatment probably have a higher likelihood of intellectual impairment. Autism and various behavioral disturbances, including hyperkinesis, aggressiveness, sleep disturbance, and frank psychosis, are sometimes noted, either as isolated problems or in combination with epilepsy or intellectual deficit.[22,23]

Some children seem to develop normally until the onset of seizures, after which their progress slows or they actually lose developmental milestones. Of those who regress, most eventually stabilize. Patients who are normal by school age are likely to remain so,[24] although some of these children will still be disabled by behavioral disorders.

Subependymal nodules (SENs) are present in 80% of TSC individuals, and subependymal giant cell astrocytomas (SEGAs) develop in 5–15% of TSC patients.[25,26] Such a tumor may be heralded by a new focal neurological deficit, unexplained behavior change, deterioration of seizure control, or other potential signs or symptoms of increased intracranial pressure. Acute or subacute onset of neurological dysfunction may result from sudden obstruction of the ventricular system by an intraventricular portion of the tumor or by hemorrhage within the tumor itself.[27] Giant cell astrocytomas most often occur in the anterior horn of the lateral ventricle (Figure 80.3).

Giant cell astrocytomas remain asymptomatic for long periods, and for this reason the best management is not always clear. Many of these lesions exhibit contrast enhancement on either cranial computed tomography (CT) or magnetic resonance imaging (MRI),[25] but contrast enhancement does not reliably predict tumor enlargement. Brain MRI every 1–3 years until age 25 years helps identify and follow giant cell astrocytomas. Individuals with an astrocytoma by that age will need continued future imaging but those who do not have these tumors identified by then can be imaged as needed.[25,28]

Neuropathology

Several types of brain lesions characterize TSC (Figure 80.4), but like the clinical manifestations, the neuropathological changes are quite variable. Not all patients have cortical tubers, whereas others have multiple. The size of these lesions varies. Histologically the tuber is a hamartoma, composed of heterogeneous neurons and glial cells with disorganized lamination (Figure 80.5).[29] A single brain hamartoma is suggestive of TSC, but multiple lesions are diagnostic.

Subependymal nodules of TSC most often are found in the anterior portions of the lateral ventricles. These nodules are made up of glial and vascular elements and, in time, accumulate the dense-calcium deposits that identify them radiographically. Some of the lesions contain large cells similar to those of the less common giant cell astrocytoma.[30]

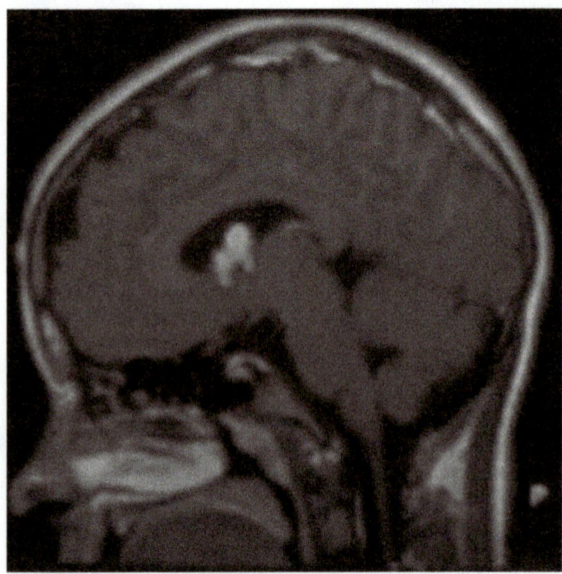

FIGURE 80.3 **Subependymal giant cell astrocytoma.** With growth, this introduces potential for obstructive hydrocephalus. Figure courtesy of Dr. E. Steve Roach, Ohio State University College of Medicine, with permission.

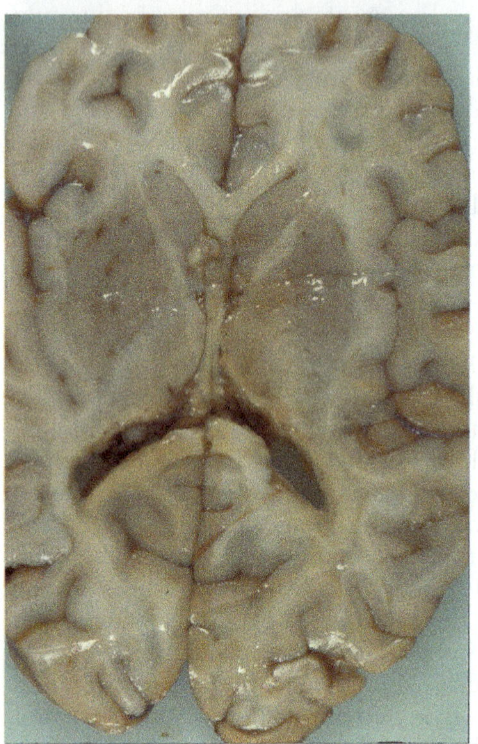

FIGURE 80.4 **A neuropathologic hallmark of tuberous sclerosis is the nodular subependymal lesion, often visible on computed tomographic scans as a calcified nodule adjacent to the ventricular system.** Reprinted with permission from Roach ES. Neurocutaneous syndromes. *Pediatr Clin North Am*. 1992;39:591–620.

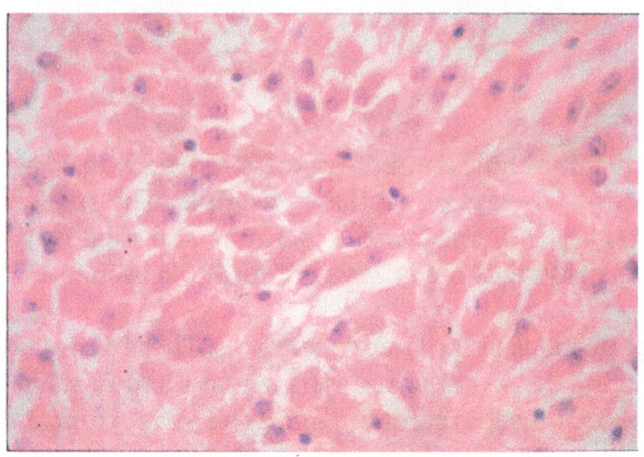

FIGURE 80.5 **Histology of a giant cell astrocytoma.** Figure courtesy of Dr. E. Steve Roach, Ohio State University College of Medicine, with permission.

Failure of proper neuronal migration during neocortical formation leads to the formation of multiple areas of cortical dysplasia that are often identifiable on imaging. Sometimes islands of heterotopic gray matter or areas with little myelin formation underlie these obvious cortical lesions. The appearance of the cortex aside from cortical defects and tubers generally is often normal, but subtle changes in the organizational pattern of the neurons and gliosis can be demonstrated even in parts of the brain that are grossly normal.[31] These findings probably account for the occasional patient with neurological symptoms even without a large number of visible cortical defects and tubers.

Neuroradiographic Findings

Neuroimaging can be an important part of making a diagnosis of TSC and then monitoring TSC. Cortical tubers appear on CT as hypodense and become hyperdense with calcification (Figure 80.6). Calcium deposits in tubers tend to become denser with time, and therefore these lesions may not be fully evident in the imaging of infants.[32] Cortical tubers and focal cortical dysplasias may be noted on CT scans, but these lesions are far more obvious on T2-weighted MRI. At times a linear tract perpendicular to the cerebral cortex is visible on MRI scans, probably a residual lesion from abnormal neuronal migration.[33] Nodular subependymal lesions that have not yet calcified produce a high-signal lesion on T2-weighted MRI studies. Large calcific lesions are seen on MRI scans as a signal void. Cerebellar lesions are found with MRI in about one-fourth of TSC patients.[34]

Multimodality imaging has been helpful in epilepsy surgery evaluations in patients without tuberous sclerosis, but the imaging abnormalities in tuberous sclerosis can be extensive whereas the epileptogenic region may be fairly localized. Interictal positron emission tomography (PET) imaging with 2-deoxy-2-(^{18}F) fluoro-D-glucose (FDG) demonstrates multiple areas of hypometabolism, but the tracer α-[^{11}C]-methyl-L-tryptophan (AMT) has been able to identify epileptogenic regions on the basis of tryptophan metabolism that may be altered by neuroinflammation affecting the kynurenine pathway.[35]

Cardiovascular Complications

Cardiac rhabdomyomas occur in approximately two-thirds of TSC patients when examined early in life.[36] Few of these individuals develop cardiac symptoms, and most of those who do manifest cardiac failure immediately after birth. Some older patients, however, develop an arrhythmia, most commonly Wolff–Parkinson–White syndrome.[37] Rhabdomyomas also may manifest as high output cardiac failure at birth. The lesions tend to be multiple, and there is evidence that their size and number diminish with age.[38] Lesions may be evident on prenatal ultrasound. In contrast, only 30% of TSC patients undergoing autopsy have a cardiac rhabdomyoma.[39] Most neonates who have symptomatic cardiac rhabdomyomas later are diagnosed with TSC.[40] Cardiac rhabdomyomas, like most other tumors associated with TSC, are considered hamartomas rather than true neoplasms.[41]

Congestive heart failure can be caused either by obstruction of blood flow by an intraluminal tumor or by inadequate normal myocardium to maintain perfusion. Surgical removal of a symptomatic intraluminal tumor may be beneficial, although the surgery is often difficult because of unstable cardiac function. When extensive, treatment of intramural rhabdomyoma may be limited to digoxin and diuretics, and many of these children do poorly,[41] while others stabilize after medical treatment and eventually improve.

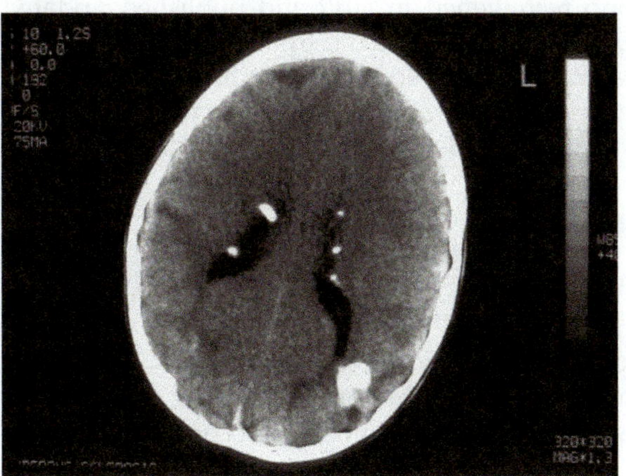

FIGURE 80.6 A computed tomographic scan of a child with tuberous sclerosis demonstrates typical subependymal calcifications. A large calcified parenchymal lesion is present on the left and a hypodense cortical tuber is visible on the right. Reprinted with permission from[33], © 2006 John Wiley and Sons.

Occasional patients develop cerebral thromboembolism,[42,43] probably because an intracardiac tumor promotes thrombus formation, although actual embolization of a tumor fragment was suggested in one patient.[43] Intracranial arteriopathy such as aneurysm occurs rarely but at a greater rate than in the general population and perhaps an earlier age but has uncertain clinical significance.[44]

Renal Involvement

Renal angiomyolipomas occur in about two-thirds of TSC patients and are probably the leading cause of death in adults with TSC.[45] Angiomyolipomas histologically are benign tumors consisting of varying amounts of vascular tissue, fat, and smooth muscle.[46] Although an angiomyolipoma is not considered specific for TSC, at least half of patients with these renal lesions have other evidence of TSC.[47] The percentage with TSC may actually be higher now that patients with subtle forms of TSC are more likely to be recognized; occasional patients with an isolated renal angiomyolipoma have children with typical TSC.[48,49] The prevalence and size of these renal hamartomas increase with age. Bilateral tumors or more than one tumor in a single kidney are common;[48] tumors larger than 4cm in diameter are more likely to enlarge and become symptomatic.[50,51]

Renal cell carcinoma and other kidney malignancies are more common in TSC patients and can occur at an earlier age compared to the general population.[52] Single or multiple renal cysts occur less often in TSC patients than angiomyolipomas but also may appear earlier in life. A subgroup of patients has adult polycystic kidney disease,[53] and it was this clue that led to the discovery of the second gene for TSC adjacent to the gene for adult polycystic kidney disease of chromosome 16. Cysts vary in size and location within the kidney. Larger cysts are easily identified with ultrasonography or CT scanning, and the combination of renal cysts and angiomyolipomas is characteristic of TSC.

Both angiomyolipomas and renal cysts can remain asymptomatic and may require no treatment. Hematuria, flank or abdominal pain, or retroperitoneal hemorrhage are the common modes of presentation. Renal failure can result from displacement of the normal renal parenchyma by tumor or from obstructive uropathy. MRI of the abdomen is the recommended mode of surveillance to identify lesions that are enlarging or are likely to enlarge. Timely identification sometimes allows these lesions to be selectively removed, sparing some of the kidney.[50,54]

Pulmonary Complications

Once considered to be uncommon, pulmonary lesions can be demonstrated in almost half of women with TSC who undergo chest CT.[55] Moreover, the risk of both pulmonary lesions and the likelihood of symptomatic lymphangioleiomyomatosis (LAM) increase steadily with age, and eventually the majority of women will develop pulmonary lesions and about two-thirds of them become symptomatic.[55] Manifestations of pulmonary TSC include spontaneous pneumothorax, dyspnea, cough, hemoptysis, and respiratory failure. Pulmonary disease is far less common in males, but it does occur.[56] Of individuals with TSC who develop symptoms, 10–12% die within 10 years of symptom onset from complications of pulmonary TSC.[55,57] Better understanding of the role of the mTOR pathway has led to treatment options with mTOR inhibitors.[58]

DIAGNOSTIC CRITERIA

The International Tuberous Sclerosis Complex Consensus Group reconvened in 2012 and published updated guidelines.[26] The diagnostic criteria remain clinically based, but genetic testing has been introduced as a parallel diagnostic criterion. Finding of a disease-causing TSC1 or TSC2 mutation can be the sole basis for making a diagnosis of TSC and implementing surveillance or therapeutic options even in the absence of threshold clinical signs. Clinical features still remain as major and minor criteria to define definite versus probable TSC; the 2012 guidelines stipulate numbers and size of varying clinical features (Table 80.1).[26]

GENETIC AND MOLECULAR BASIS

TSC is inherited as an autosomal dominant trait with variable penetrance. Even prior to gene identification, the estimated rate of spontaneous mutation varied from 56–86% depending in part on the completeness of investigation of the extended family.[59,60] Phenotypic heterogeneity was noted as the rule, as illustrated by discordance between homozygous twins.[8]

TABLE 80.1 Updated Diagnostic Criteria for Tuberous Sclerosis Complex 2012

A. GENETIC DIAGNOSTIC CRITERIA

The identification of a pathogenic mutation in either TSC1 or TSC2 is sufficient to make a definite diagnosis of TSC. A pathogenic mutation is defined as a mutation that clearly inactivates the function of the TSC1 or TSC2 proteins (e.g., out-of-frame indel or nonsense mutation), prevents protein synthesis (e.g., large genomic deletion), or is a missense mutation whose effect on protein function has been established. Other TSC1 or TSC2 variants whose effect on function is less certain do not meet these criteria and are insufficient to support a definite diagnosis of TSC. Note that 10–25% of TSC patients have no mutation identified by conventional genetic testing, so a normal result does not exclude TSC or affect the use of clinical diagnostic criteria to diagnose TSC.

B. CLINICAL DIAGNOSTIC CRITERIA

Major features:
1. Hypomelanotic macules (≥3, at least 5-mm diameter)
2. Angiofibromas (≥3) or fibrous cephalic plaque
3. Ungual fibromas (≥2)
4. Shagreen patch
5. Multiple retinal hamartomas
6. Cortical dysplasias*
7. Subependymal nodules
8. Subependymal giant cell astrocytoma
9. Cardiac rhabdomyoma
10. Lymphangioleiomyomatosis (LAM)†
11. Angiomyolipomas (≥2)†

Minor features:
1. "Confetti" skin lesions
2. Dental enamel pits (≥3)
3. Intraoral fibromas (≥2)
4. Retinal achromic patch
5. Multiple renal cysts
6. Nonrenal hamartomas

Definite diagnosis: Two major features or one major feature with ≥2 minor features
Possible diagnosis: Either one major feature or ≥2 minor features

**Includes tubers and cerebral white matter radial migration lines.*
†A combination of the two major clinical features (LAM and angiomyolipomas) without other features does not meet criteria for a definite diagnosis.
Adapted from[26].

Examination of multigenerational families with multiple affected individuals led to the initial identification of a TSC candidate gene region on the long arm of chromosome 9.[61] The 9q34 locus with the gene product hamartin later was identified as the TSC1 gene.[62] Another TSC locus at chromosome 16p13 later was confirmed as the TSC2 gene for tuberin. The discovery of this locus arose with the examination of loci close to the adult polycystic kidney disease gene—since patients with tuberous sclerosis frequently had renal cysts.[63] TSC1 mutations frequently are single-base substitutions that result in a nonsense mutation; nonsense mutations occur less frequently in the TSC2 mutations. Large deletion mutations are more likely to occur for TSC2 compared to TSC1. Detection rate for mutations in these two genes is 75–90%; this may be explained by limitations of testing, mosaicism, or the rare possibility of another locus. Sporadic mutations more commonly affect TSC2. The larger size of the TSC2 coding region may enable this. Mutations are more evenly split between TSC1 and TSC2 in familial cases.[64]

Aberrant neuronal migration had been suspected to cause much of the neurological impairment of patients with TSC, whereas abnormal cellular differentiation and proliferation was suspected to contribute to tumor formation in various organs. The identification of tuberin and hamartin as the responsible gene products did not explain how the two genes produce a common phenotype until the discovery that the two formed a single functional complex. *Drosophila* studies tied this complex to a phosphoinositide-3-kinase (PI3K) pathway. The tuberin–hamartin complex regulates the function of mTOR (initially mTOR signified "mammalian target of rapamycin," but the ubiquitous nature of the pathway across species lead to its more recent designation as "mechanistic target of rapamycin"). The PI3K and mTOR pathways regulate cell proliferation. In particular, phosphorylation of tuberin inhibits the tuberin–hamartin complex and the subsequent activation of the mTOR pathway.[65]

To target the dysfunctional upstream regulation of the mTOR pathway, rapamycin was tested *in vitro* and *in vivo* and ultimately has led to the trial and use of the oral rapamycin derivative, everolimus. This exciting bench-to-bedside process has introduced the use of everolimus as intervention for multiple features of tuberous sclerosis: SEGAs, renal angiomyolipomas, pulmonary LAM, and facial angiofibromas. The EXIST-1 trial demonstrated decreased progression of SEGA volume in a significant number of patients as a primary endpoint and a positive impact on skin findings and angiomyolipomas as a secondary endpoint and exploratory endpoint, respectively; the EXIST-2 trial demonstrated significant reduction in angiomyolipoma volume as a primary endpoint.[66] The independent impact on seizures has been under investigation.[67]

Genetic Counseling

The revised 2012 management guidelines suggest that genetic counseling should be offered to individuals affected with tuberous sclerosis when they reach reproductive age.[28] When the clinical diagnosis is unclear, genetic testing should be offered. When a mutation has been identified, family members at risk can be tested for the known mutation in their kindred if indicated for an indeterminate diagnosis in the family member or for the purposes of family planning in the family member. Different centers likely vary in their practices to perform genetic testing in unaffected family members. Taking responsibility for screening a possibly undiagnosed family member can be an ambiguous issue. Genetics clinics and many tuberous sclerosis clinics may examine adult and pediatric family members or at least screen for symptoms that might warrant further referral or investigation. Clinical examination and radiographic examination of unaffected family members can yield previously unappreciated hypopigmented patches or undetected findings of cardiac rhabdomyoma, intracranial calcification and renal cysts.

References

1. Crino PB, Nathanson KL, Henske EP. The tuberous sclerosis complex. *N Engl J Med*. 2006;355:1345–1356.
2. Marcotte L, Crino PB. The neurobiology of the tuberous sclerosis complex. *Neuromolecular Med*. 2006;8:531–546.
3. Wiederholt WC, Gomez MR, Kurland LT. Incidence and prevalence of tuberous sclerosis in Rochester, Minnesota, 1950 through 1982. *Neurology*. 1985;35:600–603.
4. Hunt A, Lindenbaum RH. Tuberous sclerosis: a new estimate of prevalence within the Oxford region. *J Med Genet*. 1984;21:272–277.
5. Shepherd CW, Beard CM, Gomez MR, Kurland LT, Whisnant JP. Tuberous sclerosis complex in Olmsted County, Minnesota, 1950–1989. *Arch Neurol*. 1991;48:400–401.
6. Baraitser M, Patton MA. Reduced penetrance in tuberous sclerosis. *J Med Genet*. 1985;22:29–31.
7. Smalley SL, Burger F, Smith M. Phenotypic variation of tuberous sclerosis in a single extended kindred. *J Med Genet*. 1994;31:761–765.
8. Gomez MR, Kuntz NL, Westmoreland BF. Tuberous sclerosis, early onset of seizures, and mental subnormality: study of discordant homozygous twins. *Neurology*. 1982;32:604–611.
9. Oppenheimer EY, Rosman NP, Dooling EC. The late appearance of hypopigmented maculae in tuberous sclerosis. *Am J Dis Child*. 1985;139:408–409.
10. Gold AP, Freeman JM. Depigmented nevi: the earliest sign of tuberous sclerosis. *Pediatrics*. 1965;35:1003–1005.
11. Nickel WR, Reed WB. Tuberous sclerosis- special reference to the microscopic alterations in the cutaneous hamartomas. *Arch Dermatol*. 1962;85:209–226.
12. Jozwiak S, Schwartz RA, Janniger CK, Michalowicz R, Chmielik J. Skin lesions in children with tuberous sclerosis complex: their prevalence, natural course, and diagnostic significance. *Int J Dermatol*. 1998;37:911–917.
13. Aronow ME, Nakagawa JA, Gupta A, Traboulsi EI, Singh AD. Tuberous sclerosis complex: genotype/phenotype correlation of retinal findings. *Ophthalmology*. 2012;119:1917–1923.
14. Nyboer JH, Robertson DM, Gomez MR. Retinal lesions in tuberous sclerosis. *Arch Ophthalmol*. 1976;94:1277–1280.
15. Thiele EA. Managing epilepsy in tuberous sclerosis complex. *J Child Neurol*. 2004;19:680–686.
16. Bebin EM, Kelly PJ, Gomez MR. Surgical treatment for epilepsy in cerebral tuberous sclerosis. *Epilepsia*. 1993;34:651–657.
17. Roach ES, Williams DP, Laster DW. Magnetic resonance imaging in tuberous sclerosis. *Arch Neurol*. 1987;44:301–303.
18. Jambaque I, Cusmai R, Curatolo P, Cortesi F, Perrot C, Dulac O. Neuropsychological aspects of tuberous sclerosis in relation to epilepsy and MRI findings. *Dev Med Child Neurol*. 1991;33:698–705.
19. Shepherd CW, Houser OW, Gomez MR. MR findings in tuberous sclerosis complex and correlation with seizure development and mental impairment. *AJNR Am J Neuroradiol*. 1995;16:149–155.
20. Winterkorn EB, Pulsifer MB, Thiele EA. Cognitive prognosis of patients with tuberous sclerosis complex. *Neurology*. 2007;68:62–64.
21. Shepherd CW, Stephenson JB. Seizures and intellectual disability associated with tuberous sclerosis complex in the west of Scotland. *Dev Med Child Neurol*. 1992;34:766–774.
22. Hunt A, Stores G. Sleep disorder and epilepsy in children with tuberosis sclerosis: a questionnaire-based study. *Dev Med Child Neurol*. 1994;36:108–115.
23. Curatolo P, Cusmai R, Cortesi F, Chiron C, Jambaque I, Dulac O. Neuropsychiatric aspects of tuberous sclerosis. *Ann N Y Acad Sci*. 1991;615:8–16.
24. Webb DW, Thomson JL, Osborne JP. Cranial magnetic resonance imaging in patients with tuberous sclerosis and normal intellect. *Arch Dis Child*. 1991;66:1375–1377.
25. Roth J, Roach ES, Bartels U, et al. Subependymal giant cell astrocytoma: diagnosis, screening, and treatment. Recommendations from the international tuberous sclerosis complex consensus conference 2012. *Pediatr Neurol*. 2013.

26. Northrup H, Krueger DA. Tuberous sclerosis complex diagnostic criteria update: recommendations of the 2012 International Tuberous Sclerosis Complex Consensus Conference. *Pediatr Neurol*. 2013;49:243–254.
27. Waga S, Yamamoto Y, Kojima T, Sakakura M. Massive hemorrhage in tumor of tuberous sclerosis. *Surg Neurol*. 1977;8:99–101.
28. Krueger DA, Northrup H. Tuberous sclerosis complex surveillance and management: recommendations of the 2012 International Tuberous Sclerosis Complex Consensus Conference. *Pediatr Neurol*. 2013;49:255–265.
29. Grajkowska W, Kotulska K, Jurkiewicz E, Matyja E. Brain lesions in tuberous sclerosis complex. Review. *Folia Neuropathol*. 2010;48:139–149.
30. Reagan TJ. Neuropathology. In: Gomez MR, ed. Tuberous Sclerosis. 2nd ed. New York: Raven Press; 1988:63–74.
31. Marcotte L, Aronica E, Baybis M, Crino PB. Cytoarchitectural alterations are widespread in cerebral cortex in tuberous sclerosis complex. *Acta Neuropathol*. 2012;123:685–693.
32. Datta AN, Hahn CD, Sahin M. Clinical presentation and diagnosis of tuberous sclerosis complex in infancy. *J Child Neurol*. 2008;23:268–273.
33. Roach ES, Kerr J, Mendelsohn D, Laster DW, Raeside C. Diagnosis of symptomatic and asymptomatic gene carriers of tuberous sclerosis by CCT and MRI. *Ann N Y Acad Sci*. 1991;615:112–122.
34. Vaughn J, Hagiwara M, Katz J, et al. MRI characterization and longitudinal study of focal cerebellar lesions in a young tuberous sclerosis cohort. *AJNR Am J Neuroradiol*. 2013;34:655–659.
35. Chugani HT, Luat AF, Kumar A, Govindan R, Pawlik K, Asano E. alpha-[11C]-Methyl-L-tryptophan–PET in 191 patients with tuberous sclerosis complex. *Neurology*. 2013;81:674–680.
36. Gibbs JL. The heart and tuberous sclerosis. An echocardiographic and electrocardiographic study. *Br Heart J*. 1985;54:596–599.
37. Nir A, Tajik AJ, Freeman WK, et al. Tuberous sclerosis and cardiac rhabdomyoma. *Am J Cardiol*. 1995;76:419–421.
38. Smythe JF, Dyck JD, Smallhorn JF, Freedom RM. Natural history of cardiac rhabdomyoma in infancy and childhood. *Am J Cardiol*. 1990;66:1247–1249.
39. Nixon JR, Miller GM, Okazaki H, Gomez MR. Cerebral tuberous sclerosis: postmortem magnetic resonance imaging and pathologic anatomy. *Mayo Clin Proc*. 1989;64:305–311.
40. Webb DW, Thomas RD, Osborne JP. Cardiac rhabdomyomas and their association with tuberous sclerosis. *Arch Dis Child*. 1993;68:367–370.
41. Fenoglio JJ, McAllister HA, Ferrans VJ. Cardiac rhabdomyoma: a clinicopathologic and electron microscopic study. *Am J Cardiol*. 1976;38:241–251.
42. Kandt RS, Gebarski SS, Goetting MG. Tuberous sclerosis with cardiogenic cerebral embolism: magnetic resonance imaging. *Neurology*. 1985;35:1223–1225.
43. Konkol RJ, Walsh EP, Power T, Bresnan MJ. Cerebral embolism resulting from an intracardiac tumor in tuberous sclerosis. *Pediatr Neurol*. 1986;2:108–110.
44. Boronat S, Shaaya EA, Auladell M, Thiele EA, Caruso P. Intracranial arteriopathy in tuberous sclerosis complex. *J Child Neurol*. 2013.
45. Shepherd CW, Gomez MR, Lie JT, Crowson CS. Causes of death in patients with tuberous sclerosis. *Mayo Clin Proc*. 1991;66:792–796.
46. Monaghan HP, Krafchik BR, MacGregor DL, Fitz CR. Tuberous sclerosis complex in children. *Am J Dis Child*. 1981;135:912–917.
47. Crosett AD. Roentgenographic findings in the renal lesion of tuberous sclerosis. *Am J Roentgenol Radium Ther Nucl Med*. 1966;98:739–743.
48. van Baal JG, Fleury P, Brummelkamp WH. Tuberous sclerosis and the relation with renal angiomyolipoma. A genetic study on the clinical aspects. *Clin Genet*. 1989;35:167–173.
49. Farrow GM, Harrison EG, Utz DC, Jones DR. Renal angiolipoma: a clinicopathological study of 32 cases. *Cancer*. 1968;22:564–570.
50. Steiner MS, Goldman SM, Fishman EK, Marshall FF. The natural history of renal angiomyolipoma. *J Urol*. 1993;150:1783–1786.
51. Oesterling JE, Fishman EK, Goldman SM, Marshall FF. The management of renal angiomyolipoma. *J Urol*. 1986;135:1121–1124.
52. Bjornsson J, Short MP, Kwiatkowski DJ, Henske EP. Tuberous sclerosis-associated renal cell carcinoma. Clinical, pathological, and genetic features. *Am J Pathol*. 1996;149:1201–1208.
53. Webb DW, Super M, Normand IC, Osborne JP. Tuberous sclerosis and polycystic kidney disease. *Br Med J*. 1993;306:1258–1259.
54. Blute ML, Malek RS, Segura JW. Angiomyolipoma: clinical metamorphosis and concepts for management. *J Urol*. 1988;139:20–24.
55. Cudzilo CJ, Szczesniak RD, Brody AS, et al. Lymphangioleiomyomatosis screening in women with tuberous sclerosis. *Chest*. 2013;144:578–585.
56. Adriaensen ME, Schaefer-Prokop CM, Duyndam DA, Zonnenberg BA, Prokop M. Radiological evidence of lymphangioleiomyomatosis in female and male patients with tuberous sclerosis complex. *Clin Radiol*. 2011;66:625–628.
57. Johnson SR, Whale CI, Hubbard RB, Lewis SA, Tattersfield AE. Survival and disease progression in UK patients with lymphangioleiomyomatosis. *Thorax*. 2004;59:800–803.
58. Henske EP, McCormack FX. Lymphangioleiomyomatosis - a wolf in sheep's clothing. *J Clin Invest*. 2012;122:3807–3816.
59. Fleury P, De Groot WP, Delleman JW, Verbeeten Jr. B, Frankenmolen-Witkiezwicz IM. Tuberous sclerosis: the incidence of sporadic cases versus familial cases. *Brain Dev*. 1980;2:107–117.
60. Geist RT, Gutmann DH. The tuberous sclerosis 2 gene is expressed at high levels in the cerebellum and developing spinal cord. *Cell Growth Differ*. 1995;6:1477–1483.
61. Fryer AED, Chalmers A, Connor JM, et al. Evidence that the gene for tuberous sclerosis is on chromosome 9. *Lancet*. 1987;1:659–661.
62. van Slegtnenhorst M, de Hoogt R, Hermans C, Nellist M, Janssen B, Verhoef S. Identification of the tuberous sclerosis gene TSC1 on chromosome 9q34. *Science*. 1997;277:805–808.
63. Kandt RS, Haines JL, Smith M, et al. Linkage of an important gene locus for tuberous sclerosis to a chromosome 16 marker for polycystic kidney disease. *Nat Genet*. 1992;2:37–41.
64. Dabora SL, Jozwiak S, Franz DN, et al. Mutational analysis in a cohort of 224 tuberous sclerosis patients indicates increased severity of TSC2, compared with TSC1, disease in multiple organs. *Am J Hum Genet*. 2001;68:64–80.
65. Manning BD, Cantley LC. United at last: the tuberous sclerosis complex gene products connect the phosphoinositide 3-kinase/Akt pathway to mammalian target of rapamycin mTOR signalling. *Biochem Soc Trans*. 2003;31:573–578.
66. Franz DN. Everolimus in the treatment of subependymal giant cell astrocytomas, angiomyolipomas, and pulmonary and skin lesions associated with tuberous sclerosis complex. *Biologics*. 2013;7:211–221.
67. Krueger DA, Wilfong AA, Holland-Bouley K, et al. Everolimus treatment of refractory epilepsy in tuberous sclerosis complex. *Ann Neurol*. 2013;745:679–687.

CHAPTER

81

Sturge–Weber Syndrome

Anne M. Comi[*,†], *Douglas A. Marchuk*[‡], *and Jonathan Pevsner*[*,†]

[*]Hugo W. Moser Research Institute at Kennedy Krieger, Baltimore, MD, USA
[†]Johns Hopkins Medicine, Baltimore, MD, USA
[‡]Duke University Medical Center, Durham, NC, USA

CLINICAL FEATURES

Historical Overview

The association between the facial birthmark and seizures was first noted in 1879 by William Allen Sturge (1850–1919). While others such as Dürk (1910), Volland (1912), Hebold (1913), and Krabbe (1932) performed seminal studies noting calcification of the brain, it was F. Parkes Weber (1863–1962) whose 1922 paper led to the naming of the syndrome as Sturge–Weber. By the 1940s, Sturge–Weber syndrome (SWS) cases were routinely characterized by radiology, with X-rays revealing typical tram track abnormalities.[1] Dr. E. Steve Roach in 1992 proposed a classification for the spectrum of "encephalotrigeminal angiomatosis."[2] While these designations have not replaced the diagnosis of Sturge–Weber syndrome it is nevertheless useful to recognize that the syndrome is a spectrum, both in terms of the structures involved and the severity of their involvement.

In 1987, Dr. Rudolf Happle hypothesized somatic mosaic mutations as a probable mechanism for neurocutaneous disorders that occur sporadically and only affect a region of the individual's body.[3] Sturge–Weber syndrome was offered as an example of such a condition: The defect always occurs sporadically, the lesions are distributed in a mosaic pattern, there is variable extent of involvement, and the sex ratio is 1:1. He hypothesized that mutation would be embryonic lethal were it to be present in all cells of a zygote. These insights were consistent with recent approaches to finding somatic mosaic mutations in several cutaneous, malformation, and vascular disorders by comparing DNA from affected regions to that of unaffected regions of the same individuals.

Historically SWS was categorized as one of the phakomatoses; others include neurofibromatosis 1, tuberous sclerosis complex, von Hippel–Lindau disease, and xeroderma pigmentosum.[4] These other disorders are congenital and hereditary diseases that are also characterized by cutaneous lesions, neuro-ophthalmic features, and tumor formation.[5,6] As discussed below, the recent discovery of somatic mutations in *GNAQ* shows that SWS too is caused by a mutation in a gene that is an oncogene in some contexts, although SWS is not typically associated with tumor formation.

Mode of Inheritance and Prevalence

SWS (Online Mendelian Inheritance in Man #185300) occurs sporadically; there is no single known validated familial case. It is the third most common neurocutaneous disorder, after neurofibromatosis (OMIM #162200) and tuberous sclerosis 1 (OMIM #191100). While no population based study has been done, estimates of prevalence range from 1 in 20–50,000 live births. Isolated port-wine birthmarks occur much more frequently; approximately 1 in 300 infants are born with a port-wine birthmark and the face, head, and neck are common locations for the birthmark.[7]

Rosenberg's Molecular and Genetic Basis of Neurological and Psychiatric Disease
http://dx.doi.org/10.1016/10.1016/B978-0-12-410529-4.00081-4

NATURAL HISTORY

When a child is born with a port-wine birthmark (PWB) on the forehead or the upper eyelid, he or she has a 10–50% risk of brain involvement. When the PWB involves the upper and the lower eyelid, the risk of eye involvement and glaucoma is highest—about 50%.[8] Glaucoma can be present from birth; therefore, performing an ophthalmological exam as soon as possible is recommended. Most newborns, however, will have otherwise normal exams and may have no clinical evidence of brain or eye involvement. Nevertheless, they are at risk for SWS and require additional follow-up and testing.

The most common presentation for SWS brain involvement is seizures; approximately 75% of patients begin having seizures by 1 year of age and 90% within the first 2 years of life.[9] Prior to the onset of symptoms, infants commonly develop normally, although a subset of patients will develop early handedness or a visual gaze preference prior to seizure onset. With the onset of seizures, it is common for infants to acutely develop hemiparesis and focal deficits. Most commonly the seizures are focal and complex partial in nature. Neurological, cognitive, and seizure control outcomes have been correlated to the extent of brain involvement and the age of seizure onset.[10]

In older patients including children under the age of 10, headaches and migraines are common. Seizures can trigger headaches and headaches can trigger seizures. Therefore, for these patients, treatment with anticonvulsants that also treat migraines such as topiramate can be helpful, and aggressively treating the migraines can help bring patients' seizures under better control. Insufficient sleep is a common trigger, as is insufficient hydration. Sumatriptan medications have been safely used and are helpful for some patients.[7] In toddlers and young children, minor head trauma from falls can commonly result in migraines and stroke-like episodes. Therefore, sports such as football and soccer should be avoided.

Strokes and stroke-like episodes are common and are frequently associated with seizures and/or migraines. Stroke-like episodes are defined as a focal neurological deficit lasting longer than 24 hours. These deficits can resolve the over a period of days to weeks. When the neurological deficit does not resolve but rather persists for months to years a patient is considered to have had a stroke. Neuroimaging performed during a stroke-like episode usually demonstrates a focal area of slowing, and occasionally perfusion deficits, although this is uncommon.[11] On the other hand, patients who are seizure- and migraine-free for a prolonged period demonstrate recovery of focal neurological deficits.[12]

A variable degree of hemiparesis is common although for many patients the weakness is slight, resulting in a functional gait, ability to run, and use of the affected hand. Occupational and physical therapy from an early age is beneficial, as is encouraging training in a musical instrument if possible. It is unclear whether vision therapy is helpful to those with a visual gaze reference. Neuropsychological testing is extremely helpful in preparing young children who have SWS for school and for optimizing their learning experience. Even children with relatively intact cognitive abilities are frequently affected by attention-deficit disorder and other learning disabilities.[13,14]

Average life expectancy in Sturge–Weber has not been well studied or determined. Morbidity from sudden unexpected death from epilepsy is very rare but occurs, as well as other comorbidities commonly seen with significant epilepsy, such as injuries from falls and accidents. Older individuals with soft and bony hypertrophy of their face underlying their birthmark can manifest sleep apnea and other upper airway issues.[15,16] Hemorrhage, both intracranial and extracranial have been reported but are rare and the factors to predict this are not well understood.[17]

MOLECULAR GENETICS

A causative somatic mosaic mutation in *GNAQ* was recently reported.[18] The mutation was identified by whole-genome sequencing of DNA from affected and unaffected tissue from three individuals with SWS, followed by extensive confirmation in affected and unaffected brain and skin tissue from individuals with SWS, isolated PWB, and controls. Both SWS and isolated PWB are caused by the same somatic mosaic mutation in *GNAQ* that produces a nonsynonymous p.Arg183Gln substitution in the encoded protein, guanine nucleotide-binding protein G(q) subunit alpha (abbreviated Gαq).[18] This discovery confirms the hypothesis offered by Dr. Happle and the understanding that SWS, and most cases of isolated PWB, are not inherited, a confirmation that is of great value and relief to families. However, it is important to recognize that because the identical somatic mosaic mutation causes both SWS and isolated PWB, the discovery at this point does not aid in the determination of whether an infant with a facial port-wine birthmark has SWS brain involvement; this remains a neuroimaging diagnosis.

Normal Gene Product

Gαq is an alpha subunit of a heterotrimeric GTP-binding protein. Proteins of this class (of which there are 24 paralogs in humans) couple to Gβ and Gγ subunits, and coordinately interact with seven transmembrane-spanning G-protein coupled receptors (GPCRs). Gα subunits are key components of signaling pathways that are characterized by specificity (based on the ligand-binding properties of GPCRs) and tremendous signal amplification (based on coupling receptor binding to second messenger pathways). While Gαq is known to couple with several GPCRs (including serotonin, glutamate, angiotensin 2 and endothelin receptors), those relevant to SWS are not yet known. The best-characterized downstream effectors of Gαq are phospholipase β (which cleaves phosphatidylinositol bisphosphate [PIP_2] into diacylglycerol and inositol triphosphate [IP_3] promoting intracellular calcium release), low molecular-weight GTPases such as Rac and Rho, and pathways mediated by activation of mitogen-activated protein kinase (MAPK). Other downstream effectors include p38, ERK, JNK, and Trio (see Figure 81.1).[19,20]

Abnormal Gene Product(s)

The Arg.183Gln mutation in *GNAQ* impacts the site of water autohydrolysis within the GTP–GDP binding site of Gαq. Therefore the mutation is predicted to decrease the efficiency of the autohydrolysis, which returns the guanine nucleotide protein to the deactivated (GDP-bound) state and complex with its GPCR. Therefore, this mutation is predicted to result in constitutive overactivation of downstream pathways. When mutant constructs were transfected into a human kidney epithelial cell line (293T cells), the R183Q mutant caused a mild but statistically significant increase in phosphorylated ERK compared to wild-type construct transfection.[18]

Mutations in *GNAQ* causing a Q209L substitution have been shown to cause uveal melanoma, a highly aggressive cancer, and blue nevi.[21] This establishes *GNAQ* as a dominant-acting oncogene in melanocytes. The same R183Q mutation in *GNAQ* (as well as R183C and R183H mutations in the paralog *GNA11*) is a relatively rare cause of uveal melanoma and blue nevi.[22] Compared to the R183Q mutation, the Q209L mutation transfected into 293T cells was more hyperactivating of downstream pathways. How the same mutation can result in cancer in one setting, and a vascular malformation in another remains to be further delineated, but this phenomenon almost certainly relates to

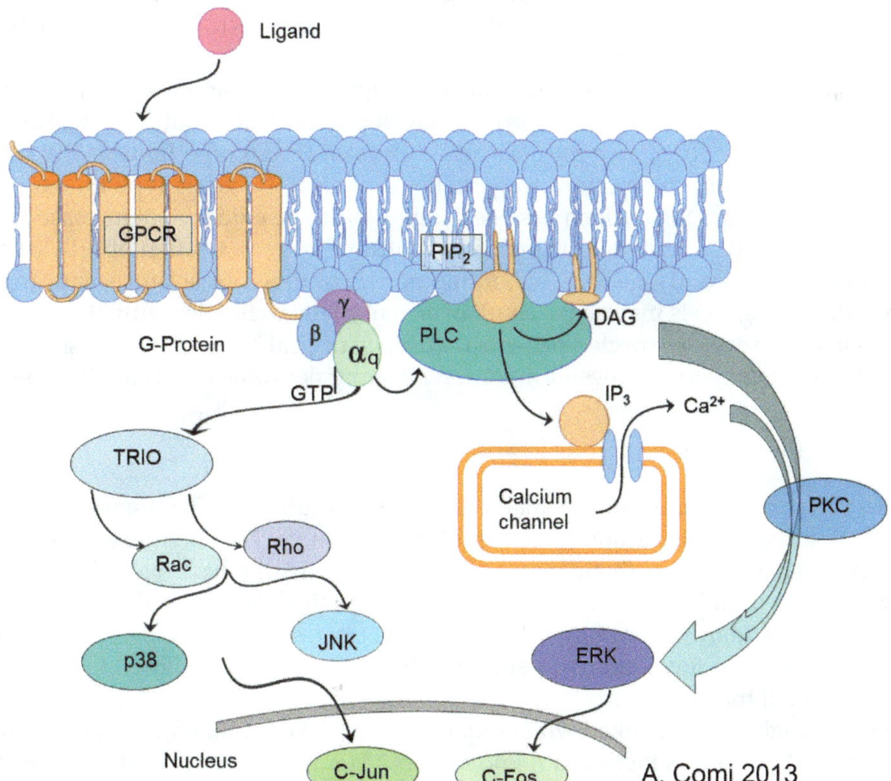

FIGURE 81.1 **Downstream effectors of Gαq.** When the ligand for the G protein-coupled receptor binds to its receptor site, GTP is bound by Gαq, displacing GDP and dissociating the activated trimeric guanine nucleotide protein. Gαq then binds and activates downstream effectors such as phospholipase β, Rac and Rho, ERK, Trio, p38 and JNK, which in turn alter gene expression.

differences in cell type mutated (melanocyte vs. vascular cell type, as yet to be determined for SWS and PWB) and developmental stage at the time the mutation occurs (adulthood vs. fetal life). Nevertheless, insights from the burgeoning field of cancer, especially as it relates to uveal melanoma, including the impacted downstream pathways and treatment with inhibitors, are very likely to have relevance to novel treatment strategies for both SWS and PWB.

Genotype–Phenotype Correlations

At this point only one somatic mosaic mutation has been associated with both SWS and with isolated PWB. Only a minority of the tested samples did not demonstrate the p.Arg183Gln *GNAQ* mutation. It is unclear whether these cases are due to a different somatic mutation in *GNAQ* or in other gene(s).

DISEASE MECHANISMS

Sturge–Weber syndrome brain involvement is characterized by an increased number of dilated leptomeningeal vessels on the surface of the brain associated with a decrease in normal cortical vessels, cortical and subcortical perivascular calcification, and, in many surgical brain samples, evidence of cortical dysgenesis is also described.[23] These vessels are abnormally innervated[24] and endothelin 1 expression is increased.[25] Furthermore, immunohistochemical studies have suggested that there is ongoing vascular remodeling of these abnormal blood vessels with increased mitotic index.

The cortical calcification develops over time as the brain atrophies and is injured. Where little or no injury has yet occurred, it is possible for no evidence of calcification to be visualized on neuroimaging. Given the association with atrophy and impaired perfusion it is likely that the calcification is secondary to anoxic injury to vascular and/or glial cells and increased permeability of cerebral vessels. Di Trapani et al. hypothesized that a putative vascular factor secreted by the vessel connective tissue was central to the formation of the calcifications.[26] In the context of the newly discovered causative genetic defect for SWS, this putative "vascular factor" may be the constitutive hyperactivation of GPCR–Gαq–ERK pathways, possibly leading to a cycle of reactive oxygen species (ROS) and intracellular calcium release, blood vessel dilatation, impaired venous drainage, seizures, ischemia with cell death, brain atrophy and calcium release and triggering additional hyperactivation of GPCR–Gαq–ERK pathways. Further study is required to identify additional important factors in this cascade.

Cortical dysgenesis has been reported by several groups in surgical brain samples and suggests that the spectrum of malformations in SWS includes dysgenesis of cortical development.[23] Cases of SWS with medically intractable epilepsy on histology have reportedly shown focal cortical dysplasia (FCD) type IIa near the region of leptomeningeal angiomatosis, cortical dysplasia, and polymicrogyria. Focal cortical dysplasia type IIa has been associated with increased phosphorylation of S6 and S6K, which has been suggested to be evidence of increased mTOR activity. ERK phosphorylation can increase mTOR activity. Congenital constitutive hyperactivation of ERK pathways has been associated with other causes of cortical dysgenesis, including models of tuberous sclerosis and fragile X syndrome.[27,28] Therefore, while the cortical dysgenesis may be the result of an early fetal ischemic insult, the presence of focal cortical malformations, at least in more severely affected patients requiring surgical intervention for seizure control, could also result from the underlying genetic somatic mosaic mutation affecting neuronal differentiation and migration.

Pathophysiology

Histopathological studies consistently report that vascular innervation of PWB and SWS cortical is abnormal and likely contributes to impaired blood flow and vascular function.[24,29] SWS malformed vessels in the brain are innervated only by noradrenergic sympathetic nerve fibers rich in tyrosine hydroxylase and neuropeptide Y. The authors hypothesized that the noradrenergic innervation of these cortical vessels could impair vascular responses to hypotension and seizures. Vascular innervation of facial PWBs noted a significant decrease in the nerve density of blood vessels, suggesting that here too abnormal neural modulation of vascular flow likely contributes to vascular remodeling and progression of the birthmark.

Endothelin 1 signals through a GPCR linked with Gαq. Expression of endothelin 1 by intracranial vascular malformations of four patients with SWS was increased.[25] Endothelin 1 is a vasoactive peptide whose expression is stimulated by high pressure and shear stress resulting from high blood flow; the effects are dependent upon the vessel type, but in pial cells the result of increased endothelin can be dilatation.[30] Study of endothelin 1 expression in port-wine stains, however, has shown normal levels of expression.[31] It has been shown that ROS increase endothelin 1 expression from endothelial cells via the Raps/Raff/ERK pathway;[32] it is possible, therefore, that endothelin expression was increased more in the brain secondary to greater increases in ROS in the studied brain tissue samples than in the skin samples. Additionally, endothelin 1 has also been shown to have effects on the normal embryonic development of neural crest-derived tissues, vascular remodeling, mitogenic actions, angiogenesis, apoptosis and the extracellular matrix; therefore, abnormal signaling of the endothelin receptor because of the mutant Gαq may have a role in the development and progression of the vascular malformations.

There is also evidence that the extracellular matrix is abnormal in and around the vascular malformations. Broader fibronectin, laminin and type IV collagen staining around dilated capillary and venous vessel walls in port-wine skin samples than in controls has been described,[33] and fibronectin expression was also increased in SWS cortical brain samples.[34] The extracellular matrix regulates angiogenesis and vasculogenesis, and is essential in the maintenance of blood vessel structure and function, as well as brain tissue responses to seizures.[35] Matrix metalloproteinases (MMP) 2 and 9 are more likely to be present in the urine of subjects with SWS and their levels correlate with severity of neurologic involvement; these results are consistent with ongoing vascular remodeling in SWS and PWB and suggest that MMPs may prove useful as vascular biomarkers. ERK phosphorylation is involved in the release of MMPs by endothelial cells suggesting that the underlying somatic mutation may have a role in their abnormal urine levels. Together this data suggests that the regulation of blood vessel structure and function, vascular innervation, extracellular matrix expression, and vasoactive molecules is altered in SWS.

Interpretation of the available histopathological studies, while greatly aided by the recent discovery of the underlying pathogenic somatic mosaic mutation in *GNAQ*, is nevertheless hindered by two important issues. First, the cell type (or types) affected by the gene mutation is at this time unknown. Second, the histopathologic changes described may, in part, be due to the chronic effects of the vascular malformation such as vasodilatation, hypoxia, seizures, and venous hypertension. It will be difficult to separate these effects from those due directly to the mutation itself. Identification of mutated cell type(s) will further advance our understanding of the relevant pathways and mechanism of disease progression. Development of relevant *in vitro* and animal models of SWS and PWB will enable studies that can distinguish the pathways and histologic changes resulting from the primary molecular defect and those aberrations that are secondary to chronic degenerative processes.

Animal Models

Animal models based on the recent discovery of the somatic mutation causing SWS and PW birthmarks are currently under development and once available will greatly enhance the understanding of how the mutation results in the vascular malformations.

Current Research

Current research has exploded in several important directions recently. Studies are ongoing to determine the cell type(s) harboring the mutation in affected tissue, to develop animal models and cell culture assays, analyze downstream proteins in SWS tissue, and develop new biomarkers and outcome measures. Inhibitors of the overactivated pathways are currently in clinical trials or being utilized clinically in the treatment of various cancers, and pilot clinical trials to assess safety and begin to evaluate efficacy of these treatments for SWS are planned for the next few years.

DIFFERENTIAL DIAGNOSIS

The main differential question is whether an individual with a facial birthmark has SWS brain and/or eye involvement. If asymptomatic and normally developing then a magnetic resonance imaging (MRI) with and without contrast, post-contrast flair, susceptibility-weighted imaging, and perfusion imaging after 1 year of age excludes SWS in most cases. Affected individuals must be followed for life, however, for evidence of glaucoma. Theoretically infants with extensive PWB, glaucoma, and macrocephaly could have either SWS or megalencephaly-capillary malformation-polymicrogyria syndrome, which is associated with a mutation in *PIK3CA* (22729224). However, in practice these syndromes are infrequently confused and the intracranial involvement is very different; the imaging findings for SWS are diagnostic.

TESTING

Molecular Diagnosis

While in principle a molecular test for SWS could be developed, in practice this test would not significantly aid the diagnosis of SWS. The diagnosis of a port-wine birthmark in most cases is easily made clinically by a dermatologist or other physician familiar with vascular birthmarks. A crucial clinical question is whether the infant born with a facial port-wine birthmark has brain or eye involvement; molecular testing cannot answer this question since the same mutation causes both isolated port-wine birthmark and SWS.

Clinical and Ancillary Testing: Imaging, Physiological, Electrodiagnostic

Diagnosis of SWS brain involvement is made with neuroimaging. MRI with and without contrast is more sensitive for visualizing the leptomeningeal vessels (Figure 81.2), and susceptibility-weighted imaging aids imaging of deep draining vessels. Atrophy, cortical dysgenesis and other vascular or brain malformations are also seen on MRI. Head computed tomography (CT), when warranted, is best for visualizing the extent of calcification. Recently the prenatal diagnosis of SWS was reported with ultrasound and MRI at 33 weeks gestation. However, MRI in the newborn period frequently is normal and does not exclude brain involvement. Electroencephalograms (EEGs) are helpful for elucidating the nature of the epilepsy when seizures develop, monitoring medication response, and for surgical planning.[36] Quantitative EEG assessment of asymmetry of power quantified from EEG can be useful for screening for brain involvement.[37]

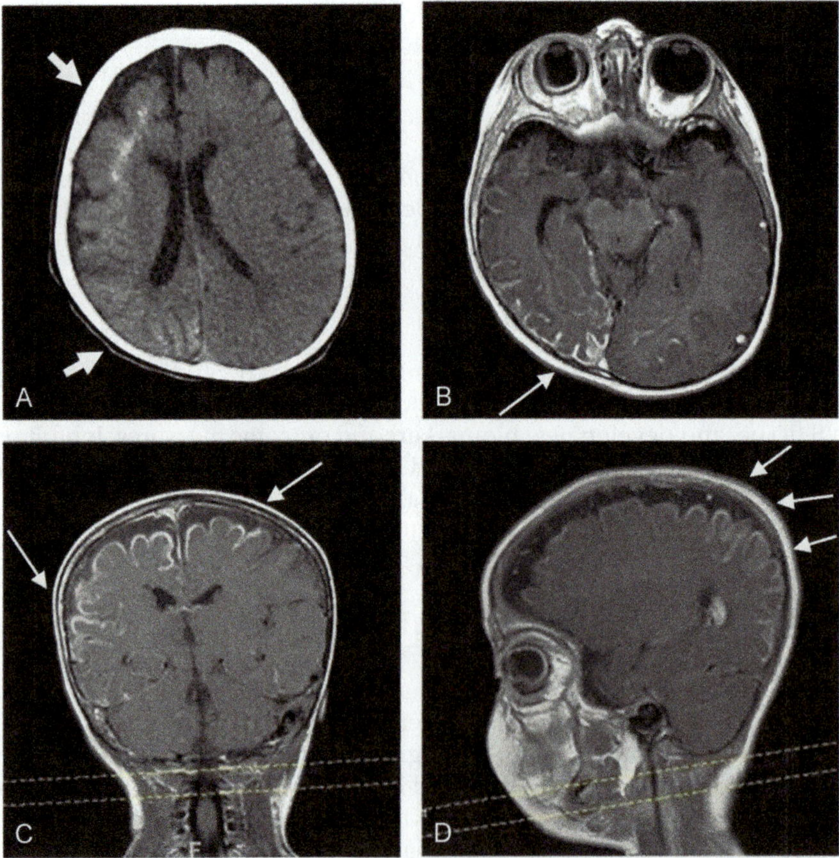

FIGURE 81.2 **Sturge–Weber syndrome brain involvement.** (**A**) Head CT from a child with bilateral Sturge–Weber syndrome brain involvement. The thick arrows point to the calcification noted and the brain atrophy seen particularly on the right side. (**B–D**) T1-weighted post-contrast enhanced images demonstrating leptomeningeal enhancement (thin arrows) bilaterally, but of the right side much more extensively than the left.

MANAGEMENT

Standard of Care

Laser treatment of the port-wine birthmark can begin early in infancy. There are good data suggesting that the birthmark responds better to laser treatment when the treatments are started in infancy.[38] Whether early laser treatment prevents later onset of soft and bony tissue hypertrophy, development of blebs and nodules, and complications due to the effects of the birthmark progression upon vision, the airway and swallowing is currently being studied. A variety of lasers have been developed to treat various birthmarks. Depending on the size, color, and thickness of the birthmark, different wavelengths of light are successfully used to lighten the birthmark by heating the hemoglobin within blood vessels and thereby obliterating them.

Treatment for glaucoma involves attempting to lower eye pressure through the use of eye drops that decrease fluid production in the eye with medications such as timolol and lantaprost.[39,40] Various combinations of eye drops may be tried. When medical management fails or if the glaucoma is fulminant then surgery to relieve the pressure is required.[41] The major complication of surgical intervention stems from the risks of relieving the eye pressure too quickly; this can result in hemorrhage and further sudden vision loss.

Treatment of the neurologic consequences of SWS primarily involves the use of anticonvulsants often combined with low-dose aspirin.[42] Given that prolonged seizures, particularly in infants and young children can result in strokes,[43] aggressive management of seizures and use of anticonvulsants is warranted. Practically speaking, this can mean some or all of the following approaches: providing parents of at-risk infants with the knowledge of how to recognize and handle seizures; the use of an abortive medication; and maintaining anticonvulsant dose in the medium-to-high range as tolerated by regularly increasing the dose for weight gain, especially in infants and young children.

There is an increased risk of hypothyroidism (commonly central) and growth hormone deficiency.[44,45] While the majority of patients with SWS do not have these conditions, endocrine condition should be screened for when appropriate symptoms are present. It is important to provide treatment for this aspect of SWS.

Failed Therapies and Barriers to Treatment Development

The rarity of Sturge–Weber syndrome and the lack of a coordinated network of research centers (until recently) able to carry out clinical trials on SWS have been barriers to the development of new treatments and the development of consensus in the approach to treatment. However, in the last 4 years, a network of Sturge–Weber syndrome clinical research groups has developed through the Brain Vascular Malformation Consortium with funding from the National Institutes of Health Office of Rare Diseases Research at the National Center for Advancing Translational Science and the National Institute of Neurological Disorders and Stroke. This network is beginning pilot clinical trials with inhibitors of the hyperactivated pathways downstream of GNAQ in the near future.

Therapies Under Investigation

Low-dose aspirin was first proposed by Dr. Stephen Roach about 20 years ago with the goal of improving sluggish blood flow through the abnormal venous leptomeningeal vessels and thereby reducing the risk of ischemic injury and preventing brain atrophy and neurologic deficits. Since then, several groups have studied the effects of low-dose aspirin (3–5 mg per kg per day) and reported evidence that it decreases the frequency and severity of stroke-like episodes and seizures.[42,46] The most commonly reported side effects are increased bruising, nosebleeds and gum bleeds; allergic reactions and rarely more seriously bleeding can occur. While many centers now use low-dose aspirin in addition to anticonvulsants, consensus on this approach has not been reached.

ACKNOWLEDGEMENTS

We acknowledge editorial assistance from Cathy Bachur. We are also grateful for funding from the National Institute of Neurological Disorders and Stroke (NINDS) (National Institutes of Health [NIH] U54NS065705, to Drs. Marchuk, Comi, and Pevsner) and from Hunter's Dream for a Cure Foundation (to Dr. Comi). The Brain Vascular Malformation Consortium (U54NS065705) is a part of the NIH Rare Disease Clinical Research Network, supported through a collaboration between the NIH Office of Rare Diseases Research at the National Center for Advancing Translational Science and the NINDS.

References

1. Norman RM. *The Sturge–Weber Syndrome*. Bristol: John Wright & Sons Ltd; 1960.
2. Roach ES. Neurocutaneous syndromes. *Pediatr Clin North Am*. 1992;39:591–620.
3. Happle R. Lethal genes surviving by mosaicism: a possible explanation for sporadic birth defects involving the skin. *J Am Acad Dermatol*. 1987;16:899–906.
4. Kurlemann G. Neurocutaneous syndromes. *Handb Clin Neurol*. 2012;108:513–533.
5. Chan JW. Neuro-ophthalmic features of the neurocutaneous syndromes. *Int Ophthalmol Clin*. 2012;52:73–85 xi.
6. Haug SJ, Stewart JM. Retinal manifestations of the phakomatoses. *Int Ophthalmol Clin*. 2012;52:107–118.
7. Powell J. Update on hemangiomas and vascular malformations. *Curr Opin Pediatr*. 1999;11:457–463.
8. Enjolras O, Riche MC, Merland JJ. Facial port-wine stains and Sturge–Weber syndrome. *Pediatrics*. 1985;76:48–51.
9. Sujansky E, Conradi S. Sturge–Weber syndrome: age of onset of seizures and glaucoma and the prognosis for affected children. *J Child Neurol*. 1995;10:49–58.
10. Kramer U, Kahana E, Shorer Z, Ben Zeev B. Outcome of infants with unilateral Sturge–Weber syndrome and early onset seizures. *Dev Med Child Neurol*. 2000;42:756–759.
11. Cakirer S, Yagmurlu B, Savas MR. Sturge–Weber syndrome: diffusion magnetic resonance imaging and proton magnetic resonance spectroscopy findings. *Acta Radiol*. 2005;46:407–410.
12. Lee JS, Asano E, Muzik O, et al. Sturge–Weber syndrome: correlation between clinical course and FDG PET findings. *Neurology*. 2001;57:189–195.
13. Chapieski L, Friedman A, Lachar D. Psychological functioning in children and adolescents with Sturge–Weber syndrome. *J Child Neurol*. 2000;15:660–665.
14. Zabel TA, Reesman J, Wodka EL, et al. Neuropsychological features and risk factors in children with Sturge–Weber syndrome: four case reports. *Clin Neuropsychol*. 2010;24:841–859.
15. Greene AK, Taber SF, Ball KL, Padwa BL, Mulliken JB. Sturge–Weber syndrome: soft-tissue and skeletal overgrowth. *J Craniofac Surg*. 2009;20(suppl 1):617–621.
16. Irving ND, Lim JH, Cohen B, Ferenc LM, Comi AM. Sturge–Weber syndrome: ear, nose, and throat issues and neurologic status. *Pediatr Neurol*. 2010;43:241–244.
17. Lopez J, Yeom KW, Comi A, Van HK. Case report of subdural hematoma in a patient with Sturge–Weber syndrome and literature review: questions and implications for therapy. *J Child Neurol*. 2013;28:672–675.
18. Shirley MD, Tang H, Gallione CJ, et al. Sturge–Weber syndrome and port-wine stains caused by somatic mutation in GNAQ. *N Engl J Med*. 2013;368:1971–1979.
19. Vaque JP, Dorsam RT, Feng X, et al. A genome-wide RNAi screen reveals a Trio-regulated Rho GTPase circuitry transducing mitogenic signals initiated by G protein-coupled receptors. *Mol Cell*. 2013;49:94–108.
20. Rozengurt E. Mitogenic signaling pathways induced by G protein-coupled receptors. *J Cell Physiol*. 2007;213:589–602.
21. Van Raamsdonk CD, Bezrookove V, Green G, et al. Frequent somatic mutations of GNAQ in uveal melanoma and blue naevi. *Nature*. 2009;457:599–602.
22. Van Raamsdonk CD, Griewank KG, Crosby MB, et al. Mutations in GNA11 in uveal melanoma. *N Engl J Med*. 2010;363:2191–2199.
23. Murakami N, Morioka T, Suzuki SO, et al. Focal cortical dysplasia type IIa underlying epileptogenesis in patients with epilepsy associated with Sturge–Weber syndrome. *Epilepsia*. 2012;53:e184–e188.
24. Sa M, Barroso CP, Caldas MC, Edvinsson L, Gulbenkian S. Innervation pattern of malformative cortical vessels in Sturge–Weber disease: an histochemical, immunohistochemical, and ultrastructural study. *Neurosurgery*. 1997;41:872–876.
25. Rhoten RL, Comair YG, Shedid D, Chyatte D, Simonson MS. Specific repression of the preproendothelin-1 gene in intracranial arteriovenous malformations. *J Neurosurg*. 1997;86:101–108.
26. Di Trapani G, Di Rocco C, Abbamondi AL, Caldarelli M, Pocchiari M. Light microscopy and ultrastructural studies of Sturge–Weber disease. *Childs Brain*. 1982;9:23–36.
27. Phoenix TN, Temple S. Spred1, a negative regulator of Ras-MAPK-ERK, is enriched in CNS germinal zones, dampens NSC proliferation, and maintains ventricular zone structure. *Genes Dev*. 2010;24:45–56.
28. Chevere-Torres I, Kaphzan H, Bhattacharya A, et al. Metabotropic glutamate receptor-dependent long-term depression is impaired due to elevated ERK signaling in the DeltaRG mouse model of tuberous sclerosis complex. *Neurobiol Dis*. 2012;45:1101–1110.
29. Rydh M, Malm M, Jernbeck J, Dalsgaard CJ. Ectatic blood vessels in port-wine stains lack innervation: possible role in pathogenesis. *Plast Reconstr Surg*. 1991;87:419–422.
30. Hahn AW, Resink TJ, Mackie E, Scott-Burden T, Buhler FR. Effects of peptide vasoconstrictors on vessel structure. *Am J Med*. 1993;94:13S–19S.
31. Katugampola GA, Lanigan SW, Rees AM. Normal distribution of endothelin in port wine stain vasculature. *Br J Dermatol*. 1997;137:323–324.
32. Grant K, Loizidou M, Taylor I. Endothelin-1: a multifunctional molecule in cancer. *Br J Cancer*. 2003;88:163–166.
33. Mitsuhashi Y, Odermatt BF, Schneider BV, Schnyder UW. Immunohistological evaluation of endothelial markers and basement membrane components in port-wine stains. *Dermatologica*. 1988;176:243–250.
34. Comi AM, Weisz CJ, Highet BH, et al. Sturge–Weber syndrome: altered blood vessel fibronectin expression and morphology. *J Child Neurol*. 2005;20:572–577.
35. Hoffman KB, Pinkstaff JK, Gall CM, Lynch G. Seizure induced synthesis of fibronectin is rapid and age dependent: implications for long-term potentiation and sprouting. *Brain Res*. 1998;812:209–215.
36. Jansen FE, van der Worp HB, van Huffelen A, van Nieuwenhuizen O. Sturge–Weber syndrome and paroxysmal hemiparesis: epilepsy or ischaemia? *Dev Med Child Neurol*. 2004;46:783–786.
37. Ewen JB, Kossoff EH, Crone NE, et al. Use of quantitative EEG in infants with port-wine birthmark to assess for Sturge–Weber brain involvement. *Clin Neurophysiol*. 2009;120:1433–1440.

38. Chapas AM, Eickhorst K, Geronemus RG. Efficacy of early treatment of facial port wine stains in newborns: a review of 49 cases. *Lasers Surg Med*. 2007;39:563–568.
39. Ray D, Mandal AK, Chandrasekhar G, Naik M, Dhepe N. Port-wine vascular malformations and glaucoma risk in Sturge–Weber syndrome. *J AAPOS*. 2010;14:105.
40. Iwach AG, Hoskins Jr. HD, Hetherington Jr. J, Shaffer RN. Analysis of surgical and medical management of glaucoma in Sturge–Weber syndrome. *Ophthalmology*. 1990;97:904–909.
41. Patrianakos TD, Nagao K, Walton DS. Surgical management of glaucoma with the Sturge Weber syndrome. *Int Ophthalmol Clin*. 2008;48:63–78.
42. Lance EI, Sreenivasan AK, Zabel TA, Kossoff EH, Comi AM. Aspirin use in Sturge–Weber syndrome: side effects and clinical outcomes. *J Child Neurol*. 2013;28:213–218.
43. Namer IJ, Battaglia F, Hirsch E, Constantinesco A, Marescaux C. Subtraction ictal SPECT co-registered to MRI (SISCOM) in Sturge–Weber syndrome. *Clin Nucl Med*. 2005;30:39–40.
44. Miller RS, Ball KL, Comi AM, Germain-Lee EL. Growth hormone deficiency in Sturge–Weber syndrome. *Arch Dis Child*. 2006;91:340–341.
45. Comi AM, Bellamkonda S, Ferenc LM, Cohen BA, Germain-Lee EL. Central hypothyroidism and Sturge–Weber syndrome. *Pediatr Neurol*. 2008;39:58–62.
46. Maria BL, Neufeld JA, Rosainz LC, et al. Central nervous system structure and function in Sturge–Weber syndrome: evidence of neurologic and radiologic progression. *J Child Neurol*. 1998;13:606–618.

CHAPTER

82

Hemangioblastomas of the Central Nervous System

Ana Metelo[*,†,‡] *and Othon Iliopoulos*[†,‡]

[*]Coimbra University, Coimbra, Portugal
[†]Harvard Medical School, Boston, MA, USA
[‡]Massachusetts General Hospital, Boston, MA, USA

INTRODUCTION

Hemangioblastomas (HB) are vascular tumors of the central nervous system (CNS) that develop in patients with von Hippel–Lindau (VHL) disease or sporadically. Loss-of-*VHL* tumor suppressor gene is a hallmark of both types of hemangioblastomas. The tumors comprise of stromal cells, proliferating vascular endothelial cells, and may contain hematopoietic components. The cell of origin and the lineage relations between the different tumor components are currently unclear. Hemangioblastomas are treated with surgical excision or radiation therapy. Currently, there is no targeted therapy available for treatment of hemangioblastomas. Future studies, based on next-generation sequencing and animal models, will discover and validate oncogenic mutations that can be targeted for medical therapy of these tumors.

CLINICAL FEATURES

Definitions

Hemangioblastomas are tumors mainly of the central nervous system (CNS). They develop most often in the cerebellum, but also in the brainstem, along the spinal cord, above the tentorium, as well as the optic nerve and the retina. HBs occur either as sporadic tumors or as part of VHL disease. Below we will compare and contrast both forms of the disease with regards to molecular genetics, their age of onset, natural history, treatment approach and surveillance.

Mode of Inheritance and Prevalence

Patients with VHL disease harbor an inactivating mutation in one of the *VHL* tumor suppressor gene alleles in their germline.[1] Their offspring has a 50% chance to inherit the mutant allele. All somatic cells therefore have one *VHL* allele inactivated. Inactivating mutations of the second *VHL* allele leads to the formation of specific tumors that give VHL patients their typical clinical presentation. Specifically, VHL patients are at a lifetime risk of developing renal cell carcinomas in either or both kidneys (60–80%), multiple and bilateral pheochromocytomas (20%) and rarely paragangliomas, pancreatic cysts and pancreatic neuroendocrine tumors (10–20%) as well as papillary cystadenomas of the middle ear, the pancreas, the epididymis and the adnexal organs (10–20%).[2] VHL patients may develop all or some of these tumors, simultaneously or sequentially. In addition to these tumors, hemangioblastomas are the commonest (80%), earliest, and the most morbid lesions developed in VHL patients.[2]

http://dx.doi.org/10.1016/10.1016/B978-0-12-410529-4.00082-6

Hemangioblastomas also develop in patients that do not suffer from VHL disease (sporadic HBs). Similarly to the VHL-related hemangioblastomas, the sporadic ones are characterized by inactivation of both VHL alleles, but the sequential inactivation of the alleles is somatic and the patients test negative for germline VHL mutations.[3,4]

Natural History

VHL Disease-Related Hemangioblastomas

Hemangioblastomas develop predominantly in the cerebellum in both sporadic and VHL-related cases. In contrast to sporadic HBs, patients with VHL disease develop HBs at a high frequency in the brainstem, the cervico-medullary junction and above the tentorium (the most classic location is hypothalamus). Supratentorial HB is almost pathognomonic of VHL disease.[5]

The optic nerve and the retina are an extension of the CNS and VHL patients often develop HBs in these organs. Retinal HBs may develop in the nasal or temporal periphery of the retina (90%), the macular retina and the papilla (8%), and less frequently in the optic nerve (1%).[6–8]

The penetrance of VHL disease is 100% by the age of 50 years old but the expressivity of the VHL disease varies widely. The average VHL patient develops multiple, synchronous or sequential HBs in multiple locations of the CNS, including the retina and the optic nerve. The CNS HB may be diagnosed as early as 8 years of age. Retinal HBs may be present and diagnosed even at birth. Annual retinal exam since infancy and brain magnetic resonance imaging (MRI) starting at 8 years of age are therefore part of the standard surveillance protocol for VHL patients.

The symptoms caused by HBs depend on the location, the size of the tumor, and the growth rate. Cerebellar and infratentorial lesions may cause cerebellar ataxia and symptoms of increased intracranial pressure (headaches, nausea, vomiting). Brainstem lesions may cause very labile blood pressure, noncoordinated swallowing, and aspiration pneumonias. And finally, spinal cord lesions may cause peripheral motor and sensory (pain, dysesthesia) symptoms.[2] HBs may also cause systemic symptoms; they may produce high erythropoietin levels that lead to secondary erythrocytosis, which in turn may generate headaches, blurred vision, abnormal sweating, and thromboembolic episodes.[9]

Retinal hemangioblastomas may present with decreased visual acuity because of retinal exudate and, if left untreated, can cause retinal hemorrhage, detachment and blindness. Optic nerve HBs present the most serious challenge because they affect visual acuity directly and cannot be treated with laser or radiation.[6]

Rare locations of HB, such as skin, liver, peripheral nerve root, peripheral nerve and "hemangioblastosis" have been reported in VHL patients.[10]

Sporadic Hemangioblastomas

The mean age of diagnosis of sporadic hemangioblastomas is higher than the one for VHL disease-related HBs. Sporadic HBs are usually single; they predominantly develop in the cerebellum and in the retina and, rarely, in the spinal cord.[11] Table 82.1 compares the clinical features of sporadic to VHL disease-related HBs.

TABLE 82.1 Comparison of the Clinical Features of VHL Disease-Related and Sporadic Hemangioblastomas

Hemangioblastomas		VHL-associated HBs	Sporadic HBs
Mean age of onset		25 years	50 years
Number of hemangioblastomas		Multiple	Single
Location	Cerebellum	75%	Common
	Spinal cord	25%	Unusual
	Brainstem	5%	Unusual
	Retina	Very common	Rare
	Rare locations	Meningeal, liver, skin	–
Associated with		Renal cell carcinoma	–
		Pheochromocytoma	–
		Pancreatic neuroendocrine tumor	–
		Papillary cystadenomas	–

MOLECULAR GENETICS

Loss-of-*VHL* tumor suppressor gene function is a hallmark of VHL disease-related and sporadic HBs.[4,12] VHL patients harbor a loss-of-function germline mutation in one *VHL* allele. The second allele is somatically inactivated in the tumor.[3] Both *VHL* alleles are somatically inactivated in sporadic HBs.[13] The *VHL* gene maps to chromosome 3p25 and encodes the VHL tumor suppressor protein (pVHL). Biochemically, pVHL serves as the substrate recognition receptor of an E3 ligase complex, which targets hypoxia inducible factors 1a and 2a (HIF1a, HIF2a) for degradation.[14,15] HIF1a/2a are transcription factors that bind specific DNA sequences and activate a large family of genes, including growth and angiogenic factors (such as vascular endothelial growth factor, VEGF; platelet derived growth factor, PDGF), erythropoietin (EPO), enzymes of the intermediary metabolism responsible for the anaerobic metabolism in cancer cells, genes that promote epithelial to mesenchymal transformation, and finally genes that are linked to stem cell proliferation.[16] HIF is undetectable in well-oxygenated cells because they bind to von Hippel–Lindau (VHL) E3 ubiquitin ligase and it is degraded with a half-life of a few seconds. The outcome of this tight control is that well-oxygenated, nonmalignant, human cells have barely any detectable activity. However loss-of-function *VHL* mutations lead to the stabilization of HIFa subunits and activation of downstream targets, rendering neoplastic cells dependent on HIF activity for growth.[17]

Hemangioblastomas are known to overexpress HIF1a and HIF2a subunits.[18] Erythropoietin, which is a *bona-fide* target of HIF2a activation, is also highly expressed in hemangioblastomas, as well as EpoR, which is activated by HB-produced Epo.[19] Both proteins contribute to a paracrine signaling pathway within the tumor microenvironment. Similarly to Epo and EpoR, other HIF1a/2a target genes are expressed and involved in autocrine/paracrine pathways that may contribute to HB growth, such as angiopoietin 1 and Tie2,[20] VEGF and Flk1,[13] and finally, SDF1 and CXCR4.[21]

Very little is known about the molecular events that "drive" the growth of HB tumors, beyond VHL inactivation. Gain of chromosome 4 and loss of chromosomes 19, 22q13.2 and 6 were detected in HBs.[22,23] However the genes that are targeted by these deletions have not been identified. It is possible that the genetic pathways that lead to sporadic and VHL disease-related HBs are overlapping but not indentical, for example loss of chromosome 6 characterizes the sporadic only HBs.

DISEASE MECHANISMS

Origin of Human Hemangioblastoma

The complex cytology of the HB stromal cells, their unique morphology, and the presence of hematopoetic tissue in HBs led to the idea that the stromal cells have an angiomesenchymal origin.[24] Other cell types have also been proposed to be the hemangioblastoma precursor (including glial, endothelial, arachnoid, embryonic choroid plexus and neuroendocrine cells).[25–30] Currently, the histogenesis of HBs remains unknown.

After the identification and characterization of the hemangioblast entity *in vivo*,[31,32] it was shown that hemangioblastoma stromal cells express proteins that are hemangioblast-specific.[20] The hemangioblast is a transient cell that forms in the embryo during gastrulation and has the ability to rapidly differentiate into an endothelial or a hematopoietic stem cell. Hemangioblast cells are identified by expressing brachyury and Flk-1 simultaneously, as well as, stem-cell leukemia factor as a consequence of their mesodermal lineage.[33] In addition, hemangioblasts are known to express the endothelial marker Tie2 and the hematopoietic markers Gata-1 and Csf-1R. Considering that all these proteins were shown to be expressed in hemangioblastoma stromal cells, it is likely that the HB stromal cell arises from a developmentally arrested hemangioblast or an embryonic progenitor with hemangioblastic differentiation potential.

Molecular studies indicate that the stromal HB cell has a developmentally complex phenotype. First, fetal hemoglobin was detected in hemangioblastoma red blood cells.[19] Second, HB stromal cells express Epo receptor (EpoR), which in adults is restricted to red blood cells.[19] Third, multiple "developmentally arrested" structures were found in the cerebellum and spinal cord of VHL patients.[34,35] And finally, the pluripotent marker Oct4 is strongly expressed in the nucleus of hemangioblastoma stromal cells.[36] Although the multi/pluripotency of stromal cells is unknown, they express markers of progenitor cells such as CD133, which labels stem cells, neural stem cells, endothelial progenitor cells, and mesenchymal stem cells.[36] Human, tumor-derived, hemangioblastoma cells may actually differentiate to both endothelial cells and red blood cells *in vitro*.[37]

Animal Models of Human HB

Currently, there is no animal model that faithfully recapitulates CNS hemangioblastomas. The *VHL* knockout (KO) mouse is embryonically lethal, and although the embryos have clear vascular abnormalities, they do not develop CNS tumors.[38] The conditional liver *VHL* KO mouse (Lox-*VHL*-Lox driven by Albumin-Cre) develops vascular liver lesions which nevertheless do not recapitulate the histology of human HBs.[39] Mice in which *VHL* was deleted in the renal proximal tubule and hepatocytes (Lox-VHL-Lox driven by PEPCK-Cre) develop kidney cysts and polycythemia, but do not form HBs.[40] And finally, to the best of our knowledge, there is not yet a reported conditional *VHL* KO mouse that targets the CNS.

The only animal model that seems to recapitulate some aspects of human HB biology is the *vhl* zebrafish model. Although these embryos do not develop CNS tumors, most likely because they only survive until 12 to 13 days postfertilization, they develop hypervascular lesions in the brain and retina; in addition they display an abnormal expansion of their embryonic caudal hematopoietic tissue.[41,42]

Future Directions

Future research will use next-generation DNA/RNA sequencing of isolated components of human HBs and HB-derived cell lines to gain insights into oncogenic drivers of these tumors. These studies will allow us to find specific therapeutic targets for both sporadic and VHL-related HBs. In addition, lineage-tracing studies will allow us to unravel the origin of these tumors.

DIAGNOSTIC TESTING

Diagnostic Imaging

The diagnostic imaging of choice for HBs is contrast-enhanced MRI. HBs are very vascular tumors that avidly uptake contrast and appear as very T1-bright lesions. Typically they form cystic lesions with a mural nodule (80%), or they appear as purely solid lesions (10%), mixed cystic and solid lesions (5%), or even as simple cysts (5%).[43] HBs may be surrounded by edema (T2-bright signal) and possibly syringomyelia, in the case of spinal HBs. Edema may be the first sign of a cyst formation.[44] Noncontrast MRI or contrast-enhanced computed tomography (CT) have very low sensitivity and they are not useful imaging approaches for diagnosis of HBs.[45]

Pathology

Hemangioblastomas, as previously mentioned, are formed by a complex network of three different cell types: large vacuolated stromal cells, endothelial cells, and pericytes. They can be divided into two different subtypes: reticular or mesenchymal hemangioblastomas, and cellular or epithelioid hemangioblastomas. The more common subtype, reticular hemangioblastoma, is characterized by individual stromal cells enmeshed in an abundant network of capillaries. In contrast, the cellular or epithelioid subtype has predominantly clusters of tumor cells.[22] The epithelioid histology is predominantly found in tumors of bigger size and extramedullary hematopoiesis is restricted to those epithelioid areas.[35] The diagnosis of HB is confirmed by pathologic examination of the surgically excised tumor. Immunohistochemical markers facilitate the diagnosis of HB: stromal cells are detected in clinical samples as D240+, Inhibin A+, CD34- cells; endothelial cells as CD34+, D240-, Inhibin A– cells; and hematopoietic components may stain positive for EPO and EPO receptor.[46–48]

Molecular Diagnosis

All patients with a pathologically confirmed HB should undergo germline testing for mutation in the *VHL* gene. The patients with multiple HBs (synchronous or metachronous) or a single HB and a history of another VHL disease-related lesion (current or in the past), are diagnosed with VHL disease on clinical grounds.[49] For these patients, germline testing is confirmatory of the VHL diagnosis and the knowledge of the specific VHL mutation can serve to screen the family members at risk. Patients with family history of VHL that present with a CNS HB can be considered as carriers of the disease. Most importantly, all patients with "seemingly sporadic" HB (i.e., a single HB, a negative family history indicative of VHL disease, and an absence of another VHL disease-associated lesion concurrently with the HB or in the past) should undergo genetic testing for VHL disease. Woodward et al. showed that 5% of these "seemingly sporadic" HB patients do harbor a germline *VHL* mutation. In addition, approximately 5% of the patients that test negative for *VHL* mutation, will develop a second VHL disease-related lesion(s) within 5 years from diagnosis of the initial HB.[50]

All patients known to have VHL disease should undergo MRI imaging of the brain, cervical spine, and thoracic spine as well as retinal exam as part of their surveillance protocol (usually annual), in order to detect early and asymptomatic lesions.

MANAGEMENT

The care of a HB patient should be coordinated in a multidisciplinary clinic where neurosurgeons, radiation therapists, and medical oncologists can contribute to an individualized treatment plan. The feasibility, success rate and outcome of surgery or radiation therapy are likely to depend on the location of the tumor, the size of the tumor, and the experience of the physician with the disease.

Surgery is the definitive treatment for HBs when it can be performed.[51,52] Location of hemangioblastomas in areas such as the hypothalamus or the brainstem may render surgery impossible.

Radiation therapy (stereotactic radiosurgery, SRS; external beam radiotherapy; or proton beam therapy) has a selective role and can be optimally applied to hemangioblastomas of a small size (usually not more than 10 mm in diameter), without cystic component and with a safe margin from sensitive normal tissue.[53,54] A recent review of long-term follow-up of VHL patients with HB that were treated with SRS strongly suggests that "SRS should not be used to prophylactically treat asymptomatic tumors and should be reserved for the treatment of tumors that are not surgically resectable."[55]

Loss-of-VHL and constitutive upregulation of HIF1a/2a is a hallmark of HBs. VEGF is one of many downstream HIF-targets and its inhibition by antibodies or small molecule inhibitors is now in routine clinical practice.[56] Renal cell cancer (RCC) is driven by loss-of-VHL and HIF2a expression and it develops, along with HBs, in VHL patients. VEGF kinase inhibitors are active against RCC in VHL patients,[57] but so far have failed to alter significantly the size or natural history of VHL disease-related HBs.[57] Small molecule inhibitors of HIF2a have been reported and they are currently under preclinical development.[58,59] It appears reasonable to target HBs with HIF2a inhibitors when they become clinically available. Lastly, next-generation sequencing of HBs is likely to highlight oncogenic mutations that drive HB growth, in addition to VHL inactivation, and provide novel targets for therapy with biologics or small molecules.

References

1. Latif F, Tory K, Gnarra J, et al. Identification of the von Hippel–Lindau disease tumor suppressor gene. *Science*. 1993;260(5112):1317–1320.
2. Lonser RR, Glenn GM, Walther M, et al. von Hippel–Lindau disease. *Lancet*. 2003;361(9374):2059–2067.
3. Gnarra JR, Tory K, Weng Y, et al. Mutations of the VHL tumour suppressor gene in renal carcinoma. *Nat Genet*. 1994;7(1):85–90.
4. Lee JY, Dong SM, Park WS, et al. Loss of heterozygosity and somatic mutations of the VHL tumor suppressor gene in sporadic cerebellar hemangioblastomas. *Cancer Res*. 1998;58(3):504–508.
5. Lonser RR, Butman JA, Kiringoda R, Song D, Oldfield EH. Pituitary stalk hemangioblastomas in von Hippel–Lindau disease. *J Neurosurg*. 2009;110(2):350–353.
6. Wong WT, Agron E, Coleman HR, et al. Clinical characterization of retinal capillary hemangioblastomas in a large population of patients with von Hippel–Lindau disease. *Ophthalmology*. 2008;115(1):181–188.
7. Wong WT, Agron E, Coleman HR, et al. Genotype–phenotype correlation in von Hippel–Lindau disease with retinal angiomatosis. *Arch Ophthalmol*. 2007;125(2):239–245.
8. Toy BC, Agron E, Nigam D, Chew EY, Wong WT. Longitudinal analysis of retinal hemangioblastomatosis and visual function in ocular von Hippel–Lindau disease. *Ophthalmology*. 2012;119(12):2622–2630.
9. Trimble M, Caro J, Talalla A, Brain M. Secondary erythrocytosis due to a cerebellar hemangioblastoma: demonstration of erythropoietin mRNA in the tumor. *Blood*. 1991;78(3):599–601.
10. Roessler K, Dietrich W, Haberler C, Goerzer H, Czech T. Multiple spinal "military" hemangioblastomas in von Hippel–Lindau (vHL) disease without cerebellar involvement. A case report and review of the literature. *Neurosurg Rev*. 1999;22(2–3):130–134.
11. Neumann HP, Eggert HR, Scheremet R, et al. Central nervous system lesions in von Hippel–Lindau syndrome. *J Neurol Neurosurg Psychiatry*. 1992;55(10):898–901.
12. Vortmeyer AO, Huang SC, Pack SD, et al. Somatic point mutation of the wild-type allele detected in tumors of patients with VHL germline deletion. *Oncogene*. 2002;21(8):1167–1170.
13. Wizigmann-Voos S, Breier G, Risau W, Plate KH. Up-regulation of vascular endothelial growth factor and its receptors in von Hippel–Lindau disease-associated and sporadic hemangioblastomas. *Cancer Res*. 1995;55(6):1358–1364.
14. Jaakkola P, Mole DR, Tian YM, et al. Targeting of HIF-alpha to the von Hippel–Lindau ubiquitylation complex by O2-regulated prolyl hydroxylation. *Science*. 2001;292(5516):468–472.

15. Schofield CJ, Ratcliffe PJ. Oxygen sensing by HIF hydroxylases. *Nat Rev Mol Cell Biol*. 2004;5(5):343–354.
16. Majmundar AJ, Wong WJ, Simon MC. Hypoxia-inducible factors and the response to hypoxic stress. *Mol Cell*. Oct 22;40(2):294–309.
17. Kaelin Jr. WG. The von Hippel–Lindau tumour suppressor protein: O_2 sensing and cancer. *Nat Rev Cancer*. 2008;8(11):865–873.
18. Shively SB, Falke EA, Li J, et al. Developmentally arrested structures preceding cerebellar tumors in von Hippel–Lindau disease. *Mod Pathol*. 2011;24(8):1023–1030.
19. Vortmeyer AO, Frank S, Jeong SY, et al. Developmental arrest of angioblastic lineage initiates tumorigenesis in von Hippel–Lindau disease. *Cancer Res*. 2003;63(21):7051–7055.
20. Glasker S, Li J, Xia JB, et al. Hemangioblastomas share protein expression with embryonal hemangioblast progenitor cell. *Cancer Res*. 2006;66(8):4167–4172.
21. Zagzag D, Krishnamachary B, Yee H, et al. Stromal cell-derived factor-1alpha and CXCR4 expression in hemangioblastoma and clear cell-renal cell carcinoma: von Hippel–Lindau loss-of-function induces expression of a ligand and its receptor. *Cancer Res*. 2005;65(14):6178–6188.
22. Rickert CH, Hasselblatt M, Jeibmann A, Paulus W. Cellular and reticular variants of hemangioblastoma differ in their cytogenetic profiles. *Hum Pathol*. 2006;37(11):1452–1457.
23. Beckner ME, Sasatomi E, Swalsky PA, Hamilton RL, Pollack IF, Finkelstein SD. Loss of heterozygosity reveals non-VHL allelic loss in hemangioblastomas at 22q13. *Hum Pathol*. 2004;35(9):1105–1111.
24. Stein AA, Schilp AO, Whitfield RD. The histogenesis of hemangioblastoma of the brain. A review of twenty-one cases. *J Neurosurg*. 1960;17:751–761.
25. Becker I, Paulus W, Roggendorf W. Histogenesis of stromal cells in cerebellar hemangioblastomas. An immunohistochemical study. *Am J Pathol*. 1989;134(2):271–275.
26. Jurco 3rd. S, Nadji M, Harvey DG, Parker Jr. JC, Font RL, Morales AR. Hemangioblastomas: histogenesis of the stromal cell studied by immunocytochemistry. *Hum Pathol*. 1982;13(1):13–18.
27. Mizuno J, Iwata K, Takei Y. Immunohistochemical study of hemangioblastoma with special reference to its cytogenesis. *Neurol Med Chir (Tokyo)*. 1993;33(7):420–424.
28. Jakobiec FA, Font RL, Johnson FB. Angiomatosis retinae. An ultrastructural study and lipid analysis. *Cancer*. 1976;38(5):2042–2056.
29. Frank TS, Trojanowski JQ, Roberts SA, Brooks JJ. A detailed immunohistochemical analysis of cerebellar hemangioblastoma: an undifferentiated mesenchymal tumor. *Mod Pathol*. 1989;2(6):638–651.
30. Tanimura A, Nakamura Y, Hachisuka H, Tanimura Y, Fukumura A. Hemangioblastoma of the central nervous system: nature of the stromal cells as studied by the immunoperoxidase technique. *Hum Pathol*. 1984;15(9):866–869.
31. Choi K, Kennedy M, Kazarov A, Papadimitriou JC, Keller G. A common precursor for hematopoietic and endothelial cells. *Development*. 1998;125(4):725–732.
32. Kennedy M, Firpo M, Choi K, et al. A common precursor for primitive erythropoiesis and definitive haematopoiesis. *Nature*. 1997;386(6624):488–493.
33. Huber TL, Kouskoff V, Fehling HJ, Palis J, Keller G. Haemangioblast commitment is initiated in the primitive streak of the mouse embryo. *Nature*. 2004;432(7017):625–630.
34. Mack FA, Rathmell WK, Arsham AM, Gnarra J, Keith B, Simon MC. Loss of pVHL is sufficient to cause HIF dysregulation in primary cells but does not promote tumor growth. *Cancer Cell*. 2003;3(1):75–88.
35. Shively SB, Beltaifa S, Gehrs B, et al. Protracted haemangioblastic proliferation and differentiation in von Hippel–Lindau disease. *J Pathol*. 2008;216(4):514–520.
36. Welten CM, Keats EC, Ang LC, Khan ZA. Hemangioblastoma stromal cells show committed stem cell phenotype. *Can J Neurol Sci*. 39(6):821–827.
37. Park DM, Zhuang Z, Chen L, et al. von Hippel–Lindau disease-associated hemangioblastomas are derived from embryologic multipotent cells. *PLoS Med*. 2007;4(2):e60.
38. Gnarra JR, Ward JM, Porter FD, et al. Defective placental vasculogenesis causes embryonic lethality in VHL-deficient mice. *Proc Natl Acad Sci U S A*. 1997;94(17):9102–9107.
39. Haase VH, Glickman JN, Socolovsky M, Jaenisch R. Vascular tumors in livers with targeted inactivation of the von Hippel–Lindau tumor suppressor. *Proc Natl Acad Sci U S A*. 2001;98(4):1583–1588.
40. Rankin EB, Tomaszewski JE, Haase VH. Renal cyst development in mice with conditional inactivation of the von Hippel–Lindau tumor suppressor. *Cancer Res*. 2006;66(5):2576–2583.
41. van Rooijen E, Voest EE, Logister I, et al. Zebrafish mutants in the von Hippel–Lindau tumor suppressor display a hypoxic response and recapitulate key aspects of Chuvash polycythemia. *Blood*. 2009;113(25):6449–6460.
42. van Rooijen E, Voest EE, Logister I, et al. von Hippel–Lindau tumor suppressor mutants faithfully model pathological hypoxia-driven angiogenesis and vascular retinopathies in zebrafish. *Dis Model Mech*. May-Jun;3(5–6):343–353.
43. Richard S, Campello C, Taillandier L, Parker F, Resche F. Haemangioblastoma of the central nervous system in von Hippel–Lindau disease. French VHL Study Group. *J Intern Med*. 1998;243(6):547–553.
44. Lonser RR, Vortmeyer AO, Butman JA, et al. Edema is a precursor to central nervous system peritumoral cyst formation. *Ann Neurol*. 2005;58(3):392–399.
45. Filling-Katz MR, Choyke PL, Oldfield E, et al. Central nervous system involvement in Von Hippel–Lindau disease. *Neurology*. 1991;41(1):41–46.
46. Hoang MP, Amirkhan RH. Inhibin alpha distinguishes hemangioblastoma from clear cell renal cell carcinoma. *Am J Surg Pathol*. 2003;27(8):1152–1156.
47. Roy S, Chu A, Trojanowski JQ, Zhang PJ. D2-40, a novel monoclonal antibody against the M2A antigen as a marker to distinguish hemangioblastomas from renal cell carcinomas. *Acta Neuropathol*. 2005;109(5):497–502.
48. Bohling T, Maenpaa A, Timonen T, Vantunen L, Paetau A, Haltia M. Different expression of adhesion molecules on stromal cells and endothelial cells of capillary hemangioblastoma. *Acta Neuropathol*. 1996;92(5):461–466.
49. Maher ER, Kaelin Jr. WG. von Hippel–Lindau disease. *Medicine (Baltimore)*. 1997;76(6):381–391.

50. Woodward ER, Wall K, Forsyth J, Macdonald F, Maher ER. VHL mutation analysis in patients with isolated central nervous system haemangioblastoma. *Brain*. 2007;130(Pt 3):836–842.
51. Jagannathan J, Lonser RR, Smith R, DeVroom HL, Oldfield EH. Surgical management of cerebellar hemangioblastomas in patients with von Hippel–Lindau disease. *J Neurosurg*. 2008;108(2):210–222.
52. Kanno H, Yamamoto I, Nishikawa R, et al. Spinal cord hemangioblastomas in von Hippel–Lindau disease. *Spinal Cord*. 2009;47(6):447–452.
53. Moss JM, Choi CY, Adler Jr. JR, Soltys SG, Gibbs IC, Chang SD. Stereotactic radiosurgical treatment of cranial and spinal hemangioblastomas. *Neurosurgery*. 2009;65(1):79–85 discussion.
54. Kano H, Niranjan A, Mongia S, Kondziolka D, Flickinger JC, Lunsford LD. The role of stereotactic radiosurgery for intracranial hemangioblastomas. *Neurosurgery*. 2008;63(3):443–450 discussion 50–51.
55. Asthagiri AR, Mehta GU, Zach L, et al. Prospective evaluation of radiosurgery for hemangioblastomas in von Hippel–Lindau disease. *Neuro Oncol*. Jan;12(1):80–86.
56. Atkins MB, Bukowski RM, Escudier BJ, et al. Innovations and challenges in renal cancer: summary statement from the Third Cambridge Conference. *Cancer*. 2009;115(10 suppl):2247–2251.
57. Jonasch E, McCutcheon IE, Waguespack SG, et al. Pilot trial of sunitinib therapy in patients with von Hippel–Lindau disease. *Ann Oncol*. Dec;22(12):2661–2666.
58. Zimmer M, Ebert BL, Neil C, et al. Small-molecule inhibitors of HIF-2a translation link its 5′UTR iron-responsive element to oxygen sensing. *Mol Cell*. 2008;32(6):838–848.
59. Scheuermann TH, Li Q, Ma HW, et al. Allosteric inhibition of hypoxia inducible factor-2 with small molecules. *Nat Chem Biol*. Apr;9(4):271–276.

CHAPTER

83

Incontinentia Pigmenti

A. Yasmine Kirkorian and Bernard Cohen

Johns Hopkins University and Johns Hopkins Hospital, Baltimore, MD, USA

INTRODUCTION

Incontinentia pigmenti (IP; Bloch–Sulzberger syndrome; OMIM 308300) is an X-linked dominantly inherited genodermatosis that affects the skin in combination with anomalies of other organs, including the central nervous system (CNS). IP results from mutations in the inhibitor of kappa beta kinase gamma (*IKBKG*; OMIM 300248), formerly known as NEMO, located on locus Xq28.[1] Absence of IKBKG results in complete inhibition of nuclear factor kappa beta (NFKB) signaling, leading to the varied manifestations seen in IP. The wide spectrum of clinical manifestations in IP results from lyonization of the X chromosome with skewed X inactivation. Therefore IP, although rare, may serve as a prototypical example of the clinical presentation of X-linked dominant disorders.

DIAGNOSTIC CRITERIA FOR IP

Diagnostic criteria for IP were first proposed by Landy and Donnai in 1993 and were based on the evaluation of clinical manifestations in 111 patients and a review of cases published in the literature.[2] These criteria were refined by Minić et al. in 2013 to reflect new knowledge of the genetic basis for IP as well as to include novel clinical features not emphasized in the original criteria such as CNS anomalies, nonretinal ocular anomalies, nondental oral anomalies, nipple anomalies, and histopathologic findings.[3]

The major criteria include any of the four IP skin stages, which are discussed below. The minor criteria include CNS anomalies, ocular anomalies, alopecia and abnormal hair, abnormal nails, oral anomalies, breast and nipple anomalies, multiple male miscarriages, and IP histopathologic findings. Eosinophilia and the evidence of skewed X-chromosome inactivation studies may be carefully considered as conditions that support the diagnosis of IP. If there is no evidence of IP in a first-degree female relative and genetic *IKBKG* mutation data are lacking, at least two or more major criteria or one major criterion and one or more minor criteria are necessary to make a diagnosis of sporadic IP. If there is no evidence of IP in a first-degree female relative and the case is confirmed for *IKBKG* mutation typical for IP, any single major or minor criterion is diagnostic. If there is evidence of IP in a first-degree female relative, any single major criterion or at least two minor criteria are sufficient.

SKIN MANIFESTATIONS OF IP

The first manifestations of IP are often dermatologic and are found in nearly all patients with IP.[4] Skin manifestations of IP occur within lines of Blaschko.[5] These lines are distributed in well-defined whorls along the body and reflect genomic or epigenetic mosaicism.[6]

There are four stages of skin manifestations in IP (Figure 83.1). These stages typically occur sequentially, but different stages may overlap and not all patients display all four stages. Some patients, especially older ones, may not display any obvious skin lesions, and this should not preclude a diagnosis of IP.

Rosenberg's Molecular and Genetic Basis of Neurological and Psychiatric Disease
http://dx.doi.org/10.1016/10.1016/B978-0-12-410529-4.00083-8

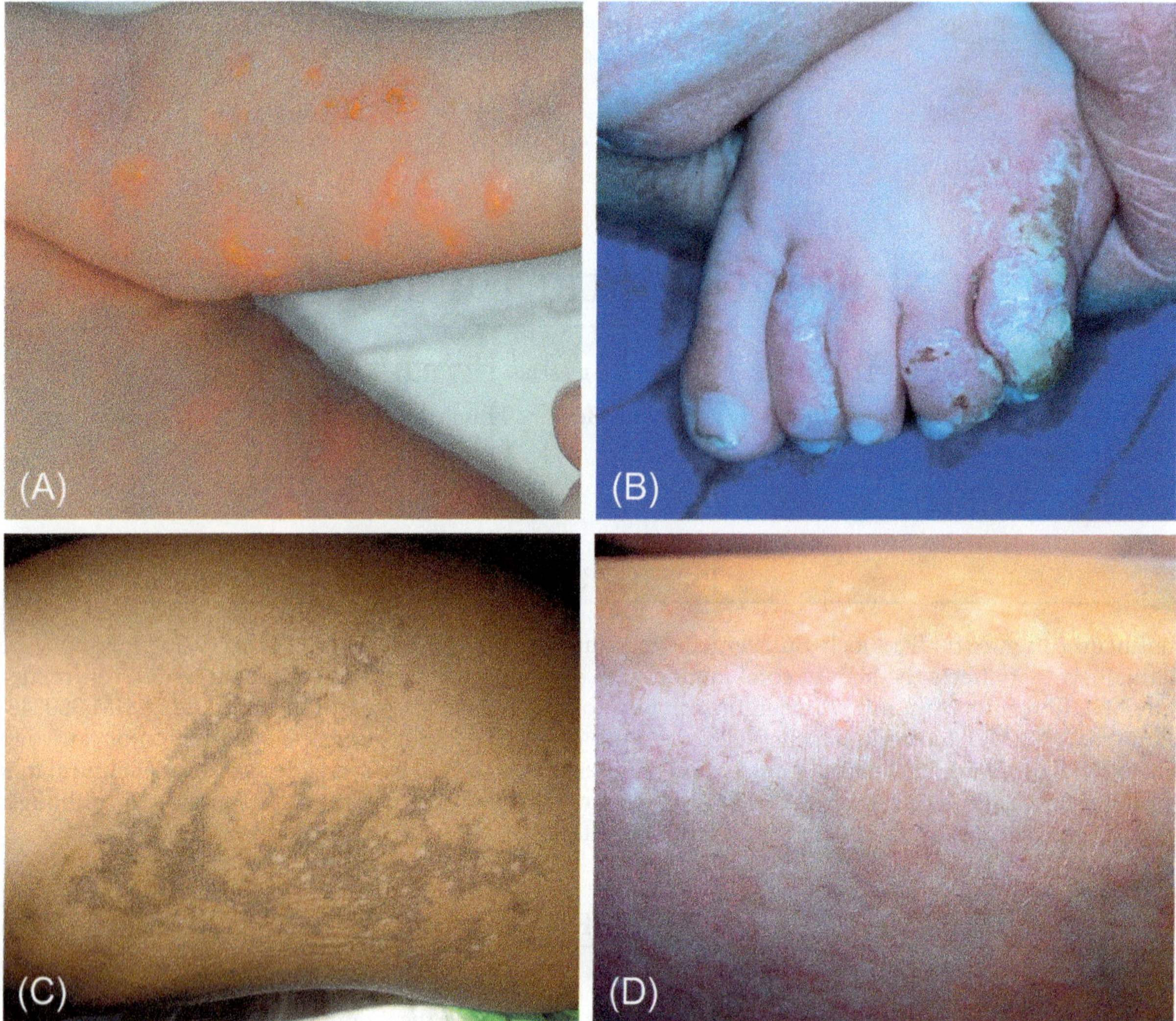

FIGURE 83.1 **Cutaneous stages of incontinentia pigmenti.** (**A**) Vesiculobullous stage; (**B**) verrucous stage; (**C**) hyperpigmented stage; (**D**) hypopigmented stage.

Stage 1 is also known as the vesiculobullous stage. This stage is seen in 90% of IP patients. In 92% of patients with an observed first stage, the lesions were noted within the first 2 weeks of life and resolved by 18 months of age.[7] Skin lesions consist of vesicles or bullae on an inflammatory urticarial base distributed within lines of Blaschko. The trunk is more commonly affected than the extremities and scalp. Accumulation of inflammatory cells within the vesicles results in the development of pustules. A skin biopsy in this stage will show eosinophilic spongiosis and intraepidermal vesicles containing eosinophils as well as many apoptotic keratinocytes in the epidermis.

Stage 2, known as the verrucous stage, occurs in approximately 70% of patients and presents within the first few weeks to months of life. The distribution of verrucous lesions usually does, but may not, correspond to the prior inflammatory vesiculobullous lesions. A skin biopsy in this stage is characterized by papillomatosis, hyperkeratosis, and acanthosis of the epidermis with many apoptotic cells in the epidermis forming squamous eddies. Major melanin incontinence is noted and reflected clinically as hyperpigmentation.

Stage 3, the hyperpigmented stage, usually begins as stage 2 resolves and it may persist into adulthood. Clinically patients demonstrate hyperpigmented grayish-brown whorled or splashed lesions referred to classically as "Chinese figures." A skin biopsy in this stage shows marked melanin incontinence with numerous melanophages in the dermis with scattered necrotic keratinocytes in the epidermis.

Stage 4, the hypopigmented or atrophic stage, usually begins as stage 3 resolves and it is clinically characterized by areas of linear hypopigmentation and atrophy along the lines of Blaschko. Skin lesions typically demonstrate an absence of pilosebaceous units and patients may note absence of hair growth or sweating in the affected area. A skin biopsy in this stage shows an atrophic epidermis, massive reduction of melanin in the basal layer, persistence of apoptotic bodies in the epidermis or papillary dermis, and the complete absence of pilosebaceous units and eccrine glands. Unusual distinctive ectatic vessels may explain the vascular appearance of certain lesions.[8]

Recurrence of stage 1 lesions has been well documented as a sporadic occurrence,[7,9] often following illness (viral infection, etc.),[10] topical application of estrogen cream,[11] vaccination,[12] or laser treatment.[13]

Skin lesions of IP at different stages may be confused with other infectious and inflammatory conditions. For example, lesions in the vesiculobullous stage have been mistaken for contact dermatitis and herpes virus infection.[14,15] The coexistence of neonatal IP and herpes simplex virus infection has also been reported.[16] Skin lesions of the verrucous stage may mimic pyoderma gangrenosum.[17] The differential diagnosis of the different stages of IP is broad and diagnosis of IP requires careful recognition of the blaschkoid distribution of skin lesions, skin histology, knowledge of affected relatives, and evaluation of other minor diagnostic criteria.

Mouse models of IP help to elucidate the mechanism by which mutation in *IKBKG* results in the skin phenotype. In mice with an *IKBKG* mutation an increase of interleukin 1 beta (IL1B) synthesis was observed in the epidermis before detecting any skin abnormalities. The reason for this elevation is uncertain, but it is hypothesized to result from cellular necrosis or dysregulated response to the bacterial environment at birth. Elevation in IL1B is then thought to induce synthesis of tumor necrosis factor alpha (TNFA) in neighboring wild type keratinocytes. This cytokine would in turn act on mutant IKBKG-expressing cells, inducing their death and clearance. The expression of TNFA is critical because IKBKG-mutated mice crossed with TNFA knockout mice did not express the IP phenotype.[18] NFKB has a wide variety of functions in the skin including protection against apoptosis, and promotion of terminal differentiation of keratinocytes therefore absence secondary to *IKBKG* mutation is expected to lead to a variety of manifestations in the skin.[19]

CNS MANIFESTATIONS OF IP

Neurologic manifestations of IP are among the most devastating consequences of this disorder. These manifestations vary from a single seizure episode to psychomotor delay, intellectual disability, learning disabilities, hemiplegia, epilepsy, cerebellar ataxia, microcephaly, neonatal encephalopathy, childhood encephalomyelitis, and neonatal and childhood ischemic stroke.[20,21] CNS sequelae in IP may occur from the first 12 hours of life through 10 years of age,[22] but are present in the first month of life in 66.9% of cases in which neurologic manifestations are documented.[23]

The frequency of such manifestations is a matter of debate in the literature. Several authors[24] have suggested a frequency of approximately 30%, which was supported by a recent review of the literature reporting 44 cases of IP and neurologic sequelae.[23] Other authors[7] report a lower frequency (10%) and question whether the reports of higher frequency might be due to confounding of early cases of IP with other syndromes such as pigmentary mosaicism (formerly hypomelanosis of Ito). A retrospective case series of 25 probands and an additional 28 female family members fulfilling the Landy and Donnai criteria demonstrated no evidence of neurologic sequelae in the 28 relatives of individuals with IP.[25] The authors suggest that there may be an underrecognition of IP in patients with a milder phenotype lacking neurological manifestations. Our experience as dermatologists supports this possibility as we have evaluated patients with only subtle cutaneous and dental manifestations of IP diagnosed as adults after giving birth to more severely affected children.

The mechanism by which mutations in *IKBKG* result in the variable neurologic sequelae seen in IP is unclear. Studies in female mice heterozygotes for *IKBKG* mutations demonstrate findings suggestive of neurologic dysfunction by day 8 of life including spasms, tremors, and locomotor difficulties.[26] In humans, IKBKG is present in neurons, astrocytes, microglia, and oligodendrocytes. An important role of NFKB is postulated in the control of neuronal death during development of the nervous system and in the modulation of synaptic function following glutamate receptor stimulation.[27] Some authors propose that neurologic dysfunction is a result of infarcts in the cerebral microvasculature. This hypothesis is not supported, however, by histopathologic studies of brain tissue from patients with IP. Those studies demonstrate focal necrosis in cerebral hemispheres in the white matter and cortex that are not associated with vascular territories.[28]

Results of brain imaging demonstrate white matter and corpus callosum abnormalities, cortical malformation, cerebral atrophy as well as cerebral and cerebellar hemorrhage. The clinical impact of these changes is not always straightforward as demonstrated by the report of a neurologically intact child with white matter changes on CNS imaging.[29]

Minić et al. analyzed the literature for patterns of CNS involvement seen in computed tomography (CT) or magnetic resonance imaging (MRI) of patients with IP.[30] The authors hypothesize that CNS lesions of IP are distributed in lines analogous to lines of Blaschko on the skin and are due to apoptosis of cells containing the *IKBKG* mutation.

Their hypothesis stems from the following observations: 1) parenchymal abnormalities were most severe in patients with severe neonatal cutaneous lesions, especially if these were located on the scalp; and 2) MRI findings do not correlate with vascular territories. They point to the radial unit model of development and the structure of cerebral cortex and the common embryonic origin of skin, CNS, and eyes and propose that CNS Blaschko line analogs, similar to those in the skin, represent the trace of the development of clones of neurons and glial cells arising from a cell with *IKBKG* mutation. Extracutaneous "lines of Blaschko" have been described in the brain, bones, teeth, and the eye, reinforcing the concept of mosaicism in presentation of X-linked disorders.[30]

OCULAR MANIFESTATIONS OF IP

In the updated criteria for IP proposed by Minić et al.,[7] ophthalmologic findings have been expanded to include extraretinal manifestations of IP. In that study, 7.24% of 831 IP patients had a variety of ocular anomalies. These included retinal anomalies, the majority of which were vascular anomalies such as telangiectasia, ectasia, hemorrhage, arteriovenous anastomoses, neovascularization, and avascularity of retina. Nonretinal anomalies included vitreous and lens anomalies. Microphthalmia and amaurotic eyes represented 2.30% and 6.91% of anomalies, respectively. Cataracts were classified as an IP specific feature. A case report of a female patient with IP described "bilateral whorl-like corneal epitheliopathy," which again might be hypothesized to represent an extracutaneous line of Blaschko.[31]

Given the potential for retinal detachment and blindness secondary to ocular involvement in IP, all infants with suspected IP should be screened immediately by ophthalmology. Early laser treatment of retinal vasculature can reverse retinopathy of IP and may be vision saving.[32,33]

DISORDERS OF SKIN APPENDAGES IN IP

In addition to absence of pilosebaceous units noted in stage 4 skin lesions of IP, 50% of patients have other hair anomalies. Alopecia is common, especially at the vertex of the scalp and often after blistering or verrucous lesions at this site. Nonalopecia hair anomalies included sparse hair, wooly hair, and anomalies of the eyebrows and eyelashes.[7]

Nail abnormalities were observed in 8.7% of 723 IP patients. These abnormalities included mild ridging or pitting to severe nail dystrophy or subungual or periungual tumors (see next section).[7] A patient with IP with 20-nail-dystrophy has been described.[34] Multiple family members were diagnosed with IP after late-onset of onychodystrophy.[35]

BONY MANIFESTATIONS OF IP

Bony anomalies noted in IP reflect the critical role of NFKB in bony morphogenesis.[36,37] The most commonly seen bony anomalies in osteolysis of the distal phalanx associated with hyperkeratotic subungual tumors that involve the fingers more than the toes.[38–42] These subungual tumors are referred to as subungual tumors of IP (STIP). They often do not present until adulthood and may be associated with pregnancy. These tumors are analogous to the cutaneous stage 2 verrucous lesions. A biopsy will show an epidermis that is hyperplastic, papillomatous and contains glassy keratinocytes. Notably these features are also shared with keratoacanthomas and squamous cell carcinoma and patients have mistakenly undergone unnecessary amputations due to misdiagnosis. Excision is curative although recurrence has been described. Other treatment options are oral and topical retinoids.

DENTAL AND ORAL MANIFESTATIONS OF IP

After skin lesions, dental anomalies associated with IP are among the most common and can be helpful in screening relatives of patients with IP. Dental anomalies will not be seen until 1 year of age when primary teeth erupt. A case series of 25 patients with IP that focused specifically on dental and oral anomalies reported anomalies in 70% of patients.[43] These included teeth-shape anomalies (cone-like or peg-like teeth), the presence of numerous cariotic teeth, early dental loss, and other oral findings, such as delayed eruption, gothic palate, and partial anodontia. In our clinical experience, dental anomalies are quite common even in those individuals with minimal skin disease and no other systemic involvement.

OTHER MINOR CRITERIA

Many other findings have been noted in IP patients and are considered minor criteria for diagnosis. These include breast and nipple anomalies such as supernumerary nipples, nipple hypoplasia, breast hypoplasia, and aplasia.

Infants with IP may have leukocytosis with eosinophilia, particularly when their skin is in stage 1 and stage 2. Investigators hypothesize that apoptosis and necrosis of keratinocytes with absent NFKB secondary to mutation in *IKBKG* results in increased secretion of inflammatory cytokines by neighboring normal keratinocytes. These inflammatory cytokines induce epidermal expression of eotaxin, which recruits eosinophils to the skin. Eosinophil degranulation and release of proteases results in intracellular edema and may lead to the blistering process of the first stage of IP.[44] Four patients with IP were found to have ultrastructurally disordered leukocytes with pseudoplatelets. The authors hypothesized that this resulted from the *IKBKG* mutation leading to misregulation of apoptosis.[45]

GENETICS OF IP

IP results from mutations in the inhibitor of kappa beta kinase gamma (*IKBKG*, OMIM 300248), formerly known as NEMO, located on locus Xq28.[46] *IKBKG* mutations are found in approximately 70% of individuals with IP. In 60–88% of IP patients harboring a mutation, a common deletion of exon 4–10 is found.[47] The rate of *de novo* mutations in *IKBKG* is 65%.[48] To date, 53 different mutations (from large deletions to single amino-acid substitutions) affecting *IKBKG* have been reported.[49] Hypomorphic *IKBKG* mutations in males lead to a different phenotype of X-linked hypohydrotic ectodermal dysplasia and immunodeficiency (HED-ID, OMIM 300291). In contrast to *IKBKG* mutations causing IP, which usually result in severe truncations of the IKBKG protein, frameshift or nonsense mutations causing HED-ID often only remove the *IKBKG* zinc finger.

IKBKG has a wide variety of functions that result from its regulation of NFKB.[50] IKBKG is a regulatory subunit of inhibitor I-kappa-B kinase (IKK) complex. NFKB is normally kept inactive in the cytoplasm due to interaction with inhibitor I-kappa-B (IKB) proteins. Phosphorylation of inhibitor molecules leads to ubiquitination of IKB and degradation by the proteasome pathway, which results in release of active NFKB. Active NFKB translocates to the nucleus resulting in transcription of genes involved in signal transduction pathways, regulating genes involved in critical developmental processes, innate and adaptive immune responses, cell adhesion, and protection against apoptosis. Absence of IKBKG results in complete inhibition of NFKB signaling reflecting its major role in activation of the IKK complex.

IP IN MALES

The *IKBKG* mutation is typically lethal in males, resulting in an increased rate of miscarriages of male fetuses in affected female patients. Surviving male patients with IP have been found to have Klinefelter syndrome,[51] XXY mosaicism,[52] or somatic mutation.[53,54]

TREATMENT AND FUTURE DIRECTIONS

To date there are no known therapeutic interventions that prevent the various manifestations of IP. Topical corticosteroids have been used to treat inflammation associated with stage 1 vesiculobullous lesions.[55] Retinal vascular changes can be addressed with laser and other interventions discussed above. The authors have reported a case of a neonate with IP whose seizures were not controlled with anticonvulsants but were controlled with use of systemic corticosteroids. Seizures recurred upon cessation of steroids and were once again controlled upon reinitiation of corticosteroids.[56] Whether such interventions affect the long-term prognosis of patients with IP is a matter for future research.

In the future specific interventions that address the NFKB pathway might be expected to be useful in treatment and prevention of the manifestations of IP. At this time, IP serves as an important model for the clinical presentation of genodermatoses with X-linked dominant inheritance.

References

1. Smah A, Courtois G, Vabres P, et al. Genomic rearrangement in NEMO impairs NF-kappaB activation and is a cause of incontinentia pigmenti. The International Incontinentia Pigmenti (IP) Consortium. *Nature.* 2000;405:466e–472e.
2. Landy SJ, Donnai D. Incontinentia pigmenti (Bloch-Sulzberger syndrome). *J Med Genet.* 1993;30(1):53–59.
3. Minić S, Trpinac D, Obradović M. Incontinentia pigmenti diagnostic criteria update. *Clin Genet.* 2013 Jun 26.
4. Berlin AL, Paller AS, Chan LS. Incontinentia pigmenti: a review and update on the molecular basis of pathophysiology. *J Am Acad Dermatol.* 2002;47(2):169–187.
5. Blaschko A. Die Nervenverteilung in der Haut in ihrer Beziehung zu den Erkrankungen der Haut. Beilage zu den Verhandlungen der Deutschen Dermatologischen Gesellschaft; 7. Congress zu Breslau im Mai 1901. Wien Leipzig: Braumuller.
6. Happle R. X-chromosome inactivation: role in skin disease expression. *Acta Paediatr Suppl.* 2006;95(451):16–23.
7. Darné S, Carmichael AJ. Isolated recurrence of vesicobullous incontinentia pigmenti in a schoolgirl. *Br J Dermatol.* 2007;156(3):600.
8. Hadj-Rabia S, Rimella A, Smahi A, et al. Clinical and histologic features of incontinentia pigmenti in adults with nuclear factor-κB essential modulator gene mutation. *J Am Acad Dermatol.* 2011;64(3):508–515.
9. Van Leeuwen RL, Wintzen M, Van Praag MCG. Incontinentia pigmenti: an extensive second episode of a "first-stage" vesiculobullous eruption. *Pediatr Dermatol.* 2000;17:70.
10. Bodak N, Hadj-Rabia S, Hamel-Teillac D, de Prost Y, Bodemer C. Late recurrence of the inflammatory first stage lesions in incontinentia pigmenti: an unusual phenomenon and a fascinating pathological mechanism. *Arch Dermatol.* 2003;139:201–204.
11. Patrizi A, Neri I, Guareschi E, Cocchi G. Bullous recurrent eruption of incontinentia pigmenti. *Pediatr Dermatol.* 2004;21(5):613–614.
12. Alikhan A, Lee AD, Swing D, Carroll C, Yosipovitch G. Vaccination as a probable cause of incontinentia pigmenti reactivation. *Pediatr Dermatol.* 2010;27(1):62–64.
13. Nagase T, Takanashi M, Takada H, Ohmore K. Extensive vesiculobullous eruption following limited ruby laser treatment for incontinentia pigmenti: a case report. *Australas J Dermatol.* 1997;58:155–157.
14. Faloyin M, Levitt J, Bercowitz E, Carrasco D, Tan J. All that is vesicular is not herpes: incontinentia pigmenti masquerading as herpes simplex virus in a newborn. *Pediatrics.* 2004;114:e270–e272.
15. Okan F, Yapici Z, Bulbul A. Incontinentia pigmenti mimicking a herpes simplex virus infection in the newborn. *Childs Nerv Syst.* 2008;24:149–151.
16. Stitt WZ, Scott GA, Caserta M, Goldsmith LA. Coexistence of incontinentia pigmenti and neonatal herpes simplex virus infection. *Pediatr Dermatol.* 1998;15(2):112–115.
17. Yoshida M, Oiso N, Kimura M, Itoh T, Kawada A. Skin ulcer mimicking pyoderma gangrenosum in a patient with incontinentia pigmenti. *J Dermatol.* 2011;38(10):1019–1021.
18. Courtois G, Israël A. IKK regulation and human genetics. *Curr Top Microbiol Immunol.* 2011;349:73–95.
19. Kaufman CK, Fuchs E. It's got you covered. NF-kappaB in the epidermis. *J Cell Biol.* 2000;149(5):999–1004.
20. Meuwissen ME, Mancini GM. Neurological findings in incontinentia pigmenti; a review. *Eur J Med Genet.* 2012;55(5):323–331.
21. Pizzamiglio MR, Piccardi L, Bianchini F, et al. Incontinentia pigmenti: learning disabilities are a fundamental hallmark of the disease. *PLoS One.* 2014;9(1):e87771.
22. Hadj-Rabia S, Froidevaux D, Bodak N, et al. Clinical study of 40 cases of incontinentia pigmenti. *Arch Dermatol.* 2003;139(9):1163–1170.
23. Minić S, Trpinac D, Obradović M. Blaschko line analogies in the central nervous system: a hypothesis. *Med Hypotheses.* 2013;81(4):671–674.
24. Carney RG. Incontinentia pigmenti. A world statistical analysis. *Arch Dermatol.* 1976;112(4):535–542.
25. Phan TA, Wargon O, Turner AM. Incontinentia pigmenti case series: clinical spectrum of incontinentia pigmenti in 53 female patients and their relatives. *Clin Exp Dermatol.* 2005;30(5):474–480.
26. Makris C, Godfrey VL, Krähn-Senftleben G, et al. Female mice heterozygous for IKK gamma/NEMO deficiencies develop a dermatopathy similar to the human X-linked disorder incontinentia pigmenti. *Mol Cell.* 2000;5(6):969–979.
27. Mattson MP, Camandola S. NF-kappaB in neuronal plasticity and neurodegenerative disorders. *J Clin Invest.* 2001;107(3):247–254.
28. Minić S, Trpinac D, Obradović M. Blaschko line analogies in the central nervous system: a hypothesis. *Med Hypotheses.* 2013;81(4):671–674.
29. Bryant SA, Rutledge SL. Abnormal white matter in a neurologically intact child with incontinentia pigmenti. *Pediatr Neurol.* 2007;36(3):199–201.
30. Rott HD. Extracutaneous analogies of Blaschko lines. *Am J Med Genet.* 1999;85(4):338–341.
31. Selvadurai D, Salomão DR, Baratz KH. Corneal abnormalities in incontinentia pigmenti: histopathological and confocal correlations. *Cornea.* 2008;27(7):833–836.
32. Batioglu F, Ozmert E. Early indirect laser photocoagulation to induce regression of retinal vascular abnormalities in incontinentia pigmenti. *Acta Ophthalmol.* 2010;88(2):267–268.
33. Nguyen JK, Brady-Mccreery KM. Laser photocoagulation in preproliferative retinopathy of incontinentia pigmenti. *J AAPOS.* 2001;5(4):258–259.
34. Scardamaglia L, Howard A, Sinclair R. Twenty-nail dystrophy in a girl with incontinentia pigmenti. *Australas J Dermatol.* 2003;44(1):71–73.
35. Bittar M, Danarti R, König A, Gal A, Happle R. Late-onset familial onychodystrophy heralding incontinentia pigmenti. *Acta Derm Venereol.* 2005;85(3):274–275.
36. Kmetz EC, Shashidhar Pai G, Burges GE. Incontinentia pigmenti with a foreshortened hand: evidence for the significance of NFkappaB in human morphogenesis. *Pediatr Dermatol.* 2009;26(1):83–86.
37. Handler MZ, Alshaiji J, Amini S, Izakovic J. Dorsal dimelia of the fourth toe in a patient with incontinentia pigmenti. *JAMA Dermatol.* 2013;149(6):761–762.
38. Simmons DA, Kegel MF, Scher RK, Hines YC. Subungual tumors in incontinentia pigmenti. *Arch Dermatol.* 1986;122(12):1431–1434.
39. Young A, Manolson P, Cohen B, Klapper M, Barrett T. Painful subungal dyskeratotic tumors in incontinentia pigmenti. *J Am Acad Dermatol.* 2005;52(4):726–729.
40. Hartman DL. Incontinentia pigmenti associated with subungual tumors. *Arch Dermatol.* 1966;94(5):632–635.
41. Lamb RC, Milne AW, Tavadia S. A subungal lesion on the finger of a young woman. *Int J Dermatol.* 2012;51(10):1177–1179.

42. Montes CM, Maize JC, Guerry-Force ML. 118. Incontinentia pigmenti with painful subungual tumors: a two-generation study. *J Am Acad Dermatol*. 2004;50(2 suppl):S45–S52.
43. Minić S, Novotny GE, Trpinac D, Obradović M. Clinical features of incontinentia pigmenti with emphasis on oral and dental abnormalities. *Clin Oral Investig*. 2006;10:343–347.
44. Jean-Baptiste S, O'Toole EA, Chen M, Guitart J, Paller A, Chan LS. Expression of eotaxin, an eosinophil-selective chemokine, parallels eosinophil accumulation in the vesiculobullous stage of incontinentia pigmenti. *Clin Exp Immunol*. 2002;127(3):470–478.
45. Minić S, Trpinac D, Obradović M, Novotny GE, Gabriel HD, Kuhn M. Incontinentia pigmenti with ultrastructurally disordered leucocytes. *J Clin Pathol*. 2010;63(7):657–659.
46. Smah A, Courtois G, Vabres P, et al. Genomic rearrangement in NEMO impairs NF-kappaB activation and is a cause of incontinentia pigmenti. The International Incontinentia Pigmenti (IP) Consortium. *Nature*. 2000;405:466e–472e.
47. Fusco F, Bardaro T, Fimiani G, et al. Molecular analysis of the genetic defect in a large cohort of IP patients and identification of novel NEMO mutations interfering with NF-kappaB activation. *Hum Mol Genet*. 2004;13:1763–1773.
48. Fusco F, Paciolla M, Napolitano F, et al. Genomic architecture at the Incontinentia Pigmenti Locus favours de novo pathological alleles through different mechanisms. *Hum Mol Genet*. 2012;21:1260–1271.
49. Fusco F, Pescatore A, Steffann J, Royer G, Bonnefont JP, Ursini MV. Clinical Utility Gene Card for: incontinentia pigmenti. *Eur J Hum Genet*. 2013;21(7).
50. Courtois G, Israël A. IKK regulation and human genetics. *Curr Top Microbiol Immunol*. 2011;349:73–95.
51. Buinauskaite E, Buinauskiene J, Kucinskiene V, Strazdiene D, Valiukeviciene S. Incontinentia pigmenti in a male infant with Klinefelter syndrome: a case report and review of the literature. *Pediatr Dermatol*. 2010;27(5):492–495.
52. Franco LM, Goldstein J, Prose NS, et al. Incontinentia pigmenti in a boy with XXY mosaicism detected by fluorescence in situ hybridization. *J Am Acad Dermatol*. 2006;55(1):136–138.
53. Fusco F, Fimiani G, Tadini G, Michele D, Ursini MV. Clinical diagnosis of incontinentia pigmenti in a cohort of male patients. *J Am Acad Dermatol*. 2007;56(2):264–267.
54. Pacheco TR, Levy M, Collyer JC, et al. Incontinentia pigmenti in male patients. *J Am Acad Dermatol*. 2006;55(2):251–255.
55. Kaya TI, Tursen U, Ikizoglu G. Therapeutic use of topical corticosteroids in the vesiculobullous lesions of incontinentia pigmenti. *Clin Exp Dermatol*. 2009;34(8):e611–e613.
56. Wolf DS, Golden WC, Hoover-Fong J, et al. High-dose glucocorticoid therapy in the management of seizures in neonatal incontinentia pigmenti: a case report. *J Child Neurol*. first published on March 28, 2014 as doi:10.1177/0883073813517509.

S E C T I O N V I I I

EPILEPSY

CHAPTER

84

The Genetic Epilepsies

Robert L. Macdonald and Martin J. Gallagher

Vanderbilt University, Nashville, TN, USA

INTRODUCTION

Definitions of Seizures and Epilepsy

The International League Against Epilepsy (ILAE) and the International Bureau for Epilepsy (IBE) have proposed definitions of the terms *epileptic seizure* and *epilepsy*.[1] "An *epileptic seizure* is a transient occurrence of signs and/or symptoms due to abnormal excessive or synchronous neuronal activity in the brain. *Epilepsy* is a disorder of the brain characterized by an enduring predisposition to generate epileptic seizures, and by the neurobiological, cognitive, psychological, and social consequences of this condition. The definition of epilepsy requires the occurrence of at least one epileptic seizure." Not everyone with seizures has epilepsy. Provocative factors that can induce a seizure, but not necessarily cause epilepsy, include acute metabolic derangements, head trauma, alcohol withdrawal, and high fever (in children).

In this chapter we will focus on genetic causes of epilepsy that do not directly alter brain structure or metabolism and thereby cause a secondary or symptomatic epilepsy. We will focus on genetic mutations that cause epilepsy as the primary phenotype.

Epidemiology

A conservative estimate of the overall incidence of epilepsy is between 31 and 57 cases per 100,000, which indicates that there are 70,000 to 129,000 new cases per year in the United States. This compares to 3 cases per 100,000 for Duchenne muscular dystrophy and 0.4 per 100,000 for myasthenia gravis. The prevalence of active epilepsy, when corrected for under-reporting, is 6.42 cases per 1000 using the 1975 Rochester data, which calculates to approximately 1.6 million people in the United States, a relatively high proportion of whom are children.

Classification

To investigate the pathophysiology, genetics, treatment, and prognosis of any class of disorders, a rational classification scheme is required. This effort has been particularly difficult regarding the epilepsies due to the variety of terms, past and present, used to characterize the phenotype. These include terms that relate to the presumed brain focus (e.g., frontal, temporal), perceived impact (e.g., grand mal, petit mal), associated electroencephalogram (EEG) pattern (e.g., spike-wave) involved, neurophysiologic system (e.g., motor, sensory, autonomic), seizure manifestation (motor, sensory, psychic, psychomotor), knowledge of etiology (idiopathic, symptomatic), and presumed pattern of onset (partial/focal/local/generalized). Modifiers such as simple, complex, minor, major, typical, and atypical were used in an idiosyncratic fashion and often had fundamentally different meanings when used by different authors.

Two types of classification systems have been in widespread use. The first is a classification based on seizure type, whereas the second is founded upon the concept of the epilepsy syndrome (i.e., a constellation of clinical, historical, and electrophysiological features that describes a patient population). Recently, the ILAE Commission on Classification and Terminology has revised concepts, terminology, and approaches for classifying seizures and forms

Rosenberg's Molecular and Genetic Basis of Neurological and Psychiatric Disease
http://dx.doi.org/10.1016/10.1016/B978-0-12-410529-4.00084-X

of epilepsy.[2] The fundamental difference among seizures as classified by this system is the presumed mode of onset (i.e., generalized vs. focal; Table 84.1).

> "Generalized epileptic seizures are conceptualized as originating at some point within, and rapidly engaging, bilaterally distributed networks. Such bilateral networks can include cortical and subcortical structures, but do not necessarily include the entire cortex. Although individual seizure onsets can appear localized, the location and lateralization are not consistent from one seizure to another. Generalized seizures can be asymmetric. They may be discretely localized or more widely distributed. There are multiple forms of generalized seizures that include tonic-clonic, absence, myoclonic, clonic, tonic and atonic seizures as well as several different forms of absence and myoclonic seizures."

Focal epileptic seizures are conceptualized as originating within networks limited to one hemisphere. They may be discretely localized or more widely distributed. They may originate in subcortical structures.

> "For each seizure type, ictal onset is consistent from one seizure to another, with preferential propagation patterns that can involve the contralateral hemisphere. In some cases, however, there is more than one network, and more than one seizure type, but each individual seizure type has a consistent site of onset. Focal seizures do not fall into any recognized set of natural classes based on any current understanding of the mechanisms involved. Descriptors of focal seizures may be used, individually or in combination with other features depending on the purpose."

Examples are presented in Table 84.2. Instead of the terms idiopathic, symptomatic, and cryptogenic, the following three terms and their associated concepts are recommended:

The underlying cause of the epilepsy has been divided into three categories, genetic, structural/metabolic and unknown.[2]

Genetic Cause

> "The concept of genetic epilepsy is that the epilepsy is, as best as understood, the direct result of a known or presumed genetic defect(s) in which seizures are the core symptom of the disorder. The knowledge regarding the genetic contributions may derive from specific molecular genetic studies that have been well replicated and even become the basis of diagnostic tests (e.g., *SCN1A* and Dravet syndrome) or the evidence for a central role of a genetic component may come from appropriately designed family studies. Designation of the fundamental nature of the disorder as genetic does not exclude the possibility that environmental factors (outside the individual) may contribute to the expression of disease. At the present time, there is virtually no knowledge to support specific environmental influences as causes of or contributors to these forms of epilepsy."

TABLE 84.1 Classification of Seizures*[2]

GENERALIZED SEIZURES
Tonic–clonic (in any combination)
Absence
Typical
Atypical
Absence with special features
Myoclonic absence
Eyelid myoclonia
Myoclonic
Myoclonic
Myoclonic atonic
Myoclonic tonic
Clonic
Tonic
Atonic
FOCAL SEIZURES
UNKNOWN
Epileptic spasms

Seizures that cannot be clearly diagnosed into one of the preceding categories should be considered unclassified until further information allows their accurate diagnosis. This is not considered a classification category, however.[2]

TABLE 84.2 Descriptors of Focal Seizures According to Degree of Impairment During Seizure[2]

Without impairment of consciousness or awareness
With observable motor or autonomic components. This roughly corresponds to the concept of "simple partial seizure." ("Focal motor" and "autonomic" are terms that may adequately convey this concept depending on the seizure manifestations)
Involving subjective sensory or psychic phenomena only. This corresponds to the concept of an aura
With impairment of consciousness or awareness. This roughly corresponds to the concept of complex partial seizure. "Dyscognitive" is a term that has been proposed for this concept
Evolving to a bilateral, convulsive seizure (involving tonic, clonic, or tonic and clonic components). This expression replaces the term "secondarily generalized seizure"

Structural/Metabolic Cause

> "Conceptually, there is a distinct other structural or metabolic condition or disease that has been demonstrated to be associated with a substantially increased risk of developing epilepsy in appropriately designed studies. Structural lesions of course include acquired disorders such as stroke, trauma, and infection. They may also be of genetic origin (e.g., tuberous sclerosis, many malformations of cortical development); however, as we currently understand it, there is a separate disorder interposed between the genetic defect and the epilepsy."

Unknown Cause

> "Unknown is meant to be viewed neutrally and to designate that the nature of the underlying cause is as yet unknown; it may have a fundamental genetic defect at its core or it may be the consequence of a separate as yet unrecognized disorder."

Electroclinical Syndromes

Not only is the EEG the major laboratory means of recording abnormal brain electrical activity, but the nature of the abnormalities has been used in several classification schemes. Furthermore, EEG traits may be present in asymptomatic relatives of patients with epilepsy and have been used as a marker of disease for pedigree analysis.

The EEG is generated by the summed excitatory and inhibitory postsynaptic potentials in the superficial layers of cortex underlying the electrode. Although the precise mechanisms are unknown, cortical neurons are stimulated and synchronized by diencephalic structures such as the thalamus. The recorded EEG has a characteristic amplitude and frequency distribution across various regions of cortex. These characteristics change dramatically as a function of state of arousal and age. Thus, identification of abnormal activity must take these factors into account. The distribution of the abnormalities is defined as generalized if both cerebral hemispheres are rapidly engaged or focal if restricted to one hemisphere; abnormalities in this latter case are typically present only in a small number of EEG electrodes. The morphology of the abnormality is described by its amplitude, sharpness, and frequency. Although significant interindividual differences exist, it is clear that the expression of both normal and abnormal patterns is at least partially under genetic control.

Some of the epilepsies have a complex of clinical features, signs, and symptoms that together define a distinctive, recognizable clinical disorder referred to as an "electroclinical syndrome".[2] These electroclinical syndromes are distinctive disorders identifiable on the basis of a typical age of onset, specific EEG characteristics, seizure types, and often other features which, when taken together, permit a specific diagnosis. The diagnosis, in turn, often has implications for treatment, management, and prognosis. The currently recognized electroclinical syndromes organized by typical age at onset are presented in Table 84.3.

DEFINING EPILEPSY GENES AND GENETIC EPILEPSY SYNDROMES

Human Epilepsy Genes

At first glance, it should be simple to produce a universally accepted definition of a "human epilepsy (hEP) gene." However, the association of epilepsy with genetic variation is complex. Genome-wide association studies enrolling thousands of patients demonstrated statistically significant effects of common genetic variations on the development of an epilepsy phenotype. Although these common variations produced statistically significant associations useful in population studies, the effect size of any individual variation was so small that the identification of such a variation in any particular individual would not provide useful diagnostic information.

TABLE 84.3 Electroclinical Syndromes Arranged by Age at Onset[2]

NEONATAL PERIOD
Benign familial neonatal epilepsy (BFNE)
Early myoclonic encephalopathy (EME)
Ohtahara syndrome (early-infantile epileptic encephalopathy, EIEE)
INFANCY
Epilepsy of infancy with migrating focal seizures (malignant migrating partial seizures of infancy, MMPSI)
West syndrome
Myoclonic epilepsy in infancy (MEI)
Benign infantile epilepsy
Benign familial infantile epilepsy
Dravet syndrome
Myoclonic encephalopathy in nonprogressive disorders
CHILDHOOD
Febrile seizures plus (FS+) (can start in infancy)
Panayiotopoulos syndrome
Epilepsy with myoclonic atonic (previously astatic) seizures
Benign epilepsy with centrotemporal spikes (BECTS)
Autosomal dominant nocturnal frontal lobe epilepsy (ADNFLE)
Late-onset childhood occipital epilepsy (Gastaut type)
Epilepsy with myoclonic absences
Lennox–Gastaut syndrome
Epileptic encephalopathy with continuous spike-and-wave during sleep (CSWS)
Landau–Kleffner syndrome (LKS)
Childhood absence epilepsy (CAE)
ADOLESCENCE–ADULT
Juvenile absence epilepsy (JAE)
Juvenile myoclonic epilepsy (JME)
Epilepsy with generalized tonic–clonic seizures alone
Progressive myoclonus epilepsies (PME)
Autosomal dominant epilepsy with auditory features (ADEAF)
Other familial temporal lobe epilepsies
LESS SPECIFIC AGE RELATIONSHIP
Familial focal epilepsy with variable foci (childhood to adult)
Reflex epilepsies

In contrast to the view that epilepsy results from the combination of multiple common variations with low effect size, epilepsy can also result from rare mutations/variations that have large effects. For the purposes of this chapter, we will define hEP genes as those in which single mutations, duplications, or deletions have been strongly associated with the development of epilepsy. It is important to emphasize that although it is easier to appreciate the association of epilepsy with hEP gene mutations than with the combinations multiple common variations, the latter syndrome is most likely a much more common means by which epilepsy is transmitted. We readily admit that classifying the strength of an association between a hEP gene and the phenotype is subjective. As is true with many disease-associated genes, hEP gene mutations are rarely 100% penetrant. In addition, family members with the same epilepsy gene mutation sometimes exhibit phenotypes with very different severities.

Because they are transmitted with high penetrance, hEP genes transmitted with Mendelian inheritance have been identified in multiple kindreds using classical linkage analysis and candidate gene sequencing. However, some hEP genes cause severe syndromes with intractable seizures, substantial cognitive impairment, and mortality, thus resulting in low reproductive fitness. Autosomal dominant mutations and copy number variants in these epilepsy genes that are associated with such severe physical impairments are usually acquired *de novo* and have been identified by high resolution chromosomal analysis, candidate gene sequencing and, recently, by whole-exome sequencing.

Genetic Epilepsy Syndromes

As discussed above, the 2010 ILAE proposed classification criteria defined epilepsies resulting from a genetic cause.[2] Therefore, guided by this classification scheme, we define genetic epilepsy syndromes as those in which there is either evidence of association with a hEP gene or strong examples of familial inheritance of the phenotype. Epilepsy syndromes are not considered genetic if they cause gross structural or metabolic abnormalities. It should be noted that there is no widely accepted list of genetic epilepsy syndromes. Some syndromes can be produced by either genetic or acquired factors. For example, approximately 75% of cases of Lennox–Gastaut syndrome (LGS) result from acquired factors such as cortical malformations or perinatal hypoxia. However, the remaining 25% have no known acquired source and their syndrome is thought to result from a genetic etiology. In this chapter, we will include as genetic epilepsy syndromes those, such as LGS, which can result from genetic factors even if they can, in other circumstances, also result from acquired factors. We should also note that although a definition of a genetic epilepsy syndrome would exclude diseases with gross structural abnormalities, it is possible that genetic epilepsy syndromes could be associated with abnormalities on high resolution imaging studies or pathological examinations. For example, although patients with the genetic epilepsy syndrome, juvenile myoclonic epilepsy (JME), have normal findings on clinical magnetic resonance imaging (MRI) studies, high-resolution research MRI studies have revealed subtle changes in cortical thickness and thalamic volume.

HUMAN GENETIC EPILEPSY SYNDROMES

Here, we will first give a clinical description of many of the known human genetic epilepsy syndromes and then provide the evidence demonstrating that the syndrome is a genetic epilepsy syndrome along with any hEP genes known to associate with the syndrome (Table 84.4). Because some hEP genes are associated with more than one genetic epilepsy syndrome, we will not provide a description of the pathophysiology of the epilepsy genes in this section, but, instead, will describe the pathophysiology of the epilepsy mutations in the next section on hEP genes.

Benign Familial Neonatal Epilepsy

Clinical

Starting in the 1960s, investigators reported different families in which several members experienced multiple daily motor seizures that began in the first week of life, remitted within 1 month of age, and were associated with normal cognitive development.[3] Because of these characteristic clinical features, this syndrome was named benign familial neonatal epilepsy (BFNE). One prospective epidemiological study estimated the incidence of BFNE at 14.4 per 100,000 births.[4] On interictal EEGs, BFNE patients typically demonstrate a relatively normal background without paroxysmal, inactive, or burst-suppression patterns and, on ictal EEG, a generalized attenuation that develops into generalized or focal spike-wave activity.[5] Despite its classical clinical characteristics, some BFNE patients exhibit variable phenotypes. For example, in some patients, the onset of seizures did not begin until 3 months of age or remit until 2 years of age. In addition, approximately 10% of BFNE patients exhibited seizures later in life.

Genetics

The first investigators that reported BFNE noted that it occurred in an autosomal pattern with high penetrance and thus BFNE has been long established as genetic epilepsy syndrome. In 1989, Leppert et al. linked BFNE in a kindred to chromosomal region 20q13.2. Subsequently, Singh et al. discovered a 20q13.2 microdeletion in one BFNE kindred and found that this region contained the *KCNQ2* gene encoding the voltage-gated potassium channel, Kv7.2.[6,7] This group also found other autosomal dominant *KCNQ2* mutations in additional BFNE kindreds. To date, there are 73 BFNE-associated *KCNQ2* mutations listed in the public Human Gene Mutation Database (HGMD, Table 84.4).

BFNE has been linked to other loci besides 20q13.2. In contrast to its typical autosomal dominant inheritance, BFNE was inherited in an autosomal recessive pattern in a consanguineous Iranian Jewish family with linkage analysis excluding 20q.[8] Moreover, an autosomal dominant syndrome of BFNE and hypotelorism was associated with a pericentric inversion of chromosome 5.[9] Finally, autosomal dominant BFNE in a large Mexican-American family was linked to region 8q.[10] This family was subsequently found to possess a mutation in a different voltage-gated potassium channel gene, *KCNQ3*.[11] There are now four *KCNQ3* mutations associated with BFNE.

TABLE 84.4 Epilepsy Genes Associated with Genetic Epilepsy Syndromes

Genetic epilepsy syndrome	Human epilepsy (hEP) gene(s)	Inheritance	Number of mutations associated with syndrome
BFNE	*KCNQ2*	AD	73
	KCNQ3	AD	4
Early epileptic encephalopathies (EIEE, MPSI, WS, LGS)	*ARX*	XR	5†
MMPSI	*CDKL5*	XD	58‡
	KCNQ2	AD	19
	KCNT1	AD	5
	PLCB1	AR	2
	SCN1A	AD	11
	SCN2A	AD	4
	SCN8A	AD	2
	SPTAN1	AD	3
	STXBP1	AD	5
Febrile Seizures*	*GABRG2*	AD	1
	SCN1A	AD	4
	SCN9A	AD	4
GEFS+	*GABRG2*	AD	5
	SCN1A	AD	41
	SCN1B	AD	1
Dravet syndrome	*GABRG2*	AD	1
	SCN1A	AD	209
CAE	*GABRA1*	AD	1
	GABRB3	AD	3
	GABRG2	AD	3
JME	*GABRA1*	AD	1
	EFHC1	AD	10
	CASR	AD	5
ADNFLE	*CHRNA2*	AD	1
	CHRNA4	AD	4
	CHRNB2	AD	4
ADEAF	*LGI1*	AD	27

Data were obtained using the public release of the Human Gene Mutation Database in conjunction with the references provided in the text of this chapter.
**FS is not a genetic epilepsy syndrome, but is included here given the strong genetic component to this seizure syndrome.*
†Includes Ohtahara syndrome and infantile spasm syndrome, X-linked.
‡Includes encephalopathy with early epilepsy, mental retardation and epilepsy, Rett syndrome, variant with infantile seizures, and encephalopathy with refractory epilepsy.
Abbreviations: AD, autosomal dominant; ADEAF, autosomal dominant epilepsy with auditory features; ADNFLE, autosomal dominant nocturnal frontal lobe epilepsy; AR, autosomal recessive; BFNE, benign familial neonatal epilepsy; CAE, childhood absence epilepsy; EIEE, early-infantile epileptic encephalopathy; GEFS+, genetic epilepsy with febrile seizures plus; JME, juvenile myoclonic epilepsy; LGS, Lennox–Gastaut syndrome; MMPSI malignant migrating partial seizures of infancy; WS, West syndrome; XD, X-linked dominant; XR, X-linked recessive.

Infantile and Childhood Epileptic Encephalopathies

Clinical

Infantile and childhood epileptic encephalopathies include early infantile epileptic encephalopathy (EIEE, also called Ohtahara syndrome), West syndrome (WS, an electroclinical syndrome in which the infants exhibit infantile spasms) and Lennox–Gastaut syndrome (LGS). In addition, malignant migrating partial seizures of infancy (MMPSI) is often considered an epileptic encephalopathy of infancy. It may seem strange to discuss these syndromes in a section on genetic epilepsy syndromes. The 1989 ILAE epilepsy classification scheme assigned these syndromes to cryptogenic or symptomatic generalized epilepsy categories.[12] The proposed 2010 ILAE classification scheme does not assign any of these syndromes to genetic, structural–metabolic, or unknown.[2] This is appropriate given that these syndromes often result from direct structural or metabolic insults, but also sometimes result from genetic mutations that cause structural or metabolic abnormalities or from mutations that are not associated with structural or metabolic abnormalities. It is this latter group that fits our definition of a genetic epilepsy syndrome and will be discussed here.

These epileptic encephalopathy syndromes share many common features: they are associated with intractable seizures and cognitive dysfunction, exhibit grossly abnormal interictal EEG tracings, and, in the majority of cases, result from anoxic, structural, or metabolic causes and not genetic etiologies.[13] The relationship among EIEE, WS, and LGS is also highlighted by the observation that 75% of EIEE cases evolve into WS (accounting for 2.6% WS) and 59% of WS cases evolve to LGS (accounting for 36% of LGS).[14] These syndromes are most readily distinguished by their age of onset, seizure types, and electrographic patterns. EIEE, WS, and LGS typically begin within the first 3 months, between 4–8 months, and between 2–5 years, respectively. The major seizure type in EIEE is tonic spasms (lasting up to 10 seconds), although patients can also exhibit partial seizures. Electrophysiological recordings in EIEE demonstrate an interictal suppression-burst pattern (2–5 second "suppression"—extremely low voltage activity—alternating with 1–3 seconds of high voltage activity) and ictal desynchronization, desynchronization with low-voltage fast activity, or hypersynchronization.[14] WS is characterized by the triad of epileptic spasms (infantile spasms), developmental delay, and the EEG hypsarrhythmia pattern (generalized, disorganized high voltage activity). LGS patients have multiple seizure types including tonic, atonic, and atypical absence seizures; focal and myoclonic seizures can also occur. Their interictal EEG typically demonstrates diffuse background slowing, multifocal spikes and slow spike-wave complexes (<2.5 Hz), and their ictal EEG is characterized by slow spike-wave complexes during atypical absence seizures, low-voltage fast activity during tonic seizures, and voltage attenuation or diffuse spike-wave complexes during atonic seizures.[13]

Genetics

Although most cases of EIEE, WS, and LGS are associated with structural or metabolic abnormalities, a small, but significant, fraction is thought to result from a direct genetic etiology. Because these diseases typically result in substantial morbidity, mortality, and cognitive delay, most of the patients do not reproduce and thus the mutations either arise *de novo* or are inherited from a parent who is only partially affected. This lack of classical Mendelian inheritance has complicated traditional genetic studies. However, recent advances in genetic association studies and whole-exome sequencing has identified several epilepsy genes involved in these infantile and childhood epileptic encephalopathy syndromes. As discussed above, these three epileptic encephalopathy syndromes are clinically interrelated and thus it is not surprising that some of the same genes confer multiple epileptic encephalopathy syndromes.

KCNQ2

Although *KCNQ2* mutations are typically associated with BFNE, some are associated with EIEE. Sequencing of the *KCNQ2* gene in a patient with EIEE as well as his mother who had BFNE revealed a mutation located at a highly conserved amino acid in the fifth transmembrane region and thus raised the possibility that *KCNQ2* mutations could increase the risk of epilepsy syndromes that were more severe than BFNE.[15] This conclusion was supported by the identification of a *KCNQ2*(761_770del10insA) mutation in a large family containing eight members with BFNE and one member with EIEE.[16] Weckhuysen et al. performed mutational analysis of the *KCNQ2* and *KCNQ3* genes in 80 unrelated EIEE patients who lacked causative structural abnormalities on MRI.[17] They found that eight patients possessed heterozygous *KCNQ2* mutations. Six of these mutations were confirmed to occur *de novo*, one mutation was expressed mosaically in the patient's unaffected father, and one was unclassified because the patient's father's DNA could not be obtained. Recently, Weckhuysen et al. screened 84 neonates (<1 month) with unexplained epileptic encephalopathy for *KCNQ2* mutations and found nine different heterozygous *de novo* mutations in 11 patients.[18] Finally, the Epi4K study group performed whole-exome sequencing of 264 patients with unexplained WS or LGS and found one *de novo* nonconservative missense *KCNQ2* mutation and one *de novo* nonconservative missense *KCNQ3* mutation.[19]

ARX

While candidate gene sequencing found the EIEE-associated *KCNQ2* mutations, several other EIEE genes were first identified characterizing sex-linked and somatic chromosomal abnormalities. In 1977, Feinberg and Leahy reported a three-generation family in which five males experienced infantile spasms and developmental delay in a clear pattern of X-linked recessive inheritance.[20] Two of the affected patients underwent an EEG that showed hypsarrhythmia. Although one patient had a porencephalic cyst, the proband had no structural lesion on brain imaging and no metabolic disorder, findings that suggested this represented a genetic (X-linked recessive) form of WS. A second family with a nonstructural/metabolic X-linked recessive WS was identified and the disease locus was linked to Xp21.3-Xp22.1.[21] Within this region, the Aristaless-related homeobox X-linked gene (*ARX*) was considered a prime candidate gene based on its expression profile in fetal, infant, and adult brain. Screening for *ARX* mutations in five families with X-linked WS and in other families with X-linked mental retardation syndromes revealed multiple *ARX* mutations including polyalanine repeats and a missense and nonsense mutation.[22] To date, there are several *ARX* mutations associated with X-linked WS and EIEE in the human genetic mutations database. In addition, there are many other *ARX* mutations associated with other mental retardation syndromes with and without epilepsy and structural abnormalities. It is interesting to note that the Epi4K study did not find disruptive *de novo ARX* mutations associated with epileptic encephalopathy syndromes.[19]

CDKL5

Mutations in the cyclin-dependent kinase-like 5 (*CDKL5*) gene cause X-linked dominant forms of EIEE and WS. High-resolution chromosome analysis studies in two unrelated girls with EIEE revealed balanced translocations between the distal short arm of chromosome X and a somatic chromosome; neither translocation was found in the parents.[23] Analyses of the translocation breakpoints revealed disruption of the serine/threonine kinase 9 (*STK9*) gene, also known as *CDKL5*. Subsequently, many *CDKL5* mutations have been associated with EIEE, and WS (Table 84.4), and many other *CDKL5* mutations have been associated with other X-linked dominant neurodevelopmental disorders that are not classified as either EIEE or WS (e.g., variants of Rett syndrome). The Epi4K study found three *CDKL5* mutations associated with epileptic encephalopathy.[19]

STXBP1 AND *SPTAN1*

Saitsu et al. evaluated a patient with EIEE (and her parents) for variations in gene copy number and found that she possessed a *de novo* 2.0-Mb microdeletion at 9q33.3-q34.11.[24] This chromosomal region contained the syntaxin-binding protein 1 gene (*STXBP1*). This same group screened for *STXBP1* mutations in 13 EIEE patients and found four *de novo* heterozygous *STXBP1* missense mutations. Two other patients with 9q33.3-q34.11 microdeletions lacked any mutations in *STXBP1* and were found to have mutations (3bp deletion and 6bp duplication) in another gene, spectrin, alpha, non-erythrocytic 1 (*SPTAN1*).[25] When expressed in cultured neurons, recombinant alpha-spectrin containing these EIEE mutations caused the formation of alpha-spectrin aggregates as well as the disruption of clustering of voltage-gated sodium channels at the axon initial segment.

PLCB1 AND *KCNT1*

A genome-wide scan of polymorphisms in a patient with EIEE born of consanguineous parents suggested the homozygous loss of a 0.5Mb region of chromosome 20. Subsequent copy number analysis of this region confirmed a homozygous deletion that contained the gene, phospholipase C-beta 1 (*PLCB1*). Subsequently, gene copy number evaluation in a patient with MMPSI (also born to consanguineous parents) identified a second homozygous *PLCB1* deletion.[26] Barcia et al. performed exome sequencing in three unrelated patients with MMPSI and found heterozygous *de novo* missense mutation of a highly conserved residue in the sodium-activated potassium channel (*KCNT1*).[27] *KCNT1* was sequenced in nine additional patients and their parents and two additional *KCNT1* missense mutations were identified. The Epi4K study found one *de novo KCNT1* mutation associated with epileptic encephalopathy.[19]

SCN1A, *SCN2A* AND *SCN8A*

There is emerging evidence that mutations in voltage-gated sodium channels are associated with some of these severe epilepsy syndromes of infancy and childhood. Carranza Rojo et al. screened 15 children with MMPSI for mutations and copy number variation (CNV) in five genes associated with EIEE and found one patient having a *de novo* missense (R862G) mutation in the sodium channel mutation, *SCN1A*.[28] In addition, Freilich et al. sequenced *SCN1A* from another child with MMPSI and found another missense mutation, A1669E.[29] Although, this patient's father did not possess this mutation, it could not be classified as *de novo* because the child was conceived *in vitro* using a donated ovum. Disruptive *de novo* mutations in the *SCN2A* gene are also associated with early-life epileptic encephalopathies.[30-32] In their whole-exome search for *de novo* mutation-associated epileptic encephalopathies, the Epi4K study found seven *SCN1A* mutations, two *SCN2A* mutations, and two *SCN8A* mutations.[19]

Febrile Seizures

Clinical

Febrile seizures (FSs) are seizures occurring in childhood between 3 months and 6 years of age associated with a febrile illness not caused by an infection of the central nervous system, without previous neonatal seizures or a previous unprovoked seizure, and not meeting the criteria for other acute symptomatic seizures. FSs are the most common seizure syndrome, affecting approximately 3% of the Caucasian population[33] with a clear genetic component that has been established through family and twin studies. However, FSs are not classified as a form of epilepsy and instead are classified as "epileptic seizures that are traditionally not diagnosed as a form of epilepsy *per se*".[2]

FSs have considerable variability in phenotype. FSs can present either as a simple FS (<10 minutes with no complex features), a complex FS (one or more of: focal features, additional seizures within 24 hours, or with the same febrile illness, duration >10 minutes and <30 minutes) or febrile status epilepticus.[34] In addition, there are several FS subsyndromes including true FSs, FSs plus (FS+) and FSs with later epilepsy. True FSs occur between 3 months and 6 years of age in association with fever. FS+ occurs outside these ages and/or in association with afebrile generalized tonic–clonic seizures.

Epidemiological studies have indicated that patients with simple FS who do not have additional risk factors (history of prior neurologic deficit or developmental problem, prior afebrile seizures, family history of afebrile seizures) do not have a substantially higher incidence of development of epilepsy than patients without febrile seizures. However, this conclusion is partly based on a large prospective study (National Collaborative Perinatal Project; NCPP) that followed children to only 7 years of age. Follow-up through the first 20–30 years of life may be necessary to firmly establish this conclusion.

Children with complex FSs are at significantly higher risk for developing FSs with later epilepsy than those with simple FS. For children with focal or prolonged seizures or repeated episodes of FSs with the same illness, the risks of additional afebrile seizures ranges from 6–8% with one of these features, to 17–22% with two of these features, to 49% with all three. The age at which the unprovoked seizures develop varies, ranging from early childhood, after the first FS, to well into young adulthood, with several individuals experiencing their first afebrile seizure between the ages of 20 and 25 years.

Genetic

In the United States and Western Europe, approximately 5% of the population experiences FSs, whereas in Japan, the estimate is approximately 8%. Males and females are equally affected. It is clear that such seizures have a familial predisposition, although the exact mode of inheritance is not known. The NCPP data indicate a positive family history of seizures in approximately 7% of individuals with FS. Other studies indicate a higher incidence: 12% in parents of children with FS and as high as 25% in siblings. Twin studies have shown somewhat variable incidences of concordance rates for monozygotic twins, ranging from 31% (with a case-wise risk of 44%) to 70%. For dizygotic twins, the concordance rate also ranges around 14–18% in all of the above studies. A community-based twin sample was used to analyze genetic factors within different FS subtypes and FS syndromes. The results suggested a strong genetic contribution to different FS subtypes and subsyndromes and support existence of distinct genetic factors for different FS subtypes and subsyndromes, especially FS+.[34]

Mutations in sodium channel and $GABA_A$ receptor genes have been reported in several large families with FS syndromes including genetic epilepsy with febrile seizures plus (GEFS+).[26,35,36] These autosomal dominant mutations with large effect are relevant for only a few families with FS since most FS syndromes likely have a polygenic basis.[37] In families where the proband had a single FS, polygenic inheritance was most likely, while in families where the proband had multiple FSs, the presence of a highly penetrant single gene was more likely.[38] This and linkage studies suggesting involvement of multiple loci supports the presence of genetic heterogeneity in FS.[39–42]

SCN1A

A mutation M145T of *SCN1A*, which codes for the α subunit of $Na_V1.1$ sodium channels, has been shown to be associated with autosomal dominant simple FSs.[43]

GABRG2

Wallace et al. identified a large Australian family in which several members had FS and/or childhood absence epilepsy (CAE).[39] Linkage analysis and subsequent gene sequencing identified a mutation (R82Q) in the gene that encodes the γ2 subunit of the $GABA_A$ receptor (*GABRG2*). Subsequent evaluation of the R82Q pedigree suggested that the mutation accounts for the FS phenotype and that an interaction of this gene with another gene or genes was required for the CAE phenotype.[44] This was confirmed in a *Gabrg2(R82Q)* knock-in (KI) mouse.[45]

Audenaert et al. in 2006 described a multigenerational family with a history of autosomal dominant pure FSs that cosegregated with the *GABRG2(R177G)* mutation. The family was relatively small, and it cannot be certain that the mutation is not associated with FS+ due to insufficient family members.

Genetic Epilepsy with Febrile Seizures Plus

Clinical

Several families have been described in which some individuals had FS+ while others had FS+ and absences, FS+ and myoclonic seizures or myoclonic-atonic epilepsy or other seizure types. These complex familial epilepsy syndromes have been termed genetic epilepsy with febrile seizures plus (GEFS+).[36,46]

Genetic

As with FSs, sodium channel and $GABA_A$ receptor subunit gene mutations have been reported in FS syndromes including GEFS+.[35,36,47]

SCN1A AND *SCN1B*

Genetic study of patients with the autosomal dominant epilepsy syndrome GEFS+ has revealed more than 20 mutations of the α subunit of $Na_V1.1$ sodium channels.[48] In addition, a mutation in the $Na_V\beta1$ subunit (C121W) has been reported that also causes GEFS+.[49]

GABRG2

Multiple mutations have been identified in *GABRG2* including missense mutations (D219N and K289M) and nonsense and frameshift mutations (Q390X, W429X and S443delC).[50]

Dravet Syndrome

Clinical

Dravet syndrome (also known as severe myoclonic epilepsy of infancy) is a debilitating infantile-onset encephalopathy with epilepsy characterized by tonic, clonic, or tonic–clonic seizures that often occur initially in the context of febrile illnesses but then occur without fever. Later, other forms of seizures occur, including myoclonic, absence, and focal seizures with and without alterations in consciousness. Children developing this form of epilepsy usually experience psychomotor retardation in the second year of life.

Genetic

Mutations in the sodium channel subunit gene *SCN1A* account for 70–80% of Dravet syndrome cases,[51] and mutations in *GABRG2* and homozygous *SCN1B* mutations can cause an encephalopathy syndrome similar to Dravet syndrome.[52–54] However, for the remaining cases, the cause remains unclear. Additional possible causes include sporadic mutations in unknown genes or a polygenic inheritance.

SCN1A

Many patients with GEFS+ possess mutations in *SCN1A*. Because patients with both GEFS+ and Dravet syndrome have febrile seizures,[55] investigated whether patients with Dravet syndrome also possess *SCN1A* mutations. They screened seven unrelated patients with Dravet syndrome for *SCN1A* mutations and found that all seven patients possessed *de novo SCN1A* mutations including four frameshift, one nonsense, one splice-donor, and one missense mutation.

GABRG2

A few patients with Dravet syndrome have been shown to have mutations in *GABRG2* including Q390X[53] and Q40X.[56]

Childhood Absence Epilepsy

Clinical

Childhood absence epilepsy (CAE) is a genetic epilepsy syndrome that occurs predominantly in developmentally normal children between the ages of 4 and 10 years, with females representing approximately 70% of cases. It has been estimated that CAE is present in between 13–17% of individuals with epilepsy. The core symptoms include multiple daily absence seizures, which are characterized as brief events (5–10 seconds) of significantly impaired consciousness associated with a generalized 3-Hz spike-wave discharge. Phenotypic variations such as myoclonic jerks of the extremities, mouth, and eyelids and irregular, fragmentary, or polyspike EEG discharges may be present. The magnitude of these movements and their timing in relation to the appearance of the absence and the EEG findings have been used to include or exclude individuals from a diagnosis of typical absence epilepsy. Furthermore, it has been suggested that the clinical course distinguishes subsyndromes with different genetic etiologies. These syndromes include children who stop having seizures of all types (66%), those with continued absences and generalized tonic–clonic seizures as adults (22%), and those who develop juvenile myoclonic epilepsy after CAE (12%). The concept of absence seizures as part of individual syndromes as opposed to a biological phenomenon common to several disorders has been previously discussed in detail. CAE needs to be distinguished from juvenile absence, a related genetic generalized epilepsy (GGE) occurring in a later age group, and atypical absence, which is frequently associated with neurological lesions and psychomotor retardation.

Genetics

CAE appears to have a clear genetic predisposition. Concordance rates for monozygotic twins with absence epilepsy have been recorded at 75% if only absence seizures are considered and as high as 84% if 3-Hz spike-wave EEGs are also included in the analysis. This would indicate a high degree of penetrance for those who inherit this particular gene (or set of genes). Several *GABR* genes have been shown to be associated with CAE including *GABRG2*, *GABRB3* and *GABRA1*.

GABRG2

A large four-generation family with FS and CAE was studied by Wallace et al. in 2001 and they reported that a *GABRG2(R82Q)* mutation was associated with, but not solely responsible for, the CAE phenotype. Since this report, additional *GABRG2* mutations have been reported to be associated with CAE and FSs including the missense mutation P83S[57] and the splice donor site mutation IVS6+2T>G.[58]

GABRA1

Maljevic et al. sequenced the gene that encodes the $GABA_A$ receptor α1 subunit (*GABRA1*) in 98 patients with GGE syndromes. They found a *de novo* frameshift mutation (S326fs328X) in one patient with CAE.[59]

GABRB3

Feucht et al. in 1999 demonstrated an association between *GABRB3* and CAE and this was followed up by Urak et al. who used the same patient group and showed an association between 45 CAE patients and 13 single-nucleotide polymorphisms (SNPs) between the exon 1a promoter and the beginning of intron 3 in *GABRB3*.[60,61] Tanaka et al., 2008, then reported two heterozygous mutations in exon 1a (P11S and F15S) and one in exon 2 (G32R) that segregated with CAE patients.

Juvenile Myoclonic Epilepsy

Clinical

JME typically arises in adolescence. Patients experience myoclonic seizures, usually in the morning upon awakening. Most of the patients also experience generalized tonic–clonic seizures (GTCS), and approximately 30% of the patients have absence seizures as well.[62] The most common EEG abnormality is 4–6-Hz bilaterally symmetrical, polyspike wave complexes and 3–4-Hz spike-wave complexes during the absence seizures. JME can be divided into several subsyndromes.[63] The most common JME subsyndrome is "classic JME" in which the myoclonic, generalized tonic–clonic, and absence (rare, if present) seizures start in adolescence. The second most common subsyndrome is "CAE evolving to JME" in which patients experience absence seizures and 3–4-Hz spike and wave before 12 years of age and then develop myoclonic and tonic–clonic seizures and 4–6-Hz polyspike wave complexes in adolescence.

Genetics

FAMILIAL INHERITANCE

JME clusters within families. Martinez-Juarez et al. prospectively studied 257 JME probands (72% classic JME, 18% CAE evolving to JME, and 10% with other JME subsyndromes).[63] For the patients with classic JME, 50 of the 480 first-degree relatives (10%) had some form of epilepsy, and 26 of these relatives (5.4%) had JME. For the patients with CAE evolving to JME, 32 of the possible 161 first-degree relatives (20%) had epilepsy; the most frequent epilepsy types were absence only (4%), generalized tonic–clonic seizures only (5%), or JME (4%).

GABRA1

Cossette et al. identified the first monogenic mutation associated with JME.[64] They performed gene linkage studies and candidate gene sequencing for 14 members of a 30-member, four-generation, French Canadian family that included 10 JME patients. They found that each patient (n=8) possessed a heterozygous missense mutation (A322D) in the gene encoding $GABA_AR$ α1 subunit (*GABRA1*). The mutation was inherited in an autosomal dominant pattern with high penetrance; the mutation was not found in any of the unaffected members of the family (n=6) or in 400 control chromosomes of French Canadian origin.

EFHC1

Genetic linkage studies performed on different families from California, Mexico, Belize, and the Netherlands identified a region on chromosome 6p12-p11 associated with JME.[65–68] Sequencing of candidate genes in this region excluded all genes except for the gene, EF-hand domain (C-terminal)-containing protein 1 (*EFHC1*). Sequencing of the *EFHC1* gene in 31 Mexican, 12 European-American, and one Belizean family revealed four heterozygous mutations (F229L, D210N, D235, and double heterozygous P77T/R221H) in 21 affected members of eight families. Thirteen of these family members had epilepsy (10 JME, three with tonic–clonic only) and eight family members had epileptiform abnormalities on EEG, but did not have a history of seizures.[69,70] These mutations were not present in 382 unrelated healthy controls. However, the F229L, D210L, R182H, and P77T/R221H mutations were present in unaffected family members and thus the combined penetrance of these mutations was 78%. In addition, they found a R182H variant in 10 affected and seven unaffected members in another family; however, this variant was also found in control chromosomes and may be a benign variant. Following these results, Annesi et al. sequenced the *EFHC1* gene in 27 Italian families that contained at least two family members with JME. They found both the previously described F229L mutation in two families as well as a novel R353W mutation in one family.[71] This finding demonstrated that the *EFHC1* mutations were not restricted to North American populations. *EFHC1* gene sequencing in 44 consecutive Mexican and Honduran JME patients and 67 consecutive Japanese JME patients revealed 9% percent of the Mexican/Honduran cohort and 3% of the Japanese patients contained *EFHC* mutations. These mutations included a truncation mutation (Q277X) that was transmitted in an autosomal dominant pattern in one Honduran family and occurred sporadically in another as well as two missense variants (T252K and T508R) found in families in which multiple members carried the mutation, but only one member was affected, and a frameshift mutation and promoter nucleotide deletion that were in both the proband and mother.[72] Similarly, Jara-Prado sequenced *EFHC1* in 41 Mexican JME patients and found three new mutations, R118C, also found in the patient's affected father and brother, R182L, also found in the patient's unaffected father, and R153Q, found in the patient's asymptomatic father.[73]

CASR

Kapoor et al. identified a multigenerational family from India in which four patients had JME and one patient had FS and two patients had complex partial seizures.[74] Linkage analysis and candidate gene sequencing identified a mutation (R898Q) in the calcium-sensing receptor gene (*CASR*) in all affected and no unaffected family members. In addition, they sequenced the *CASR* gene in 96 apparently unrelated JME patients and found four additional missense mutations, E354A, I686V, A988V, and A988G. Although mutations of other *CASR* amino acids are associated with genetic hyper- and hypocalcemia syndromes, the affected patients in this study were normocalcemic, and there is no evidence that changes in calcium levels were associated with their seizures.

Autosomal Dominant Nocturnal Frontal Lobe Epilepsy

Clinical

Autosomal dominant nocturnal frontal lobe epilepsy (ADNFLE) was first described in 1994 in six families from Australia, Canada and the United Kingdom.[75] These patients were originally misdiagnosed as having hysteria, nightmares, night terrors, or paroxysmal nocturnal dystonia. The clinical manifestations of ADNFLE were further enumerated in 1995 in five of these families that collectively contained 47 affected individuals.[76] Usually, the seizures began in childhood (range 2 months to 52 years, median 8 years) and persisted throughout adulthood. Seizures occurred in clusters averaging eight/night and were present in NREM sleep, typically while dozing or shortly before awakening. Cognitive function and neurological examination were normal in all individuals. Reports of auras were frequent with a variety of somatosensory, special sensory, psychic, and autonomic phenomena. Motor activity included tonic, clonic, and hyperkinetic movements. The interictal EEG was normal in 84%; abnormalities, when present, consisted of bifrontal, central, unitemporal, and frontotemporal discharges. Ictal EEG recording revealed bihemispheric, frontal predominant spike-wave (n=3), generalized rhythmic 9 Hz (n=1), or no ictal features (n=6). Neuroimaging (computed tomography or magnetic resonance) was reported as normal in all 14 individuals examined.

Genetics

CHRNA4

One of the originally described Australian ADNFLE families contained 27 affected individuals spanning six generations.[76] Gene linkage analysis of DNA from this family allowed chromosomal assignment to 20q13.2.[77] Subsequently, Steinlein et al. found a mutation (S248F) in the *CHRNA4* gene that encodes the α4 subunit of the neuronal nicotinic acetylcholine receptor (nAChR) in all the affected individuals and four obligate carriers, but not in 333 healthy control subjects.[78] This *CHRNA4* S248F mutation was found independently in large Norwegian and Scottish ADNFLE families.[79,80] In addition, other *CHRNA4* mutations were found in other ADNFLE families including S248L, S252F, S252L, and 259insL.[81–86]

CHRNB2 AND *CHRNA2*

In 2000, Gambardella et al. described a large Italian family with ADNFLE that linked to chromosome 1 between 1p21 to 1p24, a region that contains the gene that encodes the nAChR β2 subunit (*CHRNB2*).[87] De Fusco et al. sequenced all the *CHRNB2* exons of this family and found a V287L missense mutation.[88] In addition, Phillips et al. described a second mutation (V287M) of the same CHRNB2 amino acid in a Scottish ADNFLE family.[89] Finally, Aridon et al. described an Italian family with an epilepsy syndrome that shares many features of ADNFLE, but also presents with ictal fear. Linkage studies and gene sequencing revealed that this syndrome was associated with an I279N mutation in the *CHRNA2* gene, the gene that encodes the nAChR α2 subunit.

KCNT1

Heron et al. studied a family with ADNFLE as well as intellectual impairment and psychiatric disease.[90] They performed linkage analysis and candidate gene sequencing and identified a *KCNT1* mutation R928C. This result demonstrated that *KCNT1* mutations can cause two very different epilepsy phenotypes such as MMPSI and ADNFLE.

Autosomal Dominant Epilepsy with Auditory Features

Clinical

Autosomal dominant epilepsy with auditory features (ADEAF), also known as autosomal dominant lateral temporal lobe epilepsy (ADLTLE), is a rare epilepsy syndrome that confers focal seizures that often secondarily generalize.[91] The seizures can begin at any age, but usually start in the second or third decade of life. As the syndrome's name suggests, the majority of the patients (64%) have focal seizures that begin with an auditory component. Some have simple auditory hallucinations, such as ringing that changes in volume,[92] or a "buzzing, or humming like a machine." [93] In other patients, auditory hallucinations are formed, such as voices or singing.[92] Some patients have visual, autonomic, psychic or vertiginous focal seizures. In some pedigrees, the seizures are accompanied by a sensory aphasia with or without auditory hallucinations.[94] In most patients, the seizures are infrequent (only several times per year before starting medication) and can usually be controlled with anticonvulsant drugs.

Interictal EEG abnormalities, if present, are usually left temporal spike and sharp wave complexes. ADEAF patients do not have causative brain lesions on conventional MRI imaging. However, one diffusion tensor imaging study suggested that some patients may have subtle malformations in the left temporal cortex.[95] Finally, although their neurological exams are normal, functional imaging and magnetoencephalography studies of members of four ADEAF families were consistent with impaired language processing.[96]

Genetics

LGI1

At the time of its first description, ADEAF was linked to a 10-cM region on chromosome 10q with a 71% penetrance.[97] Linkage studies in another family narrowed the region to approximately 3 cM.[93] Kalachikov et al. sequenced all exons and intron/exon junctions from one affected patient form three different ADEAF pedigrees and then genotyped all family members from five different ADEAF pedigrees.[98] They found that all affected family members and obligate carriers possessed mutations (four frameshift/intron retention truncation mutations and one missense mutation) in the leucine-rich, glioma-inactivated 1 gene (*LGI1*). Some unaffected family members also possessed the mutations, a finding consistent with the reduced penetrance found in the gene linkage studies. There are now 27 *LGI1* mutations associated with ADEAF (Table 84.4).

Less than 50% of ADEAF families and less than 2% of sporadic ADEAF patients have *LGI1* mutations. Recently, a new ADEAF locus was found in a large Brazilian family. The DNA from 11 affected and 14 unaffected family and performed genotyping found linkage to region 19q13.11–q13.31 with incomplete penetrance.[99]

PATHOPHYSIOLOGY OF SELECTED EPILEPSY GENE MUTATIONS

As discussed in the above section, human genetic epilepsy syndromes have been associated with mutations of many genes. In some cases, mutations of different genes result in the same epilepsy syndrome (phenocopies), whereas in other instances, mutations in the same gene cause different epilepsy syndromes. Here, we will discuss the biochemical and physiological consequences of epilepsy gene mutations (Table 84.5) and, if possible, provide likely explanations for the effect of genotype on phenotype.

Potassium Channels

KCNQ2/KCNQ3

The *KCNQ2* and *KCNQ3* genes associated with BFNE and EIEE encode the voltage-gated potassium channel subunits Kv7.2 and Kv7.3, respectively. The molecular biology, biochemistry, and physiology of wild type Kv7.2 and Kv7.3 subunits have been reviewed.[100] Four subunits oligomerize to form a functional tetramer. Each subunit has intracellular *N*- and *C*-termini and six transmembrane segments, S1–S6. The voltage sensor is located in the S4 domain and the ion channel pore is formed from one S5–6 domain from each subunit in a tetramer.

The Kv7.2 subunits can form functional homotetramers or form heterotetramers with Kv7.3 subunits. Kv7.3 subunits can homotetramerize, but homotetrameric Kv7.3 channels have markedly reduced cell surface expression when expressed in heterologous cells. Not only does the formation of Kv7.2/Kv7.3 heterotetramers greatly increase the surface expression of Kv7.3, it also increases the surface expression and opening probability of Kv7.2-containing channels as well. The voltage-gated potassium channels formed from Kv7.2 and Kv7.3 subunits are partially active at rest, but activate fully by membrane depolarization. They activate slowly and do not inactivate. The outward current conducted by the Kv7.2-containing and Kv7.3-containing channels (called the "M" current because it is inhibited by muscarinic agonists) hyperpolarizes neurons. Because the M current is small and activates and deactivates so slowly, it does not shape the action potential. However, it does delay the onset and quicken the termination of action potential bursts.

There are more than 70 *KCNQ2* mutations associated with BFNE. They include missense, splice site, and large and small insertions and deletions. In contrast, there are only four missense *KCNQ3* mutations associated with BFNE. The consequences of many of these mutations have been the subject of comprehensive reviews.[101,102] Although *KCNQ2* mutations are found throughout all Kv7.2 protein domains, there are a large number of mutations in the S4 domain (voltage sensor), S5–S6 domains (ion channel pore), and intracellular *C*-terminus (facilitates cell surface expression). The *KCNQ3* mutations are located in the pore domain. *In vitro* studies performed with recombinant wild-type and mutant Kv7.2 and Kv7.3 proteins expressed in heterologous cells demonstrated that the mutations caused loss of function without producing dominant negative effects. In general, *KCNQ2* mutations in the S4 voltage sensor increased the amount of depolarization required to activate the channel whereas mutations in the *C*-terminal domain, a region that mediates interaction with other proteins, reduces the cell surface expression of the ion channels. On average, expression of mutant Kv7.2 or Kv7.3 subunits with their wild-type counterparts (to mimic heterozygous expression) reduced peak currents by 25%, a finding which suggests that only small changes in the M current are needed to produce neuronal hyperexcitability and seizures.

It is not known why some *KCNQ2* mutations cause BFNE while others cause EIEE. For example, mutation of a single positively charged residue in the S4 voltage sensor domain (R213) to tryptophan confers BFNE,[103] while mutation to glutamine is associated with EIEE.[17] Recombinant mutant Kv7.2(R213W) or Kv7.2(R213Q) cDNA were coexpressed with either wild-type Kv7.2 alone (0.5 : 0.5 molar ratio) or with Kv7.2 and Kv7.3 (0.5 : 0.5 : 1 molar ratio) in heterologous cells and patch voltage clamp studies were used to measure the effects of activating depolarizing voltage potassium current.[104] These experiments demonstrated that while both mutations reduced the voltage sensitivity, the EIEE-associated R123Q mutation had a much greater effect than the BFNE-associated R213W mutation. This result suggests that mutations that cause greater Kv7.2/7.3 dysfunction may predispose patients to EIEE rather

TABLE 84.5 Pathophysiology of Selected Epilepsy Gene Mutations

Gene class	Epilepsy gene	Epilepsy syndrome	*In vitro* characterization	Mouse model
K^+ channels	*KCNQ2*	BFNE, EIEE	Loss of function. Extent of K^+ current reduction may correlate with disease severity	No spontaneous seizures. Increased sensitivity to pharmacological- and electrical-evoked seizures
	KCNQ3	BFNE	N/A	N/A
	KCNT	EIEE (MMPSI) ADNFLE	R409Q and A913T mutations increased potassium currents	N/A
Na^+ channels	*SCN1A*	GEFS+, Dravet, EIEE, MMPSI, WS	Different mutations, even from same disease phenotype, either enhanced or reduced sodium channel currents	Knockout mouse exhibited reduced sodium current in hippocampal interneurons. Knock-in mouse showed reduced action potential bursts. Both models are consistent with disinhibition
	SCN1B	GEFS+		
$GABA_A$ receptor	*GABRA1*	CAE, JME, unspecific GGE	The S326fs328X mutation associated with CAE showed complete elimination of the mutant protein by two biochemical mechanisms. The A322D mutation caused misfolding and elimination of 80–90% of the protein	Heterozygous *Gabra1* mice had absence seizures. Reduced endocytosis of cortical $GABA_A$ receptors was associated with increased relative surface expression of the α1 subunit and increased total and surface expression of the α3 subunit
	GABRB3	CAE	P11S, S15F and G32R mutations produced abnormal hyperglycosylation and decreased GABA evoked currents	*Gabrb3* mice had generalized seizures
	GABRG2	CAE, GEFS+	The R82Q mutation reduced oligomerization of γ2(R82Q) subunits with other subunits, reducing surface receptors	Heterozygous *Gabrg2(R82Q)* mice had absence seizures and hyperthermia-induced seizures
nAChR	*CHRNA2* *CHRNA4* *CHRNB2*	ADNFLE ADNFLE ADNFLE	The mutations increase nAChR function by different mechanisms such as reducing EC50 for nicotinic agonists	*Chrna4(S252F)* and *Chrna4(+L264)* mice had spontaneous and nicotine-evoked seizures. Mutant mice had altered frequency, amplitude and decay of nicotinic-evoked mIPSCs
Homeodomain	*ARX*	EIEE	Mutant ARX had reduced gene repressor function. Expanded polyalanine repeats increased intracellular inclusions	Heterozygous polyalanine expansion mice had spontaneous myoclonus and seizures. Less ARX was localized to the nucleus and repression was reduced
EF-hand domain-containing protein	*EFHC1*	JME	Mutant EFHC1 caused less apoptosis than wild-type EFHC1 when overexpressed. Mutant EFHC1 altered interneuron migration in cultured brain slices	EFHC1 deletion mouse had spontaneous myoclonus and reduced seizure threshold. At two months of life, mutant mouse had increased frequency of mIPSCs in cortical layers II/III

Abbreviations: ADEAF, autosomal dominant epilepsy with auditory features; ADNFLE, autosomal dominant nocturnal frontal lobe epilepsy; BFNE, benign familial neonatal epilepsy; CAE, childhood absence epilepsy; EIEE, early-infantile epileptic encephalopathy; GEFS+, genetic epilepsy with febrile seizures plus; GGE, genetic generalized epilepsy; JME, juvenile myoclonic epilepsy; MMPSI, malignant migrating partial seizures of infancy; WS, West syndrome.

than BFNE. However, some BFNE cases result from heterozygous *KCNQ2* deletion (i.e. total dysfunction of that *KCNQ2* allele). Therefore, if greater Kv7.2/7.3 dysfunction resulted in EIEE rather than BFNE, why do patients with *KCNQ2* deletions exhibit BFNE? Although this was not directly addressed in the Miceli et al. study, it is possible that the Kv7.2(R213Q) mutation did cause a dominant negative effect. Although the R213 mutations reduced the sensitivity to activating voltage, they did not alter cell surface trafficking and thus the cells expressed the same number of channels, but with less sensitivity. In the heterozygous *KCNQ2* deletion cases, if the cells increased the expression of the residual wild-type *KCNQ2* allele in order to compensate for the deleted allele, the cell surface expression of the Kv7.2/7.3 channels may not be substantially decreased, and their physiology would be the same as wild-type. This possibility could be explored in *KCNQ2* and *KCNQ3* deletion and KI mice.

Kcnq2 knockout (KO) mice as well as mice expressing the Kv7.2(A306T) and Kv7.3(G111V) BFNE mutations have been studied to determine the effects of these mutations on seizures.[105,106] Homozygous *Kcnq2* KO mice died shortly after birth due to pulmonary atelectasis. When studied at 3 weeks of life, the heterozygous *Kcnq2* KO mice did not have spontaneous absence seizures but did exhibit a higher sensitivity to pentylenetetrazole-evoked seizures. Homozygous, but not heterozygous Kv7.2(A306T) and Kv7.3(G111V) KI mice exhibited spontaneous tonic–clonic seizures. As was seen with the heterozygous *Kcnq2* KO mice, heterozygous Kv7.2(A306T) and Kv7.3(G111V) KI mice were more sensitive to evoked seizures (electrical) than wild-type mice.

BFNE is an autosomal dominant disease and thus it is not known why heterozygous KO or KI mice did not exhibit spontaneous seizures. BFNE only occurs during a very specific time window in human development, a window that may be dictated by developmental changes in *KCNQ2/3* expression, other gene products or brain formation and myelination. It is conceivable that this narrow window does not exist in rodents. Possibly, the creation of a mouse model of a *Kcnq2*-mediated EIEE syndrome (i.e., one that confers a severe phenotype and does not remit with age) will have a corresponding phenotype in mice.

KCNT1

Mutations of the *KCNT1* gene are associated with EIEE and ADNFLE. *KCNT1* encodes Slo2.2 (also known as Slack), a sodium-activated potassium channel (K_{Na} channel; for reviews see [107,108]). The existence of K_{Na} channels was deduced in the 1980s from electrophysiological patch-clamp recordings of cardiac myocytes in which the application of a high concentration of sodium to the cytoplasmic side the membrane evoked a current. The high cytoplasmic sodium concentration needed to activate K_{Na} channels (half-maximal activation at 66 mM) was surprising because typical intracellular sodium concentrations (in resting neurons) range from 4–15 mM. However, stimulation substantially increases intracellular sodium concentration; the intracellular sodium concentration can exceed 100 mM in activated dendritic spines. It is now recognized that K_{Na} channels cause a prolonged afterhyperpolarization that is critical for adapting neuronal firing rates and shaping intrinsic bursting.

Like the Kv7.2 and Kv7.3 channels, Slo2.2 contains six transmembrane domains, an intracellular *N*-terminus and a large intracellular *C*-terminus. However, unlike Kv7.2 or Kv7.3, the *C*-terminus of Slo2.2 contains domains that are thought to be sodium-binding sites that regulate channel opening. In addition, unlike Kv7.2 or Kv7.3, the S4 domain does not possess positive charges and thus Slo2.2 is not voltage activated. Although they are not voltage activated, Slo2.2 channels are outwardly rectifying.

Barcia et al. performed functional studies of wild-type and two EIEE-associated mutant Slo2.2 channels.[27] They expressed the rat orthologs of the human R428Q and A934T EIEE mutations (R409Q and A913T) in *Xenopus* oocytes and recorded K_{Na} currents. Surprisingly, both mutations increased potassium currents by 2–3-fold. It is not yet understood why a gain of K_{Na} function, which would be expected to increase afterhyperpolarization, would lead to seizures and epilepsy.

Voltage-Gated Sodium Channels

Sodium channel mutations are associated with multiple genetic epilepsy syndromes including GEFS+, Dravet syndrome, and EIEE. There are several published reviews of sodium channel structure, function, and pathophysiology.[109–111] Sodium channels are composed of α and β subunits—each of which exist as multiple isoforms. It is thought that the stoichiometry of α to β subunits is 1:2.

The α subunit genes encode a ~260-kDa protein with four homologous domains (I–IV). Similar to voltage-gated potassium channels, each sodium channel domain contains six transmembrane segments (S1–6) with the S4 segment forming the voltage sensor and one loop between S5 and S6 segments of each domain lining the ion channel pore. Unlike voltage-gated potassium channels, sodium channel α subunits also contain an intracellular loop between domains III and IV that form an inactivation gate. There are nine genes encoding α subunit isoforms (*SCN1A-SCN11A*); the *SCN6A/7A* genes do not encode voltage-gated sodium channel subunits. The proteins encoded by these genes are designated $Na_V1.1$– $Na_V1.9$. The isoform of the α subunit determines where the sodium channel is expressed. The $Na_V1.1$, channels are highly expressed in the central nervous system in dendrites, cell bodies and the initial axonal segment of fast spiking parvalbumin-containing interneurons.

Many of the properties of sodium channels are mediated by the β subunits. There are four isoforms of voltage-gated sodium channel β subunits encoded by genes *SCN1B–SCN4B* and encoding proteins $Na_V\beta1$–4. The β subunit proteins have only one transmembrane segment and a large extracellular *N*-terminus as well as a large intracellular *C*-terminus. During assembly, either a β1 or β2 and either a β3 or β4 subunit associates with an α subunit and the β3 and β4 subunits are covalently linked to the α subunit by a disulfide bond.

Voltage-gated sodium channels are activated by depolarization of the cell membrane. The S4 transmembrane segment of the α subunit contains positively charged residues, which, upon depolarization, move outward and rotate to alter the conformation of the ion channel pore to the open-channel state. When the sodium channels activate, they briefly (1–2 ms) conduct an inward sodium current that then inactivates within 1–2 ms. The fast inactivation is mediated by intracellular loop between domains III and IV of the α subunit and is absolutely critical to allow the sodium channel to fire repeatedly during trains of action potentials.

SCN1A

SCN1A mutations are associated with multiple epilepsy syndromes including approximately 85% of cases of Dravet syndrome and 10% of cases of GEFS+. In addition, *SCN1A* mutations have been found in patients with EIEE, MMPSI, WS and FSs. Of the GEFS+ mutations, approximately 90% were missense mutations and only 10% were either frameshift or truncation mutations. In contrast, for the Dravet syndrome mutations, only 22% were missense mutations and 78% were frameshift, truncation, amplification, duplication or deletion mutations. This suggests that the more disruptive *SCN1A* mutations could result in the more severe Dravet syndrome rather than GEFS+.

The pathophysiological consequences of epilepsy-associated *SCN1A* mutations have been evaluated by expressing recombinant wild-type and mutant $Na_V1.1$ subunit in heterologous cells (fibroblasts or oocytes) and performing patch-clamp electrophysiology experiments to record voltage-gated sodium channel currents. These experiments demonstrated that the missense mutations caused a variety of loss-of-function and gain-of-function changes in sodium currents. The mutations R859C, T875M, W1204R, R1648H, and D1866Y are all associated with the same phenotype, GEFS+. However, when expressed in *Xenopus* oocytes, the W1204R, R1648H and D1866Y mutations increase sodium channel activity (e.g., by reducing the voltage-dependence for activation, decreasing the use-dependent inactivation, increasing the rate of recovery, and increasing the level of persistent current), while the R859C and T875M mutations decrease sodium channel activity (e.g., by increasing the voltage dependence for activation, reducing the rate of recovery from slow inactivation, or increasing the amount of slow inactivation).

A mutation M145T of *SCN1A* has been shown to be associated with autosomal dominant simple FSs. The M145T mutation is in the first transmembrane segment of domain I of the human $Na_V1.1$ channel α-subunit, and functional studies in transfected mammalian cells demonstrated that the mutation reduced current density and produced a 10-mV positive shift of the activation curve, suggesting that it is a loss-of-function mutation.[43]

In reviewing these functional studies, Escayg and Goldin suggested that the channel physiology observed in a heterologous expression system may not fully reproduce what would be observed *in vivo*.[109] Possibly, the partnering β subunits would be different in specific neurons than would be recombinantly expressed in heterologous cells. Additional differences could include alterations in processing and trafficking as well as repertoire of other α subunit isoforms present in neurons. The type of heterologous cells in which the recombinant α subunits were expressed clearly affected sodium channel physiology. For example, while the R1648H GEFS+ mutant decreased use-dependent inactivation and increased the rate of recovery from inactivation when expressed in *Xenopus* oocytes, it increased the persistent current when expressed in mammalian fibroblasts.

To address the conflicting results concerning the study of the effects of the GEFS+ mutations when overexpressed in heterologous cells, Martin et al. introduced the R1648H mutation into the mouse *Scn1a* gene.[112] Heterozygous mice experienced infrequent spontaneous seizures as well as a decreased threshold for hyperthermic seizures. Interestingly, sodium currents in heterozygous cortical pyramidal neurons were not substantially affected. However, sodium currents from cortical interneurons demonstrated that the R1648H mutation reduced the function of voltage-gated sodium channels (i.e., an opposite overall effect than was observed in heterologous cells). Cortical neurons from heterozygous R1648H mice exhibited a slower recovery from inactivation, a greater use-dependent inactivation, and reduced action potential firing. These results are consistent with the R1648H GEFS+ mutation causing disinhibition, which then results in seizures.

Scn1a mutations have been introduced into mice to model Dravet syndrome. Yu et al. produced *Scn1a* deletion mice by inserting a neomycin selection cassette to disrupt a large portion of the last coding exon.[113] Ogiwara et al. generated mice containing the *Scn1a* R1407X nonsense mutation associated with human Dravet syndrome.[114] Heterozygotes of both mouse models exhibited spontaneous seizures. In the hippocampi of *Scn1a* deletion mice, sodium currents were reduced in the interneurons, but not the pyramidal neurons. Current clamp recordings of a burst of evoked action potentials from layer II/III interneurons in the R1407X mouse revealed that the mutation substantially reduced the spike amplitude at the end of the burst. Together, these findings suggest that like the GEFS+ mutation, the Dravet syndrome-associated R1407X mediates its effect by reducing sodium currents in interneurons resulting in disinhibition.

SCN1B

A mutation in the $Na_V\beta1$ subunit (C121W) also causes GEFS+,[49] likely by impairing expression and function of $Na_V1.1$ channels.

Ligand-Gated Ion Channels

Neuronal nAChRs

The nAChRs are pentameric, ligand-gated ion channels expressed in muscle, autonomic ganglia and the central nervous system. (For a review on nAChR structure and function, see [115].) Neuronal nAChR are heteropentamers or homopentamers whose five subunits consist of α and β subunits. There are six α subunit isoforms (α2–7, encoded by genes *CHRNA2–CHRNA7*) and three β subunits (β2–4, encoded by genes *CHRNB2–CHRNB3*). The different subunit isoforms confer different physiological properties to the receptors and are expressed in different regions of the nervous system; this allows the nervous system to fine-tune nicotinic cholinergic signaling.

Each subunit is a transmembrane protein that consists of a large extracellular *N*-terminus that contributes to the agonist-binding site, four transmembrane helices (M1–M4), and a large intracellular loop between the M3 and M4 helices. The five subunits arrange pseudosymmetrically around a central depression that is the ion channel pore and one M2 segment from each of the five subunits lines the pore. When two molecules of agonist bind the nAChR, a conformational change in protein structure is transmitted from the *N*-terminal domains to the transmembrane segments to open a cation-selective ion channel. Continued exposure to agonist causes the nAChR to adopt an agonist-bound, nonconducting state called the desensitized state.

Several mutations in *CHRNA4*, *CHRNB2*, and *CHRNA3* are associated with ADNFLE. The physiological and biochemical consequences of these mutations were tested by expressing wild-type and mutant nAChR subunits in heterologous cells.[51] One consistent finding was that the ADNFLE mutations increased nAChR activity (i.e., gain of function). For example, the *CHRNA4*(S248F) reduced the EC_{50} for nicotinic agonist and the *CHRNB2*(V287I) mutation reduced the rate of nAChR desensitization.

A gain of nAChR function was also observed in mice expressing ADNFLE-associated nAChR subunit mutations. Klaassen et al. generated mice expressing the *Chrna4*(S252F) and *Chrna4*(+L264) ADNFLE-associated mutations.[116] Both genotypes of mice experienced spontaneous and nicotine-evoked seizures. Importantly, although neither mutation altered the baseline mIPSCs in cortical layer II/III pyramidal neurons, both mutations increased the mIPSC frequency, amplitude and decay time in response to nicotine. The authors concluded that the gain-of-function of the mutant nAChR expressed on the interneurons enhanced nicotine-evoked GABAergic signaling. The authors then suggested that enhanced GABAergic activity served to hypersynchronize the multiple pyramidal cells innervated by a single interneuron which thereby results in seizures.

$GABA_A$ Receptors

$GABA_A$ receptors are the primary mediators of fast inhibitory synaptic transmission in the central nervous system and reduction of $GABA_A$ receptor-mediated inhibition has been shown to produce seizures. These inhibitory receptors are heteropentamers (although some homopentamers are known to assemble from β3 and ρ subunits) formed by assembly of multiple subunit subtypes (α1–α6, β1–β3, γ1–γ3, δ, ε, π, θ, and ρ1–ρ3) that are each encoded by a different gene. The receptors form chloride ion channels and heteropentamers most commonly contain two α subunits, two β subunits and a γ or δ subunit. $GABA_A$ receptors mediate both phasic, inhibitory synaptic transmission and tonic, perisynaptic inhibition, and several antiepileptic drugs including benzodiazepines, tiagabine and barbiturates act by enhancing $GABA_A$ receptor currents.

A wide range of monogenic GGEs have been shown to be associated with $GABA_A$ receptor subunit mutations (Table 84.4) that cluster in the three hEP genes *GABRs—GABRA1*, *GABRB3* and *GABRG2*. CAE alone has been associated with *GABRB3* missense mutations located in the β3 subunit signal peptide (P11S, S15F) and the *N*-terminal region of the mature subunit (G32R),[117] as well as with a specific haplotype of the *GABRB3* promoter.[61] In addition, CAE is associated with a frameshift mutation in *GABRA1* (975delC, S326fs328X) that produces a premature translation-termination codon (PTC). The β3 subunit mutation (P11S) has also been associated with autism pedigrees with some patients who also had epilepsy. JME has been associated with a missense mutation in *GABRA1* (A322D) and a variant in *GABRD* (R220H). FSs alone have been associated with missense mutations in *GABRG2* that are located in the *N*-terminal region of the mature γ2 subunit (R82Q, R177G) and in the intron 6 splice donor site (IVS6 2T->G). GEFS+ has been associated with missense and nonsense mutations in *GABRG2* (Q40X, K328M,

Q390X, W429X) and with variants in *GABRD* (E177A, R220C). While a clear genotype–phenotype correlation has not been well established, to date mutations in *GABRB3* and *GABRA1* have been associated with CAE or JME alone while *GABRG2* mutations and other *GABRD* variants have been associated with FS, FS with CAE and GEFS+. These monogenic mutations have been shown to impair $GABA_A$ receptor function by several mechanisms.[118]

GGE-Associated GABR Missense Mutations

One GGE-associated *GABR* missense mutation *GABRG2(K328M)* causes impaired channel gating without altering intracellular trafficking. The nonsense mutation K328M reduces $GABA_A$ receptor current by decreasing the stability of channel open states, thus reducing mean channel open time.

GGE-Associated GABR Nonsense, Insertion/Deletion-Frameshift and Splice Donor Site Mutations

Eight mutations have been reported that generate PTCs among the more than 20 mutations/variants in *GABRs* that have been associated with GGEs. The PTCs are in the hEP genes *GABRA1* and *GABRG2,* and are produced by nonsense mutations (*GABRG2(Q40X)*, *GABRG2(R136X)*, *GABRG2(Q390X)* and *GABRG2(W429X)*, insertion/deletion-frameshift mutations (*GABRA1(S326fs328X)*, *GABRG2(S443delC)*, and *GABRA1(K353delins18X)*, and an intron 6 splice donor site mutation (*GABRG2(IVS6+2 T→G)*.[58,119]

GGE-ASSOCIATED GABR NONSENSE MUTATIONS THAT ACTIVATE NONSENSE-MEDIATED MRNA DECAY (NMD) AND ENDOPLASMIC RETICULUM-ASSOCIATED DECAY (ERAD)

Five mutations in the $GABA_A$ receptor subunit hEP genes that have been reported to be associated with GGEs have been shown to activate NMD to degrade mRNA coding for truncated subunits and ERAD to degrade misfolded and misassembled subunits. These mutations are in *GABRG2* (Q40X, IVS6+2T→G, and R136X) and in *GABRA1* (K353delins18X and 975delC, S326fs328X). All of the mutations produce mRNAs that are partially degraded by NMD and truncated proteins that are degraded by ERAD and likely cause epilepsy due to *GABRG2* and *GABRA1* haploinsufficiency.

Recently, the effects of heterozygous α1 subunit deletion, a model of the S326fs328X mutation, were determined in mice. Even though heterozygous loss of the α1 subunit caused absence seizures when expressed in both the DBA/2J and C57BL/6J backgrounds, it had slightly different effects in the two strains.[120] In the DBA/2J strain, heterozygous α1 subunit KO caused the same incidence in absence seizures in male and female mice and did not have any effects on viability. However, in the C57BL/6J strain, heterozygous α1 subunit deletion caused a greater incidence of absence seizures in female than male mice and reduced the rate of weight gain in females. These results in the C57BL/6J mice are consistent with the observation that CAE is more prevalent in girls than boys.

Biochemical and immunohistochemical studies revealed that cortical neurons partially compensated for the heterozygous loss of α1 subunit by two mechanisms.[121] First, neurons increased the surface expression of the residual wild-type α1 subunit driven off the wild-type allele. Second, cortical neurons increased both the total and surface expression of the α3 subunit. Electrophysiological studies revealed that synaptic currents in mutant mice had reduced peak current amplitudes and increased decay times, features consistent with α3 subunit-containing $GABA_A$ receptors and that the compensatory responses in surface α1 subunit and total and surface α3 subunit expression was at least partially driven by decreased $GABA_A$ receptor endocytosis. These studies revealed biochemical mechanisms by which altered $GABA_A$ receptor expression changed GABAergic inhibition, which may lead to disinhibition and seizures.

GGE-ASSOCIATED GABR NONSENSE MUTATIONS THAT DO NOT ACTIVATE NMD BUT DO CAUSE ER RETENTION OF TRUNCATED SUBUNITS

Two nonsense mutations in the last exon of *GABRG2* (Q390X and Q429X) have been reported to be associated with GGEs. Both of the mutations produce mRNAs that are stable and not degraded by NMD and truncated proteins that are degraded incompletely by ERAD. The GABRG2 mutation, Q390X may cause the severe epilepsy phenotype of Dravet syndrome by forming stable intracellular aggregates of the truncated γ2 (Q390X) subunit protein. The aggregated protein likely causes neuronal stress and apoptosis resulting in the severe neurological phenotype.

GGE-Associated GABR Missense Mutations that Cause ER Retention and Activate ERAD and the Unfolded Protein Response (UPR)

Nine missense mutations in the $GABA_A$ receptor subunit hEP genes and two variants in *GABRD* have been reported to be associated with GGEs. These mutations are in *GABRA1* (A322D, D219N), *GABRB3* (P11S, S15F, and G32R), and *GABRG2* (R82Q, P83S, R177G, K328M). At least five of the mutations, *GABRG2(R82Q)*, *GABRG2(P83S)*,

GABRG2(R177G), *GABRA1(D219N)*, and *GABRA1(A322D)* disrupt $GABA_A$ receptor biogenesis and trafficking, and the mutant subunits to varying extent were reported to have impaired folding and impaired oligomerization with other subunits and to be subject to ER retention, activation of the UPR and degradation by ERAD.[118] Interestingly, all three *GABRB3* mutations were shown to reduce GABA-evoked currents in HEK 293T cells and to have abnormal glycosylation.[117] For G32R, the increased glycosylation was not associated with the impaired function of the β3 subunit,[122] but the basis for the trafficking defect produced by the other two *GABRB3* is unclear.

A *Gabrg2(R82Q)* KI mouse was made,[123] and heterozygous KI mice were shown to have 6–7-Hz spike-wave discharges associated with behavioral arrest. Consistent with *in vitro* studies, the mutant γ2(R82Q) subunit had reduced cell surface trafficking associated with a small reduction in GABAergic inhibition. The mouse was later used to demonstrate that the CAE phenotype was due to *Gabrg2* haploinsufficiency.[124]

Non-Ion Channel Genes

EFHC1

EFHC1 mutations are found in 5–10% of JME patients in some populations; it has also been implicated in juvenile absence epilepsy. *EFHC1* encodes the protein, EF-hand domain-containing protein (Efhc1 protein, also known as myoclonin1). Unlike many epilepsy-associated proteins whose function was known prior to them being linked to epilepsy, *EFHC1* was first discovered in 2004 when it was associated with familial JME,[70] and thus therefore, many details of its structure and function remain unknown. Multiple theories concerning the mechanistic role of *EFHC1* mutations in JME have been reviewed.

Protein and mRNA studies demonstrated that although *Efhc1* expression in brain is highest during embryogenesis, it is present in multiple regions of adult and immature brain.[125] Mouse brains express one *Efhc1* isoform (isoform A) that contains 640 amino acids and has a mass of 74kDa. In addition to expressing isoform A, human and monkey brains also express the alternatively spliced isoform B that contains 278 amino acids and has a mass of 32kDa.[72] Analysis of the amino acid sequence of Efhc1 isoform A revealed that it contained an EF-hand domain, a 12-amino acid calcium-binding sequence that is flanked by two alpha helices. Isoform A also contains three DM10 domains, ~105 amino acid sequences of unknown function that have been identified in cilia- and flagella-containing tissues in mammals and green algae, *Chlamydomonas*.[126] DM10 domains are present in nucleoside diphosphate kinases, proteins that are tightly associated with flagellar axomes as well as Rib72, a *Chlamydomonas* flagella microtubule-associated protein that is structurally similar to Efhc1 in that it contains three repeated DM10 regions followed by two EF-hand domains.

The homology of Efhc1 to Rib72 suggests that it could be a ciliary protein. Therefore, it was not surprising Efhc1 is highly expressed in the cilia ependymal cells lining the ventricles and homozygous Efhc1 deletion mice demonstrate enlarged cerebral ventricles and a reduced ciliary beat frequency (127). Possibly, an altered flow of cerebrospinal fluid changes the migration of neural progenitor cells from the subventricular zone to the cortex and thereby changes neurodevelopment and causes seizures.

Homology between Efhc1 and Rib72 also suggested that Efhc1 functions as a microtubule-associated protein. Immunofluorescence microscopy experiments demonstrated that the *N*-terminus of Efhc1 mediates its interaction with the centrosome of the mitotic spindle.[127] Disrupting Efhc1 by expressing an *N*-terminus truncation mutant or by small interfering RNA (siRNA), inhibited proper mitotic spindle formation in cultured cells. In addition, electroporation of these constructs into developing rat brain inhibited the radial migration of neurons by affecting mitosis and cell cycle exit of neuronal progenitors and by disrupting radial migration.[128] Accordingly, overexpression of several JME-associated Efhc1 mutants, but not wild-type Efhc1, in cultured developing brain disrupted neuronal migration into the cortex.[129] Therefore, it is possible that *EFHC1* mutations act, in part, to disrupt proper cortical development.

Suzuki et al. generated an *Efhc1* deletion mouse and found that both heterozygous and homozygous mice exhibited spontaneous myoclonus starting at 7–8 months of age.[130] In addition, both heterozygous and homozygous null mice demonstrated a greater sensitivity to pentylenetetrazol-evoked seizures than wild-type mice. In support of the proposed role of Efhc1 in the ependymal cilia, the homozygous null mice demonstrated enlarged cerebral ventricles and reduced ciliary movements. However, in heterozygous mice, the model of patients in this autosomal dominant disease, there was no change in ventricular size and no reduction in ciliary movements. This finding suggested that altered cerebrospinal flow was not the predominant pathogenic mechanism in causing myoclonus. These investigators also examined the effects of *Efhc1* deletion on GABAergic transmission. They did not observe any visually apparent changes in interneuron number in the cortex of heterozygous 9–12 month mice. In addition, although *Efhc1* deletion did not cause statistically significant changes in the amplitude or time course of miniature inhibitory postsynaptic currents in layer II/III neurons, it did cause an age-dependent increase in mIPSC frequency at 2 months. These researchers did not examine the electrophysiological characteristics at time points greater than 2 months. Future experiments will help to determine if heterozygous *Efhc1* deletion causes further changes in GABAergic or glutamatergic transmission at ~7–8 months of age, the time point at which the mice experience spontaneous myoclonus.

ARX

ARX mutations are associated with EIEE, WS, and LGS. The *ARX* gene encodes the Aristaless-related homeobox protein (also abbreviated *ARX*), a 562 amino acid homeodomain-containing protein (a protein that contains a fold that binds to DNA and regulates transcription). Cloning and sequence analysis revealed substantial homology between the homeodomians of *ARX* and *Drosophila's* aristaless gene, a gene that produces mutants with altered development of the distal parts of the legs and antenna including the "arista" antenna bristles.[131–133] Besides the homeodomain, other ARX domains that are highly conserved among vertebrate species include the octapeptide (transcriptional repressor), nuclear localization, aristaless (transcriptional repressor), and the four polyalanine domains.[134] As will be discussed, expansion of the polyalanine domains is involved in many epilepsy syndromes.

Knowledge of the changes in the temporal and spatial expression of ARX protein is critical to understand how the mutations of this gene cause epilepsy (reviewed in [135]). In rodents, *ARX* is first expressed at embryonic day eight (E8) in the developing telencephalon and diencephalon. By E14.5, *ARX* is predominantly expressed in proliferating neuroblasts in the ganglionic eminence (source of developing cortical GABAergic cells) as well as the ventricular zone with trails of *ARX* positive cells migrating in the subventricular, intermediate and marginal zones. By postnatal day 14 (P14), ARX is only present in a few scattered cells in the cortex. Therefore, it is clear that ARX is critically involved in cell migration and cortical development. Because of its homology to *Drosophila* aristaless protein, *ARX* was thought to be a transcriptional repressor. This hypothesis was confirmed in gene expression profiling studies that compared transcript abundance in wild-type and heterozygous *ARX* deletion mice.[136]

ARX mutations can be classified as those that cause gross brain malformations (i.e., with MRI-apparent abnormalities such as lissencephaly that would be classified as structural epilepsy) and those without gross brain lesions (those classified as a genetic epilepsy syndromes). Interestingly, the mutations that cause complete loss of *ARX* function (e.g., nonsense, frameshift, deletion, splice-site, and some missense mutations) are much more likely to produce a structural epilepsy while the less disruptive mutations (some missense mutations and polyalanine expansions) are more likely to result in a genetic epilepsy syndrome.[135] The polyalanine expansion mutations are of particular interest. Affected members in two EIEE families possessed seven additional GCG repeats (encoding alanine) within their normal stretch of 10 GCG repeats in exon 2.[22] When overexpressed in cultured cells, recombinant *ARX* that contained the additional seven GCG repeats formed cytoplasmic aggregates and partially shifted *ARX*'s localization from the nucleus to the cytoplasm.[137] These data suggested that polyalanine expansion may have both a partial loss of function (mislocalization) and toxic effects. Genetically modified mice containing the $ARX^{(GCG)10+7}$ polyalanine expansion possess many of the phenotypic features of WS including spontaneous myoclonus and other seizure types and EEG abnormalities including multifocal spikes and electrodecremental episodes.[138] Pathological examination of the $ARX^{(GCG)10+7}$ mice demonstrated that they, like the heterologous cells used in the overexpression experiments, possessed a decreased fraction of *ARX* localized to nucleus. In addition, these mice showed a selective reduction of calbindin-containing interneurons in the cortex. These results suggest that this *ARX* polyalanine expansion causes loss of function leading to mislocalization and a reduced ability to populate and/or retain selective types of inhibitory interneuron in the cortex thus leading to disinhibition and seizures.

CONCLUSION

Our understanding of GGEs has had an explosive growth over the past two decades with the identification of "epilepsy genes" that allow diagnosis and classification of patients with inherited epilepsies. However, the work discussed in this review is only the tip of the genetic epilepsy iceberg. Virtually all of the work reviewed was based on autosomal dominant inheritance of "monogenic" epilepsies that only account for about 2% of all inherited epilepsies. The vast majority of GGEs have complex or polygenic inheritance with multiple variants with small effects or are *de novo* mutations and so most of these patients remained undiagnosed as to their causal mutation.

There have been, and are presently ongoing, several efforts to begin to develop an approach to identify the genetic disruptions in this the vast patient population. Klassen et al., 2011, used whole-exome sequencing to examine 237 channel genes including the known hEP genes in patients diagnosed with a generalized epilepsy (152) and a group of age-matched controls without epilepsy (139). The hypothesis was that they would identify combinations of nonsynonymous SNPs that would correlate with the presence of a GGE. These variants with small effect would in combination cause abnormal excitability and epilepsy. Unfortunately, the study did not identify and SNP combinations that could explain polygenic inheritance of GGEs and in fact the SNP complexity in channel genes was equally complex in affected and unaffected groups.

There are other approaches to this problem that are ongoing. The National Institutes of Health funded Epilepsy Phenome/Genome Project (EPGP) is a multi-institutional, retrospective phenotype–genotype study designed to obtain and analyze detailed phenotypic information and DNA samples on 5250 participants, including probands with specific forms of epilepsy and, in a subset, parents of probands who do not have epilepsy. The Epi4K consortium of EPGP investigators has used a screen using trios, an affected individual and both parents, to identify *de novo* mutation that could not be identified in family pedigrees. They studied patients with two epileptic encephalopathies: infantile spasms (149 patients) and Lennox–Gastaut syndrome (115 patients) and confirmed 329 *de novo* mutations.

While this represents exciting progress, much work must be done. The EPGP project should provide a rich resource with a large database that may allow informatics approaches to identify genes involved in polygenetic inheritance in the future.

References

1. Fisher RS, van Emde BW, Blume W, et al. Epileptic seizures and epilepsy: definitions proposed by the International League Against Epilepsy (ILAE) and the International Bureau for Epilepsy (IBE). *Epilepsia*. 2005;46(4):470–472.
2. Berg AT, Berkovic SF, Brodie MJ, et al. Revised terminology and concepts for organization of seizures and epilepsies: report of the ILAE Commission on Classification and Terminology, 2005–2009. *Epilepsia*. 2010;51(4):676–685.
3. Plouin P, Anderson VE. Benign familial and non-familial neonatal seizures. In: Roger J, Bureau M, Dravet C, Genton P, Tossinari CA, Wolf P, eds. *Epileptic Syndromes in Infancy, Childhood and Adolescence*. 4th ed. Montrouge, France: John Libbey Eurotext Ltd.; 2005:3–15.
4. Ronen GM, Penney S, Andrews W. The epidemiology of clinical neonatal seizures in Newfoundland: a population-based study. *J Pediatr*. 1999;134(1):71–75.
5. Plouin P. Benign familial neonatal seizures and benign idiopathic neonatal seizures. In: Engel J, Pedley TA, eds. *Epilepsy a Comprehensive Textbook*. 2nd ed. Philadelphia: Wolters Kluwer Health/Lippincott Williams & Wilkins; 2008.
6. Biervert C, Schroeder BC, Kubisch C, et al. A potassium channel mutation in neonatal human epilepsy. *Science*. 1998;279(5349):403–406.
7. Singh NA, Charlier C, Stauffer D, et al. A novel potassium channel gene, KCNQ2, is mutated in an inherited epilepsy of newborns. *Nat Genet*. 1998;18(1):25–29.
8. Schiffmann R, Shapira Y, Ryan SG. An autosomal recessive form of benign familial neonatal seizures. *Clin Genet*. 1991;40(6):467–470.
9. Concolino D, Iembo MA, Rossi E, et al. Familial pericentric inversion of chromosome 5 in a family with benign neonatal convulsions. *J Med Genet*. 2002;39(3):214–216.
10. Lewis TB, Leach RJ, Ward K, O'Connell P, Ryan SG. Genetic heterogeneity in benign familial neonatal convulsions: identification of a new locus on chromosome 8q. *Am J Hum Genet*. 1993;53(3):670–675.
11. Charlier C, Singh NA, Ryan SG, et al. A pore mutation in a novel KQT-like potassium channel gene in an idiopathic epilepsy family. *Nat Genet*. 1998;18(1):53–55.
12. Proposal for revised classification of epilepsies and epileptic syndromes. *Epilepsia*. 1989;30(4):389–399.
13. Nordli Jr. DR. Epileptic encephalopathies in infants and children. *J Clin Neurophysiol*. 2012;29(5):420–424.
14. Yamatogi Y, Ohtahara S. Early-infantile epileptic encephalopathy with suppression-bursts, Ohtahara syndrome; its overview referring to our 16 cases. *Brain Dev*. 2002;24(1):13–23.
15. Dedek K, Fusco L, Teloy N, Steinlein OK. Neonatal convulsions and epileptic encephalopathy in an Italian family with a missense mutation in the fifth transmembrane region of KCNQ2. *Epilepsy Res*. 2003;54(1):21–27.
16. Bassi MT, Balottin U, Panzeri C, et al. Functional analysis of novel KCNQ2 and KCNQ3 gene variants found in a large pedigree with benign familial neonatal convulsions (BFNC). *Neurogenetics*. 2005;6(4):185–193.
17. Weckhuysen S, Mandelstam S, Suls A, et al. KCNQ2 encephalopathy: emerging phenotype of a neonatal epileptic encephalopathy. *Ann Neurol*. 2012;71(1):15–25.
18. Weckhuysen S, Ivanovic V, Hendrickx R, et al. Extending the KCNQ2 encephalopathy spectrum: clinical and neuroimaging findings in 17 patients. *Neurology*. 2013;81(19):1697–1703.
19. Allen AS, Berkovic SF, Cossette P, et al. De novo mutations in epileptic encephalopathies. *Nature*. 2013;501(7466):217–221.
20. Feinberg AP, Leahy WR. Infantile spasms: case report of sex-linked inheritance. *Dev Med Child Neurol*. 1977;19(4):524–526.
21. Bruyere H, Lewis S, Wood S, MacLeod PJ, Langlois S. Confirmation of linkage in X-linked infantile spasms (West syndrome) and refinement of the disease locus to Xp21.3-Xp22.1. *Clin Genet*. 1999;55(3):173–181.
22. Stromme P, Mangelsdorf ME, Shaw MA, et al. Mutations in the human ortholog of Aristaless cause X-linked mental retardation and epilepsy. *Nat Genet*. 2002;30(4):441–445.
23. Kalscheuer VM, Tao J, Donnelly A, et al. Disruption of the serine/threonine kinase 9 gene causes severe X-linked infantile spasms and mental retardation. *Am J Hum Genet*. 2003;72(6):1401–1411.
24. Saitsu H, Kato M, Mizuguchi T, et al. De novo mutations in the gene encoding STXBP1 (MUNC18-1) cause early infantile epileptic encephalopathy. *Nat Genet*. 2008;40(6):782–788.
25. Saitsu H, Tohyama J, Kumada T, et al. Dominant-negative mutations in alpha-II spectrin cause West syndrome with severe cerebral hypomyelination, spastic quadriplegia, and developmental delay. *Am J Hum Genet*. 2010;86(6):881–891.
26. Poduri A, Chopra SS, Neilan EG, et al. Homozygous PLCB1 deletion associated with malignant migrating partial seizures in infancy. *Epilepsia*. 2012;53(8):e146–e150.
27. Barcia G, Fleming MR, Deligniere A, et al. De novo gain-of-function KCNT1 channel mutations cause malignant migrating partial seizures of infancy. *Nat Genet*. 2012;44(11):1255–1259.

28. Carranza RD, Hamiwka L, McMahon JM, et al. De novo SCN1A mutations in migrating partial seizures of infancy. *Neurology*. 2011;77(4):380–383.
29. Freilich ER, Jones JM, Gaillard WD, et al. Novel SCN1A mutation in a proband with malignant migrating partial seizures of infancy. *Arch Neurol*. 2011;68(5):665–671.
30. Kamiya K, Kaneda M, Sugawara T, et al. A nonsense mutation of the sodium channel gene SCN2A in a patient with intractable epilepsy and mental decline. *J Neurosci*. 2004;24(11):2690–2698.
31. Liao Y, Anttonen AK, Liukkonen E, et al. SCN2A mutation associated with neonatal epilepsy, late-onset episodic ataxia, myoclonus, and pain. *Neurology*. 2010;75(16):1454–1458.
32. Ogiwara I, Ito K, Sawaishi Y, et al. De novo mutations of voltage-gated sodium channel alphaII gene SCN2A in intractable epilepsies. *Neurology*. 2009;73(13):1046–1053.
33. Nelson KB, Ellenberg JH. Predictors of epilepsy in children who have experienced febrile seizures. *N Engl J Med*. 1976;295(19):1029–1033.
34. Eckhaus J, Lawrence KM, Helbig I, et al. Genetics of febrile seizure subtypes and syndromes: a twin study. *Epilepsy Res*. 2013;105(1–2):103–109.
35. Helbig I, Lawrence KM, Connellan MM, et al. Obstetric events as a risk factor for febrile seizures: a community-based twin study. *Twin Res Hum Genet*. 2008;11(6):634–640.
36. Scheffer IE, Berkovic SF. Generalized epilepsy with febrile seizures plus. A genetic disorder with heterogeneous clinical phenotypes. *Brain*. 1997;120(3):479–490.
37. Berkovic SF, Scheffer IE. Febrile seizures: genetics and relationship to other epilepsy syndromes. *Curr Opin Neurol*. 1998;11(2):129–134.
38. Rich SS, Annegers JF, Hauser WA, Anderson VE. Complex segregation analysis of febrile convulsions. *Am J Hum Genet*. 1987;41(2):249–257.
39. Wallace RH, Marini C, Petrou S, et al. Mutant GABA(A) receptor gamma2-subunit in childhood absence epilepsy and febrile seizures. *Nat Genet*. 2001;28(1):49–52.
40. Audenaert D, van Broeckhoven C, de Jonghe P. Genes and loci involved in febrile seizures and related epilepsy syndromes. *Hum Mutat*. 2006;27(5):391–401.
41. Kugler SL, Stenroos ES, Mandelbaum DE, et al. Hereditary febrile seizures: phenotype and evidence for a chromosome 19p locus. *Am J Med Genet*. 1998;79(5):354–361.
42. Wallace RH, Berkovic SF, Howell RA, Sutherland GR, Mulley JC. Suggestion of a major gene for familial febrile convulsions mapping to 8q13-21. *J Med Genet*. 1996;33(4):308–312.
43. Mantegazza M, Gambardella A, Rusconi R, et al. Identification of an Nav1.1 sodium channel (SCN1A) loss-of-function mutation associated with familial simple febrile seizures. *Proc Natl Acad Sci U S A*. 2005;102(50):18177–18182.
44. Marini C, Harkin LA, Wallace RH, Mulley JC, Scheffer IE, Berkovic SF. Childhood absence epilepsy and febrile seizures: a family with a GABA(A) receptor mutation. *Brain*. 2003;126(Pt 1):230–240.
45. Reid CA, Kim T, Phillips AM, et al. Multiple molecular mechanisms for a single GABAA mutation in epilepsy. *Neurology*. 2013;80(11):1003–1008.
46. Singh R, Scheffer IE, Crossland K, Berkovic SF. Generalized epilepsy with febrile seizures plus: a common childhood-onset genetic epilepsy syndrome. *Ann Neurol*. 1999;45(1):75–81.
47. Poduri A, Lowenstein D. Epilepsy genetics–past, present, and future. *Curr Opin Genet Dev*. 2011;21(3):325–332.
48. Escayg A, MacDonald BT, Meisler MH, et al. Mutations of SCN1A, encoding a neuronal sodium channel, in two families with GEFS+2. *Nat Genet*. 2000;24(4):343–345.
49. Wallace RH, Wang DW, Singh R, et al. Febrile seizures and generalized epilepsy associated with a mutation in the Na+-channel beta1 subunit gene SCN1B. *Nat Genet*. 1998;19(4):366–370.
50. Macdonald RL, Kang JQ. Molecular pathology of genetic epilepsies associated with GABAA receptor subunit mutations. *Epilepsy Current*. 2009;9(1):18–23.
51. Marini C, Guerrini R. The role of the nicotinic acetylcholine receptors in sleep-related epilepsy. *Biochem Pharmacol*. 2007;74(8):1308–1314.
52. Depienne C, Trouillard O, Saint-Martin C, et al. Spectrum of SCN1A gene mutations associated with Dravet syndrome: analysis of 333 patients. *J Med Genet*. 2009;46(3):183–191.
53. Harkin LA, Bowser DN, Dibbens LM, et al. Truncation of the GABA(A)-receptor gamma2 subunit in a family with generalized epilepsy with febrile seizures plus. *Am J Hum Genet*. 2002;70(2):530–536.
54. Patino GA, Claes LR, Lopez-Santiago LF, et al. A functional null mutation of SCN1B in a patient with Dravet syndrome. *J Neurosci*. 2009;29(34):10764–10778.
55. Claes L, Del-Favero J, Ceulemans B, Lagae L, van Broeckhoven C, de Jonghe P. De novo mutations in the sodium-channel gene SCN1A cause severe myoclonic epilepsy of infancy. *Am J Hum Genet*. 2001;68(6):1327–1332.
56. Ishii A, Kanaumi T, Sohda M, et al. Association of nonsense mutation in GABRG2 with abnormal trafficking of GABAA receptors in severe epilepsy. *Epilepsy Res*. 2014;108(3):420–432.
57. Lachance-Touchette P, Brown P, Meloche C, et al. Novel alpha1 and gamma2 GABAA receptor subunit mutations in families with idiopathic generalized epilepsy. *Eur J Neurosci*. 2011;34(2):237–249.
58. Kananura C, Haug K, Sander T, et al. A splice-site mutation in GABRG2 associated with childhood absence epilepsy and febrile convulsions. *Arch Neurol*. 2002;59(7):1137–1141.
59. Maljevic S, Krampfl K, Cobilanschi J, et al. A mutation in the GABA(A) receptor alpha(1)-subunit is associated with absence epilepsy. *Ann Neurol*. 2006;59(6):983–987.
60. Feucht M, Fuchs K, Pichlbauer E, et al. Possible association between childhood absence epilepsy and the gene encoding GABRB3. *Biol Psychiatry*. 1999;46(7):997–1002.
61. Urak L, Feucht M, Fathi N, Hornik K, Fuchs K. A GABRB3 promoter haplotype associated with childhood absence epilepsy impairs transcriptional activity. *Hum Mol Genet*. 2006;15(16):2533–2541.
62. Genton P, Thomas P, Kasteleijn-Nolst Trenite DG, Medina MT, Salas-Puig J. Clinical aspects of juvenile myoclonic epilepsy. *Epilepsy Behav*. 2013;28(suppl 1):S8–S14.
63. Martinez-Juarez IE, Alonso ME, Medina MT, et al. Juvenile myoclonic epilepsy subsyndromes: family studies and long-term follow-up. *Brain*. 2006;129(Pt 5):1269–1280.

64. Cossette P, Liu L, Brisebois K, et al. Mutation of GABRA1 in an autosomal dominant form of juvenile myoclonic epilepsy. *Nat Genet*. 2002;31(2):184–189.
65. Bai D, Alonso ME, Medina MT, et al. Juvenile myoclonic epilepsy: linkage to chromosome 6p12 in Mexico families. *Am J Med Genet*. 2002;113(3):268–274.
66. Liu AW, Delgado-Escueta AV, Serratosa JM, et al. Juvenile myoclonic epilepsy locus in chromosome 6p21.2-p11: linkage to convulsions and electroencephalography trait. *Am J Hum Genet*. 1995;57(2):368–381.
67. Liu AW, Delgado-Escueta AV, Gee MN, et al. Juvenile myoclonic epilepsy in chromosome 6p12-p11: locus heterogeneity and recombinations. *Am J Med Genet*. 1996;63(3):438–446.
68. Pinto D, de Haan GJ, Janssen GA, et al. Evidence for linkage between juvenile myoclonic epilepsy-related idiopathic generalized epilepsy and 6p11-12 in Dutch families. *Epilepsia*. 2004;45(3):211–217.
69. Suzuki T, Morita R, Sugimoto Y, et al. Identification and mutational analysis of candidate genes for juvenile myoclonic epilepsy on 6p11-p12: LRRC1, GCLC, KIAA0057 and CLIC5. *Epilepsy Res*. 2002;50(3):265–275.
70. Suzuki T, Delgado-Escueta AV, Aguan K, et al. Mutations in EFHC1 cause juvenile myoclonic epilepsy. *Nat Genet*. 2004;36(8):842–849.
71. Annesi F, Gambardella A, Michelucci R, et al. Mutational analysis of EFHC1 gene in Italian families with juvenile myoclonic epilepsy. *Epilepsia*. 2007;48(9):1686–1690.
72. Medina MT, Suzuki T, Alonso ME, et al. Novel mutations in Myoclonin1/EFHC1 in sporadic and familial juvenile myoclonic epilepsy. *Neurology*. 2008;70(22 Pt 2):2137–2144.
73. Jara-Prado A, Martinez-Juarez IE, Ochoa A, et al. Novel Myoclonin1/EFHC1 mutations in Mexican patients with juvenile myoclonic epilepsy. *Seizure*. 2012;21(7):550–554.
74. Kapoor A, Satishchandra P, Ratnapriya R, et al. An idiopathic epilepsy syndrome linked to 3q13.3-q21 and missense mutations in the extracellular calcium sensing receptor gene. *Ann Neurol*. 2008;64(2):158–167.
75. Scheffer IE, Bhatia KP, Lopes-Cendes I, et al. Autosomal dominant frontal epilepsy misdiagnosed as sleep disorder. *Lancet*. 1994;343(8896):515–517.
76. Scheffer IE, Bhatia KP, Lopes-Cendes I, et al. Autosomal dominant nocturnal frontal lobe epilepsy. A distinctive clinical disorder. *Brain*. 1995;118(Pt 1):61–73.
77. Phillips HA, Scheffer IE, Berkovic SF, Hollway GE, Sutherland GR, Mulley JC. Localization of a gene for autosomal dominant nocturnal frontal lobe epilepsy to chromosome 20q 13.2. *Nat Genet*. 1995;10(1):117–118.
78. Steinlein OK, Mulley JC, Propping P, et al. A missense mutation in the neuronal nicotinic acetylcholine receptor alpha 4 subunit is associated with autosomal dominant nocturnal frontal lobe epilepsy. *Nat Genet*. 1995;11(2):201–203.
79. McLellan A, Phillips HA, Rittey C, et al. Phenotypic comparison of two Scottish families with mutations in different genes causing autosomal dominant nocturnal frontal lobe epilepsy. *Epilepsia*. 2003;44(4):613–617.
80. Steinlein OK, Stoodt J, Mulley J, Berkovic S, Scheffer IE, Brodtkorb E. Independent occurrence of the CHRNA4 Ser248Phe mutation in a Norwegian family with nocturnal frontal lobe epilepsy. *Epilepsia*. 2000;41(5):529–535.
81. Cho YW, Motamedi GK, Laufenberg I, et al. A Korean kindred with autosomal dominant nocturnal frontal lobe epilepsy and mental retardation. *Arch Neurol*. 2003;60(11):1625–1632.
82. Hirose S, Iwata H, Akiyoshi H, et al. A novel mutation of CHRNA4 responsible for autosomal dominant nocturnal frontal lobe epilepsy. *Neurology*. 1999;53(8):1749–1753.
83. Rozycka A, Skorupska E, Kostyrko A, Trzeciak WH. Evidence for S284L mutation of the CHRNA4 in a white family with autosomal dominant nocturnal frontal lobe epilepsy. *Epilepsia*. 2003;44(8):1113–1117.
84. Saenz A, Galan J, Caloustian C, et al. Autosomal dominant nocturnal frontal lobe epilepsy in a Spanish family with a Ser252Phe mutation in the CHRNA4 gene. *Arch Neurol*. 1999;56(8):1004–1009.
85. Sansoni V, Nobili L, Proserpio P, Ferini-Strambi L, Combi R. A de novo mutation in an Italian sporadic patient affected by nocturnal frontal lobe epilepsy. *J Sleep Res*. 2012;21(3):352–353.
86. Steinlein OK, Magnusson A, Stoodt J, et al. An insertion mutation of the CHRNA4 gene in a family with autosomal dominant nocturnal frontal lobe epilepsy. *Hum Mol Genet*. 1997;6(6):943–947.
87. Gambardella A, Annesi G, De FM, et al. A new locus for autosomal dominant nocturnal frontal lobe epilepsy maps to chromosome 1. *Neurology*. 2000;55(10):1467–1471.
88. De Fusco M, Becchetti A, Patrignani A, et al. The nicotinic receptor beta 2 subunit is mutant in nocturnal frontal lobe epilepsy. *Nat Genet*. 2000;26(3):275–276.
89. Phillips HA, Favre I, Kirkpatrick M, et al. CHRNB2 is the second acetylcholine receptor subunit associated with autosomal dominant nocturnal frontal lobe epilepsy. *Am J Hum Genet*. 2001;68(1):225–231.
90. Heron SE, Smith KR, Bahlo M, et al. Missense mutations in the sodium-gated potassium channel gene KCNT1 cause severe autosomal dominant nocturnal frontal lobe epilepsy. *Nat Genet*. 2012;44(11):1188–1190.
91. Michelucci R, Pasini E, Nobile C. Lateral temporal lobe epilepsies: clinical and genetic features. *Epilepsia*. 2009;50(suppl 5):52–54.
92. Winawer MR, Ottman R, Hauser WA, Pedley TA. Autosomal dominant partial epilepsy with auditory features: defining the phenotype. *Neurology*. 2000;54(11):2173–2176.
93. Poza JJ, Saenz A, Martinez-Gil A, et al. Autosomal dominant lateral temporal epilepsy: clinical and genetic study of a large Basque pedigree linked to chromosome 10q. *Ann Neurol*. 1999;45(2):182–188.
94. Brodtkorb E, Gu W, Nakken KO, Fischer C, Steinlein OK. Familial temporal lobe epilepsy with aphasic seizures and linkage to chromosome 10q22-q24. *Epilepsia*. 2002;43(3):228–235.
95. Tessa C, Michelucci R, Nobile C, et al. Structural anomaly of left lateral temporal lobe in epilepsy due to mutated LGI1. *Neurology*. 2007;69(12):1298–1300.
96. Ottman R, Rosenberger L, Bagic A, et al. Altered language processing in autosomal dominant partial epilepsy with auditory features. *Neurology*. 2008;71(24):1973–1980.
97. Ottman R, Risch N, Hauser WA, et al. Localization of a gene for partial epilepsy to chromosome 10q. *Nat Genet*. 1995;10(1):56–60.
98. Kalachikov S, Evgrafov O, Ross B, et al. Mutations in LGI1 cause autosomal-dominant partial epilepsy with auditory features. *Nat Genet*. 2002;30(3):335–341.

99. Bisulli F, Naldi I, Baldassari S, et al. Autosomal dominant partial epilepsy with auditory features: a new locus on chromosome 19q13.11-q13.31. *Epilepsia*. 2014 Mar 1.
100. Cooper EC. Made for "anchorin": Kv7.2/7.3 (KCNQ2/KCNQ3) channels and the modulation of neuronal excitability in vertebrate axons. *Semin Cell Dev Biol*. 2011;22(2):185–192.
101. Maljevic S, Wuttke TV, Lerche H. Nervous system KV7 disorders: breakdown of a subthreshold brake. *J Physiol*. 2008;586(7):1791–1801.
102. Maljevic S, Wuttke TV, Seebohm G, Lerche H. KV7 channelopathies. *Pflugers Arch*. 2010;460(2):277–288.
103. Sadewa AH, Sasongko TH, Lee MJ, et al. Germ-line mutation of KCNQ2, p.R213W, in a Japanese family with benign familial neonatal convulsion. *Pediatr Int*. 2008;50(2):167–171.
104. Miceli F, Soldovieri MV, Ambrosino P, et al. Genotype-phenotype correlations in neonatal epilepsies caused by mutations in the voltage sensor of K(v)7.2 potassium channel subunits. *Proc Natl Acad Sci U S A*. 2013;110(11):4386–4391.
105. Watanabe H, Nagata E, Kosakai A, et al. Disruption of the epilepsy KCNQ2 gene results in neural hyperexcitability. *J Neurochem*. 2000;75(1):28–33.
106. Singh NA, Otto JF, Dahle EJ, et al. Mouse models of human KCNQ2 and KCNQ3 mutations for benign familial neonatal convulsions show seizures and neuronal plasticity without synaptic reorganization. *J Physiol*. 2008;586(14):3405–3423.
107. Bhattacharjee A, Kaczmarek LK. For K^+ channels, Na^+ is the new Ca^{2+}. *Trends Neurosci*. 2005;28(8):422–428.
108. Dryer SE. Na(+)-activated K+channels: a new family of large-conductance ion channels. *Trends Neurosci*. 1994;17(4):155–160.
109. Escayg A, Goldin AL. Sodium channel SCN1A and epilepsy: mutations and mechanisms. *Epilepsia*. 2010;51(9):1650–1658.
110. Marban E, Yamagishi T, Tomaselli GF. Structure and function of voltage-gated sodium channels. *J Physiol*. 1998;508(Pt 3):647–657.
111. Catterall WA. Voltage-gated sodium channels at 60: structure, function and pathophysiology. *J Physiol*. 2012;590(Pt 11):2577–2589.
112. Martin MS, Dutt K, Papale LA, et al. Altered function of the SCN1A voltage-gated sodium channel leads to gamma-aminobutyric acid-ergic (GABAergic) interneuron abnormalities. *J Biol Chem*. 2010;285(13):9823–9834.
113. Yu FH, Mantegazza M, Westenbroek RE, et al. Reduced sodium current in GABAergic interneurons in a mouse model of severe myoclonic epilepsy in infancy. *Nat Neurosci*. 2006;9(9):1142–1149.
114. Ogiwara I, Miyamoto H, Morita N, et al. Nav1.1 localizes to axons of parvalbumin-positive inhibitory interneurons: a circuit basis for epileptic seizures in mice carrying an Scn1a gene mutation. *J Neurosci*. 2007;27(22):5903–5914.
115. Hurst R, Rollema H, Bertrand D. Nicotinic acetylcholine receptors: from basic science to therapeutics. *Pharmacol Ther*. 2013;137(1):22–54.
116. Klaassen A, Glykys J, Maguire J, Labarca C, Mody I, Boulter J. Seizures and enhanced cortical GABAergic inhibition in two mouse models of human autosomal dominant nocturnal frontal lobe epilepsy. *Proc Natl Acad Sci U S A*. 2006 Dec 4.
117. Tanaka M, Olsen RW, Medina MT, et al. Hyperglycosylation and reduced GABA currents of mutated GABRB3 polypeptide in remitting childhood absence epilepsy. *Am J Hum Genet*. 2008;82(6):1249–1261.
118. Macdonald RL, Kang JQ. mRNA surveillance and endoplasmic reticulum quality control processes alter biogenesis of mutant GABAA receptor subunits associated with genetic epilepsies. *Epilepsia*. 2012;53(suppl 9):59–70.
119. Tian M, Macdonald RL. The intronic GABRG2 mutation, IVS6+2T->G, associated with childhood absence epilepsy altered subunit mRNA intron splicing, activated nonsense-mediated decay, and produced a stable truncated gamma2 subunit. *J Neurosci*. 2012;32(17):5937–5952.
120. Arain FM, Boyd KL, Gallagher MJ. Decreased viability and absence-like epilepsy in mice lacking or deficient in the GABA(A) receptor alpha1 subunit. *Epilepsia*. 2012;53(8):e161–e165.
121. Zhou C, Huang Z, Ding L, et al. Altered cortical GABAA receptor composition, physiology, and endocytosis in a mouse model of a human genetic absence epilepsy syndrome. *J Biol Chem*. 2013;288(29):21458–21472.
122. Gurba KN, Hernandez CC, Hu N, Macdonald RL. GABRB3 mutation, G32R, associated with childhood absence epilepsy alters alpha1beta3gamma2L gamma-aminobutyric acid type A (GABAA) receptor expression and channel gating. *J Biol Chem*. 2012;287(15):12083–12097.
123. Tan HO, Reid CA, Single FN, et al. Reduced cortical inhibition in a mouse model of familial childhood absence epilepsy. *Proc Natl Acad Sci U S A*. 2007;104(44):17536–17541.
124. Reid CA, Kim T, Phillips AM, et al. Multiple molecular mechanisms for a single GABAA mutation in epilepsy. *Neurology*. 2013;80(11):1003–1008.
125. Leon C, de Nijs L, Chanas G, Delgado-Escueta AV, Grisar T, Lakaye B. Distribution of EFHC1 or Myoclonin 1 in mouse neural structures. *Epilepsy Res*. 2010;88:196–207.
126. King SM. Axonemal protofilament ribbons, DM10 domains, and the link to juvenile myoclonic epilepsy. *Cell Motil Cytoskeleton*. 2006;63(5):245–253.
127. de Nijs L, Lakaye B, Coumans B, et al. EFHC1, a protein mutated in juvenile myoclonic epilepsy, associates with the mitotic spindle through its N-terminus. *Exp Cell Res*. 2006;312(15):2872–2879.
128. de Nijs L, Leon C, Nguyen L, et al. EFHC1 interacts with microtubules to regulate cell division and cortical development. *Nat Neurosci*. 2009;12(10):1266–1274.
129. de Nijs L, Wolkoff N, Coumans B, Delgado-Escueta AV, Grisar T, Lakaye B. Mutations of EFHC1, linked to juvenile myoclonic epilepsy, disrupt radial and tangential migrations during brain development. *Hum Mol Genet*. 2012;21(23):5106–5117.
130. Suzuki T, Miyamoto H, Nakahari T, et al. Efhc1 deficiency causes spontaneous myoclonus and increased seizure susceptibility. *Hum Mol Genet*. 2009;18(6):1099–1109.
131. Stern C, Bridges CB. The mutants of the extreme left end of the second chromosome of DROSOPHILA MELANOGASTER. *Genetics*. 1926;11(6):503–530.
132. Schneitz K, Spielmann P, Noll M. Molecular genetics of aristaless, a prd-type homeo box gene involved in the morphogenesis of proximal and distal pattern elements in a subset of appendages in *Drosophila*. *Genes Dev*. 1993;7(1):114–129.
133. Miura H, Yanazawa M, Kato K, Kitamura K. Expression of a novel aristaless related homeobox gene 'Arx' in the vertebrate telencephalon, diencephalon and floor plate. *Mech Dev*. 1997;65(1–2):99–109.
134. Gecz J, Cloosterman D, Partington M. ARX: a gene for all seasons. *Curr Opin Genet Dev*. 2006;16(3):308–316.
135. Marsh ED, Golden JA. *Developing Models of Aristaless-related homeobox mutations*. 2012.

136. Colasante G, Sessa A, Crispi S, et al. Arx acts as a regional key selector gene in the ventral telencephalon mainly through its transcriptional repression activity. *Dev Biol*. 2009;334(1):59–71.
137. Shoubridge C, Cloosterman D, Parkinson-Lawerence E, Brooks D, Gecz J. Molecular pathology of expanded polyalanine tract mutations in the Aristaless-related homeobox gene. *Genomics*. 2007;90(1):59–71.
138. Price MG, Yoo JW, Burgess DL, et al. A triplet repeat expansion genetic mouse model of infantile spasms syndrome, Arx(GCG)10+7, with interneuronopathy, spasms in infancy, persistent seizures, and adult cognitive and behavioral impairment. *J Neurosci*. 2009;29(27): 8752–8763.

SECTION IX

WHITE MATTER DISEASES

S E C T I O N I X

WHITE MATTER DISEASES

S E C T I O N I X

WHITE MATTER DISEASES

CHAPTER

85

Multiple Sclerosis

Stephen L. Hauser, Jorge R. Oksenberg, and Sergio E. Baranzini

University of California, San Francisco, CA, USA

INTRODUCTION

Multiple sclerosis (MS) is the prototypic disease of central nervous system (CNS) myelin.[1] Myelin is a multilayered insulating protein and lipid-rich membrane that surrounds and isolates axons and speeds impulse conduction by permitting action potentials to jump between naked regions of axons (nodes of Ranvier) and across myelinated segments. Myelin sheaths also contribute to the structure and stability of axons. The myelin-forming cells in the CNS are the oligodendrocytes and in the peripheral nervous system the Schwann cells. Myelin formation is a spiraling process around the axon creating multiple membrane bilayers that are tightly apposed by charged protein interactions. A single oligodendrocyte ensheaths many axons, unlike each Schwann cell, which typically myelinates a single axon. Myelin injury results in reduced support for the axons, as well as redistribution of ion channels, destabilization of axonal membrane potentials, reduced excitability, and conduction block. Axons can initially adapt and restore conduction, explaining remissions, but eventually distal and retrograde degeneration occurs. Therefore, the early promotion of remyelination and preservation of oligodendrocytes remains an important therapeutic goal.

MS is the most common cause of acquired neurological dysfunction arising during early and mid-adulthood; it affects more than 1 million people in North America and Western Europe. MS is approximately threefold more common in women than in men. The age of onset is typically between 20 and 40 years (slightly later in men than in women), but the disease can present across the lifespan. Approximately 10% of cases begin before age 18 years of age, and extremes with onset as early as 1–2 years of age or as late as the eighth decade have been described.

MS pathogenesis is complex and multifactorial with a genetic component that is not strictly Mendelian and involves the interaction, either programmed or stochastic, of multiple genes, as well as a number of postgenomic DNA changes. In addition, it is likely that interactions with infectious, nutritional, climatic, and/or other environmental influences affect susceptibility considerably. This complex array of factors results in the loss of immune homeostasis and self-tolerance and the development of abnormal and pathogenic inflammatory responses against structural components of the CNS. Myelin loss, gliosis, and varying degrees of axonal pathology culminate in progressive neurological dysfunction, including sensory loss, weakness, visual loss, vertigo, incoordination, sphincter disturbances, and altered cognition. The autoimmune model of MS pathogenesis has set the tone for immunotherapy in this disease, first by global immunosuppression using anti-inflammatory drugs, and more recently by selectively targeting specific components of the immune response.[2,3] During the past few years, however, the recognition that a neurodegenerative process unresponsive to immunosuppression was responsible for progressive neurological impairment has brought myelin biology to the forefront of MS research, emphasizing the need to better understand the axonal changes that follow demyelination, axon–myelin interactions essential to normal neuronal function and survival, and oligodendrocyte differentiation and remyelination.[1]

CLINICAL FEATURES

Symptoms of MS result from interruption of myelinated tracts in the CNS. Initial symptoms are commonly one or more of the following: weakness or diminished dexterity in one or more limbs, a sensory disturbance, monocular visual loss

Rosenberg's Molecular and Genetic Basis of Neurological and Psychiatric Disease
http://dx.doi.org/10.1016/10.1016/B978-0-12-410529-4.00085-1

(optic neuritis), double vision (diplopia), gait instability, and ataxia. As the disease evolves, bladder dysfunction, fatigue, and heat sensitivity occurs in most patients. Additional symptoms include Lhermitte sign (an electric shock-like sensation down the spine and into the limbs evoked by neck flexion), facial weakness or pain, vertigo, brief tonic spasms, and other paroxysmal symptoms (thought to represent discharges originating along demyelinated axons). Cognitive deficits are common, especially in advanced cases and include memory loss, impaired attention, problem-solving difficulties, slowed information processing, and difficulties in shifting between cognitive tasks.

Approximately 85% of patients with MS experience an abrupt onset, and early symptoms may be severe or seem so trivial that a patient may not seek medical attention for months or years. Thereafter, the clinical course may be characterized by acute episodes of worsening (exacerbations or relapses), gradual progression of disability, or combinations of both. Four clinical patterns are recognized by international consensus. Patients with relapsing-remitting MS (RRMS) experience relapses with or without complete recovery and are clinically stable between these episodes. Approximately half of patients with RRMS convert to secondary progressive MS (SPMS) within 15 years of disease onset. The secondary progressive phase is characterized by gradual progression of disability with or without superimposed relapses. In contrast, patients with primary progressive MS (PPMS) experience gradual progression of disability from onset without superimposed relapses. Approximately 15% of patients with MS experience this clinical pattern. A very small proportion of patients, perhaps 1–2%, experience gradual progression of disability from disease onset only later accompanied by one or more relapses; this clinical pattern is designated progressive relapsing MS (PRMS).

Most patients with MS ultimately experience progressive disability. Fifteen years after diagnosis, fewer than 20% of patients with MS have no functional limitation, 50–60% require assistance when ambulating, 70% are limited or unable to perform major activities of daily living, and 75% are not employed.

A number of variants of demyelinating disease exist, noteworthy among these is the Marburg variant, a rare, acute fulminant process that generally ends in death from brainstem involvement within 1–2 years. There are no remissions. Diagnosis can be established with certainty only by direct examination of tissue; widespread demyelination, axonal loss, edema, and macrophage infiltration are characteristic, and discrete plaques may also be seen. It has been suggested that acute MS may be associated with an immature form of myelin that is more susceptible to immune-mediated attack. It is unclear whether the Marburg variant represents an extreme form of MS or some other disease altogether.

Neuromyelitis optica (NMO) was previously classified as an MS spectrum disorder but is now believed to be a distinct neuroinflammatory disease. NMO is characterized by separate attacks of acute optic neuritis that may be unilateral or bilateral and severe, and by myelitis that is typically longitudinally extensive and transverse (involving >50% of the transverse width of the cord on axial sections). Optic neuritis may precede or follow an attack of myelitis by days, months, or years. Respiratory failure may result from cervical cord lesions. In contrast to patients with MS, patients with NMO do not typically experience widespread involvement, but lower brainstem lesions and also hemispheric lesions with a "cloudy" appearance by magnetic resonance imaging (MRI) may occur. In contrast to MS, histopathology of NMO lesions may reveal areas of necrosis, thickening of blood vessel walls, and an astrocytopathy. Knowledge of NMO has been transformed by the discovery that antibodies specific for the aquaporin-4 (Aqp4) water channel, a component of the dystroglycan protein complex located at astrocyte foot processes and concentrated at the blood–brain barrier, represent a specific biomarker for this condition. Anti-Aqp4 antibodies are present in the circulation of most patients with NMO and appear to be pathogenic, as they can passively transfer neuroinflammatory pathology experimentally. This syndrome is unusual in whites but appears to be more common in Asians and African Americans; nearly half of affected individuals have evidence of a systemic autoimmune disease.

DIAGNOSIS

The diagnosis of MS has been revolutionized by MRI technology, which reveals multiple, asymmetrically located white matter lesions distributed throughout the white matter of the CNS with a predilection for the corpus callosum and deep periventricular regions (Figure 85.1). New lesions are heralded by breakdown of the blood–brain barrier (BBB) associated with perivenous inflammation and detected by extrusion of the heavy metal gadolinium across the blood–brain barrier. Spinal cord lesions are frequently present and can be detected with high sensitivity using high-field MRI. Ancillary tests, used primarily in uncertain or problematic cases, consist of cerebrospinal fluid studies revealing low levels of inflammation with mononuclear cells (generally 50 cells per mm^3) and raised levels of immunoglobulin, including antibodies with restricted clonotypes (oligoclonal bands). Evoked potentials in the visual, auditory, or sensory pathways may also be helpful in identifying additional, silent, lesions. Additional laboratory

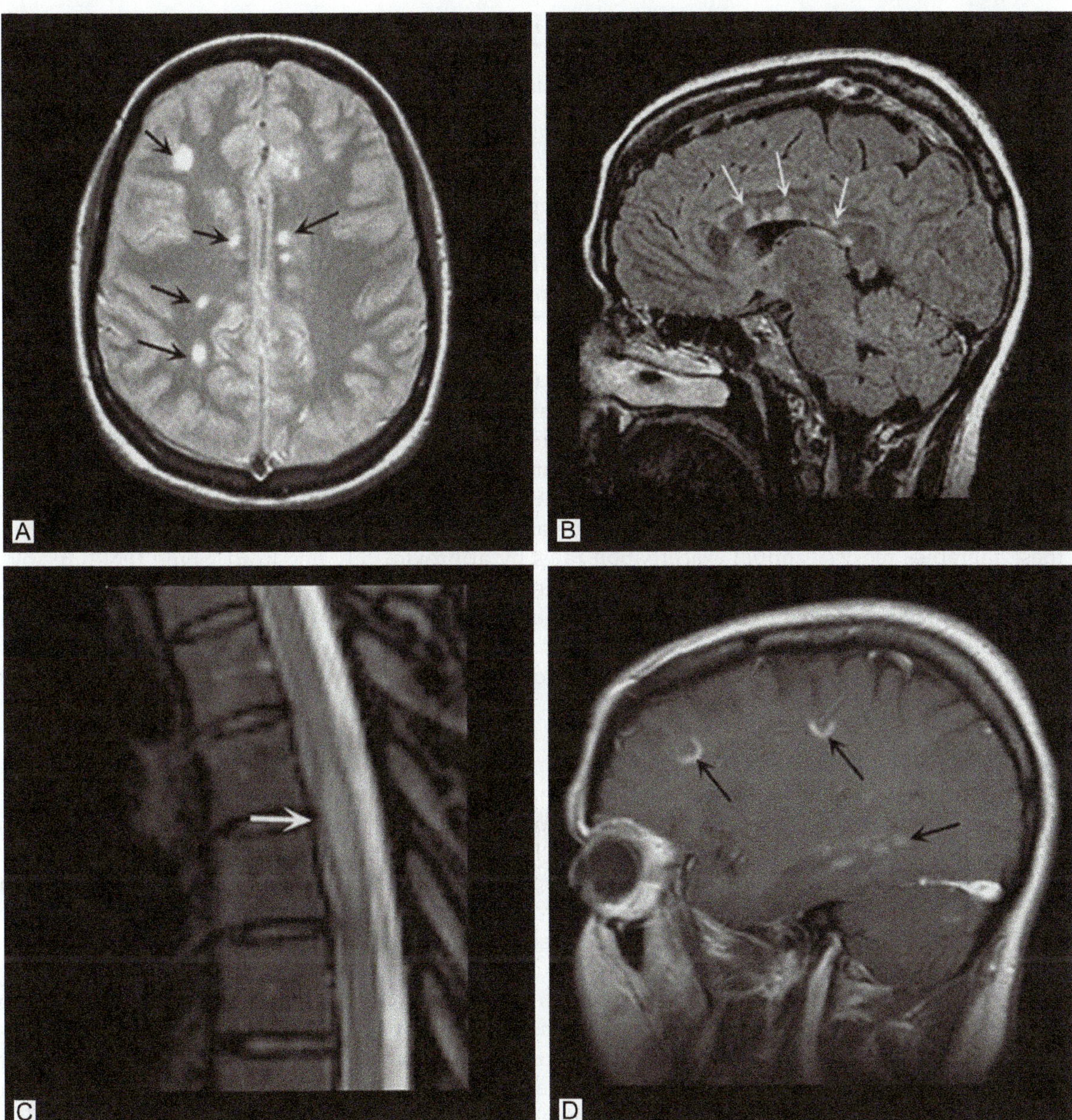

FIGURE 85.1 **MRI images in multiple sclerosis.** (**A**) Axial first-echo image from T2-weighted sequence demonstrates multiple bright signal abnormalities in white matter, typical for multiple sclerosis (MS). (**B**) Sagittal T2-weighted fluid-attenuated inversion recovery (FLAIR) image in which the high signal of cerebrospinal fluid has been suppressed. Cerebrospinal fluid appears dark, while areas of brain edema or demyelination appear high in signal as shown here in the corpus callosum (arrows). Lesions in the anterior corpus callosum are frequent in MS and rare in vascular disease. (**C**) Sagittal T2-weighted fast-spin echo image of the thoracic spine demonstrates a fusiform high signal-intensity lesion in the midthoracic spinal cord. (**D**) Sagittal T1-weighted image obtained after the intravenous administration of gadolinium-DTPA (diethyl triamine penta-acetic acid) reveals focal areas of blood–brain barrier disruption, identified as high signal-intensity regions (arrows). Reproduced from Longo et al. Harrison's Principles of Internal Medicine. 18th ed. New York: McGraw Hill; 2012.

testing that may be advisable in certain cases to exclude phenotypically similar diseases includes a serum vitamin B_{12} level, human T-cell lymphotropic virus type I (HTLV-I) titer, erythrocyte sedimentation rate, rheumatoid factor, antinuclear anti-DNA antibodies (systemic lupus erythematosus), serum VDRL (venereal disease research laboratory test), angiotensin-converting enzyme (sarcoidosis), Borrelia serology (Lyme disease), very-long-chain fatty acids (adrenoleukodystrophy), and serum or cerebrospinal fluid lactate, muscle biopsy, or mitochondrial DNA analysis (mitochondrial disorders).

DISEASE-MODIFYING TREATMENT

Therapy for MS can be divided into interventions useful for: 1) prophylaxis for relapses; and (2) progressive MS.

Prophylaxis Against Relapses

Multiple US Food and Drug Administration (FDA)-approved therapies now exist. These fall into three general categories: 1) first-generation injectable therapies with modest efficacy and a long-term outstanding safety record; 2) newer oral therapies, mostly with moderate efficacy and relatively favorable short-term safety profiles; and 3) infusion therapies that are highly effective but entail significant risks.

First-Generation Injectable Therapies

First-generation therapies for patients with relapsing forms of MS usually consist of either an interferon beta (IFN-β) preparation (Betaseron, Avonex or Rebif) or glatiramer acetate (Copaxone). All of these treatments are administered by subcutaneous or intramuscular injection. IFN-β is a class I interferon originally identified as an antiviral agent. It is likely to work against MS via immunomodulation. Glatiramer acetate is a synthetic, random polypeptide composed of four amino acids. Its mechanism of action may include inducing antigen-specific suppressor T-cells, binding to MHC molecules, or altering the balance between pro-inflammatory and regulatory cytokines. Each of these therapies reduces annual exacerbation rates by approximately one-third. Some IFN-β recipients develop neutralizing antibodies within 12 months of initiating therapy, and these patients may experience a return to their pretreatment attack rate. Treatment should probably be discontinued in all patients who continue to experience frequent attacks or gradual progression of disability for 6 or more months.[4–6] It is unknown whether patients who fail treatment with one of these interventions will respond favorably to one of the alternatives or to combination therapy.

Newer Oral Therapies

Fingolimod (Gilenya), the first oral MS therapy, is a sphingosine-1-phosphate (S1P) inhibitor that prevents the egress of lymphocytes from lymphoid organs. Its mechanism of action is probably due, in part, to blocking pathogenic lymphocytes from reaching the brain. A large head-to-head phase III randomized study demonstrated the clear superiority of fingolimod over low-dose (weekly) IFN-β. However, liver function test abnormalities develop in up to 12% of treated patients, and treatment-associated bradycardia, heart block, and retinal edema are additional risks.

Dimethyl fumarate (DMF; Tecfidera) is the most recently approved therapy for relapsing MS. Although the precise mechanisms of action of DMF are not fully understood, it seems to have anti-inflammatory effects through its modulation of the expression of pro-inflammatory and anti-inflammatory cytokines. Also, DMF inhibits the ubiquitylation and degradation of nuclear factor E2-related factor 2 (Nrf2) that induces the transcription of several antioxidant proteins. DMF reduces the attack rate and significantly improves all measures of disease severity in MS patients. However, its twice-daily oral dosing schedule is cumbersome, and gastrointestinal disturbances, skin reactions, and occasional hepatic enzyme elevations are risks. Also, several cases of progressive multifocal leukoencephalopathy (PML) were recently identified in patients receiving similar but not identical fumarate preparations.

Teriflunomide (Aubagio) is another oral therapy that was recently approved for RRMS. It is an inhibitor of pyrimidine synthesis and has only a modest effect on the relapse rate.

Infusion Therapies

Natalizumab (Tysabri) is a humanized monoclonal antibody directed against the α4 subunit of α4β1 integrin, a cellular adhesion molecule expressed on the surface of lymphocytes. It prevents lymphocytes from binding to endothelial cells, penetrating the BBB and entering the CNS. It is administered by monthly infusion. Natalizumab is approximately twice as effective as first-line therapies for relapsing forms of MS. The major concern with long-term treatment is the risk of developing progressive multifocal leukoencephalopathy (PML), a devastating neurotropic virus infection. A blood test to detect antibodies against the PML (JC) virus can identify individuals who are at risk for this complication.

Alemtuzumab (Lemtrada) is a humanized monoclonal antibody directed against the CD52 antigen, which is expressed on both monocytes and lymphocytes. It causes long-lasting lymphocyte depletion (of both B- and T-cells) and a change in the composition of lymphocyte subsets. In preliminary trials alemtuzumab markedly reduced the attack rate and MRI measures of disease activity. In two phase III trials, however, its impact on clinical disability was less convincing. The European and Canadian drug agencies have approved alemtuzumab for use in RRMS, but by contrast the FDA has thus far declined to approve. The reasons for the FDA decline relate both to concerns with trial design and potential toxicity. The toxicities of special concern to the FDA were the occurrence (during the trial or thereafter) of 1) autoimmune diseases; 2) malignancies; and 3) serious infections and infusion reactions. This decision is currently being appealed and alemtuzumab therapy may yet become available in the US.

Each of these therapies may improve the long-term outcome of MS when administered in the RRMS stage of the illness, although this key outcome goal has not yet been proven in prospective studies. Already established progressive symptoms do not appear to respond to treatment with these disease-modifying therapies.[7] Because progressive symptoms are likely to result from late effects of earlier demyelinating episodes, many experts now suggest early treatment for most MS patients.

Based on the epidemiologic data that vitamin D deficiency is associated with MS risk, and possibly also MS disease activity, many experts recommend vitamin D supplementation (up to 4000 IU daily); the vast majority of people who live in the developed world are Vitamin D deficient. Dietary supplementation with long-chain fatty acids (omega-3 fish oils, 500 mg bid) may also have immunosuppressive properties that are clinically useful.

Treatment of Progressive Disease

Developing treatments for progressive MS represents the "holy grail" of MS therapeutics and the greatest unmet need for patients.[8,9] High-dose IFN-β probably has a beneficial effect in patients with SPMS who are still experiencing acute relapses. IFN-β is probably ineffective in patients with SPMS who are not having acute attacks. It is likely that the same principle applies to other agents effective against relapsing MS, but data from randomized clinical trials have not yet addressed this question.

Chronic immunosuppression is advocated as therapy for some MS patients with progressive disease. Although mitoxantrone has been FDA approved for patients with progressive MS, this is not the population studied in the pivotal trial. Therefore no evidence-based recommendation can be made with regard to its use in this setting. The antimetabolite azathioprine given orally is relatively safe and well tolerated. But its beneficial effect is modest and must be weighed against potential risks that include hepatitis, susceptibility to infection, and a theoretical cancer risk. Methotrexate is another antimetabolite that has been used in progressive MS. Pulse therapy with the alkylating agent cyclophosphamide may be of benefit to young (<40 years) ambulatory patients with rapidly progressive MS. The side effects are considerable and include nausea, hair loss, a risk of hemorrhagic cystitis, and temporary profound immunosuppression.

No clearly effective therapies exist for PPMS. A phase III clinical trial of glatiramer acetate in PPMS was stopped because of lack of efficacy. A phase III trial of the B-cell depleting drug rituximab was also negative, although a secondary analysis suggested that patients with gadolinium (Gd) enhancement at enrollment had a favorable response; a follow-up phase III study with a similar B-cell depleting agent, ocrelizumab, is underway.

PATHOLOGY

The pathological hallmark of MS is the plaque, a well-demarcated gray or pink lesion, characterized histologically by inflammation, demyelination, proliferation of astrocytes with ensuing gliosis, and variable axonal degeneration. MS plaques are typically multiple, asymmetric, and clustered in deep white matter near the lateral ventricles, corpus callosum, floor of the fourth ventricle, deep periaqueductal region, optic nerves and tracts, corticomedullary junction, and the spinal cord. Plaques vary in size from 1 or 2 mm to several centimeters in diameter. Although early microscopic studies reported changes in the cerebral cortex of affected brains, MS is generally perceived as a white matter disease. There is, however, an increasing evidence for involvement of gray matter in MS, consisting both of demyelinating plaques within the cortex as well as a cortical neurodegenerative process, and increasing data support the concept that cortical plaques are important contributors to motor, sensory, and cognitive disability in MS.[10] Importantly, routine MRI scans are insensitive for detecting this cortical pathology.

Although cortical demyelination and progressive cortical atrophy now appear to be characteristic of MS, changes are particularly notable in progressive MS cases. In early relapsing MS, cortical changes appear to track with white matter disease, but in long-standing disease cortical changes appear to develop out of proportion to new changes in the white matter pathology. Microglia activation and diffuse axonal injury are typically associated with active lesions in the cortex, but the classical perivascular lymphocytic infiltration characteristic of the white matter plaques is absent. Overall, there is an emerging consensus that cortical lesions and atrophy are major contributors to disease burden in patients with progressive MS. In particular, recent data has highlighted previously unrecognized inflammatory changes in the meninges, especially the finding that abnormal, ectopic, lymphoid follicle-like structures rich in B-cells occur in the meninges overlying deep cortical sulci, and that these meningeal changes are associated with adjacent cortical pathology.[11–15] The development of novel animal models with demyelinating lesions in the cortex will be of great value for fully assessing the role of cortical pathology in MS.

Perivascular and parenchymal infiltration by mononuclear cells, both T-cells and macrophages, is characteristic of the acute MS white matter lesion. Parenchymal and perivascular T-cells consist of variable numbers of CD4 and CD8 cells. Although the vast majority bears the common form of the antigen receptor (i.e., the α/β heterodimer), T-cells carrying the other form, the γ/δ heterodimer, have been also identified. The observed selective accumulation of activated T-cells during certain stages of the plaque cycle indicates a specific pattern in the trafficking of T-cells to the lesion and suggests that an immune response to discrete antigenic molecules is present. Although fewer in number, B-cells also contribute to the inflammatory response. The role of this inflammatory response in MS pathogenesis is discussed later in this chapter.

At sites of inflammation, the blood–brain barrier is disrupted, but the vessel wall itself is preserved, distinguishing the MS lesion from vasculitis. In some inflammatory lesions, dissolution of the multilamellated compact myelin sheaths that surround axons occurs. As lesions evolve, axons traversing the plaque show marked irregular beading, proliferation of astrocytes occur, and lipid-laden macrophages containing myelin debris are prominent. Progressive fibrillary gliosis ensues, and mononuclear cells gradually disappear. In some MS lesions but not others, proliferation of oligodendrocytes appears to be present initially, but these cells are apparently destroyed as the gliosis progresses. In chronic MS lesions, complete or nearly complete demyelination, dense gliosis (more severe than in most other neuropathologic conditions), and loss of oligodendroglia are found. In some chronic active MS lesions, gradations in the histologic findings from the center to the lesion edge suggest that lesions expand by concentric outward growth.

MS lesions have long been known to be heterogeneous in terms of the intensity of the cellular infiltrate and antibody and complement deposition; whether or not endothelial damage is detected; in the fate of oligodendrocytes; and in the presence of remyelination, among other variables. One critically important question is whether this apparent heterogeneity observed in biopsy and autopsy tissues reflects a single underlying pathologic process observed at different stages or whether multiple, fundamentally distinct processes may be responsible in different cases.[16]

The traditional neuropathological view of MS highlights myelin loss as the prominent event occurring in the plaque, resulting in exposure to ion channels and impaired propagation of action potentials across the demyelinated region of the axon. However, even the early literature on MS described substantial axonal damage in active demyelination lesions. It is not known whether this process is independent or whether it is a consequence of demyelination, but renewed interest in this aspect of MS pathology has considerably focused attention into the neurodegenerative aspects of this disease. Contemporary high-resolution histopathological studies reveal abundant transected and dystrophic axons in sites of active inflammation and demyelination and confirm that partial or total axonal transection begins early in the disease process.[17–19] Axonal damage appears to take place in every newly formed lesion, and the cumulative axonal loss is now considered to be the reason for progressive and irreversible neurological disability in MS.[20] Furthermore, reduced axonal density is also observed in inactive and remyelinating lesions, cortical tissue, and the normal-appearing white matter in the brain and spinal cord. Pathology studies indicate that as many as 70% of axons are lost from lateral corticospinal tracts in patients with advanced paraparesis due to MS.[21] Advanced MRI techniques are an increasingly clinically useful tool and provide a quantitative window into the dynamic processes that result in progressive axonal and neuronal loss over time. Examples include accelerated rates of whole brain atrophy,[22] gradual worsening of focal (T1) lesions over time,[23] reductions in the predominantly neuronal/axonal marker *N*-acetyl aspartate by spectroscopy,[24] disruption of individual white matter tracts measured by diffusion tensor imaging,[25] and evidence of plasticity and altered functional connectivity by functional MRI.[26] These *in vivo* methods lend further support to the concept that axonal and neuronal loss is responsible for the persistent neurological dysfunction that occurs in patients with MS.

Axonal and neuronal death may result from glutamate excitotoxicity,[27,28] oxidative injury,[15,29] iron accumulation,[30] or mitochondrial failure due to free radical injury or accumulation of deletions in mitochondrial DNA.[31]

IMMUNOLOGIC BASIS

An important conceptual development in understanding MS pathogenesis has been the compartmentalization of the mechanistic process into two distinct but overlapping and connected phases, inflammatory and neurodegenerative. During the initial state of the inflammatory phase, lymphocytes with encephalitogenic potential are activated in the periphery and migrate to the CNS, become attached to receptors on endothelial cells, and then proceed to pass across the blood–brain barrier, through the endothelium and the subendothelial basal lamina directly into the interstitial matrix.[32] Remarkably, the presence of immunocompetent cells with autoimmune potential appears to be an embedded characteristic of the healthy immune system in vertebrates. These cells may provide important inflammatory signals necessary for wound healing, angiogenesis, neuroprotection, and other maintenance functions. The transition from physiological to pathological autoimmunity involves at least two factors: i) the loss of immune homeostasis, normally maintained through inhibitory signaling pathways, induction of anergy or apoptosis, and anti-idiotypic networks; and ii) the engagement and activation of lymphocytes by adjuvant signals including, conceivably, recurrent exposures to exogenous pathogens. This could occur via nonspecific polyclonal activation of T- and B-cells by bacterial or viral antigens or, alternatively, as a consequence of structural homology between a self-protein and a protein in the pathogen, a process commonly referred to as molecular mimicry. It is notable, for example, that components of the myelin sheath share amino acid homologies with proteins of measles, influenza, herpes, papilloma, adenoviruses, and other viruses.[33] These pathogens acquired sufficient homology to engage myelin-specific T-cells with the potential for a misguided response. In addition, amino acid identity may not even be required for cross-reactivity to occur between the auto-antigen and the mimic, as long as they share chemical properties at critical residues that allow anchoring to antigen-presenting molecules and interaction with the T-cell antigen receptor. An alternative hypothesis proposes that activation of autoimmune cells occurs as a consequence of viral infection of CNS cells; such infections may be asymptomatic but cause cytopathic effects to target cells in the course of an antiviral response. The prolonged release of neural antigens may then induce inflammatory responses that eventually become self-perpetuating and pathologic.

Once activated, T-cells express surface molecules called integrins, which mediate binding to the specialized capillary endothelial cells of the blood–brain barrier.[34] One such integrin, VLA-4, binds the vascular cell adhesion molecule (VCAM) expressed in the capillary endothelial cells following induction by tumor necrosis factor-α (TNF-α) and interferon-γ during an inflammatory response. As the activated T-cells migrate across the blood–brain barrier to reach the CNS parenchyma, they express gelatinases (matrix metalloproteinases) responsible for lysis of the dense subendothelial basal lamina. Metalloproteinases may act not only as mediators of cell traffic across the blood–brain barrier, but they may also increase the inflammatory reaction through TNF processing. Furthermore, a direct neurotoxic effect for metalloproteinases has been proposed; microinjection of activated matrix metalloproteinases into the cortical white matter of experimental animals results in axonal injury, even in the absence of local inflammation.

A different group of molecules involved in leukocyte migration and extravasation comprises soluble chemoattractants called chemokines and their receptors.[34] Chemokines are members of an expanding family of small serum proteins of between 7 and 16 kDa in size primarily involved in selective trafficking and migration of leukocytes to sites of infection and inflammation, leukocyte maturation in the bone marrow, tissue repair and vascularization, and hematopoiesis and renewal of circulating leukocytes. The spatial and temporal expression of chemokines correlates with disease activity in experimental autoimmune encephalomyelitis (EAE) and MS. In addition, chemokine receptors mediate entry of microorganisms into target cells and also participate in the viral-mediated induction of pro-inflammatory cytokines, both potential mediators of the encephalitogenic response.

After traversing the blood–brain barrier, pathogenic T-cells are believed to be reactivated by fragments of myelin antigens. This appears to be a two-step process.[35] Primed CD4 T-cells are first engaged by antigen-presenting cells in the perivascular space before moving into the parenchyma. Reactivation induces additional release of pro-inflammatory cytokines that stimulate microglia, further open the brain–blood barrier, and stimulate chemotaxis, resulting in additional waves of inflammatory cell recruitment and leakage of antibody and other plasma proteins into the nervous system. The activated microglial cells also contribute to the inflammatory milieu by secreting T-cell-activating factors such as interleukin-12 (IL-12), IL-23, osteopontin, and toxic mediators such as nitric oxide and oxygen radicals. Pathogenic T-cells may not be capable of producing or inducing tissue injury in the absence of the secondary leukocyte recruitment. For example, in EAE mediated by adoptive transfer of myelin-reactive encephalitogenic T-lymphocytes, these cells are among the first to infiltrate the CNS, but they constitute only a minor component of the total infiltrate in the full-blown lesion.

A large body of experimental data firmly establish that myelin-specific T-cells in MS patients are present in greater numbers than in healthy subjects, have lower thresholds of activation, and have different effector profiles. However, whereas the role of CD4 and CD8 T-cells as initiators and regulators of the CNS inflammatory response is well established, their role as direct effectors of myelin injury in MS remains uncertain. Potential T-cell–mediated mechanisms of myelin damage have been established *in vitro*; TNF-α kills myelinating cells in culture, antimyelin basic protein CD4 cells can display cytolytic functions, and CD8 cells induce cytoskeleton breaks in neurites. Axonal injury correlates better with the presence of CD8 T-cells and macrophages than it does with CD4 T-cells. It is clear, however, that the MS lesion is not exclusively T-cell mediated; rather, a synergistic cellular and humoral response is required to produce demyelination and axonal damage.

B-cell activation and antibody responses appear to be necessary for the full development of demyelination, both in humans and in experimentally induced disease. In most MS patients, an elevated level of immunoglobulins synthesized intrathecally can be detected in the cerebrospinal fluid. Although the specificity of these antibodies is mostly unknown, antimyelin basic protein specificities have been detected. Myelin-specific infiltrating B-cells have been detected in the brains of patients with MS, and in the cerebrospinal fluid and brains of affected individuals there is an elevated frequency of clonally expanded B-cells with properties of postgerminal center memory or antibody-forming lymphocytes.[36,37] Further, these B-cells are responsible for production of oligoclonal immunoglobulin in CSF.[38] The systemic administration of B-cell-depleting antibodies such as the anti-CD20 monoclonal antibodies rituximab and ocrelizumab have shown considerable promise for treating MS.[39,40] Based on recent data from animal models, it is likely that B-cells participate in MS, at least in part, by serving as antigen presenting cells for the activation of pathogenic T-cells.

Myelin-specific autoantibodies have been detected bound to the vesiculated myelin fragments, at least in some patients, and these antibodies are thought to promote demyelination.[32] Some bound antibodies appear to be directed against the CNS specific myelin protein, myelin oligodendrocyte glycoprotein (MOG).[41] More recently, antibodies against a different molecule, the potassium channel Kir4.1 expressed in the CNS, have been identified in some MS patients.[42] However, despite multiple lines of indirect evidence, convincing proof of an autoimmune etiology for MS is lacking because a specific antigen has not been convincingly demonstrated, and transfer of MS with antibodies has not been accomplished.

A third class of cells, the resident microglia, lying within the parenchyma, also become activated as a result of locally released cytokines.[43] Microglia act as scavengers that remove debris, and as antigen-presenting cells that present processed antigens to T-cells, contributing to their local clonal expansion. Mutual interactions between T-cells and macrophages induce proliferation of both cell types through mediation of such molecules as interleukin 2 (IL-2) and colony-stimulating factors. Furthermore, endothelia and T-cells provide colony-stimulating factors that maintain macrophage activation and prevent apoptosis and cell death. Microglia are also likely to directly induce myelin damage and killing of oligodendroglial cells through the release of mediators such as free radicals (nitric oxide and superoxide anion), glutamate, vasoactive amines, complement, proteases, cytokines (IL-1, TNF-α), and eicosanoids. Other factors released by cells of the innate immune system may also interact with the adaptive arm in previously unrecognized ways,[44] or even promote recovery.[45]

EPIDEMIOLOGY

Geographical gradients have been repeatedly observed in MS, with prevalence rates decreasing with decreasing latitude. The highest known prevalence for MS (250 per 100,000) occurs in the Orkney Islands, located north of Scotland, and similarly high rates are found throughout northern Europe, the northern United States, and Canada. In contrast, the prevalence is low in Japan (6 per 100,000), in other parts of Asia, in equatorial Africa, and in the Middle East. One proposed explanation for the latitude effect on MS is that there is a protective effect of sun exposure.[46] Ultraviolet radiation from sun exposure is the most important source of vitamin D in most individuals, and low levels of vitamin D are common at high latitudes where sun exposure may be low, particularly during the winter months. Prospective studies have confirmed that relative vitamin D deficiency is associated with an increase in the subsequent risk for MS. Immunoregulatory effects of vitamin D could explain this possible relationship. The prevalence of MS also appears to have steadily increased over the past century; furthermore, this increase has apparently occurred primarily in women. Recent epidemiologic data suggests that the latitude effect on MS may currently be decreasing, for unknown reasons.

Migration studies and identification of possible point epidemics provide further support for an effect of the environment on MS risk. Migration studies suggest that some MS-related exposure appears to occur in childhood, years before MS is clinically evident. In some studies, migration early in life from a low- to a high-risk area was found to increase MS risk, and, conversely, migration from a high- to a low-risk area decreased risk. With respect to possible point epidemics, the most convincing example occurred in the Faroe Islands north of Denmark after the British occupation during World War II.

MS risk also correlates with high socioeconomic status, which may reflect improved sanitation and delayed initial exposures to infectious agents.[47,48] By analogy, some viral infections (e.g., poliomyelitis and measles viruses) produce neurologic sequelae more frequently when the age of initial infection is delayed. High antibody titers against many viruses have been reported in serum and cerebrospinal fluid of MS patients, including measles, herpes simplex, varicella, rubella, Epstein–Barr, influenza C, and some parainfluenza strains. Numerous viruses and bacteria (or their genomic sequences) have been recovered from tissues and fluids of MS patients. Occasionally, reports seem to implicate a specific infectious agent such as human herpes virus type 6 (HHV-6) or *Chlamydia pneumoniae*, although, in general, the available reports have been quite inconsistent.

Most intriguing, the evidence of a remote Epstein–Barr virus (EBV) infection playing some role in MS is supported by a number of epidemiologic and laboratory studies. Infectious mononucleosis (associated with EBV infection relatively later in life) and higher antibody titers to the nuclear EB antigens are associated with MS, and, conversely, individuals never infected with EBV are at quite low MS risk. At this time, however, a causal role for EBV, or for any specific infectious agent in MS, remains uncertain.

The possibility that smoking increases MS risk was supported by recent data that cigarettes, but not chewing tobacco, were associated with MS risk,[49] and that this correlated with a genetic variant responsible for metabolism of inhaled smoke (L. Barcellos, unpublished observations). These results suggest that how the lung handles cigarette smoke could be important in MS, and correlate with experimental data indicating that the lung is a key site for activation of disease-inducing T-cells.[50]

In recent years, fundamental changes appear to have taken place in the epidemiology of MS; the disease has become more frequent, the female to male ratio has increased, and the latitude gradient has decreased. A recent observation that may relate to the increasing prevalence of MS is that a high salt diet, associated with consumption of processed foods, can activate pro-inflammatory immune responses mediated by Th17 T-lymphocytes.[51,52]

GENETIC BASIS OF MS

Classical Genetics

MS clusters with the so-called complex genetic diseases, a group of common disorders characterized by modest disease-risk heritability and multifaceted gene–environment interactions. The genetic component in MS is primarily suggested by familial aggregation of cases and the high incidence in some ethnic populations (particularly those of northern European origin) compared with others (African and Asian groups), regardless of geographic location.[46] Evidence of risk heritability in the form of familial recurrence has long been known.[53] The degree of familial aggregation can be determined by estimating the ratio of the prevalence in siblings (recurrence risk) over the disease prevalence in the population (λs). The most recent estimates of the λs for MS are between 7.1[54] and 16.8.[55] The lowest estimate comes from the analysis of 28,396 patients from the Swedish national registry and their relatives. Interestingly, despite a well-established lower prevalence of multiple sclerosis amongst males, the relative risks were equal among maternal and paternal relatives in this study. By analyzing recurrence rates in monozygotic and dizygotic twins and siblings, this study also reported the broad sense heritability of MS to be 0.64.[54] The higher estimate (λs = 16.8) comes from another recent article in which a meta-analysis of 18 twin and family studies was performed.[55] Half-sibling, adoptee,[56] and spouse[57] risk assessment studies performed in Canada appear to confirm that genetic and not environmental factors are primarily responsible for the familial clustering of cases. However, an intriguing association with month of birth was observed in the Canadian familial cases, reflecting perhaps an interaction between genes and an environmental factor operating during gestation or shortly after birth.[58] Concordant siblings tend to share age of symptom onset rather than year of onset, and second-and third-degree relatives of MS patients are also at an increased risk. Furthermore, twin studies from different populations consistently indicate pairwise concordance (20–40% in identical twin pairs compared to 2–5% in like-sex fraternal twin pairs), providing additional evidence for a genetic etiology in MS.[59,60] Twin concordance appears to exhibit, at least in the Canadian series, gender dimorphism.[61] Finally, parent-of-origin effects may also influence both disease susceptibility and outcome,[62–64] and concordance in families for early and late clinical features has been observed as well, suggesting that in addition to susceptibility, genes may influence disease severity or other aspects of the clinical phenotype.[65–67] For example, a significant difference in transmission from father to son compared with mother to son was found from a total of 13,923 father–child and 29,265 mother–child pairs from the Swedish national registry. This finding is in contrast with previous reports[64,68] and lends support to the presence of the Carter effect (preferential transmission by fathers) in MS.

Altogether, neither the recurrence familial rate nor the twin concordance supports the presence of a Mendelian trait. Modeling of the available data predicts that the MS-prone genotype results from multiple interacting polymorphic genes, each exerting a small or at most a moderate effect on the overall risk. The incomplete penetrance and moderate individual effect of such genes probably reflect epistatic interactions and postgenomic events; these may include genes that rearrange somatically to encode a vast variety of immune receptors, posttranscriptional regulatory mechanisms, and incorporation of retroviral sequences. An additional layer of difficulty is encountered when genetic heterogeneity is considered, whereby specific genes or alleles influence susceptibility and pathogenesis in some affected individuals but not in others.

The Major Histocompatibility Complex

The *HLA-DRB1* gene is the strongest genetic susceptibility factor yet identified in MS. The HLA (human leukocyte antigen) genes are located in the 3.5-Mb major histocompatibility complex (MHC) region on the short arm of chromosome 6 (6p21). HLA genes encode highly polymorphic cell surface glycoproteins, several of which play a crucial role in the self- and nonself-antigen recognition by the immune system. The earliest associations between MS and HLA were described for the class I alleles A3[69] and B7[70] in the 1970s using serological based measurements. The association with HLA-A3 was reported to be secondary to HLA-B7, which in turn was shown to be secondary to a suggested primary association with class II antigens HLA-DR2 and -DQw6.[71–73] It was originally suggested that this indirect association of HLA class I genes was due to linkage disequilibrium (LD) within the HLA region, a puzzle that would take several more decades to solve. With the advent of new technologies, the main susceptibility region was finally narrowed down to the haplotype DRB5*01:01, DRB1*15:01, and DQB1*06:02, that encode β-chains for the HLA class II molecules DR2a, DR2b and DQ6, respectively. In all ethnic groups previously studied, DRB1*15:01 and DRB5*01:01 were almost inseparable due to LD. However, this effect is less pronounced in some African populations due to their shorter haplotype block structure, thus resulting in the identification of African American MS subjects carrying the DRB1*15:01/DRB5*01:01 part of the haplotype but not the DQB1*06:01/02. These association studies thus indicate that DRB5*01:01 and DRB1*15:01 are the primary MS susceptibility genes independent of DQB1*06:02.[74] Notably, the *HLA-DRB5* gene, located at the telomeric boundary of the HLA class II region, and only expressed by HLA-DRB1*15 haplotypes is reported to attenuate MS severity.[75] Interestingly, HLA-DRB1*15:01 has a low frequency even among healthy individuals in the indigenous Sami (natives of the northern part of the Scandinavian countries) which coincides with a low MS prevalence, suggesting a general reduced genetic risk for MS in this population.[76,77] The HLA-DRB1*15:01 haplotype is carried by 28–33% of northern Caucasian MS patients compared to 9–15% of healthy controls, corresponding to an average OR of 3.08,[78] making this the single strongest susceptibility locus in MS genome-wide.

A high degree of allelic heterogeneity of the HLA-DRB1 locus exists in Caucasians, showing that variation in this region influences MS susceptibility in a complex manner and highlighting the need for in-depth follow-up studies that help refine these categories. For example, a dominant HLA-DRB1*15 dose effect has been identified as well as a modest recessive dose effect for the HLA-DRB1*03 allele (i.e., two copies of either of these susceptibility haplotypes further increases disease risk).[79] A large genome-wide screening suggested HLA-DRB1*13:03, HLA-DRB1*01:08, and HLA-DRB1*03:01 alleles as also being associated with MS.[78] In the Sardinia region of Italy, where MS prevalence is high, HLA-DRB1*04, HLA-DRB1*03:01, and HLA-DRB1*13:01 (in addition to HLA-DRB1*15:01) have been associated with MS.[80]

Whole-Genome Mapping

A fine-mapping effort using data from genome-wide studies and advanced statistical methods redefined the contribution of the HLA region to MS susceptibility to a total of 11 statistically independent effects. In addition to the well-known DRB1*15:01, significant effects were found in other 5 DRB1 genes (*03:01, *13:03, *04:04, *04:01, and *14:01). In addition, effects in the HLA-DPB1, HLA-A*02:01, HLA-B*37:01, HLA-B*38:01, and the region between MICB and LST1 (which contains several genes including *TNF*, *LTA* and *LTB*) were also identified.[81] These discoveries highlight the complexity of this region and how new laboratory techniques, novel methods of analysis, and well-powered datasets can contribute to refining the genetic map of MS.

Indeed, recent advances in technology, including the development of novel chemistries, miniaturization, and automation made possible the effective screening of thousands of single nucleotide polymorphisms (SNPs) in thousands of samples at an affordable price. These technologies spurred a series of genome-wide association studies (GWAS), capable of detecting susceptibility genes of modest effects in multifactorial, complex genetic disorders such as MS.

In 2007 a GWAS identified the first non-HLA regions with genome-wide significance ($p<10–8$) using a two-step approach.[82] After screening of 931 family trio samples using 334,923 SNPs and replication in 609 family trios, 2322 case subjects and 789 control subjects, the genes encoding the IL7Rα and IL2Rα were found to be significantly associated with relatively modest odds ratios (OR<1.35). Simultaneously, the *IL7Rα* gene was confirmed to be associated in other MS[83,84] cohorts. In the years that followed, a new series of GWAS and meta-analyses were performed in different MS cohorts, steadily adding more regions to the list of confirmed MS associated loci.

However, a low replication rate of variants identified in earlier studies suggested that larger studies were needed to identify new susceptibility alleles. Some power estimates indicated that to reach the genome-wide significance level for risk loci with OR as modest as 1.2 or less (as expected from earlier studies), the sample size should ideally include approximately 10,000 cases and controls.[85] To achieve this number of MS samples, the International MS Genetic Consortium (IMSGC) embarked on an unprecedented effort and in, collaboration with the Wellcome Trust Case Control Consortium 2 (WTCCC2),[78] performed a GWAS with 9772 MS cases and 17,376 controls. In all, this genomic screen identified 52 MS risk loci with genome-wide significance, of which 29 were novel at the time. As could be expected from previous studies, all loci (except HLA) showed OR in the range 1.1–1.2. Remarkably, the vast majority of these confirmed MS associated loci are located close or inside genes encoding immune related molecules, strongly supporting the hypothesis that MS is primarily an immune mediated disease.

These recent studies also underscored that more than one SNP may be needed to account for the risk of the same region. As an example, two of the associated SNPs (rs3118470 and rs7090512) are located 9 kb from one another in the *IL2RA* gene and independently contribute to MS risk. Allelic heterogeneity has been previously described at this locus, in which the same allele that predisposes to MS protects from type 1 diabetes.[86] Of interest, one-third of the identified loci were reportedly associated with at least one other autoimmune disease, strengthening the notion that common disease mechanism(s) may underlie most, if not all autoimmune diseases.[87,88]

In an effort to further fine-map and replicate previous candidate regions, the IMSGC conducted an association study in samples from 14,498 MS patients and 24,091 healthy controls with 161,311 SNPs mostly located in regions of immunological relevance (the "immunoChip").[89] This latest effort identified yet another 48 novel associations, bringing the total number of susceptibility variants to 128 (Figure 85.2). It is estimated that all these variants combined may only explain <30% of the sibling recurrence risk; thus, a large portion of the heritable component of MS is still unidentified. Subsequent efforts are currently underway by the IMSGC to perform a final meta-analysis of all previous studies, with an appropriate replication phase to provide the highest quality MS susceptibility map possible to obtain. It is anticipated this analysis will identify additional susceptibility regions, although with ever-smaller effects.

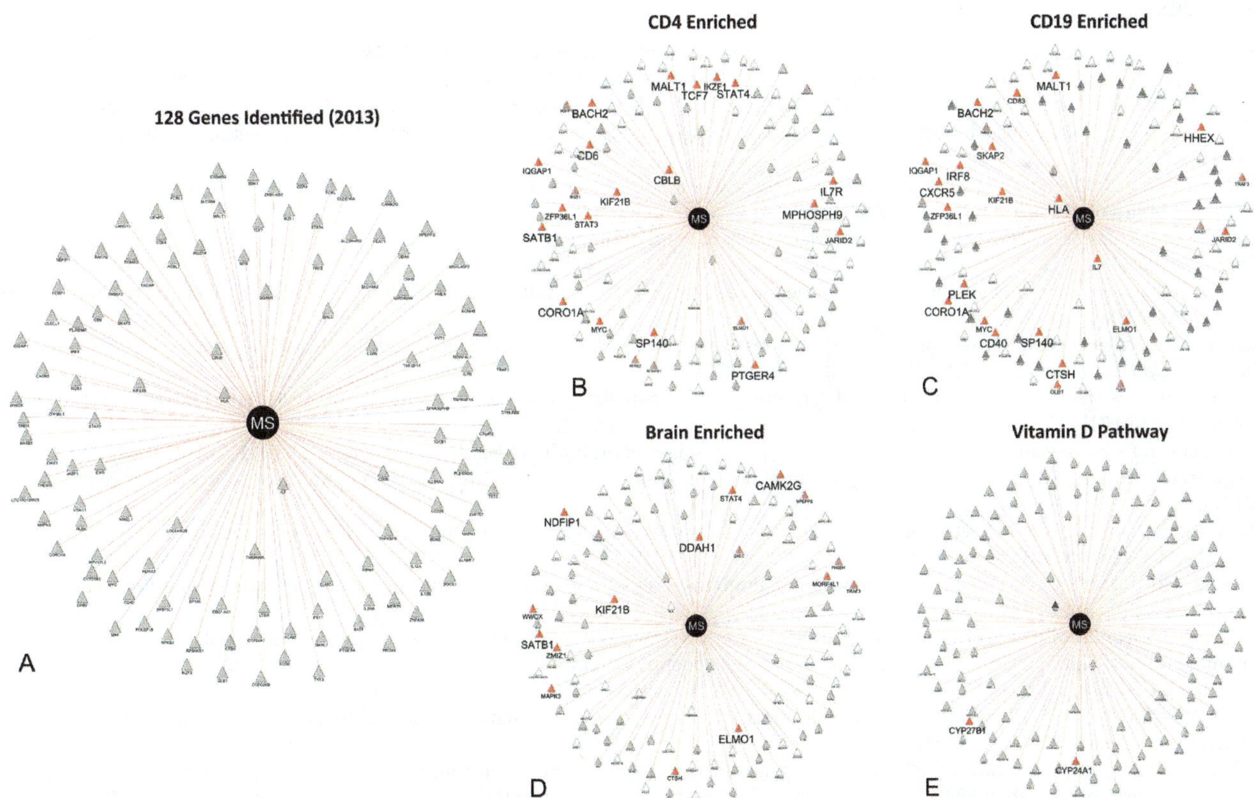

FIGURE 85.2 **The genetic landscape of multiple sclerosis.** A total of 128 multiple sclerosis (MS) risk variants have been identified to date (**A**); extragenic variants are identified here as the closest nearby gene in the vicinity. The distance from the center approximates the odds ratio of each variant on MS risk. Those genes with a known function or expression in CD4 T-cells (**B**), CD19 B-cells (**C**), brain (**D**) and vitamin D pathway (**E**) are shown.

Given the scale of the above gene mapping efforts in MS, it is unlikely that additional susceptibility regions with large effects will be found by exploring common DNA variation. Emerging studies aimed at unraveling a larger portion of the heritability in MS include those targeting rare variation by whole-genome or exome sequencing, gene–gene interactions, and the effect of epigenomics.

CONCLUSIONS

Compelling epidemiologic and molecular data indicate that genes play a critical role in determining who is at risk for developing MS. The genetic component of MS etiology is believed to result from the action of common allelic variants in several genes and a yet unidentified contribution of rare variants and complex network of molecular and functional interactions. Some loci may be involved in the initial pathogenic events, while others could influence the development and progression of the disease. The incomplete penetrance and moderate individual effect of these genes probably reflect interactions with other genes, posttranscriptional regulatory mechanisms, and significant environmental and epigenetic influences. Equally significant, it is likely that genetic heterogeneity exists, meaning that specific genes influence susceptibility and pathogenesis in some individuals with MS but not others.

The past few years have seen real progress in the development of laboratory and analytical approaches to study non-Mendelian complex genetic disorders and in defining the pathological basis of demyelination, setting the stage for the final characterization of the genes involved in MS susceptibility and pathogenesis. The careful and methodic collection of data could lead to the identification of coregulated genes and characterization of genetic networks that underlie specific cellular processes. This complex organization is what ultimately defines biological function and, therefore, the disease phenotype. With the aid of high-capacity technologies, the combined and integrative analysis of genomic, transcriptional, proteomic, and extensive phenotypic information, including environmental exposure data, will define a useful conceptual model of pathogenesis and a framework for understanding the mechanisms of action of existing therapies for this disorder, as well as the rationale for novel curative strategies.

References

1. Hauser SL, Chan JR, Oksenberg JR. Multiple sclerosis: prospects and promise. *Ann Neurol*. 2013;74(3):317–327.
2. Hauser S, Josephson SA. Harrison's Neurology in Clinical Medicine. 3rd ed. New York, NY: McGraw-Hill Education; 2013.
3. Bruck W, Gold R, Lund BT, et al. Therapeutic decisions in multiple sclerosis: moving beyond efficacy. *JAMA Neurol*. 2013;70(10):1315–1324.
4. Trojano M, Pellegrini F, Paolicelli D, et al. Real-life impact of early interferon beta therapy in relapsing multiple sclerosis. *Ann Neurol*. 2009;66(4):513–520.
5. Shirani A, Zhao Y, Karim ME, et al. Association between use of interferon beta and progression of disability in patients with relapsing-remitting multiple sclerosis. *JAMA*. 2012;308(3):247–256.
6. Greenberg BM, Balcer L, Calabresi PA, et al. Interferon beta use and disability prevention in relapsing-remitting multiple sclerosis. *JAMA Neurol*. 2013;70(2):248–251.
7. Coles AJ, Cox A, Le Page E, et al. The window of therapeutic opportunity in multiple sclerosis: evidence from monoclonal antibody therapy. *J Neurol*. 2006;253(1):98–108.
8. Rommer PS, Stuve O. Management of secondary progressive multiple sclerosis: prophylactic treatment-past, present, and future aspects. *Curr Treat Options Neurol*. 2013;15(3):241–258.
9. Castro-Borrero W, Graves D, Frohman TC, et al. Current and emerging therapies in multiple sclerosis: a systematic review. *Ther Adv Neurol Disord*. 2012;5(4):205–220.
10. Lassmann H. Pathology and disease mechanisms in different stages of multiple sclerosis. *J Neurol Sci*. 2013;333(1–2):1–4.
11. Magliozzi R, Howell OW, Reeves C, et al. A gradient of neuronal loss and meningeal inflammation in multiple sclerosis. *Ann Neurol*. 2010;68(4):477–493.
12. Magliozzi R, Howell O, Vora A, et al. Meningeal B-cell follicles in secondary progressive multiple sclerosis associate with early onset of disease and severe cortical pathology. *Brain*. 2007;130(Pt 4):1089–1104.
13. Choi SR, Howell OW, Carassiti D, et al. Meningeal inflammation plays a role in the pathology of primary progressive multiple sclerosis. *Brain*. 2012;135(Pt 10):2925–2937.
14. Prineas JW, Kwon EE, Cho ES, et al. Immunopathology of secondary-progressive multiple sclerosis. *Ann Neurol*. 2001;50(5):646–657.
15. Lassmann H, van Horssen J. The molecular basis of neurodegeneration in multiple sclerosis. *FEBS Lett*. 2011;585(23):3715–3723.
16. Lucchinetti CF, Parisi J, Bruck W. The pathology of multiple sclerosis. *Neurol Clin*. 2005;23(1):77–105, vi.
17. Bitsch A, Schuchardt J, Bunkowski S, Kuhlmann T, Bruck W. Acute axonal injury in multiple sclerosis. Correlation with demyelination and inflammation. *Brain*. 2000;123(Pt 6):1174–1183.
18. Ferguson B, Matyszak MK, Esiri MM, Perry VH. Axonal damage in acute multiple sclerosis lesions. *Brain*. 1997;120(Pt 3):393–399.
19. Trapp BD, Peterson J, Ranshoff RM, Rudick R, Mork S, Bo L. Axonal transection in the lesions of multiple sclerosis. *N Engl J Med*. 1998;338:278–285.
20. Neumann H. Molecular mechanisms of axonal damage in inflammatory central nervous system diseases. *Curr Opin Neurol*. 2003;16(3):267–273.

21. Bjartmar C, Kidd G, Mork S, Rudick R, Trapp BD. Neurological disability correlates with spinal cord axonal loss and reduced N-acetyl aspartate in chronic multiple sclerosis patients. *Ann Neurol*. 2000;48(6):893–901.
22. Miller DH, Barkhof F, Frank JA, Parker GJ, Thompson AJ. Measurement of atrophy in multiple sclerosis: pathological basis, methodological aspects and clinical relevance. *Brain*. 2002;125(Pt 8):1676–1695.
23. Bagnato F, Butman JA, Gupta S, et al. *In vivo* detection of cortical plaques by MR imaging in patients with multiple sclerosis. *AJNR Am J Neuroradiol*. 2006;27(10):2161–2167.
24. Gadea M, Martinez-Bisbal MC, Marti-Bonmati L, et al. Spectroscopic axonal damage of the right locus coeruleus relates to selective attention impairment in early stage relapsing-remitting multiple sclerosis. *Brain*. 2004;127(Pt 1):89–98.
25. Pagani E, Filippi M, Rocca MA, Horsfield MA. A method for obtaining tract-specific diffusion tensor MRI measurements in the presence of disease: application to patients with clinically isolated syndromes suggestive of multiple sclerosis. *Neuroimage*. 2005;26(1):258–265.
26. Cader S, Cifelli A, Abu-Omar Y, Palace J, Matthews PM. Reduced brain functional reserve and altered functional connectivity in patients with multiple sclerosis. *Brain*. 2006;129(Pt 2):527–537.
27. Ouardouz M, Coderre E, Basak A, et al. Glutamate receptors on myelinated spinal cord axons: I. GluR6 kainate receptors. *Ann Neurol*. 2009;65(2):151–159.
28. Ouardouz M, Coderre E, Zamponi GW, et al. Glutamate receptors on myelinated spinal cord axons: II. AMPA and GluR5 receptors. *Ann Neurol*. 2009;65(2):160–166.
29. Mahad D, Ziabreva I, Lassmann H, Turnbull D. Mitochondrial defects in acute multiple sclerosis lesions. *Brain*. 2008;131(Pt 7):1722–1735.
30. Hametner S, Wimmer I, Haider L, Pfeifenbring S, Bruck W, Lassmann H. Iron and neurodegeneration in the multiple sclerosis brain. *Ann Neurol*. 2013;74(6):848–861.
31. Campbell GR, Ziabreva I, Reeve AK, et al. Mitochondrial DNA deletions and neurodegeneration in multiple sclerosis. *Ann Neurol*. 2011;69(3):481–492.
32. Nylander A, Hafler DA. Multiple sclerosis. *J Clin Invest*. 2012;122(4):1180–1188.
33. Lang HL, Jacobsen H, Ikemizu S, et al. A functional and structural basis for TCR cross-reactivity in multiple sclerosis. *Nat Immunol*. 2002;3(10):940–943.
34. Holman DW, Klein RS, Ransohoff RM. The blood-brain barrier, chemokines and multiple sclerosis. *Biochim Biophys Acta*. 2011;1812(2):220–230.
35. Platten M, Steinman L. Multiple sclerosis: trapped in deadly glue. *Nat Med*. 2005;11(3):252–253.
36. von Budingen HC, Bar-Or A, Zamvil SS. B cells in multiple sclerosis: connecting the dots. *Curr Opin Immunol*. 2011;23(6):713–720.
37. von Budingen HC, Kuo TC, Sirota M, et al. B cell exchange across the blood-brain barrier in multiple sclerosis. *J Clin Invest*. 2012;122(12):4533–4543.
38. Obermeier B, Mentele R, Malotka J, et al. Matching of oligoclonal immunoglobulin transcriptomes and proteomes of cerebrospinal fluid in multiple sclerosis. *Nat Med*. 2008;14(6):688–693.
39. Hauser SL, Waubant E, Arnold DL, et al. B-cell depletion with rituximab in relapsing-remitting multiple sclerosis. *N Engl J Med*. 2008;358(7):676–688.
40. Kappos L, Li D, Calabresi PA, et al. Ocrelizumab in relapsing-remitting multiple sclerosis: a phase 2, randomised, placebo-controlled, multicentre trial. *Lancet*. 2011;378(9805):1779–1787.
41. Genain CP, Cannella B, Hauser SL, Raine CS. Identification of autoantibodies associated with myelin damage in multiple sclerosis. *Nat Med*. 1999;5(2):170–175.
42. Srivastava R, Aslam M, Kalluri SR, et al. Potassium channel KIR4.1 as an immune target in multiple sclerosis. *N Engl J Med*. 2012;367(2):115–123.
43. Mayo L, Quintana FJ, Weiner HL. The innate immune system in demyelinating disease. *Immunol Rev*. 2012;248(1):170–187.
44. Davidson MG, Alonso MN, Yuan R, et al. Th17 cells induce Th1-polarizing monocyte-derived dendritic cells. *J Immunol*. 2013;191(3):1175–1187.
45. Miron VE, Boyd A, Zhao JW, et al. M2 microglia and macrophages drive oligodendrocyte differentiation during CNS remyelination. *Nat Neurosci*. 2013;16(9):1211–1218.
46. Islam T, Gauderman WJ, Cozen W, Hamilton AS, Burnett ME, Mack TM. Differential twin concordance for multiple sclerosis by latitude of birthplace. *Ann Neurol*. 2006;60(1):56–64.
47. Ascherio A, Munger KL. Environmental risk factors for multiple sclerosis. Part I: the role of infection. *Ann Neurol*. 2007;61(4):288–299.
48. Ascherio A, Munger KL. Environmental risk factors for multiple sclerosis. Part II: noninfectious factors. *Ann Neurol*. 2007;61(6):504–513.
49. Hedstrom AK, Baarnhielm M, Olsson T, Alfredsson L. Exposure to environmental tobacco smoke is associated with increased risk for multiple sclerosis. *Mult Scler*. 2011;17(7):788–793.
50. Odoardi F, Sie C, Streyl K, et al. T cells become licensed in the lung to enter the central nervous system. *Nature*. 2012;488(7413):675–679.
51. Wu C, Yosef N, Thalhamer T, et al. Induction of pathogenic TH17 cells by inducible salt-sensing kinase SGK1. *Nature*. 2013;496(7446):513–517.
52. Kleinewietfeld M, Manzel A, Titze J, et al. Sodium chloride drives autoimmune disease by the induction of pathogenic TH17 cells. *Nature*. 2013;496(7446):518–522.
53. Robertson NP, Fraser M, Deans J, Clayton D, Walker N, Compston DA. Age-adjusted recurrence risks for relatives of patients with multiple sclerosis. *Brain*. 1996;119(Pt 2):449–455.
54. Westerlind H, Ramanujam R, Uvehag D, et al. Modest familial risks for multiple sclerosis: a registry-based study of the population of Sweden. *Brain*. 2014;137(Pt 3):770–778.
55. O'Gorman C, Lin R, Stankovich J, Broadley SA. Modelling genetic susceptibility to multiple sclerosis with family data. *Neuroepidemiology*. 2013;40(1):1–12.
56. Ebers GC, Sadovnick AD, Risch NJ. A genetic basis for familial aggregation in multiple sclerosis. Canadian Collaborative Study Group. *Nature*. 1995;377(6545):150–151.
57. Ebers GC, Yee IM, Sadovnick AD, Duquette P. Conjugal multiple sclerosis: population-based prevalence and recurrence risks in offspring. Canadian Collaborative Study Group. *Ann Neurol*. 2000;48(6):927–931.
58. Willer CJ, Dyment DA, Sadovnick AD, Rothwell PM, Murray TJ, Ebers GC. Timing of birth and risk of multiple sclerosis: population based study. *BMJ*. 2005;330(7483):120.

59. Sadovnick AD, Armstrong H, Rice GP, et al. A population-based study of multiple sclerosis in twins: update. *Ann Neurol*. 1993;33(3):281–285.
60. Mumford CJ, Wood NW, Kellar-Wood H, Thorpe JW, Miller DH, Compston DA. The British Isles survey of multiple sclerosis in twins. *Neurology*. 1994;44(1):11–15.
61. Willer CJ, Dyment DA, Risch NJ, Sadovnick AD, Ebers GC. Twin concordance and sibling recurrence rates in multiple sclerosis. *Proc Natl Acad Sci U S A*. 2003;100(22):12877–12882.
62. Hupperts R, Broadley S, Mander A, Clayton D, Compston DA, Robertson NP. Patterns of disease in concordant parent-child pairs with multiple sclerosis. *Neurology*. 2001;57(2):290–295.
63. Ebers GC, Sadovnick AD, Dyment DA, Yee IM, Willer CJ, Risch N. Parent-of-origin effect in multiple sclerosis: observations in half-siblings. *Lancet*. 2004;363(9423):1773–1774.
64. Kantarci OH, Barcellos LF, Atkinson EJ, et al. Men transmit MS more often to their children vs women: the Carter effect. *Neurology*. 2006;67(2):305–310.
65. Brassat D, Azais-Vuillemin C, Yaouanq J, et al. Familial factors influence disability in MS multiplex families. French Multiple Sclerosis Genetics Group. *Neurology*. 1999;52(8):1632–1636.
66. Barcellos LF, Oksenberg JR, Green AJ, et al. Genetic basis for clinical expression in multiple sclerosis. *Brain*. 2002;125(Pt 1):150–158.
67. Kantarci OH, de Andrade M, Weinshenker BG. Identifying disease modifying genes in multiple sclerosis. *J Neuroimmunol*. 2002;123(1–2):144–159.
68. Herrera BM, Ramagopalan SV, Orton S, et al. Parental transmission of MS in a population-based Canadian cohort. *Neurology*. 2007;69(12):1208–1212.
69. Naito S, Namerow N, Mickey MR, Terasaki PI. Multiple sclerosis: association with HL-A3. *Tissue Antigens*. 1972;2(1):1–4.
70. Jersild C, Svejgaard A, Fog T. HL-A antigens and multiple sclerosis. *Lancet*. 1972;1(7762):1240–1241.
71. Jersild C, Fog T, Hansen GS, Thomsen M, Svejgaard A, Dupont B. Histocompatibility determinants in multiple sclerosis, with special reference to clinical course. *Lancet*. 1973;2(7840):1221–1225.
72. Bertrams HJ, Kuwert EK. Association of histocompatibility haplotype HLA-A3-B7 with multiple sclerosis. *J Immunol*. 1976;117(5 Pt.2):1906–1912.
73. Compston DA, Batchelor JR, McDonald WI. B-lymphocyte alloantigens associated with multiple sclerosis. *Lancet*. 1976;2(7998):1261–1265.
74. Oksenberg JR, Barcellos LF, Cree BA, et al. Mapping multiple sclerosis susceptibility to the HLA-DR locus in African Americans. *Am J Hum Genet*. 2004;74(1):160–167.
75. Caillier SJ, Briggs F, Cree BA, et al. Uncoupling the roles of HLA-DRB1 and HLA-DRB5 genes in multiple sclerosis. *J Immunol*. 2008;181(8):5473–5480.
76. Harbo HF, Riccio ME, Lorentzen AR, et al. Norwegian Sami differs significantly from other Norwegians according to their HLA profile. *Tissue Antigens*. 2010;75(3):207–217.
77. Harbo HF, Utsi E, Lorentzen AR, et al. Low frequency of the disease-associated DRB1*15-DQB1*06 haplotype may contribute to the low prevalence of multiple sclerosis in Sami. *Tissue Antigens*. 2007;69(4):299–304.
78. International Multiple Sclerosis Genetics Consortium, 2 WTCCC. Genetic risk and a primary role for cell-mediated immune mechanisms in multiple sclerosis. *Nature*. 2011;10(476):214–219.
79. Barcellos LF, Sawcer S, Ramsay PP, et al. Heterogeneity at the HLA-DRB1 locus and risk for multiple sclerosis. *Hum Mol Genet*. 2006;15(18):2813–2824.
80. Coraddu F, Sawcer S, D'Alfonso S, et al. A genome screen for multiple sclerosis in Sardinian multiplex families. *Eur J Hum Genet*. 2001;9(8):621–626.
81. Patsopoulos NA, Barcellos LF, Hintzen RQ, et al. Fine-mapping the genetic association of the major histocompatibility complex in multiple sclerosis: HLA and non-HLA effects. *PLoS Genet*. 2013;9(11):e1003926.
82. Hafler DA, Compston A, Sawcer S, et al. Risk alleles for multiple sclerosis identified by a genomewide study. *N Engl J Med*. 2007;357(9):851–862.
83. Gregory SG, Schmidt S, Seth P, et al. Interleukin 7 receptor alpha chain (IL7R) shows allelic and functional association with multiple sclerosis. *Nat Genet*. 2007;39(9):1083–1091.
84. Lundmark F, Duvefelt K, Iacobaeus E, et al. Variation in interleukin 7 receptor alpha chain (IL7R) influences risk of multiple sclerosis. *Nat Genet*. 2007;39(9):1108–1113.
85. Sawcer S. Bayes factors in complex genetics. *Eur J Hum Genet*. 2010;18(7):746–750.
86. Maier LM, Lowe CE, Cooper J, et al. IL2RA genetic heterogeneity in multiple sclerosis and type 1 diabetes susceptibility and soluble interleukin-2 receptor production. *PLoS Genet*. 2009;5(1):e1000322.
87. Baranzini SE. The genetics of autoimmune diseases: a networked perspective. *Curr Opin Immunol*. 2009;21(6):596–605.
88. Cotsapas C, Voight BF, Rossin E, et al. Pervasive sharing of genetic effects in autoimmune disease. *PLoS Genet*. 2011;7(8):e1002254.
89. Beecham AH, Patsopoulos NA, Xifara DK, et al. Analysis of immune-related loci identifies 48 new susceptibility variants for multiple sclerosis. *Nat Genet*. 2013;45(11):1353–1360.

CHAPTER

86

Vanishing White Matter Disease

Orna Elroy-Stein and Raphael Schiffmann†*

*Tel Aviv University, Tel Aviv, Israel

†Baylor Research Institute, Dallas, TX, USA

CLINICAL FEATURES

Historical Overview

Disease Identification

Childhood ataxia with central nervous system (CNS) hypomyelination (CACH) was first identified in 1992.[1] Similar patients were later described as suffering from vanishing white matter disease (VWM) or myelinopathia centralis diffusa.[2–4] The initial disease identification was based on the most common early childhood-onset variety but, as often happens, the identification of the genetic defect of CACH/VWM led to the recognition of a wider clinical spectrum.[5,6]

Gene Identification

The etiology of this genetically heterogeneous disease was only found thanks to astute use of molecular genetics in the population of a limited geographical region.[7–9] Among the 120 mutations that have been described until 2009,[10] 83% (99 different mutations) are missense mutations that cause a single amino acid to be altered to another amino acid, and the remainder is a mixture of premature nonsense mutations, some causing a frameshift and others that alter splicing. All affected individuals have two altered copies of a single *EIF2B* gene (autosomal recessive inheritance).

Most of the mutations reside in *EIF2B5*, the gene encoding the catalytic subunit. Genetic screens enabled the tracking of founder effects: two mutations originated in Dutch populations,[11] whereas the mutation causing the severe "Cree" encephalopathy originated in indigenous people in Manitoba and Québec, Canada.[12] Other novel mutations were found in Poland,[13] China[14] and Japan.[15]

Mode of Inheritance and Prevalence

CACH/VWM is an autosomal recessive disorder. The prevalence of CACH/VWM is not known; it is considered one of the most common leukodystrophies. In a study of unclassified leukodystrophies in childhood, CACH/VWM was the most common.[16] In some countries, the incidence of CACH/VWM is close to that of metachromatic leukodystrophy (see Chapter 29).[17]

Natural History

Age of Onset

Phenotypes range from a congenital or early infantile form to a subacute infantile form (onset age <1 year), an early-childhood onset form (onset age 1–5 years), a late-childhood/juvenile onset form (onset age 5–15 years), and an adult-onset form.[18,19] Both the childhood and juvenile forms have been observed in sibs;[8] the infantile and juvenile/adult forms have never been observed within the same family. The rate of progression of the disease is affected by the presence of certain mutations.[20,21]

Rosenberg's Molecular and Genetic Basis of Neurological and Psychiatric Disease
http://dx.doi.org/10.1016/10.1016/B978-0-12-410529-4.00086-3

Disease Evolution

The neurologic signs include ataxia and spasticity, but optic atrophy is usually absent. In the early-onset forms, encephalopathy is severe, seizures are often a predominant clinical feature, and decline is rapid and followed quickly by death; in the later-onset forms, decline is usually slower and milder.[9] The progressive neurological motor decline is associated with the development of cystic cavitation in the white matter seen on T1-weighted images and particularly well on fluid-attenuated inversion recovery (FLAIR) magnetic resonance images (MRI) that reflect neuronal loss.[19,22] Chronic progressive decline can be exacerbated by rapid deterioration during febrile illness or following minor head trauma or fright.[19,23]

Disease Variants

While the juvenile and adult forms are often associated with primary or secondary ovarian failure, a syndrome referred to as "ovarioleukodystrophy,"[24,25] includes ovarian dysgenesis that may occur in any of the forms regardless of age of onset, as has been found at autopsy in infantile and childhood cases.[6] Because the affected individuals were prepubertal, the ovarian dysgenesis was not manifested clinically. Another infantile-onset phenotype was described as "Cree leukoencephalopathy" because of its occurrence in the native North American Cree and Chippewayan indigenous populations.[26,27] Infants typically have hypotonia followed by the sudden onset of seizures (age 3–6 months), spasticity, rapid breathing, vomiting (often with fever), developmental regression, blindness, lethargy, and cessation of head growth, with death by 2 years of age.

End of Life: Mechanisms and Comorbidities

Death typically occurs secondary to aspiration or in the clinical picture of progressive inanition.

MOLECULAR GENETICS

Normal Gene Product

eIF2B complex is a master regulator of protein synthesis. It reactivates the G protein eIF2, which, in its GTP-bound form, binds initiator methionyl–tRNA (Met–$tRNA_i^{Met}$) to form GTP–eIF2/Met–$tRNA_i^{Met}$ ternary complex (TC). The TC is delivered to the 40S small ribosomal subunit before mRNA binding. This step is fundamental for translation initiation since AUG, the start codon of the vast majority of cellular mRNAs, is recognized by Met–$tRNA_i^{Met}$. Upon AUG recognition, eIF5, a GTPase-activating protein (GAP) and a GDP dissociation inhibitor (GDI), promotes the hydrolysis of GTP, followed by release of eIF2-GDP/eIF5 complex from the ribosome. eIF2B then reactivates eIF2 by two independent consecutive activities. First, as a GDI displacement factor (GDF), it recruits eIF2 from the eIF2•GDP/eIF5 complex. Second, as a GEF, it promotes the association of GTP to eIF2. The transitions of eIF2 between inactive (GDP-bound) and active (GTP-bound) states, are thus governed by eIF5 and eIF2B[39–41] (Figure 86.1B).

eIF2B is composed of five different subunits (α–ε), encoded by five genes (*EIF2B1–5* in humans) (Figure 86.1A). The largest subunit, eIF2Bε, is the catalytic one. Whereas the extreme *C*-terminal region of eIF2Bε is sufficient for low nucleotide exchange activity *in vitro*, formation of complete five-subunit complexes is required for full eIF2B activity.[28,29] eIF2Bγ and eIF2Bε heterodimerize to form a "γε catalytic subcomplex," which elicits guanine nucleotide exchange factor (GEF) activity by itself,[30] but eIF2Bβ is crucial for binding eIF2 *in vivo*.[31] eIF2Bγ and ε share a conserved *N*-terminal region, harboring sequence and tertiary structure similarity with each other and with sugar nucleotide pyrophosphorylase enzymes. This region is critical for eIF2B five-subunits complex formation.[32,33] eIF2B-αβδ subunits also share sequence similarity[34] and form a second "αβδ regulatory subcomplex." Importantly, the regulatory subcomplex mediates inhibition of eIF2B GEF activity when its eIF2 substrate is phosphorylated in its α-subunit.[35,36] Phosphorylated eIF2 (eIF2α-P) binds to the regulatory subcomplex with higher affinity than eIF2,[30] and acts as a competitive inhibitor of eIF2B, leading to inhibition of translation initiation[37] (Figure 86.1A). Recent mass spectrometry experiments revealed that eIF2B is actually decameric, composed of a dimer of eIF2B(βγδε) tetramers stabilized by two copies of eIF2Bα. The decamers bind eIF2 more efficiently and thus are more active.[38]

Regulation of eIF2B Enzymatic Activity

eIF2B activity is tightly regulated via several molecular mechanisms, which effectively match global translation initiation to varying physiological requirements. A major mechanism of eIF2B regulation under stress conditions is via phosphorylation of Ser^{51} in the *N*-terminal domain of the α subunit of eIF2. Phosphorylation increases the binding affinity of eIF2β to eIF2B, turning it from a substrate to a potent competitive inhibitor. Four different kinases

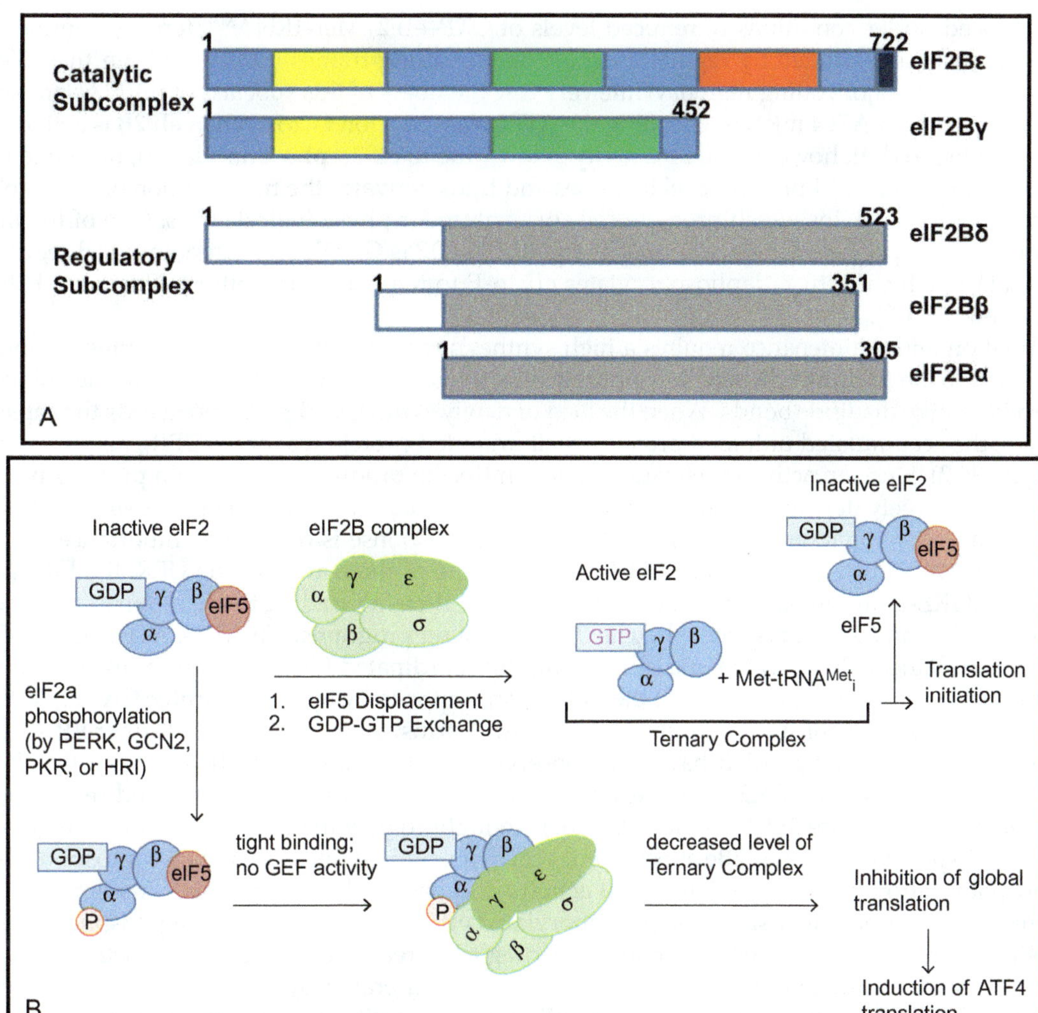

FIGURE 86.1 **Structure and function of eIF2B complex.** (**A**) Schematic illustration of the five subunits of eIF2B. Yellow, the regions of eIF2B ε and γ with similarity to nucleotidyl transferases; green, similarity to acyl transferases (the "I-patch regions"); red, the catalytic domain; black, aromatic and acidic residues-rich region; gray, the sequence similarity between the α, β and δ subunits. (**B**) Schematic presentation showing the function of eIF2B complex in promoting the exchange of GDP for GTP on eIF2, thereby generating active GTP-bound eIF2, which can bind the initiator tRNA Met–rRNA$^{Met}_{i}$ to form a ternary complex (TC). Upon start codon recognition following TC loading on the ribosome, eIF5 promotes GTP hydrolysis and inhibits GDP dissociation, thereby releasing the inactive GDP-bound form of eIF2. eIF2B displaces eIF5 and then exchanges the GDP with GTP, thereby regenerating the active GTP-bound form of eIF2. When eIF2 is phosphorylated on its α subunit (by PERK, GCN2, PKR, or HRI kinases), it binds more tightly to eIF2B, thereby blocking its guanine nucleotide exchange activity. The decreased level of active eIF2 leads to inhibition of global mRNA translation. However, under conditions of decreased TC levels, translation of specific sub-class of mRNAs, including ATF4 mRNA, is enhanced.

(HRI, GCN2, PKR and PERK) regulate eIF2B in this way. Each of them is activated by a different stress signal. HRI (heme-regulated inhibitor of translation), which is activated by heme deficiency, primarily functions in reticulocytes to balance globin synthesis with heme availability.[42] GCN2 (general control non-derepressible 2) is primarily activated by uncharged tRNAs in response to amino acid starvation, which is essential for tolerance to nutrient deprivation. GCN2 is also important for adaptation to other stresses and for development, differentiation, and normal function of mammalian organs (for review see[43]). PKR (protein kinase R) is activated by binding double-stranded RNAs, primarily as part of the cellular antiviral response. PKR functions also to control cell growth and cell differentiation, and its activity can be controlled by the action of several oncogenes (for review see[44]). PERK (PKR-like endoplasmic reticulum [ER] kinase) is an ER transmembrane protein with an ER luminal domain that responds to the stress of unfolded proteins in the ER. In response to unfolded proteins accumulation, its cytoplasmic kinase domain phosphorylates eIF2 and limits further translation of ER-destined proteins. Downregulation of eIF2B activity by eIF2 kinases enables two responses: a general reduction in the translation of most mRNAs and a simultaneous enhancement of translation of stress-responsive mRNAs encoding rescue proteins, since the translation of certain mRNAs is enhanced under conditions of reduced levels of GTP–eIF2/Met–tRNA$^{Met}_{i}$ ternary complexes (TC). This regulatory mechanism is mediated by regulatory short open reading frames (uORFs) within their 5′-untranslated region, upstream of the major coding region. While very few members of this subclass of mRNAs are known, a well-documented case is that of ATF4 mRNA, encoding the ATF4 transcription factor. When eIF2B is active, ATF4 protein can hardly be synthesized. If, however, eIF2B activity is low (due to eIF2α phosphorylation, or due to hypomorphic CACH/VWM mutations), ATF4 protein level increases and trans-activates the transcription of its target genes, most of which encode rescue proteins which promote cell survival under physiological stress. One of the target genes is *CHOP*, a second transcription factor. An important target of CHOP is GADD34, a stress-induced regulatory subunit of PP1 phosphatase, which in turn dephosphorylates eIF2α-P to relieve the inhibitory effect on eIF2B and resume translation (reviewed in [45]).

Myelin formation and maintenance requires a high synthesis rate of glycoproteins and transmembrane proteins, which are targeted to the ER lumen, where they undergo posttranslation processing events such as folding by chaperones and introduction of disulfide bonds. When the load of newly synthesized proteins exceeds the capacity of the ER to process them, the accumulated unfolded proteins elicit unfolded protein response (UPR), also known as ER-stress response (Figure 86.2). Since an active myelinating oligodendrocyte produces ~ 10^5 myelin proteins per minute,[46] the myelination process largely depends on full capacity of the ER to guarantee complete processing of all client proteins throughout the entire myelination process. Thus, the ER-stress response is of specific importance in CACH/VWM disease. Note that while PERK/eIF2/eIF2B/ATF4 represents one of three arms of the UPR, the ER-transmembrane proteins, ATF6 and IRE-1, are similarly activated in response to ER-stress, leading to activation of two additional transcription factors, ATF6 and XBP1, respectively. The three signaling programs, via ATF4, ATF6, and XBP-1 transcription factors are simultaneously activated and must be tightly coordinated for successful completion of the ER-stress response (for review see [45]). Hyperactivation of the PERK arm, supposedly by hypomorphic CACH/VWM mutations in eIF2B, may lead to lack of coordination with the ATF6/IRE-1 arms, thereby leading to further ER-stress.

Additional ways of regulating eIF2B have been observed. First, mammalian eIF2B is directly phosphorylated by four protein kinases: CK1 and CK2, DYRK (dual-specificity tyrosine phosphorylated and regulated kinase) and GSK3 (glycogen synthase kinase 3).[47,48] Of these, GSK3, is thought to play the greatest role in the regulation of eIF2B activity. DYRK phosphorylates rat eIF2Be at Ser^{539} and acts as a priming kinase that allows phosphorylation at Ser^{535} by GSK3. Phosphorylation by GSK3 inhibits eIF2B activity and has a role in apoptosis.[49] Insulin leads to inhibition of GSK3; therefore, in response to insulin, the phosphorylated eIF2Bε Ser^{535} undergoes dephosphorylation in a phosphatidyl inositol 3-kinase (PI3K)-dependent manner, leading to increased eIF2B enzymatic activity.[50]

Fusel alcohols, such as butanol and 3-methylbutan-1-ol, are generated by amino acid catabolism when amino acids are used as a nitrogen source. Their presence indicates lack of nitrogen. Interestingly, in yeast, fusel alcohols regulate translation initiation by inhibiting eIF2B in a mechanism that may involve altering the integrity and dynamics of the eIF2B body.[51]

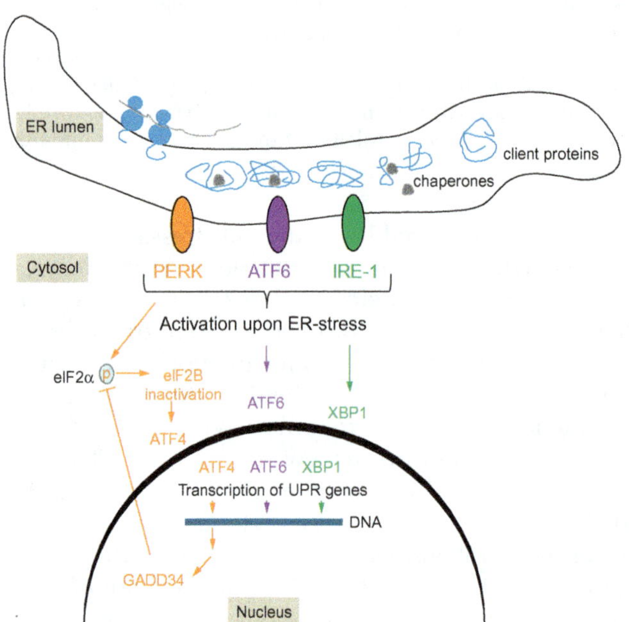

FIGURE 86.2 ER-stress response, also termed Unfolded Protein Response (UPR). Accumulation of unfolded proteins in the ER lumen leads to formation of three transcription factors, ATF4, ATF6, and XBP-1, via activation of the three ER-transmembrane proteins, PERK, ATF6 and IRE-1, respectively. Each of the three factors activates the transcription of multiple rescue genes. The PERK arm is responsible for downregulating further mRNA translation at the initial phase of the response, by phosphorylating eIF2α, which in turn inhibits eIF2B activity. GADD34, the stress-induced eIF2α phosphatase (downstream of PERK), is responsible for eIF2α-P dephosphorylation to resume translation. The three signaling programs must be tightly coordinated for successful rescue. Hypomorphic eIF2B mutations lead to hyperactivation of the PERK arm, lack of tight coordination and increased stress.

Finally, recent findings imply that heterogeneity in the cellular proportions of eIF2B(αβγδε) and eIF2B(βγδε) complexes have important implications for the regulation of translation in individual cell types.[38]

Abnormal Gene Products

Malfunction of eIF2B complex as a translation initiation factor may theoretically result from various different scenarios. Mutations may directly impair the catalytic activity, reduce the stability of the protein, reduce the ability to form a proper heteropentameric complex, or weaken the capacity of the complex to bind to its eIF2 substrate. All of the above "loss-of-function" circumstances are eventually conveyed by reduced GEF activity and decreased TC levels, thereby leading to global inhibition of protein synthesis, which is linked to enhanced ATF4 mRNA translation. Another type of mutation, so far not found in human patients, might affect the ability of eIF2B activity to be inhibited by phosphorylated eIF2 in response to physiological and environmental signals, leading to the absence of ATF4 mRNA translation.

The high evolutionary conservation of eIF2B subunits from yeast to mammals prompted the use of yeast (*Saccharomyces cerevisiae*) as a model system for the analysis of 12 CACH/VWM human mutations, selected based on sequence alignments indicating a conserved amino acid residue in both species.[52] One mutation in *EIF2B1* and seven mutations in *EIF2B5* tested in this study led to lower expression levels of the mutated protein subunit, explaining the observed outcome of reduction in global protein synthesis. In contrast, none of the mutations in eIF2B2 caused protein instability. One eIF2B2 variant (V341D) failed to recruit eIF2Bδ into *EIF2B* holocomplex, explaining the decreased global translation and the reduced growth rate of the corresponding mutant yeast strain. Five of the tested CACH/VWM mutations led to increased activity of Gcn4p, the yeast homolog of ATF4, consistent with down regulation of eIF2B function. Interestingly, four of the mutations also led to butanol-specific sensitivity of the corresponding mutant yeast strains.[52]

The effect of many additional CACH/VWM mutations was analyzed in the context of the human eIF2B complex, which offers a more authentic molecular milieu compared to the yeast model system.[14,53,54] By cotransfecting five recombinant plasmids into a human cell line, a tagged version of each of the mutant subunits was coexpressed with the rest of the four wild-type subunits, followed by pull-down of the generated recombinant eIF2B holocomplex for further biochemical analysis of its integrity, substrate binding and GEF activity. The majority of the mutations tested led to partial loss of *in vitro* GEF activity of the recombinant eIF2B, via various mechanisms. The deletion mutations causing premature translation termination (directly or via a frameshift) are effectively null mutations since they lead to the formation of truncated eIF2Bβ and eIF2Bε unstable polypeptides, which failed to form complexes with other subunits. Certain point mutations also impair eIF2B holocomplexes formation ability. For example, the G200V, P291S and G329V mutations in eIF2Bβ caused complete, or almost complete, inability to form complexes, or loss of the regulatory eIF2Bα subunit, respectively. However, other mutations caused decreased GEF activity without affecting complex formation. One mutation in eIF2Bγ (R225Q) and three in the catalytic domain of eIF2Bε (W628R, I649T, E650K) showed defective ability to bind its eIF2 substrate, and a significant drop in eIF2B GEF activity, while two mutations in eIF2Bβ (E213G and V316D) actually enhanced eIF2 binding but also led to decreased GEF activity and increased expression of enhanced green fluorescent protein (EGFP) from a reporter construct containing the regulatory features of ATF4 mRNA. The V316D mutation in eIF2Bβ showed a deficient ability to bind the other subunits, similarly to the effect observed in the yeast study.[52,53]

The effect of 40 mutations on eIF2B biochemical function is summarized in Table 86.1. The overall picture that emerged from these studies was that most of the mutations are hypomorphic, i.e., cause partial loss-of-function of eIF2B GEF activity. Thus, most of the mutations are associated with decreased levels of TC and decreased global protein synthesis. None of the mutations were found to alter the response of eIF2B to phosphorylated eIF2. However, surprisingly, several mutations, including two associated with the most severe form of the disease (A391D in eIF2Bδ and Y495C in eIF2Bε) did not affect complex formation and only very slightly decreased GEF activity. Two mutations (V73G and R269G in eIF2Bε) actually caused an increase in GEF activity.[54] Thus, some mutations may lead to CACH/VWM disease by affecting eIF2B function in as yet unknown ways, which may not be related directly to its role as a translation initiation factor.

Genotype–Phenotype Correlations

The diagnosis of CACH/VWM comprises a wide spectrum in terms of age of onset, clinical symptoms, and disease deterioration.[6,18] Given that CACH/VWM mutations reside in genes encoding eIF2B subunits, it was intuitively hypothesized that mutations affecting the biochemical performance of eIF2B as a translation initiation factor are

TABLE 86.1 The Effect of Mutations on eIF2B Biochemical Function

Subunit (gene)	Mutation	Equivalent in yeast	Mutant-subunit stability	eIF2B complex formation	Binding to eIF2	GEF activity†	Translation of GCN4/ATF4	Sensitivity to butanol
eIF2B1	N208Y	N209Y	Decreased[52]					
eIF2B2	S171F					Decreased[54]		
	G200V			Complete loss[54]				
	Del leading to M203fs*			Decreased[53]		Decreased[53]		
	E213G	E239G	Stable[52]		Increased[53]	Decreased[53]	Increased[53]	
	K273R	K300R	Stable[52]			Decreased[54]		
	P291S			Decreased[34]				
	V316D	V341D	Stable[52]	Decreased[52,53]	Increased[53]	Decreased[52,53]	Increased[52,53]	
	G329V			eIF2Bα lost [54]		Decreased[54]		
eIF2B3	Q136P					Decreased[54]		
	R225Q				Decreased[54]	Decreased[54]		
	H341Q					Decreased[54]		
eIF2B4	R357W			Decreased[54]		Decreased[54]		
	A391D					Decreased[54]		
	R483W			Decreased[54]		Decreased[54]		
eIF2B5	A16D					Decreased[54]		
	A56V					Decreased[54]		
	D62V					Decreased[14]		
	L68S					Decreased[54]		
	V73G	V57G	Decreased[52]			Increased[54]	Increased[52]	
	T91A			Decreased[53]		Decreased[53]		
	L106F	I90F	Decreased[52]					Yes[52]
	R113H					Decreased[53]		
	R136C					Decreased[54]		
	R195H					Decreased[53]		
	Del to F264fs*			Decreased[53]		Decreased[53]		
	S610-D613del		Decreased[14]	Decreased[14]		Decreased[14]		
	R269G					Increased[54]		
	R269del*		Decreased[14]	Decreased[14]		Decreased[14]		
	R299H	R284H	Decreased[52]				Increased[52]	Yes[52]
	R315G					Decreased[53]		
	D335S					Decreased[14]		
	R339P	R323P	Decreased[52]			Decreased[53]	Increased[52]	
	N376D					Decreased[14]		
	G386V	G369V	Decreased[52]					
	V430A	I413A	Decreased[52]					
	Y495C					Decreased[54]		
	W628R	W618R	Decreased[52]		Decreased[53]	Decreased[53]	Increased[52]	Yes[52]
	I649T				Decreased[54]	Decreased[54]		
	E650K				Decreased[54]	Decreased[54]		

**Mutation leading to truncated protein.*
†Extent varies depending on the study.

directly correlated with the severity of disease. The elucidation of genotype–phenotype correlation is not a trivial task due to several reasons: i) most of the patients are heterozygous for two different mutations, thus their phenotype is a combinatorial effect of both;[55] ii) the ratio of eIF2 to eIF2B varies between different cell types, leading to variations in eIF2B biochemical performance in different experimental systems; iii) the random effect of physiological and environmental stress on disease onset and deterioration may mask the primary genotype–phenotype relationship. In fact, a cohort of 83 individuals carrying 46 distinct mutations in eIF2Bα–ε subunits and representing a large clinical spectrum, from rapidly fatal infantile to adult forms, demonstrated that disease severity is correlated with age at onset but not with the type of the mutated subunit nor with the position of the mutation within the protein.[20]

The availability of blood samples and skin biopsies enabled eIF2B enzymatic activity testing in lysates of patient-derived cells. A significant decrease of 20–70% in GEF activity was observed in Epstein–Barr virus (EBV)-transformed lymphocytes (e.g., lymphoblasts) from 30 CACH/VWM patients, compared to lymphoblasts from 10 unaffected heterozygotes, and 22 controls without *EIF2B* mutations.[56] In this study, the age of disease onset inversely correlated with the degree of the decrease in eIF2B enzymatic activity.[56] A larger study, using lymphoblasts from 63 patients carrying different *EIF2B* mutations and presenting different clinical forms of the disease, provided confirmation for the decrease in eIF2B GEF activity in cells from all eIF2B-mutated patients compared to healthy controls and to patients with other leukodystrophies lacking *EIF2B* mutations.[57] This verification prompted the suggestion to use eIF2B GEF activity assay as an initial screen for CACH/VWM disease prior to sequencing of all *EIF2B1–5* genes. However, this large study also indicated the weak correlation between GEF activity and age at disease onset. While patients with a severe disease and onset at ≤2 years always had eIF2B GEF activity <55%, discrepancies in the GEF activity values were found among patients of later onset, and between siblings carrying the same mutations.[57]

Another study[58] analyzed eIF2B GEF activity in lymphoblasts derived from two patients homozygous for mutations that are significantly associated with milder forms (eIF2B5-R113H and eIF2B2-E213G),[55] and two patients homozygous for mutations associated with more severe forms (eIF2B4-R374C and eIF2B5-T91A). This study demonstrated similar profound loss of eIF2B activity in the cells of all patients, which was not correlated to disease severity of the affected individuals,[58] pointing out the lack of clear genotype–phenotype correlation. Interestingly, an intrafamilial phenotypic variability was observed for two sisters carrying the so-called "mild" eIF2B5-R113H mutation. One sister had slight gait disturbances that started at age 42, whereas her sister presented a coma episode at age 8 after febrile illness, progressive gait abnormalities at age 20, and death at 35 years of age after a moderate cranial trauma.[59] It is now accepted that eIF2B GEF activity assayed in lysates of patient-derived lymphoblasts do not reliably demonstrate robust genotype–phenotype correlations. Notably, GDP–GTP exchange on eIF2 is not rate limiting for protein synthesis in lymphoblasts. Despite the reduced eIF2B activity assayed in this cell type, labeling studies with [^{35}S]-methionine of control- and CACH/VWM-derived lymphoblasts indicated no major differences in protein synthesis under normal or hyperthermic conditions, nor a failure of the ability to recover from high temperature.[58]

Since the process of cell transformation by EBV affects the translation machinery, it was important to analyze eIF2B GEF activity in patient-derived primary cells that have not been transformed or immortalized by any exogenous agent that may affect the cellular phenotype. Analyses using lysates from primary fibroblasts that had been derived from patients carrying a range of CACH/VWM mutations demonstrated only a minor loss of basal eIF2B activity, and therefore failed to provide genotype–phenotype correlation.[54,60] Nonetheless, the heightened stress response of primary fibroblasts from CACH/VWM patients, as exhibited by the significantly greater increase in ATF4 induction in response to pharmacologically induced UPR,[60] prompted the idea to employ this feature as an initial screening tool for CACH/VWM leukodystrophy. However, no major difference in the strength of UPR was detected between eIF2B-mutated lymphocytes or lymphoblasts and the corresponding control cells.[61]

As detailed in the previous section above, some mutations do not affect any parameter related to eIF2B complex integrity and enzymatic activity, even though they cause a harsh phenotype, implying that severe disease can result from alterations in eIF2B function that are related to yet unknown mechanisms.[54]

DISEASE MECHANISMS AND PATHOPHYSIOLOGY

Observations in Human Subjects

Neuropathologic findings show a destructive cavitating orthochromatic leukodystrophy (Figure 86.3). Cerebral and cerebellar myelin is markedly diminished, whereas the spinal cord is relatively spared. Axons are hypomyelinated in areas of preserved white matter with increased numbers of oligodentrocytes, whereas astrocytes are

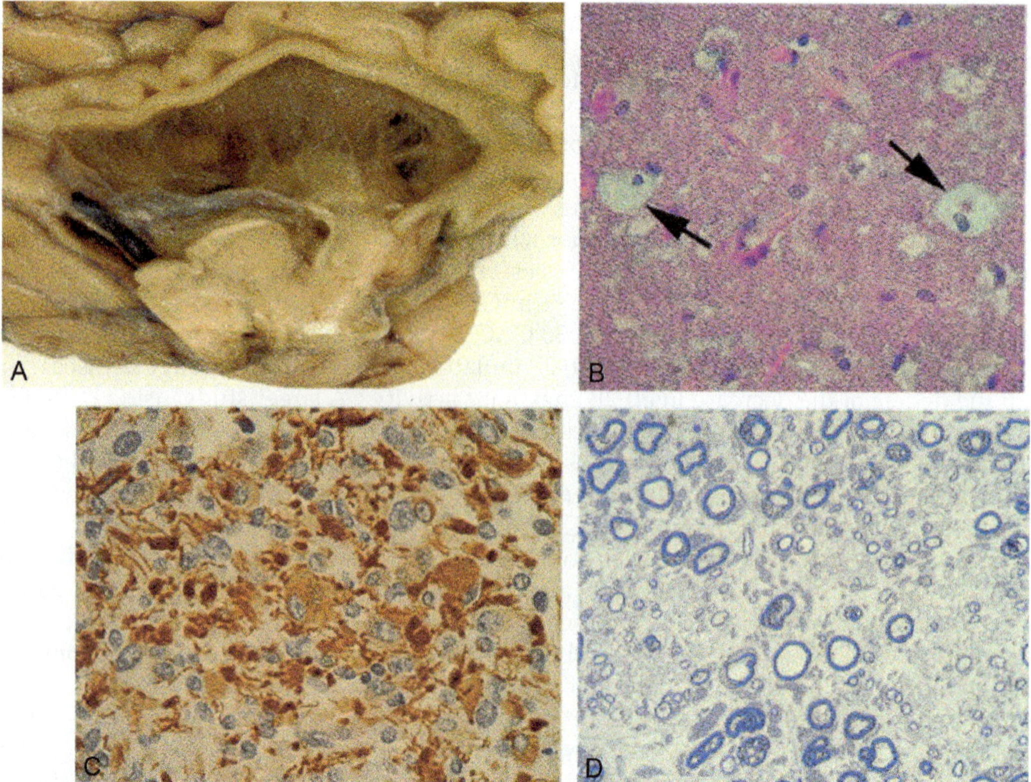

FIGURE 86.3 **Characteristic pathological finding in CACH/VWM.**[80] (**A**) Gross pathology: although the cortex looks largely intact, all that is left of the white matter is thin and fragile strands tissue. (**B**) Alcian blue–periodic acid–Schiff staining shows large "foamy" oligodendrocytes (arrows). x100. (**C**) GFAP staining showing grossly coarse and atypical astrocytes. x100. (**D**) Sural nerve of a 15-year-old male with CACH/VWM, showing mild loss of large myelinated axons with a greater loss of small myelinated axons. The unmyelinated axons appear swollen with greater than normal cross sectional diameter. There are scattered severely atrophic axons and occasional degenerating axons. All axons had a thinner than normal myelin sheath but "onion bulbs" were not identified. x100. All panels are courtesy of Kondi Wong, M.D.

decreased, especially in the severe infantile form.[62–64] The pathological hallmark is the presence of oligodendrocytes with "foamy" cytoplasm and markedly hypotrophic and sometimes atypical astrocytes (Figure 86.3B).[65] The white matter astrocytes and oligodendrocytes are immature precursor cells, explaining the lack of myelin production and scarce gliosis.[66] A pattern of glycosylation abnormality in the CNS was suggested by the finding of decreased spinal fluid asialotransferrin/sialotransferrin ratio. This finding may be a biomarker of the CACH/VWM.[67,68] Although the phenotype of at least 90% of patients with a CACH/VWM is *EIF2B*-related, the mechanism of disease in the others is different. For example, the juvenile patient in the original description of the disease turned out not to have *EIF2B* abnormality.[69]

Animal Models

Systematic analysis of the effect of *EIF2B* mutations on human brain development and performance is not feasible due to the low number of CACH/VWM patients, lack of presymptomatic clinical data, and the heterogeneous genetic background/diet/environmental conditions within the human population. Animal models not only solve these limitations, but also provide a source for brain tissue for research under defined conditions with a goal of understanding disease onset and progression. The first reported knock-in mouse model for CACH/VWM disease harbors a histidine instead of arginine residue at position 132 of the catalytic subunit of eIF2B.[70] Murine eIF2B5-R132H is homologous to the human eIF2B5-R136H mutation, which at a homozygous state in the human patient led to a classical form of CACH/VWM symptoms, followed by death at 10 years of age. In contrast to the human patient, homozygous Eif2b5$^{R132H/R132H}$ mice have a normal lifespan and mild clinical symptoms, reflecting the vast difference between human and mice white matter physiology, volume, and complexity. However, the mild symptoms, associated with a 20% reduction in eIF2B GEF activity in the mouse brain, include impaired motor functions with involvement

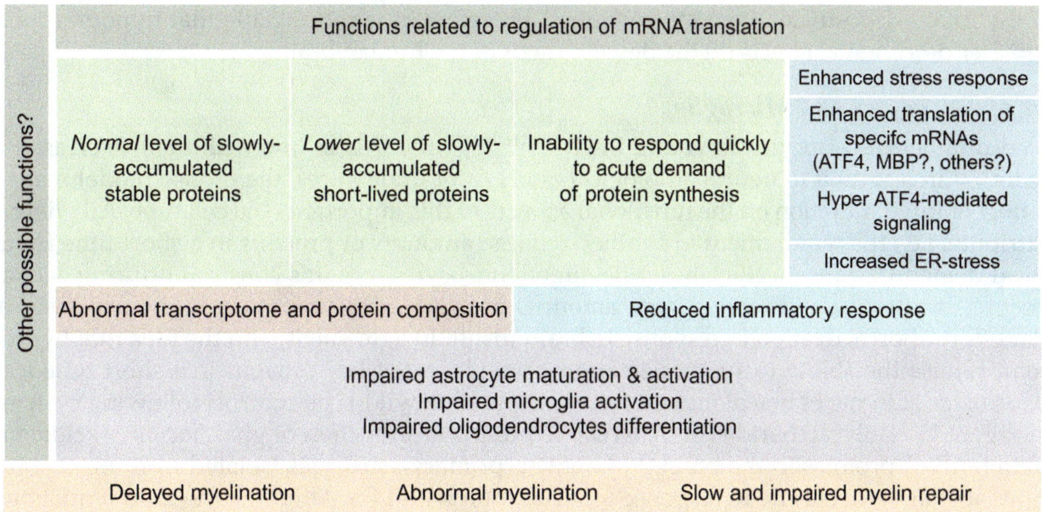

FIGURE 86.4 **Pathophysiological consequences of mutations in eIF2B.** Current model based on accumulated experimental data.

of the corpus callosum, internal capsule and brainstem, thus representing the late-onset form in humans carrying hypomorphic *EIF2B* alleles with mild mutations. Eif2b5$^{R132H/R132H}$ mice have provided fundamental insights related to the effect of a 20% reduction in eIF2B GEF activity on brain development and function under defined genetic background and environmental conditions, as detailed below and summarized in Figure 86.4.

Postnatal Brain Development

Surprisingly, the hypomorphic Eif2b5-R132H alleles affect global gene expression in the mouse brain, throughout its transition from a primarily proliferative state during the first postnatal week to a highly differentiated state later on. The unique time point-specific transcriptome signatures reflect delayed waves of gene expression and an adaptive process to cellular stress.[71] The delay is not uniform across the transcriptome and seems to affect specific subsets of genes. In particular, cell cycle-associated genes are significantly dysregulated at postnatal day 1, whereas white matter-specific genes are significantly dysregulated at postnatal day 21, at the peak of myelin formation. This phenomenon results from a differential effect of the mutated Eif2b on translation rates of mRNAs encoding key regulators, e.g., transcription factors and RNA-binding proteins, which thereby differentially affects RNA synthesis, processing, and turnover of a wide range of target mRNAs. The negative effect of hypomorphic Eif2b mutations is expected to be more detrimental to rapidly translated mRNAs. Obviously, the overall difference in gene expression patterns reflects multiple layers of outcomes that accumulate and build up as the brain develops and matures. This fact by itself provides one of the explanations for the severity of CACH/VWM disease, despite the slight reduction in eIF2B enzymatic activity associated with many hypomorphic mutations.

Time-course diffused tensor MRI scans confirmed the delayed postnatal CNS white matter development of Eif2b5$^{R132H/R132H}$ mice.[70] While the fractional anisotropy (FA) parameter seems to be pseudonormalized at an older age, the steady increase in apparent diffusion coefficient (ADC) values throughout the life of the mutant mice is indicative of a degenerative process and tissue pathology, which progresses from the lower parts of the motor pathway. Transmission electron microscopy (TEM) analysis indicated small-caliber axons in the young mutant mice, while increased abundance of demyelinated axons and axons ensheathed with split and damaged myelin in older mutant mice is indicative of a neurodegenerative process.[70] The abnormal content of the two major myelin proteins, MBP and PLP/DM20, at a young age, followed by normalization of MBP but not PLP/DM20 content at an older age, is in agreement with abnormal myelination during development.[70] Similarly, the elevated abundance of NG2-positive oligodendrocytes in young mutant mice is in agreement with the increased number of oligodendrocytes in human patients,[5,62–64] suggesting a possible difficulty in proceeding to more advanced differentiation stages. The time course MRI scans of the mutant mice began at 3 weeks of age, therefore the maldevelopment of only the slowest-developing regions (e.g., internal capsule, brainstem, and hippocampus) was followed. The latter enabled the observation that gray matter (e.g., hippocampus) is also involved, which is not surprising given the significant role of glial cells in neuronal function. The delayed brain development of Eif2b5$^{R132H/R132H}$ mice suggests a similar phenomenon in presymptomatic human patients. Maturation of white matter pathways in humans continues until late childhood,[72] compared to the first 3–4 postnatal weeks in rodents.[73] Therefore, it is conceivable that hypomorphic mutations in *EIF2B* have greater detrimental effects on the development of the human brain.

Involvement of Astrocytes and Microglia

The reduced number of astrocytes in young Eif2b5$^{R132H/R132H}$ mice[70] is in line with their decreased abundance in human patients.[64,74] In addition to neuropathological analysis of brain slices, the mouse model made it possible to address the effect of Eif2b mutation on the functional aspects of this important glial cell type. All glial cell types share a common attribute, i.e., the requirement to synthesize large amounts of proteins in a short time frame in response to physiological demand. More specifically, while oligodendrocytes face this constraint during active myelination, reactive astrocytes synthesize and secrete large amounts of proteins to support brain homeostasis, and microglia share a similar effort upon activation in response to brain insult. In agreement with the idea that hypomorphic Eif2b mutations compromise the ability to synthesize appropriate levels of key proteins in a short time frame, impaired astrogliosis was evident in the brains of mutant mice compared to wild-type controls following systemic injection of the pyrogenic agent lipopolysaccharide (LPS), which led to lower induction of glial fibrillary acidic protein (GFAP) and interleukin 6 (IL-6) proteins in the brain.[75] Moreover, primary astrocytes isolated from mutant brains failed to undergo proper activation in culture in response to incubation with LPS. Although the mild reduction in eIF2B GEF activity did not lead to decreased global protein synthesis rates under normal conditions, it did prevent the normal increase in global translation rates upon exposure to LPS. In agreement with this anomaly, lower levels of pro- and anti-inflammatory cytokines were produced despite similar induction of the encoding mRNAs.[75] Most importantly, primary astrocytes and microglia isolated from Eif2b5$^{R132H/R132H}$ mice brains suffer a similar impairment, implicating the involvement of microglia in the poor astrogliosis outcome in the mutant brains. These results were the first to suggest a role for microglia in CACH/VWM pathology, and the first to relate poor performance of neuroinflammatory signaling to the etiology of the disease.[75] Eif2b5$^{R132H/R132H}$ mice highlighted the importance of accurate protein synthesis rates to brain homeostasis restoration via microglia–astrocyte crosstalk, and underscored the significance of fully functional translation machinery for efficient cerebral inflammatory response upon insults.

The Effect of Stress

CACH/VWM disease is strongly associated with stress: a) myelin loss and associated neurological symptoms of human patients deteriorate upon exposure to various stressors; b) the UPR pathway (see Figure 86.2) was found to be activated in the brains of CACH/VWM patients;[76,77] and c) primary cultured fibroblasts isolated from CACH/VWM patients, as well as oligodendroglia derived-cell line expressing mutated Eif2b, are hypersensitive to induced ER stress.[60,78] The final stages of oligodendrocyte precursor cells differentiation into mature myelinating oligodendrocytes involve active synthesis of glycoproteins within the ER in a short time window. Involvement of the ER in myelin formation explains why myelination by itself serves as an endogenous stress agent, explaining the dependence of myelin formation on full performance of the UPR. The decreased abundance of PLP/DM20 in total hippocampus homogenates of Eif2b5$^{R132H/R132H}$ mice[70] may result from retarded ER function, in agreement with previous findings of abnormal ER stress response in eIF2B-mutated primary fibroblasts isolated from CACH/VWM patients.[60] Unbalanced expression of UPR-related genes may prevent efficient differentiation of oligodendrocyte precursor cells and may lead to apoptosis or maldifferentiation. Moreover, various stressors, e.g., head trauma, or febrile illnesses caused by systemic infections, elicit a brain response to guarantee the maintenance of homeostasis, which includes myelin maintenance and repair. In agreement with the above, Eif2b5$^{R132H/R132H}$ mice fail to recover from cuprizone-induced demyelination.[70] Moreover, they exhibit insufficient cerebral inflammatory response and poor astrogliosis upon exposure to LPS.[75] It seems that, in mutant mice, myelin generation is slow and abnormal in response to brain damage, just as it is throughout early postnatal development. Note that slow and inefficient astrogliosis, reflected by increased GFAP levels in mutant mice brains compared to back-to-normal levels in wild-types towards the end of the repair process, can be misinterpreted as increased astrogliosis. The defective astrogliosis observed in Eif2b$^{R132H/R132H}$ mice in response to cuprizone-induced demyelination reflects abnormal performance, in agreement with a range of atypical astrogliosis variations reported for CACH/VWM depending on the specific mutation, disease severity and brain region analyzed.[5,62,64,74] Experiments using Eif2b5$^{R132H/R132H}$ mice have highlighted the link between hypomorphic eIF2B mutations and limited capability to execute efficient brain repair. The mouse model enabled the understanding that two types of cellular effects are elicited by hypomorphic Eif2b5 mutations: i) defective UPR; and ii) compromised ability to synthesize sufficient amounts of proteins within a short time window in response to physiological demand. The demand for massive synthesis of myelin components by oligodendrocytes may also be referred to as "acute" demand for increased translation within a short time frame. The prolonged myelin production during early postnatal brain development of Eif2b5$^{R132H/R132H}$ mice, their failure to overcome cuprizone-induced demyelination and their prolonged astrogliosis during the delayed repair period support this concept. Eif2b5$^{R132H/R132H}$ mice provided evidence that even a mild reduction in eIF2B enzymatic activity causes slow white matter development and compromised remyelination in response to physiological stress. Therefore, it is speculated that repeated brain insults before complete myelination or remyelinating will eventually lead to neurological deterioration. Based on data obtained from the mouse model, it is hypothesized that a significant factor affecting disease onset and severity is the extent and frequency of stress events each individual experiences during critical times of postnatal brain development and during the process of brain repair. According to this idea, a patient carrying mild mutations may present a severe clinical picture if exposed to multiple and frequent stress events, whereas a patient carrying more potent mutations may be associated with milder symptoms if s(he) was fortunate to avoid them. It is suggested that brain insults in individuals carrying hypomorphic eIF2B alleles result in a more rapid breakdown of abnormal myelin, followed by inefficient remyelination, imposing a major threat to demyelinated axons and leading to eventual axonal swelling and loss.

Current Research

Current research focuses on further characterization of the involved molecular mechanisms and on development of therapeutic approaches. Both embryonic stem cells (mESC) derived from Eif2b-mutant mice and induced pluripotent stem cells (iPSC) derived from human patients are currently used for *in vitro* differentiation studies of the oligodendroglial lineage. Additional mouse models are currently constructed bearing more severe mutations in Eif2b5 and in other eIF2B subunits. Mouse models demonstrating a severe phenotype similar to VWM/CACH clinical symptoms will make it possible to test future cell therapy approaches that involve transplantation of genetically engineered iPSC. Larger sets of mouse models will enable the identification of key mRNA transcripts whose translational control is essential for normal myelin development and maintenance. The use of system biology tools for analyzing transcriptome, translatome, and proteome will enable the discovery of the molecular circuits involved in this pathology and will provide the basis for rational drug design. Recently, a genetic approach identified Gadd34, the stress-induced eIF2α-P phosphatase, as a potential drug target for VWM/CACH (Elroy-Stein, in preparation). Current experiments use this novel insight, as well as the hypersensitivity of eIF2B-mutated cells to ER stress, for screening of drug-like small molecules, which will hopefully protect *EIF2B*-mutants from stress-related deterioration. Future experiments will shed light on the implications of type and timing of stress on disease onset, severity and progression.

DIFFERENTIAL DIAGNOSIS

There are other disorders with diffuse white matter abnormalities in childhood to consider, with their distinct MRI findings: X-linked adrenoleukodystrophy, metachromatic leukodystrophy, Krabbe disease, and Canavan disease. These disorders, however, do not provoke cystic degeneration.

Alexander disease usually has a frontal predominance with cystic degeneration possible in the subcortical or deep white matter.[22] Basal ganglia and thalamic abnormalities are frequently present. Contrast enhancement of characteristic structures often facilitates the diagnosis. Megalencephalic leukoencephalopathy with subcortical cysts (MLC) is characterized by diffusely abnormal and mildly swollen cerebral hemispheric white matter that does not show signs of diffuse cystic degeneration.[22] Subcortical cysts are almost always present in the anterior temporal area and often in other regions. The cysts are best seen on proton density and FLAIR MRI. The diagnosis can usually be established with molecular genetic testing. MRI abnormalities similar to those seen in CACH/VWM with prominent and diffuse white matter rarefaction and cystic degeneration may be seen in mitochondrial disorders.[79] Diffuse hyperintensity of the white matter on T2-weighted images is also observed in leukodystrophies with primary hypomyelination, such as the PLP1-related disorders; however, these disorders have a normal or nearly normal white matter signal on T1-weighted images and computed tomography (CT) scan.[22] CADASIL, lamin B1 mutations, or acquired white matter disorders such as multiple sclerosis need to be considered in individuals with adult-onset CACH/VWM; however, the early, constant, diffuse, symmetric alteration of the white matter on MRI in *EIF2B*-related disorders is distinctive (Figure 86.5).

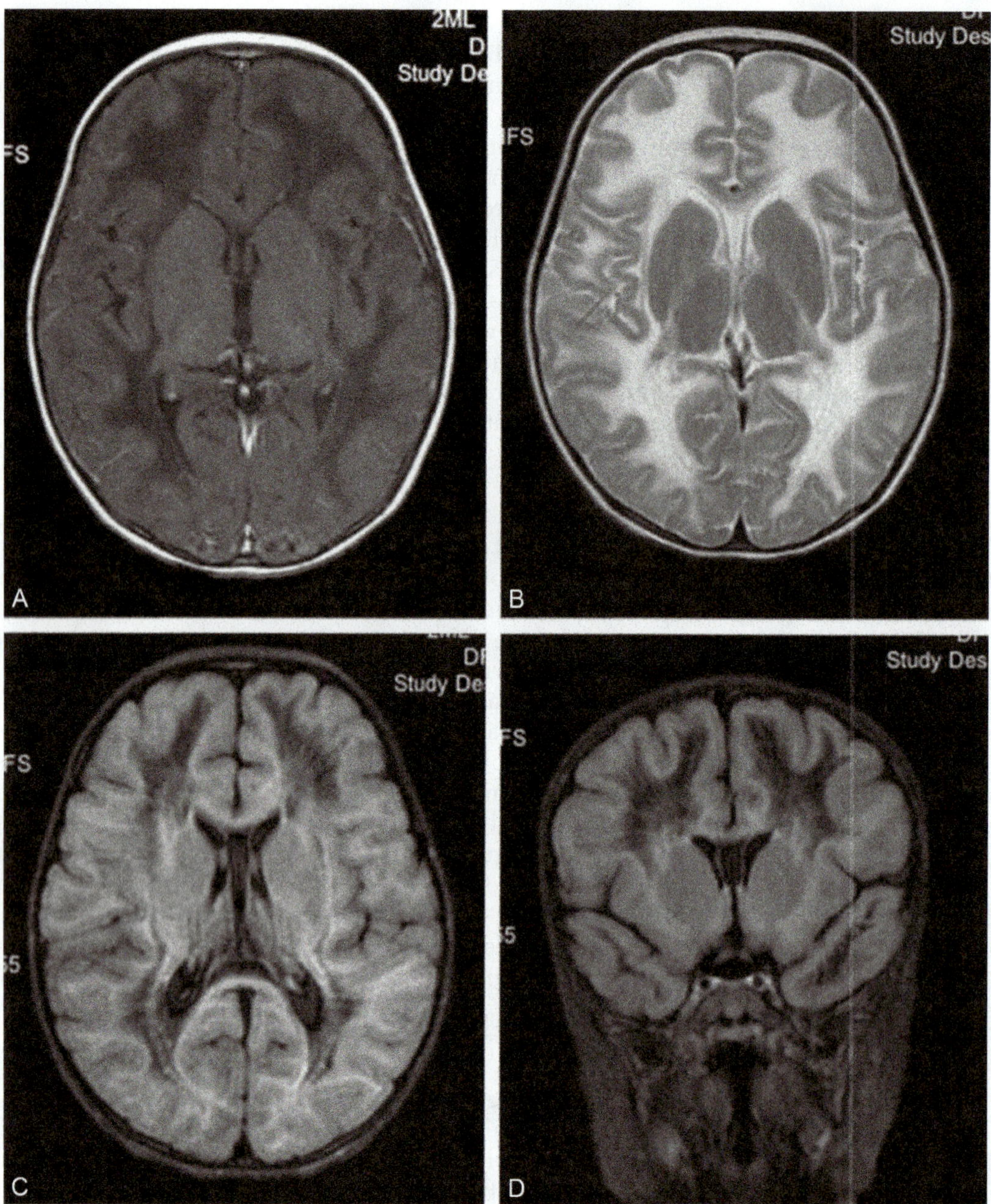

FIGURE 86.5 **Typical MRI of a 2-year-old patient with CACH/VWM, showing the white matter to be diffusely hypointense on T1-weighted images (A) and hyperintense on a T2-weighted image (B).**[80] Note the small lateral ventricles and absence of cortical atrophy. Axial (**C**) and coronal (**D**) fluid-attenuated inversion recovery (FLAIR) images in the same patient show widespread secondary breakdown of the abnormal white matter that leaves visible bundles of nerve fibers crossing the subcortical region.

TESTING

The MRI is virtually pathognomonic and must be present in order to justify molecular testing.

Molecular Diagnosis

Molecular genetic testing (in order) of *EIF2B5*, *EIF2B2*, *EIF2B4*, *EIFB3*, and *EIF2B1* is recommended. Sequencing of the coding sequence and associated splice sites must be performed.[17]

Analytical Testing

Routine blood tests are normal. Cerebrospinal fluid (CSF) asialotransferrin/total transferrin ratio may be used to identify those likely to have mutations in any of the eIF2B subunit genes.[67]

Clinical and Ancillary Testing: Imaging, Physiological, Electrodiagnostic

Brain MRI (Figure 86.5) shows symmetrically and diffusely abnormal white matter that has low signal intensity compared to the cortex on T1-weighted images and high signal on T2-weighted and on FLAIR images. On T1-weighted and FLAIR images, a typical radiating appearance of white matter bundles or dots on sagittal and coronal images is seen on transverse or coronal images (Figure 86.5).[19] Over time, white matter is replaced with CSF, particularly in infants and children, via cystic breakdown seen on proton density or FLAIR images (Figure 86.5).[19] Cerebellar vermian atrophy may develop. Importantly, the cerebral cortex does not show significant atrophy. Supratentorial corticosubcortical white matter atrophy with or without cystic degeneration can be observed in adult-onset forms with slow progression.[21] Cranial CT scan is of limited use and shows diffuse and symmetric hypodensity of the cerebral hemispheric white matter with no calcifications.[69] Evoked potential or electromyography are typically normal and are not helpful in the diagnosis of CACH/VWM.

MANAGEMENT

Standard of Care

There is no specific therapy for CACH/VWM. Standard supportive therapy such as physical therapy and rehabilitation for spasticity, ataxia or tremor may be used, as well as antiepileptic medications for seizures. Contact sports that may cause head trauma and stressful situations, including high body temperature, should be avoided when possible.

Barriers to Treatment Development

In order to be effective, future specific interventions should occur very early in the disease process, ideally before any clinical abnormalities are apparent. Therapy is unlikely to be effective once significant axonal loss has already occurred. Therefore, early recognition using various screening methods will have to be applied.

Therapies Under Investigation

Efforts to develop inhibitors of GADD34, the stress-induced eIF2α-P phosphatase, are in progress (see also above).

References

1. Schiffmann R, Trapp BD, Moller JR, et al. Chidhood ataxia with diffuse central nervous system hypomyelination. *Ann Neurol*. 1992;32(3):484.
2. Hanefeld F, Holzbach U, Kruse B, Wilichowski E, Christen HJ, Frahm J. Diffuse white matter disease in three children: an encephalopathy with unique features on magnetic resonance imaging and proton magnetic resonance spectroscopy. *Neuropediatrics*. 1993;24(5):244–248.
3. van der Knaap MS, Barth PG, Gabreels FJ, et al. A new leukoencephalopathy with vanishing white matter. *Neurology*. 1997;48(4):845–855.
4. Bruck W, Herms J, Brockmann K, Schulz-Schaeffer W, Hanefeld F. Myelinopathia centralis diffusa (vanishing white matter disease): evidence of apoptotic oligodendrocyte degeneration in early lesion development. *Ann Neurol*. 2001;50(4):532–536.
5. van der Knaap MS, Kamphorst W, Barth PG, Kraaijeveld CL, Gut E, Valk J. Phenotypic variation in leukoencephalopathy with vanishing white matter. *Neurology*. 1998;51(2):540–547.
6. van der Knaap MS, van Berkel CG, Herms J, et al. eIF2B-related disorders: antenatal onset and involvement of multiple organs. *Am J Hum Genet*. 2003;73(5):1199–1207.
7. Leegwater PA, Konst AA, Kuyt B, et al. The gene for leukoencephalopathy with vanishing white matter is located on chromosome 3q27. *Am J Hum Genet*. 1999;65(3):728–734.
8. Leegwater PA, Vermeulen G, Konst AA, et al. Subunits of the translation initiation factor eIF2B are mutant in leukoencephalopathy with vanishing white matter. *Nat Genet*. 2001;29(4):383–388.
9. van der Knaap MS, Leegwater PA, Konst AA, et al. Mutations in each of the five subunits of translation initiation factor eIF2B can cause leukoencephalopathy with vanishing white matter. *Ann Neurol*. 2002;51(2):264–270.
10. Pavitt GD, Proud CG. Protein synthesis and its control in neuronal cells with a focus on vanishing white matter disease. *Biochem Soc Trans*. 2009;37(Pt 6):1298–1310.

11. Pronk JC, Leegwater PA, van der Knaap MS. From gene to disease; a defect in the regulation of protein production leading to vanishing white matter. *Ned Tijdschr Geneeskd*. 2002;146(41):1933–1936.
12. Fogli A, Dionisi-Vici C, Deodato F, Bartuli A, Boespflug-Tanguy O, Bertini E. A severe variant of childhood ataxia with central hypomyelination/vanishing white matter leukoencephalopathy related to EIF21B5 mutation. *Neurology*. 2002;59(12):1966–1968.
13. Mierzewska H, van der Knaap MS, Scheper GC, Jurkiewicz E, Schmidt-Sidor B, Szymanska K. Leukoencephalopathy with vanishing white matter due to homozygous EIF2B2 gene mutation. First Polish cases. *Folia Neuropathol*. 2006;44(2):144–148.
14. Leng X, Wu Y, Wang X, et al. Functional analysis of recently identified mutations in eukaryotic translation initiation factor 2Bvarepsilon (eIF2Bvarepsilon) identified in Chinese patients with vanishing white matter disease. *J Hum Genet*. 2011;56(4):300–305.
15. Matsukawa T, Wang X, Liu R, et al. Adult-onset leukoencephalopathies with vanishing white matter with novel missense mutations in EIF2B2, EIF2B3, and EIF2B5. *Neurogenetics*. 2011;12(3):259–261.
16. van der Knaap MS, Breiter SN, Naidu S, Hart AA, Valk J. Defining and categorizing leukoencephalopathies of unknown origin: MR imaging approach. *Radiology*. 1999;213(1):121–133.
17. Schiffmann R, Fogli A, Van der Knaap MS, Boespflug-Tanguy O. Childhood ataxia with central nervous system hypomyelination/vanishing white matter. In: Pagon RA, Adam MP, Bird TD, Dolan CR, Fong CT, Stephens K, eds. GeneReviews. Seattle, WA: University of Washington; 1993.
18. Fogli A, Boespflug-Tanguy O. The large spectrum of eIF2B-related diseases. *Biochem Soc Trans*. 2006;34(Pt 1):22–29.
19. van der Knaap MS, Pronk JC, Scheper GC. Vanishing white matter disease. *Lancet Neurol*. 2006;5(5):413–423.
20. Fogli A, Schiffmann R, Bertini E, et al. The effect of genotype on the natural history of eIF2B-related leukodystrophies. *Neurology*. 2004;62(9):1509–1517.
21. Labauge P, Horzinski L, Ayrignac X, et al. Natural history of adult-onset eIF2B-related disorders: a multi-centric survey of 16 cases. *Brain*. 2009;132(Pt 8):2161–2169.
22. Schiffmann R, van der Knaap MS. Invited article: an MRI-based approach to the diagnosis of white matter disorders. *Neurology*. 2009;72(8):750–759.
23. Vermeulen G, Seidl R, Mercimek-Mahmutoglu S, Rotteveel JJ, Scheper GC, van der Knaap MS. Fright is a provoking factor in vanishing white matter disease. *Ann Neurol*. 2005;57(4):560–563.
24. Schiffmann R, Tedeschi G, Kinkel RP, et al. Leukodystrophy in patients with ovarian dysgenesis. *Ann Neurol*. 1997;41(5):654–661.
25. Fogli A, Rodriguez D, Eymard-Pierre E, et al. Ovarian failure related to eukaryotic initiation factor 2B mutations. *Am J Hum Genet*. 2003;72(6):1544–1550.
26. Fogli A, Wong K, Eymard-Pierre E, et al. Cree leukoencephalopathy and CACH/VWM disease are allelic at the EIF2B5 locus. *Ann Neurol*. 2002;52(4):506–510.
27. Black DN, Booth F, Watters GV, et al. Leukoencephalopathy among native Indian infants in northern Quebec and Manitoba. *Ann Neurol*. 1988;24(4):490–496.
28. Gomez E, Mohammad SS, Pavitt GD. Characterization of the minimal catalytic domain within eIF2B: the guanine-nucleotide exchange factor for translation initiation. *EMBO J*. 2002;21(19):5292–5301.
29. Mohammad-Qureshi SS, Haddad R, Hemingway EJ, Richardson JP, Pavitt GD. Critical contacts between the eukaryotic initiation factor 2B (eIF2B) catalytic domain and both eIF2beta and -2gamma mediate guanine nucleotide exchange. *Mol Cell Biol*. 2007;27(14):5225–5234.
30. Pavitt GD, Ramaiah KV, Kimball SR, Hinnebusch AG. eIF2 independently binds two distinct eIF2B subcomplexes that catalyze and regulate guanine-nucleotide exchange. *Genes Dev*. 1998;12(4):514–526.
31. Dev K, Qiu H, Dong J, Zhang F, Barthlme D, Hinnebusch AG. The beta/Gcd7 subunit of eukaryotic translation initiation factor 2B (eIF2B), a guanine nucleotide exchange factor, is crucial for binding eIF2 in vivo. *Mol Cell Biol*. 2010;30(21):5218–5233.
32. Reid PJ, Mohammad-Qureshi SS, Pavitt GD. Identification of intersubunit domain interactions within eukaryotic initiation factor (eIF) 2B, the nucleotide exchange factor for translation initiation. *J Biol Chem*. 2012;287(11):8275–8285.
33. Wang X, Wortham NC, Liu R, Proud CG. Identification of residues that underpin interactions within the eukaryotic initiation factor (eIF2) 2B complex. *J Biol Chem*. 2012;287(11):8263–8274.
34. Price NT, Mellor H, Craddock BL, et al. eIF2B, the guanine nucleotide-exchange factor for eukaryotic initiation factor 2. Sequence conservation between the alpha, beta and delta subunits of eIF2B from mammals and yeast. *Biochem J*. 1996;318(Pt 2):637–643.
35. Yang W, Hinnebusch AG. Identification of a regulatory subcomplex in the guanine nucleotide exchange factor eIF2B that mediates inhibition by phosphorylated eIF2. *Mol Cell Biol*. 1996;16(11):6603–6616.
36. Pavitt GD, Yang W, Hinnebusch AG. Homologous segments in three subunits of the guanine nucleotide exchange factor eIF2B mediate translational regulation by phosphorylation of eIF2. *Mol Cell Biol*. 1997;17(3):1298–1313.
37. Krishnamoorthy T, Pavitt GD, Zhang F, Dever TE, Hinnebusch AG. Tight binding of the phosphorylated alpha subunit of initiation factor 2 (eIF2alpha) to the regulatory subunits of guanine nucleotide exchange factor eIF2B is required for inhibition of translation initiation. *Mol Cell Biol*. 2001;21(15):5018–5030.
38. Wortham NC, Martinez M, Gordiyenko Y, Robinson CV, Proud CG. Analysis of the subunit organization of the eIF2B complex reveals new insights into its structure and regulation. *FASEB J*. 2014;28(5):2225–2237.
39. Pavitt GD. eIF2B, a mediator of general and gene-specific translational control. *Biochem Soc Trans*. 2005;33(Pt 6):1487–1492.
40. Jennings MD, Pavitt GD. eIF5 has GDI activity necessary for translational control by eIF2 phosphorylation. *Nature*. 2010;465(7296):378–381.
41. Jennings MD, Zhou Y, Mohammad-Qureshi SS, Bennett D, Pavitt GD. eIF2B promotes eIF5 dissociation from eIF2*GDP to facilitate guanine nucleotide exchange for translation initiation. *Genes Dev*. 2013;27(24):2696–2707.
42. Han AP, Yu C, Lu L, et al. Heme-regulated eIF2alpha kinase (HRI) is required for translational regulation and survival of erythroid precursors in iron deficiency. *EMBO J*. 2001;20(23):6909–6918.
43. Murguia JR, Serrano R. New functions of protein kinase Gcn2 in yeast and mammals. *IUBMB Life*. 2012;64(12):971–974.
44. Garcia MA, Meurs EF, Esteban M. The dsRNA protein kinase PKR: virus and cell control. *Biochimie*. 2007;89(6–7):799–811.

45. Walter P, Ron D. The unfolded protein response: from stress pathway to homeostatic regulation. *Science*. 2011;334(6059):1081–1086.
46. Pfeiffer SE, Warrington AE, Bansal R. The oligodendrocyte and its many cellular processes. *Trends Cell Biol*. 1993;3(6):191–197.
47. Wang X, Paulin FE, Campbell LE, et al. Eukaryotic initiation factor 2B: identification of multiple phosphorylation sites in the epsilon-subunit and their functions in vivo. *EMBO J*. 2001;20(16):4349–4359.
48. Woods YL, Cohen P, Becker W, et al. The kinase DYRK phosphorylates protein-synthesis initiation factor eIF2Bepsilon at Ser539 and the microtubule-associated protein tau at Thr212: potential role for DYRK as a glycogen synthase kinase 3-priming kinase. *Biochem J*. 2001; 355(Pt 3):609–615.
49. Pap M, Cooper GM. Role of translation initiation factor 2B in control of cell survival by the phosphatidylinositol 3-kinase/Akt/glycogen synthase kinase 3beta signaling pathway. *Mol Cell Biol*. 2002;22(2):578–586.
50. Wang X, Janmaat M, Beugnet A, Paulin FE, Proud CG. Evidence that the dephosphorylation of Ser(535) in the epsilon-subunit of eukaryotic initiation factor (eIF) 2B is insufficient for the activation of eIF2B by insulin. *Biochem J*. 2002;367(Pt 2):475–481.
51. Taylor EJ, Campbell SG, Griffiths CD, et al. Fusel alcohols regulate translation initiation by inhibiting eIF2B to reduce ternary complex in a mechanism that may involve altering the integrity and dynamics of the eIF2B body. *Mol Biol Cell*. 2010;21(13):2202–2216.
52. Richardson JP, Mohammad SS, Pavitt GD. Mutations causing childhood ataxia with central nervous system hypomyelination reduce eukaryotic initiation factor 2B complex formation and activity. *Mol Cell Biol*. 2004;24(6):2352–2363.
53. Li W, Wang X, Van Der Knaap MS, Proud CG. Mutations linked to leukoencephalopathy with vanishing white matter impair the function of the eukaryotic initiation factor 2B complex in diverse ways. *Mol Cell Biol*. 2004;24(8):3295–3306.
54. Liu R, van der Lei HD, Wang X, et al. Severity of vanishing white matter disease does not correlate with deficits in eIF2B activity or the integrity of eIF2B complexes. *Hum Mutat*. 2011;32(9):1036–1045.
55. van der Lei HD, van Berkel CG, van Wieringen WN, et al. Genotype-phenotype correlation in vanishing white matter disease. *Neurology*. 2010;75(17):1555–1559.
56. Fogli A, Schiffmann R, Hugendubler L, et al. Decreased guanine nucleotide exchange factor activity in eIF2B-mutated patients. *Eur J Hum Genet*. 2004;12(7):561–566.
57. Horzinski L, Huyghe A, Cardoso MC, et al. Eukaryotic initiation factor 2B (eIF2B) GEF activity as a diagnostic tool for EIF2B-related disorders. *PLoS One*. 2009;4(12):e8318.
58. van Kollenburg B, Thomas AA, Vermeulen G, et al. Regulation of protein synthesis in lymphoblasts from vanishing white matter patients. *Neurobiol Dis*. 2006;21(3):496–504.
59. Damon-Perriere N, Menegon P, Olivier A, et al. Intra-familial phenotypic heterogeneity in adult onset vanishing white matter disease. *Clin Neurol Neurosurg*. 2008;110(10):1068–1071.
60. Kantor L, Harding HP, Ron D, et al. Heightened stress response in primary fibroblasts expressing mutant eIF2B genes from CACH/VWM leukodystrophy patients. *Hum Genet*. 2005;118(1):99–106.
61. Horzinski L, Kantor L, Huyghe A, et al. Evaluation of the endoplasmic reticulum-stress response in eIF2B-mutated lymphocytes and lymphoblasts from CACH/VWM patients. *BMC Neurol*. 2010;10:94.
62. Rodriguez D, Gelot A, della Gaspera B, et al. Increased density of oligodendrocytes in childhood ataxia with diffuse central hypomyelination (CACH) syndrome: neuropathological and biochemical study of two cases. *Acta Neuropathol*. 1999;97(5):469–480.
63. Van Haren K, van der Voorn JP, Peterson DR, van der Knaap MS, Powers JM. The life and death of oligodendrocytes in vanishing white matter disease. *J Neuropathol Exp Neurol*. 2004;63(6):618–630.
64. Francalanci P, Eymard-Pierre E, Dionisi-Vici C, et al. Fatal infantile leukodystrophy: a severe variant of CACH/VWM syndrome, allelic to chromosome 3q27. *Neurology*. 2001;57(2):265–270.
65. Wong K, Armstrong RC, Gyure KA, et al. Foamy cells with oligodendroglial phenotype in childhood ataxia with diffuse central nervous system hypomyelination syndrome. *Acta Neuropathol*. 2000;100(6):635–646.
66. Bugiani M, Boor I, van Kollenburg B, et al. Defective glial maturation in vanishing white matter disease. *J Neuropathol Exp Neurol*. 2011;70(1):69–82.
67. Vanderver A, Hathout Y, Maletkovic J, et al. Sensitivity and specificity of decreased CSF asialotransferrin for eIF2B-related disorder. *Neurology*. 2008;70(23):2226–2232.
68. Vanderver A, Schiffmann R, Timmons M, et al. Decreased asialotransferrin in cerebrospinal fluid of patients with childhood-onset ataxia and central nervous system hypomyelination/vanishing white matter disease. *Clin Chem*. 2005;51(11):2031–2042.
69. Schiffmann R, Moller JR, Trapp BD, et al. Childhood ataxia with diffuse central nervous system hypomyelination. *Ann Neurol*. 1994;35(3):331–340.
70. Geva M, Cabilly Y, Assaf Y, et al. A mouse model for eukaryotic translation initiation factor 2B-leucodystrophy reveals abnormal development of brain white matter. *Brain*. 2010;133(Pt 8):2448–2461.
71. Marom L, Ulitsky I, Cabilly Y, Shamir R, Elroy-Stein O. A point mutation in translation initiation factor eIF2B leads to function- and time-specific changes in brain gene expression. *PLoS One*. 2011;6(10):e26992.
72. Barnea-Goraly N, Menon V, Eckert M, et al. White matter development during childhood and adolescence: a cross-sectional diffusion tensor imaging study. *Cereb Cortex*. 2005;15(12):1848–1854.
73. Noble M, Mayer-Proschel M, Miller RH. The oligodendrocyte. In: Rao MS, Jacobson M, eds. Developmental Neurobiology 4th ed. Springer, New-York. 2005:151–196.
74. Dietrich J, Lacagnina M, Gass D, et al. EIF2B5 mutations compromise GFAP+astrocyte generation in vanishing white matter leukodystrophy. *Nat Med*. 2005;11(3):277–283.
75. Cabilly Y, Barbi M, Geva M, et al. Poor cerebral inflammatory response in eIF2B knock-in mice: implications for the aetiology of vanishing white matter disease. *PLoS One*. 2012;7(10):e46715.
76. van der Voorn JP, van Kollenburg B, Bertrand G, et al. The unfolded protein response in vanishing white matter disease. *J Neuropathol Exp Neurol*. 2005;64(9):770–775.
77. van Kollenburg B, van Dijk J, Garbern J, et al. Glia-specific activation of all pathways of the unfolded protein response in vanishing white matter disease. *J Neuropathol Exp Neurol*. 2006;65(7):707–715.

78. Kantor L, Pinchasi D, Mintz M, Hathout Y, Vanderver A, Elroy-Stein O. A point mutation in translation initiation factor 2B leads to a continuous hyper stress state in oligodendroglial-derived cells. *PLoS One*. 2008;3(11):e3783.
79. de Lonlay-Debeney P, von Kleist-Retzow JC, Hertz-Pannier L, et al. Cerebral white matter disease in children may be caused by mitochondrial respiratory chain deficiency. *J Pediatr*. 2000;136(2):209–214.
80. Schiffmann R, Elroy-Stein O. Childhood ataxia with CNS hypomyelination/vanishing white matter disease–a common leukodystrophy caused by abnormal control of protein synthesis. *Mol Genet Metab*. 2006;88(1):7–15.

S E C T I O N X

NEUROPATHIES AND NEURONOPATHIES

CHAPTER

87

Amyotrophic Lateral Sclerosis

Jemeen Sreedharan and Robert H. Brown, Jr.†*

*Babraham Institute, Cambridge, UK

†University of Massachusetts Medical School, Worcester, MA, USA

INTRODUCTION

In this chapter, we will review those genes most strongly associated with classical amyotrophic lateral sclerosis (ALS). We will not review the clinical features of this disorder in detail other than to emphasize that this a uniformly fatal, paralytic disorder that begins with focal motor weakness, usually in the distal limbs but sometimes in the bulbar musculature, and then spreads to paralyze all skeletal muscles. Without respiratory assistance, death from ventilatory failure ensues, typically within 5 years. In some cases, these features are associated with frontotemporal dementia (FTD), characterized initially by disturbances of behavior and speech. At the microscopic level, the primary finding is motor neuron death often with deposition in neurons and non-neuronal cells of proteinaceous aggregates and, in some cases, intranuclear RNA deposits. Most cases at autopsy reveal both astrogliosis and microgliosis in addition to loss of motor neurons. Importantly, both upper motor neurons (UMNs) and lower motor neurons (LMNs) are affected. About 10% of cases are inherited as dominant traits (familial ALS; fALS).

The effort to identify ALS genes began in earnest in the early and mid-1980s. In a setting in which the fundamental pathology of ALS was ill understood, the premise was that insights into pathogenesis would be garnered from analysis of pathways implicated by mutant ALS genes. Not surprisingly, the technologies for gene discovery have evolved substantially over these last 30 years. The first and initially most productive phase of ALS gene discovery invoked genetic linkage methods, in which co-migration of common variants with the disorder could be used in pedigrees to define the general chromosomal addresses of the genes. This method required both detailed, multigenerational family structure with clear inheritance patterns and extensive work after linkage was defined to detect the causative mutation within the linked locus; usually that entailed laborious manual sequencing of scores of candidate genes. The second phase focused on genome-wide association studies (GWAS), which offered at least two potent advantages: the use of hundreds of thousands of single nucleotide polymorphisms across the genome and the ability to screen sporadic ALS (sALS) as well as fALS for association of variants with this disease. Several candidate genes have been identified through this technology, although in general this method has been less definitive than the linkage studies. In the third and present phase of ALS genetic investigations, next-generation high-throughput sequencing has been combined with exome capture to greatly facilitate identification of ALS-causing gene mutations.

Thirty years later it can be argued that the premise that ALS genetics would define critical molecular events in ALS pathogenesis was well founded, although there is, as yet, no effective ALS treatment. More than 100 genetic variants have been reported to correlate either with ALS susceptibility or specific phenotypes (e.g., http://alsod.iop.kcl.ac.uk/). As recently reviewed, there are several approaches to discerning which of these may be robustly described as "ALS genes," including methods using bioinformatics.[1] This chapter will discuss a set of genes that in our view may robustly be considered ALS genes, as gauged by publication records and the degree to which they illuminate aspects of ALS biology. It is useful to group these into three broad categories of molecular pathology involving: 1) altered conformational stability and turnover of critical proteins; 2) disturbances in RNA biology, implicating RNA structure and homeostasis and RNA binding proteins; and 3) perturbations in aspects of axonal biology (Figure 87.1).

http://dx.doi.org/10.1016/10.1016/B978-0-12-410529-4.00087-5

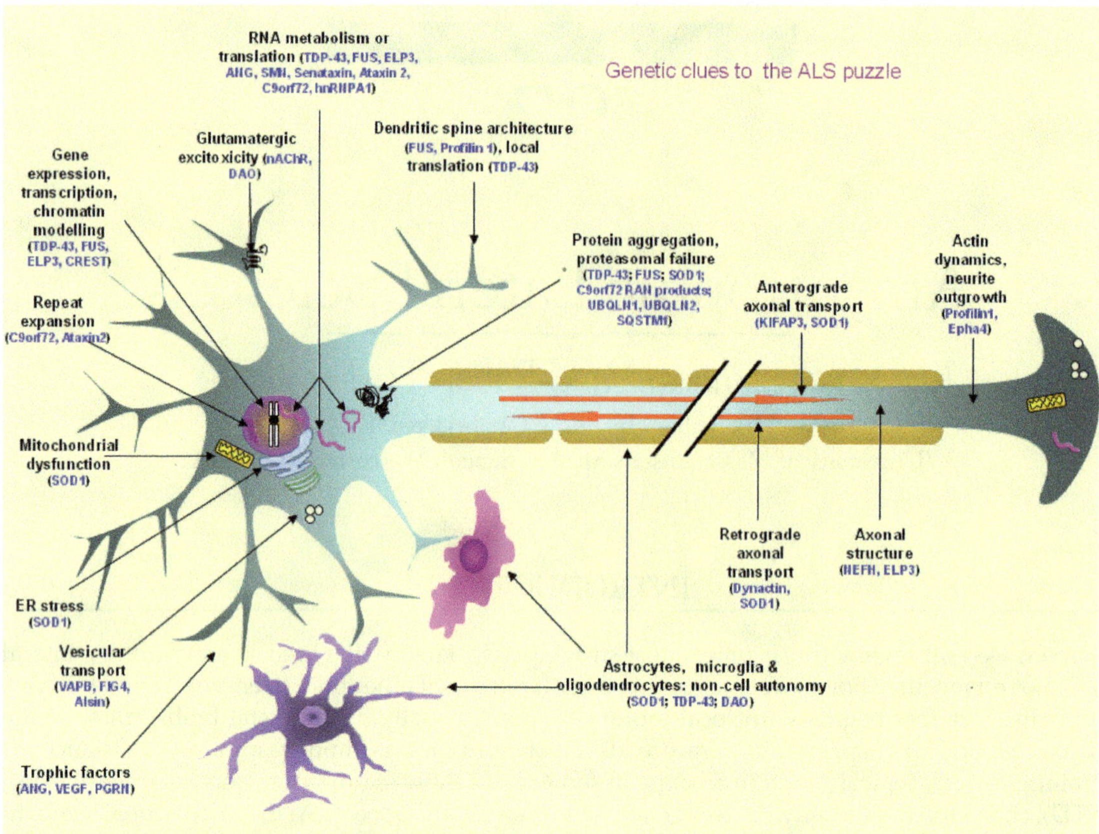

FIGURE 87.1 **Overview of ALS genes.** The complexity of motor neuron degeneration in ALS is striking, as illustrated by this diagram, which highlights processes and pathways in motor neuron death identified by analysis of ALS-linked gene mutations. Motor neurons are vulnerable in many ways, partly due to their sheer size. Motor neuron cell bodies are among the largest of all neuronal cells, and their axons among the longest. The motor neuron cell body may approach 100 μm in diameter; the axons may be several thousand-fold longer than the diameter of the motor neuron and encompass more than 90% of the volume of the cell. The length of the axon is clearly underestimated in this figure (hence the break in the axon indicated by two lines). The role of glial cells (axon ensheathing oligodendrocytes, astrocytes, which have roles in the regulation of extracellular neurotransmitter levels, and microglia, which have important neuroinflammatory functions) is also indicated, with both TDP-43 and SOD1 models showing that these proteins have an important contributory role in neurodegeneration in animal and cell models.
For the last 20 years it has been acknowledged that protein toxicity is a critical element in motor neuron death in ALS, especially in the context of intraneuronal inclusions of SOD1, TDP-43, and FUS, often concomitantly with p62 (see also Figure 87.2). An emerging theme from ALS genetics (see text) is that disturbances in RNA processing are pivotal in both fALS and sALS (see Figure 87.3). Studies of the *C9orf72* ALS gene implicate both RNA and protein toxicity via intranuclear RNA foci and cytoplasmic dipeptide repeat inclusions. Endoplasmic reticulum (ER) stress and perturbations of proteasomal and autophagosomal protein processing are also implicated, as is excessive electrical excitability of this cell. The theme of excitotoxicity is consistent with the finding of rare variants in genes that could promote glutamate excitotoxicity (and also by the observation that riluzole has modest benefit in ALS).
Downstream effects of aberrant RNA processing and protein toxicity include early disruption of the structure and function of neuronal peripheral compartments, which make up the vast majority of a neuron by volume and which are at a great distance from the cell body. Thus, axon degeneration is a critical early step in ALS and, therefore, is an important therapeutic target. Intriguingly, a role for developmental abnormalities in the pathogenesis of ALS is suggested by mutations in genes with a role in the outgrowth of neural processes, such as *Profilin1* and *Epha4*.

We recognize that our understanding of pathophysiology is incomplete and that some genetic variants are likely to be neurotoxic through diverse mechanisms. Nonetheless, this approach provides a thematic scheme for organizing these genes, a schema that one hopes will help guide the practice of future clinical trials.

Table 87.1 lists the genes most strongly linked to both familial and apparently sporadic classical ALS. Table 87.2 enumerates a larger set of genes whose mutations cause not only ALS but also other forms of motor neuron disease. In a broad overview of these tables, a few points are striking. First, despite the importance of trophic factors and their receptors in the development of motor neurons and related neural circuitry, it is striking that neither is primarily implicated in any of the major categories of motor neuron diseases. Second, it appears that the rapidly growing category of ALS gene defects involving altered RNA metabolism are not commonly implicated in other types of motor neuron disease. And, third, it also appears that motor neuron disorders that are slowly progressive and nonlethal (e.g., hereditary spastic paraplegia [HSP] and type 2 Charcot–Marie–Tooth disease [CMT]) more commonly involve

TABLE 87.1 Major Amyotrophic Lateral Sclerosis Genes

Familial (26)			Sporadic (6)
Disease variant	Locus	Gene	Gene
ALS1	21q	*SOD1*	*ELP3*
ALS2	2q	*Alsin*	*KIFAP3*
ALS4	9q	*Senataxin*	*CHGB*
ALS6	16q	*FUS*	*UNC13A*
ALS8	20q	*VAPB*	*EphA4*
ALS9	14q	*ANG*	*CREST*
ALS10	1p	*TDP-43*	
ALS11	6q	*FIG4*	
ALS12	10p	*OPTN*	
ALS13	12q	*Ataxin-2*	
ALS14	9p	*VCP*	
ALS18	17p	*PFN1*	
ALS	12q	*DAO*	
ALS	17q	*TAF15*	
ALS	19p	*NTE*	
ALS	2p	*Dynactin*	
ALS	5q	*SQSTM1/p62*	
ALS	7q	*PON1–3*	
ALS	9p	*C9orf72*	
ALS	9p	*Sigma-R1*	
ALS	9q	*Ubiquilin-1*	
ALS	X	*Ubiquilin-2*	
ALS	12q	*hnRNPA1*	
ALS	7p	*hnRNPA2B1*	

genes and proteins related to the axonal cytoskeleton than is the case in ALS. Stated differently, one lesson from the genetics of motor neuron disease is that the fulminant process of cell death in ALS is primarily somatocentric with components of axonal pathology, while the central pathology in HSP and axonal CMT is fundamentally axonal.

ALTERED CONFORMATIONAL STABILITY AND TURNOVER OF CRITICAL PROTEINS IN ALS

Cytosolic Copper/Zinc Superoxide Dismutase (SOD1)

The first successful genome-wide linkage study of fALS implicated a locus at chromosome 21q21, leading to the identification of mutations in the *SOD1* gene in fALS.[2,3] To date, nearly 180 mutations in this gene have been identified in the SOD1 protein, affecting nearly every residue (153 amino acids), with no "hotspots"; as assessed by their rarity in the National Heart, Lung, and Blood Institute (NHLBI) exome database, most of these mutations are likely to be pathogenic. Mutations in *SOD1* are the second commonest known cause of ALS, accounting for 7–20% of fALS and 1–3% of sALS cases in most populations studied (http://alsod.iop.kcl.ac.uk/Als). Of the ~180 *SOD1* gene mutations reported in fALS, the vast majority are missense mutations, with a smaller number of nonsense mutations, insertions and deletions. Most mutations are inherited in an autosomal dominant manner. In northern Sweden and Finland, the D90A missense variant occurs heterozygously as a polymorphism in the normal population; however, when homozygously present, it causes a slowly progressive form of ALS, often with prominent corticospinal features. Thus, in these populations this mutation causes ALS recessively.[4] In these populations, haplotype markers around the *SOD1* locus point to a single ancient founder for D90A cases.[5] By contrast, in other populations, the same D90A mutation is

TABLE 87.2 Summary of 109 Motor Neuron Disease Genes by Gene Category

Disorder	#	Growth factor and receptors	GEF, other signaling	DNA/RNA metabolism	Heat shock proteins	ER, Golgi vesicles	Transcription factors	Mito-chondrial	ABC transporters	Enzymes	Axonal transport, motors	Cytoskel., axonal sprouting	Surface proteins, myelin	Neuro-transmitters, receptors	Protein degradation	Membrane transporter
fALS	14		OPTN	FUS TPD43 TAF15 C9orf72 hnRNPA1		VAPB				SOD1 PON1–3 D-AO		Pfn-1			SQSTM1	
sALS	11		EphA4	EP3 ANG CREST		FIG4 CHGB					KIFAP3			UNC13A CHRNA3, A4,B4		
ALS-like	6		Alsin	SETX hnRNPA2B1		SIGMAR1				NTE	Dynactin					
PN	30	TrkA NGFβ	RhoGEF10 IKBPKAP	GARS YARS AARS	HSP22 HSP27	SIMPLE FAM134B	ERG2 MTM2 NMDRG1 IKAP SOD10	Mitofucin	ABC1	SerPT1 SerPT2 Rab7 ASA Gal-ases PhyCOase Por-deam WNK1	KIF1Bb	NFL Gigaxonin	SCN9			
LMN	16		PLEKHG5	SMN And-R IGHMBP2	HSBP1 HSBP2	PIP5K1C		SCO2 TK2		Hex A,B	KIF	SMAX3				SLC52A1 SLC52A2 SLC52A3
HSP	27		Spardin		HSP60 HSPD1	Msprdn REEP1 TFG Spartin	BSCL2 ZFYVE27	Prplegin Cytoch C REEP1	ABCD1	CYP7B1	KIF5a Atlastin NIPA1 Spatacsin Spastizin Maspardin	Stumpellin SLC33A1 SLC16A2	L1CAM PLP GJC2 CYP7B1			
ALS-FTD	5		PGRN			CHMP2B VCP						Tau			UBQL2	
TOTAL	109	2	8	16	6	13	7	6	2	17	10	8	5	4	2	3

Abbreviations: fALS, familial amyotrophic lateral sclerosis; sALS, sporadic amyotrophic lateral sclerosis; PN, peripheral neuropathy; LMN, lower motor neuron; HSP, hereditary spastic paraplegia; ALS-FTD, amyotrophic lateral sclerosis with frontotemporal dementia.

transmitted as a dominant trait. An additional recessive *SOD1* mutation (delta G27/P28) has also been described in a Filipino kindred. The unique exonic six base pair deletion results in alternative splicing and reduced expression of the mutant transcript, which may explain the reduced penetrance of this mutation.[6]

With some exceptions (notably the D90A/D90A cases in Sweden) most *SOD1* mutations cause a mixed UMN/LMN or predominantly LMN phenotype. No pedigrees have been reported with significant dementia attributed to mutant *SOD1*. Except for a few variants, nearly all *SOD1* mutations are associated with a poor prognosis. The commonest mutation in the USA, A4V, is particularly aggressive, with survival sometimes only a few months, while homozygous D90A patients have a slower progression with an average survival of 14 years. The I113T mutation has a very low penetrance and is the most frequent mutation found in sporadic ALS cases. Apart from these exceptions, there is a relatively poor genotype–phenotype correlation for *SOD1* mutations.

The identification of mutations in the gene encoding Cu/Zn superoxide dismutase (*SOD1*)[7] led rapidly on to the creation of transgenic mice overexpressing mutant SOD1 isoforms, which demonstrated striking ALS-like phenotypes.[8] Mutant SOD1 ALS mice, and subsequently ALS rats, have now been generated using several mutant *SOD1* alleles. The most extensively studied is the G93A allele; most cause focal then disseminated weakness in mice, usually starting only after a period of normal motor development. For some alleles, such as G93A, it is clearly the case that the age of onset and rate of progression are dependent on the dose of mutant protein. In these mice, as in some human instances, higher levels of the mutant protein correlate with more severe phenotypes (for review see [9]).

Mutant SOD1 disturbs cellular function in many ways (Figure 87.2), including (but not limited to) the following: 1) In ALS mice, mutant SOD1 disturbs dendritic arborization of neurons in the spinal cord. 2) It directly impairs mitochondrial function accumulating in the inner membrane space of this organelle.[10] 3) It induces endoplasmic reticulum (ER) stress and thereby exerts several deleterious influences, including activating the pro-apoptotic protein CHOP.[11] ER stress also leads to activation of the unfolded protein response (UPR), which is pro-apoptotic.[11] The UPR also reduces protein synthesis; while this may help relieve some cellular stress, prolonged shutdown of protein synthesis may lead to loss of key proteins (e.g., in the synapse[12]). 4) Mutant SOD1 promotes excessive electrical activity, probably in part by degrading glutamate transporters such as EAAT2.[13,14] 5) It impairs axonal transport by altering the phosphorylation state of kinases that modulate transport rate.[15] 6) Because SOD1 has also been detected in small amounts in the nucleus, it is possible that cytoplasmic retention of SOD1 may deplete nuclear SOD1 and thereby promote oxidative pathology in the nucleus.[16] 7) Moreover, mutant SOD1 is secreted. Upon secretion, mutant SOD1 activates microglia through the CD14/TLR receptor increasing the generation of superoxide and promoting inflammation.[17,18] 8) Another important mechanism whereby mutant SOD1 may enhance motor neuron death is through a prion-like mechanism.[19] Data suggest that mutant SOD1 is toxic, either as a monomer or through formation of toxic oligomers.[20–22] Importantly, several studies have shown that wild-type SOD1 also can misfold under conditions of cell stresses, implicating SOD1 toxicity in sALS.[23–25] Whether induced by cellular stress or germline mutations, misfolded SOD1 has the capacity to seed misfolding in normal SOD1 and, thus, in a prion-like manner, propagate the fibrillization and misfolding process both intracellularly and, after these proteins are secreted, extracellularly. It has been argued that the latter process, extracellular spread of misfolded toxic SOD1 species may underlie the spread of ALS after focal onset, although this has not been proven. Although originally proposed in the context of mutant SOD1, the prion hypothesis now encompasses other ALS genes that possess domains of low complexity thought to predispose to prion-like protein aggregation. Interestingly, both TDP-43 and FUS (described below) have also been shown in modeling studies to have prion-like potential in their low complexity domains.[26]

Ubiquilin 2

It has been recognized for more than 20 years that there are rare families in which fALS or fALS and dementia are transmitted as X-linked traits. It was therefore important that missense mutations in the X-linked gene encoding ubiquilin 2 were identified in X-linked ALS and X-linked ALS-FTD.[27] At least at this time, these missense mutations are the only known X-linked cause of ALS. In the first cases identified, the offending mutations all involve the proline-rich, so-called PXX, domain (residues 491–526).

Rare *UBQLN2* mutations have been found by several other groups in large European cohorts of both sporadic and familial ALS. Overall, *UBQLN2* mutations probably account for 1–2% of fALS.[27–30] While many of these variants also convert prolines within the PXX domain (clearly a mutation hotspot), other mutations outside (but still near to) the PXX domain have also been detected, although these have not been proven to segregate with disease in kindreds as DNA was not available from additional family members. Thus, it remains plausible that some of these mutations may not be pathogenic. Indeed, a control case has been found with an N439I mutation, while one individual with fALS with the known disease-causing TDP-43 M337V mutation was also found to have a ubiquilin 2 M446R

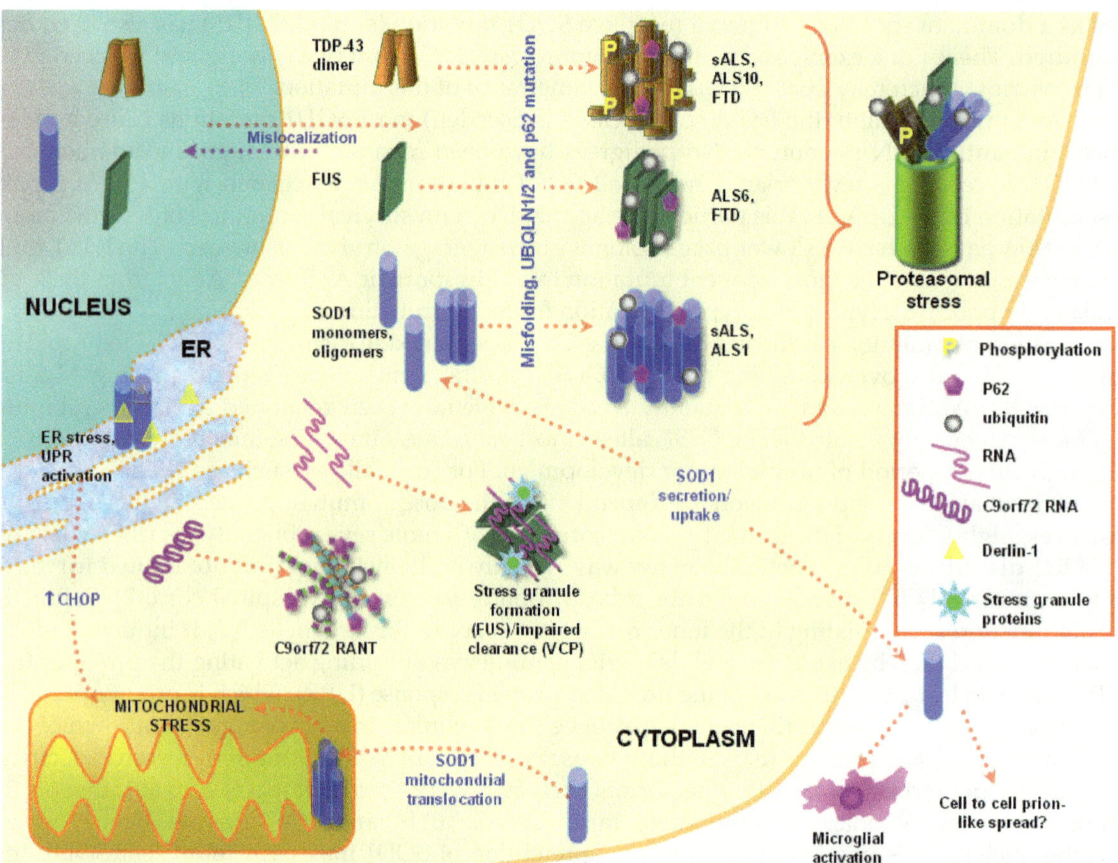

FIGURE 87.2 **Defective protein processing in ALS.** Protein toxicity in ALS can entail both loss of normal protein function and the adverse consequences of novel protein functions acquired via mutations. This figure depicts some of the proposed forms of neurotoxicity from ALS-linked mutant proteins, with a focus on SOD1, TDP-43, FUS and C9orf72. A common finding in these different forms of ALS is the presence of cytoplasmic protein aggregates, often tagged with ubiquitin and p62. To date, it remains unclear whether these accumulations in the cytoplasm (and, in the case of TDP-43, also in the nucleus) are themselves directly toxic. C9orf72 RAN (repeat associated non-ATG) translation products (RANT) are repeat dipeptides that are detected as cytoplasmic aggregates. Abnormal cytosolic proteins can compromise protein degradative pathways such as the proteasome and autophagy. Mutations in p62 (*SQSTM1*) and ubiquilin 1 and 2 (*UBQLN1* and *2*) also occur in ALS and may impair the labeling and chaperoning of misfolded proteins towards lysosomal and proteasomal degradation.
As described in the text, the deleterious effects of mutant SOD1 on cellular function are numerous and complex, involving mitochondrial impairment (accumulation of mutant SOD1 in the inner membrane space), endoplasmic reticulum (ER) stress and activation of the unfolded protein response (UPR), reduced protein synthesis, excessive electrical activity, impaired axonal transport, and possibly even reduction of intranuclear antioxidative benefits of SOD1. Additionally, secretion of mutant SOD1 activates microglia.
Mutant SOD1 may also be neurotoxic via a prion-like mechanism. Both mutant (and also wild-type) SOD1 can misfold under conditions of cellular stress and, once misfolded, recruit other molecules of SOD1 to misfold and fibrillize. Extracellular spread of prion-like SOD1 complexes may explain the clinical spread of ALS after focal onset. This prion hypothesis has been extended to encompass other ALS genes that possess domains of low complexity thought to predispose to prion-like protein aggregation, such as *TDP-43* and *FUS*. Stress granules are likely to accelerate this process by sequestering RNA binding proteins with low complexity, prion-like domains, thereby facilitating binding and seeding of fibrillization.

mutation.[30] Another fALS case with an optineurin Q314L mutation was found to have a ubiquilin 2 P533L mutation.[30] Which of these mutations in these two latter cases is disease-causing, or whether both mutations are required for pathogenesis, remains to be determined.

Ubiquilin 2 is thought to be involved in regulating proteasomal degradation of ubiquitinated proteins. The mechanisms whereby the mutations trigger ALS or ALS-FTD is not clear as functional studies of the ubiquilin 2 variants are only just emerging. It is important to note that ubiquilin 2 is detected in neuronal aggregates not only in patients with *UBQLN2* mutations, but also in sporadic ALS cases with wild-type UBQLN2. This suggests that ubiquilin 2 is implicated rather broadly in all types of ALS. Of particular interest are studies showing that ubiquilin 2 overexpression in human neuroglioma H4 cells *in vitro* leads to a fall in levels of TDP-43, another gene and protein that are important in ALS pathogenesis.[31] Ubiquilin 2 is able to bind TDP-43 and may play a role in clearing TDP-43, suggesting a link to TDP-43-mediated aggregation.

It is of interest that ubiquilin 2 is a member of a family of four ubiquilin genes. While the other ubiquilins have not yet been implicated in ALS pathogenesis, a novel mutation in the gene encoding ubiquilin 1 (*UBQLN1*) has been reported in a form of childhood-onset motor neuron disease with early bulbar involvement and sensorineural hearing loss. That mutation was shown to correlate with enhanced cytosolic mislocalization and aggregation of TDP-43 and with reduced proteasomal function.[32]

Valosin-Containing Protein (VCP)

Inclusion body myopathy with Paget disease of the bone and frontotemporal dementia (abbreviated IBMPFD; OMIM 167320) is a rare autosomal dominant disease that has an interesting clinical, pathological, and genetic overlap with ALS. Mutations in valosin-containing protein (VCP) were the first identified cause of IBMPFD.[33] More recently, exome sequencing led to the identification of a VCP mutation (R191Q) in a large Italian pedigree with typical ALS phenotype.[34] Further screens have identified other rare missense mutations of VCP in fALS and apparently sporadic ALS. So far, eight missense variants (R155H, R155C, R159G, R191Q, and D592N in fALS; and R159C, N387T, and R662C in sALS) and a number of intronic variants of uncertain significance have been described.[34–36] Studies of cohorts of fALS of European descent (amounting to around 300–400 probands) indicate that VCP mutations account for <1% of fALS. One woman of African descent has been described with VCP mutation.[37] Although VCP mutations have also been found in Korean and Chinese patients with the IBMPFD phenotype,[38,39] VCP mutation has not as yet been noted in ALS patients from the Far East.

The phenotype associated with VCP mutation may vary within and between kindreds. The same mutation may cause inclusion body myopathy (IBM) or typical ALS (with limb onset and unequivocal upper and lower motor neuron involvement), with or without Paget disease or FTD. Age of onset can vary from late 20s to mid-50s and survival appears to show some correlation with phenotype, with myopathic patients having long protracted courses over decades, while ALS patients progress more quickly (although still relatively slowly compared to sporadic ALS). Where weakness is the predominant feature, it tends to be of limb onset, usually in the legs and usually proximal. Bulbar and respiratory muscles do become affected in some patients. Interestingly, upper motor neuron signs tend to be seen in those with neurogenic as opposed to myogenic electromyogram (EMG) findings, strongly suggesting that these individuals have true ALS rather than IBM.

As always, careful clinical characterization of the individual case is important to make an accurate diagnosis. In this respect, careful note should be made of the serum creatine kinase and alkaline phosphatase, which may indicate myopathy and Paget disease, respectively. Bedside testing and formal neuropsychometry may reveal frontotemporal dysfunction. EMG may show neurogenic or myopathic changes. Intriguingly, some patients demonstrate both myopathic and neurogenic changes,[35] suggesting dual pathology with both IBM and ALS. Muscle histological studies of these patients will help answer this question.

Perhaps the most interesting pathological observation is that the brains of VCP mutation patients demonstrate TDP-43 pathology. A VCP knock-in mouse also displays TDP-43 pathology in both spinal cord motor neurons and muscle.[40] This suggests that VCP-linked ALS and IBMPFD are on the same disease spectrum as sporadic ALS and TDP-43-associated frontotemporal lobar degeneration (FTLD-TDP). Thus, understanding VCP malfunction may help us understand ALS-FTLD in the broadest sense.

VCP is a type II AAA+ (ATPase associated with multiple activities) protein, which forms hexamers. It has roles in ER-associated degradation (ERAD), ER protein export into the cytosol, nuclear envelope reconstruction, membrane fusion, postmitotic Golgi reassembly, DNA damage response, suppressing apoptosis, ubiquitin-dependent protein degradation, and the recruitment and targeting of proteins to the 26S proteasome for lysosomal degradation.[41–47] VCP deficiency causes uncoupling of mitochondrial respiration from oxidative phosphorylation, causing loss of mitochondrial membrane potential, although the mechanism is unclear.[48] VCP is a remarkably highly conserved protein in multiple species including yeast (CDC48), flies (ter94) and mice (p97). Coded by 17 exons and consisting of 806 amino acids, VCP consists of an *N*-terminal domain, followed by a linker, an ATPase domain, another linker, a second ATPase domain, and a *C*-terminal domain. VCP is extensively modified by phosphorylation, acetylation, and by lysine methylation.[49] The *N*-terminus binds different cofactors, which may explain how VCP is able to perform a variety of functions. Given that almost all of the mutations in VCP occur in this *N*-terminal region, one might therefore expect some genotype–phenotype correlation, as a consequence of mutation-specific defect in cofactor interaction. However, apart from the observation that A229E causes especially severe IBMPFD,[50] so far there does not appear to be a clear relationship between mutation and phenotype.

Optineurin

Linkage analysis of consanguineous Japanese families with ALS led to the identification of autosomal recessive mutations in the *OPTN* gene, which encodes optineurin.[51] Subsequently, rare heterozygous mutations were identified in European ALS cases.[52] Truncation and missense mutations are responsible for less than 1% of fALS. It is clear that there is clinical diversity among patients carrying *OPTN* mutations. In addition to the ALS cases, some *OPTN* mutations cause open-angle glaucoma.[53] Moreover, a recent GWAS also highlighted a role for optineurin in Paget disease of bone.[54]

Optineurin is a multifunctional transcription factor with roles in Golgi membrane trafficking, and inflammation, vasoconstriction, and apoptosis.[55] How its mutations compromise motor neuron viability is unclear. While it is likely that *OPTN* mutations mediate motor neuron disease via loss of function of the protein, an acquired toxic function cannot be excluded. Some data support the original suggestion that the mutations impair the ability of *OPTN* to inhibit activation of the nuclear factor kappa-B (NFκB) pathway.[51] The mechanistic complexity is underscored by the finding that both recessive and heterozygous mutations in *OPTN* cause ALS, although an important observation is that *OPTN* cases demonstrate TDP-43 inclusions on neuropathological examination.[51]

Sequestosome 1 (*SQSTM1*)

SQSTM1 encodes the ubiquitin-binding protein p62, which regulates ubiquitin binding and the activation of NFκB. p62 is multifunctional and figures importantly in protein homeostasis, controlling proteasomal and autophagic protein degradation. Mutations in this gene cause Paget disease of bone.[56] Missense and deletion mutants have been detected infrequently (<1%) among ALS cases.[57] However, co-migration of mutations in *SQSTM1* with ALS in a large pedigree has not been described.

DEFECTS IN RNA PROCESSING GENES AND PROTEINS IN ALS

TAR DNA Binding Protein (*TARDBP*, TDP-43)

In 2006, the transactivating response DNA binding protein of 43 kD (TDP-43, encoded by the *TARDBP* gene) was found to be the major ubiquitinated protein in the pathological inclusions that characterize sALS and FTLD.[58,59] This observation consolidated the previously established clinical and genetic link between these two disease entities. TDP-43 is a highly conserved, ubiquitously expressed heterogeneous nuclear ribonucleoprotein (hnRNP). It is present mostly in the nucleus, but shuttles in and out of the cytoplasm and along axons. It functions in gene transcription (Figure 87.3), including gene silencing,[60–62] as well as RNA binding, splicing, and transport (extensively reviewed in [63]). In disease, TDP-43 mislocalizes to the cytoplasm, is hyperphosphorylated and cleaved to yield C-terminal fragments that accumulate in insoluble aggregates. It is still not known if any of these biochemical changes are directly pathogenic *in vivo*. However, 2008 saw the first of a series of reports of mutations in *TARDBP*, the gene encoding TDP-43, in familial and apparently sporadic ALS, clearly indicating some kind of role for TDP-43 in the pathogenesis of ALS.[64–66] Now, around 40 different missense mutations in the glycine-rich C-terminal domain have been identified, almost exclusively in ALS cases. In addition, one truncation mutation (Tyr374STOP) has been found in a male sporadic patient[67] and one mutation in the first RNA recognition motif (RRM1) has been described in a female sporadic case.[65] The A90V variant in the middle of the bipartite nuclear localizing sequence has been suggested to promote disease, but has also been seen in controls and is of uncertain pathogenicity. Also of note is the fact that homozygous cases have been described.[68] One individual well into their seventh decade appears to be well with homozygous TDP-43 mutations. Notably, these cases come from closed populations, a similar scenario to that seen with D90A SOD1 mutation in Scandinavia (see above) and H517Q FUS mutation in Cape Verde (see below).

TDP-43 mutations account for around 1–4% of fALS in different populations around the world, making it the third most commonly mutated ALS gene. Like the *UBQLN2* cases, TDP-43 mutant individuals sometimes demonstrate features of FTLD. Furthermore, one patient with apparently pure FTLD and no family history of ALS has recently been found to have a TDP-43 mutation, again indicating that aberrant TDP-43 biology causes ALS-FTLD.[69]

In essence, TDP-43 mutations are seen exclusively in the context of ALS-FTD. Parkinson disease is the only other disease in which TDP-43 mutation has been described. The A382T mutation has been found in sporadic Parkinson disease (PD) cases in the closed Sardinian population mentioned above. However, these patients developed onset in their seventies and, given the apparent reduced penetrance of A382T, it is possible that this variant is not the cause

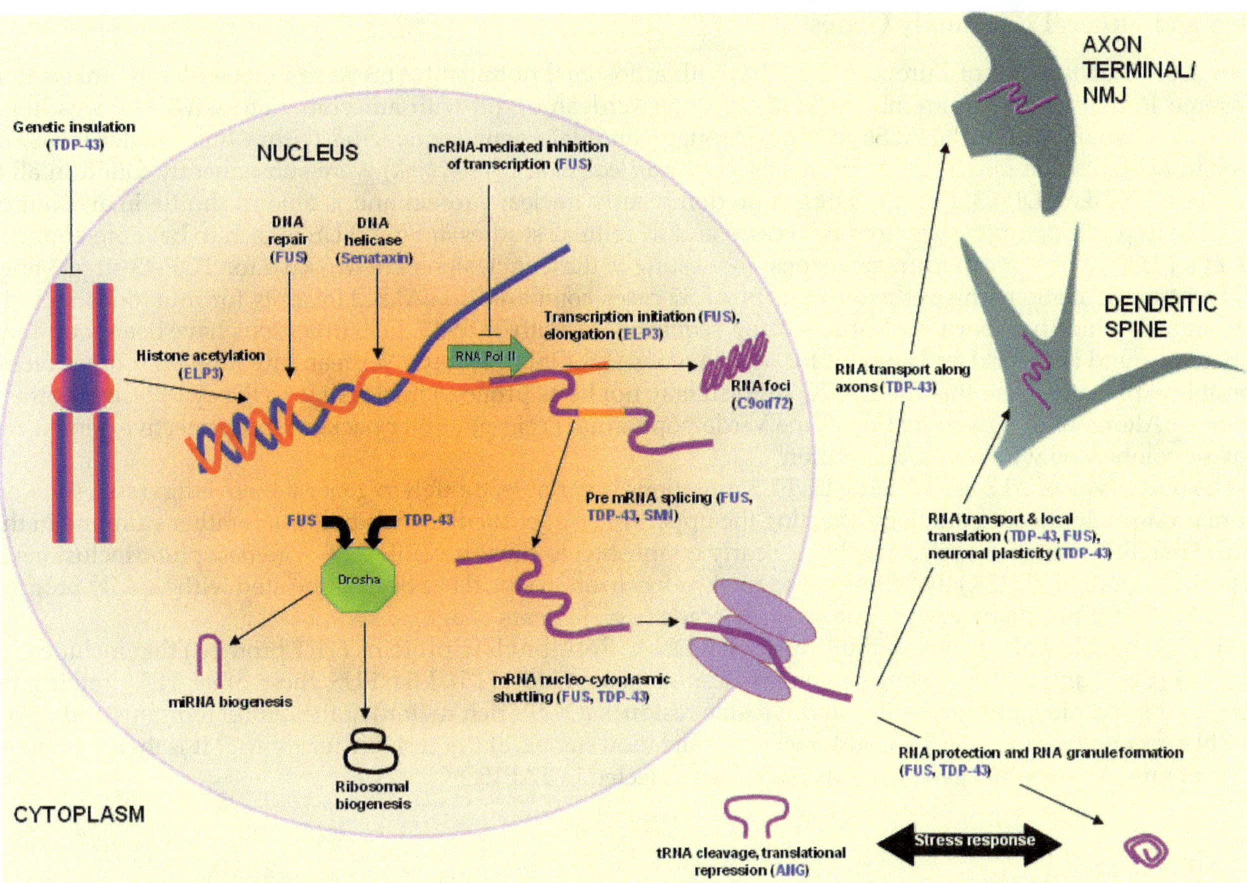

FIGURE 87.3 **Defective RNA processing in ALS.** Genetic studies implicate gene expression and RNA-processing abnormalities in ALS. TDP-43 and FUS each directly bind to RNA transcripts from thousands of genes. They are pivotal in almost all aspects of DNA and RNA metabolism including transcription, splicing, DNA repair, microRNA biogenesis, and transport of RNA-containing complexes for local, activity-dependent protein translation in dendritic spines. Senataxin, whose mutations cause early adult-onset, slowly progressive ALS, has a DNA/RNA helicase domain; although its biology remains incompletely understood it too may have a role in DNA repair as well as DNA replication, recombination and transcription, RNA processing, transcript stability, and translation initiation. Elongator protein 3 (ELP3), implicated as a susceptibility factor in sALS via a GWAS study, figures both in initiation of transcription and in histone acetylation. Survival motor neuron (SMN), the spinal muscular atrophy gene, functions in RNA splicing and in the formation of small nuclear ribonuclear proteins. The intronic hexanucleotide repeat expansions in the *C9orf72* gene (see text) may be detrimental at the DNA level, predisposing to the formation of a quadruplex structure, at the RNA level (possibly reducing levels of expression of *C9orf72* and forming intranuclear foci) and at the protein level (see Figure 87.2 and text).
TDP-43 and FUS are both able to bind and translocate mRNA molecules, while FUS (and possibly TDP-43) may be able to protect certain mRNAs in stress granules. Additionally, both FUS and TDP-43 are part of the Drosha complex, which is critical for miRNA biogenesis. Distant translocation of mRNAs is another recent observation in the studies of TDP-43, which is able to associate with other RNA binding proteins in axons *in vitro* and also transport RNA molecules along axons in fly models. This further implicates axonal and synaptic pathobiology in ALS.

of their disease. Nonetheless, another mutation (N267S) has recently been described in a different Parkinson patient, which adds weight to the hypothesis that TDP-43 mutation may rarely contribute to pathogenesis of PD.[70]

The mechanisms by which these mutations cause disease are unknown but almost certainly complex. This is in large measure a consequence of the fact that TDP-43 itself is functionally complex. TDP-43 binds to literally thousands of RNA targets and is involved in multiple components of RNA biology. Several lines of *in vitro* and *in vivo* experiments have suggested in diverse model systems (invertebrate and vertebrate) that the TDP-43 mutations exert both dominant negative and toxic gain-of-function effects. In yeast it is evident that excessive levels of wild-type TDP-43 can be as toxic as mutant TDP-43. This is distinctly different from the observation that forced expression of high levels of wild-type SOD1 is well tolerated *in vitro* and *in vivo*. Also unclear is whether any of the thousands of RNA binding targets of TDP-43 (which notably include RNAs encoding neuronal proteins, such as neurofilament light chain) have a role in neurodegeneration. One point that has received considerable attention is the finding that most TDP-43 mutations cluster at the C-terminus. This region is predicted to have low structural complexity and thus to be able to interact with many types of proteins. It is thought that this favours protein aggregation and may seed prion-like behavior of TDP-43, particularly in its mutant forms.[26]

FUS and Other TET Family Genes

In 2003, four families of European ancestry with autosomal dominant ALS were independently linked to chromosome 16q12.[71,72] In 2009, an inbred family of Cape Verdean origin with autosomal recessive ALS was linked to a narrow domain within the ALS6 locus.[73] Through candidate gene sequencing within this smaller locus, mutations in *FUS* (*fused in sarcoma*, also known as *TLS*, *translocated in liposarcoma*) were subsequently found in all these kindreds.[73,74] Like TDP-43, FUS protein is a predominantly nuclear protein and is able to shuttle in and out of the nucleus. In postmortem tissues from ALS cases and in cellular studies *in vitro*, FUS is seen to be more abundant in the cytoplasm where it can form inclusions. Interestingly, these inclusions are negative for TDP-43 suggesting that different mechanistic pathways may occur in ALS6 cases compared to sALS. Hotspots for mutations include the C-terminus, which harbors a nuclear localizing sequence. As with TDP-43, FUS mutations have been found in ALS patients around the world and account for up to 4% of fALS. Only the Cape Verdean mutation (H517Q) is recessive. Possible explanations are that the H517Q isoform may not be as prone to mislocalize to the cytoplasm as other mutations.[73] Alternatively, given that the Cape Verde population is closed, other background protective genetic factors may be coinherited with the FUS mutation.

The phenotype of ALS associated with FUS mutations is variable, though in general there is aggressive lower motor neuron-predominant pathology affecting the upper limbs in particular. Two points are rather stunning in this regard. First, FUS mutations have now been clearly established to cause juvenile ALS with basophilic inclusions. And, a specific mutation (P525K) that eliminates a nuclear localization signal has been associated with rapidly progressing ALS in children beginning early in the second decade (e.g., 11 years of age).

FUS is a member of a family of multifunctional DNA/RNA-binding proteins (TET proteins) that includes Ewing sarcoma (EWS) and TATA-binding protein-associated factor 15 (TAF15). Like FUS, these other TET proteins are enriched for glutamine, glycine, serine, and tyrosine residues (QGSY-rich region) at the amino terminus and also have an RNA-recognition motif (RRM), and nuclear localization signals at the carboxy terminus.[75] It is therefore noteworthy that rare, ALS-specific gene variants have been detected in TAF15.[76,77]

Ataxin 2

A role for the *ataxin2* (*ATXN2*) gene in neurodegeneration was firmly established by the observation that expansions of segments of CAG repeats cause spinocerebellar ataxia type 2.[78–80] Normally, there are less that 24 repeats in this gene; in SCA2, there are more than 33 repeats. That *ATXN2* is implicated in ALS was indicated by observations that an ATXN2 homolog augmented TDP-43 pathology in yeast.[81] This prompted an analysis of the relation between *ATXN2* expansions ALS, disclosing that intermediate lengths of ATXN2 (27–33 repeats) are over-represented in ALS (5.5% of ALS vs. 2.5% of controls in the initial report). Moreover, onset of ALS is earlier in the presence of the intermediate length CAG expansions. This association was consistent with an earlier clinical observation that individuals with SCA2 sometimes showed features suggestive of motor neuron disease.[78–80] First reported in the context of the aforementioned *Drosophila* study, this finding has now been robustly confirmed by multiple investigators around the world,[82–86] although it has been suggested that the expansion size linked with ALS may be limited to a range of 31–33.[87] However, with ataxin-2 the phenotype is also dependent on the "purity" of the expansion. Thus, a large expansion (34–59 repeats) composed of only CAGs results in SCA2. The ALS associated intermediate expansions of 27–33 repeats possess occasional CAA triplets.[81] By contrast, expansions of 33–49 repeats with occasional CAA motifs are associated with levodopa-responsive Parkinson disease, with or without a superimposed spinocerebellar syndrome.[88–90] Given that CAA and CAG both code for glutamine, one could speculate that the basis for the different patterns of neurodegeneration (ALS vs. parkinsonism) lies at the mRNA level rather than a polyglutamine repeat.

The mechanism of this impact of the ATXN2 expansions is not clear, although an *in vitro* study has suggested that the intermediate expansion may specifically lead to caspase activation.[91] ATXN2 protein is cytoplasmic and present in the trans Golgi network.[92] It participates normally in several aspects of RNA processing including microRNA (miRNA) synthesis, mRNA polyadenylation, stress granule formation, polyribosome formation, and RNA translation.[93–95] It may also be important in Golgi maintenance. What seems clear at this point is that ATXN2 can complex with TDP-43 and that ATXN2 is mislocalized in cytoplasmic aggregates in ALS spinal cord.[81]

Angiogenin

The concept that angiogenic factors might influence ALS susceptibility was first suggested by studies showing that, in some populations, deletions in the oxygen-sensing component of the vascular endothelial growth factor (VEGF) gene, and SNPs within that gene, may be associated with ALS.[96–98] These findings, which were not uniformly confirmed, raised the possibility that another protein that promotes angiogenesis, angiogenin (ANG), might also be implicated as a susceptibility factor.[99] This view was strongly substantiated by a report documenting that rare genetic variants in ANG are over-represented in ALS and may even be coinherited with the disease, at least in some small families.[100] A subsequent study of 6471 ALS cases, 3146 Parkinson disease cases, and 7668 controls confirmed that ANG variants are more common in both ALS (0.45%) and Parkinson disease (0.48%) than in controls (0.04%), although the variants did not appear to influence the phenotype of either disorder.[101] Initial studies suggest that the mutations impair several functions of ANG, including its neuroprotective properties. Because these variants are typically present as single copies (heterozygous with the wild-type allele), the overall implication is that haploinsufficiency of ANG is a risk factor for susceptibility to both ALS and Parkinson disease. This is in contrast with many of the dominantly inherited ALS genes (above) but is reminiscent of the relation of progranulin mutations to frontotemporal dementia.[102] Another model for these effects of ANG mutations observes that ANG cleaves tRNA during stress and that tRNA-derived stress-induced fragments (tiRNA) inhibit protein synthesis and trigger phopsho-eIF2alpha-independent assembly of stress granules (SGs), which may predispose to aggregation and neurotoxicity of RNA binding proteins.[103] Interestingly, tRNA fragments have been shown in an independent study of the CLP1 RNA kinase to promote oxidative stress and motor neuron death in mice.[104]

C9orf72

In 2006 it was reported that in families with ALS and FTD (and sometimes both in the same individual) the disease trait was coinherited with markers on chromosome 9p.[105,106] In those studies, two large kindreds with autosomal dominant inheritance of ALS, FTLD, or ALS-FTLD were linked to chromosome 9p. Despite extensive deep sequencing of this region, no mutations were found. Intriguingly, parallel GWAS in sALS implicated a replicable locus in the same region of chromosome 9p.[107–109] The answer to both the familial and sporadic questions was found in 2011 to be an extraordinary hexanucleotide (GGGGCC) intronic repeat expansion in the *chromosome 9 open reading frame 72* gene (*C9orf72*), a gene of unknown function.[110,111] The GC-rich repeat precluded straightforward identification by conventional sequencing approaches and although repeat primed polymerase chain reaction (PCR) can readily detect expansion, Southern blotting is, at present, the most commonly used method for confirming genetic diagnosis and estimating the size of the expansion. A normal length has been estimated at no more than ~25 repeats, while disease is associated with up to several thousands of repeats.

The C9orf72 intronic expansions are the most common genetic variant causing ALS. These are present in ~50% of fALS in populations of white European descent, up to 25% of familial FTD, and ~5% of apparently sporadic ALS and FTLD. These expansions are also seen in ~0.5% of controls.[112,113] Haplotype analysis indicates that a common European founder appears to be responsible for all cases.[114] Several models have been suggested to explain the neurotoxicity of these hexanucleotide expansions: 1) It is likely that one element is a toxic consequence of aggregation of the expanded RNA transcripts, now detected in both sense and antisense strands deposited in intranuclear foci in neurons (and other cell types). 2) The mutations may attenuate expression of the C9orf72 transcript, leading to haploinsufficiency of this gene. This model is supported by the finding of reduced levels of C9orf72 transcripts in ALS brains and by the parallel observation that in zebrafish reduction in C9orf72 expression also causes motoneuron degeneration.[115] Haploinsufficiency might be anticipated if an aberrant conformation of the hexanucleotide-expanded genomic DNA impaired C9orf72 transcription and/or if expanded RNA transcripts were abnormally spliced or translated. 3) Another factor in the pathobiology of the C9 expansions may be that they can form very stable G-quadruplex structures[116] and hairpin loops. DNA and RNA G-quadruplexes function in transcriptional and translational regulation, RNA transport and telomere stability, each of which are potentially relevant to maintaining homeostatic, long-term viability of cells like motor neurons.[117] Conceivably, the salutary impact of the G-quadruplexes may be blunted if the pathological C9orf72 expansions bind and sequester proteins that normally act in conjunction with binding to the normal C9orf72 G-quadruplexes. 4) The hexanucleotide repeat may excessively sequester transcription factors, by analogy with sequestration of the transcription factor muscleblind by tri- and tetranucleotide expansions in myotonic dystrophy types 1 (DM1) and 2. 5) The fifth putative mechanism for toxicity of C9orf72 expansions is a form of illegitimate protein translation termed repeat-associated, non-ATG (RAN) translation. This mechanism was first documented in myotonic dystrophy type 1 and spinocerebellar ataxia 8.[118] Two teams have described RAN translation peptides arising from C9orf72 expansions yielding RAN polypeptides that present as intracytoplasmic aggregates.[119,120] The degree of toxicity of these RNA peptides remains to be defined.

hnRNPA2B1 and hnRNPA1

Mutations in hnRNPA2B1 and hnRNPA1 were recently described as two further very rare causes of the IBMPFD phenotype in association with motor neuron disease.[121] The overarching phenotype was labeled "multisystem proteinopathy" (MSP). One family with ALS was found to harbor a D262N mutation in hnRNPA1, though clinical information was not provided. An hnRNPA2B1 D290V mutation was described in a family with an MSP phenotype in which one individual had EMG evidence of denervation, although this person had no upper motor neuron features as one would expect in ALS.[36] No other studies have found these hnRNPs to be a common cause of ALS, so they remain very rare causes of motor neuron disease. Nonetheless, they are biologically interesting given that they have a prion-like domain, whose prionogenicity increases as a consequence of the mutations. Furthermore, they both interact with TDP-43. Genetic interaction studies in flies indicate that VCP toxicity is mitigated by downregulation of fly homologs of TDP-43, hnRNPA2B1 and hnRNPA1.[121,122]

CANDIDATE ALS GENES IMPLICATED IN TRANSCRIPTIONAL REGULATION

Elongater Protein 3 (ELP3)

A microsatellite GWAS and, in a parallel, a *Drosophila* mutagenesis study both linked ALS risk with reduced levels of elongator protein 3 (ELP3), a highly conserved RNA polymerase II component.[123] This is discussed further in the section on GWAS, below.

CREST

Through analysis of ALS trios, in which cases and both unaffected parents underwent exome sequencing, 25 novel *de novo* heterozygous amino acid-altering mutations were identified.[124] Five of these were mutations in chromatin-modeling proteins. One of these genes, *CREST*, is particularly notable. *CREST* codes for a calcium-regulated transcriptional activator with a role in dendrite outgrowth. The mutation, which truncated the last nine amino acids of CREST, potentially affects the protein's interaction with the histone acetylase CBP. *In vitro* studies showed a negative effect of the mutation on activity induced dendrite outgrowth, and also suggested that CREST interacts with FUS. An additional missense mutation in CREST (I123M) was found to segregate with disease in a small kindred with ALS. Furthermore, a *de novo* truncating mutation in SRCAP, another CBP-interacting transcriptional coactivator, was also found in a case of ALS.

PERTURBATIONS IN ASPECTS OF AXONAL BIOLOGY IN ALS

Profilin-1 (PFN1)

Over the last few years, genetic analyses in ALS have strongly incriminated disturbances of axonal homeostasis and function (e.g., axonal transport) as critical pathologies in ALS. PFN1 is a protein that is essential for ATP-mediated polymerization of filamentous actin, which, in turn, is essential for many cellular events such as outgrowth of axonal growth cones and nuclear movement during mitosis. Exome capture methods have identified missense mutations that impair PFN1 function in some cases of fALS.[125] These ALS-associated PFN1 mutations impair axonal extension and growth cone elongation, leading to adult onset, predominantly lower motor neuron deterioration. It is of interest in this regard that in some experimental systems both mutant SOD1 and mutant TDP-43 can also impair neurite outgrowth.[126,127] Also of note is the observation that PFN1 can interact with VCP. It is estimated that mutations in PFN1 may account for up to 1% of fALS in some populations, although they have not been uniformly identified in all cohorts studied.

Dynactin (DCTN1)

That axonal transport may be implicated in ALS has long been intimated on general grounds (i.e., that normal transport is indispensable for a motor neuron, whose axonal length may be several 1,000-fold the diameter of the cell). A role-altered axonal transport in ALS was also suggested by early pilot studies of isolated human motor nerve axons from ALS cases that discerned defects in axonal transport.[128] Over the last 5 years, more direct evidence has been adduced, documenting that mutant SOD1 impairs axonal transport in multiple lines of transgenic ALS mice and in isolated axoplasm from squid giant axons. Additional direct evidence that altered axonal transport may have a primary role in ALS biology is a report that a mutation in the gene encoding dynactin p150^{glued} (DCTN1) causes a somewhat atypical form of slowly progressive, predominantly lower motor neuron disease that is often heralded by vocal cord weakness.[129] A different mutation in another family was associated with ALS and FTD.[130] The DCTN1 protein is a component of the dynein–dynactin motor protein complex that is essential for loading cargo onto the retrograde axonal transport motor. At least one of the ALS-related mutations (G59S) decreases binding of p150^{glued} to dynein and leads to aggregation of p150^{glued}. These DCTN1 mutations are rare in ALS but are more common in the extrapyramidal disorder Perry syndrome (PD with weight loss, depression and hypoventilation).[131]

Axoskeletal Proteins: Neurofilament (NF) and Peripherin (PRPH)

The most abundant structural proteins in axons are neurofilaments, a class of intermediate filaments that determine axonal diameter. It has long been recognized that disturbances of neurofilament (NF) architecture are present in fALS and sALS[132–135] and other motor neuron diseases and neuropathies. It is striking that forced expression of the NF heavy subunit, NF-H, prevents the normal distribution of NFs into the axon yet also prolongs survival in mutant SOD1 ALS mice.[136] A mutation in the NF light subunit gene (NF-L) causes a slowly progressive motor neuropathy. Although mutations in the genes encoding NF-H can be detected in sALS, this is quite rare.[137] Another type of intermediate filament also reportedly implicated in ALS is peripherin (PRPH). This protein is detected with neurofilaments within axonal inclusions in ALS. Moreover, a neurotoxic fragment of PRPH that has an insertion of 3 kD of coding sequence is detected in the spinal cords of ALS cases. To date two novel mutations in the peripherin gene have been detected rarely in ALS.[138,139]

Ephrin A4 (EphA4)

Recent studies by Van Hoecke and colleagues have determined that naturally occurring reductions in expression of the gene ephrin A4 (EphA4) are correlated with longer survival in ALS.[140] This finding stemmed first from studies of morpholino-induced reductions of EphA4 in zebrafish, and was then confirmed using pharmacological reduction of EphA4 in ALS mice and rats. An explicit relation to human sALS was defined by this group's finding that individuals with ALS who expressed lower levels of EphA4 (as measured in peripheral blood cells) had longer survival. Furthermore, two EphA4 gene mutations (both disrupting EphA4 function) were identified in one sporadic and one familial ALS patient; both patients had very long disease durations (the fALS patient is still alive after 12 years).[140]

EphA4 is a receptor for ephrins, a family of proteins involved in axon repulsion during development. In the context of neurodegeneration, ephrins may also have a role in adults, playing a role in synapse formation, plasticity and memory, and reduced levels have been found in postmortem brain tissues from patients with Alzheimer disease.[141] One speculation is that reductions in levels of EphA4 will permit enhanced (nonrepulsed) axon growth in individuals with ALS, thereby partially and temporarily rescuing the denervation process.

GENOME-WIDE ASSOCIATION STUDIES (GWAS) IN ALS

ALS is a complex disease in which a small minority of cases are due to very clear gene mutations (as detailed above), while the majority are apparently sporadic. Some cases of sALS will have mutations in known ALS genes. The vast majority, however, remain unexplained. In our view, truly sporadic ALS reflects a combination of unknown environmental factors, aging, and genetic susceptibility factors of low penetrance (for review see [1]). Several GWAS of increasing sizes have been conducted to identify common genetic variants that are associated with sALS. Several hits have been found, but, apart from the C9orf72 chromosome 9p21 locus, replication has been a challenge.

Association with the 9p21 locus was originally found in 2009 and replicated by several groups soon after.[107–109] This remains the most strongly associated linkage peak identified in over a dozen published ALS GWAS (as of March 2014). The subsequent identification of the *C9orf72* expansion and the common haplotype indicating a northern European common founder confirmed the validity of these robust GWAS.

Probably the second most confirmed GWAS locus in ALS is that which is in proximity with the *UNC13A* gene, with the intronic common variant rs12608932 being associated with disease susceptibility.[108] Because UNC13A regulates the release of neurotransmitters, including glutamate, this is a "plausible" candidate ALS gene. A further study of over 2000 cases and 2000 controls suggested the minor allele of rs12608932 was indeed associated with susceptibility and survival in northern Europeans.[142] A smaller Italian study was less conclusive, finding only a 1-year decrease in survival of patients with the minor allele.[143]

Other large SNP GWAS have implicated *FGGY* (*FLJ10986*), *ITPR2* and *DPP6* genes in ALS.[144–146] However, these have not uniformly been replicated across multiple studies; accordingly, doubt remains as to their validity as true ALS susceptibility factors.[147–149] The largest GWAS in ALS (as of March 2014) involved over 6000 European ALS cases.[150] It replicated previous chromosome 9p data, but also found an intriguing hit at chromosome 17q11.2, near the *SARM1* gene. SARM1 is involved in promoting axon destruction and may be a plausible hit. It remains to be determined whether the associated SNP has functional relevance.

A recent study of Han Chinese has further implicated two loci at 1q32 and 22p11, although the size of the replication cohort was small at around 500.[151] Replication studies are hopefully in the pipeline. Indeed, the study of non-European populations may be fruitful territory for GWAS approaches, as no big hit such as C9orf72 (which is essentially confined to Europeans) has yet to emerge.

Finally, a unique microsatellite GWAS identified risk alleles in the *ELP3* gene (noted above), which encodes an RNA polymerase component. Risk alleles were associated with reduced brain expression of ELP3 in postmortem studies.[123] The *ELP3* locus has been weakly associated with ALS in one separate GWAS,[149] but to date no other studies replicating this data have been published.

References

1. Al-Chalabi A, Hardiman O. The epidemiology of ALS: a conspiracy of genes, environment and time. *Nat Rev Neurol*. 2013;9:617–628. doi:10.1038/nrneurol.2013.203.
2. Siddique T, Figlewicz DA, Pericak-Vance MA, et al. Linkage of a gene causing familial amyotrophic lateral sclerosis to chromosome 21 and evidence of genetic-locus heterogeneity. *N Engl J Med*. 1991;324:1381–1384.
3. Rosen DR, Siddique T, Patterson D, et al. Mutations in Cu/Zn superoxide dismutase gene are associated with familial amyotrophic lateral sclerosis. *Nature*. 1993;362:59–62.
4. Andersen PM, Nilsson P, Ala-Hurula V, et al. Amyotrophic lateral sclerosis associated with homozygosity for an Asp90Ala mutation in CuZn-superoxide dismutase. *Nat Genet*. 1995;10:61–66.
5. Al-Chalabi A, Andersen PM, Chioza B, et al. Recessive amyotrophic lateral sclerosis families with the D90A SOD1 mutation share a common founder: evidence for a linked protective factor. *Hum Mol Genet*. 1998;7:2045–2050.
6. Zinman L, Liu HN, Sato C, et al. A mechanism for low penetrance in an ALS family with a novel SOD1 deletion. *Neurology*. 2009;72:1153–1159.
7. Rosen DR. Mutations in Cu/Zn superoxide dismutase gene are associated with familial amyotrophic lateral sclerosis. *Nature*. 1993;364:362. doi:10.1038/364362c0.
8. Gurney ME. Transgenic-mouse model of amyotrophic lateral sclerosis. *N Engl J Med*. 1994;331:1721–1722.
9. Turner BJ, Talbot K. Transgenics, toxicity and therapeutics in rodent models of mutant SOD1-mediated familial ALS. *Prog Neurobiol*. 2008;.
10. Higgins CM, Jung C, Xu Z. ALS-associated mutant SOD1G93A causes mitochondrial vacuolation by expansion of the intermembrane space and by involvement of SOD1 aggregation and peroxisomes. *BMC Neurosci*. 2003;4:16.
11. Nishitoh H, Kadowaki H, Nagai A, et al. ALS-linked mutant SOD1 induces ER stress- and ASK1-dependent motor neuron death by targeting Derlin-1. *Genes Dev*. 2008;22:1451–1464. doi:10.1101/gad.1640108.
12. Moreno JA, Halliday M, Molloy C, et al. Oral treatment targeting the unfolded protein response prevents neurodegeneration and clinical disease in prion-infected mice. *Sci Transl Med*. 2013;5:206ra138. doi:10.1126/scitranslmed.3006767.
13. Bendotti C, Tortarolo M, Suchak SK, et al. Transgenic SOD1 G93A mice develop reduced GLT-1 in spinal cord without alterations in cerebrospinal fluid glutamate levels. *J Neurochem*. 2001;79:737–746.
14. Howland DS, Liu J, She Y, et al. Focal loss of the glutamate transporter EAAT2 in a transgenic rat model of SOD1 mutant-mediated amyotrophic lateral sclerosis (ALS). *Proc Natl Acad Sci U S A*. 2002;99:1604–1609.
15. Morfini GA, Bosco DA, Brown H, et al. Inhibition of fast axonal transport by pathogenic SOD1 involves activation of p38 MAP kinase. *PLoS One*. 2013;8:e65235. doi:10.1371/journal.pone.0065235, PONE-D-13-11272 [pii].
16. Sau D, De Biasi S, Vitellaro-Zuccarello L, et al. Mutation of SOD1 in ALS: a gain of a loss of function. *Hum Mol Genet*. 2007;16:1604–1618.
17. Urushitani M, Sik A, Sakurai T, et al. Chromogranin-mediated secretion of mutant superoxide dismutase proteins linked to amyotrophic lateral sclerosis. *Nat Neurosci*. 2006;9:108–118.
18. Turner BJ, Atkin JD, Farg MA, et al. Impaired extracellular secretion of mutant superoxide dismutase 1 associates with neurotoxicity in familial amyotrophic lateral sclerosis. *J Neurosci*. 2005;25:108–117.
19. Munch C, O'Brien J, Bertolotti A. Prion-like propagation of mutant superoxide dismutase-1 misfolding in neuronal cells. *Proc Natl Acad Sci U S A*. 2011;108:3548–3553. doi:10.1073/pnas.1017275108.
20. Lindberg MJ, Normark J, Holmgren A, Oliveberg M. Folding of human superoxide dismutase: disulfide reduction prevents dimerization and produces marginally stable monomers. *Proc Natl Acad Sci U S A*. 2004;101:15893–15898.
21. Rakhit R, Crow JP, Lepock JR, et al. Monomeric Cu, Zn-superoxide dismutase is a common misfolding intermediate in the oxidation models of sporadic and familial amyotrophic lateral sclerosis. *J Biol Chem*. 2004;279:15499–15504.
22. Teilum K, Smith MH, Schulz E, et al. Transient structural distortion of metal-free Cu/Zn superoxide dismutase triggers aberrant oligomerization. *Proc Natl Acad Sci U S A*. 2009;106:18273–18278. doi:10.1073/pnas.0907387106.
23. Ezzi SA, Urushitani M, Julien JP. Wild-type superoxide dismutase acquires binding and toxic properties of ALS-linked mutant forms through oxidation. *J Neurochem*. 2007;102:170–178.

24. Bosco DA, Morfini G, Karabacak NM, et al. Wild-type and mutant SOD1 share an aberrant conformation and a common pathogenic pathway in ALS. *Nat Neurosci*. 2010;13:1396–1403. doi:nn.2660, [pii] 10.1038/nn.2660.
25. Guareschi S, Cova E, Cereda C, et al. An over-oxidized form of superoxide dismutase found in sporadic amyotrophic lateral sclerosis with bulbar onset shares a toxic mechanism with mutant SOD1. *Proc Natl Acad Sci U S A*. 2012;109:5074–5079. doi:1115402109, [pii] 10.1073/pnas.1115402109.
26. King OD, Gitler AD, Shorter J. The tip of the iceberg: RNA-binding proteins with prion-like domains in neurodegenerative disease. *Brain Res*. 2012;1462:61–80. doi:10.1016/j.brainres.2012.01.016.
27. Deng HX, Chen W, Hong ST, et al. Mutations in UBQLN2 cause dominant X-linked juvenile and adult-onset ALS and ALS/dementia. *Nature*. 2011;477:211–215. doi:10.1038/nature10353.
28. Millecamps S, Corcia P, Cazeneuve C, et al. Mutations in UBQLN2 are rare in French amyotrophic lateral sclerosis. *Neurobiol Aging*. 2012;33:839.e831–839.e833. doi:10.1016/j.neurobiolaging.2011.11.010.
29. Williams KL, Warraich ST, Yang S, et al. UBQLN2/ubiquilin 2 mutation and pathology in familial amyotrophic lateral sclerosis. *Neurobiol Aging*. 2012;33:2527.e3–2527.e10. doi:10.1016/j.neurobiolaging.2012.05.008.
30. Gellera C, Tiloca C, Del Bo R, et al. Ubiquilin 2 mutations in Italian patients with amyotrophic lateral sclerosis and frontotemporal dementia. *J Neurol Neurosurg Psychiatry*. 2013;84:183–187. doi:10.1136/jnnp-2012-303433.
31. Cassel JA, Reitz AB. Ubiquilin-2 (UBQLN2) binds with high affinity to the C-terminal region of TDP-43 and modulates TDP-43 levels in H4 cells: characterization of inhibition by nucleic acids and 4-aminoquinolines. *Biochim Biophys Acta*. 2013;1834:964–971. doi:10.1016/j.bbapap.2013.03.020.
32. González-Pérez P, Lu Y, Chian RJ, et al. Association of UBQLN1 mutation with Brown-Vialetto-Van Laere syndrome but not typical ALS. *Neurobiol Dis*. 2012;48:391–398. doi:10.1016/j.nbd.2012.06.018.
33. Watts GD, Wymer J, Kovach MJ, et al. Inclusion body myopathy associated with Paget disease of bone and frontotemporal dementia is caused by mutant valosin-containing protein. *Nat Genet*. 2004;36:377–381. doi:10.1038/ng1332.
34. Johnson JO, Mandrioli J, Benatar M, et al. Exome sequencing reveals VCP mutations as a cause of familial ALS. *Neuron*. 2010;68:857–864. doi:S0896-6273(10)00978-5, [pii] 10.1016/j.neuron.2010.11.036.
35. Abramzon Y, Johnson JO, Scholz SW, et al. Valosin-containing protein (VCP) mutations in sporadic amyotrophic lateral sclerosis. *Neurobiol Aging*. 2012;33:2231.e2231–2231.e2236. doi:S0197-4580(12)00234-5, [pii] 10.1016/j.neurobiolaging.2012.04.005.
36. Benatar M, Wuu J, Fernandez C, et al. Motor neuron involvement in multisystem proteinopathy: implications for ALS. *Neurology*. 2013;80:1874–1880. doi:WNL.0b013e3182929fc3, [pii] 10.1212/WNL.0b013e3182929fc3.
37. DeJesus-Hernandez M, Desaro P, Johnston A, et al. Novel p.Ile151Val mutation in VCP in a patient of African American descent with sporadic ALS. *Neurology*. 2011;77:1102–1103. doi:10.1212/WNL.0b013e31822e563c.
38. Kim EJ, Park YE, Kim DS, et al. Inclusion body myopathy with Paget disease of bone and frontotemporal dementia linked to VCP p.Arg155Cys in a Korean family. *Arch Neurol*. 2011;68:787–796. doi:10.1001/archneurol.2010.376.
39. Gu JM, Ke YH, Yue H, et al. A novel VCP mutation as the cause of atypical IBMPFD in a Chinese family. *Bone*. 2013;52:9–16. doi:10.1016/j.bone.2012.09.012.
40. Nalbandian A, Llewellyn KJ, Badadani M, et al. A progressive translational mouse model of human valosin-containing protein disease: the VCP(R155H/+) mouse. *Muscle Nerve*. 2013;47:260–270. doi:10.1002/mus.23522.
41. Hetzer M, Meyer HH, Walther TC, et al. Distinct AAA-ATPase p97 complexes function in discrete steps of nuclear assembly. *Nat Cell Biol*. 2001;3:1086–1091. doi:10.1038/ncb1201-1086.
42. Rabinovich E, Kerem A, Frohlich KU, Diamant N, Bar-Nun S. AAA-ATPase p97/Cdc48p, a cytosolic chaperone required for endoplasmic reticulum-associated protein degradation. *Mol Cell Biol*. 2002;22:626–634.
43. Rabouille C, Kondo H, Newman R, et al. Syntaxin 5 is a common component of the NSF- and p97-mediated reassembly pathways of Golgi cisternae from mitotic Golgi fragments in vitro. *Cell*. 1998;92:603–610.
44. Kondo H, Rabouille C, Newman R, et al. p47 is a cofactor for p97-mediated membrane fusion. *Nature*. 1997;388:75–78. doi:10.1038/40411.
45. Meyer HH, Shorter JG, Seemann J, Pappin D, Warren G. A complex of mammalian ufd1 and npl4 links the AAA-ATPase, p97, to ubiquitin and nuclear transport pathways. *EMBO J*. 2000;19:2181–2192. doi:10.1093/emboj/19.10.2181.
46. Hirabayashi M, Inoue K, Tanaka K, et al. VCP/p97 in abnormal protein aggregates, cytoplasmic vacuoles, and cell death, phenotypes relevant to neurodegeneration. *Cell Death Differ*. 2001;8:977–984. doi:10.1038/sj.cdd.4400907.
47. Kobayashi T, Tanaka K, Inoue K, Kakizuka A. Functional ATPase activity of p97/valosin-containing protein (VCP) is required for the quality control of endoplasmic reticulum in neuronally differentiated mammalian PC12 cells. *J Biol Chem*. 2002;277:47358–47365. doi:10.1074/jbc.M207783200.
48. Bartolome F, Wu HC, Burchell VS, et al. Pathogenic VCP mutations induce mitochondrial uncoupling and reduced ATP levels. *Neuron*. 2013;78:57–64. doi:10.1016/j.neuron.2013.02.028.
49. Cloutier P, Lavallee-Adam M, Faubert D, Blanchette M, Coulombe B. A newly uncovered group of distantly related lysine methyltransferases preferentially interact with molecular chaperones to regulate their activity. *PLoS Genet*. 2013;9:e1003210. doi:10.1371/journal.pgen.1003210.
50. Weihl CC, Pestronk A, Kimonis VE. Valosin-containing protein disease: inclusion body myopathy with Paget's disease of the bone and fronto-temporal dementia. *Neuromuscul Disord*. 2009;19:308–315. doi:10.1016/j.nmd.2009.01.009.
51. Maruyama H, Morino H, Ito H, et al. Mutations of optineurin in amyotrophic lateral sclerosis. *Nature*. 2010;465:223–226. doi:10.1038/nature08971.
52. Del Bo R, Tiloca C, Pensato V, et al. Novel optineurin mutations in patients with familial and sporadic amyotrophic lateral sclerosis. *J Neurol Neurosurg Psychiatry*. 2011;82:1239–1243. doi:10.1136/jnnp.2011.242313.
53. Rezaie T, Child A, Hitchings R, et al. Adult-onset primary open-angle glaucoma caused by mutations in optineurin. *Science*. 2002;295:1077–1079. doi:10.1126/science.1066901.
54. Albagha OM, Visconti MR, Alonso N, et al. Genome-wide association study identifies variants at CSF1, OPTN and TNFRSF11A as genetic risk factors for Paget's disease of bone. *Nat Genet*. 2010;42:520–524. doi:10.1038/ng.562.
55. Maruyama H, Kawakami H. Optineurin and amyotrophic lateral sclerosis. *Geriatr Gerontol Int*. 2013;13:528–532. doi:10.1111/ggi.12022.

56. Laurin N, Brown JP, Morissette J, Raymond V. Recurrent mutation of the gene encoding sequestosome 1 (SQSTM1/p62) in Paget disease of bone. *Am J Hum Genet*. 2002;70:1582–1588. doi:10.1086/340731.
57. Fecto F, Yan J, Vemula SP, et al. SQSTM1 mutations in familial and sporadic amyotrophic lateral sclerosis. *Arch Neurol*. 2011;68:1440–1446. doi:10.1001/archneurol.2011.250.
58. Neumann M, Sampathu DM, Kwong LK, et al. Ubiquitinated TDP-43 in frontotemporal lobar degeneration and amyotrophic lateral sclerosis. *Science*. 2006;314:130–133.
59. Arai T, Hasegawa M, Akiyama H, et al. TDP-43 is a component of ubiquitin-positive tau-negative inclusions in frontotemporal lobar degeneration and amyotrophic lateral sclerosis. *Biochem Biophys Res Commun*. 2006;351:602–611.
60. Ou SH, Wu F, Harrich D, Garcia-Martinez LF, Gaynor RB. Cloning and characterization of a novel cellular protein, TDP-43, that binds to human immunodeficiency virus type 1 TAR DNA sequence motifs. *J Virol*. 1995;69:3584–3596.
61. Abhyankar MM, Urekar C, Reddi PP. A novel CpG-free vertebrate insulator silences the testis-specific SP-10 gene in somatic tissues: role for TDP-43 in insulator function. *J Biol Chem*. 2007;282:36143–36154. doi:10.1074/jbc.M705811200.
62. Lalmansingh AS, Urekar CJ, Reddi PP. TDP-43 is a transcriptional repressor: the testis-specific mouse acrv1 gene is a TDP-43 target in vivo. *J Biol Chem*. 2011;286:10970–10982. doi:10.1074/jbc.M110.166587.
63. Buratti E, Baralle FE. Multiple roles of TDP-43 in gene expression, splicing regulation, and human disease. *Front Biosci*. 2008;13:867–878.
64. Sreedharan J, Blair IP, Tripathi VB, et al. TDP-43 mutations in familial and sporadic amyotrophic lateral sclerosis. *Science*. 2008;319:1668–1672.
65. Kabashi E, Valdmanis PN, Dion P, et al. TARDBP mutations in individuals with sporadic and familial amyotrophic lateral sclerosis. *Nat Genet*. 2008;40:572–574.
66. Yokoseki A, Shiga A, Tan CF, et al. TDP-43 mutation in familial amyotrophic lateral sclerosis. *Ann Neurol*. 2008;63:538–542. doi:10.1002/ana.21392.
67. Daoud H, Valdmanis PN, Kabashi E, et al. Contribution of TARDBP mutations to sporadic amyotrophic lateral sclerosis. *J Med Genet*. 2009;46:112–114. doi:10.1136/jmg.2008.062463.
68. Borghero G, Floris G, Cannas A, et al. A patient carrying a homozygous p.A382T TARDBP missense mutation shows a syndrome including ALS, extrapyramidal symptoms, and FTD. *Neurobiol Aging*. 2011;32:2327.e2321–2327.e2325. doi:10.1016/j.neurobiolaging.2011.06.009.
69. Synofzik M, Born C, Rominger A, et al. Targeted high-throughput sequencing identifies a TARDBP mutation as a cause of early-onset FTD without motor neuron disease. *Neurobiol Aging*. 2014;35:1212.e1211–1212.e1215. doi:10.1016/j.neurobiolaging.2013.10.092.
70. Rayaprolu S, Fujioka S, Traynor S, et al. TARDBP mutations in Parkinson's disease. *Parkinsonism Relat Disord*. 2013;19:312–315. doi:10.1016/j.parkreldis.2012.11.003.
71. Sapp PC, Hosler BA, McKenna-Yasek D, et al. Identification of two novel loci for dominantly inherited amyotrophic lateral sclerosis. *Am J Hum Genet*. 2003;73(2):397–403.
72. Ruddy DM, Parton MJ, Al-Chalabi A, et al. Two families with familial amyotrophic lateral sclerosis are linked to a novel locus on chromosome 16q. *Am J Hum Genet*. 2003;73:390–396.
73. Kwiatkowski Jr. TJ, Bosco DA, Leclerc AL, et al. Mutations in the FUS/TLS gene on chromosome 16 cause familial amyotrophic lateral sclerosis. *Science*. 2009;323:1205–1208. doi:323/5918/1205, [pii] 10.1126/science.1166066.
74. Vance C, Rogelj B, Hortobágyi T, et al. Mutations in FUS, an RNA processing protein, cause familial amyotrophic lateral sclerosis type 6. *Science*. 2009;323:1208–1211. doi:323/5918/1208, [pii] 10.1126/science.1165942.
75. Morohoshi F, Ootsuka Y, Arai K, et al. Genomic structure of the human RBP56/hTAFII68 and FUS/TLS genes. *Gene*. 1998;221:191–198.
76. Ticozzi N, Vance C, Leclerc AL, et al. Mutational analysis reveals the FUS homolog TAF15 as a candidate gene for familial amyotrophic lateral sclerosis. *Am J Med Genet B Neuropsychiatr Genet*. 2011;156B:285–290. doi:10.1002/ajmg.b.31158.
77. Couthouis J, Hart MP, Shorter J, et al. A yeast functional screen predicts new candidate ALS disease genes. *Proc Natl Acad Sci U S A*. 2011;108:20881–20890. doi:10.1073/pnas.1109434108.
78. Pulst SM, Nechiporuk A, Nechiporuk T, et al. Moderate expansion of a normally biallelic trinucleotide repeat in spinocerebellar ataxia type 2 [see comments]. *Nat Genet*. 1996;14:269–276.
79. Sanpei K, Takano H, Igarashi S, et al. Identification of the spinocerebellar ataxia type 2 gene using a direct identification of repeat expansion and cloning technique, DIRECT. *Nat Genet*. 1996;14:277–284. doi:10.1038/ng1196-277.
80. Imbert G, Saudou F, Yvert G, et al. Cloning of the gene for spinocerebellar ataxia 2 reveals a locus with high sensitivity to expanded CAG/glutamine repeats. *Nat Genet*. 1996;14:285–291. doi:10.1038/ng1196-285.
81. Elden AC, Kim HJ, Hart MP, et al. Ataxin-2 intermediate-length polyglutamine expansions are associated with increased risk for ALS. *Nature*. 2010;466:1069–1075. doi:10.1038/nature09320.
82. Van Damme P, Veldink JH, van Blitterswijk M, et al. Expanded ATXN2 CAG repeat size in ALS identifies genetic overlap between ALS and SCA2. *Neurology*. 2011;76:2066–2072. doi:10.1212/WNL.0b013e31821f445b.
83. Daoud H, Belzil V, Martins S, et al. Association of long ATXN2 CAG repeat sizes with increased risk of amyotrophic lateral sclerosis. *Arch Neurol*. 2011;68:739–742. doi:10.1001/archneurol.2011.111.
84. Lahut S, Ömür Ö, Uyan Ö, et al. ATXN2 and its neighbouring gene SH2B3 are associated with increased ALS risk in the Turkish population. *PLoS One*. 2012;7:e42956. doi:10.1371/journal.pone.0042956.
85. Conforti FL, Spataro R, Sproviero W, et al. Ataxin-1 and ataxin-2 intermediate-length PolyQ expansions in amyotrophic lateral sclerosis. *Neurology*. 2012;79:2315–2320. doi:10.1212/WNL.0b013e318278b618.
86. Liu X, Lu M, Tang L, et al. ATXN2 CAG repeat expansions increase the risk for Chinese patients with amyotrophic lateral sclerosis. *Neurobiol Aging*. 2013;34:2236.e2235–2236.e2238. doi:10.1016/j.neurobiolaging.2013.04.009.
87. Gellera C, Ticozzi N, Pensato V, et al. ATAXIN2 CAG-repeat length in Italian patients with amyotrophic lateral sclerosis: risk factor or variant phenotype? Implication for genetic testing and counseling. *Neurobiol Aging*. 2012;33:1847.e1815–1847.e1821. doi:10.1016/j.neurobiolaging.2012.02.004.
88. Gwinn-Hardy KA, Crook R, Lincoln S, et al. A kindred with Parkinson's disease not showing genetic linkage to established loci. *Neurology*. 2000;54:504–507.
89. Ragothaman M, Sarangmath N, Chaudhary S, et al. Complex phenotypes in an Indian family with homozygous SCA2 mutations. *Ann Neurol*. 2004;55:130–133. doi:10.1002/ana.10815.

90. Charles P, Camuzat A, Benammar N, et al. Are interrupted SCA2 CAG repeat expansions responsible for parkinsonism? *Neurology*. 2007;69:1970–1975. doi:10.1212/01.wnl.0000269323.21969.db.
91. Hart MP, Gitler AD. ALS-associated ataxin 2 polyQ expansions enhance stress-induced caspase 3 activation and increase TDP-43 pathological modifications. *J Neurosci*. 2012;32:9133–9142. doi:10.1523/JNEUROSCI.0996-12.2012.
92. Turnbull VJ, Storey E, Tarlac V, et al. Different ataxin-2 antibodies display different immunoreactive profiles. *Brain Res*. 2004;1027:103–116. doi:10.1016/j.brainres.2004.08.044.
93. Nonhoff U, Ralser M, Welzel F, et al. Ataxin-2 interacts with the DEAD/H-box RNA helicase DDX6 and interferes with P-bodies and stress granules. *Mol Biol Cell*. 2007;18:1385–1396. doi:10.1091/mbc.E06-12-1120.
94. McCann C, Holohan EE, Das S, et al. The Ataxin-2 protein is required for microRNA function and synapse-specific long-term olfactory habituation. *Proc Natl Acad Sci U S A*. 2011;108:E655–E662. doi:10.1073/pnas.1107198108.
95. Satterfield TF, Pallanck LJ. Ataxin-2 and its Drosophila homolog, ATX2, physically assemble with polyribosomes. *Hum Mol Genet*. 2006;15:2523–2532. doi:10.1093/hmg/ddl173.
96. Chen W, Saeed M, Mao H, et al. Lack of association of VEGF promoter polymorphisms with sporadic ALS. *Neurology*. 2006;67:508–510.
97. Gros-Louis F, Laurent S, Lopes AA, et al. Absence of mutations in the hypoxia response element of VEGF in ALS. *Muscle Nerve*. 2003;28:774–775.
98. Lambrechts D, Storkebaum E, Morimoto M, et al. VEGF is a modifier of amyotrophic lateral sclerosis in mice and humans and protects motoneurons against ischemic death. *Nat Genet*. 2003;34(4):383–394.
99. Greenway MJ, Alexander MD, Ennis S, et al. A novel candidate region for ALS on chromosome 14q11.2. *Neurology*. 2004;63:1936–1938.
100. Greenway MJ, Andersen PM, Russ C, et al. ANG mutations segregate with familial and 'sporadic' amyotrophic lateral sclerosis. *Nat Genet*. 2006;38:411–413.
101. van Es MA1, Schelhaas HJ, van Vught PW, et al. Angiogenin variants in Parkinson disease and amyotrophic lateral sclerosis. *Ann Neurol*. 2011;70. doi:10.1002/ana.22611.
102. Cruts M, Kumar-Singh S, Van Broeckhoven C. Progranulin mutations in ubiquitin-positive frontotemporal dementia linked to chromosome 17q21. *Curr Alzheimer Res*. 2006;3:485–491.
103. Emara MM, Ivanov P, Hickman T, et al. Angiogenin-induced tRNA-derived stress-induced RNAs promote stress-induced stress granule assembly. *J Biol Chem*. 2010;285:10959–10968. doi:10.1074/jbc.M109.077560.
104. Hanada T, Weitzer S, Mair B, et al. CLP1 links tRNA metabolism to progressive motor-neuron loss. *Nature*. 2013;495:474–480. doi:10.1038/nature11923.
105. Vance C, Al-Chalabi A, Ruddy D, et al. Familial amyotrophic lateral sclerosis with frontotemporal dementia is linked to a locus on chromosome 9p13.2-21.3. *Brain*. 2006;129:868–876.
106. Morita M, Al-Chalabi A, Andersen PM, et al. A locus on chromosome 9p confers susceptibility to ALS and frontotemporal dementia. *Neurology*. 2006;66:839–844.
107. Laaksovirta H, Peuralinna T, Schymick JC, et al. Chromosome 9p21 in amyotrophic lateral sclerosis in Finland: a genome-wide association study. *Lancet Neurol*. 2010;9:978–985. doi:10.1016/S1474-4422(10)70184-8.
108. van Es MA, Veldink JH, Saris CG, et al. Genome-wide association study identifies 19p13.3 (UNC13A) and 9p21.2 as susceptibility loci for sporadic amyotrophic lateral sclerosis. *Nat Genet*. 2009;41:1083–1087. doi:10.1038/ng.442.
109. Shatunov A, Mok K, Newhouse S, et al. Chromosome 9p21 in sporadic amyotrophic lateral sclerosis in the UK and seven other countries: a genome-wide association study. *Lancet Neurol*. 2010;9:986–994. doi:10.1016/S1474-4422(10)70197-6.
110. Renton AE, Majounie E, Waite A, et al. A hexanucleotide repeat expansion in C9ORF72 is the cause of chromosome 9p21-linked ALS-FTD. *Neuron*. 2011;72:257–268. doi:10.1016/j.neuron.2011.09.010.
111. DeJesus-Hernandez M, Mackenzie IR, Boeve BF, et al. Expanded GGGGCC hexanucleotide repeat in noncoding region of C9ORF72 causes chromosome 9p-linked FTD and ALS. *Neuron*. 2011;72:245–256. doi:10.1016/j.neuron.2011.09.011.
112. Majounie E, Renton AE, Mok K, et al. Frequency of the C9orf72 hexanucleotide repeat expansion in patients with amyotrophic lateral sclerosis and frontotemporal dementia: a cross-sectional study. *Lancet Neurol*. 2012;11:323–330. doi:10.1016/S1474-4422(12)70043-1.
113. Beck J, Poulter M, Hensman D, et al. Large C9orf72 hexanucleotide repeat expansions are seen in multiple neurodegenerative syndromes and are more frequent than expected in the UK population. *Am J Hum Genet*. 2013;92:345–353. doi:10.1016/j.ajhg.2013.01.011.
114. Smith BN, Newhouse S, Shatunov A, et al. The C9ORF72 expansion mutation is a common cause of ALS +/-FTD in Europe and has a single founder. *Eur J Hum Genet*. 2013;21:102–108. doi:10.1038/ejhg.2012.98.
115. Ciura S, Lattante S, Le Ber I, et al. Loss of function of C9orf72 causes motor deficits in a zebrafish model of Amyotrophic Lateral Sclerosis. *Ann Neurol*. 2013;74:180–187. doi:10.1002/ana.23946.
116. Fratta P, Mizielinska S, Nicoll AJ, et al. C9orf72 hexanucleotide repeat associated with amyotrophic lateral sclerosis and frontotemporal dementia forms RNA G-quadruplexes. *Sci Rep*. 2012;2:1016. doi:10.1038/srep01016.
117. Bugaut A, Balasubramanian S. 5′-UTR RNA G-quadruplexes: translation regulation and targeting. *Nucleic Acids Res*. 2012;40:4727–4741. doi:10.1093/nar/gks068.
118. Zu T, Gibbens B, Doty NS, et al. Non-ATG-initiated translation directed by microsatellite expansions. *Proc Natl Acad Sci U S A*. 2011;108:260–265. doi:10.1073/pnas.1013343108.
119. Ash PE, Bieniek KF, Gendron TF, et al. Unconventional translation of C9ORF72 GGGGCC expansion generates insoluble polypeptides specific to c9FTD/ALS. *Neuron*. 2013;77:639–646. doi:10.1016/j.neuron.2013.02.004.
120. Mori K, Weng SM, Arzberger T, et al. The C9orf72 GGGGCC repeat is translated into aggregating dipeptide-repeat proteins in FTLD/ALS. *Science*. 2013;339:1335–1338. doi:science.1232927, [pii] 10.1126/science.1232927.
121. Kim HJ, Kim NC, Wang YD, et al. Mutations in prion-like domains in hnRNPA2B1 and hnRNPA1 cause multisystem proteinopathy and ALS. *Nature*. 2013;495:467–473. doi:nature11922, [pii] 10.1038/nature11922.
122. Ritson GP, Custer SK, Freibaum BD, et al. TDP-43 mediates degeneration in a novel Drosophila model of disease caused by mutations in VCP/p97. *J Neurosci*. 2010;30:7729–7739. doi:30/22/7729, [pii] 10.1523/JNEUROSCI.5894-09.2010.
123. Simpson CL, Lemmens R, Miskiewicz K, et al. Variants of the elongator protein 3 (ELP3) gene are associated with motor neuron degeneration. *Hum Mol Genet*. 2009;18:472–481. doi:ddn375, [pii] 10.1093/hmg/ddn375.

124. Chesi A, Staahl BT, Jovičić A, et al. Exome sequencing to identify de novo mutations in sporadic ALS trios. *Nat Neurosci*. 2013;16:851–855. doi:10.1038/nn.3412.
125. Wu CH, Fallini C, Ticozzi N, et al. Mutations in the profilin 1 gene cause familial amyotrophic lateral sclerosis. *Nature*. 2012;488:499–503. doi:nature11280, [pii] 10.1038/nature11280.
126. Lee KW, Kim HJ, Sung JJ, Park KS, Kim M. Defective neurite outgrowth in aphidicolin/cAMP-induced motor neurons expressing mutant Cu/Zn superoxide dismutase. *Int J Dev Neurosci*. 2002;20:521–526.
127. Fiesel FC, Schurr C, Weber SS, Kahle PJ. TDP-43 knockdown impairs neurite outgrowth dependent on its target histone deacetylase 6. *Mol Neurodegeneration*. 2011;6:64. doi:10.1186/1750-1326-6-64.
128. Breuer AC, Atkinson MB. Fast axonal transport alterations in amyotrophic lateral sclerosis (ALS) and in parathyroid hormone (PTH)-treated axons. *Cell Motil Cytoskeleton*. 1988;10:321–330.
129. Puls I, Jonnakuty C, LaMonte BH, et al. Mutant dynactin in motor neuron disease. *Nat Genet*. 2003;33:455–456.
130. Münch C1, Rosenbohm A, Sperfeld AD, et al. Heterozygous R1101K mutation of the DCTN1 gene in a family with ALS and FTD. *Ann Neurol*. 2005;58:777–780.
131. Farrer MJ, Hulihan MM, Kachergus JM, et al. DCTN1 mutations in Perry syndrome. *Nat Genet*. 2009;41:163–165. doi:10.1038/ng.293.
132. Hirano A. Cytopathology in amyotrophic lateral sclerosis. *Adv Neurol*. 1991;56:91–101.
133. Carpenter S. Proximal axonal enlargement in motor neuron disease. *Neurology*. 1968;18:842–851.
134. Chou SM, Fakadej AV. Ultrastructure of chromatolytic motoneurons and anterior spinal roots in a case of Werdnig-Hoffmann disease. *J Neuropathol Exp Neurol*. 1971;30:368–379.
135. Hirano A, Donnenfeld H, Sasaki S, Nakano I. Fine structural observations of neurofilamentous changes in amyotrophic lateral sclerosis. *J Neuropath Exptl Neurol*. 1984;43:461–470.
136. Couillard-Després S, Zhu Q, Wong PC, et al. Protective effect of neurofilament heavy gene overexpression in motor neuron disease induced by mutant superoxide dismutase. *Proc Natl Acad Sci U S A*. 1998;95:9626–9630.
137. Al-Chalabi A1, Andersen PM, Nilsson P, et al. Deletions of the heavy neurofilament subunit tail in amyotrophic lateral sclerosis. *Hum Mol Genet*. 1999;8:157–164.
138. Gros-Louis F, Larivière R, Gowing G, et al. A frameshift deletion in peripherin gene associated with amyotrophic lateral sclerosis. *J Biol Chem*. 2004;279:45951–45956.
139. Leung CL, He CZ, Kaufmann P, et al. A pathogenic peripherin gene mutation in a patient with amyotrophic lateral sclerosis. *Brain Pathol*. 2004;14:290–296.
140. Van Hoecke A, Schoonaert L, Lemmens R, et al. EPHA4 is a disease modifier of amyotrophic lateral sclerosis in animal models and in humans. *Nat Med*. 2012;18:1418–1422. doi:nm.2901, [pii] 10.1038/nm.2901.
141. Matsui C, Inoue E, Kakita A, et al. Involvement of the gamma-secretase-mediated EphA4 signaling pathway in synaptic pathogenesis of Alzheimer's disease. *Brain Pathol*. 2012;22:776–787. doi:10.1111/j.1750-3639.2012.00587.x.
142. Diekstra FP, van Vught PW, van Rheenen W, et al. UNC13A is a modifier of survival in amyotrophic lateral sclerosis. *Neurobiol Aging*. 2012;33:630.e633–630.e638. doi:10.1016/j.neurobiolaging.2011.10.029.
143. Chiò A, Mora G, Restagno G, et al. UNC13A influences survival in Italian amyotrophic lateral sclerosis patients: a population-based study. *Neurobiol Aging*. 2013;34:357 e351–357 e355. doi:10.1016/j.neurobiolaging.2012.07.016.
144. Dunckley T, Huentelman MJ, Craig DW, et al. Whole-Genome analysis of sporadic amyotrophic lateral sclerosis. *N Engl J Med*. 2007.
145. van Es MA, Van Vught PW, Blauw HM, et al. ITPR2 as a susceptibility gene in sporadic amyotrophic lateral sclerosis: a genome-wide association study. *Lancet Neurol*. 2007;6:869–877.
146. van Es MA, van Vught PW, Blauw HM, et al. Genetic variation in DPP6 is associated with susceptibility to amyotrophic lateral sclerosis. *Nat Genet*. 2008;40:29–31.
147. Chiò A, Schymick JC, Restagno G, et al. A two-stage genome-wide association study of sporadic amyotrophic lateral sclerosis. *Hum Mol Genet*. 2009;18:1524–1532. doi:ddp059, [pii] 10.1093/hmg/ddp059.
148. Fogh I, D'Alfonso S, Gellera C, et al. No association of DPP6 with amyotrophic lateral sclerosis in an Italian population. *Neurobiol Aging*. 2011;32:966–967. doi:10.1016/j.neurobiolaging.2009.05.014.
149. Kwee LC, Liu Y, Haynes C, et al. A high-density genome-wide association screen of sporadic ALS in US veterans. *PLoS One*. 2012;7:e32768. doi:10.1371/journal.pone.0032768.
150. Fogh I, Ratti A, Gellera C, et al. A genome-wide association meta-analysis identifies a novel locus at 17q11.2 associated with sporadic amyotrophic lateral sclerosis. *Hum Mol Genet*. 2014;23(8):2220–2231.
151. Deng M, Wei L, Zuo X, et al. Genome-wide association analyses in Han Chinese identify two new susceptibility loci for amyotrophic lateral sclerosis. *Nat Genet*. 2013;45:697–700. doi:10.1038/ng.2627.

CHAPTER

88

Peripheral Neuropathies

Steven S. Scherer, Kleopas A. Kleopa*†*, and Merrill D. Benson*‡*

*The University of Pennsylvania, Philadelphia, PA, USA
†The Cyprus Institute of Neurology and Genetics, Nicosia, Cyprus
‡Indiana University School of Medicine, Indianapolis, IN, USA

INTRODUCTION

The diseases considered in this chapter are listed in Tables 88.1 and 88.2, along with their associated genetic causes and accession numbers in the Online Mendelian Inheritance in Man (OMIM; http://www.ncbi.nlm.nih.gov/omim). Owing to constraints of space, we discuss the phenotypic consequences of different mutations for a selected number of genes; in addition to OMIM, please consult the Inherited Neuropathies Consortium (http://rarediseasesnetwork.epi.usf.edu/INC/), and GeneReviews (http://www.ncbi.nlm.nih.gov/books/NBK1358/), and the Washington University Neuromuscular website (http://neuromuscular.wustl.edu/). For the original references of each mutation, see the website http://www.molgen.vib-ua.be/CMTMutations/Mutations/MutByGene.cfm.

In addition to the topics covered in this chapter, an axonal neuropathy is a feature of many other inherited diseases. Many of these appear elsewhere in this volume—in the chapters on certain mitochondrial diseases (Chapters 23 and 24), certain types of hereditary spastic paraparesis (Chapter 77), inherited ataxias (Chapter 71), including Friedreich ataxia (Chapter 72) and ataxia-telangiectasia (Chapter 73); fragile X-associated disorders (Chapter 17), Fabry disease (Chapter 38), Schindler–Kanzaki disease (Chapter 39), cerebrotendinous xanthomatosis (Chapter 52), the porphyrias (Chapter 66), inherited causes of vitamin E or B12 deficiency (Chapter 47), and disorders of lipid (Chapter 50) and amino acid (Chapter 56) metabolism, including maple sugar urine disease (Chapter 59). Similarly, a demyelinating neuropathy is a feature of Refsum disease (Chapter 64), metachromatic leukodystrophy (Chapter 29), and globoid cell leukodystrophy/Krabbe disease (Chapter 30).

CLASSIFYING INHERITED NEUROPATHIES

Inherited neuropathies have been recognized since the late 1800s, when Charcot, Marie, and Tooth (hence, the term "Charcot–Marie–Tooth" [CMT]) described families with what we now know to be a dominantly inherited neuropathy.[1] With an estimated prevalence of 1 in 2500 persons, CMT is one of the commonest kinds of neurogenetic disease. The classical features arise from length-dependent axonal loss that worsens over time. For CMT, the onset is typically insidious in the first or second decade, and the rate of progression is slow. Motor and sensory nerve function are symmetrically affected in a length-dependent manner, with distal muscle weakness and atrophy, impaired sensation (particularly "large fiber" modalities), and diminished or absent deep tendon reflexes. Patients most commonly present with signs and symptoms related to weakness, more often involving the lower legs and feet, later progressing to affect the hands and then forearms. When weakness is present, a history of abnormal (steppage) gait, as well as tripping and falling, is frequently elicited. Patients may also present because of foot deformities (pes cavus, hammer toes) even in childhood; patients may also develop "claw hand" deformities owing to weakness of intrinsic hand muscles. Despite the involvement of the sensory nerves in CMT, patients may not report sensory disturbances, including neuropathic pain.

Rosenberg's Molecular and Genetic Basis of Neurological and Psychiatric Disease
http://dx.doi.org/10.1016/10.1016/B978-0-12-410529-4.00088-7

TABLE 88.1 Nonsyndromic Inherited Neuropathies

Disease (OMIM)	Locus in OMIM	Affected gene (OMIM)
CMT1 (AUTOSOMAL OR X-LINKED DOMINANT DEMYELINATING)		
HNPP (162500)	17p12	*PMP22* (601097)
CMT1A (118220)	17p12	*PMP22* (601097)
CMT1B (118200)	1q23.3	*MPZ* (159440)
CMT1C (601098)	16p13.13	*LITAF* (603795)
CMT1D (607678)	10q21.3	*EGR2* (129010)
CMT1E (118300)	17p12	*PMP22* (601097)
CMT1F (607734)	8p21.2	*NEFL* (162280)
CMT1X (302800)	Xq13.1	*GJB1* (304040)
Slowed nerve conduction velocity (608236)	8p23.3	*ARHGEF10* (608136)
AUTOSOMAL RECESSIVE DEMYELINATING NEUROPATHY ("CMT4")		
CMT4A (214400)	8q21.11	*GDAP1* (606598)
CMT4B1 (601382)	11q21	*MTMR2* (603557)
CMT4B2 (604563)	11p15.4	*SBF2* (607697)
CMT4B3 (615284)	22q13.33	*SBF1* (603560)
CMT4C (601596)	5q32	*SH3TC2* (608206)
CMT4D (601455)1	8q24.22	*NDRG1* (605262)
CMT4E (605253)	10q21.3	*EGR2* (129010)
CMT4F (614895)	19q13.2	*PRX* (605725)
CMT4G (605285)	10q22.1	*HK1* (142600)
CMT4H (609311)	12p11.21	*FGD4* (11104)
CMT4J (611228)	6q21	*FIG4* (609390)
CMT4	17p12	*PMP22* (601097)
CMT4	1q23.3	*MPZ* (159440)
SEVERE, DEMYELINATING NEUROPATHIES; DOMINANT OR RECESSIVE		
Déjérine–Sottas neuropathy (DSN) (145900)	17p12	*PMP22* (601097)
	1q23.3	*MPZ* (159440)
	10q21.3	*EGR2* (129010)
	19q13.2	*PRX* (605725)
	12p11.21	*FGD4* (11104)
	6q21	*FIG4* (609390) *MTMR2* (603557)
	11q21	*SBF2* (607697) *SH3TC2* (608206)
	11p15.4	
	5q32	
Congenital hypomyelinating neuropathy (CHN) (605253)	17p12	*PMP22* (601097)
	1q23.3	*MPZ* (159440)
	10q21.3	*EGR2* (129010)
CMT, DOMINANT INTERMEDIATE (CMTDI)		
CMTDIA (606483)	10q24.1-q25.1	
CMTDIB (606482)	19p13.2	*DNM2* (602378)

TABLE 88.1 Nonsyndromic Inherited Neuropathies—cont'd

Disease (OMIM)	Locus in OMIM	Affected gene (OMIM)
CMTDIC (608323)	1p35.1	*YARS* (603623)
CMTDID (607791)	1q23.3	*MPZ* (159440)
CMTDIE (614455)*	14q32.33	*INF2* (610982)
CMTDIF (615185)	3q26.33	*GNB4* (610863)
CMT2 (AUTOSOMAL DOMINANT AXONAL) AND DOMINANT CMTX		
CMT2A (609260)	1p36.22	*MFN2* (608507)
CMT2B (600882)	3q21.3	*RAB7A* (602298)
CMT2C (606071)	12q24.11	*TRPV4* (605427)
CMT2D (601472)	7p14.3	*GARS* (600287)
CMT2E (607684)	8p21.2	*NEFL* (162280)
CMT2F (606595)	7q11.23	*HSPB1* (602195)
CMT2G (608591)	12q12-q13.3	
CMT2I (607677)	1q23.3	*MPZ* (159440)
CMT2J (607736)	1q23.3	*MPZ* (159440)
CMT2K (607831)	8q21.11	*GDAP1* (606598)
CMT2L (608673)	12q24.23	*HSPB8* (608014)
CMT2M (606482)	19p13.2	*DNM2* (602378)
CMT2N (613287)	16q22.1	*AARS* (601065)
CMT2O (614228)†	14q32.31	*DYNC1H1* (600112)
CMT2P (614436)	9q33.3	*LRSAM1* (610933)
CMT2Q (615025)	10p14	*DHTKD1* (614984)
HMSNP (604484)‡	3q12.2	*TFG* (602498)
CMT2	5q31.3	*HARS* (142810)
CMT2	12q13.3	*MARS* (156560)
CMTX6 (300905)	Xp22.11	*PDK3* (300906)
Giant axonal neuropathy (610100)		*DCAF8* (no OMIM)
Hereditary neuralgic amyotrophy (162100)	17q25.2-q25.3	*SEPT9* (604061)
AUTOSOMAL RECESSIVE AXONAL NEUROPATHY (ARCMT2); SOME ARE SEVERE EARLY-ONSET AXONAL NEUROPATHY (SEOAN) PHENOTYPE		
ARCMT2B1 (605588)	1q22	*LMNA* (150330)
ARCMT2B2 (605589)	19q13.33	*MED25* (610197)
ARCMT2	1p36.22	*MFN2* (608507)
ARCMT2	8p21.2	*NEFL* (162280)
"CMT2P" (614436)	9q33.3	*LRSAM1* (610933)
Neuromyotonia and axonal neuropathy (137200)	5q23.3	*HINT1* (601314)
"CMT2R" (615490)	4q31.3	*TRIM2* (614141)
Lethal congenital contractural syndrome type 2 (607598)	12q13.2	*ERBB3* (190151)
Early-onset ARCMT2	5q13.2	*SMN1* (600354)
Lethal neonatal AR axonal neuropathy (604431)		
Congenital neuropathy		5q deletion

Continued

TABLE 88.1 Nonsyndromic Inherited Neuropathies—cont'd

Disease (OMIM)	Locus in OMIM	Affected gene (OMIM)
CMT, RECESSIVE INTERMEDIATE (CMTRI)		
CMTRIA (608340)	8q21.11	*GDAP1* (606598)
CMTRIB (613641)	16q23.1	*KARS* (601421)
CMTRIC (615376)	1p36.31	*PLEKHG5* (611101)
HEREDITARY SENSORY (AND AUTONOMIC) NEUROPATHY (HSN OR HSAN)		
HSN1A (162400)	9q22.31	*SPTLC1* (605712)
HSN1B (608088)	3p24-p22	
HSN1C (613640)	14q24.3	*SPTLC2* (605713)
HSN1D (613708)	14q22.1	*ATL1* (606439)
HSN1E (614116)	19p13.2	*DNMT1* (126375)
Small fiber neuropathy (133020)	2q24.3	*SCN9A* (603415)
HSAN2A (201300)	12p13.33	*WNK1* (605232)
HSN2B (613115)	5p15.1	*FAM134B* (613114)
HSN2C (614213)	2q37.3	*KIF1A* (601255)
HSAN3 (223900)	9q31.3	*IKBKAP* (603722)
HSAN4 (256800)*†	1q23.1	*NTRK1* (191315)
HSAN5 (608654)†	1p13.2	*NGF* (162030)
HSAN6 (614653)*	6p12.1	*DST* (113810)
HSAN7 (615548)	3p22.2	*SCN11A* (604385)
Congenital indifference to pain (147430)§	2q24.3	*SCN9A* (603415)
Congenital indifference to pain (243000)§	3p22.2	*SCN11A* (604385)
Primary erythermalgia (133020)§	2q24.3	*SCN9A* (603415)
Paroxysmal extreme pain disorder (167400)§	2q24.3	*SCN9A* (603415)
FEPD 1 (615040)§	8q13.3	*TRPA1* (604775)
FEPD 2 (615551)§	3p22.2	*SCN10A* (604427)
FEPD 3 (615552)§	3p22.2	*SCN11A* (604385)
CISS1 (272430)§	19p13.11	*CRLF1* (604237)
CISS2 (610313)§	11q13.2	*CLCF1* (607672)
HEREDITARY MOTOR NEUROPATHIES (HMN OR "DISTAL SMA")		
HMN1 (182960)	7q34-q36	
HMN2A (158590)	12q24.23	*HSPB8* (608014)
HMN2B (608634)	7q11.23	*HSPB1* (602195)
HMN2C (613376)	5q11.2	*HSPB3* (604624)
HMN2D (615575)	5q32	*FBXO38* (608533)
HMN2 (no OMIM)	5q31.3	*HARS* (142810)
HMN5A (600794)	7p14.3	*GARS* (600287)
HMN5A (600794)¶	11q12.3	*BSCL2* (606158)
HMN5B/SPG31 (614751)¶	2p11.2	*REEP1* (609139)
HMN/ALS4 (602433)¶	9q34.13	*SETX* (608465)

TABLE 88.1 Nonsyndromic Inherited Neuropathies—cont'd

Disease (OMIM)	Locus in OMIM	Affected gene (OMIM)
DSMA1/SMARD1 (604320)	11q13.3	*IGHMBP2* (600502)
DSMA2 (605726)	9p21.1-p12 11q13 1p36.31	*PLEKHG5* (611101)
DSMA3 (607088)	2q35	*DNAJB2* (604139)
DSMA4 (611067)		
DSMA5 (614881)		
HMN7A (158580)	2q12.3	*SLC5A7* (608761)
HMN7B (607641)	2p13.1	*DCTN1* (601143)
SMAX3 (300489)	Xq21.1	*ATP7A* (300011)
Scapuloperoneal SMA (181405)	12q24.11	*TRPV4* (605427)
Congenital distal SMA (600175)	12q24.11	*TRPV4* (605427)
SMALED1 (158600)[†]	14q32.31	*DYNC1H1* (600112)
SMALED2 (615290)[†]	9q22.31	*BICD2* (609797)
X-LINKED (DOMINANT OR RECESSIVE) CMT OR HMN/DSMA		
CMT1X (302800)	Xq13.1	*GJB1* (304040)
CMTX6 (300905)	Xp22.11	*PDK3* (300906)
SMAX3 (300489)	Xq21.1	*ATP7A* (300011)

[*]Syndromic neuropathy that is classified as CMT or HSAN in OMIM.

[†]Static clinical conditions, more likely the result of neuronal loss during embryonic development than an ongoing peripheral neuropathy.

[‡]CMT2O has been reconsidered to be a progressive sensory neuropathy superimposed on a static congenital motor neuronopathy. Recessive TFG mutations cause a myelopathy plus neuropathy (602498).

[§]Altered peripheral sensory and/of autonomic function without known evidence of axonal loss.

[¶] These patients typically have a myelopathy and a length-dependent motor axonopathy.

Abbreviations: CISS, cold-induced sweating syndrome; FEPD, familial episodic pain disorder; HNPP, hereditary neuropathy with liability to pressure palsies; SMALED, spinal muscular atrophy, lower extremity predominant; SMARD, spinal muscular atrophy with respiratory distress; SPG, spastic paraplegia.

Nonsyndromic inherited neuropathies are grouped by their patterns of inheritance and phenotypes. Some diseases are thus listed more than once. The genes are named by the HUGO gene nomenclature (http://www.genenames.org/). The diseases, their chromosomal locus, and genes are named according to OMIM (http://www.ncbi.nlm.nih.gov/Omim/).

Dyck and colleagues subdivided CMT according to clinical, electrophysiological, and histological features, and introduced the term hereditary motor and sensory neuropathy (HMSN).[1] CMT1/HMSN-I is more common and is characterized by an earlier age of onset (first or second decade of life), nerve conduction velocities (NCVs) less than 38 m per second in upper limb nerves, and segmental demyelination and remyelination with onion bulb formations in nerve biopsies. CMT2/HMSN-II has a later onset, NCVs greater than 38 m per second, and biopsies mainly show a loss of myelinated axons. CMT1 and CMT2 are dominantly inherited.

The term "dominant-intermediate" CMT (CMTDI) was proposed for patients with dominantly inherited CMT and median motor conduction velocities between 25–45 m per second.[2] In addition to the ones shown in Table 88.1, one could add CMT1X (*GJB1*), some cases of CMT1B (a subset of *MPZ* mutations, including the one termed CMTDID in OMIM), CMT2E (*NEFL*), CMT2N (*AARS*), *ARHGEF10*, and others. The term "recessive intermediate CMT" has been used to convey the same idea for recessive forms; *GDAP1* mutations are the commonest cause.

The terms Déjérine–Sottas neuropathy (DSN), CMT3, and HMSN-III apply to children who have a severe neuropathy.[3] They have delayed motor development before 3 years of age, sometimes extending to infancy. Their motor abilities typically improve during the first decade, followed by progressive weakness to the point that many affected individuals use wheelchairs. Ventilatory failure (presumably caused by phrenic nerve involvement) can occur, even during infancy or childhood. Kyphoscoliosis, short stature, and foot deformities are common in older children. Sensory loss is often profound, especially for modalities subserved by myelinated axons, to the point that some children have a severe sensory ataxia. Tendon reflexes are absent. Occasional patients have cranial nerve involvement—miosis, reduced pupillary responses to light, ptosis, facial weakness, nystagmus, and hearing loss. Cerebrospinal fluid (CSF) protein may be elevated, and nerve roots may enhance by magnetic resonance imaging (MRI). Sensory responses are absent; motor responses have reduced amplitudes, and very slow (<10 m per second)

TABLE 88.2 Genetic Classification of the Syndromic Inherited Neuropathies

Disease (MIM)	Locus	Affected gene	Other clinical features
SYNDROMIC DEMYELINATING NEUROPATHIES			
Waardenburg syndrome type 2E (611584)	22q13.1 dominant	*SOX10* (602229)	Hirschsprung disease
PCWH (609136)	22q13.1 dominant	*SOX10* (602229)	CNS demyelination, Hirschsprung disease
Metachromatic leukodystrophy (250100)	22q13.33 recessive	*ARSA* (607574)	Optic atrophy, mental retardation, hypotonia
Globoid cell leukodystrophy (245200)	14q31.3 recessive	*GALC* (606890)	Spasticity, optic atrophy, mental retardation
Refsum Disease (266500)	10p13 recessive	*PHYH* (602026)	Deafness, retinitis pigmentosa, ichthyosis, heart failure
PBD9B (614879)	q23.3 recessive	*PEX7* (601757)	Infantile (more severe) variant
Gonadal dysgenesis with minifascicular neuropathy (607080)	12q13.12 recessive	*DHH* (605423)	Mental retardation, hypogonadism
CDG1A (212065)	16p13.2 recessive	*PMM2* (601785)	Leukodystrophy, abnormal serum glycoproteins, mental retardation, hypotonia, ataxia
MDC1A (607855)	6q22.33 recessive	*LAMA2* (156225)	Congenital muscular dystrophy, mildly slowed PNS conduction, abnormal T2 MRI signal white matter
Duchenne muscular dystrophy (310200)	Xp21.2-p21.1 X-linked	*DMD* (300377)	Muscular dystrophy
PMD (312080) SPG2 (312920)	Xq22.2 X-linked	*PLP1* (300401)	Neuropathy is associated with certain loss of function mutations
Hypomyelinating leukodystrophy 5 (610532)	7p15.3 recessive	*FAM126A* (610531)	Congenital cataracts; abnormal MRI signal in CNS white matter
ACCPN (218000)	15q14 recessive	*SLC12A6* (604878)	Seizures, malformed corpus callosum, mental retardation
CCFDN (604168)	18q23 recessive	*CTDP1* (604927)	Rudari Gypsies: congenital cataracts and microcornea, facial dysmorphism
Cockayne syndrome A (216400)	5q12.1 recessive	*ERCC8* (609412)	Dwarfism, optic atrophy, mental retardation
Cockayne syndrome B (133540)	10q11.23 recessive	*ERCC6* (609413)	Dwarfism, optic atrophy, mental retardation
SYNDROMIC DEMYELINATING NEUROPATHIES, MITOCHONDRIA-RELATED			
MTDPS1 (603041)	22q13.33 recessive	*TYMP* (131222)	MNGIE phenotype
Leigh syndrome (256000) variant	9q34.2 recessive	*SURF1* (185620)	Leigh syndrome
	17p12 recessive	*COX10* (602125)	Myopathy, premature ovarian failure, hearing loss, pigmentary maculopathy, renal tubular dysfunction
SYNDROMIC AXONAL NEUROPATHIES			
Porphyria, acute intermittent (176000)	11q23.3 dominant	*HMBS* (609806)	Abdominal pain, psychosis, depression, dementia, seizures
Coproporphyria (121300)	3q11.2-q12.1 dominant	*CPOX* (612732)	Skin photosensitivity
Porphyria, variegata (176200)	1q23.3 dominant	*PPOX* (600923)	Skin photosensitivity, mainly South African Dutch
FAP-1 (105210) **FAP-2** (115430)	18q12.1 dominant	*TTR* (176300)	Dysautonomia, cardiac disease; type 2 have carpal tunnel syndrome
FAP-3 (105200)	11q23.3 dominant	*APOA1* (107680)	Nephropathy, liver disease
FAP-4 (105120)	9q33.2 dominant	*GSN* (137350)	Corneal lattice dystrophy, cranial neuropathies

TABLE 88.2 Genetic Classification of the Syndromic Inherited Neuropathies—cont'd

Disease (MIM)	Locus	Affected gene	Other clinical features
Familial visceral amyloidosis (105200)	15q21.1 dominant	*B2M* (109700)	Orthostatic hypotension, diarrhea
Somatic and autonomic neuropathy	20p13 dominant	*PRNP* (176640)	Autonomic sensory axonal neuropathy precede cognitive decline
CFEOM1 (600638)	16q24.3 dominant	*TUBB3* (602661)	Congenital fibrosis of extraocular muscles
Giant axonal neuropathy-1 (256850)	16q23.2 recessive	*GAN* (605379)	Mental retardation, spasticity, kinky or curly hair, slowed NCVs
Giant axonal neuropathy	10q26.11 recessive	*BAG3* (603883)	Giant axons, myopathy, cardiomyopathy, scoliosis
IND1 (256600)	22q13.1 recessive	*PLA2G6* (603604)	Progressive motor and mental deterioration
NIBA2B (610217)	22q13.1 recessive	*PLA2G6* (603604)	Ataxia, spasticity
PKAN (234200)	20p13 recessive	*PANK2* (606157)	Retinopathy; deposition of iron give "eye of the tiger" on MRI
Schindler disease (609241)	22q13.2 recessive	*NAGA* (104170)	Infantile onset, neuroaxonal dystrophy of CNS and PNS
Kanzaki disease (609242)	22q13.2 recessive	*NAGA* (104170)	Adult onset— angiokeratoma, sensorineural hearing loss, vertigo
Fabry disease (301500)	Xq22.1 X-linked	*GLA* (300644)	Angiokeratoma, painful neuropathy, cardiomyopathy, renal failure
ALD (300100)	Xq28 X-linked	*ABCD1* (300371)	Spastic paraparesis, adrenal insufficiency
AAAS (231550)	12q13.13 recessive	*AAAS* (605378)	Achalasia, addisonianism, alacrima
Tangier disease (205400)	9q31.1 recessive	*ABCA1* (600046)	Orange tonsils, organomegaly; pain, paresthesias, anesthesia; cranial neuropathies
THMD4 (613710)	17q25.1 recessive	*SLC25A19* (606521)	Bilateral striatal necrosis
Tyrosinemia type 1 (276700)	15q25.1 recessive	*FAH* (613871)	Neuropathy similar to porphyria (but younger age)
MMACHC (277400)	1p34.1 recessive	*MMACHC* (609831)	Onset infancy to adulthood; hematologic, retinal problems
BVVLS2 (614707)	8q24.3 recessive	*SLC52A2* (607882)	Infantile onset of vision and hearing loss, hearing loss, bulbar dysfunction
Leukoencephalopathy with dystonia and motor neuropathy (613724)	1p32.3 recessive	*SCP2* (184755)	Myelopathy; dystonia, hypogonadism, cerebellar finding
APBD (263570)	3p12.2 recessive	*GBE1* (607839)	Cognitive impairment, MRI changes in white matter
Galactosialidosis (256540)	20q13.12 recessive	*CTSA* (613111)	Coarse facies, motor neuropathy, cerebellar findings
GACR (258870)	10q26.13 recessive	*OAT* (613349)	Chorioretinal degeneration, cataracts, and type II fiber atrophy
Cerebrotendinous xanthomatosis (213700)	2q35 recessive	*CYP27A1* (606530)	Neonatal jaundice, ataxia, myelopathy, dementia, cataracts, low cholesterol, atherosclerosis, xanthomas
CEDNIK syndrome (609528)	22q11.21 recessive	*SNAP29* (604202)	Cerebral dysgenesis, neuropathy, ichthyosis, palmoplantar keratoderma
MEDNIK (609313)	7q22.1 recessive	*AP1S1* (603531)	Mental retardation, enteropathy, deafness, neuropathy, keratoderma
Xeroderma pigmentosum (278700)	9q22.33 recessive	*XPA* (611153)	Cutaneous lesions, increased risk of skin cancers
Chediak–Higashi syndrome (214500)	1q42.3 recessive	*LYST* (606897)	Partial albinism, immunodeficient

Continued

TABLE 88.2 Genetic Classification of the Syndromic Inherited Neuropathies—cont'd

Disease (MIM)	Locus	Affected gene	Other clinical features
PNMHH (614369)	19q13.33 recessive	*MYH14* (608568)	Myopathy, hoarseness, hearing loss
EDS6 (225400)	1p36.22 recessive	*PLOD1* (153454)	Skin hyperextensibility, articular hypermobility, tissue fragility
CTRCT21 (610202)	16q23.2 recessive	*MAF* (177075)	Cataracts, diminished vibration sensation
Sensory neuropathy and deafness	1p34.3 dominant	*GJB3* (603324)	Hearing loss
Muscular dystrophy and neuropathy	1q22 dominant	*LMNA* (150330)	Muscular dystrophy, cardiac disease, leukonychia
Motor-predominant neuropathy	14q32.12 dominant	*FBLN5* (604580)	Macular degeneration, hyperelastic skin
NF2 (101000)	22q12.2 dominant	*NF2* (607379)	Schwannomas
AXONAL NEUROPATHIES ASSOCIATED WITH ATAXIA			
NARP (551500)	m8993T>G	*MT-ATP6* (516060)	Neuropathy, ataxia, retinitis pigmentosa; seizures
SANDO (607459)	15q26.1 recessive	*POLG* (174763)	Sensory ataxic neuropathy, dysarthria, ophthalmoplegia
PEOA3 (609286)	10q24.31 dominant	*C10orf2* (606075)	PEO, sensory neuropathy, hearing loss, dysphagia
MTDPS7 (271245)	10q24.31 recessive	*C10orf2* (606075)	Ataxia, ophthalmoplegia, ptosis, encepthalopathy, seizures
AXPC1 (609033)	1q32.3 recessive	*FLVCR1* (609144)	Childhood-onset retinitis pigmentosa and later onset of gait ataxia due to sensory loss
SACS (270550)	13q12.12 recessive	*SACS* (604490)	Myelopathy
Friedreich ataxia / FRDA-1 (229300)	9q21.11 recessive	*FXN* (606829)	Ataxia, cardiomyopathy
FRDA-2 (601992)	9p23-p11 recessive		Type 2 cannot be distinguished clinically from FRDA-1
VED (277460)	8q12.3 recessive	*TTPA* (600415)	Similar to Friedreich ataxia, tendon xanthomas, retinitis pigmentosa
LBSL (611105)	1q25.1 recessive	*DARS2* (610956)	Ataxia, myelopathy
Abetalipoprotein emia (200100)	4q23 recessive	*MTTP* (157147)	Likely vitamin E deficiency—retinitis pigmentosa, acanthocytosis
FXTAS (300623)	Xq27.3 X-linked	*FMR1* (309550)	Axonal neuropathy, tremor, cognitive changes, parkinsonian, MRI changes in cerebellum
Ataxia telangiectasia (208900)	11q22.3 recessive	*ATM* (607585)	Oculomotor apraxia, telangiectasias
EAOH (208920)	9p21.1 recessive	*APTX* (606350)	Cerebellar atrophy, hypoalbuminemia
SCAN1 (607250)	14q32.11 recessive	*TDP1* (607198)	Cerebellar ataxia
SCAR1 (606002)	9q34.13 recessive	*SETX* (608465)	Juvenile onset ataxia, increased α-fetoprotein, nystagmus, cerebellar and pontine atrophy; oculomotor apraxia
SCAR3 (271250)	6p23-p21 recessive		Neuropathy
SCAR4 (607317)	1p36 recessive		Saccadic intrusions
CANVAS (614575)	recessive		Cerebellar ataxia, neuropathy, vestibular areflexia syndrome
	1p36.32 recessive	*PEX10* (602859)	Progressive ataxia, motor neuropathy
	19q13.33 recessive	*PNKP* (605610)	Progressive cerebellar atrophy
SCA1 (164400)	6p22.3 dominant	*ATXN1* (601556)	Gaze paresis, slow / absent saccades, spasticity, neuropathy in 40%
SCA2 (183090)	12q24.12 dominant	*ATXN2* (601517)	Slow saccades, tremor, myoclonus, neuropathy in 80%

TABLE 88.2 Genetic Classification of the Syndromic Inherited Neuropathies—cont'd

Disease (MIM)	Locus	Affected gene	Other clinical features
SCA3/MJD (109150)	14q32.12 dominant	*ATXN3* (607047)	Gaze paresis, extrapyramidal, bulging eyes; neuropathy in 50%
SCA4 (600223)	16q22.1 dominant		Sensory neuropathy
SCA7 (164500)	3p14.1 dominant	*ATXN7* (607640)	Retinopathy and neuropathy
SCA10 (603516)	22q13.31 dominant	*ATXN10* (611150)	Neuropathy in some families
SCA12 (604326)	5q32 dominant	*PPP2R2B* (604325)	Subclinical neuropathy
SCA18 (607458)	7q22-q32 dominant		Sensory and motor neuropathy With ataxia
SCA23 (610245)	20p13 dominant	*PDYN* (131340)	Sensory neuropathy
SCA25 (608703)	2p21-p13 , dominant	*IFRD1*	Sensory neuropathy
SCA27 (609307)	13q33.1 dominant	*FGF14* (601515)	Cerebellar findings, mild neuropathy
SCA36 (614153)	20p13 dominant	*NOP56* (614154)	Distal motor neuropathy
AXONAL NEUROPATHIES ASSOCIATED WITH SPASTIC PARAPARESIS			
HSN with spastic paraplegia (256840)	5p15.2 recessive	*CCT5* (610150)	Severe mutilating sensory neuropathy with spastic paraplegia
SPOAN (609541)	11q13 recessive		Spastic paraplegia, optic atrophy, and neuropathy
	3q12.2 recessive	*TFG* (602498)	Optic atrophy and neuropathy
ALD (300100)	Xq28 X-linked	*ABCD1* (300371)	Spastic paraparesis, adrenal insufficiency
SPG2 (312920)	Xq22.2 X-linked	*PLP1* (300401)	
SPG3A (182600)	14q22.1 dominant	*ATL1* (606439)	Neuropathy is not associated with most mutations but the R495 mutation causes HSN1D
SPG7 (607259)	16q24.3 recessive	*SPG7* (602783)	Optic atrophy; neuropathy reported in some patients
SPG9 (601162)	10q23.3-q24.1 dominant		Short stature, cataracts, motor neuropathy
SPG10 (604187)	12q13.3 dominant	*KIF5A* (602821)	Adult onset; neuropathy in some patients
SPG11 (604360)	15q21.1 recessive	*SPG11* (610844)	Thin corpus callosum, neuropathy
SPG14 (605229)	3q27-q28 recessive		Mild mental retardation, distal motor neuropathy; sural nerve biopsy normal
SPG15 (270700)	14q24.1 recessive	*ZFYVE26* (612012)	Mental retardation, retinal degeneration
SPG17 (270685)	11q12.3 dominant	*BSCL2* (606158)	Silver syndrome; spasticity, motor neuropathy in arms>legs
SPG20 (275900)	13q13.3 recessive	*SPG20* (275900)	Mental retardation, pseudobulbar affect
SPG23 (270750)	1q24-q32 recessive		Axonal neuropathy; depigmentation
SPG26 (609195)	12p11.1-q14 recessive	*B4GALNT1* (601873)	Intellectual disability, ataxia, peripheral neuropathy
SPG28 (609340)	14q22.1 recessive	*DDHD1* (14603)	Subclinical axonal neuropathy
SPG36 (613096)	12q23-q24 dominant		Axonal neuropathy
SPG38 (612335)	4p16-p15 dominant		Silver syndrome
SPG39 (612020)	19p13.2 recessive	*PNPLA6* (603197)	Childhood onset of slowly progressive spastic paraplegia; progressive distal motor neuropathy beginning in early through late adolescence

Continued

TABLE 88.2 Genetic Classification of the Syndromic Inherited Neuropathies—cont'd

Disease (MIM)	Locus	Affected gene	Other clinical features
SPG46 (614409)	9p13.3 recessive	*GBA2* (609471)	Mental impairment, thin corpus callosum, ataxia, neuropathy
SPG55 (615035)	12q24.31 recessive	*C12orf65* (613541)	Mild neuropathy
SYNDROMIC, AXONAL NEUROPATHIES, MITOCHONDRIA-RELATED			
MELAS (540000)	m1095T>C	*MT-RNR1* (561000)	Parkinsonism, deafness
MELAS (540000)	m3243A>G	*MT-TL1* (590050)	
MERRF (545000)	m8313G>A m8344A>G	*MT-TK* (590060)	Bulbar weakness, severe myopathy
MERRF (545000)	m8529G>A	*MT-ATP8* (516070)	Hypertrophic cardiomyopathy
NARP (551500)	m8618insTm8993T>G m8993T>C m9176T>C m9185T>C	*MT-ATP6* (516060)	Neuropathy, ataxia, retinitis pigmentosa, cardiomyopathy
Kearns–Sayre syndrome (530000)	mDNA deletions		Ophthalmoplegia, retinitis pigmentosa, heart block, ptosis
MTDPS3 (251880)	2p13.1 recessive	*DGUOK* (601465)	Liver failure, myopathy
MTDPS4B (613662)	15q26.1 recessive	*POLG* (174763)	MNGIE
MTDPS6 (256810)	2p23.3 recessive	*MPV17* (137960)	Corneal opacification, liver disease, acromutilation
MTDPS7 (271245)	10q24.31 recessive	*C10orf2* (606075)	Infantile-onset ataxia, PEO, encephalopathy, seizures
MTDPS8B (612075)	8q22.3 recessive	*RRM2B* (604712)	PEO, MNGIE, minimal neuropathy
Multiple symmetric lipomatosis (151800)	dominant		Multiple lipomas with predilection for the neck, typically in alcoholics
PDHAD (312170)	Xp22.12 X-linked	*PDHA1* (300502)	Hypotonia, lethargy, mental retardation
Cowchock syndrome (310490)	Xq24-q26.1 X-linked	*AIFM1* (300169)	Mental retardation (60%), deafness
Adult onset sensory motor neuropathy	5p13.2 recessive	*AMACR* (604489)	Retinopathy, myelopathy
EAOH (208920)	9p21.1 recessive	*APTX* (606350)	Cerebellar atrophy, hypoalbuminemia
MSUDIb (248600)	6q14.1 recessive	*BCKDHB* (248611)	Intermittent seizures, drowsiness, ataxia
Friedreich ataxia; FRDA-1 (229300)	9q21.11 recessive	*FXN* (606829)	Ataxia, cardiomyopathy
LBSL (611105)	1q25.1 recessive	*DARS2* (610956)	Ataxia, myelopathy
TFP (609015)	2p23.3 recessive	*HADHA* (600890) *HADHB* (143450)	Episodic weakness, myalgias, and rhabdomyolysis
Syndromic optic atrophy (125250)	3q29 dominant	*OPA1* (605290)	PEO, deafness, myelopathy, MS-like
SACS (270550)	13q12.12 recessive	*SACS* (604490)	Myelopathy
CEMCOX1 (604377)	22q13.33 recessive	*SCO2* (604272)	Leigh-like syndrome, cardiomyopathy
SPG7 (607259)	16q24.3 recessive	*SPG7* (602783)	Optic atrophy; neuropathy reported in some patients
SPG20 (275900)	13q13.3 recessive	*SPG20* (275900)	Mental retardation, pseudobulbar affect
SYNDROMIC X-LINKED NEUROPATHIES (DOMINANT OR RECESSIVE)			
"CMTX2" (302801)	Xp22.2 recessive		Infantile onset of neuropathy; 2/5 patients had mental retardation

TABLE 88.2 Genetic Classification of the Syndromic Inherited Neuropathies—cont'd

Disease (MIM)	Locus	Affected gene	Other clinical features
"CMTX3" (302802)	11q12.3 , dominant	*BSCL2* (606158)	Not X-linked; family had HMN5A
Cowchock syndrome/ "CMTX4" (310490)	Xq24-q26.1 X-linked	*AIFM1* (300169)	Mental retardation, deafness
"CMTX5" (311070)	Xq22.3 recessive	*PRPS1* (311850)	Hearing loss, visual impairment
Fabry disease (301500)	Xq22.1 recessive	*GLA* (300644)	Angiokeratoma, painful neuropathy, cardiomyopathy, renal failure
ALD (300100)	Xq28 recessive	*ABCD1* (300371)	Spastic paraparesis, adrenal insufficiency
PMD (312080) SPG2 (312920)	Xq22.2 X-linked	*PLP1* (300401)	Neuropathy is associated with certain loss of function mutations
FXTAS (300623)	Xq27.3 X-linked	*FMR1* (309550)	Axonal neuropathy, tremor, cognitive changes, parkinsonian, MRI changes in cerebellum
PDHAD (312170)	Xp22.12 X-linked	*PDHA1* (300502)	Hypotonia, lethargy, mental retardation
SBMA (313200)	Xq12 recessive	*AR* (313700)	Motor neuron disease and sensory neuropathy, androgen insensitivity

Abbreviations: AAAS, achalasia-addisonianism-alacrima syndrome; ACCPN, genesis of the corpus callosum with peripheral neuropathy; ALD, adrenoleukodystrophy; AXPC1, posterior column ataxia with retinitis pigmentosa; BVVLS2, Brown–Vialetto–Van Laere syndrome 2; CANVAS, cerebellar ataxia, neuropathy, vestibular areflexia syndrome; CCFDN, congenital cataracts, facial dysmorphism, and neuropathy; CDG1A, congenital disorder of glycosylation type Ia; CEDNIK syndrome, cerebral dysgenesis neuropathy, ichthyosis, and palmoplantar keratoderma; CEMCOX1, cardioencephalomyopathy, fatal infantile, due to cytochrome C oxidase deficiency 1; CTRCT21, cataract 21, multiple types; EDS6, Ehlers–Danlos syndrome, type 6; EAOH, ataxia, early onset, with oculomotor apraxia and hypoalbuminemia; FAP, familial amyloidotic polyneuropathy; FRDA, Friedreich ataxia; FXTAS, fragile X tremor/ataxia syndrome; GACR, gyrate atrophy of choroid and retina; IND1, infantile neuroaxonal dystrophy 1; LBSL, leukoencephalopathy with brainstem and spinal cord involvement and lactate elevation; MDC1A muscular dystrophy, congenital merosin-deficient, 1A; MEDNIK, mental retardation enteropathy, deafness, peripheral neuropathy, keratoderma; MELAS, mitochondrial myopathy, encephalopathy, lactic acidosis, and stroke-like episodes; MERRF, myoclonic epilepsy associated with ragged-red fibers; MJD, Machado-Joseph disease; MMACHC, early onset methylmalonic aciduria and homocystinuria; MNGIE, mitochondrial neurogastrointestinal encephalopathy syndrome; MSUDIb, maple syrup urine disease; MTDPS, mitochondria DNA depletion syndrome; NARP, neuropathy, ataxia, retinitis pigmentosa; NF2, Neurofibromatosis type 2; NIBA2B, neurodegeneration with brain iron accumulation 2B; PCWH, peripheral demyelinating syndrome, central dysmyelinating leukodystrophy, Waardenburg syndrome, and Hirschsprung disease; PBD9B, peroxisome biogenesis disorder; PDHAD, pyruvate dehydrogenase E1α deficiency; PEO, progressive external ophthalmoplegia; PKAN, pantothenate kinase-associated neurodegeneration; PMD, Pelizaeus-Merzbacher disease; PNMHH, peripheral neuropathy, myopathy, hoarseness, hearing loss; THMD4, thiamine metabolism dysfunction syndrome 4; SACS, spastic ataxia, Charlevoix–Saguenay type; SANDO, sensory ataxic neuropathy, dysarthria, and ophthalmoparesis; SBMA, spinal and bulbar muscular atrophy, X-linked; SCA, spinocerebellar ataxia; SCAN1, spinocerebellar ataxia, autosomal recessive, with axonal neuropathy; SCAR4, spinocerebellar ataxia, autosomal recessive 4; SMNA, autosomal dominant sensory/motor neuropathy with ataxia; SPG, spastic paraplegia; SPOAN, spastic paraplegia, optic atrophy, and neuropathy; TFP, trifunctional protein deficiency; THMD4, thiamine metabolism dysfunction syndrome, type 4; VED, vitamin E, familial isolated deficiency.

Syndromic inherited neuropathies grouped by their pattern of inheritance and phenotypes. Some diseases are thus listed more than once. The genes are named according to the HUGO gene nomenclature (http://www.genenames.org/). The diseases, their chromosomal locus, and genes are named according to OMIM (http://www.ncbi.nlm.nih.gov/omim/). Diseases in bold are dominant.

NCVs, with marked temporal dispersion. Nerves are often enlarged, and biopsies reveal a complete absence of axons with normal (thick) myelin sheaths, and endoneurial edema. The remaining myelinated axons have inappropriately thin myelin sheaths for the axonal caliber. Onion bulbs are prominent, composed of Schwann cell processes or just their basal laminae. While these clinical, electrophysiological, and pathological features distinguish DSN from CMT1, the literature contains many examples of patients who are said to have CMT (often "severe CMT"), many of whom could have just as readily been labeled DSN. *PMP22* duplications, new dominant mutations in *MPZ*, *PMP22*, and *EGR2*, as well as recessive mutations in *MPZ*, *PMP22*, *EGR2*, *PRX*, *FGD4*, *FIG4*, *MTMR2*, *MTMR13*, and especially *SH3TC2*, cause DSN.[3] Thus, DSN is a convenient clinical label but lacks specificity in terms of causality.

Congenital hypomyelinating neuropathy (CHN) is a term for infants who have hypotonic weakness at birth caused by a severe neuropathy.[3] These infants may even have arthrogryposis caused by a prenatal onset, and swallowing or respiratory difficulties may result in death during infancy or later. NCVs are severely reduced (<5 m per second), and biopsies show similar features as described for DSN. New dominant mutations in *MPZ*, *PMP22*, and *EGR2* mutations are the known causes.[3] It is possible that the most severe cases of CHN, associated with arthrogryposis, have different genetic causes, as *MPZ*, *PMP22*, or *EGR2* mutations have been rarely shown to cause this phenotype.

CMT4 (HMSN-IV) was originally restricted to recessive demyelinating forms of CMT, subsuming CHN, DSN, and CMT1 phenotypes, but in recent years, some authors used this term for both demyelinating and axonal forms of CMT. OMIM uses the term "CMT2B" for recessively inherited axonal neuropathies, which is unfortunate because this term is easily confused with CMT2B, an axonal neuropathy caused by dominant *RAB7* mutations. In Table 88.1, we have used the less ambiguous, alternative term, "autosomal recessive CMT2" (ARCMT2). The apt term, "severe early-onset axonal neuropathy" has been proposed for the forms of CMT that fit this description, including cases caused by new dominant MFN2 and *NEFL* mutations, inherited dominant *TRPV4* mutations, and recessive mutations of *MFN2*, *NEFL*, and especially *GDAP1*.[3–5]

Hereditary sensory and autonomic neuropathies (HSAN) predominantly affect sensory neurons and/or their axons; in most forms the autonomic neuropathy is clinically inconspicuous so the term hereditary sensory neuropathy (HSN) has been used.[6] Hereditary motor neuropathy (HMN; also called distal spinal muscular atrophy [DSMA]) includes various disorders that affect motor axons in a length-dependent manner.

There are several problems with the nomenclature. A few neuropathies, including neuralgic amyotrophy, could have been called CMT, as neuropathy dominates the typical clinical picture. Conversely, the neuropathy of CMTDIE (*IFN2*), CMT4D (*NDRG1*), CMTX2 (gene unknown), "CMTX3" (gene unknown), "CMTX4" (*AIFM1*), "CMTX5" (*PRPS1*), HSAN4 (*NTRK1*), and HSAN6 (*DST*) is part of a larger syndrome; it is misleading to label them as CMT. In some syndromic neuropathies, the neuropathy may be the presenting issue and/or dominate the clinical picture. In some cases, specific mutations in various genes may produce a CMT phenotype (Table 88.3). Finally, some forms of CMT—CMT2O and SMALED1 (*DYNC1H1*), SMALED2 (*BICD2*), HSAN4 (*NTRK1*), and HSAN5 (*NGF*)—appear to have static neurological deficits, reflecting neuronal cell loss during development; there is little evidence of ongoing neuronal or even axonal loss.

Myelinating Schwann Cells and Demyelinating Neuropathies

The demyelinating forms are the most common kind of CMT, and are caused by cell autonomous effects of the mutation in myelinating Schwann cells.[7] Myelin is a multilamellar spiral of specialized cell membrane that ensheathes axons larger than 1 μm in diameter. By reducing the capacitance and/or increasing the resistance, myelin reduces current flow across the internodal axonal membrane, thereby facilitating saltatory conduction at nodes. Figure 88.1A depicts two internodes; one has been unrolled to reveal its trapezoidal shape. The myelin sheath itself can be divided into two domains—compact and noncompact myelin—each containing a nonoverlapping set of proteins (Figure 88.1B). Compact myelin forms the bulk of the myelin sheath. It is largely composed of lipids, mainly cholesterol and sphingolipids (including galactocerebroside and sulfatide), and the intrinsic membrane proteins, myelin protein zero (MPZ or P0) and peripheral myelin protein 22 kDa (PMP22). P0 is the molecular glue of compact myelin and is the most abundant protein component; PMP22 is a minor component of unknown function. Noncompact myelin is found in the paranodes and incisures, and contains the gap junction protein connexin32 (Cx32).

PMP22 Deletions Cause Hereditary Neuropathy with Liability to Pressure Palsies (HNPP)

Deletion or duplication of a 1.5-Mb segment of chromosome 17, containing the *PMP22* gene, causes HNPP or CMT1A, respectively, establishing a singular genetic mechanism as the cause of the two most common inherited neuropathies.[8] Two homologous DNA sequences flanking the *PMP22* gene are the molecular basis for its frequent deletion/duplication; their high degree of homology promotes unequal crossing over during meiosis, which simultaneously generates a duplicated and a deleted allele.

Given this shared molecular mechanism, HNPP should be as common as CMT1A, but far fewer HNPP patients than CMT1A patients are seen in neuropathy clinics. Thus, one suspects that most patients who harbor the *PMP22* deletion are not diagnosed, or even seen by neurologists. Part of the reason is likely to be that many patients never experience the episodic mononeuropathies at typical sites of nerve compression—the hallmark of HNPP. Over half of the patients recover completely, usually within days to months, but deficits may persist.

In addition to focal changes at common sites of nerve entrapment, genetically affected individuals with or without episodes of pressure palsies have a mild, sensory–motor polyneuropathy. Sensory velocities are mildly slowed, especially in the upper extremities. Motor NCVs are minimally slowed, but distal motor latencies are consistently prolonged. Biopsies of unpalsied nerves show focal thickenings (tomaculae) caused by folding of the myelin sheath, as well as segmental demyelination and remyelination. While it is plausible that severe focal demyelination (causing conduction block) followed by remyelination (restoring conduction) are the cellular alterations associated with the focal neuropathies, this remains to be directly demonstrated.

How reduced gene dosage affects myelinating Schwann cells is not known, but there is less PMP22 mRNA in peripheral nerves from HNPP patients, and PMP22 protein is reduced in the compact myelin from individuals with HNPP. *Pmp22* +/− mice, an animal model of HNPP, also develop a demyelinating neuropathy with many features of HNPP.

TABLE 88.3 Syndromic Neuropathies that can Present with Neuropathy

Disease (OMIM)	Affected gene (OMIM)
CMT1-LIKE (DOMINANT DEMYELINATING NEUROPATHY)	
CMT1-like (no OMIM)	*FBLN5* (604580) R373C
CMT2-LIKE (DOMINANT AXONAL NEUROPATHY)	
FAP-1 (105210)	*TTR* (176300)
Somatic and autonomic neuropathy (no OMIM)	*PRNP* (176640) Y163X
CFEOM1 (600638)	*TUBB3* (602661) D416N, E410K
CMT2-like (no OMIM)	*MT-ATP6* (516060) m9185T>C, m9176T>C
CMT4-LIKE (RECESSIVE, DEMYELINATING NEUROPATHY)	
Globoid cell leukodystrophy (245200)	*GALC* (606890)
Metachromatic leukodystrophy (250100)	*ARSA* (607574)
ACCPN (218000)	*SLC12A6* (604878)
Leigh syndrome (256000)	*SURF1* (185620)
MTDPS1 (603041)	*TYMP* (131222)
ARCMT-LIKE (RECESSIVE, AXONAL NEUROPATHY)	
GAN-1 (256850)	*GAN* (605379)
SANDO (607459)	*POLG* (174763)
SACS (270550)	*SACS* (604490)
DSMA1/SMARD (604320)	*IGHMBP2* (600502)
X-LINKED AXONAL NEUROPATHY	
ALD (300100)	*ABCD1* (300371)
Fabry disease (301500)	*GLA* (300644)
HEREDITARY MOTOR NEUROPATHY-LIKE (HMN)	
HMN-like (no OMIM)	*FBLN5* (604580) G90S, V126M
SPG4 (182601)	*SPAST* (604277) Asp321fs
HMN-like (no OMIM)	*MT-ATP6* (516060) m9185T>C

Abbreviations: ACCPN, genesis of the corpus callosum with peripheral neuropathy; ALD, adrenoleukodystrophy; CFEOM1, congenital fibrosis of extraocular muscles 1; FAP, familial amyloidotic polyneuropathy; GAN, giant axonal neuropathy; SACS, spastic ataxia, Charlevoix–Saguenay type; SANDO, sensory ataxic neuropathy, dysarthria, and ophthalmoparesis; SMARD, spinal muscular atrophy with respiratory distress; SPG, spastic paraplegia.

Listed are syndromic inherited neuropathies that can be diagnosed as nonsyndromic neuropathies. The genes are named according to the HUGO gene nomenclature (http://www.genenames.org/). The diseases, their chromosomal locus, and genes are named as in OMIM (http://www.ncbi.nlm.nih.gov/omim/). Diseases in bold are dominant.

PMP22 Duplications Cause CMT1A

CMT1A is, by far, the most common cause of CMT, and is almost always associated with the typical 17p11.2 duplication. In one large series, the age at onset ranged from 2 to 76 years, but most patients were affected by end of the second decade, but neonatal and even congenital onsets have been reported.[3] Weakness, atrophy, and sensory loss in the lower limbs, foot deformities and areflexia were invariably present. In one series, at a mean age of 40 years, 35% of the patients had mild functional disability, 25% were asymptomatic, 61% had difficulty walking or running, but were still independent, and only one patient (less than 1%) required a wheelchair. There is considerable variability in the degree of neurological deficits within families.

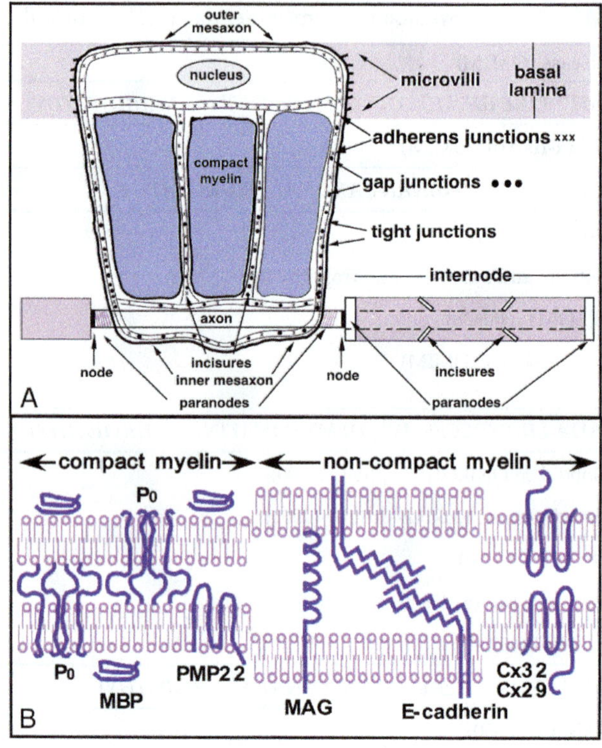

FIGURE 88.1 **The architecture of the myelinated axon in the peripheral nervous system.** (**A**) One of the Schwann cells has been "unrolled" to reveal the regions forming compact myelin, as well as paranodes and incisures, regions of noncompact myelin. (**B**) Note that P_0, PMP22, and MBP are found in compact myelin, whereas Cx29, Cx32, MAG, and E-cadherin are localized in noncompact myelin.

Sensory responses are usually absent. Motor NCVs are slowed, ranging from 5 to 35 m per second in forearm motor nerves, but most average around 20 m per second. The lack of conduction block or temporal dispersion and the high correlation between the motor NCVs in different nerves are all hallmarks. NCVs are slow in children, even well before the clinical onset of disease. Conduction velocities, like clinical disability, vary widely within families, but an early age at onset and greatly reduced median NCV are associated with a more severe course. In individual patients, motor NCVs do not change significantly even over two decades, whereas the motor amplitudes and the number of motor units decrease slowly, which correlates with clinical disability. Thus, it is axonal loss and not reduced conduction velocity *per se* that causes weakness.

How the duplication of *PMP22* causes demyelination is not settled. The amount of PMP22 protein in compact myelin appears to be increased, which could destabilize the myelin sheath. Overexpressing PMP22 in myelinating Schwann cells of rodents leads to dose-dependent alterations in myelination, with nearly complete failure to myelinate in the highest expressing lines. Decreasing PMP22 expression in myelinating Schwann cells is an obvious target for treating CMT1A, and has been pursued in high-throughput screens with cell-based assays.

PMP22 Mutations Cause DSN, CMT1A, or HNPP

In addition to the much more common *PMP22* duplications and deletions, 72 different heterozygous *PMP22* mutations have been published. Each family or affected individual typically has a "private" mutation—three are benign polymorphisms, 25 cause CHN or DSN, 22 cause CMT1, and 21 cause HNPP. Homozygous deletions of *PMP22* and compound heterozygous mutations of *PMP22* cause DSN.

PMP22 mutations that cause phenotypes that are more severe than HNPP must have gain-of-function/toxic effects. The PMP22 mutant proteins analyzed to date do not reach the compact myelin; they appear to be retained in the endoplasmic reticulum (ER) and/or an intermediate compartment between the ER and the Golgi. One thought is that PMP22 mutants that are retained in the ER have a prolonged half-life and sequester chaperones that are essential for the proper processing of glycosylated proteins, thereby causing a toxic gain of function. PMP22 accumulates intracellularly in myelinating Schwann cells of *Trembler* (which harbor a D150G mutation in *Pmp22*) and *TremblerJ* mice (which harbor a L16P mutation in *Pmp22*), and in nerve biopsies of patients with a *PMP22* missense mutation. Perhaps the mutant PMP22 interacts abnormally with its wild-type counterpart, forming aggregates that accumulate in the ER, possibly triggering a maladaptive ER "stress response."

MPZ Mutations Cause CHN, DSN, CMT1B, CMTDID, or a CMT2-Like Phenotype

MPZ mutations are the third most common cause of CMT. Each family or affected individual typically has a "private" mutation. Of the 164 published heterozygous mutations, seven are benign polymorphisms, 35 cause CHN or DSN, 10 cause severe CMT1B, 68 cause CMT1B, 27 cause CMT2, 15 cause a mild, late-onset, demyelinating neuropathy, and two are clinically uncharacterized. These *MPZ* mutations can be lumped into an early-onset, severe phenotype with very slow NCVs, and a late-onset phenotype (that includes both axonal and demyelinating forms) with mildly slowed or even normal NCVs.[9] Homozygous V102 frameshift, homozygous D224Y mutations, and compound heterozygous S62F/N131Y mutations cause DSN. Duplication of the *MPZ* gene (one or five extra copies) also causes a CMT1 phenotype.

How can one make sense of so many mutations and the diversity of phenotypes they cause? The lesson from *PMP22* mutations is to compare the effects of *MPZ* mutations to a null allele. Deletion of one *MPZ* gene in humans has not been described, but the V102 frameshift mutation may be comparable to a deletion, as the protein should not function as a cell adhesion molecule because it lacks its transmembrane domain. It was discovered in two siblings with DSN who were both homozygous for this mutation, but their asymptomatic heterozygous parents and grandparents had a mild demyelinating neuropathy. Thus, given the mild phenotypes of *MPZ*, mutations with loss-of-function, mutations that cause DSN, severe CMT1B, and typical CMT1B, must have a toxic gain-of-function.

A large family with a K96E mutation illustrates the early onset CMT1B phenotype. The youngest members were all affected before age 10, several before age 3. Delayed walking and clumsy running were common early symptoms; absent tendon reflexes, followed by distal weakness and atrophy were early clinical findings. Proximal weakness and atrophy developed over time, and motor impairment outweighed sensory findings in most patients. The disease was insidiously progressive, but did not affect longevity. The median and ulnar motor NCVs ranged from 6 to 15 m per second, changing little over time, except that distal motor responses became unobtainable, presumably due to the loss of motor axons. An autopsy of a 92-year-old affected woman revealed degeneration of fasciculus gracilis, which is consistent with severe axonal/neuronal loss caused by demyelination.

Some *MPZ* mutations cause a CMT2-like phenotype, including the mutation designated as CMTDID (Table 88.1). The best example is the T124M mutation, which initially causes unreactive pupils and progresses to a late-onset (fourth decade or later) neuropathy; dysphagia, positive sensory phenomena (including painful lancinations), and hearing loss are also common. In spite of a late onset, patients may progress to the point of using a wheelchair. Motor NCVs may be nonuniformly slowed into the "intermediate" range, and nerve biopsies show some thinly myelinated axons as well as clusters of regenerated axons.

How do different *MPZ* mutations result in such different phenotypes? The possibility that P_0 mutants have diminished adhesion has been evaluated in transfected cells. Disrupting the disulfide bond, truncating the cytoplasmic domain, abolishing *N*-linked glycosylation, or phosphorylation of an intracellular serine all reduces adhesion, sometimes even of a coexpressed wild-type *MPZ* allele (a dominant-negative effect), but it is hard to relate the diverse range of phenotypes to adhesion alone. The substitution by cysteine generates a thiol group, which could act in a dominant manner by forming aberrant disulfide bonds. Some mutants (e.g., Ser63deleted) are retained in the ER and cause an unfolded protein response, which appears to have adaptive as well as maladaptive effects.

GJB1 Mutations Cause CMT1X

Although long under-recognized, CMT1X turned out to be the second most common form of CMT1.[10] In affected males, the clinical onset is between 5 and 20 years. The initial symptoms include difficulty running and frequent ankle sprains; foot drop and sensory loss in the legs develop later. Depending on the tempo of the disease, the distal weakness may progress to involve the gastrocnemius and soleus muscles. Weakness, atrophy, and sensory loss also develop in the hands, particularly in thenar muscles. Sensory responses are absent in affected men, and the motor NCVs in the arms show "intermediate" slowing (25–40 m per second); motor amplitudes become reduced as the disease progresses. As compared to CMT1A, CMT1X patients typically have more heterogeneous NCVs and more distal axonal loss. Nerve biopsies from CMT1X patients show less demyelination and remyelination and more axonal degeneration and regeneration than do those from CMT1A/1B patients.

Some *GJB1* mutations are associated with a variety of central nervous system (CNS) manifestations. These include static features such as hearing loss and extensor plantar responses, but also transient episodes—with abnormal brain MRIs—that have been mistaken for strokes and acute disseminated encephalomyelitis.[11] One presumes that these CNS phenotypes owe to the expression of particular Cx32 mutants in oligodendrocytes, but the mechanism is unclear.

CMT1X is considered to be an X-linked dominant trait because it affects female carriers; their clinical involvement varies, likely owing to the proportion of myelinating Schwann cells that inactivate the X chromosome carrying the mutant *GJB1* allele. Affected women usually have a later onset (after the end of second decade) and a milder version of the same phenotype. Some degree of slowing is usually found in female carriers, even asymptomatic ones.

CMT1X is caused by mutations in *GJB1*, the gene that encodes the gap junction protein Cx32. More than 400 different mutations have been identified; each affected person/family typically has a "private" mutation. The mutations are predicted to affect every domain of the Cx32 protein, the promoter and 5′ untranslated region, as well as deletions of the entire gene. All mutations appear to cause a similar phenotype, indicating that loss of function is the common denominator. Expressing the mutant proteins in heterologous cells reveals that most mutants do not form functional gap junctions, indicating the importance of each amino acid in the function of this highly conserved protein. In mammalian cells, most mutants that fail to reach the cell membrane accumulate in the ER or Golgi apparatus and are degraded via endosomal and proteosomal pathways. Some mutants, however, have normal electrophysiological characteristics; how these cause demyelination remains to be determined. Some mutants have novel electrophysiological characteristics, including the ability to form abnormal hemi-channels; these characteristics may cause a more severe phenotype. *Gjb1*-null mice provide direct evidence that loss of Cx32 causes a demyelinating neuropathy.

Other Causes of CMT1

Seven different, dominant, missense mutations of *SIMPLE/LITAF* cause CMT1C. SIMPLE encodes a widely expressed gene/protein that interacts with components of the endosomal sorting complex required for transport (ESCRT). Dominant SIMPLE mutants cause prolonged activation of ERK1/2 in response to neuregulin signaling.

EGR2 is a transcription factor that is required for the normal differentiation and maintenance of myelinating Schwann cells; many myelin-related genes have EGR2 (often with SOX10) binding sites. Two recessive and 13 dominant *EGR2* mutations cause phenotypes that range from CHN to CMT1. Most dominant mutations affect the zinc finger domain, impairing DNA binding, and the amount of residual binding correlates with disease severity. A homozygous mutation in a key enhancer of *EGR2* results in a particularly severe amyelinating neuropathy associated with arthrogryposis.

Recessively Inherited Demyelinating Neuropathies (CMT4)

Autosomal recessive forms of demyelinating CMT have been collectively designated CMT4 (Table 88.1). All are rare, but are instructive because they illuminate the diversity of genes that are required for normal myelination.[12]

CMT4B1, CMT4B2, and CMT4B3 are caused by recessive mutations in *MTMR2*, *SBF2/MTMR13*, and *SBF1/MTMR5*, respectively.[13] CMT4B1–3 share a common phenotype, including a unique pathological finding—focally folded myelin sheaths. These genes belong to the family of myotubularin-related dual specific phosphatases whose substrates are phosphoinositols; MTMR13 lacks phosphatase activity but interacts directly with MTMR2 and MTMR5. MTMR2 and MTMR13 form complexes that dephosphorylate the 3′ phosphate of PI(3)P or PI(3,5)P. Gene knockouts in mice demonstrate that myelinating Schwann cells require MTMR2 and MTMR13 to maintain normal myelin sheaths. This may be related to abnormal vesicular trafficking, as well as altered ERK1/2 and AKT signaling, in myelinating Schwann cells. Further, recessive mutations in *FIG4*, a PI(3,5)P2 5′ phosphatase, also cause a demyelinating neuropathy (CMT4J), further supporting a special role for PI(3,5)P2 in myelinating Schwann cells. FIG4 has phosphatase activity like MTMR, but when complexed with Vac14 and Fab1 kinase, FIG4 activates PI(3,5)P2 production. Thus, loss of FIG4 or Vac14 function produces less, not more, PI(3,5)P2 in yeast and vertebrate cells—the opposite of what is expected in CMT4B1/B2/B3. Further, *Fig4* deficiency rescues the loss of MTMR2 function in Schwann cells—consistent with opposing effects on PI(3,5)P2 levels. Only certain *FIG4* mutations, likely those that have some residual activity, however, cause neuropathy, which often has a distinctive phenotype—"highly variable onset and severity, proximal as well as distal and asymmetric muscle weakness, electromyography demonstrating denervation in proximal and distal muscles, and frequent progression to severe amyotrophy."[14]

A variety of recessive mutations in *SH3TC2* cause CMT4C. Affected individuals have a moderate to severe, demyelinating neuropathy that begins in infancy or childhood and may result in loss of ambulation. Scoliosis is typical. SH3TC2 directly interacts with Rab11, which regulates endosomal recycling and neuregulin 1 signaling in Schwann cells.

CMT4D could be considered to be a syndromic neuropathy, as it is associated with hearing loss and dysmorphic features. It has an onset in childhood, with foot deformities and abnormal gait, and progresses to severe disability by the fifth decade. Motor NCVs are severely reduced and become unobtainable after the age of 15, indicating severe axonal loss. Biopsies show demyelination, onion bulbs, and cytoplasmic inclusions in Schwann cells. It is usually caused by a homozygous R148x mutation (a founder effect in the Roma) in *NDRG1*.

Recessive mutations in *PRX* cause CMT4F. Affected individuals have phenotypes that range from DSN (delayed motor milestones, progressing to severe distal weakness, areflexia, sensory loss, and even sensory ataxia) to CMT. The most affected individuals have severely slowed (<5m per second) MCVs, and biopsies show thinly myelinated axons, onion bulbs, and severe axonal loss. The *PRX* gene encodes periaxin, a PDZ domain protein that is expressed by myelinating Schwann cells, and links dystroglycan to the actin cytoskeleton and possibly compact myelin.

CMT4H is an autosomal recessive hypomyelinating neuropathy caused by homozygous, putative loss-of-function mutations in *FGD4*. *FGD4* encodes Frabin, a guanine nucleotide exchange factor (GEF) for Cdc42, one of the small rhoGTPases that regulate cellular morphogenesis, including myelination. Because the GTP-bound form of Cdc42 is active, loss of Frabin function should decrease Cdc42 activity. Knocking out Cdc42 in Schwann cells severely affects their ability to properly ensheathe and myelinate axons.

Syndromic Demyelinating Neuropathies

A demyelinating neuropathy has been reported to be a part of many different and apparently unrelated diseases (Table 88.2). In most cases, myelinating Schwann cells express the affected gene (Refsum disease is the exception), so that demyelination is a cell autonomous manifestation of the mutation. Of these, a few stand out. Dominant *SOX10* mutations are associated with complex phenotypes—PCWH: peripheral demyelinating neuropathy, central demyelinating leukodystrophy, Waardenburg–Shah syndrome, and Hirschsprung disease—or partial variants. PCWH phenotypes include moderate (CMT1-like) or severe (CHN/DSS-like) demyelinating neuropathies.[7] Recessive/loss-of-function mutations in *ARSA* and *GALC* cause metachromatic leukodystrophy and Krabbe disease, with accumulation of sulfatide and galactoceramide, respectively. A demyelinating neuropathy is part of mitochondrial neurogastrointestinal encephalopathy syndrome (MNGIE) caused by recessive *TYMP* mutations (mutations in other genes also cause MNGIE), and some *SURF1* mutations (usually associated with Leigh syndrome) cause a demyelinating neuropathy, indicating that abnormal mitochondrial function can affect myelination, as has been directly demonstrated in mice by deleting nuclear-encoded genes for mitochondrial proteins in Schwann cells.

Axons and Axonal Neuropathies

Inherited axonal neuropathies are usually named according to the most prominent clinical features. If both sensory and motor axons (and not CNS axons), then CMT2 and ARCMT2 are the conventional terms; if only sensory axons, then HSN; if sensory plus autonomic axons, then HSAN; if only motor axons, then HMN (Tables 88.1 and 88.2). If spasticity is the most obvious finding, then the disorder will often be called hereditary spastic paraplegia (SPG); if ataxia and spasticity are the main manifestations, then the term spinocerebellar ataxia (SCA) is often used, but there are many other eponyms and acronyms (Table 88.2). With a few exceptions, the selective vulnerability of different axonal populations is largely unexplained, as both peripheral nervous system (PNS) and CNS neurons typically express the mutated gene. Below we discuss some selected, unifying themes.

It is commonly supposed that the selective vulnerability of neurons that leads to neuropathy is the length of the axons themselves, which makes neurons the longest cells in the body. The sheer length of axons has important implications. Neurons must synthesize and maintain all of the components of axons, which may be >99.9% of the neuron's volume, and these components must be transported down the axon, which may be more than 1 meter in length. Orthograde axonal transport is mediated by a subset of kinesins, microtubule-activated ATPases that carry specific cargoes. One motor protein, dynein, and its associated activator, dynactin, mediates retrograde axonal transport. Both kinesins and dynein/dynactin use microtubules as tracks. Microtubules are formed from the association of α/β-tubulin dimers into protofilaments, which associate laterally to form a tubule.

In keeping with their biology, mutations in the genes that encode these axonal proteins can cause axonal neuropathies. Mutations of axonal transport cause a variety of axonal neuropathies. Mutations in the CAP-Gly domain of dynactin (*DCTN1*) cause HMN7B. In model systems, the mutations prevent the distal enrichment of dynactin thereby inhibiting the initiation of retrograde transport. Dominant mutations in *KIF5A* cause SPG10 (neuropathy is an associated feature), and recessive mutations of *KIF1A* cause a sensory neuropathy. Dominant mutations in *DYNC1H1*, which encodes the dynein heavy chain, were reported to cause CMT2, but later reports made a strong case that these cause a static motor neuronopathy (SMALED1).

Mutations that affect the cytoskeleton cause a variety of axonal neuropathies. Neurofilaments are the main cytoskeletal component of axons, and dominant mutations in *NEFL*, the gene encoding the light subunit of neurofilaments, cause an axonal neuropathy of variable severity (CMT2E); recessive *NEFL* mutations also cause a severe neuropathy. Further, recessive mutations in *GAN*, which encodes a protein that regulates the degradation of neurofilaments, affect both CNS and PNS axons. Some dominant mutations of *TUBB3*, which encodes a neuronal microtubule subunit, cause an axonal neuropathy, indicating that an abnormal microtubule highway leads to bad outcomes. *SPAST* encodes a microtubule severing protein, and dominant *SPAST* mutations cause SPG4, a common form of SPG (in which neuropathy is infrequently noted), although one dominant mutation causes HMN5. Dominant mutations in *HSPB1* and *HSPB8* cause motor predominant axonal neuropathies, sometimes diagnosed as CMT2 and sometimes as HMN. *HSPB1* and *HSPB8* encode heat shock protein of 28 kDa (HSP28) and HSP22, respectively, which are chaperones that interact with each other. Some HSPB1 mutants show increased binding to tubulin, enhancing the stability of microtubules; this has been suggested to be the cause of the neuropathy.

An axonal neuropathy is a cardinal feature of NARP (neuropathy, ataxia, and retinitis pigmentosa) and SANDO (sensory ataxic neuropathy, dysarthria, and ophthalmoparesis), and is a common finding in many "mitochondrial diseases" (Tables 88.1 and 88.2) that are caused by a variety of mutations of the mitochondrial genome and nuclear-encoded mitochondrial proteins.[15] The failure to produce enough local energy within axons could be a common cause of the neuropathy, but there are unexplained differences between the effects of different mutations. A subset of mitochondria DNA depletion syndromes (MTDPS), which are caused by recessive mutations of several different genes, cause an axonal neuropathy. Further, a subset of mutations in mitochondrial DNA cause neuropathy, including m9176T>C and m9185T>C (both affecting *MT-ATP6*), which can mimic CMT2 and/or a motor neuropathy (Table 88.3). Because m9176T>C and m9185T>C are more commonly associated with Leigh disease and/or NARP, it seems plausible that these more mildly affected individuals (who have CMT2/HMN) have a lower burden of mutations at least in their PNS neurons, but this remains to be shown.

MFN2 and *GDAP1* mutations are exceptions to the idea that most mitochondrial diseases are syndromes. Most *MFN2* mutations are new, dominant mutations, found in children who have a severe early-onset axonal neuropathy, and are the commonest cause of this phenotype. Recessive *MFN2* mutations also cause a severe, axonal neuropathy. Some CMT2A patients also have optic atrophy and/or myelopathy, and abnormal brain MRIs. The shared feature of optic atrophy could be related to a shared role of mitofusin-2 and OPA1, which is mutated in optic atrophy 1 (OMIM 165500; occasionally has an associated axonal neuropathy) and cooperates with mitofusin-2 to promote mitochondrial fusion. Conversely, several dominant and many recessive *GDAP1* mutations cause an axonal neuropathy, likely by impairing mitochondrial fission. Why altered mitochondrial fusion and fission results in a nonsyndromic axonal neuropathy is unclear.

Dominant mutations in several of the genes that encode the enzymes that "charge" amino acids to their respective tRNAs—*YARS* (CMTDIC), *GARS* (CMT2D and HMN5A), *AARS* (CMT2N), *HARS* (CMT2 and HMN2), and *MARS* (CMT2), as well as recessive *KARS* (CMTRIB) and *DARS2* (syndromic) mutations—cause an axonal neuropathy. Why this should be the case has yet to be determined.

Hereditary Sensory and Autonomic Neuropathies (HSAN)

The first members of this sundry group were designated as HSAN types 1–5,[16,17] but many of the subsequently discovered disorders that could fit into this category have not been classified accordingly (Table 88.1). Positive sensory phenomena (especially neuropathic pain) and especially the loss of sensory (and variably autonomic) function bind these disorders together. In HSAN4 and HSAN5 (caused by recessive mutations in the *NTRKA* and *NGFB*, respectively), the sensory and autonomic abnormalities are likely present before birth, although the secondary changes due to unheeded trauma may not be evident until late infancy or childhood. In accord, the corresponding nociceptive and sympathetic neurons are lost *in utero* in animal models of HSAN4 and HSAN5, as the interaction of nerve growth factor (NGF) with its receptor TrkA is required for the survival of nociceptive and sympathetic neurons.[18] The issue of a static versus a progressive deficit is not settled for HSAN3 (also known as Riley–Day syndrome or familial dysautonomia with congenital indifference to pain), which is usually caused by a recessive mutation that leads to missplicing of the *IKBKAP* transcript, particularly in neurons.

Dominant *SCN9A* (encodes the voltage-gated Na^+ channel, Nav1.7), *SCN11A* (encodes Nav1.9), and *TRPA1* mutations cause various neuropathic pain disorders that are not associated with neuropathy, although other dominant *SCN9A* mutations are associated with a painful, "small fiber" neuropathy (HSN2D). Conversely, recessive mutations in *SNC9A* and one dominant mutation in *SCN11A* result in "congenital indifference to pain". Nav1.7, Nav1.9, and TrpA1 are expressed in nociceptors, and the mutant channels have electrophysiological characteristics that conceptually give rise to these phenotypes.

Dominant mutations in *SPTLC1* and *SPTLC2*, the genes that encode two subunits of serine palmitoyl transferase, the enzyme that conjugates serine with fatty acids, cause HSAN1. These mutants result in the misincorporation of alanine and glycine in place of serine in novel/toxic lipids that are thought to mediate that neuropathy. Supplemental L-serine lowers the levels of these novel lipids, and improves the neuropathy in a mouse model.[19]

Hereditary Motor Neuropathies (HMN)

This group of disorders (also known as distal spinal muscular atrophy [DSMA]), present as length-dependent motor neuropathies. Seven types have been traditionally delineated, based on their clinical phenotypes and mode of inheritance,[20] and many of the gene defects have now been identified (Table 88.1). The onset is usually in childhood or early adulthood, with distal weakness typically affecting the extensor muscles of the toes and feet. Sensory responses are normal, although mild axonal loss may be present in sensory nerve biopsies. Motor responses have reduced amplitudes, with normal or moderately reduced NCVs, and needle electromyography demonstrates length-dependent denervation. Vocal cord weakness is a hallmark of HMN7. Upper limb predominance is characteristic of HMN5, and is caused by dominant mutations in a number of genes, including two *BSCL2* mutations that disrupt the glycosylation of the protein. Even in the same kindred, these *BSCL2* mutations can produce a clinical picture of a HMN5, with prominent weakness of intrinsic hand muscles, and/or spastic paraparesis (SPG17; also known as Silver syndrome). A motor neuropathy and myelopathy is also the clinical picture in HMN5B/SPG31, ALS4, and DSMA2; a distal motor neuropathy is also found in SPG14, SPG20, and some patients with the m9185T>C mutation in *MT-ATP6*.

DIAGNOSIS AND TREATMENT OF CMT AND RELATED DISORDERS

An accurate diagnosis of an inherited neuropathy is facilitated by a family history, as well as the patient's physical exam and clinical electrophysiology. The findings that should suggest various diagnoses have already been described. In the authors' opinion, a nerve biopsy should not be part of the initial diagnostic evaluation except for research and in problematic cases. Distinguishing an acquired chronic axonal neuropathy from an inherited cause, particularly in older patients, is a common and problematic example, especially because genetic testing rarely reveals a cause in late-onset CMT2. Some inherited demyelinating neuropathies (MNGIE caused by recessive *TYMP* mutations and CMT1X, in particular) can masquerade as chronic inflammatory demyelinating polyneuropathy.

The strategy for genetic testing is rapidly evolving. The classical approach—to find the common causes of particular phenotypes— is summarized by Saporta et al.;[21] this relies on detection of the deletion/duplication of *PMP22* and Sanger sequencing specific genes, and is widely available (see http://www.genetests.org/). HNPP (*PMP22* deletion), CMT1A (*PMP22* duplication), CMT1B (*MPZ* sequencing), CMT1X (*GJB1* sequencing), CMT2A (*MFN2* sequencing), and CMT4A/ARCMT2 (*GDAP1* sequencing) could be diagnosed in this way. For other cases, "next-generation sequencing" can be done for little more than the cost of Sanger sequencing a few genes,[22] and as the price drops further, this may become the preferred option. For patients who have CMT2, HSAN, and HMN, one could argue that whole-exome sequencing is the correct test, as the known genes account for less than one-half of these cases. Whole-genome sequencing is still on the horizon.

No treatment slows the progression of CMT. There are cell-based assays and even excellent animal models for some kinds of inherited neuropathy;[23] these facilitate preclinical work that provides the proof of concept of a given approach. Multicenter clinical trials have been done for CMT1A, and have highlighted the need for better, sensitive outcome measures. Current treatment is symptomatic, involving neurologists, physiatrists, orthopedic surgeons, and physical and occupational therapists, according to the needs of the patient. Daily heel cord stretching exercises can prevent shortening of the Achilles tendon, and special shoes, including those with good ankle support, may be needed. Patients often require ankle/foot orthoses to correct foot drop and aid walking, and orthopedic surgery to correct deformities. Forearm crutches or canes may be required for gait stability, and some patients will need wheelchairs. Patients who have profound sensory loss need to take precautions to avoid falls and ulcers. Exercise and weight control are desirable. When pain is present, the cause should be identified as accurately as possible. Musculoskeletal pain may respond to acetaminophen or nonsteroidal anti-inflammatory agents, whereas neuropathic pain may respond to the appropriate medications, such as tricyclic antidepressants, "mixed" (norepinephrine and serotonin) uptake inhibitors, gabapentin, pregabalin, and opioids. Whether CMT will affect career and employment options should be discussed. Patients should be made aware of their options for having children; a genetics counselor is invaluable in this regard.

Chemotherapeutic agents known to cause neuropathy should be used with great caution in patients with inherited neuropathies. In the case of vincristine, total avoidance is strongly advised for CMT1A, CMT1D, and CMT2, as serious consequences have been documented following the administration of standard oncological dosages in patients. At least one patient with CMT1X, however, received vincristine uneventfully, although another patient with CMT1X developed foot drops when treated concurrently with voriconazole and vincristine.

AMYLOID NEUROPATHIES

Neuropathy is a part of several kinds of amyloidosis—systemic diseases in which soluble plasma proteins are transformed into β-structured fibrils that are deposited in various organs and apparently cause dysfunction by their presence and magnitude.[24] While each kind is characterized by the specific protein that contributes the subunit of the fibrils, the pathologic processes leading to amyloid deposition are similar. A normally soluble protein exits the circulation, is partially proteolyzed by cell-associated mechanisms, and then assembles into proteolytically resistant β-structured fibrils in extravascular spaces. These fibril deposits progressively enlarge, displacing normal structures including cells, basement membranes, and connective tissue. The degree and distribution of organ function impairment dictates the clinical manifestations. The types of amyloidosis that may cause neuropathy are immunoglobulin (AL) amyloidosis and the hereditary amyloidoses associated with mutations of plasma transthyretin (*TTR*), gelsolin (*GSN*), apolipoprotein AI (*ApoAI*), and b2-microglobulin (*B2M*) genes (Table 88.2; http://amyloidosismutations.com/). Neuropathy is not a feature of reactive (secondary) amyloidosis nor of the hereditary amyloidoses caused by mutations in fibrinogen Aα chain, lysozyme, apolipoprotein AII genes, and most of the variant forms of ApoAI. Only the inherited forms of amyloid that cause neuropathy are considered here; all are dominant disorders.

Transthyretin Amyloidosis (FAP1 and FAP2)

The most common type of hereditary amyloid neuropathy is caused by mutations in *TTR*—at least 112 mutations to date. All are caused by a single nucleotide change that results in an amino acid substitution, except for one that deletes a codon, resulting in the loss of a valine (ΔVal122). Only two cases of *de novo* mutations have been verified; one in Japan (Gly47Arg) and one in the United States (Ala25Ser). Many mutations that have been found associated with amyloidosis have been assumed, but not proven, to be the definitive cause of the disease because of insufficient data. The associations with well characterized mutations, or well studied families of inherited amyloidosis, are probably valid.

TTR mutations have been described in most countries of the world and in many ethnic groups, but it is a rare disease overall except in certain populations. The Val30Met mutation was probably present in Portugal by AD 1400 from whence it was taken to Sweden and later to Japan and Brazil. In Skelleftea and Lulea (in northern Sweden) and among the fishermen of Pavoa de Varzim (in northern Portugal), 3–5% percent of the inhabitants have the Val30Met mutation. In the United States, the majority of mutations are found in families of European ancestry and many have been traced to their country of origin. The Thr60Ala mutation originated in Northern Ireland and was brought to the United States before 1800. The Leu58His mutation came with German migrants to Pennsylvania in the late 1700s. The Val122Ile is the most prevalent, affecting ~3% of African Americans, but is under appreciated because it usually causes clinically significant disease only after age 60, manifesting primarily as congestive heart failure that is often attributed to other more common causes such as hypertension. It is often associated with carpal tunnel syndrome, and does not usually cause a generalized neuropathy.

The typical neuropathy of Val30Met amyloidosis was described by Andrade on kindreds from northern Portugal,[25] later designated FAP1. It starts with dysesthesias in the lower extremities with or without small fiber findings such as decreased temperature sensation. Sensory loss usually progresses over several years, and may eventually involve the trunk. When sensory loss has reached the level of the knees, similar loss occurs in the upper extremities and motor loss in the lower extremities declares itself by foot drop. Progression of motor impairment may span 10–20 years and often leads to use of braces, canes, crutches, and finally a wheelchair. Autonomic dysfunction—constipation and diarrhea or impotence—often occurs early in the disease and may be the first symptom. Orthostatic hypotension occurs later, probably partly caused by autonomic dysfunction, but it is most prominent in individuals who also have restrictive cardiomyopathy. Before antibiotics and improved levels of supportive care were developed, leg ulcers, osteomyelitis, and malnutrition from loss of gastrointestinal function took their toll; now heart failure caused by restrictive cardiomyopathy has replaced neuropathy as the major direct cause of death.

The *TTR* gene (18q23) has four exons. No mutations have been found in exon 1, which encodes a signal peptide and the first three residues of the mature 127-amino acid protein. Exons 2 (amino acid residues 4–47), 3 (residues 48–92), and 4 (residues 93–127) each have similar proportion of mutations that cause (n > 100) and do not cause amyloidosis (n ~ 12). To date, neither the position of the altered amino acid, nor the type of amino acid substitution, has been implicated in amyloid formation. Nevertheless, there are many variations on the clinical theme of amyloidosis and some can be related to specific mutations. Many affected individuals have the carpal tunnel syndrome and, in some, the disease starts with this entity, especially in the Maryland/German (Leu58His) and Indiana/Swiss (Ile84Ser) kindreds; this was the justification for naming this clinical variant FAP2. Amyloid deposition in structures around the nerve roots may occur, causing spinal claudication that can magnify the deficits in patients who only have moderate neuropathy. Leptomeningeal amyloid deposition may be a major manifestation of certain variants (Leu12Pro, Asp18Gly, Val30Gly, Leu55Arg, Phe64Ser, Try69His, Tyr114Cys, ΔVal122), and can present as intracerebral hemorrhage, seizures, headaches, and in some cases, dementia. Many of the TTR variants that cause leptomeningeal deposition also cause vitreous opacities (Val30Gly, Tyr69His, Tyr114Cys). Overall, amyloid deposition in the vitreous is a variable feature, occurring in about one-quarter of the kindreds, and may be the only manifestation of the disease in some individuals.

Although TTR amyloidosis is an autosomal dominant disease, penetrance is variable and can be obscured by the late onset of manifestations. Even when the manifestations are present, TTR amyloidosis is often not diagnosed, thereby obscuring the family history and misleading the clinician. Fortunately, the availability of DNA testing has greatly facilitated the ability to diagnose suspected cases of amyloidosis. When the specific *TTR* mutation is known, this can be done for any individual in the kinship. If a *TTR* mutation is suspected, but no specific mutation is being considered, the current practice is to sequence all of the coding regions.

TTR is a plasma protein that is synthesized almost exclusively by the liver, and has a plasma half-life of only 1–2 days. The retinal pigment epithelium and choroid plexus probably contribute TTR to the vitreous of the eye and the cerebrospinal fluid, respectively. Whether the amyloid deposited in the peripheral nerve is partially derived from choroid plexus is unknown; the association of amyloid with endoneurial vessels suggests a blood (hepatic) origin. In plasma, TTR is a tetramer composed of four identical monomers, each of which is composed of a single polypeptide chain of 127 amino acids. Each monomer has extensive β-structure (four β-strands in each of two planes), which must be an important factor in the formation of amyloid fibrils, as even wild-type TTR is amyloidogenic, providing the fibrils in senile cardiac amyloidosis. How the *TTR* mutations promote amyloid formation is unknown. It has been suggested that they destabilize the tetramer, thereby favoring the aggregation of subunits and fibrillogenesis. Mutations may increase the metabolic turnover of TTR, as individuals with amyloid-associated TTR mutants have reduced plasma TTR levels (but equal amounts of normal and variant TTR mRNA in the liver). Why the delay of fibril deposition until adult life and the variation of time of disease onset between individuals with different or even the same mutation are persistent mysteries in the pathogenesis of amyloidosis.

The TTR tetramer has a central channel with two binding sites for thyroxin, and retinol/vitamin A-saturated retinol-binding protein binds to the outside of the tetramer. Although TTR was assumed to serve as a transporter of these two moieties, mice lacking TTR have low plasma levels of retinol, but normal fecundity and development, so that reducing TTR levels may be a feasible strategy for treatment.

Until recently liver transplantation was the only specific therapy for TTR amyloidosis. Because all plasma TTR is produced by the liver, after hepatic transplant, plasma levels of mutant TTR fall rapidly. As of 2010, more than 2000 patients received liver transplants for this disease, and for many individuals (especially those with the Val30Met mutation), progression of amyloidosis was stopped or at least slowed to allow prolonged survival. Unfortunately, for patients with other *TTR* mutations (including Glu42Gly and Thr60Ala), the neuropathy and cardiomyopathy progress because normal TTR still forms more amyloid. Normal TTR accounts for ~40% of cardiac and nerve amyloid prior to transplantation but may increase to 60% at the time of death in patients who survive for more than 3 years post liver transplantation.

Recently, Vyndaqel (tafamidis), a small organic compound postulated to stabilize plasma TTR and impede amyloid formation, has been approved in the European Union for treatment of mild-to-moderate peripheral neuropathy. Whether the drug affects the progression of cardiomyopathy has not been studied, and it has not yet been approved by the Federal Drug Administration (FDA) in the USA. Diflunisal (Dolobid), a nonsteroidal anti-inflammatory drug approved for treatment of arthritis, also stabilizes plasma TTR, and has been studied for altering progression of TTR polyneuropathy. Both antisense oligonucleotides and small interfering RNA (siRNA), have been shown to decrease hepatic TTR expression, and both compounds have now progressed to studies in humans. Unfortunately, none of these therapies have been shown to affect amyloid deposition in the eye or leptomeninges, where TTR synthesis is the result of expression in the choroid plexus and the retinal pigment epithelium.

Medical therapies can prolong the life and function of individuals with TTR amyloidosis, including neuropathic pain, diarrhea, and heart failure. Avoiding drugs that impair cardiac contractility (including most antiarrhythmics) can reduce the incidence and degree of heart failure. Cardiac pacing is often necessary. Renal dialysis may benefit some patients. Surgical decompression of carpal tunnel syndrome should be thorough because amyloid deposition is progressive, and recurrence is possible. A laminectomy can partially relieve spinal claudication due to amyloid infiltration.

Gelsolin Amyloidosis (FAP4)

Gelsolin amyloidosis is an autosomal dominant disease with odd features. Its main manifestations are lattice corneal dystrophy and facial paralysis, but why these are the sites of amyloid deposition is unknown. Corneal deposits appear first, usually at about 20 years of age. These appear as linear branching structures and usually do not cause serious loss of vision for several years. Progressive cranial neuropathy develops about age 40, causing bilateral facial palsies, which can be disfiguring and can cause corneal damage owing to the failure of the upper and lower lids to close. Amyloid is also deposited in other organs, including the heart, intestines, and kidneys, but usually does not shorten longevity. In some individuals who are homozygous for the mutant gelsolin, renal and cardiac disease is accelerated and life threatening.

Gelsolin is a plasma protein, and by analogy to its intracellular form, it is assumed to cleave intravascular actin. Amyloidosis in large kindreds in Finland is caused by a mutation in plasma gelsolin (Asp187Asn of the mature protein). The fibril subunit is a 71-residue peptide that contains the variant residue and, unlike TTR, no normal gelsolin peptide is found in the amyloid fibrils. Thus, the normal and variant proteolytic fragments go their separate ways prior to fibril formation. In addition to the well-characterized disease which originated in Finland, a similar disease caused by an Asp187Tyr mutation has been reported from kindreds in Denmark and the Czech Republic. Recently a third gelsolin mutation, Gly167Arg, has been identified in one person with renal gelsolin amyloidosis, but no neuropathy or lattice corneal dystrophy.

There is no specific therapy for gelsolin amyloidosis. If vision is impaired by corneal dystrophy, corneal transplantation is an accepted and successful procedure, but the condition can recur after several years and may require repeat transplantation. Plasma gelsolin is synthesized in large part by skeletal muscle and, therefore, unlike TTR, organ transplantation is not a therapeutic option. In the unusual patients with symptomatic renal, gastrointestinal, or cardiac amyloidosis, therapy is similar to that described for TTR amyloidosis.

Apolipoprotein AI Amyloidosis (FAP3)

ApoAI amyloidosis was first described by Van Allen et al.[26] They documented a progressive axonal neuropathy involving sensory, motor, and autonomic neurons. Sensory loss in the legs was usually the first sign of neuropathy, followed by motor impairment. Impotence was common, but gastrointestinal motility problems did not occur. Several affected individuals developed deafness. This family was found to have a Gly26Arg substitution of ApoAI. Subsequently, other families with this mutation have been identified: some families have renal amyloidosis but no clinical evidence of neuropathy, while other families have both the nephropathy and neuropathy. To date, 20 different *ApoAI* mutations have been identified—all cause renal, hepatic, or cardiac amyloidosis, but neuropathy is only associated with the Gly26Arg mutation. Gly26Arg amyloidosis is slowly progressive, with survival for 15–20 years after clinical presentation. Renal dialysis is the major supportive therapy because renal failure is the major cause of death. Hypertension (probably related to the renal disease) is common and should be treated early. Even though not all circulating apoAI is of hepatic origin, several patients have had liver transplantation, and early results are encouraging, so that increasing the relative amount of normal to variant ApoAI may favorably affect the disease.

ApoAI is part of plasma high-density lipoprotein (HDL) and is believed to be important in reverse cholesterol transport. How ApoAI, which has little β-structure, is altered to form amyloid-β fibrils is not clear; this property is shared with serum amyloid A (SAA), another apolipoprotein constituent of HDL (and gives rise to the amyloid fibrils in reactive amyloidosis). Only the mutant Gly26Arg protein is incorporated into amyloid fibrils, suggesting that apoAI is dissociated from HDL before the mutant protein enters the pathway to fibril formation. The organ distribution of ApoAI amyloid is similar to that seen with SAA (kidney, liver, adrenal), but only the amyloid formed by the ApoAI Gly26Arg mutant, and not by other HDL apolipoproteins (SAA and ApoAII), causes neuropathy.

Diagnosis of Amyloid Neuropathy

Amyloidosis typically affects multiple organ systems, so even if it starts with neuropathic symptoms, this will not remain the sole feature for long. The greatest deterrent to a timely diagnosis is a lack of familiarity with the disease and, therefore, the failure to search for more extensive involvement. Neurologists, like most subspecialists, tend not to look far beyond their organ system of special interest. For the patient with an axonal neuropathy, or even signs of autonomic neuropathy alone (e.g., impotence), it is essential to look for renal disease (proteinuria, elevated serum creatinine), cardiac disease (angina-type pain, electrocardiogram, echocardiogram) and gastrointestinal dysfunction (constipation, diarrhea, early satiety).

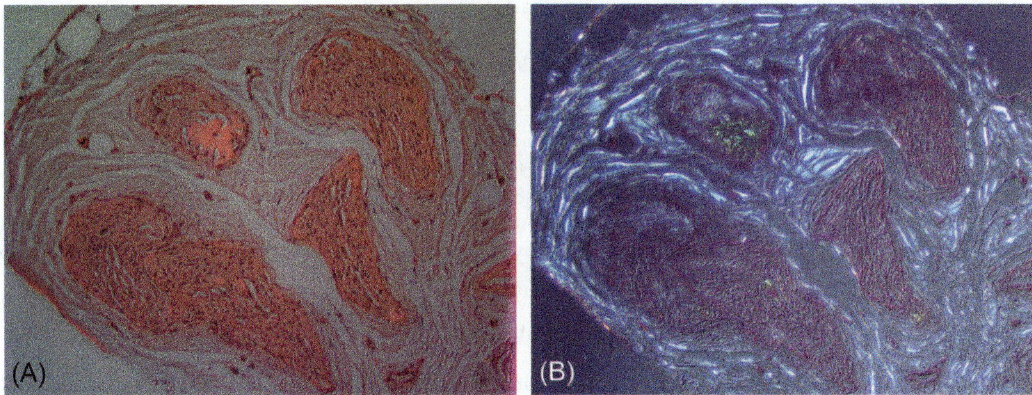

FIGURE 88.2 **Sural nerve biopsy stained with Congo red from a patient with TTR Thr49Ala amyloidosis, viewed under transmitted light (A) and crossed polarized filters (B).** In panel B, note green birefringence of Congo red stained deposits—diagnostic for amyloid deposits.

A definitive diagnosis is made by demonstrating amyloid deposits. A nerve biopsy is usually not needed because biopsies of rectal mucosa or minor salivary glands, or an abdominal fat aspirate will reveal the amyloid deposits in approximately 80% of cases. If these biopsies are negative, a nerve biopsy may yield the diagnosis, but even a negative nerve biopsy does not rule out amyloidosis owing to the sporadic nature of amyloid deposition (Figure 88.2). After biopsy diagnosis, determining the exact type of amyloidosis is essential for both counseling and treatment. A family history of amyloidosis effectively excludes AL amyloidosis (the most common cause of amyloid neuropathy), but the absence of a family history does not exclude inherited amyloidosis. The clinical syndromes of AL, TTR, and ApoAI neuropathies are similar, and distinct from gelsolin neuropathy. Several laboratories offer genetic testing for amyloid-associated mutations (http://www.genetests.org/), which facilitates genetic counseling. Expertise in this area can be found at the Amyloid Research Group at Indiana University School of Medicine (http://www.iupui.edu/~amyloid/) and the Amyloid Treatment and Research Program at Boston University Medical Center (http://www.bu.edu/amyloid/).

ACKNOWLEDGEMENTS

We thank our colleagues, especially Drs. Charles Abrams, Ueli Suter, and Larry Wrabetz, for discussion, as well as the NINDS, the National Multiple Sclerosis Society, The Muscular Dystrophy Association, and the Judy Seltzer Levenson Memorial Fund for CMT Research for supporting our work.

References

1. Shy ME, Lupski JR, Chance PF, Klein CJ, Dyck PJ. Hereditary motor and sensory neuropathies: an overview of clinical, genetic, electrophysiologic, and pathologic features. In: Dyck PJ, Thomas PK, eds. *Peripheral Neuropathy*. 4th ed. Philadelphia: Saunders; 2005:1623–1658.
2. Nicholson G, Myers S. Intermediate forms of Charcot-Marie-Tooth neuropathy. *NeuroMolec Med*. 2006;8:123–130.
3. Baets J, Deconinck T, De Vriendt E, et al. Genetic spectrum of hereditary neuropathies with onset in the first year of life. *Brain*. 2011; 134(Pt 9):2664–2676.
4. Yiu E, Ryan MM. Genetic axonal neuropathies and neuronopathies of pre-natal and infantile onset. *J Peripher Nerv Syst*. 2012;17:285–300.
5. Wilmshurst JM, Ouvrier R. Hereditary peripheral neuropathies of childhood: an overview for clinicians. *Neuromuscul Disord*. 2011;21(11):763–775.
6. Rotthier A, Baets J, Timmerman V, Janssens K. Mechanisms of disease in hereditary sensory and autonomic neuropathies. *Nat Rev Neurol*. 2012;8(2):73–85.
7. Scherer SS, Wrabetz L. Molecular mechanisms of inherited demyelinating neuropathies. *Glia*. 2008;56:1578–1589.
8. Lupski JR, Garcia CA. Charcot-Marie-Tooth peripheral neuropathies and related disorders. In: Scriver CR, Beaudet AL, Sly WS, Valle D, Childs B, Kinzler KW, eds. *The Metabolic & Molecular Basis of Inherited Disease*. 8th ed. New York: McGraw-Hill; 2001:5759–5788.
9. Shy ME, Jani A, Krajewski K, et al. Phenotypic clustering in *MPZ* mutations. *Brain*. 2004;127:371–384.
10. Scherer SS, Kleopa KA. X-linked Charcot-Marie-Tooth disease. In: Dyck PJ, Thomas PK, eds. *Peripheral Neuropathy*. 4th ed. Philadelphia: Elsevier Saunders; 2005:1791–1804.
11. Abrams CK, Scherer SS. Gap junctions in inherited human disorders of the central nervous system. *Biochim Biophys Acta*. 1818;2012:2030–2047.

12. Vallat JM, Tazir M, Magelaine C, Sturtz F, Grid D. Autosomal-recessive Charcot-Marie-Tooth diseases. *J Neuropathol Exp Neurol.* 2005;64(5):363–370.
13. Previtali SC, Quattrini A, Bolino A. Charcot-Marie-Tooth type 4B demyelinating neuropathy: deciphering the role of MTMR phosphatases. *Expert Reviews Mol Med.* 2007;9:1–16.
14. Nicholson G, Lenk GM, Reddel SW, et al. Distinctive genetic and clinical features of CMT4J: a severe neuropathy caused by mutations in the PI(3,5)P(2) phosphatase FIG4. *Brain.* 2011;134(Pt 7):1959–1971.
15. Koopman WJH, Willems PHGM, Smeitink JAM. Monogenic mitochondrial disorders. *N Engl J Med.* 2012;366:1132–1141.
16. Houlden H, King R, Blake J, et al. Clinical, pathological and genetic characterization of hereditary sensory and autonomic neuropathy type 1 (HSAN I). *Brain.* 2006;129:411–425.
17. Rotthier A, Baets J, De Vriendt E, et al. Genes for hereditary sensory and autonomic neuropathies: a genotype-phenotype correlation. *Brain.* 2009;132:2699–2711.
18. Bibel M, Barde Y-A. Neurotrophins: key regulators of cell fate and cell shape in the vertebrate nervous system. *Genes Dev.* 2000;14:2919–2937.
19. Garofalo E, Penno A, Schmidt BP, et al. Oral L-serine supplementation in mice and humans with hereditary sensory autonomic neuropathy type 1. *J Clin Invest.* 2011;121:4735–4745.
20. Rossor AM, Kalmar B, Greensmith L, Reilly MM. The distal hereditary motor neuropathies. *J Neurol Neurosurg Psychiatry.* 2012;83(1):6–14.
21. Saporta ASD, Sottile SL, Miller LJ, Feely SME, Siskind CE, Shy ME. Charcot-Marie-Tooth disease subtypes and genetic testing strategies. *Ann Neurol.* 2011;69:22–33.
22. Rossor AM, Polke JM, Houlden H, Reilly MM. Clinical implications of genetic advances in Charcot-Marie-Tooth disease. *Nat Rev Neurol.* 2013;9(10):562–571.
23. Bouhy D, Timmerman V. Animal models and therapeutic prospects for Charcot-Marie-Tooth disease. *Ann Neurol.* 2013;74(3):391–396.
24. Benson MD. Amyloidosis. In: Scriver CR, Beaudet AL, Sly WS, et al., eds. *The Metabolic and Molecular Bases of Inherited Disease.* 8th ed. New York: McGraw Hill; 2000:5345–5378.
25. Andrade C. A peculiar form of peripheral neuropathy. Familial atypical generalized amyloidosis with special involvement of the peripheral nerves. *Brain.* 1952;75:408–427.
26. Van Allen MW, Frohlich JA, Davis JR. Inherited predisposition to generalized amyloidosis. *Neurology.* 1969;19:10–25.

CHAPTER

89

Spinal Muscular Atrophy

Bakri H. Elsheikh, W. David Arnold†, and John T. Kissel†*

*Johns Hopkins Aramco Healthcare, Dhahran, Saudi Arabia

†The Ohio State University, Columbus, OH, USA

CLINICAL FEATURES

Historical Overview

The initial description of infantile proximal spinal muscular atrophy (SMA) was made in 1891 by Guido Werdnig, an Austrian neurologist who reported two patients with hereditary progressive muscular atrophy presenting like a muscular dystrophy but on a neurogenic basis.[1–4] In 1892, John Hoffmann, a German neurologist, established the spinal nature of the disease and first coined the term spinal muscular atrophy.[2] Autopsy analysis of their cases revealed atrophy of the ventral roots and loss of motor neurons in the anterior horn.[1,2] In 1956, Eric Kugelberg and Lisa Welander described the disease in older children who could walk for some time after diagnosis.[5] Subsequently, others confirmed previous reports and described a wide clinical spectrum of disease, including an adult-onset form, and the relationship between disease severity and age of onset.[6–8] Several attempts to classify the autosomal recessive proximal SMAs led to a consensus classification by an International SMA Consortium in 1992 that was based on the age at symptom onset and maximal motor milestones achieved.[9–11] This classification recognized three types of childhood SMA that ranged from infants with severe disease and early death (type 1), to children with milder weakness and normal lifespan (type 3), in addition to an intermediate form (type 2). Subsequent modifications recognized an adult-onset disease (type 4) and recognized patients with prenatal symptoms and very early death (type 0). The first breakthrough in understanding the molecular genetics of SMA occurred in 1990 with the genetic mapping of proximal SMA to chromosome 5 11.27-13.3.[12–14] Five years later, Lefebvre et al. reported homozygous mutation of the telomeric copy of the survival motor neuron gene (*SMN1*) as the causative mutation for the disease.[15]

The mainstay of SMA treatment is supportive care especially in relation to the orthopedic and respiratory complications of the disease. Advances in the understanding of the molecular pathophysiology of the disease led to exploration of several pharmacologic treatment strategies, including efforts to enhance neuroprotection; anabolic effects; energy metabolism; and, most notably, increase *SMN2* expression or stability.[16–28] Unfortunately, none of the agents tested to date in controlled trials showed a clinical benefit.[29,30] More recently, antisense oligonucleotide and gene replacement therapies have shown much promise, with human trials now underway.

Mode of Inheritance and Prevalence

SMA is an autosomal recessive disorder that is the leading inherited cause of infant mortality. The incidence is estimated to be 1 in 11,000 live births, with an overall carrier frequency of 1 in 54.[31] The carrier frequency is variable depending on the population and is as low as 1 in 72 for the African American population, compared to 1 in 47 for Caucasians. There are approximately 600 new cases of SMA yearly and about 6–7 million carriers in the United States.[32–35] There is an unexplained slight male preponderance.[36,37] The quoted prevalence of the different SMA types is as follows: approximately 45% of cases are type 1; 23% are type 2; 30% are type 3 (16% 3a and 14% 3), and only 2% are type 4. However, these figures likely underestimate the frequency of type 1 SMA given the high mortality rate of this type.[38,39]

Rosenberg's Molecular and Genetic Basis of Neurological and Psychiatric Disease
http://dx.doi.org/10.1016/10.1016/B978-0-12-410529-4.00089-9

Natural History

SMA Type 1

Type 1 SMA patients develop clinical symptoms and signs in the first 6 months of life, and by definition are never able to sit unsupported.[11] Historically 70–80% of patients die in the first 2 years of life from respiratory complications.[38–40] However, in recent years survival has increased due to advances in proactive pulmonary, nutritional, and supportive care.[41,42] There are several modifications of this phenotype that have prognostic implications.[43] SMA type 0 is a term that has been applied to a particularly severe phenotype associated with prenatal onset of symptoms, severe weakness at birth, joint contractures, a requirement for early respiratory support, and (usually) death within the first few weeks. Type 1a refers to a severe neonatal phenotype with onset before 1 month of age that also carries a poor prognosis. Type 1b represents the majority of type 1 cases, and has onset of weakness within the first 6 months; these patients never sit or achieve head control, and have an intermediate prognosis. Some type 1c patients, which are a small portion of type 1 patients, achieve head control and sit with support and have a better prognosis within this group.

Type 1 patients present with severe hypotonia, poor head control, weak cry, poor feeding, diffuse, proximal predominant weakness, and respiratory distress (Figure 89.1).[44] Patients appear alert for their degree of weakness because of absent or minimal facial involvement.[45] Eye muscles and cognitive function are spared, and children with SMA typically have a higher than normal IQ.[46] Tongue atrophy and fasciculations are noted in over half of patients.[6,45] Motor examination reveals severe areflexic hypotonia with the extremities held in a splayed out "frog-leg" like posture. A bell-shaped chest results from paradoxical breathing, i.e., flattening of the upper chest and flaring of the lower chest and abdominal protrusion during inspiration caused by weakness of the intercostal muscles and relative sparing of the diaphragm.[45] Although SMA is usually not considered a multisystem disease, there have been scattered reports of patients with severe type 1 disease associated with various cardiac structural abnormalities and digital vascular necrosis and thrombosis.[47,48]

SMA Type 2

These patients become symptomatic between the ages of 6 and 18 months, and by definition achieve the ability to sit unsupported (Figure 89.2). In the majority, this will occur within the normal expected age range of less than 9 months and in the rest by 30 months.[49] Some type 2 patients even achieve the ability to stand using a long-leg brace or standing frame but they never stand or walk unaided. Weakness is more severe in the legs and a fine finger tremor (minipolymyoclonus) is often noted.[50] Deep tendon reflexes are absent in most patients. Cranial nerves examination reveals sparing of facial and extraocular muscles but tongue atrophy and fasciculations are common. Swallowing difficulty and respiratory compromise are more common in those at the weak end of the disease spectrum, i.e., those barely able to sit unsupported compared to stronger children with better trunk control.[51–53] Additional factors that contribute to compromised respiratory function include scoliosis, recurrent aspiration, infections, and sleep disordered breathing.[54,55] Joint contractures and kyphoscoliosis often develop and worsen over time. Osteopenia and a tendency to obesity are prevalent in this population.[56,57] Overall, the rate of functional decline slows with time with many manifesting an accelerated phase of strength decline followed by a more static phase with a long plateau period.[58]

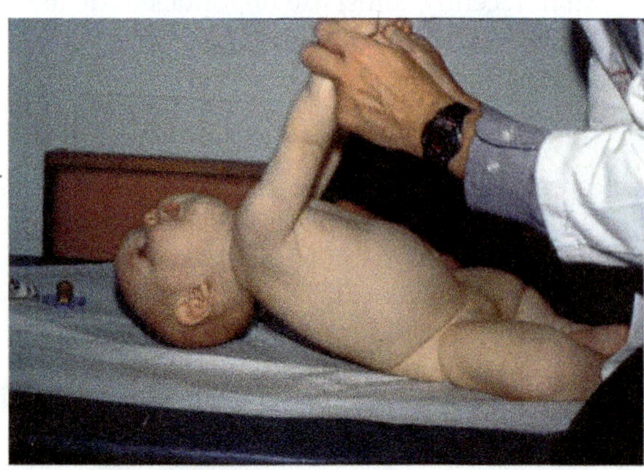

FIGURE 89.1 **SMA type 1 infant with typical marked head lag and noticeable proximal weakness when pulled to a sit.** Slight sternal retraction is also seen.

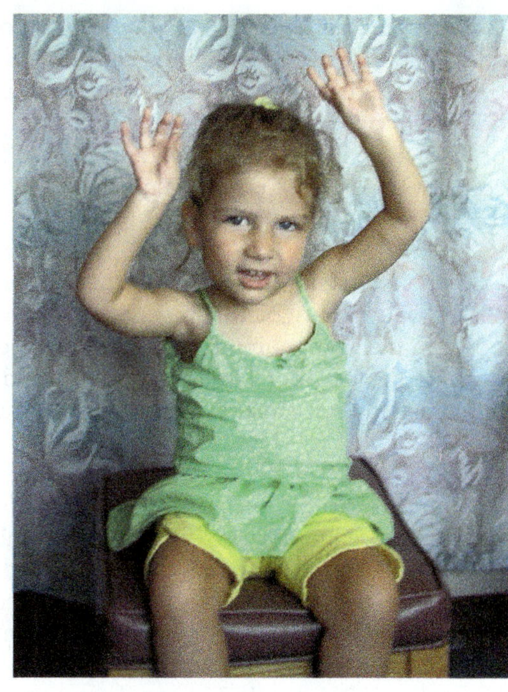

FIGURE 89.2 Three-year-old girl with SMA 2 sitting without support.

SMA Type 3

These patients start to have symptoms after 18 months of age, and by definition achieve the ability to stand and walk independently; in 10% this is delayed until after 18 months of age.[39,49] Two subtypes have been described based on severity and age at onset: SMA 3a and SMA 3b, with the latter presenting after age 3 years and having a milder disease and better prognosis. Similar to other SMA types, the weakness is mainly proximal with predominant leg involvement. Patients usually present with recurrent falls, difficulty going up stairs and trouble arising from the floor, sometimes employing a Gowers' maneuver to stand (Figure 89.3). Distance walking is limited by leg fatigue and correlates with muscle strength and disease duration.[59] Some type 3 patients maintain a remarkable ability to ambulate despite profound weakness; this is probably because of the segmental distribution of the weakness allowing for development of compensatory mechanisms.[60] These patients typically have a waddling gait due to hip abductor weakness and protuberant abdomen secondary to abdominal muscle weakness. As in other SMA subtypes, there is minimal or lack of facial and complete lack of extraocular muscle involvement, sparing of sensation, depressed or absent deep tendon reflexes, and minipolymyoclonus.[50] Fasciculations in limb muscles and calf hypertrophy are

FIGURE 89.3 Four-year-old SMA type 3 patient using Gowers' maneuver to stand from the floor.

seen in some patients.[61,62] Ventilatory function is usually normal in type 3 patients, although restrictive lung disease and sleep disordered breathing can develop in those with severe weakness. Despite the fact that SMA type 3 is a progressive disease, patients tend to have long plateau periods lasting many years.[58] Slow functional decline may be seen in type 2 and 3 patients with observation periods exceeding 1 year.[63] The separation of patients into type 3a or 3b disease is important prognostically: 58% of SMA type 3b patients are still ambulatory at the age of 40 years in contrast to just 22% of type 3a patients.[38]

SMA Type 4

In this mildest form of SMA, patients do not manifest symptoms until adulthood, with a mean age of onset in the 30s.[64–66] The clinical picture is similar to that of type 3 disease but is less severe. Patients usually report trouble getting up from the floor, rising from a chair or crouch going up stairs with or without the help of handrail. Many patients develop fasciculations in limb muscles, leading to a mistaken diagnosis of amyotrophic lateral sclerosis. Type 4 patients usually remain ambulatory and typically do not develop dysphagia, respiratory difficulties, or scoliosis.

Other SMA Variants

There are multiple other rare SMA disorders that are best classified by mode of inheritance and whether the weakness is predominantly proximal or distal (Table 89.1).[67–76] Associated features and age of onset help characterize these patients. The most common of these disorders is X-linked spinal and bulbar muscular atrophy (Kennedy disease) associated with expansion of CAG trinucleotide repeat in the androgen receptor gene.[73–75] This condition starts in the fourth decade and is characterized by a slowly progressive atrophy and weakness of the bulbar and proximal limb muscles. Characteristic features include facial and perioral fasciculations, gynecomastia, and testicular atrophy. Several types of distal SMA are characterized by slowly progressive distal predominant weakness and can be confused with the Charcot–Marie–Tooth disorders.[76] The lack of significant sensory deficits either clinically or electrophysiologically in the distal SMAs is crucial in making this distinction.

TABLE 89.1 Non-SMN-Related Types of SMA

PROXIMAL PREDOMINANT SMA VARIANTS		
Autosomal recessive	**Locus**	**Gene**
SMA 0-IV	5q	Survival motor neuron (*SMN1*)
Congenital with arthrogryposis	5q	Survival motor neuron (*SMN1*)
Congenital axonal neuropathy	5q	Survival motor neuron (*SMN1*)
SMA with pontocerebellar hypoplasia	14q32.2	Vaccina related kinase (*VRK1*)
Autosomal dominant	**Locus**	**Gene**
Adult proximal SMA (Finkle)	20q 13.22	Vesicle associated membrane protein associated protein B (*VABP*)
Scapuloperoneal syndrome	12q24	Vanilloid transient receptor protein (*TRPV4*)
SMA with lower extremity predominance (SMALED)	14q32	Dynein, cytoplasmic 1, heavy chain 1 (*DYNC1H1*)
X linked	**Locus**	**Gene**
Bulbospinal muscular atrophy (Kennedy)	Xq12	Androgen receptor
Infantile SMA with arthrogryposis	X p11.3	Ubiquitin activating enzyme 1 (UBE1)
DISTAL PREDOMINANT SMA VARIANTS		
Autosomal dominant	**Locus**	**Gene**
Distal SMA 1	7q24-q36	Unknown
Distal SMA 2A	12q24	*HSPB8/Heat shock protein 22*
Distal SMA 2B	7q11	*HSPB1/heat shock protein 27*

TABLE 89.1 Non-SMN-Related Types of SMA—cont'd

DISTAL PREDOMINANT SMA VARIANTS		
Autosomal dominant	**Locus**	**Gene**
Distal SMA 2C	5q 11	*HSPB3/Heat shock protein*
Distal SMA 5A	7p14	Glycyl tRNA Synthetase (*GARS*)
Distal SMA 5B	2p11.2	Receptor expression enhancing protein 1 gene (*REEP1*)
Distal SMA 5C	11q12	Berardinelli–Seip congenital lipodystrophy (*BSCL2*)
Distal SMA 7A	2q12.3	Solute carrier family 5, member 7 (*SLC5A7*)
Distal SMA 7B	2q13	Dynactin (*DCTN1*)
Distal SMA with UMN Signs	9q34	Sentaxin (*SETX6*)
Autosomal recessive	**Locus**	**Gene**
Distal SMA 6 (SMARD1)	11q13	Immunoglobulin mu binding protein 2 (*IGHMBP2*)
Distal SMA 3	11q13	Unknown
SMA (Jerash type)	9p21.1-p12	Unknown
Distal SMA4	1P36	*Pleckstrin homology domain containing, family G (PLEKHG5)*
Distal SMA 2B	7q11	*HSPB1/Heat shock protein 27*

End of Life: Mechanisms and Comorbidities

Without respiratory support, patients with SMA type 1 rarely live longer than 2 years because of respiratory complications.[38–40] However, recent reports have shown an increase in SMA 1 survival because of early proactive respiratory care and aggressive nutritional support.[41,77,78] An earlier decline in respiratory function is seen in type 2 patients compared to those with type 3.[79] Survival studies have shown that about 52–68% of SMA type 2 patients are alive to age 20–25 years and most type 3 and 4 patients have a normal lifespan.[39,80] For patients with severe respiratory and bulbar involvement, end of life discussions about various options should not be delayed, and resources should be available to allow parents and families to make their decisions.

MOLECULAR GENETICS

SMA is caused by a genetic defect in the survival motor neuron (*SMN*) gene on chromosome 5. This gene in humans exists in two nearly identical forms, the telomeric *SMN1* gene and the centromeric *SMN2* gene. The main functional difference between these two genes relates to a single nucleotide change at the beginning of exon 7 with cytosine in *SMN1* replaced by a thymidine in *SMN2*. This C-to-T transition alters an exonic splice modulator leading to frequent skipping of exon 7 during transcription of *SMN2*. While the majority of *SMN1* transcripts are full length and result in stable SMN protein, approximately 90% of *SMN2* transcripts are truncated and produce a truncated and unstable SMN protein that is rapidly degraded. However, due to alternative splicing, 10% of transcripts of the *SMN2* gene include exon 7 and therefore produce full-length protein (Figure 89.4).[81]

Homozygous deletion of *SMN1* is seen in 95% of all SMA 0, 1, 2, and 3 patients and 70% of type 4 patients.[15,65] The remaining 5% represent compound heterozygotes involving an *SMN1* deletion of one allele and an intragenic missense or frameshift mutation of the other allele. *SMN2* copy number is the primary determinant of SMA phenotype variability and is inversely correlated to the disease severity (Table 89.2).[82–87]Although *SMN2* copy number is the principal determinant of SMA phenotype, other disease modifiers have been reported. One *SMN2* gene variant containing a c.859G>C mutation in exon 7 is a positive disease modifier that causes a 20% increased production of in full-length SMN protein. This results in a milder phenotype; in the initial report on this mutation, three adults with only two copies of *SMN2* had a mild type 3b phenotype.[88–90] Plastin 3 has been proposed to be a possible protective

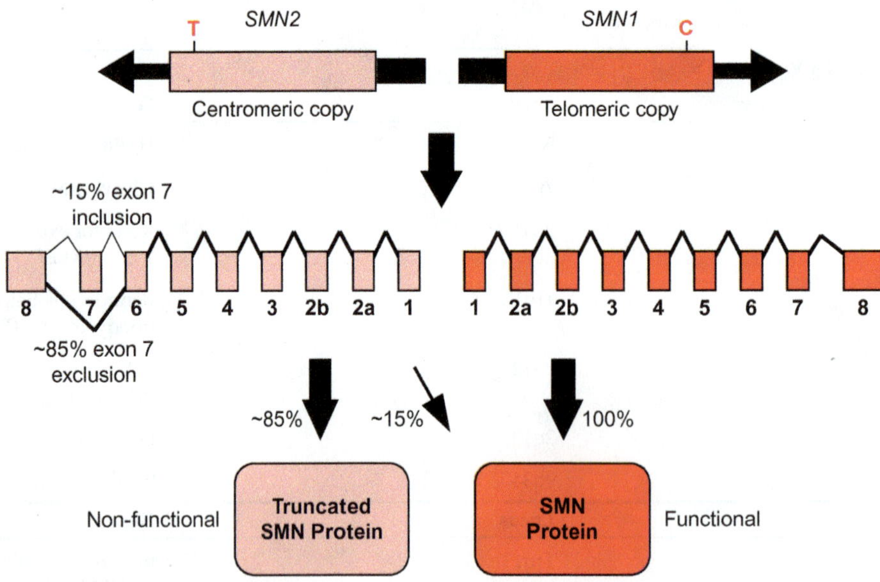

FIGURE 89.4 **Schematic of the human *SMN1* and *SMN2* gene region and pre-mRNAs produced by these genes.** All SMA patients have homozygous mutations (usually deletions) in both copies of *SMN1*. The *SMN2* gene is also expressed but most *SMN2* pre-mRNA lacks exon 7 because of a C-to-T transition at position 6 of exon 7. The resultant truncated SMN protein is unstable and nonfunctional. Due to alternative splicing, however, a small proportion of the *SMN2* pre-mRNA is full-length mRNA containing exon 7, resulting in functional full-length SMN protein.

TABLE 89.2 Classification of SMN-related SMA

Type	Age at onset	Highest function	Natural age at death	SMN 2 no.
0	Prenatal	Respiratory support	<1 month	1
1	0–6 months	Never sit	<2 years	2
2	<18 months	Never stand	>2 years	3, 4
3	>18 months	Stand & walk unaided	Adult	
3a	18 months–3 years	Stand & walk unaided	Adult	3, 4
3b	>3 years	Stand & walk unaided	Adult	4
4	>21 years	Stand & walk unaided	Adult	4–8

modifier of SMA in females.[91,92] However, the exact role of plastin 3 in SMA remains questionable given that high levels were found in female siblings with the severe phenotype.[93]

DISEASE MECHANISMS

Importantly, SMA is related to deficiency of SMN protein, not absence. Although 10–15% of the normal population lack *SMN2*, retention of at least one *SMN2* copy is necessary for survival in the setting of homozygous *SMN1* deletion. The function of the SMN protein and thus the pathogenesis of SMA are not yet fully elucidated. The SMN protein is ubiquitously expressed in essentially all organs of the body with the highest tissue levels during development. In the cell, SMN is present within the cytoplasm and nucleus; in the latter it is localized in inclusions known as "gems."[93–96] A principal function of SMN protein is assembly of Sm proteins onto small nuclear RNA, and loss of this housekeeping function provides a clear explanation why absence of SMN protein is incompatible with cellular function and survival.[97–99] Other possible SMN-related functions that might be relevant to the SMA pathogenesis include modulation of apoptosis, translational regulation, axonal growth, and neuromuscular junction formation.[100] Deficiency of the ubiquitously expressed SMN protein results in selective degeneration only of motor neurons.[101,102]

The basis for this selectivity is unclear but may involve heightened susceptibility of the motor neuron to SMN deficiency and potential roles for SMN protein at the neuromuscular junction and other sites within the motor unit. There is also some emerging evidence of possible involvement of other organ systems in SMA.[47,48]

Animal Models

Much of the understanding of SMA pathogenesis comes from the development and utilization of animal models—most notably mouse and zebrafish models.[103] In contrast to all other vertebrate species, the *SMN2* gene is unique to humans. Therefore, because SMA is a disease of SMN deficiency, animal models cannot be generated using a simple gene knockout approach. Mice have a single *SMN* gene and initial knockout models of the mouse *SMN* allele resulted in embryonic lethality.[103] The first viable mouse model was successfully produced by the introduction of two copies of a human *SMN2* transgene onto a background of mouse *SMN* knockout allele.[104] This mouse model had a severe SMA phenotype with average lifespan of 5 days.[104] Insertion of eight or more copies of *SMN2* transgene completely rescued this model.[104] The most commonly utilized and important mouse model was generated by introduction of two copies of the cDNA of the *SMN2* gene lacking exon 7 (SMNΔ7) onto the background of the severe mouse background (*SMN* knockout and 2 copies of the human *SMN2*).[105] The addition of the SMNΔ7 transgene demonstrated the beneficial effects, albeit moderate, of truncated SMN protein produced from transcripts lacking exon 7 by extending the lifespan from 5 days to 14 days.[106]

Additional SMA murine models have been created with inducible alleles and Cre-Lox P systems to investigate the role of SMN restoration at specific periods during development and in specific cellular populations.[107–111] Creating milder mouse models of SMA to mimic type 2 and 3 disease has been a challenge, seemingly due to sensitivity of the models to small increases in SMN protein. A mouse model for a milder disease was generated using the SMNΔ7 transgene manipulated to encode a modestly more functional protein referred to as SMN read-through (SMN [RT]).[112] While mouse models have been the most critical in the understanding of the pathogenesis of SMA and in preclinical development of therapeutic agents, other important models include yeast, nematode, fly, and zebrafish. A zebrafish model was generated using a transgenic zebrafish expressing the human *SMN2* gene (hSMN2), which produces a low amount of full-length SMN, and crossed onto the smn−/−background.[113] This model has proven particularly useful for studying the effects of SMN protein on axonal growth and differentiation.

Current Research

The downstream targets of SMN protein deficiency that lead to loss of motor neurons are unknown; an emerging theory postulates downstream deleterious impact of defective RNA processing on motor neuron development, survival or both.[114] Importantly, when splicing is assayed in SMA mouse spinal cord, there are relatively minimal splicing changes.[115] Similarly, changes in alternative splicing and remodeling of the motor neuron transcriptome have been investigated using laser capture microdissection to selectively obtain motor neuron tissue from SMA mice.[116] These investigations are being used in attempts to identify downstream targets of SMN deficiency that could underlie motor neuron loss in SMA. While SMN protein restoration offers promise as a therapeutic target, understanding the targets of SMN deficiency will allow for the development of specific therapeutics and a greater understanding of when and where therapeutic intervention will be required.

DIFFERENTIAL DIAGNOSIS

The differential diagnosis for SMA 1 includes all other etiologies for a hypotonic "floppy" infant, including causes related to central injury involving the cerebral hemisphere, basal ganglia, brainstem, cerebellum or spinal cord, and peripheral causes, such as congenital myopathies, congenital myotonic dystrophy and other muscular dystrophies.[117] Other considerations include congenital myasthenic syndromes; infantile botulism; metabolic myopathies, such as Pompe disease; chromosomal disorders, e.g., Prader–Willi syndrome; peroxisomal disorders, e.g., Zellweger disease; neonatal adrenoleukodystrophy; and infantile Refsum disease. Since most of these disorders are uncommon, SMA testing has become a standard initial step in infants with hypotonia felt to be of peripheral origin.

Several disorders that cause proximal weakness in older children and adolescents mimic SMA type 2 and 3, such as congenital and metabolic myopathies, dystrophinopathies, limb girdle dystrophies, inflammatory myopathies, myasthenia gravis, and immune-mediated peripheral neuropathies. Detailed history and examination to uncover associated clinical features and pertinent family history, and directed diagnostic testing such as brain and spinal cord imaging, electrodiagnostic studies, and muscle biopsy can help explore the differential diagnostic possibilities. Other motor neuron disorders such as hexosaminidase A deficiency can mimic SMA,[118] although cognitive decline, developmental regression, personality change, ataxia, spasticity, and seizures usually help distinguish this disorder.

TESTING

Molecular genetic testing is the gold standard for confirming the diagnosis in patients with suspected SMA, and has eliminated the need for electrodiagnostic testing and muscle biopsy in the vast majority of cases, especially infants. Targeted mutation analysis is commercially available and detects the exon 7 homozygous deletion present at 95% sensitivity and near 100% specificity.[15] Sequence analysis detects the 2–5% of SMA patients who are compound heterozygotes for *SMN1* deletion and *SMN1* intragenic mutations.[119] *SMN2* copy number can also be determined through commercially available testing. SMA carrier detection uses a gene dose analysis method to determine the *SMN1* copy number. The method is sensitive; however, there are variables that complicate interpretation and lead to false-negative results in the 4% of individuals who have no *SMN1* copies in one chromosome and two *SMN1* copies in one chromosome (i.e., a *cis* configuration). Furthermore, 2% of SMA subjects have a *de novo* mutation that occurs in one SMA allele and only one parent is a true carrier.[119] It is feasible in specialized centers to pursue prenatal diagnosis in high-risk pregnancies using chorionic villus sampling or amniocentesis to obtain fetal DNA. Similarly, embryonic testing can be done pre-implantation in the setting of *in vitro* fertilization. Formal genetic counseling is crucial when pursuing genetic testing, especially in regards to the implications of *SMN2* copy number because of its prognostic implications.

Electrodiagnostic testing and muscle biopsy are valuable in those with negative genetic testing or other types of SMA. As in other anterior horn cell disorders, nerve conduction studies show normal sensory responses with reduced compound muscle action potential (CMAP) amplitudes and normal or slightly reduced motor conduction velocities.[39] Electromyography reveals active and chronic denervation and reinnervation with fibrillation potentials, positive sharp waves, fasciculations, large amplitude and long duration motor unit potentials with reduced recruitment pattern. Regular spontaneous motor unit action potentials occur in SMA type 1.[120,121] Patients with SMA may also have a mild to moderate increase in creatine kinase (CK), leading to erroneous diagnosis of muscle disease.[122] Muscle biopsy is rarely indicated, except in patients with negative genetic testing but in type 1 infants reveals intermixed hypertrophic type 1 fibers and sheets of atrophic round fibers.[123,124] Biopsy findings in type 2 patients are often similar to type 1 but can also show hypertrophic type 2 fibers and fiber type grouping. In SMA type 3 secondary myopathic features, including fiber splitting, increased internal nuclei and endomysial connective tissue can be seen in addition to fiber-type grouping.[39]

MANAGEMENT

Currently, the cornerstone for SMA treatment relies on aggressive supportive care, stressing the pivotal role of a multidisciplinary team approach. The consensus statement for standard of care by the international SMA Standard of Care Committee expert panel serves as an excellent resource for health care professionals and provides recommendations for treatment.[125] Several respiratory issues, such as difficulty handling secretions, poor development of the lungs and thorax, recurrent pneumonia, and sleep disordered breathing are the main cause of morbidity and mortality in type 1 and 2 patients.[112] Early and aggressive initiation of cough assist (using devices or manual techniques), vaccinations, infection control, and gastroesophageal reflux treatment is important.[126,127] Noninvasive ventilation using bilevel positive air way pressure machines (BiPap) is beneficial and leads to less hospitalization, prolongation of life, and possible improvement in lung development and function.[77,128,129] The use of tracheotomy and chronic ventilation is always a difficult decision for families in this situation and are somewhat controversial in SMA 1.[130,131] The decision making in this regard should be a process that involves explanation and discussion with the family about the expected outcome, and quality of life as well as exploration of other options such as palliative care.

Gastrointestinal and nutritional complications in patients with SMA include: dysphagia and increased aspiration risk secondary to bulbar dysfunction; constipation; delayed gastric emptying; and reflux secondary to gastroesophgeal dysmotility.[45,125] Swallowing evaluations by speech therapy and use of videofluoroscopic swallow studies are useful. In selected patients gastrostomy tube for feeding is essential to insure adequate nutrition. H2 blocker and proton pump inhibitors for reflux and use of metoclopramide or erythromycin to help gastric motility are also beneficial. Nissen fundoplication should be reserved for those failing medical treatment of reflux and is currently routinely performed at many US centers on the vast majority of SMA type 1 infants.

There is lack of quality data regarding proper assessment and treatment of growth and nutritional status in SMA. Type 1 SMA children have a lower caloric requirement compared to their age-matched peers and thus are at risk for both over- or undernourishment.[132] SMA children have higher fat mass on dual-energy X-ray absorptiometry (DEXA) scans, and scrupulous control of excessive weight gain is crucial.[57,132] Orthopedic complications include scoliosis, kyphosis, contractures, osteopenia, and fractures.[125,133] There is an overall correlation between development and progression of spinal deformity with weakness.[117] There are numerous benefits to surgical correction, such as improvement in sitting, balance, comfort, and delays in decline of pulmonary function.[134–138] The timing for surgery should take into account the degree of curve, child growth, and pulmonary function, and the surgery should optimally be performed by surgeons experienced in managing children with neuromuscular disorders. Although human data is lacking, animal studies suggest that there may be a beneficial effect of regular exercise in patients with SMA.[139] The consensus statement encourages routine exercise such as swimming and adaptive sports to maintain and optimize fitness and endurance.[125] There is marked variability in the care provided to SMA patients and, if feasible, patients should always be referred to a center experienced in managing such patients.[140]

Failed Therapies and Barriers to Treatment Development

Although the presence of multiple animal models allows for effective screening of therapeutic agents, the success of several compounds in these models has not yet translated into successful human therapeutic trials.[29,30,141] The strategies previously tested include gabapentin, riluzole, albuterol, creatine, carnitine, valproic acid, hydroxyurea, and phenylbutyrate.[16–30] Work is ongoing to test other treatment strategies, especially in relation to pharmacologic agents that upregulate *SMN2* production of full length protein. There are several gaps in the understanding of the pathophysiology that need to be filled to allow more effective treatment discovery. The search for valid and reliable biomarkers that are clinically meaningful is also ongoing.[142,143] Similarly, and despite the progress made over the last decade, further work is needed for validation of a unified clinical outcome measure for use in future trials.[100] Use of new psychometric techniques such as Rasch analysis is helpful in the validation process and improvement of these outcomes.[144] Current evidence suggests an early intervention is likely to provide the best option for desired clinical outcome and several studies have demonstrated the feasibility of a neonatal screening program.[145–148]

Therapies Under Investigation

Ongoing research to identify therapeutic small molecules for SMA has met with some success. As a result of high throughput screening programs, quinazoline-based compounds were found to increase SMN promoter activity, SMN mRNA and SMN protein levels, and increase the lifespan of SMA mice by 21–30%.[149] Even more encouraging are small molecules that increase exon 7 into *SMN2* mRNA and efficiently correct the splicing defect of *SMN2*. Some of these compounds increase SMN protein levels in SMA patient cell lines and mouse models and extend survival 10-fold in mice.[150] Other therapies under investigation include stem cell-based interventions. Initial animal studies using neural stem cells and embryonic stem cell-derived neural cells demonstrated motor neuron and astrocyte differentiation, motor neuron protection, and growth factor production. Genetically corrected, induced pluripotent stem cells generated from skin fibroblasts of SMA patients (SMA-iPSCs) and transplanted into an SMA mouse model extended lifespan and improved the disease phenotype, most likely due to improved neural environments due to growth factor production by the transplanted stem cells.[151–153] There are, however, multiple hurdles to stem cell-based therapy such as type and amount of cell to be transplanted, delivery methodologies, and issues of rejection and the need for immunosuppression. The most promising strategies in the current pipeline being tested include antisense oligonucleotide (ASO)-mediated therapy and gene replacement therapy.[105] ASO base sequences have been developed complementary to the gene's messenger RNA (mRNA) and used as a tool to modify RNA processing or achieve gene suppression.[105] In the SMA setting, this works by converting the *SMN2* message into *SMN1* by potentiation of *SMN2* read through. There are a number of *cis*-elements involved in the splicing regulation of SMN exon 7. ASOs targeted to these splice elements have shown robust rescue of the phenotype in multiple SMA mouse models.[154–157] Human clinical trials using the delivery of such drugs intrathecally are underway. Another promising strategy is gene therapy aimed at the replacement of the *SMN1* gene. The excitement stems from the recent preclinical success story with multiple groups showing substantial and robust animal response to gene replacement using a self complementary adeno-associated virus vector (scAAV) for delivery through intravenous, intrathecal (or both) routes.[158–160] Human clinical studies are underway for this exciting therapy. Because the therapeutic window for rescuing motor neurons may be extremely small (as suggested by studies in the animal models) it will be crucial going forward to explore non-SMN based therapies as well to treat these devastating disorders.

References

1. Werdnig G. Zwei fruhinfantile hereditare falle von progressiver muskelatrophie unter dem bilde de dystrophie, ager auf neurotischer grundlage. *Arch Pyschiatr*. 1891.
2. Dubowitz V. Ramblings in the history of spinal muscular atrophy. *Neuromuscul Disord*. 2009;19:69–73.
3. Biros I, Forrest S. Spinal muscular atrophy: untangling the knot? *J Med Genet*. 1999;36:1–8.
4. Hoffmann J. Ueber chronische spinale muskelatrophie im kindesalter auf familiärer basis. *Dtsch Z Nervenheilkd*. 1892;7:395–415.
5. Kugelberg E, Welander L. Heredofamilial juvenile muscular atrophy simulating muscular dystrophy. *AMA Arch Neurol Psychiatry*. 1956;75:500–509.
6. Byers RK, Banker BQ. Infantile muscular atrophy. *Arch Neurol*. 1961;5:140–164.
7. Dubowitz V. Infantile muscular atrophy. A prospective study with particular reference to a slowly progressive variety. *Brain*. 1964;87:707–718.
8. Pearn JH, Wilson J. Acute Werdnig-Hoffmann disease: acute infantile spinal muscular atrophy. *Arch Dis Child*. 1973;48:425–430.
9. Emery AE. The nosology of the spinal muscular atrophies. *J Med Genet*. 1971;8:481–495.
10. Pearn J. Classification of spinal muscular atrophies. *Lancet*. 1980;1:919–922.
11. Munsat TL, Davies KE. International SMA consortium meeting. (26–28 June 1992, Bonn, Germany). *Neuromuscul Disord*. 1992;2:423–428.
12. Brzustowicz LM, Lehner T, Castilla LH, et al. Genetic mapping of chronic childhood-onset spinal muscular atrophy to chromosome 5q11.2-13.3. *Nature*. 1990;344:540–541.
13. Melki J, Sheth P, Abdelhak S, et al. Mapping of acute (type I) spinal muscular atrophy to chromosome 5q12-q14. The French Spinal Muscular Atrophy Investigators. *Lancet*. 1990;336(8710):271–273.
14. Gilliam TC, Brzustowicz LM, Castilla LH, et al. Genetic homogeneity between acute and chronic forms of spinal muscular atrophy. *Nature*. 1990;345(6278):823–825.
15. Lefebvre S, Burglen L, Reboullet S, et al. Identification and characterization of a spinal muscular atrophy-determining gene. *Cell*. 1995;80:155–165.
16. Miller RG, Moore DH, Dronsky V, et al. SMA Study Group. A placebo-controlled trial of gabapentin in spinal muscular atrophy. *J Neurol Sci*. 2001;191:127–131.
17. Kinali M, Mercuri E, Main M, et al. Pilot trial of albuterol in spinal muscular atrophy. *Neurology*. 2002;59:609–610.
18. Merlini L, Solari A, Vita G, et al. Role of gabapentin in spinal muscular atrophy: results of a multicenter, randomized Italian study. *J Child Neurol*. 2003;18:537–541.
19. Russman BS, Iannaccone ST, Samaha FJ. A phase 1 trial of riluzole in spinal muscular atrophy. *Arch Neurol*. 2003;60:1601–1603.
20. Chang JG, Hsieh-Li HM, Jong YJ, Wang NM, Tsai CH, Li H. Treatment of spinal muscular atrophy by sodium butyrate. *Proc Natl Acad Sci U S A*. 2001;98:9808–9813.
21. Mercuri E, Bertini E, Messina S, et al. Pilot trial of phenylbutyrate in spinal muscular atrophy. *Neuromuscul Disord*. 2004;14:130–135.
22. Weihl CC, Connolly AM, Pestronk A. Valproate may improve strength and function in patients with type III/IV spinal muscle atrophy. *Neurology*. 2006;67:500–501.
23. Pane M, Staccioli S, Messina S, et al. Daily salbutamol in young patients with SMA type II. *Neuromuscul Disord*. 2008;18(7):536–540.
24. Swoboda KJ, Scott CB, Reyna SP, et al. Phase II open label study of valproic acid in spinal muscular atrophy. *PLoS One*. 2009;4(5):e5268.
25. Swoboda KJ, Scott CB, Crawford TO, et al. Project Cure Spinal Muscular Atrophy Investigators Network. SMA CARNIVAL trial part I: double-blind, randomized, placebo-controlled trial of L-carnitine and valproic acid in spinal muscular atrophy. *PLoS One*. 2010;5(8):e12140.
26. Kissel JT, Scott CB, Reyna SP, et al. Project Cure Spinal Muscular Atrophy Investigators' Network. SMA CARNIVAL trial part II: a prospective, single-armed trial of L-carnitine and valproic acid in ambulatory children with spinal muscular atrophy. *PLoS One*. 2011;6(7):e21296.
27. Chen TH, Chang JG, Yang YH, et al. Randomized, double-blind, placebo-controlled trial of hydroxyurea in spinal muscular atrophy. *Neurology*. 2010;75:2190–2197.
28. Kissel JT, Elsheikh B, King WM, for the Project Cure Spinal Muscular Atrophy Investigators Network. SMA valiant trial: a prospective, double-blind, placebo-controlled trial of valproic acid in ambulatory adults with spinal muscular atrophy. *Muscle Nerve*. 2014;49(2):187–192.
29. Wadman RI, Bosboom WM, van den Berg LH, Wokke JH, Iannaccone ST, Vrancken AF. Drug treatment for spinal muscular atrophy types II and III. *Cochrane Database Syst Rev*. 2011;12:CD006282.
30. Wadman RI, Bosboom WM, van der Pol WL, et al. Drug treatment for spinal muscular atrophy type I. *Cochrane Database Syst Rev*. 2012;4:CD006281.
31. Sugarman EA, Nagan N, Zhu H, et al. Pan-ethnic carrier screening and prenatal diagnosis for spinal muscular atrophy: clinical laboratory analysis of >72,400 specimens. *Eur J Hum Genet*. 2012;20:27–32.
32. Burd L, Short SK, Martsolf JT, Nelson RA. Prevalence of type I spinal muscular atrophy in North Dakota. *Am J Med Genet*. 1991;41(2):212–215.
33. Mostacciuolo ML, Danieli GA, Trevisan C, Müller E, Angelini C. Epidemiology of spinal muscular atrophies in a sample of the Italian population. *Neuroepidemiology*. 1992;11(1):34–38.
34. Thieme A, Mitulla B, Schulze F, Spiegler AW. Epidemiological data on Werdnig-Hoffmann disease in Germany (West-Thuringen). *Hum Genet*. 1993;91:295–297.
35. Ludvigsson P, Olafsson E, Hauser WA. Spinal muscular atrophy. Incidence in Iceland. *Neuroepidemiology*. 1999;18:265–269.

36. Pearn J. Genetic studies of acute infantile spinal muscular atrophy (SMA type I). An analysis of sex ratios, segregation ratios, and sex influence. *J Med Genet*. 1978;15:414–417.
37. Hausmanowa-Petrusewicz I, Zaremba J, Borkowska J, Szirkowiec W. Chronic proximal spinal muscular atrophy of childhood and adolescence: sex influence. *J Med Genet*. 1984;21:447–450.
38. Zerres K, Rudnik-Schoneborn S. Natural history in proximal spinal muscular atrophy. Clinical analysis of 445 patients and suggestions for a modification of existing classifications. *Arch Neurol*. 1995;52:518–523.
39. Zerres K, Davies KE. 59th ENMC international workshop: spinal muscular atrophies: recent progress and revised diagnostic criteria 17–19 April 1998, Soestduinen, The Netherlands. *Neuromuscul Disord*. 1999;9:272–278.
40. Thomas NH, Dubowitz V. The natural history of type I (severe) spinal muscular atrophy. *Neuromuscul Disord*. 1994;4:497–502.
41. Oskoui M, Levy G, Garland CJ, et al. The changing natural history of spinal muscular atrophy type 1. *Neurology*. 2007;69:1931–1936.
42. Mannaa MM, Kalra M, Wong B, Cohen AP, Amin RS. Survival probabilities of patients with childhood spinal muscle atrophy. *J Clin Neuromuscul Dis*. 2009;10:85–89.
43. Bertini E, Burghes A, Bushby K, et al. 134th ENMC international workshop: outcome measures and treatment of spinal muscular atrophy, 11–13 February 2005. *Naarden Neuromuscul Disord*. 2005;802–816, 5d.
44. Iannaccone ST. Spinal muscular atrophy. *Semin Neurol*. 1998;18:19–26.
45. Iannaccone ST, Browne RH, Samaha FJ, Buncher CR. Prospective study of spinal muscular atrophy before age 6 years. DCN/SMA Group. *Pediatr Neurol*. 1993;9:187–193.
46. von Gontard A, Zerres K, Backes M, et al. Intelligence and cognitive function in children and adolescents with spinal muscular atrophy. *Neuromuscul Disord*. 2002;12:130–136.
47. Rudnik-Schoneborn S, Heller R, Berg C, et al. Congenital heart disease is a feature of severe infantile spinal muscular atrophy. *J Med Genet*. 2008;45:635–638.
48. Rudnik-Schoneborn S, Vogelgesang S, Armbrust S, Graul-Neumann L, Fusch C, Zerres K. Digital necroses and vascular thrombosis in severe spinal muscular atrophy. *Muscle Nerve*. 2010;42:144–147.
49. Rudnik-Schoneborn S, Hausmanowa-Petrusewicz I, Borkowska J, Zerres K. The predictive value of achieved motor milestones assessed in 441 patients with infantile spinal muscular atrophy types II and III. *Eur Neurol*. 2001;45:174–181.
50. Moosa A, Dubowitz V. Spinal muscular atrophy in childhood. Two clues to clinical diagnosis. *Arch Dis Child*. 1973;48:386–388.
51. van den Engel-Hoek L, Erasmus CE, van Bruggen HW, et al. Dysphagia in spinal muscular atrophy type II: more than a bulbar problem? *Neurology*. 2009;73:1787–1791.
52. Samaha FJ, Buncher CR, Russman BS, et al. Pulmonary function in spinal muscular atrophy. *J Child Neurol*. 1994;9:326–329.
53. Iannaccone ST, Hynan LS. Reliability of 4 outcome measures in pediatric spinal muscular atrophy. *Arch Neurol*. 2003;60:1130–1136.
54. Chng SY, Wong YQ, Hui JH, Wong HK, Ong HT, Goh DY. Pulmonary function and scoliosis in children with spinal muscular atrophy types II and III. *J Paediatr Child Health*. 2003;39:673–676.
55. Mellies U, Dohna-Schwake C, Stehling F, Voit T. Sleep disordered breathing in spinal muscular atrophy. *Neuromuscul Disord*. 2004;14:797–803.
56. Khatri IA, Chaudhry US, Seikaly MG, Browne RH, Iannaccone ST. Low bone mineral density in spinal muscular atrophy. *J Clin Neuromuscul Dis*. 2008;10:11–17.
57. Sproule DM, Montes J, Montgomery M, et al. Increased fat mass and high incidence of overweight despite low body mass index in patients with spinal muscular atrophy. *Neuromuscul Disord*. 2009;19:391–396.
58. Crawford TO, Pardo CA. The neurobiology of childhood spinal muscular atrophy. *Neurobiol Dis*. 1996;3:97–110.
59. Montes J, Dunaway S, Garber CE, Chiriboga CA, De Vivo DC, Rao AK. Leg muscle function and fatigue during walking in spinal muscular atrophy type 3. *Muscle Nerve*. 2013. doi:10.1002/mus.24081.
60. Deymeer F, Serdaroglu P, Poda M, Gulsen-Parman Y, Ozcelik T, Ozdemir C. Segmental distribution of muscle weakness in SMA III: implications for deterioration in muscle strength with time. *Neuromuscul Disord*. 1997;7:521–528.
61. Reimers CD, Schlotter B, Eicke BM, Witt TN. Calf enlargement in neuromuscular diseases: a quantitative ultrasound study in 350 patients and review of the literature. *J Neurol Sci*. 1996;143:46–56.
62. Oh J, Kim SM, Shim DS, Sunwoo IN. Neurogenic muscle hypertrophy in Type III spinal muscular atrophy. *J Neurol Sci*. 2011;308:147–148.
63. Kaufmann P, McDermott MP, Darras BT, et al. Muscle Study Group (MSG), Pediatric Neuromuscular Clinical Research Network for Spinal Muscular Atrophy (PNCR). Prospective cohort study of spinal muscular atrophy types 2 and 3. *Neurology*. 2012;79(18):1889–1897.
64. Pearn JH, Hudgson P, Walton JN. A clinical and genetic study of spinal muscular atrophy of adult onset: the autosomal recessive form as a discrete disease entity. *Brain*. 1978;101:591–606.
65. Brahe C, Servidei S, Zappata S, Ricci E, Tonali P, Neri G. Genetic homogeneity between childhood-onset and adult-onset autosomal recessive spinal muscular atrophy. *Lancet*. 1995;346:741–742.
66. Clermont O, Burlet P, Lefebvre S, Burglen L, Munnich A, Melki J. SMN gene deletions in adult-onset spinal muscular atrophy. *Lancet*. 1995;346:1712–1713.
67. Burglen L, Amiel J, Viollet L, et al. Survival motor neuron gene deletion in the arthrogryposis multiplex congenita-spinal muscular atrophy association. *J Clin Invest*. 1996;98:1130–1132.
68. Bingham PM, Shen N, Rennert H, et al. Arthrogryposis due to infantile neuronal degeneration associated with deletion of the SMNT gene. *Neurology*. 1997;49(3):848–851.
69. Renbaum P, Kellerman E, Jaron R, et al. Spinal muscular atrophy with pontocerebellar hypoplasia is caused by a mutation in the VRK1 gene. *Am J Hum Genet*. 2009;85:281–289.
70. Deng HX, Klein CJ, Yan J, et al. Scapuloperoneal spinal muscular atrophy and CMT2C are allelic disorders caused by alterations in TRPV4. *Nat Genet*. 2010;42:165–169.
71. Harms MB, Ori-McKenney KM, Scoto M, et al. Mutations in the tail domain of DYNC1H1 cause dominant spinal muscular atrophy. *Neurology*. 2012;78(22):1714–1720.
72. Ramser J, Ahearn ME, Lenski C, et al. Rare missense and synonymous variants in UBE1 are associated with X-linked infantile spinal muscular atrophy. *Am J Hum Genet*. 2008;82:188–193.

73. Kennedy WR, Alter M, Sung JH. Progressive proximal spinal and bulbar muscular atrophy of late onset. A sex-linked recessive trait. *Neurology*. 1968;18:671–680.
74. La Spada AR, Wilson EM, Lubahn DB, Harding AE, Fischbeck KH. Androgen receptor gene mutations in X-linked spinal and bulbar muscular atrophy. *Nature*. 1991;352:77–79.
75. Amato AA, Prior TW, Barohn RJ, Snyder P, Papp A, Mendell JR. Kennedy's disease: a clinicopathologic correlation with mutations in the androgen receptor gene. *Neurology*. 1993;43:791–794.
76. Irobi J, Dierick I, Jordanova A, Claeys KG, De Jonghe P, Timmerman V. Unraveling the genetics of distal hereditary motor neuronopathies. *Neuromolecular Med*. 2006;8:131–146.
77. Lemoine TJ, Swoboda KJ, Bratton SL, Holubkov R, Mundorff M, Srivastava R. Spinal muscular atrophy type 1: are proactive respiratory interventions associated with longer survival? *Pediatr Crit Care Med*. 2012.
78. Gregoretti C, Ottonello G, Chiarini Testa MB, et al. Survival of patients with spinal muscular atrophy type 1. *Pediatrics*. 2013;131(5):e1509–e1514.
79. Khirani S, Colella M, Caldarelli V, et al. Longitudinal course of lung function and respiratory muscle strength in spinal muscular atrophy type 2 and 3. *Eur J Paediatr Neurol*. 2013;17(6):552–560.
80. Farrar MA, Vucic S, Johnston HM, du Sart D, Kiernan MC. Pathophysiological insights derived by natural history and motor function of spinal muscular atrophy. *J Pediatr*. 2013;162(1):155–159.
81. Kolb SJ, Kissel JT. Spinal muscular atrophy: a timely review. *Arch Neurol*. 2011;68:979–984.
82. Mailman MD, Heinz JW, Papp AC, et al. Molecular analysis of spinal muscular atrophy and modification of the phenotype by SMN2. *Genet Med*. 2002;4:20–26.
83. Feldkotter M, Schwarzer V, Wirth R, Wienker TF, Wirth B. Quantitative analyses of SMN1 and SMN2 based on real-time lightCycler PCR: fast and highly reliable carrier testing and prediction of severity of spinal muscular atrophy. *Am J Hum Genet*. 2002;70:358–368.
84. Swoboda KJ, Prior TW, Scott CB, et al. Natural history of denervation in SMA: relation to age, SMN2 copy number, and function. *Ann Neurol*. 2005;57:704–712.
85. Wirth B, Brichta L, Schrank B, et al. Mildly affected patients with spinal muscular atrophy are partially protected by an increased SMN2 copy number. *Hum Genet*. 2006;119:422–428.
86. Tiziano FD, Bertini E, Messina S, et al. The Hammersmith functional score correlates with the SMN2 copy number: a multicentric study. *Neuromuscul Disord*. 2007;17:400–403.
87. Elsheikh B, Prior T, Zhang X, et al. An analysis of disease severity based on SMN2 copy number in adults with spinal muscular atrophy. *Muscle Nerve*. 2009;40:652–656.
88. Prior TW, Krainer AR, Hua Y, et al. A positive modifier of spinal muscular atrophy in the SMN2 gene. *Am J Hum Genet*. 2009;85:408–413.
89. Vezain M, Saugier-Veber P, Goina E, et al. A rare SMN2 variant in a previously unrecognized composite splicing regulatory element induces exon 7 inclusion and reduces the clinical severity of spinal muscular atrophy. *Hum Mutat*. 2010;31:E1110–E1125.
90. Bernal S, Alías L, Barceló MJ, et al. The c.859G>C variant in the SMN2 gene is associated with types II and III SMA and originates from a common ancestor. *J Med Genet*. 2010;47(9):640–642.
91. Oprea GE, Krober S, McWhorter ML, et al. Plastin 3 is a protective modifier of autosomal recessive spinal muscular atrophy. *Science*. 2008;320:524–527.
92. Stratigopoulos G, Lanzano P, Deng L, et al. Association of plastin 3 expression with disease severity in spinal muscular atrophy only in postpubertal females. *Arch Neurol*. 2010;67:1252–1256.
93. Bernal S, Also-Rallo E, Martinez-Hernandez R, et al. Plastin 3 expression in discordant spinal muscular atrophy (SMA) siblings. *Neuromuscul Disord*. 2011;21:413–419.
94. Cauchi RJ. SMN and Gemins: 'we are family' … or are we?: insights into the partnership between Gemins and the spinal muscular atrophy disease protein SMN. *Bioessays*. 2010;32(12):1077–1089.
95. Coovert DD, Le TT, McAndrew PE, et al. The survival motor neuron protein in spinal muscular atrophy. *Hum Mol Genet*. 1997;6:1205–1214.
96. Lefebvre S, Burlet P, Liu Q, et al. Correlation between severity and SMN protein level in spinal muscular atrophy. *Nat Genet*. 1997;16:265–269.
97. Schrank B, Gotz R, Gunnersen JM, et al. Inactivation of the survival motor neuron gene, a candidate gene for human spinal muscular atrophy, leads to massive cell death in early mouse embryos. *Proc Natl Acad Sci U S A*. 1997;94:9920–9925.
98. Cifuentes-Diaz C, Frugier T, Tiziano FD, et al. Deletion of murine SMN exon 7 directed to skeletal muscle leads to severe muscular dystrophy. *J Cell Biol*. 2001;152:1107–1114.
99. Vitte JM, Davoult B, Roblot N, et al. Deletion of murine Smn exon 7 directed to liver leads to severe defect of liver development associated with iron overload. *Am J Pathol*. 2004;165:1731–1741.
100. Tiziano FD, Melki J, Simard LR. Solving the puzzle of spinal muscular atrophy: what are the missing pieces? *Am J Med Genet A*. 2013;161A(11):2836–2845.
101. Monani UR. Spinal muscular atrophy: a deficiency in a ubiquitous protein; a motor neuron-specific disease. *Neuron*. 2005;48:885–896.
102. Burghes AH, Beattie CE. Spinal muscular atrophy: why do low levels of survival motor neuron protein make motor neurons sick? *Nat Rev Neurosci*. 2009;10:597–609.
103. Sumner CJ. Molecular mechanisms of spinal muscular atrophy. *J Child Neurol*. 2007;22(8):979–989.
104. Monani UR, Sendtner M, Coovert DD, et al. The human centromeric survival motor neuron gene (SMN2) rescues embryonic lethality in Smn(-/-) mice and results in a mouse with spinal muscular atrophy. *Hum Mol Genet*. 2000;9:333–339.
105. Arnold WD, Burghes AH. Spinal muscular atrophy: development and implementation of potential treatments. *Ann Neurol*. 2013;74(3):348–362.
106. Le TT, Pham LT, Butchbach ME, et al. SMNDelta7, the major product of the centromeric survival motor neuron (SMN2) gene, extends survival in mice with spinal muscular atrophy and associates with full-length SMN. *Hum Mol Genet*. 2005;14(6):845–857.
107. Le TT, McGovern VL, Alwine IE, et al. Temporal requirement for high SMN expression in SMA mice. *Hum Mol Genet*. 2011;20:3578–3591.
108. Lutz CM, Kariya S, Patruni S, et al. Postsymptomatic restoration of SMN rescues the disease phenotype in a mouse model of severe spinal muscular atrophy. *J Clin Invest*. Aug 1, 2011;121(8):3029–3041.
109. Lee AJ, Awano T, Park GH, Monani UR. Limited phenotypic effects of selectively augmenting the SMN protein in the neurons of a mouse model of severe spinal muscular atrophy. *PLoS One*. 2012;7(9):e46353.

110. Martinez TL, Kong L, Wang X, et al. Survival motor neuron protein in motor neurons determines synaptic integrity in spinal muscular atrophy. *J Neurosci*. 2012;32(25):8703–8715.
111. Paez-Colasante X, Seaberg B, Martinez TL, Kong L, Sumner CJ, Rimer M. Improvement of neuromuscular synaptic phenotypes without enhanced survival and motor function in severe spinal muscular atrophy mice selectively rescued in motor neurons. *PLoS One*. 2013;8(9):e75866.
112. Cobb MS, Rose FF, Rindt H, et al. Development and characterization of an SMN2-based intermediate mouse model of spinal muscular atrophy. *Hum Mol Genet*. 2013;22(9):1843–1855.
113. Hao le T, Burghes AH, Beattie CE. Generation and characterization of a genetic zebrafish model of SMA carrying the human SMN2 gene. *Mol Neurodegener*. 2011;6(1):24.
114. Kolb SJ, Sutton S, Schoenberg DR. RNA processing defects associated with diseases of the motor neuron. *Muscle Nerve*. 2010;41:5–17.
115. Baumer D, Lee S, Nicholson G, et al. Alternative splicing events are a late feature of pathology in a mouse model of spinal muscular atrophy. *PLoS Genet*. 2009;5:e1000773.
116. Zhang Z, Pinto AM, Wan L, et al. Dysregulation of synaptogenesis genes antecedes motor neuron pathology in spinal muscular atrophy. *Proc Natl Acad Sci U S A*. 2013;110(48):19348–19353.
117. Cwik V. Childhood spinal muscular atrophy. *Continuum*. 1997;3:10–32.
118. Diek A, Saunders-Pullman R. Atypical presentation of late-onset Tay-Sachs disease. *Muscle Nerve*. 2013 Dec 10. doi:10.1002/mus.24146 [Epub ahead of print].
119. Prior TW. Spinal muscular atrophy diagnostics. *J Child Neurol*. 2007;22:952–956.
120. Buchthal F, Olsen PZ. Electromyography and muscle biopsy in infantile spinal muscular atrophy. *Brain*. 1970;93:15–30.
121. Hausmanowa-Petrusewicz I, Karwanska A. Electromyographic findings in different forms of infantile and juvenile proximal spinal muscular atrophy. *Muscle Nerve*. 1986;9:37–46.
122. Rudnik-Schoneborn S, Lutzenrath S, Borkowska J, Karwanska A, Hausmanowa-Petrusewicz I, Zerres K. Analysis of creatine kinase activity in 504 patients with proximal spinal muscular atrophy types I-III from the point of view of progression and severity. *Eur Neurol*. 1998;39:154–162.
123. Dubowitz V. Chaos in classification of the spinal muscular atrophies of childhood. *Neuromuscul Disord*. 1991;1:77–80.
124. Iannaccone ST, Bove KE, Vogler CA, Buchino JJ. Type 1 fiber size disproportion: morphometric data from 37 children with myopathic, neuropathic, or idiopathic hypotonia. *Pediatr Pathol*. 1987;7:395–419.
125. Wang CH, Finkel RS, Bertini ES, et al. Consensus statement for standard of care in spinal muscular atrophy. *J Child Neurol*. 2007;22:1027–1049.
126. Schroth MK. Special considerations in the respiratory management of spinal muscular atrophy. *Pediatrics*. 2009;123(suppl 4):S245–S249.
127. Iannaccone ST. Modern management of spinal muscular atrophy. *J Child Neurol*. 2007;22:974–978.
128. Bach JR, Baird JS, Plosky D, Navado J, Weaver B. Spinal muscular atrophy type 1: management and outcomes. *Pediatr Pulmonol*. 2002;34:16–22.
129. Bach JR, Bianchi C. Prevention of pectus excavatum for children with spinal muscular atrophy type 1. *Am J Phys Med Rehabil*. 2003;82:815–819.
130. Simonds AK. Ethical aspects of home long term ventilation in children with neuromuscular disease. *Paediatr Respir Rev*. 2005;6:209–214.
131. Mitchell I. Spinal muscular atrophy type 1: what are the ethics and practicality of respiratory support? *Paediatr Respir Rev*. 2006;7 (suppl 1):S210–S211.
132. Poruk KE, Davis RH, Smart AL, et al. Observational study of caloric and nutrient intake, bone density, and body composition in infants and children with spinal muscular atrophy type I. *Neuromuscul Disord*. 2012;22(11):966–973.
133. Fujak A, Kopschina C, Forst R, Gras F, Mueller LA, Forst J. Fractures in proximal spinal muscular atrophy. *Arch Orthop Trauma Surg*. 2010;130(6):775–780.
134. Evans GA, Drennan JC, Russman BS. Functional classification and orthopaedic management of spinal muscular atrophy. *J Bone Joint Surg Br*. 1981;63B:516–522.
135. Barsdorf AI, Sproule DM, Kaufmann P. Scoliosis surgery in children with neuromuscular disease: findings from the US National Inpatient Sample, 1997 to 2003. *Arch Neurol*. 2010;67:231–235.
136. Chandran S, McCarthy J, Noonan K, Mann D, Nemeth B, Guiliani T. Early treatment of scoliosis with growing rods in children with severe spinal muscular atrophy: a preliminary report. *J Pediatr Orthop*. 2011;31:450–454.
137. McElroy MJ, Shaner AC, Crawford TO, et al. Growing rods for scoliosis in spinal muscular atrophy: structural effects, complications, and hospital stays. *Spine (Phila Pa 1976)*. 2011;36:1305–1311.
138. Fujak A, Raab W, Schuh A, Richter S, Forst R, Forst J. Natural course of scoliosis in proximal spinal muscular atrophy type II and IIIa: descriptive clinical study with retrospective data collection of 126 patients. *BMC Musculoskelet Disord*. 2013;14:283.
139. Grondard C, Biondi O, Armand AS, et al. Regular exercise prolongs survival in a type 2 spinal muscular atrophy model mouse. *J Neurosci*. 2005;25:7615–7622.
140. Bladen CL, Thompson R, Jackson JM, et al. Mapping the differences in care for 5,000 spinal muscular atrophy patients, a survey of 24 national registries in North America, Australasia and Europe. *J Neurol*. 2014;261(1):152–163.
141. Monani UR, Coovert DD, Burghes AH. Animal models of spinal muscular atrophy. *Hum Mol Genet*. 2000;9(16):2451–2457.
142. Crawford TO, Paushkin SV, Kobayashi DT, et al. Pilot study of biomarkers for spinal muscular atrophy trial group. Evaluation of SMN protein, transcript, and copy number in the biomarkers for spinal muscular atrophy (BforSMA) clinical study. *PLoS One*. 2012;7(4):e33572.
143. Finkel RS, Crawford TO, Swoboda KJ, et al. Pilot study of biomarkers for spinal muscular atrophy trial group. Candidate proteins, metabolites and transcripts in the biomarkers for spinal muscular atrophy (BforSMA) clinical study. *PLoS One*. 2012;7(4):e35462.
144. Cano SJ, Mayhew A, Glanzman AM, et al., on behalf of the International Coordinating Committee for SMA Clinical Trials Rasch Task Force. Rasch analysis of clinical outcome measures in spinal muscular atrophy. *Muscle Nerve*. 2014;49(3):422–430.
145. Rothwell E, Anderson RA, Swoboda KJ, Stark L, Botkin JR. Public attitudes regarding a pilot study of newborn screening for spinal muscular atrophy. *Am J Med Genet A*. 2013;161A(4):679–686.

146. Pyatt RE, Prior TW. A feasibility study for the newborn screening of spinal muscular atrophy. *Genet Med*. 2006;8:428–437.
147. Gitlin JM, Fischbeck K, Crawford TO, et al. Carrier testing for spinal muscular atrophy. *Genet Med*. 2010;12:621–622.
148. Prior TW, Snyder PJ, Rink BD, et al. Newborn and carrier screening for spinal muscular atrophy. *Am J Med Genet A*. 2010;152A:1608–1616.
149. Butchbach ME, Singh J, Thorsteinsdóttir M, et al. Effects of 2,4-diaminoquinazoline derivatives on SMN expression and phenotype in a mouse model for spinal muscular atrophy. *Hum Mol Genet*. 2010;19(3):454–467.
150. Naryshkin N, Narasimhan J, Dakka A, et al. T.P.7 Small molecule compounds correct alternative splicing of the SMN2 gene and restore SMN protein expression and function. *Neuromuscul Disord*. 2012;22:848.
151. Corti S, Nizzardo M, Nardini M, et al. Neural stem cell transplantation can ameliorate the phenotype of a mouse model of spinal muscular atrophy. *J Clin Invest*. 2008;118(10):3316–3330.
152. Corti S, Nizzardo M, Nardini M, et al. Embryonic stem cell-derived neural stem cells improve spinal muscular atrophy phenotype in mice. *Brain*. 2010;133(2):465–481.
153. Corti S, Nizzardo M, Simone C, et al. Genetic correction of human induced pluripotent stem cells from patients with spinal muscular atrophy. *Sci Transl Med*. 2012;4(165):165ra162.
154. Burghes AH, McGovern VL. Antisense oligonucleotides and spinal muscular atrophy: skipping along. *Genes Dev*. 2010;24:1574–1579.
155. MacKenzie A. Sense in antisense therapy for spinal muscular atrophy. *N Engl J Med*. 2012;366(8):761–763.
156. Porensky PN, Burghes AH. Antisense oligonucleotides for the treatment of spinal muscular atrophy. *Hum Gene Ther*. 2013;24(5):489–498.
157. Sivanesan S, Howell MD, Didonato CJ, Singh RN. Antisense oligonucleotide mediated therapy of spinal muscular atrophy. *Transl Neurosci*. 2013;4(1).
158. Valori CF, Ning K, Wyles M, et al. Systemic delivery of scAAV9 expressing SMN prolongs survival in a model of spinal muscular atrophy. *Sci Transl Med*. 2010;2(35):35ra42.
159. Dominguez E, Marais T, Chatauret N, et al. Intravenous scAAV9 delivery of a codon-optimized SMN1 sequence rescues SMA mice. *Hum Mol Genet*. 2011;20(4):681–693.
160. Benkhelifa-Ziyyat S, Besse A, Roda M, et al. Intramuscular scAAV9-SMN injection mediates widespread gene delivery to the spinal cord and decreases disease severity in SMA mice. *Mol Ther*. 2013;21(2):282–290.

CHAPTER

90

Pain Genetics

William Renthal

The University of Texas Southwestern Medical Center, Dallas, TX, USA

INTRODUCTION

Pain is defined by the International Association for the Study of Pain as an "unpleasant sensory and emotional experience associated with actual or potential tissue damage." The process by which pain is experienced has intrigued physicians and philosophers for millennia. Rene Descartes in the 17th century was one of the first to postulate that a peripheral pain stimulus traveled to the brain for the actual perception of this experience. There is now ample experimental evidence that pain-specific nerve fibers detect and transmit noxious stimuli at the site of injury to the brain for processing and initiating the appropriate motor response (e.g., withdrawing hand from fire). The survival benefit of an efficient pain-response pathway to damaging environmental stimuli is clear across multiple species, as well as evident from genetic mutations in humans who cannot experience pain. Many of these patients succumb to tragic injuries and burns because of their inability to sense or to process these stimuli. This chapter reviews the basic pathways through which pain perception occurs as well as the clinical implications of gene mutations in these pathways. Emphasis will be placed on the clinical characteristics, diagnosis, and molecular genetics of the Mendelian pain disorders, as well as genetic insights from patients with chronic pain and from animal pain models.

NEUROBIOLOGY OF PAIN

Nociception

Nociception is the detection of painful stimuli. Specialized neurons in the dorsal root ganglia (DRG) or the trigeminal ganglia project into skin and soft tissue to detect extremes of heat, cold, mechanical and chemical signals, and alert the body of potential dangers.[1] These neurons are pseudo-unipolar in structure, which is unique in that their axons project both distally to the skin and proximally to the spinal cord. This structure permits noxious stimuli to be detected along the length of the fiber and is clinically important because it allows pain treatments to be targeted either topically to the skin surface or directly to the epidural or intrathecal space.

Nociception occurs via membrane-bound neuronal receptors (nociceptors) at the nerve terminals that are activated by specific noxious stimuli. This is followed by the conversion of that signal into an action potential and its propagation to the dorsal horn of the spinal cord. Nociceptors are as diverse as the noxious stimuli themselves. The detection of noxious temperatures, typically above 43 °C, activate transient receptor potential (TRP)-type proteins that are located in "heat responsive" C and Aδ fibers (described below).[2] Studies in mouse pain models have shown that the TRP channel, TRPV1, is required for the detection of noxious heat. Interestingly, capsaicin, the compound that makes chili peppers "hot" also directly activates TRPV1 and is used in chronic pain management to deplete pain fibers of pronociceptive neurotransmitters. There are multiple TRP family members that are activated at distinct temperatures, so it is thought that these channels work together with TRPV1 to detect innocuous and noxious heat signals.[2–4]

Cold sensation is mediated by a distinct set of proteins that typically respond to temperatures less than 25 °C. One well-described cold sensor is TRPM8, which is thought to mediate much of cold sensation between 10 and 30 °C.

Rosenberg's Molecular and Genetic Basis of Neurological and Psychiatric Disease
http://dx.doi.org/10.1016/10.1016/B978-0-12-410529-4.00090-5

Mice lacking TRPM8, however, do not lose all cold sensation, especially at much lower temperatures, so additional cold sensors are still being characterized. Menthol also directly activates TRPM8, which is likely the mechanism by which it is perceived as "cool".[2]

Similar to heat and cold nociception, mechanical nociception is thought to be mediated by the free nerve endings of Aδ and C fibers. Innocuous mechanosensation, however, is controlled by large diameter Aβ fibers that terminate in Pacinian corpuscles, Merkle cells, and hair follicles. The molecular receptors that respond to noxious mechanical stimuli have been difficult to identify because of challenges in developing experimental systems that accurately model *in vivo* mechanosensation. Several candidate genes have been identified though, and investigations are underway to determine which are important for noxious mechanical sensation in mouse models.

Chemical stimuli encompass a variety of molecules, including foods (e.g., chili peppers, wasabi), environmental toxins (e.g., smoke, formalin), and importantly, tissue damage. TRP channels, especially TRPA1, detect a wide range of these chemical stimuli. At sites of tissue damage or inflammation, release of prostaglandins, bradykinins, neuropeptides, and neurotransmitters mediate the hyperalgesia and allodynia experienced at sites of tissue damage or inflammation. Importantly, chemical nociception not only serves as an alarm to damaging infectious or inflammatory substances, it also plays an active role in the immune response itself.[5]

Propagation of the Pain Signal

Once a noxious stimulus is detected by the nociceptor, it must be transduced to the spinal cord via an action potential (Figure 90.1). Similar to any neuron in the nervous system, this occurs primarily through the interplay between sodium, potassium, and calcium channels. Interestingly, certain isoforms of these ion channels are highly enriched in DRG neurons and thus play an essential role in pain transmission. Indeed, several of these ion channels have been implicated in human pain disorders. For example, loss-of-function mutations in sodium channel $Na_v1.7$ result in a dangerous condition in which patients are unable to recognize noxious stimuli. Conversely, gain-of-function mutations in $Na_v1.7$ result in hypersensitivity to pain. $Na_v1.8$ is another sodium channel highly enriched in C fibers and is responsible for transducing noxious cold sensation. Potassium channels, such as the K_v7 channels and the HCN cyclic nucleotide-gated channels, have been shown to play a role in pain transmission and neuropathic pain. The P/Q-type calcium channels have been implicated in familial hemiplegic migraine, suggesting a role in pain processing. These human pain syndromes will be discussed in detail below. Importantly, several of these channels are under intense investigation as targets for pharmaceutical intervention aimed at reducing DRG excitability and thus pain transmission.

Anatomy of Pain Pathways

The dorsal root and trigeminal ganglia support nociception in the body and face, respectively. The DRG are located along the vertebral column, and the trigeminal ganglia are located intracranially. In order to relay a wide variety of painful stimuli at appropriate speed to the spinal cord, these nociceptive axons form functionally distinct types of fibers, Aδ and C fibers. Aδ fibers are thinly myelinated neurons that quickly transmit (2–30 meters per second) the

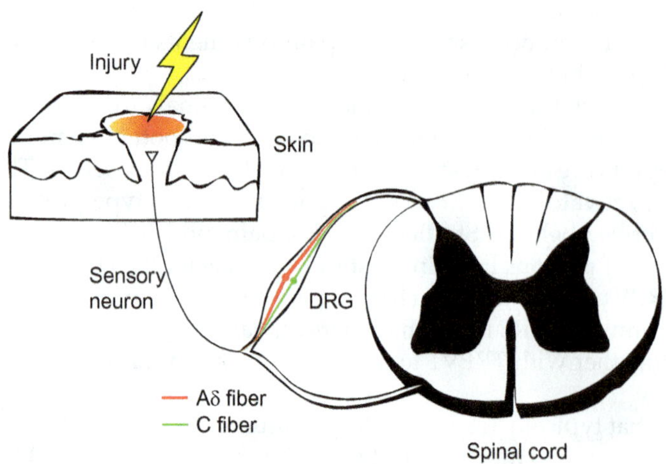

FIGURE 90.1 Pain detection. At site of skin injury, the sensory neuron transmits the pain signal via myelinated Aδ fibers (red) and unmyelinated C fibers (green) whose cell bodies originate in the dorsal root ganglion (DRG) and synapse in the dorsal horn of the spinal cord.

initial pain response to noxious mechanical or heat stimuli. C fibers are comprised of small, unmyelinated neurons with slow conduction speeds (<2 meters per second) that are responsible for the delayed burning associated with pain after the injury. Innocuous stimuli, such as light touch, are carried by large, rapidly conducting (30–70 meters per second) myelinated Aβ fibers that can also modulate the intensity of noxious stimuli at the level of the spinal cord (e.g., rubbing your elbow after hitting it on a counter).[1,6] These peripheral fibers all track into the dorsal horn of the spinal cord in an organized fashion where they synapse with second-order neurons that then ascend to the brain (Figure 90.1). The cell bodies of second-order neurons make up the gray matter of the spinal cord, which is subdivided based on microscopic morphology into laminae. These second-order neurons then transmit pain information to the brain through five distinct groups of long tracts: the spinothalamic, spinoreticular, spinomesencephalic, cervicothalamic, and spinohypothalamic tracts (Figure 90.2).

The lateral spinothalamic tract, which is made up of neurons from laminae I and V–VII, is responsible for transmitting most acute noxious stimuli to the brain. This tract crosses and ascends within a couple of spinal levels of its origin along the lateral funiculus. Lesions of this tract result in loss of pain and temperature sensation on the side contralateral to the lesion. The spinothalamic tract fibers predominantly terminate in the ventral posterolateral (VPL) nucleus of the thalamus along with smaller contributions to other thalamic nuclei. This information is then transmitted to the cortex by thalamocortical projections. The cervicothalamic tract has a similar function, but originates mainly from lamina IV cells at cervical levels C1–C4. The spinoreticular tract arises from the neurons in laminae VII and VIII and synapses in the reticular formations of the medulla and pons before reaching the thalamus. This tract is thought to transmit information important for the more primitive or automatic response to pain. The spinohypothalamic tract originates primarily from laminae I, V, VII, and X, and is thought to play a role in the autonomic, endocrine, and possibly even emotional responses to painful stimuli. These neurons project to the hypothalamus where they decussate again and continue on to the ipsilateral thalamus and brainstem. The spinomesencephalic tract is made up of laminae I and IV–VI neurons that project to multiple midbrain structures including the periaqueductal gray (PAG). This connection to the PAG is thought to contribute to the affective component of pain as well as link to the descending pain modulatory pathways.[1,7]

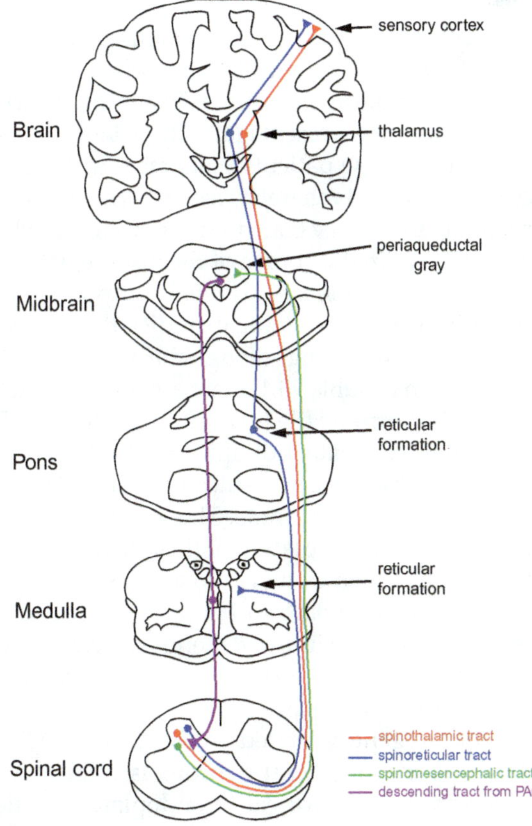

FIGURE 90.2 **Ascending and descending pain pathways.** The cell bodies of the spinothalamic (red), spinoreticular (blue), and spinomesencephalic (purple) tracts reside in the dorsal horn. Their axons cross in the spinal cord anterior commissure and then ascend to brainstem or thalamic structures as shown. A descending pain modulatory pathway begins with cell bodies in the periaqueductal gray (PAG), which receive inputs from multiple forebrain structures. This pathway synapses in the rostral ventromedial medulla where second-order neurons project back onto the dorsal horn to modulate pain. Circles = cell bodies; triangles = terminal synapses.

Descending Modulation

The PAG plays a crucial role in central pain integration. Electrical stimulation of this area causes analgesia to noxious stimuli. Moreover, microinjection of opiates into the PAG activates this region and leads directly to analgesia. This demonstrates that opiates exert their analgesic effects through potent modulation of both central and peripheral nervous systems. One mechanism by which the PAG causes analgesia is by inhibiting the ascending transmission of pain at the level of the spinal cord. The PAG accomplishes this by first synapsing in the rostral ventromedial medulla (RVM), which includes the serotonergic cells in the dorsal raphe magnus and the neighboring reticular formation. Axons from the RVM form the descending long tracts that synapse in the dorsal horn of the spinal cord and directly inhibit laminae I, II, and V neurons. Because the PAG receives ascending pain information directly from the dorsal horn as well as multiple inputs from limbic forebrain structures, such as the anterior cingulate, insular cortex, and amygdala, it is perfectly situated to integrate multiple environmental and emotional factors into the ultimate perception of pain. Indeed, it is thought the PAG plays a key role in mediating the absent or strongly attenuated pain response described by some soldiers experiencing combat-related trauma.

MENDELIAN DISORDERS OF PAIN

Many of the key breakthroughs in understanding the mechanisms of pain transduction have come from families with genetic mutations that lead to defects in these pathways. These heritable disorders of pain will be discussed with a focus on their clinical presentation, diagnosis, molecular genetics, and treatment.

Hereditary Sensory and Autonomic Neuropathies

The hereditary sensory and autonomic neuropathies (HSAN) are a set of peripheral neuropathies that are primarily characterized by axonal damage to sensory and autonomic neurons, which distinguishes them from the motor and sensory neuropathies.[1,8,9] HSANs are grouped into five subtypes based on their mode of inheritance, symptoms, and usual age of onset (Table 90.1). Loss of pain sensation is common to each of the HSANs, albeit to varying degrees.

HSAN1

Hereditary sensory and autonomic neuropathy 1 is the most common HSAN and is an autosomal dominant syndrome that usually begins in early adulthood. The mean age of onset is 29, but can begin between the ages of 12 and 70 years. The cardinal feature is loss of pain and temperature sensation in the distal extremities, which spreads proximally. This often presents initially as lancinating leg pains but progresses to painless ulcerations and Charcot joints, which make these patients vulnerable to developing lower extremity infections and limb damage. Unlike the other HSANs, HSAN1 is characterized by weakness in addition to sensory loss. This weakness is typically more distal than proximal, and is associated with muscle atrophy, which can confuse the clinical picture with Charcot–Marie–Tooth disease. Later in the course of the disease, vibration, proprioception, and light touch are affected in addition to pain, but to a lesser degree. Associated symptoms include pes cavus and loss of reflexes. The diagnosis of HSAN is supported by electrophysiologic studies that show primarily axonal sensory and often motor peripheral neuropathy. Together, these findings along with a family history, are diagnostic for HSAN1.

As shown in Table 90.1, there are several functionally distinct mutations that give rise to clinically distinct versions of HSAN1 (termed HSAN1A through D). The *SPTLC1* gene encodes for a serine palmitoyltransferase, which is a key player in the formation of lipid rafts in cell membranes that facilitate signal transduction. Mutations in *SPTLC1* and *SPTLC2* are thought to result in the accumulation of two toxic sphingolipid metabolites, 1-deoxysphinganine and 1-deoxymethylsphinganine. These metabolites are toxic to primary sensory neurons, and to a lesser extent, to motor neurons. The accumulation of these metabolites in serum inversely correlates with the serine : alanine ratio. This observation has been extended to a mouse model of HSAN1 in which L-serine diet supplementation both normalized the serine : alanine ratio and improved peripheral sensation. A small pilot of 14 HSAN1 patients supplemented with L-serine also found improvements in their serine : alanine ratio, as well as some subjective reports of improved sensation and hair growth.[10] However, these data are anecdotal as there was no placebo arm, so L-serine supplementation remains experimental.

Another gene mutated in HSAN1 is *ATL1*, which encodes the Atlastin-1 GTPase. While the function of Atlastin-1 in peripheral neurons is unclear, it is highly expressed in the endoplasmic reticulum of cells, and in *Drosophila*, it is critical to endoplasmic reticulum structure and function. In cortical neurons, this protein is

TABLE 90.1 Hereditary Sensory and Autonomic Neuropathies

Disorder	OMIM	Locus/gene	Protein	Inheritance/onset	Characteristics
HSAN1-A	162400	9q22.31/*SPTLC1*	Serine palmitoyltransferase, long chain 1	AD/adolescence	Loss of pain and temperature; ulcerations; variable distal weakness
HSAN1-C	613640	14q24.3/*SPTLC2*	Serine palmitoyltransferase, long chain 2	AD/adulthood	Loss of pain and temperature; ulcerations; variable distal weakness
HSAN1-D	613708	14q22.1/*ATL1*	Atlastin-1	AD/adulthood	Distal axonal sensory neuropathy affecting all modalities; ulcerations
HSAN1-E	614116	19p13.2/*DNMT1*	DNA methyltransferase 1	AD/adulthood	Sensory loss to all modalities; progressive hearing loss; early-onset dementia
HSAN2-A	201300	12p13.33/*WNK1*	WNK lysine deficient protein kinase 1	AR/childhood	Loss of pain, temperature, tactile sensation, ulcerations; mainly hands and feet
HSAN2-B	613115	5p15.1/*FAM134B*	Family with sequence similarity 134, member B	AR/childhood	Impaired nociception and progressive ulcerations of hands and feet
HSAN2-C	614213	2q37.3/*KIF1A*	Kinesin family member 1A	AR/childhood	Distal numbness of hand and feet; ulcerations; mild distal weakness
HSAN3	223900	9q31.3/*IKBKAP*	Inhibitor of kappaB	AR/congenital	Indifference to pain and temperature, neuropathic joints, alacrima, autonomic dysfunction
HSAN4	256800	1q23.1/*NTRK1*	Neurotrophic tyrosine kinase, type 1	AR/congenital	Insensitivity to pain and temperature; neuropathic joints; anhidrosis; hyperpyrexia, autonomic dysfunction
HSAN5	608654	1p13.2/*NGFB*	Nerve growth factor β	AR/congenital	Loss of pain and temperature, neuropathic joints, ulceration

Abbreviations: AD, autosomal dominant; AR, autosomal recessive; HSAN, hereditary sensory autonomic neuropathy; OMIM, Online Mendelian Inheritance of Man.

important in axonal elongation, which may implicate this process in HSAN1 pathophysiology. While the precise molecular genetics remain under investigation, it is interesting to note that mutations in the same gene have been described in hereditary spastic paraplegia. This disease is associated with upper motor neuron axonal degeneration as well as peripheral neuropathy; however, it remains unclear how such distinct phenotypes can arise from the same mutation.

A subtype of HSAN1 that is associated with early-onset dementia and hearing loss is caused by heterozygous mutations in the *DNMT1* gene. The *DNMT1* gene encodes a DNA methyltransferase that is ubiquitously expressed and plays an essential role in establishing DNA methylation patterns during development. These intricate patterns of DNA methylation help coordinate the correct level and timing of specific gene expression patterns in every cell. The mechanisms by which mutations in *DNMT1* result in neuropathy, hearing loss, and dementia remain unclear, but the mutation alters the way the enzyme binds to DNA and likely leads to transcriptional dysregulation that over time causes axonal damage.

HSAN2

Hereditary sensory and autonomic neuropathy 2 is also characterized by the loss of pain, temperature, pressure, and touch sensation, but has recessive inheritance and typically begins in early childhood. Clinical presentation and nerve conduction studies demonstrate that HSAN2 is predominantly a sensory neuropathy, but there may be mild hypotonia on exam or slightly reduced motor conduction velocity.[8,9]

The molecular genetics from families with HSAN2 have revealed three diverse genes, *WNK1*, *FAM134B*, and *KIF1A*. The *WNK1* gene encodes a serine-threonine kinase that is highly enriched in neurons, especially in dorsal root ganglia. A mutation within exon 8 of *WNK1* (previously thought to be its own single-exon gene named *HSN2*) results in loss of function. The WNK1 protein has been described to have multiple functions, such as regulating intracellular ion flux and limiting the expression of TrpV1, a cell surface mechanical and thermal nociceptor. Somewhat paradoxically, patients who are heterozygous for *WNK1* mutations become hypersensitive to painful stimuli. One hypothesis is that the increased levels of TrpV1 seen in *WNK1* heterozygous patients mediate pain hypersensitivity, whereas in homozygous patients the levels of TrpV1 reach excitotoxic levels and ultimately lead to peripheral sensory neuron loss. Further research is needed to clarify this mechanism.

Mutations in *FAM134B* have also been reported in families with HSAN2. The normal protein is highly expressed in peripheral sensory neurons, but its function is not well understood. Studies in dorsal root ganglion cells suggest a role in Golgi apparatus structure and cell survival, since its experimental reduction leads to apoptosis. Further research is needed to determine if an analogous process explains sensory neuron loss in patients with *FAM134B* mutations.

The third mutation known to result in HSAN2 is in the molecular transport protein, kinesin family member 1A (KIF1A). The *KIF1A* mutations reported in HSAN2 reduce the protein's ability to bind to and transport synaptic vesicles. It is thought that, over time, reduced transport of synaptic vesicles leads to neuronal loss. Interestingly, the KIF1A protein directly interacts with WNK1, but the functional significance of this relationship is still under investigation.

HSAN3

Hereditary sensory and autonomic neuropathy 3, also known as Riley–Day syndrome or familial dysautonomia, has an autosomal recessive inheritance pattern and is found almost exclusively in Ashkenazi Jews. The most pronounced symptoms of this syndrome are related to a severe autonomic neuropathy. Symptoms begin in infancy with gastrointestinal dysmotility, decreased or increased tearing, bladder dysfunction, and are progressive. Patients can experience dysautonomic storms, in which bouts of parasympathetic hyperactivity, usually involving nausea and vomiting, alternate with sympathetic hyperactivity such as tachycardia, hypertension, and irritability. Orthostatic hypotension, temperature dysregulation, and pupil dilation are often noted in these patients. In addition to dysautonomia, these patients also have severe peripheral neuropathy manifested as sensory loss of pain, temperature, and vibration. This occurs because of progressive small myelinated and unmyelinated axonal loss. The causative mutation is the *IKBKAP* gene, which encodes the IkB kinase complex-associated protein, ELP1. ELP1 is a ubiquitously expressed protein that has distinct actions in the nucleus and cytoplasm. In the nucleus it associates with RNA polymerase II and facilitates transcriptional elongation. In the cytoplasm it acts as a scaffold protein to facilitate signal transduction pathways. Cell culture studies have shown a role for ELP1 in the expression of numerous important neuronal genes, some of which control cytoskeleton dynamics and may contribute to the HSAN3 phenotype. Despite our increasing understanding of the molecular genetics of HSAN3, management of this syndrome is primarily supporting the autonomic dysregulation.

HSAN4

HSAN4 is also known as congenital insensitivity to pain with anhidrosis (CIPA), and is inherited in an autosomal recessive pattern. Symptoms begin in infancy and include complete loss of pain and temperature sensation, anhidrosis and impaired thermoregulation, inadvertent self-mutilation behavior, and sometimes mild mental retardation. Multiple families with this syndrome have mutations in the *NTRK1* gene. This gene encodes the protein TrkA, which is the primary cell surface receptor for nerve growth factor (NGF). NGF signaling plays a critical role in the survival of autonomic and peripheral nerves, and thus there is a complete loss of small unmyelinated fibers and sympathetic innervation of sweat glands. There is likely neuronal loss in the central nervous system (CNS) as well, as these patients often have cognitive deficits. Interestingly, in addition to mediating neuronal development and survival, NGF plays a key role in pain and itch sensation in normal individuals; however this is not felt to be related to the pathogenesis of HSAN4. Treatment at this time is symptomatic.

HSAN5

Hereditary sensory and autonomic neuropathy 5 is an autosomal recessive syndrome that is phenotypically similar to HSAN4/CIPA. Patients with HSAN5 can be distinguished from HSAN4 because they have hypohidrosis instead of anhidrosis and do not have signs of dementia. Nerve conduction studies are normal, but nerve biopsy shows loss of both unmyelinated and myelinated small-diameter fibers. Mutations in the *NGF* gene, which encodes the protein NGF-beta, cause HSAN5. NGF-beta is the ligand for the TrkA protein mutated in HSAN4 and thus explains the similarity between the two disorders. The mutations in NGF-beta that have been described in HSAN5 reduce overall levels of NGF-beta and also block its interaction with the TrkA co-receptor, $p75^{NTR}$. In addition to co-signaling with TrkA, $p75^{NTR}$ has an independent signaling pathway that helps mediate activation of peripheral nerve fibers in models of acute and inflammatory pain. Together, lower levels of NGF-beta and decreased TrkA/$p75^{NTR}$ signaling likely explain why HSAN5 is symptomatically similar but less severe than HSAN4, in which there is absent TrkA signaling. Similar to HSAN4, treatment is symptomatic.

Na_v1.7 Channelopathies

Voltage-gated sodium channels are fundamental elements of action potentials and neuronal transmission. These channels are made up of alpha and beta subunits, which form a voltage-sensitive pore and a modulatory domain, respectively. There are nine known isoforms of the alpha subunit, which make up the Na_v1.1–Na_v1.9 channels. Each isoform has a distinct expression pattern and electrophysiological properties. Three sodium channel isoforms, Na_v1.7, Na_v1.8, and Na_v1.9, are selectively enriched in sensory neurons.[11,12] Na_v1.7, which is encoded by the *SCN9* gene, is expressed primarily in DRG and sympathetic ganglion neurons. Several mutations in *SCN9* have been described, which, depending on the functional consequence on Na_v1.7 function, can result in complete indifference to pain or intractable attacks of debilitating pain (Table 90.2). Na_v1.8 and 1.9, which are encoded by the *SCN10A* and *SCN11A* genes, are also expressed primarily on nociceptive sensory neurons and are thought to play a key role in pain processing. Indeed, *SCN10A* gain-of-function mutations have recently been identified in patients with painful peripheral neuropathies.[13] Mutations in *SCN11A* have not yet been reported in human pain disorders, but Na_v1.9 activity is thought to play a key role in mediating inflammatory pain.

Congenital Indifference to Pain

Congenital indifference to pain (CIP) is an autosomal recessive disorder in which patients have a specific inability to perceive pain but intact sensation to other stimuli such as vibration, proprioception, and light touch. Motor, autonomic, and cognitive modalities also remain unaffected. The first described case was reported in 1932 of a circus performer who acted as a human pincushion. Other patients have been described as walking on hot coals or performing crucifixion re-enactments. Similar to the recessive HSANs, these patients often present in preschool or younger with repeated injuries, such as accidental tongue or finger biting, trauma, or burns. Later in life, skin ulcers and Charcot joints are debilitating complications of this disorder.

TABLE 90.2 Na_v1.7 Channelopathies

Disorder	OMIM	Locus/gene	Protein	Mutation	Inheritance/onset	Characteristics
CIP	243000	2q24.3/*SCN9A*	Na_v1.7 sodium channel	Loss of function	AR/childhood	Loss of pain sensation, but intact to all other modalities; prone to accidental burns or self-mutilation; no autonomic or motor deficits
PE	133020	2q24.3/*SCN9A*	Na_v1.7 sodium channel	Gain of function	AD/childhood	Episodic symmetrical vasodilation, erythema, and burning pain, often in lower extremities; exercise is trigger
PEPD	167400	2q24.3/*SCN9A*	Na_v1.7 sodium channel	Gain of function	AD/congenital	Episodic pain attacks, often involving rectal, ocular, and mandibular pain; carbamazepine often helps symptoms

Abbreviations: AD, autosomal dominant; AR, autosomal recessive; CIP, congenital indifference to pain; OMIM, Online Mendelian Inheritance of Man; PE, primary erythermalgia; PEPD, paroxysmal extreme pain disorder.

SCN9A was discovered as the gene responsible for CIP with help of three consanguineous families in Pakistan in 2006.[14] Three nonsense mutations were identified at this time (S459X, I757X, and W897X). Additional *SCN9A* mutations have been reported in several other families in Europe and Canada. Using cultured cells expressing wild-type or mutant $Na_v1.7$, all of the human mutations reported so far cause a loss of function of $Na_v1.7$.[11] Diagnosis is suspected based on clinical history, exam, and electrophysiology, but genetic testing is also available for confirmation. Treatment at this time is focused on educating these patients and their parents about taking appropriate safety precautions.

Primary Erythromelalgia

Primary erythromelalgia (PE) is an autosomal dominant disorder in which patients develop attacks of severe burning pain, erythema, and heat and burning pain in the feet and sometimes hands and parts of the face. These attacks can be precipitated by hot weather, fever, exercise, socks or tight shoes, and sometimes even spicy or hot food or drink. The attacks are variable in frequency, and can occur infrequently, multiple times a day, or even be constant as the disease progresses. Diagnosis of this disorder is made by the history of temperature sensitivity and characteristic pain attacks with swelling and erythema. Nerve electrophysiologic studies may be abnormal, but there is no characteristic finding. Skin biopsies may also be abnormal but are nonspecific. Genetic testing is available for definitive diagnosis.

The molecular genetics of PE is an elegant example of symmetry in pain biology. PE is caused by missense mutations in *SCN9A*, the same gene that is mutated in CIP.[15] The opposite phenotypes are explained by the functional result of the different mutations. The mutations in CIP result in loss of protein function, whereas in PE, they result in a gain of function. Specifically, in PE the missense mutations occur in the pore-forming domain and result in a hyperactive channel that is activated at lower levels of depolarization. This results in the transmission of pain sensation spontaneously and significant allodynia. One fascinating feature of this disorder is the pain localization and lack of autonomic involvement. Some insight into this comes from *in vitro* studies showing that the hyperexcitability of mutant $Na_v1.7$ requires a second type of sodium channel, $Na_v1.8$, to be expressed in the same neurons. Since $Na_v1.8$ is almost exclusively expressed in small nociceptive sensory neurons, this may explain the lack of autonomic involvement in PE.[12] Treatment of PE is challenging. Patients will often soak their feet in ice water during these attacks to manage the pain. Neuropathic pain medications such as gabapentin or pregabalin can be tried, but most patients do not get significant relief.

Paroxysmal Extreme Pain Disorder

Paroxysmal extreme pain disorder (PEPD) is an autosomal dominant disorder involving paroxysmal pain attacks involving skin flushing. Pain attacks most often occur in the mandible and orbit, but rectal pain is also very common. Indeed, this disorder was formerly termed familial rectal pain disorder. These paroxysms of pain are described as a sudden onset of excruciating pain that localizes to the previously described areas but can also extend to the entire body. There are also reports of tonic seizures accompanying these pain attacks, typically in children. The attacks are usually preceded by a trigger, such as cold temperature, defecation, or crying. While the attacks are more frequent and more severe in younger patients, PEPD remains a lifelong illness. Diagnosis is made based on careful clinical history, as electrophysiology, and skin biopsies are not specific. Gene testing is now available for definitive diagnosis, although there is a small cohort of PEPD patients for whom the causal mutation is still unclear.

Similar to PE, the molecular genetics of PEPD converges on *SCN9A* missense mutations that result in a hyperactive sodium channel.[16] However, functional analysis of the described *SCN9A* mutations reveals a distinct biophysical mechanism of action that may explain the clinical differences. In PEPD, the *SCN9A* mutations impair sodium channel inactivation, which prolongs the action potential and thus the downstream intensity of pain signaling. Mutations in PE, however, decrease the threshold of channel activation making them more likely to fire rather than firing for prolonged durations. Another important feature that is unique to PEPD is that many of these patients get significant relief, albeit never complete, from carbamazepine. Carbamazepine is a commonly used anticonvulsant, which acts by antagonizing sodium channels. The putative mechanism of relief in PEPD patients is thought to be via peripheral neuron sodium channel antagonism.

Familial Hemiplegic Migraine

Familial hemiplegic migraine (FHM) is an autosomal dominant headache disorder characterized by migraine with aura and reversible motor weakness that typically resolves within minutes to hours. Hemiplegia is often associated with reversible visual, sensory, or speech disturbances as well. In rare cases, patients have been reported to have loss of consciousness, coma, or seizures. A positive family history is necessary for diagnosis of FHM, but there are cases of sporadic mutations in the same causal genes leading to new families carrying the disease. Most cases of FHM are predominantly characterized by the headache disorder, but some patients also have cerebellar ataxia and seizures secondary to the same genetic mutation. Careful workup must be performed to rule out other neurological causes prior to diagnosis of FHM. FHM is very rare, with less than 200 families and 200 sporadic cases reported.

Three genes have been classically associated with FHM and a fourth was recently identified (Table 90.3).[17] FHM1 families have gain-of-function missense mutations in the *CACNA1A* gene, which codes for the alpha-1A subunit of the P/Q-type calcium channel $Ca_v2.1$. These mutations increase calcium ion influx and therefore neuronal excitability. This has been shown to reduce the threshold for cortical spreading depression, the wave of cortical depolarization thought to mediate migraine with aura. FHM2 is caused by missense mutations in *ATP1A2*, which encodes the alpha2 subunit of a glial Na-K ATPase. Interestingly, a mutation in the same gene has been reported in a family with basilar-type migraine as well. *ATP1A2* mutations are loss-of-function mutations that are thought to result in slower glial reuptake of glutamate and thus hyperexcitability. FHM3 is caused by mutations in *SCN1A*, which encodes an alpha subunit of the voltage-gated sodium channel, $Na_v1.1$. These mutations are thought to accelerate recovery from channel inactivation, making neurons hyperexcitable. Mutations in $Na_v1.1$ have also been associated with epilepsy. The penetrance for the more common FHM1 and FHM2 is 60–80%; in the five reported cases of FHM3 the penetrance is 100%. A fourth gene identified in FHM is *PRRT2*, which encodes a membrane protein that interacts with the synaptosomal-associated protein, SNAP25, and therefore may regulate calcium channel activity or synaptic vesicle release. The precise functional significance of this mutation is still under investigation. While *CACNA1A*, *ATP1A2*, *SCN1A*, and *PRRT2* are the best described causal mutations in FHM, screening for these genes in multiple families explained less than 14% of the cases. Thus, additional FHM genes remain to be discovered.

Treatment of hemiplegic migraine is similar to treatment of typical migraine with aura. Abortive agents such as NSAIDs, Tylenol, and antiemetics are the recommended initial treatment. The use of triptans is controversial, and while many physicians believe it to be safe, the manufacturers label it as contraindicated for hemiplegic migraine. This is out of concern that a pharmacologic vasoconstrictor could exacerbate the vasoconstriction already occurring in the migraine and lead to stroke. In severe cases, hemiplegic patients may need to be admitted to the hospital to provide supportive care for fever, coma, or seizures. For migraine prophylaxis, verapamil and acetazolamide are often tried first, but lamotrigine, amitriptyline, and topiramate may also be effective.

TABLE 90.3 Familial Hemiplegic Migraine

Disorder	OMIM	Locus/gene	Protein	Mutation	Inheritance/onset	Characteristics
FHM1	141500	19p13.2/*CACNA1A*	$Ca_v2.1$ calcium channel	Gain of function	AD/childhood to young adult	Hemiparesis/plegia for minutes to hours followed by migraine; severe attacks can cause LOC or coma
FHM2	602481	1q23.2/*ATP1A2*	Na-K ATPase, α2 subunit	Gain of function	AD/childhood to adolescence	Hemiplegic migraines, basilar type migraines; associated with seizures and coma complicating headaches
FHM3	609634	2q24/*SCN1A*	$Na_v1.1$ sodium channel	Gain of function	AD/childhood to adolescence	Hemiplegic migraines; associated with idiopathic epilepsy
HM	614386	16p11.2/*PRRT2*	Proline-rich transmembrane protein 2	Probably loss of function	AD/childhood to adolescence	Hemiplegic migraines; benign infantile convulsions; paroxysmal kinesigenic dyskinesia, epilepsy

Abbreviations: AD, autosomal dominant; FHM, familial hemiplegic migraine; HM, hemiplegic migraine; LOC, loss of consciousness; OMIM, Online Mendelian Inheritance of Man.

NON-MENDELIAN PAIN GENETICS

While rare, the heritable pain disorders have been amenable to classical human genetic analysis. It has been far more challenging to identify the more common genetic variance that underlies individual differences to pain and analgesics. Part of this difficulty stems from the inability to dissociate the genetics that result in the primary cause of pain (e.g., arthritis, ulcerative colitis) from the genetics that mediate the perception of pain itself. Even when attempts are made to control for the specific disease processes, there may be multiple distinct mechanisms that contribute to pain in a given individual (e.g., inflammation, mood). Thus, it can be extremely complicated to interpret the biological significance of data, even when statistically significant. Despite these challenges, progress has been made from both a candidate gene approach in focused patient populations as well as genome-wide association studies. Candidate genes are often chosen from studies in rodent models or genes functionally similar to those already implicated in pain. Genome-wide association studies take large populations of patients with a specific phenotype, perform genome sequencing or microarrays on patient DNA, and then use bioinformatic analysis on thousands of genes simultaneously to determine which genes, if any, are significantly associated with the phenotype in question. Table 90.4 displays a select list of genes that have been significantly associated with pain phenotypes in human studies.[8,18,19]

The *CYP2D6* gene encodes a cytochrome P450 that catalyzes the *O*-demethylation of codeine into morphine. Since the analgesic effect of morphine is ~200 times stronger than codeine, this metabolic conversion has a dramatic effect on dosing. It is estimated that 2.6% of Caucasians have a *CYP2D6* gene duplication and are thus sensitized to codeine, while 7% have loss-of-function mutations that render them resistant. The *OPRM1* gene encodes the μ-opioid receptor, which multiple studies have implicated in responses to analgesia and pain. Patients with *OPRM1* 188A→G single nucleotide polymorphis (SNP) who receive morphine have significantly less pain relief and pupillary miosis. However, the mechanism by which this occurs is unclear since it appears the 188A→G SNP does not directly affect morphine receptor binding. One hypothesis is that this SNP results in reduced *Oprm1* mRNA expression.

Polymorphisms in *COMT*, a methyltransferase involved in the degradation of catecholamines, have been associated with pain perception in several studies. The *COMT* SNP, val^{158}met, reduces enzyme activity 3–4 fold and is associated with increased sensitivity to painful stimuli in certain paradigms. The *SCL6A4* gene encodes the serotonin transporter (SERT), which is involved in the reuptake of serotonin from the synaptic cleft and is the target of most selective serotonin reuptake inhibitors. The *SCL6A4* gene promoter has a polymorphism that results in either a "short" or "long" form of the gene. The "short" form has been associated with higher rates of depression, anxiety, fibromyalgia, and reported pain levels after third molar extraction. The mechanism by which SERT mediates pain hypersensitivity is unclear.

TABLE 90.4 Genes Implicated in Human Pain

Gene	Protein	Function	Polymorphism phenotype
CACNG2	Voltage-dependent calcium channel, gamma 2	AMPA receptor subunit; receptor trafficking and stabilizes closed state	SNPs are associated with increased risk of neuropathic pain after mastectomy
COMT	Catechol-*O*-methyltransferase	Catecholamine degradation	SNPs resulting in higher enzyme activity may increase sensitivity to pain
CYP2D6	Cytochrome P450, family 2, subfamily D, polypeptide 6	Metabolism of drugs (e.g., codeine into morphine)	Gene duplication results in increased sensitivity to codeine
GCH1	GTP cyclohydrolase 1	Tetrahydrobiopterin biosynthesis, required for catecholamine synthesis	Decreased sensitivity to experimental pain
HTR2A	Serotonin receptor 2A	Serotonin signaling	Increased post-surgical pain and chronic widespread pain
KCNS1	Voltage-gated potassium channel, delayed-rectifier	Neuronal excitability, resting membrane potential	Increased risk for chronic back pain, sciatica or post-amputation pain
SCL6A4	Solute carrier family 6	Serotonin transporter	SNPs associated with increased sensitivity to pain and chronic widespread pain
SCN9A	Voltage-gated sodium channel, type 9	Neuronal excitability, action potential formation in nociceptive neurons	Increased risk for chronic pain (e.g., sciatica, pancreatitis)
OPRM1	Opioid receptor, μ1	Opioid signaling	Decreased analgesic response to morphine

Abbreviation: AMPA, α-amino-3-hydroxy-5-methyl-4-isoxazolepropionic.

Table 90.4 displays several additional genes that have also been associated with pain in human studies. While these examples illustrate the potential power of studying individual differences in human pain perception, it should be noted that it has been challenging for the community to reproduce the same findings across cohorts and investigators. The reasons for this problem in pain genetics are likely multifactorial, including differing study designs and populations, and the genetic and environmental heterogeneity inherent to pain perception. For example, a person's pain threshold could vary even with the duration of sleep or mood prior to testing, so sample sizes need to be large enough to overcome these obstacles.

Rodent Pain Genetics

While human genetics has provided key insights into the mechanisms underlying pain perception, pain research in animal models has been important in both novel gene identification and in mechanistic studies. Animal pain models have been designed to study acute nociception, inflammatory pain, neuropathic pain, and more recently, operant pain behaviors. Acute nociception assays of thermal, mechanical, or chemical stimuli, measure paw withdrawal reflex time from application of a stimulus. Inflammatory pain models attempt to understand the pain sensitization that occurs in inflammatory conditions such as arthritis. Formalin, capsaicin, and complete Freund's adjuvant are commonly used inflammatory compounds that are directly applied or injected into the study site. The inflammatory pathway initiated by these compounds causes profound hyperalgesia and allodnyia that can be assayed via paw withdrawal latency to tactile or thermal stimuli. Neuropathic pain models usually involve surgically induced damage to a large nerve (e.g., sciatic nerve) through physical crush, cuffing, or partial ligation. A pain syndrome develops over days to weeks, which permits the testing of drugs that have delayed effects such as antidepressants or anticonvulsants. Even though the inflammatory and neuropathic models are focused on the effects of chronic pain, the acute nociceptive reflexes to mechanical or thermal stimuli are still the behavioral readouts. This may overlook the higher-order emotional and cognitive elements related to the pain experience in the rodent. There are several experimental paradigms that attempt to model the operant effects of pain, including conditioned place preference to analgesic administration and self-administration of analgesics. In the place preference paradigm, an animal learns to associate a specific environment with the delivery of an analgesic, which enables quantification of how much an animal "prefers" to be in this environment when given a choice. Self-administration of analgesics is based on a similar principle, except that an animal is trained to push a lever for analgesic administration and the number of lever presses is recorded. Each of these pain models can be combined with neuropharmacology and/or genetic engineering to explore the role of specific genes and pathways.[20,21]

Some rodent genetics studies use quantitative trait locus (QTL) mapping of pain phenotypes in which unique pain responses are associated with chromosomal regions and ultimately specific genes. Other studies generate transgenic animals based on specific candidate genes to identify the pathways that are altered in animal models of chronic pain. Numerous genes have been implicated using these animal models, and current efforts are focusing on which genes also apply to human pain (for review see [19]). QTL mapping and genome-wide analyses attempt to identify the genotype responsible for a specific phenotype, but variations of this approach are also powerful. One example of this is the *SCN10A* gene, which encodes the $Na_v1.8$ sodium channel that is expressed in nociceptive C fibers and is involved in peripheral pain transduction to the dorsal horn. Clever mouse genetics has enabled expression of the diphtheria neurotoxin in $Na_v1.8$-positive C fiber neurons, resulting in loss of nearly all C fibers. This new transgenic animal model enables researchers to study the role of C fibers directly in various pain models.

Taken together, cutting edge genetics tools when combined with rodent pain models are significantly contributing to our understanding of pain physiology. That said, how these data translate to human pain is difficult to predict. One example of this is substance P, which is a key neuropeptide in pain processing. In animal models, substance P is clearly linked with nociception, and disruption of this signaling leads to dramatic reductions in pain sensitivity. However, when antagonists were generated for use in human trials, the results were disappointing. This has led the pharmaceutical industry in particular to take a cautious approach to the direct application of preclinical mouse models to human pain perception.

FUTURE DIRECTIONS

The future of pain genetics is bright because there are powerful new tools awaiting application to both animal models and patients, but major obstacles still remain. Difficult-to-control factors, such as the heterogeneity of pain etiologies and the inherent variance in pain scale reporting, make it challenging to generate homogeneous cohorts of patients amenable to genetic analysis. Future studies will need to control better for these factors, generate large sample sizes, and/or include novel metrics that more reliably report pain. One possibility could be to incorporate functional magnetic resonance imaging into this assessment as more is understood about these pain signatures. With the establishment of larger and better-characterized human cohorts, we can begin to combine our knowledge of mouse and human pain genetics with appropriate statistical power to draw meaningful conclusions. This will also enable the full potential of whole-genome sequencing and genome-wide association studies, which have been successfully applied to other areas of medicine.

The emerging frontiers of stem cells, genetic engineering, and epigenetics also bring promise to the pain field. New stem cell technology permits the conversion of skin cells from patients into certain types of neurons, and attempts are currently underway to generate nociceptive neurons using these techniques. This will enable electrophysiological and molecular analysis of cells taken directly from pain patients. Combined with stem cell technology, new genetic engineering techniques make it easy to introduce the mutations discovered in patients directly into stem cells to study the functional significance. One additional area of interest is in the study of epigenetics, the long-lasting alterations in gene expression caused by changes to chromatin structure rather than a genetic mutation or polymorphism. Epigenetic modifications such as DNA methylation and posttranslational modifications to histones are thought to mediate the interaction between genes and environment to result in individualized phenotypes. Indeed, some evidence for this is beginning to emerge in the study of pain transduction in the spinal cord and brain.

CONCLUDING REMARKS

Disorders in pain perception can occur at any step along the signaling pathway. Without the membrane-bound nociceptors that initially detect noxious stimuli, no pain signal is generated. Without the structural proteins, ion channels, and neurotransmitters involved in transducing pain signals to the brain, the signal is not delivered. And finally, without multiple CNS networks functioning together, pain perception either does not occur or does not result in the appropriate behavioral response. As would be expected, we know the most about heritable phenotypes for which the sequencing of multiple affected family members was the easiest. These human mutations along with rodent pain models and newly emerging genetics tools have yielded numerous breakthroughs in our understanding of pain physiology and promise many more to come.

References

1. McMahon S, Koltzenburg M. Wall and Melzack's Textbook of Pain. Textbook of Pain (*Wallach*), Vol. 5 Churchill Livingstone; 2005, 1280.
2. Basbaum AI, et al. Cellular and molecular mechanisms of pain. *Cell*. 2009;139(2):267–284.
3. Caterina MJ, et al. Impaired nociception and pain sensation in mice lacking the capsaicin receptor. *Science*. 2000;288(5464):306–313.
4. Davis JB, et al. Vanilloid receptor-1 is essential for inflammatory thermal hyperalgesia. *Nature*. 2000;405(6783):183–187.
5. Chiu IM, von Hehn CA, Woolf CJ. Neurogenic inflammation and the peripheral nervous system in host defense and immunopathology. *Nat Neurosci*. 2012;15(8):1063–1067.
6. Kandel E, Schwartz J, Jessell T. Principles of neural science. 4th ed. New York, NY: McGraw-Hill; 2000.
7. Craig AD. Pain mechanisms: labeled lines versus convergence in central processing. *Annu Rev Neurosci*. 2003;26:1–30.
8. Goldberg YP, et al. Human Mendelian pain disorders: a key to discovery and validation of novel analgesics. *Clin Genet*. 2012;82(4):367–373.
9. Rotthier A, et al. Mechanisms of disease in hereditary sensory and autonomic neuropathies. *Nat Rev Neurol*. 2012;8(2):73–85.
10. Garofalo K, et al. Oral L-serine supplementation reduces production of neurotoxic deoxysphingolipids in mice and humans with hereditary sensory autonomic neuropathy type 1. *J Clin Invest*. 2011;121(12):4735–4745.
11. Dib-Hajj SD, et al. Sodium channels in normal and pathological pain. *Annu Rev Neurosci*. 2010;33:325–347.
12. Dib-Hajj SD, et al. The Na(V)1.7 sodium channel: from molecule to man. *Nat Rev Neurosci*. 2013;14(1):49–62.
13. Faber CG, et al. Gain-of-function Nav1.8 mutations in painful neuropathy. *Proc Natl Acad Sci U S A*. 2012;109(47):19444–19449.
14. Cox JJ, et al. An SCN9A channelopathy causes congenital inability to experience pain. *Nature*. 2006;444(7121):894–898.
15. Yang Y, et al. Mutations in SCN9A, encoding a sodium channel alpha subunit, in patients with primary erythermalgia. *J Med Genet*. 2004;41(3):171–174.
16. Fertleman CR, et al. SCN9A mutations in paroxysmal extreme pain disorder: allelic variants underlie distinct channel defects and phenotypes. *Neuron*. 2006;52(5):767–774.
17. Russell MB, Ducros A. Sporadic and familial hemiplegic migraine: pathophysiological mechanisms, clinical characteristics, diagnosis, and management. *Lancet Neurol*. 2011;10(5):457–470.
18. Lacroix-Fralish ML, Mogil JS. Progress in genetic studies of pain and analgesia. *Annu Rev Pharmacol Toxicol*. 2009;49:97–121.
19. Mogil JS. Pain genetics: past, present and future. *Trends Genet*. 2012;28(6):258–266.
20. Mogil JS. Animal models of pain: progress and challenges. *Nat Rev Neurosci*. 2009;10(4):283–294.
21. Barrot M. Tests and models of nociception and pain in rodents. *Neuroscience*. 2012;211:39–50.

SECTION XI

MUSCLE AND NEUROMUSCULAR JUNCTION DISORDERS

CHAPTER

91

Dystrophinopathies

Eric P. Hoffman

Children's Research Institute and George Washington University, Washington, DC, USA

HISTORICAL OVERVIEW

Disease Identification

The most common of the muscular dystrophies is Duchenne muscular dystrophy (DMD), named after the French physician Guillaume-Benjamine Armand Duchenne, who wrote extensively about the disease in the late 1800s. However, a British physician, Edward Meryon, had elegantly described the disorder and its inheritance pattern some years earlier.[1] The disease is one of the most common inherited lethal disorders of humans (1:3500 males, although incidence has decreased to 1:5000 in developed countries due to recent preventive measures)[2] and likely affects all world populations equally. Due to the very high mutation rate of the gene, many cases are sporadic. When inherited, the disease shows an X-linked recessive pattern, with 50% risk of disease to male offspring of a carrier mother. Boys with Duchenne muscular dystrophy show delayed milestones in infancy (1–3 years), although are often not brought to a clinician's attention until early school years (4–6 years).[3]

Gene Identification

DMD was the first human inherited disorder to be clarified at the molecular level by positional cloning, where the mutated genetic sequences were identified prior to any knowledge of the protein or biochemical abnormality. The identification of the causative gene was aided by a number of early observations that narrowed the regions of the genome requiring analysis. First, it was known that the disease was X-linked, which limited the search to the 10% of the genome in the X chromosome. Second, some girls with muscular dystrophy clinically indistinguishable from DMD in boys had X:autosome translocations involving a specific region on the short arm of the X chromosome, *Xp21*. Third, a boy affected with multiple X-linked recessive diseases, including DMD, was found to harbor a small but microscopically visible deletion of the *Xp21* region of the X chromosome.[4] Genomic DNA from this boy was used to isolate small fragments of genetic material from the *Xp21* region, and one of these (*pERT87*) led to the identification of a neighboring coding sequence (exons) of the gene that was eventually characterized: Its protein product was dubbed *dystrophin*.[5–7]

The dystrophin gene consists of 79 exons, with multiple spliced isoforms and multiple tissue-specific promoters (alternative first exons). The different promoters often lead to smaller mRNA transcripts and protein isoforms derived from a subset of the 79 exons. The best-characterized mRNA transcript and protein is called full-length or muscle dystrophin (427 kDa); it derives from the muscle promoter and includes all downstream exons. The loss of this particular transcript in DMD/Becker muscular dystrophy (BMD) patients appears to be responsible for most clinical problems in both muscle and heart. Lack of other isoforms can cause abnormalities of the tissues in which they are expressed. For example, a specific dystrophin isoform is expressed in the retina, and DMD patients show abnormal electroretinograms.[8] Dystrophin is expressed in neurons by the brain promoter, and about 30% of patients show mental retardation as part of their clinical picture.[9]

http://dx.doi.org/10.1016/10.1016/B978-0-12-410529-4.00091-7

Early Treatments

Supportive therapy involves physical therapy to counteract joint contractures, particularly at the heel cords. Pharmacological therapy is typically limited to chronic glucocorticoids (prednisone or deflazacort).[10,11] Daily or chronic intermittent use of glucocorticoids increases strength and mobility, and slows disease progression. However, the efficacy of daily glucocorticoids is offset in part by the extensive side effect profiles, including mood changes, bone fragility, and stunting of growth.

MODE OF INHERITANCE AND PREVALENCE

The disease when inherited shows an X-linked recessive pattern, with affected male children and female carriers. However, as noted above, the high spontaneous mutation rate of the responsible gene leads to a high proportion of nonfamilial, sporadic (isolated) cases. In many isolated cases, the mother is not a carrier of the mutation seen in her DMD son. Prevalence is 1:3500 if no preventative measures are taken (genetic counseling, prenatal diagnosis). Where genetic counseling and genetic testing has become routine, incidence has dropped to about 1:5000.

NATURAL HISTORY

Age of Onset

Affected boys typically present at age 3–5 years with proximal muscle weakness and difficulty keeping up with peers. However, studies of 1–3 year old DMD patients have shown gross motor milestone scores to be about half those of peers.[3] Newborn screening based on elevated serum muscle enzymes (creatine kinase) is also possible, but only narrowly implemented due to lack of availability of presymptomatic therapies.

Disease Evolution

Proximal weakness manifests as difficulty keeping up with peers, awkward running, and difficulty climbing stairs. The classic Gower maneuver is pathognomonic: in rising from the floor, the child uses the hands on his knees to push the body to a standing position. Progressive proximal weakness then leads to excessive lordosis and a waddling gait. Progression is relatively stereotypical: boys have increasing difficulty with walking, and lose ability to ambulate between ages 7 and 15 years. The hip-girdle weakness is initially most pronounced, although weakness of the shoulder girdle soon follows. Some muscles, particularly those of the calves and the sartorius, are unusually large. This enlarged muscle mass has traditionally been dubbed *pseudohypertrophy* due to the concomitant muscle weakness and the connective tissue proliferation seen in muscle biopsies. However, correlations with animal models suggest that this represents true hypertrophy of muscle fibers, at least in the early stage of the disease.

Strength continuously decreases, and mobility is gradually restricted due to muscle wasting and contractures, although facial musculature and speech are relatively spared. Contractures can be disabling, with Achilles tendon contractures leading to toe walking and instability. Physical therapy and tendon releases are used to counteract the contractures. Nonambulatory patients continue to develop contractures, and spinal fusions may be indicated to arrest progressive scoliosis. Patients are often ventilated in the second decade, but typically require assistance with activities of daily living.

Patients can show other symptoms in addition to skeletal muscle involvement. The heart is affected: there is first hypertrophic, then dilated, cardiomyopathy, and progressive fibrosis is most pronounced in the basolateral free wall of the left ventricle. A minority of Duchenne dystrophy patients die of overt heart failure. Approximately 30% of patients show nonprogressive mental retardation that affects verbal more than other functions. The mental retardation may result from hypersensitivity of the central nervous system to ischemic insults, which are most likely to happen in the perinatal period.[9] Dystrophin is also deficient in smooth muscle, and may present with gastric hypomotility in some patients.

Regarding laboratory findings, striking elevations of muscle-derived creatine kinase elevations in peripheral blood (serum CK) have been strongly associated with dystrophinopathies since the 1950s. All patients show grossly elevated serum creatine kinase levels from birth (more than 50 times the upper limit of normal), and transaminases and lactate dehydrogenase are also elevated. Although the transaminases almost certainly derive from muscle tissue,

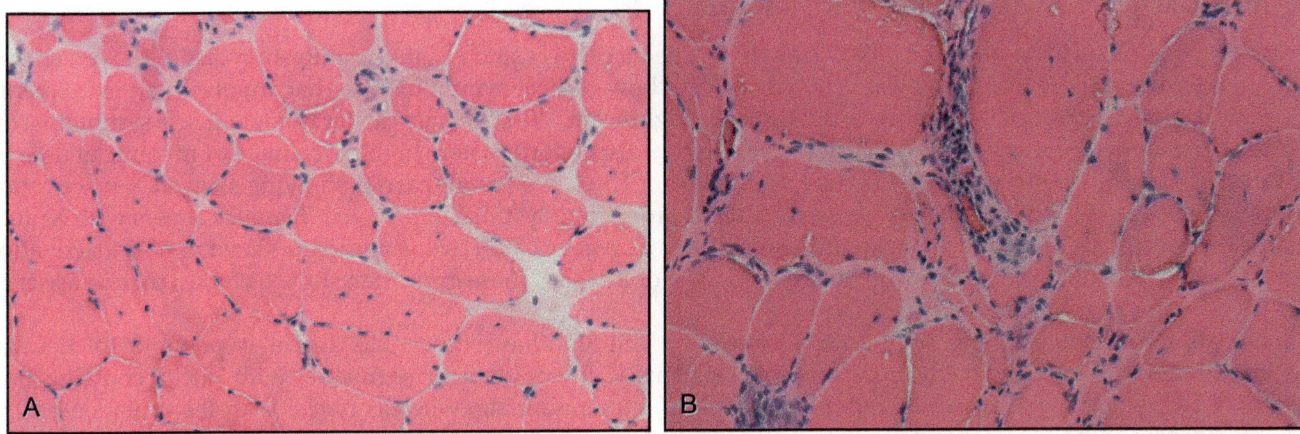

FIGURE 91.1 **Skeletal muscle histopathology of dystrophinopathies.** (**A**) Becker muscular dystrophy. (**B**) Duchenne muscular dystrophy. Panel A shows a relatively mild dystrophic process, with neighboring fascicles showing either very mild (lower) or moderate (upper) changes. Mild dystrophic changes include central nucleation of some myofibers, and mild fiber size variation. The more moderate changes in the upper fascicle include endomysial fibrosis (connective tissue proliferation between adjacent myofibers) and more extensive fiber size variation. Panel B shows a later stage more severely dystrophic muscle area, with dramatic fiber size variation, including hypertrophied fibers, and smaller angulated fibers suggestive of failed attempts at regeneration. Fibrosis is more extensive, and there is endomysial inflammation. The histopathology of dystrophinopathies is typically progressive, with age-dependent gradual replacement of muscle tissue with fibrofatty connective tissue.

occasionally these elevated serum enzyme levels mislead clinicians into investigating liver dysfunction in younger or mildly symptomatic patients. Neonatal screening for DMD is highly efficient at detecting all patients at birth, but the current lack of effective therapy has made screening programs controversial. New programs have been initiated by the US Centers for Disease Control.[12]

Muscle biopsy shows features consistent with a dystrophic myopathy, including myofiber size variation, eosinophilic hypercontracted fibers, increased endomysial connective tissue (fibrosis), and occasional areas of overt necrosis with macrophage infiltration (Figure 91.1). The histopathology changes as a function of age, with young patients showing less fibrosis and less dramatic changes in muscle, whereas end-stage muscle shows large numbers of very small fibers that have failed to fully regenerate, extensive endomysial fibrosis, and a few remaining hypertrophic fibers (Figure 91.1). No histopathologic feature is diagnostic for DMD, but muscle biopsies can be tested for dystrophin protein content, which is considered diagnostic. In the past, electromyography (EMG) was a standard diagnostic tool, but the findings are relatively nonspecific, and EMG is less commonly employed today. Increasingly, direct testing of the dystrophin gene for causative gene mutations is the primary diagnostic test utilized.

DISEASE VARIANTS

Becker Muscular Dystrophy

Becker muscular dystrophy (BMD) derives its eponym from Emil Becker, a German physician who described extensive X-linked pedigrees with a form of muscular dystrophy that was less severe than DMD. Unlike DMD, the disease was often compatible with reproduction, and all daughters of BMD males are carriers of the disease. The distribution of weakness can be similar to that of the more severe DMD, although the progression is typically slower. While severe DMD is due to absence of dystrophin, BMD is caused by present but abnormal dystrophin (altered molecular weight [quality], quantity, or both). Presentations of BMD can include proximal weakness starting at nearly any age, myoglobinuria, muscle cramps, muscle pain, weakness limited to quadriceps, heart failure, or high serum creatine kinase in the absence of other symptoms (hyperCKemia).

Disease progression can be quite variable. Some patients show progressive weakness affecting both limb girdles, whereas others can remain quite static. Intrafamilial variability is common. The cardiomyopathy present in all dystrophinopathy patients is often more pronounced in BMD, presumably because the preservation of strength relative to DMD puts greater demands on the compromised cardiac musculature.

Female Carriers (Asymptomatic or Manifesting)

Girls and women in families with DMD are at risk of being carriers, and isolated manifesting carriers have also been frequently described. Most female carriers only rarely show clinical symptoms. Cardiac involvement is frequently seen by sensitive tests, but this is only rarely clinically significant. When female carriers show muscle symptoms, this is usually the result of skewed X-inactivation, whereby a disproportionately high percentage of myonuclei uses the mutant dystrophin gene and inactivates the normal gene. Thus, in a female carrier showing skewed X-inactivation, only 10% or less of myofiber nuclei may be capable of producing dystrophin protein.[13] Female carriers with X: autosome translocations involving a breakpoint in the dystrophin gene (*Xp21*) show complete skewing of X-inactivation, thus keeping the mutant dystrophin gene active in most or all cells. These girls often show a clinical picture as severe as DMD in boys.

Dystrophin protein testing of muscle biopsies by immunofluorescence has made it possible to visualize X-inactivation patterns indirectly in female muscle. Muscle fibers (or fiber segments) with nuclei containing the normal X chromosome active show normal dystrophin immunostaining, whereas nuclei with the mutation-bearing X chromosome show dystrophin deficiency. Thus, most carrier muscle shows a mosaic pattern of dystrophin-positive and dystrophin-negative muscle fibers in cross-section. It became apparent that isolated female muscular dystrophy patients without a family history of DMD in male relatives may also have dystrophin mosaicism (female carrier state) as the underlying cause of their disease. Nearly all of these isolated female cases were diagnosed as limb-girdle muscular dystrophy before dystrophin testing became available. Large collaborative studies have shown that approximately 10% of female limb-girdle patients are instead carriers of DMD.[13]

The clinical presentation of dystrophinopathy in isolated female patients is as varied as BMD in males. Female patients can show asymmetric weakness due to varied X-inactivation patterns in different muscles. Also, the disease in female patients is generally less rapidly progressive than in male patients with similar presentations. The reason for the slower progression is the tendency of female muscle to become more dystrophin-normal with age (genetic and biochemical normalization).[14]

END OF LIFE: MECHANISMS AND COMORBIDITIES

Duchenne muscular dystrophy patients typically succumb to respiratory failure, pneumonia, or cardiac failure in the second or third decade, unless ventilated. Ventilatory support can increase lifespan considerably. Scoliosis and concomitant reductions in vital capacity (lung volume) are a serious morbidity in DMD patients, often requiring scoliosis surgery (spinal rods). However, use of glucocorticoids reduces scoliosis and the need for spinal surgery.[15,16]

Patients must be managed to compensate for the gradual loss of muscle tissue and strength. Tendon contractures affect the comfort and mobility of the patient. Achilles tendon contractures often lead to toe walking; stretching of these and other involved tendons via physical therapy can slow or alleviate contractures. Night splints and other forms of bracing are often recommended in young patients (age 4–5 years) to prevent leg contractures. Surgical lengthening of the Achilles tendon is frequently done if toe walking begins to limit mobility.

MOLECULAR GENETICS

Normal Gene Product

The dystrophin gene is by far the largest gene identified to date in any species. At about 2.3 million bp (2.3 Mbp), it is about twice as large as the next largest gene. It takes 16 hours to produce a single mRNA from this enormous gene.[17]

Abnormal Gene Products

The most common type of gene mutation in both DMD and BMD are deletions or duplications encompassing one or more exons. Because introns take up about 99.5% of the transcript unit compared to the smaller exons, nearly all deletion and duplication mutations begin and end in intron sequences. There are two deletion hotspots, one in the region of exons 44 to 50, and another near the beginning of the gene (exons 2 to 13).

The most important feature of a deletion or duplication mutation regarding prognosis is the resulting reading frame.[18] If the remaining exons can still be spliced into a RNA that maintains the triplet codon reading frame, then some semifunctional dystrophin protein is often produced in muscle, and a milder Becker muscular dystrophy phenotype (BMD) is usually observed. Deletion or duplication mutations where remaining exons are spliced into an out-of-frame mRNA (e.g., frameshift) show complete loss of dystrophin protein in muscle and clinically severe DMD. There are many exceptions to the reading frame rule, but the amount of dystrophin in muscle still most accurately predicts the clinical phenotype.[19] These exceptions occur primarily at the 5′ end of the gene, where the transcriptional machinery can override the frameshift by use of a different initiator methionine codon (ATG) or can make use of alternative splicing.

The dystrophin protein has domains with specific subfunctions, and mutations within specific domains differentially disrupt dystrophin function (Figure 91.2). The carboxyl terminus of the dystrophin protein, where the syntrophins and dystroglycan proteins bind, seems particularly important for dystrophin function. If this domain is lacking or defective, a severe Duchenne phenotype usually results, even if residual dystrophin is present in patient muscle.[20] The amino-terminal actin-binding domain also appears to be important. Many patients lacking this domain have a particularly severe BMD phenotype, although the remainder of the protein is intact. Most deletion mutations occur in the large central rod domain, which fortunately appears to be the least critical for dystrophin function; some mild

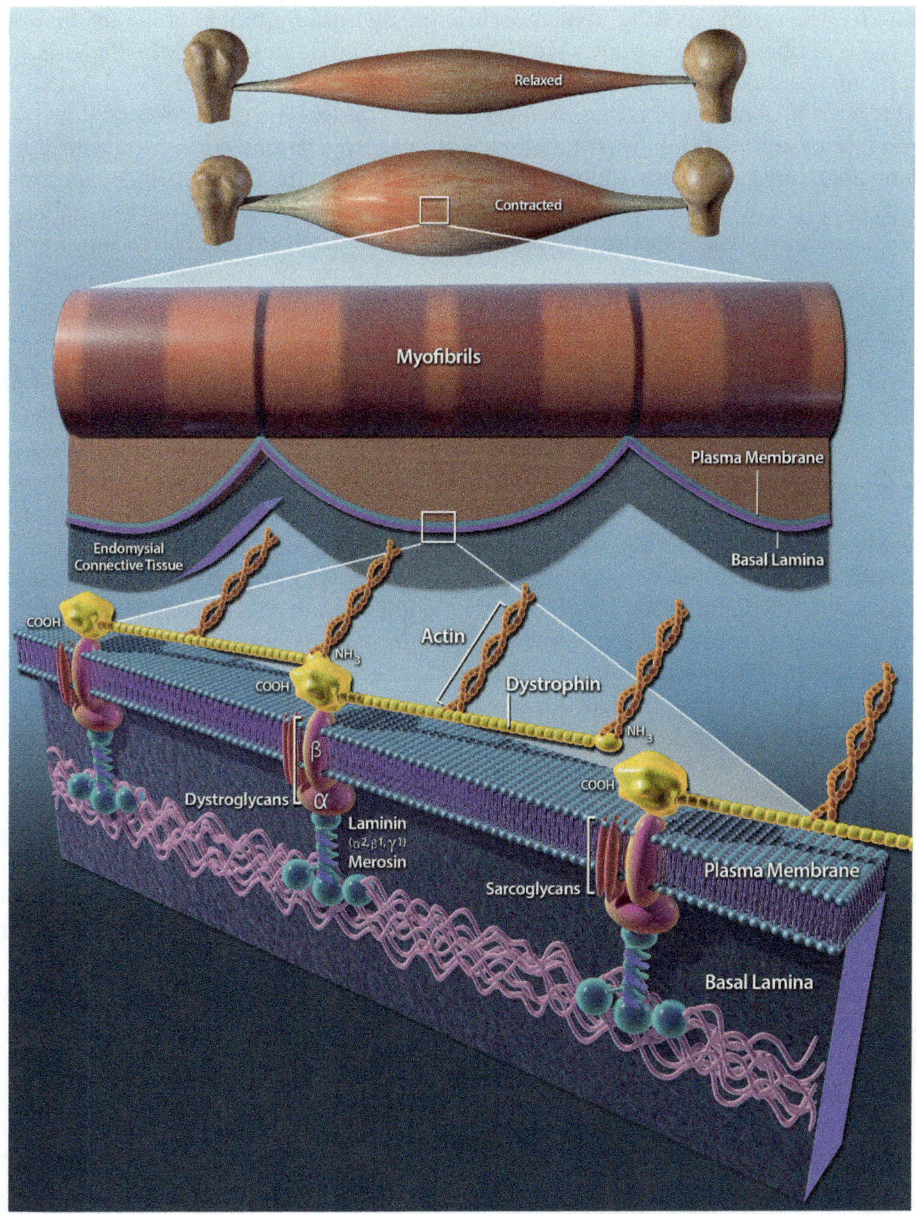

FIGURE 91.2 **Dystrophin protein localization in the muscle myofiber plasma membrane cytoskeleton.**

Becker dystrophy patients have most of this region either deleted or duplicated. The amount of abnormal protein that accumulates can affect clinical severity in BMD. However, there is interpatient variability, even when dystrophin gene and protein findings are similar.

Genotype–Phenotype Correlations

As with many genetic disorders, the relative strength of genotype–phenotype correlations is as much a matter of perspective, rather than any strict quantitative science. Considering the wide range of phenotypes seen in dystrophinopathies, in both males and females, there is a relatively good correlation with the amount of residual dystrophin in muscle, and the severity of the clinical phenotype. That said, the correlations are approximate and there are always exceptions. For example, it has been noted that isolated cases of female manifesting carriers show a strong correlation between the extent of skewed X-inactivation (e.g., percentage of dystrophin-positive myonuclei), the amount of dystrophin observed in muscle from the same patient, and the severity of the clinical phenotype. However, this same correlation was lost in *familial* cases of manifesting carriers, suggesting considerable ascertainment bias can exist for both family members and physicians caring for them (e.g., a family with multiple DMD patients may be more attuned to possible weakness in female relatives, independent of X-inactivation status).[14]

Overall, patients with undetectable dystrophin in muscle tissue by immunoblot or only rare dystrophin-positive fibers by immunostaining (<3% of fibers) show a more severe phenotype consistent with DMD. Patients with ~5–15% normal levels of dystrophin in muscle show a severe Becker phenotype (loss of ambulation at 16–25 years without steroid treatment), and patients with >20% of normal dystrophin levels show a milder Becker phenotype (loss of ambulation in third decade or later, or no loss of ambulation). Most patients show gene mutations consistent with the dystrophin protein observations, but there are many exceptions. Indeed, systematic studies at the DNA, mRNA, and protein level in a series of Becker muscular dystrophy patients showed many different molecular events taking place that were challenging to predict based on gene mutations studies alone, thus leading to many exceptions to the reading frame rule.

BIOCHEMISTRY

Dystrophin was found to be associated with the plasma membrane both by subcellular fractionation studies and by immunostaining of muscle biopsy cryosections. Immunoelectron microscopy studies showed that dystrophin was at the intracellular face of the plasma membrane, with some periodicity suggesting a network (Figure 91.2). Dystrophin associates with the membrane via a series of dystrophin-associated proteins. The carboxyl-terminal region of dystrophin binds to the transmembrane β-dystroglycan. Both the sarcoglycan complex (four proteins in stoichiometric tetramers) and α-dystroglycan stabilize the association of dystrophin with β-dystroglycan, and α-dystroglycan in turn binds to the myofiber basal lamina via laminin α2/β1/γ1 tetramer in the extracellular basal lamina (also called merosin). Dystrophin binds to filamentous actin at multiple sites along the molecule, with the strongest interacting sites in the amino-terminal actin-binding domain of the protein. In linking the intracellular actin network to the plasma membrane and to the extracellular basal lamina, the dystrophin-based membrane cytoskeleton likely protects the membrane from contraction-induced damage (Figure 91.2). Both sarcoglycans and dystroglycans show secondary reductions in protein levels in dystrophinopathy patients. Mutations in any of the four sarcoglycan genes cause autosomal recessive dystrophies that are clinically similar to DMD.

Much is known about the structure and function of the many dystrophin-associated proteins (including the syntrophins, actin, neuronal nitric oxide synthase [nNOS], aciculin, and others).[21] nNOS appears particularly important because it regulates vascular perfusion, mitochondrial function, calcium trafficking, necrosis, and myofibrillar contractility. nNOS usually associates indirectly with dystrophin via syntrophin, but when dystrophin is lost, nearly all nNOS immunostaining at the plasma membrane is secondarily lost. The loss of nNOS is probably responsible for the functional ischemia seen in dystrophin-deficient muscle.

There is a second membrane cytoskeleton system in myofibers: the vinculin/integrin/laminin cytoskeleton, which is partially redundant with the dystrophin/dystroglycan/laminin cytoskeleton. Overexpression of components of muscle integrin can rescue the dystrophin-deficient *mdx* mice.[22]

PATHOPHYSIOLOGY AND ANIMAL MODELS

Findings from a series of *in vivo* and *in vitro* studies have shown that dystrophin-deficient muscle membranes are more fragile than normal myofiber membranes. First, cultured dystrophin-negative myotubes are considerably more susceptible than normal myotubes to damage by cell volume changes caused by osmotic shock.[23] Second, dystrophin is concentrated at specialized sites on the membrane that are subjected to the greatest stress, including the myotendinous junction,[24] costameres,[25] and the neuromuscular junctions. Third, membrane instability has been directly measured in dystrophin-deficient cells, and dystrophin-deficient *mdx* myotubes showed a fourfold decrease in membrane stiffness.[26] Fourth, electron microscopic studies of human muscle biopsies have demonstrated overt breaks of the plasma membrane in dystrophin-deficient muscle.[27] Finally, dystrophin-deficient myofibers are more susceptible to damage by proteases *in vivo*.[28] There may be similarities between the cues signaling muscle hypertrophy in normal muscle and the hypertrophy seen in young patients with DMD: Sublethal damage of the myofiber membrane may be the normal physiologic signal for subsequent muscle hypertrophy. Thus, the fragility of the plasma membrane in Duchenne muscle may lead to miscued hypertrophy early in the disease, but the progressive aspects of the disease lead to fibrofatty infiltration, failed regeneration, wasting, and weakness with increasing age.

It is tempting to conclude that any disruption of the membrane cytoskeleton causes myofiber death and hence a muscular dystrophy phenotype. Considerable evidence suggests, however, that the pathophysiology of DMD is much more complicated than a simple cause-and-effect relationship between membrane instability and muscle weakness. Dystrophin and its associated proteins are fully expressed at birth, and dystrophin deficiency can be biochemically detected from fetal life onward. Thus, the static biochemical defect stands in marked contrast to the progressive clinical course. Although many animal models of DMD have dystrophin deficiency and dystrophin gene mutations, clinical progression in these animals is quite different from the human disease. Dystrophin-deficient *mdx* mice have elevated serum creatine kinase concentrations soon after birth but do not show overt histopathology until 3–4 weeks of age. A remarkable histopathologic feature of the mouse is that large groups of fibers die synchronously (grouped necrosis) at 3–6 weeks. After this initial wave of grouped necrosis, the muscle successfully regenerates. After 8 weeks of age, the murine *mdx* muscle shows rarer and less pronounced bouts of degeneration/regeneration. Mice do in fact show a relatively mild reduction in lifespan, but they show little evidence of clinical weakness earlier than 1 year of age.[29]

Sporadically occurring dystrophin-deficient cats have been found in all parts of North America, and each affected cat shows a similar phenotype.[30] Affected cats show progressive muscle hypertrophy that can itself become lethal despite the absence of muscle weakness or wasting. For example, one cat died from esophageal constricture secondary to dramatic hypertrophy of the diaphragm, and another presented with renal failure secondary to dehydration from hypertrophied lingual musculature that made lapping water difficult. Muscle histopathology looks similar to that of the *mdx* mouse.

In dogs, dystrophin deficiency has been documented in golden retrievers, wire-haired fox terriers, rottweilers, and shelties, among others.[31,32] Dystrophin-deficient dogs are clinically variable, with some not surviving the neonatal period, others showing a severely progressive course with early death, and others stabilizing with long lifespan.

Many pathogenetic models have attempted to explain the enigmatic progression of DMD and the differences in phenotypes of the different animal models. One theory suggests that other proteins can compensate for dystrophin to some extent, and it is the variability in these compensatory proteins that conditions the progression and species differences. Utrophin (also a dystrophin-related protein) is a 420-kDa protein that looks much like dystrophin at the primary amino-acid sequence level but has a very different tissue distribution and temporal expression pattern. One therapeutic avenue being pursued is to induce excess production (upregulation) of utrophin in patient muscle through pharmacologic means (see discussion of dystrophinopathies above).[33] Another theory places the burden of disease progression on the failure of regeneration, either due to loss of satellite cell (muscle stem cell) populations, or asynchronous remodeling and associated inflammation that misdirects the regenerative process.[34] It is becoming increasingly likely that all of the pathophysiologic pathways discussed have a component role in the complex downstream consequences of dystrophin deficiency. It also follows that therapeutics targeted at slowing disease progression may need pharmacologic approaches to address multiple pathophysiologic pathways at the same time.

EXPERIMENTAL THERAPY AND FUTURE RESEARCH DIRECTIONS

Experimental therapeutics for DMD are focused on either the primary genetic and biochemical defect or the multiple downstream pathways causing disease progression. Pharmacologic trials addressing the downstream pathophysiology of dystrophin deficiency and/or compensatory pathways have been done in the *mdx* mouse model. Insulin-like growth factor 1 (IGF-1), carnitine, creatine, pentoxifylline, glutamine, oxatamide, integrins, myostatin inhibitors, green tea extract, nNOS, increase in glycosylation via GalNAc transferase expression, and utrophin upregulation have all been shown to improve the strength and/or histopathology of *mdx* mice.[35–40] One approach that has shown promise in the *mdx* mouse model is the development of novel steroid drugs that appear to retain or enhance efficacy of glucocorticoids, while reducing side effect profiles.[41]

There is also considerable work on rescuing the primary defect (e.g., dystrophin gene or protein replacement). One method to replace the defective gene in DMD patients is transplant of either muscle tissue or muscle stem cells (myoblasts). Whole muscle groups in a DMD patient could conceivably be transplanted, but all muscles are eventually involved in the disease process, and whole body muscle transplantation is not envisioned as a possible therapy for DMD. Another approach is to introduce normal muscle cells from a donor into the muscle of the DMD patient. Cellular transplantation is being tried in many disorders and has also been attempted in DMD. Although this approach, known as *myoblast transfer*, showed some success in mice, barriers to use as a DMD patient therapy include the observations that relatively few cells survive after injection, efficient delivery of adequate cells to muscle myofibers is problematic, and the presence of significant immunologic barriers.[42] A second approach to delivering functional dystrophin genes to patient muscle is gene therapy using viral vectors. Clinical trials have begun, but many of the barriers regarding delivery and immunology facing stem cell approaches are shared with gene therapy approaches.[43]

A promising alternative approach is to repair the patient's own dystrophin gene, rather than providing a new one. The molecular approach that is closest to human applications is called *exon skipping*.[44] The exon skipping approach aims to force the mRNA machinery to fail to include one or more exons neighboring a DMD deletion mutation, such that the reading frame is restored. The mRNA subjected to exon skipping now maintains the appropriate translational reading frame and makes a stable mRNA and protein resembling a Becker muscular dystrophy-like dystrophin protein. This approach has shown success in restoring dystrophin protein expression systemically in *mdx* mice,[45] dogs,[46] and initial human trials.[47,48] Unfortunately, a relatively large phase II/III trial of the 2′-*O*-methyl chemistry of exon skipping failed to show significant improvement in the primary clinical outcome (6 minute walk). However, there may be more promise with the alternative morpholino chemistry.[49]

References

1. Emery AEH, Emery MLH. *The History of a Genetic Disease. Duchenne Muscular Dystrophy, or Meryon's Disease*. London: Royal Society of Medicine Press; 1995.
2. Moat SJ, Bradley DM, Salmon R, Clarke A, Hartley L. Newborn bloodspot screening for Duchenne muscular dystrophy: 21 years experience in Wales (UK). *Eur J Hum Genet*. 2013;21:1049–1053.
3. Connolly AM, Florence JM, Cradock MM, et al. MDA DMD Clinical Research Network. Motor and cognitive assessment of infants and young boys with Duchenne Muscular Dystrophy: results from the Muscular Dystrophy Association DMD Clinical Research Network. *Neuromuscul Disord*. 2013;23:529–539.
4. Kunkel LM, Monaco AP, Middlesworth W, et al. Specific cloning of DNA fragments absent from the DNA of a male patient with an X chromosome deletion. *Proc Natl Acad Sci U S A*. 1985;82:4778–4782.
5. Monaco AP, Neve RL, Colletti-Feener C, et al. Isolation of candidate cDNAs for portions of the Duchenne muscular dystrophy gene. *Nature*. 1986;323:646–650.
6. Koenig M, Hoffman EP, Bertelson CJ, et al. Complete cloning of the Duchenne muscular dystrophy (DMD) cDNA and preliminary genomic organization of the DMD gene in normal and affected individuals. *Cell*. 1987;50:509–517.
7. Hoffman EP, Brown RH, Kunkel LM. Dystrophin: the protein product of the Duchenne muscular dystrophy locus. *Cell*. 1987;51:919–928.
8. Pillers DM, Bulman DE, Weleber RG, et al. Dystrophin expression in the human retina is required for normal function as defined by electroretinography. *Nat Genet*. 1993;4:82–86.
9. Mehler MF, Haas KZ, Kessler JA, et al. Enhanced sensitivity of hippocampal pyramidal neurons from mdx mice to hypoxia-induced loss of synaptic transmission. *Proc Natl Acad Sci U S A*. 1992;89:2461–2465.
10. Ricotti V, Ridout DA, Scott E, et al. NorthStar Clinical Network. Long-term benefits and adverse effects of intermittent versus daily glucocorticoids in boys with Duchenne muscular dystrophy. *J Neurol Neurosurg Psychiatry*. 2013;84(6):698–705.
11. Bushby K, Finkel R, Birnkrant DJ, et al. DMD Care Considerations Working Group. Diagnosis and management of Duchenne muscular dystrophy, part 1: diagnosis, and pharmacological and psychosocial management. *Lancet Neurol*. 2010 Jan;9(1):77–93.
12. Mendell JR, Shilling C, Leslie ND, et al. Evidence-based path to newborn screening for Duchenne muscular dystrophy. *Ann Neurol*. 2012;71(3):304–313.
13. Hoffman EP, Arahata K, Minetti C, et al. Dystrophinopathy in isolated cases of myopathy in females. *Neurology*. 1992;42:967–975.
14. Pegoraro E, Schimke RN, Garcia C, et al. Genetic and biochemical normalization in female carriers of Duchenne muscular dystrophy: evidence for failure of dystrophin production in dystrophin competent myonuclei. *Neurology*. 1995;45:677–690.
15. Connolly AM, Kim HJ, Bridwell KH. Corticosteroids can reduce the severity of scoliosis in Duchenne muscular dystrophy. *J Bone Joint Surg Am*. 2013;95(12):e86. doi:10.2106/JBJS.M.00428, PubMed PMID: 23783217.
16. Lebel DE, Corston JA, McAdam LC, Biggar WD, Alman BA. Glucocorticoid treatment for the prevention of scoliosis in children with Duchenne muscular dystrophy: long-term follow-up. *J Bone Joint Surg Am*. 2013;95(12):1057–1061. doi:10.2106/JBJS.L.01577, PubMed PMID: 23783200.
17. Tennyson CN, Klamut HJ, Worton RG. The human dystrophin gene requires 16 hours to be transcribed and is cotranscriptionally spliced. *Nat Genet*. 1995;9:184–190.

18. Monaco AP, Bertelson CJ, Liechti-Gallati S, et al. An explanation for the phenotypic differences between patients bearing partial deletions of the DMD locus. *Genomics*. 1988;2:90–95.
19. Kesari A, Pirra LN, Bremadesam L, et al. Integrated DNA, cDNA, and protein studies in Becker muscular dystrophy show high exception to the reading frame rule. *Hum Mutat*. 2008;29(5):728–737.
20. Goldberg LR, Hausmanowa-Petrusewicz I, Fidzianska A, Duggan DJ, Steinberg LS, Hoffman EP. A dystrophin missense mutation showing persistence of dystrophin and dystrophin-associated proteins yet a severe phenotype. *Ann Neurol*. 1998;44(6):971–976.
21. Barresi R, Campbell KP. Dystroglycan: from biosynthesis to pathogenesis of human disease. *J Cell Sci*. 2006;119(Pt 2):199–207.
22. Fairclough RJ, Perkins KJ, Davies KE. Pharmacologically targeting the primary defect and downstream pathology in Duchenne muscular dystrophy. *Curr Gene Ther*. 2012;12(3):206–244.
23. Menke A, Jockusch H. Decreased osmotic stability of dystrophin-less muscle cells from the mdx mouse. *Nature*. 1991;349:69–71.
24. Tidball JG, Law DJ. Dystrophin is required for normal thin filament-membrane associations at myotendinous junctions. *Am J Pathol*. 1991;138:17–21.
25. Minetti C, Bentrame F, Marcenaro G, et al. Dystrophin at the plasma membrane of human muscle fibers shows a costameric localization. *Neuromuscul Disord*. 1992;2:99–109.
26. Pasternak C, Wong S, Elson EL. Mechanical function of dystrophin in muscle cells. *J Cell Biol*. 1995;128:355–361.
27. Mokri B, Engel AG. Duchenne dystrophy: electron microscopic findings pointing to a basic or early abnormality in the plasma membrane of the muscle fiber. *Neurology*. 1975;25:1111–1120.
28. Gorospe JRM, Tharp MD, Demitsu T, et al. Dystrophin-deficient myofibers are vulnerable to mast cell granule-induced necrosis. *Neuromuscul Disord*. 1994;4:325–333.
29. Pastoret C, Sebille A. mdx Mice show progressive weakness and muscle deterioration with age. *J Neurol Sci*. 1995;129:97–105.
30. Gaschen FP, Hoffman EP, Gorospe JRM, et al. Dystrophin deficiency causes lethal muscle hypertrophy in cats. *J Neurol Sci*. 1992;110:149–159.
31. Gorospe JRM, Nishikawa BK, Hoffman EP. Pathophysiology of Duchenne muscular dystrophy: a clinical and biological enigma. In: Lucy JA, Brown SC, eds. *Dystrophin: Gene, Protein and Cell*. London: Cambridge University Press; 1997:201–232.
32. Kornegay JN, Bogan JR, Bogan DJ, et al. Canine models of Duchenne muscular dystrophy and their use in therapeutic strategies. *Mamm Genome*. 2012;23(1–2):85–108.
33. Ljubicic V, Burt M, Jasmin BJ. The therapeutic potential of skeletal muscle plasticity in Duchenne muscular dystrophy: phenotypic modifiers as pharmacologic targets. *FASEB J*. 2014;28(2):548–568.
34. Dadgar S, Wang Z, Johnston H, et al. Asynchronous remodelling as a driver for failed regeneration in Duchenne Muscular Dystrophy. *J Cell Biol*. in press.
35. Granchelli JA, Pollina C, Hudecki MS. Pre-clinical screening of drugs using the mdx mouse. *Neuromuscul Disord*. 2000;10:235–239.
36. Barton ER, Morris L, Musaro A, et al. Muscle-specific expression of insulin-like growth factor I counters muscle decline in mdx mice. *J Cell Biol*. 2002;157:137–148.
37. Buetler TM, Renard M, Offord EA, et al. Green tea extract decreases muscle necrosis in mdx mice and protects against reactive oxygen species. *Am J Clin Nutr*. 2002;75:749–753.
38. Wehling M, Spencer MJ, Tidball JG. A nitric oxide synthase transgene ameliorates muscular dystrophy in mdx mice. *J Cell Biol*. 2001;155:123–131.
39. Nguyen HH, Jayasinha V, Xia B, et al. Overexpression of the cytotoxic T cell GalNAc transferase in skeletal muscle inhibits muscular dystrophy in mdx mice. *Proc Natl Acad Sci U S A*. 2002;99:5616–5621.
40. Rybakova IN, Patel JR, Davies KE, et al. Utrophin binds laterally along actin filaments and can couple costameric actin with sarcolemma when overexpressed in dystrophin-deficient muscle. *Mol Biol Cell*. 2002;13:1512–1521.
41. Heier CR, Damsker JM, Yu Q, et al. VBP15, a novel anti-inflammatory and membrane-stabilizer, improves muscular dystrophy without side effects. *EMBO Mol Med*. 2013;5(10):1569–1585.
42. Partridge TA. Stem cell route to neuromuscular therapies. *Muscle Nerve*. 2003;27(2):133–141.
43. Mendell JR, Campbell K, Rodino-Klapac L, et al. Dystrophin immunity in Duchenne's muscular dystrophy. *N Engl J Med*. 2010;363(15):1429–1437.
44. Koo T, Wood MJ. Clinical trials using antisense oligonucleotides in duchenne muscular dystrophy. *Hum Gene Ther*. 2013;24(5):479–488.
45. Alter J, Lou F, Rabinowitz A, et al. Systemic delivery of morpholino oligonucleotide restores dystrophin expression bodywide and improves dystrophic pathology. *Nat Med*. 2006;12:175–177.
46. Yokota T, Lu QL, Partridge T, et al. Efficacy of systemic morpholino exon-skipping in Duchenne dystrophy dogs. *Ann Neurol*. 2009;65(6):667–676.
47. Cirak S, Arechavala-Gomeza V, Guglieri M, et al. Exon skipping and dystrophin restoration in patients with Duchenne muscular dystrophy after systemic phosphorodiamidate morpholino oligomer treatment: an open-label, phase 2, dose-escalation study. *Lancet*. 2011;378(9791):595–605.
48. Mendell JR, Rodino-Klapac LR, Sahenk Z, et al. Eteplirsen Study Group. Eteplirsen for the treatment of Duchenne muscular dystrophy. *Ann Neurol*. 2013;74(5):637–647.
49. Hoffman EP, McNally EM. Exon-skipping therapy: a roadblock, detour, or bump in the road? *Sci Transl Med*. 2014;6(230):230fs14.

92

CHAPTER

92

Limb-Girdle Muscular Dystrophy

Wen-Chen Liang and Ichizo Nishino‡*

*Kaohsiung Medical University Hospital, and College of Medicine, Kaohsiung Medical University, Kaohsiung, Taiwan
‡National Institute of Neuroscience, and Translational Medical Center, National Center of Neurology and Psychiatry, Tokyo, Japan

INTRODUCTION

Limb-girdle muscular dystrophy (LGMD) is a clinically and pathologically similar, but genetically heterogeneous, disease group. It is clinically characterized by progressive weakness predominantly in shoulder and pelvic girdles with sparing of the facial muscles, and a pathologically necrotic and regenerating process with a variable extent of fibrosis. The term "LGMD" was first introduced in the 1950s.[1] This disease entity was categorized into two major groups according to the hereditary trait of autosomal dominant or recessive type, indicated by number 1 and 2, respectively. As time goes by, the number of subtypes has markedly expanded, especially in the past two decades. So far there are five type 1 (LGMD1A–1E) and 17 type 2 (LGMD2A–2Q) LGMD whose causative genes have been identified (Table 92.1).

Of the LGMD, the recessive forms (type 2) are more common than the dominant types (type 1). LGMD2A is most prevalent in many geographic areas followed by LGMD2B, but not everywhere.[2,3] Due to the founder mutations, LGMD2I (c.826C>A) and 2L (c.191dupA and c.2272C>T) appear to be almost equally common to LGMD2A and 2B in North Europe.[4–6] Sarcoglycanopathy (LGMD2C–2F) constitute about 10–15% of all LGMD2 in different populations and are relatively common in North Africa, India, and Brazil.[7–9] All LGMD1 and other LGMD2 are relatively rare.

LGMD1

LGMD1A is caused by the mutations in the *MYOT* gene, encoding myotilin.[10] Myotilin is one of the Z-disk-associated proteins, playing an important role in sarcomere assembly and the integrity of the contracting muscle cell.[11] Patients with LGMD1A clinically present with adult-onset slowly progressive proximal weakness followed by distal involvement later, and pathologically share the same features as those of myofibrillar myopathy (MFM), typically showing rimmed vacuoles with inclusions. Except for LGMD1A, the patients with *MYOT* mutations can initially show only distal predominant weakness, especially in the lower legs.

LGMD1B is due to the mutations in the *LMNA* gene, encoding A-type lamins. A-type lamins, intermediate filament proteins, form nuclear laminae and anchor inner nuclear membrane proteins, thus providing a mechanically resistant meshwork.[12] However, in addition to LGMD1B, the *LMNA* gene mutations actually can cause a dozen different clinical conditions, including Emery–Dreifuss muscular dystrophy, lipodystrophy, Charcot–Marie–Tooth disease and progeroid syndrome, etc. LGMD1B is characterized by slowly progressive proximal weakness together with cardiac conduction defects or cardiomyopathy. The age of disease onset is variable, ranging from childhood to adult. The muscle weakness is usually mild but, in contrast, the cardiac symptoms often deteriorate more rapidly.[13] Muscle pathology usually shows mild dystrophic or only chronic myopathic changes.

Rosenberg's Molecular and Genetic Basis of Neurological and Psychiatric Disease
http://dx.doi.org/10.1016/10.1016/B978-0-12-410529-4.00092-9

TABLE 92.1 Current Classification of Limb-Girdle Muscular Dystrophy

Disease	Gene	Protein	Protein properties	Typical onset	Progression/ elevated CK level	Cardiac/ pulmonary involvement	Mouse model
AUTOSOMAL DOMINANT							
LGMD1A	*MYOT*	Myotilin	Sarcomere-associated	Adulthood	Slow	– / –	Myo$^{-/-}$
LGMD1B	*LMNA*	Lamin A/C	Nuclear envelope-associated	Variable	Slow	++ / +	Lmna$^{-/-}$; Lmna$^{H222P/H222P}$
LGMD1C	*CAV3*	Caveolin-3	Sarcolemma-associated	Variable	Slow–mod / + – ++	+ / –	Cav3$^{-/-}$
LGMD1D	*DNAJB6*	DNAJB6	Sarcomere-associated, cochaperone	Adulthood	Slow	– / –	Nil
LGMD1E	*DES*	Desmin	Intermediate filament	Adulthood	Slow	++ / +	Des$^{-/-}$; D7-des
AUTOSOMAL RECESSIVE							
LGMD2A	*CAPN3*	Calpain-3	Myofibril-associated	Adolescence	Slow–mod / + – ++	Rarely	Capn3$^{-/-}$; Capn3$^{CS/CS}$
LGMD2B	*DYSF*	Dysferlin	Sarcolemma-associated	Young Adulthood	Slow / ++	Rarely	SJL/J; Dysf$^{-/-}$
LGMD2C	*SGCG*	Sarcoglycan	Sarcolemma-associated	Childhood	Rapid / ++	++ / ++	Sgcg$^{-/-}$
LGMD2D	*SGCA*	Sarcoglycan	Sarcolemma-associated	Childhood	Rapid / ++	+ / +	Sgca$^{-/-}$
LGMD2E	*SGCB*	Sarcoglycan	Sarcolemma-associated	Childhood	Rapid / ++	++ / ++	Sgcb$^{-/-}$
LGMD2F	*SGCD*	Sarcoglycan	Sarcolemma-associated	Childhood	Rapid / ++	++ / ++	BIO14.6; Sgcd$^{-/-}$
LGMD2G	*TCAP*	Telethonin	Sarcomere-associated	Adolescence	Slow / +–++	+ / +	Tcap$^{-/-}$
LGMD2H	*TRIM32*	Tripartite motif-containing 32	Sarcomere-associated	Adulthood	Slow / +–++	+ / +	Trim32$^{-/-}$
LGMD2I	*FKRP*	Fukutin related protein	Putative glycosyltransferase	Childhood	Mod / ++	++ / +–++	Fkrp$^{L276I/P448L}$
LGMD2J	*TTN*	Titin	Sarcomere-associated	Childhood	Mod–rapid / ++	? / +	Ttn$^{FINmaj/FINmaj}$
LGMD2K	*POMT1*	Protein O-mannosyltransferase 1	Glycosyltransferase	Childhood	Slow	+ / ?	Pomt1$^{-/-}$
LGMD2L	*ANO5*	Anoctamin 5	Transmembrane protein	Variable	Slow / ++	Rarely	Nil
LGMD2M	*FKTN*	Fukutin	Putative glycosyltransferase	Childhood	Mod	++ / +	myf5-Cre/Fktn (E8) KO
LGMD2N	*POMT2*	Protein O-mannosyltransferase 2	Glycosyltransferase	Childhood	Slow	+ / +	POMT2$^{f/f}$
LGMD2O	*POMGNT1*	Protein *O*-mannose β- 1,2-*N*-acetylglucosaminyltransferase	Glycosyltransferase	Childhood	Mod	+ / +	Pomgnt1$^{-/-}$
LGMD2P	*DAG1*	Dystroglycan	Sarcolemma-associated	Childhood	Slow / ++	? / ?	Dag1$^{-/-}$
LGMD2Q	*PLEC1*	Plectin 1f	Cytolinker protein	Childhood	Mod–rapid / ++	– / –	cKO-ple

Abbreviation: Mod, moderate.

LGMD1C results from the mutation in the *CAV3* gene, encoding caveolin-3. Caveolin-3 is the striated-specific form of the caveolin family, the major component of caveolae. Caveolae, localized on the cytoplasmic surface of the sarcolemmal membrane, are involved in many cellular processes, including the maintenance of plasma membrane integrity, the regulation of vesicular trafficking and signal transduction. LGMD1C can show diverse clinical manifestations. The typical feature is slow-to-moderately progressive proximal weakness with various age of onset. However, muscle pain and/or stiffness can be more remarkable than weakness and therefore be the initial presentation. Isolated hyperCKemia and hypertrophic cardiomyopathy have been reported in the patients with *CAV3* mutations.[14,15] Rippling muscle disease, showing the signs of increased muscle irritability is another phenotype associated with *CAV3* mutations.[14] Muscle pathology shows dystrophic change and immunohistochemistry (IHC) together with western blot is very helpful to identify the deficiency of caveolin-3 in the muscle. Creatine kinase (CK) level can be up to 5–10 times the normal limit, higher than that of other LGMD1.

A new causative gene for autosomal dominant LGMD, *DNAJB6* was recently reported.[16,17] However, the nomenclature for this new type of LGMD1 is still undetermined as both LGMD1D and LGMD1E were used in different reports. The *DNAJB6* gene encodes DNAJB6, a cochaperone and a member of the heat shock protein 40 (Hsp40) family. DNAJB6 has been reported to suppress protein aggregation and the toxicity of disease-associated polyglutamine and also to participate in the turnover of proteins and organelles.[18–20] Therefore, the defect of DNAJB6 may impair the clearance of misfolded proteins. The age of onset often ranges from 20 to 60 years, but can even occur in childhood. The initial presentation is usually proximal predominant weakness in the legs, and upper limbs are involved later. Distal weakness can develop and the posterior compartment of the lower legs is more severely involved compared to the anterior. Neither cardiac nor respiratory impairment has been reported. Muscle pathology can show classic dystrophic features; however, as DNAJB6 is a Z-disk-associated protein and involved in the protein clearance, the pathological findings also include autophagic vacuoles and protein aggregation, overlapping with those seen in MFM.

Another LGMD1 with ununified nomenclature (LGMD1D or LGMD1E) is caused by *DES* mutations. The *DES* gene encodes desmin, an intermediate filament protein forming a framework around the Z-disk and thus connecting the myofibrils to the subsarcolemmal cytoskeleton. The *DES* gene mutations have been known to lead to MFM over one decade, and typically present with distal myopathy. However, recently one big family previously categorized as a new type of LGMD1 with dilated cardiomyopathy was found to carry a *DES* mutation.[21,22] The patients showed slowly progressive proximal dominant weakness accompanied by cardiac conduction defect and/or cardiomyopathy. The cardiac symptoms were relatively marked and several family members expired due to sudden death. Muscle pathology reveals dystrophic change and granulofilamentous inclusions.

LGMD2

LGMD2A (calpainopathy) is the most frequent LGMD2 in many countries, as previously mentioned; it is caused by *CAPN3* mutations. Calpain-3, encoded by *CAPN3*, is the skeletal muscle-specific isoform of the calpain family, calcium-dependent enzymes. It functions as a protease and is involved in disassembly of the sarcomere and muscle cytoskeleton during muscle remodeling.[23] These patients usually become aware of the symptoms of proximal muscle weakness especially with the involvement of pelvic girdle in adolescence. The posterior compartment of the thigh, especially hip adductors is preferentially affected. Scapular winging is also often seen and sometimes calf hypertrophy. In the advanced stage, soleus and medial head of the gastrocnemius are involved in the calf.[24] Disease progression is slow to moderate; respiratory and cardiac problems are rarely reported. Isolated hyperCKemia up to more than 10 times the normal upper limit can be found in the preclinical period. Muscle pathology showed dystrophic change, often together with many lobulated fibers, indicating disorganized intermyofibrillar networks. Western blot is useful to detect calpain-3 deficiency but is not specific for LGMD2A, as the deficiency may be secondary.

LGMD2B (dysferlinopathy) is also one of the most common LGMD2. The mutations in the *DYSF* gene, encoding dysferlin, result in this disease. Dysferlin is a transmembrane protein, highly expressed in skeletal muscle, heart and kidney. It is involved in sarcolemmal resealing, muscle differentiation and regeneration, and has a role to stabilize stress-induced calcium signaling in the transverse tubule.[25–28] The age of onset of LGMD2B is usually in early adulthood, but can occur in childhood. Typical clinical features are mild and involve slowly progressive proximal muscle weakness. However, dysferlinopathy can present with distal weakness, known as Miyoshi myopathy and distal myopathy with anterior tibial onset, and mixed proximal and distal involvement. Isolated hyperCKemia up to more than 10 times the normal limit may be seen. The patients with LGMD2B do not show overt respiratory and cardiac involvement. Muscle pathology reveals typical dystrophic change often with remarkable inflammatory cell infiltration, which easily leads to misdiagnosis of myositis. The inflammatory response appears to be a secondary event.[29,30] IHC can make the final diagnosis if complete deficiency of dysferlin is present.

LGMD2C–2 F (sarcoglycanopathy) are caused by the mutations in the *SGCG, SGCA, SGCB* and *SGCD* genes, encoding γ-, α-, β- and δ-sarcoglycans respectively. In skeletal muscle, these four sarcoglycans form heterotetramers in sarcolemma, composing dystrophin–glycoprotein complex (DGC) together with other proteins. The DGC has been known to link the cytoskeleton and extracellular matrix, which participate not only in mechanical force transmission but cell signaling. Among all sarcoglycanopathies, LGMD2D is the most prevalent. The clinical phenotype thus overlaps with that seen in the muscular dystrophies caused by the defects of other DGC proteins. Similar to Duchenne muscular dystrophy, the age of onset is often in early childhood and the disease progression is rapid. However, there are also a proportion of milder cases. The patients with sarcoglycanopathy typically present with proximal muscle weakness and calf hypertrophy. Cardiac and respiratory involvement is commonly seen in the advanced stage, but seems to be relatively rare in LGMD2D.[31] Muscle pathology shows typical dystrophic change and IHC for the four types of sarcoglycan is very helpful to identify their deficiencies but is not completely specific.

LGMD2G is caused by the mutation in the *TCAP* gene, encoding titin-cap/telethonin, which connects *N*-terminus of titin with other Z-disk proteins. Telethonin can regulate the assembly of the sarcomere and is necessary for the cardiomyocytes' stretch sensor and the structural organization of the cardiac sarcomere.[32,33] So far, only a few families have been reported, mainly Brazilian. The age of onset is usually in childhood and the teenage years but the disease progression is often slow; however, some patients lose ambulation within the third or fourth decade.[34] Scapular winging, lower leg weakness, and calf hypertrophy can be observed, as well as cardiomyopathy and respiratory impairment.[32,34] However, a patient with *TCAP* mutations presenting with congenital muscular dystrophy has been reported.[35] In addition to dystrophic change, muscle biopsy often shows rimmed vacuoles and sometimes lobulated fibers.[36,37]

Mutations in *TRIM32*, a ubiquitous E3 ubiquitin ligase belonging to the tripartite motif (TRIM) protein family, have been associated with LGMD2H.[38] TRIM32 is a sarcomere-related protein, localized in the Z-disk. TRIM32 has been reported to interact with myosin IIA and ubiquitinates many proteins including actin and dysbindin, a protein associated with lysosome trafficking.[39,40] It is probably also involved in regulating the differentiation of neural progenitor cells and skeletal muscle stem cell, and adult muscle regeneration.[41,42] Recently a series of studies discussed the role of TRIM32 in regulating TNFα-induced apoptosis.[43,44] Patients with LGMD2H present with slowly progressive proximal weakness with onset usually in the second or third decades of life. Some patients complain of paresthesia and myalgia. Cardiac, respiratory and peripheral nerve involvement has also been reported.[45] It is noteworthy that the mutations in *TRIM32* also cause sarcotubular myopathy, a more early-onset and severe phenotype, even from the same mutation found in LGMD2H patients.[46] Muscle pathology shows dystrophic change or vacuolar myopathy.

LGMD2I, 2K, 2M, 2N, 2O and 2P belong to so-called α-dystroglycanopathy, allelic to several types of congenital muscular dystrophy (congenital muscular dystrophy type 1C, Fukuyama type, muscle–eye–brain disease, Walker–Warburg syndrome). They are caused by mutations in *FRKP, POMT1, FKTN, POMT2, POMGNT1* and *DAG1* respectively.[47–52] *DAG1* encodes dystroglycan (DG), which further forms two different proteins, α- and β-DG after posttranslational cleavage, while other genes encode the proteins involved in the glycosylation of α-DG. Extracellular α-DG binds to laminin via sugar chain, and transmembrane β-DG interacts with dystrophyin. These two dystroglycans are, therefore, components of DGC. Due to defects of glycosylation of α-DG or α-DG itself, binding to laminin would thus be affected, leading to the dysfunction of DGC. The clinical phenotype could be as severe as the dystrophinopathy type or as mild as adult-onset LGMD. Calf hypertrophy, cardiomyopathy, and respiratory insufficiency are commonly seen. Mental retardation has been reported in some patients. CK level can be markedly elevated to several to 10 times the normal limit. Muscle pathology shows typical dystrophic change. IHC and western blot for the sugar chains of α-DG are helpful for screening the glycosylation defect. The correlation between the reduced α-dystroglycan staining and clinical phenotype is controversial.

LGMD2J results from mutations in *TTN*, which encodes titin, the largest protein identified in nature. It stretches from the Z-disk to the M-line to keep the contractile elements of the sarcomere in place, and is responsible for muscle elasticity.[53] Titin also provides multiple ligand binding sites for many other muscle proteins, including calpain-3, α-actinin, myosin, myomesin, and telethonin,[54–56] and it interacts indirectly with myotilin via α-actinin.[57] Loss of protein interactions of C-terminal titin is thus a likely consequence of the mutations to cause the disease. LGMD2J, mostly caused by homozygous FINmaj M-line titin mutation (the Finnish founder mutation), is a severe childhood-onset disease causing proximal muscle weakness and it progresses over the next two decades to the loss of the ability to walk. Facial weakness and calf hypertrophy are sometimes seen. Cardiomyopathy has not been described in LGMD2J, but an early-onset myopathy with fatal cardiomyopathy has been reported to have homozygous titin deletion.[58] In addition to LGMD2J, the mutations in titin actually cause broad-spectrum clinical phenotypes, including tibial muscular dystrophy (TMD) and hereditary myopathy with early respiratory failure (HMERF) and familiar dilated cardiomyopathy.[59–61] However, these allelic diseases are inherited with an autosomal dominant trait. Muscle pathological findings are variable. Typically, it shows myofibrillar changes, including rimmed vacuoles and inclusions. However, core-like and centronuclear patterns have also been reported.[62] Western blot for titin and calpain-3, together and cDNA analysis, are helpful for the diagnosis.

LGMD2L has been associated with mutations in *ANO5*, encoding anoctamin-5. Anoctamin-5 belongs to a family all exhibiting eight transmembrane domains, and recently some anoctamin family members have been recognized as a calcium-activated chloride channel, although the precise function of anoctamin-5 is still unknown.[63–65] LGMD2L is characterized by adult-onset proximal weakness primarily affecting the pelvic girdle and thigh muscles, with less prominent distal leg weakness and high CK values.[4] Asymmetric involvement is often seen. Calf hypertrophy is not uncommon in the early stage but usually becomes atrophic as the disease progresses. Dilated cardiomyopathy in LGMD2L was recently reported, but no overt respiratory problems were noticed.[66] Asymptomatic hyperCKemia can also be seen. As LGMD2L overlaps some clinical features of dysferlinopathy; the differential diagnosis of these two diseases should be kept in mind. A remarkable male predominance in anoctaminopathies has been reported in several studies; females are also less severely affected than males.[5] Dominant mutations in *ANO5* are also responsible for gnathodiaphyseal dysplasia, a rare skeletal syndrome characterized by bone fragility, cement-osseous lesions and diaphyseal sclerosis.[67] Muscle pathology shows dystrophic change. So far, no suitable antibody for anoctamin-5 can be used in IHC or western blot.

Mutations in *PLEC* have recently been found to cause LGMD2Q.[68] *PLEC* encodes plectin, a multimodular protein interacting with many cytoskeletal proteins. In humans, eight different first coding exons are spliced to a common exon 2, leading to different isoforms.[69] Plectin 1, 1b, 1d and 1f are most abundantly expressed in muscle tissue. A homozygous 9-bp deletion containing the initiation codon of exon 1f, associated with DGC, is the cause of LGMD2Q. Other muscle isoforms are associated with outer nuclear/ER membrane, mitochondria and Z-disk.[70] So far, only a few patients with LGMD2Q have been reported, who presented with childhood-onset proximal muscle weakness and were confined to wheelchair in early adulthood. Neither cardiac nor respiratory defect was found. The recessive mutations in *PLEC* were initially reported to cause epidermolysis bullosa simplex with muscular dystrophy (EBS-MD).[71] Then other clinical phenotypes, EBS with pyloric atresia (EBS-PA) and EBS-Ogna, were also correlated to *PLEC* mutations.[72,73] More recently, a subtype of EBS-MD was recognized, EBS-MD with myasthenic syndrome.[74] Muscle biopsy for LGMD2Q shows dystrophic change. The analysis of plectin 1f mRNA expression may be helpful to confirm the diagnosis.

ANIMAL MODELS

Various animal models have been developed for each type of LGMD except for LGMD1D/1E (DNAJB6) and LGMD2L. Some recapitulate the human phenotype in many aspects, thus are more suitable for preclinical trial, while some mimic only the myopathological or biochemical phenotypes seen in humans, which might be more useful for probing the pathomechanism (summarized in Table 92.1).

DIFFERENTIAL DIAGNOSIS

As the key clinical feature of all types of LGMD is shoulder and pelvic girdle weakness, it may be difficult to make the differential diagnosis based on only clinical features. However, the age of onset, the presence of cardiac or respiratory involvement, and the disease course can still provide the clue to narrow the possible diagnoses as the first step.

Muscle pathology, together with IHC and western blot analysis, can further provide the evidence of protein deficiency, as previously mentioned. However, partial deficiency shown on IHC or western blot should be carefully interpreted, as it might be a secondary phenomenon, rather than a primary deficiency. For example, decreased caveolin-3 expression could be observed in inflammatory myopathy; reduced calpain-3 expression could occur in LGMD1C (caveolinopathy) and LGMD2J (titinopathy); reduced sarcoglycan or dystroglycan expression does not absolutely indicate sarcoglycanopathy (LGMD2C ~ 2 F) or α-dystroglycanopathy (LGMD2I, K, M, N, O and P) as it is usually shown in dystrophinopathy.

Recently, muscle imaging has emerged to be an additional tool used for the differential diagnosis.[75–78] For most LGMD2, the posterior compartment is usually more severely affected at both the thigh and calf levels, and the sartorius and gracilis muscles are often spared. In LGMD2A, the muscle involvement seems to start from the adductor magnus, spread to the semitendinosus and then all the hamstring muscles. In LGMD2B, the pattern is similar to that seen in LGMD2A but is usually more diffuse. Both proximal and distal leg muscles are often affected simultaneously; the calf muscles, especially the gastrocnemii and soleus muscles, are the earliest to be involved. In LGMD2C–2F, unlike other LGMD, the anterior compartment of the thigh is more severely affected. The lower leg muscles often have very mild changes. In LGMD2I, except for adductor magnus, the long head of the biceps femoris muscle is often involved early and severely. The vastus lateralis muscle is often spared. In LGMD2L, the pattern of muscle involvement is similar to that seen in LGMD2B, but more homogeneous in all patients. Compared to other LGMD2, gluteal muscle involvement is more severe in LGMD2A and 2I; asymmetric involvement is more commonly seen in LGMD2B and 2L. As LGMD1 and other types of LGMD2 are rare, there are very few muscle imaging reports available. It is worth noting that the muscle involvement of each LGMD does overlap, in part, and much depends on the disease stage.

MANAGEMENT AND FUTURE PERSPECTIVES

So far, there is no consensus on the standard care for LGMD, as this disease group is very heterogeneous in many aspects. However, cardiac conduction defects should be carefully monitored for certain types of LGMD, such as LGMD1B (laminopathy) and 1D/1E (desminopathy), as sudden death may occur.

There is no cure for any type of LGMD to date. However, variable animal models enhance the progress in the exploration of pathomechanism, which will be very helpful in providing novel avenues for therapeutic intervention. Most ongoing or recently completed clinical trials are focusing on LGMD2B, LGMD2C and LGMD2D, and involve gene therapy, stem cell therapy, steroid, myostatin inhibitor, proteosome inhibitor and other agents (www.clinicaltrials.gov). Nevertheless, there is no doubt that more and more therapeutic agents will be developed and clinical trials will be conducted with the increased understanding of clinical, biochemical, pathological and genetic features of each LGMD.

References

1. Walton JN, Nattrass FJ. On the classification, natural history and treatment of the myopathies. *Brain*. 1954;77(2):169–231.
2. Moore SA, Shilling CJ, Westra S, et al. Limb-girdle muscular dystrophy in the United States. *J Neuropathol Exp Neurol*. 2006;65(10):995–1003.
3. van der Kooi AJ, Frankhuizen WS, Barth PG, et al. Limb-girdle muscular dystrophy in the Netherlands: gene defect identified in half the families. *Neurology*. 2007;68(24):2125–2128.
4. Hicks D, Sarkozy A, Muelas N, et al. A founder mutation in Anoctamin 5 is a major cause of limb-girdle muscular dystrophy. *Brain*. 2011;134(Pt 1):171–182.
5. Sarkozy A, Hicks D, Hudson J, et al. ANO5 gene analysis in a large cohort of patients with anoctaminopathy: confirmation of male prevalence and high occurrence of the common exon 5 gene mutation. *Hum Mutat*. 2013;34(8):1111–1118.
6. Sveen ML, Schwartz M, Vissing J. High prevalence and phenotype-genotype correlations of limb girdle muscular dystrophy type 2I in Denmark. *Ann Neurol*. 2006;59(5):808–815.
7. Merlini L, Kaplan JC, Navarro C, et al. Homogeneous phenotype of the gypsy limb-girdle MD with the gamma-sarcoglycan C283Y mutation. *Neurology*. 2000;54(5):1075–1079.
8. Piccolo F, Jeanpierre M, Leturcq F, et al. A founder mutation in the gamma-sarcoglycan gene of gypsies possibly predating their migration out of India. *Hum Mol Genet*. 1996;5(12):2019–2022.
9. Vainzof M, Passos-Bueno MR, Pavanello RC, Marie SK, Oliveira AS, Zatz M. Sarcoglycanopathies are responsible for 68% of severe autosomal recessive limb-girdle muscular dystrophy in the Brazilian population. *J Neurol Sci*. 1999;164(1):44–49.
10. Hauser MA, Horrigan SK, Salmikangas P, et al. Myotilin is mutated in limb girdle muscular dystrophy 1A. *Hum Mol Genet*. 2000;9(14):2141–2147.
11. van der Ven PF, Wiesner S, Salmikangas P, et al. Indications for a novel muscular dystrophy pathway. gamma-filamin, the muscle-specific filamin isoform, interacts with myotilin. *J Cell Biol*. 2000;151(2):235–248.
12. Broers JL, Kuijpers HJ, Ostlund C, Worman HJ, Endert J, Ramaekers FC. Both lamin A and lamin C mutations cause lamina instability as well as loss of internal nuclear lamin organization. *Exp Cell Res*. 2005;304(2):582–592.
13. Muchir A, Bonne G, van der Kooi AJ, et al. Identification of mutations in the gene encoding lamins A/C in autosomal dominant limb girdle muscular dystrophy with atrioventricular conduction disturbances (LGMD1B). *Hum Mol Genet*. 2000;9(9):1453–1459.
14. Carbone I, Bruno C, Sotgia F, et al. Mutation in the CAV3 gene causes partial caveolin-3 deficiency and hyperCKemia. *Neurology*. 2000;54(6):1373–1376.
15. Hayashi T, Arimura T, Ueda K, et al. Identification and functional analysis of a caveolin-3 mutation associated with familial hypertrophic cardiomyopathy. *Biochem Biophys Res Commun*. 2004;313(1):178–184.

16. Harms MB, Sommerville RB, Allred P, et al. Exome sequencing reveals DNAJB6 mutations in dominantly-inherited myopathy. *Ann Neurol.* 2012;71(3):407–416.
17. Sarparanta J, Jonson PH, Golzio C, et al. Mutations affecting the cytoplasmic functions of the co-chaperone DNAJB6 cause limb-girdle muscular dystrophy. *Nat Genet.* 2012;44(4):450–455, S451–S452.
18. Hageman J, Rujano MA, van Waarde MA, et al. A DNAJB chaperone subfamily with HDAC-dependent activities suppresses toxic protein aggregation. *Mol Cell.* 2010;37(3):355–369.
19. Rose JM, Novoselov SS, Robinson PA, Cheetham ME. Molecular chaperone-mediated rescue of mitophagy by a Parkin RING1 domain mutant. *Hum Mol Genet.* 2011;20(1):16–27.
20. Watson ED, Geary-Joo C, Hughes M, Cross JC. The Mrj co-chaperone mediates keratin turnover and prevents the formation of toxic inclusion bodies in trophoblast cells of the placenta. *Development.* 2007;134(9):1809–1817.
21. Greenberg SA, Salajegheh M, Judge DP, et al. Etiology of limb girdle muscular dystrophy 1D/1E determined by laser capture microdissection proteomics. *Ann Neurol.* 2012;71(1):141–145.
22. Messina DN, Speer MC, Pericak-Vance MA, McNally EM. Linkage of familial dilated cardiomyopathy with conduction defect and muscular dystrophy to chromosome 6q23. *Am J Hum Genet.* 1997;61(4):909–917.
23. Kramerova I, Beckmann JS, Spencer MJ. Molecular and cellular basis of calpainopathy (limb girdle muscular dystrophy type 2A). *Biochim Biophys Acta.* 2007;1772(2):128–144.
24. Mercuri E, Bushby K, Ricci E, et al. Muscle MRI findings in patients with limb girdle muscular dystrophy with calpain-3 deficiency (LGMD2A) and early contractures. *Neuromuscul Disord.* 2005;15(2):164–171.
25. Bansal D, Campbell KP. Dysferlin and the plasma membrane repair in muscular dystrophy. *Trends Cell Biol.* 2004;14(4):206–213.
26. Chiu YH, Hornsey MA, Klinge L, et al. Attenuated muscle regeneration is a key factor in dysferlin-deficient muscular dystrophy. *Hum Mol Genet.* 2009;18(11):1976–1989.
27. de Luna N, Gallardo E, Soriano M, et al. Absence of dysferlin alters myogenin expression and delays human muscle differentiation "in vitro". *J Biol Chem.* 2006;281(25):17092–17098.
28. Kerr JP, Ziman AP, Mueller AL, et al. Dysferlin stabilizes stress-induced Ca2+ signaling in the transverse tubule membrane. *Proc Natl Acad Sci U S A.* 2013;110(51):20831–20836.
29. Confalonieri P, Oliva L, Andreetta F, et al. Muscle inflammation and MHC class I up-regulation in muscular dystrophy with lack of dysferlin: an immunopathological study. *J Neuroimmunol.* 2003;142(1–2):130–136.
30. De Luna N, Gallardo E, Sonnet C, et al. Role of thrombospondin 1 in macrophage inflammation in dysferlin myopathy. *J Neuropathol Exp Neurol.* 2010;69(6):643–653.
31. Politano L, Nigro V, Passamano L, et al. Evaluation of cardiac and respiratory involvement in sarcoglycanopathies. *Neuromuscul Disord.* 2001;11(2):178–185.
32. Hayashi T, Arimura T, Itoh-Satoh M, et al. Tcap gene mutations in hypertrophic cardiomyopathy and dilated cardiomyopathy. *J Am Coll Cardiol.* 2004;44(11):2192–2201.
33. Zhang S, Londhe P, Zhang M, Davie JK. Transcriptional analysis of the titin cap gene. *Mol Genet Genomics.* 2011;285(3):261–272.
34. Moreira ES, Wiltshire TJ, Faulkner G, et al. Limb-girdle muscular dystrophy type 2G is caused by mutations in the gene encoding the sarcomeric protein telethonin. *Nat Genet.* 2000;24(2):163–166.
35. Ferreiro A, Mezmezian M, Olive M, et al. Telethonin-deficiency initially presenting as a congenital muscular dystrophy. *Neuromuscul Disord.* 2011;21(6):433–438.
36. Paim JF, Cotta A, Vargas AP, et al. Muscle phenotypic variability in limb girdle muscular dystrophy 2G. *J Mol Neurosci.* 2013;50(2):339–344.
37. Moreira ES, Vainzof M, Marie SK, Sertie AL, Zatz M, Passos-Bueno MR. The seventh form of autosomal recessive limb-girdle muscular dystrophy is mapped to 17q11-12. *Am J Hum Genet.* 1997;61(1):151–159.
38. Frosk P, Weiler T, Nylen E, et al. Limb-girdle muscular dystrophy type 2H associated with mutation in TRIM32, a putative E3-ubiquitin-ligase gene. *Am J Hum Genet.* 2002;70(3):663–672.
39. Kudryashova E, Kudryashov D, Kramerova I, Spencer MJ. Trim32 is a ubiquitin ligase mutated in limb girdle muscular dystrophy type 2H that binds to skeletal muscle myosin and ubiquitinates actin. *J Mol Biol.* 2005;354(2):413–424.
40. Locke M, Tinsley CL, Benson MA, Blake DJ. TRIM32 is an E3 ubiquitin ligase for dysbindin. *Hum Mol Genet.* 2009;18(13):2344–2358.
41. Nicklas S, Otto A, Wu X, et al. TRIM32 regulates skeletal muscle stem cell differentiation and is necessary for normal adult muscle regeneration. *PLoS One.* 2012;7(1):e30445.
42. Sato T, Okumura F, Kano S, Kondo T, Ariga T, Hatakeyama S. TRIM32 promotes neural differentiation through retinoic acid receptor-mediated transcription. *J Cell Sci.* 2011;124(Pt 20):3492–3502.
43. Albor A, El-Hizawi S, Horn EJ, et al. The interaction of Piasy with Trim32, an E3-ubiquitin ligase mutated in limb-girdle muscular dystrophy type 2H, promotes Piasy degradation and regulates UVB-induced keratinocyte apoptosis through NFkappaB. *J Biol Chem.* 2006;281(35):25850–25866.
44. Ryu YS, Lee Y, Lee KW, et al. TRIM32 protein sensitizes cells to tumor necrosis factor (TNFα)-induced apoptosis via its RING domain-dependent E3 ligase activity against X-linked inhibitor of apoptosis (XIAP). *J Biol Chem.* 2011;286(29):25729–25738.
45. Borg K, Stucka R, Locke M, et al. Intragenic deletion of TRIM32 in compound heterozygotes with sarcotubular myopathy/LGMD2H. *Hum Mutat.* 2009;30(9):E831–E844.
46. Schoser BG, Frosk P, Engel AG, Klutzny U, Lochmuller H, Wrogemann K. Commonality of TRIM32 mutation in causing sarcotubular myopathy and LGMD2H. *Ann Neurol.* 2005;57(4):591–595.
47. Brockington M, Yuva Y, Prandini P, et al. Mutations in the fukutin-related protein gene (FKRP) identify limb girdle muscular dystrophy 2I as a milder allelic variant of congenital muscular dystrophy MDC1C. *Hum Mol Genet.* 2001;10(25):2851–2859.
48. Beltran-Valero de Bernabe D, Currier S, Steinbrecher A, et al. Mutations in the O-mannosyltransferase gene POMT1 give rise to the severe neuronal migration disorder Walker-Warburg syndrome. *Am J Hum Genet.* 2002;71(5):1033–1043.
49. van Reeuwijk J, Janssen M, van den Elzen C, et al. POMT2 mutations cause alpha-dystroglycan hypoglycosylation and Walker-Warburg syndrome. *J Med Genet.* 2005;42(12):907–912.

50. Murakami T, Hayashi YK, Noguchi S, et al. Fukutin gene mutations cause dilated cardiomyopathy with minimal muscle weakness. *Ann Neurol*. 2006;60(5):597–602.
51. Yoshida A, Kobayashi K, Manya H, et al. Muscular dystrophy and neuronal migration disorder caused by mutations in a glycosyltransferase, POMGnT1. *Dev Cell*. 2001;1(5):717–724.
52. Hara Y, Balci-Hayta B, Yoshida-Moriguchi T, et al. A dystroglycan mutation associated with limb-girdle muscular dystrophy. *N Engl J Med*. 2011;364(10):939–946.
53. Labeit S, Kolmerer B. Titins: giant proteins in charge of muscle ultrastructure and elasticity. *Science*. 1995;270(5234):293–296.
54. Young P, Ferguson C, Banuelos S, Gautel M. Molecular structure of the sarcomeric Z-disk: two types of titin interactions lead to an asymmetrical sorting of alpha-actinin. *EMBO J*. 1998;17(6):1614–1624.
55. Sorimachi H, Kinbara K, Kimura S, et al. Muscle-specific calpain, p94, responsible for limb girdle muscular dystrophy type 2A, associates with connectin through IS2, a p94-specific sequence. *J Biol Chem*. 1995;270(52):31158–31162.
56. Trinick J. Titin and nebulin: protein rulers in muscle? *Trends Biochem Sci*. 1994;19(10):405–409.
57. Salmikangas P, Mykkanen OM, Gronholm M, Heiska L, Kere J, Carpen O. Myotilin, a novel sarcomeric protein with two Ig-like domains, is encoded by a candidate gene for limb-girdle muscular dystrophy. *Hum Mol Genet*. 1999;8(7):1329–1336.
58. Carmignac V, Salih MA, Quijano-Roy S, et al. C-terminal titin deletions cause a novel early-onset myopathy with fatal cardiomyopathy. *Ann Neurol*. 2007;61(4):340–351.
59. Hackman P, Vihola A, Haravuori H, et al. Tibial muscular dystrophy is a titinopathy caused by mutations in TTN, the gene encoding the giant skeletal-muscle protein titin. *Am J Hum Genet*. 2002;71(3):492–500.
60. Ohlsson M, Hedberg C, Bradvik B, et al. Hereditary myopathy with early respiratory failure associated with a mutation in A-band titin. *Brain*. 2012;135(Pt 6):1682–1694.
61. Gerull B, Gramlich M, Atherton J, et al. Mutations of TTN, encoding the giant muscle filament titin, cause familial dilated cardiomyopathy. *Nat Genet*. 2002;30(2):201–204.
62. Ceyhan-Birsoy O, Agrawal PB, Hidalgo C, et al. Recessive truncating titin gene, TTN, mutations presenting as centronuclear myopathy. *Neurology*. 2013;81(14):1205–1214.
63. Bolduc V, Marlow G, Boycott KM, et al. Recessive mutations in the putative calcium-activated chloride channel Anoctamin 5 cause proximal LGMD2L and distal MMD3 muscular dystrophies. *Am J Hum Genet*. 2010;86(2):213–221.
64. Caputo A, Caci E, Ferrera L, et al. TMEM16A, a membrane protein associated with calcium-dependent chloride channel activity. *Science*. 2008;322(5901):590–594.
65. Almaca J, Tian Y, Aldehni F, et al. TMEM16 proteins produce volume-regulated chloride currents that are reduced in mice lacking TMEM16A. *J Biol Chem*. 2009;284(42):28571–28578.
66. Wahbi K, Behin A, Becane HM, et al. Dilated cardiomyopathy in patients with mutations in anoctamin 5. *Int J Cardiol*. 2013;168(1):76–79.
67. Tsutsumi S, Kamata N, Vokes TJ, et al. The novel gene encoding a putative transmembrane protein is mutated in gnathodiaphyseal dysplasia (GDD). *Am J Hum Genet*. 2004;74(6):1255–1261.
68. Gundesli H, Talim B, Korkusuz P, et al. Mutation in exon 1f of PLEC, leading to disruption of plectin isoform 1f, causes autosomal-recessive limb-girdle muscular dystrophy. *Am J Hum Genet*. 2010;87(6):834–841.
69. Rezniczek GA, Abrahamsberg C, Fuchs P, Spazierer D, Wiche G. Plectin 5′-transcript diversity: short alternative sequences determine stability of gene products, initiation of translation and subcellular localization of isoforms. *Hum Mol Genet*. 2003;12(23):3181–3194.
70. Paavilainen VO, Oksanen E, Goldman A, Lappalainen P. Structure of the actin-depolymerizing factor homology domain in complex with actin. *J Cell Biol*. 2008;182(1):51–59.
71. Smith FJ, Eady RA, Leigh IM, et al. Plectin deficiency results in muscular dystrophy with epidermolysis bullosa. *Nat Genet*. 1996;13(4):450–457.
72. Nakamura H, Sawamura D, Goto M, et al. Epidermolysis bullosa simplex associated with pyloric atresia is a novel clinical subtype caused by mutations in the plectin gene (PLEC1). *J Mol Diagn*. 2005;7(1):28–35.
73. Koss-Harnes D, Hoyheim B, Anton-Lamprecht I, et al. A site-specific plectin mutation causes dominant epidermolysis bullosa simplex Ogna: two identical de novo mutations. *J Invest Dermatol*. 2002;118(1):87–93.
74. Selcen D, Juel VC, Hobson-Webb LD, et al. Myasthenic syndrome caused by plectinopathy. *Neurology*. 2011;76(4):327–336.
75. Carlier PG, Mercuri E, Straub V. Applications of MRI in muscle diseases. *Neuromuscul Disord*. 2012;22(suppl 2):S41.
76. Mercuri E. An integrated approach to the diagnosis of muscle disorders: what is the role of muscle imaging? *Dev Med Child Neurol*. 2010;52(8):693.
77. Straub V, Carlier PG, Mercuri E. TREAT-NMD workshop: pattern recognition in genetic muscle diseases using muscle MRI: 25-26 February 2011, Rome, Italy. *Neuromuscul Disord*. 2012;22(suppl 2):S42–S53.
78. ten Dam L, van der Kooi AJ, van Wattingen M, de Haan RJ, de Visser M. Reliability and accuracy of skeletal muscle imaging in limb-girdle muscular dystrophies. *Neurology*. 2012;79(16):1716–1723.

CHAPTER

93

The Congenital Myopathies

Heinz Jungbluth[*,†], *Caroline Sewry*[‡,§], *and Francesco Muntoni*[‡]

[*]Evelina London Children's Hospital, Guy's & St. Thomas' Hospital NHS Foundation Trust, London, UK

[†]King's College, London, UK

[‡]University College London Institute of Child Health and Great Ormond Street Hospital for Children, London, UK

[§]RJAH Orthopaedic Hospital, Oswestry, UK

INTRODUCTION

The most common congenital myopathies (CMs)—central core disease (CCD), multi-minicore disease (MmD), myotubular/centronuclear myopathy (MTM/CNM), nemaline myopathy (NM) and congenital fiber type disproportion (CFTD)—are a heterogeneous group of inherited early-onset neuromuscular disorders defined by and named after characteristic features on muscle biopsy. Although presentation is usually nonspecific with hypotonia and generalized weakness pronounced in axial and proximal muscles, certain features such as external ophthalmoplegia, prominent distal muscle involvement or marked exertional myalgia may suggest a specific CM. Respiratory impairment is common and often out of proportion to the limb muscle weakness. Primary cardiac involvement is not usually a feature but distinct conditions with primary cardiomyopathies have to be considered in the differential diagnosis. Patients with mutations in the skeletal muscle ryanodine receptor (*RYR1*) gene, the most common identifiable cause, may be susceptible to malignant hyperthermia.

The CMs are associated with all modes of Mendelian inheritance. Many of the causative genes encode proteins that are involved in sarcomeric assembly, membrane trafficking, autophagy, or calcium homeostasis and excitation–contraction coupling. Mutations in specific genes are often associated with different histopathological features, with considerable histopathological overlap between genes primarily responsible for a well-defined pathological entity. This fact complicates the diagnostic pathway and may be due to structural association of the respective gene products, or their involvement in related physiological processes. Management is currently mainly supportive based on a multidisciplinary team approach but more specific pharmacological and genetic therapies are approaching rapidly.

CLINICAL AND HISTOPATHOLOGICAL FEATURES

The congenital myopathies are a group of inherited neuromuscular conditions associated with diverse genetic backgrounds and characterized by variable degrees of muscle weakness and the presence of characteristic structural abnormalities on muscle biopsy (for general review see [1]). The concept of the congenital myopathies as distinct entities dates back to the 1950s and 60s, when new histochemical techniques were introduced to the investigation of diseased muscle, resulting in identification of a number of structural abnormalities—central cores, multi-minicores, nemaline rods, and central nuclei—considered to be specific at the time. The abundance of any such structural abnormality on muscle biopsy resulted in the description of conditions such as central core disease (CCD), multi-minicore disease (MmD), nemaline myopathy (NM) and centronuclear myopathy (CNM). Congenital fiber type disproportion (CFTD) is another pathological entity whose status as a distinct CM has, however, not been universally acknowledged. Apart from the defining abnormality, most CMs share type 1 predominance and hypotrophy as a common feature. Very rare additional CM characterized by other structural defects such as caps, spheroid bodies,

Rosenberg's Molecular and Genetic Basis of Neurological and Psychiatric Disease
http://dx.doi.org/10.1016/10.1016/B978-0-12-410529-4.00093-0

zebra bodies, fingerprint bodies, cylindrical spirals, tubular aggregates have also been described, but in view of their rarity they will not be discussed in this chapter.

The CMs are associated with all modes of Mendelian inheritance. Autosomal dominant, recessive, or X-linked recessive mutations in genes involved in sarcomeric assembly, intracellular membrane trafficking, autophagy, calcium homeostasis, and excitation–contraction coupling, or a combination of the above, have been identified (Table 93.1). Mutations in the skeletal muscle ryanodine receptor (*RYR1*) gene are the most common cause. Recent genetic advances have both clarified and confused the concept of the CMs, as different mutations in the same gene may give rise to different and/or more than one histopathological features, while the same histopathological feature may be caused by mutations in different genes; this probably reflects structural or functional association of the respective gene products.

Presentation of the CMs is usually at birth or in early infancy with marked hypotonia and variable weakness often involving the facial, ocular, and bulbar muscles, but may occur later in childhood with motor developmental delay and mainly proximal weakness. The occurrence of selective muscle involvement (i.e., ophthalmoplegia or distal weakness) or exertional myalgia are more common in some subgroups than others, and may hint at the correct diagnosis. Scoliosis is frequently seen, corresponding to prominent axial weakness. Many patients have joint hypermobility, and contractures other than those affecting the Achilles tendon are uncommon, except the most severe neonatal cases who may present with arthrogryposis. Associated and often disproportionate respiratory and bulbar involvement is common to many forms and may necessitate ventilatory support and nasogastric feeds or gastrostomy insertion, respectively. Cardiac involvement is rare, and usually secondary to substantial respiratory impairment. Central nervous system (CNS) involvement or an associated neuropathy are not typical features, but some subgroups (e.g., with mutations in *TPM2/3, DNM2* or *RYR1*) may have an associated neuromuscular transmission defect. Apart from the most severe early-onset cases with high mortality, the typical course during childhood and adolescence is static or only slowly progressive, with the large majority of patients achieving and maintaining independent ambulation. The degree of respiratory involvement is the most important prognostic factor. Clinical and histopathological features of specific CMs are outlined in more detail below and typical muscle biopsy changes are illustrated in Figure 93.1.

Congenital Myopathies with Cores

The core myopathies, CCD and MmD, the most frequent forms of CM, are pathologically characterized by focally reduced oxidative enzyme activity of variable appearance ("central cores, multi-minicores;" for review see [2]). They are genetically heterogeneous: In *RYR1*-related core myopathies there is often marked predominance or unifomity of type 1 fibers, an increase in internal or central nucleation and, occasionally, additional nemaline rods. Both *RYR1*- and *SEPN1*-related myopathies can be associated with a disproportion in the size of type 1 versus type 2 fibers with or without cores,[3,4] and substantial increase in fat and connective tissue, which can cause pathological confusion with a congenital muscular dystrophy. Cores are highly variable, ranging from extensive areas with reduced oxidative enzyme activity running along an appreciable proportion of the muscle fiber axis on longitudinal sections ("central cores," often associated with dominant *RYR1* mutations) to multiple focal areas with reduced oxidative enzyme activity affecting only a few sarcomeres ("multi-minicores," often associated with recessive *RYR1* or *SEPN1* mutations). It is important nevertheless to acknowledge that the presence of cores of variable size is a nonspecific and a common feature that occurs in several other disorders, including neurogenic disorders, muscular dystrophies, and myasthenic conditions. Mutations in *MYH7*[5] and *TTN*[6] are also increasingly recognized in the differential diagnosis of core myopathies.

Onset of CCD due to dominant *RYR1* mutations (for review see [7]) is typically in infancy or early childhood with proximal weakness pronounced in the hip girdle. Exertional myalgia is common and may be a presenting feature in mild cases. Facial involvement is typically mild and extraocular muscles are usually spared. Bulbar and respiratory involvement is mild in most cases with the exception of the rare severe neonatal cases. Orthopedic complications, in particular congenital dislocation of the hips and scoliosis, are common. Most patients achieve independent ambulation and have a static or only slowly progressive course. Serum creatine kinase (CK) levels may only rarely be elevated (up to 5 times normal).

Clinical features of MmD (for review see [7]), typically associated with autosomal recessive inheritance, depend on the gene responsible. The first gene to be discovered was *SEPN1*,[8] and the associated myopathy is characterized by marked axial and shoulder girdle weakness, early spinal rigidity, scoliosis, and respiratory impairment. Mild facial weakness is common, but ophthalmoplegia is not typical. Recessively inherited *RYR1*-related core myopathies with multi-minicores show a distribution of weakness and wasting with some similarity to the *SEPN1*-related form but lack the severe respiratory involvement; however, they often are associated with ophthalmoplegia and bulbar

TABLE 93.1 Genetics of the Congenital Myopathies

	Gene	Locus	Inheritance	Protein
CONGENITAL MYOPATHIES WITH NEMALINE RODS				
Nemaline myopathy	*ACTA1*	1q42	AD or AR	Skeletal α-actin
	NEB	2q2	AR	Nebulin
	TPM3	1q2	AD or AR	α-tropomyosin
	TPM2	9p13	AD or AR	β-tropomyosin
	TNNT1	19q13	AR	Troponin T
	CFL2	14q12	AR	Cofilin-2
	KBTBD13	15q25	AD	Kelch 13
	KLHL40	3p21	AR	Kelch-like 40
	KLHL41	2q31	AR	Kelch-like 41
CONGENITAL MYOPATHIES WITH CORES*				
Central core disease	*RYR1*	19q13	AD or AR	Ryanodine receptor 1
Multi-minicore disease	*SEPN1*	1p36	AR	Selenoprotein N1
± cardiomyopathy	*TTN*	2q31	AR	Titin
	MYH7	14q11	AD	Slow myosin heavy chain
CONGENITAL MYOPATHIES WITH CENTRAL NUCLEI				
Myotubular myopathy	*MTM1*	Xq28	XLR	Myotubularin
Centronuclear myopathy	*DNM2*	19p13	AD	Dynamin-2
	BIN1	2q14	AR	Amphiphysin
	RYR1	19q13	AR†	Ryanodine receptor 1
	CCDC78	16p13.3	AD	Coiled-coil domain-containing protein 78
± cardiomyopathy	*TTN*	2q31	AR	Titin
CONGENITAL FIBER TYPE DISPROPORTION				
Congenital fiber type disproportion	*ACTA1*	1q42	AD	Skeletal α-actin
	SEPN1	1p36	AR	Selenoprotein N1
	TPM3	1q2	AD	α-tropomyosin
	TPM2	9q13	AD	β-tropomyosin
± cardiomyopathy	*MYH7*	14q11	AD	Slow myosin heavy chain
	RYR1	19q13	AR	Ryanodine receptor 1
	HACD1	10p12	AR	3-hydroxyacyl-CoA dehydratase 1

Only the most common genetic backgrounds and predominant modes of inheritance are indicated.

**Core-like lesions occur in association with defects in several genes (see text).*

†One isolated case with RYR1*-related CNM and a* de novo *heterozygous missense mutation has also been reported.*

Abbreviations: AD, autosomal dominant; AR, autosomal recessive; XLR, X-linked recessive.

weakness.[9] A primary cardiomyopathy is not a feature of *SEPN1*- or *RYR1*-related core myopathies and, if present in conjunction with cores on muscle biopsy, should raise the suspicion of other conditions, for example *MYH7*- or *TTN*-related myopathies.[5,6]

Malignant hyperthermia susceptibility (MHS) is a potential risk in *RYR1*-related CMs, although the association is less certain for some of the recessive compared to the dominant *RYR1* mutations.

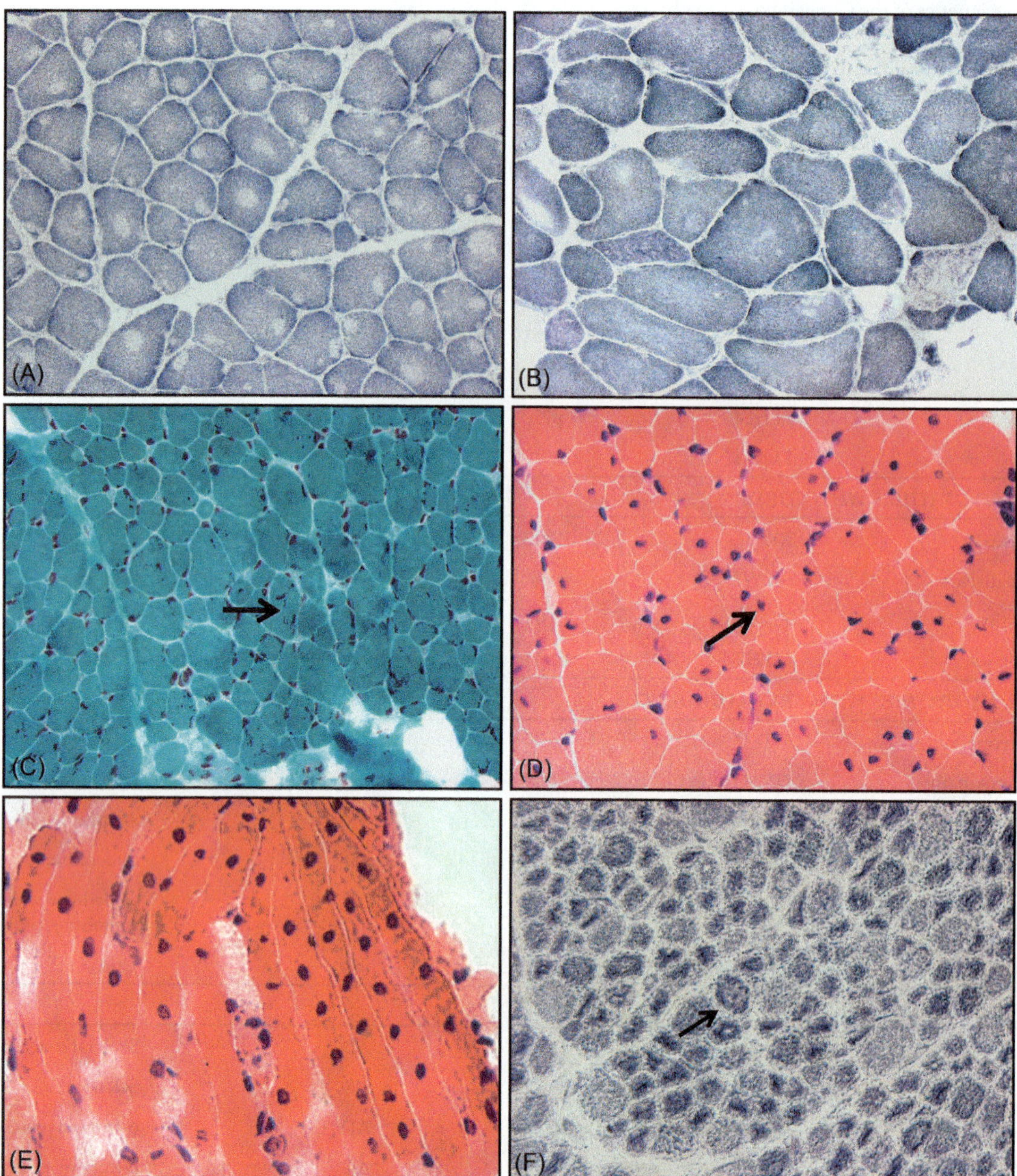

FIGURE 93.1 **Histopathological abnormalities in common congenital myopathies.** Muscle biopsies showing: (**A**) large central or peripheral cores with absence of staining for NADH-tetrazolium reductase (NADH-TR) and uniform fiber typing in a case with a dominantly inherited *RYR1* mutation; (**B**) small cores and unevenness of staining for NADH-TR in both fiber types in a case with recessively inherited *SEPN1* mutations; (**C**) red stained rods with the Gömöri trichrome technique (arrow) and a population of small diameter fibers in a case with recessively inherited *NEB* mutations; (**D, E**) hematoxylin and eosin staining showing several fibers with central nuclei (arrow) which are spaced in longitudinal section (**E**) in a case of X-linked *MTM1* myotubular myopathy; (**F**) pale peripheral halos (arrow), darkly stained centers and several small more darkly stained type 1 fibers (NADH-TR).

Congenital Myopathies with Central Nuclei

The CNMs[2,10] are characterized by a variable number of fibers with centralized nuclei, often spaced down the long axis of the fibers. Central areas of enhanced oxidative enzyme activity and a pale peripheral halo are also common. This pathological appearance is similar to that seen in congenital myotonic dystrophy. Strictly centralized nuclei are more common than multiple internalized nuclei in the *MTM1*-, *DNM2*-, and *BIN1*-related forms,[11–13] the opposite applies to *RYR1*-related cases.[14] A radial distribution of sarcoplasmic strands with staining for NADH-tetrazolium reductase (NADH-TR) and with periodic acid–Schiff (PAS) staining is seen in some, in particular, *DNM2*-related cases;[12] the latter may also exhibit excess endomysial connective tissue as seen in a muscular dystrophy but necrosis is not usually a feature. Additional histopathological features include cores evolving over time, particularly in *RYR1*-related CNM, "necklace" fibers often seen in milder *MTM1*-mutated cases or female carriers, and, in most forms, ultrastructural triad abnormalities.[15]

Extraocular muscle involvement is the most consistent clinical feature of all the CNMs (for review see [16]). X-linked myotubular myopathy (XLMTM) due to X-linked recessive myotubularin (*MTM1*) mutations is the most severe form and often associated with antenatal onset, multiple contractures, profound hypotonia and weakness, and associated bulbar and respiratory involvement almost always necessitating ventilation. In ~80% of cases the condition is fatal over weeks or months unless constant respiratory support is provided. Some long-term survivors show associated medical problems such as pyloric stenosis, gallstones, and hepatic peliosis. Neuromuscular junction abnormalities have been reported in XLMTM and other forms of CNM.[17] Some XLMTM carriers may develop mild weakness, sometimes late in life, but more severe manifestations in females are usually due to skewed X-inactivation or other X-chromosomal abnormalities. Dominantly inherited CNM associated with mutations in *DNM2* is frequently associated with a relatively mild condition,[12] but even more severe cases with neonatal onset may improve over time provided respiratory complications are managed proactively. Facial weakness, ptosis, and ophthalmoplegia are common. Muscle weakness is mainly proximal and axial, with additional distal involvement in many patients and, occasionally, marked muscle hypertrophy, particularly affecting the calves. Signs of subtle peripheral or central nervous system involvement, neutropenia or cataracts have occasionally been observed. Recessively inherited CNM due *RYR1* mutations[14] is clinically similar to other myopathies due to recessive mutations in the same gene. CNM due to recessive and, less frequently, dominant mutations in *BIN1* has only been reported in few families,[13,18] with variable severity depending on the mode of inheritance.

Congenital Myopathies with Nemaline Rods

The defining histopathological feature of nemaline myopathy (NM; for review see [2]) is the presence of numerous nemaline rod bodies, which stain red with the Gömöri trichrome stain and are prominent in semithin resin sections stained with Toluidine Blue. Confirmation of the rods with electron microscopy (EM) is needed, especially in cases in which rods are very small. The different genetic forms of NM can rarely be distinguished on histopathological grounds. Exceptions are cases with *ACTA1* mutations that have nuclear rods or accumulation of actin, or show cardiac actin in all fibres.[19] Nemaline rods are also not specific and may appear in other clinical contexts, in normal eye muscles and at myotendinous junctions.

Clinical features of NM are highly variable depending on the genetic background (for review see [20]): The most common form of NM due to recessive *NEB* mutations[21] is characterized by onset in infancy with marked hypotonia, axial and proximal weakness, and pronounced bulbar involvement, the latter often the most prominent feature at presentation. Distal involvement in the lower limb becomes prominent later in life and may be a presenting feature in some patients. Respiratory involvement is almost universal in *NEB*-related NM, may present throughout life and is disproportionate to skeletal muscle involvement. *ACTA1*-related NM often presents with profound weakness from birth, particularly if due to *de novo* mutations, but milder forms have also been reported. An associated cardiomyopathy is not a feature of *NEB*-related NM but has been reported rarely in patients with *ACTA1* mutations. More specific clinical features include the tremor and progressive contractures in *TNNT1*-related NM prevalent in the Old Order Amish,[22] and the distinct slowness of movement seen in association with *KBTBD13* mutations.[23] Recently two genes responsible for the very severe end of the NM spectrum, often not compatible with life, have been reported, *KLHL40*[24] and *KLHL41*.[25]

Congenital Fiber Type Disproportion

Congenital fiber type disproportion (CFTD) is defined by type 1 fibers being at least 25% smaller than type 2 fibers in the absence of other pathological features on muscle biopsy (for review see [26]). The concept of CFTD as a distinct myopathy, however, is controversial, as type 1 fibers are often smaller than type 2 in all congenital myopathies and specific histopathological features may evolve over time. CFTD has recently been attributed to mutations in genes implicated in several forms of CM,[3,4,27–31] and associated clinical features have been similar to previously reported patients with the same genetic background but a different histopathological presentation.

MOLECULAR GENETICS AND DISEASE MECHANISMS

Congenital Myopathies with Cores

Dominantly inherited CCD and the MHS trait are mainly associated with heterozygous missense mutations often affecting the C-terminal of the skeletal muscle ryanodine receptor (*RYR1*) gene and other *RYR1* mutational hotspots.[32] Recessively inherited MmD is often due to compound heterozygosity for truncating and missense mutation distributed throughout the *RYR1* coding sequence and associated with marked reduction of the functional RyR1 protein,[32] the principal sarcoplasmic reticulum (SR) calcium release channel with a crucial role in excitation–contraction coupling. Two models for RyR1 receptor malfunction in CCD have been proposed (for review see [7]), depletion of SR calcium stores at rest with resulting increase in cytosolic calcium levels ("leaky channel" hypothesis), and disturbance of excitation–contraction coupling (E–C uncoupling hypothesis). Disturbed E–C coupling, probably mediated by marked reduction of RyR1 and associated proteins, is also a likely mechanism in myopathies associated with recessive *RYR1* mutations.[33] A number of animal models carrying knocked-in *RYR1* mutations related to MHS and CCD provide valuable insights into the pathophysiology and the evolution of associated histopathological changes (for review see [34]). There is currently no murine model for recessive *RYR1*-related myopathies; however, the sporadic zebrafish *relatively relaxed* mutant[35] mimics one of the probable molecular mechanisms underlying recessive *RYR1*-related core myopathies closely.

Selenoprotein N encoded by *SEPN1*[8] is a member of a family of proteins that mediate the effect of selenium and are involved in antioxidant defense systems and various metabolic pathways. *SEPN1* mutations are predominantly truncating, but the missense mutations reported result in a very severe phenotype. As demonstrated in sepn1−/− mouse and zebrafish models, deficiency of selenoprotein N is associated with disturbed myogenesis,[36] aberrant redox regulation and abnormalities of calcium homeostasis and excitation–contraction coupling,[37] the latter probably explaining the clinicopathological similarities between *RYR1*- and *SEPN1*-related core myopathies.

Congenital Myopathy with Central Nuclei

CNM has been associated with X-linked recessive mutations in the *MTM1* gene encoding myotubularin,[11] autosomal dominant mutations in *DNM2* encoding dynamin 2,[12] autosomal recessive mutations in *RYR1*,[14] and *BIN1* encoding amphiphysin 2,[13] the latter recently also implicated in a few families with a possible dominant form of CNM.[18] Dominant mutations in *CCDC78*[38] and recessive mutations in *TTN*[39] have recently emerged as additional causes of congenital myopathies with central nuclei, with or without additional cores.

Models for the investigation of affected cellular pathways (for review see [18]) are now available in yeast, *Caenorhabditis elegans*, *Drosophila*, zebrafish, mouse, and dog and have suggested several potential pathogenic mechanisms, many of them common to different forms of CNM. These include alteration of triads and calcium handling, defects of the neuromuscular junction, satellite cell activation, mitochondria, and the desmin cytoskeleton, as well as the autophagy pathway. Several proof-of-concept studies aimed at therapy development have already been reported, including replacement of the myotubularin protein through either viral transfer[40] or enzyme replacement therapy,[41] the use of drugs acting on the neuromuscular junction or modifying the autophagy pathway.

Congenital Myopathies with Nemaline Rods

NM has been associated with mutations in nine genes to date, most commonly recessive mutations in the nebulin (*NEB*) gene,[21] and (*de novo*) dominant mutations in the skeletal muscle α-actin (*ACTA1*) gene.[42] Dominant mutations in the α-tropomyosin (*TPM3*)[43] and the β-tropomyosin (*TPM2*)[44] genes, as well as recessive mutations in the slow troponin T (*TNNT1*),[22] cofilin-2 (*CFL2*),[45] Kelch-repeat and BTB (POZ) domain containing 13 (*KBTBD 13*),[23] *KLHL40*[24] and *KLHL41*[25] are less common, although founder mutations in *TNNT1*,[22] *KHL40*, and *TPM3* have been reported in certain populations.

The pathogenic mechanisms concerning the most commonly mutated genes have been extensively studied (for review see [20]): Dominant *ACTA1* mutations exert a dominant negative effect on muscle function, possibly mediated by lower calcium sensitivity.[46] Recessively inherited *ACTA1* mutations do not express any functional protein, and severity of the phenotype is possibly dependent on compensatory expression of the cardiac actin isoform. *NEB* mutations are likely to affect the specific role of nebulin in thin filament regulation and force generation.[47]

Congenital Fiber Type Disproportion

CFTD has recently been attributed to mutations in genes implicated in other forms of CM, including *RYR1*, *SEPN1*, *ACTA1*, *TPM2*, *TMP3*, *MYH7*, and *HACD1*,[3,4,27–31] suggesting similar pathogenic mechanisms.

DIFFERENTIAL DIAGNOSIS

The differential diagnosis in the patient presenting with a congenital myopathy depends on the age of presentation and the disease characteristics of the specific condition.

A wide range of non-neurological, neurological, and neuromuscular causes has to be considered when confronted with the most common presentation, profound hypotonia and weakness in the neonatal period or early infancy ("the floppy infant syndrome"). Non-neurological conditions that have to be considered in the floppy infant include severe systemic illness such as sepsis, chromosomal abnormalities including Down syndrome or Prader–Willi syndrome, and primary metabolic abnormalities, in particular Pompe disease, mitochondrial conditions, or systemic disorders of glycosylation. Primary neurological causes that at least initially may give rise to profound hypotonia include severe perinatal hypoxemia and neonatal encephalopathies. Within the group of primary neuromuscular conditions presenting early in life with both hypotonia and weakness, congenital myotonic dystrophy and spinal muscular atrophy are the most common and have to be excluded by specific genetic testing. In cases where ptosis, extraocular, and bulbar involvement are prominent, congenital myasthenic syndromes (CMS) as well as autoimmune transient neonatal myasthenia gravis (MG) are important considerations. There is also some clinicopathological overlap between the congenital myopathy and some of the congenital muscular dystrophies, in particular forms due to mutations in the *SEPN1* and the *COL6A1*, *2* and *3* genes, the latter in particular often mimicking *RYR1*-related myopathies.

In the child presenting later in life with motor developmental delay and proximal weakness, milder forms of SMA and late presentations of congenital myasthenic syndromes have to be considered in the differential diagnosis.

TESTING

The process of establishing the diagnosis of a specific congenital myopathy is mainly based on a detailed clinical assessment, ancillary investigations, and a muscle biopsy to identify the characteristic histopathological abnormality. A detailed overview of the approach to the diagnosis of the congenital myopathies has been recently published by an international expert consortium.[48]

CK levels are usually normal or only mildly elevated, but moderate elevations may be seen in *RYR1*-related myopathies, particularly if muscle cramps are a feature. Neurophysiological studies including electromyography (EMG), stimulated single fiber EMG (ST-SFEMG) and nerve conduction studies (NCS) may help in the differential diagnosis with congenital myasthenic syndromes, bearing in mind however that some CMs may show a modest functional involvement of the neuromuscular junction.

Muscle imaging applying muscle ultrasound (US) and muscle magnetic resonance imaging (MRI) has become an increasingly important diagnostic modality (for review see [49]), Muscle imaging may reveal a pattern of selective muscle involvement suggestive of specific genetic backgrounds, particularly in congenital myopathies due to mutations in *RYR1*, *SEPN1*, and *NEB*.

Considering that the diagnosis rests on histopathological appearance, muscle biopsy and investigation of diseased muscle tissue with a standard panel of histological and histochemical stains has been historically the single most important investigation in the assessment of the congenital myopathies. Additional immunohistochemical stains may be of value to distinguish CMs from the (congenital) muscular dystrophies, particularly in *RYR1*- and *DNM2*-related forms where increases in fat and connective tissue may be marked. Immunoanalysis of developmental, fast and slow myosin isoforms can also be useful. EM is very helpful to clarify or confirm the pathognomonic structural abnormalities seen with light microscopy.

Diagnostic genetic testing by Sanger sequencing is now routinely available for most causative genes, although this approach remains problematic and costly when applied to very large genes such as *RYR1* and *NEB*. These difficulties may be partly overcome by the more widespread introduction of large-scale parallel sequencing into the diagnosis of the CMs.

MANAGEMENT

Management of the CMs is mainly supportive and based on a multidisciplinary team approach involving a neuromuscular neurologist, respiratory physicians, orthopedic surgeons, physiotherapists, occupational therapists, and speech language therapists. Pharmacological therapies aimed at common pathogenic mechanisms are at the evaluation stage and gene therapy for the most severe forms is at the horizon. Detailed guidelines for the management of patients with CMs have been recently published by an international expert consortium.[50]

Supportive management of the individual patient should be tailored to an understanding of the natural history of the specific CM and anticipation of associated complications. Regular physiotherapy and provision of orthotic support is of benefit to all patients in order to promote mobility and to prevent development of contractures. Dysarthria is common and will benefit from dedicated speech therapy input. Substantial bulbar involvement and poor weight gain may require at least temporary gastrostomy insertion. Regular respiratory function monitoring (including overnight oxymetry) and timely institution of (noninvasive) ventilation are essential particularly in forms where respiratory impairment is a prominent feature. Regular cardiac monitoring is relevant in individuals where the genetic defect is uncertain, and in patients with significant respiratory impairment not controlled by respiratory intervention because of the associated risk of cor pulmonale. Scoliosis correction and other orthopedic surgery should be undertaken at a center with experience in the management of neuromuscular disorders and postoperative mobilization should be considered as soon as feasible. MHS has to be anticipated in the anesthetic management of patients particularly those with *RYR1*-involvement or uncertain genetic background.

Pharmacological therapies have only been investigated in a small number of patients and include tyrosine supplementation in NM[51] and oral salbutamol in *RYR1*-related myopathies.[52] Antioxidants such as acetylcysteine have been suggested as compounds of potential benefit for *SEPN1*- and *RYR1*-related myopathies but no human data are available.[53] Recent reports of an associated neuromuscular transmission defect suggest that some CMs may at least be partly amenable to acetylcholine esterase inhibitors or other pharmacological agents used in the treatment of myasthenic conditions.[17]

Genetic and enzyme replacement therapies are approaching the clinical trial stage in XLMTM.[40,41] Upregulation of proteins with compensatory potential such as cardiac actin in *ACTA1* null mutants has been promoted with some success in animal models of NM.[54]

References

1. Wallgren-Petterson C, Jungbluth H. Congenital (structural) myopathies. In: Rimoin DL, Connor JM, Pyeritz RE, Korf B, eds. Emery and Rimoin's Principles and Practice of Medical Genetics. Vol 3. Philadelphia, PA: Elsevier Churchill Livingston; 2007:2963–3000.
2. Dubowitz V, Sewry CA, Oldfors A. Muscle Biopsy: A Practical Approach. 4th ed. Oxford: Elsevier; 2013.
3. Clarke NF, Waddell LB, Cooper ST, et al. Recessive mutations in RYR1 are a common cause of congenital fiber type disproportion. *Hum Mutat*. 2010;31:E1544–E1550.
4. Clarke NF, Kidson W, Quijano-Roy S, et al. SEPN1: associated with congenital fiber-type disproportion and insulin resistance. *Ann Neurol*. 2006;59:546–552.
5. Cullup T, Lamont PJ, Cirak S, et al. Mutations in MYH7 cause Multi-minicore Disease (MmD) with variable cardiac involvement. *Neuromuscul Disord*. 2012;22:1096–1104.
6. Chauveau C, Bonnemann CG, Julien C, et al. Recessive TTN truncating mutations define novel forms of core myopathy with heart disease. *Hum Mol Genet*. 2014;23:980–991.
7. Jungbluth H, Sewry CA, Muntoni F. Core myopathies. *Semin Pediatr Neurol*. 2011;18:239–249.
8. Ferreiro A, Quijano-Roy S, Pichereau C, et al. Mutations of the selenoprotein N gene, which is implicated in rigid spine muscular dystrophy, cause the classical phenotype of multiminicore disease: reassessing the nosology of early-onset myopathies. *Am J Hum Genet*. 2002;71:739–749.
9. Jungbluth H, Zhou H, Hartley L, et al. Minicore myopathy with ophthalmoplegia caused by mutations in the ryanodine receptor type 1 gene. *Neurology*. 2005;65:1930–1935.
10. Romero NB, Bitoun M. Centronuclear myopathies. *Semin Pediatr Neurol*. 2011;18:250–256.
11. Laporte J, Hu LJ, Kretz C, et al. A gene mutated in X-linked myotubular myopathy defines a new putative tyrosine phosphatase family conserved in yeast. *Nat Genet*. 1996;13:175–182.
12. Bitoun M, Maugenre S, Jeannet PY, et al. Mutations in dynamin 2 cause dominant centronuclear myopathy. *Nat Genet*. 2005;37:1207–1209.
13. Nicot AS, Toussaint A, Tosch V, et al. Mutations in amphiphysin 2 (BIN1) disrupt interaction with dynamin 2 and cause autosomal recessive centronuclear myopathy. *Nat Genet*. 2007;39:1134–1139.
14. Wilmshurst JM, Lillis S, Zhou H, et al. RYR1 mutations are a common cause of congenital myopathies with central nuclei. *Ann Neurol*. 2010;68:717–726.
15. Toussaint A, Cowling BS, Hnia K, et al. Defects in amphiphysin 2 (BIN1) and triads in several forms of centronuclear myopathies. *Acta Neuropathol*. 2011;121:253–266.
16. Romero NB. Centronuclear myopathies: a widening concept. *Neuromuscul Disord*. 2010;20:223–228.

17. Robb SA, Sewry CA, Dowling JJ, et al. Impaired neuromuscular transmission and response to acetylcholinesterase inhibitors in centronuclear myopathies. *Neuromuscul Disord*. 2011;21:379–386.
18. Jungbluth H, Wallgren-Pettersson C, Laporte JF. 198th ENMC international workshop: 7th workshop on centronuclear (Myotubular) myopathies, 31st May – 2nd June 2013, Naarden, The Netherlands. *Neuromus Disord*. 2013;23:1033–1043.
19. Goebel HH, Warlo I. Nemaline myopathy with intranuclear rods–intranuclear rod myopathy. *Neuromuscul Disord*. 1997;7:13–19.
20. Wallgren-Pettersson C, Sewry CA, Nowak KJ, Laing NG. Nemaline myopathies. *Semin Pediatr Neurol*. 2011;18:230–238.
21. Pelin K, Hilpela P, Donner K, et al. Mutations in the nebulin gene associated with autosomal recessive nemaline myopathy. *Proc Natl Acad Sci U S A*. 1999;96:2305–2310.
22. Johnston JJ, Kelley RI, Crawford TO, et al. A novel nemaline myopathy in the Amish caused by a mutation in troponin T1. *Am J Hum Genet*. 2000;67:814–821.
23. Sambuughin N, Yau KS, Olive M, et al. Dominant mutations in KBTBD13, a member of the BTB/Kelch family, cause nemaline myopathy with cores. *Am J Hum Genet*. 2010;87:842–847.
24. Ravenscroft G, Miyatake S, Lehtokari VL, et al. Mutations in KLHL40 are a frequent cause of severe autosomal-recessive nemaline myopathy. *Am J Hum Genet*. 2013;93:6–18.
25. Gupta VA, Ravenscroft G, Shaheen R, et al. Identification of KLHL41 mutations implicates BTB-Kelch-mediated ubiquitination as an alternate pathway to myofibrillar disruption in nemaline myopathy. *Am J Hum Genet*. 2013;93:1108–1117.
26. Clarke NF. Congenital fiber-type disproportion. *Semin Pediatr Neurol*. 2011;18:264–271.
27. Clarke NF, Ilkovski B, Cooper S, et al. The pathogenesis of ACTA1-related congenital fiber type disproportion. *Ann Neurol*. 2007;61:552–561.
28. Clarke NF, Kolski H, Dye DE, et al. Mutations in TPM3 are a common cause of congenital fiber type disproportion. *Ann Neurol*. 2008;63:329–337.
29. Laing NG, Clarke NF, Dye DE, et al. Actin mutations are one cause of congenital fibre type disproportion. *Ann Neurol*. 2004;56:689–694.
30. Ortolano S, Tarrio R, Blanco-Arias P, et al. A novel MYH7 mutation links congenital fiber type disproportion and myosin storage myopathy. *Neuromuscul Disord*. 2011;21:254–262.
31. Muhammad E, Reish O, Ohno Y, et al. Congenital myopathy is caused by mutation of HACD1. *Hum Mol Genet*. 2013;22:5229–5236.
32. Klein A, Lillis S, Munteanu I, et al. Clinical and genetic findings in a large cohort of patients with ryanodine receptor 1 gene-associated myopathies. *Hum Mutat*. 2012;33:981–988.
33. Zhou H, Rokach O, Feng L, et al. RyR1 deficiency in congenital myopathies disrupts excitation-contraction coupling. *Hum Mutat*. 2013;34:986–996.
34. Maclennan DH, Zvaritch E. Mechanistic models for muscle diseases and disorders originating in the sarcoplasmic reticulum. *Biochim Biophys Acta*. 1813;2010:948–964.
35. Hirata H, Watanabe T, Hatakeyama J, et al. Zebrafish relatively relaxed mutants have a ryanodine receptor defect, show slow swimming and provide a model of multi-minicore disease. *Development*. 2007;134:2771–2781.
36. Castets P, Bertrand AT, Beuvin M, et al. Satellite cell loss and impaired muscle regeneration in selenoprotein N deficiency. *Hum Mol Genet*. 2011;20:694–704.
37. Jurynec MJ, Xia R, Mackrill JJ, et al. Selenoprotein N is required for ryanodine receptor calcium release channel activity in human and zebrafish muscle. *Proc Natl Acad Sci U S A*. 2008;105:12485–12490.
38. Majczenko K, Davidson AE, Camelo-Piragua S, et al. Dominant mutation of CCDC78 in a unique congenital myopathy with prominent internal nuclei and atypical cores. *Am J Hum Genet*. 2012;91:365–371.
39. Ceyhan-Birsoy O, Agrawal PB, Hidalgo C, et al. Recessive truncating titin gene, TTN, mutations presenting as centronuclear myopathy. *Neurology*. 2013;81:1205–1214.
40. Childers MK, Joubert R, Poulard K, et al. Gene therapy prolongs survival and restores function in murine and canine models of myotubular myopathy. *Sci Transl Med*. 2014;6:220ra210.
41. Lawlor MW, Armstrong D, Viola MG, et al. Enzyme replacement therapy rescues weakness and improves muscle pathology in mice with X-linked myotubular myopathy. *Hum Mol Genet*. 2013;22:1525–1538.
42. Nowak KJ, Wattanasirichaigoon D, Goebel HH, et al. Mutations in the skeletal muscle alpha-actin gene in patients with actin myopathy and nemaline myopathy. *Nat Genet*. 1999;23:208–212.
43. Laing NG, Wilton SD, Akkari PA, et al. A mutation in the alpha tropomyosin gene TPM3 associated with autosomal dominant nemaline myopathy. *Nat Genet*. 1995;9:75–79.
44. Donner K, Ollikainen M, Ridanpaa M, et al. Mutations in the beta-tropomyosin (TPM2) gene–a rare cause of nemaline myopathy. *Neuromuscul Disord*. 2002;12:151–158.
45. Agrawal PB, Greenleaf RS, Tomczak KK, et al. Nemaline myopathy with minicores caused by mutation of the CFL2 gene encoding the skeletal muscle actin-binding protein, cofilin-2. *Am J Hum Genet*. 2007;80:162–167.
46. Ravenscroft G, Jackaman C, Bringans S, et al. Mouse models of dominant ACTA1 disease recapitulate human disease and provide insight into therapies. *Brain*. 2011;134:1101–1115.
47. Ochala J, Lehtokari VL, Iwamoto H, et al. Disrupted myosin cross-bridge cycling kinetics triggers muscle weakness in nebulin-related myopathy. *FASEB J*. 2011;25:1903–1913.
48. North KN, Wang CH, Clarke N, et al. Approach to the diagnosis of congenital myopathies. *Neuromuscul Disord*. 2014;24:97–116.
49. Wattjes MP, Kley RA, Fischer D. Neuromuscular imaging in inherited muscle diseases. *Eur Radiol*. 2010;20:2447–2460.
50. Wang CH, Dowling JJ, North K, et al. Consensus statement on standard of care for congenital myopathies. *J Child Neurol*. 2012;27:363–382.
51. Ryan MM, Sy C, Rudge S, et al. Dietary L-tyrosine supplementation in nemaline myopathy. *J Child Neurol*. 2008;23:609–613.
52. Messina S, Hartley L, Main M, et al. Pilot trial of salbutamol in central core and multi-minicore diseases. *Neuropediatrics*. 2004;35:262–266.
53. Arbogast S, Beuvin M, Fraysse B, Zhou H, Muntoni F, Ferreiro A. Oxidative stress in SEPN1-related myopathy: from pathophysiology to treatment. *Ann Neurol*. 2009;65:677–686.
54. Nowak KJ, Ravenscroft G, Jackaman C, et al. Rescue of skeletal muscle alpha-actin-null mice by cardiac (fetal) alpha-actin. *J Cell Biol*. 2009;185:903–915.

CHAPTER

94

The Distal Myopathies

Ami Mankodi, Bjarne Udd†, and Robert C. Griggs‡*

*National Institute of Neurological Disorders and Stroke, Bethesda, MD, USA
†University of Helsinki, Helsinki, Finland
‡University of Rochester Medical Center, Rochester, NY, USA

INTRODUCTION

Distal myopathies are a group of inherited or sporadic primary muscle disorders characterized by predominant distal muscle weakness and atrophy in hands, forearms, lower legs, and feet. In this chapter we emphasize progress in understanding disease mechanisms and describe some novel forms of distal myopathies.

Advances in molecular genetics and immunohistochemical techniques have clarified the genetic basis and potential pathophysiology of distal myopathies. Specific chromosomal localization is now known for all major disorders. The gene product has been identified for Miyoshi myopathy,[1] hereditary inclusion body myopathy (Nonaka myopathy),[2] tibial muscular dystrophy (Udd myopathy),[3] early-onset distal myopathy (Laing myopathy),[4] Welander myopathy,[5,6] matrin-3 distal myopathy,[7] VCP-mutated distal myopathy,[8] αB-crystallin-mutated myopathy,[9] distal ABD-filaminopathy,[10] KLH9-mutated distal myopathy,[11] and distal anoctaminopathy.[12] During the last few years it has been shown that the genes responsible for the pathologically defined category of myofibrillar myopathy clinically frequently show a distal myopathy phenotype.[13] This was first established for desminopathy[14] and αB-crystallinopathy,[15] and later shown also for myotilinopathy[16] and zaspopathy, as with the Markesbery–Griggs late-onset distal myopathy.[17] For some of these genes, mutations in the same gene can cause either a proximal or a distal phenotype—dysferlinopathy causing limb-girdle muscular dystrophy (*LGMD2B*) or Miyoshi myopathy,[1] myotilinopathy causing LGMD1A or a very late-onset distal myopathy.[18,19] Nonaka myopathy, hereditary inclusion body myopathy, and quadriceps-sparing vacuolar myopathy are all caused by mutations in GNE (UDP-*N*-acetylglucosamine 2-epimerase/*N*-acetylmannosamine kinase) and consequently are considered the same disease, now termed, GNE-myopathy.[20] Thus, future classifications of distal myopathies should arguably be based on the gene protein product or gene mutation rather than on the distribution of weakness. However, the current classification based on age at onset, pattern of muscle involvement, and mode of inheritance is clinically useful and facilitates further diagnostic procedures (Table 94.1).

LATE ADULT-ONSET DISTAL MYOPATHIES

Welander Myopathy

Clinical Features

Inheritance is autosomal dominant. Most reported families are of Scandinavian descent. Symptom onset is usually in the fifth decade, although some patients first notice weakness as late as in their 70s. Onset before age 30 years is unusual. Patients usually develop symptoms in distal upper extremities, typically finger and wrist extensors. Marked atrophy of thenar and intrinsic hand muscles becomes manifest after several years of disease. In 85% of patients, the weakness is moderately asymmetric. With time, the finger and wrist flexors become affected. Symptoms gradually extend to the lower extremities, with weakness predominantly in toe and ankle extensors. Proximal weakness is

Rosenberg's Molecular and Genetic Basis of Neurological and Psychiatric Disease
http://dx.doi.org/10.1016/10.1016/B978-0-12-410529-4.00094-2

TABLE 94.1 Current Classification of Distal Myopathies Based on Age at Onset, Pattern of Muscle Involvement, and Mode of Inheritance

Type	Description	Inheritance	Locus/*Gene*	Onset age (years)	Early symptoms	CK	Pathology
DEFINITE ENTITIES							
Welander myopathy	Welander 1951	AD	2p13/*TIA1*	>40	Hands, finger extensors	1–3×	Dystrophic, rimmed vacuoles
Tibial muscular dystrophy (Udd myopathy)	Udd 1993	AD	2q31/*TTN*	>35	Anterior lower leg	1–4×	Dystrophic, rimmed vacuoles in tibial anterior muscle
Distal zaspopathy	Markesbery and Griggs 1974	AD	10q22/*ZASP (LDB3)*	40–50	Clinically anterior and posterior lower leg on muscle imaging	1–4×	Dystrophic, rimmed and nonrimmed vacuoles, desmin–myotilin aggregates
Laing myopathy (MPD1)	Laing 1995	AD	14q/*MYH7*	35–60	Anterior lower leg	1–3×	Mild to moderate dystrophic
Distal myopathy with vocal cord and pharyngeal weakness (MPD2)	Feit 1998	AD	5q31/*MATR3*	35–60	Anterior lower leg, dysphonia	1–8×	Rimmed vacuoles
Miyoshi myopathy	Miyoshi 1985	AR	2p13/*DYSF*	15–30	Posterior lower leg, calf	20–150×	Dystrophic, dysferlin defect
Distal anoctaminopathy	Bolduc 2010	AD	11p14.3/*ANO5*	20–40	Posterior lower leg, calf	20–150×	Dystrophic, early scattered fiber necrosis
Nonaka myopathy (GNE myopathy)	Nonaka 1981	AR	9p1-q1/*GNE*	15–30	Anterior lower leg	3–4×	Dystrophic, prominent rimmed vacuoles
Distal nebulin myopathy	Wallgren-Pettersson 2007	AR	2q21/*NEB*	1–20	Anterior lower leg	1–3×	Myopathic, group atrophy, no rods on light microscopy
VCP-mutated distal myopathy	Palmio 2011	AD	9p13.3/*VCP*	>35	Anterior lower leg; dementia during later years	1–2×	Dystrophic, rimmed vacuoles
Distal ABD-filaminopathy	Williams 2005, Duff 2011	AD	7q32-q35/*FLNC*	20–30	Thenar, anterior forearm and posterior lower leg	1–2×	Myopathic
KLH9-mutated distal myopathy	Cirak 2010	AD	9p21.2-p22.3/ *KLH9*	8–16	Anterior lower leg; distal sensory impairment	1–8×	Myopathic
Distal caveolinopathy	Tateyama 2002	Sporadic	3p25/*CAV3*	<20	Hands and feet	20×	Myopathic, reduced caveolin-3
MYOFIBRILLAR MYOPATHIES WITH DISTAL PHENOTYPE							
Distal desminopathy	Milhorat and Wolf 1943	AD	2q35/*DES*	Variable	Distal and proximal weakness; cardiomyopathy	1–4×	Dystrophic, rimmed vacuoles, desmin aggregates
αB-crystallin-mutated distal myopathy	Reilich 2010	AD	11q/*CRYAB*	50–60	Distal leg and hands	1.5–2.5×	Myopathic, rimmed and nonrimmed vacuoles, Z-disk streaming, desmin and αB-crystallin aggregates
Distal myotilinopathy	Penisson-Besnier 1998	AD	5q31/*MYOT*	50–60	Posterior more than anterior lower leg	1–3×	Dystrophic, rimmed and nonrimmed vacuoles, desmin–myotilin aggregates

TABLE 94.1 Current Classification of Distal Myopathies Based on Age at Onset, Pattern of Muscle Involvement, and Mode of Inheritance—cont'd

Type	Description	Inheritance	Locus/*Gene*	Onset age (years)	Early symptoms	CK	Pathology
SINGLE FAMILIES WITH UNKNOWN GENETIC CAUSE							
Adult-onset distal myopathy	Felice 1999	AD	2, 9, 14 excluded	20–40	Anterior lower leg	2–6×	Mild nonspecific changes
Adult-onset distal myopathy (MPD3)	Mahjneh 2003	AD	8p-q and 12q	30	Hand thenar and hypothenar	1–4×	Dystrophic, rimmed vacuoles and eosinophilic bodies
Distal myopathy with pes cavus and areflexia	Servidei 1999	AD	19p13	15–50	Lower leg; dysphonia; dysphagia	2–6×	Dystrophic, rimmed vacuoles

Abbreviations: AD, autosomal dominant; AR, autosomal recessive.

uncommon. Muscle stretch reflexes are preserved except for ankle reflexes, which may be lost later in the disease. Sensory system examination is usually normal. The progression of disease is typically slow, and most patients continue full activities and have a normal lifespan. Cardiomyopathy does not occur. Homozygous patients are occasionally seen and have a more severe disease with early age at onset, proximal muscle weakness, and wheelchair dependency.

Laboratory Findings

Serum creatine kinase (CK) level is normal or two- to three-fold elevated. Needle electromyography (EMG) reveals small, brief motor unit potentials with early recruitment. Fibrillations and complex repetitive discharges may occur. Although routine nerve conduction studies are normal, mild abnormalities in sural nerve biopsies and deficits in vibration and temperature examination by quantitative sensory testing suggest underlying asymptomatic, length-dependent, predominantly sensory small-fiber neuropathy. Muscle biopsy shows dystrophic features with fiber size variability, increase in connective and fat tissue, central nuclei, and split fibers. Rimmed vacuoles and 15- to 18-nm cytoplasmic and nuclear filaments are seen in patients with moderate-to-severe weakness. Absence of inflammation helps distinguish Welander myopathy from inclusion body myositis. Muscle imaging shows considerable involvement of posterior calf muscles besides fatty degenerative changes in the anterior compartment.

Molecular Genetics

Welander myopathy is caused by a heterozygous mutation (c.1150G>A; p. E384K) in the C-terminal glutamine-rich prion-like domain of T-cell intracellular antigen-1 or TIA1 (MIM #603518).[5,6] TIA1 is an RNA-binding protein that regulates alternative splicing of select pre-mRNAs and promotes the assembly of stress granules, which are dynamic aggregates of stalled translation initiation complexes formed in response to environmental stress.[21,22] Initial studies have revealed that the disease-causing mutation (E384K) in TIA1 affects dynamics of stress granules.[5] Klar et al.[6] showed increased levels of spliced SMN2 (without exon 7) in skeletal muscle of patients compared to controls. Hackman et al.[5] found no difference in the splicing of known TIA1 target exons in *TIA1*, *FAS*, and *PLOD2* in skeletal muscle between patients and controls. Further research is needed to better understand the role of altered functions of TIA1 in the disease mechanism underlying the degeneration of skeletal muscle fibers in Welander myopathy.

Tibial Muscular Dystrophy (TMD; Udd Myopathy)

Clinical Features

Patients with TMD present after age 35 with weakness in ankle dorsiflexion and later visible atrophy of anterior compartment lower leg muscles. At onset symptoms and signs may be asymmetric; progression is slow. After age 70 years some patients have proximal weakness in the lower extremities, but only rarely become wheelchair bound. Sparing of short toe extensors (extensor digitorum brevis) is a typical clinical finding and hand muscles are rarely affected. A recent study of 207 mutation-confirmed patients showed phenotypic heterogeneity in 9% of the patients, including: onset of weakness and atrophy in proximal leg muscles; involvement of upper limb muscles; onset in calf muscles; generalized weakness in childhood; and persistent asymmetric and focal atrophies.[23]

Prevalence of the disease is high in Finland: more than 7/100,000.[23] TMD has been identified in Sweden, Norway, Germany, and Canada, occurring in descendants of Finnish immigrants. Moreover, TMD families without Finnish ancestry have been identified in France, Belgium and Spain.

Laboratory Findings

Serum CK is normal or mildly elevated. In affected muscles, EMG shows low-amplitude, short-duration motor unit potentials on moderate activity, frequent fibrillation potentials, and occasional high-frequency and complex repetitive discharges. In unaffected upper limbs polyphasic potentials may be recorded. Computed tomography (CT) and magnetic resonance imaging (MRI) show fatty degeneration in clinically weak anterior tibial muscles. Later lesions appear in the long toe extensor and hamstring muscles. Asymptomatic focal lesions may occur in soleus and medial gastrocnemius. The pattern of MRI involvement is different in the aberrant phenotypes.

Muscle pathology includes variation of fiber size, thin atrophic fibers, central nuclei, structural changes within the fibers, endomysial fibrosis, usually rimmed vacuoles in the tibial anterior, and fatty replacement in the end stage muscle. Necrotic fibers, some showing phagocytosis, are rare in TMD. Both major fiber types are equally involved in the pathological process. There are no neurogenic findings. Many rimmed vacuoles are acid phosphatase positive, while others are ubiquitin positive and, with rare exceptions, they are not lined by sarcolemmal membrane proteins. Congo red stains and immunohistochemistry for amyloid-β and amyloid precursor protein remain negative in contrast to inclusion body myopathies. Ultrastructural studies in TMD revealed overall well preserved sarcomere structure, even in the homozygote LGMD2J mutants. Autophagy and occasional tubulofilamentous inclusions occur in the vacuolated fibers.

Molecular Genetics

TMD is caused by mutations in C-terminal titin physically located in the M-line of the sarcomere. All Finnish patients carry one common founder mutation (FINmaj), a complex 11-bp insertion–deletion mutation changing four amino acids without frameshift.[3] A point mutation changing a lysine to proline was found in two unrelated French families in the same last exon of titin.[24] A third mutation in the same exon was found in a Belgian TMD family.[25] More recently, four other mutations in the last and second-to-last exons have been identified in Spanish, Italian and other French families.[26] In new, unrelated TMD patients, mutations should be sought by sequencing the last titin exons. Mutation testing also confirmed homozygously inherited FINmaj titin mutation in all four available LGMD phenotype patients with childhood onset in the large Larsmo kindred. Recently, adult onset distal titinopathy was also reported with recessive titin mutations.[27]

Titin is the third most abundant protein (after myosin and actin) in muscle and constitutes the third filament system of the sarcomere. Titin binds calpain-3 in the N2A-line of I-band titin and in its C-terminus, and homozygous LGMD2J patient muscle shows secondary calpain-3 deficiency.[28] In primary calpain-3 defect, LGMD2A, perturbations of the IκBα/NF-κB pathway and apoptotic myonuclei have been observed.[29] In TMD/LGMD2J similar changes with secondary calpain-3 defect, clusters of apoptotic myonuclei were also found, suggesting similar molecular pathology. Mutant titin is transcribed, translated, and incorporated in the sarcomere as shown by immunohistochemistry in homozygous muscle. However, C-terminal antibodies do not label mutant titin, indicating that the C-terminus is either markedly conformational or cleaved. The titin C-terminus contains motifs for signaling, and a catalytic kinase domain with several interacting signaling molecules.[3]

Myofibrillar Myopathies with Dominant Distal Myopathy

Myofibrillar myopathies are a heterogenous group of severe distal myopathies, often associated with cardiomyopathy.[13,30] These myopathies are characterized pathologically by findings of disintegrated myofibrillar structure and abnormal accumulation of proteins like desmin, dystrophin, vimentin, gelsolin, and β-spectrin. Desmin was the first of the proteins studied and thus the term "desmin-related myopathies" was applied in the past. The two major structural abnormalities in light as well as electron microscopy are: nonhyaline lesions consisting of foci of myofibrillar destruction; and hyaline lesions composed of compacted and degraded myofibrillar elements (cytoplasmic or spheroid bodies). The nonhyaline lesions appear as irregular, lobulated dark green areas of amorphous material with the modified Gömöri trichrome stain. These lesions include desmin, dystrophin, neural cell adhesion molecule, gelsolin, and amyloid-β precursor protein, but do not contain myosin, actin, or α-actinin. The hyaline lesions are spherical, serpentine, or pleomorphic cytoplasmic inclusions, usually eosinophilic with the hematoxylin and eosin (H&E) stain and dark blue or purple–red with the modified Gömöri trichrome stain. These lesions react variably to antibodies against desmin, whereas they react strongly to antibodies against dystrophin, gelsolin, amyloid-β precursor protein, titin, nebulin, actin, myosin and α-actinin. Staining the protein aggregations with anti-myotilin antibodies has proved to be a very sensitive marker. Indeed, pathogenic mutations have been identified in the desmin gene in some families originally described as distal myopathies,[14,31] and recently even the classic "neurogenic" scapuloperoneal syndrome of Stark–Kaeser was identified as desmin-mutated disease.[32] Exome sequencing revealed a mutation in the desmin gene in a large Swedish pedigree with autosomal dominant predominantly distal weakness and arrhythmogenic right ventricular cardiomyopathy that was previously linked to chromosome 10q22.3.[33,34] Focal accumulation of desmin is, however, a nonspecific finding. It can be seen in many neuromuscular diseases, including spinal muscular atrophy, congenital myotonic dystrophy, myotubular myopathy, and nemaline myopathy, as well as diffusely in regenerating muscle fibers of many etiologies. Desmin is a muscle protein consisting of intermediary filaments and is located at Z-disks and in subsarcolemmal regions of skeletal, cardiac, and smooth muscle, where it serves as a major scaffold structure. It appears to play an important role in maintaining the structural and functional integrity of muscle fibers by linking the Z-disks to the plasma membrane. *DES* knockout mice develop normally and are fertile. Overexpression of mutant desmin protein (L385P) in cultured cells formed intracytoplasmic *DES*-positive aggregates quite similar to the accumulations seen in muscle biopsy specimens. The majority of cells with cytoplasmic desmin aggregates died within 72 hours by apoptosis, suggesting a direct toxic effect of the mutant DES protein.[35]

In addition, a p.G154S mutation in the *CRYAB* gene encoding αB-crystallin gene was identified in a pedigree with dominantly inherited late-onset distal vacuolar myopathy.[9] αB-crystallin is a molecular chaperone and is believed to interact with DES in the assembly of intermediary filaments. Mutations in the filamin C rod and dimerization domain cause a myofibrillar myopathy phenotype,[36–38] whereas mutations in the actin-binding domain of the filamin C gene cause a distal myopathy lacking myofibrillar protein aggregation pathology (see below).[10] In contrast to typical slow progression in myofibrillar myopathies, a mutation in the *BAG3* gene was identified in a pedigree with severe childhood-onset myopathy characterized by distal contractures, rigid spine, scoliosis, cardiomyopathy, and respiratory failure.[39] Together with myotilin and ZASP (discussed below), these currently known six genes associated with myofibrillar myopathy constitute about half of the cases identified as myofibrillar myopathy. It is likely that additional genetic defects causing myofibrillar myopathies will be identified in the future.

Zaspopathy (Markesbery–Griggs Distal Myopathy)

Clinical Features

This form of late-onset, dominantly inherited distal myopathy has been reported in families of English, French, and German descent. Symptoms usually begin after age 40 years with ankle weakness. As the disease progresses, finger and wrist extensors may be affected. Much later in life, proximal weakness may also occur. In some the disease may progress rapidly. The ability to walk may be lost after 15–20 years of disease, progressing to complete incapacity after 30 years. One patient in the first-described family had cardiomyopathy with heart block requiring a pacemaker. Facial, bulbar, and respiratory muscles are not affected.[40]

Laboratory Findings

Serum CK levels can be normal or 3–4-fold elevated. EMG reveals myopathic changes in affected muscles. Muscle imaging at onset of symptoms shows changes in the posterior compartment of the legs; later, all lower leg muscles and proximal leg muscles become involved. Muscle biopsy shows prominent rimmed vacuoles and dark structures in trichrome stain compatible with myofibrillar myopathy. Immunostaining of muscle fibers reveals cytoplasmic accumulation of myotilin, desmin, ZASP, F-actin, and αB-crystallin.[40,41]

Molecular Genetics

Two mutations in ZASP (Z-disk alternatively spliced PDZ domain-containing protein, also termed the *LDB3* gene) are frequently associated with this type of distal myopathy.[42] The causative A165V mutation in the Markesbery–Griggs family was shown to be an ancient European founder mutation based on a relatively short common haplotype around the mutation in six unrelated families tested.[17] The other recurring ZASP mutation, A147T, causes an identical phenotype. These mutations are located within the actin-binding domain of ZASP and cause disruption of skeletal muscle actin filaments.[41]

Distal Myotilinopathy

Clinical Features

Myotilin mutations were first associated with dominant limb-girdle phenotype (LGMD1A).[19] Similar to zaspopathy, late-onset distal phenotype was identified in patients classified based on histopathology as having myofibrillar myopathy. In subsequent studies, myofibrillar myopathy due to myotilin mutations most frequently displayed late-onset distal myopathy as the clinical phenotype.[16] In some families the first symptom was loss of ankle dorsiflexion between ages 50 and 60 years, followed by plantar flexion weakness.[43] In others, weakness and atrophy of calf muscles was the first sign, followed by a period of pain and cramps.[18,44] Involvement of upper limbs or proximal leg muscles was moderate or severe at later stages. Respiratory and bulbar involvement do not occur.

Laboratory Findings

Serum CK levels ranged from normal to less than two-fold. EMG showed myopathic changes with fibrillations and complex repetitive discharges. Muscle imaging paralleled clinical findings with extensive fatty degenerative changes in calf muscles and milder proximal leg muscle involvement. Muscle biopsy findings were consistent with myofibrillar myopathy: frequent large nonrimmed vacuoles, and focal cytoplasmic H&E-basophilic and trichrome dark material in both fiber types. Other features included occasional rimmed vacuoles and some fiber splitting. Electron microscopy showed autophagic vacuoles and large zones of myofibrillar disorganization with IBM-type 15–18-nm tubular filaments close to the vacuoles. Early changes were widening of dark material in the Z-disk. Of the different proteins associated with myofibrillar myopathy, myotilin showed the highest degree of abnormal irregularly aggregated and mislocated protein.

Molecular Genetics

All reported mutations in myotilin are dominant missense mutations located within the serine-rich second domain, regardless of clinical phenotype. Mutations most often involve a serine and threonine residues. In our own cohort of late-onset distal myopathy patients, we have observed recurrence of the mutations S60F and S60C. The previously described spheroid body myopathy also proved to be caused by myotilin mutation.[45]

Distal Myopathy with Vocal Cord and Pharyngeal Weakness (MPD2)

Clinical Features

Feit et al.[46] described a large North American pedigree with an autosomal dominantly inherited myopathy characterized by distal upper and lower extremity weakness and development of vocal cord and pharyngeal weakness. Symptoms begin between ages 35 and 60 years with weakness in ankle dorsiflexors and toe extensors. Sometimes finger extensors may be affected before leg muscles. The weakness may be asymmetric at the onset. More recently a multigeneration Bulgarian pedigree with a typical MPD2 phenotype was reported, which facilitated the discovery of the gene defect.[7]

Laboratory Findings

Serum CK levels ranged from normal to up to eight-fold increased. Nerve conduction studies showed mild slowing of velocities. EMG showed myopathic potentials. Muscle biopsy revealed a myopathy with rimmed vacuoles.

Molecular Genetics

MPD2 is the first hereditary human disease caused by a mutation in a nuclear matrix protein, matrin 3.[7] Nuclear matrix is an intricate meshwork of proteins spread throughout the nucleus. Matrin 3 plays a role in essential nuclear functions, which include gene expression, RNA splicing, RNA export, and nuclear protein import and export. The consequences of the disease-causing mutation p.S85C on the structure and functions of matrin 3 protein are not yet known. Initial studies have indicated altered distribution of the mutant protein in myonuclei in affected muscle such that the mutant protein was exclusively confined to the insoluble fraction of nuclear matrix, whereas the wild-type protein was mostly localized to the nucleic acid-bound fraction of the matrix.[7] Overexpression studies in cultured nonmuscle cells showed a similar nuclear redistribution of the mutant protein. Based on immunofluorescence analysis of the transfected cells, the authors argued that this altered distribution did not result from the formation of insoluble nuclear aggregates.

VCP-Mutated Distal Myopathy

Clinical Features

Palmio et al.[8] reported a Finnish pedigree with nine affected members in three generations. All patients had symptom onset after age 35 years. Anterior leg muscles were most affected. Three patients developed rapidly progressive frontotemporal dementia 20 years after onset of ankle weakness, were bedridden, and died of pneumonia. Neuropsychological evaluation revealed changes in information processing and executive functions but no memory deficits.

Laboratory Findings

CK levels were normal or slightly elevated. Muscle biopsy showed myopathic changes with prominent rimmed vacuoles. Fatty degenerative change of anterior lower leg muscles was evident on CT/MRI.

Molecular Genetics

The combination of distal myopathy with rimmed vacuoles and dementia in patients provided the clue to genetic diagnosis. Previously, missense mutations in the *VCP* gene have been associated with hereditary inclusion body myopathy, Paget disease, and frontotemporal dementia.[64] Using a candidate gene approach, a novel missense mutation p.P137L in the *VCP* gene was identified in all affected members.[8] Affected members in the Finnish pedigree did not have signs of Paget disease.[8]

EARLY ADULT-ONSET DISTAL MYOPATHIES

Nonaka Distal Myopathy

Clinical Features

Initial reports of this disease came from Japan. However, subsequently the disease has been reported in patients from other parts of the world. Inheritance is autosomal recessive. Symptoms begin late in the second or third decade, and average age at onset is 26 years. Weakness initially involves ankle dorsiflexors and toe extensors, causing foot drop and a steppage gait. Mild distal upper extremity weakness may be present earlier in the illness. At later stages, patients may develop proximal weakness, although the quadriceps muscles remain relatively spared. Most patients lose ambulation 10–15 years after disease onset. Neck flexors may be affected. Cranial muscles are not involved. Cardiac arrhythmia necessitating pacemaker implant has been reported.

Laboratory Findings

Serum CK level is elevated 3–4-fold. EMG shows small, brief motor-unit potentials and fibrillation potentials. Muscle biopsy reveals prominent vacuoles and dystrophic changes. Vacuoles are lined with granular material that is basophilic with H&E staining and purple–red with modified Gömöri trichrome staining. The vacuoles exhibit acid phosphatase activity. Electron microscopy reveals 15- to 18-nm filamentous inclusions in the nucleus and cytoplasm, in addition to the vacuoles.

Families having a similar clinical presentation have been reported from Middle Eastern countries, North America, and India with the diagnosis of hereditary inclusion body myopathy or quadriceps-sparing myopathy. These patients differ from those with sporadic inclusion body myopathy by their earlier age at onset, initial symptom of foot drop, autosomal recessive inheritance, and absence of inflammation on muscle biopsy.

Molecular Genetics

Like quadriceps-sparing myopathy, Nonaka myopathy is a hereditary inclusion body myopathy linked to chromosome 9p12-13.[20] The responsible gene was first identified in patients with hereditary inclusion body myopathy,[2] and subsequently confirmed in Nonaka myopathy families.[20,47,48] Forty-seven Jewish families from Middle Eastern countries were homozygous for a missense mutation (2186 T→C, M712T) in the kinase domain of UDP-*N*-acetylglucosamine 2 epimerase/*N*-acetyl mannosamine kinase (GNE).[2] In Japanese patients, one mutation is a more frequent founder mutation V572L,[47] and families of different ethnic origins (Asian Indian, North American, and Caribbean) were heterozygous for distinct missense mutations in the kinase and epimerase domains of the GNE.[2] GNE is a bifunctional enzyme that catalyzes the first two steps in the biosynthesis of *N*-acetylneuraminic acid or sialic acid. Interestingly, dominant mutations in the epimerase domain of the gene cause sialuria. None of the patients with hereditary inclusion body myopathy had elevated sialic acid levels. GNE is exclusively shared by vertebrates and bacteria. There is no GNE ortholog in *Drosophila melanogaster*, *Caenorhabditis elegans*, or yeast. The two enzymatic activities of GNE are carried out by separate proteins in bacteria. GNE is a ubiquitous molecule, encoded by a single gene. GNE has been shown to be the rate-limiting enzyme in the sialic acid biosynthetic pathway. Sialic acid modification of glycoproteins and glycolipids expressed at the cell surface is crucial for their function in many biologic processes, including cell adhesion and signal transduction. Hyposialylation of proteins in affected muscles has been proposed in Nonaka myopathy,[49,50] but was not confirmed by others.[51] A recent study showed that prophylactic treatment with sialic acid metabolites prevented muscle atrophy and weakness in a mouse model of GNE myopathy.[50] These observations have raised a possibility that the GNE myopathy is treatable in patients. Clinical trials testing the safety and efficacy of sialic acid metabolites in patients have begun in the USA and Japan.

Miyoshi Distal Myopathy

Clinical Features

Miyoshi's original cases of myopathy were reported from Japan. Subsequently, the disease has been reported in many ethnic groups. Symptoms begin between ages 15 and 25 years. Inheritance is autosomal recessive. Initial symptoms are in the distal lower extremities, usually the gastrocnemius muscles. Patients complain that they cannot walk on their toes or climb stairs. Aching or discomfort in the calves is common. The gastrocnemius muscles become atrophic, and ankle reflexes are lost. The anterior compartment muscles are spared initially, but later become involved. The predilection for early involvement of the posterior compartment in Miyoshi myopathy distinguishes it from other distal myopathies. At presentation, calf involvement may be asymmetric, and in a small number of patients symptoms may remain confined to one leg. As the disease progresses, proximal muscles in legs and arms are affected, and the two phenotypes in dysferlinopathy, Miyoshi myopathy and *LGMD2B*, merge into one. Progression of the disease appears to correlate better with disease duration than with age at onset. About one-third of patients are confined to a wheelchair within 10 years of symptom onset. Patients with anterior tibial onset and fairly rapid progression have also been reported.[52]

Laboratory Findings

Miyoshi myopathy is characterized by a marked increase in CK levels compared to other distal myopathies. Serum CK levels range from 20 to 150 times normal. Indeed, elevated CK levels (hyperCKemia) are often detected prior to symptoms or signs in these patients. EMG shows small, brief myopathic motor-unit potentials and early recruitment. Very weak and atrophic gastrocnemius muscles may show long duration polyphasic motor-unit potentials with reduced recruitment. Muscle biopsy in severely affected gastrocnemius muscle may show "end-stage" disease with widespread fibrosis, fatty replacement, and loss of most muscle fibers. The ideal muscles to biopsy are the hamstrings, as they provide most information on muscle histology. Rimmed vacuoles are not common in these patients. About 10% of patients with dysferlinopathy have inflammatory cell infiltrates in biopsies. With identification of the gene dysferlin (*DYSF*),[1] characterization of its protein product, and development of antibodies, the diagnosis of Miyoshi myopathy can now be confirmed by a simple immunohistochemical technique. Dysferlin normally localizes to the plasma membrane of muscle fibers. In patients with Miyoshi myopathy and *LGMD2B*, dysferlin is absent in the plasma membrane, whereas scattered granular staining in the cytoplasm or nuclear membrane may be observed.

Molecular Genetics

The first gene causing any distal myopathy was identified in patients with Miyoshi myopathy. Interestingly, identical mutations in the *DYSF* gene encoding dysferlin protein on chromosome 2p13 (MMD1) can cause Miyoshi myopathy, *LGMD2B*, or distal myopathy with clinical onset in anterior tibial muscles.[1] In a single pedigree, the same mutation can cause Miyoshi myopathy in one sibling and *LGMD2B* in another. The only clinical features shared by these diseases are early age at onset, recessive inheritance, and markedly elevated serum CK levels. The patterns of muscle involvement in these allelic phenotypes are quite distinct at onset, as their names suggest. The reason for such diverse phenotypes caused by similar gene defects (phenotypic heterogeneity) is not known.

Dysferlin is a novel protein without homology to other known mammalian proteins. However, it shows homology throughout its length to the *C. elegans* spermatogenesis factor *fer-1*. In fact, the name dysferlin comes from dystrophy-associated *fer-1-like* protein. Dysferlin is expressed in many tissues, including heart, skeletal muscle, kidney, stomach, liver, spleen, lung, uterus and, to a lesser extent, brain and spinal cord.[53] Dysferlin is expressed in the embryonic tissues from the earliest time point examined.[53] As mentioned earlier, it is localized to the plasma membrane. However, dysferlin does not interact with dystrophin, sarcoglycans, or dystroglycans.

The presence of C2 domains in dysferlin suggests that it may play an important role in signaling pathways. C2 domains are believed to bind calcium and thereby trigger signal transduction and membrane trafficking events. Immunoprecipitation studies revealed that dysferlin interacts with caveolin-3.[54] (Mutations in caveolin-3 cause *LGMD1C* and a novel form of distal myopathy.[55]) Dysferlin staining was abnormal in two *LGMD1C* muscles. Possible caveolin-3 binding domains have been identified in dysferlin. Indeed, structural abnormalities of the sarcolemma, including subsarcolemmal vacuoles and papillary projections, have been reported in patients with Miyoshi myopathy.[56] Further studies on dysferlin indicate essential functions in the membrane repair mechanisms.[57]

A mouse model for dysferlin deficiency (SJL-Dysf)[58] is due to spontaneous mutations in the fourth C2 domain. These mice develop muscle weakness, atrophy, and histologic changes identical to *LGMD2B*. This animal model may help understand the mechanism of muscle degeneration due to dysferlin deficiency, and may be useful for the development of therapeutic rescue strategies.

Distal Anoctaminopathy

Not all patients with a Miyoshi-like phenotype have dysferlinopathy. There are families with the Miyoshi phenotype and families with adult and later onset of symptoms in whom linkage to the locus 2p has been excluded.[59,60] Two unrelated Dutch families with Miyoshi myopathy have been linked to a 23 cM region on chr 10p (MMD2) with a maximum LOD score of 2.578.[59] More recent studies identified a third locus on chr 11p (MMD3) in Finnish and Dutch pedigrees. Affected members were found to have homozygous mutations in the *ANO5* gene encoding anoctamin-5 protein associated with either a proximal muscle weakness (LGMD2L) or Miyoshi myopathy.[12,61,62] There were no clear phenotype–genotype correlations. Females tended to have a milder phenotype compared to males with similar mutations. Electron microscopy studies of patient muscle and *in vitro* studies using patient-derived fibroblasts suggest a defect in membrane repair as an underlying disease mechanism similar to dysferlinopathy.[12,60]

Distal ABD-Filaminopathy

Clinical Features

Muscle weakness starts with loss of firm handgrip in the early third decade, followed by thenar atrophy and calf weakness in the next decades. Disease progression is slow and despite late proximal limb weakness, patients do not lose ambulantion.[10,63]

Laboratory Findings

CK levels are normal or mildly elevated. EMG is myopathic and in contrast to the myofibrillar myopathy filaminopathies, muscle biopsy findings are nonspecific; there are no rimmed vacuoles and no myofibrillar disintegration with protein aggregates.

Molecular Genetics

The muscle specific isoform of filamins, filamin-C, is located in the periphery of the Z-disks and interacts with myotilin. Filamins are generally actin binding cytoskeletal proteins. Filamin-C also binds sarcolemmal proteins and is cleaved by calpain-3. In distal ABD-filaminopathy, the mutations are located in the *N*-terminal actin-binding domain (ABD) of filamin-C.[10]

EARLY-ONSET DISTAL MYOPATHIES

Laing Distal Myopathy (MPD1)

Clinical Features

Symptoms begin between the ages of 2 and 25 years. Inheritance is autosomal dominant. Since Laing's report of an English–Welsh family in 1995,[65] additional families have been identified.[66–68] Weakness begins in ankle dorsiflexors, toe extensors, and neck flexor weakness is prominent. Finger extensors and shoulder muscles are affected later in the disease. The course is protracted, and most patients remain ambulatory.

Laboratory Findings

Serum CK level is normal or mildly elevated (up to threefold). EMG shows short, brief myopathic potentials, and muscle biopsy reveals moderate myopathy. Vacuoles are usually not observed. On immunohistochemistry using myosin heavy chain (MyHC) antibodies, fiber types are abnormally distributed in the target tibial anterior muscle, showing very atrophic type 1 MyHC slow fibers that may be hybrids expressing also fast MyHC.

Molecular Genetics

The disease has been linked to chromosome 14q11[65] and the responsible gene is *MYH7* that encodes slow β-myosin heavy chain protein, which is the main myosin isoform in type 1 slow muscle fibers and in cardiac muscle fibers. All identified mutations causing early-onset distal myopathy are located in the tail region of the MYH7 molecule, whereas mutations in other parts of the protein may cause cardiomyopathy or hyaline body myopathy.[4]

Distal Nebulin Myopathy

Clinical Features

In a number of patients with early-onset sporadic or recessive distal myopathy referred to us for titin analysis, we have identified nebulin mutations. Extensor muscles of feet and later hands are predominantly affected. The progression is very mild and adult patients do not have major disability.

Laboratory Findings

Selective fatty degeneration in the anterior tibial muscles is shown by muscle imaging, EMG is myopathic and CK is normal or mildly elevated. Muscle biopsy may reveal peculiar scattered and grouped atrophic fibers, first interpreted as neurogenic changes. Light microscopy did not show nemaline rods. Later re-evaluation by electron microscopy disclosed some small rod-like dense material associated with Z-disks observed in some but not all patients.

Molecular Genetics

A large number of different mutations in nebulin are known to cause autosomal recessive nemaline myopathy.[69] The reason for this new clinical distal phenotype not to have been identified earlier as nebulinopathy is the absence of nemaline rods on light microscopy. The reason for different pathology is that nemaline patients have more disrupting mutations while this milder distal phenotype is caused by missense mutations in the large nebulin gene.[70]

Distal Caveolinopathy

Mutations in the *CAV3* gene encoding caveolin-3 protein cause various forms of myopathies (caveolinopathies) including *LGMD1C*, hyperCKemia, rippling muscle disease, and a novel form of distal myopathy.[55] Tateyama et al.[55] reported a sporadic patient with hyperCKemia (diagnosed at age 12 years, with a 20-fold increase), atrophy, and moderate weakness of small muscles in hands and feet. EMG at age 12 years revealed myopathic potentials in hand muscles, and muscle biopsy from the biceps brachii showed mild myopathy. Immunoreactivity for caveolin-3 and dysferlin was markedly reduced in muscle, while it was normal for dystrophin, merosin, α-sarcoglycan, and β-dystroglycan. By western blot, caveolin-3 was absent whereas dysferlin was normal in size and amount. Sequence analysis of the *CAV3* revealed a previously reported mutation (80G→A substitution).

Early-Onset Dominant Distal Myopathy Caused by Mutations in the Kelch-Like Homolog 9 (*KLHL9*) Gene

Clinical Features

Cirak et al.[11] reported an autosomal dominant distal myopathy in a German pedigree with symptom onset between ages 8 and 16 years. Ankle extensor muscles were prominently affected. All patients developed ankle contractures, intrinsic hand muscle weakness, and glove and stocking sensory loss. The disease progression was slow in all affected members. Proximal muscles were relatively spared.

Laboratory Findings

Serum CK levels were elevated (200–1400U/L). Nerve conduction studies were normal in most patients, but increased distal motor latencies in some were present. There was sural sensory action potential in one patient. EMG of leg muscles showed fibrillations and neurogenic motor-unit potentials. These changes were thought to be secondary to myopathy and not neuropathy. MRI of muscles showed symmetric fatty atrophy of the tibialis anterior, gastrocnemius and soleus muscles, with relative preservation of the tibialis posterior and peroneus longus muscles. And in the lower thigh, the most significant changes were seen in the semimembranosus muscle, in the biceps femoris muscle, as well as in the vastus intermedius muscle, whereas the lateral and medial vastus muscles and parts of the adductor muscles gracilis and sartorius were preserved. Muscle biopsy of the gastrocnemius muscle showed myopathic changes with fiber size variation, internal nuclei, and increased connective tissue. A sural nerve biopsy did not show axon loss or demyelination.

Molecular Genetics

All patients harbored a heterozygous missense mutation p.L95F in the *KLH9* gene encoding kelch-like homolog 9 protein. The disease causing mutation disrupts the formation of E3 ubiquitin ligase complex by inhibiting the association of KLH9 with Cul3. The disease mechanism underlying the degeneration of skeletal muscle has not yet been delineated.

SINGLE DISTAL MYOPATHY FAMILIES WITH UNKNOWN MOLECULAR CAUSE

Felice et al.[71] described an autosomal dominant distal myopathy that does not localize to chromosome 2, 9, or 14 loci (which have been linked to other distal myopathies).

Mahjneh et al.[72] reported a new dominant distal myopathy in a Finnish family. Symptoms started after age 30 years with weakness of the intrinsic hand muscles or with asymmetric weakness in the anterior lower leg muscles. Rimmed vacuoles were frequent, with prominent eosinophilic inclusions. Known distal myopathies were excluded by linkage and significant LOD score >3 was instead shown for two separate loci, 8p22-q11 and 12q13-q22.[73]

Servidei et al.[74] have reported one Italian family with 10 affected members in three generations with areflexia and pes cavus. Weakness in both anterior and posterior lower legs started between the second and sixth decades of life and progressed to upper limbs and proximal muscles. Dysphagia and dysphonia were early signs. The family was linked to chromosome 19p13 with a LOD score of 3.03. Rimmed vacuoles were frequent on muscle biopsy together with dystrophic changes, and features of both lysosomal and nonlysosomal degradation were reported.[75]

CONCLUSION

Although distal myopathies were initially reported in Scandinavian and Japanese populations, they are now recognized in all ethnic groups. Molecular definition has helped increase awareness of these diseases. With identification of gene products, including dysferlin and caveolin-3, it is now feasible to diagnose some of these myopathies by immunohistochemistry or western blotting. Characterization of pathologic findings has helped define the heterogeneous group of myofibrillar myopathies and early-onset distal myopathies, and has facilitated molecular diagnosis in families and in sporadic patients. The final diagnosis rests firmly on the identification of gene mutations. With better understanding of disease mechanisms, we have begun to define targets for genetic treatment strategies in the distal myopathies.

References

1. Liu J, Aoki M, Illa I, et al. Dysferlin, a novel skeletal muscle gene, is mutated in Miyoshi myopathy and limb girdle muscular dystrophy. *Nat Genet*. 1998;20:31–36.
2. Eisenberg I, Avidan N, Potikha T, et al. The UDP-*N*-acetylglucosamine 2-epimerase/*N*-acetylmannosamine kinase gene is mutated in recessive hereditary inclusion body myopathy. *Nat Genet*. 2001;29:83–87.
3. Hackman P, Vihola A, Haravuori H, et al. Tibial muscular dystrophy is a titinopathy caused by mutations in TTN, the gene encoding the giant skeletal-muscle protein titin. *Am J Hum Genet*. 2002;71:492–500.
4. Meredith C, Herrmann R, Parry C, et al. Mutations in the slow skeletal muscle fiber myosin heavy chain gene (MYH7) cause Laing early-onset distal myopathy (MPD1). *Am J Hum Genet*. 2004;75:703–708.
5. Hackman P, Sarparanta J, Lehtinen S, et al. Welander distal myopathy is caused by a mutation in the RNA-binding protein TIA1. *Ann Neurol*. 2013;73(4):500–509.
6. Klar J, Sobol M, Melberg A, et al. Welander distal myopathy caused by an ancient founder mutation in TIA1 associated with perturbed splicing. *Hum Mutat*. 2013;34:572–577.

7. Senderek J, Garvey SM, Krieger M, et al. Autosomal-dominant distal myopathy associated with a recurrent missense mutation in the gene encoding the nuclear matrix protein, matrin 3. *Am J Hum Genet*. 2009;84:511–518.
8. Palmio J, Sandell S, Suominen T, et al. Distinct distal myopathy phenotype caused by VCP gene mutation in a Finnish family. *Neuromuscul Disord*. 2011;21:551–555.
9. Reilich P, Schoser B, Schramm N, et al. The p.G154S mutation of the alpha-B crystallin gene (CRYAB) causes late-onset distal myopathy. *Neuromuscul Disord*. 2010;20:255–259.
10. Duff RM, Tay V, Hackman P, et al. Mutations in the *N*-terminal actin-binding domain of filamin C cause a distal myopathy. *Am J Hum Genet*. 2011;88:729–740.
11. Cirak S, von Deimling F, Sachdev S, et al. Kelch-like homologue 9 mutation is associated with an early onset autosomal dominant distal myopathy. *Brain*. 2010;133:2123–2135.
12. Bolduc V, Marlow G, Boycott KM, et al. Recessive mutations in the putative calcium-activated chloride channel Anoctamin 5 cause proximal LGMD2L and distal MMD3 muscular dystrophies. *Am J Hum Genet*. 2010;86:213–221.
13. Selcen D, Ohno K, Engel AG. Myofibrillar myopathy: clinical, morphological and genetic studies in 63 patients. *Brain*. 2004;127:439–451.
14. Sjoberg G, Saavedra-Matiz CA, Rosen DR, et al. A missense mutation in the desmin rod domain is associated with autosomal dominant distal myopathy, and exerts a dominant negative effect on filament formation. *Hum Mol Genet*. 1999;8:2191–2198.
15. Vicart P, Caron A, Guicheney P, et al. A missense mutation in the alphaB-crystallin chaperone gene causes a desmin-related myopathy. *Nat Genet*. 1998;20:92–95.
16. Selcen D, Engel AG. Mutations in myotilin cause myofibrillar myopathy. *Neurology*. 2004;62:1363–1371.
17. Griggs R, Vihola A, Hackman P, et al. Zaspopathy in a large classic late-onset distal myopathy family. *Brain*. 2007;130:1477–1484.
18. Penisson-Besnier I, Talvinen K, Dumez C, et al. Myotilinopathy in a family with late onset myopathy. *Neuromuscul Disord*. 2006;16:427–431.
19. Hauser MA, Horrigan SK, Salmikangas P, et al. Myotilin is mutated in limb girdle muscular dystrophy 1A. *Hum Mol Genet*. 2000;9:2141–2147.
20. Nishino I, Noguchi S, Murayama K, et al. Distal myopathy with rimmed vacuoles is allelic to hereditary inclusion body myopathy. *Neurology*. 2002;59:1689–1693.
21. Forch P, Puig O, Martinez C, Seraphin B, Valcarcel J. The splicing regulator TIA-1 interacts with U1-C to promote U1 snRNP recruitment to 5′ splice sites. *EMBO J*. 2002;21:6882–6892.
22. Gilks N, Kedersha N, Ayodele M, et al. Stress granule assembly is mediated by prion-like aggregation of TIA-1. *Mol Biol Cell*. 2004;15:5383–5398.
23. Udd B, Vihola A, Sarparanta J, Richard I, Hackman P. Titinopathies and extension of the M-line mutation phenotype beyond distal myopathy and LGMD2J. *Neurology*. 2005;64:636–642.
24. de Seze J, Udd B, Haravuori H, et al. The first European family with tibial muscular dystrophy outside the Finnish population. *Neurology*. 1998;51:1746–1748.
25. Van den Bergh PY, Bouquiaux O, Verellen C, et al. Tibial muscular dystrophy in a Belgian family. *Ann Neurol*. 2003;54:248–251.
26. Pollazzon M, Suominen T, Penttila S, et al. The first Italian family with tibial muscular dystrophy caused by a novel titin mutation. *J Neurol*. 2010;257:575–579.
27. Evilä A, Vihola A, Sarparanta J, et al. Atypical phenotypes in titinopathies explained by second titin mutations. *Ann Neurol*. 2014;75:230–240.
28. Haravuori H, Vihola A, Straub V, et al. Secondary calpain3 deficiency in 2q-linked muscular dystrophy: titin is the candidate gene. *Neurology*. 2001;56:869–877.
29. Richard I, Broux O, Allamand V, et al. Mutations in the proteolytic enzyme calpain 3 cause limb-girdle muscular dystrophy type 2A. *Cell*. 1995;81:27–40.
30. Amato AA, Kagan-Hallet K, Jackson CE, et al. The wide spectrum of myofibrillar myopathy suggests a multifactorial etiology and pathogenesis. *Neurology*. 1998;51:1646–1655.
31. Milhorat AT, Wolff HG. Studies in diseases of muscle XII. Heredity of progressive muscular dystrophy; Relationship between age at onset of symptoms and clinical course. *Arch Neurol Psychiatry*. 1943;49:641–654.
32. Walter MC, Reilich P, Huebner A, et al. Scapuloperoneal syndrome type Kaeser and a wide phenotypic spectrum of adult-onset, dominant myopathies are associated with the desmin mutation R350P. *Brain*. 2007;130:1485–1496.
33. Melberg A, Oldfors A, Blomstrom-Lundqvist C, et al. Autosomal dominant myofibrillar myopathy with arrhythmogenic right ventricular cardiomyopathy linked to chromosome 10q. *Ann Neurol*. 1999;46:684–692.
34. Hedberg C, Melberg A, Kuhl A, Jenne D, Oldfors A. Autosomal dominant myofibrillar myopathy with arrhythmogenic right ventricular cardiomyopathy 7 is caused by a DES mutation. *Eur J Hum Genet*. 2012;20:984–985.
35. Sugawara M, Kato K, Komatsu M, et al. A novel de novo mutation in the desmin gene causes desmin myopathy with toxic aggregates. *Neurology*. 2000;55:986–990.
36. Vorgerd M, van der Ven PF, Bruchertseifer V, et al. A mutation in the dimerization domain of filamin C causes a novel type of autosomal dominant myofibrillar myopathy. *Am J Hum Genet*. 2005;77:297–304.
37. Shatunov A, Olive M, Odgerel Z, et al. In-frame deletion in the seventh immunoglobulin-like repeat of filamin C in a family with myofibrillar myopathy. *Eur J Hum Genet*. 2009;17:656–663.
38. Luan X, Hong D, Zhang W, Wang Z, Yuan Y. A novel heterozygous deletion-insertion mutation (2695-2712 del/GTTTGT ins) in exon 18 of the filamin C gene causes filaminopathy in a large Chinese family. *Neuromuscul Disord*. 2010;20:390–396.
39. Selcen D, Muntoni F, Burton BK, et al. Mutation in BAG3 causes severe dominant childhood muscular dystrophy. *Ann Neurol*. 2009;65:83–89.
40. Markesbery WR, Griggs RC, Leach RP, Lapham LW. Late onset hereditary distal myopathy. *Neurology*. 1974;24:127–134.
41. Lin X, Ruiz J, Bajraktari I, et al. ZASP mutations in actin-binding domain cause disruption of skeletal muscle actin filaments in myofibrillar myopathy. *J Biol Chem*. 2014;289:13615–13626.
42. Selcen D, Engel AG. Mutations in ZASP define a novel form of muscular dystrophy in humans. *Ann Neurol*. 2005;57:269–276.
43. Olive M, Goldfarb LG, Shatunov A, Fischer D, Ferrer I. Myotilinopathy: refining the clinical and myopathological phenotype. *Brain*. 2005;128:2315–2326.

44. Penisson-Besnier I, Dumez C, Chateau D, Dubas F, Fardeau M. Autosomal dominant late adult onset distal leg myopathy. *Neuromuscul Disord*. 1998;8:459–466.
45. Foroud T, Pankratz N, Batchman AP, et al. A mutation in myotilin causes spheroid body myopathy. *Neurology*. 2005;65:1936–1940.
46. Feit H, Silbergleit A, Schneider LB, et al. Vocal cord and pharyngeal weakness with autosomal dominant distal myopathy: clinical description and gene localization to 5q31. *Am J Hum Genet*. 1998;63:1732–1742.
47. Tomimitsu H, Ishikawa K, Shimizu J, Ohkoshi N, Kanazawa I, Mizusawa H. Distal myopathy with rimmed vacuoles: novel mutations in the GNE gene. *Neurology*. 2002;59:451–454.
48. Kayashima T, Matsuo H, Satoh A, et al. Nonaka myopathy is caused by mutations in the UDP-*N*-acetylglucosamine-2-epimerase/*N*-acetylmannosamine kinase gene (GNE). *J Hum Genet*. 2002;47:77–79.
49. Nishino I, Malicdan MC, Murayama K, Nonaka I, Hayashi YK, Noguchi S. Molecular pathomechanism of distal myopathy with rimmed vacuoles. *Acta Myol*. 2005;24:80–83.
50. Malicdan MC, Noguchi S, Hayashi YK, Nonaka I, Nishino I. Prophylactic treatment with sialic acid metabolites precludes the development of the myopathic phenotype in the DMRV-hIBM mouse model. *Nat Med*. 2009;15:690–695.
51. Salama I, Hinderlich S, Shlomai Z, et al. No overall hyposialylation in hereditary inclusion body myopathy myoblasts carrying the homozygous M712T GNE mutation. *Biochem Biophys Res Commun*. 2005;328:221–226.
52. Illa I, Serrano-Munuera C, Gallardo E, et al. Distal anterior compartment myopathy: a dysferlin mutation causing a new muscular dystrophy phenotype. *Ann Neurol*. 2001;49:130–134.
53. Anderson LV, Davison K, Moss JA, et al. Dysferlin is a plasma membrane protein and is expressed early in human development. *Hum Mol Genet*. 1999;8:855–861.
54. Matsuda C, Hayashi YK, Ogawa M, et al. The sarcolemmal proteins dysferlin and caveolin-3 interact in skeletal muscle. *Hum Mol Genet*. 2001;10:1761–1766.
55. Tateyama M, Aoki M, Nishino I, et al. Mutation in the caveolin-3 gene causes a peculiar form of distal myopathy. *Neurology*. 2002;58:323–325.
56. Selcen D, Stilling G, Engel AG. The earliest pathologic alterations in dysferlinopathy. *Neurology*. 2001;56:1472–1481.
57. Bansal D, Campbell KP. Dysferlin and the plasma membrane repair in muscular dystrophy. *Trends Cell Biol*. 2004;14:206–213.
58. Bittner RE, Anderson LV, Burkhardt E, et al. Dysferlin deletion in SJL mice (SJL-Dysf) defines a natural model for limb girdle muscular dystrophy 2B. *Nat Genet*. 1999;23:141–142.
59. Linssen WH, de Visser M, Notermans NC, et al. Genetic heterogeneity in Miyoshi-type distal muscular dystrophy. *Neuromuscul Disord*. 1998;8:317–320.
60. Jaiswal JK, Marlow G, Summerill G, et al. Patients with a non-dysferlin Miyoshi myopathy have a novel membrane repair defect. *Traffic*. 2007;8:77–88.
61. Hicks D, Sarkozy A, Muelas N, et al. A founder mutation in Anoctamin 5 is a major cause of limb-girdle muscular dystrophy. *Brain*. 2011;134:171–182.
62. Penttila S, Palmio J, Suominen T, et al. Eight new mutations and the expanding phenotype variability in muscular dystrophy caused by ANO5. *Neurology*. 2012;78:897–903.
63. Williams DR, Reardon K, Roberts L, et al. A new dominant distal myopathy affecting posterior leg and anterior upper limb muscles. *Neurology*. 2005;64:1245–1254.
64. Watts GD, Wymer J, Kovach MJ, et al. Inclusion body myopathy associated with Paget disease of bone and frontotemporal dementia is caused by mutant valosin-containing protein. *Nat Genet*. 2004;36:377–381.
65. Laing NG, Laing BA, Meredith C, et al. Autosomal dominant distal myopathy: linkage to chromosome 14. *Am J Hum Genet*. 1995;56:422–427.
66. Voit T, Kutz P, Leube B, et al. Autosomal dominant distal myopathy: further evidence of a chromosome 14 locus. *Neuromuscul Disord*. 2001;11:11–19.
67. Zimprich F, Djamshidian A, Hainfellner JA, Budka H, Zeitlhofer J. An autosomal dominant early adult-onset distal muscular dystrophy. *Muscle Nerve*. 2000;23:1876–1879.
68. Lamont PJ, Udd B, Mastaglia FL, et al. Laing early onset distal myopathy: slow myosin defect with variable abnormalities on muscle biopsy. *J Neurol Neurosurg Psychiatry*. 2006;77:208–215.
69. Wallgren-Pettersson C, Pelin K, Nowak KJ, et al. Genotype-phenotype correlations in nemaline myopathy caused by mutations in the genes for nebulin and skeletal muscle alpha-actin. *Neuromuscul Disord*. 2004;14:461–470.
70. Wallgren-Pettersson C, Lehtokari VL, Kalimo H, et al. Distal myopathy caused by homozygous missense mutations in the nebulin gene. *Brain*. 2007;130:1465–1476.
71. Felice KJ, Meredith C, Binz N, et al. Autosomal dominant distal myopathy not linked to the known distal myopathy loci. *Neuromuscul Disord*. 1999;9:59–65.
72. Mahjneh I, Haravuori H, Paetau A, et al. A distinct phenotype of distal myopathy in a large Finnish family. *Neurology*. 2003;61:87–92.
73. Haravuori H, Siitonen HA, Mahjneh I, et al. Linkage to two separate loci in a family with a novel distal myopathy phenotype (MPD3). *Neuromuscul Disord*. 2004;14(3):183–187.
74. Servidei S, Capon F, Spinazzola A, et al. A distinctive autosomal dominant vacuolar neuromyopathy linked to 19p13. *Neurology*. 1999;53:830–837.
75. Di Blasi C, Moghadaszadeh B, Ciano C, et al. Abnormal lysosomal and ubiquitin-proteasome pathways in 19p13.3 distal myopathy. *Ann Neurol*. 2004;56:133–138.

CHAPTER

95

Hereditary Inclusion-Body Myopathies

Aldobrando Broccolini and Massimiliano Mirabella

Catholic University School of Medicine, Rome, Italy

INTRODUCTION

The term "hereditary inclusion-body myopathies" (HIBMs) was originally proposed by Askanas and Engel to identify different muscle disorders with autosomal recessive or dominant inheritance and muscle pathology similar to that of inclusion-body myositis (s-IBM), except for the presence of lymphocytic inflammation.[1] Although clinical presentation varies among different forms, they all have a progressive course leading to severe disability. Typical abnormalities of HIBM muscle biopsy include i) myopathic changes with increased scatter of muscle fiber diameter and centralization of myonuclei, ii) muscle fibers containing various-sized and mainly rimmed vacuoles, and iii) cytoplasmic and occasionally nuclear 15–21-nm diameter tubulofilaments (Figure 95.1A, B and C).[2]

GNE MYOPATHY

The most common form of HIBM was originally described in Persian–Jewish (PJ) families (MIM# 600737)[3] and is due to mutations of the UDP-*N*-acetylglucosamine 2-epimerase/*N*-acetylmannosamine kinase (*GNE*) gene on chromosome 9, hence the name GNE myopathy.[4] *GNE* codes for a bifunctional enzyme (GNE/MNK), with both epimerase and kinase activities, that has a central role in sialic acid biosynthesis.[5] A homozygous mutation converting methionine to threonine at codon 712 (p.M712T) has been found in all Middle Eastern patients of both PJ and non-PJ descent, thus suggesting a common founder effect.[4,6,7] Shortly after, *GNE* mutations were found in patients with distal myopathy with rimmed vacuoles, also known as Nonaka myopathy (DMRV; MIM# 605820),[8] therefore confirming that GNE myopathy and DMRV are indeed the same entity. In the same way as in Middle Eastern patients, strong linkage disequilibrium has been demonstrated in the majority of Japanese patients with the homozygous missense mutation p.V572L.[9] Conversely, patients of other ethnic origins are usually compound heterozygous for mutations in different regions of the gene.[4,7,10]

Clinical and Pathologic Features

The onset of GNE myopathy is usually in late teenage or early adulthood years, with weakness and atrophy of the distal muscles of the lower extremities and later proximal progression involving iliopsoas, gluteal, hamstring, and adductor muscles. Peculiarly, the quadriceps muscles often retain a normal or close-to-normal strength despite the severe involvement of other hip muscles. Upper limb muscles usually become affected later in the disease. Serum creatine kinase (CK) level is normal or slightly elevated and electromyography (EMG) may show the presence of mixed myopathic and neurogenic features.[3] However, the identification of the causative gene defect has allowed the recognition of phenotypic variants. These include patients lacking distal weakness or with quadriceps involvement, as well as patients with facial weakness.[6] On the contrary, it has been shown that sparing of the quadriceps is not a unique feature of GNE myopathy as it has been described also in patients with a muscle disorder unrelated to the *GNE* gene.[11] Furthermore, the age at onset of symptoms may be sometimes postponed even to late adulthood, as isolated patients carrying either the p.M712T or the p.A578T mutations who are still asymptomatic in their sixth

Rosenberg's Molecular and Genetic Basis of Neurological and Psychiatric Disease
http://dx.doi.org/10.1016/10.1016/B978-0-12-410529-4.00095-4

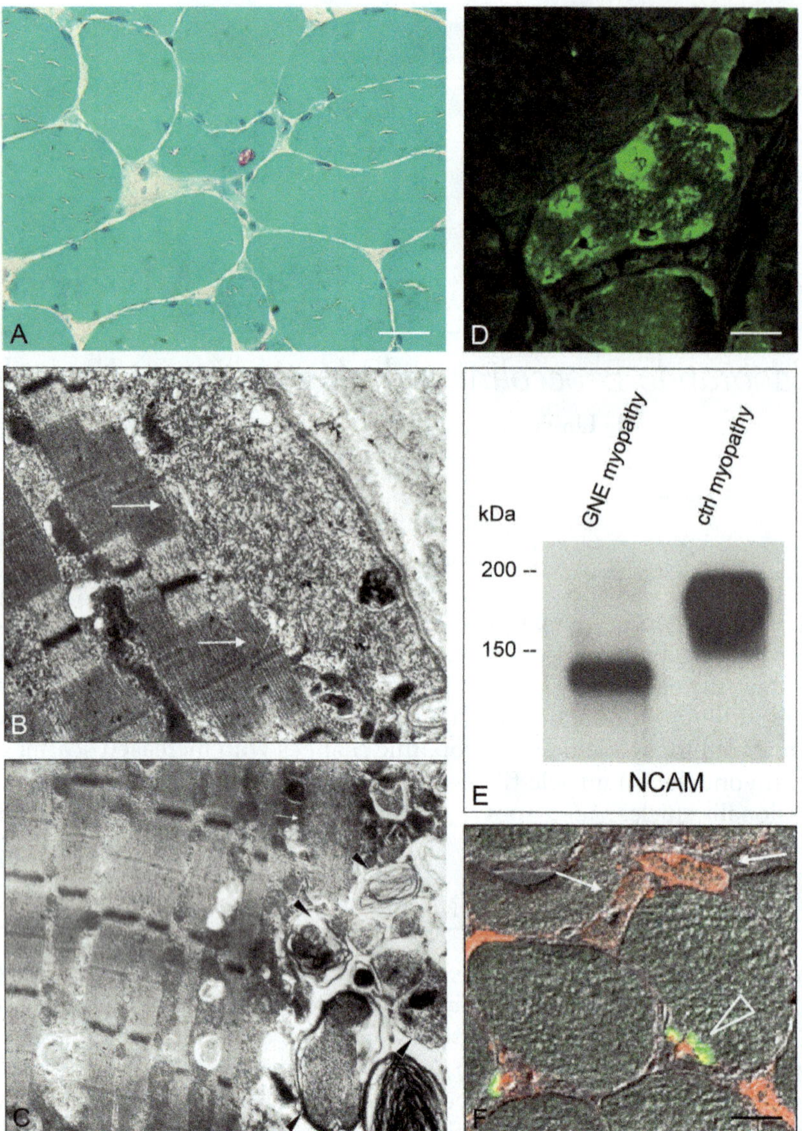

FIGURE 95.1 **Morphological, ultrastructural and biochemical abnormalities of GNE myopathy.** (**A**) Gomori stain of a representative muscle biopsy of GNE myopathy shows myopathic changes and a fiber with a cytoplasmic rimmed vacuole. (Scale bar = 20 μm.) (**B**,**C**) Electron microscopy micrographs showing subsarcolemmal collection of typical 15–21 tubulofilaments (B, arrows), myeloid bodies, membranous material and partially degraded organelles (C, arrow heads). Original magnification ×16000 (B) and ×10000 (C). (**D**) Immunohistochemistry for amyloid-β. Vacuolated muscle fibers showing accumulation of amyloid-β. (Scale bar = 40 μm.) (**E**) Western blot analysis for NCAM. In GNE myopathy hyposialylated NCAM has a lower molecular weight and therefore migrates as a discrete band of 130 kDa, whereas in a control myopathy the protein migrates as a broad band of 150–200 kDa. (**F**) Immunohistochemistry for polysialylated NCAM. In GNE myopathy muscle polysialylated NCAM (red) is expressed in isolated regenerating fibers (arrows) and along the postsynaptic side of a neuromuscular junction identified by α-bungarotoxin labeling (arrow head, green). (Scale bar = 40 μm.)

or seventh decade of life have been described.[6,9] Therefore, degree of progression of symptoms and distribution of muscle weakness may be variable but it is not known whether this is because different *GNE* mutations are not functionally equivalent or other factors modulate the clinical phenotype. Moreover, the selective involvement of skeletal muscle is overtly peculiar as this tissue expresses relatively low levels of the enzyme in comparison to other tissues (i.e., liver, lung, and kidney) that remain unaffected. This suggests the existence of putative susceptibility factors of skeletal muscle to a generalized metabolic impairment.

Muscle pathology of GNE myopathy is similar to that of s-IBM, the most frequent muscle disorder occurring in elderly patients. In fact, in both disorders there is the abnormal accumulation of proteins commonly associated with Alzheimer disease brain pathology including amyloid-β (Aβ) (Figure 95.1D) and paired helical filaments containing hyperphosporylated tau.[2] The similarities between GNE myopathy and s-IBM muscles suggest that, despite different etiologies, both disorders possibly share some common downstream pathogenic mechanisms.

Considerations about Pathogenesis

The cellular pathogenic mechanism(s) activated by mutations of the *GNE* gene and leading to muscle fiber degeneration are for the most part undefined. Sialic acid is normally present on the distal ends of *N*- and *O*-glycans and is involved in many biological functions including cellular adhesion, formation or masking of recognition determinants, stabilization of glycoprotein structure, and signal transduction.[5] Whether mutations of the *GNE* gene lead to hyposialylation of muscle glycoproteins (which has a role in the pathogenic cascade) is still debated. However, it has been shown that transgenic mice expressing the human *GNE* gene with the p.D176V mutation on a *Gne* knockout background (*Gne*$^{(-/-)}$h*GNE*D176V-Tg) have reduced sialic acid levels in serum and other tissues and develop a myopathy with a similar molecular phenotype. Indeed, the muscle of these mice shows rimmed vacuoles and cytoplasmic inclusions within the fibers. Interestingly, abnormal muscle fibers of *Gne*$^{(-/-)}$h*GNE*D176V-Tg mice accumulate Aβ and this seems to precede other abnormalities. In this animal model, total sialic acid content is reduced in liver, spleen, and kidney to a greater extent than that of skeletal muscle, although the latter is the only tissue that becomes affected.[12] This evidence provides further strength to the hypothesis that skeletal muscle may be more sensitive to perturbations in sialic acid metabolism.

In an attempt to identify glycoproteins whose abnormal sialylation may be relevant to the pathogenesis of this disorder, initial attention has been given to α-dystroglycan (α-DG), a structural protein that provides connection between the cellular cytoskeleton and proteins of the extracellular matrix such as laminin. An abnormal glycosylation of α-DG, as seen in some congenital muscular dystrophies, results in impairment of its laminin-binding activity and possibly promotes muscle fiber degeneration.[13] A similar defect was initially hypothesized as a possible pathogenic mechanism of GNE myopathy. However, in GNE myopathy α-DG can be variably hyposialylated but the laminin-binding capacity is never impaired, thus undermining the involvement of the protein in the pathogenesis of this disorder.[14]

The neural cell adhesion molecule (NCAM) is a member of the immunoglobulin superfamily of adhesion molecules and physiologically binds long linear homopolymers of sialic acid residues, thus forming polysialylated NCAM. Therefore, NCAM can be considered a sensitive indicator of the level of cellular sialic acid.[15] Accordingly, in GNE myopathy muscle NCAM is, for the most part, hyposialylated, as evidence by its increased electrophoretic mobility when the protein is studied by western blot (Figure 95.1E).[11,16] However, the possible pathogenic role of hyposialylated NCAM in GNE myopathy is not known. Moreover, our later studies have shown a heterogeneous sialylation of NCAM within muscle fibers. Indeed, the NCAM protein that is present postsynaptically at the neuromuscular junctions appears to be normally sialylated, whereas the amount of protein accumulated in the cytoplasm of abnormal fibers (that probably represents the majority of NCAM expressed by the diseased muscle) is not. Likewise, the NCAM that is expressed by the rare regenerating muscle fibers of GNE myopathy muscle appears to be normally sialylated (Figure 95.1F). The mechanism underlying such difference in sialylation of NCAM needs further explanation. In addition, the presence of sialylated NCAM in regenerating muscle fibers is congruent with the evidence that primary cultured GNE myopathy myotubes are capable of producing normally sialylated NCAM (Broccolini and Mirabella, personal observation 2009). Therefore, GNE myopathy muscle is still able to produce some amount of sialic acid but mainly during its developmental/regenerative stage and, to a lesser extent, in the mature muscle fiber. A plausible explanation of such difference between GNE myopathy adult muscle and cultured myotubes is that the latter express a higher level of *GNE* mRNA, and possibly of the corresponding protein, which compensates for the enzyme impairment. Alternatively, in immature fibers, sialic acid can be synthesized also by other metabolic pathways that are progressively shut down as muscle differentiation progresses. Despite the lack of evidence of hyposialylated NCAM having a role in GNE myopathy pathogenesis, such a peculiar feature can be used as a pre-genetic cellular marker in the routine diagnostic workup of muscle biopsy. This can be helpful especially when encountering patients with uncommon clinical or pathologic features. Indeed, we have shown that out of a cohort of 84 patients with a yet uncharacterized muscle disorder, in three patients the western blot analysis disclosed an NCAM protein with increased electrophoretic mobility, thus suggesting abnormal sialylation. The subsequent genetic study demonstrated that they all carried pathogenic mutations of the *GNE* gene.[16]

Aβ accumulates within abnormal fibers of GNE myopathy muscle and this possibly plays a central role in the pathogenic cascade leading to muscle degeneration, as previously shown for s-IBM.[17] However, the functional relationship between *GNE* defect, sialic acid metabolism and the accumulation of Aβ is still elusive. A possible explanation is an impairment of cellular mechanisms involved in Aβ processing. For example, Neprilysin (NEP), a zinc metallopeptidase known to cleave Aβ at multiple sites, is known to contain a large amount of sialic acid and changes

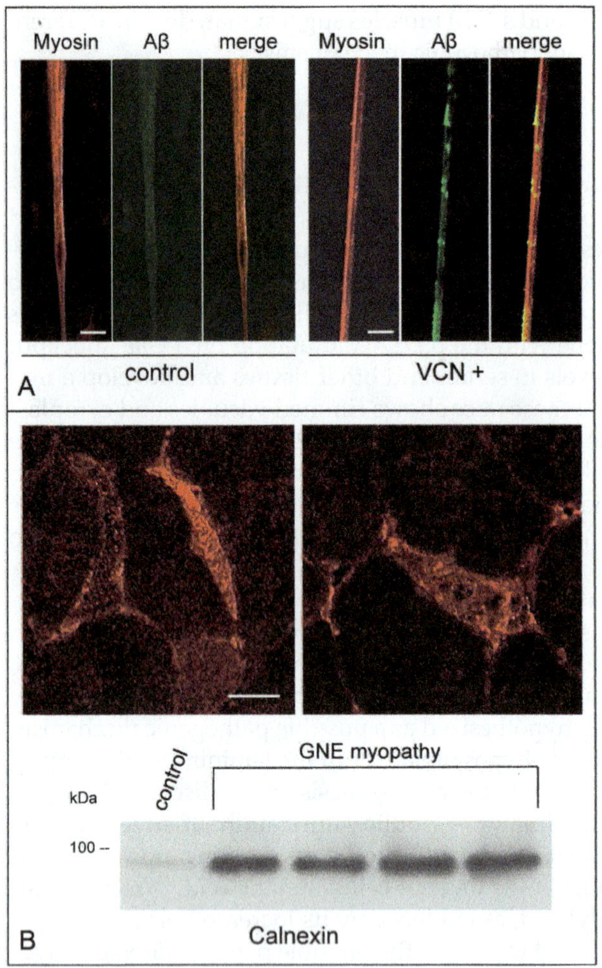

FIGURE 95.2 **GNE myopathy: biochemical clues of possible pathogenic mechanisms.** (**A**) Immunocytochemistry for amyloid-β (Aβ) in normal cultured myotubes treated with vibrio cholerae neuraminidase (VCN) for the enzymatic removal of sialic acid from glycoproteins. In control myotubes Aβ is detected as a faint cytoplasmic signal. Myosin heavy chain staining is used to identify myotubes. On the contrary, the experimental desialylation of cultured myotubes results in the appearance of distinct foci strongly immunoreactive for Aβ within the cytoplasm of myosin-positive myotubes. (Scale bar = 10 μm.) (**B**) Immunochemical analysis of calnexin, an endoplasmic reticulum (ER) chaperone, in GNE myopathy muscle. Abnormal muscle fibers show overexpression of calnexin in the cytoplasm (Scale bar = 20 μm). Accordingly, by western blot analysis there is an overall increased level of the protein compared to control muscle, thus suggesting activation of ER stress.

in the sugar moieties of the protein can affect its stability and enzymatic activity.[18] We have shown that NEP is hyposialylated in GNE myopathy muscle and this is associated with a significant reduction of its expression and enzymatic activity.[19] The link between abnormal sialylation and functional impairment of NEP is provided by the evidence that, *in vitro*, the enzymatic removal of sialic acid from glycoproteins of cultured human normal myotubes results in reduced NEP activity and this is associated with the appearance of Aβ cytoplasmic inclusions (Figure 95.2A).[19] We do not know whether this functional defect of NEP is *per se* sufficient to trigger Aβ accumulation, but in the complex and still undisclosed abnormal milieu of GNE myopathy muscle, it is possible that hyposialylated and dysfunctional NEP has a role in hampering the cellular Aβ clearing system. How hyposialylation of NEP affects its stability is not known, although interference with the correct processing of the protein in the endoplasmic reticulum (ER) leading to increased degradation can be hypothesized. In general terms, the possibility exists that hyposialylation of glycoproteins may perturb their proper folding and trafficking through the ER and Golgi network and their translocation to the plasma membrane. This would activate a mechanism of ER stress that is intended to manage the accumulation of abnormal proteins. Once an ER stress condition is established, the misfolded and unfolded proteins trapped in the ER are retrotranslocated to the cytoplasm and degraded by either the ubiquitin–proteasome system or the autophagic process.[20] Indeed, there is evidence that ER stress and the unfolded protein response, two cellular mechanisms intended to cope with abnormal folding and accumulation of proteins, are activated in GNE myopathy muscle (Figure 95.2B) (Broccolini and Mirabella, personal observation 2010, and [21]).

Therapeutic Perspectives

The identification of the key pathogenic mechanism(s) triggered by mutations of the *GNE* gene is a fundamental prerequisite to design a therapeutic strategy. If hyposialylation of glycoproteins plays a pivotal role in the relentless muscle degeneration observed in GNE myopathy, it is plausible that the early restoration of a normal sialylation status within the cellular environment may help to recover homeostasis of muscle fibers. The first short-term open-label pilot study of administration of sialic acid has been performed on four GNE myopathy patients who received sialic acid via intravenous immune globulin G (IVIG), a glycoprotein that contains 8 μmoles of this sugar per gram, for 1 month. Although there was no evidence of increased sialylation of specific muscle glycoproteins at the end of the treatment, the systemic supplementation of sialic acid via IVIG was associated with a temporary improvement of objective and subjective measures of muscle strength and function.[22] With all the limitations connected with a small and short-term study, the authors reasoned that the provision of sialic acid may hold therapeutic promise and suggested that the use of its precursors *N*-acetyl-D-mannosamine (ManNAc) could represent a suitable alternative source of sialic acid. Indeed, previous reports have shown that the exogenous supplementation of ManNAc is able to increase in a dose-dependent manner the sialylation level of a subclone of B lymphoma cell line lacking GNE/MNK activity.[23] This is because ManNAc enters the sialic acid biosynthetic pathway immediately beyond the metabolic block determined by mutated GNE/MNK and is phosphorylated by a different enzyme with ManNAc kinase activity (most likely *N*-acetylglucosamine kinase). Then ManNAc-6-phosphate can be further metabolized to sialic acid. Indeed, Malicdan et al. have demonstrated that the oral prophylactic supplementation of ManNAc in *Gne*$^{(-/-)}$h*GNE*D176V-Tg mice results in an increase of sialic acid in muscle to a nearly normal level and prevents the development of the muscle phenotype. In particular, ManNAc-treated mice have increased strength, muscle mass, body weight, and overall survival compared to untreated control littermates. This is associated with increase of the mean muscle fiber cross-sectional area and a reduction of vacuolated fibers containing Aβ inclusions.[24] A phase I clinical trial of sialic acid supplementation therapy for human patients was carried out between October 2010 and June 2011 in Japan and September 2011 and April 2012 in the US, the latter using slow-release tablets of sialic acid; a phase II trial will be conducted in the near future.

HEREDITARY INCLUSION-BODY MYOPATHY WITH PAGET DISEASE OF THE BONE AND FRONTOTEMPORAL DEMENTIA

Hereditary inclusion-body myopathy associated with Paget disease of bone (PDB) and frontotemporal dementia (IBMPFD, MIM# 167320) is a rare multisystem degenerative autosomal dominant disorder due to mutations of the valosin-containing protein gene (*VCP*, MIM# 601023).[25] Patients do not always express the all three variably penetrant phenotypic features. In fact, while 90% of patients develop muscle weakness, approximately half have PDB and one-third have frontotemporal dementia (FTD). Also, IBMPFD patients can express only one phenotypic component, as approximately 30% of patients present with a pure myopathy. This implies that, in the presence of an isolated myopathy, the clinician should enquire for the presence of dementia or PDB in other members of the family. Other phenotypic features have been reported as well, including dilated cardiomyopathy, hepatic steatosis, cataracts, sensory motor axonal neuropathy, pyramidal tract dysfunction, sphincter disturbance, sensorineural hearing loss, and amyotrophic lateral sclerosis-like features.[26]

Clinical and Pathologic Features

Muscle involvement is usually variable, although the majority of patients display a limb-girdle distribution of weakness and scapular winging. Axial weakness with head drop and weakness of the distal muscle of limbs have also been described.[27] Serum CK level is within normal values or slightly elevated. EMG reveals myopathic changes often associated with fibrillation potentials and positive sharp waves. Nerve conduction studies are normal.[27] Muscle biopsy features include myopathic changes with increased scatter of fiber diameter, fibers with cytoplasmic rimmed vacuoles, and patchy increase of endomysial connective tissues. Spread out small angulated fibers are also observed. Immunohistochemistry for Aβ, ubiquitin and VCP highlights focal cytoplasmic accumulation of these proteins in the majority of cases.[28] Of note, accumulation of VCP is not specific for IBMPFD as it has been also demonstrated in other muscle disorders with prominent multi-protein aggregate pathology.[29] Electron microscopy (EM) analysis of skeletal muscle shows tubulofilamentous inclusions within myofibers and myonuclei (more frequently than in GNE myopathy) along with disrupted myofibrils, vacuolization, and myelin-like whorls.[28]

Clinical features of cognitive decay may be variable among different patients with IBMPFD. However, in many cases an early symptom is language impairment with preserved episodic memory. Later in the disease, auditory comprehension deficits for one-step commands, alexia, and agraphia may be documented. Not uncommonly, a behavioral disorder or psychotic features are described as initial symptoms. Brain magnetic resonance imaging

(MRI) mainly shows atrophy of the frontal and temporal lobes with ventricular enlargement. FTD usually lacks the typical pathology of Alzheimer disease brain.[26] Differently from other familial variants of FTD due to mutation of tau protein, a key feature of VCP-related frontotemporal dementia is the presence of ubiquitinated inclusions within neuronal nuclei, thus defining IBMPFD as a subtype of frontotemporal degeneration with ubiquitinated inclusions (FTLD-U).[30,31] Ubiquitin-positive inclusions colocalize with TDP-43, an RNA binding protein involved in exon skipping of the pre-mRNA.[32] Interestingly, TDP-43 also accumulates in abnormal muscle fibers of IBMPFD. However, differently from the brain that presents only intranuclear ubiquitinated and TDP-43 inclusions, affected muscle shows both myonuclear and sarcoplasmic inclusions while TDP-43 is diminished in the nuclei of inclusion-bearing fibers.[33] TDP-43 redistribution from nucleus to the cytosol is not exclusive of this disorder as it is observed in other neurodegenerative diseases and in muscle disorders with rimmed vacuoles such as s-IBM and GNE myopathy.[32,34]

PDB is caused by overactive osteoclasts and is associated clinically with pain, elevated serum alkaline phosphatase, and X-ray findings of coarse trabeculation and sclerotic lesions. PDB is often focal and the bones most commonly involved are the skull, vertebrae, and pelvis. Other laboratory features include increased urine concentrations of pyridinoline and deoxypyridinoline, whereas radionuclide total-body scan using Tc-99m shows focally increased bony uptake of the tracer.

Considerations about Pathogenesis

VCP, a member of the AAA-ATPase superfamily, is characterized by a tripartite structure with an *N*-terminal domain, involved in ubiquitin binding, and two central domains that bind and hydrolyze ATP.[35] VCP regulates critical steps in ubiquitin-dependent protein quality control and intracellular signaling pathways and participates in ubiquitin-labeled proteins processing for recycling or degradation by the proteasome.[36] Other known functions comprise membrane fusion, transcription activation, nuclear envelope reconstruction, DNA repair, post-mitotic organelle reassembly, cell cycle control and apoptosis.[36–39] Approximately 20 different missense mutations have been identified in patients with IBMPFD, the majority of these involving the *N*-terminal domain of the protein.[40,41] Interestingly, codon 155 has been identified as a mutation hot spot, as at least five different mutations involve this codon and an identical mutation has been described in patients with an apparently different genetic background.[41] The underlying pathogenic mechanism of IBMPFD is largely indefinite, but increasing evidence points to impaired protein processing and turnover having a role.[26,41] Because VCP participates in protein degradation and autophagy, mutations hampering the protein functions result in abnormal accumulation of ubiquitinated proteins and activation of ER stress response. Indeed, *in vitro*, the expression of IBMPFD VCP mutations results in increase of ER-associated and cytosolic ubiquitinated inclusions.[42] Moreover, co-expression of the aggregate-prone expanded polyglutamine in IBMPFD mutant cells leads to increased aggregated protein compared to cells expressing the wild type VCP. Overall, it appears that VCP has a role in autophagosome maturation, as *in vitro* it has been shown that these subcellular structures are formed but then fail to mature and be degraded.[42] Interestingly, TDP-43 accumulates in the cytosol upon autophagy inhibition, similar to what is seen after IBMPFD mutation expression, highlighting the possible role of impaired autophagy in the pathologic changes seen in IBMPFD muscle.[42] To date there are no promising therapeutic strategies or compounds for IBMPFD. One can speculate that IBMPFD being a disorder where abnormal trafficking of misfolded and aggregated proteins leads to impaired autophagy, then molecules that promote a more efficient transfer of protein aggregates into cellular areas of autophagy may hold a therapeutic promise.

HEREDITARY INCLUSION-BODY MYOPATHY WITH CONGENITAL JOINT CONTRACTURES AND EXTERNAL OPHTHALMOPLEGIA

This form of autosomal dominant HIBM was first described in a large multigeneration family from Sweden (MIM# 605637).[43] Typical clinical features consist of early joint contractures, external ophthalmoplegia, and predominantly proximal muscle weakness. Joint contractures normalize during early childhood and overall the myopathy has a mild course in young individuals. A mild upgaze paresis can be evidenced. Accordingly, when a muscle biopsy is performed during this stage, histopathologic abnormalities are usually minor. During adulthood the disease has a more aggressive course, with weakness and atrophy mainly affecting the proximal muscles of upper and lower extremities. The quadriceps are usually severely affected. Complete external ophthalmoplegia is a consistent finding and a hand tremor is common. Serum CK level can be elevated to up to tenfold the normal value and EMG shows myopathic changes. Muscle biopsy shows more pronounced abnormalities consisting in dystrophic changes and presence of rimmed vacuoles. Structural alterations initially affect mainly type 2A fibers, where EM investigation reveals focal disorganization of contractile filaments, although in older patients all fiber types have morphological changes including rimmed vacuoles and 15–21-nm tubulofilamentous inclusions.[43,44]

This disorder is associated with a mutation of the myosin heavy chain IIa gene (*MyHC-IIa*) on chromosome 17, converting glutamic acid to lysine at codon 706 (p.E706K). This affects the SH1 helix of the protein, the core of myosin head, and therefore causes a dominant negative effect through interference with filaments assembly, impaired myosin–actin interaction or functional defect in ATPase activity. Because type IIa MyHC is the main constituent of type 2A fibers, this probably justifies the fact that these fibers are predominantly affected.[44]

OTHER VARIANTS OF HEREDITARY INCLUSION-BODY MYOPATHY

Various and less common forms of hereditary myopathies have been originally included among the HIBM myopathy group based on muscle biopsy findings. For example, the autosomal dominant myopathy identified as HIBM1 (MIM# 601419) is an adult-onset disorder affecting the distal muscles of the lower extremities and the quadriceps. The identification of mutations of *desmin* as the causative gene defect has moved this disorder into the group of myofibrillar myopathies.[45] Likewise, the hereditary myopathy with early respiratory failure (MIM# 607569), also initially considered a form of HIBM, is an autosomal dominant disorder that has been later shown to be a titinopathy.[46] Various reports have described forms of myopathies in isolated families worldwide with pathologic abnormalities resembling that of HIBM but molecular studies are still insufficient to warrant a precise nosological classification.

References

1. Askanas V, Engel WK. New advances in inclusion-body myositis. *Curr Opin Rheumatol*. 1993;5(6):732–741.
2. Askanas V, Engel WK. Sporadic inclusion-body myositis and hereditary inclusion-body myopathies: current concepts of diagnosis and pathogenesis. *Curr Opin Rheumatol*. 1998;10(6):530–542.
3. Argov Z, Yarom R. "Rimmed vacuole myopathy" sparing the quadriceps. A unique disorder in Iranian Jews. *J Neurol Sci*. 1984;64(1):33–43.
4. Eisenberg I, Avidan N, Potikha T, et al. The UDP-*N*-acetylglucosamine 2-epimerase/*N*-acetylmannosamine kinase gene is mutated in recessive hereditary inclusion body myopathy. *Nat Genet*. 2001;29(1):83–87.
5. Hinderlich S, Stasche R, Zeitler R, Reutter W. A bifunctional enzyme catalyzes the first two steps in N-acetylneuraminic acid biosynthesis of rat liver. Purification and characterization of UDP-*N*-acetylglucosamine 2-epimerase/*N*-acetylmannosamine kinase. *J Biol Chem*. 1997;272(39):24313–24318.
6. Argov Z, Eisenberg I, Grabov-Nardini G, et al. Hereditary inclusion body myopathy: the Middle Eastern genetic cluster. *Neurology*. 2003;60(9):1519–1523.
7. Eisenberg I, Grabov-Nardini G, Hochner H, et al. Mutations spectrum of GNE in hereditary inclusion body myopathy sparing the quadriceps. *Hum Mutat*. 2003;21(1):99.
8. Kayashima T, Matsuo H, Satoh A, et al. Nonaka myopathy is caused by mutations in the UDP-*N*-acetylglucosamine-2-epimerase/*N*-acetylmannosamine kinase gene (GNE). *J Hum Genet*. 2002;47(2):77–79.
9. Nishino I, Noguchi S, Murayama K, et al. Distal myopathy with rimmed vacuoles is allelic to hereditary inclusion body myopathy. *Neurology*. 2002;59(11):1689–1693.
10. Broccolini A, Ricci E, Cassandrini D, et al. Novel GNE mutations in Italian families with autosomal recessive hereditary inclusion-body myopathy. *Hum Mutat*. 2004;23(6):632.
11. Ricci E, Broccolini A, Gidaro T, et al. NCAM is hyposialylated in hereditary inclusion body myopathy due to GNE mutations. *Neurology*. 2006;66(5):755–758.
12. Malicdan MC, Noguchi S, Nonaka I, Hayashi YK, Nishino I. A Gne knockout mouse expressing human GNE D176V mutation develops features similar to distal myopathy with rimmed vacuoles or hereditary inclusion body myopathy. *Hum Mol Genet*. 2007;16(22):2669–2682.
13. Michele DE, Barresi R, Kanagawa M, et al. Post-translational disruption of dystroglycan-ligand interactions in congenital muscular dystrophies. *Nature*. 2002;418(6896):417–422.
14. Broccolini A, Gliubizzi C, Pavoni E, et al. alpha-Dystroglycan does not play a major pathogenic role in autosomal recessive hereditary inclusion-body myopathy. *Neuromuscul Disord*. 2005;15(2):177–184.
15. Hong Y, Stanley P. Lec3 Chinese hamster ovary mutants lack UDP-*N*-acetylglucosamine 2-epimerase activity because of mutations in the epimerase domain of the Gne gene. *J Biol Chem*. 2003;278(52):53045–53054.
16. Broccolini A, Gidaro T, Tasca G, et al. Analysis of NCAM helps identify unusual phenotypes of hereditary inclusion-body myopathy. *Neurology*. 2010;75(3):265–272.
17. Askanas V, Engel WK, Alvarez RB. Light and electron microscopic localization of beta-amyloid protein in muscle biopsies of patients with inclusion-body myositis. *Am J Pathol*. 1992;141(1):31–36.
18. Howell S, Nalbantoglu J, Crine P. Neutral endopeptidase can hydrolyze beta-amyloid(1-40) but shows no effect on beta-amyloid precursor protein metabolism. *Peptides*. 1995;16(4):647–652.
19. Broccolini A, Gidaro T, De Cristofaro R, et al. Hyposialylation of neprilysin possibly affects its expression and enzymatic activity in hereditary inclusion-body myopathy muscle. *J Neurochem*. 2008;105(3):971–981.
20. Ni M, Lee AS. ER chaperones in mammalian development and human diseases. *FEBS Lett*. 2007;581(19):3641–3651.

21. Li H, Chen Q, Liu F, et al. Unfolded protein response and activated degradative pathways regulation in GNE myopathy. *PLoS One.* 2013;8(3):e58116.
22. Sparks S, Rakocevic G, Joe G, et al. Intravenous immune globulin in hereditary inclusion body myopathy: a pilot study. *BMC Neurol.* 2007;7:3.
23. Hinderlich S, Berger M, Keppler OT, Pawlita M, Reutter W. Biosynthesis of N-acetylneuraminic acid in cells lacking UDP-*N*-acetylglucosamine 2-epimerase/*N*-acetylmannosamine kinase. *Biol Chem.* 2001;382(2):291–297.
24. Malicdan MC, Noguchi S, Hayashi YK, Nonaka I, Nishino I. Prophylactic treatment with sialic acid metabolites precludes the development of the myopathic phenotype in the DMRV-hIBM mouse model. *Nat Med.* 2009;15(6):690–695.
25. Watts GD, Wymer J, Kovach MJ, et al. Inclusion body myopathy associated with Paget disease of bone and frontotemporal dementia is caused by mutant valosin-containing protein. *Nat Genet.* 2004;36(4):377–381.
26. Weihl CC. Valosin containing protein associated fronto-temporal lobar degeneration: clinical presentation, pathologic features and pathogenesis. *Curr Alzheimer Res.* 2011;8(3):252–260.
27. Watts GD, Thomasova D, Ramdeen SK, et al. Novel VCP mutations in inclusion body myopathy associated with Paget disease of bone and frontotemporal dementia. *Clin Genet.* 2007;72(5):420–426.
28. Hubbers CU, Clemen CS, Kesper K, et al. Pathological consequences of VCP mutations on human striated muscle. *Brain.* 2007;130(Pt 2): 381–393.
29. Greenberg SA, Watts GD, Kimonis VE, Amato AA, Pinkus JL. Nuclear localization of valosin-containing protein in normal muscle and muscle affected by inclusion-body myositis. *Muscle Nerve.* 2007;36(4):447–454.
30. Schroder R, Watts GD, Mehta SG, et al. Mutant valosin-containing protein causes a novel type of frontotemporal dementia. *Ann Neurol.* 2005;57(3):457–461.
31. Cairns NJ, Bigio EH, Mackenzie IR, et al. Neuropathologic diagnostic and nosologic criteria for frontotemporal lobar degeneration: consensus of the Consortium for Frontotemporal Lobar Degeneration. *Acta Neuropathol.* 2007;114(1):5–22.
32. Neumann M, Sampathu DM, Kwong LK, et al. Ubiquitinated TDP-43 in frontotemporal lobar degeneration and amyotrophic lateral sclerosis. *Science.* 2006;314(5796):130–133.
33. Weihl CC, Temiz P, Miller SE, et al. TDP-43 accumulation in inclusion body myopathy muscle suggests a common pathogenic mechanism with frontotemporal dementia. *J Neurol Neurosurg Psychiatry.* 2008;79(10):1186–1189.
34. Salajegheh M, Pinkus JL, Taylor JP, et al. Sarcoplasmic redistribution of nuclear TDP-43 in inclusion body myositis. *Muscle Nerve.* 2009;40(1):19–31.
35. Dai RM, Li CC. Valosin-containing protein is a multi-ubiquitin chain-targeting factor required in ubiquitin-proteasome degradation. *Nat Cell Biol.* 2001;3(8):740–744.
36. Jarosch E, Geiss-Friedlander R, Meusser B, Walter J, Sommer T. Protein dislocation from the endoplasmic reticulum–pulling out the suspect. *Traffic.* 2002;3(8):530–536.
37. Rabinovich E, Kerem A, Frohlich KU, Diamant N, Bar-Nun S. AAA-ATPase p97/Cdc48p, a cytosolic chaperone required for endoplasmic reticulum-associated protein degradation. *Mol Cell Biol.* 2002;22(2):626–634.
38. Rabouille C, Kondo H, Newman R, Hui N, Freemont P, Warren G. Syntaxin 5 is a common component of the NSF- and p97-mediated reassembly pathways of Golgi cisternae from mitotic Golgi fragments in vitro. *Cell.* 1998;92(5):603–610.
39. Yamanaka K, Sasagawa Y, Ogura T. Recent advances in p97/VCP/Cdc48 cellular functions. *Biochim Biophys Acta.* 2012;1823(1):130–137.
40. Gonzalez-Perez P, Cirulli ET, Drory VE, et al. Novel mutation in VCP gene causes atypical amyotrophic lateral sclerosis. *Neurology.* 2012;79(22):2201–2208.
41. Nalbandian A, Donkervoort S, Dec E, et al. The multiple faces of valosin-containing protein-associated diseases: inclusion body myopathy with Paget's disease of bone, frontotemporal dementia, and amyotrophic lateral sclerosis. *J Mol Neurosci.* 2011;45(3):522–531.
42. Ju JS, Fuentealba RA, Miller SE, et al. Valosin-containing protein (VCP) is required for autophagy and is disrupted in VCP disease. *J Cell Biol.* 2009;187(6):875–888.
43. Darin N, Kyllerman M, Wahlstrom J, Martinsson T, Oldfors A. Autosomal dominant myopathy with congenital joint contractures, ophthalmoplegia, and rimmed vacuoles. *Ann Neurol.* 1998;44(2):242–248.
44. Martinsson T, Oldfors A, Darin N, et al. Autosomal dominant myopathy: missense mutation (Glu-706→Lys) in the myosin heavy chain IIa gene. *Proc Natl Acad Sci U S A.* 2000;97(26):14614–14619.
45. Goldfarb LG, Vicart P, Goebel HH, Dalakas MC. Desmin myopathy. *Brain.* 2004;127(Pt 4):723–734.
46. Pfeffer G, Elliott HR, Griffin H, et al. Titin mutation segregates with hereditary myopathy with early respiratory failure. *Brain.* 2012;135 (Pt 6):1695–1713.

CHAPTER

96

The Myotonic Dystrophies

Richard T. Moxley, III, James E. Hilbert*, and Giovanni Meola†*

*University of Rochester Medical Center, Rochester, NY, USA

†University of Milan, Milan, Italy

INTRODUCTION

Disease Characteristics

The myotonic dystrophies are the most common adult muscular dystrophies.[1] Their core characteristics are dominant inheritance, myotonia, cataracts, and muscle weakness, but they also have a broad spectrum of multisystem manifestations.[1] Myotonic dystrophy type 1 (DM1) results from an unstable trinucleotide repeat expansion (CTG) in the dystrophia myotonica-protein kinase (*DMPK*) gene located on chromosome 19q13.3.[2–4] Clinical presentation of DM1 can occur from infancy through adulthood. Myotonic dystrophy type 2 (DM2) results from an unstable four-nucleotide repeat expansion (CCTG) at intron 1 of the *CNBP/ZNF9* gene located on chromosome 3q21.3.[5,6] DM2 patients usually develop symptoms during early to mid-adulthood. DM1 and DM2 can have similar multisystem manifestations involving the brain, eyes, smooth muscle, cardiovascular and endocrine systems; however, DM2 patients have greater proximal than distal weakness, less muscle wasting, and less severe cardiorespiratory complications.[1,7,8] A small percentage of DM1 patients present at birth with congenital myotonic dystrophy (CDM), a severe illness including respiratory failure, hypotonia, loss of swallowing, and other complications. DM2 does not have a congenital form.

Diagnosis/Testing

A high level of clinical suspicion is necessary early in the disease to establish the diagnosis of DM1 or DM2, especially in adults. Initial symptoms in adults are often variable. Delay in obtaining a correct diagnosis is frequent.[9] Family history is very helpful, and, if positive, can prompt specific diagnostic testing. Sensitive and reliable leukocyte DNA testing is available for DM1 and DM2. Early detection curtails a potential diagnostic odyssey, reduces cost of care, and permits timely family planning.

Current Research

Investigations in animal models and in patients have accelerated our understanding of the molecular pathology responsible for DM1 and DM2. The findings indicate that there is a toxic gain-of-function of mutant RNA generated from the microsatellite repeat expansions in the non-coding regions of the genes involved.[10] The nuclear accumulation of this mutant RNA sequesters nuclear regulatory proteins and, in turn, leads to altered regulation of pre-mRNA splicing of the message for specific proteins, such as the skeletal muscle chloride channel (causing myotonia) and the insulin receptor (causing insulin resistance).[10] This discovery has stimulated the rapid development of therapeutic strategies to neutralize or degrade the toxic mutant RNA. Recent studies of a mouse model of DM1 using antisense oligonucleotides have offered encouraging results and have shown a reversal of muscle pathology and correction of abnormal mRNA splicing.[11] In response to these and other promising preclinical study results, clinical researchers

Rosenberg's Molecular and Genetic Basis of Neurological and Psychiatric Disease
http://dx.doi.org/10.1016/10.1016/B978-0-12-410529-4.00096-6

are pursuing studies in DM1 patients to identify optimal biomarkers and clinical endpoints to assess efficacy of treatment for use in future therapeutic trials.

CLINICAL FEATURES

Historical Overview: Disease and Gene Identification

The classical manifestations of Steinert disease, myotonic dystrophy type 1 (DM1), were first described clinically in 1909 by Hans Steinert (Leipzig, Germany), including descriptions of the typical pattern of muscle weakness and myotonia in DM1. Subsequent contributions reported dominant inheritance and multisystem manifestations (skeletal and smooth muscle, heart, brain, eye, respiratory and endocrine systems).[1] In 1992, researchers discovered that DM1 results from an unstable trinucleotide repeat expansion (CTG) in the dystrophia myotonica-protein kinase (*DMPK*) gene located on chromosome 19q13.3.[2–4]

In 1994, clinical researchers in the United States and Germany published descriptions of patients without abnormal expansion of the CTG repeat in the *DMPK* gene with dominant inheritance of myotonia, cataracts, and weakness and whose clinical manifestations were similar to but distinct from Steinert disease. They proposed two names, one name being myotonic dystrophy type 2 (DM2), and the other, proximal myotonic myopathy (PROMM).[7,8] Genetic mapping linked both disorders to chromosome 3q21,[5] and in 2001 researchers discovered that DM2/PROMM resulted from an unstable CCTG repeat expansion in intron 1 of the *CNBP/ZNF9* gene at locus 3q21.3.[6]

Mode of Inheritance and Prevalence

DM1 and DM2 are autosomal dominant. Anticipation (earlier onset of more severe symptoms in successive generations) occurs primarily in DM1.[12] CDM occurs almost exclusively through maternal transmission from a mother with DM1. In contrast, both mothers and fathers with DM1 can have children with childhood-onset of DM1.[1] However, only 5–10% of DM1 patients have onset before their mid-teens.[1] DM2 occurs neither in infancy or childhood, although teenage onset of myotonic symptoms occurs in some DM2 patients carrying concomitant mutations in the gene for the skeletal muscle chloride channel.[13]

Determining the exact prevalence of DM1 and DM2 is a challenge because DNA confirmation of DM1 has only become widely available since the mid-1990s[2–4] and for DM2 after 2003.[5,6,14] Previous estimates of prevalence of myotonic dystrophy relied solely upon clinical diagnosis. Study populations are likely to have included a mixture of symptomatic patients with DM1 and DM2. Patients with minimal or no symptoms are unlikely to have participated. Within these limitations, previous publications report a prevalence range for myotonic dystrophy of 5–20 per 100,000.[1] Higher prevalence of DM1 has been reported in the Basque region of Spain, northern Sweden,[15] Istria region of Croatia,[16] and the Saguenay-Lac St. Jean region of Quebec, Canada.[17,18] The remarkably high incidence in Quebec (158–189 per 100,000) traces to a common ancestor from France.[19] Another recent study of DM1 in Canada calculated the prevalence of CDM to be approximately 2.1/100,000 live births, based upon incident reports and birth data from Canada's Central Statistical Office.[20] This estimate of prevalence of CDM is lower than prevalence estimates in previous non-population based studies in Sweden, Spain, and Britain.

Until recently no estimates of prevalence for DM2 were available. In 2011, clinical researchers in Finland published findings indicating a higher prevalence of DM2 in Europeans than most clinicians suspected. In a large study of over 5000 anonymous blood donors, they observed a higher prevalence for DM2 than DM1.[21] The higher prevalence of DM2 identified in this Finnish population emphasizes the wide spectrum of phenotypes and points out the diagnostic challenge that exists due to the variation in different symptoms that occur in DM2. For example, clinical findings, such as myotonia on physical examination or even on electromyographic testing may be absent, limiting the ability of myotonia screening alone as a means to identify individuals with DM2.[22,23] Whole blood DNA testing in individuals at known risk for DM1 or DM2 or in individuals presenting with one of the common manifestations (Tables 96.1 and 96.2) is needed to determine more exactly the actual prevalence of individuals with abnormal repeat expansions of *DMPK* and *CNBP/ZNF9* genes.

NATURAL HISTORY

Tables 96.1 and 96.2 give a summary of the genetics, core clinical features, and multisystem manifestations typically observed in the initial evaluation of adults with either mild or moderate signs and symptoms of DM1 and DM2.

TABLE 96.1 Summary of Genetics and Core Clinical Features of Myotonic Dystrophy (DM1 and DM2) in Mild to Moderately Affected Patients

	Adult onset myotonic dystrophy type 1 (DM1)	**Adult onset myotonic dystrophy type 2 (DM2)**
Inheritance	Autosomal dominant	Autosomal dominant
Anticipation	Pronounced	Exceptionally rare
Congenital form	Yes	No
Chromosome	19q13.3	3q21.3
Locus	*DMPK*	*CNBP*
Expansion mutation	$(CTG)_n$	$(CCTG)_n$
Location of the expansion	3′ untranslated region	Intron 1
CORE FEATURES: MILD TO MODERATELY AFFECTED PATIENTS		
Clinical myotonia	Usually occurs in the distal limb (grip and intrinsic hand muscles) Can manifest in the tongue, jaw, eye, and smooth muscles Percussion myotonia in thenar and forearm muscles often present Myotonia may be absent in some patients	Not as prominent as in DM1 Often first appears intermittently with hand grip May be limited mainly to percussion of wrist extensor muscles May be absent in some patients
Muscle weakness	Forearm finger flexors; foot dorsi- and plantar flexors; and facial muscles Ptosis Laryngeal (dysarthria) Occasional extraocular weakness Weakness of shoulder girdle and proximal upper limb, neck (flexors > extensors), and knee flexors and extensors more prominent in later stages as disease progresses	Involvement of distal and facial muscles is usually absent Initial weakness is in proximal hip girdle and neck (flexors > extensors) muscles Occasional mild ptosis less common than in DM1 Calf muscle hypertrophy is common
Muscle wasting	Significant atrophy is typically apparent in patients affected moderately Anterior neck Masseter and temporalis Distal forearm Intrinsic hand Distal lower extremity	Significant muscle atrophy is typically mild and involves anterior neck and hip and shoulder girdle
Cataracts	Iridescent, posterior capsular opacities, typically before age 50 years	Similar in appearance to DM1 but less commonly seen on initial exam

DM1 (Myotonic Dystrophy Type 1; Steinert Disease)

Age of Onset

DM1 can present from birth to late adult life. Teenage or later onset is typical for 90% of individuals with DM1. One of the core features (Table 96.1) is often the initial sign or symptom; however, one of the multisystem manifestations (Table 96.2) can be the first complaint. Because of the variable nature of these multisystem features, such as cataracts, gastrointestinal complaints, and sleep disturbances, care providers often do not consider DM1 as the cause and a delay in diagnosis occurs. Early onset in infants or children often shortens the time for diagnostic testing for DM1. Symptoms of CDM appear in the newborn period, can be life threatening, and manifestations include generalized weakness and hypotonia, respiratory failure/insufficiency, feeding difficulty, and clubfoot deformity.[20,24] Childhood myotonic dystrophy (ChDM) may be more difficult to diagnose if there is no diagnosis of an affected parent and/or no positive family history of DM1. ChDM is often defined as having onset of DM symptoms < 10 years; an uneventful prenatal and neonatal history; normal development within the first year of life; and increasing manifestations in the toddler years.[25] ChDM typically causes intellectual deficiency, difficulty with speech or hearing, delayed bowel–bladder training, clumsiness, or rarely cardiac dysrhythmia or postoperative apnea.[26,27]

TABLE 96.2 Summary of Multisystem Manifestations of Myotonic Dystrophy (DM1 and DM2) in Mild to Moderately Affected Patients

	Adult onset myotonic dystrophy type 1 (DM1)	**Adult onset myotonic dystrophy type 2 (DM2)**
MULTISYSTEM MANIFESTATIONS: MILD TO MODERATELY AFFECTED PATIENTS		
Brain	Visual spatial deficits Reduced executive function (trouble organizing and staying on task, reduced goal directed action) Apathy	Similar visual–spatial and executive function deficits to those present in DM1
Heart	Conduction defects, particularly heart block and arrhythmias	Significant disturbances in conduction much less common than in DM1
Respiratory	Reduced forced vital capacity Obstructive and central sleep apnea Weak cough Microatelectasis Frequent respiratory infections	Obstructive sleep apnea
Anesthesia	Increased frequency of respiratory insufficiency/failure following general anesthesia Paradoxical reaction to depolarizing muscle relaxant medications Hypersensitivity to opioids and sedative medications Neuroleptic malignant syndrome Failed respiratory wean Arrhythmias	Limited information is available to determine if there is a significant and increased risk of general anesthesia Recommend careful monitoring in postoperative period until more information is published
Hypersomnia and fatigue	Excessive daytime sleepiness Obstructive and central sleep apnea CNS and muscle related fatigue	Excessive daytime sleepiness is not as prominent as in DM1 Obstructive sleep apnea CNS and muscle related fatigue
Endocrine	Gonadal insufficiency Low testosterone Erectile dysfunction Insulin resistance Hyperlipidemia Hypothyroidism Decreased growth hormone release	Gonadal insufficiency Low testosterone Erectile dysfunction Insulin resistance Hyperlipidemia Hypothyroidism
Gastrointestinal	Dysphagia Gastroesophageal reflux Gastroparesis Small intestinal dysmotility (diarrhea, abdominal cramps, bloating, pain) Colonic dysmotility (pain, "spastic colon," constipation) Pseudo-obstruction Anorectal dysfunction (anal incontinence, constipation, fecal impaction)	Limited information is available Manifestations often less severe than DM1 and can include: Dysphagia Gastroesophageal reflux Abdominal pain Constipation
Muscle pain	Posterior neck, trapezius, anterior lateral chest wall; and upper and lower back Can worsen with exercise and cold temperature	Often a major symptom, especially the arms and upper and lower back Fluctuates in duration, location and intensity Can worsen with exercise and cold temperature Aches and stiffness Pain on muscle palpitation
Pregnancy	Vital capacity declines and shortness of breath occurs often in the third trimester Frequent miscarriage Failed maintenance of pregnancy Polyhydramnios Prolonged first stage of labor Placenta previa Postpartum hemorrhage	Limited information is available to determine if there is significant risk of complications during pregnancy and delivery Weakness and stiffness may worsen during pregnancy and improve following delivery

Disease Evolution

DM1, particularly in adults, may have the most variable spectrum of clinical manifestations of any inherited disease. It can present in late life with only cataracts, baldness, or occasionally with heart block. Weakness and myotonia may be absent or very mild. In the second and third decades, DM1 can present with variable grip myotonia, trouble holding items, ankle weakness, or intermittent slurring of speech. Mild weakness of the face, long finger flexors, intrinsic hand and foot dorsiflexor muscles is usually present, as is myotonia. The myotonia typically occurs following a strong grip or after percussion of the thenar and forearm wrist extensor muscles. Grip and percussion myotonia are not present in infancy and early childhood in DM1, and are sometimes difficult to detect on clinical examination in adults with marked wasting of forearm and intrinsic hand muscles. Dysarthria, difficulty swallowing, gastrointestinal dismotility (frequent bowel movements, intermittent constipation), hypersomnia, cognitive–behavioral problems, sleep apnea, decreased vital capacity, and cardiac conduction disturbances develop over subsequent years.

DM2 (Myotonic Dystrophy Type 2; Proximal Myotonic Myopathy)

Age of Onset and Disease Evolution

DM2 presents in adult life, usually with myotonic stiffness or weakness (proximal legs or long finger flexors).[7,8,14,28,29] Tables 96.1 and 96.2 summarize the core features and multisystem manifestations of DM2. Frequently, the myotonia fluctuates, being prominent only on "bad days," and the muscle pain similarly can vary in severity.[14,30–32] The muscle pain typically appears to be independent of exercise and unrelated to the severity of myotonia on clinical examination. The pain varies in severity over days to weeks. The cause for the variation is unknown. In the initial stages of DM2, there is often mild weakness of long finger flexors, thigh flexors, and hip extensors along with grip myotonia.[14,30] The myotonia is most apparent following percussion of forearm extensor and thenar muscles. The myotonia that occurs after a powerful grip often has a jerky, tremulous quality that differs from the grip myotonia observed in DM1. Patients with this jerky myotonia usually have brisk tendon reflexes and often have mild sustention tremor.

The prominent facial weakness and the wasting of facial, forearm, and distal leg muscles that are typically seen in DM1 are not observed in DM2. Even in the late stages, typical DM2 patients have relatively mild muscle wasting. Occasionally, patients with DM2 have calf muscle hypertrophy.[7,8,14,28] Early onset of cataracts that require surgery is common. Later in life, patients may present with complaints of difficulty climbing stairs or arising from a squat. Myotonia and muscle pain are less prominent when onset is late. Patients typically ascribe their complaints to "old age" or arthritis. In contrast to DM1, respiratory failure and cardiac conduction disturbances are less common in DM2. Difficulty swallowing, gastrointestinal dysfunction, and cognitive disturbances do occur in DM2, but their frequency seems less than in DM1. Additional investigation of the natural history of DM2 is necessary to establish the actual frequency of these complaints and their relationship, if any, to the severity of muscle weakness and/or myotonia.

End of Life: Mechanisms and Comorbidities

DM1

In CDM, estimated mortality ranges between 16–41% in the neonatal period.[20] Contributors to reduced survival include respiratory failure and elective withdrawal of care.

In adult DM1, reports published during the past 2 years have further defined mortality. In a study of 838 patients with DM1 in Quebec, there were 321 deaths with a median age of death of 56 years during 1985–2010.[18] This study showed significant difference between median age of death between childhood-onset (median = 48.5 years), adult-onset (median = 54 years) and mild phenotypes (median = 73 years). In another study of 406 DM1 patients in the US, the authors reported 118 deaths with a median age of 54 years,[33] and the majority resulted from respiratory and cardiac complications,[33] similar to previous reports.[17] A smaller study in Belgrade showed a median age of death of 57 years in 15 of 101 DM1 patients from 1983–2002.[34] The overall death rates in these studies compare to a previous study of DM1 patients in the Netherlands (83 deceased with a median age of death of 56 years).[35] Going forward it will be important to document how mortality rates for DM1 change over the next decade as care providers implement improvements in the treatment of respiratory and cardiac manifestations.

DM2

There are no large-scale studies evaluating mortality in DM2. Future research is necessary to determine if the causes of death in DM2 differ significantly from the general population.

MOLECULAR GENETICS

DM1

Normal Gene Product

Table 96.1 summarizes information on the dystrophia myotonica-protein kinase (*DMPK*) gene and the associated mutation that leads to DM1. The protein encoded by this gene is a serine-threonine kinase that is closely related to other kinases that interact with members of the Rho family of small GTPases. Substrates for this enzyme include myogenin, the beta-subunit of the L-type calcium channels, and phospholemman. The serine-threonine kinase appears to function normally in patients with DM1. The 3′ untranslated region of the *DMPK* gene in normal leukocyte DNA contains 5–37 copies of a CTG trinucleotide. This CTG repeat is unstable in DM1 and in circulating blood may range in length from 50 to 4,000 repeats in different patients. In asymptomatic or "pre-mutation carriers" of the unstable CTG repeat in the *DMPK* gene, the expansion size is 30–50 repeats.[1,10] The repeat size becomes unstable once it exceeds 50 CTG repeats.[1,10]

DM2

Normal Gene Product

Table 96.1 summarizes information on the *CNBP* gene (often termed the zinc finger 9 protein gene, *ZNF9* gene) that is responsible for DM2. *CNBP* encodes cellular nucleic acid-binding protein. Its role in humans is not clearly understood, but it influences embryonic development. It appears to function normally in DM2. In DM2 there is an unstable expansion of the CCTG repeat in intron 1 of the *CNBP* gene. The normal CCTG repeat structure is approximately 10–20 repeats. In carriers of the premutation there are 22–33 uninterrupted CCTG repeats. These premutation carriers are asymptomatic and are unlikely ever to show symptoms. Symptomatic DM2 patients have CCTG repeat lengths ranging from 75 to 11, 000 repeats (mean 5000 CCTG repeats).[1,10]

DM1 and DM2

Genotype–Phenotype Correlations

In DM1 there is a rough correlation between the length of the abnormal CTG expansion in circulating leukocyte DNA and the severity of disease, with exceptions occurring frequently in clinical practice.[1] Age of onset of DM1 shows a correlation. Individuals having 50–100 CTG repeats in leukocyte DNA typically develop mild symptoms later in life. Patients with 150–1000 CTG repeats often have manifestations typical of adult-onset DM1 as summarized in Tables 96.1 and 96.2. Individuals with over 1000 repeats are likely to have congenital or childhood forms of DM1.

Most DM1 patients have a continuous uninterrupted abnormally expanded length of CTG repeats in the 3′ untranslated region of the *DMPK* gene. Published reports of genotype–phenotype correlations make this assumption. However, recent reports point out the limitations in this assumption.[36–38] The most recent report estimates that approximately 4.8% of DM1 patients have CCG interruptions within the 3′ end of the CTG motif at the *DMPK* gene locus.[36] This study used bidirectional triplet primed polymerase chain reaction (TP-PCR) and sequencing to identify the variant expansions. Of the five patients identified, three tested negative for an abnormal CTG expansion using a standard DNA analysis of the *DMPK* gene by routine long-range PCR. All five patients with the variant CTG expansion structure containing the CCG interruptions showed no significant clinical differences compared to the DM1 patients with continuous CTG expansions except for one difference. Those with interrupted repeat tracts did not have significant cognitive impairment.[36]

In another report of DM1 patients with repeat tract interruptions investigators describe a Dutch family co-segregating DM1, Charcot–Marie–Tooth neuropathy, encephalopathic attacks, and early hearing loss caused by a complex variant repeat at the *DMPK* gene locus.[38] The mutation comprised an expanded CTG tract at the 5′ end and a complex array of CTG repeats interspersed with multiple CCG and GGC repeats at the 3′ end. The researchers concluded that their findings implicated a *cis*-acting modifier of mutational dynamics in the 3′ flanking DNA, and they felt that the variant repeats very likely contributed to the unusual symptoms in this Dutch family. More investigations of variant mutations affecting the CTG repeat motif of the *DMPK* gene locus are necessary before making firm conclusions about their contribution to phenotypic variation in DM1. However, these findings support the use of bidirectional TP-PCR in the detection of "variant" DM1 CTG repeat expansions and its inclusion in the routine diagnostic DNA testing to screen for DM1.

In DM2 the abnormal length of the CCTG repeat in circulating blood spans a wider range of size than in DM1 and does not correlate with the clinical severity.

Both DM2 and DM1 show somatic instability of the mutant alleles and studies indicate that the size of the repeat expansion increases over time, across generations, and shows somatic mosaicism, even in postmitotic tissue.[6,39–44] For example, CTG repeat lengths in skeletal muscle, heart and brain of DM1 patients often are 5–10-fold longer than in their circulating leukocyte DNA.[1] Intergenerational repeat variations and somatic mosaicism result from contractions and expansions due to defects in DNA replication, DNA repair, and recombination.[45–48] Studies from population-based mathematical modeling of DM1 mutant alleles from leukocyte DNA demonstrate a bias toward expansion and the bias appears to be due to expansion and contractions that are coupled to DNA repair or transcription rather than dependent on DNA replication.[49]

Disease Mechanisms for DM1 and DM2: RNA-Mediated Diseases

While the fundamental mechanisms that account for the abnormal expansion of the CTG repeats in the *DMPK* gene and CCTG repeats in the *CNBP* gene that cause DM1 and DM2 remain elusive, there is strong support of the hypothesis that both disorders are RNA-mediated diseases with many similarities in their molecular pathomechanism. The comments that follow come largely from two selected reviews that describe the disease mechanisms of DM in more detail.[10,50]

DM1 results from the abnormally expanded CTG repeat in exon 15. Expression of the mutated allele produces mRNA containing expanded CUG repeats. The mutant RNA containing the expanded CUG repeats forms a double-stranded hairpin structure that is highly stable and retained in the nucleus.

An expansion of the CCTG repeat in intron 1 of the *CNBP* gene causes DM2. As occurs with the normal sized CCUG repeat, editing of mutant pre-mRNA removes the expanded CCUG repeat from the mutant *CNBP* pre-mRNA allele. Expanded CCUG repeats are absent from the *CNBP* mRNA. However, the expanded CCUG remnant of intron 1 is highly resistant to degradation due to formation of a double-stranded hairpin structure similar to that adopted by the expanded CUG repeat in mutant DM1 mRNA.

In both DM1 and DM2, the abnormally expanded pathogenic RNA remains in the nucleus and binds nuclear regulatory protein, especially proteins in the MBNL family, such as MBNL1, and forms ribonuclear inclusions. Sequestration of MBNL1 by this expanded mutant RNA occurs in both DM1 and DM2 skeletal muscle and other target tissues, hampering the normal function of MBNL1 and leading to misregulation of alternative splicing of a large number of pre-mRNAs.[10,50] A number of studies in patients with DM1 and DM2 and in animal models support the toxic RNA gain-of-function model summarized above. However, several other factors contribute to complex pathomechanisms that likely participate in one or both forms of myotonic dystrophy. These factors include repeat-associated non-ATG translation (RAN),[51–53] bidirectional transcription,[36,48,54,55] aberrant DNA methylation,[56,57] microRNA (miRNA) dysregulation,[58–61] abnormal regulation of protein translation,[62] and other alterations.[63] Figure 96.1 provides a schematic of the different currently hypothesized pathomechanisms.[10]

Human Studies Supporting RNA-Mediated Pathomechanism for DM1 and DM2

Studies of skeletal muscle, heart, and brain from patients with DM1 and DM2 have identified abnormal ribonuclear inclusions containing mutant mRNA with abnormally expanded CUG repeats that have sequestered nuclear regulatory proteins, specifically, MBNL1, MBNL2, and CELF (CUG BP) proteins.[10,50] Initial studies in skeletal muscle of DM1 patients demonstrated toxic effects of the mutant mRNA on the regulation of alternative splicing of pre-mRNAs of the insulin receptor[64] and skeletal chloride channel.[65,66] The defect in splicing led to an imbalance in the synthesis of the correct isoforms of the insulin receptor and chloride channel with an excess of the fetal and a deficiency of adult isoforms.[10,50] This imbalance resulted primarily from sequestration of MBNL1, causing a preponderance of the less responsive form of the insulin receptor (IR-A lacking exon 11) leading to insulin resistance and in the overproduction of a non-adult form of the skeletal muscle chloride channel (edited mRNA that contains a portion of intron 7 called 7a) leading to a marked reduction in chloride conductance and myotonia.[65–68]

Alterations in splicing that may contribute to the muscle weakness in DM1 patients include: a) expression of an inactive form of BIN1 (active form required for biogenesis of muscle T tubules);[69] b) abnormal skipping of exon 29 of voltage gated calcium channel 1.1;[70] and c) missplicing of dystrobrevin exons 11A and 12.[71] All these MBNL1-dependent splicing alterations have occurred in muscle biopsies from DM1 patients and not in controls. However, conclusive correlation of these splicing defects with muscle weakness in DM1 needs further confirmation. Other recent investigations of muscle biopsies point out that MBNL1 undergoes ectopic relocation in the muscle nuclei of DM patients, similar to that observed in sarcopenia of aging, and that this maldistribution of nuclear regulatory protein caused by toxic mutant RNA may have a role in muscle weakness.[72]

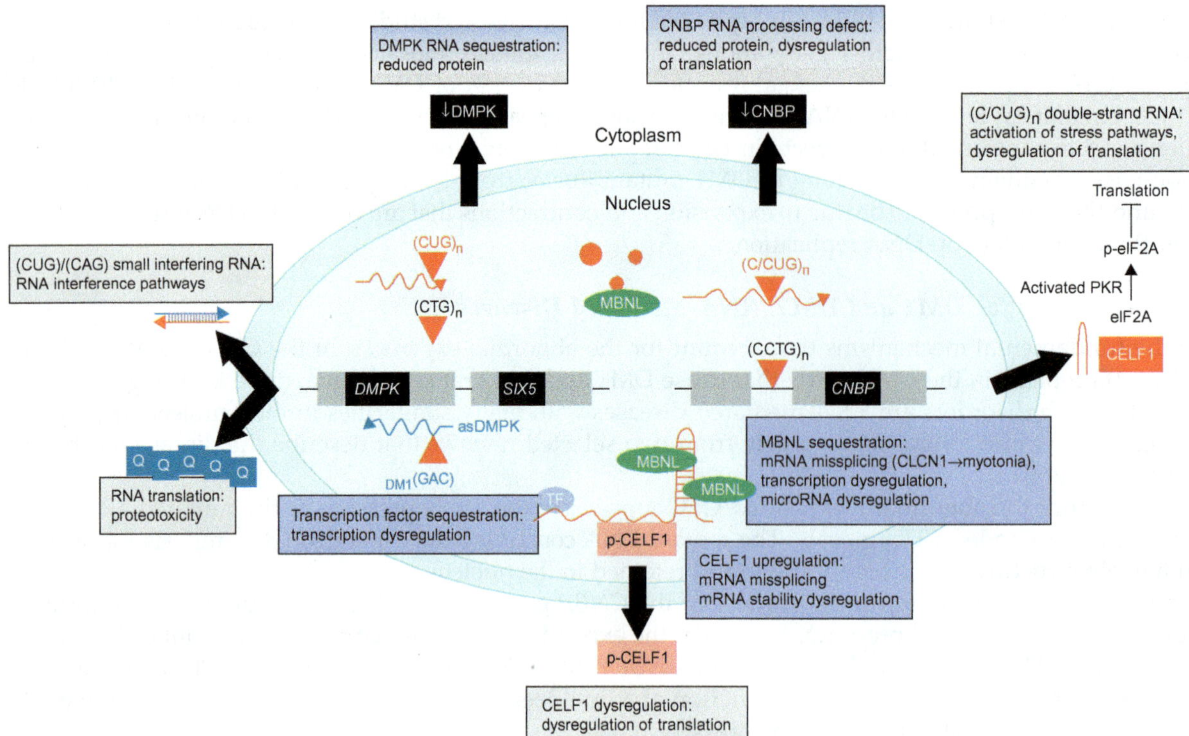

FIGURE 96.1 **Postulated pathological mechanisms underlying myotonic dystrophy types 1 and 2.** Most mechanisms are consistent with the prevailing toxic RNA gain-of-function model and affect multiple cellular aspects in the nucleus (blue boxes), the cytoplasm (gray boxes), or both (blue and gray boxes), including reduced protein function of the mutated genes, missplicing, dysregulation of transcription, protein translation and turnover, and activation of cellular stress pathways. Both mutant $(CTG)_n$ and $(CCTG)_n$ expansions in non-coding regions of *DMPK* and *CNBP* give rise to C/CUG-containing transcripts (red) that form stable secondary structures detectable as RNA foci (red circles) in the nucleus and the cytoplasm, and result in reduced amounts of *DMPK* and *CNBP* protein because of DMPK mRNA sequestration or a CNBP pre-mRNA processing defect. Members of the MBNL (green) protein family, such as MBNL1, are sequestered in ribonuclear foci leading to loss of function and dysregulation of MBNL splice and transcription targets and microRNA metabolism. Hyperphosphorylation of nuclear and cytoplasmic CELF1 (p-CELF1; salmon) by several protein kinases, including RNA-dependent protein kinase (PKR), leads to stabilization of different isoforms, which in turn affects alternative splicing, translation, and protein turnover. Inactivation of the translation initiation factor eIF2A leads to general attenuation of translation. Sequestration of transcription factors and other nuclear factors also contributes to dysregulation of gene expression. Generation of small interfering RNA from sense (s) and antisense (as) DMPK transcripts might activate RNA interference pathways that lead to wider dysregulation of mRNA and protein amounts. sDMPK and asDMPK transcripts are subject to repeat-associated non-ATG translation in the polyglutamine (poly$[Q]_n$) reading frame and possibly others, giving rise to toxic homopolymeric polypeptides that accumulate in the cytoplasm. Although some mechanisms have only been reported in myotonic dystrophy type 1, theoretically they could also operate in type 2 disease. TF, transcription factor. Reprinted from[10] with permission from Elsevier.

Animal Models

Researchers have developed a number of cell and animal models for DM1 and DM2 since the discoveries of the *DMPK* and *ZNF9* expansion mutations.[10,73,74] While there are no animal models that recapitulate the full spectrum of disease manifestations of DM1 or DM2, the available model systems play a major role in our understanding of these disorders.[10,73,74] Cell models have demonstrated that CTG and CCTG repeats accumulate in intranuclear RNA foci together with an assortment of cellular factors including MBNL1, MBNL2, hnRNP H, and Sp1. Mouse models have included transgenic mice expressing either ubiquitously expressed or tissue-specific constitutive as well as inducible CTG and CCTG repeat.[10,73,74] Knockout models of Mbnl1 and Mbnl2 recapitulate many features of the DM-associated spliceopathy and other characteristic disease manifestations, including myotonia, subcapsular ocular cataracts, and cardiac arrhythmias observed in patients. Another key pathological feature in patients with DM1, severe muscle wasting, has been observed in inducible CTG960 and CELF1 overexpression mice.[75] All the models noted above have provided insights into to our understanding of relevant pathogenic mechanisms including providing support for RNA toxicity as a major contributor. For more information describing animal and cell model systems, current research, recent treatment trials in animal models, and likely future research with different models, please refer to recent reviews.[10,76,77]

Current Human Research

Much of the current research involving patients with DM1 and DM2 focuses upon the identification of reliable, feasible measures of disease progression that are meaningful to patients and are appropriate for use in future therapeutic trials (clinicaltrials.gov). These measures include assessments of: a) strength using manual muscle testing and quantitative myometry; b) muscle mass (whole body and regional); c) myotonia (quantitative isometric grip relaxation, voluntary hand opening time, needle electromyography); d) timed functional tests (6 minute walk distance, time to travel 30 feet); e) forced vital capacity; f) whole body dual energy X-ray absorptiometry (DEXA); g) patient reported quality of life measures; and h) tissue biomarkers. One very recent study of patients with DM1 and DM2 included a cohort of 50 DM1 patients in whom investigators found 42 defects in alternative splicing of specific pre-mRNAs in small needle biopsy specimens of their tibialis anterior muscles. Twenty of these splicing defects showed graded changes that correlated with the degree of weakness of foot dorsiflexion.[78] Monitoring of defects in alternative splicing offers promise as a surrogate measure to assess early therapeutic response in future treatment trials in DM1. For example, as the promising treatment with antisense oligonucleotides transitions from the mouse model[77] to patients with DM1 using small needle biopsy analysis of alterations in splicing, alterations in the skeletal muscle chloride channel along with assessments of myotonia, may prove especially useful in detecting an early beneficial therapeutic response.

DIFFERENTIAL DIAGNOSIS AND TESTING

The major problem in making the diagnosis of DM1 and DM2 is the tendency of clinicians to ignore these entities in their practice. DM1 can present as a floppy infant or as child with delayed motor and cognitive development (see Natural History, p. 1152). There are long lists of much more common causes for these presentations than DM1. If a clinician has a low diagnostic suspicion, the family history and examination of the parents may not have probed sufficiently to reveal characteristic features of DM1 (Tables 96.1 and 96.2). The failure to consider DM1 is the principal contributor to the diagnostic odyssey and delays DNA testing, family counseling, and appropriate management and treatment.[9]

DM1 and DM2 present primarily in adult life. The core clinical features in Table 96.1 and the multisystem manifestations in Table 96.2 can occur in isolation or combination and usually develop gradually (see Natural History). Muscle stiffness, muscle pain, intermittent trouble with clear speech and swallowing, dropping items, gait instability, and trouble seeing are frequent initial symptoms. Early-onset cataracts (typically iridescent posterior capsular cataracts) and cardiac arrhythmias are examples of multisystem manifestations that can occur early in DM1 and DM2. In contrast, patients with other hereditary myotonic disorders, such as chloride or sodium channel myotonias, do not have cardiac arrhythmias or posterior capsular cataracts. DM1 patients occasionally present with prolonged apnea or delayed-onset apnea following general anesthesia. These complications can occur in individuals with only mild clinical signs of DM1. The risk of complications after general anesthesia in DM2 patients appears to be low, but until more long-term data are available, it is prudent to monitor both DM2 and DM1 patients closely during and 24 hours following general anesthesia.[79–81]

Figure 96.2 shows a diagnostic flow chart for identifying DM1 and DM2 in an adult with muscle weakness, stiffness, or pain but without clear diagnostic findings on clinical evaluation. The current gold standard to establish the diagnosis is DNA testing to identify the abnormally expanded CTG repeat in the *DMPK* gene for DM1 and the abnormally expanded CCTG repeat in the *CNBP* gene for DM2 (Table 96.1; see Molecular Genetics). In a small percentage of DM1 patients evaluated with current standard DNA testing, affected individuals have a false negative DNA test due to an interruption in the tract of CTG repeats by other nucleotide repeats (see Genotype–Phenotype Correlations). These DM1 patients have an abnormally expanded total CTG repeat length that requires the use of bidirectional TP-PCR to detect the "variant" interruptions of the CTG repeat expansion. Figure 96.2 points out that if a patient suspected of having DM1 or DM2 has negative DNA testing, performance of a muscle biopsy may help establish the diagnosis of another myotonic disorder. However, if the muscle biopsy findings are typical for DM1 (increased central nuclei, atrophy of type 1 fibers, ringed fibers, subsarcolemmal masses) or DM2 (very small fibers, type 2 nuclear clumps, type 2 fiber atrophy), further genetic analysis, including bidirectional TP-PCR may provide evidence of a mutation variant.

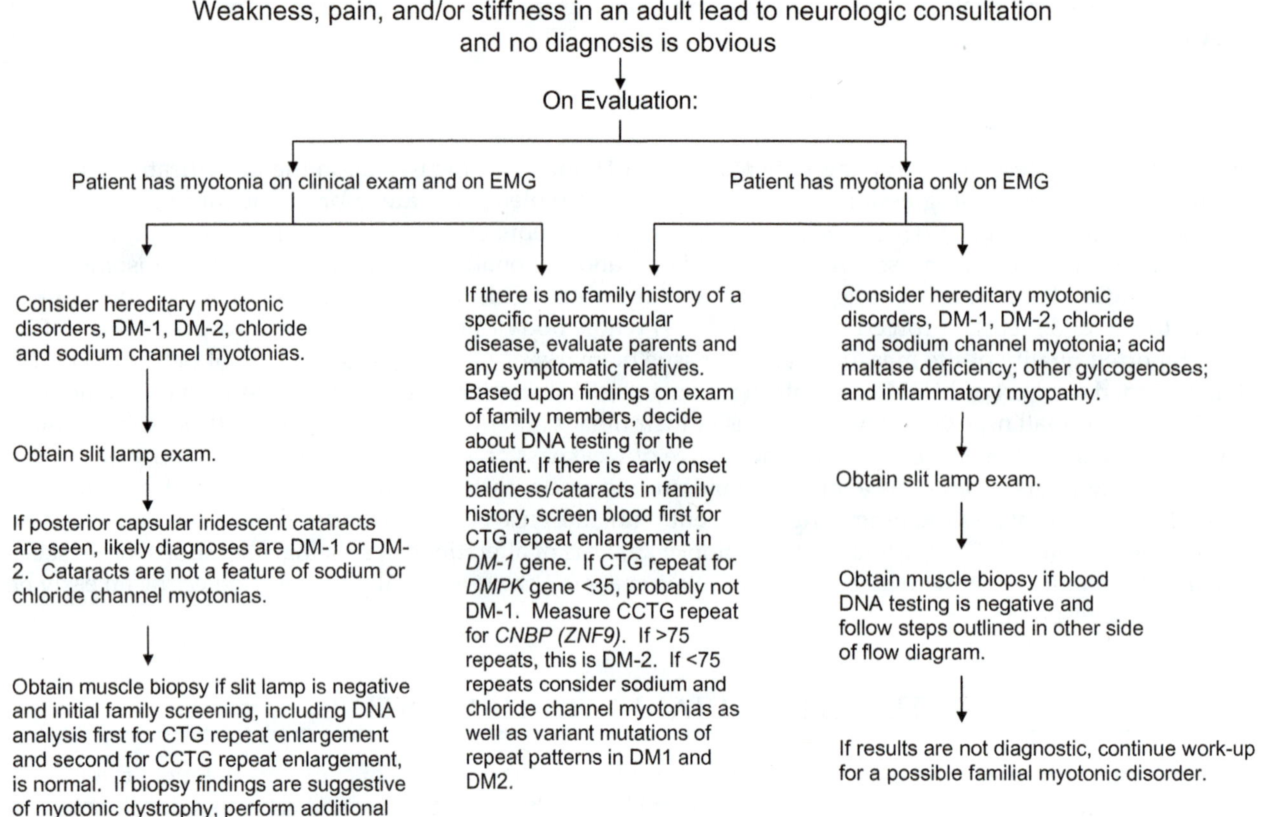

FIGURE 96.2 Diagnostic approach to myotonic dystrophy in adults.

MANAGEMENT

Table 96.2 summarizes the multisystem manifestations and Table 96.3 outlines the general guidelines for management of patients with DM1 and DM2. Detailed discussion of each disease manifestation is beyond the scope of this review. More specific information about guidelines for care is available in textbooks focused primarily on myotonic dystrophy (see Harper[1,82] and the recent evidence-based guideline being developed by the American Academy of Neurology and the American Association of Neuromuscular & Electrodiagnostic Medicine).

Standards of Care

Symptomatic treatment remains the mainstay of therapy for DM1 and DM2. Table 96.3 outlines the general approach to evaluation and approaches to care of several multisystem manifestations. We have highlighted references below to accompany Table 96.3 to provide more details related to guidelines for care involving skeletal muscle,[83–85] myotonia,[86–90] heart,[91–95] brain,[96–99] respiratory system,[100–102] sleep disturbances and fatigue,[102–105] endocrine dysfunction,[106,107] gastrointestinal system,[108,109] muscle pain,[31,32,110] and pregnancy.[111–114]

Therapies Under Investigation

As of January 2013 clinicaltrials.gov (NCT01225614; NCT01530841; NCT01406873) lists three treatment trials that are actively recruiting DM1 patients. One is a multicenter study of DM1 patients with respiratory insufficiency randomized to receive early initiation of nighttime noninvasive bi-level pressure ventilation compared to usual annual clinical monitoring. The primary outcome is the number of pulmonary complications observed over 5 years. The second is a single-center randomized trial of DM1 patients with respiratory insufficiency comparing nocturnal home ventilation with bi-level pressure support with average volume assured pressure support (AVAPS) to nocturnal home ventilation without AVAPS. The primary outcome is to compare differences at day 7 on arterial pCO_2 under

TABLE 96.3 General Guidelines for Care in Adults with Myotonic Dystrophy (DM1 and DM2)

System	Evaluation	Approaches to care
Brain	Psychological, educational, and counseling evaluations as needed Structural imaging as required Routinely assess for sleep disturbances and respiratory insufficiency	Psychological, educational, and other counseling treatment and services CNS medications (for example, stimulants) as necessary under close supervision of care providers
Heart	Yearly electrocardiograms Cardiology consultation for symptomatic patients and long-term follow-up care	Prompt pacemaker placement as needed
Respiratory	Serial monitoring of sitting and supine respiratory function; including forced vital capacity Polysomnography and pulmonary medicine consultation as required	Yearly immunizations Noninvasive or invasive ventilation as required Serial evaluation by pulmonary medicine and sleep consultation as required
Anesthesia	Before elective surgery, have anesthesia consultation and pulmonary medicine evaluation ECG reviewed by cardiology consult Discuss known risks and any previous anesthesia related problems	Use of regional anesthesia over general when appropriate Use of non-depolarizing muscle relaxants Reduce use of opioids In general anesthesia, protection of the airway and minimizing aspiration, careful cardiac monitoring, and extensive postoperative monitoring (at least 24 hours)
Hypersomnia and fatigue	Polysomnograms Metabolic and endocrine screens Psychological, educational, and sleep consultant evaluations	Use of continuous positive airway pressure (CPAP) or bi-level positive airway pressure (BiPAP) Use of CNS stimulants
Endocrine	Symptomatic assessment of testosterone deficiency Yearly lipid profile, thyroid screening, diabetes screening Monitor sleep disturbances	Hormone replacement as required Dietary intervention Medications for lipid and glucose control as needed Treatment for sleep disturbances as required
Gastrointestinal	Occupational and physical therapy consultation (dysphagia) Metabolic and endocrine screens Dietician, gastrointestinal consultations Careful assessment of bloating and signs of pseudo-obstruction	Gastroesophageal reflux may be treated with avoidance of late-night meals, elevation of the head of the bed, and medications Constipation, diarrhea, abdominal pain, and bloating may be treated with modifying the diet to small, low-fat meals Surgery as appropriate for gall bladder disease Use of cholestyramine may help alleviate diarrhea Small intestine and colon dysmotility may respond to anti-myotonia medications
Muscle pain	Psychological, educational, and pain counseling Pain treatment consultants as required	Anti-myotonia agents; NSAIDs Avoidance of opioids Aquatic therapy Other pain medication as coordinated between pain treatment consult and primary care provider
Pregnancy	Obtain obstetrician and genetic consultation prior to pregnancy as appropriate Discuss possible complications Coordinate monitoring of pregnancy with other care providers, including a neonatal pediatric specialist Closely monitor respiratory function during the third trimester	During delivery, monitor mother's ECG Use regional anesthesia Notify consultants of mother's status and request urgent evaluations as necessary

For more information, please see reviews by.[1,82]

ventilation. The third is a single-center double-blind randomized 6 month trial of mexiletine 150 mg three times daily compared to placebo. The primary outcome is the change in distance walked in 6 minutes and the degree of grip myotonia comparing baseline to 6 months of treatment.

Preclinical experimental therapies for myotonic dystrophy involve animal models and have focused on the use of antisense oligonucleotides (ASOs), small interfering RNAs, small molecules, ribozymes, and engineered small nuclear RNAs.[77]

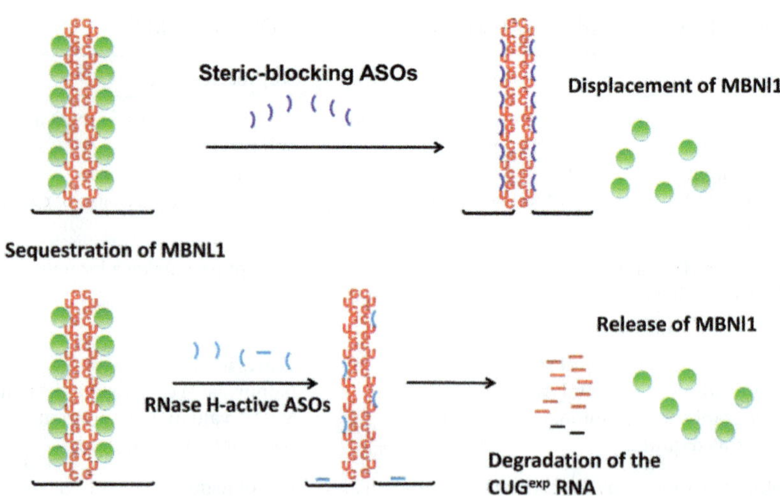

FIGURE 96.3 **Application of antisense oligonucleotides (ASOs) to eliminate RNA toxicity in myotonic dystrophy type 1 (DM1).** Steric blocking ASOs bind to CUG^{exp} RNA with high affinity and disrupt the RNA–MBNL1 (muscleblind-like 1) interactions, thereby displacing MBNL1 from the toxic RNA and reversing some features of DM1. In contrast, RNase H-active ASOs bind to the CUG^{exp} RNA and directly induce the degradation of the toxic RNA, resulting in release of MBNL1, restoration of its normal function, and reversal of disease features. Reprinted from[77] with permission from Mary Ann Liebert, Inc.

Especially promising are ASO therapies (Figure 96.3). ASOs are short synthetic analogs of DNA or RNA that bind to target RNA and act by blocking RNA function or by cleaving RNA, for example through the enzyme, RNase H.[115] In 2009, two studies showed that ASO treatments disrupted the toxic effects of mutant RNA in a mouse model of DM1.[116,117] These treatments were shown to reverse many of the symptoms. However, there are concerns that preventing the toxic RNA effects on splicing of mRNAs by MBNL may cause other, non-MBNL1-mediated toxicities associated with DM1.[118] New approaches using ASO therapies focus on degrading the mutant RNA. Two studies have shown promising results to target and cleave toxic RNA via the enzyme RNase H in mouse models of DM1.[11,119] One of these studies used a combination ASO therapy that involved both blockage and degradation (through RNase H) of toxic RNA.[119] The ASOs in this study were delivered by intramuscular injections to DM1 mice. The ASOs had high efficacy and potency, but resulted in local muscle damage.[119] The other study evaluated subcutaneous ASO injections (twice weekly for 4 weeks) and showed encouraging results.[11] The ASOs had 2′-*O*-methoxyethyl (MOE) modifications to maximize biostability and a central gap of 10 unmodified nucleotides to support RNase H activity.[11] After a total of eight injections, DM1 mice showed release of MBNL proteins from nuclear aggregates, correction of measured RNA splicing defects, reduction in myotonia, and reversal of alterations even 1 year after cessation of treatment.[11] These results are very promising, but challenges remain to assure that ASO therapies go to the tissues primarily affected in DM1 (skeletal muscle, brain, and heart) and that there are no serious off-target effects. There is also an urgent need for clinical researchers to develop and validate the most appropriate measures to use in treatment trials to assess the progression of each of the different disease manifestations, particularly those manifestations that are most important to patients.

Several questions will need to be asked by patients and care providers with the development of promising new treatments. How do we manage expectations about the distribution of these promising treatments to care providers and their patients? What patients are "first in line" to participate in early phase clinical trials? Will treatments be safe and available for children? Do we need to have further discussions of genetic testing of at-risk family members and newborn screening? The future for treatment of DM1 and DM2 holds great promise.

References

1. Harper PS. *Myotonic Dystrophy*. London: W.B. Saunders Company; 2001.
2. Fu YH, Pizzuti A, Fenwick Jr. RG, et al. An unstable triplet repeat in a gene related to myotonic muscular dystrophy. *Science*. 1992;255(5049):1256–1258.
3. Mahadevan M, Tsilfidis C, Sabourin L, et al. Myotonic dystrophy mutation: an unstable CTG repeat in the 3′ untranslated region of the gene. *Science*. 1992;255(5049):1253–1255.

4. Brook JD, McCurrach ME, Harley HG, et al. Molecular basis of myotonic dystrophy: expansion of a trinucleotide (CTG) repeat at the 3′ end of a transcript encoding a protein kinase family member. *Cell*. 1992;68(4):799–808.
5. Ranum LP, Rasmussen PF, Benzow KA, et al. Genetic mapping of a second myotonic dystrophy locus. *Nat Genet*. 1998;19(2):196–198.
6. Liquori CL, Ricker K, Moseley ML, et al. Myotonic dystrophy type 2 caused by a CCTG expansion in intron 1 of ZNF9. *Science*. 2001;293(5531):864–867.
7. Thornton CA, Griggs RC, Moxley 3rd RT. Myotonic dystrophy with no trinucleotide repeat expansion. *Ann Neurol*. 1994;35(3):269–272.
8. Ricker K, Koch MC, Lehmann-Horn F, et al. Proximal myotonic myopathy: a new dominant disorder with myotonia, muscle weakness, and cataracts. *Neurology*. 1994;44(8):1448–1452.
9. Hilbert JE, Ashizawa T, Day JW, et al. Diagnostic odyssey of patients with myotonic dystrophy. *J Neurol*. 2013;260(10):2497–2504.
10. Udd B, Krahe R. The myotonic dystrophies: molecular, clinical, and therapeutic challenges. *Lancet Neurol*. 2012;11(10):891–905.
11. Wheeler TM, Leger AJ, Pandey SK, et al. Targeting nuclear RNA for *in vivo* correction of myotonic dystrophy. *Nature*. 2012;488(7409):111–115.
12. Harper PS, Harley HG, Reardon W, et al. Anticipation in myotonic dystrophy: new light on an old problem. *Am J Hum Genet*. 1992;51(1):10–16.
13. Cardani R, Giagnacovo M, Botta A, et al. Co-segregation of DM2 with a recessive CLCN1 mutation in juvenile onset of myotonic dystrophy type 2. *J Neurol*. 2012;259(10):2090–2099.
14. Day JW, Ricker K, Jacobsen JF, et al. Myotonic dystrophy type 2: molecular, diagnostic and clinical spectrum. *Neurology*. 2003;60(4):657–664.
15. Rolander A, Floderus S. Dystrophia myotonica in the Norbotten District. *Sven Lakartidn*. 1961;58:648–652.
16. Medica I, Logar N, Mileta DL, et al. Genealogical study of myotonic dystrophy in Istria (Croatia). *Ann Genet*. 2004;47(2):139–146.
17. Mathieu J, Allard P, Potvin L, et al. A 10-year study of mortality in a cohort of patients with myotonic dystrophy. *Neurology*. 1999;52(8):1658–1662.
18. Mathieu JPC. Epidemiological surveillance of myotonic dystrophy type 1: A 25-year population-based study. *Neuromuscul Disord*. 2012;22(11):974–979.
19. Mathieu J, De Braekeleer M, Prevost C. Genealogical reconstruction of myotonic dystrophy in the Saguenay-Lac-Saint-Jean area (Quebec, Canada). *Neurology*. 1990;40(5):839–842.
20. Campbell C, Levin S, Siu VM, et al. Congenital myotonic dystrophy: Canadian population-based surveillance study. *J Pediatr*. 2013;163(1):120–125.
21. Suominen T, Bachinski LL, Auvinen S, et al. Population frequency of myotonic dystrophy: higher than expected frequency of myotonic dystrophy type 2 (DM2) mutation in Finland. *Eur J Hum Genet*. 2011;19(7):776–782.
22. Milone M, Batish SD, Daube JR. Myotonic dystrophy type 2 with focal asymmetric muscle weakness and no electrical myotonia. *Muscle Nerve*. 2009;39(3):383–385.
23. Young NP, Daube JR, Sorenson EJ, et al. Absent, unrecognized, and minimal myotonic discharges in myotonic dystrophy type 2. *Muscle Nerve*. 2010;41(6):758–762.
24. Echenne B, Rideau A, Roubertie A, et al. Myotonic dystrophy type I in childhood long-term evolution in patients surviving the neonatal period. *Eur J Paediatr Neurol*. 2008;12(3):210–223, 1090–3798.
25. Koch MC, Grimm T, Harley HG, et al. Genetic risks for children of women with myotonic dystrophy. *Am J Hum Genet*. 1991;48(6):1084–1091.
26. Harper PS. Congenital myotonic dystrophy in Britain. I. clinical aspects. *Arch Dis Child*. 1975;50(7):505–513.
27. Sinclair JL, Reed PW. Risk factors for perioperative adverse events in children with myotonic dystrophy. *Paediatr Anaesth*. 2009;19(8):740–747.
28. Meola G, Moxley III RT. Myotonic dystrophy type 2 and related myotonic disorders. *J Neurol*. 2004;251(10):1173–1182.
29. Meola G, Sansone V, Marinou K, et al. Proximal myotonic myopathy: a syndrome with a favourable prognosis? *J Neurol Sci*. 2002;193(2):89–96.
30. Udd B, Meola G, Krahe R, et al. Myotonic dystrophy type 2 (DM2) and related disorders report of the 180th ENMC workshop including guidelines on diagnostics and management 3–5 December 2010, Naarden, the Netherlands. *Neuromuscul Disord*. 2011;21(6):443–450.
31. Suokas KI, Haanpaa M, Kautiainen H, et al. Pain in patients with myotonic dystrophy type 2: a postal survey in Finland. *Muscle Nerve*. 2012;45(1):70–74.
32. George A, Schneider-Gold C, Zier S, et al. Musculoskeletal pain in patients with myotonic dystrophy type 2. *Arch Neurol*. 2004;61(12):1938–1942.
33. Groh WJ, Groh MR, Shen C, et al. Survival and CTG repeat expansion in adults with myotonic dystrophy type 1. *Muscle Nerve*. 2011;43(5):648–651.
34. Mladenovic J, Pekmezovic T, Todorovic S, et al. Survival and mortality of myotonic dystrophy type 1 (Steinert's disease) in the population of Belgrade. *Eur J Neurol*. 2006;13(5):451–454.
35. de Die-Smulders CE, Howeler CJ, Thijs C, et al. Age and causes of death in adult-onset myotonic dystrophy. *Brain*. 1998;121:1557–1563.
36. Santoro M, Masciullo M, Pietrobono R, et al. Molecular, clinical, and muscle studies in myotonic dystrophy type 1 (DM1) associated with novel variant CCG expansions. *J Neurol*. 2013;260(5):1245–1257.
37. Braida C, Stefanatos RK, Adam B, et al. Variant CCG and GGC repeats within the CTG expansion dramatically modify mutational dynamics and likely contribute toward unusual symptoms in some myotonic dystrophy type 1 patients. *Hum Mol Genet*. 2010;19(8):1399–1412.
38. Musova Z, Mazanec R, Krepelova A, et al. Highly unstable sequence interruptions of the CTG repeat in the myotonic dystrophy gene. *Am J Med Genet A*. 2009;149A(7):1365–1374.
39. Thornton CA, Johnson K, Moxley RT. Myotonic dystrophy patients have larger CTG expansions in skeletal muscle than in leukocytes. *Ann Neurol*. 1994;35:104–107.
40. Ashizawa T, Dubel JR, Harati Y. Somatic instability of CTG repeat in myotonic dystrophy. *Neurology*. 1993;43(12):2674–2678.
41. Martorell L, Monckton DG, Gamez J, et al. Progression of somatic CTG repeat length heterogeneity in the blood cells of myotonic dystrophy patients. *Hum Mol Genet*. 1998;7(2):307–312.
42. Anvret M, Ahlberg G, Grandell U, et al. Larger expansions of the CTG repeat in muscle compared to lymphocytes from patients with myotonic dystrophy. *Hum Mol Genet*. 1993;2(9):1397–1400.

43. Zatz M, Passos-Bueno MR, Cerqueira A, et al. Analysis of the CTG repeat in skeletal muscle of young and adult myotonic dystrophy patients: when does the expansion occur? *Hum Mol Genet*. 1995;4(3):401–406.
44. Bachinski LL, Udd B, Meola G, et al. Confirmation of the type 2 myotonic dystrophy $(CCTG)_n$ expansion mutation in patients with proximal myotonic myopathy/proximal myotonic dystrophy of different European origins: a single shared haplotype indicates an ancestral founder effect. *Am J Hum Genet*. 2003;73(4):835–848.
45. Goula AV, Stys A, Chan JP, et al. Transcription elongation and tissue-specific somatic CAG instability. *PLoS Genet*. 2012;8(11):e1003051.
46. Liu G, Chen X, Leffak M. Oligodeoxynucleotide binding to (CTG). (CAG) microsatellite repeats inhibits replication fork stalling, hairpin formation, and genome instability. *Mol Cell Biol*. 2013;33(3):571–581.
47. Kurosaki T, Ueda S, Ishida T, et al. The unstable CCTG repeat responsible for myotonic dystrophy type 2 originates from an AluSx element insertion into an early primate genome. *PLoS One*. 2012;7(6):e38379.
48. Nakamori M, Pearson CE, Thornton CA. Bidirectional transcription stimulates expansion and contraction of expanded (CTG)*(CAG) repeats. *Hum Mol Genet*. 2011;20(3):580–588.
49. Higham CF, Morales F, Cobbold CA, et al. High levels of somatic DNA diversity at the myotonic dystrophy type 1 locus are driven by ultra-frequent expansion and contraction mutations. *Hum Mol Genet*. 2012;21(11):2450–2463.
50. Wheeler TM, Thornton CA. Myotonic dystrophy: RNA-mediated muscle disease. *Curr Opin Neurol*. 2007;20(5):572–576.
51. Echeverria GV, Cooper TA. RNA-binding proteins in microsatellite expansion disorders: mediators of RNA toxicity. *Brain Res*. 2012;1462:100–111.
52. Pearson CE. Repeat associated non-ATG translation initiation: one DNA, two transcripts, seven reading frames, potentially nine toxic entities! *PLoS Genet*. 2011;7(3):e1002018.
53. Zu T, Gibbens B, Doty NS, et al. Non-ATG-initiated translation directed by microsatellite expansions. *Proc Natl Acad Sci U S A*. 2011;108(1):260–265.
54. Moseley ML, Zu T, Ikeda Y, et al. Bidirectional expression of CUG and CAG expansion transcripts and intranuclear polyglutamine inclusions in spinocerebellar ataxia type 8. *Nat Genet*. 2006;38(7):758–769.
55. Huguet A, Medja F, Nicole A, et al. Molecular, physiological, and motor performance defects in DMSXL mice carrying >1,000 CTG repeats from the human DM1 locus. *PLoS Genet*. 2012;8(11):e1003043.
56. Lopez CA, Nakamori M, Tome S, et al. Expanded CTG repeat demarcates a boundary for abnormal CpG methylation in myotonic dystrophy patient tissues. *Hum Mol Genet*. 2011;20(1):1–15.
57. Lopez Castel A, Nakamori M, Thornton CA, et al. Identification of restriction endonucleases sensitive to 5-cytosine methylation at non-CpG sites, including expanded (CAG)n/(CTG)n repeats. *Epigenetics*. 2011;6(4):416–420.
58. Furling D. Misregulation of alternative splicing and microRNA processing in DM1 pathogenesis. *Rinsho Shinkeigaku*. 2012;52(11):1018–1022.
59. Greco S, Perfetti A, Fasanaro P, et al. Deregulated microRNAs in myotonic dystrophy type 2. *PLoS One*. 2012;7(6):e39732.
60. Rau F, Freyermuth F, Fugier C, et al. Misregulation of miR-1 processing is associated with heart defects in myotonic dystrophy. *Nat Struct Mol Biol*. 2011;18(7):840–845.
61. Perbellini R, Greco S, Sarra-Ferraris G, et al. Dysregulation and cellular mislocalization of specific miRNAs in myotonic dystrophy type 1. *Neuromuscul Disord*. 2011;21(2):81–88.
62. Meola G, Jones K, Wei C, et al. Dysfunction of protein homeostasis in myotonic dystrophies. *Histol Histopathol*. 2013;28(9):1089–1098.
63. Ravel-Chapuis A, Belanger G, Yadava RS, et al. The RNA-binding protein Staufen1 is increased in DM1 skeletal muscle and promotes alternative pre-mRNA splicing. *J Cell Biol*. 2012;196(6):699–712.
64. Savkur RS, Philips AV, Cooper TA. Aberrant regulation of insulin receptor alternative splicing is associated with insulin resistance in myotonic dystrophy. *Nat Genet*. 2001;29(1):40–47.
65. Mankodi A, Takahashi MP, Jiang H, et al. Expanded CUG repeats trigger aberrant splicing of ClC-1 chloride channel pre-mRNA and hyperexcitability of skeletal muscle in myotonic dystrophy. *Mol Cell*. 2002;10(1):35–44.
66. Charlet-B N, Savkur RS, Singh G, et al. Loss of the muscle-specific chloride channel in type 1 myotonic dystrophy due to misregulated alternative splicing. *Mol Cell*. 2002;10(1):45–53.
67. Lueck JD, Mankodi A, Swanson MS, et al. Muscle chloride channel dysfunction in two mouse models of myotonic dystrophy. *J Gen Physiol*. 2007;129(1):79–94.
68. Wheeler TM, Lueck JD, Swanson MS, et al. Correction of ClC-1 splicing eliminates chloride channelopathy and myotonia in mouse models of myotonic dystrophy. *J Clin Invest*. 2007;117(12):3952–3957.
69. Fugier C, Klein AF, Hammer C, et al. Misregulated alternative splicing of BIN1 is associated with T tubule alterations and muscle weakness in myotonic dystrophy. *Nat Med*. 2011;17(6):720–725.
70. Tang ZZ, Yarotskyy V, Wei L, et al. Muscle weakness in myotonic dystrophy associated with misregulated splicing and altered gating of ca(V)1.1 calcium channel. *Hum Mol Genet*. 2012;21(6):1312–1324.
71. Nakamori M, Kimura T, Kubota T, et al. Aberrantly spliced alpha-dystrobrevin alters alpha-syntrophin binding in myotonic dystrophy type 1. *Neurology*. 2008;70(9):677–685.
72. Malatesta M, Giagnacovo M, Costanzo M, et al. Muscleblind-like1 undergoes ectopic relocation in the nuclei of skeletal muscles in myotonic dystrophy and sarcopenia. *Eur J Histochem*. 2013;57(2):e15.
73. Sicot G, Gomes-Pereira M. RNA toxicity in human disease and animal models: from the uncovering of a new mechanism to the development of promising therapies. *Biochim Biophys Acta*. 2013;1832(9):1390–1409.
74. Gomes-Pereira M, Cooper TA, Gourdon G. Myotonic dystrophy mouse models: towards rational therapy development. *Trends Mol Med*. 2011;17(9):506–517.
75. Ward AJ, Rimer M, Killian JM, et al. CUGBP1 overexpression in mouse skeletal muscle reproduces features of myotonic dystrophy type 1. *Hum Mol Genet*. 2010;19(18):3614–3622.
76. Sicot G, Gourdon G, Gomes-Pereira M. Myotonic dystrophy, when simple repeats reveal complex pathogenic entities: new findings and future challenges. *Hum Mol Genet*. 2011;20(R2):R116–R123.
77. Gao Z, Cooper TA. Antisense oligonucleotides: rising stars in eliminating RNA toxicity in myotonic dystrophy. *Hum Gene Ther*. 2013;24(5):499–507.

78. Nakamori M, Sobczak K, Puwanant A, et al. Splicing biomarkers of disease severity in myotonic dystrophy. *Ann Neurol*. 2013;74(6):862–872.
79. Veyckemans F, Scholtes JL. Myotonic dystrophies type 1 and 2: anesthetic care. *Paediatr Anaesth*. 2013;23(9):794–803.
80. Campbell N, Brandom B, Day JW, Moxley III RT. *Practical Suggestions for the Anesthetic Management of a Myotonic Dystrophy Patient*. 73–80 http://www.myotonic.org/sites/default/files/pages/files/Anesthesia%20Guidelines.pdf. Accessed 06.04.13.
81. Kirzinger L, Schmidt A, Kornblum C, et al. Side effects of anesthesia in DM2 as compared to DM1: a comparative retrospective study. *Eur J Neurol*. 2010;17(6):842–845.
82. Harper PS, Engelen BG, Eymard B, et al., eds. *Myotonic Dystrophy: Present Management, Future Therapy*. 1st ed. New York, NY: Oxford University Press; 2004.
83. Johnson ER, Abresch RT, Carter GT, et al. Profiles of neuromuscular diseases myotonic dystrophy. *Am J Phys Med Rehabil*. 1995;74(5):S104–S116.
84. Wiles CM, Busse ME, Sampson CM, et al. Falls and stumbles in myotonic dystrophy. *J Neurol Neurosurg Psychiatry*. 2006;77(3):393–396.
85. Voet NB, van der Kooi EL, Riphagen II, et al. Strength training and aerobic exercise training for muscle disease. *Cochrane Database Syst Rev*. 2013;7: CD003907.
86. Logigian EL, Martens WB, Moxley RT, et al. Mexiletine is an effective antimyotonia treatment in myotonic dystrophy type 1. *Neurology*. 2010;74(18):1441–1448.
87. Logigian EL, Blood CL, Dilek N, et al. Quantitative analysis of the "warm-up" phenomenon in myotonic dystrophy type 1. *Muscle Nerve*. 2005;32(1):35–42.
88. Trip J, Drost G, van Engelen BG, et al. Drug treatment for myotonia. *Cochrane Database Syst Rev*. 2006;(1), 1469–493. CD004762.
89. de Swart BJ, van Engelen BG, Maassen BA. Warming up improves speech production in patients with adult onset myotonic dystrophy. *J Commun Disord*. 2007;40(3):185–195.
90. Heatwole CR, Statland JM, Logigian EL. The diagnosis and treatment of myotonic disorders. *Muscle Nerve*. 2013;47(5):632–648.
91. Groh WJ, Bhakta D. Arrhythmia management in myotonic dystrophy type 1. *JAMA*. 2012;308(4):337–338, 1538–3598; 0098–7484.
92. Bhakta D, Shen C, Kron J, et al. Pacemaker and implantable cardioverter-defibrillator use in a US myotonic dystrophy type 1 population. *J Cardiovasc Electrophysiol*. 2011;22(12):1369–1375.
93. Duboc D, Wahbi K. What is the best way to detect infra-hisian conduction abnormalities and prevent sudden cardiac death in myotonic dystrophy? *Heart*. 2012;98(6):433–434.
94. Ha AH, Tarnopolsky MA, Bergstra TG, et al. Predictors of atrio-ventricular conduction disease, long-term outcomes in patients with myotonic dystrophy types I and II. *Pacing Clin Electrophysiol*. 2012;35(10):1262–1269.
95. Wahbi K, Meune C, Porcher R, et al. Electrophysiological study with prophylactic pacing and survival in adults with myotonic dystrophy and conduction system disease. *JAMA*. 2012;307(12):1292–1301.
96. Meola G, Sansone V. Cerebral involvement in myotonic dystrophies. *Muscle Nerve*. 2007;36(3):294–306.
97. Ekstrom AB, Hakenas-Plate L, Tulinius M, et al. Cognition and adaptive skills in myotonic dystrophy type 1: a study of 55 individuals with congenital and childhood forms. *Dev Med Child Neurol*. 2009;51(12):982–990.
98. Cup EH, Kinebanian A, Satink T, et al. Living with myotonic dystrophy; what can be learned from couples? A qualitative study. *BMC Neurol*. 2011;1, 86-2377-11-86.
99. Gagnon C, Chouinard MC, Laberge L, et al. Health supervision and anticipatory guidance in adult myotonic dystrophy type 1. *Neuromuscul Disord*. 2010;20(12):847–851.
100. Monteiro R, Bento J, Goncalves MR, et al. Genetics correlates with lung function and nocturnal ventilation in myotonic dystrophy. *Sleep Breath*. 2013;17(3):1087–1092.
101. Campbell C, Sherlock R, Jacob P, et al. Congenital myotonic dystrophy: assisted ventilation duration and outcome. *Pediatrics*. 2004;113(4):811–816.
102. Bhat S, Sander HW, Grewal RP, et al. Sleep disordered breathing and other sleep dysfunction in myotonic dystrophy type 2. *Sleep Med*. 2012;13(9):1207–1208.
103. Annane D, Moore DH, Barnes PR, et al. Psychostimulants for hypersomnia (excessive daytime sleepiness) in myotonic dystrophy. *Cochrane Database Syst Rev*. 2006;(3), CD003218.
104. Laberge L, Gagnon C, Dauvilliers Y. Daytime sleepiness and myotonic dystrophy. *Curr Neurol Neurosci Rep*. 2013;13(4):340.
105. Shepard P, Lam EM, St Louis EK, et al. Sleep disturbances in myotonic dystrophy type 2. *Eur Neurol*. 2012;68(6):377–380.
106. Cruz Guzman Odel R, Chavez Garcia AL, Rodriguez-Cruz M. Muscular dystrophies at different ages: metabolic and endocrine alterations. *Int J Endocrinol*. 2012;2012:485376.
107. Orngreen MC, Arlien-Soborg P, Duno M, et al. Endocrine function in 97 patients with myotonic dystrophy type 1. *J Neurol*. 2012;259(5):912–920.
108. Bellini M, Biagi S, Stasi C, et al. Gastrointestinal manifestations in myotonic muscular dystrophy. *World J Gastroenterol*. 2006;12(12): 1821–1828.
109. Tieleman AA, van Vliet J, Jansen JB, et al. Gastrointestinal involvement is frequent in myotonic dystrophy type 2. *Neuromuscul Disord*. 2008;18(8):646–649.
110. Jensen MP, Hoffman AJ, Stoelb BL, et al. Chronic pain in persons with myotonic dystrophy and facioscapulohumeral dystrophy. *Arch Phys Med Rehabil*. 2008;89(2):320–328.
111. Awater C, Zerres K, Rudnik-Schoneborn S. Pregnancy course and outcome in women with hereditary neuromuscular disorders: comparison of obstetric risks in 178 patients. *Eur J Obstet Gynecol Reprod Biol*. 2012;162(2):153–159.
112. Martorell L, Cobo AM, Baiget M, et al. Prenatal diagnosis in myotonic dystrophy type 1. Thirteen years of experience: implications for reproductive counseling in DM1 families. *Prenat Diagn*. 2007;27(1):68–72.
113. Rudnik-Schoneborn S, Schneider-Gold C, Raabe U, et al. Outcome and effect of pregnancy in myotonic dystrophy type 2. *Neurology*. 2006;66(4):579–580.
114. Newman B, Meola G, O'Donovan DG, et al. Proximal myotonic myopathy (PROMM) presenting as myotonia during pregnancy. *Neuromuscul Disord*. 1999;9(3):144–149.
115. Bennett CF, Swayze EE. RNA targeting therapeutics: molecular mechanisms of antisense oligonucleotides as a therapeutic platform. *Annu Rev Pharmacol Toxicol*. 2010;50:259–293.

116. Mulders SA, van Den Broek WJ, Wheeler TM, et al. Triplet-repeat oligonucleotide-mediated reversal of RNA toxicity in myotonic dystrophy. *Proc Natl Acad Sci U S A*. 2009;106(33):13915–13920.
117. Wheeler TM, Sobczak K, Lueck JD, et al. Reversal of RNA dominance by displacement of protein sequestered on triplet repeat RNA. *Science*. 2009;325(5938):336–339.
118. Mahadevan MS. Myotonic dystrophy: is a narrow focus obscuring the rest of the field? *Curr Opin Neurol*. 2012;25(5):609–613.
119. Lee JE, Bennett CF, Cooper TA. RNase H-mediated degradation of toxic RNA in myotonic dystrophy type 1. *Proc Natl Acad Sci U S A*. 2012;109(11):4221–4226.

CHAPTER

97

Facioscapulohumeral Dystrophy

Rabi Tawil

University of Rochester Medical Center, Rochester, NY, USA

CLINICAL FEATURES

Historical Overview

The first description of FSHD was in 1884 by Landouzy and Dejerine and it came to be known as the Landouzy–Dejerine syndrome for a number of years.[1] The most comprehensive and complete description of the clinical manifestations of FSHD was published in 1982 by George Padberg.[2] In 1990, FSHD was linked to chromosome 4q35, where subtelomeric rearrangements were identified.[3] These rearrangements consist of deletions of an integral number of copies of a 3.3-kb macrosatellite repeat named D4Z4.[4] Whereas normal individuals have 11 or more 3.3-kb repeats, individuals with FSHD have 1–10 repeats. Whereas repeat contractions account for about 95% of patients with FSHD (FSHD1), another 5% of patients with a clinically identical phenotype (FSHD2) have no contractions of the 4q35 D4Z4 repeats.[5,6]

Mode of Inheritance and Prevalence

Estimates of the prevalence of FSHD vary from 1:15,000–1:20,000.[2,7] Inheritance in FSHD1 is autosomal dominant with as many as 30% *de novo* mutations. Inheritance pattern in FSHD2 is more complicated as it is transmitted as a digenic disease.[2,8,9]

Natural History

Disease onset is typically late in the second decade of life for most but with a wide spectrum of age at onset. Typically, men present earlier than women and, as a group, are more severely affected. A rare but severe form of the disease has symptom onset in infancy and early childhood. Penetrance is said to be about 95% by the age of 20 but recent data suggests that it is much lower.[2,8] Certainly, many patients with minimal clinical involvement remain asymptomatic. FSHD is defined by a characteristic distribution of weakness with a generally descending pattern of involvement, starting with usually asymptomatic facial weakness followed sequentially by scapular fixator, humeral, truncal, and lower extremity weakness. The rate of progression is generally slow and steady.[10,11] Many patients report a relapsing course with long periods of quiescence interrupted by periods of rapid deterioration involving a particular muscle group, and sometimes heralded by pain in the affected limb. Eventually, 20% of patients above the age of 50 become wheelchair-bound.[2,12] A subpopulation of patients with infantile-onset FSHD is severely disabled at an early age.

The most commonly presenting symptom is scapular winging secondary to weakness of the scapular fixators, specifically the serratus, rhomboid, supraspinatus, and infraspinatus muscles. Facial weakness is almost invariably present, but typically without perceived symptoms. Although the initial symptoms are often described as recent in onset, a history of long-standing difficulties climbing rope, doing push-ups, whistling, or drinking through a straw is often elicited. Parents often describe their children as having always slept with their eyes partially open, indicating long-standing unnoticed facial weakness. Striking bifacial weakness can be the presenting symptom in infantile

Rosenberg's Molecular and Genetic Basis of Neurological and Psychiatric Disease
http://dx.doi.org/10.1016/10.1016/B978-0-12-410529-4.00097-8

FSHD and may be misdiagnosed as Mobius syndrome. Some patients first present with foot drop. These individuals invariably have coexisting, asymptomatic facial and scapular fixator weakness. Leg involvement is initially distal, but proximal weakness is present in many patients and can be more severe than distal weakness. Indeed, the availability of specific molecular diagnostic testing for FSHD is leading to an expansion of the clinical spectrum of the disease, as some sporadic patients with limb-girdle patterns of weakness are found to have the characteristic 4q35 FSHD deletion.[13]

On inspection, FSHD patients have widened palpebral fissures and decreased facial expression. The lips are pouted and dimples are often evident at the corners of the mouth. Examination confirms bifacial, typically asymmetric, weakness with a transverse smile, inability to purse lips, and inability to bury eyelashes (Figure 97.1C and D). Extraocular, eyelid, and bulbar muscles are spared. Neck flexors remain strong and are relatively spared compared to the neck extensors. The anterior view of the shoulders shows a typical appearance with straight clavicles, forward sloping, and rounding. The pectoral muscles are atrophied and occasionally absent. Pectoralis atrophy results in the appearance of an axillary crease (Figure 97.1B). The scapulae are laterally displaced, and winging is evident either at rest or with attempted forward arm flexion or abduction (Figure 97.1A). The selective involvement of the lower trapezius muscles results in a characteristic upward jutting of the scapulae when the arms are abducted. The biceps and triceps muscles are typically involved with remarkable sparing of the deltoid. With exception of the wrist extensors, the forearm and hand muscles are spared early in the course of FSHD. This pattern of intact deltoids, forearm muscles, and wasted arm muscles results in a "Popeye" arm appearance. Abdominal muscle weakness is prominent, resulting in a protuberant abdomen and secondary lumbar lordosis. The lower (occasionally upper) abdominal wall muscles are selectively involved causing a striking upward movement of the umbilicus on neck flexion (Beevor sign), a sign uncommon in other myopathic conditions.[14–16] In the lower extremities, the peroneal muscles are involved with foot dorsiflexor weakness, usually sparing the calf muscles. Although proximal leg muscles are thought to become involved only late in the disease, studies with quantitative myometry have shown early involvement of knee extensors and flexors.[11] Contractures are rarely present even in profoundly weak muscles. One of the striking features of FSHD is the degree of side-to-side asymmetry in muscle weakness that seems to be independent of limb dominance.[17]

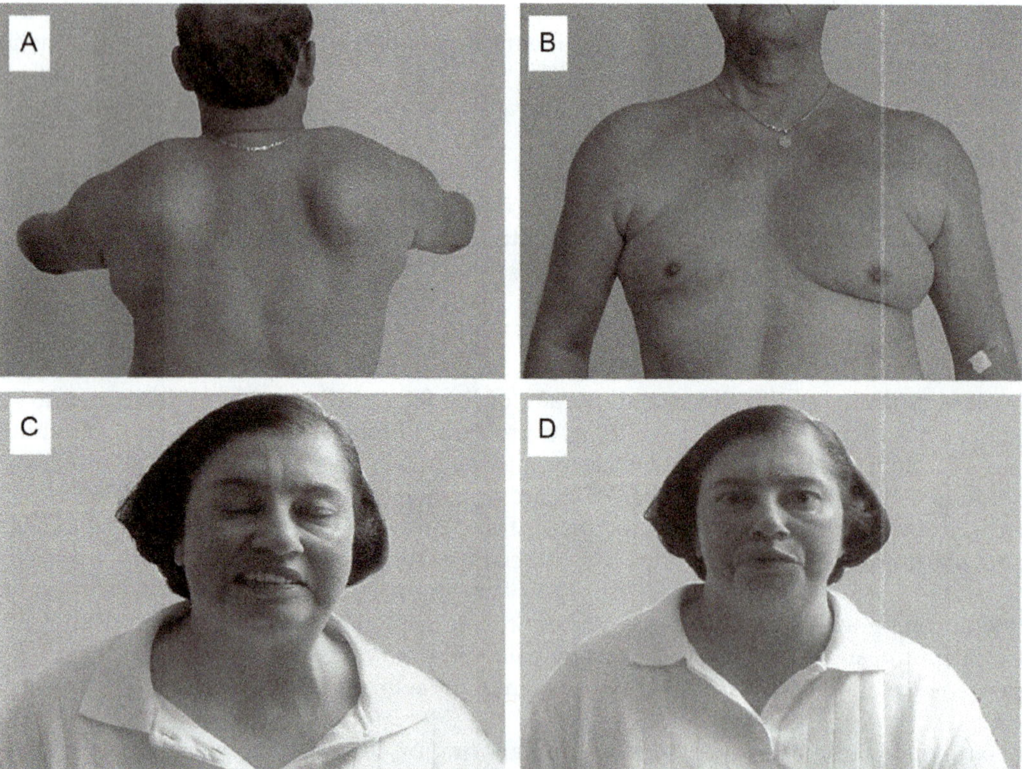

FIGURE 97.1 (A) Limitation of forward arm flexion with scapular winging due to weakness of scapular stabilizers; (B) typical appearance of shoulders with straight clavicles and bilateral axillary creases indicative of wasting of pectoral muscles; (C) weakness of forced eye closure and asymmetric lower facial muscle weakness with a transverse smile; (D) symmetric weakness of orbicularis oris muscle demonstrated by inability to pucker lips and whistle.

Extramuscular manifestations are common in FSHD, although are typically mild or asymptomatic. The most common are high-frequency hearing loss and retinal telangiectasias, occurring in 75% and 60% of affected individuals, respectively.[18,19] Only about 0.6% of patients develop a symptomatic exudative retinopathy (Coats disease) and invariably they are patients with infantile onset disease and very large deletions.[20] Coats disease can lead to visual loss and is potentially preventable if detected early. Like retinal vascular disease, hearing loss, especially symptomatic hearing, occurs most frequently in early-onset patients with large deletions. If not detected early in infants, hearing loss can lead to speech and language delay. Systematic studies of heart involvement in FSHD have shown no cardiomyopathies, but about 5% had cardiac involvement with conduction abnormalities and a predilection for supraventricular tachyarrhythmias.[21] Restrictive pattern of lung involvement occurs in a minority of patients with FSHD and only about 1% will require ventilator assistance.[22] The central nervous system is typically uninvolved in FSHD. However, rarely, patients with severe, early-onset disease can present with cognitive impairment and even seizures.[23,24]

MOLECULAR GENETICS

FSHD Type 1 (FSHD1)

As previously described, FSHD1 is the result of contraction of an integral number of D4Z4 repeats on the tip of chromosome 4q35 for >10 repeats to 1–10 repeats.[4] D4Z4 repeats contain sequences that suggest they are within heterochromatic domains of the genome, suggesting that under normal conditions, there is no transcription from within the repeats.[25] Each D4Z4 repeat contains a single open reading frame encoding a double homeobox gene designated *DUX4*.[26] The D4Z4 repeat contraction is associated with chromatin relaxation, a change making the repeats more amenable to transcription.[27] However, a contraction of the repeats alone does not result in FSHD unless it occurs on one of two (A or B) sequence variants distal to the last repeat.[28] Contraction of the repeats resulting in 1–10 residual repeats resulted in FSHD only when it occurred on the A variant. The A variant contains a polyadenylation signal critical in stabilizing *DUX4* mRNA (Figure 97.2B).[29] A repeat contraction in the setting of a permissive background results in the production of stable *DUX4* mRNA and DUX4 protein from the most distal copy of the D4Z4 repeats. *DUX4*, a retrogene that codes for a transcriptional regulator, is normally expressed only in the germline and its expression is suppressed in somatic tissue.[30]

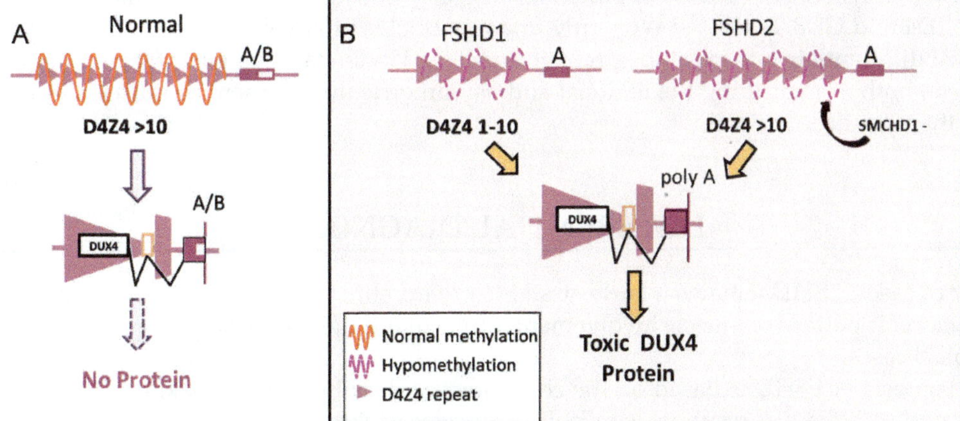

FIGURE 97.2 **Schematic of the chromosome D4Z4 region on the tip of 4q35 showing the location of the D4Z4 repeats and the two distal allelic variants of 4q35 distal to the repeats: 4qA or 4qB. Both occur in equal frequencies in the populations.** **(A)** Shows a normal individual with >10 D4Z4 repeats and highly methylated DNA (solid line) which is transcriptionally inactive resulting in no *DUX4* expression. **(B)** In FSHD1, contraction of the D4Z4 repeats to 1–10 repeats results in a more permissive chromatin structure allowing DUX4 expression from the last D4Z4 repeat but only in the presence of the distal A variant providing a polyadenylation signal does it result in a stable *DUX4* mRNA and protein expression. In FSHD2, both D4Z4 repeats are in the normal range but a permissive chromatin configuration results, in 80% of cases, from a mutation in the *SMCHD1* gene. Here again, stable *DUX4* mRNA and toxic DUX5 protein is produced in the presence of a distal variant.

FSHD Type 2 (FSHD2)

A subgroup of patients with typical FSHD phenotypes were noted to have no contraction in the number of D4Z4 repeats on 4q35.[31] However, despite the presence of a normal number of D4Z4 repeats, these patients had chromatin changes at 4q35 similar to FSHD1.[6] Whereas in FSHD1 D4Z4, chromatin relaxation, as measured by the degree of DNA methylation, was noted in the contracted allele, in FSHD2 chromatin relaxation occurred on both D4Z4 copies, as demonstrated by the presence of profound hypomethylation, and in the presence of a normal complement of repeats. Moreover, just as in FSHD1, at least one copy of 4q35 contained the permissive A sequence (Figure 97.2B).[6] These findings suggested that although FSHD1 and 2 have distinct molecular signatures, the end result, the aberrant derepression of *DUX4* expression, is the same. Whereas in FSHD1 the contraction of the repeats resulting in chromatin relaxation was the primary genetic lesion, the presence of chromatin changes selectively on both copies of the repeats suggested an independent genetic factor that predisposed to FSHD2. Exome sequencing of multiple FSHD2 kindreds has now identified mutations in about 80% of these kindreds in the *SMCHD1* gene.[9] The structural maintenance of chromosomes flexible hinge domain gene (*SMCHD1*), codes for a protein involved in X inactivation but is also involved in expression of some autosomal genes.[32]

DISEASE MECHANISMS

Aberrant expression of DUX4 protein in somatic cells is thought to be the primary cause of disease in both FSHD1 and 2. Whether *SMCHD1* mutations associated with FSHD2 result in deleterious epigenetic alterations on chromosomal regions other than 4q35 remains to be seen.[9] To date, based on evaluation of relatively few FSHD2 kindreds, there are no phenotypic differences between FSHD2 and FSHD1.[31] Relaxation of the D4Z4 chromatin structure allows *DUX4* mRNA transcription from the distal copy of *DUX4*, which is then spliced to the polyadenylation signal,[33] resulting in variegated expression of DUX4 protein in a small percentage of myonuclei.[29,33,34] DUX4 is a transcription factor that is normally expressed only in the human germline but its role in germline biology is not established.[33] There are several potential mechanisms through which DUX4 expression can result in skeletal muscle damage.

Expression of DUX4 in several cell lines demonstrate that DUX4 protein is highly toxic resulting in caspase-3-mediated apoptosis and can negatively affect myogenesis.[35–39] Moreover, overexpression of DUX4 in zebrafish and mice resulted in p53-mediated toxicity.[40] DUX4-induced apoptosis could be the result of expression of the germline gene program in postmitotic skeletal muscle cells.[41] DUX4 also activates a number of genes involved in atrophy and protein degradation as well as genes involved in innate immunity.[41] Another potential mechanism is that the germline is immune-privileged and proteins normally expressed only by the germline and misexpressed in somatic cells may induce an immune response.[42] This is of particular interest, since FSHD muscle pathology is characterized by an infiltration of CD4 and CD8 T-cells.[43,44] Why only few myonuclei, at any one time, express high levels of DUX4 protein as myoblast differentiate into myotubes, remains unclear. Understanding what factors trigger this variegated expression is crucial both in explaining the regional and asymmetric involvement of muscles in FSHD and in suggesting potential therapeutic interventions.

DIFFERENTIAL DIAGNOSIS

In the majority of cases, FSHD remains a fairly straightforward clinical diagnosis if the examining physician recognizes the characteristic pattern of muscle involvement. Molecular diagnosis is helpful in confirming the diagnosis, especially in atypical cases.

The bedside diagnosis of FSHD is based on the characteristic distribution of weakness. The presence of prominent contractures and weakness of respiratory or bulbar muscles excludes the diagnosis of FSHD. In typical FSHD, disease progression occurs in a descending fashion. Therefore, shoulder girdle muscles are typically more severely involved than leg muscle in the early stages. This gradient becomes less obvious in more advanced cases. To confirm the diagnosis in individuals with atypical presentations, careful examination of other family members is essential. Several conditions can mimic the clinical presentation of FSHD but have characteristic muscle histopathology. These include: desmin storage (myofibrillar) myopathy, polymyositis, late-onset acid maltase deficiency, inclusion-body myositis, mitochondrial myopathy, and congenital nemaline and centronuclear myopathy.[45]

TESTING

Molecular Genetic Diagnosis

Molecular diagnosis for FSHD1 is widely available whereas diagnosis for FSHD2 is restricted at this point mainly to research laboratories. FSHD1 molecular diagnosis is fairly straightforward with a highly sensitive and specific assay. However, as outlined below, because of the complexity of the molecular defect, false positive results and false negative results do occur. Molecular diagnosis of FSHD is performed by Southern blot analysis with leukocyte-derived DNA using probe p13E-11 following digestion with *Eco*RI. However, p13E-11 also hybridizes to a region on 10q26 homologous to 4q35 and detects fragments in the size range diagnostic for FSHD (potentially complicating molecular diagnosis).[46] The accuracy of FSHD molecular diagnosis was significantly improved with the finding of a restriction site (*Bln*I) that is unique to the 3.3-kb repeats of 10q26 origin.[47] Double digestion with *Eco*RI/*Bln*I results in complete digestion of 10q26 alleles and confirms the 4q35 origin of the remaining alleles (Figure 97.3). Molecular diagnosis using the double-digestion technique identifies the FSHD-associated 4q35 deletion in more than 95% of cases with the disease. Normal individuals have two alleles greater than 50 kb in size, whereas affected individuals have one allele in the normal range and one containing the deletion and thus of reduced size (10–38 kb).[48] Therefore, the detection of a deletion in affected family members or its appearance *de novo* in sporadic cases confirms a diagnosis of FSHD. Translocations between 4q35 and 10q26 creating hybrid 3.3-kb repeats can at times complicate the interpretation of the results of conventional molecular diagnosis. Optimal analysis of such cases requires the use of an additional restriction enzyme XapI, to which 10-type repeats are resistant, and the use of pulse field gel electrophoresis (PFGE).[47]

Another complicating factor is that there are two variants of 4qter known, as 4qA and 4qB, occurring with equal frequency in the general population.[28] FSHD alleles, in contrast, are exclusively 4qA type; in other words, contractions of the 3.3-kb repeats on a 4qB variant do not result in FSHD. Widespread availability of FSHD genetic testing and indiscriminate use of this test has resulted in rare false positive tests in patients who do not have FSHD.[49] A source of false negative tests is the rare occurrence of D4Z4 contraction that extends proximally to include the region of the p13E-11 probe.[50–52] In fact, some of these patients account for the families with typical FSHD previously reported as not being linked to 4q35.[51] Most laboratories will determine deletion size but will only do 4qA/4qB testing on request. In patients with a typical clinical presentation and an autosomal dominant family history, demonstrating the presence of a contracted repeat is sufficient. In atypical cases and in sporadic cases, determination of the A/B variant is necessary to avoid the possibility of a false positive test.

Clinical and Ancillary Testing

Laboratory studies other than molecular diagnostic testing are helpful only in ruling out other conditions. Creatine kinase (CK) levels are normal to slightly elevated (3–5 times normal); CK levels elevated more than 10 times normal suggest an alternative diagnosis. Electromyography shows nonspecific changes of active and chronic myopathy. The presence of myotonia or changes of chronic denervation rule out the diagnosis of FSHD. Muscle biopsy in FSHD

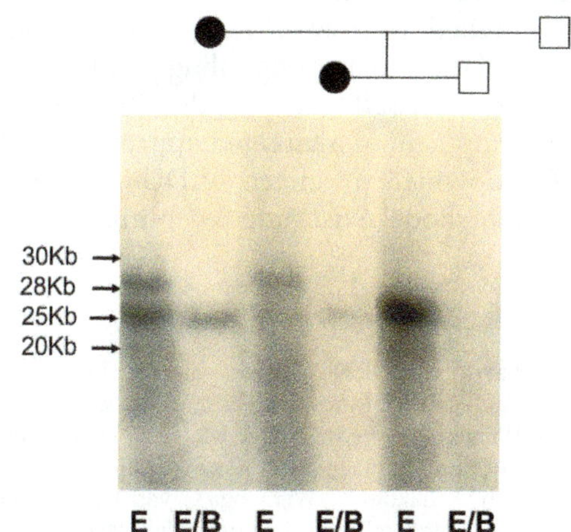

FIGURE 97.3 **Double digestion with *Eco*RI/*Bln*I enhances the accuracy of FSHD molecular diagnosis.** Pulse field gel electrophoresis (PFGE) with *Eco*RI digestion (lanes labeled E) suggest the presence of two fragments in the range associated with FSHD (28 kb and 25 kb). Double digestion with *Eco*RI/*Bln*I (lanes labeled E/B) shows the 28-kb fragment reduced to 25 kb with double digestion, indicating chromosome 4q origin; the 25-kb fragment is totally digested, indicating chromosome 10q origin (nonpathogenic).

shows nonspecific myopathic changes. As many as one-third of muscle biopsies have variable degrees of inflammation. If molecular confirmation cannot be obtained or molecular testing does not show a typical FSHD-associated deletion, a muscle biopsy should be performed to rule out potential FSHD mimickers.

MANAGEMENT

Currently there are no disease-modifying treatments in FSHD. Nevertheless, management of the patients' increasing function limitations and prevention of potential complications is crucial.[53] Management should focus on the patients' motor functional impairments and screening for potential nonmuscular manifestations.

In general, because FSHD is slowly progressive, affected individuals develop effective adaptive strategies to compensate for their disabilities, often without necessitating assistive devices. Ankle–foot orthoses are useful for foot drop, particularly early in the course of the disease. If significant quadriceps weakness sets in, however, ankle–foot orthoses can hinder ambulation. In these situations, anterior, carbon fiber, floor reaction ankle–foot orthoses may prove more useful for patients. Inability to lift the arms overhead, due to weakness of the periscapular muscles, is one of the most common disabilities in patients with FSHD. However, there are no effective ways to improve this function with any type of bracing. One option is surgical scapular fixation. In a select group of patients, scapular fixation can significantly enhance arm mobility.[54] In addition to the potential surgical complications, including breaks in the surgical wires and rarely brachial plexus injury, several factors must be considered before recommending surgery. One is whether fixation will actually result in greater arm movement in a given patient. This is easily tested at the bedside by assessing the degree of functional gain when the examiner manually fixes the scapula. Another factor is the rate of progression in the individual patient. If the disease is rapidly progressive, the benefit of surgery may be short-lived. Finally, although scapular fixation can improve arm abduction, shoulder range of motion becomes more restricted. Consequently, bilateral fixation should be avoided if possible.

Close consideration should also be given to the management of extramuscular manifestations in FSHD. Two of the most common, yet rarely symptomatic, are hearing loss and retinal vascular disease. Clinical experience, now supported by recent studies, show that in both instances only patients with very large deletions are likely to get symptomatic hearing loss or retinal vascular diseases.[20,55] Large deletions predict severe disease and typically childhood onset disease. This group of patients should therefore be carefully screened for these complications. Screening audiograms should be performed on all patients with infantile FSHD to assess the need for hearing aids. Similarly early detection of retinal telangiectasias is critical to prevent the visual loss that can result from Coats syndrome.[53] Restrictive respiratory insufficiency can be detected in a minority of patients with FSHD but the exact frequency is not well determined.[56,57] Nevertheless, at least 1% of affected individuals develop significant respiratory insufficiency.[22] Respiratory insufficiency is typically reported in wheelchair-bound patients, patients with severe truncal weakness, and those with pectus excavatum. Regular monitoring of respiratory function in severely affected individuals is essential.

Until recently, the molecular mechanism of FSHD was not known and therefore pharmacologic interventions to date were nontargeted interventions aimed at slowing disease progression. Three randomized controlled trials of albuterol in FSHD were completed, all of which failed to show an improvement in strength or function.[58–60] Additionally, a trial of a myostatin inhibitor, MYO-029, in FSHD, Becker muscular dystrophy, and limb-girdle muscular dystrophy also failed to show efficacy, although the study, a phase II study, was not powered to show efficacy.[61] Myostatin is part of the TGF beta signaling pathway and is a negative inhibitor of muscle.[62] The recent establishment of a unifying theory for FSHD and the identification of DUX4 aberrant expression as a therapeutic target, opens the way for the development of a targeted approach to FSHD therapeutics. However, obstacles to therapeutic development remain. The downstream effects of DUX4 misexpression in muscle are not yet fully characterized and although an FSHD animal model exists and recapitulates the molecular markers of FSHD, it does not have an overt muscle phenotype.[63]

References

1. Landouzy LDJ. De la myopathie atrophic progressive. *C R Acad Sci*. 1884;98:53–55.
2. Padberg GW. *Facioscapulohumeral Disease*. Leiden: University of Leiden; 1982.
3. Wijmenga C, Padberg GW, Moerer P, et al. Mapping of facioscapulohumeral muscular dystrophy gene to chromosome 4q35-qter by multipoint linkage analysis and in situ hybridization. *Genomics*. 1991;9(4):570–575.
4. van Deutekom JC, Wijmenga C, van Tienhoven EA, et al. FSHD associated DNA rearrangements are due to deletions of integral copies of a 3.2 kb tandemly repeated unit. *Hum Mol Genet*. 1993;2(12):2037–2042.

5. de Greef JC, Wohlgemuth M, Chan OA, et al. Hypomethylation is restricted to the D4Z4 repeat array in phenotypic FSHD. *Neurology*. 2007;69(10):1018–1026.
6. de Greef JC, Lemmers RJ, van Engelen BG, et al. Common epigenetic changes of D4Z4 in contraction-dependent and contraction-independent FSHD. *Hum Mutat*. 2009;30(10):1449–1459.
7. Flanigan KM, Coffeen CM, Sexton L, Stauffer D, Brunner S, Leppert MF. Genetic characterization of a large, historically significant Utah kindred with facioscapulohumeral dystrophy. *Neuromuscul Disord*. 2001;11(6–7):525–529.
8. Scionti I, Greco F, Ricci G, et al. Large-scale population analysis challenges the current criteria for the molecular diagnosis of fascioscapulohumeral muscular dystrophy. *Am J Hum Genet*. 2012;90(4):628–635.
9. Lemmers RJ, Tawil R, Petek LM, et al. Digenic inheritance of an SMCHD1 mutation and an FSHD-permissive D4Z4 allele causes facioscapulohumeral muscular dystrophy type 2. *Nat Genet*. 2012;44(12):1370–1374.
10. Statland JM, McDermott MP, Heatwole C, et al. Reevaluating measures of disease progression in facioscapulohumeral muscular dystrophy. *Neuromuscul Disord*. 2013;23(4):306–312.
11. A prospective, quantitative study of the natural history of facioscapulohumeral muscular dystrophy (FSHD): implications for therapeutic trials. The FSH-DY group. *Neurology*. 1997;48(1):38–46.
12. Statland JM, Tawil R. Risk of functional impairment in facioscapulohumeral muscular dystrophy. *Muscle Nerve*. 2013.
13. Felice KJ, Moore SA. Unusual clinical presentations in patients harboring the facioscapulohumeral dystrophy 4q35 deletion. *Muscle Nerve*. 2001;24(3):352–356.
14. Awerbuch GI, Nigro MA, Wishnow R. Beevor's sign and facioscapulohumeral dystrophy. *Arch Neurol*. 1990;47(11):1208–1209.
15. Eger K, Jordan B, Habermann S, Zierz S. Beevor's sign in facioscapulohumeral muscular dystrophy: an old sign with new implications. *J Neurol*. 2010;257(3):436–438.
16. Shahrizaila N, Wills AJ. Significance of Beevor's sign in facioscapulohumeral dystrophy and other neuromuscular diseases. *J Neurol Neurosurg Psychiatry*. 2005;76(6):869–870.
17. Tawil R, Forrester J, Griggs RC, et al. Evidence for anticipation and association of deletion size with severity in facioscapulohumeral muscular dystrophy. The FSH-DY group. *Ann Neurol*. 1996;39(6):744–748.
18. Padberg GW, Brouwer OF, de Keizer RJ, et al. On the significance of retinal vascular disease and hearing loss in facioscapulohumeral muscular dystrophy. *Muscle Nerve*. 1995;2:S73–S80.
19. Fitzsimons RB, Gurwin EB, Bird AC. Retinal vascular abnormalities in facioscapulohumeral muscular dystrophy. A general association with genetic and therapeutic implications. *Brain*. 1987;110(Pt 3):631–648.
20. Statland JM, Sacconi S, Farmakidis C, Donlin-Smith CM, Chung M, Tawil R. Coats syndrome in facioscapulohumeral dystrophy type 1: frequency and D4Z4 contraction size. *Neurology*. 2013;80(13):1247–1250.
21. Laforet P, de Toma C, Eymard B, et al. Cardiac involvement in genetically confirmed facioscapulohumeral muscular dystrophy. *Neurology*. 1998;51(5):1454–1456.
22. Wohlgemuth M, van der Kooi EL, van Kesteren RG, van der Maarel SM, Padberg GW. Ventilatory support in facioscapulohumeral muscular dystrophy. *Neurology*. 2004;63(1):176–178.
23. Grosso S, Mostardini R, Di Bartolo RM, Balestri P, Verrotti A. Epilepsy, speech delay, and mental retardation in facioscapulohumeral muscular dystrophy. *Eur J Paediatr Neurol*. 2011;15(5):456–460.
24. Funakoshi M, Goto K, Arahata K. Epilepsy and mental retardation in a subset of early onset 4q35-facioscapulohumeral muscular dystrophy. *Neurology*. 1998;50(6):1791–1794.
25. Hewitt JE, Lyle R, Clark LN, et al. Analysis of the tandem repeat locus D4Z4 associated with facioscapulohumeral muscular dystrophy. *Hum Mol Genet*. 1994;3(8):1287–1295.
26. Ding H, Beckers MC, Plaisance S, Marynen P, Collen D, Belayew A. Characterization of a double homeodomain protein (DUX1) encoded by a cDNA homologous to 3.3 kb dispersed repeated elements. *Hum Mol Genet*. 1998;7(11):1681–1694.
27. van Overveld PG, Lemmers RJ, Sandkuijl LA, et al. Hypomethylation of D4Z4 in 4q-linked and non-4q-linked facioscapulohumeral muscular dystrophy. *Nat Genet*. 2003;35(4):315–317.
28. Lemmers RJ, de Kievit P, Sandkuijl L, et al. Facioscapulohumeral muscular dystrophy is uniquely associated with one of the two variants of the 4q subtelomere. *Nat Genet*. 2002;32(2):235–236.
29. Lemmers RJ, van der Vliet PJ, Klooster R, et al. A unifying genetic model for facioscapulohumeral muscular dystrophy. *Science*. 2010;329(5999):1650–1653.
30. Geng LN, Yao Z, Snider L, et al. DUX4 activates germline genes, retroelements, and immune mediators: implications for facioscapulohumeral dystrophy. *Dev Cell*. 2011.
31. de Greef JC, Lemmers RJ, Camano P, et al. Clinical features of facioscapulohumeral muscular dystrophy 2. *Neurology*. 2010;75(17):1548–1554.
32. Mould AW, Pang Z, Pakusch M, et al. Smchd1 regulates a subset of autosomal genes subject to monoallelic expression in addition to being critical for X inactivation. *Epigenetics Chromatin*. 2013;6(1), 19-8935-6-19.
33. Snider L, Geng LN, Lemmers RJ, et al. Facioscapulohumeral dystrophy: incomplete suppression of a retrotransposed gene. *PLoS Genet*. 2010;6(10):e1001181.
34. Gabriels J, Beckers MC, Ding H, et al. Nucleotide sequence of the partially deleted D4Z4 locus in a patient with FSHD identifies a putative gene within each 3.3 kb element. *Gene*. 1999;236(1):25–32.
35. Kowaljow V, Marcowycz A, Ansseau E, et al. The DUX4 gene at the FSHD1A locus encodes a pro-apoptotic protein. *Neuromuscul Disord*. 2007;17(8):611–623.
36. Snider L, Asawachaicharn A, Tyler AE, et al. RNA transcripts, miRNA-sized fragments and proteins produced from D4Z4 units: new candidates for the pathophysiology of facioscapulohumeral dystrophy. *Hum Mol Genet*. 2009;18(13):2414–2430.
37. Bosnakovski D, Lamb S, Simsek T, et al. DUX4c, an FSHD candidate gene, interferes with myogenic regulators and abolishes myoblast differentiation. *Exp Neurol*. 2008;214(1):87–96.
38. Bosnakovski D, Xu Z, Gang EJ, et al. An isogenetic myoblast expression screen identifies DUX4-mediated FSHD-associated molecular pathologies. *EMBO J*. 2008;27(20):2766–2779.

39. Wuebbles RD, Long SW, Hanel ML, Jones PL. Testing the effects of FSHD candidate gene expression in vertebrate muscle development. *Int J Clin Exp Pathol*. 2010;3(4):386–400.
40. Wallace LM, Garwick SE, Mei W, et al. DUX4, a candidate gene for facioscapulohumeral muscular dystrophy, causes p53-dependent myopathy in vivo. *Ann Neurol*. 2011;69(3):540–552.
41. Geng LN, Yao Z, Snider L, et al. DUX4 activates germline genes, retroelements, and immune mediators: Implications for facioscapulohumeral dystrophy. *Dev Cell*. 2012;22(1):38–51.
42. van der Maarel SM, Miller DG, Tawil R, Filippova GN, Tapscott SJ. Facioscapulohumeral muscular dystrophy: consequences of chromatin relaxation. *Curr Opin Neurol*. 2012;25(5):614–620.
43. Arahata K, Ishihara T, Fukunaga H, et al. Inflammatory response in facioscapulohumeral muscular dystrophy (FSHD): immunocytochemical and genetic analyses. *Muscle Nerve*. 1995;2:S56–S66.
44. Frisullo G, Frusciante R, Nociti V, et al. CD8(+) T cells in facioscapulohumeral muscular dystrophy patients with inflammatory features at muscle MRI. *J Clin Immunol*. 2011;31(2):155–166.
45. Tawil R. Facioscapulohumeral muscular dystrophy. *Neurotherapeutics*. 2008;5(4):601–606.
46. van Deutekom JC, Bakker E, Lemmers RJ, et al. Evidence for subtelomeric exchange of 3.3 kb tandemly repeated units between chromosomes 4q35 and 10q26: implications for genetic counselling and etiology of FSHD1. *Hum Mol Genet*. 1996;5(12):1997–2003.
47. Deidda G, Cacurri S, Piazzo N, Felicetti L. Direct detection of 4q35 rearrangements implicated in facioscapulohumeral muscular dystrophy (FSHD). *J Med Genet*. 1996;33(5):361–365.
48. Orrell RW, Tawil R, Forrester J, Kissel JT, Mendell JR, Figlewicz DA. Definitive molecular diagnosis of facioscapulohumeral dystrophy. *Neurology*. 1999;52(9):1822–1826.
49. Sacconi S, Salviati L, Bourget I, et al. Diagnostic challenges in facioscapulohumeral muscular dystrophy. *Neurology*. 2006;67(8):1464–1466.
50. Lemmers RJ, Osborn M, Haaf T, et al. D4F104S1 deletion in facioscapulohumeral muscular dystrophy: phenotype, size, and detection. *Neurology*. 2003;61(2):178–183.
51. Tim RW, Gilbert JR, Stajich JM, et al. Clinical studies in non-chromosome 4-linked facioscapulohumeral muscular dystrophy. *J Clin Neuromuscul Dis*. 2001;3(1):1–7.
52. Deak KL, Lemmers RJ, Stajich JM, et al. Genotype-phenotype study in an FSHD family with a proximal deletion encompassing p13E-11 and D4Z4. *Neurology*. 2007;68(8):578–582.
53. Tawil R, van der Maarel S, Padberg GW, van Engelen BG. 171st ENMC international workshop: standards of care and management of facioscapulohumeral muscular dystrophy. *Neuromuscul Disord*. 2010;20(7):471–475.
54. Orrell RW, Copeland S, Rose MR. Scapular fixation in muscular dystrophy. *Cochrane Database Syst Rev*. 2010;(1), CD003278.
55. Lutz KL, Holte L, Kliethermes SA, Stephan C, Mathews KD. Clinical and genetic features of hearing loss in facioscapulohumeral muscular dystrophy. *Neurology*. 2013;81(16):1374–1377.
56. Stubgen JP, Schultz C. Lung and respiratory muscle function in facioscapulohumeral muscular dystrophy. *Muscle Nerve*. 2009;39(6):729–734.
57. Kilmer DD, Abresch RT, McCrory MA, et al. Profiles of neuromuscular diseases. Facioscapulohumeral muscular dystrophy. *Am J Phys Med Rehabil*. 1995;74(5 suppl):S131–S139.
58. Kissel JT, McDermott MP, Mendell JR, et al. Randomized, double-blind, placebo-controlled trial of albuterol in facioscapulohumeral dystrophy. *Neurology*. 2001;57(8):1434–1440.
59. van der Kooi EL, Vogels OJ, van Asseldonk RJ, et al. Strength training and albuterol in facioscapulohumeral muscular dystrophy. *Neurology*. 2004;63(4):702–708.
60. Payan CA, Hogrel JY, Hammouda EH, et al. Periodic salbutamol in facioscapulohumeral muscular dystrophy: a randomized controlled trial. *Arch Phys Med Rehabil*. 2009;90(7):1094–1101.
61. Wagner KR, Fleckenstein JL, Amato AA, et al. A phase I/II trial of MYO-029 in adult subjects with muscular dystrophy. *Ann Neurol*. 2008;63(5):561–571.
62. Smith RC, Lin BK. Myostatin inhibitors as therapies for muscle wasting associated with cancer and other disorders. *Curr Opin Support Palliat Care*. 2013;7(4):352–360.
63. Krom YD, Thijssen PE, Young JM, et al. Intrinsic epigenetic regulation of the D4Z4 macrosatellite repeat in a transgenic mouse model for FSHD. *PLoS Genet*. 2013;9(4):e1003415.

CHAPTER

98

Muscle Channelopathies: Periodic Paralyses and Nondystrophic Myotonias

Jeffrey Ralph and Louis Ptáček

University of California San Francisco, San Francisco, CA, USA

INTRODUCTION

Disease Characteristics

Ion channel activities determine the maintenance of the cellular resting membrane potential and, in excitable tissues, the generation and propagation of action potentials across cell membranes. In muscle, action potentials allow for the rapid, efficient, whole-cell activation of the chemical processes underlying contraction. It makes sense, therefore, that mutations in the genes encoding ion channels may lead to either electrical hyperexcitation and excessive contraction (myotonia) or hypoexcitation and paralysis. Muscle ion channel disorders are prototypic channelopathies. They share many of the classic clinical features of channelopathies, including: episodicity, emergence of symptomatology at young age and amelioration later in life, and modulation of severity by environmental factors. Many neurologists consider these disorders among the most fascinating and rewarding in all of neurology to diagnose and treat. The patient's perspective is often different. Owing to the rareness and periodicity of some of these disorders, many patients have dealt with misdiagnosis and long delays in proper management.

Diagnosis/Testing

The diagnosis is clinical, supported by laboratory, electrodiagnostic, and genetic testing. Serum potassium levels during a paralytic attack and provocative testing are helpful in the diagnostic work-up of periodic paralysis. The role of provocative testing has narrowed somewhat given the wider availability of genetic testing. However, a significant minority of patients do not have any of the recognized ion channel mutations, so ancillary testing is still helpful in the evaluation of these individuals.

Current Research

Research has focused on amelioration of myotonia as well as the prevention and treatment of paralytic attacks. For the rarer muscle channelopathies, for example Andersen–Tawil syndrome, the emphasis has been on disease characterization and natural history studies.

CLINICAL FEATURES

Chloride Channelopathies: Myotonia Congenita

Myotonia is the hallmark feature of the chloride channelopathies. Myotonia refers to involuntary muscle contractions caused by aberrant sarcolemmal action potentials. In general, myotonia does not occur spontaneously in resting muscle. Rather, voluntary activity or mechanical perturbation of the muscle (e.g., inserting an electromyography [EMG] needle) triggers runs of the abnormal, excessive electrical activity (electrographic myotonia) and the resulting

Rosenberg's Molecular and Genetic Basis of Neurological and Psychiatric Disease
http://dx.doi.org/10.1016/10.1016/B978-0-12-410529-4.00098-X

muscle contractions. If the motor neurons stop firing and the muscle is undisturbed, the myotonia fades away rapidly and completely in less than 10 seconds.

The physician may assess for myotonia by looking for delayed relaxation of muscles after forceful contraction (e.g., grip myotonia). Percussion of muscles may also produce rapid contractions followed by delayed relaxations and indentations of the overlying skin (Figure 98.1).

The autosomal dominant form of myotonia congenita (MC), Thomsen disease, was described in the late 1800s. Members of Thomsen's own family were afflicted with this condition. Symptom onset is usually in early childhood. All striated muscle groups, including the facial muscles and tongue, are affected. The presenting complaint is muscle stiffness or difficulty letting go of objects. The stiffness is most pronounced when a forceful movement occurs after 10–15 minutes of rest. Patients may have difficulty getting out of chairs or vehicles. The myotonia may lead to safety issues. A military cadet may fail to let go of the trigger to his automatic weapon. Repeated muscle contraction may reduce the stiffness, the so-called "warm-up phenomenon." Despite some limitations, patients have normal or superior strength and may be competitive in some sports. The muscles are often well developed.

Becker's descriptions of the more common autosomal recessive form of MC were published a century after Thomsen's. In contrast to Thomsen disease, the myotonia in Becker patients is not present until the age of 10–14 years. Symptom severity may slowly increase until approximately age 30. Men are affected more than women. Compared to Thomsen disease, the myotonia is more severe. In addition to muscle hypertrophy, patients tend to toe walk and have a compensatory lordosis. Becker patients frequently have transient weakness that occurs a few seconds into exercise. This weakness resolves rapidly if the attempt to exercise continues. Electrodiagnostic testing protocols, as discussed later, exploit this transient weakness phenomenon for making the diagnosis. A few Becker patients show permanent muscle weakness, including in distal muscle groups.[1] The fixed muscle weakness along with elevated creatine kinase (CK) values may make differentiating this condition from myotonic dystrophy difficult.

Mutations of *CLCN1*, which encodes the voltage-gated chloride channel of the skeletal muscle fiber membrane, cause both Thomsen and Becker MC.[2]

The prevalence of MC in northern Scandinavia is estimated to be 1:10,000.[3,4] The worldwide prevalence has been estimated at 1:100,000.[5]

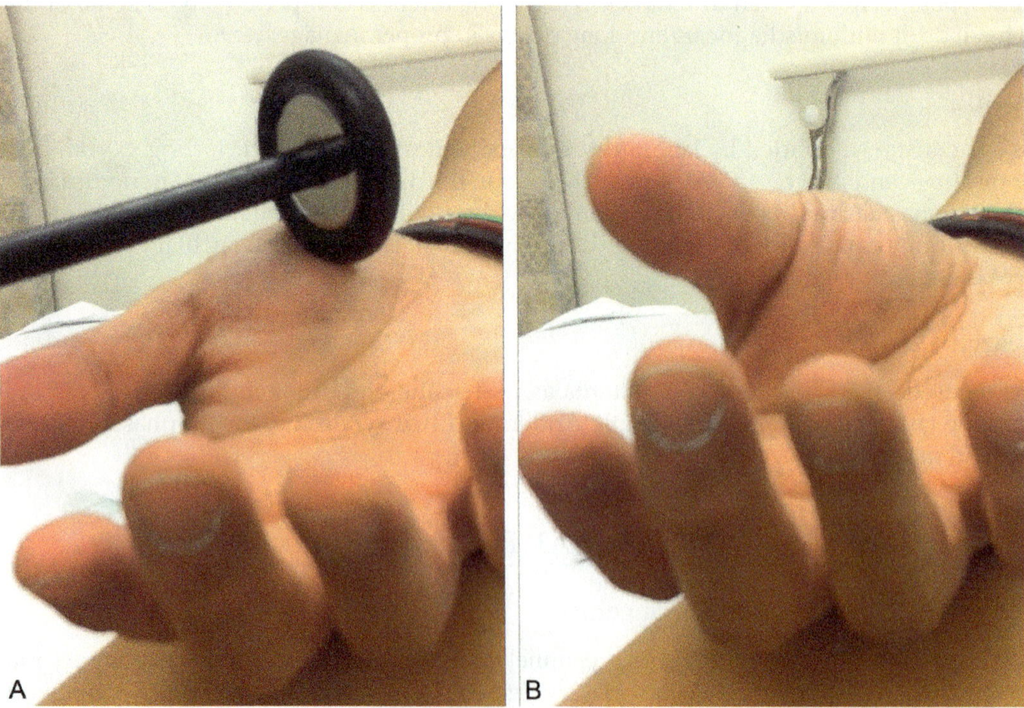

FIGURE 98.1 (A) Percussion of the right thenar eminence with a reflex hammer. (B) Thumb abduction persists long after the strike, which is driven by ongoing myotonic discharges in the thenar muscles.

Sodium Channelopathies: Disorders with Myotonia or Paralysis or Both

All of the sodium channelopathies are inherited in an autosomal dominant fashion with manifestation of symptoms before age 20. Mutations in the gene encoding the α-subunit of the muscle voltage-gated sodium channel cause these disorders. Hyperkalemic periodic paralysis (hyperKPP) was the first channelopathy to be characterized genetically.[6]

Potassium-Aggravated Myotonia

Patients with potassium-aggravated myotonia (PAM) develop severe muscle stiffness after the ingestion of potassium-rich foods or after prolonged exercise. These patients do not have bouts of muscle weakness. Some individuals with this disorder were originally suspected of having myotonia congenita, but were subsequently found to have mutations in *SCN4A*. Unlike the warm-up phenomena seen in MC, these patients get more symptomatic as the exercise continues. There is a wide range of disease severity, with some patients only experiencing mild and transient myotonia whereas others experience debilitating, constant myotonia. Cold does not significantly impact symptoms.

Paramyotonia Congenita

The hallmark features of paramyotonia congenita (PC) are: 1) exacerbations of myotonia with prolonged exercise; and 2) cold-induced myotonia. There may be considerable overlap between this entity and PAM and hyperKPP. Paramyotonic symptoms are present at birth and change little over time. A baby with this disorder cannot open the eyes after a cold washcloth is placed over the face. Repeated forceful eyelid closures in older patients leads to the same problem. Figure 98.2 illustrates this phenomenon plus the dramatic effects of cold on the severity of the myotonia. Myotonia of the extraocular muscles may lead to delayed relaxation of the upper eyelid upon downgaze (lid lag) as well as intermittent diplopia in various directions of gaze. Drinking cold fluids may impair swallowing. With intense or very prolonged cooling or exercise, stiffness may eventually give way to paralysis.

Stress, hypothyroidism, hunger, or pregnancy may increase symptoms.

The mild nature of PC symptoms prevents making estimates on prevalence because many individuals do not seek medical care. Seventy-one percent of PC patients have mutations in *SCN4A*.[7]

Acetazolamide-responsive myotonia in which patients experience exercise-induced muscle pain may also belong to this group.[8]

Hyperkalemic Periodic Paralysis (HyperKPP)

HyperKPP was first described by Tyler and colleagues in 1951. In the 1950s, other significant contributions were made by Helweg-Larsen and Gamstrop, among others.

Attacks of weakness usually begin in the first decade of life. A rise of serum potassium from any cause, including eating potassium-rich foods, can precipitate an attack. Once a paralytic attack is triggered, an outflow of potassium from muscle often occurs, which can lead to even higher serum potassium concentrations. Although the potassium levels can become quite high, cardiac arrhymthias are uncommon. In between attacks, the serum potassium level is usually normal. A typical attack begins before breakfast, lasts for 30 minutes to an hour, and then resolves. If the attack progresses to paralysis, the deep tendon reflexes disappear. A patient may become tetraplegic but ocular muscles are spared; it is uncommon for respiratory failure to occur. Sustained mild exercise after vigorous exercise may block a full-blown attack. In the interictal status, most patients have mild myotonia; some patients have all of the clinical features of PC. When myotonia is present, it points to hyperKPP over the other forms of periodic paralysis (PP). Male patients generally have more frequent and severe attacks.[9]

Rest after exercise, cold, hunger, emotional stress, and pregnancy may provoke or worsen attacks. The attacks become more frequent and severe over time until patients reach middle age, after which the frequency and severity may decrease. Many patients develop a chronic progressive myopathy that preferentially affects the pelvic girdle and thigh muscles.

In one series, 64% of patients with hyperKPP were found to have mutations in *SCN4A*. The average age at onset is younger for those with mutations (2 years) than those without (14 years). The prevalence of hyperKPP is around 1:200,000. The penetrance is around 90%.

Other Cation Channelopathies: Paralytic Syndromes without Myotonia

Hypokalemic Periodic Paralysis (HypoKPP)

The first descriptions of this disorder were in the 19th century. In 1934, hypokalemia was recognized as a feature of the paralytic attacks. The inheritance pattern is autosomal dominant with a reduced penetrance in women. Severe

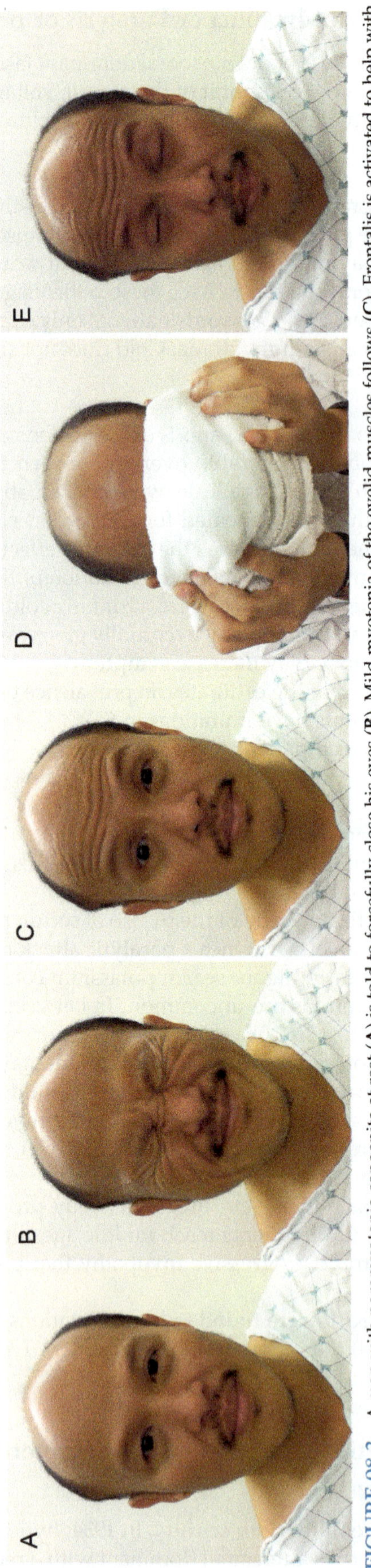

FIGURE 98.2 A man with paramyotonia congenita at rest **(A)** is told to forcefully close his eyes **(B)**. Mild myotonia of the eyelid muscles follows **(C)**. Frontalis is activated to help with eye opening. He then applies a cold cloth to the face for 30 seconds **(D)**. This produces severe myotonia of the eyelid **(E)**, much worse than with just forceful closure. He has difficulty opening his eyes for 30 seconds, even with the help of frontalis.

cases present in early childhood, although mild cases may present in the third decade of life. Paralytic attacks occur in the second half of the night or in the early morning hours. Upon awakening, the patient may be unable to move her arms, legs, or trunk. The muscles of the face are spared. There is a wide range of attack frequencies, with some patients having daily attacks and others only having a few attacks during their lifetime. Many patients report fewer attacks after age 30; some become attack-free later in life. Recovery in between attacks may be incomplete. Strength may also show diurnal variations, with increased strength later in the day. Milder attacks may affect only isolated muscle groups. As in hyperKPP, significant weakness of the extraocular muscles and facial muscles is uncommon as is respiratory failure. The symptoms may vary widely in a family. There is no clinical or electrographic evidence of myotonia.

An attack may be triggered by a fall in serum potassium levels. The triggers include strenuous exercise and a carbohydrate-rich meal. Other triggers may include illness, table salt, alcohol ingestion, fasting, medications, dehydration, lack of sleep, and prolonged bed rest. As the attack proceeds, there is an additional shift of potassium into muscle cells. The average potassium level during an attack is 2.4 mEq per L.[7]

Compared to hyperKPP patients, hypoKPP patients have later onset of disease and less frequent attacks but the attacks last much longer and are more severe. Over time, most patients developed fixed weakness affecting proximal muscles in the lower greater than in the upper extremities.

HypoKPP is the most common of the periodic paralyses with a prevalence of 1:100,000.[10] Sixty-four percent of patients carry one of the known mutations.[7] The age of onset is earlier for patients with mutations (10 vs. 22 years). Most cases of hypoKPP are caused by mutations in the L-type voltage-gated calcium channel α_1-subunit (hypoKPP1). Less commonly, mutations in *SCN4A* may also cause a very similar syndrome (hypoKPP2). Compared to hypoKPP1 patients, hypoKPP2 patients have shorter attack durations and younger ages of disease onset; fixed muscle weakness may also be less common in hypoKPP.[7,11]

Andersen–Tawil Syndrome

First described in 1971, Andersen–Tawil syndrome (ATS) is defined by the triad of periodic paralysis, ventricular ectopy, and dysmorphic features.[12,13] The paralytic attacks usually begin before age 20. The weakness may occur spontaneously or may be triggered by prolonged rest or rest after exercise. Fixed proximal weakness often develops. Ictal potassium levels may be reduced, normal, or elevated. Electrocardiogram (ECG) abnormalities include a prolonged QT interval, prominent U waves, premature ventricular contractions, ventricular bigeminy, and polymorphic ventricular tachycardia. Some have a unique form of ventricular tachycardia, characterized by beat-to-beat alternating QRS axis polarity. Some patients are asymptomatic, whereas others present with palpitations, syncope, or even cardiac arrest.[14] There seems to be a lower risk for syncope and sudden death in ATS compared to the other long QT syndromes.

ATS patients may have any of the following dysmorphic features: toe clinodactyly and syndactyly, small hands and feet, broad forehead, high arched or cleft palate, mild facial asymmetry, microcephaly, and missing teeth or persistent primary dentition.[15] ATS patients also have impaired executive function and abstract reasoning.[16] In about ~70% of cases, ATS is caused by mutation in *KCNJ2*, which encodes the strong inward-rectifier potassium channel Kir2.1.[17,18]

Thyrotoxic Hypokalemic Periodic Paralysis

Thyrotoxic hypokalemic periodic paralysis (TPP) is a sporadic disorder characterized by episodic paralysis and hypokalemia in genetically susceptible, thyrotoxic individuals. It is most prevalent in young Asian and Latin American men, and much rarer in Caucasians and Africans. Up to 10% of Chinese men who are thyrotoxic have TPP, whereas only 0.1% of thyrotoxic Caucasian Americans will have it.[19,20] The clinical features of the paralytic attacks, including exacerbating factors, are very similar to those of hypoKPP. The disorder resolves once the thyrotoxicosis is treated. In 2010, it was discovered that mutations in the potassium channel Kir2.6 cause susceptility to TPP.[21]

MOLECULAR GENETICS AND DISEASE MECHANISMS

Chloride Channelopathies

As mentioned earlier, mutations in *CLCN1* cause both dominant and recessive MC. The protein encoded by this gene, ClC-1, is comprised of symmetric homodimers (Figure 98.3) that produce two ion-conducting pores; its two-pore structure is distinct from cation channels. In addition to playing key roles in establishing the resting membrane potential, chloride channels effectively raise the threshold for action potential generation by providing a current shunt.

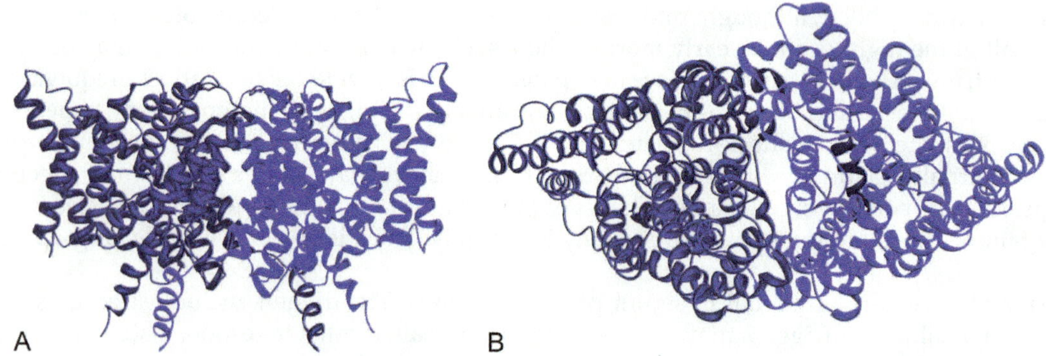

FIGURE 98.3 **Crystal structure of the ClC chloride channel from *Escherichia coli*.** (**A**) Side view of the two halves (gray and black) with the cytosol on the bottom and extracellular space on the top. (**B**) Top-down view of this same channel complex.

Mutations of *CLCN1* produce myotonia by reducing chloride conductance. The reduced chloride conductance markedly increases the depolarizing effects of potassium accumulation in the muscle T-tubule system after repeated action potentials. With wild-type ClC-1, tubular potassium accumulation leads to a depolarization of about 0.1 mV per action potential. In MC, the same potassium accumulation may cause a 10-fold greater depolarization.[22] These effects increase chances for afterdepolarizations and clinical myotonia.

Over 100 mutations in *CLCN1* may cause autosomal recessive MC, more than 15 mutations cause dominant MC, and approximately 10 mutations may cause either recessive or dominant disease. The recessive mutations are widely scattered across the gene. Dominant *CLCN1* mutations tend to cluster at the dimer interface, enabling the mutant protein to impair the normal one, thereby producing a dominant negative effect.[23] Recessive mutations produce disease because of haploinsufficiency imparted by each allele. Individuals with the F428S mutation may have a phenotype resembling PC. Patients with the G230E and T310M mutations may have a fluctuating phenotype triggered by pregnancy.[24,25] T550M mutations may be associated with proximal weakness; distal myopathy has been associated with the P932L mutation. The penetrance is variable.

An animal model of MC with a striking phenotype is the "fainting goat" of Tennessee. When startled, the limbs extend and stiffen, leading to collapse. Fallen animals look like toy animals that have been toppled over by a careless child. It goes without saying that wild animals having such mutations would face significant evolutionary hurdles. The low chloride-conductance theory was developed for the myotonic goat long before *CLCN1* was cloned and characterized; curiously, the goat homolog of *CLCN1* was cloned after the human's. At least two strains of mice spontaneously developed mutations that produced a MC phenotype.

The mechanism of the warm-up phenomenon remains unclear.

Sodium Channelopathies

SCN4A encodes the pore-forming, α-subunit of the skeletal muscle sodium channel (Nav1.4). Four identical domains (DI–DIV), each with six transmembrane α helices (S1–S6) surround a single pore (Figure 98.4). The S4 segment has voltage sensitivity. It moves in response to depolarization, resulting in channel opening. The small D3–4 intracellular loop is the site of the fast inactivation gate. This structure is poised to enter the intracellular portion of the ion-conducting pore and obstruct ion passage (fast inactivation). Other mechanisms underlie slow inactivation.

Mutations of *SCN4A* cause disease (PAM, PC, and hyperKPP) by producing abnormally persistent sodium currents. If the activities from other ion channels compensate for this aberrant current and the muscle can eventually restore stable repolarization, myotonia but not weakness occurs. However if this cannot be achieved and the cell membrane remains persistently depolarized, normal sodium channels become inactivated. Once this occurs, the muscle is rendered electrically inexcitable. Compounding the problem, the prolonged depolarization leads to potassium efflux from the muscle cell. The resulting hyperkalemia has an additional depolarizing effect.

Most of the gain-of-function mutations of *SCN4A* lead to abnormal fast inactivation of the sodium channel.[26,27] A few mutations produce enhanced activation. Some mutations may also impair slow inactivation as well.

Some mutations (L689I, T704M, I693T, A1156T, I1495F, and the double mutant F1490L+M1493I) produce pure hypoKPP and no stiffness; others produce both weakness and stiffness. Rare families have a hybrid phenotype of hyperKPP and PC.[28] Patients with the T704M mutation do not have interictal myotonia. Approximately 50% of these

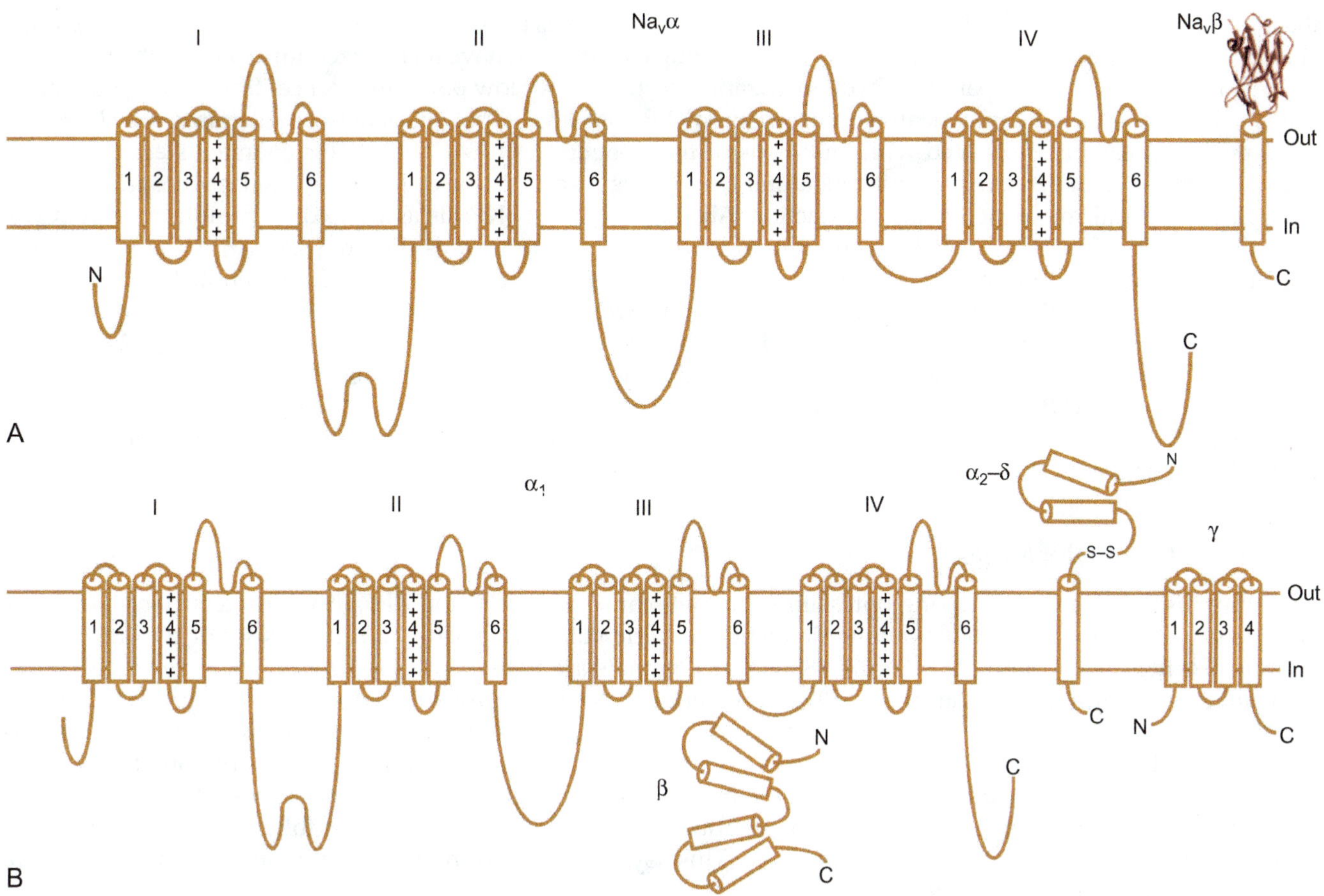

FIGURE 98.4 **Schematic structure of voltage-gated sodium and calcium channel subunits.** (**A**) The sodium channel α-subunit (left) is comprised of four domains, labeled I through IV. Each domain consists of six membrane-spanning helices (S1 through S6) with a reentrant pore-forming loop between the fifth and sixth of these segments. The S4 segment of each domain contains a series of positively charged residues (+ symbols) at every third residue. The β-subunit (right) is composed of a single transmembrane helix with an extracellular *N*-terminus and intracellular *C*-terminus. The *N*-terminus bears homology to immunoglobulin-like folds, for which an immunoglobulin-like crystal structure is shown. (**B**) The calcium channel α_1-subunit (left) is schematically identical to that of the sodium channel α-subunit. The α_2-δ-subunit is comprised of two segments joined by a disulfide bond. The α_2-segment is extracellular and composed of two α-helices. The δ-segment forms a membrane-spanning α-helix with an intracellular *C*-terminus and an extracellular *N*-terminus. It is the *N*-terminus that forms a disulfide bond with the *C*-terminus of the α_2-segment. Its intracellular β-subunit (bottom) is formed by four α-helices and interacts with the D1-2 linker region of the α_1-subunit. Palmitoylization of this subunit can lead to membrane anchoring. The γ-subunit (right) is comprised of four membrane-spanning helices, numbered S1 through S4 with an intracellular *N*- and *C*-terminus.

individuals develop a permanent myopathy. Individuals previously labeled as having "normokalemic periodic paralysis" also had this mutation.

An equine animal model of hyperKPP exists. The syndrome is similar to the human disease, but the manifestations may be more severe. A single point mutation in the fourth domain of the equine muscle sodium channel is responsible for all cases. In the breed, the American Quarter Horse, 4.4% are affected. The mutation has been traced back to a sire named *Impressive*.[22]

Other Cation Channelopathies

Hypokalemic Periodic Paralysis (HypoKPP)

As in hyperKPP, the muscle during a hypoKPP attack is in a depolarized, inexcitable state. How this happens is not completely understood. The mechanisms underlying the ictal hypokalemia are particularly puzzling—potassium ions during paralysis enter the muscle against their electrical and chemical gradients.

Some recent work sheds some light on another peculiar aspect of hypoKPP attacks: the paradoxical depolarization of the sarcolemma in the setting of hypokalemia. It turns out that the low extracellular potassium concentrations reduce the conductance of the inward rectifier potassium (Kir) channels that are responsible for establishing the normal resting potential.[29] This change in conductance uncouples the resting membrane potential from the potassium equilibrium potential. Once this occurs, depolarizing currents can have an outsized influence on the membrane potential. In normal muscle, paradoxical depolarization occurs at very low potassium concentrations (approximately 1 mM). How paradoxical depolarizaton occurs in hypoKPP with less severe levels of hypokalemia is not clear.

HypoKPP is usually associated with mutations of the α_1-subunit of *CACNA1S*. This channel is attached to the ryanodine receptor of the sarcoplasmic reticulum. It serves as a voltage sensor for the ryanodine receptor, which in turn plays a critical role in activating contraction. About 10% of hypoKPP is associated with mutations in *SCN4A* (hypoKPP2). The Nav1.4 and Cav1.1 α-subunits are structurally homologous (Figure 98.4). All of the mutations in this channel that lead to hypoKPP are located in the voltage-sensitive S4 segment of domain 2. These mutations lead to altered inactivation properties. In addition, recent work suggests that the mutations may produce an abnormal gating-pore current that would tend to depolarize the cells.[30,31] A proposed model for hypoKPP2 from these mutations, then, is that intracellular sodium levels gradually increase in muscle cells that are unable to clear using sodium–potassium ATPases due to hypokalemia (extracellular potassium ions being required for this mechanism). This build-up of intracellular sodium leads to a gradual depolarization of the membrane and a shift of voltage-gated channels to inactivated states.

Potassium Channelopathies

Mutations in inwardly rectifying potassium channels cause both ATS and TPP. Inwardly rectifying potassium channels are comprised of four homomeric or heteromeric subunits. Each subunit consists of two membrane-spanning α-helices (M1 and M2) (Figure 98.5). Kir2 channels are gated primarily in two ways: rectification and phosphatidylinositol 4,5-bisphosphate (PIP_2) interaction. Inward rectification, due to binding by intracellular polyamines and magnesium, is an asymmetric current voltage relationship where the channel is able to pass more inward than outward current. Polyamines bind to residues on the C-terminus and M2 region of the channel subunits and block potassium ions from traversing the channel.[32] Potential gradients across the cell membrane keep these interactions low during hyperpolarization but allow the polyamines to follow potassium ions outward through the channel during depolarization. This asymmetric current–voltage relationship is important for

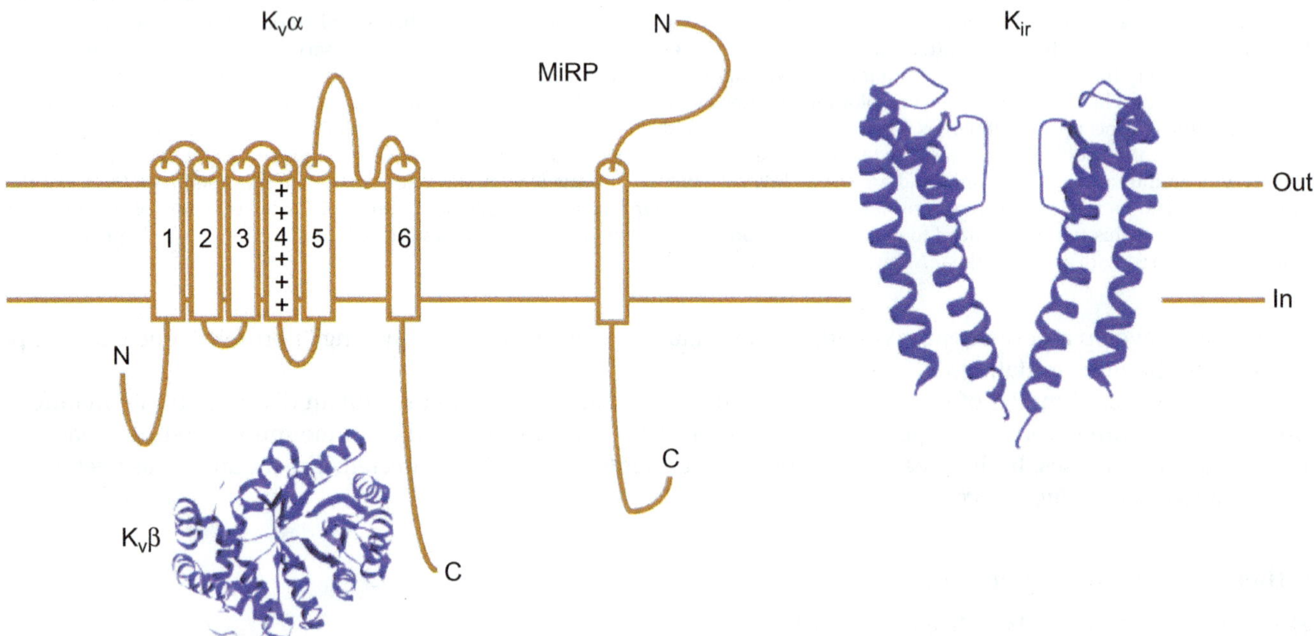

FIGURE 98.5 **Structure of voltage-gated and inwardly rectifying potassium channels.** Left: The Kv channel α-subunit is structured similarly to that of the sodium and calcium channel. It consists of six membrane-spanning α-helices (S1 through S6) with a reentrant loop between S5 and S6. As with other voltage-gated channels, the S4 segment contains positively charged residues at every third residue (+ symbols). The α-subunit associates with both the minK/MiRP family (middle) and β-subunits (bottom). MiRP subunits are constructed from a single membrane-spanning helix. The crystal structure of the Kvβ_2-subunit is also shown (bottom). The inwardly rectifying potassium channel (right) is constructed of two membrane-spanning helices, M1 and M2. The crystal structure of the bacterial KCSA protein is shown. The pore is in the center, formed by the P-loop and the M2 helix. The *N*- and *C*-termini are not shown but are both intracellular.

setting the resting potential of cells while not leading to current shunting during depolarization. PIP_2 mediated gating is due to the interaction with numerous residues throughout the *N*- and *C*-termini of the channel. The binding of these residues likely causes a change of structure of the channel, increasing the diameter of the inner-pore (specifically the G-loop).[33–35]

Early on, the importance of proper Kir2.1 function in cardiac function and skeletal development was demonstrated by knockout of this gene in mice, leading to cardiac abnormalities.[36] Numerous mutations located throughout Kir2.1 cause ATS, all seeming to lead to dominant-negative loss of Kir2.1 current. Some mutations are at sites directly (R312C) or indirectly (V302M) interacting with membrane PIP_2. PIP_2 binding is required for channel opening, so it is unsurprising that these mutant channels with decreased PIP_2 affinity show decreased current.[34] Instead of directly interacting with PIP_2, V302M is involved in coupling PIP_2 interaction with other residues to movement of the G-loop and, thereby, channel opening.[37] Another common change caused by ATS mutations is altered trafficking. Two such mutations, 95_98del and 314_315del, are translated but do not traffic to the surface, decreasing whole-cell current by 94%.[38] Regardless of the underlying mechanism, these mutations lead to membrane depolarization in a dominant negative manner and an inability of voltage-gated channels to recover from inactivation. While this explains cardiac and skeletal muscle abnormalities, it is unclear how ATS mutations lead to skeletal dysmorphisms.

There are no mouse models for ATS; the Kir2.1 knockout is lethal.

Thyrotoxic Periodic Paralysis

Mutations of *KCNJ18* are associated with susceptibility for TPP. *KCNJ18* was serendipitously discovered during the search for candidate genes for this disorder. The investigators screened for mutations in ion channels known to be expressed in muscles that contained thyroid response elements in their promoter regions. While screening *KCNJ12*, they noted a number of polymorphisms that violated the Hardy–Weinberg equilibrium, suggesting the existence of another gene (*KCNJ18*). It encodes Kir2.6, which is expressed primarily in skeletal muscle. The amino acid sequence of Kir2.6 is 96–98% homologous to Kir2.2. Unlike Kir2.2, the amino acid sequence of Kir2.6 prevents the efficient transport of it from the endoplasmic reticulum to the sarcolemma. Kir2.6 can coassemble with Kir2.1 and Kir 2.2. In the context of hyperthyroidism, it is possible that mutations in Kir2.6 have a dominant negative effect by blocking the normal cellular trafficking of Kir2.1 and Kir2.2 to the cell surface.[39]

DIFFERENTIAL DIAGNOSIS

Myotonias

When trying to sort out a patient with a suspected myotonic syndrome, it is important to consider the possibility of myotonic dystrophy (DM). DM type 1 is a dominantly inherited multisystem disease with skeletal muscle, cardiac, brain, eye, and endocrine manifestations caused by expansion of an unstable CTG trinucleotide repeat in the 3′ untranslated region of the DM protein kinase (*DMPK*) gene. The expansion leads to alternate splicing of the mRNA transcript of the chloride channel, creating a secondary chloride channelopathy. Characteristic craniofacial atrophy, prominent distal weakness, and early-onset cataracts help establish the diagnosis. For DM2, which is caused by an expansion of an unstable CCTG tetranucleotide repeat in intron 1 of the *ZNF9* gene, early-onset cataracts and the presence of proximal weakness help to make the diagnosis. It should be remembered that some Becker MC patients may also have fixed weakness in either proximal or distal muscle groups. Clinical and electrographic myotonia is more conspicuous in MC than in DM1 or DM2. In DM2, clinical and electrographic myotonia may be scarce.

Patients with mutations of the caveolin-3 gene may be misdiagnosed as having a myotonic disorder. These patients may complain of muscle stiffness. Some patients have weakness in proximal muscle groups of the limbs. Some patients have percussion-induced rapid muscle contractions (PIRC), which mimic percussion myotonia. However, the relaxation phase is not as delayed in PIRCs as it is in myotonia. Finally, EMG analysis may show increased insertional activity; these discharges may be misconstrued as myotonic runs but lack the waxing and waning qualities of myotonic discharges.

Periodic Paralysis

In considering the differential diagnosis for familial PP, it is important to understand that severe, sustained hyperkalemia or hypokalemia, from a variety of medical causes, can itself lead to periodic bouts of weakness.

Hyperkalemic paralysis may occur when the serum potassium concentration exceeds 7 mmol per L. The weakness spares the facial, ocular, and bulbar muscles. Rest after exercise may provoke an attack. Patients may complain of paresthesias. The diagnosis is suspected by the exceedingly high potassium levels and hyperkalemia in-between attacks. Causes for hyperkalemia include renal or adrenal insufficiency, potassium-retaining medications, and angiotensin-converting enzyme inhibitors.

Acquired hypokalemic periodic paralysis may occur in individuals with excessive renal or intestinal losses of potassium. The hypokalemia is more marked than that seen in hypoKPP. It is not known whether the ion shifts are similar to those of hypoKPP attacks. Measurement of serum and urinary concentrations of electrolytes along with associated symptoms can differentiate this disorder from hypoKPP. In potassium-depleted animals, the muscle fiber becomes paradoxically depolarized. The physiology of this phenomenon is incompletely understood. Some specific potassium-deficiency syndromes that may cause PP include Bartter syndrome, Liddle syndrome, and distal renal tubular acidosis.

TESTING

An accurate channelopathy diagnosis can often be made based solely on the history and physical examination. Bedside testing can be useful. For example, transient weakness of the finger flexors after squeezing the physician's hand for a few seconds suggests MC. Difficulty opening the eyelids after brief cooling of the eyes with an ice pack suggests PC. Particularly helpful ancillary test results include: 1) the potassium levels during weakness attacks (if there are any), and 2) the results of electrodiagnostic testing (presence or absence of myotonia or of a decrement of the compound muscle action potential after exercise testing). Of course, molecular genetic testing is confirmatory. Often, it's reasonable to skip laboratory tests and proceed directly to genetic testing.

Laboratory Testing

SERUM POTASSIUM As previously mentioned, ictal potassium levels over 5 mEq per L would favor hyperKPP. A level of 2.4 mEq per L is typical for hypoKPP attacks.[7] Potassium concentrations of <2 or >7 mEq per L suggest that the paralysis is secondary to medical illness. Potassium levels are normal in between attacks for all forms of PP.

CREATINE KINASE (CK) CK levels are usually mildly elevated in the familial periodic paralysis and in MC.

THYROID FUNCTION Distinguishing TPP from hypoKPP depends on checking thyroid-stimulating hormone and free thyroxine levels.

ELECTROCARDIOGRAMS For hyperKPP, hypoKPP, and TPP, the ECGs show the expected changes associated with derangements of the serum potassium level. ECG monitoring is recommended when inpatients with hypoKPP are receiving potassium repletion; these patients are at a risk for rebound hyperkalemia. As discussed before, a prolonged QT segment in a patient with paralytic attacks would suggest ATS. Cardiac conduction defects may suggest one of the myotonic dystrophies.

PROVOCATIVE TESTING Provocative testing is generally reserved for patients with negative results from genetic and other ancillary testing. It should be performed with continuous cardiac monitoring and serial serum potassium measurements. Patients with pre-existing cardiac or renal disease should not undergo this testing.

For hyperKPP an attack can be triggered with a 2–10-g potassium load. An attack is usually induced within 1 hour; it typically lasts 30–60 minutes. An alternate option is having the patient exercise on a bicycle ergometer for 30 minutes followed by bed rest. The potassium level should rise during exercise, decline after exercise, and then rise again 20 minutes after exercise.

An attack of hypoKPP may be triggered with an oral glucose load (2 g per kg) and/or insulin administration (10 units subcutaneously). An exercise test could also be performed.

ELECTRODIAGNOSTIC EVALUATION The presence or absence of myotonic discharges on needle EMG is a key piece of information for narrowing the differential diagnosis. Myotonic potentials are seen in association with the chloride and sodium channelopathies, but not with mutations of the calcium or potassium channels. An electrodiagnostic study of 51 patients with a variety of channelopathies revealed five distinct patterns of findings based on the results of needle EMG and short and long exercise test protocols. If a patient matches one of these patterns, the specific genetic diagnosis can be predicted.[40] More recently, the same authors have added a cooling protocol which helps to differentiate various sodium channelopathies.[41]

MUSCLE BIOPSY Muscle biopsies are generally not needed in the work-up of suspected channelopathies. The findings are usually nonspecific and do not change management. Patients with long-standing hyperKPP may develop a vacuolar myopathy. Specimens from hypoKPP patients may have vacuoles or tubular aggregates. Muscle tissue is usually normal in MC, although an absence of type 2B fibers is sometimes noted.

Molecular Diagnosis

Genetic testing for MC involves sequence analysis of *CLCN1*. Distinguishing between the autosomal dominant and recessive MC depends on the family history more than the genetic analysis.

Targeted mutation analysis for the common mutations for hyperKPP and hypoKPP may be followed by full-sequence analysis of the relevant genes if none of the common mutations are found. The yield of the genetic testing has been discussed previously.

MANAGEMENT

Myotonias

Many patients with MC and PC do not need pharmacologic treatment. Patients with MC should avoid rapid movements after periods of rest. However, if stiffness does occur, continued exercised mitigates the symptoms. Patients with PC should avoid strenuous exercise and cold environments. However, if the myotonia limits routine activities, pharmacological intervention is indicated. Use-dependent sodium channel blockers are effective in suppressing myotonia. Mexilitine starting at 150mg twice daily and increasing to 200–300mg three times daily may be used. Procainamide (125–1000mg per day) and phenytoin (300–400mg per day) can also be used. Acetazolamide has been helpful to some individuals.

Depolarizing muscle relaxants, including suxamethonium, should be avoided in MC patients because they may cause prolonged weakness. Nondepolarizing muscle relaxants are safe.

Periodic Paralysis

HyperKPP attacks may be blunted with mild exercise. Ingestion of carbohydrates, inhalation of β-agonists, or intravenous calcium gluconate can also help.

Attacks may be prevented by frequent small meals containing carbohydrates, use of potassium-wasting diuretics, acetazolamide, and avoidance of potassium-rich foods. In terms of surgical planning, depolarizing agents such as suxamethonium should be avoided.

Patients with hypoKPP should limit sodium and carbohydrates and eat foods rich in potassium. Oral intake of potassium salts and acetazolamide in some individuals is also helpful. Patients should avoid overexerting themselves, sweets, alcohol, and glucose infusions. Corticosteroids should be avoided if possible.

Paralytic hypoKPP attacks should be treated with intravenous or oral potassium to normalize the serum potassium and to shorten the episode. Ideally, the ECG and serum potassium levels should be monitored during repletion.

Although studies suggest the benefit of acetazolamide and the more potent carbonic anhydrase inhibitor dichlorphenamide for preventing hyperKPP and hypoKPP attacks, a Cochrane Database Systemic Review did not provide a guideline on management because of insufficient evidence.[42] A randomized study of dichlorphenamide for hyperKPP and hypoKPP is in progress.

The cardiac issues associated with ATS lead to some special management recommendations. Potassium-wasting diuretics, β-agonist inhalers, and other medications known to prolong QT intervals should all be avoided because of increased arrhythmia risk. Cardiac screening involves annual 12-lead ECGs and 24-hour Holter monitoring. Flecainide treatment may be considered for significant, frequent ventricular arrhythmias in the setting of reduced ventricular function.[43–45] Patients with cardiogenic syncope should receive implanted cardioverter–defibrillators. The management of the paralytic attacks is similar to the other periodic paralysis; it should be tailored to the serum potassium level. Carbonic anhydrase inhibitors may help to prevent attacks.

The paralytic attacks of TPP may be treated with potassium. Oral (3 mg per kg) or intravenous propanolol may also help by decreasing the insulin response to glucose. Acetazolamide is ineffective or harmful. The definitive treatment for this disorder involves restoring the euthyroid state, after which the paralytic attacks stop.

ACKNOWLEDGEMENT

This chapter contains contributions by Devon Ryan, PhD, who wrote the text that appeared in the previous edition of this book.

References

1. Nagamitsu S, Matsuura T, Khajavi M, et al. A "dystrophic" variant of autosomal recessive myotonia congenita caused by novel mutations in the CLCN1 gene. *Neurology*. 2000;55(11):1697–1703.
2. Koch MC, Steinmeyer K, Lorenz C, et al. The skeletal muscle chloride channel in dominant and recessive human myotonia. *Science*. 1992;257(5071):797–800.
3. Papponen H, Toppinen T, Baumann P, et al. Founder mutations and the high prevalence of myotonia congenita in northern Finland. *Neurology*. 1999;53(2):297–302.
4. Sun C, Tranebjaerg L, Torbergsen T, Holmgren G, Van Ghelue M. Spectrum of CLCN1 mutations in patients with myotonia congenita in northern Scandinavia. *Eur J Hum Genet*. 2001;9(12):903–909.
5. Emery AE. Population frequencies of inherited neuromuscular diseases—a world survey. *Neuromuscul Disord*. 1991;1(1):19–29.
6. Ptacek LJ, George Jr AL, Griggs RC, et al. Identification of a mutation in the gene causing hyperkalemic periodic paralysis. *Cell*. 1991;67(5):1021–1027.
7. Miller TM, Dias da Silva MR, Miller HA, et al. Correlating phenotype and genotype in the periodic paralyses. *Neurology*. 2004;63(9):1647–1655.
8. Ptacek LJ, Tawil R, Griggs RC, et al. Sodium channel mutations in acetazolamide-responsive myotonia congenita, paramyotonia congenita, and hyperkalemic periodic paralysis. *Neurology*. 1994;44(8):1500–1503.
9. Renner DR, Ptacek LJ. Periodic paralyses and nondystrophic myotonias. *Adv Neurol*. 2002;88:235–252.
10. Fontaine B. Periodic paralysis. *Adv Genet*. 2008;63:3–23.
11. Sternberg D, Maisonobe T, Jurkat-Rott K, et al. Hypokalaemic periodic paralysis type 2 caused by mutations at codon 672 in the muscle sodium channel gene SCN4A. *Brain*. 2001;124(Pt 6):1091–1099.
12. Tawil R, Ptacek LJ, Pavlakis SG, et al. Andersen's syndrome: potassium-sensitive periodic paralysis, ventricular ectopy, and dysmorphic features. *Ann Neurol*. 1994;35(3):326–330.
13. Sansone V, Griggs RC, Meola G, et al. Andersen's syndrome: a distinct periodic paralysis. *Ann Neurol*. 1997;42(3):305–312.
14. Tristani-Firouzi M, Jensen JL, Donaldson MR, et al. Functional and clinical characterization of KCNJ2 mutations associated with LQT7 (Andersen syndrome). *J Clin Invest*. 2002;110(3):381–388.
15. Yoon G, Oberoi S, Tristani-Firouzi M, et al. Andersen-Tawil syndrome: prospective cohort analysis and expansion of the phenotype. *Am J Med Genet A*. 2006;140(4):312–321.
16. Yoon G, Quitania L, Kramer JH, Fu YH, Miller BL, Ptacek LJ. Andersen–Tawil syndrome: definition of a neurocognitive phenotype. *Neurology*. 2006;66(11):1703–1710.
17. Plaster NM, Tawil R, Tristani-Firouzi M, et al. Mutations in Kir2.1 cause the developmental and episodic electrical phenotypes of Andersen's syndrome. *Cell*. 2001;105(4):511–519.
18. Donaldson MR, Jensen JL, Tristani-Firouzi M, et al. PIP2 binding residues of Kir2.1 are common targets of mutations causing Andersen syndrome. *Neurology*. 2003;60(11):1811–1816.
19. Kelley DE, Gharib H, Kennedy FP, Duda Jr. RJ, McManis PG. Thyrotoxic periodic paralysis. Report of 10 cases and review of electromyographic findings. *Arch Intern Med*. 1989;149(11):2597–2600.
20. Dias Da Silva MR, Cerutti JM, Arnaldi LA, Maciel RM. A mutation in the KCNE3 potassium channel gene is associated with susceptibility to thyrotoxic hypokalemic periodic paralysis. *J Clin Endocrinol Metab*. 2002;87(11):4881–4884.
21. Ryan DP, da Silva MR, Soong TW, et al. Mutations in potassium channel Kir2.6 cause susceptibility to thyrotoxic hypokalemic periodic paralysis. *Cell*. 2010;140(1):88–98.
22. Lehmann-Horn F, Rudel R, Jurkat-Rott K. Altered excitability of the cell membrane. In: Engel AG, Franzini-Armstrong C, eds. 3rd ed. Myology. vol. 2. 2004:1257.
23. Duffield M, Rychkov G, Bretag A, Roberts M. Involvement of helices at the dimer interface in ClC-1 common gating. *J Gen Physiol*. 2003;121(2):149–161.
24. Lacomis D, Gonzales JT, Giuliani MJ. Fluctuating clinical myotonia and weakness from Thomsen's disease occurring only during pregnancies. *Clin Neurol Neurosurg*. 1999;101(2):133–136.
25. Wu FF, Ryan A, Devaney J, et al. Novel CLCN1 mutations with unique clinical and electrophysiological consequences. *Brain*. 2002;125(Pt 11): 2392–2407.
26. Cannon SC, Brown Jr. RH, Corey DP. A sodium channel defect in hyperkalemic periodic paralysis: potassium-induced failure of inactivation. *Neuron*. 1991;6(4):619–626.
27. Lehmann-Horn F, Kuther G, Ricker K, Grafe P, Ballanyi K, Rudel R. Adynamia episodica hereditaria with myotonia: a non-inactivating sodium current and the effect of extracellular pH. *Muscle Nerve*. 1987;10(4):363–374.

28. de Silva SM, Kuncl RW, Griffin JW, Cornblath DR, Chavoustie S. Paramyotonia congenita or hyperkalemic periodic paralysis? Clinical and electrophysiological features of each entity in one family. *Muscle Nerve*. 1990;13(1):21–26.
29. Geukes Foppen RJ, van Mil HG, van Heukelom JS. Effects of chloride transport on bistable behaviour of the membrane potential in mouse skeletal muscle. *J Physiol*. 2002;542(Pt 1):181–191.
30. Jurkat-Rott K, Weber MA, Fauler M, et al. K+-dependent paradoxical membrane depolarization and Na+overload, major and reversible contributors to weakness by ion channel leaks. *Proc Natl Acad Sci U S A*. 2009;106(10):4036–4041.
31. Sokolov S, Scheuer T, Catterall WA. Gating pore current in an inherited ion channelopathy. *Nature*. 2007;446(7131):76–78.
32. Lu Z. Mechanism of rectification in inward-rectifier K+channels. *Annu Rev Physiol*. 2004;66:103–129.
33. Soom M, Schonherr R, Kubo Y, Kirsch C, Klinger R, Heinemann SH. Multiple PIP_2 binding sites in Kir2.1 inwardly rectifying potassium channels. *FEBS Lett*. 2001;490(1–2):49–53.
34. Lopes CM, Zhang H, Rohacs T, Jin T, Yang J, Logothetis DE. Alterations in conserved kir channel-PIP_2 interactions underlie channelopathies. *Neuron*. 2002;34(6):933–944.
35. Du X, Zhang H, Lopes C, Mirshahi T, Rohacs T, Logothetis DE. Characteristic interactions with phosphatidylinositol 4,5-bisphosphate determine regulation of kir channels by diverse modulators. *J Biol Chem*. 2004;279(36):37271–37281.
36. Zaritsky JJ, Redell JB, Tempel BL, Schwarz TL. The consequences of disrupting cardiac inwardly rectifying K(+) current (I(K1)) as revealed by the targeted deletion of the murine Kir2.1 and Kir2.2 genes. *J Physiol*. 2001;533(Pt 3):697–710.
37. Ma D, Tang XD, Rogers TB, Welling PA. An Andersen-Tawil syndrome mutation in Kir2.1 (V302M) alters the G-loop cytoplasmic K+conduction pathway. *J Biol Chem*. 2007;282(8):5781–5789.
38. Bendahhou S, Donaldson MR, Plaster NM, Tristani-Firouzi M, Fu YH, Ptacek LJ. Defective potassium channel Kir2.1 trafficking underlies Andersen–Tawil syndrome. *J Biol Chem*. 2003;278(51):51779–51785.
39. Dassau L, Conti LR, Radeke CM, Ptacek LJ, Vandenberg CA. Kir2.6 regulates the surface expression of Kir2.x inward rectifier potassium channels. *J Biol Chem*. 2011;286(11):9526–9541.
40. Fournier E, Arzel M, Sternberg D, et al. Electromyography guides toward subgroups of mutations in muscle channelopathies. *Ann Neurol*. 2004;56(5):650–661.
41. Fournier E, Viala K, Gervais H, et al. Cold extends electromyography distinction between ion channel mutations causing myotonia. *Ann Neurol*. 2006;60(3):356–365.
42. Sansone V, Meola G, Links TP, Panzeri M, Rose MR. Treatment for periodic paralysis. *Cochrane Database Syst Rev*. 2008;(1) doi(1):CD005045.
43. Bokenkamp R, Wilde AA, Schalij MJ, Blom NA. Flecainide for recurrent malignant ventricular arrhythmias in two siblings with Andersen–Tawil syndrome. *Heart Rhythm*. 2007;4(4):508–511.
44. Fox DJ, Klein GJ, Hahn A, et al. Reduction of complex ventricular ectopy and improvement in exercise capacity with flecainide therapy in Andersen–Tawil syndrome. *Europace*. 2008;10(8):1006–1008.
45. Pellizzon OA, Kalaizich L, Ptacek LJ, Tristani-Firouzi M, Gonzalez MD. Flecainide suppresses bidirectional ventricular tachycardia and reverses tachycardia-induced cardiomyopathy in Andersen–Tawil syndrome. *J Cardiovasc Electrophysiol*. 2008;19(1):95–97.

CHAPTER

99

Congenital Myasthenic Syndromes

Andrew G. Engel

Mayo Clinic, Rochester, MN, USA

INTRODUCTION

Congenital myasthenic syndromes (CMSs) are a heterogeneous group of disorders in which the safety margin of neuromuscular transmission is compromised by one or more specific mechanisms.[1] The CMSs can be identified by generic criteria that include myasthenic symptoms since birth or early childhood involving ocular and other cranial, axial and limb muscles, history of similarly affected relatives, decremental electromyographic (EMG) response of the compound muscle action potential (CMAP) on low-frequency (2–3 Hz) stimulation, and negative tests for antibodies against the acetylcholine receptor (AChR), the muscle-specific tyrosine kinase (MuSK), and the P/Q type calcium channels. There are exceptions, however. Some CMS are sporadic or present in later life; a decremental EMG response may not be present in all muscles or at all times, or is elicited only at higher rates of stimulation; and the weakness may not involve cranial muscles.[2]

In normal neuromuscular transmission, activation of AChRs trigger an endplate potential (EPP) that activates the $Na_v1.4$ voltage-dependent sodium channels to give rise to a propagated action potential. A high concentration of AChRs on the terminal expansions of the synaptic folds and of $Na_v1.4$ in the depth of the folds ensures that excitation is propagated beyond the endplate (EP). Thus, the safety margin of neuromuscular transmission is a function of the difference between the depolarization caused by the EPP and the depolarization required to activate $Na_v1.4$.[3]

The safety margin of neuromuscular transmission can be resolved into the following major categories: factors that affect the number of acetylcholine (ACh) molecules per synaptic vesicle; factors that affect quantal release mechanisms (i.e., the quantal content of EPP); and factors that affect the efficacy of individual quanta. Quantal efficacy is affected by endplate geometry, the density of acetylcholinesterase (AChE) in the synaptic space, the density and distribution of AChRs on the postsynaptic membrane, and the kinetic properties of single AChR and $Na_v1.4$ channels. Electrophysiologic and morphologic tests can probe the involvement of different mechanisms in CMS. These provide deeper understanding of newly identified disease genes or mutations, or can provide further clues for mutation analysis. Table 99.1 lists the currently identified CMS disease genes and their frequency in the Mayo CMS cohort.

A genetic diagnosis of a specific CMS is greatly facilitated when clinical, physiologic, or morphologic studies point to a candidate gene (Table 99.2). If these are lacking, mutation analysis can be based on frequencies of the heretofore identified mutations in different EP proteins, as shown in Table 99.1. If a sufficient number of informative relatives is available, linkage analysis can be performed. If successful, it will point to a candidate chromosomal locus. This approach seldom works for CMS because large informative CMS kinships are seldom available; however, it has been successful in inbred populations with multiple consanguineous families.[4]

Testing for CMS mutations in previously identified CMS genes is now commercially available. Because it is expensive, it is best used in a targeted manner based on specific clinical clues. In recent years, whole-exome sequencing has been used to identify CMS mutations. This approach presently captures only ~97% of the entire exome but reads only 75% of the exome with more than 20 times coverage. The enormous amount of generated data needs to be filtered against previously identified variants deemed nonpathogenic and scrutinized for mutations in genes encoding CMS related genes. The putative mutations must be confirmed by Sanger sequencing and the nontruncating mutations examined by expression studies. The cost of exome sequencing with the required bioinformatics analysis is still high. Other pitfalls in this approach are that 1) disease-causing variants in noncoding regions and some large deletions

http://dx.doi.org/10.1016/10.1016/B978-0-12-410529-4.00099-1

TABLE 99.1 Classification of the CMS Based on Site of Defect*

Defect	Index cases observed at Mayo Clinic	%
PRESYNAPTIC CMS (7%)		
Choline acetyltransferase deficiency†	18	5
Paucity of synaptic vesicles and reduced quantal release	1	0.3
Lambert–Eaton syndrome-like	1	0.3
Other presynaptic defects	2	0.6
SYNAPTIC BASAL LAMINA-ASSOCIATED DEFECTS (14%)		
Endplate AChE deficiency†	45	13
β2-laminin deficiency†	1	0.3
POSTSYNAPTIC DEFECTS (73%)		
Primary kinetic defect in AChR with/without AChR deficiency†	62	18
Primary AChR deficiency with/without minor kinetic abnormality†	118	33
OTHER MYASTHENIC SYNDROMES		
Kinetic defect in $Na_v1.4$†	1	0.3
Plectin deficiency†	2	0.6
Myasthenic syndrome associated with centronuclear myopathy	1	0.3
CMS CAUSED BY DEFECTS IN ENDPLATE DEVELOPMENT AND MAINTENANCE		
Agrin deficiency†	1	0.3
MuSK deficiency†	1	0.3
Dok-7 myasthenia†	36	10
Rapsyn deficiency†	51	14
CMS CAUSED BY A CONGENITAL DEFECT OF GLYCOSYLATION		
GFPT1 myasthenia†	11	3
DPAGT1 myasthenia†	2	0.6
ALG2 myasthenia†	0	0
ALG14 myasthenia†	0	0
OTHER MYASTHENIC SYNDROMES		
Kinetic defect in Nav1.4†	1	0.3
Plectin deficiency†	2	0.6
Myasthenic syndrome associated with centronuclear myopathy	1	0.3
Total (100%)	354	100

*Classification based on cohort of CMS patients investigated at the Mayo Clinic between 1988 and 2012.[72] The ALG2 and ALG14 myasthenias were reported elsewhere.[1]

†Gene defects identified.

Abbreviations: AChE, acetylcholinesterase; AChR, acetylcholine receptor; ChAT, choline acetyltransferase; MuSK, muscle specific protein kinase; Dok-7, docking protein 7; GFPT1, glutamine-fructose-6-phosphate transaminase; DPAGT1, dolichyl-phosphate (UDP-N-acetylglucosamine) N-acetyl-glucosaminephosphotransferase; ALG2, alpha-1,3/1,6-mannosyltransferase; ALG14, UDP-N-acetylglucosaminyltransferase subunit.

or duplications can be missed, 2) pathogenic variants causing rare diseases may have previously been entered in dbSNP, 3) synonymous variants in candidate genes that might affect a splice enhancer/repressor or can cause exon skipping are often filtered out, and 4) novel disease genes might be overlooked. Exome sequencing is most efficient when large and/or multiple kinships are available for analysis. Large deletion or duplication mutations can be missed by either Sanger sequencing or whole-exome sequencing. Although rare, they can be identified by array based comparative genomic hybridization.[5]

TABLE 99.2 Phenotypic Clues to CMS Disease Proteins or Syndromes

Dominant inheritance: SCCMS
Refractory or worsened by cholinesterase inhibitors: ColQ, Dok-7, SCCMS
Congenital contractures: Rapsyn, AChR δ- or γ-subunit, ChAT
Repetitive CMAP: SCCMS, ColQ
Defect induced by subtetanic stimulation (10 Hz for 5 min) followed by slow recovery: ChAT
Sudden apneic episodes provoked by fever or stress: ChAT, rapsyn, Na^{2+} channel myasthenia
Predominantly limb-girdle distribution: Dok-7, GFPT1, DPAGT1, ALG2, ALG14; occasionally rapsyn and ColQ
Nephrotic syndrome and ocular abnormalities: β2-laminin

Abbreviations: ChAT, choline acetyltransferase; CMAP, compound muscle action potential; ColQ, triple-stranded collagenic tail; SCCMS, Slow-channel congenital myasthenic syndromes.

If a novel CMS disease gene is discovered, then expression studies with the genetically engineered mutant protein can be used to confirm pathogenicity and to analyze the properties of the abnormal protein.[6]

This presentation will discuss only those CMSs with an established gene defect.

PRESYNAPTIC CMS

Defects in Choline Acetyltransferase (ChAT)

ChAT catalyzes the synthesis of ACh by transfer of an acetyl group from acetyl-CoA to choline in cholinergic neurons. Pathogenic mutations, alone or in combination, alter expression, catalytic efficiency, or structural stability of the enzyme.[7,8] The decreased rate of ACh resynthesis progressively depletes the ACh content of the synaptic vesicles, and hence the amplitude of the miniature EP potential (MEPP), when neuronal impulse flow is increased (Figure 99.1).

Some patients present with hypotonia, bulbar paralysis, and apnea at birth. Other patients are normal at birth and develop apneic attacks during infancy or childhood precipitated by infections, excitement, or no apparent cause.[8–12] However, apneic episodes also can occur in Na-channel myasthenia,[13] in some patients harboring mutations in rapsyn[14,15] or in the AChR δ-subunit.[16] In some children, an acute attack is followed by respiratory insufficiency that lasts

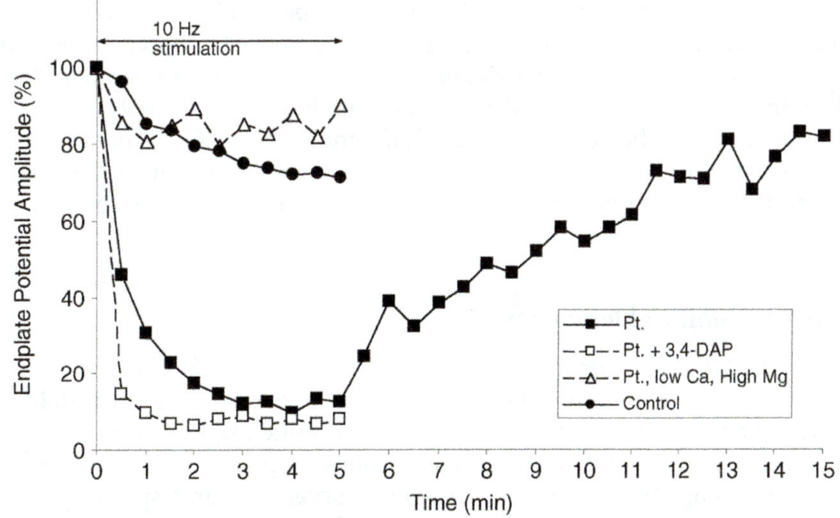

FIGURE 99.1 **CMS caused by EP choline acetyltransferase (ChAT) deficiency.** 10-Hz stimulation for 5 minutes causes rapid decline of the endplate potential (EPP), which then recovers slowly over more than 10 minutes. 3,4-Diaminopyridine (3,4-DAP), which accelerates acetylcholine (ACh) release, accelerates decline of the EPP, whereas a low Ca^{2+}–high Mg^{2+} solution, which reduces ACh release, prevents it. (From Engel AG, Ohno K, Sine SM. Congenital myasthenic syndromes: progress over the past decade. *Muscle Nerve.* 2003 Jan;27(1):4-25.)

for weeks.[17] Few patients are apneic, respirator dependent, and paralyzed since birth,[8] and some develop cerebral atrophy after episodes of hypoxemia.[8,12] Others improve with age, but still have variable ptosis, ophthalmoparesis, fatigable weakness, and recurrent cyanotic episodes, or complain only of mild to moderately severe fatigable weakness. The symptoms are worsened by exposure to cold, likely because this further reduces the catalytic efficiency of the mutant enzyme.[10] Between apneic attacks, a decremental EMG response on 2-Hz stimulation is usually absent in rested muscles, but appears after a conditioning train of 10-Hz stimuli for 5 minutes.

Mutations with severe kinetic effects are located in the active-site tunnel of the enzyme, or adjacent to its substrate-binding site, or exert their effect allosterically. Some mutations have no kinetic effects and express well but impair the thermal stability of ChAT by introducing a proline residue into an α-helix. In three very severely affected patients with life-long apnea, permanent paralysis, and failure to respond to pyridostigmine, one of the two mutations was near the active-site of ChAT.[8]

Although some patients fail to respond to AChE inhibitors, these agents should be tried in the initial management of all patients. Therapy should be continued prophylactically even when patients are asymptomatic. The parents should be provided with a portable ventilatory device, instructed in the intramuscular injection of neostigmine methylsulfate, and advised to install an apnea monitor in the home.

DEFECTS IN BASAL LAMINA PROTEINS

Defects in EP Species of AChE Due to Mutations in ColQ Protein

With few exceptions, EP AChE deficiency causes a severe CMS. The clinical clues are a repetitive CMAP evoked by single nerve stimulus and absence of AChE from the EPs. Electron microscopy shows small nerve terminals that are often isolated from the synaptic cleft by an intruding Schwann cell process, and degeneration of the junctional folds with loss of AChR. The small size and encasement by Schwann cells of the nerve terminals restricts evoked release of ACh. The absence of AChE prolongs the lifetime of ACh in the synaptic space; this increases the duration of the synaptic potentials and currents so that they outlast the refractory period of the muscle fiber and excite a second, or repetitive, CMAP. The prolonged EP currents also overload the postsynaptic region with cations and instigate an EP myopathy causing loss of AChR. The safety margin of neuromuscular transmission is compromised by decreased quantal release, loss of AChR, and desensitization of AChR at physiologic rates of stimulation.[18]

The molecular form of AChE at the endplate is an asymmetric enzyme composed of 1, 2, or 3 homotetramers of globular catalytic subunits ($AChE_T$) attached to a triple-stranded collagenic tail (ColQ) forming A_4, A_8, and A_{12} species of AChE (Figure 99.2A and B).[19] After the sequence the human *COLQ* gene was determined,[20] the molecular basis of endplate AChE deficiency was traced to recessive mutations in *COLQ*.[20,21] Figure 99.2B shows some of the mutations identified to date. When wild-type $ACHE_T$ (Figure 99.2C) or wild-type $ACHE_T$ and *COLQ* (Figure 99.2D) are transfected into COS cells, the density gradient sedimentation profile of the cell extract reveals the abundance of the globular and asymmetric species of AChE. The consequences of some mutations are illustrated in density gradient profiles of extracts of COS cells: transfection only with wild-type $AChE_T$ is shown in Figure 99.2C; with wild-type *ColQ* and $AChE_T$ in Figure 99.2D, and with the indicated wild-type $ACHE_T$ and mutant *COLQ* constructs in Figure 99.2F–H.

Therapy is still unsatisfactory, but ephedrine has had a significant beneficial effect in some patients.[22–24] More recently, albuterol was found to be as or more effective than ephedrine[25] but long-term therapy is required to obtain a reasonable benefit.

CMS Associated with β2-Laminin Deficiency

β2-Laminin, encoded by *LAMB2*, is a component of the basal lamina of different tissues and is highly expressed in kidney, eye, and the neuromuscular junction. Synaptic β2-laminin governs the appropriate alignment of the axon terminal with the postsynaptic region and, hence, pre- and postsynaptic trophic interactions. Defects in β2-laminin result in Pierson syndrome with renal and ocular malformations. A patient carrying heteroallelic missense and frameshift mutations in *LAMB2* had Pierson syndrome as well as severe ocular, respiratory, and proximal limb muscle weakness.[26] *In vitro* microelectrode studies revealed decreased quantal release by nerve impulse and a reduced MEPP amplitude. Electron microscopy showed the nerve terminals were abnormally small and often encased by Schwann cells, accounting for the decreased quantal release. The synaptic space was widened and the junctional folds were simplified, accounting for the decreased MEPP amplitude.

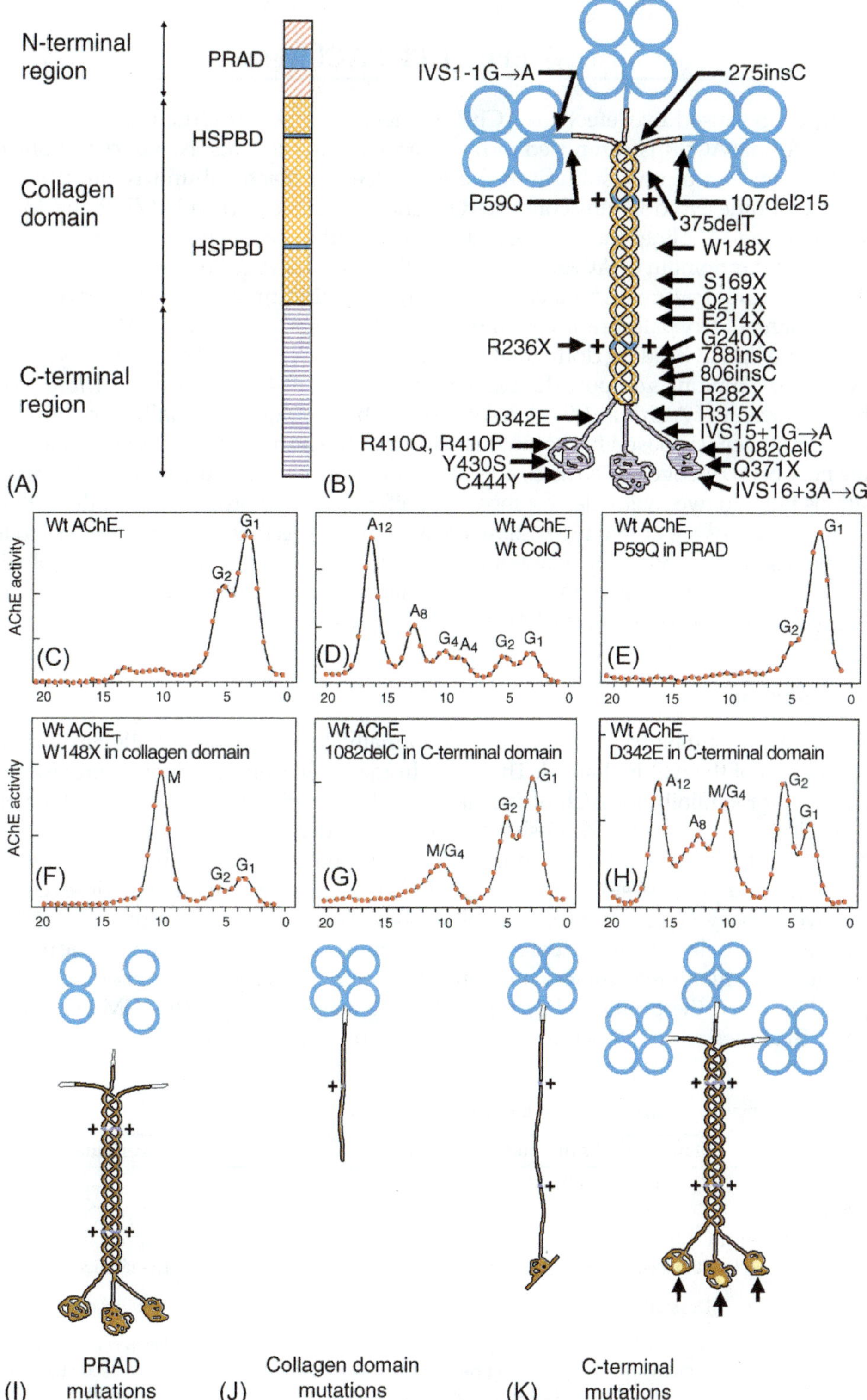

FIGURE 99.2 **CMS caused by defects in ColQ protein.** (**A**) Schematic diagram showing domains of a ColQ strand. (**B**) Schematic diagram showing the A12 species of asymmetric AChE with 24 identified ColQ mutations. **C–H:** Density gradient profiles of acetylcholinesterase (AChE) extracted from COS cells transfected with wild-type *ACHE* and different types of *COLQ* mutants. **I–K:** Schematic diagrams of the abnormal species of AChE formed in transfected HEK cells. In (**E**) and (**I**) note that disruption of the PRAD domain produces a sedimentation profile identical with that obtained after transfection with *ACHE*$_T$ alone. Thus, *ACHE* fails to attach to ColQ and no asymmetric AChE is formed. In (**F**) and (**J**), note that the asymmetric A_4, A_8, or A_{12} moieties are absent and there is a prominent -10.5S mutant (M) peak, representing a G_4 tetramer of the catalytic subunit linked to the truncated ColQ peptide. In (**G**) and the left diagram in (**K**), note presence of a small M peak but absence of peaks corresponding to triple-stranded asymmetric enzymes. In (**H**) and the right diagram in (**K**), note that both an M peak and asymmetric AChE are present. PRAD, proline-rich attachment domain; HSPBD, heparan sulfate proteoglycan binding domain. (Reproduced with permission from Engel AG, et al. Congenital myasthenic syndromes. In: Engel AG, Franzini-Armstrong C, eds. Myology, 3rd ed. New York: McGraw-Hill, 2004.)

DEFECTS IN AChR

Before discussing CMS caused by defects in AChR we need to consider structural features of the receptor. The muscle form of the nicotinic AChR is composed of five homologous subunits: two of α and one of β, δ, and ε in the adult receptor, and a γ instead of an ε subunit in the fetal receptor. Each subunit is composed of an extracellular domain, four transmembrane (M) domains connected by short linkers between M1/M2 and M2/M3 and a long cytoplasmic M3/M4 linker, and M2 lines the ion channel. The subunits are arranged like staves in a barrel in the order of α, β, δ, α, ε. The extracellular domain of AChR has two binding sites for ACh located at interfaces of the α/ε and α/δ subunits as well as structural features which ensure that agonist binding result in fast and efficient opening of the transmembrane ion channel. Crystallography studies of the ACh binding protein (ACHBP) of invertebrate species provided a model for the ligand-binding domain of AChR at 2.1–2.7-Å resolution,[27,28] and cryoelectron microscopy of *Torpedo* AChR revealed the binding and pore domains of the receptor at 4-Å resolution.[29] More recent crystallography studies image the α1-subunit of AChR at 1.94-Å resolution.[30] These reports rationalized previous structure–function studies and enabled mutagenesis-based homology modeling of the human AChR ligand-binding domain.[31]

Mutations have now been discovered in the α, β, δ, and ε subunits of adult AChR, as well as in the γ subunit of fetal AChR. The mutations fall into two major classes: those that alter its kinetic properties and those that reduce expression of AChR. The kinetic mutations are of two types: slow-channel mutations that increase, and fast-channel mutations that decrease the synaptic response to ACh. The slow- and fast-channel mutations represent physiologic opposites in both their phenotypic consequences and alterations of fundamental steps underlying receptor activation (Table 99.3). Some fast-channel mutations also reduce expression of AChR, which contributes to, or even dominates, the clinical phenotype.

Slow-Channel Syndromes

The name slow-channel syndrome originates from the abnormally slow decay of synaptic currents caused by abnormally prolonged opening events of the AChR channel. These syndromes arise from dominant gain-of-function mutations that either enhance the receptor's affinity for ACh or increase its gating efficiency by increasing channel-opening rate or by slowing the channel-closing rate.[32] Figure 99.3A shows some of the identified slow-channel mutations. The clinical phenotypes vary. Some slow-channel patients present in early life and have severe disability by the end of the first decade; others present later in life and progress slowly, resulting in little disability even in the sixth or seventh decade of life. Most patients show selectively severe involvement of cervical and of wrist and finger extensor muscles. Except for severely affected patients, the cranial muscles tend to be spared except for mild and sometimes asymmetric ptosis. Progressive spinal deformities and respiratory embarrassment are common complications during evolution of the illness.

As in EP AChE deficiency, the prolonged synaptic currents trigger a repetitive CMAP. The prolonged synaptic currents (Figure 99.3B) as well as spontaneous openings of the unliganded closed channel overload the postsynaptic

TABLE 99.3 Kinetic Abnormalities of the Acetylcholine Receptor (AChR)

	Slow-channel syndromes	Fast-channel syndromes
Endplate currents	Slow decay	Fast decay
Channel opening events	Prolonged	Brief
Open states	Stabilized	Destabilized
Closed states	Destabilized	Stabilized
Mechanisms*	Increased affinity for ACh Increased channel opening rate Decreased channel closing rate	Decreased affinity for ACh Decreased channel opening rate Increased channel closing rate Mode-switching kinetics
Pathology	Endplate myopathy from cationic overloading	No anatomic footprint
Genetic background	Dominant gain-of-function mutations	Recessive loss-of-function mutations
Response to therapy	Long-lived open channel blockade of AChR with quinidine or fluoxetine	3,4-DAP and AChE inhibitors

**Different combinations of mechanisms operate in the individual slow- and fast-channel syndromes.*

Abbreviation: 3,4-DAP, 3,4-diaminopyridine.

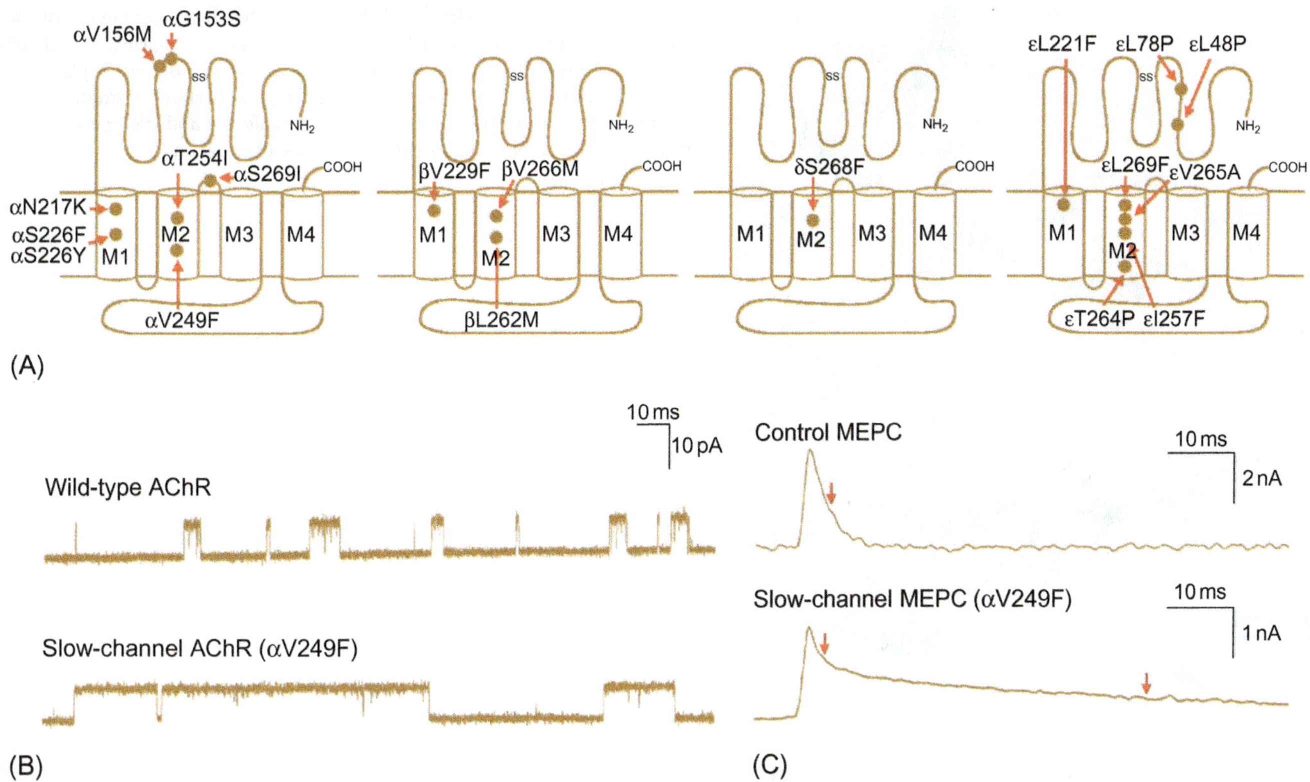

FIGURE 99.3 **Slow-channel currents and mutations.** (**A**) Schematic diagram of acetylcholine receptor (AChR) subunits positions of slow-channel mutations. (**B**) Single-channel currents from wild-type and slow-channel (αV249F) AChRs expressed in HEK cells. (**C**) Miniature endplate currents (MEPC) recorded from endplates (EPs) of a control subject and a patient harboring the αV249F slow-channel mutation. The slow-channel MEPC decays biexponentially due to expression of both wild-type and mutant AChRs at the EP, so one decay time constant is normal and the other is markedly prolonged. (Reproduced with permission from Engel AG, et al. Congenital myasthenic syndromes. In: Engel AG, Franzini-Armstrong C, eds. Myology, 3rd ed. New York: McGraw-Hill, 2004:1801–1844.)

region with cations, and especially with Ca^{2+}, and cause an EP myopathy manifested by degeneration of the junctional folds, loss of AChR from the folds, widening of synaptic space that alters the EP geometry (Figure 99.4A), vacuolar change in muscle fiber regions underlying the EP, and apoptosis of nearby nuclei (Figure 99.4B).

Slow-channel mutations in the ε-subunit are more deleterious than in other subunits because the ε-subunit enhances the Ca^{2+} permeability of the receptor more than slow-channel mutations in other subunits.[33,34] The safety margin of neuromuscular transmission is compromised by: 1) destruction of junctional folds with loss of AChR, 2) altered EP geometry, 3) staircase summation of markedly prolonged EP potentials, causing a depolarization block at physiologic rates of stimulation, and 4) an increased propensity of the mutant receptor to become desensitized.

The slow-channel syndromes are refractory to, or are worsened by, cholinergic agonists but are improved by long-lived open-channel blockers of the AChR, such as quinine[35] or quinidine, or fluoxetine.[36–38] The long-term side effects of relatively high doses of fluoxetine may prevent its use in a minority of patients[38] but in the author's experience these patients can be managed by using lower doses of fluoxetine in combination with small doses of quinine or quinidine.

Fast-Channel Syndromes

The fast-channel syndromes are physiologic opposites of the slow-channel syndromes (Table 99.3). They are recessively inherited disorders caused either by: decreased affinity for ACh; reduced gating efficiency; destabilization of the channel kinetics; or by a combination of these mechanisms; and they leave no anatomic footprints. The clinical consequences vary from mild to severe.

The name "fast-channel syndrome" originates from abnormally fast decay of the synaptic current (Figure 99.5A) caused by abnormally brief channel opening events (Figure 99.5B). Some fast-channel mutations documented to date are shown in Figure 99.5C. Fast-channel mutations cause loss of function and are typically recessive, but an αF256L mutation in the M2 domain of the AChR α-subunit has a dominant negative effect.[39] In most cases, the fast-channel

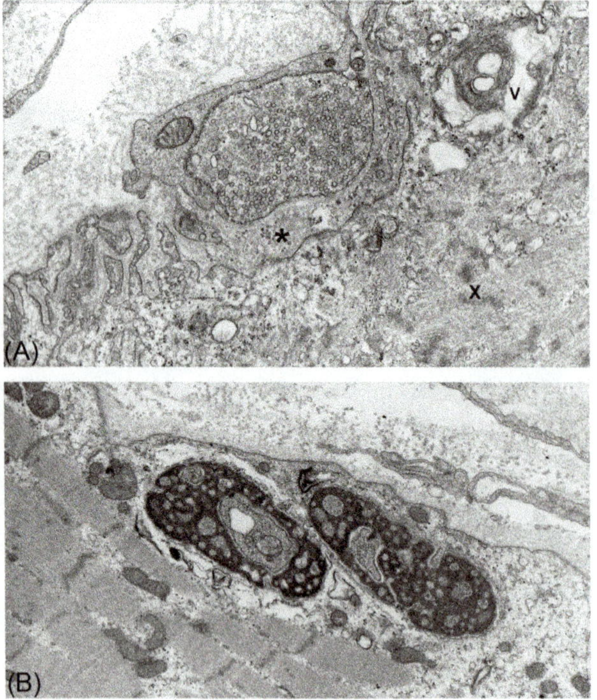

FIGURE 99.4 **Ultrastructural findings in the slow-channel syndrome.** (**A**) Postsynaptic endplate (EP) region of patient harboring the εL269F mutation. Note destruction of the junctional folds and their replacement by globular debris (*). The postsynaptic region harbors a small vacuole containing a degenerate membranous organelle (V) and disorganized myofibrils (X). (**B**) Apoptotic nuclei near a slow-channel endplate. (**A**) and (**B**), ×26,000.

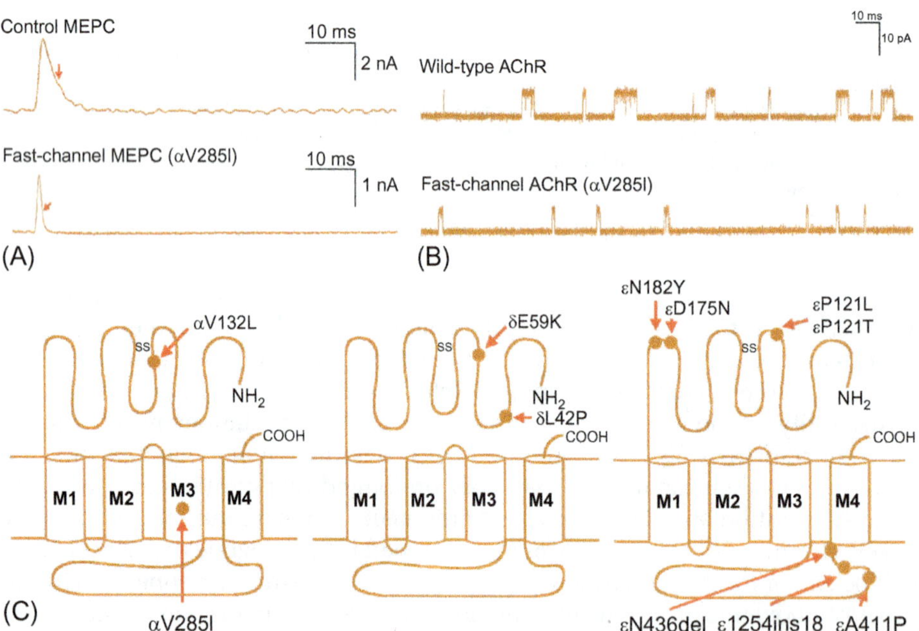

FIGURE 99.5 **Fast-channel currents and mutations.** (**A**) Miniature endplate (EP) currents (MEPC) recorded from EPs of a control subject and from a patient harboring the αV285I fast-channel mutation. Arrows indicate decay time constants. (**B**) Single-channel currents from wild-type and fast-channel (αV285I) acetylcholine receptors (AChRs) expressed in HEK cells. (**C**) Schematic diagram of fast-channel mutations in the AChR α-, β-, and δ-subunits. (Reproduced with permission from Engel AG, et al. Congenital myasthenic syndromes. In: Engel AG, Franzini-Armstrong C, eds. Myology, 3rd ed. New York: McGraw-Hill, 2004:1801–1844.)

mutation in one allele is accompanied by a null mutation in the second allele so that the fast-channel mutation dominates the clinical phenotype. The mutation can reduce gating efficiency, or decrease affinity for ACh, or impair the fidelity of gating.

Fast-channel mutations in the extracellular domain of AChR at or near the agonist binding sites impede channel activation by decreasing agonist affinity. A highly lethal fast-channel CMS located the at the interface between the ε- and α-subunit binding sites exerts its effect by reducing agonist affinity by 30-fold (Figure 99.6).[40] Other fast-channel

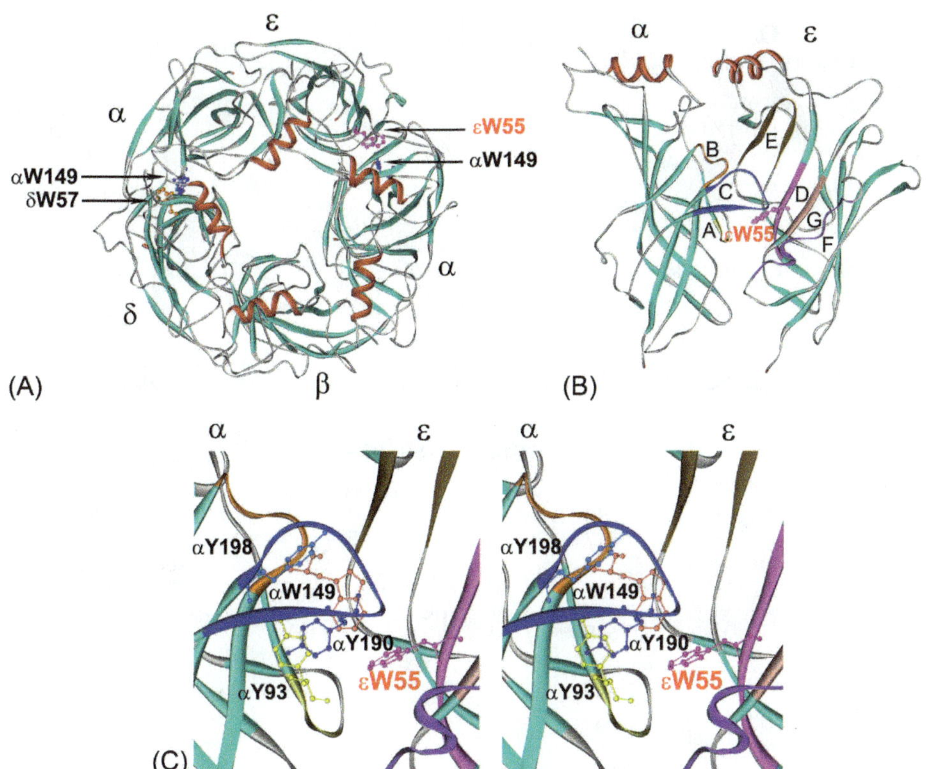

FIGURE 99.6 **The extracellular domain of acetylcholine receptor (AChR) and the εW55R fast-channel mutation at the α/ε ACh binding site.** (**A**) Structural model of extracellular domains of human AChR viewed from the synaptic space indicating positions of tryptophan residues at the α/δ and α/ε binding sites. (**B**) Side-view of the α- and ε-subunits showing loops E, D, G, and F in the ε-subunit, and loops A, B, and C in the α-subunit that define the two ACh binding sites. (**C**) Stereo view of the α/ε binding site showing positions of the anionic aromatic residues shrouding the binding pocket. In each panel the mutated εTrp55 is highlighted in red. (Based on the crystal structure of the ACh-binding protein [PDB 1I9B] and lysine-scanning mutagenesis delineating the structure of the human AChR-binding domain.)[31] (Reproduced with permission from [40].)

mutations in the extracellular domain reduce gating efficiency by exerting their effect in the fleeting transitional state during which the liganded receptor isomerizes from the closed to the open state.[41–44] Finally, mutations near the C-terminal of the long cytoplasmic loop between M3 and M4 of the ε-subunit corrupt the fidelity of gating so that the channels open and close with irregular kinetics and with a reduced overall probability of opening[45,46] (Figure 99.5C).

Mutations Causing EP AChR Deficiency

CMS with severe EP AChR deficiency result from homozygous or more often heterozygous recessive mutations in AChR subunit genes. These mutations are concentrated in the ε-subunit are the commonest cause of the CMS and some of these are shown in Figure 99.7. The likely reason for this is that expression of the fetal type γ-subunit, although at a low level, can partially compensate for absence of the ε-subunit from the adult receptor,[46–48] whereas patients harboring null mutations in non-ε subunit genes might not survive for lack of a substituting subunit.

Different types of recessive low-expressor mutations have been identified. Some cause premature termination of the translational chain; these mutations are frameshifting,[44,47–53] occur at a splice site,[49,51] or produce a stop codon directly.[48] An important mutation in this group is the 1369delG in the ε-subunit that results in loss of a C-terminal cysteine, C470, crucial to both maturation and surface expression of the adult receptor.[54] Thus, any mutation that truncates the ε-subunit upstream of C470 is predicted to inhibit expression of the ε-subunit. Homozygous low-expressor ε-subunit mutations are endemic in Mediterranean or other Near Eastern countries,[51,55] and the frameshifting ε1267delG mutation occurring at homozygosity is endemic in Gypsy families,[49,50,52] where it derives from a common founder.[50]

A second type of recessive mutation is point mutation in the Ets binding site, or N-box, of the promoter region of the ε-subunit gene: ε-154G>A,[56] ε-155G>A,[57] and ε-156C>T.[58] The N-box represents the end point of a signaling cascade driven by neuregulin through ERBB receptors. ERBB receptors phosphorylate mitogen-activated protein (MAP) kinases, which in turn phosphorylate GABPα and GABPβ (members of the Ets family of transcription factors) that bind to the N-box.[59–61] That these mutations impair AChR expression is direct evidence that the neuregulin

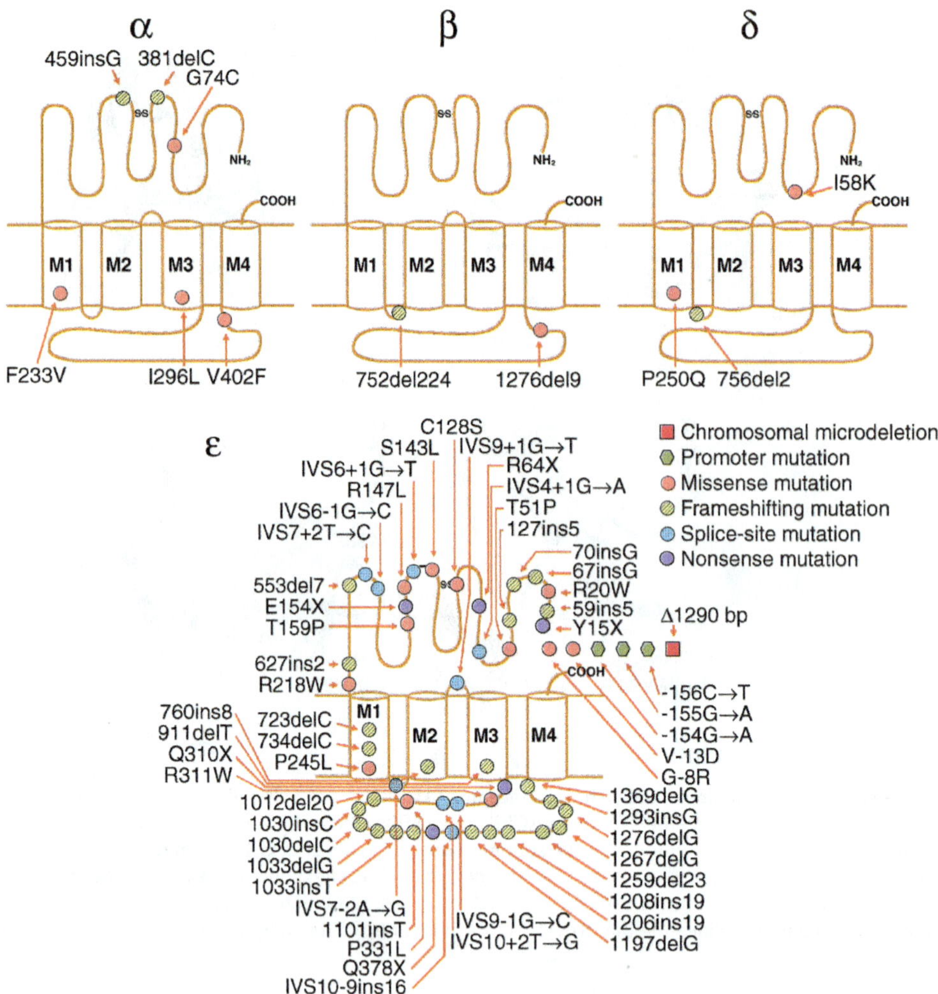

FIGURE 99.7 **Schematic diagram of low-expressor and null mutations reported in the α-, β-, δ-, and ε-subunits of the acetylcholine receptor (AChR).** Note that most mutations appear in the ε-subunit and are highly concentrated in the long cytoplasmic loop between the third and fourth transmembrane domains (M). Splice-site mutations point to *N*-terminal codons of the predicted skipped exons. (Reproduced with permission from Engel AG, et al. Congenital myasthenic syndromes. In: Engel AG, Franzini-Armstrong C, eds. Myology, 3rd ed. New York: McGraw-Hill, 2004:1801–1844.)

signaling pathway participates in regulating synapse-specific transcription at the human EP. There are also missense mutations in a signal peptide region (εG-8R[62] and εV-13D[51]), and of residues essential for assembly of the pentameric receptor located at an *N*-glycosylation site (εS143L),[62] in the C128-C142 disulfide (εC128S),[46] and in the long cytoplasmic loop of the β-subunit causing a three-codon deletion.[63]

Prenatal Syndromes Caused by Mutations in AChR Subunits and Other EP-Specific Proteins

The first identified prenatal myasthenic syndrome were caused by recessive, nonsense, frameshift, splice-site, or missense mutations in the fetal AChR γ-subunit. In humans, γ-AChR appears on myotubes around the ninth developmental week and becomes concentrated at nascent nerve–muscle junctions around the sixteenth developmental week. Subsequently, the γ-subunit is replaced by the adult ε-subunit and disappears from the fetal endplates after the thirty-first developmental week.[64] Thus, pathogenic mutations of the γ-subunit result in hypomotility *in utero*, mostly between the sixteenth and thirty-first developmental weeks. The clinical consequences at birth are multiple joint contractures, small muscle bulk, multiple pterygia (webbing of the neck, axilla, elbows, fingers, or popliteal fossa), camptodactyly rocker-bottom feet with prominent heels, and characteristic facies with mild ptosis and a small mouth with downturned corners. Myasthenic symptoms are absent after birth because by then the normal ε-subunit is expressed at the endplates.[65] More recently, lethal fetal akinesia syndromes arising from biallelic null mutations in the AChR α, β, and δ subunits as well as in rapsyn[66,67] and Dok-7[68] were also identified.

DEFECTS IN MECHANISMS GOVERNING EP DEVELOPMENT AND MAINTENANCE

Mutations in four proteins essential for EP development and maintenance, MuSK, agrin, Dok-7, and rapsyn are presently known to cause CMS. Agrin secreted into the synaptic space by the nerve terminal, binds to the lipoprotein-related protein LRP4 in the postsynaptic membrane. The Agrin–LRP4 complex binds to and activates MuSK, a single-pass postsynaptic membrane protein. Association of MuSK with LRP4 in the postsynaptic membrane enhances MuSK phosphorylation, which leads to clustering of LRP4 and MuSK. Activated MuSK in concert with postsynaptic Dok-7 and other postsynaptic proteins acts on rapsyn to concentrate AChR in the postsynaptic membrane, increases synapse-specific gene expression by postsynaptic nuclei, and promotes postsynaptic differentiation. Clustered LRP4, in turn, signals to motor axons to promote their differentiation. The agrin–LRP4–MusK–Dok-7 signaling system is also essential for maintaining the structure of the adult neuromuscular junction (for review see [69]).

Mutations in Agrin

Agrin is a multidomain proteoglycan. The muscle isoform of agrin harbors A and B regions near its *C*-terminal. Two reports describe a CMS caused by defects in agrin. In one report, two siblings with eyelid ptosis but normal ocular ductions had only mild facial and hip-girdle muscle weakness.[70] They carry a homozygous missense mutation in *AGRN* at codon 1709 (G1709R). The mutation is in the laminin G-like 2 domain, upstream of the A and B inserts of agrin required for MuSK activation and EP formation. AChR and agrin expression at the EP were normal. Structural studies showed EPs with misshaped synaptic gutters partially filled by nerve endings and formation of new EP regions. The postsynaptic regions were preserved. Expression studies revealed no effect on MuSK activation by agrin or binding of agrin to α-dystroglycan. Forced expression of a mutant mini-agrin gene in mouse soleus muscle induced changes similar to those at patient EPs. Thus, the observed mutation perturbs the maintenance of the EP without preventing agrin to initiate postsynaptic differentiation. Cholinesterase inhibitors and 3,4-DAP were ineffective.

The second report describes a severe form of CMS caused by a nonsense (Q353X) mutation located at the *N*-terminal and a missense (V1727F) mutation in the second laminin G-like domain of agrin.[71] The synaptic contacts were dispersed and fragmented, the postsynaptic regions were simplified, the nerve terminals were reduced in size, and membranous debris was present in the synaptic space and in the subsynaptic cytoplasm. Expression studies revealed that the nonsense mutation abolished agrin expression and the missense mutation markedly reduced AChR clustering in C2 muscle cells. The index patient failed to respond to a cholinesterase inhibitor and 3,4-DAP, but responded partially to ephedrine.

Mutations in MuSK

This CMS has been reported in three kinships. In one kinship, a brother and sister carried two heteroallelic mutations: a frameshifting null mutation (c.220insC), and a missense mutation (V790M) that reduced MuSK expression and stability and AChR aggregation. The EPs consisted of multiple small regions linked by nerve sprouts. AChR expression per endplate was reduced to approximately 45% of normal.[72] When the missense mutation was overexpressed in mouse muscle by electroporation, it reduced synaptic AChR expression and resulted in aberrant axonal outgrowth, similar to that observed in the patient.[73] The safety margin in this CMS is likely compromised by the AChR deficiency.

A second report describes heterozygous M605I and A727V mutations in a patient with severe myasthenic symptoms since early life that improved after puberty but worsened after menstrual periods. The synaptic response to ACh was reduced to about 30% and quantal release to 50% of normal. Synaptic contacts were small and electron microscopy showed simplified postsynaptic regions with sparse secondary synaptic clefts. The patient failed to respond to pyridostigmine, ephedrine or 3,4-DAP, but responded partially to albuterol.[74]

A third kinship harbored a homozygous P344R missense mutation in the cysteine-rich extracellular MuSK domain.[75] The clinical course was progressive. Low doses of pyridostigmine and 3,4-DAP led to gradual improvement; ephedrine or higher doses of pyridostigmine were not tolerated.

Mutations in Dok-7

After the discovery in 2006 of Dok-7 as a muscle-intrinsic activator of MuSK,[76] numerous CMS-related mutations were identified in *DOK7*,[77–81] and Dok-7 myasthenia is now recognized as a common cause of CMS. The first report-described *DOK7* mutations were in 19 patients.[77] All had limb-girdle weakness, with lesser face, jaw or neck muscle weakness. The EPs were small and simplified EPs on light microscopy. The synaptic response to ACh was half-normal.[82]

A subsequent study of 16 other patients harboring mutations in *DOK7* showed a wider clinical spectrum.[79] All patients experienced short-term fatigability on exertion. Neonatal onset was evident in eight, severe bulbar weakness in five, and significant oculoparesis in three. The clinical course and disability was mild to severe. Type 1 fiber preponderance, type 2 fiber atrophy, isolated necrotic and regenerating fibers, and pleomorphic oxidative enzyme decreases or target formations were variable associated features. All EPs consisted of one to multiple small synaptic contacts. Consistent with the known function of Dok-7 to participate in the maintenance of the structure of neuromuscular junction, electron microscopy studies revealed ongoing destruction and remodeling of the EPs (Figure 99.8). Neuromuscular transmission was compromised by decreased quantal release and reduced response to released ACh. Each patient harbored two heteroallelic mutations: four previously reported[77] and 12 novel. No consistent phenotype–genotype correlations emerged.[83] Importantly, this CMS is worsened by cholinergic agonists but responds to ephedrine[84] and albuterol.[83]

Mutations in Rapsyn

Rapsyn, under the influence of agrin, LRP4, MuSK, and Dok7 concentrates AChR in the postsynaptic membrane and links it to the subsynaptic cytoskeleton through dystroglycan. (Figure 99.9). The clue to discovery of mutations in *RAPSN* came from CMS patients with demonstrated EP AChR deficiency with no mutation in any AChR subunit.[14] Since then, numerous rapsyn mutations have been identified, and some of these are shown in Figure 99.9. Defects in rapsyn are now the second commonest cause of CMS after those caused by mutations in the AChR ε-subunit. With few exceptions,[85] Indo-European patients with mutations in the translated region of *RAPSN* carry the N88K mutation on at least one allele; other mutations in the translated region are dispersed over different rapsyn domains. There is evidence for an ancient Indo-European founder for N88K,[86,87] but not all patients with the N88K mutation carry the same haplotype.[86,88] Mutations also occur in the E-box of the *RAPSN* promoter (Figure 99.9A). One of these mutations, -38A>G, was observed at homozygosity and shown to arise from a common founder in Near-Eastern Jews with marked jaw and other facial malformations.[89] Apart from the mild course of patients carrying -38A>G at homozygosity, no distinct genotype–phenotype correlations have emerged to date.

Most patients present at birth or during infancy. The few patients who present in the second or even third decade of life have a milder clinical course than those with an earlier onset of symptoms. Cranial muscles can be affected but the ocular ductions tend to be spared. AChR expression at the EP is usually mild, ranging from about 20–50% of normal. The EMG decrement on 2-Hz stimulation can be mild or absent, and single-fiber EMG may be required to reveal the defect of neuromuscular transmission. Arthrogryposis is present at birth in about 20–30% of the patients. Febrile illnesses worsen the weakness in most patients and can precipitate episodes of respiratory failure or even apnea.[14,15,90,91]

The morphologic features of the EP and the factors that impair the safety margin of neuromuscular transmission are the same as in primary AChR deficiency but the EP AChR deficiency is relatively mild in most patients. In some patients single-fiber EMG is required to demonstrate a defect in neuromuscular transmission.

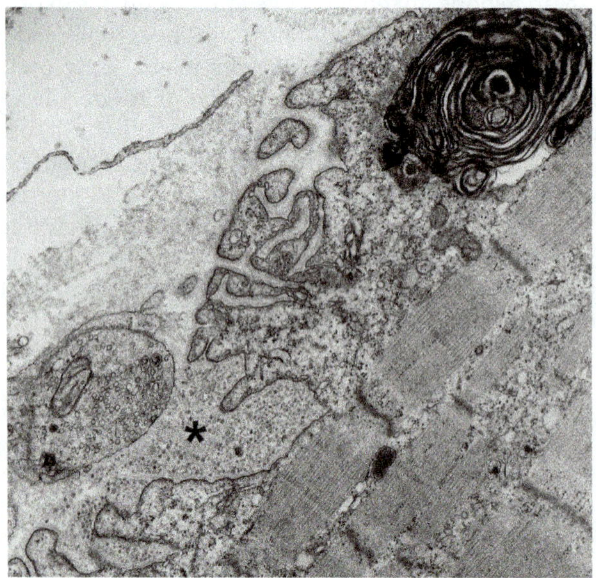

FIGURE 99.8 **Abnormal endplate (EP) in Dok-7 CMS.** Lower left, widened synaptic space harbors globular residues of degenerate junctional fold (*). Upper right, large myeloid structure replaces part of the junctional sarcoplasm. ×20,000.

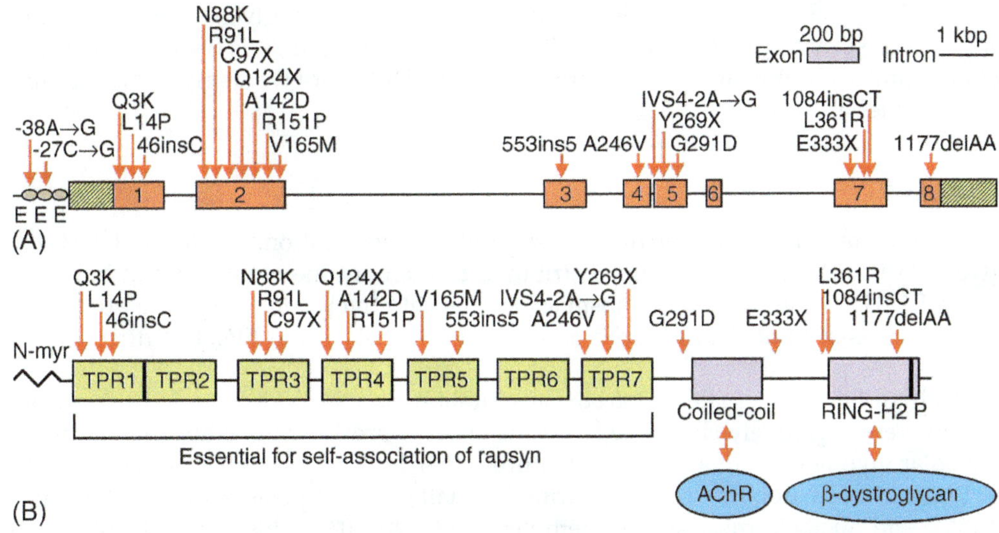

FIGURE 99.9 **Schematic diagram showing structure of the *RAPSN* gene (A) and domains of rapsyn (B) with identified mutations.** Shaded areas in (**A**) indicate untranslated regions. Rapsyn carries a myristoylation signal at the *N*-terminus required for membrane association, has seven tetratricopeptide repeats (TPRs) that subserve self-association, a coiled-coil domain that interacts with AChR, a RING-H2 domain that binds to the cytoplasmic domains of β-dystroglycan and mediates the MuSK-induced phosphorylation of AChR, and a serine phosphorylation site at codon 406. Transcription of rapsyn in muscle is under the control of helix–loop–helix myogenic determination factors that bind to the *cis*-acting E-box sequence in the *RAPSN* promoter. (Reproduced from [90].)

Most patients respond well to AChE inhibitors; some derive additional benefit from the use of 3,4-diaminopyridine (3,4-DAP)[92] and some observed by the author benefited from the added use of ephedrine or albuterol.

DEFECTS OF GLYCOSYLATION

Glycosylation increases solubility, folding, stability, assembly, and intracellular transport of nascent peptides. *O*-glycosylation occurs in the Golgi apparatus with addition of sugar residues to hydroxyl groups of serine and threonine; *N*-glycosylation occurs in the endoplasmic reticulum in sequential reactions that decorate amino group of asparagine with a core glycan composed of two glucose, nine mannose and two *N*-acetylglucosamine (GlcNAc) residues. To date, four enzymes subserving glycosylation are known to be associated with a CMS: GFPT1 (glutamine fructose-6-phosphate transaminase),[4,93] DPAGT1 (dolichyl-phosphate [UDP-*N*-GlcNAc], *N*-GlcNAc phosphotransferase 1),[94] ALG2 (alpha-1,3-mannosyltransferase), and ALG14 (UDP-*N*-acetylglucosaminyltransferase subunit).[95] The three distinguishing features of these syndromes are a predominantly limb-girdle distribution of the weakness, tubular aggregates of the sarcoplasmic reticulum in type 2 muscle fibers, and responsiveness to cholinergic agonists. However, in some patients with GFPT1- and DPAGT1-CMS there is also facial and bulbar involvement and the tubular aggregates may be difficult to find or are absent.

Mutations in GFPT1

GFPT1 controls the flux of glucose into the hexosamine pathway, and thus the formation of hexosamine products and the availability of precursors for *N*- and *O*-linked protein glycosylation. A CMS caused by defects in GFPT1 was reported in 2011 in 16 patients.[4] The effects of the different mutations on EP fine structure and the extent to which they alter parameters of neuromuscular transmission were not determined.

A subsequent report of 11 patients harboring mutations in *GFPT1* demonstrated grape-like synaptic contacts.[93] Histochemical studies in nine patients revealed tubular aggregates in six and rimmed vacuoles in three. Electron microscopy showed abnormally small EP regions and many with poorly developed junctional folds. Microelectrode studies of intercostal muscle EPs in five patients revealed reduced synaptic response to ACh in three. One patient with a nonsense mutation and second mutation that disrupts the muscle specific exon of *GFPT1* never moved *in utero*, was arthrogrypotic at birth, and is bed-fast and tube-fed at age 9 years. She has a severe myopathy with numerous dilated and degenerating vesicular profiles, autophagic vacuoles, and bizarre apoptotic nuclei.[93] This patient was not improved by pyridostigmine or 3,4-DAP.

That many EPs in this CMS are underdeveloped likely stems from hypoglycosylation and altered function of EP-specific glycoproteins, such as MuSK, agrin, and dystroglycans. However, it is unclear why mutations in *GFPT1* selectively affect neuromuscular transmission and muscle fiber architecture when glycosylated proteins are widely distributed in many tissues and organs.

Mutations in DPAGT1

DPAGT1 catalyzes the first committed step of *N*-linked protein glycosylation. A defect in DPAGT1 predicts failure of asparagine glycosylation of multiple proteins distributed throughout the organism, but in the first five patients with mutations in DPAGT1, neuromuscular transmission was selectively affected.[94] The adverse effect on neuromuscular transmission was attributed to reduced expression of AChR at the EPs, but direct EP studies were not performed.

A subsequent study of two other patients harboring mutations in DPAGT1 extended the phenotypic spectrum of the disease.[96] Both were cognitively challenged and the affected brother of one also had autistic features. These patients had a vacuolar myopathy with autophagic features, fiber type disproportion, as well as tubular aggregates in muscle fibers. Electron microscopy revealed abnormally small pre- and postsynaptic regions with poorly developed junctional folds and microelectrode studies demonstrated a 40–60% reduction of quantal release, of synaptic response to ACh, and of EP AChR content. Immunoblots of muscle extracts with two different antibodies showed reduced to absent protein glycosylation. One patient responded modestly and one poorly to cholinergic agonists.

Mutations in ALG2 and ALG14[1]

ALG2 catalyzes the second and third committed steps of *N*-glycosylation. In one kinship, four affected siblings were homozygous for an insertion/deletion mutation; and another patient was homozygous for a low-expressor V68G mutation. ALG14 forms a multiglycosyltransferase complex with ALG13 and DPAGT1 and catalyzes the first committed step of *N*-glycosylation. In one family, two affected siblings carried heteroallelic P65L and V68G mutations. EP ultrastructure and parameters of neuromuscular transmission were not investigated.

OTHER MYASTHENIC SYNDROMES

Sodium Channel Myasthenia

Only one patient with this type of CMS has been observed to date.[13] This was the case of a 20-year-old normokalemic woman with eyelid ptosis, marked generalized fatigable weakness, and recurrent attacks of respiratory and bulbar paralysis since birth that caused anoxic brain injury. The clue that the defect resides in $Na_v1.4$, the skeletal muscle sodium channel encoded by *SCN4A*, came from the observation that EPPs depolarizing the muscle fiber resting potential to −40mV failed to excite action potentials. *SCN4A* harbored two heteroallelic mutations: S246L in the S4/S5 cytoplasmic linker in domain I, and V1442E in the S3/S4 extracellular linker in domain IV (Figure 99.10). Expression studies on the observed mutations in HEK cells revealed that V1442E markedly enhanced fast inactivation close to the resting potential and also enhanced use-dependent inactivation on high frequency stimulation; S246L caused only minor kinetic abnormalities, suggesting that it is a benign polymorphism. $Na_v1.4$ expression at the EPs and over the sarcolemma was normal by immunocytochemical criteria.[13]

This disease differs from periodic paralyses caused by other mutations in *SCN4A*. The onset is neonatal, the disorder is normokalemic, the attacks selectively involve bulbar and respiratory muscles, physiologic rates of stimulation decrement the CMAP abnormally, and the muscle fiber membrane potential is normal when action potential generation fails. Periodic paralyses stemming from other mutations in *SCN4A* present later in life, the attacks typically spare cranial, bulbar, and respiratory muscles, the serum potassium level increases or declines during attacks in

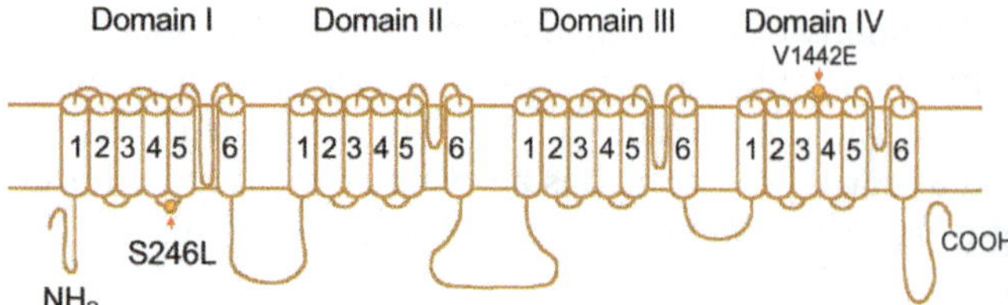

FIGURE 99.10 Scheme of skeletal muscle sodium channel $Na_v1.4$ encoded by *SCN4A* and identified mutations causing CMS. S246L in the S4/S5 cytoplasmic linker of domain I is likely a benign mutation. V1442E in the S3/S4 extracellular linker of domain IV markedly enhances fast inactivation of $Na_v1.4$ near the resting membrane potential. (Reproduced with permission from [13].)

most cases, mild exercise for brief periods does not decrement the CMAP, and the resting membrane potential of the muscle fiber is decreased when action potential generation fails.

Mutations in Plectin

Plectin, encoded by *PLEC*, is a highly conserved and ubiquitously expressed intermediate filament-linking protein with tissue and organelle-specific transcript isoforms that serves as a versatile linker of cytoskeletal components to target organelles in cells of different tissues.[97–99] It is concentrated at sites of mechanical stress, such as the postsynaptic membrane lining junctional folds, the sarcolemma, Z-disks in skeletal muscle, hemidesmosomes in skin, and intercalated disks in cardiac muscle. In skeletal muscle, multiple alternatively spliced transcripts of the exon preceding a common exon 2 link cytoskeletal intermediate filaments to specific targets: the outer nuclear membrane (isoform 1), the outer mitochondrial membrane (isoform 1b), Z-disks (isoform 1d), and to sarcolemmal costameres (isoform 1f).[99] Pathogenic mutations in plectin result in epidermolysis bullosa simplex, a progressive muscular dystrophy in many patients, and a myasthenic syndrome in some patients (for review see [100,101]). Heteroallelic nonsense, frameshift, or splice-site mutations in *PLEC* were recently reported in four unrelated patients with documented defects in neuromuscular transmission.[101–103] In two patients investigated by the author, microelectrode studies of intercostal muscle EPs showed low-amplitude MEPPs. Morphologic studies revealed dislocated and degenerating muscle fiber organelles, plasma membrane defects resulting in Ca^{2+} overloading of the muscle fibers as in Duchenne dystrophy, and extensive degeneration of the junctional folds, all attributable to lack of cytoskeletal support.[101] The patients responded poorly to pyridostigmine and 3,4-DAP. Another patient harbored homozygous frameshift mutations in both *PLEC* and in *CHRNE*.[103] Finally, a recent study identified a homozygous deletion mutation in isoform 1f that caused limb-girdle muscular dystrophy, but neither epidermolysis bullosa simplex nor myasthenia.[104]

PHARMACOTHERAPY OF THE CMS

Table 99.4 shows treatment of the currently recognized CMS. Three important principles should be noted. First, agents that benefit one type of CMS can be ineffective or harmful in another type of CMS. This underlines the need for an accurate molecular diagnosis before start of therapy. Second, because different mutations can have different

TABLE 99.4 Pharmacotherapy of the CMS

CMS subtype	Pharmacotherapy
ChAT deficiency	Pyridostigmine
AChE deficiency	Albuterol Avoid pyridostigmine
Simple AChR deficiency	Pyridostigmine 3,4-DAP also helps in 30–50%
Slow-channel CMS	Quinidine or fluoxetine (long-lived open channel blockers of AChR channel) Avoid pyridostigmine
Fast-channel CMS	Pyridostigmine and 3,4-DAP
Rapsyn deficiency	Pyridostigmine; 3,4-DAP; albuterol
Na^{2+}-channel myasthenia	Pyridostigmine and acetazolamide
Plectin deficiency	Refractory to pyridostigmine; 3,4-DAP may help
Dok-7 myasthenia	Albuterol Avoid pyridostigmine
Agrin deficiency	No response to pyridostigmine or 3,4-DAP in one patient[70] Some response to pyridostigmine in another patient[71]
MuSK deficiency	Variable responses to low doses of pyridostigmine or to 3,4-DAP or albuterol[72,74,75]
GFPT1, DPAGT1, ALG2 and ALG14 myasthenia	Pyridostigmine, 3,4-DAP

Abbreviations: 3,4-DAP, 3,4-diaminopyridine; AChE, acetylcholinesterase; AChR, acetylcholine receptor; ChAT, choline acetyltransferase.

functional consequences, patients harboring different mutations in a given gene may respond differently to a given form of therapy. Third, the cholinergic agonists pyridostigmine and 3,4-DAP exert their effect as soon as the medication is absorbed. In contrast, the beneficial effects of fluoxetine, quinine, and quinidine become fully effective only after weeks or months. Similarly, the beneficial effects of albuterol and ephedrine, which likely act through a second messenger, appear slowly over weeks or months.

References

1. Engel AG. The investigation of congenital myasthenic syndromes. *Ann N Y Acad Sci*. 1993;681:425–434.
2. Engel AG, Ohno K, Sine SM. Congenital myasthenic syndromes. In: Engel AG, ed. Myasthenia Gravis and Myasthenic Disorders. New York: Oxford University Press; 1999:251–297.
3. Wood SJ, Slater CP. Safety factor at the neuromuscular junction. *Prog Neurobiol*. 2001;64:393–429.
4. Senderek J, Muller JS, Dusl M, et al. Hexosamine biosynthetic pathway mutations cause neuromuscular transmission defect. *Am J Hum Genet*. 2011;88:162–172.
5. Shinawi MCS. The array CGH and its clinical applications. *Drug Discov Today*. 2008;13:760–770.
6. Engel AG. The investigation of congenital myasthenic syndromes. *Ann N Y Acad Sci*. 1993;681:425–434.
7. Ohno K, Tsujino A, Shen XM, et al. Choline acetyltransferase mutations cause myasthenic syndrome associated with episodic apnea in humans. *Proc Natl Acad Sci U S A*. 2001;98(4):2017–2022.
8. Shen X-M, Crawford TO, Brengman J, et al. Functional consequences and structural interpretation of mutations in human choline acetyltransferase. *Hum Mutat*. 2011;32:1259–1267.
9. Byring RF, Pihko H, Shen X-M, et al. Congenital myasthenic syndrome associated with episodic apnea and sudden infant death. *Neuromuscul Disord*. 2002;12:548–553.
10. Maselli RA, Chen D, Mo D, et al. Choline acetyltransferase mutations in myasthenic syndrome due to deficient acetylcholine resynthesis. *Muscle Nerve*. 2003;27:180–187.
11. Mallory LA, Shaw JG, Burgess SL, et al. Congenital myasthenic syndrome with episodic apnea. *Pediatr Neurol*. 2009;41:42–45.
12. Schara U, Christen H-J, Durmus H, et al. Long-term follow-up in patients with congenital myasthenic syndrome due to ChAT mutations. *Eur J Paediatr Neurol*. 2010;14:326–333.
13. Tsujino A, Maertens C, Ohno K, et al. Myasthenic syndrome caused by mutation of the *SCN4A* sodium channel. *Proc Natl Acad Sci U S A*. 2003;100(12):7377–7382.
14. Ohno K, Engel AG, Shen X-M, et al. Rapsyn mutations in humans cause endplate acetylcholine receptor deficiency and myasthenic syndrome. *Am J Hum Genet*. 2002;70(4):875–885.
15. Burke G, Cossins J, Maxwell S, et al. Rapsyn mutations in hereditary myasthenia. Distinct early- and late-onset phenotypes. *Neurology*. 2003;61:826–828.
16. Müller JS, Baumeister SK, Schara U, et al. *CHRND* mutation causes a congenital myasthenic syndrome by impairing co-clustering of the acetylcholine receptor with rapsyn. *Brain*. 2006;129:2784–2793.
17. Kraner S, Lufenberg I, Strassburg HM, et al. Congenital myasthenic syndrome with episodic apnea in patients homozygous for a *CHAT* missense mutation. *Arch Neurol*. 2003;60:761–763.
18. Engel AG, Lambert EH, Gomez MR. A new myasthenic syndrome with end-plate acetylcholinesterase deficiency, small nerve terminals, and reduced acetylcholine release. *Ann Neurol*. 1977;1:315–330.
19. Bon S, Coussen F, Massoulié J. Quaternary associations of acetylcholinesterase. II. The polyproline attachment domain of the collagen tail. *J Biol Chem*. 1997;272:3016–3021.
20. Ohno K, Brengman JM, Tsujino A, et al. Human endplate acetylcholinesterase deficiency caused by mutations in the collagen-like tail subunit (ColQ) of the asymmetric enzyme. *Proc Natl Acad Sci U S A*. 1998;95(16):9654–9659.
21. Donger C, Krejci E, Serradell P, et al. Mutation in the human acetylcholinesterase-associated gene, *COLQ*, is responsible for congenital myasthenic syndrome with end-plate acetylcholinesterase deficiency. *Am J Hum Genet*. 1998;63(4):967–975.
22. Bestue-Cardiel M, de-Cabazon-Alvarez AS, Capablo-Liesa JL, et al. Congenital endplate acetylcholinesterase deficiency responsive to ephedrine. *Neurology*. 2005;65:144–146.
23. Brengman JM, Capablo-Liesa JL, Lopez-Pison J, et al. Ephedrine treatment of seven patients with congenital endplate acetylcholinesterase deficiency. *Neuromuscul Disord*. 2006;16(suppl 1):S129.
24. Mihaylova V, Muller JS, Vilchez JJ, et al. Clinical and molecular genetic findings in COLQ-mutant congenital myasthenic syndromes. *Brain*. 2008;131:747–759.
25. Liewluck T, Selcen D, Engel AG. Beneficial effects of albuterol in congenital endplate acetylcholinesterase deficiency and DOK-7 myasthenia. *Muscle Nerve*. 2011;44:789–794.
26. Maselli RA, Ng JJ, Andreson JA, et al. Mutations in *LAMB2* causing a severe form of synaptic congenital myasthenic syndrome. *J Med Genet*. 2009;46:203–208.
27. Brejc K, van Dijk WV, Schuurmans M, et al. Crystal structure of ACh-binding protein reveals the ligand-binding domain of nicotinic receptors. *Nature*. 2001;411(6835):269–276.
28. Celie PHN, Klaassen RV, van Rossum-Fikkert SE, et al. Crystal structure of acetylcholine-binding protein from *Bulinus truncatus* reveals the conserved structural scaffold and sites of variation in nicotinic acetylcholine receptors. *J Biol Chem*. 2005;280:26457–26466.
29. Unwin N. Refined structure of the nicotinic acetylcholine receptor at 4 Å resolution. *J Mol Biol*. 2005;346(4):967–989.
30. Dellisanti CD, Yao Y, Stroud JC, et al. Crystal structure of the extracellular domain of the nAChR α1 bound to α-bungarotoxin at 1.94 Å resolution. *Nat Neurosci*. 2007;10(8):953–962.
31. Sine SM, Wang H-L, Bren N. Lysine scanning mutagenesis delineates structure of nicotinic receptor binding domain. *J Biol Chem*. 2002;277:29210–29223.
32. Engel AG, Ohno K, Sine SM. Sleuthing molecular targets for neurological diseases at the neuromuscular junction. *Nat Rev Neurosci*. 2003;4(5):339–352.

33. Fucile S, Sucapane A, Grassi A, et al. The human adult subtype AChR channel has high Ca^{2+} permeability. *J Physiol (London)*. 2006;573:35–43.
34. Di Castro A, Martinello K, Grassi F, et al. Pathogenic point mutations in a transmembrane domain of the ε subunit increase the Ca^{2+} permeability of the human endplate ACh receptor. *J Physiol (London)*. 2007;579:671–677.
35. Harper CM, Engel AG. Quinidine sulfate therapy for the slow-channel congenital myasthenic syndrome. *Ann Neurol*. 1998;43:480–484.
36. Harper CM, Fukudome T, Engel AG. Treatment of slow channel congenital myasthenic syndrome with fluoxetine. *Neurology*. 2003;60:170–173.
37. Colomer J, Muller JS, Vernet A, et al. Long-term improvement of slow-channel myasthenic syndrome with fluoxetine. *Neuromuscul Disord*. 2006;16:329–333.
38. Chaouch A, Muller JS, Guergueltcheva V, et al. A retrospective clinical study of the treatment of slow-channel congenital myasthenic syndromes. *J Neurol*. 2012;259:478–481.
39. Webster R, Brydson M, Croxen R, et al. Mutation in the AChR channel gate underlies a fast channel congenital myasthenic syndrome. *Neurology*. 2004;62:1090–1096.
40. Shen XM, Brengman J, Edvardson S, et al. Highly fatal fast-channel congenital syndrome caused by AChR ε subunit mutation at the agonist binding site. *Neurology*. 2012;79:449–454.
41. Shen X-M, Fukuda T, Ohno K, et al. Congenital myasthenia-related AChR δ subunit mutation interferes with intersubunit communication essential for channel gating. *J Clin Invest*. 2008;118(5):1867–1876.
42. Sine SM, Engel AG. Recent advances in Cys-loop receptor structure and function. *Nature*. 2006;440(7083):448–455.
43. Shen X-M, Brengman J, Sine SM, et al. Myasthenic syndrome AChRα C-loop mutant disrupts initiation of channel gating. *J Clin Invest*. 2012;122:2613–2621.
44. Shen X-M, Ohno K, Tsujino A, et al. Mutation causing severe myasthenia reveals functional asymmetry of AChR signature Cys-loops in agonist binding and gating. *J Clin Invest*. 2003;111(4):497–505.
45. Wang H-L, Ohno K, Milone M, et al. Fundamental gating mechanism of nicotinic receptor channel revealed by mutation causing a congenital myasthenic syndrome. *J Gen Physiol*. 2000;116:449–460.
46. Milone M, Wang H-L, Ohno K, et al. Mode switching kinetics produced by a naturally occurring mutation in the cytoplasmic loop of the human acetylcholine receptor ε subunit. *Neuron*. 1998;20:575–588.
47. Engel AG, Ohno K, Bouzat C, et al. End-plate acetylcholine receptor deficiency due to nonsense mutations in the ε subunit. *Ann Neurol*. 1996;40:810–817.
48. Ohno K, Quiram P, Milone M, et al. Congenital myasthenic syndromes due to heteroallelic nonsense/missense mutations in the acetylcholine receptor ε subunit gene: identification and functional characterization of six new mutations. *Hum Mol Genet*. 1997;6:753–766.
49. Ohno K, Anlar B, Özdirim E, et al. Myasthenic syndromes in Turkish kinships due to mutations in the acetylcholine receptor. *Ann Neurol*. 1998;44:234–241.
50. Abicht A, Stucka R, Karcagi V, et al. A common mutation (ε1267delG) in congenital myasthenic patients of Gypsy ethnic origin. *Neurology*. 1999;53:1564–1569.
51. Middleton L, Ohno K, Christodoulou K, et al. Congenital myasthenic syndromes linked to chromosome 17p are caused by defects in acetylcholine receptor ε subunit gene. *Neurology*. 1999;53:1076–1082.
52. Croxen R, Newland C, Betty M, et al. Novel functional ε-subunit polypeptide generated by a single nucleotide deletion in acetylcholine receptor deficiency congenital myasthenic syndrome. *Ann Neurol*. 1999;46:639–647.
53. Shen X-M, Ohno K, Fukudome T, et al. Congenital myasthenic syndrome caused by low-expressor fast-channel AChR δ subunit mutation. *Neurology*. 2002;59:1881–1888.
54. Ealing J, Webster R, Brownlow S, et al. Mutations in congenital myasthenic syndromes reveal an ε subunit C-terminal cysteine, C470, crucial for maturation and surface expressions of adult AChR. *Hum Mol Genet*. 2002;11:3087–3096.
55. Ohno K, Anlar B, Ozdemir C, et al. Frameshifting and splice-site mutations in acetylcholine receptor ε subunit gene in 3 Turkish kinships with congenital myasthenic syndromes. *Ann N Y Acad Sci*. 1998;841:189–194.
56. Abicht A, Stucka R, Schmidt C, et al. A newly identified chromosomal microdeletion and an N-box mutation of the AChR ε gene cause a congenital myasthenic syndrome. *Brain*. 2002;125:1005–1013.
57. Ohno K, Anlar B, Engel AG. Congenital myasthenic syndrome caused by a mutation in the Ets-binding site of the promoter region of the acetylcholine receptor ε subunit gene. *Neuromuscul Disord*. 1999;9:131–135.
58. Nichols PR, Croxen R, Vincent A, et al. Mutation of the acetylcholine receptor ε-subunit promoter in congenital myasthenic syndrome. *Ann Neurol*. 1999;45:439–443.
59. Duclert A, Savatier N, Schaeffer L, et al. Identification of an element crucial for the sub-synaptic expression of the acetylcholine receptor epsilon-subunit gene. *J Biol Chem*. 1996;271:17433–17438.
60. Schaeffer L, Duclert N, Huchet-Dymanus M, et al. Implication of a multisubunit Ets-related transcription factor in synaptic expression of the nicotinic acetylcholine receptor. *EMBO J*. 1998;17:3078–3090.
61. Fromm L, Burden SJ. Synapse-specific and neuregulin-induced transcription require an Ets site that binds GABPα/GAPBβ. *Genes Dev*. 1998;12:3074–3083.
62. Ohno K, Wang H-L, Milone M, et al. Congenital myasthenic syndrome caused by decreased agonist binding affinity due to a mutation in the acetylcholine receptor ε subunit. *Neuron*. 1996;17(1):157–170.
63. Quiram P, Ohno K, Milone M, et al. Mutation causing congenital myasthenia reveals acetylcholine receptor β/δ subunit interaction essential for assembly. *J Clin Invest*. 1999;104:1403–1410.
64. Hesselmans LFGM, Jennekens FGI, Vand Den Oord CJM, et al. Development of innervation of skeletal muscle fibers in man: relation to acetylcholine receptors. *Anat Rec*. 1993;236:553–562.
65. Morgan NV, Brueton LA, Cox P, et al. Mutations in the embryonal subunit of the acetylcholine receptor (*CHNRG*) cause lethal and Escobar variants of the multiple pterygium syndrome. *Am J Hum Genet*. 2006;79:390–395.
66. Vogt J, Harrison BJ, Spearman H, et al. Mutation analysis of CHRNA1, CHRNB1, CHRND, and RAPSN genes in multiple pterygium syndrome/fetal akinesia patients. *Am J Hum Genet*. 2008;82:222–227.
67. Michalk A, Stricker S, Becker J, et al. Acetylcholine receptor pathway mutations explain various fetal akinesia deformation sequence disorders. *Am J Hum Genet*. 2008;82:464–476.

68. Vogt J, Morgan NV, Marton T, et al. Germline mutation in DOK7 associated with fetal akinesia deformation sequence. *J Med Genet.* 2009;46:338–340.
69. Burden SJ, Yumoto N, Zhang W. The role of MuSK in synapse formation and neuromuscular disease. *Cold Spring Harb Perspect Biol.* 2013;5(5), a009167.
70. Huze C, Bauche S, Richard P, et al. Identification of an agrin mutation that causes congenital myasthenia and affects synapse function. *Am J Hum Genet*. 2009;85(2):155–167.
71. Maselli RA, Fernandez JM, Arredodndo J, et al. LG2 agrin mutation causing severe congenital myasthenic syndrome mimics functional characteristics of non-neural agrin (z-) agrin. *Hum Genet*. 2012;131:1123–1135.
72. Chevessier F, Faraut B, Ravel-Chapuis A, et al. MUSK, a new target for mutations causing congenital myasthenic syndrome. *Hum Mol Genet*. 2004;13:3229–3240.
73. Chevessier F, Girard E, Molgo J, et al. A mouse model for congenital myasthenic syndrome due to MuSK mutations reveals defects in structure and function of neuromuscular junctions. *Hum Mol Genet*. 2008;17:3577–3595.
74. Maselli R, Arredondo J, Cagney O, et al. Mutations in *MUSK* causing congenital myasthenic syndrome impair MuSK-Dok-7 interaction. *Hum Mol Genet*. 2010;19:2370–2379.
75. Mihaylova V, Salih MA, Mukhtar MM, et al. Refinement of the clinical phenotype in Musk-related congenital myasthenic syndromes. *Neurology*. 2009;73:1926–1928.
76. Okada K, Inoue A, Okada M, et al. The muscle protein Dok-7 is essential for neuromuscular synaptogenesis. *Science*. 2006;312:1802–1805.
77. Beeson D, Higuchi O, Palace J, et al. Dok-7 mutations underlie a neuromuscular junction synaptopathy. *Science*. 2006;313(5795):1975–1978.
78. Muller JS, Herczegfalvi A, Vilchez JJ, et al. Phenotypical spectrum of DOK7 mutations in congenital myasthenic syndromes. *Brain*. 2007;130:1497–1506.
79. Selcen D, Milone M, Shen X-M, et al. Dok-7 myasthenia: phenotypic and molecular genetic studies in 16 patients. *Ann Neurol*. 2008;64:71–87.
80. Anderson JA, Ng JJ, Bowe C, et al. Variable phenotypes associated with mutations in *DOK7*. *Muscle Nerve*. 2008;37:448–456.
81. Ammar AB, Petit F, Alexandri K, et al. Phenotype-genotype analysis in 15 patients presenting a congenital myasthenic syndrome due to mutations in *DOK7*. *J Neurol*. 2010;257:754–766.
82. Slater CR, Fawcett PRW, Walls TJ, et al. Pre- and postsynaptic abnormalities associated with impaired neuromuscular transmission in a group of patients with "limb-girdle myasthenia". *Brain*. 2006;127:2061–2076.
83. Selcen D, Milone M, Shen X-M, et al. Dok-7 myasthenia: clinical spectrum, endplate electrophysiology and morphology, 12 novel DNA rearrangements, and genotype-phenotype relations in a Mayo cohort of 13 patients. *Neurology*. 2007;68(suppl. 1):A106–A107.
84. Schara U, Barisic N, Deschauer M, et al. Ephedrine therapy in eight patients with congenital myasthenic syndrome due to *DOK7* mutations. *Neuromuscul Disord*. 2010;19:828–832.
85. Müller JS, Baumeister SK, Rasic VM, et al. Impaired receptor clustering in congenital myasthenic syndrome with novel RAPSN mutations. *Neurology*. 2006;67:1159–1164.
86. Richard P, Gaudon K, Andreux F, et al. Possible founder effect of rapsyn N88K mutation and identification of novel rapsyn mutations in congenital myasthenic syndromes. *J Med Genet*. 2003;40:81e.
87. Müller JS, Abicht A, Burke G, et al. The congenital myasthenic syndrome mutation RAPSN N88K derives from an ancient Indo-European founder. *J Med Genet*. 2004;41(8):e104.
88. Ohno K, Engel AG. Lack of founder haplotype for the rapsyn mutation: N88K is an ancient founder mutation or arises from multiple founders. *J Med Genet*. 2004;41(1):e8.
89. Ohno K, Sadeh M, Blatt I, et al. E-box mutations in *RAPSN* promoter region in eight cases with congenital myasthenic syndrome. *Hum Mol Genet*. 2003;12:739–748.
90. Engel AG, Sine SM. Current understanding of congenital myasthenic syndromes. *Curr Opin Pharmacol*. 2005;5:308–321.
91. Cossins J, Burke G, Maxwell S, et al. Diverse molecular mechanisms involved in AChR deficiency due to rapsyn mutations. *Brain*. 2006;129:2773–2783.
92. Banwell BL, Ohno K, Sieb JP, et al. Novel truncating *RAPSN* mutation causing congenital myasthenic syndrome responsive to 3,4-diaminopyridine. *Neuromuscul Disord*. 2004;14:202–207.
93. Selcen D, Shen X-M, Milone M, et al. GFPT1-myasthenia: clinical, structural, and electrophysiologic heterogeneity. *Neurology*. 2013;81: 370–378.
94. Belaya K, Finlayson S, Slater C, et al. Mutations in DPAGT1 cause a limb-girdle congenital myasthenic syndrome with tubular aggregates. *Am J Hum Genet*. 2012;91:1–9.
95. Cossins J, Belaya K, Hicks D, et al. Congenital myasthenic syndromes due to mutations in ALG2 and ALG14. *Brain*. 2013;136:944–956.
96. Selcen D, Shen X-M, Li Y, et al. DPAGT1 myasthenia and myopathy. Genetic, phenotypic, and expression studies. *Neurology*. May 2014; In press.
97. Elliott CE, Becker B, Oehler S, et al. Plectin transcript diversity: identification and tissue distribution of variants with distinct first coding exons and rodless isoforms. *Genomics*. 1997;42:115–125.
98. Fuchs P, Zorer M, Rezniczek GA, et al. Unusual 5′ transcript complexity of plectin isoforms: novel tissue-specific exons modulate actin binding activity. *Hum Mol Genet*. 1999;8:2461–2472.
99. Konieczny P, Fuchs P, Reipert S, et al. Myofiber integrity depends on desmin network targeting to Z-disks and costameres via distinct plectin isoforms. *Neurology*. 2014;82:1822–1830.
100. Banwell BL, Russel J, Fukudome T, et al. Myopathy, myasthenic syndrome, and epidermolysis bullosa simplex due to plectin deficiency. *J Neuropathol Exp Neurol*. 1999;58:832–846.
101. Selcen D, Juel VC, Hobson-Webb LD, et al. Myasthenic syndrome caused by plectinopathy. *Neurology*. 2011;76(4):327–336.
102. Forrest K, Mellerio JE, Robb S, et al. Congenital muscular dystrophy, myasthenic symptoms and epidermolysis bullosa simplex (EBS) associated with mutations in the *PLEC1* gene encoding plectin. *Neuromuscul Disord*. 2010;20:709–711.
103. Maselli R, Arredondo J, Cagney O, et al. Congenital myasthenic syndrome associated with epidermolysis bullosa caused by homozygous mutaions in *PLEC1* and *CHRNE*. *Clin Genet*. 2011;80:444–451.
104. Gundesli H, Talim B, Korkusuz P, et al. Mutation in exon 1f of PLEC, leading to disruption of plectin isoform 1f, causes autosomal-recessive limb-girdle muscular dystrophy. *Am J Hum Genet*. 2010;87:834–841.

SECTION XII

STROKE

CHAPTER

100

Cerebral Vasculopathies

Michael M. Dowling

The University of Texas Southwestern Medical Center, Dallas, TX, USA

INTRODUCTION

There is a clear genetic component underlying the major causes of stroke in adults, including hypertension, diabetes, and disorders of cholesterol metabolism and coagulation. Heritable differences in inflammation and response to infection likely also play a role in stroke risk. The pathogenesis of stroke in most cases is multifactorial. Less commonly, stroke can occur secondary to a single gene mutation or a chromosomal disorder affecting the cerebral vasculature. Here we will review the less common cerebral vasculopathies associated with either monogenic or chromosomal disorders (Tables 100.1 and 100.2). These disorders are important to identify for their implications for prevention and treatment but also to allow identification of other family members at risk.

These disorders include inborn errors of metabolism such as Fabry disease, homocystinuria, and the mitochondrial disorders, as well as the genetic disorders leading to early atherosclerosis including the hyperlipidemias and Tangier disease. Collagen vascular disorders such as Ehlers–Danlos syndrome type IV and Marfan syndrome can increase the risk of arterial dissection. Moyamoya syndrome, a severe cerebrovascular disorder, has been described in a variety of monogenic and chromosomal disorders including sickle cell anemia, neurofibromatosis type I, and also Down and Williams syndrome. Small vessel vasculopathies include CADASIL and CARASIL.

INBORN ERRORS OF METABOLISM WITH CEREBROVASCULAR INVOLVEMENT

Some inborn errors of metabolism, such as the organic acidurias or urea cycle disorders, can increase stroke risk through a mechanism of an acute metabolic decompensation, or "metabolic stroke," leading to acute focal neurologic deficits, often during an illness or other episode of metabolic stress when toxic metabolites can accumulate in vulnerable areas of the brain. Others, including the mitochondrial disorders can cause infarction though local failures of energy supply or the accumulation of oxygen free radicals that can also affect the vasculature. The mitochondrial disorders and the role of the vasculature in these disorders will be discussed elsewhere. Other metabolic disorders can cause stroke through a vascular mechanism (Table 100.1) and will be reviewed below.

Homocystinuria

Classic homocystinuria is due to mutations in the enzyme cystathionine beta synthase (CBS), leading to failure of conversion of homocysteine to cystathionine. Serum homocysteine is greatly increased in plasma (up to 100-fold) as is plasma methionine, and homocysteine is excreted in the urine.[1] The disorder is inherited in an autosomal recessive pattern and patients have skeletal, ocular, and neurologic complications, as well as vascular disease. Patients typically are tall, with increased arm span, pes cavus, high arched palate, lens dislocation (ectopia lentis), and cognitive deficits.[2] The body habitus and lens dislocation can lead to misdiagnosis as Marfan syndrome.

The elevated homocysteine can directly damage the vascular endothelium with subsequent platelet activation and thrombus formation as well as causing smooth muscle proliferation.[3–5] There is a high risk of thrombotic or thromboembolic events, including stroke in about one-third. There is also increased risk of sinovenous thrombosis.[6]

Rosenberg's Molecular and Genetic Basis of Neurological and Psychiatric Disease
http://dx.doi.org/10.1016/10.1016/B978-0-12-410529-4.00100-5

TABLE 100.1 A Partial List of Genetic Causes of Cerebral Vasculopathies

Inborn Errors of Metabolism with Cerebrovascular Involvement
Homocystinuria Fabry disease Menkes disease Mitochondrial encephalopathies
Genetic disorders with early atherosclerosis
Familial hyperlipidemias Familial hypercholesterolemia Tangier disease
Genetic disorders with increased prevalence of dissection
Alpha-1-antitrypsin deficiency ICAM-1 variants MTHFR variants Osteogenesis imperfecta Loeys–Dietz syndrome Autosomal dominant polycystic kidney disease
Connective tissue disorders
Ehlers–Danlos syndrome type IV Marfan Syndrome Pseudoxanthoma elasticum Williams–Beuren syndrome
Moyamoya
Genetic causes of small vessel disease
CADASIL CARASIL COL4A1 mutations Cerebroretinal vasculopathy and HERNS Cerebral amyloid angiopathy

About half of the patients with homocystinuria due to CBS deficiency will respond to treatment with pyridoxine at doses of 100–200 mg daily. The plasma methionine will normalize and the total plasma homocysteine will fall dramatically, and in these pyridoxine-responsive patients, risk of further complications is significantly reduced. Folate supplementation should also be provided.

In those patients not responsive to pyridoxine, treatment includes restriction of dietary methionine and addition of betaine, a methyl-donor agent. All CBS deficiency patients need an emergency regimen at times of physical stress, illness, or surgery including adequate hydration and consideration of antiplatelet agents.[1,7]

Less common causes of homocystinuria include the "remethylation defects," inborn errors of cobalamin metabolism (methionine synthase), and tetrahydrofolate metabolism (5,10-methylenetetrahydrofolate reductase; MTHFR). In these disorders, the plasma homocysteine levels are elevated to approximately 10-fold with a reduced or low–normal plasma methionine. Treatment is difficult and includes cobalamin, folinic acid, and betaine.[1] Other mutations in the MTHFR gene and heterozygotes for CBS deficiency can have more modest elevations of plasma homocysteine without homocystinuria. Elevated homocysteine has been identified as risk factor for stroke, premature atherosclerosis, and thromboembolism.[8,9]

Fabry Disease

Fabry disease is a lysosomal storage disorder due to deficiency of α-galactosidase, which leads to the accumulation of the sphingolipid trihexosylceramide causing a diffuse vasculopathy. This X-linked recessive disorder commonly affects young boys and presents with characteristic punctuate angiokeratomas in a "bathing suit" distribution, agonizing burning pain in the palms and soles, lens and corneal opacities, renal and cardiac involvement, and later stroke. This disorder is discussed in more detail elsewhere. Recognition is important, as it is treatable by enzyme replacement therapy.[10]

TABLE 100.2 A Partial List of Conditions Associated with Moyamoya*

Metabolic disease	Homocystinuria Diabetes Hyperphosphatemia
Anemias	Sickle cell disease Hereditary spherocytosis β-thalassemia Fanconi anemia Paroxysmal nocturnal hemoglobinuria Hemolytic-uremic syndrome
Disorders of thrombosis	Protein C or S deficiency Factor XII deficiency Lupus anticoagulant Antiphospholipid antibodies Factor V Leiden Hemophilia A
Neurocutaneous disorders	Neurofibromatosis type I Hypomelanosis of Ito Tuberous sclerosis Livido reticularis and Sneddon syndrome
Tumors	Brain tumors Wilm tumor Leukemia
Autoimmune disease	Systemic lupus erythematosus Sjogren syndrome Juvenile rheumatoid arthritis Ulcerative colitis
Infection	Leptospirosis Tuberculus meningitis Sarcoidosis Epstein–Barr virus Varicella zoster/herpes zoster opthalmicus
Immunodeficiencies	HIV IgA deficiency
Miscellaneous genetic disorders	Down syndrome Prader–Willi syndrome Turner syndrome Noonan syndrome Marfan syndrome Ehlers–Danlos syndrome, type IV Williams syndrome Alagille syndrome Osteogenesis imperfecta Robinow syndrome
Other conditions	Cranial irradiation Cocaine dependency Pituitary gigantism CREST syndrome Hirschsprung disease Bechet disease Hemiplegic migraine
Cardiac disease	Aortic stenosis Coronary artery disease Dilated cardiomyopathy Coarctation of the aorta
Vascular associations with moyamoya	Cerebral aneurysm Cerebral arteriovenous malformation Internal carotid artery dissection Fibromuscular dysplasia Persistent primitive trigeminal or hypoglossal arteries Sagittal sinus thrombosis Renal artery stenosis Renovascular hypertension Primary pulmonary hypertension Peripheral artery disease Raynaud phenomenon

*Adapted from Ganesan V, Kirkham F, eds. *Stroke and Cerebrovascular Disease in Childhood*. London: Mac Keith Press; 2011.

Menkes Disease

Menkes disease (trichopoliodystrophy or kinky hair disease) is an X-linked recessive disorder of copper transport. Mutations in the *ATP7A* gene, which is essential for the translocation of metal cations across the cell membrane, lead to impaired intestinal absorption of copper, which affects the function of a variety of cuproenzymes (including copper-zinc superoxide dismutase, cytochrome c oxidase, dopamine β-hydroxylase, and others).[11] Affected young boys are typically normal at birth but suffer a progressive neurodegenerative course at about 6–12 weeks of age with developmental regression, lethargy, poor feeding, and seizures. The classic diagnostic feature is the abnormal hair. It is short, easily broken, and often white, with a brush or wool-like appearance. Under the microscope, the individual hair shafts appear twisted (torti pili). There are other abnormalities including scurvy-like bone lesions and elongated tortuous cerebral (and visceral) arteries, which are subject to occlusion.[12] On magnetic resonance imaging (MRI), there are progressive white matter abnormalities and diffuse atrophy as well as multifocal areas of infarction. Vessels show areas of localized narrowing and dilatation with a defect in the elastic fibers of the vessel wall.[13] The copper deficiency leads to reduced activity of the copper-dependent enzyme lysyl oxidase, causing defective cross-linking of elastin and collagen.[14] This may explain the increased risk of stroke although there may also be a contribution from reduced cytochrome c oxidase activity.[15]

Diagnosis depends on recognition of the clinical syndrome, microscopic examination of the hair, and by low plasma copper and low serum ceruloplasmin levels. Defective copper export can be evaluated in cultured fibroblasts and DNA testing can be performed.[16,17]

Affected children typically die between 3 and 12 months, with some surviving until 3 years or beyond, likely depending on the severity of the copper transport deficiency.[18] Therapy with parenteral copper can normalize serum copper and ceruloplasmin levels as well as normalize liver copper, but transport into the brain is still impaired.[19] There have been successful reports of treatment with copper-histidinate started soon after birth or even *in utero*.[20] In one study of 12 patients, survival was increased with early treatment (within 22 days of birth) and in two boys with partial ATPase activity, normal neurological development and myelination was observed.[11]

GENETIC DISORDERS WITH EARLY ATHEROSCLEROSIS

The development of an apparent atherosclerotic disease in a child or young adult is suggestive of an underlying genetic disorder such as one of the familial hyperlipidemias. One of the most common, familial hypercholesterolemia, with severely elevated low-density lipoprotein cholesterol levels, is primarily due to autosomal dominant mutations in one of three genes (LDLR, low-density lipoprotein receptors; APOB, apolipoprotein B-100; and PCSK9, proprotein convertase subtilisin/kexin type 9).[21] Familial hypercholesterolemia is characterized by cholesterol deposits in the tendons (xanthomas), or around the eyes (xanthelasmas), a white, gray, or blue ring in the corneal margin (corneal arcus), coronary artery disease, and increased risk of stroke. Other familial disorders, including familial hyperlipoproteinemias and homocystinuria (see above), can lead to early atherosclerotic vasculopathy.

Tangier Disease

Tangier disease is a rare autosomal recessive disease with low levels of high-density lipoproteins (HDL) due to mutations in the ATP-binding cassette-1 gene. Cholesterol esters accumulate in the tissues resulting in large yellow–orange tonsils, as well as enlargement of the liver, spleen, and lymph nodes. Patients have low cholesterol levels and chylomicron remnants.[22]

Affected patients have a peripheral neuropathy with widespread loss of pain and temperature sensation, and muscle wasting and weakness in the hands and arms in what is described as a "syringomylelia-like" syndrome,[23] with "relapsing" painful peripheral neuropathy in some. Corneal clouding, early coronary artery, and carotid artery disease is also noted.[24]

GENETIC DISORDERS WITH INCREASED PREVALENCE OF DISSECTION

Arterial dissection occurs when there is a tear or splitting of the intimal layer of the artery, typically the carotid or vertebral, with arterial blood penetrating the subintimal layer and causing a longitudinal extension of the intramural hematoma. The radiographic appearance is that of a tapering of the vessel lumen with focal stenosis or occlusion. The area of the dissection can serve as a nidus for thromboembolism. Coincident hypercoagulability or other factors may play a role in the symptomatic expression of a dissection as stroke.

Identified risk factors for cervicocephalic arterial dissections include tobacco use, oral contraceptives, migraine, alpha-1-antitrypsin deficiency, elevated homocysteine, MTHFR variants, and hypertension.[25-29] Infection may also play a role, as there is a seasonal pattern, with a higher incidence of nontraumatic dissection in winter and infection has been associated with multiple dissections.[30,31] An inflammatory role is suggested by observations of an association with autoimmune thyroid disease and variants of the pro-inflammatory gene ICAM-1.[27b,32]

Dissections are usually attributed to blunt trauma or stretching injury to the vessels. However, dissection after trivial trauma or multiple dissections raises concern for an underlying connective tissue disorder. Dissection and other neurovascular complications have been described in Marfan syndrome, fibromuscular dysplasia,

Ehlers–Danlos syndrome type IV, Loeys–Dietz syndrome osteogenesis imperfecta, pseudoxanthoma elasticum, and autosomal dominant polycystic kidney disease.[33–35]

Dissection has also been associated with certain forms of congenital heart disease where there may also be an underlying genetic predisposition. These include coarctation of the aorta, bicuspid aortic valve, and other disorders.[36–38]

Ehlers–Danlos Syndrome Type IV

Ehlers–Danlos syndrome type IV (or the vascular type) is an autosomal dominant disorder resulting from mutations in the collagen type III gene (*COL3A1*). Patients with this form typically lack the dramatic skin hyperelasticity and joint hypermobility found in types I–III, but have thin, translucent, and fragile skin, with joint laxity that is limited to the small joints in the hand. The clinical features include easy bruising, translucent skin with visible veins, with a characteristic facial appearance of expressive eyes, thin nose and lips, and lobeless ears.[35,39] Spontaneous mutations are common and about half of the cases have no reported family history of the disorder.

Patients are at risk of rupture of the arteries, heart, uterus, liver, spleen, or intestines. Cardiac complications are common and include prolapse of the mitral and tricuspid valves, septal defects, and dilatation of the aortic root and pulmonary arteries. Cerebrovascular lesions include intracranial aneurysms and arterial dissection or rupture, and subarachnoid hemorrhage.[35,40] Aneurysms can be saccular or fusiform and are common in the cavernous sinus where rupture can lead to spontaneous carotid-cavernous fistula. In a series of 220 patients and 199 of their affected relatives, 10.5% had central nervous system arterial complications at a mean age of onset of 33 years with cavernous-carotid fistulae in 2.4%, aneurysm in 1.2%, and arterial rupture in 0.5%.[39]

The main cause of death is typically arterial dissection or rupture of the thoracic or abnormal artery, with additional deaths from organ rupture.[39] Angiography and angioplasty should be avoided if possible due to a high rate of complications and surgery is complicated by the delicate friable arteries. Women with Ehlers–Danlos type IV are at high risk of uterine or arterial rupture during pregnancy.

Marfan Syndrome

Marfan syndrome is an autosomal dominant condition caused by mutations in the fibrillin-1 gene (*FBN1*) on chromosome 15. Mutations are spontaneous in up to 30% of cases. There is significant clinical variability in this syndrome with skeletal, ocular, and cardiovascular abnormalities, but many patients have a characteristic appearance including tall stature, long, thin limbs (dolichostenomelia) or digits (arachnodactyly), prominent jaw (prognathism), high-arched palate, kyphoscoliosis, chest deformity (pectus excavatum or carinatum), and joint hypermobility. Eye findings include myopia, lens dislocation (ectopia lentis), and retinal detachment. Cardiac manifestations include aortic and mitral value insufficiency and aneurysmal dilatation of the aortic root with risk of aortic dissection. There can also be dural ectasia, inguinal hernia, and spontaneous pneumothorax. There is also vascular fragility, which can complicate surgery.

The neurovascular complications of Marfan syndrome are primarily related to the cardiac and aortic abnormalities with dissection of the aorta extending into the common carotid or spinal arteries leading to stroke. Spontaneous dissection of the common carotid artery, or the extracranial internal carotid and vertebral arteries has been reported.[41,42] However, the prevalence of intracranial abnormalities may have been overestimated or related to misclassification of patients in earlier studies.

No clear association with intracranial dissection was noted in a retrospective study of 513 Marfan syndrome patients. "Neurovascular complications" were reported in 18 patients (3.5%), primarily transient ischemic attack (TIA; 2%), cerebral infarction (0.4%), spinal cord infarction (0.4%), subdural hematoma (0.4%), and spinal subarachnoid hemorrhage (0.2%).[43] However, a cardioembolic source was identified in 12/13 patients with cerebral ischemia, with prosthetic heart valves in nine of the patients. Thus intracranial dissection may play a much smaller role in the neurovascular complications of Marfan syndrome than embolic events from the heart or aorta. Cerebral emboli were reported in 25/675 (3.7%) of Marfan syndrome patients undergoing aortic root replacement surgery.[44]

Similarly, while intracranial aneurysms have also been reported in Marfan syndrome, larger studies have not shown an association.[45] A small autopsy study of 25 patients with Marfan syndrome found no increased prevalence of intracranial aneurysms and a review of 750 neurosurgical patients with intracranial aneurysms found no patients with Marfan syndrome.[46] No symptomatic or ruptured intracranial aneurysms were noted in a 12-year retrospective study of 135 Marfan syndrome patients followed in a disease-specific clinic.[47]

Loeys–Dietz Syndrome

Loeys–Dietz syndrome is another connective tissue disorder with a clinical presentation that can be similar to Marfan syndrome. Patients can have long limbs and fingers with joint laxity as well as other characteristics including widely spaced eyes, cleft palate, bifid uvula, and chest wall abnormalities (pectus excavatum or carinatum). There is an increased prevalence of congenital heart abnormalities, including patent ductus arteriosis and atrial septal defects, and increased risk of aortic and other arterial aneurysms and dissections. Arteriography can demonstrate irregularity and increased tortuosity of the large vessels. It is caused by dominantly inherited mutations in either of two genes encoding the transforming growth factor beta receptor (*TGFBR1* or *TGFBR2*).[48,49]

Pseudoxanthoma Elasticum

In this connective tissue disorder, there is calcification of the elastic fibers in the skin, eyes, heart, and vasculature.[50] The skin is thin and has an appearance described as that of a "peau d'orange" or "plucked chicken" and there are characteristic yellow–orange papules that form in the areas of joint flexure typically beginning in the neck. Similar lesions can occur in the mucus membranes or in the intestinal mucosa. Upper gastrointestinal hemorrhage was reported in 17/94 patients and there can be acute or chronic anemia from gastrointestinal bleeding.[51] Eye findings include retinal mottling and angioid streaks in the retina due to breakdown of the elastic lamina of Bruch membranes. Optic disk drusen are common as well.[52] Vision loss can occur, either gradually from secondary macular degeneration or acutely from retinal hemorrhages that are either spontaneous or follow minor trauma.[53] Cardiovascular complications result from calcification of the internal elastic lamina of mid-sized arteries including the coronary and cerebral arteries as well as peripheral vessels. This can result in myocardial infarction, stroke, hypertension, and intermittent claudication.

Inheritance is autosomal recessive but dominant forms are reported. The disorder is due to mutations in the *ABCC6* gene on chromosome 16p13.1, a member of the ATP-binding cassette (ABC) gene, subfamily C. This gene encodes the multidrug resistance-associated protein 6 (MRP6).[54] It is highly expressed in the liver but only at low levels in the skin, retina, and blood vessels, and appears to serve as a transmembrane transporter of polyanionic, glutathione-conjugated molecules, but the transported substrates that lead to calcification (or other mechanisms of action) are unclear. However, other pseudoxanthoma elasticum-like lesions have been noted in hepatic disorders that have effects on calcification-regulating plasma proteins such as fetuin.[55] Similar skin findings and angioid streaks have also been reported in patients with thalassemia[56] and have been found to accompany deficiency of the vitamin-K-dependent clotting factors.[57] Thus, the absence or decreased function of MRP6 in the liver could result in a deficiency of circulating factors that inhibit arterial calcification.[58] This is supported by transplant experiments in mice where skin from wild-type mice was grafted onto ABCC6-deficient mice and mineralization was observed in the transplanted tissue.[59] Discovery of the substrate transported by MRP6 could have therapeutic implications.

Interestingly, another rare disorder with pathological arterial calcifications appears to have considerable phenotypic and genotypic overlap with pseudoxanthoma elasticum. Generalized arterial calcification of infancy (GACI) is a rare autosomal recessive disorder with calcification of the internal elastic lamina, intimal proliferation, and arterial stenosis with early severe congestive heart failure, hypertension, and myocardial ischemia leading to death in the neonatal period unless treated with bisphosphonates. This disorder is due to mutations in ecto-nucleotide pyrophosphatase/phosphodiesterase 1 (*ENPP1*). Mutations in *ABCC6* were found in a significant subset of GACI patients and some children with *ENPP1* mutations developed pseudoxanthomas and angioid streaks in early childhood, suggesting that the two disorders might represent two ends of a spectrum of pathological calcification disorders.[58]

MOYAMOYA

Moyamoya disease is a progressive, nonatherosclerotic, noninflammatory intracranial vasculopathy typically affecting the large vessels of the anterior circulation. Stenosis of the distal internal carotid arteries, as well as the anterior and middle cerebral arteries (and rarely the posterior cerebral arteries), leads to the development of extensive small collateral blood vessels at the base of the brain and basal ganglia that on cerebral angiography resemble a "puff of smoke" or *moyamoya* in Japanese. This occlusive disease leads to transient ischemic attacks, stroke, or intracranial hemorrhage from rupture of the collaterals.

The term "moyamoya" alone is used to describe the characteristic findings on angiography of this spontaneous occlusion of the circle of Willis with collateral formation. "Moyamoya disease" describes the presence of the bilateral arteriopathy in the absence of other associated disorders, while "moyamoya syndrome" is a more general term referring to cases of unilateral involvement or those occurring in conjunction with other disorders (Table 100.2) such as other congenital malformations, genetic disorders, or environmental factors such as cranial irradiation.[60]

Moyamoya disease is typically idiopathic, and while more common in East Asians (including Japanese and Korean), it has been reported in all racial and ethnic groups. Among Japanese, the annual incidence is 0.35/100,000 with a female preponderance of 1.8 : 1[61] with a higher annual incidence of 1.7–2.3/100,000 noted in a recent study in Korea.[62] There appear to be two distinct periods of symptom onset, one at about 5–10 years of age and another in the 40s.[61,63] The incidence in Europe is only one-tenth of that in Japan.[64]

There is evidence of a strong genetic component with 7–12% of cases being familial and an 80% concordance in monozygotic twins.[65] Familial moyamoya disease has an autosomal dominant inheritance pattern with variable penetrance. Genomic imprinting may be involved as daughters of affected females are more likely to develop moyamoya.[66] Linkage studies in families with moyamoya disease identified multiple candidate genetic loci on chromosomes 3p24-26, 6q25, 8q13-24, 12p12-13, and 17q25.[67–70]

Recently, using genome-wide and locus-specific association studies, as well as exome sequencing, mutations in the *RNF213* gene at the 17q25 locus were found to be highly associated with moyamoya disease.[71,72] The initial variant identified in the *RNF213* gene was present in Japanese, Korean and, to a lesser extent, Chinese patients with moyamoya disease, but not in Caucasians with the disease, consistent with a founder effect for this variant. This could explain why moyamoya disease is more prevalent in Japan, where the carrier frequency of the founder variant is high at 1/72, compared to Western countries, where no carriers were identified in 400 Caucasian controls.[71] Subsequent sequencing identified additional "private" mutations in the *RNF213* gene in other Asian and non-Asian patients with moyamoya disease.

In a study of 204 patients with moyamoya disease, the founder variant was identified in 95.1% of patients with familial moyamoya disease and only 1.8% of controls, and homozygosity for the variant was observed only in patients, not in controls. Age of symptom onset correlated with genotype with earlier onset for homozygotes, compared to heterozygotes or wild-types (median age of onset was 3, 7, and 8 years, respectively).[73] Infarction at presentation, as well as posterior circulation involvement, both signs of more severe disease, were also significantly more prevalent in homozygotes. The variant also appeared to have an effect on the clinical phenotype in one sibling pair. The child who was homozygous for the founder variant had earlier onset (age 2 vs. 17 years) with more severe disease and more extensive vascular changes compared to the heterozygous sibling.[74]

This *RFN213* variant may also have more far reaching effects. In a case control study, it was found to be associated with non-moyamoya related stenosis or occlusion of major intracranial arteries, with the variant identified in 9 of 41 patients. There was no association with the variant found for patients with either cerebral aneurysm or cervical carotid stenosis or occlusion.[75] As for the distinction between unilateral "moyamoya syndrome" and bilateral "moyamoya disease," the founder variant was identified in four of six patients with unilateral disease suggesting the unilateral and bilateral forms may have the same genetic predisposition.[71]

The susceptibility gene *RFN213* encodes a 5256 amino acid, ring finger protein 213. The RING (Really Interesting New Gene) finger motif has been associated with function as an E3 ubiquitin ligase. There is also a Walker motif, an ATPase domain characteristic of energy-dependent unfoldases. It is ubiquitously expressed, with the highest levels of expression in immune tissues such as the spleen and leukocytes.[71] There is no allele-specific expression identified in moyamoya patients.[72]

It is still not clear how variants of *RNF213* result in moyamoya disease. When expression of the *RNF213* gene was suppressed in a zebrafish model, there was severely abnormal sprouting of vessels in the head region, especially from the optic vessels. This supports the role of *RNF213* in intracranial angiogenesis.[72] However, *RNF213* deficient mice did not spontaneously develop moyamoya disease suggesting that simple functional loss of the protein activity is insufficient to induce the disorder.[76] Other genetic or environmental factors are likely to be important as the founder variant carrier frequency is high (approximately 1%) while the disease is still rare in those carriers (1/10,000).

These other factors could include infection, inflammation, autoimmunity, or exposure to radiation.[77–82] Interestingly, a transforming growth factor beta (TGF-β) polymorphism has been associated with moyamoya disease and TGF-β is elevated in cerebrospinal fluid, blood, and arteries of patients with moyamoya disease.[83,84] In addition, the same angiographic pattern can be seen in association with a variety of other disorders, including several genetic syndromes, such as sickle cell disease, hereditary spherocytosis, neurofibromatosis type I, homocystinuria, Marfan syndrome, as well as chromosomal disorders such as Down syndrome and Williams syndrome. Moyamoya syndrome has also been reported in a host of other disorders ranging from metabolic conditions, to anemias, disorders of thrombosis, other neurocutaneous disorders, autoimmune disorders, infections, and other genetic syndromes as well as cardiac disease (Table 100.2).

There is little class I or class II evidence to guide therapy in patients with moyamoya. Medical therapy with aspirin is typically recommended for stroke prevention, but the efficacy is not well established. Anticoagulation is reserved to patients who continue to have TIA or stroke despite aspirin but must be used cautiously given increased risk of hemorrhage. Optimal therapy after age 18 years is also unclear as the risk of hemorrhage is increased in adulthood. The most common clinical problems in patients with moyamoya, aside from TIA or stroke, are headache and hypertension. In general, triptans are not recommended for acute headache treatment as they are associated with vasoconstriction, but calcium channel blockers (with careful attention to maintaining adequate perfusion), and the anticonvulsants topiramate and sodium valproate, as well as amitriptyline, have been recommended.[85,86]

Treatment of hypertension in affected patients is problematic. In some, cerebral perfusion is dependent on elevated systemic blood pressure and about 10% will have renal artery stenosis. The treatment goal is then finding a balance between cerebral hypoperfusion (leading to symptomatic confusion or TIA) versus potential end organ damage from hypertension. Blood pressure tends to drop after cerebral revascularization surgery in children without renovascular disease. Sudden drops in blood pressure should be avoided as this will increase the risk of stroke and extra care must be taken to maintain fluid status during illnesses with vomiting and diarrhea as well as during the perioperative period.[86] Of note, the founder variant of the *RNF213* gene that is associated with moyamoya is also associated with elevated systolic blood pressure in a Japanese population.[87]

As the external carotid artery branches are not affected by the moyamoya arteriopathy, they can provide an alternate blood supply to the brain, but the choice of surgical treatment is controversial. The two main options are a direct or indirect approach. In the direct approach, the superficial temporal artery is anastomosed to a cerebral artery, typically the middle cerebral artery. Indirect techniques involve placing the superficial temporal artery or other vascularized tissues (such as the temporalis muscle, dura mater, or omentum) on the surface of the brain to promote collateral formation. While there is debate over the individual treatment choice, there is considerable data that the surgical outcomes, at high volume centers, are excellent compared to the natural history of the disorder.[85] In a review of 57 studies of revascularization surgery for pediatric moyamoya syndrome involving 1448 patients, the indirect procedure was performed alone in 73%, and used in combination with direct procedures in 23%.[88] The perioperative stroke rate was 4.4% and 6.1%, respectively, in these groups. Eighty-seven percent had symptomatic benefit, reported as reduction or disappearance in symptomatic cerebral ischemia with no significant difference between the indirect versus the direct/combined groups. Good collateral formation was significantly more frequent in the direct/combined group compared to the indirect group but this was not associated with differences in symptomatic outcome. Even though the progressive moyamoya arteriopathy is not arrested by revascularization surgery, there is a 96% probability of remaining stroke free in the 5 years postoperatively.[85]

GENETIC CAUSES OF SMALL VESSEL DISEASE

CADASIL

Cerebral autosomal dominant arteriopathy with subcortical infarcts and leukoencephalopathy (CADASIL) is a highly descriptive name for a disorder caused by mutations in the *NOTCH3* gene on chromosome 19.[89] This is perhaps the most common heritable cause of stroke and vascular dementia in adults, with a clinical presentation including migraine headaches and neuropsychiatric changes, including mood disturbances and dementia. There is also a high incidence of small deep and subcortical infarction on MRI with increased T2-weighted signal in the deep white matter consistent with demyelination. There is sparing of the cortex and the U-fibers. The distribution of white matter changes of CADASIL is distinct from that of hypertension. White matter changes in the anterior temporal pole and the external capsule are more common in CADASIL.[90] Two MRI studies have shown that the volume of lacunar infarction, and not the white matter abnormalities, were correlated with the degree of cognitive dysfunction.[91,92]

Clinically, the first symptom is that of recurrent migraine-like headaches or behavioral abnormalities such as manic or depressive episodes.[93] This is followed by recurrent subcortical ischemic strokes in the fifth decade, typically in the absence of other risk factors. There is progression to pseudobulbar palsy and dementia, although some may have vascular parkinsonism as a prominent feature.[94] Pediatric cases have been reported, the youngest being aged 3 years,[95] but it is more commonly reported in adolescents, where migraine and memory or attention deficits are the most common presenting signs although stroke may be the presenting symptom.[96] Often there is an unusual headache pattern with prolonged auras, or auras without headache, or complicated migraine, and the white matter lesions can be noted on MRI.[97,98]

Symptoms are primarily neurologic, although the vasculopathy of CADASIL is systemic. Electron microscopy of skin biopsies demonstrates granular osmophilic material adjacent to the basement membrane of smooth muscle cells of arterioles with a sensitivity of only 45% but a specificity of 100%. Immunostaining with a NOTCH3 monoclonal antibody can increase the sensitivity to 100%.[99] DNA testing is commercially available.

Notch3 is a transmembrane receptor with 34 tandem repeats of an epidermal growth factor-like ligand-binding extracellular domain. Notch3 signaling involves release of the cytoplasmic domain from the cell surface, which is translocated to the nucleus to function as a transcriptional activator.[100] Its expression is restricted to the vascular smooth muscle cells and in CADASIL the mutated extracellular domain forms aggregates and sequesters extracellular matrix proteins and disrupts their function.[101] Defective NOTCH3 signaling may also be involved in the disease process. In a mouse model of CADASIL, transgenic mice develop Notch3 aggregates and granular osmophilic deposits in brain vessels, white matter abnormalities, and cerebral hypoperfusion, as well as enhanced "cortical spreading depression," which is associated, in humans, with migraine aura.[100,102]

There is no specific treatment. There is a report of a high complication rate (69%) of cerebral angiography in CADASIL patients, which suggests that angiography might be contraindicated.[103] Most patients are treated with antiplatelet or anticoagulant medications but there is no evidence that treatments change the natural history of the disease and patients with CADASIL can have increased risk of brain hemorrhage.[104] Control of other risk factors for lacunar infarction, such as blood pressure and glucose, may be beneficial.[105] A clinical trial of donepezil to enhance cholinergic signaling did not improve cognition, but did show some benefit in secondary study endpoints for executive function and processing speed.[106] Acetazolamide may be helpful for migraine prophylaxis in CADASIL.[107]

CARASIL

Cerebral autosomal recessive arteriopathy with subcortical infarcts and leukoencephalopathy (CARASIL) is a small vessel vasculopathy caused by mutations in the *HTRA1* gene, which encodes HtrA serine peptidase/protease 1.[108] It has been described in several families of Japanese background. Like CADASIL, there are small vessel infarcts with an associated vascular dementia with diffuse white matter changes, but headache and migraine are not a prominent feature. Instead, the neurologic features appear at a younger age, in the 20s to 40s, and the condition is more prevalent in males (M:F=3:1). There is also alopecia, which develops after the neurologic signs, along with attacks of low back pain from disk herniation and with other bony abnormalities including kyphosis, degenerative arthropathy, and elbow deformity. Vessel walls have concentric thickening with narrowing of the lumen and fragmentation of the internal elastic lamina, but no deposition of granular osmophilic material as is seen in CADASIL. There is no known treatment and the disease is slowly progressive with stepwise deterioration and death within 10–20 years from onset of symptoms.[109,110]

COL4A1 Mutations

Autosomal dominant mutations in the alpha1 (IV)-chain of type IV collagen are associated with familial perinatal intracerebral hemorrhage with porencephaly as well as with small vessel strokes, leukoaraiosis, and microhemorrhages in adults.[111] There is also increased tortuosity of the retinal arteries, cataracts, glaucoma, and dysgenesis of the ocular anterior segment. In addition to cerebral vessel disease, there can be more widespread systemic disease with aneurysm of larger vessels, muscle cramps, Raynaud phenomena, kidney defects and cardiac arrhythmias.[112]

COL4A1 mutations should be considered in children presenting with stroke if there is a family history suggestive of autosomal dominant inheritance.[113] Outside of pre- or perinatal strokes, there can also be strokes in young adulthood. MRI shows diffuse white matter changes with involvement of the posterior periventricular areas, subcortical infarcts, cerebral microbleeds, and dilated perivascular spaces. The temporal lobe and arcuate fibers are spared, with the highest lesion load typically found in the frontal and parietal white matter.

References

1. Surtees R, Bowron A, Leonard J. Cerebrospinal fluid and plasma total homocysteine and related metabolites in children with cystathionine beta-synthase deficiency: the effect of treatment. *Pediatr Res*. 1997;42(5):577–582.
2. Mudd SH, Skovby F, Levy HL, et al. The natural history of homocystinuria due to cystathionine beta-synthase deficiency. *Am J Hum Genet*. 1985;37(1):1–31.
3. Harker LA, Slichter SJ, Scott CR, et al. Homocystinemia: vascular injury and arterial thrombosis. *N Engl J Med*. 1974;291(11):537–543.
4. Harker LA, Ross R, Slichter SJ, et al. Homocystine-induced arteriosclerosis: the role of endothelial cell injury and platelet response in its genesis. *J Clin Invest*. 1976;58(3):731–741.

5. Tsai JC, Perrella MA, Yoshizumi M, et al. Promotion of vascular smooth muscle cell growth by homocysteine: a link to atherosclerosis. *Proc Natl Acad Sci U S A*. 1994;91(14):6369–6373.
6. Buoni S, Molinelli M, Mariottini A, et al. Homocystinuria with transverse sinus thrombosis. *J Child Neurol*. 2001;16(9):688–690.
7. Yap S. Classical homocystinuria: vascular risk and its prevention. *J Inherit Metab Dis*. 2003;26(2–3):259–265.
8. Welch GN, Loscalzo J. Homocysteine and athero-thrombosis. *N Engl J Med*. 1998;338:1042–1050.
9. Selhub J, Jacques PF, Bostom AG, et al. Association between plasma homocysteine concentrations and extracranial carotid-artery stenosis. *N Engl J Med*. 1995;332(5):286–291.
10. Wang RY, Bodamer OA, Watson MS, Wilcox WR. Lysosomal storage diseases: diagnostic confirmation and management of presymptomatic individuals. *Genet Med*. 2011;13:457–484.
11. Kaler SG, Holmes CS, Goldstein DS, et al. Neonatal diagnosis and treatment of Menkes disease. *N Engl J Med*. 2008;358:605–614.
12. Menkes JH. Menkes disease and Wilson disease: two sides of the same copper coin. Part I: Menkes disease. *Eur J Paediatr Neurol*. 1999;3(4):147–158.
13. Danks DM, Campbell PE, Mayne V, et al. Menkes's kinky hair syndrome. An inherited defect in copper absorption with widespread effects. *Pediatrics*. 1972;50:188.
14. Peltonen L, Kuivaniemi H, Palotie A, et al. Alterations in copper and collagen metabolism in the Menkes syndrome and a new subtype of the Ehlers-Danlos syndrome. *Biochemistry*. 1983;22(26):6156–6163.
15. Barkovich AJ, Good WV, Koch TK, et al. Mitochondrial disorders: analysis of their clinical and imaging characteristics. *AJNR Am J Neuroradiol*. 1993;14(5):1119–1137.
16. Goldstein DS, Holmes CS, Kaler SG. Relative efficiencies of plasma catechol levels and ratios for neonatal diagnosis of Menkes disease. *Neurochem Res*. 2009;34(8):1464–1468.
17. Tumer Z, Birk Moller L, Horn N. Screening of 383 unrelated patients affected with Menkes disease and finding of 57 gross deletions in ATP7A. *Hum Mutat*. 2003;22:457–464.
18. Gerdes AM, Tønnesen T, Pergament E, et al. Variability in clinical expression of Menkes syndrome. *Eur J Pediatr*. 1988;148(2):132–135.
19. Williams DM, Atkin CL, Frens DB, et al. Menkes' Kinky hair syndrome: studies of copper metabolism and long term copper therapy. *Pediatr Res*. 1977;11(7):823–826.
20. Kaler SG, Miller RC, Wolf EJ, et al. In utero treatment of menkes disease. *Pediatr Res*. 1993;123:828.
21. Youngblom E, Knowles JW. Familial hypercholesterolemia. 2014 Jan 2. In: Pagon RA, Adam MP, Bird TD, et al., eds. *GeneReviews™ [Internet]*. Seattle (WA): University of Washington, Seattle; 1993–2014. Available from: http://www.ncbi.nlm.nih.gov/books/NBK174884/.
22. Brooks-Wilson A, Marcil M, Clee SM, et al. Mutations in ABC1 in Tangier disease and familial high-density lipoprotein deficiency. *Nat Genet*. 1999;22:336–345.
23. Pietrini V, Rizzuto N, Vergani C, Zen F, Ferro Milone F. Neuropathy in Tangier disease: a clinicopathologic study and a review of the literature. *Acta Neurol Scand*. 1985;72:495–505.
24. Schippling S, Orth M, Beisiegel U, et al. Severe Tangier disease with a novel ABCA1 gene mutation. *Neurology*. 2008;71:1454–1455.
25. Vila N, Millan M, Ferrer X, et al. Levels of alpha 1-antitrypsin in plasma and risk of spontaneous cervical artery dissections: a case-control study. *Stroke*. 2003;34:E168.
26. Pezzini A, Del Zotto E, Archettti S, et al. Plasma homocysteine concentration, C677T MTHFR genotype, and 844ins68bp CBS genotype in young adults with spontaneous cervical artery dissection and atherothrombotic stroke. *Stroke*. 2002;33:664.
27a. Pezzini A, Caso V, Zanferrari C, et al. Arterial hypertension as risk factor for spontaneous cervical artery dissection. A case–control study. *J Neurol Neurosurg Psychiatry*. 2006;77:95–97.
27b. Pezzini A, Del Zotto E, Mazziotti G, et al. Thyroid autoimmunity and spontaneous cervical artery dissection. *Stroke*. 2006;37(9):2375–2377.
28. Rubinstein SM, Peerdeman SM, van Tulder MW, et al. A systematic review of the risk factors for cervical artery dissection. *Stroke*. 2005;36:1575.
29. Pezzini A, Granella F, Grassi M, et al. History of migraine and the risk of spontaneous cervical artery dissection. *Cephalalgia*. 2005;25:575.
30. Paciaroni M, Georgiadis D, Arnold M, et al. Seasonal variability in spontaneous cervical artery dissection. *J Neurol Neurosurg Psychiatry*. 2006;77(5):677–679.
31. Dziewas R, Konrad C, Dräger B, et al. Cervical artery dissection-clinical features, risk factors, therapy and outcome in 126 patients. *J Neurol*. 2003;250(10):1179–1184.
32. Longoni M, Grond-Ginsbach C, Grau AJ, et al. The ICAM-1 E469K gene polymorphism is a risk factor for spontaneous cervical artery dissection. *Neurology*. 2006;66(8):1273–1275.
33. Baumgartner RW, Bogousslavsky J, Caso V, et al., eds. *Handbook on Cerebral Artery Dissection*. Basel: Karger; 2005.
34. Caplan LR. Dissections of brain-supplying arteries. *Nat Clin Pract Neurol*. 2008;4(1):34–42.
35. Schievink WI, Michels VV, Piepgras DG. Neurovascular manifestations of heritable connective tissue disorders. A review. *Stroke*. 1994;4:889–903.
36. Ganesan V, Kirkham FJ. Stroke due to arterial disease in congenital heart disease. *Arch Dis Child*. 1997;76(2):175.
37. Schievink WI, Mokri B, Piepgras DG, et al. Intracranial aneurysms and cervicocephalic arterial dissections associated with congenital heart disease. *Neurosurgery*. 1996;39(4):685–689.
38. Schievink WI, Mokri B. Familial aorto-cervicocephalic arterial dissections and congenitally bicuspid aortic valve. *Stroke*. 1995;26(10):1935–1940.
39. Pepin M, Schwarze U, Superti-Furga A, Byers PH. Clinical and genetic features of Ehlers-Danlos syndrome type IV, the vascular type. *N Engl J Med*. 2000;342(10):673–680.
40. North KN, Whiteman DA, Pepin MG, Byers PH. Cerebrovascular complications in Ehlers-Danlos syndrome type IV. *Ann Neurol*. 1995;38(6):960–964.
41. Austin MG, Schaefer RF. Marfan's syndrome, with unusual blood vessel manifestations. *Arch Pathol*. 1957;64(2):205–209.
42. Youl BD, Coutellier A, Dubois B, et al. Three cases of spontaneous extracranial vertebral artery dissection. *Stroke*. 1990;21(4):618–625.
43. Wityk RJ, Zanferrari C, Oppenheimer S. Neurovascular complications of Marfan syndrome: a retrospective, hospital-based study. *Stroke*. 2002;33(3):680–684.

44. Gott VL, Greene PS, Alejo DE, et al. Replacement of the aortic root in patients with Marfan's syndrome. *N Engl J Med*. 1999;340(17):1307–1313. PMID:10219065.
45. Schievink WI. Genetics of intracranial aneurysms. *Neurosurgery*. 1997;4:651–662.
46. Conway JE, Hutchins GM, Tamargo RJ. Marfan syndrome is not associated with intracranial aneurysms. *Stroke*. 1999;8:1632–1636.
47. van den Berg JS, Limburg M, Hennekam RC. Is Marfan syndrome associated with symptomatic intracranial aneurysms? *Stroke*. 1996;27(1):10–12.
48. Loeys BL, Chen J, Neptune ER, et al. A syndrome of altered cardiovascular, craniofacial, neurocognitive and skeletal development caused by mutations in TGFBR1 or TGFBR2. *Nat Genet*. 2005;37(3):275–281.
49. Loeys BL, Schwarze U, Holm T, et al. Aneurysm syndromes caused by mutations in the TGF-beta receptor. *N Engl J Med*. 2006;355(8):788–798.
50. Uitto J, Bercovitch L, Terry SF, Terry PF. Pseudoxanthoma elasticum: progress in diagnostics and research towards treatment: summary of the 2010 PXE International Research Meeting. *Am J Med Genet A*. 2011;155A:1517–1526.
51. van den Berg JS, Hennekam RC, Cruysberg JR, et al. Prevalence of symptomatic intracranial aneurysm and ischaemic stroke in pseudoxanthoma elasticum. *Cerebrovasc Dis*. 2000;10(4):315–319.
52. Coleman K, Ross MH, Mc Cabe M, et al. Disk drusen and angioid streaks in pseudoxanthoma elasticum. *Am J Ophthalmol*. 1991;112(2):166–170.
53. Carlbourg U, Ejrup B, Gronblad E, et al. Vascular studies in pseudoxanthoma elasticum and angioid streaks. *Acta Med Scand*. 1959;166 (suppl):1–68.
54. Hendig D, Knabbe C, Götting C. New insights into the pathogenesis of pseudoxanthoma elasticum and related soft tissue calcification disorders by identifying genetic interactions and modifiers. *Front Genet*. 2013;4:114. doi: 10.3389/fgene.2013.00114. eCollection 2013. PMID:23802012[PubMed] PMCID:PMC3685813.
55. Hendig D, Schulz V, Arndt M, Szliska C, Kleesiek K, Götting C. Role of serum fetuin-A, a major inhibitor of systemic calcification, in pseudoxanthoma elasticum. *Clin Chem*. 2006;52:227–234.
56. Hamlin N, Beck K, Bacchelli B, Cianciulli P, Pasquali-Ronchetti I, Le Saux O. Acquired pseudoxanthoma elasticum-like syndrome in beta-thalassaemia patients. *Br J Haematol*. 2003;122:852–854.
57. Rongioletti F, Bertamino R, Rebora A. Generalized pseudoxanthoma elasticum with deficiency of vitamin K-dependent clotting factors. *J Am Acad Dermatol*. 1989;21:1150–1152.
58. Nitschke Y, Baujat G, Botschen U, et al. Generalized arterial calcification of infancy and pseudoxanthoma elasticum can be caused by mutations in either ENPP1 or ABCC6. *Am J Hum Genet*. 2012;90(1):25–39.
59. Jiang Q, Endo M, Dibra F, Wang K, Uitto J. Pseudoxanthoma elasticum is a metabolic disease. *J Invest Dermatol*. 2009;129:348–354.
60. Scott RM, Smith ER. Moyamoya disease and moyamoya syndrome. *N Engl J Med*. 2009;360(12):1226–1237.
61. Wakai K, Tamakoshi A, Ikezaki K, et al. Epidemiological features of moyamoya disease in Japan: findings from a nationwide survey. *Clin Neurol Neurosurg*. 1997;99:S1–S5.
62. Ahn IM, Park DH, Hann HJ, Kim KH, Kim HJ, Ahn HS. Incidence, prevalence, and survival of moyamoya disease in Korea: a nationwide, population-based study. *Stroke*. 2014 Mar 4, [Epub ahead of print] PMID:24595588.
63. Baba T, Houkin K, Kuroda S. Novel epidemiological features of moyamoya disease. *J Neurol Neurosurg Psychiatry*. 2008;79:900–904.
64. Yonekawa Y, Ogata N, Kaku Y, Taub E, Imhof HG. Moyamoya disease in Europe, past and present status. *Clin Neurol Neurosurg*. 1997;99:S58–S60.
65. Kuriyama S, Kusaka Y, Fujimura M, et al. Prevalence and clinicoepidemiological features of moyamoya disease in Japan: findings from a nationwide epidemiological survey. *Stroke*. 2008;39:42–47.
66. Mineharu Y, Takenaka K, Yamakawa H, et al. Inheritance pattern of familial moyamoya disease: autosomal dominant mode and genomic imprinting. *J Neurol Neurosurg Psychiatry*. 2006;77(9):1025–1029.
67. Ikeda H, Sasaki T, Yoshimoto T, Fukui M, Arinami T. Mapping of a familial moyamoya disease gene to chromosome 3p24.2-p26. *Am J Hum Genet*. 1999;64:533–537.
68. Inoue TK, Ikezaki K, Sasazuki T, Matsushima T, Fukui M. Linkage analysis of moyamoya disease on chromosome 6. *J Child Neurol*. 2000;15:179–182.
69. Sakurai K, Horiuchi Y, Ikeda H, et al. A novel susceptibility locus for moyamoya disease on chromosome 8q23. *J Hum Genet*. 2004;49:278–281.
70. Yamauchi T, Tada M, Houkin K, et al. Linkage of familial moyamoya disease (spontaneous occlusion of the circle of Willis) to chromosome 17q25. *Stroke*. 2000;31:930–935.
71. Kamada F, Aoki Y, Narisawa A, et al. A genome-wide association study identifies RNF213 as the first Moyamoya disease gene. *J Hum Genet*. 2011;56:34–40.
72. Liu W, Morito D, Takashima S, et al. Identification of RNF213 as a susceptibility gene for moyamoya disease and its possible role in vascular development. *PLoS One*. 2011;6:e22542.
73. Miyatake S, Miyake N, Touho H, et al. Homozygous c. 14576G>A variant of RNF213 predicts early-onset and severe form of moyamoya disease. *Neurology*. 2012;78:803–810.
74. Miyatake S, Touho H, Miyake N, et al. Sibling cases of moyamoya disease having homozygous and heterozygous c.14576G>A variant in RNF213 showed varying clinical course and severity. *J Hum Genet*. 2012;57(12):804–806.
75. Miyawaki S, Imai H, Takayanagi S, Mukasa A, Nakatomi H, Saito N. Identification of a genetic variant common to moyamoya disease and intracranial major artery stenosis/occlusion. *Stroke*. 2012 Dec;43(12):3371–3374.
76. Sonobe S, Fujimura M, Niizuma K, et al. Temporal profile of the vascular anatomy evaluated by 9.4-T magnetic resonance angiography and histopathological analysis in mice lacking RNF213: a susceptibility gene for moyamoya disease. *Brain Res*. 2014;1552:64–71.
77. Ueno M, Oka A, Koeda T, Okamoto R, Takeshita K. Unilateral occlusion of the middle cerebral artery after varicella-zoster virus infection. *Brain Dev*. 2002;24:106–108.
78. El Ramahi KM, Al Rayes HM. Systemic lupus erythematosus associated with moyamoya syndrome. *Lupus*. 2000;9:632–636.
79. Ogawa K, Nagahiro S, Arakaki R, et al. Antialpha-fodrin autoantibodies in Moyamoya disease. *Stroke*. 2003;34:e244–e246.

80. Czartoski T, Hallam D, Lacy JM, Chun MR, Becker K. Postinfectious vasculopathy with evolution to moyamoya syndrome. *J Neurol Neurosurg Psychiatry*. 2005;76:256–259.
81. Tanigawara T, Yamada H, Sakai N, et al. Studies on cytomegalovirus and Epstein-Barr virus infection in moyamoya disease. *Clin Neurol Neurosurg*. 1997;99(suppl):S225–S228.
82. Bitzer M, Topka H. Progressive cerebral occlusive disease after radiation therapy. *Stroke*. 1995;26:131–136.
83. Hojo M, Hoshimaru M, Miyamoto S, et al. Role of transforming growth factor-beta1 in the pathogenesis of moyamoya disease. *J Neurosurg*. 1998;89:623–629.
84. Phillips 3rd. JA, Poling JS, Phillips CA, et al. Synergistic heterozygosity for TGFbeta1 SNPs and BMPR2 mutations modulates the age at diagnosis and penetrance of familial pulmonary arterial hypertension. *Genet Med*. 2008;10:359–365.
85. Smith ER, Scott RM. Spontaneous occlusion of the circle of Willis in children: pediatric moyamoya summary with proposed evidence-based practice guidelines. *J Neurosurg Pediatrics*. 2012;9:353–360.
86. Ganesan V. Moyamoya: to cut or not to cut is not the only question. A paediatric neurologist's perspective. *Dev Med Child Neurol*. 2010;52:10–13.
87. Koizumi A, Kobayashi H, Liu W, et al. P.R4810K, a polymorphism of RNF213, the susceptibility gene for moyamoya disease, is associated with blood pressure. *Environ Health Prev Med*. 2013;18:121–129.
88. Fung LW, Thompson D, Ganesan V. Revascularisation surgery for paediatric moyamoya: a review of the literature. *Childs Nerv Syst*. 2005;21:358–364.
89. Joutel A, Corpechot C, Ducros A, et al. Notch3 mutations in CADASIL, a hereditary adult-onset condition causing stroke and dementia. *Nature*. 1996;383:707–710.
90. O'Sullivan M, Jarosz JM, Martin RJ, Deasy N, Powell JF, Markus HS. MRI hyperintensities of the temporal lobe and external capsule in patients with CADASIL. *Neurology*. 2001;56:628–634.
91. Liem MK, van der Grond J, Haan J, et al. Lacunar infarcts are the main correlate with cognitive dysfunction in CADASIL. *Stroke*. 2007;38(3):923–928.
92. Viswanathan A, Gschwendtner A, Guichard JP, et al. Lacunar lesions are independently associated with disability and cognitive impairment in CADASIL. *Neurology*. 2007;69(2):172–179.
93. Bousser MG, Tournier-Lasserve E. Summary of the proceedings of the first international workshop on CADASIL. *Stroke*. 1994;25:704–707.
94. Ragno M, Berbellini A, Cacchio G, et al. Parkinsonism is a late, not rare, feature of CADASIL: a study on Italian patients carrying the R1006C mutation. *Stroke*. 2013;44(4):1147–1149.
95. Benabu Y, Beland M, Ferguson N, Maranda B, Boucher RM. Genetically proven cerebral autosomal-dominant arteriopathy with subcortical infarcts and leukoencephalopathy (CADASIL) in a 3-year-old. *Pediatr Radiol*. 2013;43(9):1227–1230.
96. Granild-Jensen J, Jensen UB, Schwartz M, Hansen US. Cerebral autosomal dominant arteriopathy with subcortical infarcts and leukoencephalopathy resulting in stroke in an 11-year-old male. *Dev Med Child Neurol*. 2009;51(9):754–757.
97. Cleves C, Friedman NR, Rothner AD, Hussain MS. Genetically confirmed CADASIL in a pediatric patient. *Pediatrics*. 2010;126(6):e1603–e1607.
98. Dichgans M, Mayer M, Muller-Mysok B, Straube A, Gasser T. Identification of a key recombinant narrows the CADASIL gene region to 8 cM and argues against allelism of CADASIL and familial hemiplegic migraine. *Genomics*. 1996;32:151–154.
99. Markus HS, Martin RJ, Simpson MA, et al. Diagnostic strategies in CADASIL. *Neurology*. 2002;59:1134–1138.
100. Joutel A, Monet-Lepretre M, Gosele C, et al. Cerebrovascular dysfunction and microcirculation rarefaction precede white matter lesions in a mouse genetic model of cerebral ischemic small vessel disease. *J Clin Invest*. 2010;120:433–445.
101. Monet-Lepretre M, Haddad I, Baron-Menguy C, et al. Abnormal recruitment of extracellular matrix proteins by excess Notch3 ECD: a new pathomechanism in CADASIL. *Brain*. 2013;136(Pt 6):1830–1845.
102. Eikermann-Haerter K, Yuzawa I, Dilekoz E, et al. Cerebral autosomal dominant arteriopathy with subcortical infarcts and leukoencephalopathy syndrome mutations increase susceptibility to spreading depression. *Ann Neurol*. 2011;69:413–418.
103. Dichgans M, Petersen D. Angiographic complications in CADASIL. *Lancet*. 1997;349:776–777.
104. Kotorii S, Goto H, Kondo T, Matsuo H, Takahashi K, Shibuya N. Case of CADASIL showing spontaneous subcortical hemorrhage with a novel mutation of Notch3 gene. *Rinsho Shinkeigaku*. 2006;46(9):644–648.
105. Viswanathan A, Guichard JP, Gschwendtner A, et al. Blood pressure and haemoglobin A1c are associated with microhaemorrhage in CADASIL: a two-centre cohort study. *Brain*. 2006;129(Pt 9):2375–2383.
106. Dichgans M, Markus HS, Salloway S, et al. Donepezil in patients with subcortical vascular cognitive impairment: a randomised double-blind trial in CADASIL. *Lancet Neurol*. 2008;7(4):310–318.
107. Donnini I, Nannucci S, Valenti R, et al. Acetazolamide for the prophylaxis of migraine in CADASIL: a preliminary experience. *J Headache Pain*. 2012;13:299–302.
108. Fukutake T. Cerebral autosomal recessive arteriopathy with subcortical infarcts and leukoencephalopathy (CARASIL):from discovery to gene identification. *J Stroke Cerebrovasc Dis*. 2011;20(2):85–93.
109. Fukutake T, Hirayama K. Familial young-adult-onset arteriosclerotic leukoencephalopathy with alopecia and lumbago without arterial hypertension. *Eur Neurol*. 1995;35(2):69–79.
110. Yanagawa S, Ito N, Arima K, Ikeda S. Cerebral autosomal recessive arteriopathy with subcortical infarcts and leukoencephalopathy. *Neurology*. 2002;58(5):817–820.
111. Gould DB, Phalan FC, Breedveld GJ, et al. Mutations in COL4A1 cause perinatal cerebral hemorrhage and porencephaly. *Science*. 2005;308:1167–1171.
112. Sibon I, Coupry I, Menegon P, et al. COL4A1 mutation in Axenfeld-Rieger anomaly with leukoencephalopathy and stroke. *Ann Neurol*. 2007;62:177–184.
113. Shah S, Kumar Y, McLean B, et al. A dominantly inherited mutation in collagen IV A1 (COL4A1) causing childhood onset stroke without porencephaly. *Eur J Paediatr Neurol*. 2010;14(2):182–187.

CHAPTER

101

Coagulopathies

Fenella J. Kirkham

Institute of Child Health, University College London, London, UK

INTRODUCTION

There is a long list of inherited conditions that predispose to cerebrovascular disease and stroke, but our understanding of the precise mechanisms has been limited until recently. Several conditions with cerebral vascular manifestations are now being intensively studied and there has been some initial exploration of related areas within neurology and psychiatry, including migraine and dementia. The distribution of candidate genes for stroke risk associated with coagulopathy is very different in diverse populations, which makes pooling the results of small single-center studies fraught with difficulty. The advent of genome-wide studies will lead to an explosion of literature on this subject over the next few years. This chapter deals with a few of the conditions that have been extensively studied in adults and children, with enough patients studied for appropriate meta-analysis, as models for the study of associations. Some conditions for which there are few large studies in adults, or genetic disorders for which there is an important mechanism involving lipid metabolism[1] or the vessel wall, such as ADAMSTS13[2] and von Willebrand factor[3] or potentially acquired acute abnormality, such as protein S and C[1,4] and antithrombin deficiencies and increased factor VIII, are not discussed.

Although considered less important in etiology of stroke in adults, recognized disorders of coagulation occur acutely in up to 38% of children with childhood stroke,[5] but up to half resolve within 3 months of the ictus, and the prevalence of known inherited coagulopathies is around 10% in previously well patients.[6] Williams and colleagues quoted a higher figure of 25%,[7] but this series included a high proportion of children with sickle cell disease, in which prothrombotic abnormalities are known to be common, but are not necessarily associated with stroke and cerebrovascular disease.[8]

Β-FIBRINOGEN ON CHROMOSOME 4q28

Historical Overview and Disease Identification

Fibrinogen is a glycoprotein of 340,000 Da consisting of three polypeptide chains: α, β and γ. Genes encoding the polypeptides are colocated on chromosome 4q28. Fibrinogen levels are determined by environmental factors, e.g., smoking and obesity, particularly in men,[9] although also in children,[10] and a genetic component that may account for up to half of the variability, although currently known variants explain only a small proportion.[11] The research focus has mainly been on genetic variability in β-fibrinogen.

Molecular Genetics

- Normal gene product: fibrinogen.
- Abnormal gene product(s): increased fibrinogen level.

Rosenberg's Molecular and Genetic Basis of Neurological and Psychiatric Disease
http://dx.doi.org/10.1016/10.1016/B978-0-12-410529-4.00101-7

Clinical Features and Genotype–Phenotype Correlations

Cerebral Venous Thrombosis

Variation in the α and γ chains may be important in venous thromboembolism[12–16] although there are few data on cerebral venous sinus thrombosis (Figures 101.1 and 101.2).

Arterial Ischemic Stroke

There is evidence for an association of high fibrinogen levels with stroke (Figures 101.3 and 101.4) in adults,[17,18] with evidence for an effect in large vessel disease and lacunar stroke/silent cerebral infarction.[19,20] A recent case control study suggests that high fibrinogen levels are also a risk factor for stroke in children.[21] High fibrogen levels also appear to predict poor outcome[22] and symptomatic recurrence in adults.[23]

Large studies in Asia suggest associations between polymorphisms in the β-fibrogen gene (-148C/T, -448G/A and -455G/A) and stroke risk[24–30] and in some European populations[9,31] but not others.[32] All stroke subtypes have been reported as associated.[31,33] There are few data in children.

Migraine

There is evidence for an increase in fibrinogen levels in patients with migraine,[34] although not all studies have found an association in adults[35] or children.[36] Levels do not appear to be higher during attacks[34] but are higher in young women with migraine and lacunar stroke than in those with either migraine or lacunar stroke.[37]

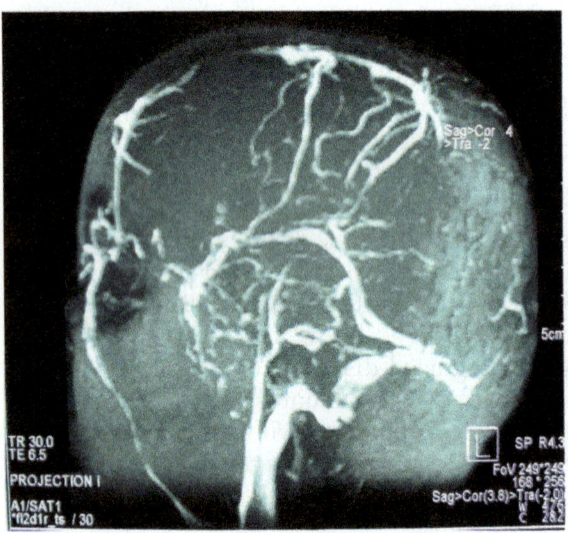

FIGURE 101.1 Magnetic resonance venogram showing chronic occlusion of the posterior portion of the sagittal sinus in a child with a hemolytic anemia and a psychiatric presentation who had systemic lupus erythematosus.

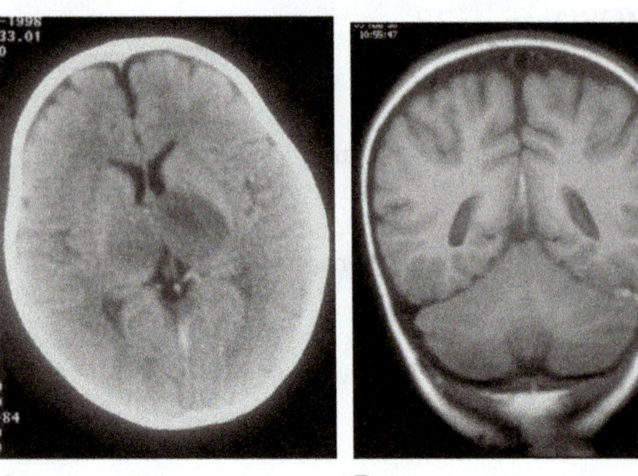

FIGURE 101.2 Bilateral thalamic infarction (A) in a child with severe iron deficiency anemia (hemoglobin 4 g/dL) and transverse sinus thrombosis (B).

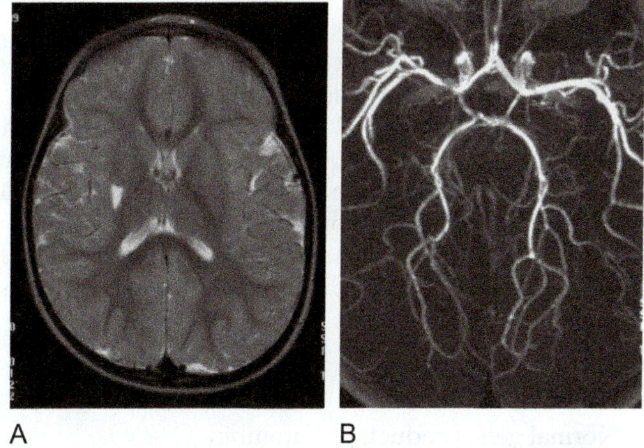

FIGURE 101.3 Small basal ganglia infarct (A) in a child with focal cerebral arteriopathy involving the middle cerebral artery on magnetic resonance arteriography (B).

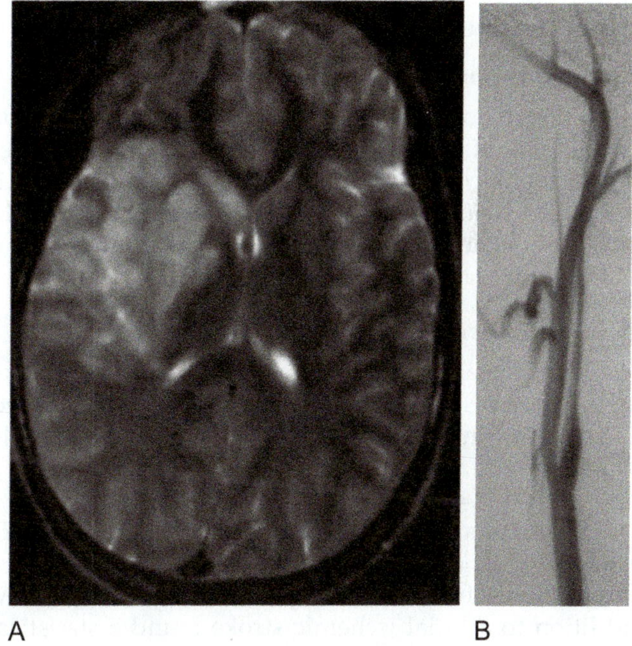

FIGURE 101.4 Large basal ganglia infarct (A) in a child with a conventional arteriogram showing a "rat's tail" appearance consistent with an internal carotid artery dissection (B).

Dementia

Increased fibrinogen levels have been associated with vascular dementia,[38] Alzheimer disease[39–41] and dementia in Parkinson disease.[42] Plasma concentrations of five proteins, including γ-fibrinogen, together with age and sex, explained more than 35% of variance in whole brain volume in Alzheimer disease patients and is consistent with the possibility that coagulation plays a role in pathogenesis.[43] Peptide fragments of β-fibrinogen may be related to a useful biomarker in Alzheimer disease.[44]

FACTOR II (PROTHROMBIN) GENE 20210G>A ON 11p11-q12

Historical Overview and Disease Identification

Prothrombin (factor II) is a proenzyme for thrombin, a serine protease that catalyses the conversion of fibrinogen to fibrin. Prothrombin levels are higher in patients with venous thrombosis than in controls.[45] Prothrombin is encoded by the factor II gene, 11p11-q12. A single-nucleotide polymorphism in position 20210G/A is associated with an increased prothrombin level and thrombin formation.

Gene Identification

The prothrombin gene was examined as a candidate gene for venous thrombosis in patients with a documented familial history by Poort and colleagues.[45] All the exons and the 5′- and 3′-UT region of the prothrombin gene were analyzed by polymerase chain reaction and direct sequencing in 28 probands. No deviations were found in the coding regions and the 5′-UT region known polymorphic sites. Only one nucleotide change (a G to A transition) at position 20210 was identified in the sequence of the 3′-UT region[45] in 18% of patients with venous thrombosis and 1% of controls. The incidence of the 20210A allele in a group of 164 healthy controls was 1.2% (allele frequency 0.61%, 95% CI 0.08–2.19).[46] Further studies have confirmed that heterozygosity is a risk factor for venous and arterial thrombosis[47] while homozygotes have a high risk[48] and a wide variety of presentations.

Molecular Genetics

- Normal gene product: prothrombin.
- Abnormal gene product(s): increased prothrombin level.

Clinical Features and Genotype–Phenotype Correlations

Cerebral Venous Thrombosis

Although not commonly found in cerebral venous sinus thrombosis,[49,50] and not in excess compared with control populations of children in relatively small studies,[51] the factor II (prothrombin) G20210A mutation appears to be a risk factor for cerebral venous thrombosis[52] (Figures 101.1 and 101.2) and is more prevalent in this group than in those with deep venous thrombosis in the lower extremity.[53] In pooled data from several European studies, this mutation was the only one to be associated with increased risk of recurrence.[54] There was no evidence of increased prevalence of this mutation in dural arteriovenous fistulae.[55]

Arterial Ischemic Stroke

There appears to be an association with arterial stroke (Figures 101.3 and 101.4) in young adults and children.[48,56] In the Genetics of Early-Onset Stroke study, adults aged 15–42 years with ischemic stroke were significantly more likely than controls to have the mutation (odds ratio [OR] = 5.9; 95% confidence intervals [CI] = 1.2–28.1; P = 0.03) and in a meta-analysis of 17 studies, the mutation was associated with significantly increased stroke risk in adults ≤55 years (OR = 1.4; 95% CI = 1.1–1.9; P = 0.02).[57]

In a meta-analysis confined to stroke in an arterial territory in children, for which 18 publications fulfilled entry criteria, factor II G20210A was more common in children with arterial ischemic stroke than controls.[58] A full meta-analysis looking at cerebral venous sinus thrombosis (CVST) and perinatal stroke in addition to arterial ischemic stroke found a statistically significant association with stroke onset for factor II G20210A (OR = 2.52; 95% CI = 1.71–3.72).[59] However, not all studies have found an association and there may be geographical differences.[60,61]

FACTOR V LEIDEN

Factor V is a pro-enzyme cleaved to active factor Va under the influence of thrombin, which is important for the conversion of prothrombin to thrombin. Factor V is encoded by a gene on chromosome 1q23. Activated protein C resistance was discovered in a man with a strong personal and family history of multiple thrombotic events[62] who did not have prolongation of clotting time with activated protein C in activated partial thromboplastin time, factor IXa- or Xa-based assays. The activated protein C (APC) limits clot formation by proteolytic inactivation of factors Va and VIIIa. Although there are other genes underlying APC resistance, the Leiden mutation in the position 1691G/A of the factor V gene has been studied most comprehensively. After activation, abnormal factor V is less efficiently degraded by activated protein C, leading to increased thrombin generation. Factor V Leiden is a highly prevalent hereditary risk factor for venous thromboembolism in Europe but less so in Asia and Africa,[63] with the risk of thrombosis increased by up to 10-fold in heterozygotes and up to 100-fold in homozygotes.

Molecular Genetics

- Abnormal gene product(s): activated protein C resistance.

Genotype–Phenotype Correlations

Cerebral Venous Sinus Thrombosis

Although the proportion of patients with cerebral venous sinus thrombosis (Figures 101.1 and 101.2) who have factor V Leiden is relatively small,[49] there is evidence for an association in adult[64,65] and pediatric case control studies,[51] and meta-analyses involving mainly European populations.[59] However, there is no evidence for an increased risk in Indian women during the puerperium.[66] In adults the mutation predicts recurrence[64] but there is no evidence that this is the case in children.[54]

Arterial Ischemic Stroke

Activated protein C resistance or the factor V Leiden mutation has been linked to stroke in adults, particularly women and those aged <55 years,[67,68] as well as neonates[69–71] and children, in populations of mainly Caucasian extraction.[58,59,72] For South America, one study found a link[73] while others in adults[74] and children[75] did not, probably representing differences in ethnicity in this continent. Factor V Leiden was not found in any child with stroke in an Indian series[76] and is rare in Afro-Caribbeans, and at present, there is little evidence for a link with stroke for this or any other genetically determined prothrombotic disorder in patients with sickle cell disease.[8]

In fact, large case control studies of stroke secondary to all causes have found little evidence for an association with arterial ischemic stroke in adults.[56,77–81] Although there is little evidence that the prevalence of the factor V Leiden mutation is greater in adults with stroke specifically secondary to atrial fibrillation,[82,83] there is some evidence for an effect on early and total recurrence risk (Figure 101.5).[83] For neonatal stroke there is evidence for an effect on outcome.[84]

Migraine

An excess of factor V Leiden has been found in patients with migraine with aura in some populations[85,86] but not others.[87,88] There is little evidence for any association with migrainous infarction.[89,90]

Dementia

There is evidence for an association of the factor V Leiden mutation with dementia, specifically vascular dementia in some populations[91] but not others.[92]

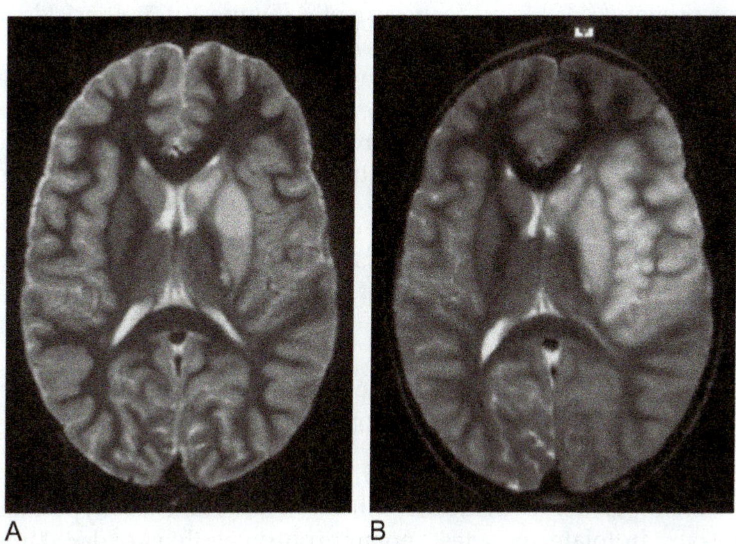

FIGURE 101.5 **Magnetic resonance imaging (MRI) scans showing enlargement of a stroke in the middle cerebral artery territory in an 8-year-old boy with middle cerebral artery stenosis on MR and conventional arteriography who was heterozygous for factor V Leiden and prothrombin 20210 and homozygous for thermolabile methylene tetrahydrofolate reductase mutations.**

THERMOLABILE METHYLENE TETRAHYDROFOLATE REDUCTASE POLYMORPHISM

Historical Overview and Disease Identification

Homocysteine is a sulphur-containing amino acid synthesized during the metabolism of methionine, which in excess has adverse effects on the vessel wall rather than on blood clotting, although it is often considered with the prothrombotic disorders discussed above. Classical homocystinuria (deficiency of cystathionine β-synthase) has long been recognized as an important cause of arterial vascular disease and infarction, and may present with stroke in infancy, childhood, and adolescence. The large number of mutations of cystathionine β-synthase are individually relatively rare.

Gene Identification

Reduction in the activity of the 5,10-methylene tetrahydrofolate reductase gene also results in reduction of 5-methyl tetrahydrofolate for the conversion of homocysteine to methionine and therefore also in hyperhomocysteinemia. A common thermolabile variant of the methylene tetrahydrofolate reductase gene was identified in 1995,[93] for which about 60% of Caucasians are heterozygous (CT) and 5–15% homozygous (TT).

Molecular Genetics

- Normal gene product: homocysteine.
- Abnormal gene product(s): high homocysteine.

Genotype–Phenotype Correlations

Cerebral Venous Sinus Thrombosis

The thermolabile methylene tetrahydrofolate reductase mutation is a risk factor for cerebral venous sinus thrombosis (Figures 101.1 and 101.2) in adults,[52] but the available data have not shown a clear association in children.[59] There is no evidence for an effect on recurrence risk.[54]

Arterial Ischemic Stroke

In adults, stroke risk is linked to homocysteine levels[94–96] and also appears to be a risk factor for childhood[97,98] and neonatal[99] stroke. Homocysteine levels appear to be higher in patients with sickle cell disease who have stroked than in those who have not, and are inversely related to red cell folate, despite the relatively low prevalence of the thermolabile methylene tetrahydrofolate reductase mutation in Afro-Caribbeans.[100] High homocysteine level is a risk factor for recurrent stroke (Figure 101.5) and all-cause mortality in adults.[101]

Homozygosity for the thermolabile methylene tetrahydrofolate reductase polymorphism appears to be a risk factor for primary[72,98,102,103] and for recurrent[102] (Figure 101.5) nonsickle stroke in childhood in Northern Europe, but not in Turkey[104] or India.[76] Meta-analysis has confirmed the association.[58,59,105] There is a trend for an effect in neonatal stroke.[106] Studies of the association of ischemic stroke in adults with thermolabile methylene tetrahydrofolate reductase status are conflicting, perhaps because of the confounding effect of acquired risk factors (e.g., smoking, poor diet, type 2 diabetes, metabolic syndrome) in this age group as well as geographical variations in the prevalence of the allele.[27,56,77,94,107–112] There are other common genes influencing homocysteine metabolism, some of which appear to influence ischemic stroke.[113–116]

Hemorrhagic Stroke

The thermolabile methylene tetrahydrofolate reductase polymorphism also appears to be a risk factor for hemorrhagic stroke (Figure 101.6) in adults.[117–120]

Migraine

Several polymorphisms of the methylene tetrahydrofolate reductase gene, including the thermolabile variant, have been reported to be more common in patients with migraine with aura than controls.[121,122]

Dementia

There is some evidence for an association of hyperhomocysteinemia and the thermolabile methylene tetrahydrofolate reductase polymorphism with vascular dementia but little evidence of any association with Alzheimer disease.[123–125]

FIGURE 101.6 Computed tomography scan showing intracerebral hemorrhage (A) in a child with an arteriovenous malformation (B).

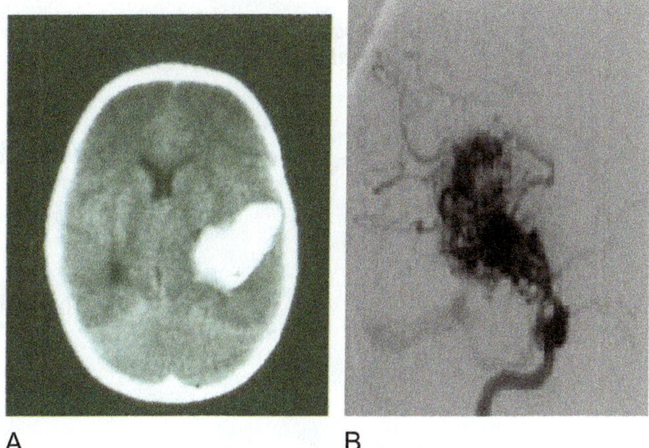

DISEASE MECHANISMS

Specific Vascular Stroke Syndromes

Venous Sinus (Sinovenous) Thrombosis

Genetic causes of thrombophilia, including factor V Leiden, prothrombin 20210, and the thermolabile variant of the methylene tetrahydrofolate reductase gene are risk factors for cerebral venous sinus (sinovenous) thrombosis (Figures 101.1 and 101.2),[52] together with environmental triggers such as systemic illness or infection, dehydration, and iron deficiency anemia.

Paradoxical Embolus from the Venous Circulation through a Patent Foramen Ovale

The mechanism of increased arterial ischemic stroke risk may involve paradoxical embolism through a patent foramen ovale (Figure 101.7), particularly in cryptogenic stroke.[126–128] This implies that the thrombosis associated with this polymorphism occurs in the venous circulation even when the stroke is in an arterial territory. It is possible that the link between stroke and migraine involves this mechanism as there is little evidence for atherosclerosis of large vessels.[129]

Arterial Stenosis

Although high fibrinogen levels may be associated with large vessel disease,[31] there is little evidence that any of the polymorphisms discussed is a risk factor for cervical artery stenosis in adults[130,131] or intracranial stenosis in children (Figure 101.3). However, one study suggested that homozygotes for the thermolabile variant of the methylene tetrahydrofolate reductase gene were more likely to have carotid atherosclerosis.[132]

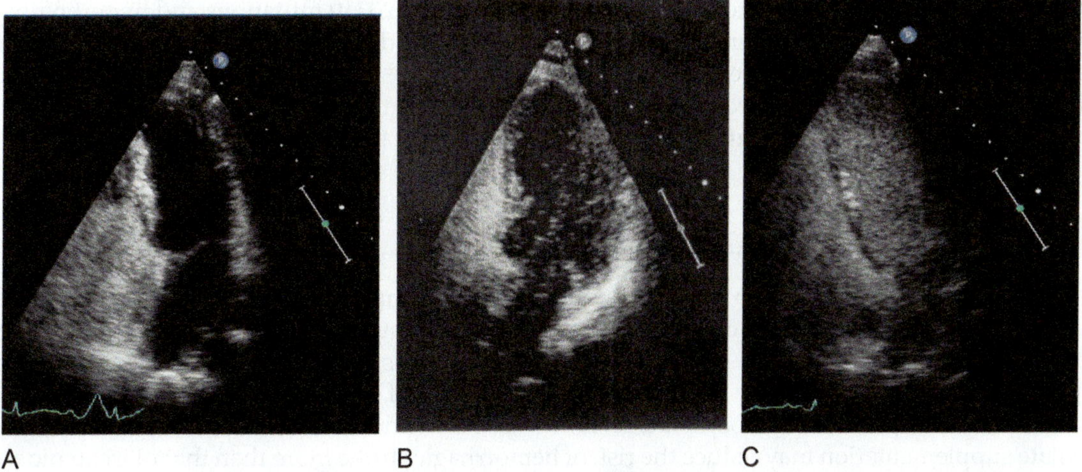

FIGURE 101.7 Echocardiograms showing: (A) an intact interatrial septum; (B) a few microemboli crossing a patent foramen ovale; (C) many microemboli crossing a patent foramen ovale.

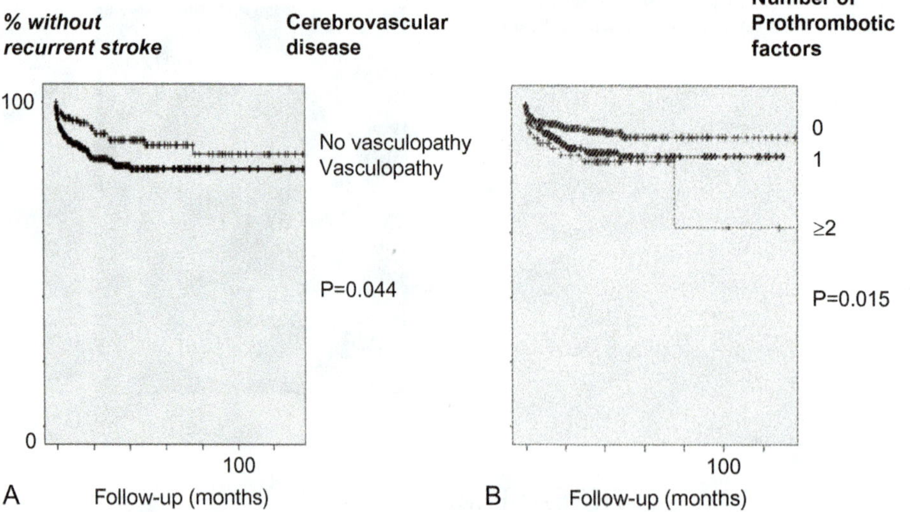

FIGURE 101.8 **Kaplan–Meier curves showing time to recurrent stroke for children with and without vasculopathy (A) and with no, 1 or ≥2 prothrombotic disorders identified in data pooled from stroke databases from Münster, Toronto and London (B).** P values are for comparisons using the Log rank test.

Arterial Dissection

Dissection of the extracranial and intracranial vasculature (Figure 101.4) is certainly an important cause of stroke, but there is little evidence for an important link with prothrombotic disorders.[133] Homocysteine levels were higher in adults with arterial dissection than in controls in some series,[134] but there is controversy over whether the thermolabile variant of the methylene tetrahydrofolate reductase gene polymorphism is associated,[135] with or without migraine with aura.[136]

Moyamoya

Primary moyamoya disease is relatively common in Asia, particularly in Japan, and sporadic primary cases are also relatively common in this population. In the USA and Europe, however, most cases are of moyamoya syndrome secondary to other conditions with Mendelian inheritance. There is also evidence for a role of prothrombotic disorders, particularly lupus anticoagulant.[137] The relative importance of genetic and environmental factors remains controversial and although the radiological appearances are similar, there may be important differences in pathology.

Interactions

A significant proportion of patients with stroke have multiple prothrombotic disorders and there is evidence for interaction, for example between the factor V Leiden and prothrombin 20210 mutations and hyperhomocysteinemia in determining risk of stroke and dementia. Multiple prothrombotic disorders appear to predict recurrent stroke (Figure 101.8),[138] particularly in cryptogenic arterial ischemic stroke.[139] Other risk factors, including hypertension, type 2 diabetes, obesity, metabolic syndrome, infection, trauma, exposure to hypoxia (e.g., during obstructive sleep apnea), smoking and air pollution appear to interact with genetic risk factors in complex ways.

MANAGEMENT

There is little evidence that long-term anticoagulation reduces primary or secondary stroke or dementia risk in any of these prothrombotic disorders. A case can be made for not investigating patients with stroke or dementia for prothrombotic disorders as appropriate advice on aspirin prophylaxis, healthy diet, discontinuing smoking, etc., would be given whatever the laboratory diagnosis.[140–142] High intake of fruit and vegetables is unlikely to do harm. Apart from genetic predisposition, homocysteine levels are also influenced by intake of folate, vitamin B_{12} and vitamin B_6.[143] Folate supplementation may reduce the risk of hemorrhagic stroke more than that of ischemic stroke.[117] The data on whether supplementation of these vitamins reduces stroke risk is conflicting, at least in part because of the geographical differences in universal supplementation in bread flour.[144]

SUMMARY AND FUTURE DIRECTIONS

Although conditions with a clear genetic basis are commonly associated with childhood stroke, we are a long way from understanding the mechanisms. A very clear description of the phenotype will be important, as, for example, the genetic basis of cerebrovascular disease with the angiographic appearances of moyamoya may be very different from that associated with venous sinus thrombosis, even if the underlying sickle cell disease is the same. The relative importance of intermediate factors, such as prothrombotic disorders, abnormalities of homocysteine metabolism, and factors affecting endothelial function may be very different in different populations. To date, most studies have used a candidate gene approach and a case control design, which has limitations, but international collaboration has recently allowed more sophisticated approaches. It is likely that stroke occurs secondary to interactions between several genes and the environment and teasing out the relative importance of each component will be complex. The hope is that an understanding of the genetic basis of stroke and dementia will lead to better management of this important group of patients.

References

1. Strater R, Becker S, von Eckardstein A, et al. Prospective assessment of risk factors for recurrent stroke during childhood—a 5-year follow-up study. *Lancet*. 2002;360:1540–1545.
2. Lambers M, Goldenberg NA, Kenet G, et al. Role of reduced ADAMTS13 in arterial ischemic stroke: a pediatric cohort study. *Ann Neurol*. 2013;73:58–64.
3. Della-Morte D, Beecham A, Dong C, et al. Association between variations in coagulation system genes and carotid plaque. *J Neurol Sci*. 2012;323:93–98.
4. Douay X, Lucas C, Caron C, Goudemand J, Leys D. Antithrombin, protein C and protein S levels in 127 consecutive young adults with ischemic stroke. *Acta Neurol Scand*. 1998;98:124–127.
5. deVeber G, Monagle P, Chan A, et al. Prothrombotic disorders in infants and children with cerebral thromboembolism. *Arch Neurol*. 1998;55:1539–1543.
6. Ganesan V, McShane MA, Liesner R, Cookson J, Hann I, Kirkham FJ. Inherited prothrombotic states and ischaemic stroke in childhood. *J Neurol Neurosurg Psychiatry*. 1998;65:508–511.
7. Williams LS, Garg BP, Cohen M, Fleck JD, Biller J. Subtypes of ischemic stroke in children and young adults. *Neurology*. 1997;49:1541–1545.
8. Liesner R, Mackie I, Cookson J, et al. Prothrombotic changes in children with sickle cell disease: relationships to cerebrovascular disease and transfusion. *Br J Haematol*. 1998;103:1037–1044.
9. Carter AM, Catto AJ, Bamford JM, Grant PJ. Gender-specific associations of the fibrinogen B beta 448 polymorphism, fibrinogen levels, and acute cerebrovascular disease. *Arterioscler Thromb Vasc Biol*. 1997;17:589–594.
10. Nowak-Gottl U, Langer C, Bergs S, Thedieck S, Strater R, Stoll M. Genetics of hemostasis: differential effects of heritability and household components influencing lipid concentrations and clotting factor levels in 282 pediatric stroke families. *Environ Health Perspect*. 2008;116:839–843.
11. Sabater-Lleal M, Huang J, Chasman D, et al. Multiethnic meta-analysis of genome-wide association studies in >100,000 subjects identifies 23 fibrinogen-associated loci but no strong evidence of a causal association between circulating fibrinogen and cardiovascular disease. *Circulation*. 2013;128:1310–1324.
12. Nowak-Gottl U, Weiler H, Hernandez I, et al. Fibrinogen alpha and gamma genes and factor V Leiden in children with thromboembolism: results from two family-based association studies. *Blood*. 2009;114:1947–1953.
13. Carter AM, Catto AJ, Kohler HP, Ariens RA, Stickland MH, Grant PJ. Alpha-fibrinogen Thr312Ala polymorphism and venous thromboembolism. *Blood*. 2000;96:1177–1179.
14. Ko YL, Hsu LA, Hsu TS, et al. Functional polymorphisms of FGA, encoding alpha fibrinogen, are associated with susceptibility to venous thromboembolism in a Taiwanese population. *Hum Genet*. 2006;119:84–91.
15. Uitte de WS, de Visser MC, Houwing-Duistermaat JJ, Rosendaal FR, Vos HL, Bertina RM. Genetic variation in the fibrinogen gamma gene increases the risk for deep venous thrombosis by reducing plasma fibrinogen gamma' levels. *Blood*. 2005;106:4176–4183.
16. Reiner AP, Lange LA, Smith NL, Zakai NA, Cushman M, Folsom AR. Common hemostasis and inflammation gene variants and venous thrombosis in older adults from the Cardiovascular Health Study. *J Thromb Haemost*. 2009;7:1499–1505.
17. Beg M, Nizami A, Singhal KC, Mohammed J, Gupta A, Azfar SF. Role of serum fibrinogen in patients of ischemic cerebrovascular disease. *Nepal Med Coll J*. 2007;9:88–92.
18. Zhu YC, Cui LY, Hua BL, Pan JQ. Correlation between fibrinogen level and cerebral infarction. *Chin Med Sci J*. 2006;21:167–170.
19. Lang Q, Zhou M, Feng H, Guo J, Chen N, He L. Research on the relationship between fibrinogen level and subtypes of the TOAST criteria in the acute ischemic stroke. *BMC Neurol*. 2013;13:207.
20. Aono Y, Ohkubo T, Kikuya M, et al. Plasma fibrinogen, ambulatory blood pressure, and silent cerebrovascular lesions: the Ohasama study. *Arterioscler Thromb Vasc Biol*. 2007;27:963–968.
21. Kopyta I, Sarecka-Hujar B, Emich-Widera E, Marszal E, Zak I. Association between lipids and fibrinogen levels and ischemic stroke in the population of the Polish children with arteriopathy and cardiac disorders. *Wiad Lek*. 2010;63:17–23.
22. Swarowska M, Janowska A, Polczak A, et al. The sustained increase of plasma fibrinogen during ischemic stroke predicts worse outcome independently of baseline fibrinogen level. *Inflammation*. 2014.
23. Tohgi H, Chiba K, Takahashi H, Tamura K, Sasaki K, Suzuki H. Comparison of symptomatic and asymptomatic reinfarctions after small subcortical stroke. *Eur Neurol*. 1994;34:140–146.

24. Chen XC, Xu MT, Zhou W, Han CL, Chen WQ. A meta-analysis of relationship between beta-fibrinogen gene -148C/T polymorphism and susceptibility to cerebral infarction in Han Chinese. *Chin Med J (Engl)*. 2007;120:1198–1202.
25. Guo X, Zhang D, Zhang X. Fibrinogen gene polymorphism Bbeta-148C/T. in Uygur patients with cerebral infarction. *Neurol Res*. 2009;31:381–384.
26. Chen XC, Xu MT, Zhou W, Han CL, Chen WQ. A meta-analysis of beta-fibrinogen gene-455G/A polymorphism and plasma fibrinogen level in Chinese cerebral infarction patients. *Biomed Environ Sci*. 2007;20:366–372.
27. Xu X, Li J, Sheng W, Liu L. Meta-analysis of genetic studies from journals published in China of ischemic stroke in the Han Chinese population. *Cerebrovasc Dis*. 2008;26:48–62.
28. Gu L, Liu W, Yan Y, et al. Influence of the beta-fibrinogen-455G/A polymorphism on development of ischemic stroke and coronary heart disease. *Thromb Res*. 2014; S0049-3848:00010-3.
29. Gu L, Wu G, Su L, et al. Genetic polymorphism of beta-fibrinogen gene-455G/A can contribute to the risk of ischemic stroke. *Neurol Sci*. 2014;35:151–161.
30. Lee SH, Kim MK, Park MS, et al. Beta-fibrinogen gene -455 G/A polymorphism in Korean ischemic stroke patients. *J Clin Neurol*. 2008;4:17–22.
31. Kessler C, Spitzer C, Stauske D, et al. The apolipoprotein E and beta-fibrinogen G/A-455 gene polymorphisms are associated with ischemic stroke involving large-vessel disease. *Arterioscler Thromb Vasc Biol*. 1997;17:2880–2884.
32. Golenia A, Chrzanowska-Wasko J, Jagiella J, et al. The beta-fibrinogen -455G/A gene polymorphism and the risk of ischaemic stroke in a Polish population. *Neurol Neurochir Pol*. 2013;47:152–156.
33. Martiskainen M, Pohjasvaara T, Mikkelsson J, et al. Fibrinogen gene promoter -455 A allele as a risk factor for lacunar stroke. *Stroke*. 2003;34:886–891.
34. Yucel Y, Tanriverdi H, Arikanoglu A, et al. Increased fibrinogen, D-dimer and galectin-3 levels in patients with migraine. *Neurol Sci*. 2014;35:545–549.
35. Kurth T, Ridker PM, Buring JE. Migraine and biomarkers of cardiovascular disease in women. *Cephalalgia*. 2008;28:49–56.
36. Herak DC, Antolic MR, Krleza JL, et al. Inherited prothrombotic risk factors in children with stroke, transient ischemic attack, or migraine. *Pediatrics*. 2009;123:e653–e660.
37. Salobir B, Sabovic M, Peternel P, Stegnar M, Grad A. Classic risk factors, hypercoagulability and migraine in young women with cerebral lacunar infarctions. *Acta Neurol Scand*. 2002;105:189–195.
38. Gallacher J, Bayer A, Lowe G, et al. Is sticky blood bad for the brain?: Hemostatic and inflammatory systems and dementia in the Caerphilly Prospective Study. *Arterioscler Thromb Vasc Biol*. 2010;30:599–604.
39. Cortes-Canteli M, Zamolodchikov D, Ahn HJ, Strickland S, Norris EH. Fibrinogen and altered hemostasis in Alzheimer's disease. *J Alzheimers Dis*. 2012;32:599–608.
40. Cortes-Canteli M, Paul J, Norris EH, et al. Fibrinogen and beta-amyloid association alters thrombosis and fibrinolysis: a possible contributing factor to Alzheimer's disease. *Neuron*. 2010;66:695–709.
41. Wood H. Alzheimer disease: fibrinogen links amyloid with vascular dysfunction. *Nat Rev Neurol*. 2010;6:413.
42. Slawek J, Roszmann A, Robowski P, et al. The impact of MRI white matter hyperintensities on dementia in Parkinson's disease in relation to the homocysteine level and other vascular risk factors. *Neurodegener Dis*. 2013;12:1–12.
43. Thambisetty M, Simmons A, Hye A, et al. Plasma biomarkers of brain atrophy in Alzheimer's disease. *PLoS One*. 2011;6:e28527.
44. Noguchi M, Sato T, Nagai K, et al. Roles of serum fibrinogen alpha chain-derived peptides in Alzheimer's disease. *Int J Geriatr Psychiatry*. 2013.
45. Poort SR, Rosendaal FR, Reitsma PH, Bertina RM. A common genetic variation in the 3'-untranslated region of the prothrombin gene is associated with elevated plasma prothrombin levels and an increase in venous thrombosis. *Blood*. 1996;88:3698–3703.
46. Cumming AM, Keeney S, Salden A, Bhavnani M, Shwe KH, Hay CR. The prothrombin gene G20210A variant: prevalence in a U.K. anticoagulant clinic population. *Br J Haematol*. 1997;98:353–355.
47. Young G, Manco-Johnson M, Gill JC, et al. Clinical manifestations of the prothrombin G20210A mutation in children: a pediatric coagulation consortium study. *J Thromb Haemost*. 2003;1:958–962.
48. de Stefano V, Chiusolo P, Paciaroni K, et al. Prothrombin G20210A mutant genotype is a risk factor for cerebrovascular ischemic disease in young patients. *Blood*. 1998;91:3562–3565.
49. Sebire G, Tabarki B, Saunders DE, et al. Cerebral venous sinus thrombosis in children: risk factors, presentation, diagnosis and outcome. *Brain*. 2005;128:477–489.
50. Heller C, Becker S, Scharrer I, Kreuz W. Prothrombotic risk factors in childhood stroke and venous thrombosis. *Eur J Pediatr*. 1999;158(suppl 3):S117–S121.
51. Heller C, Heinecke A, Junker R, et al. Cerebral venous thrombosis in children: a multifactorial origin. *Circulation*. 2003;108:1362–1367.
52. Marjot T, Yadav S, Hasan N, Bentley P, Sharma P. Genes associated with adult cerebral venous thrombosis. *Stroke*. 2011;42:913–918.
53. Wysokinska EM, Wysokinski WE, Brown RD, et al. Thrombophilia differences in cerebral venous sinus and lower extremity deep venous thrombosis. *Neurology*. 2008;70:627–633.
54. Kenet G, Kirkham F, Niederstadt T, et al. Risk factors for recurrent venous thromboembolism in the European collaborative paediatric database on cerebral venous thrombosis: a multicentre cohort study. *Lancet Neurol*. 2007;6:595–603.
55. Kraus JA, Stuper BK, Muller J, et al. Molecular analysis of thrombophilic risk factors in patients with dural arteriovenous fistulas. *J Neurol*. 2002;249:680–682.
56. Hamzi K, Tazzite A, Nadifi S. Large-scale meta-analysis of genetic studies in ischemic stroke: five genes involving 152,797 individuals. *Indian J Hum Genet*. 2011;17:212–217.
57. Jiang B, Ryan KA, Hamedani A, et al. Prothrombin G20210A mutation is associated with young-onset stroke: the genetics of early-onset stroke study and meta-analysis. *Stroke*. 2014;45:961–967.
58. Haywood S, Liesner R, Pindora S, Ganesan V. Thrombophilia and first arterial ischaemic stroke: a systematic review. *Arch Dis Child*. 2005;90:402–405.
59. Kenet G, Lutkhoff LK, Albisetti M, et al. Impact of thrombophilia on risk of arterial ischemic stroke or cerebral sinovenous thrombosis in neonates and children: a systematic review and meta-analysis of observational studies. *Circulation*. 2010;121:1838–1847.

60. Kopyta I, Sarecka-Hujar B, Sordyl J, Sordyl R. The role of genetic risk factors in arterial ischemic stroke in pediatric and adult patients: a critical review. *Mol Biol Rep*. 2014.
61. Munshi A, Aliya N, Jyothy A, Kaul S, Alladi S, Shafi G. Prothombin gene G20210A mutation is not a risk factor for ischemic stroke in a South Indian Hyderabadi Population. *Thromb Res*. 2009;124:245–247.
62. Dahlback B, Carlsson M, Svensson PJ. Familial thrombophilia due to a previously unrecognized mechanism characterized by poor anticoagulant response to activated protein C: prediction of a cofactor to activated protein C. *Proc Natl Acad Sci U S A*. 1993;90:1004–1008.
63. Rees DC, Cox M, Clegg JB. World distribution of factor V Leiden. *Lancet*. 1995;346:1133–1134.
64. Ludemann P, Nabavi DG, Junker R, et al. Factor V Leiden mutation is a risk factor for cerebral venous thrombosis: a case-control study of 55 patients. *Stroke*. 1998;29:2507–2510.
65. Zuber M, Toulon P, Marnet L, Mas JL. Factor V Leiden mutation in cerebral venous thrombosis. *Stroke*. 1996;27:1721–1723.
66. Dindagur N, Kruthika-Vinod TP, Christopher R. Factor V gene A4070G mutation and the risk of cerebral veno-sinus thrombosis occurring during puerperium. *Thromb Res*. 2007;119:497–500.
67. Kim RJ, Becker RC. Association between factor V Leiden, prothrombin G20210A, and methylenetetrahydrofolate reductase C677T mutations and events of the arterial circulatory system: a meta-analysis of published studies. *Am Heart J*. 2003;146:948–957.
68. Hamedani AG, Cole JW, Mitchell BD, Kittner SJ. Meta-analysis of factor V Leiden and ischemic stroke in young adults: the importance of case ascertainment. *Stroke*. 2010;41:1599–1603.
69. Thorarensen O, Ryan S, Hunter J, Younkin DP. Factor V Leiden mutation: an unrecognized cause of hemiplegic cerebral palsy, neonatal stroke, and placental thrombosis. *Ann Neurol*. 1997;42:372–375.
70. Lynch JK, Nelson KB, Curry CJ, Grether JK. Cerebrovascular disorders in children with the factor V Leiden mutation. *J Child Neurol*. 2001;16:735–744.
71. Simchen MJ, Goldstein G, Lubetsky A, Strauss T, Schiff E, Kenet G. Factor V Leiden and antiphospholipid antibodies in either mothers or infants increase the risk for perinatal arterial ischemic stroke. *Stroke*. 2009;40:65–70.
72. Nowak-Gottl U, Strater R, Heinecke A, et al. Lipoprotein A and genetic polymorphisms of clotting factor V, prothrombin, and methylenetetrahydrofolate reductase are risk factors of spontaneous ischemic stroke in childhood. *Blood*. 1999;94:3678–3682.
73. de Paula SA, Ribeiro DD, Carvalho M, Cardoso J, Dusse LM, Fernandes AP. Factor V Leiden and increased risk for arterial thrombotic disease in young Brazilian patients. *Blood Coagul Fibrinolysis*. 2006;17:271–275.
74. Rey RC, de Larranaga G, Lepera S, et al. Activated protein C resistance in patients with arterial ischemic stroke. *J Stroke Cerebrovasc Dis*. 2001;10:128–131.
75. Bonduel M, Sciuccati G, Hepner M, et al. Factor V Leiden and prothrombin gene G20210A mutation in children with cerebral thromboembolism. *Am J Hematol*. 2003;73:81–86.
76. Konanki R, Gulati S, Saxena R, et al. Profile of prothrombotic factors in Indian children with ischemic stroke. *J Clin Neurosci*. 2014; pii: S0967-5868(14)00029-0.
77. Shi C, Kang X, Wang Y, Zhou Y. The coagulation factor V Leiden, MTHFRC677T variant and eNOS 4ab polymorphism in young Chinese population with ischemic stroke. *Clin Chim Acta*. 2008;396:7–9.
78. Zunker P, Hohenstein C, Plendl HJ, et al. Activated protein C resistance and acute ischaemic stroke: relation to stroke causation and age. *J Neurol*. 2001;248:701–704.
79. Juul K, Tybjaerg-Hansen A, Steffensen R, Kofoed S, Jensen G, Nordestgaard BG. Factor V Leiden: the Copenhagen City Heart Study and 2 meta-analyses. *Blood*. 2002;100:3–10.
80. Cushman M, Rosendaal FR, Psaty BM, et al. Factor V Leiden is not a risk factor for arterial vascular disease in the elderly: results from the Cardiovascular Health Study. *Thromb Haemost*. 1998;79:912–915.
81. Hamedani AG, Cole JW, Cheng Y, et al. Factor V leiden and ischemic stroke risk: the Genetics of Early Onset Stroke GEOS study. *J Stroke Cerebrovasc Dis*. 2013;22:419–423.
82. Go AS, Reed GL, Hylek EM, et al. Factor V Leiden and risk of ischemic stroke in nonvalvular atrial fibrillation: the Anticoagulation and Risk Factors in Atrial Fibrillation ATRIA Study. *J Thromb Thrombolysis*. 2003;15:41–46.
83. Berge E, Haug KB, Sandset EC, Haugbro KK, Turkovic M, Sandset PM. The factor V Leiden, prothrombin gene 20210GA, methylenetetrahydrofolate reductase 677CT and platelet glycoprotein IIIa 1565TC mutations in patients with acute ischemic stroke and atrial fibrillation. *Stroke*. 2007;38:1069–1071.
84. Mercuri E, Cowan F, Gupte G, et al. Prothrombotic disorders and abnormal neurodevelopmental outcome in infants with neonatal cerebral infarction. *Pediatrics*. 2001;107:1400–1404.
85. D'Amico D, Moschiano F, Leone M, et al. Genetic abnormalities of the protein C system: shared risk factors in young adults with migraine with aura and with ischemic stroke? *Cephalalgia*. 1998;18:618–621.
86. Maitrot-Mantelet L, Horellou MH, Massiou H, Conard J, Gompel A, Plu-Bureau. Should women suffering from migraine with aura be screened for biological thrombophilia?: Results from a cross-sectional French study. *Thromb Res*. 2014; pii: S0049-3848(14)00049-8.
87. Soriani S, Borgna-Pignatti C, Trabetti E, Casartelli A, Montagna P, Pignatti PF. Frequency of factor V Leiden in juvenile migraine with aura. *Headache*. 1998;38:779–781.
88. Ferrara M, Capozzi L, Bertocco F, Ferrara D, Russo R. Thrombophilic gene mutations in children with migraine. *Hematology*. 2012;17:115–117.
89. Haan J, Kappelle LJ, de Ronde H, Ferrari MD, Bertina RM. The factor V Leiden mutation R506Q. is not a major risk factor for migrainous cerebral infarction. *Cephalalgia*. 1997;17:605–607.
90. Iniesta JA, Corral J, Gonzalez-Conejero R, Rivera J, Vicente V. Prothrombotic genetic risk factors in patients with coexisting migraine and ischemic cerebrovascular disease. *Headache*. 1999;39:486–489.
91. Bots ML, van Kooten F, Breteler MM, et al. Response to activated protein C in subjects with and without dementia. The Dutch Vascular Factors in Dementia Study. *Haemostasis*. 1998;28:209–215.
92. Chapman J, Wang N, Treves TA, Korczyn AD, Bornstein NM. ACE, MTHFR, factor V Leiden, and APOE polymorphisms in patients with vascular and Alzheimer's dementia. *Stroke*. 1998;29:1401–1404.
93. Frosst P, Blom HJ, Milos R, et al. A candidate genetic risk factor for vascular disease: a common mutation in methylenetetrahydrofolate reductase. *Nat Genet*. 1995;10:111–113.

94. Li Z, Sun L, Zhang H, et al. Elevated plasma homocysteine was associated with hemorrhagic and ischemic stroke, but methylenetetrahydrofolate reductase gene C677T polymorphism was a risk factor for thrombotic stroke: a Multicenter Case-Control Study in China. *Stroke*. 2003;34:2085–2090.
95. Salem-Berrabah OB, Mrissa R, Machghoul S, et al. Hyperhomocysteinemia, C677T MTHFR polymorphism and ischemic stroke in Tunisian patients. *Tunis Med*. 2010;88:655–659.
96. Biswas A, Ranjan R, Meena A, et al. Homocystine levels, polymorphisms and the risk of ischemic stroke in young Asian Indians. *J Stroke Cerebrovasc Dis*. 2009;18:103–110.
97. van Beynum IM, Smeitink JA, den Heijer M, te Poele Pothoff MT, Blom HJ. Hyperhomocysteinemia: a risk factor for ischemic stroke in children. *Circulation*. 1999;99:2070–2072.
98. Cardo E, Monros E, Colome C, et al. Children with stroke: polymorphism of the MTHFR gene, mild hyperhomocysteinemia, and vitamin status. *J Child Neurol*. 2000;15:295–298.
99. Hogeveen M, Blom HJ, Van AM, Boogmans B, van Beynum IM, Van De Bor M. Hyperhomocysteinemia as risk factor for ischemic and hemorrhagic stroke in newborn infants. *J Pediatr*. 2002;141:429–431.
100. Andrade FL, Annichino-Bizzacchi JM, Saad ST, Costa FF, Arruda VR. Prothrombin mutant, factor V Leiden, and thermolabile variant of methylenetetrahydrofolate reductase among patients with sickle cell disease in Brazil. *Am J Hematol*. 1998;59:46–50.
101. Zhang W, Sun K, Chen J, et al. High plasma homocysteine levels contribute to the risk of stroke recurrence and all-cause mortality in a large prospective stroke population. *Clin Sci (Lond)*. 2010;118:187–194.
102. Prengler M, Sturt N, Krywawych S, Surtees R, Liesner R, Kirkham F. Homozygous thermolabile variant of the methylenetetrahydrofolate reductase gene: a potential risk factor for hyperhomocysteinaemia, CVD, and stroke in childhood. *Dev Med Child Neurol*. 2001;43:220–225.
103. Zak I, Sarecka-Hujar B, Kopyta I, et al. The T allele of the 677C>T polymorphism of methylenetetrahydrofolate reductase gene is associated with an increased risk of ischemic stroke in Polish children. *J Child Neurol*. 2009;24:1262–1267.
104. Akar N, Akar E, Ozel D, Deda G, Sipahi T. Common mutations at the homocysteine metabolism pathway and pediatric stroke. *Thromb Res*. 2001;102:115–120.
105. Sarecka-Hujar B, Kopyta I, Pienczk-Reclawowicz K, Reclawowicz D, Emich-Widera E, Pilarska E. The TT genotype of methylenetetrahydrofolate reductase 677C>T polymorphism increases the susceptibility to pediatric ischemic stroke: meta-analysis of the 822 cases and 1,552 controls. *Mol Biol Rep*. 2012;39:7957–7963.
106. Gunther G, Junker R, Strater R, et al. Symptomatic ischemic stroke in full-term neonates : role of acquired and genetic prothrombotic risk factors. *Stroke*. 2000;31:2437–2441.
107. Banerjee I, Gupta V, Ganesh S. Association of gene polymorphism with genetic susceptibility to stroke in Asian populations: a meta-analysis. *J Hum Genet*. 2007;52:205–219.
108. Sabino A, Fernandes AP, Lima LM, et al. Polymorphism in the methylenetetrahydrofolate reductase C677T. gene and homocysteine levels: a comparison in Brazilian patients with coronary arterial disease, ischemic stroke and peripheral arterial obstructive disease. *J Thromb Thrombolysis*. 2009;27:82–87.
109. Sun JZ, Xu Y, Lu H, Zhu Y. Polymorphism of the methylenetetrahydrofolate reductase gene association with homocysteine and ischemic stroke in type 2 diabetes. *Neurol India*. 2009;57:589–593.
110. Semmler A, Moskau S, Stoffel-Wagner B, Weller M, Linnebank M. The effect of MTHFR c.677C>T on plasma homocysteine levels depends on health, age and smoking. *Clin Invest Med*. 2009;32:E310.
111. Isordia-Salas I, Barinagarrementeria-Aldatz F, Leanos-Miranda A, et al. The C677T polymorphism of the methylenetetrahydrofolate reductase gene is associated with idiopathic ischemic stroke in the young Mexican-Mestizo population. *Cerebrovasc Dis*. 2010;29:454–459.
112. Li P, Qin C. Methylenetetrahydrofolate reductase MTHFR. gene polymorphisms and susceptibility to ischemic stroke: a meta-analysis. *Gene*. 2014;535:359–364.
113. Arsene D, Gaina G, Balescu C, Ardeleanu C. C677T and A1298C methylenetetrahydropholate reductase MTHFR. polymorphisms as factors involved in ischemic stroke. *Rom J Morphol Embryol*. 2011;52:1203–1207.
114. Zhao X, Jiang H. Quantitative assessment of the association between MTHFR C677T polymorphism and hemorrhagic stroke risk. *Mol Biol Rep*. 2013;40:573–578.
115. Kang S, Wu Y, Liu L, Zhao X, Zhang D. Association of the A1298C polymorphism in MTHFR gene with ischemic stroke. *J Clin Neurosci*. 2014;21:198–202.
116. Giusti B, Saracini C, Bolli P, et al. Early-onset ischaemic stroke: analysis of 58 polymorphisms in 17 genes involved in methionine metabolism. *Thromb Haemost*. 2010;104:231–242.
117. Van GB, Hultdin J, Johansson I, et al. Folate, vitamin B12, and risk of ischemic and hemorrhagic stroke: a prospective, nested case-referent study of plasma concentrations and dietary intake. *Stroke*. 2005;36:1426–1431.
118. Sazci A, Ergul E, Tuncer N, Akpinar G, Kara I. Methylenetetrahydrofolate reductase gene polymorphisms are associated with ischemic and hemorrhagic stroke: dual effect of MTHFR polymorphisms C677T and A1298C. *Brain Res Bull*. 2006;71:45–50.
119. Hultdin J, Van GB, Winkvist A, et al. Prospective study of first stroke in relation to plasma homocysteine and MTHFR 677C>T and 1298A>C genotypes and haplotypes - evidence for an association with hemorrhagic stroke. *Clin Chem Lab Med*. 2011;49:1555–1562.
120. Kang S, Zhao X, Liu L, Wu W, Zhang D. Association of the C677T polymorphism in the MTHFR gene with hemorrhagic stroke: a meta-analysis. *Genet Test Mol Biomarkers*. 2013;17:412–417.
121. Scher AI, Terwindt GM, Verschuren WM, et al. Migraine and MTHFR C677T genotype in a population-based sample. *Ann Neurol*. 2006;59:372–375.
122. Stuart S, Cox HC, Lea RA, Griffiths LR. The role of the MTHFR gene in migraine. *Headache*. 2012;52:515–520.
123. Gorgone G, Ursini F, Altamura C, et al. Hyperhomocysteinemia, intima-media thickness and C677T MTHFR gene polymorphism: a correlation study in patients with cognitive impairment. *Atherosclerosis*. 2009;206:309–313.
124. Jin P, Hou S, Ding B, et al. Association between MTHFR gene polymorphisms, smoking, and the incidence of vascular dementia. *Asia Pac J Public Health*. 2013;25:57S–63S.
125. Liu H, Yang M, Li GM, et al. The MTHFR C677T polymorphism contributes to an increased risk for vascular dementia: a meta-analysis. *J Neurol Sci*. 2010;294:74–80.

126. Lichy C, Reuner KH, Buggle F, et al. Prothrombin G20210A mutation, but not factor V Leiden, is a risk factor in patients with persistent foramen ovale and otherwise unexplained cerebral ischemia. *Cerebrovasc Dis*. 2003;16:83–87.
127. Karttunen V, Hiltunen L, Rasi V, Vahtera E, Hillbom M. Factor V Leiden and prothrombin gene mutation may predispose to paradoxical embolism in subjects with patent foramen ovale. *Blood Coagul Fibrinolysis*. 2003;14:261–268.
128. Botto N, Spadoni I, Giusti S, Ait-Ali L, Sicari R, Andreassi MG. Prothrombotic mutations as risk factors for cryptogenic ischemic cerebrovascular events in young subjects with patent foramen ovale. *Stroke*. 2007;38:2070–2073.
129. Stam AH, Weller CM, Janssens AC, et al. Migraine is not associated with enhanced atherosclerosis. *Cephalalgia*. 2013;33:228–235.
130. Oh SH, Kim NK, Kim HS, Kim WC, Kim OJ. Plasma total homocysteine and the methylenetetrahydrofolate reductase 677C>T polymorphism do not contribute to the distribution of cervico-cerebral atherosclerosis in ischaemic stroke patients. *Eur J Neurol*. 2011;18:491–496.
131. Chutinet A, Suwanwela NC, Snabboon T, Chaisinanunkul N, Furie KL, Phanthumchinda K. Association between genetic polymorphisms and sites of cervicocerebral artery atherosclerosis. *J Stroke Cerebrovasc Dis*. 2012;21:379–385.
132. They-They TP, Nadifi S, Rafai MA, Battas O, Slassi I. Methylenehydrofolate reductase C677T. polymorphism and large artery ischemic stroke subtypes. *Acta Neurol Scand*. 2011;123:105–110.
133. Jara-Prado A, Alonso ME, Martinez RL, et al. MTHFR C677T, FII G20210A, FV Leiden G1691A, NOS3 intron 4 VNTR, and APOE epsilon4 gene polymorphisms are not associated with spontaneous cervical artery dissection. *Int J Stroke*. 2010;5:80–85.
134. Arauz A, Hoyos L, Cantu C, et al. Mild hyperhomocysteinemia and low folate concentrations as risk factors for cervical arterial dissection. *Cerebrovasc Dis*. 2007;24:210–214.
135. McColgan P, Sharma P. The genetics of carotid dissection: meta-analysis of a MTHFR/C677T common molecular variant. *Cerebrovasc Dis*. 2008;25:561–565.
136. Pezzini A, Grassi M, Del ZE, et al. Migraine mediates the influence of C677T MTHFR genotypes on ischemic stroke risk with a stroke-subtype effect. *Stroke*. 2007;38:3145–3151.
137. Bonduel M, Hepner M, Sciuccati G, Torres AF, Tenembaum S. Prothrombotic disorders in children with moyamoya syndrome. *Stroke*. 2001;32:1786–1792.
138. Nowak-Gottl U, Kirkham FJ, Chan AK, et al. Recurrent stroke: the role of prothrombotic disorders. *J Thromb Haemost*. 2005;(3 suppl 1):OR355.
139. Ganesan V, Prengler M, Wade A, Kirkham FJ. Clinical and radiological recurrence after childhood arterial ischemic stroke. *Circulation*. 2006;114:2170–2177.
140. Blinkenberg EO, Kristoffersen AH, Sandberg S, Steen VM, Houge G. Usefulness of factor V Leiden mutation testing in clinical practice. *Eur J Hum Genet*. 2010;18:862–866.
141. Bradbeer P, Teague L, Cole N. Testing for heritable thrombophilia in children at Starship Children's Hospital: an audit of requests between 2004 and 2009. *J Paediatr Child Health*. 2012;48:921–925.
142. Morris JG, Singh S, Fisher M. Testing for inherited thrombophilias in arterial stroke: can it cause more harm than good? *Stroke*. 2010;41:2985–2990.
143. Holmes MV, Newcombe P, Hubacek JA, et al. Effect modification by population dietary folate on the association between MTHFR genotype, homocysteine, and stroke risk: a meta-analysis of genetic studies and randomised trials. *Lancet*. 2011;378:584–594.
144. Husemoen LL, Skaaby T, Jorgensen T, et al. MTHFR C677T genotype and cardiovascular risk in a general population without mandatory folic acid fortification. *Eur J Nutr*. 2014.

CHAPTER

102

Sickle Cell Disease

Fenella J. Kirkham

Institute of Child Health, University College London, London, UK

INTRODUCTION

People with sickle cell disease, a chronic hemolytic anemia, present with a wide variety of neurological syndromes, including ischemic and hemorrhagic stroke, anterior and posterior territory transient ischemic attacks (TIAs), "soft neurological signs," seizures, headache, coma, visual loss, and altered mental status. There is a peak for ischemic stroke in childhood, typically associated with stenosis or occlusion of the distal internal carotid and proximal middle cerebral arteries diagnosable using magnetic resonance angiography or transcranial Doppler ultrasound (TCD). For hemorrhagic stroke, the peak age is early adulthood, when aneurysms are common. Covert infarction may be detected on magnetic resonance imaging (MRI) in around 25%. Cognitive difficulties, characteristically affecting attention, executive function, memory, arithmetic and processing speed, are also common. Indefinite transfusion is standard care for secondary prevention (and primary in those with TCD velocities >200 cm per second). In addition to the possibility of cure from transplantation, hydroxyurea, aspirin and overnight respiratory support are under investigation as strategies to prevent neurological complications.

DISEASE CHARACTERISTICS

Sickle cell disease (SCD) is an autosomal recessive disorder of hemoglobin, which is one of the most common single-gene disorders and particularly affects those of African origin.[1] It is the most common cause of childhood stroke worldwide[2] and has a wide range of neurological complications in adults and children.[3]

Hallmark Manifestations

The clinical features of SCD are caused by a chronic hemolytic anemia combined with intermittent vaso-occlusion.[1] Although disease severity is variable, many suffer repeated episodes of bone pain[4] and progressive organ damage, resulting in a poor quality of life and reduced life expectancy.[5] Pain is a cardinal feature and typically restricts normal activities and requires strong analgesia. Other common complications include dactylitis, chest crisis, splenic sequestration, aplastic anemia, skin ulceration, and priapism.[1] At least 10% are at risk of stroke in childhood,[6,7] and many have educational difficulties as a result of memory and attentional problems, decoding difficulties, and abnormalities in visual motor integration.[8–10]

Inheritance: Autosomal Recessive Inheritance of Sickle Hemoglobin (HbS)

Hemoglobin S is unstable and has an increased propensity to polymerize under certain physical conditions including exposure to hypoxia.

Disease Variants:

- Homozygotes have homozygous sickle cell anemia (HbSS);
- Heterozygotes have sickle cell trait (HbAS);

Rosenberg's Molecular and Genetic Basis of Neurological and Psychiatric Disease
http://dx.doi.org/10.1016/10.1016/B978-0-12-410529-4.00102-9

- Complex heterozygotes:
  - Hemoglobin SC disease (HbSC);
  - Sickle cell-beta thalassemia (HbSβthal).

Diagnosis/Testing

- Hemoglobin electrophoresis;
- Chromatography;
- Mass spectrometry;
- DNA analysis.

CLINICAL FEATURES

Historical Overview and Disease Identification

Although widely recognized as a cause of painful crisis in Africa for centuries, the condition was first recognized as a chronic anemia with abnormal sickle-shaped cells in a dental student from Grenada by Herrick in 1910.[11] The distinction between hemoglobin S and hemoglobin A was made using electrophoresis by Linus Pauling in 1949 and he claimed this as the "first molecular disease."[12] Polymerization of sickle hemoglobin as a key pathophysiological mechanism was described by Perutz in 1951.[13]

Gene Identification

An abnormal beta globin gene with a substitution of thymine for adenine codes the substitution of the amino acid valine for glutamine in the beta globin chain.

Early Treatments

Blood transfusion was the mainstay of treatment for acute anemic events and was then used to prevent recurrence of stroke from the 1970s.[14] Penicillin prophylaxis to prevent life-threatening pneumococcal infection in the context of poor splenic function was introduced for children in the 1980s.[15] Hydroxycarbamide was shown to reduce the frequency of painful crisis in the 1990s.[16]

Mode of Inheritance and Prevalence

- Heterozygotes (HbAS): selective advantage by virtue of a relative resistance to falciparum malaria but may have complications, e.g., splenic infarction on exposure to altitude.
- Homozygotes (HbSS): serious life-threatening disease, sickle cell anemia.
- Complex heterozygotes:
  - Hemoglobin SC disease (HbSC)
  - Sickle cell-beta thalassemia (HbSβthal).

NATURAL HISTORY

Age of Onset

Most cases in developed countries are now identified through neonatal screening. Sickle cell anemia often presents in the first year of life with dactylitis or splenic sequestration.

Disease Evolution

There is a very wide spectrum of severity, with some people not presenting until adulthood even with the homozygous form, although this is more common with SC disease or S$\beta_0$thalassemia. Others have frequent neurological and non-neurological complications, often requiring acute hospital admission.[1]

End of Life: Mechanisms and Comorbidities

Mortality in childhood has substantially reduced in the past few decades[17,18] but the median age at death is in middle age.[5,19,20] The main causes of death in childhood are sepsis, acute splenic sequestration, chest syndrome, and stroke, while chronic organ damage, including cardiopulmonary problems such as pulmonary hypertension and diastolic dysfunction,[21,22] stroke, and renal failure are associated with mortality in adults.

MOLECULAR GENETICS

- Normal gene product: hemoglobin A.
- Abnormal gene product(s): hemoglobin S.

Genotype–Phenotype Correlations

Overt Stroke

Stroke is 250 times more common in children with SCD than in healthy children,[2,7] with an incidence similar to that for a general population of elderly adults. Stroke and transient ischemic attacks (TIA) occur with an incidence of 0.61 and 0.52–0.74/100 patient years in children and young adults, respectively, with homozygous sickle cell anemia (SCA, HbSS).[7] The prevalence of stroke in children aged 2–19 years is 2–6% and in those over 30 years is 7–9%. In patients with the other genotypes (HbSC and HbSβthal), stroke is less common than in those with HbSS, but is still much more common than in the general childhood population.[7] For those with SCD who survive into adult life, ischemic and hemorrhagic stroke are common. Approximately 25% of patients with SCA and 10% of those with HbSC will have had a stroke by the age of 45 years.[7]

PATHOLOGY

Angiography shows that the majority of SCD patients with stroke have narrowing of the arteries of the Circle of Willis at the base of the brain, usually involving the distal internal carotid (ICA) and proximal middle cerebral (MCA) arteries (Figures 102.1, 102.2B, 102.3B, and 102.4B).[23] Pathological examination of these arteries shows endothelial proliferation, fibroblastic reaction, hyalinization and fragmentation of the internal elastic lamina. In some cases, recent thrombus may be seen either within the narrowed lumen or occasionally without associated arterial wall changes.[24] In a proportion, fibrous occlusion of these arteries is associated with the development of very thin collateral vessels which bypass the occlusion. The appearance of these collaterals is known on angiography as moyamoya (Figure 102.4B), from the Japanese expression describing the angiogram appearing like a "puff of smoke."[25,26] The arterial occlusion (Figure 102.2B) and stenosis (with or without moyamoya; Figures 102.3B and 102.4B) leads to cerebral infarction either in the middle cerebral artery territory (Figure 102.2A) or more characteristically in the superficial and deep borderzones between the anterior and middle cerebral artery territories (Figure 102.3A).[27,28] Pathological changes in the small vessels are also well described and both infarction and hemorrhage are documented in the absence of intracranial arteriographic changes. Carotid and vertebral stenosis, occlusion, dissection and tortuosity occur in sickle cell disease (Figure 102.4).[29] Cardioembolic embolus may play a role in some cases,[30] and there is evidence for an excess of patent foramen ovale (see Chapter 101, Figure 101.7) in children with stroke,[31] which is under investigation in the PFAST study. Subarachnoid and intracerebral hemorrhage may occur (Figure 102.5A),[32] sometimes as a result of rupture of the fragile moyamoya vessels or of aneurysms usually located at the bifurcations of major vessels, most commonly in the posterior circulation (Figure 102.5B).[33] Sinovenous thrombosis (see Chapter 101, Figures 101.1 and 101.2) may be associated with infarction or hemorrhage (Figure 102.5A).[34]

CLINICAL PRESENTATION

There is a broad spectrum of acute presentation with cerebrovascular accident (CVA) and other neurological complications in patients with SCD. In addition to clinical stroke, with focal signs lasting >24 hours, patients with SCD also have transient ischemic attacks (TIAs) with symptoms and signs resolving within 24 hours, although many of these individuals are found to have cerebral infarction or atrophy on imaging.[3] The insidious onset of "soft

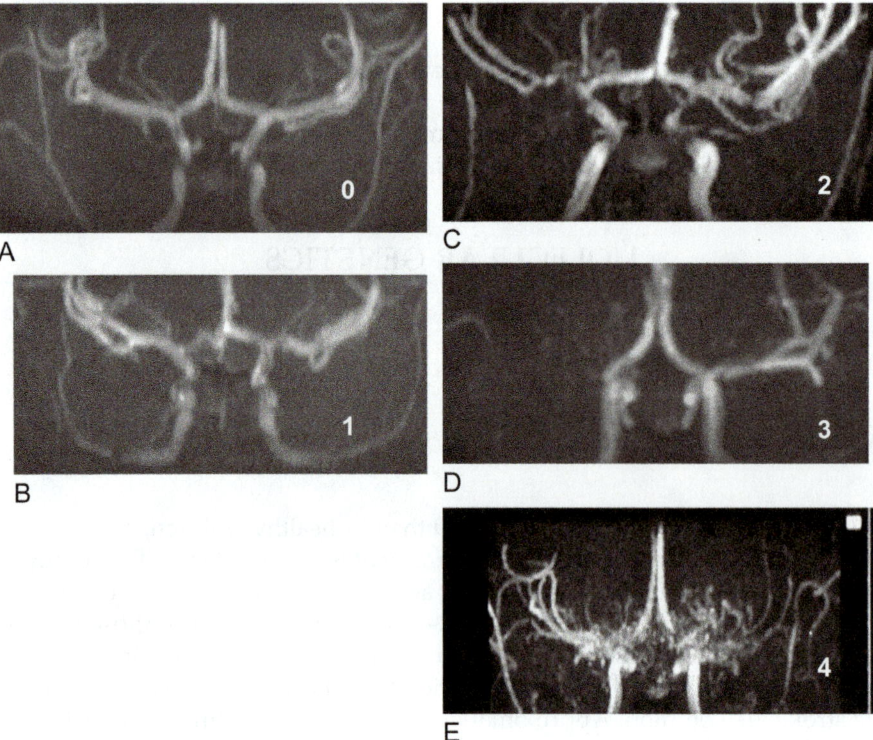

FIGURE 102.1 **Magnetic resonance angiography of children with sickle cell disease.** (**A**) Normal (grade 0); (**B**) mild (grade 1) turbulence; (**C**) moderate (grade 2) turbulence; (**D**) severe (grade 3) turbulence; (**E**) moyamoya (grade 4). From[60].

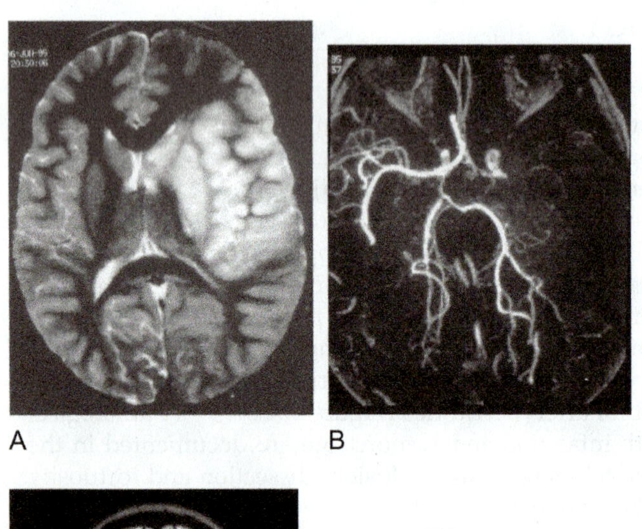

FIGURE 102.2 **Large middle cerebral artery stroke (A) in a child with sickle cell disease and middle cerebral artery occlusion (B).**

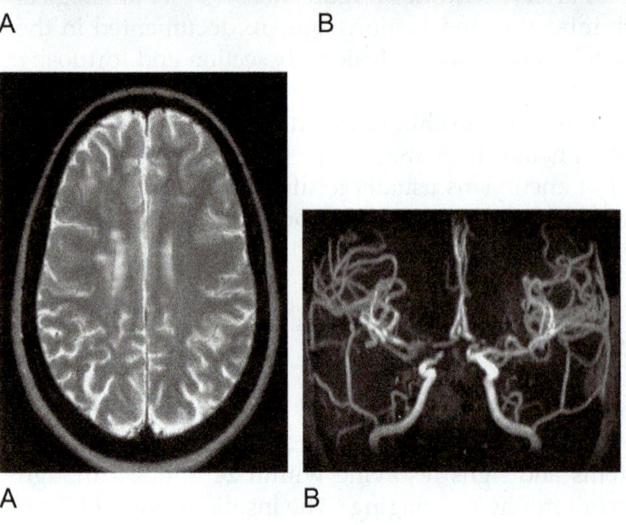

FIGURE 102.3 **"Silent" cerebral infarction in the deep white matter (A) in a child with sickle cell disease and middle cerebral artery stenosis (B).**

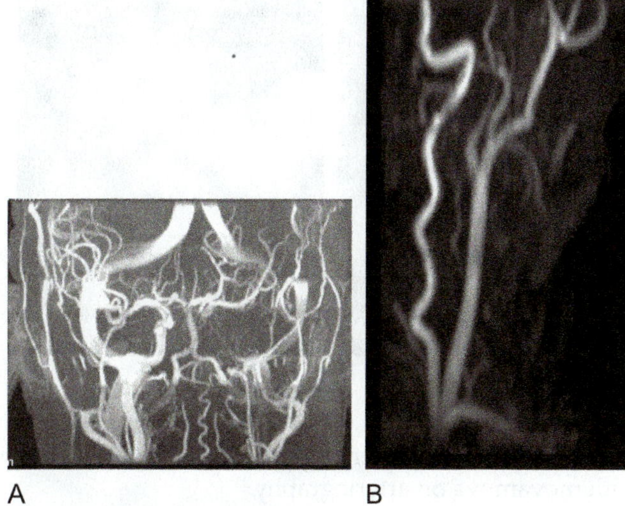

FIGURE 102.4 (A) Occlusion of the internal carotid artery in the neck. (B) Tapering occlusion compatible with internal carotid dissection.

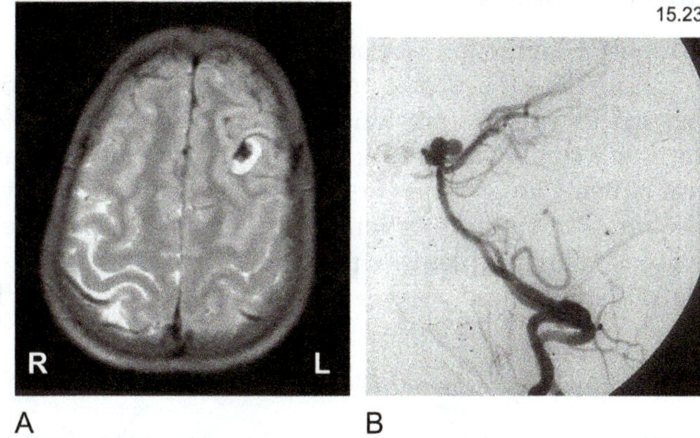

FIGURE 102.5 (A) Hemorrhage in a child with sickle cell disease and acute headache. (B) Posterior circulation aneurysm in sickle cell disease.

neurological signs," such as difficulty in tapping quickly, is usually associated with cerebral infarction. Seizures are also more common in the SCD population, affecting around 10%.[35,36] In addition, coma and headache are common presentations of infarctive and hemorrhagic stroke and cerebrovascular disease in children with SCD.[37] Altered mental status—with or without reduced level of consciousness, headache, seizures, visual loss or focal signs—can occur in numerous contexts, including infection, acute chest syndrome (ACS) and acute anemia, as well as apparently spontaneously. These patients are classified clinically as having had a CVA, although there is a wide differential of focal and generalized vascular and nonvascular pathologies, often distinguished using acute MR techniques, with important management implications.[3]

Risk Factors for Acute Neurological Complications

High white-cell count, low hemoglobin and oxyhemoglobin desaturation predict neurological complications.[7,38] In addition, risk factors for overt ischemic stroke include hypertension and previous TIA or chest crisis.[7] For hemorrhagic stroke, aneurysms are common in adults but not children, who often present with hypertension after transfusion or corticosteroids.[32] Seizures are more common in males and those with dactylitis in infancy, and those with cerebrovascular disease and covert infarction.[35,36]

Recurrence

Around two-thirds of children with sickle cell disease presenting with their first stroke will have a further episode if untreated (Figure 102.6).[14] Regular simple, manual or automated (erythrocytapheresis) transfusion to a hemoglobin S of <30% reduces the recurrence rate over the longer term to around 10%.[39] The risk of ischemic stroke recurrence is

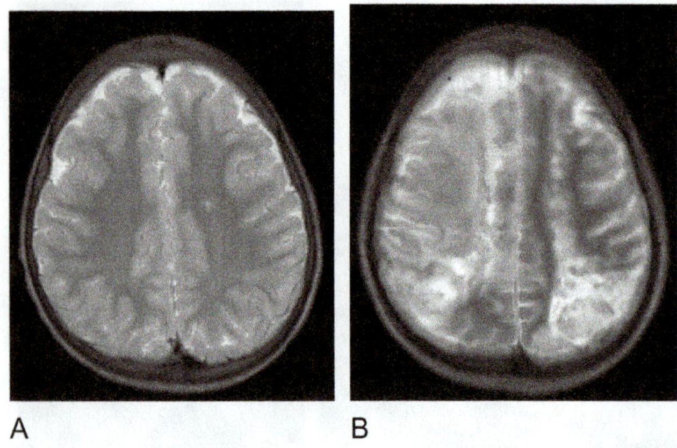

FIGURE 102.6 (A) Small silent infarct in the deep white matter of a 7-year-old boy with sickle cell anemia H. (B) Extensive bilateral stroke in the same boy with sickle cell anemia 1 year later after a chest crisis.

higher for patients presenting spontaneously,[40] those who underwent simple transfusion at presentation[41] and those with moyamoya on arteriography.[25]

Covert or Silent Cerebral Infarction

Cerebral infarction can be demonstrated on magnetic resonance imaging (MRI) in patients with SCD without symptoms of either stroke or TIA (Figure 102.3).[42] This covert or "silent" infarction affects around 20–35% of children and adolescents,[42] characteristically in the anterior and/or posterior borderzones. From a diffusion-weighted imaging study conducted during screening for the Silent Infarct Transfusion trial (SIT), the incidence of acute silent cerebral ischemic events appears to be 47.3 per 100 patient years.[43] Infarction may occur in the context of acute anemia, e.g. after *Parvovirus* infection.[44] Risk factors include male gender, low hemoglobin, relative hypertension, laboratory evidence of hyposplenism, previous seizures and relatively infrequent pain.[42,45,46] Neurological examination is normal, although these patients may have had subtle TIAs, headaches or seizures.

Cognitive Sequelae

By mid-childhood there is a reduction in the IQ of patients with SCD compared with siblings and other controls,[8,9,47] which may be progressive,[48] particularly in those with moyamoya syndrome.[26] Impaired cognitive function appears to be related to covert stroke, poor growth, low hemoglobin level, and high platelet count,[49] as well as lowered level of oxyhemoglobin saturation.[9,50] Arithmetic may be particularly affected and subtle deficits in attention and executive function have also been demonstrated in association with covert infarction.[42,47,51]

DISEASE MECHANISMS AND PATHOPHYSIOLOGY

Human Observations

SCD is characterized by both acute and reversible, as well as chronic and irreversible, changes in the properties and deformability of sickle red cells. Intracellular hemoglobin polymerization, combined with abnormal membrane properties, results in abnormal rheology of the blood. The primary determinant of the amount of polymer formation within the sickle red blood cell is the degree of oxyhemoglobin saturation.[1] Hypoxic exposure, whether chronic sustained, intermittent or acute, has a wide range of effects to which the human body adapts but which may alter the cerebrovascular circulation.[52–54] Nitric oxide resistance in the context of hemolysis may play a role in the development of cerebrovascular disease.[55]

Animal Models

There are mouse models of stroke in sickle cell disease which suggest that the mechanism of cerebrovascular disease involves inflammation, neutrophils, and thrombin generation,[56] with relative protection from atherosclerosis at least, in part, related to upregulation of heme oxygenase.[57] There are alterations in the cerebral hemodynamic response to oxygen which may be important in stroke risk.[58]

TESTING

Molecular Diagnosis

Neonatal screening, combined with prophylactic penicillin and immunization, has had a dramatic impact on deaths due to sepsis.

Analytical Testing: Clinical and Ancillary Testing for Neuropsychiatric Complications

Brain Imaging in Acute Neurological Complications

Computed tomography (CT) may be performed rapidly to exclude hemorrhage, although MRI is usually required to demonstrate the extent of infarction as early as possible, as well as potentially reversible pathologies such as posterior leukoencephalopathy.[3] In patients presenting with clinical stroke, likely findings include large infarct in the distribution of the middle cerebral artery, or smaller lesions in the basal ganglia or deep white or gray matter of the borderzones.[27,28] Occipitoparietal or thalamic involvement should suggest sinovenous thrombosis.[34]

ARTERIAL IMAGING

Between 60–90% of children with SCD and arterial ischemic stroke in an arterial distribution have abnormal findings on conventional angiography or magnetic resonance angiography (MRA; Figures 102.1–102.4). Typical abnormalities include stenosis or occlusion of the distal internal carotid or middle cerebral arteries.[3] Moyamoya syndrome,[25] vertebral or internal carotid dissection,[29] and small vessel vasculitis have also been documented. Sinovenous thrombosis is probably underdiagnosed because many patients do not undergo acute vascular imaging; if emergency MRA is available and the results are found to be normal, MR or CT venography should be considered.[34]

Screening for Preventable Pathology

MRI

Covert or "silent" infarction can be detected using T2-weighted clinical MRI at 1.5 Tesla (Figure 102.2) but cannot be recommended unless the Silent Infarct Transfusion trial publishes a positive result during 2014.[59] Focal abnormalities in perfusion demonstrated using MRI (arterial spin-tagging or gadolinium as tracer) or positron emission tomography are common, particularly in those with cerebrovascular disease, and may be associated with cognitive deficits.[60–62] Compared with controls, using voxel-based morphometry there is evidence of damage in the white matter of the borderzones, even in patients with SCA and normal T2-weighted MRI, but this cannot currently be used for diagnosis in individual patients.[63] Diffusion tensor imaging has confirmed earlier work showing abnormality in the anterior corpus callosum.[64–66] There are also reductions in the volume of deep gray matter structures which may impact on cognitive function.[67,68]

TRANSCRANIAL DOPPLER FOR CEREBROVASCULAR SCREENING

Stenosis may be diagnosed before stroke occurs as high time-averaged maximum velocity on transcranial Doppler ultrasound (TCD) (Figure 102.7). Adams et al.[69] categorized the highest time-averaged maximum velocity on either side into three groups:

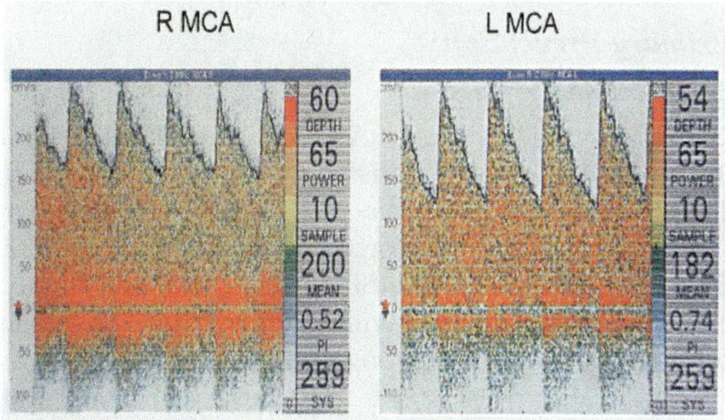

FIGURE 102.7 **Transcranial Doppler showing abnormal cerebral blood velocity of 200 cm/second on the right and conditional cerebral blood velocity >170 <200 cm/second on the left.**

1. those with normal TCD, defined as ICA/MCA velocity <170 cm per second;
2. those with conditionally abnormal ICA/MCA velocities, i.e., between 170 cm and 200 cm per second, who had a 7% risk of stroke over the next 3 years;
3. those with abnormal ICA/MCA velocities, i.e., >200 cm per second, who had a 40% risk of stroke over the next 3 years.

Transcranial Doppler screening is now widespread in developed countries and regular prophylactic blood transfusion is recommended for those with ICA/MCA velocities > 200 cm per second (see below) as this strategy has been associated with reduction in stroke risk.[18,70–72]

Additional Genetic Risk Factors for Neurological Complications

In African Americans, there is a high incidence of individuals with only two or three β-globin genes (rather than the normal complement of four). This α-thalassemia appears to reduce the risk of stroke, probably in relation to the increased hemoglobin levels.[73,74] High hemoglobin F levels appear to ameliorate the risk of overt stroke and silent infarction,[75–77] at least in childhood. The β-globin haplotypes might also alter risk, probably by influencing hemoglobin F levels, although the data are conflicting. Increasing hemoglobin F levels provides the rationale for the use of hydroxyurea to ameliorate disease severity. The role of glucose-6-phosphate dehydrogenase is controversial.[78,79] There are a number of other genes which appear to modify the risk of stroke in sickle cell disease.[80–83]

MANAGEMENT

Standard of Care

Emergency Management of Stroke in Sickle Cell Disease

This therapy has evolved through hematological clinical experience[39] rather than being subject to rigorous evaluation by randomized controlled trials. If available, exchange rather than top-up transfusion is recommended by most hematologists, and appears to be associated with reduced risk of recurrence,[41] although there are no randomized data. Although published cases are rare, once diagnosed on appropriate neuroimaging, there is no known contraindication to management of hemorrhage, sinovenous thrombosis, or dissection as outlined in the protocols for nonsickle stroke. Interventional neuroradiology with coils has been successfully employed for the management of aneurysms in sickle cell disease,[33] and represents a reasonable alternative to surgery.

Prevention of Recurrent Stroke

Although there has never been a controlled trial, several cohort studies have shown that recurrent stroke is reduced to around 10% in those on a chronic transfusion regime, at least 6 weekly, to reduce the sickle cell hemoglobin (HbS) level below 30%, compared with a recurrence rate of 66–90% in untreated patients.[3,84] Until further evidence is available, children with sickle cell disease who have had a stroke should be transfused long-term to achieve a hemoglobin S below 30% for 3 years and below 50% subsequently, as there is considerable, albeit anecdotal, evidence that this is an effective method of reducing recurrence. Hydroxyurea, bone marrow transplantation and revascularization for moyamoya are options for some patients, particularly those who cannot tolerate chronic transfusion or experience recurrent events despite reduction of the hemoglobin S percentage to <30%.[3]

Primary Prevention

BLOOD TRANSFUSION

Current policy in the USA and Europe arising from the STOP I and STOP II randomized controlled trials of regular prophylactic blood transfusion is that children with SCA (HbSS) or HbSβ_0 thalassemia should be screened with TCD to find those with velocities > 200 cm per second and those children should then be transfused indefinitely;[3,85–87] this policy has been endorsed in official guidelines in the USA and UK. The number needed to treat is six. The Silent Infarct Transfusion (SIT) Study commenced late 2004,[59,88] and comprises 29 sites in the United States, Canada, England and France and will report in 2014. The main goal is to determine the efficacy of blood transfusion therapy as a treatment for preventing silent infarctions.

Failed Therapies and Barriers to Treatment Development

Studies of drugs that might modify mechanisms considered important in laboratory studies, such as the tendency to polymerization, the relative deficit in endothelial nitric oxide, and the activation of white cells, platelets, and the endothelium have not involved endpoints involving the central nervous system. Reasons include the relative rarity of central nervous system events, such as stroke and seizures, and the expense of accurately documenting less obvious pathology, such as silent infarction and cognitive deficit. Trials of Gardos channel inhibitors, arginine, and specific inhibitors of components of the inflammatory pathway, e.g., e-selectin, have not yet produced any convincing positive results.[89–91]

Therapies Under Investigation

Bone Marrow Transplantation

Although there has not been a randomized controlled trial, it is clear that successful bone marrow transplantation cures sickle cell disease.[3] National centers quote survival rates of >90% whilst >80% are disease-free. However, there is a possible risk of a fatal cerebral hemorrhage in the acute phase. Approximately 20% of patients will have seizures despite phenytoin prophylaxis and supplementary magnesium. Although recurrent clinical stroke is rare after transplantation, even if transplantation is performed for the indication of a previous stroke, individuals who have already had a silent infarction are at risk of an extension. Until recently, relatively few patients were eligible, but stem cell and unrelated donor transplantation[92,93] may open up this cure to a larger number of patients, although the community remains concerned about the mortality.

Hydroxyurea

Hydroxurea increases fetal hemoglobin, which has a higher oxyhemoglobin affinity, and probably also reduces inflammation. In preliminary studies there is evidence for a beneficial effect on TCD abnormality[94] and cognition.[95] However, many patients are reluctant to embark on this drug because of the uncertainty of the long-term risk of cancer. For patients who have abnormal TDD there is currently a trial recruiting which compares long-term blood transfusion treatment with switching to hydroxyurea after several years of blood transfusion treatment (TWITCH). Examining whether this drug has a role in primary prevention of neurocognitive abnormality in the BABY HUG trial[96] has proved difficult because of the requirement to sedate very young children to undergo MRI.

Aspirin

Aspirin has been shown to reduce secondary stroke in adults and has been investigated as a method of primary prevention, although the risk of bleeding may outweigh any benefit, as has been shown for the general population with a low risk of stroke. Two small trials in the early 1980s showed that aspirin had no effect on the number of painful episodes in SCD; however, importantly, no major adverse effects were reported. Data from pilot trials and cohorts of children with sickle cell disease and stroke suggest that aspirin is probably safe.[87,97]

Management of Sleep Disordered Breathing

Young people with sickle cell disease are at high risk of sleep disordered breathing and there is some evidence for an effect on stroke risk[38] and cognitive function.[50] Adenotonsillectomy appears to be of benefit in terms of reducing stroke risk.[98] In a pilot trial of auto-adjusting continuous positive airways pressure, adherence was excellent and there was evidence for an improvement in cancellation, a test of attention and processing speed.[99] A phase II study including adults as well as children is underway.

Infection, Blood Pressure, Nutrition and Family Support

Children with sickle cell disease should be vaccinated against *Pneumococcus* and *Haemophilus* and should take regular penicillin. Blood pressure is lower in the majority of patients with sickle cell disease, but is relatively increased in those with overt and silent stroke;[7,46] there is no evidence on whether control of hypertension reduces recurrence risk. A good diet, including at least five portions of fruit and vegetables a day, and regular exercise within the limits of tolerance, may also reduce the risk of stroke.[3] Support for families and appropriate role models for good parenting is likely to be of benefit, as there is evidence that a "learned helplessness" parenting style is associated with cognitive problems[100] and that behavior problems are more likely if there is family conflict. Trials of educational remediation are under way.[101]

DISCUSSION

This is an exciting time for those caring for patients with sickle cell disease, as greater understanding of the pathophysiology has led to evidence-based treatments for the primary and secondary prevention of neurological and cognitive morbidity, and these are starting to become widely available in parallel with evidence of benefit to the population. However, many have side effects or are an unacceptable burden to the child, and further research using our improved understanding of the molecular basis of sickle cell disease is urgently required. Large-scale randomized controlled trials will need to involve multiple sites across continents, but the studies conducted to date have shown that this is an achievable aim.

ACKNOWLEDGEMENTS

This work has been funded by Action Research, the Wellcome Trust, the Stroke Association, the National Institutes for Health (USA), and the National Institutes for Health Research (UK). Research at UCL Institute of Child Health and Great Ormond Street Hospital for Children NHS Foundation Trust benefits from R&D funding received from the National Health Service.

References

1. Rees DC, Williams TN, Gladwin MT. Sickle-cell disease. *Lancet*. 2010;376:2018–2031.
2. Earley CJ, Kittner SJ, Feeser BR, et al. Stroke in children and sickle-cell disease: Baltimore–Washington Cooperative Young Stroke Study. *Neurology*. 1998;51:169–176.
3. Kirkham FJ. Therapy insight: stroke risk and its management in patients with sickle cell disease. *Nat Clin Pract Neurol*. 2007;3:264–278.
4. Platt OS, Thorington BD, Brambilla DJ, et al. Pain in sickle cell disease. Rates and risk factors. *N Engl J Med*. 1991;325:11–16.
5. Platt OS, Brambilla DJ, Rosse WF, et al. Mortality in sickle cell disease. Life expectancy and risk factors for early death. *N Engl J Med*. 1994;330:1639–1644.
6. Balkaran B, Char G, Morris JS, Thomas PW, Serjeant BE, Serjeant GR. Stroke in a cohort of patients with homozygous sickle cell disease. *J Pediatr*. 1992;120:360–366.
7. Ohene-Frempong K, Weiner SJ, Sleeper LA, et al. Cerebrovascular accidents in sickle cell disease: rates and risk factors. *Blood*. 1998;91:288–294.
8. Schatz J, Finke RL, Kellett JM, Kramer JH. Cognitive functioning in children with sickle cell disease: a meta-analysis. *J Pediatr Psychol*. 2002;27:739–748.
9. Hogan AM, Pit-ten Cate IM, Vargha-Khadem F, Prengler M, Kirkham FJ. Physiological correlates of intellectual function in children with sickle cell disease: hypoxaemia, hyperaemia and brain infarction. *Dev Sci*. 2006;9:379–387.
10. Berkelhammer LD, Williamson AL, Sanford SD, et al. Neurocognitive sequelae of pediatric sickle cell disease: a review of the literature. *Child Neuropsychol*. 2007;13:120–131.
11. Herrick JB. Peculiar elongated and sickle-shaped red blood corpuscles in a case of severe anemia. *Ann Inter Med*. 1910;6:517–521.
12. Pauling L, Itano HA. Sickle cell anemia a molecular disease. *Science*. 1949;110:543–548.
13. Perutz RR, Liquori AM, Eirich F. X-ray and solubility studies of the haemoglobin of sickle-cell anaemia patients. *Nature*. 1951;167:929–931.
14. Powars D, Wilson B, Imbus C, Pegelow C, Allen J. The natural history of stroke in sickle cell disease. *Am J Med*. 1978;65:461–471.
15. Gaston MH, Verter JI, Woods G, et al. Prophylaxis with oral penicillin in children with sickle cell anemia. A randomized trial. *N Engl J Med*. 1986;314:1593–1599.
16. Charache S, Terrin ML, Moore RD, et al. Effect of hydroxyurea on the frequency of painful crises in sickle cell anemia. Investigators of the Multicenter Study of Hydroxyurea in Sickle Cell Anemia. *N Engl J Med*. 1995;332:1317–1322.
17. Quinn CT, Rogers ZR, McCavit TL, Buchanan GR. Improved survival of children and adolescents with sickle cell disease. *Blood*. 2010;115:3447–3452.
18. Telfer P, Coen P, Chakravorty S, et al. Clinical outcomes in children with sickle cell disease living in England: a neonatal cohort in East London. *Haematologica*. 2007;92:905–912.
19. Wierenga KJ, Hambleton IR, Lewis NA. Survival estimates for patients with homozygous sickle-cell disease in Jamaica: a clinic-based population study. *Lancet*. 2001;357:680–683.
20. Makani J, Cox SE, Soka D, et al. Mortality in sickle cell anemia in Africa: a prospective cohort study in Tanzania. *PLoS One*. 2011;6:e14699.
21. Gladwin MT, Sachdev V, Jison ML, et al. Pulmonary hypertension as a risk factor for death in patients with sickle cell disease. *N Engl J Med*. 2004;350:886–895.
22. Sachdev V, Machado RF, Shizukuda Y, et al. Diastolic dysfunction is an independent risk factor for death in patients with sickle cell disease. *J Am Coll Cardiol*. 2007;49:472–479.
23. Stockman JA, Nigro MA, Mishkin MM, Oski FA. Occlusion of large cerebral vessels in sickle-cell anemia. *N Engl J Med*. 1972;287:846–849.
24. Rothman SM, Fulling KH, Nelson JS. Sickle cell anemia and central nervous system infarction: a neuropathological study. *Ann Neurol*. 1986;20:684–690.
25. Dobson SR, Holden KR, Nietert PJ, et al. Moyamoya syndrome in childhood sickle cell disease: a predictive factor for recurrent cerebrovascular events. *Blood*. 2002;99:3144–3150.

26. Hogan AM, Kirkham FJ, Isaacs EB, Wade AM, Vargha-Khadem F. Intellectual decline in children with moyamoya and sickle cell anaemia. *Dev Med Child Neurol*. 2005;47:824–829.
27. Adams RJ, Nichols FT, McKie V, McKie K, Milner P, Gammal TE. Cerebral infarction in sickle cell anemia: mechanism based on CT and MRI. *Neurology*. 1988;38:1012–1017.
28. Pavlakis SG, Bello J, Prohovnik I, et al. Brain infarction in sickle cell anemia: magnetic resonance imaging correlates. *Ann Neurol*. 1988;23:125–130.
29. Telfer PT, Evanson J, Butler P, et al. Cervical carotid artery disease in sickle cell anemia: clinical and radiological features. *Blood*. 2011;118:6192–6199.
30. Dowling MM, Quinn CT, Rogers ZR, Journeycake JM. Stroke in sickle cell anemia: alternative etiologies. *Pediatr Neurol*. 2009;41:124–126.
31. Dowling MM, Lee N, Quinn CT, et al. Prevalence of intracardiac shunting in children with sickle cell disease and stroke. *J Pediatr*. 2010;156:645–650.
32. Strouse JJ, Hulbert ML, DeBaun MR, Jordan LC, Casella JF. Primary hemorrhagic stroke in children with sickle cell disease is associated with recent transfusion and use of corticosteroids. *Pediatrics*. 2006;118:1916–1924.
33. Vicari P, Choairy AC, Siufi GC, Arantes AM, Fonseca JR, Figueiredo MS. Embolization of intracranial aneurysms and sickle cell disease. *Am J Hematol*. 2004;76:83–84.
34. Sebire G, Tabarki B, Saunders DE, et al. Cerebral venous sinus thrombosis in children: risk factors, presentation, diagnosis and outcome. *Brain*. 2005;128:477–489.
35. Ali SB, Reid M, Fraser R, MooSang M, Ali A. Seizures in the Jamaica cohort study of sickle cell disease. *Br J Haematol*. 2010;151:265–272.
36. Prengler M, Pavlakis SG, Boyd S, et al. Sickle cell disease: ischemia and seizures. *Ann Neurol*. 2005;58:290–302.
37. Hines PC, McKnight TP, Seto W, Kwiatkowski JL. Central nervous system events in children with sickle cell disease presenting acutely with headache. *J Pediatr*. 2011.
38. Kirkham FJ, Hewes DK, Prengler M, Wade A, Lane R, Evans JP. Nocturnal hypoxaemia and central-nervous-system events in sickle-cell disease. *Lancet*. 2001;357:1656–1659.
39. Kirkham FJ, DeBaun MR. Stroke in children with sickle cell disease. *Curr Treat Options Neurol*. 2004;6:357–375.
40. Scothorn DJ, Price C, Schwartz D, et al. Risk of recurrent stroke in children with sickle cell disease receiving blood transfusion therapy for at least five years after initial stroke. *J Pediatr*. 2002;140:348–354.
41. Hulbert ML, Scothorn DJ, Panepinto JA, et al. Exchange blood transfusion compared with simple transfusion for first overt stroke is associated with a lower risk of subsequent stroke: a retrospective cohort study of 137 children with sickle cell anemia. *J Pediatr*. 2006;149:710–712.
42. DeBaun MR, Armstrong FD, McKinstry RC, Ware RE, Vichinsky E, Kirkham FJ. Silent cerebral infarcts: a review on a prevalent and progressive cause of neurological injury in sickle cell anemia. *Blood*. 2012.
43. Quinn CT, McKinstry RC, Dowling MM, et al. Acute silent cerebral ischemic events in children with sickle cell anemia. *JAMA Neurol*. 2013;70:58–65.
44. Dowling MM, Quinn CT, Plumb P, et al. Acute silent cerebral ischemia and infarction during acute anemia in children with and without sickle cell disease. *Blood*. 2012;120:3891–3897.
45. Kinney TR, Sleeper LA, Wang WC, et al. Silent cerebral infarcts in sickle cell anemia: a risk factor analysis. The Cooperative Study of Sickle Cell Disease. *Pediatrics*. 1999;103:640–645.
46. DeBaun MR, Sarnaik SA, Rodeghier MJ, et al. Associated risk factors for silent cerebral infarcts in sickle cell anemia: low baseline hemoglobin, gender and relative high systolic blood pressure. *Blood*. 2011.
47. Watkins KE, Hewes DK, Connelly A, et al. Cognitive deficits associated with frontal-lobe infarction in children with sickle cell disease. *Dev Med Child Neurol*. 1998;40:536–543.
48. Wang W, Enos L, Gallagher D, et al. Neuropsychologic performance in school-aged children with sickle cell disease: a report from the Cooperative Study of Sickle Cell Disease. *J Pediatr*. 2001;139:391–397.
49. Bernaudin F, Verlhac S, Freard F, et al. Multicenter prospective study of children with sickle cell disease: radiographic and psychometric correlation. *J Child Neurol*. 2000;15:333–343.
50. Hollocks MJ, Kok TB, Kirkham FJ, et al. Nocturnal oxygen desaturation and disordered sleep as a potential factor in executive dysfunction in sickle cell anemia. *J Int Neuropsychol Soc*. 2012;18:168–173.
51. DeBaun MR, Schatz J, Siegel MJ, et al. Cognitive screening examinations for silent cerebral infarcts in sickle cell disease. *Neurology*. 1998;50:1678–1682.
52. Setty BN, Stuart MJ, Dampier C, Brodecki D, Allen JL. Hypoxaemia in sickle cell disease: biomarker modulation and relevance to pathophysiology. *Lancet*. 2003;362:1450–1455.
53. Kirkham FJ, Datta AK. Hypoxic adaptation during development: relation to pattern of neurological presentation and cognitive disability. *Dev Sci*. 2006;9:411–427.
54. Quinn CT, Dowling MM. Cerebral tissue hemoglobin saturation in children with sickle cell disease. *Pediatr Blood Cancer*. 2012.
55. Kato GJ, Hsieh M, Machado R, et al. Cerebrovascular disease associated with sickle cell pulmonary hypertension. *Am J Hematol*. 2006;81:503–510.
56. Gavins FN, Russell J, Senchenkova EL, et al. Mechanisms of enhanced thrombus formation in cerebral microvessels of mice expressing hemoglobin-S. *Blood*. 2011;117:4125–4133.
57. Wang H, Luo W, Wang J, et al. Paradoxical protection from atherosclerosis and thrombosis in a mouse model of sickle cell disease. *Br J Haematol*. 2013;162:120–129.
58. Kennan RP, Suzuka SM, Nagel RL, Fabry ME. Decreased cerebral perfusion correlates with increased BOLD hyperoxia response in transgenic mouse models of sickle cell disease. *Magn Reson Med*. 2004;51:525–532.
59. Vendt BA, McKinstry RC, Ball WS, et al. Silent Cerebral Infarct Transfusion (SIT) trial imaging core: application of novel imaging information technology for rapid and central review of MRI of the brain. *J Digit Imaging*. 2009;22:326–343.
60. Kirkham FJ, Calamante F, Bynevelt M, et al. Perfusion magnetic resonance abnormalities in patients with sickle cell disease. *Ann Neurol*. 2001;49:477–485.

61. Strouse JJ, Cox CS, Melhem ER, et al. Inverse correlation between cerebral blood flow measured by continuous arterial spin-labeling (CASL) MRI and neurocognitive function in children with sickle cell anemia (SCA). *Blood*. 2006;108:379–381.
62. Hales PW, Kawadler JM, Aylett SE, Kirkham FJ, Clark CA. Arterial spin labeling characterization of cerebral perfusion during normal maturation from late childhood into adulthood: normal 'reference range' values and their use in clinical studies. *J Cereb Blood Flow Metab*. 2014.
63. Baldeweg T, Hogan AM, Saunders DE, et al. Detecting white matter injury in sickle cell disease using voxel-based morphometry. *Ann Neurol*. 2006;59:662–672.
64. Schatz J, Buzan R. Decreased corpus callosum size in sickle cell disease: relationship with cerebral infarcts and cognitive functioning. *J Int Neuropsychol Soc*. 2006;12:24–33.
65. Sun B, Brown RC, Hayes L, et al. White matter damage in asymptomatic patients with sickle cell anemia: screening with diffusion tensor imaging. *AJNR Am J Neuroradiol*. 2012;33:2043–2049.
66. Balci A, Karazincir S, Beyoglu Y, et al. Quantitative brain diffusion-tensor MRI findings in patients with sickle cell disease. *AJR Am J Roentgenol*. 2012;198:1167–1174.
67. Kawadler JM, Clayden JD, Kirkham FJ, Cox TC, Saunders DE, Clark CA. Subcortical and cerebellar volumetric deficits in paediatric sickle cell anaemia. *Br J Haematol*. 2013;163:373–376.
68. Mackin RS, Insel P, Truran D, et al. Neuroimaging abnormalities in adults with sickle cell anemia: associations with cognition. *Neurology*. 2014;82:835–841.
69. Adams RJ, McKie VC, Carl EM, et al. Long-term stroke risk in children with sickle cell disease screened with transcranial Doppler. *Ann Neurol*. 1997;42:699–704.
70. Fullerton HJ, Adams RJ, Zhao S, Johnston SC. Declining stroke rates in Californian children with sickle cell disease. *Blood*. 2004;104:336–339.
71. Lehman LL, Fullerton HJ. Changing ethnic disparity in ischemic stroke mortality in US children after the STOP trial. *JAMA Pediatr*. 2013;167:754–758.
72. Cherry MG, Greenhalgh J, Osipenko L, et al. The clinical effectiveness and cost-effectiveness of primary stroke prevention in children with sickle cell disease: a systematic review and economic evaluation. *Health Technol Assess*. 2012;16:1–129.
73. Hsu LL, Miller ST, Wright E, et al. Alpha Thalassemia is associated with decreased risk of abnormal transcranial Doppler ultrasonography in children with sickle cell anemia. *J Pediatr Hematol Oncol*. 2003;25:622–628.
74. Belisario AR, Rodrigues CV, Martins ML, Silva CM, Viana MB. Coinheritance of alpha-thalassemia decreases the risk of cerebrovascular disease in a cohort of children with sickle cell anemia. *Hemoglobin*. 2010;34:516–529.
75. Powars DR, Weiss JN, Chan LS, Schroeder WA. Is there a threshold level of fetal hemoglobin that ameliorates morbidity in sickle cell anemia? *Blood*. 1984;63:921–926.
76. Adekile AD, Yacoub F, Gupta R, et al. Silent brain infarcts are rare in Kuwaiti children with sickle cell disease and high Hb F. *Am J Hematol*. 2002;70:228–231.
77. Marouf R, Gupta R, Haider MZ, Adekile AD. Silent brain infarcts in adult Kuwaiti sickle cell disease patients. *Am J Hematol*. 2003;73:240–243.
78. Miller ST, Milton J, Steinberg MH. G6PD deficiency and stroke in the CSSCD. *Am J Hematol*. 2011;86:331.
79. Thangarajh M, Yang G, Fuchs D, et al. Magnetic resonance angiography-defined intracranial vasculopathy is associated with silent cerebral infarcts and glucose-6-phosphate dehydrogenase mutation in children with sickle cell anaemia. *Br J Haematol*. 2012;159:352–359.
80. Sebastiani P, Ramoni MF, Nolan V, Baldwin CT, Steinberg MH. Genetic dissection and prognostic modeling of overt stroke in sickle cell anemia. *Nat Genet*. 2005;37:435–440.
81. Hoppe C, Klitz W, D'Harlingue K, et al. Confirmation of an association between the TNF-308. promoter polymorphism and stroke risk in children with sickle cell anemia. *Stroke*. 2007;38:2241–2246.
82. Flanagan JM, Sheehan V, Linder H, et al. Genetic mapping and exome sequencing identify 2 mutations associated with stroke protection in pediatric patients with sickle cell anemia. *Blood*. 2013;121:3237–3245.
83. Cox SE, Makani J, Soka D, et al. Haptoglobin, alpha-thalassaemia and glucose-6-phosphate dehydrogenase polymorphisms and risk of abnormal transcranial Doppler among patients with sickle cell anaemia in Tanzania. *Br J Haematol*. 2014; in press.
84. Hulbert ML, McKinstry RC, Lacey JL, et al. Silent cerebral infarcts occur despite regular blood transfusion therapy after first strokes in children with sickle cell disease. *Blood*. 2011;117:772–779.
85. Adams RJ, McKie VC, Hsu L, et al. Prevention of a first stroke by transfusions in children with sickle cell anemia and abnormal results on transcranial Doppler ultrasonography. *N Engl J Med*. 1998;339:5–11.
86. Adams RJ, Brambilla D. Discontinuing prophylactic transfusions used to prevent stroke in sickle cell disease. *N Engl J Med*. 2005;353:2769–2778.
87. Kirkham FJ, Lerner NB, Noetzel M, et al. Trials in sickle cell disease. *Pediatr Neurol*. 2006;34:450–458.
88. Jordan LC, McKinstry III RC, Kraut MA, et al. Incidental findings on brain magnetic resonance imaging of children with sickle cell disease. *Pediatrics*. 2010;126:53–61.
89. Bartolucci P, Galacteros F. Clinical management of adult sickle-cell disease. *Curr Opin Hematol*. 2012;19:149–155.
90. Morris CR. Mechanisms of vasculopathy in sickle cell disease and thalassemia. *Hematology Am Soc Hematol Educ Program*. 2008;177–185.
91. Kato GJ. Novel small molecule therapeutics for sickle cell disease: nitric oxide, carbon monoxide, nitrite, and apolipoprotein A-I. *Hematology Am Soc Hematol Educ Program*. 2008;186–192.
92. Dedeken L, Le PQ, Azzi N, et al. Haematopoietic stem cell transplantation for severe sickle cell disease in childhood: a single centre experience of 50 patients. *Br J Haematol*. 2014.
93. Kharbanda S, Smith AR, Hutchinson SK, et al. Unrelated donor allogeneic hematopoietic stem cell transplantation for patients with hemoglobinopathies using a reduced-intensity conditioning regimen and third-party mesenchymal stromal cells. *Biol Blood Marrow Transplant*. 2014;20:581–586.
94. Zimmerman SA, Schultz WH, Burgett S, Mortier NA, Ware RE. Hydroxyurea therapy lowers transcranial Doppler flow velocities in children with sickle cell anemia. *Blood*. 2007;110:1043–1047.
95. Puffer E, Schatz J, Roberts CW. The association of oral hydroxyurea therapy with improved cognitive functioning in sickle cell disease. *Child Neuropsychol*. 2007;13:142–154.

96. Wang WC, Ware RE, Miller ST, et al. Hydroxycarbamide in very young children with sickle-cell anaemia: a multicentre, randomised, controlled trial (BABY HUG). *Lancet*. 2011;377:1663–1672.
97. Majumdar S, Miller M, Khan M, et al. Outcome of overt stroke in sickle cell anaemia, a single institution's experience. *Br J Haematol*. 2014.
98. Tripathi A, Jerrell JM, Stallworth JR. Cost-effectiveness of adenotonsillectomy in reducing obstructive sleep apnea, cerebrovascular ischemia, vaso-occlusive pain, and ACS episodes in pediatric sickle cell disease. *Ann Hematol*. 2011;90:145–150.
99. Marshall MJ, Bucks RS, Hogan AM, et al. Auto-adjusting positive airway pressure in children with sickle cell anemia: results of a phase I randomized controlled trial. *Haematologica*. 2009;94:1006–1010.
100. Thompson Jr. RJ, Armstrong FD, Link CL, Pegelow CH, Moser F, Wang WC. A prospective study of the relationship over time of behavior problems, intellectual functioning, and family functioning in children with sickle cell disease: a report from the Cooperative Study of Sickle Cell Disease. *J Pediatr Psychol*. 2003;28:59–65.
101. King AA, White DA, McKinstry RC, Noetzel M, DeBaun MR. A pilot randomized education rehabilitation trial is feasible in sickle cell and strokes. *Neurology*. 2007;68:2008–2011.

SECTION XIII

PSYCHIATRIC DISEASE

CHAPTER

103

Depression

Steven T. Szabo[*] *and Charles B. Nemeroff*[†]

[*]Duke University Medical Center, Durham, NC, USA

[†]Leonard M. Miller School of Medicine, University of Miami, Miami, FL, USA

INTRODUCTION

Depression affects approximately 121 million people worldwide and is the most common of the psychiatric illnesses.[1] Symptoms of major depressive disorder are first characterized by clinicians as representing either unipolar or bipolar depression (for a review on neurobiology of bipolar depression see [2]). Although there are differences in prevalence rates of major depressive disorders and bipolar depression around the world,[3,4] in industrialized nations the lifetime prevalence of depression from epidemiologic and family studies remains at approximately 5–12% in adult males and 10–22% in adult females using retrospective designs. The results of prospective studies have suggested that prevalence rates of depression may actually be twice that previously reported,[5] with approximately 1 in 10 children also suffering with a depressive disorder.[6]

It is not surprising that depression is the second most common cause of disability-adjusted life years (DALYs) lost from ages 15–44, and by 2030 it is anticipated to be the single biggest health burden, amounting for 15% of all health conditions worldwide, outranking heart attacks.[7] In addition to the morbidity associated with depression, suicide occurs in 8–15% of patients previously hospitalized for depression, highlighting the dire need for more effective and more rapidly acting treatments. This is of major concern because approximately 37,000 deaths and an estimated 1.4 million years of potential life lost occurs in the United States due to suicide alone.[8] Depression is also now understood to be a systemic disease, a syndrome that impacts global body function, as evidenced by its role as an independent risk factor for coronary artery disease and stroke, with worse outcomes following myocardial infarction, cancer, and diabetes. Reducing the burden of depression will lead to decreased morbidity and mortality and appreciable gains in quality of life.

CLINICAL FEATURES

Diagnosis of Depressive Disorders

The ability to fully understand the pathophysiology of the major psychiatric disorders has been largely hampered by poor clinical access to the target tissue organ "the brain." Albeit inherently limited, peripheral and indirect measures of altered biochemical events taken from biologic fluids of patients with depression have shed light on possible neural mechanisms underlying this disease. Unlike many other fields of medicine, the lack of biochemical or genetic markers of high diagnostic specificity and sensitivity have rendered the process of psychiatric diagnosis to remain as a purely clinical endeavor based on observation, interview, history, and collateral information. The Diagnostic and Statistical Manual of Mental Disorders (DSM) is currently in its fifth edition (DSM-5) and provides a diagnostic framework for all of the psychiatric disorders; this new edition has recently been released with some notable changes from the prior publication (Tables 103.1A and B). The mood disorders section in the DSM-5 is partitioned into three different subsections with varying scope. The first subsection lists the various syndromes of mood episodes as either

http://dx.doi.org/10.1016/10.1016/B978-0-12-410529-4.00103-0

TABLE 103.1A DSM-5 Diagnostic Criteria: Major Depressive Disorder

A. Five or more of the following symptoms have been present during the same 2-week period and represent a change from previous functioning; at least one of the symtoms is either 1) depressed mood or 2) loss of interest or pleasure:

Note: Do not include symptoms that are clearly attributable to another medical condition

1. Depressed mood most of the day, nearly every day, as indicated by either subjective report (e.g., feels sad, empty, hopeless) or observation made by others (e.g., appears fearful). (Note: in children and adolescents, can be irritable mood)
2. Markedly diminished interest or pleasure in all, or almost all, activities most of the day, early every day (as indicated by either subjective account or observation)
3. Significant weight loss when not dieting or weight gain (e.g., a change in more than 5% of body weight a month) or decrease or increase in appetite nearly every day. (Note: in children, consider failure to make expected weight gain)
4. Insomnia or hypersomnia nearly every day
5. Psychomotor agitation or retardation nearly every day (observable by others, not merely subjective feelings of restlessness or being slowed down)
6. Fatigue or loss of energy nearly every day
7. Feelings of worthlessness or excessive or inappropriate guilt (which may be delusional) nearly every day (not merely self-reproach or guilt about being sick)
8. Diminished ability to think or concentrate, or indecisiveness, nearly every day (either by subjective account or as observed by others)
9. Recurrent thoughts of death (not just fear of dying), recurrent suicidal ideation without a specific plan, or suicide attempt or a specific plan for committing suicide

B. The symptoms cause clinically significant distress or impairment in social, occupational, or other important areas functioning

C. The episode is not attributable to the physiological effects of a substance or to another medical condition

Note: Criteria A–C represent a major depressive episode

Note: Responses to a significant loss (e.g., bereavement, financial ruin, losses from a natural disaster, a serious medical illness or disability) may include feelings of intense sadness, rumination about the loss, insomnia, poor appetite, and weight loss noted in Criterion A, which may resemble a depressive episode. Although such symptoms may be understandable or considered appropriate to the loss, the presence of a major depressive episode in addition to the normal response to a significant loss should also be carefully considered. This decision inevitably requires the exercise of clinical judgment based on the person's past history of major depressive episodes, whether the symptoms are disproportionately severe given the nature of the loss, and the individual's cultural norms for the expression of distress in the context of their loss

D. The occurrence of the major depressive episode is not better explained by schizoaffective disorder, schizophrenia, schizophreniform disorder, delusional disorder, or other specified and unspecified schizophrenia spectrum and other psychotic disorders

E. There has never been a manic episode or a hypomanic episode

Note: This exclusion does not apply if all the manic-like or hypomanic-like episodes are substance-induced or are attributable to the physiological effects of another medical condition

TABLE 103.1B DSM-5 Diagnostic Changes to Mood Disorders

Bipolar Disorder	Inclusion of increased energy/activity as a core symptom of hypomania/mania Addition of mixed features specifier for manic, hypomanic, and major depressive episodes
Major depressive episode	Elimination of the bereavement exclusion Addition that the presence of a major depressive episode can be considered in addition to the normal response to a significant loss Explanatory information is included about the difference between bereavement and major depressive episode Addition of anxious distress specifier for major depressive episode
Disruptive mood dysregulation disorder	New disorder in DSM-5 Addresses presentations of severe, nonepisodic irritability that has contributed to upsurge of pediatric bipolar diagnoses

being major depressive, manic, hypomanic, or mixed episode. The next subheading describes necessary criteria that must be met for establishing a diagnosis. Once a diagnosis is reached, the third subheading allows for a "specifier" to be added such as mild moderate, severe with psychotic features, etc. (Table 103.2A) and describes the most recent episode or course of recurrent episodes the patient suffers (Table 103.2B).

Major depressive disorder is primarily differentiated from bipolar disorder as being devoid of characteristic features of mania or hypomania (i.e., elation for a period of time, decreased need for sleep, etc.). Distinct diagnostic

TABLE 103.2A DSM-5 Severity and Course: Major Depressive Disorder

Severity/course specifier	Single episode	Recurrent episode*
Mild (pp. 188)	296.21 (F32.0)	296.31 (F33.0)
Moderate (pp. 188)	296.22 (F32.1)	296.32 (F33.1)
Severe (pp. 188)	296.23 (F32.2)	296.33 (F33.2)
With psychotic features (pp. 186)	296.24 (F32.3)	296.34 (F33.3)
In partial remission (pp. 188)	296.25 (F32.4)	296.35 (F33.4)
In full remission (pp. 188)	296.26 (F32.5)	296.36 (F33.5)
Unspecified	296.20 (F32.9)	296.30 (F33.9)

**For an episode to be considered recurrent there must be an interval of at least 2 consecutive months between separate episodes in which criteria are not met for a major depressive episode. The definitions of specifiers are found on the indicated pages.*

TABLE 103.2B DSM-5 Specifiers: Major Depressive Disorder

Specify:

If the full criteria are currently met for a major depressive episode, specify its current clinical status and/or features:
With anxious distress (pp. 184)
With mixed features (pp. 184–185)
With melancholic features (pp. 185)
With atypical features (pp. 185–186)

In distinguishing grief from a major depressive episode (MDE), it is useful to consider that in grief the predominant affect is feelings of emptiness and loss, while in MDE it is persistent depressed mood and the inability to anticipate happiness or pleasure. The dysphoria in grief is likely to decrease in intensity over days to weeks and occurs in waves, the so-called pangs of grief. These waves tend to be associated with thoughts or reminders of the deceased. The depressed mood of MDE is more persistent and not tied to specific thoughts or preoccupations. The pain of grief may be accompanied by positive emotions and humor that are uncharacteristic of the pervasive unhappiness and misery characteristic of MDE. The thought content associated with grief generally features a preoccupation with thoughts and memories of the deceased, rather than the self-critical or pessimistic ruminations seen in MDE. In grief, self-esteem is generally preserved, whereas in MDE feelings of worthlessness and self-loathing are common. If self-derogatory ideation is present in grief, it typically involves perceived failings vis-à-vis the deceased (e.g., not visiting frequently enough, not telling the deceased much he or she was loved). If a bereaved individual thinks about death and dying, such thoughts are generally focused on the deceased and possibly about "joining" the deceased, whereas in MDE such thoughts are focused on ending ones own life because of feelings of worthlessness, undeserving of life, or unable to cope with the pain of depression.

With mood-congruent psychotic features (pp. 186)
With mood-incongruent psychotic features (pp. 186)
With catatonic features (pp. 186) Coding note: use additional code 781.99 (R29.818)
With peripartum onset (pp. 186–187)
With seasonal pattern (recurrent episode only) (pp. 187–188)

Specify current or most recent episode:
Single episode
Recurrent episode: defined as the presence of two or more lifetime major depressive episodes. To be considered separate episodes, there must be an interval of at least 2 consecutive months in which criteria are not met for a major depressive episode

Specify current severity:
Mild (pp. 188)
Moderate (pp. 188)
Severe (pp. 188)

Specify:
Level of concern for suicide in the current assessment period regardless of current episode or remission status (pp. 182)

criteria for classifying an individual as having a depressive disorder necessitates that five or more of the below listed definable symptoms has occurred over a 2-week period (Criterion A, Table 103.1A). Depressed mood and loss of interest or pleasure (also known as anhedonia) must be one of the identifiable symptoms, as well as four of these other symptoms must also be present: 1) fatigue or loss of energy nearly every day; 2) feelings of worthlessness, or excessive/inappropriate guilt; 3) significant weight gain or weight loss apart from dieting (5% change in body weight); 4) decrease or increase in appetite nearly every day; 5) insomnia or increased desire to sleep nearly every day; 6) trouble making decisions, thinking or concentrating; 7) restlessness or slowed behavior; or 8) recurrent thoughts of death or suicidal behaviors. These symptoms need not only be based on the patient's own feeling, but can be based on observations by others.

Subtypes of major depression, based on symptom cluster or course, by which specifiers noted above added to a patient's diagnosis help more accurately document the patient's illness. Thus, those making a diagnosis of depression can comment on the severity, presence of psychosis (often seen in late-life depression), and level of remission (Table 103.2B). This helps detail patient differences in major depressive disorder, which may inform clinicians of the appropriate treatment and course of illness trajectory. Changes in DSM-5 for major depression have included *removal of the bereavement exclusion*, which will allow clinicians to diagnose major depression in an individual suffering from a complicated and extended grief over loss a loved one, as well as elucidation of chronic depression. DSM-5 now also includes *mixed anxiety/depression* and *disruptive mood dysregulation disorder* as diagnosable depressive entities. Introducing the former entity of mixed features in the DSM-5 has been criticized by some who feel that a lack of information on mixed anxiety/depression exists and it fails to warrant a specific category.[9–11] The inherent subjectivity of DSM diagnostics taken together with widespread burden of depression that is often comorbid with numerous medical disorders, places increasing importance on clinicians to be competent in diagnosing mood disorders. Objective biology-based diagnostics are needed to augment the clinical evaluation of patients in the same way that electroencephalography augments the diagnosis of epilepsy. Such advancements will also aid in decreasing the stigma of mood disorders, improve accuracy in diagnosis, and provide more personalized treatment.

Course

The course of major depression can vary considerably in individuals and may, in part, reflect differences in epigenetic factors, vulnerability and resiliency genes, gene–environment interactions and stress-load effects. Because approximately 80% of patients with depression have comorbid anxiety, common pathophysiologic responses to stress are implicated in the diathesis of both mood and anxiety disorders. Anxiety and mild depressive symptoms are often first observed during a prodromal period, which can predate the full depressive episode by weeks to months. Following this, depressive symptoms generally will progress over days to weeks. If the depressive episode is left untreated, symptoms may last for 6 months or longer, but in general a complete remission of symptoms occur over time. However, upwards of 30% of patients will not recover completely from a depressive episode for months or longer (regardless of age of onset), and it is well known that patients who have experienced one major depressive episode will most likely have a syndromal recurrence later on in life (approximately a 90% chance of having another episode if the patient has >3 prior episodes). Three factors place an individual at greatest risk for more severe and refractory episodes, with less likelihood of remission: 1) time to treatment; 2) number of previous episodes; and 3) a history of childhood abuse and neglect. Lastly, approximately 5–10% of patients meet criteria for a major depressive disorder during 2 or more years without remission, so called "chronic depression."

Syndromes Secondary to Medical and Neurological Disease

Depressive symptoms can occur in the presence of numerous systemic medical illnesses, and as such are diagnosed as mood disorder due to a general medical condition (if the medical condition is deemed to be leading to the depressed mood). Primary care physicians are a major access point for evaluating patients with depression and initiating treatment early. Unfortunately, successful clinical resolution of depression in the primary care setting is often uncommon. It has recently been estimated that only half of patients suffering from major depression are identified in the primary care setting, and only a quarter of these patients receive treatment, with less than 10% obtaining adequate treatment or achieving remission.[12] Diabetes, both type I and type II, is associated with a 33% lifetime prevalence of depression; the latter further predisposes individuals to coronary artery disease (CAD) and congestive heart failure (CHF). Because the cardiac morbidity and mortality in depressed patients is more severe when compared to matched cardiac patients without depression, this is of great clinical significance. Indeed, antidepressant treatment in patients with stable coronary artery disease results in lower mental stress-induced myocardial ischemia.[13] Patients who are depressed 1 week post-myocardial infarction are 3–4 times more likely to die within the next 6 months than patients without depression.[14,15]

In addition to cardiac disease and diabetes, other medical disorders are associated with depressed mood. Patients with Cushing disease have up to a 67% lifetime prevalence of syndromal depression and anxiety disorders. Various types of cancers, such as pancreatic cancer and breast cancer, are also well known to be associated with higher rates of depression than those seen in the general population, and these syndromal mood symptoms are independent of prognosis. Recent work on pro-inflammatory cytokine mediators as mediators of depression have been the focus of intense scrutiny and a subset of patients with increased C-reactive protein (CRP) concentrations and depression respond to cytokine antagonist treatment.[16] This is concordant with the well-known "depressogenic" effects of interferon-α used in the treatment of malignant melanoma and hepatitis C.

It is not surprising that patients with certain neurological disorders are especially prone to depressive disorders. Notably, Parkinson disease patients have roughly a 50% lifetime prevalence of depression, thought to relate primarily to dopamine signaling deficits, although reduced serotonin (5-HT) metabolites in the cerebrospinal fluid (CSF) have also been reported in these patients. Huntington disease patients have a very high lifetime prevalence rate of depression and approximately 40% show depressive symptoms prior to neurological symptoms. Cerebrovascular disease, stroke, and Alzheimer disease represent very common causes of late-life depression and account for 30–50% of the lifetime prevalence of depression. Patients with multiple sclerosis exhibit depression rates as high as 50%, but they may also experience manic or hypomanic symptoms, at times exacerbated by corticosteroid therapy, which complicates their treatment. Although physicians often attribute depression in these patients to an inability to cope with debilitating disease, epidemiologic and biological data suggest that this is not the case. Many biologic mechanisms leading to systemic and neurological disease are also antecedents to depression (i.e., loss of monoamine neurons in degenerative disorders, paraneoplastic syndromes). In the next section we describe the neurochemical substrates of depression.

PATHOLOGY: BIOCHEMICAL ALTERATIONS

Monoamine Neurotransmitter Abnormalities

The discovery, initially driven largely by serendipity, that most effective antidepressant agents were found to biochemically target monoamine neurotransmitter systems (5-HT, norepinephrine [NE], and dopamine [DA]), led to their extensive study in depression (for reviews see [1,17] and Figure 103.1). Numerous studies have contributed to the notion that depression is, at least in part, the result of monoamine dysregulation in the brain; however, monoamine metabolites in blood products, urine, and CSF, as well as postmortem studies from brains of patients with mood disorders lack consistency in demonstrating monoamine deficiencies. This likely reflects heterogeneity in the depressed patients studied, limitations and uniformity of experimental techniques, and possibly nonspecific factors. Because much of our understanding of depression and its treatment have centered upon monoamine mechanisms, it is important to summarize these findings.

SEROTONIN 5-HT neurons project widely throughout the central nervous system (CNS), modulating virtually every part of the neuroaxis. L-Tryptophan, an amino acid that is derived primarily from diet, is actively transported into 5-HT-containing neurons and is the precursor to 5-HT. The availability of L-tryptophan is rate limiting and tryptophan hydroxylase (TrpH) converts L-tryptophan to 5-hydroxytryptophan (5-HTP). This reaction can be effectively inhibited by the drug *p*-chlorophenylalanine (PCPA). PCPA is associated with symptoms of depression, likely due to serotonin depletion. Mice transfected with a human-occurring brain mutation of TrpH (Tph2 [C1473G]) exhibit reduced 5-HT synthesis and a depressive-like phenotype.[18] Experimental depletion of 5-HT in humans can be accomplished by orally administering a drink containing all aromatic amino acids except tryptophan, and this prevents endogenous tryptophan from entering the brain, rapidly depleting the CNS of 5-HT. Interestingly, 5-HT depletion reliably causes a relapse of depression in patients successful treated with a selective 5-HT reuptake inhibitor (SSRI), but not a norepinephrine reuptake inhibitor, and does not induce depression in healthy subjects. 5-HT depletion also does not alter the response rate of patients treated with electroconvulsive therapy (ECT). These findings suggest that

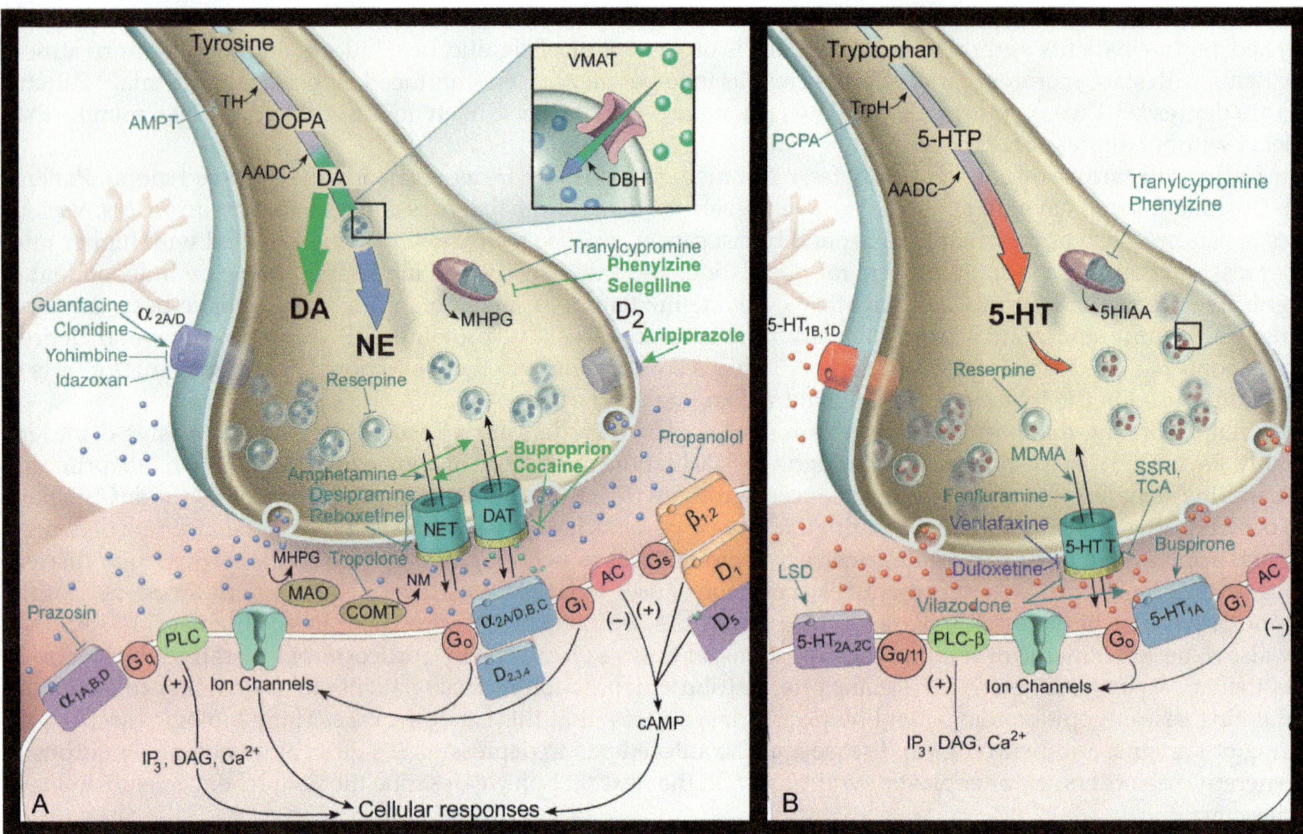

FIGURE 103.1 (**A**) Schematic representation of a catecholamine neuron and biosynthetic pathway to production of dopamine (DA) and norepinephrine (NE). This catecholamine neuron is impinging on a postsynaptic target that has DA (D_1, D_2, D_3, D_4, D_5) and NE (α_1, α_2, β_1, β_2, β_3) receptors linked to second-messenger signaling and intracellular events. The amino acid L-tyrosine is actively transported into presynaptic catecholamine nerve terminals and converted to catecholamines. The rate-limiting step is conversion of L-tyrosine to L-dihydroxyphenylalanine (levodopa; a drug used in treatment of Parkinson disease) by the enzyme tyrosine hydroxylase (TH). Notably, α-methyl-*p*-tyrosine (AMPT) is a competitive inhibitor of TH and results in reduced catecholaminergic function in clinical studies of mood disorders. Aromatic amino acid decarboxylase (AADC) takes levodopa to DA down the synthetic pathway. DA is then taken up from the cytoplasm and packaged into vesicles by vesicle monoamine transporters (VMATs) to await release into the synapse. However, DA can also be hydroxylated by dopamine-β-hydroxylase (DBH) in the presence of O_2 and ascorbate to form NE. Normetanephrine (NM) is formed by the action of catechol-*O*-methyltransferase (COMT) on NE and is then further metabolized by monoamine oxidase (MAO) and aldehyde reductase to 3-methoxy-4-hydroxyphenylglycol (MHPG). Tropolone is an agent that can block COMT. Phenylzine, selegiline, and tranylcypromine are antidepressants that inhibit MAO. Reserpine causes a depletion of monoamines from vesicles by interfering with uptake and storage mechanisms (depressive-like symptoms have been reported). Once released from the presynaptic terminal, DA can interact with a variety of presynaptic and postsynaptic receptors. Presynaptic regulation of NE and DA neuron firing activity and release occurs through somatodendritic (not shown) and nerve-terminal α_2-adrenoreceptors and D_2 receptors, respectively. Yohimbine potentiates NE neuronal firing and NE release by blocking these presynaptic α_2-adrenoreceptors, thereby disinhibiting these neurons from a negative feedback influence. This agent can also induce panic attacks in individuals that are prone to this disorder. Conversely, clonidine attenuates NE neuron firing and release by activating presynaptic α_2-adrenoreceptors. This antihypertensive agent can help with reducing opiate withdrawal symptoms and is often used in children with attention-deficit hyperactivity disorder (ADHD). Idazoxan is a relatively selective α_2-adrenoreceptor antagonist primarily used for pharmacological purposes and guanfacine is used in treatment of ADHD. Aripiprazole is a non-selective partial D_2 receptor agonist and antipsychotic agent used in treatment of mood disorders. The binding of NE to G protein receptors (Go, Gi, etc.) coupled to adenylyl cyclase (AC) and phospholipase C-β (PLC-β) produces a cascade of second messenger and cellular effects (see diagram and later sections of the text). Propranolol (β_2-receptor antagonist) has beneficial effects in situational anxiety and prazosin (α_1-adrenoceptor antagonist) reduces nightmares associated with posttraumatic stress disorder (PTSD; both agents are antihypertensives). NE has its action terminated in the synapse by rapidly being taken back into the presynaptic neuron via NE transporters (NETs). Once inside the neuron it can either be repackaged into vesicles for reuse or undergo enzymatic degradation. The selective NE reuptake inhibitor antidepressant reboxetine and older-generation tricyclic antidepressant desipramine are able to interfere/block the reuptake of NE. On the other hand, amphetamine is able to facilitate NE release by altering NET function. Cocaine and buproprion exert effects, in part, through nonselective blockade of DA transporters (DAT). Note: green spheres represent DA neurotransmitters; blue spheres represent NE neurotransmitters. (**B**) Schematic representation of a serotonin (5-HT) neuron and its postsynaptic target. 5-HT is produced from L-tryptophan, an amino acid actively transported into presynaptic 5-HT-containing terminals. It is converted to 5-hydroxytryptophan (5-HTP) by the rate-limiting enzyme tryptophan hydroxylase (TrpH). This enzyme is effectively inhibited by the drug *p*-chlorophenylalanine (PCPA), a depressogenic agent. Aromatic amino acid decarboxylase (AADC) converts 5-HTP (a 5-HT enhancer) to 5-HT. Once released from the presynaptic terminal, 5-HT can interact with a variety (i.e., 15 different types) of presynaptic and postsynaptic receptors. Presynaptic regulation of 5-HT neuron firing activity and release occurs through somatodendritic 5-HT_{1A} (not shown) and $5\text{-HT}_{1B/1D}$ autoreceptors, respectively, located on nerve terminals. Buspirone is a partial 5-HT_{1A}

FIGURE 103.1 receptor agonist that activates both pre- and postsynaptic receptors and is the non-benzodiazepine receptor agonist used in treatment of anxiety. Vilazodone is a newer selective serotonin reuptake inhibitor (SSRI) and partial 5-HT$_{1A}$ receptor agonist. The binding of 5-HT to G-protein receptors (Go, Gi, etc.) that are coupled to adenylyl cyclase (AC) and phospholipase C-β (PLC-β) will result in the production of a cascade of second-messenger and cellular effects. Lysergic acid diethylamide (LSD) likely interacts with numerous 5-HT receptors to mediate its effects. 5-HT has its action terminated in the synapse by rapidly being taken back into the presynaptic neuron through 5-HT transporters (5-HTT). Once inside the neuron it can either be repackaged into vesicles for reuse or undergo enzymatic catabolism. The SSRIs and older-generation tricyclic antidepressants (TCAs) are able to interfere/block the reuptake of 5-HT (agents such as fluoxetine, paroxetine, sertraline, and citalopram are examples). Venlafaxine and duloxetine are SNRIs, i.e., serotonin and norepinephrine reuptake inhibitors, which enhance the availability of both these monoamines. 5-HT is then metabolized to 5-hydroxyindoleacetic acid (5-HIAA) by monoamine oxidase (MAO), located on the outer membrane of mitochondria or sequestered and stored in secretory vesicles by vesicle monoamine transporters (VMATs). Reserpine causes a depletion of 5-HT in vesicles by interfering with uptake and storage mechanisms (depressive-like symptoms have been reported with this agent). Tranylcypromine and phenylzine are MAO inhibitors (MAOIs) and effective antidepressants. Fenfluramine (an anorectic agent) and MDMA ("ecstasy") are able to facilitate 5-HT release by altering 5-HTT function. DAG, diacylglycerol; 5-HTT, serotonin transporter; IP3, inositol-1,4,5-triphosphate. Adapted with permission from Szabo ST, Gould TD, Manji HK. Introduction to neurotransmitters, receptors, signal transduction, and second messengers. In: Schatzberg AF, Nemeroff CB, eds. The American Psychiatric Publishing Textbook of Psychopharmacology. 4th ed. Washington, DC: American Psychiatric Publishing; 2009.

modulating the 5-HT system is only one way to modulate affect and treat depression.[19] Genetic vulnerability to stress has also been associated with variant alelles of the 5-HT transporter and may predispose individuals to depression and differential responses to antidepressant treatment.[20]

NOREPINEPHRINE The locus coeruleus contains the cell bodies of NE neurons that project most of the NE in the brain (90% of NE in the forebrain and 70% of total NE in the brain; see [21] for review). The amino acid L-tyrosine is actively transported into presynaptic NE nerve terminals, but unlike the 5-HT system, the rate-limiting step in NE production is conversion of L-tyrosine to L-dihydroxyphenylalanine (levodopa) catalyzed by the enzyme tyrosine hydroxylase (TH). Aromatic amino acid decarboxylase (AADC) converts levodopa to dopamine (DA), and dopamine-β-hydroxylase (DβH) converts DA to NE. Depletion of NE by α-methyl-*p*-tyrosine (a TH inhibitor) has contributed to the evidence that catecholamines play a seminal role in the pathophysiology and treatment of mood and anxiety disorders.[22,23] Because antidepressants produce a downregulation/desensitization of β_1-receptors in rat forebrain, these effects have been suggested to play a role in their therapeutic efficacy. Interestingly, β-receptors also regulate emotional memories, leading to the proposal that β antagonists may have utility in the treatment of posttraumatic stress disorder (PTSD).[24,25] The neuroanatomical and chemical interactions between 5-HT and NE neurons have led to detailing a brainstem circuit accounting for monoamine circuit modulation with relevance to treatment of anxiety and depression.[26–28]

DOPAMINE DA is largely produced in midbrain nuclei (substantia nigra and ventral tegmental area) having a biochemical synthesis pathway near identical to that of NE (see above and Figure 103.1). Essentially, the only difference is that NE neurons possess DβH to convert DA to NE. The early observations that patients with Parkinson disease frequently have depressive symptoms and reserpine (a depleter of monoamine vesicles) induces depression in some individuals led to the hypothesis that decreased activation of DA pathways may underlie symptoms of psychomotor retardation and melancholic depression in some patients.[29] More specifically, mesocortical DA pathways are associated with regulation of motivation–emotion, mesolimbic DA pathway signal aspects of reward, and psychomotor changes are implicated through abnormalities in nigrostriatal DA circuits.[30] Postmortem and imaging studies support a pre-eminent role for DA circuit dysfunction in the pathogenesis of depression.[29] Monoamine oxidase inhibitors (MAOI) were some of the first available antidepressants and increase the availability of DA, 5-HT and NE. Complex interaction between 5-HT, NE and DA circuits occur, and alterations in one monoamine system impact the others.[31] This concentration of findings implicates dysfunction of DA pathways as contributing to the pathophysiology of depression.[32]

Glutamate and Excitatory Amino Acids

Glutamate (L-glutamic acid) and aspartate are excitatory amino acids that play a prominent role in synaptic plasticity, learning, and memory. Glutamate at high concentrations is a potent neuronal excitotoxin, which triggers rapid or delayed neuronal death. Unlike the monoamines, which require transport of precursor amino acids across the blood–brain barrier, glutamate and aspartate cannot adequately penetrate the brain from the periphery and are produced locally. The metabolic and synthetic enzymes responsible for the formation of these nonessential amino acids are located both in glial cells and neurons. Glutamine is the precursor molecule for synthesis of both glutamate and the major inhibitory neurotransmitter γ-aminobutyric acid (GABA). Once glutamate is formed, it can be packaged and released from presynaptic nerve terminal vesicles to mediate transsynaptic effects, taken up by the surrounding glial cells through glutamate transporters, and converted back to glutamine (Figure 103.2).

Although evidence has accrued suggesting that glutamatergic dysregulation occurs in mood disorders, no therapeutic agents targeting this system have gained US Food and Drug Administration (FDA) approval.[33–35] Antidepressants that primarily target monoamine systems have been reported to reduce the function of glutamate receptors (*N*-methyl-D-aspartate [NMDA], α-amino-3-hydroxy-5-methyl-4-isoxazolepropionic acid [AMPA], and metabotropic glutamate receptors [mGluR]) after sustained treatment, and these receptors have known importance in modulating synaptic plasticity, a process that may be defective in depression. Ketamine is a high-affinity, non-competitive NMDA receptor antagonist (i.e., anesthetic agent with abusive potential, "Special K") reported to be a rapidly acting antidepressant (i.e., within 72 hours of administration), with effects that persist from a few days to 1–2 weeks (for a recent review see [36]). Clinical trial designs that incorporate an appropriate active placebo agent are needed because the dissociative effects of ketamine are well documented and difficult to "blind" to patients and investigators. Notwithstanding the inherent pitfalls of using ketamine in the treatment of depression, NMDA receptors and other glutamate receptor agents (i.e., AMPA modulators) hold promise as novel agents in the treatment of mood

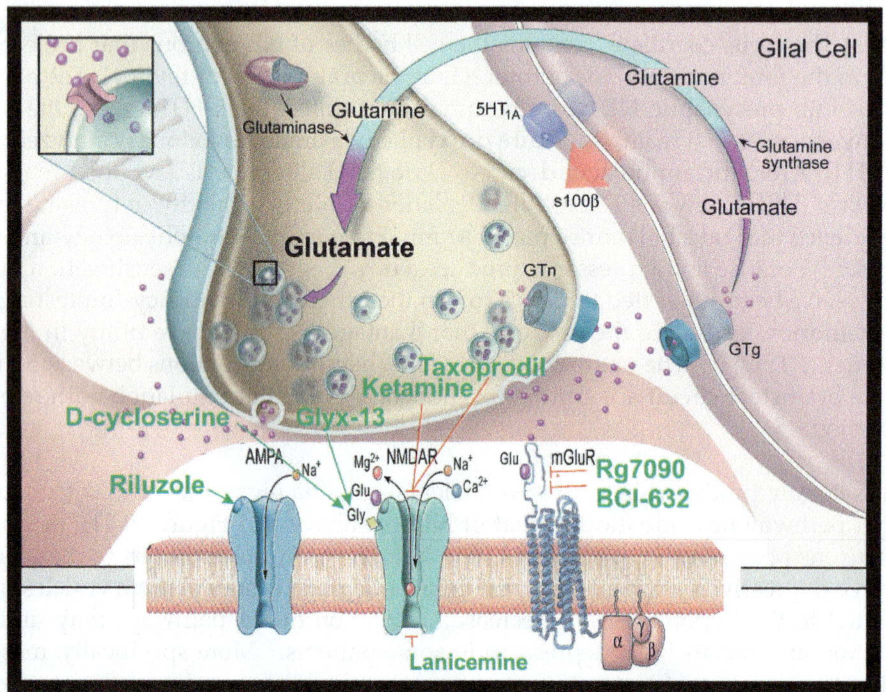

FIGURE 103.2 **The glutamate system has now presented itself as a viable target in antidepressant action.** This figure schematically represents various processes involved in glutamatergic regulation and neurotransmission. Synthesis of glutamate derives from glucose and undergoes transamination to α-ketoglutarate; however, a small proportion of glutamate can be formed from glutamine by glutamine synthetase. Conversion of glutamine to glutamate occurs in glia cells and can then be transported to neurons, with mitochondria glutaminase being needed to complete the formation of glutamate from its precursors. Glutamine can also undergo oxidation to yield α-ketoglutarate in astrocytes and transported to neurons for glutamate synthesis and vesicular storage. Calcium-dependent excitotoxic processes control glutamate release and this neurotransmitter can impact a variety of postsynaptic receptors (*N*-methyl-D-aspartate [NMDA], α-amino-3-hydroxy-5-methyl-4-isoxazolepropionic acid [AMPA], and metabotropic [mGluR]). The $mGluR_2$ and $mGluR_3$ localization on glutamate neurons confer autoreceptor capabilities and targeting these receptors or reuptake transporters (found on neurons [GTn] and astrocytes [GTg]) have been proposed to be viable antidepressant targets. Reductions in astrocytes are documented in mood disorders and possibly contribute to glutamatergic deregulation, with glial-derived neurotrophic factor (GDNF) and S100β secreted from these cells being important in maintaining synapse integrity. 5-HT_{1A} receptors are important in antidepressant action and can modulate the release of S100β, potentially bridging 5-HT regulation of the glutamate system and synapse plasticity. Listed in this figure are agents that biochemically target glutamate receptors and have the potential to mediate antidepressant effects. However, none as of yet are FDA approved for clinical use. Glu, glutamate; Gly, glycine; GTg, glutamate transporter glial; GTn, glutamate transporter neuronal; mGluR, metabotropic glutamate receptor; purple sphere, glutamate; green arrow, agonist; red line, antagonist. Adapted with permission from Szabo ST, Gould TD, Manji HK. Introduction to neurotransmitters, receptors, signal transduction, and second messengers. In: Schatzberg AF, Nemeroff CB, eds. The American Psychiatric Publishing Textbook of Psychopharmacology. 4th ed. Washington, DC: American Psychiatric Publishing; 2009.

disorders[37] (Figure 103.2). Recently, traxoprodil (an NMDA receptor antagonist) and lanicemine (a nonselective "low trapping" NMDA receptor antagonist) exhibited beneficial effects in treatment-resistant depression.[38,39] However, these were small studies with intravenously infused agents, and similar to ketamine, produced dissociative effects in some subjects. However, as in the ketamine trials, dissociation did not correlate with whether the individual exhibited an antidepressant response. Caution in using NMDA receptor antagonists as potential antidepressants is needed because neuronal vacuolization and other potentially toxic enduring receptor changes have been observed following sustained administration of dizocilpine and other NMDA antagonists during animal studies.[40,41]

Glycine is an excitatory amino acid that can modulate NMDA receptors and the glycine-binding site has been posited as a novel avenue for antidepressant treatment.[42] Thus, D-cycloserine binds to the glycine B site of NMDA receptors, where it acts as either a partial agonist or full agonist depending on the composition of the subunits comprising the NMDA receptor (i.e., GluN2A–2D).[43] A placebo-controlled add-on of high dose D-cycloserine (1000 mg per day) led to gradual improvement in depressive symptoms when administered to treatment resistant patients kept on their antidepressant regimen for 6 weeks.[44] Furthermore, GLYX-13 (a glycine receptor partial agonist) has been shown to possess antidepressant effects without the dissociative components observed with the abovementioned NDMA receptor antagonists tested in depression.[45] Treatment with inhibitors of glutamate release and ions such as Zn^{2+} and Mg^{2+} can alter NMDA receptor function, and their ability to target allosteric modulation sites of NMDA receptors have been suggested as a worthwhile approach to novel antidepressant development.[35]

Metabotropic glutamate receptors (mGluR) are found on postsynaptic neurons ($mGluR_5$) and localized directly on glutamate neurons ($mGluR_{2/3}$), the latter serving as autoreceptors to regulate neuron firing activity and glutamate release profiles. Two mGluR antagonists are currently under investigation for the treatment of depression: RG7090 is a mGluR5 antagonist in phase II clinical trials and BCI-632 is a mGluR2/3 antagonist in phase I (see [46] for review). Additionally, AMPA receptor potentiators have exhibited antidepressant properties in preclinical studies and may represent a viable target in developing glutamate-based antidepressants.[46] For instance, riluzole is an AMPA receptor modulator used in treatment of amyotrophic lateral sclerosis (ALS) that has documented preclinical antidepressant properties;[47,48] however, following a single ketamine infusion in patients with depression, add-on treatment with riluzole was not more effective then placebo.[49] Interestingly, the antidepressant effects of $mGluR_{2/3}$ antagonists are negated by AMPA receptor antagonism. Chronic treatment with monoamine-based antidepressants enhance AMPA receptor function by altering the phosphorylation state of specific AMPA receptor subunits,[50] leading to altered membrane expression of AMPA receptors,[51] an interesting potential mechanism of antidepressant action and mood stabilization.[52–54]

Neuroendocrine Factors

Hypothalamic–Pituitary–Adrenal Axis (CRF and Neuroactive Steroids)

Stressful life events and chronic stress can produce major CNS alterations as assessed by structural and functional brain imaging as well as neurodevelopmental sequelae, i.e., hyperglutamatergic state[54] and chronic and persistent changes in cortisol release from the adrenal cortex. Increased corticotropin releasing factor (CRF) secretion from the hypothalamus and adrenocorticotrophin (ACTH) from the anterior pituitary gland drives the aberrant cortisol responses to stress. Hyperactivity of the hypothalamic–pituitary–adrenal (HPA) axis represents one of the most consistent alterations in patients with mood disorders, but can also be observed in other psychiatric and nonpsychiatric illnesses. As many as 75% of patients with major depression exhibit overactivity of the HPA axis. Depressed patients and suicide victims have been reported to exhibit: 1) elevation of CRF concentrations in CSF; 2) increased hypothalamic and cerebrocortical CRF immunoreactivity; 3) increased CRF mRNA expression in postmortem tissue; and 4) increased ACTH and cortisol levels during depressive episodes; 5) decreased CRF_1 receptor binding and decreased $CRFR_1$ mRNA expression have also been reported in postmortem brain tissue of suicide victims. Sensitive tests of HPA function such as the dexamethasone/CRF suppression test confirms hyperactivity of the HPA axis in depressed patients. Preclinical animal models and clinical studies have contributed to the development of glucocorticoid receptor antagonists, now reported to be effective in severe types of depression, including depression with psychotic features. Lastly, cross-sectional area and volume of both the pituitary and adrenal glands are increased in individuals with depression compared to nondepressed individuals; the adrenal gland is likely enlarged due to chronically elevated ACTH release[55,56] (Figure 103.3).

Neuroactive steroids are neurohormones that are produced by neurons and glial cells in the brain, as well as outside the nervous system (i.e., adrenals and gonads). Although numerous neurosteroids and metabolites are produced in the nervous system, allopregnanolone, dehydroepiandrosterone (DHEA), dehydroepiandrosterone sulfate

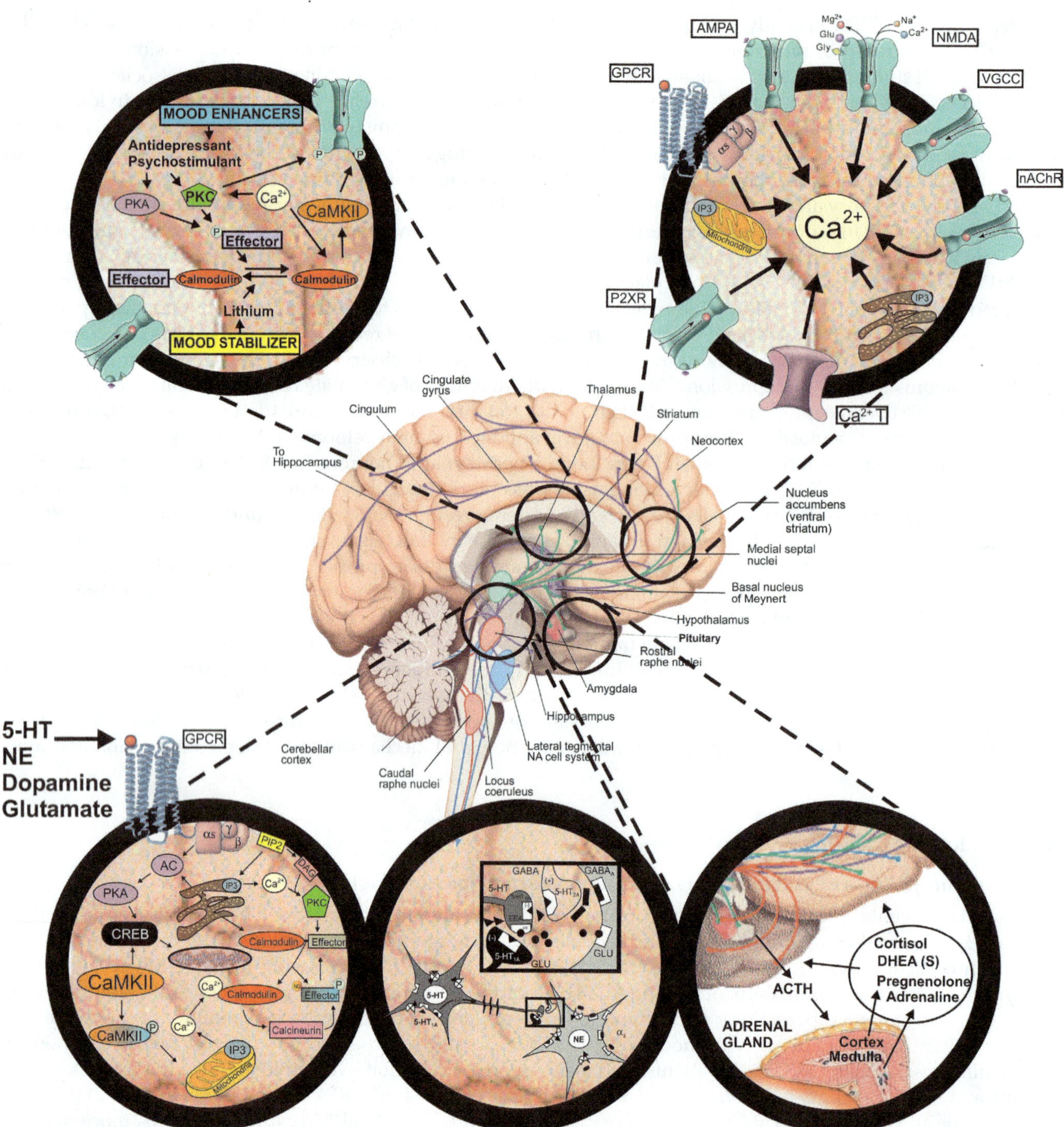

FIGURE 103.3 **Schematic representation of a midsagittal section of the human brain (center).** Circular blowout representation of neurotransmitter–receptor transduction mechanisms to biogenic amines and glutamate signaling implicated in antidepressant action (bottom left). Specifically, activation of G protein-coupled receptors can mediate changes in second messengers, such as adenylate cyclase and calcium (Ca^{2+}) transduction. This impacts various intracellular components deemed important in antidepressant action, such as changes in cAMP response element-binding protein (CREB) and brain-derived neurotrophic factor (BDNF) regulation. Circular blowout representation of the reciprocal interaction between serotonin (5-HT) neurons (raphe nuclei) and norepinephrine (NE) neurons (locus coeruleus). 5-HT neurons recruit other neural elements that ultimately impinge on NE neuron activity (i.e., through glutamate and γ-aminobutyric acid [GABA] terminals). Conversely, activation of α_1-adrenoceptors directly on 5-HT neurons mediates NE effects in the dorsal raphe (not shown). As most antidepressants biochemically target 5-HT and/or NE neuron elements, their reciprocal interactions and neural changes to enhance monoamine neurotransmission may lead to different regulatory set-points of forebrain activation. Circular blowout representation of the hypothalamic-pituitary adrenal (HPA) axis with notable neurohormones implicated in mood and anxiety (bottom right). These neurohormones can shape neural circuits and the stress–anxiety–depression axis. Circular blowout representation of glutamate receptors (*N*-methyl-D-aspartate [NMDA], α-amino-3-hydroxy-5-methyl-4-isoxazolepropionic acid [AMPA]), nicotinic receptor (nAChR), adenosine receptor (P_2X), voltage gated Ca^{2+} channel (VGCC), and calcium transporter ($Ca^{2+}T$), with intracellular mitochondria, endoplasmic reticulum, and membrane bound receptors able to change Ca^{2+} dynamics

(DHEA-S), 5α-dihydroprogesterone, pregnenolone, and progesterone are considered by many to be the major neurosteroids. Neuroactive steroids rapidly alter neuronal excitability by acting at ligand-gated ion channels, such as $GABA_A$ and NMDA receptors, and are increased during stress. For example, DHEA has been reported to possess antidepressant properties, antiglucocorticoid effects, and when exogeneously administered can elevate plasma levels of allopregnanolone (a neurosteroid that modulates $GABA_A$ receptors).[57] It should be noted that neurosteroids induce intracellular effects similar to those of other cholesterol-based neurohormones, which may be related to their mechanism of action in mood and anxiety disorders.[58] The interaction between neurosteroids and HPA axis regulation in the etiology and treatment of depression is an ongoing avenue of investigation,[59] with novel neurosteroid agents currently undergoing investigation.[58]

Hypothalamic–Pituitary–Thyroid Axis

The hypothalamic–pituitary–thyroid (HPT) axis is organized hierarchically, with thyrotropin-releasing hormone (TRH), a tripeptide, the first of the releasing factors to be characterized, synthesized in neurons in the hypothalamus to act on receptors in the anterior pituitary, triggering release of thyroid-stimulating-hormone (TSH). This in turn stimulates the release of the thyroid hormones thyroxine (T_4) and tri-iodothyronine (T_3) from the thyroid gland into the general circulation. Thyroid abnormalities are associated with altered mood states; *hypo*thyroidism is characterized by symptoms of depression and *hyper*thyroidism with anxiety and mania. Higher than expected rates of autoimmune thyroiditis are found in patients with depression. Both T_4 and T_3 are transported across the blood–brain barrier into the brain, but the brain contains only T_3 receptors; T_4 in the brain must be converted to T_3 by a deiodinase enzyme. Thyroid hormones, like steroids, are capable of diffusing through cellular membranes and impacting gene transcription directly. In addition to mood symptoms of patients with thyroid disease, alterations of the HPT axis in patients with depression has been reported,[60] including increased CSF TRH concentrations and a blunted TSH response to intravenously administered TRH in euthyroid depressed patients. There is considerable evidence that T_3 treatment augments the effects of antidepressants in nonresponders. Results from the STAR*D clinical trial revealed T_3 addition to be well tolerated, and led to a 24.7% improvement in patients with depression failing to remit after two adequate trials of an antidepressant treatment.[61] Furthermore, TRH administration given nocturnally to bipolar depressive patients induces a rapid antidepressant response. Moreover, the beneficial effects of repetitive transmagnetic stimulation treatment (rTMS), an FDA approved treatment for depression, correlates with changes in blood TSH levels.[62,63] Neurogranin is a gene expressed almost exclusively in the brain as a postsynaptic neural protein, in high quantities in the cortex, hippocampus, striatum, and amygdala (areas of the brain implicated in depression). Genetic mutations of neurogranin have been documented in schizophrenia and bipolar disorder,[64] and interestingly thyroid hormones bind to the first intron of neurogranin. It has been recently proposed that mood-regulating psychotropics and sleep deprivation (i.e., a known mood enhancing strategy) differentially modulates the phosphorylation state of neurogranin. Thus, neurogranin represents a novel underlying pathophysiology in the potential treatment for mood disorders.[50] When evaluating a patient with mood symptoms for the first time, thyroid function (i.e., TSH) should be assessed to rule out hypothyroidism.

Intracellular and Second-Messenger Signaling Changes

Signals generated from a variety of receptors can activate associated G proteins and stimulate single second-messenger pathways to alter neuron activity. The ability of intracellular and second-messenger pathways to interact with multiple receptors in a cell provides for elegant mechanisms to organize neurotransmitter effects. Abnormalities in a variety of human diseases arise from primary perturbations in G-protein subunits and G-protein signaling cascades, but direct evidence for the involvement of G proteins in depression is still lacking. Postmortem studies have demonstrated that individuals with depression and suicide victims exhibit changes in numerous second-messenger and

FIGURE 103.3, cont'd and neuronal function. These processes have been implicated in depression as either viable treatment targets or mutations that may contribute to mood disorders (top right). Circular blowout representing an example of a unified mechanism by which mood regulating agents can change intracellular components, such as PKC and calmodulin (CaM) mediated signaling. Phosphorylation states of proteins and enzymes, which can alter cellular activity profiles, may be important intermediates in antidepressant action. For instance, effector proteins that are bound to CaM (i.e., presynaptic localized GAP_{43} and postsynaptic localized neurogranin) can regulate Ca^{2+} signaling set-points through their phosphorylation state and lead to altered activation of CaMKII (calmodulin dependent protein kinase II), circuit activity, and potentially mood regulation. Presynaptic GAP_{43} and postsynaptic neurogranin are major phosphorylation targets of PKC and together can shape neurotransmitter release profiles and postsynaptic events to numerous psychotropic agents.[50,127] Adapted with permission from Szabo ST, Gould TD, Manji HK. Introduction to neurotransmitters, receptors, signal transduction, and second messengers. In: Schatzberg AF, Nemeroff CB, eds. The American Psychiatric Publishing Textbook of Psychopharmacology. 4th ed. Washington, DC: American Psychiatric Publishing; 2009.

intracellular signaling cascades.[65] The effects of antidepressants on these systems are helping shape our understanding of the mechanism of action of these medications. Lithium is a monovalent cation that can permeate the cell membrane and, in addition to its efficacy in mania, is a well-established antidepressant augmentation strategy.[61] Chronic administration of lithium increases 5-HT signaling, but attenuates both Gs and Gi function, resulting in elevated basal cAMP levels and dampened receptor-mediated effects.[66] Effects of cAMP due to G-protein activation are then mediated largely by protein kinase A (PKA) and its major downstream target cAMP response element-binding protein (CREB). The action of PKA on CREB phosphorylation is thought to exert a major role in neuroplasticity and antidepressant action. Phosphorylated CREB binds to the cAMP response element (CRE), which is a sequence found in the promoter region of certain genes, and antidepressants activate CREB and thereby enhance expression of a major target gene, *BDNF*. In short, receptors can both positively (e.g., α-adrenergic, D_1) or negatively (e.g., 5-HT_{1A}, D_2) impact various intracellular signaling cascades (i.e., ability of adenylate cyclase to regulate cAMP levels and CREB phosphorylation/BDNF production) modulating brain circuitry and mood. Because the scope of this review does not allow for a detailed discussion of all the various mechanistic potentials of second-messenger and intracellular regulation in depression, or pathophysiology, see [67] for a recent review.

PATHOLOGY: FUNCTIONAL NEUROBIOLOGY

Mood disorder patients have abnormalities in brain morphology and variability between patients is likely due to the existence of different subtypes. For example, elderly patients with late-onset depression exhibit white matter hyperintensities and other imaging correlates of cerebrovascular disease as well as lateral ventricle enlargement compared to elderly depressed patients with early-onset depression.[68] Brain volume changes in the prefrontal cortex (orbital, ventrolateral, dorsal anterolateral), cingulate, and temporal structures appear to prevail in the CNS of depressed patients, with decreased left anterior cingulate cortex (gray matter) and subgenual portion of the corpus callosum (white matter) being the most consistent findings in younger individuals with a family history of depression and/or depression with psychotic features.[69] Hippocampal abnormalities are notable in depression, but inconsistent findings may reflect clinical heterogeneity of this illness with distinct neuroanatomical–clinical associations (i.e., time spent depressed, women with early-life trauma). After correcting for brain volume deficits, changes in cerebral blood flow occur in patients with depression (i.e., anterior cingulate, prefrontal cortex, and amygdala).[70] In general, activity of the dorsal prefrontal cortex is decreased, and ventral prefrontal cortex and orbital prefrontal cortex increased, in depression. Various effective antidepressant treatments (i.e., psychotherapy, medications, deep brain stimulation (DBS), magnetic stimulation therapy) are capable of producing a decrease in cerebral blood flow, whereas monoamine-depletion paradigms that induce depressive symptoms in individuals enhance cerebral blood flow in these brain areas.[71–74]

Advancements in brain imaging techniques (i.e., magnetoencephalography, optical imaging) are providing increasing ability to appreciate brain circuits that are compromised and linked by symptoms germane to depression, while shedding light into etiology and treatment. Aberrant connections of networks that subserve evaluation and regulation of emotional states have been implicated in depressed patients, including the orbital and medial prefrontal cortex, amygdala, hippocampal subiculum, ventromedial striatum, mediodorsal and midline thalamic nuclei, and ventral pallidum.[75] Although pertubations of monoamine neuron activity can lead to symptoms of depression, and pharmacologic treatments that biochemically target these systems are effective, functional brain imaging techniques have not focused on these nuclei, in part due to their relative small size. Studies assessing glucose metabolism with positron emission tomography (PET) and resting state blood flow have reported hypofrontality in patients with depression and brain circuit models of depression implicate reciprocal modulation between ventral limbic and dorsal cortical regions of negative mood states.[76] Effective treatments of depression including ECT, DBS, and cognitive–behavioral therapy (CBT) can change the activation patterns of these brain areas in patients that show clinical improvement, consistent with the notion that neural plasticity is essential for response to treatment. The ability to further characterize distinct subtypes and aspects of depression with targeted treatments will help to better optimize patient care and provide a greater understanding of mood disorders.

PATHOLOGY: NEURAL PLASTICITY AND RESILIENCE

Various brain regions have been reported to be decreased in size as assessed by structural magnetic resonance imaging (MRI) in depressed patients. Theories of neuronal atrophy and synaptic loss have been proposed as a consequence of defective neuronal plasticity in mood disorders. Reduction in connectivity between the hippocampus and prefrontal cortex in the presence of increased connectivity of other brain areas in some studies preclude simplistic mechanisms to account for depression pathophysiology. Postmortem studies have highlighted reduced pyramidal cell size of the dorsolateral prefrontal cortex, as well as deficient GABA interneurons, glia, and synaptic connections. Similarly, the hippocampus is reportedly reduced in neuronal cell size and number. Mechanism(s) underlying the cellular compromise in depression may relate, in part, to altered cortisol/neurosteroid production and signaling. The backbone of neurosteroid production is cholesterol, which has been shown to be decreased in the plasma and brain of patients suffering from depression. HPA axis activation due to chronic stress results in atrophy of neurons in the prefrontal cortex and hippocampus in laboratory animal studies, and patients with depression have aberrant negative feedback of the HPA axis and increased adrenal gland volume, likely in part due to ACTH hypersecretion. Changes to adrenal gland size and activity can chronically increase cortisol levels, as well as production of other neurosteroids, which represents an area of novel treatment approaches in mood disorders.[77] For example, DHEA has been reported to possess antidepressant effects in patients with depression,[59] and increased levels of DHEA-S: cortisol may be associated with resiliency against stress.[78]

A major intracellular target postulated to be involved in deficiencies of synaptic neural plasticity in depression is brain-derived neurotrophic factor (BDNF). BDNF is essential for neuronal viability and axon growth, both of which are integral to synaptic plasticity, and its availability is reportedly reduced by stress and increases in cortisol secretion. Receptor-mediated activation in CREB occurs during chronic antidepressant administration, which binds to promoter regions on the *BDNF* gene to facilitate transcription. Moreover, 5-HT_{1A}-receptor activation, which accompanies chronic SSRI treatment, is important in transcribing BDNF, and the subsequent induction of neurogenesis makes for a unified mechanism between facilitating neurotransmission and brain circuit remodeling. Postmortem studies of depression indicate that levels of phosphorylated CREB and total CREB are reduced in cortical areas, and suicide victims have been reported to exhibit decreased BDNF levels in the hippocampus. Alterations in genetic transcription of BDNF can also occur through mutations of the *BDNF* gene, with the hippocampus being smaller in individuals carrying the Val66Met *BDNF* polymorphism.

TREATMENT: MECHANISM OF ANTIDEPRESSANT ACTION

Antidepressant Agents

As noted above, the vast majority of FDA-approved antidepressants target monoamine neuronal circuits, predominantly regarded as being neuromodulatory. It has been suggested that the delay in antidepressant action of these agents is associated with the need for monoamine systems to adapt to their biochemical targeted effects, as well as in subsequent regulation of limbic and cortical brain function. Neurohormonal changes through monoamine regulation likely also exert profound effects onto antidepressant action. Although it is not precisely known why antidepressants take weeks to work, have variable response rates, and different side-effect profiles in individuals, much research has focused on neurotransmitter changes. The ability of antidepressants to enhance monoamine availability through either inhibition of enzymatic destruction of neurotransmitter amount, such as with the MAOIs, or blockade of monoamine reuptake, as with the tricyclic antidepressants (TCAs), are the mechanisms believed to be most germane to their therapeutic action. Most newer antidepressants have similar but more restricted mechanisms of action, as seen with SSRIs (fluoxetine, paroxetine, sertraline, escitalopram, citalopram, fluoxetine), norepinephrine reuptake inhibitor (NRI; reboxetine), and dual 5-HT and NE reuptake inhibitors (SNRIs; venlafaxine, *O*-desmethylvenlafaxine, duloxetine, milnacipran). The tetracyclic antidepressant mirtazapine does not block monoamine reuptake or inhibit catabolism of monoamines, but rather enhances monoamine neurotransmission after a sustained treatment through blockade of α_2 and 5-HT_2 receptors. All of the aforementioned monoamine-targeted antidepressants require weeks to exert beneficial effects, in spite of the fact that their ability to enhance the concentration of monoamines in the synaptic cleft is nearly immediate, but a delay in enhanced monoamine transmission occurs. The discrepancy may relate to autoinhibitory neural mechanisms that are in place to modulate neurotransmission, which over time become deactivated. Monoamine receptor and transporter desensitization/downregulation-mediated effects have largely prevailed as a hypothesis behind the delayed onset of antidepressant action, with a role for other neurotransmitter systems and intracellular events gaining momentum. Numerous therapeutic strategies that biochemically impact monoamine elements have been targeted to increase the onset of antidepressant action without avail. Figure 103.1 schematically summarizes the biochemical targets of antidepressants that clinicians have at their disposal in treatment of depression.

Ketamine, an NMDA receptor antagonist, can induce dendritic sprouting (with a time course of about 24 hours), mirroring the short delay of its reported therapeutic action in treatment-resistant patients with depression.[43,79] As with monoamine antidepressants, it appears that the antidepressant effects of ketamine also relies on BDNF activation of Trk_B receptors. Individuals with certain (BDNF) Val66Met polymorphisms appear resistant to the antidepressant effects of ketamine treatment[80] and this mutation when introduced in mice demonstrates synaptic deficits and negates the antidepressant actions of ketamine.[81] Rapid-acting antidepressants, such as ketamine and scopolamine, may be effective, in part, as a result of their ability to recruit Akt and mTOR signaling to trigger expression of AMPA glutamate receptor subunits and activity-regulated cytoskeleton-associated protein (Arc), which is necessary for actin polymerization and dendritic spine outgrowth and stabilization.[82,83] However, GSK_3 inhibition can also occur through stimulation of Akt or by blockade of NMDA receptors and it is of interest that lithium is also an indirect inhibitor of GSK-3, and possesses antidepressant and clear antisuicidal properties. Animal studies suggest that using low doses of ketamine combined with lithium together are capable of modulating synaptogenic and antidepressant responses.

Device-Based Treatments

Electroconvulsive Therapy

One of the most highly regarded treatments in depression is electroconvulsive therapy (ECT). Although ECT is generally considered to be the most effective of all of the available antidepressant treatments, it is also surely the most stigmatized, and as such, is mainly reserved for treatment refractory depression.[84] With the modern use of anesthesia and determining optimal stimulus intensity in patients undergoing ECT, the side-effect profile of this modality has been considerably reduced and its application is being heralded by patients and families (and insurance companies) due to decreased inpatient admissions and days spent in hospital. Although ECT is clearly effective, its exact mechanism of action (as with all antidepressants) remains largely unknown. Application of electricity to the brain induces widespread neurotransmitter release, which, in part, is in line with the monoamine hypothesis of depression. The ability of mossy fiber sprouting, changes in intracellular elements related to synaptic remodeling, desensitization of presynaptic autoreceptors on monoamine neurons with enhanced sensitivity to postsynaptic $5\text{-}HT_{1A}$ receptors (latter phenomenon being notably associated with ECT, TCAs, and lithium), may explain why these treatments are thought by many to possess superior efficacy.

Vagal Nerve Stimulation

Vagal nerve stimulation (VNS) can be likened to an implantable pacemaker used for regulation of heart rate, but instead delivers electrical stimulation directly to the vagus nerve. VNS first gained approval in the treatment of pharmacoresistant epilepsy with improvement in mood observed in epilepsy patients receiving VNS independent of changes in seizure frequency.[85] This led to development of VNS as an adjunctive therapy for patients with treatment-resistant depression, including those who failed to respond to ECT. Although the effect size for VNS was small (31% response rate and a 15% remission rate at 3 months), given the severity of the treatment-resistant population, the FDA approved the use of VNS as an adjunctive therapy for treatment resistant depression (i.e., individuals failing four or more medications).[86,87] Newer data suggests that longer treatments may lead to more higher response rates and decreased suicide; thus, there is a need for an enhanced understanding of its neurobiologic mechanism(s) of action.

The vagus nerve (i.e., cranial nerve 10) is comprised of parasympathetic fibers regulating autonomic functions as well as a large number of sensory afferent fibers (from the head, neck, and body). The limited efferent fibers of this nucleus, which projects to motor areas, have been linked to some of the VNS side-effect profile (i.e., hoarseness due to activation of laryngeal nerve). The mechanism by which VNS is thought to work is primarily through monoamine neuron modulation. NE neurons in the locus coeruleus receive innervation from the nucleus of the solitary tract, a brainstem nucleus that is innervated by vagus nerve afferents. Interestingly, lesioning the locus coeruleus in animal studies blocks the anticonvulsant effects of VNS,[88] with increased activity of these NE neurons to VNS occuring seconds to minutes after stimulation.[89] However, no significant change in CSF metabolites of NE and 5-HT was seen in VNS patients treated for 3 months compared with pretreatment levels.[90] In animal experiments that model VNS in patients, sustained 3-month treatment enhanced firing activity of both 5-HT and NE neurons, consistent with the slow but significant reduction in depression rating scale scores over time.[91] The increase in 5-HT neuron firing activity is not mediated by changes in $5\text{-}HT_{1A}$ autoreceptors, as have been observed after chronic SSRI treatment, indicating a different mechanism of action.[92]

Transcranial Magnetic Stimulation and Magnetic Seizure Therapy

Transcranial magnetic stimulation (TMS) is a technique to noninvasively stimulate the cerebral cortex by applying a strong magnetic field. TMS can be used to probe neural circuits in humans for neuroscience research[93] and pulses when given in a repetitive fashion (rTMS) have been used in the treatment of mood disorders.[94] High-frequency rTMS (>1 Hz) can activate cortical areas, whereas low-frequency rTMS (<1 Hz) has been shown to inhibit activity. Efficacy of rTMS in mood disorders is an active avenue of investigation, with side-effect profiles and restrictions being less problematic than ECT. It is now FDA approved for the treatment of depressed patients who have failed one adequate trial with SSRIs. Clinical improvement in depression using rTMS has been associated with changes in cerebral blood flow in the prefrontal and paralimbic areas. Responders to rTMS demonstrate two patterns of change in regional blood flow, with treatment most characteristic of reduced orbitofrontal blood flow and/or decreased anterior cingulate blood flow. rTMS can alter cortical glutamate/glutamine levels (both close to the site of stimulation and in distant structures), implicating cortical glutamate neuron regulation as a mechanism of action (as well as monoamine changes). Interestingly, rTMS also increases TSH levels and can normalize the dexamethasone suppression test (DST) in depressed patients.[62,95]

Magnetic seizure therapy (MST) uses transcranial magnetic stimulation (TMS) at convulsive settings to induce therapeutic seizures under general anesthesia. After its introduction in 2000, only a few case reports described successful treatment of depressed patients using MST, and it is not yet established that MST has antidepressant efficacy. MST seizures may cause fewer cognitive side effects than ECT given its ability to deliver focused seizures and possibly sparing brain regions associated with memory loss. The limited clinical data does suggest that MST possesses antidepressant properties and fewer cognitive side effects than ECT, with patients able to orient themselves more quickly following the procedure, having fewer attention difficulties or retrograde amnesia as compared to ECT.

Deep Brain Stimulation

As noted earlier, cortical (subgenual cingulate) and limbic systems are dysregulated in depression and effects of antidepressants and certain psychotherapies may lead to normalization of these circuits and function. By monitoring brain activity through PET scans, cerebral blood flow abnormalities in depression (i.e., decreased blood flow in the prefrontal structures and increased blood flow in the subgenual cingulate [Cg25]) can be normalized in treatment-responsive patients using deep brain stimulation (DBS). Effectiveness of DBS as an antidepressant may rest on returning neurons to their regular firing frequency and correcting deficient brain circuit function. DBS is thought to lead to release of inhibitory neurotransmitters from nearby axons (i.e., GABA) and reduce neuron activity, as well as, when stimulation is rapid enough (i.e., tetanic), neurotransmitter depletion can lead to overactivation of voltage-gated ion channels and depolarization block. Although depression is probably a disorder of multiple brain areas, neuronal pathways, and neurotransmitters, with genetic and environmental predispositions, the fact that DBS works through stimulating a single brain area and leads to beneficial antidepressant effects is impressive. Encouraging results have been obtained by Mayberg and colleagues who have targeted the subgenual cingulate (Broadman area 25). More recently, Schlaepfer and colleagues have reported positive results with DBS of the medial forebrain bundle. Furthermore, stimulation of other brain areas may impact circuits that can also lead to decreased mood and suicidality,[96] truly suggesting that depression is a manifestation of abnormal circuit function, and when compromised or normalized, leads to exacerbation or treatment of symptoms, respectively.

Psychotherapeutic Modalities

There are numerous psychotherapeutic modalities that have been shown through controlled clinical trials to be as effective as antidepressant agents. Regretfully, many therapeutic modalities have not undergone rigorous scientific testing and this lack of standardization limits the ability to evaluate their efficacy. Nonetheless, cognitive–behavioral therapy (CBT) is a standardized therapy that repeatedly has been shown to be effective in the treatment of depression. It has also been demonstrated that the combination of CBT with antidepressant treatment exerts additive effects and provides greater efficacy to patients, particularly those with chronic depression.[97] One obstacle of such psychotherapeutic interventions is expense and another is the limited availability to many patients. Although antidepressant agents and device-based treatments have well-known side effects that are associated with their use in some patients, therapy also comes with limitations including a malalignment with the therapy style or therapist; it does however have a reduced side-effect profile.

GENETICS: UNIPOLAR DEPRESSION AS A HERITABLE DISEASE

Family Association Studies

Depression fails to exhibit classic Mendelian disease transmission patterns of genetic inheritance, but it is well established that a strong genetic "familial" component to depression exists. The relationship between familial genetics and susceptibility to depression has been well documented through countless observations of families that have been devastated by multiple suicides, and clinically family history is an important question to ask patients with mood disorders. Severe forms of depression (i.e., early onset-before the age of 30, recurrent, psychotic features) are considered more heritable. From twin and epidemiologic studies, the concordance rate of major depression between monozygotic and dizygotic twins is approximately 37%, which is substantial but considerably lower than the heritability of bipolar disorder (≅65%). First-degree relatives of individuals with major depression have a 15% increased prevalence of the disease as compared to the general population (5.4%),[98] with polygenetic and multifactorial components to major depression becoming increasingly clear. Major depressive disorder is a genetically complex disease that is virtually a prototypical gene–environment disorder. Simply stated, the major psychiatric disorders are not caused by single gene abnormalities, but represent complex genetic profiles conferring individual susceptibility after exposure to environmental perturbations such as severe early life stress. Genome-wide association studies (GWAS) have largely been unsuccessful in pinpointing specific genetic underpinnings for major depression. However, a small proportion of gene variants may account for depression in certain individuals, with these genes also having a low predictive power in doing so. In contrast, a number of single nucleotide polymorphisms (SNPs) have been identified that mediate the depressogenic and anxiogenic effects of child abuse and neglect.[99,100]

Adoption and Twin Studies

Because of the focus on gene–environment interactions in the diathesis of depression, adoption studies are helpful in providing the framework to determine if factors related to familial transmission of depression involve environmental or biologic factors. In the presence of methodologic constraints and limited investigations, these studies provide evidence that a significant increase in risk of developing depression occurs if the biological parents suffered with depression (regardless of the adoptive environment). Twin studies have nonetheless proven to be a very powerful tool for determining the relative contribution of heritability to major depression. The comparison between monozygotic (who share virtually all their genes) and dizygotic twins (who share only 50% of genes) allows for an analysis of the relative contributions of gene versus environment. However, monozygotic twins are treated much more similarly than dizygotic twins by their family members, and the assumption that differences between the two groups is completely genetic may not be entirely correct. Studies using a sample of over 7000 twins have identified environmental risk factors for major depression, being predominantly stressful life events, humiliation, and loss.[101]

GENETIC STUDIES: THE SEARCH FOR QUANTITATIVE TRAITS

Numerous genetic and environmental factors can contribute to major depressive disorders. Several SNPs with potentials for predicting vulnerability for major depression have been reported, but lack of replication likely reflects heterogeneity of the disease and inclusion of diverse subjects in the analysis. Although it is believed that these disorders are not transmitted in a Mendelian fashion, there are certain chromosomal regions associated with inheritance or likelihood of disease (LOD) score related to candidate genes. For instance, areas on chromosome 11 and 17 have genes that code for tyrosine hydroxylase and the serotonin transporter (5-HTT), respectively, have been implicated in major depression. Also, a region on chromosome 2 codes for CREB1 and is associated with early-onset major depression in women.

A meta-analysis of genetic association studies in major depression has uncovered five genes that may be related to major depression, including *APOE* (implicated in cognitive disorders), *GN3* (β3-subunit of the G protein complex), *MTHFR* (5,10-methylenetetrahydrofolate reductase), *SCL6A3* (DAT) and *SCL6A4* (5-HTT).[102] Taking *SCL6A4* as an example, polymorphisms in the promoter region of this gene produce two allelic variants, 484 and 528, representing the short and long allele, respectively. The short 484 allele compromises gene expression and is associated with reduced 5-HTT availability.[103] Studies have linked this polymorphism to susceptibility in developing major depression in individuals exposed to traumatic events during childhood, and may represent more severe types of depression associated with melancholia and suicide. The work of Caspi and colleagues[20] also exemplifies this avenue of investigation as depressed adults with the short (S) allele of a 5-HTT promoter polymorphism (5-HTTLPR) are more vulnerable to developing depression if exposed to stressful early life events than long-allele carriers. Because the presence of childhood abuse in the S allele carriers actually predicted adult depression and suicide attempts, how early life trauma epigenetically interacts with promoter regions of the 5-HT transporter continues to be an important area of research.

GWAS using genotyping platforms for genetic variability (on upwards of 1 million genetic markers) are transforming genetic risk factors of complex diseases without the requisite need for hypothesis-driven exploration. In the three largest GWAS for major depression, no polymorphism reached significance and these findings suggest the factors responsible for the lack of success are: 1) the heterogeneity of depression; 2) the likely importance of several genes each with a small effect; and 3) this technique precludes an appreciation of genetic risk profiles.[104] The incorporation of genetic risk profiles with functional brain imaging and phenotypic characteristics associated with depression will likely be more informative and is currently ongoing. For instance, elevation of left amygdala activity during 5-HT depletion was restricted to homozygotes for the long allele of the 5-HTT,[105] and changes in amygdala function may represent familial risk and temperamental variation in individuals at high risk for bipolar disorder.[106,107]

EPIGENETICS: ENVIRONMENTAL INFLUENCE AT THE GENETIC LEVEL

Epigenetics correspond to gene expression changes that occur independent of direct modifications to the genomic code. Epigenetics in the traditional sense signifies events that occur outside the gene coding sequence proper (i.e., adenine, guanine, thymine, and cytosine), but can impact their function. Research now convincingly demonstrates that there are epigenetic transgenerational inheritance patterns[108,109] that influence germline transmission of epigenetic modifications between generations to alter genome activity and phenotype. Thus, epigenetic factors that change expression of preprogrammed cellular genetic products are also inherited and are now a major focus in psychiatric research given the preeminent role of environment effects in the pathogenesis of major mental illness. One such example is the impact of epigenetic factors (stress, early-life trauma, environmental contaminants) that lead to "methylated DNA" and in epigenetic terms refers to modification of gene promoter elements that can impact transcription of its adjacent gene product (i.e., usually to silence expression).

Laboratory animal studies have demonstrated that maternal influences represent epigenetic factors when imparted onto rat pups in the form of maternal care profiles. Essentially "high licking" mothers are purportedly "nurturing" and provide for less fearful offspring, whereas more fearful offspring were raised by low-tactile stimulating mothers. The epigenomic changes relate, in part, to modulation of the gene encoding for glucocorticoid receptors in the hippocampus, which occur through differential histone acetylation and DNA methylation events. Interestingly, these epigenetic changes can be reversed by cross-fostering or infusion of histone deacetylase (HDAC) inhibitors. This may be particularly applicable to the therapeutic effects of lithium and valproate, both well known HDAC inhibitors, and their effects on epigenetic mechanisms may control glucocorticoid receptor regulation.[110]

In humans, adverse childhood experiences (i.e., child abuse or early separation from parents) adds a 2–3-fold risk for developing major depression as an adult, and mirrors results obtained from animal experiments. Childhood trauma is also a strong predictor of pathologic responses to psychosocial stressors, CRF stimulation, combined dexamethasone suppression/CRF stimulation,[111–114] and altered CSF CRF concentrations in adults.[115–117] Thus, CRF acts in the nucleus accumbens to increase DA release through coactivation of $CRFR_1$/$CRFR_2$; and stress disables the capacity of CRF to regulate DA release and may be an early pathophysiologic mechanism of depression.[118] Early adverse events can also influence the 5-HT system with a number of studies demonstrating that SSRIs reverse the stress-effects of CRF administration,[119] and depressed patients treated with SSRIs exhibit decreased CSF CRF concentrations.[120,121] Furthermore, $CRFR_1$-mediated increases in 5-HT_2 receptor signaling are dependent on receptor internalization and receptor recycling that increases expression of 5-HT_2 receptors.[122] Thus, the ability of sustained administration of antidepressants to downregulate 5-HT_2 receptors is in keeping with $CRFR_1$ antagonists under study for the treatment of depression;[123] thus, the negative studies of CRF antagonists in depression likely relate to the heterogeneity of this disorder and highlight the need for biomarker development.

As mentioned previously, HPA axis dysfunction and glucocorticoid resistance are among the most robust biological findings in major depression moderated by genetic factors. One such mechanism is through polymorphisms in the human gene for FK506 binding protein 5 (*FKBP5*). *FKBP5* is involved in glucocorticoid receptor sensitivity and variants of this gene are associated with vulnerability to mood and anxiety disorders. The mechanism is rather complex, but FKBP5 is a cochaperone of heat shock protein 90 (hsp90) and this complex binds to glucocorticoid receptors to regulate its sensitivity. Upon glucocorticoid receptor activation, changes in FKBP5-receptor binding and subsequent translocation to the nucleus occurs, allowing FKBP5 to impart effects as a transcription factor in glucocorticoid-related gene transcription. Interestingly, rare homozygous genotype SNPs of *FKBP5* are associated with stronger transcription of FKBP5 by cortisol activation with less corticotropin release to combined dexamethasone–CRH test in depressed patients.[124] Thus, not only are genetic polymorphisms of FKBP5 related to depression, but also early childhood (as well as *in utero*) trauma can induce epigenetic changes on FKBP5 to influence regulation of the HPA axis.

In short, epigenetic programming due to early life stress in individuals is likely capable of being attenuated or reversed during adulthood by effective psychiatric treatments, a mechanism whereby psychotherapy and environmental influences can also exert profound neurobiological effects in concert with pharmacologic and device-based treatments.

CONCLUSION: REDUCING BURDEN BY INCREASING THERAPEUTIC EFFECT

The impact of mental illness and particularly mood disorders are clearly devastating not only to the affected individual, but globally to societal function. Recognition of symptoms, access and adherence to treatment, with increased therapeutics and research to fuel our understanding thereof, needs to be continued. Although not entirely understood, current treatment modalities for depression provide a wealth of benefit and should be widely available. Greater emphasis through resource allocation by governmental and nongovernmental entities must provide the foundation to stop this growing epidemic and its medical consequences. Indeed, depression is now undisputed to be a disorder that impacts several organ systems, rendering individuals susceptible to a variety of medical disorders.[125,126] Better outcomes in prevention of disease and personalized treatments of depression will reduce morbidity and the mortality secondary to suicide and death from comorbid mental disorders. Given the numerous mechanisms whereby a person may become depressed, with the multitude of interactions that can take place at the psychosocial and neurobiologic interface, a more personalized therapeutic approach to treatment of depression is dictated. The "one size fits all" approach to categorizing depression and its treatment leads to tremendous variability in treatment response, but also ambiguity as to which treatments would benefit individuals the most. Physicians at present have adopted the framework of therapy plus medications for treatment of depression, as it has been shown to be the most powerful combination that will lead to remission of symptoms. The inherent subjectivity of our descriptions of depressive symptoms must be augmented by the addition of objective biomarkers. Given that the neuroscience techniques used to understand the brain are evolving at an exponential rate, application of these techniques in patients with depression will undoubtedly lead to advances in diagnosis and treatment.

FINANCIAL DISCLOSURES

At the time of publication, STS does not have any conflicts of interest or financial disclosures. CBN was a consultant for Xhale, Takeda, SK Pharma, Shire, Roche, Lilly, Allergan, Lundbeck; had received grant/support from the NIH; was a stock shareholder of CeNeRx BioPharma, PharmaNeuroBoost, Revaax Pharma, Xhale, NovaDel Pharma; was on the Board of Directors of the American Foundation for Suicide Prevention (AFSP), NovaDel (2011), Skyland Trail, Gratitude America, ADAA; sat on the Scientific Advisory Board for AFSP, CeNeRx BioPharma, the Brain and Behavior Research Foundation (BBRF), Xhale, PharmaNeuroBoost, the Anxiety Disorders Association of America (ADAA), Skyland Trail; holds a patent for method and devices for transdermal delivery of lithium (US 6,375,990B1), and for a method of assessing antidepressant drug therapy via transport inhibition of monoamine neurotransmitters by *ex vivo* assay (US 7,148,027B2); and had equity or other financial interests in PharmaNeuroBoost, CeNeRx BioPharma, NovaDel Pharma, Reevax Pharma, American Psychiatric Publishing and Xhale.

References

1. Belmaker RH, Agam G. Mechanisms of disease: major depressive disorder. *N Engl J Med*. 2008;358(1):55–68.
2. Newberg AR, Catapano LA, Zarate CA, Manji HK. Neurobiology of bipolar disorder. *Expert Rev Neurother*. 2008;8(1):93–110.
3. Merikangas KR, Lamers F. The 'true' prevalence of bipolar II disorder. *Curr Opin Psychiatry*. 2012;25(1):19–23.
4. Bromet E, Andrade LH, Hwang I, et al. Cross-national epidemiology of DSM-IV major depressive episode. *BMC Med*. 2011;9:90.
5. Patten SB. Accumulation of major depressive episodes over time in a prospective study indicates that retrospectively assessed lifetime prevalence estimates are too low. *BMC Psychiatry*. 2009;9:19.
6. Costello EJ, Foley DL, Angold A. 10-year research update review: the epidemiology of child and adolescent psychiatric disorders: II. Developmental epidemiology. *J Am Acad Child Adolesc Psychiatry*. 2006;45(1):8–25.
7. Mathers CD, Loncar D. Projections of global mortality and burden of disease from 2002 to 2030. *PLoS Med*. 2006;3(11):e442.

8. O'Connor E, Gaynes BN, Burda BU, Soh C, Whitlock EP. Screening for and treatment of suicide risk relevant to primary care: a systematic review for the U.S. Preventive Services Task Force. *Ann Intern Med*. 2013;158(10):741–754.
9. Batelaan NM, Spijker J, de Graaf R, Cuijpers P. Mixed anxiety depression should not be included in DSM-5. *J Nerv Ment Dis*. 2012;200(6):495–498.
10. Cassidy F. Anxiety as a symptom of mixed mania: implications for DSM-5. *Bipolar Disord*. 2010;12(4):437–439.
11. Koukopoulos A, Sani G, Ghaemi SN. Mixed features of depression: why DSM-5 is wrong (and so was DSM-IV). *Br J Psychiatry*. 2013;203:3–5.
12. Pence BW, O'Donnell JK, Gaynes BN. The depression treatment cascade in primary care: a public health perspective. *Curr Psychiatry Rep*. 2012;14(4):328–335.
13. Jiang W, Velazquez EJ, Kuchibhatla M, et al. Effect of escitalopram on mental stress-induced myocardial ischemia: results of the REMIT trial. *JAMA*. 2013;309(20):2139–2149.
14. Frasure-Smith N, Lesperance F. Depression and cardiac risk: present status and future directions. *Heart*. 2010;96(3):173–176.
15. Frasure-Smith N, Lesperance F, Talajic M. Depression following myocardial infarction. Impact on 6-month survival. *JAMA*. 1993;270(15):1819–1825.
16. Raison CL, Rutherford RE, Woolwine BJ, et al. A randomized controlled trial of the tumor necrosis factor antagonist infliximab for treatment-resistant depression: the role of baseline inflammatory biomarkers. *JAMA Psychiatry*. 2013;70(1):31–41.
17. Blier P, Haddjeri N, Szabo ST, Dong J. Enhancement of serotoninergic function - a sometimes insufficient cause of antidepressant action. *Hum Psychopharmacol*. 2001;16(1):23–27.
18. Jacobsen JP, Siesser WB, Sachs BD, et al. Deficient serotonin neurotransmission and depression-like serotonin biomarker alterations in tryptophan hydroxylase 2 (Tph2) loss-of-function mice. *Mol Psychiatry*. 2012;17(7):694–704.
19. Cassidy F, Weiner RD, Cooper TB, Carroll BJ. Combined catecholamine and indoleamine depletion following response to ECT. *Br J Psychiatry*. 2010;196(6):493–494.
20. Caspi A, Sugden K, Moffitt TE, et al. Influence of life stress on depression: moderation by a polymorphism in the 5-HTT gene. *Science*. 2003;301(5631):386–389.
21. Szabo ST, Blier P. Response of the norepinephrine system to antidepressant drugs. *CNS Spectr*. 2001;6(8):679–684.
22. Coupland NJ. Social phobia: etiology, neurobiology, and treatment. *J Clin Psychiatry*. 2001;62(suppl 1):25–35.
23. McCann UD, Thorne D, Hall M, et al. The effects of L-dihydroxyphenylalanine on alertness and mood in alpha-methyl-para-tyrosine-treated healthy humans. Further evidence for the role of catecholamines in arousal and anxiety. *Neuropsychopharmacology*. 1995;13(1):41–52.
24. Cahill L, Prins B, Weber M, McGaugh JL. Beta-adrenergic activation and memory for emotional events. *Nature*. 1994;371(6499):702–704.
25. Przybyslawski J, Roullet P, Sara SJ. Attenuation of emotional and nonemotional memories after their reactivation: role of beta adrenergic receptors. *J Neurosci*. 1999;19(15):6623–6628.
26. Blier P, Szabo ST. Potential mechanisms of action of atypical antipsychotic medications in treatment-resistant depression and anxiety. *J Clin Psychiatry*. 2005;66(suppl 8):30–40.
27. Szabo ST, Blier P. Serotonin (1A) receptor ligands act on norepinephrine neuron firing through excitatory amino acid and GABA(A) receptors: a microiontophoretic study in the rat locus coeruleus. *Synapse*. 2001;42(4):203–212.
28. Szabo ST, Blier P. Functional and pharmacological characterization of the modulatory role of serotonin on the firing activity of locus coeruleus norepinephrine neurons. *Brain Res*. 2001;922(1):9–20.
29. Dunlop BW, Nemeroff CB. The role of dopamine in the pathophysiology of depression. *Arch Gen Psychiatry*. 2007;64(3):327–337.
30. Heinz A, Schlagenhauf F. Dopaminergic dysfunction in schizophrenia: salience attribution revisited. *Schizophr Bull*. 2010;36(3):472–485.
31. Guiard BP, El Mansari M, Blier P. Prospect of a dopamine contribution in the next generation of antidepressant drugs: the triple reuptake inhibitors. *Curr Drug Targets*. 2009;10(11):1069–1084.
32. Ruhe HG, Mason NS, Schene AH. Mood is indirectly related to serotonin, norepinephrine and dopamine levels in humans: a meta-analysis of monoamine depletion studies. *Mol Psychiatry*. 2007;12(4):331–359.
33. Krystal JH. Ketamine and the potential role for rapid-acting antidepressant medications. *Swiss Med Wkly*. 2007;137(15–16):215–216.
34. Maeng S, Zarate Jr. CA. The role of glutamate in mood disorders: results from the ketamine in major depression study and the presumed cellular mechanism underlying its antidepressant effects. *Curr Psychiatry Rep*. 2007;9(6):467–474.
35. Skolnick P, Popik P, Trullas R. Glutamate-based antidepressants: 20 years on. *Trends Pharmacol Sci*. 2009;30(11):563–569.
36. Murrough JW. Ketamine as a novel antidepressant: from synapse to behavior. *Clin Pharmacol Ther*. 2012;91(2):303–309.
37. Zarate Jr. CA, Manji HK. The role of AMPA receptor modulation in the treatment of neuropsychiatric diseases. *Exp Neurol*. 2008;211(1):7–10.
38. Preskorn SH, Baker B, Kolluri S, Menniti FS, Krams M, Landen JW. An innovative design to establish proof of concept of the antidepressant effects of the NR2B subunit selective N-methyl-D-aspartate antagonist, CP-101,606, in patients with treatment-refractory major depressive disorder. *J Clin Psychopharmacol*. 2008;28(6):631–637.
39. Zarate Jr. CA, Mathews D, Ibrahim L, et al. A randomized trial of a low-trapping nonselective N-methyl-d-aspartate channel blocker in major depression. *Biol Psychiatry*. 2013;74(4):257–264.
40. Jentsch JD, Redmond Jr. DE, Elsworth JD, Taylor JR, Youngren KD, Roth RH. Enduring cognitive deficits and cortical dopamine dysfunction in monkeys after long-term administration of phencyclidine. *Science*. 1997;277(5328):953–955.
41. Olney JW, Labruyere J, Price MT. Pathological changes induced in cerebrocortical neurons by phencyclidine and related drugs. *Science*. 1989;244(4910):1360–1362.
42. Szakacs R, Janka Z, Kalman J. The "blue" side of glutamatergic neurotransmission: NMDA receptor antagonists as possible novel therapeutics for major depression. *Neuropsychopharmacol Hung*. 2012;14(1):29–40.
43. Krystal JH, Sanacora G, Duman RS. Rapid-acting glutamatergic antidepressants: the path to ketamine and beyond. *Biol Psychiatry*. 2013;73(12):1133–1141.
44. Heresco-Levy U, Gelfin G, Bloch B, et al. A randomized add-on trial of high-dose D-cycloserine for treatment-resistant depression. *Int J Neuropsychopharmacol*. 2013;16(3):501–506.
45. Burgdorf J, Zhang XL, Nicholson KL, et al. GLYX-13, a NMDA receptor glycine-site functional partial agonist, induces antidepressant-like effects without ketamine-like side effects. *Neuropsychopharmacology*. 2013;38(5):729–742.
46. Russo SJ, Charney DS. Next generation antidepressants. *Proc Natl Acad Sci U S A*. 2013;110(12):4441–4442.

47. Pittenger C, Coric V, Banasr M, Bloch M, Krystal JH, Sanacora G. Riluzole in the treatment of mood and anxiety disorders. *CNS Drugs*. 2008;22(9):761–786.
48. Zarate CA, Manji HK. Riluzole in psychiatry: a systematic review of the literature. *Expert Opin Drug Metab Toxicol*. 2008;4(9):1223–1234.
49. Ibrahim L, Diazgranados N, Franco-Chaves J, et al. Course of improvement in depressive symptoms to a single intravenous infusion of ketamine vs add-on riluzole: results from a 4-week, double-blind, placebo-controlled study. *Neuropsychopharmacology*. 2012;37(6):1526–1533.
50. Szabo ST, Machado-Vieira R, Yuan P, et al. Glutamate receptors as targets of protein kinase C in the pathophysiology and treatment of animal models of mania. *Neuropharmacology*. 2009;56(1):47–55.
51. Martinez-Turrillas R, Del Rio J, Frechilla D. Neuronal proteins involved in synaptic targeting of AMPA receptors in rat hippocampus by antidepressant drugs. *Biochem Biophys Res Commun*. 2007;353(3):750–755.
52. Du J, Gray NA, Falke CA, et al. Modulation of synaptic plasticity by antimanic agents: the role of AMPA glutamate receptor subunit 1 synaptic expression. *J Neurosci*. 2004;24(29):6578–6589.
53. Du J, Quiroz JA, Gray NA, Szabo ST, Zarate Jr. CA, Manji HK. Regulation of cellular plasticity and resilience by mood stabilizers: the role of AMPA receptor trafficking. *Dialogues Clin Neurosci*. 2004;6(2):143–155.
54. Du J, Szabo ST, Gray NA, Manji HK. Focus on CaMKII: a molecular switch in the pathophysiology and treatment of mood and anxiety disorders. *Int J Neuropsychopharmacol*. 2004;7(3):243–248.
55. Krishnan KR, Doraiswamy PM, Lurie SN, et al. Pituitary size in depression. *J Clin Endocrinol Metab*. 1991;72(2):256–259.
56. Rubin RT, Phillips JJ, Sadow TF, McCracken JT. Adrenal gland volume in major depression. Increase during the depressive episode and decrease with successful treatment. *Arch Gen Psychiatry*. 1995;52(3):213–218.
57. Butterfield MI, Stechuchak KM, Connor KM, et al. Neuroactive steroids and suicidality in posttraumatic stress disorder. *Am J Psychiatry*. 2005;162(2):380–382.
58. Zorumski CF, Mennerick S. Neurosteroids as therapeutic leads in psychiatry. *JAMA Psychiatry*. 2013;70(7):659–660.
59. Schmidt PJ, Daly RC, Bloch M, et al. Dehydroepiandrosterone monotherapy in midlife-onset major and minor depression. *Arch Gen Psychiatry*. 2005;62(2):154–162.
60. Sullivan PF, Wilson DA, Mulder RT, Joyce PR. The hypothalamic-pituitary-thyroid axis in major depression. *Acta Psychiatr Scand*. 1997;95(5):370–378.
61. Nierenberg AA, Fava M, Trivedi MH, et al. A comparison of lithium and T(3) augmentation following two failed medication treatments for depression: a STAR*D report. *Am J Psychiatry*. 2006;163(9):1519–1530, quiz 665.
62. Szuba MP, O'Reardon JP, Rai AS, et al. Acute mood and thyroid stimulating hormone effects of transcranial magnetic stimulation in major depression. *Biol Psychiatry*. 2001;50(1):22–27.
63. Szuba MP, Amsterdam JD, Fernando AT 3rd, Gary KA, Whybrow PC, Winokur A. Rapid antidepressant response after noctumal TRH administration in patients with bipolar type I and bipolar type II major depression. *J Clin Psychopharmacol*. 2005;25(4):325–330.
64. Williams HJ, Craddock N, Russo G, et al. Most genome-wide significant susceptibility loci for schizophrenia and bipolar disorder reported to date cross-traditional diagnostic boundaries. *Hum Mol Genet*. 2011;20(2):387–391.
65. Pandey GN, Dwivedi Y. Neurobiology of Teenage Suicide. Boca Raton, FL: CRC Press; 2012.
66. Haddjeri N, Szabo ST, de Montigny C, Blier P. Increased tonic activation of rat forebrain 5-HT(1A) receptors by lithium addition to antidepressant treatments. *Neuropsychopharmacology*. 2000;22(4):346–356.
67. Niciu MJ, Ionescu DF, Mathews DC, Richards EM, Zarate CA. Second messenger/signal transduction pathways in major mood disorders: moving from membrane to mechanism of action, part I: major depressive disorder. *CNS Spectr*. 2013;1–10.
68. Drevets WC. Neuroplasticity in mood disorders. *Dialogues Clin Neurosci*. 2004;6(2):199–216.
69. Drevets WC, Price JL, Furey ML. Brain structural and functional abnormalities in mood disorders: implications for neurocircuitry models of depression. *Brain Struct Funct*. 2008;213(1–2):93–118.
70. Murray EA, Wise SP, Drevets WC. Localization of dysfunction in major depressive disorder: prefrontal cortex and amygdala. *Biol Psychiatry*. 2011;69(12):E43–E54.
71. Mayberg HS, Lozano AM, Voon V, et al. Deep brain stimulation for treatment-resistant depression. *Neuron*. 2005;45(5):651–660.
72. Drevets WC, Bogers W, Raichle ME. Functional anatomical correlates of antidepressant drug treatment assessed using PET measures of regional glucose metabolism. *Eur Neuropsychopharmacol*. 2002;12(6):527–544.
73. Neumeister A, Nugent AC, Waldeck T, et al. Neural and behavioral responses to tryptophan depletion in unmedicated patients with remitted major depressive disorder and controls. *Arch Gen Psychiatry*. 2004;61(8):765–773.
74. Hasler G, Fromm S, Carlson PJ, et al. Neural response to catecholamine depletion in unmedicated subjects with major depressive disorder in remission and healthy subjects. *Arch Gen Psychiatry*. 2008;65(5):521–531.
75. Ongur D, Ferry AT, Price JL. Architectonic subdivision of the human orbital and medial prefrontal cortex. *J Comp Neurol*. 2003;460(3):425–449.
76. McGrath CL, Kelley ME, Holtzheimer PE, et al. Toward a neuroimaging treatment selection biomarker for major depressive disorder. *JAMA Psychiatry*. 2013;1–9.
77. Sripada RK, Marx CE, King AP, et al. DHEA enhances emotion regulation neurocircuits and modulates memory for emotional stimuli. *Neuropsychopharmacology*. 2013;38(9):1798–1807.
78. Morgan 3rd. CA, Southwick S, Hazlett G, et al. Relationships among plasma dehydroepiandrosterone sulfate and cortisol levels, symptoms of dissociation, and objective performance in humans exposed to acute stress. *Arch Gen Psychiatry*. 2004;61(8):819–825.
79. Duman RS, Li N, Liu RJ, Duric V, Aghajanian G. Signaling pathways underlying the rapid antidepressant actions of ketamine. *Neuropharmacology*. 2012;62(1):35–41.
80. Laje G, Lally N, Mathews D, et al. Brain-derived neurotrophic factor Val66Met polymorphism and antidepressant efficacy of ketamine in depressed patients. *Biol Psychiatry*. 2012;72(11):e27–e28.
81. Liu RJ, Lee FS, Li XY, Bambico F, Duman RS, Aghajanian GK. Brain-derived neurotrophic factor Val66Met allele impairs basal and ketamine-stimulated synaptogenesis in prefrontal cortex. *Biol Psychiatry*. 2012;71(11):996–1005.
82. Duman RS, Aghajanian GK. Synaptic dysfunction in depression: potential therapeutic targets. *Science*. 2012;338(6103):68–72.

83. Voleti B, Navarria A, Liu RJ, et al. Scopolamine rapidly increases mammalian target of rapamycin complex 1 signaling, synaptogenesis, and antidepressant behavioral responses. *Biol Psychiatry*. 2013;74(10):742–749.
84. Weiner R, Lisanby SH, Husain MM, et al. Electroconvulsive therapy device classification: response to FDA advisory panel hearing and recommendations. *J Clin Psychiatry*. 2013;74(1):38–42.
85. Harden CL, Pulver MC, Ravdin LD, Nikolov B, Halper JP, Labar DR. A pilot study of mood in epilepsy patients treated with vagus nerve stimulation. *Epilepsy Behav*. 2000;1(2):93–99.
86. Cusin C, Dougherty DD. Somatic therapies for treatment-resistant depression: ECT, TMS, VNS, DBS. *Biol Mood Anxiety Disord*. 2012;2(1):14.
87. Nahas Z, Marangell LB, Husain MM, et al. Two-year outcome of vagus nerve stimulation (VNS) for treatment of major depressive episodes. *J Clin Psychiatry*. 2005;66(9):1097–1104.
88. Krahl SE, Clark KB, Smith DC, Browning RA. Locus coeruleus lesions suppress the seizure-attenuating effects of vagus nerve stimulation. *Epilepsia*. 1998;39(7):709–714.
89. Groves DA, Bowman EM, Brown VJ. Recordings from the rat locus coeruleus during acute vagal nerve stimulation in the anaesthetised rat. *Neurosci Lett*. 2005;379(3):174–179.
90. Carpenter LL, Moreno FA, Kling MA, et al. Effect of vagus nerve stimulation on cerebrospinal fluid monoamine metabolites, norepinephrine, and gamma-aminobutyric acid concentrations in depressed patients. *Biol Psychiatry*. 2004;56(6):418–426.
91. Dorr AE, Debonnel G. Effect of vagus nerve stimulation on serotonergic and noradrenergic transmission. *J Pharmacol Exp Ther*. 2006;318(2):890–898.
92. Sackeim HA, Rush AJ, George MS, et al. Vagus nerve stimulation (VNS) for treatment-resistant depression: efficacy, side effects, and predictors of outcome. *Neuropsychopharmacology*. 2001;25(5):713–728.
93. Luber B, Lou HC, Keenan JP, Lisanby SH. Self-enhancement processing in the default network: a single-pulse TMS study. *Exp Brain Res*. 2012;223(2):177–187.
94. Lisanby SH, Kinnunen LH, Crupain MJ. Applications of TMS to therapy in psychiatry. *J Clin Neurophysiol*. 2002;19(4):344–360.
95. Pridmore S. Rapid transcranial magnetic stimulation and normalization of the dexamethasone suppression test. *Psychiatry Clin Neurosci*. 1999;53(1):33–37.
96. Rodrigues AM, Rosas MJ, Gago MF, et al. Suicide attempts after subthalamic nucleus stimulation for Parkinson's disease. *Eur Neurol*. 2010;63(3):176–179.
97. Keller MB, McCullough JP, Klein DN, et al. A comparison of nefazodone, the cognitive behavioral-analysis system of psychotherapy, and their combination for the treatment of chronic depression. *N Engl J Med*. 2000;342(20):1462–1470.
98. Gershon ES, Rieder RO. Major disorders of mind and brain. *Sci Am*. 1992;267(3):126–133.
99. Ressler KJ, Bradley B, Mercer KB, et al. Polymorphisms in CRHR1 and the serotonin transporter loci: gene x gene x environment interactions on depressive symptoms. *Am J Med Genet*. 2010;153B(3):812–824.
100. Labonte B, Suderman M, Maussion G, et al. Genome-wide epigenetic regulation by early-life trauma. *Arch Gen Psychiatry*. 2012;69(7):722–731.
101. Kendler KS, Hettema JM, Butera F, Gardner CO, Prescott CA. Life event dimensions of loss, humiliation, entrapment, and danger in the prediction of onsets of major depression and generalized anxiety. *Arch Gen Psychiatry*. 2003;60(8):789–796.
102. Lopez-Leon S, Janssens AC, Gonzalez-Zuloeta Ladd AM, et al. Meta-analyses of genetic studies on major depressive disorder. *Mol Psychiatry*. 2008;13(8):772–785.
103. Collier DA, Stober G, Li T, et al. A novel functional polymorphism within the promoter of the serotonin transporter gene: possible role in susceptibility to affective disorders. *Mol Psychiatry*. 1996;1(6):453–460.
104. Mitjans M, Arias B. The genetics of depression: what information can new methodologic approaches provide? *Actas Esp Psiquiatr*. 2012;40(2):70–83.
105. Neumeister A, Hu XZ, Luckenbaugh DA, et al. Differential effects of 5-HTTLPR genotypes on the behavioral and neural responses to tryptophan depletion in patients with major depression and controls. *Arch Gen Psychiatry*. 2006;63(9):978–986.
106. Whalley HC, Sussmann JE, Chakirova G, et al. The neural basis of familial risk and temperamental variation in individuals at high risk of bipolar disorder. *Biol Psychiatry*. 2011;70(4):343–349.
107. Hariri AR. The highs and lows of amygdala reactivity in bipolar disorders. *Am J Psychiatry*. 2012;169(8):780–783.
108. Anway MD, Skinner MK. Epigenetic transgenerational actions of endocrine disruptors. *Endocrinology*. 2006;147(6 suppl):S43–S49.
109. Skinner MK, Mohan M, Haque MM, Zhang B, Savenkova MI. Epigenetic transgenerational inheritance of somatic transcriptomes and epigenetic control regions. *Genome Biol*. 2012;13(10):R91.
110. Zhou R, Gray NA, Yuan P, et al. The anti-apoptotic, glucocorticoid receptor cochaperone protein BAG-1 is a long-term target for the actions of mood stabilizers. *J Neurosci*. 2005;25(18):4493–4502.
111. Heim C, Nemeroff CB. The role of childhood trauma in the neurobiology of mood and anxiety disorders: preclinical and clinical studies. *Biol Psychiatry*. 2001;49(12):1023–1039.
112. Heim C, Newport DJ, Heit S, et al. Pituitary-adrenal and autonomic responses to stress in women after sexual and physical abuse in childhood. *JAMA*. 2000;284(5):592–597.
113. Heim C, Mletzko T, Purselle D, Musselman DL, Nemeroff CB. The dexamethasone/corticotropin-releasing factor test in men with major depression: role of childhood trauma. *Biol Psychiatry*. 2008;63(4):398–405.
114. Tyrka AR, Wier L, Price LH, Ross NS, Carpenter LL. Childhood parental loss and adult psychopathology: effects of loss characteristics and contextual factors. *Int J Psychiatry Med*. 2008;38(3):329–344.
115. Carpenter LL, Tyrka AR, McDougle CJ, et al. Cerebrospinal fluid corticotropin-releasing factor and perceived early-life stress in depressed patients and healthy control subjects. *Neuropsychopharmacology*. 2004;29(4):777–784.
116. Lee R, Geracioti TD, Kasckow JW, Coccaro EF. Childhood trauma and personality disorder: positive correlation with adult CSF corticotropin-releasing factor concentrations. *Am J Psychiatry*. 2005;162(5):995–997.
117. Lee RJ, Gollan J, Kasckow J, Geracioti T, Coccaro EF. CSF corticotropin-releasing factor in personality disorder: relationship with self-reported parental care. *Neuropsychopharmacology*. 2006;31(10):2289–2295.

118. Lemos JC, Wanat MJ, Smith JS, et al. Severe stress switches CRF action in the nucleus accumbens from appetitive to aversive. *Nature*. 2012;490(7420):402.
119. Lowry CA, Hale MW, Plant A, et al. Fluoxetine inhibits corticotropin-releasing factor (CRF)-induced behavioural responses in rats. *Stress*. 2009;12(3):225–239.
120. Debellis MD, Gold PW, Geracioti TD, Listwak SJ, Kling MA. Association of fluoxetine treatment with reductions in Csf concentrations of corticotropin-releasing hormone and arginine vasopressin in patients with major depression. *Am J Psychiatry*. 1993;150(4):656–657.
121. Nikisch G, Mathe AA, Czernik A, et al. Long-term citalopram administration reduces responsiveness of HPA axis in patients with major depression: relationship with S-citalopram concentrations in plasma and cerebrospinal fluid (CSF) and clinical response. *Psychopharmacology (Berl)*. 2005;181(4):751–760.
122. Magalhaes AC, Holmes KD, Dale LB, et al. CRF receptor 1 regulates anxiety behavior via sensitization of 5-HT2 receptor signaling. *Nat Neurosci*. 2010;13(5):622–629.
123. Paez-Pereda M, Hausch F, Holsboer F. Corticotropin releasing factor receptor antagonists for major depressive disorder. *Expert Opin Investig Drugs*. 2011;20(4):519–535.
124. Binder EB, Salyakina D, Lichtner P, et al. Polymorphisms in FKBP5 are associated with increased recurrence of depressive episodes and rapid response to antidepressant treatment. *Nat Genet*. 2004;36(12):1319–1325.
125. Whiteford HA, Baxter AJ. The global burden of disease 2010 study: what does it tell us about mental disorders in Latin America? *Rev Bras Psiquiatr*. 2013;35(2):111–112.
126. Zajecka J, Kornstein SG, Blier P. Residual symptoms in major depressive disorder: prevalence, effects, and management. *J Clin Psychiatry*. 2013;74(4):407–414.
127. Einat H, Yuan P, Szabo ST, Dogra S, Manji HK. Protein kinase C inhibition by tamoxifen antagonizes manic-like behavior in rats: implications for the development of novel therapeutics for bipolar disorder. *Neuropsychobiology*. 2007;55(3–4):123–131.

CHAPTER

104

Bipolar Disorder

Scott C. Fears[†] and Victor I. Reus[*]

[†]University of California, Los Angeles, CA, USA
[*]University of California, San Francisco, CA, USA

INTRODUCTION

Bipolar disorder (BP), also known as manic-depressive illness, affects approximately 1–2% of the adult population. The disorder is characterized by cyclic disturbances of mood, energy pattern, and behavior, and is associated with high morbidity and mortality. Epidemiologic studies have shown that BP is caused by a complex mix of environmental and genetic factors, but very little is known about the specific biological mechanisms underlying BP. Environmental risk factors, such as loss of social support or disruption of the sleep/wake cycle, contribute to the manifestation of symptoms. Family, twin, and adoption studies have shown that genetic factors play an important role in its pathogenesis, and recent genome-wide association studies have started to identify confirmed risk loci. These recent findings have provided new insights into the cellular and molecular pathways that may be involved in the disorder. However, the complex nature of the disorder, the variability of the phenotype, and the lack of good animal models continue to challenge our progress in understanding the pathophysiology of mood disorders such as BP. In this chapter, we discuss some current strategies for overcoming difficulties in finding the underlying causes of BP.

CLINICAL FEATURES

The predominant symptom in BP is a cyclic disturbance of mood. Episodes of elevated or irritable mood, also known as mania, alternate with episodes of depression.[1] Mixed states, in which symptoms of mania and depression co-occur can be present as well. Mania, the cardinal feature of the disorder, is required for a diagnosis of BP and consists of an abnormally elevated mood, characterized by euphoria, expansiveness, and/or irritability. Other classic features include inflated self-esteem, grandiosity, low frustration tolerance, increased need for immediate gratification, decreased impulse control, an increase in provocative behavior, and an increase in goal-directed activity (social, work-related, or creative). In addition, increased energy, decreased need for sleep, sexual or religious preoccupation, and racing thoughts are often present. Psychotic symptoms occur in about 50% of patients.[2]

Depression consists of a pervasive or consistently low mood in conjunction with neurovegetative and cognitive features such as difficulty sleeping or excessive sleep, low energy, anhedonia, social isolation, and changes in appetite. Depressive episodes in BP are similar to those in unipolar major depressive disorder (MDD), but atypical depression and psychotic depression occur more often in BP than in MDD. Depression is a characteristic part of BP, and in fact individuals with BP spend more time depressed than in elevated states,[3] but it is not a necessary component; approximately 10% of patients with BP never have a clear depressive episode.

Mixed states, also called dysphoric mania, are defined as the simultaneous co-occurrence of all or some of the full symptoms of mania and depression and are more common than previously recognized.[4] Between 15–70% of patients will experience a mixed episode, and about 50% of inpatient hospitalizations for BP are for mixed states. Research into mixed states has been limited by the high symptom thresholds required in the fourth edition of the Diagnostic

http://dx.doi.org/10.1016/10.1016/B978-0-12-410529-4.00104-2

and Statistical Manual of Mental Disorders (DSM-IV) that have excluded many patients with substantial evidence of sub-threshold hypomanic symptoms and has resulted in more relaxed criteria in the fifth edition (DSM-5).

The lifetime prevalence of BP I is between 0.5 and 1.5% using a narrow diagnostic classification and up to 4.4% using a broader diagnostic scheme that includes BP II and related spectrum disorders.[5] Epidemiologic investigations have been complicated by the fact that the expression of symptoms as well as the diagnosis itself may be influenced by factors such as class, race, culture, and even climate. For example, schizophrenia (SZ) is more frequently diagnosed in ethnic minority populations than BP, even when the presenting symptoms are the same. When such biases are accounted for, however, it appears that BP is approximately equally prevalent in all ethnic populations. In addition, BP occurs with equal frequency in men and women, in contrast with unipolar depressive disorders that are at least twice as prevalent in women.[6] The typical age of onset in BP is around 25 years of age; however, many patients experience symptoms during adolescence. Childhood-onset BP used to be considered rare. Although this diagnosis is becoming increasingly common, it remains more difficult to establish with confidence than adult-onset BP, as affected children demonstrate predominantly behavioral disturbances rather than emotional symptomatology.[7–9] Attention deficit hyperactivity disorder (ADHD), in particular, shares many features in common with childhood BP and may exist as a comorbid or independent condition.

The burden for the individual affected with BP and for society is high. According to the Global Burden of Disease Study, a 5-year study initiated by the World Health Organization, MDD is the leading cause of premature mortality, disease, and injury worldwide, and BP is ranked sixth.[10,11] BP is often accompanied by psychiatric and medical conditions that complicate the disease process. Comorbid eating disorders, for example, as well as anxiety and panic disorders, are frequent in patients with BP, as are high rates of medical conditions, such as cardiovascular and metabolic disorders.[12,13] Substance abuse and dependence occur in about 50% of patients with BP, and these comorbidities play an important role in the occurrence of suicide events.[14,15] BP patients have a high risk of suicide, with a rate 20–30 times greater than the general population.[16]

Differential Diagnosis

Currently accepted diagnostic criteria for BP specify a minimum duration for mood episodes. However, significant differences in symptomatology and course of illness are observed even among patients who meet these criteria. Patients vary substantially in the course of the disease, especially in the number, frequency, regularity, and severity of episodes, and the ratio of manic to depressive episodes, as well as their response to treatment.[17] These factors contribute to the notorious difficulty diagnosing BP as evidenced by the finding that only 20% of patients with BP presenting with depressive symptoms are correctly diagnosed during the first year of treatment, with the average delay between illness onset and diagnosis of up to 10 years.[18]

BP may be differentiated from other disorders that are considered to form a spectrum of mood disorders (Figure 104.1). The diagnosis of BP I, based on DSM-5, requires at least one episode each of mania and major depression, whereas BP II requires one full episode of major depression and at least one episode of elevated mood (hypomania) without full mania. Major differences between hypomanic and manic episodes include relatively less occupational and/or social dysfunction and the absence of psychotic symptoms in BP II. The distinction between BP I and BP II is probably the most challenging differential diagnosis in the mood disorder spectrum because functional impairment is often difficult to assess in an objective manner.[19]

Cyclothymia consists of episodic mood alterations that fail to meet the duration or the severity requirements of either major depression or mania. Unipolar MDD is characterized by episodes of sadness, often associated with somatic symptoms, such as weight loss, disturbed sleep, and decreased energy, without a history of manic or hypomanic episodes. Schizoaffective disorder is frequently viewed as an intermediate form between BP and SZ, in which episodes of clear mania and/or depression occur, but in which chronic psychotic symptoms persist even in the absence of mood symptoms. Often a subtype with predominantly affective symptoms (schizoaffective, BP type) is distinguished from a subtype with predominantly schizophrenic symptoms (schizoaffective, SZ type). Epidemiologic studies support this classification and suggest different genetic susceptibility of these subtypes.[20]

There are no routine laboratory tests for BP. However, identification of valid biomarkers is an active area of research and changes in oxidative stress markers, such as glutathione, vitamin E, and zinc, chemokines, such as interleukins IL-4,IL-6,IL-8, tumor necrosis factor alpha (TNF-α) and soluble interleukin-2 receptor, and serum brain-derived neurotrophic factor (BDNF) have been frequently reported in association with a BP diagnosis (discussed in more detail below).[21]

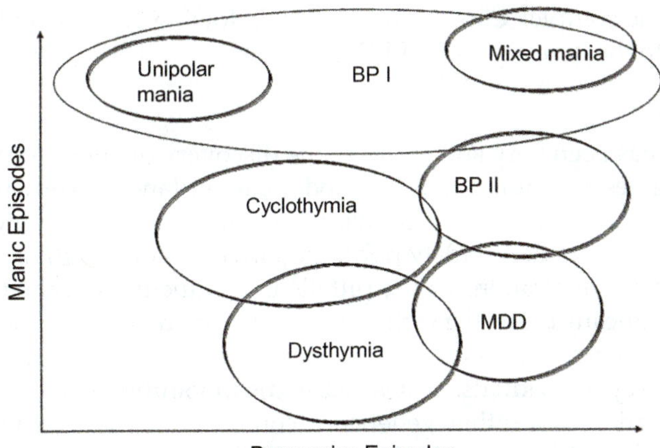

FIGURE 104.1 **The spectrum of mood disorders.** Bipolar disorder (BP) is generally thought of as being part of a group of affective illnesses, characterized by episodes of mania and depression, and that includes BP type I (BP I), BP type II (BP II), cyclothymia, dysthymia, and major depressive disorder (MDD). Unipolar mania and mixed mania are considered subtypes of BP I. These conditions must be differentiated from mood disorders that are due to a general medical condition or that are secondary to substance abuse. This figure summarizes the relationship between types of episodes and Diagnostic and Statistical Manual of Mental Disorders, Fifth Edition (DSM-5) disorders. Psychotic features can be found in MDD, BP I, and mixed mania.

GENETIC EPIDEMIOLOGY

Bipolar disorder shows strong evidence of genetic influence. Siblings of BP probands have a 5–10 times greater risk of developing the disorder compared to the general population and twin studies estimate the heritability for BP to be 80–90%.[22] Recent genome-wide association studies in population samples have generated the first confirmed risk loci associated with BP (described in more detail below); however, the genetic architecture of BP remains largely unresolved. It is hypothesized that the disorder may be due to a combination of small effect alleles that are relatively common in the population (>1%) and rare alleles that have a large effect at the individual level, but do not influence disease risk at the population level.[23,24] The challenges related to characterizing the genetic architecture of BP have motivated innovations in phenotyping and genetic analysis. Combined with rapidly evolving molecular technology (e.g., next-generation sequencing), psychiatric genetics appears poised for fruitful discoveries in the near future. In the following section, a brief history and overview of progress and challenges in the field is provided.

A major challenge for the genetic investigation of psychiatric disorders is defining the best phenotype for subsequent analysis. Psychiatric disorders are defined categorically using checklists that broadly capture similar patterns of symptoms but are widely believed to cluster heterogeneous disorders under a single categorical diagnosis. Within a sample of individuals meeting criteria for BP, there is likely significant heterogeneity regarding genetic influences on the disorder, which reduces the power of genetic analysis to identify genotype–phenotype correlations. A related problem of categorical diagnosis based on symptom checklists is the growing recognition that different disorders can share common genetic risks. For example, some families ascertained for SZ probands also aggregate BP.[25] Direct evidence for shared genetic influence among psychiatric disorders was demonstrated in several recent genome-wide association studies in population studies of samples including multiple diagnostic groups.[26,27] These studies found that a high proportion of loci that predict disease risk are shared between BP and SZ. Additionally, both BP and SZ show genetic overlap with depression, although not to the same extent as each other. Testing known SZ risk loci in BP samples shows that some variants contribute risk to both disorders (e.g., *CACNA1C*), whereas other loci appear specific to SZ.[28] Detailed analysis of the clinical characteristics of patients across disorders suggests that shared genetic risk may be linked to common clinical symptoms, like mania.

Gene Mapping Studies

Two general approaches are used to map genetic variants that contribute to disease risk. Linkage analysis maps disease variants by identifying cosegregation of genetic loci of known location (genotypes) and disease traits in families. The alternative approach is association analysis, which maps variants by identifying loci that are enriched in clinical samples compared to healthy controls. The two approaches are complementary and in combination have started to shed some light on the complex etiology of BP. The following sections will provide an overview of each approach and its contribution to the investigation of BP.

Genetic Linkage Approaches

Parametric linkage analysis has been very successful in the discovery of genes for rare Mendelian disorders. This method is useful in large pedigrees. It requires that the mode of inheritance is known or can be estimated relatively accurately and the penetrance (the degree to which symptoms are expressed in individuals carrying the disease genotype) of the disease gene is specified. A widely publicized investigation of BP in an extended Old Order Amish pedigree provided one of the first illustrations of the pitfalls in conducting parametric linkage studies of complex traits. An initial report of a significant LOD (logarithm of odds) score on chromosome 11[29] could not be confirmed in a reanalysis of this pedigree, mainly based on an extension of the pedigree to include new branches and reclassification of the phenotypes of key individuals.[30] These early disappointments highlighted the fact that psychiatric diseases are not Mendelian disorders, but rather, genetically complex disorders for which the mode of transmission and the penetrance are generally not known. Mapping a complex disorder requires addressing factors such as: i) incomplete penetrance; ii) presence of phenocopies (forms of the disease that are not caused by genetic factors); iii) locus heterogeneity (different genes associated with similar phenotypes in different families or populations); iv) polygenic inheritance (multiple genes acting together to cause disease); and v) multifactorial inheritance (genes and environment interacting to cause a trait or disease).

The lack of success of many subsequent parametric linkage studies of BP (and other complex traits) led to a shift to nonparametric linkage strategies and association approaches. Nonparametric linkage methods rely on the principle that affected relatives should share DNA sequences inherited from their ancestors (identical by descent; IBD) in the region of the disease-causing gene more frequently than would be expected by chance. These methods are less powerful for detecting linkage than a correctly specified parametric linkage approach but are considered more robust than parametric approaches in the presence of genetic complexity. However, a major limitation of linkage investigations is the difficulty replicating results. Significant findings in small studies are often not replicated in samples of different families. Meta-analyses attempt to combine the numerical results of several individual studies to estimate the average effect of interest[31] and thereby overcome the low power of small linkage scans. An analysis combining the results of 18 genome scans for BP found no loci at a genome-wide level of significance; the most promising finding was for a location on chromosome 9p21-22 under a narrow definition of the affected phenotype.[32] Another meta-analysis that combined the genotyping data from 11 BP scans found linkage signals on chromosomes 6q21-25 and 8q24 that some have interpreted as meeting genome-wide levels of significance.[33] The largest meta-analysis to date consisted of almost 1000 families,[34] but it is still debatable whether even this relatively large study achieved any unequivocally replicated linkage findings. Many explanations have been proposed for the disappointing results of such studies; broadly speaking it is most likely that they have been statistically underpowered. If multiple interacting genes with small to moderate effects cause BP, then these genes may be difficult to detect with linkage approaches, especially in the presence of genetic heterogeneity. In the extreme situation that each family is segregating a unique set of genetic variants, linkage analysis would have little power to identify risk loci. A more likely scenario is that the genetic architecture of BP includes a mix of common and rare alleles. Much of the recent effort in psychiatric genetics has shifted to identifying common risk loci of small effect. Compared to linkage approaches, these investigations are starting to appreciate considerable success (described in the next section). However, it has become increasingly clear that the variants identified in population-based association approaches explain only a small proportion of disease risk (~1–2%). Therefore, family-based linkage approaches have a potentially important role identifying rare risk variants that occur on the background of common risk variants and determine which individuals develop the disorder. An additional advantage of family-based linkage analysis is that they have more power to identify quantitative trait loci that contribute to brain and behavioral traits associated with BP (see "Intermediate Phenotypes," page 1279)

Association Studies

The lack of success of linkage studies for common disorders such as BP has cast doubt on the idea that a few alleles of large effect are solely responsible for the genetic susceptibility to such disorders. Rather, it has been argued that the genetic contribution to disease risk derives, in large part, from multiple variants that are common in the population. Association studies are much more powerful than linkage designs for detecting such common variants. In these studies, polymorphisms are tested for association with a disease or trait in a sample of individuals drawn from the population. Association studies can roughly be divided into two categories: candidate gene studies, in which polymorphisms in a particular gene or set of genes are selected based on *a priori* hypothesis that these genes are involved in the disease process, and whole-genome studies, in which there is no such hypothesis. Interest in association studies has been heightened by several recent developments that have made it feasible to conduct genome-wide association studies to identify common susceptibility variants, most of which are expected to be single nucleotide polymorphisms (SNPs). In particular, the International HapMap project has generated a catalog of several million SNPs positioned across the genome, and advances in genotyping technology have made it feasible to assay these SNPs in large samples of cases and controls in a cost-effective manner.

Prior to the availability of sufficiently large samples with genome-wide genotypes, investigators implemented the candidate gene approach, in which, polymorphisms in specific genes are selected for association testing based on the hypothesis that the known biological functions of these genes make them relevant to the trait being investigated.[35] The monoamine hypothesis has driven the preponderance of candidate association studies of BP. Examples of genes that have been investigated for association with BP on the basis of this hypothesis include the serotonin transporter, monoamine oxidase A (MAOA), the dopamine D3 receptor gene (DRD3), and the tyrosine hydroxylase gene.[35] For each of these genes, there have been some nominally significant associations reported, but the results have not been significant when appropriate corrections have been made for multiple testing. In addition, a high proportion of association studies to such genes has been negative.[36] The equivocal results of candidate association studies conducted to date, together with increased recognition of the low prior probability of association to BP of any of the tested genes, has turned the field toward genome-wide association strategies, which do not require prior knowledge about the underlying disease process.

In contrast to the disappointing results from linkage analysis and candidate gene investigations, genome-wide association studies (GWAS) of BP have begun to yield compelling findings, albeit more slowly compared to non-psychiatric complex disorders.[22] The first GWAS of BP was conducted in 2007 by the Wellcome Trust Case Control Consortium (WTCCC) in 2000 BP cases and 3000 controls genotyped at ~500,000 SNPs.[37] In contrast to the six other common diseases that were investigated in the same study, no genome-wide significant risk variants were identified in the BP sample. A similarly sized GWAS was performed in 2008, using samples from the Systematic Treatment Enhancement Program for Bipolar Disorder (STEP-BD) and University College of London (UCL),[38] and also failed to identify any genome-wide significant risk loci. A combined analysis of the WTCCC, STEP, and UCL data, with 4387 BP cases and 6209 controls, finally reported the first genome-wide significant association for BP in the *ANK3* (ankyrin G) gene.[39] Subsequent investigations have replicated the *ANK3* finding and extended the list of risk loci such that by the end of 2013, about a dozen loci had been established as confirmed BP risk alleles (Table 104.1).[22] The effect sizes of the risk alleles are low, with odds ratios generally less than 1.2, indicating that these common variants only contribute to a moderate increase in disease risk.

Pathway analysis is an adaptation of the GWAS method that assesses the significance of gene clusters with similar biological function instead of individual genetic variants. These methods improve statistical power by reducing the overall number of tests, and also leveraging the growing body of evidence in gene ontology databases to identify specific biological pathways implicated in BP. Such approaches have provided evidence for the involvement of a range of cellular processes in BP including ion channelopathies, endocrine regulation, and posttranscriptional modification.[40]

Although genome-wide association studies are potentially very powerful for detecting variants of even relatively small effect, they do have some important limitations.[41] They are sensitive to false-positive findings if cases and controls are not well matched in terms of their ethnic origin; in population genetics terms this is referred to as sensitivity to population stratification and population substructure. In addition, because millions of statistical tests are performed in genome-wide association studies, it is important to correct appropriately for multiple testing. It is generally accepted that multiple replications are required to consider the variants identified in association studies as validated findings. It should also be noted that although GWAS have better spatial resolution compared to linkage methods, the causal genetic variants for the GWAS findings have not been unequivocally established. An additional caveat concerning genome-wide association studies is that they have little power to detect rare variants and their identification might depend on the development of inexpensive whole-genome sequencing technologies.

Intermediate Phenotypes

Despite recent success in identifying *bona fide* risk variants for BP, the mechanisms by which risk variants are linked to BP pathophysiology is not clear. The investigation of intermediate phenotypes, also called endophenotypes, has been proposed as a strategy for overcoming the obstacles presented by the complexity of psychiatric disorders.[42] The endophenotype concept assumes that neurophysiological, biochemical, endocrine, neuroanatomical, cognitive, or neuropsychological variations underlie clinical features of a given disease and are more amenable to genetic mapping than the disease phenotypes described in our current diagnostic systems.[43,44] Pedigree-based

TABLE 104.1 Bipolar Disorder: Significant Genome-Wide Association Study Findings

SNP	Chromosomal location	Nearest gene	p-value	Odds ratio	Clinical sample	Study
rs12576775	11q14.1	*ODZ4*	4.4×10^{-8}	1.14	BP	PGC-BP[100]
rs4765913	12p13.3	*CACNA1C*	1.5×10^{-8}	1.14	BP & SZ	PGC-BP[100]
rs1064395	19p12	*NCAN*	2.1×10^{-9}	1.17	BP	Cichon et al.[101]
rs7296288	12p13.1	*RHEBL1, DHH*	9.0×10^{-9}	0.90	BP	Green et al.[102]
rs3818253	20q11.22	*TRPC4AP*	3.9×10^{-8}	1.16	BP	Green et al.[102]
rs9371601	6q25	*SYNE1*	2.9×10^{-8}	1.10	BP	Green et al.[103]
rs1344706	2q32.1	*ZNF804A*	4.1×10^{-13}	1.11	BP & SZ	O'Donovan et al.[104]
rs2239547	3p21.1	*ITIH3/4*	7.8×10^{-9}	1.12	BP & SZ	PGC-SZ[105]
rs10994359	10q21	*ANK3*	2.4×10^{-8}	1.22	BP & SZ	PGC-SZ[105]
rs4583255	16p11.2	*MAPK3*	6.6×10^{-11}	1.08	BP & SZ	Steinberg et al.[106]
rs2251219	3p21	*PBRM1*	3.6×10^{-8}	0.87	BP & MDD	McMahon et al.[107]

Abbreviations: BP, bipolar disorder, SZ, schizophrenia; MDD, major depressive disorder; PGC-BP, Psychiatric Genome-Wide Association Study Consortium Bipolar Disorder Working Group; PGC-SZ, Schizophrenia Psychiatric Genome-Wide Association Study Consortium.

linkage analysis may have greater power to genetically map quantitative endophenotypes than categorical disease diagnoses. Criteria for considering a trait useful as an endophenotype for a particular disease includes the following: i) the trait is associated with the disease, i.e., individuals who display the disease phenotype are more likely to demonstrate extreme values of the trait than control subjects; ii) trait values are not state-dependent (only present during illness episodes); and iii) the trait is heritable. Because endophenotypes represent only a fraction of the disease phenotype, they may be present in some unaffected relatives of patients and absent in some of the affected individuals. Endophenotypes may also be easier to model, measure, and map in animals. Traits that have been proposed as endophenotypes in BP include measures of attention and cognitive function, circadian rhythm instabilities, alteration in motivation and reward, brain structural abnormalities and white matter hyperintensities, stress response, abnormalities of calcium metabolism, reduced serotonin function and tryptophan depletion, and increased sensitivity to stimulants.[45] Individuals who are later diagnosed with BP have been characterized as having distinctive antecedent personality profiles, with elevations in dimensions of novelty seeking, harm avoidance, and self-transcendence, and lower self-directedness and cooperativeness than people with other psychiatric diagnoses and the population at large; such assessments have been found to be replicable, heritable, and stable over time, suggesting that BP may be constructed as a dimensional disturbance on a continuum with normal function as opposed to a categorical diagnosis, and that these personality traits provide tractable targets for genetic investigation.[46] A recent study implementing the endophenotype approach provides support for its potential for the genetic dissection of BP. The study ascertained BP subjects and their first-degree relatives from genetically isolated populations in Latin America and assessed a multidimensional range of brain and behavioral measures using structural magnetic resonance imaging (MRI), computer-based cognitive tests, and self-report questionnaires of personality and temperament traits and found that a majority of the measures were genetically influenced and many of the traits were different between BP subjects and non-BP family members (Figure 104.2).[47]

DISEASE MECHANISMS

Pathophysiology

The pathophysiology of BP is poorly understood. Our current knowledge rests on a composite of findings from various areas of research, including neuropsychological assessments, brain imaging, histological and histochemical studies on postmortem brain tissue, and pharmacological studies. The following section summarizes these findings separately by fields of study. Following this section, an overview of the molecular biology investigations of BP is presented.

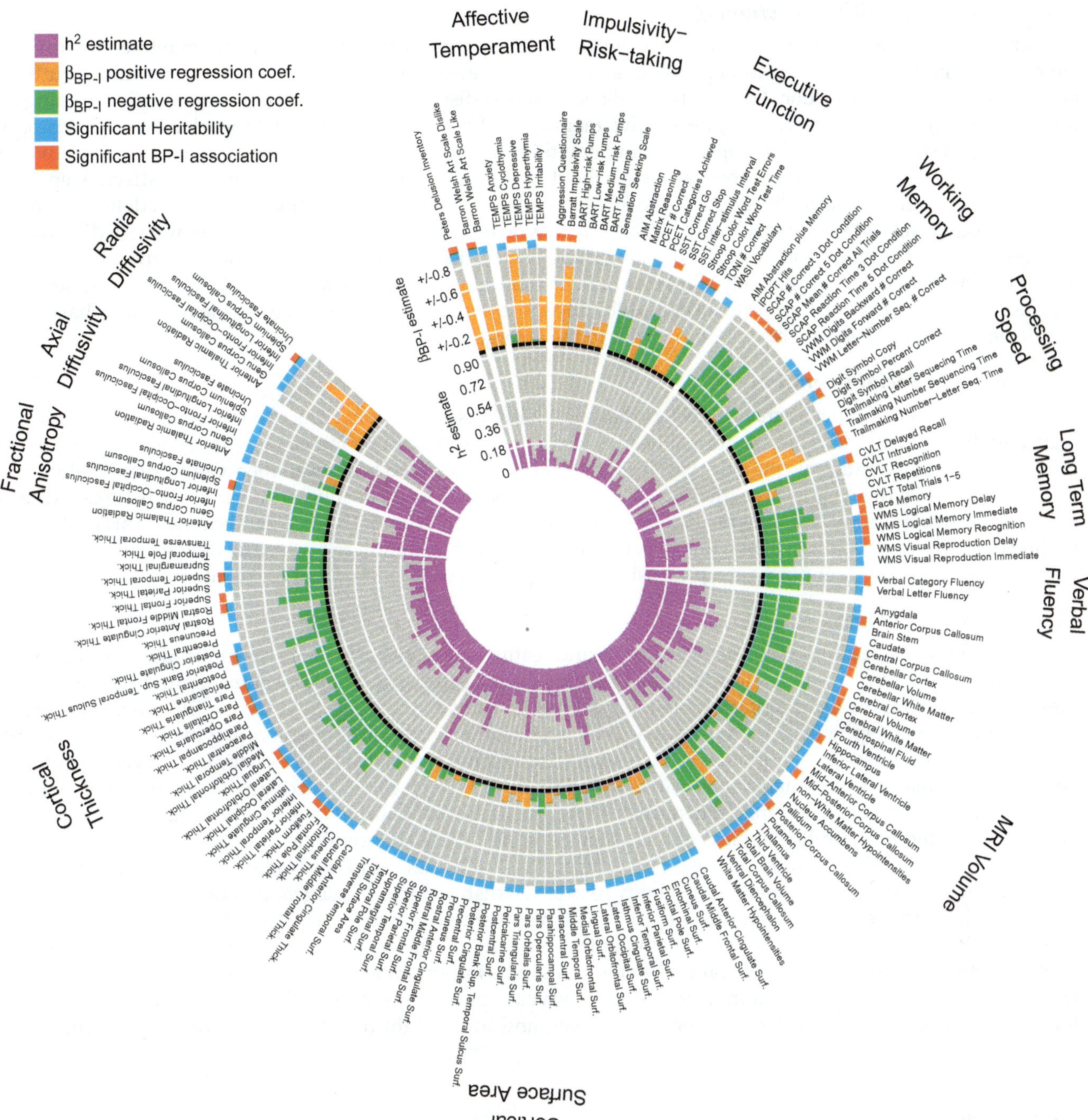

FIGURE 104.2 **Example of intermediate phenotype (endophenotype) approach for the genetic investigation of bipolar disorder (BP).** A broad range of neuroimaging and behavioral phenotypes were acquired from more than 700 members of families heavily loaded for BP in genetically isolated regions of Latin America.[47] Each phenotype was tested for two criteria: familial aggregation (i.e., heritability) and association with BP. The results of analyses of heritability and of association with BP I are shown as two histograms stacked on top of each other. Inner histogram purple bars show the magnitude of the heritability estimate for each component phenotype and the blue box next to the trait name at the outer edge of the plot indicates estimates that passed the significance threshold. Outer histogram shows the magnitude of the estimated regression coefficient for the BP I association test. In orange are positive coefficients representing traits that are higher in BP I subjects compared to non-BP I family members. In green are negative coefficients representing traits that are lower in BP I subjects. A red box at the outer edge of the circle indicates traits that exceeded the significance threshold for association with BP I. Abbreviations: PCET, Penn Conditional Exclusion Test; SST, Stop Signal Task; TONI, Test of Nonverbal Intelligence; AIM, Abstraction Inhibition and Memory test; IPCPT, Identical Pairs Continuous Performance Test; VWM, verbal working memory; CVLT, California Verbal Learning Test; WMS, Wechsler Memory Scale; BART, Balloon Analog Risk Task; TEMPS, Temperament Evaluation of Memphis, Pisa, Paris and San Diego; WASI, Wechsler Abbreviated Scale of Intelligence; SCAP, Spatial Capacity Delayed Response Test. Reproduced with permission from[47]. © American Medical Association.

Disturbance in Cognitive Functioning

Even though cyclic and dramatic mood shifts dominate the clinical presentation of BP, cognitive abnormalities are observed both in mania and in depression, and emerging evidence of cognitive impairment during euthymic periods is highlighting functional disability unrelated to mood disturbance. Executive dysfunction in planning and problem-solving tasks, verbal fluency, declarative memory disturbances, and deficits in working memory, processing speed, and sustained attention appear in about one-third of euthymic patients and their nonaffected relatives.[48,49] The severity of these disturbances correlates with the duration of the illness and the number of affective episodes. These neurocognitive abnormalities seem to be more prevalent in patients with hallucinations and delusions than in those without psychotic symptoms,[50] and also appear to worsen with age,[51] but it is unclear how these disturbances contribute to the pathophysiology of the disease. There is significant heterogeneity in the cognitive deficits found in BP subjects, which are qualitatively similar to deficits found in schizophrenia but tend to be much less severe. In contrast to schizophrenia, BP subjects do not exhibit the same degree of prodromal impairment but may have subtle deficits in visuospatial and verbal reasoning prior to the emergence of the full syndrome.

Disturbance in Social Cognition

Emerging evidence indicates that BP subjects have impairments in emotional processing/regulation and functions related to theory of mind.[52] These abnormalities are generally exacerbated by manic or depressive states, but appear present even during euthymic periods. Social cognition constructs have been examined using functional imaging methods and taken together indicate that BP patients show overactivation of subcortical limbic structures indicative of emotional hyperarousal and underactivation of prefrontal cortical regions that regulate emotional arousal.[53] These findings suggest a combination of baseline hyperarousal to emotional stimuli and a general deficit in inhibitory modulation of the limbic system during the processing and regulation of emotional/social cognition.

Disturbance of Biological Rhythms

Cyclic alterations in biological functions are striking features of manic and depressive episodes,[54] and recent research has shifted from viewing circadian changes like sleep disturbance as epiphenomena, towards recognizing these changes as part of the multifactorial etiological mechanisms underlying the disorder. BP patients experience disturbance of sleep, appetite, energy, sexual activity, and diurnal secretion of hormones, typically during the prodromal and acute phase of the illness. In fact, such disruptions, especially in the sleep/wake cycle, can initiate episodes of mania and hypomania in sensitive individuals, and sleep disturbances are often the initial symptom of mania. However, even euthymic patients can show interepisode abnormalities in their daily activity patterns, demonstrating more variability and less stability than controls;[55] these abnormalities have been associated with cognition, emotional regulation, substance use, and general physical health,[56] and may indicate an intrinsic instability in circadian rhythms and an increased vulnerability to external disturbances. A subgroup of BP patients meets clinical criteria for insomnia even between mood episodes.[57] This phenomenon may be related to other characteristics of euthymic BP patents, notably inability to maintain regular daily routines, increased cognitive arousal, and reduced daytime activity. Seasonal influences on symptoms seem to be important as well, with some individuals appearing to be particularly sensitive to environmental effects, such as exposure to sunlight. These individuals develop symptoms predominantly during certain seasons of the year, and bright-light treatment can reduce their symptoms of depression.[58]

Neuroimaging

A rich literature on structural and functional neuroimaging in BP suggests abnormalities in prefrontal cortical areas, striatum, and amygdala early during the course of the disease.[59] Taken as a whole, neuroimaging findings suggest that abnormalities during brain development lead to altered white matter connectivity and pruning, especially in limbic and prefrontal areas involved in cognitive and emotional control. Gray matter changes in BP patients as compared to healthy controls are not consistently observed across studies, but generally point to gray matter reductions in the prefrontal and possible temporal association cortices (Figure 104.3). In addition, changes in the volume of subcortical structures including, the amygdala, hippocampus, and basal ganglia have been reported, but are complicated by the possibility that treatment with lithium and other neurotrophic drugs may mask or even reverse regional changes in brain volume.[60] Neuroimaging changes identified in BP II subjects appear parallel to those noted in BP I.[61] A particularly appealing feature of brain image measures is that they tend to be highly heritable (Figure 104.3) and therefore are compelling intermediate phenotypes for genetic analysis.

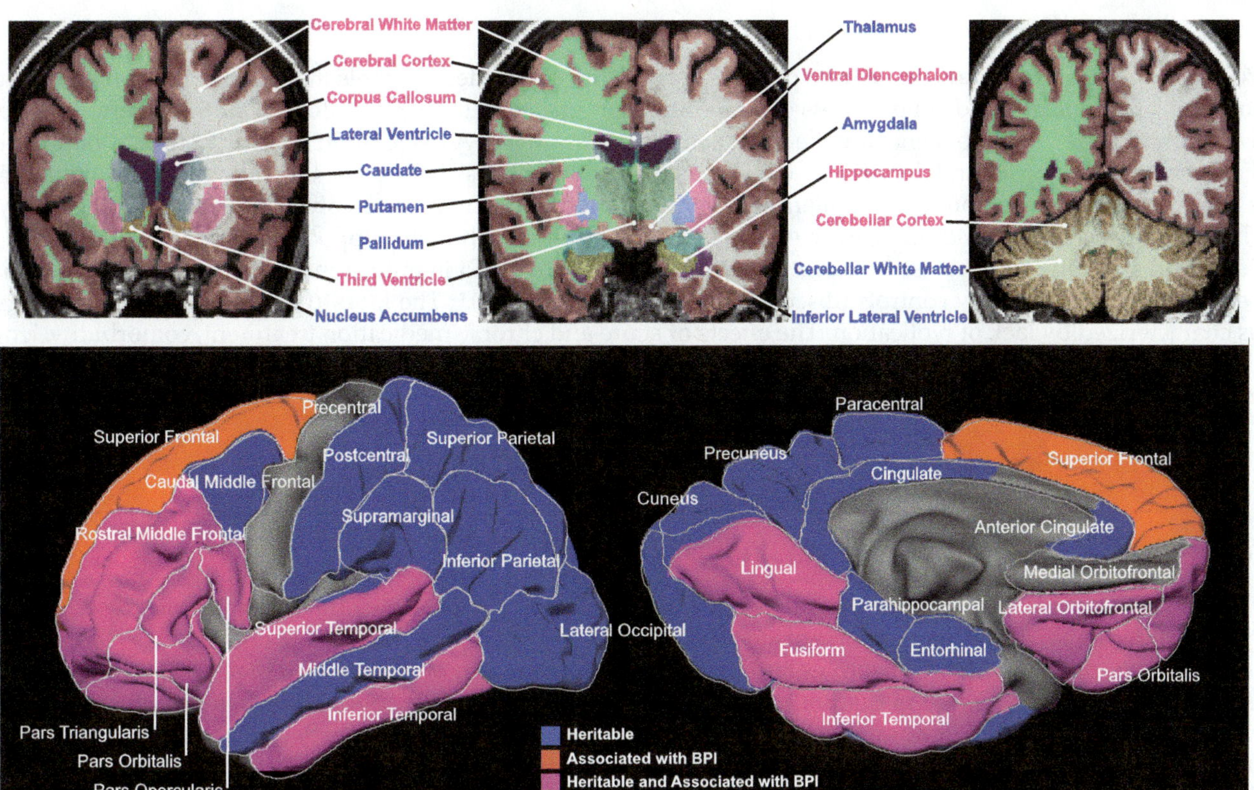

FIGURE 104.3 **Structural neuroimaging phenotypes in bipolar disorder (BP).** Results of a neuroimaging analysis in a large sample of BP individuals. This data was obtained from the same families discussed in Figure 104.2 and shows the results of heritability and BP-association analysis on brain measures generated from high-resolution brain images from 527 individuals.[47] The upper panel shows results of the heritability and BP I association analyses of volumetric MRI phenotypes. The three representative T1-weighted MRI coronal images depict the results of the Freesurfer segmentation overlaid as colored masks selected to better distinguish the anatomy. Mask colors are not related to the results. The colors of the text labels indicate structures that showed significant evidence of familial aggregation (blue) and structures that were both heritable and associated with BP I (magenta). The lower panel depicts cortical thickness phenotypes and shows the results of the heritability and BP I association analysis for cortical gray matter thickness. Heritable cortical regions are colored in blue, BP I-associated regions are shown in red and regions that were both heritable and associated with BP I are colored in magenta. The medial surface is rotated upwards by 60° to provide a view of the ventral surface. Reproduced with permission from[47]. © American Medical Association.

White matter abnormalities, suggestive of reduced integrity in longitudinal and interhemispheric white matter tracts, have been reported in patients with BP, compared to their unaffected relatives.[62] Together with early studies showing changes in patterns of resting-state functional connectivity,[63] these studies support the role of abnormal connectivity in the pathogenesis of BP. These changes increase with age and are often associated with an ischemic or degenerative process, but they can be found in children with BP as well.[64] Even though white matter hyperintensities are among the most replicated neuroimaging abnormalities in BP, they are found in only a minority of patients, and it is unclear how these findings relate to the cognitive or emotional abnormalities found in BP.

Numerous functional MRI studies have been performed during the past decade and have revealed complex abnormalities in BP.[59] Abnormalities in several major neural systems involved in emotional regulation have been identified; a subcortical system involved in emotion and reward processing in the amygdala and ventral striatum, a medial prefrontal cortical circuit related to implicit emotional regulation, and a lateral prefrontal cortical system required for cognitive control and "top-down" effortful control of emotion. At least some of the altered patterns of brain activity are seen in other psychiatric disorders including depression and anxiety; current work is focused on determining which abnormalities are core features common to multiple disorders and which findings are specific to BP and may be able to distinguish it from other mood and anxiety disorders.

Neuropathology (Postmortem Studies)

Postmortem studies on brain tissue from patients with BP have provided valuable insight into abnormalities of cell counts, receptor binding, and gene expression. Alterations in glia cell number and density in the anterior cingulate and prefrontal cortex of BP patients are the most consistent findings.[65] A decrease in specific neuronal populations and alterations in synapses and dendrites are found as well, although less frequently. Overall, these findings suggest that abnormal neurodevelopment and alterations in cellular plasticity may be involved in the pathophysiology of BP. Immunohistochemical and microarray gene-expression studies on human hippocampal tissue from BP and SZ patients find a significant decrease in expression levels of genes involved in mitochondrial energy metabolism in BP patients compared to healthy controls (discussed in more detail below).[66] The literature on postmortem findings in BP is characterized by lack of replication. The effects of acute and chronic medication treatment, comorbidity, length of the postmortem interval, premortem processes, such as death involving anoxia, and the generally small sample sizes of the studies are variables that contribute to this problem.

MOLECULAR GENETICS

Neurotransmitter Receptors and Intracellular Signaling Pathways

The monoamine system has been the subject of extensive investigations since the serendipitous discovery that tricyclic antidepressants and monoamine-oxidase inhibitors are effective treatments for depression.[67] Monoamines function as neurotransmitters in a complex and highly integrated system of neurons. The cell bodies of presynaptic monoamine neurons are located in midbrain structures, particularly in the limbic system, which is involved in the regulation of sleep, appetite, arousal, sexual functions, and emotional stages, such as fear and rage. The cell dendrites, however, are widely distributed over the entire brain and form a complex network involved in modulation and integration of stimuli and response.

Disturbances in neurotransmitter release and reuptake seem to explain some of the findings in BP. Norepinephrine and dopamine turnover appear to decrease in depression and increase in mania, while serotonergic changes are more variable. Drugs with antidepressive effects in patients inhibit the reuptake of norepinephrine and serotonin, whereas agents that significantly increase catecholamine release can precipitate mania.[68] Limitations of these models are that the effect on reuptake is an immediate one, whereas the antidepressant effect in patients usually occurs only after chronic administration. Another conundrum with the monoamine hypotheses is that up to 50% of patients have an inadequate response to antidepressants, which has led to the search for alternative etiologic mechanisms including other transmitter systems and downstream processes related to intracellular signaling.

Non-amine neurotransmitters may also play a significant role in the pathogenesis of BP.[69] Glutamate, for example, is involved in neuronal plasticity and is one of the major excitatory neurotransmitters in circuits that are presumed to be affected in recurrent mood disorder. The effect of glutamate is transmitted through postsynaptic receptors, including α-amino-3-hydroxy-5-methyl-4-isoxazolepropionic acid (AMPA), kainate, and *N*-methyl-D-aspartate (NMDA) receptors. These receptors belong to the class of ion channel receptors and their trafficking, expression, and internalization are regulated through phosphorylation at various receptor sites.[70] It has been suggested that excessive glutamate-induced excitation plays a major role in mood disorders,[69] a theory that has received partial support from the fact that antiglutamatergic agents like lamotrigine and riluzole have therapeutic properties. Additionally, lithium and valproate have been shown to regulate the expression and the phosphorylation of the AMPA receptor subunit glutamate receptor 1 (GluR1) after chronic treatment with therapeutically relevant concentrations, and this effect appears to be AMPA-receptor specific.[71] The involvement of AMPA receptor transmission in BP is further supported by the fact that receptor specific antagonists reduce manic-like symptoms induced through amphetamine administration. Molecules in this postsynaptic signaling cascade appear to be promising targets for therapeutic intervention.

The discovery of highly complex and integrated postsynaptic second messenger systems has shifted focus from neurotransmitters and their receptors to modulators of the intracellular response (Figure 104.4).[72] Three major signaling cascades, activated by cell surface receptors in the brain, have received considerable attention: the cyclic adenosine monophosphate (cAMP) generating pathway, the phosphoinositide (PI) pathway, and the Wnt signaling cascade (Figure 104.4).[73,74] Lithium and valproate can deplete the inositol second messenger pool through inhibition of the enzymatic breakdown of inositol phosphates to inositol. Lithium and valproate also appear to inhibit the enzyme glycogen synthase kinase-3β (GSK3β), an enzyme in the Wnt pathway involved in cell apoptosis. More generally, intracellular signaling cascades are linked to a variety of cellular processes implicated in the etiology of mood disorders including neurogenesis, synaptogenesis, and long-term potentiation.

Neuroinflammation

Immune regulation in the brain involves a complex interaction between the peripheral immune system, the blood–brain barrier and the specialized immune networks within the central nervous system. It has long been appreciated that autoimmune disorders like systemic lupus erythematosus, Hashimoto encephalopathy, and autoimmune synaptic encephalitis can cause BP-like syndromes.[75] The pathogenic mechanisms leading to neuropsychiatric symptoms in autoimmune disorders include the breakdown of the blood–brain barrier and production of neurocytotoxic antibodies.[76] More recently, there is growing evidence that more general features of immune dysregulation and inflammation are linked to pathological processes in BP.[77] Investigations of the peripheral immune system in BP patients have identified an imbalance of immunomodulatory mediators with upregulation of pro-inflammatory mediators, including TNF-α, C-reactive protein, and interleukin-6, and downregulation of anti-inflammatory mediators like interleukin-10.[76] In the brain, microglia are the first line of defense and interact with internal and peripheral signals to modulate immune responses. Postmortem studies in BP have found evidence of microglia dysregulation with elevated levels of inflammatory and excitotoxic markers. Additionally, BP patients show a downregulation of antioxidant genes and an up regulation of proapoptotic genes inducing vulnerability to cell damage and suppression of neurogenesis. In combination, immune dysregulation may create conditions for a vicious cycle between cytokines, oxidative stress, mitochondrial dysfunction, and neural damage.[77] The impact of these dysregulated processes impact cognitive and emotional function at multiple biological levels. For example, at the cellular level, cytokines can alter synaptic behavior by upregulating serotonin transporter activity.[78] At the level of neural circuits, neuroinflammation may be linked to white matter abnormalities that are found even early in BP and may contribute to the hypothesized disruption of cortical connections involved in emotional regulation.[79]

Channelopathies

Channelopathies refer to disorders resulting from defective ion channel functioning and include neurological, cardiovascular, and muscle disorders.[80] Ion channels are complex macromolecular structures that span cellular membranes and regulate the flow of ion current. The identification of altered calcium concentrations in cerebrospinal fluid of bipolar patients suggested that ion imbalances might be involved in the pathophysiology of depression and mania.[81] The finding of elevated calcium levels in platelets and lymphoblasts from BP patients added additional support to this hypothesis.[82] Most recently, the identification of confirmed BP risk loci in the *CACNA1C* and *ANK3* gene regions in genome-wide association studies has highlighted their potential role in neuropsychiatric disorders. *CACNA1C* encodes the alpha 1C subunit of the L-type voltage-gated calcium channel and is involved in regulation of gene expression, neurotransmitter release, and neuronal plasticity.[83] The *ANK3* gene encodes a protein that is required to direct the localization of potassium channels to the axons and nodes of Ranvier.[84] Ion channels provide a tractable target for treatment and in fact, numerous investigations have suggested that the therapeutic action of lithium may be related to ion flow modulation.[85,86] Additionally, some preliminary evidence suggests that calcium channel antagonists, especially the dihydropyridine compounds, have mood stabilizing and antidepressant effects.[85]

Mitochondrial Dysfunction

Mitochondrial dysfunction results in impaired energy metabolism and can be particularly damaging to energy-dependent organs like the brain. Psychiatric symptoms including mood, anxiety, psychosis and mania can be the presenting feature for primary mitochondrial disorders caused by gene mutations in nuclear or mitochondrial genes required for normal mitochondrial functioning.[87] In these cases, psychiatric symptoms often precede other physical symptoms by up to a decade, adding support to the fact that the brain is particularly sensitive to energy imbalances. Evidence from several lines of investigation is converging on the possibility that mitochondrial dysfunction may be a more general feature of many BP cases, not only those linked to a known mitochondrial disorder. Immunohistochemical and gene-expression studies on postmortem hippocampal tissue from BP patients identified a decrease in expression levels of genes involved in mitochondrial energy metabolism in BP patients compared to healthy controls.[88] Another postmortem study found altered mitochondrial morphology in the prefrontal cortex of BP patients.[89] Investigations of the enzyme complexes involved in oxidative phosphorylation in brain and peripheral

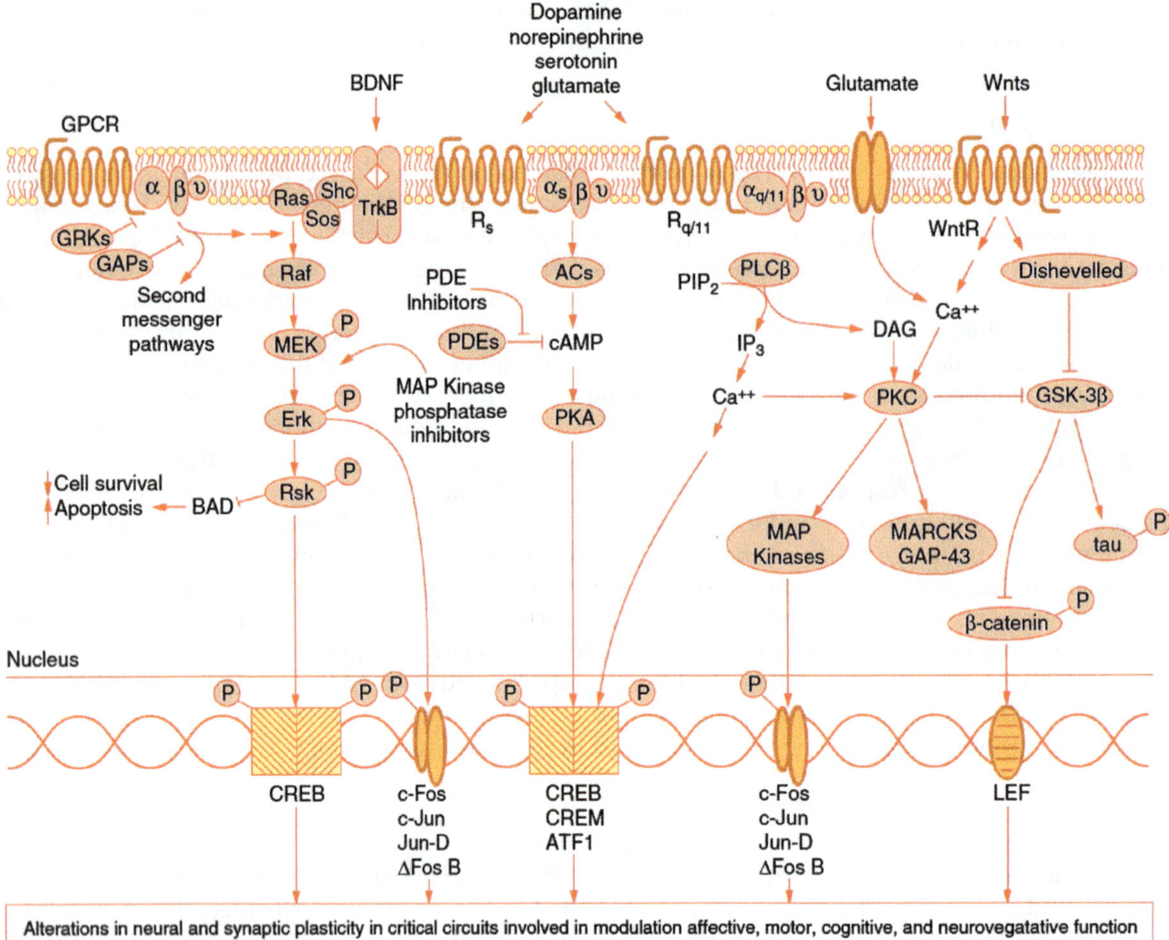

FIGURE 104.4 **Major intracellular signaling pathways in the brain.** The figure depicts some of the major intracellular signaling pathways involved in neural and behavioral plasticity; it has been proposed that these pathways may be perturbed in bipolar disorder. Cell-surface receptors transduce extracellular signals such as neurotransmitters and neuropeptides into the interior of the cell. Most neurotransmitters and neuropeptides communicate with other cells by activating seven transmembrane-spanning G-protein-coupled receptors (GPCRs). GPCRs activate selected G-proteins, which are composed of α and βγ subunits. Two families of proteins turn off the GPCR signal (and may therefore represent attractive targets for new medication development). G-protein-coupled receptor kinases (GRKs) phosphorylate GPCRs and thereby uncouple them from their respective G-proteins. GTPase-activating proteins (GAPs, also called RGS or regulators of G-protein signaling proteins) accelerate the G-protein turn-off reaction (an intrinsic GTPase activity). Two major signaling cascades activated by GPCRs are the cyclic adenosine monophosphate (cAMP)-generating second messenger system, and the phosphoinositide (PI) system. cAMP activates protein kinase A (PKA), a pathway that has been implicated in the therapeutic effects of antidepressants. Among the potential targets for the development of new antidepressants are certain phosphodiesterases (PDEs). PDEs catalyze the breakdown of cAMP; thus, PDE inhibitors would be expected to sustain the cAMP signal, and they may represent an antidepressant augmenting strategy. Activation of receptors coupled to PI hydrolysis results in the breakdown of phosphoinositide 4,5-biphosphate (PIP_2) into two second messengers, inositol 4,5-triphosphate (IP_3) and diacylglycerol (DAG). IP_3 mobilizes Ca^{2+} from intracellular stores, whereas DAG is an endogenous activator of protein kinase C (PKC), which is also directly activated by Ca^{2+}, PKC, PKA, and other Ca^{2+}-dependent kinases, directly or indirectly activating several important transcription factors, including CREB, CREM, ATF-1, c-Fos, c-Jun, Jun-D, and ΔFos-B. Endogenous growth factors, such as brain-derived neurotrophic factor (BDNF) use different types of signaling pathways. BDNF binds to and activates its tyrosine kinase receptor (TrkB); this facilitates the recruitment of other proteins (SHC, SOS), which results in the activation of the ERK-MAP kinase cascade (via sequential activation of Ras, Raf, MEK, Erk, and Rsk). In addition to regulating several transcription factors, the ERK-MAP kinase cascade, via Rsk, downregulates BAD, a proapoptotic protein. Enhancement of the ERK-MAP kinase cascade may have effects similar to those of endogenous neurotrophic factors; one potential strategy is to use inhibitors of MAP kinase phosphatases (which would inhibit the turn-off reaction) as potential drugs with neurotrophic properties. In addition to using GPCRs, many neurotransmitters (e.g., glutamate and γ-aminobutyric acid [GABA]) produce their responses via ligand-gated ion channels. Although these responses are very rapid, they also bring about more stable changes via regulation of gene transcription. One pathway gaining increasing recent attention in adult mammalian neurobiology is the Wnt signaling pathway. Wnts are a group of glycoproteins active in development, but they are now known to play important roles in the mature brain. Binding of Wnts to the Wnt receptor (WntR) activates an intermediary protein, dishevelled, which regulates a glycogen synthase kinase (GSK-3β). GSK-3β exerts many cellular effects; it regulates cytoskeletal proteins, including tau, and also plays an important role in determining cell survival/cell death decisions. GSK-3β has recently been identified as a target for the actions of lithium. GSK-3β also regulates phosphorylations of β-catenin, a protein that when dephosphorylated acts as a transcription factor at LEF (lymphoid enhancer factor) sites. Abbreviations: CREB, cAMP response element binding protein; R_q and R_s, extracellular GPCRs coupled to stimulation or inhibition of adenylyl cyclases (ACs), respectively; $R_{q/11}$, GPCR coupled to activation of phospholipase C (PLC); MARCKS, myristoylated alanine-rich C kinase substrate, a protein associated with several neuroplastic events; MAP, mitogen-activated protein; SHC, (Sre hormone 2 domain containing) transforming protein; SOS, Son of sevenless (*Drosophila melanogaster*). Reprinted from Einat H, Manji HK. Cellular plasticity cascades: genes-to-behavior pathways in animal models of bipolar disorder. *Biol Psychiatr*. 2006;59:1160–1171, with permission.

tissue has shown lower enzyme activity in BP patients relative to controls.[87] Additional support for mitochondrial dysfunction comes from magnetic resonance spectroscopy (MRS), a neuroimaging method that can noninvasively assay *in vivo* brain chemistry.[90] These studies have found reduced *N*-acetyl aspartate (NAA) in BP patients compared to controls. NAA is synthesized in the mitochondria and may serve as an energy substrate during periods of high demand,[91] and lithium treatment appears to reverse the NAA findings. MRS imaging has also found decreased intracellular pH and increased concentrations of lactate in the neurons of BP subjects, suggesting that there is a shift away from energy production via oxidative phosphorylation towards glycolysis, a lower efficiency process. Except in cases of mutations known to cause mitochondrial disorders, it is unclear whether mitochondrial dysfunction observed in most BP patients is a cause or consequence of the disorder. On the one hand, factors not directly related to energy metabolism like changes in calcium concentrations may lead to secondary effects on mitochondrial functioning; on the other hand, it has been hypothesized that energy dysregulation is primary and compensatory mechanisms, including the overactivation of monoaminergic systems, a process that has been demonstrated in rodent hippocampal neurons,[92] then leads to the emergence of mood symptoms.

ANIMAL MODELS

Animal models of mania and depression offer the opportunity to model certain aspects of symptoms and to evaluate treatment effects.[93] Depressive episodes have been mimicked using the primate separation model and models of learned helplessness.[72] Other frequently used models for symptoms of depression in rodents include the forced swim test and the tail suspension test, both of which are sensitive to amelioration by antidepressant treatment. Animal models of mania focus principally on features of hyperactivity and irritability that can be precipitated by repeated administration of dopamine agonists, such as amphetamines, and reversed by treatment with lithium, neuroleptics, or cholinergic agents. Although amphetamine-induced behavior provides a clear target for experimental assays, these models are limited because of the ambiguous link between BP and stimulant-induced cellular and molecular changes. More recently, genetic engineering methods have been used to manipulate specific genes hypothesized to play a role in BP pathophysiology. Based on the knowledge that lithium targets GSK3β (see above), a transgenic mouse overexpressing the protein was generated and showed BP-like behavioral changes including hypophagia, hyperactivity, and reduced anxiety-like behavior.[94] In another model, mice with loss-of-function mutations in the *Clock* gene display BP-like behavior including hyperactivity, decreased sleep, lowered depression-like behavior, and increased reward-seeking behavior.[95] Interestingly, these behavioral differences were attenuated with lithium administration. Given the possible role of circadian rhythm disturbances in BP, these findings are particularly intriguing. However, there is little evidence for the direct involvement of *GSK3*β or circadian rhythm gene variants in BP cases; therefore, the direct relevance of these genetically engineered mice to understanding the pathophysiology of BP remains unclear.

MANAGEMENT

Treatment of BP is symptomatic, primarily directed at mood stabilization, and involves different categories of psychopharmacologic agents, depending on the prevailing symptoms and the phase of the illness.[96] In the acute manic phase, standard treatment includes lithium carbonate or an anticonvulsant agent such as valproate or lamotrigine, in combination with an atypical antipsychotic agent. Other anticonvulsants, such as oxcarbazepine and zonisamide also may have some efficacy in the treatment of BP, although there are still few well-controlled trials of these medications. Gabapentin and tiagabine have been shown to be without specific benefit. Atypical antipsychotic drugs possess specific antimanic effects and possibly acute antidepressant efficacy as well. The US Food and Drug Administration has approved several atypical antipsychotic agents for the treatment of acute mania and quetiapine for the treatment of bipolar depression. In treatment of refractory patients or when there is concern about the safety of standard treatments, electroconvulsive therapy (ECT) can be indicated. In addition, specific psychoeducational interventions, such as regulating lifestyle, are often used as adjunct treatments.

An important element in the long-term treatment of BP is interepisode prophylaxis. Lithium, carbamazepine, valproate, and lamotrigine have been shown to decrease risk of recurrence of a manic episode, while lithium and lamotrigine have the added benefit of prophylaxis against depressive episodes as well. Lithium appears to specifically decrease suicide risk and lamotrigine may also be advantageous in patients with rapid mood cycling.[96]

The use of antidepressants remains controversial in the treatment of BP. The STEP-BD collaborative trial found that antidepressants added to mood stabilizers were no more effective than placebo for bipolar depression.[97] Monotherapy with tricyclic or serotonin targeting antidepressants is not recommended in BP, as they may accelerate cycle frequency and precipitate progression to mania. However, in combination with olanzapine, fluoxetine may be more effective for bipolar depression compared to placebo or olanzapine alone.[98] Some studies have reported that individuals with BP may be more likely to respond beneficially to a monoamine oxidase inhibitor (MAOI), but concurrent treatment with a mood stabilizer is again recommended.

Investigations into pharmacogenetic predictors of pharmacologic response have thus far been conflicting, but lithium response has been shown to be a heritable trait and both genome-wide linkage scans and GWAS have identified nominally significant loci in *GRIA2*, *SYN2*, and *ACCN1* genes.[99]

CONCLUSION

The investigation of BP has entered an exciting phase. Like other complex disorders, BP appears to be a heterogeneous disorder caused by a mix of genetic and environmental factors. Advances in approaches to the genetic investigation of complex disorders have lead to the identification of a number of risk loci for BP. The findings so far indicate that the long-held hypotheses regarding dysfunction in monoamine systems are unlikely to explain much of the etiologic mechanisms underlying BP. Instead, alterations in ion channels, neuroplasticity, inflammation, and energy metabolism appear to associate in a complex network of interactions that impacts the biology of individual neurons. At the level of neural circuits, these pathophysiologic mechanisms lead to impairments in cognition and emotional reactivity and regulation. Although current treatments are still based on largely serendipitous findings that are decades old, recent findings regarding the molecular and cellular biology of the disorder suggest new pathways for targeted intervention.

References

1. American Psychiatric Association. *DSM-IV-TR: Diagnostic and Statistical Manual of Mental Disorders*. 4th ed. Text Revision Washington: American Psychiatric Association; 2004.
2. Morgan VA, Mitchell PB, Jablensky AV. The epidemiology of bipolar disorder: sociodemographic, disability and service utilization data from the Australian National Study of Low Prevalence (Psychotic) Disorders. *Bipolar Disord*. 2005;7:326–337.
3. Hirschfeld RM, Lewis L, Vornik LA. Perceptions and impact of bipolar disorder: how far have we really come? Results of the national depressive and manic-depressive association 2000 survey of individuals with bipolar disorder. *J Clin Psychiatry*. 2003;64:161–174.
4. Marneros A. Origin and development of concepts of bipolar mixed states. *J Affect Disord*. 2001;67:229–240.
5. Merikangas KR, Akiskal HS, Angst J, et al. Lifetime and 12-month prevalence of bipolar spectrum disorder in the National Comorbidity Survey replication. *Arch Gen Psychiatry*. 2007;64:543–552.
6. Sherazi R, McKeon P, McDonough M, Daly I, Kennedy N. What's new? The clinical epidemiology of bipolar I disorder. *Harv Rev Psychiatry*. 2006;14:273–284.
7. Costello EJ, Pine DS, Hammen C, et al. Development and natural history of mood disorders. *Biol Psychiatry*. 2002;52:529–542.
8. Faust DS, Walker D, Sands M. Diagnosis and management of childhood bipolar disorder in the primary care setting. *Clin Pediatr (Phila)*. 2006;45:801–808.
9. Berk M, Berk L, Dodd S, et al. Stage managing bipolar disorder. *Bipolar Disord*. 2013;16(5):471–7.
10. Murray CJL, Lopez AD, eds. *The global burden of disease: a comprehensive assessment of mortality and disability from diseases, injuries and risk factors in 1990 and projected to 2020*. Cambridge, Harvard School of Public Health on behalf of the World Health Organization and the World Bank, 1996.
11. Kessler RC, Akiskal HS, Ames M, et al. Prevalence and effects of mood disorders on work performance in a nationally representative sample of U.S. workers. *Am J Psychiatry*. 2006;163:1561–1568.
12. Conus P, Macneil C, McGorry PD. Public health significance of bipolar disorder: implications for early intervention and prevention. *Bipolar Disord*. 2013.
13. McElroy SL, Keck PE. Metabolic syndrome in bipolar disorder: a review with a focus on bipolar depression. *J Clin Psychiatry*. 2014;75:46–61.
14. Karch DL, Barker L, Strine TW. Race/ethnicity, substance abuse, and mental illness among suicide victims in 13 US states: 2004 data from the National Violent Death Reporting System. *Inj Prev*. 2006;12(suppl 2):ii22–ii27.
15. Di Florio A, Craddock N, van den Bree M. Alcohol misuse in bipolar disorder. A systematic review and meta-analysis of comorbidity rates. *Eur Psychiatry*. 2014;29:117–124.
16. Pompili M, Gonda X, Serafini G, et al. Epidemiology of suicide in bipolar disorders: a systematic review of the literature. *Bipolar Disord*. 2013;15:457–490.
17. Phillips ML, Kupfer DJ. Bipolar disorder diagnosis: challenges and future directions. *Lancet*. 2013;381:1663–1671.
18. Baldessarini RJ, Tondo L, Baethge CJ, Lepri B, Bratti IM. Effects of treatment latency on response to maintenance treatment in manic-depressive disorders. *Bipolar Disord*. 2007;9:386–393.

19. Vieta E, Reinares M, Bourgeois M, Goodwin F, Marneros A, eds. *Bipolar I and Bipolar II: A Bichotomy?* Cambridge: Cambridge University Press: 88–108.
20. Laursen TM, Labouriau R, Licht RW, Bertelsen A, Munk-Olsen T, Mortensen PB. Family history of psychiatric illness as a risk factor for schizoaffective disorder: a Danish register-based cohort study. *Arch Gen Psychiatry*. 2005;62:841–848.
21. Goldstein BI, Young LT. Toward clinically applicable biomarkers in bipolar disorder: focus on BDNF, inflammatory markers, and endothelial function. *Curr Psychiatry Rep*. 2013;15:425.
22. Craddock N, Sklar P. Genetics of bipolar disorder. *Lancet*. 2013;381:1654–1662.
23. Craddock N, Sklar P. Genetics of bipolar disorder: successful start to a long journey. *Trends Genet*. 2009;25:99–105.
24. Purcell SM, Wray NR, Stone JL, et al. Common polygenic variation contributes to risk of schizophrenia and bipolar disorder. *Nature*. 2009;460:748–752.
25. Shih RA, Belmonte PL, Zandi PP. A review of the evidence from family, twin and adoption studies for a genetic contribution to adult psychiatric disorders. *Int Rev Psychiatry*. 2004;16:260–283.
26. Genetic Risk Outcome of Psychosis (GROUP) Consortium. Identification of risk loci with shared effects on five major psychiatric disorders: a genome-wide analysis. *Lancet*. 2013;381:1371–1379.
27. Lee SH, Ripke S, Neale BM, et al. Genetic relationship between five psychiatric disorders estimated from genome-wide SNPs. *Nat Genet*. 2013;45:984–994.
28. Ruderfer DM, Fanous AH, Ripke S, et al. Polygenic dissection of diagnosis and clinical dimensions of bipolar disorder and schizophrenia. *Mol Psychiatry*. 2013.
29. Egeland JA, Gerhard DS, Pauls DL, et al. Bipolar affective disorders linked to DNA markers on chromosome 11. *Nature*. 1987;325:783–787.
30. Kelsoe JR, Ginns EI, Egeland JA, et al. Re-evaluation of the linkage relationship between chromosome 11p loci and the gene for bipolar affective disorder in the Old Order Amish. *Nature*. 1989;342:238–243.
31. Levinson DF. Meta-analysis in psychiatric genetics. *Curr Psychiatry Rep*. 2005;7:143–151.
32. Segurado R, Detera-Wadleigh SD, Levinson DF, et al. Genome scan meta-analysis of schizophrenia and bipolar disorder, part III: bipolar disorder. *Am J Hum Genet*. 2003;73:49–62.
33. McQueen MB, Devlin B, Faraone SV, et al. Combined analysis from eleven linkage studies of bipolar disorder provides strong evidence of susceptibility loci on chromosomes 6q and 8q. *Am J Hum Genet*. 2005;77:582–595.
34. Badner JA, Koller D, Foroud T, et al. Genome-wide linkage analysis of 972 bipolar pedigrees using single-nucleotide polymorphisms. *Mol Psychiatry*. 2012;17:818–826.
35. Lohmueller KE, Pearce CL, Pike M, Lander ES, Hirschhorn JN. Meta-analysis of genetic association studies supports a contribution of common variants to susceptibility to common disease. *Nat Genet*. 2003;33:177–182.
36. Seifuddin F, Mahon PB, Judy J, et al. Meta-analysis of genetic association studies on bipolar disorder. *Am J Med Genet B Neuropsychiatr Genet*. 2012;159B:508–518.
37. Wellcome Trust Case Control Consortium. Genome-wide association study of 14,000 cases of seven common diseases and 3,000 shared controls. *Nature*. 2007;447:661–678.
38. Sklar P, Smoller JW, Fan J, et al. Whole-genome association study of bipolar disorder. *Mol Psychiatry*. 2008;13:558–569.
39. Ferreira MA, O'Donovan MC, Meng YA, et al., Wellcome Trust Case Control Consortium. Collaborative genome-wide association analysis supports a role for ANK3 and CACNA1C in bipolar disorder. *Nat Genet*. 2008;40:1056–1058.
40. Holmans P, Green EK, Pahwa JS, et al. Gene ontology analysis of GWA study data sets provides insights into the biology of bipolar disorder. *Am J Hum Genet*. 2009;85:13–24.
41. Hirschhorn JN, Daly MJ. Genome-wide association studies for common diseases and complex traits. *Nat Rev Genet*. 2005;6:95–108.
42. Gould TD, Gottesman II. Psychiatric endophenotypes and the development of valid animal models. *Genes Brain Behav*. 2006;5:113–119.
43. Freimer N, Sabatti C. The human phenome project. *Nat Genet*. 2003;34:15–21.
44. Freimer N, Sabatti C. The use of pedigree, sib-pair and association studies of common diseases for genetic mapping and epidemiology. *Nat Genet*. 2004;36:1045–1051.
45. Hasler G, Drevets WC, Gould TD, Gottesman II, Manji HK. Toward constructing an endophenotype strategy for bipolar disorders. *Biol Psychiatry*. 2006;60:93–105.
46. Greenwood TA, Badner JA, Byerley W, et al. Heritability and linkage analysis of personality in bipolar disorder. *J Affect Disord*. 2013.
47. Fears SC, Service SK, Kremeyer B, et al. Multisystem component phenotypes of bipolar disorder for genetic investigations of extended pedigrees. *JAMA Psychiatr*. 2014;71:375–387.
48. Arts B, Jabben N, Krabbendam L, van Os J. Meta-analyses of cognitive functioning in euthymic bipolar patients and their first-degree relatives. *Psychol Med*. 2008;38:771–785.
49. Bora E, Yucel M, Pantelis C. Cognitive endophenotypes of bipolar disorder: a meta-analysis of neuropsychological deficits in euthymic patients and their first-degree relatives. *J Affect Disord*. 2009;113:1–20.
50. Bora E, Yücel M, Pantelis C. Neurocognitive markers of psychosis in bipolar disorder: a meta-analytic study. *J Affect Disord*. 2010;127:1–9.
51. Samamé C, Martino DJ, Strejilevich SA. A quantitative review of neurocognition in euthymic late-life bipolar disorder. *Bipolar Disord*. 2013;15:633–644.
52. Samamé C. Social cognition throughout the three phases of bipolar disorder: a state-of-the-art overview. *Psychiatry Res*. 2013;210:1275–1286.
53. Cusi AM, Nazarov A, Holshausen K, Macqueen GM, McKinnon MC. Systematic review of the neural basis of social cognition in patients with mood disorders. *J Psychiatry Neurosci*. 2012;37:154–169.
54. Mansour HA, Monk TH, Nimgaonkar VL. Circadian genes and bipolar disorder. *Ann Med*. 2005;37:196–205.
55. Jones SH, Hare DJ, Evershed K. Actigraphic assessment of circadian activity and sleep patterns in bipolar disorder. *Bipolar Disord*. 2005;7:176–186.
56. Harvey AG. Sleep and circadian functioning: critical mechanisms in the mood disorders? *Annu Rev Clin Psychol*. 2011;7:297–319.
57. Harvey AG, Schmidt DA, Scarnà A, Semler CN, Goodwin GM. Sleep-related functioning in euthymic patients with bipolar disorder, patients with insomnia, and subjects without sleep problems. *Am J Psychiatry*. 2005;162:50–57.

58. Hakkarainen R, Johansson C, Kieseppä T, et al. Seasonal changes, sleep length and circadian preference among twins with bipolar disorder. *BMC Psychiatry*. 2003;3:6.
59. Strakowski SM, Adler CM, Almeida J, et al. The functional neuroanatomy of bipolar disorder: a consensus model. *Bipolar Disord*. 2012;14:313–325.
60. Hafeman DM, Chang KD, Garrett AS, Sanders EM, Phillips ML. Effects of medication on neuroimaging findings in bipolar disorder: an updated review. *Bipolar Disord*. 2012;14:375–410.
61. Elvsåshagen T, Westlye LT, Bøen E, et al. Bipolar II disorder is associated with thinning of prefrontal and temporal cortices involved in affect regulation. *Bipolar Disord*. 2013;15:855–864.
62. Skudlarski P, Schretlen DJ, Thaker GK, et al. Diffusion tensor imaging white matter endophenotypes in patients with schizophrenia or psychotic bipolar disorder and their relatives. *Am J Psychiatry*. 2013;170:886–898.
63. Townsend J, Altshuler LL. Emotion processing and regulation in bipolar disorder: a review. *Bipolar Disord*. 2012;14:326–339.
64. Beyer JL, Young R, Kuchibhatla M, Krishnan KR. Hyperintense MRI lesions in bipolar disorder: a meta-analysis and review. *Int Rev Psychiatry*. 2009;21:394–409.
65. Gigante AD, Young LT, Yatham LN, et al. Morphometric post-mortem studies in bipolar disorder: possible association with oxidative stress and apoptosis. *Int J Neuropsychopharmacol*. 2011;14:1075–1089.
66. Anglin RE, Mazurek MF, Tarnopolsky MA, Rosebush PI. The mitochondrial genome and psychiatric illness. *Am J Med Genet B Neuropsychiatr Genet*. 2012;159B:749–759.
67. Preskorn SH, Drevets WC. Neuroscience basis of clinical depression: implications for future antidepressant drug development. *J Psychiatr Pract*. 2009;15:125–132.
68. Sanacora G. Reviewing medications for bipolar disorder: understanding the mechanisms of action. *J Clin Psychiatry*. 2009;70:e02.
69. Kugaya A, Sanacora G. Beyond monoamines: glutamatergic function in mood disorders. *CNS Spectr*. 2005;10:808–819.
70. Huganir RL, Nicoll RA. AMPARs and synaptic plasticity: the last 25 years. *Neuron*. 2013;80:704–717.
71. Du J, Gray NA, Falke CA, et al. Modulation of synaptic plasticity by antimanic agents: the role of AMPA glutamate receptor subunit 1 synaptic expression. *J Neurosci*. 2004;24:6578–6589.
72. Nestler EJ, Gould E, Manji H, et al. Preclinical models: status of basic research in depression. *Biol Psychiatry*. 2002;52:503–528.
73. Zarate CA, Singh J, Manji HK. Cellular plasticity cascades: targets for the development of novel therapeutics for bipolar disorder. *Biol Psychiatry*. 2006;59:1006–1020.
74. Chen G, Manji HK. The extracellular signal-regulated kinase pathway: an emerging promising target for mood stabilizers. *Curr Opin Psychiatry*. 2006;19:313–323.
75. Najjar S, Pearlman DM, Alper K, Najjar A, Devinsky O. Neuroinflammation and psychiatric illness. *J Neuroinflammation*. 2013;10:43.
76. Rege S, Hodgkinson SJ. Immune dysregulation and autoimmunity in bipolar disorder: Synthesis of the evidence and its clinical application. *Aust N Z J Psychiatry*. 2013;47:1136–1151.
77. Berk M, Kapczinski F, Andreazza AC, et al. Pathways underlying neuroprogression in bipolar disorder: focus on inflammation, oxidative stress and neurotrophic factors. *Neurosci Biobehav Rev*. 2011;35:804–817.
78. Zhu CB, Carneiro AM, Dostmann WR, Hewlett WA, Blakely RD. p38 MAPK activation elevates serotonin transport activity via a trafficking-independent, protein phosphatase 2A-dependent process. *J Biol Chem*. 2005;280:15649–15658.
79. Fries GR, Pfaffenseller B, Stertz L, et al. Staging and neuroprogression in bipolar disorder. *Curr Psychiatry Rep*. 2012;14:667–675.
80. Felix R. Calcium channelopathies. *Neuromolecular Med*. 2006;8:307–318.
81. Carman JS, Wyatt RJ. Calcium: pacesetting the periodic psychoses. *Am J Psychiatry*. 1979;136:1035–1039.
82. Emamghoreishi M, Schlichter L, Li PP, et al. High intracellular calcium concentrations in transformed lymphoblasts from subjects with bipolar I disorder. *Am J Psychiatry*. 1997;154:976–982.
83. Striessnig J, Bolz HJ, Koschak A. Channelopathies in Cav1.1, Cav1.3, and Cav1.4 voltage-gated L-type Ca2+ channels. *Pflugers Arch*. 2010;460:361–374.
84. Pan Z, Kao T, Horvath Z, et al. A common ankyrin-G-based mechanism retains KCNQ and NaV channels at electrically active domains of the axon. *J Neurosci*. 2006;26:2599–2613.
85. Casamassima F, Hay AC, Benedetti A, Lattanzi L, Cassano GB, Perlis RH. L-type calcium channels and psychiatric disorders: a brief review. *Am J Med Genet B Neuropsychiatr Genet*. 2010;153B:1373–1390.
86. Judy JT, Zandi PP. A review of potassium channels in bipolar disorder. *Front Genet*. 2013;4:105.
87. Anglin RE, Tarnopolsky MA, Mazurek MF, Rosebush PI. The psychiatric presentation of mitochondrial disorders in adults. *J Neuropsychiatry Clin Neurosci*. 2012;24:394–409.
88. Konradi C, Eaton M, MacDonald ML, Walsh J, Benes FM, Heckers S. Molecular evidence for mitochondrial dysfunction in bipolar disorder. *Arch Gen Psychiatry*. 2004;61:300–308.
89. Cataldo AM, McPhie DL, Lange NT, et al. Abnormalities in mitochondrial structure in cells from patients with bipolar disorder. *Am J Pathol*. 2010;177:575–585.
90. Stork C, Renshaw PF. Mitochondrial dysfunction in bipolar disorder: evidence from magnetic resonance spectroscopy research. *Mol Psychiatry*. 2005;10:900–919.
91. Madhavarao CN, Chinopoulos C, Chandrasekaran K, Namboodiri MA. Characterization of the N-acetylaspartate biosynthetic enzyme from rat brain. *J Neurochem*. 2003;86:824–835.
92. Smith GA, Brett CL, Church J. Effects of noradrenaline on intracellular pH in acutely dissociated adult rat hippocampal CA1 neurones. *J Physiol*. 1998;512(Pt 2):487–505.
93. Nestler EJ, Hyman SE. Animal models of neuropsychiatric disorders. *Nat Neurosci*. 2010;13:1161–1169.
94. Prickaerts J, Moechars D, Cryns K, et al. Transgenic mice overexpressing glycogen synthase kinase 3beta: a putative model of hyperactivity and mania. *J Neurosci*. 2006;26:9022–9029.
95. Roybal K, Theobold D, Graham A, et al. Mania-like behavior induced by disruption of CLOCK. *Proc Natl Acad Sci U S A*. 2007;104:6406–6411.
96. Geddes JR, Miklowitz DJ. Treatment of bipolar disorder. *Lancet*. 2013;381:1672–1682.

97. Peters AT, Nierenberg AA. Stepping back to step forward: lessons from the Systematic Treatment Enhancement Program for Bipolar Disorder (STEP-BD). *J Clin Psychiatry*. 2011;72:1429–1431.
98. Benazzi F, Berk M, Frye MA, Wang W, Barraco A, Tohen M. Olanzapine/fluoxetine combination for the treatment of mixed depression in bipolar I disorder: a post hoc analysis. *J Clin Psychiatry*. 2009;70:1424–1431.
99. Severino G, Squassina A, Costa M, et al. Pharmacogenomics of bipolar disorder. *Pharmacogenomics*. 2013;14:655–674.
100. Psychiatric GWAS Consortium Bipolar Disorder Working Group. Large-scale genome-wide association analysis of bipolar disorder identifies a new susceptibility locus near ODZ4. *Nat Genet*. 2011;43:977–983.
101. Cichon S, Mühleisen TW, Degenhardt FA, et al. Genome-wide association study identifies genetic variation in neurocan as a susceptibility factor for bipolar disorder. *Am J Hum Genet*. 2011;88:372–381.
102. Green EK, Hamshere M, Forty L, et al. Replication of bipolar disorder susceptibility alleles and identification of two novel genome-wide significant associations in a new bipolar disorder case–control sample. *Mol Psychiatry*. 2013;18:1302–1307.
103. Green EK, Grozeva D, Forty L, et al. Association at SYNE1 in both bipolar disorder and recurrent major depression. *Mol Psychiatry*. 2013;18:614–617.
104. O'Donovan MC, Craddock N, Norton N, et al., Molecular Genetics of Schizophrenia Collaboration. Identification of loci associated with schizophrenia by genome-wide association and follow-up. *Nat Genet*. 2008;40:1053–1055.
105. Schizophrenia Psychiatric Genome-Wide Association Study (GWAS) Consortium. Genome-wide association study identifies five new schizophrenia loci. *Nat Genet*. 2011;43:969–976.
106. Steinberg S, de Jong S, Mattheisen M, et al. Common variant at 16p11.2 conferring risk of psychosis. *Mol Psychiatry*. 2012.
107. McMahon FJ, Akula N, Schulze TG, et al., Bipolar Disorder Genome Study (BiGS) Consortium. Meta-analysis of genome-wide association data identifies a risk locus for major mood disorders on 3p21.1. *Nat Genet*. 2010;42:128–131.

CHAPTER

105

Schizophrenia

David W. Volk and David A. Lewis

University of Pittsburgh, Pittsburgh, PA, USA

INTRODUCTION

Schizophrenia is a chronic, severe psychiatric illness characterized by: 1) positive (or psychotic) symptoms, such as delusions and hallucinations; 2) disorganized speech and behavior; 3) negative symptoms such as social withdrawal, affective flattening, and amotivation; and 4) cognitive dysfunction. The cognitive symptoms, which include impairments in executive function and working memory, are the most disabling aspects of the disorder and contribute to a lifetime of distress.[1] The clinical features of the disorder frequently become apparent in the late second to third decade of life, with the average age of onset generally about 5 years earlier in males than females. The wide range of signs and symptoms that can be present in schizophrenia are indicative of disturbances in cognitive, emotional, perceptual and motor processes.[1] This clinical heterogeneity, in concert with differences across individuals in the presence and magnitude of various brain abnormalities, supports the idea that what we recognize clinically as schizophrenia is likely to encompass a set of disorders that differ with respect to their underlying causes and pathophysiological mechanisms.

This complexity has made the identification of specific etiological factors in schizophrenia a great challenge. Available data indicate that genetic, environmental and developmental factors all contribute to the risk of developing schizophrenia, although the specific combinations of factors that may give rise to the illness remain elusive. Thus, as is the case for many human disease states, the clinical syndrome termed schizophrenia may represent the endpoint of many different pathogenetic paths.

CLINICAL FEATURES

Historical Overview and Disease Identification

The initial formal descriptions of schizophrenia over a century ago highlighted the substantial differences across affected individuals in specific constellations of clinical features, illness severity, and course. Emil Kraeplin (1856–1926) first used the term "dementia praecox" to identify an illness that affected cognitive functioning (dementia) and had a premature onset (praecox). Kraeplin noted that this illness generally had a chronic and progressive course, was associated with prominent psychotic symptoms such as hallucinations and delusions, and was distinguishable from the more episodic psychosis associated with manic depression. Later, Eugen Bleuler (1857–1939) noted that not all patients with the disorder had a deteriorating course and instead coined the term "schizophrenia" to highlight the observed splitting of cognition and affect seen in the patients with the disorder. Bleuler also identified the primary symptoms of schizophrenia to be loosening of associations, affective problems, ambivalence, and autism (the "four A's" of the disorder), and the psychotic symptoms were identified as secondary symptoms. Kurt Schneider (1887–1967) then provided additional clinically useful descriptions of "first-rank symptoms" of psychosis, such as auditory hallucinations, in which voices discuss the patient or provide a running commentary, thought insertion, thought withdrawal, thought broadcasting, and somatic passivity.

Rosenberg's Molecular and Genetic Basis of Neurological and Psychiatric Disease
http://dx.doi.org/10.1016/10.1016/B978-0-12-410529-4.00105-4

Modes of Inheritance and Prevalence

Genetic Liability

Several convergent lines of evidence support a critical role of inheritance in the etiology of schizophrenia. For example, family, twin, and adoption studies have demonstrated that the risk of developing schizophrenia in the relatives of affected individuals increases with the percentage of shared genes.[2] In contrast to the 1% incidence of schizophrenia in the general population, schizophrenia occurs in approximately 2% of the third-degree relatives (e.g., first cousins) of an individual with schizophrenia, in 2–6% of the second-degree relatives (e.g., nieces/nephews), and in 6–17% of the first-degree relatives (e.g., parents, siblings, or children).[2] In addition, if one member of a twin pair has the illness, the risk of schizophrenia in the other twin is approximately 17% for fraternal twins and approaches 50% for identical twins.[2] Other lines of evidence indicate that the familial nature of schizophrenia is not due to shared environmental factors. For example, when the biological children of individuals with schizophrenia are adopted away, their risk of developing schizophrenia is elevated, as expected for first-degree relatives, and is much higher than the general population rates of schizophrenia present in their adoptive families.[3,4] Furthermore, the offspring of identical twins discordant for schizophrenia have elevated rates of the disorder, regardless of whether the parent was affected or unaffected.[5]

Thus, the etiology of schizophrenia clearly involves genetic factors, with the heritability of the disorder estimated at 65–80%. However, about 60% of all persons with schizophrenia have neither a first- nor second-degree relative with the disorder.[6] In addition, given that the degree of concordance for schizophrenia among monozygotic twins is only about 50%,[2] genetic liability alone is not sufficient for the clinical appearance of the disorder. These observations suggest both that the genetic predisposition to schizophrenia is complex, and that additional, nongenetic factors are required for the inherited risk to become manifest.

Environmental Risk

Because inheritance alone cannot account for schizophrenia, substantial effort has been expended to identify the potential etiological role of a variety of environmental factors. The importance of these factors is demonstrated by the fact that in twin studies, the nonshared environment accounts for almost all of the liability for schizophrenia attributable to environmental effects.[7] A variety of adverse events, especially during early development, have been associated with an increase in the relative risk of schizophrenia. For example, a history of *in utero* events, such as maternal exposure to viral or parasitic infections and associated elevations in maternal cytokine levels, severe maternal malnutrition, and complications during labor or delivery, have been found with increased frequency in individuals with schizophrenia.[8] Environmental exposures that occur later during childhood and adolescence, such as cannabis use, growing up in an urban environment, trauma, and minority group position, have also been linked to higher rates of schizophrenia.[9,10] However, it is important to note that most individuals with these exposures do not develop schizophrenia.

Developmental Processes

Based on these findings, most models of the etiology of schizophrenia propose additive and/or interactive effects between multiple susceptibility genes and environmental factors. Furthermore, the temporal delay between environmental events of possible etiological significance (e.g., *in utero* viral exposure) and the appearance of clinical illness during late adolescence has played a major role in the idea that schizophrenia may be a disorder of neural development. In addition, numerous observations of subtle disturbances in motor, cognitive, and social functions during childhood and adolescence of individuals who later manifest schizophrenia have been viewed as evidence of disturbances in neurodevelopmental processes.

MOLECULAR GENETICS

The advent of large-scale genome-wide association study (GWAS) data sets of hundreds of thousands of single nucleotide polymorphisms (SNPs) collected in thousands of cases of schizophrenia and control subjects has allowed a more in-depth investigation into the high heritability of schizophrenia to proceed at the single nucleotide level. These studies have identified a mix of common, rare, and even *de novo* structural mutations that contribute to the genetic liability of schizophrenia. For example, recent GWAS studies have identified common (i.e., frequently present in the control population) SNPs that confer a relatively small increased risk (odds ratios ~1.1) for schizophrenia. These SNPs have been particularly commonly seen, for example, in the major histocompatibility complex region, microRNA 137, and the transcription factor ZNF804A, among other genes.[11–14] These GWAS studies have also provided estimates that 23–32% of the variance in liability for schizophrenia can be attributed to SNPs.[12,15,16]

Rare but important structural genomic variants, such as large (i.e., >100kb) deletions or duplications of DNA, have also been identified that convey a high degree (>5×) genotypic relative risk for the disorder.[17–19] However, these rare copy number variants (CNVs), such as deletions at chromosomes 22q11, 1q21, or 5q13, are only present in a small fraction (<1%) of individuals with schizophrenia and are also found in other disorders such as autism, intellectual disability, and epilepsy. More recent studies have also identified an overrepresentation of rare, but disruptive, single nucleotide variants that result in nonsense, frameshift, or splice site alterations in schizophrenia.[20] In addition, *de novo* CNVs (i.e., genetic variants that are not present in either parent of an individual with schizophrenia) have also been reported in 5–10% of individuals with schizophrenia, which is 5–8 times more frequently than *de novo* CNVs are seen in control subjects.[21,22] Furthermore, exome sequencing technology has also revealed an overrepresentation of small *de novo* mutations, those that involve only a few nucleotides, in schizophrenia. These *de novo* mutations, which frequently involve postsynaptic proteins, can often disrupt the function of the allele.[23] Importantly, the advent of whole-genome sequencing promises to expand the discovery of genetic variants relevant for schizophrenia in the future.

To summarize, many risk alleles for schizophrenia have been identified that range in frequency from rare to common, with corresponding genotypic relative risks that range from large to small; however, each structural variant, when considered individually, explains only a small proportion of the total variance in liability to schizophrenia.

DISEASE MECHANISMS

In terms of revealing the molecular basis of this illness, the existing literature is limited by the fact that any particular finding: 1) may be restricted to the specific subset of subjects studied; 2) may represent a secondary process that is associated with the illness, but that does not reflect the primary causal mechanisms; or 3) may, given the number of different molecules that have been studied, represent a false-positive observation. For these and other reasons, it is not possible at this time to present a comprehensive, definitive review of the molecular basis of schizophrenia. Thus, we have chosen to highlight studies illustrating different and complementary approaches that have yielded interesting data regarding altered molecular systems in schizophrenia.

Alterations in GABA Neurotransmission

Cognitive deficits are a particularly disabling feature of schizophrenia. Since cognitive impairments are present prior to the onset of psychotic symptoms[24] and are associated with long-term functional impairment in the illness,[25] cognitive deficits are considered to be a core feature of schizophrenia. Cognitive control, or the ability to regulate thoughts and behaviors to achieve goals, is thought to be subserved, at least in part, through gamma frequency (30–80Hz) oscillations that tightly coordinate neural activity in different brain regions, including the prefrontal cortex.[26] Individuals with schizophrenia show disturbances in cognitive control,[27] altered activation of the prefrontal cortex,[28] and reduced gamma oscillation power in the frontal cortex.[29,30] Thus, understanding the underlying disturbances in prefrontal cortex circuitry that normally supports gamma oscillations may assist in uncovering the pathophysiology of cognitive dysfunction in schizophrenia.

Cortical inhibitory circuitry plays a critical role in enabling synchronized firing of pyramidal neurons at the gamma frequency. For example, the subpopulation of cortical γ-aminobutyric acid (GABA) neurons that expresses the calcium-binding protein parvalbumin and which provides inhibitory input onto the cell bodies and proximal dendrites of pyramidal neurons (termed parvalbumin basket neurons) largely drives cortical gamma oscillations.[31,32] Interestingly, several lines of evidence indicate altered functioning of cortical parvalbumin basket neurons in schizophrenia. For example, approximately 50% of prefrontal parvalbumin neurons express undetectable mRNA levels for the GABA-synthesizing enzyme glutamate decarboxylase (GAD67) in the disorder.[33] Prefrontal parvalbumin basket neurons also have a large reduction in GAD67 protein in their axon terminals in schizophrenia.[34] Furthermore, parvalbumin neuron inhibitory inputs onto pyramidal neurons are largely mediated by $GABA_A$ receptors that contain the α1 subunit.[35,36] Lower mRNA levels for the $GABA_A$ receptor α1 subunit have been consistently reported in the prefrontal cortex in schizophrenia,[37–39] and the deficits in $GABA_A$ receptor α1 mRNA are primarily found in pyramidal neurons, which are the primary target of parvalbumin basket neurons.[40] Taken together, these data suggest the presence of deficient presynaptic (reduced GABA synthesis) and postsynaptic (reduced availability of $GABA_A$ receptors) GABA function of prefrontal parvalbumin basket neurons in schizophrenia. Consequently, these disturbances in prefrontal GABA neurotransmission may contribute to disturbances in gamma oscillations and, consequently, cognitive functioning in schizophrenia and may represent a pathophysiologically informed target for further therapeutic intervention.

Alterations in Dopamine Neurotransmission

The dopamine hypothesis, which posits that the psychotic features of schizophrenia result from excessive activity of the subcortical dopamine system, was initially based on two clinical observations. First, dopamine agonists, such as amphetamine, induce and exacerbate psychotic symptoms in healthy subjects and subjects with schizophrenia, respectively. Second, the potency of antipsychotic medications is proportional to the degree of dopamine D2 receptor antagonism (for review, see[41] and see Management, below). Direct evidence of elevated subcortical dopamine function was subsequently provided by imaging studies demonstrating that amphetamine induces a greater amount of dopamine release in the striatum in drug-free schizophrenia subjects than in control subjects.[42] A recent meta-analysis of a large number of positron emission tomography (PET) and single-photon emission computed tomography (SPECT) studies of striatal dopamine signaling further confirmed the presence of elevated presynaptic dopamine function, which includes an increased capacity for dopamine synthesis and release in schizophrenia.[43] Furthermore, this finding was not attributable to treatment with D2 receptor antagonists. However, while the primary mechanism of antipsychotic medication treatment for schizophrenia for decades has involved blockade of the D2 receptor (discussed under Management, below), evidence for altered D2 receptor function in the striatum is equivocal.[43] These data suggest that while reducing postsynaptic dopamine signaling through D2 receptor antagonism may be beneficial in the illness, a potentially more direct approach may involve addressing the upstream problem of excessive presynaptic dopamine activity in the striatum.

DIFFERENTIAL DIAGNOSIS

The differential diagnosis for an individual presenting with psychotic symptoms can include a wide variety of illnesses including substance-induced psychotic disorder (e.g., hallucinogens, stimulants, cannabis), psychotic disorder due to a general medication condition (e.g., head injury, encephalitis, encephalopathy), temporal lobe epilepsy, dementia, delirium, delusional disorder, paranoid or schizotypal personality disorders, and others. These disorders can generally be ruled out with additional laboratory testing (e.g., urine drug screen, brain imaging, electroencephalogram, blood cell count, blood chemistries; see Testing, below) and, most importantly, a thorough clinical history with collateral information provided by relatives and friends. However, the disorders that present with clinical features most closely resembling schizophrenia, and that can be difficult to distinguish from schizophrenia using only clinical history, include schizoaffective disorder and bipolar disorder with psychotic features. Schizoaffective disorder presents with all of the features of schizophrenia and also the symptoms of a mood disorder (e.g., a major depressive, manic, or mixed episode) that are present for a substantial duration of the illness. However, in schizoaffective disorder, psychotic symptoms are also present in the absence of the mood disorder symptoms.

Bipolar disorder, in which individuals have had at least one manic episode (e.g., a period of at least 1 week with elevated or irritable mood, decreased need for sleep, grandiosity, pressured speech, risky behavior), can also be difficult to distinguish from schizophrenia. For example, individuals with bipolar disorder can also present with psychotic features and cognitive impairments[44] that are highly similar to schizophrenia. Interestingly, genetics studies have also found similarities between schizophrenia and bipolar disorder. For example, large-scale GWAS data have found evidence of a high degree of coheritabilities of commonly found SNPs in schizophrenia and bipolar disorder.[45] However, rare CNVs (those present in less than 1% of the general population) appear to be more commonly found in schizophrenia, but not in bipolar disorder, relative to healthy subjects.[46,47] Furthermore, neuropathological studies have revealed some similarities in findings between the two disorders. For example, disturbances in the inhibitory circuitry of the prefrontal cortex seen in schizophrenia[33,34,48–55] (described earlier in Disease Mechanisms) have also been reported to varying degrees in the prefrontal cortex in bipolar disorder.[50,56,57] Taken together, these clinical, genetic, and pathological similarities may indicate a degree of shared etiopathogenesis in schizophrenia and bipolar disorder.

TESTING

Clinical history is primarily used to make the diagnosis of schizophrenia, and molecular diagnosis protocols are not yet available for schizophrenia. However, some laboratory testing is often conducted around the time of illness onset to rule out other disorders (described in Differential Diagnosis).

MANAGEMENT

Antipsychotic Treatment

The first line of treatment for schizophrenia is the use of antipsychotic medications, which are categorized as either typical or atypical. Typical antipsychotics (e.g., haloperidol, perphenazine, fluphenazine) primarily block the D2 subtype of dopamine receptors. Antagonism of D2 receptors in the mesolimbic dopamine pathway contributes to the beneficial antipsychotic efficacy of typicals. However, blockade of D2 receptors in the nigrostriatal dopamine pathway can result in the development of side effects including extrapyramidal symptoms (e.g., dystonia, akathisia, cogwheel rigidity, bradykinesia), while D2 receptor antagonism in the tuberoinfundibular dopamine pathway can raise serum prolactin levels, which may lead to side effects of abnormal lactation and irregular menstrual periods in females and gynecomastia in males. Long-term exposure to typical antipsychotics is associated with a 25–50% risk of developing tardive dyskinesia, which involves repetitive involuntary movements such as chewing, protruding the tongue, facial grimacing, and abnormal movements of the limbs.

In contrast, atypical antipsychotics (e.g., clozapine, risperidone, olanzapine, quetiapine) generally have a lower affinity for the D2 receptor than typicals and also block the 5HT2A subtype of serotonin receptors, although the overall receptor antagonism profile also differs across atypicals. Consequently, atypicals are less likely to produce extrapyramidal symptoms and tardive dyskinesia and might have greater efficacy for negative symptoms than typical antipsychotics. Clozapine has the highest antipsychotic efficacy relative to all other antipsychotic medications, but also carries a more substantial burden of side effects. For example, clozapine is associated with a 1% risk of agranulocytosis, a potentially life-threatening decrease in white blood cells that increases risk of infection and necessitates regular monitoring of white blood cell levels. In addition, clozapine blocks: 1) muscarinic acetylcholine receptors, which can produce constipation, dry mouth, blurred vision and drowsiness; 2) histamine receptors, which can produce sedation and weight gain; and 3) $\alpha 1$ norepinephrine receptors, which can produce orthostatic hypotension. In contrast, risperidone has a higher affinity for D2 receptors than other atypicals and thus has a higher risk for extrapyramidal symptoms and elevated prolactin levels. Meanwhile, the high affinity of olanzapine and quetiapine for histamine receptors leads to greater risk for sedation and weight gain. Antipsychotic medications, and in particular the atypicals, also increase the risk of developing the metabolic syndrome, which consists of obesity, elevated blood glucose and cholesterol levels, and high blood pressure, which requires regular monitoring as well. Because the effectiveness of many atypical antipsychotics (excluding clozapine) is generally similar to that of typical antipsychotics, the prescribing patterns of psychiatrists are generally more informed by matching the side effect profiles of antipsychotic medications with the clinical situation of individual patients.

Psychosocial Treatment Interventions

Multiple forms of psychosocial treatment interventions are employed in the treatment of schizophrenia. For example, social skills training in the areas of employment, activities of daily living, and relationships can help improve social functioning. Individual, as well as group, psychotherapy assists in enhancing medication compliance, self-esteem and coping strategies for life stressors. In addition, by focusing on reducing expressed emotion, such as criticism and hostility, in the family setting, family therapy can help lower the risk of relapse and rehospitalization. Optimal management of schizophrenia involves the integration of psychosocial interventions with antipsychotic medications and service coordination such as through an assertive community treatment (ACT) team. An ACT team typically consists of a psychiatrist, a therapist, a nurse and licensed social worker who together provide a comprehensive treatment approach and assist the patient in accessing community resources. The primary goal of the integrated treatment approach is to reduce the risk of relapse and rehospitalization, enhance medication adherence, and lead to a higher level of functioning.

References

1. Lewis DA, Lieberman JA. Catching up on schizophrenia: natural history and neurobiology. *Neuron*. 2000;28:325–334.
2. Gottesman II. *Schizophrenia Genesis: The Origins of Madness*. New York: W.H. Freeman; 1991.
3. Gottesman II, Shields J. *Schizophrenia: The Epigenetic Puzzle*. Cambridge, UK: Cambridge University Press; 1982.
4. Ingraham LJ, Kety SS. Adoption studies of schizophrenia. *Am J Med Genet*. 2000;97:18–22.
5. Gottesman II, Bertelsen A. Confirming unexpressed genotypes for schizophrenia. Risks in the offspring of Fischers's Danish identical and fraternal discordant twins. *Arch Gen Psychiatry*. 1989;46:867–872.
6. Gottesman II, Erlenmeyer-Kimling L. Family and twin strategies as a head start in defining prodromes and endophenotypes for hypothetical early-interventions in schizophrenia. *Schizophr Res*. 2001;51:93–102.

7. Tsuang MT, Stone WS, Faraone SV. Genes, environment and schizophrenia. *Br J Psychiatry*. 2001;40S:18–24.
8. Brown AS, Derkits EJ. Prenatal infection and schizophrenia: a review of epidemiologic and translational studies. *Am J Psychiatry*. 2010;167:261–280.
9. van Os J, Kenis G, Rutten BP. The environment and schizophrenia. *Nature*. 2010;468:203–212.
10. Casadio P, Fernandes C, Murray RM, Di Forti M. Cannabis use in young people: the risk for schizophrenia. *Neurosci Biobehav Rev*. 2011;35:1779–1787.
11. O'Donovan MC, Craddock N, Norton N, et al. Identification of loci associated with schizophrenia by genome-wide association and follow-up. *Nat Genet*. 2008;40:1053–1055.
12. Purcell SM, Wray NR, Stone JL, et al. Common polygenic variation contributes to risk of schizophrenia and bipolar disorder. *Nature*. 2009;460:748–752.
13. Stefansson H, Ophoff RA, Steinberg S, et al. Common variants conferring risk of schizophrenia. *Nature*. 2009;460:744–747.
14. Ripke S, Sanders AR, Kendler KS, et al. Genome-wide association study identifies five new schizophrenia loci. *Nat Genet*. 2011;43:969–976.
15. Lee SH, DeCandia TR, Ripke S, et al. Estimating the proportion of variation in susceptibility to schizophrenia captured by common SNPs. *Nat Genet*. 2012;44:247–250.
16. Ripke S, O'Dushlaine C, Chambert K, et al. Genome-wide association analysis identifies 13 new risk loci for schizophrenia. *Nat Genet*. 2013;45:1150–1159.
17. Stone JL, O'Donovan MC, Gurling H, et al. Rare chromosomal deletions and duplications increase risk of schizophrenia. *Nature*. 2008;455:237–241.
18. Levinson DF, Duan J, Oh S, et al. Copy number variants in schizophrenia: confirmation of five previous findings and new evidence for 3q29 microdeletions and VIPR2 duplications. *Am J Psychiatry*. 2011;168:302–316.
19. Sullivan PF, Daly MJ, O'Donovan M. Genetic architectures of psychiatric disorders: the emerging picture and its implications. *Nat Rev Genet*. 2012;13:537–551.
20. Purcell SM, Moran JL, Fromer M, et al. A polygenic burden of rare disruptive mutations in schizophrenia. *Nature*. 2014;506:185–190.
21. Xu B, Roos JL, Levy S, van Rensburg EJ, Gogos JA, Karayiorgou M. Strong association of de novo copy number mutations with sporadic schizophrenia. *Nat Genet*. 2008;40:880–885.
22. Malhotra D, McCarthy S, Michaelson JJ, et al. High frequencies of de novo CNVs in bipolar disorder and schizophrenia. *Neuron*. 2011;72:951–963.
23. Fromer M, Pocklington AJ, Kavanagh DH, et al. De novo mutations in schizophrenia implicate synaptic networks. *Nature*. 2014;506:179–184.
24. Reichenberg A, Caspi A, Harrington H, et al. Static and dynamic cognitive deficits in childhood preceding adult schizophrenia: a 30-year study. *Am J Psychiatry*. 2010;167:160–169.
25. Green MF. Cognitive impairment and functional outcome in schizophrenia and bipolar disorder. *J Clin Psychiatry*. 2006;67(suppl 9):3–8.
26. Howard MW, Rizzuto DS, Caplan JB, et al. Gamma oscillations correlate with working memory load in humans. *Cereb Cortex*. 2003;13:1369–1374.
27. Lesh TA, Niendam TA, Minzenberg MJ, Carter CS. Cognitive control deficits in schizophrenia: mechanisms and meaning. *Neuropsychopharmacology*. 2011;36:316–338.
28. Minzenberg MJ, Laird AR, Thelen S, Carter CS, Glahn DC. Meta-analysis of 41 functional neuroimaging studies of executive function in schizophrenia. *Arch Gen Psychiatry*. 2009;66:811–822.
29. Cho RY, Konecky RO, Carter CS. Impairments in frontal cortical gamma synchrony and cognitive control in schizophrenia. *Proc Natl Acad Sci U S A*. 2006;103:19878–19883.
30. Minzenberg MJ, Firl AJ, Yoon JH, Gomes GC, Reinking C, Carter CS. Gamma oscillatory power is impaired during cognitive control independent of medication status in first-episode schizophrenia. *Neuropsychopharmacology*. 2010;35:2590–2599.
31. Cardin JA, Carlen M, Meletis K, et al. Driving fast-spiking cells induces gamma rhythm and controls sensory responses. *Nature*. 2009;459:663–667.
32. Sohal VS, Zhang F, Yizhar O, Deisseroth K. Parvalbumin neurons and gamma rhythms enhance cortical circuit performance. *Nature*. 2009;459:698–702.
33. Hashimoto T, Volk DW, Eggan SM, et al. Gene expression deficits in a subclass of GABA neurons in the prefrontal cortex of subjects with schizophrenia. *J Neurosci*. 2003;23:6315–6326.
34. Curley AA, Arion D, Volk DW, et al. Cortical deficits of glutamic acid decarboxylase 67 expression in schizophrenia: clinical, protein, and cell type-specific features. *Am J Psychiatry*. 2011;168:921–929.
35. Nusser Z, Sieghart W, Benke D, Fritschy J-M, Somogyi P. Differential synaptic localization of two major γ-aminobutyric acid type A receptor α subunits on hippocampal pyramidal cells. *Proc Natl Acad Sci U S A*. 1996;93:11939–11944.
36. Doischer D, Hosp JA, Yanagawa Y, et al. Postnatal differentiation of basket cells from slow to fast signaling devices. *J Neurosci*. 2008;28:12956–12968.
37. Hashimoto T, Arion D, Unger T, et al. Alterations in GABA-related transcriptome in the dorsolateral prefrontal cortex of subjects with schizophrenia. *Mol Psychiatry*. 2008;13:147–161.
38. Beneyto M, Abbott A, Hashimoto T, Lewis DA. Lamina-specific alterations in cortical GABAA receptor subunit expression in schizophrenia. *Cereb Cortex*. 2011;21:999–1011.
39. Hoftman GD, Volk DW, Bazmi HH, Li S, Sampson AR, Lewis DA. Altered cortical expression of GABA-related genes in schizophrenia: Illness progression vs. developmental disturbance. *Schizophr Bull*. 2013, Epub ahead of print.
40. Glausier JR, Lewis DA. Selective pyramidal cell reduction of GABA(A) receptor alpha1 subunit messenger RNA expression in schizophrenia. *Neuropsychopharmacology*. 2011;36:2103–2110.
41. Seeman P. Targeting the dopamine D2 receptor in schizophrenia. *Expert Opin Ther Targets*. 2006;10:515–531.
42. Laruelle M, Abi-Dargham A, van Dyck CH, et al. Single photon emission computerized tomography imaging of amphetamine-induced release in drug-free schizophrenic subjects. *Proc Natl Acad Sci U S A*. 1997;93:9235–9240.
43. Howes OD, Kambeitz J, Kim E, et al. The nature of dopamine dysfunction in schizophrenia and what this means for treatment. *Arch Gen Psychiatry*. 2012;69:776–786.

44. Zanelli J, Reichenberg A, Morgan K, et al. Specific and generalized neuropsychological deficits: a comparison of patients with various first-episode psychosis presentations. *Am J Psychiatry*. 2010;167:78–85.
45. Lee SH, Ripke S, Neale BM, et al. Genetic relationship between five psychiatric disorders estimated from genome-wide SNPs. *Nat Genet*. 2013;45:984–994.
46. Grozeva D, Kirov G, Ivanov D, et al. Rare copy number variants: a point of rarity in genetic risk for bipolar disorder and schizophrenia. *Arch Gen Psychiatry*. 2010;67:318–327.
47. Grozeva D, Kirov G, Conrad DF, et al. Reduced burden of very large and rare CNVs in bipolar affective disorder. *Bipolar Disord*. 2013;15:893–898.
48. Akbarian S, Kim JJ, Potkin SG, et al. Gene expression for glutamic acid decarboxylase is reduced without loss of neurons in prefrontal cortex of schizophrenics. *Arch Gen Psychiatry*. 1995;52:258–266.
49. Volk DW, Austin MC, Pierri JN, Sampson AR, Lewis DA. Decreased glutamic acid decarboxylase67 messenger RNA expression in a subset of prefrontal cortical gamma-aminobutyric acid neurons in subjects with schizophrenia. *Arch Gen Psychiatry*. 2000;57:237–245.
50. Guidotti A, Auta J, Davis JM, et al. Decrease in reelin and glutamic acid decarboxylase67 (GAD67) expression in schizophrenia and bipolar disorder. *Arch Gen Psychiatry*. 2000;57:1061–1069.
51. Straub RE, Lipska BK, Egan MF, et al. Allelic variation in GAD1 (GAD67) is associated with schizophrenia and influences cortical function and gene expression. *Mol Psychiatry*. 2007;12:854–869.
52. Morris HM, Hashimoto T, Lewis DA. Alterations in somatostatin mRNA expression in the dorsolateral prefrontal cortex of subjects with schizophrenia or schizoaffective disorder. *Cereb Cortex*. 2008;18:1575–1587.
53. Fung SJ, Webster MJ, Sivagnanasundaram S, Duncan C, Elashoff M, Weickert CS. Expression of interneuron markers in the dorsolateral prefrontal cortex of the developing human and in schizophrenia. *Am J Psychiatry*. 2010;167:1479–1488.
54. Mellios N, Huang HS, Baker SP, Galdzicka M, Ginns E, Akbarian S. Molecular determinants of dysregulated GABAergic gene expression in the prefrontal cortex of subjects with schizophrenia. *Biol Psychiatry*. 2009;65:1006–1014.
55. Volk DW, Matsubara T, Li S, et al. Deficits in transcriptional regulators of cortical parvalbumin neurons in schizophrenia. *Am J Psychiatry*. 2012;169:1082–1091.
56. Woo TU, Kim AM, Viscidi E. Disease-specific alterations in glutamatergic neurotransmission on inhibitory interneurons in the prefrontal cortex in schizophrenia. *Brain Res*. 2008;1218:267–277.
57. Sibille E, Morris HM, Kota RS, Lewis DA. GABA-related transcripts in the dorsolateral prefrontal cortex in mood disorders. *Int J Neuropsychopharmacol*. 2011;14:721–734.

CHAPTER

106

Obsessive–Compulsive Disorder

Michael H. Bloch, Jessica B. Lennington, Gabor Szuhay, and Paul J. Lombroso
Yale University, New Haven, CT, USA

INTRODUCTION

Disease Characteristics, Hallmark Manifestations and Inheritance

Obsessive–compulsive disorder (OCD) is characterized by recurrent and intrusive thoughts or images (obsessions) that are often accompanied by intentional repetitive behaviors (compulsions).[1] Genetic studies among OCD subjects suggest that both genetic and environmental factors play a critical role in the etiology and expression of symptoms.

Diagnosis and Testing

A diagnosis of OCD is currently made solely based on a clinical evaluation using diagnostic criteria.[1] Rating scales such as the Yale–Brown Obsessive Compulsive Scale (Y–BOCS) for adults, or the Children's Yale–Brown Obsessive Compulsive Scale (CY–BOCS) for children are useful in measuring symptom severity and monitoring response to treatment.[2–4]

Current Research

Genetic studies in OCD patients implicate several neurotransmitter systems, including serotonin and glutamate signaling, yielding promising treatment targets.[5] Neuroimaging studies in humans have revealed alterations in OCD in several brain regions including the orbitofrontal cortex, anterior cingulate, and basal ganglia.

CLINICAL FEATURES

OCD is characterized by recurrent and intrusive thoughts or images (obsessions) that are often accompanied by intentional repetitive behaviors (compulsions).[1] The compulsions may be performed to relieve the anxiety associated with the obsession or to achieve a sense of completion.[6] In order to meet the current diagnostic criteria for OCD, the obsessions and/or compulsions must be time consuming and cause marked distress and impairment.[1]

In an effort to understand the apparent heterogeneity of OCD, several investigators used results from factor analysis to propose a dimensional approach to classifying OCD subtypes. These include: 1) forbidden thoughts (aggression, sexual and religious obsessions, and checking compulsions); 2) symmetry (symmetry obsessions and ordering, arranging and counting compulsions); 3) cleaning (contamination obsessions and cleaning compulsions); and 4) hoarding.[7]

Clinical investigations into the heterogeneity of OCD have led to the delineation of a new specifier in the Diagnostic and Statistical Manual of Mental Disorders, Fifth Edition (DSM-5) recognizing tic-related OCD as an important subtype of OCD patients. The term "tic-like" compulsion refers to touching, tapping, rubbing, blinking in patterns, and repeating routine activities such as opening and closing a door, setting down an object and picking it back up again and again. These tic-like compulsions are more common in OCD patients with a history of tics, and in some cases,

Rosenberg's Molecular and Genetic Basis of Neurological and Psychiatric Disease
http://dx.doi.org/10.1016/10.1016/B978-0-12-410529-4.00106-6

may be difficult to distinguish from complex tics. In other cases, however, it is clear that these repetitive habits are performed to relieve the distress caused by an obsession rather than prompted by a perceptional urge. OCD patients with comorbid tics have a worse response to selective serotonin re-uptake inhibitor (SSRI) pharmacotherapy (but a similar response to behavioral treatment) when compared to OCD patients without tics.[8,9] Additionally, OCD with comorbid tics have an improved response to antipsychotic augmentation.[10]

Mode of Inheritance and Prevalence

Considerable evidence exists that OCD contains a hereditable component. The strongest evidence supporting the hereditability of OCD comes from family studies and twins studies. Four family studies, each examining between 300–500 first-degree relatives of OCD patients have estimated the rate of OCD to be 5–10-fold greater in first-degree relatives of OCD patients compared to the relatives of psychiatrically healthy controls.[11,12] Twin studies in OCD have suggested an 87% concordance rate among monozygotic twins with the disorder compared to 47% for dizygotic twins.[13] Since monozygotic and dizygotic twins share similar environmental influences but different degrees of genetic relatedness, a higher rate of an illness among monozygotic twins is seen as strong evidence of a hereditary component to illness. The heritability of OCD was estimated at 0.41 in a recent large genome-wide association study involving 1061 cases and 4236 controls.[14] Childhood-onset OCD cases (onset before 16 years) demonstrated an increased heritability compared to adult-onset cases. In contrast to Tourette syndrome (TS), variants with minor allele frequency did not contribute to the overall heritability of OCD.[14]

Results from the Epidemiological Catchment Area study examining over 18,500 adults and using structured clinical interviews estimated the lifetime prevalence of OCD among adults to be between 1.9–3.3%.[15] Epidemiological studies of adolescents provide similar estimates in the range of 1.9–3.6%.[16,17] The sex distribution of these epidemiologic studies suggests that OCD affects males and females equally after puberty. However, males typically have an earlier age of onset than females with the first presentation occurring well before puberty.

Natural History

Age of Onset

Obsessive–compulsive disorder is generally considered to have a bimodal distribution in age of onset. One peak of incidence occurs around the peripubertal years and the other occurs in early adulthood. Across the life cycle, symptoms of OCD are similarly expressed.[18,19] However there are several important differences between pediatric and adult-onset OCD. Pediatric-onset OCD cases have a male predominance (unlike adult-onset OCD cases, which are female predominant), have a stronger family history of OCD and have higher rates of comorbid attention-deficit hyperactivity disorder (ADHD) and tic disorders. The specific content of obsessions and compulsions of OCD also differ across age ranges.[18] Children with OCD have much higher rates of aggression obsession (such as fear of a catastrophic event, fear of harm coming to self or others) and have OCD with poor insight (be unable to recognize their obsessions and compulsions as excessive and unreasonable). Adolescents with OCD have a higher proportion of obsessions with sexual and religious themes, whereas cleaning and contamination related symptoms were quite prevalent across age ranges. Patients with OCD and comorbid tics have a significantly higher rate of intrusive violent or aggressive thoughts and images, sexual and religious preoccupations, concerns with symmetry and exactness, hoarding and counting rituals, and touching and tapping compulsions compared to patients with nontic OCD, who often suffer primarily from contamination worries and cleaning compulsions. Compulsions designed to eliminate a perceptually tinged mental feeling of unease, coined in the literature as "Just Right" perceptions, are typically seen in patients with OCD and comorbid tics.[19]

Disease Evolution

Evidence suggests that a large portion of children with OCD achieve a remission of OCD symptoms with evidence-based treatment. Meta-analysis suggests that the long-term persistence of OCD into adulthood is less than 50%.[20,21] Adults with OCD are likely to achieve a significant improvement in their OCD symptoms with evidence-based treatments.[22] However, a large proportion of adults with OCD still experience a subclinical or clinically significant degree of OCD symptoms even after undergoing successful treatment.[23,24]

MOLECULAR GENETICS

Linkage analysis has been used to isolate genes responsible for various neurological disorders. Traditional linkage analyses are less informative when more than one gene is involved and have been less successful for identifying candidate genes for psychiatric disorders such as OCD. Two studies have used a genome-wide scan for OCD and reported a candidate region on 9p24 that met criteria for trending significance.[25,26] Based on the identified linkage peak, epithelial excitatory amino acid transporter (ECCA-1, also known as solute carrier family 1 or SLC1A1), a transporter that acts at the terminals of the excitatory syapses to regulate synaptic glutamate concentration, has been studied extensively in OCD.[27,28] SLC1A1 was found to be associated with the transmission of OCD in male offspring; however, there were no functional mutations in the families of seven OCD probands included in a genome scan.[29]

Studies examining potential genetic mechanisms have yet to identify unequivocally any genes that are involved in the phenotypic expression of OCD. A recent meta-analysis examining published genetic association studies in OCD reported a significant but small association between multiple different genes in the serotinergic system and a diagnosis of OCD. A serotonin receptor gene, *HTR2A* rs6311 and rs6313 single-nucleotide polymorphism (SNP) was weakly associated with an OCD diagnosis (OR = 1.2, ES = 0.1). Additionally the serotonin transporter gene, *5-HTTLPR*, when classified into trialleic variants (La vs. Lg or S), but not into biallelic variants, was also weakly associated with a diagnosis of OCD (OR = 1.3, ES = 0.1). OCD within males (but not females) was also associated with polymorphisms involved in catecholamine modulation (catechol-*O*-methyltransferase [COMT], overall OR = 1.2; males only OR = 1.5; and monoamine oxidase A [MAOA], overall OR = 1.3; males only OR = 2.9).[30]

A large genome-wide association study in OCD involving over 20 international sites and including 1465 cases, 5557 controls, and 400 trios failed to demonstrate a single SNP that reached genome wide significance. Top candidate genes from the study included *DLGAP1*, a member of the neuronal postsynaptic density complex, and *BTBD3*, a transcription factor.[31] Genetic studies of OCD have been hampered by the fact that OCD is a complex disorder with a large degree of heterogeneity in its clinical phenotypes and our current lack of understanding into the molecular mechanisms underlying the disorder. It is hoped that the development of quantitative endophenotypes using clinical, neuropsychological, imaging or electrophysiological measures, will advance the field.

DISEASE MECHANISMS: PATHOPHYSIOLOGY AND CURRENT RESEARCH

Human Observations

Neuroimaging studies of OCD patients have supported the hypothesis that cortico-striatal-thalamo-cortical circuits (CSTC), particularly orbitofrontal-subcortical dysfunction, are (circuits) central to the pathogenesis of the disorder. In CSTC circuits, the basal ganglia form an essential crossroad for neural pathways projecting from the cerebral cortex to the thalamus and back to cortex. The pathways are somatotopically organized and arranged in minimally overlapping circuits, regulating motor, somatosensory, cognitive, and emotional functions. The somatotopic arrangement provides a rational functional and structural organizational plan and current understanding of the pathophysiology of OCD suggests that a disruption of one or more of these pathways underlies the symptoms of these disorders.

Positron emission tomography (PET) symptom provocation studies demonstrate hypermetabolism in the anterior-lateral orbitofrontal cortex, caudate nucleus and paralimbic areas of OCD patients when they are exposed to situations that provoke OCD symptoms compared to neutral stimulus periods. PET studies also implicate hypermetabolism in the orbitofrontal cortex and caudate nucleus in subjects with OCD compared to normal controls during resting conditions.[32] Hypermetabolism in these areas reverses with successful pharmacotherapy with SSRIs or cognitive–behavioral therapy (CBT).[32] Recently, studies have suggested the possibility of atypical perhaps less mature cortical interconnectivity existing in OCD as is observed in patients with other psychiatric disorders such as Tourette syndrome.[33–35] They showed that brain networks in OCD patients had less efficient small world properties, leading to at times desynchronized performance and compensatory repetitive signals/behaviors and atypical modularity properties of the effected OCD brain's default-mode networks (DMN) and sensorimotor pathways, and more developed connections between executive function and spatial cognitive control pathways, possibly trying to compensate for the others' deficits. These findings were also observed in unaffected siblings of OCD patients.[36]

A large, multisite structural magnetic resonance imaging (MRI) study in adults with OCD demonstrated smaller volumes of frontal gray and white matter bilaterally, including the dorsomedial prefrontal cortex, the anterior cingulate cortex, and the inferior frontal gyrus extending to the anterior insula in 412 OCD patients and 368 healthy controls.[37] Meta-analysis of previous imaging studies in OCD describes increased regional gray matter volumes in bilateral lenticular and caudate nuclei, and decreased gray matter volumes in bilateral dorsal medial frontal/anterior cingulate gyri in OCD patients compared to healthy controls.[38] MRI studies examining psychotropic-naïve children with OCD found increased anterior cingulate gray matter and globus pallidus volumes and reduced striatal volumes compared to normal controls.[39,40]

Animal Models

An animal model of OCD would be of considerable value.[41] With this tool, researchers could examine environmental and genetic interactions, confirm neurological substrates, and search for more effective medical treatments. This effort has largely been limited to animals exhibiting repetitive behaviors.[42] A quinpirole rat preparation has been suggested as a model for checking behaviors, as rats exposed to this dopamine agonist perform considerably more checking than normal rats.[43] In addition, the acral lick dermatitis model, a self-injurious grooming behavior found in dogs, has been proposed as a good animal model for OCD. Support for this model comes from the amelioration of symptoms after SSRI treatment.[44] However, due to the difficulties in analyzing the meaning of these animal behaviors, these models continue to remain controversial. Furthermore, repetitive behavior models may be better suited for TS rather than OCD.

More biologically meaningful animal models have emerged over recent years that take advantage of the importance of CSTC circuit dysfunction in the pathogenesis of OCD. A mouse model with a genetic deletion of the SAP90/PSD95-associated protein 3 (*Sapap3*; also known as Dlgap3) exhibits an anxiety phenotype with compulsive grooming behavior.[45] Sapap3 is a postsynaptic scaffolding protein at excitatory synapses and is highly expressed in the striatum. *Sapap3* knockout mice have deficits in cortico-striatal synapses. Furthermore, the compulsive grooming behaviors are reduced with acute fluoxetine administration and the phenotype can be rescued by restoring Sapap3 in the striatum by using lentiviral-mediated expression techniques.[45]

A recent paper used optogenetics to stimulate CSTC hyperaction in mice.[46] Chronic but not acute orbitofrontal cortex-ventromedial striatum stimulation produced repetitive grooming in mice that persisted at least 2 weeks following the cessation of stimulation. Furthermore, chronic fluoxetine administration reversed the increased grooming behaviors.[46] It remains to be determined how much these emerging animal models of OCD will inform on novel treatments for the disorder.

DIFFERENTIAL DIAGNOSIS

Recurrent thoughts, images, and impulses similar to the obsessions present in OCD characterize many other mental disorders. For instance, in major depressive episodes it is not uncommon for patients to have persistent thoughts about unpleasant circumstances, personal worthlessness, or about possible alternative actions. These ruminative thoughts of depression can be distinguished from OCD by the fact that they are a mood-congruent aspect of depression, ego-syntonic compared to obsessions of OCD, which are ego-dystonic. There is a high comorbidity between depression and OCD; thus it is common for both ruminative thoughts of depression and OCD to be present in the same individual. The symptoms of other anxiety disorders can also mimic the symptoms of OCD. Generalized anxiety disorder is characterized by excessive worry, but such worries can be distinguished from obsessions by the fact that the person experiences them as excessive concerns about real-life circumstances, whereas obsessions in OCD are generally experienced as excessive and unreasonable. Obsessions of OCD must also be distinguished from mental disorders where individuals have excessive worries about their appearance (body dysmorphic disorder, anorexia, and bulimia), a specific situation or circumstance (specific phobia), or of serious illness due to misinterpretation of normal bodily signals (hypochondrias).

In children, obsessions in OCD concerning the fear of harm coming to self or others must be distinguished from those typical of separation anxiety disorder. These two conditions can often be distinguished by the observation that obsessions in OCD are usually accompanied by stereotyped and specific compulsive rituals (i.e., specific checking behavior, counting), whereas the compulsive actions of separation anxiety disorder are less stereotyped. Furthermore, the presence of other OCD symptoms besides fear of harm can aid in diagnostic clarification. The repetitive stereotypies of children with autism spectrum disorders, mental retardation, and pervasive developmental disorders can resemble the compulsions of OCD. However, these stereotypies can usually be easily distinguished from OCD based on the child's accompanying symptoms as well as the fact that stereotypies are usually experienced as soothing or pleasurable whereas compulsions are ego-dystonic.

TESTING

Currently no biological test exists to diagnose OCD. Diagnosis is based on clinical information in accordance with DSM-5 criteria.[1] The Y-BOCS is the standard clinical rating scale used to assess symptom severity in adults with OCD.[3,4] The Y-BOCS is a 10-item ordinal scale (0–4) that rates the severity separately for both obsessions and compulsions of OCD according to the time occupied, degree of interference, subjective distress, internal resistance, and degree of control. The CY-BOCS is an identically designed scale for use in children with OCD.[2] CY-BOCS and Y-BOCS scales differ only according to the accompanying symptom screening checklist, with the CY-BOCS being more developmentally appropriate for children. Both scales have been validated for use in their representative patient populations and are sensitive to changes in symptom severity with treatment. Both scales rate OCD symptoms on a scale with a range from 0 (no symptoms) to 40 (severe OCD), with a score less than 8 considered as subclinical symptomatology, over 16 as clinically significant symptoms, and over 24 as moderate to severe OCD. Generally a reduction in Y-BOCS score of 25% or 35% with a final Y-BOCS is considered the criteria for response to treatment.[47] The Y-BOCS rating scale takes approximately 5 minutes to complete once the patient has completed the initial symptom checklist and serves as a good measure to assess symptom fluctuation in a clinical setting.

MANAGEMENT

Standard of Care

Standard of care treatment for OCD in both adults and children involves evidence-based treatment with cognitive–behavioral therapy or pharmacological treatment with SSRIs.[48–50]

Cognitive–Behavioral Therapy

Cognitive–behavioral therapy (CBT) involves psychoeducation, cognitive therapy, and exposure and response prevention. Exposure and response prevention is the core of CBT for OCD and involves gradual, systematic exposure to distress-producing stimuli without engaging in associated rituals or avoidance (i.e., response prevention). The putative mechanism of exposure is extinction, whereby repeated presentations of a conditioned stimulus (CS) in the absence of a previously paired unconditioned stimulus (US), lead to reductions in the conditioned response (CR). Methodologically rigorous studies among children and adolescents with OCD have established the superiority of CBT to placebo/waitlist, attention-control conditions, and serotonin reuptake inhibitor (SRI) medications.[48,51–54] A recent meta-analysis demonstrated that the effect size for CBT with SRI therapy (d = 1.704) and without (d = 1.203) was superior to SRI treatment alone (d = 0.745).[55]

Serotonin Reuptake Inhibitors

Pharmacotherapy with selective serotonin reuptake inhibitors and clomipramine has demonstrated efficacy in the treatment of both children and adults with OCD.[56] There is no evidence that any particular SSRI is any more effective than any other agent within the class for treating OCD. Meta-analytic studies of placebo-controlled trials show clomipramine to have significantly greater effect size than SSRIs when compared to placebo,[57,58] although head-to-head trials have not demonstrated superiority.[59] SSRIs are generally still used as the first-line treatment in OCD because of the better side-effect profile.

Roughly 25% of OCD patients whose symptoms fail to respond to an initial SRI trial of adequate dose and duration will respond to a second SRI trial with a different agent.[60,61]

SSRIs often need to be used at high doses—often higher than the US Food and Drug Administration (FDA)-approved dose range—to be efficacious in OCD; increasing the dose of a well-tolerated SSRI is therefore a valuable strategy in treatment of refractory OCD. A multicenter, double-blind trial supported the use of high-dose SSRIs. Sixty-six nonresponders to 16 weeks of sertraline at a standard dose (200 mg per day) were randomized to receive continued sertraline treatment at either 200 or 400 mg per day.[62] Twelve weeks of high-dose sertraline treatment resulted in a significantly greater improvement in symptoms than continued treatment at the standard dose, though not in the number of patients categorized as responders (40% vs. 33%).[62] Furthermore, meta-analysis of fixed-dose trials in OCD have demonstrated that higher doses of SSRIs have greater efficacy when compared to low doses of the same agents.

Barriers to Treatment Development

OCD remains an excellent candidate psychiatric illness to study in randomized controlled trials. Advantages to studying OCD compared to many other psychiatric illnesses are: 1) the low placebo-response rate in OCD (typically under 10% in trials); 2) OCD patients typically have good insight into their symptoms and rating scales measuring OCD severity (e.g., Y-BOCS) are relatively consistent and sensitive to change; and 3) there exists a substantial population of OCD patients who have not responded to evidence-based treatments for OCD and are motivated to help advance treatment research.

The main challenge in OCD research has been the relative dearth of industry- or federally-funded multisite trials in OCD. Pilot trials examining agents for treatment-refractory OCD have been small and severely underpowered. Thus even relatively large pilot trials in treatment-refractory OCD ($n \approx 50$) have been underpowered to detect clinically meaningful treatment benefits of promising agents. Furthermore, positive small pilot trials in OCD are at least as likely to have false-positive findings related to bias than related to actual efficacy of the agent being studied.[63] The major challenge for the field moving forward is to champion fewer, but larger, multisite trials of interventions with strong preclinical rationale and to find funding for these efforts.

Therapies Under Investigation

Antipsychotic Augmentation

Double-blind trials in the late 1990s demonstrated efficacy of augmentation of SSRI pharmacotherapy with low-dose typical and atypical antipsychotics in OCD.[64,65] A meta-analysis of nine double-blind, placebo-controlled trials of augmentation with typical or atypical antipsychotics demonstrated their efficacy compared to placebo.[10] Approximately one-third of treatment refractory OCD patients will respond to antipsychotic augmentation. OCD patients with comorbid tic disorders appear to respond particularly well to antipsychotic augmentation. Additional trials have subsequently suggested the efficacy of aripiprazole augmentation.[66,67] In general, antipsychotic augmentation should not be considered until two SRI trials of adequate dose and duration have been attempted, because of the more benign side effect profile of the SRIs and the reasonable likelihood of response to extended treatment or a switch to a second agent.

Glutamatergic Agents

Convergent evidence suggests that dysregulation of glutamate neurotransmission is involved in the pathogenesis of OCD.[68] Magnetic resonance spectroscopy (MRS) has demonstrated increased Glx, a composite measure of glutamate, glutamine, γ-aminobutyric acid (GABA), and related molecules, in the striatum of patients with OCD; this elevation normalizes with response to SRI pharmacotherapy.[69,70] Cerebrospinal fluid studies have demonstrated elevated glutamate levels in treatment-naïve OCD patients.[71] Finally, variants in the *SLC1A1* gene, which encodes a glutamate transporter, has been associated with the diagnosis of OCD.[27,28]

Riluzole (Rilutek®, Sanofi-Aventis) is an antiglutamatergic agent that is approved by the FDA for neuroprotection in the treatment of amyotrophic lateral sclerosis. Riluzole is proposed to affect glutamate neurotransmission by reducing glutamate neurotransmitter release[72,73] and increasing astrocytic uptake of glutamate.[74] A case series of 13 treatment-refractory OCD patients demonstrated that seven (59%) of these patients exhibited a treatment response to riluzole.[75] Overall, completing subjects achieved a 42% reduction in Y-BOCS score during 12 weeks of riluzole treatments.[75] We are currently conducting a double-blind study of riluzole to more definitively evaluate the efficacy of this compound.

Other agents targeting glutamate neurotransmission also hold promise in the treatment of refractory OCD. *N*-acetylcysteine (NAC) is converted to cysteine, a substrate for the glutamate/cysteine antiporter located on glial cells, and attenuates synaptic glutamate release by an indirect mechanism.[76] A randomized, placebo-controlled trial in 48 adults with OCD that failed to respond to 12 weeks of SSRI monotherapy compared the efficacy of NAC (2400 mg per day) compared to placebo.[77] NAC was superior to placebo after 12 weeks of treatment with 53% of subjects responding in the NAC group compared to 15% in the placebo group.

Ketamine is a potent *N*-methyl-D-aspartate (NMDA) receptor antagonist that is FDA-approved as an anesthetic agent at higher doses. Numerous human studies have also demonstrated that ketamine given intravenously at a dose of 0.5 mg per kg over 40 minutes leads to potent antidepressant effects that peak 1–3 days following infusion and dissipate 1–2 weeks following initial infusion. Remission rates of depressive symptoms in these trials are consistently above 50% within the first week of treatment.[78]

Two trials examined the acute effects of ketamine in adults with treatment-refractory OCD. The first open trial examined the effects of intravenous ketamine (0.5 mg per kg over 40 minutes) in 10 adults with treatment-refractory OCD as an add-on to their existing medications.[79] An extremely short-lived benefit of ketamine was observed for OCD during the first 3 hours following infusion. None of the 10 patients exhibited a response to ketamine 1–7 days following infusion. Additionally, in the seven patients with comorbid depression, ketamine significantly reduced comorbid depressive symptomatology to a greater extent than OCD symptoms from 1–7 days following infusion. Four of seven patients were judged to be responders to ketamine in terms of their depression symptomatology.[79] A second, saline-controlled crossover trial in 15 adult OCD subjects washed off their OCD medications as outpatients demonstrated a significant benefit of ketamine compared to placebo during the first week after ketamine infusion.[80] Fifty percent of OCD patients (compared to 0% on placebo) responded to ketamine infusion.[80] Additionally, there were significant carryover effects with ketamine such that the benefits of treatment lasted beyond 1 week. The differing results between these two studies may be attributable to: 1) the differences between concomitant medications received by subjects between studies; 2) difference in underlying severity between patient populations (the former involved inpatients with refractory OCD, whereas the later involved refractory OCD patients who could be washed off current medications as outpatients); and 3) issues with blinding inherent with the short-term psychotomimetic side-effects of ketamine infusion. Future studies will be needed to definitely establish the efficacy of ketamine in adult populations.

Memantine is a noncompetitive NMDA receptor antagonist that is FDA approved for the treatment of Alzheimer disease. A randomized, double-blind, placebo-controlled trial in 38 adults with OCD examined the efficacy of concurrent addition of memantine (20 mg per day titrated up over 1 week) or placebo to fluvoxamine 200 mg per day.[81] Fluvoxamine + memantine (89% response rate) was demonstrated to be superior to fluvoxamine + placebo (32% response rate) in this 8-week trial. No placebo-controlled augmentation trials have examined the efficacy of memantine in treatment-refractory OCD.

Topiramate is an FDA-approved anticonvulsant that works via activation of $GABA_A$ receptors and antagonism of specific glutamate receptors.[82] A double-blind, placebo-controlled trial in 49 adults with OCD who failed to respond to at least one SSRI trial demonstrated efficacy of topiramate.[83] Fifty percent of subjects receiving topiramate had a treatment response compared to 0 out of 25 patients in the placebo group. A second randomized, placebo-controlled study of topiramate (up to 400 mg per day) in 36 adults with treatment-refractory OCD failed to detect a significant difference between topiramate and placebo.[84] In secondary analysis, topiramate was superior to placebo in reducing Y-BOCS compulsions. However, topiramate was poorly tolerated in this trial with 28% of subjects receiving topiramate discontinuing due to tolerability issues and 39% requiring a dose reduction.

Neurosurgery

Neurosurgical techniques have been used for decades in treatment of adults with treatment-refractory OCD that causes severe impairment and distress. The ablative procedures practiced in psychosurgery for OCD include anterior cingulatomy, capsulotomy, subcaudate tractotomy, and limbic leucotomy. Currently, no published controlled studies have demonstrated the efficacy of neurosurgery for the treatment of OCD. Reviews of open studies of neurosurgical procedures in OCD have suggested 50–60% response rates observed after 6–24 months.[85]

Deep brain stimulation (DBS) involves the surgical implantation of electrodes and introducing targeted electrical stimulation to specific brain regions through the implanted electrodes. DBS in OCD typically targets the ventral capsule/striatum or subthalamic nucleus.[86] Trials using crossover trials comparing when the implanted electrodes are on compared to off have demonstrated the efficacy of DBS for both brain regions.

CONCLUSION

Research in recent years has improved our understanding of the underlying neurobiology of OCD. Future research will likely focus on how the heterogeneity in presenting OCD symptoms affects the neurobiology of OCD. Pharmacological research in OCD will focus on examining the efficacy of candidate agents targeting the glutamate system. Treatment research will additionally focus on confirming whether circuit-based interventions (e.g., DBS, neurosurgery and repetitive transcranial magnetic stimulation [rTMS]) are indeed effective in treating OCD and, if so, how to optimize them.

References

1. American Psychiatric Association. Diagnostic and Statistical Manual of Mental Disorders: DSM-5. Arlington, VA: American Psychiatric Association; 2013.
2. Scahill L, Riddle MA, McSwiggin-Hardin M, et al. Children's Yale–Brown obsessive compulsive scale: reliability and validity. *J Am Acad Child Adolesc Psychiatry*. 1997;36(6):844–852.
3. Goodman WK, Price LH, Rasmussen SA, et al. The Yale–Brown obsessive compulsive scale. II. Validity. *Arch Gen Psychiatry*. 1989;46(11):1012–1016.
4. Goodman WK, Price LH, Rasmussen SA, et al. The Yale–Brown obsessive compulsive scale. I. Development, use, and reliability. *Arch Gen Psychiatry*. 1989;46(11):1006–1011.
5. Grados MA, Specht MW, Sung HM, et al. Glutamate drugs and pharmacogenetics of OCD: a pathway-based exploratory approach. *Expet Opin Drug Discov*. 2013;8(12):1515–1527.
6. Miguel EC, do Rosário-Campos MC, Shavitt RG, et al. The tic-related obsessive–compulsive disorder phenotype and treatment implications. *Adv Neurol*. 2001;85:43–55.
7. Bloch MH, Landeros-Weisenberger A, Rosario MC, et al. Meta-analysis of the symptom structure of obsessive–compulsive disorder. *Am J Psychiatry*. 2008;165(12):1532–1542.
8. Geller DA, Biederman J, Stewart SE, et al. Impact of comorbidity on treatment response to paroxetine in pediatric obsessive–compulsive disorder: is the use of exclusion criteria empirically supported in randomized clinical trials? *J Child Adolesc Psychopharmacol*. 2003;13(suppl 1):S19–S29.
9. March JS, Franklin ME, Leonard H, et al. Tics moderate treatment outcome with sertraline but not cognitive-behavior therapy in pediatric obsessive–compulsive disorder. *Biol Psychiatry*. 2007;61(3):344–347.
10. Bloch MH, Landeros-Weisenberger A, Kelmendi B, et al. A systematic review: antipsychotic augmentation with treatment refractory obsessive–compulsive disorder. *Mol Psychiatry*. 2006;11(7):622–632.
11. Nestadt G, Samuels J, Riddle M, et al. A family study of obsessive–compulsive disorder. *Arch Gen Psychiatry*. 2000;57(4):358–363.
12. Pauls DL, Alsobrook 2nd. JP, Goodman W, et al. A family study of obsessive–compulsive disorder. *Am J Psychiatry*. 1995;152(1):76–84.
13. Taylor S. Etiology of obsessions and compulsions: a meta-analysis and narrative review of twin studies. *Clin Psychol Rev*. 2011;31(8):1361–1372.
14. Davis LK, Yu D, Keenan CL, et al. Partitioning the heritability of Tourette syndrome and obsessive compulsive disorder reveals differences in genetic architecture. *PLoS Genet*. 2013;9(10):e1003864.
15. Robins LN, Helzer JE, Croughan J, et al. National Institute of Mental Health Diagnostic Interview Schedule. Its history, characteristics, and validity. *Arch Gen Psychiatry*. 1981;38(4):381–389.
16. Zohar AH, Ratzoni G, Pauls DL, et al. An epidemiological study of obsessive–compulsive disorder and related disorders in Israeli adolescents. *J Am Acad Child Adolesc Psychiatry*. 1992;31(6):1057–1061.
17. Flament MF, Whitaker A, Rapoport JL, et al. Obsessive compulsive disorder in adolescence: an epidemiological study. *J Am Acad Child Adolesc Psychiatry*. 1988;27(6):764–771.
18. Geller D, Biederman J, Jones J, et al. Is juvenile obsessive–compulsive disorder a developmental subtype of the disorder? A review of the pediatric literature. *J Am Acad Child Adolesc Psychiatry*. 1998;37(4):420–427.
19. Leckman JF, Grice DE, Barr LC, et al. Tic-related vs. non-tic-related obsessive compulsive disorder. *Anxiety*. 1994;1(5):208–215.
20. Stewart SE, Geller DA, Jenike M, et al. Long-term outcome of pediatric obsessive–compulsive disorder: a meta-analysis and qualitative review of the literature. *Acta Psychiatr Scand*. 2004;110(1):4–13.
21. Bloch MH, Craiglow BG, Landeros-Weisenberger A, et al. Predictors of early adult outcomes in pediatric-onset obsessive–compulsive disorder. *Pediatrics*. 2009;124(4):1085–1093.
22. Foa EB, Liebowitz MR, Kozak MJ, et al. Randomized, placebo-controlled trial of exposure and ritual prevention, clomipramine, and their combination in the treatment of obsessive–compulsive disorder. *Am J Psychiatry*. 2005;162(1):151–161.
23. Eisen JL, Pinto A, Mancebo MC, et al. A 2-year prospective follow-up study of the course of obsessive–compulsive disorder. *J Clin Psychiatry*. 2010;71(8):1033–1039.
24. Skoog G, Skoog I. A 40-year follow-up of patients with obsessive–compulsive disorder [see comments]. *Arch Gen Psychiatry*. 1999;56(2):121–127.
25. Hanna GL, Veenstra-VanderWeele J, Cox NJ, et al. Genome-wide linkage analysis of families with obsessive–compulsive disorder ascertained through pediatric probands. *Am J Med Genet*. 2002;114(5):541–552.
26. Willour VL, Yao Shugart Y, Samuels J, et al. Replication study supports evidence for linkage to 9p24 in obsessive–compulsive disorder. *Am J Hum Genet*. 2004;75(3):508–513.
27. Arnold PD, Sicard T, Burroughs E, et al. Glutamate transporter gene SLC1A1 associated with obsessive–compulsive disorder. *Arch Gen Psychiatry*. 2006;63(7):769–776.
28. Dickel DE, Veenstra-VanderWeele J, Cox NJ, et al. Association testing of the positional and functional candidate gene SLC1A1/EAAC1 in early-onset obsessive–compulsive disorder. *Arch Gen Psychiatry*. 2006;63(7):778–785.
29. Stewart SE, Mayerfeld C, Arnold PD, et al. Meta-analysis of association between obsessive–compulsive disorder and the 3′ region of neuronal glutamate transporter gene SLC1A1. *Am J Med Genet B Neuropsychiatr Genet*. 2013;162B(4):367–379.
30. Taylor S. Molecular genetics of obsessive–compulsive disorder: a comprehensive meta-analysis of genetic association studies. *Mol Psychiatry*. 2013;18(7):799–805.
31. Stewart SE, Yu D, Scharf JM, et al. Genome-wide association study of obsessive–compulsive disorder. *Mol Psychiatry*. 2013;18(7):788–798.
32. Saxena S, Rauch SL. Functional neuroimaging and the neuroanatomy of obsessive–compulsive disorder. *Psychiatr Clin North Am*. 2000;23(3):563–586.
33. Hou J, Song L, Zhang W, et al. Morphologic and functional connectivity alterations of corticostriatal and default mode network in treatment-naive patients with obsessive–compulsive disorder. *PLoS One*. 2013;8(12):e83931.
34. Worbe Y, Malherbe C, Hartmann A, et al. Functional immaturity of cortico-basal ganglia networks in Gilles de la Tourette syndrome. *Brain*. 2012;135(Pt 6):1937–1946.

35. Zhang T, Wang J, Yang Y, et al. Abnormal small-world architecture of top-down control networks in obsessive–compulsive disorder. *J Psychiatry Neurosci*. 2011;36(1):23–31.
36. Peng ZW, Xu T, He QH, et al. Default network connectivity as a vulnerability marker for obsessive compulsive disorder. *Psychol Med*. 2013;1–10.
37. de Wit SJ, Alonso P, Schweren L, et al. Multicenter voxel-based morphometry mega-analysis of structural brain scans in obsessive–compulsive disorder. *Am J Psychiatry*. 2014;171(3):340–349.
38. Radua J, Mataix-Cols D. Voxel-wise meta-analysis of grey matter changes in obsessive–compulsive disorder. *Br J Psychiatry*. 2009;195(5):393–402.
39. Rosenberg DR, Keshavan MS, O'Hearn KM, et al. Frontostriatal measurement in treatment-naive children with obsessive–compulsive disorder. *Arch Gen Psychiatry*. 1997;54(9):824–830.
40. Szeszko PR, MacMillan S, McMeniman M, et al. Brain structural abnormalities in psychotropic drug-naive pediatric patients with obsessive–compulsive disorder. *Am J Psychiatry*. 2004;161(6):1049–1056.
41. Korff S, Harvey BH. Animal models of obsessive–compulsive disorder: rationale to understanding psychobiology and pharmacology. *Psychiatr Clin North Am*. 2006;29(2):371–390.
42. Eilam D, Zor R, Szechtman H, et al. Rituals, stereotypy and compulsive behavior in animals and humans. *Neurosci Biobehav Rev*. 2006;30(4):456–471.
43. Szechtman H, Eckert MJ, Tse WS, et al. Compulsive checking behavior of quinpirole-sensitized rats as an animal model of obsessive–compulsive disorder (OCD): form and control. *BMC Neurosci*. 2001;2:4.
44. Wynchank D, Berk M. Fluoxetine treatment of acral lick dermatitis in dogs: a placebo-controlled randomized double blind trial. *Depress Anxiety*. 1998;8(1):21–23.
45. Welch JM, Lu J, Rodriguiz RM, et al. Cortico-striatal synaptic defects and OCD-like behaviours in Sapap3-mutant mice. *Nature*. 2007;448(7156):894–900.
46. Ahmari SE, Spellman T, Douglass NL, et al. Repeated cortico-striatal stimulation generates persistent OCD-like behavior. *Science*. 2013;340(6137):1234–1239.
47. Pallanti S, Quercioli L. Treatment-refractory obsessive–compulsive disorder: methodological issues, operational definitions and therapeutic lines. *Prog Neuropsychopharmacol Biol Psychiatry*. 2006;30(3):400–412.
48. Pediatric OCD Treatment Study (POTS) Team. Cognitive-behavior therapy, sertraline, and their combination for children and adolescents with obsessive–compulsive disorder: the Pediatric OCD Treatment Study (POTS) randomized controlled trial. *JAMA*. 2004;292(16):1969–1976.
49. Practice parameter for the assessment and treatment of children and adolescents with obsessive–compulsive disorder. *J Am Acad Child Adolesc Psychiatry*. 2012;51(1):98–113.
50. American Psychiatric Association. Practice guideline for the treatment of patients with obsessive–compulsive disorder. Arlington, VA: American Psychiatric Association; 2007.
51. Barrett P, Healy-Farrell L, March JS. Cognitive-behavioral family treatment of childhood obsessive–compulsive disorder: a controlled trial. *J Am Acad Child Adolesc Psychiatry*. 2004;43(1):46–62.
52. Freeman JB, Garcia AM, Coyne L, et al. Early childhood OCD: Preliminary findings from a family-based cognitive-behavioral approach. *J Am Acad Child Adolesc Psychiatry*. 2008;47(5):593–602.
53. Piacentini J, Bergman RL, Chang S, et al. Controlled comparison of family cognitive behavioral therapy and psychoeducation/relaxation training for child obsessive–compulsive disorder. *J Am Acad Child Adolesc Psychiatry*. 2011;50(11):1149–1161.
54. van Balkom AJ, de Haan E, van Oppen P, et al. Cognitive and behavioral therapies alone versus in combination with fluvoxamine in the treatment of obsessive compulsive disorder. *J Nerv Ment Dis*. 1998;186(8):492–499.
55. Sánchez-Meca J, Rosa-Alcázar AI, Iniesta-Sepúlveda M, et al. Differential efficacy of cognitive-behavioral therapy and pharmacological treatments for pediatric obsessive–compulsive disorder: a meta-analysis. *J Anxiety Disord*. 2013;28(1):31–44.
56. Soomro GM, Altman D, Rajagopal S, et al. Selective serotonin re-uptake inhibitors (SSRIs) versus placebo for obsessive compulsive disorder (OCD). *Cochrane Database Syst Rev*. 2008;1: CD001765.
57. Abramowitz JS. Effectiveness of psychological and pharmacological treatments for obsessive–compulsive disorder: a quantitative review. *J Consult Clin Psychol*. 1997;65(1):44–52.
58. Geller DA, Biederman J, Stewart SE, et al. Which SSRI? A meta-analysis of pharmacotherapy trials in pediatric obsessive–compulsive disorder. *Am J Psychiatry*. 2003;160(11):1919–1928.
59. Freeman CP, Trimble MR, Deakin JF, et al. Fluvoxamine versus clomipramine in the treatment of obsessive compulsive disorder: a multicenter, randomized, double-blind, parallel group comparison. *J Clin Psychiatry*. 1994;55(7):301–305.
60. Treatment of obsessive–compulsive disorder. The Expert Consensus Panel for obsessive-compulsive disorder. *J Clin Psychiatry*. 1997;58 (suppl 4):2–72.
61. Denys D, van Megen HJ, van der Wee N, et al. A double-blind switch study of paroxetine and venlafaxine in obsessive–compulsive disorder. *J Clin Psychiatry*. 2004;65(1):37–43.
62. Ninan PT, Koran LM, Kiev A, et al. High-dose sertraline strategy for nonresponders to acute treatment for obsessive–compulsive disorder: a multicenter double-blind trial. *J Clin Psychiatry*. 2006;67(1):15–22.
63. Ioannidis JP. Why most published research findings are false. *PLoS Med*. 2005;2(8):e124.
64. McDougle CJ, Epperson CN, Pelton GH, et al. A double-blind, placebo-controlled study of risperidone addition in serotonin reuptake inhibitor-refractory obsessive–compulsive disorder. *Arch Gen Psychiatry*. 2000;57(8):794–801.
65. McDougle CJ, Goodman WK, Price LH. Dopamine antagonists in tic-related and psychotic spectrum obsessive compulsive disorder. *J Clin Psychiatry*. 1994;55(suppl):24–31.
66. Sayyah M, Sayyah M, Boostani H, et al. Effects of aripiprazole augmentation in treatment-resistant obsessive–compulsive disorder (a double blind clinical trial). *Depress Anxiety*. 2012;29(10):850–854.
67. Muscatello MR, Bruno A, Pandolfo G, et al. Effect of aripiprazole augmentation of serotonin reuptake inhibitors or clomipramine in treatment-resistant obsessive–compulsive disorder: a double-blind, placebo-controlled study. *J Clin Psychopharmacol*. 2011;31(2):174–179.

68. Pittenger C, Krystal JH, Coric V. Glutamate-modulating drugs as novel pharmacotherapeutic agents in the treatment of obsessive–compulsive disorder. *NeuroRx*. 2006;3(1):69–81.
69. Rosenberg DR, MacMaster FP, Keshavan MS, et al. Decrease in caudate glutamatergic concentrations in pediatric obsessive–compulsive disorder patients taking paroxetine. *J Am Acad Child Adolesc Psychiatry*. 2000;39(9):1096–1103.
70. Moore GJ, MacMaster FP, Stewart C, et al. Case study: caudate glutamatergic changes with paroxetine therapy for pediatric obsessive–compulsive disorder. *J Am Acad Child Adolesc Psychiatry*. 1998;37(6):663–667.
71. Chakrabarty K, Bhattacharyya S, Christopher R, et al. Glutamatergic dysfunction in OCD. *Neuropsychopharmacology*. 2005;30(9):1735–1740.
72. Urbani A, Belluzzi O. Riluzole inhibits the persistent sodium current in mammalian CNS neurons. *Eur J Neurosci*. 2000;12(10):3567–3574.
73. Wang SJ, Wang KY, Wang WC. Mechanisms underlying the riluzole inhibition of glutamate release from rat cerebral cortex nerve terminals (synaptosomes). *Neuroscience*. 2004;125(1):191–201.
74. Frizzo ME, Dall'Onder LP, Dalcin KB, et al. Riluzole enhances glutamate uptake in rat astrocyte cultures. *Cell Mol Neurobiol*. 2004;24(1):123–128.
75. Coric V, Taskiran S, Pittenger C, et al. Riluzole augmentation in treatment-resistant obsessive–compulsive disorder: an open-label trial. *Biol Psychiatry*. 2005;58(5):424–428.
76. Moran MM, McFarland K, Melendez RI, et al. Cystine/glutamate exchange regulates metabotropic glutamate receptor presynaptic inhibition of excitatory transmission and vulnerability to cocaine seeking. *J Neurosci*. 2005;25(27):6389–6393.
77. Afshar H, Roohafza H, Mohammad-Beigi H, et al. N-acetylcysteine add-on treatment in refractory obsessive–compulsive disorder: a randomized, double-blind, placebo-controlled trial. *J Clin Psychopharmacol*. 2012;32(6):797–803.
78. Aan Het Rot M, Zarate Jr. CA, Charney DS, et al. Ketamine for depression: where do we go from here? *Biol Psychiatry*. 2012;72(7):537–547.
79. Bloch MH, Wasylink S, Landeros-Weisenberger A, et al. Effects of ketamine in treatment-refractory obsessive–compulsive disorder. *Biol Psychiatry*. 2012;72(11):964–970.
80. Rodriguez CI, Kegeles LS, Levinson A, et al. Randomized controlled crossover trial of ketamine in obsessive–compulsive disorder: proof-of-concept. *Neuropsychopharmacology*. 2013;38(12):2475–2483.
81. Ghaleiha A, Entezari N, Modabbernia A, et al. Memantine add-on in moderate to severe obsessive–compulsive disorder: randomized double-blind placebo-controlled study. *J Psychiatr Res*. 2013;47(2):175–180.
82. White HS1, Brown SD, Woodhead JH, et al. Topiramate modulates GABA-evoked currents in murine cortical neurons by a nonbenzodiazepine mechanism. *Epilepsia*. 2000;41(suppl 1):S17–S20.
83. Mowla A, Khajeian AM, Sahraian A, et al. Topiramate augmentation in resistant OCD: a double-blind placebo-controlled clinical trial. *CNS Spectr*. 2010.
84. Berlin HA, Koran LM, Jenike MA, et al. Double-blind, placebo-controlled trial of topiramate augmentation in treatment-resistant obsessive–compulsive disorder. *J Clin Psychiatry*. 2011;72(5):716–721.
85. Greenberg BD, Rauch SL, Haber SN. Invasive circuitry-based neurotherapeutics: stereotactic ablation and deep brain stimulation for OCD. *Neuropsychopharmacology*. 2010;35(1):317–336.
86. Blomstedt P, Sjöberg RL, Hansson M, et al. Deep brain stimulation in the treatment of obsessive–compulsive disorder. *World Neurosurg*. 2013;80(6):e245–e253.

CHAPTER

107

Tourette Syndrome

Jessica B. Lennington, Michael H. Bloch*, Lawrence D. Scahill†, Gabor Szuhay*, Paul J. Lombroso*, and Flora M. Vaccarino**

*Yale University School of Medicine, New Haven, CT, USA

†Emory University School of Medicine, Atlanta, GA, USA

SUMMARY

Disease Characteristics, Hallmark Manifestations, and Inheritance

Tourette syndrome (TS) is a childhood-onset psychiatric disorder characterized by chronic motor and phonic tics.[1] Symptoms typically vary in nature and frequency over time and decrease with age.[2,3] There is evidence of genetic contribution; however, candidates revealed by family studies do not appear to be significant risk genes for the majority of the population, and association studies have not identified common variants above threshold significance.[4–10]

Diagnosis and Testing

TS is diagnosed when both motor and vocal tics are present for more than 12 months, with onset before age 18, and is estimated to occur in 0.1–1% of children.[11] As there are currently no physiological or genetic tests, diagnosis occurs via clinical evaluation.[12]

Current Research

Several brain regions have been implicated in TS including the cortex, hippocampus, cerebellum, and basal ganglia.[13–18] Research includes evaluation of neurotransmitter systems, postmortem brain studies, and TS-associated gene mutations.[7,19–22] The presence of antineuronal antibodies in patients, high rates of maternal autoimmune disorders, and a subcategory of children in which infection is coincident with tic onset have prompted investigations of neuroimmune interactions.[23–33]

CLINICAL FEATURES

Historical Overview

TS was first classified as a neurobiological disorder in the mid-1880s by a Gilles de la Tourette and Jean-Martin Charcot.[33,34] In the subsequent century, clinical classification continued to be refined leading to the current criteria—childhood onset of chronic motor and vocal tics.[34]

Rosenberg's Molecular and Genetic Basis of Neurological and Psychiatric Disease
http://dx.doi.org/10.1016/10.1016/B978-0-12-410529-4.00107-8

Disease Identification

General tic disorders affect 4–20% of children and are characterized by sudden, repetitive motor or vocal tics, which can be transient (2 weeks to 12 months) or chronic (more than 12 months).[36] The first symptoms are often simple motor tics, such as eye blinking and facial movements, which later progress to involve additional muscles of the face, neck, and shoulders. Vocal tics, such as throat clearing, grunting, coughing, or sniffling, typically appear several years after the motor symptoms. A minority of patients have complex vocalizations, such as swearing (coprolalia), echolalia, and the utterance of words or word parts. Patients have described a premonitory sensation that occurs immediately prior to tics, describing tic expression as an "involuntary" way to relieve the premonitory urge.[35] Tics typically occur in bouts, vary over time with the worst period occurring when the child is 10–12 years, and reduce with age.[3,37] A minority of patients are chronically disabled with severe tics that continue into adulthood.[38]

Gene Identification

To date, no gene has emerged as definitively causal. In a recent genome-wide association study of 1496 patients, no common gene variants above threshold significance were identified; yet, given the estimated population required to detect low frequency risk alleles, limited discovery of genetic candidates using these methods may not be surprising.[5,10] Significant association of TS with chromosome 2 was reported, and the strength of this association increased when diagnosis of TS and chronic tic disorder were combined, suggesting that symptoms of both emerge from common underlying genetic vulnerability.[39,40] Additional regions on chromosome 9,[41]18,[42] and 22[43–45] have been implicated. Given the difficulty of localizing genes via traditional linkage approaches, family-based association strategies have also been used and have found several genes to associate with risk of developing TS, including the dopamine receptor, DRD2; the dopamine transporter gene, *DAT1*; serotonin receptor, HTR2C; and gain-of-function alleles in the serotonin transporter gene, *SLC6A4*.[46–50]

Rare mutations were recently identified that appear to confer high risk of developing TS in particular families. Mutations in Slit and Trk-like 1 (SLITRK1) were identified after characterization of a chromosomal inversion in a child with TS.[6] SLITRK1 was found to be expressed in brain regions implicated in TS, and *in vitro* assays revealed that wild-type SLITRK1 enhanced dendritic growth compared to protein generated by a mutated version.[6,7] Rare mutations have also been described in histidine decarboxylase (HDC), which catalyzes the production of histamine; *IMMP2L*, a mitochondrial pepsidase gene; and CNTNAP2, a neurexin superfamily member.[6,7,9,51,52] It is presently unclear whether such rare mutations segregate with TS, versus comorbid disorders such as OCD,[53] or contribute to TS in the majority of patients.[51,52]

Early Treatments

Neuroleptics became the earliest, and remain the most frequently used, pharmacological treatment for TS, although with varying degrees of efficacy.[54,55] Due to frequent comorbidity of TS with obsessive–compulsive disorder (OCD) and attention-deficit hyperactivity disorder (ADHD), other early treatments included the use of adrenergic receptor agonists such as clonidine.[55] Standards of treatment continue to be refined and are described on page 1315.

Mode of Inheritance and Prevalence

Early segregation analyses indicated that inheritance is transmitted vertically as a single gene of major effect. However, while some data are consistent with an autosomal dominant locus with sex-specific penetrance, there is also evidence for intermediate inheritance, and many favor a multifactorial or polygenic model, suggesting that TS is the result of multiple genes acting together. The original description of TS included a familial expression pattern, and the rate of TS in relatives of probands is 10–15%, 10-fold higher than in the general population. Inheritance is strongly genetic: tics occur in a second twin in 77% of monozygotic, but only 23% of dizygotic twins.[4] Yet environmental factors do contribute, as monozygotic twin concordance is not 100%. Retrospective analyses revealed lower birth weights in the more severely affected twin in monozygotic pairs, suggesting that prenatal factors contribute to symptom expression later in life.[46]

Natural History

Age of Onset

Based on current classification standards, onset of TS symptoms begin in childhood before age 18. The mean age of onset is between 5 and 7 years, with simple motor tics typically preceding the onset of verbal tics by several years.[56]

Disease Evolution

TS symptoms typically vary in frequency and intensity, increase during periods of fatigue or stress, and may be affected by stimulant medications, steroids and, proposed more recently, recurrent infections. Symptoms may also wax and wane spontaneously, independent from triggers or medication.[57] Lifetime severity may be influenced by perinatal exposures that presumably affect early brain development. These observations have led to a stress-diathesis model in which the interaction of genetic vulnerability factors with environmental stressors during critical periods of development affects disease onset and severity.[35]

Disease Variants

A potential subcategory of TS has recently emerged. Initially an increased prevalence of OCD was observed in Sydenham chorea patients with a late manifestation of rheumatic fever associated with a prior streptococcal infection. After an epidemic of streptococcal pharyngitis in Providence, Rhodes Island, tic disorders increased in affected children. Swedo and colleagues then proposed the existence of a subgroup of children with tics and/or OCD now known by its acronym PANDAS (pediatric autoimmune neuropsychiatric disorder associated with streptococcal infections).[58] PANDAS is defined by five criteria: 1) the presence of a tic disorder and/or OCD; 2) prepubertal onset of neuropsychiatric symptoms; 3) history of a sudden onset of symptoms and/or an episodic course with abrupt symptom exacerbation interspersed with periods of partial or complete remission; 4) evidence of a temporal association between onset or exacerbation of symptoms and a streptococcal infection; and 5) adventitious movements (e.g., motoric hyperactivity and choreiform movements) during symptom exacerbation.[59] In an epidemiological study, patients were more likely than controls to have had streptococcal infection in the 3 months preceding onset; having multiple group A beta-hemolytic streptococcal infections (GABHS) in the preceding 12 months was associated with an increased risk for TS; and overall, 12% of TS cases appeared to be GABHS-related.[60] The proposed pathophysiology is based on the molecular mimicry hypothesis, wherein antibodies produced to fight infection cross-react with epitopes in previously unaffected tissues. This model suggests that neuronal function is compromised in a manner analogous to the sequence of events proposed for Sydenham chorea.

Dramatic therapeutic benefit has been described in TS patients treated with plasmapheresis;[61] multiple groups have shown evidence for autoantibodies in the blood serum of a subgroup of patients,[62] and additional studies have examined the involvement of cytokines, reporting decreased numbers of regulatory T cells.[23,24] However, other studies have not supported the PANDAS hypothesis; one report found no significant difference in the *in vitro* neuronal binding capacity of immunoglobulin G (IgG) isolated from PANDAS versus TS or control cases,[63] and in another study PANDAS cases had lower than expected exacerbation of symptoms and GABHS infections compared to non-PANDAS cases.[31] Further investigation is needed to elucidate the relationship between GABHS and TS, and to predict when interventions such as antibiotic treatment or IgG replacement might be of benefit.

Comorbidities

Clinically ascertained cases of TS are associated with a variety of other disorders. More than 60% of patients have problems with inattention, impulsiveness, and hyperactivity, and may warrant a separate diagnosis of ADHD.[11,64] While the high frequency of co-occurring features (aggressiveness, depression, learning disabilities, and anxiety disorders) may reflect an ascertainment bias, surveys in community samples have also observed a higher than expected frequency of ADHD and disruptive behavior among children with TS.[11] As many as 30% of individuals with TS have sufficiently severe symptoms to warrant diagnosis of OCD, and research suggests that the two may be etiologically related.[65–69]

DISEASE MECHANISMS: PATHOPHYSIOLOGY AND CURRENT RESEARCH

Several brain regions have been implicated in TS including the basal ganglia, cortex, hippocampus, and cerebellum.[13–18,70–72] The basal ganglia have long been presumed to be central for the pathophysiology of TS, based on anatomical and functional studies, and the effectiveness of D2 blockers for the treatment of tics. The structures of the basal ganglia form a crossroad of converging and diverging routes for neural pathways projecting from the cerebral cortex to the thalamus and back to cortex.[73] These pathways are somatotopically organized in parallel circuits regulating motor, somatosensory, cognitive, and emotional functions. When a movement is initiated, the inhibitory output of the direct pathway medium spiny neurons which project from the striatum to the internal globus pallidus

(GPi) and substantia nigra pars reticulata is enhanced, releasing the inhibitory GPi brake to the precise thalamic target.[74] Disruption of one or more of these pathways is currently proposed to underlie TS symptoms.

In the striatum, there are four main classes of striatal interneurons, identified by expression of calcium binding proteins and cotransmitters: three subclasses of γ-aminobutyric acid (GABA)ergic interneurons: 1) parvalbumin; 2) calretinin; 3) somatostatin/nitric oxide synthase/neuropeptide Y; and 4) cholinergic interneurons.[75–77] Glutamatergic, cholinergic, GABAergic, and dopaminergic inputs converge on the spines and dendrites of the medium spiny neurons to regulate their output. Dopamine input and the GABAergic and cholinergic activities within the striatum appear to regulate the excitatory glutamatergic input from the cortex, and ultimately the information processing to the thalamus.[74,78,79] Inhibitory parvalbumin interneurons are directly stimulated by the cortex and convey strong feedforward and widespread inhibition to the striatal medium spiny neurons, focusing their activity on the most relevant inputs. A deficiency in striatal interneurons was reported in cases of severe unremitting TS suggesting that these cells may play an important role in the genesis or persistence of tics.[17,18,80]

Human Observations

Neuroimaging of TS patients revealed larger brain volumes in prefrontal regions and parieto-occipital cortical regions, smaller inferior occipital volumes, and smaller caudate volumes,[15] and caudate volume in childhood was found to inversely correlate with tic severity in adulthood.[16] Functional studies have found alterations in cortical-basal ganglia circuits, suggesting increased activity in motor, premotor, sensorimotor cortical regions, and striatum in TS subjects.[81,82] Compared to controls, TS subjects displayed heightened activity in the left caudate that correlated with tic severity during prepulse inhibition to startle tasks.[83] In addition, increased metabolic activity in the right caudate was found during tic suppression, which negatively correlated with tic severity.[84] As tic severity increased, the signal changes decreased, suggesting that tics are due to a failure of inhibitory systems.[84,85]

Positron emission tomography (PET) and single-photon emission computed tomography (SPECT) studies have suggested increased density of striatal dopamine receptors in TS subjects.[81,82] Involvement of dopamine signaling is also indicated by considerable clinical pharmacological data: 1) the most consistently effective drugs for suppressing tics are dopamine D2 receptor blockers; 2) tics can worsen following exposure to stimulants known to increase central dopaminergic activity; 3) withdrawal from neuroleptic treatment is often accompanied by a rebound in tic symptoms, presumably due to heightened receptor sensitivity and/or upregulation of postsynaptic dopamine receptors; and 4) elevated levels of a dopamine metabolite, homovanillic acid, have been found in the cerebrospinal fluid of TS subjects.[86,87]

Other candidate neurotransmitter systems include noradrenergic and opioid systems, and serotonin, GABA, glutamate, and acetylcholine that are known to play a role in cortico-striato-thalamo-cortical circuitry.[78] Reductions of three classes of interneurons in the basal ganglia in TS subjects have been found, based on immunohistochemical labeling of postmortem tissue with antibodies to parvalbumin, choline acetyltransferase,[17,18] and nitric oxide synthase (Table 107.1[88]). Parvalbumin$^+$-GABAergic interneurons form a widespread inhibitory network responsible for maintaining inhibition within the striatum, particularly in sensorimotor territories.[89,90] In addition to decreases of parvalbumin interneurons in the caudate, a relative "excess" was found in the GPi.[17] While it is unknown whether this could be due to changes in cell migration, cell genesis, or cell death, it is plausible that discharge patterns from the caudate, the GPi, or both could be disrupted, leading to aberrant inhibition of discrete thalamic neurons and their cortical targets, resulting in tics.[17]

TABLE 107.1 Reductions of Three Classes of Interneurons in the Basal Ganglia in Tourette Syndrome (TS) Subjects

Neuronal composition of the caudate and putamen	Density in TS subjects
Medium spiny neurons (96%)	No change
Cholinergic interneurons (1%)	Reduced by 51% (Cd), 37% (Pt)
GABAergic interneurons (~4%):	
Parvalbumin +	Reduced by 52% (Cd), 40% (Pt)
Calretinin +	No change
Nitric oxide synthase/neuropeptide Y/somatostatin triple +	Reduced by 38% (Cd), 42% (Pt)

Abbreviations: Cd, caudate; Pt, putamen.

Animal Models

Experimental animal models may yield critical insights, and currently include genetic modeling to understand the roles of SLITRK1 and HDC in mice.[6,9,22] Models of striatal interneuron ablation are being evaluated for their potential to evoke stereotypic behaviors. Stereotypy can be induced through stimulant exposure, which affects expression of immediate early genes and generates an imbalance of striatal activity within striosome versus matrix compartments.[91,92] In rodents, low doses of agents that activate the dopamine pathways (i.e., amphetamine, apomorphine, and cocaine) increase locomotion while elevated doses produce stereotypies.

Neuroimmune modeling has revealed induction of stereotypies in rodents following striatal infusion of serum from TS patients with high levels of antineural autoantibodies, but not after infusing serum from controls or TS patients with low levels of antineural autoantibodies.[93,94] Subsequent studies were not able to reproduce these results, and additional research is necessary to determine whether autoantibodies play a role in some TS cases.[95] Interestingly, *in vitro* studies have shown that immune modulation is sufficient to alter signaling mechanisms critical for parvalbumin neuron function.[96,97]

DIFFERENTIAL DIAGNOSIS

Many diseases can manifest with tics, such as Huntington disease and stroke, but do not typically have pediatric onset. Others diseases such as choreas (i.e., Sydenham chorea), dystonias, and some rare genetic conditions may require more careful evaluation.[98,99] Events such as carbon monoxide poisoning are germane to children and may present with transient tics; however, these are distinguishable as TS involves chronic symptoms.[99]

TESTING

There are presently no molecular, neuroimaging, or physiological mechanisms that can reliably aid in TS diagnosis. Instead, diagnosis relies on clinical evaluation, using rating systems such as the the Yale Global Tic Severity Score (YGTSS).[12] The YGTSS rates multiple features of symptoms including the number, frequency, intensity, and complexity of motor and vocal tics, as well as the level to which tics interfere with daily function.

MANAGEMENT

Standard of Care

The goal of treatment is to achieve a balance between adequate tic control and minimizing side effects, rather than tic eradication. Several medications (Table 107.2) have been used for the treatment of tics, although few have been adequately tested in placebo-controlled studies.[100] The antipsychotics haloperidol and pimozide have been considered the most effective medications for tics. Haloperidol is a potent postsynaptic D2 receptor blocker often effective at low doses but associated with a high frequency of adverse effects. Pimozide is also effective, but at high doses, or in combination with drugs that inhibit CYP3A4, QT prolongation is a potential consideration and cardiac monitoring is warranted.[100]

Atypical neuroleptics have also been used for tic treatment. Tiapride and sulpiride, unavailable in the United States, have shown positive results in controlling tics; however, both require more study to confirm their efficacy. The D2- and 5HT2-blocking agents risperidone and ziprasidone were found to be effective in placebo-controlled studies.[101,102] Ziprasidone has a unique receptor profile including 5-HT1A-agonist and modest norepinephrine- and serotonin-reuptake blocking properties, suggesting that it may have additional anxiolytic and antidepressant effects.[100]

Adrenergic alpha-2 agonists are frequently used in the treatment of tics or ADHD that is commonly associated with TS. Clonidine was introduced as a treatment for tics in the 1970s and has shown modest benefits for ADHD. Although unlikely to be as effective as D2 blocking antipsychotics in reducing tics, it is commonly used in children with TS due to concerns about long-term exposure to neuroleptics. In the 1990s another alpha-2 agonist, guanfacine, was introduced for the treatment of ADHD in children with TS, and in a randomized clinical trial was superior to placebo for ADHD and tic symptoms.[103]

TABLE 107.2 Current Pharmacological Treatments for Tourette Syndrome (TS)

Class/mode of action	Agent	Typical dosage	Note
Anti-psychotics (D2 blockers)	Haloperidol	0.5–2.0 mg/day	Many side effects including sedation, dysphoria, cognitive dulling, weight gain, parkinsonism, dystonia, dyskinesia, and akathisia
	Pimozide	0.5–4 mg/day	QT prolongation at high doses or in combination with CYP3A4 inhibitors
Atypical neuroleptics (D2 blockers)	Tiapride	*Unavailable in USA*	Rhabdomyolysis, QT prolongation
	Sulpiride	0.1–0.8 g/day; *not available in the US*	Low frequency of neurological side effects
	Risperidone[101]	1.0–3.0 mg/day in two divided doses	Low frequency of neurological side effects; drowsiness and weight gain are common; also 5HT2 blocker
	Ziprasidone[100]		Low frequency of side effects and no weight gain; monitor for QTc prolongation; also 5HT2 blocker
Adrenergic alpha-2 agonists	Clonidine[118,119]	0.1–0.3 mg/day in 3–4 divided doses	Minimal side effects; modest benefit for ADHD
	Guanfacine[102]	1.5–3 mg/day in three divided doses	Minimal side effects; benefit for ADHD

Abbreviation: ADHD, Attention-deficit hyperactivity disorder.

Several other medications have been tried with varying success in their capacity to suppress tics.[100] These include stimulants typically used to treat ADHD, although use of stimulants requires close monitoring by the treating therapist;[104] the dopamine agonist, pergolide,[105,106] antagonist, metoclopramide,[107] and depleting agent, tetrabenazine; the serotonin antagonist, ondansetron;[108] the muscle relaxant, baclofen;[109] the anticonvulsants levetiracetam[110] and topiramate;[111] and the antidystonia injection, botulinum toxin.[112] Of these, metoclopramide, topiramate, botulinum toxin, and pergolide improved symptoms superior to placebo,[104–106,111,112] although pergolide is associated with serious adverse effects and has fallen out of use. Other dopamine agonists such as ropinirole warrant investigation.[113]

In addition to pharmacological strategies, there is growing interest in behavioral interventions such as parent training, anger control training, and habit reversal training.[114,115] Clinical trials have found behavioral therapy to be effective, with some patients demonstrating continued improvement 6 months post-treatment.[115–117]

Failed Therapies and Barriers to Treatment Development

Treatment of TS is challenging, as its natural course is variable. Even without treatment, symptoms reduce or resolve by young adulthood in the majority of patients. While emerging studies continue to support a standard of care for TS, these have also revealed an unusually robust placebo response.[118–121] This presents a potential barrier to identifying effective therapies, and has been noted in other disorders that involve the basal ganglia.[123]

Therapies Under Investigation

Immunotherapies including antibiotic treatments and immunoglobulin G serum exchanges are under investigation for their efficacy in reducing tic symptoms in patients with putative GABHS-related onset.[124] Preliminary results in severe, refractory TS cases suggest that procedures such as neurofeedback, transcranial magnetic stimulation, and deep brain stimulation may have a place in the treatment of TS.[74,124–127] Although offering great promise for the treatment of severe TS cases, consensus has not yet been achieved, for instance, on the optimal placement of the deep brain electrodes, and much needs to be learned about how best to apply these technologies.

CONCLUSIONS

Research over the next decade is likely to build on the recent discoveries in neurobiology, epidemiology, genetics, and neuroimaging, and lead to better treatments. New psychopharmacological agents based on a better understanding of the neurobiology will be used in combination with behavioral interventions. Moreover, longitudinal clinical investigations into the phenomenology and natural history may provide additional clues on the biological processes involved. The cellular abnormalities that have been detected in patients' brains will serve as a guide for the development of animal models, which will allow investigators to explore specific hypotheses. In addition, induced pluripotent stem cells can be generated directly from patients, which may allow reproduction of aspects of TS neurobiological abnormalities *in vitro*, facilitating the generation of large scale genetic and drug screens to find patient-tailored treatments.[129] Through these combined efforts, we will eventually come to a more complete understanding of the neurobiological mechanisms and treatment of TS.

ACKNOWLEDGEMENTS

This chapter is dedicated to the memory of our colleague and friend, Marcos Mercadente. The work was supported in part by NIMH grants U10MH66764, R01MH070802, R01MH069874, and CDC contract 504IPA05360 to Dr. Scahill; R01 NS054994, Tourette's Syndrome Association (TSA) awards and a BRAIN (NARSAD) Distinguished Investigator Award to Dr. Vaccarino.

References

1. Centers for Disease Control and Prevention. Prevalence of diagnosed Tourette syndrome in persons aged 6–17 years – United States, 2007. *MMWR Morb Mortal Wkly Rep*. 2009;58(21):581–585.
2. Robertson MM. Tourette syndrome, associated conditions and the complexities of treatment. *Brain*. 2000;123(3):425–462.
3. Bloch MH, Leckman JF. Clinical course of Tourette syndrome. *J Psychosom Res*. 2009;67(6):497–501.
4. Price RA, Kidd KK, Cohen DJ, et al. A twin study of Tourette syndrome. *Arch Gen Psychiatry*. 1985;42(8):815–820.
5. The Tourette Syndrome Association International Consortium for Genetics. A complete genome screen in sib pairs affected by Gilles de la Tourette syndrome. *Am J Hum Genet*. 1999;65(5):1428–1436.
6. Abelson JF, Kwan KY, O'Roak BJ, et al. Sequence variants in SLITRK1 are associated with Tourette's syndrome. *Science*. 2005;310(5746):317–320.
7. Stillman AA, Krsnik Z, Sun J, et al. Developmentally regulated and evolutionarily conserved expression of SLITRK1 in brain circuits implicated in Tourette syndrome. *J Comp Neurol*. 2009;513(1):21–37.
8. O'Rourke JA, Scharf JM, Yu D, et al. The genetics of Tourette syndrome: a review. *J Psychosom Res*. 2009;67(6):533–545.
9. Ercan-Sencicek AG, Stillman AA, Ghosh AK, et al. L-histidine decarboxylase and Tourette's syndrome. *N Engl J Med*. 2010;362(20):1901–1908.
10. Scharf JM, Yu D, Mathews CA, et al. Genome-wide association study of Tourette's syndrome. *Mol Psychiatry*. 2013;18(6):721–728.
11. Scahill L, Sukhodolsky D, Williams S, et al. The public health importance of tics and tic disorders. *Adv Neurol*. 2005;96:240–248.
12. Leckman JF, Riddle MA, Hardin MT, et al. The Yale Global Tic Severity Scale: initial testing of a clinician-rated scale of tic severity. *J Am Acad Child Adolesc Psychiatry*. 1989;28(4):566–573.
13. Morshed SA, Parveen S, Leckman JF, et al. Antibodies against neural, nuclear, cytoskeletal, and streptococcal epitopes in children and adults with Tourette's syndrome, Sydenham's chorea, and autoimmune disorders. *Biol Psychiatry*. 2001;50(8):566–577.
14. Stern E, Silbersweig DA, Chee KY, et al. A functional neuroanatomy of tics in Tourette syndrome. *Arch Gen Psychiatry*. 2000;57(8):741–748.
15. Peterson BS, Thomas P, Kane MJ, et al. Basal ganglia volumes in patients with Gilles de la Tourette syndrome. *Arch Gen Psychiatry*. 2003;60(4):415–424.
16. Bloch MH, Leckman JF, Zhu H, et al. Caudate volumes in childhood predict symptom severity in adults with Tourette syndrome. *Neurology*. 2005;65(8):1253–1258.
17. Kalanithi PS, Zheng W, Kataoka Y, et al. Altered parvalbumin-positive neuron distribution in basal ganglia of individuals with Tourette syndrome. *Proc Natl Acad Sci U S A*. 2005;102(37):13307–13312.
18. Kataoka Y, Kalanithi PS, Grantz H, et al. Decreased number of parvalbumin and cholinergic interneurons in the striatum of individuals with Tourette syndrome. *J Comp Neurol*. 2010;518(3):277–291.
19. Katayama K, Yamada K, Ornthanalai VG, et al. Slitrk1-deficient mice display elevated anxiety-like behavior and noradrenergic abnormalities. *Mol Psychiatry*. 2010;15(2):177–184.
20. Kajiwara Y, Buxbaum JD, Grice DE. SLITRK1 binds 14-3-3 and regulates neurite outgrowth in a phosphorylation-dependent manner. *Biol Psychiatry*. 2009;66(10):918–925.
21. Proenca CC, Gao KP, Shmelkov SV, et al. Slitrks as emerging candidate genes involved in neuropsychiatric disorders. *Trends Neurosci*. 2011;34(3):143–153.
22. Krusong K, Ercan-Sencicek AG, Xu M, et al. High levels of histidine decarboxylase in the striatum of mice and rats. *Neurosci lett*. 2011;495(2):110–114.
23. Leckman JF, Katsovich L, Kawikova I, et al. Increased serum levels of interleukin-12 and tumor necrosis factor-alpha in Tourette's syndrome. *Biol Psychiatry*. 2005;57(6):667–673.

24. Kawikova I, Leckman JF, Kronig H, et al. Decreased numbers of regulatory T cells suggest impaired immune tolerance in children with Tourette syndrome: a preliminary study. *Biol Psychiatry*. 2007;61(3):273–278.
25. Bos-Veneman NG, Olieman R, Tobiasova Z, et al. Altered immunoglobulin profiles in children with Tourette syndrome. *Brain Behav Immun*. 2011;25(3):532–538.
26. Murphy TK, Storch EA, Turner A, et al. Maternal history of autoimmune disease in children presenting with tics and/or obsessive–compulsive disorder. *J Neuroimmunol*. 2010;229(1–2):243–247.
27. Martino D, Dale RC, Gilbert DL, et al. Immunopathogenic mechanisms in Tourette syndrome: a critical review. *Mov Disord*. 2009;24(9):1267–1279.
28. Leonard HL, Swedo SE, Garvey M, et al. Postinfectious and other forms of obsessive–compulsive disorder. *Child Adolesc Psychiatr Clin N Am*. 1999;8(3):497–511.
29. Swedo SE, Leonard HL, Mittleman BB, et al. Identification of children with pediatric autoimmune neuropsychiatric disorders associated with streptococcal infections by a marker associated with rheumatic fever. *Am J Psychiatry*. 1997;154(1):110–112.
30. Brilot F, Merheb V, Ding A, et al. Antibody binding to neuronal surface in Sydenham chorea, but not in PANDAS or Tourette syndrome. *Neurology*. 2011;76(17):1508–1513.
31. Leckman JF, King RA, Gilbert DL, et al. Streptococcal upper respiratory tract infections and exacerbations of tic and obsessive–compulsive symptoms: a prospective longitudinal study. *J Am Acad Child Adolesc Psychiatry*. 2011;50(2):108–118 e3.
32. Swedo SE, Baird G, Cook Jr. EH, et al. Commentary from the DSM-5 Workgroup on Neurodevelopmental Disorders. *J Am Acad Child Adolesc Psychiatry*. 2012;51(4):347–349.
33. Teive HA, Chien HF, Munhoz RP, et al. Charcot's contribution to the study of Tourette's syndrome. *Arq Neuropsiquiatr*. 2008;66(4):918–921.
34. Felling RJ, Singer HS. Neurobiology of Tourette syndrome: current status and need for further investigation. *J Neurosci*. 2011;31(35):12387–12395.
35. Leckman JF. Tourette's syndrome. *Lancet*. 2002;360(9345):1557–1586.
36. Cubo E. Review of prevalence studies of tic disorders: methodological caveats. Tremor Other Hyperkinet Mov (NY). 2012; 2. pii:tre-02-61-349-1. Epub 2012 May 18. http://www.ncbi.nlm.nih.gov/pubmed/23440028.
37. Peterson BS, Leckman JF. The temporal dynamics of tics in Gilles de la Tourette syndrome. *Biol Psychiatry*. 1998;44(12):1337–1348.
38. Bloch MH, Peterson BS, Scahill L, et al. Adulthood outcome of tic and obsessive–compulsive symptom severity in children with Tourette syndrome. *Arch Pediatr Adolesc Med*. 2006;160(1):65–69.
39. Tourette Syndrome Association International Consortium for Genetics [TSAIG]. Genome scan for Tourette disorder in affected-sibling-pair and multigenerational families. *Am J Hum Genet*. 2007;80(2):265–272.
40. Simonic I, Nyholt DR, Gericke GS, et al. Further evidence for linkage of Gilles de la Tourette syndrome (GTS) susceptibility loci on chromosomes 2p11, 8q22 and 11q23-24 in South African Afrikaners. *Am J Hum Genet*. 2001;105(2):163–167.
41. Taylor LD, Krizman DB, Jankovic J, et al. 9p monosomy in a patient with Gilles de la Tourette's syndrome. *Neurology*. 1991;41(9):1513–1515.
42. Cuker A, State MW, King RA, et al. Candidate locus for Gilles de la Tourette syndrome/obsessive compulsive disorder/chronic tic disorder at 18q22. *Am J Med Genet A*. 2004;130A(1):37–39.
43. Boghosian-Sell L, Comings DE, Overhauser J. Tourette syndrome in a pedigree with a 7;18 translocation: identification of a YAC spanning the translocation breakpoint at 18q22.3. *Am J Hum Genet*. 1996;59(5):999–1005.
44. Robertson MM, Shelley BP, Dalwai S, et al. A patient with both Gilles de la Tourette's syndrome and chromosome 22q11 deletion syndrome: clue to the genetics of Gilles de la Tourette's syndrome? *J Psychosom Res*. 2006;61(3):365–368.
45. Clarke RA, Fang ZM, Diwan AD, et al. Tourette syndrome and Klippel–Feil anomaly in a child with chromosome 22q11 duplication. *Case Rep Med*. 2009;2009:361518.
46. Pauls DL. An update on the genetics of Gilles de la Tourette syndrome. *J Psychosom Res*. 2003;55(1):7–12.
47. Herzberg I, Valencia-Duarte AV, Kay VA, et al. Association of DRD2 variants and Gilles de la Tourette syndrome in a family-based sample from a South American population isolate. *Psychiatr Genet*. 2010;20(4):179–183.
48. Yoon DY, Rippel CA, Kobets AJ, et al. Dopaminergic polymorphisms in Tourette syndrome: association with the DAT gene (SLC6A3). *Am J Med Genet B Neuropsychiatr Genet*. 2007;144B(5):605–610.
49. Moya PR, Wendland JR, Rubenstein LM, et al. Common and rare alleles of the serotonin transporter gene, SLC6A4, associated with Tourette's disorder. *Mov Disord*. 2013;28(9):1263–1270.
50. Dehning S, Müller N, Matz J, et al. A genetic variant of HTR2C may play a role in the manifestation of Tourette syndrome. *Psychiatr Genet*. 2012;20(1):35–38.
51. Bertelsen B, Debes NM, Hjermind LE, et al. Chromosomal rearrangements in Tourette syndrome: implications for identification of candidate susceptibility genes and review of the literature. *Neurogenetics*. 2013;14(3–4):197–203.
52. Paschou P. The genetic basis of Gilles de la Tourette Syndrome. *Neurosci Biobehav Rev*. 2013;37(6):1026–1039.
53. Zuchner S, Cuccaro ML, Tran-Viet KN, et al. SLITRK1 mutations in trichotillomania. *Mol Psychiatry*. 2006;11(10):887–889.
54. Rickards H, Hartley N, Robertson MM. Seignot's paper on the treatment of Tourette's syndrome with haloperidol. Classic Text No. 31. *Hist Psychiatry*. 1997;8(31 Pt 3):433–436.
55. Schwabe MJ, Konkol RJ. Treating Tourette syndrome with haloperidol: predictors of success. *Wis Med J*. 1989;88(10):23–27.
56. Leckman JF, Bloch MH, Scahill L, et al. Tourette syndrome: the self under siege. *J Child Neurol*. 2006;21(8):642–649.
57. Lin H, Yeh CB, Peterson BS, et al. Assessment of symptom exacerbations in a longitudinal study of children with Tourette syndrome or obsessive–compulsive disorder. *J Am Acad Child Adolesc Psychiatry*. 2002;41(9):1070–1077.
58. Swedo SE, Leonard HL, Garvey M, et al. Pediatric autoimmune neuropsychiatric disorders associated streptococcal infections: clinical description of the first 50 cases. *Am J Psychiatry*. 1998;155(2):264–271.
59. Lombroso PJ, Scahill L. Tourette syndrome and obsessive–compulsive disorder. *Brain Dev*. 2008;30(4):231–237.
60. Mell LK, Davis RL, Owens D. Association between streptococcal infection and obsessive–compulsive disorder, Tourette's syndrome, and tic disorder. *Pediatrics*. 2005;116(1):56–60.
61. Allen AJ, Leonard HL, Swedo SE. Case study: a new infection-triggered, autoimmune subtype of pediatric OCD and Tourette's syndrome. *J Am Acad Child Adolesc Psychiatry*. 1995;34(3):307–311.

62. Pavone P, Parano E, Rizzo R, et al. Autoimmune neuropsychiatric disorders associated with streptococcal infection: Sydenham chorea, PANDAS, and PANDAS variants. *J Child Neurol*. 2006;21(9):727–736.
63. Brilot F, Merheb V, Ding A, et al. Antibody binding to neuronal surface in Sydenham chorea, but not in PANDAS or Tourette syndrome. *Neurology*. 2011;76(17):1508–1513.
64. Sukhodolsky DG, Scahill L, Zhang H, et al. Disruptive behavior in children with Tourette's syndrome: association with ADHD comorbidity, tic severity, and functional impairment. *J Am Acad Child Adolesc Psychiatry*. 2003;41(1):98–105.
65. Grados MA, Riddle MA, Samuels JF, et al. The familial phenotype of obsessive–compulsive disorder in relation to tic disorders: the Hopkins OCD family study. *Biol Psychiatry*. 2001;50(8):559–565.
66. Miguel EC, Rosário-Campos MC, Shavitt RG, et al. The tic-related obsessive–compulsive disorder phenotype. In: Cohen DJ, Jankovic J, Goetz CG, eds. *Tourette Syndrome*. Philadelphia: Lippincott, Williams & Wilkins; 2001:43–55.
67. Hooper SD, Johansson AC, Tellgren-Roth C, et al. Genome-wide sequencing for the identification of rearrangements associated with Tourette syndrome and obsessive–compulsive disorder. *BMC Med Genet*. 2012;13:123.
68. Ozomaro U, Cai G, Kajiwara Y, et al. Characterization of SLITRK1 variation in obsessive–compulsive disorder. *PLoS One*. 2013;8(8):e70376.
69. Denys D, de Vries F, Cath D, et al. Dopaminergic activity in Tourette syndrome and obsessive–compulsive disorder. *Eur Neuropsychopharmacol*. 2013;23(11):1423–1431.
70. Sowell ER, Kan E, Yoshii J, et al. Thinning of sensorimotor cortices in children with Tourette syndrome. *Nat Neurosci*. 2008;11(6):637–639.
71. Yoon DY, Gause CD, Leckman JF, et al. Frontal dopaminergic abnormality in Tourette syndrome: a postmortem analysis. *J Neurol Sci*. 2007;255(1–2):50–56.
72. Tobe RH, Bansal R, Xu D, et al. Cerebellar morphology in Tourette syndrome and obsessive–compulsive disorder. *Ann Neurol*. 2010;67(4):479–487.
73. Vaccarino F, Kataoka Y, Lennington J. Cellular and molecular pathology in tourette syndrome. In: Martino D, Leckman JF, eds. *Tourette Syndrome*. Oxford University Press; 2013.
74. Mink JW. Neurobiology of basal ganglia and Tourette syndrome: basal ganglia circuits and thalamocortical outputs. *Adv Neurol*. 2006;99:89–98.
75. Gurney K, Prescott TJ, Wickens JR, et al. Computational models of the basal ganglia: from robots to membranes. *Trends Neurosci*. 2004;27(8):453–459.
76. Alexander GE, Crutcher MD, DeLong MR. Basal ganglia-thalamocortical circuits: parallel substrates for motor, oculomotor, "prefrontal" and "limbic" functions. *Prog Brain Res*. 1990;85:119–146.
77. Middleton FA, Strick PL. Basal ganglia and cerebellar loops: motor and cognitive circuits. *Brain Res Brain Res Rev*. 2000;31(2–3):236–250.
78. Harris K, Singer HS. Tic disorders: neural circuits, neurochemistry, and neuroimmunology. *J Child Neurol*. 2006;21(8):678–689.
79. Saka E, Graybiel AM. Pathophysiology of Tourette's syndrome: striatal pathways revisited. *Brain Dev*. 2003;25(suppl 1):S15–S19.
80. Leckman JF, Vaccarino FM, Kalanithi PS, et al. Annotation: Tourette syndrome: a relentless drumbeat–driven by misguided brain oscillations. *J Child Psychol Psychiatry*. 2006;47(6):537–550.
81. Frey KA, Albin RL. Neuroimaging of Tourette syndrome. *J Child Neurol*. 2006;21(8):672–677.
82. Butler T, Stern E, Silbersweig D. Functional Neuroimaging of Tourette syndrome: advances and future directions. *Adv Neurol*. 2006;99:115–129.
83. Zebardast N, Crowley MJ, Bloch MH, et al. Brain mechanisms for prepulse inhibition in adults with Tourette syndrome: initial findings. *Psychiatry Res*. 214(1):33–41.
84. Peterson BS, Skudlarski P, Anderson AW, et al. A functional magnetic resonance imaging study of tic suppression in Tourette syndrome. *Arch Gen Psychiatry*. 1998;55(4):326–333.
85. Wylie SA, Claassen DO, Kanoff KE, et al. Impaired inhibition of prepotent motor actions in patients with Tourette syndrome. *J Psychiatric Neurosci*. 2013;38(5):349–356.
86. Singer HS, Minzer K. Neurobiology of Tourette syndrome: concepts of neuroanatomical localization and neurochemical abnormalities. *Brain Dev*. 2003;25(suppl 1):S70–S84.
87. Lombroso PJ, Leckman JF. Tourette's syndrome and tic related disorders in children. In: Charney DS, Nestler EJ, eds. *Neurobiology of Mental Illness*. New York: Oxford University Press; 2004:968–978.
88. Lennington JB, Coppola G, Kataoka-Sasaki Y, et al. Transcriptome analysis of the human striatum in Tourette syndrome. *Biological Psychiatry*. In Press.
89. Kita H, Kosaka T, Heizmann CW. Parvalbumin-immunoreactive neurons in the rat neostriatum: a light and electron microscopic study. *Brain Res*. 1990;536(1–2):1–15.
90. Kawaguchi Y, Wilson CJ, Augood SJ, et al. Striatal interneurones: chemical, physiological and morphological characterization. *Trends Neurosci*. 1995;18(12):527–535.
91. Canales JJ, Graybiel AM. A measure of striatal function predicts motor stereotypy. *Nat Neurosci*. 2000;3(4):377–383.
92. Graybiel AN, Canales JJ. The neurobiology of repetitive behaviors: clues to the neurobiology of Tourette syndrome. In: Cohen DJ, Jankovic J, Goetz CG, eds. *Tourette Syndrome*. Philadelphia: Lippincott, Williams & Wilkins; 2001:123–134.
93. Hallett JJ, Harling-Berg CJ, Knopf PM, et al. Anti-telencephalic antibodies in Tourette syndrome cause neuronal dysfunction. *J Neuroimmunol*. 2000;111(1–2):195–202.
94. Taylor JR, Morshed S, Parveen S, et al. An animal model of Tourette's syndrome. *Am J Psychiatry*. 2002;159(4):657–660.
95. Singer HS, Mink JW, Loiselle CR, et al. Microinfusion of antineuronal antibodies into rodent striatum: failure to differentiate between elevated and low titers. *J Neuroimmunol*. 2005;163(1–2):8–14.
96. Tong L, Prieto GA, Kramár EA, et al. Brain-derived neurotrophic factor-dependent synaptic plasticity is suppressed by interleukin-1β via p38 mitogen-activated protein kinase. *J Neurosci*. 2012;32(49):17714–17724.
97. Gibney SM, McGuinness B, Prendergast C, et al. Poly I:C-induced activation of the immune response is accompanied by depression and anxiety-like behaviours, kynurenine pathway activation and reduced BDNF expression. *Brain Behav Immun*. 2013;28:170–181.
98. Peterson BS, Cohen DJ. The treatment of Tourette's syndrome: multimodal, developmental intervention. *J Clin Psychiatry*. 1998;59 (suppl 1):62–72, discussion 73–74.

99. Bagheri MM, Kerbeshian J, Burd L. Recognition and management of Tourette's syndrome and tic disorders. *Am Fam Physician*. 1999;59(8):2262–2272, 2274.
100. Scahill L, Erenberg G, Berlin CM, et al. Contemporary assessment and pharmacotherapy of Tourette syndrome. *NeuroRx*. 2006;3(2):192–206.
101. Sallee FR, Kurlan R, Goetz CG, et al. Ziprasidone treatment of children and adolescents with Tourette's syndrome: a pilot study. *J Am Acad Child Adolesc Psychiatry*. 2000;39(3):292–299.
102. Scahill L, Leckman JF, Schultz RT, et al. A placebo-controlled trial of risperidone in Tourette syndrome. *Neurology*. 2003;60(8):1130–1135.
103. Cummings DD, Singer HS, Krieger M, et al. Neuropsychiatric effects of guanfacine in children with mild Tourette syndrome: a pilot study. *Clin Neuropharmacol*. 25(6):325–332.
104. Allen AJ, Kurlan RM, Gilbert DL, et al. Atomoxetine treatment in children and adolescents with ADHD and comorbid tic disorders. *Neurology*. 2005;65(12):1941–1949.
105. Gilbert DL, Sethuraman G, Sine L, et al. Tourette's syndrome improvement with pergolide in a randomized, double-blind, crossover trial. *Neurology*. 2000;54(6):1310–1315.
106. Gilbert DL, Dure L, Sethuraman G, et al. Tic reduction with pergolide in a randomized controlled trial in children. *Neurology*. 2003;60(4):606–611.
107. Nicolson R, Craven-Thuss B, Smith J, et al. A randomized, double-blind, placebo-controlled trial of metoclopramide for the treatment of Tourette's disorder. *J Am Acad Child Adolesc Psychiatry*. 2005;44(7):640–646.
108. Toren P, Weizman A, Ratner S, et al. Ondansetron treatment in Tourette's disorder: a 3-week, randomized, double-blind, placebo-controlled study. *J Clin Psychiatry*. 2005;66(4):499–503.
109. Singer HS, Wendlandt J, Krieger M, et al. Baclofen treatment in Tourette syndrome: a double-blind, placebo-controlled, crossover trial. *Neurology*. 2001;56(5):599–604.
110. Smith-Hicks CL, Bridges DD, Paynter NP, et al. A double blind randomized placebo control trial of levetiracetam in Tourette syndrome. *Mov Disord*. 2007;22(12):1764–1770.
111. Jankovic J, Jimenez-Shahed J, Brown LW. A randomised, double-blind, placebo-controlled study of topiramate in the treatment of Tourette syndrome. *J Neurol Neurosurg Psychiatry*. 2010;81(1):70–73.
112. Porta M, Maggioni G, Ottaviani F, et al. Treatment of phonic tics in patients with Tourette's syndrome using botulinum toxin type A. *Neurol Sci*. 2004;34(6):420–423.
113. Anca MH, Giladi N, Korczyn AD. Ropinirole in Gilles de la Tourette syndrome. *Neurology*. 2004;62(9):1626–1627.
114. Himle MB, Woods DW, Piacentini JC, et al. Brief review of habit reversal training for Tourette syndrome. *J Child Neurol*. 2006;21(8):719–725.
115. Scahill L, Sukhodolsky D, Bearss K, et al. A randomized trial of parent management training in children with tic disorders and disruptive behavior. *J Child Neurol*. 2006;21(8):650–656.
116. Piacentini JC, Woods DW, Scahill L, et al. Behavior therapy for children with Tourette disorder: a randomized controlled trial. *JAMA*. 2010;303(19):1929–1937.
117. Wilhelm S, Peterson AL, Piacentini JC, et al. Randomized trial of behavior therapy for adults with Tourette syndrome. *Arch Gen Psychiatry*. 2012;69(8):795–803.
118. Scahill L, Woods DW, Himle MB, et al. Current controversies on the role of behavior therapy in Tourette syndrome. *Mov Disord*. 2013;28(9):1179–1183.
119. Goetz CG, Tanner CM, Wilson RS, et al. Clonidine and Gilles de la Tourette's syndrome: double-blind study using objective rating methods. *Ann Neurol*. 1987;21(3):307–310.
120. Leckman JF, Hardin MT, Riddle MA, et al. Clonidine treatment of Gilles de la Tourette's syndrome. *Arch Gen Psychiatry*. 1991;48(4):324–348.
121. Kurlan R, Como PG, Deeley C, et al. A pilot controlled study of fluoxetine for obsessive–compulsive symptoms in children with Tourette's syndrome. *Clin Neuropharmacol*. 1993;16(2):167–172.
122. Newcorn JH, Sutton VK, Zhang S, et al. Characteristics of placebo responders in pediatric clinical trials of attention-deficit/hyperactivity disorder. *J Am Acad Child Adolesc Psychiatry*. 2009;48(12):1165–1172.
123. de la Fuente-Fernández R, Schulzer M, Stoessl AJ. The placebo effect in neurological disorders. *Lancet Neurol*. 2002;1(2):85–91.
124. Hoekstra PJ, Dietrich A, Edwards MJ, et al. Environmental factors in Tourette syndrome. *Neurosci Biobehav Rev*. 2013;37(6):1040–1049.
125. Gilbert DL. Motor cortex inhibitory function in Tourette syndrome: studies using transcranial magnetic stimulation. *Adv Neurol*. 2006;99:107–114.
126. Hariz MI, Robertson MM. Gilles de la Tourette syndrome and deep brain stimulation. *Eur J Neurosci*. 2010;32(7):1128–1134.
127. Kuhn J, Janouschek H, Raptis M, et al. In vivo evidence of deep brain stimulation-induced dopaminergic modulation in Tourette's syndrome. *Biol Psychiatry*. 2012;71(5):e11–e13.
128. Zhuo C, Li L. The application and efficacy of combined neurofeedback therapy and imagery training in adolescents with Tourette syndrome. *J Child Neurol*. 2013.
129. Vaccarino FM, Stevens HE, Kocabas A, et al. Induced pluripotent stem cells: a new tool to confront the challenge of neuropsychiatric disorders. *Neuropharmacology*. 2011;60(7–8):1355–1363.

CHAPTER

108

Addiction

Scott D. Philibin* and John C. Crabbe†

*The University of Texas Southwestern Medical Center, Dallas, TX, USA

†Veterans Affairs Medical Center, Oregon Health & Science University and Portland Alcohol Research Center, Portland, OR, USA

DISEASE CHARACTERISTICS, CLINICAL FEATURES AND DIAGNOSTIC EVALUATION

Addiction is a challenging and enigmatic term to define for medical diagnosis. Diagnostic criteria may need to employ terms and constructs from entirely different languages, descriptions of biological factors and intrapsychic events whose basis is unknown. Most clinicians would agree that the main feature of addiction is compulsive, out-of-control behavior despite adverse consequences. The American Psychiatric Association's (APA) Diagnostic and Statistical Manual of Mental Disorders (DSM-IV) lacked the term entirely. *Addiction* now appears in the DSM-V, replacing "substance abuse and dependence" with "addictions and related disorders." This is fortunate as *abuse* can reflect cultural norms outside medical pathology and *dependence* emphasizes drug withdrawal symptoms, which occurs with prescription antidepressants and beta-blockers that do not cause a loss of control. A new "behavioral addictions" category has been added that contains one disorder: gambling. There are, however, many compulsive behaviors that could be considered addictions, such as bulimia nervosa, Internet addiction, sexual addiction, etc.[1,2]

This review will focus on drug addiction because most of our understanding of addiction *per se* stems from the pharmacology of drugs of abuse. Drug addiction is a complex trait, influenced by genetic and environmental factors, which have been extensively characterized using genetic animal models. Drugs of abuse induce neuroadaptations via molecular and cellular mechanisms, which are being characterized in behavioral neuroscience. In the following sections, we will point the reader to reviews of: 1) human molecular genetic evidence associated with addiction (primarily related to alcoholism, as it is one of the longest-studied complex traits in human genetics); 2) neurobiological basis of addictive drug effects; 3) genetic animal models related to addiction; and 4) therapeutic strategies.

HUMAN MOLECULAR AND GENETIC DATA

Classical Genetic Methods—Twins, Families, Adoptees, and Genetic Epidemiology

Given the complex interplay of genetic and environmental factors predisposing individuals to addiction, much work has concentrated on dissociating the two. Twin and family studies can identify traits with common genetic and/or environmental influence, and assess gene–environment interaction and correlation.[3] Twin studies support a 50% genetic contribution to risk for alcoholism, leaving 50% accounted for by environment.[4] There is substantial comorbidity across addiction disorders with different drug classes and other behavioral disorders. For example, an impulsivity trait may coincide with high-risk behaviors, such as alcohol abuse.[5] These studies implicate some common genes influencing multiple addiction disorders, but each disorder also reflects its own unique genetic influences.

Molecular Genetic Studies—Association and Linkage

As the Human Genome Project gained information about the specific DNA sequences in the human genome, the power and precision of genetic mapping greatly increased.[6] Genetic mapping attempts to correlate genetic markers

http://dx.doi.org/10.1016/10.1016/B978-0-12-410529-4.00108-X

(i.e., localized DNA sequences), with a phenotypic trait being mapped across groups of individuals. Co-occurrence of specific markers with the trait suggests a gene included in the marker sequence, or a proximal gene on the same chromosome (i.e., linked), is directing synthesis of a gene product influencing risk.

These kinds of analyses make hundreds of comparisons, so there is a high risk of detecting false positive associations. Also, populations that appear homogeneous may vary in gene frequencies for genes in the associated region. This problem of *stratification* is prevalent in these types of studies, as the control and addicted groups may actually be drawn from two genetically distinct populations.[7,8] Also, as addictions are multigenic or polygenic traits; each gene mapped exerts only a small influence, so there is low statistical power to detect linkage. Extremely strong association and linkage data for markers are required for even a single-gene disorder, such as Huntington disease.[9] Therefore, it has been strongly argued for seeking converging evidence from multiple and varied genetic and non-genetic techniques.[10–14]

Despite the intrinsic difficulties, hundreds of linkage and association studies pursue addiction-relevant genes. Linkage and association genome scans for addiction risk have provided converging data implicating several chromosomal regions likely to harbor allelic variants contributing to drug addiction.[15] However, it is not currently known whether the candidate genes nominated by linkage and association studies in fact contribute to the disorders being mapped. This inferential problem is shared by studies using animal models, and because the solutions are conceptually similar, we will discuss the next steps toward proving a functional role for a candidate gene in a later section.

Enhancing the Power of Human Genetic Mapping Studies

There are several approaches to solving the problems of low signal-to-noise in gene mapping studies. Population sizes can obviously be increased, but this is very expensive and may not increase power very much. The use of "endophenotypes" has become popular, as they are postulated to reflect the effects of closely linked genes on a more tractable trait. Endophenotypes can be selectively measured components of complex neuropsychiatric disorders that represent neurophysiological, biochemical, endocrinological, neuroanatomical, cognitive or neuropsychological aspects[16] and can be used to develop preclinical models,[17] which take a reductionist approach to the study of addiction.

In recognition of the comorbidity of many traits, and the pleiotropic effects of genes on multiple traits, it has been suggested that some endophenotypes may be "closer to the gene." One example is the abnormal P300 sensory evoked potential and its relation to substance use disorders.[18] On the assumption that the collection of genes contributing to abnormal P300 responses will be simpler than that contributing to the complex addiction trait, mapping studies have identified genetic regions (i.e., quantitative trait loci [QTLs], discussed later) where a gene or genes influencing P300 must reside. Similarly, a long-term epidemiological study demonstrated nearly 30 years ago that college-age men with a positive family history for alcoholism were less sensitive to the effect of an acute dose of ethanol to induce body sway and subjective intoxication.[19] Many years later, family history predicted alcoholism diagnosis, but low sensitivity to acute ethanol was an even better predictor, accounting for most individual differences. The trait conferring low sensitivity to ethanol was then mapped to specific genetic markers.[20]

The discovery of single nucleotide polymorphisms (SNPs) led to a new generation of "markers for addiction." Genome-wide linkage and association studies have been applied to the study of drug addiction with varying degrees of success.[21] For example, γ-aminobutyric acid (GABA) transmission is largely understood to mediate some of the pharmacological effects of alcohol in the brain. Despite multiple studies implicating SNPs in the genes encoding the GABRA2 and GABRG1 subunits of the GABA-A receptor in risk for alcohol-related behaviors, specific functional alleles underlying these associations have yet to be identified. Conversely, genome-wide association studies of smoking behavior have reported more reliable genetic evidence for association, indicating variants within the nicotinic acetylcholine receptor (nAChR) subunit genes on human chromosomes 15q (CHRNA5-CHRNA3-CHRNB4) and 8p (CHRNA6-CHRNB3) that influence risk for nicotine dependence. Analyzing the currently abundant number of SNP markers has been facilitated by the availability of high throughput database technologies.[22] However, the use of SNPs is not without its disadvantages, primary among which is that the map of SNPs in the genome is less well articulated than that for standard multi-base DNA markers and the sheer number of SNPs is overwhelming. Moreover, there are discrepancies among many case-controlled studies, occurring by chance, due to population stratifications, and different definitions of "cases" and/or improper control groups.

More powerful designs, such as the transmission disequilibrium test (TDT) and its variants, can increase detection power. The TDT compares marker frequencies among family members with shared traits of interest versus family members who do not, which reduces the probability of false positive associations.[23,24] Statistical genetic methods are also being developed to deal with issues in quantitative genomics, such as gene–gene interactions.[25–28] By statistically conditioning a gene's effects on the genotype at another locus, genetic signal strength can be enhanced. Such methods are computationally difficult and lead to great increases in the number of comparisons made, but they may reveal genetic effects masked by the presence of other genes.

A Case of Protective Genes—ADH and ALDH Polymorphisms

The discovery of the genes encoding alcohol dehydrogenase (*ADH*), which oxidizes ethyl alcohol to acetylaldehyde, and aldehyde dehydrogenase (*ALDH*), which oxidizes the toxic metabolite acetlyaldehyde to the innocuous products acetate and carbon dioxide, led to the most well-established genetic factors associated with alcoholism.[29] Most humans possess variant alleles for the aldehyde dehydrogenase (*ALDH*) gene that lead to synthesis of an efficient *ALDH* protein. There is a variant allele for the *ALDH* gene that leads to a slow-metabolizing form of the protein, *ALDH*2*2; this variant is present at high frequencies in Southeast Asian gene pools. When an individual heterozygous (or, more strikingly, homozygous) for *ALDH*2*2 drinks alcohol, he or she rapidly accumulates acetaldehyde and frequently experiences its adverse effects (e.g., facial flushing, nausea, dizziness, and headaches). Disulfiram, an *ALDH* inhibitor, has long been used as a treatment for alcoholism under the trade name Antabuse®. An individual taking this drug chronically experiences the flushing, nausea, etc., after drinking alcohol, just as if a variant enzyme genetically protected him or her from alcoholism. Unfortunately, treatment is limited due to a lack of patient compliance, but it is not entirely without therapeutic efficacy.

BEHAVIORAL NEUROSCIENCE FRAMEWORKS

Drug-Induced Neuroplasticity

Candidate genes are nominated based on previously existing biological data, and much of addiction research focuses on the effects of drugs of abuse on the malleable brain. Repeated administration of drugs of abuse induces progressive neuroadaptations. These adaptations can include varied aspects of neuroplasticity, such as changes in molecular and cellular processes including the modulation of receptors, ion channels, intracellular signaling proteins and gene expression. Such neuroadaptations are revealed by the development of *tolerance, sensitization* and *dependence*.

Tolerance, a diminishing pharmacological effect over time, can develop at different rates depending on the specific effect. For example, tolerance develops more frequently to analgesic, euphoric, and respiratory depressant effects of opiates, compared to pupillary constriction. However, tolerance to the respiratory depressant effects of heroin may develop more slowly and less completely than tolerance to the euphoric effects, contributing to overdose in long-term users.[30] Sensitization, an escalating pharmacological effect over time that is pronounced with psychostimulants such as cocaine and methamphetamine, may reflect neuroadaptations underlying craving and relapse.[31] Drug dependence is unmasked upon abrupt cessation of drug use and manifested as symptoms of withdrawal. Symptoms of withdrawal can be physical and/or psychological (i.e., emotional–motivational). While some addictive drugs, such as ethanol and opiates, induce severe physical symptoms of withdrawal, other drugs, such as cocaine and amphetamine, do not.

Dysregulation of Opponent Processes Theory

One theory of drug addiction is based on the dysregulation of brain reward and stress systems.[32,33] Repeated bouts of dopaminergic stimulation by drugs of abuse important for reward and positive reinforcement properties eventually compromise dopaminergic activity, while sustained CRF-CRF1 receptor activity during drug withdrawal progressively exacerbates the stress response. This combination of neuroadaptations encourages the escalation of drug use, enabling drugs of abuse with negative reinforcing properties via the temporary alleviation of this negative affective state, a "negative reinforcement" process. The ventral striatum, which includes the nucleus accumbens, is suggested to be critical for the rewarding properties of drugs and the extended amygdala is associated with the stress response and negative reinforcement.

Synaptic Plasticity and Associative Learning Theory

Drug addiction has also been conceptualized as an aberrant form of learning.[34–36] Mechanisms of natural reward learning crucial for survival during evolution appear to be usurped by drugs of abuse, which produce overriding and sustained stimulation. Dopaminergic receptor activation, critical for the rewarding properties of drugs of abuse, appears to interact with glutamatergic receptor signaling, important for reward-based learning. Complex networks of intracellular signal transduction mechanisms downstream of these receptors modulates functional and structural plasticity important for associative learning processes, such as those underlying the association of the conditioned stimulus effects of drugs of abuse with the stimuli that predict them.[37] While reward processed in the nucleus accumbens is critical for acquisition of goal-directed behaviors, these behaviors repeated over time develop into more automatic repertoires characteristic of stimulus–response or habit learning, which is largely understood to involve more progressed neuroplasticity in dorsal striatal circuits.[38]

GENETIC ANIMAL MODELS

Based on neurobiological concepts of drug addiction, such as those just reviewed, genetic underpinnings can be hypothesized and tested using genetic animal models. Why study nonhuman animals? Humans are a messy species for genetic studies. They have few offspring, small families, an inconveniently long generation time; they breed with whomever they please; and ascertaining the genetic specifics of their ancestry is difficult. The advantages of nonhuman species are obvious—on every feature named above, rodents are far more desirable subjects than humans. Unfortunately, rodent behavior can never model the full spectrum of any complex behavioral human trait. Species-specific differences in physiology, behavior, social structure, and their interactions ensure an animal model can never be isomorphic with the human psychiatric trait (or diagnosis) it targets. Fortunately, the protein-coding regions of human and mouse genomes are 85% identical based on regions evolutionarily conserved due to shared ancestry. Therefore, identifying a DNA region in the mouse genome will predict where that particular region is in the human genome about 85% of the time, and many genes and their functions are also conserved. It is possible to test the effects of DNA alterations in human diseases and carefully study the effects in the laboratory mouse, including their potential influence on response to therapeutic agents. Thus, genetic animal models have taken a reductionist approach to the study of complex traits by modeling various aspects separately.[39–41]

Classical Genetic Animal Models

Selective Breeding

Artificial selection or selective breeding is the one of the oldest and most powerful methods in behavioral genetics. In the late 1940s, high and low preferring lines of rats were bred to drink alcohol solutions in preference to water at the University of Chile. Now there are a variety of rat and mouse lines selectively bred to differ in various responses to ethanol, including drinking preference, tolerance and withdrawal severity.[42] One of the features of selective breeding is that selecting for one trait leads to the development of correlated pleiotropic differences in many other traits. For example, lines bred for differences in ethanol preference also differ in tolerance.[40] Recently, some attempts have also been made to selectively breed lines differing in traits that are comorbid with addiction disorders. Anxiety, impulsivity, antisocial behavior, and depression can be modeled in rat and mouse behavioral assays, although some of these behavioral assays have a bit more than the usual level of difficulty in convincing nonbelievers they possess face validity. Given the intrinsic power of this method to assess genetic pleiotropy, it might be worthwhile developing additional lines of mice or rats that differ in some of the other traits correlated with drug abuse susceptibility in humans.

Inbred Strains

One of the most straightforward genetic animal models is to examine existing genetic variation. There exist well over 100 inbred strains of mice and as many of rats. All same-sex members of such a strain (e.g., C57BL/6 mouse) are somewhat like monozygotic twins, essentially genetic clones reproduced through many generations of brother–sister matings, which possess two copies of a single allele at all genes. If many strains are compared for a trait in a controlled environment, mean differences among the strains can be attributed to genetic differences. C57BL/6 strain mice were shown early on to prefer to drink alcohol solutions, while other strains, such as DBA/2, were nearly complete avoiders.[41] Because the genotypes of each strain remain nearly invariant over time, studies conducted over

50 years ago are still informative today. A historical comparison found that in inbred strains, characteristic alcohol preference has remained stable over time, like brain weight data.[43] Some phenotypes, however, show less comparability over time (e.g., anxiety-like behavior in the elevated plus maze).[43] The specific genes that lead to the strain differences have remained elusive, in part, because the large numbers of alleles for any gene represented in a multistrain survey make genotyping difficult and expensive.

One boost to strain surveys came with the Mouse Phenome Project launched over a decade ago to complement the mouse genome sequencing effort.[44] By supplying up to 40 commonly used and genetically diverse inbred strains to researchers, and assembling a relational database containing the resulting behavioral and physiological data, centralized access to strain data has and continues to vastly improve. The database allows for appropriate strains to be chosen for specific behavioral tasks modeling human diseases, and has improved inferences about the influence of environment on genotype. One very powerful use of this database is correlational analyses of the strain's average phenotypes, which have already taught us much about codetermination of genetic influence.

Quantitative Trait Locus Mapping

A quantitative trait is a measurable phenotype emerging from genetic and environmental factors that is distributed in magnitude in a population rather than all or none. A quantitative trait locus (QTL) is a specific chromosomal region or genetic locus in which particular sequences of bases in DNA markers are statistically associated with variation in the trait. Several polymorphic genes and environmental conditions often influence these quantitative traits and one or many QTL(s) can influence a phenotypic trait.[45] Inbred strains, selected lines, and other genetically specified populations have been used in studies analogous to the human population association and linkage studies described above.

The goals are first to locate a QTL harboring a gene or genes affecting the trait to be mapped, and then refine that genomic map until a single gene or genes can be implicated in the effect on the trait. Currently, QTL fine mapping usually involves the development of congenic strains. In a congenic strain, a very small sequence of DNA on a chromosome is moved from one genetic background to another inbred strain background. An excellent discussion of QTL mapping methods discusses, in depth, the trait of alcohol withdrawal severity.[46] Of course, each QTL generally accounts for only a small proportion of the variability in a complex behavioral trait like addiction, so this is a difficult task and cautious interpretation is warranted. The probability of success in QTL mapping depends on: 1) the heritability of the trait; 2) whether the underlying quantitative trait gene (QTG) is dominant, recessive or additive; 3) the number of genes that affect the trait; 4) whether or not their effects are interactive; and (5) most importantly, the number of subjects that can be tested (i.e., the statistical power of the mapping effort).

Many addiction-related traits have been targeted for QTL mapping studies, although very few of these QTLs have been reduced to QTGs or quantitative trait nucleotides (QTNs). The recent discovery of the addiction-relevant QTG, *Mpdz*, which possesses pleiotropic effects on the predisposition to severe alcohol and barbiturate withdrawal, demonstrates the power of this approach.[47] Further studies have shown that variation in the human *MPDZ* gene is related to alcohol drinking.[46,48] Unfortunately, QTL studies have yet to resolve to a QTG for drinking, in part, due to problems discussed above.[49] However, three candidate genes, neuropeptide Y, α-synuclein, and *CRFR2* have been associated with ethanol-seeking.[50] Encouraging evidence shows some consistencies for alcohol and other substance-dependence phenotypes in humans and mice. The more long-term goals of QTL mapping projects are then to move to human populations for studies of the homologous or orthologous gene, and use information about the biological effects of the gene's product to help design therapeutic agents or other therapies.[15]

MOLECULAR APPROACHES

Targeted Mutagenesis

The advent of genomic manipulation has greatly increased our ability to understand the molecular mechanisms and cellular pathways underlying disease states.[51] The targeted mutation of genes is a powerful approach that has integrated molecular genetics with behavioral neuroscience. The advance in genetic animal models pivotal to this was establishing the ability to manipulate the genome directly during the 1980s. This dramatic demonstration consisted of inserting the metallothionein-growth hormone gene into the pronucleus of mouse eggs to develop very large adult mice.[52] This is an example of a knock-in mouse, in which alterations of a specific genetic locus consist of DNA substitution of, or addition to, the endogenous genetic locus. Knockout mice result when a genetic locus has been targeted and rendered nonfunctional either by the insertion of irrelevant DNA to disrupt the expression of the encoding locus or by the deletion of DNA from the targeted locus. A newer generation of knockout mice is conditional and/or inducible. These are produced by inserting recombinase recognition elements flanking the gene sequence (i.e., region of interest) to facilitate its deletion from specific cells (e.g., dopamine D1 receptor-containing neurons) and/or at a specific time of development (e.g., adult). Lists of targeted genes are available online (e.g., http://www.knockoutmouse.org/) and use of these mutant mouse models will continue to be invaluable, especially as the tissue- and temporally-specific genetic mutations engineered in mice become more widely used. A recent review discusses the use of many such mouse mutants to study alcohol-related traits.[53]

Regulation of Gene Expression

Association and linkage studies in humans and QTL mapping studies in rodents were traditionally based on differences in DNA sequence. However, the regulation of gene expression has been suggested as a mechanism underlying the transition to the drug addicted state. A mapped QTL can turn out to affect the trait because the sequence identified harbors transcription factors, proteins that bind to regulatory regions of specific genes to modulate and stabilize their expression. Chronic administration of drugs of abuse may induce neuroadaptations expressed via several transcription factors, including ΔFosB, cAMP response element-binding protein (CREB) and nuclear factor kappa B (NFκB).[54] These nuclear changes could underlie persistent alterations in the expression of specific target genes, resulting in drug-induced neuroplasticity underlying maladaptive behaviors characteristic of addiction.

For example, ΔFosB has been proposed as a "molecular switch" for the initiation and maintenance of the addicted state, as it is persistently increased by repeated administration of various drugs of abuse in addiction-relevant brain areas (e.g., nucleus accumbens and dorsal striatum). Overexpression of ΔFosB can be induced exclusively in the nucleus accumbens and dorsal striatum of mutant mice, which exhibit augmented behavioral responses to cocaine, a phenotype that suggests vulnerability to addiction.[54] A downstream target of ΔFosB is the neuronal protein kinase, cyclin-dependent kinase 5 (Cdk5). Cdk5 mRNA, protein and activity are increased in response to repeated cocaine. Cdk5 appears to serve a homeostatic function that dampens the effects of D1 dopamine receptor/cAMP/PKA signaling activated by chronic cocaine. Cdk5 is also implicated in the modulation of cocaine-induced plasticity in dendritic spine morphology. Moreover, conditional knockout potentiates many of the behavioral effects of cocaine.[55] These converging lines of evidence implicate Cdk5 in cocaine-induced neuroplasticity relevant to the transition to the addicted state.[56]

The most recent studies of gene expression have taken the approach of seeking the influence of groups of genes rather than single genes. Many genes show correlated patterns of expression, and when expression is surveyed genome-wide, different gene clusters are identified. Within these clusters, key genes can be identified as those whose expression is highly correlated to many other genes. One such analysis recently explored mouse lines selectively bred for binge-like drinking and identified several networks, including those containing genes previously implicated in alcohol drinking such as *Gabarg1* encoding a GABA-A receptor.[57]

THERAPY

Existing Therapy

Addiction treatment strategies are varied and often applied in combination. These include the familiar twelve-step self-help programs, psychosocial therapies, brief interventions, behavior therapies, psychodynamic therapies, cognitive–behavioral therapy, and multiple forms of couples/family/group/community-targeted therapies. Most medical disorders are currently treated with drugs, and the addictions are no exception.

Three drugs are approved by the US Food and Drug Administration (FDA) for treatment of alcoholism in the USA: the aldehyde dehydrogenase blocker disulfiram (discussed previously), the opioid antagonist naltrexone, and the functional glutamate antagonist acamprosate (a drug with multiple effects including on NMDA receptors).

A number of other agents are under investigation and have shown promise, including selective serotonin reuptake inhibitors (SSRIs) and other drugs affecting serotonin function (e.g., fluoxetine, sertraline, ondansetron). Opioid agonists (methadone, levo-alpha-acetyl-methadol [LAAM]) and nicotine replacement are used for their respective addictions. The use of the long-acting opioid agonists (methadone and certain of its derivatives) to blunt heroin's reinforcing and physical withdrawal effects is a well-established case where targeting a drug's receptor binding sites has been therapeutically efficacious. Many years have been spent in pursuit of a drug that would "antagonize" cocaine and/or methamphetamine's effects, with little success (e.g., tricyclic antidepressants, SSRIs, disulfiram, dopaminergic agents and $GABA_A$-$GABA_B$ ligands). Potential new treatments for stimulant abuse, especially cocaine, focus on vaccine technologies, including gene transfer of highly optimized monoclonal antibodies.[58] Addiction research on the therapeutic effects of classic hallucinogens was halted in the early 1970s, but clinical trials have recently begun.[59] Ayahuasca is a hallucinogenic botanical mixture that contains *N,N*-dimethyltryptamine (DMT), a hallucinogen, and harmine, a monoamine oxidase inhibitor (MAOI), which attenuates the breakdown of DMT. Intriguingly, ayahuasca appears to have therapeutic effects in the treatment of addiction to alcohol and cocaine.[60]

In addition, there are many complementary or alternative medical treatments applied to the addictions, whether by practitioners or by the patient, including acupuncture, yoga, and herbal therapies. These alternative therapies are often administered in conjunction with routine medical treatment and are increasingly used. As they are generally not studied in a scientifically controlled design, their efficacy is hard to evaluate. To an increasing extent, different treatment modalities are being offered in combination; for example, multiple drugs with different actions combined with psychotherapy, or using vouchers in a community reinforcement approach in patients on methadone maintenance. Potentially problematically, self-treatment with herbal preparations can influence metabolism of medically prescribed drugs. The authors are neither clinicians nor involved in human subjects research, so opinion stated here is derived from the literature rather than experience. It appears to the authors that nearly all of the conventional therapies will help a percentage of the individuals who try them, on the order of 20–25% in most large controlled studies. It has proven difficult to predict who will be helped by which therapy, and the majority of addicted patients seeking therapy of any sort will not succeed in remaining abstinent for at least a year. Thus, while there is clear value to the existing therapies, there is also ample room and need for improvement. The potential value in combining these treatments is largely unknown.

Prospects for Genomically Driven Therapies

The term pharmacogenomics has been used to emphasize the use of genome-wide approaches to identify genes conferring risk for or protection against disease or drug response. Pharmacogenomics is an emerging field that seeks to use genetic information to discover new targets for drug development and to individualize treatment based on an individual's genetic fingerprint. It has assumed the knowledge from the field of "pharmacogenetics," which has long sought to identify polymorphisms in drug enzymes and receptors that result in genetically based, individual differences in drug metabolism, and builds upon that knowledge. A useful review of the field[61] describes the central feature of pharmacogenomics as its focus on groups of genetic polymorphisms used to predict both efficacy and toxicity of a drug for an individual. We know of no cases to date where this method has been applied to addiction-related traits, but the enterprise is a logical extension of the wealth of accumulating Human Genome Project-driven genomic data. The general idea is that this approach will elucidate targets other than those already established by hypothesis-driven research in neurobiology.

Major pharmaceutical companies are avidly pursuing the characterization of many complex disease traits using this strategy. However, there has been a major lack of interest by the companies in pursuing the addictions that is historical, which we find very puzzling. Given the very high prevalence of addiction, and the fact that most of the afflicted are gainfully employed and could thus pay for their drugs, it is troubling that addiction disorders have not been a focus of greater interest for the pharmaceutical industry.

CONCLUSIONS AND FUTURE DIRECTIONS

Molecular and genetic studies of complex diseases have experienced spectacular progress in the last few years, much of it enabled by knowledge of and the ability to manipulate the genome. Nonetheless, progress has been greatest for those disorders with simpler genetic structure, i.e., those where a gene or a few genes are of great importance to risk. Progress in the treatment of pathological complex traits, such as addiction, will require substantially more effort to succeed. This will require identifying the interactions across the levels of genes, their proteins, and environmental influences. We will need to confirm brain regional differences in expression and functional significance of those changes. The interaction of these factors should be explored systematically at the whole-organism level, whose most integrated expression is behavior (i.e., behavioral genomics). The potential of a behavioral genomics perspective that will discern the pathways leading from genetic predisposition to addiction is there. This is a large challenge for scientists, but the basic data are in place and the research tools are rapidly being improved.

ACKNOWLEDGEMENTS

Preparation of this chapter was supported by grants from the National Institute on Drug Abuse, the National Institute on Alcohol Abuse and Alcoholism, the Brain & Behavior Research Foundation (Formerly NARSAD), the Department of Veterans Affairs, and the National Institute on Alcohol Abuse and Alcoholism and the Department of the Army/DoD-TATRC.

References

1. Grant JE, Brewer JA, Potenza MN. The neurobiology of substance and behavioral addictions. *CNS Spectr*. 2006;11(12):924–930.
2. Grant JE, Potenza MN, Weinstein A, Gorelick DA. Introduction to behavioral addictions. *Am J Drug Alcohol Abuse*. 2010;36(5):233–241.
3. Lachman HM. An overview of the genetics of substance use disorders. *Curr Psychiatry Rep*. 2006;8(2):133–143.
4. Enoch MA. The influence of gene-environment interactions on the development of alcoholism and drug dependence. *Curr Psychiatry Rep*. 2012;14(2):150–158.
5. Birkley EL, Smith GT. Recent advances in understanding the personality underpinnings of impulsive behavior and their role in risk for addictive behaviors. *Curr Drug Abuse Rev*. 2011;4(4):215–227.
6. Williams M. The genome: five years on. *Curr Opin Investig Drugs*. 2006;7(1):14–17.
7. Lichtermann D, Franke P, Maier W, Rao ML. Pharmacogenomics and addiction to opiates. *Eur J Pharmacol*. 2000;410(2–3):269–279.
8. Ott J, Kamatani Y, Lathrop M. Family-based designs for genome-wide association studies. *Nat Rev Genet*. 2011;12(7):465–474.
9. Gusella JF, MacDonald ME. Huntington's disease: seeing the pathogenic process through a genetic lens. *Trends Biochem Sci*. 2006;31(9):533–540.
10. Wong CC, Schumann G. Review. Genetics of addictions: strategies for addressing heterogeneity and polygenicity of substance use disorders. *Philos Trans R Soc Lond B Biol Sci*. 2008;363(1507):3213–3222.
11. Belknap JK, Hitzemann R, Crabbe JC, Phillips TJ, Buck KJ, Williams RW. QTL analysis and genomewide mutagenesis in mice: complementary genetic approaches to the dissection of complex traits. *Behav Genet*. 2001;31(1):5–15.
12. Jacob T, Sher KJ, Bucholz KK, et al. An integrative approach for studying the etiology of alcoholism and other addictions. *Twin Res*. 2001;4(2):103–118.
13. Agrawal A, Verweij KJ, Gillespie NA, et al. The genetics of addiction – a translational perspective. *Transcult Psychiatry*. 2012;2:e140.
14. Ho MK, Goldman D, Heinz A, et al. Breaking barriers in the genomics and pharmacogenetics of drug addiction. *Clin Pharmacol Ther*. 2010;88(6):779–791.
15. Uhl GR, Drgon T, Johnson C, et al. "Higher order" addiction molecular genetics: convergent data from genome-wide association in humans and mice. *Biochem Pharmacol*. 2008;75(1):98–111.
16. Gottesman II, Gould TD. The endophenotype concept in psychiatry: etymology and strategic intentions. *Am J Psychiatry*. 2003;160(4):636–645.
17. Gould TD, Gottesman II. Psychiatric endophenotypes and the development of valid animal models. *Genes Brain Behav*. 2006;5(2):113–119.
18. Singh SM, Basu D. The P300 event-related potential and its possible role as an endophenotype for studying substance use disorders: a review. *Addict Biol*. 2009;14(3):298–309.
19. Schuckit MA. Ethanol-induced changes in body sway in men at high alcoholism risk. *Arch Gen Psychiatry*. 1985;42(4):375–379.
20. Schuckit MA, Edenberg HJ, Kalmijn J, et al. A genome-wide search for genes that relate to a low level of response to alcohol. *Alcohol Clin Exp Res*. 2001;25(3):323–329.
21. Wang JC, Kapoor M, Goate AM. The genetics of substance dependence. *Annu Rev Genomics Hum Genet*. 2012;13:241–261.
22. Teufel A, Krupp M, Weinmann A, Galle PR. Current bioinformatics tools in genomic biomedical research (Review). *Int J Mol Med*. 2006;17(6):967–973.
23. Ewens WJ, Li M, Spielman RS. A review of family-based tests for linkage disequilibrium between a quantitative trait and a genetic marker. *PLoS Genet*. 2008;4(9):e1000180.
24. Ewens W, Li M. Comments on the entropy-based transmission/disequilibrium test. *Hum Genet*. 2008;123(1):97–100.
25. McKinney BA, Reif DM, Ritchie MD, Moore JH. Machine learning for detecting gene-gene interactions: a review. *Appl Bioinformatics*. 2006;5(2):77–88.
26. Moore JH. Detecting, characterizing, and interpreting nonlinear gene-gene interactions using multifactor dimensionality reduction. *Adv Genet*. 2010;72:101–116.
27. Gui J, Moore JH, Kelsey KT, Marsit CJ, Karagas MR, Andrew AS. A novel survival multifactor dimensionality reduction method for detecting gene-gene interactions with application to bladder cancer prognosis. *Hum Genet*. 2011;129(1):101–110.
28. Gui J, Andrew AS, Andrews P, et al. A robust multifactor dimensionality reduction method for detecting gene-gene interactions with application to the genetic analysis of bladder cancer susceptibility. *Ann Hum Genet*. 2011;75(1):20–28.
29. Liu J, Zhou Z, Hodgkinson CA, et al. Haplotype-based study of the association of alcohol-metabolizing genes with alcohol dependence in four independent populations. *Alcohol Clin Exp Res*. 2011;35(2):304–316.
30. Warner-Smith M, Darke S, Lynskey M, Hall W. Heroin overdose: causes and consequences. *Addiction*. 2001;96(8):1113–1125.
31. Hearing MC, Zink AN, Wickman K. Cocaine-induced adaptations in metabotropic inhibitory signaling in the mesocorticolimbic system. *Rev Neurosci*. 2012;23(4):325–351.
32. Koob GF. Theoretical frameworks and mechanistic aspects of alcohol addiction: alcohol addiction as a reward deficit disorder. *Curr Top Behav Neurosci*. 2013;13:3–30.
33. George O, Le Moal M, Koob GF. Allostasis and addiction: role of the dopamine and corticotropin-releasing factor systems. *Physiol Behav*. 2012;106(1):58–64.
34. Torregrossa MM, Corlett PR, Taylor JR. Aberrant learning and memory in addiction. *Neurobiol Learn Mem*. Mar 3 2011.

35. Hyman SE, Malenka RC, Nestler EJ. Neural mechanisms of addiction: the role of reward-related learning and memory. *Annu Rev Neurosci.* 2006;29:565–598.
36. Milton AL, Everitt BJ. The persistence of maladaptive memory: addiction, drug memories and anti-relapse treatments. *Neurosci Biobehav Rev.* 2012;36(4):1119–1139.
37. Philibin SD, Hernandez A, Self DW, Bibb JA. Striatal signal transduction and drug addiction. *Front Neuroanat.* 2011;5:60.
38. Pierce RC, Vanderschuren LJ. Kicking the habit: the neural basis of ingrained behaviors in cocaine addiction. *Neurosci Biobehav Rev.* 2010;35(2):212–219.
39. Sher KJ, Dick DM, Crabbe JC, Hutchison KE, O'Malley SS, Heath AC. Consilient research approaches in studying gene x environment interactions in alcohol research. *Addict Biol.* 2010;15(2):200–216.
40. Crabbe JC, Kendler KS, Hitzemann RJ. Modeling the diagnostic criteria for alcohol dependence with genetic animal models. *Curr Top Behav Neurosci.* 2013;13:187–221.
41. Crabbe JC, Phillips TJ, Belknap JK. The complexity of alcohol drinking: studies in rodent genetic models. *Behav Genet.* 2010;40(6):737–750.
42. Crabbe JC. Translational behaviour-genetic studies of alcohol: are we there yet? *Genes Brain Behav.* 2012;11(4):375–386.
43. Wahlsten D, Bachmanov A, Finn DA, Crabbe JC. Stability of inbred mouse strain differences in behavior and brain size between laboratories and across decades. *Proc Natl Acad Sci U S A.* 2006;103(44):16364–16369.
44. Maddatu TP, Grubb SC, Bult CJ, Bogue MA. Mouse phenome database (MPD). *Nucleic Acids Res.* Jan 2012;40(Database issue):D887–D894.
45. Abiola O, Angel JM, Avner P, et al. The nature and identification of quantitative trait loci: a community's view. *Nat Rev Genet.* 2003;4(11):911–916.
46. Milner LC, Buck KJ. Identifying quantitative trait loci (QTLs) and genes (QTGs) for alcohol-related phenotypes in mice. *Int Rev Neurobiol.* 2010;91:173–204.
47. Shirley RL, Walter NA, Reilly MT, Fehr C, Buck KJ. Mpdz is a quantitative trait gene for drug withdrawal seizures. *Nat Neurosci.* 2004;7(7):699–700.
48. Ehlers CL, Walter NA, Dick DM, Buck KJ, Crabbe JC. A comparison of selected quantitative trait loci associated with alcohol use phenotypes in humans and mouse models. *Addict Biol.* 2010;15(2):185–199.
49. Crabbe JC. Review. Neurogenetic studies of alcohol addiction. *Philos Trans R Soc Lond B Biol Sci.* 2008;363(1507):3201–3211.
50. Spence JP, Liang T, Liu L, et al. From QTL to candidate gene: a genetic approach to alcoholism research. *Curr Drug Abuse Rev.* 2009;2(2):127–134.
51. Doyle A, McGarry MP, Lee NA, Lee JJ. The construction of transgenic and gene knockout/knockin mouse models of human disease. *Transgenic Res.* 2012;21(2):327–349.
52. Palmiter RD, Brinster RL, Hammer RE, et al. Dramatic growth of mice that develop from eggs microinjected with metallothionein-growth hormone fusion genes. *Nature.* 1982;300(5893):611–615.
53. Bilbao A. Advanced transgenic approaches to understand alcohol-related phenotypes in animals. *Curr Top Behav Neurosci.* 2013;13:271–311.
54. Nestler EJ. Transcriptional mechanisms of drug addiction. *Clin Psychopharmacol Neurosci.* 2012;10(3):136–143.
55. Benavides DR, Quinn JJ, Zhong P, et al. Cdk5 modulates cocaine reward, motivation, and striatal neuron excitability. *J Neurosci.* 2007;27:1967–12976.
56. Benavides DR, Bibb JA. Role of Cdk5 in drug abuse and plasticity. *Ann N Y Acad Sci.* 2004;1025:335–344.
57. Iancu OD, Overbeck D, Darakjian P, et al. Selection for drinking in the dark alters brain gene coexpression networks. *Alcohol Clin Exp Res.* 2013;37(8):1295–1303.
58. Brimijoin S, Shen X, Orson F, Kosten T. Prospects, promise and problems on the road to effective vaccines and related therapies for substance abuse. *Expert Rev Vaccines.* 2013;12(3):323–332.
59. Bogenschutz MP, Pommy JM. Therapeutic mechanisms of classic hallucinogens in the treatment of addictions: from indirect evidence to testable hypotheses. *Drug Test Anal.* 2012;4(7–8):543–555.
60. Brierley DI, Davidson C. Developments in harmine pharmacology–implications for ayahuasca use and drug-dependence treatment. *Prog Neuropsychopharmacol Biol Psychiatry.* 2012;39(2):263–272.
61. Eichelbaum M, Ingelman-Sundberg M, Evans WE. Pharmacogenomics and individualized drug therapy. *Annu Rev Med.* 2006;57:119–137.

SECTION XIV

A NEUROLOGIC GENE MAP

CHAPTER

109

A Neurologic Gene Map

Saima N. Kayani, Kathleen S. Wilson, and Roger N. Rosenberg

The University of Texas Southwestern Medical Center, Dallas, TX, USA

TABLE 109.1 Genes Associated with Neurologic Disorders

Genetic disorder	Gene	Locus	Gene product	Inheritance (if established) and phenotype
ACAD9 deficiency	*ACAD9*	3q26	Acyl-CoA dehydrogenase family, member 9	AR. Mitochondrial complex 1 deficiency due to ACAD 9 deficiency
Achondroplasia	*FGFR3*	4p16.3	Fibroblast growth factor receptor 3	AD. Achondroplasia
Action myoclonus-renal failure syndrome	*SCARB2*	4q13-q21	Scavenger receptor class B, member 2	AR. Action myoclonus-renal failure syndrome
Acyl-CoA dehydrogenase, medium chain, deficiency of	*ACADM*	1p31	Acyl-CoA dehydrogenase, C-4 to C-12 straight chain	AR. MCAD. Impaired fatty acid β-oxidation. Fasting intolerance
Adenylosuccinase deficiency	*ADSL*	22q13.1	Adenylosuccinate lyase	AR. Psychomotor delays, epilepsy, autistic features
Adrenoleukodystrophy, neonatal	*PEX5*	12p13.3	Peroxisomal biogenesis factor 5	AR. Peroxisome biogenesis disorder
Adrenoleukodystrophy, neonatal	*PEX26*	22q11.21	Peroxisomal biogenesis factor 26	AR. Peroxisome biogenesis disorder
Adrenoleukodystrophy/adrenomyeloneuropathy	*ABCD1*	Xq28	ATP-binding cassette, subfamily D (ALD), member 1	AR. Adrenoleukodystrophy/adrenomyeloneuropathy
AGAT deficiency/cerebral creatine deficiency syndrome 3	*GATM*	15q15.3	Glycine amidinotransferase (L-arginine: glycine amidinotransferase)	AR. Creatine/phosphocreatine depletion in brain causing psychomotor delays
Agenesis of the corpus callosum with peripheral neuropathy	*SLC12A6*	15q13-q14	Solute carrier family 12 (potassium/chloride transporters), member 6	AR. Agenesis of the corpus callosum with motor/sensory neuropathy
Aicardi syndrome	*AIC*	Xp22	Aicardi syndrome	X-dominant. Aicardi syndrome
Aicardi–Goutieres syndrome 1, dominant and recessive	*TREX1*	3p21.3-p21.2	3′ repair exonuclease 1	AD/AR. Aicardi–Goutieres syndrome 1, dominant and recessive
Aicardi–Goutieres syndrome 2	*RNASEH2B*	13q14.1	Ribonuclease H2, subunit B	AR. Aicardi–Goutieres syndrome 2
Aicardi–Goutieres syndrome 3	*RNASEH2C*	11q13.2	Ribonuclease H2, subunit C	AR. Aicardi–Goutieres syndrome 3
Aicardi–Goutieres syndrome 4	*RNASEH2A*	19p13.13	Ribonuclease H2, subunit A	AR. Aicardi–Goutieres syndrome 4
Aicardi–Goutieres syndrome 5	*SAMHD1*	20q11.2	SAM domain and HD domain 1	AR. Aicardi–Goutieres syndrome 5
Aicardi–Goutieres syndrome 6	*ADAR1*	1q21	Adenosine deaminase RNA specific	AR. Aicardi–Goutieres syndrome 6
Alexander disease	*GFAP*	17q21	Glial fibrillary acidic protein	AD. Alexander disease

Continued

http://dx.doi.org/10.1016/10.1016/B978-0-12-410529-4.00109-1

TABLE 109.1 Genes Associated with Neurologic Disorders—cont'd

Genetic disorder	Gene	Locus	Gene product	Inheritance (if established) and phenotype
Allan–Herndon–Dudley syndrome	*SLC16A2*	Xq13.2	Solute carrier family 16, member 2 (monocarboxylic acid transporter 8)	X-linked recessive. Allan–Herndon–Dudley syndrome
Alopecia with mental retardation syndrome 1	*APMR1*	3q26.3-q27.3	Alopecia-mental retardation syndrome	AR. Alopecia with mental retardation syndrome 1
Alopecia with mental retardation syndrome 2	*APMR2*	3q26.2-q26.31	Alopecia with mental retardation syndrome 2	AR. Alopecia with mental retardation syndrome 2
Alopecia, neurologic defects, and endocrinopathy syndrome	*RBM28*	7q32.1	RNA-binding motif protein 28	AR. Alopecia, neurologic defects, and endocrinopathy syndrome
α-Thalassemia myelodysplasia syndrome (somatic). α-thalassemia/mental retardation syndrome. X-linked mental retardation-hypotonic facies syndrome	*ATRX*	Xq21.1	*ATR-X* gene, helicase 2 X-linked	X-linked. α-Thalassemia and mental retardation syndrome. Myelodysplasia
α-Ketoglutarate dehydrogenase deficiency	*OGDH*	7p14-p13	Oxoglutarate (α-ketoglutarate) dehydrogenase (lipoamide)	AR. Axial hypotonia, lactic acidosis, hypoglycemia during infections
α-Thalassemia/mental retardation syndrome, type 1	*HBHR*	16pter-p13.3	α-Thalassemia/mental retardation syndrome, type 1	Contiguous gene deletion syndrome at 16pter
Alport syndrome	*COL4A5*	Xq22.3	Collagen, type IV, α5	X-linked. Chronic nephritis and sensorineural deafness
Alport syndrome, autosomal recessive	*COL4A3*	2q36-q37	Collagen, type IV, α3 (Goodpasture antigen)	AR. Chronic nephritis and sensorineural deafness
Alport syndrome, autosomal recessive	*COL4A4*	2q36-q37	Collagen, type IV, α4	AR. Chronic nephritis and sensorineural deafness
Alport syndrome, mental retardation, midface hypoplasia, and elliptocytosis	*AMMEC*	Xq22.3	Contiguous gene deletion syndrome	Contiguous gene deletion syndrome. Chronic nephritis and sensorineural deafness with mental retardation, midface hypoplasia, and elliptocytosis
Alzheimer disease, familial cerebral amyloid angiopathy with Dutch, Italian, Iowa, Flemish and Arctic variants	*APP*	21q21.3	Amyloid-β A4 precursor protein	Alzheimer disease
Alzheimer disease, type 2	*APOE*	19q13.31	Apolipoprotein E4	Late-onset familiar Alzheimer disease
Alzheimer disease, type 3	*PSEN1*	14q24.3	Presenilin 1	Early-onset familial Alzheimer disease
Alzheimer disease-4	*PSEN2*	1q31-q42	Presenilin 2 (Alzheimer disease 4)	AD. Early-onset Alzheimer disease
Alzheimer disease-5	*AD5*	12p11.23-q13.12	Alzheimer disease 5	Alzheimer disease
Alzheimer disease-6	*AD6*	10q24	Susceptibility locus on 10, genes not established	Late-onset Alzheimer disease
Alzheimer disease-7	*AD7*	10p13	Alzheimer disease 7	Familial Alzheimer disease
Alzheimer disease-8	*AD8*	20p	Alzheimer disease 8	Familial Alzheimer disease
Alzheimer disease-9, late onset, susceptibility to	*AD9*	19p13.2	Alzheimer disease 9	Late-onset Alzheimer disease
Alzheimer disease-10	*AD10*	7q36	Alzheimer disease 10	Familial Alzheimer disease
Alzheimer disease-11	*AD11*	9p22.1	Alzheimer disease 11	Alzheimer disease
Alzheimer disease 12	*AD12*	8p12-q22	Alzheimer disease 12	Dominant Alzheimer disease
Alzheimer disease-13	*AD13*	1q21	Alzheimer disease 13	Alzheimer disease
Alzheimer disease-14	*AD14*	1q25	Alzheimer disease 14	Alzheimer disease—isolated population in southwestern area of the Netherlands

TABLE 109.1 Genes Associated with Neurologic Disorders—cont'd

Genetic disorder	Gene	Locus	Gene product	Inheritance (if established) and phenotype
Alzheimer disease-15	*AD15*	3q22-q24	Alzheimer disease 15	Alzheimer disease—isolated population in southwestern area of the Netherlands
Alzheimer disease-16	*AD16*	Xq21.3	Alzheimer disease 16	Late-onset Alzheimer disease
Alzheimer disease-17	*AD17*	6p21.2	Alzheimer disease 17	Late-onset Alzheimer disease
Alzheimer disease	*APBB2*	4p14	Amyloid-β (A4) precursor protein-binding, family B, member 2	Alzheimer disease, late onset
Alzheimer disease, late-onset, susceptibility to	*PLAU*	10q24	Plasminogen activator, urokinase	Alzheimer disease, susceptibility to
Alzheimer disease, pathogenesis, association with	*SORL1*	11q23.2-q24.2	Sortilin-related receptor, L(DLR class) A repeats containing	Alzheimer disease, susceptibility to
Alzheimer disease, susceptibility to	*BLMH*	17q11.2	Bleomycin hydrolase	Alzheimer disease, susceptibility to
Alzheimer disease, susceptibility to	*PAXIP1*	7q36	PAX interacting (with transcription-activation domain) protein 1	Alzheimer disease, susceptibility to
Amish infantile epilepsy syndrome	*ST3GAL5*	2p11.2	ST3 β-galactoside α-2,3-sialyltransferase 5	AR. Infantile onset epilepsy syndrome
Amyloidosis, Finnish type	*GSN*	9q33	Gelsolin	AD. Cranial/sensory peripheral neuropathy, corneal lattice dystrophy, cutis laxa
Amyloidosis, hereditary, transthyretin-related	*TTR*	18q11.2-q12.1	Transthyretin	AD. Familial amyloid neuropathy and cardiomyopathy
Amyloidosis, secondary, susceptibility to	*APCS*	1q21-q23	Amyloid P component, serum	Alzheimer disease, susceptibility to
Amyotrophic lateral sclerosis, due to SOD1 deficiency	*SOD1*	21q22.1	Superoxide dismutase 1, soluble	Amyotrophic lateral sclerosis, due to SOD1 deficiency
Amyotrophic lateral sclerosis 2, juvenile	*ALS2*	2q33	Alsin	AR. Juvenile amyotrophic lateral sclerosis
Amyotrophic lateral sclerosis 3	*ALS3*	18q21	Amyotrophic lateral sclerosis 3 (autosomal dominant)	AD. Amyotrophic lateral sclerosis 3
Amyotrophic lateral sclerosis 4, juvenile	*SETX*	9q34.13	Senataxin	AR. Juvenile amyotrophic lateral sclerosis
Amyotrophic lateral sclerosis 5, juvenile	*ALS5*	15q15.1-q21.1	Amyotrophic lateral sclerosis 5	AR. Amyotrophic lateral sclerosis 5, juvenile recessive
Amyotrophic lateral sclerosis 6, with or without frontotemporal dementia	*FUS*	16p11.2	Fused in sarcoma	AR and AD reported. Amyotrophic lateral sclerosis 6, with or without frontotemporal dementia
Amyotrophic lateral sclerosis 7	*ALS7*	20p13	Amyotrophic lateral sclerosis 7	Amyotrophic lateral sclerosis
Amyotrophic lateral sclerosis 8	*VAPB*	20q13.3	VAMP (vesicle-associated membrane protein)-associated protein B and C	AD. Amyotrophic lateral sclerosis 8
Amyotrophic lateral sclerosis 9	*ANG*	14q11.2	Angiogenin, ribonuclease, RNase A family, 5	AD. Amyotrophic lateral sclerosis 9
Amyotrophic lateral sclerosis 10, with or without failure to thrive	*TARDBP*	1p36.2	TAR DNA binding protein	AD. Amyotrophic lateral sclerosis 10, with or without failure to thrive
Amyotrophic lateral sclerosis 11	*FIG4*	6q21	*Saccharomyces cerevisiae* homolog	AD and sporadic. Amyotrophic lateral sclerosis

Continued

TABLE 109.1 Genes Associated with Neurologic Disorders—cont'd

Genetic disorder	Gene	Locus	Gene product	Inheritance (if established) and phenotype
Amyotrophic lateral sclerosis 12	*OPTN*	10p	Optineurin	AR/AD/sporadic. Amyotrophic lateral sclerosis
Amyotrophic lateral sclerosis 13, susceptibility to	*ATXN2*	12q24.12	Ataxin 2	Intermediate length polyglutamine repeat expansion. Amyotrophic lateral sclerosis
Amyotrophic lateral sclerosis 14	*VCP*	9p13-12	Valosin-containing protein	AD. Amyotrophic lateral sclerosis
Amyotrophic lateral sclerosis 15	*UBQLN2*	Xp11.23-p11.1	Ubiquilin 2	AD. Amyotrophic lateral sclerosis
Amyotrophic lateral sclerosis 16, juvenile	*SIGMAR1*	9p13.3	Sigma nonopiod intracellular receptor 1	AR. Juvenile amyotrophic lateral sclerosis
Amyotrophic lateral sclerosis 17	*CHMP2B*	3p11	CHMP family, member 2B	AD. Amyotrophic lateral sclerosis
Amyotrophic lateral sclerosis 18	*PFN1*	17p13.3	Profilin 1	AD. Amyotrophic lateral sclerosis
Amyotrophic lateral sclerosis and/or frontotemporal dementia	*C9orf72*	9p21.2	Chromosome 9 open reading frame 72	AD. Amyotrophic lateral sclerosis and/or frontotemporal dementia
Amyotrophic lateral sclerosis, susceptibility to	*PRPH*	12q12-q13	Peripherin	Heterozygous deletion causes susceptibility to amyotrophic lateral sclerosis
Amyotrophic lateral sclerosis, susceptibility to	*NEFH*	22q12.2	Neurofilament, heavy polypeptide	Deletion causes susceptibility to amyotrophic lateral sclerosis
Amyotrophic lateral sclerosis, susceptibility to	*DCTN1*	2p13.1	Dynactin	Heterozygous deletion causes susceptibility to amyotrophic lateral sclerosis
Spinal and bulbar muscular atrophy of Kennedy	*AR*	Xq12	Androgen receptor	X-linked CAG repeat disorder. Spinal and bulbar muscular atrophy of Kennedy. Various other phenotypes described including androgen insensitivity and X-linked hypospadias
Aneurysm, intracranial berry, 1	*ANIB1*	7q11.2	Aneurysm, intracranial berry, 1	AD. Intracranial berry aneurysm
Aneurysm, intracranial berry, 2	*ANIB2*	19q13	Aneurysm, intracranial berry, 2	Intracranial berry aneurysm
Aneurysm, intracranial berry, 3	*ANIB3*	1p36.13-p34.3	Aneurysm, intracranial berry, 3	AD/AR both reported. Intracranial berry aneurysm
Aneurysm, intracranial berry, 4	*ANIB4*	5p15.2-p14.3	Aneurysm, intracranial berry, 4	Intracranial berry aneurysm
Aneurysm, intracranial berry, 5	*ANIB5*	Xp22	Aneurysm, intracranial berry, 5	X-linked dominant. Intracranial berry aneurysm
Aneurysm, intracranial berry, 6	*ANIB6*	9p21	Aneurysm, intracranial berry, 6	Intracranial berry aneurysm
Aneurysm, intracranial berry, 7	*ANIB7*	11q24-q25	Aneurysm, intracranial berry, 7	Intracranial berry aneurysm
Aneurysm, intracranial berry, 8	*ANIB8*	14q23	Aneurysm, intracranial berry, 8	Intracranial berry aneurysm
Aneurysm, intracranial berry, 9	*ANIB9*	2q33.1	Aneurysm, intracranial berry, 9	Intracranial berry aneurysm
Aneurysm, intracranial berry, 10	*ANIB10*	8q12.1	Aneurysm, intracranial berry, 10	Intracranial berry aneurysm
Aneurysm, intracranial berry, 11	*ANIB11*	8q12.1	Aneurysm, intracranial berry, 11	AD. Intracranial berry aneurysm
Aneurysmal bone cysts	*ANBC*	16q22	Aneurysmal bone cysts	Aneurysmal bone cyst
Angelman syndrome	*UBE3A*	15q11-q13	Ubiquitin protein ligase E3A	*De novo* maternal deletions involving Angelman syndrome region, 2–3% imprinting defect, 2% paternal uniparental disomy. Severely delayed motor development, mental retardation, speech impairment, gait ataxia, epilepsy with abnormal EEG, as well as physical anomalies such as microcephaly, characteristic facial dysmorphology

TABLE 109.1 Genes Associated with Neurologic Disorders—cont'd

Genetic disorder	Gene	Locus	Gene product	Inheritance (if established) and phenotype
Angelman syndrome-like	*CDKL5*	Xp22.13	Cyclin dependent kinase-like 5	X-linked dominant. Angelman syndrome-like epileptic encephalopathy
Aplasia cutis congenital, reticulolinear, with microcephaly, facial dysmorphism and other congenital anomalies	*COX7B*	Xq21.1	Cytochrome C oxidase subunit VIIb	X-linked dominant. *De novo* mutations. Aplasia cutis congenital, reticulolinear, with microcephaly, facial dysmorphism and other congenital anomalies
Argininemia	*ARG1*	6q23	Arginase, liver	AR. Argininemia
Arthrogryposis, distal, type 2A	*MYH3*	17p13.1	Myosin, heavy chain 3, skeletal muscle, embryonic	Congenital joint contractures
Aspartylglucosaminuria	*AGA*	4q32-q33	Aspartylglucosaminidase	AR. Lysosomal storage disorder leading to progressive intellectual disability
Asperger syndrome susceptibility 1	*ASPG1*	3q25-q27	Susceptibility locus	Asperger syndrome, susceptibility to, 1
Asperger syndrome susceptibility 2	*ASPG2*	17p13	Susceptibility locus	Asperger syndrome, susceptibility to, 2
Asperger syndrome susceptibility 3	*ASPG3*	1q21-q22	Susceptibility locus	Asperger syndrome, susceptibility to, 3
Asperger syndrome susceptibility 4	*ASPG4*	3p24-p21	Susceptibility locus	Asperger syndrome, susceptibility to, 4
Ataxia with isolated vitamin E deficiency	*TTPA*	8q13.1-q13.3	Tocopherol (α) transfer protein	AR. Ataxia, speech difficulties and titubations, low vitamin E levels
Ataxia, early onset, with oculomotor apraxia and hypoalbuminemia	*APTX*	9p13.3	Aprataxin	AR. Onset around age 7 years with ataxia, ocular aparaxia, neuropathy, and chorea. Low albumin levels
Ataxia-oculomotor apraxia-2	*SETX*	9q34	Senataxin	AR. Onset approximately 15 years of age with ataxia, ocular aparaxia, neuropathy, and chorea. High levels of α-fetoprotein
Ataxia-telangiectasia	*ATM*	11q22.3	Ataxia telangiectasia mutated	AR. Hereditary ataxia, ocular apraxia, immune dysfunction, telangiectasias
Ataxia-telangiectasia-like disorder	*MRE11A*	11q21	MRE11 meiotic recombination 11 homolog A (*S. cerevisiae*)	AR. Phenotypic features of Ataxia-telangiectasia and Nijmegen breakage syndrome
Ataxia, cerebellar, Cayman type	*ATCAY*	19p13.3	Ataxia, cerebellar, Cayman type	AR. Psychomotor retardation, nonprogressive cerebellar dysfunction, hypotonia
Ataxia, posterior column, with retinitis pigmentosa	*FLVCR1*	1q31.3	Feline leukemia virus subgroup C cellular receptor 1	AR. Posterior column ataxia with retinitis pigmentosa
Ataxia, sensory, autosomal dominant	*RNF170*	8p12-q12.1	Ring finger protein 170	AD. Sensory ataxia
Ataxia, spastic, 3, autosomal recessive	*SPAX3*	2q33-q34	Ataxia, spastic, 3, autosomal recessive	AR. Hereditary ataxia with leukoencephalopathy
Ataxia, spastic, 4	*MTPAP*	10p11.23	Mitochondrial poly(A) polymerase	AR. Spastic ataxia
Attention deficit hyperactivity disorder (ADHD)	*ADHD1*	16p13	Attention deficit hyperactivity disorder, susceptibility to, 1	Susceptibility locus for ADHD

Continued

TABLE 109.1 Genes Associated with Neurologic Disorders—cont'd

Genetic disorder	Gene	Locus	Gene product	Inheritance (if established) and phenotype
Attention deficit hyperactivity disorder	*ADHD2*	17p11	Attention deficit hyperactivity disorder, susceptibility to, 2	Susceptibility locus for ADHD
Attention deficit hyperactivity disorder	*ADHD3*	6q12	Attention deficit hyperactivity disorder, susceptibility to, 3	Susceptibility locus for ADHD
Attention deficit hyperactivity disorder	*ADHD4*	5p13	Attention deficit hyperactivity disorder, susceptibility to, 4	Susceptibility locus for ADHD
Attention deficit hyperactivity disorder, susceptibility to, 5	*ADHD5*	2q21.1	Attention deficit hyperactivity disorder, susceptibility to, 5	Susceptibility locus for ADHD
Attention deficit hyperactivity disorder, susceptibility to, 6	*ADHD6*	13q12.11	Attention deficit hyperactivity disorder, susceptibility to, 6	Susceptibility locus for ADHD
Attention deficit hyperactivity disorder, susceptibility to	*SLC6A3*	5p15.3	Solute carrier family 6 (neurotransmitter transporter, dopamine), member 3	Susceptibility locus for ADHD
Autism	*CHD8*	14q11.2	Chromodomain helicase DNA-binding protein 8	Autistic spectrum disorder
Autism susceptibility 1	*AUTS1*	7q22	Autism susceptibility to, 1	Susceptibility to autistic spectrum disorder
Autism susceptibility 3	*AUTS3*	13q14.2-q14.1	Autism, susceptibility to, 3	Susceptibility to autistic spectrum disorder
Autism susceptibility 5	*AUTS5*	2q	Autism, susceptibility to, 5	Susceptibility to autistic spectrum disorder
Autism susceptibility 6	*AUTS6*	17q11	Autism, susceptibility to, 6	Susceptibility to autistic spectrum disorder
Autism susceptibility 7	*AUTS7*	17q21	Autism, susceptibility to, 7	Susceptibility to autistic spectrum disorder
Autism susceptibility 8	*AUTS8*	3q25-q27	Autism, susceptibility to, 8	Susceptibility to autistic spectrum disorder
Autism susceptibility 10	*EN2*	7q36	Engrailed homeobox 2	Susceptibility to autistic spectrum disorder
Autism susceptibility 11	*AUTS11*	1q41-q42	Autism, susceptibility to, 11	Susceptibility to autistic spectrum disorder
Autism susceptibility 12	*AUTS12*	21p13-q11	Autism, susceptibility to, 12	Susceptibility to autistic spectrum disorder
Autism susceptibility 13	*AUTS13*	12q14.2	Autism, susceptibility to, 13	Susceptibility to autistic spectrum disorder
Autism susceptibility 14	*AUTS14*	16p11.2	Autism, susceptibility to, 14	Susceptibility to autistic spectrum disorder
Autism susceptibility 16	*SLC9A9*	3q24	Solute carrier family 9 (sodium/hydrogen exchanger), member 9	Susceptibility to autistic spectrum disorder
Autism susceptibility 17	*SHANK2*	11q13.3-q13.4	SH3 and multiple ankyrin repeat domains 2	Susceptibility to autistic spectrum disorder
Autism susceptibility, X-linked 1	*NLGN3*	Xq13	Neuroligin 3	X-linked. Autistic spectrum disorder
Autism susceptibility, X-linked 2	*NLGN4X*	Xp22.33	Neuroligin 4, X-linked	X-linked. Autistic spectrum disorder
Autism susceptibility, X-linked 4	*AUTSX4*	Xp22.11	Autism, X-linked, susceptibility to, 4	X-linked. Autistic spectrum disorder
Autistic behavior	*CADPS2*	7q31.32	Calcium-dependent activator protein for secretion 2	Autistic spectrum disorder

TABLE 109.1 Genes Associated with Neurologic Disorders—cont'd

Genetic disorder	Gene	Locus	Gene product	Inheritance (if established) and phenotype
Autosomal recessive Hyper-IgE recurrent infection syndrome, Autosomal dominant mental retardation	*DOCK8*	9p24.3	Dedicator of cytokinesis 8	AR. Hyper-IgE recurrent infection syndrome; AD. Mental retardation
Bardet–Biedl syndrome 1	*BBS1*	11q13	Bardet–Biedl syndrome 1	AR. Retinal dystrophy, polydactyly, mental retardation, and mild obesity
Bardet–Biedl syndrome 2	*BBS2*	16q21	Bardet–Biedl syndrome 2	AR. Retinal dystrophy, polydactyly, mental retardation, and mild obesity
Bardet–Biedl syndrome 3	*ARL6*	3p12-q13	ADP-ribosylation factor-like 6	AR. Retinal dystrophy, polydactyly, mental retardation, and mild obesity
Bardet–Biedl syndrome 4	*BBS4*	15q22.3-q23	Bardet–Biedl syndrome 4	AR. Retinal dystrophy, polydactyly, mental retardation, and mild obesity
Bartter syndrome, type 4a; sensorineural deafness with mild renal dysfunction	*BSND*	1p32.3	*BSND* gene, BARTTIN	AR. Bartter syndrome, type 4a; sensorineural deafness with mild renal dysfunction
Bardet–Biedl syndrome 5	*BBS5*	2q31	Bardet–Biedl syndrome 5	AR. Retinal dystrophy, polydactyly, mental retardation, and mild obesity
Bardet–Biedl syndrome 7	*BBS7*	4q27	Bardet–Biedl syndrome 7	AR. Retinal dystrophy, polydactyly, mental retardation, and mild obesity
Bardet–Biedl syndrome 8	*TTC8*	14q32.1	Tetratricopeptide repeat domain 8	AR. Retinal dystrophy, polydactyly, mental retardation, and mild obesity
Bardet–Biedl syndrome 9	*BBS9*	7p14	Bardet–Biedl syndrome 9	AR. Retinal dystrophy, polydactyly, mental retardation, and mild obesity
Bardet–Biedl syndrome 10	*BBS10*	12q21.2	Bardet–Biedl syndrome 10	AR. Retinal dystrophy, polydactyly, mental retardation, and mild obesity
Bardet–Biedl syndrome 12	*BBS12*	4q27	Bardet–Biedl syndrome 12	AR. Retinal dystrophy, polydactyly, mental retardation, and mild obesity
Bardet–Biedl syndrome 14	*CEP290*	12q21.32	Centrosomal protein 290-kd	Multiple phenotypes: Joubert syndrome 5; Leber congenital amaurosis 10; Meckel syndrome 4; Senior–Loken syndrome 6
Bardet–Biedl syndrome 15	*WDPCP*	2p15	WD repeat containing planar cell polarity effector	AR. Retinal dystrophy, polydactyly, mental retardation, and mild obesity
Basal ganglia calcification, idiopathic	*IBGC1*	14q	Idiopathic basal ganglia calcification 1	AD. Cerebral calcifications involving basal ganglia
Basal ganglia disease, biotin-responsive	*SLC19A3*	2q36.3	Solute carrier family 19, member 3	AR. Biotin or thiamine responsive basal ganglia disease, childhood presentation with subacute encephalopathy, may present with seizures, neurologic findings and basal ganglia MRI lesions
Bethlem myopathy	*COL6A1*	21q22.3	Collagen, type VI, α1	AD. Benign muscular dystrophy, slow progression and more prominent involvement of proximal muscles
Bethlem myopathy	*COL6A2*	21q22.3	Collagen, type VI, α2	AD. Benign muscular dystrophy, slow progression and more prominent involvement of proximal muscles
Bethlem myopathy	*COL6A3*	2q37	Collagen, type VI, α3	AD. Benign muscular dystrophy, slow progression and more prominent involvement of proximal muscles

Continued

TABLE 109.1 Genes Associated with Neurologic Disorders—cont'd

Genetic disorder	Gene	Locus	Gene product	Inheritance (if established) and phenotype
Biotinidase deficiency	*BTD*	3p25	Biotinidase	AR. Alopecia, dermatitis, hypotonia, developmental regression
Birk–Barel mental retardation dysmorphism syndrome	*KCNK9*	8q24.1-q24.3	Potassium channel, subfamily K, member 9	Imprinting defect, maternal transmission with silencing of paternal allele. Features include, intellectual disability, hypotonia, facial dysmorphism
Blepharospasm, primary benign	*DRD5*	4p16.1-p15.3	Dopamine receptor D5	Multifactorial inheritance. Focal dystonia involving orbicularis oculi muscle
Borjeson–Forssman–Lehmann syndrome	*PHF6*	Xq26.3	PHD finger protein 6	X-linked intellectual disability, epilepsy, marked obesity and hypometabolism
Bosley–Salih–Alorainy syndrome	*HOXA1*	7p15.3	Homeobox A1	AR. Bilateral Duane syndrome, deafness, ear defects, delayed motor development and autistic features
Brachydactyly-mental retardation syndrome	*HDAC4*	2q37.2	Histone deacetylase 4	AD. Short stature, intellectual disability, brachymetaphalangia and eczema
Brody myopathy	*ATP2A1*	16p12	ATPase, Ca^{2+} transporting, cardiac muscle, fast twitch 1	AR. Muscle pain and cramping with exercise, pseudomyotonia
Brunner syndrome	*MAOA*	Xp11.23	Monoamine oxidase A	X-linked. Impulsive behavior, aggressiveness, mild mental retardation
Canavan disease	*ASPA*	17pter-p13	Aspartoacylase	AR. A form of leukodystrophy with macrocephaly, elevated *N*-acetyl-L-aspartic acid peak on MR spectroscopy, severe hypotonia followed by spasticity, seizures, severe developmental delay, feeding difficulties, speech and hearing problems
Carbamoyl phosphate synthetase I deficiency	*CPS1*	2q35	Carbamoyl-phosphate synthase 1, mitochondrial	AR. Urea cycle disorder presents with hyperammonemia and protein intolerance
Carnitine-acylcarnitine translocase deficiency	*SLC25A20*	3p21.31	Solute carrier family 25 (carnitine/acylcarnitine translocase), member 20	AR. Features of fatty acid oxidation disorder, e.g., hypoglycemia, cardiac disease, hepatomegaly and hepatic dysfunction
CDAGS syndrome	*CDAGS*	22q12-q13	CDAGS	AR. Craniosynostosis, anal anomalies, and porokeratosis syndrome
Central hypoventilation syndrome, congenital	*PHOX2B*	4p12	Paired-like homeobox 2b	Polyalanine repeat expansion. Abnormal control of respiration in absence of cardiopulmonary disease or clear brain stem lesion
Centronuclear myopathy, autosomal, modifier of	*MTMR14*	3p25.3	Myotubularin related protein 14	AD. Progressive muscular weakness that starts early in life
Centrotemporal epilepsy	*ECT*	11p13	Centralopathic epilepsy	AD. Benign epilepsy of childhood with centrotemporal spikes/rolandic epilepsy
Cerebellar ataxia and mental retardation with or without quadrupedal locomotion 3	*CA8*	8q11-q12	Carbonic anhydrase VIII	AR. Cerebellar ataxia and mental retardation with or without quadrupedal locomotion 3
Cerebellar atrophy with progressive microcephaly	*CLAM*	7q11-q21	Cerebellar atrophy with progressive microcephaly	Proposed AR. Pontocerebellar hypoplasia, severe developmental delays, seizures, small cerebellum and brainstem. Clinical features may vary

TABLE 109.1 Genes Associated with Neurologic Disorders—cont'd

Genetic disorder	Gene	Locus	Gene product	Inheritance (if established) and phenotype
Cerebellar atrophy, ataxia, and mental retardation	*SCN8A*	12q13	Sodium channel, voltage gated, type VIII, α subunit	AD. Clinical features vary. Delayed psychomotor development, ataxia, and ADHD
Cerebellar hypoplasia and mental retardation with or without quadrupedal locomotion 1	*VLDLR*	9p24	Very low density lipoprotein receptor	AR. Nonprogressive cerebellar ataxia and mental retardation
Cerebral amyloid angiopathy	*CST3*	20p11.2	Cystatin C	AD. Progressive deposition of amyloid protein in cerebral blood vessels with spontaneous cerebral hemorrhage
Cerebral amyloid angiopathy, familial British/Danish variants	*ITMB2*	13q14.2	Integral membrane protein 2B	AD. Progressive deposition of amyloid protein in cerebral blood vessels. Clinical signs of neurodegeneration, dementia and strokes
Cerebral amyloid angiopathy, Dutch, Italian, Iowa, Flemish, Arctic variants	*APP*	21q21	Amyloid-β (A4) precursor protein	AD. Progressive deposition of amyloid protein in cerebral blood vessels. Variable neurologic manifestations including dementia, ischemic lesion, and intracranial hemorrhage
Cerebral arteriopathy with subcortical infarcts and leukoencephalopathy	*NOTCH3*	19p13.2-p13.1	Notch 3	AD. Cerebral arteriopathy with subcortical infarcts and leukoencephalopathy
Cerebral cavernous malformations-1	*KRIT1*	7q11.2-q21	KRIT1, ankyrin repeat containing	AD. Two hit mechanism. Cerebral venous malformation can cause seizures or bleeding. Can be asymptomatic
Cerebral cavernous malformations-2	*CCM2*	7p13	Cerebral cavernous malformation 2	AD. Two hit mechanism. Cerebral venous malformation can cause seizures or bleeding. Can be asymptomatic
Cerebral cavernous malformations-3	*PDCD10*	3q26.1	Programmed cell death 10	AD. Two hit mechanism. Cerebral venous malformation can cause seizures or bleeding. Can be asymptomatic
Cerebral dysgenesis, neuropathy, ichthyosis, and palmoplantar keratoderma syndrome	*SNAP29*	22q11.2	Synaptosomal-associated protein, 29 kDa	AR. Microcephaly, severe neurologic impairment, psychomotor delays, neuropathy, ichthyosis, and palmoplantar keratoderma syndrome
Cerebral infarction, susceptibility to	*PRKCH*	14q22-q23	Protein kinase C, eta	Susceptibility to stroke
Cerebral palsy, ataxic	*CPAT1*	9p12-q12	Cerebral palsy, ataxic 1	AR. Cerebral palsy, ataxia
Cerebral palsy, spastic quadriplegic	*KANK1*	9p24.3	KN motif and ankyrin repeat domains 1	Imprinting disorder, paternally inherited deletion, maternally imprinted. Cerebral palsy, spastic quadriplegia
Cerebral palsy, spastic quadriplegic, 3	*AP4M1*	7q22.1	Adaptor-related protein complex 4, μ1 subunit	AR. Spastic paraplegia
Cerebral palsy, spastic, symmetric	*GAD1*	2q31	Glutamate decarboxylase 1 (brain, 67 kDa)	AR. Spastic cerebral palsy
Cerebrooculofacioskeletal syndrome 4	*ERCC1*	19q13.2-q13.3	Excision repair cross-complementing rodent repair deficiency, complementation group 1 (includes overlapping antisense sequence)	AR. Dysmorphic features arthrogryposis and neurologic abnormalities
Cerebrotendinous xanthomatosis	*CYP27A1*	2q33-qter	Cytochrome P450, family 27, subfamily A, polypeptide 1	AR. Severe neurologic deterioration, premature atherosclerosis and cataracts

Continued

TABLE 109.1 Genes Associated with Neurologic Disorders—cont'd

Genetic disorder	Gene	Locus	Gene product	Inheritance (if established) and phenotype
Ceroid-lipofuscinosis, neuronal-1, infantile	*PPT1*	1p34.2	Palmitoyl-protein thioesterase 1	AR. Santavuori–Haltia disease. Progressive dementia, myoclonic seizures, microcephaly, vision loss and ataxia. Intracellular fine granular osmiophilic deposits by electron microscopy
Ceroid-lipofuscinosis, neuronal-2, classic late infantile	*TPP1*	11p15.5	Tripeptidyl peptidase I	AR. Jansky–Bielschowsky disease. Progressive dementia, myoclonic seizures, microcephaly, vision loss and ataxia. Intracellular deposit of eosinophilic curvilinear material
Ceroid-lipofuscinosis, neuronal-3, juvenile	*CLN3*	16p12.1	Ceroid-lipofuscinosis, neuronal 3	AR. Onset at age 4–10 years. Progressive dementia, ataxia, cerebellar signs, progressive vision loss. Lipopigment in extra neuronal cells/fingerprint profiles, curvilinear profiles on ultrastructure exam
Ceroid-lipofuscinosis, neuronal-4A, adult onset	*CLN6*	13q21.1-q32	Ceroid-lipofuscinosis, neuronal	AR. Kufs disease. Onset in adulthood. Type A with progressive myoclonic epilepsy; type B with dementia, motor disturbances, and facial dyskinesias
Ceroid-lipofuscinosis, neuronal-4B, adult onset	*DNAJC5*	20q13.3	Ceroid-lipofuscinosis, neuronal	AD. Kufs disease. Onset in adulthood. Type B with dementia, motor disturbances, and facial dyskinesias. Ultrastructurally finger print or curvilinear bodies
Ceroid-lipofuscinosis, neuronal-5, variant late infantile	*CLN5*	13q22.3	Ceroid-lipofuscinosis, neuronal 5	AR. Onset at 4–7 years; late onset also seen. Progressive dementia, ataxia, cerebellar signs, progressive vision loss. Lipopigment in extraneuronal cells/fingerprint profiles, curvilinear profiles on ultrastructure exam
Ceroid-lipofuscinosis, neuronal-6, variant late infantile	*CLN6*	15q21-q23	Ceroid-lipofuscinosis, neuronal 6, late infantile, variant	AR. Progressive dementia, myoclonic seizures, microcephaly, vision loss, and ataxia. Intracellular deposit of eosinophilic curvilinear material and fingerprint profile ultrastructurally
Ceroid-lipofuscinosis, neuronal-7	*MFSD8*	4q28.1-q28.2	Major facilitator superfamily domain containing 8	AR. Onset in childhood and rapidly progressive disease. Fingerprint profile and rectilinear profile on ultrastructural analysis
Ceroid-lipofuscinosis, neuronal-8	*CLN8*	8p23	Ceroid-lipofuscinosis, neuronal 8 (epilepsy, progressive with mental retardation)	AR. Clinically similar to late onset ceroid-lipofuscinosis 2 and 5; however, more severe seizures with progressive course and intracellular fingerprint and curvilinear profile on ultrastructural analysis
Ceroid-lipofuscinosis, neuronal-10	*CTSD*	11p15.5	Cathepsin D	AR. Congenital form, symptoms present at birth
Ceroid-lipofuscinosis, neuronal-11	*GRN*	17q21-31	Granulin precursor	AD and AR reported. Onset in young adulthood
Charcot–Marie–Tooth disease, type 1A	*PMP22*	17p11.2	Peripheral myelin protein 22	AD. Demyelinating hereditary motor sensory neuropathy IA
Charcot–Marie–Tooth disease, type 1B	*MPZ*	1q23.3	Myelin protein zero	AD. Demyelinating hereditary motor sensory neuropathy IB

TABLE 109.1 Genes Associated with Neurologic Disorders—cont'd

Genetic disorder	Gene	Locus	Gene product	Inheritance (if established) and phenotype
Charcot–Marie–Tooth disease, type 1C	*LITAF*	16p13.3-p12	Lipopolysaccharide-induced TNF factor	AD. Demyelinating hereditary motor sensory neuropathy IC
Charcot–Marie–Tooth disease, type 1D	*EGR2*	10q21.3	Early growth response 2	AD. Demyelinating hereditary motor sensory neuropathy ID
Charcot–Marie–Tooth disease, type 1F	*NEFL*	8p21.2	Neurofilament protein light polypeptide	AD. Demyelinating hereditary motor sensory neuropathy IF. Early onset and severe neuropathy
Charcot–Marie–Tooth disease, type 2A1	*KIF1B*	1p36.2	Kinesin family member 1B	AD. Hereditary motor sensory neuropathy, axonal type 2A1
Charcot–Marie–Tooth disease, type 2A2	*MFN2*	1p36.2	Mitofusin 2	AD. Hereditary motor sensory neuropathy, axonal type 2A2
Charcot–Marie–Tooth disease, type 2B	*RAB7A*	3q21	RAB7A, member RAS oncogene family	AD. Hereditary motor sensory neuropathy, axonal type 2B
Charcot–Marie–Tooth disease, type 2B1	*LMNA*	1q22	Lamin A/C	AR. Hereditary motor sensory neuropathy, axonal type 2B1
Charcot–Marie–Tooth disease, type 2B2	*MED25*	19q13.3	Mediator complex subunit 25	AR. Hereditary motor sensory neuropathy, axonal type 2B2
Charcot–Marie–Tooth disease, type 2C	*TRPV4*	12q24.11	Transient receptor potential cation channel, subfamily V, member IV	AD. Hereditary motor sensory neuropathy, axonal type 2C, also have respiratory failure, cranial nerve palsy, and sensorineural hearing loss
Charcot–Marie–Tooth disease, type 2D	*GARS*	7p14.3	Glycyl-tRNA synthetase	AD. Hereditary motor sensory neuropathy, axonal type 2D. Onset around 18 years with slow progression
Charcot–Marie–Tooth disease, type 2E	*NEFL*	8p21	Neurofilament, light polypeptide	AD. Hereditary motor sensory neuropathy, axonal type 2E
Charcot–Marie–Tooth disease, axonal, type 2F	*HSPB1*	7q11.23	Heat shock-27 kd-protein	AD. Hereditary motor sensory neuropathy, axonal type 2F. Variable onset 15–40 years
Charcot–Marie–Tooth disease, axonal, type 2G	*CMT2G*	12q12-q13.3	Charcot–Marie–Tooth disease, axonal, type 2G	AD. Hereditary motor sensory neuropathy, axonal type 2G. Peak age of onset in second decade
Charcot–Marie–Tooth disease, axonal, type 2H	Undetermined	Undetermined	Undetermined	AR. Hereditary motor sensory neuropathy, axonal type 2H with pyramidal features
Charcot–Marie–Tooth disease, axonal, type 2I	*MPZ*	1q23.3	Myelin protein zero	AD. Hereditary motor sensory neuropathy, axonal type 2I
Charcot–Marie–Tooth disease, axonal, type 2J	*MPZ*	1q23.3	Myelin protein zero	AD. Hereditary motor sensory neuropathy, axonal type 2J with sensorineural hearing loss, slow pupillary response, and dysphagia
Charcot–Marie–Tooth disease, type 2K	*GDAP1*	8q21.1	Ganglioside-induced differentiation associated protein 1	AD/AR. Hereditary motor sensory neuropathy, axonal type 2K. Kyphoscoliosis is also seen
Charcot–Marie–Tooth disease, axonal, type 2N	*AARS*	16q22	Alanyl-tRNA synthetase	AD. Hereditary motor sensory neuropathy, axonal type 2N, sensorineural deafness
Charcot–Marie–Tooth disease, dominant intermediate B	*DNM2*	19p13.2	Dynamin 2	AD. Progressive muscle weakness and atrophy

Continued

TABLE 109.1 Genes Associated with Neurologic Disorders—cont'd

Genetic disorder	Gene	Locus	Gene product	Inheritance (if established) and phenotype
Charcot–Marie–Tooth disease, dominant intermediate C	*YARS*	1p35	Tyrosyl-tRNA synthetase	AD. Progressive muscle weakness and atrophy
Charcot–Marie–Tooth disease, recessive intermediate B	*KARS*	16q22.2-q22.3	Lysyl-tRNA synthetase	AR. Progressive muscle weakness and atrophy
Charcot–Marie–Tooth disease, type 4A	*GDAP1*	8q13-q21.1	Ganglioside-induced differentiation associated protein 1	AR. Hereditary motor sensory neuropathy demyelinating type, axonal features may exist
Charcot–Marie–Tooth disease, type 4B1	*MTMR2*	11q22	Myotubularin-related protein 2	AR. Hereditary motor sensory neuropathy demyelinating type
Charcot–Marie–Tooth disease, type 4B2	*SBF2*	11p15	SET-binding factor 2	AR. Hereditary motor sensory neuropathy demyelinating type
Charcot–Marie–Tooth disease, type 4C	*SH3TC2*	5q32	SH3 domain and tetratricopeptide repeats 2	AR. Hereditary motor sensory neuropathy demyelinating type, cranial nerve involvement and sensorineural deafness can be present
Charcot–Marie–Tooth disease, type 4D	*NDRG1*	8q24.3	N-myc downstream regulated 1	AR. Hereditary motor sensory neuropathy demyelinating type
Charcot–Marie–Tooth disease, type 4F	*PRX*	19q13.2	Periaxin	AR. Hereditary motor sensory neuropathy demyelinating type
Charcot–Marie–Tooth disease, type 4G	Undetermined	Undetermined	Undetermined	AR. Hereditary motor sensory neuropathy demyelinating type
Charcot–Marie–Tooth disease, type 4H	*FGD4*	12p11.2	FYVE, RhoGEF and PH domain containing 4	AR. Hereditary motor sensory neuropathy demyelinating type
Charcot–Marie–Tooth disease, type 4J	*FIG4*	6q21	FIG4 homolog, SAC1 lipid phosphatase domain containing	AR. Hereditary motor sensory neuropathy demyelinating type
Charcot–Marie–Tooth neuropathy, X-linked dominant, 1	*GJB1*	Xq13.1	Gap junction protein, β1, 32kDa	X-linked dominant. Hereditary sensory and motor neuropathy. Can have both demyelinating and axonal features
Charcot–Marie–Tooth neuropathy, X-linked recessive, 2	*CMTX2*	Xp22.2	Charcot–Marie–Tooth neuropathy, X-linked 2	X-linked recessive, type 2 hereditary sensory and motor neuropathy. Can have both demyelinating and axonal features. Early onset in infancy and more severe course
Charcot–Marie–Tooth neuropathy, X-linked recessive, 3	*CMTX3*	Xq26	Charcot–Marie–Tooth neuropathy, X-linked 3	X-linked recessive hereditary sensory and motor neuropathy. Can have both demyelinating and axonal features
Charcot–Marie–Tooth neuropathy, X-linked recessive, 3	*MTX3*	Xq26	Metaxin 3	X-linked recessive hereditary sensory and motor neuropathy
Charcot–Marie–Tooth neuropathy, X-linked recessive, 4/Cowchock syndrome	*AIFM1*	Xq24-q26.1	Apoptosis-inducing factor, mitochondrion-associated, 1	X-linked recessive. Neuropathy, axonal, motor–sensory with deafness and mental retardation (Cowchock syndrome)
Charcot–Marie–Tooth neuropathy, X-linked recessive, 5	*PRSP1*	Xq21-24	Phosphoribosylpyrophosphate synthetase 1 gene	X-linked recessive. Neuropathy, optic atrophy, and deafness
Chondrodysplasia punctata, X-linked dominant	*EBP*	Xp11.23-p11.22	Emopamil-binding protein (sterol isomerase)	X-linked dominant. Chondrodysplasia punctata
Chondrodysplasia punctata, X-linked recessive	*ARSE*	Xp22.3	Arylsulfatase E (chondrodysplasia punctata 1)	X-linked recessive. Chondrodysplasia punctata

TABLE 109.1 Genes Associated with Neurologic Disorders—cont'd

Genetic disorder	Gene	Locus	Gene product	Inheritance (if established) and phenotype
Choreoacanthocytosis	*VPS13A*	9q21	Vacuolar protein sorting 13 homolog A (*S. cerevisiae*)	Progressive neurodegeneration and acanthocytosis with onset in fourth and fifth decade of life
Choreoathetosis/spasticity, episodic	*CSE*	1p31	Choreoathetosis/spasticity, episodic (paroxysmal choreoathetosis/spasticity)	AD. Childhood onset of paroxysmal choreoathetosis and progressive spastic paraplegia
Cockayne syndrome, type A	*ERCC8*	5q12	Excision repair cross-complementing rodent repair deficiency, complementation group 8	AR. Growth restriction, microcephaly, severe intellectual disability
Cockayne syndrome, type B	*ERCC6*	10q11	Excision repair cross-complementing rodent repair deficiency, complementation group 6	AR. Severe mental retardation, failure to thrive, dwarfism, joint contractures
Coenzyme Q10 deficiency D1	*COQ2*	4q21-q22	Coenzyme Q2 homolog, prenyltransferase (yeast)	AR. Various phenotypes including an encephalomyopathic form with seizures and ataxia; a multisystem infantile form with encephalopathy, cardiomyopathy, and renal failure; a predominantly cerebellar form with ataxia and cerebellar atrophy; Leigh syndrome with growth retardation; and an isolated myopathic form
Coenzyme Q10 deficiency D2	*PDSS1*	10p12.1	Prenyl (decaprenyl) diphosphate synthase, subunit 1	AR. Various phenotypes including an encephalomyopathic form with seizures and ataxia; a multisystem infantile form with encephalopathy, cardiomyopathy, and renal failure; a predominantly cerebellar form with ataxia and cerebellar atrophy; Leigh syndrome with growth retardation; and an isolated myopathic form
Coenzyme Q10 deficiency D3	*PDSS2*	6q21	Prenyl (decaprenyl) diphosphate synthase, subunit 2	AR. Various phenotypes including an encephalomyopathic form with seizures and ataxia; a multisystem infantile form with encephalopathy, cardiomyopathy and renal failure; a predominantly cerebellar form with ataxia and cerebellar atrophy; Leigh syndrome with growth retardation; and an isolated myopathic form
Coenzyme Q10 deficiency D4	*ADCK3*	1q42.2	AarF domain-containing kinase 3	AR. Various phenotypes including an encephalomyopathic form with seizures and ataxia; a multisystem infantile form with encephalopathy, cardiomyopathy, and renal failure; a predominantly cerebellar form with ataxia and cerebellar atrophy; Leigh syndrome with growth retardation; and an isolated myopathic form
Coenzyme Q10 deficiency D5	*COQ9*	16q13	Coenzyme Q9 homolog (*S. cerevisiae*)	AR. Various phenotypes including an encephalomyopathic form with seizures and ataxia; a multisystem infantile form with encephalopathy, cardiomyopathy, and renal failure; a predominantly cerebellar form with ataxia and cerebellar atrophy; Leigh syndrome with growth retardation; and an isolated myopathic form

Continued

TABLE 109.1 Genes Associated with Neurologic Disorders—cont'd

Genetic disorder	Gene	Locus	Gene product	Inheritance (if established) and phenotype
Coenzyme Q10 deficiency D6	*COQ6*	14q24.3	Coenzyme Q6 homolog (*S. cerevisiae*)	AR. Various phenotypes including an encephalomyopathic form with seizures and ataxia; a multisystem infantile form with encephalopathy, cardiomyopathy and renal failure; a predominantly cerebellar form with ataxia and cerebellar atrophy; Leigh syndrome with growth retardation; and an isolated myopathic form
Coffin–Lowry syndrome	*RPS6KA3*	Xp22.2-p22.1	Ribosomal protein S6 kinase, 90 kDa, polypeptide 3	X linked dominant. Intellectual disability with distinct facial features. Short tapering fingers. Microcephaly and short stature can be present
Cohen syndrome	*VPS13B*	8q22-q23	Vacuolar protein sorting 13 homolog B (yeast)	AR. Hypotonia, psychomotor retardation, microcephaly, characteristic facial features
Combined oxidative phosphorylation deficiency 1	*GFM1*	3q25.1-q26.2	Elongation factor, mitochondrial 1	AR. Multisystem fatal disorder resulting from deficiency of mitochondrial oxidative phosphorylation
Combined oxidative phosphorylation deficiency 2	*MRPS16*	10q22.1	Mitochondrial ribosomal protein S16	AR. Multisystem fatal disorder resulting from deficiency of mitochondrial oxidative phosphorylation
Combined oxidative phosphorylation deficiency 3	*TSFM*	12q13-q14	Ts translation elongation factor, mitochondrial	AR. Multisystem fatal disorder resulting from deficiency of mitochondrial oxidative phosphorylation
Combined oxidative phosphorylation deficiency 4	*TUFM*	16p11.2	Tu translation elongation factor, mitochondrial	AR. Multisystem fatal disorder resulting from deficiency of mitochondrial oxidative phosphorylation
Combined oxidative phosphorylation deficiency 5	*MRPS22*	3q23	Mitochondrial ribosomal protein S22	AR. Multisystem fatal disorder resulting from deficiency of mitochondrial oxidative phosphorylation
Combined oxidative phosphorylation deficiency 6	*AIFM1*	Xq25-q26	Apoptosis-inducing factor, mitochondrion-associated, 1	X-linked. Multisystem fatal disorder resulting from deficiency of mitochondrial oxidative phosphorylation
Combined oxidative phosphorylation deficiency 7	*C12orf65*	12q24.31	Chromosome 12 open reading frame 65	AR. Multisystem fatal disorder resulting from deficiency of mitochondrial oxidative phosphorylation
Combined oxidative phosphorylation deficiency 8	*AARS2*	6p21	Mitochondrial alanine tRNA synthetase	AR. Multisystem fatal disorder resulting from deficiency of mitochondrial oxidative phosphorylation
Combined oxidative phosphorylation deficiency 9	*MRPL3*	3q22	Mitochondrial ribosomal protein L3	AR. Multisystem fatal disorder resulting from deficiency of mitochondrial oxidative phosphorylation
Combined oxidative phosphorylation deficiency 10	*MTO1*	6q13	Mitochondrial translation optimization, *S. cerevisiae* homolog of	AR. Multisystem fatal disorder resulting from deficiency of mitochondrial oxidative phosphorylation
Combined oxidative phosphorylation deficiency 11	*RMND1*	6q25	Required for meiotic nuclear division, *S. cerevisiae* homolog of	AR. Multisystem fatal disorder resulting from deficiency of mitochondrial oxidative phosphorylation
Combined oxidative phosphorylation deficiency 12	*EARS2*	16p13	Glutamate tRNA synthetase 2	AR. Multisystem fatal disorder resulting from deficiency of mitochondrial oxidative phosphorylation

TABLE 109.1 Genes Associated with Neurologic Disorders—cont'd

Genetic disorder	Gene	Locus	Gene product	Inheritance (if established) and phenotype
Combined oxidative phosphorylation deficiency 13	*PNPT1*	2p16	Polyribonucleotide nucleotidyltransferase 1	AR. Multisystem fatal disorder resulting from deficiency of mitochondrial oxidative phosphorylation
Combined oxidative phosphorylation deficiency 14	*FARS2*	6p25	Phenylalanine tRNA synthetase 2, mitochondrial	AR. Multisystem fatal disorder resulting from deficiency of mitochondrial oxidative phosphorylation
Combined oxidative phosphorylation deficiency 15	*MTFMT*	15q	Methionyl tRNA formyltransferase, mitochondrial	AR. Multisystem fatal disorder resulting from deficiency of mitochondrial oxidative phosphorylation
Congenital cataracts, facial dysmorphism, and neuropathy	*CTDP1*	18q23	CTD (carboxy-terminal domain, RNA polymerase II, polypeptide A) phosphatase, subunit 1	AR. Congenital cataracts, facial dysmorphism, and neuropathy
Congenital disorder of glycosylation, type Ia	*PMM2*	16p13.3-p13.2	Phosphomannomutase 2	AR. Severe encephalopathy, psychomotor retardation, cerebellar hypoplasia, retinitis pigmentosa. Lipodystrophy. All congenital disorders of glycosylation have abnormal isoelectric focusing of serum transferrin
Congenital disorder of glycosylation, type Ib	*MPI*	15q22-qter	Mannose phosphate isomerase	AR. Failure to thrive, protein-losing enteropathy, hypoglycemia
Congenital disorder of glycosylation, type Ic	*ALG6*	1p22.3	Asparagine-linked glycosylation 6, α-1,3-glucosyltransferase homolog (*S. cerevisiae*)	AR. Severe encephalopathy, psychomotor retardation, cerebellar hypoplasia, motor and speech delays, and protein-losing enteropathy
Congenital disorder of glycosylation, type Id	*ALG3*	3q27	Asparagine-linked glycosylation 3, α-1,3-mannosyltransferase homolog (*S. cerevisiae*)	AR. Genetically heterogeneous group of disorders caused by defective glycosylation of glycoconjugates. Usually presents in infancy with hypotonia, psychomotor delays, peripheral neuropathy, eye abnormalities
Congenital disorder of glycosylation, type Ie	*DPM1*	20q13.13	Dolichyl-phosphate mannosyltransferase polypeptide 1, catalytic subunit	AR. Genetically heterogeneous group of disorders caused by defective glycosylation of glycoconjugates
Congenital disorder of glycosylation, type If	*MPDU1*	17p13.1-p12	Mannose-P-dolichol utilization defect 1	AR. Genetically heterogeneous group of disorders caused by defective glycosylation of glycoconjugates
Congenital disorder of glycosylation, type Ig	*ALG12*	22q13.33	Asparagine-linked glycosylation 12, α-1,6-mannosyltransferase homolog (*S. cerevisiae*)	AR. Genetically heterogeneous group of disorders caused by defective glycosylation of glycoconjugates
Congenital disorder of glycosylation, type Ih	*ALG8*	11pter-p15.5	Asparagine-linked glycosylation 8, α-1,3-glucosyltransferase homolog (*S. cerevisiae*)	AR. Genetically heterogeneous group of disorders caused by defective glycosylation of glycoconjugates
Congenital disorder of glycosylation, type Ii	*ALG2*	9q22	Asparagine-linked glycosylation 2, α-1,3-mannosyltransferase homolog (*S. cerevisiae*)	AR. Genetically heterogeneous group of disorders caused by defective glycosylation of glycoconjugates
Congenital disorder of glycosylation, type Ij	*DPAGT1*	11q23.3	Dolichyl-phosphate (UDP-*N*-acetylglucosamine) *N*-acetylglucosaminephosphotransferase 1 (GlcNAc-1-P transferase)	AR. Genetically heterogeneous group of disorders caused by defective glycosylation of glycoconjugates
Congenital disorder of glycosylation, type Ik	*ALG1*	16p13.3	Asparagine-linked glycosylation 1, β-1,4-mannosyltransferase homolog (*S. cerevisiae*)	AR. Genetically heterogeneous group of disorders caused by defective glycosylation of glycoconjugates

Continued

TABLE 109.1 Genes Associated with Neurologic Disorders—cont'd

Genetic disorder	Gene	Locus	Gene product	Inheritance (if established) and phenotype
Congenital disorder of glycosylation, type Il	*ALG9*	11q23	Asparagine-linked glycosylation 9, α-1,2-mannosyltransferase homolog (*S. cerevisiae*)	AR. Genetically heterogeneous group of disorders caused by defective glycosylation of glycoconjugates
Congenital disorder of glycosylation, type Im	*DOLK*	9q34.11	Dolichol kinase	AR. Genetically heterogeneous group of disorders caused by defective glycosylation of glycoconjugates
Congenital disorder of glycosylation, type In	*RFT1*	3p21.1	RFT1 homolog (*S. cerevisiae*)	AR. Genetically heterogeneous group of disorders caused by defective glycosylation of glycoconjugates
Congenital disorder of glycosylation, type Io	*DPM3*	1q12-q21	Dolichyl-phosphate mannosyltransferase polypeptide 3	AR. Genetically heterogeneous group of disorders caused by defective glycosylation of glycoconjugates
Congenital disorder of glycosylation, type Ip	*ALG11*	13q14.3	Asparagine-linked glycosylation 11, α-1,2-mannosyltransferase homolog (yeast)	AR. Genetically heterogeneous group of disorders caused by defective glycosylation of glycoconjugates
Congenital disorder of glycosylation, type Iq	*SRD5A3*	4q12	Steroid 5 α-reductase 3	AR. Genetically heterogeneous group of disorders caused by defective glycosylation of glycoconjugates
Congenital disorder of glycosylation, type IIa	*MGAT2*	14q21	Mannosyl (α-1,6-)-glycoprotein β-1,2-*N*-acetylglucosaminyltransferase	AR. Genetically heterogeneous group of disorders caused by defective glycosylation of glycoconjugates
Congenital disorder of glycosylation, type IIb	*MOGS*	2p13-p12	Mannosyl-oligosaccharide glucosidase	AR. Genetically heterogeneous group of disorders caused by defective glycosylation of glycoconjugates
Congenital disorder of glycosylation, type IIc	*SLC35C1*	11p11.2	Solute carrier family 35, member C1	AR. Genetically heterogeneous group of disorders caused by defective glycosylation of glycoconjugates
Congenital disorder of glycosylation, type IId	*B4GALT1*	9p13	UDP-Gal:βGlcNAc β-1,4-galactosyltransferase, polypeptide 1	AR. Genetically heterogeneous group of disorders with multisystem involvement. Multiple forms have been identified
Congenital disorder of glycosylation, type IIe	*COG7*	16p12.2	Component of oligomeric Golgi complex 7	AR. Genetically heterogeneous group of disorders with multisystem involvement. Multiple forms have been identified
Congenital disorder of glycosylation, type IIf	*SLC35A1*	6q15	Solute carrier family 35 (CMP-sialic acid transporter), member A1	AR. Genetically heterogeneous group of disorders with multisystem involvement. Multiple forms have been identified
Congenital disorder of glycosylation, type IIg	*COG1*	17q25.1	Component of oligomeric Golgi complex 1	AR. Genetically heterogeneous group of disorders with multisystem involvement. Multiple forms have been identified
Congenital disorder of glycosylation, type IIh	*COG8*	16q22.1	Component of oligomeric Golgi complex 8	AR. Genetically heterogeneous group of disorders with multisystem involvement. Multiple forms have been identified
Congenital disorder of glycosylation, type IIi	*COG5*	7q31	Component of oligomeric Golgi complex 5	AR. Genetically heterogeneous group of disorders with multisystem involvement. Multiple forms have been identified

TABLE 109.1 Genes Associated with Neurologic Disorders—cont'd

Genetic disorder	Gene	Locus	Gene product	Inheritance (if established) and phenotype
Congenital disorder of glycosylation, type IIj	*COG4*	16q22.1	Component of oligomeric Golgi complex 4	AR. Genetically heterogeneous group of disorders with multisystem involvement. Multiple forms have been identified
Convulsions, benign familial infantile	*BFIC*	19q13	Benign familial infantile convulsions	AD. Benign familial partial seizures, controlled easily on medications, typically resolve after 1 year of age
Convulsions, benign familial infantile, 2	*PRRt2*	16p12-q12	Proline-rich transmembrane protein 2	Benign familial infantile, convulsions-2
Convulsions, benign familial infantile, 3	*SCN2A*	2q23-q24.3	Sodium channel, voltage-gated, type II, α subunit	AD. Benign form of epilepsy. Onset can vary from 2–7 months and resolves typically after 1 year of life
Convulsions, familial febrile, 1	*FEB1*	8q13-q21	Febrile convulsions 1	AD. Febrile seizures, typically start at 6 months and resolve after 5 years
Convulsions, familial febrile, 2	*FEB2*	19p13.3	Febrile convulsions 2	AD. Febrile seizures, typically start at 6 months and resolve after 5 years
Convulsions, familial febrile, 3A	*SCN1A*	2q24	Sodium channel, voltage-gated, type I, α subunit	AD. Febrile seizures, typically start at 6 months and resolve after 5 years
Convulsions, familial febrile, 3B	*SCN9A*	2q24	Sodium channel, voltage-gated, type IX, α subunit	AD. Febrile seizures, typically start at 6 months and resolve after 5 years
Convulsions, familial febrile, 4	*GPR98*	5q14.3	G protein-coupled receptor 98	AD. Febrile seizures, typically start at 6 months and resolve after 5 years
Febrile convulsions, familial, 5	*FEB5*	6q22-q24	Febrile convulsions 5	AD. Febrile convulsions, familial, 5
Febrile convulsions, familial, 6	*FEB6*	18p11.2	Febrile convulsions 6	AD. Febrile convulsions, familial, 6
Febrile convulsions, familial, 7	*FEB7*	21q22	Febrile convulsions 7	AD. Febrile convulsions, familial, 7
Convulsions, familial febrile, 8	*GABRG2*	5q34	γ-Aminobutyric acid (GABA) receptor	AD. Febrile seizures, typically start at 6 months and resolve after 5 years. Same gene implicated in generalized epilepsy febrile seizure plus syndrome
Febrile convulsions, familial, 9	*FEB9*	3p24.2-p23	Febrile convulsions, familial, 9	AD. Febrile convulsions and absence seizures
Febrile convulsions, familial, 10	*FEB10*	3q26.2-q26.33	Febrile convulsions, familial, 10	AD. Febrile convulsions, familial, 10
Convulsions, familial febrile, 11	*CAP6*	5q34	Carboxypeptidase 6	AD. Febrile seizures, typically start at 6 months and resolve after 5 years
Convulsions, infantile and paroxysmal choreoathetosis	*PRRt2*	16p12-q12	Proline-rich transmembrane protein 2	AD. Seizures, paroxysmal choreoathetosis and dystonia
Cornelia de Lange syndrome 1	*NIPBL*	5p13.1	Nipped-B homolog (*Drosophila*)	AD. Multisystem malformation disorder, with characteristic facial dysmorphism and varying degree of limb anomalies and intellectual disability
Cornelia de Lange syndrome 2	*SMC1A*	Xp11.22-p11.21	Structural maintenance of chromosomes 1A	X-linked. Multisystem malformation disorder, with characteristic facial dysmorphism and varying degree of limb anomalies and intellectual disability
Cornelia de Lange syndrome 3	*SMC3*	10q25	Structural maintenance of chromosomes 3	AD. Multisystem malformation disorder, with characteristic facial dysmorphism and varying degree of limb anomalies and intellectual disability

Continued

TABLE 109.1 Genes Associated with Neurologic Disorders—cont'd

Genetic disorder	Gene	Locus	Gene product	Inheritance (if established) and phenotype
Cortical dysplasia-focal epilepsy syndrome	*CNTNAP2*	7q35-q36	Contactin associated protein-like 2	AR. Cortical dysplasia-focal epilepsy syndrome; Pitt–Hopkin-like syndrome; autism susceptibility
Cowchock syndrome	*AIFM1*	Xq24-q26.1	Apoptosis-inducing factor, mitochondrion-associated, 1	X-linked recessive. Neuropathy, axonal, motor–sensory with deafness and mental retardation (Cowchock syndrome)
Cowden disease	*PTEN*	10q23.31	Phosphatase and tensin homolog	AD. Multiple hamartomatous lesions involving skin, thyroid gland, breast. Cerebellar tumor and megalencephaly. Hamartomatous gastrointestinal tract polyps
CPT deficiency, hepatic, type IA	*CPT1A*	11q13	Carnitine palmitoyltransferase 1A (liver)	AR. Hypoketotic hypoglycemia, hypotonia, lethargy, hepatomegaly and cardiomegaly
Craniosynostosis, Adelaide type	*CRSA*	4p16	Craniosynostosis, Adelaide type	AD. Craniosynostosis and hand bone anomalies, including hypoplasia of middle phalanges
Craniosynostosis, type 2	*MSX2*	5q34-q35	Msh homeobox 2	AD. Craniosynostosis
Creatine deficiency syndrome, X-linked/cerebral creatine deficiency syndrome 1	*SLC6A8*	Xq28	Solute carrier family 6 (neurotransmitter transporter, creatine), member 8	X-linked. Cerebral creatine deficiency. Intellectual disability, speech delays, behavioral abnormalities and seizures
Creutzfeldt–Jakob disease (CJD)	*PRNP*	20pter-p12	Prion protein	AD. Neurologic manifestation with vision problems, supranuclear gaze palsy, ataxia, extrapyramidal muscular rigidity, dementia, confusion, behavioral abnormalities. Pathology reveals diffuse nerve cell degeneration with spongiform changes
Creutzfeldt–Jakob disease, variant, resistance to	*HLA-DQB1*	6p21.3	Major histocompatibility complex, class II, DQ β1	Reduced frequency of this allele in patients with variant CJD
Crouzon syndrome	*FGFR2*	10q26	Fibroblast growth factor receptor 2	AD. Craniosynostosis causing craniofacial dysmorphism with frontal bossing, proptosis, atretic external auditory canals
Cytochrome c oxidase deficiency	*COX6B1*	19q13.1	Cytochrome c oxidase subunit VIb polypeptide 1 (ubiquitous)	AR. Weakness/pain, progressive ataxia and neurologic deterioration
D-2-hydroxyglutaric aciduria	*D2HGDH*	2q37.3	D-2-hydroxyglutarate dehydrogenase	AR. Macrocephaly, variable phenotypes exist from milder to severe form of disease. Severe disease form has early onset of encephalopathy
D-2-hydroxyglutaric aciduria 2	*IDH2*	15q26.1	Isocitrate dehydrogenase 2 (NADP+), mitochondrial	AR. Macrocephaly, variable phenotypes exist from milder to severe form of disease. Severe disease form has early onset of encephalopathy
Dandy–Walker malformation	*ZIC1*	3q24	Zic family member 1	AR/heterogeneous. Brain malformation consisting of partial or complete absence of cerebellar vermis with cystic dilation of fourth ventricle
Dandy–Walker malformation	*ZIC4*	3q24	Zic family member 4	AR/heterogeneous. Brain malformation consisting of partial or complete absence of cerebellar vermis with cystic dilation of fourth ventricle

TABLE 109.1 Genes Associated with Neurologic Disorders—cont'd

Genetic disorder	Gene	Locus	Gene product	Inheritance (if established) and phenotype
De Sanctis–Cacchione syndrome	*ERCC6*	10q11.23	Excision repair, cross-complementing group 6	AR. Xeroderma pigmentosum, intellectual disability, neurologic deterioration and microcephaly
Dejerine–Sottas neuropathy, autosomal recessive	*PRX*	19q13.1-q13.2	Periaxin	AR. Demyelinating peripheral neuropathy with onset in infancy
Dementia, familial British	*ITM2B*	13q14	Integral membrane protein 2B	AD. Dementia, spasticity, hyperreflexia
Dementia, familial, nonspecific	*CHMP2B*	3p11.2	Charged multivesicular body protein 2B	AD. Frontotemporal dementia
Dementia, frontotemporal	*PSEN1*	14q24.2	Presenilin 1	AD. Frontotemporal dementia
Dementia, frontotemporal, with or without parkinsonism	*MAPT*	17q21.1	Microtubule-associated protein tau	AD. Frontotemporal dementia with or without parkinsonism
Dementia, Lewy body	*SNCB*	5q35	Synuclein, β	Dementia, Lewy body
Dentatorubro-pallidoluysian atrophy	*ATN1*	12p13.31	Atrophin 1	CAG expansion in *ATN1* gene. Neurodegenerative disorder, dementia, intellectual disability, and seizures
DiGeorge syndrome/ velocardiofacial syndrome	*TBX1*	22q11.2	DiGeorge syndrome	Hemizygous microdeletion. Variable clinical features, craniofacial dysmorphism, hypoparathyroidism, immune dysfunction, cardiac anomalies
DiGeorge syndrome/ velocardiofacial syndrome complex-2	*DGS2*	10p14-p13	DiGeorge syndrome chromosome region-2	Microdeletion. Another locus for DiGeorge syndrome
Down syndrome	*DCR*	21q22.3	Down syndrome chromosome region	Critical region for trisomy 21
Duane retraction syndrome 1	*DURS1*	8q13	Duane retraction syndrome 1	AD. Congenital eye movement disorder due to failure of normal development of cranial nerve 6 with restricted horizontal eye movements with globe retraction and small palpebral fissure
Duane retraction syndrome 2	*CHN1*	2q31-q32.1	Chimerin (chimaerin) 1	AD. Congenital eye movement disorder, see above
Duchenne muscular dystrophy	*DMD*	Xp21.2	Dystrophin	X-linked recessive muscular dystrophy. Weakness with pseudohypertrophy of calf muscles, Gowers sign present, respiratory insufficiency, dilated cardiomyopathy. Mild form is Becker muscular dystrophy
Dyggve–Melchior–Clausen disease	*DYM*	18q21.1	Dymeclin	AR. Hurler syndrome like disorder, Multisystem disorder with severe neurologic deterioration, coarse facial features and skeletal findings
Dysautonomia, familial	*IKBKAP*	9q31	Inhibitor of kappa light polypeptide gene enhancer in B-cells, kinase complex-associated protein	AR. Hereditary sensory and autonomic neuropathy
Dyslexia, susceptibility to, 1	*DYX1C1*	15q21	Dyslexia susceptibility 1 candidate 1	Susceptibility locus for dyslexia 1
Dyslexia, susceptibility to, 2	*DYX2*	6p22.2	Dyslexia susceptibility 2	Susceptibility locus for dyslexia 2

Continued

TABLE 109.1 Genes Associated with Neurologic Disorders—cont'd

Genetic disorder	Gene	Locus	Gene product	Inheritance (if established) and phenotype
Dyslexia, susceptibility to, 2	*KIAA0319*	6p22.2	KIAA0319	Susceptibility locus for dyslexia 2
Dyslexia, susceptibility to, 3	*DYX3*	2p16-p15	Dyslexia susceptibility 3	Susceptibility locus for dyslexia 3
Dyslexia, susceptibility to, 5	*DYX5*	3p12-q13	Dyslexia susceptibility 5	Susceptibility locus for dyslexia 5
Dyslexia, susceptibility to, 6	*DYX6*	18p11.2	Dyslexia susceptibility 6	Susceptibility locus for dyslexia 6
Dyslexia, susceptibility to, 8	*DYX8*	1p36-p34	Dyslexia susceptibility 8	Susceptibility locus for dyslexia 8
Dyslexia, susceptibility to, 9	*DYX9*	Xq27.3	Dyslexia susceptibility 9	Susceptibility locus for dyslexia 9
Dystonia 1, torsion	*TOR1A*	9q34	Torsin family 1, member A (torsin A)	AD. Primary torsion dystonia. Variable phenotypes
Dystonia 2, torsion	Undetermined	Undetermined	Undetermined	AR. Primary torsion dystonia. Early onset in childhood
Dystonia 3, torsion; Dystonia-parkinsonism, X-linked	*TAF1*	Xq13	TAF1 RNA polymerase II, TATA box binding protein (TBP)-associated factor, 250 kDa	X-linked recessive. Dystonia and parkinsonism, onset in fourth decade
Dystonia 4, torsion	*TUBB4A*	19p13.3	Tubulin, β4A	AD. Neurodegenerative disorder, also called whispering dysphonia
Dystonia 6, torsion	*THAP1*	8p11.21	THAP domain containing, apoptosis associated protein 1	AD. Dystonia with early involvement of craniofacial muscles, often involves arms and laryngeal muscles
Dystonia 7, torsion	*DYT7*	18p	Dystonia 7, torsion	AD. Focal dystonia
Dystonia 11, myoclonic	*SGCE*	7q21	Sarcoglycan, epsilon	AD. Dystonia and myoclonus with behavioral and psychiatric symptoms. Upper limb affected more than lower limb
Dystonia, myoclonic	*DRD2*	11q23.1	Dopamine receptor D2	AD. Rare form of myoclonic dystonia
Dystonia 13, torsion	*DYT13*	1p36.32-p36.13	Dystonia 13, torsion	AD. Focal onset in craniofacial area or upper limbs
Dystonia 12	*ATP1A3*	19q12-q13.2	ATPase, Na^+/K^+ transporting, α3 polypeptide	AD. Craniofacial dystonia with parkinsonism and emotional lability and depression
Dystonia 15, myoclonic	*DYT15*	18p11	Dystonia 15, myoclonic	AD. Dystonia and myoclonus. Upper limb affected more than lower limb
Dystonia 16	*PRKRA*	2q31.3	Protein kinase, interferon-inducible double-stranded RNA dependent activator	Inheritance not established. Early-onset dystonia-parkinsonism
Dystonia 17, primary torsion	*DYT17*	20p11.2-q13.12	Dystonia 17	AR. Described in a Lebanese family. Focal or generalized dystonia seen
Dystonia, dopa-responsive, with or without hyperphenylalanemia	*GCH1*	14q22.1-q22.2	GTP cyclohydrolase 1	AD/AR. Dystonia with diurnal variation, defect in BH4 syntheses, dopa-responsive, clinical and genetic heterogeneity
Dystonia, dopa-responsive, due to sepiapterin reductase deficiency	*SPR*	2p14-p12	Sepiapterin reductase (7,8-dihydrobiopterin:NADP+ oxidoreductase)	AR/possible AR. Dystonia, dopa-responsive, due to sepiapterin reductase deficiency
Dystonia, juvenile-onset	*ACTB*	7p22-p12	Actin, β	AD. Juvenile onset dopa-unresponsive dystonia and neurodegenerative disorder
Emanuel syndrome	*DER22T11-22*	22q11.2	Emanuel syndrome	Chromosomal imbalance. Multiple congenital anomalies, craniofacial dysmorphism, developmental delays

TABLE 109.1 Genes Associated with Neurologic Disorders—cont'd

Genetic disorder	Gene	Locus	Gene product	Inheritance (if established) and phenotype
Emery–Dreifuss muscular dystrophy 1	*EMD*	Xq28	Emerin	X-linked recessive. Contractures of elbows and achilles. Slowly progressive weakness. Cardiac arrhythmias
Emery–Dreifuss muscular dystrophy, 2, AD	*LMNA*	1q21.2	Lamin A/C	AD. Dilated cardiomyopathy, spine rigidity, and progressive weakness and atrophy. AR form is termed Emery–Dreifuss muscular dystrophy 3
Emery–Dreifuss muscular dystrophy 4, AD	*SYNE1*	1q21.2	Synaptic nuclear envelope protein 1	AD. Emery–Dreifuss muscular dystrophy
Emery–Dreifuss muscular dystrophy 5	*SYNE2*	14q23	Spectrin repeat containing, nuclear envelope 2	AD. Emery–Dreifuss muscular dystrophy
Emery-Dreifuss muscular dystrophy 6	*FHL1*	Xq26.3	Four and a half LIM domain	AD. Muscular dystrophy with progressive proximal weakness, joint contractures and cardiac conduction defects
Emery–Dreifuss muscular dystrophy 7	*TMEM43*	3p25.1	Transmembrane protein 43	AD. Muscular dystrophy with progressive proximal weakness, joint contractures, and cardiac conduction defects
Encephalopathy, lethal, due to defective mitochondrial peroxisomal fission	*DNM1L*	12p11.21	Dynamin 1-like	Inheritance not established. Microcephaly, abnormal brain development, optic atrophy and hypoplasia, and lactic acidemia
Encephalopathy, acute necrotizing 1	*RANBP2*	2q11-q13	RAN binding protein 2	Inheritance not established. Acute necrotizing encephalopathy
Encephalopathy, familial, with neuroserpin inclusion bodies	*SERPINI1*	3q26	Serpin peptidase inhibitor, clade I (neuroserpin), member 1	Inheritance not established. Familial encephalopathy with neuroserpin inclusion bodies
Encephalopathy, progressive mitochondrial, with proximal renal tubulopathy due to cytochrome c oxidase deficiency	*COX10*	17p12-p11.2	COX10 homolog, cytochrome c oxidase assembly protein, heme A: farnesyltransferase (yeast)	AR. Phenotypic heterogeneity that can include sensorineural hearing loss, anemia, hypertrophic cardiomyopathy
Endplate acetylcholinesterase deficiency	*COLQ*	3p25	Collagen-like tail subunit (single strand of homotrimer) of asymmetric acetylcholinesterase	AR. Infantile respiratory and feeding difficulties, muscle weakness, ophthalmoplegia
Epilepsy, benign neonatal, 1	*KCNQ2*	20q13.3	Potassium voltage-gated channel, KQT-like subfamily, member 2	AD. Epilepsy by 2–8 days of life, typically remits by 6 weeks with myokymia
Epilepsy, benign neonatal, type 2	*KCNQ3*	8q24	Potassium voltage-gated channel, KQT-like subfamily, member 3	AD. Epilepsy by 2–8 days of life, typically remits by 2 months
Epilepsy, childhood absence, 1	*ECA1*	8q24	Epilepsy, childhood absence 1	AD. Childhood absence seizures
Epilepsy, childhood absence, 2	*GABRG2*	5q34	GABA receptor γ-2	AD. Childhood absence seizures
Epilepsy, familial adult myoclonic, 1	*FAME1*	8q24	Epilepsy, familial adult myoclonic, 1	AD. Adult-onset myoclonic epilepsy and tremors
Epilepsy, familial adult myoclonic, 2	*FAME2*	2p11.1-q12.2	Epilepsy, familial adult myoclonic, 2	AD. Adult-onset myoclonic epilepsy
Epilepsy, familial adult myoclonic, 3	*FAME3*	5p15.31-p15.1	Epilepsy, familial adult myoclonic, 3	AD. Adult-onset myoclonic epilepsy

Continued

TABLE 109.1 Genes Associated with Neurologic Disorders—cont'd

Genetic disorder	Gene	Locus	Gene product	Inheritance (if established) and phenotype
Epilepsy, familial adult myoclonic, 4	*FAME4*	3q26.3-q28	Familial cortical myoclonic tremor with epilepsy	AD. Familial cortical myoclonic tremor with epilepsy 4
Epilepsy, familial mesial temporal lobe	*FMTLE*	4q13.2-q21.3	Epilepsy, familial mesial temporal lobe	AD. Familial mesial temporal lobe epilepsy
Epilepsy, familial temporal lobe	*ETL2*	12q22-q23.3	Epilepsy, familial temporal lobe	AD. Simple and complex partial seizures with temporal lobe-related auras
Generalized epilepsy with febrile seizures plus	*SCN1B*	19q13.1	Sodium channel, voltage-gated, type I, β subunit	AD. Generalized epilepsy with febrile seizures plus, type 1
Epilepsy, generalized, with febrile seizures plus, type 2	*SCN1A*	2q24	Sodium channel, voltage-gated, type I, α subunit	AD. Generalized epilepsy with febrile seizures plus, type 2
Epilepsy, generalized, with febrile seizures plus, type 3	*GABRG2*	5q31.1-q33.1	γ-Aminobutyric acid (GABA) A receptor, γ2	AD. Generalized epilepsy with febrile seizures
Epilepsy, generalized, with febrile seizures plus, type 4	*GEFSP4*	2p24	Epilepsy, generalized, with febrile seizures plus, type 4	AD. Generalized epilepsy with febrile seizures
Generalized epilepsy with febrile seizures plus, type 5, susceptibility to	*GABRD*	1p36.3	γ-Aminobutyric acid (GABA) A receptor, δ	AD. Generalized epilepsy with febrile seizures
Generalized epilepsy with febrile seizures plus, type 6	*GEFSP6*	8p23-p21	Generalized epilepsy with febrile seizures plus, type 6	AD. Generalized epilepsy with febrile seizures
Epilepsy, hot water, 1	*HWE1*	10q21.3-q22.3	Epilepsy, hot water, 1	AD. Partial seizures with secondary generalization provoked by hot/warm water
Epilepsy, hot water, 2	*HWE2*	4q24-q28	Epilepsy, hot water, 2	AD. Partial seizures with secondary generalization provoked by hot/warm water
Epilepsy, idiopathic generalized, susceptibility to	*ME2*	18q21	Malic enzyme 2, NAD(+)-dependent, mitochondrial	Susceptibility locus for idiopathic generalized epilepsy
Epilepsy, idiopathic generalized, susceptibility to, 1	*EGI*	8q24	Epilepsy, generalized, idiopathic	Susceptibility locus for idiopathic generalized epilepsy
Epilepsy, idiopathic generalized, susceptibility to, 2	*EIG2*	14q23	Epilepsy, idiopathic generalized, susceptibility to, 2	Susceptibility locus for idiopathic generalized epilepsy
Epilepsy, idiopathic generalized, susceptibility to, 3	*EIG3*	9q32-q33	Epilepsy, idiopathic generalized, susceptibility to, 3	Susceptibility locus for idiopathic generalized epilepsy
Epilepsy, idiopathic generalized, susceptibility to, 4	*EIG4*	10q25-q26	Epilepsy, idiopathic generalized, susceptibility to, 4	Susceptibility locus for idiopathic generalized epilepsy
Epilepsy, idiopathic generalized, susceptibility to, 5	*EIG5*	10p11.22	Epilepsy, idiopathic generalized, susceptibility to, 5	Susceptibility locus for idiopathic generalized epilepsy
Epilepsy, idiopathic generalized, susceptibility to, 6	*CACNA1H*	16p13.3	Calcium channel, voltage-dependent, T type, α 1H subunit	Susceptibility locus for idiopathic generalized epilepsy
Epilepsy, idiopathic generalized, susceptibility to, 7	*EJM2*	15q14	Epilepsy, juvenile myoclonic 2	Susceptibility locus for idiopathic generalized epilepsy
Epilepsy, juvenile myoclonic 3	*EJM3*	6p21	Epilepsy, juvenile myoclonic 3	Inheritance not established. Juvenile myoclonic epilepsy
Epilepsy, juvenile myoclonic, susceptibility to, 5	*GABRA1*	5q34-q35	γ-Aminobutyric acid (GABA) A receptor, α1	Susceptibility locus for juvenile myoclonic epilepsy
Epilepsy, juvenile myoclonic, susceptibility to, 6	*CACNB4*	2q22-q23	Calcium channel, voltage-dependent, β4 subunit	Susceptibility locus for juvenile myoclonic epilepsy

TABLE 109.1 Genes Associated with Neurologic Disorders—cont'd

Genetic disorder	Gene	Locus	Gene product	Inheritance (if established) and phenotype
Epilepsy, juvenile myoclonic, susceptibility to, 8	*CLCN2*	3q26	Chloride channel, voltage-sensitive 2	Susceptibility locus for juvenile myoclonic epilepsy
Myoclonic epilepsy, juvenile, 4	*EJM4*	5q12-q14	Myoclonic epilepsy, juvenile, 4	AD. Juvenile myoclonic epilepsy
Myoclonic epilepsy, juvenile, susceptibility to, 1	*EFHC1*	6p12-p11	EF-hand domain (C-terminal) containing 1	Susceptibility to myoclonic epilepsy
Myoclonic epilepsy, juvenile, susceptibility to, 1	*EJM1*	6p12-p11	Epilepsy, juvenile myoclonic 1	Susceptibility to myoclonic epilepsy
Epilepsy, nocturnal frontal lobe, 1	*CHRNA4*	20q13.2-q13.3	Cholinergic receptor, nicotinic, α4 (neuronal)	AD. Nocturnal frontal lobe motor seizures
Epilepsy, nocturnal frontal lobe, type 2	*ENFL2*	15q24	Epilepsy, nocturnal frontal lobe, type 2	AD. Nocturnal frontal lobe motor seizures
Epilepsy, nocturnal frontal lobe, 3	*CHRNB2*	1q21	Cholinergic receptor, nicotinic, β2 (neuronal)	AD. Nocturnal frontal lobe motor seizures
Epilepsy, nocturnal frontal lobe, type 4	*CHRNA2*	8p21	Cholinergic receptor, nicotinic, α2 (neuronal)	AD. Familial epilepsy with nocturnal wandering and ictal fear
Epilepsy, occipitotemporal lobe, and migraine with aura	*EPOLM*	9q21-q22	Epilepsy, occipitotemporal lobe, and migraine	AD. Familial temporal lobe epilepsy 4. Epilepsy, occipitotemporal lobe, and migraine with aura
Epilepsy, partial, with auditory features	*LGI1*	10q24	Leucine-rich, glioma inactivated 1	AD. Epilepsy, partial, with simple auditory sounds or complex auditory sounds
Epilepsy, partial, with pericentral spikes	*EPPS*	4p15	Epilepsy, partial, with pericentral spikes	AD. Epilepsy, partial, with pericentral spikes
Epilepsy, partial, with variable foci	*DEPDC5*	22q11-q12	DEP-Dopamine containing protein 5	AD. Epilepsy, partial, with variable foci
Epilepsy, progressive myoclonic 1/Unverricht–Lundborg disease	*CSTB*	21q22.3	Cystatin B (stefin B)	AR. Epilepsy, progressive myoclonic 1/Unverricht–Lundborg disease. Neurodegenerative disease
Epilepsy, progressive myoclonic 1B	*PRICKLE1*	12q12	Prickle homolog 1 (*Drosophila*)	AR. Progressive neurologic disease. Onset of ataxia in early childhood followed by myoclonic seizures
Epilepsy, progressive myoclonic 2A (Lafora)	*EPM2A*	6q24	Epilepsy, progressive myoclonus type 2A, Lafora disease (laforin)	AR. Progressive myoclonic epilepsy with neurologic degeneration often starts with headaches, myoclonic epilepsy, and visual hallucinations
Epilepsy, progressive myoclonic 2B (Lafora)	*NHLRC1*	6p22.3	NHL repeat containing 1	AR. Progressive myoclonic epilepsy with neurologic degeneration often starts with headaches, myoclonic epilepsy and visual hallucinations
Epilepsy, progressive myoclonic 3	*KCTD7*	7q11.21	Potassium channel tetramerization domain containing 7	AR. Progressive neurodegenerative disease with severe myoclonic epilepsy and cognitive decline
Epilepsy, pyridoxine-dependent	*ALDH7A1*	5q31	Aldehyde dehydrogenase 7 family, member A1	AR. Pyridoxine-dependent epilepsy
Epilepsy, rolandic, with paroxysmal exercise-induced dystonia and writer's cramp	*EPRPDC*	16p12-p11.2	Epilepsy, rolandic, with paroxysmal exercise-induced dystonia and writer's cramp	Possible AD. Epilepsy, rolandic, with paroxysmal exercise-induced dystonia and writer's cramp

Continued

TABLE 109.1 Genes Associated with Neurologic Disorders—cont'd

Genetic disorder	Gene	Locus	Gene product	Inheritance (if established) and phenotype
Epilepsy, X-linked, with variable learning disabilities and behavior disorders	*SYN1*	Xp11.4-p11.2	Synapsin I	X-linked. Epilepsy, with variable learning disabilities and behavior disorders
Epileptic encephalopathy	*ARX*	Xp22.13	Aristaless-related homeobox	Epileptic encephalopathy, early infantile, 1
Epileptic encephalopathy, hydrancephaly with abnormal genital, X-linked lissencephaly, X-linked mental retardation 29, Partington syndrome, Proud syndrome	*ARX*	Xp21.3	Aristaless-related homebox, X-linked	X-linked recessive. Epileptic encephalopathy, hydrancephaly with abnormal genital, X-linked lissencephaly, X-linked mental retardation 29, Partington syndrome, Proud syndrome
Epileptic encephalopathy, early infantile, 2	*CDKL5*	Xp22	Cyclin-dependent kinase-like 5	X-linked dominant. Epileptic encephalopathy, early infantile, 2
Epileptic encephalopathy, early infantile, 3	*SLC25A22*	11p15.5	Solute carrier family 25 (mitochondrial carrier: glutamate), member 22	AR. Epileptic encephalopathy, early infantile, 3
Epileptic encephalopathy, early infantile, 4	*STXBP1*	9q34.1	Syntaxin-binding protein 1	AD. Epileptic encephalopathy, early infantile, 4
Epileptic encephalopathy, early infantile, 5	*SPTAN1*	9q33-q34	Spectrin, α, nonerythrocytic 1 (α-fodrin)	AD. Epileptic encephalopathy, early infantile, 5
Epileptic encephalopathy, early infantile, 6. Dravet syndrome	*SCN1A*	2q24	Sodium channel, voltage-gated, type I, α subunit	AD. Febrile seizures, progressive disorder with myoclonic seizures and epileptic encephalopathy
Epileptic encephalopathy, early infantile, 7	*KCNQ2*	20q13.3	Potassium voltage-gated channel, KQT-like subfamily, member 2	AD. Epileptic encephalopathy, early infantile, 7
Epileptic encephalopathy, early infantile, 8	*ARHGEF9*	Xq22.1	Cdc42 guanine nucleotide exchange factor (GEF) 9	X-linked recessive. Epileptic encephalopathy, early infantile, 8
Epileptic encephalopathy, early infantile, 9	*PCDH19*	Xq22	Protocadherin 19	X-linked. Epileptic encephalopathy, early infantile, 9
Epileptic encephalopathy, early infantile, 10	*PNKP*	19q13.4	Polynucleotide kinase 3′-phosphatase	AR. Epileptic encephalopathy, early infantile, 10
Epileptic encephalopathy, early infantile, 11	*SCN2A*	2q24	Sodium channel, voltage-gated, type 2, α subunit	AD. Progressive epileptic encephalopathy
Epileptic encephalopathy, early infantile, 12	*PLCB1*	20p12	Phospholipase C, β1 (phosphoinositide-specific)	AR. Epileptic encephalopathy, early infantile, 12
Epileptic encephalopathy, early infantile, 13	*SCN8A*	2q24	Sodium channel, voltage-gated, type VIII, α subunit	AD. Progressive epileptic encephalopathy
Epileptic encephalopathy, early infantile, 14	*KCNT1*	9q34.3	Potassium channel, subfamily T, member 1	AD. Progressive epileptic encephalopathy
Epileptic encephalopathy, early infantile, 15	*ST3GAL3*	1p34.1	ST3 β-galactoside α-2,3-sialyltransferase 3	AR. Progressive epileptic encephalopathy
Epileptic encephalopathy, early infantile, 16	*TBC1D24*	16p13.3	TBC1 domain family, member 24	AR. Epileptic encephalopathy, early infantile, 16
Epileptic encephalopathy, Lennox–Gastaut type	*MAPK10*	4q21.3	Mitogen-activated protein kinase 10	AR. Epileptic encephalopathy, Lennox–Gastaut type
Episodic ataxia/myokymia syndrome	*KCNA1*	12p13	Potassium voltage-gated channel, shaker-related subfamily, member 1	AD. Episodic ataxia with myokymia, type 1
Episodic ataxia, type 2	*CACNA1A*	19p13	Calcium channel, voltage-dependent, P/Q type, α1A subunit	AD. Episodic ataxia with vertigo, hemiplegic migraines

TABLE 109.1 Genes Associated with Neurologic Disorders—cont'd

Genetic disorder	Gene	Locus	Gene product	Inheritance (if established) and phenotype
Episodic ataxia, type 3	*EA3*	1q42	Episodic ataxia, type 3	AD. Episodic ataxia with vertigo and tinnitus
Episodic ataxia, type 4	*EA4*	?	Episodic ataxia, type 4	AD. Episodic ataxia with vertigo and tinnitus
Episodic ataxia, type 5	*CACNB4*	2q22-q23	Calcium channel, voltage-dependent, β4 subunit	AD. Episodic ataxia with vertigo and tinnitus
Episodic ataxia, type 6	*SLC1A3*	5p13	Solute carrier family 1 (glial high affinity glutamate transporter), member 3	AD. Episodic ataxia with vertigo and tinnitus
Episodic ataxia, type 7	*EA7*	19q13	Episodic ataxia, type 7	AD. Episodic ataxia with vertigo
Episodic muscle weakness, X-linked	*EMWX*	Xp22.3	Episodic muscle weakness, X-linked	X-linked episodic muscle weakness
Ethylmalonic encephalopathy	*ETHE1*	19q13.32	Ethylmalonic encephalopathy 1	AR. Developmental regression failure to thrive, retinal lesions
Fabry disease	*GLA*	Xq22	Galactosidase, α	X-linked multisystem disorder, can cause cardiac and renal failure, corneal lesions, strokes, and skin lesions, among others
Facial paresis, hereditary congenital, 1	*MBS2*	3q21-q22	Möbius syndrome 2	AD. Unilateral or bilateral facial paralysis/Möbius syndrome 2
Facioscapulohumeral muscular dystrophy-1A	*FSHMD1A*	4q35	Facioscapulohumeral muscular dystrophy 1A	AD. Progressive facial and proximal weakness and atrophy, dysphagia, and wheelchair dependence
FG syndrome 1. Opitz–Kaveggia syndrome	*MED12*	Xq13	Mediator complex subunit 12	X-linked intellectual disability, peculiar dysmorphic features, hypotonia, sensorineural hearing loss, constipation
FG syndrome 2	*FLNA*	Xq28	Filamin A	X-linked intellectual disability, peculiar dysmorphic features, hypotonia, sensorineural hearing loss, constipation
FG syndrome 3	*FGS3*	Xp22.3	FG syndrome 3	X-linked. See above
FG syndrome, mental retardation with microcephaly and pontine and cerebellar hypoplasia, mental retardation with or without nystagmus	*CASK*	Xp11.4	Calcium/calmodulin-dependent serine protein	X-linked. FG syndrome, mental retardation with microcephaly and pontine and cerebellar hypoplasia, mental retardation with or without nystagmus
FG syndrome 5	*FGS5*	Xq22.3	FG syndrome 5	X-linked. See above
Fibrosis of extraocular muscles, congenital, 3A	*TUBB3*	16q24.3	Tubulin, β3, class III	AD. Congenital fibrosis of extraocular muscles
Folate malabsorption, hereditary	*SLC46A1*	17q11.1	Solute carrier family 46 (folate transporter), member 1	AR. Megaloblastic anemia, diarrhea, immune deficiency, and neurologic deficits
Forsythe–Wakeling syndrome	*FWS*	1p33-p31.1	Forsythe–Wakeling syndrome	AR. One family reported. Global developmental delay, onset of nephrotic syndrome, and thrombocytopenia

Continued

TABLE 109.1 Genes Associated with Neurologic Disorders—cont'd

Genetic disorder	Gene	Locus	Gene product	Inheritance (if established) and phenotype
Fragile X syndrome	*FMR1*	Xq27.3	Fragile X mental retardation 1	X-linked recessive caused by trinucleotide (CGG) repeat expansion greater than 200. Phenotype is macroorchidism, mental retardation. Premutations can cause fragile-X ataxia syndrome
Friedreich ataxia	*FXN*	9q13	Frataxin	AR. Progressive neurologic disease with gait and limb ataxia, hypertrophic cardiomyopathy
Friedreich ataxia 2	*FRDA2*	9p23-p11	Friedreich ataxia 2	AR. Progressive neurologic disease with gait and limb ataxia, hypertrophic cardiomyopathy
Frontotemporal lobar degeneration with ubiquitin-positive inclusions	*GRN*	17q21.32	Granulin	AD. Type of frontotemporal dementia
Fucosidosis	*FUCA1*	1p34	Fucosidase, α-L- 1, tissue	AR. Lysosomal storage disease with progressive psychomotor retardation, angiokeratoma, dysostosis multiplex
GABA-transaminase deficiency	*ABAT*	16p13.3	4-Aminobutyrate aminotransferase	AR. Psychomotor retardation, hypotonia, hyperreflexia
Galactokinase deficiency with cataracts	*GALK1*	17q24	Galactokinase 1	AR. Disorder of galactose metabolism
Galactose epimerase deficiency	*GALE*	1p36-p35	UDP-galactose-4-epimerase	AR. Disorder of galactose metabolism
Galactosemia	*GALT*	9p13	Galactose-1-phosphate uridylyltransferase	AR. Disorder of galactose metabolism. Classic galactosemia
Galactosialidosis	*CTSA*	20q13.1	Cathepsin A	AR. Lysosomal storage disease with visceromegaly, short stature, coarse facial features, neurologic signs
GAMT deficiency/cerebral creatine deficiency syndrome 2	*GAMT*	19p13.3	Guanidinoacetate *N*-methyltransferase	AR. Creatine/phosphocreatine depletion in brain causing psychomotor delays
Gaucher disease, type 1	*GBA*	1q21	Glucosidase, β, acid	AR. Non-neuronopathic form, hepatosplenomegaly, bone marrow infiltration
Gaucher disease, types 2 and 3	*GBA*	1q21	Glucosidase, β, acid	AR. Neuropathic form, hepatosplenomegaly, bone marrow infiltration, and nervous system involvement
Gaze palsy, horizontal, with progressive scoliosis	*ROBO3*	11q23-q25	Roundabout, axon guidance receptor, homolog 3 (*Drosophila*)	AR. Gaze palsy, horizontal, with progressive scoliosis
Generalized epilepsy and paroxysmal dyskinesia	*KCNMA1*	10q22.3	Potassium large conductance calcium-activated channel, subfamily M, α member 1	AD. Generalized epilepsy and paroxysmal dyskinesia
Gerstmann–Straussler disease	*PRNP*	20pter-p12	Prion protein	Inheritance not established. Cerebral amyloid angiopathy
Giant axonal neuropathy 1	*GAN*	16q24.1	Gigaxonin	AR. Central and peripheral nervous system involvement
Glioma susceptibility 4	*GLM1*	15q23-q26.3	Glioma, familial, 1	Susceptibility locus for glioma
Glioma susceptibility 5	*GLM5*	9p21.3	Glioma susceptibility 5	Susceptibility locus for glioma
Glioma susceptibility 6	*GLM6*	20q13.33	Glioma susceptibility 6	Susceptibility locus for glioma

TABLE 109.1 Genes Associated with Neurologic Disorders—cont'd

Genetic disorder	Gene	Locus	Gene product	Inheritance (if established) and phenotype
Glioma susceptibility 8	*GLM8*	5p15.33	Glioma susceptibility 8	Susceptibility locus for glioma
Gliosis, familial progressive subcortical	*GPSC*	17q21-q22	Gliosis, familial progressive subcortical	Susceptibility locus for glioma
GLUT1 deficiency syndrome 1	*SLC2A1*	1p35-p31.3	Solute carrier family 2 (facilitated glucose transporter), member 1	AD. Rarely AR. Static neurologic dysfunction with absence or myoclonic epilepsy, movement disorders or anemia
Glutamine deficiency, congenital	*GLUL*	1q31	Glutamate-ammonia ligase	AR. Craniofacial dysmorphism, bradycardia, apnea, encephalopathy, hypotonia, seizures
Glutaric aciduria, type I	*GCDH*	19p13.2	Glutaryl-CoA dehydrogenase	AR. Glutaric aciduria
Glutaric aciduria, type IIA	*ETFA*	15q23-q25	Electron-transfer flavoprotein, α polypeptide	AR. Glutaric aciduria
Glutaric aciduria, type IIB	*ETFB*	19q13.3	Electron-transfer flavoprotein, β polypeptide	AR. Glutaric aciduria
Glutaric aciduria, type IIC	*ETFDH*	4q32-qter	Electron-transferring-flavoprotein dehydrogenase	AR. Glutaric aciduria
Glutathionuria	*GGT6*	17p13.2	γ-Glutamyltransferase 6	AR. Intellectual disability
Glycine encephalopathy	*GCSH*	16q24	Glycine cleavage system protein H (aminomethyl carrier)	AR. Encephalopathy, seizures, hypotonia, behavioral problems
Glycine encephalopathy	*AMT*	3p21.2-p21.1	Aminomethyltransferase	AR. Encephalopathy, seizures, hypotonia, behavioral problems
Glycine encephalopathy	*GLDC*	9p22	Glycine dehydrogenase (decarboxylating)	AR. Encephalopathy, seizures, hypotonia, behavioral problems
Glycogen storage disease 0, muscle	*GYS1*	19q13.3	Glycogen synthase 1 (muscle)	AR. Hypoglycemia, occasional muscle cramping
Glycogen storage disease Ib	*SLC37A4*	11q23	Solute carrier family 37 (glucose-6-phosphate transporter), member 4	AR. Hypoglycemia, hepatomegaly, hyperlipidemia, lactic acidosis, hyperuricemia
Glycogen storage disease Ic	*SLC17A3*	6p21.3	Solute carrier family 17 (sodium phosphate), member 3	AR. Hypoglycemia, hepatomegaly, hyperlipidemia, lactic acidosis, hyperuricemia
Glycogen storage disease II	*GAA*	17q25.2-q25.3	Glucosidase, α; acid	AR. Hepatomegaly, muscle weakness, heart failure
Glycogen storage disease IIb	*LAMP2*	Xq24	Lysosomal-associated membrane protein 2	AR. Hepatomegaly, muscle weakness, heart failure
Glycogen storage disease IIIa	*AGL*	1p21	Amylo-α-1,6-glucosidase, 4-α-glucanotransferase	AR. Hypoglycemia, hepatomegaly, hyperlipidemia, myopathy
Glycogen storage disease IV	*GBE1*	3p12	Glucan (1,4-α-), branching enzyme 1	AR. Hepatomegaly with cirrhosis, failure to thrive, death at age ~5 years
Glycogen storage disease IXc	*PHKG2*	16p12.1-p11.2	Phosphorylase kinase, γ2 (testis)	AR. Hypoglycemia, hepatomegaly, hyperlipidemia, delayed motor development, growth retardation
Glycogen storage disease VII	*PFKM*	12q13.3	Phosphofructokinase, muscle	AR. Exercise-induced muscle cramps and weakness, growth retardation, hemolytic anemia
Glycogen storage disease X	*PGAM2*	7p13-p12.3	Phosphoglycerate mutase 2 (muscle)	AR. Exercise-induced cramps, occasional myoglobinuria

Continued

TABLE 109.1 Genes Associated with Neurologic Disorders—cont'd

Genetic disorder	Gene	Locus	Gene product	Inheritance (if established) and phenotype
Glycogen storage disease XI	*LDHA*	11p15.4	Lactate dehydrogenase A	AR. Hypoglycemia, hepatomegaly
Glycogen storage disease XII	*ALDOA*	16p11.2	Aldolase A, fructose-bisphosphate	AR. Exercise intolerance, cramps
Glycogen storage disease XIII	*ENO3*	17pter-p12	Enolase 3 (β, muscle)	AR. Exercise-induced myalgias
Glycogen storage disease XIV	*PGM1*	1p31	Phosphoglucomutase 1	AR. Exercise intolerance, myoglobinuria
Glycogen storage disease XV	*GYG1*	3q24-q25.1	Glycogenin 1	AR. Muscle weakness and cardiac arrhythmias
Glycogen storage disease, type 0	*GYS2*	12p12.2	Glycogen synthase 2 (liver)	AR. Hypoglycemia
Glycogen storage disease, type IXa1	*PHKA2*	Xp22.2-p22.1	Phosphorylase kinase, α2 (liver)	AR. Hypoglycemia, hepatomegaly, hyperlipidemia, delayed motor development, growth retardation
G_{M1}-gangliosidosis, type I	*GLB1*	3p21.33	Galactosidase, β1	AR. Coarse facial features, intellectual disability, cerebral degeneration, hepatosplenomegaly, dilated cardiomyopathy
G_{M2}-gangliosidosis, AB variant	*GM2A*	5q31.3-q33.1	G_{M2} ganglioside activator	AR. Encephalopathy, seizures, dementia, hypotonia with late hypertonia, increased startle response
Greig cephalopolysyndactyly syndrome	*GLI3*	7p13	GLI family zinc finger 3	AD. Frontal bossing, scaphocephaly and hypertelorism, normal intellect
Griscelli syndrome, type 1	*MYO5A*	15q21	Myosin VA (heavy chain 12, myoxin)	AR. Hypomelanosis with primary neurologic manifestations and no immune deficits
Griscelli syndrome, type 2	*RAB27A*	15q21	RAB27A, member RAS oncogene family	AR. Hypomelanosis with primary neurologic manifestations and immune deficits
Growth retardation with deafness and mental retardation due to IGF1 deficiency	*IGF1*	12q22-q24.1	Insulin-like growth factor 1 (somatomedin C)	AR. Growth retardation with deafness and mental retardation
Growth retardation, developmental delay, coarse facies, and early death	*FTO*	16q12.2	Fat mass and obesity associated	AR. Growth retardation, developmental delay, coarse facies, and early death
Gustavson syndrome	*GUST*	Xq26	Gustavson syndrome	X-linked. Gustavson mental retardation syndrome (with microcephaly, optic atrophy and blindness)
Hartnup disorder	*SLC6A19*	5p15.33	Solute carrier family 6 (neutral amino acid transporter), member 19	AR. Seizure, cognitive delays, hypertonia, short stature, dermatitis, and atrophic glossitis. Neutral aminoaciduria
Hemiplegic migraine, familial	*CACNA1A*	19p13	Calcium channel, voltage-dependent, P/Q type, α1A subunit	AD. Familial hemiplegic migraine
Hemorrhagic destruction of the brain, subependymal calcification, and cataracts	*JAM3*	11q25	Junctional adhesion molecule 3	AR. Microcephaly, with seizures, hypertonia, hyperreflexia, cataracts with renal problems and hepatomegaly
Holoprosencephaly-1	*HPE1*	21q22.3	Holoprosencephaly 1, alobar	AR. Holoprosencephaly
Holoprosencephaly-2	*SIX3*	2p21	SIX homeobox 3	AD. Holoprosencephaly
Holoprosencephaly-3	*SHH*	7q36	Sonic hedgehog	AD. Holoprosencephaly

TABLE 109.1 Genes Associated with Neurologic Disorders—cont'd

Genetic disorder	Gene	Locus	Gene product	Inheritance (if established) and phenotype
Holoprosencephaly-4	*TGIF1*	18p11.3	TGFB-induced factor homeobox 1	AD. Holoprosencephaly
Holoprosencephaly-5	*ZIC2*	13q32	Zic family member 2	AD. Holoprosencephaly
Holoprosencephaly-6	*HPE6*	2q37.1-q37.3	Holoprosencephaly 6	Haploinsufficiency. Holoprosencephaly
Holoprosencephaly-7	*PTCH1*	9q22.32	Patched, *Drosophila* homolog of	AD. Holoprosencephaly
Holoprosencephaly-8	*HPE8*	14q13	Holoprosencephaly 8	Haploinsufficiency. Holoprosencephaly
Holoprosencephaly-9	*HPE9*	2q14.3	Holoprosencephaly 9	AD. Holoprosencephaly
Holoprosencephaly-10	*HPE10*	1q41-42	Holoprosencephaly 10	Deletion. Holoprosencephaly
Holoprosencephaly-11	*CDON*	11q24.2	Cell adhesion molecule related-downregulated by oncogene	AD. Holoprosencephaly
Homocystinuria due to MTHFR deficiency	*MTHFR*	1p36.3	Methylenetetrahydrofolate reductase (NAD(P)H)	AR. Microcephaly, with seizures, intellectual disability, strokes. Psychiatric disturbances
Homocystinuria-megaloblastic anemia, cbl E type	*MTRR*	5p15.3-p15.2	5-Methyltetrahydrofolate-homocysteine methyltransferase reductase	AR. Failure to thrive, nystagmus, psychomotor retardation with seizures with homocystinuria and megaloblastic anemia
Homocystinuria, B6-responsive and nonresponsive types	*CBS*	21q22.3	Cystathionine-β-synthase	AR. Homocystinuria
Homocystinuria, cblD type, variant 1	*MMADHC*	2q23.2	Methylmalonic aciduria (cobalamin deficiency) cblD type, with homocystinuria	AR. Homocystinuria
HSN I with cough and gastroesophageal reflux	*HSN1B*	3p24-p22	Hereditary sensory neuropathy, type IB	AD. Sensory axonal neuropathy with distal sensory loss. Cough and cough syncope, sensorineural hearing loss
Huntington disease	*HTT*	4p16.3	Huntington	AD. Neurodegenerative disease with chorea, dystonia, incoordination, cognitive decline and behavioral difficulties
Huntington disease-like 1	*PRNP*	20pter-p12	Prion protein	AD. Clinical phenotype similar to Huntington
Huntington disease-like 2	*JPH3*	16q24.3	Junctophilin 3	AD. Clinical phenotype similar to Huntington
Huntington disease-like 3	*HLN2*	4p15.3	Huntington-like neurodegenerative disorder 2	AR. Clinical phenotype similar to Huntington
Hydrocephalus due to aqueductal stenosis	*L1CAM*	Xq28	L1 cell adhesion molecule	X-linked recessive. Hydrocephalus, aqueductal stenosis and agenesis of corpus callosum and septum pellucidum
Hyperekplexia/startle disease-1	*GLRA1*	5q32	Glycine receptor, α1	AD/AR. Exaggerated startle response to external stimuli
Hyperekplexia-2	*GLRB*	4q31.3	Glycine receptor, β	AR. Exaggerated startle response to external stimuli
Hyperekplexia-3	*SLC6A5*	11p15.2-p15.1	Solute carrier family 6 (neurotransmitter transporter, glycine), member 5	AD/AR. Exaggerated startle response to external stimuli
Hyperkalemic periodic paralysis, type 2	*SCN4A*	17q23.1-q25.3	Sodium channel, voltage-gated, type IV, α subunit	AD. Episodes of flaccid paralysis and weakness, hyperkalemia during attacks

Continued

TABLE 109.1 Genes Associated with Neurologic Disorders—cont'd

Genetic disorder	Gene	Locus	Gene product	Inheritance (if established) and phenotype
Hyperlysinemia	*AASS*	7q31.3	Aminoadipate-semialdehyde synthase	AR. Impaired sexual development, intellectual disability, ectopia lentis, lax ligaments and joints
Hyperornithinemia-hyperammonemia-homocitrullinemia syndrome	*SLC25A15*	13q14	Solute carrier family 25 (mitochondrial carrier; ornithine transporter) member 15	AR. Failure to thrive, psychomotor retardation with seizures, spasticity, liver dysfunction, episodic vomiting
Hyperphenylalaninemia, BH4-deficient A	*PTS*	11q22.3-q23.3	6-Pyruvoyltetrahydropterin synthase	AR disorder with hyperphenylalaninemia with depletion of serotonin and dopamine, progressive cognitive and motor delays
Hyperphenylalaninemia, BH4-deficient B	*GCH1*	14q22.1	GTP cyclohydrolase 1	AR disorder with hyperphenylalaninemia with depletion of serotonin and dopamine, progressive cognitive and motor delays
Hyperphenylalaninemia, BH4-deficient C	*QDPR*	4p14.32	Dihydropteridine reductase	AR disorder with hyperphenylalaninemia with depletion of serotonin and dopamine, progressive cognitive and motor delays
Hyperphenylalaninemia, BH4-deficient, D	*PCBD1*	10q22	Pterin-4 α-carbinolamine dehydratase/dimerization cofactor of hepatocyte nuclear factor 1 α	AR disorder with hyperphenylalaninemia with depletion of serotonin and dopamine, progressive cognitive and motor delays
Hypoceruloplasminemia, hereditary	*CP*	3q23-q24	Ceruloplasmin (ferroxidase)	AR. Blepharospasm, retinal degeneration, extrapyramidal signs
Hypokalemic periodic paralysis, type 1	*CACNA1S*	1q32	Calcium channel, voltage-dependent, L type, α1S subunit	AD. Episodes of flaccid paralysis and weakness, hypokalemia during attacks
Hypomyelination, global cerebral	*SLC25A12*	2q24	Solute carrier family 25 (mitochondrial carrier, Aralar), member 12	AR. Severe psychomotor delays, hypotonia with later hypertonia, MRI with diffuse hypomyelination
Inclusion body myopathy with early-onset Paget disease and frontotemporal dementia	*VCP*	9p13-p12	Valosin-containing protein	AD. Proximal weakness, rimmed vacuoles seen on biopsy, gait abnormalities, and expressive dysphasia
Inclusion body myopathy-2	*GNE*	9p13.3	UDP-*N* acetylglucosamine 2 epimerase	AR. Progressive ascending weakness, distal muscle atrophy, limb girdle weakness. Rimmed vacuoles on biopsy
Inclusion body myopathy-3	*MYH2*	17p13.1	Myosin, heavy chain 2, skeletal muscle, adult	AD. Joint contractures, external ophthalmoplegia and proximal weakness. Rimmed vacuoles on biopsy
Incontinentia pigmenti, type II	*IKBKG*	Xq28	Inhibitor of kappa light polypeptide gene enhancer in B-cells, kinase γ	X-linked dominant. Skin pigmentation that follows Blaschko lines. Microcephaly, seizures, intellectual disability, retinal changes
Infantile neuroaxonal dystrophy 1	*PLA2G6*	22q13.1	Phospholipase A2, group VI (cytosolic, calcium-independent)	AR. Psychomotor delays, hypotonia, generalized weakness, ataxia and spasticity

TABLE 109.1 Genes Associated with Neurologic Disorders—cont'd

Genetic disorder	Gene	Locus	Gene product	Inheritance (if established) and phenotype
Insomnia, fatal familial	*PRNP*	20p13	Prion protein	AD. Refractory insomnia, dementia, neurodegeneration, diplopia, dysphagia, dysautonomia
Isobutyryl-CoA dehydrogenase deficiency	*ACAD8*	11q25	Acyl-CoA dehydrogenase family, member 8	AR. Cardiomyopathy, anemia, and secondary carnitine deficiency
Isovaleric acidemia	*IVD*	15q14-q15	Isovaleryl-CoA dehydrogenase	AR. Isovaleric acidemia
Jacobsen syndrome	*JBS*	11q23	Jacobsen syndrome	Haploinsufficiency. Craniofacial dysmorphism, genital anomalies, hypotonia, spasticity, and mental retardation
Johanson–Blizzard syndrome	*UBR1*	15q15-q21.1	Ubiquitin protein ligase E3 component n-recognin 1	AR. Growth restriction, intellectual disability and dysmorphic features
Joubert syndrome 1	*INPPE5*	9q34.3	Inositol polyphosphate-5-phosphatase, 72 kDa	AR. Joubert syndrome: Complex brainstem malformation including dysgenesis or agenesis of cerebellar vermis, hypoplasia of brainstem; macrocephaly; failure to thrive, nephronophthisis
Joubert syndrome 2	*TMEM216*	11q13	Transmembrane protein 216	AR. Joubert syndrome
Joubert syndrome 3	*AHI1*	6q23.3	Abelson helper integration site 1	AR. Joubert syndrome
Joubert syndrome 4	*NPHP1*	2q13	Nephrocystin 1	AR. Joubert syndrome
Joubert syndrome 5	*CEP290*	12q21.3	Centrosomal protein, 290 kDa	AR. Joubert syndrome
Joubert syndrome 5	*LCA10*	12q21.3	Lung carcinoma-associated protein 10	AR. Joubert syndrome
Joubert syndrome 6	*TMEM67*	8q21	Transmembrane protein 67	AR. Joubert syndrome
Joubert syndrome 7	*RPGRIP1L*	16q12.2	RPGRIP1-like	AR. Joubert syndrome
Joubert syndrome 8	*ARL13B*	3q11.2	ADP-ribosylation factor-like 13B	AR. Joubert syndrome
Joubert syndrome 9	*CC2D2A*	4p15.3	Coiled-coil and C2 domain containing 2A	AR. Joubert syndrome
Joubert syndrome 10	*OFD1*	Xp22.2	Chromosome X open reading frame 5	X-linked recessive. Joubert syndrome
Joubert syndrome 12	*KIF7*	15q26.1	Kinesin family member	AR. Joubert syndrome
Joubert syndrome 13	*TCTN1*	12q24.1	Tectonic family member 1	AR. Joubert syndrome
Joubert syndrome 14	*TMEM237*	2q33.1	Transmembrane protein 237	AR. Joubert syndrome
Joubert syndrome 15	*CEP41*	7q32.2	Centrosomal protein, 41 kDa	AR. Joubert syndrome
Joubert syndrome 16	*TMEM138*	11q12.2	Transmembrane protein 138	AR. Joubert syndrome
Joubert syndrome 17	*C5orf42*	5p13.2	Chromosome 5 open reading frame 42	AR. Joubert syndrome
Joubert syndrome 18	*TCTN3*	10q24.1	Tectonic family member 3	AR. Joubert syndrome
Joubert syndrome 19	*ZNF423*	16q12.1	Zinc finger protein 423	AD. Joubert syndrome
Joubert syndrome 20	*TMEM231*	16q23	Transmembrane protein 231	AR. Joubert syndrome
Kabuki syndrome	*MLL2*	12q13.12	Myeloid/lymphoid or mixed-lineage leukemia 2	AD. Intellectual disability, epilepsy, specific facial features resembling make-up of Kabuki

Continued

TABLE 109.1 Genes Associated with Neurologic Disorders—cont'd

Genetic disorder	Gene	Locus	Gene product	Inheritance (if established) and phenotype
Kabuki syndrome 2	*KDM6A*	Xp11.3	Lysine specific demethylase 6A	X-linked dominant. Intellectual disability, epilepsy, specific facial features resembling make-up of Kabuki
Kahrizi syndrome	*KRIZI*	4p12-q12	Kahrizi syndrome	AR. Intellectual disability, coloboma, coarse facial features
Kallmann syndrome 1	*KAL1*	Xp22.31	Kallmann syndrome 1	X-linked. Characterized by hypogonadotropic hypogonadism and anosmia, maybe associated with renal agenesis of olfactory lobes and ataxia, synkinesia, nystagmus, visual defects, clubfoot, cleft palate, including rare cases of sporadic isolated GNRH deficiency
Kleefstra syndrome	*EHMT1*	9q34.3	Euchromatic histone-lysine *N*-methyltransferase 1	AD. 9q subtelomeric deletion syndrome with intellectual disability, microcephaly/brachycephaly. Epilepsy, synophrys, anteverted nares
Knobloch syndrome, type 1	*COL18A1*	21q22.3	Collagen, type XVIII, α1	AR. Myopia, retinal detachment, occipital skull defects
Krabbe disease	*GALC*	14q31	Galactosylceramidase	AR. Leukodystrophy with severe intellectual disability and psychomotor regression
Kuru, susceptibility to	*PRNP*	20p13	Prion protein	Susceptibility to Kuru
Lactic acidemia due to PDX1 deficiency	*PDHX*	11p13	Pyruvate dehydrogenase complex, component X	AR. The clinical spectrum can range from fatal lactic acidosis in the newborn to chronic neurologic dysfunction with structural abnormalities in the central nervous system without systemic acidosis
Legius syndrome	*SPRED1*	15q13.2	Sprouty-related EVH1 domain-containing protein 1	AD, multiple café-au-lait spots, macrocephaly, attention deficit, and learning problems
Legius syndrome	*SPRED1*	15q13.2	Sprouty-related, EVH1 domain containing 1	AD, multiple café-au-lait spots, macrocephaly, attention deficit and learning problems
Leigh syndrome	*NDUFS3*	11p11.11	NADH dehydrogenase (ubiquinone) Fe–S protein 3, 30 kDa (NADH-coenzyme Q reductase)	AR/mitochondrial. Progressive neurodegenerative disease with focal bilateral lesions in brain stem, thalamus, basal ganglia or spinal cord. Clinical features depend upon the area of involvement. The most common underlying defect is defect in oxidative phosphorylation
Leigh syndrome	*NDUFS8*	11q13	NADH dehydrogenase (ubiquinone) Fe–S protein 8, 23 kDa (NADH-coenzyme Q reductase)	AR/mitochondrial. Progressive neurodegenerative disease with focal bilateral lesions in brain stem, thalamus, basal ganglia or spinal cord. Clinical features depend upon the area of involvement. The most common underlying defect is defect in oxidative phosphorylation

TABLE 109.1 Genes Associated with Neurologic Disorders—cont'd

Genetic disorder	Gene	Locus	Gene product	Inheritance (if established) and phenotype
Leigh syndrome	*NDUFV1*	11q13	NADH dehydrogenase (ubiquinone) flavoprotein 1, 51 kDa	AR/mitochondrial. Progressive neurodegenerative disease with focal bilateral lesions in brain stem, thalamus, basal ganglia or spinal cord. Clinical features depend upon the area of involvement. The most common underlying defect is defect in oxidative phosphorylation
Leigh syndrome	*NDUFS7*	19p13	NADH dehydrogenase (ubiquinone) Fe–S protein 7, 20 kDa (NADH-coenzyme Q reductase)	AR/mitochondrial. Progressive neurodegenerative disease with focal bilateral lesions in brain stem, thalamus, basal ganglia or spinal cord. Clinical features depend upon the area of involvement. The most common underlying defect is defect in oxidative phosphorylation
Leigh syndrome	*SDHA*	5p15	Succinate dehydrogenase complex, subunit A, flavoprotein (Fp)	AR/mitochondrial. Progressive neurodegenerative disease with focal bilateral lesions in brain stem, thalamus, basal ganglia or spinal cord. Clinical features depend upon the area of involvement. The most common underlying defect is defect in oxidative phosphorylation
Leigh syndrome	*NDUFS4*	5q11.1	NADH dehydrogenase (ubiquinone) Fe–S protein 4, 18 kDa (NADH-coenzyme Q reductase)	AR/mitochondrial. Progressive neurodegenerative disease with focal bilateral lesions in brain stem, thalamus, basal ganglia or spinal cord. Clinical features depend upon the area of involvement. The most common underlying defect is defect in oxidative phosphorylation
Leigh syndrome due to mitochondrial complex I deficiency	*NDUFA2*	5q31.2	NADH dehydrogenase (ubiquinone) 1 α subcomplex, 2, 8 kDa	AR/mitochondrial. Progressive neurodegenerative disease with focal bilateral lesions in brain stem, thalamus, basal ganglia or spinal cord. Clinical features depend upon the area of involvement. The most common underlying defect is defect in oxidative phosphorylation
Leigh syndrome due to mitochondrial complex I deficiency	*C8orf38*	8q22.1	Chromosome 8 open reading frame 38	AR/mitochondrial. Progressive neurodegenerative disease with focal bilateral lesions in brain stem, thalamus, basal ganglia, or spinal cord. Clinical features depend upon the area of involvement. The most common underlying defect is defect in oxidative phosphorylation
Leigh syndrome, due to COX deficiency	*SURF1*	9q34	Surfeit 1	AR/mitochondrial. Progressive neurodegenerative disease with focal bilateral lesions in brain stem, thalamus, basal ganglia or spinal cord. Clinical features depend upon the area of involvement. The most common underlying defect is defect in oxidative phosphorylation

Continued

TABLE 109.1 Genes Associated with Neurologic Disorders—cont'd

Genetic disorder	Gene	Locus	Gene product	Inheritance (if established) and phenotype
Leigh syndrome, French-Canadian type	*LRPPRC*	2p21-p16	Leucine-rich pentatricopeptide repeat containing	AR/mitochondrial. Progressive neurodegenerative disease with focal bilateral lesions in brain stem, thalamus, basal ganglia or spinal cord. Clinical features depend upon the area of involvement. The most common underlying defect is defect in oxidative phosphorylation
Lesch–Nyhan syndrome	*HPRT1*	Xq26-q27.2	Hypoxanthine phosphoribosyltransferase 1	AR. Intellectual disability, spastic cerebral palsy, choreoathetosis, uric acid urinary stones, and self-destructive biting of fingers and lips
Leukodystrophy, adult-onset, autosomal dominant	*LMNB1*	5q23.3-q31.1	Lamin B1	AD. Adult-onset leukodystrophy
Leukodystrophy, hypomyelinating, 2	*GJC2*	1q41-q42	Gap junction protein, γ2, 47 kDa	AR. Hypotonia, developmental delays, ataxia, titubation, spastic paraplegia
Leukodystrophy, hypomyelinating, 5	*FAM126A*	7p15.3	Family with sequence similarity 126, member A	AR. Cataracts, progressive neurodegeneration progressive scoliosis, hypomyelination
Leukoencephalopathy with brainstem and spinal cord involvement and lactate elevation	*DARS2*	1q25.1	Aspartyl-tRNA synthetase 2, mitochondrial	AR. Nystagmus, joint contractures, muscle weakness and wasting
Leukoencephalopathy with dystonia and motor neuropathy	*SCP2*	1p32	Sterol carrier protein 2	AR. Leukoencephalopathy with dystonia and motor neuropathy
Leukoencephalopathy with metaphyseal chondrodysplasia	*LKMCD*	Xq25-q27	Leukoencephalopathy with metaphyseal chondrodysplasia	X-linked. Leukoencephalopathy with metaphyseal chondrodysplasia
Leukoencephalopathy with vanishing white matter	*EIF2B1*	12q24.31	Eukaryotic translation initiation factor 2B, subunit 1α, 26 kDa	AR. Leukoencephalopathy with vanishing white matter
Leukoencephalopathy with vanishing white matter	*EIF2B2*	14q24	Eukaryotic translation initiation factor 2B, subunit 2β, 39 kDa	AR. Leukoencephalopathy with vanishing white matter
Leukoencephalopathy with vanishing white matter	*EIF2B3*	1p34.1	Eukaryotic translation initiation factor 2B, subunit 3γ, 58 kDa	AR. Leukoencephalopathy with vanishing white matter
Leukoencephalopathy with vanishing white matter	*EIF2B4*	2p23.3	Eukaryotic translation initiation factor 2B, subunit 4δ, 67 kDa	AR. Leukoencephalopathy with vanishing white matter
Leukoencephalopathy with vanishing white matter	*EIF2B5*	3q27	Eukaryotic translation initiation factor 2B, subunit 5ε, 82 kDa	AR. Leukoencephalopathy with vanishing white matter
Leukoencephalopathy, cystic, without megalencephaly	*RNASET2*	6q27	Ribonuclease T2	AR. Cystic leukoencephalopathy
Li–Fraumeni syndrome	*CHEK2*	22q12.1	Checkpoint kinase 2	AD. Increased risk for multiple cancers including hematologic, breast, soft tissue sarcomas and uterine
Limb-girdle muscular dystrophy, type 1G	*LGMD1G*	4q21	Limb-girdle muscular dystrophy 1G (autosomal dominant)	AD. Limb-girdle muscular dystrophy
Lissencephaly 1	*PAFAH1B1*	17p13.3	Platelet-activating factor acetylhydrolase 1b, regulatory subunit 1 (45 kDa)	Isolated/AR. Lissencephaly
Lissencephaly 3	*TUBA1A*	12q12-q14	Tubulin, α1a	Isolated/AR. Lissencephaly
Lissencephaly syndrome, Norman–Roberts type	*RELN*	7q22	Reelin	Isolated/AR. Lissencephaly
Lissencephaly, X-linked	*DCX*	Xq22.3-q23	Doublecortin	X-linked. Lissencephaly

TABLE 109.1 Genes Associated with Neurologic Disorders—cont'd

Genetic disorder	Gene	Locus	Gene product	Inheritance (if established) and phenotype
Lysinuric protein intolerance	*SLC7A7*	14q11.2	Solute carrier family 7 (amino acid transporter light chain, y+L system), member 7	AR. Intermittent hyperammonemia, failure to thrive and cognitive delays
Machado–Joseph disease	*ATXN3*	14q24.3-q31	Ataxin 3	AD. Spinocerebellar ataxia type 3
Major affective disorder 1	*MAFD1*	18p	Major affective disorder 1	Inheritance not established. Major affective disorder 1
Major affective disorder 2	*MAFD2*	Xq28	Major affective disorder 2	Inheritance not established. Major affective disorder 2
Major affective disorder 3, early onset	*MAFD3*	21q22.13	Major affective disorder 3, early onset	Inheritance not established. Major affective disorder 3, early onset
Major affective disorder 4	*MAFD4*	16p12	Major affective disorder 4	Inheritance not established. Major affective disorder 4
Major affective disorder 5	*MAFD5*	2q22-q24	Major affective disorder 5	Inheritance not established. Major affective disorder 5
Major affective disorder 6	*MAFD6*	6q23-q24	Major affective disorder 6	Inheritance not established. Major affective disorder 6
Major affective disorder 7, susceptibility to	*XBP1*	22q12	X-box binding protein 1	Susceptibility to major affective disorder 7
Major affective disorder 8, susceptibility to	*MAFD8*	10q21	Major affective disorder 8, susceptibility to	Susceptibility to major affective disorder 8
Major affective disorder 9, susceptibility to	*MAFD9*	12p13.3	Major affective disorder 9, susceptibility to	Susceptibility to major affective disorder 9
Major depressive disorder 1	*MDD1*	12q22-q23.2	Major depressive disorder	Inheritance not established. Major depressive disorder 1
Major depressive disorder 2	*MDD2*	15q25.3-q26.2	Major depressive disorder 2	Inheritance not established. Major depressive disorder 2
Major depressive disorder and accelerated response to antidepressant drug treatment	*BP51*	6p21.3-p21.2	Blood pressure QTL 51	Inheritance not established. Major depression with rapid response to antidepressant therapy
Major depressive disorder and accelerated response to antidepressant drug treatment	*FKBP5*	6p21.3-p21.2	FK506 binding protein 5	Inheritance not established. Major depression with rapid response to antidepressant therapy
Malignant hyperthermia susceptibility 1	*RYR1*	19q13.1	Ryanodine receptor 1 (skeletal)	Susceptibility to malignant hyperthermia
Malignant hyperthermia susceptibility 2	*MHS2*	17q11.2-q24	Malignant hyperthermia susceptibility 2	Susceptibility to malignant hyperthermia
Malignant hyperthermia susceptibility 3	*MHS3*	7q21-q22	Malignant hyperthermia susceptibility 3	Susceptibility to malignant hyperthermia
Malignant hyperthermia susceptibility 4	*MHS4*	3q13.1	Malignant hyperthermia susceptibility 4	Susceptibility to malignant hyperthermia
Malignant hyperthermia susceptibility 6	*MHS6*	5p	Malignant hyperthermia susceptibility 6	Susceptibility to malignant hyperthermia
Malonyl-CoA decarboxylase deficiency	*MLYCD*	16q24	Malonyl-CoA decarboxylase	AR. Malonic aciduria with variable phenotypic features including language and psychomotor delay

Continued

TABLE 109.1 Genes Associated with Neurologic Disorders—cont'd

Genetic disorder	Gene	Locus	Gene product	Inheritance (if established) and phenotype
Mannosidosis, α-, types I and II	*MAN2B1*	19cen-q12	Mannosidase, α, class 2B, member 1	AR. Lysosomal storage disorder with coarse facial features, intellectual disability, skeletal abnormalities, hearing problems
Mannosidosis, β	*MANBA*	4q22-q25	Mannosidase, βA, lysosomal	AR. Developmental delays and intellectual disability. Different levels of severity
Marfan syndrome	*FBN1*	15q21.1	Fibrillin 1	AD. Disorder of fibrous connective tissue affecting skeletal, ocular, and cardiovascular systems
Marinesco–Sjögren syndrome	*SIL1*	5q31	SIL1 homolog, endoplasmic reticulum chaperone (*S. cerevisiae*)	AR. Congenital cataracts, cerebellar ataxia, progressive weakness due to myopathy, and delayed psychomotor development
Martin–Probst deafness-mental retardation syndrome	*MRXSMP*	Xq22.1	Martin–Probst deafness-mental retardation syndrome	X-linked recessive. Sensorineural hearing loss, mental retardation, short stature, umbilical hernia, facial dysmorphism, abnormal teeth, widely spaced nipples, and abnormal dermatoglyphics
Mast syndrome	*SPG21*	15q21-q22	Spastic paraplegia 21 (autosomal recessive, Mast syndrome)	AR, spastic paraplegia 21
McArdle disease	*PYGM*	11q13	Phosphorylase, glycogen, muscle	AR. Exercise intolerance and muscle cramps presenting in childhood or adolescence
Meckel syndrome, type 1	*MKS1*	17q23	Meckel syndrome 1	AR. Cystic renal disease, central nervous system malformation, most commonly occipital encephalocele and postaxial polydactyly
Meckel syndrome, type 2	*TMEM216*	11q13	Transmembrane protein 216	AR. Cystic renal disease, central nervous system malformation, most commonly occipital encephalocele and postaxial polydactyly
Meckel syndrome, type 3	*TMEM67*	8q	Transmembrane protein 67	AR. Cystic renal disease, central nervous system malformation, most commonly occipital encephalocele and postaxial polydactyly
Meckel syndrome, type 4	*CEP290*	12q	Centrosomal protein, 290 kDa	AR. Cystic renal disease, central nervous system malformation, most commonly occipital encephalocele and postaxial polydactyly
Meckel syndrome, type 5	*RPGRIP1L*	16q12	RPGRIP1-like	AR. Cystic renal disease, central nervous system malformation, most commonly occipital encephalocele and postaxial polydactyly
Meckel syndrome, type 6	*CC2D2A*	4p15	Coiled-coil and C2 domains-containing protein 2A	AR. Cystic renal disease, central nervous system malformation, most commonly occipital encephalocele and postaxial polydactyly
Meckel syndrome, type 7	*NPHP3*	3q22	Nephrocystin 3	AR. Cystic renal disease, central nervous system malformation, most commonly occipital encephalocele and postaxial polydactyly

TABLE 109.1 Genes Associated with Neurologic Disorders—cont'd

Genetic disorder	Gene	Locus	Gene product	Inheritance (if established) and phenotype
Meckel syndrome, type 8	*TCTN2*	12q24.31	Tectonic family, member 2	AR. Cystic renal disease, central nervous system malformation, most commonly occipital encephalocele and postaxial polydactyly
Meckel syndrome, type 9	*B9D1*	17p11.2	B9 domain-containing protein 1	AR. Cystic renal disease, central nervous system malformation most commonly occipital encephalocele and postaxial polydactyly
Meckel syndrome, type 10	*B9D2*	19p13	B9 domain-containing protein 2	AR. Cystic renal disease, central nervous system malformation, most commonly occipital encephalocele and postaxial polydactyly
Medulloblastoma	*PTCH2*	1p32	Patched 2	Somatic mutation. Medulloblastoma and basal cell nevus
Medulloblastoma and involved in reciprocal translocation with BCR gene in Philadelphia chromosome	*ABR*	17p13.3	Active BCR-related gene	Chromosomal translocation. Medulloblastoma
Medulloblastoma, desmoplastic	*SUFU*	10q24-q25	Suppressor of fused homolog (*Drosophila*)	Somatic mutation. Medulloblastoma
Megalencephalic leukoencephalopathy with subcortical cysts	*MLC1*	22q13.33	Megalencephalic leukoencephalopathy with subcortical cysts 1	Inheritance not established. Megalencephalic leukoencephalopathy with subcortical cysts
MEHMO syndrome	*MEHMO*	Xp22.13-p21.1	Mental retardation, epileptic seizures, hypogonadism, microcephaly and obesity syndrome	Inheritance not established. Mental retardation, epileptic seizures, hypogonadism, microcephaly and obesity
Melkersson–Rosenthal syndrome	*MROS*	9p11	Melkersson–Rosenthal syndrome	Inheritance not established. Recurrent peripheral facial nerve palsy
Memory impairment, susceptibility to	*BDNF*	11p13	Brain-derived neurotrophic factor	Susceptibility to memory impairment
Meningioma	*MN1*	22q12.3-qter	Meningioma (disrupted in balanced translocation) 1	Chromosomal translocation. Meningioma
Meningioma, radiation-induced	*MNRI*	1p11	Meningioma, radiation-induced	Somatic mutation. Meningioma
Meningioma, SIS-related	*PDGFB*	22q12.3-q13.1	Platelet-derived growth factor β polypeptide	Chromosomal deletion. Meningioma
Menkes disease, occipital horn syndrome, X-linked spinal muscular atrophy, distal	*ATP7A*	Xq21.1	ATPase, Cu($^{2+}$) transporting, α polypeptide	X-linked recessive. Copper deficiency resulting in growth retardation, peculiar hair, and focal cerebral and cerebellar degeneration
Mental retardation, X-linked Fraxe type	*AFF2*	Xq28	AF4, FMR2 family, member 2; AFF2	X-linked. Intellectual disability
Mental retardation, X-linked syndromic, Fried type	*AP1S2*	est	Adaptor-related protein complex 1, sigma 2 subunit	X-linked. Intellectual disability
Mental retardation syndrome, X-linked, Abidi type	*MRXSAB*	Xq13.2	Abidi X-linked mental retardation syndrome	X-linked. Intellectual disability
Mental retardation syndrome, X-linked, Armfield type	*MRXSA*	Xq28	Armfield X-linked mental retardation syndrome	X-linked. Intellectual disability
Mental retardation syndrome, X-linked, Lubs type	*MRXSL*	Xq28	Lubs X-linked mental retardation syndrome	X-linked. Intellectual disability

Continued

TABLE 109.1 Genes Associated with Neurologic Disorders—cont'd

Genetic disorder	Gene	Locus	Gene product	Inheritance (if established) and phenotype
Mental retardation syndrome, X-linked, Siderius type	*PHF8*	Xp11.2	PHD finger protein 8	X-linked. Intellectual disability
Mental retardation with language impairment and autistic features	*FOXP1*	3p14.1	Forehead box P1	Inheritance not established. Mental retardation
Mental retardation-skeletal dysplasia	*MRSD*	Xq28	Mental retardation-skeletal dysplasia	X-linked. Intellectual disability
Mental retardation, anterior maxillary protrusion, and strabismus	*SOBP*	6q21	Sine oculis binding protein homolog (*Drosophila*)	Inheritance not established. Mental retardation, anterior maxillary protrusion, and strabismus
Mental retardation, autosomal dominant 1	*MBD5*	2q23.1	Methyl-CpG binding domain protein 5	AD. Intellectual disability
Mental retardation, autosomal dominant 2	*DOCK8*	9p24	Dedicator of cytokinesis 8	AD. Intellectual disability
Mental retardation, autosomal dominant 3	*CDH15*	16q24.3	Cadherin 15, type 1, M-cadherin (myotubule)	AD. Intellectual disability
Mental retardation, autosomal dominant 4	*KIRREL3*	11q24.2	Kin of IRRE like 3 (*Drosophila*)	AD. Intellectual disability
Mental retardation, autosomal dominant 5	*SYNGAP1*	6p21.3	Synaptic Ras GTPase activating protein 1	AD. Intellectual disability
Mental retardation, autosomal dominant 7	*DYRK1A*	21q22.3	Dual-specificity tyrosine phosphorylation-regulated kinase 1A	AD. Intellectual disability
Mental retardation, autosomal recessive 1	*PRSS12*	4q25-q26	Protease, serine, 12 (neurotrypsin, motopsin)	AR. Intellectual disability
Mental retardation, autosomal recessive 2A	*CRBN*	3p26.2	Cereblon	AR. Intellectual disability
Mental retardation, autosomal recessive 3	*CC2D1A*	19p13.12	Coiled-coil and C2 domain containing 1A	AR. Intellectual disability
Mental retardation, autosomal recessive, 4	*MRT4*	1p21.1-p13.3	Mental retardation, nonsyndromic, autosomal recessive, 4	AR. Intellectual disability
Mental retardation, autosomal recessive, 5	*MRT5*	5p15-p14	Mental retardation, nonsyndromic, autosomal recessive, 5	AR. Intellectual disability
Mental retardation, autosomal recessive, 6	*GRIK2*	6q21	Glutamate receptor, ionotropic, kainate 2	AR. Intellectual disability
Mental retardation, autosomal recessive 7	*TUSC3*	8p22	Tumor suppressor candidate 3	AR. Intellectual disability
Mental retardation, autosomal recessive, 8	*MRT8*	10q22	Mental retardation, nonsyndromic, autosomal recessive, 8	AR. Intellectual disability
Mental retardation, autosomal recessive, 9	*MRT9*	14q11.2-q12	Mental retardation, nonsyndromic, autosomal recessive, 9	AR. Intellectual disability
Mental retardation, autosomal recessive, 10	*MRT10*	16p12-q12	Mental retardation, nonsyndromic, autosomal recessive, 10	AR. Intellectual disability
Mental retardation, autosomal recessive, 11	*MRT11*	19q13.2-q13.3	Mental retardation, nonsyndromic, autosomal recessive, 11	AR. Intellectual disability
Mental retardation, autosomal recessive, 12	*MRT12*	1p34-p33	Mental retardation, nonsyndromic, autosomal recessive, 12	AR. Intellectual disability

TABLE 109.1 Genes Associated with Neurologic Disorders—cont'd

Genetic disorder	Gene	Locus	Gene product	Inheritance (if established) and phenotype
Mental retardation, autosomal recessive 13	*TRAPPC9*	8q24.3	Centrosomal protein, 290 kDa	AR. Intellectual disability
Mental retardation, FRA12A type	*DIP2B*	12q13.12	DIP2 disco-interacting protein 2 homolog B (*Drosophila*)	Inheritance not established. Mental retardation
Mental retardation, joint hypermobility and skin laxity, with or without metabolic abnormalities	*ALDH18A1*	10q24.3	Aldehyde dehydrogenase 18 family, member A1	Inheritance not established. Mental retardation
Mental retardation, profound	*RAB40AL*	Xq22.2	RAB40A, member RAS oncogene family-like	X-linked. Intellectual disability
Mental retardation, severe, with spasticity and tapetoretinal degeneration	*MRST*	15q24	Mental retardation, severe, with spasticity and tapetoretinal	Inheritance not established. Mental retardation
Mental retardation, stereotypic movements, epilepsy, and/or cerebral malformations	*MEF2C*	5q14	Myocyte enhancer factor 2C	Inheritance not established. Mental retardation, stereotypic movements, epilepsy, and/or cerebral malformations
Mental retardation, truncal obesity, retinal dystrophy, and micropenis	*INPP5E*	9q34.3	Inositol polyphosphate-5-phosphatase, 72 kDa	Inheritance not established. Mental retardation, truncal obesity, retinal dystrophy, and micropenis
Mental retardation, X-linked 1	*IQSEC2*	Xp11.22	IQ motif and Sec7 domain 2	X-linked. Intellectual disability
Mental retardation, X-linked 14	*MRX14*	Xp11.3-q13.3	Mental retardation, X-linked 14	X-linked. Intellectual disability
Mental retardation, X-linked 20	*MRX20*	Xp11-q21	Mental retardation, X-linked 20	X-linked. Intellectual disability
Mental retardation, X-linked 23, nonspecific	*MRX23*	Xq23-q24	Mental retardation, X-linked 23	X-linked. Intellectual disability
Mental retardation, X-linked 27, nonspecific	*MRX27*	Xq23-q24	Mental retardation, X-linked 27	X-linked. Intellectual disability
Mental retardation, X-linked 35, nonspecific	*MRX35*	Xq23-q24	Mental retardation, X-linked 35	X-linked. Intellectual disability
Mental retardation, X-linked 80, nonspecific	*MRX80*	Xq23-q24	Mental retardation, X-linked 80	X-linked. Intellectual disability
Mental retardation, X-linked 30	*PAK3*	Xq21.3-q24	p21 protein (Cdc42/Rac)-activated kinase 3	X-linked. Intellectual disability
Mental retardation, X-linked 45	*ZNF81*	Xp22.1-p11	Zinc finger protein 81	X-linked. Intellectual disability
Mental retardation, X linked 46	*ARHGEF6*	Xq26.3	Rho guanine nucleotide exchange factor 6	X-linked. Intellectual disability
Mental retardation, X-linked 52	*MRX52*	Xp11.21-q22.3	Mental retardation, X-linked 52	X-linked. Intellectual disability
Mental retardation, X-linked 58	*TSPAN7*	Xq11	Tetraspanin 7	X-linked. Intellectual disability
Mental retardation, X-linked 59	*AP1S2*	Xp22	Adaptor-related protein complex 1, sigma 2 subunit	X-linked. Intellectual disability
Mental retardation, X linked 63	*ACSL4*	Xq23	Acyl-CoA synthase long chain family, member 4	X-linked. Intellectual disability
Mental retardation, X-linked 77	*MRX77*	Xq12-q21.3	Mental retardation, X-linked 77	X-linked. Intellectual disability
Mental retardation, X-linked 78	*MRX78*	Xp11.4-p11.23	Mental retardation, X-linked 78	X-linked. Intellectual disability
Mental retardation, X-linked 81	*MRX81*	Xp11.2-q12	Mental retardation, X-linked 81	X-linked. Intellectual disability
Mental retardation, X-linked 82	*MRX82*	Xq24-q25	Mental retardation, X-linked 82	X-linked. Intellectual disability

Continued

TABLE 109.1 Genes Associated with Neurologic Disorders—cont'd

Genetic disorder	Gene	Locus	Gene product	Inheritance (if established) and phenotype
Mental retardation, X-linked 84	*MRX84*	Xp11.3-q22.3	Mental retardation, X-linked 84	X-linked. Intellectual disability
Mental retardation, X-linked 93	*BRWD3*	Xq13	Bromodomain and WD repeat domain containing 3	X-linked. Intellectual disability
Mental retardation, X-linked 94	*GRIA3*	Xq25-q26	Glutamate receptor, ionotropic, AMPA 3	X-linked. Intellectual disability
Mental retardation, X-linked 95	*MAGT1*	Xq13.1-q13.2	Magnesium transporter 1	X-linked. Intellectual disability
Mental retardation, X-linked nonspecific	*GDI1*	Xq28	GDP dissociation inhibitor 1	X-linked. Intellectual disability
Mental retardation, X-linked nonspecific, 42	*MRX42*	Xq26	Mental retardation, X-linked 42	X-linked. Intellectual disability
Mental retardation, X-linked nonspecific, 63	*ACSL4*	Xq22.3	Acyl-CoA synthetase long-chain family member 4	X-linked. Intellectual disability
Mental retardation, X-linked nonspecific, type 46	*ARHGEF6*	Xq26	Rac/Cdc42 guanine nucleotide exchange factor (GEF) 6	X-linked. Intellectual disability
Mental retardation, X-linked nonspecific, type 50	*MRX50*	Xp11.3-p11.21	Mental retardation, X-linked 50	X-linked. Intellectual disability
Mental retardation, X-linked syndromic	*UBE2A*	Xq24-q25	Ubiquitin-conjugating enzyme E2A	X-linked. Intellectual disability
Mental retardation, X-linked syndromic, Christianson type	*SLC9A6*	Xq26.3	Solute carrier family 9 (sodium/hydrogen exchanger), member 6	X-linked. Intellectual disability
Mental retardation, X-linked syndromic, Turner type	*HUWE1*	Xp11.2	HECT, UBA and WWE domain containing 1, E3 ubiquitin protein ligase	X-linked. Intellectual disability
Mental retardation, X-linked 49	*MRX49*	Xp22.3	Mental retardation, X-linked 49	X-linked. Intellectual disability
Mental retardation, X-linked 53	*MRX53*	Xq22.2-q26	Mental retardation, X-linked 53	X-linked. Intellectual disability
Mental retardation, X-linked 72	*RAB39B*	Xq28	RAB39B, member RAS oncogene family	X-linked. Intellectual disability
Mental retardation, X-linked 88	*AGTR2*	Xq22-q23	Angiotensin II receptor, type 2	X-linked. Intellectual disability
Mental retardation, X-linked 89	*ZNF41*	Xp22.1-cen	Zinc finger protein 41	X-linked. Intellectual disability
Mental retardation, X-linked 9	*FTSJ1*	Xp11.23	FtsJ homolog 1 (*Escherichia coli*)	X-linked. Intellectual disability
Mental retardation, X-linked 90	*DLG3*	Xq13.1	Discs, large homolog 3 (*Drosophila*)	X-linked. Intellectual disability
Mental retardation, X-linked 91	*ZDHHC15*	Xq13.3	Zinc finger, DHHC-type containing 15	X-linked. Intellectual disability
Mental retardation, X-linked 92	*ZNF674*	Xp11	Zinc finger protein 674	X-linked. Intellectual disability
Mental retardation, X-linked, 1	*MRX1*	Xp11.3-q21.1	Mental retardation, X-linked 1 (non-dysmorphic)	X-linked. Intellectual disability
Mental retardation, X-linked, 2	*MRX2*	Xp22.3	Mental retardation, X-linked 2 (nondysmorphic)	X-linked. Intellectual disability
Mental retardation, X-linked, 21/34	*IL1RAPL1*	Xp22.1-p21.3	Interleukin 1 receptor accessory protein-like 1	X-linked. Intellectual disability
Mental retardation, X-linked, FRAXE type	*FRAXE*	Xq28	Fragile site, folic acid type, rare, fra(X)(q28) E	X-linked. Intellectual disability
Mental retardation, X-linked, FRAXE type	*AFF2*	Xq28	AF4/FMR2 family, member 2	X-linked. Intellectual disability

TABLE 109.1 Genes Associated with Neurologic Disorders—cont'd

Genetic disorder	Gene	Locus	Gene product	Inheritance (if established) and phenotype
Mental retardation, X-linked, Shashi type	*MRXS11*	Xq26-q27	Mental retardation, X-linked, syndromic 11	X-linked. Intellectual disability
Mental retardation, X-linked, Shashi type	*S11*	Xq26-q27	Surface antigen (X-linked) 2	X-linked. Intellectual disability
Mental retardation, X-linked, Snyder–Robinson type	*SMS*	Xp22.1	Spermine synthase	X-linked. Intellectual disability
Mental retardation, X-linked, syndromic 12	*MRXS12*	Xp11	Mental retardation, X-linked, syndromic 12	X-linked. Intellectual disability
Mental retardation, X-linked, syndromic 12	*S12*	Xp11	Surface antigen (X-linked) 3	X-linked. Intellectual disability
Mental retardation, X-linked, syndromic 14	*UPF3B*	Xq25-q26	UPF3 regulator of nonsense transcripts homolog B (yeast)	X-linked. Intellectual disability
Mental retardation, X-linked, syndromic 15 (Cabezas type)	*CUL4B*	Xq24	Cullin 4b	X-linked. Intellectual disability
Mental retardation, X-linked, syndromic 6, with gynecomastia and obesity	*WTS*	Xp21.1-q22	Wilson–Turner X-linked mental retardation syndrome	X-linked. Intellectual disability
Mental retardation, X-linked, syndromic 7	*MRXS7*	Xp11.3-q22	Mental retardation, X-linked, syndromic 7	X-linked. Intellectual disability
Mental retardation, X-linked, syndromic 2, with dysmorphism and cerebral atrophy	*PRS*	Xp11-q21	Prieto X-linked mental retardation syndrome	Mental retardation, X-linked, syndromic 2, with dysmorphism and cerebral atrophy
Mental retardation, X-linked, syndromic 4, with congenital contractures and low fingertip arches	*MCS*	Xq13-q22	Miles–Carpenter X-linked mental retardation syndrome	Mental retardation, X-linked, syndromic 4, with congenital contractures and low fingertip arches
Mental retardation, X-linked, syndromic 5, with Dandy–Walker malformation, basal ganglia disease, and seizures	*MRXS5*	Xq25-q27	Mental retardation, X-linked, syndromic 5	Mental retardation, X-linked, syndromic 5, with Dandy–Walker malformation, basal ganglia disease, and seizures
Mental retardation, X-linked, syndromic, JARID1C-related	*KDM5C*	Xp11.22-p11.21	Lysine (K)-specific demethylase 5C	Mental retardation, X-linked, syndromic, JARID1C-related
Mental retardation, X-linked, with cerebellar hypoplasia and distinctive facial appearance	*HN1*	Xq12	Hematological and neurological expressed 1	Mental retardation, X-linked, with cerebellar hypoplasia and distinctive facial appearance
Mental retardation, X-linked, with cerebellar hypoplasia and distinctive facial appearance	*OPHN1*	Xq12	Oligophrenin 1	Mental retardation, X-linked, with cerebellar hypoplasia and distinctive facial appearance
Mental retardation, X-linked, with epilepsy	*ATP6AP2*	Xp11.4	ATPase, H^+ transporting, lysosomal accessory protein 2	Mental retardation, X-linked, with epilepsy
Mental retardation, X-linked, with isolated growth hormone deficiency	*SOX3*	Xq26.3	SRY (sex determining region Y)-box 3	Mental retardation, X-linked, with isolated growth hormone deficiency
Mental retardation, X-linked, with or without epilepsy	*SYP*	Xp11.23-p11.22	Synaptophysin	Mental retardation, X-linked, with or without epilepsy
Mental retardation, X-linked, with short stature	*MRSS*	Xq24	Mental retardation, X-linked, with short stature	Mental retardation, X-linked, with short stature
Mental retardation, X-linked, ZDHHC9-related	*ZDHHC9*	Xq26.1	Zinc finger, DHHC-type containing 9	Mental retardation, X-linked, ZDHHC9-related

Continued

TABLE 109.1 Genes Associated with Neurologic Disorders—cont'd

Genetic disorder	Gene	Locus	Gene product	Inheritance (if established) and phenotype
Mental retardation, X-linked 4, with scoliosis and spotty abdominal hypopigmentation, slight facial asymmetry, and clinodactyly; may be several genes in the interval	*MRX4*	?	Mental retardation, X-linked 4	Mental retardation, X-linked 4, with scoliosis and spotty abdominal hypopigmentation, slight facial asymmetry, and clinodactyly; may be several genes in the interval
Metachromatic leukodystrophy	*ARSA*	22q13.31-qter	Arylsulfatase A	AR. Leukodystrophy, progressive neurodegenerative disorder with attention deficit, cognitive regression, blindness
Metachromatic leukodystrophy due to SAP-b deficiency	*PSAP*	10q22.1	Prosaposin	AR. Metachromatic leukodystrophy
Methylmalonic aciduria and homocystinuria, cblC type	*MMACHC*	1p34.1	Methylmalonic aciduria (cobalamin deficiency) cblC type, with homocystinuria	Inheritance not established. Methylmalonic aciduria and homocystinuria
Methylmalonic aciduria due to transcobalamin receptor defect	*CD320*	19p13.2	CD320 molecule	AR. Methylmalonic aciduria
Methylmalonic aciduria, mut(0) type	*MUT*	6p21	Methylmalonyl CoA mutase	AR. Methylmalonic aciduria
Methylmalonic aciduria, vitamin B_{12}-responsive	*MMAA*	4q31.1-q31.2	Methylmalonic aciduria (cobalamin deficiency) cblA type	AR. Methylmalonic aciduria
Methylmalonyl-CoA epimerase deficiency	*MCEE*	2p13.3	Methylmalonyl CoA epimerase	AR. Methylmalonic aciduria
Mevalonic aciduria	*MVK*	12q24	Mevalonate kinase	AR. Mevalonic aciduria
Mitochondrial phosphate carrier deficiency	*SLC25A3*	12q23	Solute carrier family 25 (mitochondrial carrier; phosphate carrier), member 3	AR. Lactic acidosis, hypertrophic cardiomyopathy, and muscular hypotonia
Microcephalic osteodysplastic primordial dwarfism, type II	*PCNT*	21q22.3	Pericentrin	AD. Microcephaly
Microcephaly with digital anomalies	*JAWAD*	18p11.22-q11.2	Microcephaly with digital anomalies	AD. Microcephaly with digital anomalies
Microcephaly, Amish type	*MCPHA*	17q25.3	Microcephaly, Amish	AD. Microcephaly
Microcephaly, Amish type	*SLC25A19*	17q25.3	Solute carrier family 25 (mitochondrial thiamine pyrophosphate carrier), member 19	AD. Microcephaly
Microcephaly, cortical malformations, and mental retardation	*WDR62*	19q13.12	WD repeat domain 62	Inheritance not established. Microcephaly, cortical malformations, and mental retardation
Microcephaly, mental retardation, and distinctive facies, with cardiac and genitourinary malformations	*MMRFCGU*	16p13.3	Microcephaly, mental retardation, and distinctive facies, with cardiac and genitourinary malformations	Inheritance not established. Microcephaly, mental retardation, and distinctive facies, with cardiac and genitourinary malformations
Microcephaly, postnatal progressive, with seizures and brain atrophy	*MED17*	11q21	Mediator complex subunit 17	Inheritance not established. Microcephaly, seizures and brain atrophy
Microcephaly, primary autosomal recessive, 1	*MCPH1*	8p23	Microcephalin 1	AR. Primary microcephaly
Microcephaly, primary autosomal recessive, 3	*CDK5RAP2*	9q33.3	CDK5 regulatory subunit associated protein 2	AR. Primary microcephaly
Microcephaly, primary autosomal recessive, 4	*CEP152*	15q21.1	Centrosomal protein 152kDa	AR. Primary microcephaly

TABLE 109.1 Genes Associated with Neurologic Disorders—cont'd

Genetic disorder	Gene	Locus	Gene product	Inheritance (if established) and phenotype
Microcephaly, primary autosomal recessive, 5, with or without simplified gyral pattern	*ASPM*	1q31	Asp (abnormal spindle) homolog, microcephaly associated (*Drosophila*)	AR. Primary microcephaly
Microcephaly, primary autosomal recessive, 6	*CENPJ*	13q12.2	Centromere protein J	AR. Primary microcephaly
Microcephaly, primary autosomal recessive, 7	*STIL*	1p32	SCL/TAL1 interrupting locus	AR, Microcephaly
Microphthalmia, syndromic 1	*MAA*	Xq27-q28	Microphthalmia or anophthalmia and associated anomalies	Inheritance not established. Microphthalmia with abnormal brain development
Microphthalmia, syndromic 3	*SOX2*	3q26.3-q27	SRY (sex determining region Y)-box 2	Inheritance not established. Microphthalmia with congenital anomalies of brain, cognitive and motor delays and seizures, hearing loss and other congenital anomalies
Microphthalmia, syndromic 4	*ANOP1*	Xq27-q28	Anophthalmos 1	Inheritance not established. Mental retardation, without limb anomalies or dental or urogenital abnormalities
Microphthalmia, syndromic 4	*COPS4*	Xq27-q28	COP9 constitutive photomorphogenic homolog subunit 4 (*Arabidopsis*)	Inheritance not established. Microphthalmia
Microphthalmia, syndromic 5	*OTX2*	14q22.3	Orthodenticle homeobox 2	Inheritance not established. Microphthalmia with congenital anomalies of brain, cognitive and motor delays and seizures
Microphthalmia, syndromic 6, orofacial cleft 11	*BMP4*	14q22.2	Bone morphogenetic protein 4	Inheritance not established. Microphthalmia with congenital anomalies of brain and digital anomalies
Microphthalmia, syndromic 7	*HCCS*	msl2	Holocytochrome c synthase	X-linked dominant. Unilateral or bilateral microphthalmia and linear skin defects with areas of aplastic skin that heal with age to form hyperpigmented areas in affected females. *In utero* lethality for males
Microphthalmia, syndromic 9	*STRA6*	15q24.1	Stimulated by retinoic acid gene 6 homolog (mouse)	Inheritance not established. Anophthalmia with mild facial dysmorphism and variable anomalies of lungs, heart and diaphragm
Migraine, familial typical, susceptibility to, 1	*MGR2*	Xq	Migraine, familial typical, susceptibility to	Susceptibility to migraine
Migraine, familial hemiplegic, 2	*ATP1A2*	1q21-q23	ATPase, Na^+/K^+ transporting, $\alpha 2$ polypeptide	Inheritance not established. Hemiplegic migraines
Migraine with or without aura, susceptibility to, 3	*MGR3*	6p21.1-p12.2	Migraine, familial, with or without aura, susceptibility to	Susceptibility to migraines
Migraine without aura, susceptibility to, 4	*MGR4*	14q21.2-q22.3	Migraine, susceptibility to, 4	Susceptibility to migraines
Migraine with or without aura, susceptibility to, 5	*MGR5*	19p13	Migraine with or without aura, susceptibility to, 5	Susceptibility to migraines
Migraine, familial hemiplegic	*MGR*	1q31	Migraine, several forms	Inheritance not established. Migraines

Continued

TABLE 109.1 Genes Associated with Neurologic Disorders—cont'd

Genetic disorder	Gene	Locus	Gene product	Inheritance (if established) and phenotype
Migraine with aura, susceptibility to, 7	*MGR7*	15q11.2-q12	Migraine with aura, susceptibility to, 7	Susceptibility to migraines
Migraine, susceptibility to, 8	*MGR8*	5q21	Migraine, susceptibility to, 8	Susceptibility to migraines
Migraine with aura, susceptibility to, 9	*MGR9*	11q24	Migraine with aura, susceptibility to, 9	Susceptibility to migraines
Migraine, with or without aura, susceptibility to, 12	*MGR12*	10q22-q23	Migraine, with or without aura, susceptibility to, 12	Susceptibility to migraines
Migraine, with or without aura, susceptibility to, 13	*KCNK18*	10q25.3	Potassium channel, subfamily K, member 18	Susceptibility to migraines
Mitochondrial complex I deficiency	*NDUFV1*	11q13.2	NADH dehydrogenase (ubiquinone) flavoprotein 1, 51 kDa	AR. Mitochondrial complex 1 deficiency. Severe encephalopathy and neurodegenerative disease
Mitochondrial complex I deficiency	*NDUFV2*	18p11.31-p11.2	NADH dehydrogenase (ubiquinone) flavoprotein 2, 24 kDa	AR. Mitochondrial complex 1 deficiency. Severe encephalopathy and hypertrophic cardiomyopathy
Mitochondrial complex I deficiency	*NDUFS1*	2q33-q34	NADH dehydrogenase (ubiquinone) Fe–S protein 1, 75 kDa (NADH-coenzyme Q reductase)	AR. Mitochondrial complex 1 deficiency. Severe encephalopathy and neurodegenerative disease
Mitochondrial complex I deficiency	*NDUFS2*	1q23	NADH dehydrogenase (ubiquinone) Fe–S protein 2, 49 kDa (NADH-coenzyme Q reductase)	AR. Mitochondrial complex 1 deficiency. Severe encephalopathy and neurodegenerative disease
Mitochondrial complex I deficiency	*NDUFS3*		NADH dehydrogenase (ubiquinone) Fe–S protein 3, 30 kDa	AR. Mitochondrial complex 1 deficiency. Leigh syndrome due to mitochondrial complex 1 deficiency. Severe encephalopathy and neurodegenerative disease
Mitochondrial complex I deficiency	*NDUFS4*	5q11.2	NADH dehydrogenase (ubiquinone) Fe–S protein 4, 18 kDa	AR. Mitochondrial complex 1 deficiency. Leigh syndrome due to mitochondrial complex 1 deficiency. Severe encephalopathy and neurodegenerative disease
Complex I, mitochondrial respiratory chain, deficiency of	*NDUFS6*	5pter-p15.33	NADH dehydrogenase (ubiquinone) Fe–S protein 6, 13 kDa (NADH-coenzyme Q reductase)	AR. Mitochondrial complex 1 deficiency. Severe encephalopathy and neurodegenerative disease
Complex I, mitochondrial respiratory chain, deficiency of	*NDUFS7*	19p13.3	NADH dehydrogenase (ubiquinone) Fe–S protein 7, 20 kDa (NADH-coenzyme Q reductase)	AR. Leigh syndrome
Complex I, mitochondrial respiratory chain, deficiency of	*NDUFS8*	11q13.2	NADH dehydrogenase (ubiquinone) Fe–S protein 8, 23 kDa (NADH-coenzyme Q reductase)	AR. Leigh syndrome
Mitochondrial complex I deficiency	*NDUFAF2*	5q12.1	NADH dehydrogenase (ubiquinone) 1α subcomplex, assembly factor 2	AR. Leigh syndrome
Mitochondrial complex I deficiency	*NDUFA11*	19p13.3	NADH dehydrogenase (ubiquinone) 1α subcomplex, 11, 14.7 kDa	AR. Mitochondrial complex 1 deficiency. Severe encephalopathy and neurodegenerative disease
Mitochondrial complex I deficiency	*NDUFB3*	2q33.1	NADH dehydrogenase (ubiquinone) 1β subcomplex, 3	AR. Mitochondrial complex 1 deficiency. Severe encephalopathy and neurodegenerative disease

TABLE 109.1 Genes Associated with Neurologic Disorders—cont'd

Genetic disorder	Gene	Locus	Gene product	Inheritance (if established) and phenotype
Mitochondrial complex I deficiency	*FOXRED1*	11q24.2	FAD-dependent oxidoreductase domain containing 1	AR. Mitochondrial complex 1 deficiency. Severe encephalopathy and Leigh syndrome
Mitochondrial complex I deficiency	*ACAD9*	3q21.3	Acyl-CoA dehydrogenase family member 9	AR. Mitochondrial complex 1 deficiency due to ACAD 9 deficiency
Mitochondrial complex I deficiency	*NUBPL*	14q12	Nucleotide binding protein-like	AR. Mitochondrial complex 1 deficiency. Severe encephalopathy and neurodegenerative disease
Mitochondrial complex I deficiency	*NDUFAF1*		NADH dehydrogenase (ubiquinone) 1α subcomplex, assembly factor 1	AR. Mitochondrial complex 1 deficiency. Severe encephalopathy and hypertrophic cardiomyopathy
Mitochondrial complex I deficiency	*NDUFAF2*	5q12.1	NADH dehydrogenase (ubiquinone) 1α subcomplex, assembly factor 2	AR. Mitochondrial complex 1 deficiency. Severe encephalopathy and neurodegenerative disease
Mitochondrial complex I deficiency	*NDUFAF3*	3p21.31	NADH dehydrogenase (ubiquinone) 1α subcomplex, assembly factor 3	AR. Mitochondrial complex 1 deficiency. Severe encephalopathy and neurodegenerative disease
Mitochondrial complex I deficiency	*NDUFAF4*	6q16.1	NADH dehydrogenase (ubiquinone) 1α subcomplex, assembly factor 4	AR. Mitochondrial complex 1 deficiency. Severe encephalopathy and neurodegenerative disease
Mitochondrial complex 1 deficiency	*C20orf7*	20p12.1	Chromosome 20 open reading frame 7	AR. Mitochondrial complex 1 deficiency. Severe encephalopathy and neurodegenerative disease
Mitochondrial complex I deficiency	*NDUFA10*	2q37.2	NADH dehydrogenase (ubiquinone) 1α subcomplex, 10, 7.5 kDa	AR. Leigh syndrome
Mitochondrial complex I deficiency	*MTND1*	Mitochondrial	Complex 1 subunit ND1	Mitochondrial. Leber optic atrophy, nonsyndromic deafness, MELAS
Mitochondrial complex I deficiency	*MTND2*	Mitochondrial	Complex 1 subunit ND2	Mitochondrial. Leber optic atrophy, Leigh syndrome, mitochondrial complex 1 deficiency
Mitochondrial complex I deficiency	*MTND3*	Mitochondrial	Complex 1 subunit ND3	Mitochondrial. Leigh syndrome, mitochondrial complex 1 deficiency
Mitochondrial complex I deficiency	*MTND4*	Mitochondrial	Complex 1 subunit ND4	Mitochondrial. Leber optic atrophy, encephalopathy due to complex 1 deficiency
Mitochondrial complex I deficiency	*MTND5*	Mitochondrial	Complex 1 subunit ND5	Mitochondrial. Leber optic atrophy, MELAS
Mitochondrial complex I deficiency	*MTND6*	Mitochondrial	Complex 1 subunit ND6	Mitochondrial. Leber optic atrophy with dystonia, Leigh syndrome with dystonia, MELAS
Mitochondrial complex I deficiency	*NDUFA1*	Xq24	NADH dehydrogenase (ubiquinone) 1α subcomplex, 1, 7.5 kDa	X-linked. Complex 1 deficiency, milder symptoms in females
Mitochondrial complex II deficiency	*SDHA*	5p15.33	Succinate dehydrogenase complex, subunit A, flavoprotein	AR. Leigh syndrome, dilated cardiomyopathy
Mitochondrial complex II deficiency	*SDHAF1*	19q12-q13.2	Succinate dehydrogenase complex assembly factor 1	AR. Progressive neurodegenerative disorder with dementia and myoclonic epilepsy

Continued

TABLE 109.1 Genes Associated with Neurologic Disorders—cont'd

Genetic disorder	Gene	Locus	Gene product	Inheritance (if established) and phenotype
Mitochondrial complex III deficiency	*BCS1L*	2q33	BCS1-like (*S. cerevisiae*)	AR. Progressive encephalomyopathy due to mitochondrial complex III deficiency. Cataracts, failure to thrive and hepatorenal dysfunction
Mitochondrial complex III deficiency	*TTC19*	17p12	Tetratricopeptide repeat domain-containing protein 19	AR. Progressive encephalomyopathy due to mitochondrial complex III deficiency. Cataracts, failure to thrive, and hepatorenal dysfunction
Mitochondrial complex III deficiency	*UQCRB*	8q22	Ubiquinol-cytochrome c reductase binding protein	AR. Described in a Turkish patient with mild disease presenting with intermittent hypoglycemia and lactic acidosis
Mitochondrial complex III deficiency	*UQCRQ*	5q31.1	Ubiquinol-cytochrome c reductase, complex III subunit VII, 9.5 kDa	AR. Progressive neurodegenerative disease due to mitochondrial complex III deficiency. Encephalopathy and pyramidal signs
Mitochondrial complex III deficiency	*UQCRC2*	16p12.2	Ubiquinol-cytochrome c reductase core protein II	AR. Intermittent hypoglycemia and lactic acidosis
Mitochondrial complex III deficiency	*CYC1*	8q24.3	Cytochrome C1	AR. Intermittent hypoglycemia and lactic acidosis
Mitochondrial complex IV deficiency	*FASTKD2*	2q33.3	FAST kinase domains 2	AR. Variable clinical presentation with myopathy to severe neurodegenerative disease
Mitochondrial complex V deficiency 1	*ATPAF2*	17p11.2	ATP synthase, mitochondrial F1 complex, assembly factor 2	AR. Presents with neonatal hypotonia, failure to thrive, hypertrophic cardiomyopathy, psychomotor delays
Mitochondrial complex V deficiency 2	*TMEM70*	8q21.11	Transmembrane protein 70	AR. Encephalocardiomyopathy, neonatal, mitochondrial, due to ATP synthase deficiency
Mitochondrial complex V deficiency 3	*ATP5E*	20q13.2	ATP synthase, H^+ transporting, mitochondrial F1 complex, epsilon subunit	AR. Mitochondrial complex 5 deficiency
Mitochondrial complex V deficiency 4	*ATPAF2*	18q21.1	ATP synthase, H^+ transporting, mitochondrial F1 complex, epsilon subunit 1	AR. Mitochondrial complex 5 deficiency
Mitochondrial DNA depletion syndrome 1	*TYMP, ECGF1*	22q13.33	Thymidine phosphorylase	AR. Mitochondrial DNA depletion syndrome 1—MNGIE type
Mitochondrial DNA depletion syndrome 2	*TK2*	16q21	Thymidine kinase 2, mitochondrial	AR. Mitochondrial DNA depletion syndrome 2—myopathic type
Mitochondrial DNA depletion syndrome 3	*DGUOK, DGK*	2p13.1	Deoxyguanosine kinase	AR. Hepatocerebral type—mitochondrial DNA depletion syndrome 3
Mitochondrial DNA depletion syndrome 4A	*POLG, POLG1, POLGA,*	15q26.1	Polymerase (DNA directed), γ	AR. Mitochondrial DNA depletion syndrome 4A—Alpers type
Mitochondrial DNA depletion syndrome 4B	*POLG, POLG1, POLGA*	15q26.1	Polymerase (DNA directed), γ	AR. Mitochondrial DNA depletion syndrome 4B—MNGIE type
Mitochondrial DNA depletion syndrome 5	*SUCLA2*	13q14.2	Succinate-CoA ligase, ADP-forming, β subunit	AR. Mitochondrial DNA depletion syndrome 5—encephalomyopathic with or without methylmalonic aciduria

TABLE 109.1 Genes Associated with Neurologic Disorders—cont'd

Genetic disorder	Gene	Locus	Gene product	Inheritance (if established) and phenotype
Mitochondrial DNA depletion syndrome 6	*MPV17*	2p23.3	Mouse homolog of MpV17. Mitochondrial inner membrane protein	AR. Hepatocerebral type—mitochondrial DNA depletion syndrome 6
Mitochondrial DNA depletion syndrome 7	*C10orf2, TWINKLE*	10q24.31	Chromosome 10 open reading frame 2	AR. Mitochondrial DNA depletion syndrome 7—hepatocerebral type
Mitochondrial DNA depletion syndrome 8A	*RRM2B, P53R2,*	8q22.3	Ribonucleotide reductase M2 B (TP53 inducible)	AR. Mitochondrial DNA depletion syndrome 8A (encephalomyopathic type with renal tubulopathy)
Mitochondrial DNA depletion syndrome 8B	*RRM2B, P53R2*	8q22.3	Ribonucleotide reductase M2 B (TP53 inducible)	AR. Mitochondrial DNA depletion syndrome 8B
Mitochondrial DNA depletion syndrome 9	*SUCLG1, SUCLA1*	2p11.2	Succinate-CoA ligase, α subunit	AR. Encephalomyopathic type—mitochondrial DNA depletion syndrome with methylmalonic aciduria; fatal in infancy
Mitochondrial DNA depletion syndrome 11	*MGME1, C20orf72*	20p11.23	Mitochondrial genome maintenance exonuclease 1	AR. Mitochondrial DNA depletion syndrome 11
Mitochondrial DNA depletion syndrome 12	*SLC25A4*	4q35.1	Solute carrier family 25 (mitochondrial carrier; adenine nucleotide translocator), member 4	AR. Mitochondrial DNA depletion syndrome 12—cardiomyopathic type
Progressive external ophthalmoplegia, autosomal recessive	*POLG*	15q25	Polymerase (DNA directed), γ	AR. Progressive external ophthalmoplegia
Mitochondrial myopathy and sideroblastic anemia 1	*PUS1*	12q24.33	Pseudouridylate synthase 1	AR. Oxidative phosphorylation disorder affecting skeletal muscle and bone marrow
Miyoshi muscular dystrophy 1	*DYSF1*	2p13.2	Dysferlin	AR. Late-onset distal muscular dystrophy, sparing intrinsic hand muscles
Miyoshi muscular dystrophy 2	*MMD2*	8q22.3	Miyoshi muscular dystrophy 2	AR. Late-onset distal muscular dystrophy, sparing intrinsic hand muscles
Miyoshi muscular dystrophy 3	*ANO5*	11p14.3	Anoctamin 5	AR. Late-onset distal muscular dystrophy, sparing intrinsic hand muscles
Molybdenum cofactor deficiency, type A	*MOCS1*	6p21.2	Encodes for enzyme in the first step of molybdenum cofactor (Moco) biosynthesis	AR. Severe intellectual disability, epilepsy, dislocated lenses, and excretion of urinary xanthine stones
Molybdenum cofactor deficiency, type B	*MOCS2*	5q11.2	Encodes for enzyme in the first step of molybdenum cofactor (Moco) biosynthesis	AR. Severe intellectual disability, epilepsy, dislocated lenses, and excretion of urinary xanthine stones
Molybdenum cofactor deficiency, type C	*GPHN*	14q24	Gephyrin	AR. Severe intellectual disability, epilepsy, dislocated lenses and excretion of urinary xanthine stones
Motor neuropathy, distal hereditary, with vocal cord paralysis	*HMN7A*	2q14	Motor neuropathy, distal hereditary, type VIIA	AD. Hereditary distal motor neuropathy
Mowat–Wilson syndrome	*ZEB2*	2q22	Zinc finger E-box binding homeobox 2	AD. Delayed motor development, intellectual disability and epilepsy
Moyamoya disease	*MYMY1*	3p26-p24.2	Moyamoya disease 1	Susceptibility to moyamoya disease
Moyamoya disease 2	*RNF213*	17q25	Ring finger protein/Alk lymphoma oligomerization partner on chr 17	Susceptibility to moyamoya disease

Continued

TABLE 109.1 Genes Associated with Neurologic Disorders—cont'd

Genetic disorder	Gene	Locus	Gene product	Inheritance (if established) and phenotype
Moyamoya disease 3	*MYMY3*	8q23	Moyamoya disease 3	Moyamoya disease 3/studied in Japanese population
Moyamoya disease 4	*MMY4*	Xq28	Moyamoya disease 4	X-linked. Moyamoya disease 4 with short stature, hypergonadotropic hypogonadism, and facial dysmorphism
Moyamoya disease 5	*MMY5*	10q	Moyamoya disease 5	AD. Moyamoya disease 5
Mucolipidosis	*GNPTAB*	16p	*N*-Acetylglucosamine-1-phosphotransferase,α/β subunit	AR. Mucolipidosis II α/β (or I-cell disease) and mucolipidosis III α/β are autosomal recessive disorders, characterized clinically by short stature, skeletal abnormalities, cardiomegaly, and developmental delay
Mucolipidosis III α/β	*GNPTAB*	12q23.3	*N*-Acetylglucosamine-1-phosphate transferase, α and β subunits	AR. Short stature, developmental delays, cardiomegaly
Mucolipidosis III γ	*GNPTAG*	16p	*N*-Acetylglucosamine-1-phosphotransferase,γ subunit	AR. Mucolipidosis III γ is characterized clinically by short stature, skeletal abnormalities, cardiomegaly, and developmental delay
Mucopolysaccharidisis type IIIA (Sanfilippo A)	*SGSH*	17q25.3	*N*-Sulfoglucosamine sulfohydrolase	AR. Coarse facial features, joint contractures, cardiomyopathy
Mucopolysaccharidosis IH	*IDUA*	4p16.3	Iduronidase, α-L-	AR. Hurler syndrome, multisystem disorder with severe neurologic deterioration, coarse facial features, corneal clouding, skeletal deformities, and dysostosis multiplex
Mucopolysaccharidosis Ih/S	*IDUA*	4p16.3	Iduronidase, α-L-	AR. Hurler–Scheie syndrome
Mucopolysaccharidosis II	*IDS*	Xq28	Iduronate 2-sulfatase	X-linked recessive. Hunter syndrome
Mucopolysaccharidosis IS	*IDUA*	4p16.3	Iduronidase, α-L-	AR. Scheie syndrome. Caused by deficiency of IDUA, but milder clinical phenotype than Hurler syndrome
Mucopolysaccharidosis IVA	*GALNS*	16q24.3	Galactosamine (*N*-acetyl)-6-sulfate sulfatase	AR. Skeletal and dental findings with corneal clouding. No central nervous system involvement
Mucopolysaccharidosis type IIIA (Sanfilippo A)	*SGSH*	17q25.3	*N*-Sulfoglucosamine sulfohydrolase	AR. Lysosomal storage disease with major nervous system involvement. Macrocephaly and intellectual disability
Mucopolysaccharidosis type IIIB (Sanfilippo B)	*NAGLU*	17q21	*N*-Acetylglucosaminidase, β	AR. Lysosomal storage disease with major nervous system involvement. Macrocephaly and intellectual disability
Mucopolysaccharidosis type IIIC (Sanfilippo C)	*HGSNAT*	8p11.1	Heparan-α-glucosaminide *N*-acetyltransferase	AR. Lysosomal storage disease with major nervous system involvement. Macrocephaly and intellectual disability
Mucopolysaccharidosis type IIID	*GNS*	12q14	Glucosamine (*N*-acetyl)-6-sulfatase	AR. Lysosomal storage disease with major nervous system involvement. Macrocephaly and intellectual disability

TABLE 109.1 Genes Associated with Neurologic Disorders—cont'd

Genetic disorder	Gene	Locus	Gene product	Inheritance (if established) and phenotype
Mucopolysaccharidosis type VI	*ARSB*	5q14	Arylsulfatase B	AR. Maroteaux–Lamy. Skeletal findings, facial dysmorphism, minimal neurologic involvement
Mucopolysaccharidosis VII	*GUSB*	7q21.11	Glucuronidase, β	AR. Lysosomal storage disease with skeletal abnormalities, craniofacial dysmorphism and various degrees of nervous system involvement. Phenotype can vary in severity
Multiple sclerosis, susceptibility to	*CD24*	6q21	CD24 molecule	Susceptibility locus for multiple sclerosis
Multiple sclerosis, susceptibility to, 2	*MS2*	10p15.1	Multiple sclerosis, susceptibility to, 2	Susceptibility locus for multiple sclerosis
Multiple sclerosis, susceptibility to, 3	*MS3*	5p13.2	Multiple sclerosis, susceptibility to, 3	Susceptibility locus for multiple sclerosis
Multiple sclerosis, susceptibility to, 4	*MS4*	1p36	Multiple sclerosis, susceptibility to, 4	Susceptibility locus for multiple sclerosis
Multiple sulfatase deficiency	*SUMF1*	3p26	Sulfatase modifying factor 1	AR. Psychomotor delays, hydrocephalus, hepatosplenomegaly, dysostosis multiplex
Muscle glycogenosis	*PHKA1*	Xq13	Phosphorylase kinase, α1 (muscle)	X-linked. Exercise-induced weakness and stiffness
Muscular dystrophy with epidermolysis bullosa simplex	*PLEC*	8q24	Plectin	AR. Childhood onset progressive muscular dystrophy and bullous skin changes
Muscular dystrophy with rimmed vacuoles	*MDRV*	19p13.3	Muscular dystrophy, with rimmed vacuoles	AD. Muscular dystrophy, with rimmed vacuoles
Muscular dystrophy-dystroglycanopathy (congenital with brain and eye anomalies), type A, 1	*POMT1*	9q34.1	Protein-*O*-mannosyltransferase 1	AR. Muscular dystrophy with brain and eye anomalies. Walker–Warburg syndrome
Muscular dystrophy-dystroglycanopathy (congenital with brain and eye anomalies), type A, 2	*POMT2*	14q24.3	Protein-*O*-mannosyltransferase 2	AR. Muscular dystrophy with brain and eye anomalies. Walker–Warburg syndrome
Muscular dystrophy-dystroglycanopathy (congenital with brain and eye anomalies), type A, 3	*POMGNT1*	1p34-p33	Protein *O*-linked mannose β1,2-*N*-acetylglucosaminyltransferase	AR. Muscular dystrophy with brain and eye anomalies. Walker–Warburg syndrome
Muscular dystrophy-dystroglycanopathy (congenital with brain and eye anomalies), type A, 4	*FKTN*	9q31	Fukutin	AR. Muscular dystrophy with brain and eye anomalies. Fukuyama congenital muscular dystrophy
Muscular dystrophy-dystroglycanopathy (congenital with brain and eye anomalies), type A, 5	*FKRP*	19q13.3	Fukutin-related protein	AR. Muscular dystrophy with brain and eye anomalies. Walker–Warburg syndrome
Muscular dystrophy-dystroglycanopathy (congenital with brain and eye anomalies), type A, 6	*LARGE*	22q12.3-q13.1	Like-glycosyltransferase	AR. Muscular dystrophy with brain and eye anomalies. Walker–Warburg syndrome

Continued

TABLE 109.1 Genes Associated with Neurologic Disorders—cont'd

Genetic disorder	Gene	Locus	Gene product	Inheritance (if established) and phenotype
Muscular dystrophy, congenital merosin-deficient	*LAMA2*	6q22-q23	Laminin, α2	AR. Congenital muscular dystrophy—merosin deficient
Muscular dystrophy, congenital, 1B	*MDC1B*	1q42	Muscular dystrophy, congenital, 1B	AR. Congenital muscular dystrophy
Muscular dystrophy, congenital, due to ITGA7 deficiency	*ITGA7*	12q13	Integrin, α7	AR. Congenital muscular dystrophy
Muscular dystrophy, limb-girdle, type 1A	*MYOT*	5q31	Myotilin	AD. Muscular dystrophy primarily affecting proximal muscles
Muscular dystrophy, limb-girdle, type 1D	*LGMD1D*	7q	Limb-girdle muscular dystrophy 1D (autosomal dominant)	AD. Muscular dystrophy primarily affecting proximal muscles
Muscular dystrophy, limb-girdle, type 1F	*LGMD1F*	7q32.1-q32.2	Limb-girdle muscular dystrophy 1F (autosomal dominant)	AD. Muscular dystrophy primarily affecting proximal muscles
Muscular dystrophy, limb-girdle, type 1H	*LGMD1H*	3p25.1-p23	Limb-girdle muscular dystrophy 1H (autosomal dominant)	AD. Muscular dystrophy primarily affecting proximal muscles
Muscular dystrophy, limb-girdle, type 2A	*CAPN3*	15q15.1-q21.1	Calpain 3, (p94)	AR. Muscular dystrophy primarily affecting proximal muscles
Muscular dystrophy, limb-girdle, type 2B	*DYSF*	2p13.3-p13.1	Dysferlin, limb-girdle muscular dystrophy 2B (autosomal recessive)	AR. Muscular dystrophy primarily affecting proximal muscles
Muscular dystrophy, limb-girdle, type 2C	*SGCG*	13q12	Sarcoglycan, γ (35 kDa dystrophin-associated glycoprotein)	AR. Muscular dystrophy primarily affecting proximal muscles
Muscular dystrophy, limb-girdle, type 2D	*SGCA*	17q12-q21.33	Sarcoglycan, α (50 kDa dystrophin-associated glycoprotein)	AR. Muscular dystrophy primarily affecting proximal muscles
Muscular dystrophy, limb-girdle, type 2E	*SGCB*	4q12	Sarcoglycan, β (43 kDa dystrophin-associated glycoprotein)	AR. Muscular dystrophy primarily affecting proximal muscles
Muscular dystrophy, limb-girdle, type 2F	*SGCD*	5q33	Sarcoglycan, δ (35 kDa dystrophin-associated glycoprotein)	AR. Muscular dystrophy primarily affecting proximal muscles
Muscular dystrophy, limb-girdle, type 2G	*TCAP*	17q12	Titin-cap (telethonin)	AR. Muscular dystrophy primarily affecting proximal muscles
Muscular dystrophy, limb-girdle, type 2H	*TRIM32*	9q31-q34.1	Tripartite motif containing 32	AR. Muscular dystrophy primarily affecting proximal muscles
Muscular dystrophy, limb-girdle, type IC	*CAV3*	3p25	Caveolin 3	AR. Muscular dystrophy primarily affecting proximal muscles
Muscular dystrophy, rigid spine, 1	*SEPN1*	1p36-p35	Selenoprotein N, 1	AR. Muscular dystrophy
Myasthenia, familial infantile, 1	*FIMG1*	17p13	Myasthenia gravis, familial infantile, 1	Inheritance not established. Myasthenia gravis
Myasthenia, limb-girdle, familial	*DOK7*	4p16.2	Docking protein 7	AR. Congenital myasthenic syndrome with proximal weakness
Myasthenia, limb-girdle, familial	*AGRN*	1pter-p32	Agrin	AR. Congenital myasthenic syndrome with proximal weakness
Myasthenic syndrome, congenital, associated with acetylcholine receptor deficiency	*RAPSN*	11p11.2-p11.1	Receptor-associated protein of the synapse	AR. Congenital myasthenic syndrome
Myasthenic syndrome, congenital, associated with acetylcholine receptor deficiency	*MUSK*	9q31.3-q32	Muscle, skeletal, receptor tyrosine kinase	AR. Congenital myasthenic syndrome
Myasthenic syndrome, congenital, associated with episodic ataxia	*CHAT*	10q11.2	Choline *O*-acetyltransferase	AR. Congenital myasthenic syndrome with episodic ataxia

TABLE 109.1 Genes Associated with Neurologic Disorders—cont'd

Genetic disorder	Gene	Locus	Gene product	Inheritance (if established) and phenotype
Myasthenic syndrome, slow-channel congenital	*CHRNB1*	17p12-p11	Cholinergic receptor, nicotinic, β1 (muscle)	AR. Congenital myasthenic syndrome
Myasthenic syndrome, slow-channel congenital	*CHRNE*	17p13-p12	Cholinergic receptor, nicotinic, ε (muscle)	AR. Congenital myasthenic syndrome
Myasthenic syndrome, slow-channel congenital	*CHRNA1*	2q24-q32	Cholinergic receptor, nicotinic, α1 (muscle)	AR. Congenital myasthenic syndrome
Myasthenic syndrome, slow-channel congenital	*CHRND*	2q33-q34	Cholinergic receptor, nicotinic, δ (muscle)	AR. Congenital myasthenic syndrome
Myoadenylate deaminase deficiency	*AMPD1*	1p21-p13	Adenosine monophosphate deaminase 1	AR. Myopathy
Myoclonic epilepsy, infantile, familial	*TBC1D24*	16p13.3	TBC1 domain family, member 24	AR. Infantile myoclonic epilepsy. Early infantile epileptic encephalopathy is an allelic form
Myoglobinuria, acute recurrent, autosomal recessive	*LPIN1*	2p21	Lipin 1	AR. Acute recurrent myoglobinuria
Myopathy due to CPT II deficiency	*CPT2*	1p32	Carnitine palmitoyltransferase 2	AR. Weakness, cramps and stiffness after prolonged exercise
Myopathy with lactic acidosis, hereditary	*ISCU*	12q24.1	Fe–S cluster scaffold homolog (*E. coli*)	AR. Intolerance to exercise, weakness, cramps, palpitation with lactic acidosis during exercise
Myopathy, cardioskeletal, desmin-related, with cataract	*CRYAB*	11q22.3-q23.1	Crystallin, αB	AR/AD. Hypertrophic cardiomyopathy, congenital myopathy
Myopathy, centronuclear	*MYF6*	12q21	Myogenic factor 6 (herculin)	AD. Facial and limb weakness
Myopathy, centronuclear, autosomal recessive	*BIN1*	2q14	Bridging integrator 1	AR. Centronuclear myopathy
Myopathy, congenital, Compton–North	*CNTN1*	12q11-q12	Contactin 1	AR. Congenital myopathy
Myopathy, congenital, with fiber-type disproportion, X-linked	*CFTDX*	Xq13.1-q22.1	Myopathy, congenital, with fiber-type disproportion, X-linked	X-linked congenital myopathy
Myopathy, distal 1	*MYH7*	14q11.2	Myosin heavy chain 7	AD. Distal myopathy
Myopathy, distal 2	*MATR3*	5q31.2	Matrin 3	AD. Distal myopathy
Myopathy, distal 3	*MPD3*	8p22-q11	Myopathy, distal 3	AD. Distal myopathy
Myopathy, hyaline body	*MHB*	3p22.2-p21.32	Myopathy, hyaline body, autosomal recessive	AR. Hyaline body myopathy
Myopathy, lactic acidosis, and sideroblastic anemia 2	*YARS2*	12p11.21	Tyrosyl-tRNA synthetase 2, mitochondrial	AR. Myopathy, sideroblastic anemia, and lactic acidosis
Myopathy, mitochondrial progressive, with congenital cataract, hearing loss, and developmental delay	*GFER*	16p13.3-p13.12	Growth factor, augmenter of liver regeneration	AR. Combined mitochondrial complex deficiency
Myopathy, myofibrillar, BAG3-related	*BAG3*	10q25.2-q26.2	BCL2-associated athanogene 3	AD. Cardioskeletal myopathy
Myopathy, myofibrillar, filamin C-related	*FLNC*	7q32	Filamin C, γ	AD. Cardioskeletal myopathy
Myopathy, myofibrillar, ZASP-related	*LDB3*	10q22.2-q23.3	LIM domain binding 3	AD. Cardioskeletal myopathy

Continued

TABLE 109.1 Genes Associated with Neurologic Disorders—cont'd

Genetic disorder	Gene	Locus	Gene product	Inheritance (if established) and phenotype
Nemaline myopathy 1, autosomal dominant	*TPM3*	1q22-q23	Tropomyosin 3	AD/AR. Nemaline myopathy
Nemaline myopathy 2, autosomal recessive	*NEB*	2q22	Nebulin	AR. Nemaline myopathy
Myopathy, nemaline, 3	*ACTA1*	1q42.1	Actin, α1, skeletal muscle	AR. Nemaline myopathy
Myopathy, X-linked, with excessive autophagy	*VMA21*	Xq28	VMA21 vacuolar H^+-ATPase homolog (*S. cerevisiae*)	X-linked myopathy
Myotonia congenital	*CLCN1*	7q34	Chloride channel, voltage-sensitive 1	AD. Myotonia congenital
Myotonia congenita, recessive	*CLCN1*	7q35	Chloride channel, voltage-sensitive 1	AR. Myotonia congenita
Myotonic dystrophy	*DMPK*	19q13.2-q13.3	Dystrophia myotonic-protein kinase	AD. Myotonic dystrophy
Myotonic dystrophy, type 2	*CNBP*	3q13.3-q24	CCHC-type zinc finger, nucleic acid binding protein	AD. Myotonic dystrophy
Myotubular myopathy, X-linked	*MTM1*	Xq28	Myotubularin 1	X-linked myotubular myopathy
Narcolepsy 1	*HCRT*	17q21	Hypocretin (orexin) neuropeptide precursor	AD. Narcolepsy
Narcolepsy 2	*NRCLP*	4p13-q21	Narcolepsy, HLA-associated	Susceptibility locus for narcolepsy
Narcolepsy 3	*NRCLP3*	21q11.2	Narcolepsy 3	AD. Narcolepsy
Narcolepsy 4	*NRCLP4*	22q13	Narcolepsy 4	Susceptibility locus for narcolepsy
Narcolepsy 5	*NRCLP5*	14q11.2	Narcolepsy 5	Susceptibility locus for narcolepsy
Nasu–Hakola disease	*TREM2*	6p21.2	Triggering receptor expressed on myeloid cells 2	AR. Presenile dementia with bone cysts
Nasu–Hakola disease	*TYROBP*	19q13.1	TYRO protein tyrosine kinase binding protein	AR. Presenile dementia with bone cysts
Neural tube defects, susceptibility to	*T*	6q27	T, brachyury homolog (mouse)	Susceptibility to neural tube defects
Neuroblastoma, susceptibility to, 3	*ALK*	2p23	Anaplastic lymphoma receptor tyrosine kinase	Susceptibility to neuroblastoma
Neurodegeneration due to cerebral folate transport deficiency	*FOLR1*	11q13.3-q13.5	Folate receptor 1 (adult)	AR. Neurodegeneration due to cerebral folate transport deficiency
Neurodegeneration with brain iron accumulation 1	*PANK2*	20p13-p12.3	Pantothenate kinase 2	AR. Neurodegenerative disease due to brain iron accumulation
Neurofibromatosis, type 1	*NF1*	17q11.2	Neurofibromin 1	AD. Neurofibromatosis
Neurofibromatosis, type 2	*NF2*	22q12.2	Neurofibromin 2 (merlin)	AD. Neurofibromatosis
Niemann–Pick disease, type A	*SMPD1*	11p15.4-p15.1	Sphingomyelin phosphodiesterase 1, acid lysosomal	AR. Lysosomal storage disease due to sphingomyelinase deficiency. Neurodegenerative disease with early death
Niemann–Pick disease, type C1	*NPC1*	18q11-q12	Niemann–Pick disease, type C1	AR. Lysosomal storage disease. Defect in intracellular trafficking of lipids. Progressive neurodegenerative disease with early death

TABLE 109.1 Genes Associated with Neurologic Disorders—cont'd

Genetic disorder	Gene	Locus	Gene product	Inheritance (if established) and phenotype
Niemann–Pick disease, type C2	*NPC2*	14q24.3	Niemann–Pick disease, type C2	AR. Lysosomal storage disease. Defect in intracellular trafficking of lipids. Progressive neurodegenerative disease with early death
Nijmegen breakage syndrome	*NBS1/NBN*	8q21	Nibrin	AR, chromosomal instability syndrome with nonsyndromic microcephaly, growth restriction, immunodeficiency, and predisposition to cancer
Nijmegen breakage syndrome-like disorder	*RAD50*	5q31	RAD50 homolog (*S. cerevisiae*)	AR. Nijmegen breakage syndrome-like disorder
Norrie disease	*NDP*	Xp11.4	Norrie disease (pseudoglioma)	Retinal blindness, sensorineural hearing loss and psychotic features
Nystagmus 1, congenital, X-linked	*FRMD7*	Xq26.2	FERM domain containing 7	X-linked congenital nystagmus
Nystagmus 3, congenital	*NYS3*	7p11.2	Nystagmus 3, congenital autosomal dominant	AD. Congenital nystagmus
Nystagmus 4, congenital	*NYS4*	13q31-q33	Nystagmus 4, congenital autosomal dominant	AD. Congenital nystagmus
Nystagmus 5, congenital, X-linked	*NYS5*	Xp11.4	Nystagmus 5, infantile periodic alternating	X-linked congenital nystagmus
Nystagmus-2, autosomal dominant	*NYS2*	6p12	Nystagmus 2, congenital autosomal dominant	AD. Congenital nystagmus
Oculopharyngeal muscular dystrophy	*PABPN1*	14q11.2-q13	Poly(A) binding protein, nuclear 1	AD. Oculopharyngeal muscular dystrophy
Ornithine transcarbamylase deficiency	*OTC*	Xp21.1	Ornithine carbamoyltransferase	X-linked recessive. Urea cycle disorder
Parkinson disease 1	*SNCA*	4q22.1	Synuclein, α	AD. Parkinson disease
Parkinson disease, juvenile, type 2	*PARK2*	6q25.2-q27	Parkinson protein 2, E3 ubiquitin protein ligase (parkin)	AR. Parkinson disease, early onset
Parkinson disease 3	*PARK3*	2p13	Parkinson disease 3 (autosomal dominant, Lewy body)	AD. Parkinson disease
Parkinson disease 4	*SNCA*	4q21	Synuclein, α (non-A4 component of amyloid precursor)	AD. Parkinson disease
Parkinson disease 5, susceptibility to	*UCHL1*	4p14	Ubiquitin carboxyl-terminal esterase L1 (ubiquitin thiolesterase)	AD. Parkinson disease
Parkinson disease 6, early onset	*PINK1*	1p36	PTEN-induced putative kinase 1	AR. Parkinson disease, early onset
Parkinson disease 7, autosomal recessive early onset	*DJ1*	1p36	Oncogene DJ1	AR. Parkinson disease, early onset
Parkinson disease 8	*LRRK2*	12q12	Leucine-rich repeat kinase 2	AD. Parkinson disease
Parkinson disease 9	*ATP13A2*	1p36	ATPase type 13A2	AR. Parkinson disease. Also known as Kufor–Rakeb disease
Parkinson disease 10	*PARK10*	1p32	Parkinson disease 10 (susceptibility)	Parkinson disease
Parkinson disease 11	*GIGYF2*	2q37.1	GRB10 interacting GYF protein 2	AD. Parkinson disease
Parkinson disease 12	*PARK12*	Xq21-q25	Parkinson disease 12 (susceptibility)	X-linked Parkinson disease susceptibility locus

Continued

TABLE 109.1 Genes Associated with Neurologic Disorders—cont'd

Genetic disorder	Gene	Locus	Gene product	Inheritance (if established) and phenotype
Parkinson disease 13	*HTRA2*	2p12	HtrA serine peptidase 2	AD. Parkinson disease
Parkinson disease 15, autosomal recessive	*FBXO7*	22q12-q13	F-box protein 7	AR. Parkinson disease
Parkinson disease 16	*PARK16*	1q32	Parkinson disease 16 (susceptibility)	Parkinson disease
Parkinson disease 17	*VPS35*	16p11.2	Vacuolar protein sorting 35, yeast homolog of	AD. Parkinson disease
Parkinson disease 18	*EIF4G1*	3q27.1	Eukaryotic translation initiation factor 4	AD. Parkinson disease
Parkinson disease 19	*DNAJC6*	1p31.3	DNAJ/HSP40 homolog, subfamily C, member 6	AR. Parkinson disease
Parkinson disease 20	*SYNJ1*	21q22.1	Synaptojanin	AR. Parkinson disease
Paroxysmal kinesigenic choreoathetosis	*DYT10*	16p11.2-q12.1	Dystonia 10	AD. Episodic dyskinesias, dystonia, and choreoathetosis
Paroxysmal nonkinesigenic dyskinesia	*PNKD*	2q35	Paroxysmal nonkinesigenic dyskinesia	AD. Episodic choreoathetosis, dystonia, myokymia, dysphagia
Paroxysmal nonkinesigenic dyskinesia 2	*PNKD2*	2q31	Paroxysmal nonkinesigenic dyskinesia 2	AD. Episodic choreoathetosis, dystonia, myokymia, dysphagia
Pelizaeus–Merzbacher disease	*PLP1*	Xq22	Proteolipid protein 1	X-linked recessive. Neurologic disease with hypomyelination and abnormal eye movements. Hypotonia progressing to hypertonia and diffuse weakness and spasticity
Periventricular heterotopia with microcephaly	*ARFGEF2*	20q13.13	ADP-ribosylation factor guanine nucleotide-exchange factor 2 (brefeldin A-inhibited)	AR. Periventricular heterotopia with microcephaly
Periventricular nodular heterotopia 3	*PVNH3*	5p15.1	Periventricular nodular heterotopia 3	Periventricular nodular heterotopia
Periventricular nodular heterotopia 5	*PVNH5*	5q14.3-q15	Periventricular nodular heterotopia 5	Periventricular nodular heterotopia
Peroxisomal acyl-CoA oxidase deficiency	*ACOX1*	17q25	Acyl-CoA oxidase 1, palmitoyl	Clinical symptoms similar to Zellweger syndrome
Phenylketonuria	*PAH*	12q24.1	Phenylalanine hydroxylase	AR. Associated with excessive levels of phenylalanine that can cause nervous system toxicity and developmental delays, microcephaly, psychomotor delays
Phosphoglycerate kinase 1 deficiency	*PGK1P1*	Xq13	Phosphoglycerate kinase 1, pseudogene 1	X-linked. Variable clinical symptoms consisting of hemolytic anemia, myopathy, and neurologic involvement
Phosphorylase kinase deficiency of liver and muscle, autosomal recessive	*PHKB*	16q12-q13	Phosphorylase kinase, β	AR. Glycogen storage disease
Phosphoserine aminotransferase deficiency	*PSAT1*	9q21.31	Phosphoserine aminotransferase 1	AR. Low plasma and cerebrospinal fluid (CSF) concentrations of serine and glycine and clinically by intractable seizures, acquired microcephaly, hypertonia, and psychomotor retardation

TABLE 109.1 Genes Associated with Neurologic Disorders—cont'd

Genetic disorder	Gene	Locus	Gene product	Inheritance (if established) and phenotype
Photoparoxysmal response 1	*PPR1*	6p21.1	Photoparoxysmal response 1	Inheritance not established. Abnormal occurrence of cortical spikes or spike and wave discharges on electroencephalogram (EEG) in response to intermittent light stimulation
Photoparoxysmal response 2	*PPR2*	13q31.3	Photoparoxysmal response 2	Inheritance not established. Abnormal occurrence of cortical spikes or spike and wave discharges on electroencephalogram (EEG) in response to intermittent light stimulation
Photoparoxysmal response 3	*PPR3*	7q32	Photoparoxysmal response 3	Inheritance not established. Abnormal occurrence of cortical spikes or spike and wave discharges on electroencephalogram (EEG) in response to intermittent light stimulation
Pitt–Hopkins syndrome 1	*TCF4*	18q21.2	Transcription factor 4	Inheritance not established. Mental retardation, wide mouth and distinctive facial features, and intermittent hyperventilation followed by apnea
Pitt–Hopkins-like syndrome 2	*NRXN1*	2p16.3	Neurexin 1	Inheritance not established. Mental retardation, wide mouth and distinctive facial features, and intermittent hyperventilation followed by apnea
Polyhydramnios, megalencephaly, and symptomatic epilepsy	*STRADA*	17q23.3	STE20-related kinase adaptor α	Inheritance not established. Polyhydramnios, megalencephaly, and symptomatic epilepsy
Polymicrogyria with optic nerve hypoplasia	*TUBA8*	22q11	Tubulin, α8	AR. Polymicrogyria with optic nerve hypoplasia
Polymicrogyria, symmetric or asymmetric	*TUBB2B*	6p25.2	Tubulin, β2A class IIa	AD. Polymicrogyria, symmetric or asymmetric
Polymicrogyria, bilateral frontoparietal	*GPR56*	16q13	G protein-coupled receptor 56	AR. Polymicrogyria, bilateral frontoparietal
Polyneuropathy, hearing loss, ataxia, retinitis pigmentosa, and cataract	*ABHD12*	20p11.21	Abhydrolase domain containing 12	AR. Polyneuropathy, hearing loss, ataxia, retinitis pigmentosa, and cataract
Pontocerebellar hypoplasia type 1	*VRK1*	14q32	Vaccinia-related kinase 1	AR. Severe intellectual disability, poor feeding, hypotonia and psychomotor delays with associated pontocerebellar hypoplasia
Pontocerebellar hypoplasia type 2A	*TSEN54*	17q25.1	tRNA splicing endonuclease 54 homolog (*S. cerevisiae*)	AR. Pontocerebellar hypoplasia with progressive microcephaly, dyskinesia, chorea and epilepsy
Pontocerebellar hypoplasia type 2B	*TSEN2*	3p25.1	tRNA splicing endonuclease 2 homolog (*S. cerevisiae*)	AR. See pontocerebellar hypoplasia
Pontocerebellar hypoplasia type 2C	*TSEN34*	19q13.4	tRNA splicing endonuclease 34 homolog (*S. cerevisiae*)	AR. Pontocerebellar hypoplasia with progressive microcephaly, dyskinesia, chorea and epilepsy
Porencephaly	*COL4A1*	13q34	Collagen, type IV, α1	AD. Porencephaly

Continued

TABLE 109.1 Genes Associated with Neurologic Disorders—cont'd

Genetic disorder	Gene	Locus	Gene product	Inheritance (if established) and phenotype
Porphyria variegata	*PPOX*	1q22	Protoporphyrinogen oxidase	AD. Cutaneous manifestations and acute exacerbation with abdominal pain and neuropsychiatric symptoms
Porphyria, acute hepatic	*ALAD*	9q34	Aminolevulinate dehydratase	AR. Failure to thrive, abdominal colic, vomiting, neuropathy
Porphyria, acute intermittent	*HMBS*	11q23.3	Hydroxymethylbilane synthase	AD. Recurrent attacks of abdominal pain, gastrointestinal dysfunction, and neurologic disturbances
Potocki–Lupski syndrome	*PTLS*	17p11.2	Potocki–Lupski syndrome	Chromosomal microduplication. Developmental disorder characterized by hypotonia, failure to thrive, mental retardation, pervasive developmental disorders, and congenital anomalies
Potocki–Shaffer syndrome	*PSS*	11p11.2	Potocki–Shaffer syndrome	Contiguous gene syndrome. Multiple exostoses, craniofacial dysostosis, and intellectual disability
Prader–Willi syndrome	*NDN*	15q11-q13	Necdin homolog (mouse)	AD. Loss of part of paternally inherited long arm of 15. Hypotonia, short stature, hyperphagia, obesity, small hands and feet
Prader–Willi syndrome	*SNRPN*	15q12	Small nuclear rib nucleoprotein polypeptide N	Chromosomal microdeletion, uniparental disomy. Prader–Willi syndrome
Primary lateral sclerosis, adult, 1	*PLSA1*	4p16	Primary lateral sclerosis, adult, 1	AD. Neurodegenerative disorder with paralysis and limb spasticity
Progressive external ophthalmoplegia with mitochondrial DNA deletions 3	*SLC25A4*	4q35	Solute carrier family 25 (mitochondrial carrier; adenine nucleotide translocator), member 4	AD. Progressive external ophthalmoplegia with mitochondrial DNA deletions
Progressive external ophthalmoplegia, autosomal dominant, 3	*C10orf2*	10q24	Chromosome 10 open reading frame 2	AD. Progressive external ophthalmoplegia
Progressive external ophthalmoplegia with mitochondrial DNA deletions, autosomal dominant 4	*POLG2*	17q23-q24	Polymerase (DNA directed), $\gamma 2$, accessory subunit	AD. Progressive external ophthalmoplegia with mitochondrial DNA deletions
Progressive external ophthalmoplegia, autosomal recessive	*POLG*	15q25	Polymerase (DNA directed), γ	AR. Progressive external ophthalmoplegia
Proliferative vasculopathy and hydranencephaly-hydrocephaly syndrome	*FLVCR2*	14q24.3	Feline leukemia virus subgroup C cellular receptor family, member 2	AR. Proliferative vasculopathy and hydranencephaly-hydrocephaly syndrome
Ptosis, congenital	*ZFHX4*	8q21.12	Zinc finger homeobox 4	AD. Congenital ptosis
Ptosis, hereditary congenital 2	*PTOS2*	Xq24-q27.1	Ptosis, hereditary congenital 2	X-linked congenital ptosis
Pyridoxamine 5′-phosphate oxidase deficiency	*PNPO*	17q21.32	Pyridoxamine 5′-phosphate oxidase	AR. Intractable seizures starting after birth, myoclonus, abnormal eye movements; seizures respond to pyridoxal phosphate

TABLE 109.1 Genes Associated with Neurologic Disorders—cont'd

Genetic disorder	Gene	Locus	Gene product	Inheritance (if established) and phenotype
Pyruvate carboxylase deficiency	*PC*	11q13.4-q13.5	Pyruvate carboxylase	AR. North American phenotype (lactic acidemia, psychomotor retardation); French form (lactic acidemia, intracellular redox disturbance, relatively decreased survival); benign form
Pyruvate dehydrogenase deficiency	*PDHA1*	Xp22.2-p22.1	Pyruvate dehydrogenase (lipoamide) α1	X-linked dominant. Pyruvate dehydrogenase deficiency
Pyruvate dehydrogenase E1-β deficiency	*PDHB*	3p13-q23	Pyruvate dehydrogenase (lipoamide) β	AR. Pyruvate dehydrogenase E1-β deficiency
Pyruvate dehydrogenase E2 deficiency	*DLAT*	11q23.1	Dihydrolipoamide *S*-acetyltransferase	AR. Decreased activity of pyruvate dehydrogenase complex
Pyruvate dehydrogenase phosphatase deficiency	*PDP1*	8q22.1	Pyruvate dehydrogenase phosphatase catalytic subunit 1	AR. Decreased activity of pyruvate dehydrogenase complex
Refsum disease	*PHYH*	10pter-p11.2	Phytanoyl-CoA 2-hydroxylase	AR. Sensorineural deafness, cardiomyopathy, peripheral sensorimotor neuropathy
Renpenning syndrome	*PQBP1*	Xp11.23	Polyglutamine binding protein 1	X-linked intellectual disability
Restless legs syndrome 1	*RLS*	12q12-q21	Restless legs syndrome, susceptibility to	Susceptibility to restless leg syndrome
Restless legs syndrome 2	*RLS2*	14q13-q21	Restless legs syndrome 2	Susceptibility to restless leg syndrome
Restless legs syndrome 3	*RLS3*	9p24-p22	Restless legs syndrome 3	Susceptibility to restless leg syndrome
Restless legs syndrome 4	*RLS4*	2q33	Restless legs syndrome 4	Susceptibility to restless leg syndrome
Restless legs syndrome 7	*RLS7*	2p14-p13	Restless legs syndrome 7	Susceptibility to restless leg syndrome
Restless legs syndrome, susceptibility to, 5	*RLS5*	20p13	Restless legs syndrome, susceptibility to, 5	Susceptibility to restless leg syndrome
Restless legs syndrome, susceptibility to, 6	*RLS6*	6p21	Restless legs syndrome, susceptibility to, 6	Susceptibility to restless leg syndrome
Rett syndrome	*MECP2*	Xq28	Methyl CpG binding protein 2 (Rett syndrome)	X-linked dominant. Progressive neurodegenerative disease with psychomotor impairment and arrest of development at around 6 months of age, stereotypic movement, acquired microcephaly and seizures. Seen in females, lethal in males
Rett syndrome, congenital variant	*FOXG1*	14q13	Forkhead box G1	Inheritance not established. Isolated, features of Rett syndrome with earlier onset
Ribose 5-phosphate isomerase deficiency	*RPIA*	2p11.2	Ribose 5-phosphate isomerase A	AR. Optic atrophy, with psychomotor delays, seizures, and cerebellar ataxia
Rippling muscle disease-1	*RMD1*	1q41	Rippling muscle disease 1	AD. Muscle cramps and pain with exercise
Rubinstein–Taybi syndrome	*CREBBP*	16p13.3	CREB-binding protein	AD. Multiple congenital anomalies including intellectual disability, broad thumbs and great toes
Salla disease	*SLC17A5*	6q14-q15	Solute carrier family 17 (anion/sugar transporter), member 5	AR. Phenotype can vary from a milder adult-onset form to a severe infantile form. Neurodegenerative features with excretion of sialic acid in urine

Continued

TABLE 109.1 Genes Associated with Neurologic Disorders—cont'd

Genetic disorder	Gene	Locus	Gene product	Inheritance (if established) and phenotype
Sandhoff disease, infantile, juvenile, and adult forms	*HEXB*	5q13	Hexosaminidase B (β polypeptide)	AR. Progressive neurodegenerative disease, clinically similar to Tay–Sachs disease
Sarcoidosis, susceptibility to, 2	*BTNL2*	6p21.3	Butyrophilin-like 2 (MHC class II associated)	Susceptibility to sarcoidosis
Sarcoidosis, susceptibility to, 3	*SS3*	10q22.3	Sarcoidosis, susceptibility to, 3	Susceptibility to sarcoidosis
Scapuloperoneal myopathy, X-linked dominant	*FHL1*	Xq27.2	Four and a half LIM domains 1	X-linked dominant scapuloperoneal dystrophy
Schindler disease, type I	*NAGA*	22q13.2	*N*-acetyl-α-D-galactosaminidase	AR. Neuroaxonal dystrophy
Schizencephaly	*EMX2*	10q26.1	Empty spiracles homeobox 2	AD. Schizencephaly
Seckel syndrome 1	*ATR*	3q22-q24	Ataxia telangiectasia and Rad3 related	AR. Primordial dwarfism with microcephaly and intellectual disability
Seckel syndrome 2	*SCKL2*	18p11.31-q11.2	Seckel syndrome 2	AR. Growth restriction, intellectual disability and characteristic facial features
Seckel syndrome 3	*SCKL3*	14q21-q22	Seckel syndrome 3	AR. Primordial dwarfism with microcephaly and intellectual disability
Segawa syndrome, recessive	*TH*	11p15.5	Tyrosine hydroxylase	AR. Infantile onset dopa-responsive dystonia
Seizures, benign familial infantile, 3	*SCN2A*	2q23-q24.3	Sodium channel, voltage-gated, type II, α subunit	AD. Benign form of epilepsy. Onset can vary from 2–7 months and resolve typically after 1 year of life
Seizures, benign neonatal, 1	*KCNQ2*	20q13.3	Potassium voltage-gated channel, KQT-like subfamily, member 2	AD. Epilepsy by 2–8 days of life, typically remits by 6 weeks; + myokymi
Septooptic dysplasia	*HESX1*	3p21.2-p21.1	HESX homeobox 1	AD/AR. Septooptic dysplasia
SESAME syndrome	*KCNJ10*	1q23.2	Potassium inwardly-rectifying channel, subfamily J, member 10	AR. Seizures, sensorineural deafness, ataxia, intellectual disability
Severe combined immunodeficiency with microcephaly, growth retardation, and sensitivity to ionizing radiation	*NHEJ1*	2q35	Nonhomologous end-joining factor 1	AR. Severe combined immunodeficiency with microcephaly, growth retardation, and sensitivity to ionizing radiation
Sialidosis, type I	*NEU1*	6p21.3	Sialidase 1 (lysosomal sialidase)	AR. Progressive neurodegenerative disorder. Progressive myoclonic epilepsy, ataxia, blindness, and sensorineural hearing loss
Sialuria	*GNE*	9p13.3	Glucosamine (UDP-*N*-acetyl)-2-epimerase/*N*-acetylmannosamine kinase	AD. Visceromegaly and coarse facial features. Developmental delay and seizures
Smith–Lemli–Opitz syndrome	*DHCR7*	11q13.4	7-Dehydrocholesterol reductase	AR. Multiple congenital anomalies and intellectual disability
Smith–Magenis syndrome	*RAI1*	17p11.2	Retinoic acid induced 1	AD/isolated cases. Various congenital anomalies, developmental delay, brachycephaly and craniofacial dysmorphism, self-destructive behavior

TABLE 109.1 Genes Associated with Neurologic Disorders—cont'd

Genetic disorder	Gene	Locus	Gene product	Inheritance (if established) and phenotype
Sotos syndrome	*NSD1*	5q35	Nuclear receptor binding SET domain protein 1	Isolated cases, macrocephaly and developmental delay
Spastic ataxia 1, autosomal dominant	*SAX1*	12p13	Spastic ataxia 1 (autosomal dominant)	AD. Spastic ataxia
Spastic ataxia, Charlevoix–Saguenay type	*SACS*	13q12	Spastic ataxia of Charlevoix–Saguenay (sacsin)	Inheritance not established. Spastic ataxia
CRASH syndrome	*L1CAM*	Xq28	L1 cell adhesion molecule gene	X-linked recessive. Corpus callosum hypoplasia, retardation, adducted thumbs, spastic paraplegia, and hydrocephalus
MASA syndrome	*L1CAM*	Xq28	L1 cell adhesion molecule gene	X-linked recessive. Mental retardation, adducted thumbs, shuffling gait, and aphasia
Spastic paraplegia 2	*PLP1, PMD, HLD1, SPG2*	Xq22.2	Proteolipid protein 1	X-linked spastic paraplegia
Spastic paraplegia 3A	*ATL1, SPG3A, HSN1D*	14q22.1	Atlastin GTPase 1	AD. Spastic paraplegia
Spastic paraplegia 4	*SPAST, SPG4*	2p22.3	Spastin	AD. Spastic paraplegia
Spastic paraplegia 5A	*CYP7B1, CBAS3, SPG5A*	8q12.3	Cytochrome P450, family 7, subfamily B, polypeptide 1; bile acid synthesis defect, congenital, 3; spastic paraplegia 5a	AR. Spastic paraplegia
Spastic paraplegia 6	*NIPA1, SPG6*	15q11.2	Nonimprinted in Prader–Willi/Angelman syndrome 1	AD. Spastic paraplegia
Spastic paraplegia 7	*PGN, SPG7, CMAR, CAR*	16q24.3	Spastic paraplegia 7 (pure and complicated autosomal recessive)	AR. Spastic paraplegia
Spastic paraplegia 8	*KIAA0196, SPG8*	8q24.13	KIAA0196	AD. Spastic paraplegia
Spastic paraplegia 9	*SPG9*	10q23.3-q24.1	Spastic paraplegia 9	AD. Spastic paraplegia with amyopathy, cataracts, and gastroesophageal reflux
Spastic paraplegia 10	*KIF5A, NKHC, SPG10*	12q13.3	Kinesin 5A	AD. Spastic paraplegia without peripheral neuropathy
Spastic paraplegia 11	*SPG11, KIAA1840, FLJ21439*	15q21.1	Spastic paraplegia 11	AR. Spastic paraplegia with cognitive impairment and thin corpus callosum
Spastic paraplegia 12	*RTN2, NSPL1, SPG12*	19q13.32	Neuroendocrine specific protein-like	AD. Spastic paraplegia
Spastic paraplegia 13	*HSPD1, SPG13, HSP60, HLD4*	2q33.1	Heat shock 60 kDa protein 1 (chaperonin)	AD. Spastic paraplegia
Spastic paraplegia 14	*SPG14*	3q27-q28	Spastic paraplegia 14	AR. Spastic paraplegia
Spastic paraplegia 15	*ZFYVE26, KIAA0321, SPG15*	14q24.1	Zinc finger, FYVE domain containing 26	AR. Spastic paraplegia
Spastic paraplegia 16	*SPG16*	Xq11.2	Spastic paraplegia 16	X-linked. Spastic paraplegia
Silver spastic paraplegia syndrome	*BSCL2, SPG17, HMN5*	11q12.3	Spastic paraplegia 17	AD. Spastic paraplegia

Continued

TABLE 109.1 Genes Associated with Neurologic Disorders—cont'd

Genetic disorder	Gene	Locus	Gene product	Inheritance (if established) and phenotype
Spastic paraplegia 18	*ERLIN2, SPFH2, C8orf2, SPG18*	8p11.23	Endoplasmic reticulum lipid raft-associated protein 2	AR. Spastic paraplegia with onset in early childhood and severe clinical course and psychomotor retardation
Spastic paraplegia 19	*SPG19*	9q	Spastic paraplegia 19	AD. Spastic paraplegia
Mast syndrome	*ACP33, MAST, SPG21*	15q22.31	Acidic cluster protein, 33kDa	AR. Spastic paraplegia 21
Spastic paraplegia 24	*SPG24*	13q14	Spastic paraplegia 24	AR. Spastic paraplegia. Consanguineous family from Saudi described with age of onset in early childhood
Spastic paraplegia 25	*SPG25*	6q23-q24.1	Spastic paraplegia 25	AR. Consanguineous Italian family described with adult onset spastic paraplegia
Spastic paraplegia 26	*SPG26*	12p11.1-q14	Spastic paraplegia 26	AR. Spastic paraplegia with age of onset in first or second decade
Spastic paraplegia 27	*SPG27*	10q22.1-q24.1	Spastic paraplegia 27	AR. Spastic paraplegia
Spastic paraplegia 28	*DDHD1, PAPLA1, KIAA1705, SPG28*	14q22.1	Spastic paraplegia 28	AR. Spastic paraplegia
Spastic paraplegia 29	*SPG29*	1p31.1-p21.1	Spastic paraplegia 29	AD. Spastic paraplegia with sensorineural hearing loss
Spastic paraplegia 30	*KIF1A, ATSV, UNC104, SPG30, HSN2C, MRD9*	2q37.3	Kinesin family member 1A	AR. Slowly progressive spastic paraplegia
Spastic paraplegia 31	*REEP1, C2ORF23, SPG31, HMN5B*	2p11.2	Receptor expression-enhancing protein-1 gene	AD. Spastic paraplegia
Spastic paraplegia 32	*SPG32*	14q12-q21	Spastic paraplegia 32	AR. Spastic paraplegia, described in consanguineous Portuguese family
Spastic paraplegia 33	*ZFYVE27, SPG33*	10q24.2	Zinc finger, FYVE domain containing 27	AD. Later-onset spastic paraplegia
Spastic paraplegia 34	*SPG34*	Xq24-q25	Spastic paraplegia 34	X-linked spastic paraplegia
Spastic paraplegia 35	*FA2H, FAAH, FAXDC1, FAH1, SCS7, SPG35*	16q23.1	Fatty acid 2-hydroxylase	AR. Childhood onset with gait difficulties progressing to paraplegia and associated with cognitive delays and leukodystrophy on brain imaging
Spastic paraplegia 36	*SPG36*	12q23-q24	Spastic paraplegia 36	AD. Spastic paraplegia with demyelinating sensory disturbances in lower limbs
Spastic paraplegia 37	*SPG37*	8p21.1-q13.3	Spastic paraplegia 37	AD. Spastic paraplegia with decreased vibration sense and sensory disturbances
Spastic paraplegia 38	*SPG38*	4p16-p15	Spastic paraplegia 38	AD. Spastic paraplegia with onset described in second decade

TABLE 109.1 Genes Associated with Neurologic Disorders—cont'd

Genetic disorder	Gene	Locus	Gene product	Inheritance (if established) and phenotype
Spastic paraplegia 39	*PNPLA6, NTE, SPG39, NTEMND*	19p13.2	Patatin-like phospholipase domain containing 6	AR. Spastic paraplegia with onset in childhood and distal wasting of lower and upper limbs
Spastic paraplegia 41	*SPG41*	11p14.1-p11.2	Spastic paraplegia 41	AD. Spastic paraplegia
Spastic paraplegia 42	*SLC33A1, ACATN, AT1, SPG42, CCHLND*	3q25.31	Solute carrier family 33 member 1	AD. Spastic paraplegia
Spastic paraplegia 43	*C19orf12, NBIA4, SPG43*	19q12	Chromosome 19 open reading frame 12	AR. Spastic paraplegia
Spastic paraplegia 44	*GJC2, GJA12, CX47, PMLDAR, HLD2, SPG44, LMPH1C*	1q42.13	Gap junction protein γ2	AR. Spastic paraplegia
Spastic paraplegia 45	*SPG45*	10q24.3-q25.1	Spastic paraplegia 45	AR. Spastic paraplegia, complicated with cognitive impairment. Nystagmus also described
Spastic paraplegia 46	*GBA2, KIAA1605, SPG46*	9p13.3	Glucosidase β acid 2	AR. Spastic paraplegia with cognitive impairment and cerebellar signs with brain imaging abnormalities
Spastic paraplegia 47	*AP4B1, SPG47, CPSQ5*	1p13.2	Adaptor related protein complex 4, β-1 subunit	AR. Neonatal onset with hypotonia progresses to hypertonia and severe intellectual disability
Spastic paraplegia 48	*AP5Z1, KIAA0415, SPG48*	7p22.1	Adaptor related protein complex 5, zeta-1 subunit	AR. Spastic paraplegia
Spastic paraplegia 49	*TECPR2, KIAA0329, SPG49*	14q32.31	Tectonin β-propeller repeat-containing protein 2	AR. Spastic paraplegia complicated form with severe intellectual disability
Spastic paraplegia 50	*AP4M1, SPG50, CPSQ3*	7q22.1	Adaptor related protein complex 4, μ-1 subunit	AR. Spastic paraplegia with neonatal onset and severe intellectual disability
Spastic paraplegia 51	*AP4E1, SPG51, CPSQ4*	15q21.2	Adaptor related protein complex 4, ε-1 subunit	AR. Neonatal onset with hypotonia progresses to hypertonia and severe intellectual disability
Spastic paraplegia 52	*AP4S1, CPSQ6, SPG52*	14q12	Adaptor related protein complex 4, σ-1 subunit	AR. Neonatal onset with hypotonia progresses to hypertonia and severe intellectual disability
Spastic paraplegia 53	*VPS37A, HCRP1, SPG53*	8p22	Vacuolar protein sorting 37, yeast homolog of	AR. Spastic paraplegia with cognitive impairment
Spastic paraplegia 54	*DDHD2, KIAA0725, SPG54*	8p11.23	DDHD domain containing protein 2	AR. Spastic paraplegia with intellectual disability and early-onset spasticity of lower limbs
Spastic paraplegia 55	*C12orf65, COXPD7, SPG55*	12q24.31	Chromosome 12 open reading frame 65	AR. Spastic paraplegia. Optic atrophy has been described
Spastic paraplegia 56	*CYP2U1, SPG56*	4q25	Cytochrome P450, family 2, subfamily U, polypeptide 1	AR. Spastic paraplegia
Troyer syndrome	*SPG20*	13q13.3	*SPG20* gene	

Continued

TABLE 109.1 Genes Associated with Neurologic Disorders—cont'd

Genetic disorder	Gene	Locus	Gene product	Inheritance (if established) and phenotype
Spastic paraplegia, optic atrophy, and neuropathy	*SPOAN*	11q13	Spastic paraplegia, optic atrophy, and neuropathy	AR. Spastic paraplegia, optic atrophy, and neuropathy
Specific language impairment 4	*SLI4*	7q35-q36	Specific language impairment 4	Inheritance not established. Language impairment
Specific language impairment QTL, 1	*SLI1*	16q	Specific language impairment QTL, 1	Inheritance not established. Language impairment
Specific language impairment QTL, 2	*SLI2*	19q	Specific language impairment QTL, 2	Inheritance not established. Language impairment
Specific language impairment QTL, 3	*SLI3*	13q21	Specific language impairment QTL, 3	Inheritance not established. Language impairment
Speech-language disorder-1	*FOXP2*	7q31	Forkhead box P2	Inheritance not established. Speech disorder
Speech-sound disorder	*SSD*	3p12-q13	Speech-sound disorder	Inheritance not established. Speech disorder
Spina bifida, folate-sensitive, susceptibility to	*MTHFD1*	14q24	Methylenetetrahydrofolate dehydrogenase (NADP+ dependent) 1, methenyltetrahydrofolate cyclohydrolase, formyltetrahydrofolate synthetase	Susceptibility to folate-sensitive spina bifida
Spinal muscular atrophy-1	*SMN1*	5q12.2-q13.3	Survival of motor neuron 1, telomeric	AR. Spinal muscular atrophy
Spinal muscular atrophy, chronic distal, autosomal recessive	*SMAR*	11q13	Spinal muscular atrophy, chronic distal, autosomal recessive	AR. Spinal muscular atrophy
Spinal muscular atrophy, congenital nonprogressive, of lower limbs	*SMAL*	12q23-q24	Spinal muscular atrophy, congenital nonprogressive, of lower limbs	AR. Spinal muscular atrophy
Spinal muscular atrophy, distal, autosomal recessive, 4	*PLEKHG5*	1p36	Pleckstrin homology domain containing, family G (with RhoGef domain) member 5	AR. Spinal muscular atrophy
Spinal muscular atrophy, lower extremity, autosomal dominant	*SMALED*	14q32	Spinal muscular atrophy, lower extremity, autosomal dominant	AD. Spinal muscular atrophy
Spinal muscular atrophy, type III, modifier of	*SMN2*	5q12.2-q13.3	Survival of motor neuron 2, centromeric	AD. Spinal muscular atrophy
Spinal muscular atrophy, X-linked 2, infantile	*UBA1*	Xp11.23	Ubiquitin-like modifier activating enzyme 1	X-linked recessive. Spinal muscular atrophy, infantile
Spinocerebellar ataxia 1	*6p22.3*	ATXN1, ATX1	Ataxin 1	AD. Cerebellar degenerative disorder presenting with cerebellar ataxias
Spinocerebellar ataxia 2	*12q24.12*	ATXN2, ATX2	Ataxin 2	AD. Cerebellar degenerative disorder presenting with cerebellar ataxias
Machado–Joseph disease	*14q32.12*	ATXN3, MJD	Atxin 3	AD. Cerebellar degenerative disorder presenting with cerebellar ataxia, spasticity and ocular abnormalities
Spinocerebellar ataxia 4	*16q22.1*	SCA	Spinocerebellar ataxia 4	AD. Cerebellar degenerative disorder presenting with cerebellar ataxias
Spinocerebellar ataxia 5	*11q13.2*	SPTBN2	Spectrin, β, nonerythrocytic 2	AD. Cerebellar degenerative disorder presenting with cerebellar ataxias
Spinocerebellar ataxia 6	*19p13.2*	CACNA1A	Calcium channel, voltage-dependent, P/Q type, α1A subunit	AD. Cerebellar degenerative disorder presenting with cerebellar ataxias

TABLE 109.1 Genes Associated with Neurologic Disorders—cont'd

Genetic disorder	Gene	Locus	Gene product	Inheritance (if established) and phenotype
Spinocerebellar ataxia 7	*3p14.1*	ATXN7	Ataxin 7	AD. Cerebellar degenerative disorder presenting with cerebellar ataxias
Spinocerebellar ataxia 8	*13q21*	ATXN8	Ataxin 8	AD. Cerebellar degenerative disorder presenting with cerebellar ataxias
Spinocerebellar ataxia 8	*13q21.33*	ATXN8OS	ATXN8 opposite strand (non-protein coding)	AD. Cerebellar degenerative disorder presenting with cerebellar ataxias
Spinocerebellar ataxia 10	*22q13.31*	ATXN10	Ataxin 10	AD. Cerebellar degenerative disorder presenting with cerebellar ataxias
Spinocerebellar ataxia 11	*15q15.2*	TTBK2	Tau tubulin kinase 2	AD. Cerebellar degenerative disorder presenting with cerebellar ataxias
Spinocerebellar ataxia 12	*5q32*	PPP2R2B	Protein phosphatase 2, regulatory subunit B, β	AD. Cerebellar degenerative disorder presenting with cerebellar ataxias
Spinocerebellar ataxia 13	*19q13.33*	KCNC3	Potassium voltage-gated channel, Shaw-related subfamily, member 3	AD. Cerebellar degenerative disorder presenting with cerebellar ataxias
Spinocerebellar ataxia 14	*19q13.42*	PRKCG, PKCC, PKCG,	Protein kinase C, γ	AD. Cerebellar degenerative disorder presenting with cerebellar ataxias
Spinocerebellar ataxia 15	*3p26.1*	ITPR1	Inositol 1,4,5-trisphosphate receptor, type 1	AD. Cerebellar degenerative disorder presenting with cerebellar ataxias
Spinocerebellar ataxia 17	*6q27*	TBP	TATA box binding protein	AD. Cerebellar degenerative disorder presenting with cerebellar ataxias
Spinocerebellar ataxia 18	*7q22-q32*	SCA18	Spinocerebellar ataxia 18 (sensory with neurogenic muscular atrophy)	AD. Cerebellar degenerative disorder presenting with cerebellar ataxias
Spinocerebellar ataxia 19	*1p13.2*	KCND3, KCND3S, KCND3	Spinocerebellar ataxia 19	AD. Cerebellar degenerative disorder presenting with cerebellar ataxias
Spinocerebellar ataxia 20	*11q12*	DUP11q12, C11DUPq12	Spinocerebellar ataxia 20	AD. Cerebellar degenerative disorder presenting with cerebellar ataxias
Spinocerebellar ataxia 21	*7p21.3-p15.1*	SCA21	Spinocerebellar ataxia 21	AD. Cerebellar degenerative disorder presenting with cerebellar ataxias
Spinocerebellar ataxia 23	*20p13*	PDYN	Prodynorphin	AD. Cerebellar degenerative disorder presenting with cerebellar ataxias
Spinocerebellar ataxia 25	*2p21-p13*	SCA25	Spinocerebellar ataxia 25	AD. Cerebellar degenerative disorder presenting with cerebellar ataxias
Spinocerebellar ataxia 26	*19p13.3*	EEF2, EF2	Spinocerebellar ataxia 26	AD. Cerebellar degenerative disorder presenting with cerebellar ataxias
Spinocerebellar ataxia 27	*13q33.1*	FGF14, FHF4	Fibroblast growth factor 14	AD. Cerebellar degenerative disorder presenting with cerebellar ataxias
Spinocerebellar ataxia 28	*18p11.21*	AFG3L2	AFG3 ATPase family gene 3-like 2 (*S. cerevisiae*)	AD. Cerebellar degenerative disorder presenting with cerebellar ataxias
Spinocerebellar ataxia 29, congenital nonprogressive	*3p26.1*	ITPR1	Spinocerebellar ataxia 29	AD. Cerebellar degenerative disorder presenting with cerebellar ataxias
Spinocerebellar ataxia 30	*4q34.3-q35.1*	SCA30	Spinocerebellar ataxia 30	AD. Cerebellar degenerative disorder presenting with cerebellar ataxias
Spinocerebellar ataxia 31	*16q21*	BEAN	Brain expressed, associated with NEDD4, 1	AD. Cerebellar degenerative disorder presenting with cerebellar ataxias
Spinocerebellar ataxia 32	*7q32-q33*	SCA32	Spinocerebellar ataxia 32	AD. Cerebellar degenerative disorder presenting with cerebellar ataxias

Continued

TABLE 109.1 Genes Associated with Neurologic Disorders—cont'd

Genetic disorder	Gene	Locus	Gene product	Inheritance (if established) and phenotype
Spinocerebellar ataxia 34	*6p12.3-q16.2*	SCA34	Spinocerebellar ataxia 34	AD. Cerebellar degenerative disorder presenting with cerebellar ataxias
Spinocerebellar ataxia 35	*20p13*	TGM6, TG6	Transglutaminase 6	AD. Cerebellar degenerative disorder presenting with cerebellar ataxias
Spinocerebellar ataxia 36	*20p13*	NOP56	NOP56, *S. cerevisiae*, homolog of	AD. Cerebellar degenerative disorder presenting with cerebellar ataxias
Spinocerebellar ataxia, autosomal recessive 2	*SCAR2*	9q34-qter	Spinocerebellar ataxia, autosomal recessive 2	AR. Spinocerebellar ataxia
Spinocerebellar ataxia, autosomal recessive 3	*SCAR3*	6p23-p21	Spinocerebellar ataxia, autosomal recessive 3	AR. Spinocerebellar ataxia
Spinocerebellar ataxia, autosomal recessive 4	*SCASI*	1p36	Spinocerebellar ataxia with saccadic intrusions	AR. Spinocerebellar ataxia
Spinocerebellar ataxia, autosomal recessive 5	*ZNF592*	15q25.3	Zinc finger protein 592	AR. Spinocerebellar ataxia
Spinocerebellar ataxia, autosomal recessive 6	*CLA3*	20q11-q13	Cerebellar ataxia 3 (cerebellar parenchyma disorder 1)	AR. Spinocerebellar ataxia
Spinocerebellar ataxia, autosomal recessive 6	*SCAR6*	20q11-q13	Spinocerebellar ataxia, autosomal recessive 6	AR. Spinocerebellar ataxia
Spinocerebellar ataxia, autosomal recessive 7	*SCAR7*	11p15	Spinocerebellar ataxia, autosomal recessive 7	AR. Spinocerebellar ataxia
Spinocerebellar ataxia, autosomal recessive 8	*SYNE1*	6q25	Spectrin repeat containing, nuclear envelope 1	AR. Spinocerebellar ataxia
Spinocerebellar ataxia, autosomal recessive 10	*PPBP*	4q12-q13	Pro-platelet basic protein (chemokine (C-X-C motif) ligand 7)	AR. Spinocerebellar ataxia
Spinocerebellar ataxia, autosomal recessive with axonal neuropathy	*TDP1*	14q31-q32	Tyrosyl-DNA phosphodiesterase 1	AR. Spinocerebellar ataxia with neuropathy
Spinocerebellar ataxia, X-linked 1	*SCAX1*	Xp11.21-q21.3	Spinocerebellar ataxia, X-linked 1	X-linked spinocerebellar ataxia
Spinocerebellar ataxia, X-linked 5	*SCAX5*	Xq25-q27.1	Spinocerebellar ataxia, X-linked 5	X-linked spinocerebellar ataxia
Stocco dos Santos syndrome	*SHROOM4*	Xp11.2	Shroom family member 4	X-linked intellectual disability
Stroke, susceptibility to	*ALOX5AP*	13q12	Arachidonate 5-lipoxygenase-activating protein	Susceptibility to stroke
Stroke, susceptibility to, 1	*PDE4D*	5q12	Phosphodiesterase 4D, cAMP-specific	Susceptibility to stroke
Succinic semialdehyde dehydrogenase deficiency	*ALDH5A1*	6p22	Aldehyde dehydrogenase 5 family, member A1	AR. Developmental delays, seizures, ataxia and aggression.
Supranuclear palsy, progressive, 1	*PSNP1*	17q21.31	Microtubule-associated protein tau	AD. Parkinsonian features, supranuclear gaze palsy and postural instability
Supranuclear palsy, progressive, 2	*PSNP2*	1q31.1	Supranuclear palsy, progressive, 2	AD. Parkinsonian features, supranuclear gaze palsy and postural instability
Supranuclear palsy, progressive, 3	*PSNP3*	11p12-p11	Supranuclear palsy, progressive, 3	AD. Parkinsonian features, supranuclear gaze palsy, and postural instability
Tay–Sachs disease	*HEXA*	15q23-q24	Hexosaminidase A (α polypeptide)	AR. Severe neurodegenerative disease, increased startle response and cherry-red spot on fundoscopic examination
Telangiectasia, hereditary hemorrhagic, type 1	*ENG*	9q34.11	Endoglin	AD. Multiple telangiectasia involving skin, mucosa, and visceral organs

TABLE 109.1 Genes Associated with Neurologic Disorders—cont'd

Genetic disorder	Gene	Locus	Gene product	Inheritance (if established) and phenotype
Telangiectasia, hereditary hemorrhagic, type 2	*ACVRL1*	12q11-q14	Activin A receptor type II-like 1	AD. Multiple telangiectasia involving skin, mucosa, and visceral organs
Telangiectasia, hereditary hemorrhagic, type 3	*HHT3*	5q31.3-q32	Hereditary hemorrhagic telangiectasia, type 3	AD. Multiple telangiectasia involving skin, mucosa and visceral organs
Telangiectasia, hereditary hemorrhagic, type 4	*HHT4*	7p14	Hereditary hemorrhagic telangiectasia, type 4	AD. Multiple telangiectasia involving skin, mucosa, and visceral organs
Thyrotoxic periodic paralysis, susceptibility to, 2	*KCNJ18*	17p11.2	Potassium inwardly rectifying channel, subfamily J, member 18	Susceptibility focus for thyrotoxic periodic paralysis
Timothy syndrome	*CACNA1C*	12p13.3	Calcium channel, voltage-dependent, L type, α1C subunit	AD, congenital heart defects, lethal arrhythmias, webbing of fingers, immune deficiency, and cognitive defects
Treacher Collins syndrome 2	*POLR1D*	13q12.2	Polymerase (RNA) I polypeptide D, 16 kDa	AD. Mandibulofacial dysostosis
Treacher Collins syndrome 3	*POLR1C*	6p22.3	Polymerase (RNA) I polypeptide C, 30 kDa	AD. Mandibulofacial dysostosis
Trifunctional protein deficiency	*HADHB*	2p23	Hydroxyacyl-CoA dehydrogenase/3-ketoacyl-CoA thiolase/enoyl-CoA hydratase (trifunctional protein), β subunit	AR. Neonatal form presents with SIDS, infantile form presents with Reye-like syndrome and late-onset form with skeletal myopathy
Troyer syndrome	*SPG20*	13q12.3	Spartin	AR. Spastic paraplegia 20 (Troyer syndrome)
Tuberous sclerosis 1	*TSC1*	9q34	Tuberous sclerosis 1	AD. Multisystem disorder characterized by hamartomas in various organ systems
Tuberous sclerosis 2	*TSC2*	16p13.3	Tuberous sclerosis 2	AD. Multisystem disorder characterized by hamartomas in various organ systems
Tyrosinemia I, hepatorenal	*FAH*	15q25.1	Fumarylacetoacetate hydrolase (fumarylacetoacetase)	AR. Tyrosinemia type 1
Tyrosinemia, type II	*TAT*	16q22.1-q22.3	Tyrosine aminotransferase	AR. Tyrosinemia, type II, keratitis, painful palmoplantar hyperkeratosis, mental retardation, and elevated serum tyrosine levels
Tyrosinemia, type III	*HPD*	12q24-qter	4-hydroxyphenylpyruvate dioxygenase	AR. Tyrosinemia, type III. Mild mental retardation with absence of liver damage
Usher syndrome, type 1B	*MYO7A*	11q13.5	Myosin VIIA	AR. Congenital hearing loss, speech problems, early retinitis pigmentosa
Usher syndrome, type 1C	*USH1C*	11p15.1	Harmonin	AR. Usher syndrome 1C
Usher syndrome, type 1D	*USH1D*	10q22.1	Cadherin 23	AR. Digenic recessive. Usher syndrome 1D. Hearing loss, retinitis pigmentosa, and vestibular dysfunction
Usher syndrome, type 1E	*USH1E*	21q21	Usher syndrome 1E (autosomal recessive)	AR. Usher syndrome 1E
Usher syndrome, type 1F	*PCDH15*	10q21-q22	Protocadherin-related 15	AR. Usher syndrome 1F
Usher syndrome, type 1G	*USH1G*	17q24-q25	Usher syndrome 1G (autosomal recessive)	AR. Usher syndrome 1G

Continued

TABLE 109.1 Genes Associated with Neurologic Disorders—cont'd

Genetic disorder	Gene	Locus	Gene product	Inheritance (if established) and phenotype
Usher syndrome, type 1H	*USH1H*	15q22-q23	Usher syndrome 1H (autosomal recessive)	AR. Usher syndrome 1H
Usher syndrome, type 2A	*USH2A*	1q41	Usher syndrome 2A (autosomal recessive, mild)	AR. Usher syndrome, mild
Usher syndrome, type 3	*CLRN1*	3q21-q25	Clarin 1	AR. Usher syndrome
von Hippel–Lindau syndrome	*VHL*	3p26-p25	von Hippel–Lindau tumor suppressor, E3 ubiquitin protein ligase	AD. Inherited cancer predisposition to various benign and malignant neoplasms
Waardenburg syndrome, type 1	*PAX3*	2q35	Paired box 3	AD. Sensorineural hearing loss with pigmentary abnormalities of hair, skin, and eyes. Type 1 has dystopia canthorum
Waardenburg syndrome, type 2B	*WS2B*	1p21-p13.3	Waardenburg syndrome, type 2B	AD. Sensorineural hearing loss with pigmentary abnormalities of hair, skin, and eyes. Type 2 is characterized by absence of dystopia canthorum
Waardenburg syndrome, type 2C	*WS2C*	8p23	Waardenburg syndrome, type 2C	AD. See above
Waardenburg syndrome, type 2D	*SNAI2*	8q11	Snail homolog 2 (*Drosophila*)	AD. See above
Waardenburg syndrome, type 3	*PAX3*	2q35	Paired box 3	AD. Sensorineural hearing loss with pigmentary abnormalities of hair, skin, and eyes. Dystopia canthorum and upper limb anomalies
Waardenburg syndrome, type 4B	*EDN3*	20q13.2-q13.3	Endothelin 3	AD. Sensorineural hearing loss with pigmentary abnormalities of hair, skin, and eyes. Also has Hirschsprung disease
Waisman parkinsonism-mental retardation syndrome	*WSN*	Xq28	Waisman syndrome	X-linked. Early-onset parkinsonism and intellectual disability syndrome
Warburg micro syndrome 1	*RAB3GAP1*	2q21.3	RAB3 GTPase activating protein subunit 1 (catalytic)	AR. Characterized by microcephaly, micro cornea, optic atrophy, congenital brain anomalies
Warsaw breakage syndrome	*DDX11*	12p11	DEAD/H (Asp-Glu-Ala-Asp/His) box helicase 11	AR. Developmental delay, hypotonia, facial dysmorphism
Welander distal myopathy	*WDM*	2p13	Welander distal myopathy	AD. Late-onset distal muscular dystrophy
Wieacker–Wolff syndrome	*WWS*	Xq13-q21	Wieacker–Wolff syndrome	X-linked recessive. Congenital contractures of the feet at birth, a slowly progressive predominantly distal muscle atrophy, dyspraxia of the eyes, face, and tongue muscles, and mild mental retardation
Wilson disease	*ATP7B*	13q14.3-q21.1	ATPase, Cu^{2+} transporting, β polypeptide	AR. Abnormalities in copper metabolism with subsequent damage to neurologic system
Wolf–Hirschhorn syndrome	*WHCR*	4p16.3	Wolf–Hirschhorn syndrome chromosome region	Microdeletion syndrome, characterized by developmental delay and multisystem involvement
Wolfram syndrome	*WFS1*	4p16.1	Wolfram syndrome 1 (wolframin)	AR, neurodegenerative disease characterized by diabetes mellitus, optic atrophy, diabetes insipidus, and deafness. Additional features include renal anomalies, ataxia, dementia

TABLE 109.1 Genes Associated with Neurologic Disorders—cont'd

Genetic disorder	Gene	Locus	Gene product	Inheritance (if established) and phenotype
Wolfram syndrome 2	*CISD2*	4q22-q24	CDGSH iron–sulfur domain 2	AR, neurodegenerative disease characterized by diabetes mellitus, optic atrophy, diabetes insipidus, and deafness. Additional features include renal anomalies, ataxia, dementia
Wolman disease	*LIPA*	10q23.31	Lipase A, lysosomal acid, cholesterol esterase	AR. Massive infiltration of organs by macrophages filled with cholesterol esters. Early death
Woodhouse–Sakati syndrome	*DCAF17*	2q22.3-q35	DDB1 and CUL4 associated factor 17	AR, multisystem disorder
X-linked mental retardation	*DLG3*	Xq13.1	Discs large, *Drosophila*, homolog of, 3 neuroendocrine DLG	X-linked intellectual disability
X-linked lissencephaly, X-linked subcortical laminal heterotopia	*DCX*	Xq23	Doublecortin	X-linked lissencephaly, X-linked subcortical laminal heterotopia
Zellweger syndrome-1	*PEX1*	7q21-q22	Peroxisomal biogenesis factor 1	AR. Progressive multisystem disorder characterized by severe neurologic symptoms, liver failure and craniofacial dysmorphism. Genetic heterogeneity exists. See below
Zellweger syndrome-2	*ABCD3*	1p22-p21	ATP-binding cassette, subfamily D (ALD), member 3	AR. See Zellweger syndrome
Zellweger syndrome-3	*PEX2*	8q21.1	Peroxisomal biogenesis factor 2	AR. See Zellweger syndrome
Zellweger syndrome	*PEX14*	1p36.2	Peroxisomal biogenesis factor 14	AR. See Zellweger syndrome
Zellweger syndrome	*PEX10*	1p36.32	Peroxisomal biogenesis factor 10	AR. See Zellweger syndrome
Zellweger syndrome	*PEX19*	1q22	Peroxisomal biogenesis factor 19	AR. See Zellweger syndrome
Zellweger syndrome	*PEX13*	2p15	Peroxisomal biogenesis factor 13	AR. See Zellweger syndrome
Zellweger syndrome, complementation group 9	*PEX16*	11p12-p11.2	Peroxisomal biogenesis factor 16	AR. See Zellweger syndrome
Zellweger syndrome, complementation group G	*PEX3*	6q23-q24	Peroxisomal biogenesis factor 3	AR. See Zellweger syndrome
17-β-Hydroxysteroid dehydrogenase X deficiency	*HSD17B10*	Xp11.2	Hydroxysteroid (17-β) dehydrogenase 10	X-linked mental retardation 17/31. Progressive psychomotor degeneration, epilepsy, microduplication
2-Methylbutyrylglycinuria	*ACADSB*	10q25-q26	Acyl-CoA dehydrogenase, short/branched chain	AR. Severe developmental delay, microcephaly, epilepsy
3-Methylglutaconic aciduria, type I	*AUH*	9q22.3	3-Methylglutaconyl-CoA hydratase	AR. 1) Childhood-onset psychomotor retardation. 2) Adult-onset ataxia, spasticity and dementia
3-Methylglutaconic aciduria, type III	*OPA3*	19q13.32	OPA3A and OPA3B	AR. Optic atrophy 3 (autosomal recessive, with chorea and spastic paraplegia); AD optic atrophy with cataract
46,XY partial gonadal dysgenesis, with minifascicular neuropathy	*DHH*	12q13.1	Desert hedgehog-signaling molecules	AR. 46,XY partial gonadal dysgenesis, with minifascicular neuropathy

Abbreviations: AR, autosomal recessive; AD, autosomal dominant.

Method Description: *The genes in this list were identified through searching databases available in the public domain (The International Standards for Cytogenomic Arrays Consortium database [www.iscaconsortium.org], which generates this information using NCBI's database of genomic structural variation [dbVar, www.ncbi.nlm.nih.gov/dbvar/]; The Human Gene Compendium Gene Cards [http://www.genecards.org/]) as well genes of interest to neurologists at U.T. Southwestern. This list was curated from approximately 5000 to 1200 genes following review and annotation utilizing two other databases also in the public domain, OMIM and HUGO.*[1,2]

References

1. Online Mendelian Inheritance in Man, OMIM®. *McKusick-Nathans Institute of Genetic Medicine*. Baltimore, MD: Johns Hopkins University; 2014. http://omim.org/.
2. Gray KA, Daugherty LC, Gordon SM, Seal RL, Wright MW, Bruford EA. genenames.org: the HGNC resources in 2013. *Nucleic Acids Res.* 2013;41(Database issue):D545–D552. doi:10.1093/nar/gks1066 Epub 2012 Nov 17 PMID:23161694.

Index

Note: Page numbers followed by *f* indicate figures and *t* indicate tables.

B

E

F

G

I

R

T

U

V